Microbes in Practice

Microbes in Practice

Prakash S. Bisen
Emeritus Professor
School of Studies in Biotechnology
Jiwaji University
Gwalior
&
Research and Development Centre
Tropilite Foods Pvt. Ltd.
Davar Campus, Tansen Road
Gwalior

ISBN: 978-93-89307-33-7

Edition: 2019

Printed at: Rekha Printers

Dedicated to
my Beloved Parents

Preface

Microbes in Practice impinges on almost every aspect of human existence. An attempt to cover the whole of the subject would have been an impossible task. I have confined myself to a consideration of the aspects of microbiology in which I have research experience: bacteriology, mycology, immunology and virology. I have also tried to concentrate on the fundamental problems in the subject. What constitute the microbes? How do they differ from higher organisms and from each other? How can microbes be controlled, visualized, enumerated and cultured? How are they classified? The answer to these and other could be found in this book with illustrations. With this book, I hope to provide a firm framework of microbiological concepts for students whose requirements are more modest. There are several excellent textbooks available to students of Microbiology. However, there is a lack of concise presentation that gives an overview of the field to the students of microbiology and also presents the required basic knowledge of general microbiology to students of zoology, botany, chemistry, agriculture, nutrition, pharmacy, biotechnology, etc. This book is designed to appeal to such a wide spectrum of readers. Its aim is to convey a general view as well as specific knowledge and to stimulate.

Microbiology deals mostly with the large group of fungi, bacteria and viruses which equal in diversity and physiological phenomena the more traditional group of organisms within the disciplines of botany and zoology. During recent years the study of microorganisms has contributed important insights into the basic problems of biology. Because of their ease of manipulation, rapid growth, highly developed capacity for adaptation, and other properties, microorganisms are one of the preferred objects of research in biochemistry and genetics. They are employed successfully for varied purpose as studying the pathogenic nature of an organism, growing auxotrophs, gene transfer, gene expression, experiments in molecular biology, protoplast fusion, gene therapy, modification of industrial strains, food, dairy, beverages industries, pharmaceutical industries, freeing plants of pathogens and other innumerable works for the welfare of the mankind. The topics and the techniques described in the book reflect this diversity.

I believe that many biologists would like to use microbiological techniques in their research, but hesitate to do so because of lack of information. In spite of several excellent books on the physiology, biochemistry and molecular biology of microorganisms there exists no single book which can explain the metabolism of tiny creatures with a practical approach and which includes a brief text on the microbiological experiments useful for the workers in the allied fields at various levels. A constant need for such a text material was widely felt which describes the methods in simple language supplemented with sufficient illustrations and various media composition. The chapters have been made complete by giving the expected results and flow charts to make experiment easier and enjoyable. The subject matter and their results are illustrated in line diagrams in lucid manner.

Beginning with the characteristic difference between prokaryotes and eukaryotes and classification of microorganisms the book depicts the various methods employed to characterize and cultivate different forms of microbes, different media used and various procedures involved with the isolation and characterization of microorganisms. Several staining procedures, chemical compositions of media, biochemical tests, medical microbiology and immunology, microbes in relation to plants, waste management, industrial application, aquatic environment and soil has been compiled at one place in different chapters. Separate chapters have been formulated for dairy and food microbiology and medical microbiology and immunology with application of membrane filter technique, standard plate count at several places. Mushroom cultivation technology has been described in depth.

Every care has been taken to include the minute details. Author claims no originality for the methods described in the book, except for the fact that each method has been carried out in the laboratories before being included. It is therefore; presumed that there would be no difficulty for any moderately equipped laboratory in the use of these methods. The exercises have been formulated by keeping in mind the hazards of chemicals and only those exercises are given which minimize the exposure of students to potentially pathogenic organisms and hazardous chemicals.

This book is written during a period of rapid expansion in higher and technical education and when several institutions (Public funded and Private players) was embracing modularization and transforming their basic courses into vocational education competing with other professional courses. Writing this text provided an enjoyable relief. To all these and other I owe a debt of gratitude to my friends and students. Their advice has been invaluable, but any mistake in the text is now all mine.

Prakash S Bisen
School of Studies in Biotechnology
Jiwaji University
Gwalior
&
Research and Development Center
Tropilite Foods Pvt. Ltd.
Davar Campus, Tansen Road
Gwalior

Acknowledgements

During the preparation of this book, I have sought the help and valuable advice of a number of persons to whom I am indebted. My mentor Professor G.P. Agarwal, Former Chairman and Dean, Department of Postgraduate Studies and Research in Biological Sciences, Rani Durgawati Vishwavidyalaya, Jabalpur, India has been a constant source of inspiration and supportive throughout. Without his unfaltering belief, mind-easing humour, and comforting support, I would not be who I am, and where I am today.

I have to thank colleagues and students for stimulating and critical comments. At various stages during book writing and during my academic tenure, my former colleagues and students have commented and influenced me on different aspects of the text. These have included Dr. Vidya P. Sahni, Professor S. K. Hasija, Professor D.P. Tewari, Professor Ms. D. Kulhara, Professor Ms. Indu Gupta, Professor R. C. Rajak, Professor Ms. Anjana Sharma, Professor Ms. Divya Bagchi, Professor S. S. Sandhu and Mr. Saurabh Bundela from Department of Postgraduate Studies and Research in Biological Sciences, Rani Durgawati Vishwavidyalaya, Jabalpur, Professor N.D. Sharma, Jawaharlal Nehru Agriculture University, Jabalpur, Professor Ms. Surabhi Shrivastava, Department of Microbiology, Gragi College, Delhi University, Delhi, Professor Surendra Singh, Department of Botany, Banaras Hindu University, Varanasi, Professor Brij Bhan Singh, Department of Biochemistry and Molecular Biology, School of Medicine and Health Sciences, University of North Dakota, Grand Forks, ND, USA, Dr. Ms. Harshini Mukundan, Los Alamos National Laboratory, Los Alamos, New Mexico, USA, Dr. Sanjay K. Garg, University of Michigan Medical School, Ann Arbor, MI, USA, Dr. Zakir Khan, Cedars-Sinai Medical Center University of California, Los Angeles, USA, Dr. Anshul Bhardwaj, Department of Biochemistry and Molecular Biology, Thomas Jefferson University, Philadelphia, Dr Ms. Archana Singh Sikarwar International Medical University, Kuala Lumpur, Malaysia, Professor Ms. Dolly Malhotra, Principal, Government Gitanjali Postgraduate Autonomous College, Bhopal, Professor D. P. Singh, School of Environmental Sciences, Babasaheb Bhimrao Ambedkar Central University, Lucknow, Dr. Ms. Mousumi Debnath, CEG Test House and Research Centre Pvt. Ltd., Jaipur, Professor Ms. Shanthy Sundaram, Biotechnology Programme, Allahabad Central University, Allahabad, Dr. Ms. Ragini Gothalwal, Department of Biotechnology, Barkatullah University, Bhopal and Dr. Ms. Namita Singh, Department of Bio- and Nanotechnology, Guru Jambheswar University of Science and Technology, Hisar, Haryana.

I very well acknowledge the lively day-to-day interactions I used to have with my research colleagues and express my sincere thanks to Prof. G.B.K.S. Prasad, Dean, Faculty of Life Sciences, Jiwaji University, Gwalior and Dr. Anish Zacharia J., School of Studies in Biochemistry, Jiwaji University, Gwalior; Professor Ishan K. Patro and Dr. Ms. Nisha Patro, School of Studies in Neurosciences, Jiwaji University, Gwalior; Dr. Ms. Divya

Shrivastava, Ms. Meenal Dixit, Mr. Amit Kumar, Mr. Sumit Govil, and Mr. Shilesh Kumar from Jaipur National University, Jaipur, Dr. Ms. Shruti Mathur, Amity University, Rajasthan, Jaipur, Dr. Ram P. Tiwari, Rapid Diagnostic Group of Companies, New Delhi, Dr. Anubhav Jain, and Dr. Pankaj Krishna from Diagnostic R&D Division, RFCL Limited, New Delhi, Dr. Ms. Ruchika Raghuvanshi, Mr. Bhagwan S. Sanodiya, Dr. Gulab S. Thakur, Dr. Rakesh K. Baghel, Mr. Rohit Sharma and Dr. Mithilesh K. Jaiswal, Ms. Pallavi Jaiswal from Tropilite Foods Pvt. Ltd., Davar Campus, Tansen Road, Gwalior for their invaluable assistance in preparation of this book.

I am indebted to Mr. Devendra Singh and Mr. Rahul Jha for careful and faithful execution of the drawings, shaping the tables in the presentable form and for computational help. The cover page of the book was designed and conceived by Mr. Rahul Jha. I acknowledge my thanks to many in my life who in their splendid individual ways have encouraged, inspired and cheered for me to introduce to me to microbial world especially my elder brother Late Shri Onkar Nath Pande and my very old friend Mrs Vijayalakshmi Sharma from Mumbai.

Finally. my family, my wife Shashi and children Ankur, Anand, Sonia, Parul and Angad, have learned to live with microbes and have shown full understanding for the priorities. For this. I acknowledge my heartfelt thanks.

Prakash S. Bisen
School of Studies in Biotechnology
Jiwaji University
Gwalior
&
Research and Development Center
Tropilite Foods Pvt. Ltd.
Davar Campus, Tansen Road
Gwalior

Contents

Chapter 1

Microbial World

1.1 INTRODUCTION

Microbes include several distinct groups of organisms, differing widely in form and life cycle but resembling one another in their small size and relatively simple structure, the way they live and multiply, their form and classification and the role they play in the varied habitats they occupy. Molecular phylogenies divide all living organisms into three domains Bacteria, Archaea, and Eukarya (eukaryotes: protists, fungi, plants, animals).

This has been described in the three-kingdom classification proposed by Woese and Fox in 1977, but the place of viruses in the phylogenetic tree of life is still uncertain. Though, some attention has been devoted to viruses and the problems posed by them in certain biological contexts. The domains of Bacteria and Archaea encompass a huge diversity of organisms that differ in their sources of carbon or nitrogen, their metabolic pathways, the end products of their metabolism, and their ability to attack various naturally occurring organic compounds. These bacteria and archaea have adapted to available climate and microenvironment on earth depending upon their specific requirements, for example; halophilic microorganisms grow in brine ponds encrusted with salt; thermophilic microorganisms grow on smoldering coal piles or in volcanic hot springs, and barophilic microorganisms prefer enormous pressure available in the depths of the seas. Bacteria do have different living patterns as some are symbionts of plants while others live as intracellular parasites inside mammalian cells or form stable consortia with other microorganisms. The seemingly limitless diversity of the microorganisms provides an immense pool of raw material for applied microbiology. Taxonomic systems for biological organisms are hierarchical. The most inclusive unit of classification is a kingdom (or domain), followed by phylum (or division), class, order, family, genus, species, and subspecies. An additional rank below the subspecies level pathovar, serovar, or biotype is added when a strain needs to be distinguished by its special character. For example, the rank of a pathovar (or pathotype) is applied to an organism with pathogenic properties for a certain host or hosts, as exemplified by *Xanthomonas campestris* pv *vesicatoria*, the causal agent of bacterial spot of pepper and tomato. Serovar (or serotype) refers to distinctive antigenic properties, and biovar (or biotype) is applied to strains with special biochemical or physiological properties. Principally, organisms can be classified according to any set of criteria, as long as the scheme results in reproducible identification of new strains. However, a classification scheme based on totally arbitrary criteria is likely to be of very limited practical use. Thus taxonomists may group together apparently similar, presumably related species into a genus and related genera into a family in the hope that this classification accurately reflects the evolutionary or phylogenetic relationships among various organisms.

1.2 MICROORGANISMS

Microorganisms are ubiquitous microscopic organisms (exception is multicellular organisms like *Thiomargarita namibiensis*) found in almost every habitat in nature. They are very diverse including bacteria, fungi, archaea, green algae and animals also, such as plankton and planarian. They are actually the true masters of life on earth. Microorganisms are the most versatile and adaptable forms of life on earth as they have existed here for some 3.5 billion years (Schopf, 2006; Altermann and Kazmierczak, 2003). Indeed for the first 2 billion years of their existence, prokaryotes alone ruled the biosphere, colonizing every accessible ecological niche, from glacial ice to the hydrothermal vents of the deep-sea bottoms. As these early prokaryotes evolved, they developed the major metabolic pathways characteristic of all living organisms today, as well as various other metabolic processes, such as nitrogen fixation which is still restricted to prokaryotes alone. Over their long period of global dominance, prokaryotes also changed the earth, transforming its anaerobic atmosphere to one rich in oxygen and generating massive amounts of organic compounds (Figure 1.1 (a)). Eventually, they created an environment suited to the maintenance of more complex forms of life (Figure 1.1 (b)).

Viruses are also considered as microorganisms by some microbiologists, but others consider them as non-living. Microorganisms reside in all parts of the biosphere including water, soil, hot springs, on the ocean floor, in the atmosphere and deep inside rocks within the earth's crust. Microorganisms are critical to nutrient recycling in ecosystems as they act as decomposers. As some microorganisms can fix nitrogen, they are a vital part of the nitrogen cycle, and studies indicate that airborne microbes may play a role in precipitation and weather (Christner et al., 2008). Most microorganisms can reproduce rapidly and some of them as bacteria can freely exchange genes by methods of genetic recombination including conjugation, transformation and transduction (Figure 1.2) between widely divergent species (Wolska, 2003).

Single-celled microorganisms were the first forms of life to develop on earth, approximately 3–4 billion years ago. Further evolution was slow, and for about 3 billion years in the Precambrian eon, all organisms were microscopic (Stanley, 1973). So, for most of the history of life on earth the only forms of life were microorganisms. Microbes are also exploited by people in fermentation, both in traditional food and beverage preparation, and in modern technologies based on genetic engineering (Table 1.1 and Table 1.2).

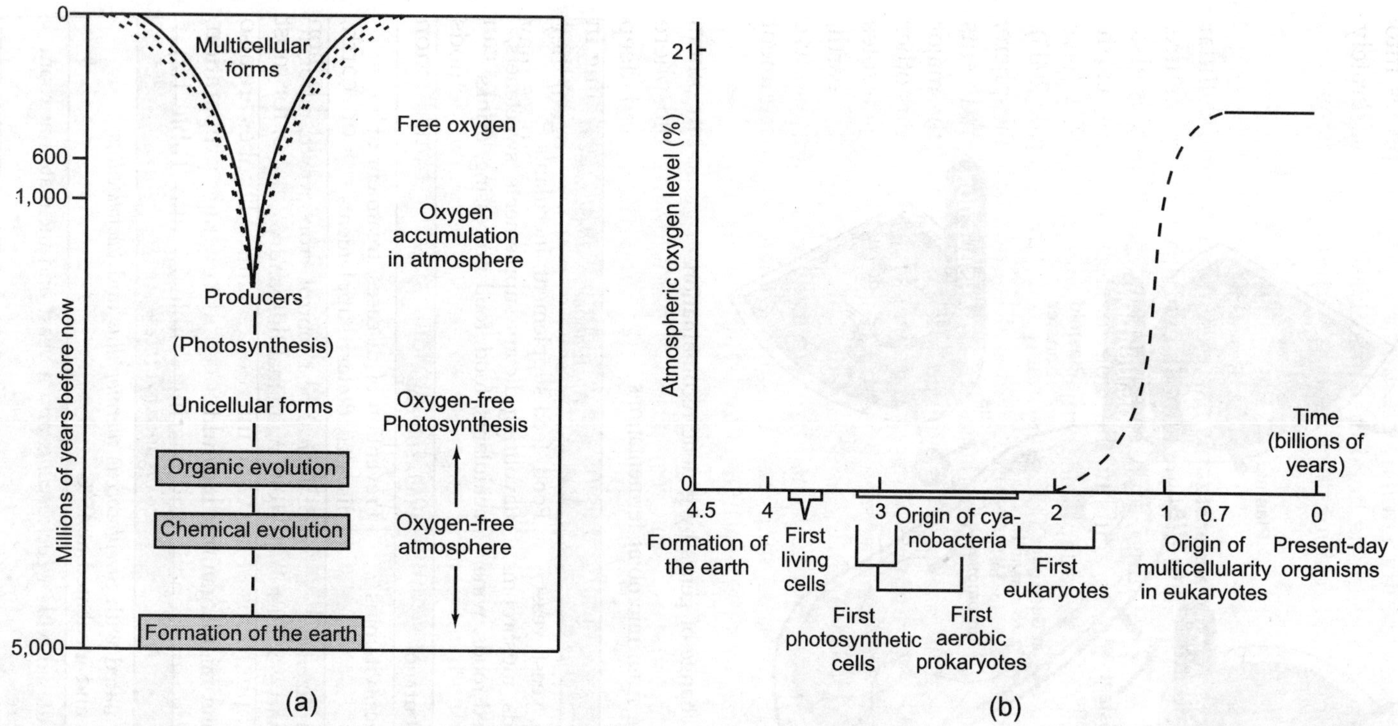

Figure 1.1 (a) Diagrammatic repersentations of probable events in the history of life on earth (b) A putative scenario of the main steps in cellular evolution.

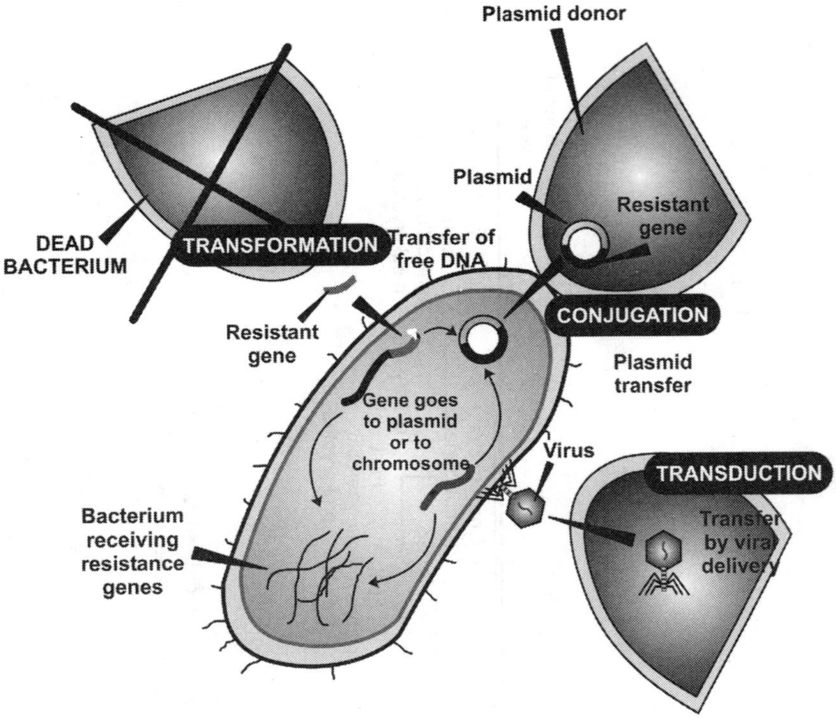

Figure 1.2 Exchange of genes by genetic recombination.

Table 1.1 Some important products from microbial fermentations

Area/Industry	*Product*	*Applications food/beverages*
Food/beverages	Single cell protein, yeast, yeast extract, amino acids, mushrooms, enzymes, fermented foods, wine, beer	Food/feed supplement, inoculants, flavouring dietary supplements, sweeteners, acidulants, food, food processing, drinks
Chemicals	Ethanol, acetone, butanol	Fuel, solvent
Health	Vaccines, antibiotics, vitamins	Prevention of diseases, treatment of diseases, dietary supplements
Agriculture	Insecticides, biofertilizers, antibiotics, veterinary vaccines	Biological pest control, crop growth, nitrogen fixation, animal husbandry

Table 1.2 Some foods and feeds and microorganisms involved

Products	*Microorganisms/Process*
Yogurt	Milk inoculated with *Streptococcus thermophilus* and *Lactobacillus bulgaricus* and incubated at 45°C
Russian kefir	Milk inoculated with a *Lactobacillus* and a yeast and incubated at room temperature

Contd...

Contd...

Indian idli	Polished rice and dehulled *mung* beans are soaked separately, grounded and incubated. The principal fermenting agent involved is *Leuconostoc mesenterodies,* and *Streptococcus faecalis*
African acid porridges	*Corynebacterium manihot* and *Geotrichum candidum* hydrolyze starch. In cassava porridge and produce lactic and formic acids.
Acidic cereal porridges	Produced generally at temperature higher than 50°C *Lactobacillus delbrueckii* the principal fermenting microbe involved.
Beers, Wines	Yeasts (e.g., *Saccharomyces cerevisiae, Hansenula, Endomycopsis*) ferment starch/sugars in germinating (malted) grains, to ethyl alcohol.
Shoyu, Soy Sauce, Miso	Soybean paste, rice wheat, sorghum, or maize are (soybeen paste), Koji fermented by *Aspergillus soyae, A. oryzae, Saccharomyces rouxii, Lactobacillus* spp.
Indonesian tempeh	Made by fermenting soaked soybeans with *Rhizopus* sp. Rich in vitamin B12

However, some microorganisms are pathogenic and very harmful to living beings, since they invade and grow within other organisms, causing diseases that kill people, other animals and plants.

Phylogenetically, this diverse microbial community can be classified into prokaryotes including archaea and bacteria and eukaryotes including fungi, plants and animals (Figure 1.3).

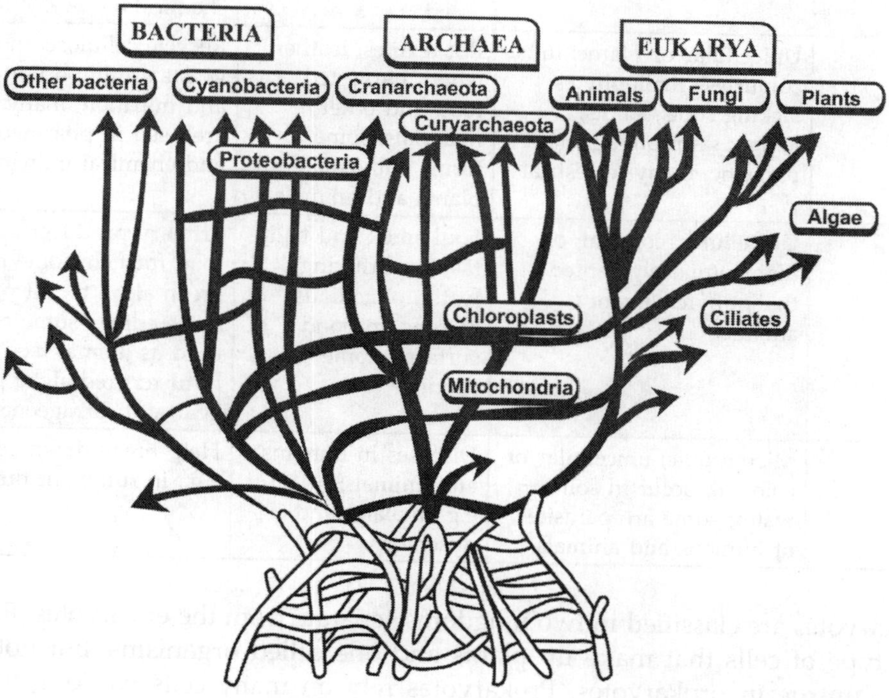

Figure 1.3 Phylogenetic classification of microorganisms.

1.3 PROKARYOTES AND EUKARYOTES

All living microorganisms can be classified into one of two groups depending on the fundamental structure of their cells, these are prokaryotes and eukaryotes and the distinction between prokaryotes and eukaryotes is considered to be the most important distinction among various groups of organisms. The major difference is the presence or absence of membrane bound organelles. Eukaryotic cells contain membrane-bound organelles as the nucleus, while prokaryotic cells do not. Other differences in cellular structure of prokaryotes and eukaryotes include the presence of mitochondria and chloroplasts, the cell wall, and the structure of chromosomal DNA (Table 1.3).

Table 1.3 Main characteristics of major groups of microorganisms

Microorganisms	*Chief characteristics*	*Problems caused*	*Important uses*
Viruses	Nucleoproteins; infective; multiply in living cells	Several diseases in humans, animals, and plants	Carry genetic information and transducer characters in bacteria; some cause diseases in pests and insects and are useful in pest control.
Bacteria	Mostly unicellular; spherical, rod shaped, or spiral; mostly saprophytic	Some are pathogenic to humans, animals and plants	Degrade organic matter and promote soil fertility; aid waste disposal and biogas production; some yield antibiotics and other chemicals.
Fungi	Unicellular or filamentous, colourless thallophytes lacking roots, stems, leaves; saprophytic; parasitic; many terrestrial	Rot textiles, leather, wood, harvested food and other materials; some forms pathogenic for plants and animals	Recycle cellulose and lignin; yeasts are important in food and nutrition; many forms are used in pharmaceutical and chemical industries.
Algae	Unicellular, colonial, or filamentous pigmented thallophytes; mainly aquatic	Foul smell and bad taste in drinking water; obnoxious blooms on pond surfaces; some produce toxins	Primary producers; chief food for fish; some blue-green algae fix nitrogen; some edible; some are used as animal feed or bio-fertilizer; red algae produce agar and carrageenan.
Protozoa	Microscopic; unicellular or colonial; occur in soil or water; some are parasites of humans and animals	Diseases in humans and animals, e.g., malaria and dysentery	Help break down cellulose, e.g., in ruminant nutrition

Prokaryotes are classified in two kingdoms separate from the eukaryotes. Eukaryotes are the type of cells that make up plants and one celled organisms, but not animals, which is untrue in prokaryotes. Prokaryotes rely on many cells working together to function. While eukaryotes are radically different from one another, they do have three general parts that allow them to carry out these processes of life. These are the cell membrane, the nucleus, and the third part comprises other organelles. The organelles are very important to the cell's functioning. One of these organelles includes mitochondria, which transfer energy from organic compounds to ATP. The ribosomes organize the synthesis of proteins (which is used to get energy). The rough endoplasmic reticulum prepares proteins for export. The smooth endoplasmic reticulum regulates calcium levels, breaks down toxic substances, and synthesizes steroids. The Golgi body processes and packages substances produced by the cell. The lysosomes digest molecules. There are other little parts to the cells which aid in all of this. These include microfilaments, cilia, flagella (those two assist in transportation), and of course the nucleus. In a plant, there are also the cell wall, the vacuole, and the plastid. All of these organelles are used to carry out the life processes (Figure 1.4).

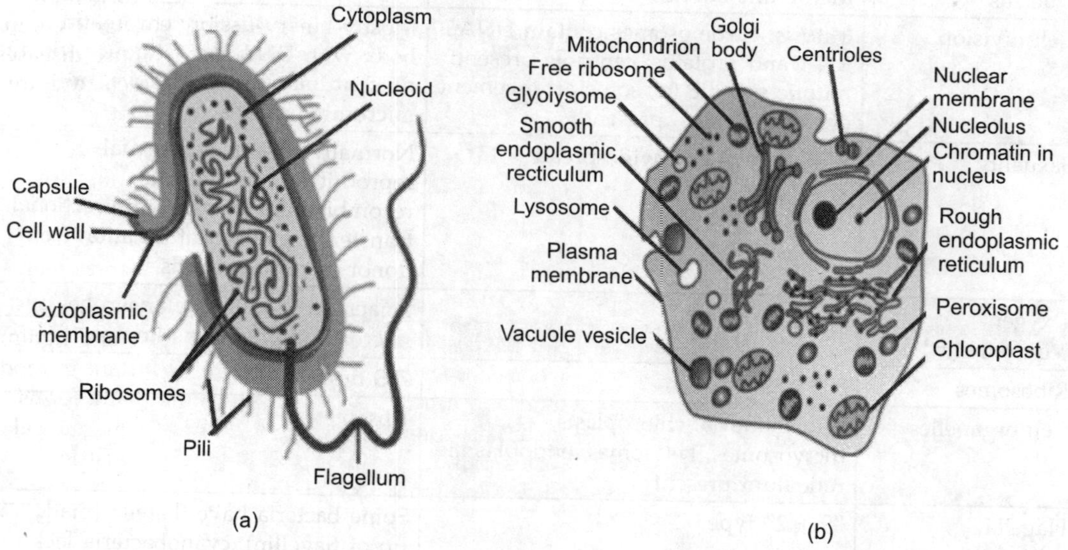

Figure 1.4 Cell structure of (a) Prokaryote (b) Eukaryote.

1.4 PROKARYOTES

In prokaryotic organisms, the chromosome (nucleoid) is a closed circular DNA molecule, lies in the cytoplasm, is not surrounded by a nuclear membrane, and contains all of the information necessary for cell reproduction. Bacteria and archaea are prokaryotes, whereas, fungi are eukaryotes. The choice of a fungus (such as the yeast *Saccharomyces cerevisiae*) or a bacterium (such as *Escherichia coli*) for a particular application is often dictated by the basic genetic, biochemical, and physiological differences between prokaryotes and eukaryotes.

The prokaryotes are dependent on other similar cells and different cells in order to carry out life processes, while each eukaryotic cell has all the organelles it needs to carry out the processes of life (Table 1.4 and Table 1.5). It is thought that at one time, only prokaryotes existed on earth. They 'ate' organic compounds and reproduced, slowly forming more complicated structures. After a while, some of them evolved into eukaryotes. This is how different organisms managed to come about on earth. Of course, this is only a scientific theory, not an absolute proven fact.

Table 1.4 Distinctions between eukaryotes and prokaryotes

Parameters	*Eukaryotes*	*Prokaryotes*
Size	Mostly large and macroscopic but a few are microbes larger than 10 m	Most are microscopic (1-10 μm)
Cell wall	Cellulosic (green plants); chitinous (fungi); mucopeptides absent	Have mucopeptides as a major component
Nucleus	Membrane-bound	No distinct membrane-bound nucleus
Cell division	Mitosis; chromosomes contain DNA, RNA and proteins; centriole present; mitotic spindle (or some arrangement of microtubules) present	Mostly binary fission; chromatin body with DNA but without protein; no centriole, mitotic spindle, or microtubules present
Sexuality	Well-known and widespread	Normally absent; parasexual reproduction and genetic recombination involve unidirectional transfer of genetic information from donor to recipient cells.
Oxygen requirement	Mostly aerobic	Anaerobic; facultative anaerobes and microaerophilic forms known
Ribosomes	80S type	70S type
Cell organelles	Mitochondria, chloroplasts, dictyosomes, lysosomes, endoplasmic reticulum present	Absent
Flagella	"9 + 2" type	Some bacteria have flagella (made up of flagellin), cyanobacteria lack flagella

Table 1.5 Comparison of selected microorganisms

Parameters	*Mycoplasma*	*Rickettsia*	*Bacteria*	*Virus*	*Protozoa*
Growth in cell-free medium	+	–	+	–	+
Cell wall	No	Yes	Yes	No	Yes
Smallest size 200 mμ or less	+	–	–	+	–
Possess both RNA and DNA	+	+	+	–	+
Growth inhibition by antibiotics	+	+	+	–	+
Sterol requirement	+	–	–	–	+
Growth inhibition by antibodies	+	+	–	+	–

1.4.1 The Bacteria

Bacteria are successful and ancient form of life. They are small cells, found in the environment as either individual cells or aggregated together as clumps, and their intracellular structure is far simpler than eukaryotes. Bacteria have a single circular DNA chromosome that is found within the cytoplasm of the cell as they do not have a nucleus. Indeed they lack any of the intracellular organelles so characteristic of eukaryotic cells, such that they do not have the Golgi apparatus, endoplasmic reticulum, lysosomes nor mitochondria. However, they are generally capable of 'free-living' and therefore possess all the necessary biosynthetic machinery including 70S ribosomes (as opposed to the larger 80S forms found in eukaryotes) distributed throughout the cytoplasm (Table 1.6).

Table 1.6 Major differences in genome organization between prokaryotes and eukaryotes

Characteristics of genome organization	*Prokaryotes*	*Eukaryotes*
Genome size	600 kb-9.5 Mb	3 Mb-14000 Mb
Average gene size/number	950 bp/4300	2500 bp/19000
Operon-like regulatory unit	General	No
Horizontal gene transfer	Significant	Negligible
Rate of non-coding sequences	Low	High
Introns	Rare	General
Redundant gene number	Rare	General
Ploidy level	Haploidy	Haploidy-Polyploidy
Chromosome number	One	More than one
Heterozygosity	No	Yes

The most complex region of the cell is often the cell surface. The cell wall/outer membrane, but in addition some bacteria may secrete a polysaccharide capsule onto their outer surface, some may have flagella, which they require for mobility and some may have external projections such as fimbriae and pili which are useful for adherence in their chosen habitat. Although bacteria are generally far simpler than eukaryotic cells, they are extremely efficient within their own little niche, and this may include the ability to cause human infections. Bacteria multiply by binary fission and there is no sexual interaction.

1.4.2 Rickettsiae

The rickettsiae are named after H.T. Ricketts, who discovered American 'Rocky Mountain spotted fever'. This group is also referred to as the 'spotted' fever (typhus fever) group after the disease produced by the best known of these pathogens (*Rickettsia prowazekii*). Their natural distribution is by host vectors such as lice, fleas, ticks, mites, etc., in which they exist as harmless parasites or even symbionts. On being transferred to other animal hosts or humans by bites, scratches, or inhalation, they produce very serious disease symptoms. Although the size of rickettsiae is of the same order of magnitude as pox viruses, they can be differentiated quite unequivocally from any viruses. They contain DNA as well as RNA in a ratio of 1:3.5. The cells are surrounded by a cell wall that contains muramic acid and is lysozyme sensitive. Nuclear regions and cell walls can also be distinguished in electron micrographs of thin section.

Many rickettsiae have never been cultured except in living cells, but they can be grown in incubating chicken eggs or in experimental animals; about 10^9 cells can be produced in the yolk sac of a chicken egg. A few enzymes of intermediary metabolism can be demonstrated in isolated cells, but the metabolic activity of the cells declines on storage; respiration can be restored by addition of ATP, organic acids and amino acids. It appears that the rickettsiae are able to develop their own metabolism, but they are apparently unable to control uptake and excretion of metabolites because of the permeability characteristics of their cell envelope.

The best known pathogens among the rickettsiae are those of the typhus group *R. prowazekii* is the causal agent of typhus fever. The reservoir is humans, and the organism is transmitted by lice (head lice and clothes lice). The lice are rapidly (in a few days) killed by the organism, and infection is via their faeces. *R. typhi* is the causal organism of endemic and murine typhus, which produces a similar but less severe syndrome. The organism is spread by rats which remain symptomless and is transmitted from rat to rat to humans by fleas.

Whilst most rickettsiae are relatively sensitive to heat and dehydration, *Coxiella burnetii*, which causes Q-fever, can survive outside the host. It is transmitted by ticks to sheep, goats and cattle and can infect humans by tick bites and also via animal dust, infected soil, and consumption of milk. The usual pasteurization of milk (heating at 60°C for 30 min) does not kill *Coxiella*, thus 72.5°C for 15 seconds in now used.

1.4.3 Chlamydiae

Chlamydiae are human pathogens. *Chlamydiae trachomatis* causes trachoma, the Egyptian eye disease that starts as conjunctivitis and leads to blindness, and the venereal disease lymphogranuloma venereum. In both cases transmission is by contact. *C. psittaci* is the causative agent of ornithoses, the best known of which is psittacosis, a feverish pneumonia. The main hosts of chlamydiae are birds.

On the basis of their biochemical characteristics, chlamydiae belong to the prokaryotes (bacteria) and not, as formerly thought, to the viruses. They contain RNA and DNA in a ratio of characteristic of bacteria, and they synthesize substances which eukaryotic cells are unable to produce, such as muramic acid, diaminopimelic acid, D-alanine, and folic acid. These properties are in accord with their sensitivity to penicillin and sulphonamides. Their genome is very small (relative size, 0.66×10^9) and corresponds to only about one quarter of the genetic information of *E. coli*. Chlamydiae grow only in living cells and are cultured in chicken eggs and tissue cultures. Their dependence on the metabolism of host cells is apparently due to absence of an ATP-generating system of their own. They are unable to phosphorylate glucose or to metabolise it. On the other hand, they are extraordinarily permeable to ATP and CoA. They can, therefore, be regarded as 'energy parasites'. The obligate cell parasites among the bacteria are apparently the result of a regressive development. The adaptation to their host cells has been accompanied by loss of various synthetic capacities.

1.4.4 Mycoplasma Group

The members of the Mycoplasma group (class **Mollecutes**) are the smallest independently replicating prokaryotes. They do not have any cell walls. Since the protoplasts are surrounded only by a cytoplasmic membrane, they are osmotically extremely labile. The genome size of several mycoplasmas (*Mycoplasma* and *Ureaplasma*) corresponds to only a fraction of that of *E. coli*. With a chromosome size of 500-900 kb, they have the smallest genome of any prokaryotes capable of autoreplication. The order **Mycoplasmatales** has therefore been elevated to a special class of bacteria with the name **Mollicutes** (soft skinned) to express the phylogenetic differentiation of this group from all other bacteria (Figure 1.5).

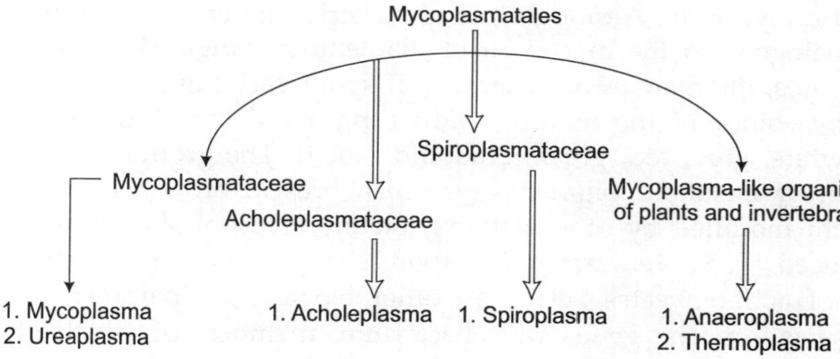

Figure 1.5 Biology of Mycoplasma.

The first representative of the mycoplasmas described was the causal organism of pleuropneumonia, a lung infection of cattle. This organism grows on agar media containing serum in small colonies that have a 'fried egg' appearance. Similar growth forms have therefore been designated as pleuropneumonia-like organisms (PPLO). Mycoplasmas have been recognized as causative organisms for a number of diseases, as contaminants in tissue cultures, and as harmless saprophytes. The colonies consist of cells and fragments of various sizes and can be described as cocci, filaments, discs and rosettes. They multiply by binary fission, by fragmentation of filaments and rings into coccoid cells, and by a kind of sprouting. In liquid media they are often found as very irregular and sometimes branched forms that can pass through membrane filters, rather like viruses.

1.4.5 Spirochetes

Spirochetes are not a large group, there are only six genera, having tremendous impact on our lives. Both syphilis and Lyme disease are caused by these bacteria, and other species are important symbionts in the stomachs of cows and other ruminants. The main spirochete cluster contains the genera *Spirochaeta, Borrelia* and *Treponema.*

These organisms may be free-living or host associated and many are well known pathogens belonging to Eubacteria. The genus *Leptospira* forms a distinct branch within the spirochete group. Spirochetes have traditionally been grouped on the basis of their unique morphology and mode of motility, and these traits turn out to be valid phylogenetic indicators as well.

1.4.6 Actinomycetes

The actinomycetes are a large group of filamentous bacteria (earlier recognized as filamentous fungi) sometime referred as Actinobacteria. Phylogenetically, the filamentous actinomycetes form a coherent group; thus the mycelia spore-forming habit is of both phylogenetic as well as taxonomic importance (Figure 1.6). Most members of the group are soil inhabitants, aerobic to anaerobic mostly saprophytic but few members are human pathogenic. The members of this group have got tremendous medical, agricultural and industrial importance. They are usually Gram-positive and form branching filaments. As a result of successful growth and branching, a ramifying network of filaments is formed, called a mycelium. Although it is of bacterial dimensions, the mycelium is in some ways analogous to the mycelium of filamentous fungi. Most members of this group form spores; the manner of formation of spore varies and is used in separating subgroups. The colour of the mature sporulating aerial mycelium is recorded in a simple way (white, grey, red, green, blue and violet). The grouping is made also on the production of melanoid pigments (greenish brown, brownish black or distinct brown, pigment modified by other colours) on the medium shown as produced (+) and not produced (-). Spore surface and spore chain is another important feature in identification of actinomycetes beside all other biochemical parameters followed in bacterial classification. The genus Mycobacterium, members of which often show a tendency to form branches, is also placed in the actinomycetes by some taxonomists. DNA base composition of all members of the actinomycete falls within a relatively narrow range of 63 to 78% GC. Organisms at the upper end of this range have the highest GC percentage of any bacteria known.

The preservation methods are similar to that of bacteria such as subculturing, freezing especially in liquid nitrogen, freeze-drying and maintenance of strains in mineral oil.

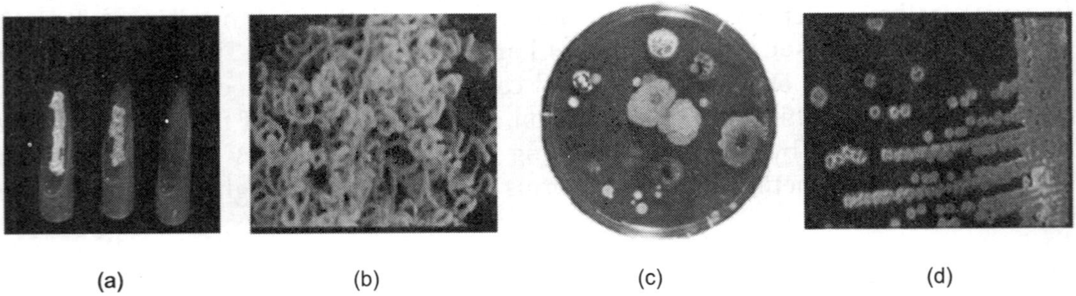

Figure 1.6 Member of Actinomycetes (a) Pigment production in agar slants (b) Electron micrograph view *Streptomyces* sp. (c) Members of Actinomycetes growing on Petri plates (d) *Streptomyces griseus* mycelia and conidiospores.

1.4.7 Cyanobacteria

Cyanobacteria are aquatic and photosynthetic, that is, they live in water, and can manufacture their own food. As the suffix "bacteria", they are quite small and usually unicellular, though they often grow in colonies which can be seen by naked eye. They have the distinction of being the oldest known fossils, more than 3.5 billion years old; they are one of the largest and most important groups of bacteria on earth. Many Proterozoic oil deposits are attributed to the activity of cyanobacteria. They are also important providers of nitrogen fertilizer in the cultivation of rice and beans. The cyanobacteria have also been tremendously important in shaping the course of evolution and ecological change throughout earth's history. The oxygen atmosphere that we depend on was generated by numerous cyanobacteria during the Archaean and Proterozoic eras. Before that time, the atmosphere had a very different chemistry, unsuitable for life as we know it today (Figure 1.7).

The other great contribution of the cyanobacteria is the origin of plants. The chloroplast with which plants make food for themselves is actually a cyanobacterium living within the plant's cells. Sometime in the late Proterozoic, or in the early Cambrian, cyanobacteria began to take up residence within certain eukaryote cells, making food for the eukaryote host in return for a home. This event is known as endosymbiosis, and is also the origin of the eukaryotic mitochondrion.

Because of their photosynthetic and aquatic behaviour, cyanobacteria are often called "blue-green algae". Cyanobacteria resemble the bacteria, more than eukaryotic alga, and it is only the chloroplast in eukaryote which relates them. Cyanobacteria include unicellular and colonial species. Colonies may form filaments, sheets or even hollow balls. Some filamentous colonies show the ability to differentiate into several different cell types: vegetative cells, the normal, photosynthetic cells that are formed under favourable growing conditions; akinetes, the climate-resistant spores that may form when environmental conditions become harsh; and thick-walled heterocysts, which contain the enzyme nitrogenase, vital for nitrogen fixation. Heterocysts may also form under the appropriate environmental conditions (anoxic) when fixed nitrogen is scarce. Heterocyst-forming species are specialized for nitrogen fixation and are able to fix nitrogen gas into ammonia (NH_3), nitrites (NO_2) or nitrates (NO_3) which can be absorbed by plants and converted to protein and nucleic acids (atmospheric nitrogen cannot be used by plants directly).

The rice paddies of Asia, which produce about 75% of the world's rice could not do so well were it not for healthy populations of nitrogen-fixing cyanobacteria in the rice paddy fertilizer. Many cyanobacteria also form motile filaments, called hormogonia, that travel away from the main biomass to bud and form new colonies elsewhere. The cells in a hormogonium are often thinner than in the vegetative state, and the cells on either end of the motile chain may be tapered. In order to break away from the parent colony, a hormogonium often must tear apart a weaker cell in a filament, called a necridium.

Each individual cell of a cyanobacterium typically has a thick, gelatinous cell wall. They differ from other Gram-negative bacteria in that the quorum sensing molecules

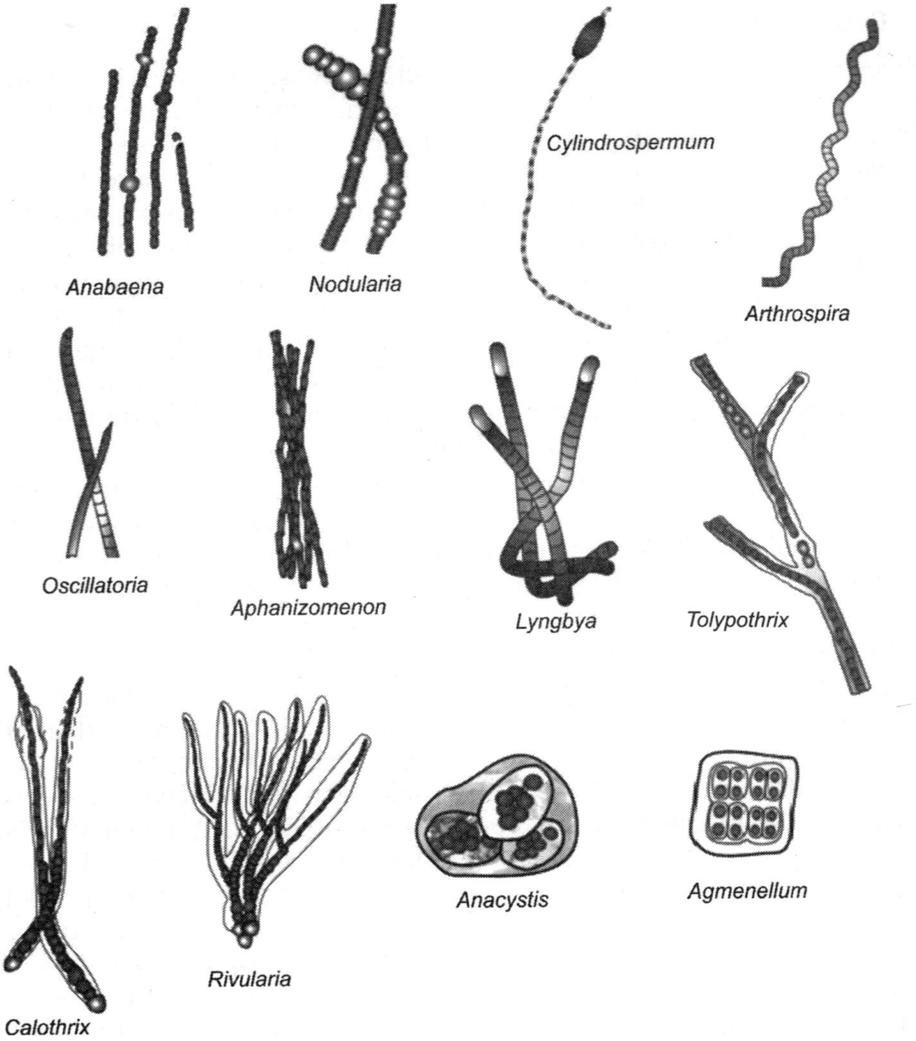

Figure 1.7 Microscopic view of some common members of cyanobacteria.

autoinducer-2 and acyl-homoserine lactones are absent. They lack flagella, but possess hormogonia and some species may move about by gliding along surfaces. Many of the multicellular filamentous forms of *Oscillatoria* are capable of a waving motion; the filament oscillates back and forth. In water columns some cyanobacteria float by forming gas vesicles, like in archaea. These vesicles are not organelles as such as they are not bounded by lipid membranes but by a protein sheath. Some of these organisms contribute significantly to global ecology and the oxygen cycle. The tiny marine cyanobacterium *Prochlorococcus* was discovered in 1986 and accounts for more than half of the photosynthesis of the open ocean. Many cyanobacteria even display the circadian rhythms that were once thought to exist only in eukaryotic cells.

1.5 THE ARCHAEA

Most Archaea live in extreme environments called extremophiles. Other Archaea species may live in ordinary temperatures and salinities. Some even live in your guts. Some extremophile species are heat loving; they prefer to live in boiling water, like the geysers, and inside volcanoes hence earned the nickname "thermophile", which means " heat loving", and it would probably freeze to death at ordinary room temperature. Other extremophile Archaea prefer to live in very salty, called hypersaline, environments. They are able to survive in these extreme places where other organisms cannot. These salt-loving Archaea are called halophiles (Figure 1.8).

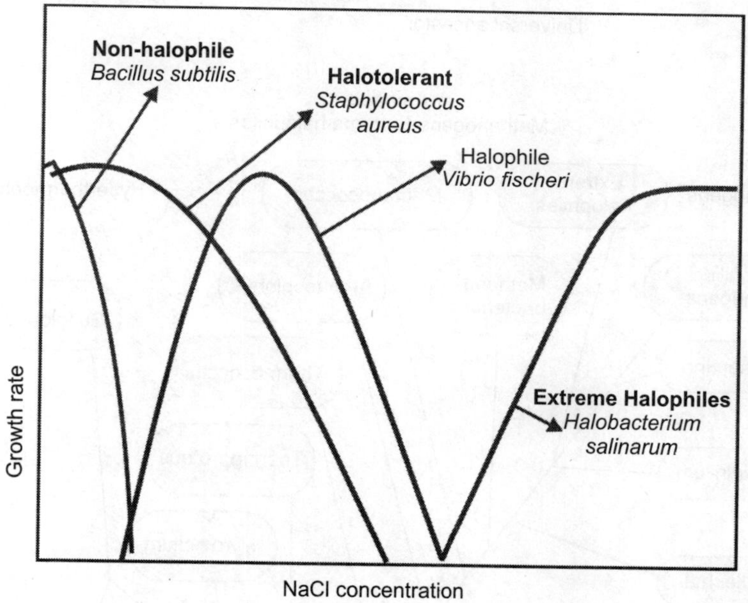

Figure 1.8 Growth pattern of different salt loving bacteria.

Archaea was originally thought to be just like bacteria, but Archaea is a much different and simpler form of life. It is also considered the oldest form of life on earth. Archaea requires neither sunlight for photosynthesis as do plants nor oxygen. Archaea absorbs CO_2, N_2, or H_2S and gives off methane gas as a waste product the same way humans breathe in oxygen and breathe out carbon dioxide. Planets which contain an environment wherein archaea might survive include Venus, the past environment of Mars, Jupiter, Saturn, and Jupiter's moon. The relationship between archaea and eukaryotes remains problematic. Aside from the similarities in cell structure and function, many genetic trees group the two as shown in Figure 1.9, whereas, Figure 1.10 shows the morphological diversity in microbes.

Complicating factors include claims that the relationship between Eukaryotes and the archaeal phylum Euryarchaeota is closer than the relationship between the Euryarchaeota and the phylum Crenarchaeota and the presence of archaean-like genes in certain bacteria, such as *Thermotoga maritima,* from horizontal gene transfer. The leading

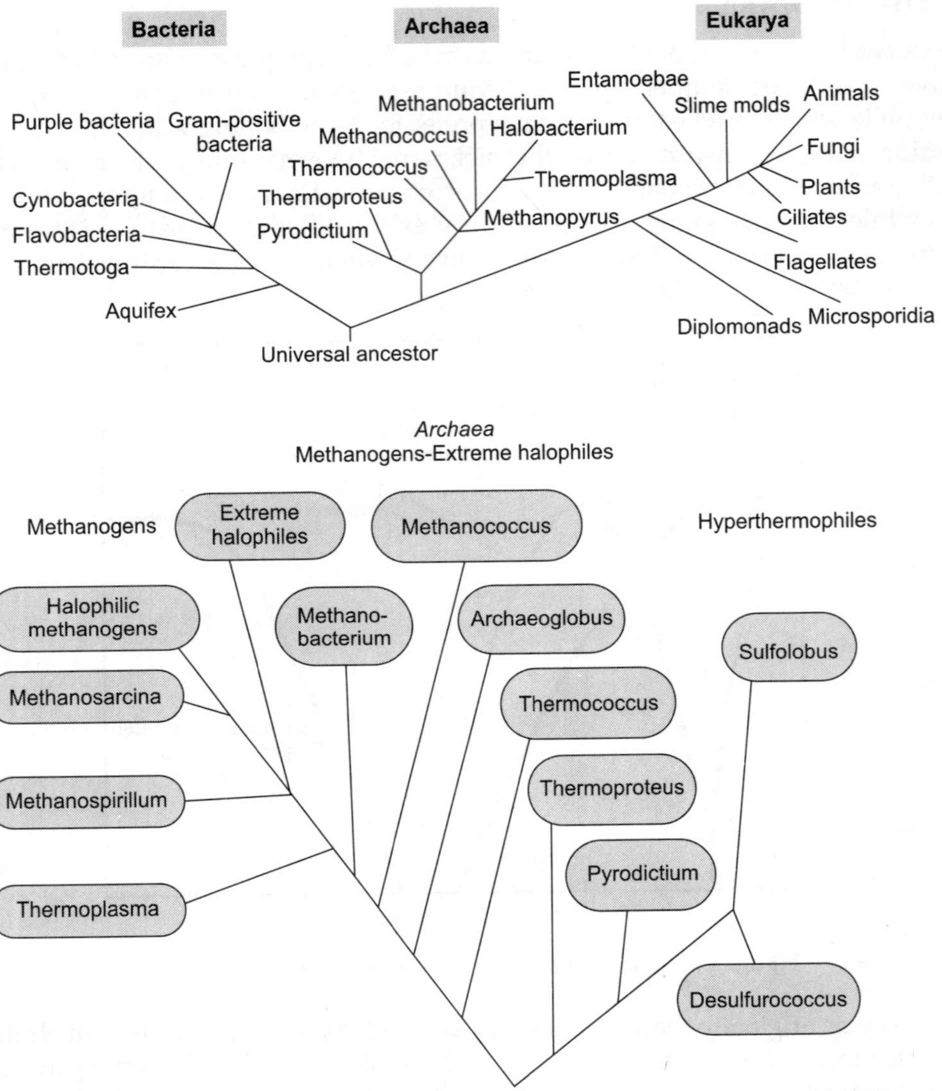

Figure 1.9 Phylogenetic trees representing major groups in the Archaea commonly found in extreme enviroments: methanogens, halophiles, and thermaphiles (Ogunseitan 2005).

hypothesis is that the ancestor of the eukaryotes diverged early from the Archaea, and that Eukaryotes arose through fusion of an archaean and eubacterium, which became the nucleus and cytoplasm; this accounts for various genetic similarities but runs into difficulties explaining cell structure.

1.6 EUKARYA

The Domain Eukarya arose from the first prokaryotic organisms more than 1.7 biliion years ago. It includes all of the organisms with eukaryotic cells--that is, those with membranous organelles (including mitochondria and chloroplasts) (Figure 1.11). The organisms in this domain represent the vast majority of organisms we see each day. In terms of classification, members of the domain have the following characteristics:

- Eukaryotic cells
- Unicellular (some Protists and yeasts), colonial (some Protists) or multicellular (most Fungi, Plantae, and Animalia) organisms
- Cell division by mitosis, with a variety of cycles of reproduction and recombination
- A variety of modes of access to energy and carbon, varying with kingdom.

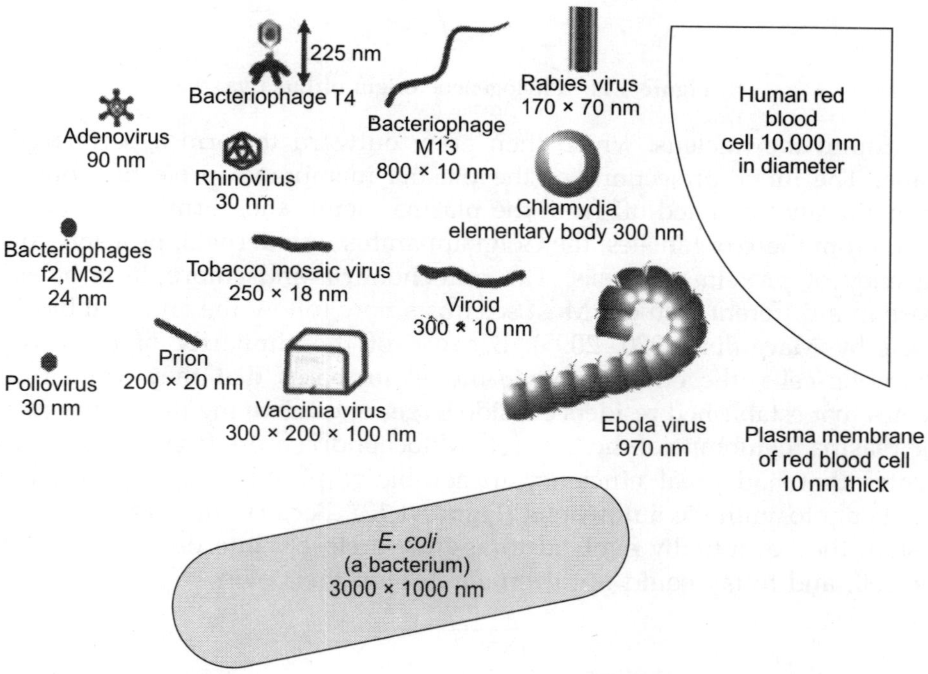

Figure 1.10 Morphological diversity in microbes.

All of the membranous organelles except for the mitochondria and chloroplasts are ultimately continuous with the plasma membrane and with each other. They comprise the endomembrane system, and are highly dependent on each other for the partitioning of the cell for metabolic activity. The organelles of the endomembrane system include the nucleus (surrounded by a double membrane), rough and smooth endoplasmic reticulum, Golgi apparatus, lysosomes, peroxisomes, vacuoles, and transport vesicles that move between the components.

The eukaryotic endomembrane system arose by a process known as membrane infolding. The plasma membranes of the original prokaryotic cells that gave rise to the first eukaryote ones folded inward. Originally, this membrane may have formed small

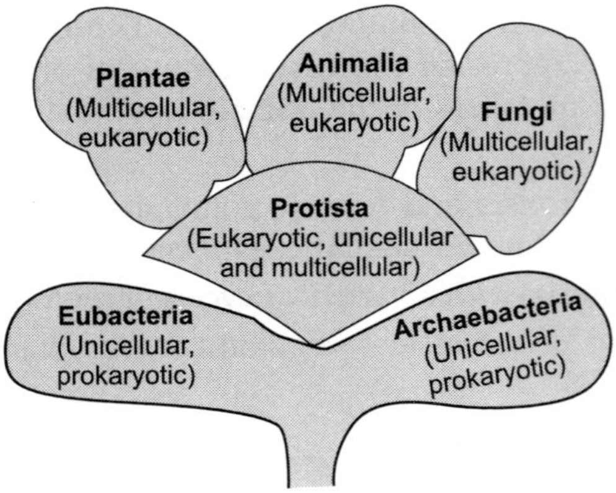

Figure 1.11 Phylogenetic origin of Eukarya.

vesicles around the nucleus, which then grew outward to form a perforated double membrane. The inner projections of the plasma membrane (projecting out from the nuclear membrane) pinched off from the plasma membrane, forming the endoplasmic reticulum. From these organelles, the Golgi apparatus was derived, probably to improve the efficiency of protein synthesis. The mitochondria and chloroplasts of eukaryotic cells arose in a different fashion. Most scientists now follow the endosymbiosis model, developed by Margulis (1970, 2005). Because of the similarity of these organelles to prokaryotic cells, the endosymbiotic model proposes that they arose from small prokaryotes that established residence inside larger ones, deriving their outer membranes from the plasma membrane of the host cell. Mitochondria arose from small heterotrophic prokaryotes that had great efficiency in aerobic respiration, and chloroplasts arose from small photosynthetic autotrophs (Figure 1.12). Because the organelles could not be digested, they eventually evolved to assume a closely interdependent relationship with the cell, and today could not even survive on their own.

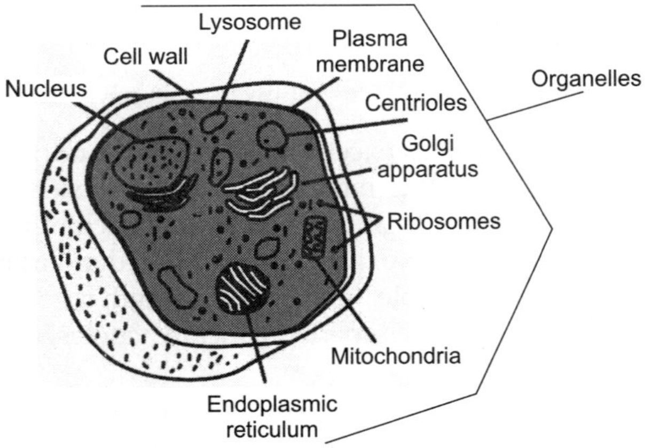

Figure 1.12 Microscopic view of some common members of fungi.

1.6.1 The Fungi

The kingdom Fungi encompasses an extraordinary diversity of organisms: bread molds, yeasts, powdery mildews, cup and sponge fungi, smuts, rusts, puffballs, and mushrooms. Some are invisible to the naked eye. Others grow to over two feet in diameter. Whatever their differences, however, all fungi have certain important properties in common. They are eukaryotic, reproduce by means of sexual and asexual reproduction, and grow as hyphae exhibiting apical growth or as yeasts. They are heterotrophic and do not perform photosynthesis. Most fungi are saprophytes or symbionts, but some are parasites of plants, man and other animals. They absorb nutrients through their cell membranes known as phagotrophy and ingestion of solid food particles is a very rare property among the fungi. To utilize particulate or high molecular weight substrates, the fungi secrete various degradative, extracellular enzymes. They generally have rigid, polysaccharide-rich cell walls.

They are cosmopolitan and most of them are inconspicuous because of their smaller structure, and their cryptic lifestyles in soil, on dead matter, and as symbionts of plants, animals, or other fungi. They may become noticeable when fruiting, either as mushrooms or molds (Figure 1.13). Fungi perform an essential role in the decomposition of organic matter and have fundamental roles in nutrient cycling and exchange. They have long been used as a direct source of food, such as mushrooms and truffles, as a leavening agent for bread, and in fermentation of various food products, such as wine, beer, and soy sauce. Fungi can survive the intense UV and cosmic radiation encountered during space travel.

Classification of fungi is based on morphological characteristics and mode of reproduction. The biological species concept discriminates species based on their ability to mate. The application of molecular tools, such as DNA sequencing and phylogenetic analysis, to study diversity has greatly enhanced the resolution and added robustness to estimates of genetic diversity within various taxonomic groups. Phylogenetic analyses based upon 18S rDNA and protein sequences indicate that the fungi are more closely related to animals than to plants or to the algae. An eleborated description of fungi has been discussed separately in chapter 11.

1.6.2 Algae

The subkingdom Algae include a large and diverse assemblage of all the photosynthetic eukaryotic organisms in the protista that contain *chlorophyll a* which is necessary to carry out oxygenic photosynthesis. Algae are organisms that use light energy to convert CO_2 and H_2O to carbohydrates and other cellular products with the release of oxygen. Algae should not be confused with cyanobacteria, which are also oxygenic phototrophs, but which are eubacteria and thus evolutionarily quite distinct from algae. The algae may be unicellular, colonial, or filamentous and most of them are microscopic hence are clearly microorganisms, a number of forms are macroscopic, some seaweed growing to over several feet in length (Figure 1.14). The undifferentiated algal structure is often referred to as thallus. They lack the stem, root, leaf structure that result from tissue specialization. They contain chlorophyll in well organized membranous chloroplast and

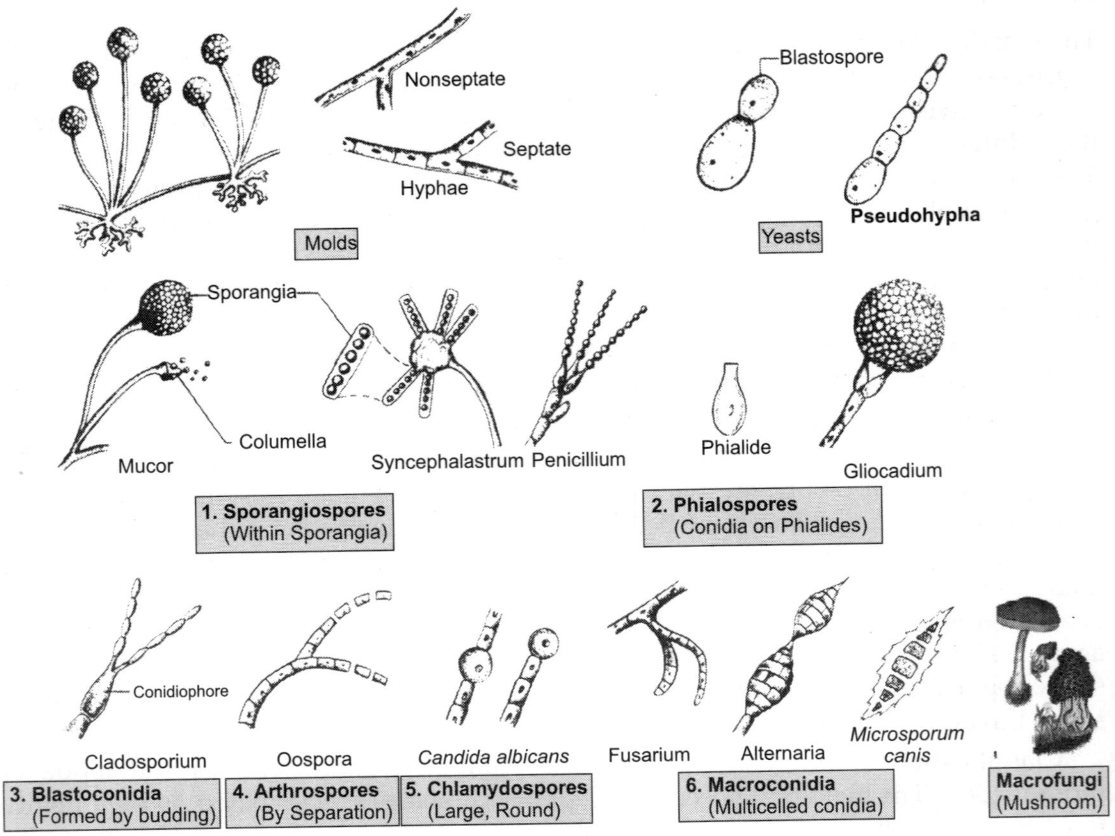

Figure 1.13 Microscopic view of some common members of fungi.

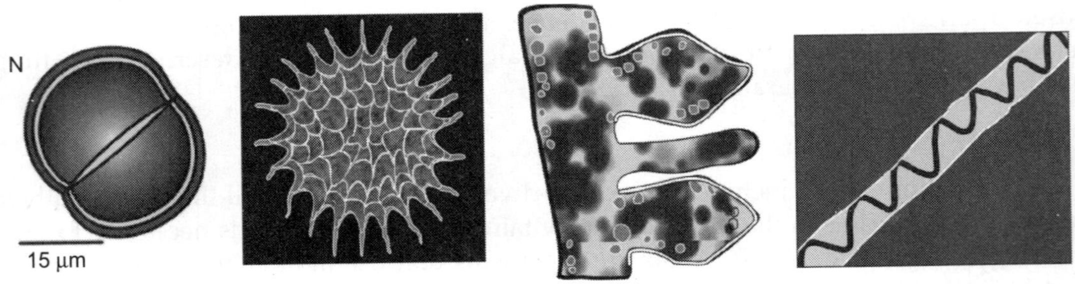

Figure 1.14 Schematic diagram of different forms of algae from unicellur to filamentous and microscopic.

visible nuclei, and other membranous structures (Figure 1.15). Photosynthesis takes place within these bodies. The size, shape, distribution and number of chloroplasts vary considerably from species to species. In some instances a single chloroplast may occupy most of the cell space.

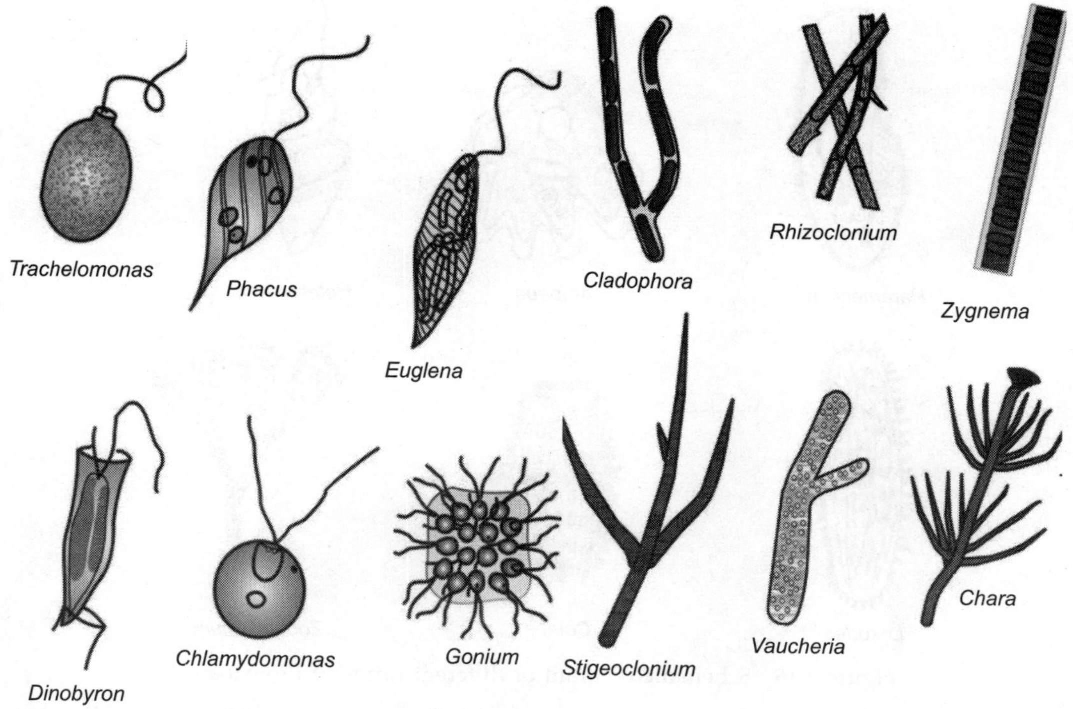

Figure 1.15 Microscopic view of some common members of algae.

However, a few kinds of common algae are not green but appear brown or red because in addition to chlorophyll, other pigments such as phycocyanin, xanthocyanin, phycoerythrin, and carotenoids, etc., are present which mask the green colour and extend the range of light waves that can be used by these organisms for photosynthesis. In Whittakers five kingdom system, some of the algae have been included with protozoan.

Algae, however, lack a well organised vascular system. Algae do not directly infect humans but some produce toxins that cause paralytic shell fish poisoning. Some of these toxins do not cause illness in the shellfish that feed on the algae but accumulate in their tissues and when eaten by man cause nerve damage. As one of the primary producers of carbohydrate and other cellular products, the algae are essential in the food chains. In addition, they produce a large proportion of the oxygen in the atmosphere.

1.6.3 Protozoa

The protozoa are one-celled animals and the smallest of all animals. Most of them can only be seen under a microscope (Figure 1.16). They do breathe, move, and reproduce like multicelled animals. They live in water or at least where it is damp. Animals in this group include the *Paramecium*, the *Euglena* and the *Amoeba*. Nearly 200,000 Protozoans are known as per Global Biodiversity Estimate for the UN.

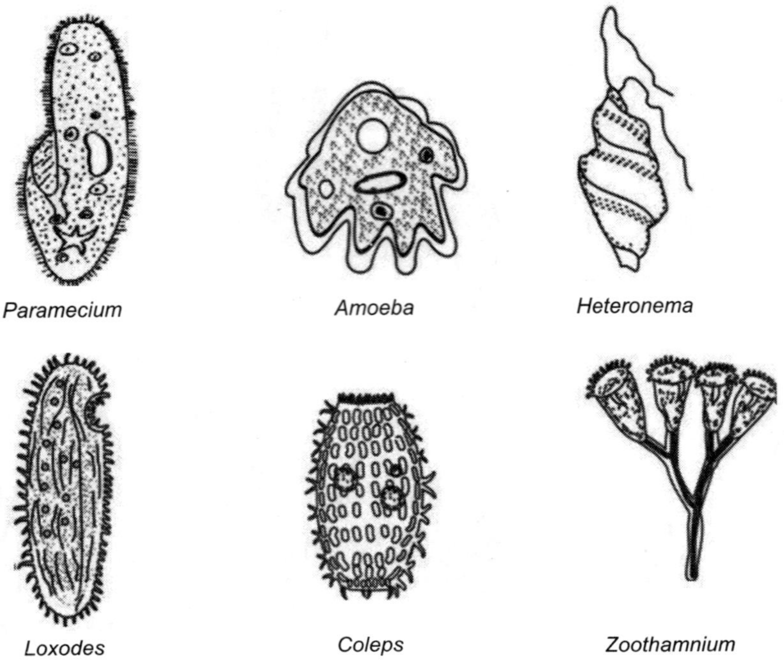

Figure 1.16 Schematic diagram of different forms of Protozoans.

They are cosmopolitan in distribution and include aquatic (fresh water and marine) forms too. Phylum protozoa includes free living animals like *Euglena, Paramecium, Amoeba, Noctiluca* and *Elphidium* and parasitic animals like *Plasmodium* species, *Monocystic, Entamoeba, Trypanosoma* and *Giardia*. The organisms of this phylum greatly vary in shape, size, locomotory organelles, and method of reproduction.

The subkingdom protozoa include single celled organisms; however, some of them do form colonial aggregates. Externally, the cells are covered with cell membranes, or a pellicle; cell walls are absent, and distinct nuclei with nuclear membranes are present. Specialized organelles such as contractile vacuoles, cytostomes, mitochondria, ribosomes, flagella and cilia may also be present (Figure 1.17).

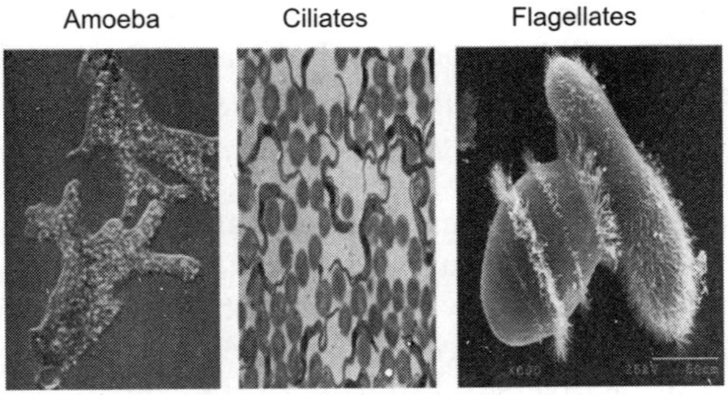

Figure 1.17 Different forms of Protozoans.

Some Protozoans are harmful to man as they can cause serious diseases. Others are helpful because they eat harmful bacteria and are food for fish and other animals. As components of the micro- and meiofauna, protozoa are an important food source for micro-invertebrates. Thus, the ecological role of protozoa in the transfer of bacterial and algal production to successive trophic levels is important. As predators, they prey upon unicellular or filamentous algae, bacteria, and microfungi. Protozoa are both herbivores and consumers in the decomposer link of the food chain. They also control bacteria populations and biomass to some extent. Protozoa such as the malaria parasites (*Plasmodium* spp.), trypanosomes and *Leishmania* are also important as parasites and symbionts of multicellular animals. All the protozoa produce cysts, which are resistant dormant stages that enable them to survive drought, heat and freezing. They produce asexually by cell division and exhibit various degrees of sexual reproduction. The subkingdom Protozoa is divided into three phyla: Sarcomastigophora, Ciliophora and Apicomplexa (Chapter 11). Type of locomotion plays an important role in classifiying them. Protozoa can reproduce by binary fission or multiple fission. Some protozoa reproduce sexually, some asexually, while some use a combination, (e.g. Coccidia). An individual protozoon is hermaphroditic.

1.6.4 Nematodes

The members of this phylum are the roundworms. They are commonly known as Nemas or nematodes. They are characteristically round in cross section, have an external cuticle without cilia, lack eyes, and have a tubular digestive system complete with the mouth, intestine and anus. The males are generally much smaller than the females and have a hooked posterior end. The number of the named species is only a fraction of the total nematodes in existence. Species identification of these invertebrates requires very detailed study of the many minute anatomical features, which requires complete knowledge of anatomy.

1.6.5 Microscopic Invertebrates

They are large, transparent, complex multicellular moving "monsters" microorganisms or rotifers with organ systems. If organ systems are present, then the organisms cannot be protists, because organs indicate the presence of tissue differentiation. Collectively, these microscopic forms are designated as "invertebrates." A few invertebrates, such as *Dugesia* and *Hydra,* are macroscopic in adult form but microscopic when immature. Be sure to judge size differences by reading the scale of each organism. The phyla are listed according to the degree of complexity, the simplest first (Figure 1.18).

Many tissues from multicellular invertebrates do not require supplemental oxygen or complex media, unlike vertebrate tissues. Their central nervous systems are not as well-developed as those of vertebrates; therefore invertebrates are probably not self-aware.

Studies on the phylogeny of the oligochaete bacteria have revealed a remarkable diversity of the symbionts, with up to 5 - 6 different 16S rRNA phylotypes co-occurring in the symbiont-containing region. In all host species, the primary endosymbionts are

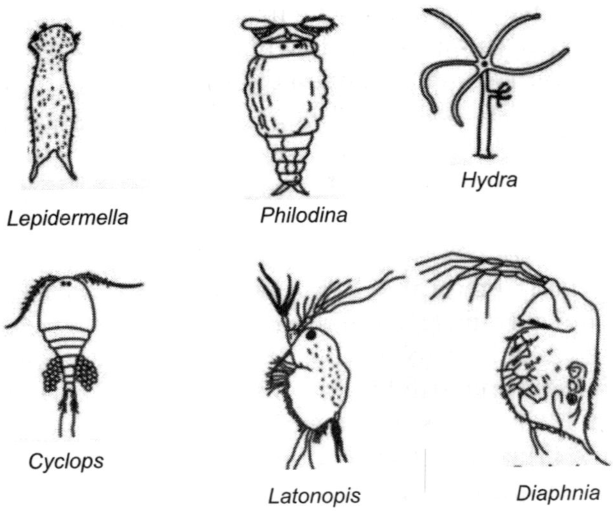

Figure 1.18 Microscopic invertebrates.

chemoautotrophic sulfide-oxidizers that belong to the gamma subclass of Proteobacteria and are related to free-living sulfide-oxidizers such as *Allochromatium vinosum*. In addition to the primary symbionts, up to 5 other bacterial phylotypes belonging to the Alphaproteobacteria (Blazejak *et al.*, 2005), Deltaproteobacteria, or the Spirochaetes can co-exist in the same host species (Dubilier *et al.*, 2006). In the case of chloroplasts and mitochondria, examples of recent and ancient gene transfer to the chromosomes of their host cell illustrate the process of genetic merger in the wake of organelle origins. Symbiotic microorganisms, both beneficial and potentially pathogenic reside in the gastrointestinal tract of humans. Imbalances in the bacterial composition, known as dysbiosis, are thought to be a major determinant in inflammatory bowel diseases such as Crohn's disease in humans. It has long been known that intestinal microorganisms are important for the development of intestinal tissues. Studies have demonstrated that a human symbiotic microorganism called *Bacteroides fragilis* protects animals from experimental colitis induced by *Helicobacter hepaticus*. Polysaccharide A (PSA) from this bacteria single-handedly protects humans from the inflammatory bowel diseases.

Microbial symbioses are interesting in their own right and also serve as exemplary models to help biologists to understand two important symbioses in the evolutionary past of eukaryotic cells: the origins of chloroplasts and mitochondria. Most, if not all, microbial symbioses have a chemical basis: compounds produced by one partner are useful for the other. But symbioses can also entail the transfer of genes from one partner to the other, which in some cases cements two cells into a bipartite, co-evolving unit. Here, we discuss some microbial symbioses in which progress is being made in uncovering the nature of symbiotic interactions: anaerobic methane-oxidizing consortia Archaea in red, Sulfate Reducing Bacteria in green. Microscopic image shows a consortium after fluorescent in situ hybridization (CARD-FISH). Marine worms that possess endosymbionts instead of a digestive tract, amino acid producing endosymbionts of aphids, prokaryotic endosymbionts living within a prokaryotic host within mealybugs, endosymbionts of an insect vector of human disease and a photosynthetic sea slug that steals chloroplasts from algae (Figure 1.19).

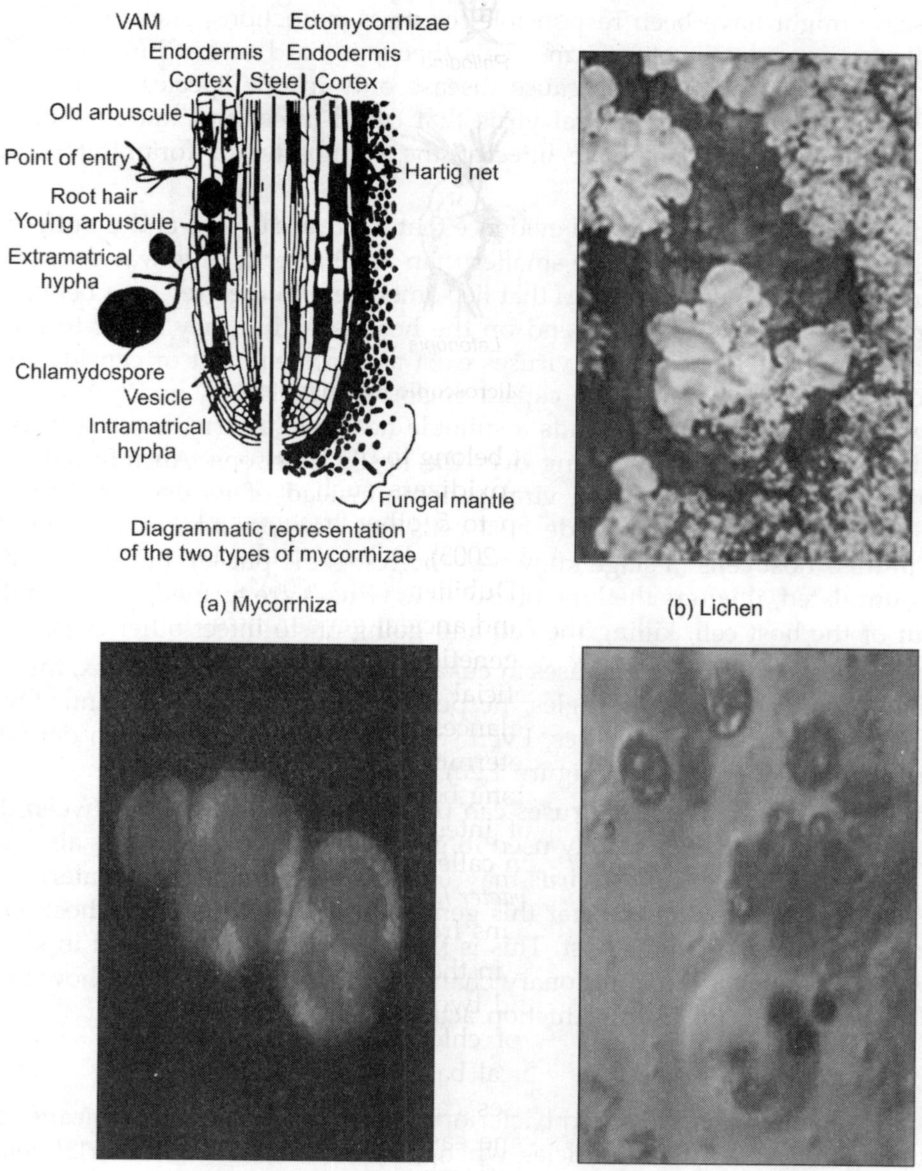

(a) Mycorrhiza (b) Lichen

(c) Archaea (d) Purple sulphur bacteria

Figure 1.19 Microbial interrelationship.

1.7 VIRUSES

Viruses themselves do not have any fossil record, but it is quite possible that they have left traces in the history of life. It has been hypothesized that viruses may be responsible for some of the extinctions seen in the fossil record, and the outbreaks of viral disease might have been responsible for mass extinctions, such as the extinction of the dinosaurs and other life forms. This theory is hard to test but seems unlikely, since a given virus can typically cause disease only in one species or in a group of related species. Even a hypothetical virus that could infect and kill all dinosaurs, 65 million years ago, could not have infected the ammonites or foraminifera that also went extinct at the same time.

Loeffler and Frosch (1898) found evidence that the cause of foot-and-mouth disease in livestock was an infectious particle smaller than any bacteria. This was the first clue to the nature of viruses, genetic entities that lie somewhere in the grey area between living and non-living states. Viruses depend on the host cells that they infect to reproduce. When found outside of host cells, viruses exist as a protein coat or capsid, sometimes enclosed within a membrane. The capsid encloses either DNA or RNA which codes for the virus elements. When it finds a suitable host cell, a virus can insert its genetic material into its host, literally taking over the host's functions. An infected cell starts producing more viral protein and viral genome instead of its usual products. Some viruses may remain dormant inside host cells for long periods, causing no obvious change in their host cells (a stage known as the lysogenic phase). But when a dormant virus is stimulated, it enters the lytic phase: new viruses are formed, self-assemble, and burst out of the host cell, killing the cell and going on to infect other cells.

Viruses cause a number of diseases in eukaryotes. In humans, smallpox, the common cold, chickenpox, influenza, shingles, herpes, polio, rabies, Ebola, hanta fever, and AIDS are examples of viral diseases. Even some types of cancer though definitely not all have been linked to viruses (Figure 1.20).

On the other hand, because viruses can transfer genetic material between different species of host, they are extensively used in genetic engineering. Viruses also carry out natural "genetic engineering": a virus may incorporate some genetic material from its host as it is replicating, and transfer this genetic information to a new host, even to a host unrelated to the previous host. This is known as transduction, and in some cases it may serve as a means of evolutionary change although it is not clear how important an evolutionary mechanism transduction actually is.

1.7.1 Bacterial Viruses

Viruses infecting bacteria are called bacteriophage or just phage (*phage* means "to eat"). Bacteriophage exists as inert particles when they are outside of bacterial cells. They possess complex protein coats with defined structure and may also have tail structures (Figure 1.21). The protein coat, or capsid, surrounds the deoxyribonucleic acid (DNA) or ribonucleic acid (RNA) molecules that make up the bacteriophage genome. Phage genomes can be single stranded or double stranded, and are either circular or linear. Different bacteriophages can encode as few as four proteins or as many as one hundred in their genome.

Families of Viruses that Affect Humans

Characteristics/ Dimensions	Viral Family	Important genera	Clinical of special features
Single stranded DNA nonenveloped 18-25 nm	Parvoviridae	Human parvovirus B19	Fifth disease; anaemia in immunocompromised patients
Double-stranded DNA nonenveloped 70-90 nm	Adenoviridae	Mastadenovirus	Medium-sized viruses that cause various respiratory infections in humans; some cause humans in animals
40-57 nm	Papovaviridae	*Papillomavirus* (human wort virus) *Polyomavirus*	Small viruses that induce tumours; the human wort virus (papilloma) and certain viruses that produce cancer in animals (polyoma and simian) belong to this family.
Double-stranded DNA enveloped 200-350 nm	Poxviridae	*Orthopoxvirus* (vaccinia and smallpox viruses) Molluscipoxvirus	Very large, complex, brick-shaped viruses that cause diseases such as smallpox (variola), molluscum contagiosum (wortlike skin lesion), and cowpox.
150-200 nm	Herpesviridae	*Simplexvirus* (HHV-1 and 2) *Varicellovirus* (HHV-3) *Lymphocrytovirus* (HHV-4) *Cytomegalovirus* (HHV-5) *Roseolovirus* (HHV-6) HHV-7 *Kaposi's sarcona* (HHV-8)	Medium-sized viruses that cause various human diseases, such as fever blisters, chickenpox, shingles, and infectious mononucleosis; implicated in a type of human cancer called Burkit's lymphoma.
42 nm	Hepadnoviridae	Hepachavirus (hepatitis B virus)	After protein synthesis, hepatitis B virus uses reverse transcriptase to produce its DNA from mRNA; causes hepatitis B and liver tumours.

Figure 1.20 Human viral families.

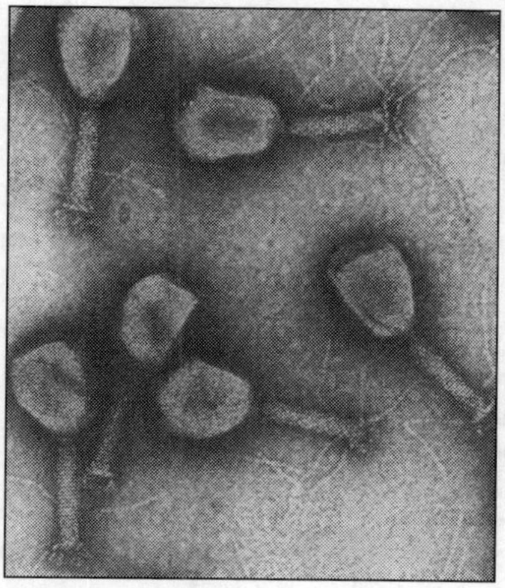

Figure 1.21 Electronmicrograph of Bacteriophage.

Phages are similar to animal and plant viruses in that they are not alive, since they cannot replicate themselves or conduct metabolic processes. All phages require bacterial cells for reproduction, but each phage type exhibits a defined host range. Some phages are very specific for one or two closely related bacterial species, while others can infect and replicate in a variety of bacterial cells. The host cell functions required for bacteriophage reproduction define host range. These include attachment to specific molecules on the bacterial cell surface, injection of the bacteriophage DNA into the bacterial cell cytoplasm, avoidance of host cell defences, proper expression and regulation of bacteriophage genes, production and assembly of the capsid, replication of the phage nucleic acid (DNA or RNA), packaging the nucleic acid into the capsid, and exit from the bacterial cell.

The bacteriophage that infects the bacterium *Escherichia coli* can be used to illustrate many of the properties of different bacteriophages. One of the well-studied bacteriophages is bacteriophage λ. The λ genome exists as a linear, double-stranded DNA molecule in the bacteriophage particle. There are 48,514 base pairs of DNA that encode about 50 genes that define the λ genome. λ phage binds to a receptor on the *E. coli* cell surface that includes a protein involved in transporting the sugar molecule. maltose (It is common for viruses to use cellular molecules designed for another function for their own ends.)

The λ genome is injected through the cell envelope into the cytoplasm, where it is converted from a linear to circular form. At this point a choice is made between two different programs: a lytic or lysogenic state (Figure 1.22). When λ phage undergoes lytic growth, replication produces hundreds of copies of its genome and phage genes produce the proteins that make up the capsid, in which the phage genome is inserted to make the mature phage particles. These phage particles are released by enzymes that break open the bacterial cell. Lysogeny is a dormant state, where the λ genome becomes part of the bacterial genome and is inherited by the bacterial offspring as a prophage. Bacteriophages that produce lysogeny, like λ, are called temperate since they do not harm the bacteria, while those that can only replicate are called virulent, since they commonly kill the host cell (Chapter 12).

1.7.2 Viroids and Prions

Viroids are a unique class of pathogens which differ from viruses in the absence of a dormant phase (virions) and in the having much smaller genomes. They lack the protein

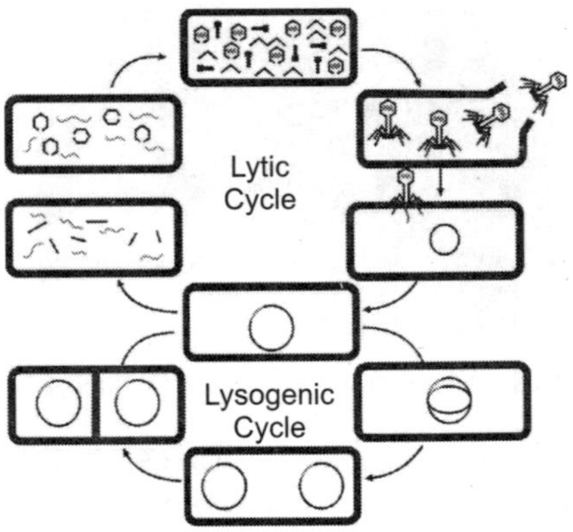

Figure 1.22 Lytic and lysogenic cycle of bacteriophage infection cycle of T_4 bacterophage on *E. coli.*

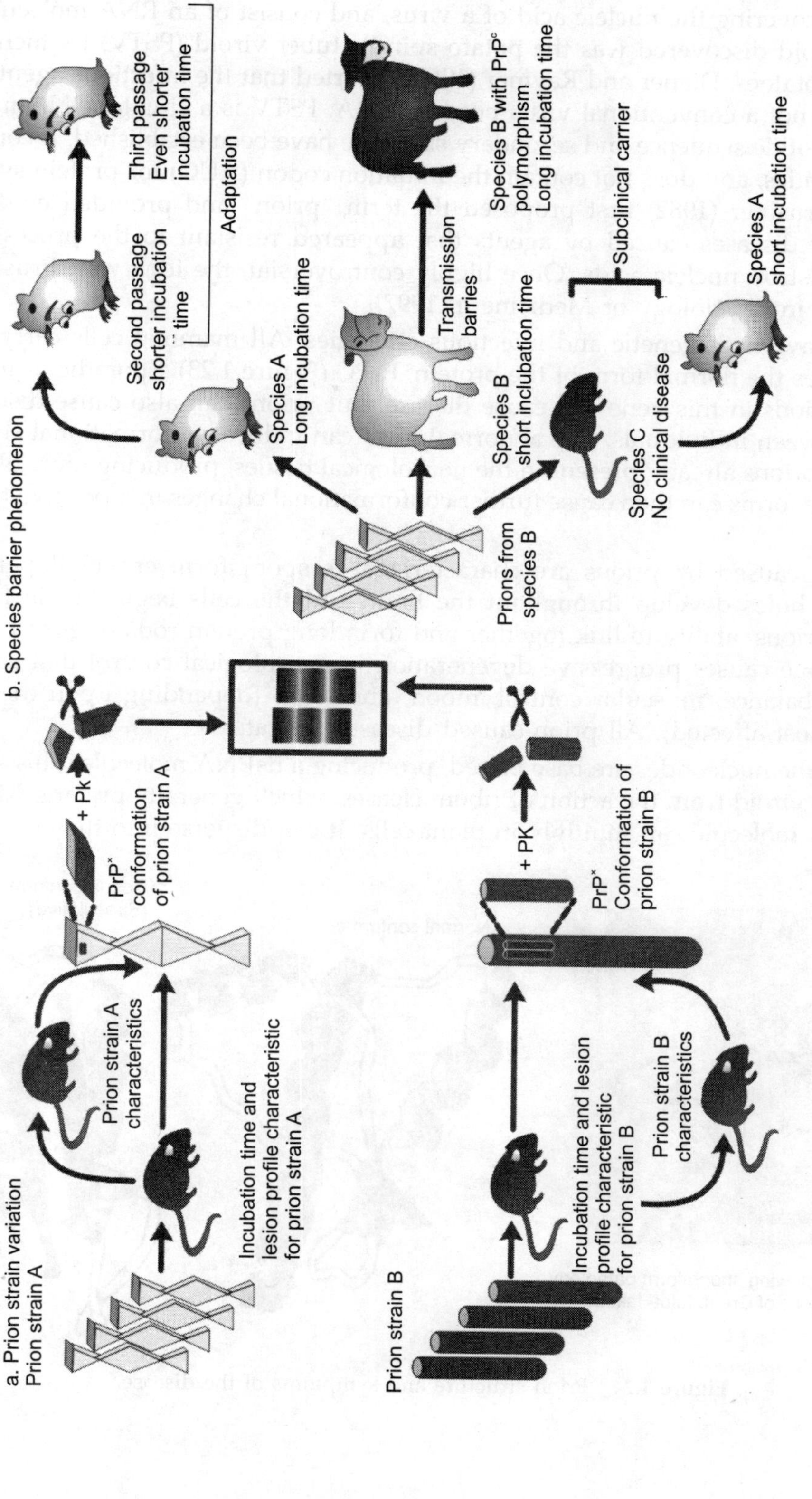

Figure 1.23 Prion development cycle.

component covering the nucleic acid of a virus, and consist of an RNA molecule only. The first viroid discovered was the potato spindle tuber viroid (PSTV) I which causes disease in potatoes. Diener and Raymer (1967) reported that the infectious agent for the disease was not a conventional virus but free RNA. PSTV is a circular RNA molecule whose nucleotide sequence and secondary structure have been established. It consists of only nucleotides, and does not contain the initiation codon (AUG) for protein synthesis. Stanley B Prusiner (1982) first proposed the term "prion" and provided evidence of neurological diseases caused by agents that appeared resistant to the processes that normally destroy nucleic acids. Once highly controversial, the idea won Prusiner the Nobel Prize in Physiology or Medicine in 1997.

Prions have both genetic and infectious etiologies. All mammal cells carry a gene that produces the normal form of the protein, PrPC (Figure 1.23). Sporadic or inherited point mutations in this gene can cause disease. But prions can also cause diseases by passing between individuals. The abnormal form can induce conformational change in the normal prions already present in the neurological tissues, producing more abnormal forms. These forms can then cause further conformational changes in a positive feedback loop.

Diseases caused by prions are characterized as spongiform encephalopathies. In the disease holes develop throughout the brain and the cells begin to die from the abnormal prions' ability to link together and form long protein rods (Figure 1.24). The tissue damage causes progressive degeneration in neurological control that can affect things like balance, muscular control, mood, and sleep (depending upon the area of the brain most affected). All prion-caused diseases are fatal.

Most of the nucleotides are base paired, producing a dsRNA molecule. This structure protects the viroid from the action of ribonucleases, which generally prefer ssRNA: The naked RNA molecule can multiply in plant cells. It can disperse into the environment

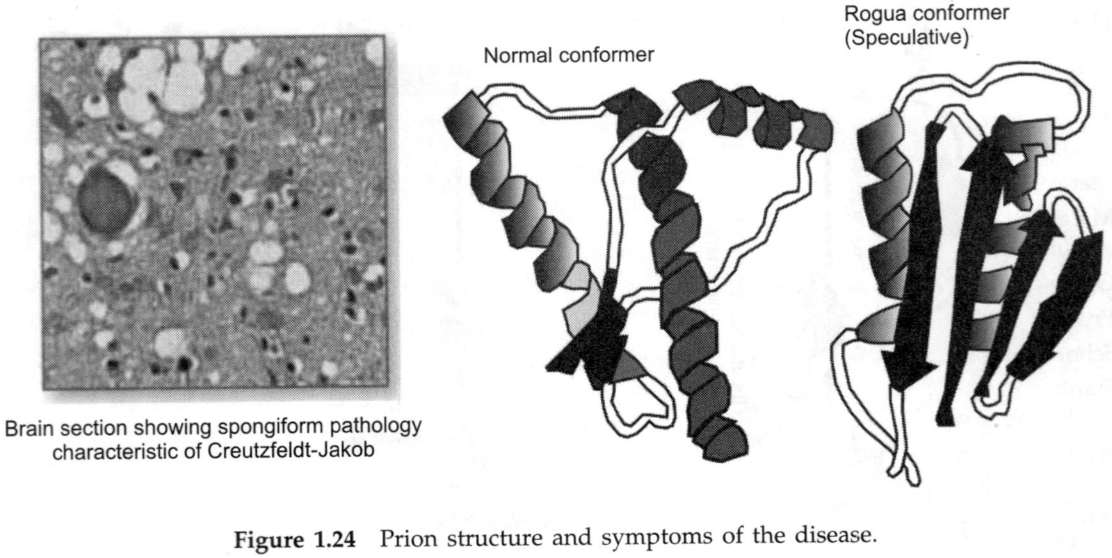

Figure 1.24 Prion structure and symptoms of the disease.

and produce diseases. The molecular weight of PST V RNA is just sufficient to code for a very small protein of 70- 80 amino acids. This has led to the suggestion that PSTV is not a single RNA species, but a population of several small RNA molecules of the same size but with different nucleotide sequences. These molecules were together supposed to constitute a viral genome of a size comparable to that of a typical viral genome. Existing knowledge, however, does not support this model, and the current view is that the PSTV genome is very small.

Another suggestion was that PSTV may be associated with a helper virus. However, it was experimentally found that when low molecular weight PSTV was simultaneously inoculated into different species of solanaceous plants, all showed signs of infection.

Suggested Readings

Ainsworth GC. 1971. *Ainsworth and Bislays's dictionary of the fungi.* Commonwealth Mycological Institute, Kent, Surrey, 662p.

Altermann W, Kazmierczak J. 2003. Archean microfossils: a reappraisal of early life on Earth. *Res Microbiol* **154**, 611–817.

Blazejak A, Erseus C, Amann R, Dubilier N. 2005. Coexistence of bacterial sulfide oxidizers, sulfate reducers, and spirochetes in a gutless worm (Oligochaeta) from the Peru margin. *Appl Environ Microbiol* **71**, 1553–1561.

Christner BC,_Skidmore ML, Priscu JC, Tranter M, Foreman C. 2008. Geographic, seasonal, and precipitation chemistry influence on the abundance and activity of biological ice nucleators in rain and snow. *Proc Natl Acad Sci USA* **105**, 18854–18859.

Diener TO, Raymer WB. 1967. Potato spindle tuber virus: a plant virus with properties of a free nucleic acid. *Science* **158**, 378–381.

Dubilier N, Bergin C, Lott C, 2008. Symbiotic diversity in marine animals: the art of harnessing chemosynthesis. *Nature Reviews Microbiology* **6**, 725-740.

Handelsman J. 2004. Metagenomics: Application of Genomics to uncultured microorganisms. *Microbiol Mol Biol Rev* **68**, 669- 685.

Loeffler F, Frosch P. 1898. Berichte der Kommission zur Erforschung der Maul- und Klauenseuche bei dem Institut fur Infektionskrankheiten in Berlin. *Centralblatt fur Bakteriologie, Parasitenkunde und Infektionskrankheiten, Abt. I* **23**: 371-391.

Margulis L. 1970. *Origin of Eukaryotic Cells,* Yale University Press, Malden, MA, USA, 312 p.

Margulis L. 2005. Hans Ris (1914-2004). Genophore, chromosomes and the bacterial origin of chloroplasts. *International microbiology : The official journal of the Spanish Society for Microbiology* **8**: 145–148.

Ogunseitan O. 2005. *Microbial diversity: Form and function in Prokaryotes.* Blackwell, New York.

Prusiner SB. 1982. Novel proteinaceous infectious particles cause scrapie. *Science* **216**, 136–144.

Schopf J. 2006. Fossil evidence of Archaean life. *Philos Trans R Soc Lond B Biol Sci* **361**, 869–885.

Stanley S. 1973. An Ecological Theory for the Sudden Origin of Multicellular Life in the Late Precambrian. *Proc Natl Acad Sci USA* **70**, 1486–1489.

Woese, CR, Fox GE .1977. Phylogenetic structure of the prokaryotic domain: the primary kingdoms. *Proc Natl Acad Sci USA* **74** , 5088–5090.

Wolska K . 2003. Horizontal DNA transfer between bacteria in the environment. *Acta Microbiol Pol* **52**, 233–243.

Chapter 2

Preservation and Culture Collection of Microorganisms

2.1 INTRODUCTION

Man has accomplished to choose and select thousands of useful *microorganisms* which form the platform for strong **BIOTECHNOLOGICAL** processes and they are extensively utilized as research tool in foundational and basic methods of molecular biology. These microorganisms require special preservation methods in order to ensure their optimal long-term viability and genetic stability. The culture collection and preservation is essential as pure culture of microorganisms is not only necessary for research purposes but also many of these strains are involved in the commercial production of various products including dairy industry, alcohol production, vaccines and antibiotics production, enzymes, vinegar, etc. The need for the preservation of strains also arises to examine their vitality, specificity, activity, immunogenicity and other properties in laboratory conditions. The production standard and quality depend on the right choice of strain preservation methods in order to maintain the vitality and authenticity of the cultivated strains. The method opted for maintenance and preservation of an organism should conserve all the specific desirable characteristics of the organism. Microbial Culture Collection Centres are fundamental biological resource centres which are able to supply more than just strains but essential data related to the genome, transcriptome, proteome and metabolome of the microorganisms collected (De vero and Giudici, 2013).

2.2 PRESERVATION

Microbial culture collections focus on the acquisition, authentication, production, preservation, cataloguing and distribution of viable cultures of standard reference microbes, cell lines and other materials for research in microbial systematic (Madigan, 2012). Once a microorganism has been isolated and grown in pure culture, it becomes necessary to maintain its viability and purity by keeping the pure cultures free from contamination. Normally, to allow continuous growth and viability of microorganisms, the pure cultures are transferred periodically onto or into a fresh medium by sub-culturing methods in laboratories. The transfer is always subject to aseptic conditions to avoid contamination. Since repeated subculturing is time consuming, it is quite difficult to maintain a large number of pure cultures successfully for longer time period. Additionally, there might be the risk of genetic changes leading to morphological and physiological changes in the organism being subcultured as well probability of techniques that do not need frequent subculturing. These methods include refrigeration, paraffin method, cryopreservation, and lyophilization (freeze drying).

These preservation techniques can either be categorized as metabolically inactive preservation techniques including cryopreservation and drying or as metabolically active preservation techniques including periodic transfer on agar or in liquid medium and keeping agar cultures under mineral oil. These metabolically inactive preservation techniques are CABRI (Common Access to Biological Resources and Information) accepted methods. A strain can only be maintained by metabolically active methods in case it cannot be preserved by one of the metabolically inactive methods, or in addition to one of these methods. Based on the duration required for preservation, these techniques have also been divided into the following two categories:

A. Short-Term Preservation

1. Refrigeration
2. Paraffin/Mineral Oil method

B. Long-Term Preservation

1. Lyophilization
2. Cryopreservation

2.2.1 Short-Term Preservation

Short-term methods preserve and maintain cultures for up to one year. Most of the fungal cultures can be maintained for that period by serial transfer. It is a simple, inexpensive, and widely used method, but is time consuming and labour intensive. Periodic transfer is a prfteferred option for small collections with cultures in constant use for short periods (less than one year).

Short-term preservation techniques also have some disadvantages as the preserved cultures must be frequently checked for contamination by mites or other microorganisms and for drying. Additionally, there might be the change in the morphology and physiology of a fungal culture during preservation over time to time. Particularly, the cultures may lose their ability to sporulate or to infect a host on repeated subculturing. The technique is, therefore, not considered appropriate for long-term preservation. Short-term preservation involves the transfer of inoculum from an actively growing fungus culture to test tubes (screw cap or plugged with cotton or foam) or Petri dishes (wrapped with Para film to reduce drying) containing an agar medium of choice. Alternating growth in nutrient-rich with nutrient-poor media at each transfer helps to maintain healthy cultures. Following two practices are normally employed for short-term preservation:

2.2.1.1 Refrigeration

This method involves the low temperature preservation of fungal cultures on the slanting surface of an appropriate medium which has been solidified in a test tube inclined from the perpendicular so as to give a greater area than that of the lumen of the tube. Pure culture is preserved after successful incubation at 0-4°C either in refrigerators or in cold-rooms. Some as endophytic and entomopathogenic species of fungi, have specific media requirements (Bacon 1990; Singleton *et al.*, 1992; Humber 1994). While short-term preservation, the cultures must be checked periodically for contamination and desiccation. Fungi such as oomycetes and some basidiomycetes (e.g., *Boletus, Coprinus, Cortinarius,* and *Mycena)* should be transferred monthly if kept at 16°C (von Arx and Schipper 1978). Most of the filamentous fungi can survive around 1-2 years at 4°C. Vigorously sporulating cultures also can be sealed tightly and preserved by storing in a freezer at -20°C (Carmichael 1956, 1962) or could be stored at −70°C (Pasarell and McGinnis 1992) to enhance survival and increase the interval between required transfers. This method of preservation has a disadvantage as could be applied for short duration only (2-3 weeks for bacteria and 3-4 months for fungi) because during storage, the metabolic activities of the microorganisms are although slowed down but not completely stopped which leads to slow growth thereby utilizing nutrients of the preservation medium. These organisms also release metabolic by-products in the medium which finally leads to death of preserved organism.

2.2.1.2 Paraffin method/mineral oil method

The mineral oil method is originally developed to conserve bacterial cultures. It has also been used to preserve filamentous fungi and has now been established as a simple and the most economical method of maintaining pure cultures of bacteria and fungi. This method involves pouring of sterile liquid paraffin over the slant (slope) of culture which is stored upright at room temperature. This layer of paraffin ensures anaerobic conditions and prevents dehydration of the medium, which helps microorganisms or pure culture to remain in a dormant state and, therefore, the culture can be preserved from months to years (depending upon the species). For fungal strain preservation, cultures grown on agar slants are covered with about 10 mm of mineral oil or paraffin. The entire agar surface and fungal culture should be submerged completely in the oil and stored at room temperature. Preservation by this method requires periodic checking of oil level in the tubes or vials, and adding of more oil when necessary so that the culture can be preserved for several months to several years.

2.2.2 Long-Term Preservation

Long-term preservation of biological cultures requires reduction in the metabolism to a bare minimum preserves stock cultures. The biological processes of microorganisms which lead to their inactivation are slowed down, and the biological materials can be maintained in inactive state for a longer time period. Organisms could be preserved longer by any one of the following methods.

2.2.2.1 Lyophilization/freeze-drying/cryodesiccation

Lyophilization or freeze-drying is a dehydration process which removes the water from the sample material and is typically used to preserve a perishable material and making the material more convenient for transport purposes. Freeze-drying is based on a physical phenomenon called sublimation which works by freezing the material and then reducing the surrounding pressure and adding enough heat to allow the frozen water in the material to sublime directly from the solid phase to the gas phase and the water in the frozen sample is removed as water vapour without thawing the sample. It has been used as a preservation or processing technique for a wide variety of products. Some of the applications include the processing of pharmaceuticals, diagnostic kits, restoration of water damaged documents, river bottom sludge prepared for hydrocarbon analysis, ceramics used in the semiconductor industry, viral or bacterial cultures, tissues prepared for analysis, the production of synthetic skins and restoration of historic/reclaimed boat hulls. Lyophilization is also used in the biotechnology and biomedical industries to preserve vaccines, blood samples, purified proteins, and other biological materials. These lyophilized materials can be revived by opening the vials, adding liquid medium, and transferring the rehydrated culture to a suitable growth medium. Freeze-dried formulations not only have the advantage of better stability, but also easier to handle as small vials can be sent conveniently to other microbiology laboratories when packaged in a special sealed mailing containers. Freeze-drying is a time- and energy-intensive process that could take days or even weeks to finish if the freeze-drying cycle is not optimized. The stability of the biological material during freeze-drying and storage and the duration of the cycle are the two major considerations for freeze-drying process optimization (Figure 2.1).

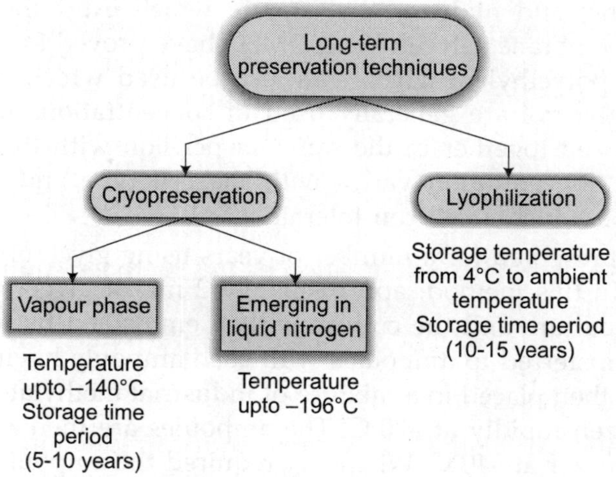

Figure 2.1 Different types of cryopreservation techniques.

2.2.2.2 Cryopreservation/cryoconservation

Scientists require genetically stabilizing living cells to ensure reproducible results and continuity in research and biomedical processes. Serial subculturing has always been time consuming and can lead to contamination or genetic drift. Cryopreservation or cryoconservation is a process in which cells, tissues or other biological substances which are susceptible to damage caused by chemical reactivity or time can be preserved by cooling to sub-zero temperatures or cryogenic temperatures, which cease any enzymatic or chemical reactivity which might cause damage to the material. Advances in cryopreservation technology have led to techniques which are available for the preservation of microorganisms, isolated tissue cells, small multicellular organisms, and even more complex organisms such as embryos.

Water is the major component of all living cells and it should be available for the chemical processes of life to occur. Cellular metabolism stops when all water in the system is converted to ice during cryopreservation. Rapid cooling minimizes the solute concentration effects as ice forms uniformly, but leads to more intracellular ice. During slow cooling, freezing occurs external to the cell before forming intracellular ice. As ice forms, water is removed from the extracellular environment. An osmotic imbalance occurs across the cell membrane leading to water migration out of the cell resulting increase in solute concentration extra- and intracellular which can be detrimental to cell survival. If too much water remains inside the cell, damage due to ice crystal formation and recrystallization during warming can occur. Cryoprotective additives or chemicals protect the cells during freezing. They can also minimize the detrimental effects of increased solute concentration and ice crystal formation. Cryoprotectants are of two types: penetrating agents such as glycerol and Dimethyl sulfoxide (DMSO), which readily pass through the cell membrane and protect intracellularly and extracellularly, and non-penetrating agents such as sucrose, lactose, glucose, mannitol, sorbitol, dextran, polyvinyl-pyrrolidone, and hydroxyethyl starch, which exert their protective effect external to the cell membrane. Glycerol and DMSO have proved to be the most effective for fungi, although polyethylene glycol can also be used which is other penetrating agent. DMSO and glycerol are generally used in concentrations ranging from 5-10% (v/v), and are not used together in the same suspension with the exception of plant cells. The optimum concentration varies with the cell type and we should use the highest concentration which a cell can tolerate.

Cultures can be preserved for a number of years using glycerol at a temperature of –40°C in a freezer. In this method, approximately 2 mL of glycerol solution is added onto the agar slant culture and the culture is then emulsified by shaking the culture. Such emulsion is transferred to ampoules with each ampoule having 5 mL of culture. These ampoules are then placed in a mixture of industrial methylated spirit and carbon dioxide and are frozen rapidly at –70°C. The ampoules are then removed and placed directly in a deep freezer at –40°C. Whenever required the ampoules are kept in water bath at 45°C for about a few seconds and then could be used directly.

For the preservation of spore-forming fungi, there is a requirement that harvesting and suspension preparation of the spores should be done in fresh growth medium containing the cryoprotective agent. When freezing fungal spores, care must be taken as freezing process should not be too long to avoid germination prior to freezing. In case of non-sporulating fungi, special procedures for harvesting mycelia must be employed prior to freezing. For fungi with tough mycelia, the culture is harvested from agar growth by cutting and removing agar plugs containing the mycelia and placing the plugs into fresh growth medium containing the cryoprotective agent. Tough mycelia that do not adhere well to agar cultures are grown in broth culture and the mycelial mass is blended prior to freezing. The viability and an estimate of recovery should be determined both before and after freezing the culture. A comparison of the counts prior to and after freezing gives an indication of the degree of recovery or the success of the preservation procedure.

2.3 STORAGE DURING PRESERVATION

Smaller packaging for storage is preferred as it is convinient to use and also minimizes viability loss. A variety of small containers such as flame-sealed glass ampoules and screw-cap polypropylene vials can be used for storing cells at ultra-low temperatures. The most commonly used sizes are 1.2 to 2.0 mL vials which maximize storage capacity and are easy for handling. Generally, 0.5-1.0 mL of the cell suspension is placed into each vial. As the sample has been frozen for 48 hours, a vial should be checked by thawing its content to determine whether the cells are viable and able to establish a cell population or not, i.e., it should survive the freezing procedure. The temperature at which frozen preparations are stored should be proportional to the duration of storage. The lower the storage temperature, the longer will be the viable storage period.

Biological materials can be stored by two ways through the emersion of those materials in the liquid nitrogen for ultimate security at approximately –196°C such as animal viruses infected cells at –150°C or plasmids at –150°C. However, there are risks in storing vials directly in liquid nitrogen, as improperly sealed glass ampoules may have micro channels leading to liquid nitrogen penetration over time. When these are retrieved from liquid nitrogen to ambient temperature, rapid conversion of the liquid nitrogen to vapour inside the ampoule may result in explosion of the ampoule. Plastic vials with screw top closures are also susceptible to liquid nitrogen penetration, but do minimize the explosion potential. However, during warming of plastic vials liquid can spray from the cap/vial interface with potential dissemination of the vial contents. Another concern is that cultures in leaky vials may be contaminated with bacteria or spores that may be present in the liquid-nitrogen freezer. Therefore, various alternative techniques such as using plastic straws instead of vials or tubes are already available in the market for liquid nitrogen storage. Another way for storage is the storage in the vapour phase of the liquid nitrogen which provides all vapour phase storage at the temperature range of approximately –140°C to –130°C. It is ideal for the storage of plant viruses, fungal spores and some bacterial species.

This method requires that liquid nitrogen freezer should maintain the proper working temperature, to assure this the volume of liquid nitrogen in the unit be adjusted to a level which results in a temperature of –150°C just above the stored material. This working temperature can be attained in most of the liquid nitrogen freezers.

During preservation improper handling of material that has maintained at cryogenic temperatures can have detrimental effect on the viability of frozen cells. Therefore, the storage systems should be designed so as to minimize exposure of stored material to warmer temperatures, as well as minimize prolonged exposure of personnel during specimen retrieval. When stainless steel racks are used, maintain a small number of vials of each preparation in the top box of the rack and store the remaining vials of each preparation in lower boxes. By doing this, a vial of one preparation can be retrieved without exposing all vials of any particular culture or lot. To maximize the available space in liquid nitrogen freezers and minimize exposure of material during retrieval, use small storage boxes or aluminium canes.

2.4 RETRIEVAL OF STORAGE CELLS

The cryopreserved biological materials are subjected to reconstitution or thawing prior to retrieval. For most of the cells, warming from the frozen state should occur as rapidly as possible until complete thawing is achieved. To achieve rapid warming, place the frozen vial into a 37°C water bath. There might be the risk of contamination during reconstitution, which can be minimized by disinfecting the external surface of the vial by wiping with alcohol-soaked gauze prior to opening. For most cultures, the entire contents of the vial may be placed into fresh media. However, further dilution may be necessary for cell lines. It is recommended that the suspension be centrifuged at 100 × g for 10 minutes after initial dilution, the supernatant material should be removed, and the cells are then re-suspended into fresh growth medium to remove residual cryoprotective agents. Safety precautions must be observed throughout the preservation and maintenance process. Proper record keeping should be maintained during cryopreservation process.

The choice of preservation method depends on the species of concern, the resources available, and the goal of the project. Some low-cost methods of preservation, such as storage in distilled water and the silica gel method are good, but not considered permanent. The maximum duration of storage varies with each method and with the species being preserved, but it generally is 10 years or less. Fungal strains should be preserved with one of the permanent methods (lyophilization, cryopreservation) whenever possible. Permanent preservation is essential for strains with critically important characteristics and for type specimens. Cultures that are permanently preserved in metabolically inactive states now can serve as type specimens, according to Article 8.4 of the International Code of Botanical Nomenclature (Greuter *et al.* 2000).

As a whole there is no single appropriate preservation procedure which can be uniformly applied for all the organisms. Instead, there are some basic methods that differ in cost and convenience, which can be utilized in combination for proper preservation of microbial cultures.

2.5 PRESERVATION OF FUNGAL CULTURES

Preservation methods for fungal cultures vary depending on the type and their degree of sporulation. Spore-forming fungal strains can usually be freeze-dried with the exception of zoosporic fungi as compared to the non-sporulating strains. Both fungal types can be frozen and stored for longer periods in liquid nitrogen. All plasmid-containing or mutant strains should also be frozen directly from the original material particularly to prevent any alteration or loss of desired characteristics. Fungi are such a diverse group thereby require several methods of cultivation and preservation to ensure the viability and morphological, physiological, and genetic integrity of the cultures over time. Secondly, the cost and convenience of each method, however, also must be considered. Maintenance and preservation of fungal cultures are essential elements of systematic and biodiversity studies. The primary methods of culture preservation are continuous growth, drying, and freezing.

2.5.1 Continuous Growth Method

This method employs the growth of cultures on agar, this is typically used for short-term storage. Such cultures are stored at temperatures of from 5°C to 20°C, or they may be frozen to increase the interval between subcultures. The methods are simple and inexpensive because they do not require specialized equipment.

2.5.2 Drying

Drying is the most useful method of preservation for cultures that produce spores or other resting structures. Silica gel, glass beads, and soil are substrata commonly used in drying. Fungi have been stored successfully on silica gel for up to 11 years (Smith and Onions 1983a, 1983b, 1994; Smith and Thomas 1998; Smith *et al.*, 2001; Smith and Ryan, 2004). Drying methods are technically simple and also do not require expensive equipment.

2.5.3 Freezing Methods

Cryopreservation is a versatile and widely applicable method. Most fungi can be preserved by this method, with or without cryoprotectants, in liquid nitrogen or in standard home procedures (Figure 2.2).

As microorganisms are required for different biotechnological processes which further have the requirement of reliable sources of pure, authenticated cultures. There are over 500 culture collections worldwide that make strains of bacteria and fungi available, generally for a modest fee. These collections obtain most of their strains from microbiologists working in universities or research institutes; other strains come from industries that no longer have a use for them. Moreover, the law now requires that if a process that uses a microorganism is to be patented, a culture of the microorganism must be deposited with a recognized culture collection. In England alone, the national culture collections hold more than 27,000 strains of bacteria and fungi. The American Type Culture Collections include more than 35,000 strains of bacteria, fungi, yeasts, viruses, and plasmids. No single preservation procedure is appropriate for all organisms. Instead, there are four basic methods that differ in cost and convenience. Microbial cells can be maintained for short periods on slants and stabs of appropriate nutrient-containing agar or stored for longer periods in freeze-dried or other frozen form. For organisms that produce spores, the latter can be preserved in dry form on solid supports.

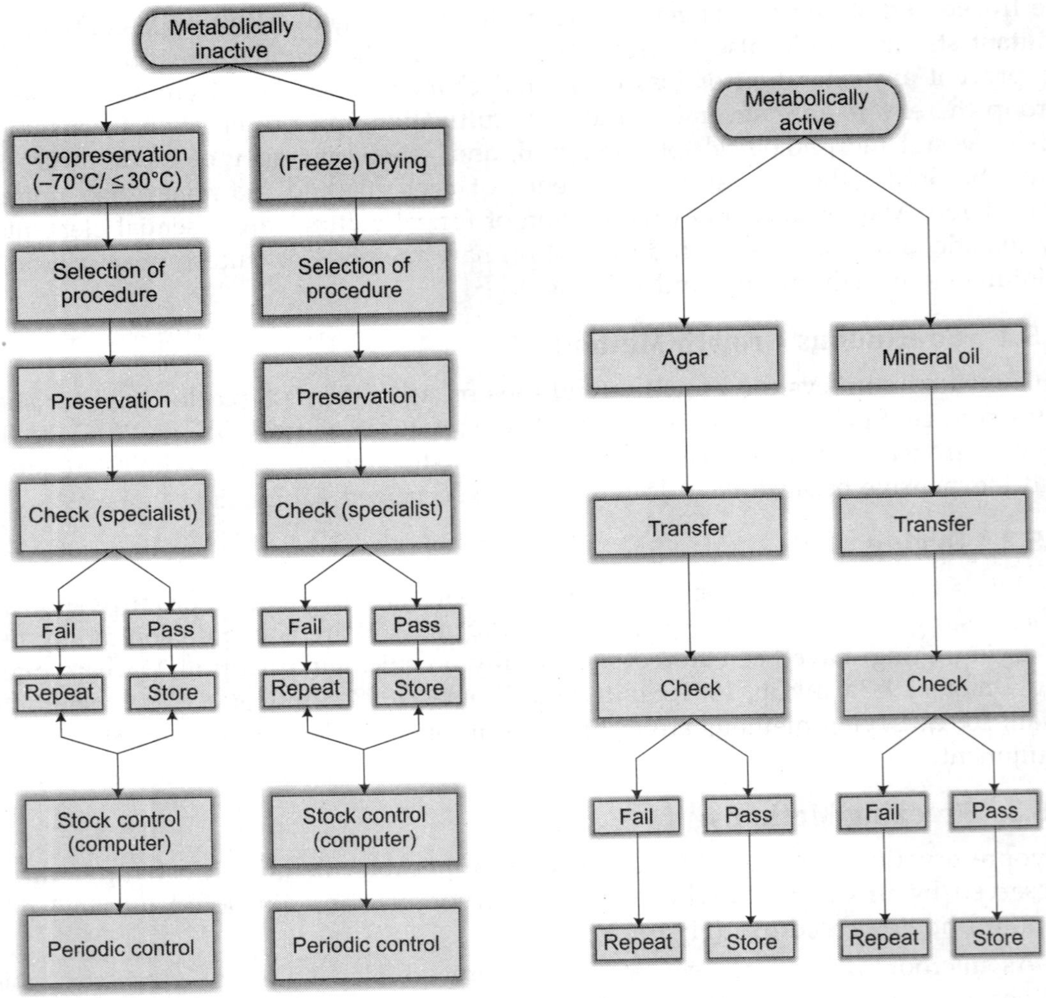

Figure 2.2 Freezing methods used for different types of organisms.

- The simplest procedure is to transfer cultures periodically to fresh solid slants of agar in the appropriate medium and to incubate them at a suitable growth temperature. Once the slant cultures are well established, they are kept in a refrigerator at 5°C enclosed in a container to avoid desiccation. This is the least expensive procedure and keeps cells viable up to months, but there is a disadvantage that mutants or contaminants may accumulate in such cultures.
- Lyophilization (freeze-drying) described above is a convenient preservation method. Microbial cells are mixed with a medium containing skim milk powder (at 20% w/v) or sucrose (at 12% w/v) and frozen, after which the water is removed from them by sublimation under partial vacuum. Lyophilized samples remain viable for many years and can be shipped without refrigeration.

- Cells can be stored for prolonged periods of time at liquid nitrogen temperature. In this procedure, the cells are placed in ampules with media containing (by volume) either 10% glycerol or 5% dimethylsulfoxide and are slowly frozen; their temperature is decreased by 1°C to 2°C per minute until it reaches about -50°C. The ampules are then stored at –156°C to –196°C in a liquid nitrogen refrigerator. Additives such as skim milk, sucrose, glycerol, and di-methyl-sulfoxide minimize damage to the cells by preventing ice crystals from forming during the freezing process. Many spore-forming fungi can be preserved by slowly air drying the spores at ambient temperature on the surface of sterilized soil, silica gel, or glass beads.

2.6 MICROBIAL TYPE CULTURE COLLECTIONS AND BANK

Microorganisms are not only required for the production of useful substances; they also play unique roles in element cycles with plants and animals. Microorganisms are the significant gene pools, and these gene pools must be conserved. They can be regarded as the cultural heritage and the cultural property, and must be transferred to the next generation in a normal and healthy condition. Therefore, reliable culture collections are needed as the depository and for the further study and application of the cultures (Komagata, 1999).

Implementation of quality measures, compliance with the Convention on Biological Diversity (CBD), and adoption of latest bioinformatics tools are various main steps to be taken by microbial culture collections in order to provide resources for the emerging area of the knowledge-based bio-economy. These measures have to be introduced side by side with the deposition of increasingly phylogenetically and physiologically diverse microbiological organisms. However, the necessary expansion of human resources and infrastructure is moving slowly, if at all. Furthermore, considering that the vast majority of microbial isolates do not find their way into public collections, a strategy should be devised to encourage researchers to deposit a higher fraction of strains. It appears obvious that in order to make available an even broader range of diversity to users and researchers, collections will have to decide whether to diversify on a broad taxon spectrum of the hierarchic system, holding a small number of representatives per species, or to follow the route of focusing on in-depth holdings of selected groups of organisms, depending on existing taxonomic expertise. These decisions require a worldwide coordinated activity with the outcome to be made transparent to users in an emerging global network (Stackebrandt, 2010).

A survey showed more than 300 species of leaf-feeding insects on a single tropical tree and about half of the species among them were unidentified yet. In addition, ants belonging to 43 species and 26 genera were recognized on a single leguminous tree, the number of species found being comparable to that in the whole of England. These findings reflect close association among plants, animals, and microorganisms, and the richness of biological diversity of tropical rain forests. Microorganisms form an important part of ecosystems on the earth, and are responsible for various environmental processes as nitrogen fixation, bioremediation of pollutants, and soil building, and the life of plants and animals, and even human depends on the activities of microorganisms. Food chains in nature cannot be completed without microorganisms. The development of modern biotechnology aspires for isolation of new microorganisms and improvement of their attributes. Since the early days of microbiology, astronomical numbers of microorganisms were isolated from a wide variety of natural sources, and used for the scientific research and for the industrial fermentation. However, large numbers of microorganisms had already extinct in the past, and they are no longer available now. Microbiologists often lose microbial cultures that they studied because of the change of their interests and difficulties in keeping the cultures. It is probably due to the absence of reliable culture collections in which microorganisms could be maintained properly and supply them promptly on demand. Through the study of microbial cultures conserved in the culture collection, potential properties of microorganisms have been developed, and the future perspective of microbiology will be presumed. Effective research needs adequate and reliable sources of properly preserved cultures. In the near future, uncountable numbers of microbial strains will be isolated through the study of microbial diversity, and the attributes of a large number of strains will be improved. Therefore, reliable and well-organized culture collections are needed as the depository and for the promotion of research and application of the strains. In fact, culture collections play a role of the depository of the type strains in bacteriology, and the study of bacterial systematics will not be completed without culture collections consequently. Major contribution of culture collections are the collection and maintenance of important and useful microbial cultures, supply of these cultures on demand, and preparation of their informative documents. Social needs for culture collections are increasing year by year, and effective and smooth management is required for better services of culture collections. In this point of view, exchange of information and cooperation among culture collections are the most important feature to achieve the purpose of the culture collection. According to the data of World Data Center for Microorganisms (WDCM), culture collections are distributed in 58 countries in the world, and the number of all the holdings is recorded to be 815,568. However, major public culture collections, which preserve a wide variety of microorganisms and supply them to the public, are estimated to be about 30 collections. In addition, bacterial cultures account for 343,253 (42%), filamentous fungi for 372,304 (46%), viruses for 14,376 (1.8%), cell lines for 5,156 (0.6%), and others for 80,485 (9.9%) by the data of WDCM. South East Asian countries are rich in microbial diversity, but the exploitation of microorganisms in such areas has not been completed. Large number of microbial strains will be isolated from a wide variety of sources in South East Asia, and hundreds of thousands of improved strains in biotechnology will be obtained in the near future. Thus improvement of culture collections is critical and crucial for the further development of microbiology, microbial industry, and biotechnology. In addition, good operation and management of culture collection is the need of the day in the area of biological sciences beside American Type Culture Collection and National Culture Collection Centre, England (Table 2.1).

Table 2.1

Abbreviation	Name	Location
ATCC	American Type Culture Collection	Rockville, MD USA
CBS	Centraalbureau voor Schirnmelcultur	Baarm, Netherlands
CCM	Czechoslovak Collection of Microorganisms	J.E. Purkyne university, Brno
CDDA	Canadian Department of Agriculture	Ottawa, Canada
ClP	Collection of the Institut Pasteur	Paris, France
CMI	Commonwealth Mycological Institute	Kew, UK
DSM	Deutsche Sammlung von Mikroorganismen	Gottingen, Germany
FAT	Faculty of Agriculture, Tokyo University	Tokyo, Japan
lAM	Institute of Applied Microbiology	University of Tokyo, Japan
NCIB	National Collection of Industrial Bacteria	Aberdeen, Scotland
NCTC	National Collection of Type Cultures	London, UK
NRRL	Northern Regional Research Laboratory	Peoria, IL USA
MMCC	Medical Microbiology Culture Collection	Aarhus, Denmark
IMT	Institute of Microbial Technology	Chandigarh, India
NCIM	National Chemical Laboratory	Pune, India
UON	University of'Neuchatel	Neuchatel, Switzerland.
MCBL	Indian Institute of Science	Bangalore, India.
MICRO	National University of Singapore	Singapore

Cryobank of Microorganisms, Research Institute of Antibiotics and Enzymes, St. Petersburg, Russia

National Bank for Industrial Microorganisms and Cell Cultures, 125 Tsarigradsko chaussee blvd., B1.2, BG-III3 Sofia, Bulagaria.

National Culture Bank, Universita di Udine, Dip. BiologiaApplicata Difesa Piante, Area Rizzi, Via delle Scienze 208, 1-331 00 Udine, Italy.

Research Institute of Fermentation, Yamanashi University, Kofu, Japan.

Collection of Cultures, Institute of Hygiene, Warsow, Poland.

Food and Nutrition Culture Collection Center, Centre for Food and Nutrition Studies, Gadjah Mada University, Indonesia.

Listed here are just a few of the general culture collections. Many universities and research laboratories maintain collections of specific microbial groups.

Suggested Readings

Arx J. A. von and M. A. A. Schipper. 1978. The CBS fungus collection. *Adv Applied Microbiol* **24**: 215-236.

Bacon C. W. 1993. Abiotic stress tolerances (moisture, nutrients) and photosynthesis in endophyte-infected tall fescue. *Agri Ecosys Environ* **44**: 123-141.

Carmichael J. W. 1962. Viability of mold cultures stored at –20°C. *Mycologia* **54**:432-436.

Carmichael J. W. 1956. Frozen storage for stock culture of fungi. *Mycologia* **48**: 378-381.

De Vero L and Giudici P. 2013. Significance and management of acetic acid bacteria culture collections. *Acetic Acid Bacteria* volume 2(s1):e9

Dworkin M, F Stanley, E Rosenberg, K-H **SCHLEIFER**, and E Stackebrandt. 2006. *The Prokaryotes: A Handbook on the Biology of Bacteria*, Volume 5. Springer Dordrecht.

Greuter W, J McNeill, F R Barrie, H M Burdet, V Demoulin, T S Filgueiras, D H Nicolson, P C Silva, J E Skog, et al. 2000. *International Code of Botanical Nomenclature (Saint Louis Code)*. Königstein, Germany: Koeltz Scientific Books.

Humber R A. 1984. Foundations for an evolutionary classification of the Entomophthorales (Zygomycetes). In: *Fungus-insect relationships: perspectives in ecology and evolution*. Wheeler Q, Blackwell M, eds. Columbia University Press, New York: 166-183.

Komagata K. 1990. Systematics of methanol-utilizing bacteria. *FEMS Microbio Lett* **87**: 291-296.

Madigan, M T. 2012. *Brock biology of microorganisms* (13th ed.). San Francisco: Benjamin Cummings. ISBN 9780321649638.

Pasarell L , and M R McGinnis. 1992. Viability of fungal cultures maintained at −70°C. *J Clin Microbiol* **30:***1000-1004.*

Singleton L L, J D Mihail, and C M Rush eds. 1992. *Methods for research on soil-borne phytopathogenic fungi*. St Paul, MN, USA, American Phytopathological Society Press. 264 pp.

Smith D, and A H S Onions 1983a. *The Preservation and Maintenance of Living Fungi*. Kew: CAB Mycological Institute.

Smith D, and A H S Onions 1983b. A comparison of some preservation technques for fungi. *Transactions of the British Mycological Society* **81**, 535-540.

Smith D, and A H S Onions 1994. *The Preservation and Maintenance of Living Fungi*. Second edition. IMI Technical Handbooks No. 2, pp 122. Wallingford, UK: CAB INTERNATIONAL.

Smith D, and M J Ryan 2004. Current status of fungus culture collections and their role in biotechnology. *In: Handbook of Biotechnology, Mycology*, Second edition, Ed. D.K. Arora. New York: Marcel Dekker. 527-538.

Smith D, and V E Thomas. 1998. Cryogenic light microscopy and the development of cooling protocols for the cryopreservation of filamentous fungi. *World Journal of Microbiology & Biotechnology* **14**: 49-57.

Smith D, M J Ryan, and J D Day, Eds 2001. *The UK National Culture Collection Biological Resource:* Properties, maintenance and management. pp 382. UK National Culture Collection, Egham.

Stackebrandt E. 2012. Pre-and post-publication deposition of Biological Resources, *BRIF Workshop*, Toulouse, Seventh Framework Programme, Research area, INFRA-2008-1.1.2 (2009-2012) 22.10.2012.

Chapter

3

Culture Medium

3.1 INTRODUCTION

Nutritional requirement of microorganisms vary greatly to which there are vast differences in the chemical composition of the media used in the laboratory. They exhibit differences with respect to the physical conditions favouring the growth. Knowledge of microbial nutrition allows the microbiologists to grow microorganisms in the laboratory effectively aseptically and to study their physiological, biochemical and genetical mode. In a general sense, the nutrient requirements of all organisms are the same, that is, all organisms require the macro- and micronutrients. But more specifically, certain organisms require certain nutrients in specific forms. Thus, the proportion and type of each nutrient in a culture medium can vary dramatically for growth of different microorganisms.

3.2 CULTURE MEDIUM

Literally, thousands of different recipes for culture media have been published, and most experienced microbiologists pay careful attention to the nutritional aspects of the organism or organisms they are growing. The common chemical by which many of the nutrients are supplied in culture media are listed in Table 3.1.

Table 3.1 Broad ingredients of a culture medium

Ingredients	*Nutrient supplied*	*Function*
Peptone	Amino acids and peptides	Serve as sources of C, N, and energy
Meat extract	Amino acids, peptides, nucleotides, organic acids, vitamins, minerals	Serve as sources of C, N and energy
Yeast extract	Vitamin of B group, growth factor, minerals, nucleotides amino acids	Serve as the sources of C, N and growth regulators.
Glucose	Simple sugar	Energy
Fructose	Sugar	Energy
Cellulose	Sugar	Energy
Blood	Vitamins, organic acids minerals, amino acids, peptides	Serve as the sources of C, N and growth regulators
Sodium chloride	Na^+ and Cl^- ions	Export and import of ions
Ammonium phosphate	Nitrogen and phosphate	Serve as sources of C and N

Contd...

Contd...

Magnesium sulphate	Magnesium and sulphate	Serve as the sources of Mg^{++} and SO_4^-
Potassium phosphate	Potassium and phosphate	Serve as sources of potassium and phosphate
Sodium carbonate	Sodium and carbonate	Serve as sources of sodium and CO_2
Bile salt	Growth inhibitor	Inhibiting the growth of Gram-positrve bacteria and favour the growth of Gram-negative bacteria
Crystal violet	Growth inhibitor	Do
Basic fuchine	Growth inhibitor	Do
Phenol red	Dye	pH indicator
Methyl red	Dye	pH indicator

Media that contain some ingredients of unknown chemical composition are called complex media. Such media are very useful for routine cultivation of bacteria under controlled laboratory conditions. The main undefined components of complex media are peptone, beef extract and yeast extract. They may be modified according to requirements, by adding supplements like blood, serum and specific chemicals to enhance the growth of fastidious bacteria and isolation of selective bacteria.

Two classes of culture media are used in microbiology: **chemically defined (synthetic) or undefined (complex)**. Chemically defined media are prepared by adding precise amounts of pure inorganic or organic chemicals to distilled water. Therefore, the exact chemical composition of a defined medium is known. In many cases, however, knowledge of the exact composition is not critical. In these instances undefined media may suffice, or may even be advantageous. Complex media employ crude digests of substances such as casein (milk protein), beef, soybeans, yeast cells, or a number of other highly nutritious, yet nutritionally undefined, substances. Such digests are available commercially in powdered form and can be weighed out rapidly and added to culture media. However, an important disadvantage of using complex media is the loss of control over the precise nutrient specifications of the medium. They must be cultured in the laboratory in order to understand the various functions of microorganism. Culture media are prepared with essential nutrients in forms that the organism can utilize it. In synthetic culture media, all of the elements are provided from known chemical compounds. In complex culture media, undefined materials such as plant or animal extract provide the necessary ingredients (Table 3.2).

Table 3.2 Nutrient requirements of microorganisms and common means to satisfy them in culture

Nutrient	*Chemical form supplied in culture media*
Carbon	Organic defined media: glucose, acetate, pyruvate, malate, hundreds of other compounds; complex media: yeast extract, beef extract, peptone, many other digests, inorganic CO_2, HCO_3^-
Nitrogen	Organic amino acids, nitrogenous NH_4CI, $(NH_4)_2SO_4$, KNO_3, N_2

Contd...

Contd...

Phosphorus	KH_2PO_4, Na_2HPO_4
Sulfur	Na_2SO_4, H_2S
Potassium	KCI, K_2HPO_4
Magnesium	$MgCI_2$, $MgSO_4$
Sodium	NaCI
Iron	$FeCI_3$, $Fe(NH_4)(SO_4)_2$, iron chelates
Micronutrients	$CoCI_2ZnCI_2$, Na_2MoO_4, $CuCI_2$, $MnSO_4$, $NiCI_2$, Na_2SeO_4, Na_2WO_4, Na_2VO_4

Culture media can be prepared for use either in a liquid state, or in a gel (semisolid) state. A liquid culture medium is converted to the semisolid state by addition of a gelling agent. **Agar** is the most commonly used gelling agent. It is manufactured from certain seaweeds and is not a nutrient for most microorganisms. Culture media containing agar are dispensed in flat dishes called Petri plates, where microbial cells can grow and form visible masses called colonies (Figure 3.1).

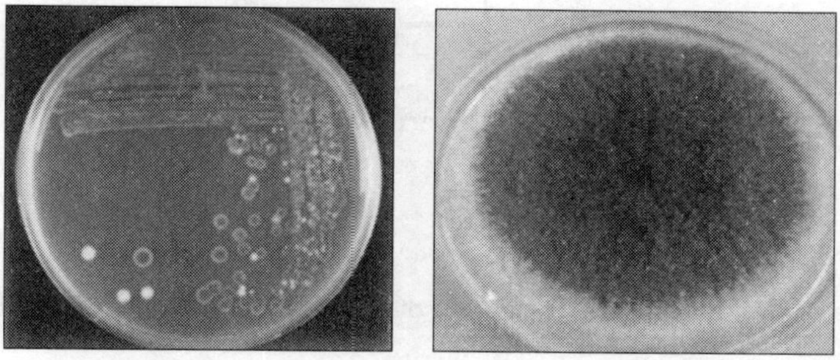

Figure 3.1 Agar-containing media in Petri plates showing bacterial and fungal growth.

3.3 TRACE ELEMENTS

The micronutrients (trace elements) play special role in the preparation of culture media. Most of these minerals are required in extremely small amounts. In chemically defined media, trace elements are usually added in small amounts from a stock solution containing a mixture of the metals whereas, trace elements are not needed in undefined media.

Iron plays a major role in cellular respiration, being a key component of the cytochromes and iron sulfur proteins involved in electron transport. It is considered as a trace mineral but required in larger amounts than other trace minerals. Mostly inorganic iron salts are highly insoluble, providing an adequate supply of iron to a culture medium.

One way of supplying iron is to complex it with an organic chemical called a **chelating agent.** Two common chelating agents used to provide iron to culture media are ethylene diamine tetraacetic acid (EDTA) and nitrilotriacetic acid (NTA).

Many organisms produce specific iron chelators **ironophores**, which solubilize iron salts and transport iron into the cell. Once the iron-hydroxamate complex has passed into the cell, the iron is released and the hydroxamate can pass out of the cell and be utilized again for iron transport (Figure 3.2). In some becteria, the iron-binding compounds are not hydroxamates but phenolic acids. Enteric bacteria such as *Escherichia coli* and *Salmonella* spp. produce complex phenolic derivatives called **enterobactins** or **enterochelins**.

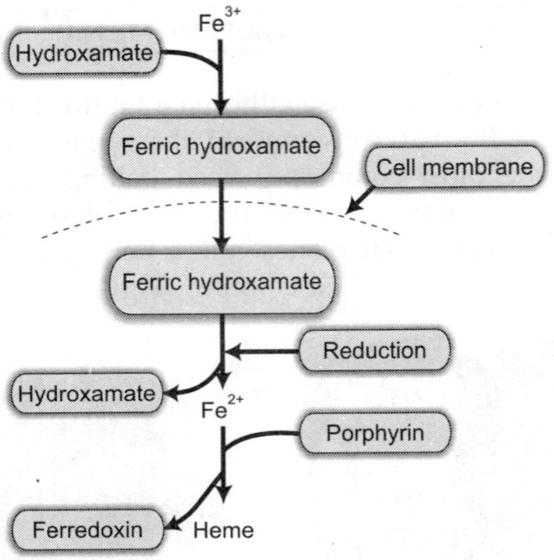

Figure 3.2 Role of iron-binding hydroxamate of an organism living in low iron environment.

3.4 GROWTH FACTORS

Growth factors are specific organic compounds that are required in very small amounts and that cannot be synthesized by some cells. Substances frequently serving as growth factors are vitamins, amino acids, purines, and pyrimidines. Although most organisms are able to synthesize all of these compounds, certain others require them preformed from the environment. Hence, they must be added to the culture medium. If a complex organic material such as yeast extract or peptone is used in a culture medium, most or all of the potential growth factors will already be supplied, but in synthetic media, provision of proper growth factors is often of great importance.

Vitamins are the most commonly needed growth factors. Vitamins are defined as organic compounds required in small amounts for growth and function that do not serve as either energy sources or building blocks of macromolecules. The first growth factors to be discovered and studied in any detail were the vitamins. Most vitamins function as parts of coenzymes and these are summarized in Table 3.3. Many microorganisms are able to synthesize all of the components of their coenzymes, but some microorganisms are unable to do so and must be provided with certain parts of these coenzymes in the form of vitamins. The lactic acid bacteria, which include the genera Streptococcus and

Lactobacillus, are renowned for their complex vitamin requirements, which are even greater than those of humans. The vitamins most commonly required by microorganisms are thiamine (vitamin B_1), biotin, pyridoxine (vitamin B_6), and cobalamin (vitamin B_{12}).

Table 3.3 Vitamins and their functions

Vitamin	*Function*
p-Aminobenzoic acid	Precursor of folic acid
Folic acid	One-carbon metabolism; methyl group transfer
Biotin	Fatty acid biosynthesis; α-decarboxylations; CO_2 fixation
Cobalamin (B_{12})	Reduction of and transfer of single carbon fragments; synthesis of deoxyribose
Lipoic acid	Transfer of acyl groups in decarboxylation of pyruvate and α-ketoglutarate
Nicotinic acid (niacin)	Precursor of NAD; electron transfer in oxidation-reduction reactions
Pantothenic acid	Precursor of coenzyme A; activation of acetyl and other acyl derivatives
Riboflavin	Precursor of FMN, FAD in flavoproteins involved in electron transport
Thiamin (B_1)	α-Decarboxylations; Transketolase
Vitamins B_6 (pyridoxal-Pyridoxamine group)	Amino acid and keto acid transformations
Vitamin K group; quinines	Electron transport; synthesis of sphingolipids
Hydroxamates (terregens factor, coprogen, mycobactin, etc.)	Iron-binding compounds; solubilization of iron and transport into cell
Coenzyme M (Co-M)	Required by certain methanogens; plays a role in methanogenesis

3.5 ENVIRONMENTAL FACTORS

In addition to appropriate nutrients, culture of microorganisms requires appropriate environmental conditions. Suitable temperature, pH, and oxygen conditions must be adjusted appropriately for each organism to be cultured. One of the most important environmental factors that must be adjusted is **temperature.** Each microorganism has a defined temperature range over which it is capable of growing. Satisfactory growth will not be obtained if temperature is too low or too high for that organism. **Optimum** temperature is the temperature at which the organism grows the fastest. Some microorganisms have optimum temperatures as low as 5°C, and others are known with optima as high as 105°C. However, no one organism spans more than a small part of this temperature range. Microorganisms capable of causing disease in warm-blooded animals grow best near body temperature, which is about 37°C. Microorganisms found in soil surface and water usually grow best at temperature 25°C, whereas microorganisms from excessively heated environments require higher temperature. As a general rule, it can be stated that the most appropriate temperature of culture of a microorganism is near the temperature of the habitat in which that microorganism is growing.

A proper acidity or alkalinity must be provided for each organism. Acidity or alkalinity of a solution is expressed by its pH value on a scale in which neutrality is pH 7. Those pH values that are less than 7 are acidic and those greater than 7 are alkaline. Microorganisms vary in their need for, or tolerance of, **oxygen.** They can be divided into several groups, depending on the effect of oxygen. Those which are able to use oxygen are called **aerobes.** Some organisms are unable to use oxygen and may actually be harmed by it. Organisms unable to use oxygen are called **anaerobes** and those harmed by it are called **obligate anaerobes**. Oxygen is to be excluded to culture the anaerobes. Oxygen is ubiquitous in the air and to culture microorganisms under anaerobic conditions, bottles or tubes are to be filled completely up to the top with culture medium provided with tightly fitting stoppers so as to maintain the anaerobic conditions for organisms that are not too sensitive to small amounts of oxygen. It is also possible to add a chemical to the medium which reacts with oxygen and therefore removes it from the culture medium. Such a substance is called a reducing agent. A number of organisms, called **facultative aerobes,** are able to grow in the presence or absence of oxygen.

Yeast is a good example of a facultative aerobe. They can grow anaerobically by fermentation, producing alcohol and CO_2 and can also grow aerobically by respiration, producing only CO_2 Whether yeast grows aerobically or anaerobically depends, in the first instance on the availability of O_2. If O_2 is present, yeast will grow aerobically, producing a large amount of ATP, and consequently a large amount of cell mass. If O_2 is absent and a fermentable energy source such as glucose is present, yeast can grow by fermentation, producing much less cell mass, but produce alcohol and CO_2.

3.6 STERILIZATION AND ASEPTIC

This is the most essential and important aspect for any microbiological experiments.

3.6.1 Technique

Microbes are everywhere and they are easily dispersed in the air and on surfaces (such as the human skin) because of their small size. Therefore, we must **sterilize** the culture medium soon after its preparation to eliminate microorganisms already contaminating it. It is equally important to take precautions during the subsequent handling of a sterile culture medium to exclude undesired organisms from the medium. Therefore, other materials which come into contact with culture media must also be sterilized. It is also necessary to sterilize contaminated materials in order to control the spread of harmful bacteria.

3.6.2 Sterilization of Media

The most common way of sterilizing culture media is by heat. However, since heat may also cause damage and change the chemical nature in many of the ingredients of culture media, it is desirable to keep the heating time as short as possible. The best procedure is to heat under pressure, since at pressures above atmospheric, the temperature can be increased above 100°C, thus decreasing the time necessary to make the medium sterile. Devices for heating under pressure are called autoclaves. The conditions used to sterilize in an autoclave are generally 15 minutes at 121°C at 15 psi.

The technique used in the prevention of contamination during manipulations of cultures and sterile culture media is called aseptic technique. Experience is required for success in the microbiology laboratory. Air-borne contaminants are the most common problem, since air always contains dust particles that generally have a population of microorganisms. They must be handled in such a way that the air loaded with contaminant does not enter when containers are opened. This is best done by keeping the containers at an angle so that most of the opening is not exposed to the air.

Manipulations should preferably be carried out in a dust-free room in which air currents are absent. Aseptic transfer of a culture from one tube of medium to another is usually accomplished with an inoculating loop or needle which has been sterilized by incineration in a flame (Figure 3.3).

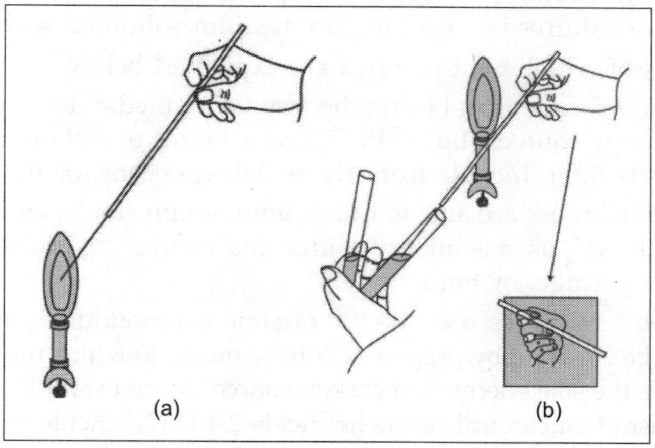

Figure 3.3 Aseptic transfer, loop is heated until rehot (a) tube is uncapped and loop is cooled in air briefly, sample is removed and tube is recapped. Sample is transferred to a sterile tube (b) loop is reheated before being taken out of service.

3.7 PROCEDURES FOR ISOLATION

Different procedures are adopted in order to isolate the microbes from different sources.

3.7.1 Microorganisms from Nature

In nature, microorganisms always live in mixture. The organism must first be isolated in pure culture before most characteristics of a particular microorganism can be determined. The general procedure to isolate an organism in pure culture is to use a culture medium. Incubation conditions which encourage the growth of desired organism and discourage other organisms are very important. Once the desired organism has grown in appreciable amount in relation to the others, it is purified and the isolated culture contains only single type microorganism desired.

3.7.2 Enrichment Culture

Microbes vary widely in their nutritional and environmental requirements. It is possible, by choosing the appropriate growth conditions, to select from a mixture a microbe of interest. The procedure of adjusting culture conditions to select a particular microbe called **enrichment culture** and is a procedure of considerable importance in microbiology.

The usual practice in enrichment culture procedures is to prepare a culture medium that will favour the growth of the microbe of interest. The medium is then inoculated with a sample of material thought to contain the microbe, such as soil, water, blood or tissue. The appropriate conditions of aeration, temperature, and pH are provided and the inoculated medium is then incubated. After an appropriate incubation time, usually several days, the enrichment culture is examined visually and microscopically for microbial growth. Cultures in which growth has taken place are then used further for isolation of pure cultures by streaking on medium solidified with agar (Figure 3.4).

Some examples of enrichment cultures are explained below:

1. Phototrophic microbes can be enriched by using media devoid of any organic or inorganic energy sources (but with CO_2 as a source of carbon) and incubating the cultures in the light. Inocula from rivers, lakes, oceans, or the soil may be used.
2. Lithotrophic microbes are able to utilize ammonium can be enriched by preparing medium with NH_4^+ as sole energy source and incubating in the dark. Inocula can be from soil, sewage, or mud.
3. Organotrophic microbes use specific organic compounds as sole energy source. Culture can be enriched by preparing culture media in which the organic compound in question is the sole energy and carbon source. As an example, if we are interested in an organism that can utilize the herbicide 2,4-D (2,4-dichlorophenoxyaceic acid), we could prepare a culture medium in which 2,4-D is the sole carbon and energy source.

   Loopful of inoculums is removed from tube, streak is made over a sterile agar plate, spreading the organisms, appearance of the streaked plate after incubation. Note the presence of isolated colonies. It is from such well-isolated colonies that pure cultures can usually be obtained.
4. If required to isolate a microbe capable to grow at high temperatures, enrichment cultures can be prepared using high temperatures for incubation.
5. If anaerobic organisms are desired, enrichment cultures devoid of O_2 can be prepared.

   Above examples are just a few of many that could be tried using appropriate enrichment culture procedures. Most of our knowledge on microbial diversity has come from a study of pure cultures derived by enrichment procedures.

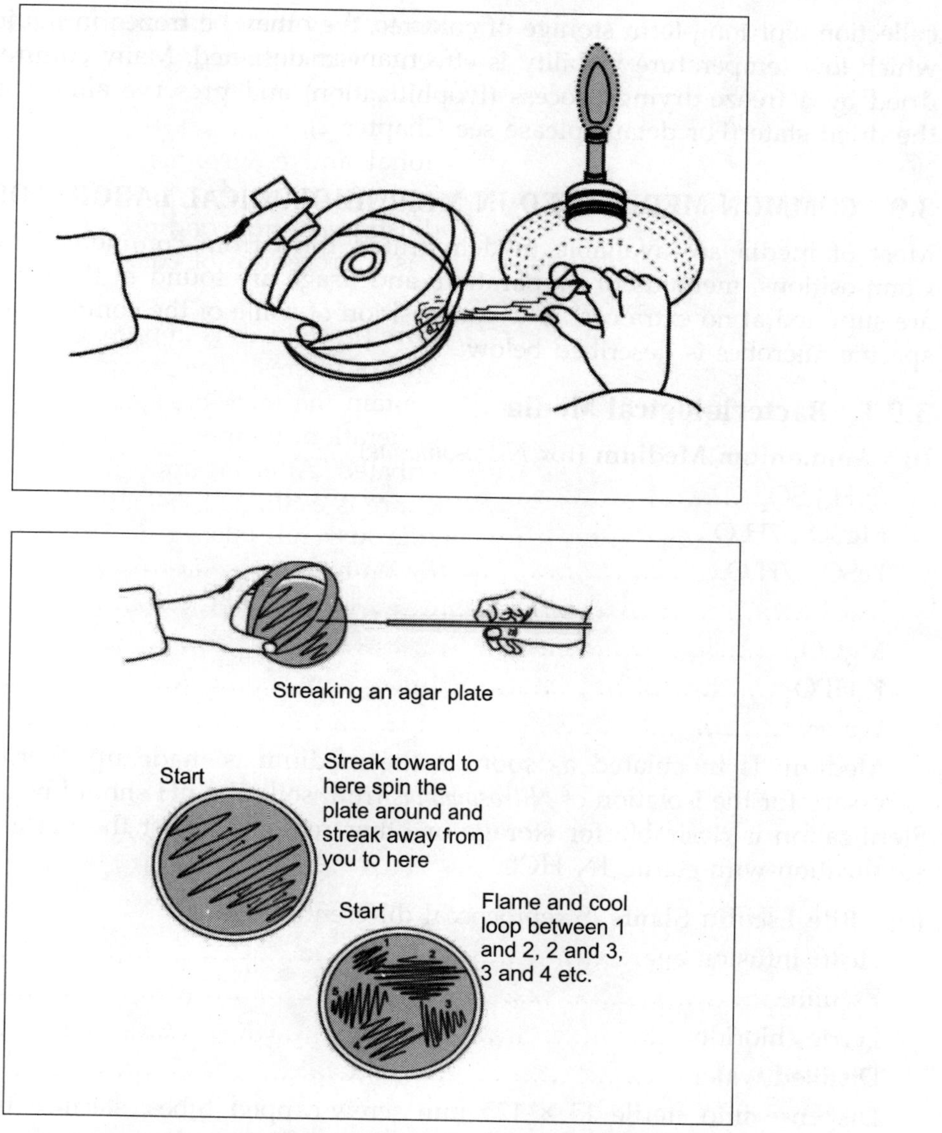

Figure 3.4 Method of pouring plates and making a streak plate to obtain pure cultures.

3.8 MAINTENANCE OF PURE CULTURES

One of the most frequent ways in which erroneous results and conclusions are obtained in microbiology is by the use of contaminated cultures. Obtained/isolated pure culture must be kept pure. Pure culture of microbes of interest is maintained in the laboratory for study and reference is called stock cultures. The stock culture must be maintained to retain viability should be free from contaminants and genetically homogeneous. Cultures which are infrequently used can be purchased from a commercial culture collection. For long-term storage of cultures, they may be frozen in liquid nitrogen, at which low temperature viability is effectively maintained. Many cultures can also be dried by a freeze-drying process (lyophilization) and preserve almost indefinitely in the dried state (For details please see Chapter 2).

3.9 COMMON MEDIA USED IN MICROBIOLOGICAL LABORATORIES

Most of media are available in dehydrated form from commercial manufacturers. Compositions, methods of preparation, and usage are found in their manuals, which are supplied at no extra cost. The composition of some of the common media used for specific microbes is described below:

3.9.1 Bacteriological Media

(i) Ammonium Medium (for *Nitrosomonas*)

$(NH_4)_2SO_4$........ 2.0 g
$MgSO_4 \cdot 7H_2O$........ 0.5 g
$FeSO_4 \cdot 7H_2O$........ 0.03 g
NaCl........ 0.3 g
$MgCO_3$........ 10.0 g
K_2HPO_4........ 1.0 g
Water........ 1000 mL

Medium is inoculated as soon as the medium is made up. Sterilization is not necessary for the isolation of *Nitrosomonas* from soil. The pH should be adjusted to 7.3. Sterilization is desirable for storage or other reasons, adjust the pH aseptically after sterilization with sterile 1N HCL.

(ii) Bile Esculin Slants (Streptococcal differentiation)

Heart infusion agar........ 4.0 g
Esculin........ 1.0 g
Ferric chloride........ 0.5 g
Distilled water........ 1000 mL

Dispense into sterile 15 × 125 mm screw-capped tubes, sterilize in autoclave at 121°C for 15 minutes, and slant during cooling.

(iii) Blood Agar

Trypticase soy agar powder........ 40 g
Distilled water........ 1000 mL
Final pH of 7.3
Defibrinated sheep or rabbit blood........ 50 mL

Liquefy and sterilize 1000 mL of Trypticase soy agar (TSA) in a large Erlenmeyer flask. While the TSA is being sterilized, warm up 50 mL of defibrinated blood to 50°C. After cooling the TSA to 50°C, aseptically transfer the blood to the flask and mix by gently rotating the flask (cold blood may cause lumpiness). Pour 10-12 mL of the mixture into sterile Petri plates. If bubbles form on the surface of the medium, flame the surface gently with a Bunsen burner before the medium solidifies.

(iv) Bromthymol Blue Carbohydrate Broths

Make up stock indicator solution:

Bromthymol blue.. 8 g
95% ethyl alcohol.. 250 mL
Distilled water.. 250 mL

Indicator is dissolved first in alcohol and then water is added.

Make up broth

Sugar base (lactose, sucrose, glucose, etc.).. 5 g
Tryptone.. 10 g
Yeast extract.. 5 g
Indicator solution.. 2 mL
Distilled water.. 1000 mL
Final pH 7.0

(v) Chlorobium Agar Shake Deeps (Photosynthetic bacteria)

The incompatibility of the ingredients of this medium at autoclave temperatures requires that the medium be made up of in five different portions that can be sterilized separately and mixed at a lower temperature. Ideally, the medium is made up the day it is to be used and kept liquid in a water bath until the inoculations are made.

Solutions A
(Make in 2000 mL flask)

NH_4Cl.. 1.0 g
KH_2PO_4.. 1.0 g
$MgCL_2$.. 0.5 g
Agar.. 15.0 g
Water.. 700 mL

Solutions B

$NaHCO_3$.. 2.0 g
Water.. 100 mL

Solutions C

$Na_2S.\ 9H_2O$.. 1.0 g
Water.. 100 mL

Solutions D

$FeCl_3.\ 6H_2O$.. 5.0 g
Water.. 100 mL

Solutions E

(1N H_3PO_4)

After autoclaving all five solutions at 121°C for 15 minutes, cool them to 50°C and pour solutions B, C and D into solution A flask. Use solution E to adjust the pH of the medium to 7.3.

Dispense aseptically to sterile 16 mm dia. Soft glass test tube (10 mL per tube). Keep the tubes of media liquefied in a 50°C water bath until inoculations are completed.

(vi) Chlorobium Enrichment Medium

It is best to prepare it without sterilization just before it is to be used. It is unstable in nature also precludes storage of the medium for any length of time in the refrigerator. Ingredients are incompatible at high temperature.

NH_4Cl	1.0 g
KH_2PO_4	1.0 g
$MgCl_2$	.0.5 g
$NaHCO_3$	2.0 g
$Na_2S.\ 9H_2O$	1.0 g
$FeCL_3.6H_2O$	0.0005 g
Water	1000 mL

(vii) Chromatium Agar Shake Deeps

This medium is similar in composition to that used for Chlorobium. Thus, the problems are similar. The medium must be made up in three different portions that can be sterilized separately and mixed at a lower temperature. It is best to make up the medium the same day it is to be used. It must be kept liquid in a water bath until the inoculations are made.

Solutions A

(Make in 2000 mL flask)

NH_4Cl	1.0 g
KH_2PO_4	1.0 g
$MgCl_2$	0.5 g
Agar	20.0 g
Water	780 mL

Solutions B

$NaHCO_3$	2.0 g
Water	100 mL

Solutions C

$Na_2S.\ 9H_2O$	1.0 g
Water	100 mL

After sterilizing each of the three solutions in the autoclave at 121°C for 15 minutes, cool them to 50°C and pour solutions B, and C into flask A.

Dispense aseptically to sterile 16 mm dia. Soft glass test tube (10 mL per tube). Keep the tubes of media liquefied in a 50°C water bath until inoculations are completed.

(viii) Chromatium Enrichment Medium

This medium is similar in composition to the enrichment medium used for Chlorobium. Thus, the problems are similar. It is best to prepare this medium without sterilization and use it immediately, it does not store well, even in the refrigerator, because of its unstable nature

NH_4Cl	1.0 g
KH_2PO_4	1.0 g
$MgCl_2$	0.05 g
$NaHCO_3$	2.0 g
Na_2S. $9H_2O$	1.0 g
Water	1000 mL

(ix) Deca-Strength Phage Broth (DSPB) for Bacteriophage

Peptone	100 g
Yeast extract	50 g
NaCl	25 g
K_2HPO_4	80 g
Distilled water	1000 mL

Final pH 7.6

(x) Glucose-Minimal Salts Agar (Media used for Bacterial Genetics)

This medium is made from glucose, agar, and Vogel-Bonner medium E (50x)

VOGEL-Bonner medium E (50x)

Distilled water	670 mL
$MgSO_4.7H_2O$	10 g
Citric acid monohydrate	100 g
K_2HPO_4 (anhydrous)	500 g
Sodium ammonium phosphate ($NaHN_4PO_4.4H_2O$)	175 g

Add salt in the order indicated to warm water (45°C) in a 2 L beaker or flask placed on a magnetic stirring hot plate. Allow each salt to dissolve completely before adding the next. Adjust the volume to 1 L. Distribute into two 1 L Glass bottle. Autoclave, loosely cap, for 20 minutes at 121°C.

(xi) Plates of Glucose Minimal Salts Agar

Agar	15 g
Distilled water	930 mL

50 × V-B salts.. 20 mL
40% glucose.. 50 mL

Add 15 g of agar to 930 mL of distilled water in 2 L flask. Autoclave for 20 minutes using slow exhaust. When the solution has cooled slightly, add 20 mL of sterile 50 × V-B salts and 50 mL of sterile 40% glucose. For mixing, a large magnetic stir bar can be added to the flask before autoclaving. After all the ingredients have been added, the solutions should be stirred thoroughly. Pour 30 mL into each Petri plate. **Important** The 50 × V-B salts and 40% glucose should be autoclave separately.

(xii) Glucose Peptone Acid Agar

Glucose.. 10 g
Peptone.. 5 g
Monopotassium phosphate.. 1 g
Magnesium sulfate ($MgSO_4.7H_2O$).. 0.5 g
Agar.. 15 g
Water.. 1000 mL

While still liquid after sterilization, add sufficient sulfuric acid to bring the pH down to 4.0.

(xiii) Glycerol Yeast Extract Agar

Glycerol.. 5 mL
Yeast extract.. 2 g
Dipotassium phosphate.. 1 g
Agar.. 15 g
Water.. 1000 mL

(xiv) Endo MF Broth (Bacterial count with Membrane filter)

This medium is extremely hygroscopic in the dehydrated form and oxidizes to cause deterioration of the medium after the bottle has been opened. Once a bottle has been opened it should be dated and discarded after one year. If the medium becomes hardened within that time it should be discarded. Storage of the bottle inside a larger bottle that contains silica gel will extend its shelf life. It is best to make up the medium the day it is to be used. It should not be stored over 96 hours prior to use.

1. Into a 250 mL screw-cap Erlenmeyer flask place the following:

   Distilled water.. 50 mL
   95% ethyl alcohol.. 2 mL
   Dehydrated medium (m Endo MF broth).. 4.8 g

   Shake the above mixture by swirling the flask until the medium is dissolved and then add another 50 mL of distilled water.

2. Cap the flask loosely and immerse it into a pan of boiling water. As soon as the medium begins to simmer, remove the flask from the water bath. Do not boil the medium any further.

3. Cool the medium to 45°C, and adjust the pH to between 7.1 and 7.3.
4. If the medium be stored for a few days, place it in the refrigerator at 2° – 10°C, with screw-cap tightened securely.

(xv) Milk Salt Agar (15% NaCl)

Prepare three separate beakers of the following ingredients:

1. Beaker containing 200 g of sodium chloride.
2. Large beaker (2000 mL size) containing 50 g of skim milk powder in 500 mL of distilled water.
3. Glycerol-peptone agar medium.

$MgSO_4.7H_2O$	5.0 g
$MgNO_3.6H_2O$	1.0 g
$FeCl_3.7H_2O$	0.025 g
Difco protecose-peptone #3	5.0 g
Glycerol	10.0 g
Agar	30.0 g
Distilled water	500 mL

Sterilize the above three beakers separately. The milk solution should be sterilized at 113°C – 115°C (10 psi) in autoclave for 20 minutes. The salt and glycerol-peptone agar can be sterilized at conventional pressure and temperature. After the milk solution has cooled to 55°C. Add the sterile salt, which should also be cooled down to a moderate temperature. If the salt is too hot, coagulation may occur. Combine the milk salt and glycerol-peptone agar solutions by gently swirling with a glass rod. Dispense aseptically into Petri plates.

(xvi) Minimal Agar (Bacterial Conjugation Experiments)

Glucose	2 g
$(NH_4)_2SO_4$	1 g
K_2HPO_4	7 g
$MgSO_4$	0.5 g
Agar	15 g
Distilled water	1000 mL

(xvii) Nitrite Medium (for *Nitrobacteria*)

$NaNO_2$	1.0 g
$MgSO_4.7H_2O$	0.5 g
$FeSO_4.\ 7H_2O$	0.03 g
NaCl	0.3 g
Na_2CO_3	1.0 g
K_2HPO	1.0 g
Water	1000 mL

Medium is inoculated as soon as the medium is made up. Sterilization is not necessary for the isolation of *Nitrobacteria* from soil. The pH should be adjusted to 7.3. Sterilization is desirable for storage or other reasons, adjust the pH aseptically after sterilization with sterile 1N HCl.

(xviii) Nitrogen-Free Glucose Agar (Free living Nitrogen Fixer Microorganisms)

Add 15 g of agar to the basal salts portion of the above recipe without nitrate, bring to a boil, and sterilize in the autoclave at 121°C for 15 minutes. 1 gm glucose is dissolved in 100 mL of water and sterilized separately in similar manner mix the two solutions aseptically, and dispense into sterile Petri plates.

(xix) Nitrogen-Free Medium (*Azotobacter*)

K_2HPO_4	1.0 g
$MgSO_4.7H_2O$	0.2 g
$FeSO_4$. $7H_2O$	0.05 g
$CaCl_2.2H_2O$	0.01 g
Na_2MoO_4. $2H_2O$	0.001 g
*Glucose	10.0 g
Distilled water	1000 mL

*Sterilize Separately.

Adjust pH to 7.2

Medium is inoculated as soon as the medium is made up. Sterilization is not necessary for the isolation of *Azotobacter* from soil. The pH should be adjusted to 7.2. Sterilization is desirable for storage or other reasons, adjust the pH aseptically after sterilization with sterile 1N HCl.

(xx) Phage Growth Medium (Isolation of Bacteriophages)

KH_2PO_4	1.5 g
Na_2HPO_4	3.0 g
NH_4Cl	1.0 g
$MgSO_4.4H_2O$	0.2 g
Glycerol	10.0 g
Acid-hydrolyzed casein	5.0 g
dl-Tryptophan	0.01 g
Gelatin	0.02 g
Tween-80	0.2 g
Distilled water	1000 mL

Sterilize in autoclave at 121°C for 20 minutes.

(xxi) Phage Lysing Medium (Lytic Phage)

Add sufficient sodium cyanide (NaCN) to the above growth medium to bring the concentration up to 0.02 M. For 1 L of lysing medium this will amount to about 1 g (actually 0.98 gm) of NaCN. When an equal amount of this lysing medium is added to the growth medium during the last 6 hours of incubation, the concentration of NaCN in the combined medium is 0.01 M.

(xxii) Russell Double Sugar Agar (Gram-negative Intestinal Pathogens)

Beef extract	1 g
Proteose peptone No. 3 (Difco)	12 g
Lactose	10 g
Dextrose	1 g
Sodium chloride	5 g
Agar	15 g
Phenol Red (Difco)	0.025 g
Distilled water	1000.0 mL

Final pH 7.5 at 25°C

Dissolve ingredients in water, and bring to boiling. Cool to 50°C- 60°C and dispense about 8 mL per tube (16 mm dia tubes). Slant tubes to cool. Butt depth should be about 1/2″.

(xxiii) Skim Milk Agar

Skim milk powder	100 g
Agar	15 g
Distilled water	1000 mL

Dissolve 15 g of agar into 700 mL of distilled water by boiling. Pour into a flask and sterilize at 121°C, 15 lb pressure.

In a separate container, dissolve 100 g of skim milk powder into 300 mL of water heated to 50°C. Sterilize this milk solution at 113°C-115°C (10 psi) for 20 minutes.

After the two solutions have been sterilized, cool to 55°C and combine in one flask, swirling gently to avoid bubbles. Dispense into sterile Petri plates.

(xxiv) Sodium Chloride (6.5%) Tolerance Broth (*Streptococci* differentiation)

Heart infusion broth	25 g
NaCl	60 g
Indicator (1.6 g bromcresol purple) in 100 mL 95% ethanol)	1 mL
Dextrose	1 g
Distilled water	1000 mL

Add all reagents together up to 1000 mL (final volumes). Dispense in 15 × 125 mm screw-capped tubes and sterilize in autoclave for 15 minutes at 121°C.

A positive reaction is recorded when the indicator changes from purple to yellow or when growth is obvious even though the indicator does not change.

(xxv) Sodium Hippurate Broth (Streptococci differentiation)

Heart infusion broth........ 25 g
Sodium hippurate........ 10 g
Distilled water........ 1000 mL

Sterilize in autoclave at 121°C for 15 minutes after dispensing in 15 × 15 mm screw-capped. Tighten caps to prevent evaporation.

(xxvi) Soft Nutrient Agar (for bacteriophage)

Dehydrated nutrient broth........ 8 g
Agar........ 7 g
Distilled water........ 1000 mL

Sterilize in autoclave at 121°C for 20 minutes.

(xxvii) Spirit Blue Agar (Bacterial Hydrolysis)

This medium is used to detect lipase production by bacteria. Lipolytic bacteria cause the medium to change from pale lavender to deep blue.

Spirit blue agar (Difco)........ 35 g
Lipase reagent (Difco)........ 35 mL
Distilled water........ 1000 mL

Dissolve the spirit blue agar in 1000 mL of water by boiling. Sterilize in autoclave for 15 minutes at 15 psi (121°C). Cool to 55°C and slowly add the 35 mL of lipase reagent, agitating to obtain even distribution. Dispense into sterile Petri plates.

(xxviii) Streptomycin Agar (Replica Plating Mutant Isolation)

To 1000 mL sterile liquid nutrient agar (50°C), aseptically add 100 mg of streptomycin sulfate. Pour directly into sterile Petri plates.

(xxix) Top Agar (Bacterial Mutagenicity Test)

Tubes containing 2 mL of top agar made up just prior to using from bottles of top agar base and his/bio stock solution.

His/Bio Stock Solution

D-Biotin (F.W.247.3)........ 30.9 mg
L-Histidine HCl (F.W.191.7)........ 24.0 mg
Distilled water........ 250 mL

Dissolve by heating the water to the boiling point. This can be done in a micro-wave oven. Sterilize by filtration through 0.22 μm membrane filter or autoclave for 20 minutes at 121°C. store in a glass bottle at 4°C.

Top Agar Base

Agar........ 6 g
Sodium chloride (NaCl)........ 5 g
Distilled water........ 1000 mL

The agar may be dissolved in a steam bath or microwave oven or by autoclaving briefly. Mix thoroughly and transfer 100 mL aliquots to 250 mL glass bottles with screw caps. Autoclave for 20 minutes with loosened caps. Slow exhaust. Cool the agar and tighten caps.

Just before using, add 10 mL of the his/bio stock solution to bottles of 100 mL of liquefied top agar base (45°C). After thoroughly mixing, distribute, aseptically, 2 mL of this mixture to sterile tubes (13 mm × 100 mm). Hold tubes at 45°C until used.

(xxx) Tryptone Agar

Tryptone.. 10 g
Agar.. 15 g
Distilled water.. 1000 mL

Tryptone Broth

Tryptone.. 10 g
Distilled water.. 1000 mL

(xxxi) Tryptone Yeast Extract Agar

Tryptone.. 10 g
Yeast extract.. 5 g
Dipotassium phosphate.. 3 g
Sucrose.. 50 g
Agar.. 15 g
Water.. 1000 mL

pH 7.4

(xxxii) Nutrient Broth/Agar (Routine Bacteriological Test)

Peptone.. 5 g
Beef extract.. 3 g
Sodium Chloride.. 5 g
Agar.. 15 g
Water.. 1000 mL

pH 7.2

Prepare plates/slants/broth (without agar) after routine steam sterilization by autoclaving at 121°C

3.9.2 Media for Fungal Growth

3.9.2.1 Non-synthetic media (natural)

(i) Corn meal Agar

Com meal	20 g
Peptone (if desired)	20 g
Dextrose (if desired)	15 g

Agar	15 g
Distilled water	1000 mL
(ii) Oat meal Agar	
Oats	100 g
Agar	15 g
Distilled water	1000 mL
(iii) Potato Dextrose Agar	
Peeled potato	250 g
Glucose	20 g
Agar	15 g
Distilled water	1000 mL
pH	6-6.5
(iv) Soil Extract Agar	
Soil extract	100 mL
Glucose	1 g
K_2HPO_4	0.5 g
Agar	15 g
Tap water	900 mL
pH	7-7.2

Preparation of Soil Extract

Steam 1000 g garden soil in one L of tap water in an autoclave at 15 psi for 30 min. Add a small quantity of calcium carbonate and filter the soil suspension through double layers of filter papers. The filtrate is sterilised in a flask and stocked for future use.

3.9.2.2 Synthetic media

(i) Malt Extract Agar

Medium is useful for the growth of wood destroying fungi.

Malt extract	25 g
Agar	15 g
Distilled water	1000 mL

(ii) Martin's Rose Bengal Streptomycin Agar

Dextrose	10 g
Peptone	5 g
KH_2PO_4	1 g
$MgSO_4$	0.5 g
Rose Bengal (1 part in 30,000 parts of medium)	
Agar	20 g

Streptomycin	0.03 g
Distilled water	1000 mL

(iii) Waksman's medium

Specially used for counting soil actinomycetes

Glucose	10 g
Peptone	5 g
K_2HPO_4	1 g
$MgSO_4$	0.5 g
Agar	20 g
Distilled water	1000 mL

Adjust pH to 4

(iv) Sabraud's Medium

Medium is suitable for many fungi including human or animal pathogens.

Glucose (or Maltose)	40 g
Peptone	10 g
Agar	15 g
Distilled water	1000 mL

(v) Modified Sabouraud's glucose agar (Emmons Culture Medium For Fungi)

C.W. Emmons developed the following recipe as an improvement over Sabouraud's glucose agar for the cultivation of fungi. Its principal advantage is that a neutral pH does not inhibit certain molds that have difficulty growing on Sabouraud's agar (pH 5.6). Instead of relying on a low pH to inhibit bacteria, it contains chloramphenicol, which does not adversely affect the fungi.

Glucose	20 g
Neopeptone	10 g
Agar	20 g
Chloramphenicol	40 mg
Distilled water	1000 mL

After the glucose, peptone, and agar are dissolved, heat to boiling, add the chloramphenicol which has been suspended in 10 mL of 95% alcohol and remove quickly from the heat. Autoclave for only 10 minutes.

(vi) Czapek's Medium (for soil inhabiting fungi)

Sucrose	30 g
Sodium nitrate	2 g
K_2HPO_4	1 g
$MgSO_4$	0.5 g
KCI	0.5 g
$FeSO_4$	0.01 g
Distilled water	1000 mL

(vii) Coon's Medium (for *Fusarium*)

Sucrose	7.2 g
Dextrose	3.6 g
$MgSO_4$	1.23 g
K_2HPO_4	2.72 g
KNO_3	2.02 g
Distilled water	1000 mL

Add 25 ppm of malachite green or 40 ppm Gention Violet.

(viii) Richards' Medium (for *Aspergillus*)

KNO_3	10 g
KH_2PO_4	5 g
$MgSO_4$	2.5 g
$FeCI_3$	0.02 g
Sucrose	50 g
Distilled water	1000 mL
pH	6.6-7.2

3.9.3 Cyanobacteria

Cyanobacteria can easily be grown in culture under laboratory conditions. Some of the simple techniques for isolation include direct inoculation into minimal media, soil dilution and moist culture techniques. Rice-field soils when dry can be used for preparation of moist culture plates and many taxa in the reproductive phase as spores or akinetes come up in the plates within a short time. By repeatedly streaking, selected colonies on agar plates mixed cultures can be made unicyanobacterial. Preparation of axenic culture is known to be achieved through various treatments. Some of the methods are: (a) washing with chlorine water (b) treatment with detergent and phenol (c) treatment with antibiotics (d) exposure to ultraviolet light (e) exposure to gamma radiation (f) temperature treatment and (g) picking up individual hormogoia. Among these methods antibiotic treatment has been successfully used for preparation of axenic cultures. Axenic cultures are obtained making full use of one's skill and intelligence.

(i) Modified Chu-10 medium (Gerloff *et al.*, 1950)

$Ca(NO_3)_2.4H_2O$	------	40 mg
$K_2HPO_4.3H_2O$	------	10 mg
$MgSO_4.7H_2O$	------	25 mg
$NaCO_3$	------	20 mg
$NaSiO_3.9H_2O$	------	25 mg
Ferric citrate	------	3 mg
Citric acid	------	3 mg
Glass distilled water	------	1 L

1 mL/L of Fogg's micronutrient solution is to be added in to the above medium. The composition of the solution is:

$MnCl_2.4H_2O$	------	10 mg
$Na_2MoO_{4.}2H_2O$	------	10 mg
H_3BO_3	------	10 mg
$CuSO_4.5H_2O$	------	1 mg
$ZnSO_4.7H_2O$	------	1 mg
Distilled Water	------	100 mL

pH of this medium should be adjusted to 7.5 before autoclaving.

(ii) Allen and Arnon's medium (Allen and Arnon, 1955)

KNO_3H	------	2.020 g
$K_2HPO_4.3H_2O$	------	456 mg
$MgSO_4.7H_2O$	------	246 mg
NaCl	------	232 mg
$CaCl_2.2H_2O$	------	74 mg

1 mL/L Fogg's micronutrient solution and 4 mg/L of Fe (as Fe EDTA complex) should be added.

(iii) BG-11 medium (Rippka et al., 1979)

$NaNo_3$	------	1.5 mg
$K_2HPO_4.3H_2O$	------	40 mg
$MgSO_4.7H_2O$	------	75 mg
$CaCl_2.2H_2O$	------	36 mg
Citric acid	------	6 mg
Ferric ammonium citrate	------	6 mg
EDTA (Na_2Mg salt)	------	1 mg
Na_2CO_3	------	20 mg
Glass distilled water	------	1 L
Trace metal solution$^+$	------	1 mL

For preparing N-free medium $NaNO_3$ may be omitted.

$^+$Composition of trace metal solution "A_5+C"

H_3BO_3	------	2.86 g/L
$MnCl_2.4H_2O$	------	1.81 g/L
$ZnSO_4.7H_2O$	------	0.222 g/L
$Na_2MoO_42H_2O$	------	2.86 g/L
$CuSO_4.5H_2O$	------	1.81 g/L
$Co(NO_3)_2.6H_2O$	------	0.0494 g/L

After autoclaving and cooling the pH of the medium is 7.1

(iv) Allen (1968) Modified from Hughes *et al.* (1958)

Composition	g/L
$NaNO_3$	15
K_2HPO_4	0.03
$MgSO_4.7H_2O$	0.075
Na_2CO_3	0.02
$CaCI_2$	0.027
$Na_2SiO_3.9H_2O$	0.058
EDTA	0.005
Citric acid	0.006
Ferrous citrate	0.006
Microelements 1 mL/L of a solution having g/L	
H_3BO_3	2.86
$MnCl_2.4\ H_2O$	1.81
$ZnSO_4.7H_2O$	0.222
$Na_2MoO_4.2H_2O$	0.391
$CuSO_4.5H_2O$	0.079
$Co\ (NO_3)2.6H_2O$	0.0494
pH	7.8

Used for unicellular and colonial form, viz., *Anacystis nidulans; Gleocapsa alpieola.* Extremely valuable for surface culture and assists the growth of single celled colony when 1.5% agar is used.

(v) Cannon *et al.* (1971) Modified from Safferman and Morris (1964)

Composition	g/ L
$NaNO_3$	0.17
$Ca\ (NO_3)2\ .4\ H_2O$	0.01
K_2HPO_4	0.01
$MgSO_4.7H_2O$	0.025
Na_2CO_3	0.02
$Na_2SiO_3.5H_2O$	0.044
Ferric citrate	0.35
Citric acid	0.35
$CuCI_2.2H_2O$	0.001
$ZnCI_2$	0.04
H3BO3	0.24
$CoCI_2.6H_2O$	0.02
$MnCI_2.4H_2O$	0.014

(vi) Employed for *Plectonema boryanum*

Composition	g /L
Nitrilotriacetic acid	0.1
$CaSO_4.2H_2O$	0.06
$MgSO_4.7H_2O$	0.1
NaCI	0.008
KNO_3	0.103
NaNO3	0.689
Na2HPO4	0.111
$FeCl_3$ solution	0.2905 g (1 mL)*
Microelement solution 0.5 mL	

Microelement Solution

Composition	g /L
$MnSO_4.H_2O$	2.28
$ZnSO_4.7H_2O$	0.50
H_3BO_3	0.50
$CuSO_4.2H_2O$	0.025
$Na_2MoO_4.2H_2O$	0.045
Cone H_2SO_4	0.50

*Values in parentheses indicate the amount of solution in 1000 mL of media.
pH before autoclaving 8.2 and 7.5-7.6 after sterilization.

(vii) KM modified after Kratz and Myers (1955)

Composition	g /L
$MgSO_4.7H_2O$	0.25
K_2HPO_4	1.00
Ca $(NO_3)_2.4H_2O$	0.025
$NaNO_3$	1.00
Na_2 EDTA	0.031
Fe $(SO_4).6H_2O$	0.004
NH_4CI	0.1
Microelement solution	0.6 mL

Composition	g/L
$ZnSO_4.7H_2O$	8.82
$MnCi_2.4H_2O$	1.44
MoO_3	0.71
$CuSO_4.5H_2O$	1.57
$CO(NO_3)_2.6H_2O$	0.49
H_3BO_3	2.86

KM medium is employed for unicellular and/or coccoid cyanobacteria especially for the quantitative surface plating of *A. nidulans* to accurate estimation of viable counts.

(viii) Medium D

Composition	**g/L**
$MgSO_4.7H_2O$	0.15
K_2HPO_4	0.01
$NaNO_3$	1
EDTA	0.05
$Fe_2(SO_4)_3.\ 6H_2O$	0.004
NH_4Cl	0.1
Microelement solution 0.6 mL	

Composition	**g/L**
$ZnSO_4.\ 7H_2O$	8.82
$MnCl_2.4H_2O$	1.44
MoO_3	0.71
$CuSO_4$	1.57
$Co\ (NO_3)_2.6H_2O$	0.49

Medium D is employed for unicellular coccoid especially for the quantitative surface plating of *A. nidulans* for accurate estimation of viable counts.

(ix) Medium 'Z' modified after Zenader

Composition	**g/L**
$NaNO_3$	0.467
$Ca\ (NO_3)_2.4H_2O$	0.059
K_2HPO_4	0.031
$MgSO_4\ .\ 7H_2O$	0.025
Na_2CO_3	0.021
Microelement solution	10 mL

Composition	**g/L**
EDTA	0.08
$FeSO_4.7H_2O$	0.01

Suggested Readings

Alexopoulos, C. J., C.W. Mims and M. Blackwell. 1996. *Introductory Mycology*, 4th ed. New York, John Wiley and Sons.

Allen M. B. and D. I. Arnon. 1955. Studies on nitrogen-fixing blue green algae I. Growth and nitrogen fixing by Anabaena cylindrical Lemm. *Plant Physiol* **30**, 366-372.

Aneja K. R. 2011. *Experiments in microbiology and Plant Pathology and Biotechnology*. Pp 607, New Delhi. New Age International Publishers.

Benson H. J. 1998. *Microbiological applications: Laboratory manual in General microbiology*. Pp 462, Boston, WCB McGraw-Hill.

Bisen P. S. 2014. *Laboratory Protocols in Applied Life Sciences*, Boca Ratan, Florida, CRC Taylor and Francis.

Bisen P. S. and K Verma. 2004. *Handbook of Microbiology*, pp. 199 New Delhi, CBS Publishers

Bisen, P.S., G. B. K. S. Prasad and M. Debnath. 2012. *Microbes: Concepts and Applications*, 699. 1st ed. New York, John Wiley and Sons.

Cappuccino, J.G. and N. Sherman, N. 1991. *Microbiology: A Laboratory Manual*. The Amsterdam, The Netherlands Benjamin/Climmings Publishing Company,

Carr N G. and B A Whitton. 1982. Cyanobacteria current perspectives. In: *The Biology of Cyanobacteria*. Carr, N. G. and Whitton, B. A. (eds.) Blackwell Scientific Publications, Oxford, 1-8.

Cochran, W.G. 1950. Estimation of bacterial densities by means of the most probable number *Biometrics* 6: 105-116.

Dubey R. C. and D. K. Maheshwari. 2002. Practical microbiology, pp. 411, New Delhi, S. Chand & Company.

Hawker L. E. and A. H. Linton. 1971. *Mico-organisms: Function, form and environment*. London, Edward Arnold (Publishers) Ltd.

Kratz W. A. and J. Myers 1955. Nutrition and growth of several blue green algae. *Am J Bot* **42**, 282-287.

Pollack R. A., L. Findlay, W. Mondschein, R. Ronald Modesto. 2004. *Laboratory Exercises in Microbiology*, 280 pages New York, John Wiley and Sons.

Seeley, H.W., Jr. and P. J. Denmark. 1975. *Microbes in action: A Laboratory manual of microbiology*, 361. Mumbai, India, D.B. Taraporevala Sons & Co. Pvt. Ltd.

Chapter 4

Sterilization

4.1 INTRODUCTION

Killing all the bacteria or other **microorganisms** in or on objects is a process called sterilization in microbiological terms and the procedures that Pasteur and others used were eventually refined and carried over into microbiological research. Theory of spontaneous generation could be disapproved only due to the development of effective streilisation procedures without which modern-day microbiology as a science could not have developed and established. It was later shown that flasks and other vessels could be protected from contamination by cotton stoppers permitting the exchange of air but not particulate matter. The handling of sterile objects so as to maintain their sterility is termed aseptic technique. Although Pasteur was successful in sterilizing materials with simple boiling, some workers found that boiling was insufficient. The failure resulted from the presence of unusually heat-resistant structures called endospores.

4.2 STERILIZATION

It is essential for all microbiological experiments to ensure to allow growth of the desired microorganisms. All the components of experiments need to be sterilized, including the nutrient medium and other ingredients. Glassware and all other accessory equipment must also be sterilized to ensure pristine conditions and virtually no contamination. Different techniques of streilisation are available based on the properties of the component being sterilized. Most commonly used techniques include: heat sterilization, moist heat sterilization, dry heat sterilization, incineration, and boiling; sterilization by chemicals; sterilization by filtration; and sterilization by radiation (Figure 4.1). Type of sterilization used is based on the following conditions:

1. **The type of microorganism present:** Some microorganisms are resistant and it is very difficult to kill them compared to others.
2. **The number of microorganisms present:** It is much easier to kill lesser number of organisms compared to larger.
3. **The amount and type of organic material that protects the microorganisms:** Blood or tissue remaining on poorly cleaned labware/instruments acts as a shield to microorganisms during the sterilization process.
4. **The number of cracks and crevices on an instrument that might harbour microorganisms:** Microorganisms collect in, and are protected by, scratches, cracks and crevices such as the serrated jaws of tissue forceps.

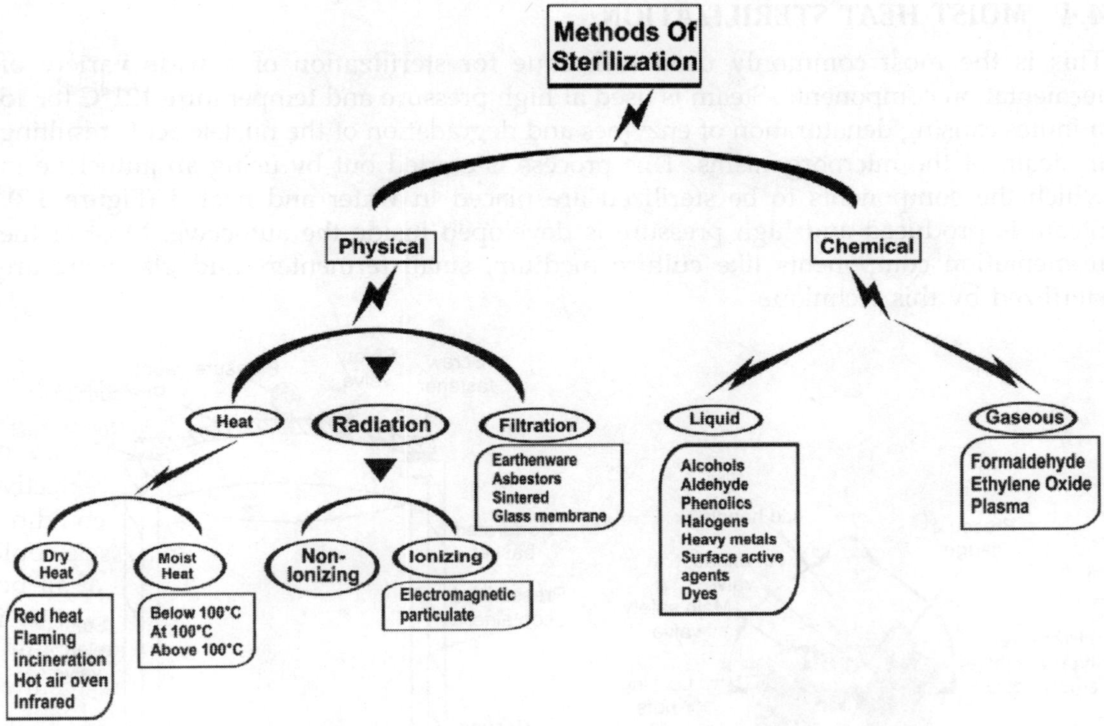

Figure 4.1 Various methods of sterilization.

4.3 HEAT

Heat is considered to be the most reliable method of sterilization of articles that can withstand heat. Heat acts by oxidative effects as well as denaturation and coagulation of proteins. Those articles that cannot withstand high temperatures can still be sterilized at lower temperature by prolonging the duration of exposure.

4.3.1 Factors Affecting Sterilization by Heat

- Nature of heat: Moist heat is more effective than dry heat
- Temperature and time: temperature and time are inversely proportional. As temperature increases the time taken decreases.
- Number of microorganisms: More the number of microorganisms, higher the temperature or longer the duration required.
- Nature of microorganism: Depends on species and strain of microorganism, sensitivity to heat may vary. Spores are highly resistant to heat.
- Type of material: Articles that are heavily contaminated require higher temperature or prolonged exposure. Certain heat sensitive articles must be sterilized at lower temperature.
- Presence of organic material: Organic materials such as protein, sugars, oils and fats increase the time required.

4.4 MOIST HEAT STERILIZATION

This is the most commonly used technique for sterilization of a wide variety of fermentation components. Steam is used at high pressure and temperature 121°C for 15 minutes causing denaturation of enzymes and degradation of the nucleic acids resulting in death of the microorganisms. This process is carried out by using an autoclave in which the components to be sterilized are placed in water and heated (Figure 4.2). Steam is produced and high pressure is developed inside the autoclave. Most of the fermentation components like culture medium, small fermenters and glassware are sterilized by this technique.

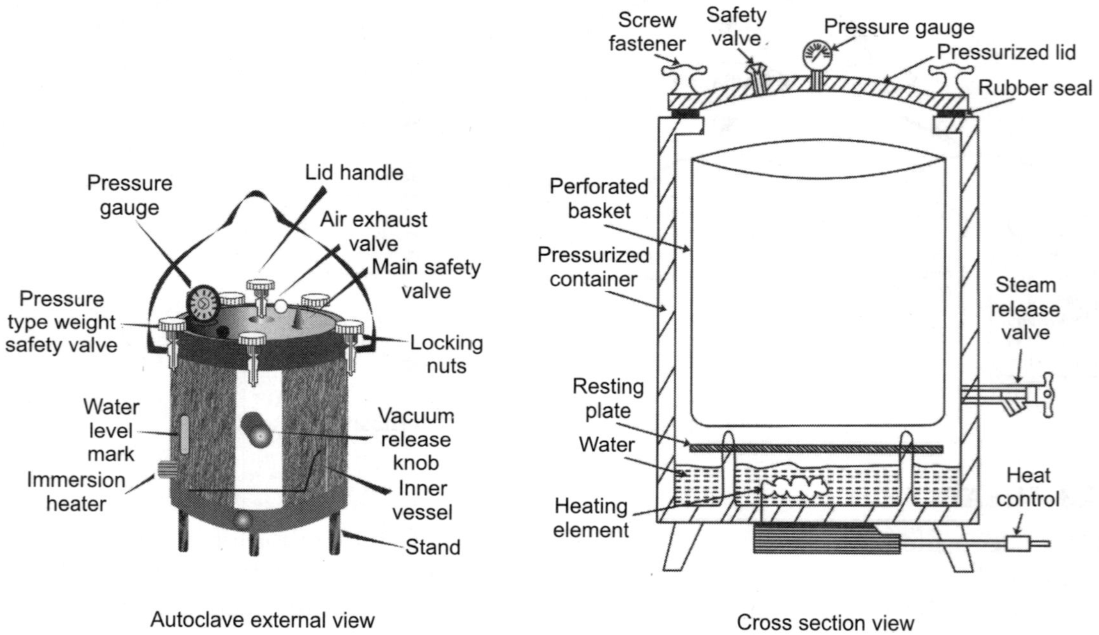

Figure 4.2 Autoclave.

Steam sterilization requires four conditions: **adequate contact, sufficiently high temperature, correct time** and **sufficient moisture**. Although all are necessary for sterilization to take place. Sterilization failures in clinics and hospitals are most often caused by lack of steam contact or failure to attain adequate temperature. Moist heat acts to coagulate and denature proteins of microorganisms.

4.4.1 BELOW 100°C

4.4.1.1 Pasteurization

This process was originally employed by Louis Pasteur. Currently, this procedure is employed in food and dairy industry. There are two methods of pasteurization, the holder method (heated at 63°C for 30 minutes) and flash method (heated at 72°C for 15 sec) followed by quickly cooling to 13°C. Other pasteurization methods include Ultra-High Temperature (UHT), 140°C for 15 sec and 149°C for 0.5 sec. This method is suitable to destroy most milk-borne pathogens like *Salmonella, Mycobacteria, Streptococci, Staphylococci* and *Brucella,* however *Coxiella* may survive pasteurization. Efficacy is tested by phosphatase test and methylene blue test.

4.4.1.2 Vaccine bath

The contaminating bacteria in a vaccine preparation can be inactivated by heating in a water bath at 60°C for one hour. Only vegetative bacteria are killed and spores survive. Serum bath: The contaminating bacteria in a serum preparation can be inactivated by heating in a water bath at 56°C for one hour on several successive days. Proteins in the serum will coagulate at higher temperature. Only vegetative bacteria are killed and spores survive.

4.4.1.3 Inspissation

This is a technique to solidify as well as disinfect egg and serum containing media. The medium containing serum or egg are placed in the slopes of an inspissator and heated at 80-85°C for 30 minutes on three successive days. On the first day, the vegetative bacteria would die and those spores that germinate by next day are then killed the following day. The process depends on germination of spores in between inspissation. If the spores fail to germinate then this technique cannot be considered for sterilization.

4.4.2 AT 100°C

Boiling: Boiling water (100°C) kills most vegetative bacteria and viruses immediately. Certain bacterial toxins such as Staphylococcal enterotoxin are also heat resistant. Some bacterial spores are resistant to boiling and survive; hence this is not a substitute for sterilization. The killing activity can be enhanced by addition of 2% sodium bicarbonate. When absolute sterility is not required, certain metal articles and glasswares can be disinfected by placing them in boiling water for 10-20 minutes. The lid of the boiler must not be opened during the period.

Steam at 100°C: Instead of keeping the articles in boiling water, they are subjected to free steam at 100°C. Traditionally, Arnold's and Koch's steamers were used. An autoclave (with discharge tap open) can also serve the same purpose. A steamer is a metal cabinet with perforated trays to hold the articles and a conical lid. The bottom of steamer is filled with water and heated. The steam that is generated sterilizes the articles when exposed for a period of 90 minutes. Media such as TCBS, DCA and selenite broth are sterilized by steaming. Sugar and gelatin in medium may get decomposed on autoclaving; hence they are exposed to free steaming for 20 minutes for three successive days. This process is known as tyndallisation (after John Tyndall) or fractional sterilization or intermittent sterilization. The vegetative bacteria are killed in the first exposure and the spores that germinate by next day are killed in subsequent days. The success of process depends on the germination of spores.

4.4.3 ABOVE 100°C

4.4.3.1 Autoclave

Sterilization can be effectively achieved at a temperature above 100°C using an autoclave. Water boils at 100°C at atmospheric pressure, but if pressure is raised, the temperature at which the water boils also increases. In an autoclave the water is boiled in a closed chamber. As the pressure rises, the boiling point of water also raises. At a pressure of 15 lbs inside the autoclave, the temperature is said to be 121°C. Exposure of articles to this temperature for 15 minutes sterilizes them. To destroy the infective agents associated with spongiform encephalopathies (prions), higher temperatures or longer times are used; 135°C or 121°C for at least one hour are recommended (Figure 4.2).

Advantages of steam: It has more penetrative power than dry air, it moistens the spores (moisture is essential for coagulation of proteins), condensation of steam on cooler surface releases latent heat, and condensation of steam draws in fresh steam.

Advantage: Very effective way of sterilization, quicker than hot air oven.

Disadvantages: Drenching and wetting or articles may occur, trapped air may reduce the efficacy, takes long time to cool.

Sterilization control: Physical method includes automatic process control, thermocouple and temperature chart recorder. Chemical method includes Browne's tube No. 1 (black spot) and succinic acid (whose melting point is 121°C) and Bowie Dick tape. Bowie Dick tape is applied to articles being autoclaved. If the process has been satisfactory, dark brown stripes will appear across the tape. Biological method includes a paper strip containing 10^6 spores of *Geobacillus stearothermophilus*.

4.5 DRY HEAT STERILIZATION

Dry heat acts by protein denaturation, oxidative damage and toxic effects of elevated levels of electrolytes. The moist heat acts by coagulation and denaturation of proteins. Moist heat is superior to dry heat in action. Temperature required to kill microbe by dry heat is more than the moist heat. **Thermal death time** is the minimum time required to kill a suspension of organisms at a predetermined temperature in a specified environment.

4.5.1 Red Heat

Articles such as bacteriological loops, straight wires, tips of forceps and searing spatulas are sterilized by holding them in Bunsen flame till they become red hot. This is a simple method for effective sterilization of such articles, but is limited to those articles that can be heated to redness in flame.

4.5.2 Flaming

This is a method of passing the article over a Bunsen flame, but not heating it to redness. Articles such as scalpels, mouth of test tubes, flasks, glass slides and cover slips are passed through the flame a few times. Even though most vegetative cells are killed, there is no guarantee that spores too would die on such short exposure. This method too is limited to those articles that can be exposed to flame. Cracking of the glassware may occur.

4.5.3 Hot Air Oven

This method was introduced by Louis Pasteur. Articles to be sterilized are exposed to high temperature (160°C) for duration of one hour in an electrically heated oven. Since air is poor conductor of heat, even distribution of heat throughout the chamber is achieved by a fan. The heat is transferred to the article by radiation, conduction and convection. The oven should be fitted with a thermostat control, temperature indicator, meshed shelves and must have adequate insulation.

Articles sterilized: Metallic instruments (like forceps, scalpels, and scissors), glassware (such as Petri dishes, pipettes, flasks, all-glass syringes), swabs, oils, grease, petroleum jelly and some pharmaceutical products.

Sterilization process: Articles to be sterilized must be perfectly dry before placing them inside to avoid breakage. Articles must be placed at sufficient distance so as to allow free circulation of air in between. Mouths of flasks, test tubes and both ends of pipettes must be plugged with cotton wool. Articles such as Petri dishes and pipettes may be arranged inside metal canisters and then placed. Individual glass articles must be wrapped in kraft paper or aluminium foils.

4.5.4 Sterilization Cycle

This takes into consideration the time taken for the articles to reach the sterilizing temperature, maintenance of the sterilizing temperature for a defined period (holding time) and the time taken for the articles to cool down. Different temperature-time relations for holding time are 60 minutes at 160°C, 40 minutes at 170°C and 20 minutes at 180°C. Increasing temperature by 10 degrees shortens the sterilizing time by 50 per cent. The hot air oven must not be opened until the temperature inside has fallen below 60°C to prevent breakage of glassware.

4.6 INCINERATION

This is a method of destroying contaminated material by burning them in incinerator. Articles such as soiled dressings; animal carcasses, pathological material and bedding, etc., should be subjected to incineration. This technique results in the loss of the article, hence is suitable only for those articles that have to be disposed. Burning of polystyrene materials emits dense smoke, and hence they should not be incinerated.

It is a waste treatment process that involves the combustion of organic substances contained in waste materials. Incineration and other high temperature waste treatment systems are described as "thermal treatment". Incineration of waste materials converts the waste into ash, flue gas, and heat. The ash is mostly formed by the inorganic constituents of the waste, and may take the form of solid lumps or particulates carried by the flue gas. The flue gases must be cleaned of gaseous and particulate pollutants before they are dispersed into the atmosphere. In some cases, the heat generated by incineration can be used to generate electric power.

Incineration with energy recovery is one of several waste-to-energy (WtE) technologies such as gasification, plasma arc gasification, pyrolysis and anaerobic digestion. Incineration may also be implemented without energy and materials recovery. In several countries, there are still concerns from experts and local communities about the environmental impact of incinerators. In some countries, incinerators built just a few decades ago often did not include a materials separation to remove hazardous, bulky or recyclable materials before combustion. These facilities tended to risk the health of the plant workers and the local environment due to inadequate levels of gas cleaning and combustion process control. Most of these facilities did not generate electricity. Incinerators reduce the solid mass of the original waste by 80–85% and the volume (already compressed somewhat in garbage trucks) by 95-96%, depending on composition and degree of recovery of materials such as metals from the ash for recycling. This means that while incineration does not completely replace land-filling, it significantly reduces the necessary volume for disposal. Garbage trucks often reduce the volume of waste in a built-in compressor before delivery to the incinerator. Alternatively, at landfills, the volume of the uncompressed garbage can be reduced by approximately 70% by using a stationary steel compressor, albeit with a significant energy cost. In many countries, simpler waste compaction is a common practice for compaction at landfills.

4.7 CHEMICALS

Disinfectants are those chemicals that destroy pathogenic bacteria from inanimate surfaces. Some chemical have very narrow spectrum of activity and some have very wide. Those chemicals that can sterilize are called chemisterilants. Those chemicals that can be safely applied over skin and mucus membranes are called antiseptics.

An ideal antiseptic or disinfectant should have the following properties:

- wide spectrum of activity
- be able to destroy microbes within practical period of time
- be active in the presence of organic matter
- make effective contact and be wettable
- be active in any pH
- be stable
- have long shelf life
- be speedy
- have high penetrating power
- be non-toxic, non-allergenic, non-irritative or non-corrosive
- not have bad odour

- not leave non-volatile residue or stain
- not lose efficacy on reasonable dilution
- not be expensive and must be available easily

Such an ideal disinfectant is not yet available. The level of disinfection achieved depends on contact time, temperature, type and concentration of the active ingredient, the presence of organic matter, the type and quantum of microbial load. The chemical disinfectants at working concentrations rapidly lose their strength on standing.

4.7.1 Alcohols

Mode of action: Alcohols dehydrate cells, disrupt membranes and cause coagulation of protein.

Examples: Ethyl alcohol, isopropyl alcohol and methyl alcohol

Application: A 70% aqueous solution is more effective at killing microbes than absolute alcohols. 70% ethyl alcohol (spirit) is used as antiseptic on skin. Isopropyl alcohol is preferred to ethanol. It can also be used to disinfect surfaces. It is used to disinfect clinical thermometers. Methyl alcohol kills fungal spores, hence is useful in disinfecting inoculation hoods.

Disadvantages: Skin irritant, volatile (evaporates rapidly), inflammable.

4.7.2 Aldehydes

Mode of action: Acts through alkylation of amino-, carboxyl- or hydroxyl group, and probably damages nucleic acids. It kills all microorganisms, including spores.

Examples: Formaldehyde, Gluteraldehyde

Application: 40% Formaldehyde (formalin) is used for surface disinfection and fumigation of rooms, chambers, operation theatres, biological safety cabinets, wards, sick rooms, etc. Fumigation is achieved by boiling formalin, heating paraformaldehyde or treating formalin with potassium permanganate. It also sterilizes bedding, furniture and books. 10% formalin with 0.5% tetraborate sterilizes clean metal instruments. 2% gluteraldehyde is used to sterilize thermometers, cystoscopes, bronchoscopes, centrifuges, anasethetic equipment, etc. An exposure of at least 3 hours at alkaline pH is required for action by gluteraldehyde. 2% formaldehyde at 40°C for 20 minutes is used to disinfect wool and 0.25% at 60°C for six hours to disinfect animal hair and bristles.

Disadvantages: Vapours are irritating (must be neutralized by ammonia), has poor penetration, leaves non-volatile residue, activity is reduced in the presence of protein. Gluteraldehyde requires alkaline pH and only those articles that are wettable can be sterilized.

4.7.3 Phenol

Mode of action: Act by disruption of membranes, precipitation of proteins and inactivation of enzymes.

Examples: 5% phenol, 1-5% Cresol, 5% Lysol (a saponified cresol), hexachlorophene, chlorhexidine, chloroxylenol (Dettol).

Applications: Joseph Lister used it to prevent infection of surgical wounds. Phenols are coal-tar derivatives. They act as disinfectants at high concentration and as antiseptics at low concentrations. They are bactericidal, fungicidal, mycobactericidal but are inactive against spores and most viruses. They are not readily inactivated by organic matter. The corrosive phenolics are used for disinfection of ward floors, in discarding jars in laboratories and disinfection of bedpans. Chlorhexidine can be used in an isopropanol solution for skin disinfection, or as an aqueous solution for wound irrigation. It is often used as an antiseptic handwash. 20% Chlorhexidine gluconate solution is used for pre-operative hand and skin preparation and for general skin disinfection. Chlorhexidine gluconate is also mixed with quaternary ammonium compounds such as cetrimide to get stronger and broader antimicrobial effects (e.g., Savlon). Chloroxylenols are less irritant and can be used for topical purposes and are more effective against Gram-positive bacteria than Gram-negative bacteria. Hexachlorophene is chlorinated diphenyl and is much less irritant. It has marked effect over Gram-positive bacteria but poor effect over Gram-negative bacteria, mycobacteria, Fungi and viruses. Triclosan is organic phenyl ether with good activity against Gram-positive bacteria and effective to some extent against many Gram-negative bacteria including Pseudomonas. It also has fair activity on fungi and viruses.

Disadvantages: It is toxic, corrosive and skin irritant. Chlorhexidine is inactivated by anionic soaps. Chloroxylenol is inactivated by hard water.

4.7.4 Halogens

Mode of action: They are oxidizing agents and cause damage by oxidation of essential sulfydryl groups of enzymes. Chlorine reacts with water to form hypochlorous acid, which is microbicidal.

Examples: Chlorine compounds (chlorine, bleach, hypochlorite) and iodine compounds (tincture iodine, iodophores).

Applications: Tincture of iodine (2% iodine in 70% alcohol) is an antiseptic. Iodine can be combined with neutral carrier polymers such as polyvinylpyrrolidone to prepare iodophores such as povidone-iodine. Iodophores permit slow release and reduce the irritation of the antiseptic. For hand washing iodophores are diluted in 50% alcohol. 10% Povidone Iodine is used undiluted in pre- and postoperative skin disinfections. Chlorine gas is used to bleach water. Household bleach can be used to disinfect floors. Household bleach used in a stock dilution of 1:10. In higher concentrations chlorine is used to disinfect swimming pools. 0.5% sodium hypochlorite is used in serology and virology. Used at a dilution of 1:10 in decontamination of spillage of infectious material. Mercuric chloride is used as a disinfectant.

Disadvantages: They are rapidly inactivated in the presence of organic matter. Iodine is corrosive and staining. Bleach solution is corrosive and will corrode stainless steel surfaces.

4.7.5 Heavy Metals

Mode of action: Act by precipitation of proteins and oxidation of sulfydryl groups. They are bacteriostatic.

Examples: Mercuric chloride, silver nitrate, copper sulfate, organic mercury salts (e.g., mercurochrome, merthiolate)

Applications: 1% silver nitrate solution can be applied on eyes as treatment for opthalmia neonatorum (Crede's method). This procedure is no longer followed. Silver sulphadiazine is used topically to help to prevent colonization and infection of burn tissues. Mercurials are active against viruses at dilution of 1:500 to 1:1000. Merthiolate at a concentration of 1:10000 is used in preservation of serum. Copper salts are used as a fungicide.

Disadvantages: Mercuric chloride is highly toxic, and readily inactivated by organic matter.

4.7.6 Surface Active Agents

Mode of actions: They have the property of concentrating at interfaces between lipid containing membrane of bacterial cell and surrounding aqueous medium. These compounds have long chain hydrocarbons that are fat soluble and charged ions that are water-soluble. Since they contain both of these, they concentrate on the surface of membranes. They disrupt membrane resulting in leakage of cell constituents.

Examples: These are soaps or detergents. Detergents can be anionic or cationic. Detergents containing negatively charged long chain hydrocarbon are called anionic detergents. These include soaps and bile salts. If the fat-soluble part is made to have a positive charge by combining with a quaternary nitrogen atom, it is called cationic detergents. Cationic detergents are known as quaternary ammonium compounds (or quat). Cetrimide and benzalkonium chloride act as cationic detergents.

Application: They are active against vegetative cells, mycobacteria and enveloped viruses. They are widely used as disinfectants at dilution of 1-2% for domestic use and in hospitals.

Disadvantages: Their activity is reduced by hard water, anionic detergents and organic matter. Pseudomonas can metabolise cetrimide, using them as a carbon, nitrogen and energy source.

4.7.7 Dyes

Mode of action: Acridine dyes are bactericidal because of their interaction with bacterial nucleic acids.

Examples: Aniline dyes such as crystal violet, malachite green and brilliant green. Acridine dyes such as acriflavin and aminacrine. Acriflavine is a mixture of proflavine and euflavine. Only euflavine has effective antimicrobial properties. A related dye, ethidium bromide, is also germicidal. It intercalates between base pairs in DNA. They are more effective against Gram-positive bacteria than Gram-negative bacteria and are more bacteriostatic in action.

Applications: They may be used topically as antiseptics to treat mild burns. They are used as paint on the skin to treat bacterial skin infections. The dyes are used as selective agents in certain selective media.

4.7.8 Hydrogen Peroxide

Mode of action: It acts on the microorganisms through its release of nascent oxygen. Hydrogen peroxide produces hydroxyl-free radical that damages proteins and DNA.

Application: It is used at 6% concentration to decontaminate the instruments, equipment such as ventilators. 3% hydrogen peroxide solution is used for skin disinfection and deodorising wounds and ulcers. Strong solutions are sporicidal.

Disadvantages: Decomposes in light, broken down by catalase, proteinaceous organic matter drastically reduces its activity.

4.7.9 Ethylene Oxide (EO)

Mode of action: It is an alkylating agent. It acts by alkylating sulfydryl-, amino-, carboxyl- and hydroxyl groups.

Properties: It is a cyclic molecule, which is a colourless liquid at room temperature. It has a sweet ethereal odour, readily polymerizes and is flammable.

Application: It is a highly effective chemisterilant, capable of killing spores rapidly. Since it is highly flammable, it is usually combined with CO_2 (10% CO_2 + 90% EO) or dichlorodifluoromethane. It requires presence of humidity. It has good penetration and is well absorbed by porous material. It is used to sterilize heat labile articles such as bedding, textiles, rubber, plastics, syringes, disposable Petri dishes, and complex apparatus like heart-lung machine, respiratory and dental equipment. Efficiency testing is done using *Bacillus subtilis* var *niger*.

Disadvantages: It is highly toxic, irritating to eyes, skin, highly flammable, mutagenic and carcinogenic.

4.7.10 Beta-Propiolactone (BPL)

Mode of action: It is an alkylating agent and acts through alkylation of carboxyl- and hydroxyl-groups.

Properties: It is a colourless liquid with pungent to slightly sweetish smell. It is a condensation product of ketone with formaldehyde.

Application: It is an effective sporicidal agent, and has broad-spectrum activity. 0.2% is used to sterilize biological products. It is more efficient in fumigation that formaldehyde. It is used to sterilize vaccines, tissue grafts, surgical instruments and enzymes

Disadvantages: It has poor penetrating power and is a carcinogen.

4.8 FILTRATION

Filtration does not kill microbes, it separates them out. Membrane filters with pore sizes between 0.2-0.45 µm are commonly used to remove particles from solutions that can't be autoclaved. It is used to remove microbes from heat labile liquids such as serum, antibiotic solutions, sugar solutions, and urea solution. Various applications of filtration include removing bacteria from ingredients of culture media, preparing suspensions of viruses and phages free of bacteria, measuring sizes of viruses, separating toxins from culture filtrates, counting bacteria, clarifying fluids and purifying hydatid fluid. Filtration is aided by using either positive or negative pressure using vacuum pumps. The older filters made of earthenware or asbestos are called depth filters.

4.8.1 Type of Filters

4.8.1.1 Earthenware filters

These filters are made up of diatomaceous earth or porcelain. They are usually baked into the shape of candle. Different types of earthenware filters are:

- Pasteur-Chamberland filter: These candle filters are from France and are made up of porcelain (sand and kaolin). Similar filter from Britain is Doulton. Chamberland filters are made with various porosities, which are graded as L1, L1a, L2, L3, L5, L7, L9 and L11. Doulton filters are P2, P5 and P11.
- Berkefeld filter: These are made of Kieselguhr, a fossilized diatomaceous earth found in Germany. They are available in three grades depending on their porosity (pore size); they are V (veil), N (normal) and W (wenig). Quality of V grade filter is checked using culture suspension of *Serrtia marcescens* (0.75 µm).
- Mandler filter: This filter from America is made of kieselguhr, asbestos and plaster of Paris.

4.8.1.2 Asbestos filters

These filters are made from chrysotile type of asbestos, chemically composed of magnesium silicate. They are pressed to form disc, which is to be used only once. The disc is held inside a metal mount, which is sterilized by autoclaving. They are available in the following grades; HP/PYR (for removal of pyrogens), HP/EKS (for absolute sterility) and HP/EK (for clarifying).

4.8.1.3 Sintered glass filters

These are made from finely ground glass that are fused sufficiently to make small particles adhere to each other. They are usually available in the form of disc fused into a glass funnel. Filters of Grade 5 have average pore diameter of 1-1.5 µm. They are washed in running water in reverse direction and cleaned with warm concentrated H_2SO_4 and sterilized by autoclaving.

4.8.1.4 Membrane filters

These filters are made from a variety of polymeric materials such as cellulose nitrate, cellulose diacetate, polycarbonate and polyester. The older type of membrane, called gradocol (graded colloid ion) membrane was composed of cellulose nitrate. Gradocol membranes have average pore diameter of 3-10 µm. The newer ones are composed of cellulose diacetate. These membranes have a pore diameter ranging from 0.015 µm to 12 µm. These filters are sterilized by autoclaving. Membrane filters are made in two ways, the capillary pore membranes have pores produced by radiation while the labyrinthine pore membranes are produced by forced evaporation of solvents from cellulose esters.

The disadvantages of depth filters are migration of filter material into the filtrate, absorption or retention of certain volume of liquid by the filters, pore sizes are not definite and viruses and mycoplasma could pass through. The advantages of membrane filters are known porosity, no retention of fluids, re-usable after autoclaving and compatible with many chemicals. However, membrane filters have little loading capacity and are fragile.

4.8.1.5 *Air filters*

Air can be filtered using HEPA (High Efficiency Particle Air) filters. They are usually used in biological safety cabinets. HEPA filters are at least 99.97% efficient for removing particles >0.3 µm in diameter. Examples of areas where HEPA filters are used include rooms housing severely neutropenic patients and those operating rooms designated for orthopaedic implant procedures. HEPA filter efficiency is monitored with the dioctylphthalate (DOP) particle test using particles that are 0.3 µm in diameter.

4.9 RADIATION

Two types of radiation are used, ionizing and non-ionizing. Non-ionizing rays are low-energy rays with poor penetrative power while ionizing rays are high-energy rays with good penetrative power. Since radiation does not generate heat, it is termed "cold sterilization". In some parts of Europe, fruits and vegetables are irradiated to increase their shelf life up to 500 per cent.

4.9.1 Non-ionizing Rays

Rays of wavelength longer than the visible light are non-ionizing. Microbicidal wavelength of UV rays lie in the range of 200-280 nm, with 260 nm being most effective. UV rays are generated using a high-pressure mercury vapour lamp. It is at this wavelength that the absorption by the microorganisms is at its maximum, which results in the germicidal effect. UV rays induce formation of thymine-thymine dimers, which ultimately inhibits DNA replication. UV readily induces mutations in cells irradiated with a non-lethal dose. Microorganisms such as bacteria, viruses, yeast, etc., that are exposed to the effective UV radiation are inactivated within seconds. Since UV rays don't kill spores, they are considered to be of use in surface disinfection. UV rays are employed to disinfect hospital wards, operation theatres, virus laboratories, corridors, etc. Disadvantages of using UV rays include low penetrative power, limited life of the UV bulb, some bacteria have DNA repair enzymes that can overcome damage caused by UV rays, organic matter and dust prevents its reach, rays are harmful to skin and eyes. It doesn't penetrate glass, paper or plastic.

4.9.2 Ionizing Rays

Ionizing rays are of two types, particulate and electromagnetic rays.

- Electron beams are particulate in nature while gamma rays are electromagnetic in nature. High speed electrons are produced by a linear accelerator from a heated cathode. Electron beams are employed to sterilize articles like syringes, gloves, dressing packs, foods and pharmaceuticals. Sterilization is accomplished in few seconds. Unlike electromagnetic rays, the instruments can be switched off. Disadvantage includes poor penetrative power and requirement of sophisticated equipment.
- Electromagnetic rays such as gamma rays emanate from nuclear disintegration of certain radioactive isotopes (Co^{60}, Cs^{137}). They have more penetrative power than electron beam but require longer time of exposure. These high-energy radiations damage the nucleic acid of the microorganism. A dosage of 2.5 megarads kills all bacteria, fungi, viruses and spores. It is used commercially to sterilize disposable petri dishes, plastic syringes, antibiotics, vitamins, hormones, glassware and fabrics. Disadvantages include; unlike electron beams, they can't be switched off, glassware tend to become brownish, loss of tensile strength in fabric. Gamma irradiation impairs the flavour of certain foods. *Bacillus pumilus* E601 is used to evaluate sterilization process.

Suggested Readings

Anonymous. 2002. Quality Control of Culture Media. *In* : Culture Media Manual LAB M. (IDG Ltd., United Kingdom); pp. 16-18.

Arora DR. 2004. Quality Assurance in Microbiology. *Indian J Med Microbiol* **22** :81-86.

Bridson EY, Brecker A. 1970. Design and Formulation of Microbial Culture Media. In: Norris SR, Ribbon DW. ed., *Methods in Microbiology* **39**, 414-422. Academic Press, New York.

Bridson EY. 1998. Culture Media *In*: The Oxoid Manual 8th edn. OXOID England: Limited Hampshire; pp. 2-8.

Duguid JP, Collee JG, Fraser AG, Aikman KW. 1996. Organization of the clinical bacteriology laboratory; quality assurance *In* : Mackie & McCartney Practical Medical Microbiology 14th edn. Churchill Livingstone, London; pp. 1-16.

Evans GL, Bell RH, Cunningham LV, *et al* .1996. Quality Assurance for Commercially Prepared Microbiological Culture Media; Approved standards- 2nd edn. *NCCLS* M22-A2:16.

Ferraro MJ, Wikler MA, Craig WA, *et al* .2003. Performance Standards for Antimicrobial Disk Susceptibility Tests; Approved Standards 8th edn. *NCCLS* M2-A8:23.

Gardner JF, Peel MM. 1986. Introduction to Sterilization and Disinfection. Churchill Livingstone.

Jones RN, Krisher K, Bird DS. 2003. Results of the survey of the quality assurance for the commercially prepared microbiology media. *Arch Pathol Lab Med* **127**: 661-665.

Krisher K, Callihan DR, Jones RN, *et al*. 2004. Quality Control for Commercially prepared Microbiological Culture Media; Approved Standards-3rd edn. *NCCLS* M22-A3:24.

Krisher K. 2001. Survey results: the use of the NCCLS standard M22-A2 and Quality assurance for commercially prepared microbiologic culture media by clinical laboratories. *Clin Microbiol Newsletter* **23**: 59-63.

Lewis IM. 1930. The inhibition of phytomonas malvaceara in culture media containing sugars. ***J BACTERIOL* 19**, 423–433.

MacLowry JD, Edson DC, Dreskin R. 1985. CAP Microbiology Resource Committee survey of commercially prepared media. *In*: *The role of Clinical Microbiology in Cost Effective Health Care.* Skokie III. Smith JW. pp. 555-559.

Mossel DAA, Trees MG, Laarhoven BV, et al. 1983. Quality assurance of selective culture media for bacteria, molds and yeast: An attempt to standardization at international level. *J Appl Bact* **54**: 313-327.

Page B. 1996. ISO Standard redefines limits for ETO residuals. *MDDI* archive.

Perkins JJ. 1969. Sterilizer control, Sterilization indicators and culture tests. *In*: Principles and methods of Sterilization in health sciences. Illinois: Charles C. Thomas Publishers; pp. 483-500.

Stumbo CR. 1973. Thermobacteriology in Food Processing. 2nd Edition. Academic Press, New York.

Weenk GH, vd Brink J, Meeuwissen J, van Oudenallen A, van Schievan RR. 1992. A standard protocol for the quality control of microbiology media. *Int J Food Microbiol* **17**: 183-198.

Zoutman D, Fleming C, Richardson H. 2002. Compliance with NCCLS approved standard M22-A2 for the bacteriologic media quality assurance: a survey of 124 Ontario microbiology laboratories. *Diagn Microbiol Infect Dis* **42**: 29-34.

Chapter 5

Microbial Growth

5.1 INTRODUCTION

Growth of a biological system or of a living organism or part of one may, however, be defined as an increase in mass or size (in any direction) accompanied by the synthesis of macromolecules, leading to the production of new organized structure. Increase in size of a macromolecule in a living cell by the addition of molecular groups from the neighbouring pool of intermediate substances as a result of the activity of cell enzyme is growth. When such macromolecules are integrated into the cell organelles, thus producing new organized structure, the organelles grow in size. The range of compounds that support the growth of a given organism, the way these compounds are degraded, and the nature of the end products (including the involvement of oxygen in the process) represent an important set of phenotypic characters. Because the word 'growth' is applied to a number of analogous biological phenomena at different levels of organization, it is often advisable to use it preceded by a qualifying descriptive word, such as population, mass volume, linear, localized, etc., to avoid ambiguity.

5.2 MICROBIAL GROWTH PATTERN

Microbial growth is defined as an increase in number of microbial cell in a population, which can also be measured as an increase in microbial mass. Growth rate is the change in cell number or mass per unit time. During this cell-division cycle, all the structural components of the cell double. The interval for the formation of two cells from one is called a generation, and the time required for this to occur is called the generation time. The generation time is thus the time required for the cell number to double (Figure 5.1).

Some time also it is referred as the doubling time. Growth is measured by following changes in number of cell or weight of cell mass by counting cell number, by observing turbidity in liquid broth or estimating cell mass suited to different organisms or different problems. Microorganisms are cultured in water with appropriate nutrients is called as culture medium. The nutrients present in the culture medium provide the microbial cell with those ingredients required for cell to produce more cells like itself. Besides an energy source, which could be an organic or inorganic chemicals, or light, a culture medium must have a source of carbon, nitrogen and other macro- and micronutrients. Microbial control by chemical and physical means the use of antiseptics, disinfectants, antibiotics, ultraviolet light, and many other agents. The experiments explained on

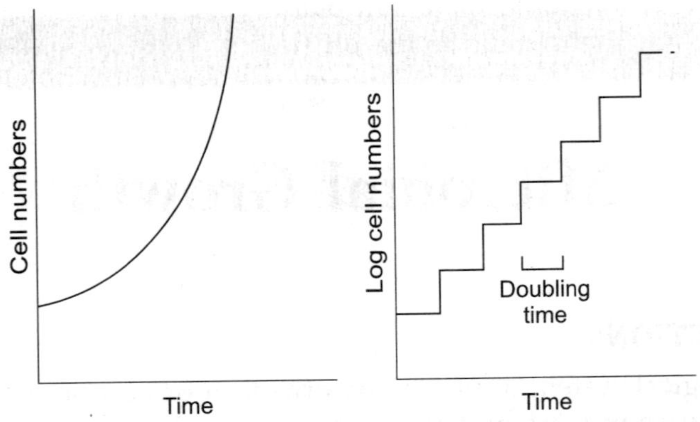

Figure 5.1 Microbial growth in terms of doubling time and cell number.

growth are related to these aspects primarily, to demonstrate methods of measurement with in-depth evaluation. The complete killing of the microorganism is called sterilization and is brought about by use of heat, radiation or chemicals.

5.2.1 Bacterial Growth

Microbiologists involved with industries are concerned, on the one hand, with providing optimum growth conditions to optimize growth for biomass or industrial products. The physician, nurse, and other members of the medical profession, on the other hand, are concerned with the limitation of microbial populations in disease prevention and treatment. An understanding of one of these facets of microbial existence enhances the other. Selection of media for microbial growth that contain all the essential nutritional needs is very important for any microbiological experiment. Temperature, oxygen, and/or hydrogen ion concentration are other limiting factors. An organism provided with all its nutritional needs may fail to grow if one or more be sustained to achieve the desired growth of microorganism. Some techniques are qualitative in their intent. That is, they provide a "yes or no" answer. Other techniques are quantitative in their intent. These techniques provide numerical information about a sample.

Bacterial growth is defined as the division of one bacterium into two cells, taking place by a process called binary fission. One initial bacterium cell eventually becomes millions. Each cell divides into two cells in minutes, making for the exponential rate at which bacteria population increases. Binary fission is the asexual method of bacterial reproduction. The DNA of each cell is first replicated then attaches itself to the plasma membrane. Once the DNA is attached to the membrane, the cell elongates by growing inward and splits into two separate entities and this process is referred to as cytokinesis. Bacterial growth undergoes four stages. The first two stages are the lag phase, where growth is slow at first, bacteria are becoming "acclimatized" to the new environmental conditions to which they have been introduced (pH, temperature, nutrients, etc). There is no significant increase in numbers with time. The log phase is defined as the phase where the amount of bacteria double every few minutes. The final two stages of growth are the stationary phase, where growth stabilizes due to competition for nutrients among a high number of bacteria, and the death phase, where waste builds up and nutrients run out, allowing for the bacteria to die off (Figure 5.2).

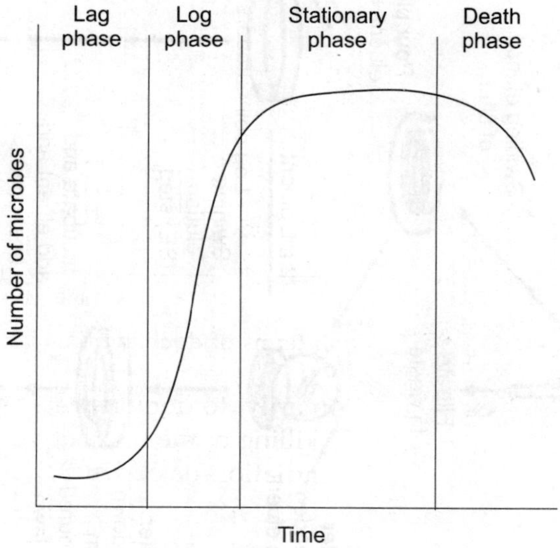

Figure 5.2 Bacterial growth pattern.

5.3 METHODS OF MICROBIAL GROWTH MEASUREMENT

Several different methods are available for population counts. The selection of method depends on the purpose of the study. It is possible to do a population count by preparing the dilution and counting the organisms in a number of microscopic fields on a slide without involvement of much of the equipment. Direct examination of milk samples with this technique can be performed very quickly, and the results obtained are quite reliable. A technique similar to this can be performed on a Petrof-Hauser counting chamber.

Bacterial counts of gas-forming bacteria can be made by inoculating a series of tubes of lactose broth and using statistical probability tables to estimate bacterial numbers. This method, which we use in exercise to estimate numbers of coliform bacteria in water samples, is easy to use, works well in water testing, but is limited to water, milk, and food testing. **Quantitative plating** (standard plate count, or SPC) and **turbidity measurements** to determine the numbers of bacteria in a culture sample are being commonly used. The two methods are somewhat parallel in the results they yield, however, there are distinct differences. The SPC reveals information only as related to viable organisms; that is, colonies that are seen on the plates after incubation represent only living organisms, not dead ones. Turbidimetry results, on the other hand, reflect the presence of all organisms in a culture, dead and living.

5.3.1 Standard Plate Count: A Quantitative Test

This determines the total number of bacteria in a specified amount of milk, generally a millilitre (mL). This is used for grading of milk. Under aseptic conditions, one mL of

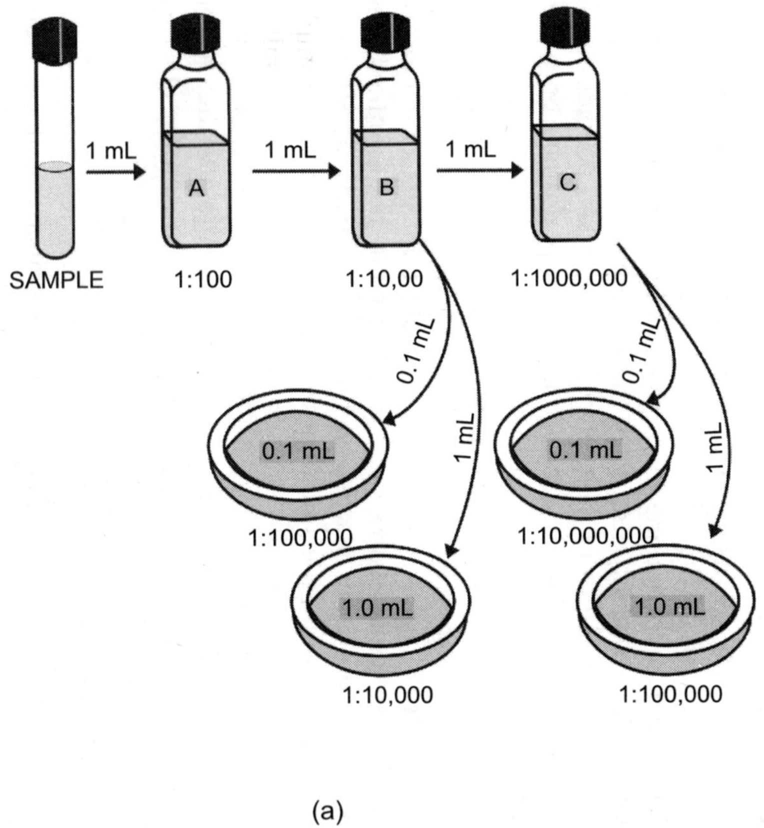

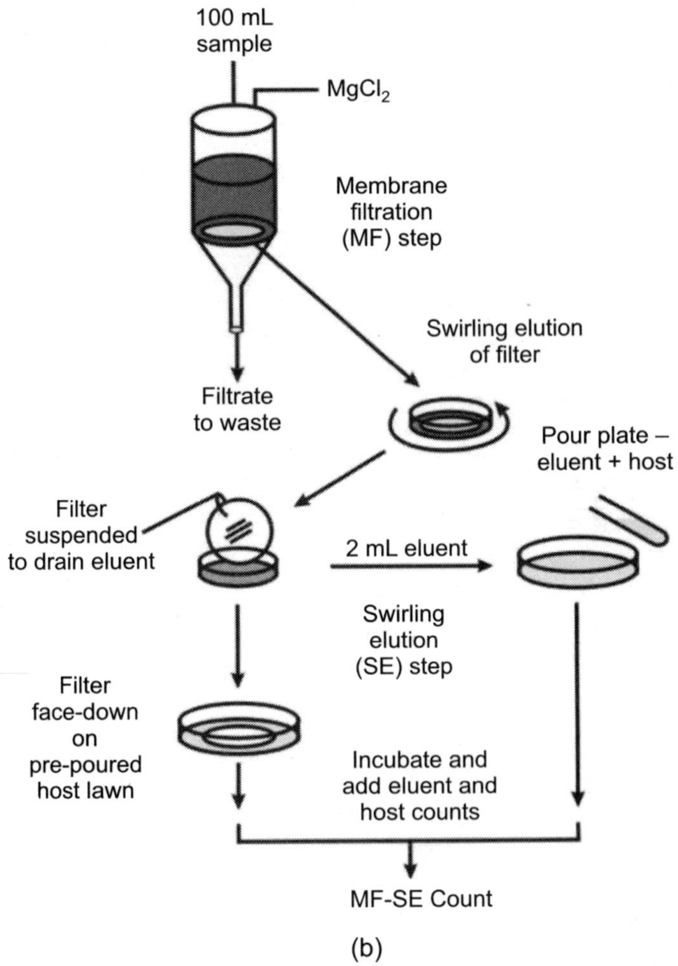

Figure 5.3 (a) Standard plate count after dilution of the samples (b) Membrane filtration.

milk is added to 99 mL of distilled water or buffer. One mL and one-tenth mL samples are then transferred to sterile Petri dishes and prepare 1:100 and 1: 1000 dilutions of the milk, respectively. Other dilutions may also be prepared successively (Figure 5.3 (a) and (b)).

A growth medium like plate count agar or tryptone glucose yeast extract agar is then added, and the milk samples are mixed with the medium. The dishes are incubated at 37°C for 24-48 hours. The plates are then placed on a counting device like a Quebec colony counter and the number of bacterial colonies is recorded. The colony count falling between 30 and 300 is selected and multiplied by the reciprocal of the dilution factor to obtain the bacterial count per mL of milk. If 248 colonies appeared on the 1: 100 plate and 16 colonies on, the 1:1000 plate, 248 would be selected and multiplied by 100, giving 24,800 total bacteria per mL of milk sample.

The standard plate count (SPC), also referred to as the aerobic plate count or the total viable count, is one of the most common tests applied to indicate the microbiological quality of food. The significance of SPCs, however, varies markedly according to the type of food product and the processing it has received. When SPC testing is applied on a regular basis it can be a useful means of observing trends by comparing SPC results over time.

Three levels of SPC are listed below based on food type and the processing/handling the food.

Level 1 applies to ready-to-eat foods in which all components of the food have been cooked in the manufacturing process/preparation of the final food product and, as such, microbial counts should be low.

Level 2 applies to ready to eat foods which contain some components that have been cooked and then further handled (stored, sliced or mixed) prior to preparation of the final food or where no cooking process has been used.

Level 3 SPCs not applicable. This applies to foods such as fresh fruits and vegetables (including salad vegetables), fermented foods and foods incorporating these (such as sandwiches and filled rolls). It would be expected that these foods would have an inherent high plate count because of the normal microbial flora present.

Note: An examination of the microbiological quality of a food should not be based on SPCs alone. The significance of high (unsatisfactory) SPCs cannot truly be made without identifying the microorganisms that predominate or without other microbiological testing.

5.3.2 Turbidity/OD Assay

Following inoculation the bacterial culture is stored or incubated in an environment suitable for growth in liquid broth or semisolid agar slants. The mass of daughter cells are visible to naked eye either as cloudiness or turbidity in liquid broth or as an isolated colony on solid media. The quantitative plate count method becomes a rather cumbersome tool while you are handling a large number of cultures. It is time-consuming beside takes a considerable amount of glassware and media. Measurement of turbidity of the culture is a much faster with a spectrophotometer and translate this into the number of organisms. To accomplish this, however, the plate count must be used to establish the count for one culture of known turbidity. The bacterial cells act as a colloidal suspension, which will intercept the lights as it passes through the spectrophotometer. The amount of light that is absorbed is directly proportional to the concentration of cells within certain limits.

Figure 5.4 illustrates the path of light through a spectrophotometer. A beam of white light passes through two lenses and an entrance slit into horizontal beams of all colours of the spectrum. Short wavelengths (violet and ultraviolet) are at one end and long wavelengths (red and infrared) are at the other end. The spectrum of lights falls on a dark screen with a slit (exit slit) cut in it. Only that portion of the spectrum that happens to fall on the slit goes through into the sample. It will be a monochromatic beam of light. By turning a wavelength control Knob on the instrument, the diffraction grating can be reoriented to allow the slit. The light that passes through the culture activates a phototube, which, in turn, registers **percent transmittance (1%T)** on a galvanometer. The higher the percent transmittance, the fewer are cells in suspension.

There should be a direct proportional relationship between the concentration of bacterial cells and the absorbance (optical density, O.D.) of the culture. These values will be converted to O.D. and plotted on a graph as a function of culture dilution. There is a linear relationship between concentration of cells and O.D. values the relationship may not be linear. That is, for a doubling in cell concentration, there may be less than a doubling in O.D.

5.3.3 Membrane Filter Method

The MF technique is an effective, accepted technique for testing fluid samples for microbiological contamination. It involves less preparation than many traditional methods, and is one of a few methods that will allow the isolation and enumeration of microorganisms. The MF technique also provides presence or absence information within 24 hours. The Membrane Filter (MF) technique was introduced in the late 1950s as an alternative to the Most Probable Number (MPN) procedure for microbiological analysis of water samples. The MF technique offers the advantage of isolating discrete colonies of bacteria (Figure 5.3 (b)), whereas the MPN procedure only indicates the presence or absence of an approximate number or organisms. The various steps are described below.

- Collect the sample and make any necessary dilutions.
- Select the appropriate nutrient or culture medium. Dispense the broth into a sterile Petri dish, evenly saturating the absorbent pad.
- Flame the forceps, and remove the membrane from the sterile package.
- Place the membrane filter into the funnel assembly.
- Flame the pouring lip of the sample container and pour the sample into the funnel.
- Turn on the vacuum and allow the sample to draw completely through the filter.

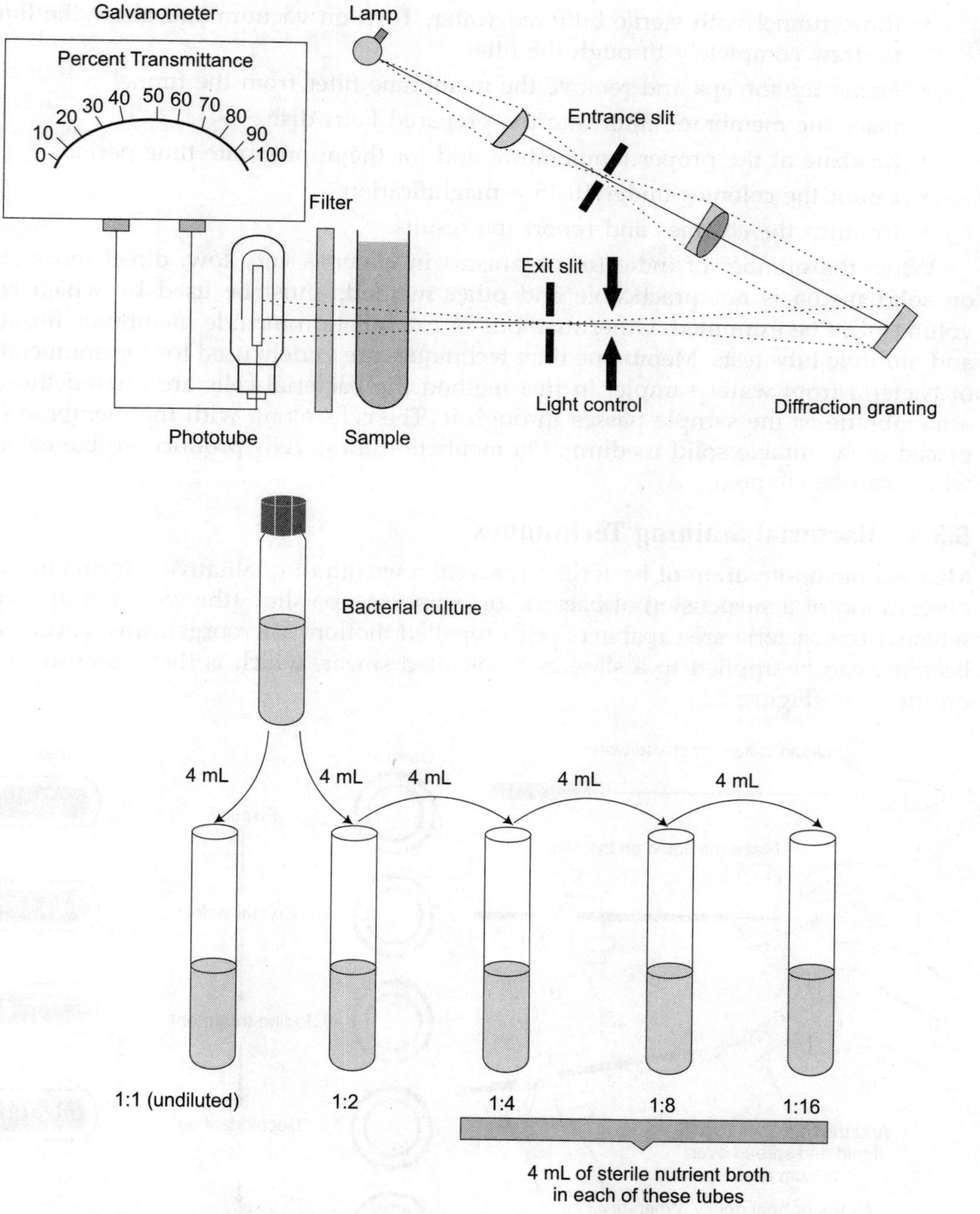

Figure 5.4 Colorimetric measurement of bacterial/turbidity growth.

- Rinse funnel with sterile buffered water. Turn on vacuum and allow the liquid to draw completely through the filter.
- Flame the forceps and remove the membrane filter from the funnel.
- Place the membrane filter into the prepared Petri dish.
- Incubate at the proper temperature and for the appropriate time period.
- Count the colonies under 10-15 × magnification.
- Confirm the colonies and report the results.

When the number of indicator organisms in water is very low, direct inoculation on solid media is not practicable and other methods must be used by which large volumes can be examined. Other methods of examination include membrane filtration and multiple tube tests. Membrane filter techniques are widely used for the enumeration of bacteria from water sample. In this method the bacterial cells are filtered through a membrane as the sample passes through it. The cells along with the membrane are placed on a suitable solid medium. On incubation these cells produce visible colonies which can be counted.

5.3.4 Bacterial Staining Techniques

Microscopic observation of bacteria can reveal a wealth of qualitative information. The observation of a suspension of bacteria on a microscope slide (the wet mount) reveals whether the bacteria are capable of self propelled motion. Microorganisms, particularly bacteria, can be applied to a slide as a so-called smear, which is then allowed to dry on the slide (Figure 5.5).

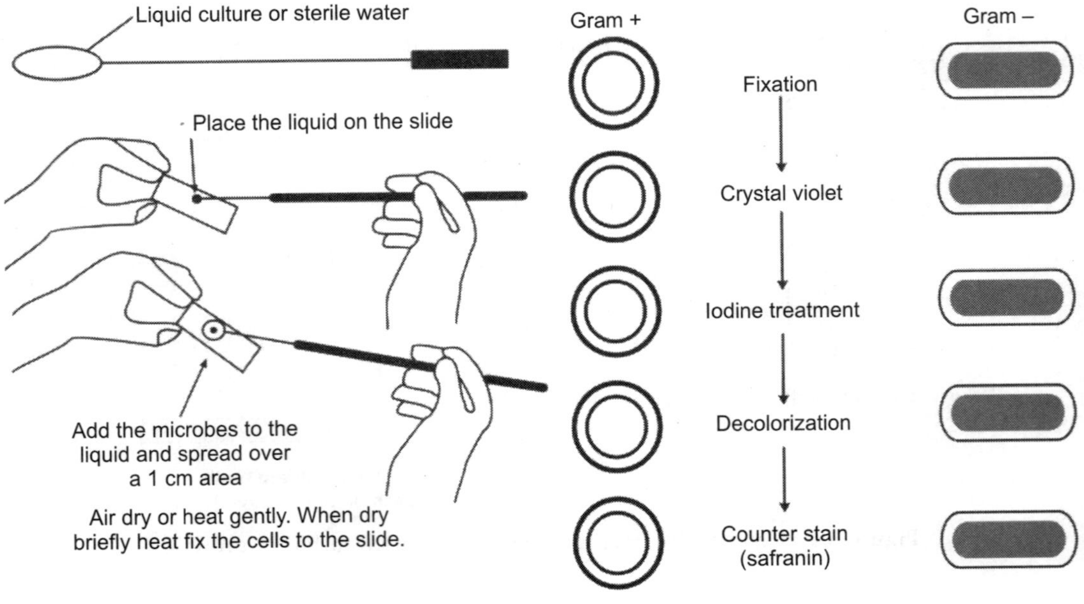

Figure 5.5 Method of making a bacterial smear and Gram stain.

The dried bacteria can be stained to reveal, for example, whether they retain the primary stain in the Gram stain protocol (Gram-positive) or whether that stain is washed out of the bacteria and a secondary stain retained (Gram-negative). Examination of such smears will also reveal the shape, size, and arrangement (singly, in pairs, in chains, in clusters) of the bacteria. These qualitative attributes are important in categorizing bacteria.

5.3.5 Fluorescence Staining Qualitative Technique

Microscopy can be extended to provide qualitative information. The incorporation of antibodies to specific components of the sample can be used to calculate the proportion of the samples in a population that possess the target of interest. Fluorescent-labelled antibodies, or antibodies combined with a dark appearing molecule such as ferritin, are useful in such studies (Figure 5.6).

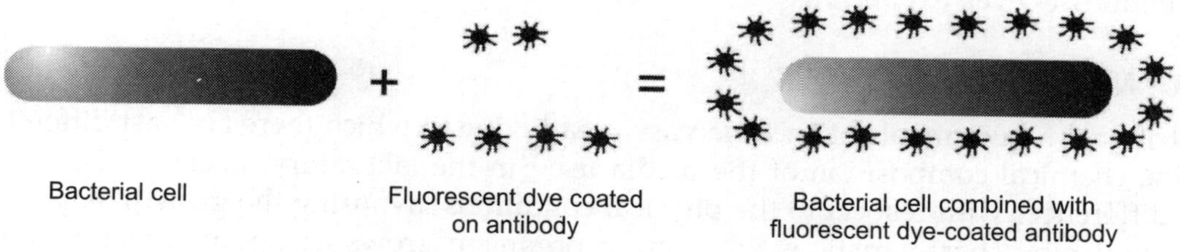

Figure 5.6 Qualitative fluorescene staining technique.

The scanning confocal microscope is tremendously useful in this regard. The optics of the microscope allows visual data to be obtained at various depths through a sample (typically the sample is an adherent population of microorganisms). These optical thin sections can be reconstructed via computer imaging to produce a three-dimensional image of the specimen. The use of fluorescent tagged antibodies allows the location of protein within the living biofilm to be assessed.

The self-propelled movement of living microorganisms, a behaviour that is termed motility, can also provide quantitative information. For example, recording a moving picture image of the moving cells is used to determine their speed of movement, and whether the presence of a compound acts as an attractant or a repellent to the microbes.

Bacterial growth is another area that can yield qualitative or quantitative information. Water analysis for the bacterium *Escherichia coli* provides an example. A specialized growth medium allows the growth of only *Escherichia coli*. Another constituent of the growth medium is utilized by the growing bacteria to produce a by-product that fluoresces when exposed to ultraviolet light. If the medium is dispensed in bottles, the presence of growing *Escherichia coli* can be detected by the development of fluorescence. However, if the medium is dispensed in smaller volumes in a grid-like pattern, then the number of areas of the grid that are positive for growth can be related to a mathematical formula to produce a most probable number of living *Escherichia coli* in the water sample. Viable bacterial counts can be determined for many other bacteria by several other means. The ability of bacteria to grow or not to grow on a media containing controlled amounts and types of compounds yields quantitative information about the nutritional requirements of the microbes.

The advent of molecular techniques has expanded the repertoire of quantitative information that can be obtained. For example, a technique involving reporter genes can show whether a particular gene is active and can indicate the number of copies of the gene product that is manufactured. Gene probes have also been tagged to fluorescent or radioactive labels to provide information as to where in a population a certain metabolic activity is occurring and the course of the activity over time.

Many other qualitative and quantitative techniques exist in microbiological analysis. A few examples include immunoelectrophoresis, immunoelectron microscopy, biochemical dissection of metabolic pathways, the molecular construction of cell walls and other components of microorganisms, and mutational analysis. The scope of the techniques is ever expanding.

5.4 NUTRITIONAL TYPES

Nutritional requirements of bacteria vary greatly due to which there are vast differences in the chemical composition of the media used in the laboratory. Bacteria also exhibit the differences with respect to the physical conditions favouring the growth. A primary distinguishing characteristic is whether an organism grows aerobically, anaerobically, facultatively (i.e., in either the presence or absence of oxygen), or microaerobically (i.e., in the presence of a less than atmospheric partial pressure of oxygen) (Figure 5.7).

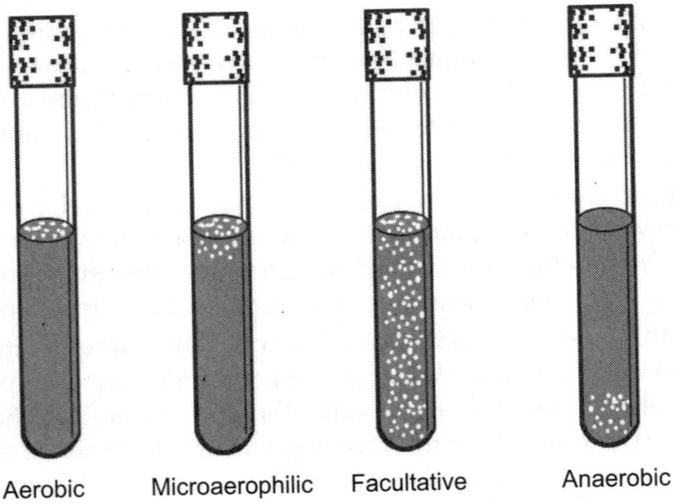

Figure 5.7 Schematic illustration of the growth of bacteria in deep agar tubes showing differences response to atmospheric oxygen.

The proper atmospheric conditions are essential for isolating and identifying bacteria. Other important growth assessments include the incubation temperature, pH, nutrients required, and resistance to antibiotics. For example, one diarrheal disease agent, *Campylobacter jejuni*, grows well at 42°C in the presence of several antibiotics; another, *Y. enterocolitica*, grows better than most other bacteria at 4°C. *Legionella, Haemophilus*, and some other pathogens require specific growth factors, whereas *E. coli* and most other Enterobacteriaceae can grow on minimal media. Bacteria can be classified into different nutritional types (Table 5.1).

Table 5.1 Nutritional types of bacteria in terms of energy sources and hydrogen donors

	Energy Source and Hydrogen Donor	*Nutritional Type*
A	**Energy Source**	
	(i) Radiation (light)	Phototrophs
	(ii) Oxidation of organic or inorganic compounds	Chemotrophs
B	**Hydrogen donor**	
	(i) Reduced inorganic H donors (NO_2, NO_3, H_2S, S_2O_3, Fe^{2+}, CO, H_2)	Lithotrophs
	(ii) Organic compounds	Organotrophs

Addition of a dried polysaccharide agar, derived from a red algae is made when solid media is required. Agar has the property to go into solution (1.5-2%) at 100°C and solidify at 45°C.

Many special purpose media have been designed which facilitate recognition, enumeration and isolation of certain bacterial types.

(i) *Selective media:* It provides nutrients that enhance the growth and predominance of a particular type of bacterium and may inhibit other cell types.

(ii) *Differential Media:* Certain reagents or supplements when incorporated into culture media, may allow differentiation of several bacterial kinds.

5.4.1 Preparation of Media

1. Each ingredient or the completely dehydrated medium is dissolved in the appropriate volume of distilled water.
2. The pH of the fluid medium is determined by a pH meter and adjusted if necessary.
3. If a solid medium is required, agar is added and the medium is boiled to dissolve the agar.
4. The medium is dispensed into tubes or flasks.
5. Media is sterilised by autoclaving (15 lbs pressure maintained for 20 mins). Some ingredients which are heat labile are sterilised by filtration.

5.5 ROLE OF OXYGEN

A very important requirement of cellular growth and metabolism is the requirement for oxygen. Microbes can be grouped into five different categories based on their requirement for molecular oxygen. Strict or **obligate aerobes** must have O_2 for growth because O_2 is used as the terminal electron acceptor for oxidative phosphorylation.

In contrast, strict or **obligate anaerobes** cannot grow in the presence of O_2 and may be killed by trace amounts of O_2. **Microaerophilic** organisms need a small amount of O_2 for growth but too much O_2 will kill them. A subset of anaerobes can tolerate O_2 but they do not use it. They are called **aerotolerant anaerobes**. **Facultative anaerobes** grow best when O_2 is available, but they can also grow, though not as well, if O_2 is not present (Figure 5.8).

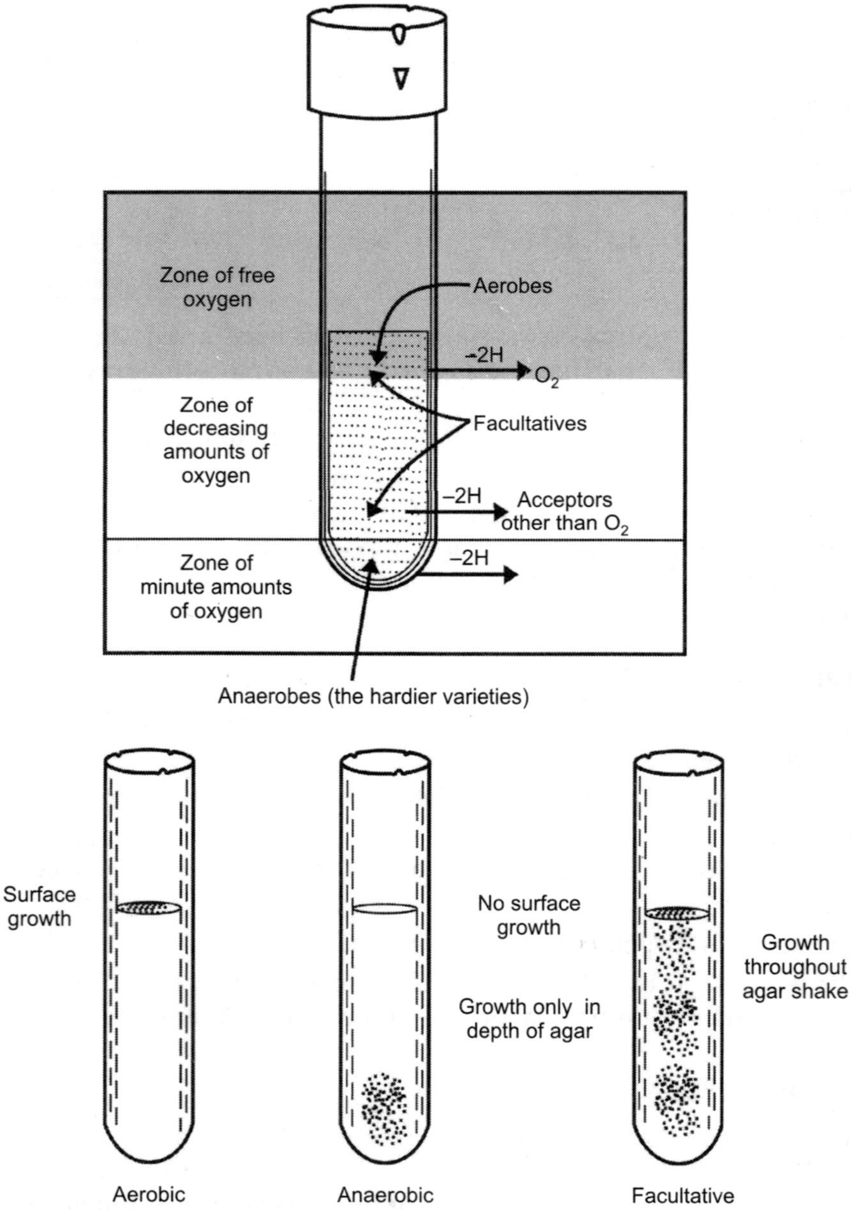

Figure 5.8 Relation of oxygen to microbial growth in a solid medium.

5.5.1 Mode of Action of Oxygen

The strong oxidising nature of oxygen, makes it a suitable terminal electron acceptor for the energy yielding process—the *respiration.* However, it may act as a toxic substance as in anaerobic organisms. Aerobic and facultative ones have developed protective mechanisms that greatly mitigate this toxicity. In anaerobes following factors are among those that have been implicated in oxygen toxicity (Figure 5.7).

(a) *Inactivation of enzymes by oxygen:* Molecular oxygen oxidises reduced groups (e.g., thiol-SH, or enzymes) thus rendering them inactive.

(b) *Derivatives of Oxygen:* Derivatives like Super-oxide (O_2^-) radicals can inactivate vital cell components.

$$O_2 + O^- \rightarrow O^-_2 \qquad (1)$$

Hydrogen peroxide generated by (O^-_2) is another poison.

$$2O^-_2 + 2H^+ \rightarrow O_2 + H_2O_2 \qquad (2)$$

The energized derivative of oxygen known as singlet oxygen (Δ_8), is a powerful oxidising agent highly toxic to cells.

In aerobes and facultative organisms, an enzyme *Superoxide dismutase* eliminates super oxide radicals as in reaction (2). The hydrogen peroxide produced by this reaction can in turn be dissipated by *Catalase* and *Peroxidase* enzymes.

$$2H_2O_2 \xrightarrow{\text{Catalase}} 2H_2O + O_2$$

$$H_2O_2 + \text{reduced substrate} \xrightarrow{\text{Peroxidase}} 2H_2O_2 + \text{oxidized substrate}$$

5.6 ROLE OF TEMPERATURE

Temperature is probably the most important environmental factor affecting growth. If temperature is too hot or too cold microorganisms will not grow. The minimum and maximum temperatures for microbial growth vary widely among microorganisms and are usually a reflection of the temperature range and average temperature of their habitat. Temperature affects living organisms in two opposing ways:

1. As temperature rises, chemical and enzymatic reactions precede at a faster rate and the growth rate increases.
2. Above a certain temperature proteins are irreversibly damaged.

As temperature is increased within a certain range therefore, growth and metabolic activity increases up to a point where inactivation reactions set in. Above this temperature, cell functions fall sharply to zero. Each microorganism thus has:

- A minimum temperature below which no growth occurs.
- An optimum temperature at which growth is most rapid.
- A maximum temperature, above which growth is not possible.

The optimum temperature is always closer to the maximum rather than the minimum.

These three temperatures, called the cardinal temperatures, are usually characteristic of each type of organism. They are not completely fixed however, because they can be modified by other environmental factors especially the chemical composition of the medium. The cardinal temperature for the growth of some common bacteria is shown in Table 5.2.

Table 5.2 Cardinal temperature for many bacteria

	Temperature of Growth °C		
	Minimum	*Optimum*	*Maximum*
Vibrio psychoerythrus	0	15	19
Pseudomonas fluorescence	4	25-30	40
Staphylococcus aureus	6.5	30-37	46
Corynebacterium diptheriae	15	37	40
Streptococcus thermophilus	20	40-45	50
Thermus aquaticus	40	70-72	79

5.6.1 Factors Affecting Maximum and Minimum Temperatures

The maximum growth temperature usually reflects the inactivation of one or more key proteins in the cell. The factors affecting minimum temperatures are less clear. It may result from the "freezing" of the cytoplasmic membrane, impairing its ability to transport nutrients or form proton gradients. Experiments have shown that adjustment in membrane lipid composition can cause changes in minimum temperature. Cardinal temperatures vary greatly throughout the microbial world. Temperature classes of microorganisms can be broadly categorized under four broad categories

1. Psychrophiles – these have low temperature optima usually at 0°C or lower.
2. Mesophiles – these have mid range temperature optima usually at 25 to 40°C.
3. Thermophiles – these have high temperature optima usually above 45°C.
4. Hyperthermophiles – these have very high temperature optima usually above 55°C.

However, it is important to note that a bacterial species may not manifest the same characteristics in every detail when grown at different temperatures. For example: *Serratia marcescens* forms a blood red to orange pigment when cultured at 25°C but produces little or no pigment when cultured at 37°C (Figure 5.9). Similarly, *Lactobacillus planatarum* does not require amino acid phenylamine for growth when cultured at 25°C but does require it at 37°C.

Table 5.3 Minimum, maximum and optimum pH for growth of certain prokaryotes

Organism	*Maximum pH*	*Optimum pH*	*Maximum pH*
Thiobacillus thiooxidans	0.5	2.0-2.8	4.0-6.0
Sulfolobus acidocaldarius	1.0	2.0-3.0	5.0

Contd...

Contd...

Bacillus acidocaldarius	2.0	4.0	6.0
Zymomonas lindneri	3.5	5.5-6.0	7.5
Lactobacillus acidophilus	4.0-4.6	5.8-6.6	6.8
Staphylococcus aureus	4.2	7.0-7.5	9.3
Escherichia coli	4.4	6.0-7.0	9.0
Clostridium sporogenes	5.0-5.8	6.0-7.6	8.5-9.0
Erwinia caratovora	5.6	7.1	9.3
Pseudomonas aeruginosa	5.6	6.6-7.0	8.0
Thiobacillus novellus	5.7	7.0	9.0
Streptococcus pneumonia	6.5	7.8	8.3
Nitrobacter sp	6.6	7.6-8.6	10.0

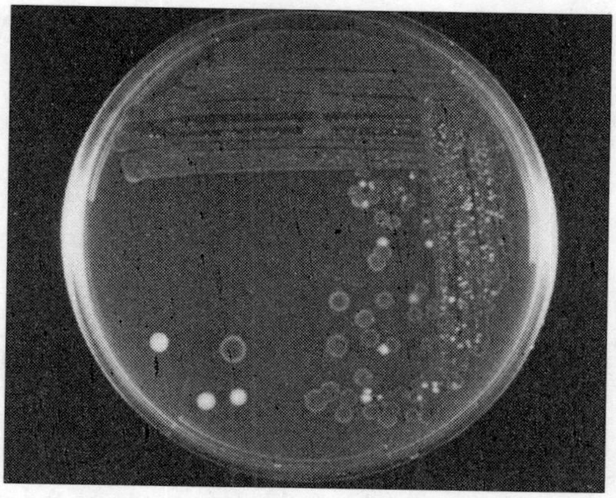

Figure 5.9 Chromogenic Bacterium *Serratia marcescens* producing orange pigments at 25°C.

5.7 ROLE OF pH

Microbes need a physiological pH inside their cells, just like all other living organisms (Table 5.3). Their ability to survive in extreme pH (either high or low) depends on their ability to correct for the difference between inside and out. One example of a bacterium that can live in acidic environments is *Helicobacter pylori* which live in the stomach. It produces high amounts of urease which is an enzyme that degrades urea, and by doing so decreases the acidity (raises the pH). Imagine the bacteria produce a 'cloud' of neutral pH around them to protect them from the acidic environment. There are other bacteria that are specialized to live in basic pH, for instance near black smokers, geological fountains of minerals that shoot highly alkaline minerals into the ocean. In conclusion, which pH is lethal for the bacteria depends on the species. Their defence is to keep the protons or OH^- ions out. Would they not succeed, then their proteins would rapidly denature. That is the lethal toxicity of non-physiological pH. pH affects bacteria the same way it affects all living things.

Extremes of pH affect the function of enzyme systems by denaturing them. However, bacteria become adapted over time to their surroundings. Just like we have enzymes that are adapted to the pH of our stomachs (very acidic) or to the small intestine (basic), bacteria that live in acid conditions are adapted to them. If they are moved to an environment that is neutral or basic they will probably die. Bacteria that are human or animal pathogens are generally adapted to a pH of about 7.4 which is slightly basic.

Each organism has a pH range within which growth is possible, and most have well defined pH optima (Figure 5.10). Most natural environments have pH values between 5–9 and most organisms have pH optima in this range. Very few species can grow at pH values below 2 or above 10. Organisms capable of living at low pHs are called acidophiles. Those capable of living at very high pHs are called alkaliphiles.

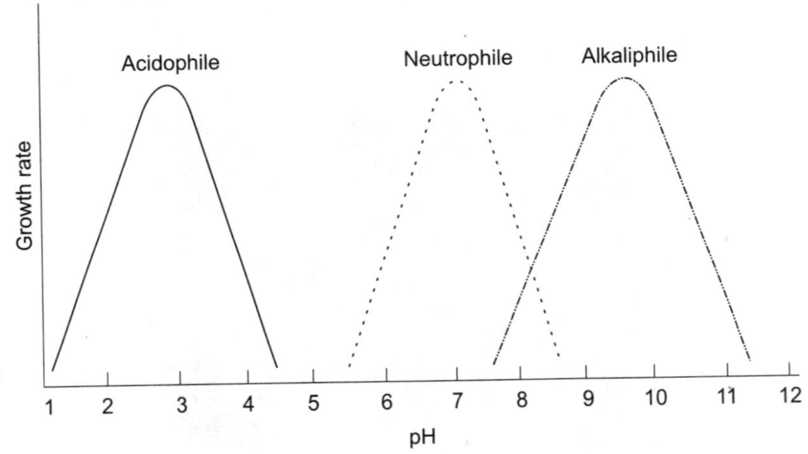

Figure 5.10 Growth rate vs pH for three enviromental classes of procaryotes. Most free-living bacteria grow over a pH range of about three units. Note the symmetry of the curves below and above the optimum pH for growth.

As a group, fungus tend to be more acid tolerant than bacteria. Many grow optimally at pH 5 or below, and a few grow well at pH values as low as some bacteria are also acidophilic and some cannot grow at neutral pH (obligate acidophiles). These include several species of *Thiobacillus*, and several genera of the archaea including sulfolobus. A high concentration of hydrogen ions is required for the stability of the cytoplasmic membrane of obligate acidophiles. This dissolves if pH is raised to neutrality. Despite the pH requirements of particular organisms for growth, the optimal growth pH represents the pH of the extracellular environment only. The intracellular pH must remain near neutrality to prevent destruction of acid or alkali labile macromolecules in the cell. In extreme acidophiles and extreme alkalinophiles the intracellular pH may vary by several units from neutrality.

5.8 OSMOTIC PRESSURE

Osmotic pressure is another limiting factor in the growth of bacteria. Bacteria contain about 80-90% water; they require moisture to grow because they obtain most of their nutrients from their aqueous environment. Cell walls protect prokaryotes against changes in osmotic pressure over a wide range. However, sufficiently hypertonic media at concentrations greater than those inside the cell (such as 20% sucrose) cause water loss from the cell by osmosis. Fluid leaves the bacteria causing the cell to contract, which, in turn, causes the cell membrane to separate from the overlying cell wall. This process of cell shrinkage is called plasmolysis. Because plasmolysis inhibits bacterial cell growth, the addition of salts or other solutes to a solution inhibits food spoilage by bacteria, as occurs when meats or fish is salted.

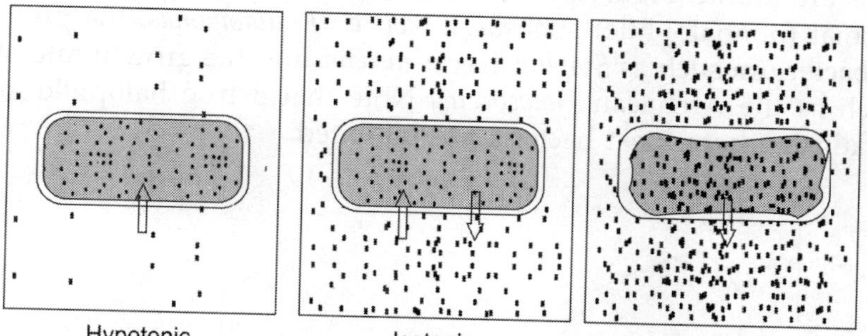

Figure 5.11 Osmotic variabilities.

Osmolarity is inversely related to water activity (A_w), which is more like a measure of the concentration of water (H_2O) in a solution. Osmolarity is determined by solute concentration in the environment. Increased solute concentration means increased osmolarity and decreased A_w. Some of the limiting water activities (A_w) for growth of certain prokaryotes are shown in Table 5.4. Some types of bacteria, called extreme or obligate halophiles, are adapted to and require high salt concentrations, such as found in the Dead Sea, where salt concentrations can reach 30%. Facultative halophiles do not require high salt environments to survive, but are capable of tolerating these conditions. Halophiles can grow in salt concentrations up to 2%, a level that would inhibit the growth of other bacteria. However, some facultative halophiles, such as *Halobacterium halobium* grow in salt lakes, salt flats, and other environments where the concentration of salts is up to seven times greater than that of the oceans. When bacteria are placed in hypotonic media with concentrations weaker than the inside of the cell, water tends to enter by osmosis. The accumulation of this water causes the cell to swell and then to burst, a process called osmotic lysis (Figure 5.11).

Table 5.4 Limiting water activities (A_w) for growth of certain prokaryotes

Organism	*Minimum A_w for growth*
Caulobacter	1.00
Spirillum	1.00
Pseudomonas	.91
Salmonella/E.coli	.91

Contd...

Contd...

Lactobacillus	.90
Bacillus	.90
Staphylococcus	.85
halococcus	.75

When the osmolarity of the different prokaryotes are plotted various types of growth curve patterns are found (Figure 5.12). From left to right the graph shows the growth rate of a normal (nonhalophile) such as *E. coli* or *Pseudomonas,* the growth rate of a halotolerant bacterium such as *Staphylococcus aureus,* and the growth rate of an extreme halophile such as the Archaean *Halococcus.* Note that a true halophile grows best at salt concentrations where most bacteria are inhibited.

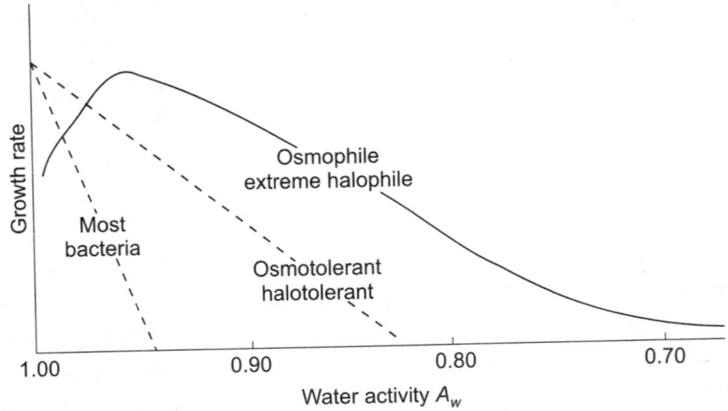

Figure 5.12 Growth rate vs. osmolarity for different classes of prokaryotes.

5.9 UV LIGHT

UV light is nonionizing short wavelength radiation that falls between 4 nm and 400 nm in the visible spectrum. Most bacteria are killed by the effects of UV light (Figure 5.13) and it is routinely used to sterilize surfaces, such as work areas of transfer hoods used for the inoculation of cultures. The primary lethal effects of UV are due to its mutagenic properties. When DNA absorbs UV light, it causes the formation of pyrimidine dimers. These dimers deform the DNA molecule, so that DNA polymerase is unable to replicate the strand and genes can no longer be transcribed. Cells have evolved various repair mechanisms to deal with the damage (including photoreactivation, nucleotide excision repair and the SOS system). However, if sufficiently large numbers of dimers form in the DNA, the systems cannot cope and begin making errors by inserting incorrect bases for the damaged bases, eventually resulting in cell death.

5.10 OLIGODYNAMIC ACTION

The inhibiting or killing of microorganisms by use of very small amounts of a chemical substance is oligodynamic. The activity of heavy metals on microorganisms is termed

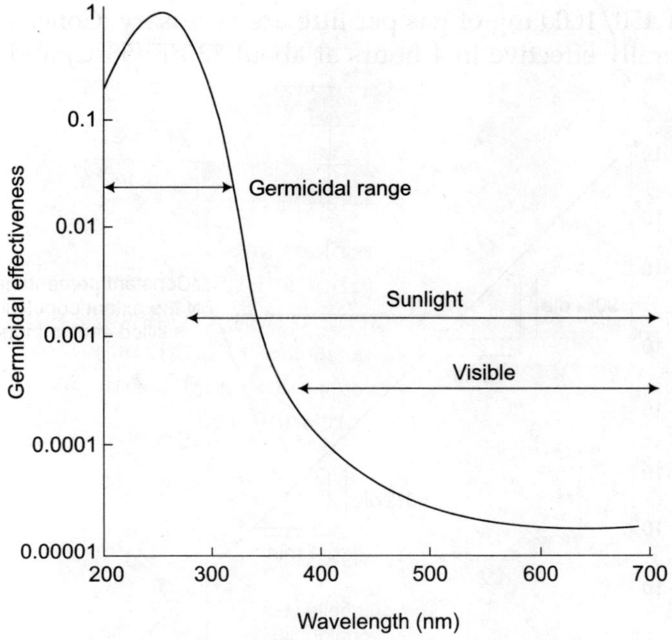

Figure 5.13 Effect of various wavelengths of light on the germicidal effectiveness.

oligodynamic action (Figure 5.14). Metals as silver, mercury and copper are used. Mercury is an older antiseptic, used as mercuric chloride ($HgCl_2$). In products like mercurochrome, merthiolate and metaphen, mercury is combined with organic carrier compounds, that reduces its toxicity to skin. Copper is particularly active against algae. It is used as copper sulphate in swimming pools and municipal water supplies. Silver, as silver nitrate is used as an antiseptic and disinfectant. One drop of 1% $AgNO_3$, solution is placed in eyes of newborn to protect against infection by the gonococcus, *Neisseria gonorrhoeae*. Silver nitrate can also be combined with an antimicrobial drug for use in treatment of burns. Ethylene oxide has been used to sterilize many substances, including spices, biological preparations, soil, plastics, certain medical preparations, and contaminated lab equipment. It has been used to decontaminate certain spacecraft components. Beside its good penetration, this gas has a wide spectrum of activity (meaning it is effective against many different kinds of microbes). It is also effective at relatively low temperatures, and it does not damage materials exposed to it. One advantage is its relatively slow action upon microorganisms. The mode of action of ethylene oxide is believed to be alkylation reactions with organic compounds such as enzymes and other proteins. Alkylaton consists in the replacement of an active hydrogen atom in an organic compound with an alkyl group. In this reaction, the ring in the ethylene oxide molecule splits and attaches itself where the hydrogen was originally placed. The reaction would for instance, inactivate an enzyme. Ethylene oxide is usually measured in terms of milligrams of the pure gas per litre of space. For sterilization, concentrations of 450/1000 mg of gas per litre are necessary. Concentrations of 500 mg per litre are generally effective in 4 hours at about 130°F (58°C) and relative humidity of about 40%.

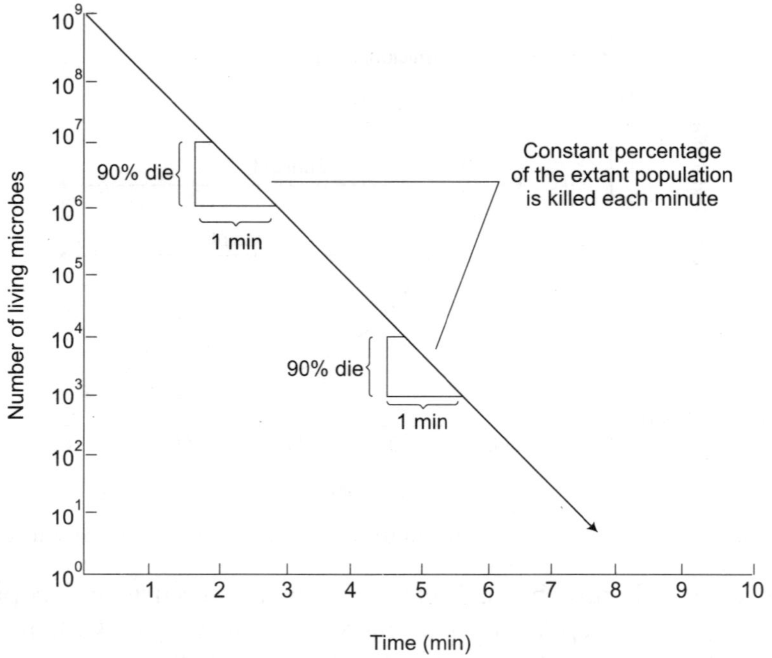

Figure 5.14 Oligodynamic property.

Beta-propiolactone is a colourless liquid at room temperature with a high boiling point (162.3°C). It has a sweet but irritating odour. Although unstable at room temperature, it may be stored at 4°C (refrigerated) for months without deterioration. Although beta-propiolactone is not flammable and does not penetrate and, it is much more active than ethylene oxide. Aqueous solutions effectively inactivate some viruses, including polio and rabies and also kill bacteria and bacterial spores. The vapour, in a concentration of about 1.5 mg of the lactone per litre of air with a high relative humidity (75-80%) at about 25°C kills spores in a few minutes. Beta-propiolactone is not a substitute for ethylene oxide (because of its low penetrating power) but is used in place of formaldehyde for surface disinfection or sterilization. It is used to sterilize large enclosed spaces. Advantages over formaldehyde include more rapid antimicrobial action and faster removal from enclosure after application. Its mode of action is thought to be similar to that of ethylene oxide.

Formaldehyde gas is stable only in high concentrations and at elevated temperatures. At room temperature it polymerizes, forming a solid substance. The important polymer is paraformaldehyde, a colourless solid which rapidly yields formaldehyde upon heating. Formaldehyde is also marketed in aqueous solution as formalin, which contains 37-40% formaldehyde. Formalin and paraformaldehyde are the two main sources of formaldehyde for gaseous sterilization. Vaporization of formaldehyde from either of these sources into an enclosed space for an adequate time will effect sterilization. Humidity and temperature have a pronounced effect on the microbicidal action of formaldehyde; in order to sterilize an enclosure, the temperature should be about room temperature (22°C) and the relative humidity between 60-80%. One disadvantage is the limited ability of these vapours to penetrate covered surfaces.

5.11 BACTERIAL CULTIVATION

In nature, microbial populations do not segregate themselves by species but exist with a mixture of many other cell types. In laboratory, these populations can be separated into pure cultures. Many different microbial species normally inhabit various parts of our bodies, such as oral cavity, intestinal tract and the skin. A single sneeze may therefore disperse from 10,000 to 100,000 bacteria. One gram of faeces contains 10^{11} bacteria. Studies of microorganisms in these habitats require knowledge of the specific microbes present. This, in turn, requires unravelling of mixed population into pure culture of different species.

Three kinds of selective methods are employed to obtain pure cultures.

(i) Chemical
(ii) Physical
(iii) Biological

5.11.1 Chemical Methods of Selection

(a) *Use of specific carbon or nitrogen source:* All the microbes have the same basic requirements, but there is diversity in the use of organic and inorganic compounds. Some media contain solutions of inorganic salts and supplemented with organic compounds while other media contain complex ingredients hence all media should provide carbon, nitrogen sources, in addition to minerals and other growth factors. There are no media available that can support the growth of all microorganisms. The method is an example of enrichment media preparation. Here, for example, if we wish to, select nitrogen fixing organism, N_2 gas can be supplied as sole nitrogen source. One classical example deals with the isolation of *Pseudomonas* where supply of sodium benzoate as sole carbon source ensures the development of only desired bacterial colonies.

(b) *Use of dilute media:* Certain bacteria like *Caulobacter* are capable of growing with very low levels of carbon and nitrogen source. In order to isolate such types, a media containing 0.01% peptone is employed since other bacteria are unable to grow in such dilute conditions (Figure 5.15).

(c) *Addition of toxic chemicals:* Gram-negative bacteria can grow in presence of low concentrations of various dyes that inhibit the growth of Gram-positive species. A medium containing crystal violet dye plus sodium-deoxy cholate allows Gram-negative intestinal bacteria to grow but will inhibit most other kinds. Use of Mac Conkey agar which is used to select for Gram-negative intestinal pathogens such as *Salmonella* and *Shigella* is based on above principle (Figures 5.16 and 5.17).

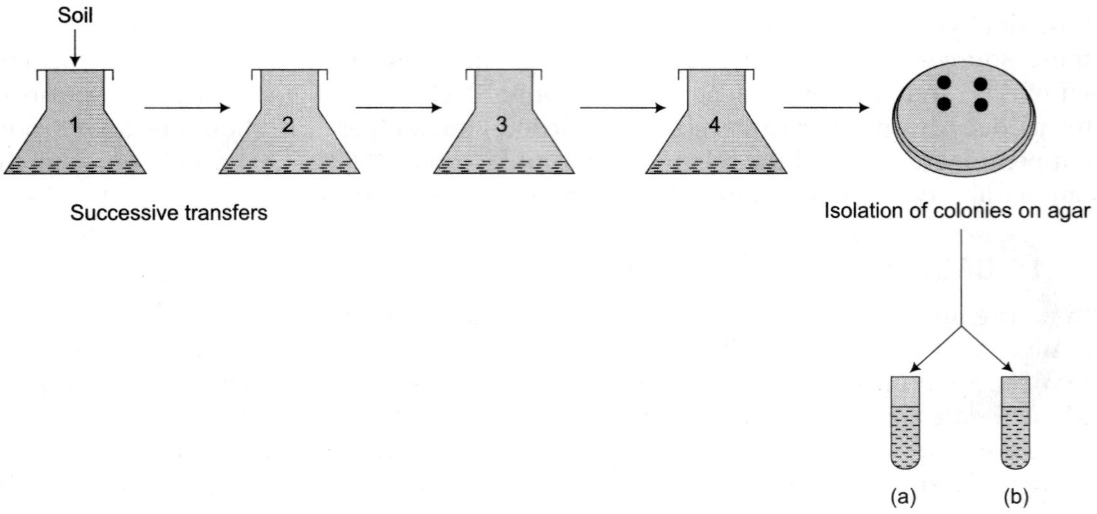

Figure 5.15 Isolation of pure colonies and verification of ability to utillize substrates (enrichment methods), +/– indicates presence or absence of specific substrate.

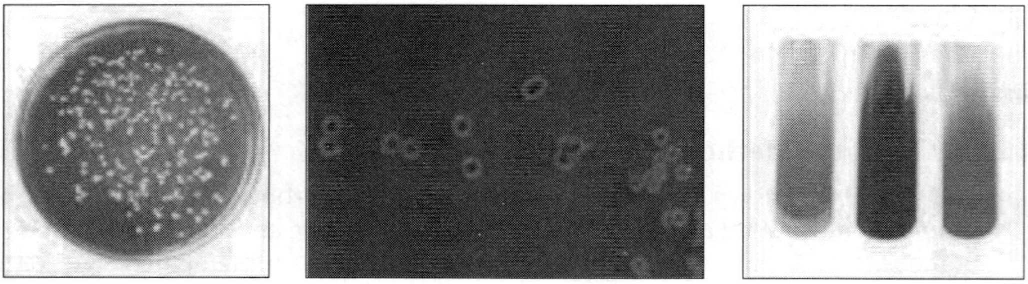

(a) (b) (c)

Figure 5.16 Culture of Salmonella typhosa (a) Mac Conkey agar (b) Colonial growth pattern (c) Tripple sugar iron.

5.11.2 Physical Methods of Selection

(a) *Heat treatment:* To select for endospore forming bacteria, e.g., *Bacillus, Clostridium* a mixed culture is heated to 80°C, 10 min before being used to inoculate culture media. Vegetative cells will be killed by this treatment but endospores will survive (Figure 5.17).

(b) *pH and incubation temperature:* The selection of psychrophillic microbe cultures are incubated at low temperature, e.g., 0 to 5°C. On the other hand, a hunt for thermophiles, is achieved at an incubation temperature of 55°C. Similarly, to select for acid tolerant bacteria a low pH medium can be used, e.g., *Lactobacilli* (pH 5.35) while alkali tolerant like *Vibrio cholerae* obtained from stool sample thrives at pH 8.5.

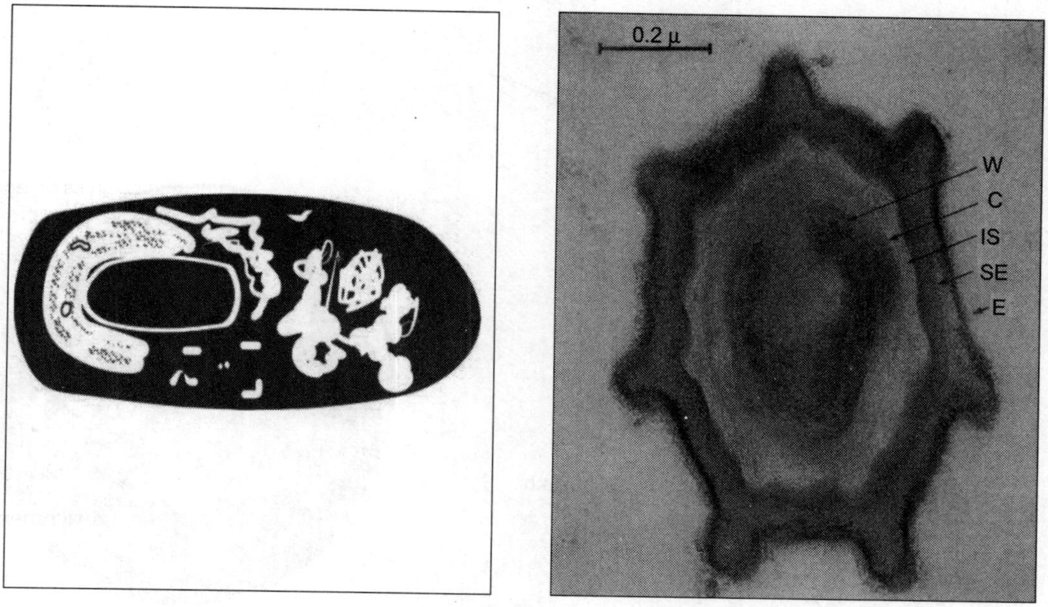

(a) (b)

Figure 5.17 (a) Diagrammatic sketch of electron micrograph of an endospore of *Bacillus cereas* (b) EM of a thin section of bacterial endospores of *Bacillus circulans* circulans where SE denotes for spore envelope; IS layered inner space envelops: C cortex; W cell well of sopre; E endospore.

(c) *Cell size and motility:* Use of small cell diameter and ability to migrate helps smooth identification in some bacterial sp., e.g., *Treponema* sp., membrane filter having a pore size of 0.15 µm is placed on the surface of an agar plate and gingival scrapings are placed on the filter. The small *Treponemes* penetrate the pores of filter and reach the underlying surface. The motility feature help them to migrate away from filter (Figures 5.18 and 5.19).

5.11.3 Biological Methods of Selection

A disease-producing species occurring in a mixed culture can often be selected by taking advantage of its pathogenic properties. For example, sputum having *Streptococcus pneumoniae* is ordinarily contaminated by many bacterial species. However, laboratory mouse, are extremely susceptible to infection by S. *pneumoniae,* and if the sputum sample is injected into a mouse the pathogen will multiply extensively and the culture is isolated in pure form.

5.12 ISOLATION AND PURIFICATION

Before commencing to the techniques related with pure culture, knowledge of certain microbiological terms is essential. Essentially, all culture media are of two forms: liquid or broth media and solid or agar media. Agar enters into solution at 97-100°C and solidifies at about 45°C. It is not included as food, but only as a hardening agent.

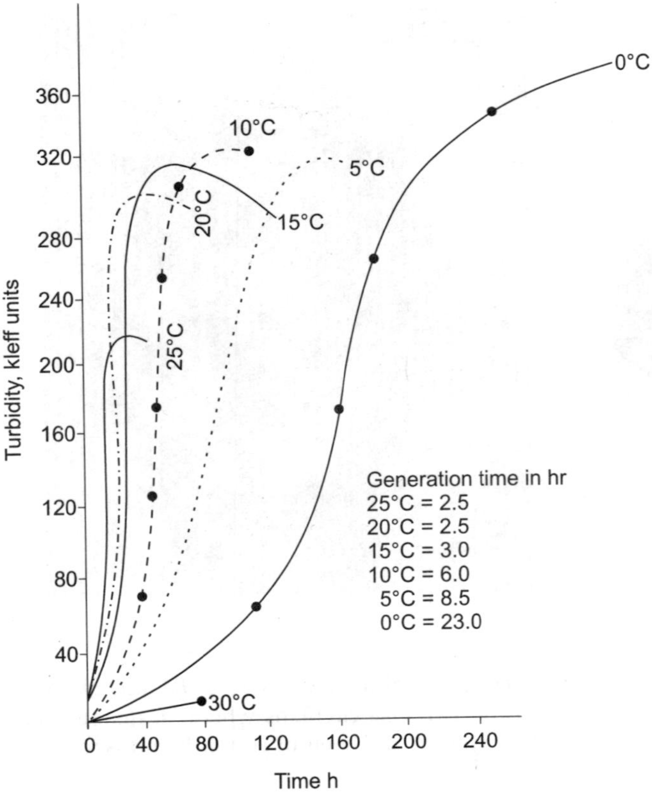

(a)

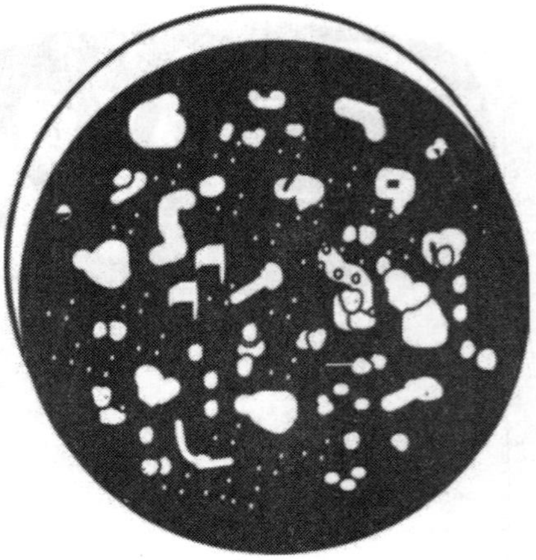

(b)

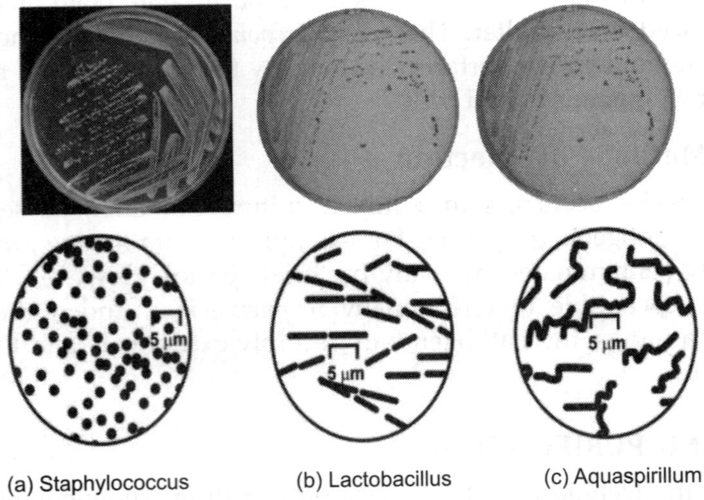

(a) Staphylococcus (b) Lactobacillus (c) Aquaspirillum

Figure 5.18 (a) Effect of temperature on bacterial growth, (b) Mixed culture.

Figure 5.19 Motality test specimen.

The technique which ensures prevention of the entry of undesired microbes is aseptic technique which is followed at every step of cultivation. Since any medium immediately after preparation contains microorganisms from its ingredients and from the surfaces of the utensils for glassware, it must be sterilized by heating or other methods. Prior to sterilization a tube containing the medium is plugged with cotton or plastic closures. This prevents the entry of new contaminants but permits free interchange of gases or air.

5.12.1 Inoculation

Inoculation is the method by which the desired microbe is transferred to the sterilized media. This is achieved by heating the inoculation loop to redness by flaming immediately before and after making the transfer. The flaming destroys any living forms on the surface of needle or loop (Figure 5.20).

Following inoculation the bacterial culture is stored or incubated in an environment suitable for growth. The mass of daughter cells are visible to naked eye either as cloudiness or turbidity in liquid broth or as an isolated colony on solid media. The turbidity in liquid broth in terms of growth is measured by using a colorimeter/ spectrocolorimeter measuring light scattering (Figure 5.4).

5.12.2 Pure Culture Techniques

(1) *Streak Plate:* It is a method of applying microbial culture on to agar surface with loop or a bent glass rod (spreader). This is called streaking and a plate so prepared is streak plate (Figures 5.21 and 5.22).

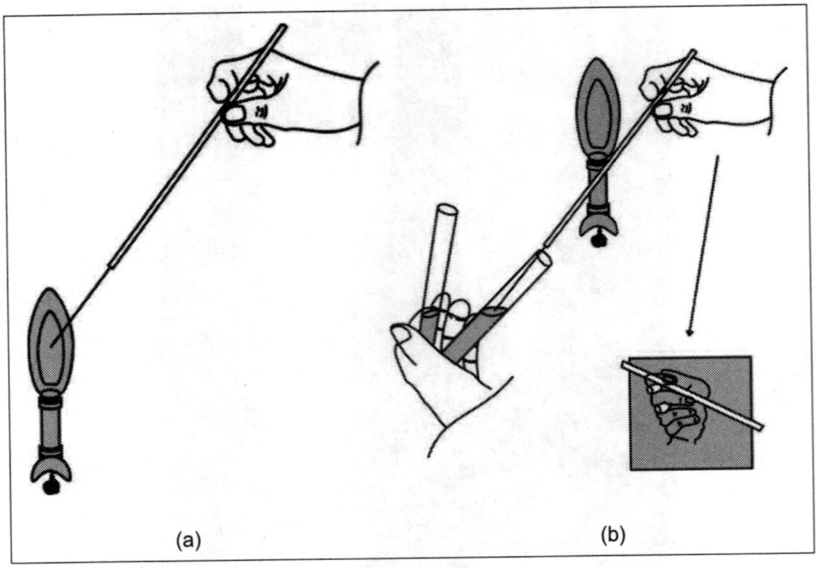

Figure 5.20 (a) Flame the inoculation loop (b) Correct method for holding tubes before and after inoculation transferring cultures.

During inoculation the closely packed cells of the start form colonies that run together. However as the inoculum becomes thin towards the end, the colonies obtained here are isolated.

(2) *Pour Plate*: Here the agar medium is inoculated while it is liquid (but cool) and so the colonies develop throughout the medium. Before inoculating a culture it is preferable to dilute the sample through loop-dilution technique or serial dilution technique. Dilution ensures isolated colonies (Figures 5.23, 5.24 and 5.25).

5.12.2.1 *Preparation of serial dilution*

1. Withdraw 1 mL sample or the bacterial culture and add it to 9 mL of 1 N saline solution.
2. Again take 1 mL of above inoculated saline and add to 9 mL of sterilized 1 N saline. This dilutes the bacterial sample to 10^{-2} times.
3. Again withdraw 1 mL of bacterial sample and add to 9 mL of saline.
4. This raises the dilution to 10^{-3} times.
5. In this way by repeated inoculation prepare dilution of 10^{-6}.
6. Now inoculate 10 μL of above sample to the plate by pour plate method in triplicate.

After sufficient incubation, the bacteria grow so as to form a mixed culture each having characteristic colony morphology of its own. The nature of whole colony, the edge, the elevation and the surface fixture in itself provides maximum information about the type of bacterium. These characteristics are made use during identification of bacteria (Figure 5.26).

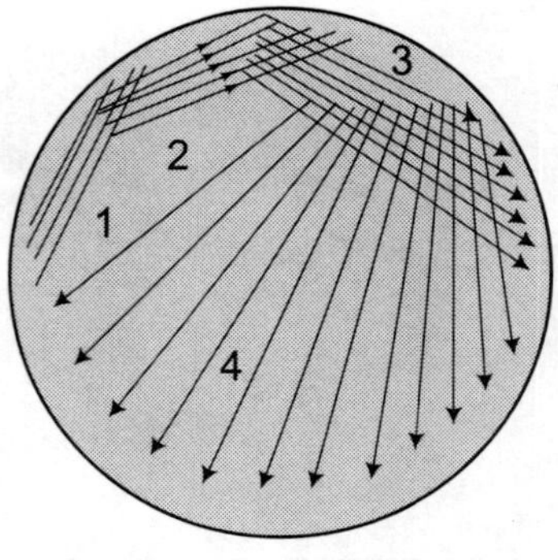

Quadrant streak

1. Streak one loopful of organisms over Area 1 near edge of the plate. *Apply the loop lightly*. Don't gouge into the medium.
2. Flame the loop, cool 5 seconds, and make 5 or 6 streaks from Area 1 through Area 2.
3. Flame the loop again, cool it, and make 6 or 7 streaks from Area 2 through Area 3.
4. Flame the loop again and make as many streaks as possible from Area 3 into Area 4, using up the remainder of the plate surface.
5. Flame the loop before putting it aside.

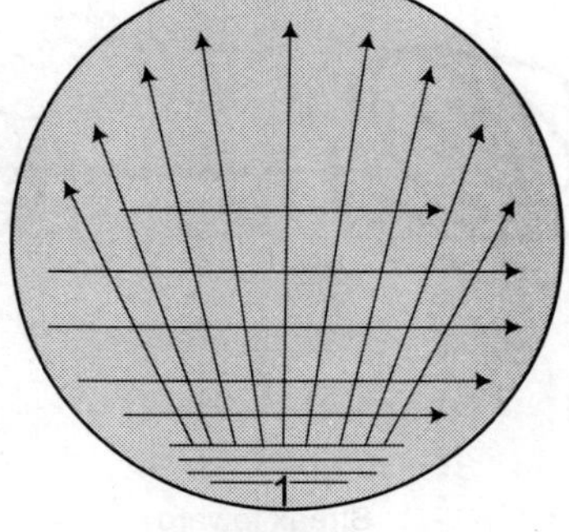

Radiant streak

1. Spread a loopful of organisms in small area near edge of plate (Area 1).
2. Flame the loop and allow it to cool for 5 seconds.
3. From the edge of area 1 make 7 or 8 straight streakes to the opposite side of the plate.
4. Flame the loop again and cross-streak over the last streaks, starting near Area 1.
5. Flame the loop again before putting it aside.

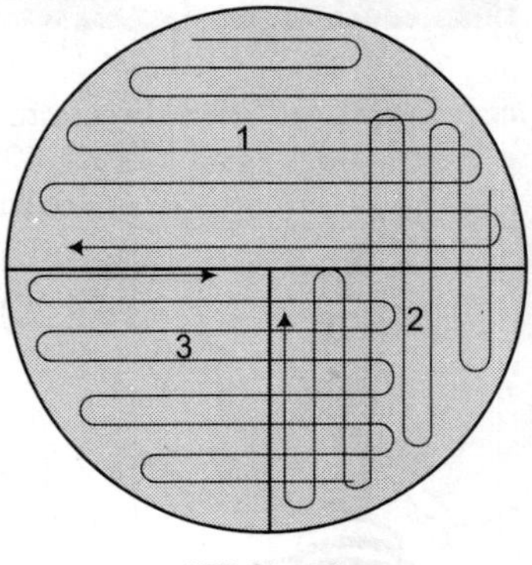

"T" Streak

1. With marking pencil draw a "T" on bottom of plate, dividing the plate into one half and two quarters.
2. Inoculate the half portion with one loopful of organisms in the continuous line from edge of the plate to the midline of the plate. Don't gouge the medium.
3. After flaming the loop, cross-streak from Area 1 into Area 2 with one continuous streak, filling Area 2.
4. After flaming loop again, cross-streak from Area 2 to Area 3 with a single continuous streak.

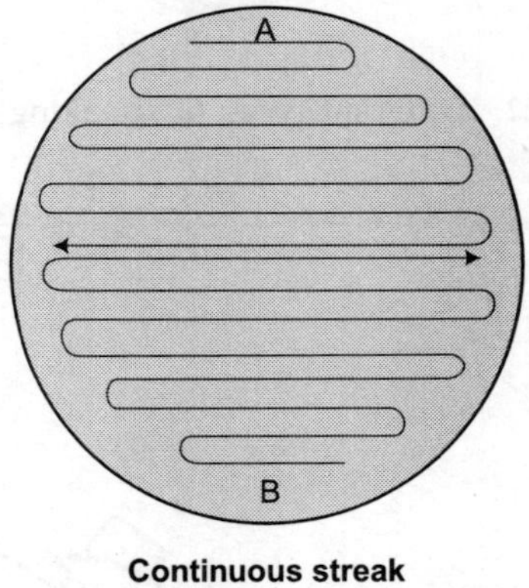

Continuous streak

1. Starting at the edge of the plate (Area A) with a loopful of organisms, spread the organisms in a single continuous movement to the cenre of the plate.
2. Rotate the plate 180 degrees so that the uninoculated portion of the plate is away from you.
3. Without flaming loop, and using the same face of the loop, continue streaking the other half of the plate by starting at Area B and working toward the centre.
4. Flame your loop before setting it aside.

Figure 5.21 Streak plate.

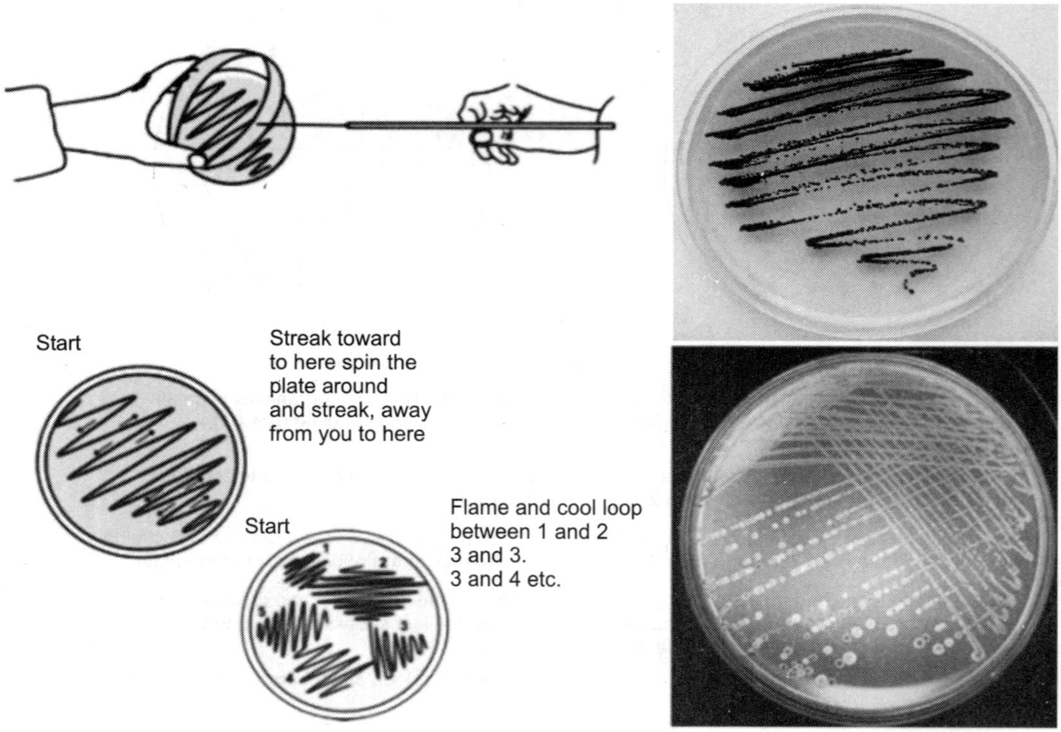

Figure 5.22 Different types of streaking and motion of loop in one and two direction.

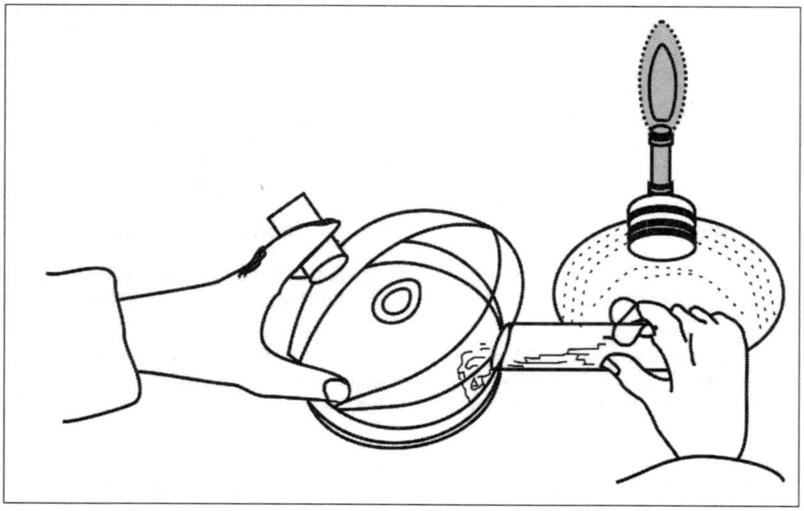

Figure 5.23 Pouring an agar plate.

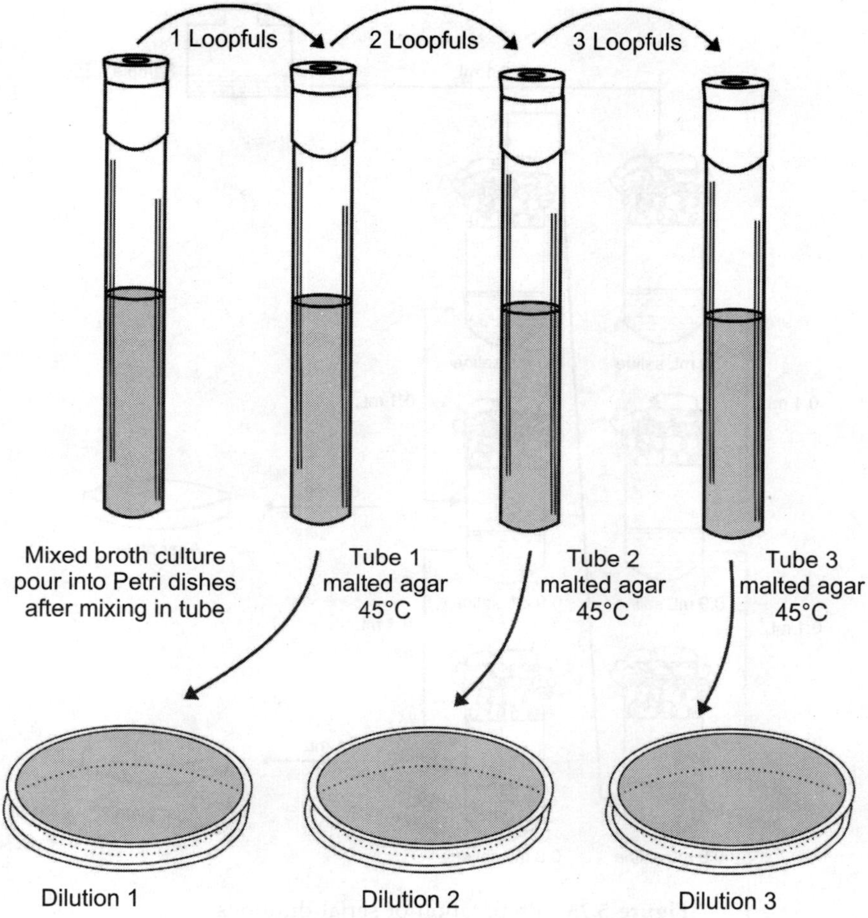

Figure 5.24 The loop-dilution technique.

5.12.2.2 Mixed to pure culture

From a mixed culture, study the colony morphology and then transfer each of the colony type to different plates having the same isolation media as used before. Now watch if there is the development of only single colony type on the plate. If it is so the culture can be said to be pure culture. There are various methods to store a pure culture (Figure 5.27). However, the most common one which is followed generally is the transfer of pure culture from plate to an agar slant (a slant is a test tube 2/3rd filled with agar media and kept at an angle during solidification). The inoculated slant is kept in the incubator and after proper growth it is transferred to the refrigerator.

5.13 CULTIVATION OF ANAEROBES

There are many bacteria that find oxygen toxic or at least inhibitory to their existence. An environment is to be created for developing anaerobic condition. Using special

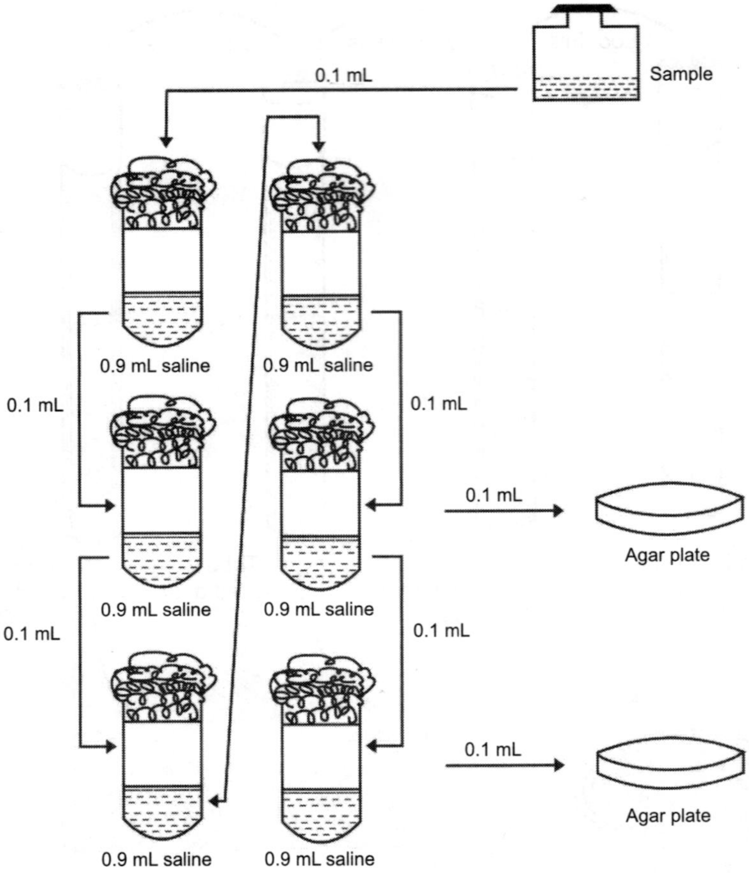

Figure 5.25 Preparation of serial dilutions.

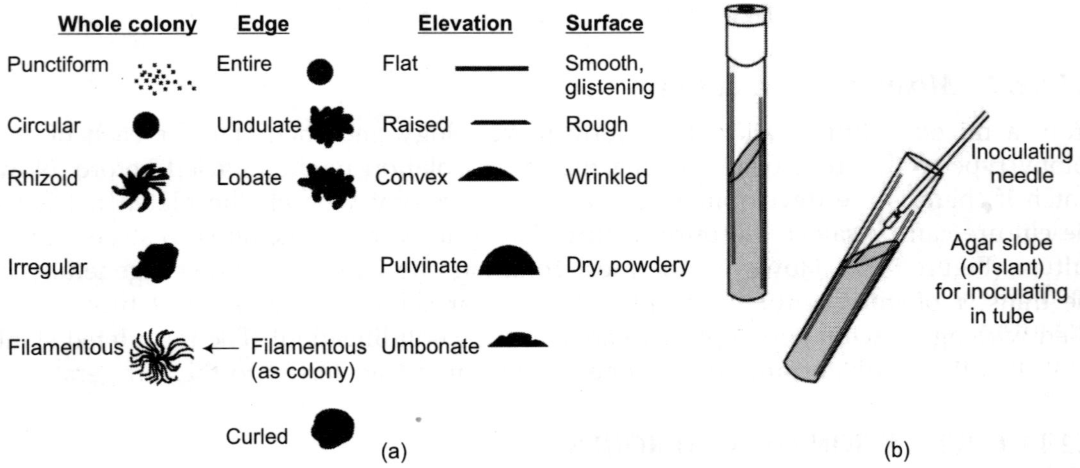

Figure 5.26 (a) Guide to terms used in colony descriptions (b) Different types of growth on agar slopes.

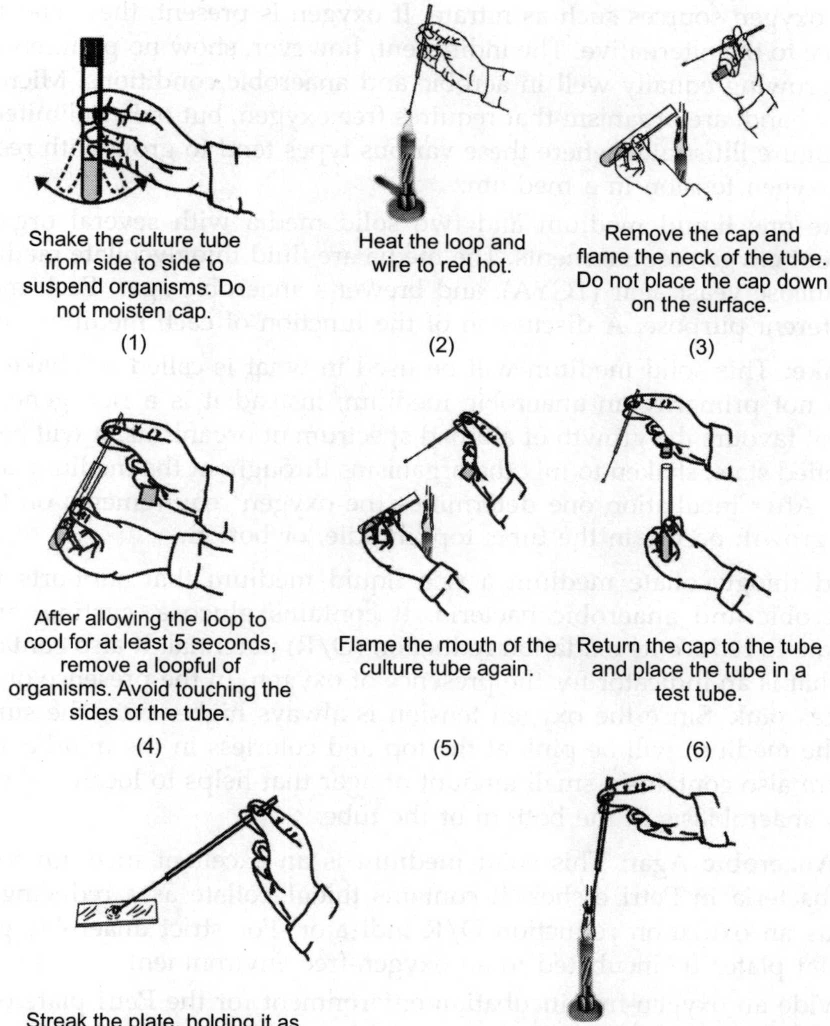

Figure 5.27 Illustration of asceptic procedure for organism removal.

media deficient or lacking in oxygen and containers that are free from oxygen. Several organisms of different cultural requirements are inoculated to different special media to evaluate their oxygen needs.

The oxygen requirements of bacteria range from strict (obligate) aerobes that cannot exist without this gas to the strict (obligate) aerobes that die in its presence. In between these extremes are the facultative, indifferent, and microaerophilics. The facultative are bacteria that have enzyme systems enabling them to utilize free oxygen or some alternative oxygen sources such as nitrate. If oxygen is present, they tend to utilize it in preference to the alternative. The indifferent, however, show no preference for either condition, growing equally well in aerobic and anaerobic conditions. Microaerophiles on the other hand, are organism that requires free oxygen, but only in limited amounts. Following figure illustrates where these various types tend to grow with respect to the degree of oxygen tension in a medium.

Inoculate one liquid medium and two solid media with several organisms that have different oxygen requirements. The media are fluid thioglycollate medium (FTM), tryptone glucose yeast agar (TGYA), and brewer's anaerobic agar. Each medium will serve a different purpose. A discussion of the function of each medium follows:

TGYA Shake: This solid medium will be used in what is called a "shake tube." The medium is not primarily an anaerobic medium; instead it is a rich general purpose medium that favours the growth of a broad spectrum of organisms. It will be inoculated in the liquefied state, shaken to mix the organisms throughout the medium, and allowed to solidify. After incubation one determines the oxygen requirements on the basis of where the growth occurs in the tube: top, middle, or bottom.

FTM: Fluid thioglycollate medium a rich liquid medium that supports the growth of both aerobic and anaerobic bacteria. It contains glucose, cystine, and sodium thioglycollate to reduce its oxidation reduction (O/R) potential. It also contains the dye resazurin that is an indicator for the presence of oxygen. In the presence of oxygen the dye becomes pink. Since the oxygen tension is always higher near the surface of the medium, the medium will be pink at the top and colorless in the middle and bottom. The medium also contains a small amount of agar that helps to localize the organisms and favors anaerobiasis in the bottom of the tube.

Brewer's Anaerobic Agar: This solid medium is an excellent medium for culturing anaerobic bacteria in Petri dishes. It contains thioglycollate as a reducing agent and resazurin as an oxidation reduction O/R indicator. For strict anaerobic growth it is essential that plates be incubated in an oxygen-free environment.

To provide an oxygen-free incubation environment for the Petri plate of anaerobic agar Gas Pak anaerobic jar will be used hydrogen is generated in the jar, which removes the oxygen by forming water. Palladium pellets catalyze the reaction at room temperature (Figure 5.28). The generation of hydrogen is achieved by adding water to a plastic envelope of chemicals. CO_2 is produced, which is a requirement for the growth of many fastidious bacteria. To make certain that anaerobic condition actually exist in the jar, an indicator strip of Methylene blue becomes colorless in the total, absence of oxygen. If the strip is not reduced (decolorized) within 2 hours, the jars has not been sealed properly, or the chemical reaction has failed to occur.

In addition a microscopic study of the types of endospores formed by spore-formers are described below:

The tubes and all of the plates placed in a Gaspak jar and incubate at 37°C in incubator.

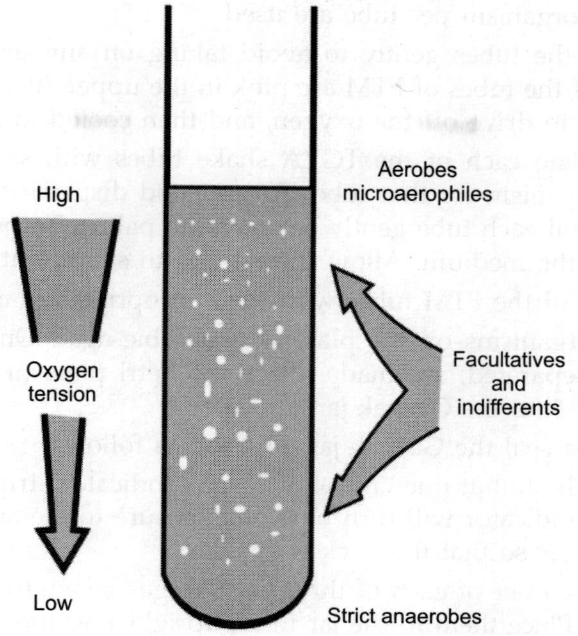

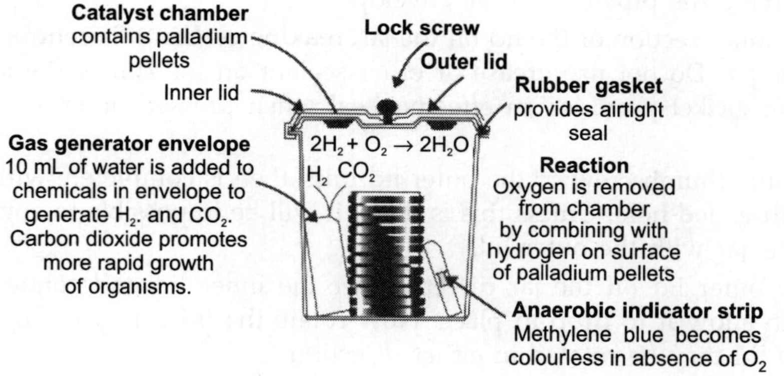

Figure 5.28 Gas pack anaerobic chamber and anaerobic enviroment.

5.13.1 Procedure

Three tubes of fluid thioglycollate medium, 3 TGYA shake tubes (liquefied), 1 Petri plate of Brewer's anaerobic agar. Broth cultures of *Staphylococcus aureus, Streptococcus, faecalis, Clostridium sporogenes, Bacillus subtilis, Escherichia coli,* and *Blostridium rubrum* are used. Gaspak anaerobic jar, Gaspak generator envelopes, Gaspak anaerobic generator strip, scissors and one 10 mL pipette water baths.

1. Water bath is set up at 45°C close to workplace in to keep tubes of TGYA shakes from solidifying.

2. Six tubes one organism per tube are used.

   Note: Handle the tubes gently to avoid taking on any unwanted oxygen into the medium. If the tubes of FTM are pink in the upper 30%, they must be boiled a few minutes to drive off the oxygen, and then cooled to inoculate.
3. Heavily inoculate each of the TGYA shake tubes with several loopfuls of the appropriate organism for that tube. To get good dispersion of the organisms in the medium, roll each tube gently between the palms. To prevent uptake do not overly agitate the medium. Allow these tubes to solidify at room temperature.
4. Inoculate each of the FTM tubes with the appropriate organisms.
5. Streak three organisms on the plate of anaerobic agar. Only three straight-line streaks, well separated, are made. Place the Petri plate (inverted) in a canister and then place into the Gaspak jar.
6. To activate and seal the Gaspak jar, proceed as follows:
   - Peel apart the foil at one end of a Gaspak indicator strip and pull it halfway down. The indicator will turn blue on exposure to the air. Place the indicator strip in the jar so that the wick is visible.
   - Cut off the corner of each of three Gaspak gas generator envelopes with pair of scissors. Place them in the jar in an upright position.
   - Pipette 10 mL of tap or distilled water into the open corner of each envelope. Avoid forcing the pipette into the envelope.
   - Place the inner section of the lid on the jar, making certain it is centred on the top of the jar. Do not use grease or other sealant on the rim of the jar since the O-ring gasket provides an effective seal when pressed down on a clean surface.
   - Unscrew the thumbscrew or the outer lid until then is completely withdrawn into the threaded hole. Unless this is done, it will be impossible to engage the lugs of the jar with the outer lid.
   - Place the outer lid on the jar directly over the inner lid and rotate the lid slightly to allow it to drop in place. Now rotate the lid firmly to engage the lugs. The lid may be rotated in either direction.
   - Tighten the thumbscrew by turning clockwise. If the outer lid raises up, the plugs are not properly engaged.
7. Place the jar in a 37°C incubator. After 2 or 3 hours check the jar to note if the indicator strip has lost its blue colour. If depolarization has not occurred, replace the palladium pellets and repeat the entire process.
8. Incubate the tubes and plates for 24 to 48 hours.

5.13.2 Culture Evaluations and Spore Staining

Remove the lid from the Gaspak jar. If vacuum holds the inner lid firmly in place, break the vacuum by sliding the lid to the edge. Take care not to agitate the FTM tubes during transportation. The position of growth in the medium can be easily changed if handled carelessly. Prepare a combinsed slide with three separate smears of three spore-formers, using either one of the two spore-staining methods.

5.14 PHOTOSYNTHETIC BACTERIA

A considerable number of pigmented bacteria exist that derive energy by photosynthesis. At present three families are recognized that contain around 20 genera other than cyanobacteria. These phototrophic bacteria are a diverse group of cocci, rods, vibrio-, and spiral-shaped forms. All of them are Gram-negative. Although cell suspensions of various species occur as purple, red, orange-brown, brown, or green, the genera are subdivided into two main groups: those that are purple and those that are green. All of them are aquatic, being abundant in the ooze of ponds.

The purple photosynthetic bacteria (suborder Rhodospirillineae) include the families Rhodospirillineae and Chromatiaceae. The green species (suborder Chromatiaceae) belong to one family, the Chlorobiaceae.

The photosynthetic bacteria differ from the algae and green plants in that the bacteria are anaerobic and utilize bacteriochlorophyll instead of chlorophyll-a for photosynthesis. While the algae produce oxygen as a by-product of photosynthesis, the Chromatiaceae and Chlorobiaceae produce sulfur instead. The following chemical reaction of photosynthesis in these organisms indicates how hydrogen sulfide is the hydrogen donor.

$$CO_2 + 2H_2S \longrightarrow C(H_2O) + 2S + H_2O$$

In addition to differing in colour, Chromatiaceae and Chlorobiaceae also differ in that the former produce intracellular sulfur granules (one exception) and the Chlorobiaceae produce extra cellular sulfur granules. The Rhodospirillaceae, incidentally, are nonsulfur phototrophs and were formerly called the Athiorhodaceae.

Species of *Chromatium* (family Chromatiaceae) and *Chlorobium* (family Chlorobiaceae) are isolated from the ooze of stagnant ponds. Bacterial decomposition in the pond ooze provides the hydrogen sulfide essential to their photosynthesis. Specified enrichment media are first inoculated with ooze collected from various ponds. These bottles of inoculated media are then incubated at room temperature for 7 days under illumination with incandescent lamps 24 hours a day. Isolated colonies of phototrophic bacteria, agar plates from shake deeps are made from bottles by streak plate methods and green, red, brown, or purple are picked up after 7 days.

5.14.1 Enrichment

The enrichment media for these two genera contain essentially the same ingredients with differences being only in amounts. The Chromatium medium will turn red in 4 to 7 days if any chromatia are present. The Chlorobium medium should turn green in about the same length of time, but it may become overgrown by red chromatia in a short period of time; thus, subculture timing is more critical with the latter (Figure 5.29).

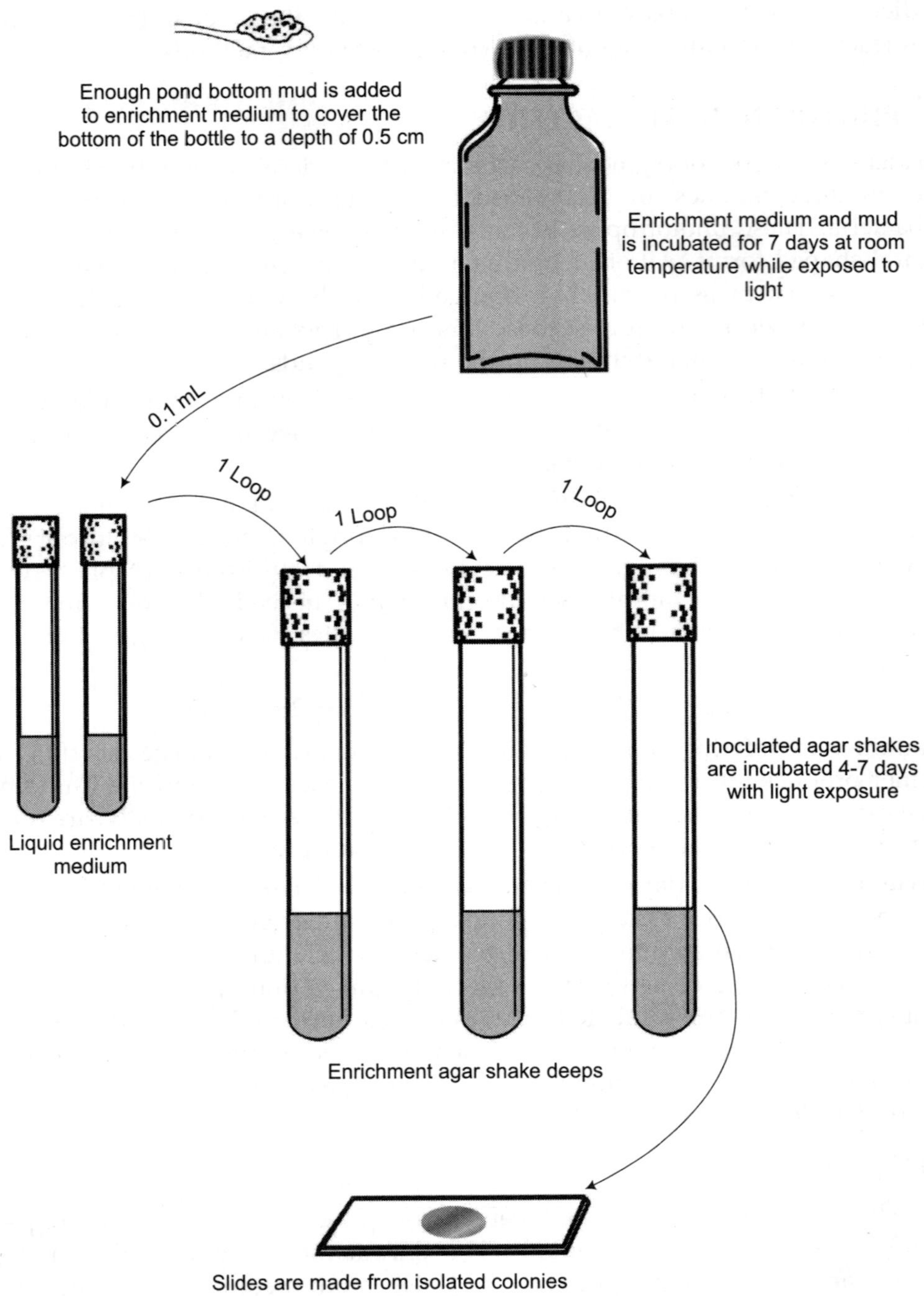

Figure 5.29 Photosynthetic bacterial isolation method by enrichment.

5.14.2 Materials

Samples of pond mud, Chromatium enrichment medium (nonsterile, in 3 oz prescription bottle), Chlorobium enrichment medium (nonsterile, in 3 oz prescription bottle) and Desk lamp with 60 watt bulb.

5.14.3 Procedure

1. Place enough pond mud into medium containing bottles with a minimum depth of 0.5 cm exists in the bottom. Fill the bottles to maximum capacity with medium and twist the lids securely. Ideally, there should be no air bubble in the bottle several times to mix the mud and medium.
2. Place the bottles about 1-1.5 feet away from a desk lamp incubate the bottles at room temperature for 7 days. The lamp should remain on 24 hours per day. Examine the bottles each day to look for colour changes. The Chromatium medium will become reddish if species of this genus are present in the pond mud. The Chlorobium medium will become green if the mud contains species of *Chlorobium.*
3. Prepare a hanging drop slide of organisms from both bottles and examine it with phase optics or bright field oil immersion.

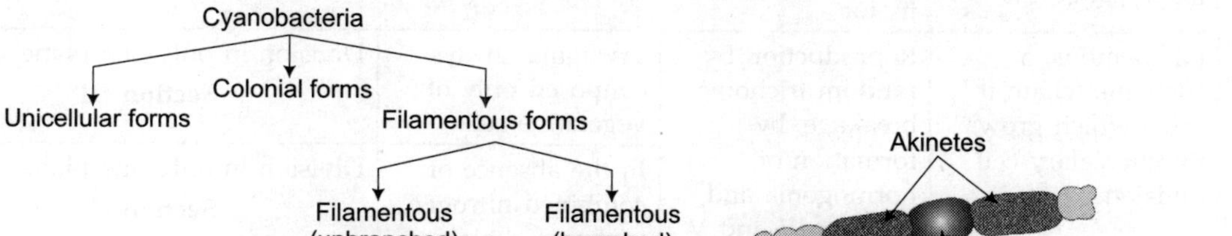

Figure 5.30 Classification of cyanobacteria with filament showing akinetes and heterocyst.

5.15 CYANOBACTERIA

Cyanobacteria include unicellular and colonial species. Colonies may form filaments, sheets or even hollow balls. Some filamentous colonies show the ability to differentiate into several different cell types: vegetative cells, the normal, photosynthetic cells that are formed under favourable growing conditions; akinetes, the climate-resistant spores that may form when environmental conditions become harsh; and thick-walled heterocysts, which contain the enzyme nitrogenase, vital for nitrogen fixation (Figure 5.30 and Table 5.5). Heterocysts may also form under the appropriate environmental conditions (anoxic) when fixed nitrogen is scarce. Heterocyst-forming species are specialized for nitrogen fixation and are able to fix nitrogen gas into ammonia (NH_3), nitrites (NO_2) or nitrates (NO_3), which can be absorbed by plants and converted to protein and nucleic acids [atmospheric nitrogen cannot be used by plants directly].

Table 5.5 Broad classification of cyanobacteria

Unicellular; cells single or forming colonical aggregates held together by additional outer cell wall layers	Reproduction by binary fission or by budding		**Section I**
	Reproduction by multiple fission giving rise to small daughter cells (baeocytes), or by both multiple fission and binary fission		**Section II**
Filamentous; a trichome (chain if cells) which grows by intercalary cell division	Reproduction by random trichome breakage, by formation of hormogonia and (Section IV and V only) sometimes by germination of akinetes	Trichome always conposed only of vegetative cells	Division in only one plane **Section III**
		In the absence of combined nitrogen, trichome contains heterocysts; some also produce akinetes	Division in only one plane **Section IV**
			Division in more than one plane **Section V**

Often initial isolations of cyanobacteria from the natural environment give mixed cultures. Therefore, it is essential to purify the individual types of cyanobacteria from the mixture. This could be achieved easily by streak or pour plate techniques. These are essentially dilution techniques resulting in the physical separation of individual cyanobacterium from the mixture allowing them to form distinct colonies, which can then be picked up to make pure cultures. Also, for obtaining clonal cultures of filamentous forms, these techniques prove extremely useful.

5.15.1 Isolation and Purification

5.15.1.1 Streak plate

A loop containing varying numbers of cyanobacteria, when touched to the agar surface in a Petri dish, allows to adhere maximum number of cyanobacteria, as the streak progresses further on the surface of the medium and towards the end of streak the number gets so much reduced that on incubation form separate colonies.

5.15.1.1.1 Protocol

1. Prepare agar plates containing suitable medium.
2. Flame the inoculation needle and cool it by jabbing it into the edge of the agar medium.
3. With the drop on the edge away from the body, streak the culture back and forth, edge to edge in parallel lines, moving towards the body.
4. When the needle reaches the centre of the plate, spin it around 180° and continue streaking, now moving away from the body. This reversal avoids interference with the needle.

5.15.1.2 Spread plate

In this method, the cyanobacteria in liquid medium is directly spread over the entire surface of the solid medium resulting in separation at many places.

5.15.1.2.1 Protocol

1. Prepare suitable agar medium.
2. Sterilize the medium, Petri plates, and tissue grinder, etc.
3. Grind a small portion of cyanobacteria with the help of tissue grinder homogenously.
4. Take 1 mL of ground material and make serial dilutions.
5. Pour the sterilized agar medium and allow it to solidify.
6. Inoculate 0.1 mL of diluted cyanobacterial sample on to the surface of agar and spread it thoroughly with the L-shaped glass rod.
7. Label the Petri plate before incubation.

5.15.1.3 Pour plate

In pour plate technique, there is direct dilution of cyanobacteria while being suspended in molten agar medium, resulting in separation at the time of plating. Solidifying substance traps the individual cells in place and produces a fixed colony of the cell or filament.

5.15.1.3.1 Procedure

1. Prepare a set of sterile 90 mL water blanks in 250 mL conical flasks.
2. Add 10 mL of cyanobacterial sample in water blank and shake the contents thoroughly. Label it as 10^{-1} dilution.
3. Transfer 10 mL from 10^{-1} dilution to 90 mL water blank and mix thoroughly. Label it as 10^{-2} dilution.
4. Transfer 0.1 or 1 mL of the diluted suspension into sterile Petri dish and pour molten agar medium cooled to 47°C.
5. Mix the contents by swirl action and allow solidification.

6. Prepare 10, 10 and 10 dilutions following similar procedure and transfer 0.1 mL suspension from each dilution in separate Petri dish and pour the medium to prepare poured plates from each dilution.

Comments: There is no accurate way of predicting the number of viable cells in a given sample by this method. Therefore, one must always make several dilutions of the sample and pour several plates. However, experience with a certain type of samples will minimize wasteful dilutions. The success of this method depends upon thorough mixing and even distribution of cells.

5.15.2 Inoculation Procedure

In order to grow an organism in a sterilized medium, a number of cells or filaments (inoculum) are transferred (inoculated) into the medium with special precautions to maintain the purity of the culture. In the inoculation procedure the needle or loop, when used to transfer cyanobacteria, should be heated to redness by flaming immediately, before and the lower part of the handle. The flaming destroys any living forms on the surface of needle or loop. During transfer, hold the tube in the left hand and hold the plug between the fingers of the right hand. Never lay a plug down. Hold the tube as nearly horizontal as feasible during transfer and do not leave it open longer than necessary. The mouths of the tubes from which cultures are taken and into which they are transferred should also be passed through the burner flame immediately before and after the needle is introduced and removed. In addition to destroying any organism on the tip of the tube, flaming tends to create outward convection currents, thus decreasing the chance of contamination.

5.15.2.1 Incubation conditions

The cultures are normally incubated at 29 ± 2°C under continuous illumination (2,000-3,000 lux) unless otherwise required. Because of the high concentration of water in agar, condensation of water may occur in Petri dishes during incubation and moisture is likely to drip from the cover into the surface of the agar and spread out resulting in a confluent mass of growth and ruining individual colony formation. To avoid this, the Petri dishes are incubated bottom side up. The slant cultures should be placed in a slanting position, with the streak towards the light source. Always label all Petri dishes, tubes, flasks and culture bottles with name, date and identification of contents.

5.15.2.2 Culture room

A culture room (growth room or incubation chamber) for cyanobacteria can be designed depending upon the space and nature of work. In general, a growth room should have a provision of controlled illumination, temperature and aeration facilities for the culture vessels.

5.15.3 Isolation Procedures

5.15.3.1 Unicellular cyanobacteria

1. Transfer sample from primary enrichment tubes to fresh liquid medium and incubate.
2. From the secondary tubes, streak the cyanobacteria, onto the mineral agar plates and incubate.
3. After incubation, isolate colonies with a sterile loop and inoculate in fresh medium.
4. Successive transfers from liquid to solid media usually result in unicyanobacterial cultures.

5.15.3.2 Filamentous cyanobacteria

Pure cultures of filamentous forms usually can be obtained simply by repeated liquid transfer of small amounts of material. Phototactic response of the cyanobacteria is also utilized to separate them from other cyanobacteria.

1. Place a small mass of cells on a slide in a liquid medium.
2. Cover with cover glass and seal with vaseline, paraffin or similar compounds.
3. Incubator slide in unidirectional light for 2-10 hours. The filament moves from the centre so that single filament is observed at the outer edge.
4. Isolate with a sterile Pasteur-type capillary pipette, and transfer it to a tube of liquid medium or on the surface of solidified medium (1%) in a Petri dish (a low agar concentration less than 1.5% is important for obtaining good growth and the filament rapidly migrates through 1% agar).
5. Incubate in unidirectional light.
6. Isolate filaments that are visibly free of bacterial contamination from the rapidly moving front on the illuminated side of the plate. By several passages through agar, axenic filamentous strains can be obtained (Figure 5.31).

5.15.4 Purification Methods

It is often possible to obtain axenic cultures from field material by performing isolation and the subsequent purification steps on plates as described. Sometimes, however, this is not the case, particularly if the natural sample is highly contaminated by bacteria and cyanobacterial species are immobile and incapable of self-purification by gliding away from their contaminants, or if the single filaments or micro colonies do not grow on solid media. Purification of such cyanobacteria should be attempted after the establishment of unicyanobacterial isolates in liquid cultures.

5.15.4.1 Repeated liquid subculture

This technique has been successfully used when a natural collection is particularly rich in a specific cyanobacterium.

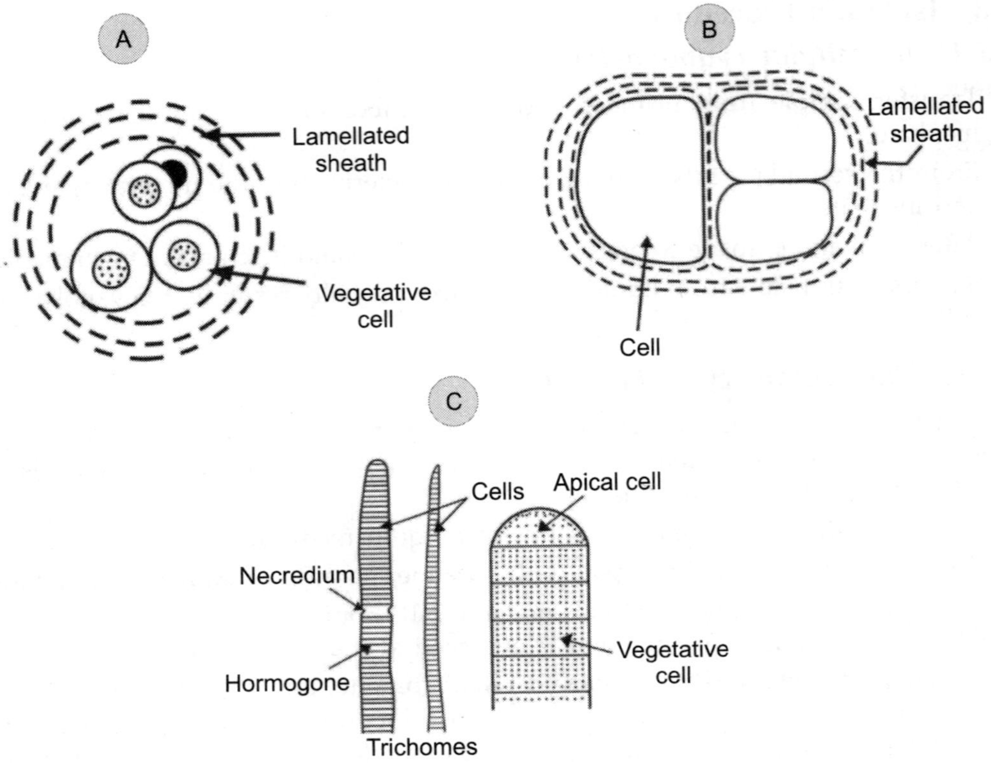

Figure 5.31 Diagrammatic view of some common cyanobacteria (a) *Gloeocapsa* sp. (b) *Chroococcus* sp. (c) *Oscillatoria*.

5.15.4.1.1 Protocol

1. Prepare suitable liquid medium and distribute it to flask and sterilize.
2. Make frequent subcultures to the fresh medium to get rid of the other contaminating bacteria.

5.15.4.2 Fragmentation

5.15.4.2.1 Protocol

1. Homogenize filamentous forms with a glass homogenizer for 5-10 min or using a sonicator to break mucilagenous colonies allowing short filaments (4-8 cells) long to be obtained.
2. Then follow either streak/spread/pour plate method to obtain individual colonies.

5.15.4.3 Antibiotic treatments

It may be difficult to remove certain contaminants by repeated subculturing and/or ultrasonic treatment. In such instances, use of a chemical method rather than a physical method is preferred, but physical methods such as washing and ultrasonic treatment may proceed. One such chemical method uses antibiotics, singly or in combination to kill or inhibit the growth of tenaciously attached contaminants. Purification though antibiotic treatment has been adopted by many workers.

5.15.4.3.1 Protocol

1. Dissolve 100 mg penicillin G (K or Na salt) and 50-mg streptomycin sulphate, together in 10 mL distilled water: add 10-mg chloramphenicol (dissolved in 1 mL 95% ethanol) to the penicillin-streptomycin solution and mix well.
2. Filter the triple antibiotic solution quickly, using a membrane or Seitz filter.
3. Place 1 mL of cyanobacterial suspension to be purified in each of six 125 mL Erlenmeyer flasks, each containing 50 mL culture medium.
4. Add one of the following volumes of antibiotic solution to each of the flasks: 0.125, 0.25, 0.5, 1.0, 2.0, and 3.0 mL. This provides penicillin levels ranging from approximately 20-500 mg and corresponding levels of two other antibiotics.
5. Place the culture flasks in dark conditions suitable for bacterial growth.
6. After 12 to 24 hrs aseptically transfer some cyanobacterial cells from each flask to tubes of sterile, antibiotic free culture medium. Prepare tubes in triplicates at specific intervals and incubate.
7. Check culture tubes for bacterial contamination after 2-3 days using nutrient agar plates or nutrient broth.
8. Some of the bacteriostatic compounds like potassium tellurite (K_2IeO_3 –10 µg mL^{-1}) has also been used to obtain bacteria-free cyanobacterial algal cultures. Purification can also be done by treatment with cycloserine (5 µg mL^{-1}).

5.15.4.4 Ultraviolet (UV) irradiation

This method has been widely used to obtain pure cultures of cyanobacteria.

5.15.4.4.1 Protocol

Place a dilute cyanobacterial suspension in the quartz-windowed chamber and irradiate for 20-30 min. at 275 nm. Agitate the suspension by continuous stirring during the irradiation period. Remove the cyanobacterial samples from the irradiation chamber at 5 min. intervals and prepare a large number of dilution cultures from each. Inoculate the diluted cyanobacterial suspension into fresh medium and incubate in a culture room. Usually, cultures from long exposure times fail to grow while those from shorter exposure are contaminated. However, certain subcultures, particularly from samples with intermediate exposures, will be free of bacteria.

5.15.5 Test for Purity

Cyanobacterial cultures grown in liquid medium under stationary phase are examined critically under the microscope using phase contrast objectives and oil immersion. For additional macroscopic mineral medium supplemented with casamino acids (0.02-0.05% w/v) and glucose (0.5% w/v).

The plates are then incubated in the dark for several days at a temperature typical for growth of the cyanobacteria to be tested since any contaminants still present should grow at that temperature. The contaminants will then be easily detected by their superior growth, colour and other typical appearances (e.g., fungal hyphae). If the culture is heavily contaminated the entire area of the test drop will be covered with bacterial or fungal growth: a low degree of contamination will reveal itself in the form of micro colonies on the cyanobacterial lawn., Some contaminants (e.g., gliding bacteria) do not form discrete colonies on the surface of the agar. Their presence (or, its hoped absence) can only be confirmed by additional microscopy, after screening small sections from the agar occupied by the cyanobacterial deposit.

5.15.6 Culture Media

To isolate and maintain the cyanobacteria, appropriate media are used, and congenial environmental conditions are maintained. Phosphate, nitrate, magnesium and calcium are the macronutrients generally required by them. The essential micronutrients are iron, zinc, manganese, copper and molybdenum. Good growth is obtained at 2000-3000 lux light intensity. The optimal temperature required for their growth is 29 ± 1°C. A number of culture media for growing cyanobacteria are available. For the isolation and maintenance of heterocystous cyanobacteria, the source of combined nitrogen ($NaNO_3$ or KNO_3) in the synthetic medium should be omitted and possibly be replaced by the corresponding chlorides.

5.15.7 Isolation of Cyanobacteria from Soil/Water

5.15.7.1 Introduction

Cyanobacteria are photoautotrophic, nitrogen fixing prokaryotes occurring everywhere in a diversified habitat. Some of the members of cyanobacteria are being commercially exploited for mass production of biofertilizer, pharmaceutical and value added products. They have been classified on the basis of their morphology from unicellular to filamentous and differentiated structures.

5.15.7.2 Materials

Soil sample, 10 mL sterile pipettes, Erlenmeyer's flasks (250 mL capacity), illuminated growth chamber, inoculation needle, polythene bags, Bunsen burner, sterile distilled water.

Modified Chu-10 medium: $Ca(NO_3)_2.4H_2O$ 40 mg, $K_2HPO_4.3H_2O$ 10 mg, $MgSO_4.7H_2O$ 25 mg, Na_2CO_3 20 mg, $NaSiO_3.9H_2O$ 25 mg, ferric citrate 3 mg, citric acid 3 mg, glass distilled water to 1 L.

To this medium 1 mL/L of **Fogg's micronutrient solution** is to be added. The composition of the solution:

$MnCl_2.4H_2O$10 mg, $Na_2MoO_42H_2O$ 10 mg, H_3BO_3 10 mg, $CuSO_4.5H_2O$ 1 mg, $ZnSO_4.7H_2O$ 1 mg, distilled water 100 mL, pH of this medium should be adjusted to 7.5 before autoclaving.

5.15.7.3 *Protocol*

1. Collect the sample in a polythene bag or colonies of cyanobacteria floating on the water surface in the rice fields in a glass bottle and bring it to the laboratory.
2. Prepare Chu-10 medium, transfer 20 mL into 100 mL Erlenmeyer flasks and autoclave at 15 psi for 15 min.
3. Serially dilute the soil/water sample (10^{-2} to 10^{-7}) in sterile distilled water.
4. Pour aliquots of 1 mL of various dilution of soil/algal sample into 20 mL of sterilized medium in flasks.
5. Incubate the flasks at 30 ± 1°C for 2-3 weeks in an alternate light and dark regime of 12 fours. Cyanobacterial colonies appear in broth medium.
6. Soon pick up individual colonies and transfer onto fresh agar slants and incubate as above.

5.15.7.4 *Results and observations*

Several colonies of cyanobacteria grow on slants. They secrete copious amount of mucilage. Observe them under the microscope and identify following literature. Filaments of various colour and shape are observed under the microscope. The filament may be heterocystous or non-heterocystous, branched or unbranched.

5.15.8 Measurement of Specific Growth Rate and Generation Time in a Given Cyanobacterium

5.15.8.1 *Introduction*

Microorganism in a culture will increase exponentially until an essential nutrient is exhausted. Typically the first organism splits into two daughter organisms, which is then each split to form four, who split to form eight, and so on. Consequently, the organisms continuously add in their growth and biomass. There are several parameters to measure the growth of a given cyanobacterium such as cell number, chlorophyll content, DNA and RNA content, protein contents, absorbance (optical density), and dry weight. Of all the growth process, a sigmoidal curve (s-shaped growth) is a typical one (Figure 5.32). The cyanobacteria show four distinct types of growth phases as are observed in bacteria. In addition, the generation time is the time required by any of the initial population to double its population. It is designed as G. Any two points are selected in the log phase of growth and generation time can be calculated by using the following formula:

$$G = \frac{T_2 - T_1}{2.302\,(\log N_2 - \log N_1)}$$

where, G = Generation time, T_1 = Initial time of growth, T_2 = final time of growth, N_1 = initial number of cell mass or any other parameter, N_2 = Final cell number or cell mass or any other parameter.

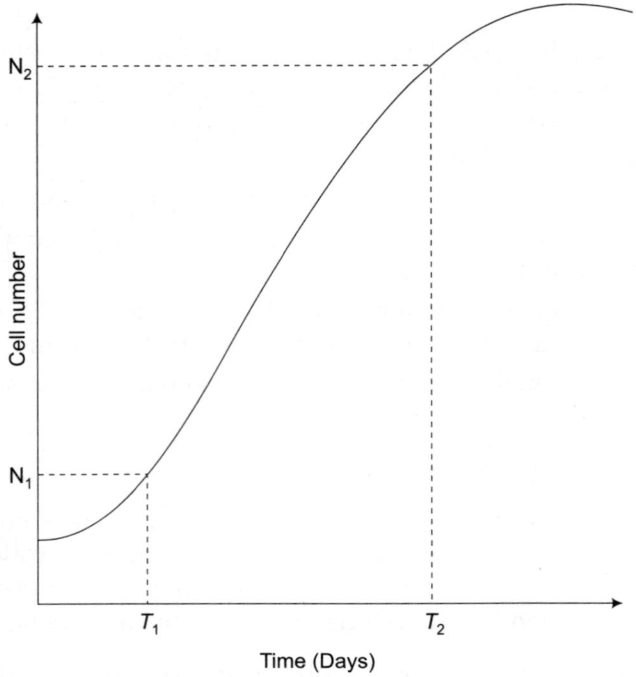

Figure 5.32 A typical sigmoidal growth curve.

5.15.8.2 Materials

Cyanobacterial culture, Erlenmeyer flasks, pipettes, culture tubes, spectrocolorimeter, water bath,

BG-11 growth medium (g/L) in distilled water: $NaNO_3$ 1.5, $K_2HPO_4.3H_2O$ 0.04, $MgSO_4.7H_2O$ 0.75, $CaCl_2.2H_2O$ 0.036, Citric acid 0.006, Ferric ammonium citrate 0.006, EDTA (Na$_2$Mg salt) 0.001, Na_2CO_3 0.02.

Micronutrient solution (g/L) in distilled water: H_3BO_3 2.86, $MnCl_2.4H_2O$ 1.81, $ZnSO_4.7H_2O$ 0.222, $Na_2MoO_42H_2O$ 0.390, $CuSO_4.5H_2O$ 0.79, $Co(NO_3)_2.6H_2O$ 0.0494.

1 mL of the micronutrient solution is added to BG-11 growth medium.

Note: $K_2HPO_4.3H_2O$ is autoclaved separately and prepared fresh when added to the medium as it precipitates when autoclaved with other ingredients.

After autoclaving and cooling the pH of the medium is 7.1

5.15.8.3 Protocol

1. Autoclave the BG-11 medium at 121°C for 30 min, inoculate the medium with actively growing pure culture of the cyanobacterium and incubate at 30-35°C for 10-15 days, harvest the exponentially growing cells by centrifugation at 5000 × g.
2. Wash the pellets twice with sterile distilled water, re-suspend the pellets in BG-11 medium and measure immediately OD at 663 nm after resuspending the culture in fresh medium.

3. Again take the sample after each day interval and measure OD at 663 nm.
4. Plot the graph between incubation time and OD (Figure 5.32).

5.15.8.4 Growth measurement by measuring Chlorophyll-a

3 mL of culture is centrifuged 3000 × g for 5 min, pellet is washed and suspended in equal volume (3 mL) of methanol 90% v/v. After thorough mixing, the resulting suspension is heated at 60°C for 4-5 min in a water bath. Thereafter, it is centrifuged to collect the cell-free supernatant, and the pellet is discarded. Absorbance of the methanol extract is estimated at 665 nm with a spectrocolorimeter against 90% methanol as a blank. Quantification of chlorophyll-a is made by employing the absorption coefficient 12.5.

Concentraion of chlorophyll-a (ug/mL) = 12.5 × A_{665} where A_{665} = Absorption at 665 nm

5.15.8.5 Results and observations

Calculate the specific growth rate (*K*) by using the following formula

$$K = \frac{T_2 - T_1}{2.302 (\log N_2 - \log N_1)}$$ where, N_1 = OD at time T_1, N_2 = OD at time T_2

5.16 MEASUREMENT OF FUNGAL GROWTH BY COLONY DIAMETER METHOD

The fungi grow on semi-synthetic or synthetic media forming colony with radial growth of different size, shape and colour. Size of colonies enhances every day attaining the full growth at a definite period. However, the fungal growth can be estimated by measuring the size of the colonies per day (Figure 5.33).

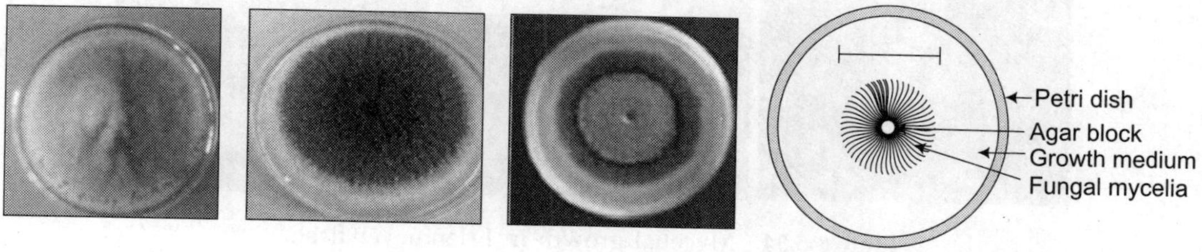

Figure 5.33 Measurement of fungal growth by colony diameter method.

5.16.1 Materials

Potato dextrose medium (PDA), culture tube of a test organism, sterile Petri dishes, cork borer (95 mm diameter), measuring scale or vernier's caliper, para film, incubator, Bunsen burner.

5.16.2 Protocol

1. Grow the fungus on the PDA plates in three replicates, cut agar blocks of 5 mm diameter from actively growing margin of test fungus and transfer block in the centre of agar plates.

2. Seal the margin of the plates with para film to avoid contamination and incubate the inoculated plates at 28 ± 1°C for 5 days in inverted position.
3. At regular intervals of 24 hours, measure colony diameter of the fungus from the four different corners growing on the medium and tabulate the results.

5.16.3 Results and Observations

Take the average value of the three readings and calculate radial growth of the test fungus/day by dividing this value by total number of days incubated. For example, A fungus attained 60 mm of its colony in 5 days, then its mean growth rate will be 12 mm/day.

5.17 BIOMASS MEASUREMENT OF FUNGAL GROWTH (DRY WEIGHT METHOD)

Radial growth measurement method does not give the correct picture of growth of some of the fungi. Several fungi form mycelia on the upper surface of agar showing colony, whereas a few penetrate inside the agar and form mycelia network. Therefore, those growing inside semisolid agar could not be taken into account. In addition, growing edges of hyaline mycelia are not readily observed with naked eyes. Therefore, mycelia dry weight method may be used to estimate fungal growth.

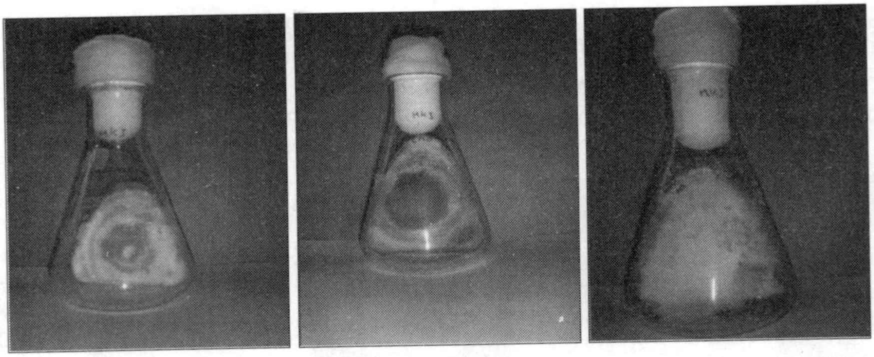

Figure 5.34 Mycelial growth in Erlenmeyer flask.

5.17.1 Materials

Czepek Dox broth medium, pure culture of the test fungus, cork borer (5 mm diameter), inoculation needle, Whatman No. 42 filter paper, incubator.

Czapek-Dox agar: $NaNO_3$ 2 g, K_2HPO_4 1 g, $MgSO_4.7H_2O$ 0.5 g, KCl 0.5 g, FeSO4 0.01 g, sucrose 30 g, agar agar 15- 20 g (for slants and Petri plates) and distilled water to 1 L.

If glass distilled water is used, add 1 mL of 1% $ZnSO_4$ and 0.5% $CuSO_4$. Heat all chemical solution with sucrose in a water bath for 15 minutes. After cooling add sucrose and agar agar and autoclave.

5.17.2 Protocol

1. Prepare Czepek Dox broth medium and sterilize at 15 psi/inch², cut agar block of 5 mm diameter from actively growing margin of the test fungus by using a sterile cork borer, and transfer one block into each flask with the help of a sterile inoculation needle, and incubate the flask at 28 ± 1°C in an incubator for 15 days depending on the fungal growth.
2. Filter the mycelia mat through pre weighed Whatman No. 42 filter paper and dry at 80°C for 24 hours in an oven, weigh dry mycelia mat and record dry weight (Figure 5.34).

5.17.3 Results and Observations

Measure average mycelia dry weight (average of three readings) and estimate growth rate of the test fungus by dividing the value by the total number of the incubation period.

5.18 EFFECT OF INCUBATION TEMPERATURE ON GROWTH

5.18.1 Materials

Culture of Curvularia lunata, Trichoderma viridie, Escherichia coli, Bacillus globissporus and Bacillus hermophilus.

Nutrient agar slants, Czapek–Dox agar plates, Bunsen burner, inoculating loop, inoculating needle. For incubators set at 15, 35, 55 and 75°C.

5.18.2 Protocol

1. Label each of the nutrient agar slants with the test organisms (bacteria) and the temperature of incubation (0, 4, 15, 35, 55 and 75°C).
2. Inoculate six nutrient agar slants (labelled 0, 4, 15, 35, 55 and 75°C) with *E. coli* by streaking the surface of each slope.
3. Repeat step 2 for inoculation of the remaining cultures.
4. Label each of the Czapek–Dox agar plates with *Curvularia lunata* and *Trichoderma viridie* and the temperature of incubation (0, 5, 15, 35, 55 and 75°C).
5. Inoculate six Czapek–Dox agar plates with *Curvularia lunata* and other six plates with *Trichoderma viridie* by transferring a disc on the center of each plate.
6. Incubate one inoculated slant and plate from each set at 0°C (ice box of refrigerator), 4°C (refrigerator) for 10-14 days and other slants and plates at 15, 35, 55 and 75°C (incubators set at these temperatures) for 24-48 hours (bacterial) and 7-14 days (fungal cultures).

5.18.3 Results and Observations

Examine the cultures after two days incubation (incubated at 15, 35, 55, and 75°C) and after 10-14 days incubation (at 0°C and 4°C) for the presence (+) or absence (–) of growth and the degree of growth (i.e., minimal growth (+), moderate growth (2^+), heavy growth (3^+), very heavy (maximum growth (4^+). Express the results in the form of a table and determine the minimum, maximum and optimum temperature ranges for each microorganism and classify each organism as psychrophile, mesophile and thermophile (Table 5.6).

Table 5.6 Growth observed in four test organisms at various incubation temperatures

Species	*Growth at °C* 0 4 15 35 55 75	*Cardinal temperature* *Min Max Opt*	*Tern. Classification* *T M P*
C. lunata			
T. viridie			
E. coli			
B. globissporus			
B. hermophilus			

Growth is indicated as –. +, 2+, 3+, or 4+, T = thermophile, M =mesophile and Pr = psychrophile

5.19 SOIL DECOMPOSITION AND MICROBIAL COMMUNITY STRUCTURE (WINOGRADSKY TECHNIQUE)

Soil decomposition is a necessary and important process in every ecosystem. It is the process of recycling nutrients that all organisms need for survival. Bacteria and fungi are the main components of this process that involves the breaking down of detritus and dead organic materials. Humus (decomposed material) is the final step in this process that supply soil with nutrient such as calcium, phosphate, potassium, and other ions.

In any given condition, natural environment teems with microorganisms to provide a specific combination of nutrients and oxygen that allows only certain microorganisms to survive. One can use their own designed composed media and inoculum from any water body to create and study a mini-ecosystem, called a Winogradsky column, in the laboratory (Figure 5.35).

5.19.1 Procedure

Place about 200 mL compost soils (agriculture or forest soil) to a mixing glass or plastic clear container; add pond water and stir until it is about the consistency of apple sauce; add 5 g grass cutting for the agriculture soil and 5 g forest plants cutting for the forest soil having lignocellulose as a carbon source; add equal amount of calcium carbonate and calcium sulfate and mix until the mixture become drier as a source of carbon and sulfur.

Pour or spoon this mixture into the decomposition chamber to approximately 3-4 cm in depth; mix it well with a spoon or stirring rod to remove any air pockets; add plain agriculture or forest soil to the respective chamber until the depth of the soil mixture reaches between 6 and 8 cm. DO NOT STIR! Add pond water, leaving about 4 cm of headspace; insert a thermometer into the soil and record the starting temperature.;

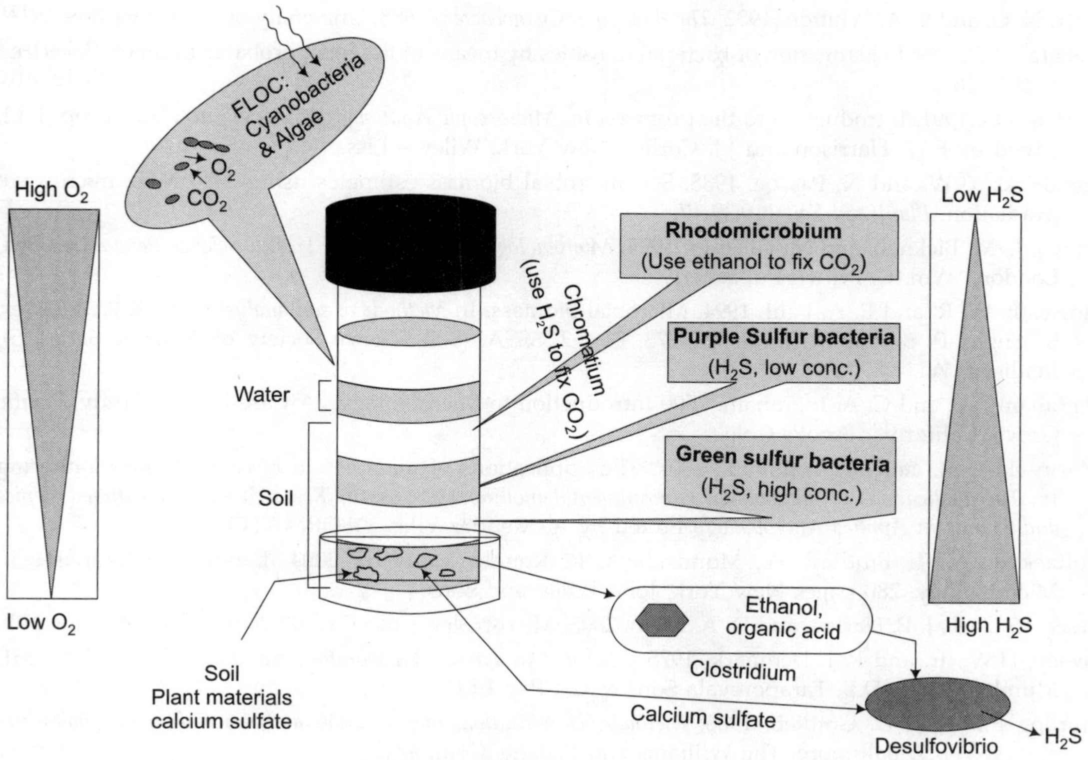

Figure 5.35 Illustration of Winogradsky technique.

the soil column in each decomposition chamber should be covered with aluminium foil, seal with parafilm, and placed next to window. Observe the developed soil layers weekly.

Suggested Readings

Alexander, M. 1961. *Introduction to Soil Microbiology*,314. John Wiley and Sons, New York and London,

Alexopoulos, C.J., C.W. Mims and M. Blackwell. 1996. *Introductory Mycology*, 4th ed. New York, John Wiley and Sons.

Benson, H.J. 1990. *Microbiological Applications*-A *Laboratory Manual in General Microbiology*. Iowa, Win. C. Brown Publishers,

Bisen, P.S. and A. Sharma. 2012. Introduction to instrumentation in Life Sciences, 3553. 1st ed. Boca Raton:, CRC Press.

Bisen, P.S., G. B. K. S. Prasad and M. Debnath. 2012. *Microbes: Concepts and Applications*, 699. 1st ed. New York, John Wiley and Sons.

Cappuccino, J.G. and N. Sherman, N. 1991. *Microbiology: A Laboratory Manual*. The Amsterdam, The Netherlands Benjamin/Climmings Publishing Company,

Cappuccino, J.G. and N. Sherman, N. 1991. *Microbiology: A Laboratory Manual*. 462. The Benjamin/ Climmings Publishing Company, Amsterdam, the Netherlands.

Carr, N. G. and B. A. Whitton. 1982. *The Biology of Cyanobacteria*, 688. University of California Press, CA,

Cochran, W.G. 1950. Estimation of bacterial densities by means of the most probable number. *Biometrics* 6: 105-116.

Corliss, J.O. 1991. Introduction to the protozoa *In*: *Microscopic Anatomy of Invertebrates*. Vol. 1, pp. 1-12, edited by F.W. Harrison and J.J. Corliss. New York. Wiley - Liss.

Hendricks, C.W. and N. Pascoe. 1988. Soil microbial biomass estimates using 2450 MHz microwave irradiation. *Plant and Soil* 110:39-47.

Hleyn, J., M. Bicknell, and M. Gilstrap. 1995. *Microbiology Experiments: A Health Science Perspective*. 354. London., Wm. C. Brown Publishers.

Horwath, W. R. and E. A. Paul. 1994. Microbial biomass. In *Methods of soil analysis*, eds. R.W. Weaver, S. Angle, P. Bottomley, et al. 753-773. Part 2. SSSA. (Soil Science Society of America Series, 5), Madison, WI.

Ingraham J. L. and C. A. Ingraham. 2000 Introduction to Microbiology : A Case-History Study, Pacific Grove, California, Brooks Cole.

Marinšek, L. R., and M. Vodovnik. 2007. The applications of microbes in environmental monitoring *In: The applications of microbes in environmental monitoring. Current Research and Educational Topics and Trends in Applied Microbiology* Edited by A. Méndez-Vilas, ©FORMATEX.

Pollack R. A., L. Findlay, W. Mondschein, R. Ronald Modesto. 2004. Laboratory Exercises in Microbiology, 280 pages New York, John Wiley and Sons.

Prsecott, L. M., J. P. Harley, and D. A. Klein. 2005. Microbiology. 6th Ed. 992. New York, McGraw Hill.

Seeley, H.W., Jr. and P. J. Denmark. 1975. *Microbes in action: A Laboratory manual of microbiology*, 361. Mumbai, India, D.B. Taraporevala Sons & Co. Pvt. Ltd.

Shirling E.B., and D. Gottlieb. 1966. *Methods, classification, identification and description of genera and species. Vol. 2,* Baltimore, The Williams and Wilkins Company.

Tortora, G.J., B.R. Funke and C. L. Case. 2009 Microbiology: An Introduction, Eighth 10th Ed. 960. San Fransisco, C.A., Benjamin Cummings.

Chapter 6

Microbial Staining

6.1 INTRODUCTION

The correct identification of microorganisms is of fundamental importance to microbial systematists as well as to scientists involved in many other areas of applied research and industry (e.g., agriculture, clinical microbiology and food production). Most bacteria are quite colourless and transparent and have a refractive index similar to that of the aqueous fluids in which they are suspended. Unless the diaphragm is carefully adjusted usually there is considerable difficulty in bringing the organisms into focus. Owing to the small size of bacteria little structural details can be seen with the ordinary light microscope unless the organisms are stained. The majority stain readily with aniline dyes. Some staining techniques, such as the Gram and Ziel Neelsen stains, although of great diagnostic value because of their differential staining properties for specific bacteria, reveal little internal structure. Other such as Feulgen stain for nuclear bodies, demonstrates specific structure. Because of its importance, different important stains are described in some detail.

6.2 HANGING DROP TECHNIQUE

The techniques employed are meant for microscopic observation of living bacteria.

The study of bacterial morphology is performed in two ways:

1. Observing unstained cells live by hanging drop preparation.
2. Observing dead cells by making use of chemical nature of their unicellular, body. This is achieved by staining.

Hanging drop technique enables viewing of size, shape, arrangement and motility of live microorganisms in fluid media. It requires the use of special ground slides. In this technique a loopful of bacterial suspension is placed in the centre of a cover slip. In the four corners tiny droplets of mineral oil are placed. The hollow ground slide is placed over the cover slip with the depression side down and the slide is inverted quickly so that the water cannot run off to one side. However, the lack of contrast yields limited though valuable information. For pathogens one tube one plate method can be used. Each method has its advantage and limitations. The method depends on which one is most suitable for the situation at hand.

6.2.1 Hanging Drop Preparation or Motility Test

1. Apply vaseline around the depression of the hanging-drop slide.
2. Using the inoculation loop, aseptically transfer one drop of the culture to the centre of a clean cover slip.
3. Invert the hanging-drop slide and centre its well over the drop of the culture. Press down on the edge of the cover slip so that the vaseline makes a firm seal.
4. Quickly and carefully turn the slide right side up so as to suspend the hanging drop in the well. Don't let the drop fall or touch the bottom of the well.
5. To examine, first locate its edge in centre of the microscopic field with low power objective and markedly lower the light. The edge will be visible, as a bright wavy line against a dark background. Now the slide can be focused under oil immersion (Figure 6.1).

When working with pathogenic microorganisms such as the typhoid bacillus, it is too dangerous to attempt to determine motility with slide techniques. A much safer method is to culture the organisms in a special medium that can demonstrate the presence of motility. The procedure is to inoculate a tube of semisolid or SIM medium that can demonstrate the presence of motility. Both media have a very soft consistency that allows motile bacteria to migrate readily through them causing cloudiness. Figure 6.2 illustrates the inoculation procedure.

6.3 MICROBIAL STAINING

It is a chemical or a physical union between the dye and like component of a cell. If it is a chemical reaction a new compound is formed and a simple washing with water does not liberate the bound dye but if purely physical it is easy to decolorize such stained organism. Usually it is a combination of chemical and physical reactions.

The main advantages of staining are that it

(i) provides contrast between microorganisms and their backgrounds, permitting differentiation among various morphological types;
(ii) permits study of internal structures of the bacterial cell, such as the cell wall, vacuoles or nuclear bodies and other cellular structures; and
(iii) enables the bacteriologist to use higher magnifications.

6.3.1 Fixing

Before staining it is essential to fix the bacterial sample on to the slide. Smear is prepared in the following way:

(i) With a wire loop place a small drop of the broth culture or a loop full of bacteria on a clean slide.
(ii) Place a drop of water over it.
(iii) Spread the culture so as to form a thin film.
(iv) Allow slide to dry in the air or by holding it above a bunsen flame.
(v) Avoid excess heating.

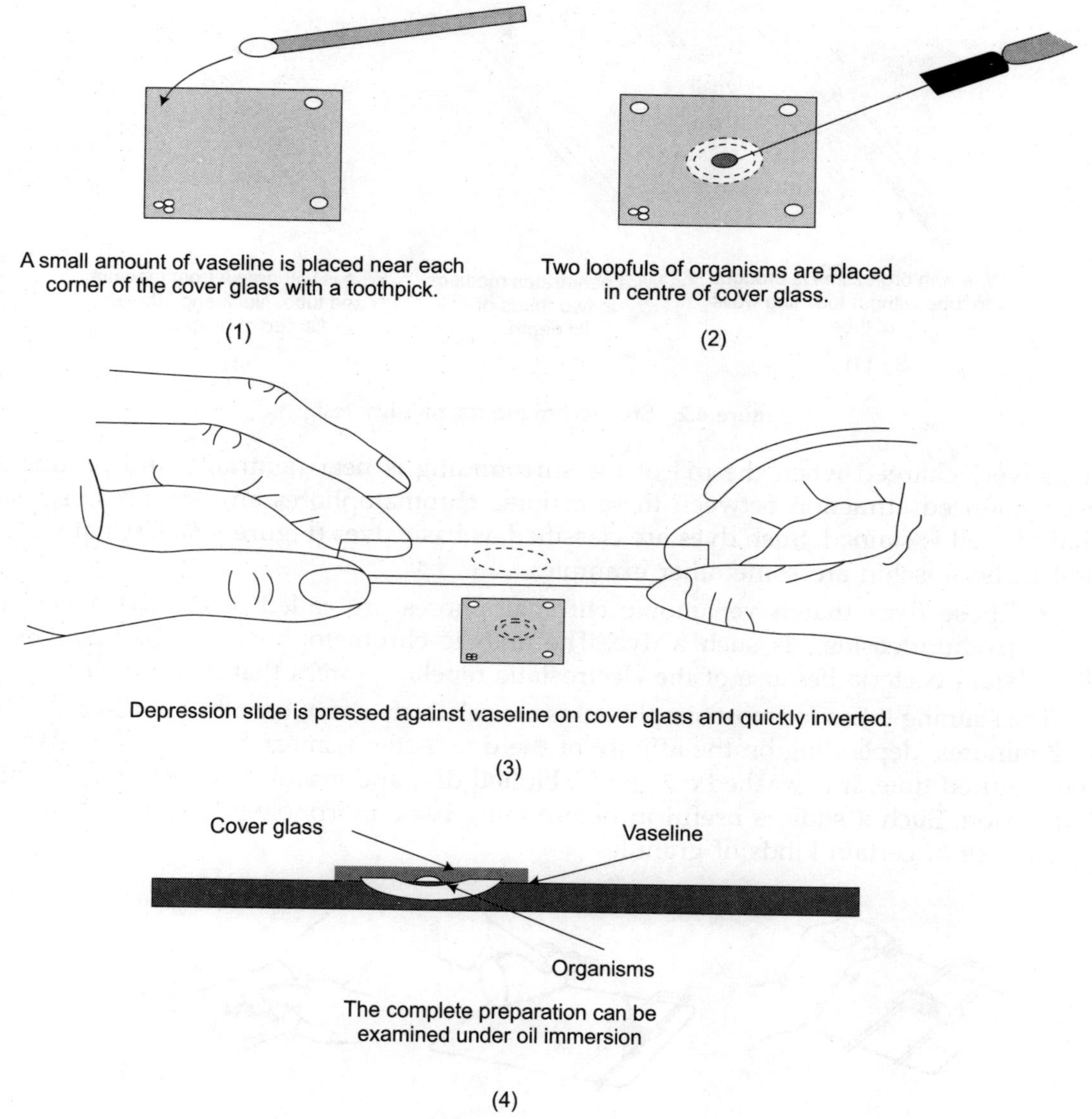

Figure 6.1 Hanging drop preparation.

The purpose of fixation is to kill the microorganisms, coagulate the protoplasm of the cell and cause it to adhere to the slide.

6.3.2 Type of Stains

6.3.2.1 Simple staining

The use of a single stain to colour a bacterial organism is commonly referred as simple staining. All these dyes work well on bacteria as they have colour bearing ions (chromatophores) and are positively charged. The fact that bacteria are slightly

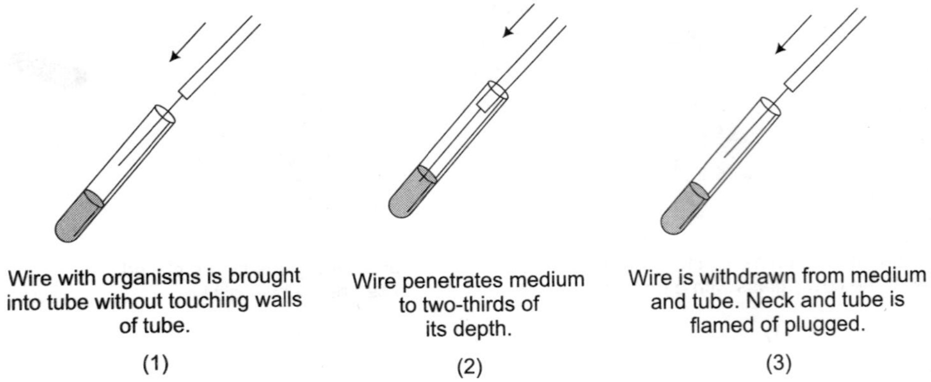

Figure 6.2 Stab technique for motility test.

negatively charged when the pH of the surrounding is near neutrality and produces a pronounced attraction between these cationic chromatophores and the organism so that the cell is stained. Such dyes are classified as basic dyes (Figure 6.3). Crystal violet and carbol fuschin are some other examples.

- Those dyes that have anionic chromatophores are called acidic dyes. Eosine (sodium$^+$eosine$^-$) is such a dye. The anionic chromatophores, eosine$^-$, will not stain bacteria because of the electrostatic repelling forces that are involved.

The staining times for most simple stains are relatively short, usually from 30 seconds to 2 minutes, depending on the affinity of the dye. After a smear has been stained for the required time, it is washed off gently, blotted dry, and examined directly under oil immersion. Such a slide is useful in determining basic morphology and the presence or absence of certain kinds of granules.

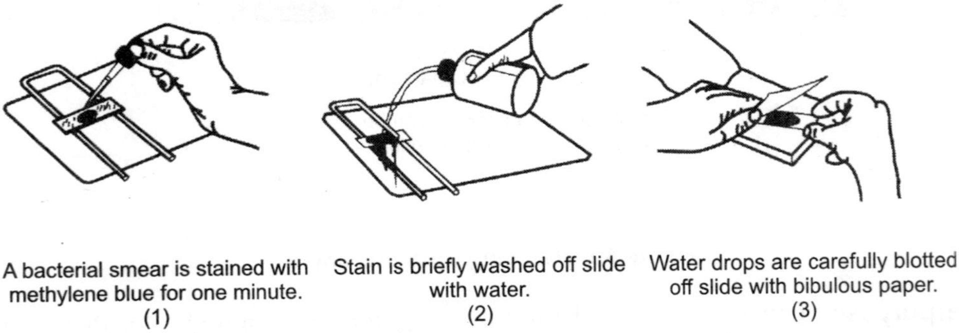

Figure 6.3 Demonstration of staining procedure.

6.3.2.2 Negative staining

A better way to observe bacteria for the first time is to prepare a slide by a process called negative or background staining. This method consists of mixing the microorganisms in a small amount of nigrosine or India ink and spreading the mixture over the surface of the slide (nigrosine is far superior to India ink). Since these two pigments are not really bacterial stains, they do not penetrate the microorganisms; instead they obliterate the background, leaving the organisms transparent and visible in a darkened field. Although this technique has a limitation, it can be useful for determining cell morphology and size. Since no heat is applied to the slide, there is no shrinkage of the cell and consequently more accurate cell size determination result than with some other methods. This method is also useful for studying spirochaetes that does not stain readily with ordinary dyes.

Negative staining can be performed by one of the following methods. Figure 6.4 illustrates the more commonly used method in which the organisms are mixed in a drop of nigrosine and spread over the slide with another slide in order to prepare a smear that is thick at one end and feather thin at the other end. Somewhere between the too thick and too thin areas will be an ideal spot to study the organisms.

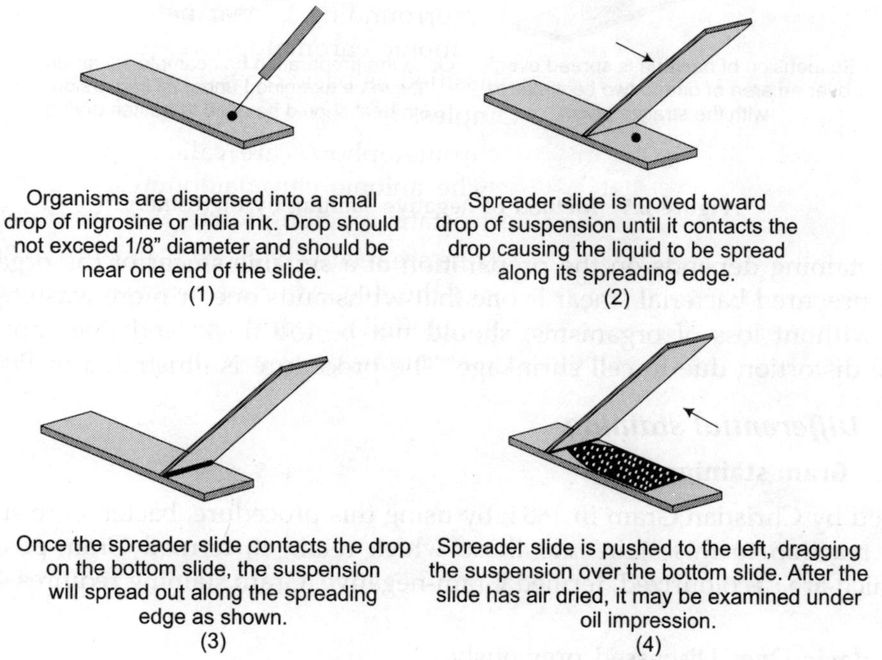

Figure 6.4 Demonstration of method of negative staining.

In the second method, the organisms are mixed in only a loopful of nigrosine instead of a full drop. The organisms are spread over a smaller area in the centre of the slide with an inoculating needle. No spreader slide is used in this method. It gives more accurate view of the bacterial cell (Figure 6.5).

While negative staining is a simple enough process to make bacteria more visible with a brightfield microscope, it is of little help when one attempts to observe anatomical microstructures such as flagella, granule, and enodospore. Only by applying specific bacteriological stains to organisms one can see such organelles. However, success in

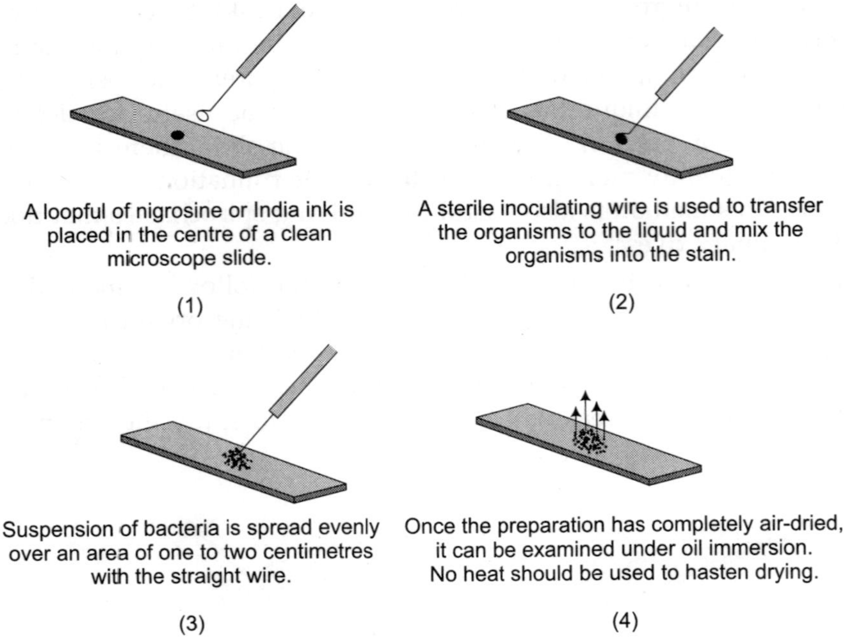

Figure 6.5 Method of negative staining by nigrosine.

bacterial staining depends on the preparation of a suitable smear of the organisms. A properly prepared bacterial smear is one that withstands one or more washings during staining without loss of organisms; should not be too thick; and does not result in excessive distortion due to cell shrinkage. The procedure is illustrated in Figure 6.6.

6.3.2.3 Differential staining

6.3.2.3.1 Gram staining

Developed by Christian Gram in 1884; by using this procedure, bacteria are subdivided by their reaction to this stain into those which retain it, termed Gram-positive, and those which are decolourized, termed Gram-negative. Gram staining requires 4 different solutions.

(i) **A Basic Dye:** Discussed previously.

(ii) **A Mordant:** It increases the affinity or attraction between the cell and the dye, e.g., Iodine.

(iii) **A Decolorising agent:** It removes the dye from a stained cell, e.g., alcohol, acetone or ether.

(iv) **Counter stain:** It is a basic dye of a different colour than the initial one, e.g., Safranin.

Stain Preparation

(i) **Crystal Violet:** Dissolve 2 g of crystal violet in 20 mL of 95% ethanol. Dissolve 0.8 g of ammonium oxalate in 80 mL of distilled water. Mix these 2 solutions; stand for 24 h and filter.

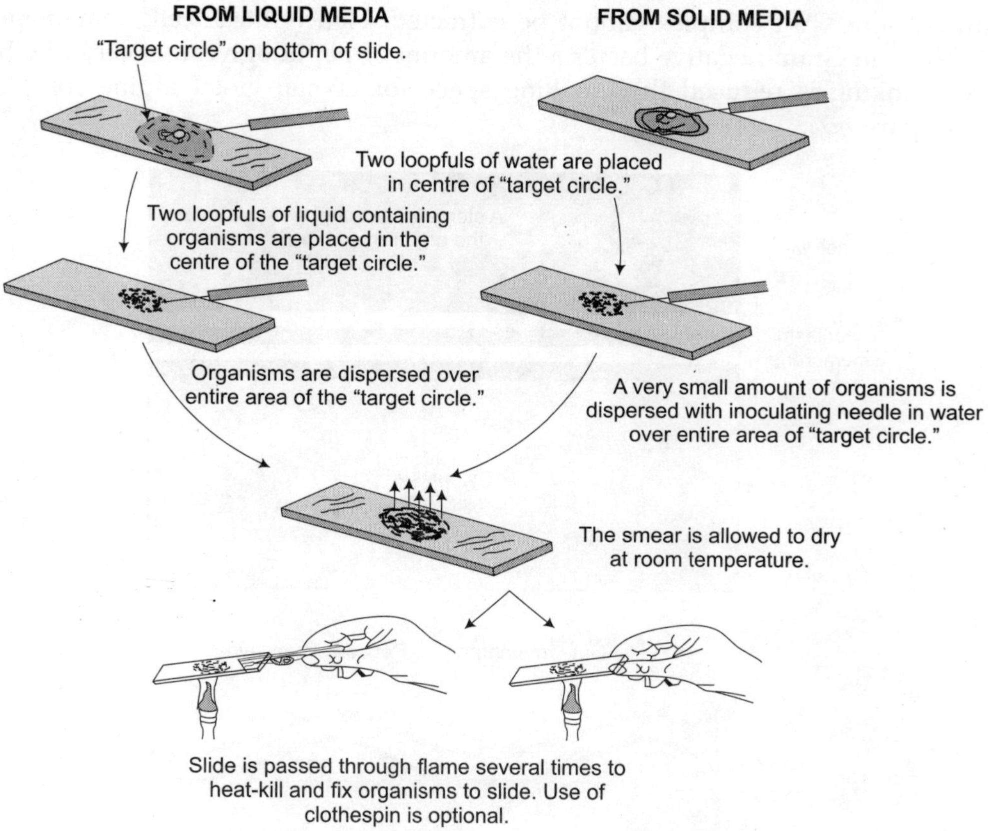

Figure 6.6 Complete ideal staining procedure from two different sources.

(ii) **Gram's Iodine:** Dissolve 2 g of potassium iodide and 1 g of iodine in 300 mL of distilled water.

(iii) **Safranin:** Grind 0.25 g of Safranin in a mortar with 10 mL of 95% ethanol.

Principle

The differential response towards Gram reaction is attributed to the difference in the cell wall of these bacteria. Gram-negative bacteria have cell wall generally thinner than those of Gram-positive bacteria. Gram-negative Ones have higher lipid content than the Gram-positive bacteria. During staining of Gram-positive bacteria, the alcohol treatment extracts the lipid, which results in increased porosity or permeability of the cell wall. The CV-I (crystal violet iodine) complex can be extracted and the Gram-negative organism is decolorized. These cells subsequently take the colour of the safranin a *Counter stain.* The cell walls of Gram-positive bacteria, due to their different composition (lower lipid content) become dehydrated during treatment with alcohol presumably causes diminution in the pore diameter of the walls of peptioglycan and CV-I complex is trapped in the wall following ethanol treatment. The pore size decreases, permeability is reduced, and CV-I complex cannot be extracted. Hence, these cells remain purple violet. While in Gram-negative bacteria the amount of peptidoglycan is very low hence the cross linking is reduced thus making space for crystal violet iodine complex to escape (Figure 6.7).

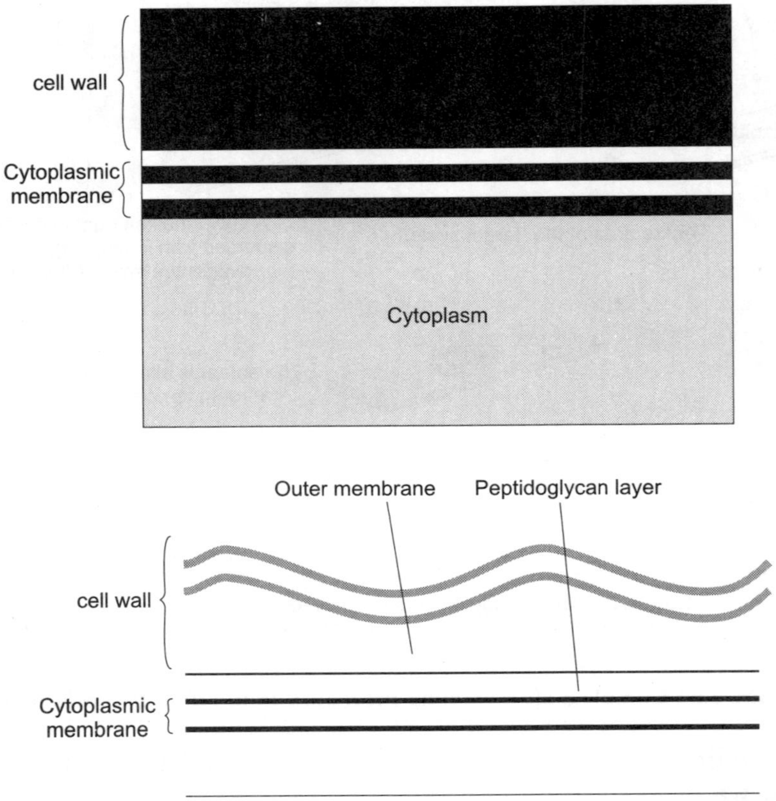

Figure 6.7 Cell wall of Gram-positive and Gram-negative bacteria.

Procedure

1. Prepare smear on a clean slide.
2. Stain with crystal violet for 30 seconds.
3. Rinse with water.
4. Flood the film with Grams iodine and allow it to act for 30 sec.
5. Rinse with water.
6. Decolorise with 95% alcohol.
7. Rinse with water.
8. Counterstain with safranin for 20-30 sec.
9. Rinse with water and blot dry.
10. Examine under oil immersion objective.

Table 6.1 Steps in Gram stain

Step	*Process*	*Results*	
		Gram+	*Gram–*
Initial stain Mordant	CV for 30 sec.	Stains purple	Stains purple
Decolorisation	12 for 30 sec.	Remains purple	Remains purple
Counterstain	95% alcohol for 10-20 sec.	Remains purple	Becomes colourless
	Safranin for 20-30 sec.	Remains purple	Stains pink

6.3.2.3.2 Acid fast stain

The Ziel-Nehlson method employs carbol fuchsin, acid alcohol and a blue or green counter stain.

Acid fastness is a phenomenon that is found in certain types of bacteria where they resist the process of de-colourization that occurs when acid is used to wash a sample that contains these bacteria. These bacteria also resist staining and it may require heat and concentrated staining to colourize them. Once this colorization has taken place, it becomes difficult to decolorize using acid, or stain them with another colour unless the heat and concentration technique is used. This is why the term given to them is 'acid fast' just like one would use the term 'colour fast' for cloth that does not leak colour when washed (Figure 6.8).

The most common type of acid fast bacteria is mycobacteria. A strain of mycobacteria is responsible for tuberculosis. In diagnostic medicine, the acid fast test may be used on a person in order to detect the existence of the bacteria that can cause tuberculosis. It is therefore used as a diagnosis of the disease because other symptoms of tuberculosis are not useful as they overlap with other conditions. Acid fast stain results can confirm the presence of the bacteria known as mycobacteria tuberculosis which is the bacteria responsible for causing tuberculosis. In laboratories where large number of sputum, gastric washings, urine, and other body fluid samples are tested for pathogenic mycobacteria, fluorochrome acid fast staining is used in conjunction with the Ziel-Nehlson method. The advantage of using a fluorescence method is that fluorochrome stained slides can be scanned under lower magnification, while a Ziel-Nehlson prepared slide must be examined under oil immersion (1000× magnification), fluorochrome stained slides can be examined with 60× or 100× magnification.

Stain Preparation

(i) Carbol fuchsin

1. Basic fuchsin 58 g
2. Crystalline phenol 25 g
3. 95% alcohol 50 mL
4. Distilled water to 500 mL

Dissolve the fuchsin and phenol in alcohol over a warm water bath, and then add water. Filter, before use.

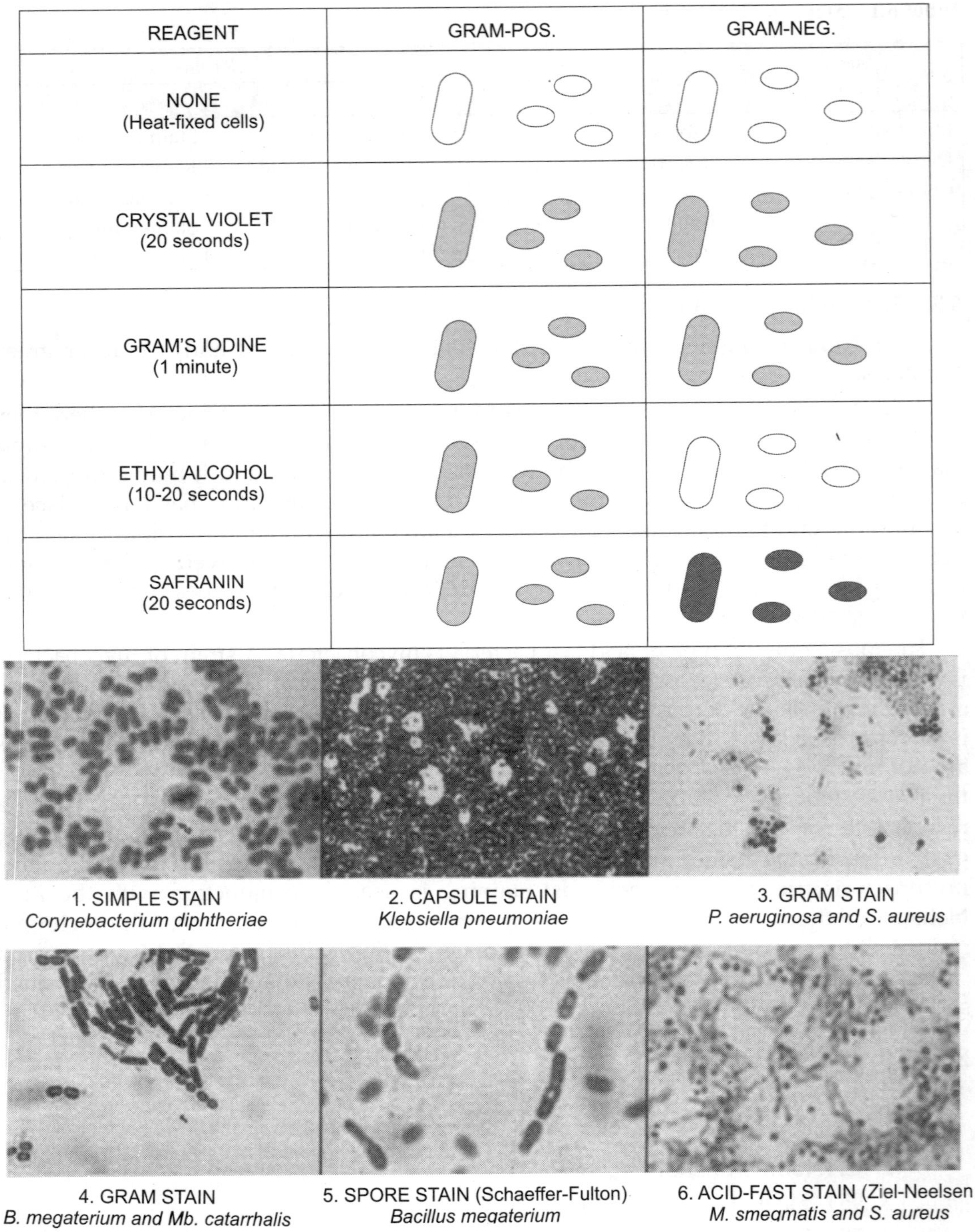

Figure 6.8 Demonstration of the process of Gram staining and acid fast stain.

(ii) 95% ethyl alcohol 970 mL
Concentrated HCl 30 mL

(iii) 0.5% methylene blue or malachite green in distilled water.

Procedure

Pour carbol fuchsin on the slide and heat carefully until steam rises. Stain for 3-5 min but do not allow to dry. Wash well with water. Decolorize using acid alcohol for 10-20 s. changing twice and counter stain for 2 min with methylene blue or malachite green. Acid fast organisms, e.g., *Mycobacterium tuberculosis and Mycobacterium leprae* stain well (Figure 6.9).

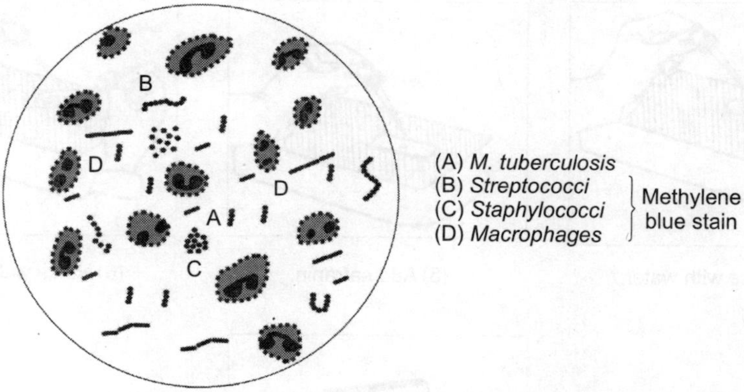

Figure 6.9 Carbon Fuchsin staining of acid fast bacteria.

6.3.2.4 *Structural staining*

Structural staining allows you to observe certain structures on bacteria. This is important because certain structures on bacteria can be antigenic or act as an endotoxin. Structural stains are more complex than simple ones and use more than one stain to differentiate cellular components. They are used to examine structural differences between bacterial groups or to provide contrast to different structures within the same organism. Some of these stains are explained below.

6.3.2.4.1 Endospore stain

Species of bacteria, belonging principally to the genera *Bacillus and Clostridium* produce extremely heat resistant structures called endospores. In addition to being heat-resistant, they are also resistant to many chemicals that destroy non-spore forming bacteria. This resistance to heat and chemicals is due primarily to a thick, tough spore coat. They resist staining and, once stained they resist decolorisation and counter staining. Several methods are available that employ heat to provide stain penetration. However, the malachite green-Schaeffer and Fulton and Dorner methods are commonly used by most bacteriologists. Thus endospore stains green but rest of the cell or a cell without endospore stains light red (Figures 6.10 and 6.11).

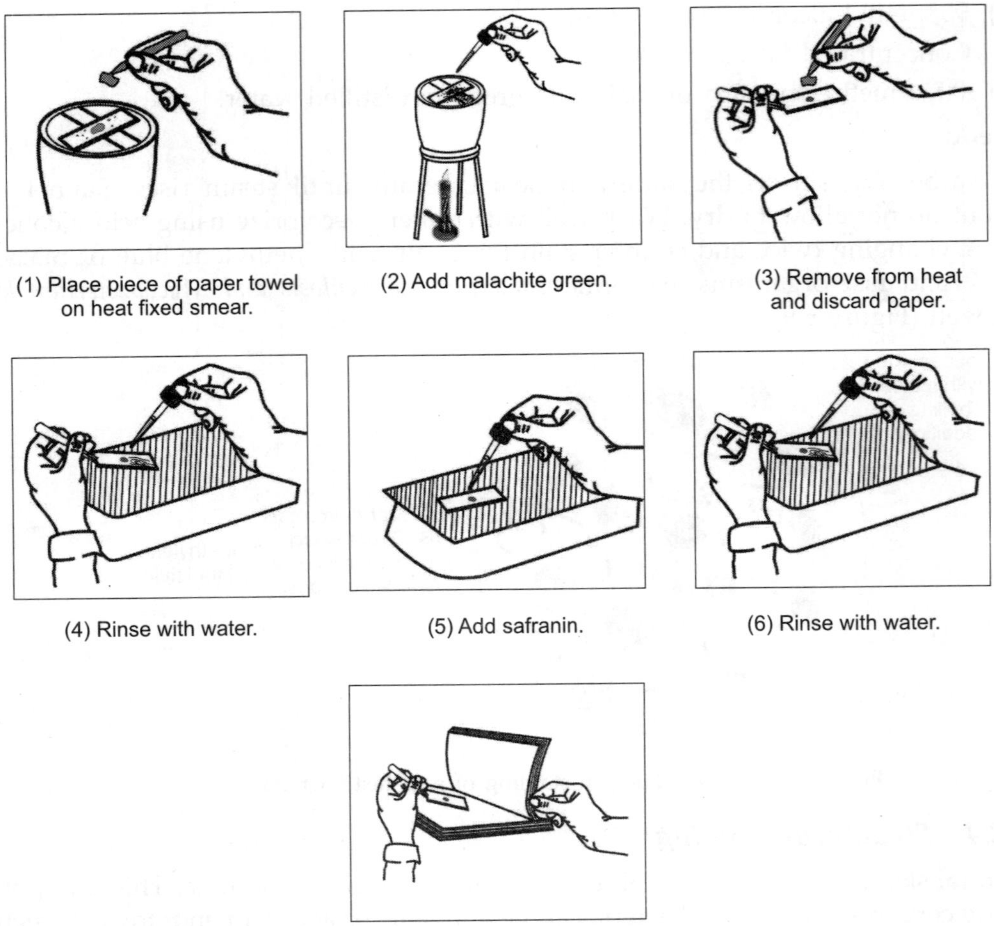

(1) Place piece of paper towel on heat fixed smear.
(2) Add malachite green.
(3) Remove from heat and discard paper.
(4) Rinse with water.
(5) Add safranin.
(6) Rinse with water.
(7) Blot dry.

Figure 6.10 Steps in the endospore stain.

(i) The malachite green-Schaeffer and Fulton Method

Procedure

1. Prepare a thin smear on a clean slide.
2. Place the slide on staining rack above boiling water.
3. Cover the smear with small pieces of paper towel, keep saturated with malachite green (5% aqueous solution) and continue heating for 5 min.
4. Wash gently with water.
5. Counterstain with safranin for 30 seconds.
6. Wash with water and blot dry.
7. Examine under oil immersion objective.

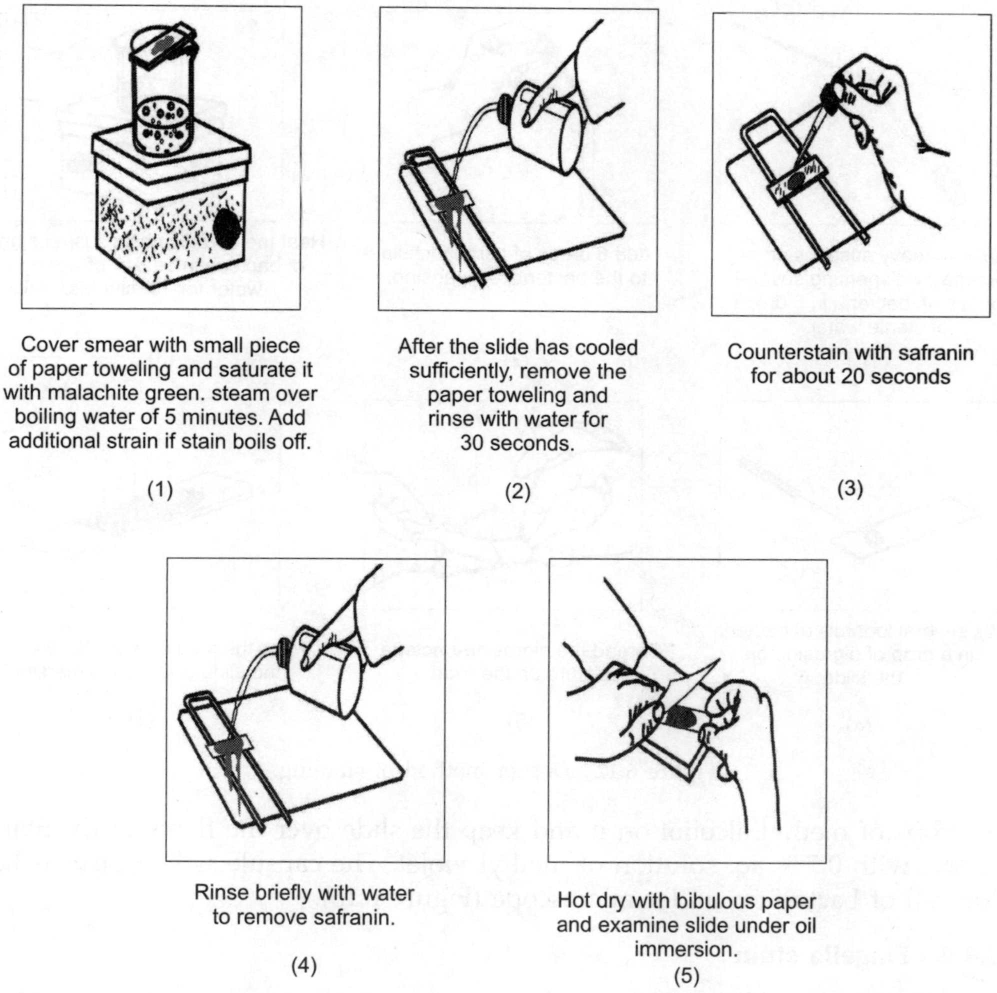

Figure 6.11 Malachite green-Schaeffer and Fulton Stain method.

6.3.2.4.2 Dorner method

The Dorner method for staining endospore produces a red spore within a colourless sporangium. Nigrosine is used to provide a dark background for contrast.The sporangium and endospore are stained during boiling in step 3, however, the sporangium is decolourized by the diffusion of safranin molecules into the nigrosine. The six steps involved in this technique are illustrated in Figure 6.12.

6.3.2.4.3 Capsule stain

Place a small drop of India ink on a clean slide. Mix into it a small loopful of bacterial culture. Drop a cover glass and blot off excess ink. Examine.

For dry preparations mix one loopful of India ink with one loopful of suspension of organism in 5% dextrose solutions at one end of a slide. Allow to dry and pour

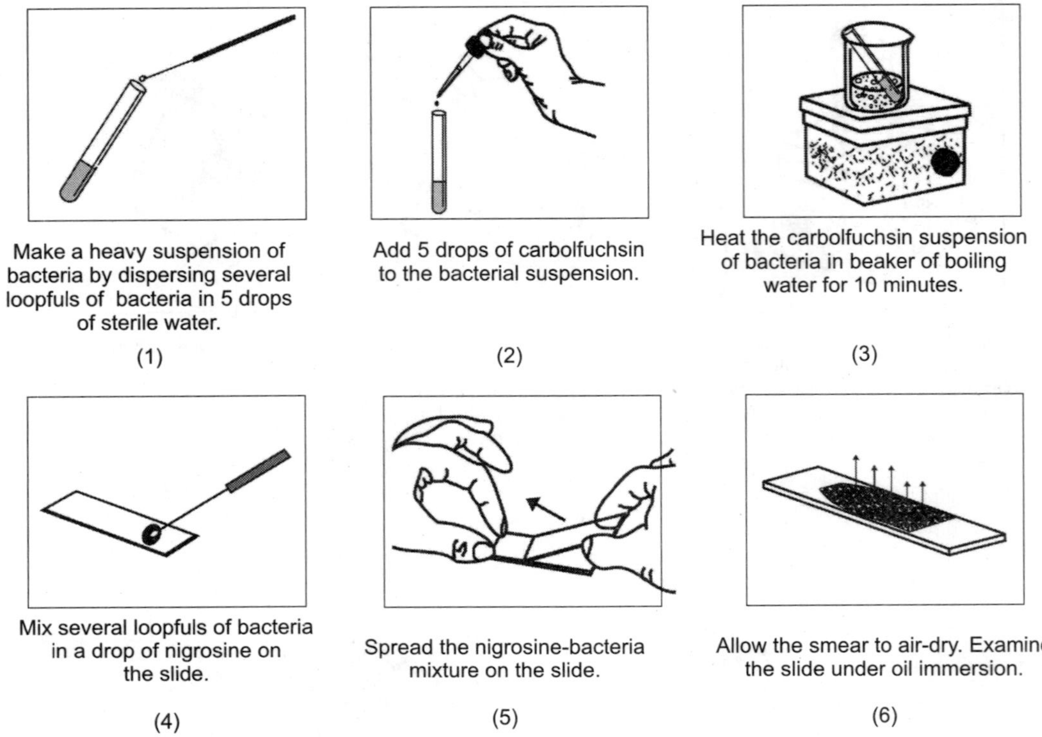

Figure 6.12 Dorner method of staining.

a few drops of methyl alcohol on it and keep the slide over the flame to fix. Stain for a few sec. with 0.5.% aq. solution of methyl violet. The capsule will appear as haloes in blue cell of bacterium under microscope (Figure 6.13).

6.3.2.4.4 Flagella stain

(i) Mordant

20% tannic acid solution	3 parts
5% tartar emetic solution	2 parts
Distilled water	5 parts

Mix and redissolve the heavy precipitate by boiling.

(ii) Silver solution: Dilute a saturated solution of silver sulphate with an equal volume of distilled water. Add 33% monoethylamine till the precipitate formed is redissolved. Solution can be stored for longer time, heat gently before using (Figure 6.14).

Procedure

1. Take a clean slide and transfer a loopful of bacterial suspension holding the slide vertically. Allow the drop to run to the other end. Dry the film in air or in incubator.

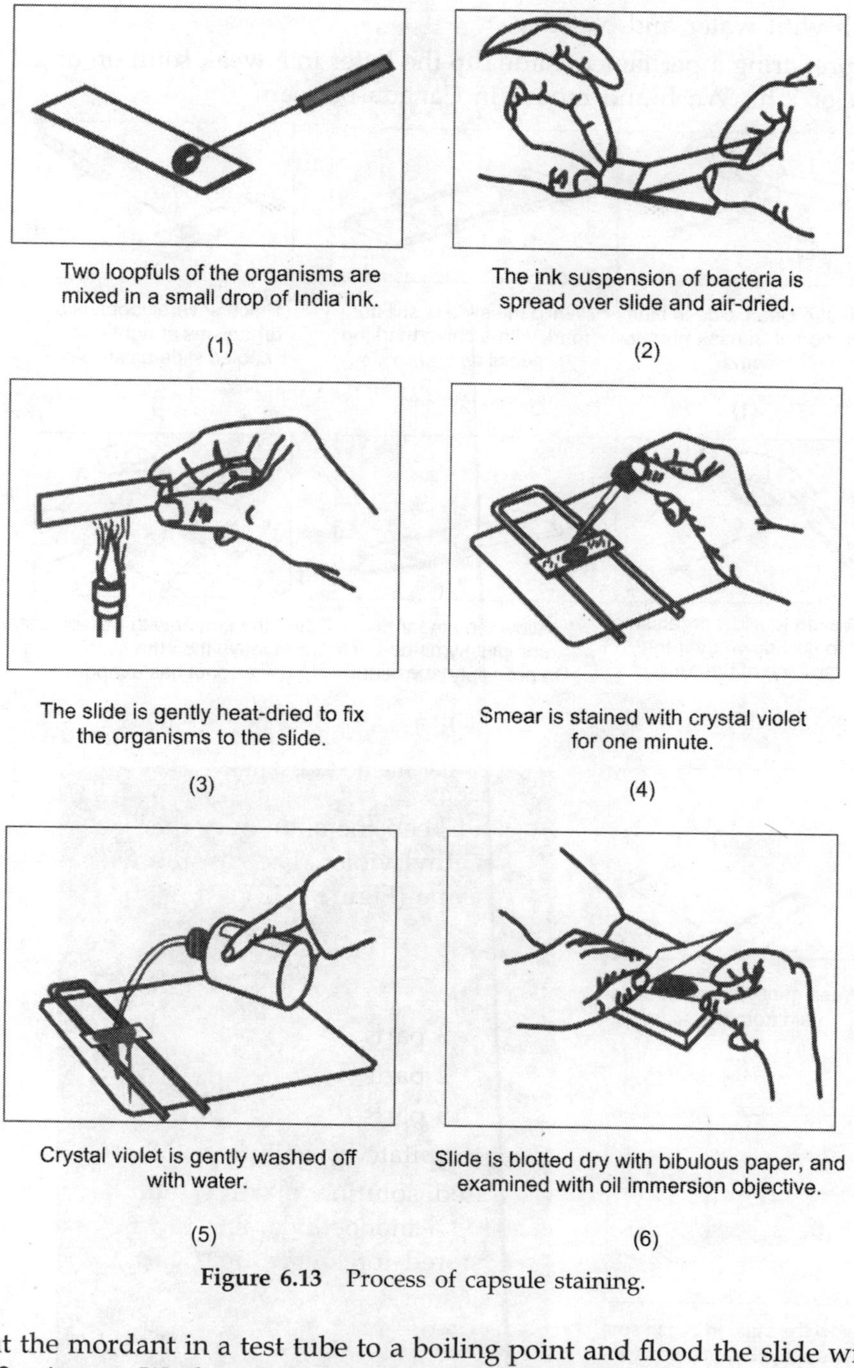

Two loopfuls of the organisms are mixed in a small drop of India ink.

(1)

The ink suspension of bacteria is spread over slide and air-dried.

(2)

The slide is gently heat-dried to fix the organisms to the slide.

(3)

Smear is stained with crystal violet for one minute.

(4)

Crystal violet is gently washed off with water.

(5)

Slide is blotted dry with bibulous paper, and examined with oil immersion objective.

(6)

Figure 6.13 Process of capsule staining.

2. Heat the mordant in a test tube to a boiling point and flood the slide with it wait for 2 minutes. Wash well in hot water and finally in cold water. Allow to dry.
3. Cover the film with silver solution till thick portion of the film becomes dark to brown and metallic sheen appears on the edge.

4. Wash with water and blot dry.
5. For preparing a permanent slide dip the latter in a weak solution of gold chloride for 1 or 2 hr. Wash and mount in Canada balasam.

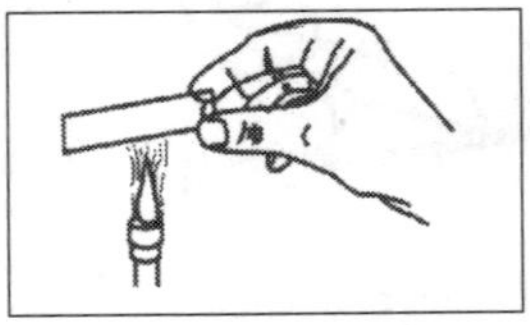

Heat a clean slide in blue portion of Bunsen burner flame.

(1)

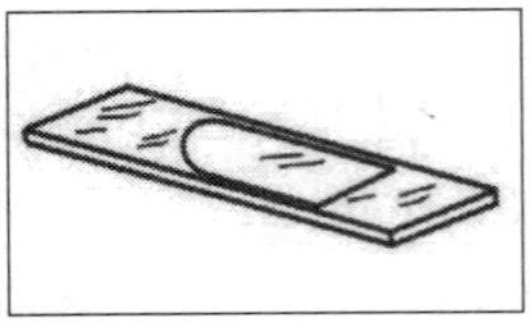

While the slide is still hot, mark with a china marking pencil as shown.

(2)

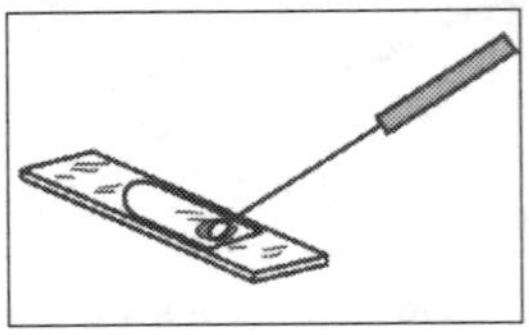

Place several loopfuls of organisms at right end of cooled slide as shown.

(3)

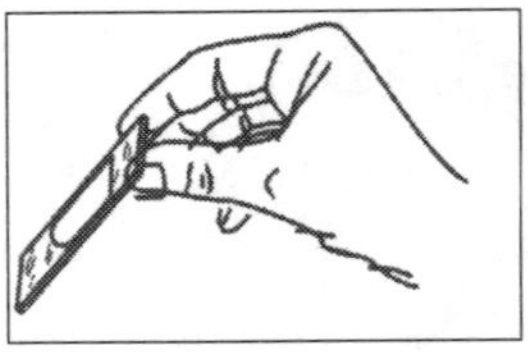

Tilt slide to allow organisms to flow down over the surface of the slide.

(4)

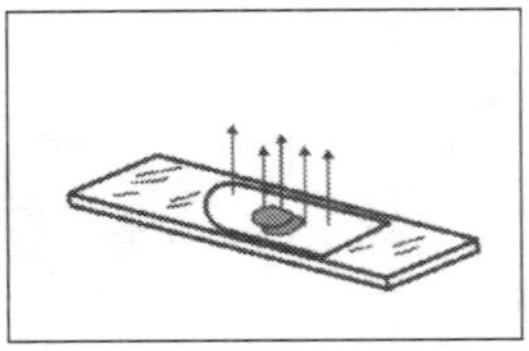

Allow the smear to completely air-dry. Do not apply any heat.

(5)

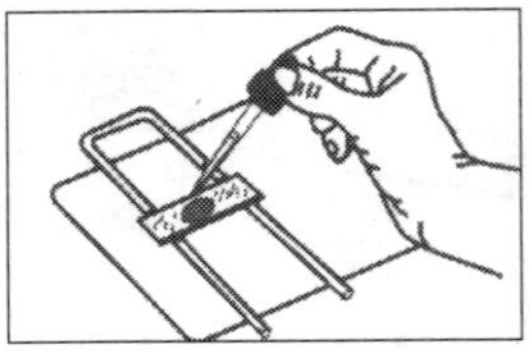

Cover the smear with Leifson's stain and leave it on the slide until all the alcohol has evaporated.

(6)

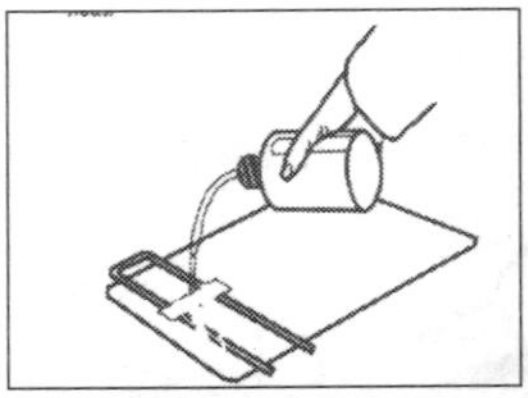

Wash gently to remove the stain from the slide.

(7)

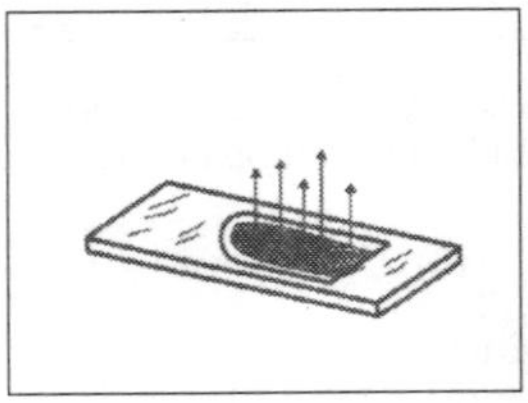

Allow the stained organisms to air-dry. Examine under oil immersion.

(8)

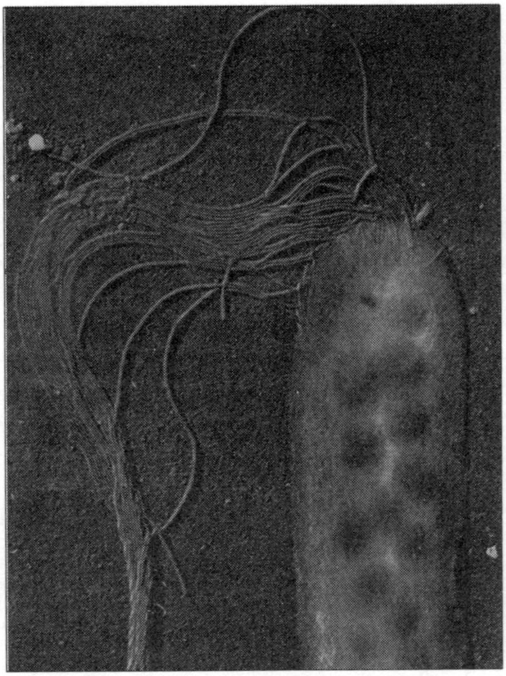

Figure 6.14 Demonstration of flagella staining.

Suggested Readings

Anderson NL, Noble MA, Weissfeld AS, Zabransky RJ. 2005. *Cumitech 3B, Quality Systems in the Clinical Microbiology Laboratory*. Coordinating ed., Sewell DL. ASM Press, Washington, D.C.

Beveridge TJ, Graham LL. 1991. Surface Layers of Bacteria. *Microbiol Rev* **55**, 684-705.

Brown E. 2004. *Alfred, Benson's Microbiological Applications*, ninth edition, McGraw Hill Publication, New York.

Cappuccino GJ, Sherman N. 2013. *Microbiology A laboratory manual*, tenth edition, Benjamin Cummings, London.

Carlone GM, Valadez MJ, Pickett MJ. 1983. Methods for Distinguishing Gram-Positive from Gram-Negative Bacteria. *J Clin Microbiol* **16**, 1157-1159.

Cerny G. 1976 Method for distinction of gram negative from gram positive bacteria. *Eur J Appl Microbiol* **3**, 223-225.

Davies JA, Anderson GK, Beveridge TJ, H C Clark HC. 1983. Chemical Mechanism of the Gram Stain and Synthesis of a New Electron-opaque Marker for Electron Microscopy Which Replaces the Iodine Mordant of the Stain. *J Bacteriol* **156**, 837-845.

Farmer T. 2005. A Definitive, Rapid Alternative to the Gram Stain Assay. *Rapid Microbiology Newsletter* **3**, 2-3.

Gram C. 1884. The Differential Staining of Schizomycetes in Tissue Sections and in Dried Preparations. *Fortschitte der Medicin* **2**, 185-189.

Gregersen T. 1978. Rapid Method for Disitinction of Gram-negative From Gram-positive Bacteria. *Appl Microbiol Biotech* **5**, 123-127.

Halebian S, Harris B, Finegold SM, Rolfe RD. 1981. Rapid Method That Aids in Distinguishing Gram-Positive from Gram-Negative Anaerobic Bacteria. *J Clin Microbiol* **13**, 444-448.

Holm C, Jespersen L. 2003. A Flow-Cytometric Gram-Staining Techniques for Milk-Associated Bacteria. *Appl Environ Microbiol* **69**, 2857 – 2863.

Ingrahm JL, Ingrahm CA. 2000. *Introduction to Microbiology*. II Ed. Brooks/Cole, Thompson Learning division. Pacific Grove, California.

Iwanaga S. 1987. Chromogenic Substrate for Horseshoe Crab Clotting Enzyme: Its Application of the Assay of Bacterial Endotoxins. *Haemostasis* **7**, 183.

McClelland R. 2001. Gram's Stain: the Key to Microbiology. *Med Lab Observer* April pp. 20-31.

Murray, RGE, Doetsch RN, Robinow CF. 1994. Determinative and Cytological Light Microscopy. In: Gerhardt P. ed., *Methods for General and Molecular Bacteriology*, ASM Press, Washington, DC. pp21-41.

Pelczar JM, Chan ECS, Krieg RN. 2001. *Microbiology*, fifth Edition, Tata McGraw-Hill Publishing Company Limited, New York.

Powers EM. 1995. Efficacy of the Ryu Nonstaining KOH Technique for Rapidly Determining Gram Reactions of Food-Borne and Waterborne Bacteria and Yeasts. *Appl Environ Microbiol* **61**, 3756-3758.

Prescott ML, Harley PJ, Klein AD. 2004. *Microbiology*, sixth edition, McGraw-Hill Higher Education, New York.

Romero S, Schell RF, Pennell DR. 1988. Rapid Method for the Differentiation of Gram-Positive and Gram-Negative Bacteria on Membrane Filters. *J Clin Microbiol* **26**, 1378-1382.

Salle AJ. 1992. *Fundamental Principles of Bacteriology*. VII Ed. McGraw Hill Publishing Co. Ltd. New York.

Sizemore RK, Caldwell JJ, Kendrick AS. 1990. Alternate Gram Staining Technique Using a Fluorescent Lectin. *Appl Environ Microbiol* **56**, 2245-2247.

Stanier YR, Ingrahm JL, Wheelis ML, Painter PR. 2003. General Microbiology. fifth edition. MacMillan Press Ltd. New Jersey.

Von Gravenitz A, Bucher C. 1983. Accuracy of the KOH and Vancomycin Tests in Determining the Gram Reaction of Non-Enterobacterial Rods. *J Clin Microbiol* **16**, 983-985.

Chapter 7

Biochemical Tests

7.1 INTRODUCTION

Bacteria are very small and are rather difficult to see the differences between different species under the microscope. There are a variety of biochemical tests that can be used to identify bacteria and their ability/lack of ability to perform certain biochemical transformations. Culture media can be non-selective or selective. Non-selective media allow a wide variety of bacteria to grow (e.g., nutrient agar or blood agar). Selective media allow only certain organisms to grow because they have specific inhibitors added to the media (e.g., the bile salts in MacConkey agar). Additionally, culture media may also be differential, which allows group of biochemically related bacteria to be distinguished from other group of bacteria. In many distinct areas of microbiology, the ability to identify microorganisms has important application. For example, in food microbiology it is important to be able to accurately identify food spoilage contaminants. In microbial ecology, the identification of microorganism helps us characterize biodiversity. In the field of medical microbiology, a branch of microbiology that investigates pathogenic microorganisms, the primary focus is to isolate, identify, and study microorganisms responsible for infectious disease. Classical biochemical tests are often used to identify microorganisms and the results are seen by colour change. In most cases, detection is based on the reaction of an enzyme with a certain substrate.

7.2 BIOOXIDATION

Morphological and cultural characteristics of bacteria divide them into 11 seperate groups of genera. Determination of morphological and cultural characteristics is essential in getting to the genus and species determination requires a good deal of more information. Since all physiological (biochemical) reactions in organisms are catalyzed by enzymes, and since each enzyme is expressed by individual genes, a genetic profile is essential to confirm the identification. The enzymic reactions used for the identification criteria belong to respiration and fermentation.

Biooxidation reactions in bacteria pertain to the manner in which they get their energy; some are oxidative and others are fermentative. Strict aerobes that oxidize organic substances to produce the end products carbon dioxide and water are said to be oxidative. By utilizing organic compounds as electron donors, with oxygen as the ultimate electron (and hydrogen) acceptor, they produce CO_2 and water as end products to release energy. The ability to utilize free oxygen is accomplished by a cytochrome enzyme system.

Fermentative bacteria are organisms that also utilize organic compounds for energy but lack a cytochrome system. Instead of producing only CO_2 and H_2O, they produce complex end products, such as acids, aldehydes, and alcohols that are oxidizable and reducible. In these organisms oxygen is not the ultimate electron acceptor, and the reactions occur under anaerobic conditions. Various gases, such as carbon dioxide, hydrogen and methane, are also produced. In fermentative bacteria the organic compounds act both as electron donors and electron acceptors.

Sugars particularly glucose, are the compounds most widely used by fermenting organisms. Other substances such as organic acids, amino acids, purines, and pyrimidines also can be fermented by some bacteria. The end product of a particular fermentation are determined by the nature of the organism, the characteristics of the substrate, and environment conditions such as temperature and pH.

Although fermentation and respiration represent two different types of energy-yielding biooxidation, they can both be present in the same organism, as is true of the facultative anaerobes. An exception, however, is seen in the lactic acid bacteria where fermentation occurs in the presence of air (O_2).

Six types of biooxidations are normally studied in bacterial identification and isolation procedures (1) Durham tube sugar fermentation, (2) Mixed acid fermentation, (3) Butanediol fermentation, (4) Catalase production, (5) Oxidase production, and (6) Nitrate reduction.

The performance of all above tests involves a considerable number of inoculations because a set of positive test controls will also be needed. In actual practice physiological tests are used very selectively. The experiments are illustrated diagrammatically as line diagramme in Figure 7.1.

7.2.1 Durham Tube Sugar Fermentation Tests

The organism is able to utilize various sugars added in the broth with Durham tubes. If an organism is able to ferment a particular sugar, acid will be produced and gas may be produced. The presence of acid is detectable with the colour change of a pH indicator in the medium. Gas production is revealed by the formation of a void in the inverted vial of the Durham tube. The sugar broth used contains 0.5% of the specific carbohydrate plus sufficient amounts of beef extract and peptone to satisfy the nitrogen and mineral needs of the bacteria. The pH indicator phenol red is included for acid detection. This indicator is red when the pH is above 7 and yellow below this point.

7.2.2 Mixed Acid Fermentation (Methyl Red Test)

Several Gram-negative intestinal (enteric) bacteria can be differentiated on the basis of end products produced when they ferment the glucose in MR-VP medium. Genera of the bacteria such as *Escherichia*, *Salmonella*, *Proteus*, and *Aeromonas* ferment glucose to produce large amounts of lactic, acetic, succinic, and formic acids, plus CO_2, H_2 and ethanol. The accumulation of these acids lowers the pH of the medium to 5.0 and less. If Methyl red is added to such a culture, the indicator turns red, an indication that the organism is a mixed acid fermenter. These organisms are generally great gas producers,

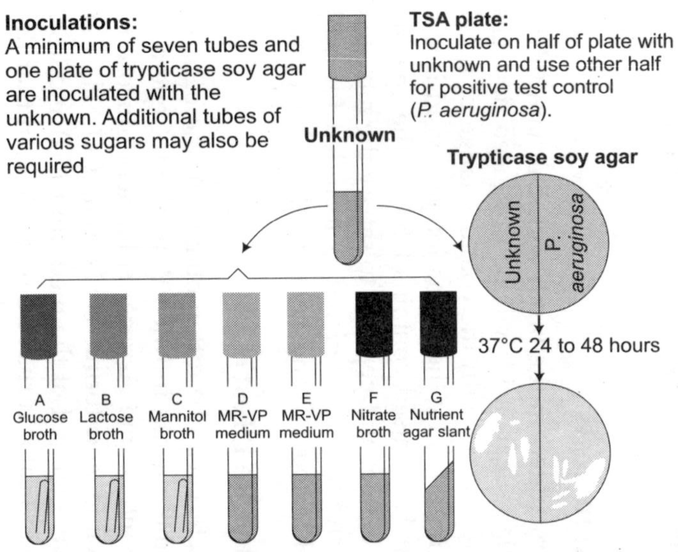

Incubation	Evaluation and tests	Positive test results
Except for tube E, incubate all tubes for 24-48 hrs at the optimum temperature for the unknown. Tube E should be incubated for 3-5 days at the optimum temperature.	Tubes A, B and C: If broth turns yellow, acid has been produced. If a bubble is present in the inverted vial, gas has been produced. Tube D: Do methyl red test. Tube E: Do Voges-Proskauer test. Tube F: Do nitrite test. Tube G: Do catalase test. TSA plate: Do oxidase test.	Tube D: Red colour is positive for mixed acids production. Tube E: Pink or Red colour is positive for butanediol production. Tube F: Red colour is positive for nitrate reduction to nitrite. Tube G: Bubbles effervescing from streak indicate that catalase is produced. TSA plate: Black colonies indicate that oxidase is produced.

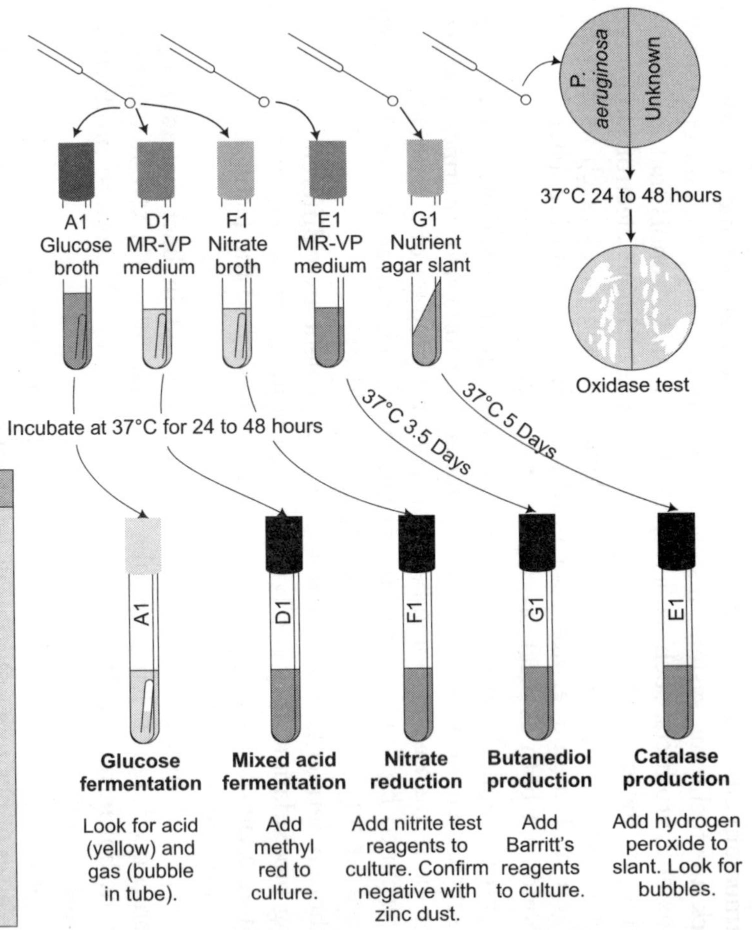

Figure 7.1 Demonstration of different biooxidation tests.

too, because they produce the enzyme formic hydrogenyl, which splits formic acid into equal parts of CO_2, and H_2.

$$HCOOH = \xrightarrow{\text{formic hydrogenylase}} CO_2 + H_2$$

MR-VP medium is essentially a glucose broth with some buffered peptone and dipotassium phosphate.

7.2.3 Butanediol Fermentation (Voges-Proskauer Test)

A negative methyl red test may indicate that the organism being tested produced a lot of 2,3 butanediol and ethanol instead of acids. All species of *Enterobacter* and *Serratia,* as well as some species of *Erwinia, Bacillus,* and *Aeromonas,* do just that. The production of these non-acid end products results in less lowering of the pH in MR-VP medium, causing the methyl red test to be negative. There is no satisfactory test for 2, 3 butanediol; however, acetoin (acetylmethylcarbino), a precursor of 2,3 butanediol, is easily detected with Barritt's reagent.

Barritt's reagent consists of alpha-naphthol and KOH. When added to a 3- to 5-day culture of MR-VP medium, and allowed to stand for some time, the medium changes to pink or red in the presence of acetoin. Since acetoin and 2,3 butanediol are always simultaneously present, the test is valid. This indirect method of testing for 2,3 butanediol is called the *Voges-Proskauer* test.

7.2.4 Catalase Production

Most aerobes and facultatives that utilize oxygen produce hydrogen peroxide, which is toxic to their own enzyme systems. Their survival in the presence of this antimetabolite is possible because they produce an enzyme called catalase, which converts hydrogen peroxide to water and oxygen.

$$2H_2O_2 = \xrightarrow{\text{catalase}} 2H_2O + O_2$$

The death of the strict anaerobes in the presence of oxygen may be due to the suicidal act of H_2O_2 production in the absence of catalase production. The presence or absence of catalase production is an important means of differentiation between certain groups of bacteria.

Catalase is measured by placing few drops of 3% hydrogen peroxide on the organisms of a slant culture. If the hydrogen peroxide effervesces, the organism is catalase positive.

7.2.5 Oxidase Production

Oxidase production is one of the most significant tests for differentiating certain groups of bacteria. All the members of Enterobacteriaceae are oxidase-negative and most species of *Pseudomonas* are oxidase positive. Another important group, the *Neisseria,* are oxidase producers. Two methods are normally employed to perform the test. The first method utilizes the entire TSA plate; the second method is less demanding to that only a loopful of organisms from the plate is used.

Unknown and *Pseudomonas aeruginosa* are streaked on to the TSA plates, pour the oxidase test reagent covering the colonies of both the organisms. Oxidase test reagent is prepared with 1% solution of dimethyl-*p*-phenylenediamine hydrochloride). *Pseudomonas aeruginosa* colonies first become pink, then change to maroon, dark red and finally black. If unknown follows the same colour sequence, it, too, is oxidase-positive. In other method, place several drops of oxidase test reagent on a piece of Whatmaan No. 2 filter paper. Remove a loopful of the organisms from one of the colonies and smear the organisms over a small area of the paper. The positive colour reaction will show up within 10-15 seconds.

7.2.6 Nitrate Reduction

Many facultative bacteria are able to use the oxygen in nitrate as a hydrogen acceptor in anaerobic respiration, thus converting nitrate to nitrite. This enzymatic reaction is controlled by an inducible enzyme called nitrase.

$$NO_3^- + 2e^- + 2H^+ \xrightarrow{nitrase} NO_2^- + H_2O$$

Since the presence of free oxygen prevents nitrate reduction, actively multiplying organisms will use up the oxygen first and then utilize the nitrate. In culturing some organisms, it is desirable to use anaerobic methods to ensure nitrate reduction.

The nitrate broth used to culture the organism consists of beef extract, peptone, and potassium nitrate. To test for nitrate after incubation, two reagents designated as A and B are used. Reagent A contains sulfanilic acid and reagent B contains dimethyl-alpha-naphthylamine. In the presence of nitrite, these reagents cause the culture to turn red. Negative results must be confirmed as negative with zinc dust. Add two or three drops of nitrite test solution A (sulfanilic acid) and an equal amount of solution B (dimethyl-alpha-naphthylamine) to the nitrate broth culture. Red colour should appear almost immediately indicating that nitrate reduction has occurred.

7.3 PHYSIOLOGICAL TESTS FOR HYDROLYSIS OF SUBSTRATES

Hydrolysis of polysaccharides, protein, fat, casein, starch, tryptophan and urea by extracellular hydrolytic enzymes play an important role in the identification of certain bacteria. The entire protocol for performing the hydrolysis of unknown bacteria is described in Figure 7.2.

7.3.1 Starch Hydrolysis

Many bacteria are able to hydrolyze starch which is a large molecule consisting of two constituents: amylase, a straight chain polymer of 200 to 300 glucose units, and amylopectin, a larger branched polymer with phosphate groups. Bacteria that hydrolyze starch produce amylases that yield molecules of maltose, glucose, and dextrins.

Grams/s iodine solution is an indicator of starch. When iodine comes in contact with a medium containing starch, it turns blue. If starch is hydrolyzed and is no longer present, the medium will have a clear zone next to growth. By pouring the

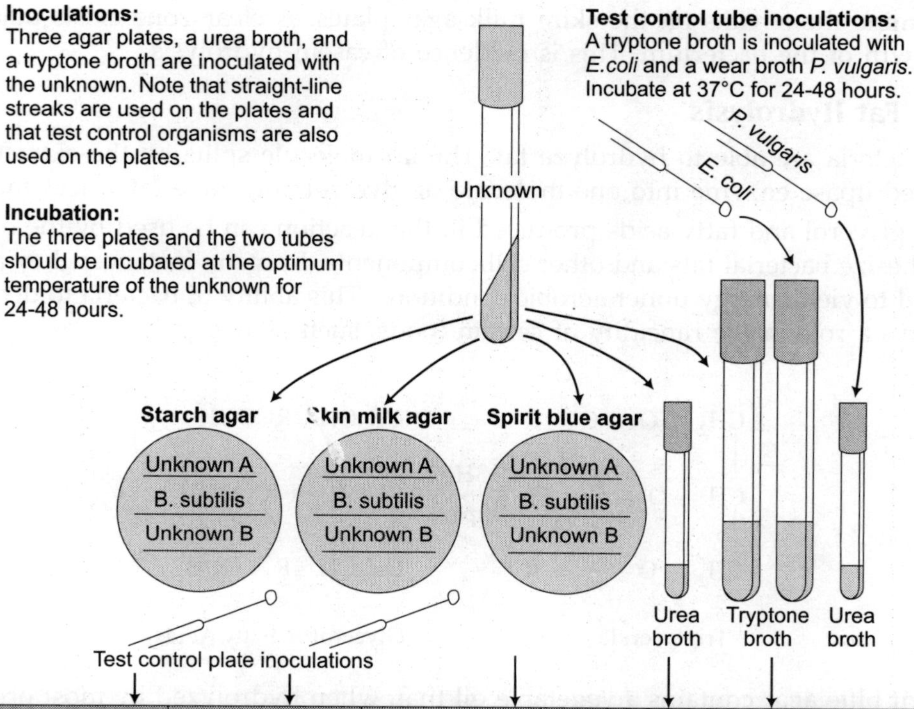

Starch hydrolysis	Casein hydrolysis	Fat hydrolysis	Urea hydrolysis	Tryptophan hydrolysis
Add several drops of Gram's iodine to growth on plate. Clear areas next to growth indicated that starch is hydrolyzed.	If clear zone is seen adjacent to growth on medium, casein is hydrolyzed.	If growth streak exhibits a dark blue precipitate, the organism is positive for fat hydrolysis.	If broth exhibits a red or cerise colour, the organism can hydrolyze urea.	If after adding Kovacs' reagent, a red ring appears on the surface of the broth, the organisms are able to hydrolyze tryptophan (indole-positive).

Figure 7.2 Hydrolysis tests.

Gram's iodine over the growth on the medium, one can see clearly where starch has been hydrolyzed. If the area immediately adjacent to the growth is clear, amylase is produced. Pour enough iodine over each streak to completely wet the entire surface of the plate. Rotate and tilt the plate gently to spread the iodine. Compare with unknown with the positive result.

7.3.2 Casein Hydrolysis

Casein is the prominent milk protein. Its presence causes milk to have its characteristic white appearance. Many bacteria produce extracellularly *caseinase,* which hydrolyzes casein to produce more soluble, transparent derivatives. Protein hydrolysis also is referred to as proteolysis, or *peptonization.*

Examine the streaks on the skim milk agar plates. A clear zone exists adjacent to the growth of the bacterium. This is evidence of casein hydrolysis.

7.3.3 Fat Hydrolysis

Some bacteria are able to hydrolyze fat. The fat molecule splits by the extracellularly produced lipase enzyme into one molecule of glycerol and three fatty acid molecules.

The glycerol and fatty acids produced in this reaction can be used by the organism to synthesize bacterial fats and other cell components. In many instances they are even oxidized to yield energy under aerobic conditions. This ability of bacteria to decompose fats plays a role in the rancidity of certain foods, such as margarine.

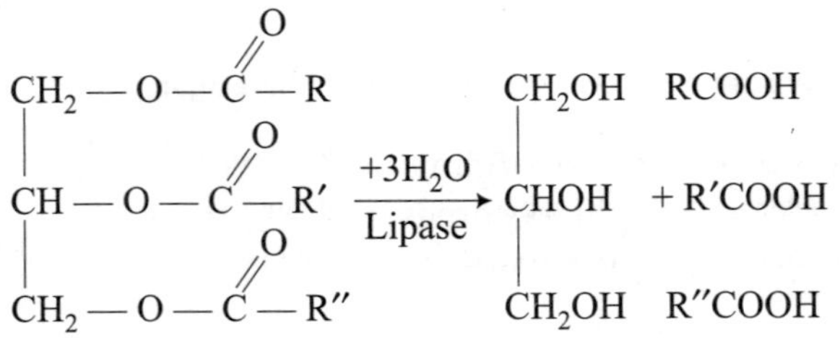

Triglyceride Glycerol Fatty Acids

Spirit blue agar contains a vegetable oil that, when hydrolyzed by most organisms, results in the lowering of the pH sufficiently to produce a dark blue precipitate. Unfortunately, the hydrolytic action of some organisms on this medium does not produce a blue precipitate because the pH is not lowered sufficiently.

7.3.4 Tryptophan Hydrolysis

Certain bacteria have the ability to split the amino acid tryptophan into indole and pyruvic acid. The enzyme that causes this hydrolysis is tryptophanase. Indole can be easily detected with Kovac's reagent. This test is particularly useful in differentiating *E. coli* from some closely related enteric bacteria.

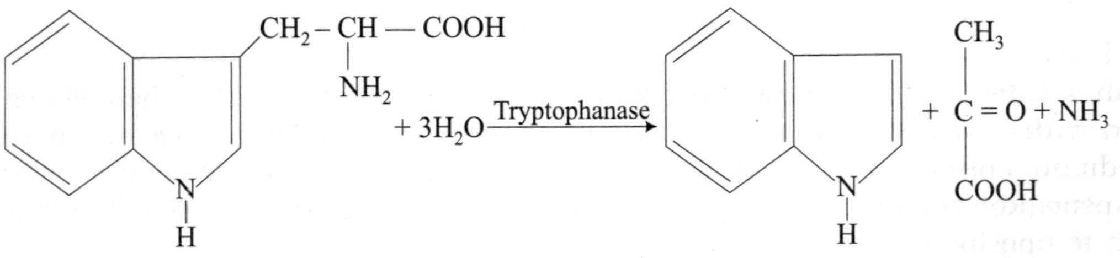

Tryptophane Indole Pyruvic acid

Tryptophan broth (1%) is used for this test because it contains a great deal of tryptophan. Tryptophan is a peptone derived from casein by pancreatic digestion. To test for indole add 10-12 drops of Kovac's reagent to the actively growing culture in tryptone broth. A red layer should form at the top of the culture.

7.3.5 Urea Hydrolysis

The differentiation of Gram-negative bacteria is greatly helped in demonstrating the production of urease from unknown sample. This enzyme splits off ammonia from the urea molecule. Urea broth is a buffered solution of yeast extract and urea. It also contains phenol red as a pH indicator. Since urea is unstable and breaks down in the autoclave at 15 psi steam pressure, it is often sterilized by filtration. It is tubed in small amounts to hasten the visibility of the reaction.

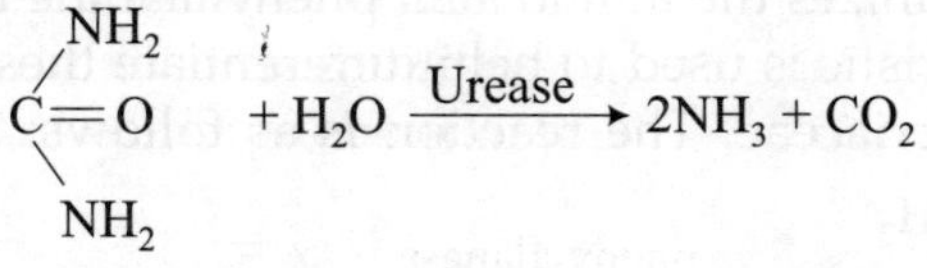

Urea Ammonia

When urease is produced by an organism in this medium, the ammonia that is released raises the pH. As the pH becomes higher, the phenol red changes from a yellow colour (pH 6.8) to a red or cerise colour (pH 8.1 or more).

7.4 MISCELLANEOUS TESTS

7.4.1 Hydrogen Sulfide Production

Some bacteria, such as *Proteus vulgaris,* produce hydrogen sulfide from the amino acid cysteine. These organisms produce the enzyme *cysteine desulfurase,* which works in conjunction with the coenzyme pyridoxyl phosphate. The production of H_2S is the initial step in the deamination of cysteine as indicated below:

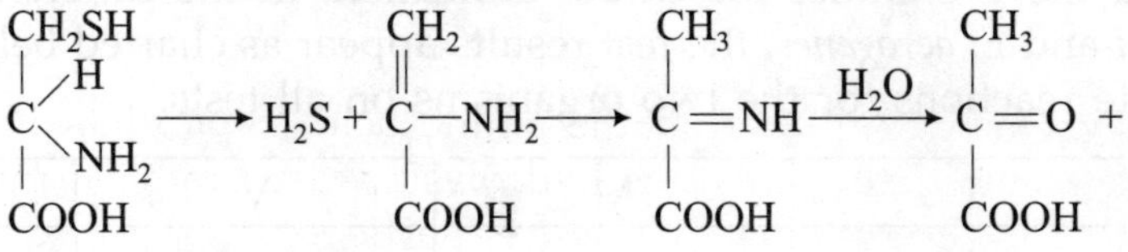

cysteine α amino crylic acid imino acid pyruvic acid

Kligler's iron agar or SIM medium is used here to detect hydrogen sulfide production. Both of these media contain iron salts that react with H_2S to form a dark precipitate of iron sulfide. Kligler's iron agar also contains glucose, lactose, and phenol red. When this medium is used in slants it is an excellent medium for detecting glucose and lactose fermentation. SIM medium, on the other hand, can also be used for determining motility and testing for indole production. Examine the tube of one of these media that was inoculated with *P. vulgaris,* if it is Kligler's iron agar positive tubes will have black precipitate.

7.4.2 Citrate Utilization

Some bacteria are able to utilize citrate as a sole source of carbon can be a very useful differentiation characteristic in working with intestinal bacteria. Koser's citrate medium and Simmons citrate agar are two media that are used to detect this ability in bacteria. In both of these synthetic media sodium citrate is the sole carbon source; nitrogen is supplied by ammonium salts instead of amino acids. The distinct Prussian blue colour change is the positive indication of citrate utilization.

7.4.3 Phenylalanine Deamination

Few bacteria, such as *Proteus, Morganella,* and *Providencia,* produce the deaminase phenylalanase, that deaminizes the amino acid phenylalanine to produce phenylpyruvic acid (PPA). This characteristic is used to help differentiate these three genera from other genera of the Enterobacteriaceae. The reaction is as follows:

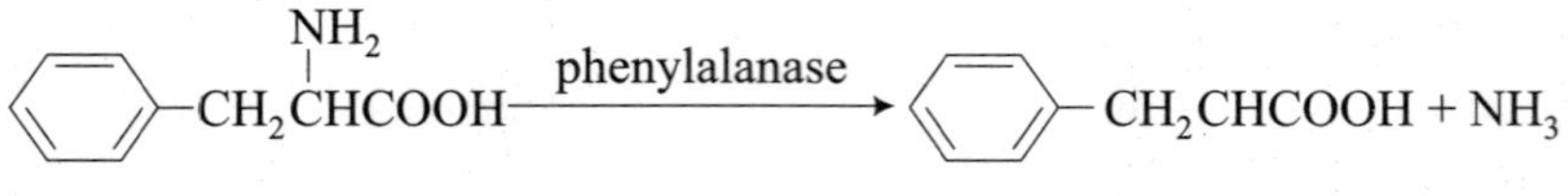

Phenylalanine Phenylpyruvic acid

The production of phenylpyruvic acid is evidence of the production of phenylalanase.

Allow 5 – 10 drops of 10% ferric chloride to flow down over the slants of the test control. To hasten the reaction, use a loop to emulsify the organisms into solution. A deep green colour should appear on the test slant in 1 – 5 minutes.

7.4.4 IMViC Tests

Four physiological tests have been grouped together into IMViC Tests. The I stands for indoel; the M and V stand for methyl red and Voges-Proskauer tests; i simply facilitates pronunciation; and the C signifies for citrate utilization. In the differentiation of the two coliform *E. coli* and *E. aerogenes,* the test results appear as charted below, revealing completely opposite reactions for the two organisms on all tests.

	I	M	V	C
E.coil	+	+	–	–
E. aerogenes	–	–	+	+

The significance of these tests is that when testing drinking water for the presence of the sewage indicator *E. coli,* one must be able to rule out *E. aerogenes,* which has many of the morphological and physiological characteristics of *E. coli.* Since *E. aerogenes* is not always associated with sewage, its presence in water would not necessarily indicate sewage contamination. Above key help in identifying a Gram-negative, facultative, rod-shaped bacterial organism, group these series of tests together in this manner to see how your unknown fits this combination of tests.

7.4.5 Litmus Milk Reaction

Litmus milk contains 10% powdered skim milk and a small amount of litmus as a pH indicator. When the medium is made up, its pH is adjusted to 6.8. It is an excellent growth medium for many organisms and can be very helpful in unknown characterization. In addition to revealing the presence or absence of fermentation, it can detect certain proteolytic characteristic in bacteria. A number of facultative bacteria with strong reducing power are able to utilize litmus as an alternative electron acceptor to render it colourless.

Some of the reaction takes 4-5 days to occur, the cultures should be incubated for at least this period of time; they should be examined every 24 hours and observe for the following reactions:

- ***Acid Reaction*** Litmus becomes pink. Typical of fermentative bacteria.
- ***Alkaline Reaction*** Litmus turns blue or purple. Many proteolytic bacteria cause this reaction in the first 24 hours.
- ***Litmus Reaction*** Culture becomes white, actively reproducing bacteria reduce the oxidation/reduction potential of the medium.
- ***Coagulation*** Curd formation. Solidification is due to protein coagulation. Tilting tube at 45°C will indicate whether or not this has occurred.
- ***Peptonization*** Medium becomes translucent. It often turns brown at this stage. Caused by proteolytic bacteria.
- ***Ropiness*** Thick, slimy residue in the bottom of the tube. Ropiness can be demonstrated with sterile loop.

Experiments

7.5 AMYLASE ACTIVITY

Amylases, glycoside hydrolases, break starch into glucose, maltose, maltotriose and dextrin by hydrolysis of glycosidic bonds. Starch is broken down in the intestinal tract of mammalian animal species by amylase-producing bacteria. This is presumably accomplished through the production of amylase. Therefore they are also called digestive enzymes. Amylase is an enzyme confined to the starch hydrolysing organisms. It can be tested using Lugol's iodine.

Starch is made up of two polysaccharides:

(i) Amylose, soluble in water and a polymer of 200-1000 D-glucose molecules linked by 1-4 glycosidic linkages (Figure 7.3 and Figure 7.4). It is dissolved in hot water giving blue colour with iodine.

(ii) Amylopectin is insoluble in hot water and is a polymer of 100 chains of D-glucose molecules linked by 1-4 glucosidic linkages. The chains are united by 1-6 glycosidic linkage.

Mode of Action of Amylase: It is a complex of α and β amylases. Both degrade amylopectin to maltose unit; the latter is acted upon by β-glucosidase to give glucose. While α-amylase alone degrades amylose to yield monomer glucose (Figure 7.3).

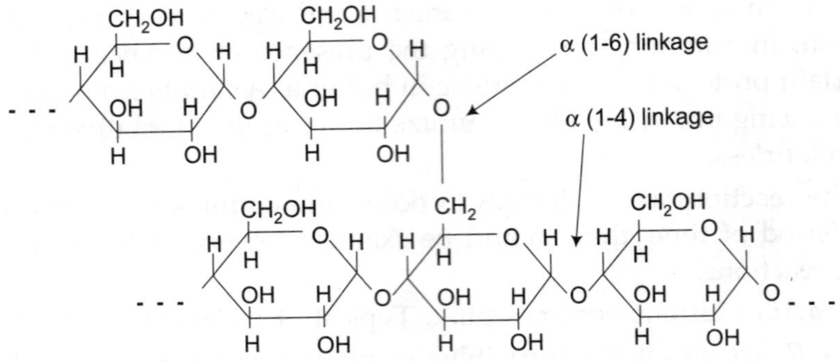

Figure 7.3 An α (1 → 6) branch point in amylopectin.

7.5.1 Procedure

7.5.1.1 Reagent and media preparation

Starch agar

1. Prepare a 10% solution of soluble starch in water and steam for 1 h.
2. Add 20 mL of this solution to 100 mL of melted nutrient agar and pour plates.

Composition of nutrient agar

Beef extract	3 g
Peptone	5 g
NaCI	5 g
$NaNO_3$	5 g
Distilled water	100 mL
pH	7.0

Lugol's Iodine

Iodine	1 g
Potassium iodide	2 g
Distilled water	300 mL

Inoculate starch agar and incubate for 24 hrs. Flood the plate with Lugol's iodine and observe for the clear zones.

7.6 CASEINASE

Casein is the principal protein of milk. It exists as a colloidal suspension that gives milk its opaque whiteness. Many bacteria are equipped with caseinase enzyme that hydrolyses this protein. This reaction of peptonisation is useful for the identification of microorganisms.

Media Preparation

1. Reconstitute milk powder and mix equal volumes of this and double strength nutrient agar melted at 55°C mix and pour plates.

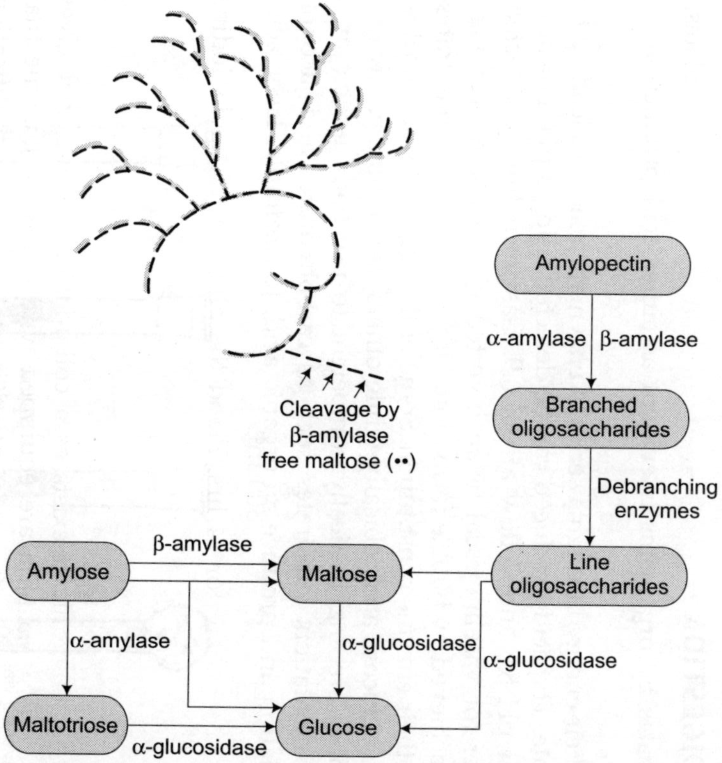

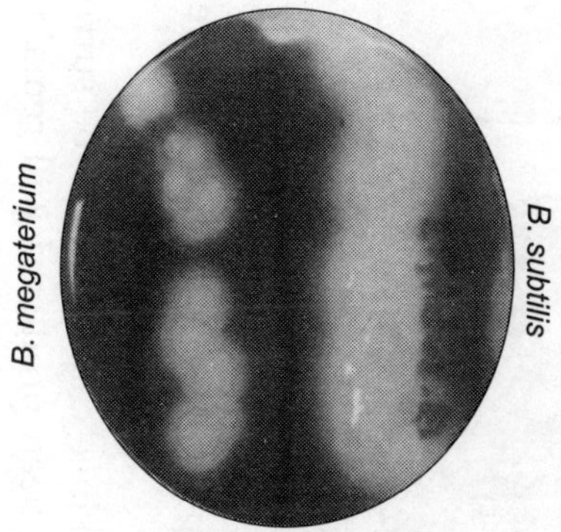

Starch hydrolysis

Figure 7.4 Enzyme conversion of starch to glucose initiated by α- and β-amylase.

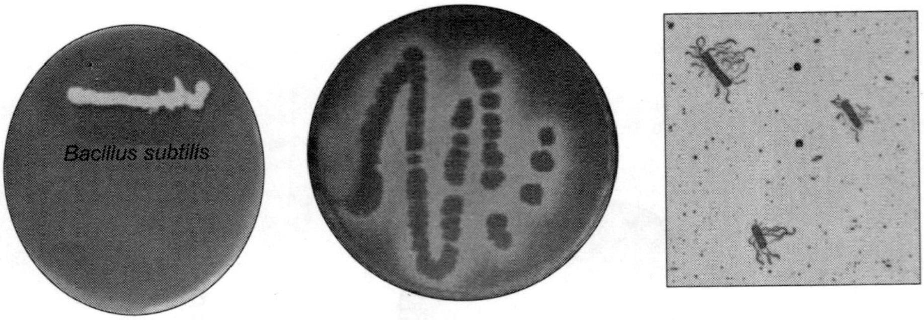

Figure 7.5 Caseinase production by *Bacillus subtilis*.

2. Inoculate the plates with a loop to place cells at the centre only.
3. Incubate at 37°C for 24-48 hrs.
4. Colonies of organisms that produce caseinase will be surrounded by clear zone. Areas where caseinase has not been produced will remain opaque (Figure 7.5).

7.7 CARBOHYDRATE DIGESTION

The digestion of carbohydrates by organisms is evident as either acid production or gas production (Figure 7.6).

Microorganisms obtain their energy through a series of enzymatic reactions leading to the oxidation of carbohydrate. The acid produced can be detected using a pH indicator phenol red, which is red at pH 8.5 and yellow at pH 6.9. Gas formation is detected through the use of a vasper (or vaseline) seal or an inverted vial Durham tube. Gas formation in agar media is achieved by bubble formation. Organisms use carbohydrates differently depending upon their enzyme constitution. Some of the organisms are capable of fermenting sugars such as glucose anaerobically while others use aerobic pathways. Still, the facultative anaerobes are enzymatically competent to use both aerobic and anaerobic pathways. In fermentation, substrates such as carbohydrates and alcohol undergo anaerobic dissimilation and produce an organic acid, for example, lactic acid

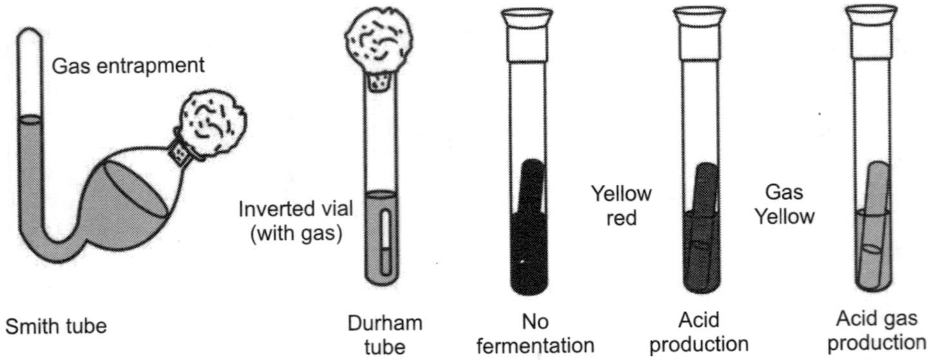

Figure 7.6 Illustrations of carbohydrate utilization with or without gas formation.

and may be accompanied by the production of gases like hydrogen or carbon dioxide. Following incubation, carbohydrates that have been fermented and produce acid wastes turn the phenol from red to yellow, thereby indicating the production of acids. In some cases, acid production is accompanied by the evolution of gas (CO_2) that is visible as bubbles in the inverted (Durham) tube. The acid produced can be detected using a pH indicator, phenol red, which is red at pH 8.5 and yellow at pH 6.9. Gas formation is detected through the use of a vaspar (or vaseline) seal or an inverted vial Durham tube. Gas formation in agar media is achieved by bubble formation. Carbohydrate is digested either by oxidation or fermentation, which in Gram-negative rods is an important identification feature. Three simple methods for carbohydrate digestion either by oxidation or fermentation are described below.

7.7.1 Inverted Tube Method

(a) Media preparation: (Peptone water broth)

Peptone	0.8 g
NaCl	1.4 g
$NaHCO_3$	0.02g
KCI	0.04 g
$CaCI_2$	0.04 g
KH_2PO_4	0.24 g
$Na_2HPO_4.2H_2O$	0.88 g
Distilled water	1000 mL

Steam the ingredients at 100°C for 20 min, and filter through double fold Whatman No. 1 filter paper. The pH is adjusted to 7.2.

Addition of Sugar: Prepare 1% solution of the desired sugar. Autoclave at low pressure and add to above solution in order to make to a final concentration of 10% by volume.

Addition of Indicator (Phenol Red): Dissolve 0.1 g of phenol sulphonaphthalein in 14.1 mL of 0.02 N NaOH and dilute to 250 mL with water, autoclave and add to basal medium making a final concentration of 0.4% by volume.

Phenol red is yellow at pH 6.8 (acid production) and red at pH 8.4 (no acid).

(i) Transfer media in test tubes and invert a Durham tube in each of these tubes.
(ii) Inoculate each tube with the culture to be identified.
(iii) Incubate at 37°C.
(iv) Observe the change in colour, if any (Figure 7.7)

7.7.2 Disc Method

It differs from the above method in that the sugar is not added to medium directly but sterilised filter paper discs are impregnated with the sugar solution and then placed on the inoculated agar plates aseptically. Observe after 48 hrs (Figure 7.8).

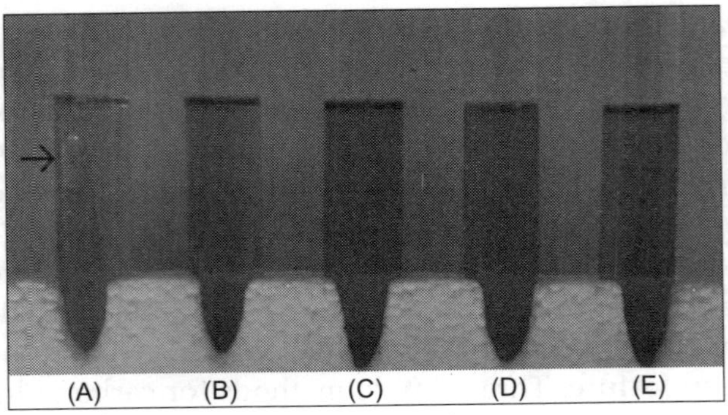

Figure 7.7 Possible phenol red tube results include: (A) formation of acid and gas (bubble is indicated by arrow), (B) formation of acid, (C) Uninoculated control, (D) alkaline by products, and (E) no acid or gas formation.

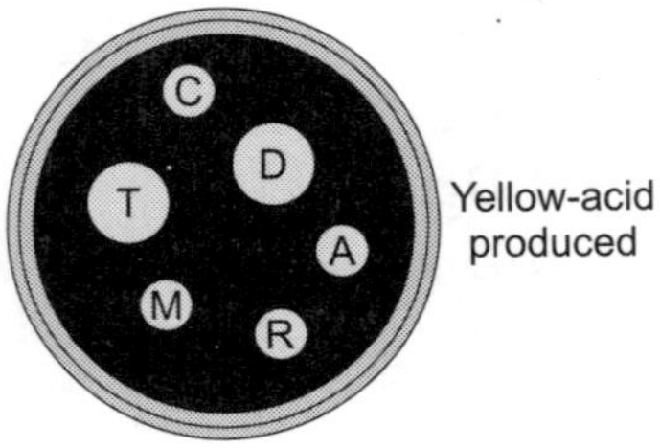

Figure 7.8 The disc method for determining carbohydrate fermentation. Letters indicate type of carbohydrate.

7.7.3 Oxidative/Fermentation Test

Test illustrates whether the carbohydrate is utilised oxidatively or fermentatively. Here oxidation fermentation medium (OF) is used with bromothymol blue as a pH indicator.

The purpose of this test is to determine whether an organism attacks sugars (in this case glucose is used) by fermentation or oxidation. Two tubes of Hugh and Leifon's medium are used. In one tube the medium is covered with vaseline. If the organism is an oxidizer it will produce acid only in the open tube (without vaseline), if it is a fermenter it will produce acid in the vaseline covered tube and in the open tube. Some aerobic bacteria may use the peptone in the medium, producing ammonia, with resulting alkalinity (blue) in the top part of the open tube. The indicator used is bromthymol blue. Motile bacteria will show diffuse growth away from the stab line. Non-motile bacteria will only show growth confined to the stab line. In the examples given here *E. coli* is motile and *Klebsiella* is non-motile (Figure 7.9).

Tubes are inoculated by stabbing. One tube is incubated aerobically while the other is kept in anaerobic condition. Fermented cultures will turn both tubes yellow while oxidative culture will turn only one tube yellow. In some oxidative culture tubes no acid end product will be detected. If the colour production is homogeneous, it also confirms the motility of the organism.

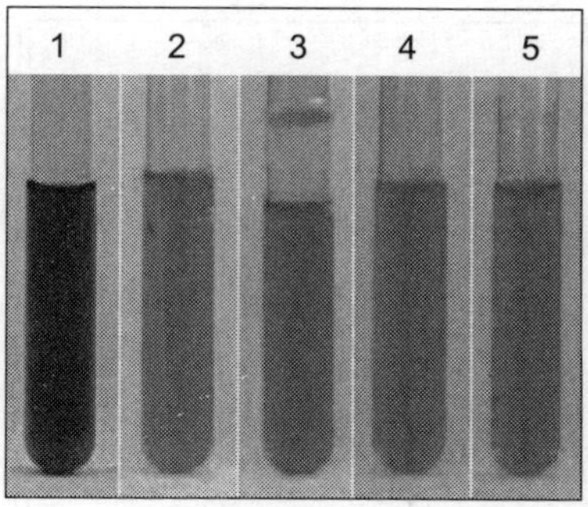

Figure 7.9 1 uninoculated medium, 2 and 3 fermentative reaction with gas production in closed tubes *Escherichia coli* and 4 and 5 fermentative reaction with gas production in the closed tube: *Klebsiellae.*

OF Agar Medium/Oxidation-Fermentation Medium

Peptone	2 g
Sodium chloride	1.5 g
K_2HPO_4	1 g
Distilled water	1000 mL
Agar	3 g

Heat to dissolve, adjust pH to 7.2 and add bromothymol blue.

Bromothymol blue: Dissolve 0.1 g dibromo-cresol sulphnaphthalein in 8 mL of 0.02 N NaOH and make up to 250 mL with water. This is yellow at pH 6 and blue at pH 7.6. Sterilize and add in OF medium making a final concentration of bromothymol blue solution as 0.4%, dextrose sugar is added (dextrose sugar can be replaced by any other desired sugar) which is autoclaved separately or filter sterilized, if possible. Make final concentration of sugar as 0.1% by volume.

7.7.3.1 Procedure

1. Inoculate the slants of OF medium by stabbing once (Figure 7.10).
2. Cover one tube with a thin layer of vaspar (vaseline and paraffin are mixed in equal amount after melting)
3. Incubate for 48 hrs at 37°C.
4. Observe for production of acid and gas from dextrose. Fermentative cultures turn both tubes yellow, while oxidative cultures will turn only the open tube yellow.

7.8 CATALASE ACTIVITY

Some bacteria contain flavoproteins that reduce oxygen (O_2), resulting in the production of hydrogen peroxide (H_2O_2) and, in some cases, an extremely toxic superoxide (O_2^-). Accumulation of these substances will result in death of the organism as they are powerful oxidizing agents and destroy cellular constituents very rapidly unless they can

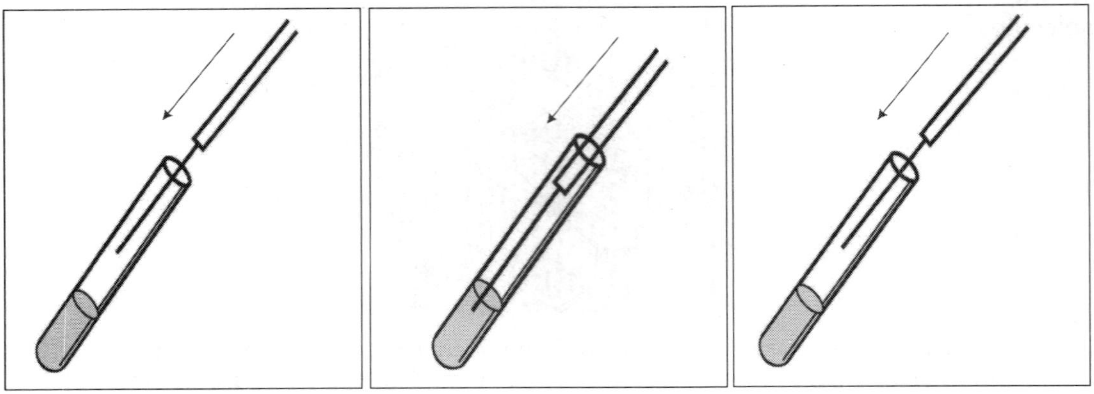

Wire with organisms is brought into tube without touching walls of tube.

(1)

Wire penetrates medium to two-thirds of its depth.

(2)

Wire is withdrawn from medium and tube. Neck of tube is flamed and plugged.

(3)

Figure 7.10 Preparing an agar-stab culture.

be enzymatically degraded. These substances are produced when aerobes, facultative anaerobes, and microaerophiles use the aerobic respiratory pathway in which oxygen is the final electron acceptor, during degradation of carbohydrates for energy production.

A bacterium must be able to protect itself against such O_2 products or it will be killed. Many bacteria possess enzymes that afford protection against toxic O_2 products. Facultative anaerobes and obligate aerobes usually contain the enzymes superoxide dismutase, which has the ability to catalyze the destruction of superoxide, and either catalase or peroxidase, which catalyze the destruction of hydrogen peroxide as follows:

$$2O_2 — \quad + \quad 2H+ \xrightarrow{\text{Superoxide dismutase}} \underset{\text{Oxygen}}{O_2} \quad + \quad \underset{\text{Hydrogen peroxide}}{H_2O_2}$$

$$2\,H_2O_2 \xrightarrow[\text{Peroxidase}]{\text{Catalase}} \underset{\text{Water}}{2\,H_2O} \quad + \quad \underset{\text{Free oxygen}}{O_2}$$

The inability of strict anaerobes to synthesize catalase, peroxidase, or superoxide dismutase may explain why oxygen is poisonous to these microorganisms. In the absence of these enzymes, the toxic concentration of H_2O_2 cannot be degraded when these organisms are cultivated in the presence of oxygen.

Organisms capable of producing catalase rapidly degrade hydrogen peroxide which is a tetramer containing four polypeptide chains, which are usually 500 amino acids long. It also contains four porphyrin heme groups (i.e., iron groups) that will allow the enzyme to react with the hydrogen peroxide (Figure 7.11). The enzyme catalase is present in most cytochrome-containing aerobic and facultative anaerobic bacteria. Catalase is the enzyme which has one of the highest turnover numbers compared to all other enzymes; one molecule of catalase has the ability to convert millions of molecules of hydrogen peroxide to water and oxygen in each second.

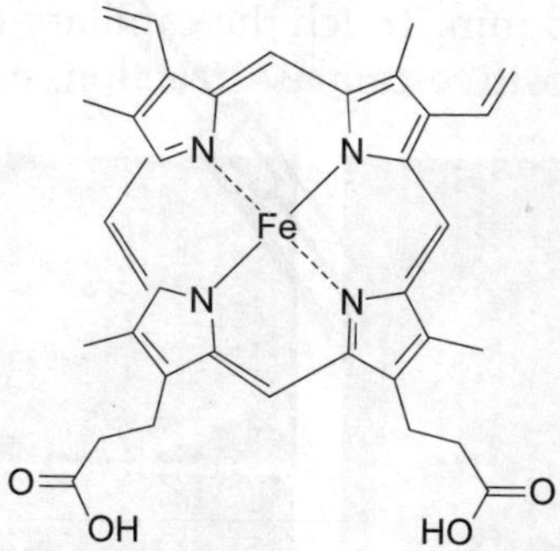

Figure 7.11 Heme, an example of an iron chelate porphyrin.

Catalase production and activity can be detected by adding the substrate H_2O_2 to an appropriately incubated (18- to 24-hour) tryptic soy agar slant culture. Organisms which produce the enzyme break down the hydrogen peroxide, and the resulting O_2 production produces bubbles in the reagent drop, indicating a positive test. Organisms lacking the cytochrome system also lack the catalase enzyme and are unable to break down hydrogen peroxide, into O_2 and water and are catalase negative.

Catalase activity is very useful in differentiating between groups of bacteria. For example, the morphologically similar *Enterococcus* (catalase negative) and *Staphylococcus* (catalase positive) can be differentiated using the catalase test.

7.8.1 Procedure

7.8.1.1 Method I

Add a few drops of 3% H_2O_2 to the slope and broth cultures and observe closely for the appearance of oxygen bubbles and a surface-broth accumulation (Figure 7.12).

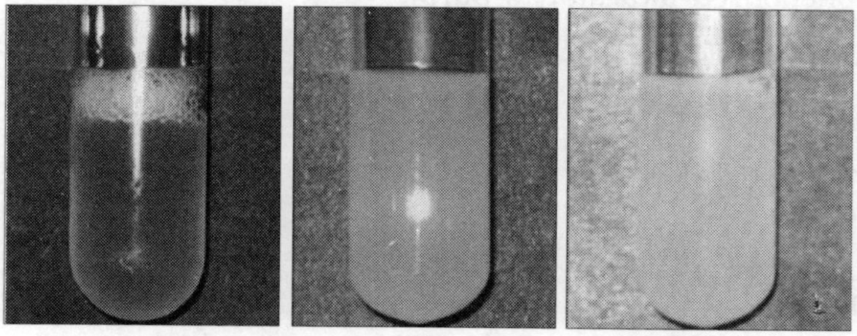

A B C

Figure 7.12 Catalase activity test A positive, B negative and C control.

7.8.1.2 Method II

Dip a glass capillary in the beaker containing 3% H_2O_2 and allow capillary action to tilt the tube to a height of about 20 mm. Touch the capillary on one of the colonies. Observe the bubbles in capillary for positive catalase reaction, e.g., *Lactobacillus* (Figure 7.13).

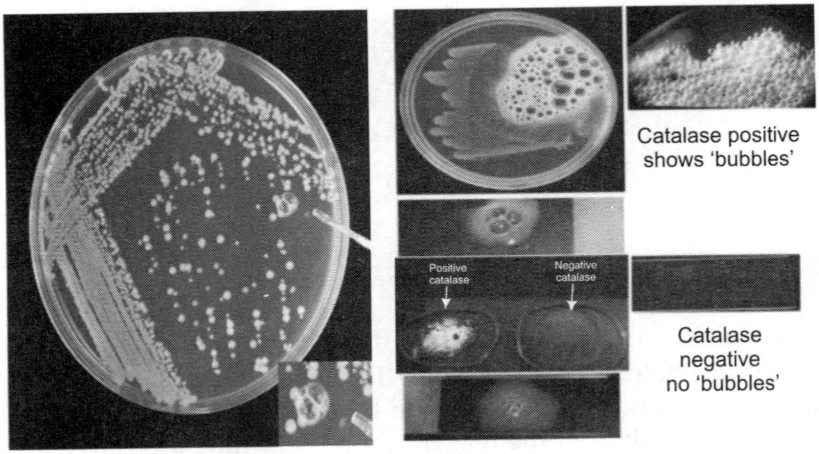

Figure 7.13 Catalase activity.

7.9 UREASE

Some bacteria split the urea molecule in two, releasing carbon dioxide and ammonia. This reaction is mediated by the enzyme urease, which can be seen in the culture medium prepared with urea added as the substrate. Phenol red is also added as a pH indicator. When bacterial cells that produce urease are grown in this medium, urea is degraded, ammonia is released, and the pH becomes alkaline. This pH shift is detected by a change in the indicator colour from orange-red to dark pink. Since the NH_3 production raises pH of the medium, urease activity can be detected by the change of the phenol red indicator to purple.

7.9.1 Procedure

Urea agar slant

Peptone	1 g
Glucose	1 g
NaCl	5 g
K_2HPO_4	5 g
Phenol red	6 mL of 1 : 500 aqueous solution
Agar	20 g
Distilled water	1000 mL
pH	6.8-6.9

1. Prepare agar base and sterilize.
2. Prepare urea solution 40% separately. Sterilize using bacteriological filter.
3. Add urea solution to agar base so as to make a final concentration of 20% of urea. Dispense in tubes to prepare slopes.
4. Inoculate and incubate.
5. Observe for colour change of phenol red indicator to purple (Figure 7.14), for example, *P. vulgaris*.

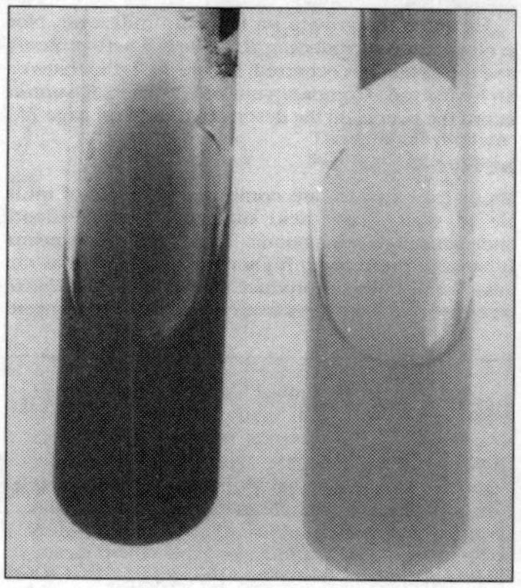

Figure 7.14 Urease test.

7.10 H_2S PRODUCTION

H_2S production in iron rich medium is an indication of breakdown of sulphur amino acids or reduction of thiosulphate. The test helps in identifying *Salmonella, Alcaligens, Edwardsiella* and *Proteus*. The activity of some bacteria on S-containing amino acids frequently results in the liberation of H_2S. This is best observed in a triple sugar iron agar slant (Figure 7.15).

7.10.1 Procedure

TSI Agar

Trypticase	10 g
Peptone	10 g
Sodium chloride	5 g
Lactose	10 g
Sucrose	10 g
Dextrose	1 g
Ferrous ammonium sulphate	0.2 g
Sodium thiosulfate	0.2 g
Phenol red	0.025 g

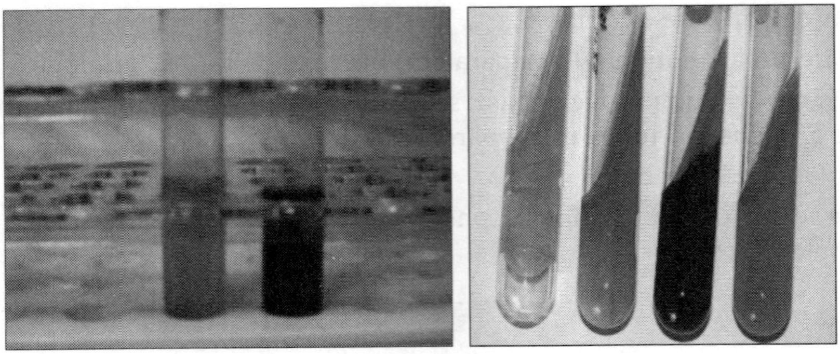

Figure 7.15 Urease test.

Agar	13 g
Distilled water	1000 mL
pH	7.3

Table 7.1 Key reaction for TSI agar

Reactions butts and slants	*Sugar fermention*	*Possible organism*
A and G in butt; A on slant H_2S not produced °1	Glucose – AG Lactose and/pr Sucrose – AG	*Escherichia, Klebsiella, intermediate coliform bacteria; Proteus, Paracolon*
A and G in butt Alk on slant: H_2S not produced	Glucose – AG Lactose and Sucrose – not fermented	*Salmonella, Proteus, Paracolon*
A-only in butt Alk. on slant: H_2S not produced °2	Glucose – AG Lactose and Sucrose – not fermented	*Salmonella, Shigella, Paracolon*
Alk. or neutral on butt AlK. on slant: H_2S not produced	None	*Alcaligenes, Pseudomonas*
No change in butt in 24 hours A on butt in 48 hours A on slant: H_2S not produced	Glucose and Sucrose – slight A; Lactose – not fermented	*Pasteurella°3*

Acid (yellow), Alk = Alkaline (red), G = Gas, °1 = *Cir. freundii* produced H_2S, °2 = *Slam. typli* produced a small amount of H_2S in the butt, and occasional strain may be encountered to give a small amount of gas in the butt; °3 = *Pasteurella multocida* is rarely encountered. *Enterococci* may give a similar type of reaction.

7.11 IMViC REACTIONS

The IMViC tests were developed to help distinguish morphologically and physiologically similar members of the **Enteric** group of bacteria, which includes many common pathogens of the intestine. IMViC is an acronym for the names of the four tests: I = Indole, M = Methyl red, Vi = Voges-Proskauer (the "i" facilitates euphony), C = Citrate.

7.11.1 Indole Test

This test reveals the presence of the enzyme **tryptophanase**, which cleaves the amino acid tryptophan into indole, pyruvic acid and ammonia, as shown in the reaction below (Figure 7.16). When amino acids are needed as a carbon source and tryptophan is present, some bacteria will begin to produce this enzyme and convert tryptophan to indole. Indole is a nitrogenous compound and a degradation product of amino acid tryptophan of various bacteria.

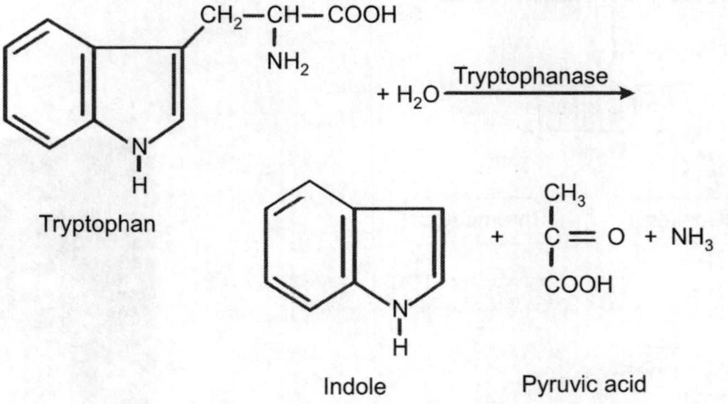

Figure 7.16

7.11.1.1 Procedure

Composition of peptone water broth medium g/L in distilled water: Peptone 0.8, NaCl 1.4, $NaHCO_3$ 0.02, KCl 0.04, $CaCl_2$ 0.04, KH_2PO_4 0.24, $Na_2HPO_4.2H_2O$ 0.88.

Steam the ingredients at 100°C for 20 min, and filter through double fold Whatman No. 1 filter paper. The pH is adjusted to 7.2. Autoclave at 121°C for 15 min.

Bacterium to be tested is inoculated in peptone water, which contains amino acid tryptophan and incubated overnight at 37°C. Following incubation few drops of Kovac's reagent are added. Kovac's reagent consists of para-dimethyl aminobenzaldehyde, isoamyl alcohol and con. HCl. Ehrlich's reagent is more sensitive in detecting indole production in anaerobes and non-fermenters. Formation of a red or pink coloured ring at the top is taken as positive.

Example: *Escherichia coli*: Positive; *Klebsiella pneumoniae*: Negative

(a) **Put a strip of filter paper (soaked in a saturated solution of oxalic acid and dried)** inside the overnight grown culture tube. Keep the filter paper suspended by the cotton plug. The paper becomes pink, when the test is positive.

(b) **Kovac's Reagent method**: p-dimethyl amino benzaldehyde 1 g, iso-amyl alcohol 15 mL.

To an overnight peptone medium grown culture, add Kovac's reagent to layer it, shake and allow the reagent to float. Indole is detected by the appearance of pink colour at the top (Figure 7.17).

7.11.2 Methyl-Red Test (MR Test)

It detects organisms that do not convert acidic products to neutral products and produce a final pH lower than that of organisms producing neutral products. Because of the

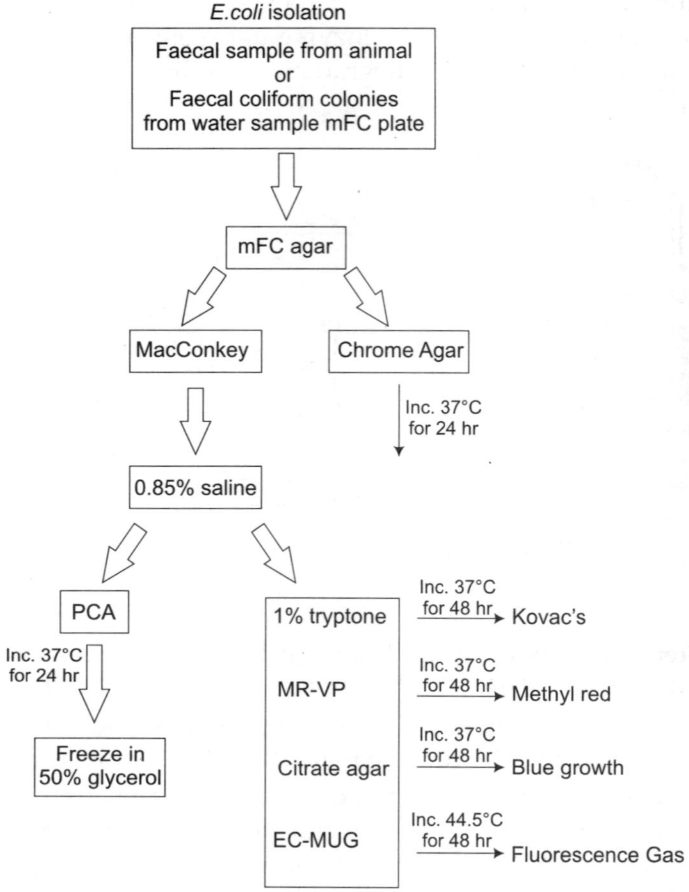

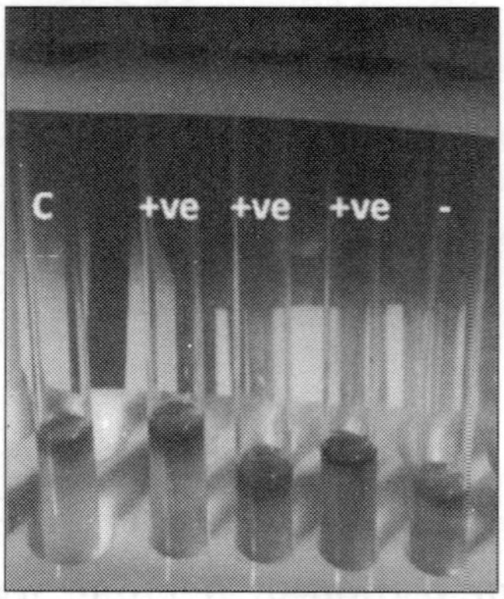

Indole test

Figure 7.17 Protocol for indole test for *E. coli.*

lower pH, the methyl-red indicator changes to a red colour, which is a positive test (Figure 7.18 and Figure 7.19).

7.11.2.1 Procedure

Composition of glucose-phosphate broth

Glucose	5 g
Proteose peptone	5 g
K_2HPO_4	5 g
Distilled water	1000 mL

1. Inoculate a tube of glucose phosphate broth with unknown culture.
2. Incubate for 48 hrs at 37°C.
3. Test for acid production by adding a few drops of an alcoholic solution of methyl red to the rest of the culture. A distinct red colour is considered to be a positive test; yellow is negative.

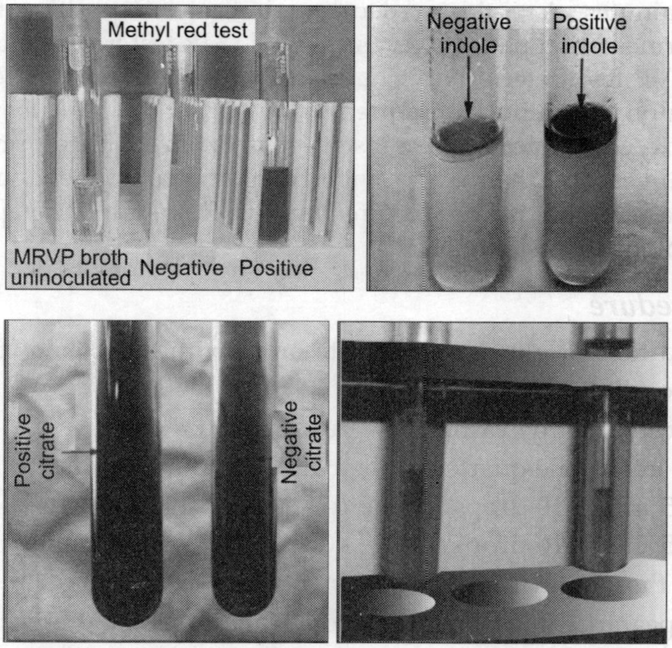

Figure 7.18 Complete demonstration for IMViC test.

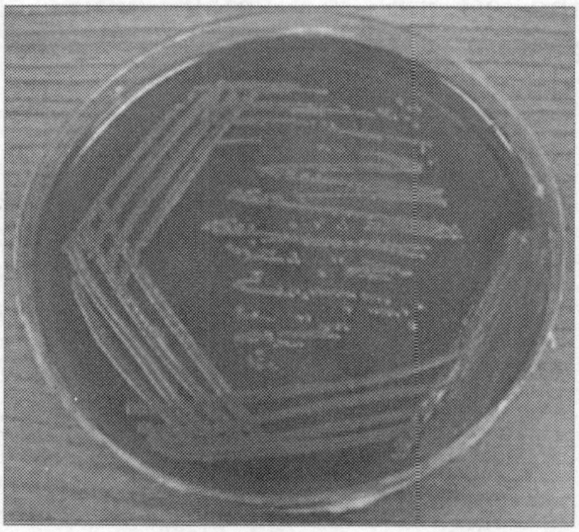

Figure 7.19 *Escherichia coli* produce red colony on MacConkey agar.

Methyl Red: 0.1 g of methyl red in 300 mL alcohol. Dissolve and make volume up to 500 mL.

7.11.3 Voges-Proskauer Test (VP Test)

The VP test uses alpha naphthol and potassium hydroxide to test for the presence of acetylmethylcarbinol (acetoin), an intermediate of the 2,3-butanediol fermentation pathway. After adding both reagents, the tube is shaken vigorously then allowed to sit for 5-10 minutes. A pinkish-red colour indicates a positive test, meaning the 2,3-butanediol fermentation pathway is used. While MR test is useful in detecting mixed acid producers, VP test detects butylene glycol producers. In this test, two reagents, 40% KOH and alpha-naphthol, are added to test broth after incubation and exposed to atmospheric oxygen. If acetoin is present, it is oxidized in the presence of air and KOH to diacetyl. Diacetyl then reacts with guanidine components of peptone, in the presence of alpha-naphthol, to produce red colour. Role of alpha-naphthol is that of a catalyst and a colour intensifier.

7.11.3.1 Procedure

1. Inoculate a tube of glucose-phosphate broth with the unknown culture.
2. Incubate for 48 hrs at 37°C.
3. Test for acetyl methyl carbinol (Figure 7.20).
   (a) Decant about one-quarter of the culture into a clean test tube.
   (b) Add 0.5 mL (8-10 drops) of the α-naphthol solution (5% solution in alcohol).
   (c) Add 0.5 mL (8-10 drops) of the 40% KOH solution containing 0.3% creatine.
   (d) Shake thoroughly and allow to stand for 5-30 minutes.
   (e) The appearance of a pink to red colour indicates the presence of acetyl methyl carbinol.

Note: The tubes of medium contain only 5 mL solution. This establishes a ratio of a large volume of air in the test tube to a small volume of medium a condition that promotes acetyl-methyl carbinol formation.

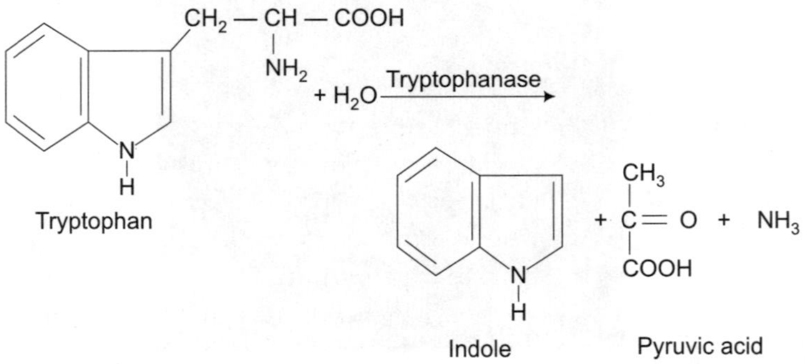

Figure 7.20 Bacterial cleavage of tryptophan into indole, pyruvic acid and ammonia.

7.11.4 Citrate Utilization

The ability to utilize citrate as a sole source of carbon and energy can be used to distinguish certain Gram-negative rods. Simmons' citrate agar contains citrate as its only carbon and energy source. Growth on this medium is a positive test for citrate utilization.

7.11.4.1 Procedure

Simmons' Citrate Agar

$MgSO_4$	0.2 g
$(NH_4)_2HPO_4$	1.0 g
K_2HPO_4	1.0 g
Sodium citrate	2.0 g
NaCI	5.0 g
Agar	15 g
Bromothymol blue	0.08 g
Distilled water	1000 mL

Prepare slants and inoculate.

An inoculum from a pure culture is transferred aseptically to a sterile tube of Simmons citrate agar. The inoculated tube is incubated at 35-37°C for 24 hours and the results are determined. Abundant growth on the slant and a change from green to blue in the medium indicates a positive test for growth using citrate (Figure 7.21).

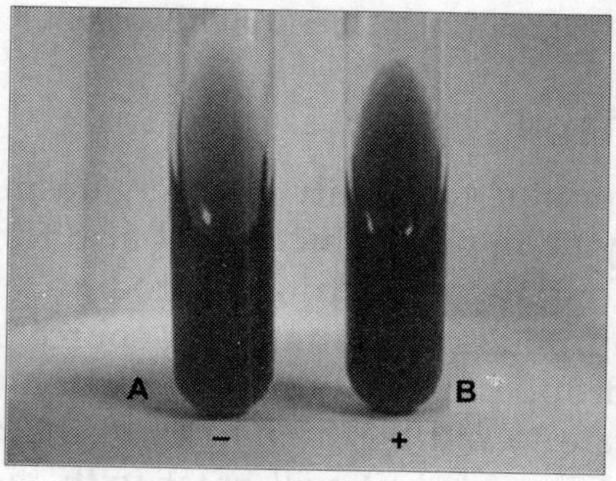

Figure 7.21 Citrate utilization test.

7.12 LITMUS-MILK REACTION

Milk is a complex nutritional source that contains proteins (mainly casein) in an aqueous solution of lactose and minerals. Bacterial enzymes alter the media and may bring about various changes. Litmus is added to the medium to detect pH changes that may occur as a result of these enzymatic reactions. Above a pH of 8.3 litmus is blue, while below a pH of 4.5 litmus is red. Here litmus powder is added to detect the reaction after fermentation of lactose, when it becomes red or may show clotting or peptonisation of milk. By producing acid from the fermentation of lactose present in milk a bacterium may also cause the milk to curdle or clot in the bottom of the tube. Litmus may also act as an electron acceptor thus becoming reduced by bacterial metabolism.

7.12.1 Procedure

Preparation of litmus milk media

(1) Dissolve 150 g of skimmed milk powder to one litre of water. Add sufficient amount of 2.5 percent aqueous solution of azolitmin.

(2) Inoculate and observe the change in the reaction of milk. It will turn red if acid is produced and blue if alkali is produced.

(3) Clotting is followed by peptonisation of curd (Figure 7.22).

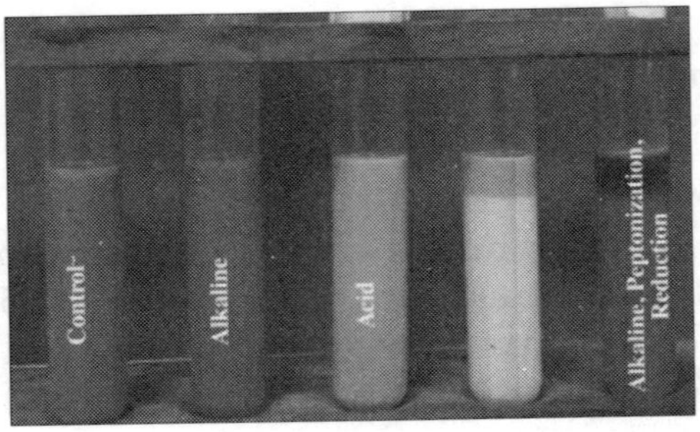

Figure 7.22 Litmus milk reaction.

7.13 NITRATE REDUCTION

Nitrate broth is used to determine the ability of an organism to reduce nitrate (NO_3) to nitrite (NO_2) using the enzyme nitrate reductase. It also tests the ability of organisms to perform nitrification on nitrate and nitrite to produce molecular nitrogen. Nitrate broth contains nutrients and potassium nitrate as a source of nitrate. After incubating the nitrate broth, add a dropperful of sulfanilic acid and α-naphthylamine. If the organism has reduced nitrate to nitrite, the nitrites in the medium will form nitrous acid. When sulfanilic acid is added, it will react with the nitrous acid to produce diazotized sulfanilic acid. This reacts with the α-naphthylamine to form a red-coloured compound. Therefore, if the medium turns red after the addition of the nitrate reagents, it is considered a positive result for nitrate reduction. If the medium does not turn red after the addition of the reagents, it can mean that the organism was unable to reduce the nitrate, or it could mean that the organism was able to denitrify the nitrate or nitrite to produce ammonia or molecular nitrogen. Therefore, another step is needed in the test. If the medium does not turn red after the addition of the nitrate reagents, add a small amount of powdered zinc. **Be careful, as powdered zinc is hazardous!** If the tube turns red after the addition of the zinc, it means that unreduced nitrate was present. Therefore, a red colour on the second step is a negative result. The addition of the zinc reduced the nitrate to nitrite, and the nitrite in the medium formed nitrous acid, which reacted with sulfanilic acid. The diazotized sulfanilic acid that was thereby produced reacted with the α-naphthylamine to create the red complex. If the medium does not turn red after the addition of the zinc powder, then the result is called a positive complete. If no red colour forms, there was no nitrate to reduce. Since there was no nitrite present in the medium, either, that means that denitrification took place and ammonia or molecular nitrogen were formed. The reduction of nitrate is brought about by several soil inhabiting bacteria – nitrate, ammonia, nitrous oxide, nitrogen gas, etc., are end products of nitrate reduction. They serve as H_2 acceptors for these species in absence of oxygen.

7.13.1 Procedure

Nitrate broth

Nutrient broth plus 5 g/L KNO_3 or $NaNO_3$

Nutrient Broth

Beef extract	3 g
Peptone	5 g
NaCl	5 g
$NaNO_3$	5 g
Distilled water	1000 mL
pH	7.0

1. Inoculate one tube of nitrate broth having a Durham tube with unknown culture.
2. Incubate at 37°C for 24 h.
3. Observe for gas formation if there is no gas incubate further.
4. Add 1 mL of α-naphthylamine reagent and 1 mL of sulfanilamide reagent to the culture tubes after incubation.
5. A red colour within 80 second shows nitrate reduction (Figure 7.23).

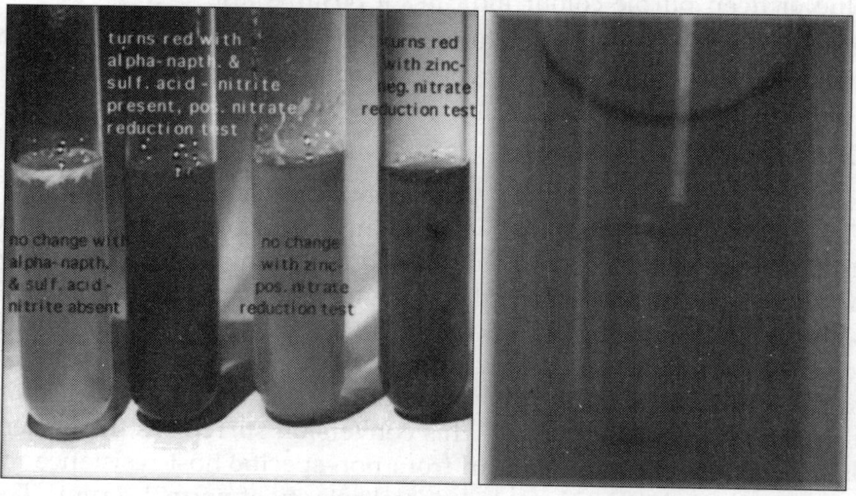

Figure 7.23 Nitrate reduction test.

Reagent Preparation

(a) **Sulfanilamide:** Sulfanillic acid (0.8%) dissolves 8 g of reagent in 1000 mL of 5 N acetic acid store in a glass stoppered bottle.

(b) **α-Napthylamine (0.5%):** Dissolve 5 g of reagent in less than 1000 mL of 5 N acetic acid (30%) by heating gently. Filter with through washed adsorbent cotton.

Store in a glass stoppered brown bottle in a refrigerator.

$$NO_3 \rightarrow NO_2 \rightarrow NH_3 \rightarrow N_2O$$

7.14 OXIDASE TEST

The oxidase test is used to identify bacteria that produce *cytochrome c oxidase,* an enzyme of the bacterial electron transport chain. When present, the *cytochrome c oxidase* oxidizes the reagent (tetramethyl-ρ-phenylenediamine) to (indophenols) purple colour end product (Figure 7.24). When the enzyme is not present, the reagent remains reduced and is colourless. Oxidation on one hand correlates with high cytochrome oxidase activity of some bacteria, e.g., *Pseudomonas* and on the other a negative oxidase reaction shows presence of the enteric bacteria, e.g., *Escherichia coli.*

7.14.1 Procedure

Reagent 1% solution of tetramethyl-ρ-phenylenediamine hydrochloride.

(1) Using a sterilized loop, pick growing well-isolated colony and rub into area 1 of the test strip (a filter paper disc, impregnated with oxidase reagent).
(2) Wait for 30 seconds and observe the inoculated area.
(3) A blue or deep purple colour indicates a positive test.
(4) Instead of using paper strips, solution can directly be poured onto the agar plate. However, a test strip keeps the reagent stable for longer time.

7.15 COAGULASES

Coagulases are enzymes that clot blood plasma by a mechanism that is similar to normal clotting. The coagulase test identifies whether an organism produces this exoenzyme. This enzyme clots the plasma component of blood. The only significant disease causing bacteria of humans that produce coagulase enzyme are *Staphylococcus aureus.* Thus this enzyme is a good indicator of the pathogenic potential of *S. aureus.* In human host, the action of coagulase enzyme produces clotting of the plasma by converting fibrinogen to fibrin in the immediate vicinity of the bacterium as a means of protection by itself. The fibrin meshwork that is formed by this conversion surrounds the bacterial cells or infected tissues, protecting the organism from non-specific host resistance mechanisms such as phagocytosis and the anti-staphylococcal activity of normal serum. This senables the bacterium to persist in the presence of a host immune response, which can lead to the establishment of infection. Thus, coagulase is described as a virulence factor (disease-causing factor) of *Staphylococcus aureus.* Citrate and EDTA are usually added to act as anticoagulants and prevent false-positive results. Most strains of *S. aureus*

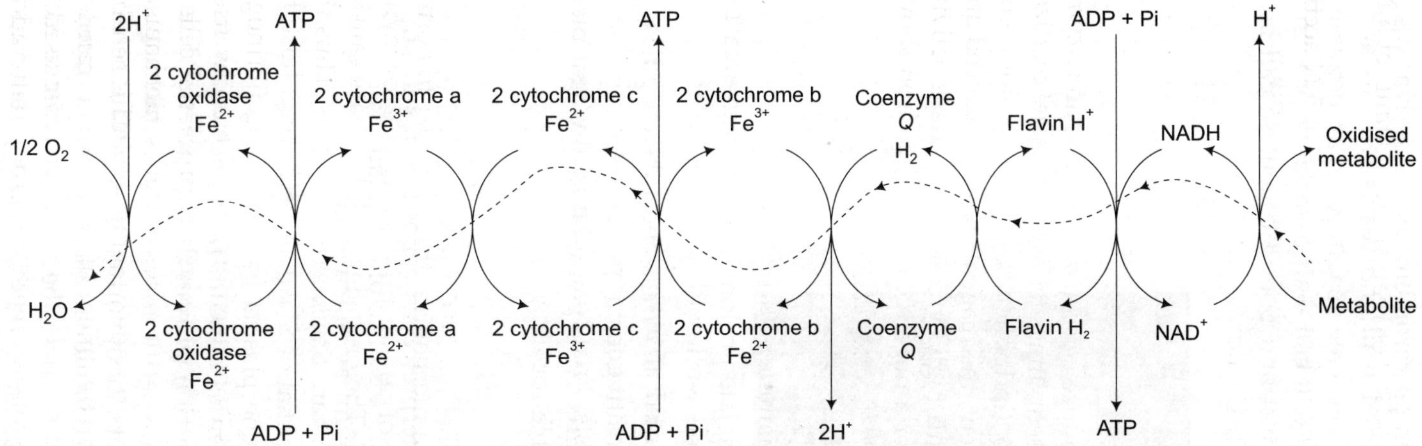

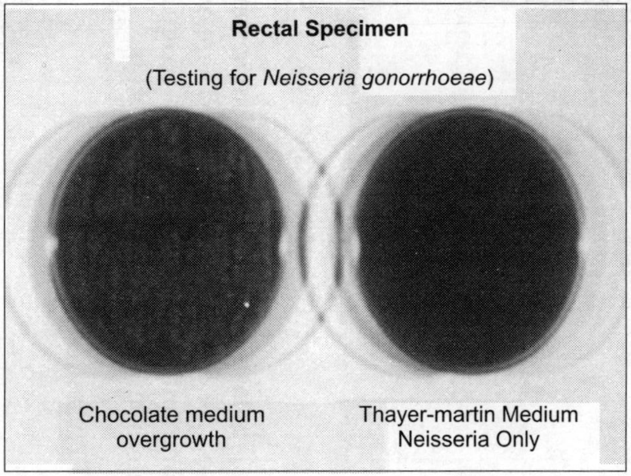

Figure 7.24 The electron transport system; the dashed line indicates the flow of electrons in *Neisseira* sp.

produce one or two types of coagulase; free coagulase and bound coagulase. Free coagulase is an extracellular enzyme which reacts with prothrombin and its derivatives. Bound coagulase is localized on the surface of the cell wall and reacts with α- and β-chains of the plasma fibrinogens to form a coagulate. Free coagulase is an enzyme that is secreted extracellularly and bound coagulase is a cell wall associated protein. Free coagulase can be detected in tube coagulase test and bound coagulase can be detected in slide coagulase test (Figure 7.25).

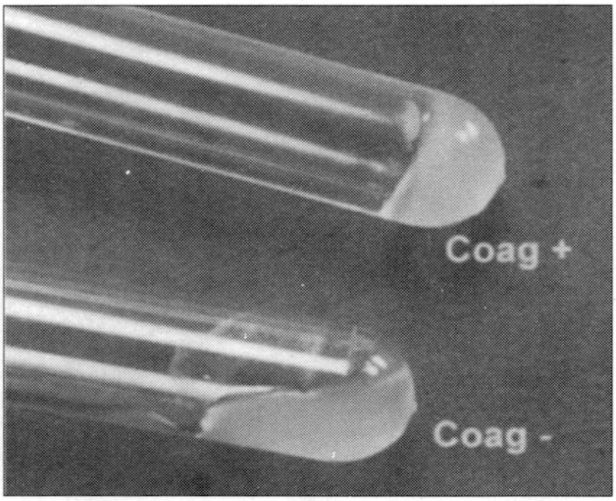

Figure 7.25 Coagulase test.

Slide coagulase test may be used to screen isolates of *S. aureus* and tube coagulase may be used for further confirmation. There are seven antigenic types of free coagulase, but only one antigenic type of bound coagulase exists. Free coagulase is always heat labile while bound coagulase is heat stable.

7.15.1 Procedure

In the test, the sample is added to rabbit plasma and held at 37°C for a specified period of time. Clot formation occurs within 4 hours is interpreted as a positive result and indicative of a virulent *Staphylococcus aureus* strain. The absence of coagulation after 24 hours of incubation is a negative result, indicative of an avirulent strain.

7.16 PECTINASE

The biotechnological potential of pectinolytic enzymes from microorganisms has drawn a great deal of attention for use as biocatalysts in a variety of industrial processes. The role of acidic pectinases in extraction and clarification of fruit juices is well established and have emerged as important industrial enzymes with wide-ranging applications in textile processing, degumming of plant base fibres, pretreatment of pectic wastewater, papermaking, and coffee and tea fermentation. Pectin methyl esterase an enzyme used in hydrolysis of pectin which is expressed in micro equivalent methoxyl group removed by the enzyme sample.

7.16.1 Procedure

To 5 mL of pectin (1% aqueous solution) adjusted to pH 6 add 2 mL of enzyme preparation (culture broth) mix and allow to react at 30°C. Note the pH and again adjust to 6 at zero hour with the help of 0.02 N NaOH. Use the boiled sample of enzyme as control. Allow the enzyme substrate to react for about 10 minutes and readjust pH to 6 using 0.02 N NaOH. Express the amount of activity in terms of unit mL^{-1}. Micro equivalent of NaOH required to maintain the original pH hr^{-1} mL^{-1} enzyme sample. The above method gives a quantitative estimation of pectinase.

Note: About 1% of pectin should be added to culture broth so as to induce pectinase synthesis.

7.17 GELATINASE

Gelatin was the first gelling agent used to solidify culture media. The protein gelatin is obtained by the hydrolysis of collagen a component of connective tissue and tendons of animals. It also demonstrates a test for proteolytic enzymes. The advantages of solid media over liquid media include isolation of pure cultures and the ability to perform plate counts. The disadvantages of gelatin include incubation at 20°C, a temperature that is lower than optimum for growing many microorganisms, and the fact that many organisms metabolize (liquefy) gelatin. Agar later replaced gelatin as a solidifying agent. Water solutions of gelatin of the concentration to be used here are liquid at room temperature and solidify in an acid bath. If the gelatin would have been hydrolysed by the action of microorganism being tested, the medium will remain liquid.

Identifying fermentative and non-fermentative Gram-negative bacilli include testing for gelatin liquefaction. If the proteolytic enzyme gelatinase is present, gelatin is hydrolyzed and loses its gelling characteristic.

7.17.1 Procedure

Nutrient Gelatin medium

Add 4% of gelatin in nutrient broth.

Nutrient broth

Beef extract	3 g
Peptone	5 g
NaCI	3 g
Distilled water	1000 mL
pH	7.0

(1) Inoculate the tubes of nutrient gelatin with the unknown culture.

(2) Incubate at 37°C together with a sterilised tube of nutrient gelatin that will serve as a control. Test after 2 days and upto 7 days or until a positive reaction is obtained.

(3) To examine hydrolysis, chill the tubes in ice water. The control tubes and tubes in which no hydrolysis has taken place will solidify. Hydrolysed gelatin will remain fluid (Figure 7.26).

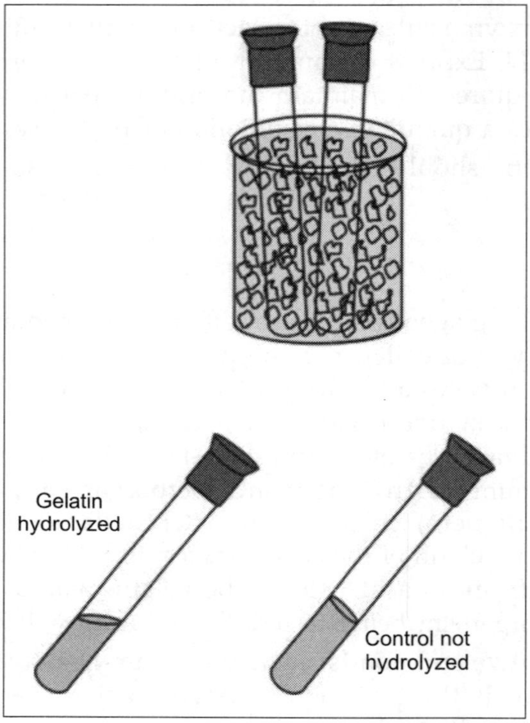

Figure 7.26 Gelatin hydrolysis.

Results

Positive: Medium remains liquefied after refrigeration.

Negative: Medium becomes solid after refrigeration.

Uninoculated control tube: Medium becomes solid after refrigeration.

7.18 LIPASE

Microbial lipases are known for their vast potential regarding their usage in different industries. Phospholipids are major components of all cell membranes, the ability to hydrolyse host cell pospholipids is an important factor in the spread of virulent microorganisms. The interest in microbial lipase production has increased in the last decades, because of its large potential in industrial applications as additives for foods (flavour modification), fine chemicals (synthesis of esters), waste water treatment (decomposition and removal of oil substances), cosmetics (removal of lipids), pharmaceuticals (digestion of oil and fats in foods), leather (removal of lipids from animal skins) and medical (blood triglyceride assay). It can be tested by growing the culture in egg yolk agar, the cleavage of phosphate ester bonds form a water insoluble lipid. The enzymatic activity is detected by a zone of opalescence in the medium surrounding the cell mass.

Two common lipids decomposed by microorganisms are triglycerides and phospholipids. The triglycerides are esters of glycerol and fatty acids.

7.18.1 Procedure

Egg Yolk Agar

Egg yolk	100 mL (5%)
Plate count agar medium	900 mL

Plate Count Agar medium

Tryptone	5 g
Yeast extract	2.5 g
Glucose	1 g
Agar	15 g
Distilled water	1000 mL

(1) Inoculate egg-yolk agar with the unknown culture.

(2) Incubate at 37°C for 48 hours.

(3) Observe for an opalescent precipitate (Figure 7.27).

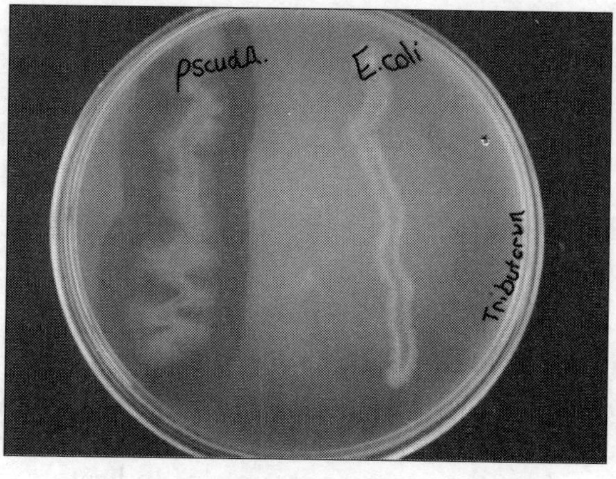

Figure 7.27 Lipid hydrolysis test.

7.19 GLUCONATE TEST

Gluconic acid is a mild organic acid derived from glucose by a simple oxidation reaction. The reaction is facilitated by the enzyme glucose oxidase (fungi) and glucose dehydrogenase (bacteria such as *Gluconobacter*). Microbial production of gluconic acid is the preferred method and it dates back to several decades. The most studied and widely used fermentation process involves the fungus *Aspergillus niger*. Gluconic acid and its derivatives, the principal being sodium gluconate, have wide applications in food and pharmaceutical industries. The test is employed to differentiate enterobacteria.

7.19.1 Media and Reagents

Gluconate Broth

Yeast extract	1.0 g
Peptone	1.5 g
K_2HPO_4	1.0 g
Potassium gluconate	40.0 g
or Sodium gluconate	37.25 g
Distilled water	1000 mL

Dissolve by heating, adjust pH to 7.0. Filter, if necessary.

Benedict's reagent (Commercially available) or

Copper sulfate	18 g
Sodium carbonate	100 g
Sodium citrate	200 g
KCNS	125 g
Potassium ferrocyanide solution	(5%) 5.0 mL
Distilled warter	100 mL

7.19.2 Procedure

(1) Inoculate gluconate broth.

(2) Incubate for 48 h at 37°C.

(3) Add an equal volume of Benedict's reagent and place in a boiling water bath for 10 min.

(4) An orange or brown precipitate indicates gluconate oxidation. A deep blue colour is negative response.

Appendix

Indicators, Stains and Reagents

All the indicators used can be made by dissolving a measured amount of the indicators in 95% ethanol, adding a measured amount of water, and filtering with filter paper. The following chart provides the correct amounts of indicators, alcohol, and water for various indicator solutions.

Indicator solution	*Indicator (g)*	*95% Ethanol (mL)*	*Distilled H_2O (mL)*
Bromcresol green	0.4	500	500
Bromcresol purple	0.4	500	500

Contd...

Contd...

Bromthymol blue	0.4	500	500
Cresol red	0.2	500	500
Methyl red	0.2	500	500
Phenolphthalein	1.0	50	50
Phenol red	0.2	500	500
Thymol blue	0.4	500	500

Stains and reagents

Acid-Alcohol (for Ziehl-Neelsen stain)

3 mL concentrated hydrochloric acid in 100 mL of 95% ethyl alcohol.

Acid-Alcohol (for Fluor chrome staining)

HCl.. 2.5 mL

Ethyl alcohol 70%.. 500 mL

NaCl.. 2.5 g

Alcohol, 70% (from 95%)

Alcohol, 95%.. 368 mL

Distilled water.. 132 mL

Auramine-Rhodamine Stain (for Mycobacteria)

Auramine 0.. 3 g

Rhodamine B.. 1.5 g

Glycerol.. 150.0 mL

Phenol crystals (liquefied at 50° C).. 20.0 mL

Distilled water.. 100 mL

Clarify by filtration with glass wool or Whatman No. 1 filter paper. Do not heat, store at room temperature.

Barritt's Reagent (Voges-Proskauer test)

Solution A: 6 g α-naphthol in 100 mL 95% ethyl alcohol.

Solution B: 16 g potassium hydroxide in 100 mL water.

Carbolfuchsin Stain (Ziehl's)

Solution A: Dissolve 0.3 g of basic fuchsia (90% dye content) in 10 mL 95% ethyl alcohol.

Solution B: Dissolve 5 g of phenol in 95 mL of water.

Mix solution A and B.

Crystal Violet Stain (Hucker modification)

Solution A: Dissolve 2 g of crystal violet (85% dye content) in 20 mL of 95% ethyl alcohol.

Solution B: Dissolve 0.8 g ammonium oxalate

In 80 mL distilled water.

Mix solution A and B.

Diphenylamine Reagent (nitrate test)

Dissolve 0.7 g diphenylamine in a mixture of 60 mL of concentrated sulfuric acid and 28.8 mL of distilled water.

cool and add slowly 11.3 mL of concentrated hydrochloric acid. After the solution has stood for 12 hours some of the base separates, showing that the reagent is saturated.

Ferric Chloride Reagent

$Fecl_3.6H_2O$	12 g
2% Aqueous HCL	100 mL

Make up the 2% aq. HCL by adding 5.4 mL of concentrated HCL (37%) to 94.6 mL H_2O. Inoculate with two or three colonies of beta haemolytic streptococci, incubate at 35°C for 20 or more hours. Centrifuge the medium to pack the cells, and pipette 0.8 mL of the clear supernate into a Kahn tube. Add 0.2 mL of the ferric chloride reagent to the Kahn tube and mix well. If a heavy precipitate remains longer than 10 minutes, the test is positive.

Gram's Iodine (Lugol's)

Dissolve 2 g of potassium iodide in 300 mL of distilled water and then add 2 g iodine crystals.

Kovac's Reagent (indole test)

N-amyl alcohol	75 mL
Hydrochloric acid (conc.)	25 mL
P-dimethylamine benzaldehyde	5 g

Lactophenol Cotton Blue Stain (Fungal stain)

Phenol crystals	20 g
Lactic acid	20 mL
Glycerol	40 mL
Cotton blue	0.05 g

Dissolve the phenol crystals in the order ingredients by heating the mixture gently under a hot water tap.

Leifson's Flagellar Stain

Prepare three separate solutions as follows:

Solution A: Dissolve 0.9 g of pararosaniline acetate and 0.3 g of pararosaniline hydrochloride in 100 mL of 95% ethyl alcohol. Let it stand overnight at room temperature to complete dissolution.

Solution B: Dissolve 3 g of tannic acid in 100 mL of distilled water.

Solution C: Dissolve 1.5 g of sodium chloride in 100 mL of distilled water.

Mix equal volumes of solutions A, B, and C and let stand for 2 hours. Store in a stopper bottle in a refrigerator (up to 2 months). Disregard precipitate that forms in bottom of bottle. Do not filter, store indefinitely, if frozen. Frozen stain solution must be thoroughly mixed after thawing since the water separates from the alcohol. After mixing, the precipitate should be allowed to settle to the bottom.

Note: Pararosaniline compounds should be certified for flagella staining.

Malachite Green Solution (spore stain)

Dissolve 5 g malachite green oxalate in 100 mL distilled water.

Methylene Blue (Loffler's)

Solution A: Dissolve 0.3 g of Methylene blue (90% dye content) in 30 mL ethyl alcohol (95%).

Solution B: Dissolve 0.01 g potassium hydroxide in 100 mL distilled water. Mix solution A and B.

Naphthol, alpha

5% alpha-naphthol in 95% ethyl alcohol.

Caution: Avoid all contact with human tissues. Alpha-naphthol is considered to be carcinogenic.

Nessler's Reagent (ammonia test)

Dissolve about 5 g of potassium iodine in 35 mL of cold ammonia-free distilled water. Add a saturated solution of mercuric chloride until a slight precipitate persists. Add 400 mL of a 50% solution of potassium hydroxide and dilute to 1 L

Allow to settle, and decant the supernatant for use.

Nitrosine Solutions (Dorner's)

Nigrosine, water soluble	10 g
Distilled water	100 mL

Boil for 30 minutes. Add as a preservative 0.5 mL formaldehyde (40%). Filter twice through double filter paper and store under aseptic conditions.

Nitrate Test Reagent (see Diphenylamine)

Nitrite Test Reagents

Solution A: Dissolve 8 g sulfanilic acid in 1000 mL 5N acetic acid (1 part glacial acetic acid to 205 parts water).

Solution B: Dissolve 5 g dimethyl-alpha-naphthylamine in 1000 mL 5N acetic acid. Do not mix solutions

Cautions: Although at this time it is not known for sure. There is a possibility that dimethyl-.α-naphthylamine in solution B may be carcinogenic. For reasons of safety, avoid all contact with tissues.

Oxidase Test Reagent

Mix 1 g of dimethyl-phenylenedianmine hydrochloride m 100 mL of distilled water. Preferably, the reagent should be made up fresh daily. It should not be stored longer than one week in the refrigerator. Tetramethyl-p-phenylenediamine dihydrochloride

(1%) is even more sensitive, but is considerably more expensive and more difficult to obtain.

Phenolized Saline

Dissolve 8.5 g sodium chloride in 1 L distilled water.

Potassium permanganate (for Fluor chrome staining)

$KMnO_4$	2.5 g
Distilled water	500 mL

Safranin (for Gram stain)

Safranin O (2.5% sol'n in 95% ethyl alcohol)	10 mL
Distilled water	100 mL

Trommsdorf''s Reagent (nitrite test)

Add slowly, with constant stirring, 100 mL of a 20% aqueous zinc chloride solution to a mixture of 4.0 g of starch in water. Continue heating until the starch in water is dissolved as much as possible, and the solution is nearly clear. Dilute with water and add 2 g of potassium iodine. Dilute to 1 L, filter and store in amber bottle.

Vaspar

Melt together in equal amount of vaseline paraffin and store in small bottles.

Voges-Proskauer Test reagent (see Barrett's)

White blood cell (WBC) Diluting Fluid

Hydrochloric acid	5 mL
Distilled water	495 mL

Add 2 small crystals of thymol as a preservative.

Suggested Readings

Association of Official Analytical Chemists. 1995. Bacteriological analytical manual, 8th ed. AOAC International, Gaithersburg, MD.

Brown AE. 2007. *Bensons's microbiological applications: laboratory manual in general microbiology.* (10th ed.). New York: Mc Graw Hill.

Cappuccino GJ, Sherman N. 2013. *Microbiology: A laboratory manual,* 10th ed., Benjamin Cummings, London.

Clarke PH, Cowan ST. 1952. Biochemical methods for bacteriology. *J Gen Microbiol* **6**,187–197.

Ewing WH. 1986. Edwards and Ewing's identification of *Enterobacteriaceae,* 4th ed. Elsevier Science Publishing Co. Inc. New York, NY.

Farmer JJ, 3rd, Davis BR, Hickman-Brenner FW, et al. 1985. Biochemical identification of new species and biogroups of Enterobacteriaceae isolated from clinical specimens. *J Clin Microbiol* **21**:46–76.

Ingrahm JL, Ingrahm CA. 2000. *Introduction to Microbiology.* 2nd ed. Brooks/Cole, Thompson Learning division. Pacific Grove, California.

Isenberg HD, Sundheim LH. 1958. Indole reactions in bacteria. *J Bacteriol* **75**, 682–690.

Isenberg HD. (ed.) 1994. Clinical microbiology procedures handbook, Sup. 1. American Society for Microbiology, Washington, D.C.

ISO 7218. 2007. Microbiology of food and animal feeding stuffs. *General requirements and guidance for microbiological examinations.*

ISO/IEC 17025. 2005. *General requirements for the competence of testing and calibration laboratories.*

Draft ISO/FDIS .2003. 11133-2. Microbiology of food and animal feeding stuffs. *Guidelines on preparation and production of culture media. Part 2. Practical guidelines on performance testing on culture media.*

ISO 11133-1. 2000. Microbiology of food and animal feeding stuffs. *Guidelines on preparation and production of culture media – Part 1: General guidelines on quality assurance for the preparation of culture media in the laboratory.*

ISO 13843. 2000. Water quality. *Guidance on validation of microbiological methods.*

ISO 6887-1. 1999. Microbiology of food and animal feeding stuffs. *Preparation of test samples, initial suspension and decimal dilutions for microbiological examination – Part 1: General rules for the preparation of the initial suspension and decimal dilutions.*

ISO Draft /DIS 16140. 1999. Food microbiology. *Protocol for the validation of alternative methods.*

Johnson TR, Case CL. 2007. *Laboratory experiments in microbiology.* (8th ed.). San Francisco: Pearson Education.

Larry McKane L, Kandel J. 1996. *Microbiology: Essentials and applications.* (2nd ed.). New York: McGraw Hill.

Murray PR, Baron EJ, Pfaller MA, Tenover FC, R. H. Yolken RH (eds.). 1995. *Manual of Clinical Microbiology,* 6th ed. American Society for Microbiology, Washington, D.C.

Payne WJ. 1973. Reduction of nitrogenous oxides by microorganisms. *Bacteriol Rev* **37**,409–452.

Salle AJ. 1992. *Fundamental Principles of Bacteriology.* 7th ed. McGraw Hill Publishing Co. Ltd. New York.

Stanier YR, Ingrahm JL, Wheelis ML, Painter PR. 2003. *General Microbiology.* 5th ed. MacMillan Press Ltd. New Jersey.

Sutter VL, Carter WT. 1972. Evaluation of media and reagents for indole-spot tests in anaerobic bacteriology. *Am J Clin Pathol* **58**, 335–338.

WHO. 2010. Good practices for pharmaceutical quality control laboratories. In: *WHO Expert Committee on Specifications for Pharmaceutical Preparations. Fortyfourth report.* Geneva, World Health Organization. WHO Technical Report Series, No. 957, Annex 1.

WHO. 2007. General guidelines for the establishment, maintenance and distribution of chemical reference substances. Revision. In: *WHO Expert Committee on Specifications for Pharmaceutical Preparations. Forty-first report.* Geneva, World Health Organization. WHO Technical Report Series, No. 943, Annex 3.

Chapter **8**

Microbial Taxonomy

8.1 INTRODUCTION

Taxonomy is an area of **biological** science which comprises three distinct, but highly interrelated disciplines that include classification, nomenclature and identification. Applied to all-living entities taxonomy provides a consistent means to classify name and identify organisms. This consistency allows biologists worldwide to use a common label for every organism they study within their particular disciplines. The common language that taxonomy provides minimizes the confusion about names and allows attention to centre on more important scientific issues and phenomena. In diagnostic microbiology, classification, nomenclature and identification of microbes play a central role in providing accurate and timely diagnosis of infection.

Classification is the organization of organisms that share similar morphologic, physiologic and genetic traits into specific groups or taxa. Nomenclature, the naming of microorganisms according to established rules and guidelines provides the accepted labels by which organisms are universally recognized. The classification of microbes is based on how they look and what they can do.

Molecular techniques for characterizing microbial genotypes provide a possible basis of defining a microbial species. Nucleic acid amplification technology has opened new avenues of microbial detection and characterization, such that growth is no longer required for microbial identification. Methods of microbial identification can be broadly delimited into genotypic techniques based on profiling an organism's genetic material (primarily its DNA) and phenotypic techniques based on profiling either an organism's metabolic attributes or some aspect of its chemical composition. Classification of microbes can be made on the basis of phenotypic characteristics and on genotypic characteristics.

8.2 CLASSIFICATION

Classification is one of the fundamental concerns of biology. Facts and objects must be arranged in an orderly fashion before their unifying principles can be discovered and used as the basis for prediction. The development of high speed electronic computers has had a profound impact on the methods of classification in biological fields. The rapidity of the computer's operation has made it possible for the first time to consider large numbers of characteristics in classifying microbes. Most approaches to the problems of microbial taxonomy have arisen from either of the two viewpoints, one derived from phylogenetic and other from practical consideration. The former viewpoint too frequently arises from some major premise, which has little practical connotation. The latter viewpoint often leads to the submergence of large groups of organisms not known to be of economic importance, because of an attitude of impatience towards any system which does not reflect the methods used in the specialized laboratory where steps in the identification of an unknown organism must be measured in terms of utility and speed. It must, therefore, be realized that the precise delineation of species cannot be the primary aim of the microbial taxonomy at present. It is seldom possible and often it may not even be desirable to compromise by recognizing the necessity for the organization within a taxonomic system of a selected body of knowledge of important differential characters which may be applied when practical consideration that demands that phylogenetically related organisms be distinguished one from the other. This implies that taxonomic systems must undergo periodic revision with the advent of new knowledge.

Classification means the act of arranging a number of objects (of any sort) into groups (or *taxa)* in relation to attributes possessed by those objects. The word classification is also applied to the result of any such arrangement. Taxonomy is concerned, *inter alia,* with definition of the aims of classification, the design of rules by which arrangements may be achieved, and with the evaluation of the end results. In biological classifications, the primary objects (microorganisms, plants, animals) are usually arranged in groups which are themselves members of larger groups (and so on) in such a way that any item or any group appears as a member of only one larger grouping, i.e., the groups are non-overlapping. This method of classification is the familiar *hierarchical* system which can be conveniently represented by a 'family tree' or *dendrogram.* The units at each level (taxonomic rank) of a hierarchical system are given distinctive names or label a branch of taxonomy known as *nomenclature.* In biology, the system of nomenclature is normally used for living organisms, which is derived from that used by the great eighteenth-century taxonomist Linnaeus (Carl von Linne). In this system, the basic unit (the *species)* is given two names one denoting its membership of a taxon at the rank that we label *genus* (generic name) followed by a second denoting the particular species (specific name). These names are written in a Latinized form and constitute the so called *Latinized binomial* (e.g., *Aspergillus niger, Bacillus subtilis, Clostridium tetani).* Taxa of higher rank (families, orders, etc.) are given single Latinized names with characteristic endings (e.g., Pseudomonadaceae, family; Pseudomonadales, order). The naming of newly discovered organisms or of newly proposed taxa of higher ranks is governed by rigid rules standardized by international agreement. It is perhaps worth emphasizing that it is by no means the only possible one. The simplest system would be merely to label the different types of organism with some sort of catalogue number which referred to a listed description. A much more useful approach might be similar to one proposed for the naming of viruses, *viz.,* the virus is given a group name (probably Latinized) which is followed by a descriptive formula akin to that used by botanists in floral diagrams or to the antigenic formula of a *Salmonella* species. The naming of the units are defined and delineated by the classification. This latter method is, in fact, reminiscent of that often used by Linnaeus, who sometimes followed his Latinized generic name with up to a dozen descriptive 'specific' epithets. Ideally, the coining of new names is contrived to convey as much information as possible about the organism or taxon. Unfortunately, both the restriction to Latinized binomials and, often, the rules of precedence make this aim difficult to achieve.

8.3 IDENTIFICATION

It can be done through various methods either by physical methods or by methods based on phylogeny. Identification simply involves the comparison of an 'unknown' object (e.g., a newly isolated bacterium, a collected microorganism, plant or animal) with all similar objects that are already known. If the 'unknown' object matches up with a 'known' then the former has been identified; if not, it may be considered to be a 'new' species, variety, or strain and, when adequately described, is added to the list of known objects. In practice this act of comparison is normally carried out not between two actual objects but between the 'unknown' *isolate* and a recorded description of previously discovered microorganisms, plants, animals, etc. The inadequacy of recorded descriptions of many microbial species can sometimes make accurate identification very difficult, if not impossible. It is not always appreciated that neither *identification* nor *nomenclature* need necessarily be connected with classification.

These three facets, or the trinity that is taxonomy, are to some extent interdependent, but in an orthodox scheme they are considered in the order given above. It is arguable whether the hen or the egg came first, but since the end of the nineteenth century microbiological ethics have demanded that we should not name a microbe before allotted it to a unit in an orderly classificatory system (Figure 8.1).

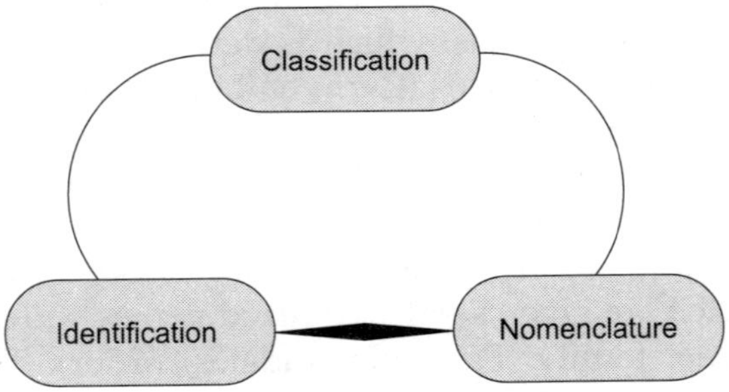

Figure 8.1 Three facets of taxonomy.

Genotypic and phenotypic criteria are based on observable physical or metabolic characteristics of microorganisms, that is, identification is through analysis of gene products. The phenotypic approaches are the classic approach of identification, and most identification strategies are still based on phenotype. The most commonly used phenotypic criteria include:

- Microscopic evaluation of microbial cellular morphology.
- Macroscopic (colony) morphology includes colony size, shape, colour (pigment) surface appearance, and any changes in the colony growth produced in the surrounding agar medium.
- Environmental conditions required for growth can be used to supplement other identification criteria.
- The enzyme-based tests are designed to measure the presence of one specific enzyme or a complete metabolic pathway that may contain several different enzymes.
- Molecular methods like Multiplex-PCR, Nested-PCR, RAPD-PCR, ARDRA, different hybridization techniques, microarrays, protein-profiling, zymographic analysis, multilocus enzyme electrophoresis, pulsed field gel electrophoresis, N-terminal sequencing, riboprinter technique and chromatographic technique have revolutionized the area of identification and characterization.

The correct identification of microorganisms is of fundamental importance to microbial systematists as well as to scientists involved in many other areas of applied research and industry (e.g., agriculture, clinical microbiology and food production). Increased use of automation and user-friendly software make these technologies more widely available. In all, the detection of infectious agents at the nucleic acid level represents a true synthesis of clinical chemistry and clinical microbiology techniques. Accurate identification requires a sound classification or system of ordering organisms into groups, as well as an unequivocal nomenclature for naming them.

Microbial taxonomy can create much order from the plethora of microorganisms. For example, the American Type Culture Collection maintains the following, which are based on taxonomic characterization (the numbers in brackets indicate the number of individual organisms in the particular category): algae (120), bacteria (14400), fungi (20200), yeast (4300), protozoa (1090), animal viruses (1350), plant viruses (590), and bacterial viruses (400). The actual number of microorganisms in each category will continue to change as new microbes are isolated and classified. The general structure, however, of this classical, so-called phenetic system will remain the same.

8.4 PRINCIPLES OF TAXONOMY

Taxonomy has 2 functions: the first is to describe as completely as possible the basic taxonomic units, or species; the second, to devise an appropriate way of arranging and cataloguing these units. The notion of species consists of assemblage of individuals that share a high degree of phenotypic similarity, coupled with an appreciable dissimilarity from other assemblage of the same general kind.

Every assemblage of individuals shows some degree of internal phenotypic diversity because of genetic variation. Ideally, species should be characterized by complete description of their phenotypes and genotypes. The influence of evolutionary criteria (*phyletic* classifications) on taxonomy during the post-Darwinian period is often thought to be the sole aim of the taxonomist. It is therefore necessary to consider whether other possible aims are valid and, indeed, whether any other approach might lead to classifications of greater value than the purely phyletic. To do so we must first make the distinction between *special* (or *artificial)* classifications and *natural* classifications. A special classification is one made for a single, defined purpose: it assists in finding the answer to a specific question. A well-known example is the classification of enteric bacteria according to the biochemical differential tests, as used by the water bacteriologists. The purpose of this classification is to group together those organisms which may indicate recent faecal pollution of a water supply and to separate these from other similar bacteria which do not have this significance. When a bacterial isolate is identified as falling within a particular group of this system an answer to the question of possible faecal pollution is obtained. A further example is the system of classification, used by medical bacteriologists, which places great weight on the pathogenicity of an organism in separating it from otherwise very similar bacteria, e.g., the anthrax bacillus from 'anthracoid' bacilli such as *Bacillus cereus;* the diphtheria bacillus from other 'diphtheroid' Corynebacteria. The question answered here is whether a fresh isolate is likely to cause disease—a question of paramount importance to the medical bacteriologist.

Such classifications are perfectly valid and perform an important function, but they make no pretence to be natural systems. In special classifications an organism may be separated from its fellows by differing in a single *key* attribute (e.g., toxigenicity) whereas the residue may be grouped under a common taxonomic title (e.g., species name) and yet differ between themselves in several attributes.

The taxonomic logic for guiding during the pre-Darwinian period for natural classification can be traced back to the ideas of Aristotle; in particular to his Logical Division Theory, which governed the ideas of Linnaeus and held sway up to the beginning of the present century. The basic notion was that organisms (or any other items) should be classified according to their *essential nature,* i.e., according to 'what they really are'. This idea is linked to the Aristotelian notion of the *species infimae* the ultimate unit of classification which became the basis of the Linnaean *species.* The *species infimae* was rather analogous to the atom of classical chemistry: it was the smallest unit into which more complex groupings could be broken down by repeated division into components. A classification based on such principles would be *sensu stricto* 'natural' but it is easily applied only to the classification of items which are clearly defined, e.g., geometrical shapes: one could construct a *genus* 'triangle' as a plane figure bounded by three straight lines and subdivide this genus into scalene, isoscales, and equilateral *species.* Here the 'essential natures' are known by definition. When attempts were made to apply this logic to the classification of living organisms taxonomists were faced with two connected difficulties which were really impossible to overcome. The first and most fundamental of these is that Aristotle's principle is one of *deductive* logic and yet taxonomists tried to apply it to situations where only *induction* is possible. we cannot deduce that cats are different from rats, we can only recognize that they differ on the basis of our observations (because we do not know the essential nature of 'cat' or 'rat').

The second difficulty is that of biological variation which makes the decision of which attributes are more 'essential' than others even more likely to be arbitrary. Following the publication of Darwin's works on the origin of species, the earlier approach to classification was replaced by one that was thought to be at least equally 'natural', viz., the phyletic system. Once the doctrine of evolution had been accepted it seemed reasonable to argue that organisms of similar 'essential nature' would have shared common lines of descent. The great advantage to taxonomists of the phyletic approach was that speculation about which attributes reflected most accurately the essential nature of organisms was replaced by decisions based on more tangible evidence such as fossil records. Even so, difficulties still remain. To mention only three: (1) fossil records are seldom adequate; (2) biological variation (both phenotypical and genetical) still poses the problem of the taxonomic level at which organisms are to be separated from each other; and (3) the *homology* of various structures or other attributes is often in doubt.

The problem of convergent evolution and homology raises a question of fundamental importance to the formulation of the aims of natural classifications. The lack of fossil evidence makes it much more difficult, if not impossible, to decide whether apparently similar microorganisms have evolved from a common ancestral organism or whether convergence, due perhaps to the selective pressures of sharing a similar habitat, has been responsible.

For the sake of illustration, let us suppose that we have two bacterial strains that share a large number of *what appear to be* similar attributes. Let us further suppose that we also know that the lines of evolution of these strains converged from very different origins. Would the objective of a natural classification be best achieved by grouping these strains together on the basis of their mutual overall similarity *(phenetic* classification) or by separating them so as to reflect their different origins *(phyletic* classification)? An argument for the phyletic approach might be that this best reflects the 'essential natures' of the two strains, to which the counter argument might be that, because of convergent evolution, their 'essential natures' have become similar.

A natural classification should have good predictive value (information content). In contrast, a special or artificial classification yields particular information to the specialized user. If we accept this distinction, it is clear that the phenetic classification would allow the most *general* predictive properties, whereas the phyletic system would offer information that is primarily of use to evolution, i.e., it is a special classification. It is possible to see a resemblance between the grouping of organisms on the basis of phenetic classification and the use of statistical parameters in characterizing sets of data. Again, if the range of bacterial variation were so great that between each 'typical' or modal strain there was an almost continuous gradation of 'intermediate' strains a phenetic classification would still have practical use in much the same way that a histogram may allow us to group and so handle what is in fact a continuous spectrum of data.

8.5 MONOTHETIC AND POLYTHETIC CLASSIFICATIONS: THE CONCEPT OF WEIGHT

Classification based on one or only a few characters are generally called 'monothetic', which means that all the objects allocated to one class must share the character or characters under consideration. Thus, the members of the class of "soluble substances" must in fact be soluble. Classification based on many characters, on the other hand are called 'polythetic'. They do not require any one character or property to be universal for a class. Thus, there are birds that lack wings, vertebrates that lack red blood cells and so on. In such cases a given "taxon", or class, is established because it contains a substantial portion of the characters employed in the classification. Assignment to the taxon is not on the basis of a single property but on the aggregate of properties , and any pair of members of the class will not necessarily share every character. The best phenetic classification is one built on comparisons based upon as many attributes as possible. Organisms which share a large number of attributes would *cluster* together to form a 'natural' group and such groups would separate from each other at 'points of rarity', i.e., at combinations of attributes which never, or very rarely, occur. If 'points of rarity' are absent it means that a continuous spectrum of 'intermediate' types of organism exists and the classification is then arbitrary (but could still be useful). A phenetic classification based on overall similarity is termed *polythetic.*

Monothetic classification is much used in the construction of artificial dichotomous keys for identification of both higher organisms and microorganisms. The essence of such a system is that certain *key characters* are selected, the possession of which automatically places the organism to be identified into a group which is itself subdivided according to the presence (or absence) of other key characters. Once a key character is selected it assumes great *weight* (importance) in determining the classificatory position of an unknown organism and we should therefore inquire whether we are justified in giving some characters more weight than others. It is obvious that, in principle, the use of key characters could nullify the aims of 'natural' classification. For example, if a new strain of bacterium were discovered that differed in a single key character from bacteria already classified together in a group, and yet had a large number of characters in common with that group, we should be forced to place the strain in a separate group according to the monothetic system, whereas it would obviously join the existing group in a polythetic system.

It is easy to justify the use of certain key characters in artificial classifications, since they may reflect the very criteria that were used in setting up the classification. For example, a special classification based on the criterion of *pathogenicity* would justifiably separate *Corynebacterium diphtheriae* from closely related 'diphtheroid' bacilli on the sole key character, toxigenicity, which thus takes on overriding weight. In the case of natural (phenetic) classifications the justification of weighting certain characters is less easy. One possible justification is in cases where we know that certain characters are homologous whereas we are unsure about others. Here we may logically argue that greater weight should be given to the homologous characters in deciding the classification. A second possibility is to argue that more weight should be given to those characters that are strongly correlated with others, a single one of these could then be used as a key character; e.g., a Gram-positive reaction in bacteria usually shows correlation with cell-wall structure, penicillin-sensitivity, sensitivity to basic dyes, etc. Two things follow from this example: First, that the same weighting would be obtained by giving *equal* weight to each of the individual correlated characters, which would then act in concert in influencing the classificatory position. Secondly, that if we eventually found that all of these correlated characters stemmed from a single genetical feature then their weight would disappear since all would be expressions of the same thing. However, we are usually in doubt about the homology of apparently similar characters in microorganisms, nor do we at present know the precise genetical reasons for observed correlations between characters. There is, therefore, an increasing trend in microbial taxonomy towards the idea that, in our position of ignorance, the best natural classification is one based upon comparison of microorganisms with respect to as many characters as possible, *each character being given equal weight* in contributing to the grouping and separation of different organisms (i.e., a polythetic system). Once such a classification has been made.

It is then possible to search for key characters which may be of use in a method of *identification.* It is, however, still unlikely that single key characters could be used as in the familiar dichotomous system, rather a set of such characters would have to be examined together in order to narrow down the possible classificatory location of an unknown organism. The idea of phenetic classification based on characters of equal weight is not new and it is now common to apply the term *Adansonian* to such classifications.

8.6 NUMERICAL TAXONOMY

Numerical taxonomy aims at a more objective system of classification. Numerical taxonomy typically invokes a number of criteria at once. The reason for this is that if only one criterion was invoked at a time there would be a huge number of taxonomic groups, each consisting of only one of a few microorganisms. The purpose of grouping would be lost. By invoking several criteria at a time, fewer groups consisting of larger number of microorganisms result. The groupings result from the similarities of the members with respect to the various criteria. A so-called similarity coefficient can be calculated. At some imposed threshold value, microorganisms are placed in the same group.

Numerical taxonomy owes much to the availability of high-speed digital computers and different kinds of software available, and interest in its application to bacterial classification. Normally, the term *Numerical Taxonomy* is applied to systems of classification which are basically Adansonian but in which the degree of similarity of organisms is assessed in quantitative, rather than merely qualitative terms. There are many advantages in having some numerical estimate of the degree of phenetic similarity or difference between a pair of organisms, of which the most obvious is that it can provide a rational basis for deciding the levels of taxonomic rank. There is at least as much difference between the 'species' of certain genera as there is between the 'genera' themselves.

This started originally with the adoption of the Adamson principle that all properties used for classification should be given equal weight. As many diagnostic characters as possible are used for numerical analysis, and these are formulated as yes or no alternatives (given (+) and (–) signs). Multiple correlations are worked out by computer; every diagnostic character of each strain is compared with every diagnostic character of all other strains. The degree of relatedness between strains is a function of the number of similar characters in proportion to the total number of characters examined. The similarities between pairs of strains is then expressed by a similarity coefficient (*S* value), which is defined as

$$S = \frac{a+d}{a+b+c+d}$$

where *a* and *d* are the sums of the character which are common to strains A and B (*a*, both positive, *d*, both negative), *b* is the sum of the characters in which A is positive and B is negative, and *c* is the sum of the characters in which A is negative and B is positive. The calculations yield values between 1 and 0; $S = 1$ means 100% similarity, i.e., identity, and $S < 0.02$ means complete unrelatedness. The values are entered on a similarity matrix, or they can be expressed as a dendogram (similar to a phylogenetic tree). Numerical taxonomy, however, is not related to phylogeny.

Microbiologists, particularly bacteriologists, have long felt the state of microbial taxonomy to be unsatisfactory. The widely used classification of bacteria (embodied in *Bergey's Manual of Determinative Bacteriology)* is a mixture of phenetic classification (but based on very different numbers of character comparisons in the different groups) and a quasi-evolutionary approach (e.g., the type of flagellation is used in this way by analogy with the classification of protozoa). Moreover, the classification is arranged in the familiar Linnaean hierarchical system and yet it is obvious that the criteria applied to what constitutes a *species* are very different in the different' genera' (e.g., the serotypes of *Salmonella* are given specific rank, whereas those of the pneumococcus are described as types of a single species *(Diplococcus pneumoniae)*. Again, the weighting of certain features results in the classification of some organisms in groups with which they have very little overall similarity (e.g., *Corynebacterium pyogenes)*.

These criticisms do not indicate that the present system is useless (which is certainly not true indeed) but rather that a more uniform approach based on Adansonian principles would almost certainly be more self-consistent and therefore a better natural classification. One disturbing aspect of the present system is that if a group of bacteria is re-examined across a set of criteria (characters) completely different from those already employed in making the existing classification, it is possible that the classification may have to be radically altered in order to accommodate the new information. This instability is unlikely to be a feature of the Adansonian approaches.

8.7 THE TECHNIQUES OF NUMERICAL TAXONOMY

There are several distinct approaches to Numerical Taxonomy, but all start by:

1. Collecting the organisms, or groups of organisms, to be compared, which are now known as *Operational Taxonomic Units* (OTUs).

2. Observing these OTUs for presence or absence (or quantity) of a large set of *characters*.
3. Drawing up of a table of OTUs *versus* characters.

A character is usually defined as an attribute about which a single statement can be made, e.g., 'present' or 'absent' or some quantitative measurement. It is important to give careful thought to what constitutes a single character before drawing up the OTU × character table. Some attributes are obviously not proper characters, e.g., the number of the OTU in the collection. Other apparent characters may not be permissible because they are redundant, i.e., are expressions of an already listed character. For instance, if an OTU ferments both glucose and sucrose with the formation of acid and gas this may generate three distinct characters, viz., acid from glucose; gas from glucose; sucrose fermented. It is improper to score 'gas from sucrose' as a separate character if we know that the fermentation of sucrose involves an initial hydrolysis to glucose, which is subsequently fermented to acid and gas.

Furthermore, it is essential to the principle of Numerical Taxonomy that each of the OTUs should be examined across the complete set of characters, so that true comparisons may be made. Care must be taken, however, not to make comparisons that are illogical. Suppose that one OTU ferments glucose to acid and gas, whereas a second OTU does not ferment glucose at all. In the case of the first OTU we may score a positive character for each of the attributes; acid from glucose; gas production. However, with regard to the second OTU we may score a negative character for lack of production of acid from glucose, but it is now illogical to score a result for 'gas production' since this depends on the prior formation of formic acid which we have already noted as absent. We therefore score 'No Comparison' (NC) for gas production by OTU number 2, which means that this character cannot be used when comparing the similarity of OTU number 2 with any other OTU.

Further questions are prompted by practical considerations, such as:

1. Since observation of characters is necessarily carried out under the artificial conditions of the laboratory, can we make a true comparison of microorganisms which might behave differently in their natural environment?
2. If we have among the OTUs some organisms that can carry out certain reactions at one temperature of incubation but not at a higher in comparison with organisms that can carry out the same reaction only at the higher temperature, should we then use different temperatures of incubation in order to correctly characterize the different OTUs?

The answer to the first question is that a comparison of microorganisms under laboratory conditions (1) is the 'best we can do' and (2) according to our practical definition of a 'natural' classification, is satisfactory because other investigators will be observing the microorganisms under similar conditions. The answer to the second question is more difficult. If there are many temperature-sensitive reactions, we may bias the comparison of OTUs, compared under standard conditions, towards an emphasis of dissimilarity when the temperature-sensitivity may be due to only a few underlying causes (which we do not know).

In our position of ignorance of the complete genetical and biochemical bases of observed characters it is generally considered best to compare OTUs over a *rigidly standardized* set of tests. Although it is almost inevitable that certain of these conditions will introduce bias when measuring the degree of similarity between pairs of OTUs, this course of action is adopted for two chief reasons: (1) practical expediency; (2) if sufficient characters are observed the bias should be 'diluted out' in much the same way that an arithmetic mean is not greatly affected by a few aberrant data, especially when they occur on both sides of the mean. Of course, tests should not be used to generate characters when it is *known* that bias is inherent in the test condition. For example, we may adjust the sensitivity of a test for urease production so that it is read as positive only with those *Enterobacteriaceae* that we call *Proteus* spp. To use this test in a phenetic comparison of *Enterobacteriaceae* would obviously introduce bias, since we have prejudged the issue by distinguishing certain species as urease positive beforehand. Such a test, however, would be perfectly valid if applied to unrelated organisms. The kinds of morphological, structural, and metabolic attributes commonly used as classificatory characters in descriptions of the various microorganisms. Other potentially valuable sources of characters include:

(1) cell-wall chemistry,
(2) electrophoretic studies on esterases and other soluble proteins,
(3) infra-red adsorption spectra,
(4) DNA base composition, and
(5) gas chromatography of cell pyrolysis products, etc.

It is clear that comparisons of OTUs based on a large number of characters are likely to be more accurate (free from bias) than comparisons based on only a few characters. *How many* characters should we observe? Guidelines to the answer may be obtained from elementary probability theory, which tells us that we are most likely to succeed in distinguishing different organisms when the number of characters is of the same order as the number of OTUs, and that we should have limited confidence in an S type Similarity Coefficient calculated on the basis of less than 50 characters.

A special difficulty may exist when an attempt is made to compare organisms that have very different growth rates under standardized conditions (e.g., pathogenic and saprophytic *Mycobacteria*). There is clearly the possibility of bias due to comparison of characters that depend on metabolic rate when similarities are calculated after an incubation period that is suboptimal for the slower growing strains. We may either incubate all strains so that the reactions of the slowest grower are realized; when difficulties may arise due, for example, to alkaline reversion in carbohydrate fermentation tests with the fast growing strains, or we may have recourse to special methods of calculation that attempt to separate effects due to *Vigour* (growth rate) from that due to *Pattern*.

8.8 METHODS OF COMPARING SIMILARITIES

After an OTU × character table has been compiled, all possible pairs of OTUs are compared and their similarities computed. There are three basic methods by which measures of similarity may be computed, only one of which has been much applied to microorganisms. These are:

1. Correlation coefficients.
2. Measures of taxonomic distance.
3. Similarity coefficients (S).

The first two methods have the advantage that characters which are expressed as quantitative data may be more or less directly incorporated into the calculations of similarity.

The correlation coefficients are closely related to the commonly used statistic *r*, which expresses the degree of correlation between two sets of bivariate data and can vary from +1 (absolute correlation), through 0 (no correlation at all), to –1 (absolute negative correlation). Thus, two organisms that were absolutely identical in all characters studied would generate a coefficient of +1, two organisms that were absolutely opposite in every character (if this were possible) would generate a coefficient of –1, whereas a coefficient of 0 would indicate no correlation of the characters of the first organism with those of the second.

Measures of taxonomic distance attempt to plot the relative positions of the OTUs in multi-dimensional space (one dimension for each character studied) in such a way that if two OTUs were identical their *mean taxonomic distance* would be 0 whereas if they were absolutely dissimilar their mean taxonomic distance would be +1. However, it is the similarity coefficient (S) that have found most application in studies of microbial classification mainly owing to the ease with which they can be computed and the results handled in subsequent stages of the classification. These 8 coefficients require that the character data must be coded in binary form, i.e., 1 (+) for the possession of a character, 0 (–) for the absence of a character, and NC for 'No Comparison'. It follows that quantitative data must be broken down into a set of single characters, and there are two chief methods of doing so, viz., the additive and the non-additive methods. Suppose that we have three OTUs one of which produces no penicillinase, a second produces a small quantity of the enzyme, and the third a large amount under comparable conditions, i.e.,

OUT	Penicillinase
A	–
B	+
C	++

In the additive method of coding we may decide as follows:

OTC	Character		
	a	b	c
A	–	NC	NC
B	+	+	–
C	+	+	+

Here character *a* codes for presence or absence of the enzyme, *b* codes for production of a small amount, and *c* for an *additional* amount. However, because we cannot distinguish *a*+*b*+ from merely *a*+ we should probably delete character *b* altogether since it contributes no additional information. The same data coded by the non-additive method gives:

OTC	Character		
	a	b	c
A	–	NC	NC
B	+	+	–
C	+	NC	+

Here character *a* codes for 'production of penicillinase', *b* codes for 'production of + penicillinase', and *c* codes for 'production of + + penicillinase' in a non-additive fashion. OTU C must therefore be scored NC for *b* since production of *a* + + quantity would mask production of a lesser amount. Here again character *b* does not give any additional information to that provided by character *a;* accordingly, character *b* would be deleted with the result that, in this simple example, the results of codings are identical by the two methods, viz.

OTC	Character	
	a	c
A	–	NC
B	+	–
C	+	+

However, if we consider a fourth OTU (D) that produces an even larger amount of enzyme (+ + +) we should obtain:

OTC (D)	Character		
	a	c	d
Additive method	+	+	+
Non-additive method	+	NC	+

In general, the difference between the two methods increases as the number of characters allotted to the quantitative data increases. Since the additive method generates a greater number of comparisons

It tends to overemphasize differences which could be due to differences in growth rate, etc. (i.e., vigour), and so tends to bias the *S* value in the direction of dissimilarity. For this reason the nonadditive approach is generally preferred.

Once the OTU × character table has been drawn up it is possible to represent the comparison of a pair of OTUs thus:

		OTU B: 1	OTU B: 0
OTU A	1	α	β
	0	γ	δ

where *a* represents the total number of characters for which both A and B are scored +, β represents the total number of characters for which A is scored + but B is scored –, and so on. Thus *a* and δ represent the number of characters on which A and B are scored similarly, whereas β and γ represent the number of unmatched characters. 'No Comparisons' are ignored in making these entries. Such tables can be drawn up for all possible pairs of OTUs.

There are two chief ways in which similarity coefficients have been calculated for application to microbial classification. One, known as S_{SM}, includes both positive and negative matches in calculating the degree of similarity, thus:

$$S_{SM} = (\alpha + \beta)/(\alpha + \beta + \gamma + \delta)$$

The other, known as S_J, bases the comparison only on the positive matches, thus:

$$S_J = \alpha/(\alpha + \beta + \gamma)$$

The point at issue in the choice between the two methods is whether two 'absences' is a valid criterion of similarity. In general, S_{SM} is currently favoured on the grounds that for many qualitative characters the coding as '+' or '–' is arbitrary. For example, penicillin-sensitivity may be scored as either '+' or , '–' according to whether one thinks of resistance as an active or passive phenomenon. The danger in including negative matches is that it is possible to bias values of *S* towards excess similarity by choosing a large number of features which the organisms do not possess. However, this applies also to some positive characters and here again it is hoped that introduced biases are 'smoothed out' by observing a sufficiently large number of characters. It is usual to delete as redundant any character which is *uniformly* positive or negative (apart from NC entries) for all OTUs under study, otherwise bias towards excess similarity would certainly occur.

It is obvious that both forms of *S* may vary from 0.000 (absolutely no matches) to 1.000 (100 per cent matches). Moreover, the dependence of *S* on the number of matches is absolutely linear, e.g., if on the basis of 100 characters two OTUs were 100 per cent similar (*S* = 1.000) a third OTU which had a single mismatch with either of the former would drop its value by 1 per cent (*S* = 0.990). This feature constitutes one of the large advantages of the similarity coefficient (particularly S_{SM}) over the other methods of comparison outlined above: it is possible to grasp the meaning of differences between S-values very easily.

When S-values have been calculated for all possible pairs of OTUs (and here the contribution of the high-speed computer is evident) they are tabulated in a *similarity matrix*. This is a table of OTUs × OTUs, which is symmetrical about its principal diagonal, since the *S* value between OTUs A and B is obviously the same as that between B and A. The values on the principal diagonal are all 1.000, since these consist only of self-comparisons. The similarity matrix is therefore usually recorded in a triangular form, omitting these redundant entries.

At this point it may be helpful to introduce a very simple hypothetical example where five OTUs are compared over only ten characters.

OTU X Character Table

	OTU				
Character	**A**	**B**	**C**	**D**	**E**
1	1	1	0	0	0
2	1	1	0	0	0
3	1	0	0	0	0
4	0	0	1	0	1
5	1	1	1	1	1
6	0	0	1	1	1
7	1	1	0	0	1
8	1	1	0	1	1
9	0	0	1	0	1
10	0	0	0	1	1

Similarity Matrix

Values of S_{SM}

OTU	A	B	C	D	E
A					
B	0.9				
C	0.2	0.3			
D	0.4	0.5	0.6		
E	0.3	0.4	0.7	0.7	

8.8.1 Cluster Analysis

After numerical estimates of the degrees of similarity between all possible pairs of OTUs have been generated, the next step is to form the groups (or clusters) which are the basis of the final classification. When using *S* coefficients there are three main ways in which this operation, known as cluster analysis, may be tackled:

1. Single linkage
2. Average linkage
3. Total linkage.

The method that has been most applied to microbial classification is that of *single linkage.* Although it has certain disadvantages (see below) its ease of computation and manipulation makes the method eminently suitable, at least for preliminary studies. Its use may be illustrated by reference to our simple example.

First, the similarity matrix is scanned at a high level of *S* and the pairs of OTUs that have mutual S-values *at least as great* as the scan level are listed. Suppose we begin by scanning at a level of $S = 1.000$ (absolute similarity), no such values appear in our example above. We next decrease the scan level by an arbitrarily selected amount that has to be chosen by reference to the scatter of S values actually obtained (or to some other criterion). In our example a decrement of 0.2 (20 per cent) would seem suitable. Thus, the next scan level becomes $S = 0.8$ and we obtain a single pair of OTUs.

Level	OTU-pairs
$S = 0.8$	A, B

Decreasing by a further amount of 0.2 we list further entries:

Level	OTU-pairs
$S = 0.6$	A.B; C, D; C,E; D,E

At this level of scan the principle of clustering by single linkage can be applied; i.e., OTU-pairs are fused to form a single cluster if anyone OTU of one pair has an S value at least as great as the scan level with anyone OTU of a second pair (or of an already existent cluster). To return to the example, we see that the last three OTU pairs satisfy this criterion and fuse into a single cluster, whereas the pair A, B remains isolated:

Level	Clusters
$S = 0.6$	A,B; C, D, E

Proceeding, we obtain:

Level	Clusters already formed	New OTU pairs
S = 0.4	A,B; C,D,E;	A,D; B,D; B,E;

The new OTU pairs fuse into a single cluster (A, B, D, E;) by the criterion of single linkage, but this cluster has elements in common (at 8 = 0.4) with the two existing clusters. Therefore the five OTUs form into a single group at 8 = 0.4 and the clustering process ends.

It is now possible to represent the results of clustering by means of a *dendrogram*, or 'family tree', resembling that of the usual hierarchical classifications.

Dendrogram

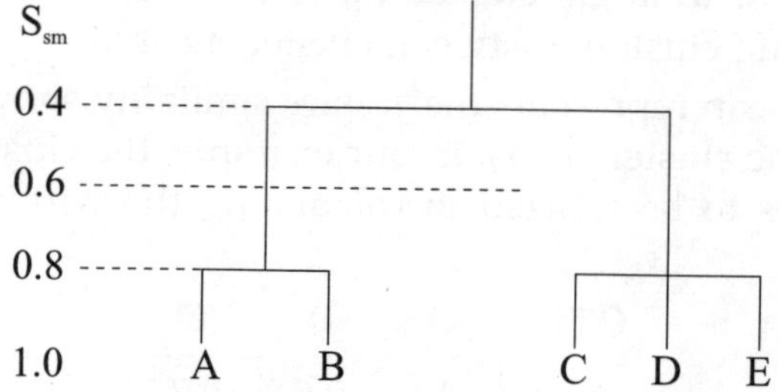

Although this form of representing the results of a cluster analysis is exceedingly useful, it is relevant to point out two distortions inherent in it. One is the fact that the points of fusion of branches of the dendrogram are shown as occurring at single levels of S, whereas the actual S values causing the fusion occur anywhere between the limits set by the arbitrarily chosen decrement. The second is that a true spatial representation of the relations between the various OTUs and clusters would require multi-dimensional space; distortion is therefore inevitable in a two-dimensional dendrogram.

Nevertheless, the method allows a tentative classification of the OTUs having the great advantage of being based on numerical estimates of the levels at which differences and similarities appear. At what level we decide to label members of a cluster 'strains', 'species', 'genera', and so on (or to abandon these terms) is still a matter of choice and agreement, but we now have a numerical 'yardstick' to guide us in this decision.

The method of clustering by single linkage has an inbuilt disadvantage which could make for grouping. Suppose the cluster A, B, C, D formed because A linked with B, B with C, and C with D. It is evident that A might be quite dissimilar from D and yet would still be clustered with it. In fact, it is easy to show that if we know $S_{A,B}$ and $S_{A,C}$ (where these are S_{SM} values) then $S_{B,C}$ may have a *minimum* value equal to

$$1 - (S_{A,B} + S_{A,C})$$

When $S_{A,B} = S_{A,C} = 0.5$, $S_{B,C}$ can be as low as zero, as is obvious from the following example:

Character	OTU A	OTU B	OTU C
1	1	1	0
2	1	1	0
3	1	1	0
4	1	0	1
5	1	0	1
6	1	0	1

Fortunately, in practice good results are commonly obtained in spite of this potential snag and a method is available that allows a check on the occurrence of serious distortion due to single linkage. In order to understand the nature of this check it is necessary to consider what is meant by *mean similarity*. Mean similarity may be computed either between the members of a single cluster (i.e., within-cluster mean) or between the members of two separate clusters (between-cluster mean).

The within-cluster mean represents the *average* similarity shown between all possible pairs of OTUs within the cluster. Thus, in our example, the cluster C, D, E was formed at $S = 0.6$. The S-values to be utilized in calculating the within-cluster mean for this example are:

OTU	C	D	E
C	1	0.6	0.7
D	0.6	1.0	0.7
E	0.7	0.7	1.0

Two forms of the within-cluster mean may be obtained. The 'square' mean (Γ mean) is the average of all 9 values in the square matrix shown above, i.e., $\Gamma = 7/9 = 0.7$. The 'triangle' mean (Δ mean) ignores the redundant comparisons and the self-comparisons, and is therefore the mean of the 3 values in the triangle, i.e., $\Delta = 2.0/3 = 0.6$. The two sorts of within-cluster mean bear a simple relation to each other: Δ is less than Γ, but the two become similar as the number of OTUs in the cluster increases.

If we compare the mean values obtained above with the level of *S* at which the cluster was formed ($S = 0.6$) we see that the means are greater than the clustering level. This indicates that the cluster is homogeneous with respect to the mutual similarities between the individual members. If OTUs had been included by single linkage that showed low levels of *S* with some existing members of the cluster, i.e., if the cluster had become heterogeneous, then the within cluster mean would have been depressed below the clustering level by an amount dependent upon the degree of heterogeneity. It is this feature that provides a check on the validity of single linkage methods of analysis.

The between-cluster mean has only one form of computation. Here each OTU in the first cluster must be compared with each OTU in the second cluster. In the example two clusters exist at $S = 0.6$:

1. A, B;
2. C, D, E;

The between-cluster mean is obtained from the rectangular matrix of *S* values:

		Cluster (1)	
		A	**B**
	C	0.2	0.3
Cluster (2)	**D**	0.4	0.5
	E	0.3	0.4

Here there are no redundancies and the between-cluster mean is $\frac{2.1}{6} = 0.35$; an average measure of the degree of similarity between the two clusters.

Between-cluster means may themselves be used as a basis for clustering: the so-called method of *average linkage* referred to above. The essence of this approach is that, at each level of clustering, individual OTUs join existing clusters, and existing clusters fuse together, only if the *mean* similarity between the OTU and its potential cluster, or the *mean* similarity between two clusters, is at least as great as the chosen level of S. This approach largely removes the danger, inherent in the single-linkage method, of creating clusters which appear to be more homogeneous than they really are; the check on the within-cluster mean may be incorporated as an additional safeguard.

There are a number of different techniques that have been used to apply the method of average linkage to classification studies but all of them require more labour, and more skilful computer programming, than does the method of single linkage-often without producing a very different result.

The method of *total linkage* represents a further extension of the attempt to ensure homogeneous clusters. In this approach the criterion of linkage is that an OTU is allowed to join a cluster only if it has the required level of S with *each* existing member, and two clusters fuse only if *each* member of the first cluster has the required level with *each* member of the second. This approach has been little used in microbial classification.

8.8.2 The Matches Hypothesis

The advantage of having a *numerical* estimate of similarity for use as a guide in making decisions on classification has already been stressed. Numerical (Adansonian) Taxonomy offers a second substantial advantage over methods that rely on qualitative, or on arbitrarily weighted, judgments. This is embodied in the *matches hypothesis,* which supposes that there is some true measure of similarity which could be computed if every possible character could be taken into account, and that the deviation from it of an actual calculated S value (based on a 'sample' of all the possible characters) will be accounted for by sampling error. Thus, a second estimate of S made between the same pair of OTUs, but based on an independent set of characters, should tend to give a value similar to that first obtained, i.e., estimates should be self-consistent. This notion is similar to that used in mathematical statistics where estimates of the true mean (μ) of a normally distributed population, obtained from the observed means (x) of randomly selected samples, cluster around μ in a manner that is predicted by the sampling error (variance).

With regard to S values the matches hypothesis seems to be borne out in practice, and the sampling error is approximated by the prediction of the Binomial Distribution:

$$\text{Standard deviation of } S = \sqrt{[S\,(1 - S)/N]}$$

Here S is taken as the probability of occurrence of a 'match' and N is the member of comparisons (characters) observed.

The advantage of self-consistency is that further studies carried out on groups of organisms already classified according to the principles outlined above are unlikely to necessitate radical changes in classification; a property that is not true for a number of existing classificatory schemes, where a new study may dictate substantial re-arrangement of taxa.

8.9 APPLICATIONS OF NUMERICAL TAXONOMY TO MICROORGANISMS

During the past decade various investigators have applied Numerical Taxonomic methods to different groups of microorganisms. These include: *Chromobacterium, Bacillus, Micrococci, Streptococci, Corynebacteria, Mycobacteria, Basidiomycetes,* and root-nodule bacteria, to mention but a few.

The results of these studies tend, in general, to confirm the prediction of the matches hypothesis, i.e., where the existing classification has been largely phenetic and based on many characters it is confirmed, with minor deviations, by the numerical study. However, even in these cases the great advantage of having some sort of quantitative criterion on which to base points of separation and combination is evident. In examples where the existing classification has been biased by reliance on a few weighted characters the numerical studies have shown up discrepancies. For instance, in a study of pigmented bacteria, it was found that the S-value clusters with the Gram-positive cocci more closely than with *Corynebacterium diphtheriae; Proteus* is as different from the *Salmonella-Escherichia* group as it is from *Bacillus.*

It will be obvious from the outline of Numerical Taxonomy given above that an overall classificatory study on microorganisms in general can be carried out only by actually comparing representative organisms over a wide range of characters. The problems of data collection and of computation make this a formidable task and the studies so far have been largely confined to more or less well-defined groups of micro-organisms. It is not entirely satisfactory to use the usual recorded descriptions as a source of data for Numerical Taxonomic studies. Often the characters recorded for the different organisms—even within a classificatory group—either do not belong to the same set, or are incomplete for any one organism, or have been obtained under different conditions. Moreover, the descriptions often record a result as 'variable' or show a range when it is the actual responses of representative organisms that are important.

Attempts have been made to gain an idea of how Numerical Taxonomy compares existing wide classifications by using published data. An example for bacteria is shown, in the form of a dendrogram, in Figure 8.2. Here, three main groups can be distinguished; the Gram-positive cocci, the Gram-negative rods, and the 'Actinomycetales'. When examined in detail, however, various examples of divergence from accepted classification become evident, e.g., *Corynebacterium pyogenes.* Although at present microbiologists will continue to use existing classifications in order to make possible communication of information, nevertheless the increasing interest that is being shown in Numerical Taxonomic studies gives promise of a more consistent and more rational (and, therefore, more generally useful) scheme of microbial classification.

8.10 STRATEGIES USED TO IDENTIFY MICROBES

Over the past century microbiologists have searched for more rapid and efficient means of microbial identification. The identification and differentiation of microorganisms has principally relied on microbial morphology and growth variables. Advances in molecular biology over the past 25 years have opened new avenues for microbial identification and characterization.

The traditional methods of microbial identification rely solely on the phenotypic characteristics of the organism. Bacterial fermentation, fungal conidiogenesis, parasitic morphology, and viral cytopathic effects are a few phenotypic characteristics commonly used. Some phenotypic characteristics are sensitive enough for strain characterization;

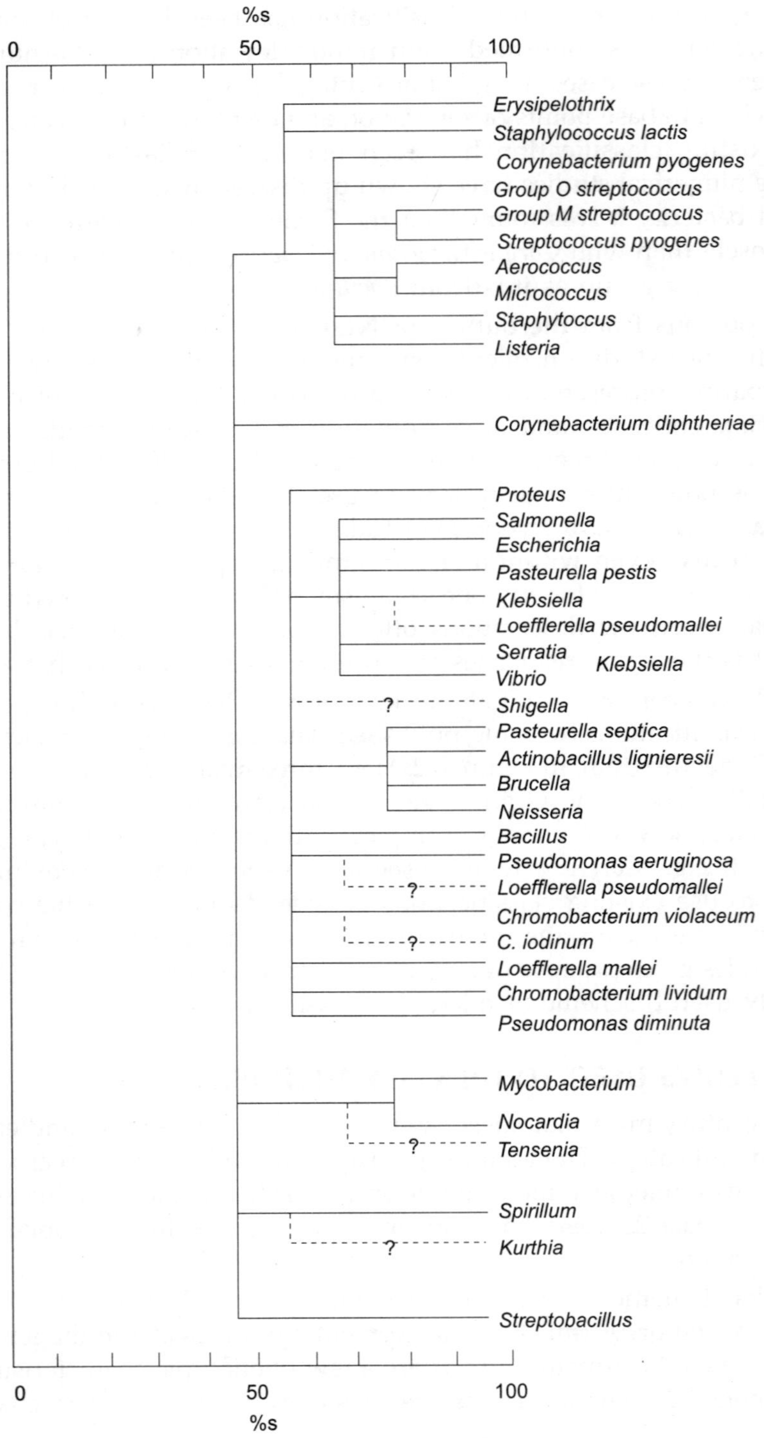

Figure 8.2 Taxonomic dendrogram of a wide range of bacteria.

these include isoenzyme profiles, antibiotic susceptibility profiles, and chromatographic analysis of cellular fatty acids. However, most phenotypic variables commonly observed in the microbiology laboratory are not sensitive enough for strain differentiation. When methods for microbial genome analysis became available, a new frontier in microbial identification and characterization was opened.

Early DNA hybridization studies were used to demonstrate relatedness amongst bacteria. This understanding of nucleic acid hybridization chemistry made possible nucleic acid probe technology. Advances in plasmid and bacteriophage recovery and analysis have made possible plasmid profiling and bacteriophage typing, respectively. Both have proven to be powerful tools for the epidemiologist investigating the source and mode of transmission of infectious diseases. These technologies, however, like the determination of phenotypic variables, are limited by microbial recovery and growth.

Nucleic acid amplification technology has opened new avenues of microbial detection and characterization, such that growth is no longer required for microbial identification. In this respect, molecular methods have surpassed traditional methods of detection for many fastidious organisms. The polymerase chain reaction (PCR) and other recently developed amplification techniques have simplified and accelerated the *in vitro* process of nucleic acid amplification. The amplified products, known as amplicons, may be characterized by various methods, including nucleic acid probe hybridization, analysis of fragments after restriction endonuclease digestion, or direct sequence analysis. Rapid techniques of nucleic acid amplification and characterization have significantly broadened the microbiologists' diagnostic arsenal.

8.11 METHODS FOR BACTERIAL IDENTIFICATION

Methods of bacterial identification can be broadly delimited into genotypic techniques based on profiling an organism's genetic material (primarily its DNA) and phenotypic techniques based on profiling either an organism's metabolic attributes or some aspect of its chemical composition (Figure 8.3). Genotypic techniques have the advantage over phenotypic methods that they are independent of the physiological state of an organism; they are not influenced by the composition of the growth medium or by the organism's phase of growth.

Phenotypic techniques, however, can yield more direct functional information that reveals what metabolic activities are taking place to aid the survival, growth, and development of the organism. These may be embodied, for example, in a microbe's adaptive ability to grow on a certain substrate, or in the degree to which it is resistant to a cohort of antibiotics. Genotypic and phenotypic approaches are complementary and use different techniques. However, this division is historical; we predict that as molecular-based identification matures, there will be more and more overlap in the information obtained using different methodologies.

Genotypic microbial identification methods can be broken into two broad categories: (1) pattern- or fingerprint-based techniques and (2) sequence-based techniques. Pattern-based techniques typically use a systematic method to produce a series of fragments

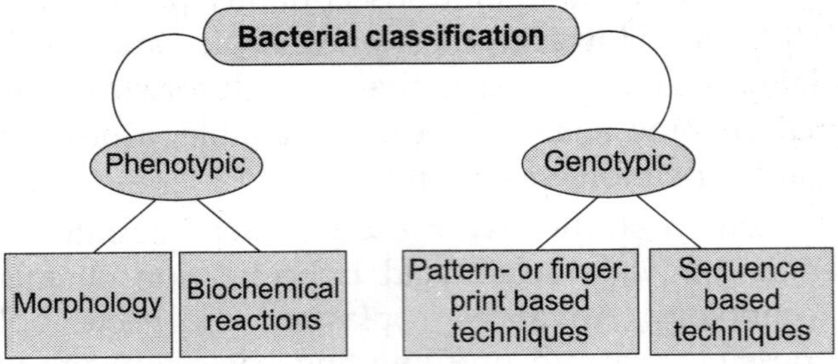

Figure 8.3 Basis of bacterial classification.

from an organism's chromosomal DNA. These fragments are then separated by size to generate a profile, or fingerprint that is unique to that organism and its very close relatives. With enough of this information, one can create a library, or database, of fingerprints from known organisms, to which test organisms can be compared. When the profiles of two organisms match, they can be considered very closely related, usually at the strain or species level.

8.12 PHENOTYPIC CHARACTERISTICS TO IDENTIFY MICROBES

Phenotypic characters of bacteria include morphology and biochemical reactions carrying out by bacteria whose results can be viewed. Morphological characteristics include colony morphology such as colour, size, shape, opacity, elevation, margin surface texture, consistency, etc. These characters are observed after the incubation period on the cultures on the solid media. In liquid cultures, we can observe the pellicle formation and sediment formation. Biochemical characteristics include enzyme production, utilization of particular sugar, aerobic or anaerobic reactions, etc.

Limited information exists on the phenotypic characteristics of bacteria found in biofilm. Both wet-mounted and properly stained bacterial cell suspensions can yield a great deal of information. These simple tests can indicate the Gram reaction of the organism; whether it is acid-fast; its motility; the arrangement of its flagella; the presence of spores, capsules, and inclusion bodies; and, of course, its shape. This information often can allow identification of an organism to the genus level, or can minimize the possibility that it belongs to one or another group. Colony characteristics and pigmentation are also quite helpful. For example, colonies of several Porphyromonas species autofluorescence under long-wavelength ultraviolet light, and Proteus species swarm on appropriate media.

A primary distinguishing characteristic is whether an organism grows aerobically, anaerobically, facultatively (i.e., in either the presence or absence of oxygen), or microaerobically (i.e., in the presence of a less than atmospheric partial pressure of oxygen). The proper atmospheric conditions are essential for isolating and identifying bacteria. Other important growth assessments include the incubation temperature, pH, nutrients required, and resistance to antibiotics. For example, one diarrheal disease agent, *Campylobacter jejuni*, grows well at 42°C in the presence of several antibiotics; another,

Y. enterocolitica, grows better than most other bacteria at 4°C. *Legionella, Haemophilus,* and some other pathogens require specific growth factors, whereas *E. coli* and most other Enterobacteriaceae can grow on minimal media. Most bacteria are identified and classified largely on the basis of their reactions in a series of biochemical tests. Some tests are used routinely for many groups of bacteria (oxidase, nitrate reduction, amino acid degrading enzymes, fermentation or utilization of carbohydrates); others are restricted to a single family, genus, or species (coagulase test for staphylococci, pyrrolidonyl arylamidase test for Gram-positive cocci).

Both the number of tests needed and the actual tests used for identification vary from one group of organisms to another. Therefore, the lengths to which a laboratory should go in detecting and identifying organisms must be decided in each laboratory on the basis of its function, the type of population it serves, and its resources. Clinical laboratories today base the extent of their work on the clinical relevance of an isolate to the particular patient from which it originated, the public health significance of complete identification, and the overall cost-benefit analysis of their procedures. For example, the Centers for Disease Control and Prevention (CDC) reference laboratory uses at least 46 tests to identify members of the Enterobacteriaceae, whereas most clinical laboratories, using commercial identification kits or simple rapid tests, identify isolates with far fewer criteria.

8.13 SEROLOGY

The protein and polysaccharides that make up a bacterium are sometimes characteristic enough to be considered identifying markers. The most useful of these are the molecules that make up surface structures including the cell wall, glycocalyx, flagella and pili. For example, some species of *Streptococcus* contains a unique carbohydrate molecule as a part of their cell wall that can be used to distinguish them from other species. These carbohydrates, as well as any distinct protein or polysaccharide can be detected using techniques that rely on the specificity of interaction between antibodies and antigens. Methods that exploit such interactions are called serology.

Highly specific identification of microorganisms can be obtained by serological techniques. *In vitro* (that is, outside the body and in an artificial environment, such as a test tube), antigens and antibodies react together in certain visible ways. The chemical composition of antigens differ, and therefore, the reactions are highly specific; that is, each antigen provokes an antibody response with that antibody only. When it provokes an antibody response, the antigen is known as an immunogen.

The cell wall of Gram-negative bacteria consists of several layers of various polysaccharides. The periplasm contains peptidoglycan, a copolymer of polysaccharide and short peptides, and a class of β-glucans. In Gram-negative bacilli, the carbohydrate antigens within the wall of the organism are called somatic (associated with the soma, that is, the body of the cell) or "O" antigens (Figure 8.4). Each species has a different array of O antigens that can detect in serological tests. In like manner, those bacilli that are motile also contain characteristic flagellar protein components called "H" antigens

(H is from the German word hauch, which refers to motility). In streptococci, the carbohydrate wall antigens are used to group the organisms by alphabetic designations A through V. Many bacteria also contain antigenic carbohydrate capsules that can be used for identification, the primary example being the pneumococci, whose capsules permit them to be differentiated into more than 80 different types. Exotoxins and other protein metabolites of bacterial cells are also antigenic. The interaction of antibody with antigen may be demonstrated in several ways. Examples of these are latex agglutination, coagglutination, and enzyme-linked assays.

These tests depend on linking antibody to a particle or enzyme in order for a positive reaction to be observed. The fluorescent antibody test is similar to the enzyme immunoassay except that the antibody is linked to a dye that fluoresces when it is reviewed microscopically under an ultraviolet light source. Fluorescent antibody tests can provide rapid diagnosis of infections caused by pathogens that are difficult to grow in culture, or that grow slowly. Thus, they have become popular for detecting such organisms as *Legionella pneumophilia* (the agent of Legionnaires disease), *Bordetella pertussis, Chlamydia trachomatis* and several viruses, directly in patient specimens. A portion of the specimen dried on a microscope slide is treated with the fluorescent antibody reagent, rinsed to remove unbound antibody, and then viewed under a fluorescence microscope with an ultraviolet light source.

In a positive test, bacteria or viral inclusions fluoresce apple green. This test is used in a similar way to identify microorganisms isolated on culture plates or in cell cultures. Bacterial agglutination test is a simpler test which detects O and H antigens of Gram-negative enteric bacilli, (usually *Salmonella* and *Shigella* species and *Escherichia coli*). When the unknown organism isolated in culture is mixed with an antiserum (prepared in animals) that contains antibodies specific for its antigenic makeup, agglutination (clumping) of the bacteria occurs. If the antiserum does not contain specific antibodies, no clumping is seen. A control test in which saline is substituted for the antiserum must always be included to be certain that the organism does not clump in the absence of the antibodies.

Commercially available antibodies are routinely used to specifically identify antigenic proteins from a wide variety of organisms. In some instances, the test may be used only to identify the genus and species of an organism. Examples of this include the cryptococcal antigen agglutination assay and the exoantigen assay for *Histoplasma capsulatum*. Other immunoassays are designed to subtype microbes. Monoclonal antibodies directed against the major subtypes of the influenza virus, as well as the various serotypes of *Salmonella,* are commonly used in speciation. Specific antigenic proteins may be detected by antibodies directed against these proteins in immunoblot methods (Figure 8.5).

Electrophoretic typing techniques have been used to examine outer membrane proteins, whole-cell lysates, and particular enzymes. Several electrophoretic methods are available to examine the protein profile of an organism. Generally, outer membrane proteins and proteins from cell lysates are examined by sodium dodecyl sulfate–

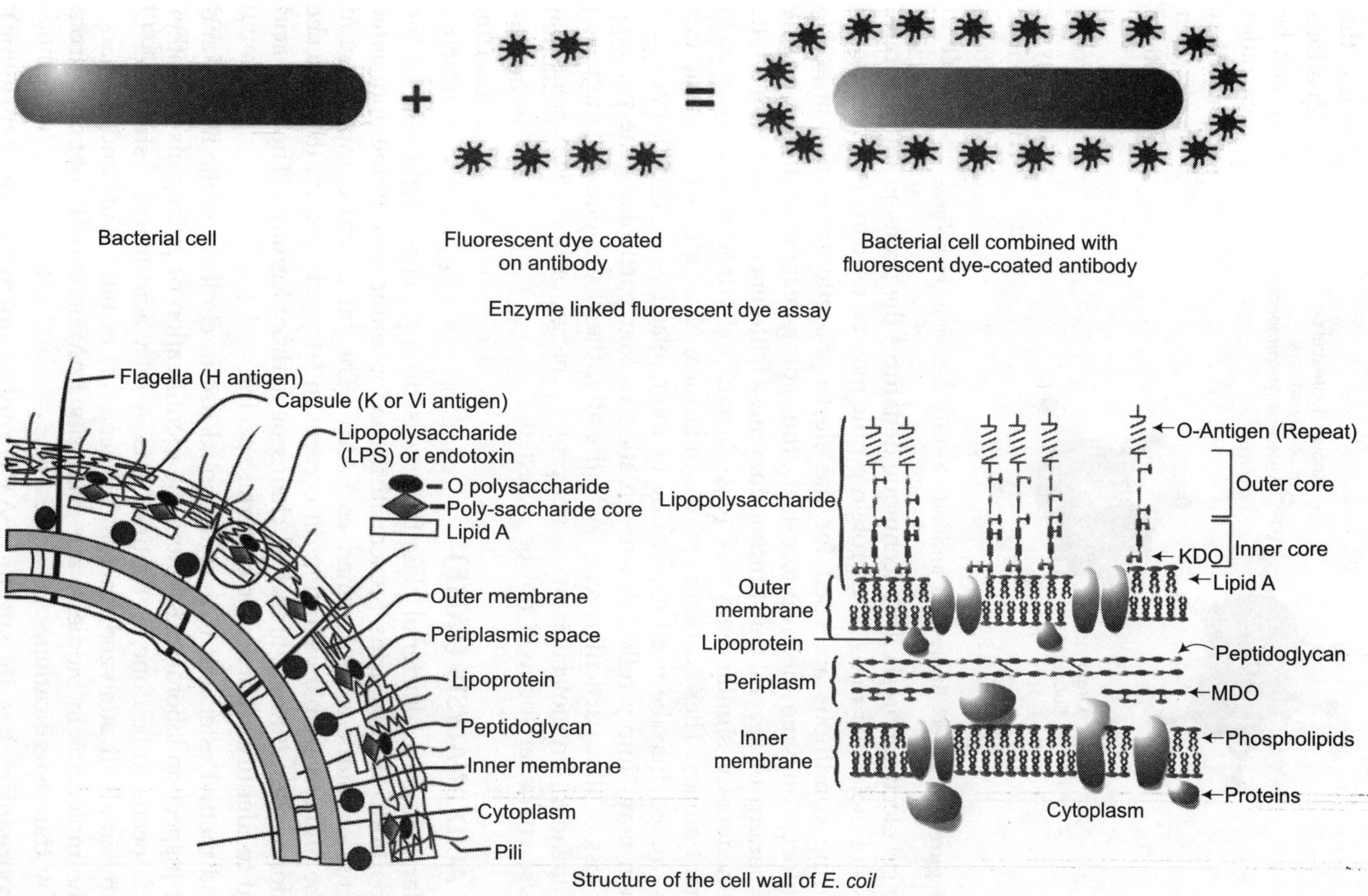

Figure 8.4 Enzyme linked fluorescent dye assay and structure of bacterial cell wall.

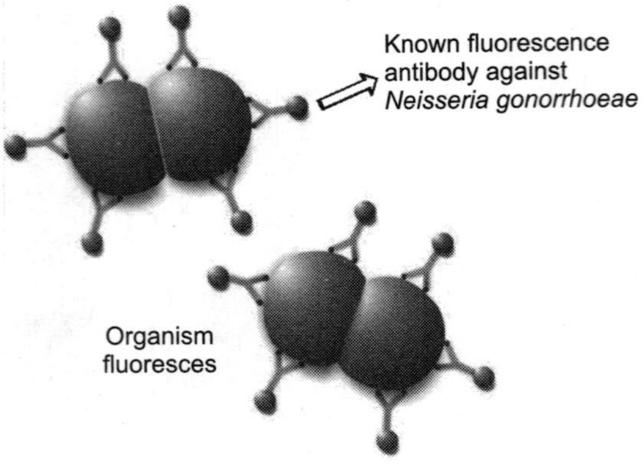

Figure 8.5 Direct fluorescent antibody test for *Neisseria gonorrhoeae.*

polyacrylamide gel electrophoresis. This technique denatures the proteins and separates them on the basis of molecular mass. The protein profile may be used to compare strains.

Non-denaturing conditions are used for the electrophoretic separation of active enzymes. Multilocus enzyme electrophoresis is the typing technique based on the electrophoretic pattern of several constitutive enzymes. Differences in electrophoretic migration of functionally similar enzymes (e.g., lactate dehydrogenase isoenzymes) represent different alleles. These differences or similarities, especially when numerous enzymes are examined, may be used to exclude or infer relatedness. The absence of a particular protein may simply reflect downregulation of that particular gene product, rather than the loss of that particular gene. Additionally, the electrophoretic migration of proteins is dependent on molecular mass, net protein charge, or both. Mutations that do not alter these characteristics will not be detected.

8.14 FATTY ACID ANALYSIS (FAME)

Another popular method of bacterial classification is through characterization of the types and proportions of fatty acids present in the cytoplasmic membrane and outer membranes. This technique is nicknamed as FAME. The fatty acid composition of prokyotes can be highly variable including differences in fatty acid length, the presence or absence of double bond, rings, branched chains or hydroxyl groups. The fatty acid profile can help to identify a particular bacterial species.

For fatty acid methyl ester and is in widespread use in clinical, public health, and food and water inspection laboratories where the identification of pathogens and other bacterial hazards needs to be done on routine basis. A fatty acid methyl ester (FAME) can be created by an alkali catalyzed reaction between fats or fatty acids and methanol (Figure 8.6). The molecules in biodiesel are primarily FAMEs, usually obtained from vegetable oils by transesterification.

Every microorganism has its specific FAME profile (microbial fingerprinting), therefore, it can be used as a tool for microbial source tracking (MST). The types and proportions of fatty acids present in cytoplasm membrane and outer membrane (Gram-negative) lipids of cells are major phenotypic trains.

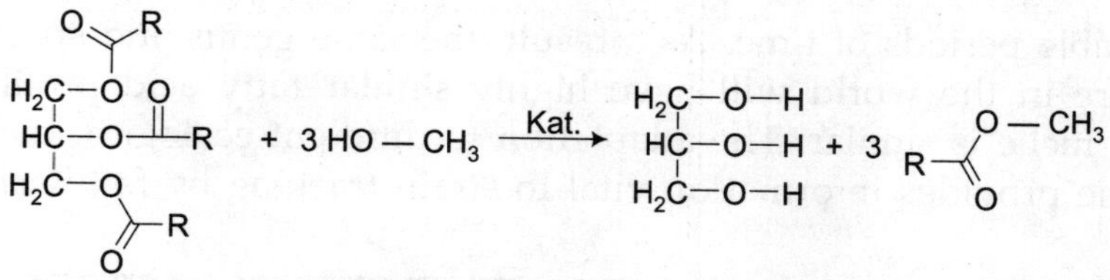

Figure 8.6 Fatty acid methyl ester is formed by an alkali catalyzed reaction.

Clinical analysis can determine the lengths, bonds, rings and branches of the FAME. To perform this analysis, a bacterial culture is taken, and the fatty acids extracted and used to form methyl esters. The volatile derivatives are then introduced into a gas chromatagraph, and the patterns of the peaks help to identify the organism. This is widely used in characterizing new species of bacteria, and is useful for identifying pathogenic strains.

More than 300 fatty acids and related compounds are found in bacteria. The wealth of information contained in these compounds is both in the qualitative differences (usually at genus level) and quantitative differences (commonly at species level). As the biochemical pathways for creating fatty acids are known, various relationships can be established. Thus 16:0 → 16:1 through action of a desaturase enzyme and is a mole-for-mole conversion. Following this, as the bacterial cell becomes physiologically mature, the shift of 16:1 → 17:0 cyclopropane is again a mole-for-mole conversion.

This information suggests that use of the cells in an actively growing stage minimizes the differences between cultures. Use of a 24 + 2 hour culture and harvesting from a rapidly growing quadrant of a quadrant streak plate reduces the differences. Controlled growth temperature and use of standardized commercially available media also contribute to the reproducibility of the fatty acid profile. Branched chain fatty acids (iso and anti-iso acids) are common in many Gram-positive bacteria, while Gram-negative bacteria are composed of predominately straight chain fatty acids. The presence of lipopolysaccharide (LPS) in Gram-negative bacteria gives rise to the presence of hydroxy fatty acids in those genera. Thus, the presence of 10:0 3OH, 12:0 3OH, and/or 14:0 3OH fatty acids indicates that the organism is Gram-negative and conversely, the absence of the LPS and hydroxy fatty acids indicates that the organism is Gram-positive. As a result, it is not necessary to perform the traditional Gram stain prior to FAME analysis. Fatty acid profiles are quite unique for *B. anthracis*, compared with other Bacillus species.

As bacteria frequently exchange plasmids, the system would not work well if such changes did cause alterations in the fatty acid composition. Similarly, treatment with ultraviolet light (a frame-shift mutagen) or point-mutagens such as nitrosoguanidine and ethyl methanesulfonate at levels that kill 99.999% of the cells and create large numbers of auxotrophic and/or motility mutants did not affect the fatty acid profile, as long as the growth rate was relatively normal. This suggests that the fatty acid composition is highly conserved genetically and that significant changes take place only over considerable periods of time. As a result, the same genus and species of bacteria from anywhere in the world will have highly similar fatty acid profiles as long as the ecological niche is similar. The adaptation to different ecological niches over long periods of time provides information vital to strain tracking by fatty acid profiling.

8.15 USING GENOTYPIC CHARACTER TO IDENTIFY MICROBES

The classification of microbes is based on not only how they look but also what they can do. These molecular techniques for characterizing microbial genotypes provide a possible basis of defining a bacterial species (Table 8.1). Molecular microbial taxonomy relies upon the generation and inheritance of genetic mutations that is the replacement of a nucleotide building block of a gene by another nucleotide. Sometimes the mutation confers no advantage to the microorganism and so is not maintained in subsequent generations. Sometimes the mutation has an adverse effect, and so is actively suppressed or changed. But sometimes the mutation is advantageous for the microorganism. Such a mutation will be maintained in succeeding generations.

Table 8.1 Molecular techniques for characterising bacterial genotypes

Approach	*Target*	*Outcome experiment*	*Main limitations*
Sequencing of rRNA			
16s rRNA gene sequencing	16s rRNA gene	16s rRNA gene sequence collection	Bias in NA extraction, PCR and cloning; laborious
RT-PCR	mRNA	Specific gene expression	Bias in NA extraction and PCR
Fingerprinting DDGE, TGGE, TTGE, T-RFLP,	mRNA	Diversity profiles	Bias in NA extraction and PCR
Non-16s rRNA gene fingerprinting	GC fractionation of DNA, cellular fatty acid	Diversity profiles	16s rRNA approaches required for identification
Quantification of 16s r-RNA and its encoding genes			
Dot-blot hybridization	l6s rRNA	Relative abundance of l6s rRNA	Laborious at species level; requires l6s rRNA gene sequence
qReal-time PCR	16s rRNA gene	Relative abundance of 16s rRNA	Laborious and expensive in early stages of development
Fish	16s rRNA	Enumeration of bacterial populations	Laborious at species level; requires 16s rRNA gene sequence data

Contd...

Contd...

DNA microarray technology			
Diversity arrays	16s rRNA genes; antibiotic resistance genes	Diversity profiles	Laborious; expensive; in early stages of development
DNA microarray	mRNA	Transcriptional fingerprints	Bias in NA extraction and NA labeling; expensive
Parallel sequencing technologies	Hundreds to thousands of genes	Sequence of microbial genome	Cost; sequences to compare and identify correctly

Because mutations occur randomly, the divergence of two initially genetically similar microorganisms will occur slowly over evolutionary time (millions of years). By sequencing a target region of genetic material, the relatedness or dissimilarity of microorganisms can be determined. When enough microorganisms have been sequenced, relationships can be established and a dendrogram constructed.

For a meaningful genetic categorization, the target of the comparative sequencing must be carefully chosen. Molecular microbial taxonomy of bacteria relies on the sequence of ribonucleic acid (RNA), dubbed 16S RNA, that is present in a subunit of prokaryotic ribosomes. Ribosomes are complexes that are involved in the manufacture of proteins using messenger RNA as the blueprint. Given the vital function of the 16S RNA, any mutation tends to have a meaningful, often deleterious, effect on the functioning of the RNA. Hence, the evolution (or change) in the 16S RNA has been very slow, making it a good molecule to compare microorganisms that are billions of years old.

The use of the chain reaction has produced a so-called bacterial phylogenetic tree. The structure of the tree is even now evolving. But the current view has the tree consisting of three main branches. One branch consists of the bacteria. There are some 11 distinct groups within the bacterial branch. Three examples are the green non-sulfur bacteria, Gram-positive bacteria, and cyanobacteria **divided** on the basis of ribosomal RNA analysis (16rRNA). Most groups (similar to phylum's) contain a variety of physiological and morphological types of bacteria. This reinforces the idea that phenotypic characteristics are inadequate to define evolutionary relationships between microbial species. Evidence to date places the Archae a bit closer on the tree to bacteria than to the final branch (the Eucarya). There are three main groups in the archae: halophiles (salt-loving), methanogens, and the extreme thermophiles (heat loving). This last group is composed of extreme thermopiles that require elemental sulfur for optimal growth (Chapter 9).

For most members, the sulfur serves as an electron acceptor in anaerobic respiration. Evolution of the eukaryotic line was characterized by periods of rapid evolution interspersed with eras of slow evolution. The accumulation of O_2 in the atmosphere about 1.5 billion years ago seems to correspond to a period of rapid evolution. Small subunit ribosomal DNA sequences were determined for 17 strains belonging to the genera *Alteromonas, Shewanella, Vibrio,* and *Pseudomonas,* and their sequences were analyzed by phylogenetic methods. The resulting data confirmed the existence of the genera *Shewanella* and *Moritella,* but suggested that the genus *Alteromonas* should be split into two genera.

In conventional taxonomy, some characteristics are given special emphasis. These include the Gram stain, cell morphology, and the presence of cell structures such as endospores. In numerical taxonomy, all phenotypic characteristics are given equal weight in classifying strains. Bergey's Manual of Systematic Bacteriology contains the phenotypic characteristics used to classify bacteria by conventional taxonomy, and keys that can be used to identify unknown strains from their phenotypic characters. Some analyses of nucleic acids have been used in conventional taxonomy. These include measurements of DNA base composition and nucleic acid hybridization.

The tools that have been developed for identifying microbes and analyzing their activity can be divided into those based on nucleic acids and other macromolecules and approaches directed at analyzing the activity of complete cells. The nucleic acid–based tools are more frequently used because of the high throughput potential provided by using PCR amplification or *ex situ* or *in situ* hybridization with DNA, RNA, or even peptide nucleic acid probes. These methods involve the study of the microbial DNA, the chromosome and plasmid, their composition, homology and presence or absence of specific genes. Application of genome-scale analysis like DNA microarray technology has revolutionized multiple scientific disciplines. Diagnostic evaluation using genotypic methods like PCR of the species-specific ligase and glycopeptide resistance genes helps to identify four *Enterococcus* species and 16S RNA sequencing, the "gold standard" for identification of enterococci-confirmed the results obtained by the FT-IR classification.

Approaches based on complete or partial genomes include DNA arrays that can be used in comparative genomics or genome-wide expression profiling. These omics approaches have now become feasible for probiotic bacteria after the recent realization of the complete genome sequences of human isolates of *Bifidobacterium longum* and *Lactobacillus plantarum.*

8.15.1 Nucleic Acid Probes to Detect Specific Nucleotide Sequence

DNA gene probes may become extremely useful in studying gene transfer and adaptation mechanisms in natural bacterial communities, and in the laboratory. This technology allows the detection of specific gene sequence(s) in bacterial species, and can be used to find and monitor recombinant DNA clones in microorganisms being considered for release into the natural environment. It may provide a new generation of highly specific tests that offers advantages over the classical approaches for identifying specific organisms. Single-stranded DNA from an organism of interest is allowed to attach itself to a membrane. A single-stranded DNA probe binds to its immobilized complementary strand. This binding can be detected by labelling the probes with radioisotopes or with non-radioactive reporter molecules, such as the biotin-streptavidin-enzyme complex (http://www.ilri.org; Figure 8.7).

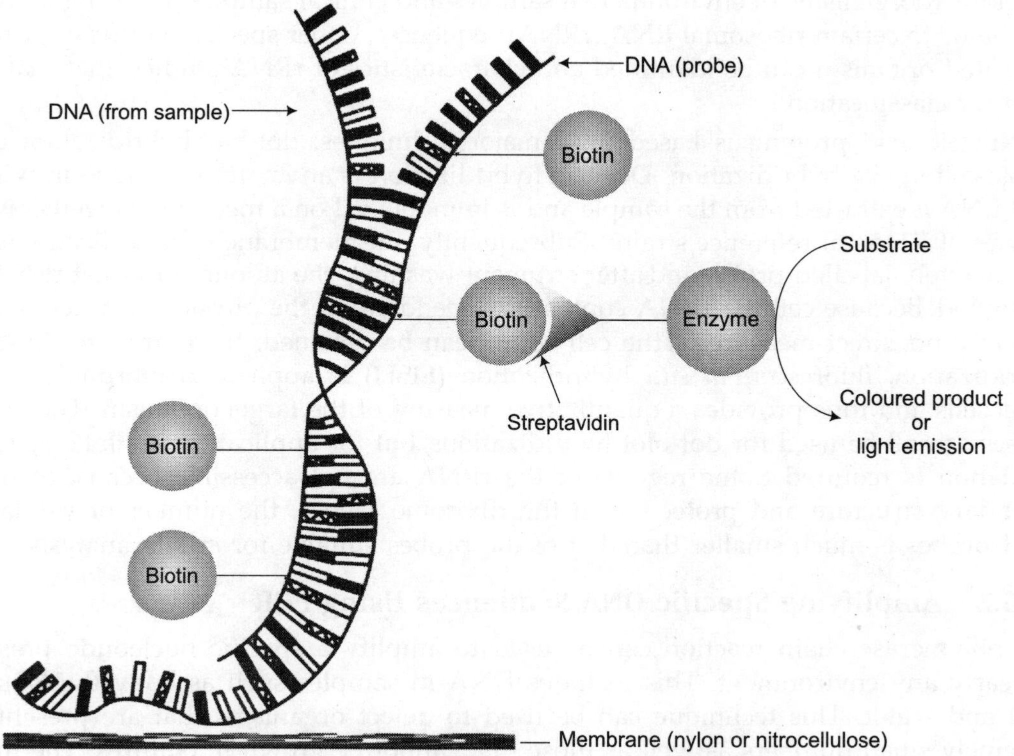

Figure 8.7 Nucleic acid hybridization process.

To adapt DNA probe methodology for use in soils, the following features of a protocol needed to be improved or developed: (i) a procedure was needed which would allow processing of more samples simultaneously and in a shorter period of time for analysis of the number of treatments and replicates needed for ecological studies (ii) the isolated DNA had to be of sufficient purity and size for use in experiments involving digestion with restriction endonucleases, transfer to cellulose nitrate membranes, and hybridization to DNA probes. If contaminants are not removed, reduction in the efficiency of digestion by restriction endonucleases and the specificity of hybridization will be seen (iii) it was also necessary to develop probes both sensitive and specific enough to detect the presence of a particular sequence of low frequency in the complex mixture of DNAs isolated from the soil bacterial community. The standard method of labelling probes by nick translation did not appear to be sensitive or specific enough for probing natural populations.

A probe is a single stranded nucleic acid that has been labelled with a detectable tag, such as radioisotope or a florescent dye. It is complementary to the sequence of interest. Floresecent *in situ* hybridisation is increasingly used to observe and identify intact microorganisms in environmental samples and clinical samples. By using a probe that binds to certain ribosomal RNA (rRNA) sequences, either species specific or groups of related organism can be identified and characteristics of rRNA studied that make it ideal for classification.

Nucleic acid probing is based on 2 major techniques: dot-blot hybridization and whole-cell *in situ* hybridization. Dot-blot hybridization is an *ex situ* technique in which total RNA is extracted from the sample and is immobilized on a membrane together with a series of RNAs of reference strains. Subsequently, the membrane is hybridized with a radioactively labelled probe, and after stringent washing, the amount of target rRNA is quantified. Because cellular rRNA content is dependent on the physiological activity of the cells, no direct measure of the cell counts can be obtained. In contrast to dot-blot hybridization, fluorescent *in situ* hybridization (FISH) is applied to morphologically intact cells and thus provides a quantitative measure of the target organism. The listed probes can all be used for dot-blot hybridizations, but for application in FISH, specific validation is required. Some regions of the rRNA are not accessible because of their secondary structure and protection in the ribosome. Hence, the number of validated FISH probes is much smaller than that of the probes suitable for *ex situ* analysis.

8.15.2 Amplifying Specific DNA Sequences Using PCR

The polymerase chain reaction can be used to amplify a specific nucleotide present in nearly any environment. This includes DNA in samples such as body fluids, soil, food and water. This technique can be used to detect organisms that are present in extremely small numbers as well as those that cannot be grown in cultures. The most commonly used DNA sequence for bacterial phylogenetics is the highly conserved 16S rRNA gene sequence (Figure 8.8), and primers have been designed to selectively amplify bacterial 16S rRNA genes.

To use PCR to detect a microbe of interest, a sample is first treated to release and denature the DNA. Specific primers and other ingredients are then added to the denatured DNA forming the components of the PCR reaction. Some information about the nucleotide sequence of the organism must be known in order to select the appropriate primers. After approximately 30 cycles of PCR, the DNA region flanked by the primer will be amplified a billionfold. In most of the cases the results in a sufficient quantity for the amplified fragment can be readily visible as a discrete band on the gel after staining with ethidium bromide.

In such situations the DNA markers most commonly used have been restriction fragment length polymorphisms (RFLPs). Fragments are usually generated by

A simplified map of the 16S r RNA molecule

Primer:	Primer sequence:	Primer Length	*E.coli* Position
Univ Bact F	5' GAG TTT GAT YMT GGC TC	17-mer	9-25
Univ Back R	5' GYT ACC TTG TTA CGA CTT	18-mer	1509-1492

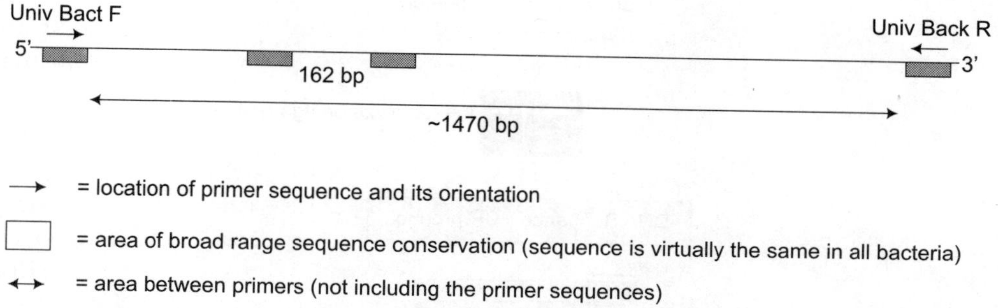

Figure 8.8 A simplified map of the 16S RNA molecule.

frequent-cutting enzymes and separated by conventional agarose gel electrophoresis, but occasionally rare-cutting enzymes are used and larger fragments are separated by pulsed-field gel electrophoresis. RFLPs have been used successfully to generate numerous microbial typing systems, but for some organisms discrimination is suboptimal because there is a tendency for one or two genetic types to predominate amongst an apparently heterogeneous population. Better discrimination between isolates can be achieved by the secondary step of Southern blot hybridisation with radio labelled probes recognising repetitive DNA sequences. However, this adds a rather laborious, expensive second step which is incompatible with large-scale epidemiological studies.

The PCR profiles obtained were unique for unrelated strains whereas similar patterns were observed for epidemiologically related strains isolated from members of the same family. In some studies, such as that carried out on human herpes virus 6 with primers from known viral DNA sequences, the amplified products were analysed by a combination of Southern blot hybridisation, digestion with restriction endonucleases and partial nucleotide sequencing. For many organisms genetic maps are not available and relatively little is known of their molecular biology.

8.15.3 Sequencing Ribosomal RNA Genes

Full and partial 16S rRNA gene sequencing methods have emerged as useful tools for identifying phenotypically aberrant microorganisms. Hence, 16S rRNA gene sequencing is also performed (Figure 8.9). In a particular case it was found that all three patients had endocarditis, and conventional methods identified isolates from patients A, B, and C as a *Facklamia* sp., *Eubacterium tenue*, and a *Bifidobacterium* sp. But when 16S rRNA gene sequencing was performed, the isolates were identified as *Enterococcus faecalis*, *Cardiobacterium valvarum*, and *Streptococcus mutans*, respectively.

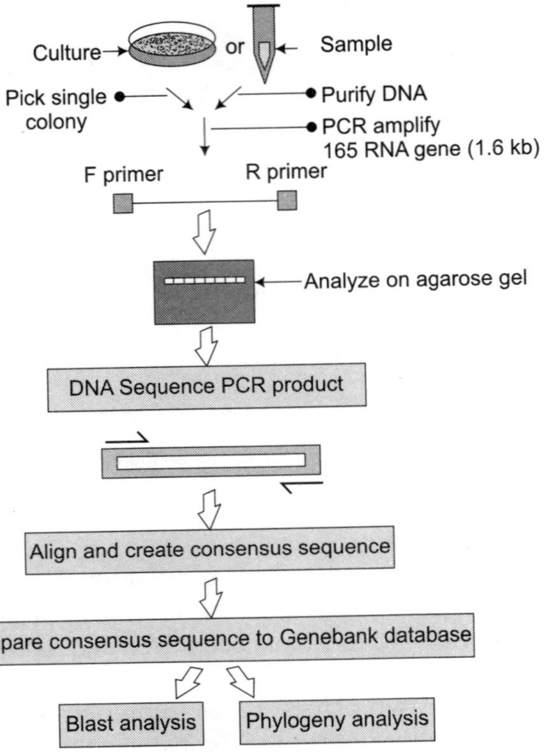

Figure 8.9 Bacterial identification by 16S rRNA gene sequencing.

Technologist bias or inexperience with an unusual phenotype or isolate may similarly compromise identification when results of biochemical tests are interpreted to fit expectations. Although not perfect, genotypic identification of microorganisms by 16S rRNA gene sequencing has emerged as a more objective, accurate, and reliable method for bacterial identification, with the added capability of defining taxonomical relationships among bacteria.

Phenotypic methods have numerous strengths but often fail because the phenotype is inherently mutable and subject to biases of interpretation. 16S rRNA gene sequencing is a more accurate and objective method of identification of microorganisms with particular utility in the clinical laboratory. It also reduces the interpretive bias and shows the need for a "pre-test" probability regarding a microorganism's classification to direct workup and database selection. Medical technologists may pursue an erroneous identification algorithm based on their phenotypic "intuition," such that when unusual microorganisms are encountered, they are made to "fit" with technologist expectations, or when common microorganisms with atypical phenotypes are encountered, they are made to "fit" characteristics of extremely unusual pathogens. Conventional automated identification systems often rely on technologists' interpretations of a microorganism's Gram stain morphology (e.g., RapID-ANA) or oxidase result (e.g., Biolog) for selecting the correct reference database. This case series demonstrates that seemingly simple biochemical or Gram reactions are not unquestionably foolproof and may lead to inappropriate use of comparative databases. Such exhaustive phenotypic testing potentially delays turnaround time without the added benefit of accuracy.

The nucleotide sequence of the ribosomal RNA (rRNA) may be used to identify prokaryotes, particulary those that are difficult or currently impossible to grown cultures. The prokaryotic 70S ribosome, which plays an indispensable role in protein synthesis is composed of proteins and 3 different rRNAs (5S, 16S and 23S). Because of its highly constrained and essential function, the nucleotide sequence changes that can occur in the rRNAs yet still allow the ribosome to operate. This is why it is proved to be so important in classification and more recently in identification.

Of the different rRNAs, the 16s molecule has proved most useful in taxonomy because of its moderate size (approximately 1500) nucleotides. The 5S molecules lacks the critical amount of information because of its small size (120 nucleotides), wheres the larger size of 23 S molecule (approx. 3000 nucleotides) has made it more difficult to sequence in the past.

Some regions in the prokaryotes are virtually the same in all prokaryotes, whereas others are variable. It is the variable region that is used to identify an organism. Once the nucleotide sequence is determined, it can be compared with 16S region of known organisms by searching extensive databases of the huge databases of rRNA sequences exists. For example, the Ribosomal Database Project (RDP) contains a large collection of such sequences, now numbering over 100,000. The RDP can be assessed electronically (http://rdp.cme.msu.edu/html/) and besides sequences contains phylogenetic tutorials, reference citations, previews of new release of sequences and a host of other features.

The methods for obtaining ribosomal RNA sequences and generating phylogenetic trees are now quite routine (Figure 8.10). Newly generated sequences can be compared with sequences in the RDP and other genetic databases such as Gen Bank (USA), DDBS (Japan), or EMBL (Germany). Then, using a treeing algorithm, a phylogentic tree is produced describing the evolutionary information inherent in the sequences.

The separation of the microorganisms is typically represented by what is known as a dendrogram. Essentially, a dendrogram appears as a tree oriented on a horizontal axis. The dendrogram becomes increasingly specialized. The similarity coefficient increases as the dendrogram moves from the left to the right. The right hand side consists of the branches of the trees. Each branch contains a group of microorganisms.

The dendrogram depiction of relationships can also be used for another type of microbial taxonomy. In the second type of taxonomy, the criterion used is the shared evolutionary heritage. This heritage can be determined at the genetic level. This is termed molecular taxonomy.

To begin with the process, the polymerase chain reaction is used to amplify the gene encoding 16S ribosomal RNA from the genomic DNA. Following this, the PCR product is sequenced by the dideoxy DNA sequencing method. Using PCR primers

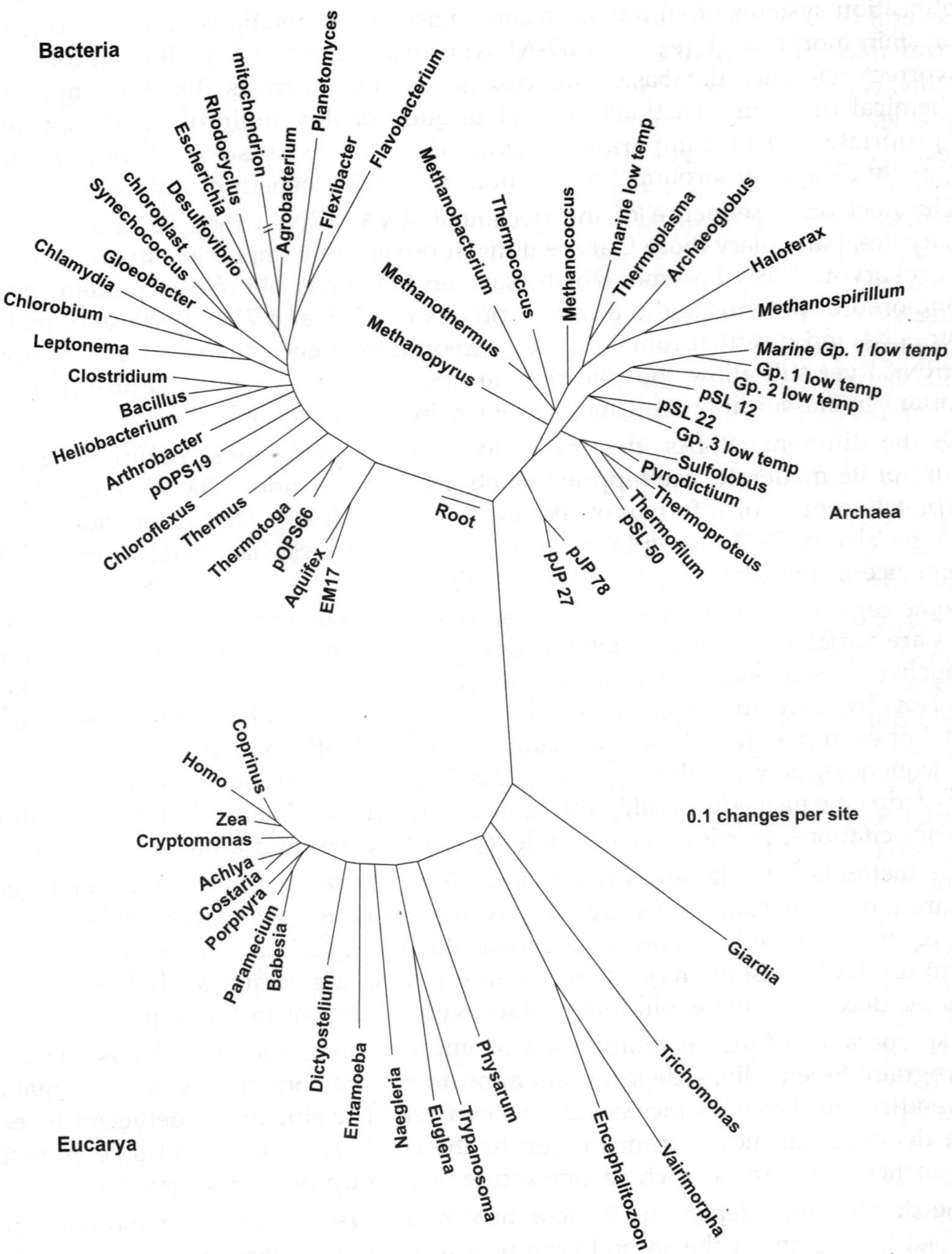

Figure 8.10 Phylogenetic relationships sequencing 16s rDNA from individual organisms.

complementary to the conserved sequences in the small unit of ribosomal RNAs, only a tiny amount of the cell material can yield a huge amount of DNA product for sequencing purposes. Once sequencing is done, it is ready for computer analysis.

Several different algorithms for sequence analysis and phylogenetic tree formation are available for comparative ribosomal sequencing (Figure 8.10).

However, regardless of which program to be used, the raw data must first be aligned with the previous aligned sequences using a sequence editor. Not all the rRNAs are exactly the same length. Thus, during alignment, gaps can be inserted wherever necessary in regions where one sequence can be shorter than the other. The aligned sequences are then imported into a treeing programme and comparative analysis is done. Two widely used treeing algorithms are distance and parsimony, using distance method, sequences are aligned and then an evolutionary distance (ED) is calculated by having the computer record every position in the dataset in which there is a difference in the sequence. From these datasets a data matrix can be constructed that shows that the ED between two sequences in the dataset. Following this, a statistical correlation is factored into the ED that considers the possibility that more than one change can occur at a given site. Once this is accounted for, a phylogenetic tree is generated in which the lengths of the lines in the tree are proportional to the evolutionary distances (Figure 8.10).

8.16 CHARACTERIZING STRAIN DIFFERENCES

If a species of bacteria is isolated and cultivated in the laboratory it is known as a strain. A single isolate with distinctive characteristic(s) may also represent a strain. Members of the same species that have small differences between them can be distinguished by additional methods. These species are then subdivided into subspecies, subgroups, biotypes, serotypes, variants, etc. Methods of bacterial strain identification can be broadly delimited into genotypic techniques based on profiling an organism genetic material (primarily its DNA) and phenotypic techniques based on profiling either an organism's metabolic attributes or some aspect of its chemical composition. Genotypic techniques have the advantage over phenotypic methods that they are independent of the physiological state of an organism; they are not influenced by the composition of the growth medium or by the organism's phase of growth. The process of differentiating strains based on their phenotypic and genotypic differences is known as 'typing'. These typing methods are useful to understand typability, reproducibility, discriminatory power, ease of performance, and ease of interpretation. Two methods of typing are found. Phenotypic techniques detect characteristics expressed by the microorganism like shape, size, staining properties, biochemical properties, antigenic properties that can be measured without reference to the genome and genotypic techniques involve direct DNA-based analysis of chromosomal or extra chromosomal genetic elements.

Molecular diagnostics provide outstanding tools for the detection, identification and characterisation of microbial strains. The application of these and other related techniques, along with the development of molecular markers for bacterial strains, greatly facilitates understanding of the ecological interactions of microbial strains, their roles, succession, competition and prevalence in food fermentations and allows the correlation of these features to desirable quality attributes of the final product. Several strains of microorganisms have been selected or genetically modified to increase the efficiency with which they produce enzymes.

8.16.1 Phenotypic Typing Methods

Traditional methods for microbial identification require the recognition of differences in morphology, growth, enzymatic activity, and metabolism to define genera and species. Phenotypic identification often suggests unusual organisms not typically associated with the submitted clinical diagnosis. Phenotypic profiles including Gram stain results, colony morphologies, growth requirements, and enzymatic and/or metabolic activities are generated, but these characteristics are not static and can change with stress or evolution. Thus, when common microorganisms present with uncommon phenotypes, when unusual microorganisms are not present in reference databases, or when databases are out of date, reliance on phenotypes can compromise accurate identification.

8.16.1.1 Biochemical typing

Traditional microbial identification methods typically rely on phenotypes, such as morphological features, growth variables, and biochemical utilization of organic substrates. The biological profile of an organism is termed a biogram. The determination of relatedness of different organisms on the basis of their biograms is termed biotyping. Investigators must determine which profile variables have the greatest differentiating capabilities for a given organism. For example, Gram stain characteristics, indole positivity, and the ability to grow on MacConkey medium do not aid in the differentiation of non-enterohaemorrhagic *Escherichia coli* from *E. coli* O157:H7. However, sorbitol fermentation has proven to be an extremely useful characteristic of the biochemical profile used to differentiate these strains.

Biograms that are identical have been used to infer relatedness between strains in epidemiological investigations. The biograms of organisms are not entirely stable, and several isotypes may exist from a single isolate. Biograms may be influenced by genetic regulation, technical manipulation, and the gain or loss of plasmids. In many instances, biotyping is used in conjunction with other methods to more accurately profile microorganisms.

Biotyping makes use of the pattern of metabolic activities expressed by an isolate, colonial morphology and environmental tolerances. Strains are referred to as "biotypes". Biochemical tests are used to identify many bacteria and also used to distinguish strains. If the biochemical variation is uncommon; it can be used for tracing the source of certain disease outbreaks. A strain has a characteristic biochemical pattern (Figure 8.11) and is called a biovar or biotype. They performed western blot analysis of the H-type BSE zebu (Charly-04) with a) a core-binding antibody (Sha31, b) an amino-terminal binding antibody (12B2) and c) a carboxy-terminal binding antibody (SAF84). Samples are assigned to the lanes as follows: negative control (N), L-type BSE (L), C-type BSE (C) and for the zebu medulla oblongata (lane 1, 15 mg tissue equivalent), cerebellar cortex (lane 2, 15 mg), hippocampus (lane 4, 0.75 mg), piriform lobe (lane 5, 15 mg), basal ganglia (lane 7, 1.5 mg), frontal cortex (lane 8, 15 mg), occipital cortex (lane 9,

15 mg) and temporal cortex (lane 10, 15 mg). The dashed line indicates the molecular mass of the unglycosylated C-type PrP and helps to visualize differences compared to the H-type BSE zebu. The same samples, but deglycosylated are shown in (d) with a carboxy-terminal binding antibody (SAF84). A molecular mass marker (in kDa) is indicated on the left in Figure 8.11.

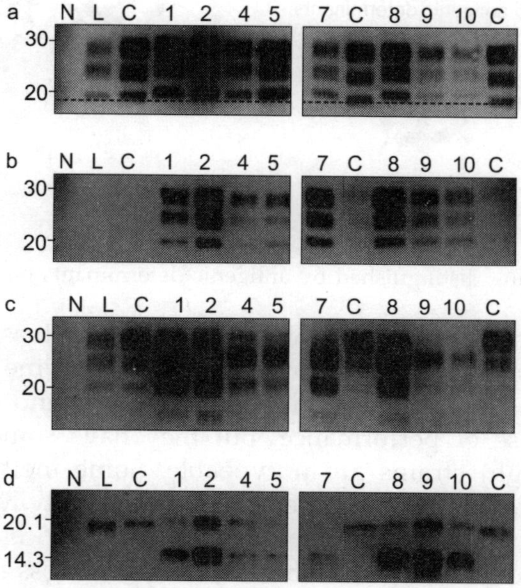

Figure 8.11 Biochemical typing of different brain regions in H-type BSE.

Biotyping may be performed manually or using automated systems. Sugar fermentation, amino acid decarboxylation/deamination, standard enzymatic tests such as IMViC, citrate, urease, tolerance to pH, chemicals and dyes, hydrolysis of compounds, haemagglutination, and hemolysis are some examples of biotyping methods. They offer some advantages as most strains are typeable. The techniques are reproducible with relatively ease in performance and interpretation. But the main disadvantages are that they have poor discriminatory power. Variation in gene expression is the most common reason for isolates that represent single strain to differ in one or more biochemical reactions. Point mutation too contributes to this problem.

8.16.1.2 Serological typing

Serologinal typing or serotyping is based on fact that strains of same species can differ in the antigenic determinants expressed on the cell surface (Figure 8.12). Surface structures such as lipopolysaccharides, membrane proteins, capsular polysaccharides, flagella and fimbriae exhibit antigenic variations. Strains differentiated by antigenic differences are known as 'serotypes'. Serotyping is used in several Gram-negative and Gram-positive bacteria.

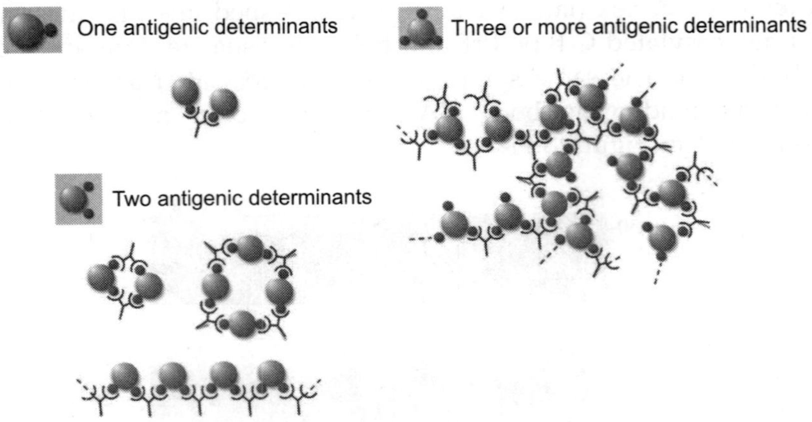

Figure 8.12 Serotyping distinguished by antigenic determinants on the surface of the cell

Serotyping is performed using several serologic tests such as bacterial agglutination, latex agglutination, co-agglutination, fluorescent and enzyme labelling assays. Most strains are typeable. They have good reproducibility and ease of interpretation though some have ease of performance. But they have some disadvantages. Some autoagglutinable (rough) strains are untypeable. Some methods of serotyping are technically demanding. There is dependency on good quality reagent from commercial sources. In-house preparation of reagents is a difficult process. Serotyping has poor discriminatory power due to large number of serotypes, cross reaction of antigens and untypeable nature of some strains.

The invention of serological typing concerns a method for typing antibodies in a sample liquid by means of type-specific antigens and in particular a method for typing antibodies to the hepatitis C virus and peptide antigens suitable for this. A further possibility of serological type differentiation of infections with the HCV types 1, 2 and 3 can be carried out by means of an indirect ELISA using peptide antigens of the amino acid regions. For this type-specific peptide antigens can be immobilized separately according to their type in individual wells of a microtitre plate and each was contacted with separate aliquots of a plasma sample from HCV-infected blood donors. The typing was carried out according to the reactivity of the serum sample with the individual peptide antigens. However, this method is relatively inaccurate and, moreover, does not allow the determination of individual viral subtypes, i.e., individual virus strains whose immunogenicity only differs to a slight extent.

8.16.1.3 *Genomic typing*

Currently, genomic typing of microorganisms is widely used in several major fields of microbiological research (Table 8.2). Taxonomy, research aimed at elucidation of evolutionary dynamics or phylogenetic relationships, population genetics of microorganisms, and microbial epidemiology all rely on genetic typing data for discrimination between genotypes. Apart from being an essential component of

Table 8.2 Survey of the characteristics of several currently used microbial typing methods

Typing method	*Typeability*	*Reproducibility*	*Discriminatory power*	*Ease of performance*	*Ease of interpre-tation*	*General availability*	*Cost*
Phenotypic							
PI Antimicrobial susceptibility	Good	Good	Poor	Excellent	Excellent	Excellent	Low
P2. manual biotyping	Good	Poor	Poor	Excellent	Excellent	Excellent	Low
P3. Automated biotyping	Good	Good	Poor	Good	Good	Variable	Medium
P4. Serotyping	Variable	Good	Variable	Good	Good	Variable	Medium
P5. Bacteriophage typing	Variable	Fair	Variable	Poor	Poor	Excellent	Medium
P6. PAGE	Excellent	Good	Good	Excellent	Fair	Good	Medium
P7. Immunoblotting	Excellent	Good	Good/excel lent	Good	Fair	Variable	Medium
P8. MLEE	Excellent	Excellent	Good	Good	Excellent	Variable	High
Genotypic							
Gl. Plasmid profiles	Variable	Fair	Variable	Fair	Good	Excellent	Medium
G2. Plasmid REA	Variable	Excellent	Good	Good	Excellent	Excellent	Medium
G3. Chromosomal REA	Excellent	Variable	Variable	Good	Fair	Variable	Medium
G4. Ribotyping	Excellent	Excellent	Good	Good	Good	Variable	High

these fundamental sciences, microbial typing clearly affects several areas of applied microbiological research. The epidemiological investigation of outbreaks of infectious diseases and the measurement of genetic diversity in relation to relevant biological properties such as pathogenicity, drug resistance, and biodegradation capacities are obvious examples. The diversity among nucleic acid molecules provides the basic information of genomic typing. However, researchers in various disciplines tend to use different vocabularies, a wide variety of different experimental methods to monitor genetic variation, and sometimes widely differing modes of data processing and interpretation.

In a unique example, minor histocompatibility antigen (HA-1) genomic typing by RSCA is easy to perform and that could be used as a routine typing method for The Kidd (JK) blood group system that is clinically important in transfusion medicine. In another example, the genetic relationship between isolates of *Listeria monocytogenes* belonging to different serotypes was determined and the suitability of automated laser fluorescent analysis (ALFA) of amplified fragment length polymorphism (AFLP) fingerprints was assessed by genomic typing of 106 L.

8.16.1.4 Phage typing

Phage typing is a method used for detecting single strains of bacteria. It is used to trace the source of outbreaks of infections. The viruses that infect bacteria are called bacteriophages ("phages" for short) and some of these can only infect a single strain of bacteria. These phages are used to identify different strains of bacteria within a single species. They help to characterize bacteria, extending to strain differences, by demonstration of susceptibility to one or more (a spectrum) races of bacteriophage; widely applied to staphylococci, typhoid bacilli, etc., for epidemiological purposes Phage typing requires the use of a standard collection of dissimilar phages. In the process of developing a phage typing set, numerous phages are first isolated and tests are undertaken to determine if they are different and useful in delineating the types of organisms under study.

For many years phage typing has been a useful epidemiologic tool for studying outbreaks of *S. typhi* and *S. typhimurium* (Figure 8.13). Ten types of phages (podoviruses) were found morphologically identical to Salmonella phage P22. Two phages are siphoviruses and identical to flagella-specific phage chi. This system was particularly useful for differentiating a group of animal strains that had a number of diverse phage types. Strains can be characterised by their pattern of resistance or susceptibility to a standard set of bacteriophages. This relies on the presence or absence of particular receptors on the bacterial surface that are used by the virus to bind to the bacterial wall. This method is used to type isolates of *Staphylococcus aureus* and *Salmonella* spp. Such stains are referred as 'phage types'. The susceptibility of an organism to a particular type of phage can be readily demonstrated in the laboratory. Firstly, a culture of the test organism is inoculated into melted, cooled nutrient agar and poured onto the surface of an agar plate thus creating an uniform layer of cells, then drops of different types of bacteriophage are carefully placed on the surface of the agar. During incubation the bacteria will multiply, forming a visible haze of cells. A clear zone will be formed at each spot where bacteriophage has been added in case the organism is susceptible to the type of phage. The pattern of clearing indicated the susceptibility to different phage and can be compared to determine the strain differences. Using phages to differentiate bacteria is justified the term "phage typing".

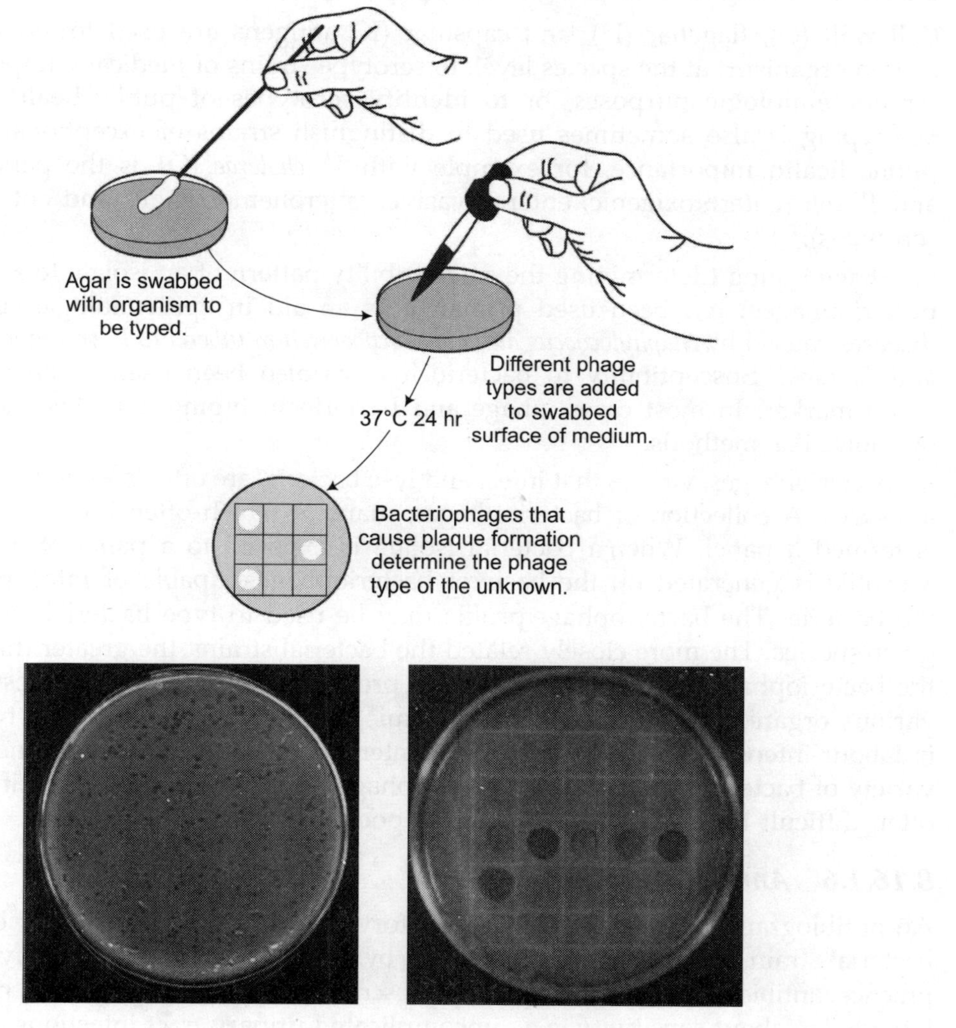

Figure 8.13 Phage typing plaque formation for *Staphylococcus aureus*.

Phage typing can be extremely important in many health situations because it can identify random, unrelated organisms as well as the isolates that are actually responsible for a given problem. Aside from relating an organism to an outbreak, this laboratory method can also be used for surveillance, assessing strain distribution, and ascertaining the effectiveness of therapeutic measures. This technique has fair amount of reproducibility, discriminatory power and ease of interpretation. But this technique also requires maintenance of biologically active phages and hence is available only at reference centres. Even for the experienced worker, the technique is demanding. Many strains are non-type able.

8.16.1.5 Antigen and phage susceptibility

Cell wall (O), flagellar (H), and capsular (K) antigens are used to aid in classifying certain organisms at the species level, to serotype strains of medically important species for epidemiologic purposes, or to identify serotypes of public health importance. Serotyping is also sometimes used to distinguish strains of exceptional virulence or public health importance, for example with *V. cholerae* (O1 is the pandemic strain) and *E. coli* (enterotoxigenic, enteroinvasive, enterohemorrhagic, and enteropathogenic serotypes).

Phage typing (determining the susceptibility pattern of an isolate to a set of specific bacteriophages) has been used primarily as an aid in epidemiologic surveillance of diseases caused by *Staphylococcus aureus, Mycobacterium tuberculi, P. aeruginosa, V. cholerae,* and *S. typhi.* Susceptibility to bacteriocins has also been used as an epidemiologic strain marker. In most cases, phage and bacteriocin typing have been supplemented by molecular methods.

Bacteriophages, viruses that infect and lyse bacteria, are often specific for strains within a species. A collection of bacteriophages, many of which often infect similar bacteria, is termed a panel. When a bacterial isolate is exposed to a panel of bacteriophages, a profile is generated on the basis of bacteriophages capable of infecting and lysing the bacteria. The bacteriophage profile may be used to type bacterial strains within a given species. The more closely related the bacterial strains, the greater the similarity of the bacteriophage profiles. Bacteriophage profiles have been used successfully to type various organisms associated with epidemic outbreaks. However, this typing method is labour-intensive and requires the maintenance of bacteriophage panels for a wide variety of bacteria. Additionally, bacteriophage profiles may fail to identify isolates, are often difficult to interpret, and may give poor reproducibility.

8.16.1.6 Antibiograms

An antibiogram is the result of a laboratory testing for the sensitivity of an isolated bacterial strain to different antibiotics. It is by definition an *in vitro* sensitivity. In clinical practice, antibiotics are most frequently prescribed on the basis of general guidelines and knowledge about sensitivity, e.g., uncomplicated urinary tract infections can be treated with a first generation quinolone, etc. This is because *Escherichia coli* is the most likely causative pathogen, and it is known to be sensitive to quinolone treatment. Infections that are not acquired in the hospital, are called "community acquired" infections.

However, many bacteria are known to be resistant to several classes of antibiotics, and treatment is not so straightforward. This is especially the case in vulnerable patients, such as patients in the intensive care unit. When these patients develop "hospital-acquired" or "nosocomial" pneumonia, more hardy bacteria like *Pseudomonas aeruginosa* are potentially involved. Treatment is then generally started on the basis of surveillance data about the local pathogens probably involved. This first treatment, based on statistical information about former patients, and aimed at a large group of potentially involved microbes, is called "empirical treatment".

Before starting this treatment, the physician will collect a sample from a suspected contaminated compartment: a blood sample when bacteria possibly have invaded the bloodstream, a sputum sample in the case of ventilator associated pneumonia, and a urine sample in the case of a urinary tract infection. These samples are transferred to the microbiology laboratory, which looks at the sample under the microscope, and tries to culture the bacteria (Figure 8.14). This can help in the diagnosis.

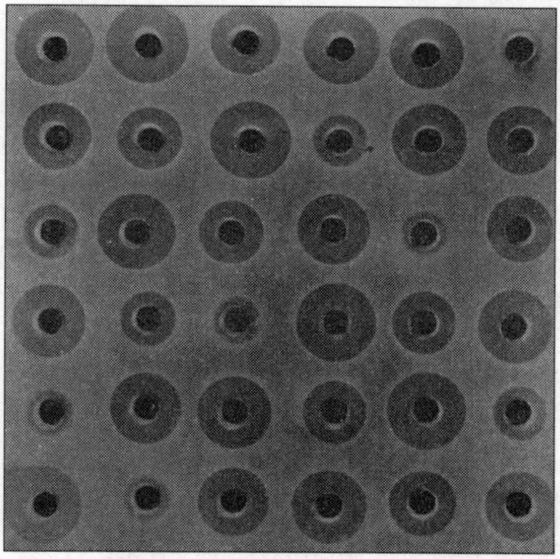

Figure 8.14 Antibiogram of *Klebsiella pneumonia* isolates, a streptomycin assay plate using the agar diffusion techniques.

Once a culture is established, there are two possible ways to get an antibiogram:

- A semi-quantitative way based on diffusion (Kirby-Bauer method); small discs containing different antibiotics, or impregnated paper discs, are dropped in different zones of the culture on an agar plate, which is a nutrient-rich environment in which bacteria can grow. The antibiotic will diffuse in the area surrounding each tablet, and a disc of bacterial lysis will become visible. Since the concentration of the antibiotic was the highest at the centre, and the lowest at the edge of this zone, the diameter is suggestive for the Minimum Inhibitory Concentration, or MIC, (conversion of the diameter in millimeter to the MIC, in μg/ml, is based on known linear regression curves).
- A quantitative way based on dilution: a dilution series of antibiotics is established (this is a series of reaction vials with progressively lower concentrations of antibiotic substance). The last vial in which no bacteria grow contains the antibiotic at the Minimal Inhibiting Concentration.

Once the MIC is calculated, it can be compared to known values for a given bacterium and antibiotic: e.g., a MIC > 0.06 µg/mL may be interpreted as a penicillin-resistant *Streptococcus pneumoniae*. Such information may be useful to the clinician, who can change the empirical treatment, to a more custom-tailored treatment that is directed only at the causative bacterium.

Antibiograms are an important resource for healthcare professionals involved in deciding and prescribing empiric antibiotic therapy. Appropriate empiric therapy is essential in attempting to treat infections correctly and quickly in an effort to decrease mortality. The use of antibiograms is also helpful in identifying trends in antibiotic resistance. Basic components of an antibiogram include: antibiotics tested, organisms tested, number of isolates for each organism, percentage susceptibility data for each drug/pathogen combination, specimen sites notations (e.g., blood, urine, catheters) and specific area or unit being tested.

It is important to tailor antibiotics as soon as sensitivities are known. This is the best way to avoid drug resistance and new/emerging organisms that are resistant. The goal to minimizing infection is to prescribe broad-spectrum antibiotics based on unit specific antibiograms.

The susceptibility or resistance of an organism to a possibly toxic agent forms the basis of the following typing techniques. The antibiogram is the susceptibility profile of an organism to a variety of antimicrobial agents, whereas the resistogram is the susceptibility profile to dyes and heavy metals. Bacteriocin typing is the susceptibility of the isolate to various bacteriocins, i.e., toxins that are produced by a collected set of producer strains. These three techniques are limited by the number of agents tested per organism.

By far, the antibiogram is the most commonly used susceptibility/resistance typing technique, most probably because the data required for antibiogram analysis are available routinely from the antimicrobial susceptibility testing laboratory. Antibiograms have been used successfully to demonstrate relatedness with limitations. Organisms with similar antibiograms may be related, such is not necessarily the case. The antibiogram of an organism is not always constant. Selective pressure from antimicrobial therapy may alter an organism's antimicrobial susceptibility profile in such a way that related organisms show different resistance profiles. These alterations may result from chromosomal point mutations or from the gain or loss of extra-chromosomal DNA such as plasmids or transposons.

This typing technique involves comparison of different isolates to a set of antibiotics. Isolates differing in their susceptibilities are considered different strains. The identification of new or unusual pattern of antibiotic resistance among isolates cultured from multiple patients is often the first indication of an outbreak. The technique has ease of performance and interpretation with fair amount of reproducibility.

As a consequence of various genetic mechanisms, different strains may develop similar resistance pattern thus reducing the discriminating power. The susceptibility pattern of isolates taken over a period of time that represents the same strain may differ for one or more antibiotics due to acquisition of resistance.

8.16.1.7 Protein typing

Protein typing relies on major or minor differences in the range of proteins made by different strains. Variations in the types and structures of the proteins expressed by bacteria can be detected by several methods. The proteins, glycoproteins or polysaccharides are extracted from a culture of the strain, separated by sodium dodecyl sulfate-polyacrylamide gel electrophoresis and stained to compare with those of other strains. More similar organisms display more similar protein patterns. In another method termed immunoblotting, the electrophoresed products are transferred to nitrocellulose membrane and then exposed to antisera raised against specific strain. The bound antibodies are then detected by enzyme labelled anti-immunoglobulins. These methods are currently employed for epidemiological studies of *Staphylococcus aureus* and *Clostridium difficile*. All strains are typeable and techniques have good reproducibility and ease of interpretation. Yet as the patterns detected are very complex, comparisons among multiple strains are difficult and the interpretation becomes difficult. Methods employed are technically demanding and equipment are costly and hence are not available in all laboratories.

8.16.1.8 Multilocus enzyme electrophoresis (MLEE)

Here, the isolates are analysed for differences in the eletrophoretic mobilities of a set of metabolic enzymes. Cell extracts containing soluble enzymes are electrophoresed in starch gels. Variations in the electrophoretic mobility of an enzyme, referred to as 'electromorph', typically reflect amino acid substitution that alter the charge of the protein. But this method is only moderately discriminatory for the epidemiological analysis of clinical isolates. It requires techniques and equipments that are not available in most laboratories.

8.16.1.9 Molecular typing methods

Genotypic characterization is becoming more widely practised and standard method for characterizing and identifying bacteria. The technique is universally applicable as all bacterial genera and species become uniformly defined according to genotypic uniqueness. The results of the phenotypic tests will correlate with the genotypic characteristics and bring about accurate and useful identification of organism. Several molecular typing techniques have been developed during the past decade for the identification and classification of bacteria at or near the strain level. The most powerful of these are genetic-based molecular methods known as DNA fingerprinting techniques, e.g., pulsed-field gel electrophoresis (PFGE) of rare-cutting restriction fragments, ribotyping, randomly amplified polymorphic DNA (RAPD), and amplified fragment length polymorphism (AFLP), which have been applied extensively for the infraspecific identification and genotyping (McCartney, 2002). Basically, these methods rely on the detection of DNA polymorphisms between species or strains and differ in their dynamic range of taxonomic discriminatory power, reproducibility, ease of interpretation, and standardization.

8.16.1.10 Plasmid analysis

The number and sizes of plasmids carried by an isolate can be determined by preparing a plasmid extract and subjecting it to gel electrophoresis. But reproducibility of this method suffers due to the existence of plasmid in different molecular forms such as super coiled, nicked or linear, each of which migrates differently on electrophoresis. Since plasmids can be spontaneously lost or readily acquired, related strains can exhibit different plasmid profiles. Clinical isolates lacking plasmids are untypeable. Those strains with one or two plasmids provide poor discriminatory powers.

8.16.1.11 Restriction endonuclease analysis (REA) of chromosomal DNA

A restriction endonuclease enzymatically cuts DNA at a specific nucleotide recognition sequence (Figure 8.15). The number and sizes of restriction fragments are influenced by the recognition sequence of enzyme and composition of DNA. Bacterial DNA is digested with endonucleases that have relatively frequent restriction sites, thereby generating hundreds of fragments ranging from ~0.5 to 50 kb in length. Such fragments can be separated by size using agarose gel electrophoresis. The pattern stained by ethidium bromide and examined under UV light. Different strains of the same species have different REA profiles because of variations in their DNA sequences. The complex profile consists of hundreds of bands that may be unresolved or overlapping thus making comparison difficult. The pattern may consist of bands generated from digestion of plasmids too. These reduce the ease of interpretation and discriminatory power.

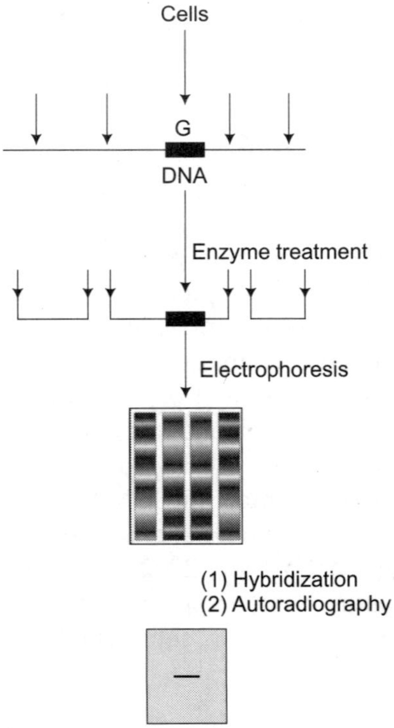

Figure 8.15 Restriction endonudease analysis.

8.16.1.12 Pulse field gel electrophoresis (PFGE) of chromosomal DNA

Pulse field gel electrophoresis technique overcomes the limitations of REA. It is a variation of agarose gel electrophoresis in which the orientation of the electric field across the gel is changed periodically. This modification enables large fragments to be effectively separated by size. Restriction fragment length polymorphism (RFLP) analysis of bacterial DNA involves the digestion of genomic DNA with rare-cutting restriction enzymes to yield a few relatively large fragments. The restriction fragments are then size-fractionated using PFGE that allows separation of large genomic fragments. The generated DNA fingerprint obtained depends on the specificity of the restriction enzyme used and the sequence of the bacterial genome and is therefore characteristic of a particular species or strain of bacteria (Figure 8.16).

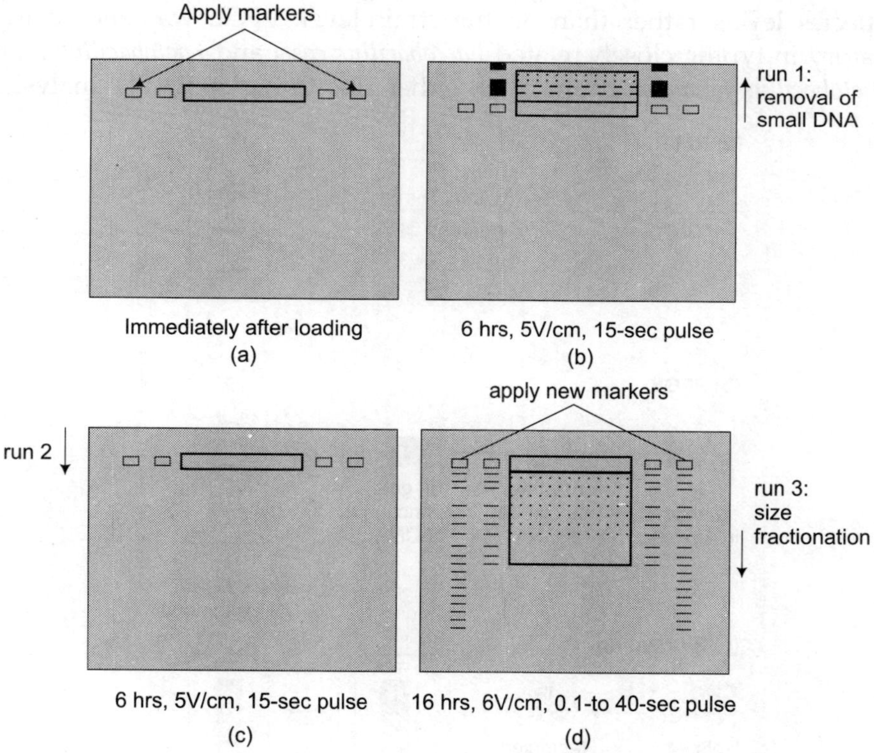

Figure 8.16 Isolation of insert DNA from agarose blocks by PFGE.

This fingerprint represents the complete genome and thus can detect specific changes (DNA deletion, insertions, or rearrangements) within a particular strain over time. Its high discriminatory power has been reported for the differentiation between strains of important probiotic bacteria, such as *Bifidobacterium longum* and *B. animalis*, *Lactobacillus casei* and *Lb. rhamnosus*, *Lb. acidophilus* complex, *Lb. helveticus*, and *Lb. johnsonii*. A new approach combining RFLP with DNA fragment sizing by flow cytometry for bacterial strain identification has been developed. DNA fragment sizing by flow cytometry is found to be faster and more sensitive than PFGE, and this technique is also amenable to automation.

8.16.1.13 Ribotyping

Ribotyping is a variation of the conventional RFLP analysis (Figure 8.17). It combines Southern hybridization of the DNA fingerprints, generated from the electrophoretic analysis of genomic DNA digests, with rDNA-targeted probing. The probes used in ribotyping vary from partial sequences of the rDNA genes or the intergenic spacer regions to the whole rDNA operon. Ribotyping has been used to characterize strains of *Lactobacillus* and *Bifidobacterium* from commercial products as well as from human faecal samples. However, ribotyping provides high discriminatory power at the species and subspecies levels rather than on the strain level. PFGE was shown to be more discriminatory in typing closely related *Lactobacillus casei* and *Lactobacillus rhamnosus* as well as *Lactobacillus johnsonii* strains than either ribotyping or RAPD analysis.

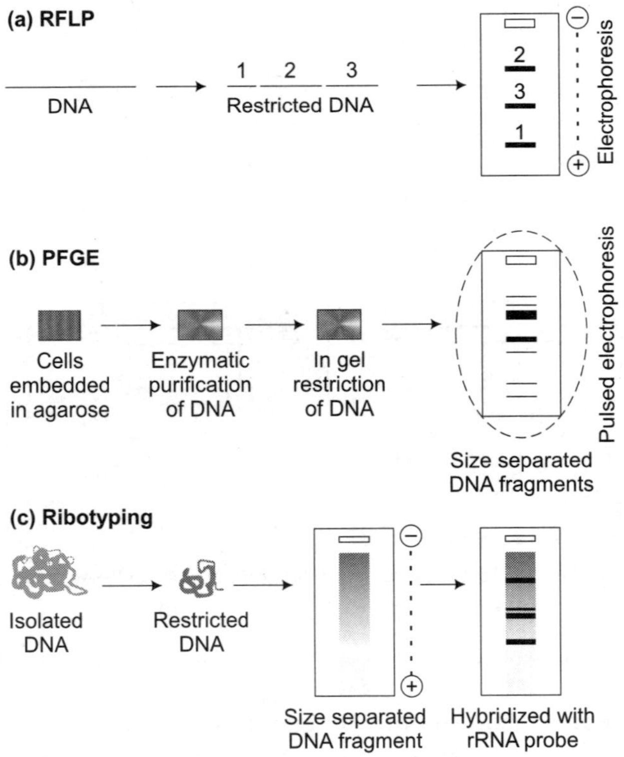

Figure 8.17 A ribotype is essentially an RFLP but differs from PFGE and RFLP.

8.16.1.14 Randomly amplified polymorphic DNA

Arbitrary amplification, also known as RAPD, has been widely reported as a rapid, sensitive, and inexpensive method for genetic typing of different strains of LAB and bifidobacteria. This PCR-based technique makes use of arbitrary primers that are able to bind under low stringency to a number of partially or perfectly complementary sequences of unknown location in the genome of an organism. If binding sites occur in a spacing and orientation that allow amplification of DNA fragments, fingerprint patterns are generated that are specific to each strain. RAPD profiling has been applied to distinguish between strains of *Bifidobacterium* and between strains of the *Lb. acidophilus* group and related strains. Several factors have been reported to influence the reproducibility and discriminatory power of the RAPD fingerprints, i.e., annealing temperature, DNA template purity and concentration, and primer combinations. The use of 5 single-primer reactions under optimized conditions improved the resolution and accuracy of the RAPD method for the characterization of dairy related bifidobacteria including *B. adolescentis, B. animalis, B. bifidum, B. breve, B. infantis,* and *B. longum.*

8.16.1.15 Amplified restriction length polymorphism

AFLP combines the power of RFLP with the flexibility of PCR-based methods by ligating primer-recognition sequences (adaptors) to the digested DNA (Figure 8.18). Total genomic DNA is digested using 2 restriction enzymes, 1 with an average cutting frequency and a second with higher cutting frequency. Double-stranded nucleotide adapters are usually ligated to the DNA fragments serving as primer binding sites for PCR amplification. The use of PCR primers complementary to the adapter and the restriction site sequence yields strain-specific amplification patterns. At present, AFLP has mostly been employed in clinical studies, but its successful application for strain typing of the *Lactobacillus acidophilus* group and *Lactobacillus johnsonii* isolates has been reported.

8.16.1.16 Other PCR approaches

PCR-based approaches other than RAPD and AFLP have been used for molecular typing, such as amplified ribosomal DNA restriction analysis (Figure 8.18). Repetitive extragenic palindromic PCR (Rep-PCR), and triplicate arbitrary primed PCR (TAP-PCR) have shown to offer a high discriminatory power for the identification.

8.16.1.17 Southern blot analysis of RFLP

In contrast to REA of DNA, southern blot analyses detect only the particular restriction fragment. The DNA is digested by endonuclease, the fragments are separated by gel electrophoresis and the fragments transferred to nitrocellulose membranes (Figure 8.19). The fragments containing specific sequences are then detected by labelled DNA probes. Variations in the number and sizes of the fragments detected are referred to as restriction fragment length polymorphism (RFLP).

There are a number of taxonomic criteria that can be used. For example, numerical taxonomy differentiates microorganisms, typically bacteria, on their phenotypic characteristics. Phenotypes are the appearance of the microbes or the manifestation of the genetic character of the microbes. Examples of phenotypic characteristics include the Gram stain reaction, shape of the bacterium, size of the bacterium, where or not

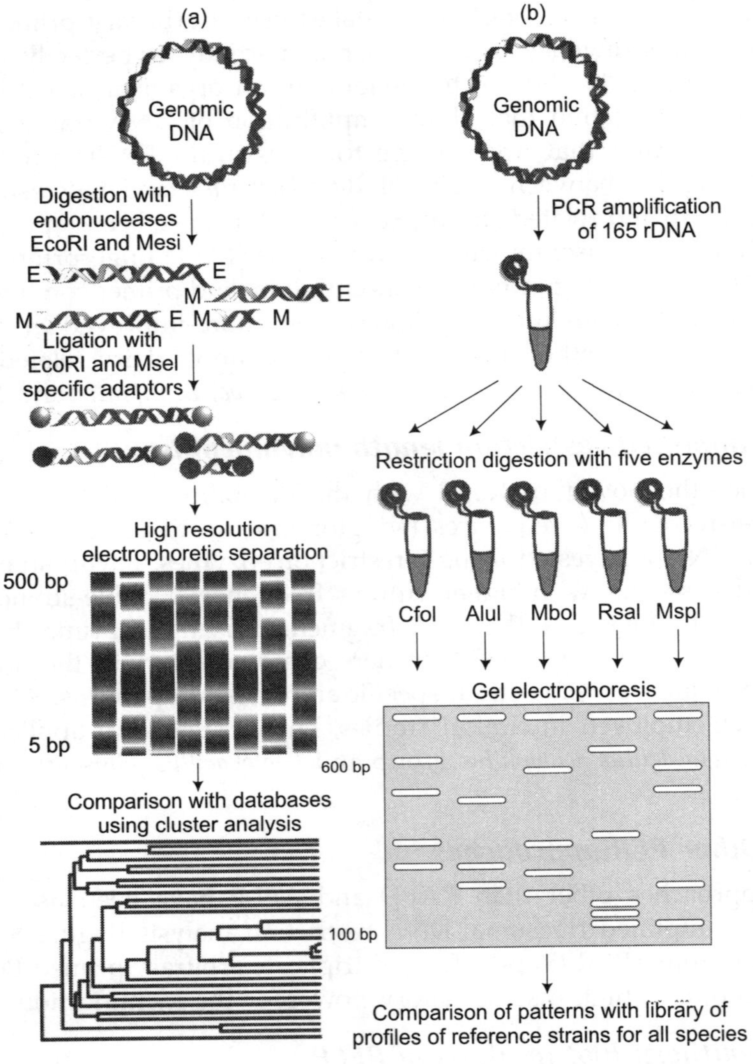

Figure 8.18 The genotypic identification of *Acinetobacter* species can be achieved by whole-genome fingerprinting, restriction analysis of a particular DNA sequence and DNA-sequence determination (a) AFLP comprises the following steps: digestion of cellular DNA with two restriction enzymes (a frequent and a rare cutter); ligation of adaptors to the restriction fragments; selective amplification of the fragments; electrophoresis of fragments on a sequencing machine; and visualization of the restriction profiles. The profiles are compared by cluster analysis to those of a reference library of AFLP fingerprints using dedicated software (b) ARDRA (see Figure, part b) ARDRA comprises amplification of 16S rDNA sequences, followed by restriction of the amplified fragments with five restriction enzymes and electrophoresis. The resulting profiles are compared with a library of restriction profiles.

the bacterium can propel itself along, the capability of the microbes to grow in the presence or absence of oxygen, types of nutrients used, chemistry of the surface of the bacterium, and the reaction of the immune system to the bacterium.

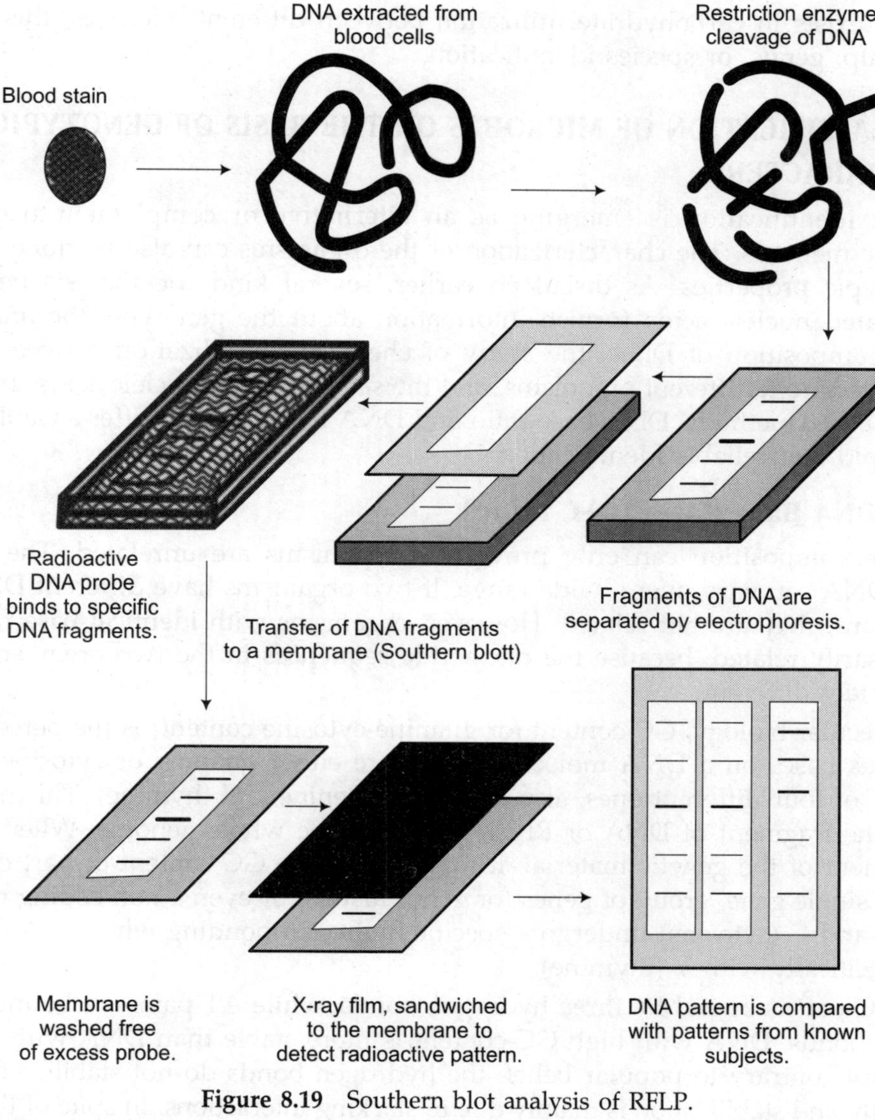

Figure 8.19 Southern blot analysis of RFLP.

Bacterial taxonomy relies on phenotypic characteristics to classify organisms, and is useful for the practical identification of unknown strains. The primary taxonomic unit is the species, which is defined by the phenotypic characteristics of a collection of similar strains. Culture collections contain type strains to serve as standards of the characteristics attributed to a particular species.

Microorganisms can be classified, or distinguished from one another, by the ability to (1) grow on different substrates and/or production of different end products, (2) produce specific enzymes, (3) use oxygen, or (4) be motile. For example, certain microbes can use different carbohydrates as sources of energy and/or carbon. Because such variability exists in carbohydrate utilization between different microbes, this can aid in the group, genus, or species identification.

8.17 CLASSIFICATION OF MICROBES ON THE BASIS OF GENOTYPIC CHARACTERS

Genotypic identification is emerging as an alternative or complement to establish phenotypic methods. The characterization of the organisms can also be done utilizing the genotypic properties. As discussed earlier, several kinds of analysis performed upon isolated nucleic acids furnish information about the genotype, the analysis of the base composition of DNA, the study of chemical hybridization between nucleic acids isolated from different organisms, and the sequencing of nucleic acids. 16S rRNA sequence-based methods, DNA base ratio and DNA hybridization offer a viable option for the rapid and reliable identification.

8.17.1 DNA Base Ratio (G+C Ratio)

DNA base composition can only prove that organisms are unrelated. The ratio of bases in DNA can vary over a wide range. If two organisms have different DNA base compositions, they are not related. However, organisms with identical base ratios are not necessarily related, because the nucleotide sequences in the two organisms could be completely different.

In molecular biology, GC-content (or guanine-cytosine content) is the percentage of nitrogenous bases on a DNA molecule which are either guanine or cytosine (from a possibility of four different ones, also including adenine and thymine). This may refer to a specific fragment of DNA or RNA, or that of the whole genome. When it refers to a fragment of the genetic material, it may denote the GC content of part of a gene (domain), single gene, group of genes (or gene clusters) or even a non-coding region. G (guanine) and C (cytosine) undergo a specific hydrogen bonding whereas A (adenine) bonds specifically with T (thymine).

The GC pair is bound by three hydrogen bonds, while AT pairs are bound by two hydrogen bonds. DNA with high GC-content is more stable than DNA with low GC-content, but contrary to popular belief, the hydrogen bonds do not stabilize the DNA significantly and stabilization is mainly due to stacking interactions. In spite of the higher thermostability conferred to the genetic material, it is envisaged that cells with DNA with high GC-content undergo autolysis, thereby reducing the longevity of the cell *per se*. Due to the robustness endowed to the genetic materials in high GC organisms it was commonly believed that the GC content played a vital role in adaptation temperatures, a hypothesis which has recently been refuted.

In PCR experiments, the GC-content of primers is used to predict their annealing temperature to the template DNA. A higher GC-content level indicates a higher melting temperature.

GC content is usually expressed as a percentage value, but sometimes as a ratio (called G+C ratio or GC-ratio). GC-content percentage is calculated as

$$\frac{G+C}{A+T+G+C} \times 100$$

whereas the AT/GC ratio is calculated as

$$\frac{A+T}{G+C}.$$

The GC-content percentages as well as GC-ratio can be measured by several means but one of the simplest methods is to measure what is called the melting temperature of the DNA double helix using spectrophotometry. The absorbance of DNA at a wavelength of 260 nm increases fairly sharply when the double-stranded DNA separates into two single strands when sufficiently heated. The most commonly used protocol for determining GC ratios uses flow cytometry for large number of samples.

GC content is found to be variable with different organisms, the process of which is envisaged to be contributed to by variation in selection, mutational bias and biased recombination-associated DNA repair. The species problem in prokaryotic taxonomy has led to various suggestions in classifying bacteria and the adhoc committee on reconciliation of approaches to bacterial systematics has recommended use of GC ratios in higher level hierarchical classification.

For example, the Actinobacteria are characterised as "high GC-content bacteria". In *Streptomyces coelicolor*, GC content is 72%. The GC-content of Yeast (*Saccharomyces cerevisiae*) is 38%, and that of another common model organism Thale Cress (*Arabidopsis thaliana*) is 36%. Because of the nature of the genetic code, it is virtually impossible for an organism to have a genome with a GC-content approaching either 0% or 100%. A species with an extremely low GC-content is *Plasmodium falciparum* (GC% = ~20%), and it is usually common to refer to such examples as being AT-rich instead of GC-poor.

Physical methods of analysis also provide an indication of the molecular homogeneity of a DNA sample if every molecule of DNA had the same G+C content, both the thermal transition in a melting curve and the band position.

The GC content is often measured by determining the temperature at which the double stranded DNA denatures (Figure 8.20). Because three hydrogen bonds occur between G and C base pairs, and only two hydrogen bonds hence, high GC content melts at a higher temperature.

The temperature at which the double stranded DNA melts can readily be determined by monitoring the absorbance of UV light by the solution of DNA as it is heated. The absorbance readily increases as double stranded DNA denatures. In a typical melting curve (Figure 8.21), the increase in UV absorbance can be measured as the temperature increases. This tracks the unwinding and denaturation of DNA. The melting point (T_m) is the temperature at which half the DNA is unwound.

DNA that consists entirely of AT base pairs melts at about 70°C and DNA that has only G/C base pairs melts at over 100°C. The T_m of any DNA molecule can be calculated if you know the base composition. The simplest formulas just take the overall composition into account and they are not very accurate. More accurate formula will use

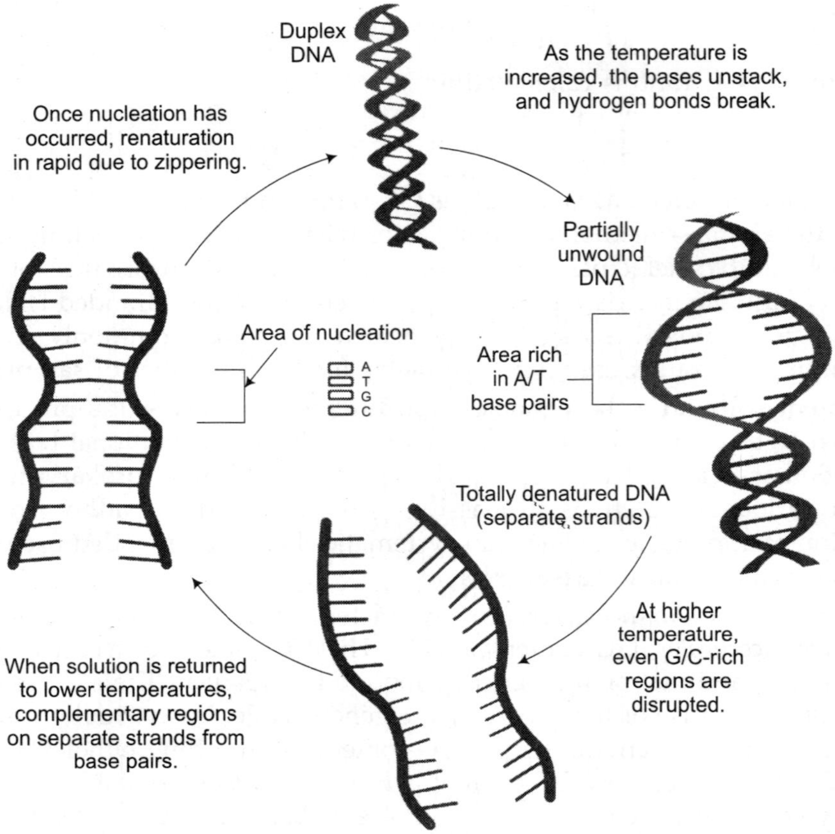

Figure 8.20 Effect of raising the temperature of DNA.

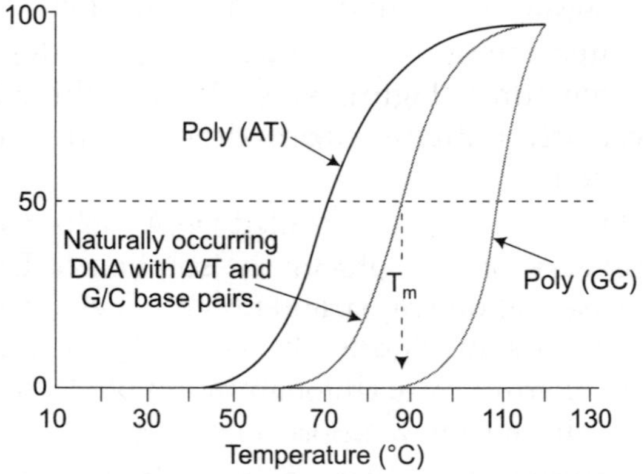

Figure 8.21 Melting curve of DNA.

the stacking interactions of each base pair to predict the melting temperature. The GC content varies among the different kinds of bacteria, with numbers ranging from 28% to 78%. Organisms that are related by other criteria have DNA base composition that are similar or identical. Thus, if the GC content of two organisms differ by more than small percent, they cannot be closely related. However, similarity does not necessarily mean that the organism is related, since many arrangements of the bases are possible. The genome size and the actual nucleotide sequences also differ greatly.

8.17.2 DNA Hybridization

DNA-DNA hybridization generally refers to a molecular biology technique that measures the degree of genetic similarity between pools of DNA sequences. It is usually used to determine the genetic distance between two species. When several species are compared that way, the similarity values allow the species to be arranged in a phylogenetic tree; it is therefore one possible approach to carrying out molecular systematics.

Hybridization between the total DNA of two organisms is useful for detecting relationships between closely related organisms. The extent of nucleotide sequence similarity between two organisms can be determined by measuring how completely single strands of their DNA will hybridize to one another. Just as two complementary strands of DNA from one organism will base pair or anneal, so will the similar DNA of the different organism. The degree of hybridization will reflect the degree of sequence similarity. DNA from organism that share many sequences will hybridize more completely the DNA from those that do not.

Upon rapid cooling of the solution of thermally denatured DNA, the single strands remain separated. However, if the solution is held at a temperature from 10°C to 30°C below the *Tm* value, specific reassociation (annealing) of the complementary strands to form double stranded molecules occur. There is always random pairing, but since a randomly matched duplex contains many mismatched base pairs, its thermal stability is low and its strands separate very rapidly at temperatures near the *Tm*. In contrast, pairing of the complementary strands forms duplexes that are quite stable because each base participates in interstrand hydrogen bonding. Thus, at temperatures near the *Tm*, only duplexes between the strands with high degree of complementarily persist; the closer that the temperature of incubation is to the *Tm*, the more stringent is the requirement of base pairing.

Shortly after the discovery of this phenomenon, it was shown that when DNA preparations from two related strains of bacteria are mixed and treated in this manner, hybrid DNA molecules are formed (Figure 8.22). The discovery of the reassociation of stranded DNA molecules from different biological sources to from hybrid duplexes laid the foundations of an entirely new approach to the study of genetic relatedness in bacteria. *In vitro* experiments of DNA-DNA associations permit an assessment of the overall degree of genetic homology between the bacteria. Since duplexes can also be formed between single stranded DNA and complementary RNA strands, analogous DNA-RNA reassociations can be performed. If the RNA preparations consist of either tRNAs or rRNAs, such experiments permit an assessment of the genetic homology

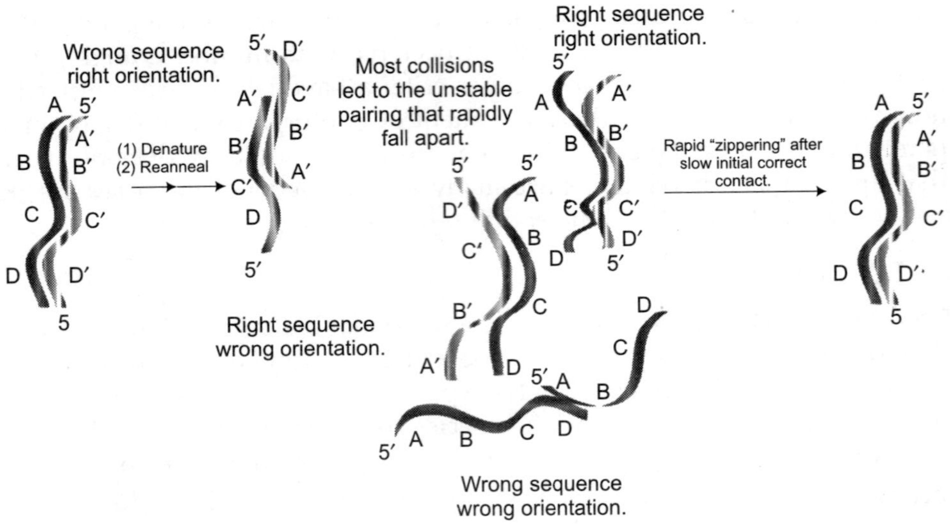

Figure 8.22 DNA-DNA hybridization.

between two bacteria with respect to specific, relatively small segments of chromosome: those that code the base sequences either of the transfer RNAs or of the ribosomal RNA. The range of organisms among which genetic homology is detectable can greatly extended by parallel studies on DNA – rRNA reassociation, because the relatively small portion of the bacterial genome that codes for the ribosomal RNA has a much more conserved sequence than the bulk of the chromosomal DNA. As a result it is frequently possible to detect the DNA – rRNA reassociation relatively high homology between the genomes of the two bacteria which shows no specific homology by DNA – DNA reassociation. The rates of the reassociation is inversely proportion to the length of the reassociating DNA (Figure 8.23).

In a bacterial group the value of nucleic acid reassociation studies is directly related to the number of strains and species that have been compared. Extensive comparative data are available for several major bacterial groups.

Whole genomic DNA-DNA hybridization has been a cornerstone of bacterial species determination but is not widely used because it is not easily implemented. Cluster analysis of the hybridization profiles revealed taxonomic relationships between bacterial strains tested at species to strain level resolution, suggesting that this approach is useful for the identification of bacteria as well as determining the genetic distance among bacteria. Since arrays can contain thousands of DNA spots, a single array has the potential for broad identification capacity.

8.17.3 Nucleotide Sequence Analysis

Genotype information at highest precision may be determined as DNA (or RNA) nucleotide-base sequences. RNA's are often sequenced either by converting the RNAs into DNA or by sequencing the DNA gene that gives rise to the RNA. By using

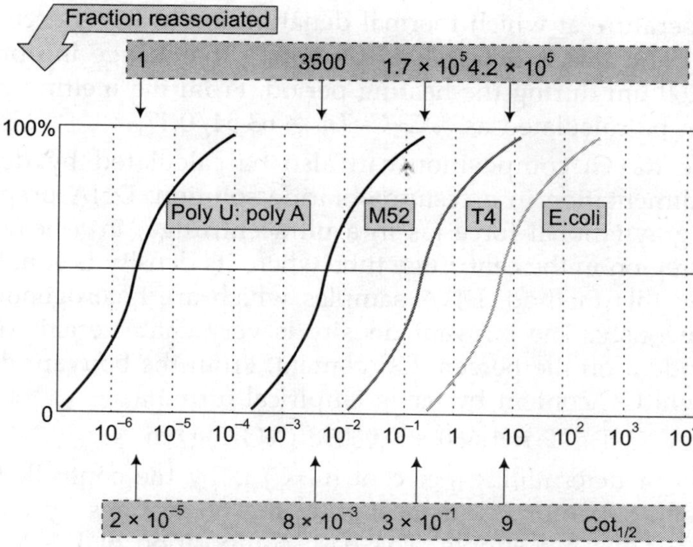

Figure 8.23 Rate of reassociation.

Polymerase Chain Reaction (PCR) to amplify a known DNA segment and automated techniques to sequence the amplified product, it is possible to compare multiple isolates.

One is the analysis of the base composition of DNA, i.e., to determine the mole per cent of guanine and cytosine in DNA (%G+C). The second is to determine the degree of similarity between two DNA samples by hybridization between DNA and DNA or DNA and RNA.The basis of this test is that the degree of hybridization would be an indication of the degree of relationship (homology). The relative percentage of guanine and cytosine (G+C/A+T+G+C) × 100 varies widely with different bacteria. The composition of chromosomal DNA is a fixed property of each cell and is independent of age and other external influences. The per cent (G+C) of chromosomal DNA can be determined by extracting DNA from cells by rupturing carefully. The DNA is then purified to remove non-chromosomal DNA. Since no preparation shows absolute molecular homogeneity, the G+C content is always a mean value and represent the peak in the normal distribution curve. Each bacterial species have DNA with a characteristic mean G+C content; this can be considered one of the important specific characters. Mean DNA base composition is a character of taxonomic value among bacteria, since the range for the group as a whole is so wide.

The base composition can then be determined either by subjecting the purified DNA to increasing temperature and determining the increase in hypochromicity or by centrifugation of the DNA in cesium chloride density gradients. The basis of the first method, i.e., the melting point method, is that when double stranded DNA is subjected to increasing temperature, the two DNA strands separate at a characteristic temperature. The melting temperature depends upon the G+C content of the DNA. Higher the G+C content, higher will be the melting point.

The mean temperature at which thermal denaturation of DNA occurs is called the melting point (*Tm*) and this is determined by noting the change in optical density of DNA solution at 260 nm during the heating period. From the melting point, the mole per cent (G+C) can be calculated as %G+C=*Tm* × 63.54/0.47.

The percentage (G+C) composition can also be calculated by determining the relative rate of sedimentation in a cesium chloride solution. DNA preparations when subjected to high gravitational force (as in a ultracentrifuge) in a heavy salt solution will sediment at a region in the centrifuge tube where its density is equal to the density of the medium. By this method, DNA samples which are heterogenous can also be separated simultaneously. The buoyant density is very characteristic of each type of DNA and is dependent on the percent GC content, From the bouyant density one can calculate the percent GC content by using empirical formula

$$P = 1.660 + 0.00098\ (\%G.C)g.cm^3$$

A third method of determining per cent (G+C) is by the controlled hydrolysis of DNA with acids and separating and measuring the nucleotides by chromatography. This method is laborious but simple. The base composition of DNA from a variety of organisms determined by these procedures. The genetic relatedness can also be determined by measuring the extent of hybridization between denatured DNA molecules between single stranded DNA and RNA species. The degree of homology is determined by mixing two kinds of single stranded DNA or single stranded DNA with RNA under appropriate conditions and then measuring the extent to which they associate to form double stranded structures. This can be precisely measured by making either the DNA or RNA radioactive. The degree of relatedness of different bacteria as determined by DNA-RNA hybridization. Although genetic relatedness can be determined by DNA-RNA hybridization, the DNA-DNA hybridization is most accurate provided precautions are taken to ensure that hybridization between two strands is uniform. The technique is advantageous as it can be applied on all strains; results are reproducible with ease in interpretation. But the process requires costly reagents and equipment besides being labour intensive.

Early in the chemical study of DNA preparation from different organisms and subsequent work has revealed that the base composition of DNA is a character of profound taxonomic importance, particularly among microorganisms.

8.17.4 Comparing the Sequence of 16S Ribosomal Nucleic Acid

Many of the modern molecular tools are based on 16S ribosomal DNA sequence, complete or partial genomes or specific fluorescent probes that monitor the physiological activity of microbial cells (Table 4.2). The tools that have been developed for identifying microbes and analyzing their activity can be divided into those based on nucleic acids and other macromolecules and approaches directed at analyzing the activity of complete cells. The nucleic acid–based tools are more frequently used because of the high throughput potential provided by using PCR amplification or *ex situ* or *in situ* hybridization with DNA, RNA, or even peptide nucleic acid probes. Notably, these include 16S rDNA sequences that can be used to place diagnostics into a phylogenetic framework and can be linked to databases providing up to 100,000 sequences (Amann and Ludwig, 2000). These 16S rDNA–based methodologies are robust and superior to traditional methods based on phenotypic approaches, which are often unreliable and lack the resolving power to analyze the microbial composition and activity of bacterial populations. In addition, a panoply of approaches that are based on DNA sequences other than rDNAs have been applied frequently to probiotic bacteria. These have been shown to be particularly useful for strain identification.

A promising method for simultaneous and selective detection of both culturable and nonculturable bacteria of defined taxonomic groups is the amplification of 16S ribosomal DNA (rDNA) or ribosomal RNA (rRNA) sequences using PCR. The primary structures are well investigated and conserved, and variable regions have been determined. Primers located in highly conserved regions have been published, allowing the amplification of 16S rDNA and subsequent sequence analysis. Certain signatures in the nucleotide sequence can be unique for particular phylogenetic groups, offering the opportunity to design genus specific probes, whereas the variable regions can be used to assign organisms to lower taxonomic groups (Mehling *et al.*, 1995). The determination of full-length 16S rDNA sequences, as opposed to partial gene sequences, of streptomycete and some other actinomycete strains has provided data which may be useful in elucidating taxonomic levels or detecting chimeric PCR-products. The design of PCR primers with potential for the differentiation of strains at the genus, species and strain levels was made possible by sequence analysis of the complete 16S rDNA sequences. The possible combinations of genus and strain specific primers permit diverse assays, such as multiplex PCR or PCR with nested primers, lessening the likelihood of false-positive identification of streptomycetes and thus increasing the fidelity of the assay.

DNA-based technology for the identification of bacteria typically uses only the 16S rRNA gene as the basis for identification. This technique has the advantage of being able to identify difficult to cultivate strains, and is growth and operator independent. As the 16S rRNA gene is highly conserved at the species level, speciation is commonly quite good, but as a result, subspecies and strain level differences are not shown. Some problems with the 16S rRNA technology are that it requires a high level of technical proficiency, and the costs per sample, as well as equipment costs are high. As a result, the technology is not well suited for routine microbial quality control (QC), but rather is best used for direct product failures (Sutton and Cundell, 2004). Technology that uses information from both the 16S rRNA and 23S rRNA genes is also used in pharmaceutical QC, but primarily to aid in strain tracking.

Sequence comparisons of small subunit rRNA have been used as a source for determining phylogenetic and evolutionary relationships among organisms of the three kingdoms Archaea, Eucarya and Bacteria. The present compilation of complete genes for the small subunit rRNA contains over 2200 16S and 16S-like sequences (Gutell *et al.*, 1994). The 16S rRNAs are highly conserved, sharing common three-dimensional structural elements of similar function. To facilitate the differential identification of the genus Streptomyces, the 16S rRNA genes of 17 actinomycetes were sequenced and screened for the existence of Streptomycete-specific signatures. The 16S rDNA of the Streptomyces strains and *Amycolatopsis orientalis* subsp lurida exhibited 95-100% similarity, while that of the 16S rDNA of *Adnoplanes utahensis* showed only 88% similarity to the streptomycete 16S rDNAs. Potential genus specific sequences were found in regions located around nucleotide positions 120,800 and 1100. Several sets of primers derived from these characteristic regions were investigated as to their specificity in PCR-mediated amplifications. Most sets allowed selective amplification of the streptomycete rDNA sequences studied. RFLPs in the 16S rDNA permitted all strains to be distinguished.

Over the last decade, hybridizations with ribosomal RNA (rRNA)-targeted probes have provided a unique insight into the structure and spatiotemporal dynamics of complex microbial communities. Nucleic acid probes can be designed to specifically target taxonomic groups at different levels of specificity (from species to domain) by virtue of variable evolutionary conservation of the rRNA molecules. Appropriate software environments such as the ARB package, a software environment for sequence data (http://www.arb-home.de/) and availability of large databases (http://rdp.cme.msu.edu/html/), or the online resource for oligonucleotide probes probeBase (http://www.microbial-ecology.de/probebase/index.html) offer powerful platforms for a rapid probe design and in silico specificity profiling. Oligonucleotide probes that are complementary to regions of 16S or 23S rRNA have been successfully used for the identification of lactic acid bacteria, and hence, they offer the potential to be used as reliable and rapid diagnostic tools.

Suggested Readings

Bosshard PP, Abels S, Altwegg M, Bottger EC, Zbinden R. 2004. Comparison of conventional and molecular methods for identification of aerobic catalase-negative gram-positive cocci in the clinical laboratory. *J Clin Microbiol* **42,** 2065-2073.

Gevers D, Cohan FM, Lawrence JG, Spratt BG, Coenye T, Feil EJ, Stackebrandt E, Van De Peer Y, Vandamme P, Thompson FL, Swings J. 2005. Defining prokaryotic species Re-evaluating prokaryotic species. *Nature Rev Microbiol* **3,** 733-739.

Goris J, Konstantinidis KT, Klappenbach JA, Coenye T, Vandamme P, Tiedje JM. 2007. DNA-DNA hybridization values and their relationship to whole-genome sequence similarities. *Int J Syst Evol Microbiol* **57,** 81-91.

Karlin S, Burge C. 1995. Dinucleotide relative abundance extremes: a genomic signature. *Trends Genet* **11,** 283-290.

Konstantinidis KT, Stackebrandt E. 2013. **Defining Taxonomic Ranks.** In *The Prokaryotes (4th edition): Prokaryotic Biology and Symbiotic Associations.* pp229, 4th edition. Edited by Rosenberg E, DeLong EF, Lory S, Stackebrandt E, Thompson FL. Springer, New York.

Konstantinidis KT, Tiedje JM. 2005. Towards a genome-based taxonomy for prokaryotes. *J Bacteriol* **187,** 6258-6264.

Kunitsky C, Osterhout G, Sasser M. 2005. Identification of microorganisms using fatty acid methyl ester (fame) analysis and the midi Sherlock microbial identification system. *In Encyclopedia of Rapid Microbiological Methods* **3,** 1-18.

Lapage SP, Sneath PHA, Lessel EF, Skerman VBD, Seeliger HPR, Clark WA. 1992. *International Code of Nomenclature of Bacteria: Bacteriological Code, 1990 Revision*. ASM Press, Washington (DC).

Márquez MC, Ventosa A, Ruiz-Berraquero F. 1987. A taxonomic study of heterotrophic halophilic and non-halophilic bacteria from a solar saltern. *J Gen Microbiol* 133, 45-46

Nakamura S, Nakaya T, Iida T. 2011 Metagenomic analysis of bacterial infections by means of high-throughput DNA sequencing. *Exp Biol Medi (Maywood, NJ)* **236,** 968-971.

Neimark HC. 1986. Origin and evolution of wall-less prokaryotes. In *The bacterial L-Forms*. 21-42. Edited by Madoff S. Marcel Dekkar Inc, New York.

Partensky F, Hess WR, Vaulot D. 1999. Prochlorococcus, a marine photosynthetic prokaryote of global significance. *Microbiol Mol Biol Rev* **63,** 106-127.

Polz MF, Alm EJ, Hanage WP. 2013. Horizontal gene transfer and the evolution of bacterial and archaeal population structure. *Trends Genet* **29,** 170-175.

Skerman VBD, McGowan V, Sneath PHA. 1980 Approved lists of bacterial names. *Int J Syst Evol Microbiol* **2,** 3-4.

Sneathp HA. 1972. Computer taxonomy. In *Methods in Microbiology,* vol. 7A, pp. 29-98. Edited byJ. R. Norris & D. W. Ribbons, London: Academic Press.

Thompson CC, Luciane Chimetto L, Edwards RA, Swings J, Stackebrandt E, Thompson LF. 2013. Microbial genomic taxonomy *BMC Genomics* **14,**913 doi:10.1186/1471-2164-14-913.

Thompson CC, Vieira NM, Vicente A, Thompson F. 2011. Towards a genome based taxonomy of Mycoplasmas**.** *Infect Genet Evol* **11,** 1798-1804.

Vandamme P, Pot B, Gillis M, de Vos P, Kersters K, Swings J. 1996. Polyphasic taxonomy, a consensus approach to bacterial systematics. *Microbiol Rev* **60,** 407-438.

Whittaker RH.1959. On the broad classification of organisms. *Quart Rev Biol* **34,** 210-226.

Willems A, Doignon-Bourcier F, Goris J, Coopman R, de Lajudie P, De Vos P, Gillis M. 2001. DNA-DNA hybridization study of *Bradyrhizobium* strains. *Int J Syst Evol Microbiol* **51**(Pt 4), 1315-1322.

Chapter 9

Microbial Phylogeny

9.1 INTRODUCTION

Microbes include several distinct groups of organisms, differing widely in form and life cycle but resembling one another in their small size and relatively simple structure, the way they live and multiply, their form and classification and the role they play in the varied habitats they occupy. The question of whether or not microorganisms are clustered in groups with certain likenesses is pivotal. If microbial diversity is organized in some way, then taxonomy needs merely to find a way to reflect that diversity. If microorganisms do not form groups with meaningful traits and evolutionary history, then organizing microbial diversity will be more difficult, if not impossible (http://www.envismadrasuniv.org). It is also possible that some microorganisms fit nicely into clusters while others do not. In this event, microbial taxonomy may have to embrace multiple approaches to classification. Moving forward in the fields of microbial taxonomy and systematic, research needs to address more fully the issues of clustering, community genomics, and the development of novel approaches for classification and characterization. The importance of this contribution is not only realized in phylogeny (the evolutionary history of organisms) but also virtually every other discipline of biology including microbiology.

9.2 PHYLOGENY

Phylogeny is the study of evolutionary relationships among groups of organisms (e.g., species, populations), which are discovered through molecular sequencing data and morphological data matrices. The result of phylogenetic studies is a hypothesis about the evolutionary history of taxonomic groups. The classification, identification, and naming of organisms is usually richly informed by phylogenetics, but remains a methodologically and logically distinct discipline. The degree to which taxonomies depend on phylogenies differs depending on the school of taxonomy: numerical taxonomy ignores phylogeny altogether, trying to represent the similarity between organisms instead; phylogenetic systematics tries to reproduce phylogeny in its classification without loss of information; evolutionary taxonomy tries to find a compromise between them in order to represent stages of evolution.

Molecular phylogenies divide all living organisms into three domains Bacteria, Archaea, and Eukarya (Figure 9.1) (eukaryotes: protists, fungi, plants, animals).

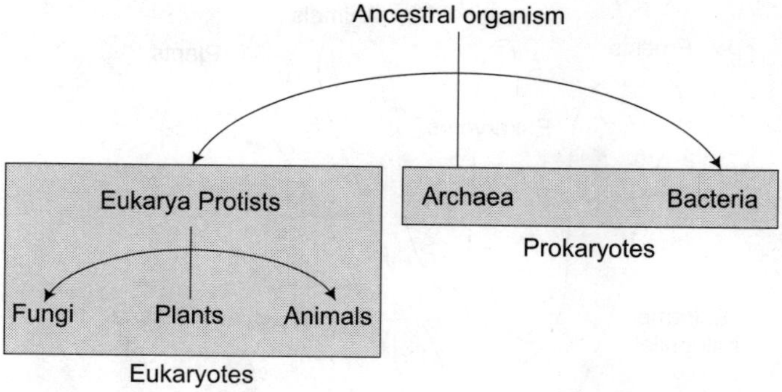

Figure 9.1 Interrelation between different groups of organisms.

Eukarya members include algae, fungi and protozoa. Algae and protozoa are also referred to as protists. Like prokaryotes there exists a diverse array of eukaryotic microorganisms.

Some of these such as algae are phototropic; they contain chlorophyll rich organelles called chloroplast and live in environment that contains only a few minerals, carbohydrate and CO_2 and light.The Eucarya are usually classified into four kingdoms: plants, animals, fungi and protists. Kingdom under domain Archaea is Archaeabacteria and kingdom under domain Bacteria is Eubacteria. Eukarya is the one which has 4 kingdoms under, namely Plantae, Animalia, Protista and Fungi. The first three of these correspond to phylogenetically coherent groups as well. However, the eukaryotic protists do not form a group, but rather are comprised of many phylogenetically disparate groups (including slime molds, multiple groups of algae, and many distinct groups of protozoa). Just as molecular analyses were required to see the natural relationships among prokaryotes, they are also allowing us to infer relationships among these non-plant, non-animal, non-fungus eukaryotes (Figure 9.2).

Many of the groups of fungi are important. In addition to providing yeast for making bread and fermenting sugar to alcohol, the fungi include the causative agents of such conditions as blight, ergot, rot, rust, smut, wilt, ringworm and athlete's foot. Although it might seem that it is straight forward to identify an organism as a fungus, this is not always the case. The unambiguous identification of the opportunistic pathogen, *Pneumocystis* (which has not been grown in culture), as a fungus required analysis of its ribosomal RNA. The "eukaryotic protists" also provide many organisms of clinical importance, including *Crithidia, Trypanosoma, Plasmodium, Amoeba,* and *Giardia.*

The Eukarya differ fundamentally from the Bacteria and Archaea in terms of the organization of the genetic machinery with multiple chromosomes being housed within a nucleus bounded by a double membrane, orgenells (Figure 9.3). This unique and complex organization must be explained by any theory on the origin of the Eukarya. The theory presented here delivers an explanation in addition to solving the lipid problem. It explains the eukaryal genetic machinery as an automatic, irreversible, almost deterministic consequence of the endosymbiosis, whereby a pre-cell was converted into

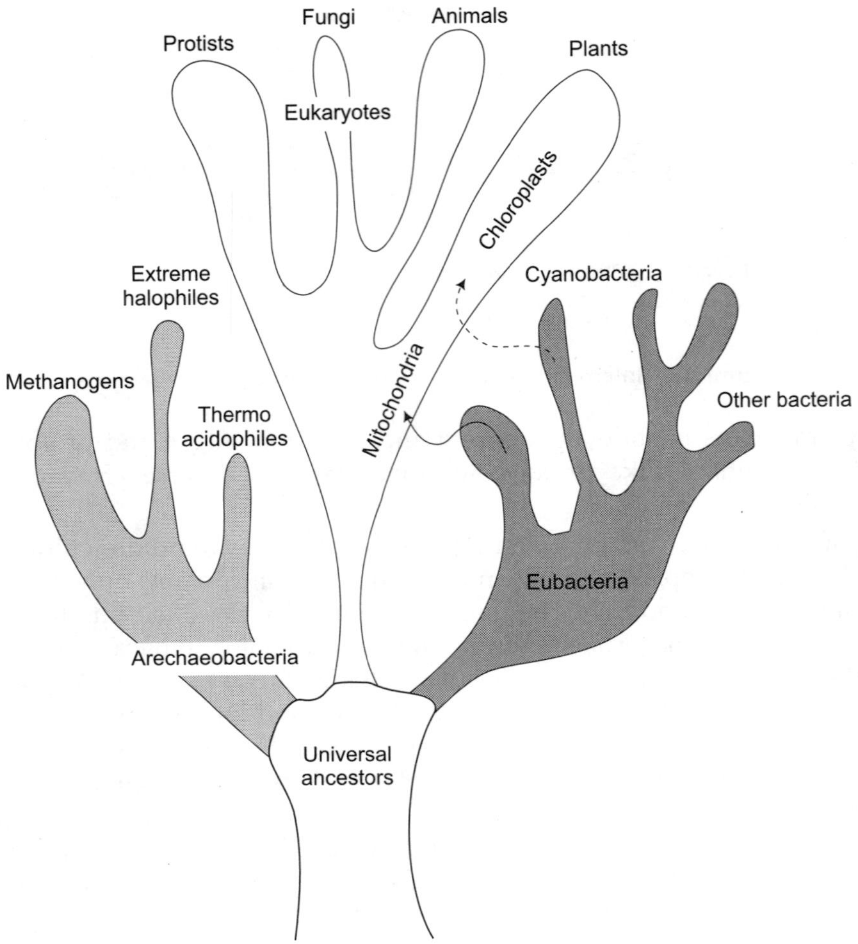

Figure 9.2 Phylogenetic evolution of Eukarya (Eukaryotes).

the nucleus. The following 11 features that are shared by all extant Eukarya may be explained by the historic origin of the nucleus in an endosymbiotic pre-cell within a bacterial host.

(i) The non-organellar genome is polychromosomal and the chromosomes are housed inside the nucleus. This can be explained by the hypothesis that a polychromosomal pre-cell turned into the nucleus, whereby the pre-cell chromosomes became trapped and separated from the outer cell membrane by the nuclear double membrane. Therefore, their replication could not become linked with the outer cell membrane during cell division. This means that there was no selective advantage in their unification into one large circular chromosome as was the case with the Bacteria and Archaea.

(ii) The nuclear membrane is endowed with pores and connected with an endoplasmic reticulum.

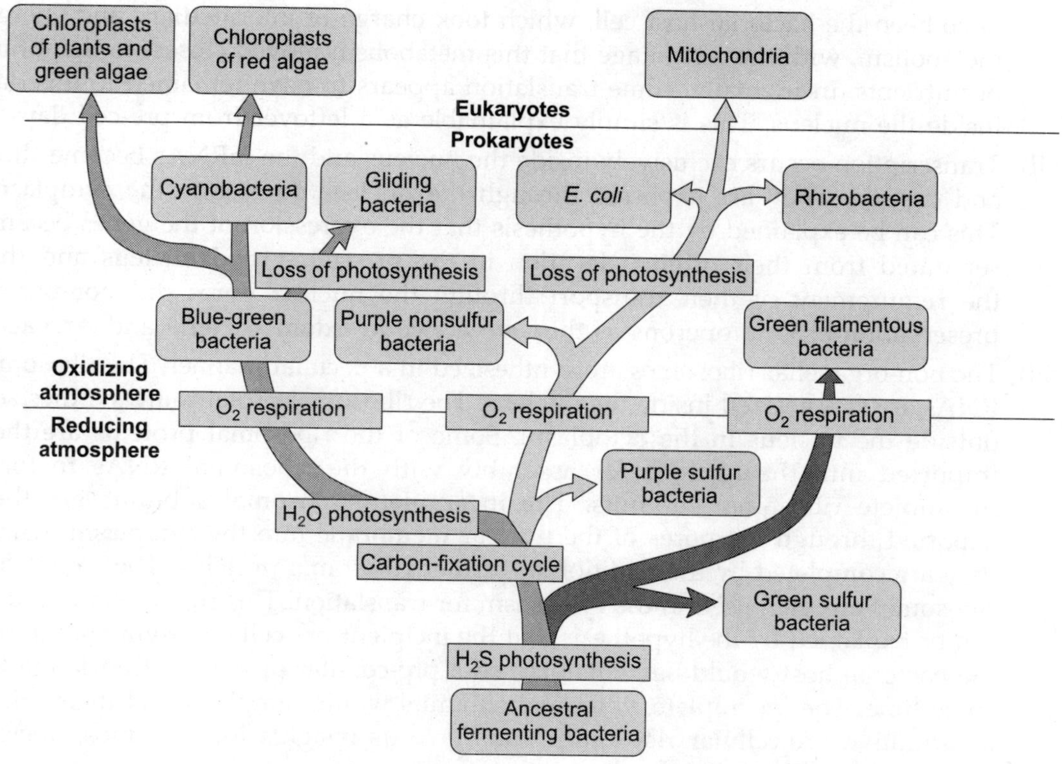

Figure 9.3 Phylogenetic evolution of organelles in Eukarya.

(iii) The non-organellar chromosomes are linear as opposed to the circular bacterial and archaeal chromosomes. This can be explained by their possible origin in the linear intermediates of the hypothesized rolling circle replication of the pre-cell chromosomes, perhaps with a multiplication in length.

(iv) Reproduction is based on a process of mitosis. The higher Eukarya, like plants and animals, have an open mitosis, while all others have a closed mitosis. This finds a simple explanation in the hypothesis that the nucleus originated from a polychromosomal pre-cell, which means that closed mitosis is original and open mitosis is derived.

(v) The nuclear chromosomes contain genes of bacterial origin. Approximately 75% of these bacterial genes in the nuclear chromosomes are clearly of non-organellar vintage. This can be explained with an evolutionary process, whereby the outer bacterial chromosomes were lost and their indispensable genes became relocated into the nuclear chromosomes.

(vi) The intermediary and energy metabolism is restricted to the outer cytoplasm. As a corollary protein synthesis by translation is also restricted to the outer cytoplasm, i.e., to exist in close quarters with the metabolism, where the enzymes are needed. This can be explained by the hypothesis that in the endosymbiosis, it would have been the bacterial host cell, which took charge of intermediary and energy metabolism, with the advantage that this metabolism moved closest to the influx of nutrients. Incidentally, some translation appears to have remained to this day inside the nucleus. This is simply explainable as a leftover from pre-cell days.

(vii) Transcription occurs exclusively inside the nucleus and the mRNAs become short and capped. They are exported through the nuclear pores into the cytoplasm. This can be explained by the hypothesis that the expression of the genes became separated from their original location in the pre-cell turned nucleus and that the requirement of their transport through the nuclear pores did not permit preservation of long operons as they are active in extant Bacteria and Archaea.

(viii) The non-organellar ribosomes are synthesized in a peculiar manner. The ribosomal RNAs are synthesized inside the nucleus. The ribosomal proteins are synthesized outside the nucleus in the cytoplasm. Some of the ribosomal proteins are then imported into the nucleus for assembly with the ribosomal RNAs to form incomplete ribosomal subunits. The incomplete ribosomal subunits are then exported through the pores of the nuclear membrane into the cytoplasm, where they are completed by the addition of further ribosomal proteins. The completed ribosomes are then used in the cytoplasm for translation. This roundabout pattern can be explained by the hypothesis that the incipient pre-cell endosymbiont inside the bacterial host would have maintained a pre-cellular process of translation for some time. The incomplete ribosomal subunits would simply reflect their origin in primitive pre-cellular ribosomes, thus give us insights into the most ancient ribosomes and their early evolution.

(ix) Most genes in the nuclear chromosomes are infested with introns of unknown origin. This infestation with introns is not lethal or even particularly deleterious due to a peculiar processing (maturation) of the primary mRNAs inside the nucleus. The introns are spliced out and the remaining mRNA segments are interconnected to form a mature, functional mRNA. This can be explained by the hypothesis that a physical isolation of the intra-nuclear, pre-cell-derived space from the extra-nuclear bacteria-derived cytoplasmic space would have ensued with endosymbiosis. This means that the intra-nuclear space would have become a safe compartment for intron splicing so that immature intron-laden mRNAs would have been prevented from entering the cytoplasm, thereby the metabolism would have been protected from being messed up with products of a translation of immature mRNAs.

(x) The genes for the intermediary metabolism are most closely related to the bacteria while the genetic machinery is most closely related to and actually more complex than the genetic machinery of the Archaea. This can now be explained with reference to the hypothetical phylogeny. The Bacteria are shown as diverging first from relatively primitive pre-cells. The Archaea are shown as diverging later from evolutionarily more advanced pre-cells. The Eukarya are shown as diverging last (by endosymbiosis) from pre-cells that would have been still farther evolved than the pre-cells, from which the Archaea originated. This rendered the pre-cell-derived genetic machinery of the Eukarya similar to, but more evolved than the genetic machinery of the Archaea. The close relationship of the metabolic genes of the Eukarya to those of the Bacteria is simply due to their bacterial origin.

(xi) The genetic machinery inside the nucleus has somewhat 'primitive' features with numerous small RNAs being involved in replication, transcription and mRNA maturation. This can now be explained by the hypothesis that these 'primitive' features would date back as leftovers to the workings of the 'primitive' pre-cells that were involved in endosymbiosis.

With these 11 explanatory points, the presentation of theory of the origin of the Eukarya by endosymbiosis of pre-cells and wall-less bacteria is completed.

The analysis and comparisons of highly conserved phylogenetic markers enabled to develop a phylogenetic tree of the microorganisms. The ribosomes, as the sites of protein synthesis, are present in all cells. Functionally, they must be very conservative. This is especially true of ribosomal RNA (rRNA) because its base sequence is not affected by the degeneracy of the genetic code, nor by suppressor mutations. rRNA, therefore, provides the properties appropriate to a general phylogenetic marker.

The 16S rRNA from the 30S ribosomes of all the organisms that are to be compared has now been isolated. For analysis the RNA is divided into large pieces (oligonucleotides) by use of specific restriction enzymes. The oligonucleotides are separated by electyrophoresis and their base sequences determined. A catalogue of the characteristic base sequences for individual organisms can then be subjected to comparative analysis with the assistance of appropriate computer programs. This yields, on the one hand, higly conserved sequences, which are largely the same in all bacteria, but differ significantly from those of the eukaryotes and those of other large groups of bacteria. On the other hand, differences for some oligonucleotides sequences are found even within genera and species. Certain 'signature' sequences have been identified which seem to be characteristic for groups of organsisms.

These analyses have produced dendograms which can be considered parts of a phylogenetic tree. They have revealed hitherto unrecognized relationships. One of the most exciting results from such 16S rRNA sequence analyses was the discovery that there was a much higher sequence homology between two such morphologicalally and physiologically distinct bacteria as *Escherichia coli* and a *Cyanobacterium,* than between *E. coli* and a morphologically very similar *Methanobacterium.* This led to the recognition of *Methanobacteria* as belonging to a large group of fundamentally very different prokaryotes. It has to be assumed that the prokaryotes split very early into two large groups, the so called **Archaebacteria** (to which the Methanobacteria belong) and the other prokarytotes designated as **Eubacteria**. It is envisaged that the very early cells – the progenotes gave rise to the Archaebacteria and Eubacteria. The third group that arose was the *Eukaryotes.* Their precursors must also have developed from the common root. Comparison of the representative Eubacteria leads to the recognition of large groups (Figure 9.4). One of these is the group of anoxygenic phototrophic purple bacteria. Further detailed comparisons of the members of this group of Gram-negative bacteria with numerous chemotrophic Gram-negative bacteria led to another surprising result: the purple bacteria can be divided into four groups which have been designated as alpha, beta, gamma and delta purple bacteria. Each of these except the delta purple bacteria contains a few well known representatives of the purple bacteria as well as a large number of heterotrophic or chemolithotrophic non-photosynthetic bacteria. Phylogenetically, this suggests that at least four groups of purple bacteria developed during the course of evolution. The heterotrophic respiratory or fermentative bacteria, as well as the chemolithotrophs and other bacteria with anaerobic respiratory systems, must then have developed from the representatives of these four branches by loss of the photosynthetic apparatus and metabolic specialization. These non-photosynthetic bacteria, which are phylogenetically related to the purple bacteria, are collectively known as Proteobacteria. This leads to the general hypothesis that the majority of the soil and water bacteria, as well as many pathogenic organisms, are derived from the purple bacteria.

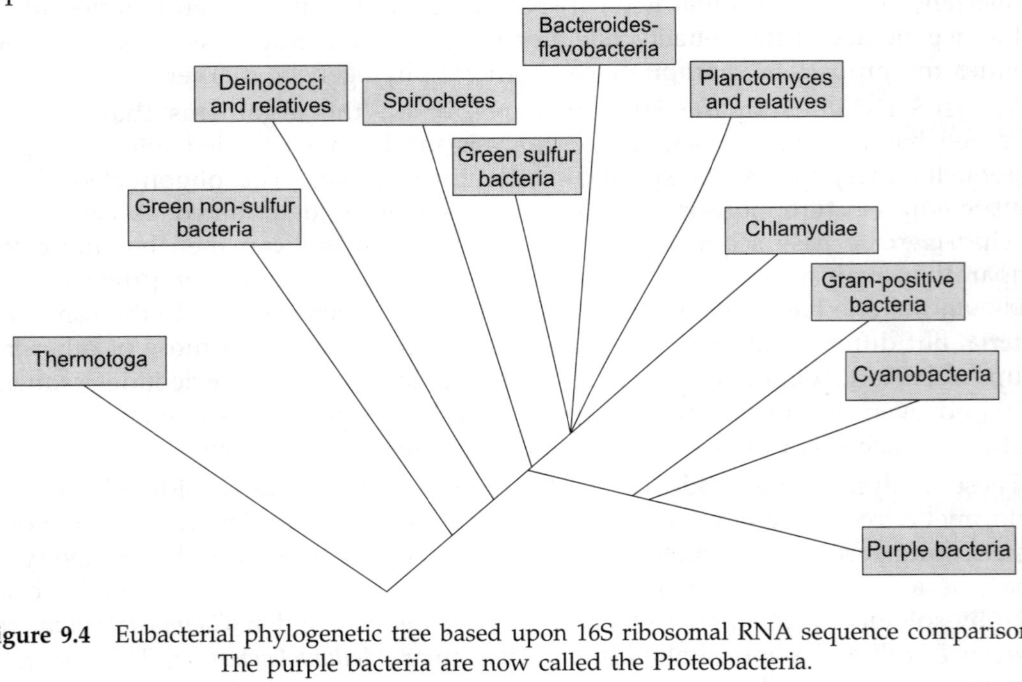

Figure 9.4 Eubacterial phylogenetic tree based upon 16S ribosomal RNA sequence comparisons. The purple bacteria are now called the Proteobacteria.

It can be readily appreciated that the taxonomic system, which is based mainly on phenotypic characters shape, flagellation, Gram staining behaviours, etc., is very different from the phylogenetic system. Only a few groups which have been classified in families or higher units by classical taxonomic criteria appear to be phylogenetically related. This is the case, for example, in the sulphate-reducing bacteria of the delta group and in case of the Eneterobacteriaceae. In contrast, members of the pseudomonads, are found in alpha, beta and gamma groups (of the phylogenetic branches) of the purple bacteria. It seems, therefore, that the cell type of Pseudomonas must be the result of convergent development.

A phylogenetically based classification system, however, does not replace the system based on the traditional taxonomy, or make it superfluous. The latter still has an important role in the identification, and in *de novo* description of bacteria. The phylogenetic system, though, conveys a wealth of heuristic principles for evolutionary studies as well as for applied microbiology.

9.3 EUBACTERIALES

The phylogeny of eubacteria, based on nucleotide catalogue analyses, suggests that eleven groups exist. The members of some groups are phylogenetically highly related. However, most of the groups are phylogenetically diverse, although coherent when compared with other groups. Some groups previously established by phenotypic criteria such as physiology or morphology, for example, green nonsulphur bacteria or spirochetes from distinct phylogenetic groups, however, are a collage of physiological and morphological types of bacteria, emphasizing the fact that phenotypic markers are often poor phylogenetic markers (Figure 9.4).

9.3.1 Purple Phototrophic Bacteria and their Relatives (Proteobacteria)

This group is the largest and physiologically most diverse of all the eubacteria. The group comprises four distinct subdivisions referred to as the *alpha, beta, gamma* and *delta* purple bacterial groups. Phototrophic purple phototrophic bacteria form the heart of the three subdivisions, but organotrophic and lithotrophic representatives are distributed across the group as well. The *delta* group contains only the organotrophic bacteria. This major grouping is now also called Proteobacteria. Major features of the each group are shown in Table 9.1.

Table 9.1 Major genera grouped under Purple Photosthetic bacteria

Major Groups	*Major Genera grouped in the purple photosynthetic bacteria*
Alpha Group	*Rhodospirillum*, Rhodopsudomonas*, Rhodobacter*, Rhodomicrobium", Rhodopila*, Rhizobium, Nitrobacter, Agrobacterium, Aquaspirillum, Hyphomicrobium, Acetobacter; Beijerinckia, Para coccus, Pseudomonas (some species)*
Beta Group	*Rhodocyclus*, Spirillum, Nitrosomonas, Sphaerotilus, Thiobacillus, Alcaligens, Pseudomonas, Bordetella, Neisseria*
Gamma Group	*Chromatium*, Thiospirillum**, and other purple sulphur bacteria *Beggiatoa, Leucothrix,* enteric bacteria, *Legionella, Azotobacter, fluorescent Pseudomonas species, Vibrio*
Delta Group	*Myxococcus, Bdellovibrio, Desulfovibrio,* and other sulfate reducing bacteria, *Desulfuronionas*

*Despite the name for the group, most of the bacteria listed are nonphotosynthetic. The major grouping is also now called Proteobacteria. Phototrophic representatives are shown by an asterisk after the name.

They are especially rich in intracytoplasmic membrane systems. In ultrathin sections these can be seen as tubules, vesicles, and plates. It is likely that the complex membrane systems and photopigments required for bacterial photosynthesis did not arise in so broad a phylogenetic group as the purple bacteria by convergent evolution. Instead, it has been proposed that the common ancestor of the purple bacterial group was itself phototrophic, and that the physiological diversity characteristic of the purple bacterial group arose as the result of the exchange of photosynthetic capacity for other physiological processes (such as sulfate reduction and lithotrophy) better adapted to new and expanding ecological niches (Figure 9.5). Such a hypothesis nicely accounts for the specific phylogenetic relatedness of organisms like Rhodopseudomonas (a budding, phototrophic bacterium containing polar stacks of lamellar membranes) and Nitrobacter (a budding, lithotrophic bacterium containing polar stacks of lamellar membranes), or the specific phylogenetic relatedness of phototrophic and certain organotrophic spirilla.

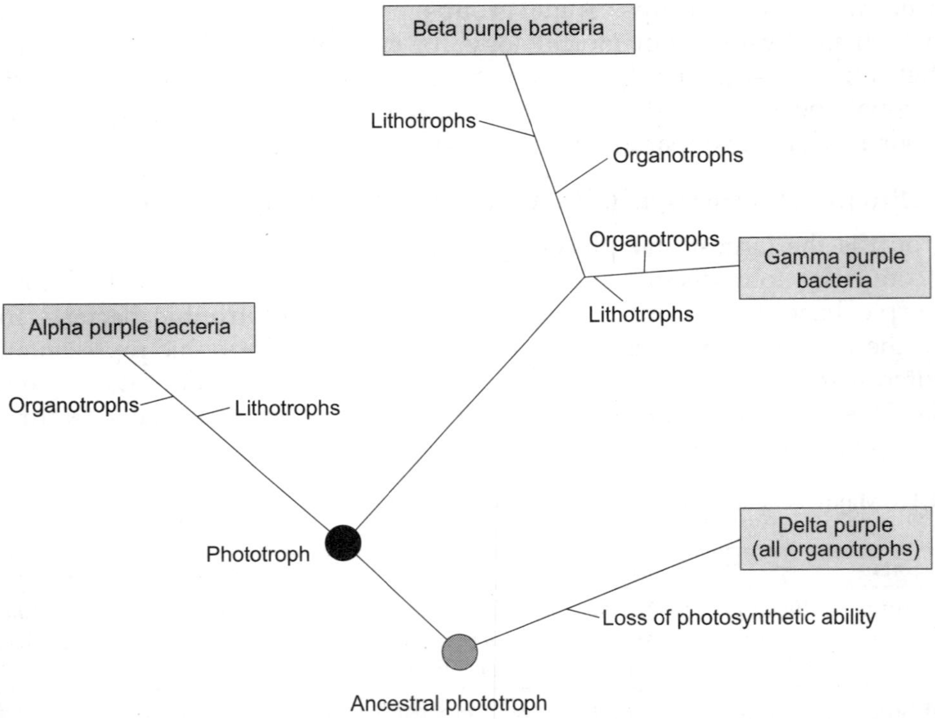

Figure 9.5 Divergence of the 'purple' bacteria (proteobacteria) to give rise to nonphotrophic relatives.

Nonsulfur purple bacteria are found in the alpha and beta subdivisions of the purple bacterial group, while purple bacteria are found in the gamma subdivisions (Table 9.1). Certain lithotrophic sulfide oxidizing bacteria which store sulfur intracellularly, for example, members of the genus *Beggiatoa* are also found within the gamma purple bacteria group. Sulfur lithotrophs carry out the oxidation of sulfide to sulfur and sulfate just as purple sulfur bacteria do, but do so as a means of obtaining energy rather than simply as a source of reducing power for CO_2 reduction. If the storage of sulfur intracellularly is a unique biochemical property, then one could imagine how the process of phototrophic sulfur storage could have been recruited as a mechanism to store sulfur for use as an energy source lithotrophically by *Beggiatoa* and its relatives. A variety of phototrophic sulfur bacteria are capable of lithotrophic (dark, microaerobic) growth at the expense of reduced sulfur compounds. Hence, the phylogenetic relationships between phototrophic and lithotrophic sulfide oxidizers are not surprising.

The *delta* group of purple bacteria is an extremely broad one that consists of physiologically diverse organsisms that appear to have no recognizable phenotypic thread among them. The major subgroups include (a) the sulfate and sulfur-reducing eubacteria, (b) the fruiting myxobacteria, and (c) the prokaryotic predator, *Bdellovibrio*. The sulfate/sulfur reducers are strictly anaerobic organotrophs that oxidize a variety of organic electron donors and reduce sulfate or sulfur to H_2S. The fruiting myxobacteria are obligate aerobes that move by gliding motility and which aggregate under certain nutritional conditions to form complex arrangements of cells called fruiting bodies. *Bdellovibrio* is a small, spiral-shaped Gram-negative bacterium which acts as a predator of other Gram-negative bacteria, attaching and eventually moving into the periplasmic space of its prey and destroying it.

Finally, all of the enteric bacteria, most *Pseudomonas* species, the free living and symbiotic nitrogen fixers, and most lithotrophs are also phylogenetically related to the purple bacteria. Hence, the purple bacteria lineage appears to have given rise to a wide diversity of ecologically and physiologically significant bacteria. The oligonucleotide sequence AAAUUCG is diagnostic of *alpha* purple bacteria, CYUUACACAUG (where Y is any base), is characteristic of the *beta* group, and ACUAAAACUCAAAG is present in the 16S RNA of most delta purple bacteria; no specific signature has been identified for the *gamma* group.

9.3.1.1 Green sulfur bacteria

The phototrophic green sulfur bacteria Chlorobium is related to other eubacteria (including purple phototrophic bacteria) at only the deepest phylogenetic levels. This lack of close relationship is not too surprising in light of the unique pigments, light harvesting apparatus (chlorosome), and autotrophic mechanisms used by these organisms. However, the finding that photosynthesis once again has deep evolutionary roots further supports the hypothesis that photosynthesis was an early invention that spread among several groups of eubacteria before the major lines of eubacterial descent were defined. Included in the green bacterial group besides Chlorobium is the genus Chloroherpeton, a gliding green bacterium; all species of Chlorobium are immotile. The sequence AUACAAUG provides a signature for green bacteria.

9.3.1.2 Green nonsulfur bacteria

The *Chloroflexus* group-A close relationship exists to green sulphur bacteria (*Chlorobium*). Phylogenetically, however, nothing could be further from the truth. The *Chloroflexus* group, which includes two nonphototrophic genera, *Herpetosiphon*, and *Thermomicrobium*, represents a distinct line of eubacterial descent. *Chloroflexus* is clearly an evolutionary puzzle because of its lack of phylogenetic relationship to green sulfur bacteria with which it shares a number of phenotypic properties. Because most of the members of the *Chloroflexus* group are thermophiles, the thermophilic property may be of more significance in defining the phylogeny of *Chloroflexus* and its relatives than is their relationship to green bacteria. The sequence CCUAAUG provides a signature for *Chloroflexus*.

9.3.2 Cyanobacteria

The taxonomy of the cyanobacteria is presently weighted toward morphological criteria, only a few morphologically defined cyanobacterial groups have firm phylogenetical standing. Both the branching and the heterocystous cyanobacteria appear to form rather tight phylogenetic group; unicellular cyanobacteria by contrast are phylogenetically diverse. The eubacterial branch leading to the cyanobacteria is also the branch within which the precursor of the green plant chloroplast arose. The nucleotide sequence AAUUUUYCG is a signature sequence for cyanobacteria.

9.3.3 Planctomyces/Pirella Group

This eubacterial group contains budding organisms which lack peptidoglycan, contain a proteinaceous cell wall, are obligately aerobic, and require very dilute culture media for growth. Many species are found in nature attached to surfaces by stalks of holdfasts. The *Planctomyces/ pirella* line form a major branch of eubacteria, albeit a branch quite distinct from the other groups. The oligonucleotide CUUAAUUCG provides a signature for members of the *Planctomyces/Pirella* group.

9.3.4 Spirochetes

The main spirochete cluster contains the genera *Spirochaeta, Borrelia,* and *Treponema*. These organisms may be free-living or host associated and many are well known pathogens. The genus *Leptospira* forms a distinct branch within the spirochete group. Spirochetes have traditionally been grouped on the basis of their unique morphology and mode of motility, and these traits turn out to be valid phylogenetic indicators as well. The signature sequence AAUCUUG is highly diagnostic for one subgroup of spirochetes; the sequence UCACACYAYCYG (where Y is any base) is diagnostic for most but not all spirochetes.

9.3.5 Bacteroides/Flavobacterium Group

This group consists of a number of genera forming a major phylogenetic line of Gram-negative eubacteria. The member of the genus Bacteroides form one subline within the group, while Cytophaga and Flavobacterium define a second. A mixture of physiological types show up here, obligate anaerobes (Bacteroides) along with obligate aerobes (Sporocytophaga). However, the two subgroups are themselves relatively homogeneous. Bacteroides and Flavobacterium are the major genera in one subgroup; both are obligately anaerobic, primarily fermentative rods. The Flavobacterium subgroup includes Cytophaga, Flexibacter, Sporocytophaga, and Saprospira, all rod-shaped or filamentous organisms, primarily respiratory in physiology, but which move by gliding motility. The sequence UUACAAUG is the signature for members of the Bacteroides group; the sequence CCCCCACACUG is the signature for the gliding group.

9.3.6 Chlamydia

This group consists thus far only of species of the obligate intracellular parasite, *Chlamydia*. Homology at the 95% level exists between the 16S rRNA sequences of the psittacosis organism, *C. psittaci*, and *C. trachomitis*, the causative agent of a variety of venereal infections and a form of acquired blindness called *trachoma*. On the basis of signature sequence comparisons, the Chlamydia; group shows a distant relationship to members of the *Planctomyces* group. It is interesting to note that members of both of these groups have cell walls that lack peptidoglycan.

9.3.7 Deinococcus-Thermus Group

This group consists of two well characterized genera, the Gram-positive, highly radiation-resistant *Deinococcus* and the Gram-negative organotrophic thermophile *Thermus*. Both *Deinococcus* and *Thermus* have an atypical cell wall that contains ornithin in place of diaminopimelic acid in its peptidoglycan (*Chloroflex* also contains ornithin in its peptidoglycan). Sequence analyses indicate that the group is tightly clustered and probably of relatively recent origin. The oligonucleotide CUUAAG is a signature for the *Deinococcus-Thermus* group.

9.3.8 Gram-positive Eubacteria

The Gram–positive eubacteria, both rods and cocci, form a coherent phylogenetic cluster. Included here are the endospore formers, lactic acid bacteria, most anaerobic and aerobic cocci, the coryneform bacteria, the actinomycetes, and even the majority of mycoplasmas, which phylogenetically appear to be clostridia that lack cell walls. Surprisingly, certain phototrophic bacteria are related to these Gram-positive bacteria. *Heliobacteria*, a group consisting of *Heliobacterium* and *Heliobacillus* appear to be distant relatives of the clostridia. This finding is significant because the phototrophic bacteria and their relatives are distributed among a number of other eubacterial groups as well. This once again points out the probably importance of photosynthesis in divergence of eubacterial lineage.

Two main classes of Gram-positive eubacteria are recognized on the basis of rRNA sequencing, those containing relatively low GC base ratios (clostridia, bacilli, lactic acid bacteria), and those with high GC ratios (actinomycete group). The oligonucleotide signature CUAAAACUCAAAG is highly diagnostic for the high GC group. No low GC group oligonucleotide signature has been found. The major genera placed in Gram-positive eubacterial group are shown in Table 9.2.

Table 9.2 Major genera of phylogenetically related Gram-positive eubacteria

GC Group	Major Genera of phylogenetically-reJated Gram-positive eubacteria
Low GC Group	Clostridium and relatives *Clostridium, Bacillus, Thermoactinomyces, Sporosarcina, Acetobacterium, Streptococcus, Peptococcus, Lactobacillus, Staphylococcus, Ruminococcus, Planococcus, Mycoplasma, Acholeplasma, Spiroplasma*
High GC Group	Actinomycetes *Actinomvces, Bifidobacterium, Propionibactyerium, Streptomyces, Nocardia, Actinoplanes, Arthrobacter, Corynebacterium, Mycobacterium, Micromonospora, Frankia, Cellulomonas, Brevibacterium* Phototrophic subdivision *Heliobacterium*

9.3.9 Thermotoga/Thermosipho

The final group of eubacteria consists of only two genera, *Thermotoga* and *Thermosipho*. Isolated thus far only from geothermally heated marine sediments, *Thermotoga* is a strictly anaerobic fermentative organism, capable of growth at temperature from 55-90°C, with an optimal of about 80°C. *Thermotoga* produces unique lipids containing extremely long-chain fatty acids and its cell wall contains peptidoglycan. *Thermpmosipho* differs from *Thermotoga* primarily in morphology and its lower GC content. Although clearly eubacteria, the *Thermotoga* group is phylogenetically positioned far from other eubacteria (Figure 9.4) and in the three kingdom tree (Figure 9.1), seems to occupy an intermediate phylogenetic position between the remaining eubacteria and the archaebacteria. Because of their extremely thermophilic nature, the discovery of *Thermotoga* and *Thermosipho* lends support to the hypothesis that early organisms were thermophilic.

9.4 ARCHAEBACTERIA

Phylogenetically, archaebacteria can be divided into four relatively broad groups, one consisting of some methanogens and the extremely halophilic bacteria, one consisting of methanogens only, one containing the extremely thermophilic bacteria, and a final "group" consisting of the genus *Thermoplasma*. The methanogenic/halophilic group is a physiological enigma. The methanoghens are a large group of obligate anaerobes unified by their ability to make methane. By contrast, the extreme halophiles are generally obligate aerobes and grow only at NaCl concentrations near saturation. The extremely thermophilic archaebacteria are mainly anaerobic organotrophs that are also extreme thermophiles. The overall phylogeny of archaebacteria, as defined by a distance-matrix tree, is shown in Figure 9.6.

It is obvious from Figure 9.6, that several subgroups of methanogens exist; this is not surprising considering the phenotypic diversity of methanogenic bacteria. The extremely halophiles prokaryotes (such as *Halobacterium*) and their relatives, group near the methanogens, more specifically near the genus *Methanobacterium*. *Thermoplasma acidophilum*, an acidophilic, thermophilic mycoplasma is also found near the methanogen branch of the archaebacteria. The sequence AYUAAG is a signature for the methanogens while the sequence AAUUAG provides a signature for extreme halophiles. The sequences AAAACUG and ACCCCA are signatures for *Thermoplasma*.

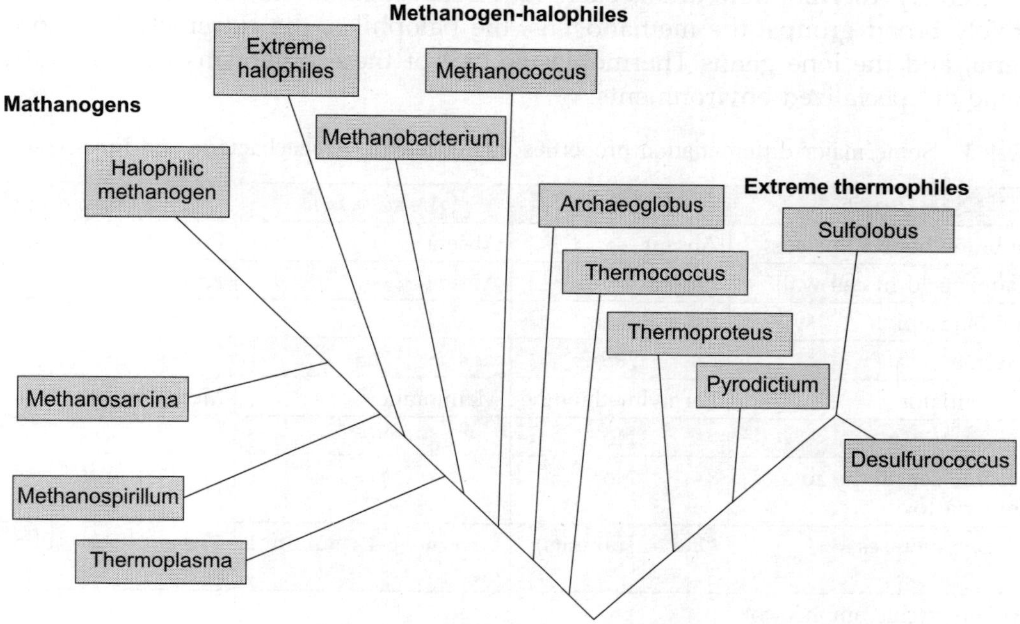

Figure 9.6 Phylogenetic tree of archaebacteria.

The other branch of the archaebacteria is the extremely thermophilic (sometimes called sulfur-dependent) archaebacteria. Phenotypically, the group is united by a variety of striking properties. First, they mostly have growth temperature optima above 80°C including the extremely thermophilic organism, *Pyrodictium*. The latter organisms has a temperature optima of 105°C. Second, most of these archaebacteria require elemental sulfur for optimal growth. This is because these organisms are for the most part anaerobes that use sulfur as an electron *acceptor*, oxidizing various organic compounds or H_2 as energy sources. One extremely thermophilic archaebacterium, *Sulfolobus*, uses elemental sulfur not as an electron acceptor but as an electron donor with oxygen or ferric iron serving as an electron acceptor. *Archaeoglobus*, an extremely thermophilic sulfate-reducing achaebacterium, may represent a phenotypic bridge between the extremely thermophilic group and the methanogens.

Sulfur plays an intimate role in the biochemistry of extremely thermophilic archaebacteria. This fact, coupled with the thermophilic character of all members of this group, has led to the suggestion that these archaebacteria represent an ancient metaboilic phenotype. Since these extreme thermophiles appear to have evolved more slowly than other archaebacteria and are closer to the root of the tree (Figure 9.6), their phenotype may resemble more closely the phenotype of ancient organisms than does that of other organisms. Such conjecture is consistent with what we know of early Earth Geochemistry, and the probable environmental conditions within which life arose. The sequence UAACACCAG and CACCACAAG are oligonucleotide signatures for extremely thermophilic archaebacteria. Archaebacteria can be divided into four relatively broad groups, the methanogens, the halophiles, the extremely thermophilic bacteria, and the lone genus Thermoplasma. All of these organisms live in relatively extreme or specialized environments.

Table 9.3 Some major differentiation properties in Eubacteria, Archaebacteria and Eukaryotes

Characteristic	*Eubacteria*	*Archaebacteria*	*Eukaryotes*
Membrane bound nucleus	Absent	Absent	Present
Muramic acid in cell wall	Present	Absent	Absent
Membrane lipids	Ester linked	Ether linked	Ester linked
Ribosome	70S	70S	80S
tRNA initiator	Formylmethionine	Methionine	Methionine
Inrons in tRNA genes	No	Yes	Yes
Ribosome sensitivity to diphtheria toxin	No	Yes	Yes
RNA polymerases	One (4 subunits)	Several (8-12 subunits each)	Three (12-14 subunits each)
Sensitivity to chloramphenicol, streptomycin, and kenamycin	Yes	No	No

9.5 EUCARYOTES

Evolutionary relationship between eukaryotic macroorganisms has traditionally been deduced from an examination of the fossil record and comparisons with extant plants and animals. Because an abundance of morphological markers are present in a macro-organisms, their phylogeny, as deduced from this comparative approach, has turned out to be a fairly accurate reflection of their true phylogeny as determined from molecular sequencing studies. However, for eukaryotic microorganisms, many of the same problems which delayed discovery of the natural phylogeny of bacteria apply here as well. Thus, molecular sequencing has greatly improved the construction of evolutionary trees for the lower eukaryotes (Table 9.3).

A eukaryotic phylogenetic tree based on ribosomal RNA sequence comparisons is shown in Figure 9.7. Because eucaryotes contain 18S rRNA instead of 16S rRNA in their small ribosomal subunits, the tree shown in Figure 9.7 is constructed from sequence comparisons of this slightly larger, but functionally equivalent, RNA species. The eukaryotic tree suggests that evolution along the nuclear line of descent was not a continuous process, but instead occurred in major epochs. It is clear with the comparison of evolutionary trees that microsporidia, a little known group of highly unusual parasitic eucaryotes, are modern relatives of one of the first major eucaryotetic cell lines, and that following the development of microsporidium-like organisms, several groups of eucaryotes radiated off the nuclear line of descent, culminating in the major eukaryotic lineages of plants and animals. Molecular sequencing suggests that microsporidia are an ancient group of eucaryotes. Microsporidia are obligate parasites which live in association with representatives of virtually all groups of eucaryotes, from microorganisms to humans.

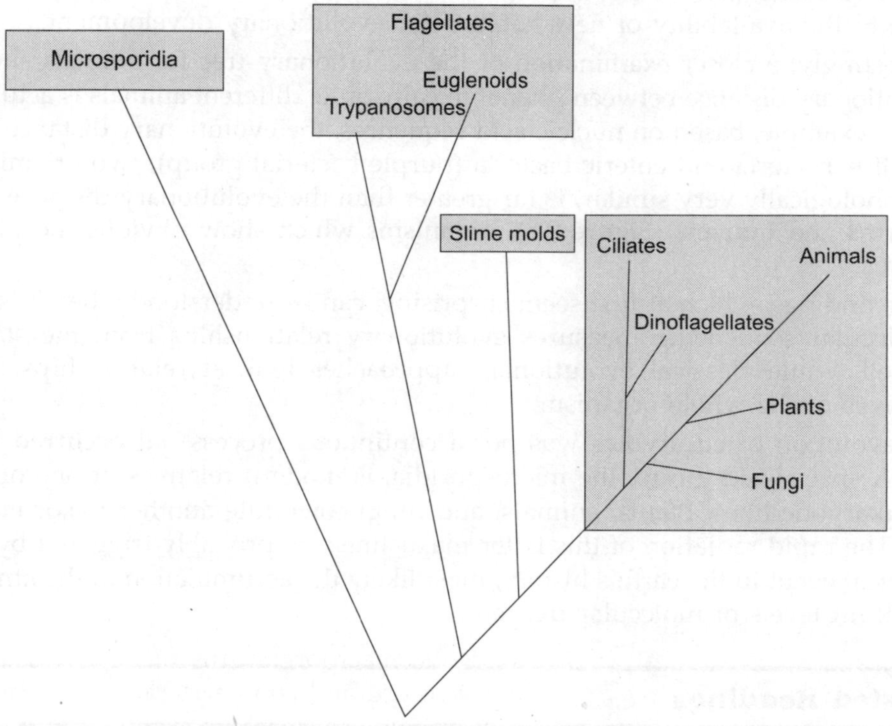

Figure 9.7 Eucaryotic phylogenetic tree based upon ribosomal RNA sequence comparisons.

Although they contain a membrane-bound nucleus, microsporidia lack mitochondria; in this connection microsporidia resemble the type of cell that first accepted stable endosymbionts. Besides lacking organelles, microsporidia differ from modern eucaryotes in many other ways. Microsporidia lack 5.8S rRNA, the small RNA species found in the large subunit of the ribosome of all other eukaryotic organisms examined. In addition, the small ribosomal subunit of microsporidia contains an RNA of about 16S rather than the 18S rRNA of all other eucaryotes. In sum, it appears that microsporidia represent an evolutionary relic of the nuclear line of descent, the latter of which is now dominated by organisms containing mitochondria, chloroplasts, and other internal structures.

The radial branches of the eukaryotic tree represent later-evolving organisms like the plants and animals. The rapid radiation observed in this region of the nuclear lineage was likely triggered by some major geochemical or geophysical event. When the fossil record is compared with the eukaryotic phylogenetic tree derived from molecular sequencing, the rapid evolutionary radiation can be dated to about 1.5 billion years ago. Geochemical evidence suggests that this is about the period in the earth's history that significant oxygen levels were accumulating in the oceans, and perhaps the rapid evolutionary radiation of eucaryotes was catalyzed by the onset of oxic conditions on the planet. The accumulation of significant O_2 levels allowed the development of an ozone shield which would have greatly expanded the number of habitats available for colonization. The explosive development of eucaryotes could then simply have been a response to the availability of new habitats for evolutionary development.

Interestingly, a closer examination of the evolutionary tree for animals shows that the evolutionary distance between phenotypically quite different animals is actually very small. For example, based on nucleic acid sequences, the evolutionary distance between green sulfur bacteria and enteric bacteria (purple bacterial group), two organisms that are morphologically very similar, is far greater than the evolutionary distance between earthworms and humans (Figure 9.8), organisms which show obvious morphological differences.

These findings, which at first seem surprising, can be understood when it is recalled that molecular sequencing measures evolutionary relationships from the standpoint of the cell, while classical evolutionary approaches look at relationships from the perspectives of the whole organism.

The evolution of eucaryotes was not a continuous process but occurred in major epochs. A specialized group, the microsporidia, is modern relatives of one of the first major eukaryotic lines. Plants, animals, and fungi constitute another major eukaryotic lineage. The rapid radiation of this latter major line was probably triggered by a major geochemical event in the earth's history, most likely the accumulation in the atmosphere of significant levels of molecular oxygen.

Suggested Readings

Beiko RG, Harlow TJ, Ragan MA. 2005. Highways of gene sharing in prokaryotes. *Proc. Natl. Acad. Sci. USA* **102,** 14332-14337.

Boucher Y, Douady CJ, Papke RT, Walsh DA, Boudreau, MER, Nesbo, CL, Case, RJ, Doolittle, WF. 2003. Lateral gene transfer and the origins of prokaryotic groups. *Annu Rev. Gent* **37**. 282 -328.

Doolittle. 1999. Phylogenetic classification and the universal tree. *Science* **284,** 2124-2129.

Dopson M, Baker-Austin C, Koppineedi PR, Bond PL. 2003. Growth in sulfidic mineral environment : metal resistance mechanisms in acidophilic microorganisms. *Microbiology* **149,** 1959-1970.

Embley TM, van der Giezen M, Horner DS. Dyal PL, Foster P, Embley TM. 2003.Mitochondria and hydrogenosomes are two forms of the same fundamental organelle. *Phil Trans Roy Soc Lond B Biol Sci* **358**, 191-201.

Maniloff J. 2002. Phylogeny and evolution. In *Molecular Biology and Pathogenicity of Mycoplasmas*. 31-43. Edited by Razin RH. New York: Kluver Academic/Plenum Publishers, New York.

Neimark HC. 1986. Origin and evolution of wall-less prokaryotes. In *The bacterial L-Forms*. 21-42. Edited by Madoff S. Marcel Dekkar Inc, New York.

Partensky F, Hess WR, Vaulot D. 1999. Prochlorococcus, a marine photosynthetic prokaryote of global significance. *Microbiol Mol Biol Rev* **63,** 106-127.

Polz MF, Alm EJ, Hanage WP. 2013. Horizontal gene transfer and the evolution of bacterial and archaeal population structure. *Trends Genet* **29,** 170-175.

Rittmann BE, Krajmalnik-Brown R, Halden RU. 2008. Pre-genomic, genomic and post-genomic study of microbial communities involved in bioenergy. *Nature Reviews Microbiology* **6,** 604-612.

Suzuki T, Yamasato K. 1994. Phylogeny of spore-forming lactic acid bacteria based on 16S rRNA gene sequences. *FEMS Microbiol Lett* **115,** 13-17.

Wächtershäuser G. 2006. From volcanic origins of chemoautotrophic life to Bacteria, Archaea and Eukarya. *Philos Trans R Soc Lond B Biol Sci* **361,** 1787-1808.

Whitman WB, Coleman DC, Wiebe WJ. 1998. Prokaryotes: The unseen majority. *Proc Nat Acad Sci USA* **95,** 6578-6583.

Woese CR, Fox GE. 1977. Phylogenetic structure of the prokaryotic domain: the primary kingdoms. *Proc. Natl. Acad. Sci.USA* **74,** 5088–5090.

Woese CR, Maniloff J, Zablen LB. 1980. Phylogenetic analysis of the mycoplasmas. *Proc Nat Acad Sci USA* **77,** 494-498.

Chapter 10

Prokaryotes

10.1 INTRODUCTION

The prokaryotic cell is a more complex structure than viruses but is generally less complex than the eukaryotic cell. It differs from the latter not only by lacking a perforated nuclear membrane surrounding the nucleoplasm, but also in the lack of complex organelles, such as mitochondria and chloroplast, the absence of cytoplasmic streaming and in the structure and composition of cell walls, flagella, and chromatophores where present. None of these characters offers such a satisfactory and precise means of differentiating between the two groups as does nuclear structure. Nevertheless, considered together and in conjunction with the latter they afford impressive support for this concept of two fundamentally different groups of organisms. The existence of a definite reproductive cycle in most, but not all, eukaryotes and the general absence of any such precise cycle among prokaryotes is additional evidence for the more primitive nature of the latter groups. Differences in fine structures frequently confirm systems of classification based on other characters but occasionally lead to an entirely new treatment of a particular group. The prokaryotes include the large and varied group of the bacteria together with cyanobacteria.

10.2 PRINCIPLES OF PROKARYOTE CLASSIFICATION

Classification of microbes can be made on the basis of phenotypic and genotypic characteristics. The microscopic characters of the bacteria remain unchanged and the relationship of bacteria to oxygen and Gram reaction properties are also relatively easy to determine. The bacteria are grouped according to their (1) shape, as cocci, rods, or curved rods, (spirallla); (2) ability to grow aerobically or anaerobically; and (3) Gram staining reaction. Metabolic diversity allows differentiating bacteria present in diverse environment into distinct categories as chemolithotrophs, chemoorganotrophs, phototrophs, and lactic acid bacteria. These are further classified according to their relationship to oxygen and for their special properties (Figure 10.1).

Phenotypic characters of bacteria include morphology and biochemical reactions carrying out by bacteria whose results can be viewed (Table 10.1). Morphological characteristics include colony morphology such as colour, size, shape, opacity, elevation, margin surface texture, consistency, etc. These characters are observed after the incubation period on the cultures on the solid media. In liquid cultures, pellicle formation and sediment formation can be observed. Biochemical characteristics include

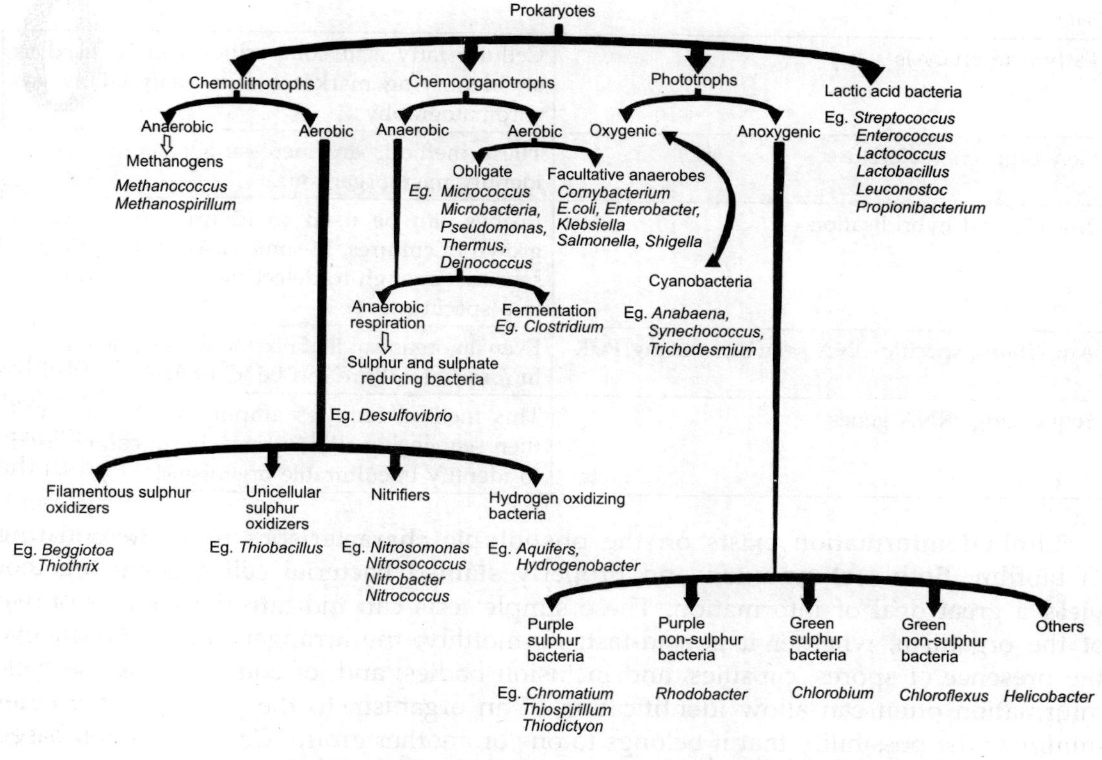

Figure 10.1 Metabolic diversity of prokaryotes.

enzyme production, utilization of particular sugar, aerobic or anaerobic reactions, etc. (Figure 10.2).

Table 10.1 Methods used to characterise genotypic and phenotypic characteristics

Method	*Comments*
Phenotypic characteristics	Most of these methods do not require sophisticated equipment and can be done easily anywhere in the world.
Microscopic morphology	Size, shape and staining characteristics such as Gram stain can give suggestive information as to the identify the organism. Further testing is needed to confirm the organism.
Metabolic differences	Culture characteristics can give suggestive information. A battery of biochemical tests can be used to confirm the identification.
Serology	Proteins and polysaccharides that make up a prokaryote are sometimes characteristics enough to be considered as a identifying marker. These can be detected using specific antibodies.

Contd...

Contd...

Fatty acid analysis	Cellular fatty acid composition can be used as an identifying marker and is analysed by gas chromatography.
Genotypic characteristics	These methods are increasingly being used to identify microorganisms.
Nucleic acid hybridisation	Probes can be used to identify prokaryotes grown in cultures. In some cases. The method is sensitive enough to detect the organism directly in a specimen.
Amplifying specific DNA sequences using PCR	Even an organism that occurs in very low number in a mixed culture can be identified
Sequencing rRNA genes	This method requires amplifying, cloning and then sequencing rRNA genes, but it can be used to identify unculturable organisms.

Limited information exists on the phenotypic characteristics of bacteria found in biofilm. Both wet-mounted and properly stained bacterial cell suspensions can yield a great deal of information. These simple tests can indicate the Gram reaction of the organism; whether it is acid-fast; its motility; the arrangement of its flagella; the presence of spores, capsules, and inclusion bodies; and, of course, its shape. This information often can allow identification of an organism to the genus level, or can minimize the possibility that it belongs to one or another group. Colony characteristics and pigmentation are also quite helpful. For example, colonies of several *Porphyromonas* species auto fluoresce under long-wavelength ultraviolet light, and *Proteus* species swarm on appropriate media. The taxonomic system, which is mainly based on phenotypic characters – shape, flagellation, gram staining behavious, etc., is very different from the phylogenetic system (Figure 10.3).

A primary distinguishing characteristic is whether an organism grows aerobically, anaerobically, facultatively (i.e., in either the presence or absence of oxygen), or microaerobically (i.e., in the presence of a less than atmospheric partial pressure of oxygen). The proper atmospheric conditions are essential for isolating and identifying bacteria. Other important growth assessments include the incubation temperature, pH, nutrients required, and resistance to antibiotics. For example, one diarrheal disease agent, *Campylobacter jejuni*, grows well at 42°C in the presence of several antibiotics; another, *Yersinia enterocolitica*, grows better than most other bacteria at 4°C. *Legionella, Haemophilus*, and some other pathogens require specific growth factors, whereas *E. coli* and most other Enterobacteriaceae can grow on minimal media. Most bacteria are identified and classified largely on the basis of their reactions in a series of biochemical tests (Figure 10.4). Some tests are used routinely for many groups of bacteria (oxidase, nitrate reduction, amino acid degrading enzymes, fermentation or utilization of carbohydrates); others are restricted to a single family, genus, or species (coagulase test for staphylococci, pyrrolidonyl arylamidase test for Gram-positive cocci) (Figure 10.5).

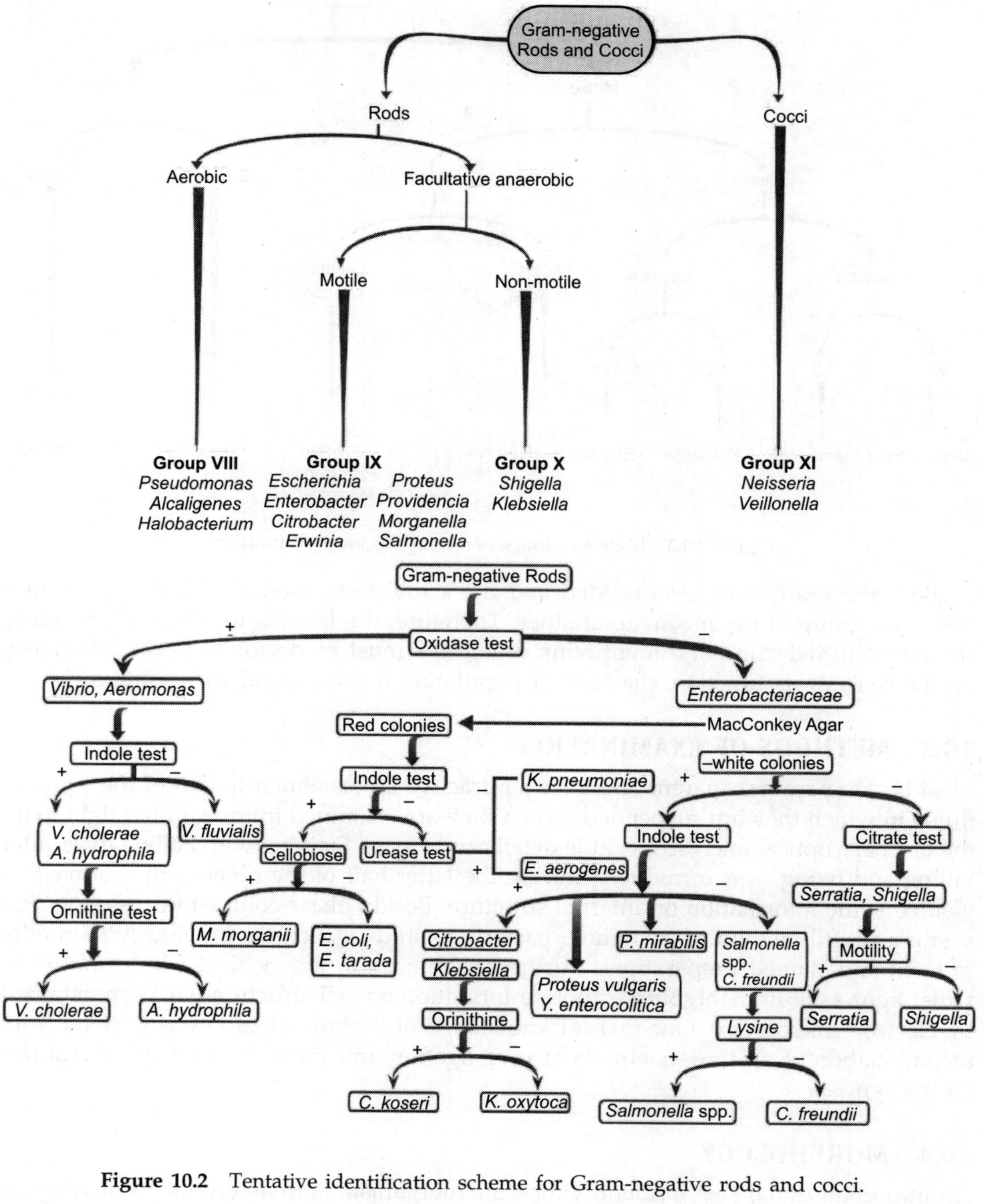

Figure 10.2 Tentative identification scheme for Gram-negative rods and cocci.

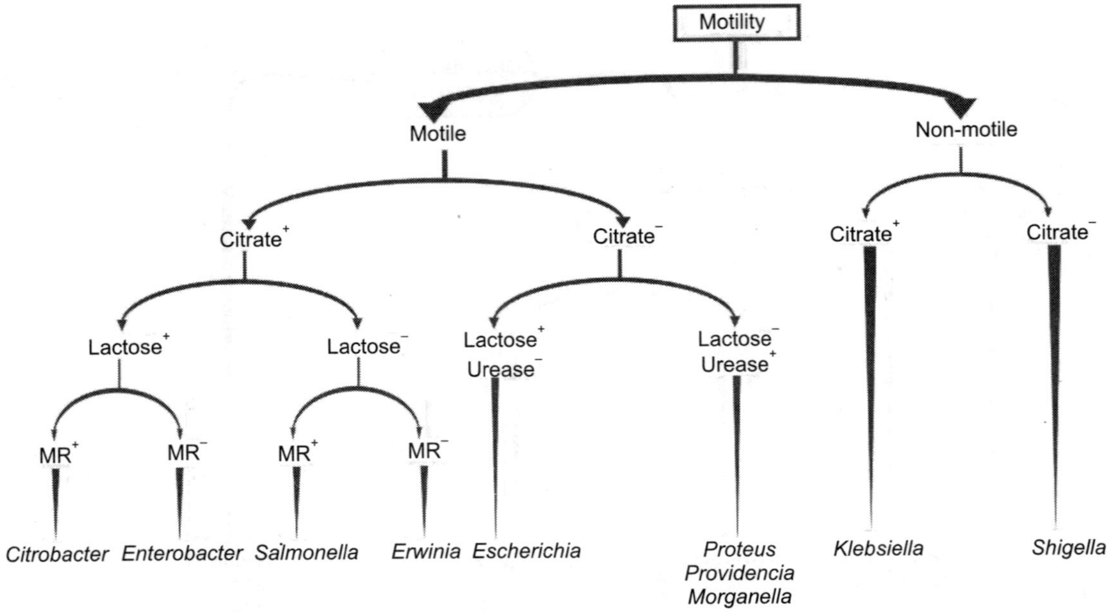

Figure 10.3 Identification scheme on the basis of motility.

Both the number of tests needed and the actual tests used for identification vary from one group of organisms to another. Therefore, the lengths to which a laboratory should go in detecting and identifying organisms must be decided in each laboratory on the basis of its function, the type of population it serves, and its resources.

10.3 METHODS OF EXAMINATION

Most bacteria are transparent and have a refractive index similar to that of the aqueous fluids in which they are suspended. When they are examined in transmitted light, with the ordinary optical microscope, little detail can be seen. Staining with aniline dyes, after killing and fixing, was introduced during the latter half of the nineteenth century and yielded some information on internal structure. Beside phase contrast microscopy, this is still a useful method of examining bacteria to study structures previously seen only in fixed and stained preparations. Further magnification was achieved by using ultra-violet light as illuminant, but far more information on cell structure has been obtained by electron microscopy. Our present knowledge of bacterial anatomy is derived from a combination of all these methods of investigation and the main anatomical features are recognized.

10.4 MORPHOLOGY

An important initial step in identifying a microorganism is to determine its shape, size and staining characteristics. Microscopic examination gives information very quickly and is sometimes enough to make a presumptive identification. Bacteria are unicellular

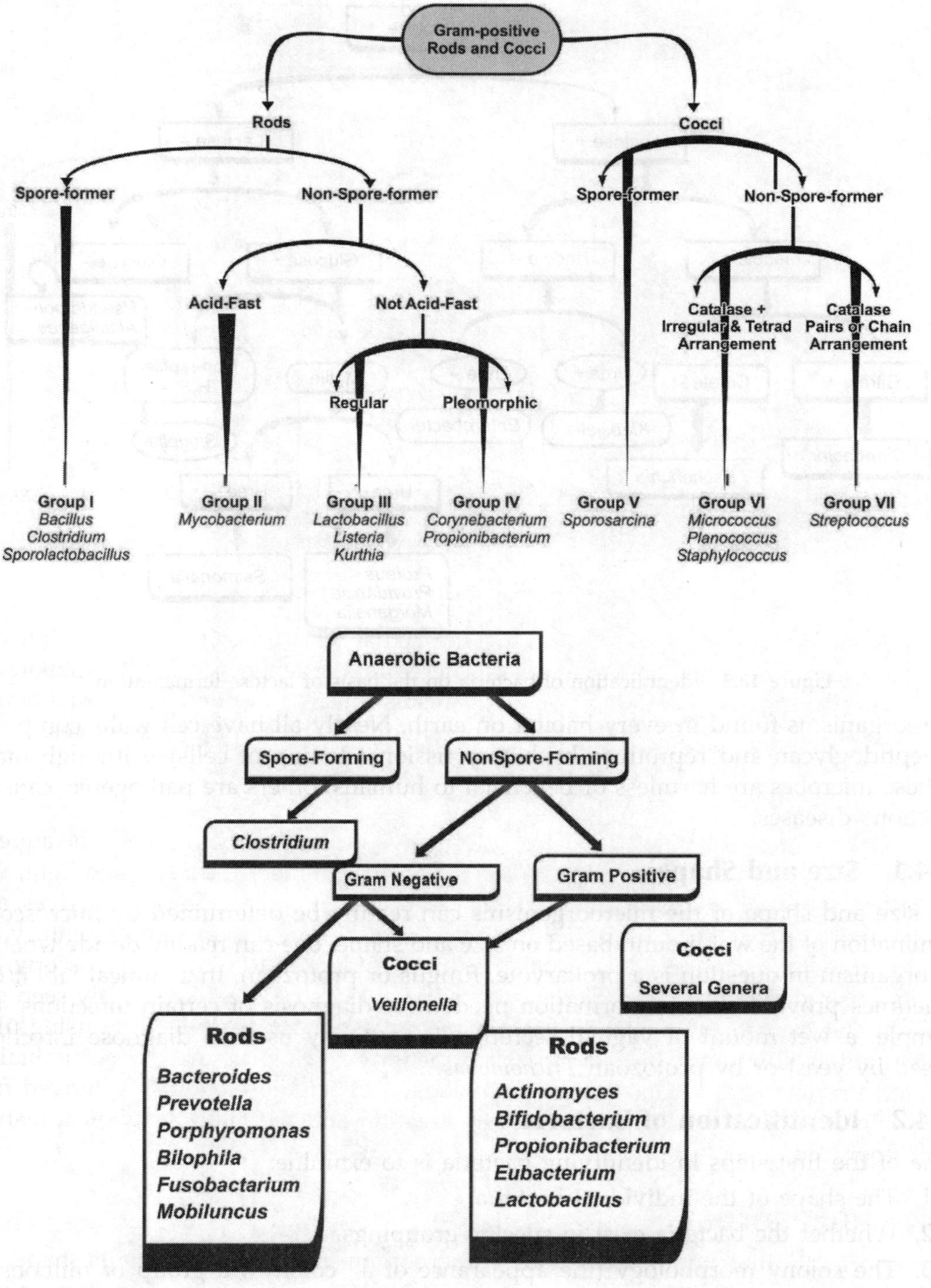

Figure 10.4 Tentative identification scheme of Gram-positive rods and coccus.

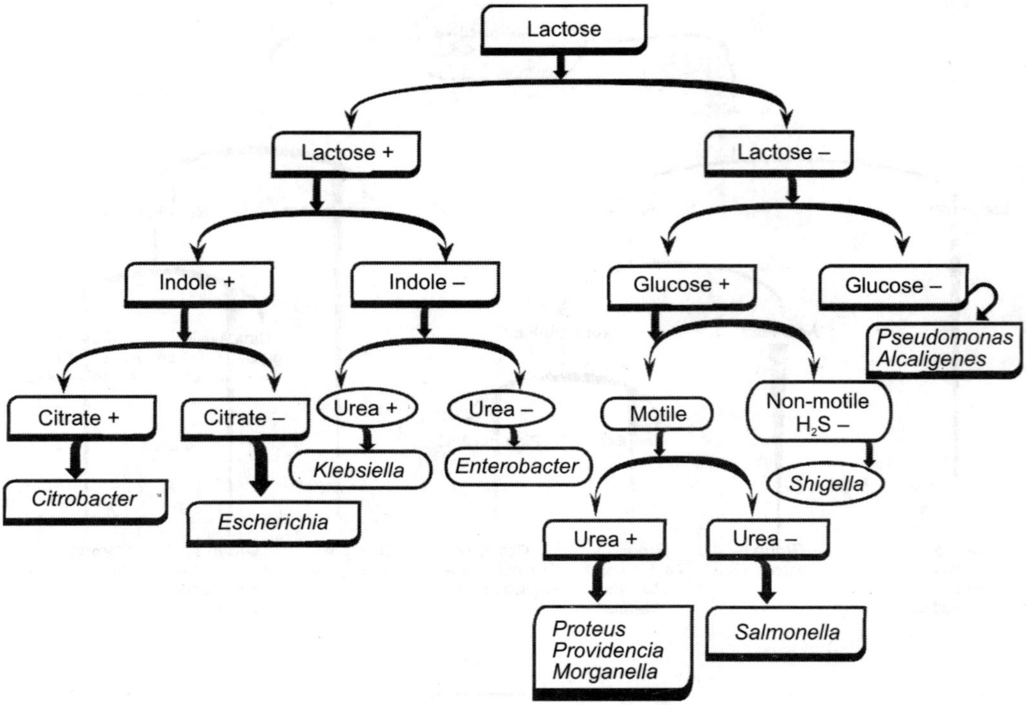

Figure 10.5 Identification of bacteria on the basis of lactose fermentation.

microorganisms found in every habitat on earth. Nearly all have cell walls composed of peptidoglycan and reproduce by binary fission (cloning of cells). Although many of these microbes are harmless or beneficial to humans, others are pathogenic, causing infectious diseases.

10.4.1 Size and Shape

The size and shape of the microorganisms can readily be determined by microscopic examination of the wet mount. Based on size and shape, one can readily decide whether the organism in question is a prokaryote, fungus or protozoan. In a clinical lab, it can sometimes provide all the information needed for diagnosis of certain infections. For example, a wet mount of vaginal secretion is routinely used to diagnose infections caused by yeast or by protozoan *Trichomonas*.

10.4.2 Identification of Bacteria

Some of the first steps in identifying bacteria is to examine:

1. The shape of the individual bacteria.
2. Whether the bacteria exist in specific groupings.
3. The colony morphology (the appearance of a "colony"; a group of millions of bacteria that arose from one single parent cell).

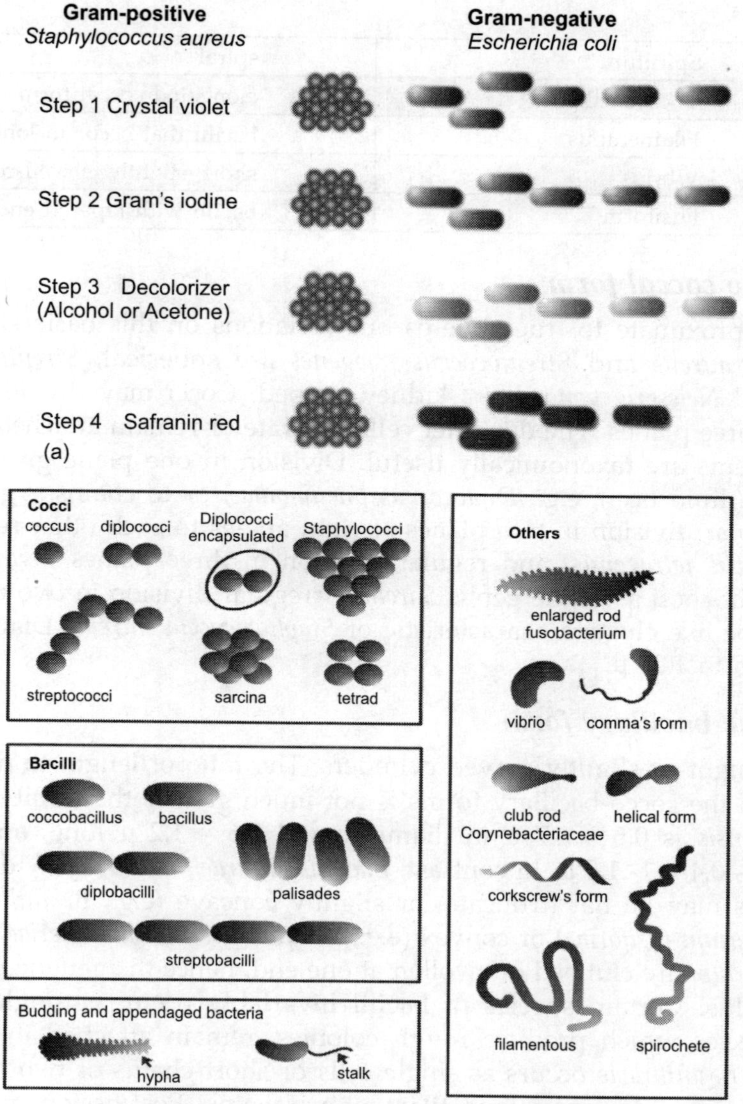

Figure 10.6 Different bacterial morphologies.

Bacterial Shapes: Most bacteria are classified according to shapes given in Table 10.2. Thus we find a range of varied bacteria ranging from rod shaped, spherical, spiral, elongated, bacilli, curved etc (Figure 10.6).

Table 10.2 Bacterial types according to the shapes

Bacteria	*Shape*
Bacillus	rod-shaped
Coccus	spherical

Contd...

Contd...

Spirillum	spiral
Coccobacilli	elongated coccal form
Filamentous	bacilli that occur in long threads
Vibrios	short, slightly curved rods
Fusiform	bacilli with tapered ends

10.4.2.1 The coccal form

Most cocci approximate to true spheres but variations on this basic structure occur. *Staphylococcus aureus* and *Streptococcus pyogenes* are spherical, *Streptococcus faecalis* ellipsoidal and *Neisseria gonorrhoeae* kidney shaped. Cocci may divide irregularly in one, two, or three planes. The daughter cells separate or remain attached; the resulting cell arrangements are taxonomically useful. Division in one plane gives rise to pairs of organisms (diplococci), e.g., *Diplococcus pneumoniae,* or to chains, e.g., *Streptococcus pyogenes.* Regular division in two planes at right angles to each other result in tetrads (e.g., *Micrococcus tetragenus)* and regular division in three planes gives rise to cubic packets of eight cocci as in the genus *Sarcina.* Irregular division in two or three planes gives the grape-like clusters characteristic of *Staphylococcus aureus.* Diameters of cocci range from 0.5 to 1.25 μ.

10.4.2.2 The bacillary form

Bacilli are straight or slightly curved cylinders. The ratio of length to diameter varies greatly and in the cocco-bacillary forms is not much greater than unity. For example, *Brucella melitensis* is 0.5 – 0.7 μ in diameter and 0.6 – 1.2 μ long, and *Haemophilus influenzae* 0.3 – 0.4 × 1- 1.5 μ. In contrast, *Bacillus anthracis* is 1-1.2 μ wide × 3-8 μ long. The end walls may be flat (truncate) or slightly concave (e.g., *B. anthracis*), tapering (e.g., *Fusobacterium fusiforme)* or convex (e.g., *Corynebacterium diphtheriae).* Some species of *Corynebacterium* are clubbed or swollen at one end, hence the generic name, meaning clubbed bacillus. Certain species of bacilli invariably occur as single cells, others, especially species which produce rough colonies, remain attached in chains. In the animal body, *B. anthracis* occurs as single cells or short chains of two or three bacilli; in culture it invariably forms long filamentous chains. Post-fission movements may give rise to characteristic arrangements as, for example, in *Corynebacterium* in which transverse fission is followed by a snapping movement, the cells remaining attached at one point, the daughter cells consequently producing angular or palisade arrangements.

10.4.2.3 The spirillar form

These are curved, twisted rods. *Spirillum* spp. are long, and cork-screw-shaped; *Vibrio* spp. are short and appear comma-shaped. *Vibrio cholerae* is 0.2 - 0.4 μ × 1.5 – 4 μ long; *Spirillum* spp. may be much longer, up to 50 μ.

10.4.2.4 Involution forms

The action of autolytic enzymes, normally restricted in actively metabolizing cells, may be responsible for the frequent occurrence in old cultures of organisms of abnormal morphology often characteristic of the bacterial species which produce them, e.g., *Pasteurella pestis* and *Neisseria* species. These are known as involution forms. Such cells are mostly dead, but those still viable, give rise to normal cells again when introduced into favourable medium.

10.4.2.5 Large bodies and L-forms

Many bacteria, under certain conditions such as cold shock, the presence of specific antiserum, bacteriophage attack or the presence of antibiotics, such as the penicillins which interfere with cell wall synthesis, tend to develop an unusual form of pleomorphism involving the formation of large spherical or distorted cells, the so-called 'large bodies'. The cells revert to the normal bacillary form if the inducing stimulus is removed after only a brief exposure. If the stimulus persists for a long time, the cells either perish or may become stabilized in the so-called *L*-form. *L*-forms consist of minute filterable bodies and nonrigid, fragile, large globules of cytoplasm and resemble the pleuropneumonia and the pleuropneumonia-like organisms. They break up readily when touched but may be examined *in situ*, by dark field microscopy from fluid cultures or in impression smears stained by Giemsa. The smears are prepared by allowing a fixative, such as Bouin's fluid, to diffuse through the agar medium, thereby fixing the colonies to cover slips placed over them. Usually, serum or blood is required in the medium for their growth. They may then breed true for many generations. *L*-forms are probably spheroplasts of bacteria which have lost their ability to produce typical cell walls or in which cell wall production is inhibited. They were originally discovered in *Streptobacillus moniliformis*, and have been subsequently reported in many other organisms including *Proteus vulgaris, Escherichia coli*, and various species of *Haemophilus, Salmonella*, and *Streptococcus*.

10.4.2.6 Cell arrangement

Bacteria sometimes occur in groups, rather than singly, and the single cell's shape influences the cell arrangements that they form as the bacterial cells divide. Bacilli divide along a single axis, and are sometimes seen in pairs or chains. Since they only divide along one axis, one will not find bacilli in clusters, such as those formed by Staphylococcal bacteria. Cocci divide on one or more planes, producing cells in pairs (diplococci), chains (streptococci), packets (sarcinae), clusters (staphylococci) etc (Figure 10.7). Size, shape and arrangement of cells are often the first clues in identification of bacteria. However, since there are many "look-alikes", methods other than microscopy must be used to determine the genus and species of an organism.

Bacterial populations grow extremely fast when they supplied with the nutrients and environmental conditions that allow them to thrive. Through this growth, different types

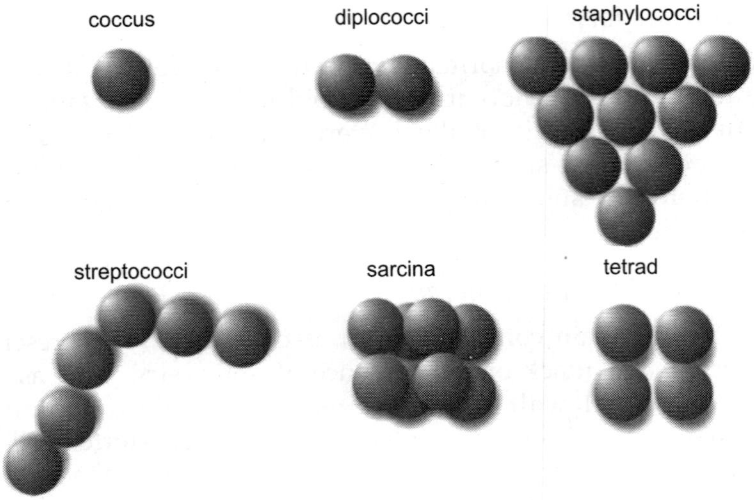

Figure 10.7 Various morphologies of bacterial coccus.

of bacteria will sometimes produce colonies that are distinctive in appearance. Some colonies may be coloured, some are circular in shape, while others are irregular. The characteristics of a colony (shape, size, color, etc.) are termed the "colony morphology".

10.4.2.7 Bacterial colonies

Bacteria are inoculated on to gelled nutrient media, single organisms, by repeated division, grow into a population of progeny producing colonies visible to the naked eye on the surface of the medium. The colonial form is usually fairly constant for a given species and is therefore of taxonomic value. Colony diameter is however readily affected by crowding, the size of the colonies then being limited by the metabolic activities of others. More constant characteristics are form (e.g., whether raised, flat, dome shaped, or swarming, as a continuous film over the surface of the medium), colour, surface appearance (e.g., smooth or rough, matt or glossy) and the nature of the colony margin (e.g., entire, lobed or crenated, rhizoid, echinulate, effuse, arborescent, beaded) (Figure 10.8).

Colony form may alter with the nature of the nutrients available or as a result of a genetic change. A frequent type of mutation is from a smooth colony, with an almost circular outline and a glistening appearance, to a rough form, with an irregular or toothed edge and a dull wrinkled surface. Such a change is known as a Smooth to Rough or S $\rightarrow$ R variation. It is often accompanied by other changes such as loss of the cell surface antigens characteristic of the smooth form, partial or complete loss of virulence, altered sensitivity to bacteriophages, or agglutination of suspended cells in the presence of electrolytes and certain dyes, such as acriflavine. Thus, the visible change from smooth to rough colony form is a valuable indicator of other fundamental changes. The opposite change from rough to smooth is less common. Loss of capsules

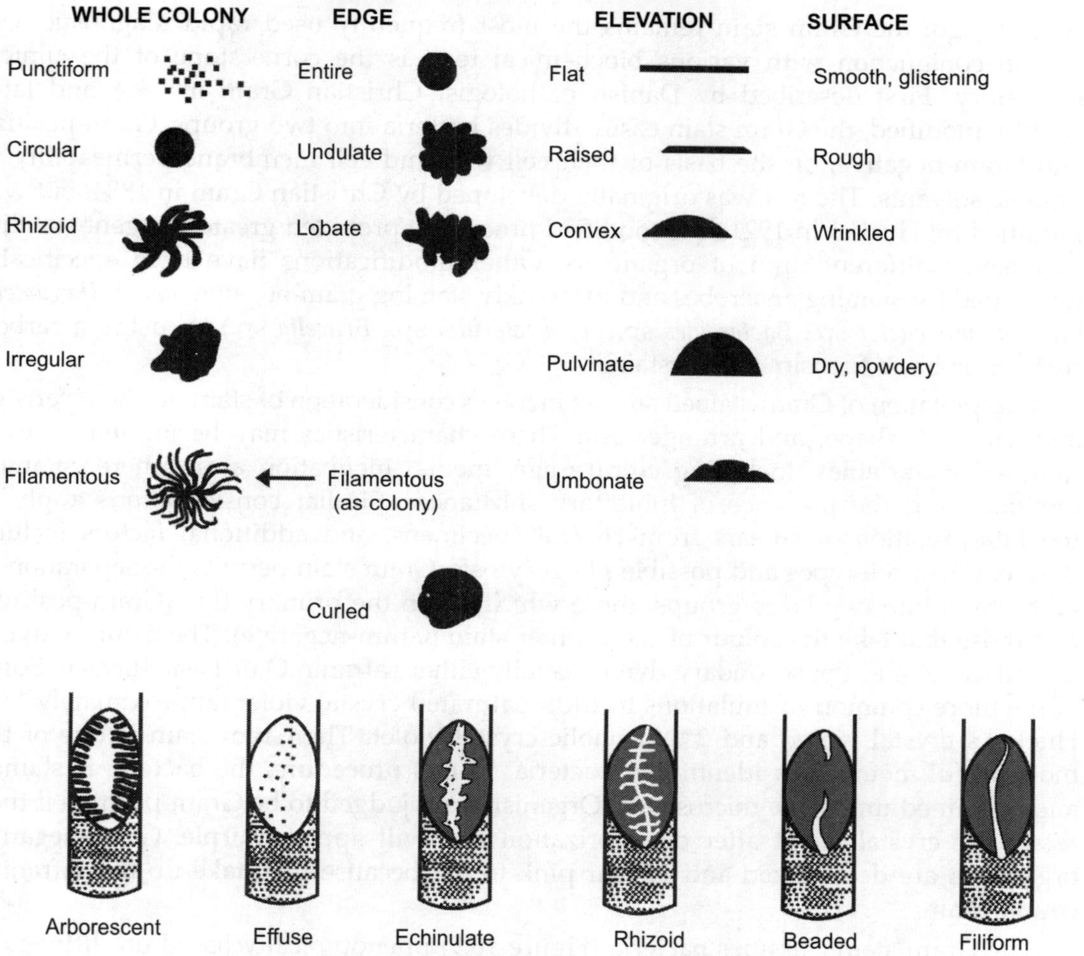

Figure 10.8 Different forms of bacterial growth on agar slants.

often results in loss of the smooth character. For instance, capsulated (virulent) strains of pneumococci produce smooth colonies, and noncapsulated (avirulent) strains rough ones.

The precise mode of cell division is of great importance in determining the form of many colonies; the slightly asymmetric division of *Bacillus cereus* var. *mycoides* results in long twisted chains of cells turning to left or right, giving the colony a rhizoidal structure. A few motile bacteria of the genus *Bacillus* give rise to colonies that migrate across the surface of solid media leaving a well-defined track of growth.

10.4.3 Gram Staining

Effective utilization and understanding of the clinical bacteriology laboratory can greatly aid in the diagnosis of infectious diseases. Although described more than a century ago, the Gram stain remains the most frequently used rapid diagnostic test, and in conjunction with various biochemical tests is the cornerstone of the clinical laboratory. First described by Danish pathologist Christian Gram in 1884 and later slightly modified, the Gram stain easily divides bacteria into two groups, Gram-positive and Gram-negative, on the basis of their cell wall and cell membrane permeability to organic solvents. The test was originally developed by Christian Gram in 1884, but was modified by Hucker in 1921. The modified procedure provided greater reagent stability and better differentiation of organisms. Other modifications have been specifically developed for staining anaerobes and for weakly staining gram-negative bacilli (*Legionella* sp., *Campylobacter* sp., *Bacteroides* sp. *Fusobacterium* sp., *Brucella* sp.) by using a carbol-fuchsin or basic fuchsin counterstain.

Interpretation of Gram-stained smears involves consideration of staining characteristics and cell size, shape, and arrangement. These characteristics may be influenced by a number of variables, including culture age, media, incubation atmosphere, staining methods, and the presence of inhibitory substances. Similar considerations apply to the interpretation of smears from clinical specimens, and additional factors include different host cell types and possible phagocytosis. Gram stain permits the separation of all bacteria into two large groups, those which retain the primary dye (Gram-positive) and those that take the colour of the counter stain (Gram-negative). The primary dye is crystal violet and the secondary dye is usually either safranin O or basic fuchsin. Some of the more common formulations include: saturated crystal violet (approximately 1%), Hucker's crystal violet, and 2% alcoholic crystal violet. The Gram stain is one of the most useful methods in identifying bacteria. In this procedure, the bacteria is stained and examined under the microscope. Organisms are judged to be Gram-positive if they retain the crystal violet after de-colorization and will appear purple. Gram-negative organisms are decolorized and appear pink to red because they take up the safranine counterstain.

The Gram stain classifies bacteria (Figure 10.9) phenotypically based on differences in cell wall thickness with differing glycosaminopeptide and lipoprotein compositions: Gram-positive bacteria have a peptidoglycan layer 10–15 times thicker than Gram-negative bacteria. The cell wall, synonymous with the peptidoglycan layer, is a rigid framework of cross linked peptidoglycan forming the outermost component of the cell. The more complex Gram-negative bacteria also have an outer membrane beyond the peptidoglycan layer that consists of lipopolysaccharide (endotoxin), lipoprotein, and phospholipids. In some Gram-negative species there also exists a periplasmic space between the outer membrane and the inner cytoplasmic membrane with β-lactamases that degrade β-lactam antibiotics.

Gram stains perform well only on fresh culture (24 hr incubation). Using older cultures often yields unexpected results. Be sure to stop the application of decolorizer immediately after it runs clear to avoid over-decolorizing. Always run positive and negative controls from known cultures.

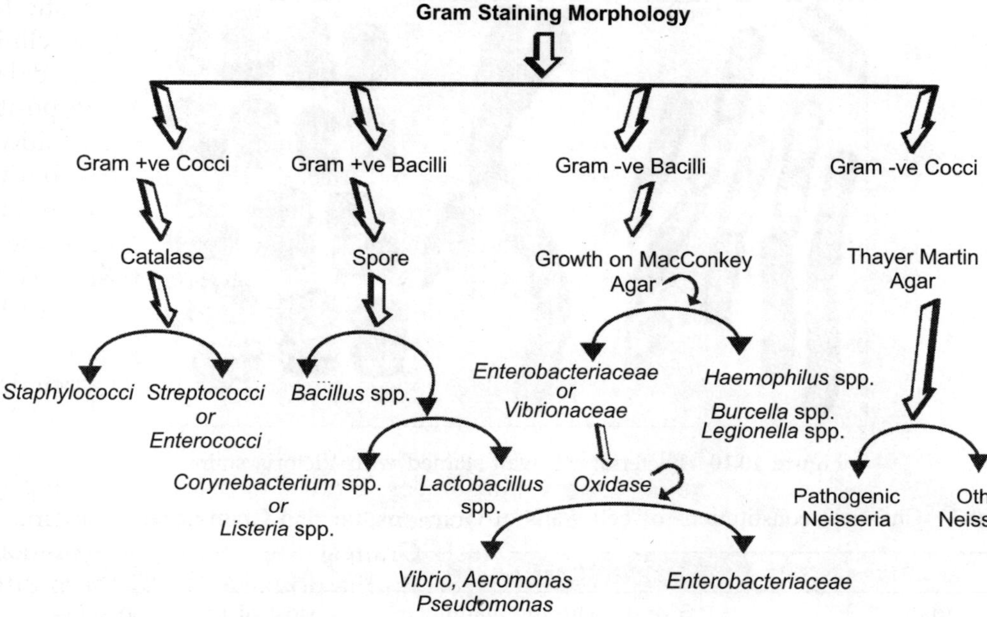

Figure 10.9 Key for tentative identification of bacteria on Gram staining.

10.4.4 The Cell Wall

The cell envelope usually consists of two components, an outer cell wall and beneath it a cytoplasmic or plasma membrane which is actually the outermost layer of the protoplast. A few bacteria, including the L-forms and the Mycoplasma have no cell wall.

The cell wall is a rigid structure which gives shape to the bacterium. Its presence can be demonstrated by plasmolysis, by electron microscopy of cell wall residues, which retain their shape when freed of their contents, and by the fact that the cellular contents assume a spherical shape when the cell wall is enzymically dissolved. The wall is chemically fairly inert and does not stain by normal techniques but special cell wall stains are available (Victoria blue stain for *Bacillus megaterium* and Hele's boundary stain for *Tetracoccus canadensis*) (Figure 10.10). The cell walls of a representative bacterium, *Staphylococcus aureus,* account for 20 per cent of the dry weight of the cell.

Pure preparations of bacterial cell walls can be obtained by mechanical disintegration of the organism followed by differential centrifugation. Autolysis, osmotic lysis, and heat shock have also been used for some cell wall preparations. Chemical and enzymic methods are not used because these alter the chemical composition. Traces of contaminating protein and nucleic acids are removed by washing or by treatment with appropriate enzymes such as trypsin or ribonuclease. The preparations may be examined with the electron microscope or subjected to chemical analysis. As stated in Table 10.3, the cell walls of Gram-positive and Gram-negative bacteria differ mainly in their amino acid and lipid content.

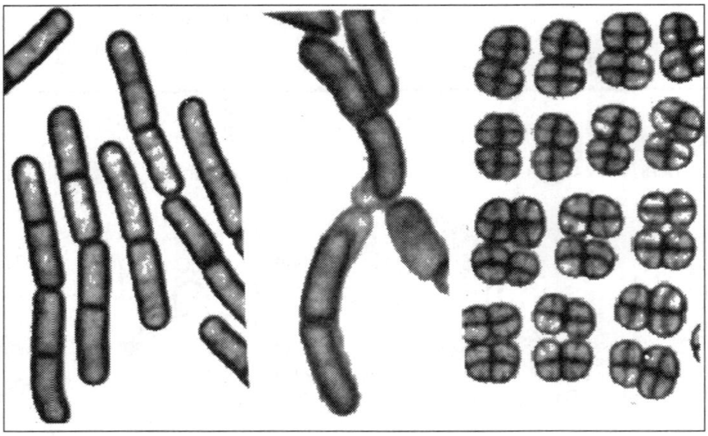

Figure 10.10 Bacterial cell wall stained with Victoria stain.

Table 10.3 Chemical constitutions of cell walls of Gram-positive and Gram-negative bacteria

Constituents	*Gram positive*	*Gram negative*
Amino acids	3 or 4 including alanine, glutamic acid, and lysine or diaminopimelic acid (no aromatic or sulphur containing amino acids)	Most of the amino acids commonly found in proteins; with diaminopimelic acid in addition.
Muramic acid	Present	Present
lipids	0-2%	10-20%
Polysaccharides	35-60%	Up to 50%
Teichoic acids	15-20%	none

The rigidity of the cell wall of Gram-positive bacteria is attributed to a mucocomplex, a polymer consisting of repeating units of the amino sugars N-acetylglucosamine and N-acetyl-muramic acid (Latin: murus, a wall) linked together by 1-4 β-inkages (Figure 10.11).

To each carboxyl group on the muramic acid residues is linked a peptide chain of alanine, glutamic acid, and either diaminopimelic acid (absent from the walls of Gram-positive cocci and related organisms) or lysine. The polysaccharide chains are linked together by peptide branches and these are themselves further interlinked by peptide bridges, the latter being more extensive in the walls of Gram-positive then in those of Gram-negative bacteria.

This results in a three-dimensional, rigid, multi-layered network around the organism, for which several possible macromolecular structures have been proposed (Figure 10.12).

Within this open meshwork are found a number of matrix materials, chiefly polysaccharides. They include the immunologically active substances which determine the antigenic specificity of certain organisms, e.g., the C polysaccharide and M protein

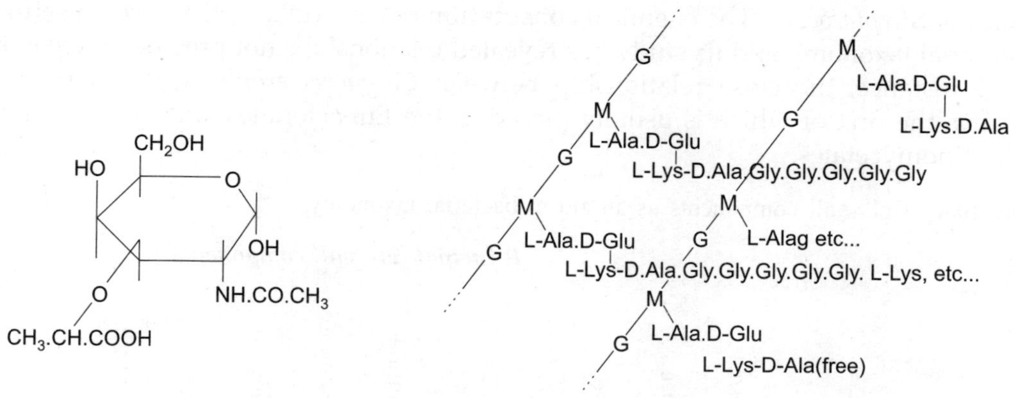

Figure 10.11 The structure of *N*-acctyl-muramic acid and over all molecular arrangement of the cross links in small pieces of mucopeptide from *Staphylococcus aureus*.

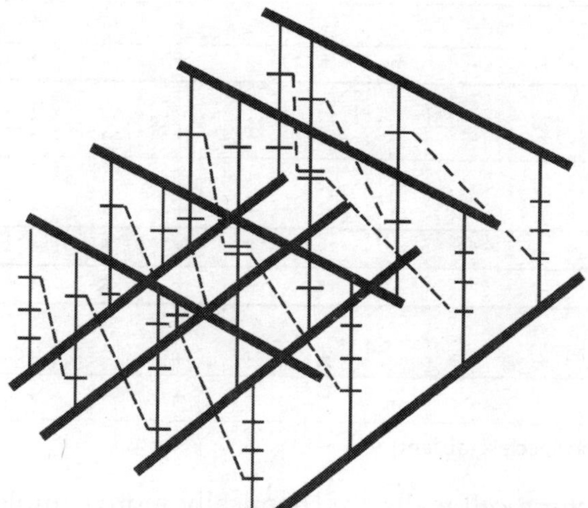

Figure 10.12 Diagrammatic representation of three-dimentional interchain links in bacterial mucopeptides. The solid horizontal bars represent the *N*-acetyl-muramic acid/*N*-acetyl glucosamine chains, the vertical crossed bars the polypeptide sidechains on the acetyl muramic acid residues and the dotted lines the penta-glycine cross bridges.

of species of *Streptococcus*. One group of these matrix materials is the teichoic and teichuronic acids. These are highly acidic substances present only in the cell walls of Gram-positive bacteria in which they may account for up to 50 per cent of the dry weight of the cell wall. They consist of either ribitol or glycerol units joined by phosphodiester linkages. Contrary to earlier views these substances are now thought to be specifically fixed to the structural components of the cell wall by covalent bonds.

Many monosaccharides have been identified in cell wall preparations following acid hydrolysis. The distribution of these sugars varies from group to group but is constant within a particular group (Table 10.4). Rhamnose, for instance, is almost unique to species of *Streptococcus*. The chemical constitution of cell walls is therefore a useful tool in bacterial taxonomy and its study has revealed relationships not previously recognized as, for instance, the close relationship between *Corynebacterium, Mycobacterium,* and *Nocardia,* the first of which is usually placed in the Eubacteriales and the other two in the Actinomycetales.

Table 10.4 Cell wall components as an aid to bacterial taxonomy.

	Principal cell wall components											
	Arabinose	Rhamnose	Galactose	Glucose	Mannose	Glucosamine	Galactosamine	Muramic acid	Alanine	Glutamic acid	Lysine	Diaminopimelic acid
Actinomyces bovis	–	–	+	–	–	+	–	+	+	+	+	–
Nocardia asteroids	+	–	+	+	–	+	–	+	+	+	–	+
Mycoberculosis tuberculosis	+	–	+	–	–	+	–	+	+	+	–	+
Corynebacterium diphtheriae	+	–	+	–	+	+	–	+	+	+	–	+
C. pyogenes	–	+	–	+	±	+	+	+	+	+	+	–
Streptococcus pyogenes	–	+	–	–	–	+	±	+	+	+	+	+
Lactobacillus Acidophilus	+	–	–	±	–	+	–	+	+	+	+	+
Staphylococcus aureus	–	–	–	–	–	+	–	+	+	+	+	–

(+ present; ± present in some species; -absent)

Gram-negative bacteria cell walls are chemically more complex than those of Gram-positive bacteria (Figure 10.13). They contain mucopeptides, lipids, and a full range of amino acids indicating the presence of protein, but less amino sugars than do the cell walls of Gram-positive bacteria.

The distinction between Gram-positive and Gram-negative species is based, not only on differences in chemical composition but also on the fine structure of their walls. Certain polysaccharides of the cell walls of Gram-negative bacteria have been correlated with antigenic specificities (Figure 10.14).

Walls of Gram-negative bacteria are relatively thin, those of Gram-positive bacteria are thicker and usually amorphous. Cells walls of *Escherichia coli* are about 8 mμ thick; those of Gram-positive bacteria vary in thickness, e.g., 15 mμ in *Staphylococcus aureus,* 22 mμ in *Bacillus megaterium* and 80 mμ in *Lactobacillus acidophilus.* In shadowed preparations of *Spirillum* spp. the inner surface consists of spherical macromolecules ranging from 10 to 14 mμ in diameter, arranged in a texagonal pattern. Such a regular macromolecular pattern is absent in the walls of all but a few of the Gram-negative bacteria examined.

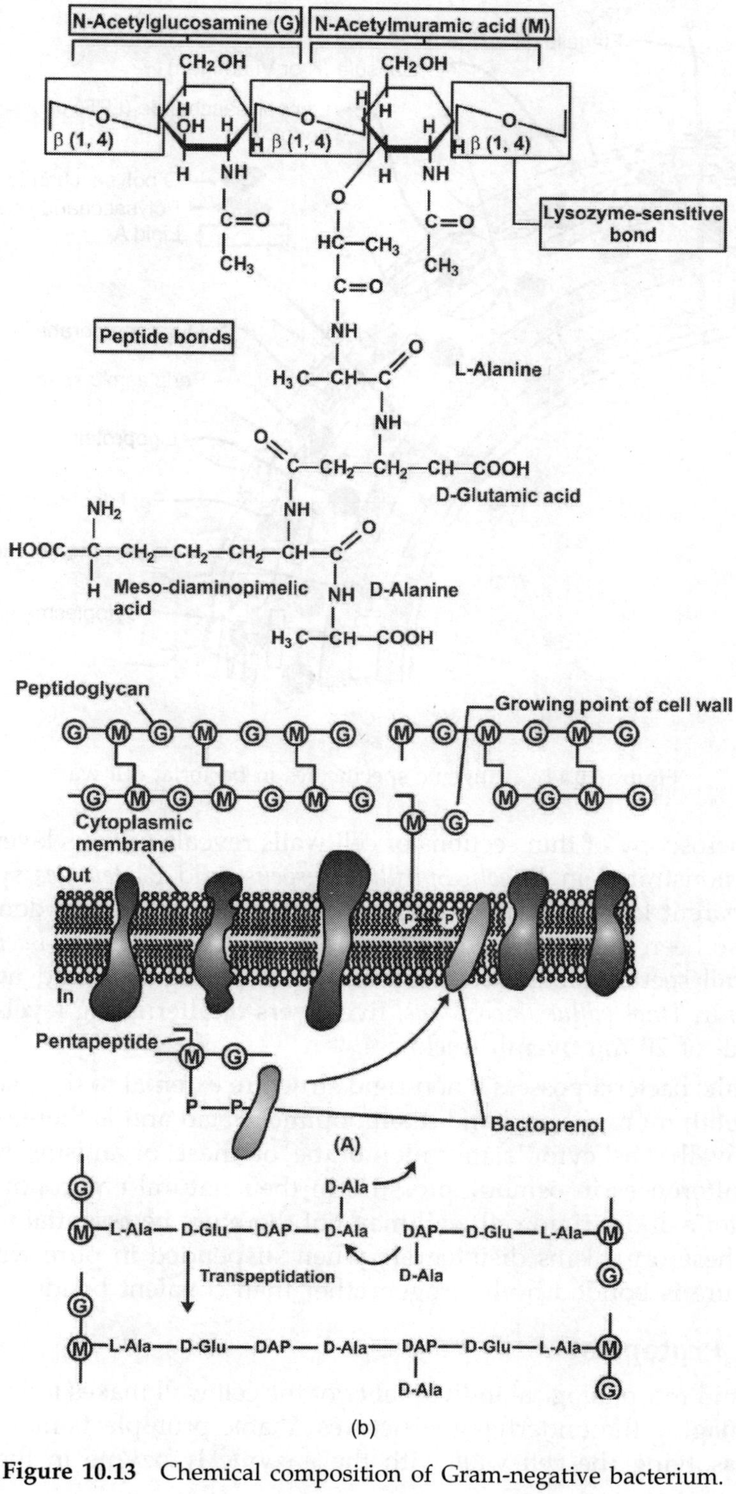

Figure 10.13 Chemical composition of Gram-negative bacterium.

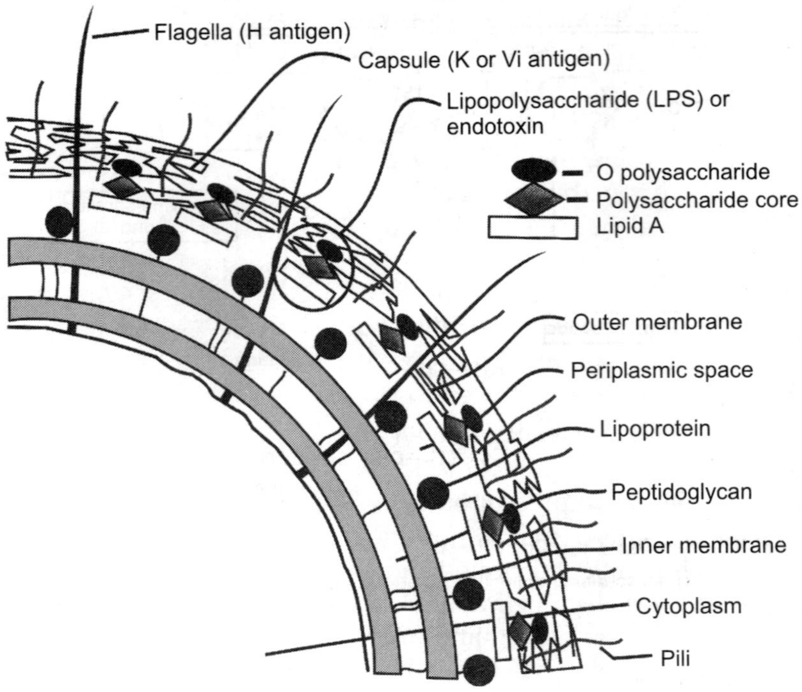

Figure 10.14 Antigenic specificities in bacterial cell wall.

Electron microscopy of thin sections of cell walls reveals various layers. Three layers have been demonstrated in *E. coli, Spirillum serpens,* and *Bacteroides* sp., in which an electron-transparent layer is enclosed on either side by an electron dense layer. Three layers have also been demonstrated in the Gram-positive organism *Bacillus megaterium* but only in wall sections fixed with osmium or permanganate and not stained with uranyl acetate. In *Thiobacillus thiooxidans,* five layers of alternating levels of density are found in a wall of 20 mμ overall thickness.

The halophilic bacteria possess a non-rigid structure external to the plasma membrane but this lacks both muramic acid and diaminopimelic acid and is therefore not a typical bacterial cell wall. The cytoplasmic membrane of these organisms is probably not subjected to differences in osmotic pressure in their natural environment as are non-halophilic bacteria and a rigid cell wall may not therefore be essential to their stability. The walls of these organisms disintegrate when suspended in pure water, suggesting that the structure is bonded by hydrogenrather than covalent bonds.

10.4.5 The Protoplast

The chemical and morphological individuality of the cell wall makes it possible to remove it without damaging the underlying structures. Viable protoplasts may be obtained by completely dissolving the cell wall with the enzyme lysozyme in the presence of a stabilizing solution such as 0.2 M sucrose which protects the naked protoplast from osmotic shock. It is also possible to prevent wall formation altogether by inclusion of penicillin in the medium or, with auxotrophic strains, by depriving the organisms of the essential diaminopimelic acid. When part of the cell wall remains attached to the protoplast the whole structure is referred to as a spheroplast. Naked protoplasts always assume a spherical form even when derived from rod-shaped organisms, indicating that the cell wall is responsible for the maintenance of the characteristic shape of the organism. The wall also prevents osmotic rupture of the protoplast in the dilute solutions in which most bacteria usually live and may protect them to some extent against harmful external agents. Isolated protoplasts are not susceptible to bacteriophage infection the cell wall being essential for this. However, protoplasts prepared from cells already infected with phage are able to support phage replication. Similarly, protoplasts prepared from cells about to produce endospores proceed to sporulation.

The protoplast accounts for between 60-80 per cent of the dry weight of the whole cell; it does not contain diaminopimelic acid, a substance which is found in many cell walls. Any flagella which were present on the original cell are retained by the separated protoplast although the latter itself is non-motile. Isolated protoplasts are capable of metabolizing many oxidizable substrates and increase in size with time although at a slower rate than do whole cells. Their ability to reproduce or regenerate cell walls is still in doubt. Spheroplasts on the other hand are able to synthesize cell walls, suggesting that the residual cell wall material either has its own synthetic mechanisms or may contain a primer essential for cell wall synthesis.

10.4.6 The Plasma Membrane

The protoplast is bounded by a plasma membrane which is distinct from the enveloping cell wall due to its shrinkage from the latter under high osmotic pressure. It is largely lipoprotein in nature and has marked affinity for dyes. In electron micrographs, the plasma membrane in most bacteria appears as two dense lines between the cytoplasm and the cell wall. This suggests that the plasma membrane consists of three layers, outer and inner electron-dense layers each about 2 to 4 mμ thick and an inner electron transparent space about 3 to 5 mμ thick. This composite structure is called a 'unit membrane' and resembles similar membranes of eukaryotes, but it may not be the only type of membrane structure to be found in bacteria cells.

The plasma membrane is the effective permeability barrier of the cell regulating, the inflow and outflow of metabolites to and from the protoplast. Soluble substances must first pass through the pores of the cell wall. This is not solely a function of their molecular size since large molecules may pass and smaller ones be retained. Certain components of the cell wall, such as the teichoic acids, may act as an ion exchange resin regulating the passage of positively charged ions. The passage of soluble substances through the plasma membrane is more complex. A few substances of low molecular weight can cross the plasma membrane by diffusion but the majority depend on specific proteins (permeases) associated with the plasma membrane. The solute is picked up by the carrier molecule, transported across the membrane and released into the cytoplasm. Some of these carrier mechanisms require metabolic energy for transport.

Relatively pure membrane fractions can be obtained by the lysis of protoplasts. Their chemical composition is similar to that of mammalian cell membranes and includes 40 per cent lipid, 60 per cent protein, and small amounts of carbohydrate. Sterols, however, are absent.

10.4.7 Cytoplasm

The cytoplasm stains homogeneously with basic stains in young cultures but become increasingly granular with age. Unlike that of many eukaryotes, bacterial cytoplasm is relatively immobile, i.e., it does not show streaming.

The fine structure of the cytoplasm of prokaryotes reveal important differences from that of eukaryotes. The cytoplasm consists of a complicated three-dimensional network of fine fibrils stretching from the plasma membrane to the nuclear mass, but lacks the elaborate endoplasmic reticulum characteristic of the eukaryotic cell.

In *Bacillus* species, typical of Gram-positive bacteria, structures continuous with the plasma membrane penetrate into the cytoplasm as tubular or vesicular invaginations. These are called mesosomes or chondrioids. Evidence at present available ascribes to them a number of possible functions. They have been shown to be the principal sites of the cellular respiratory enzymes and probably play a similar role in bacteria to that of the mitochondria of eukaryotes. In intact bacteria they are often situated at the site of the synthesis of cross-wall septa and are also attached to the chromatin material. They are therefore suitably located to perform a co-ordinating role for nuclear division and cell growth under genetic control. Mesosomes have not been seen in those Gram-negative bacteria which have been examined under the electron microscope but rosette-like structures, including rods and lamellar structures without a limiting membrane, have been located near the plasma membrane. In the absence of typical mesosomes these rosettes are considered to be the sites of respiratory enzymes, but they may also be concerned with flagella synthesis. Isolated protoplasts and *L*-forms lack mesosomes. If bacteria are treated with lysozyme the mesosomes originally present are soon extruded into the space between the cell wall and the plasma membrane and are later released along with, but separate from, the bulk of the protoplast and the cell wall fragments. In such an isolated protoplast the synthetic capacity continues, but the cell has lost its ability to synthesize new cell wall material and to divide, properties which presumably depend on the presence of mesosomes.

In photosynthetic bacteria, the pigment-containing structures, known as chromatophores, are associated with membrane structures similar to mesosomes. Under the light microscope the pigments appear to be uniformly distributed through the cell but the electron microscope shows them to be present in distinct chromatophores, which are, however, less complex than the plastids of eukaryotes. In some species the chromatophores are homogeneous particles (e.g., *Rhodospirillum rubrum,* where they measure 60 mμ in diameter), in a few (e.g., *Rhodospirillum vannielli)* lamellar structures are present and resemble those of the cyanobacteria. Some of these lamellar structures are arranged in pairs at the periphery of the cell; others are in clusters. The pigments in photosynthetic bacteria differ from those of the cyanobacteria and from the chlorophyll of plants, but are able to catalyze photophosphorylation.

The chief centres of protein synthesis in the cell are the ribosomes. By ultracentrifugation, particles with a sedimentation rate constant of 70 have been separated from disrupted cells of *E. coli*. These have been calculated to measure 14-17 mμ in width by 19 mμ in length, and since they account for 30 per cent of the cell's dry weight they would be expected to form a considerable part of the visible fine structure of the cytoplasm. Usually individual particles are not distinguishable; instead the ribosomes are seen to be integrated in long linear complexes or polysomes, closely associated with the complicated fibrillar structures of the cytoplasm. These complexes are often joined together and are possibly continuous with the plasma membrane and with mesosomes if present.

10.4.8 Nuclear Apparatus

Bacteria lack an organized nucleus of the type seen in cells of higher organisms but they possess DNA; a body of electron density less than that of the surrounding cytoplasm which corresponds to a nucleus; this is commonly referred to as a nuclear apparatus. Chromosomes of the type seen in cells of higher organisms are not found in bacteria but a single large molecules of DNA constitutes a single bacterial chromosome which contains a complete genome. This is generally considered to be a continuous molecule in the form of a circle which may open under certain circumstances to become a simple double-ended molecule. In addition some bacterial cells possess a short length of extrachromosomal DNA in the form of a plasmid. This is readily transmissible to recipient cells by conjugation or transduction and codes for characters such as drug resistance, colicin production and toxin production.

In most stained preparations the bacterial cytoplasm, being extremely basophilic, masks the presence of chromatin which also stains readily by basic stains. Chromatin, however, can be differentially stained by the Feulgen technique. This stain is a specific colour test for DNA in higher cells of animals and plants and for bacterial chromatin. The dye, basic fuchsin, is first decolorized, by reduction to the leucoform, in the presence of sulphurous acid, to give Schiff's reagent, which is used to detect the presence of aldehyde groups. Hydrolysis of smears of bacteria with hydrochloric acid removes the cytoplasmic RNA and liberates aldehyde groups on the DNA which reacts with reduced Schiff's reagent to give a purplish colour to the chromatin material. Chromatin bodies may also be demonstrated by phase contrast microscopy, and by electron microscopy of thin sections of bacteria in which the nuclear region is less electron dense than is the cytoplasm.

The fine structure of the nuclear body in thin sections is characterized by the absence of cell components such as ribosomes. The appearance of DNA in the nuclear body depends on the techniques of fixing and staining used; in some preparations numerous small aggregates of DNA fibres are seen whilst, in others, the nuclear body appears almost homogeneous; here the DNA fibres are too fine to be resolved (diameter of DNA helix is 20 Å). The fibres of DNA have been shown to be orientated in the direction of the long axis of the cell; cross sections of cells reveal that aggregates of these fibres are folded backwards and forwards into bundles which may be twisted around each other.

The shape and size of the nuclear body varies according to the physiological state of the cell. In resting cells it is spherical, oval or dumbbell-shaped and usually only one body is present in a single cell. In the non-resting cell it is almost permanently in a state of division, changing its shape within minutes from a compact to a more tape-like appearance prior to division. Division is not by mitosis with spindle formation as in eukaryotes but by controlled replication of the DNA. The migration and anchoring of the divided DNA as two separate masses at either end of the dividing cell is considered to be one of the functions of mesosomes. Division of the nuclear body occurs more or less synchronously with cell division but not infrequently more rapid division results in two or four chromatin bodies being present in one cell.

10.4.9 Inclusion Bodies

These are non-living bodies deposited in the cytoplasm of bacteria. They are found in specific organisms and their presence usually depends on the growth conditions of the culture. Volutin, i.e., metachromatic granules, are strongly basophilic, measure up to 0.5 μ in diameter and are widely distributed among bacteria and yeasts. Their presence is used as a basis for the rapid laboratory diagnosis of *Corynebacterium diphtheriae,* the causal pathogen of diphtheria. Metaphosphate is usually but not exclusively present in the granules; the volutin of *Aerobacter aerogenes,* for instance, occupies 20 per cent of the total cell volume of which only 1 per cent is metaphosphate. In many species polysaccharide (including glycogen and bacterial starch) may be detected by staining with iodine. Glycogen stains red-brown and starch, blue. Poly-β-hydroxybutyric acid granules are found in a large number of different types of bacteria. This substance can be detected with Sudan Black and acts as a storage material, being used by the cell for endogenous metabolism under appropriate conditions. Protein crystals are sometimes found, a notable example being in sporing cultures of *Bacillus thuringiensis* where a single large crystal is present in each cell. Vacuoles containing gas occur in halophilic species, such as *Halobacterium halobium* and a few other genera. Intracellular sulphur is frequently deposited by photosynthetic and chemosynthetic bacteria able to utilize hydrogen sulphide.

10.4.10 Endospores and Cysts

Endospores are produced by members of the family Bacillaceae and some species of *Vibrio, Spirillum, Metabacterium,* and *Sporosarcina.* With the exception of *Metabacterium polyspora* and certain gut bacilli, one endospore is formed in each bacterial cell, which ultimately germinates to give rise to a single vegetative cell; the process is not therefore one of multiplication. Endospores may be examined as highly refractile bodies in unstained preparations, by special staining techniques, in which hot concentrated stains, such as malachite green, are used to penetrate the highly resistant spore envelope, or by electron microscopy.

The spore position in the vegetative cell, its shape and size which may or may not distend the sporangium, are characteristic of particular species of the aerobic genus *Bacillus* and the anaerobic *Clostridium.* The spores of *Clostridium tetani,* for which instance, are spherical and terminal; those of *Clostridium septicum,* oval and sub-terminal. In both species the spores distend the vegetative cell (Figure 10.15).

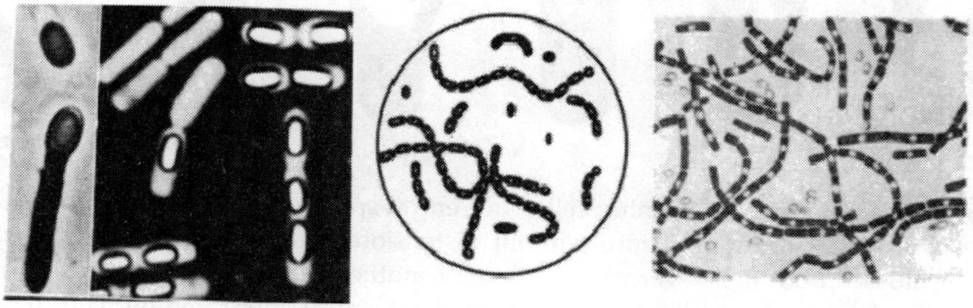

Figure 10.15 Endospores of *Clostridium pectinovorum* and *Bacillus cereus.*

10.4.10.1 Endospore formation

Spore formation is initiated and completed in culture within a relatively short time from inoculation. Whilst the exact conditions inducing spore formation are not known it generally occurs in physiologically vigorous organisms containing a full complement of enzymes after a period of rapid vegetative growth and consequent depletion of exogenous nutrients. This depletion may produce changes from exogenous to endogenous metabolism which may be the stimulus for spore formation. Sporulation involves the synthesis of new materials from low molecular weight compounds, which must be available to the cell, as well as the condensation of existing cell materials. For instance, dipicolinic acid (2,6-pyridinedicarboxylic acid) is unique to spores. Other substances, antigenically distinct from the vegetative cell, also may be found in spores. The materials necessary for spore synthesis may be provided, at least in part, by the autolytic breakdown of vegetative cell components which accompanies sporulation and which may occur before visible degeneration of the vegetative cell. Other events leading to sporulation, such as an increased requirement for cations, also take place before visible changes can be observed. Once the cell is committed to sporulation it cannot revert to vegetative proliferation (Figure 10.16).

The first visible evidence of spore formation is the appearance of a faint, clear area, later to become the spore. This gradually becomes more opaque to form the forespore, and at this stage, stains intensely. Finally, the spore envelope is laid down after which the spore no longer stains readily, owing to the impermeable nature of the wall, and becomes highly refractile. The sporangium continues to be metabolically active for a time but eventually loses its Gram-positivity, autolyses and liberates the spore.

Electron microscopy reveals the stages of the formation of the spore envelope which differs in detail between species. In *B. megaterium,* the electron dense forespore is separated from the cytoplasm by a less electron dense ring. Later distinct layers are

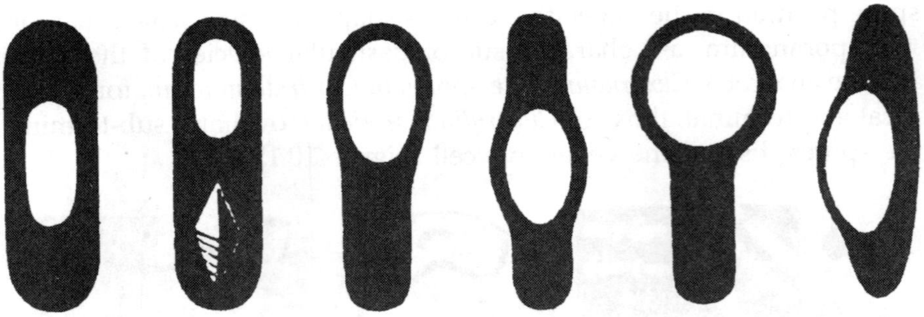

Figure 10.16 Typical Endospore-forming cells 1 = Central spore without distension of the mother cell (*Bacillus megaterium*), 2 = Terminal spore without distension of the mother cell (*Bacillus thuringiensis* with protein inclusion body), 3 = Terminal endopore mother cell distended vally (tennis racquet to spindle shpae Clostridium form (*Bacillus* macerans), 4 = Central spore, mother cell distended ovally (tennis raequet shaped) (*Bacillus polymyxa*), 5 = Terminal spherical spore, mother cell distended to drumstick shape; plectridium form (*Bacillus sphaericus*), 6 = Lateral spore mother cell distended to spindle shape (*Bacillus laterosporus*).

formed. The central core which contains the nuclear body is surrounded by the spore wall, a delicate membrane from which the cell wall of the future vegetative organism will develop. Outside this is the cortex, a thicker but less electron dense layer which, in turn, is enclosed by several layers forming the spore coat, the structure which gives the spore its high refractivity. The spore coat may be smooth, grooved or raised in geometrically-patterned ridges. In some species an additional outer structure called the exosporium surrounds the spore (Figure 10.17).

Nuclear changes accompany spore formation. In the earlier stages a part of the nuclear material, in the form of an axial filament, becomes attached by a mesosome to the adjacent plasma membrane at the site of the spore formation. Subsequently, the mesosome guides or accompanies this chromatin material into the future spore. The outer part of the nuclear material is anchored at the opposite end of the cell and ultimately decays with the parent cell or sporangium.

10.4.10.2 Endospore germination

Usually, endospores do not germinate in the mother culture and exogenous energy-supplying nutrients are essential. Germination occurs a few hours after endospores are transferred to media conducive to vegetative growth. In some species the spore first loses its optical density, refractivity, and heat resistance whilst demonstrating an increase in stainability and respiration. It next swells to about twice its volume, the rate of respiration increases, and the spore wall gradually disappears leaving only a thin outer spore coat. Eventually the germ cell breaks through this spore coat, usually at one pole. In other species little swelling and loss of refractivity occurs, the germ cell rupturing the spore coat either at one pole or along the equatorial plane leaving the parts of the coat attached to the germ cell. Finally, respiration increases to the rate characteristic of the vegetative state and the germ cell elongates to the full size of the

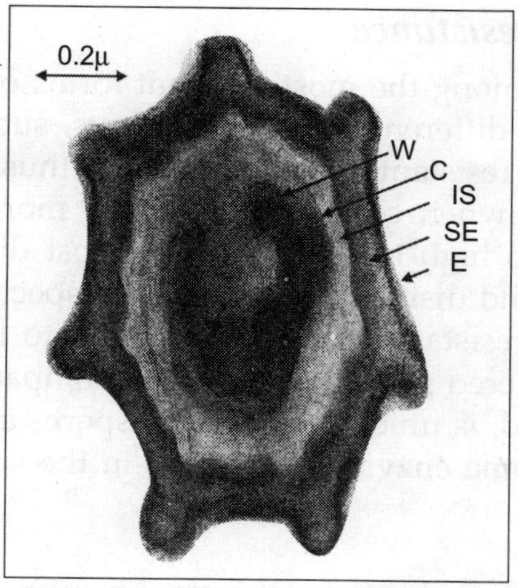

Figure 10.17 Electron micrographs of thin section of bacterial endospore of *Bacillus circulans* where SE denotes for spore envelope; IS layered inner space envelops; C cortex; W cell wall of spore.

vegetative cell prior to fission. Upon germination the nuclear material, which is not easy to demonstrate in the endospore, once more assumes a normal structure. Early in germination mesosomes develop in the germ cell, DNA replication commences and later the nuclear body is seen to be attached to the plasma membrane at the site of the first transverse septum (Figure 10.18).

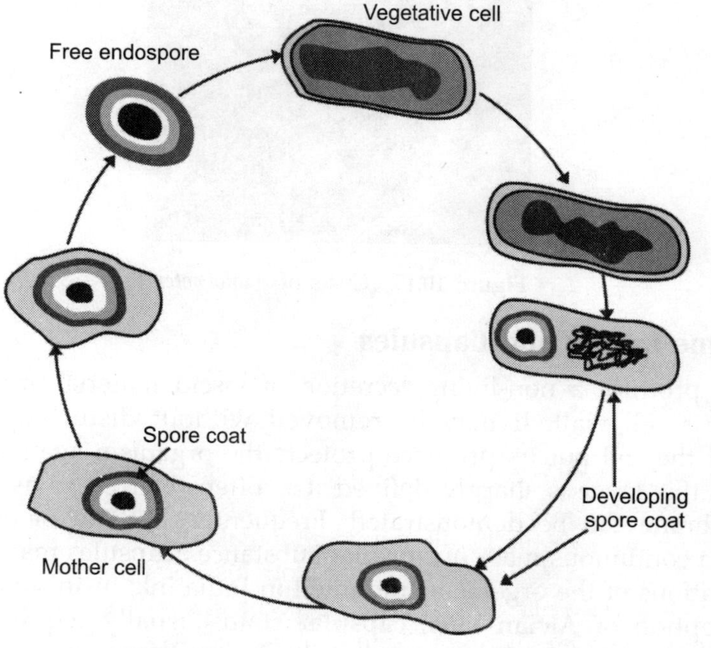

Figure 10.18 Germination of endospores.

10.4.10.3 *Endospore resistance*

Bacterial endospores are among the most resistant forms of life known but within the same species considerable differences to lethal agents, such as high temperature, are found. They are extremely resistant to desiccation and thus are able to survive for long periods and to germinate when conditions are once more favourable for vegetative growth. Their resistance to high temperature and most disinfectants enables them to survive heat sterilization and disinfection processes respectively which are adequate to kill vegetative cells. This resistance may be attributed to lower water content, higher calcium content, and an altered protein composition compared with the parent cell. The amino acid, dipicolinic acid, is unique to bacterial spores and is thought to play a role in enzyme inactivation. Some enzymes are active in the spore but most are dormant.

10.4.11 Cysts

Some species of *Azotobacter* are able to form cysts. An entire cell rounds off and develops a thick-walled, highly refractile cell. They have a low endogenous metabolism and have a resistance to environmental influences greater than that of vegetative cells but not as great as that of bacterial spores. Like the latter, a cyst can remain dormant for a long period and subsequently germinate to give rise to a single vegetative cell (Figure 10.19).

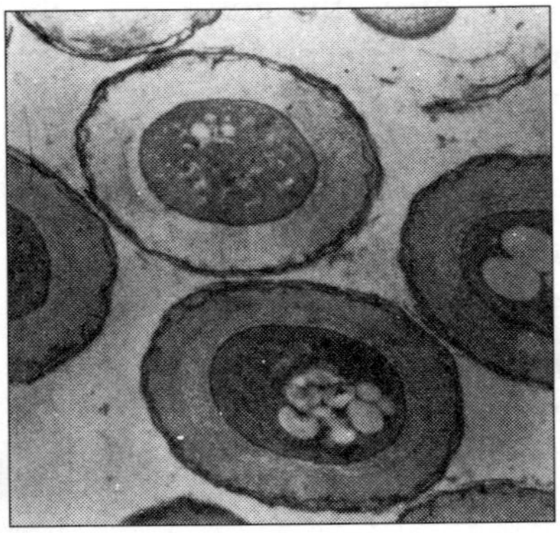

Figure 10.19 Cysts of *Azotobacter*.

10.4.12 Slime Layers and Capsules

Most bacteria produce a non-living secretion of viscid material around the external surface of their cell wall. It may be removed without disturbing the viability or metabolism of the cell but its presence protects the organism against certain external factors. When this layer is sharply defined it is often referred to as a capsule but no limiting membrane can be demonstrated. Frequently, a chain of organisms is seen embedded in a continuous mass of capsular substance. Capsules may be demonstrated in wet preparations of the organisms mounted in India ink, or in stained preparations. With the exception of Alcian blue, capsular stains usually require copper salts as mordants. Capsules may also be rendered visible by treatment with specific anticapsular antiserum which increases their refractivity. No structural detail is revealed by electron micrographs.

Slime layers usually, but not always, consist of a single chemical component, the nature of which varies with the species. For instance, the slime layers in *Diplococcus pneumoniae* and species of *Streptococcus* consist of polysaccharide containing glucosamine and hyaluronic acid respectively, in *Leuconostoc mesenteroides* of dextran (i.e., α 1-6 linked polymer of glucose), and in *Acetobacter xylinum* of cellulose (i.e., β 1-4 linked polymer of glucose). Some bacteria produce capsules of polypeptides (e.g., *Bacillus anthracis,* composed of 1-D-glutamic acid), of protein (e.g., *Pasteurella pestis),* or DNA (in certain halophilic species grown in sub-optimal salt concentrations). Occasionally, more than one component may be produced by the same species under different growth conditions.

Thus in *Streptococcus salivarius* the slime layer consists of a levan and an antigenically distinct polysaccharide; in *Bacillus megaterium,* polysaccharide is laid down in disks surrounding the sites of cell division and polypeptide material between these disks (Figure 10.20). Differences in the chemical nature of capsular polysaccharides in strains of *Diplococcus pneumoniae* are responsible for serological differences of which 70 antigenically different types are recognized.

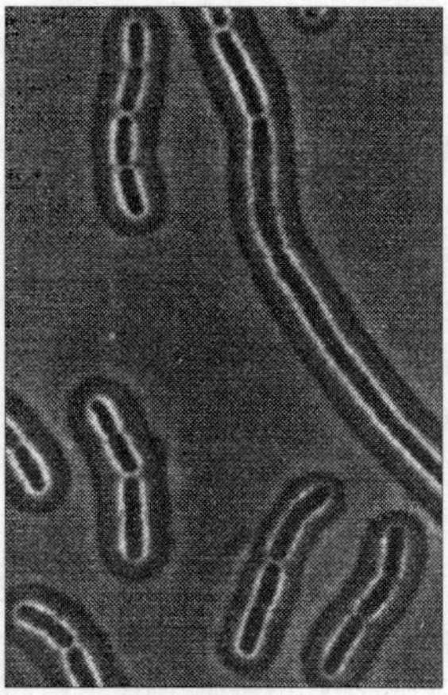

Figure 10.20 Slime layer consisting of distinct polysaccharides in *Bacillus megaterium.*

Capsule production, although primarily under genetic control, is commonly also determined by the constitution of the growth medium. For instance, when grown in the presence of sucrose, *Streptococcus salivarius* produces a dextran of high molecular weight, the quantity of the polymer being proportional, within certain limits, to the concentration of sucrose present. Capsule formation may be responsible for considerable economic loss is in dairy and other food industries. Carbohydrate-containing materials become 'ropy' when encapsulated organisms grow in them. However, some organisms, e.g., *Leuconostoc* species, are employed commercially for the production of dextran as a plasma 'extender' used in the treatment of shock resulting from blood loss. The presence of a capsule is often associated with the virulence of pathogenic organisms. When these organisms enter the animal body they are able to resist phagocytosis by the white cells of the blood, i.e., they are not readily engulfed by these scavenger cells or, if they do become engulfed, they resist digestion by the intracellular enzymes and continue to multiply. Loss of capsule, which sometimes occurs following a mutation, not only leads to loss of virulence but is often accompanied by other changes, such as alteration of the colony structure on the surface of solid media. In some organisms, e.g., *B. anthracis,* capsulation is only one of many factors involved in the pathogenicity of the organism.

10.4.13 Flagella

Not all true bacteria are motile but motility, when it does occur, is effected by means of flagella (singular flagellum; from Latin meaning 'whip'). Flagella are unbranched, sinuous filaments of uniform thicknsss throughout their length which may be many times greater than that of the bacterium.

Their small size precludes their being seen in their natural state under the light microscope but considerable thickening by means of metal deposition or special staining techniques, such as fluorescent dyes, renders them visible. Their motion can be observed by dark field microscopy or by the agitation of India ink particles in the vicinity of flagellate bacteria mounted in soft agar to slow down their rate of movement. Dark field examinations can be greatly improved by suspending organisms in 0.5 per cent methyl cellulose which not only reduces the rate of movement but also induces the aggregation of the flagella in bundles. In stained preparations the wavelength and amplitude of each wave of an individual flagellum can be determined. Flagella of different wavelength may be present on the same organism. The presence or absence of flagella and their arrangement and number is a species characteristic. A single bacterium may possess from one to a hundred flagella. Vibrios have a single flagellum (monotrichate-from Greek trichos meaning 'hair'), *Spirillum* a tuft of flagella at one or both ends (lophotrichate) (Figure 10.21). In rod-shaped organisms the flagella may be single or numerous distributed around the whole organism (peritrichate) or near one or both poles. Flagella can aggregate into a bundle and function as a single propulsive unit. Only a few spherical bacteria are flagellate (Figure 10.22).

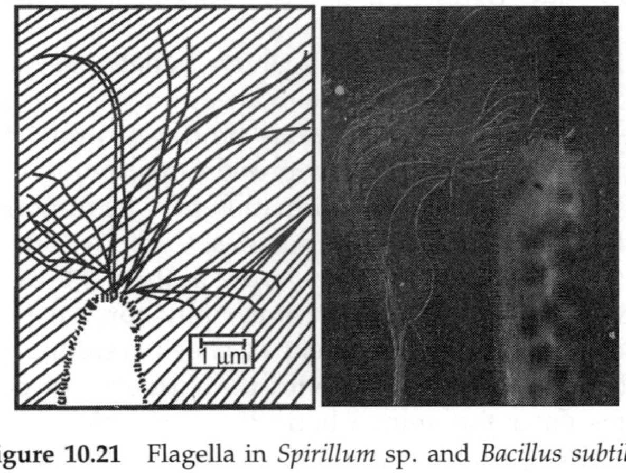

Figure 10.21 Flagella in *Spirillum* sp. and *Bacillus subtilis.*

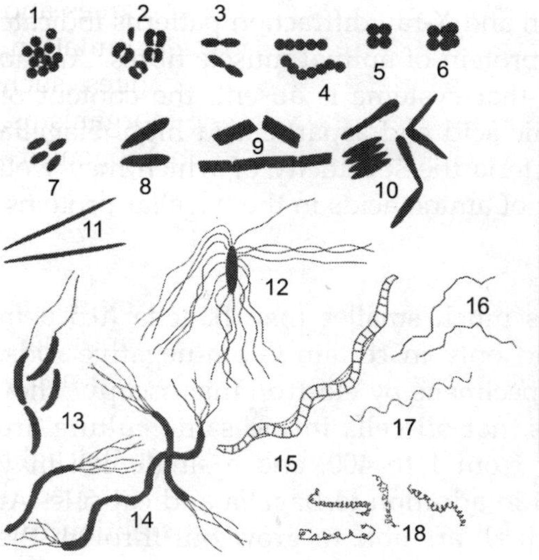

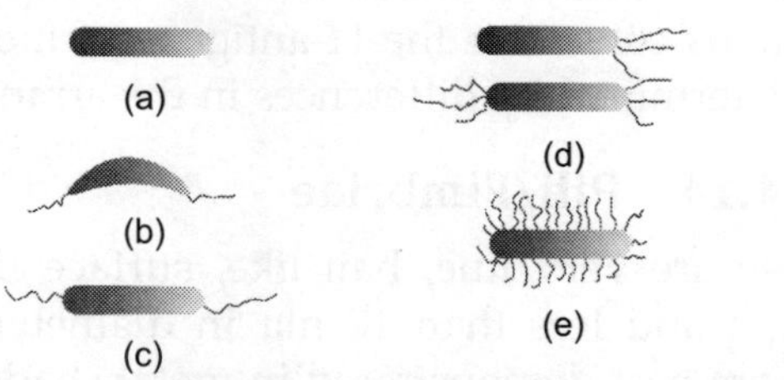

Arrangement of flagella: (a) atrichate, (b) monotrichate, (c) amphitrichate, (d) lophotrichate, (e) peritrichate.

Figure 10.22 Various morphological forms of typical bacteria drawn to approximately the same size 1- *Staphylococcus*, 2- *Nesseria diplococcic*, 3- *Diplococcus pneumonia*, 4- *Streptococcus micrococci*, 5- *Micrococcus tetragenus*, 6- *Sarcina*, 7- *Serratia marcescens* (small rods), 8- *Escherichia coli* larger, round-ended rods, 9- *Bacillus anthacis* flat ended rods both single and in chains, 10- *Corynebacterium diphtheria* clubbed bacilli arranged singly, in palisades and at various angles to one another, 11- *Fusobacterium diphtheria* clubbed bacilli arranged singly, in palisades and at various angles to one another, 12- *Salmonella typhi* flagellate rod, 13- *Vibrio cholerae* vibrios with single polar flagellum, 14- *Spirillum* spirilla with tuffs of polar flagella, 15-*Cristispira* 16- *Borrelia*, 17- *Treponema*, 18- *Leptospira*.

Electron micrographs of specimens shadowed by heavy metals or negatively stained with phosphotungstate demonstrate that a single flagellum consists of several fibrils, usually three, arranged in a helix. Flagella have an average diameter of 12 mμ. They are thus of the same order of magnitude as a single protein molecule. They are thought to arise from a basal cytoplasmic granule, and to pass through the cell membrane and the cell wall. The size of the granule depends on the type of bacterium; that of *Vibrio comma* is between 150 – 200 mμ. Some workers have expressed doubt about the size of these granules in various organisms and possibly even their existence, suggesting that they may be artifacts. In protoplasts of *Bacillus stearothermophilus* the flagella are attached by means of hooks to infoldings of the cytoplasmic membrane. Flagella are retained by protoplasts prepared from motile organisms but under these conditions lose their power to produce motility. Flagella grow very rapidly, at approximately the rate of 0.5 μ per minute, attaining full length in 10 to 20 minutes. They are thought to propel the organism, not by a lashing action, but by waves of contraction passing from base to tip, imparting a spiral motion to the flagellum and an opposite rotation to the whole cell. The speed of cell movement is very high but varies with the species; speeds of 25 μ per second for peritrichous flagellate cells and 200 μ per second for the monoflagellate *Vibrio cholerae* have been recorded.

Pure preparations of flagella can be obtained by prolonged shaking of bacterial suspensions followed by differential centrifugation. Analysis shows that at least 98 per cent of the total weight is a protein called flagellin and X-ray diffraction patterns indicate a marked resemblance of this to the contractile protein of animal muscle fibres. Amino acid analysis of a number of flagellins reveals that cysteine is absent, the content of aromatic amino acids is low and that of glutamic acid and aspartic acid high. Flagella proteins constitute the H-antigens of motile bacteria the specificity of which may well be determined by differences in the arrangement of amino acids in the flagellar proteins.

10.4.14 Pili/Fimbriae

These are very fine, hair-like, surface filaments much smaller than flagella (0.5 μ in length and less than 10 mμ in diameter), found only in certain Gram-negative rods. Pili are best demonstrated in metal-shadowed specimens by electron microscopy. They are frequently found in freshly isolated strains; not all cells in the same culture are uniformly piliate; the number per cell varying from 1 to 400, and a single cell may have one or more morphologically different pili in addition to flagella and capsule. At least eight different types of pili are recognized; all are able to grow out through the cell wall without lysing the cell.

Pili cause bacteria to adhere both to one another and to foreign cells, such as red blood cells. Piliate colonies may be detected by their more compact colony form, but more readily by their ability to haemagglutinate red blood cells when suspensions of these are poured over a plate culture. Antigenic differences occur; for instance, all strains of piliate *Shigella flexneri* are found to have pili of the same antigenic composition; in contrast, at least five antigenically distinct serotypes of piliate *Salmonella* spp. are known.

Apart from their adhesive properties the function of the majority of pili is not known. Most work has been done on sex pili which are found only on strains of certain Gram-negative bacteria, and are capable of transferring self-replicating molecules of DNA (F, colicinogenic, and R factors) from donor to recipient bacteria. Only 1 to 4 of these pili occur on each cell corresponding to the number of factors present in the cell. The donor cell synthesizes the units of pilus protein called pilin; these are assembled into hollow, hair-like structures with an axial hole of 2.5 mμ diameter, and this structure is thought to be laid down around the DNA molecule being transferred. This complex of pilus and DNA passes through the cell wall. The distal end makes contact with, and attached itself to, a recipient cell which receives the DNA through the hollow pilus (Figure 10.23).

Sex pili also act as sites for male-specific phages. The male specific RNA phages attach to the outside of the projecting pilus, enter through the pilus wall and pass along the hollow canal. The male-specific, rod-shaped DNA phages attach to the pilus end to end, the pilus again acting as a canal along which the nucleic acid enters the cell (Figure 10.24).

10.4.15 Cell Division

The normal mode of cell division in bacteria is by binary fission into two similar daughter cells. This is intimately associated with synthesis of new cell wall material, cell growth, and DNA replication. The cell first elongates beyond its normal length

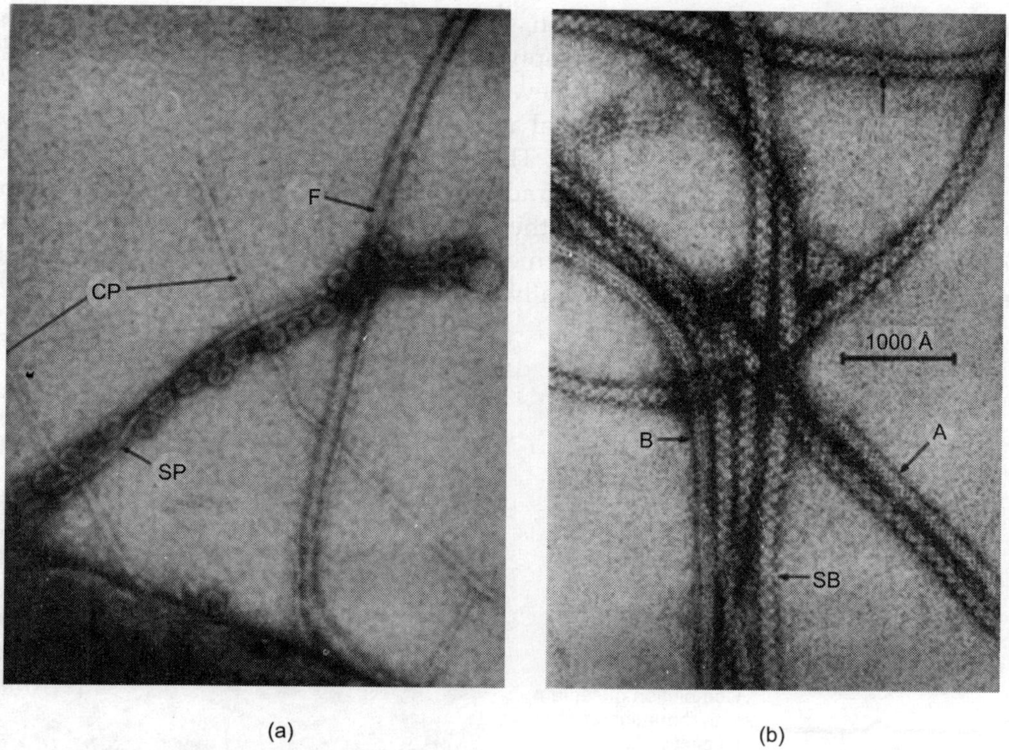

Figure 10.23 (a) Electron micrograph of appandages of *Escherichia coli* where F = flagellum, CP = common pili, SP = sex pilus (b) Molecule architecture of flagella A = one shaped like a helix, B = the other with a predominantly longitudinal arrangement C = intermediate shape = flagellum surrounded by a sheath.

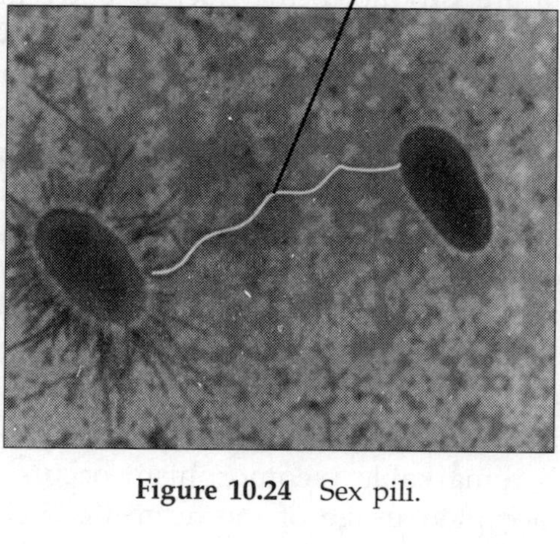

Figure 10.24 Sex pili.

prior to division. During this elongation, new cell wall material is laid down in an equatorial zone at the sites of future septum formation in Gram-positive bacteria; in Gram-negative species the addition of new cell wall material is not limited to one site but is built into any part of the original wall. Division of the cell is initiated by an ingrowth of the cytoplasmic membrane. This is followed by ingrowth of the cell wall like a slowly closing diaphragm, which gradually closes the aperture to form a complete transverse septum which is thicker than the normal cell wall. Both processes seem to be under the direct influence of associated mesosomes. Finally, separation is achieved by a median splitting of the cross wall usually from the periphery inwards (Figure 10.25).

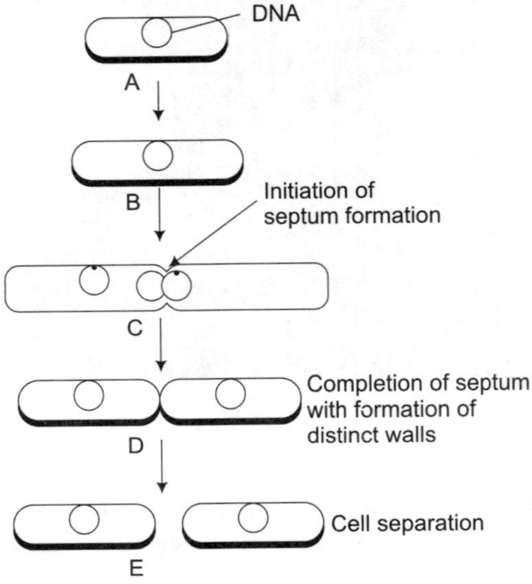

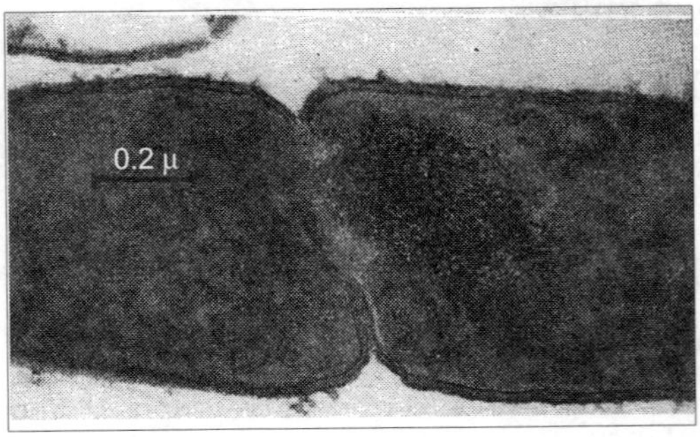

Figure 10.25 Formation of transverse wall by membrane bodies in *Bacillus megaterium*.

Before cross-wall septa are produced, the nuclear body divides. When both these processes are synchronized, the resulting daughter cells each contains a single nuclear body. Nuclear division often occurs faster than cell division and multinucleate cells are then produced; when ingrowth of cross-wall septa takes place faster than separation of daughter cells, multicellular rods and cocci result. Difference in detail in the process of separation of daughter cells is responsible for the characteristic post-fission shape and arrangement of cells in individual genera and species. When separation is incomplete, chains of cocci or filaments of bacilli result. The production of cross wall septa at different planes in regular sequence accounts for groups of four or eight cocci, as in *Micrococcus tetragenus* and *Sarcina* species respectively; irregular division results in clusters as in *Staphylococcus* species.

Complex life cycles have been described for some bacteria. While it is undoubtedly true that in some species remarkable pleomorphism occurs, evidence for a true life cycle is incomplete. The accepted usage of the term life cycle implies the occurrence of a cyclic series of phases in an obligatory sequence and that the completion of this sequence is essential for the fulfilment of some important activity, such as reproduction, genetic recombination or dispersal. Such cycles are common in higher organisms but it is most unlikely that they occur in bacteria.

10.4.16 Stalked Bacteria

Several genera of bacteria are stalked but the stalk varies considerably both in its size and nature. *Caulobacter* has a slender, relatively short protoplasmic stalk arising from the pole of the rod-shaped cell. The distal end of the stalk has an adhesive holdfast by which the organism attaches itself to the substratum, either other microorganisms or inanimate substrates, or to the holdfasts of other *Caulobacter* cells, thereby forming rosettes of cells (Figure 10.26). The stalk of *Gallionella* is usually extremely long, often spirally twisted, and has been variously considered to consist entirely of ferric hydroxide, or of organic matter impregnated with ferric hydroxide or even to be the basic structure of the cell on which inorganic material is deposited. The first of these alternatives is most unlikely and there is dispute as to which of the other suggestions is correct.

10.4.17 Bacterial Variation

Extensive variation within bacterial species occurs. Most of these have a genetic basis and are considered in Chapter on bacterial genetics.

10.5 CLASSIFICATION

The following is an outline modified from Bergey's manual. Those bacteria that cannot be assigned easily to the three basic categories are listed here as 'Large Special Groups'.

1. Cocci (spherical bacteria)

A. Gram-positive cocci

AEROBES *Micrococcus, Staphylococcus, Streptococcus, Lecuconostoc, Pediococcus, Lactococcus*

ANEROBES *Peptococcus, Peptostreptococcus, Ruminococcus, Sarcina*

B. Gram-negative cocci

AEROBES *Neisseria, Moraxella, Acinetobacter, Paracoccus, Lampropedia*

ANEROBES *Veillonella, Acidaminococcus, Megasphaera*

2. Rods (straight, cylindrical bacteria)

A. Gram-positive, non-spore forming rods

AEROBES *Lactobacillus, Listeria, Erysipelothrix, Caryophanon*

Coyneform bacteria and actinomycetes

AEROBES *Corynebacterium, Arthrobacter, Brevibacterium, Cellulomonas, Propinibacterium, Eubacterium, Bifidobacterium, Mycobacterium, Nocardia, Actinomyces, Frankia, Actinoplane, Dermatophilus, Micromonospora, Microbispora, Streptomyces, Streptosporangium*

Endospora forming rods and cocci

AEROBES *Bacillus, Sporolactobacillus, Sporosarcina, Thermoactinomyces*

ANEROBES *Clostridium, Desulfotomaculum, Oscillospira*

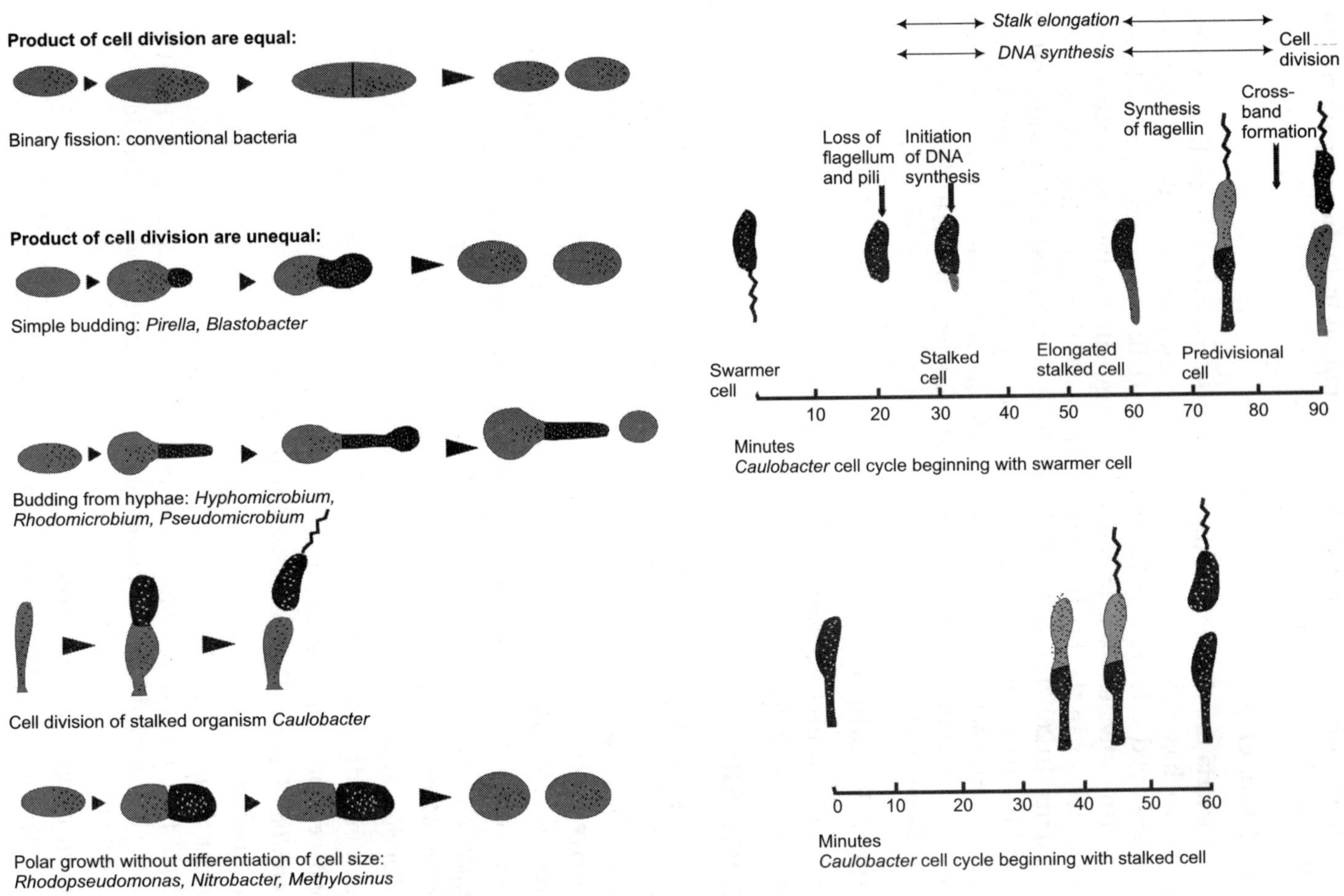

Figure 10.26 Contrast between cell division in conventional bacteria and in budding and stalked bacteria.

B. Gram-negative bacteria

Gram-negative aerobic rods and cocci

AEROBES *Pseudomonas, Xanthomonas, Zoogloea, Gluconobacter, Acetobacter, Azotobacter, Azomonas, Beijerincia, Derxia, Rhizobium, Agrobacterium, Alcaligenes, Brudella, Legionella, Thermus*

Gram-negative aerobic chemolithotrophic bacteria

AEROBES *Nitrobacter, Nitrospira, Nitrococcus, Nitrosomonas, Nitrospira, Nitrosococcus, Nitrosolobus, Thiobacillus, Thiobacterium, Thiovulum*

Sheathed bacteria

AEROBES *Sphaerotilus, Leptothrix, Streptothrix, Crenothrix*

Gram-negative facultative anaerobic

FACULTATIVE ANAEROBES *Escherichia, Klebsiella, Enterobacter, Salmonella, Shigella, Proteus, Serratia, Erwinia, Yersinia, Vibrio, Aeromonas, Photobacterium*

Gram-negative anaerobic

STRICT ANAEROBES *Bacteroides, Fusobacterium, Leptotrichia, Fibrobacter*

Archaebacteria

STRICT ANAEROBES *Methanobacterium, Methanothermus, Methanosarcina, Methanothrix, Methanococcus*

AEROBES *Halobacterium, Haloferax, Halococcus, Sulfolobus, Thermoplasma, Thermus*

ANAEROBES *Thermoproteus, Pyrodictium, Desulfurococcus, Pyrococcus, Thermococcus, Thermodiscus*

3. Curved rods and flexible cells

Gram-negative spirillar and curved bacteria

AEROBES *Spirillum, Aquaspirillum, Azospirillum, Oceanospirillum, Campylobacter, Helicobacter, Bdellovibrio, Microcyclus, Pelosigma*

Gram-negative curved anaerobic bacteria

ANAEROBES *Desulfovibrio, Succinivibrio, Butyrivibrio, Selenomonas*

Spirochaetes

AEROBES AND ANAEROBES *Spirochaeta, Cristispira, Treponema, Borrelia, Leptospira*

4. Large Special group

Gliding bacteria (always Gram-negative)

Myxococcus, Archangium, Cystobacter, Melittangium, Stigmatella, Polyangium, Nannocystis, Chondromyces, Cytophaga, Sporocytophaga, Flexibacter, Herpetosphon, Saprospira, Beggiatoa, Thiothrix, Thioploca, Achromatum, Leucothrix, Vitreoscilla, Simonsiella, Alysiella

Bacteria with appendices, prosthecate bacteria, and budding bacteria

Hyphomicrobium, Hyphomonas, Caulobacter, Asticcacaulis, Planctomyces, Ancalomicrobium, Prosthecomicrobium, Blastobacter, Seliberia, Gallionella, Nevskia

Obligate parasitic bacteria: rickettsiae and chlamydiae

Rickettisa, Coxiella, Chlamydia

Mycoplasma group (Mollicutes)

Mycoplasma, Acholeplasma, Spiroplasma, Ureoplasma, Metallogenium

Anaerobic, anoxygenic phototrophic bacteria

Rhodospirillum, Rhodopseudomonas, Rhodobacter, Rhodomicrobium, Rhodocyclus, Rhodopilus, Chromatium, Thiocystis, Thiosarcina, Thiocapsa, Thiospirillum, Thioppedia, Amoebobacter, Ectothiorhodospira, Lamprocystis, Thiodictyon, Chlorobium, Prosthecochloris, Pelodictyon, Chloroherpeton, Chloroflexus, Chloronema, Oscillochloris

Aerobic, oxygenic phototrophic bacteria: Cyanobacteria

Synechococcus, Gloeocapsa, Gloeothece, Gloeobacter, Pleurocapsa, Dermocarpa, Myxosarcina, Oscillatoria, Spirulina, Lyngbya, Phormidium, Plectonema, Anabaena, Nostoc, Calothrix, Fischerella, Scytonema

10.5.1 Gram-positive cocci

Gram-positive cocci include coccoid of lactic acid bacteria, viz., *Sterptococcus. Leuconostoc* and *Pediococcus.* These bacteria depend on fermentation reactions for their energy, are microaerotolerant, and most contain no haem pigments. Apart from these, there are the aerobic and facultative anaerobic genera *Micrococcus* and *Staphylococcus,* and the obligately anaerobic genera *Sarcina, Peptococcus* and *Ruminococcus.* The genus *Micrococcus* contains those pigmented organisms whose yellow and orange colonies are often found on 'air plates'. These were formerly assigned to the genus *Sarcina* because they form packets of cells or tetrads; the bacteria once known as *Sarcina lutea, S. flava, S. aurantiaca,* etc. are now classified together as *Micrococcus luteus.* This species also includes the organism that was named *Micrococcus lysodeikticus* because of its high lysozyme sensityivity. *Micrococcus* species are obligately bnaerobic and characterized by a high GC content (66-72%).

The name of the genus *Staphylococcus* is based on the appearance of its cells under the microscope. The cells appear to be arranged in a grape-like cluster, which results from an irregular pattern of cell division in different planes. *Staphylococcus* is facultatively anaerobic, forms cytochromes only under aerobic conditions, and is relatively resistant to dehydration. *S. aureus* is pathogenic, producing toxins and exoenzymes, and is a pus former. Other strains caused food poisoning by excreting enterotoxins during their growth on non-refrigerated food.

Sarcina ventriculi is a microaerotolerant, anaerobic organism which can be isolated easily from soil, but it is also found in the stomach contents of patients with gastric disease. It is unusual in its large size (diameter of 4 μm), in its formation of large cell clusters (more than 64 cells may be held together by cellulose), in its pH tolerance (it can grow in pH 0.9 to 9.8), and in its ability to produce endospores.

10.5.2 Gram-negative cocci

This group contains some cocci and very short rods that are Gram-negative and non-motile. Included among these are aerobes and anaerobes, pathogens, soil bacteria, and some that inhabit the intestinal tract and mucous membranes of various mammals.

10.5.2.1 Aerobic cocci

The genus *Neisseria gonorrhoeae* is oxidase positive, Gram-negative, kidney-shaped cocci occurring characteristically in pairs; some non-pathogens will grow on nutrient agar but pathogenic species require media enriched with blood or serum; some species commensal in the nasopharynx of warm-blooded animals, others pathogenic for man. *Neisseria gonorrhoeae* is the causal organism of gonorrhea, a venereal disease which had been almost eradicated in many countries because of the sensitivity of the organism to penicillin (1 µg penicillin is effective bacteriostatically); *N. meningitidis* is one of the causal organisms of meningitis. *Neisseria gonorrhoeae* is difficult to culture in the laboratory; it needs aerobic conditions with 10% CO_2 and is extremely sensitive to light and dehydration, these being the reasons why infection is limited to direct intimate contact. *N. meningitides* inhabits the nasopharyngeal tract but may penetrate into the vascular system and produce inflammation of the meninges. *Moraxella* is also oxidase-positive and very sensitive to penicillin. These are coccobacilli that are unable to utilize carbohydrates. *Acinetobacter* is an oxidase negative soil and aquatic bacterium that is easily obtained and isolated by enrichment culture in a medium containing 0.2% acetate at pH 5.5-6.0. It does not normally utilize glucose, disaccharides, or polysaccharides, but it otherwise resembles the genus *Pseudomonas* with respect to its versatility in the utilization of substrates. Within the species *A. calcoaceticus*, genetic markers can be exchanged by transformation. Since the bacterium is able to store a lot of polyphosphate it is hoped to use it for the removal of excess phosphates from effluents. The oxidase reaction that is used for differential diagnosis in this group is based on the formation of a coloured product from dimethyl-phenylene diamine (*N-N-*dimethyl-1,4-diaminobenzene) in the presence of *cytochrome c* (which is water soluble); a few drops of the reagent are applied to agar colonies, and production of a red colour formation is the diagnostic features.

10.5.2.2 Anaerobic cocci

The best known Gram-negative anaerobic cocci are *Veillonella alcalescens* (formerly *Micrococcus lactilyticus*) and *Megasphaera elsdenii* (formerly *Peptostreptococcus elsdenii*). Both are unable to ferment carbohydrates. *Veillonella alcalescens* is found in the saliva of humans and animals and in the rumen of ruminants. It ferments organic acids, especially lactate, to propionate, acetate, CO_2 and H_2. *Megasphaera elsdenii* ferments glutamic acid and other amino acids and is apparently a normal inhabitant of the intestinal tract of many animals.

10.5.3 Gram-positive, Non-Spore-Forming Rods

Gram-positive, non-spore-forming rods that can grow in the presence of oxygen and form mainly lactate by the fermentation of carbohydrates (glucose, lactose) are known collectively as 'lactic acid bacteria'. The genera *Lactobacillus,* and *Bifidobacterium* belong to these. *Bifidobacterium* is one of the few anaerobic lactic acid producers. *Listeria monocytogens* causes listeriosis in soft cheese.

Gram-positive bacteria can be arranged in a sequence according to their cell shape and other morphological and physiological properties. The rod-shaped lactic acid bacteria are at the beginning of this sequence. Most of the lactic acid bacteria are regular rods and the tendency to form club-shaped rods and slightly branched cells is not strong. This tendency is the norm in the coryneform bacteria, which include the genera *Corynebacterium* and *Arthrobacter*. The tendency to cell branching is even more strongly expressed in the mycobacterium. These are followed by the proactinomycetes, which form transient mycelia, and the sequence is completed by the actinomycetes, Many members of the above groups can be recognized easily by their cell shape, colonial morphology, and physiological properties. Others can only be identified by various biochemical characters.

Caryophanon latum can be included here also. It is a trichome-forming, flagellated bacterium that can be isolated regularly from cow dung. The Gram-positive, filamentous rods are exceptionally large (3 × 15 µm).

10.5.4 Coryneform Bacteria

10.5.4.1 Corynebacterium

These organisms are characterized by a 'snapping apart' of the cells during cell division; the connecting walls on either side of the newly made cell wall appear to separate at different rates, so that the two cells seem to twist away in opposite direction. Gram-positive, usually non-motile rods, often barred, some containing metachromatic granules; some species pleomorphic; mainly aerobic; occurring in dairy products, soil and as parasites and pathogens of animals and plants. It is unrepresentative of the genus in its physiological properties, since it grows as a microaerophile or even as an anaerobe, whilst most corynebacteria are aerobic. Stain irregularly with methylene blue to produce characteristic barring; metachromatic granules present in many species; often clubshaped; following transverse division the daughter cells remain attached but orientate themselves into angular or palisade arrangements; non-motile except for a few plant pathogens; *C. diphtheriae* produces a powerful exotoxin causing diphtheria in man; *C. renale, C. equi, C. pyogenes,* and *C. ovis* are animal pathogens causing pyelonephritis in cows, pneumonia in foals, general pyogenic infections in many animals and chronic lesions resembling tuberculosis in sheep respectively. Its pathogenicity is due to its invasion of the larynx and tonsils, and production of an exotoxin that circulates in the blood and attacks cells of the cardiac muscle, kidneys, and nerves, giving rise to post-diptherial paralysis. The only strain that has been lysogenised with phage produces the exotoxin. Thus, toxin production is a consequence of a genetic alteration of the bacterium by the phage. This is known as **'phage conversion'**. In addition, the toxin is only produced if the supply of iron to the cells is sub-optimal.

The genus Corynebacterium also contains several other animal pathogens and many are widely distributed plant pathogens (*C. michiganense, C. poinsettiae, C. fascians*). *C. mediolaneum* was the first bacterium used for biological conversion of steroids.

Listeria monocytogenes (the only species), small rods producing characteristic tumbling motility; catalase positive; produces cerebral lesions in sheep and cattle resulting in incoordinated movement referred to as 'circling disease'; produces monocytosis (i.e., an increase in blood monocytes) in rabbits.

10.5.4.2 Arthrobacter

They are characterized by a marked tendeny of the cells to produce branched and coccoid forms. Some members are motile and flagellated and all are aerobic. Usually, rods in young culture becoming coccoid later; usually non-motile; occur in soil, where they form an important part of the autochthonous flora, on plants, in cheese, and in activated sludge. Arthrobacter can be isolated easily from dry soils; together with bacilli, it can withstand several months of dry storage (in soil) whilst most non-spore-forming bacteria are killed under such conditions. Arthrobacter is pleomorphic, i.e., it assumes various shapes; in young colonies and liquid media it grows as irregularly shaped, long rods, whilst only coccoid forms are found in old cultures. Under unfavourable conditions, *Arthrobacter* can grow slowly in its coccoid form. Other strains and species of the genus are found on plants and in the active sludge of sewage plants.

Another group of coryneform bacteria in soil consists of several cellulose – utilizing species of the genus Cellulomonas. *Cellulomonas are* small pleomorphic rods or cocci, usually motile; commonly attacking cellulose; of soil or plant origin.

Several coryneform bacteria excrete glutamic acid and other amino acids into the medium and, because of this property; they have become important in industry. Other typical forms grow on soft cheese for example Brevibacterium linens, which is mildly proteolytic, plays a role in the ripening of cheese. The propionibacteria were also assigned to the coryneform bacteria group.

Corynebacterium group can be regarded as an intermediate group between the lactic acid bacteria and the mycobacteria in morphological and physiological properties. The tendency for branched cell formation increases in the series propionibacteria, coryneform bacteria, mycobacteria. In this same series, there is a general transition from anaerobic to strictly aerobic metabolism. Despite many efforts, it has not yet been possible clearly to differentiate the genera belonging to the coryneform bacteria. Knowledge of GC content, murein composition, and other properties and the application of numerical taxonomy are needed for a new classification of this extremely important and ecologically differentiated group of microorganisms.

10.5.5 Actinomycetes

10.5.5.1 Classification

Early attempts to classify Actinomycetes on morphological characters resulted in confusion owing to the tendency for these organisms to alter during periods of artificial culture. The increased use of the chemistry of the organism as a source of taxonomic criteria promises to give a more satisfactory system, but at present reliance must still be placed on morphological characteristics.

10.5.5.1.1 Mycobacteriaceae

They contains two genera, *Mycobacterium* and *Mycococcus,* the members of which range from saprophytes to obligate parasites.

Mycobacteria

Mycobacteria are invariably aerobic. Morphologically they are intermediate between the corynebacteria and proactinomycetes (Nocardia). They are characterized by the capacity to form a rudimentary mycelium under at least some conditions of cultivation. This may consist of branched vegetative cells or may approach the level of development seen in *Nocardia*. They do not form mycelia but grow in the form of irregularly shaped, slightly branched cells. They are non-motile and Gram-positive. One way in which they differ from corynebacteria is that they are **'acid fast'**. Mycobacteria and Nocardia are not decolourised by the acid treatment and are designated as **'acid fast'** (Figure 10.27). Few saprophytic mycobacteria can be decolorised with HCl alcohol, buit not with aqueous HCl. The resistance to acid is due to the high levels of mycolic acid in the cell wall, which make the cells of mycobacteria wax-like and strongly hydrophobic (Figure 10.28). The presence in the cell walls of large amounts of amorphous lipid (possibly lipoprotein) together with arabinose, galactose, and *meso-DAP*, and serological resemblances suggest a relationship not only with *Nocardia* but with *Corynebacterium* normally classified in the Eubacteria but sometimes included here. *Mycobacterium* has been distinguished on the property of acid-fastness but this is not a sound distinction since many species of *Corynebacterium* and *Nocardia* are acid-fast under particular nutritional conditions. Their murein skeleton resembles that of the Gram-negative bacteria, but it is complexed with an arabinogalactan, a polysaccharide consisting of arabinose and galactose. This is bound to lipids, namely, mycolic acids. Mycolic acids are branched hydroxyacids carrying aliphatic chain substituents in position 2 and 3. The aliphatic chains vary in length: in corynebacteria they have 32-64 carbon atoms, in nocardia, 45-58 and in mycobacteria, 79-85.

Table 10.5 Extensive keys for the identification of actinomycetes

Taxa of Actinomycetes	*Identification keys*
MYCOBACTERIACEAE *(Mycobacterium, Mycococcus)*	Mycelium rudimentary or absent, no spores
ACTINOMYCETACEAE *(Actinomyces)*	True mycelium, fragmenting into bacillary elements; anaerobic or microaerophilic
DERMATOPHILACEAE *(Dermatophilus)*	True mycelium, dividing in all planes to produce motile cocci
ACTINOPLANACEAE *(Microellobosporia, Streptosporangium, Spirillospora, Actinoplanes, Ampullariella, Amorphosporangium)*	True mycelium, dividing only transversely: Sporangia formed, sometimes conidia also

Contd...

Contd...

MICROMONOSPORACEAE *(Micromonospora)*	No sporangia: Aerial mycelium absent at ordinary temperatures; conidia borne singly on lateral branches of fine substrate mycelium; glycine and meso-DAP are major constituents of walls
STREPTOMYCETACEAE *(Streptomyces)*	Aerial mycelium bearing chains of conidia; glycine and L-DAP are major constituents of walls
NOCARDIACEAE	With or without aerial mycelium; mycelium sometimes fragmenting; conidia absent, or single, or paired, or in chains; contain meso-DAP. but lack major amounts of glycine in walls
Nocardia, Micropolyspora	Mycelium fragmenting into bacillary elements, aerobic
Thermomonospora, Thermoactinomyces	Mycelium not fragmenting; aerial mycelium
Microbispora	or in pairs
Pseudonocardia	or in chains

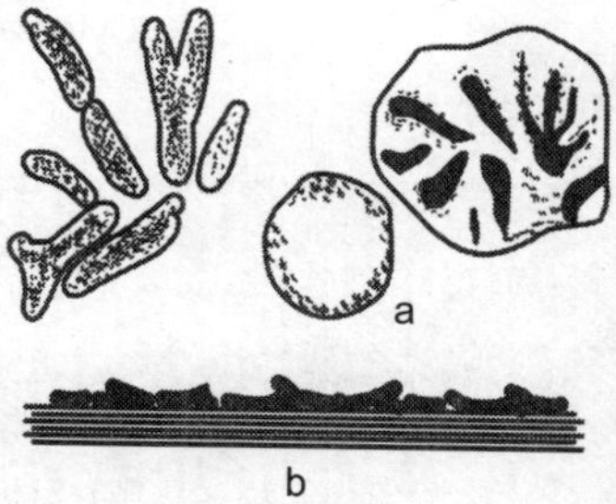

Figure 10.27 Mycobacterium (a) Representative colonial forms (b) Cross sections through growth on agar surface.

Only the long-chain aliphatic substituents render the cells acid fast *M. tuberculosis* exists in a number of strains causing tuberculosis in man, cattle, pigs, birds, or fish; *M. leprae* and *M. lepraemurium* cause leprosy in man and rats respectively; *M. johnei* causes chronic enteritis in cattle and sheep; *M. rubiacearum* is the N-fixing bacterium causing leaf-nodules in some shrubby members of the Rubiaceae.

10.5.5.1.2 Actinomycetaceae

Usually, restricted to single genus *Actinomyces,* contains true mycelium which readily fragments. Microaerophilic or anaerobic; cell wall contains lysine in place of the DAP characteristic of all other actinomycetes and lacks arabinose; best known species are

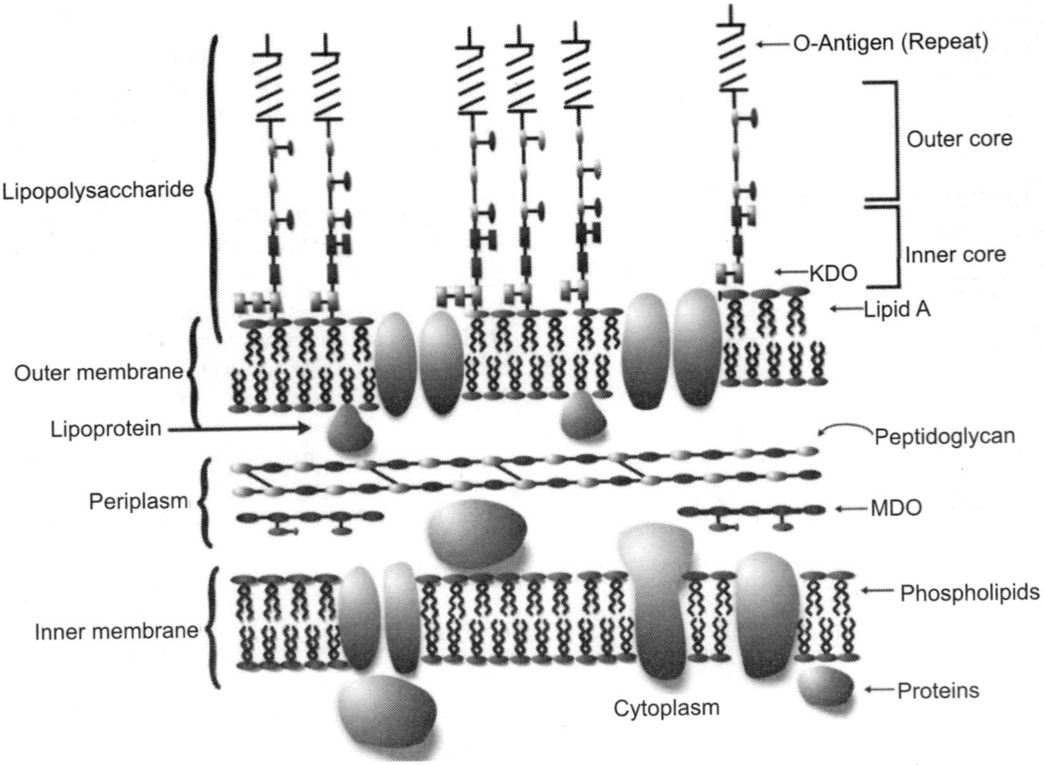

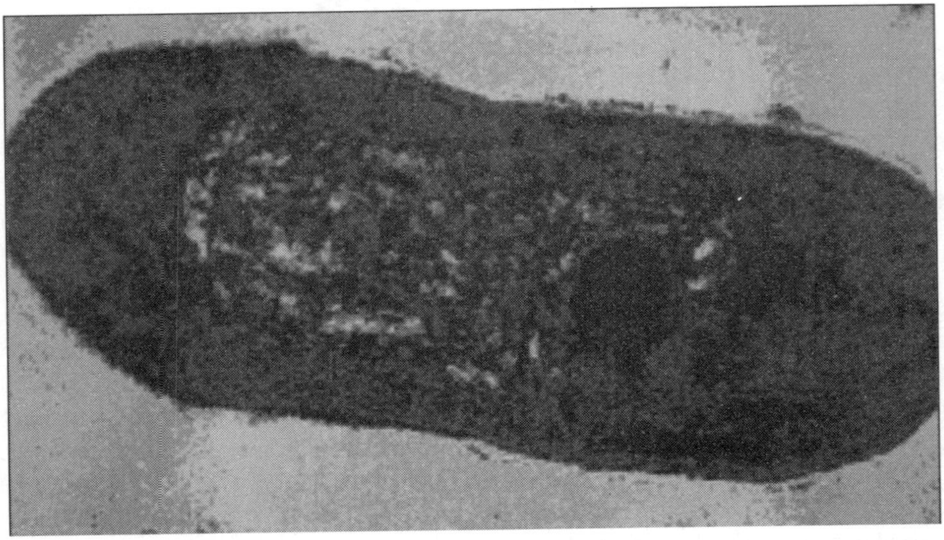

Figure 10.28 Electron micrograph of *M. tuberculosis* and *M. tuberculosis* (Acid Fast) cell wall Actinomycetes.

parasitic but serologically related ones have been isolated from soil water. Actinomycosis, a chronic disease of animals, including man, is caused by *A. israeli* and *A. bovis. A. israeli* is frequently isolated from the crypts of the normal human tonsil, where it can survive in the microaerophilic conditions prevailing, and from this site is thought to invade mouth wounds, e.g., those caused by tooth extraction. The lungs may also be infected. *A. bovis* causes 'lumpy jaw' of cattle, in which the jawbone is invaded and enlarged by the formation of new bone. Other tissues, including the udder, may be invaded and a similar infection sometimes occurs in cows. *A. baudetii* is present in certain lesions in cats and dogs.

In tissues affected by *Actinomyces* and in the resulting pus, small yellow granules (the so-called sulphur granules) occur. These are compact calcified colonies of the organism, consisting of a central Gram-positive mass of mycelium with rods and cocci derived from fragmentation and a peripheral zone of radiating Gram-negative club-shaped structures. These latter, which give the name 'ray fungus' to these organisms, consist mainly of lipid substances and are thought to be a host reaction to invasion. Actinomycoses are susceptible to penicillin treatment.

10.5.5.1.3 Nocardiaceae

They contain several genera, some of which were formerly included in Actinomycetaceae but differing from *Actinomyces* in cell wall composition, plus others formerly placed in Micromonosporaceae.

Nocardia

The mycelium at first non-septate, later becoming septate and fragmenting into rod-shaped cells (occasionally motile by one to four polar or lateral flagella); aerobic; potentially acid-fast; cell walls containing lipids (as in *Mycobacterium* but differing from those of that genus); meso-DAP (in contrast to the L-DAP of the streptomycetes) and, in most species, arabinose and galactose. Most species are soil saprophytes but several cause diseases (nocardioses) of animals including man. *N. asteroides* causes mammary and cutaneous lesions in cattle; dogs and man may also be affected. *N. farcinica* attacks lungs and lymph nodes in cattle, causing a chronic disease known as bovine farcy, inducing the development of nodules in usually one limb and resulting in death; *N. madurae* is the principal cause of madura foot, a tropical or subtropical disease of man. The organism gains entry through abrasions, caused by the habit of walking barefoot on soil, and induces swelling and malformation of the foot. Small granules of varied colour develop in the pus. Other genera differ from *Nocardia* in details of cell wall composition and in the degree of fragmentation of the hyphae. Further subdivisions of the family may follow further investigations. The sporophores in streptomycetes may sometimes be grouped to form structures resembling the coremia or pycnidia of true fungi. One family of actinomycetes (the Actinoplanaceae) is characterized by the formation of 'sporangia' which, however, are quite unlike those of the fungi. A sporangium is formed as a terminal vesicle on a hyphae, and into it grows one or more filaments which then segment into chains of spores which in some species become flagellate and motile. In the genus *Dermatophilus* the mycelium breaks up into motile cocci, and motility has also been recorded for a number of species of *Nocardia,* but is lacking in most actinomycetes (Figure 10.29).

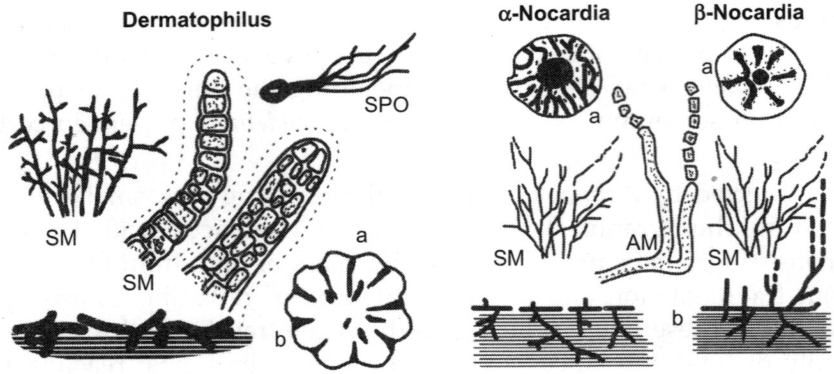

Figure 10.29 *Dermatophilus* and *Nocardia* where a = representative colonial form, b = cross section through growth on agar medium, SM = typical growth forms of the substrate mycelium, AM = typical growth forms of the aerial mycelium SPO = non-flagellate and flagellated spores.

10.5.5.1.4 Dermatophilaceae

Dermatophilaceae single genus *Dermatophilus* (syn. *Polysepta).*

Dermatophilus

They are aerobic parasites; slender hyphae (1-2 μ diam.) giving rise to wider ones (3-5 μ) which divide transversely into narrow segments and then longitudinally two or three times, the individual cells then separating as motile cocci each with a tuft of long flagella; cell walls contain major amounts of meso-DAP (cf. *Nocardia). D. congolensis* and *D. dermatonomus* cause acute local infections of the skin of cattle, sheep and other animals, following damage from other causes, and sometimes prove fatal. The damage caused to hides or wool is of economic importance. *D. pedis* is thought to cause strawberry foot rot, an extensive inflammation of the feet of sheep. Genus *Geodermatophilus* has been assigned to this family to include organisms resembling *Dermatophilus* in their complex life cycles and wall composition, but differing in having only rudimentary mycelium and in being soil inhabitants (Figure 10.29).

10.5.5.1.5 Micromonosporaceae

Contains single genus, *Micromonospora.*

Micromonospora

Slow growing; mycelium extremely fine (0.3-0.6 μ diam) profusely branched, aerial mycelium lacking at ordinary temperatures; aerial mycelium is, however, abundant on hot decomposing masses of plant material and is normally produced by the thermophilic *M. vulgaris* at 60°C; bearing spores singly at apices of short lateral branches; cell walls contain major amounts of glycine and meso-DAP. L-DAP have also been reported.

10.5.5.1.6 Streptomycetaceae

Contains the well-known genus *Streptomyces* and a few other less known ones (often of relatively complex morphology, e.g., forming sclerotia or producing spores on verticillate branches or in pycnidia).

Streptomyces

Aerobic; abundant and widespread in soils; aerial mycelium giving rise to chains of spores; in *S. griseus* and allies chains arise on substrate mycelium also; cell walls containing major amounts of glycine and L-DAP (distinguished from those of *Nocardia* by absence of arabinose and *meso-DAP)*. Sporulating streptomycetes can be identified on such characteristics as random or whorled arrangement of sparophores and whether spore chains are straight or coiled. Electron microscope studies reveal that the spores are smooth, warty, spiny, or hairy, and that the ornamentation of a given species is more or less independent of conditions of cultivation. The smooth form is commonest. Species with straight or flexuous spore chains are smooth spored. Species with ornamented spores always produce spiral spore chains, but the converse is not always true. Strains with blue to blue-green spore masses seem always to have spiny spores. The ornamentation promises to be a useful diagnostic feature in certain sections of the genus. Production of hydrogen sulphide in culture is another useful diagnostic feature. A codified system of taxonomic criteria has been proposed to simplify the identification of streptomycetes by providing a working classification which characterizes isolates which may not have been fully named. It is based on capacity to pigment the medium darkly, spore ornamentation, form and colour of aerial mycelium, and utilization of selected carbon sources. *S. scabies* causes common scab of potato. Many saprophytic species are active producers of antibiotics (including several clinically useful ones (Figure 10.30).

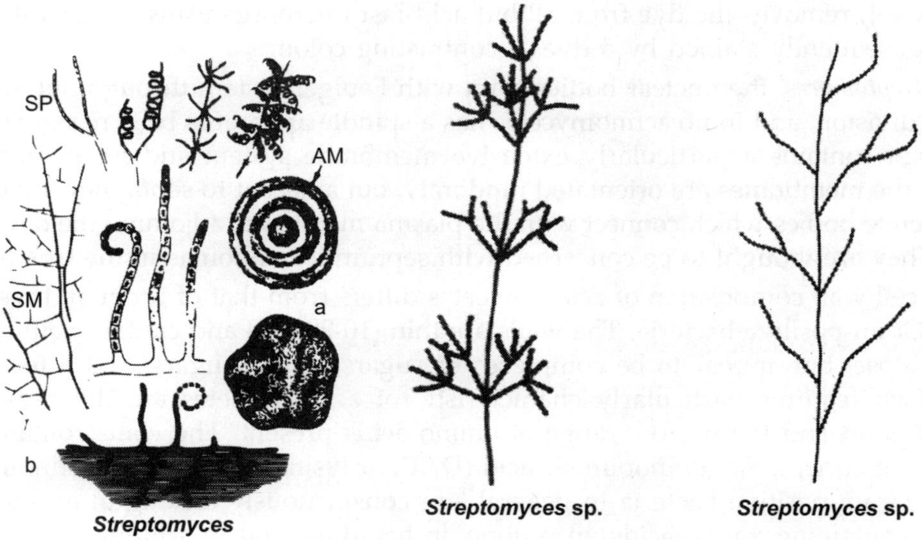

Figure 10.30 *Streptomyces* sp. growth, branching and sporulation where a = representative colonial form, b = cross section through growth on agar medium, SM = typical growth forms of the substrate mycelium, AM = typical growth forms of the aerial mycelium, Sp = sporophores.

The sporophores in streptomycetes may sometimes be grouped to form structures resembling the coremia or pycnidia of true fungi. One family of actinomycetes (the Actinoplanaceae) is characterized by the formation of 'sporangia' which, however, are quite unlike those of the fungi. A sporangium is formed as a terminal vesicle on a hyphae, and into it grows one or more filaments which then segment into chains of spores which in some species become flagellate and motile.

10.5.5.1.7 Actinoplanaceae

Family contains a few genera all of which produce spores in sporangia. These organisms are common and widespread in soil from which they are isolated by baiting with pollen of *Liquidamhar* and similar materials. The spores of *Actinoplanes, Ampullariella,* and *Spirillospora* are motile; those of other genera are not. Genera are distinguished by form of sporangia and spores; some produce chains of conidia in addition to sporangia. In culture, most species produce brilliant pigments, usually orange-coloured, and usually lack aerial mycelium. Observed differences in wall composition suggest that this family may not be a natural group.

Most actinomycetes stain relatively easily with basic dyes and are Gram-positive. A few, principally the pathogenic mycobacteria, stain only with difficulty but once stained, resist decolorization by mineral acid. This property of acid fastness varies between species and depends to some extent upon the conditions of growth. It is attributed to a cellular lipid component (mycolic acid) in the morphologically intact cell, the degree of acid-fastness being proportional to the amount of mycolic acid present. The Ziehl Neelsen technique is used for staining these organisms. The preparation is stained for 10 minutes with a hot, concentrated solution of fuchsin in 5 per cent phenol (carbol-fuchsin). Prolonged decolorization with 20 per cent mineral acid (or a mixture of acid and alcohol) removes the dye from all but acid-fast microorganisms, nonacid-fast ones being subsequently stained by a dye of contrasting colour.

In *Streptomyces,* the nuclear bodies stain with Feulgen's stain through out all stages in their division, and in no actinomycetes has a spindle apparatus been recognized. The cytoplasm contains a particularly extensive membrane system and many ribosomes. Most of the membranes are orientated randomly, but adjacent to septa they are ordered into discrete bodies which connect with the plasma membrane adjoining the developing septa. They are thought to be concerned with septum formation as in the true bacteria.

The cell wall composition of actinomycetes differs from that of fungi and resembles that of Gram-positive bacteria. The walls are thin (10-20 mμ) and contain neither chitin nor cellulose, but appear to be complexes of sugars, amino-sugars, and a few amino acids. Two features particularly characteristic of actinomycetes are the presence of muramic acid and the narrow range of amino acids present. The walls contain major amounts of either 2,6-diaminopimelic acid (DAP) or lysine. The walls of actinomycetes, as with Gram-positive bacteria in general, are conspicuously lacking in aromatic and sulphur-containing amino acids, in proline, in histidine, and in arginine.

Actinomycetes have a considerable economic importance. In nature they contribute to the mineralization of organic residues. Only a few are plant pathogens but at least one of these causes a plant disease of great importance, i.e., common scab of potato. A number cause debilitating or even lethal infections of animals and man. On the credit side, they produce the majority of antibiotics of medical importance, including substances active against bacteria, fungi, protozoa, rickettsiae, larger viruses, and neoplasms.

10.5.5.2 Structure

The actinomycetes are a group of Gram-positive bacteria with a capacity for mycelial growth. For long it was uncertain whether their phylogenetic affinities lay with the bacteria or the true fungi. However, their prokaryotic cellular organization, the chemistry of their cell walls, their nitrogen metabolism and their sensitivity to antibiotics and phages, more specifically termed actinophages, leave little doubt that they are indeed bacteria.

The production of a very fine mycelium is the primary characteristic of actinomycetes. The individual hyphae are usually not more than 1 μ in diameter. The hyphae branch and possibly in all species become septate by the inward growth of rings of wall material. Within each compartment occur one or more chromatin bodies, often of irregular form. The mycelium of some species lies wholly within or in contact with the substrate, but in others there are aerial parts too (Figure 10.31).

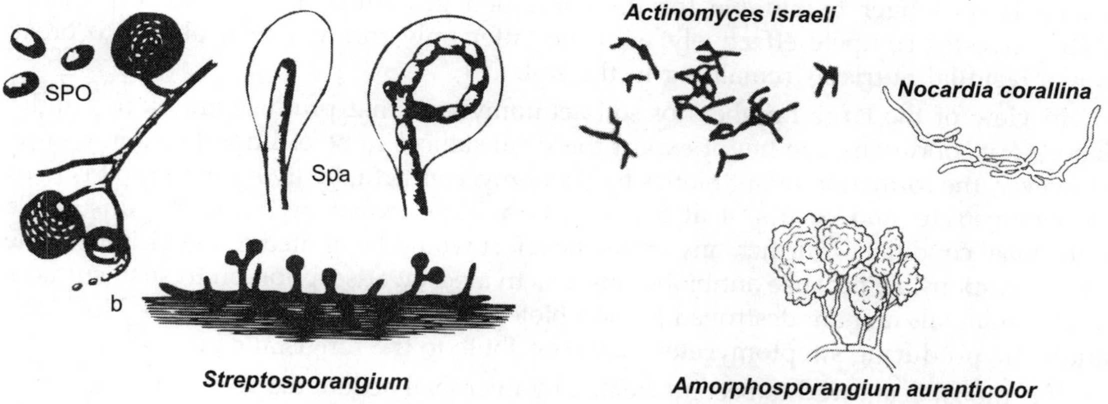

Figure 10.31 Actinomycetes b = cross section through growth on agar medium, SM = typical growth forms of the substrate mycelium, AM = typical growth forms of the aerial mycelium, Sp = sporophores, Spa = sporangia, SPO = non-flagellate spores.

Many actinomycetes produce spores which differ from the endospores of bacilli not only in method of formation but in being only mildly resistant to heat. When they germinate they form one or more germ-tubes whose walls are continuous with the spore's own walls. The combination of aerial growth and sporulation usually confers a cottony or powdery texture on the surface of the colony, while colonies which lack aerial mycelium are either glossy or matt.

In the genus *Dermatophilus* the mycelium breaks up into motile cocci, and motility has also been recorded for a number of species of *Nocardia,* but is lacking in most actinomycetes.

10.5.5.3 Natural occurrence

Most actinomycetes are soil inhabitants. Indeed the characteristic odour of soil after it is ploughed or wetted by rain is largely due to these organisms, many of which produce strong earthy smells in pure culture. The numbers of actinomycetes in soils as revealed by the dilution plate method are second only to the numbers of bacteria, but there is in fact no very accurate way of estimating their abundance in the soil. The large count on dilution plates suggests that actinomycetes are present there very largely as spores. Actinomycetes are worldwide in their distribution. Their numbers generally increase with the warmth of the climate and decrease with depth in the soil. The difficulty with which streptomycete spores are wetted may enable them to be dispersed in the air-water interfaces in the upper layers of the soil rather than to sink to lower regions. Like most other bacteria their growth is favoured by alkaline conditions. Cultivation practices which improve soil aeration often lead to increased numbers of streptomycetes which are the commonest of the soil-inhabiting actinomycetes. In acid and waterlogged soils they are relatively scarce.

The important and unique role of actinomycetes in the soil is attributable to their ability to flourish after faster growing microorganisms have transformed organic residues into a dark homogeneous mass, rich in lignin, certain hemicelluloses, and proteins; which is no longer favourable for the continued growth of these other organisms. Actinomycetes compete effectively with the latter only through their ability to break down residual nutrients remaining in the soil.

In view of the large numbers of soil actinomycetes that produce antibiotics under laboratory conditions one might expect these substances to be of importance in nature. However, the formation of antibiotics by actinomycetes actually living in the soil is hard to demonstrate, and occurs, if at all, to only a slight extent owing to the unsuitable nutritional conditions. Further, any antibiotic effect would be limited to the vicinity of the colony responsible because antibiotics are inactivated by adsorption on to such surfaces as clay minerals and are destroyed by non-biological and biological reactions. However, antibiotic-producing streptomycetes may contribute to the fungistatic properties of soil.

If compost is allowed to become heated by microbial action, the surviving microflora which continues the decomposition contains many thermophilic actinomycetes. Several species of *Streptomyces* and *Micromonospora* are thermophiles, as are all species of *Thermoactinomyces* and *Thermomonospora*. Species of *Micromonospora* sometimes form a lime-like coating on hot decomposing masses of plant material. The inhalation of dust from mouldy hay which has spontaneously heated during its maturation may produce an allergic response known as Farmer's Lung' in persons who have become hypersensitive to antigens in the dust. Spores of thermophilic actinomycetes, notably *Thermopolyspora polyspora* and *Micromonospora vulgaris*, are the principal sources of this antigen. Because these spores are small they can penetrate deeply into the respiratory tract, in the peripheral parts of which the allergic symptoms develop (Figure 10.32).

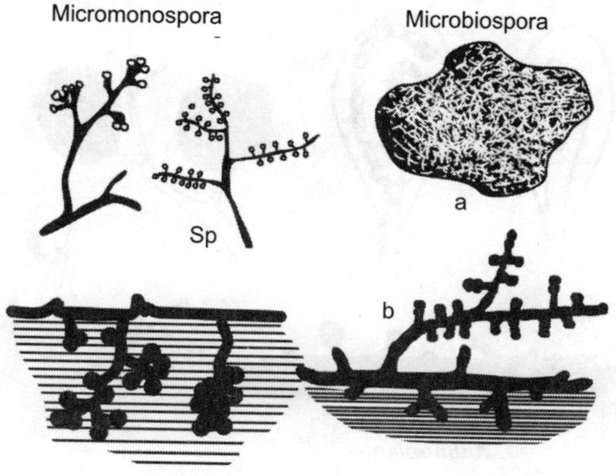

Figure 10.32 Micromonospora and Microbiospora where a = representative colonial form, b = cross section through growth on agar medium, Sp = sporophores.

A number of actinomycetes are responsible for infectious diseases of man and animals. Leprosy and tuberculosis are caused by species of *Mycobacterium*, certain species of *Actinomyces, Nocardia, Dermatophilus,* and *Streptomyces* cause superficial or even deep seated lesions.

Although actinomycetes are so common in the soil, few cause disease in plants. *Streptomyces scabies* causes the extremely important scab disease of potato, and other streptomycetes cause scab of mangel and sugar beet, and pox or soft rot of sweet potato. Nevertheless other relationships with higher organisms may exist. Actinomycetes are probably the endosymbionts within the 'nodule clusters' (rhizothamnia) that are found on the roots of a variety of dicotyledons (e.g., members of the Casuarinaceae, Myricaceae, Betulaceae, Elaeagnaceae, Coriariaceae, and Rhamnaceae) and of which a number have been shown to fix atmospheric nitrogen. Although actinomycetes have been isolated from such nodules, and electron microscope studies reveal a morphological resemblance of the endosymbionts to actinomycetes, there are no conclusive reports of successful reinfection of symbiont-free plants with the isolated organism. Undoubted actinomycetes have been found living symbiotically with insects. Usually, arthropods contain symbiotic microorganisms only if they feed on nutritionally incomplete diets during their entire life cycle. Certain blood sucking hemipterans, for example, contain *Nocardia* in crypts produced by folding of the midgut epithelium. It is suspected that such symbionts furnish their insect partners with vitamins. Organisms interpreted as actinomycetes are present even inside cells of certain insects.

Spores of streptomycetes are common in the air, and strains of *Streptomyces, Micromonospora,* and *Actinoplanes* have been demonstrated in fresh and salt water. The activity of actinomycetes in reservoirs may impart unpleasant odours to water supplies. The structure of those members of the family Actinoplanaceac that produce motile sporangiospores is clearly suited to an aquatic existence, and soil water is a rich source of these organisms (Figure 10.33).

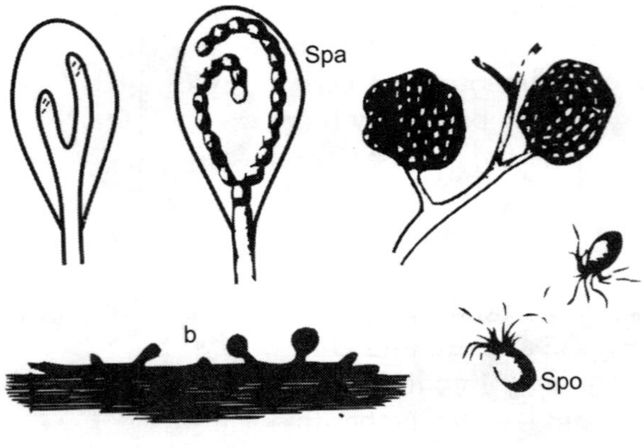

Figure 10.33 Actinoplanes: Various stages of growth where b = cross section through growth on agar medium, Spa = sporangia, Spo = non-flagellate and flagellated spores.

10.5.5.4 Physiology

The germinating spore that initiates the new actinomycete colony normally swells and then puts out one or two germ-tubes. Some spores fail to germinate if they are washed unless provided with dextrose. The germ-tube typically develops into a system of filaments which grow apically and branch monopodially. In *Actinomyces* and *Nocardia* the mycelium soon fragments into pieces which grow at both ends. Such fragmentation in *Nocardia corallina* appears to be induced by agents which accumulate in the culture medium. In *Dermatophilus,* the filaments become septate both transversely and longitudinally and separate into motile coccoid spores. In other genera the substrate mycelium typically remains intact, though in the Actinoplanaceae it is prone to fragment when grown on certain media which include potato dextrose agar. The mycelium of colonies which do not fragment grows almost entirely in the apical regions, the older parts either remaining quiescent or dying. The mycelium at first contains an optically homogeneous cytoplasm but this becomes vacuolated as it matures and fat and volutin granules may be found in the vacuoles. Anastomosis of filaments has been claimed but the evidence for this is equivocal. In *Streptomyces* and some other genera the substrate mycelium soon gives rise to aerial mycelium on which the spores are formed. The onset of sporulation in streptomycetes is related to a depletion of nutrients in the medium or to other unfavourable circumstances. The ability to form aerial mycelium implies a capacity for growth at a distance from the nutritive substrate and hence for an internal translocation of nutrients. This is a fundamentally different biological organization from that encountered in other bacteria. Growth in well-aerated submerged culture is rapid and is followed by lysis of the mycelium and even of the spores. In many streptomycetes the lytic phase is accompanied by increased liberation of antibiotics. Although some species are thermophilic and grow best at 50-56°C; the majority of actinomycetes are mesophilic. Isolates from soil and water grow best at about 25-30°C; those from warm-blooded host animals grow best at about 37°C.

The basic nutrition of the majority of saprophytic actinomycetes is simple and resembles that of a large number of other bacteria and a majority of fungi. There is a requirement for an organic source of carbon and energy but beyond this all other nutritional needs can generally be met by inorganic salts, such as nitrate or ammonium salts for nitrogen, and salts of the necessary mineral nutrients among which phosphorus, potassium, and sulphur are needed in greatest amounts. Some actinomycetes have additional requirements for particular vitamins, or for particular amino acids. Special nutrients may also be necessary to sustain particular metabolic activities especially in industrial applications of actinomycetes. For example, cobalt needs to be present for synthesis of vitamin B_{12} (cobalamin) and chlorine for the synthesis of chloramphenicol. Although nitrogen fixation by free-living actinomycetes has sometimes been claimed there is still much uncertainty as to whether this really does occur. The general consensus of opinion is that it does not. Although the minimal nutrition of actinomycetes is generally simple, the total pattern of nutrition is complicated firstly by the diversity of complex macromolecular nutrients which can be digested by extracellular enzyme action; secondly by the breadth of the range of organic nutrients that can enter the metabolic pathways of the cell; and thirdly by the development in some species of considerable degrees of specificity for particular organic nutrients. The pathogenic mycobacteria require very complex media for artificial culture. *Mycobacterium tuberculosis* grows only in the presence of serum or coagulated egg to which amino acids, particularly asparagine, are added. At blood temperature, cultures take up to 4 weeks to develop under aerobic conditions. *Mycobacterium johnei* requires, in addition, extracts of mycobacteria, which contain substances similar to vitamin K and even on such complex media grows more slowly than do tubercle bacilli.

The majority of actinomycetes inhabit the soil and these as a group are characterized by an extensive ability to digest complex materials and to use unusual molecules. This is the principal reason for their success as soil inhabitants. Enzymes able to degrade dextrins, starch, cellulose, hemicelluloses, such as mannans and xylans, pectin, chitin, fats, peptones, proteins, including keratin, humic acids, and possibly even lignin are commonly produced. Laminarin, alginates, and even agar may be digested by some species. Substances which lyse fungal hyphae and bacteria are also formed. Many soil actinomycetes, and *Nocardia* species in particular, utilize unusual molecules as sources of energy and carbon. These include long-chain fatty acids, hydrocarbons, benzene, various aromatic compounds, steroids, and heterocyclic nitrogen compounds. Species of *Streptomyces, Nocardia,* and *Mycobacterium* able to oxidize ethane have been isolated from various soils. Most have no perceptible specificity for hydrocarbons for they grow well on a variety of other media. *Mycobacterium paraffinicum,* however, grows poorly on the usual protein and carbohydrate media and grows well only with gaseous hydrocarbons such as ethane, propane, or butane. Some strains will utilize only ethane. For this reason, *M. paraffinicum* is considered to have potential value for petroleum prospecting. Among the simpler nutrients utilized as sources of carbon and energy are glucose, maltose, lactose, cellobiose, glycerol, alcohols, and various organic acids including fatty and amino acids. The latter are also particularly suitable nitrogen sources. A minority of actinomycetes utilize sucrose. The use of oxalic acid by some species of *Streptomyces* and *Nocardia* suggests they may have an important natural role in the detoxifiction of calcium oxalate in plant residues.

Most actinomycetes are aerobic organisms but obligate anaerobes occur in the genera *Actinomyces, Micromonospora* and *Microbiospora* (Figure 10.34). Anaerobic actinomycetes possess a fermentative metabolism which yields acid but not gas from a variety of carbohydrates. Even aerobic forms may produce considerable quantities of acid, largely lactic, but the acid is later consumed. Indeed, in well-aerated cultures as much as 25-30 per cent of the substrate carbon may be converted to cellular material. The intermediary metabolism of actinomycetes is still not well known, but in several streptomycetes enzymes characteristic of the tricarboxylic acid cycle have been demonstrated. A feature of the metabolism of nitrogen in those actinomycetes in which it has been studied is that lysine is synthesized through the α, ε-diaminopimelic acid pathway as it is in other bacteria, and not through the alternative α-aminoadipic acid pathway. Actinomycetes, like other bacteria, do not contain sterols.

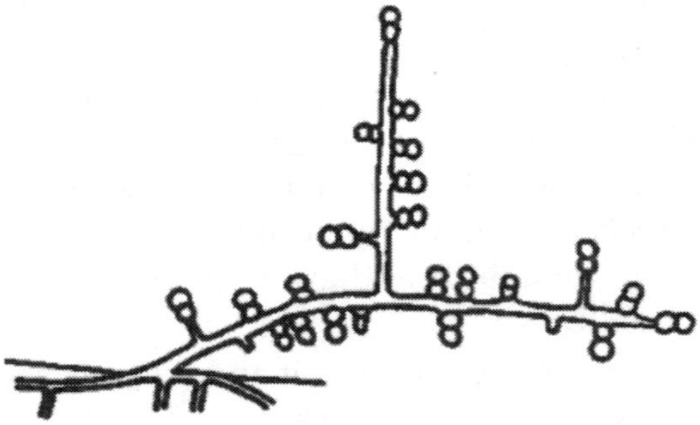

Figure 10.34 Microbiospora with spore attachment.

The smells that are characteristic of many actinomycetes are attributable to a variety of metabolites which includes acetic acid, acetaldehyde, ethanol, isobutanol, and isobutyl acetate. Earthy smelling neutral hydrocarbon oil, named geosmin, has been isolated from several species including *Streptomyces griseus.* Actinomycetes generate a great variety of substances in their environment, and these include vitamins, pigments, and antibiotics. The biosynthesis of some of these has been developed on an industrial scale. For example vitamin B_{12} is produced by *Streptomyces aureofaciens* and other species as a byproduct of the aureomycin fermentation. Aureomycin, streptomycin, and neomycin (produced by S. *aureofaciens,* S. *griseus,* and *S. fradiae* respectively) are just three examples of the many medically important antibiotics produced. Nearly eight hundred different antibiotics that are formed by actinomycetes had been described; only a few are of clinical value. The vast majority were from species of *Streptomyces,* while a few were derived from species of *Nocardia, Micromonospora, Thermoactinomyces,* and certain of the Actinoplanaceae. Antibiotic production appears to be a manifestation of the secondary metabolism of the organism. Given the appropriate precursors and enzymes systems an actinomycete begins to synthesize secondary metabolites in response to a slowing down of primary metabolic activity at the end of its growth phase or in response to derangement of normal growth activity. Antibiotics are thus produced only when the organism is grown under certain specific physical and chemical conditions. Antibiotic production is maximal near the stage of maximum growth of the actinomycete, but good growth can proceed without antibiotic production because antibiotics are not inevitable by-products of growth. The ability to produce particular antibiotics is not correlated with taxonomy, for the same substance may be produced by members of different species or different genera, while sometimes different strains of the one species or even the one organism produce several different antibiotics. Further, a strain may lose or gain the ability to produce an antibiotic as a result of mutation.

Genetical change in actinomycetes may result in the development of sectors in colonies, in changes in nutrition such as the acquisition of special organic requirements, or in alterations in pigment and antibiotic production, or pathogenicity. The ultimate source is mutation. Mutagenic agents have sometimes been used to obtain strains of altered antibiotic activity. Genetical variation by mutation is augmented by the occurrence of a form of heterokaryosis and genetical recombination. Heterokaryosis implies the transfer of a complete 'nucleus' from one cell to another in the manner familiar in fungi, though of course actinomycetes are not eukaryotic organisms, and the active migration of nuclei commonly occurring in fungi would not be expected to occur. Such a modified heterokaryosis has been reported in a number of streptomycetes. The simplest mechanism that may be invoked to account for the establishment of heterokaryons is that nuclear bodies are transferred by way of occasional hyphal fusions in the substrate, but the existence of such anastomoses has not been proven. Genetical recombination occurs in many streptomycetes apparently without the participation of any specialized sexual structures. It is not known whether this is a regular part of the life cycle. In S. *coeficolor* the possible sequence of events commences with the transfer of probably incomplete nuclear bodies by a process of conjugation between substrate hyphae. This yields heterozygous 'nuclei' which multiply and undergo some chromosomal loss and terminal deletion, and possibly mitotic crossing over. Eventually, by crossing over and reduction, haploid recombinants emerge. In *S. coeficolor* the genes are located on two linkage groups, of which the simplest physical counterpart would be a pair of chromosomes. This, if true, would contrast with the single chromosome that exists in *Escherichia coli*. Recombination has even been reported as occurring between different species. Cytoplasmic inheritance has been reported in respect of the ability of *Streptomyces scabies* to produce tyrosine.

10.5.5.5 Actinophage (Viruses Attacking Actinomycetes)

Particular viruses attack actinomycetes, and so far have been found in several streptomycetes, *Actinomyces bovis,* and *Nocardia farcinica.* Such viruses 'actinophages' are abundant in the soil. It has been possible experimentally to separate the life cycle of actinophage into phases of attachment, insertion, growth, maturation, and lysis.

Phages attack mycelium only at certain stages of its development. Aerial mycelium and spores are not lysed. The harbouring of phage without lysis occurring ('lysogeny') has also been demonstrated. Some actinophages possess a tail, which in several at least is unlike that of the T even coli phages in that it lacks a visible contractile sheath, tail fibres, and a base plate. A terminal knob or group of prongs possibly assumes the function of the base plate, though the mechanism of attachment is obscure. In other actinophages there is no tail at all.

10.5.6 Endospore-forming Rods and Cocci

The ability to produce more or less heat-resistant spores is restricted, with few exceptions, to a group of Gram-positive, motile rods with peritrichous flagella. The aerobic and facultative anaerobic rods belong to the genus *Bacillus, Sporolactobacillus,* and *Sporosarcina,* and the anaerobic rods belong to the genera *Clostridium* and *Desulfotomaculum.* Many spore formers are widely known because of their biochemical capacities. Only some representatives of the large genera *Bacillus* and *Clostridium* are described below.

10.5.6.1 Bacillus

Aerobic spore-bearing bacilli, spores are usually no wider than the vegetative cells; frequently produce rough colonies sometimes with a hair-like periphery resulting from long chain formation; active producers of enzymes such as penicillinase, collagenase, crutinase; widespread; many thermophils, e.g., *B. stearothermophilus,* cause spoilage of canned foods; others are pathogenic to insects, e.g., *B. larvae,* the cause of American foul brood of bees and one species, *B. anthracis,* is the cause of anthrax in many warm-blooded animals.

They can be differentiated into the different groups according to shape of their spores and vegetative cells.

1. The spores of the majority of bacilli are oval or cylindrical in shape and no wider than the vegetative cells (*Bacillus megaterium, B. cereus, B. subtilis, B. licheniformis, B. anthracis, B. thuringiensis*). The organisms can derive energy from aerobic respiration but also from fermentation and nitrate respiration.
2. The oval spores are wider than the vegetative cells; during sporulation the cells become distended (*B. polymyxa, B. macerans, B. stearothermophilus, B. circulans*). They produce profuse slime and the star shaped cross section of its barrel-shaped spores. It produces 2,3-butanediol, and is decidedly thermophilic organism with an optimum growth at 50-65°C.
3. Almost spherical spores in terminally distended vegetative cells (*B. pasteurii*). They are urea degrading organisms with production of CO_2 and ammonia with a capacity to adopt at high pH. *Sporosarcina urea* is physiologically similar; although it resembles sarcina morphologically, it is reckoned among the bacilli because of its physiological characteristics, i.e., thermoresistance, dipicolinic acid-containing spores, and aerobic metabolism.

10.5.6.2 Clostridium

These organisms do not require oxygen for growth. Anaerobic or aerotolerant, spore-bearing rods; cells usually swollen by the spore which may be either spherical or oval, terminal, subterminal or equatorial; commonly found in soil and in the intestinal tract; powerful exotoxins are produced by many pathogenic species, e.g., *Clostridium tetani* and *Clostridium botulinum,* the causal organisms of tetanus and botulism respectively; in addition to producing exotoxins other pathogens are saccharolytic and proteolytic, e.g., *Clostridium perfringens (welchii),* the causal organism of gas gangrene in man, lamb dysentery and enterotoxaemia in sheep, *Clostridium chauvoei* the causal organism of blackleg in cattle, *Clostridium septicum* the causal organism of gas gangrene in man, malignant oedema in cattle and braxy in sheep; saprophytic species may be thermophilic and cause spoilage of canned foods, e.g., *Clostridium thermosacharolyticum,* others are of particular importance in industrial processes, e.g., *Clostridium acetobutylicum* used in the manufacture of butyl alcohol from carbohydrates. Several species are active in the fixation of nitrogen in soil, e.g., *Clostridium pasteurianum.*

10.5.6.3 Pseudomonad and other Gram-negative rods

It has become common to refer to all polarly flagellated, Gram-negative rods as 'Pseudomonads'. They may be straight, curved or spiral, rigid, rod-shaped usually polar-flagellate at one or both ends of the cell. Frequently such physiologically highly specialized eubacteria as *Nitrosomonas, Methylomonas,* the thiobacilli, and even phototrophic bacteria (*Rhodopseudomonas*) have been designated in this way. However, such usage has only morphological significance and in no way defines a taxonomic unit (Table 10.6).

Table 10.6 Differentitation of *Acetobacter, Gluconobacter,* and *Pseudomonas*

Character	*Acetobacter*	*Gluconobacter*	*Pseudomonas*
Flagellation	Peritrichous	Polar	Polar
Growth at pH 4.5	+	+	–
Oxidation of ethanol to acetic acid at pH 4.5	+	+	–
Complete tricarboxylic acid cycle	Present	Absent	Present
DNA(mole % GC)	51-65	56-64	58-70
i umber of species	4	1	27
16S rR 'A group'	Alpha purple bacteria	Alpha purple bacteria	Alpha, beta, or gamma purple bacteria

Members of the Pseudomonadaceae obtain their energy by aerobic respiration and, in some cases, by anaerobic respiration (nitrate respiration, denitrification) but never by fermentation. The Pseudomonadaceae are chemo-organotrophs, though some are facultative chemolithotrophs. The genus Pseudomonas is the prototype of this family and is characterised by the metabolic and physiological properties of using a wide spectrum of substrates. They even use to utilize a large number of heterocyclic and aromatic compounds that are not attacked by other bacteria. They generally metabolise sugars via Entner-Doudoroff pathway. Some *Pseudomonas* species oxidize sugars incompletely and excrete sugar acids (gluconate, 2-oxogluconate). Because of their simple requirement, pseudomonads are ubiquitous. They are present in soil, water, sewage and air. When media containing mineral salts and organic acids or sugars are exposed to air, pseudomonads are usually the first colonizers. Many can be identified by their production of water soluble pigments, i.e., the blue green derivative of phenazine, pyocyanin, and the yellow green fluorescing pigments. Some of the excreted fluorescent pigments function as siderophores.

Pseudomonas species

Pseudomonas aeruginosa is an aquatic bacterium that can cause opportunistic infections in man, giving rise to inflammation of the middle ear, wound infections producing greenish blue pus, and in compromised patients, even septicaemia. *P. fluorescens* and *P. putida* are other soil and aquatic species that can oxidize a very large number of organic compounds. Many strains showing plant pathologenicity have recently been unified in the species *P. syringae*. The aquatic bacterium *P. saccharophila* was used in the discovery of the Enter-Doudoroff pathway of sugar dissimilation. 16S rRNA sequence analysis has redefined the taxonomy of many bacterial species. As a result, the genus *Pseudomonas* includes strains formerly classified in the genera *Chryseomonas* and *Flavimonas*. Other strains previously classified in the genus *Pseudomonas* are now classified in the genera *Burkholderia* and *Ralstonia*. The complete genome sequence of a *Pseudomonas* species was determined; more recently, the sequence of other strains has been determined, including *P. aeruginosa* strains PAO1, *P. putida* KT2440, *P. protegens* Pf-5, *P. syringae* pathovar tomato DC3000, *P. syringae* pathovar syringae B728a, *P. syringae* pathovar phaseolica 1448A, *P. fluorescens* PfO-1, and *P. entomophila* L48. *Pseudomonas* is reported to be the most common nucleator of ice crystals in clouds, thereby being of utmost importance to the formation of snow and rain around the world.

The yellow pigmented plant pathgogens of the Pseudomonaceae family have been unified in the genus *Xanthomonas*. The yellow pigment is a polyene compound containing bromine. Certain strains of *Xanthomonas campestris* secrete exopolysaccharides (xanthan) that are resistant to enzymic degradation. These polysaccharides are produced industrially and used to increase the viscosity of aqueous solutions such as printing ink, diet soups, and packet desserts. *Xanthomonas* spp. causes bacterial spots and blights of leaves, stems, and fruits on a wide variety of plant species (Figure 10.35). Pathogenic species show high degrees of specificity and some are split into multiple pathovars, a species designation based on host specificity. Citrus canker, caused by *Xanthomonas citri* subsp. *citri* is an economically important disease of many citrus species (lime, orange, lemon, pamelo, etc.). Multiple genomes of *Xanthomonas* have been sequenced and additional data sets/tools are available at http://www.xanthomonas.org/

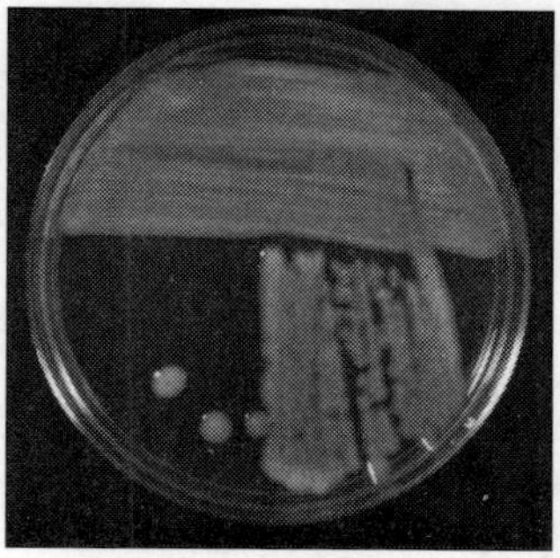

Figure 10.35 *Xanthomonas* spp. growing on sucrose peptone agar showing yellow pigment.

10.5.6.4 Other Gram-negative aerobic rods

Genera that resemble the pseudomonads metabolically are *Alcaligens, Agrobacterium* and *Rhizobium,* the acetic acid bacteria *Acetobacter* and *Gluconobacter* and the free living nitrogen fixers *Azotobacter, Beijerinckia* and *Derxia.*

Alcaligenes, Rhizobium, Agrobacterium

These genera include bacteria that resemble the aerobic pseudomonads in their energy metabolism (respiratory system), but they are not polarly flagellated. They have either subpolar flagella or only a few (2-6) peritrichous flagella. *Alcaligenes* includes the facultative autotrophic hydrogen bacterium *A. eutrophus. Rhizobium* is characterized by its ability to fix atmospheric nitrogen in endosymbiosis with leguminous plants. *Agrobacterium tumefaciens* can produc tumours in roots, leaves, and stems of plants (Table 10.7).

Table 10.7 Differential characteristics of species of hydrogen oxidizing bacteria*

Genus/species	*Denitrification*	*Growth on fructose*	*Motility*	*Other characteristics*
Gram-negative				
Alcaligenes eutrophus	+	+	+	Membrane-bound and cytoplasmic hydrogenases
Alcaligenes ruhlandii	–	–	+	Membrane-bound and cytoplasmic hydrogenases
Aquaspirillum autotrophicum	–	–	+	Only membrane-bound hydrogenases present
Pseudomonas carboxydohydrogena	–	–	+	Only membrane-bound hydrogenases present also oxidizes CO
Pseudomonas flava	–	+	+	Colonies arc bright yellow
Seliberia carboxydohydrogena	–	?	+	Also oxidizes CO
Paracoccus denitrificans	+	+	–	Only membrane-bound hydrogenase present Strong denitrifier

Contd...

Contd...

Gram-positive				
Bacillus schlegelli	–	–	+	Produces endospores
Arthrobacter sp.	–	+	–	Only membrane-bound hydrogenase present
Mycobacterium gordonae	–	?	–	Acid fast, colonies yellow 1.0 orange

*Phylogenetically, most H_2 oxidizing bacteria are α or β purple bacteria, except for the Gram-positive Organisms

10.5.6.5 Chemolithotrophic bacteria

Aerobic chemolithotrophic bacteria are characterized by their ability to use inorganic ions or compounds as electron or hydrogen donors. They can use carbon dioxide as their carbon source and carry out CO_2 fixation via the ribulose - bisphosphate pathway. Most of these bacteria are facultative autotrophs and can also utilize organic substrates. Autotrophic bacteria belong to many genera, viz., *Pseudomonas, Alcaligenes, Aquaspirillum, Xanthomonas, Mycobacterium, Bacillus and Nocardia.*

10.5.6.6 Sheathed bacteria

They are also referred as sewage 'molds'. *Sphaerotilus natans* is the best known example of this group. They grow in contaminated streams and waters, in the drainage from sugar refineries, in weirs, and in various trickling filters. It can form threads and aggregates or, when adhering to solid substrates, it can produce long pellicles and extensive surface covers. Pipes, irrigation systems and ditches can easily become plugged.

Sphaerotilus natans is Gram-negative unicellular bacterium with *polytrichously polar* flagellation. It could be regarded as a relative of the pseudomonads except for its specific growth characteristics. It grows as long chains or filaments and consist of chains of individuals held together by a thin tubular sheath. This sheath consists of a heteropolysaccharide and can be regarded as a capsule. The bacteria multiply within the sheath by division and are able to leave the sheath as motile individual cells. On agar plates two types of colonies are often found: rough colonies that contain mainly filamentous forms, and smooth colonies consisting of single cells. Cell size, filament diameter and other secondary characters have been used to differentiate a number of species.

Filamentous bacteria that are widely distributed in iron-containing waters, ditches, wells, drainage pipes, and swamps were formerly called 'ochre bacteria' (*Leptothrix ochracea*). Their natural habitat is usually deficient in utilizable organic materials but rich in iron. Their sheaths are therefore streaked and covered by iron oxide precipitates. These are also known as iron-oxidizing autotrophic bacteria. However, because neither physiological nor biochemical studies can support this assumption, *Leptothrix ochracea* and also *Cladothrix dichotoma*, which differs from *L. ochracea* in its growth form, are now regarded as variants of *Sphaerotilus*, distinguished only by their habitat.

10.5.6.7 Gram-negative facultative anaerobic bacteria

They are characterized by their fermentation products they produce. Under anaerobic condition, they gain energy for growth from fermentation and excrete several organic acids, including formic acid. Representative members of this group including *Escherichia coli, Samonella, Shigella* and others, inhibit the intestinal tract, the whole group has been called Enterobacteriaceae.

10.5.6.7.1 Enterobacteriaceae

Gram-negative, aerobic rods; active fermenters of carbohydrates, often producing CO_2 and H_2 ; many genera serologically related; many species are saprophytes; others inhabit the gut of animals, frequently causing intestinal disturbance; some are parasitic on plants, causing blights and soft rots.

Escherichia

Ferment many sugars, including lactose, to produce acids and gases (H_2 and CO_2, ratio 1:1); acetyl-methyl-carbinol not produced, methyl red test positive. *E. coli* is a parasite in the gut of many animals where it is generally non-pathogenic, consequently its presence in water supplies is used as an indicator of faecal pollution. It occasionally causes cystitis and peritonitis when gaining access to the bladder and peritoneum respectively; specific serotypes cause enteritis and septicaemia in newborn calves, bowel oedema in piglets and enteritis in babies. Other species, e.g., *E. freundii,* may also be found in the gut of animals, but also have a wider distribution in water and soil.

Aerobacter

Ferments many sugars, including lactose, to produce acids and gases (H_2 and CO_2, ratio 1: 2); produces acetyl-methyl-carbinol but methyl red test is negative; can utilize citrate as sole carbon source; widely distributed in nature, including the gut of man and other warm-blooded animals.

Klebsiella

Sugar fermentation variable, but usually a number of sugars, including lactose, are fermented; non-motile; acetyl-methyl-carbinol may or may not be produced; grows on certain media as mucoid colonies owing to excessive polysaccharide-capsule production; one cause of lobar pneumonia and infection of the urinogenital tract in man; one cause of mastitis in cattle and metritis in mares.

Erwinia

Motile rods which normally do not require organic nitrogen compounds for growth; plant pathogens causing dry necroses, galls or wilts, e.g., *E. amylovora,* the cause of fireblight or blossom-blight of pear and twig-blight of apple; unable to liquefy pectate gels or produce gas from sugars.

Pectobacterium

Motile rods, normally not requiring organic nitrogen compounds for growth; plant pathogens causing soft rots, e.g., *P. carotivorum* causing soft rot of the fleshy (parenchyma) including tissues of carrots, turnips, potatoes, bulbs, and mushrooms; able to liquefy pectate gels and many can produce gas from sugars.

Serratia

Small rods; produce various amounts of red pigment (prodigiosin), often depending upon the fermentable carbohydrate present. *S. marcescens* (syn. *Chromobacterium prodigiosum)* readily produces white variants; occurs in water, soil and milk; in the latter produces blood-like discoloration.

Proteus

Motile, straight rods with many flagella; some species swarm over the surface of solid media devoid of bile salts; pleomorphism is marked: glucose invariably fermented and usually other sugars but not lactose; urea usually decomposed; phenylalanine oxidatively deaminated to phenyl pyruvic acid; occurs in faeces or putrefying materials; *Salmonella* often present in wounds.

Salmonella

Motile rods (except *S. gallinarum* and *S. pullorum);* many sugars fermented, but not lactose; gas is usually formed (except by *S. typhi* and *S. gallinarum);* indole not produced, urea not split, acetyl-methyl-carbinol not produced and methyl red test positive; pathogenic to man and animals. *S. typhi* and *S. paratyphi* cause enteric fevers in man; other species cause infective food poisoning in man, and paratyphoid and chronic infections in animals and birds, animals may be symptomless carriers. Genus subdivided serologically, somatic antigens being characteristic of sub-groups and certain flagella antigens being species specific.

Shigella

Non-motile rods; sugars fermented without gas formation, lactose may be fermented late; possess distinctive antigenic structures; some cause bacillary dysentery in man and occasionally in primates; a few species are nonpathogenic inhabitants of the gut of warm-blooded animals.

10.5.6.8 Gram-negative anaerobic bacteria

Pleomorphic, usually Gram-negative rods, with rounded or pointed ends, varying in length to long filamentous forms; normally strict anaerobes; found primarily in the intestinal tract and in the oral cavity of warm blooded animals; sometimes pathogenic, often associated with gangrenous conditions in man and other animals. They are widely distributed in anaerobic ecosystems.

Bacteroides fragilis, has been isolated from cases of acute appendicitis, pulmonary gangrene and abscesses of the urinary tract in man; the species of *Bacteroides* have a purely fermentative metabolism. They ferment glucose to succinate, acetate, formate, lactate and other acids. *Fusobacterium* gets its name from its spindle shape; the rods are relatively long. The genus is well characterized by its chief fermentation product, butyrate. Several species occur in the oral cavity, intestinal tract, and faeces.

Sphaerophorus necrophorous is the cause of calf diphtheria; related organisms cause liver abscesses in cattle, labial necrosis in rabbits and kangaroos and foot rot in sheep.

10.5.6.9 Curved rods: spirilla and vibrios

Rigid, curved or spirally twisted cells, frequently attached after division thus forming chains, motile by tuft of polar flagella in *Spirillum* and single flagellum in other genera; usually aerobic but a few anaerobic strains known (e.g., *Desulphovibrio);* mainly water organisms but some parasites of higher animals.

Spirilla

Long screw-shaped cells, motile by tufts of flagella at one or both ends; volutin granules, i.e., polyphosphates usually present; aerobes; in fresh and salt water; one species, *Spirillum minus,* causes rat-bite fever in man. They are characterized by their helical form and bipolar polytrichous flagellation. Hey obtain energy by respiration. There are several genera. Most spirilla belong to the genus *Aquaspirillum* (*A. itersonii, A. serpens*). They also have a marked tendency not to tolerate the partial pressures of O_2 found in normal air.

Vibrio

Cells single, comma-shaped or remaining attached to form spirals; usually grow rapidly on laboratory media; widely distributed as saprophytes in salt and fresh water and soil, some parasitic in animals, including a few important pathogens, e.g., *V. cholerae,* cause of Asiatic cholera in man; *V. fetus,* cause of infertility and abortion in cattle and sheep; *V. coli,* cause of swine dysentery; and *V. jejuni,* cause of dysentery in cattle.

Desulphovibrio

Cells, single, curved or joined in short chains; strictly anaerobic; reduce sulphates to hydrogen sulphide; occurring in sea water, marine mud, fresh water, and soil. *Selenomonas, Succinivibrio,* and *Butyrivibrio* are other representative of this group. *Selenomonas* is inhabitant of oral cavity and dental mucus of humans and *Succinivibrio ruminantium* is a rumen bacterium. Both species are characterized by lateral flagellation and utilize the fermentation of carbohydrates to propionate and acetate for growth.

Bdellovibrio

Bdellovibrio bacteriovorus is an aerobic organism that parasitizes other bacteria. The cells are small and highly motile by virtue of a thick flagellum, 50 μm in diameter; its swimming velocity is 100 μm/sec which corresponds to about 70 body lengths/sec. When this parasite encounters a suitable host bacterium, it adheres with its unflagellated pole to the host's cell wall (Figure 10.36) and may rotate around its longitudinal axis. Soon afterwards the host cell rounds off and assumes a spherical form, resembling a spheroplast, whilst the *Bdellovibrio* penetrates the cell wall and settles in the periplasmic space. The *Bdellovibrio bacteriovorus* cell continues to elongate into a long cylinder until the nutrients obtained from the gradually shrinking host protoplast have been exhausted. The cylinder then undergoes multiple divisions to give rise to cells of uniform size; eventually, the host cell wall lyses and the progeny of the parasite are liberated into the medium ready to attack other host bacteria. The progressive attack and lysis of host bacteria can be seen macroscopically as the appearance of 'holes', or lytic zones, on a bacterial lawn of the host, or as a decrease in turbidity in the case of liquid cultures. In contrast to bacteriophages, which can multiply only in growing bacteria, *Bdellovibrio* can attack and lyse non-growing cultures. The several strains of *Bdellovibrio* that have been isolated from different soils are differentiated according to their host range. They lyse mainly Gram-negative bacteria and seem to prefer pseudomonads and enterobacteria. Whilst the wild type is an obligate bacterial parasite dependent on host bacteria for its nutrition, a few mutants are able to grow saprophytically outside other bacterial cells on complex nutrient media. It is generally assumed that the parasitic lifestyle of *Bdellovibrio* represents adaptation to environments with low nutrient concentrations (Table 10.8).

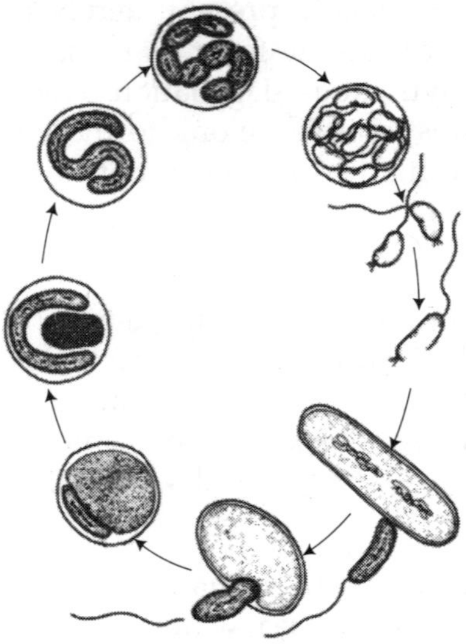

Figure 10.36 Life cycle of *Bdellovibrio bacteriovorus* in the host bacterium.

Table 10.8 Characteristics of the genera of spiral-shaped bacteria*

Genus	*165 rRNA group***	*Characteristics*
Spirillum *Aquaspirillum* *Oceanospirillum*	Beta Alpha or beta Gamma	Cell diameter 1.7 µm; microaerophilic; fresh water Cell diamter 0.2-1.5 µm; aerobic; fresh water Cell diameter 0.3-1.2 µm; aerobic; marine (require 3% NaCl)
Azospirillum *Campylobacter*	Alpha –	Cell diameter 1 µm; microaerophilic; fixes N_2 Cell diameter 0.25-0.8 µm; microaerophilic to anaerobic; pathogenic or commensal in human and animals; single polar flagellum
Bdellovibrio	Delta	Cell diameter 0.25-0.4 µm; aerobic; predatory on other bacteria; single polar sheathed flagellum
Microcyclus	–	Cell diameter 0.5 mm; curved rods forming rings; nonmotile; aerobic; sometimes gas vesicuolate

**All are Gram-negative and respiratory but never fermentative.*

***All incidicated genera are members of the purple bacteria (Protebacteria).*

10.5.6.10 Spirochaetes

The spirochaetes are long, slender, motile organisms, coiled around their long axis in the form of a helix (Figure 10.37). They differ from spiral Eubacteria, such as *Spirillum,* in being flexible, in lacking true flagella and, with the exception of *Borrelia,* in being difficult to stain with aniline dyes. They are best examined in the living state by dark-field illumination, by negative staining with nigrosin after fixing or by electron microscopy. Pathogens may be demonstrated in tissue sections after staining by Levaditi's technique which involves the deposition of silver on the organism's surface. They are divided into two families; the Spirochaetaceae which are relatively large, varying in length from 30 to 500 μ, often showing cross striations and various inclusion bodies, and the smaller Treponemataceae which rarely exceeds 16 μ in length and in which the protoplasm shows no obvious structural features.

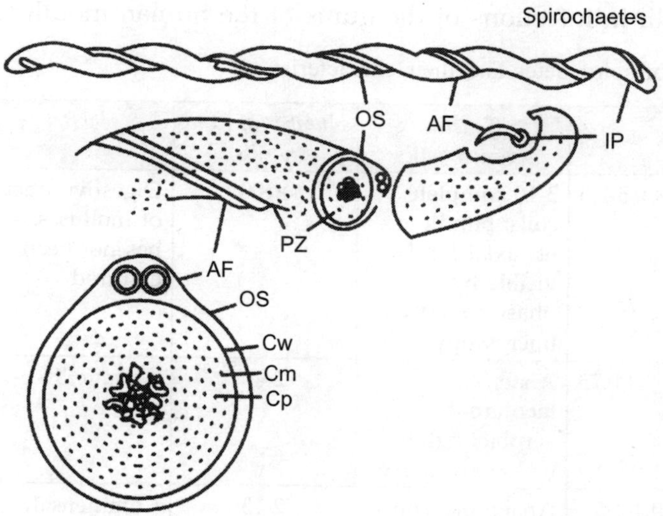

Figure 10.37 The cell of Spirochetes where PZA = Protoplasmic cylinder is surrounded by AF = hylical axial filament, each fibrilla is inserted in the protoplasmic cyclinder at one end IP = insertion pore, surrounded by OS = membranous outer sheath, Cw = Cell wall, Cm = Cytoplasmic membrane, Cp = Cytoplasm.

10.5.6.10.1 Habitat

The Spirochaetaceae are mainly free-living saprophytes, and the parasitic members cause little damage to their hosts. Free living strains are found in fresh water, marine slime, sewage, and stagnant waters. Others are parasitic on fresh water and marine shellfish including oysters and other molluscs.

The Treponemataceae are mainly parasites of birds and mammals, including man, and cause a number of serious diseases. *Treponema pallidum,* the cause of syphilis, can survive for only a short time outside the host and hence nonvenereal transmission is a very rare occurrence. The primary lesion is usually confined to the genitalia; secondary lesions are found at other sites including the mouth. In the tertiary phase vital organs, particularly of the circulatory and central nervous system, are attacked. Yaws, an ulcerative disease of man in tropical countries, is caused by the serologically related species, *Treponema pertenue; Treponema cuniculi* is transmitted venereally in rabbits; other species of *Treponema,* apparently harmless, have been found in the human mouth and in the gut of termites, cockroaches, and toads. Species of *Borrelia* are also pathogenic for a number of hosts. *Borrelia anserina* is responsible for a septicaemic disease of fowls, commonly known as fowl spirochaetosis. It is mechanically transmitted by the chicken mite *Dermanyssus gallinae* and possibly also by ingestion of faecal material from infected ticks. *Borrelia recurrentis,* the causal organism of relapsing fever in man, is also a blood-borne infection transmitted from person to person by the body louse, *Pediculus humanus* var. *corporis.* Species of *Borrelia* are frequently found in association with fusiform bacilli in infections of the gums of the human mouth (Table 10.9).

Table 10.9 Genera of spirochetes and their characteristics*

Genus	*Dimensions* μ	*General characteristics*	*Number of axial fibrills*	*Habitat*	*Diseases*
Cristispira	30-150 × 0.5-3.0	3-10 complete coils; bundle of axial fibrils visible by phase-contrast microscopy	>100	Digestive tract of molluscs; has not been cultured	None Known
Spirochaeta	5-500 × 0.2-0.75	Anaerobic or facultatively aerobic; tightly or loosely coiled	2-40	Aquatic, free-living	None Known
Treponema	5-15 × 0.1-1.5	Anaerobic, coil amplitude up to 0.511 μm;	2-15	Commensal or parasitic in humans, other animals	Syphilis, yaws
Borrelia	3-15 × 0.2-0.5	Anaerobic; 5-7 coils of approx, 1 μm: amplitude	Unknown	Humans and other mammals, arthropods	Relapsing fever
Leptospira	6-20 × 0.1	Aerobic; tightly coiled, with bent or hooked ends	2	Free-living or parasitic of humans, other mammals	Leptospirosis

**Phylogenetically, spirochetes form a distinct line of eubacteria*

The genus *Leptospira* includes many species pathogenic to various hosts and other harmless species, e.g., *L. biflexa,* living in fresh water, which are often morphologically in distinguishable from the pathogens (Figure 10.38). Pathogenic species of *Leptospira* differ from those of *Treponema* and *Borrelia* in being able to survive much longer outside their hosts and in having a much wider host range. These organisms are blood-borne but concentrate in the kidneys and are excreted in the urine. Three serotypes are known to be of particular significance to human and animal health: *L. icterohaemorrhagiae,* a natural infection of rats causes leptospiral jaundice or Weil's disease in man.

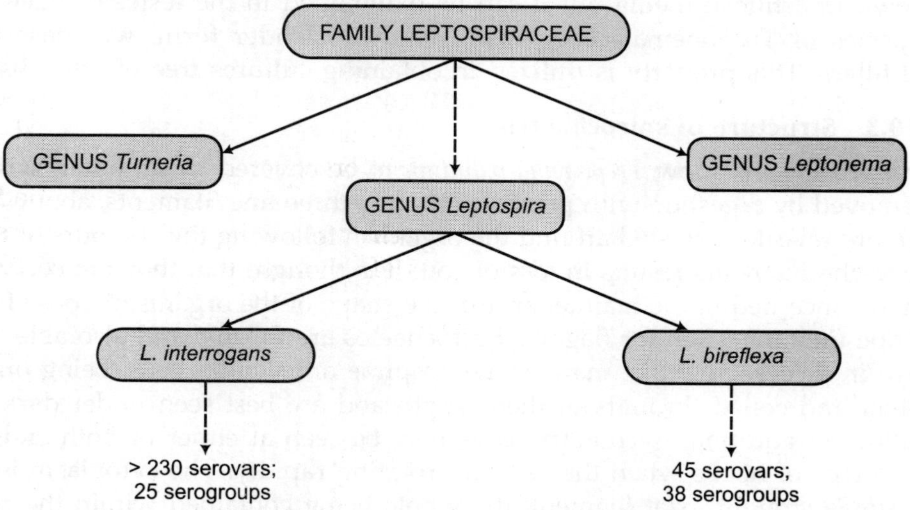

Figure 10.38 Serological classification of Genus *Leptospira.*

L. canicola, primarily causes leptospirosis in dogs but is transmissible to man; *L. pomona* causes leptospirosis in pigs from which man can contract swine-herd's disease. Many other serotypes have been isolated from both wild and domestic animals (Figure 10.39).

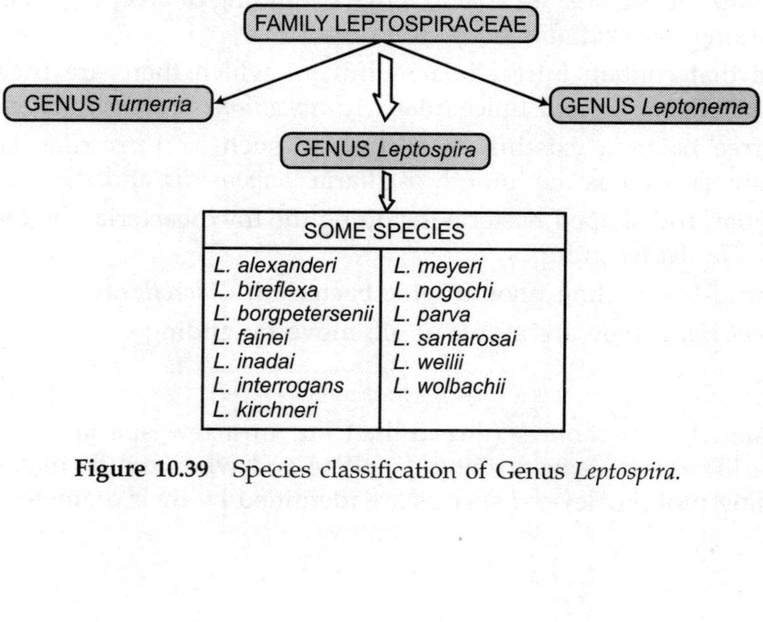

Figure 10.39 Species classification of Genus *Leptospira.*

10.5.6.10.2 Nutrition

A few saprophytic and parasitic species have been grown in pure culture on complex media. Leptospirae are grown relatively easily in fluid media containing salts and fresh rabbit serum. Anaerobic conditions are needed for *Treponema* and *Borrelia;* the other genus is aerobic. It is generally agreed that the pathogen *Treponema pallidum* has not been grown in artificial media but it can be maintained in the testes of male rabbits. Many species of Treponemataceae, owing to their slender form, will pass through bacterial filters. This property is utilized in obtaining cultures free of other bacteria.

10.5.6.10.3 Structure of spirochaetes

Electron micrographs show *Treponema pallidum* to be covered by an outer periplast. If this is removed by digestion with pepsin or trypsin three fine filaments, about 10 mμ in diameter, are revealed, twisted around the organism following the contours of the coils. Damage to the filaments results in loss of coil. It is thought that they are contractile in nature and concerned in the maintenance of the shape of the organism. Loose filaments were at one time mistaken for flagella. Spirochaetes are flexible and also able to rotate and move in a corkscrew like manner. Leptospirae are slender cells, being only 0·1 μ in diameter and coiled throughout their length and are best seen under dark-ground illumination. In addition, 'secondary' coils may be seen at either or both ends giving the appearance of hooks when the cells are rotating rapidly. The cytoplasm is wound about a single straight axial filament, the whole being contained within the cell wall. *Borrelia* are thicker than *Treponema* and *Leptospira* and consist of loose, open coils.

Genera of the family Spirochaetaceae have characteristic structures. In *Spirochaeta* and *Saprospira* the centre of the helix is also occupied by an axial filament and in *Cristispira* a membrane, the crista, spirals along the organism.

10.5.6.11 Gliding bacteria

Only a few groups of bacteria are able to move by gliding or creeping. They are grouped together as gliding bacteria and subdivided into:

1. Bacteria that contain intracellular sulfur, of which there are trichome forming (*Beggiatoa, Thiothrix*) and unicellular (*Achromatium*) representatives.
2. Sulfur-free bacteria existing as trichomes, such as *Vireoscilla, Leucothrix,* and *Saprospira,* as well as the 'mouth oscillaria' *Simonsiella* and *Alysiella.*
3. Unicellular, rod-shaped bacteria including the myxobacteria, the *Cytophaga* group and the *Flexibacter* group.
4. The thread-like gliding phototrophic bacterium *Chloroflexus*
5. *Cyanobacteria,* if they are motile at all, move by gliding.

Beggiatoa

Cells are arranged in trichomes (thread-like) but in a few species they occur singly and colourless. The organisms are non-flagellate and when motile move by a gliding, jerking or rolling motion. Several species are identified by their diameter (1.5 – 3.5 μm).

The cells are usually filled with sulphur droplet, which makes the colourless threads appear white. Those occurring as trichomes are very similar to cyanobacteria *Oscillatoria* in motion. The major difference being that they lack the pigments occurring in the latter. They are found in aquatic and terrestrial environments usually in association with decaying organic material, often of algal origin, where hydrogen sulphide is present. In the past, *Beggiatoa* has been regarded as an autotroph deriving its energy from oxidation of sulphides to sulphur which is deposited as granules within the cell. However, this genus has now been shown to be heterotrophic and does not have an obligatory requirement for H_2S although it is stimulated by this substance. The possession of intracellular sulphur granules is an important difference between *Beggiatoa* and *Vitreoscilla.* In addition to sulphur granules poly-β-hydroxybutyrate granules have been identified in *Beggiatoa.* The trichomes of *Beggiatoa* are fragile structures, but individual cells from different trichomes can exchange with one another as has been claimed for *Vitreoscilla* when two gliding trichomes cross. Some *Beggiatoa* strains found in natural habitats require organic nutrients (Figure 10.40).

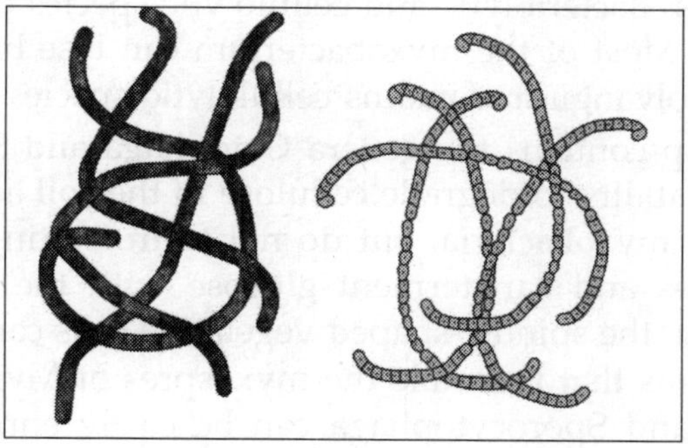

Figure 10.40 *Beggiatoa.*

Vitreoscilla

They are colourless, aerobic, multicellular, filamentous bacterium which moves by gliding and multiplies by fragmentation of the filament and can be isolated from cow dung.

Leucothrix

They grow epiphytically on marine algae, and in its manner of growth it is the organotrophic equivalent of Thiothrix. The filaments grow in bunches or tufts and adhere with their bases to a solid surface with a holdfast. They liberates gliding gonidia from the trichome trip and when cultured in a rich medium the trichome tends to form knots of varying complexity. Another difference between Leucothrix and Beggiatoa is that the former deposits sulphur externally when sulphides are oxidized.

Achromatium is a single-celled organism larger than 2 μ wide and some strains described, but not isolated in pure culture, are said to be 35-100 μ in diameter. The cells divide by constriction and under appropriate conditions contain substantial intracellular calcium carbonate sphaerules and sulphur granules.

All these organisms occur in terrestrial and aquatic environments where there is organic material and/or hydrogen sulphide. Their presence tends to prevent the hydrogen sulphide concentration building up.

10.5.6.12 Myxobacteria

Members are strictly aerobic chemoheterotrophic with gliding motility. They are soil borne and are characterized in their natural habitat by the formation of fruiting bodies that are very small (less than 1 mm). Such fruiting bodies are found in decaying plant material, rotting wood, tree bark, and on the faeces of harbivores. They can be isolated from the latter, after incubating the faeces for several weeks in damp soil. On solid media myxobacteria form very extensive flat colonies. When the vegetative cells inside the colonies aggregate, they differentiate into fruiting bodies that differ in shape, size, and pigmentation for the various genera and species of the myxobacteria.

The cells inside the fruiting bodies become dormant, i.e., they are converted to myxospores during maturation. They can be spherical (Myxococcus) or rod shaped and may be enclosed in cysts. Bacteriolytic and cellulolytic species can be distinguished by their mode of nutrition. Most of the myxobacterium can lyse bacteria by exoenzymes, whilst only the genus Polyangium contains cellulolytic species (Figure 10.41).

The Cytophaga group contains the genera Cytophaga and Sporocytophaga, which became known by their ability to degrade cellulose in the soil aerobically. They exhibit gliding motility like the myxobacteria, but do not form fruiting bodies. Some species are facultative anaerobes and can ferment glucose with the production of organic acids. In Sporocytophaga, the spindle-shaped vegetative cells can change into spherical, encapsulated lasting forms that resemble the myxospres of Myxococcus but are called microcysts. Cytophaga and Sporocytophaga can be easily enriched on cellulose, i.e., filter paper.

The Flexibacter species are aquatic bacteria. They consist of long, very flexible cells that are not considered multicellular. They become shorter and shorter and eventually fragmented into coccoid cells during continuous cultivation. Many forms contain carotenoids and show yellow, pink, or orange pigmentation. The causative agent of the columnaris fish disease (*Chondrococcus columnaris*) is now assigned to the genus Flexibacter as *F. columnaris* (Figure 10.42).

10.5.6.12.1 Prosthecate and budding bacteria

This group contains organisms whose form differs markedly froma typical bacteria either by the presence of protheca or buds or hypha like protrusions, or by the formation of extracellular stalks or masses of slime (Figure 10.43).

Budding or sprouting is the term applied to the mode of multiplication, characteristic for the budding yeasts. Contrary to binary fission, it is an unequal cell division, which proceeds via localized growth. The daughter cell (bud) is usually smaller than the mother cell and reaches normal size only after it has become separated from the mother cell. Various aquatic and soil organisms belong to the budding bacteria. *Hyphomicrobium*

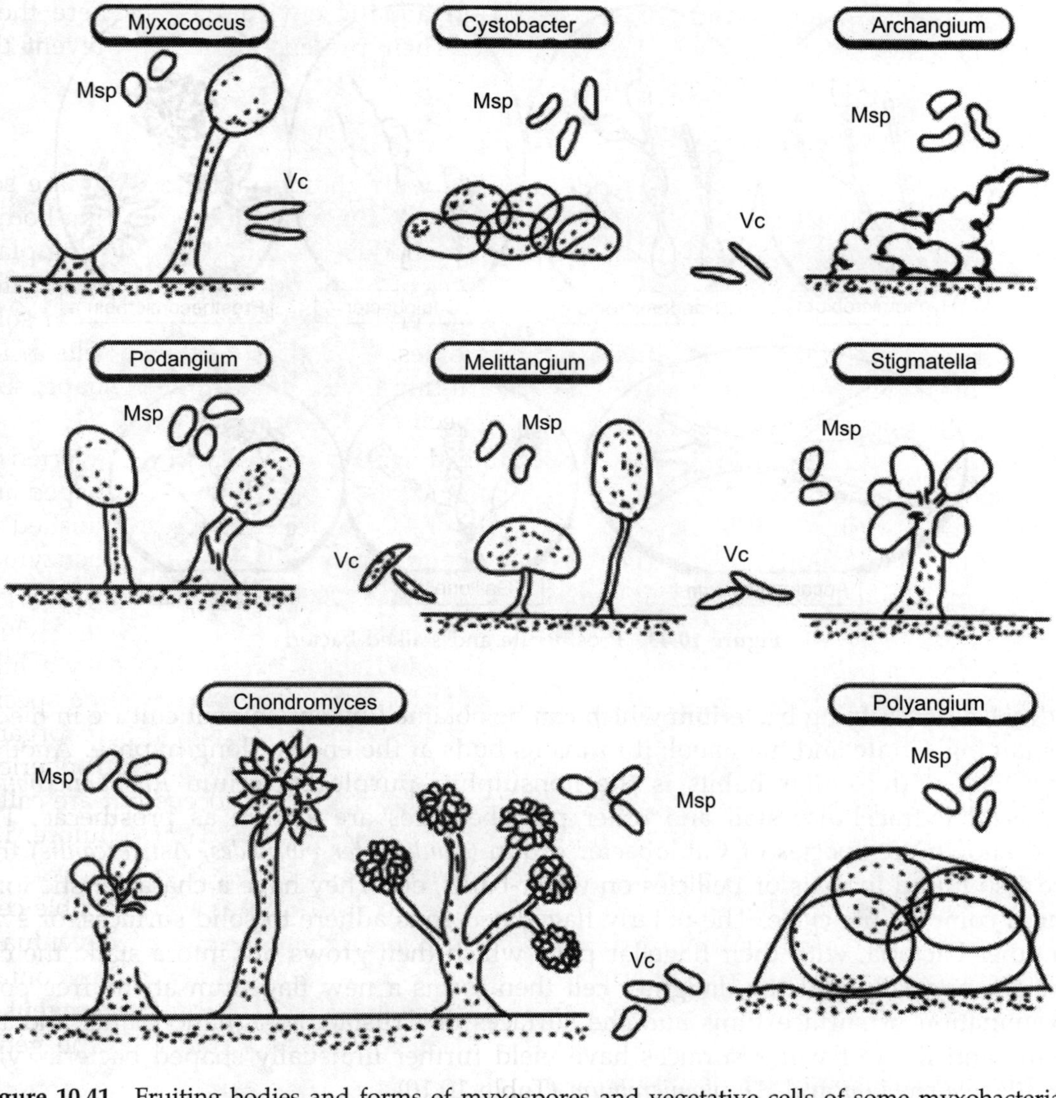

Figure 10.41 Fruiting bodies and forms of myxospores and vegetative cells of some myxobacteria where Msp = myxospores, Vc = vegetative cells.

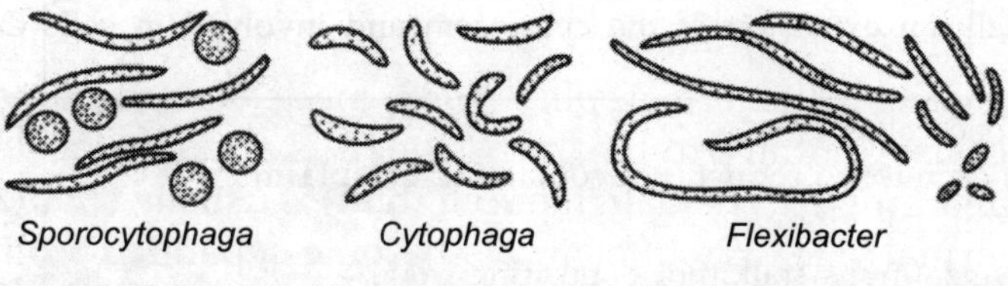

Figure 10.42 Cell forms of the genera Sporocytophaga, Cytophaga and Flexibacter.

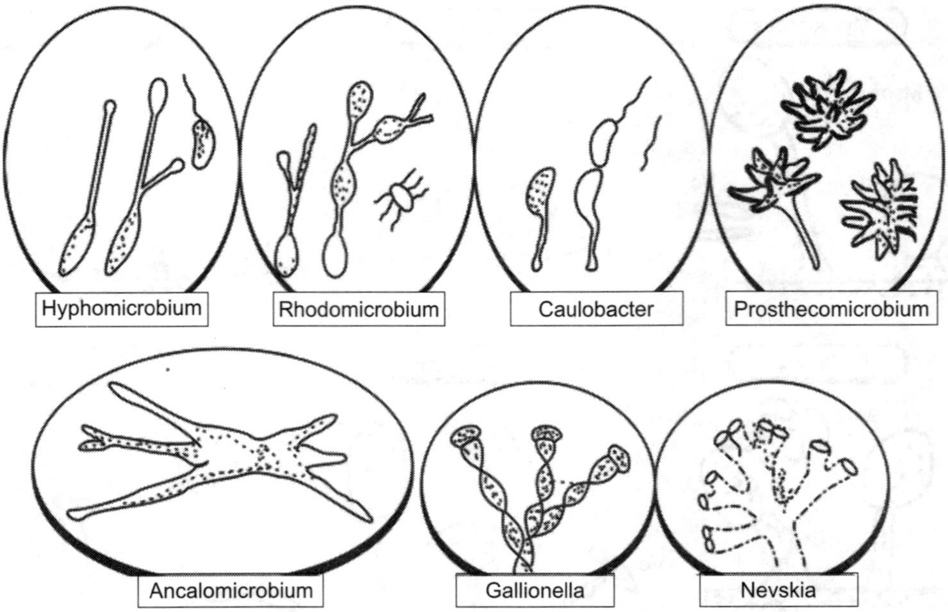

Figure 10.43 Prosthecate and stalked bacteria.

vulgare, a denitrifying bacterium which can be obtained by enrichment culture in media containing nitrate and methanol; it forms its buds at the ends of long hyphae. Another organism with similar habits is the nonsulphur purple bacterium *Rhodomicrobium vannielii*. Extracellular stall and other protuberances are known as prosthecae. The best known are species of Caulobacter group (*Caulobacter vibrioides, Asticcacaulis*) that are also found in mats or pellicles on water-baths, etc. They have a characteristic form and a complex life cycle. The polarly flagellated rods adhere to solid surfaces, or even to other bacteria, with their flagellar pole, which then grows out into a stalk; the cell divides normally and the daughter cell then forms a new flagellum at the free pole. Examination of surface films and the surfaces of Crustaceae, aquatic plants and the fauna and flora of water surfaces have yield further unusually shaped bacteria, viz., *Prosthecomicrobium* and *Ancalomicrobium* (Table 10.10).

Table 10.10 Characteristics of stalked, appendaged (prosthecate), and budding bacteria

Characteristics	*Genus*
Stalked bacteria, stalk an extension of the cytoplasm and involved in cell division	*Caulobacter*
Salked fusiform-shaped cells	*Prosthecobacter*
Stalked but stalk an excretory product, not containing cytoplasm:	
Stalk depositing iron, cell vibrioid	*Gallionella*
Laterally excreted gelatinous stalk, not depositing iron	*Nevskia*
Pear-shaped or globular cells with long stalk	*Planctomyces*

Contd...

Contd...

Appended (prosthecate) bacteria:	
Multiple prosthecae Short prosthecae, multiply by fission Flat, Star-shaped cells Long prosthecae, multiply by budding	 *Asticcacaulis* *Prosthecomicrobium* *Stella* *Ancalomicrobium*
Phtotrophic With gas vesicles	*Prosthechochloris* *Ancalochloris*
Budding bacteria:	
Phototrophic, produce hyphae	*Rhodomicrobium*
Phototrophic, budding without hyphae	*Rhodopseudomonas*
Organotrophic, pear-Shaped, stalks lacking	*Pirella*
Organotrophic, rod-Shaped cells	*Blastobacter*
Organotrophic, buds on tips of slender hyphae: Single hyphae from parent cell Multiple hyphae from parent cell	 *Hyphomicrobium* *Pedomicrobium*

**Phylogentically, Planctomyces and Pirella form a distinct line of eubacteria; most other budding/ stalked organisms are members of the gamma section of the purple bacteria (Proteobacteria)*

10.5.6.12.2 Stalked bacteria

Some widely distributed bacteria are sessile by means of a stalk consisting of slime. *Gallionella ferruginosa*, for example, is a bean-shaped cell which excretes slime on its concave side. This concave side can be seen microscopically to consist of a helical band encrusted with ferrous hydroxide. *Gallionella*, the best known iron bacterium, grows profusely, especially in the spring, in iron-containing waters such as streams, drainage pipes, etc. *Nevskia ramose* has been isolated from the pellicles it forms on ponds, ditches, and water baths. Its characteristic polar secretion of slime apparently occurs only under conditions of nutrient deficiency.

10.5.7 Obligate Cellular Parasites

The fact that some bacteria are absolutely dependent on living cells for their multiplication. Many bacteria that occur as endosymbions in insects and ciliates are obligately restricted to these locations. The basis of such obligate cell parasitic forms of life is usually attributed to degenerative changes in metabolism. Among the best studied of the obligately intracellular bacteria are the reckettsiae and chlamydiae. Both are pathogens of animals and humans and their cell envelope structure resembles with Gram-negative organisms.

10.5.7.1 Rickettsiae

The rickettsiae are named after H. T. Ricketts, who discovered American 'Rocky Mountain spotted fever'. This group is also referred to as the 'spotted' fever (typhus fever) group after the disease produced by the best known of these pathogens (*Rickettsia prowazekii*). Their natural distribution is by host vectors such as lice, fleas, ticks, mites, etc., in which they exist as harmless parasites or even as symbionts. On being transferred to other animal hosts or humans by bites, scratches, or inhalation, they produce very serious disease symptoms. Although the size of rickettsiae is of the same order of magnitude as pox viruses, they can be differentiated quite unequivocally from any viruses. They contain DNA as well as RNA in a ration of 1:3.5. The cells are surrounded by a cell wall that contains muramic acid and is lysozyme sensitive. Nuclear regions and cell walls can also be distinguished in electron micrographs of thin section.

Many rickettsiae have never been cultured except in living cells, but they can be grown in incubating chicken eggs or in experimental animals; about 10^9 cells can be produced in the yolk sac of a chicken egg. A few enzymes of intermediary metabolism can be demonstrated in isolated cells, but the metabolic activity of the cells declines on storage; respiration can be restored by addition of ATP, organic acids and amino acids. It appears that the rickettsiae are able to develop their own metabolism, but they are apparently unable to control uptake and excretion of metabolites because of the permeability characteristics of their cell envelope.

The best known pathogens among the rickettsiae are those of the typhus group *R. prowazekii* is the causal agent of typhus fever. The reservoir is humans, and the organism is transmitted by lice (head lice and clothes lice). The lice are rapidly (in a few days) killed by the organism, and infection is via their faeces. *R. typhi* is the causal organism of endemic and murine typhus, which produces a similar but less severe syndrome. The organism is spread by rats which remain symptomless and is transmitted from rat to rat to humans by fleas.

Whilst most rickettsiae are relatively sensitive to heat and dehydration, *Coxiella burnetii*, which causes Q-fever, can survive outside the host. It is transmitted by ticks to sheep, goats and cattle and can infect humans by tick bites and also via animal dust, infected soil, and consumption of milk. The usual pasteurization of milk (heating at 60°C for 30 min) does not kill *Coxiella*, thus 72.5°C for 15 seconds is now used.

10.5.7.2 Chlamydiae

Chlamydiae are human pathogens. *Chlamydiae trachomatis* causes trachoma, the Egyptian eye disease that starts as conjunctivitis and leads to blindness, and the venereal disease lymphogranuloma venereum. In both cases transmission is by contact. *C. psittaci* is the causative agent of ornithoses, the best known of which is psittacosis, a feverish pneumonia. The main hosts of chlamydiae are birds.

On the basis of their biochemical characteristics, chlamydiae belong to the prokaryotes (bacteria) and not, as formerly thought, to the viruses. They contain RNA and DNA in a ratio of characteristic of bacteria, and they synthesize substances which eukaryotic cells are unable to produce, such as muramic acid, diaminopimelic acid, D-alanine, and folic acid. These properties are in accord with their sensitivity to penicillin and sulphonamides. Their genome is very small (relative size, 0.66×10^9) and corresponds to only about one quarter of the genetic information of *E. coli*. Chlamydiae grow only in living cells and are cultured in chicken eggs and tissue cultures. Their dependence

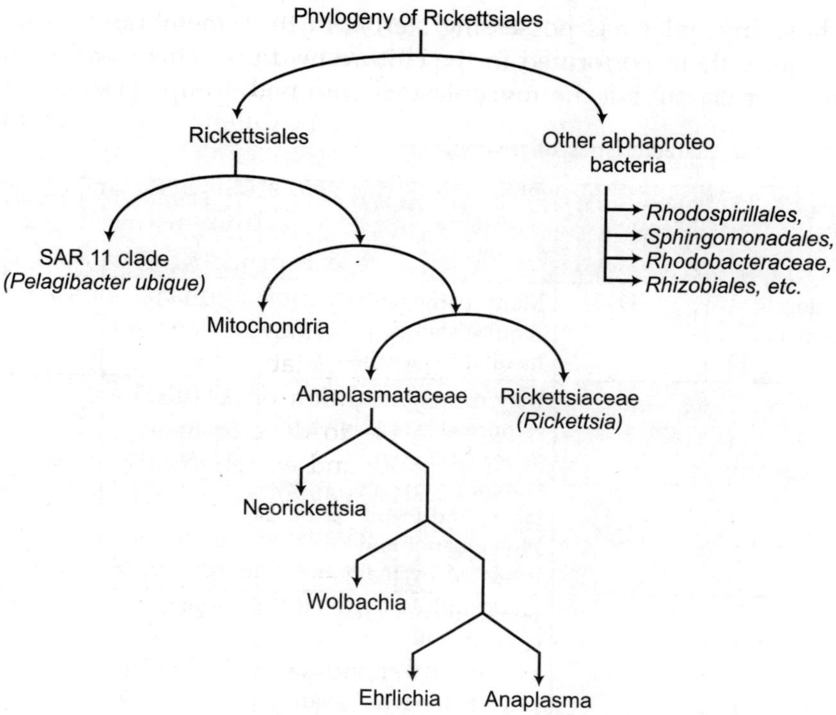

on the metabolism of host cells is apparently due to absence of an ATP-generating system of their own. They are unable to phosphorylate glucose or to metabolise it. On the other hand, they are extraordinarily permeable to ATP and CoA. They can, therefore, be regarded as 'energy parasites'.

The obligate cell parasites among the bacteria are apparently the result of a regressive development. Their adaptation to their host cells has been accompanied by loss of various synthetic capacities.

10.5.7.3 Mycoplasma group

Mycoplasma

Mycoplasmas are the organisms without cell wall that do not revert to walled organisms. They are probably the smallest organisms capable of autonomous growth and are of special evolutionary interest because of their extremely simple cell structure. Mycoplasma resemble protoplasts in their lack of a cell wall, but they are more resistant to osmotic lysis and are able to survive conditions under which protoplast lyse. The ability to resist osmotic lysis is at least partially determined by the nature of the mycoplasma cytoplasmic membrane, which is more stable than other prokaryotes. In one group of mycoplasmas, the membrane contains sterols that seem to be responsible for stability, whereas in other mycoplasmas, carotenoids or other compounds may be involved instead. Those mycoplasmas possessing sterols in their membrane do not synthesize them but require them performed in the culture medium. This sterol requirement is a current basis for classifying the mycoplasmas into two groups (Table 10.11).

Table 10.11 Major characteristics of mycoplasmas

Genus	*Number of recognized species*	*Properties*	*DNA (mole% GC)*	*Genome size*	*Presence of lipoglycans*
I. Require sterols mycoplasma	69	Many pathogenic; require sterols; facultative aerobes	23-41	5 × 10^8	+
Anaeroplasma	2	May or may not require sterols; obligate anaerobes; degrade starch, producing acetic, lactic, and formic acids plus ethanol and co.; inhibited by thallium	29 (sterol-requiring strains)	-	+
Spiroplasma	3	Spiral to corkscrew-shaped cells; associated with various phytopathogenic (plant disease) conditions	25-31	10^9	-
Ureaplasma	2	Coccoidcell; occasional clusters and shorts chanins; growth optimal at pH 6; strong urease reaction; associated with certain urinary tract infections in humans; inhibited by thallium acetate	27-30	5 × 10^8	-
II. Do not require sterols Achloeplasma	8	Facultative aerobes	27-36	10^9	+
Thermoplasma	1	Thermophilic, acidophilic archaebacterium found in heated coal refuse sites; no sterol requirement; DNA contains histones	46	10^9	+

Thermoplasma is phylogenetically unrelated to all other mycoplasmas because it is an archaebacterium. They are acidophilic, aerobic organotroph, and is also thermophilic growing optimally at 55°C and pH 2. To survive the osmotic stresses of life without a cell wall and to withstand the dual environmental extremes of low pH and high temperature, Thermoplasma has evolved a cell membrane of chemically unique structures. The membrane contains lipopolysaccharide consisting of tetraether lipid with mannose and glucose units. This molecule constitutes a major fraction of the total lipid composition of Thermoplasma. The membrane also contains glycoproteins but not sterols. Together these and other molecules render the Thermoplasma stable to hot acid conditions. The exact phylogenetic status of Thermoplasma is thus unclear.

The first representative of the mycoplasmas described was the causal organism of pleuropneumonia, a lung infection of cattle. This organism grows on agar media containing serum in small colonies that have a 'fried egg' appearance. Similar growth forms have therefore been designated as pleuropneumonia like organisms (PPLO). Mycoplasmas have been recognized as causative organisms for a number of diseases, as contaminants in tissue cultures, and as harmless saprophytes. The colonies consist of cells and fragments of various sizes and can be described as cocci, filaments, discs and rosettes. They multiply by binary fission, by fragmentation of filaments and rings into coccoid cells, and by a kind of sprouting. In liquid media they are often found as very irregular and sometimes branched forms that can pass through membrane filters, rather like viruses.

Distribution and species

The members of the Mycoplasma group (genera Mycoplasma, Acholeplasma, and Spiroplasma) are parasitic bacteria. They do not kill their hosts, but they produce predominantly chronic infections, and in this sense they are very successful parasites.

In animals, they may appear as harmless parasites on the serous epithelia of the respiratory and genital tracts (in mammals and birds). They are membrane parasites in that they adhere tightly to the epithelial cells of the serous membranes. They do not excrete toxins but because of the intimate contact between the parasite, which does not have a cell wall, and the host cell, even weakly toxic metabolites like ammonium ions and hydrogen peroxide may have toxic effects on the host cells. The animal parasites of the mycoplasma group are divided into two genera. The members of the genus Mycoplasma require cholesterol or addition of complex steroids for growth (i.e., media containing blood serum). For the genera that do not require cholesterol or similar materials, a new genus Acholeplasma has been designated. Whilst infections with mycoplasmas can be symptomless in some animals, inflammation of the respiratory system, the lungs or udder occur in others. The host specificities of Mycoplasma species are expressed in the species designation, as in *M. canis, M. gallisepticum* and *M. hominis.*

In plants, mycoplasmas cause yellowing diseases. They are predominantly localized in the phloem part of the vascular system, and because of their morphological similarity to a spirillum, they have been grouped in the genus Spiroplasma, *Spiroplasma citri* causes a yellowing disease in citrus trees. Similar forms have been demonstrated in other plants (maize, rice, Bermuda grass etc.). Spiroplasma has also been found in bees and grasshoppers and it may be assumed that insects are not only carriers but also hosts of Spiroplasma species.

Biochemical Properties

The mycoplasma group is differentiated from all other bacteria not only by the lack of a cell wall, but also by biochemical characters. The organisms grow only in isotonic or hypertonic media (with sorbitol or sucrose) and require purines and pyrimidines, as well as lipids and steroids. The lack of quinines and cytochrome suggests a very limited respiratory chain.

Relation to L-forms

A strain of *Streptobacillus moniliformis* that grew as irregular protoplasts was isolated in 1934. These cells were called L-forms after the Lister Institute of London where they were isolated. Naked protoplasts that are able to grow can also be isolated from *Salmonella*, *E. coli* and *Proteus* as well as from other bacteria by cultivation on serum agar with penicillin (100 μg/mL). They produce 'fried egg' type colonies which resemble those of *Mycoplasma* species above. Two types of L-forms have been isolated: labile forms, which revert to normal cells with complete cell walls on cultivation without penicillin; and stable forms, which do not form cell walls even in the absence of penicillin. It was at first assumed that the Mycoplasma types arose by mutation of normal bacteria to stable L-forms. However, genome size and GC content of mycoplasmas contradict their origin from eubacteria and suggest that they constitute a separate class.

10.5.8 Anaerobic Anoxygenic Phototrophic Bacteria

The anaerobic phototrophic bacteria (Rhodospirillales) are characterized by their possession of photosynthetic pigments and by their dependence on light as an energy source. Four families are distinguished on the basis of important physiological properties : purple sulphur bacteria (Chromatiaceae), non sulphur purple bacteria (Rhodospirillaceae), green sulphur bacteria (Chlorobiaceae), and the Chloroflexus group (Chloroflexaceae) and constitute a natural phylogenetic group.

10.5.9 Aerobic Oxygenic Phototrophic Bacteria

Cyanobacteria

Cyanobacteria, the largest, most widely distributed, group of photosystems are akin to those found in alage and higher plants. They occupy a wide range of oxygenic phototrophic prokaryotes, which have long been recognised as having potential in biotechnology. Their oxygen evolving habitats, which vary considerably in mean temperature and salinity and include soil surface, fresh water and seawater habitats. Free living diazotrophic cyanobacteria enter into symbiotic association with plants ranging from algae, bryophytes, fungi, gymnosperms and to angiosperm. Cyanobacteria have also been found to be associated with certain protozoa, where they are called cyanellae. Current theories suggest that archaic cyanobacteria gave rise to the plant cell chloroplast (endosymbiont theory) and that they may be responsible for the original conversion of the reducing atmosphere of the prebiotic earth to one rich in free oxygen. During evolution, the cyanobacteria have survived under many environmental stresses through the development of specialized adaptation mechanisms. These adaptations include the ability to fix atmospheric nitrogen and the development of cell types which facilitate dispersion and survival, e.g., harmogonia and akinetes. However, further adaptation was required when the free oxygen content of the atmosphere is increased (Figure 10.44). Because the nitrogenase complex is oxygen labile, a mean of excluding oxygen from the enzyme was required and one such response in filamentous species has been the evolution of a specialized cell type called heterocyst (Figure 10.45). They exhibit an enormous diversity of shapes and arrangements, from unicellular cocci or rods to long trichomes. Broad features of the classification of cyanobacteria are

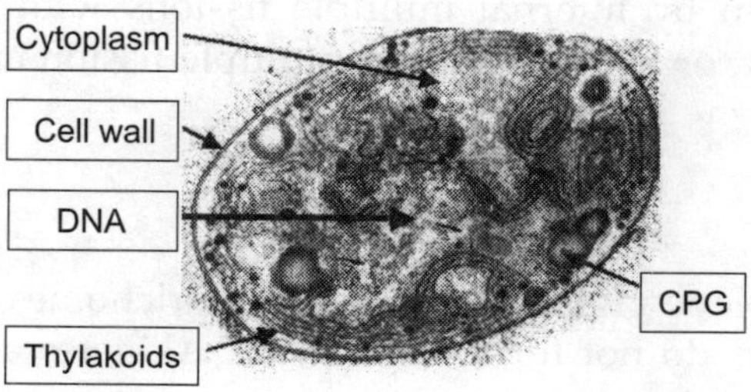

Figure 10.44 Section of a cyanobacterium *Anabaena azollae*. Most of the photosynthetic lamellae (thylakoids) are peripheral but some extend into the mid portion of the cell CPG denotes for Cynophycin granule.

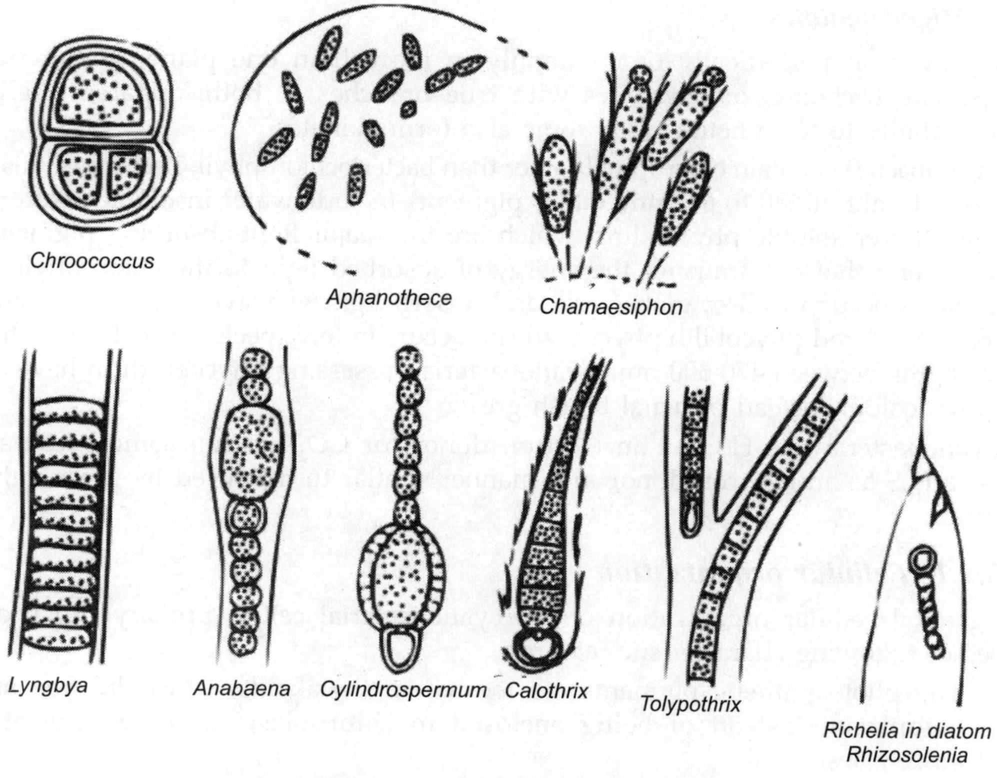

Figure 10.45 Some common cyanobacteria.

Non-filamentous

Order Chroococcales

Unicellular or non-filamentous aggregates of cells held together by outer wall or get like matrix; binary division in one, two or three planes, symmetric; or by budding. Rarely form akinetes.

Order Pleurocapsales

Unicellular or non-filamentous aggregates of cell held together by outer wall or gel like matrix; reproduction by internal multiple fissions with production of daughter cells smaller than parent; or by a mixture of multiple fission and binary fission. Rarely form akinetes.

Filamentous

Order Oscillatoriales

Binary division in one plane giving rise to 1-seriate trichomes, though sometimes with false branches; trichomes do not form heterocysts. Akinetes apparently not reported.

Order Nostocales

Binary division in one plane giving rise to 1 seriate, though sometimes with false branches, one or more cells per trichome differentiate into a heterocyst, at least when concentration of combined nitrogen is low. Some also produce akinetes.

Order Stigonematales

Binary division periodically or commonly in more than one plane, giving rise to multiseriate trichome, or trichomes with true branches or both; apparently always possess ability to form heterocysts; some also form akinetes.

Cyanobacteria contain chlorophyll rather than bacteriochlorophyll. Because of this cells absorb red light of 680 to 683 nm. Other pigments include water insoluble carotenoids and also water soluble phycobilins, which are the major light-absorbing pigments in cyanobacteria that can transmit the energy of absorbed light to the chlorophyll. Blue phycobilins occur in all cyanobacteria and absorb light at wavelengths between 500 and 650 nm. A red phycobilin phycoerythrin, occurs in few species and absorbs shorter wavelengths between 470-600 nm. Cyanobacteria possessing phycoerythrin have a red or brown colour instead of usual bluish green.

Cyanobacteria use H_2O as an electron donor for CO_2 fixation some bacteria can also use H_2S as an electron donor in a manner similar to that used by green sulphur bacteria.

10.5.9.1 Cellular organization

The general cellular organization of the cyanobacterial cell is prokaryotic. The cell possesses following characteristic features:

(i) The photosynthetic pigments are located in thylakoids, which lie free in the cytoplasm, instead of being enclosed in chloroplasts as in photosynthetic eukaryotes.

(ii) The thylakoids contain chlorophyll-a, but chlorophylls-b and c are absent.
(iii) A diversity of characteristic cytoplasmic granular inclusions.
(iv) Cell wall consist of murein (peptidoglycan), as in other eubacteria, outside which there is a lipopolysaccharide layer.
(v) A fibrous sheath.

In some species of cyanobacteria (e.g., *Synechococcus lividus*), the protoplast can be divided into two distinct regions: a peripheral pigmented region called the chromatoplasm, containing the photosynthetic apparatus, and a central region known as the centroplasm, containing DNA fibrils.

The photosynthetic apparatus is similar to that in higher plants where the two photosystems (PS I and PS II) are situated within flattened vesicles called thylakoids, which are located throughout the outer regions of the cyanobacterial cell. While, there is much evidence to support the endosymbiotic theory that cyanobacteria were the progenitors of plant chloroplast. Cyanobacteria do not contain the chlorophyll *b* pigment, which harvests light in photosynthetic eukaryotes. Instead they contain unique multimolecular structures known as phycobilisomes. These are situated on the outer surface of the thylakoids and are comprised of phycobiliprotein subunits each containing an ordered array of covalently attached bilin prosthetic groups and linker polypeptides. Phycocyanin and phycoerythrin are the two most common bilins and they efficiently funnel more than 95% of the excitation energy from light to the reaction centre of PS II.

Cyanobacteria contain a large variety of intracellular inclusions, some of which are likely to serve as reserves of metabolites for use in conditions of nutrient imbalance. For example, cyanophycin granules are high molecular mass proteinaceous granules consisting of non-ribosomally synthesised polymers containing equimolar quantities of arginine and aspartate; the arginyl residues are attached to all the free carbonyl groups of a polyaspartic acid core . Thus, cyanophycin can serve as a source of nitrogen and carbon. Polyglucose (glycogen bodies) and poly b-hydroxybutyrate granules (in *Chlorogloea*) are reserves of carbon and a source of energy, while polyphosphate bodies are stores of phosphate. Carboxysomes are polyhedral bodies containing carbonic anhydrase and ribulose-1, 5-biphosphate carboxylase oxygenase (Rubisco, the key enzyme required for the net assimilation of inorganic carbon), as well as a few additional enzymes related to carbon metabolism. They are the major sites of carbon-fixation, except for carboxysomes. However, little attention has been paid to the synthesis and role of these inclusions. They may increase the chance of survival in a changing and possible stressful environment.

10.5.9.2 Vegetative cells

Vegetative cells have both photosystems (PS I and PS II) similar to the higher plants. The photosynthetic pigments are located in thylakoid in the outer regions of the cells. In addition to chlorophyll *a* (chl *a*), carotenes and xanthophylls, they have characteristic accessory pigments called phycobiliproteins.

They have a role as light receptors, but appear to function also as readily usable nitrogen store. Vegetative cells may also contain polyhedral bodies called carboxysomes. These are molecular aggregates of the calvin-cycle enzyme, ribulose biphosphate carboxylase/oxygenase (Rubisco). The presence of rubisco in vegetative cells and its absence from heterocyst was shown first time by using immunogold localization technique. While heterocysts import the fixed carbon from vegetative cells, polyglucoside bodies may be found among the thylakoids. Excess nitrogen is stored in structured granules of copolymers of aspartic acid and arginine known as cyanophycin. In addition, vegetative cells may also contain gas vacuoles, polyphosphate and lipid bodies. Thus, they are well equipped to accumulate essential metabolites.

10.5.9.3 Heterocyst

Heterocysts are differentiated cells and the site of nitrogen fixation in cyanobacteria. In filamentous cyanobacteria, heterocyst differentiate from vegetative cells which have lost the capacity of oxygenic photosynthesis and have evolved a modified thick cell envelop. The development of heterocysts is perhaps one of the most important characteristics of cyanobacteria because no other prokaryotes possess an equivalent degree of differentiation (Figure 10.46).

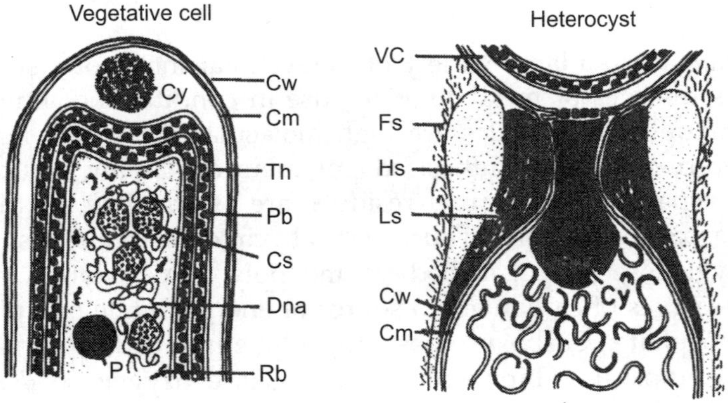

Figure 10.46 Longitudinal section of cyanobacterial cell where Cw = cell wall, Cm = cell membrane, Th = thyalokoid, Pb = phycobilisome, Cs = Carboxysomes, Cy = cynophycin bodies, Fs = fibrous layer, Hs = homogeneous layer, Ls = lamellar layer, P = polyphophate, Rb = ribosomes, VC = vegetative cell.

Heterocyst usually develops only when the concentration of combined nitrogen source is lowered. Heterocysts are terminally differentiated cells that develop in a pattern from certain vegetative cells in the filament in response to limitation in fixed nitrogen. Nitrogen fixing cyanobacteria become more important as the availability of combined nitrogen in the environment becomes a limiting factor for growth. The thick cell wall of heterocyst contains special lipopolysaccharides and forms a diffusion barrier for gases, limiting the entry of oxygen is evolved in these cells. Therefore, the heterocyst is virtually anoxic and provides an excellent environment for the oxygen sensitive nitrogenase. PS I mediated conversion of light energy in the heterocyst provides ATP to nitrogenase. However, for reducing equivalents nitrogenase depends on the transportation of carbohydrate from the neighbouring vegetative cells. Nitrogen fixing cyanobacteria have a major advantage of times when source of combined inorganic nitrogen have been depleted from the water. Cyanobacteria have high competitive ability for ammonium under nitrogen limiting conditions, but a low competitive ability for nitrate in comparison with eukaryotic algae.

10.5.9.4 Akinetes

Akinetes are 'survival' forms. They are recognized by their size, intense pigmentation and thick cell walls full of reserve materials, which enable cyanobacteria to survive when environmental conditions are not favourable for growth. They are morphologically differentiated cells, analogous to the heterocysts, and are either intercalary (*Anabaena*) or terminally arranged (*Cylindrospermum*) on the trichome. They are much larger than vegetative cells, although their size can vary, even within a single culture and its cytoplasm is more granular than the vegetative cells. They also and lack the photosynthetic pigments. Chlorophyll was largely replaced by pheophytin and contained larger proportion of xanthophylls than the intact filaments (Figure 10.47).

The akinetes contain four times as much RNA as the vegetative cells with few exceptions as well as protein content was also three times higher than vegetative cells. The increased endogenous respiration in isolated akinetes indicates the reorganization of enzyme activity prior to germination. A deficiency of fixed nitrogen, iron and phosphate with exception of planktonic species is reported to induce akinetes formation and addition of nitrogen (nitrate, nitrite and ammonia) inhibited the heterocyst differentiation induced akinetes formation in *Anabaena* and *Nostoc* 7524. Sodium glutamate, sodium chloride and various amino acids (tryptophan, aspartic acid, phenylamine, proline and isoleucine) have been shown to induce akinetes formation. While the addition of calcium gluconate resulted in the formation of longer chain of akinetes with *Cylindrospermum licheniforme* kütz as exception in which only increase in akinetes production was reported. Akinetes do permit survival of the organisms under adverse conditions, although they are not always necessary since many strains (including those capable of producing akinetes) can survive in the vegetative state. They certainly ensure the long term survival of the organisms. Isolated viable akinetes have been isolated from sediments as old as 64 years.

10.5.9.5 Hormogonia

Hormogonia are short sections of the trichome formed by breakage of longer filaments. Their formation promotes propagation. In *Oscillatoria*, their formation entails the sacrifice of at least one cell of the trichome, as the cells are connected by a common peptidoglycan layer that is not divisible. This is referred to as 'transcellular division'.

Baeocytes are small spherical reproductive cells of the plurocapsulated cyanobacteria. They are formed by multiple fission within a considerably enlarged cell that is surrounded by a thick fibrous exopolysaccharide layer. In *Dermocarpa*, for example, rapid binary divisions lead to the formation of from 4 to 1000 baeocytes.

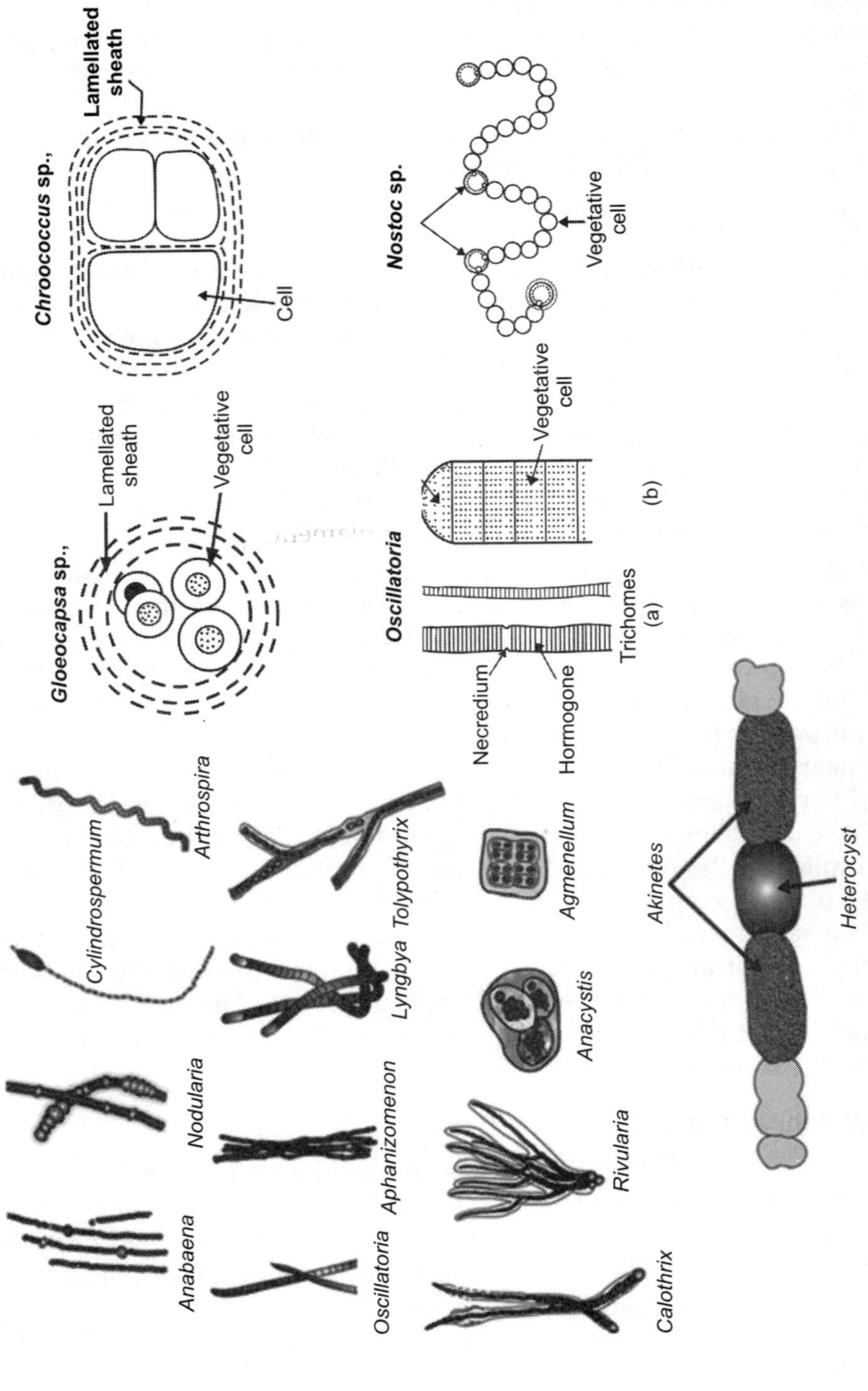

Figure 10.47 Cyanobacterial representatives.

10.5.9.6 Nitrogen metabolism in cyanobacteria

Cyanobacteria are a diverse and widely distributed group of prokaryotes which are bound to carry out oxygenic photosynthesis, making a large contribution to global nitrogen economy by unicellular, filamentous nonheterocystous and filamentous heterocystous forms.

Nitrogen constitutes as much as about 10-11% of the total dry weight of an exponentially growing cyanobacterial cell. Availability of nitrogen has been identified in many instances as the factor limiting productivity in natural habitats. Cyanonbacteria mainly use the inorganic species of nitrogen such as dinitrogen, nitrate and ammonium, in decreasing order of availability. In general, they can use nitrate or ammonium as the sole N source for growth, and many have additional ability to fix N_2. However, urea and other organic sources of N such as some amino acids, can also be assimilated by some cyanobacteria.

Under autotrophic conditions, the utilization of any form of nitrogen by cyanobacteria is strictly dependent on light and CO_2 requirement for reductants and ATP in the assimilatory process involved. The CO_2 requirement results from a set of interactions between carbon and nitrogen metabolism, some of which are regulatory in nature and determine that the assimilation of carbon and nitrogen are balanced.

The fact that inorganic nitrogen sources other than NH_4^+ are first metabolized to NH_4^+ prior to assimilation of their nitrogen atoms into nucleic acid, cell wall, porphyrins, polyamines and proteins. Some amino acids, like arginine, are assimilated by the production of NH_4^+ among their catabolic products. Glutamine synthetase and glutamate synthase (GS/GOGAT) play a central role in cyanobacterial nitrogen metabolism particularly in NH_4^+ assimilation. Glutamate produced in GS/GOGAT pathway is not only the major nitrogen donor for the biosynthesis of other nitrogen containing metabolites but is itself a precursor for some amino acids and 5-aminolevulinate, the immediate precursor for porphyrin, phycobilin and chlorophyll biosynthesis. Glutamine also donates nitrogen to some metabolites. It is evident that amino acids arginine and aspartate together make up cyanophycin, a unique storage reservoir of nitrogen (and carbon) in many cyanobacteria.

Utilization of different nitrogen sources requires their entry through the cell wall with the aid of specialized transport system followed by their hydrolysis reduction. Nitrogen-fixing cyanobacteria possess a central position in the nutrient cycling largely due to their inherent capacity to fix atmospheric nitrogen directly into ammonium, making it available to higher plant. According to one assumption cyanobacteria fix over 35 million tonnes of nitrogen annually.

NO_3^- transport is energy dependent with ATP requirement. In addition to transporting NO_3^-, it also supports NO_2^- transport and that NO_3^- and NO_2^- share a common transport system (Figures 10.48 and 10.49).

10.5.9.7 Nitrogen fixation

The majority of cyanobacteria are diazotrophic capable of carrying out simultaneously two incompatible reactions, viz., oxygenic photosynthesis and O_2 sensitive N_2-fixation.

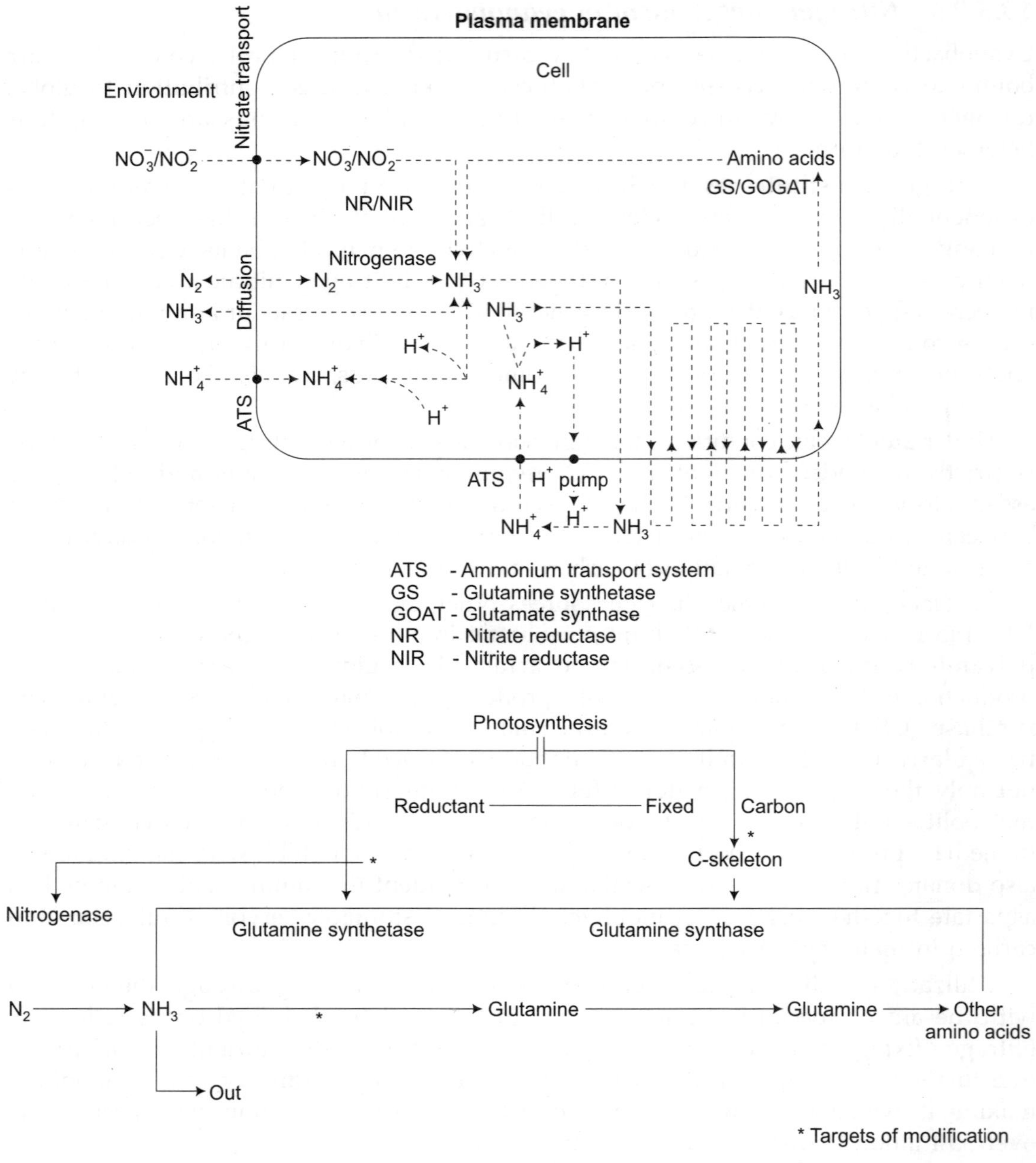

Figure 10.48 Import of exogenous inorganic nitrogen and its fate and cyanobacterial cellular metabolism.

In addition, they exhibit photo-production of hydrogen and photo-assimilation of nitrate, CO_2 and sulphate. The uniqueness of diazotrophic cyanobacteria lies in the fact that they are never energy limited and hence can continue producing H_2 from H_2O and NH_4^+ from N_2 and NO_3^-. Their ability to grow over a wide range of ecological

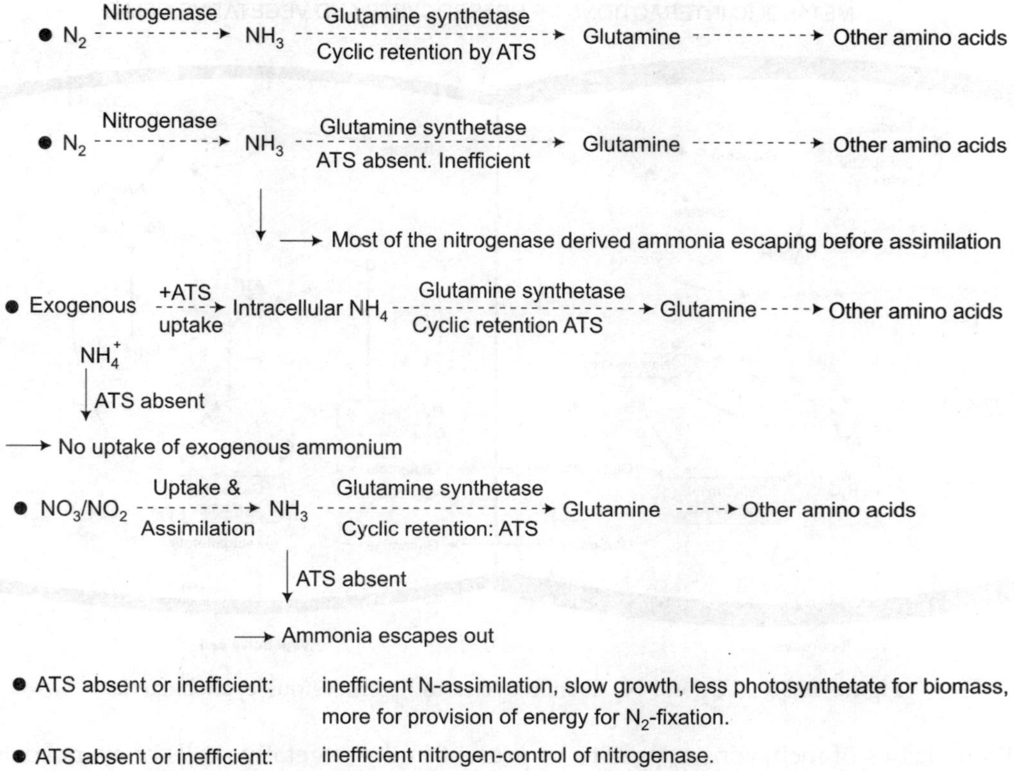

Figure 10.49 Cyanobacterial photobiological nitrogen metabolism.

situations such as light, temperature, salinity, alkalinity and pollution makes them a model photosynthetic prokaryote and study the fundamental process that participates in regulating their adaptation to varying degree of ecological amplitudes. In other words, they must have developed information processing system that enables them to cope with changing environmental conditions and thus make them survive and grow in the given situations (Figure 10.50).

The photosynthetic apparatus (i.e., chlorophyll, carotenoid pigments, photochemical reaction centres and photosynthetic electron chain) is contained in thylakoids – flattened membrane sacs located within the cell.

10.5.9.8 Photosynthetic apparatus

The common nutritrional properties of cyanobacteria have been recognized for many decades. All members of this group are aerobic photoautotrophs; the reactions of oxygenic photosynthesis are used to drive biosynthesis at the expense of inorganic nutrients. In the dark, cyanobacteria display an endogenous respiratory metabolism that is subject to immediate and severe inhibition in the light. The phenomenon of obligate photoautotrophy is widespread among cyanobacteria. However, photochemoheterotrophy and photoheterotrophy has been observed in few strains of cyanobacteria.

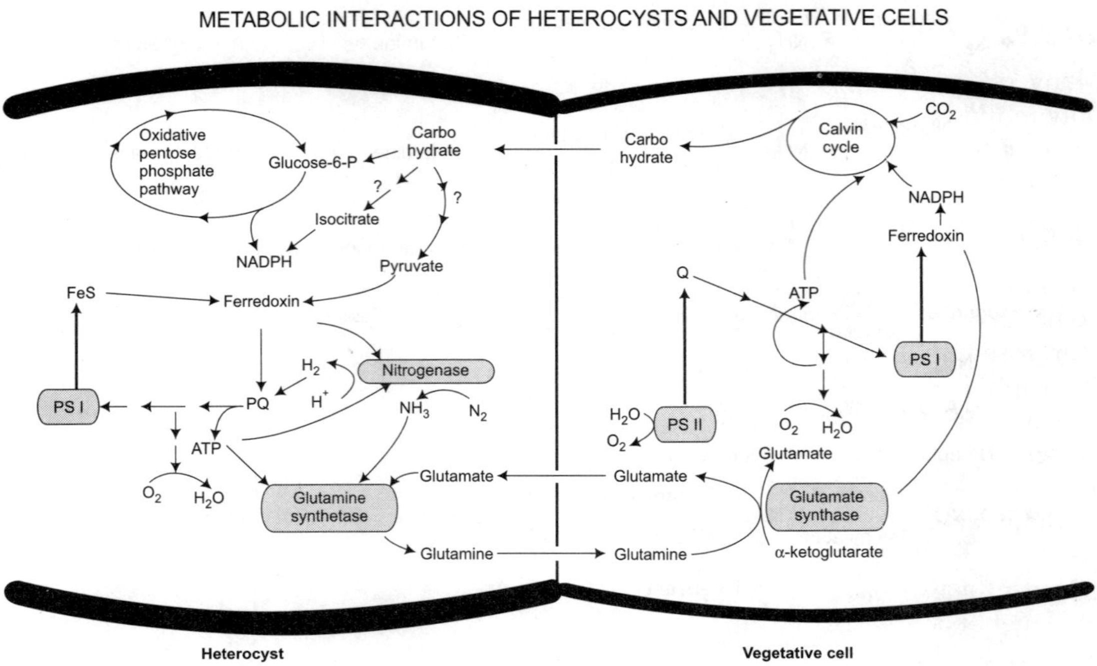

Figure 10.50 CO_2 and N_2 fixation cycles in heterocystous cyanobacteria.

Four classes of inclusions appear to be present in the vegetative cells of cyanobacteria: polyphosphate granules; glycogen granules; cyanophycin granules and carboxysomes. Other inclusions present in the vegetative cells of some cyanobacteria are poly-β-hydroxybutyrate granules and gas vesicles. Among organic source of nitrogen, urea has been found to support growth of certain cyanobacteria . Various amino acids, amides, and amino acids mixtures can serve as a source of nitrogen for other strains.

The cyanobacterial photosynthetic apparatus is located in photosynthetic thylakoids and consists of photosystem I (PS I) and photosystem II (PS II) interconnected with photosynthetic electron transport chain comprising plastoquinone, cytochrome b/f and plastocyanin. The characteristic accessory photosynthetic pigment of the group is predominantly phycocyanin, although in some forms both phycocyanin and phycoerythrin are found, where they play an indispensable role in chromatic adaptation.

10.5.9.9 Anaerobic metabolism

Several cyanobacteria are found in waters that have significant concentration of H_2S (about 5 m mol/L). *Oscillatoria limnetica* have shown that photosystem II is inactive in the presence of H_2S and that an anaerobic, anoxygenic photostnthesis takes place, similar to that is known to occur in the anaerobic purple sulphur bacteria.

$$CO_2 + 2H_2S \rightarrow (CH_2O) + H_2O + 2S$$

The ability of many cyanobacteria to function as facultative anaerobes is probably important in only a few habitats.

10.5.9.10 Obligate photoautotrophy

Many of the cyanobacteria are apparently obligate photoautotrophs; they can grow only in the light. Only a few strains can grow chemo-organotrophically by oxidation of sugars in the dark. In these cases the growth rates are always much lower than under photo-autotrophic conditions.

10.5.10 Prochlorophytes

Prochlorophytes are prokaryotic phototrophs that contain chlorophyll *a* and *b* but that do not contain phycobilins, cyanophycin and poly-β-hydroxybutyrate. Prochlorophytes therefore resemble both cyanobacteria (having prokaryotic cell structure with chlorophyll *a*) and the green plant chloroplast (having chlorophyll *b* instead of phycobilins). Prochloron, the first prochlorophyte discovered, generated obvious excitement and stimulated attempts to trace the evolutionary origin of the chloroplast using molecular approaches to phylogeny. Phenotypically, Prochloron is the best candidate yet discovered for the type of organism that led, following endosymbiotic events, to the green plant chloroplast. Prochloron is really a eubacterium due to the presence of muramic acid in the cell walls, indicating that peptidoglycan is present. DNA genome size is of 3.6×10^9.

Two members of the group are known which are closely related to cyanobacteria. *Prochloron* grows as an exosymbiont on several members of the Ascidia, but hardly grows on any synthetic media. *Prochlorothrix hollandica* is a thread-like organism, free living in fresh water, and can be cultured on mineral medium, in the light.

10.6 ARCHAEBACTERIA

Archaeabacteria was originally thought to be just like bacteria, but archaea is a much different and simpler form of life. They live in extreme environments. These are called extremophiles. Other Archaea species are not extremophiles and live in ordinary temperatures and salinities. It may also be the oldest form of life on earth. Some even live in our guts. Some extremophile species love the heat; they like to live in boiling water, like the geysers of Yellowstone Park, and inside volcanoes. They like the heat so much that it has earned the nickname "thermophile", which means "loving heat", and it would probably freeze to death at ordinary room temperature. Other extremophile Archaea love to live in very salty, called hypersaline, environments. They are able to survive in these extreme places where other organisms cannot. These salt loving Archaea are called halophyles. Archaea requires, neither sunlight for photosynthesis as do plants, or oxygen. Archaea absorbs CO_2, N_2, or H_2S and gives off methane gas as a waste product the same way humans breath in oxygen and breath out carbon dioxide. The relationship between Archaea and eukaryotes remains problematic.

Complicating factors include claims that the relationship between eukaryotes and the archaeal phylum Euryarchaeota is closer than the relationship between the Euryarchaeota and the phylum Crenarchaeota and the presence of archaean like genes in certain bacteria, such as *Thermotoga maritima,* from horizontal gene transfer. The leading hypothesis is that the ancestor of the eukaryotes diverged early from the Archaea, and that eukaryotes arose through fusion of an archaean and eubacterium, which became the nucleus and cytoplasm; this accounts for various genetic similarities but runs into difficulties explaining cell structure.

Phylogenetically, archaebacteria can be divided into four relatively broad groups, one consisting of some methanogens and the extremely halophilic bacteria, one consisting of methanogens only, one containing the extremely thermophilic bacteria, and a final "group" consisting of the genus Thermoplasma. The methanogenic/halophilic group is a physiological enigma. The methanoghens are a large group of obligate anaerobes unified by their ability to make methane. By contrast, the extreme halophiles are generally obligate aerobes and grow only at NaCl concentrations near saturation. The extremely thermophilic archaebacteria are mainly anaerobic organotrophs that are also extreme thermophiles. The overall phylogeny of archaebacteria, as defined by a distance-matrix tree, is shown in Figure 10.51.

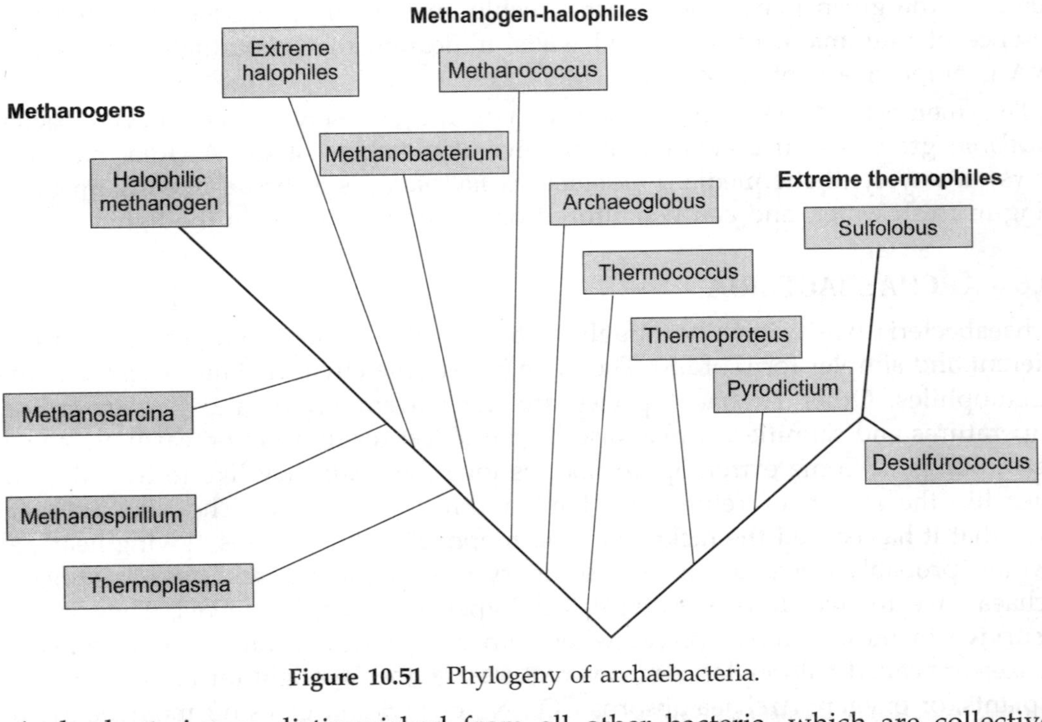

Figure 10.51 Phylogeny of archaebacteria.

Archaebacteria are distinguished from all other bacteria, which are collectively known as the eubacteria by profound differences and are found in rather extreme habitats. These 'extreme condition seem to resemble those supposed to have prevailed in the earliest times of the earth's development, i.e., archaeic times. The archaebacteria include lithotrophic and heterotrophic aerobes and anaerobes. The members of this group have many properties in common that distinguish them from the eubacteria. Within the group, however, there are differences in cell shape, cell components, and metabolism, similar to the differences found among the eubacteria.

According to the present status, the archaebacteria can be divided into three groups: methanogenic bacteria, halophilic bacteria and extreme thermo-acidophilic bacteria.

It is obvious from Figure 10.51, that several subgroups of methanogens exist; this is not surprising considering the phenotypic diversity of methanogenic bacteria. The extremely halophiles prokaryotes (such as *Halobacterium*) and their relatives, group near the methanogens, more specifically near the genus *Methanobacterium*. *Thermoplasma acidophilum*, an acidophilic, thermophilic mycoplasma is also found near the methanogen branch of the archaebacterium. The sequence AYUAAG is a signature for the methanogens while the sequence AAUUAG provides a signature for extreme halophiles. The sequences AAAACUG and ACCCCA are signatures for *Thermoplasma*.

The other branch of the archaebacteria is the extremely thermophilic (sometimes called sulfur-dependent) archaebacteria. Phenotypically the group is united by a variety of striking properties. First, they mostly have growth temperature optima above 80°C including the extremely thermophilic organism, *Pyrodictium*. The latter oorganism has a temperature optima of 105°C. Second, most of these archaebacteria require elemental sulfur for optimal growth. This is because these organisms are for the most part anaerobes that use sulfur as an electron *acceptor*, oxidizing various organic compounds or H_2 as energy sources. One extremely thermophilic archaebacterium, *Sulfolobus*, uses elemental sulfur not as an electron acceptor but as an electron donor with oxygen or ferric iron serving as an electron acceptor. *Archaeoglobus*, an extremely thermophilic sulfate-reducing archaebacterium, may represent a phenotypic bridge between the extremely thermophilic group and the methanogens.

Sulfur plays an intimate role in the biochemistry of extremely thermophilic archaebacteria. This fact, coupled with the thermophilic character of all members of this group, has led to the suggestion that these archaebacteria represent an ancient metaboilic phenotype. Since these extreme thermophiles appear to have evolved more slowly than other archaebacteria and are closer to the root of the tree (Figure 10.51), their phenotype may resemble more closely the phenotype of ancient organisms than does that of other organisms. Such conjecture is consistent with what we know of early earth geochemistry, and the probable environmental conditions within which life arose. The sequence UAACACCAG and CACCACAAG are oligonucleotide signatures for extremely thermophilic archaebacteria. Archaebacteria can be divided into four relatively broad groups, the methanogens, the halophiles, the extremely thermophilic bacteria, and the lone genus Thermoplasma. All of these organisms live in relatively extreme or specialized environments.

Common Properties

The characters common to the archaebacteria concern their cell walls, lipids, transcription and translation apparatus, coenzymes and prosthtic groups, their mechanism of autotrophic CO_2 fixation, and energy supply. Although, there are still many gaps in our knowledge (Table 10.12).

Table 10.12 Some major differentiation properties in Eubacteria, Archaebacteria and Eukaryotes

Characteristic	*Eubacteria*	*Archaebacteria*	*Eukaryotes*
Membrane bound nucleus	Absent	Absent	Present
Muramic acid in cell wall	Present	Absent	Absent
Membrane lipids	Ester linked	Ether linked	Ester linked
Ribosome	70 S	70 S	80 S
tRNA initiator	Formylmethionine	Methionine	Methionine
Inrons in tRNA genes	No	Yes	Yes
Ribosome sensitivity to diphtheria toxin	No	Yes	Yes
RNA polymerases	One (4 subunits)	Several (8-12 subunits each)	Three (12-14 subunits each)
Sensitivity to chloramphenicol, streptomycin, and kenamycin	Yes	No	No

The archaebacteria cell wall does not contain a peptidoglycan skeleton; only proteins and polysaccharides are present and, at best, a 'pseudomurein'. This explains why archaebacteria are not sensitive to antibiotics such as penicillin, cephalosporin, and D-cycloserine that act on the eubacterial cell wall.

The cytoplasmic cell membrane of archaebacteria contains glycerol ethers with C_{20} (phytanyl) and C_{40} (biphytanyl) alkyl isoprenoids in place of the fatty acid glycerol esters. Natural lipids are also found in the form of free C_{15} and C_{30} isoprenoid hydrocarbons.

The DNA-dependent RNA polymerase of archaebacteria differ from those of the eubacteria in that they consist of more than four subunits and are resistant to the antibiotics rifampicin and streptolydigin. The nucleotide sequences of archaebacterial 16S and 5S RNA differ markedly from those of the eubacteria. Translation in archaebacteria is insensitive to chloramphenicol but it inhibited by diphtheria toxin, which is ineffective in eubacteria, diphtheria toxin also inhibits translation in eukaryotes.

Some of the prosthetic groups and coenzymes of the arcaebacteria are distinct, although they may resemble those of the eubacteria or eukaryotes. Some examples are F_{420}, a 5-deazoriboflavin derivative, the nickel-tetrapyrrol factor F_{420} tetrahydromethanopterin, and coenzyme M. These factors were discovered in the methanogenic bacteria. It is not yet possible to make more general statements.

Autotrophic CO_2 fixation does not occur via the ribulose bisphosphate cycle. Methanogenic bacteria fix CO_2 via the acetyl-CoA pathway (Figure 10.52), which is also used by some eubacteria (Table 10.13).

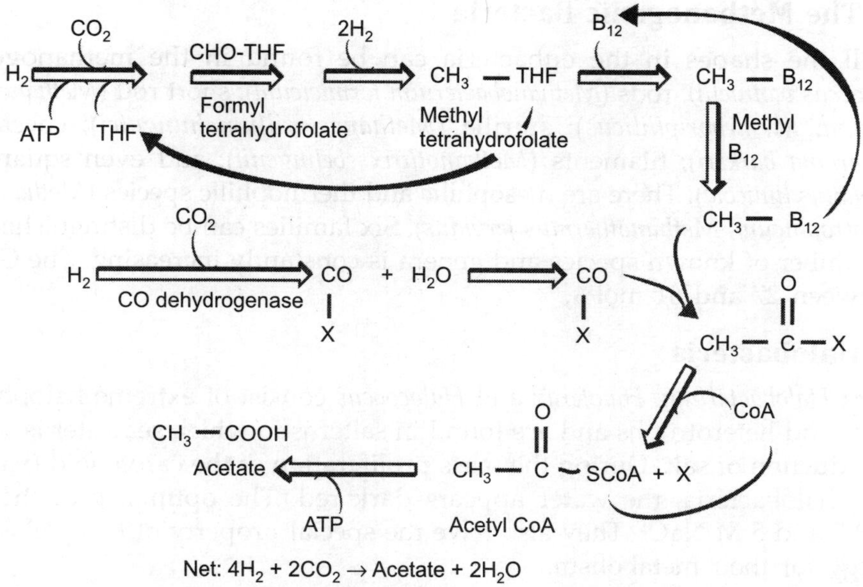

Figure 10.52 Reaction of acetyl-CoA pathway, the mechanism of autotrophy in homoacetogenic, sulfate reducing and methanogenic bacteria THF stands for tetrahydrofolate and B_{12} is vitamin $_{B12}$ in an enzyme bound intermediate. X also represents an enzyme bound form of the compound.

Table 10.13 Organisms employing the acetyl-CoA pathway of CO_2 fixation

I. The energy metabolism of Acetate *Acetobacterium noterae* *Acetobacterium wieringae* *Acetogenium kivui* *Clostridium thermoaceticum* *Clostridium formicoaceticum* *Desulfotomaculum orientis* *Sporomusa paucivorans*
II. Acetate synthesis in autotrophic metabolism Autotrophic homoacetogenic bacteria Autotrophic mcthanogenic bacteria Autotrophic sulfate-reducing bacteria
III. Acetate oxidation in energy metabolism *Reaction:* Acetate + $2H_2O$ $2CO_2$ + 8H Group II sulfate reducers (other than Desulfobacter) *Reaction:* Acetate CO_2 + CH_2 Acetoclastic methanogens (Methanosarcina, Methanothrix)

The process of energy conversion (regeneration of ATP) probably involves the formation of a proton potential gradient and the function of an ATP synthase, and this process could be regarded as a primitive anaerobic respiration. Carbon dioxide, sulphur, and in a few archaebacteria, oxygen function as the electron acceptor.

10.6.1 The Methanogenic Bacteria

Almost all the shapes in the eubacteria can be found in the methanogens: cocci (*Methanococcus vannielii*); rods (*Methanobacterium formicicum*); short rod (*Methanobrevibacter ruminantium, M. arboriphilicus*); spirilla (*Methanospirillum hungatei*); coccal packets (*Methanosarcina barkeri*); filaments (*Methanothrix soehngenii*); and even square bacteria (*Methanoplanus limicola*). There are mesophilic and thermophilic species (*Methanobacterium thermoautotrophicum, Methanothermus fervidus*). Six families can be distinguished already, and the number of known species and genera is constantly increasing. The GC content varies between 27 and 61 mol%.

10.6.2 Halobacteria

The genera *Halobacterium, Haloferax* and *Halococcus* consist of extreme halophiles. They are aerobes and heterotrophs and are found in salterns in which seawater is evaporated for the production of salt. During the mass proliferation of the carotenoid (haloruberin) containing halobacteria, the water appears dark red. The optimal growth range lies between 3.5 and 5 M NaCl. They also have the special property of being able to utilize light energy for their metabolism.

10.6.3 Thermo-acidophilic Bacteria

This group contains nonmethanogenic thermophilic archaebacteria, which do not seem to have many common features. They include autotrophs and heterotrophs, extreme acidophiles and neutrophiles, and aerobes and anaerobes.

Sulfolobus acidocaldarius is found in hot acid springs and oxidizes sulphur to sulphate. The position of *Thermoplasma acidophilus* seems quite isolated. It lacks a cell wall, as do the mycoplasma, but grows optimally at 59°C and pH 1-2. Its usual habitat is in heat-generating coal slag-heaps, but it has also been found in a hot spring. It has the smallest genome so far known in non-parasitic bacteria (1×10^9). It can grow heterotrophically under aerobic conditions in the presence of yeast extract. The above names species collected under the name of Thermoproteales. They were isolated from hot springs, volcanoes, and from the sea floor. They are extremely thermophilic with temperature optima of 85 – 105°C, and their metabolism is of the type described as 'sulphur respiration'. They oxidize hydrogen and reduce elemental sulphur to hydrogen sulphide. The group includes facultative autotrophs (*Thermoproteus tenax*), obligate autotrophs (*Thermoproteus neutrophilus, Pyrodictium accultum*), and heterotrophs (*Desulfurococcus, Thermococcus, Thermodiscus*).

It is now becoming more apparent that the archaebacteria are early deviants from the eubacteria with respect to their cell components. Most of them are probably the progeny of those 'primordial' bacteria that had 'discovered' the utilization of the inorganic hydrogen donors (H_2) and acceptors (CO_2 and sulphur) which were available in the earliest era of development. They probably contributed to the deposition of reduced carbon in sedimentary rocks that occurred some 3×10^9 years ago.

10.6.4 Thermoplasma

Thermopasma belongs to the Thermoplasmata, which thrive in acidic and high-temperature environments. *Thermoplasma* are facultative anaerobes and respire using sulfur and organic carbon. They do not contain a cell wall but instead contain a unique membrane composed mainly of a tetraether lipoglycan containing atypical archaeal tetraether lipid attached to a glucose and mannose-containing oligosaccharide. This lipoglycan is presumably responsible for the acid and thermal stability of the *Thermoplasma* membrane. Currently the genus *Thermoplasma* contains two species, *T. acidophilum* and *T. volcanium.*

Suggested Readings

Alanis AJ. 2005. Resistance to antibiotics: are we in the post-antibiotic era? *Arch Med Res* **36,** 697-705.

Albrecht M, Sharma CM, Reinhardt R, Vogel J, Rudel T. 2010. Deep sequencing-based discovery of the Chlamydia trachomatis transcriptome. *Nucleic Acids Res* 38, 866-877.

Badger JH, Hoover TR, Brun YV, Weiner RM, Laub MT, Alexandre G, Mrázek J, Ren Q, Paulsen IT, Nelson KE, Khouri HM, Radune D, Sosa J, Dodson RJ, Sullivan SA, Rosovitz MJ, Madupu R, Brinkac LM, Durkin AS, Daugherty SC, Kothari SP, Giglio MG, Zhou L, Haft DH, Selengut JD, Davidsen TM, Yang Q, Zafar N, Ward NL. 2006. Comparative Genomic Evidence for a Close Relationship between the Dimorphic Prosthecate Bacteria *Hyphomonas neptunium* and *Caulobacter crescentus. Jour Bacteriol* **188,** 6841-6850.

Baron EJ. 1996. Classification. In: S. Baron et al., eds. *Baron's Medical Microbiology*, 4th edition. University of Texas Medical Branch. ISBN 0963117211

Bisen, P.S., G. B. K. S. Prasad and M. Debnath. 2012. *Microbes Concepts and Applications*, 699. 1st ed. New York, John Wiley and Sons.

Bonde GJ. 1963. Bacterial indicators of water pollution. A study of quantitative estimation. 2nd ed. Teknish Forlog, Copengagen.

Clarridge J. 2004. Impact of 16S rRNA gene sequence analysis for identification of bacteria on clinical microbiology and infectious diseases. *Clin Microbiol Rev* **17,** 840-862.

Delong EF, Pace NR. 2001. Environmental Diversity of Bacteria and Archaea. *Syst Biol* **50,** 470-478.

Emerson D, Weiss JV, Megonigal JP. 1999. Iron-Oxidizing Bacteria Are Associated with Ferric Hydroxide Precipitates (Fe-Plaque) on the Roots of Wetland Plants. *Appl Environ Microbiol* **65,** 2758-2761.

Friedrich B, Schwartz E. 1993. Molecular biology of hydrogen utilization in aerobic chemolithotrophs. *Annu Rev Microbiol* **47,** 351-383.

Kirschner C, Maquelin K, Pina P, Ngo Thi NA, Choo-Smith LP, Sockalingum GD, Sandt C, Ami D, Orsini F, Doglia SM, Allouch P, Mainfait M, Puppels GJ, Naumann D. 2001. Classification and identification of enterococci: a comparative phenotypic, genotypic, and vibrational spectroscopic study. *J Clin Microbiol* **39,** 1763-1770.

Korn-Wendisch F, Kutzner HJ. 1992. The family Streptomycetaceae. In: *Balows A, Truper HG, Dworkin M, Harder W, Schleifer KH (eds), The prokaryotes, Springer-Verlag, New York,* 921-995.

Maslow JN, Brecher SM, Adams KS, Durbin A, Loring S, Arbeit RD. 1993. Relationship between indole production and differentiation of *Klebsiella* species: indole-positive and negative isolates of *Klebsiella* determined to be clonal. *J Clin Microbiol* **31,** 2000-2003.

Montesinos E. 2003. Plant associated microorganisms: A view from the scope of Microbiology. *Int Microbiol* **6,** 221-223.

Munn CB 2004. Marine Microbiology Ecology and Applications, Garland Science/BIOS Scientific Publishers, UK, 85-114.

Murdoch DA. 1998. Gram-positive anaerobic cocci. *Clin Microbiol Rev* **11,** 81-120.

O'Hara CM. 2005. Manual and automated instrumentation for identification of Enterobacteriaceae and other aerobic Gram-negative bacilli. J Clin Microbiol **18,** 147-162.

O'Sullivan DJ. 2000. Methods for Analysis of the Intestinal Microflora. *Curr Issues Intest Microbiol* **1,** 39-50.

Rennie RJ. 1980. Dinitrogen-fixing bacteria: computerassisted identification of soil isolates. *Can J Microbiol* **26**, 1275-1283.

Rittmann BE, Krajmalnik-Brown R, Halden RU. 2008. Pre-genomic, genomic and post-genomic study of microbial communities involved in bioenergy. *Nature Reviews Microbiology* **6,** 604-612.

Shirling EB, Gottlieb D. 1966. Methods, classification, identification and description of genera and species. *Vol. 2, The Williams and Wilkins Company, Baltimore,* 61-292.

Truper HG, Schleifer KH. 2006. Prokaryote characterization and identification. In The Prokaryotes: a Handbook on the Biology of Bacteria, 3rd edn, vol. 1, pp. 58–79. Edited by M. Dworkin,S. Falkow, E. Rosenberg, K. H. Schleifer & E. Stackebrandt. New York: Springer.

Turnbaugh PJ, Ley RE, Hamady M, Fraser-Liggett CM, Rob Knight R, Gordon JI. 2007. The Human Microbiome Project. *Nature* **449,** 804-810.

Waksman SA. 1954. The Actinomycetes. *1st ed, Watham, MASS, USA,* 185-191.

Weindling R, Ingols R. 1956. Microbial Associations and Antagonisms. *Ind Eng Chem* **48**, 1407-1410.

Whitman WB, Coleman DC, Wiebe WJ. 1998. Prokaryotes: The unseen majority. *Proc Nat Acad Sci USA* **95,** 6578-6583.

Woods GL. 2001. Molecular techniques in mycobacterial detection. *Arch Pathol Lab Med* **125,** 122-126.

Young JM, Takikawa Y, Gardan L, Stead DE. 1992. Changing Concepts in the Taxonomy of Plant Pathogenic Bacteria. *Annu Rev Phytopathol* **30,** 67-105.

Chapter **11**

Eukaryotes

11.1 INTRODUCTION

Eukaryotic organisms have complex structures with a highly developed internal membrane and microtubular structure, together with subcellular organelles. Such structures are absent from many prokaryotes, although some prokaryotes do have specialized internal membranes associated with particular metabolic functions. The domain Eukaryota appears to be monophyletic, and so makes up one of the three domains of life. The two other domains, Bacteria and Archaea, are prokaryotes and have none of the above features. Eukaryotes first developed approximately 1.6–2.1 billion years ago. Their collective worldwide biomass is estimated at about equal to that of prokaryotes due to their much larger size.

11.2 FINE STRUCTURE OF THE EUKARYOTIC CELL

The eukaryotic cells are characterized by the possession of one or more discrete nuclei each of which is enclosed by a perforated nuclear membrane unlike prokaryotes. In addition to this important difference eukaryotic cells are more complex than prokaryotic ones in a number of other ways, notably in the possession of mitochondria and vacuoles. Fungi, algae, protozoa and slime molds are all eukaryotic and although the cells of some of them may be less complex than those of higher plants and animals, considerable differentiation may occur in others (Figure 11.1).

11.2.1 Cell Ultrastructure

For the purpose of this account of the ultrastructure of eukaryotic protists in general, it is convenient to consider the different groups of organisms simply in terms of cells. Thus protozoa, slime mold swarm cells, fungal and algal zoospores may be defined as unicellular while the vegetative phases of fungi and many algae are multicellular. Although a large number of organelles are common to all cells, some are found in only a few.

11.2.1.1 Cell walls and the plasmalemma

Cells of the thalli of algae and fungi possess a true cell wall, which is more or less rigid like that of bacteria but is of different chemical composition and architectural structure. Generally, the wall consists of articulate microfibrillar skeleton or framework embedded

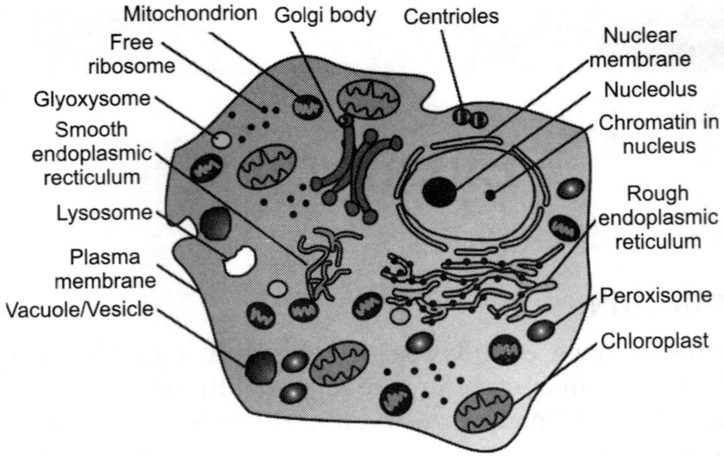

Figure 11.1 Diagrammatic representation of ultrastructure of Eukaryotic cell.

in an amorphous matrix; cellulose has been found to constitute the microfibril layer in green algae.

In most fungi the fibrillar layer is composed of chitin, though cellulose does occur in some groups. Considerable variation is found in the wall structare of cells from different stages in the life cycle of an organism. For example, some fungal spore walls possess complex ornamentations of several layers, and are often pigmented. Similarly, some algal groups have what might be termed as 'wall layer' composed of scales. Scales and other inorganic deposits are common in many protozoa and more typical cell walls are found enclosing the oocyst stages of some of them. Slime layers and filamentous layers are formed at the cell surface in some protozoa but such layers are far removed from the rigid structures of fungi and algae.

Within the cell wall the living protoplast is bound by a membrane, the plasmamembrane or plasmalemma (Figure 11.2). Many protist cells do not possess a true cell wall and are merely enclosed by the plasmalemma. Such cells are often referred to as naked cells.

The electron microscope has shown the chemicaly fixed plasmalemma to be a tripartite structure composed of two electron dense layers separated by an electron transparent layer. Each layer measures approximately 25-30Å in width and the total width of the plasmalemma is 75-100Å. This tripartite structure which has been found to occur in many biological membranes is termed a unit membrane. Numerous pieces of indirect evidence led to the concept of the plasmalemma as a bimolecular leaflet of lipid on to the outer side of which is adsorbed an extended layer of protein.

The bimolecular lipid phase is considered by many workers to consist of molecules with polar (hydrophilic) and non-polar (hydrophobic) ends, the former facing outwards and therefore associated with the protein phase. Detailed studies employing electron microscopy of both chemically fixed and freeze-etched material, together with a consideration of the physical and chemical properties of biological membranes have

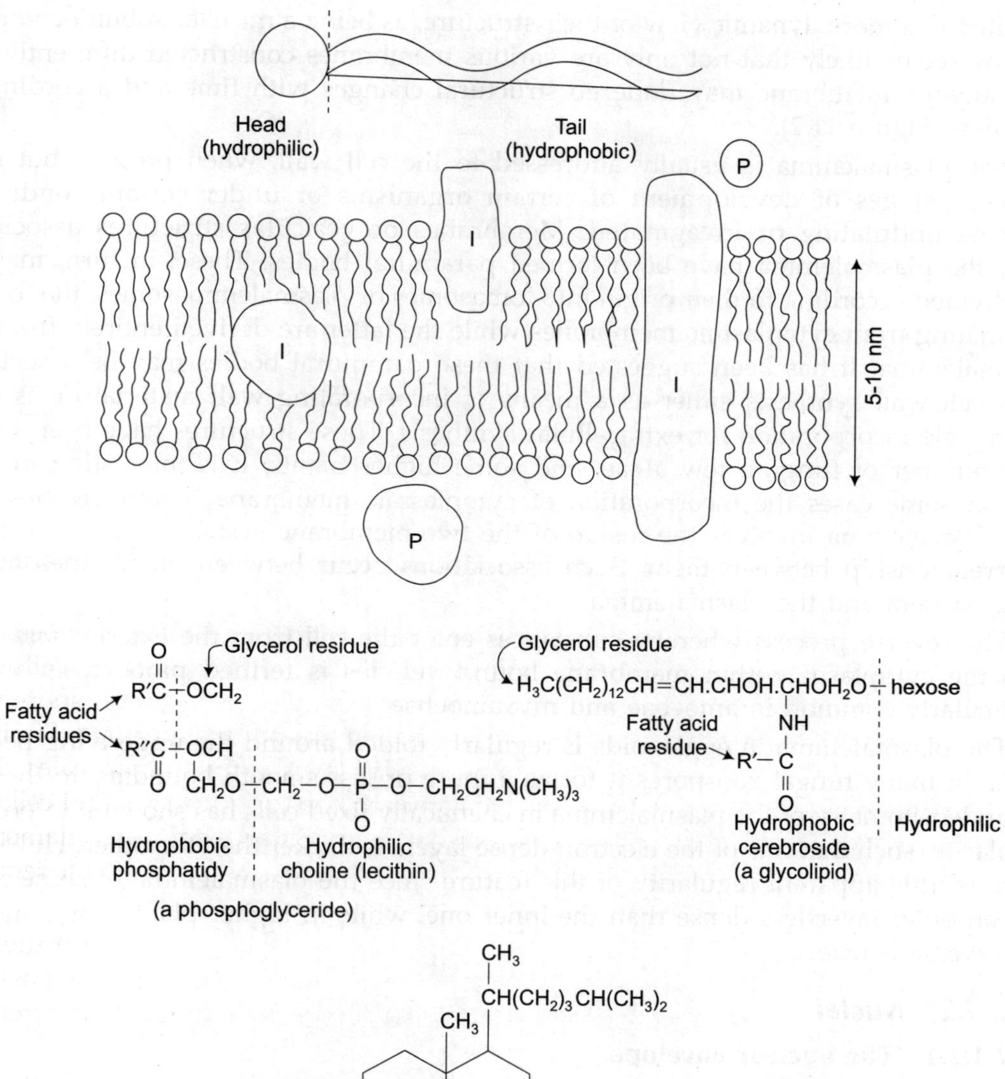

Figure 11.2 Association of protein with the phospholipid bilayer in membrane (P = peripheral protein and I = internal protein) important.

resulted in a more dynamic view of their structure, as being a micellar subunit complex. It now seems likely that not only are various membranes constructed differently but that anyone membrane may undergo structural changes with time and according to function (Figure 11.2).

The plasmalemma is usually addressed to the cell wall, when present, but may at some stages of development of certain organisms or under certain conditions become undulating or invaginated. Membranes or vesicular structures associated with the plasmalemma have been termed paramural bodies. These, in turn, may be subdivided according to their origin into lomasomes or plasmalemmasomes, the former originating from cytoplasmic membranes while the latter are derived entirely from the plasmalemma. It has been suggested that these paramural bodies may be associated with cell wall synthesis either as a means of incorporating wall material or as sites of enzyme incorporation for extracellular synthesis. These structures have been found in a number of fungi, a few algae, and some higher plants. It is interesting to note that in some cases the incorporation of cytoplasmic membrane bound vesicles into the plasmalemma involves the fusion of the two-membrane systems and suggests an interrelationship between them. Such associations occur between membranes of the Golgi system and the plasmalemma.

The reverse process whereby substances enter the cell from the exterior and pass into the cytoplasm within membrane bound vesicles is termed pinocytosis and is particularly common in amoebae and myxamoebae.

The plasmalemma in euglenoids is regularly folded around the underlying pellicle while in many fungal zoospores it forms a more or less smooth boundary to the cell. Microdensitometry of the plasmalemma in chemically fixed cells has shown it to possess a polarity, such that one of the electron-dense layers is thicker than the other. However, there is little apparent regularity of this feature since the plasmalemma of some fungi has an outer layer less dense than the inner one, while in higher plants, for example, the reverse is true.

11.2.1.2 Nuclei

11.2.1.2.1 The nuclear envelope

The nuclear envelope of most eukaryotic protist cells consists of two membranes. These are separated by a space termed the perinuclear space. Each membrane measures 70–90 Å thick and is perforated by a number of pores at the edges of which the two membranes are continuous. Additional membranes are found inside the nuclear envelope in some amoebae.

Freeze etched preparations of yeast nuclei show both open and sealed pores in the envelope and it has been suggested that a regulatory mechanism may exist to control the flow of specific nuclear substances from the nucleoplasm to the cytoplasm.

In some fungal zoospores a double membrane continuous with the outer membrane of the nuclear envelope gives rise to an enclosed structure surrounding part of the nucleus. This structure is the nuclear cap, within which densely packed ribosomes are found. Pores are absent from the nuclear cap membrane.

11.2.1.2.2 Chromosomes

The electron microscope has in general been unsuccessful in contributing to our knowledge of chromosome fine structure. Following glutaraldehyde or osmium tetroxide fixation chromosomes are usually seen as rather amorphous electron-dense structures. In some of the better preparations obtained so far, fine fibrillar components have been observed. In *Amoeba proteus* and *Pelomyxa* the nucleus contains fine threads in the form of regular helices. However, the dimensions of these structures are considerably greater than the double helix of deoxyribose nucleic acid (DNA). The fine structure of chromosomes in the dinoflagellate *Woloszynskia micra* has shown them to be electron-dense banded structures. This banding is apparently due to the arrangement of the DNA fibrils. The division into chromatids begins at one end and proceeds along the length of the chromosome with the intermediate formation of V- or Y-shaped configurations. Chromosomes appear as irregular dense masses in fungi.

11.2.1.2.3 Nucleoli

Nucleoli are seen in suitably prepared cells usually as compact, granular bodies consisting of electron-dense ribosomes.

Spindles and centrioles

Nuclear division in eukaryotic protist cells usually involves a spindle apparatus composed of tubular structures measuring 200-240 Å in diameter. These microtubules are often enclosed within a persistent nuclear envelope which does not break down until the telophase separation of daughter nuclei. This is the case in many fungi, slime molds, and certain flagellate and ciliate protozoa. In many instances, no connections with the chromosomes can be detected with microtubules while in others, some microtubules appear to terminate at or within the chromosome. Usually, continuous microtubules are found passing apparently between the chromosomes and linking the polar ends of the nucleus. The mitotic nucleus of many eukaryotic cells is associated at its polar regions with an astral ray-centriole complex. Astral rays are microtubular elements often seen radiating from the cylindrical centriole. All centrioles have a system of fibres running along their length, which, when viewed in transverse section, are seen to be arranged in a peripheral ring of nine, each composed of a triplet of subfibres. Each triplet is radially tilted in a clockwise direction when viewed from the proximal end. The cylinder of triplets is a structure characteristic of at least part of the kinetosome at the base of flagella in motile cells. Although many eukaryotic cells possess centrioles or their structural equivalents, in some, these organelles are not apparently functionally involved in the polarization of nuclei during mitosis. In addition, other organisms have what is termed anastral and acentric mitosis where neither astral fibres nor centrioles are present at all. However, when centrioles do occur and are actively involved in mitosis, they undergo division at some stage in the mitotic cycle.

In the aquatic fungus *Catenaria angumulae* a single centriole lies at either end of the prophase nucleus. These then divide some time before metaphase so that a pair of centrioles can be seen at the poles. One of each pair lies at right angles to its partner.

Although little is known of the mechanism involved in replication, DNA has been reported to occur in centrioles. It has been suggested that the centriole in certain protozoa, which is associated with complex fibrous structures, may be the organizing centre for protein molecules that are synthesized in the cytoplasm.

11.2.1.3 Flagella and associated organelles

The electron microscope has shown that in all eukaryotic protists possessing motile cell stages, flagella are composed of an axial core, enclosed within a membranous sheath continuous with the plasmalemma. With a few possible exceptions the axial core, or axoneme, consists of two sets of fibres, a peripheral ring of nine and a central pair. Transverse sections show the peripheral fibres to consist of a doublet of sub-fibres, and both these and the central fibres have a dense outer layer enclosing a less dense core.

Cilia of protozoa have an identical axial component to flagella. In some protists, for example dinoflagellates and euglenoids, varying degrees of additional differentiation within the sheath have been found. Dinoflagellates possess two flagella, one posterior, and the other transverse and often situated within a groove running round the cell. Electron microscopy has shown the transverse flagellum to be composed of a fibrous strand of periodic structure in addition to the usual 9 + 2 axoneme and linked with it by dense packing material. Euglenoid flagella in transverse section have their diameters considerably increased by the inclusion of an amorphous or crystalline paraflagellar material within the sheath. An electron dense body at the base of one of the flagella and within the sheath membrane is also found in euglenoid flagellates. This is the flagellar swelling, considered by many to be associated in some way with photoreception. Electron microscope studies on whole cells shadowed with heavy metals have shown that a variety if ornamentations may occur along the length of some flagella. Thus, it has been possible to distinguish between the so-called whiplash flagellum (Peitschengeissel), which is devoid of appendage, and the hairy flagellum (Flimmergeissel) possessing lateral hairs. These lateral hairs have been found to be bipartite in that the distal end is very much more slender than the proximal region. Some flagellates possess both types of flagella on the one cell. In such cases the posterior flagellum is of the whiplash type, while the hairy flagellum is situated anteriorly. This arrangement is termed heterokont flagellation. Scales are found covering the flagella of some algal cells and flagellar spines occur on the motile cells of some brown algae.

A further variation to the structures described above is found in the algal flagellates belonging to the family Haptophyceae . Here a short filamentous projection arises from a position within the cell close to the base of the true flagella and associated with it by striated fibrous connections. This structure, the haptonema, is a tube of three concentric membranes surrounding an inner core of seven fibres. Bridges occur locally between the central core and the haptonema wall. Transverse sections of the basal region of the haptonema show an increase in the number of central core fibres to nine, though in view of the fact that the whole structure grows out from the base, then developmentally there are nine original fibres which decrease with growth until only seven remain.

Associated with flagella and cilia is a number of organelles which lie inside the cell proper, usually at or near the base of these motile appendages. The basal body, or kinetosome, is a cylinder composed of a ring of nine fibres, each fibre being made up of three sub-fibres. This structure is consistent with that of the centriole. However, the kinetosome does not retain this regular structure throughout its length, since a central 'cartwheel' is present usually towards the lower end of the cylinder, while in the kinetosome-flagellum transition region a complex stellate fibre pattern is often seen. In some fungal zoospores this region is marked by the gradual transition from the peripheral ring of nine doublet fibres (as in the axoneme) to nine triplet fibres. The third subfibre arises as an outgrowth of one of the doublet subfibres. Distally, the kinetosome is delimited from the flagellum or cilium by a 'crosswall' of dense material, the terminal plate.

Bands of fibrous material are closely associated with the kinetosome in many motile cells. A network of fibres linking the basal bodies is found in ciliates. This network, the kinetodesma, has been extensively studied in *Paramecium.* In many flagellates, fibres, termed rhizoplasts, link the kinetosome with the nucleus; others, known as parabasal fibres, link the kinetosome with the Golgi apparatus.

11.2.1.4 Cytoplasmic membranes

11.2.1.4.1 Rough and smooth membranes

A variety of membranes is found in most eukaryotic cells. Some of these membranes are studded with dense granules (ribosomes) similar to those found in the nucleolus, while others are smooth. These are commonly referred to as rough and smooth endoplasmic reticulum respectively. The concept of the endoplasmic reticulum (ER) is held by many cell biologists to infer a continuity and interrelationship between all the internal cell membranes. Connections found between the outer membrane of the nuclear envelope and both types of ER suggest that these membranes are part of one interrelated system.

11.2.1.5 Vacuoles

Eukaryotic cells in contrast to those of prokaryotes possess a number of vacuoles. These are usually bound by a single unit membrane termed the tonoplast. Vacuoles of varying shapes and sizes perform a number of functions according to cell type and state. In some fungal and algal cells they serve to localize storage products; in others they contain water. Many protozoa possess a more or less complex vacuole system, the contractile vacuole, which is sometimes connected to a system of smart vesicles by tubular ducts. Food vacuoles are formed by invagination of the plasmalemma; the process of pinocytosis also results in the production of vesicles within the cytoplasm. Such processes are often associated with considerable membrane synthesis.

11.2.1.6 The Golgi apparatus

The Golgi apparatus is the term given to the complex system of smooth membranes present in a wide range of eukaryotes and seen with the electron microscope as stacks of elongated vesicles from the ends of which smaller spherical vesicles may be budded. Golgi system predominantly performs probably secretory function. In certain algal cells vesicles arising from the Golgi system have been found to contain scales which are subsequently deposited around the outside of the cell. In many eukaryotes the Golgi body exhibits a polarity and is associated with the nuclear envelope, ER, or with fibres of the parabasal filament. Since a large number of organisms apparently lack Golgi systems (most higher fungi and ciliates) they cannot be essential cell components. It is conceivable that a simpler system of isolating secretory products within membrane packets may operate in these groups, and the so-called multivesicular bodies found in many eukaryotes may be part of such a system.

11.2.1.7 Ribosomes

Ribosomes are small electron-dense granules preserved in cells fixed with glutaraldehyde or osmium tetroxide. They measure approximately 170-230 Å in diameter and consist of two sub-units which in the functional ribosome must be joined together. By means of fractional centrifugation, ribosomes can be isolated from the cell and subsequent chemical analysis has shown them to contain about 40 per cent protein and 60 per cent ribonucleic acid (RNA). Ribosomes of both prokaryotes and eukaryotes are the sites of protein synthesis and their occurrence and number may be related to the rate of synthesis within the cell. Eukaryotic ribosomes are larger than those of prokaryotes and are situated along the surface of rough ER, the nuclear membrane, inside the nucleus and free in the cytoplasm. Ribosomes are also found in chloroplasts, where they appear to be of the smaller 'prokaryotic type'.

11.2.1.8 Mitochondria

Typical mitochondria are characterized by a double delimiting membrane and an internal differentiation in the form of flattened membranous sacs (cristae) or tubules (tubuli). It is within these organelles that the enzymes of the respiratory system occur. Morphologically, mitochondria undergo considerable changes reflecting not only their functional state but also the metabolic state of the cell as a whole. In many motile cells, mitochondria are often intimately associated with the flagellar apparatus and in the alga *Prymnesium parvum* not only is there a relationship between the haptonema base and a mitochondrion, but also the cristae within this mitochondrion exhibit a marked polarity. Mitochondria are apparently absent from some eukaryotes living in anaerobic environments. Many ideas have been suggested regarding the ontogeny of mitochondria, perhaps the most plausible being that they arise from membrane bound initials derived from the nuclear membrane and containing the necessary genetic information for their subsequent morphogenesis.

11.2.1.9 Plastids, pyrenoids, and eyespots

Plastids are found in algal thallus cells and their motile stages (phytoflagellates). These plastids may be differentiated into chloroplasts, chromoplasts, or amyloplasts containing chlorophyll, carotenoids, and starch respectively. Chloroplasts in phytoflagellates appear to replicate by division of pre-existing ones rather than to develop from undifferentiated proplastids as in higher plant cells. The pattern of replication is probably controlled to some extent by existing chloroplasts. However, there is some evidence to suggest that the initial formation is from the nuclear membrane, during which process the genetic information necessary for the synthesis of RNA and DNA (known to occur within chloroplasts) is included. Pyrenoids are regions of granular, dense material embedded within the chloroplast or arising as an outgrowth of it, and usually associated with the deposition of reserve materials derived from photosynthates. In many chloroplasts, discrete membrane-bound starch granules are found between the lamellae. However, in euglenoids, the reserve material paramylon is formed outside the chloroplasts. Following aldehyde or osmium tetroxide fixation, eyespots are seen with the electron microscope as dense osmiophilic granules. They are considered to be associated with photoreception.

11.2.1.10 Microbodies and storage products

A variety of membrane bound vesicles associated with ER is found in a number of eukaryotic cells. Often crystalline inclusions appear in these vesicles. Their function is obscure but storage of proteinaceous material or of enzymes has been suggested. Glycogen is stored in many mature fungal cells often in the form of dense stellate granules. Lipid droplets are also found in many fungal cells, particularly mature spores.

11.2.1.11 Cytoplasmic microtubules and miscellaneous organelles

Microtubules are found in the cytoplasm of suitably fixed preparations of many eukaryotic protists. They are associated with the cell wall in many algae, while in motile cells they often occur close to the kinetosome. In motile cells of fungi belonging to the order Blastocladiales, there is apparently a close link between the kinetosome fibre pattern and the form and arrangement of microtubules which lies close to the membrane surrounding the nuclear cap. Trichocysts are found in many eukaryotic ciliates and flagellates. When discharged, they consist of a striated fibrous shaft with a spine at the tip. Some trichocysts have a fibre pattern in their tips, similar to that found in cilia and flagella, suggesting their origin from a basal body or kinetosome. Contractile structures or myonemes are found in some protozoa. These are usually bands of fibrous material running around the cell beneath the pellicle. Bands of contractile fibres occur in those diatoms which extrude mucilage which is passed in streams along the outside of the cell. Active transport of this mucilage to one end of the cell by means of these contractile elements, provides the propulsive force necessary for cell movement. One of the most striking features of the eukaryote is the development of membrane systems which effectively compartmentalize the cell, provide sites for enzyme location and chemical reactions, and a1ong which active transport can occur. It is not unusual that many of these membrane bound organelles are similarly constructed and common to a large number of eukaryotes. It is a reflection on the efficiency of the system and the limited range of molecular components involved.

11.3 FUNGI

11.3.1 Introduction

The fungi comprise large group of eukaryotic nonphotosynthetic organisms that include such diverse forms as slime molds, water molds, mushrooms, puffballs, bracket fungi, yeasts and molds. Fungi belong to Kingdom Mycota. The study of fungi is called mycology. It is the oldest discipline in microbiology. In contrast to the algae, the fungi lack chlorophyll. Fungi can be differentiated from bacteria by the fact that fungal cells are usually much larger and contain a nucleus, vacuoles and mitochondria, typical for eukaryotic cells. Although the fungi are a large and diverse group of eukaryotic microorganisms, three groups of fungi have major practical importance: the molds, yeasts and mushrooms sometimes referred as microfungi and macrofungi. Fungi could digest and (re-)mineralize natural organic substances (be they solid, liquid or gaseous, be they dead or part of a living organism) do it with high efficiency and essentially without temporal and spatial limit and also possess flexibility to adapt to new substances, one would end up with a mycelia fungus.

11.3.2 Habit and Habitat

Fungal members are omnivorous, in that they can consume a huge variety of organic compounds. They are capable of unlimited growth so much so that the mycelium of a fungus (*Armillaria bulbosa* growing in forest soil) is capable of yielding amongst the largest biomass of any single living organisms. Their hyphae is capable of transporting nutrients accumulated from rich sources to sites where the same nutrients are in short supply such that the hyphae can continue to extend or differentiate into reproductive structure. Fungi have the means to continue nutrient digestion and absorption during reproduction. They are often remarkably tolerant to external factors like pH, temperature, and dryness. They are able to colonize natural organic substrates on high mountains, sea, heavy metal and pesticide contaminated soil. The effective way of their multiplication is by air dispersal of spores. Overall, fungi are thalloid achlorophyllous, usually filamentous and multicellular, with definite cell wall. They are usually non-motile although reproductive structures are motile.

A prominent differentiation exists between reproductive and somatic structures. They are classified into parasites and saprophytes on the basis of their nutrition. They live on dead matter and are incapable of infecting organisms are called obligate saprobes. Facultative parasites are capable of causing diseases on living or dead organic matter, according to circumstances, they are facultative parasites (or facultative saprobes). Obligate parasites include those group of organisms which essentially require living protoplasm. Apart from saprophytic and parasitic attitude of fungus an interesting relationship exists between fungus and roots of higher plants. These structures are called *Mycorrhiza*.

11.3.3 Thallus Structure

Fungi usually store excess food in the form of glycogen or oil. Unlike bacteria, they prefer acidic medium for growth, usually a pH of 5 to 6. The fungal thallus consists of

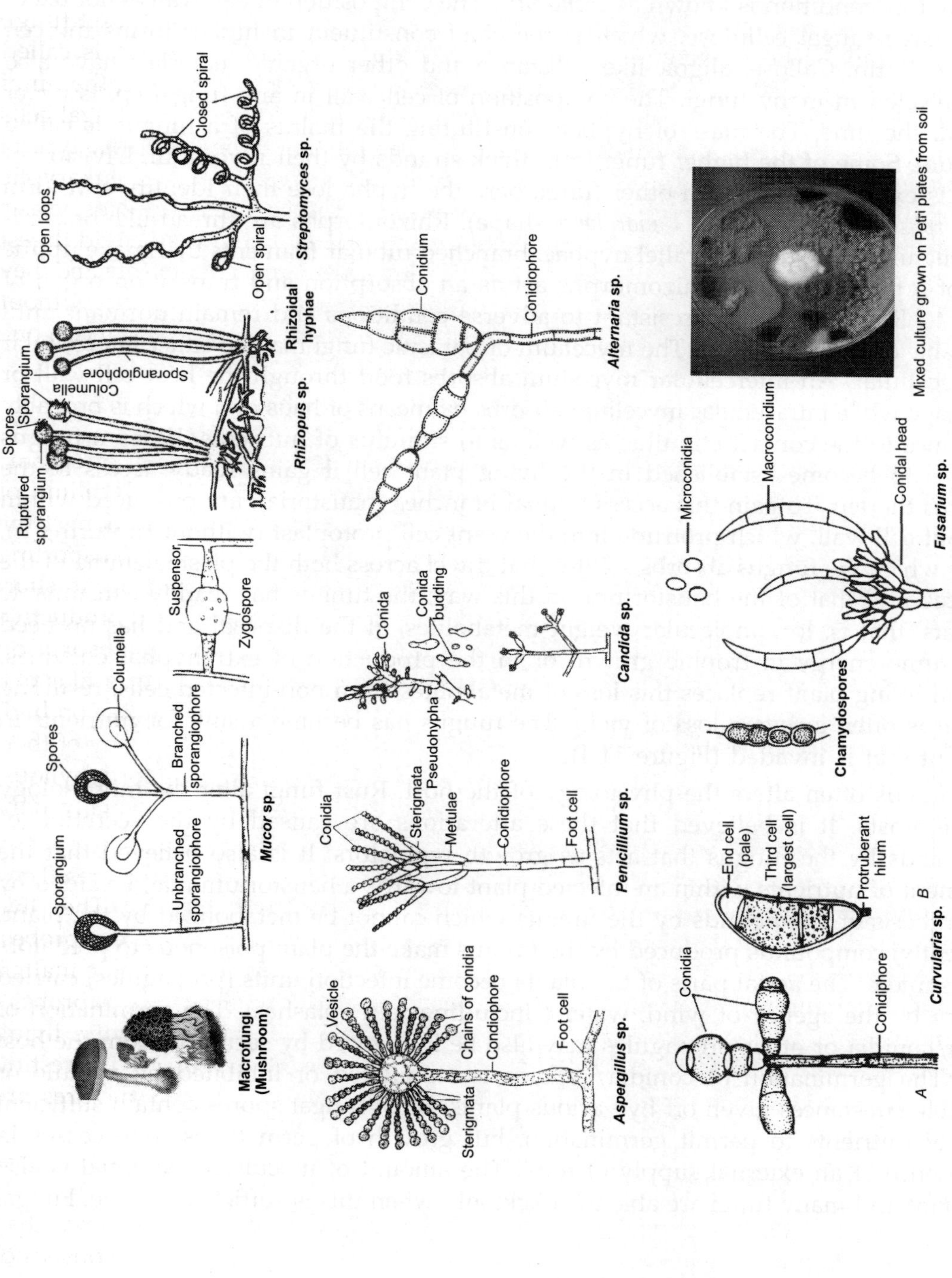

Figure 11.3 Some common soil fungi.

microscopic threads the hypha. A hyphae is of a thin, transparent, tubular wall filled with a layer of protoplasm varying in thickness.

The protoplasm is either continuous, when there are several nuclei due to merging of septa, this condition is known as *coenocytic*. The composition of cell wall is not fixed, it may have fungal cellulose, which is the chief constituent in higher forms the cell wall has chitin. Callose, alignic-like substance and other organic materials have also been detected in many fungi. The composition of cell wall in any fungal sp. is never same all the time. The mass of hyphae constituting the thallus of a fungus is called mycelium. Some of the higher fungi form thick strands by their mycelium. Division of labour is exhibited by certain other fungi, here the hypha lose their identity and form *rhizomorphs* (Gr. *rhiza* = root + *morphe* = shape). Rhizomorphs are thread-like or cord-like structure made up of parallel hyphae, branched tubular filaments that make up the body of a typical fungus. Rhizomorphs act as an absorption and translation organ of nutrients. Rhizomorphs are resistant to adverse conditions and remain dormant until favourable conditions return. The mycelium of parasitic fungi may be either intercellular or intracellular. An intercellular mycelium absorbs food through the host cell wall or membrane while intracellular mycelium absorbs by means of haustoria which is probably a response to the contact stimulus as well as to stimulus of nutrients. When a fungus manages to become established in the living plant cell it gains ready access to the nutrients therein. To gain this access hyphal branches (haustoria), are produced within the plant cell wall, which protrude into the plant cell protoplast (without rupturing it), and by which the fungus absorbs solutes that travel across both the plasmalemma of the plant cell and that of the haustorium. In this way, the fungus has readily consumable nutrients, that is, low molecular weight metabolites, at the disposal and has no need to consume energy in trophic growth or in the production of extrahyphal enzymes. The still living plant replaces this loss of metabolites from non-infected cells, resulting in what is often a severe loss of yield. The fungus has become a sink for nutrients in the plant that is invaded (Figure 11.4).

A fungus often alters the physiology of the host. Rust fungi alter the morphology of their hosts. It is believed that these alterations are caused by the secretion of compounds by the fungus that acts as growth regulators. It is also believed that the movement of nutrients within an infected plant towards a haustorium may be aided by the synthesis of compounds by the fungus which cannot be metabolized by the plant. Frequently, compounds produced by the fungus make the plant poisonous to predators or consumers. The aerial parts of the plants become infection units (propagules) carried to them by the agency of wind, water (including rain splashes). The germination of spores/conidia or other propagules may also be influenced by secretion from the host plant. The germination of conidia/ spores is stimulated or inhibited by volatile or diffusible substances given off by various plants. Most fungal spores contain sufficient essential nutrients to permit germination, but growth of germ tubes soon ceases in the absence of an external supply of food. The amount of inoculums potential is also important and many fungi are able to attack only when this is sufficiently large. Fungal

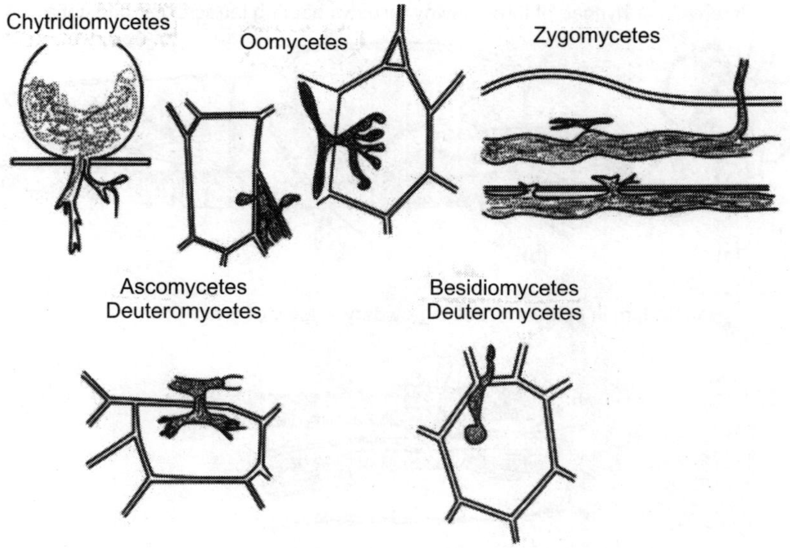

Figure 11.4 Form of haustoria in the various fungal classes.

cell enters the host only through natural openings, such as stomata or lenticels or through wounds. Fungi attach themselves firmly to any hard surface and may form an appressorium at the point of contact. From this fine penetration hyphae grow out and bore through the cuticle and epidermal cell wall (Figure 11.5).

11.3.4 Cell Wall

The vegetative cells and spores of nearly all fungi are surrounded by a definite cell wall. The zoospores and gametes of some primitive species lack such a wall and the vegetative thallus is naked in a group of primitive plant parasites, the Plasmodiophorales, in a number of internal parasites of other fungi, and in species of Coelomomyces, an internal parasite of mosquitos.

The structure of the fungus cell wall is like algae and green plants, and consists of a network of microfibrils with matrix substances filling the spaces between the fibrils. The cell wall of most of the fungi examined consists of 80-90% polysaccharides, with proteins, lipids and sometimes also pigments, polyphosphates and inorganic ions. In at least one yeast *Candida albicans,* the amount of protein is as much as 40% but this is not characteristic of yeasts as a group. The microfibrillar skeleton is frequently chitin, occasionally cellulose, or, in yeasts, non-cellulosic glucans. The cementing matrix usually contains protein, lipids and various polysaccharides. Some of the protein may be enzymic, some may be due to cytoplasmic contamination, but some must be firmly bound to the microfibrils since it is removed only by drastic treatment. The structure of the cell wall lipids is not identical with that of cytoplasmic lipids. The nature and proportion of the polysaccharides vary (Table 11.1 and Table 11.2).

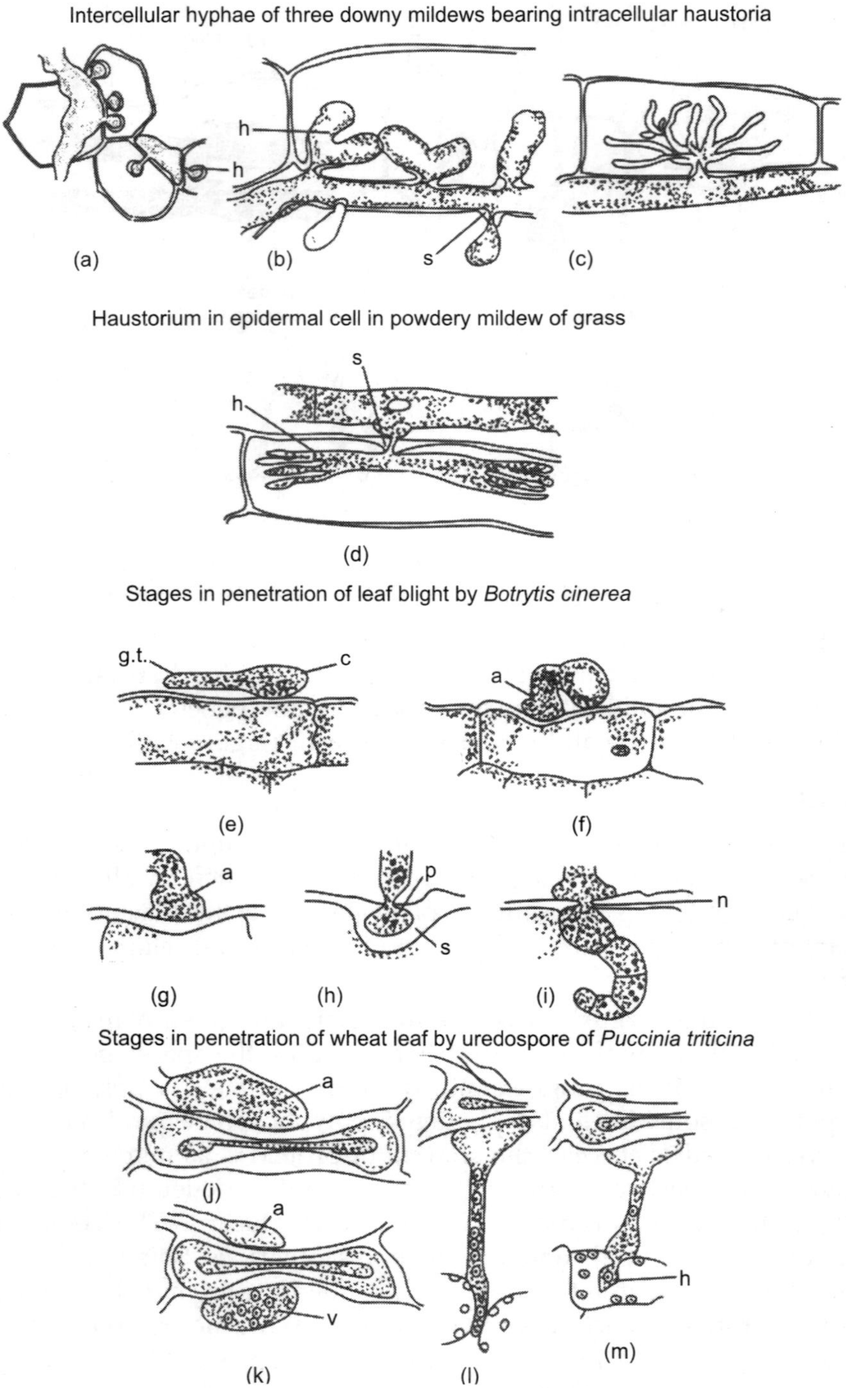

Figure 11.5 Mode of penetration of hyphae in different plant diseases caused by different groups of fungi where h indicates for haustoria, s swelling of host wall at point of penetration, c conidium germination on the surface of host tissue, g.t. germ tube, a appressorium, p penetrating hyphae, n narrow neck of hyphae, v sub-stomatal vesicles.

Table 11.1 Cell wall composition in relation to taxonomy of fungi (Bartnicki-Garcia, 1968)

Chemical category	*Taxonomic group*
Cellulose-glycogen	Acrasiales
Cellulose-glucan	Oomycetes*
Cellulose-chitin	Hyphochytridiomycetes
Chitosan-chitin	Zygomycetes**
Chitin-glucan	Chytridiomycetes Ascomycetes (except yeasts) Deuteromycetes (except asporogenous yeasts) Basidiomycetes (except mirror yeasts)
Mannan-glucan	Saccharomycetaceae
Mannan-chitin	Sporobolomycetaceae Rhodotorulaceae
Polygalactosamine galactan	Trichomycetes

*Unique among filamentous fungi in possessing a cellulosic wall.

** Differ from the filamentous higher fungi in the absence of glucan in the cell walls of vegetative hyphae; glucan has, however, been demonstrated in the spores of *Mucor rouxii* and the mannan content increases with the assumption of the yeast form by thes fungus.

Table 11.2 Cell well compostion and taxonomic classfication of representative medically important fungi (McGinnis and Tyring, 1996)

Principal cell wall polymers	*Taxonomic group*	*Examples*
Chitin-chitosan	Zygomycetes	*Rhizopus arrhizus*
Chitin-glucan	Ascomycetes (mycelia) Basidiomycetes (mycelia) Fungi Imperfecti	*Pseudallescheria boydii* *Schizophyllm commune* *Phialophora verrucose*
Glucanmannan	Ascomycetes (yeasts) Fungi Imperfecti (yeasts)	*Saccharomyces cerevisiae* *Condida albicans*
Chitin-mannan	Basidiomycetes (yeasts)	*Filobasidiella neoformans*

11.3.5 The Protoplast or Cytoplast

The protoplasm is either continuous, when there are several nuclei due to merging of septa, this condition is known as coenocytic. The protoplasts of fungi resemble, but in many ways are less complex than those of animals or green plant. They are, however, definitely eukaryotic in structure and thus considerably more complex than those of the bacteria.

11.3.6 Cell Membrane

The cell membrane of plasmalemma is similar to that of other eukaryotic cells. It is pressed more or less closely to the cell wall, except for certain areas where it curves inwards enclosing a mass of minute tubules or vesicle the lomasome, between it and the wall. Lomasomes were at first thought to be confirmed to fungi but have since been reported in algae and higher green plants. They are variable in form and have been interpreted in various ways. They have not been identified in living cells and could thus be artefacts.

11.3.7 Endoplasmic Reticulum (ER)

In vegetative hyphae, they are usually sparse, compared with that of other eukaryotes, and consist of scattered vesicles or cisternae, seldom showing a complex lamellar structure or well-defined Golgi bodies. Local aggregations of ER occur in some fungi at sites of particularly intense metabolic activity. Ribosomes are usually scattered in the endoplasmic matrix but in some fungi they are associated with the ER as with the granulated or rough ER of higher plants and animals. In the tips of growing hyphae the endoplasm contains no vacuoles, but further back from the tip small stellate vacuoles are present. These coalesce in older parts of the hyphae, the cytoplasm being reduced to a thin peripheral layer lining the cell wall. The coalescence of vacuoles has been suggested as a possible mechanism involved in the initiation and maintenance of cytoplasmic streaming. The presence of vacuoles and the occurrence of streaming are in contrast to their absence in bacterial cell.

11.3.8 Mitochondria

They are present in all actively growing fungal cells but are absent or deformed in some starved or anaerobically treated ones. They often have fewer cristae than have the mitochondria of higher plants and animals, but their shape and complexity varies with the state of activity of the cell.

11.3.9 Nuclei

Interphase nuclei of fungi resemble those of other eukaryotes. Light microscope investigations of stained preparations have demonstrated chromosomes in a large number of genera but although typical mitosis has been reported for many of these, in others the mechanism of division remains obscure.

Despite the increasing fine structural evidence for chromosomes and spindles no clear pattern has emerged as to exactly how the chromosomes separate during anaphase, whether they are attached to the spindle by discrete kinetochores and are pulled apart by contraction of polar fibres, or whether they are pushed apart by elongation of the interzonal microtubules of the central portion of the spindle. It is yet not possible to lay down any rules regarding the state of the nuclear membrane during division of the fungal nuclei. In the majority of the fungi so examined the spindle is intranuclear, the persistent nuclear envelope only breaking down at the telophase separation of the daughter nuclei. In some, however (notably Basidiomycetes), the nuclear envelope breaks down at metaphase leaving a spindle stretching between amorphous densely stained polar regions.

Centrioles have been readily demonstrated at the poles of dividing nuclei in fungi with motile stages in their life cycles, namely, the aquatic Phycomycetes. Ascomycetes, which lack any form of motile stage, have been found to possess what are termed centriolar plaques. These are electron-dense thickened areas closely associated with the nuclear membrane, and lying at the poles of the spindle during division. Similar but somewhat less regular structures have been demonstrated in the Basidiomycetes. Differential staining shows that glycogen and lipid are the commonest reserve foods in the majority of fungi and at least some of the members of Oomycetes differ from other fungi in their lack of glycogen.

11.3.10 Flagella

In contrast to the relatively simple ones of the bacteria, the flagellar apparatus is very complex consisting of a blepharoplast which is the base of the flagellum inside the motile cell, and a rhizoplast through which the flagellum is attached to the nucleus of the cell, the flagellum is composed of eleven parallel fibrils, nine peripheral fibrils forming a cylinder around two central ones. Each fibril is composed of subfibrils. The 9 + 2 pattern of flagellum is followed by algae, moss, ferns, protozoa and sperm cells of higher plants and animals as well. Some lower fungi are motile unicells or have motile phase in their life cycles. The flagella of these are of the complex type typical of the motile eukaryotic cells.

Nutrition

All fungi require a relatively large amount of a carbon source and most use carbohydrates more readily than other carbon compounds. Species and often different strains of a particular species differ in the range of carbohydrates and other carbon sources that they can utilize, but in general glucose and fructose are suitable sources of carbon; glucose is usually included in artificial media for the culture of fungi. In common with other living organisms fungi require a source of nitrogen, potassium, sulphur, magnesium, and a number of micronutrient minerals (trace elements). No fungus is known so far to fix atmospheric nitrogen. Many grow and sporulate well on a glucose salt medium from which they can synthesize all the complex organic substances they require. Others require external supplies of one or more vitamins or growth substances, which they are unable to synthesize themselves. The commonest requirement is for thiamine, but many fungi are deficient for biotin or other growth substances. Some can synthesize enough of such compounds to permit vegetative growth but need an external supply for the initiation of sporulation.

Other environmental factors such as H ion concentration of the substrate, temperature, aeration, light, water content of the substrate, and humidity, may limit fungal growth and/or reproduction. The requirements of different species and of different phases in the life cycle of a given species differ widely, but in general fungi are favoured by a slightly acid reaction (pH 5 to 6) contrary to many bacteria, temperature around 20 to 25°C, good aeration, a moist substratum and a humid atmosphere.

11.3.11 Reproduction

The formation of new individuals having all the characteristics typical of the species is accomplished in fungi by asexual and sexual reproduction. Asexual reproduction does not involve the union of nuclei, sexual reproduction, on the other hand, is characterised by the union of two nuclei. Broad classification of fungi is based on reproductive criteria (Table 11.3).

Table 11.3 Broad classification of fungi

	Criteria for Classification	
Phyla	*Sexual**	*Asexual**
Chytridiomycota	Oospores	Spores and condida
Zygomycota	Zygospores	Spores and condida
Ascomycota	Ascospores	Condida
Basidiomycota	Basidiospores	Condida
Fungi imperfecti	Absent*	Condida

*Sexual propagules (meiosis)

* Asexual propagules (mitosis)

Reproductive organs are formed either in holocarpic (where entire thallus forms reproductive organs) or eucarpic fashion (here the reproductive structures, are formed from only a portion of thallus, the remainder, continues its normal somatic activities.

11.3.11.1 Asexual reproduction

It occurs by the following means:

1. **Fragmentation of hyphae** into their component cells, **oidia** (sing. **oidium**). If cell developes a thick wall before they separate from each other or from the parent hyphae then such spores are called *chlamydospores*. Fragmentation may also be due to external force.
2. **Fission of somatic cells** into daughter cells. This is achieved by constriction and the formation of a cell wall.
3. **Budding of somatic cells or spores:** Here the mycelium forms an outgrowth, the nucleus of the parent cell divides and one daughter nucleus migrates into the bud. The bud increases in size when still attached to the parent cell and eventually breaks off and forms a new individual. At times, chains of bud may be produced (Figure 11.7).
4. **Production of spores:** This is the most common method of asexual reproduction in fungi. Fungal spores occur in infinite variety. Fungal spores produced asexually are either borne at the tips or on the sides of hyphae such spores are referred to as *conidia* (Gr. *konis* = dust + idion). Spores produced on sporangia (Gr. *sporos* = seed, spore + angion = vessel) are called *sporangiospores,* which are either motile or nonmotile.

Motile sporangiospores are called *zoospores (zoon* = animal + *sporos* = seed, spore). If nonmotile, they are called *apalanospores* (Gr. *a* = not + *planetes* = wanderer + *sporos* = seed. spore). Zoospores bear one or two flagella, either of whiplash (lower rigid and

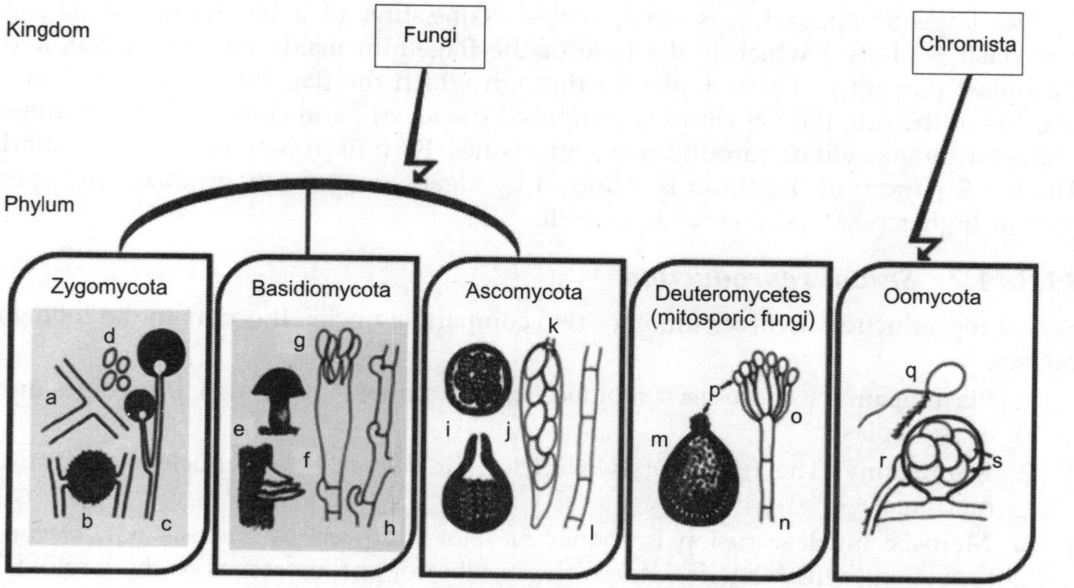

Figure 11.6 Developments in fungal taxonomy.

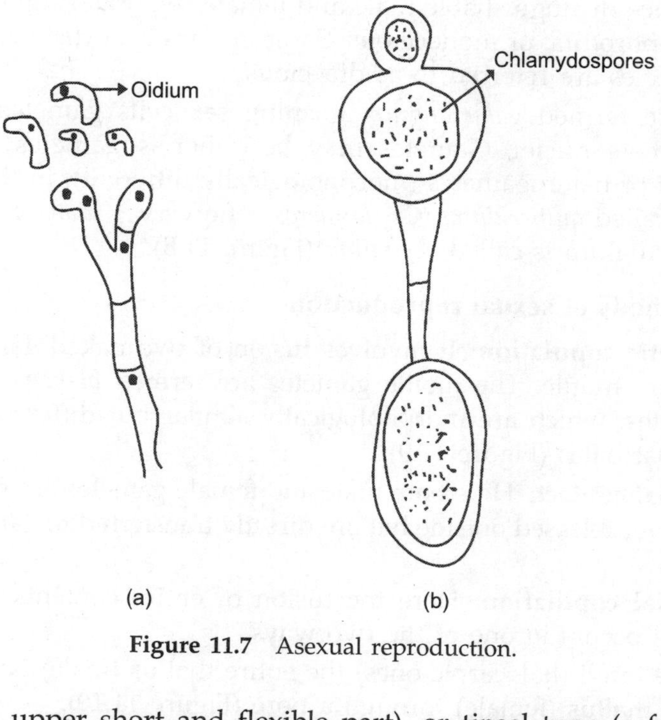

Figure 11.7 Asexual reproduction.

stronger part and upper short and flexible part), or tinsel type (a feathery structure consisting of long rachis with lateral hair like projections on all sides along its entire length.

The flagellar apparatus is very complex consisting of a blepharoplast *(blepharis* = eyelash + plastid) which is the base of the flagellum inside the motile cell, and a rhizoplast (Gr. *rhiztl* = root + plastic) through which the flagellum is attached to the nucleus of the cell, the flagellum is composed of eleven parallel fibrils, nine peripheral fibrils forming a cylinder around two central ones. Each fibril is composed of subfibrils. The 9 + 2 pattern of flagellum is followed by algae, moss, ferns, protozoa and sperm cells of higher plants and animals as well.

11.3.11.2 Sexual reproduction

Sexual reproduction involves union of two compatible nuclei. It occurs in the following phases:

1. **Plasmogamy:** (Gr. *plasma* = molded object + gamos = marriage). It involves union of protoplasts.
2. **Karyogamy:** The fusion of two nuclei (Gr. *karyon* = nut, nucleus + gamos = marriage).
3. **Meiosis:** Nuclear fusion is sooner or later followed by meiosis (Gr. *meiosis* = reduction), which again reduces the number of chromosomes to the haploid.

11.3.11.2.1 Sex organs

Some fungi produce distinguishable male and female sex organs on each thallus. Such species are hermaphroditic or monoceious. Some species bear distinct male and female thallus. These species are referred to as dioecious.

Sex organs are termed *gametangium* forming sex cells gametes. It may contain one or more gamete nuclei. Gametes may be either isogametes (morphologically indistinguishable) or heterogametes (morpnologically different). In the latter, the male gametangium is called *antheridium* (Gr. *antheros* = flowery + *idion,* dinium, suffix) and the female gametangium is called *oogonium* (Figure 11.8).

11.3.11.2.2 Methods of sexual reproduction

1. **Planogametic copulation**: It involves fusion of two naked gametes one or both of which are motile. The motile gametes are termed *planogametes. Anisogamous* planogametes, which are morphologically similar but different in size occur in monoblepharidales (Figure 11.9).
2. **Gametangial contact:** Here both male and female gametes are reduced to nucelus they are never released outside but are directly transferred to gametangium (Figure 11.9).
3. **Gametangial copulation:** Here the fusion of entire contents of two contacting gametangia occurs in one of the two ways.
   (i) In some fungi (halocarpic ones) the entire thallus (male) pours its contents to another thallus (female) through a pore (Figure 11.10).
   (ii) Two gametangial cells fuse directly to form one, the process takes place by dissolution of the walls which are in contact of two gametangia.

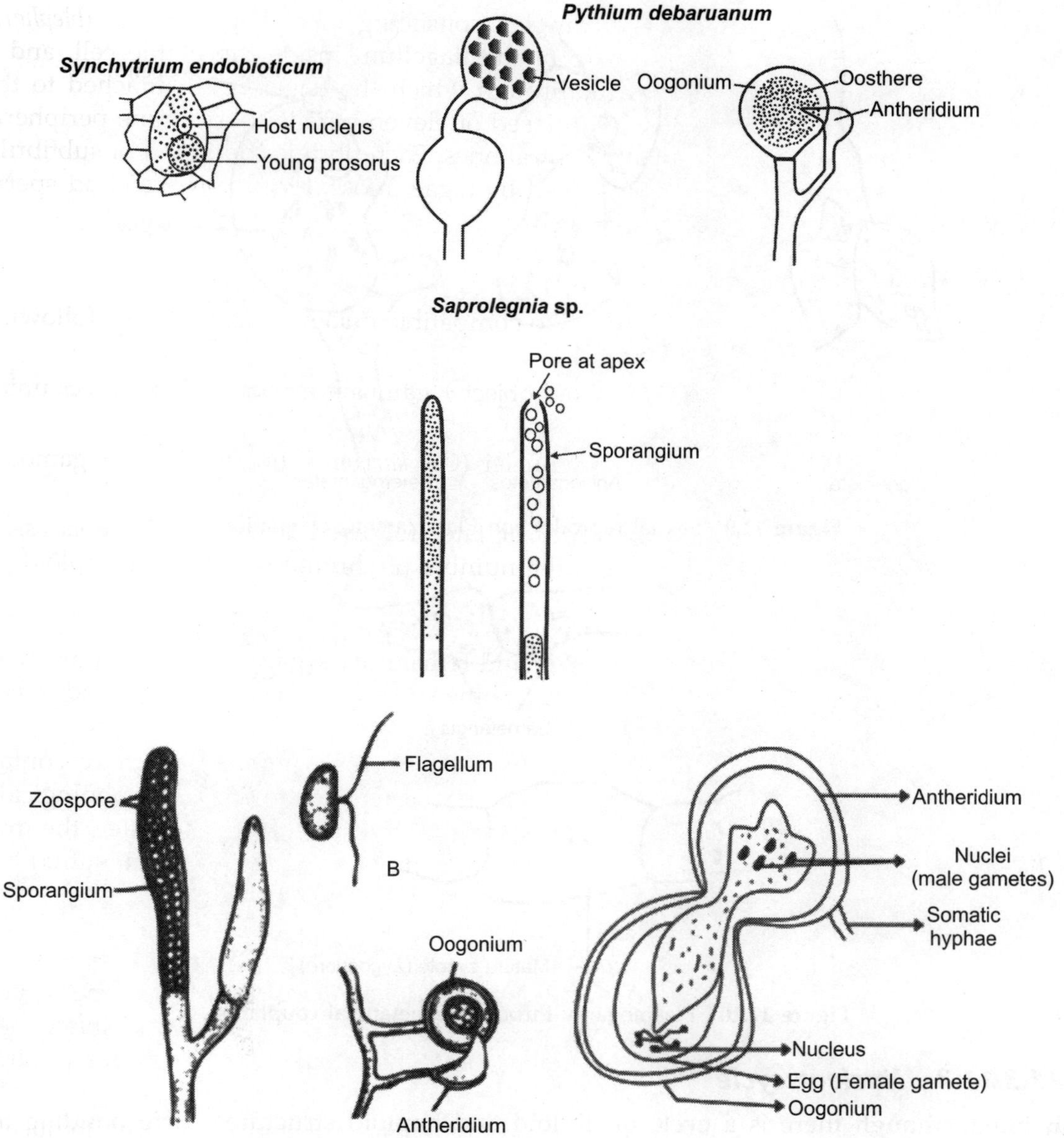

Fig. 11.8 Chytrids and oomycetes members.

4. **Spermatisation:** Few fungi bear numerous, minute, uninucleate, spore-like male structures the spermatia. The spermatia are carried by insects to other receptive hyphae where a pore develops to let the contents of spermatia enter the hyphae (Figure 11.11).
5. **Somatogamy:** It does not involve any sex organ whatsoever is produced by the fungi acts as the sexual organ.

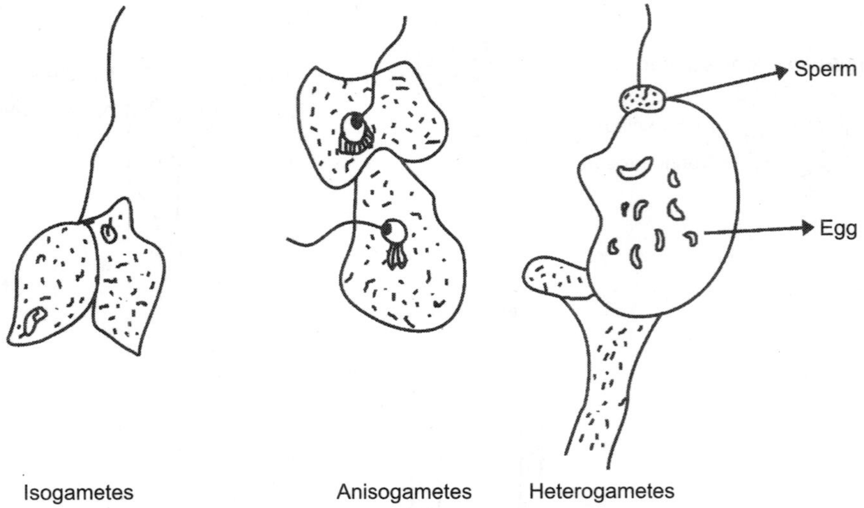

Figure 11.9 Sexual reproduction Planogametic couplation.

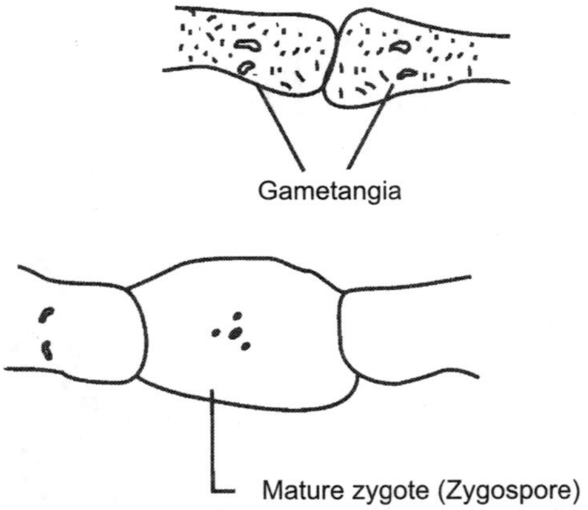

Figure 11.10 Plasmogamy through Gametangial couplation.

11.3.11.3 Nuclear cycle

In fungi, though there is a cycle of diploid and haploid structures corresponding to gametophyte and soporophyte in the green plants, still alteration in generations is not very well demonstrated. The diploid phase occupies very little of the life cycle than the haploid phase.

Heterokaryosis (Gr. *herero* = different, *karyon* = nut, nucleus). The phenomenon involves existence of different kinds of nuclei within the same individual. It may originate in 4 different ways.

On this basis most fungi are classified into 3 categoies.

(a) **Hermaphroditic:** Each thallus bears both male and female sex organs.
(b) **Dioecious:** Some thalli bear only male while others bear only female sex organs.

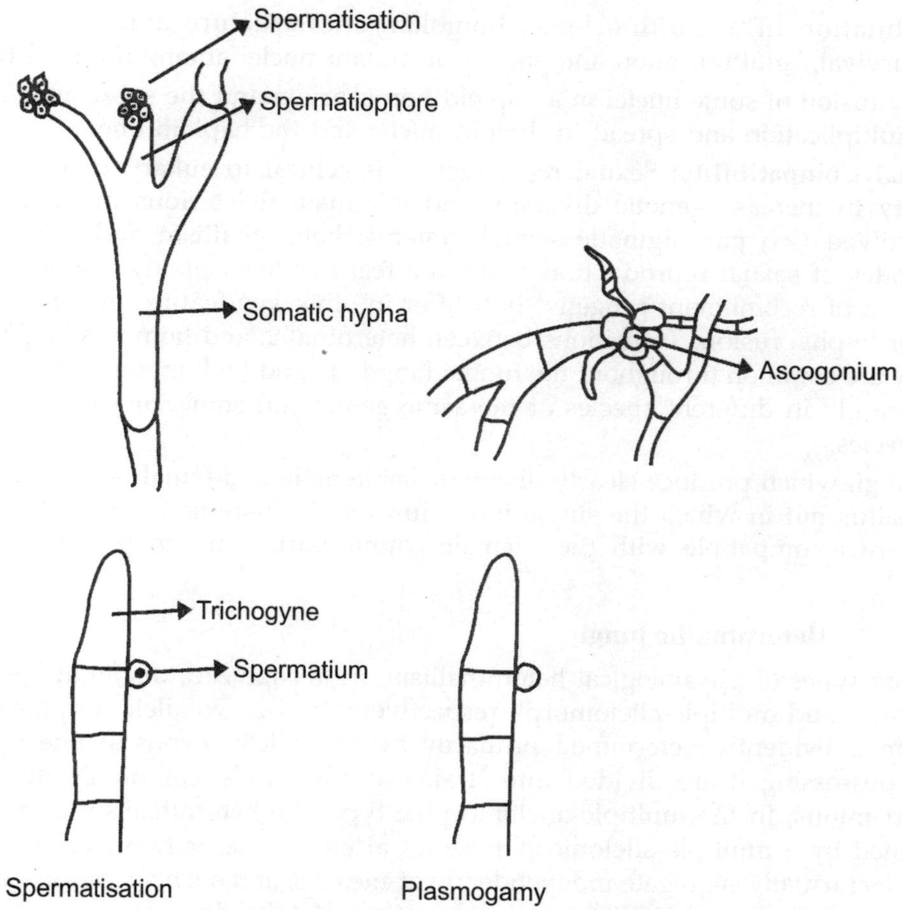

Figure 11.11 Plasmogamy by means of spermatisation.

(c) **Sexually undifferentiated:** Here, sexual functional structure are produced, which are morphologically undifferentiated as male and female sex organs.

Fungi, of the above sex categories, may be homothallic, heterothallic or secondary heterothallic depending on the sexual compatibility.

11.3.11.3.1 Homothallic fungi

Here, every thallus is sexually self-fertile and therefore, reproduces sexually by itself without the aid of another thallus, therefore no dioecious fungus is homothallic. Homothallic species carry both *MAT* genes in a single nucleus, usually closely linked or fused, in contrast to heterothallic species, which have alternate *MAT* genes in different nuclei.

(i) Germination of a heterokaryotic spore, which produces a heterokaryotic soma.

(ii) Introduction of a genetically different nuclei in a homokaryon.

(iii) Mutation in a multinucleate, homokaryotic structure and the subsequent survival, multiplication and spread of mutant nuclei among the wild type.

(iv) By fusion of some nuclei in a haploid homokaryon and the subsequent survival, multiplication and spread of diploid nuclei and the haploid one.

Sexual Compatibility: Sexual reproduction is central to eukaryotic evolution via its ability to increase genetic diversity and eliminate deleterious mutations. Fungi have evolved two paradigmatic sexual systems: heterothallism and homothallism. Both modes of sexual reproduction share key features (e.g., ploidy changes, meiosis, production of recombinant progeny) but differ in other key features involving aspects of cell or hyphal fusion. Transitions between heterothallic and homothallic patterns of sexuality are common throughout the fungal kingdom, and both modes can be observed concomitantly in different species of the same genus and sometimes even within the same species.

In fungi, which produce clearly distinguishable male and female sex organs on the same thallus but in which the single individuals are self-sterile because the male sex organs are incompatible with their female counterparts thus no plasmogamy takes place.

11.3.11.3.2 Heterothallic fungi

Two main types of physiological heterothallism are recognized, distinguished as two-allelomorph and multiple-allelomorph respectively. In the two-allelomorphic type, the condition is evidently determined primarily by two allelomorphs at one locus, and species possessing it are divided into strains of two kinds, commonly described as plus and minus. In the multiple-allelomorphic type, the heterothallism appears to be determined by a multiple-allelomorphic series at either one or two loci. In the latter, the two loci usually segregate independently at meiosis, and such species are described as tetrapolar, since each individual bears spores of four different mating types, in contradistinction to the bipolar condition of species with one locus for heterothallism. In bipolar and tetrapolar forms, the incompatibility factors control similar, if not identical, morphogenetic processes. Differentiation involves: the formation of fusion cells by two mates, and then, if the mates are compatible, the reciprocal exchange of nuclei, migration and proliferation of the exchanged nuclei within the pre-existing mycelia, nuclear pairing of migrant and resident nuclei in a 1 : 1 ratio within cells, and a complex synchronous process of nuclear and cell division involving clamp-connections.

11.3.11.3.3 Secondary homothallic fungi

In some bipolar heterothallic fungi, during spore formation two nuclei of opposite mating type are incorporated regularly in each spore. Secondary homothallism depends upon a mechanistic modification of heterothallism due to the parcelling of compatible nuclei into individual spores. Two fundamentally different means of achieving self-fertility in basically heterothallic species have been described in yeasts. In species of the genus Saccharomyces, several investigations suggest that self-fertility arises in clones of heterothallic cells through mutation of one mating type allele to the other (only two mating type alleles are known in Saccharomyces spp.). After the mutation occurs, compatible cells exist within the clone, and sexual interaction ensues.

11.3.11.3.4 Parasexuality

(Gr. *para* = besides + sex) a process peculiar to fungi and single-celled organisms, is a nonsexual mechanism for transferring genetic material without meiosis or the development of sexual structures. A parasexual cycle is initiated by the fusion of hyphae (anastomosis) during which nuclei and other cytoplasmic components occupy the same cell (heterokaryosis and plasmogamy). Fusion of the unlike nuclei in the cell of the heterokaryon results in formation of a diploid nucleus (karyogamy), which is believed to be unstable and can produce segregants by recombination involving mitotic crossing-over and haploidization. Mitotic crossing-over can lead to the exchange of genes on chromosomes; while haploidization probably involves mitotic nondisjunctions which randomly reassort the chromosomes and result in the production of aneuploid and haploid cells. Like a sexual cycle, parasexuality gives the species the opportunity to recombine the genome and produce new genotypes in their offspring. Unlike a sexual cycle, the process lacks coordination and is exclusively mitotic. Here plasmogamy, karyogamy and haploidisation (reduction division) take place, but not at specified points in the thallus or in the life cycle.

11.3.12 Life Cycle

11.3.12.1 Chytridiomycota

Chytrids are a morphologically diverse group with approximately 1000 known species. Many are saprophytic. For example, *Rhizophlyctis rosea* is a commonly encountered decomposer of cellulose in soils. Others are parasites on plants, insects, and certain amphibians. The Mastigomycotina fungi in which the spores and gametes are motile; in some systems placed in the Phycomycetes group with the Zygomycota. They may form branched chains of cells that attach to their substrate by a root-like structure called a rhizoid. Many are soil saprophytes where they are found as important decomposers. Alternatively, they are found in freshwater habitats, and may be associated with water that is polluted with sewage. Some species are found as parasites of plants or algae, and a few are parasites of insects or fish. The downy mildews are obligate parasites unable to grow in standard laboratory cultures. Asexually, Chytridiomycota reproduce through the use of zoospores. In asexual reproduction, zoospores will swim until a desireable substrate is located. The zoospore attaches itself, feeds off its host; the cytoplasm grows, meiotic divisions occur, and a cell wall forms around the original zoospore. Protoplasm increases as the cell continues to develop. Finally, cleavage of the protoplasm occurs, which produces individual zoospores that are released through a pore. Sexual reproduction is haploid dominant. It also depends on the isomorphic alternation of generations. The haploid thallus, called the gametothallus, produces female and male gametes. These occur in pairs and are terminal and subterminal. Male gametes are orange coloured, while female gametes are colourless. In addition, female gametes are much larger than male gametes. Males are attracted to females when they produce the hormone sirenin, and females are attracted to males when they produce the hormone parisin.

The diploid thallus is called the sporothallus. The sporothallus produces two types of zoosporangia: zoosporgangium (meitosporangium) and resistant sporangium (meiosporangium). Zoosporangia produce diploid zoospores, which can function as a means of asexual reproduction. Sexual reproduction may be isogamous, anisogamous, or oogamous. One species, *Allomyces macrogynus*, has a sporic life cycle, something that occurs in plants but is rare to fungi. *A. macrogynus* is an example of an anisogamous species (Figure 11.12).

So what do chytrids tell us about the origin of fungi? First of all, chytrids are predominantly **aquatic**, and not terrestrial. This means that fungi probably got their start in the water, as did plants and vertebrates.

Secondly, chytrids have **flagellated gametes** their reproductive cells have a flagellum that allows them to swim. No other fungi have flagella, which suggests that the other fungi lost this trait at some point in their evolutionary history. This is also consistent with what we know about the closest relatives to the fungi, which also have flagellae.

Finally, like other fungi, chytrids have **chitin** strengthening their cell walls, and one subgroup (Hyphochytrids) have cellulose as well, a trait unique among living fungi. The presence of chitin is thus an important defining feature of the fungi.

There is considerable variation in the morphology and ecology of chytrids. Some are freshwater, some marine; some are parasites on plants and dipterans, while others live on decaying plants and insect parts. Some are unicellular, some coenocytic, and still others produce a **mycelium** much like other fungi. Few have any noticeable impact on humans, with the exception of a few that parasitize algae, cause potato wart (*Synchytrium endoboticum*), and those used in experimental research (e.g., *Allomyces*).

The oldest fossil fungi so far known are chytrid-like forms from the Vendian of northern Russia. Older fossils of Precambrian "fungi" are now usually considered to be empty sheaths of filamentous cyanobacteria, or else are not distinct enough to be placed in any taxon with certainty. Amazingly fine chytrid fossils are known from the Devonian Rhynie Chert, where they occur alongside representatives of other major fungal groups. Some Devonian forms are parasitic on rhyniophytes and closely resemble the modern genus *Allomyces* (Blastocladiales), both in form and in reproductive features of the life cycle. Other Rhynie chytrids resemble members of the Spizellomycetales, suggesting that chytrids had diversified by this early date.

The group includes important plant pathogens, such as *Phytophthora infestans.* This is the cause of potato blight. In the nineteenth century, Irish peasants used the potato as a primary food supply. When potato blight afflicted the potato crop in Ireland, the disease was responsible for the death of more than two million people during the Irish Potato Famine.

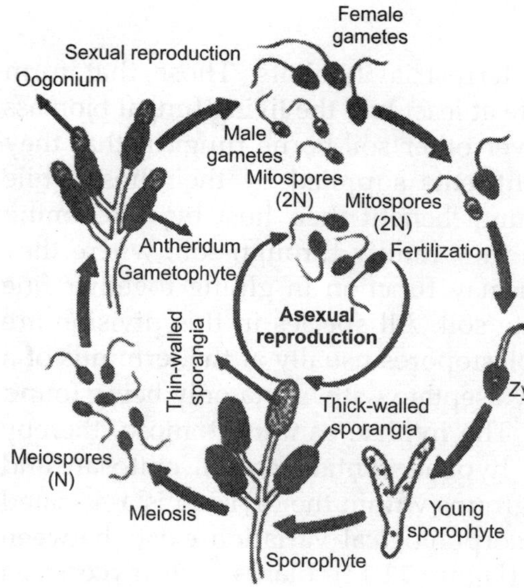

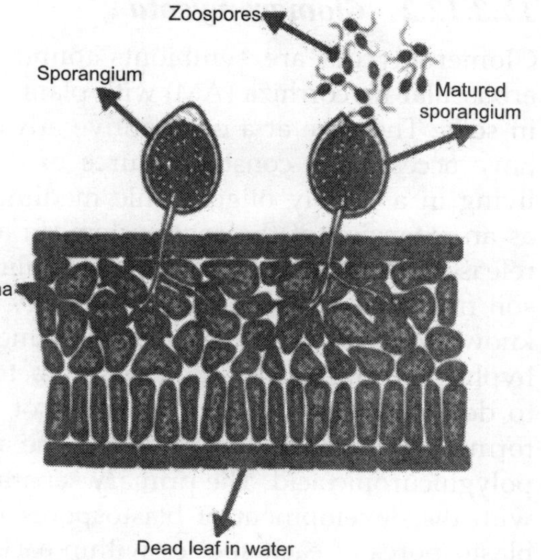

Chytridiomycetes

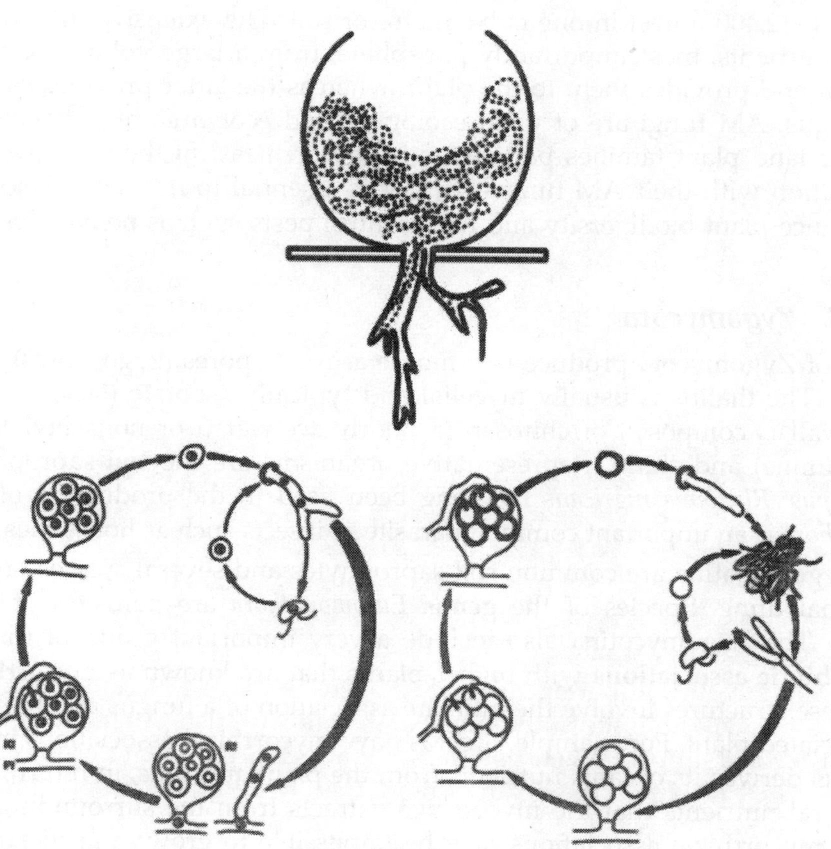

Figure 11.12 Life cycle of a member of Chytrides in which zoospores are formed.

11.3.12.2 Glomeromycota

Glomeromycota are symbionts found only in terrestrial habitats. Those that form arbuscular mycorrhiza (AM) with plants constitute at least half the living fungal biomass in soils. They are at a competitive advantage over other soil-borne fungi in that they have access to a constant source of organic nutrients supplied by their host while living in a largely oligotrophic medium. The fungi benefit their host by functioning as an extension of the root system. In addition, they ramify through soil, where they release glycoprotein, including glomalin, which may function in gluing together fine soil fragments. Hyphae also physically aggregate soil. All species in this division are known to reproduce asexually. The fungi form blastopores usually at the terminus of a hypha. Mycelia of AM fungi are characteristically aseptate with, septa only being found to delimit spores or dying portions of hyphae. The hyphae can anastomose, thereby forming interconnected mycelia. The walls of hyphae contain chitin, chitosan and polyglucuronic acid. The primary separation of groups within the division is associated with the development of blastospores. Minor morphological variation exists between blastospores of each species within each family (Figure 11.13). Plants with mycorrhiza are particularly successful in infertile soils. A mycelial network may extend to a length 20,000 km (~12,400 miles) in one cubic metre of soil. The extensive mycelium absorbs inorganic nutrients, most importantly phosphate, from a large volume of soil or other substratum and provides them to the plant, whereas the latter provides carbohydrates to the fungus. AM fungi are of great ecological and economic importance. Over 80% of vascular land plant families participate in mycorrhizas. In the case of some plants, the association with their AM fungal partner is essential to normal development. AM fungi enhance plant biodiversity and help control pests such as nematodes and fungal pathogens.

11.3.12.3 Zygomycota

Members of Zygomycota produce non-motile asexual spores (zygospores) formed in a sporocarp. The thallus is usually mycelial and typically aseptate (lacking cross-walls). The cell wall is composed of chitosan (a poorly acetylated or nonacetylated polymer of glucosamine) and chitin. Representative organisms are the soil saprophytes *Mucor* and *Rhizopus*. *Rhizopus nigricans* has long been used in the production of citric acid. *Entomophthora* is an important common parasite of insects such as house flies and aphids.

The Zygomycotina are common soil saprophytes and several species are associated with animal dung. Species of the genus *Entomophthora* are parasites of aphids and houseflies. The Zygomycotina also include a very important group of fungi that can form **symbiotic associations** with higher plants that are known as **mycorrhiza** (Figure 11.14). These structures involve the intimate association of a fungus and the root system of its associated plant. For example, orchids have mycorrhizal associations in their roots. The fungus derives its organic nutrients from the plant and thus, in return, is provided with mineral nutrients that the mycorrhiza extracts from the surrounding soil. Fungi that form mycorrhizal associations may be impossible to grow in artificial culture.

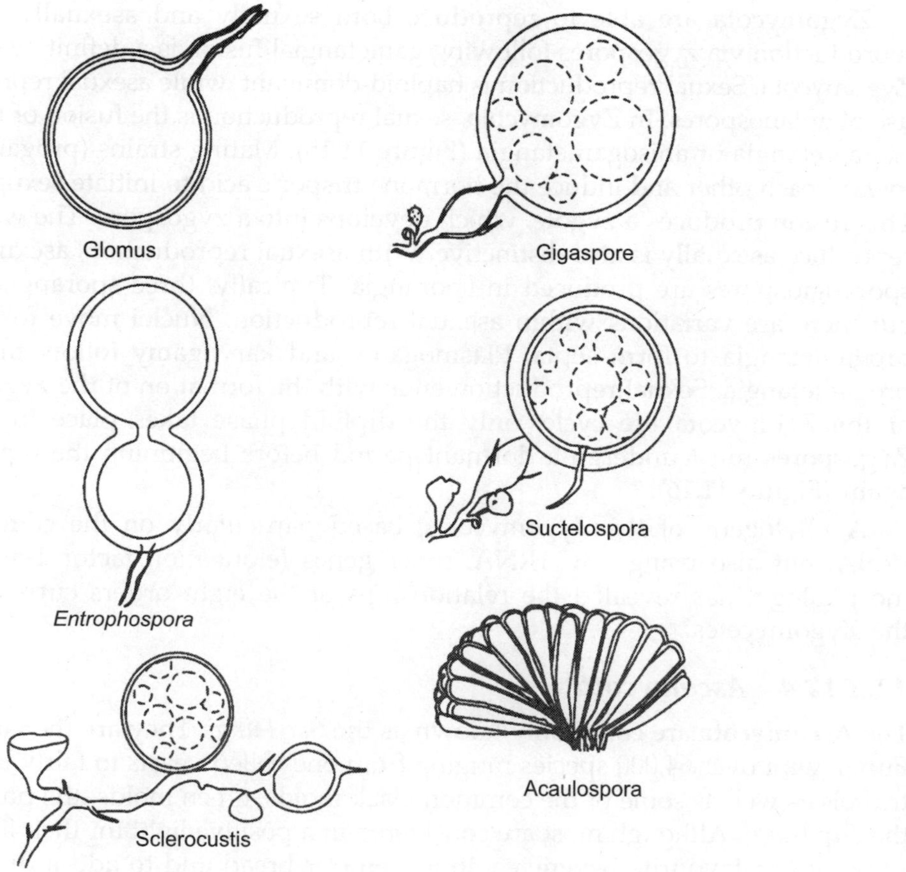

Figure 11.13 Diagrammatic view of AM fungi.

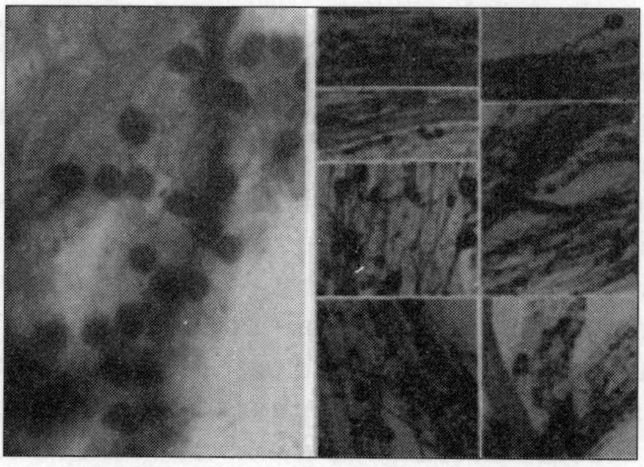

Figure 11.14 Symbiotic mycorrhizal association of Zygomycotina fungus with vascular plant.

Zygomycota are able to reproduce both sexually and asexually. In fact, sexual reproduction via zygospores following gametangial fusion is a definitive characteristic of Zygomycota. Sexual reproduction is haploid-dominant, while asexual reproduction makes use of aplanospores. In Zygomycota, sexual reproduction is the fusion of undifferentiated isogametangia or anisogametangia (Figure 11.15). Mating strains (progametangia) grow toward each other and induce the hormone trisporic acid to initiate sexual development. This fusion produces a zygote, which develops into a zygospore. The way Zygomycota reproduce asexually is also distinctive. With asexual reproduction, asexual spores called sporangiospores are produced in sporangia. Typically, three sporangia are produced, but there are variations within asexual reproduction. Nuclei move to the ends of the progametangia to form septa. Plasmogamy and karyogamy follow the fusion of the progametangia. Sexual reproduction ends with the formation of the zygospore. Because of the Zgomycota life cycle, only the diploid phase takes place in the zygospore. Zygospores must undergo a dormant period before beginning the reproductive cycle again (Figure 11.16).

A phylogeny of the Zygomycetes, based particularly on the comparison of SSU rRNA, but also using LSU rRNA, other genes (elongation factor 1-alpha [EF-1a]) or morphology, has revealed the relationships of the eight orders currently accepted in the Zygomycetes.

11.3.12.4 Ascomycota

The **Ascomycota** are commonly known as the *Sac Fungi*. They are the largest phylum of Fungi, with over 64,000 species ranging from one-celled yeasts to fairly large morels and truffels as well as some of the common black molds, green molds, the powdery mildews, the cup fungi. Although most are considered in a positive light for their ability to ferment some of our favourite beverages, to leaven our bread and to add a distinctive ping to some cheeses, others are more notorioius as disease promoting organisms (chestnut blight, Dutch Elm disease, forming rots of a number of fruit trees, powdery mildews etc.). The mycelium, when present, is composed of septate hyphae that are multinucleate with pores within the septata.

The defining feature of this fungal group is the "ascus" (from Greek: ασκός (*askos*), meaning "sac" or "wineskin"), a microscopic sexual structure in which nonmotile spores, called ascospores, are formed. However, some species of the Ascomycota are asexual, meaning that they do not have a sexual cycle and thus do not form asci or ascospores. Previously placed in the Deuteromycota along with asexual species from other fungal taxa, asexual (or anamorphic) ascomycetes are now identified and classified based on morphological or physiological similarities to ascus-bearing taxa, and by phylogenetic analyses of DNA sequences. The **dermatophyte** fungi that cause diseases such as ringworm and athlete's foot are classified as members of the Ascomycotina if the sexual reproductive stage has been identified. However, many dermatophytes are classified as Deuteromycotina because their sexual reproductive cycle has not been observed and described.

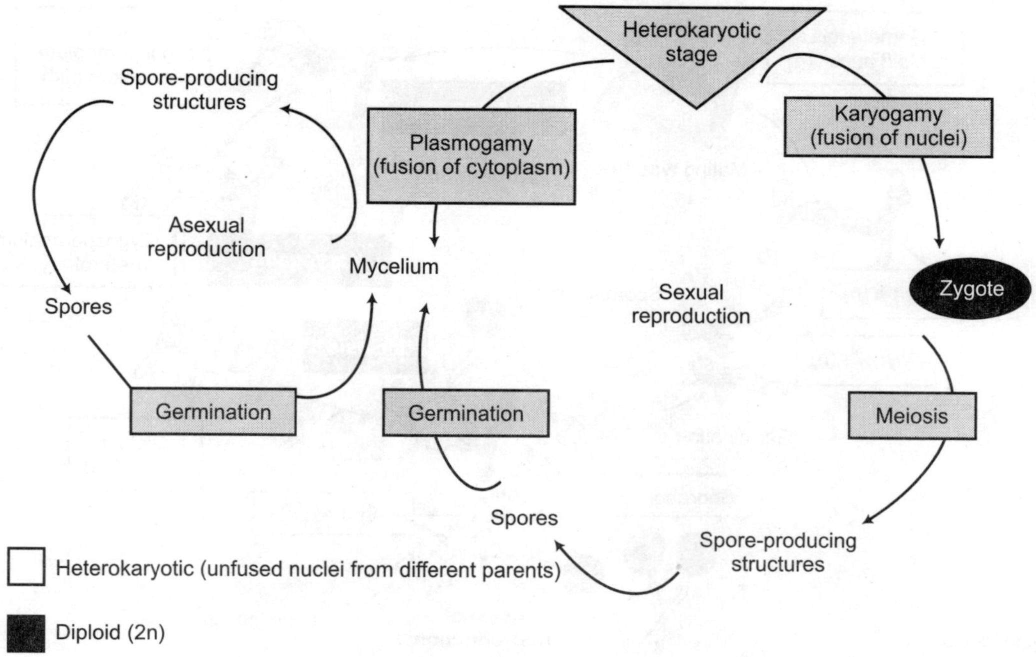

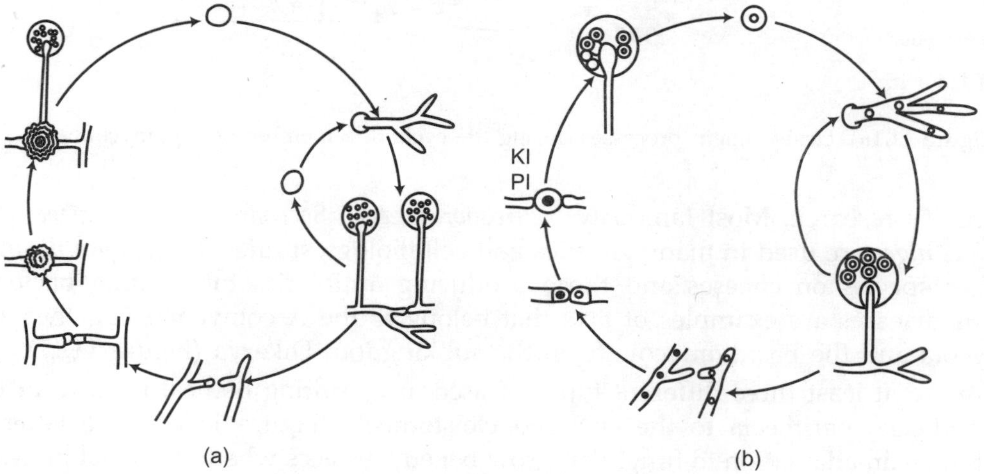

Figure 11.15 Life cycle of a saprotrophic zygomycete showing large haploid (a) and short diploid phase (b).

The ascomycetes are a monophyletic group, i.e., all of its members trace back to one common ancestor. This group is of particular relevance to humans as sources for medicinally important compounds, such as antibiotics and for making bread, alcoholic beverages, and cheese, but also as pathogens of humans and plants. The fungal symbionts in the majority of lichens (loosely termed "ascolichens") such as *Cladonia* belong to the Ascomycota. Several species of ascomycetes are biological model organisms

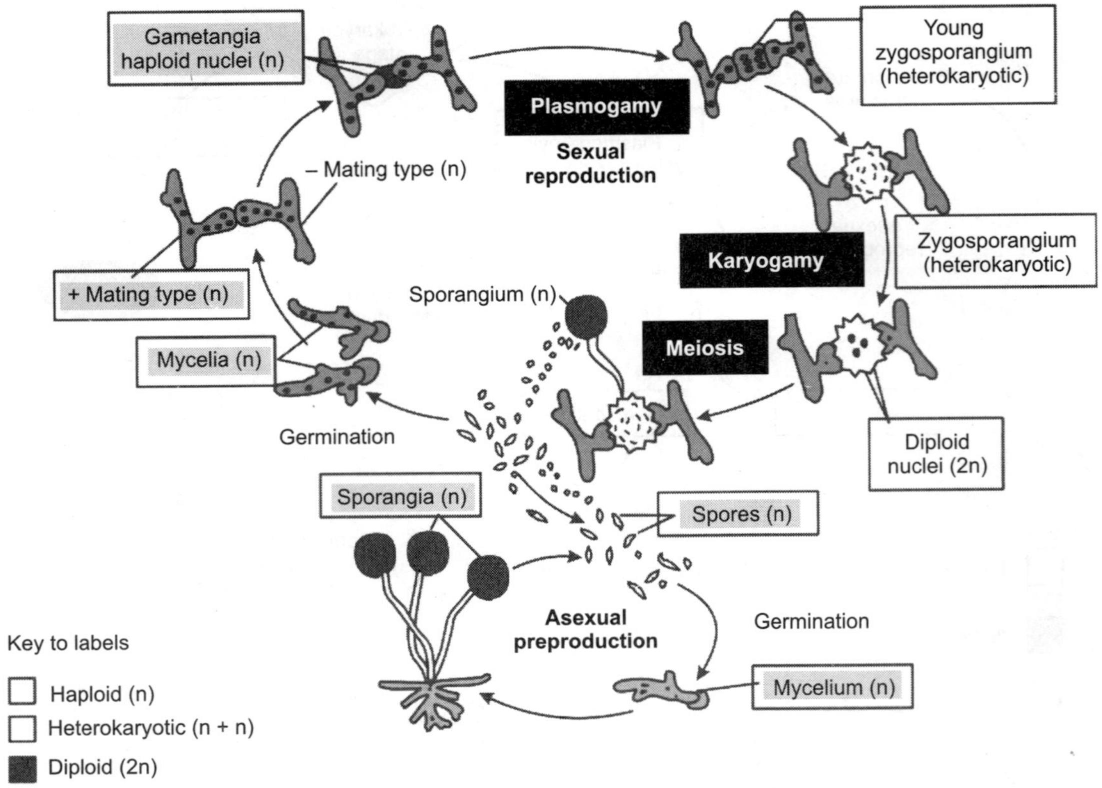

Figure 11.16 Development processes during life cycle of a member of zygomycotina.

in laboratory research. Most famously *Neurospora crassa*, *Saccharomyces cerevisiae*, and *Aspergillus niger* are used in many genetics and cell biology studies. *Penicillium notatum* and other species on cheeses and those producing antibiotics for treating bacterial infectious diseases are examples of taxa that belong to the Ascomycota. Together, the Ascomycota and the Basidiomycota form the subkingdom Dikarya (Figure 11.17).

There are at least three different types of ascocarp (fruiting bodies) form from the open apothecia, perithecia to the enclosed cleistothecia (Figure 11.19). The latter is thought to be an adaptation to fungi that grow beneth surfaces where dispersal by wind would be prohibited. In this case, an insect or motile vector may carry this packet of spores to another location where the spores can be released.

11.3.12.4.1 Asexual reproduction

Asexual reproduction is the dominant form of propagation in the Ascomycota, and is responsible for the rapid spread of these fungi into new areas. It occurs through vegetative reproductive spores, the conidia (Figure 11.18). The conidiospores commonly contain one nucleus and are products of mitotic cell divisions and thus are sometimes call mitospores, which are genetically identical to the mycelium from which they

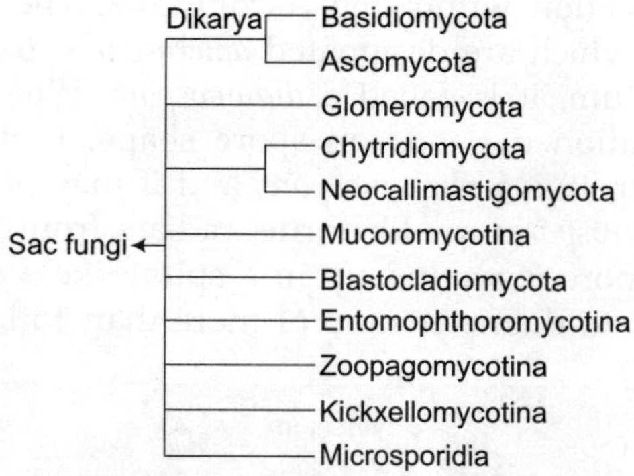

Figure 11.17 Various groups of sac fungi (Diakarya).

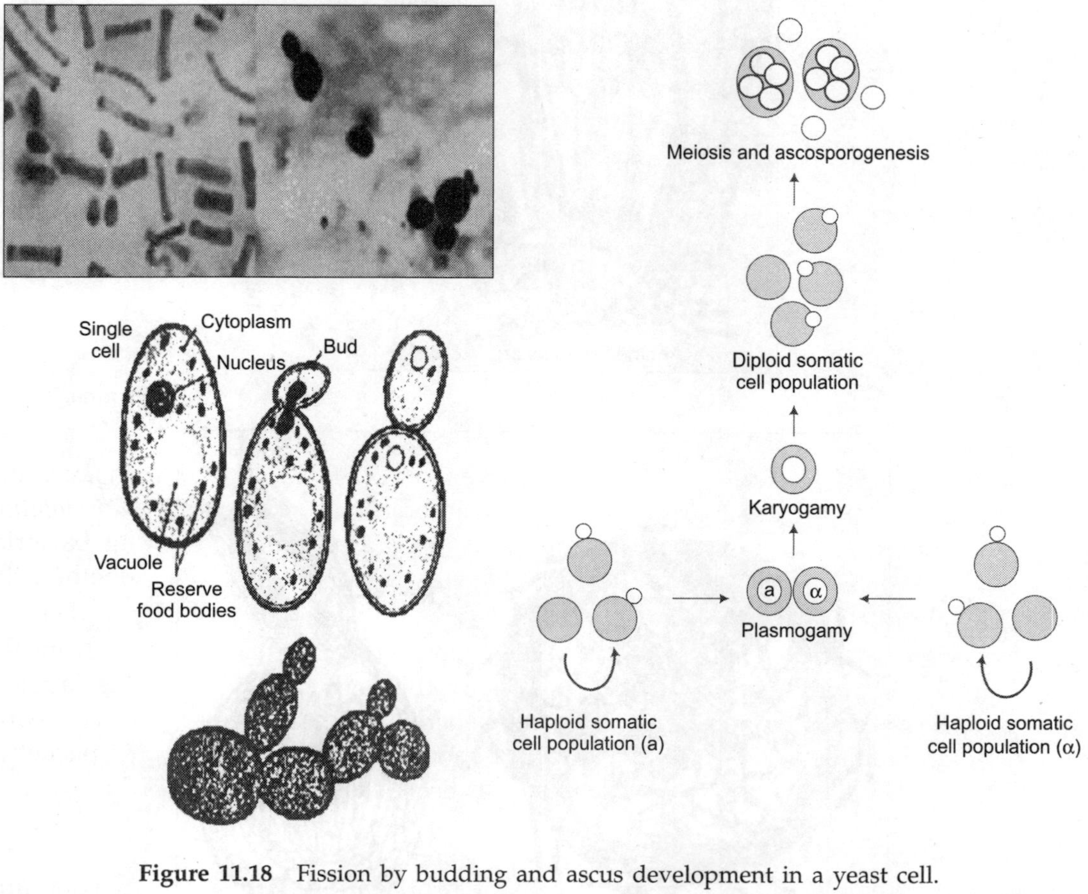

Figure 11.18 Fission by budding and ascus development in a yeast cell.

originate. They are typically formed at the ends of specialized hyphae, the *conidiophores*. Depending on the species they may be dispersed by wind or water, or by animals.

Asexual spores: Different types of asexual spores can be identified by colour, shape, and how they are released as individual spores. Spore types can be used as taxonomic characters in the classification within the Ascomycota. The most frequent types are the single-celled spores, which are designated *amerospores*. If the spore is divided into two by a cross-wall septum, it is called a *didymospore*. When there are two or more cross-walls, the classification depends on spore shape. If the septae are *transversal*, like the rungs of a ladder, it is a *phragmospore*, and if they possess a net-like structure it is a *dictyospore*. In *staurospores* ray-like arms radiate from a central body; in others (*helicospores*) the entire spore is wound up in a spiral like a spring. Very long worm-like spores with a length-to-diameter ratio of more than 15:1, are called *scolecospores*.

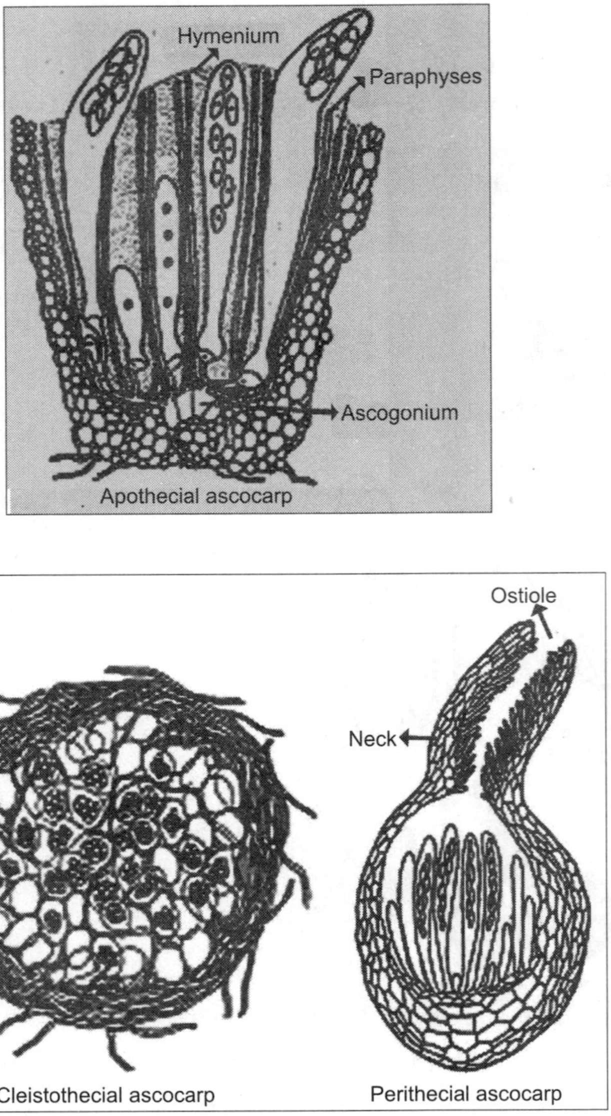

Figure 11.19 Three different types of ascocarps.

11.3.12.4.2 Heterokaryosis and parasexuality

Several Ascomycota species are not known to have a sexual cycle. Such asexual species may be able to undergo genetic recombination between individuals by processes involving *heterokaryosis* and *parasexual* events. **Parasexuality** refers to the process of heterokaryosis, caused by merging of two hyphae belonging to different individuals, by a process called *anastomosis*, followed by a series of events resulting in genetically different cell nuclei in the mycelium. The merging of nuclei is not followed by meiotic events, such as gamete formation and results in an increased number of chromosomes per nuclei. *Mitotic crossover* may enable recombination, i.e., an exchange of genetic material between homologous chromosomes. The chromosome number may then be restored to its haploid state by nuclear division, with each daughter nuclei being genetically different from the original parent nuclei. Alternatively, nuclei may lose some chromosomes, resulting in aneuploid cells.

11.3.12.4.3 Formation of sexual spores

The sexual part of the life cycle commences when two hyphal structures mate. In the case of *homothallic* species, mating is enabled between hyphae of the same fungal clone, whereas in *heterothallic* species, the two hyphae must originate from fungal clones that differ genetically, i.e., those that are of a different mating type. Mating types are typical of the fungi and correspond roughly to the sexes in plants and animals; however one species may have more than two mating types, resulting in sometimes complex vegetative incompatibility systems.

Gametangia are sexual structures formed from hyphae, and are the generative cells. A very fine hypha, called trichogyne emerges from one gametangium, the *ascogonium*, and merges with a gametangium (the *antheridium*) of the other fungal isolate. The nuclei in the antheridium then migrate into the ascogonium, and plasmogamy—the mixing of the cytoplasm—occurs. Plasmogamy is not immediately followed by the merging of the nuclei (called *karyogamy*). Instead, the nuclei from the two hyphae form pairs, initiating the *dikaryophase* of the sexual cycle, during which time the pairs of nuclei synchronously divide. Fusion of the paired nuclei leads to mixing of the genetic material and recombination and is followed by meiosis. From the fertilized ascogonium, *dinucleate* hyphae emerge in which each cell contains two nuclei. These hyphae are called *ascogenous* or fertile hyphae. They are supported by the vegetative mycelium containing uni (or mono) nucleate hyphae, which are sterile. The mycelium containing both sterile and fertile hyphae may grow into fruiting body, the *ascocarp*, which may contain millions of fertile hyphae.

The sexual structures are formed in the fruiting layer of the ascocarp, the hymenium. At one end of ascogenous hyphae, characteristic U-shaped hooks develop, which curve back opposite to the growth direction of the hyphae. The two nuclei contained in the apical part of each hypha divide in such a way that the threads of their mitotic spindles run parallel, creating two pairs of genetically different nuclei. One daughter nucleus migrates close to the hook, while the other daughter nucleus locates to the basal part of the hypha. The formation of two parallel cross-walls then divides the hypha into three sections: one at the hook with one nucleus, one at the basal of the original hypha that contains one nucleus, and one that separates the U-shaped part, which contains the other two nuclei. Fusion of the nuclei (karyogamy) takes place in the U-shaped cells in the hymenium, and results in the formation of a diploid zygote. The zygote grows into the ascus, an elongated tube-shaped or cylinder-shaped capsule. Meiosis then gives rise to four haploid nuclei, usually followed by a further mitotic division that results in eight nuclei in each ascus. The nuclei along with some cytoplasma become enclosed within membranes and a cell wall to give rise to ascospores that are aligned inside the ascus like peas in a pod.

Upon opening of the ascus, ascospores may be dispersed by the wind, while in some cases the spores are forcibly ejected from the ascus; certain species have evolved spore cannons, which can eject ascospores up to 30 cm away. When the spores reach a suitable substrate, they germinate, form new hyphae, which restarts the fungal life cycle (Figure 11.20 and Figure 11.21).

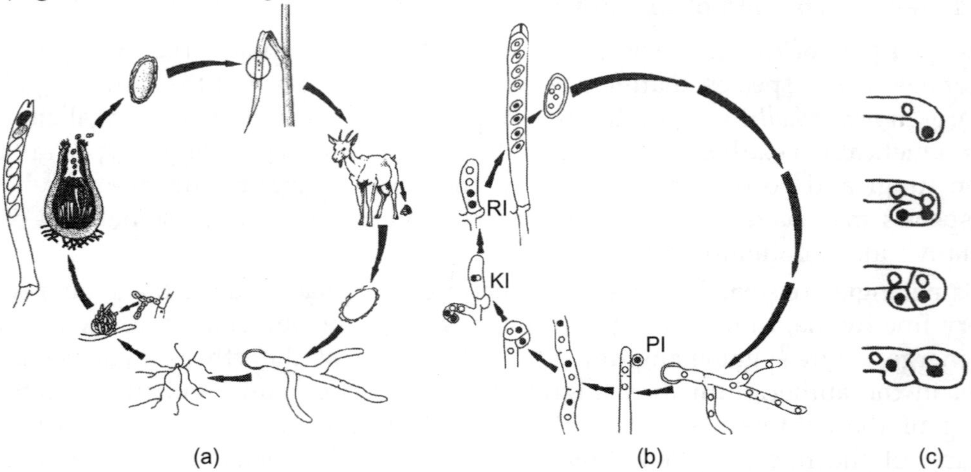

Figure 11.20 (a) Life cycle a corprophilous ascomycetes *Peziza* sp., (b) Plasmogamy and karyogemy and (c) nuclear fusion

11.3.12.5 Basidiomycota

Basidiomycetes form sexual spores on a special cell known as the basidium. As in the ascomycota, the vegetative structure is either unicellular (yeasts) or a septate mycelium. The cell walls are composed of glucans and chitin. Representatives of this subdivision include *Serpula lacrymans*, a dry-rot fungus causing wood decay; rust and smut fungi such as *Puccinia graminis*, the cause of black stem rust in grasses and cereals; and *Ustilago maydis*, which afflicts corn plants. The *Agaricus* species, the mushrooms most commonly cultivated for human consumption, are included among the Basidiomycota. Mushrooms are grown commercially on organic composts. Fungi such as those of the genus *Agaricus* form fairy rings (Figure 11.22).

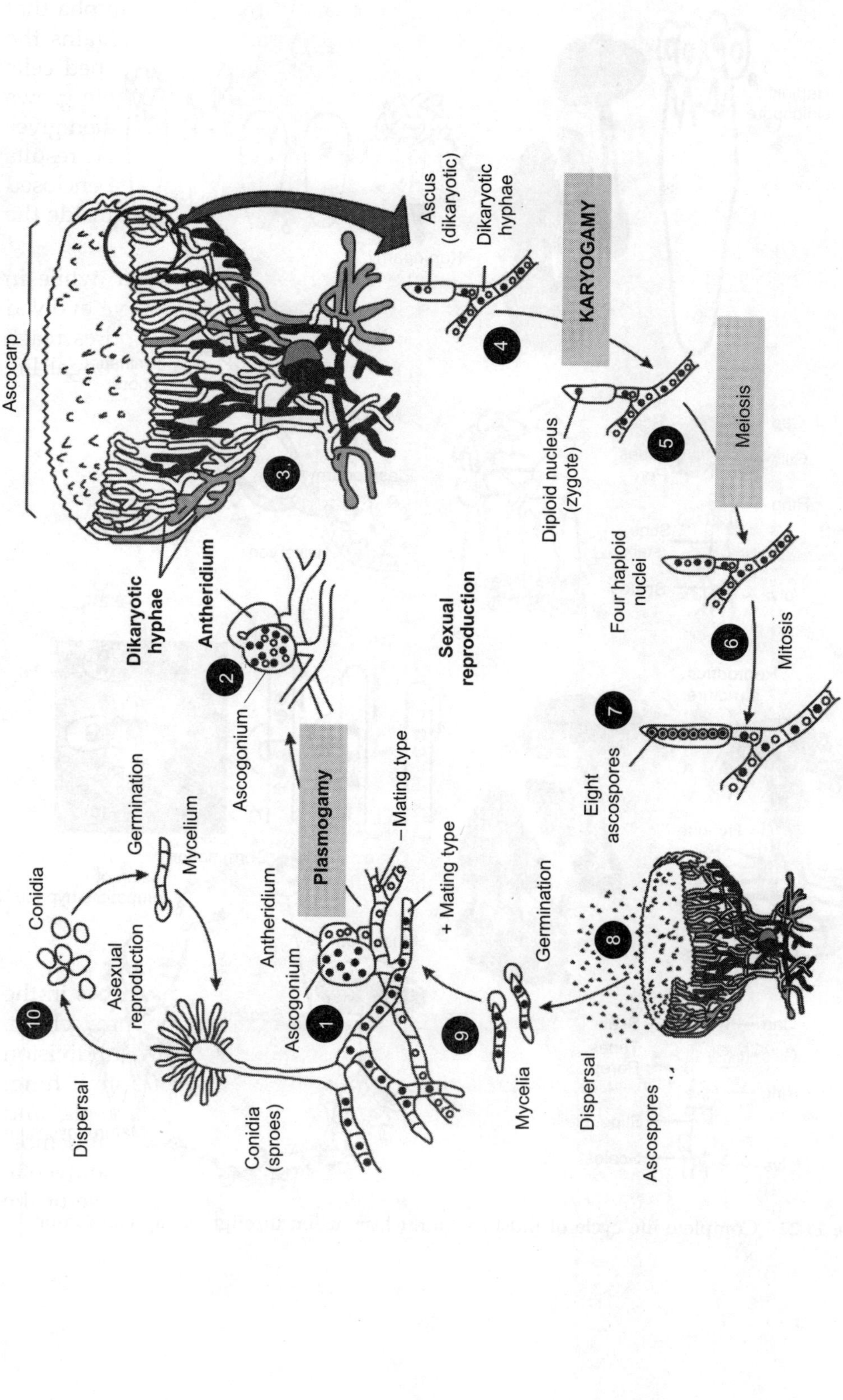

Figure 11.21 Complete life cycle of a member of Ascomycota.

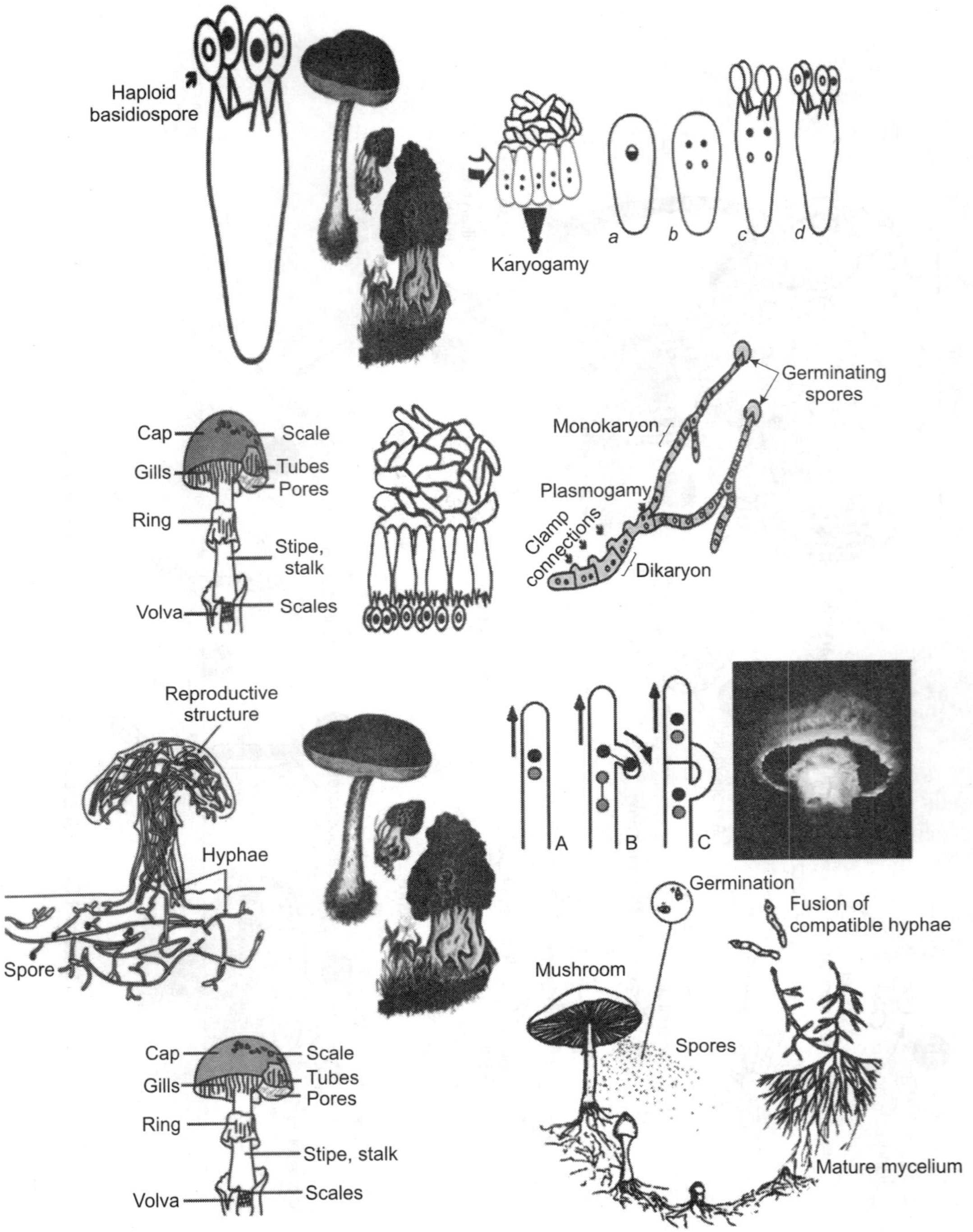

Figure 11.22 Complete life cycle of mushroom, nuclear fusion through clamp connection.

The Basidiomycotina includes many fungi that live in association with plants. Some cause disease, but most are saprophytes that grow in leaf litter, composts, soil or dung. Many of these fungi form mycorrhizas with trees. *Merulius lacrymans* is the cause of dry-rot in timber, and this single fungus is responsible for several million pounds sterling of economic loss each year in UK. Basidiomycotina of the class Teliomycetes include fungi that are responsible for plant rusts or smuts, and these are also very important economically, since they frequently affect cereal crops. Those of the order Gasteromycetes, as the name implies, include the edible fungi.

11.3.12.5.1 Rust fungi puccinia graminis

As a group, the basidiomycota have some highly characteristic features, which separate them from other fungi. They are the most evolutionarily advanced fungi, and even their hyphae have a dinstinctly "cellular" composition. The **basidiospores** usually have a single haploid nucleus. When these spores germinate they produce hyphae with usually a single nucleus in each compartment. Because these hyphae have only one nuclear type they are termed **monokaryons** (from the Greek, mono = one; karyos = kernel or nucleus). At some stage in their growth, two monokarons of different compatibility groups (mating types) fuse. This can occur either by simple hyphal fusions or by fusion of a hypha with a small spore termed an **oidium**. The fusion event is termed **plasmogamy**. Then the nuclei divide and the daughter nuclei pair so that each hyphal compatment comes to contain two nuclei—one of each mating type. At this stage the fungus is termed a **dikaryon** (i.e., with two nuclear types).

Many basidiomycota grow for most of their lives as dikaryons, until environmental signals induce them to produce fruiting bodies, such as a toadstool. All the hyphae that make up the toadstool are dikaryotic. At a late stage of development, some of these hyphae produce special cells termed **basidia** (singular, basidium). For example, the cells that line the gills of the common mushroom are basidia. Finally, the two haploid nuclei in each basidium fuse – (a process termed **karyogamy**) to form a diploid nucleus. This then undergoes meiosis to produce four haploid nuclei, and these haploid nuclei migrate into the basidiospores, which develop on small stalks (termed sterigmata) from each basidium. The dispersal of these monokaryotic spores completes the life cycle (Figure 11.23).

11.3.12.5.2 Smut fungi *Ustilago maydis*

The **smut fungi** (Basidiomycota – Order Ustilaginales) are a group of about 1000 species of pathogens that parasitise flowering plants - most commonly the grass family (Gramineae) and the sedge family (Cyperaceae). They are termed smut fungi because they produce a mass of black, powdery spores (**teliospores**) that often develop in place of the grain in cereal crops or in place of other organs such as the anthers of some flowering plants (e.g. the wild plant *Silene alba*). These teliospores are released before the normal grains would mature, and they can be carried by wind and germinate on

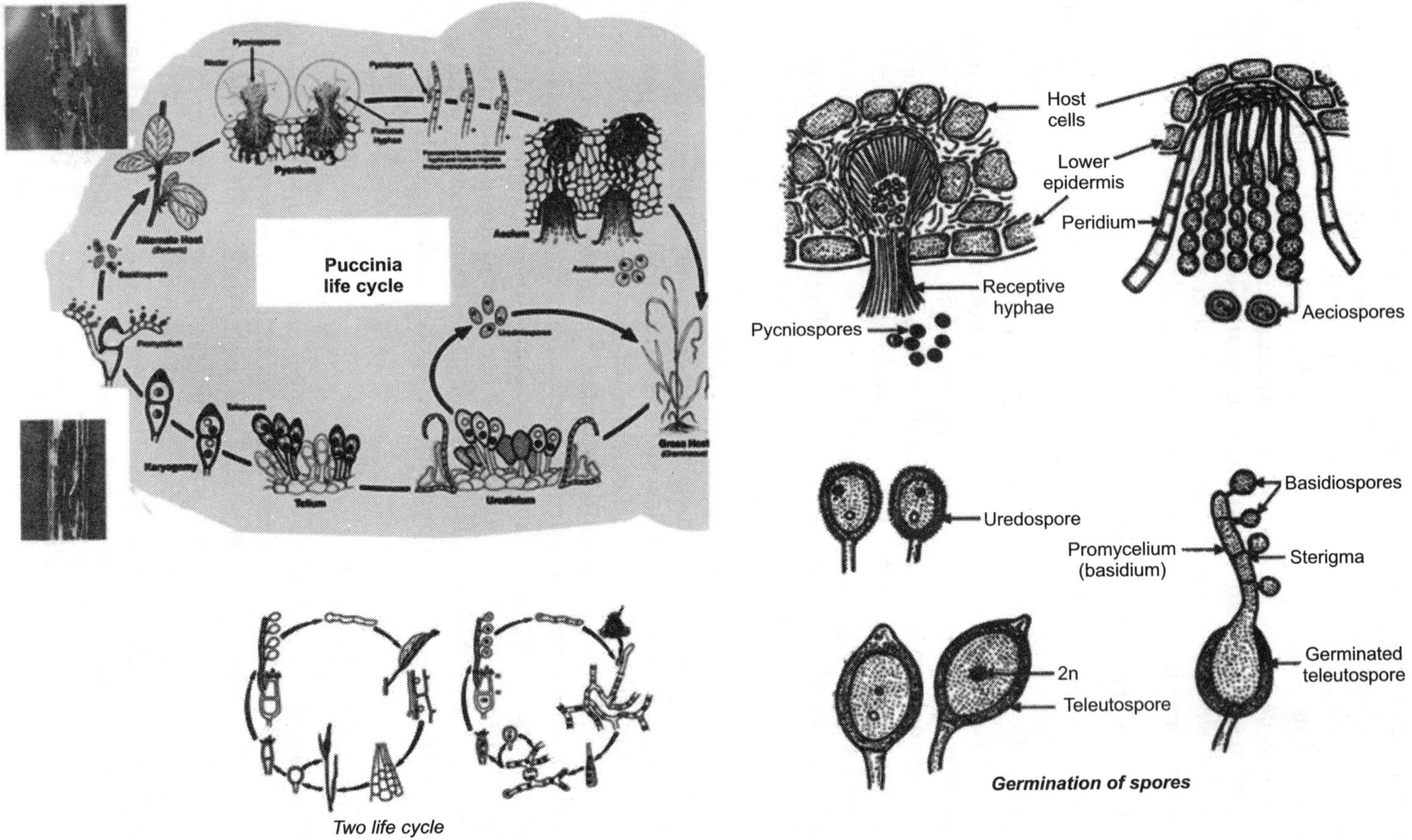

Figure 11.23 Life cycle of a Basidiomycotina rust fungus *Puccinia graminis* showing dimorphic growth on two different hosts (a) with germination of uredospore, teleutospore, basidiospores (b).

a suitable host. The teliospores of smut fungi are dikaryotic (each hyphal compartment contains 2 nuclei – one of each mating compatibility group). These spores arise from dikaryotic hyphae that grow systemically within the plant host without causing any sign of disease. But as the plant starts to mature the dikaryotic hyphae proliferate, then the 2 nuclei of each compartment fuse to form diploid cells, and the walls thicken to produce the mature spores, which are dispersed by wind. The smut spores germinate in various ways, but in *U. avenae* they produce a germ tube. The diploid nucleus enters the germ tube and undergoes meiosis, producing 4 compartments each with a single haploid nucleus. The four compartments then produce buds (sporidia) which can give rise to further budding cells while the fungus grows in a yeast-like growth form. However, the haploid yeast-like form cannot give rise to infective hyphae until the sporidia of compatible mating types fuse to establish a dikaryon. In loose smut of cereals (*Ustilago nuda*) the smutted heads arise on plants that are already systemically colonised by the fungus. The embryos of the seeds are infected, so that the infection is repeated in the following season. Infection by *Ustilago avenae* is somewhat different, because smut spores can land on the flowering heads and initiate infection of the seed as it germinates.

There are several other variations in the development of different smut fungi, but the most notable general point is that all the smut fungi live in intimate balance with their host plants. They grow slowly and progressively as the plant develops, without unduly impeding the plant's development, and in natural conditions (unlike agricultural conditions of monoculture) the degree of infection is likely to be self-regulating.

In many basidiomycota wood decomposing basidiomycete (bracket fungi) there is a rather elaborate mechanism for ensuring that the dikaryotic condition is maintained during growth of the hyphae. As shown in (Figure 11.24) below, the two nuclei in each tip cell divide at the same time, but one divides along the axis of the hypha, and the other divides so that a daughter nucleus enters a small, backward projecting branch. Then a septum forms to separate the original apical cell into two cells, and the branch fuses with the sub-terminal cell so that the nucleus in the branch migrates into this cell. The small branches at each septum are termed **clamp connections**. They can be seen at high magnification of a normal compound microscope (Figure 11.24 and Figure 11.25).

11.3.12.6 Deuteromycetes

This artificial group was created to accommodate those fungi that were known only in their asexual stage. Because their sexual state is absent, unknown, or lost and only vegetative reproduction carried out by asexual reproductive structures known as conidia is present, these fungi are also known as "fungi imperfecti." Their vegetative structure is either unicellular (yeasts) or a septate mycelium similar to those of ascomycetes and basidiomycetes. The polysaccharides of the cell wall are glucans and chitin. With molecular sequencing data, the closest sexual relatives of many of these asexual forms have been found. Members of genera of considerable economic importance, such as *Aspergillus* and *Penicillium*, are included in the deuteromycetes. Reproduction may

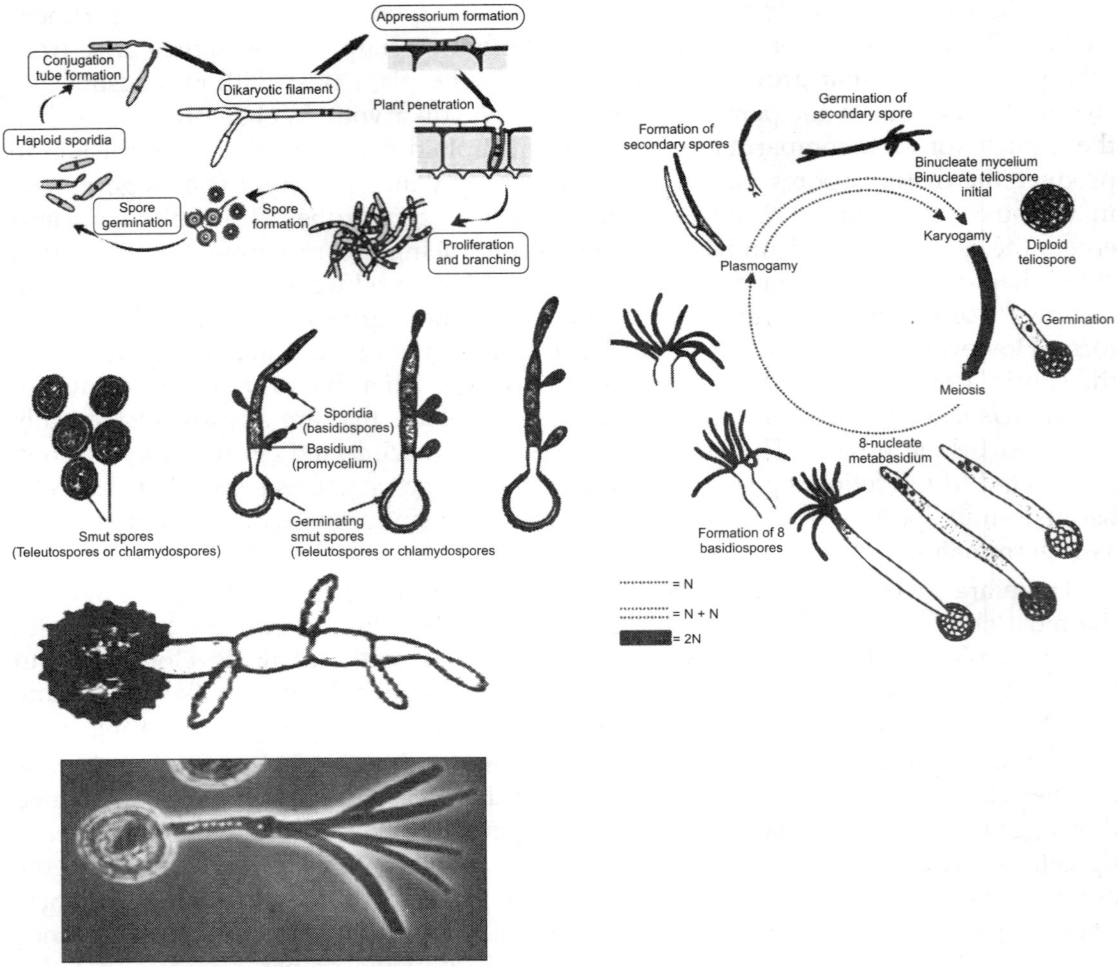

Figure 11.24 Life cycle of a smut *Ustilago maydis.*

occur from mycelium fragments, budding or from spores of various types. Budding is common. Spores produced by budding are often called blastospores. They may be oidia, arthospores, chlmydospores or conidia which are specialized non-motile asexual propagules, usually not developing by cytoplasmic cleavage or free cell formation.

The Deuteromycotina or Fungi Imperfecti, of necessity, include a wide variety of saprophytic and parasitic fungi. Many, like those of the genera *Aspergillus, Cladosporium* and *Penicillium* are important food **spoilage fungi.** *Aspergillus flavus* and related species of fungus are responsible for **aflatoxin** production. The presence of aflatoxin in foodstuffs is of great concern since aflatoxins are among the most powerful **carcinogens** so far discovered. *Aspergillus fumigatus* is responsible for the human disease aspergillosis, some forms of which cause serious and often fatal infection in immunocompromised individuals such as transplant patients. However, *Aspergillus niger* is of economic benefit,

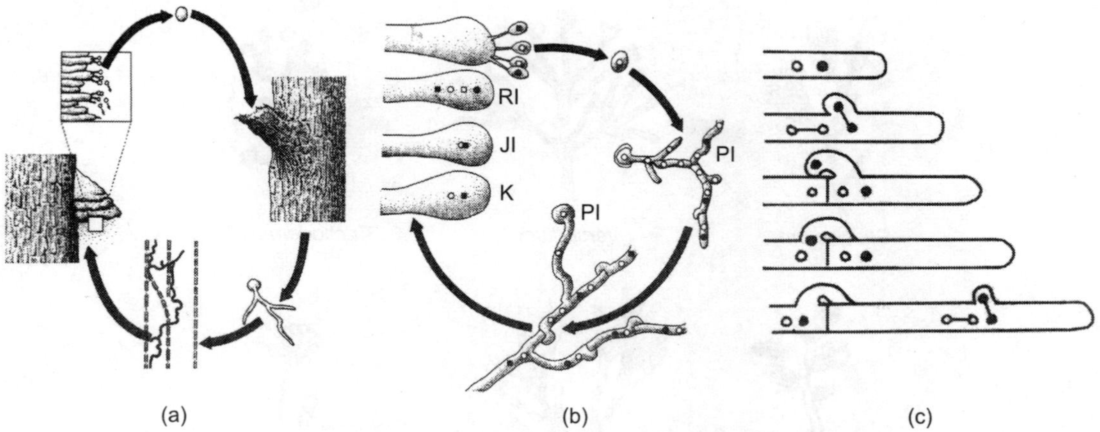

Figure 11.25 Life cycle of a wood decomposing basidiomycete (bracket fungi) (a) Life cycle on dead wood, (b) Germination of spores and (c) Nuclear fusion.

since it is used in the industrial production of citric acid. Similarly, members of the genus *Penicillium* are important in the production of antibiotics. *Penicillium chrysogenum* is used in the industrial production of the antibacterial penicillin family of antibiotics, and *Penicillium griseofulvum* is used to produce the antifungal agent griseofulvin. They are also known as conidial fungi (Figure 11.26).

11.3.13 Classification and Identification of Fungi

11.3.13.1 Introduction

Classification and identification of fungi are two independent but interrelated processes. Classification is the division of organisms into hierarchical groups. These are based upon the degree of relatedness of organisms. Identification is the process whereby an isolate is assigned to a group within a classification scheme.

Mycelial fungi, or molds, are classified according to both their macro- and micro-morphology. Yeasts are structurally more simple and therefore display a limited range of morphologies. This creates difficulties in the classification of yeasts, and this group is subdivided partly on the basis of their reactions in biochemical tests.

There is no universally accepted classification scheme for fungi. Figure 11.27 outlines a commonly used scheme, modified from that originally proposed by G. C. Ainsworth in 1971. This permits the division of fungi into broad groupings and is the most widely accepted classification scheme available at present.

The fungi are classified as Kingdom, Division, Class, Family, Genus, and Species. In accordance with the recommendations of the Committee of the International Rules on Botanical Nomenclature; Name of division of fungi ends in *mycota,* of sub-divisions in *mycotina,* of classes in *mycetes* and of sub-classes in *mycetidae.* Names of orders end in *ales,* and of families in *acae.* Genera (Sing. genus; Latin *genus* = race} and species have no standard endings. The name of an organism is in accordance with the binomial

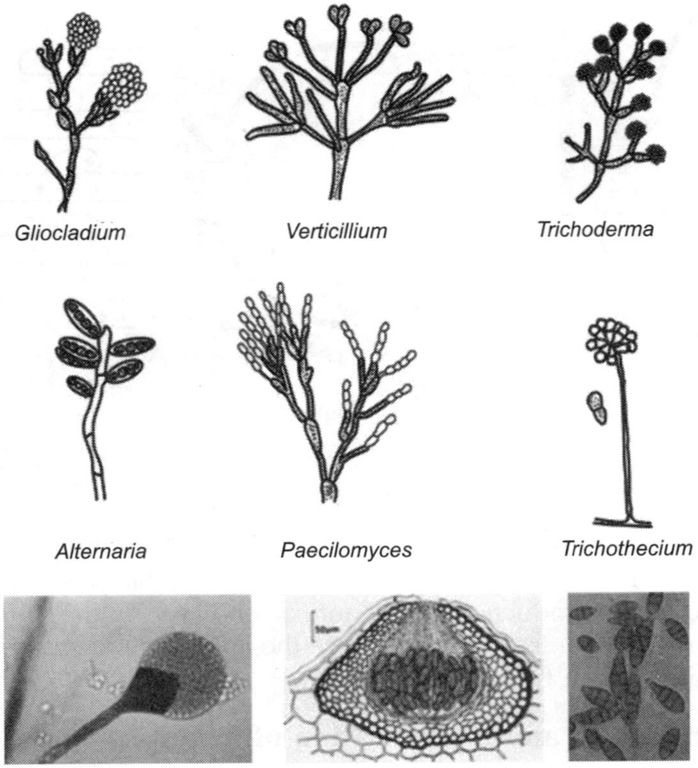

Figure 11.26 Microscopic views of different members of Deuteromycotina.

nomenclature. Taxonomists have divided the plant kingdom into many divisions. In mycology, fungi are classified on the basis of their ability to reproduce sexually, asexually, or by a combination of both. Asexual reproductive structures, which are referred to as anamorphs, are the basis for one of the sets of criteria. Because the criteria are based upon asexual morphologic forms, this system does not reflect phylogenetic relationships. It exists so that we can communicate in a simple and consistent manner by using names based upon similar morphologic structures. The second set of criteria is based upon sexual reproductive structures, which are referred to as teleomorphs. Ascospores, basidiospores, oospores, and zygospores, as well as any specialized structures associated with their development, are the basis of the second set of criteria. These criteria reflect phylogenetic relationships because they are based upon structures that form following meiosis. The term holomorph is used to describe the whole fungus, which consists of its teleomorph and anamorphs. Analyses of rRNA and rDNA sequences constitute an important complement of the morphological criteria needed to allow fungi to be more easily identified and placed on a single phylogenetic tree (Figure 11.28).

In some classification schemes the fungal divisions are referred to as phyla. Also, some have realigned the fungi grouped in the Phycomycetes. Members of the

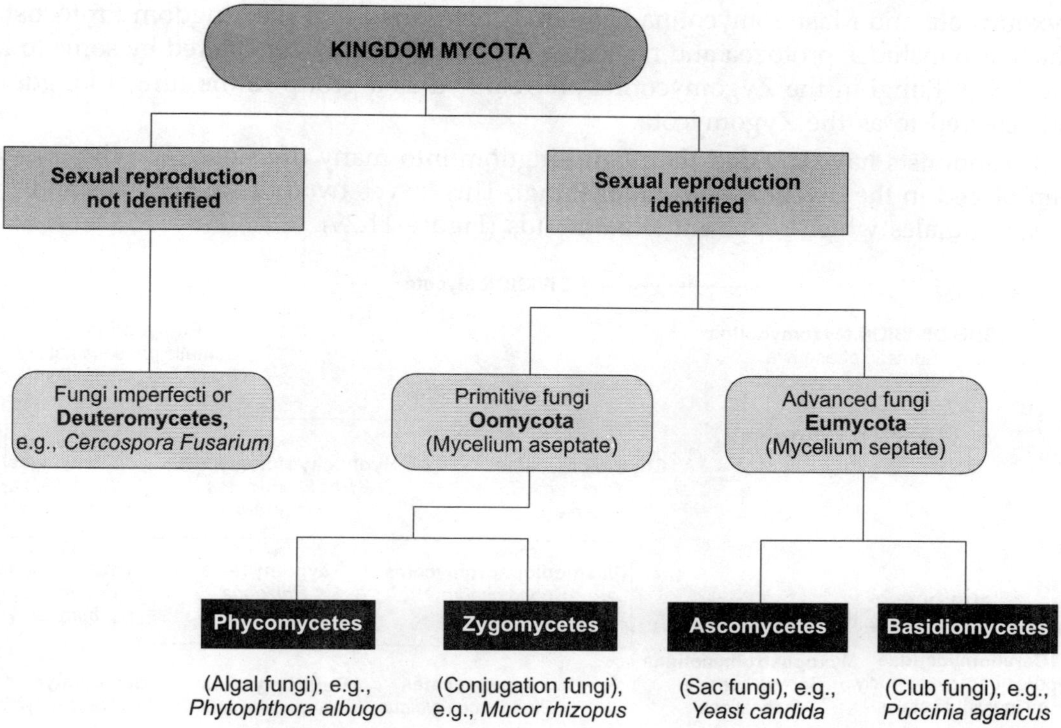

Figure 11.27 Classification of fungi as proposed by Ainsworth.

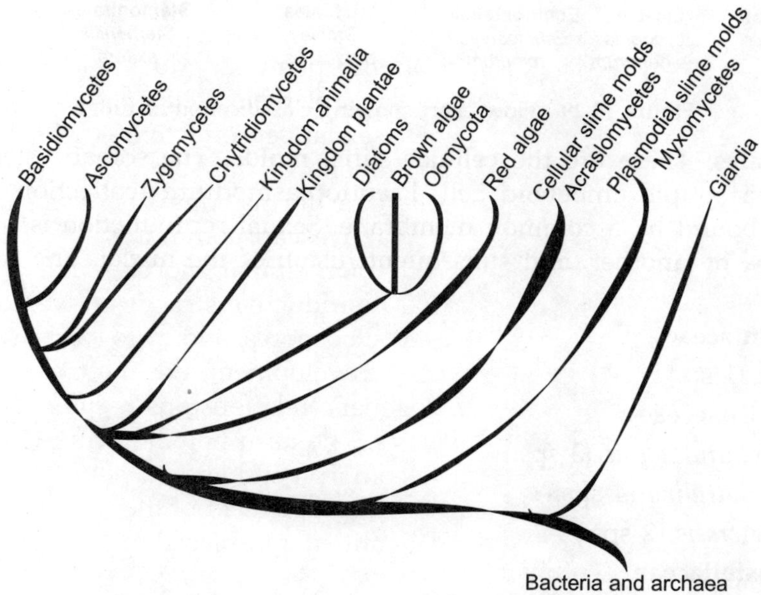

Figure 11.28 Phylogenetic tree of fungi.

Myxomycota and Mastigomycotina have been regrouped into the kingdom Protoctista, which also includes protozoa and nucleated algae, and are not considered by some to be true fungi. Fungi in the Zygomycotina are retained as a group in the fungal kingdom and referred to as the Zygomycota.

Taxanomists have divided the plant kingdom into many divisions. The fungi have been placed in the division Mycota, or Fungi. This leaves two orders Acrasials and the Labyrinthulales which represent slime molds (Figure 11.29).

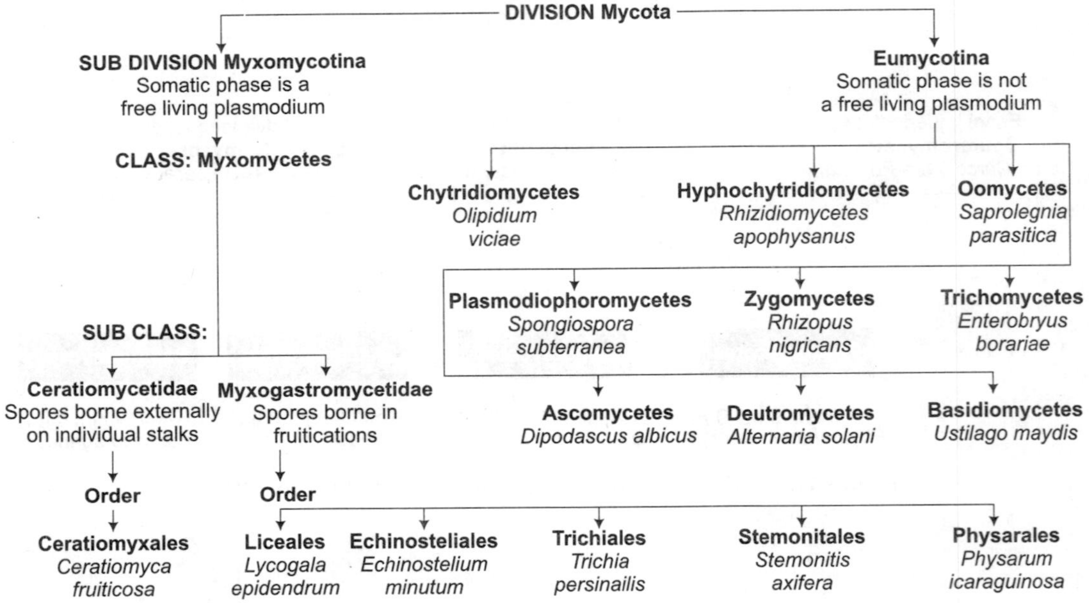

Figure 11.29 Flow chart showing classification of fungi.

Order Acrasiales: These are the cellular slime molds. The somatic structure of the Acrasiales is a simple amoeboid cell. Pseudoplasmodium (collection of amoebae). Spores never bound by a common membrane. Sexual reproduction is by engulfment of one amoeba by another, and subsequent fusion of the nuclei. The fruiting bodies are sorocarps.

Family Sappiniaceac

Genus *Sapinia* (1 sp.)

Family Guttulinaeceae

1. Genus *Guttulinopsis* (3 sps.)
2. Genus *Guttulina* (4 sps.)
3. Genus *Acrasis* (2 sps.)

Family Acytosteliaceae

1. Genus *Protostelium* (1 sp.)
2. Genus *Actostelium* (1 sp.)

Family Dictyosteliaceae

1. Genus *Dictyostelium* (6 sps.)
2. Genus *Polyspondylium*
3. Genus *Coellonia*

Order Labyrinthulales: It includes marine or freshwater parasites or saprobes. The structure is simple oval or spindle shaped. Cells secrete mucous filaments which unite to form a fine network on which the cells glide. Sexual reproduction is not known.

Genus *Labyrinthula:* It includes all marine sp. where the slime tracks are of non-living material secreted by the cells which do not feed on solid particles through phagocytosis.

Genus *Labyrinthorhiza:* It contains the freshwater species where the slime tracks are of living filaments. Some of these obtain nourishment through phagocytosis.

Division Mycota

These are the fungi. The soma varies from a unicell to an extensive mycelium. Reproduction is typically both asexual and sexual, spores being propagative units. It comprises of 2 sub-divisions - **Myxomycotina** or true slime molds, and the **Eumycotina**, or true fungi.

1. **Sub-division Myxomycotina:** Here the somatic phase is a free living plasmodium. It contains a **single class - Myxomycetes**, which is itself divided into **2 sub-classes Ceratiomycetidae** and **Myxogastromycetidae**.
   (i) **Ceratiomycetidae:** Spores are borne externally on individual stalks. It contains a single **order Ceratiomyxales**, **family Ceratiomyxaceae** e.g., *Ceratiomyxa fruticulosa* (Coral slime).
   (ii) **Myxogastromycetidae:** Spores are borne in fruitfications. It is divided into 5 orders.
      1. **Liceales:** True capitulum or columella is lacking, e.g., *Lycogala epidendrum.*
      2. **Echinosteliales:** Stalked or columella is present, e.g., *Echinostelium minutum.*
      3. **Trichiales**: Columella sessile, e.g., *Trichia scabra.*
      4. **Stemonitales**: Neither peridium (pouch bearing spores) nor the capitulum is calcareous, e.g., *Stemonitis axifera.*
      5. **Physarales:** Peridium or capitulum or both calcareous, e.g., *Physarum nicarguinasa.*
2. **Sub-division Eumycotina:** Somatic phase is not a free living plasmodium.
   (i) **Class Chytridiomycetes:** Flagella is posterior and whiplash. It is divided into 3 orders.
      1. **Chytridiales:** True mycelium lacking; rhizomycelium present in some species. It is divided into **4 families**.
         (a) **Olipidiaceae**: Sexual reproduction by copulation of isoplanogametes, e.g., *Olipidium viciae.*
         (b) **Synchytriaceae**: Reproductive organs enveloped in common membrane, e.g., *Synchytrium endobioticum.*

(c) **Phyciticiaceae:** Sexual reproduction by gametangial copulation, e.g., *Physophidium couchii.*

(d) **Megachytriaceae:** Aquatic, sporangia and resting spores, e.g., *Nowakowskiella ramosa.*

2. **Blastocladiales:** Sexual reproduction by fusion of planogametes, thick walled, resistant, sporangia characteristically formed. It contains single family **Blastocladiaceae,** e.g., *Allomyces macrogynus.*
3. **Monoblepharidiales:** Sexual reproduction by male and female apalanogamete, no resistant sporangia. **Family Monoblepharidiaceae,** e.g., *Monoblepharis, Polymorpha.*

(ii) **Class Hyphochytridiomycetes**: Flagellum anterior, tinsel type. **Family Rhizoidiomycetaceae,** e.g., *Rhizoidiomyces apophysatus.*

(iii) **Class Oomycetes**: Reproduction asexual, zoospore biflagellate, each bearing one tinsel flagellum directed forward and one whisplash directed backward. Zoopores in sporangia of several types. It is divided into four orders.

1. **Saprolegniales:** Zoospores are borne in the sporangium. Hyphae, if present, then alocarpideucarpic and without constrictions. Family *Saprolegniaceae.*
2. **Leptomitales:** Zoosporia either in sponmgium or in evanescent vesicle; hypae with eucarpic constrictions, e.g., *Apodachyla pyrifera.*
3. **Lagemdiales:** Hyphae holocarpic. Family Lagnenidiaceae, e.g., *Lagenidium rabenhostii.*
4. **Pernosporales:** Eucarpic hyphae, sporangiophore is either similar to somatic hyphae or different from it.

(a) **Family Phythiaceae:** Sporangiophore similar to somatic hyphae, e.g., *Phytophthora infestans.*

(b) **Family Albuginaceae:** Sporangia are produced in chains, e.g., *Albugo candida*

(c) **Family Perosporanceae:** Sporangia in clusters, e.g., *Plasmospora viticola*

(iv) **Class Plasmodiophoromycetes:** These obligate endcparasites have their somatic phase as a plasmodium, which develops with in the host cells. Plasmodia produce zoosporangia containing zoospores produced by clevage of plasmodium. Single **family Plasmodiophoromycetes,** e.g., *Spongospora subterranea.*

(v) **Class Zygomycetes:** Sexual reproduction by gametangial copulation resulting in the formation of zygospore. Asexual reproduction by means of non-motile spores in the form of sporangiospores or conidia. It includes 3 orders.

1. **Order Mucorales:** Chiefly saprobic, weakly parasitic, asexual reproduction by sporangia having one to many aplanospores, at times, conidia, e.g., *Rhizopus oryzae.* Mucorales have developed six lines of evolution. The families which represent each line may be grouped as follows:

(a) **Mucoraceae**
(b) Chonephoraceae
(c) Thamnidiaceae
(d) Pilobolaceae
(e) Mortierellaceae
(f) Syncephalastraceae

2. **Order Entomophthorales:** Parasitic, rarely saprobic, asexual reproduction by sporangia functioning as conidia forcibly discharged, e.g., *Entomophthora muscae.*
3. **Order Zoopagales:** Conidia not forcibly discharged; e.g., *Cochlonema verrucosum.*

(vi) **Class Trichomycetes**: Soma filamentous, attached to digestive tract or external cuticle of living arthropods, mycelium not immersed in host tissue. **Order Ecrinales** e.g., *Enterobryus borariae.*

(vii) **Class Ascomycetes**: Spores resulting from karyogamy and meiosis, borne in asci naked 3 **sub-classes**

1. **Hemiascomycetidae**
2. **Euascomycetideae**
3. **Loculoascomycetidae**

1. **Sub-class Hemiascomycetidae**: No ascogenous hyphae or ascocarps. It includes 2 orders.
   1. **Order Endomycetales**: Asci arise directly from zygotes, each derived from the copulation of two cells or parthogenetically from single cell. **4 families**
      (a) **Ascosphaeriaceae** Asci in spore balls, e.g., *Ascosphaera apis*
      (b) **Endomycetaceae** Asci 1-8 spored, mycelium aboundant, plasmogamy by copulation of uninucleate gametangia, e.g., *Endomyces*
      (c) **Spermophthoraceae** Plasmogamy by copulation of non-motile gametes released from gametangia, e.g., *Spermophthora.*
      (d) **Saccharomycetaceae** Mycelium scanty, e.g., *Saccharomyces cerevisiae.*
   2. **Order Taphrinales** Asci arise from binucleate ascogenous cells which develop simsimilarly to chlamydospores from binucleate mycelium, e.g., *Taphrina deformans.*
   3. **Sub-class Euascomycetideae** Asci typically unitunicate, if bitunicate then borne in apotheciurn. It includes **4 series.**

I. **Series Plectomycetes:** Asci scattered within the ascocarp or united in spore balls within a sporcyst. **3 orders**

(i) **Order Eurotiales** Ascocarps sessile and without ostiole **3 families.**

(a) **Assosphaeriaceae** Asci in spore balls, e.g., *Ascosphaera spis.*

(b) **Gymnoascaceac** Asci not in spore balls, peridium when present a lax network of hyphae, e.g., *Ctenomyces serratus.*

(c) **Eurotiaceae** Peridium of closely interwoven hyphae usually forming a pseudoparenchyma, e.g., *Aspergillus niger* and *Penicillium notatum*

(ii) **Order Microascales** Ascocarps ostiolate. **Family Ophistomataceae**. e.g., *Ceratocystis fimbriata.*

(iii) **Order Onygenales** Ascocarps stalked.

II. **Series Pyrenomycetes** Asci in basal or peripheral layer forming a hymenium, typically Persistent. Ascocarp closed. **8 orders.**

1. **Erysiphales** Mycelium white, asci in cleistothecia. **Family Erysiphaceas**, e.g., *Erysiphe chichoracearum.*
2. **Meliolales** Mycelium dark ascocarps ostiolate. e.g., *Meliola garnsii*
3. **Chaetomiales** Mycelium within the substratum, asci evanescent, e.g., *Chaetomium.*
4. **Sphaeriales** Ascocarps and stromata, if present are dark, membranous or carbonous ascal bases not gelatenising, mature asci remaining attached to the perithecial wall.
   (a) **Family Sordariaceae** Perithesia without an easily discernible stroma, perithecia superficial, e.g., *Sordaria fimicola.*
   (b) **Family Phyllachoraceae** Perithecia immersed in the substratum, e.g., *Phyllachora graminis.*
   (c) **Family Diatrypaceae** Stroma consisting of both host and fungal tissue.
   (d) **Family Xylariaceae** Stroma entirely fungal, e.g., *Xylaria polymorpha.*
5. **Diaporthales** Ascal bases gelatinising, mature asci loose in the perithecial cavity and exuding through the ostiole.
   (a) **Family Gnomoniaceae** Perithecia in substratum, e.g., *Gnomonia erythrostoma.*
   (b) **Family Diaporthaceae** Perithecium immersed in a stroma with only their beaks protruding out, e.g., *Glomerella cingulata.*
6. **Hypocreales** Ascocarps and stromata if present brightly coloured, soft, fleshy or waxy 3 families.
   (a) **Family Nectriaceae** Perithecia superficial, e.g., *Nectria coccinea.*
   (b) **Family Hypocreaceae** Perithecia immersed completely, e.g., *Hypomyces.*
   (c) **Family Melanosporaceae** Perithecia transparent, e.g., *Melanospora.*
7. **Clavicipitales** Perithicia in stroma composed of fungal tissue.
   Family Clavicipitaceae e.g., *Claviceps purpurea.*
8. **Cronophorales** Ascocarp opening irregularly by an ostiole, but ostiole is not funnel shaped.

III. **Series Discomycetes** Ascocarp an open apothecium as a modified form.

1. **Order Ostropales** Ascocarps aboveground and asci inoperculate.

2. **Order Helotiales** Ascospores not thread like, sometimes needle shaped 4 families.
   (a) **Family Phacidiaceae** Transitional stage between perithecia and apothecia, e.g., *Rhytisma acerinum.*
   (b) **Family Sclerotiniaceae** Apothecial initials arise from stromata and sclerotia, e.g., *Monilinia fructicola.*
   (c) **Family Geoglossaceae** Apothecia variously shaped, e.g., *Geoglossum.*
   (d) **Family Cyttariaceae** Ascocarps embedded in fleshy, globose stromata produced in large clusters, e.g., *Nothofagus.*
3. **Order Pezizales** Operculate or sub operculate. 3 families.
   (a) **Family Sacroscyphaceae** Asci sub-operculate, e.g., *Sacrosoma globosum.*
   (b) **Family Pezizaceae** Apothecia not differentiated into stipe and pileus, e.g., *Peziza vesiculosa.*
   (c) **Family Halvellaceae** Apothecia differentiated clearly into stipe and pileus, e.g., *Helvella esculenta.*
4. **Order Tuberales** Ascocarps below the ground, e.g., *Tuber aestivum.*

IV. **Series Laboulobeniomycetes** Specialised exoparasites of insects or arachinids, thallus usually limited. **Order Laboulobeniales**, e.g., *Laboulbenia formicarum.*

3. **Sub-class Loculoascomycetidae** Asci bituniate, ascocarp and ascostroma present. It is divided into 5 orders.
1. **Order Myriangiales** Locules uniascal, usually scattered at various levels in ascostroma. 2 families.
   (a) **Family Elsinocaceae** Four celled hyaline ascospores mature in globose asci, e.g., *Elsinoe veneta.*
   (b) **Family Myriangiaceae** The tissues surrounding the cavities in the centre of the ascocarp is of ascogenous hyphae, e.g., *Piedraia hortai.*
2. **Order Dothideales** Absence of interascal threads in locule and formation of bitunicate asci. 3 families.
   (a) **Family Dothideaceae** Spermatia borne in spermogonia, e.g., *Mycosphaerella fragariae.*
   (b) **Family Pseudosphaeriaceae** Ascostroma is a pseudothecium in which asci are formed single with in the ascocarp and remain separated by stromatal tissue, e.g., *Pseudoplea gaumanii.*
   (c) **Family Capxodiaceae** Individual ascostroma resemble perithecia.
3. **Order Pleosporales** Stroma not boat shaped. Pseudoparaphyses present. 3 families.
   (a) **Family Venturiaceae** Plant parasites, two celled ovoid/ellipsoid ascospores, e.g., *Venturia inequalis.*
   (b) **Family Pleosporaceae** Plant parasite, e.g., *Pleospora sp.*
   (c) **Family Lophiostomataceae** Ascocarp bears the general appearance of a perithercium with a prominent neck.

4. **Order Microthyriales** Locules polyascal arranged in a basal layer, stroma flattened and dimidiate, e.g., *Microthyrium microscopicum.*
5. **Order Mysteriales** Boat shaped or branched ascocarps with a long slit-like opening along the entire length of fruiting body, e.g., *Glonium.*

(viii) **Class Deuteromycetes** These fungi lack a sexual phase (perfect stage) hence are popularly known as Fungi Imperfecti. These may be considered as conidial stages of Ascomycetes whose sexual stages do not exist. 3 orders.

1. **Order Sphaeropsidales** Reproduction by means of conidia borne in pycnidia. 2 families.
   (a) **Sphaeropsidaceae** Small, ostiolar pycnidia sunk in the substratum very short conidiophores and hyaline spherical or oval spores, e.g., *Phyliosticta acericola* and *Phoma* sp.
   (b) **Zythiaceae** Pycnidia are light or brightly coloured and their walls are soft or waxy, e.g., *Zythia fragariae.*
2. **Order Melanconiales** Conidia when produced, borne in acervuli, e.g., *Gloeosporium musarum.*
3. **Order Moniliales** Reproduction by means of conidia borne otherwise, by oidia or by budding.
   (a) **Family Cryptococceae** Yeast like fungi
   (b) **Family Moniliaceae** Unorganised, conidia on hyaline conidiophores or hyaline hyphae, e.g., *Trichophyton megninii.*
   (c) **Family Dematiaceae** : Hyphae and conidia are typically dark, e.g., *Alternaria solani*
   (d) **Family Stibellaceae** Condiophores are united into synnemats, e.g., *Ceratiocystis ulmi.*
   (e) **Family Tuberculariaceae** Sporodochium or stromata covered with conidiophores is shaped like mushroom or is crescent shaped, e.g., *Tubercularia vulgaria.*

(ix) **Class Basidiomycetes** Spores resulting from karyogamy and meiosis borne on basidia. 2 sub-classes.
   (a) Heterobasidiomycetes
   (b) Homobasidiomycetes

(a) **Sub-class Heterobasidiomycetes** Basidium septate or deeply divided consisting of a teleutospore germinating into a promycelium, basidiospores usually capable of germinating by repetition.

1. **Order Tremellales** Basidiocarp well developed, mostly saprobic, some species parasitic on plants or scale insects. **9 families**.
   (a) **Ceratiobasidiaceae**
   (b) **Tulasnellaceae**
   (c) **Sirobasidiaceae**

(d) **Decrymycetaeaee**

(e) **Tremellaceae**

(g) **Hyaloriaceae**

(h) **Phleogenaoeae**

(i) **Auriculariaceae**

(j) **Septobasidiaceae**

2. **Order Uredinales** Teleutospores present, plant parasites, basidiospores discharged forcibly.
3. **families**

(a) **Family Puccinaceae** Teleutospores free or variously united, but never in layers or crusts, e.g., *Puccnia graminis.*

(b) **Family Melanosporaceae** Teleutospores laterally united into layers, crusts or columns *Cronatium rificola.*

(c) **Family Coleosporiaceae** Teleutospores become septate during germination without the formation of an external promycelium, e.g., *Coleosporium solidaginis.*

3. **Order Ustillaginales** (Smuts): Basidispores sessile, not forcibly discharged. **3 families.**

(a) **Family Ustillaginaceae** Promycelium septate; basidiospores produced laterally from each cell of the promycelium, e.g., *Ustilago maydis.*

(b) **Family Tilletiaceae** Promycelium non-septate, basidiospores on a terminal cluster, e.g., *Tilletia caries.*

(c) **Family Graphiolaceae** Plant Pathogenic Teleutospores are produced in cup-like fruiting body

(b) **Sub-class Homobasidiomycotes** Basidium not septate or deeply divided; basidiospores usually germinating by germ tube.

1. **Order Exobasidiales** Basidiocarp lacking, hymenium covering the surface of the parasitized plant tissues, e.g., *Exobasidium vaccinii.*

I. **Series Hymenomycetes** Hymenium present and exposed before the spores are mature 2 orders.

1. **Order Polyporales** Hymenium borne in various ways, but if basidia line pores or gills, texture of basidiocarp not soft and putrescenet. 6 families.

(a) **Thelephoraceae** Basidiocarp cobwebby membranous and hard, e.g., *Streum frustlilatum.*

(b) **Clavariaceae** Basidiocarp club shaped. Usually erect, e.g., *Clavariadelphus pistillasis.*

(c) **Cantharellaceae** Basidiocarp mushroom like or funnel shaped, e.g., *Cantharellus cibarius.* .

(d) **Hydnaceae** Hymenium covering pendant warts, spines or teeth, e.g., *Hericium coralloides.*

(e) **Meruliaceae** Hymenium lining interior of pits or tubes, pits shallow, fertile. on ridges.

(f) **Polyporaceae** Tubes deep, if shallow, sterile on ridges, texture not soft and putrescent, e.g., *Polyporus sulphureus.*

2. **Order Agaricales** Hymenium borne on gills, or if lining the interior of pores, then basidiocarp soft and putrescent. 5 families.

(a) **Boletaceae** Hymenophore made up of tubes e.g., *Boletus edulis.*

(b) **Paxillaceae** Hymenophore of lamellae which are often united by veins.

(c) **Russulaceae** Basidiocarp fleshy content of sporophore containing nests of sphaerocysts, e.g., *Russula emetica.*

(d) **Hygrophoraceae** Gills, waxy, broadly triangular in section.

(e) **Agaricaceae** Gills not waxy, narrow in section *Amanita verna.*

II. **Series Gasteromycetes** Basidiocarps remain closed at least till the spores have been released from the basidia. **5 orders**.

1. **Order Hymenogastrales** Gleba fleshy to cartillaginous, or if slimy then not exposed to maturity, e.g., *Endoprychum agaricoides.*
2. **Order Lycoperdales** Hymenium present in early stages, spores lightly coloured and small, e.g., *Clavatia gigantea.*
3. **Order Sclerodermatales** Hymenium lacking or indistinct, spores mostly dark large, e.g.,, *Astraeus hygrometricus.*
4. **Order Phallales** Gleba exposed on a receptacle, e.g., *Diclyophora duplicata.*
5. **Order Nidulariales** Glebal chambers forming waxy peridioles, or entire gleba separating as a unit from peridium, e.g., *Cyalhus striatus.*

11.3.14 Phylogenetic Approach

The correct identification of fungi is of great practical importance in the clinical setting, plant pathology, biodeterioration, biotechnology, and environmental studies. An enormous number of species of fungi are already known, and so taxonomists are being kept very busy with recognizing and describing new species and grouping taxa. Hence, most species have received only limited study, so that classification has been mainly traditional rather than numerical and has been based on readily observable morphological features. However, some groups of fungi, because of their economical or pathological importance, have been studied more extensively. Other features beside morphology, such as susceptibilities to yeast killer toxins, susceptibilities to chemicals and antifungal drugs, the use of morphograms, molecular techniques, physiological and biochemical tests, secondary metabolites, ubiquinone systems, fatty acid composition, cell wall composition, and protein composition, have been used in classification and also in identification. The increased use and availability of modern techniques have opened up many new areas within systematics and have enabled more traditional ones to be developed further. The biological concept, which was developed before the advent of modern phylogenetic analysis, emphasizes gene exchange (i.e., sexual and parasexual reproduction) within species and the presence of barriers that prevent the cross-breeding of species. A species is considered to be an actual or potential interbreeding population isolated by intrinsic reproductive barriers. However, application of the biological-species concept to fungi is complicated by the difficulties in mating and in assessing its outcome. Also, whether a cross is considered fertile or sterile depends on the frame of reference. It is applied only to sexual fungi, whereas asexual ones need similar characteristics to each other. In asexual fungi, genetic exchange through somatic hybridization is a theoretical possibility, although it is limited by vegetative incompatibility. Two recent and important developments have greatly influenced and caused significant changes in the traditional concepts of systematics. These are the phylogenetic approaches and incorporation of molecular biological techniques, particularly the analysis of DNA nucleotide sequences, into modern systematics. Molecular techniques, which were previously used only in research laboratories, are now commonplace. These developments have provided new information that has caused the biological-species concept to come under criticism in favor of the phylogenetic-species concept. This new concept has been found particularly appropriate for fungal groups in which no sexual reproduction has been observed (deuteromycetes). Hence, new concepts, specifically formulated within the field of phylogenetics, are becoming familiar to mycologists. Population studies and molecular data are increasingly showing that many widely used morphospecies actually comprise several biological or phylogenetic species. One of the problems for morphologists involves deciding how many base differences are required for strains to be considered different species. This has been partly solved by the phylogenetic-species concept, especially when based on cladistic analysis of molecular characteristics, which offers consistency in the delineation of species. Cladogram topology indicates the existence of monophyletic groups, which may represent species or supra- or subspecific taxa. Biological-species concept has been correlated with the phylogenetic-species concept by comparing the fertility of genetic crosses among heterothallic yeasts and found that the D_2 region, a variable region of the 25S rDNA gene, is sufficiently variable to recognize biological species of yeasts and that conspecific species generally show less than 1% nucleotide substitution. Molecular biology techniques, especially the analysis of rRNA sequences, are currently used for reliable phylogenetic studies, which enable a more natural classification system to be established. However, despite the effective application of these techniques in PCR-mediated identification systems, they are not yet currently available in the routine mycology environment. The definition of the phylogenetic-species concept is also complex, and several different definitions have been proposed. There appear to be two main approaches. The character-based concept, or diagnostic approach, defines a species as a group of organisms that have common observable attributes or combinations of attributes. In contrast, the history-based concept insists that organisms must be historically related before they can be considered members of any given species. Some of the different species concepts currently in use are difficult to distinguish, and so mycologists should be familiar with them and clearly define which they are using to recognize species (Figure 11.28).

11.4 PROTOZOA

11.4.1 Habit and Habitat

Protozoa are the eukaryote Protista which have the combination of locomotion and heterotrophic nutrition that is characteristic of animals, together with some other organisms that clearly have close relations with them. By definition the Protozoa are simple animals, but, since the evolutionary divergence that led to 'higher' animals and plants probably involved flagellated eukaryotes, among which there is still abundant evidence of adaptive evolution, it is not surprising that some organisms which are indisputable plants show clear evidence of relationship with genuine animals.

It is often stated that the Protozoa are unicellular, but while there are many species within the group that have the characteristics of single cells, there are many protozoan species with two or more nuclei enclosed within a single membrane and many more species which have a colonial organization of interconnected individuals. The size of protozoan individuals ranges from a few μ to several mm. Protozoa frequently contain a range of specialized and complex organelles and may have complex life cycles in which the individual passes through a number of specialized stages. The organization and life of the protozoan individual is different in character from that of a cell of a multicellular organism, and many protozoologists prefer not to refer to an individual protozoan as a cell. Multicellular animals must be assumed to have evolved from protozoan ancestors in the distant past, but there is a lack of evidence to support speculations about the forms of Protozoa which may have been the starting points for these evolutionary developments.

The major sub-groups of Protozoa are separated on the basis of their organelles of locomotion and characteristics of their life cycles. All members of two of the groups are parasitic, with limited powers of locomotion and complex life cycles; they have spores of characteristic types. In the Conidospora the complex spores contain one or several amoeboid sporoplasms (the infective individuals) and possess one or several polar filaments which are ejected before emergence of the sporoplasms from the spore. The spores of the Sporozoa contain infective sporozoites of characteristic shape; they are simpler and do not have polar filaments. Members of the Ciliophora (ciliate Protozoa) are characterized by the possession of numbers of cilia (organelles of unilateral beat), and (normally) two types of nucleus. The flagellate Protozoa (Mastigophora), which have only one type of nucleus and move by means of flagella (organelles with an undulatory beat), are now grouped with the amoeboid Protozoa (Sarcodina), whose organization characteristically depends on pseudopodia for locomotion and food capture. The occurrence in some organisms of both flagella and pseudopodia, either simultaneously or at different parts of the life cycle, led to the taxonomic linking of the flagellates and amoebae to form the Sarcomastigophora.

11.4.2 Nutrition

The nutrition of some of the pigmented flagellates may be exclusively photoautotrophic, but the majority of Protozoa are heterotrophs, dependent upon organic molecules as sources of carbon and nitrogen as well as for energy. Many protozoans (phagotrophs) eat other organisms or large fragments of these by phagocytosis, taking the organic material into food vacuoles for digestion. Such vacuoles may be formed at cell mouth or cytostome in ciliates and some flagellates, or at many parts of the body surface in amoeboid organisms and some others. Numerous Protozoa (osmotrophs) feed saprozoically, absorbing soluble organic materials, with or without pinocytosis; in many species the relative importance of this method of feeding has not been evaluated, but in some forms, including many parasites, it is probably the only mode of nutrition. Since the Protozoa include both autotrophic and heterotrophic forms, it is not surprising that dietary requirements range from a simple selection of inorganic salts to a complex mixture containing many amino acids, vitamins and other growth factors. The waste products of metabolic activities of Protozoa are believed to escape by diffusion, although in some organisms crystalline deposits may accumulate, internally or in skeletal structures. The contractile vacuole is believed to serve an osmoregulatory rather than an excretory function.

11.4.3 Ultrastructure

Studies of protozoan ultrastructure indicate that the basic form of a species can be related to a characteristic cyto-architecture which is primarily fibrous, but which also includes membrane elements and secreted structures. The structure and formation of this architecture has been fully studied in relatively few species, but the organization of some skeletal structures and of the ciliate and flagellate pellicles shows abundant evidence of the utilization of microtubular fibres, bundles of thin filaments and membranous vesicles, both as components of the complex structures and as morphogenetic organizing systems.

11.4.4 Reproduction

In most protozoan species the ability to reproduce is retained by all individuals although they may have a complex body organization. Usually, reproduction involves binary fission of an organism to form two more or less equal daughters, so that no parent remains. Such reproduction involves not only the replication of the nucleus, but also the production of new organelles; in many cases existing organelles may be broken down and a new set produced in each of the daughters. Fission of an organism to produce a larger number of daughters is not uncommon; in the best-known example, called schizogony, the nucleus of an organism divides many times in succession, and then each nucleus separates from the main mass with a small amount of cytoplasm and many daughter organisms are formed.

Some sort of sexual process occurs in all major groups of Protozoa, although there are species in several groups which are not known to have any form of sexual activity, and there are also forms with modified sexual phases which have lost the full genetic potentialities of meiosis and cross-fertilization. Some flagellates and all sporozoans are haploid and show a meiotic division of the zygote nucleus, while other flagellates, the ciliates, and some sarcodines are diploid with meiosis in the formation of the gametic pronuclei. A life cycle involving intermediary meiosis and an alternation of a haploid generation with a diploid generation (comparable in development with that of most cryptogamic land plants) is shown by several species of foraminiferan sarcodines and may be normal throughout that group. The occurrence of sexual phases may be restricted to certain parts of the life cycle, as in foraminiferans and sporozoans, or may involve unspecialized individuals. Examples are known in both flagellates and ciliates where the likelihood of fertilization between clones (rather than within clones) is increased by mating-type systems, in which only individuals of different mating-types show sexual fusion. These mating types are analogous to the cross-pollination devices of some higher plants. They bear no direct connection with sexuality, since, in ciliates at least, each individual is monoecious and produces gametes of both types.

In some Protozoa, the life cycles are regular and include a consistent sequence involving divisions, sexual fusion, cyst formation, etc.; such regularity is characteristic of parasitic forms. The life cycles of other Protozoa are more dependent upon environmental changes of a physico-chemical or biological nature for the determination of cyst formation or characteristic types of reproduction, e.g., resistant cysts are commonly produced by soil Protozoa which are subjected to periods of drought, or predatory Protozoa may encyst when the supply of food organisms is exhausted. The alternation of a motile larva and sessile adult might be regarded as a simple sort of cycle, but such organisms may also show sexual and cyst phases of irregular frequency.

Protozoa may occupy any trophic level in the food chains of natural communities where free water is present. Autotrophic flagellates are abundant in the sea and in fresh waters as well as in symbiotic associations with animals at various levels of organization, including other Protozoa. Herbivorous members of the sarcodine and ciliate groups in particular form important items in the diet of a large number of animals including carnivorous ciliates and sarcodines. Of particular importance in the economy of many communities, in damp terrestrial habitats as well as in truly aquatic ones, are the saprozoic and bacteria-feeding Protozoa which make use of the substances and organisms involved in the final 'decomposition' level of food chains and themselves form the food of organisms at a higher trophic level, thereby recirculating organic matter.

11.4.5 Classification

11.4.5.1 Mastigophora

Mastigophora includes flagellates, sarcodines and the taxonomically enigmatic opalinids. They are forms moving by flagella or pseudopodia, or both, and have only one type of nucleus. The flagellates and opalinids can further be distinguished from ciliates by the form of binary fission, which in the flagellated organisms is basically longitudinal, producing two mirror-image daughters, while in ciliates anterior and posterior daughters are produced by a transverse division running across the kineties. Many of the simpler flagellates possess chlorophyll and other pigments, or are heterotrophic derivatives of the autotrophic forms. These flagellates are classed by protozoologists as Phytamastigophorea, many of which are described in the section on Algae. The class Zoomastigophorea contains mainly parasitic forms with fairly complex structure as well as a small number of simple heterotrophic forms for which no relationship with any group of the Phytamastigophorea has yet been established.

Many pigmented flagellates feed osmotrophically or phagotrophically and at the same time use photosynthesis. Colourless, heterotrophic species occur in most of the major groups of algal flagellates; many of these are frequently classed as protozoa, and receive scant treatment in discussions on algae. Amongst the chrysomonads (= Chrysophyta) there are pigmented forms whose nutrition is supplemented with phagotrophy by means of pseudopodia, often aided by the collection of particles by flagellar activity. One of these flagellates, *Ochromonas malhamensis,* has a dietary requirement for vitamin B_{12}, and has been used in bioassay for this vitamin. Among the heterotrophic euglenoids are *Peranema,* whose diet includes algae and protozoa which are taken in through a cytostome which opens separately beside the apical invagination from which the flagella emerge, *Copromonas,* which feeds on bacteria, and osmotrophic species like *Astasia* and *Euglena.* One of the commonest of all freshwater flagellates is the osmotrophic cryptomonad *Chilomonas* (Figure 11.29). The predominantly marine dinoflagellate group (part of Pyrrophyta) includes many species that are parasitic in animals or plants as well as free-living phototrophic forms and many phagotrophic forms with diverse body forms. The majority of these plant flagellates have few flagella associated with simple root structures, and the body surface is often strengthened with cellulose or proteinaceous pellicular thickenings or scales. By contrast, members of the Zoomastigophorea frequently have large numbers of flagella whose bases are associated with complex fibrous organelles, and the body surface is seldom stiffened with anything more than sub-pellicular microtubules. The collar-flagellates form a distinctive group of zooflagellates; the characteristic form of cell (*Codonosiga botrytis*), with its single flagellum surrounded by a fine collar of microvilli, is also found in the multicellular sponges, and the common possession of this unique structure suggests a phylogenetic connection between these two groups (Figure 11.30). Many collar-flagellates are sessile, with or without stalks, and some of them secrete a lorica (a loose case within which the animal lives). These loricate forms particularly are liable to be confused with the bicosoecid flagellates, which however, lack the collar, and which have two flagella, the shorter of which attaches the flagellate to the base of the lorica. Organisms of both groups use the flagella to collect bacteria and organic particles from the water.

A number of forms showing features of both amoebae and flagellates may frequently be classed with the zooflagellates; these include species which are capable of transformation from an amoeboid form to a flagellate, like *Noegleria,* which is common in soil, and forms which possess flagella and pseudopodia simultaneously, in which case the pseudopodia may be lobed, e.g., *Mastigamoeba* or fine axopodia, e.g., *Actinomonas* and *Dimorpha* (Figure 11.30). The most important group of flagellate parasites are *Trypanosoma* and its relatives. They share with *Bodo* and some related forms the possession of a DNA rich mitochondrial organelle called the kinetoplast, which occurs in the region of the flagellar base. *Bodo* has two flagella, one anterior and one trailing posteriorly, and is commonly found in water rich in organic matter, when it feeds on bacteria which are taken in through a tubular channel at the cytostome. Some related biflagellate forms

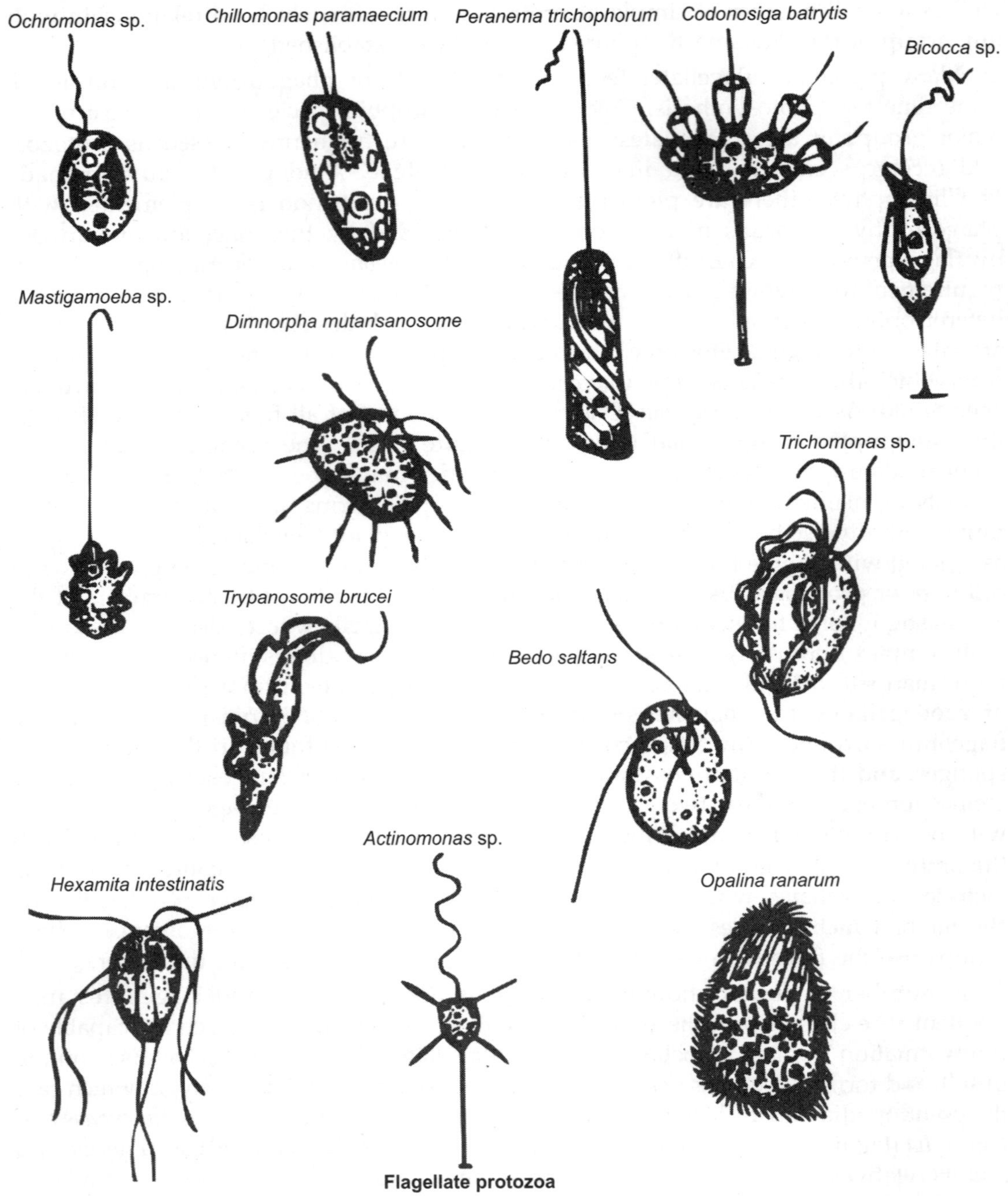

Figure 11.30 Flagellate and Opalinid protozoa.

are parasitic. The body form of *Trypanosoma* varies at different stages of the life cycle; in the familiar form from the blood plasma of vertebrates, the elongate flagellate has a single flagellum which runs forward from its basal body near the posterior end of the organism, and is connected to the body along much of its length by an extension of the flagellar membrane which forms an undulating membrane. In other stages of the life cycle the flagellum may originate near the middle or at the anterior end of an elongated body, or the body may be rounded and lack an emergent flagellum; these stages are known respectively as crithidial, leptomonad, and leishmanial stages, since their characteristic forms are found in the related genera *Crithidia* and *Leptomonas,* parasites in insects and other invertebrates, and *Leishmania,* which is responsible for kala azar and other diseases of mammals in which white blood corpuscles are entered by the flagellate. All forms may not be present in the life cycle of all trypanosomes.

There are many species of *Trypanosoma* which spend part of their life cycle in the circulatory systems of vertebrate hosts and part of the cycle in blood sucking invertebrates, such as insects and leeches. *Trypanosoma gambiense* and *T. rhodesiense* which cause sleeping sickness in man in Africa, are transmitted by tsetse flies; parasites sucked by the fly from an infected man enter the stomach of the insect, where they multiply, undergoing changes in form, and then migrate to the salivary glands, whence they may be injected into the next man bitten by the insect. There are non-pathogenic species as well as the well-known pathogens which cause diseases of man and domestic animals; it seems that severe disease is caused only by trypanosomes which leave the blood plasma and enter other tissues (Figure 11.31).

The more complex zooflagellates are sometimes referred to collectively as metamonads, and it is convenient, if artificial, to subdivide these into polymastigote and hypermastigote flagellates. Several species of the polymastigote *Trichomonas* occur as parasites in man and domestic animals, and this type illustrates several features of these metamonad flagellates, although it is one of the simplest. There is a complex of structures called the karyomastigont formed by an association of the nucleus, the flagella and internal organelles originating near the flagellar basal bodies. Many species of *Trichomonas* have three free flagella and one recurrent flagellum attached to the body by an undulating membrane; the basal bodies of all four flagella lie close together (Figure 11.31). Of the three large organelles which arise near the basal bodies, the axostyle, which runs down the middle of the body and may project posteriorly, is a hyaline rod enclosed in a sheath of microtubules, and the costa and parabasal fibre are striated. The thick fibre which curves round the side of the body beneath the undulating membrane of the recurrent flagellum is the costa, which is thought to be contractile and responsible for changes in body shape, and the slender parabasal fibre runs near the nucleus and is associated with parabasal (Golgi) bodies. *Trichomonas* is generally believed to be saprozoic, but there are some reports of phagotrophy, mostly involving bacteria.

In some other polymastigote flagellates there may be two simpler karyomastigont systems, e.g., in *Hexamita* and *Giardia* (Figure 11.32), which are parasitic in vertebrates, or one to several hundred karyomastigonts in some species symbiotic in wood-eating

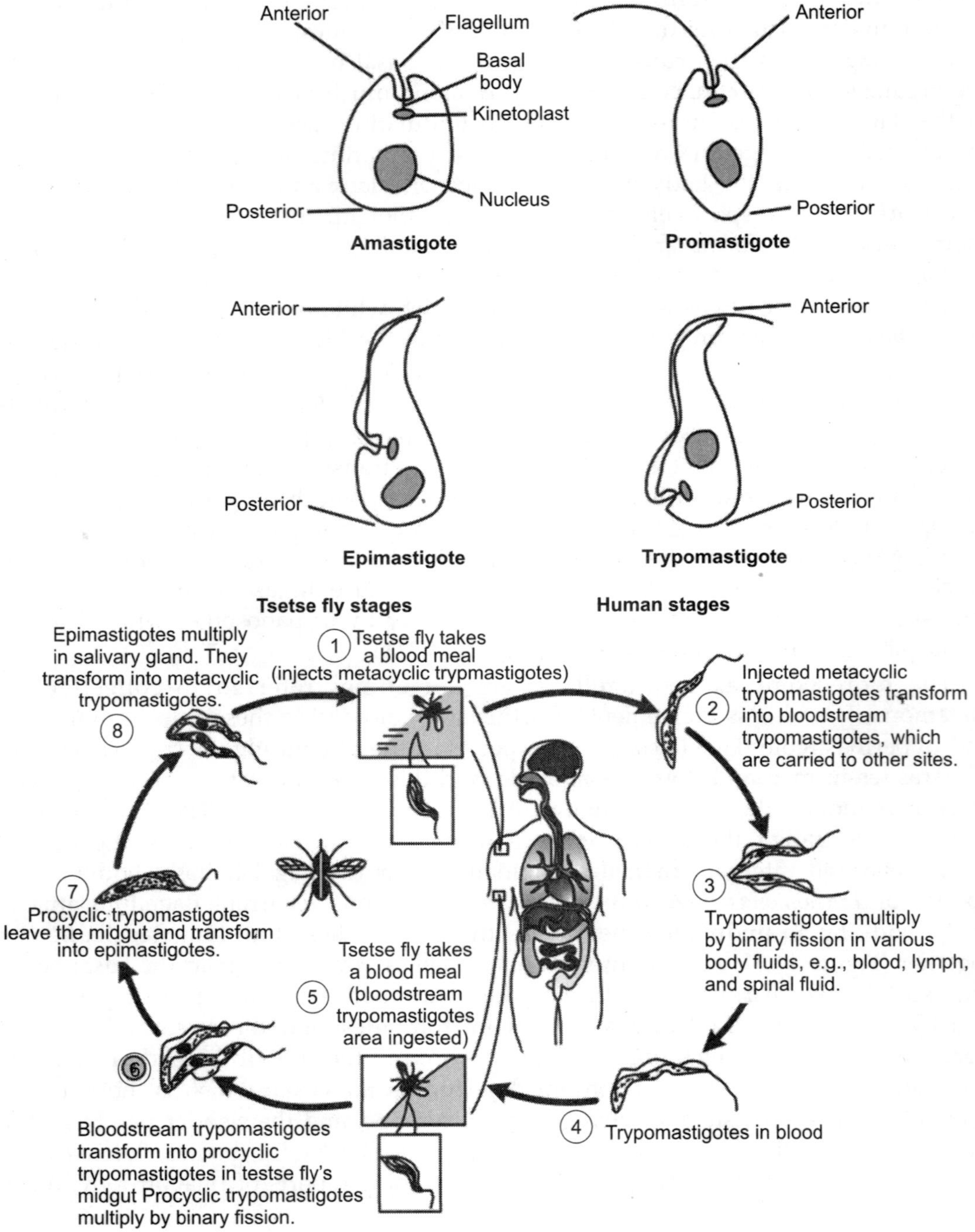

Figure 11.31 Life cycle of *Trypanosoma brucei* and major stages of *T. brucei*.

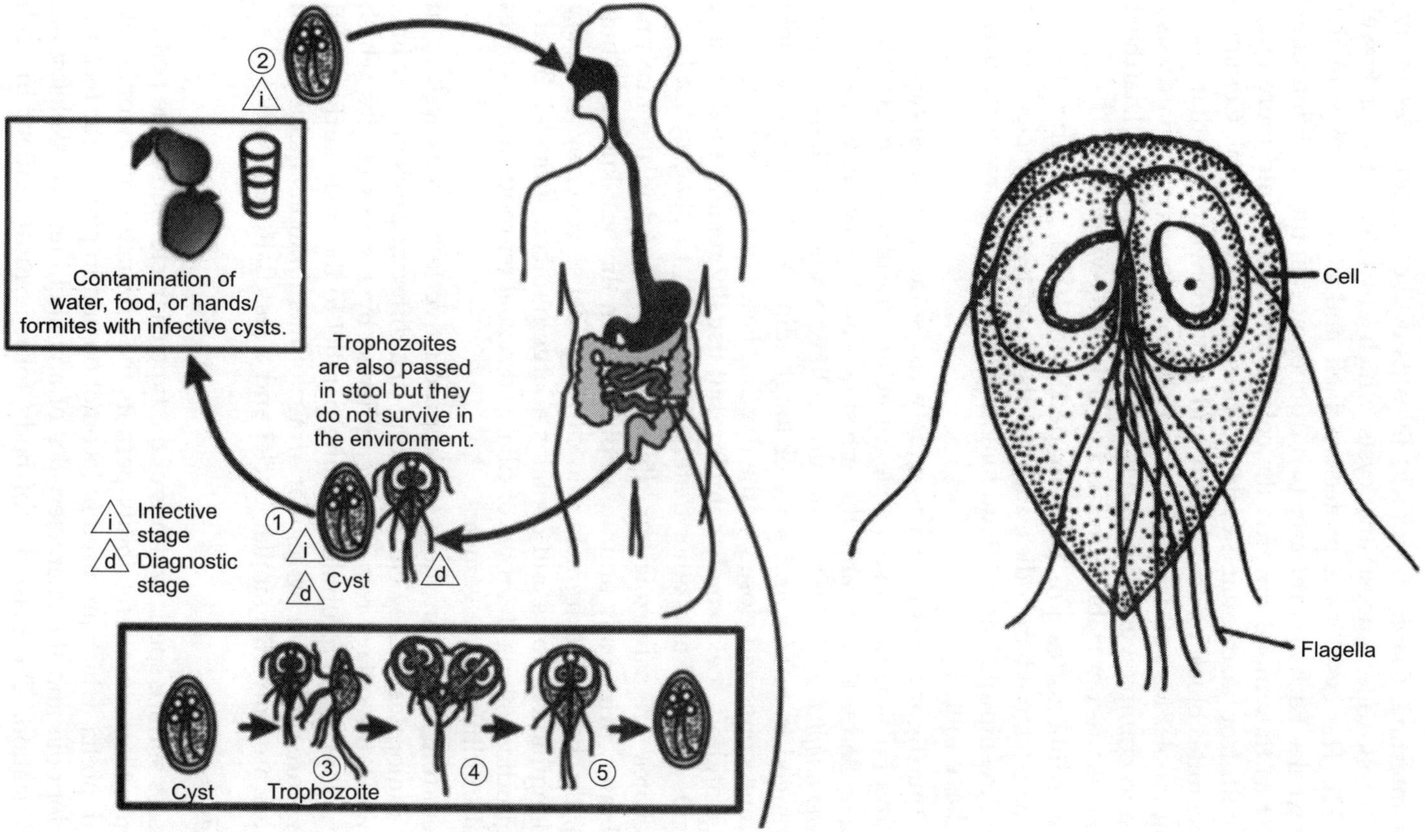

Figure 11.32 Life cycle of Giardia.

insects. *Giardia intestinalis* also known as *Giardia lamblia* or *Giardia duodenalis* causes giardiasis in the small intestine. *Giardia intestinalis* lives as active trophozoites in the small intestine. Some trophozoites encyst into cysts which are released in a bowel movement (Figure 11.32). The cyst has a protective shell and it can survive in the environment for many weeks. Each cyst releases two trophozoites in the small intestine. The trophozoites encyst as they move towards the colon. Cysts are found more often in firm stool whereas both trophozoites and cysts are present in loose stool. Common giardiasis symptoms include: bloating, bad breath and farts, dehydration, diarrhoea, fatigue, greasy floating stools, loss of appetite, malabsorption (in chronic giardiasis), nausea, stomach ache and cramps, vomiting (rarely), weakness, weight loss. Diarrheal can be fatal. Another not so recognizable effect is the lack of B_{12}-vitamin.

Giardia intestinalis trophozoites are pear-shaped and 10-20 μm long. Other characteristics include: flagella, median bodies, sucking disks and two big nuclei. *Giardia intestinalis* cysts are oval to ellipsoid and 8-19 μm long. Immature cysts have two nuclei, whereas mature cysts have four.

The details of the mastigont system vary widely; in some the axostyle is an extensive array of microtubules and is responsible for active undulations of the body. The hypermastigote flagellates also inhabit the intestines of insects: they are often xylophagous, e.g., *Trichonympha*. Members of this group have a single nucleus, a large number of flagella arranged in tufts, spiral rows or longitudinal rows, and a variable pattern of associated fibres connected closely to the flagellar bases or more remote from them. The body is large and often extremely complex, but usually retains a naked area through which fragments of wood and other materials are taken in phagocytically.

The opalinid 'flagellates' are saprozoic parasites which are found in the intestine of cold-blooded vertebrates, principally anuran amphibians. In the best-known genus, *Opalina* (Figure 11.30), the body is extremely flat, carries a dense covering of ciliary organelles in oblique longitudinal rows and contains a large number of nuclei. Binary fission is typically longitudinal, along the rows of cilia, and sexual reproduction involves the fusion of two small multiflagellate gametes.

The opalinids differ from ciliates in these two reproductive features as well as in the absence of nuclear dimorphism; they clearly belong to a different phylogenetic line. The body surface between the ciliary rows is strengthened by a series of ribs which contain a single vertical row of microtubules running parallel to the body surface, and the ribs and ciliary rows are underlain by regular arrays of membranous inclusions. In two related genera there are only two similar nuclei and a less flattened body.

11.4.5.2 Sarcodines

Sarcodines do not possess surface specializations of the type characteristic of the pellicle of ciliates and flagellates; their surface flexibility permits the production of cytoplasmic extensions called pseudopodia, which are used in locomotion and in phagocytic feeding. The form of the pseudopodia and the characteristics of any shell, test, or skeleton are features used in classification. The plasticity of body form is most evident in such naked forms as *Amoeba* (Figure 11.33), where the pseudopodia are broadly lobed. The structure is simple, perhaps because of the use of the whole body in movement, so that large permanent organelles could not be retained; smaller structures like the nucleus, contractile and other vacuoles, mitochondria, and granular inclusions move around with the more fluid interior cytoplasm. Contractile activity of the surface layer of gelled cytoplasm is held to cause the cytoplasmic streaming that is the most obvious feature of the movement of the organism. These lobose amoebae range in size from a few microns to several millimeters in diameter, and may have one, two, or many nuclei. While the majority of them are free-living forms, there are also some important parasities, e.g.,

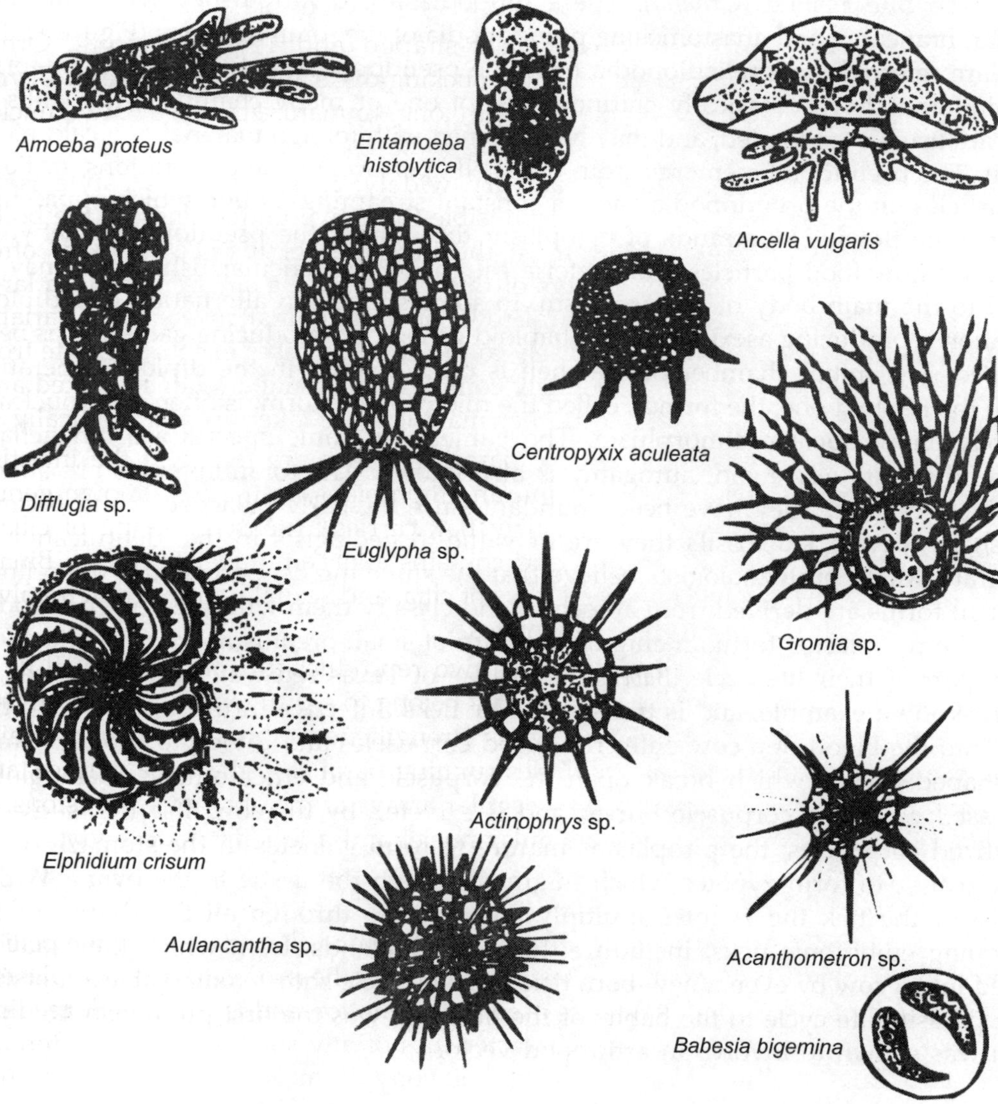

Figure 11.33 Amoeboid protozoa.

Entamoeba histolytica (Figure 11.33), which is responsible for amoebic dysentery in man, and which passes as a cyst from the faeces of one host to the mouth of another.

Some testaceous (shelled) forms have lobed pseudopodia which emerge from the aperture of the test to move the organism and collect food (Figure 11.33). The basis of the test is a chitinous membrane which may be thickened with further chitin *(Arcella)*, secreted siliceous plates *(Euglypha)*, or encrusted with sand grains *(Difflugia)* or diatom frustules *(Centropyxis)*. These organisms are free-living inhabitants of moss, soil, and fresh water; they can retreat within the shell in adverse conditions.

Slender branching pseudopodia called filopodia are a characteristic of some forms, both naked and testate (*Gromia*). These filopodia are distinguished from the fine, granular, branching and anastomosing pseudopodia of the foraminiferans (Figure 11.33), which are referred to as reticulopodia from the pseudopodial network that they form. Foraminiferans have a basically chitinous test of one or many chambers, the walls of which are usually perforated, and may be thickened with foreign material or a calcareous deposit. The pseudopodia emerge from the shell aperture, or the perforations, or both. The particles on the pseudopodia show a constant streaming motion which apparently indicates the flowing of strands of cytoplasm throughout the pseudopodial network, and shows how food particles and bacteria caught by the reticular fishing net may be carried to the main body of the organism. In some species an alternation of a diploid generation reproducing asexually and a haploid generation producing gametes has been described. The initial chamber of the shell is often smaller in the diploid generation than in the haploid one; the former, called the microspheric form, is often multinucleate, and may show nuclear dimorphism. The gametes in some species are biflagellate, and in others are amoeboid; autogamy is also known. The foraminiferans are almost exclusively marine. They have been abundant since the early palaeozoic, and because their shells leave good fossils they are of value to geologists in the identification of rock strata. Many protozoologists believe that the slime molds, particularly the cellular (acrasian) forms are derived from amoebae, and classify them with these pseudopodial forms. The piroplasms form an enigmatic group of small organisms that are amoeboid during part of their life cycle. *Babesia,* the cause of Texas red-water fever in cattle, is the best-known example, and is transmitted by ticks. Infective forms injected with tick saliva into the blood of a cow enter red blood corpuscles and divide to produce small pear-shaped bodies which break open the corpuscle and enter others; haemoglobin is released when the corpuscle bursts, and is excreted by the cow. If a tick sucks up parasitized corpuscles, the piroplasms mature to form gametes in the stomach of the tick, and fuse to form zygotes which migrate through the tissue to the ovary. Within the ova of the tick the zygotes multiply and migrate through all the tissues of the developing embryonic ticks, including the salivary glands, from which they may be injected into a cow by even a new-born tick. This life cycle shows remarkable adaptation of the parasite life cycle to the habits of the tick, and was the first protozoan life cycle which was shown to involve in arthropod vector.

The spherical floating amoebae (Actinopodea) form a distinct class of sarcodines. The pseudopodia, which are slender and radiate all round the body, are usually supported by an axial structure, when they are called axopodia; some forms have filopodia or reticulopodia. Many species produce some sort of skeletal material, either internally or externally.

The marine Radiolaria are best known from their skeletons of silica, which are abundant in the ooze of certain parts of the ocean floor, and are common fossils. The skeleton is deposited around a membranous central capsule with pores that allow communication between an inner region, containing one or more nuclei and other bodies, and an outer region in which food vacuoles occur. Many species have symbiotic algae in the outer region, but may also feed on small animals and plants caught by the filose or reticular pseudopodia. Some of the best-known species are colonial forms. Reproduction is not well known, but may include binary and multiple fission, and a sexual phase involving flagellate gametes is said to occur in certain radiolarians. In the Acantharia (Figure 11.33), which are also marine and which have often been grouped with Radiolaria, there is a skeleton of radial spines impregnated with strontium sulphate which support the pseudopodia. Their body organization appears broadly similar to that of radiolarians, with fine reticulate pseudopodia in addition to the axopodia; symbiotic algae may occur within the capsule. Some acantharians have been shown to reproduce by multiple fission which takes place in an encysted stage, and leads to the production of numerous flagellate spores.

A central capsule is not present in the Heliozoa (Figure 11.33), which are predominantly inhabitants of fresh water. They generally have axopodia supported by bundles of microtubular fibres around which the pseudopodial cytoplasm streams, and some forms have siliceous scales or spines, or a reticulate skeleton. Ciliates or other prey organisms are caught by the axopodia and surrounded by cytoplasm. One or many nuclei may be present in the central part of the body, and the surface cytoplasm between the basal parts of the axopodia is vacuolated. Heliozoans divide by binary or multiple fission, and may also reproduce sexually, in some cases autogamously (e.g., *Actinophrys*).

11.4.5.3 Sporozoa

All members of this subphylum are parasites, often with complex life cycles (Figure 11.34). In the main class the Telosporea, the infective stage is a vermiform sporozoite which may be injected naked into a host by an insect vector in a few forms or, more usually, may hatch in the host intestine from spores which are eaten by the host. Since they do not have ciliary organelles, even the most active sporozoites show only weak motility, gliding or performing slow flexural movements. The sporozoites migrate to specific host cells and feed saprozoically or sometimes phagocytically within the cell. At this stage parasites of the sub-class Gregarinia leave the host cell and continue to grow in an organ cavity of the host body, feeding saprozoically, e.g., in the seminal vesicle of the earthworm in the case of *Monocystis* (Figure 11.33). When the feeding stage (trophozoite) of the parasite is mature, it encysts with another mature trophozoite, and in both individuals nuclei divisions are followed by cytoplasmic separation to form numerous gametes. These fuse in pairs and each zygote becomes a spore within which a number of sporozoites develop. Reproduction in this case occurs only at spore formation (sporogony); in a few gregarines the trophozoite also undergoes multiple fission (schizogony) to produce a new generation of feeding individuals.

In the sub-class Coccidia the trophozoites do not grow very large since they remain intracellular. After growth they undergo schizogony to produce many new infective individuals (merozoites) which enter other cells, so that a very heavy infection can be built up by repeated schizogonies. Trophozites ultimately develop into gametocytes, some of which give rise to single macrogametes and others to several microgametes, which may be flagellate. Fusion of gametes in pairs produces zygotes, each of which develops into a few or many sporozoites, either naked or enclosed in a spore. Naked spores are characteristic of the Haemospotina (e.g., *Plasmodium),* and in the Eimeriina (e.g., *Eimeria)* the sporozoites are enclosed in a walled spore (or sporocyst). Members of the genus *Eimeria* cause disease (coccidiosis) in domestic mammals and birds as well as in other animals. Spores that are eaten hatch in the intestine and sporozoites enter the epithelial cells of the intestine or associated glands. In these cells the parasites grow and undergo repeated schizogonies, each one producing hundreds of minute merozoites; severe tissue damage or even death is caused.

Certain trophozoites mature to form gametes, and the small biflagellate microgamete seeks out a macrogamete and fuses with it to form a zygote. Each zygote produces a resistant cyst wall (oocyst) within which meiotic division of the nucleus takes place and a number of sporocysts develop, each containing a number of sporozoites. The number of sporocysts per oocyst and the number of sporozoites per sporocyst are generic features of diagnostic importance, e.g., in *Eimeria* these numbers are 4 and 2 respectively, while in *Isospora* (a species of which occurs occasionally in man) the oocyst contains 2 sporocysts, each with 4 sporozoites. The whole of the life cycle in these forms takes place in a single host individual.

Malaria, which is probably the most important human disease caused by a pathogenic protozoan, is caused by species of *Plasmodium* which are transmitted from man to man by *Anopheles* mosquitoes (Figure 11.34). Many other species of *Plasmodium* occur in other terrestrial vertebrates.

Slender sporozoites some 10 μ long, are injected with the saliva of the insect into the blood of the vertebrate. The sporozoites first enter cells around the blood vessels in the liver, and after growing for several days they undergo schizogony to produce hundreds of very small merozoite which enter red blood corpuscles. Inside the erythrocytes the trophozoites grow as intracellular phagotrophs, maturing in about 2 days. The full-grown trophozoite undergoes schizogony, bursting the erythrocyte and liberating merozoites and toxic products into the blood. The merozoites may enter other erythrocytes and continue a sequence of schizogonies, releasing toxic products into the blood at more or less regular intervals and producing the recurrent fever characteristic of malaria. Sooner or later some of the trophozoites develop into gametocytes which remain dormant within corpuscles in the bloodstream until sucked by a mosquito. In the stomach of the insect the gametocytes develop into gametes, producing either

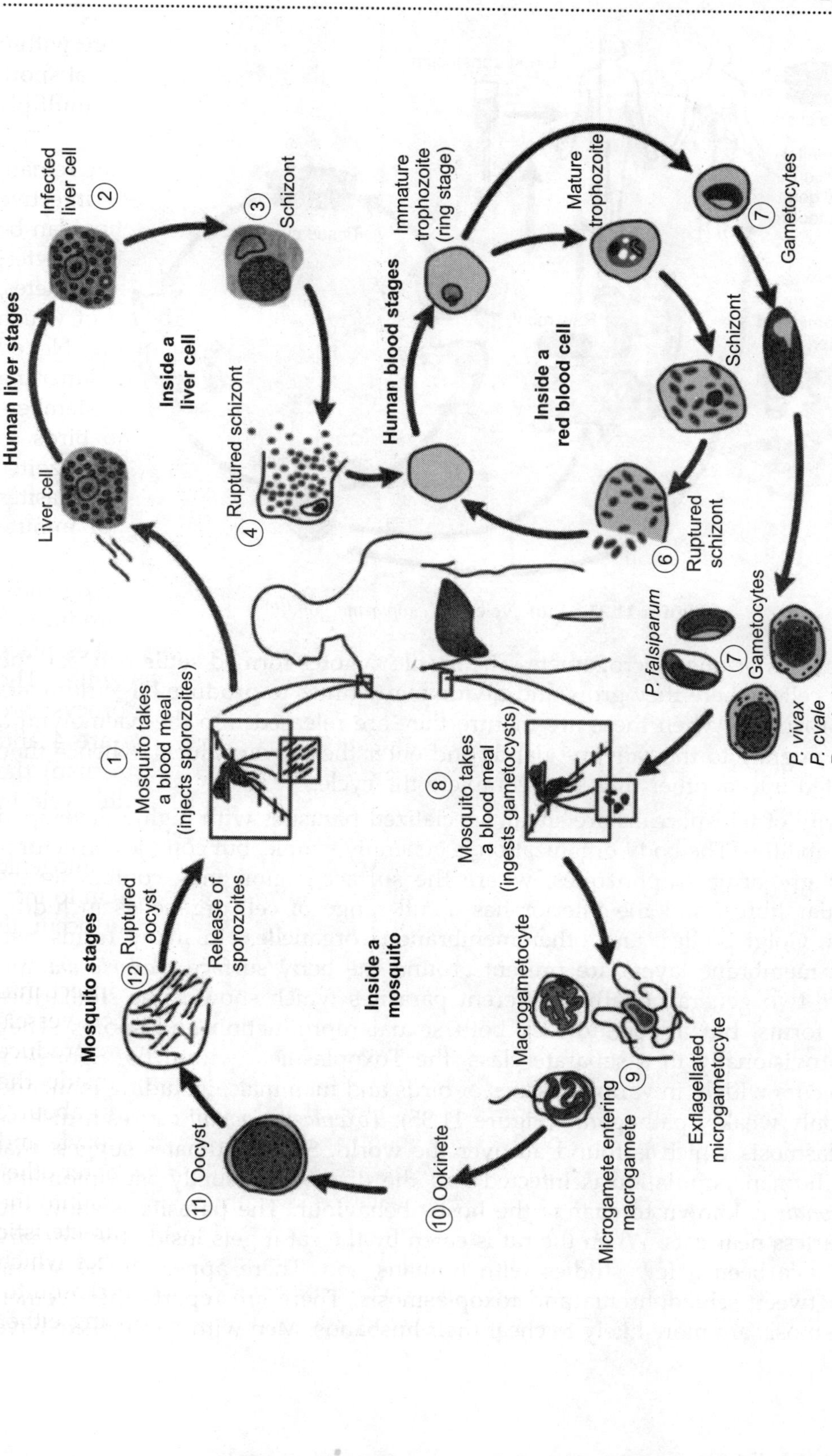

Figure 11.34 Life cycle of the *Plasmodium.*

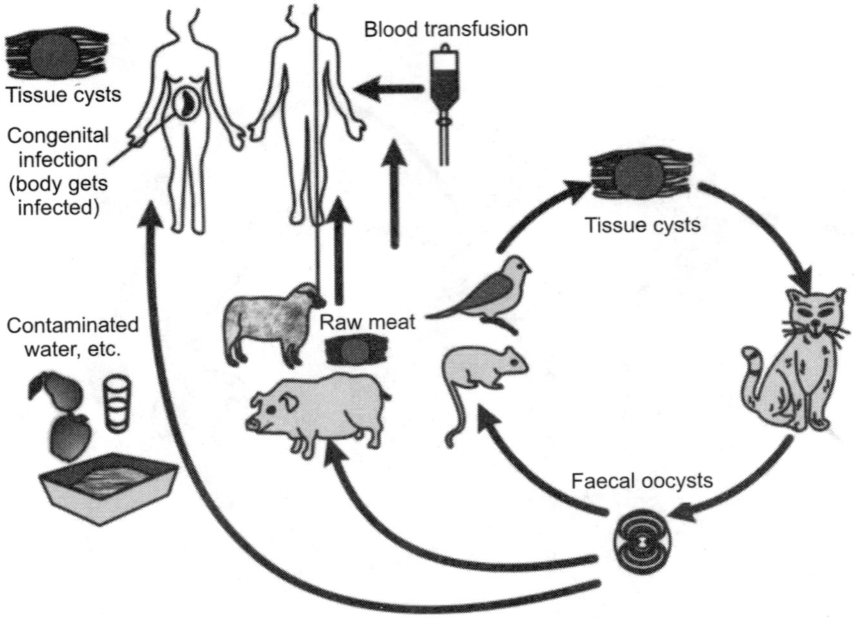

Figure 11.35 Life cycle of *Toxoplasma gondii.*

four microgametes or one macrogamete. The motile zygotes formed settle between the stomach wall cells, where they grow and divide many times to produce large numbers of naked sporozoites. When these are mature they are released into the haemolymph of the insect, migrate to the salivary glands and enter the salivary ducts, whence they may be injected into another man and complete the cycle.

The majority of telosporeans are small, specialized parasites with highly developed reproductive abilities. The body organization is generally simple, but complex structures are found in gregarine trophozoites, where the surface region may contain arrays of microtubular fibres, and the interior has a full range of cell organelles including mitochondria, Golgi bodies, and other membranous organelles. In many forms two or three unit membrane layers are present around the body surface. *Toxoplasma* and *Sarcocystis* are two genera of rather different parasites which show some similarities to coccidian forms, but appear to lack both sexual reproduction and spores. They are placed provisionally in a separate class, the Toxoplasmea, within the Sporozoa. *Toxoplasma* occurs widely in various tissues of birds and mammals, including man, and is probably only weakly pathogenic (Figure 11.35). *Toxoplasma gondii* causes a disease called toxoplasmosis which is found all over the world. Some estimates suggest that over 30% of human population is infected and the disease is usually asymptomatic. *Toxoplasma gondii* is known to change the host's behaviour. The parasite is shown to make rats fearless near cats. When the rat is eaten by the cat it gets inside the primary host. There have been a few studies with humans, too. There appear to be strong correlation between schizophrenia and toxoplasmosis. There are reports that women with toxoplasmosis are more likely to cheat their husbands. Men with the parasite have shown to be more aggressive. The trophic individuals are 5 to 10 μ long and are usually found intracellularly, where they multiply by binary fission or perhaps schizogony. The means of transmission is unknown. There is no resistant spore. *Sarcocystis* is found in the muscles of terrestrial vertebrates, including man. Tubular bodies found in the muscles contain vast numbers of banana-shaped parasites, some 10 μ long. These may multiply by binary fission, but little is known of their biology or of their means of transmission. One report claims that the spores produce septate mycelia of a fungal type.

Another curious small group of parasites is the Haplosporea (e.g., *Haplosporidium*). These have simple resistant spores, but do not show sexual reproduction. They are normally parasites of invertebrates, but have been reported from fish. The amoeboid infective individual which emerges from the spore enters cells of the intestinal epithelium and grows to produce a multinucleate form which subdivides into uninucleate bodies, each of which may develop into a spore. Spore formation involves more stages in some forms. This group forms a link with the Conidospora, since both have an amoeboid sporoplasm; and the spores are similar, except for the absence of a polar filament in spores of the Haplosporea (Figure 11.36).

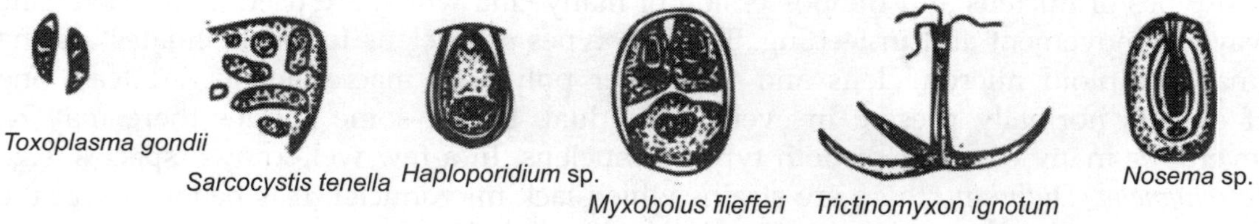

Figure 11.36 Representative members of Sporozoa.

11.4.5.4 Conidospora

Parasites of this subphylum have complex spores surrounded by a single membrane or a two- or three-valved structure, they contain one to six polar filaments, usually coiled in polar capsules, and one to many sporoplasms. The more complex spores found in the class Myxosporidea are formed from several cells during sporulation, and there is some doubt that these organisms are really protozoans. They occur in lower vertebrates and many types of invertebrates. In the host intestine, where the polar filaments are said to be extruded and anchor the spore, the amoeboid sporoplasms hatch from the spore and migrate through the gut epithelium to the organs characteristic of the species. Within the tissues the parasites grow, become multinucleate and split up into multinucleate masses, in each of which one or two spores develop. In *Myxobolus,* which causes the fatal boil disease of such fish as the carp, the parasite grows in the muscle or connective tissue producing large bulges at the surface of the fish which burst to release vast numbers of spores, each with two polar capsules and a single sporoplasm. The freshwater worm *Tubifex* is often parasitized by *Triactinomyxon,* whose spore has three valves, three polar capsules and eight sporoplasms. It is often stated that nuclear fusion occurs in the formation of the sporoplasms of myxosporidians, but the complete cytogenetic cycle remains controversial.

The simpler spores of members of the Microsporidea are of unicellular origin. The chitinous membrane of the spore has a single valve which encloses a single tubular polar filament, through which the single amoeboid sporoplasm emerges on hatching. The parasites occur in animals of most phyla, but are most commonly found in insects and fishes, where they multiply within host cells and cause hypertrophy of the host tissues. The spores hatch in the gut of the host and the sporoplasms migrate through the gut wall and around the body to enter specific cells within which they multiply. Later the trophozoites mature to form spores. The parasite remains uninucleate throughout its life cycle, except for the final stages of spore production in some species. The best-known microsporidian diseases are caused by species of *Nosema* which parasitize intestinal cells of honeybees and the cells of various tissues of the silkworm (Figure 11.36).

11.4.5.5 Ciliophora

The most distinctive features of typical members of the Ciliophora are the possession of two types of nucleus and the possession of many cilia which are used in characteristic ways in movement and in feeding. The two types of nucleus found in ciliates are the smaller diploid micronucleus and the larger polyploid macronucleus. At least one of each is normally present in every individual, but in some species there may be several or many of either or both types of nucleus. In a few well-known species, e.g., *Tetrahymena, Didinium,* there are strains which lack micronuclei. It is believed that the macronucleus has a somatic function, providing for the routine synthetic activities of the cell, while the micronucleus is not concerned with these activities. Both types of nucleus divide during binary fission of a ciliate, and nuclei of both types pass to each daughter. However, during the sexual process, which in ciliates is of a type called conjugation, only the micronucleus provides genetic continuity.

Conjugation involves the exchange of genetic material between two ciliates which come together with a temporary, local, cytoplasmic fusion; a zygote nucleus is normally formed in both ciliates during the process. At conjugation the micronucleus divides meiotically and two of the products of this division form the haploid pronuclei. A pronucleus from each conjugant ciliate crosses the cytoplasmic bridge to the body of the other ciliate and fuses with the stationary pronucleus to form a syncaryon. The two conjugants separate and undergo a process of nuclear reorganization; only later does binary fission occur, so that increase in numbers is not an essential part of the sexual process.

During conjugation the old macronucleus of each conjugant ciliate begins to disintegrate, and one of the products of the first mitotic division of the micronuclear syncaryon develops to form a new polypoid macronucleus by repeated replication of nuclear material. The details of conjugation show some variation in different species, e.g., in some forms one conjugant is much smaller than the other, the macroconjugant absorbs both the cytoplasm and pronucleus of the microconjugant, and mutual fertilization cannot occur. A form of self-fertilization may occur, in which the two pronuclei fuse with each other within the single individual; this process of autogamy does not require the pairing of ciliates, but may occur in ciliates that have paired, when it is called 'selfing'. Autogamy is more common in aged clones, while conjugation is characteristic of younger clones.

The ciliary organelles, together with their associated basal fibril systems, are a dominant feature of the pellicular organization of most ciliates. The basic arrangement of these cilia in characteristic, longitudinal, meridional rows called kineties can be seen in most ciliates, except those in which the ciliature is restricted to certain body regions or to certain parts of the life cycle. The kinety is made up of a row of kinetids, which are regarded as the units of pellicular organization.

Each kinetid consists of a cilium arising from a basal body (kinetosome), to which are attached fibres which run off into the cytoplasm beneath the surface membrane. The ciliary base is surrounded by a pair of pellicular alveoli, so that over most of the body surface there are three unit membranes, the inner two being separated by a fluid-filled space. In some ciliates two cilia occur at the centre of each kinetid. The largest fibres of the kinetid in such species as *Paramecium* and *Tetrahymena* are the kinetodesmal fibres (striated aggregates of micro-filaments) which pass forward for a number of microns from their attachment to the basal body, and run with similar fibres from other kinetids of the same row to form a 'cable' of fibres called the kinetodesma. This kinetodesma can be seen in the light microscope, and is known to occur along the (animal's) right side of each kinety. Small groups of microtubular fibrils also arise near the basal body and may terminate in close association with other fibril systems. Sub-pellicular fibrils, usually microtubular, are common in the surface layers of ciliate protozoa, and, while many are associated with ciliary bases, it is likely that others are not. In some ciliates (e.g., *Stentor, Spirostomum)* large arrays of microtubular fibrils seem to have replaced the striated microfilamentous kinetodesmata; they occur in almost the same position, but may have a different function in these highly contractile ciliates. These fibre systems of the pellicle are believed to function in maintaining the shape and rigidity of the body, and in the alignment of the kinetids; there is no evidence that they co-ordinate the activity of the cilia. Fibrous systems with an active contractile function occur in the stalk of *Vorticella* and in the subpellicular 'M' bands of *Stentor* and *Spirostomum;* they seem to form a specialized addition internal to the basic pellicular structures.

Associated with the pellicle are often arrays of organelles like mitochondria and specialized structures like trichocysts, which are capable of explosively ejecting a thread which may have toxic properties. There are also specialized regions of the pellicle, particularly at the cytostome, but also at the cytopyge (cell anus) and at the site of opening of the contractile vacuole.

Specialization of the ciliature for feeding or locomotion is a feature of most ciliate groups. Frequently, this specialization takes the form of aggregation of cilia into bands or compound organelles, and is well illustrated by the mouth organelles of *Tetrahymena.* Along the (animal's) right side of the mouth is a single row of cilia which lie so close together that they appear to form a membrane, the undulating membrane (not to be confused with the quite different structure in flagellates). At the left side of the mouth are three compound cilia in the form of flat plates called membranelles, each one based on a rectangle of basal bodies. These four organelles together function to bring food particles to the cytostome which lies between them. In many ciliates both the undulating membrane and the membranelle row may show increased development, e.g., *Condylostoma* (Figure 11.37), while in many others only the long row of membranelles (the adoral zone of membranelles) persists. Such ciliates as *Euplotes* possess compound cilia with a specialized locomotory function; these are called cirri, and they take the form of an elongate cone, being based on a more or less circular patch of basal bodies. The component cilia of compound organelles of the membranelle or cirrus type are not fused together, but normally move together as a single unit that is much more powerful than a single cilium.

The classification of ciliates reflects the distribution and specialization of their ciliary organelles. Among the holotrich ciliates, which characteristically have a fairly uniform covering of cilia over the whole body, there is a progressive development of mouth ciliature. In the simplest holotrichs, the gymnostome ciliates, the feeding is usually macrophagous, through a mouth that may be capable of extreme dilation, e.g., as a slit *(Loxophyllum)* or as a proboscis reinforced by fibres *(Didinium)*, or may be specialized to take in algal filaments or diatoms through a 'basket' of fibres *(Nassula)*; in these forms cilia are not actively used in passing food to the mouth, although the fibrous organelles may originate from ciliary roots. Trichostome ciliates frequently have abnormally long cilia around the mouth, which may be used in feeding, e.g., collection of bacteria by *Colpoda*, but these cilia are not formed into compound mouth organelles, and they represent only a slightly modified region of body cilia in a depression (vestibulum) around the cytostome. In the hymenostome ciliates compound feeding cilia occur in a region around the mouth called the buccal cavity, as in *Tetrahymena*, where they collect food particles of the size of bacteria. The development of specialized buccal cilia in hymenostomes has taken diverse forms, including the complex arrangement in the tubular buccal cavity of *Paramecium* and the large sail-like undulating membrane of *Cyclidium*. Among the holotrichs are classified the chonotrichs, an enigmatic group of ciliates which lack body cilia and live attached to crustaceans, and several specialized parasitic groups whose members have well-developed body ciliature but may have small mouths or no mouth at all.

A long adoral zone of membranelles is found throughout the spirotrich subclass. This group shows specialization from such forms as *Blepharisma* or *Condylostoma* with complete mouth ciliature and complete body ciliation, firstly by loss of the undulating membrane in, e.g., *Stentor* and *Spirostomum* and then by reduction of the body ciliature, e.g., in *Halteria* which has a few long body cilia, in the entodiniomorph ciliates (symbiotic in the rumen of herbivorous mammals) which have only one or a few tufts of membranelles, in the hypotrich ciliates like *Oxytricha, Euplotes* and *Aspidisca* with dorsoventrally flattened bodies whose ventral cilia are formed into cirri which are involved in complex locomotory behaviour, and in the pelagic loricate tintinnids.

Many of the spirotrichs are bacterial feeders, but among the large forms they are omnivores, quite a number of which may eat fragments of organic detritus as well as live food, and active carnivores, e.g., *Stentor,* which may eat multicellular rotifers or other stentors as well as a wide range of smaller ciliates and flagellates. The origins and relationships of the peritrich ciliates are uncertain, but it seems likely that they may have been derived from fairly advanced holotrichs. In feeding individuals of sessile peritrichs like *Vorticella* and *Carchesium* cilia are restricted to a band which leads to the mouth, although they develop an aboral ciliary ring in the migratory phase, and the stalk originates from a tuft of ciliary structures called the scopula. Mobile peritrichs like *Trichodina* have aboral bands of cilia for locomotion as well as the oral bands for feeding. The majority of peritrichs probably feed on bacteria. Species of *Trichodina* and related genera are ectoparasitic on aquatic organisms, and may kill their hosts, vast numbers on fish gills may suffocate the host.

The suctorian ciliates lack cilia in the adult feeding stage, although the larval form is provided with kineties or bands of locomotory cilia. Suctorians are usually sessile organisms which feed on ciliate protozoa by means of tentacles through which the prey cytoplasm is sucked into the body of the predator. The tentacles have special adhesive bodies which enable them to keep hold of a ciliate which swims into them. There is no evidence of any homology between the tentacles and cilia, although the tentacle has an internal organization of microtubular fibres. Many suctoria have stalks that are developed from a scopula of ciliary organelles (Figure 11.37).

11.4.6 Genome

The completion of five new protozoan parasite genomes lead to an improved understanding of their biology and through comparative genomics, shed light on protozoan genome diversity and evolution. *Entamoeba histolytica* is a human gut parasite that causes amoebiasis, a significant health problem in developing countries. The 23.7 Mb draft *E. histolytica* genome contains 9,938 predicted genes and a striking abundance of tandemly repeated transferRNA containing arrays. *Entamoeba histolytica* infects large intestine and causes internal inflammation. About 50 million people are infected worldwide, mostly in tropical countries in areas of poor sanitation. In humans, *E. histolytica* lives and multiplies as a trophozoite. Trophozoites are oblong and about 15-20 μm in length. The trophozoites encyst and exit the body in the feces. Both cysts and trophozoites may be seen in the faeces. The life cycle of *E. histolytica* does not require any intermediate host. Another human can get infected by ingesting faecally contaminated water, food or hands. If the cysts survive the acidic environment of stomach, they transform back into trophozoites in the small intestine. Trophozoites migrate to the large intestine where they live and multiply by binary fission. Cysts are usually confined to firm stool, whereas trophozoites are found in loose stool. Only cysts can survive longer periods and infect other humans (Figure 11.38).

The genomes of *E. histolytica* and the amitochondrial protist pathogens *Giardia lamblia* and *Trichomonas vaginalis* share several metabolic adaptations. These include reduced or eliminated mitochondrial metabolic pathways. Indeed, the genome data are consistent

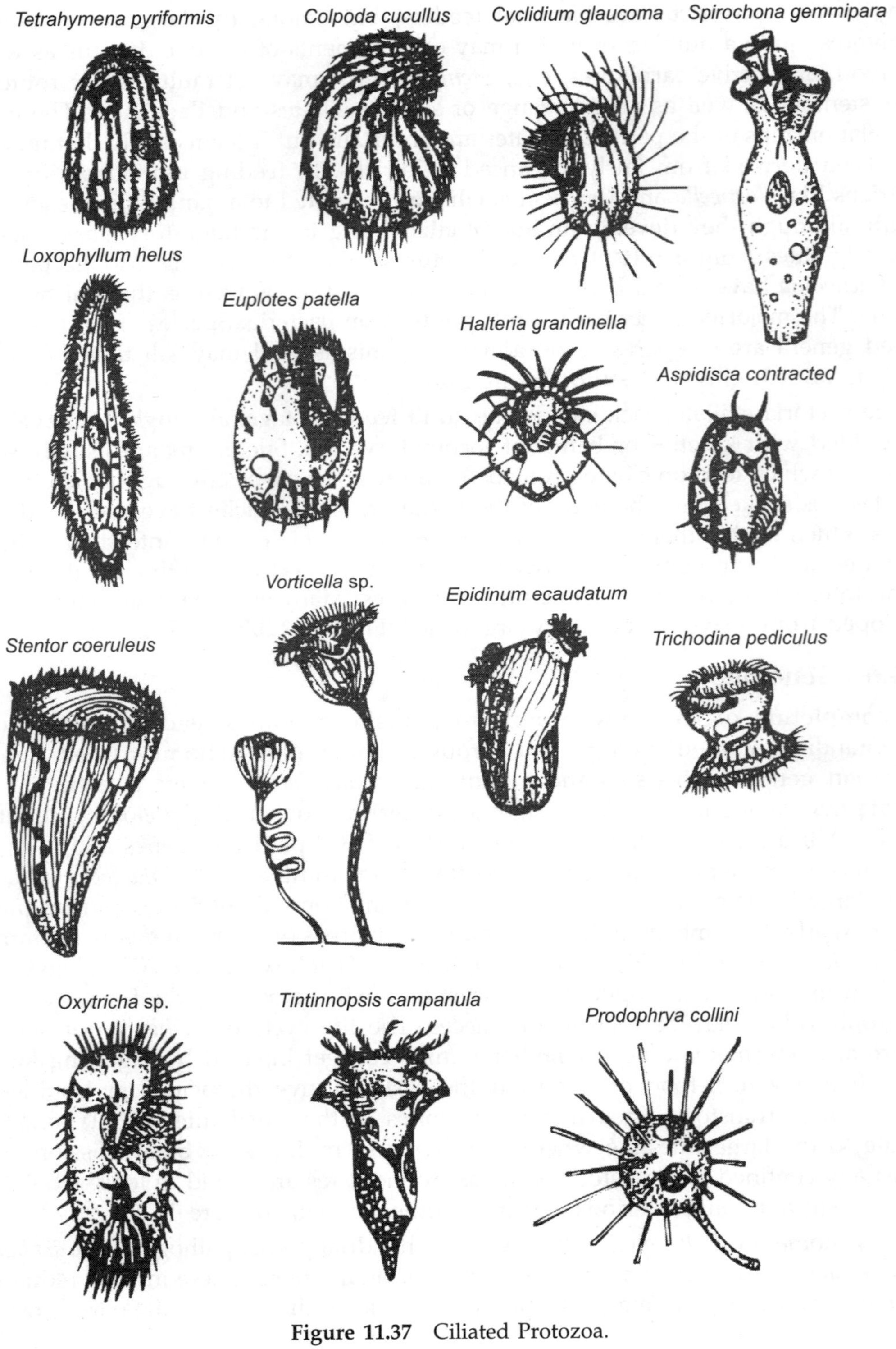

Figure 11.37 Ciliated Protozoa.

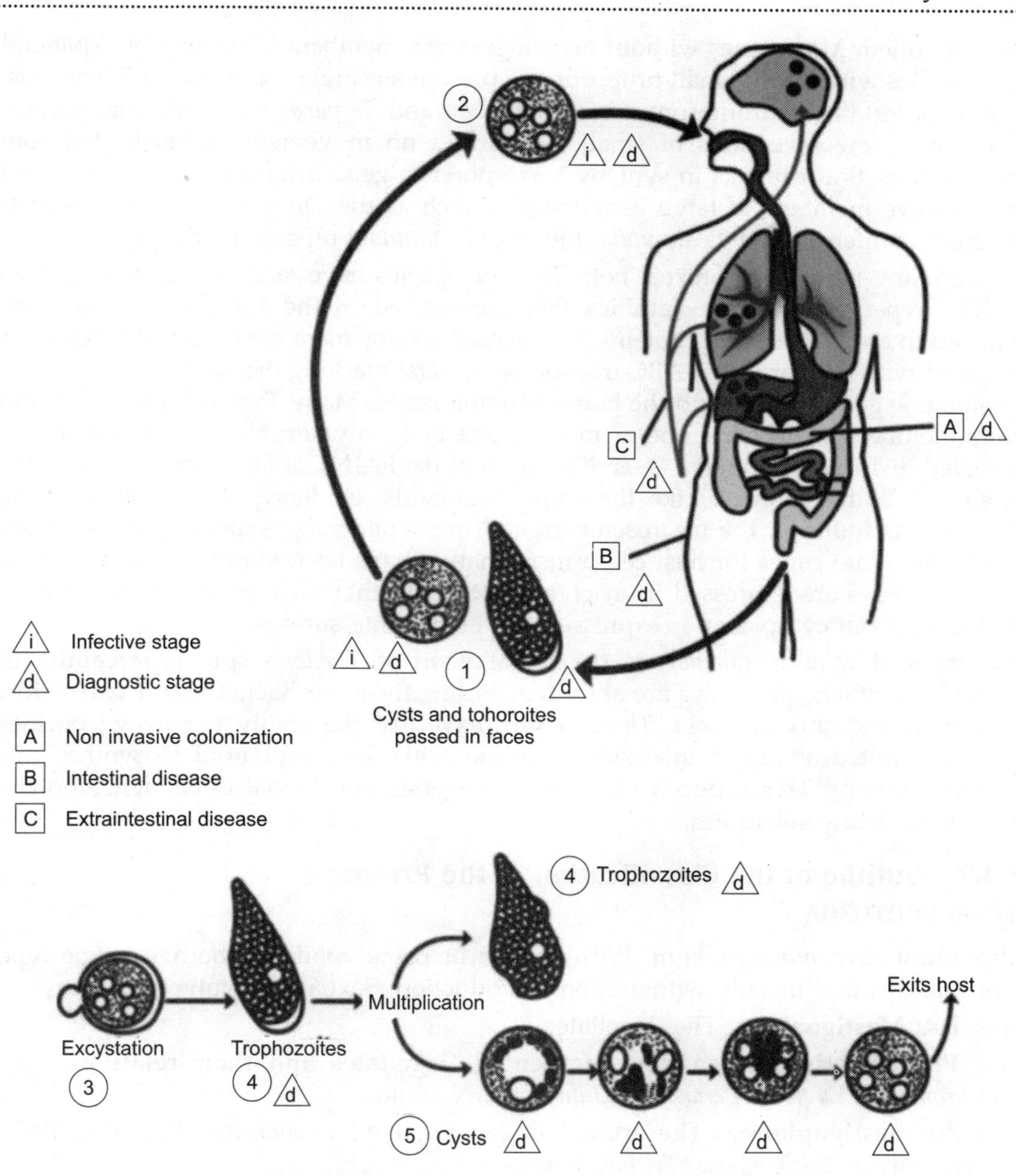

Figure 11.38 Life cycle of *Entamoeba histolytica.*

with the lack of a mitochondrial genome. Tricarboxylic-acid-cycle and mitochondrial-electron-transport-chain enzymes are lacking. Secondary gene loss and lateral gene transfer, mainly from prokaryotes, seem to have shaped *E. histolytica* metabolism. Moreover, *T. parva* and *T. annulata* have a higher gene density (4,035 and 3,792 genes, respectively) and more spliced genes. Despite conserved gene sequences and synteny, 3,265 orthologous gene pairs between the two species, several species-specific genes were identified. Most genes without orthologues are members of unequally expanded gene families with only a small proportion is present as single copies (34 in *T. annulata,* 60 in *T. parva*). The chromosomes of *T. annulata* and *T. parva* have syntenic regions with only a few inversions of small blocks and no movement of blocks between chromosomes. Short breaks in synteny correspond to gene insertions or deletions and often involve members of large gene families such as the *Tpr* genes (*T. parva* repeat) and their counterparts in *T. annulata,* the *Tar* (*T. annulata* repeat) genes.

Like many parasitic protozoa, both *Theileria* species have tandem arrays of genus-specific, hypervariable gene families that are located in the subtelomeres and are predicted to encode secreted proteins. The overall arrangement of subtelomeric genes is conserved with one (or more) ABC-transporter gene(s) marking the boundaries between subtelomeric gene families and the house-keeping genes. Many *Theileria*-specific protein family members incorporate one or more copies of a polymorphic FAINT (frequently associated in *Theileria*) domain. Over 900 copies of the FAINT domain are present in 166 proteins of both genomes. Like the trypanosomatids, evidence of positive immune selection was found in the macroschizont and merozoite stage expressed genes. Also, several candidate genes for host cell transformation have been identified, by assuming that these genes are expressed in macroschizonts and that their products are released into the host cell cytoplasm or expressed on the parasite surface.

Compared with *P. falciparum,* the metabolism of *Theileria* spp. is streamlined. Several biosynthetic pathways are absent, including those for haem, type II fatty acids, polyamines and shikimic acid. *Theileria* spp. have lost the ability to salvage purines and have limited ability to interconvert amino acids, but isoprenoid biosynthesis is, however, present. This reduced metabolism suggests substantial dependence on the host cell for many substrates.

11.4.7 Outline of the Classification of the Protozoa

Phylum PROTOZOA

Subphylum *Sarcomastigophora.* With flagella or pseudopodia or both, a single type of nucleus, and usually without spore production. Sexuality involves syngamy.

Superclass **Mastigophora**. The flagellates.

Class **Phytamastigophorea**. The pigmented flagellates and their relatives, e.g., *Ochromonas, Euglena, Peranema, Chilomonas, Noctiluca.*

Class **Zoomastigophorea**. The animal flagellates, e.g., *Codonosiga, Bicosoeca, Bodo, Trypanosoma, Trichomonas, Trichonympha.*

Superclass **Opalinata**. With numerous ciliary organelles and 2 to many nuclei, parasitic.

Class **Opalinidea**. e.g., *Opalina.*

Superclass **Sarcodina.** Pseudopodial forms, body at least partly naked, but shell or skeleton often present.

Class **Rhizopodea**. With lobed, filose or reticular pseudopodia, e.g., *Amoeba, Arcella, Gromia, Elphidium, Dictyostelium.*

Class Piroplasmea. Small piriform or amoeboid parasites in vertebrate erythrocytes, transmitted by ticks, e.g., *Babesia.*

Class Actinopodea. Spherical floating forms with axopodia or slender pseudopodia or both, often with skeleton or spicules, e.g., *Aulacantha, Sphaerozoum, Acanthometra, Actinophrys.*

Subphylum *Sporozoa.* Usually with simple spores containing sporozoites, parasitic.

Class Telosporea. With spores and characteristic sporozoites, e.g., *Monocystis, Eimeria, Plasmodium.*

Class Toxoplasmea. No spores, but cysts with many naked trophozoites, e.g., *Toxoplasma, Sarcocystis.*

Class Haplosporea. With simple spores and amoeboid sporoplasm, e.g., *Haplosporidium.*

Subphylum *Cnidospora*. Spores with polar filaments and amoeboid sporoplasms, parasitic.

Class Myxosporidea. Spore of multicellular origin, e.g., *Myxobolus, Triactinomyxon.*

Class Microsporidea. Spore of unicellular origin, single tubular polar filament, e.g., *Nosema.*

Subphylum *Ciliophora*. With cilia and two types of nucleus. Sexuality involves conjugation. **Class Ciliatea**

Subclass HOLOTRICHIA. Body ciliature usually simple and uniform, mouth ciliature simple, e.g., *Loxophyllum, Didinium, Nassula, Colpoda, Paramecium, Tetrahymena, Pleuronema.*

Sublass PERITRICHIA. Body cilia absent in mature form, oral cilia in rows winding to mouth, often attached to substrate by stalk or basal disk, e.g., *Vorticella, Carchesium, Trichodina.*

Subclass SUCTORIA. No cilia in mature stage which feeds by tentacles, larva ciliated, e.g., *Podaphrya, Dendrocometes.*

Subclass SPIROTRICHIA. Mouth cilia composed of many membranelles, forming a long adoral zone, body cilia usually reduced or compounded to form cirri, e.g., *Blepharisma, Stentor, Spirostomum, Halteria, Euplotes, Aspidisca, Epidinium.*

11.5 SLIME MOLDS (MYXOMYCOTA)

11.5.1 Introduction

The slime molds are a heterogeneous assemblage of protists requiring organic nutrients (organotrophic). All at one stage in their life-cycles consist of motile single vegetative cells, which may be amoeboid or fusiform, with or without flagella, and show what generally are considered to be animal characteristics, notably no cell wall, a capacity usually for phagocytosis, and motility. One or more of three characteristic sorts of development thereafter occur in the lifecycle; (1) formation of a motile multinucleate mass of protoplasm known as the plasmodium, (2) aggregation of individual cells as motile masses of cells, known as pseudoplasmodia, which show a greater or lesser degree of organization and coordinated behaviour or (3) transformation of single cells, plasmodia, or pseudoplasmodia into stationary spore-bearing fructifications which tend to be rather plant-like in substances which give mechanical support and which possibly include cellulose are produced.

Some authorities classify the slime molds as primitive fungi, or as protozoa, and others as a separate group. Four main subdivisions have been recognized. The status of these subdivisions (and hence the endings of the names used) varies in different taxonomic treatments of the slime molds.

11.5.2 Myxomycetes (True Slime Molds)

Myxomycetes, of which there are several orders, have a distinctive life cycle. The spore liberates one or more uninucleate protoplasts which feed and multiply and eventually may fuse in pairs. The resulting zygotes develop into plasmodia which are wall-less, multinucleate masses of cytoplasm that are capable of locomotion. Eventually, the plasmodia transform into fruit-bodies in which cellulose or chitin or both have been reported. The Myxomycetes are a homogeneous group and not likely to be related closely to the Acrasiales from which they differ in possessing flagellated stages in their life-cycle and a true plasmodium. They comprise two subclasses-Myxogasteromycetidae (or Endosporeae) and Ceratiomyxomycetidae (or Exosporeae) of which the latter contains only a single species.

Most of the four hundred odd species are probably cosmopolitan. The plasmodia generally occur on damp decaying twigs and logs, but some can live indefinitely beneath water. Most appear to be independent of the substrate on which eventually they fruit, but nevertheless some prefer either hard or soft woods or leaves for this.

The unicellular, usually uninucleate, haploid spores are enclosed in smooth or sculptured walls. Germination is influenced by factors similar to those that affect this process in true fungi. Some spores under favourable conditions germinate after only 20 minutes, others take up to 1-2 weeks, while in some species germination has proved very difficult to obtain. On germination, 1-4 naked cells escape through a crack or irregular pore in the spore coat. These cells may be myxamoebae or anteriorly flagellated swarm cells and in many species the one can change into the other. Such transformations are accompanied by changes in the exterior layers of the plasmalemma. Microelectrophoretic investigations on *Physarum nudum* suggest that the surfaces of swarm cells are predominantly protein, but those of myxamoebae are predominantly acid protein or mucoprotein, while those of the eventually formed plasmodia are predominantly fats. Though swarm cells often appear to be uniflagellate, they are usually unequally biflagellate, or at least are potentially so. Two basal bodies are consistently present. The shorter of the flagella may be visible only after special fixation. Both flagella are of the whiplash type and have the 9 + 2 fibrillar sub-structure characteristic of eukaryotic cells. Flagellum like pseudopodia also has sometimes been reported. The cells feed by ingesting solid particles such as bacteria, and can multiply rapidly, but under unfavourable conditions may encyst until favourable conditions return. Contractile vacuoles are active in swarm cells.

In almost every species that has been critically examined, fusion between pairs of haploid cells is a prerequisite of plasmodium production. Nevertheless there are reports that plasmodia in some species have arisen without prior gametic fusion. Some species are homothallic, but in others plasmodia arise only when appropriate clones of cells are paired. In some heterothallic myxomycetes there appear to be multiple alleles which govern mating type so that either mating type in one race of a species may fuse with either type of another race. Even among heterothallic species plasmodia are known sometimes to have arisen from single spores. There is evidence for three causes of this: (1) that the spore in question was binucleate, (2) that at least one of the cells it produced had mutated to a genotype that allowed mating to occur, (3) that the myxamoebae sometimes can actually self-fertilize (i.e., be apogamous).

Cell fusion is closely followed by fusion of nuclei. Thereafter nuclear divisions take place without cell division so that a multinucleate diploid plasmodium forms; further nuclear divisions are synchronized. Sometimes the plasmodium coalesces with nearby zygotes and small plasmodia, while even large plasmodia may fuse together completely if there is genetical identity at a particular locus on their diploid genomes. However, fusions between particular plasmodia of *Physarum polycephalum* are followed by their death. Plasmodia sometimes cut off large numbers of myxamoebae and swarm cells of unknown ploidy. This probably represents a mean of reproduction.

There are several types of plasmodia. The simplest, the *Proto plasmodium,* is a minute sheet of undifferentiated granular protoplasm with barely detectable cytoplasmic streaming. When eventually it fruits it forms a single sporangium. The *aphanoplasmodium* is a rather inconspicuous, very flat sheet of transparent, non-granular protoplasm which virtually lacks a sheath of gelled ectoplasm but exhibits streaming. The *phaneroplasmodium* is the most complex type. When young it resembles the protoplasmodium, but it grows to become a conspicuous, thick, fan-shaped mass of granular cytoplasm organized into veins of gelled ectoplasm within which flow currents of endoplasm. Studies of the fine structure of plasmodia have revealed mitochondria with tubular cristae, and in some plasmodia fibrils about 15 mμ diameter. The interest in this latter observation lies in the possibility that contractile proteins are involved in the mechanisms of cytoplasmic streaming and plasmodial locomotion.

Plasmodia from the field are often cluttered with relics of ingested food particles, but there is evidence that plasmodia may be nourished by soluble foods also. They are sometimes coloured, often in shades of yellow, orange, or buff, but the chemical nature of the pigments is unknown. Plasmodia which are pigmented need exposure to light before they will fruit. Studies on spectral sensitivity and light absorption suggest that a photoreceptor may be associated with fruiting which is common to a number of species. Other pigments may serve protective functions.

Protoplasmic streaming and locomotion occurring most conspicuously in phaneroplasmodia are intimately linked processes. Two contractile proteins, viz., myxomyosin and myosin B, and possibly others also are implicated in streaming, but streaming may not be primarily responsible for locomotion. Its function may be to circulate materials, and myxomyosin may serve less to motivate streaming than to control the pattern of flow. The flow follows a network of veins at speeds as great as 1 mm/sec., and reverses its direction periodically. A plasmodium under unfavourable conditions such as sudden drought may rapidly become sclerotic. Streaming ceases, the cytoplasm gels, and cell walls develop which divide it into a large number of multinucleate macrocysts. These are long-lived, and on rehydration rapidly regenerate the plasmodium.

Vegetative plasmodia eventually give rise to spore-bearing fructifications. The plasmodium usually differentiates into small clumps which become sporangia that may remain sessile or develop stalks. Sporangia are usually a millimetre or so high, and may be borne directly on the substrate or on a membranous hypothallus. Nuclear divisions occur within, and the formation of cell walls gives rise to large numbers of spores. In some species the plasmodia do not cleave into discrete primordia. Instead the poorly differentiated fructification resembles a prostrate strand of the plasmodium, the 'plasmodiocarp'. In others, the large (e.g., upto 30 cm across) and which may or may not manifest its composite nature at maturity. Within the sporangium there may be a system of threads, the capillitium, which is concerned with spore release. This consists of irregular tubes or filaments, with or without intermingled nodular masses, and sometimes contains calcium carbonate. In some genera the threads are beautifully and regularly ornamented. Most sporangia at maturity have a double wall, the outer one tough and sometimes impregnated with calcium carbonate, the inner thin and transparent.

The development of fruiting bodies has been critically studied in only a small number of genera. Generally, a membrane (the *hypothallus)* is produced on the upper surface of the plasmodium and sporangial primordia enlarge by welling up beneath it. The upgrowth of the primordium produces a sporangial region borne on a narrower support within which a central core of material is laid down. The whole of the developing sporangium is ensheathed in a membrane which is continuous with the hypothallus and which forms the peridium and the outer wall of the stalk. The development of members of the order Stemonitales differs from this. Their hypothallus if deposited *beneath* the plasmodium, and within the young sporangial primordia stalks are produced which grow apically as the protoplasm of the primordia flows up their outsides and forms the sporangial heads. The eventual peridium is not continuous with the hypothallus and stalk, for it is a separate structure formed on the outer surface after other parts of the fruit-body have matured (Figure 11.40).

The capillitium usually arises by the deposition of material in or on a system of vacuoles which develops within the sporangial protoplasm. About the same time, the nuclei in the sporangia divide. Most evidence supports the view that reduction division precedes spore formation, but reports that meiosis is delayed even until the spores are maturing may indicate some variation in this respect. The spores, which arise by cleavage of the cytoplasm, are typically uninucleate but nuclear division within the spore has been reported.

Ceratiomyxa fruticulosa, the single species in the Exosporeae, differs from the foregoing account in a number of respects. The plasmodium when about to fruit emerges from the rotting wood in which it lives and produces a minute cushion-like body from which arise digitate processes that may branch and sometimes anastomose. On their surface the plasmodium takes on a network appearance which becomes a mosaic of cells, each of which produces a single spore on a slender stalk. When first delimited the spores are diploid, but as they mature a reduction division yields four haploid nuclei. The so-called spore is in fact widely interpreted as a much reduced sporangium. It germinates by solution of its wall and releases haploid swarm cells. These multiply and eventually fuse in pairs. Single zygote appears to form plasmodia directly without coalescence with other zygotes.

Only a few myxomycetes have been induced to complete their entire life cycle in artificial culture, and so far none has done so in the absence of other microorganisms which have served as food. *Physarum polycephalum* has been cultured in the plasmodial state in a defined liquid medium, showing that at least vegetative growth is possible with solely soluble nutrients.

The classification of myxomycetes is based principally on features of the fructifications, such as whether spores are borne externally or internally, the colour of spores in the mass, the presence of calcium carbonate in various parts of the fruit body, the structure of the stipe if present, and the form, ornamentation, and location of the capillitium.

11.5.3 Acrasiomycetes or Acrasiales (Cellular Slime Molds)

A second group, the **cellular slime molds**, spend most of their lives as separate single-celled amoeboid protists, but upon the release of a chemical signal, the individual cells aggregate into a great swarm (Figure 11.39). Cellular slime molds are thus of great interest to cell and developmental biologists, because they provide a comparatively simple and easily manipulated system for understanding how cells interact to generate a multicellular organism. There are two groups of cellular slime molds, the Dictyostelida and the Acrasida, which may not be closely related to each other.

There are three families; Dictyosteliaceae, Guttulinaceae, and Sappiniaceae, though the latter are somewhat doubtful members of this order and will receive no further consideration. The general life cycle follows the sequence: germination of the spore to yield an amoeba, multiplication of amoebae, aggregation, migration of the entire aggregate, and finally transformation of the aggregate into a *sorocarp* containing spores. Throughout the sequence there is a steady increase in organization and in the precision with which shape is defined. There is no true plasmodial stage.

Differences between genera and species are most obvious at the stage of fruiting. Members of the Dictyosteliaceae form delicately proportioned stalked sorocarps; members of the Guttulinaceae form somewhat irregular and amorphous sorocarps with ill-defined stalks and spore-masses. About twenty species of Acrasiales are known and all are free-living within the soil. They appear to prefer forest to grassland soil, and occur chiefly in the surface humus and in mouldering leaf litter. For their isolation a diluted suspension of soil may be mixed with a suspension of bacteria, such as *Escherichia*

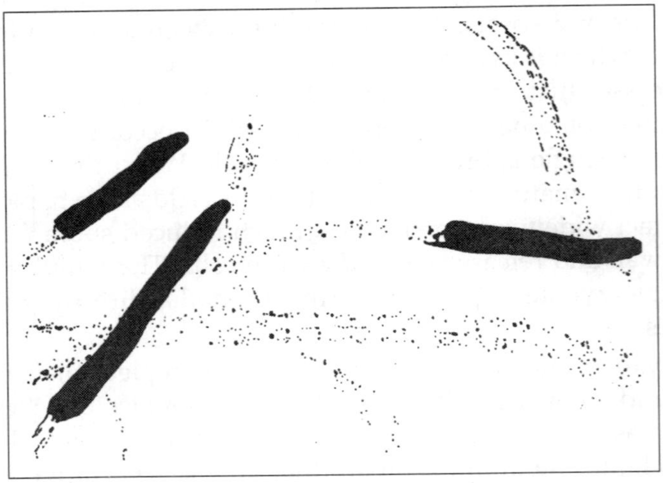

Figure 11.39 Migration of cellular slime mold slugs on an agar surface.

coli, which serve as food, and be poured over a dilute hay infusion agar buffered with phosphate to pH 6. Clones of amoebae should appear in 3-4 days. The spores of Acrasiales are firmly walled. In an aqueous medium they split and each releases an amoeba. The amoebae are small, uninucleate, and nonflagellate and live in water films in which they feed by ingesting bacteria. If nourished properly they divide every few hours. They can be grown in two membered cultures in association with a wide variety of bacteria, and myxamoebae of a *Polysphondylium* species have been cultured in a defined liquid medium. Their cellular organization is typically eukaryotic. Food vacuoles are abundant and contractile vacuoles continually empty water to the exterior. When eventually the food supply is depleted, the amoebae stop multiplying, their food vacuoles progressively disappear, and after a pause the cells begin to aggregate.

Details of subsequent behaviour vary with different species but have been well studied in members of the Dictyosteliaceae. The amoebae elongate and orient themselves radially about a number of centres towards which they begin to migrate by amoeboid movements. This may readily be observed in plate cultures. The converging amoebae form streams in which the cells adhere end to end. Aggregation streams sometimes break up and reaggregate elsewhere, and centres may form only to melt away again, but eventually the pattern of aggregation becomes confirmed. Since feeding has ceased, all further development and movement are sustained only by endogenous reserves.

The way in which aggregation is controlled and accomplished is complex. The process involves chemotaxis governed by a substance named acrasin that is produced by the amoebae themselves. Acrasin is cyclic 3′,5′-adenosine monophosphate (cyclic AMP) for some of the Acrasiales. Chemotaxis requires that a cell in some way perceives a gradient in concentration of the chemotactic principle. There is, however, no single gradient of acrasin concentration associated with an aggregation pattern of amoebae. The system that operates appear to depend on acrasin having more than a single effect.

Firstly, an amoeba orients itself in the direction of a source of acrasin. Secondly, acrasin induces responsive amoebae to produce acrasin themselves. Thirdly, it induces amoebae to become adhesive one to another. The probable mechanism of aggregation is that a particular cell or group of cells produces acrasin briefly.

The pulse diffuses outwards and induces nearby cells to orient themselves so that they are moving up the gradient and to produce a further pulse of acrasin. By repetition of this process a wave of orientation and acrasin production is propagated centrifugally. Successive pulses of acrasin production are initiated at the centre as the cells there recover from the preceding pulse. Acrasin is very labile as produced, for it is accompanied by an inactivator, which may be the enzyme phosphodiesterase that hydrolyses cyclic 3′5′-AMP to 5′AMP. The simultaneous production of an inactivator may be biologically advantageous for it effectively increases the concentration gradient and also prevents the development of a high and uniform background concentration in confined environments.

The nature of the cells that initiate aggregation centres has aroused controversy. It has been suggested that there is a small fixed proportion of special initiator cells in a population, but current opinion favours the view that there are simply some cells or groups of cells that are slightly ahead of others in their development and so seize and maintain an initiative. The number and distribution of aggregation centres has been found to be affected by diffusible and volatile factors produced by the organisms themselves.

As long as cells that are aggregating remain separate they appear to be guided only by chemotaxis, but once they begin to meet, a process known as 'contact following' can operate. An amoeba contacting a moving chain of cells is indifferent to its direction of movement until the back end of the cell that it is touching moves past. It is to this that it then adheres. Short chains of cells thus lengthen and join up with others, and eventually continuous centripetal streams form. The behaviour of cells on joining a stream may be understood in the light of what is known of the mechanism of movement in amoebae of Acrasiales. This is thought to involve the active contraction of cell surface towards the posterior rim of the roughly cylindrical cell and its active enlargement around the anterior rim. The result is that cytoplasm moves forward within a stationary surface membrane which is continuously being extended in the forward direction and continuously being removed at the back. The anterior and posterior cell surfaces are perhaps stabilized through their contact with contiguous cells, and it is to a stable end that the newly arriving amoeba adheres.

The conical aggregate of cells eventually acquires a fairly definite shape, topples on to its side, and proceeds to migrate over the surface of its substrate as if it were a single multicellular organism. This allows the organisms in the conditions best for sporulation. The phase of migration may however be eliminated in dry conditions or if the aggregate of cells is small. The migratory stage has been termed the 'grex'; the term pseudoplasmodium, though widely used, has little to commend it. In some species, for example *Dictyostelium mucoroides* and species of *Polysphondylium,* the grex as it migrates

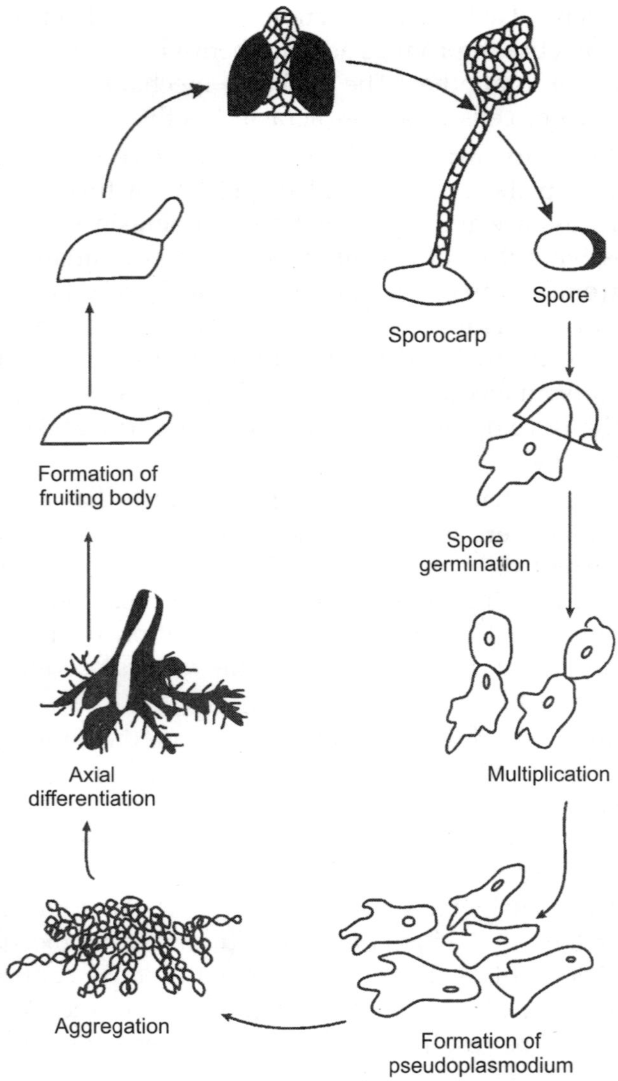

Figure 11.40 Life cycle of a slime mold.

continually secretes a postrate cellular stalk which is left behind, but in other species the grex forms no stalk and may then be termed a 'slug'. The slug of *Dictyostelium discoideum* ranges from about a fifth to five millimeters long and consists of a few hundred to hundreds of thousands of cells. Its size depends on the number of amoebae entering the aggregation and the density of amoebae in the culture. A high density leads to a large grex. The slug has a distinct polarity and becomes increasingly differentiated inside as it migrates. Its anterior third comprises cells which differ histochemically, serologically, and in size from the rest. These anterior cells will eventually form the stalk of the sorocarp, while the posterior ones will form the spores. Those amoebae that arrived first at the aggregation centre become the anterior pre-stalk cells; those that arrive later become the posterior pre-spore cells. In *Dictyostelium discoideum* those that are the very last to arrive eventually form the basal disc that is characteristic of the sorocarp of this particular species. Up to a certain stage, the differentiation of cells within a slug is reversible, for slugs cut in half will give rise to normal sorocarps containing all three types of cell, i.e., stalk, spore, and basal disc. Further, cells removed from aggregates, slugs, or immature sorocarps will multiply and produce clones of cells able eventually to form normal sorocarps. The differentiation of these cells is therefore only phenotypic (Figure 11.40).

The grex migrates at about 0.25-2 mm/hour by a slow gliding movement, all the while leaving a collapsed slime sheath behind. The larger the slug the greater is its velocity. Its direction of movement is influenced by illumination and extremely small gradients of temperature, but appears to be indifferent to acrasin concentration. There is much evidence that cells in the interior of the grex advance by their own amoeboid movement, so that they contribute to the movement of the grex as a whole, and are not simply carried along by the amoebae that are in contact with the substrate. The means by which the interior cells are able to obtain traction is an intriguing problem. It has been suggested that the slime sheath may extend throughout the grex and thereby provide traction for all the cells, but a more satisfactory explanation has emerged with the discovery that individual amoebae of Acrasiales move as a result of the forward transport of the granular cytoplasm within an apparently stationary cell surface which is being constantly augmented near the front of the cell and contracted at the back. A cell surrounded by others can thus advance independently because the surfaces of cells that adjoin it are stationary. Eventually, the grex undergoes the process known as culmination by which the erect sorocarp is produced. Culmination is hastened by a degree of desiccation, a rise in temperature, and possibly the removal of a volatile self-produced inhibitor. In the Dictyosteliaceae the sorocarp consists of a spore mass raised on a stalk built of vacuolated parenchymatous cells within a non-cellular cellulose sheath. The transformation of a grex into a sorocarp involves extensive and well ordered movements of the constituent cells and their differentiation into specialized types. The culmination proceeds in *Dictyostelium discoideum* is illustrated in Figure 11.41.

The grex halts, becomes rounded, and its tip rises to become a papilla on top of the cell mass. A stalk sheath initial then appears beneath the papilla as a more or less cylindrical, hyaline membrane, and the cells that are within this sheath become large and vacuolated. More pre-stalk cells migrate into the open upper end of the stalk sheath which is growing in length by the deposition of sheath material by the pre-stalk cells as they ascend its outside prior to entering its open end. The lengthening of the stalk is accompanied by its downward extension through the supporting mass of cells until its first-formed basal part contacts the substrate. Thereafter further extension elevates the growing end of the stalk. The pre-spore cells of the grex follow the pre-stalk cells in their ascent of the outside of the stalk, develop cellulose walls, and become a lemon-shaped or spherical mass of spores. In *D. discoideum,* but not other species, the very last cells of the grex remain around the base of the stalk where they develop into the

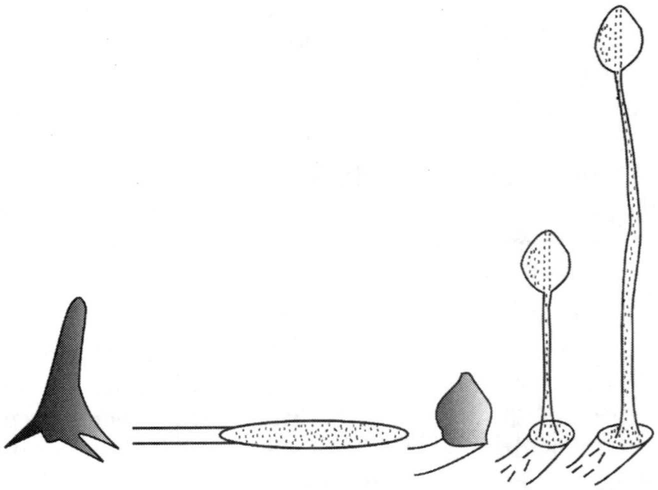

Figure 11.41 Development of cellular slime mold *Dictyostelium discoideum.*

basal disc. In *Polysphondylium* rings of cells are successively pinched off from the hind end of the grex as it ascends the erect portion of the stalk, and each forms a whorl of small sorocarps, each essentially like that of *Dictyostelium.*

The number of amoebae which enter an aggregate governs the size of the eventual sorocarp. Normally, that of *D. discoideum* is 0.5-5 mm long. The remarkable capacity for regulation of differentiation and form of the organized grex and sorocarp is revealed by the fact that though a normal sorocarp may consist of hundreds of thousands of cells, those of a mutant of the same species consist of as few as ten or twelve cells. Regardless of the number of cells, the proportion of pre-spore to pre-stalk cells remains constant.

The stalks of sorocarps always tend to be perpendicular to the substrate whatever its plane, so cannot be oriented by gravity. An unidentified gas formed by the fruit-bodies themselves and to which they are sensitive is instead involved in orientation.

Not all Acrasiales have sorocarps as precisely organized as those in the Dictyosteliaceae. In *Guttulinopsis vulgaris* the amoebae become grouped into membranous compartments in both sorocarps and aggregates. Almost all of the amoebae in the upper part of the sorocarp become spores but in the lower part and in the aggregate itself some become spores and some degenerate.

The life-cycle of members of the Acrasiales may be completed in as little as about three days. Sexuality has not been demonstrated unequivocally, though it may exist.

11.5.4 Protostelida

The organisms classified under this group as an order of slime molds, among which it probably holds a primitive position. The organisms are widespread and common in soil, dung, and plant remains from which they may be isolated and grown in two-membered culture with an appropriate bacterium, yeast or mold to serve as food. Protostelida are found in temperate forests and grasslands, where they can be located at both ground litter and aerial litter sites. They are also found on soil, bark, and dung. Some Protostelida species have been found in tundra ecosystems and boreal forests in Alaska. These organisms were found on ground litter sites. Diversity of Protostelida species is lower at higher altitudes and are also capable of growing in Tropical Wet Forests, although two species common in temperate forests were rare in this environment.

Protostelida are flagellated. Individual cells join to form a fruiting body. Their fruiting bodies consist of an acellular stalk with a small number of spores at the tip. The species *Protostelium mycophaga* has microtubule centers (MTCs) on opposite sides of the nucleus. Many other Protostelida contain MTCs that are remote from the nucleus. Some species, such as *Schizoplasmodiopsis reticulata,* have cell walls with irregular thickenings. In the ameboid phase of this and other *Schizoplasmodiopsis* species, organisms lack electron-dense bodies and mitochondrial surfaces covered with ribosomes.

The spore liberates a single amoeboid protoplast which proceeds to divide. In one genus alone the protoplasts are anteriorly flagellate. Cell and nuclear divisions are not always closely associated, and protoplasts with several nuclei may appear. In some species multinucleate plasmodia arise, while coalescence of plasmodia is common. Plasmodia at sporulation break into smaller protoplasts whose further development is essentially like that of single amoebae of non-plasmodial species. The individual protoplast becomes domed and then hat-shaped. A sheath develops over its upper surface and on withdrawal of the protoplast from beneath the rim the latter makes contact with the substrate. The protoplast then differentiates a basal core, the 'steliogen' which proceeds to secrete a non-cellular, tubular stalk. This is progressively lengthened, enveloped closely by the original sheath, so that the protoplast and steliogen become raised above the substrate. The remnant of the steliogen is usually then walled-off as the apophysis, a slight swelling which lies immediately beneath the spore which forms from the elevated protoplast. Occasionally the latter forms two spores. The spores of species which have an apophysis are usually deciduous, and in one genus are discharged forcibly. Histochemical evidence suggests that the spore wall and stalk may contain cellulose.

11.5.5 Labyrinthulales or Labyrinthulidae

This is a small group of aquatic organisms. The genus *Labyrinthula* contains nearly ten described species, most of which are parasites, chiefly of algae, but possibly some of the names will eventually prove to be synonyms. The best known species is *L. macrocystis* which causes a destructive disease of *Zostera marina* (eel-grass) along Atlantic shore lines. Its predominant growth phase is the 'net plasmodium'. This is a lacy network of filaments on which spindle-shaped uninucleate cells (average size 18 × 4 μ) move by an obscure mechanism. These cells are neither flagellate nor amoeboid. They glide along the outside of the elastic and glutinous filaments at speeds up to 150 μ/minute, but in certain species, such as *L. roscoffensis,* it has been reported that the filaments are tubes and the cells move inside them. The filaments of *L. macrocystis* are produced by the cells themselves. At the advancing edge of an active colony there are clumps of cells, and from the tips of individual cells single, fine filaments suddenly shoot out. These are 8 to 10 times longer than the cells themselves, and are apparently endowed with a capacity for independent motion. They meet and fuse to yield the track on which the cells will continue their advance. The cells divide by fission. They are nourished by absorption of dissolved nutrients; in only one species has ingestion of particles been reported. Individual cells may encyst, and sometimes 5-100 cells mass together, acquire a pellicle, and form a sorus. In due course the pellicle ruptures and the many emergent cells become spindle-shaped and resume normal development. Heterokont zoospores have been reported in one species. Cells have also been seen to form large pseudoplasmodial clumps that slowly move. There is no evidence of sexuality in these organisms. Within eel-grass, aggregates of cells lie on filaments in air spaces in the leaves, from which position they invade the cells whose walls they are able to penetrate.

Almost all the species are marine. It has been claimed that two fresh-water species have tracks that are probably composed of filamentous pseudopodia, and for these the genus *Labyrinthorhiza* has been proposed. Another genus, *Labyrinthomyxa,* includes *L. marina* (formerly *Dermocystidium marinum)* that causes heavy mortality of the commercial oyster *(Crassostrea virginica)* of North America, and another species that parasitizes clams.

11.6 ALGAE

11.6.1 Introduction

The subkingdom Algae include a large and diverse assemblage of all the photosynthetic eukaryotic organisms in the protista that contain *chlorophyll a* which is necessary to carry out oxygenic photosynthesis. They do, however, possess a wider range of accessory pigments some of which are involved in photosynthesis and although starch is the reserve product of some algal groups (as with higher plants), various other polysaccharides, fats, etc., are the main reserve substances.

Algae are organisms that use light energy to convert CO_2 and H_2O to carbohydrates and other cellular products with the release of oxygen. Algae should not be confused with cyanobacteria, which are also oxygenic phototrophs, but which are eubacteria and thus evolutionarily quite distinct from algae. The algae may be unicellular, colonial, or filamentous and most of them are microscopic hence are clearly microorganisms, a number of forms are macroscopic, some seaweed growing to over several feet in length. The undifferentiated algal structure is often referred to as thallus. They lack the stem, root, leaf structure that result from tissue specialization. They contain chlorophyll in well organized membranous chloroplast and visible nuclei, and other membranous structures. Photosynthesis takes place within these bodies. The size, shape, distribution and number of chloroplasts vary considerably from species to species. In some instances a single chloroplast may occupy most of the cell space. However, a few kinds of common algae are not green but appear brown or red because in addition to chlorophyll, other pigments such as phycocyanin, xanthocyanin, phycoerythrin, and carotenoids, etc., are present which mask the green colour and extend the range of light waves that can be used by these organisms for photosynthesis (Table 11.4). In Whittakers five kingdom system, some of the algae have been included with protozoan.

Algae, however, lack a well organised vascular system. Algae do not directly infect humans but some produce toxins that cause paralytic shell fish poisoning. Some of these toxins do not cause illness in the shellfish that feed on the algae but accumulate in their tissues and when eaten by man and cause nerve damage. As one of the primary producers of carbohydrate and other cellular products, the algae are essential in the food chains. In addition, they produce a large proportion of the oxygen in the atmosphere.

These microorganisms are universally present where ample moisture, favourable temperature and sufficient sunlight exists. Although a great majority of them live submerged in water, some grow on soil; others grow on the bark of trees or on the surface of the rocks.

11.6.2 Habit and Habitat

Algae require moisture for the movement of their reproductive cells but like most other plants many can exist for long periods in habitats which are subject to intense desiccation. Thus, they are common as epiphytes on trees especially in moist tropical regions but also in temperate zones, on bare rock surfaces especially when moistened by sea-spray, and on soils, many of which are extensively colonized by diatoms, flagellates, and non-motile coccoid and filamentous algae. In all freshwater and marine situations algae thrive down to depths to which photosynthetically usable light penetrates. Moving from the land into any body of water there is an intermediate zone where the soil (or rock) algal flora intermingles with that of the underwater soil. On the permanently submerged silt or rock surfaces characteristic algal floras occur; the former tend to be composed of motile and the latter of attached forms. Similar but not identical species coat the larger aquatic plants (algae, Bryophytes, or Angiosperms) and also occur on many animals, particularly molluscs and hydroids. The boundary between the land and the sea has its very characteristic algal flora, visually dominated by the large brown and red algae but these also are richly coated with microscopic species. In the open water a selected number of species of many algal groups are maintained in circulation by the turbulent motion of the water; this community extends over the whole ocean surface as a thin veneer of carbon fixing species. A further marine algal community exists in and on the under surface of sea ice and is of considerable importance in polar waters.

11.6.3 Fine Structure of Algal Cells

The main features of algal cells resemble those of other eukaryotic protists, but some organelles show characteristic features and others, such as chloroplasts are found among microorganisms only in algae.

11.6.3.1 Plastids

Both pigmented (chloroplasts, chromoplasts) and non-pigmented (amyloplasts) plastids occur in algal cells, though the latter are so far known only in certain siphonaceous Chlorophyta. Whatever the pigmentation, all eukaryotic algal plastids are membrane

Table 11.4 Properties of major groups of algae

Algal group	*Common name*	*Morphology*	*Pigments*	*Typical representative*	*Carbon reserve materials*	*Cell wall*	*Major habitats*
Chlorophyta	Green Algae	Unicellular to leafy	Chlorophylls a, b	Chlamy-domonas	Starch (α-I, 4-glican), sucrose	Cellulose	Freshwate, moist soils, few Marine
Euglenophyta	Euglenids	Unicellular	Chlorophylls a, b	Euglena	Paramylon (β-I, 2-glucan)	No wall present	Freshwater
Chrysophyta	Golden-Brown algae, diatoms	Unicellular	Chlorophylls a, c, e	Navicula	Lipids	Many have two over-lapping components made of silica	Freshwater, marine, soil
Phaeophyta	Brown algae	Filamentous to leafy, occasionally massive and Plantlike	Chlorophylls A, c, xantho-Phylls	Laminaria	Laminarian ($-1,3-glucan) mannitol	Cellulose	Marine
Pyrrophyta	Dino-flagellates	Unicellular flagellated	Chlorophylls a, c	Gonyaulax	Starch (α-1, 4-glucan)	Cellulose	Freshwater, marine
Rhodophyta	Red algae	Unicellular, filamentous to leafy	Chlorophylls a,d, Phycocy-anin, phy-coerythrin	Polisiphonia	Floridean starch (α-I, 4, and α-1 ,6, glucan), fluoridoside (glycerol-galactoside)	Cellulose	Marine

bound and composed of a granular matrix through which is dispersed a varying number of elongated flattened vesicles; in disrupted dinoflagellate chromatophores they are seen to be circular. These vesicles are termed thylakoids. In addition, the plastids may contain pyrenoids, starch grains, osmiophilic granules and pigment-containing granules grouped to form the stigma (eyespot). The bounding membrane is usually a double structure corresponding to two closely appressed unit membranes and in some algae it is continuous with elements of the endoplasmic reticulum. Indeed the endoplasmic reticulum may form additional layers around the plastid, e.g., in the Haptophyceae where there is a double endoplasmic reticulum membrane outside the double plastid one. However, in dinoflagellates, euglenoids, and Xanthophyta the plastid membrane is distinctly three layered. The thylakoids are usually stacked into lamellae which run along part of or the whole length of the plastids. Material of the matrix does not penetrate inside the flattened thylakoids. The arrangement of varying numbers of thylakoids into lamellae (bands or stacks) is characteristic of different algal groups. In the Red Algae the thylakoids run singly through the matrix whilst in all other algal divisions they are grouped, e.g., two thylakoids in the Cryptophyta, two to six thylakoids in Euglenophyta and Chlorophyta, and three in dinoflagellates, diatoms, Chrysophyta, and Phaeophyta.

Highly characteristic 'girdle' lamellae encircle the inner lamellae in Phaeophyta, Chrysophyta, and Xanthophyta and some diatoms; in the latter they are distinctive features even of the male gametes *(Lithodesmium)*. Most algae do not possess the characteristic stacks of disk-like vesicles, known as grana, found in the chloroplasts of Angiosperms but invagination and folding of the thylakoids result in 'pseudograna' in *Carteria* and perhaps even true grana in some *Acetabularia* species.

Colourless strains of some *Euglena* species can be induced by growth at high temperatures since chloroplast replication is then inhibited whilst cell division is not. Cell division in these circumstances produces some cells containing only a single chloroplast; each of these on further division yields one without any plastids. These cells cannot then be induced to form chloroplasts a fact which is further evidence for the theory that chloroplasts cannot arise *de novo* but only from existing chloroplasts.

Fine arrays of microtubules have been found in the chloroplasts, vegetative cells, zoospores, and eggs of *Oedogonium*. They are obliquely striate in longitudinal section. Comparable microtubules have not been found in any other group of green algae.

Pyrenoids are either immersed in the plastid or appear as outgrowths from it, e.g., in some dinoflagellates and Phaeophyta. There are areas of fine-grained dense material without bounding membranes except in some diatoms where a distinct boundary layer has been seen. In all algal groups the thylakoids are displaced by the aggregation of the pyrenoid matrix, but one or two thylakoids of the lamellar system may continue into and sometimes through the matrix. In some Chlorophyta the normally flattened thylakoids reduce to tubules as they pass into the pyrenoid and in cross section these tubules contain peripheral rod-like structures. These rod structures may be the reduced inner thylakoids of the lamellae and the tube itself may thus be derived from the outer membrane of the group of thylakoids forming the lamella. Starch is formed usually as discrete plates around the pyrenoid, though in the diatoms which do not form starch there is no apparent reserve product associated with this structure. Light microscopy suggests that the projecting pyrenoids are unconnected with starch formation, but the electron microscope reveals that some of them have a thin layer of some reserve substance surrounding the matrix, e.g., in Dinoflagellates where in the species with stalked pyrenoids the plastid membrane also covers the pyrenoid and the polysaccharide material is deposited outside this membrane. Similarly, in the euglenoid algae the paramylon granules lie outside the chloroplasts. The starch grains in *Scenedesmus* are cut off from the pyrenoid by intrusion of lamellae. In this genus the pyrenoid disappears completely at cell division and arises *de novo* in the daughter chloroplasts. In *Platymonas* and *Prasinocladus* of the Prasinophyceae the pyrenoid is unusual in that it is penetrated by branching canaliculi entering from a direction facing the nucleus and lined by the chloroplast membrane in the former and by this and also projections of the nuclear membrane in the latter.

Eyespots are composed of collections of osmiophilic granules in which carotenoid material is dissolved. When they occur within the plastid, they form either a single or multiple row(s) of granules located between thylakoids usually adjacent to the outer chloroplast membrane. Slight modifications of the chloroplast membrane and the plasmalemma in the region over the eyespot have been reported. The eyespots in the gametes of fucoid Phaeophyta appear as a series of chambers containing the pigment rather than as separate granules and it is noteworthy that in these organisms the hind flagellum is swollen as it passes over the eyespot and although not actually attached to the gamete in this position it is in some way pressed close against the cell membrane. The eyespot in euglenoid flagellates is situated free in the cytoplasm usually around the base of the anterior invagination but again consists of clusters of osmiophilic granules and some microtubular elements (also seen in some Chlorophyta). The eyespot granules are clustered in groups of 2-3 surrounded by a membrane of unit membrane dimensions, but the eyespot as a whole has no bounding membrane. Osmiophilic granules are common in the chloroplasts of many algae including euglenoid genera and there is some evidence that they can be channelled between the thylakoids towards the eyespot when this also is present.

Colourless amyloplasts occur in some siphonaceous Chlorophyta and resemble chloroplasts but contain a single large starch grain together with a small concentric mass of lamellae. It is interesting that this small concentric aggregation of lamellae also occurs in the chloroplasts of those species containing amyloplasts but not in those siphonaceous species without amyloplasts.

11.6.3.2 Mitochondria

Though basically similar to those of other plants, algal mitochondria show subtle variations between groups, particularly in the development of the cristae (tube-like, plate-like, extending to various degrees into the lumen, etc.). In *Micromonas pusilla* there is only a single mitochondrion but in most cells they are numerous but often precisely located within the cell, e.g., around the nucleus and between the lobes of the chloroplast *(Chlamydomonas eugamatos)*, between the plasmalemma and the chloroplast (other

Chlamydomonas spp. *Carteria,* and colonial species). They are often closely associated with flagella bases (kinetosomes) and in the Prasinophycean genus *Heteromastix* a single mitochondrion lies pressed against the inside of the 'C' shaped chloroplast.

11.6.3.3 Dictyosomes

Dictyosomes (Golgi bodies) are very distinctive components of algal cells and are often precisely located. They consist of stacks of elongate vesicles dilated at the periphery. In some colonial Chlorophyta they surround the nucleus and are contained within a series of 'amplexi' formed of endoplasmic reticulum which is continuous with the outer layer of the nuclear membrane. Associated with the dictyosomes are vesicles which are budded off from the cisternae and in the Haptophyceae and Prasinophyceae it is within these 'vesicles that the characteristic scales of the body and flagella are formed. In the dinoflagellate *Gonyaulax* the dictyosomes form a 'spherical shell' closely associated with the concave side of the 'C' shaped nucleus and within this area vesicles, presumably formed from the dictyosomes, give rise to the trichocysts. The dictyosomes in *Chlorella* pair prior to cell division to form corresponding 'mirror' images between which the new partition membrane is formed. In the diatoms special membrane-limited vesicles appear during cell division and within these the new silica valves are deposited. The surrounding membrane has been termed the silicalemma and it is quite distinct from the plasmalemma.

11.6.3.4 Vacuoles

Contractile vacuoles appear as empty membrane-bounded spaces often with subsidiary cisternal elements around them and fusing to form the main vacuole. In dinoflagellates more complex vacuoles (pusules) are associated one with each of the flagellar canals. These are in clusters of approximately 40 vesicles, each with a double membrane and opening by means of a narrow pore, either into a central vesicle and then into the flagellar canal or individually into the canal. These pusules are permanent vacuoles unlike the temporary contractile vacuoles which enlarge and discharge and enlarge again.

11.6.3.5 Miscellaneous organelles

Structures comparable to the lomasomes of the fungi are found in the green algal genera *Chara* and *Nitella* of the Charophyta and in *Platymonas* of the Prasinophyceae but have not been seen in other groups. Muciferous bodies are often located beneath the plasma membrane. They may discharge through it or into microtubules, e.g., in euglenoids.

Trichocysts of the dinoflagellates are complex organelles which eject fine, banded, probably proteinaceous threads from the cell. The threads are extruded through thin areas in the wall after the mature trichocyst has migrated from the central dictyosome area. The trichocyst has a single membrane around a central square core of amorphous material. The enclosing membrane is thickened by hoops or spirals of material. Small backwardly directed tubules occur along the upper end of the core and attached to the apex of the core is a group of twisted fibrils ending in even finer ones which just touch the closing membrane. Discharge is accompanied by a change in hydration of the core which takes place in a matter of milliseconds. Trichocysts occur also in some other algal groups.

11.6.4 Cell walls

Wall structure in the algae is extremely variable, though usually based on polysaccharides within or on which other substances are deposited. Mucilages are common but these may have structural microfibrillar elements within them, e.g., in colonial volvocalean genera. In many Chlorophyta the microfibrillar material is cellulose and in the classic examples *Cladophora* and *Chaetomorpha* it is deposited in layers of microfibrils running at different angles to the long axis of the cells. Shorter microfibrillar units of xylose or mannose form the wall material in some of the siphonaceous green algae.

Two groups of algae which have received considerable attention in the last decade, the Haptophyceae and Prasinophyceae have a 'wall' layer composed of scales which are formed within the cells in the vesicles arising from the dictyosomes. In the former group they consist of organic scales, in which mannose has been detected, and which have a fibrous structure with an inner face with radiating fibres and an outer one of randomly arranged fibres. In fact the two faces are really two sides of a flattened disk and in one subgroup, comprising the Coccolithophorids, it is within such scales that the calcium carbonate is deposited. Simple scales without the calcified part (coccolith) are also found on the body of the cell. Thus in the coccolithophorids at least two scale types can be demonstrated on anyone cell and in the other section of the group containing the genus *Chrysochromulina* several precisely arranged purely organic scale types can be found on any cell. In the Prasinophyceae the situation is even more complex, for here it is common to find two scale types on the flagella and three types over the body of the cell (e.g., *Pyramimonas*). In yet another genus *viz Platymonas* fine scale-like structures, is also produced in the vesicles derived from the dictyosomes, are liberated and coalesce outside the cell to form the theca which surrounds the cell. The dinoflagellates have long been known to have an 'armoured' cell wall consisting of precisely arranged plates joined together by sutures. These plates at least in some species are sandwiched between outer and inner membranes and the inner membrane connects at the edge of the plates with the underlying plasmalemma.

11.6.5 The Range of Vegetative Structure

The tremendous range of vegetative morphology can be seen to a series of types:

(l) Unicells which may be non-motile *(Chlorella)*, motile by means of flagella *(Chlamydomonas, Euglena, Ceratium)*, motile by extrusion of mucilage or other means (desmids and diatoms)

(2) Aggregations of cells into colonies motile *(Volvox, Synura)*, non-motile *(Pediastrum)*

(3) Filaments unbranched *(Ulothrix)*, branched *(Ectocarpus)*,

(4) Complex thalli based on filamentous construction (Rhodophyta and some Phaeophyta)

(5) Parenchymatous, i.e., without distinct filamentous structure in the mature stages *(Ulva, Laminaria, Fucus)* ,

(6) Coenocytic--either simple unicellular *(Protosiphon)* or branched unicellular *(Bryopsis)* or divided into multinucleate segments *(Cladophora)*.

11.6.6 Life Cycles of Algae

Algae have characteristic life histories; the simplest is shown by *Fucus,* where the spermatozoids and eggs are the only haploid phase in the life history. In others there are varying degrees of isomorphic/heteromorphic development of gametophytes and sporophytes culminating in the most complex of all plant life histories which characterize the Rhodophyta. They all have in common is the formation of gametangia which lack a sterile wall layer around the fertile tissue; this distinguishes them from all higher plant groups. The sexual process may involve simple fusion of nucleated cytoplasmic masses or of flagellated gametes and non-flagellated egg cells. Accessory reproduction of either or both gametophyte and sporophyte may be achieved either by nonmotile spores or by motile zoospores.

11.6.7 Classification of Algae

The classical concept of algae as consisting of six basic groups distinguished on the chemical basis of pigmentation and flegellation into Chlorophyta (Green), Euglenophyta (Euglenoids), Chrysophyta (Golden Brown), Phaeophyta (Brown), Pyrrophyta (Dianoflagellate) and red Rhodophyta (Red). Detailed studies particularly the increasing use of electron microscope studies has led to a clearer segregation of eukaryotic algae into a series of divisions (phyla) each of which requires consideration to give a balanced

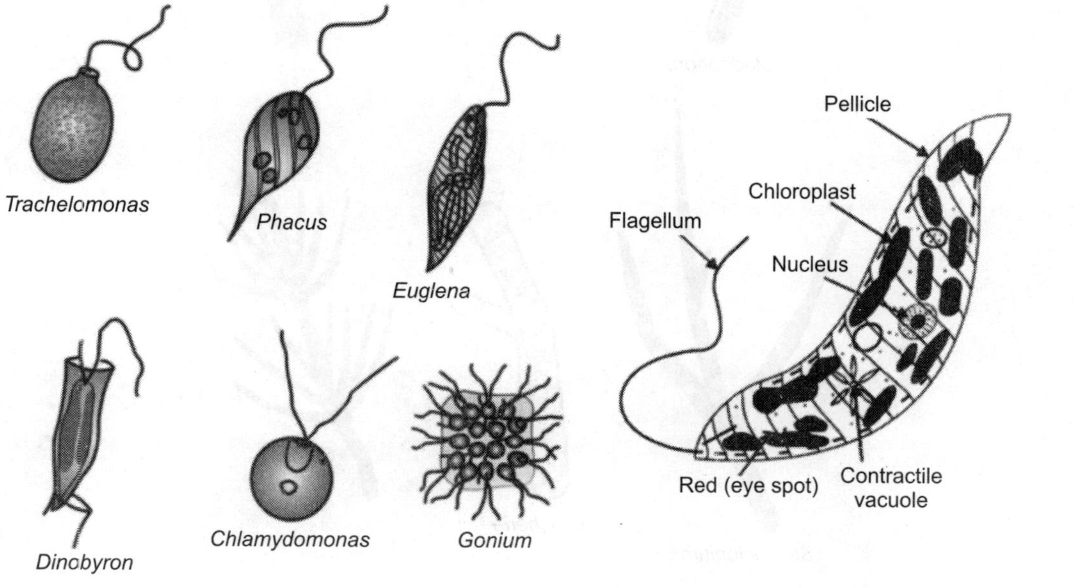

Figure 11.42 Microscopic views of unicellular flagellated algae.

view of the group. In this classification great stress has always been placed on the characteristics of the motile cells and of the reproductive processes which tend to be relatively constant throughout each division; indeed many anomalies which existed in the older schemes have now been shown to be due to unfortunate juxtaposition of unlike algal groups.

11.6.7.1 Chlorophyta (Green Algae)

The green algae are present in ponds. They may be unicellular, colonial (Figure 11.44), or filamentous They are grass green in colour in having chlorophyll *a* and *b*. They synthesize starch during photosynthesis instead of paramylum for food storage (Figure 11.42).

It is the consensus that all the filamentous algae have evolved from this flagellated form are chlorophyta. A unique group of green algae are the desmids. With the exceptions of a few species, the cells of desmids consists of two similar halves or semicells. The two halves usually are separated by a constriction, the isthmus (Figure 11.43).

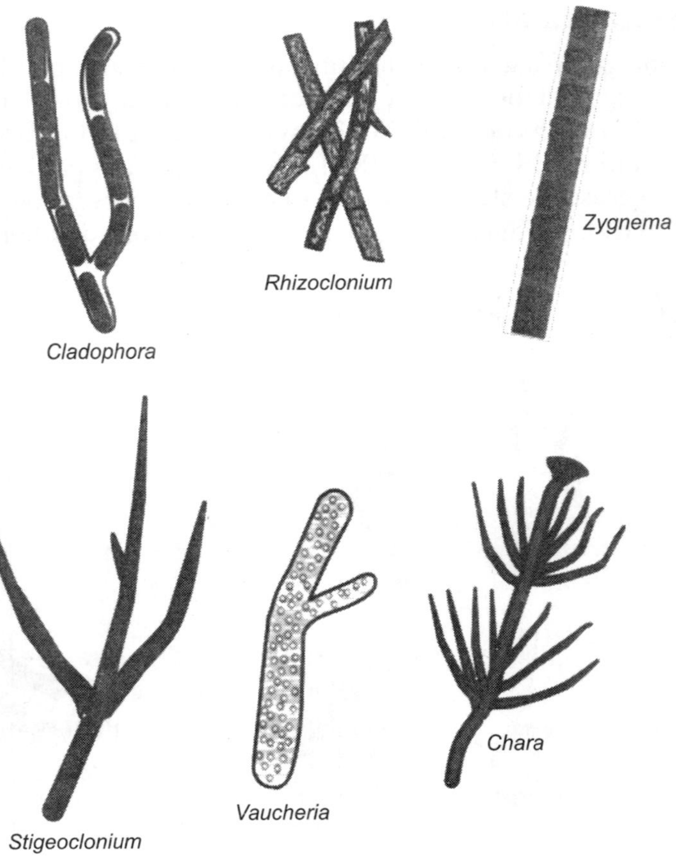

Figure 11.43 Microscopic views of filamented algae.

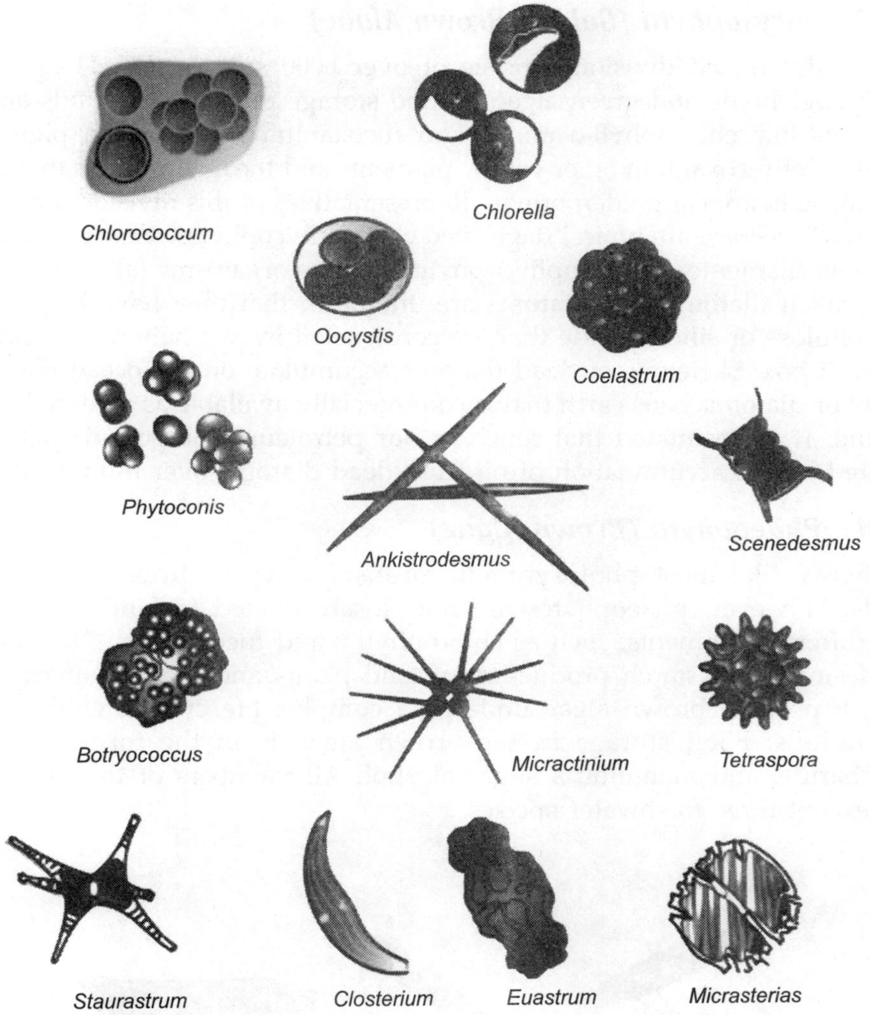

Figure 11.44 Microscopic views of non-flagellated colonial algae.

11.6.7.2 Euglenophyta (Euglenoids)

They are flagellated and appear to be intermediate between the algae and protozoa. Protozoan like characteristics seen in euglenoids are the absence of cell wall, presence of gullet, ability to assimilate organic substances and absence of chloroplast in some species. The absence of cell wall makes these protists very flexible. Instead of cell wall they possess a semirigid outer pellicle, which gives the organism a definite form. Photosynthesis types contain chlorophyll *a* and *b*, and they always have a red stigma or eyespot that is light-sensitive. Their characteristic food storage is a lipopolysaccharide, paramylum the photosynthetic euglenoids can be bleached experimentally by various means in the laboratory. The colourless forms that develop, however, cannot be induced to revert to phototrophy.

11.6.7.3 Chrysophyta (Golden Brown Algae)

This large diversified division consists of over 6000 species (Fig 11.45). They differ from the euglenoids and green algae in food storage in the form of oils and leucosin, in pigments like chlorophyll-a and c and fucoxanthin, a brownish pigment. It is a combination of fucoxanthin other yellow pigments and the chlorophylls that causes most of these algae to appear golden brown. Representatives of this division are *Chrysococcus, Synura* and *Dinobryon* are typical flagellated chrysophycophytes. *Vaucheria* and *Tribonema* are the only filamentous chrysophycophytes. All the organisms fall in special category of algae called diatoms. The diatoms are unique in that they have hard cell walls of pectin, cellulose or silicon oxide that are constructed in two halves. The two halves fit like lid and box. Skeletons of dead diatoms accumulate on the ocean bottom to form diatomite or diatomaceous earth that is commercially available as an excellent polishing compound. It is postulated that much of our petroleum compounds may have been formulated by the accumulation of oil from dead diatoms over millions of years.

11.6.7.4 Phaeophyta (Brown Algae)

Phaeophytes, like most photosynthetic protists, have traditionally been classified as plants. However, phaeophytes are not closely related to land plants; their cells contain different pigments, such as chlorophyll *c* and fucoxanthin. They also lack the plasmodesmata and starch production of land plants and their relatives. Like plants and many protists, brown algae undergo a complex life cycle involving alternation of generations. Food storage in the brown algae is in the form of laminarin, a polysaccharide, and mannitol, a sugar alcohol. All members of the Phaeophytes are marine except three freshwater species.

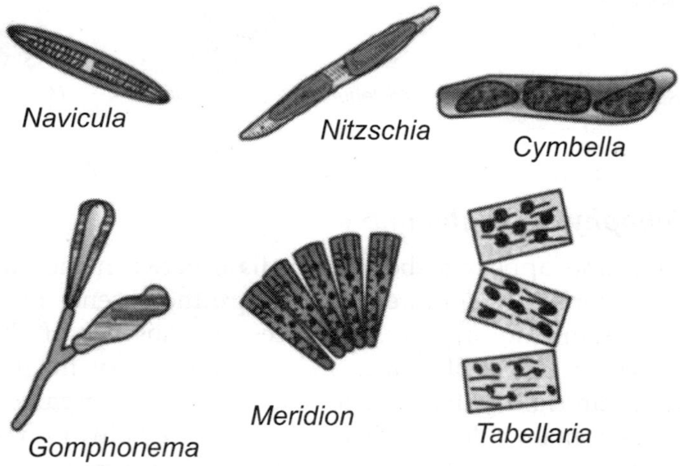

Figure 11.45 Microscopic views of Diatoms (Chrysophyta).

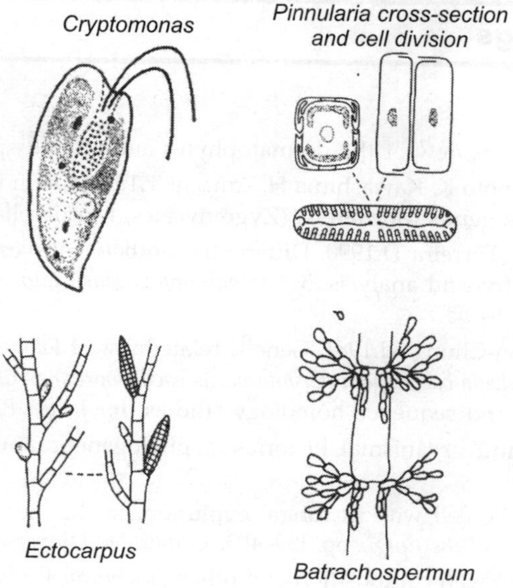

Figure 11.46 Members of the Cryptophyta algae.

11.6.7.5 *Pyrrophyta (Fire Algae)*

The division Pyrrophyta comprises a large number of unusual algal species of many shapes and sizes. Majority of these protists are marine. They are free-living "dinoflagellates" refers to the forward- spiraling swimming motion of these organisms. They are free-swimming protists (unicellular eukaryotic microorganisms) with two flagella, a nucleus with condensed chromosomes, chloroplasts, mitochondria, and Golgi bodies. Biochemically, photosynthetic species possess green pigments, chlorophylls *a* and *c*, and golden brown pigments, including peridinin. Food is variously stored in the form of starch, fats and oils. Dinoflagellates primarily exhibit asexual cell division; some species reproduce sexually, while others have unusual life cycles. Their nutrition varies from autotrophy (photosynthesis; in nearly 50% of the known species) to heterotrophy (absorption of organic matter) to mixotrophy (autotrophic cells engulf other organisms, including other dinoflagellates). Many species are found in numerous habitats, living in the plankton or attached to sediments, sand, corals, or to macroalgal surfaces or to other aquatic plants. Some species are present as parasites in marine invertebrates and fish. Some even serve as symbionts, known as zooxanthellae, providing organic carbon to their hosts: reef-building corals, sponges, clams, jellyfish, anemones and squid.

11.6.7.6 *Rhodophyta*

Red algae are red because of the presence of the pigment **phycoerythrin**; this pigment reflects red light and absorbs blue light. Because blue light penetrates water to a greater depth than light of longer wavelengths, these pigments allow red algae to photosynthesize and live at somewhat greater depths than most other "algae". Some rhodophytes have very little phycoerythrin, and may appear green or bluish from the chlorophyll and other pigments present in them. They are all marine and macroscopic.

Suggested Readings

Fungi

Ajello L. 1968. A taxonomic review of the dermatophytes and related species. *Sabouraudia* **6**, 147–159.

Amano N, Shinmen Y, Akimoto K, Kawashima H, Amachi T.1992. Chemotaxonomic significance of fatty acid composition in the genus Mortierella (Zygomycetes, Mortierellaceae). *Mycotaxon* **44**, 257–265.

Augustyn OP.H, Kock JFL, Ferreira D.1990. Differentiation between yeasts species and strains within a species by cellular fatty acid analysis. *3. Saccharomyces sensu lato, Arxiozyma* and *Pachytichospora. Syst Appl Microbiol* **13**, 44–55.

Aulakh HS, Straus SE, Kwon-Chung KJ.1981. Genetic relatedness of *Filobasidiella neoformans (Cryptococcus neoformans)* and *Filobasidiella bacillispora (Cryptococcus bacillisporus)* as determined by deoxyribonucleic acid base composition and sequence homology studies. *Int J Syst Bacteriol* **31**, 97–103.

Avise J.1989. Gene trees and organismal histories: a phylogenetic approach to population biology. *Evolution* **43**, 1192–1208.

Bartnicki-Garcia S. 1987. The cell wall in fungal evolution. In: Rayner ADM, Brasier CM, Moore D. (ed) *Evolutionary biology of the fungi,* pp 389–403. *Cambridge University Press, New York, N.Y*

Bartnicki-Garcia S.1970. Cell wall composition and other biochemical markers in fungal phylogeny. In: Harbone JB. (ed) Phytochemical Phylogeny. pp 81–103, Academic Press, Ltd. London.

Begerow D, Bauer R, Oberwinkler F.1997. Phylogenetic studies on nuclear large subunit ribosomal DNA sequences of smut fungi and related taxa. *Can J Bot 75, 2045–2056.*

Benny G. L.*1994.* Classical morphology in zygomycete taxonomy. *Can J Bot* **73** *(Suppl. 1)* S725–S730.

Benny GL, Kimbrough JW.1980. A synopsis of the orders and families of Plectomycetes with keys to genera. *Mycotaxon* **12:**1–91.

Benny GL. 2001. The Zygomycota: Zygomycetes, pp. 147-160. In: McLaughlin DJ, McLaughlin EG, Lemke PA. (eds.). *The Mycota* 7A, *Systematics and Evolution*. Springer-Verlag, Berlin, Germany.

Berbee ML, Taylor JW. 1992. Detecting the morphological convergence in true fungi using 18S RNA sequence data. *BioSystems* **28,** 117–125.

Berbee ML, Taylor JW.1992. Two ascomycete classes based on fruiting-body characters and ribosomal DNA sequence. *Mol Biol Evol* **9**, 278–284.

Berbee ML, Yoshimura A, Sugiyama J, Taylor JW.1995. Is Penicillium monophyletic? an evaluation of phylogeny in the family Trichocomaceae from 18S, 5.8S and ITS ribosomal sequence data. *Mycologia* **87**, 210–222.

Bruns TD, White TJ, Taylor JW.1991. Fungal molecular systematics. *Annu Rev Ecol Syst* **22,** *525–564.*

Cain RF. 1972. Evolution of the fungi. *Mycologia* **64,** 1–14.

Carlile MJ, Watkinson SC.1994. *The fungi.* Academic Press, Ltd. London.

Cole GT, Samson RA.1979. *Patterns of development in conidial fungi*, Pittman, London.

Eriksson OE.1982. Outline of the Ascomycetes *Mycotaxon* **15**, 203–248.

Eriksson OE.1995. DNA and ascomycete systematics. *Can J Bot* **73** (Suppl. 1) S784–S789.

Fuller MS.1976. Mitosis in fungi. *Int Rev Cytol* **45**,113–153.

Gams W, Nirenberg H. I.1989. A contribution to the generic definition of Fusarium. *Mycotaxon* **35**, 407–416.

Guého E., Leclerc M. C., de Hoog G. S., Dupont B.1997. Molecular taxonomy and epidemiology of Blastomyces and Histoplasma species. *Mycoses* **40**, 69–81.

Hawksworth DL.1991. The fungal dimension of biodiversity: magnitude, significance, and conservation. *Mycol. Res.* **95,** 641–655.

Hennig W.1966. *Phylogenetic systematics.* University of Illinois Press, Urbana.

Hibbett DS.1992. Ribosomal RNA and fungal systematics. *Trans Mycol Soc Jpn* **33,** 533–556.

Kawasaki M, Ishizaki H, Nishimura K, Miyaji M.1993. Mitochondrial DNA analysis of *Exophiala moniliae. Mycopathologia* **121,**7–10.

Kurtzman CP.1993. Systematics of the ascomycetous yeasts assessed from ribosomal RNA sequence divergence. *Antonie Leeuwenhoek Int J Genet* **63,** 165–174.

Kurtzman CP.1994. Molecular taxonomy of the yeasts. *Yeast* **10,** 1727–1740.

Meyer RJ.1991. Mitochondrial DNAs and plasmids as taxonomic characteristics in *Trichoderma viride. Appl Environ Microbiol* **57,** 2269–2276.

Nakada M, Tanaka C, Tsunewaki K, Tsuda M.*1994.* RFLP analysis for species separation in the genera *Bipolaris* and *Curvularia. Mycoscience* **35,** 271–278.

Ni Min, Feretzaki M, Sun S, Wang X, Heitman J. 2011. Sex in Fungi. *Annu Rev Gent* **45,** 405- 430.

O'Donnell K, Lutzoni FM, Ward TJ, Benny GL. 2001. Evolutionary relationships among mucoralean fungi (Zygomycota): evidence for family polyphyly on a large scale. *Mycologia* **93,**286-297.

Paterson RRM, Bridge PD. 1994. Biochemical techniques for filamentous fungi.CAB International, Wallingford, UK.

Savage JM.1995. Systematics and the biodiversity crisis. *BioScience* **45,** 673–679.

Schuessler A, Schwarzott D, Walker C. 2001. A new fungal phylum, the Glomeromycota: phylogeny and evolution. *Mycological Research* **105,** 1413-1421.

Sugiyama J, Rahayu ES, Chang J, Oyaizu H.1991. Chemotaxonomy of *Aspergillus* and associated teleomorphs. *Jpn J Med Mycol* **32,** 39–60.

Tanabe Y, Nagahama T, Saikawa M, Sugiyama J. 1999. Phylogenetic relationship of *Cephaliophora* to nematophagus hyphomycetes including taxonomic and nomenclatural emendations of the genus *Lecophagus. Mycologia* **91,** 830-835.

Tanabe Y, O'Donnell K, Saikawa M, Sugiyama J. 2000. Molecular phylogeny of parasitic Zygomycota (Dimargaritales, Zoopagales) based on nuclear small subunit ribosomal DNA sequences. *Molecular Phylogenetics and Evolution* **16,** 253-262.

Voigt K, Cigelnik E, O'Donnell K. 1999. Phylogeny and PCR identification of clinically important Zygomycetes based on nuclear ribosomal-DNA sequence data. *Journal of Clinical Microbiology* **37,** 3957-3964.

Windels CE.1991. Current status of Fusarium taxonomy. *Phytopathology* **81,** 1048–1051.

Yun Sung-Hwan, Berbee ML, Yoder OC, Turgeon BG. 1999. Evolution of the fungal self-fertile reproductive lifestyle from self-sterile ancestors. *Proc Natl Acad Sci (USA)* **96,** 5592-5597.

Protozoa

Corliss JO. 1991. Introduction to the protozoa in: Microscopic anatomy of invertebrates. Vol. 1. Pp. 1-12, (ed. FW. Harrison, J.J.Corliss). Wiley - Liss, New York.

Englund PT, Sher A (eds). 1988. *The Biology of Parasitism.* A Molecular and Immunological Approach. Alan R. Liss, New York.

Goldsmith R, Heyneman D (eds) 1989. *Tropical Medicine and Parasitology.* Appleton and Lange, East Norwalk, CT.

Hausmann K, N. Hulsmann N. 1996. *Protozoology.* Thieme Verlag; New York.

Honigberg BM, Balamuth W, Bovee EC, Corliss JO, Gojdics M, Hall RP, Kudo RR, Levine ND, Lobblich AR, Weiser J. 1964. A Revised Classification of the Phylum Protozoa. *Journal of Eukaryotic Microbiology* **11**, 7–20.

JahnTL, Bovee, EC, Jahn FF. 1979. *How to Know the Protozoa*. Wm. C. Brown Publishers, Div. of McGraw Hill, Dubuque, Iowa.

Kotler DP, Orenstein JM. 1998. Prevalence of Intestinal Microsporidiosis in HIV-infected individuals referred for gastrointestinal evaluation. *J Gastroenterol 1994*, 89.

Lee JJ, Hutner SH, Bovee EC (eds). 1985. *An Illustrated Guide to the Protozoa*. Society of Protozoologists, Lawrence, KS.

Neva FA, Brown H 1994. *Basic Clinical Parasitology*, 6th edition, Appleton & Lange, Norwalk, CT .

Roger AO, Druger M, eds. 1997.*Explore the World Using Protozoa*. National Science Teachers Association and the Society of Protozoologists. Arlington, VA.

Sleigh M, Arnold E. 1981. *The Biology of Protozoa*. London.

Slime Molds (Myxomycota)

Alexopolous CJ. 1962. second edition. Introductory Mycology. John Wiley and Sons, New York p. 78.

Bonner JT. 2003. The evolution of development in the cellular slime molds. *Evolution and Development* **5**, 305–313.

Deasey MC, Olive LS. 1981. Role of Golgi Apparatus in Sorogenesis by the Cellular Slime Mold Fonticula alba. *Science* **213**, 561–563.

Häder DP, Burkart U. 1983. Optical properties of Dictyostelium discoideum pseudoplasmodia responsible for phototactic orientation. *Exp Mycol* **7**, 1–8.

Matthew W, Spiegel FW, Silberman JD. 2009. Phylogeny of the "Forgotten" Cellular Slime Mold, Fonticula alba, Reveals a Key Evolutionary Branch within Opisthokonta. *Molecular Biology and Evolution* **26**, 2699.

Poinar GO, Waggoner BM. 1992. A fossil myxomycete plasmodium from Eocene-Oligocene amber of the Dominican Republic. *Journal of Protozoology* **39**, 639-642.

Ponte E, Bracco E, Faix J, Bozzoro S. 1998. Detection of subtle phenotypes: The case of the cell adhesion molecule csA in Dictyostelium. *Proc Natl Acad Sci USA* **95**, 9360–9365.

Swanson AR, Vadell EM, Cavender JC. 1999. Global distribution of forest soil dictyostelids. *J Biogeography* 26, 133–148.

Algae

Cavalier-Smith T. 2007. Evolution and relationships of algae: major branches of the tree of life. In: *Unravelling the algae: the past, present and future of algal systematics*, eds J. Brodie and J. Lewis, pp 21-55. CRC Press, Boca Raton, London and New York.

De Reviers B, Rousseau F, Draisma SGA. 2007. Classification of the Phaeophyceae from past to present and current challenges. In: *Unravelling the algae: the past, present and future of algal systematics*, eds J. Brodie and J. Lewis, pp. 267-284. CRC Press, Boca Raton, London and New York.

Lewis LA, McCourt RM. 2004. Green algae and the origin of land plants. *American Journal of Botany*, **91**, 1535-1556.

Maggs MA, Verbruggen H, De Clerck 2007. Molecular systematics of red algae: building future structures of firm foundations. In: *Unravelling the algae: the past, present and future of algal systematics*, eds J. Brodie and J. Lewis, pp. 103-121. CRC Press, Boca Raton, London and New York.

Saunders GW, Hommersand MH. 2004. Assessing red algal supraordinal diversity and taxonomy in the context of contemporary systematic data. *American Journal Botany* **91**, 1494-1507.

Stam W, Olsen JL, Zaleski SF, Murray SN, Brown KR. Walters LJ. 2006. A forensic and phylogenetic survey of Caulerpa species (Caulerpales, Chlorophyta) from the Florida coast, local aquarium shops and e-commerce: establishing a proactive baseline for early detection. *Journal of Phycology* **42,** 1113-1124.

Verlaque M, Durand C, Huisman JM, Boudouresque CF, Parco Y. 2003. On the identity and origin of the Mediterranean invasive Caulerpa racemosa (Caulerpales, Chlorophyta). *European Journal of Phycology* **38**, 325–339.

Yoon HS, Müller KM, Sheath RG, Ott FD, Bhattacharya D. 2006. Defining the major lineages of red algae. *Journal Phycology* **42,** 482-492.

Chapter **12**

Biology of Viruses

12.1 INTRODUCTION

By the beginning of the twentieth century much was known about the causes of the infectious diseases of animals and plants. Many parasitic bacteria and fungi had been isolated and their life cycles and relations with their hosts studied. It was felt to be only a matter of time before these types of microorganisms would be shown to be the causal agents of all infectious diseases. This state of optimism was soon shattered and the causes of many important diseases, e.g., smallpox, foot and mouth, influenza, tobacco mosaic, remained unknown. Iwanowski in 1892 showed that the symptoms of tobacco mosaic could be reproduced in healthy plants by rubbing their leaves with juice from infected ones, even after the juice had been passed through a filter capable of holding back particles of bacterial size. Unwittingly, Iwanowski had demonstrated the existence of an infectious agent entirely different from any other known at that time. However, he was sceptical of his results and clung to the idea of a bacterial aetiology of tobacco mosaic. Six years later Beijerinck confirmed these filtration experiments and realized that an infectious agent different from bacteria was responsible. However, his idea of a '*contagium vivum fluidum*' implied the idea of a nonparticulate agent and he used the term 'virus' derived from the Latin meaning of a 'poison'. In the same year as Beijerinck's work on tobacco mosaic, Loeffler and Frosch showed that foot and mouth disease of cattle could be induced by bacteria-free filtrates of the fluid from vesicles characteristic of the disease, and Sanarelli obtained comparable results with myxomatosis of rabbits. Other plant and animal diseases were soon added to the list of those caused by agents able to pass through bacterial filters.

These agents were termed 'filtrable' viruses but this description is no longer used since it is now possible to make filters fine enough to retain the smallest virus particles. It also became clear that viruses propagate only in living cells, i.e., they are obligate intracellular parasites. This explained the repeated failure of the early microbiologists to grow viruses in normal bacteriological media.

Twort (1915) and D'Herelle (1917) respectively described cultures of staphylococci and dysentery bacilli showing clear areas due to lysis of bacterial cells. From such cultures bacteria-free filtrates were obtained which were able to produce the effects in new cultures. It was concluded that this lysis was due to a virus, now termed bacteriophage. The actinomycetes are also attacked by viruses (actinophages) and phycoviruses (cyanophages) have been discovered in cyanobacteria and mycophages attacking fungi. Indeed it seems that most, if not all, types of organisms are attacked by viruses.

12.2 THE NATURE OF VIRUSES AND THEIR PLACE AS MICROORGANISMS

The development of several physical techniques (filtration with filters of graded pore size, ultracentrifugation and electron microscopy) made it clear that viruses were particulate, virus particles being termed virions. Stanley (1935) was able to purify and crystallize tobacco mosaic virus (TMV). In the following year Bawden and Pirie showed that this virus is a nucleo-protein. Many viruses have since been purified by the use of similar techniques. The ability of some viruses to crystallize emphasizes their simplicity and regularity of structure.

As knowledge about viruses increased, it became clear that they differed radically in many respects from other infectious agents. Indeed, the ability of some viruses to crystallize, called into question the concept of viruses as living microorganisms. Subsequently the uniqueness of viruses was further emphasized when, by using gentle extraction methods, it was found possible to separate the protein and nucleic acid components of some simple plant and animal viruses (e.g., TMV and poliovirus) and to show that the nucleic acid alone was able to initiate infection.

The debate on the nature of viruses as living organisms continues although most of the disputes are purely semantic and revolve around usage of the words 'living' and 'organism'. Owing to their ultraparasitic nature, viruses are intimately connected with a wide range of living organisms. Moreover, the study of bacteriophage has done much to add our knowledge of genetics both of the pages themselves and of their host bacteria. It is clear too that viruses unlike, say, the transforming DNA of some bacteria, have developed distinct means of transferring their genome from one host cell to another.

12.3 VIRUS-LIKE MICROORGANISMS

Considering the great diversity of microorganisms and the varying degrees of parasitism that have developed during the course of evolution, it seems likely that there may exist organisms which resemble viruses in some ways and bacteria in other. Such organisms are now recognised as belonging to two groups, namely the Rickettsiae and the *Bedsoniae,* also known as the Chlamydozoaceae. Organisms belonging to both these groups are small (approximately 300 mμ in diameter—a size only a little larger than the largest of the viruses) and, like the viruses, they are obligate intracellular parasites.

For these reasons, they were classified for many years as viruses and many virological techniques, e.g., methods of culture, have proved useful in their study. However, the Rickettsiae and *Bedsoniae* agents are now known to possess properties more closely related to the bacteria (Table 12.1). The bacterium-like properties of the Rickettsiae and *Bedsoniae* are as follows:

(1) particles contain both RNA and DNA, whereas viruses contain only one of these,
(2) multiplication occurs by binary fission or budding,
(3) particles possess bacterial cell wall constituents and some of their associated metabolic pathways (consequent to this is the susceptibility of these agents to antibiotics like penicillin).

Table 12.1 Comparison of the properties of viruses with some other microorganisms

Properties	*Bacteria*		*Mycoplasma*	*Rickettsiae*	*Besdoniae*	*Viruses*
	+ve	*–ve*				
Size	1 μ	1 μ	< 150 mμ	300 mμ	300 mμ	Largest 300 mμ, majority < 150 mμ
Nucleic acid	RNA and DNA		RNA and DNA	RNA and DNA	RNA and DNA	Either RNA and DNA
Growth on artificial media	+	+	+	–	–	–
Intracellular-replication	–[a]	–[b]	+	+	+	+
Replication mode	Binary fission		Binary fission	Binary fission and budding	Binary fission	Intracellular assembly of constituents (nucleic acid and protein subunits etc.)
Muramic acid	+	+++	–	+	+	–
Antibiotic sensitivity	+	+	+	+	+	–

[a]*Mycobacterium tuberculosis replicate intracelluarly*
[b]*Brucella abortus replicate intracellularly*

For the above reasons the Rickettsiae and *Bedsoniae* are no longer considered viruses but as very small bacteria which have developed an ultra-parasitic mode of life.

12.4 STRUCTURE AND CHEMICAL COMPOSITION OF VIRUSES

Efficient methods of preparing virus particles free of host cell constituents are necessary to determine the chemical and physical characteristic of viruses. The development of such purification methods (e.g., chromatography, electrophoresis, density gradient centrifugation, x-ray diffraction and electron microscopy), has greatly facilitated the study of virus structure and composition.

The simplest of the viruses (e.g., poliovirus, tobacco mosaic virus), possess two main chemical constituents, protein and nucleic acid. Viruses, unlike other microorganisms, are characterized by the presence of only one type of nucleic acid in the infectious particle. Thus viruses may be of two types, containing either RNA or DNA. Until recently all higher plant viruses examined were RNA-containing but a DNA containing virus infecting cauliflowers and several other plants have now been described. The viruses of other organisms may be of either type. The extremely small size of viruses means that the particles are capable of containing only minute amounts of nucleic acid. The extreme smallness of viral genomes has considerable influence upon viral fine structure.

It is now well established that DNA and RNA generally occur as double and single-stranded molecules respectively. The nucleic acids of some viruses, however, are atypical. For example, although most DNA viruses contain the double-stranded molecule, some phages (e.g., ∅ × 174, S13), have single-stranded DNA. In RNA viruses the nucleic acid is usually single stranded but double stranded RNA has been demonstrated in the reoviruses (a group of animal viruses), the fungal viruses and the wound tumour virus of plants. Some of the common viruses that affect human are described in Table 12.2.

Table 12.2 Families of viruses that affect humans

Characteristics/ Dimensions	*Viral family*	*Important genera*	*Clinical of special features*
Single-stranded DNA non-enveloped 18-25 nm	Parvoviridae 	Human parvovirus 819	Fifth disease; anaemia in immunocompromised patients
Double-stranded DNA non-envoloped 70-90 nm	Adenoviridae 	Mastadenvirus	Medium-sized virues that cause various respiratory infections in humans; some cause humans animals
40-57 nm	Papovaviridae 	Papillomavirus (human wort virus Polyomavirus	Small viruses that induce tumours; the human wort virus (papilloma) and certain viruses that produce cancer in animals (polyoma and simian) belong to this family.
Double-stranded DNA enveloped 200-350 nm	Poxviridae 	Orthopoxvirus (vaccina and smallpox viruses Molluscipoxvirus	Very large, complex, brick-shaped viruse that cause diseases such as smallpox (various), molluscum contagiosum (wortlike skin lesion), and cowpox.
150-200 nm	Herpesviridae 	*Simplexirus* (HHV-1 and 2) *Varicellovirus* (HHV-3) *Lymphocrytovirus* (HHV-4) *Cytomegalovirus* (HHV-5) *Roseolovirus* (HHV-6) HHV-7 *Kaposl's sarcons* (HHV-8)	Medium-sized viruses that cause various human diseases, such as fever blisters chickenpox, shingles, and infectious mononucleosis; implicated in a type of human cancer called Burkit's lymphoma.

Contd...

Contd...

42 nm	Hepadnoviridae 	*Hepachavirus* (hepatitis B virus)	After protein synthesis, hepatitis B virus uses reverse transcriptase to produce its DNA from mRNA; causes hepatitis B and liver tumours.

Most of the proteins of the virion occur as a coat, termed a capsid, surrounding the nucleic acid. Such proteins have two main functions. They protect the viral genome from adverse environmental conditions and also aid in the penetration of the virus into the host cell. The latter function may be merely the attachment of the capsid on to the host cell membrane (as with many of the animal viruses) or it may be concerned with active injection of the viral nucleic acid through the host cell wall (Figure 12.1). Beside structural proteins, essential for the morphological integrity of the particle, some viruses also contain proteins with enzymatic functions, e.g., the phages contain the enzyme lysozyme which facilitates entry of the phage nucleic acid through the host cell wall; the myxoviruses (a group of animal viruses including influenza virus) contain the mucolytic enzyme neuraminidase which may be involved in penetration of the virus into the host cell or release of progeny virus after infection. The total amount of protein present in viruses differs from virus to virus (Table 12.3).

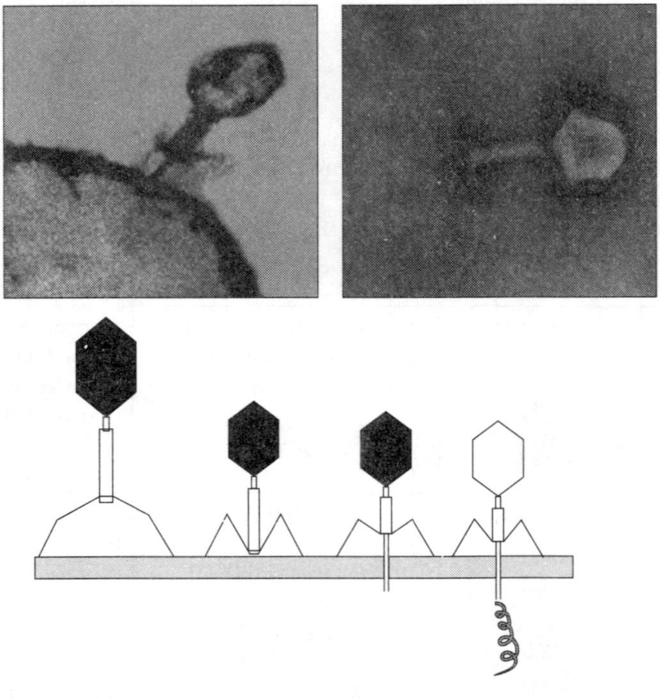

Figure 12.1 Demonstrating bacteriophage infecting bacteria.

Table 12.3 The approximate composition of some viruses

Virus	*%RNA*	*% DNA*	*% Protein*	*Molecular Weight*	
				Nucleic acid	*Virion*
Poliovirus	25-30	0	70 -75	2.2×10^6	3.6×10^6
Influenza[a]	0.8-1	0	60- 75	2.0×10^6	2.1×10^6
TMV	6	0	94	2.1×10^6	40×10^6
Vaccinia[b]	0	5	89	150×10^6	2000×10^6
Adenovirus	0	13	87	23×10^6	177×10^6
Coliophage T2	0	50	50	120×10^6	300×10^6

[a]Also contain approximately 30% lipid and 6% non-nucleic acid carbohydrate

[b]Also contain approximetely 6% lipid

Besides the two basic constituents, protein and nucleic acid, some viruses also contain lipids, e.g., the myxoviruses, herpes viruses (including herpes simplex virus of man), the arboviruses (including the virus of yellow fever), the poxviruses (including the virus of smallpox) and the phage PM2 which infects a species of *Pseudomonas.* The amount of lipid present varies from approximately 30 per cent in the case of the myxoviruses to 6 per cent in the case of the poxviruses.

Some viruses also contain a small percentage of non-nucleic acid carbohydrate, e.g., influenza virus contains approximately 6 per cent. The function of this component is unknown.

Finally, some viruses, viz., phages and plant viruses, are known to contain polyamines.

The basic properties of these compounds serve to neutralize the acidic phosphate groups of the nucleic acids. It is thought that this facilitates the folding up of the nucleic acid molecule within the small volume of the virion. Similar substances may occur in animal viruses. For many years after the discovery of viruses, little was known about their fine structure. Techniques such as filtration and ultracentrifugation gave estimates of the sizes of virions (Table 12.4) but viral fine structure only began to be elucidated following the development of electron microscopy and X-ray *diffraction.* Electron microscopy, with the aid of metal shadowing demonstrated clearly that some viruses like TMV are rod shaped while others like poliovirus and tomato bushy stunt virus are nearly spherical (isometric). However, it was X-ray diffraction which first demonstrated the structure of viruses at the molecular level. This technique can only be applied fruitfully to those viruses which, when purified, form crystals or paracrystals e.g., poliovirus and TMV, respectively. The virus, first and most extensively investigated by X-ray diffraction, was TMV. From such studies it is now clear that the capsid of TMV is a hollow rod consisting of identical protein molecules (structure units) arranged in the form of a helix around the long axis of the virion (Figure 12.2). Furthermore, the RNA of the virion also exists as a helix, within and closely associated with, the helix of protein subunits. The virions of other rod-shaped plant viruses (e.g., white clover mosaic virus), although less rigid than those of TMV, have a similar helical organization.

Table 12.4 Important features of some animal, plant and bacterial viruses

Virus	*Nucleic Acid*	*Capsid symmetry*	*Presence of envelope*	*Size of virion*
Vaccinia	DNA	?	-	230 × 300 mμ
Herpes simplex	DNA	Icosahedral	+	Non enveloped 110 mμ, enveloped 180-250 mμ
Tipula iridescent virus (infects larvae of the crane fly), Tipula paludosa	DNA	Icosahedral	-	130 mμ
Adeno virus	DNA	Icosahedral	-	75 mμ
T4 Phage	DNA	Icosahedral and Helical	-	Head 95 × 65 mμ Tail 110 × 25 mμ
Ø × 174 virus	DNA	Icosahedral	-	30 mμ
TMV	RNA	Helical	-	Rigid rod 300 mμ
Potato X	RNA	Helical	-	Flexible rod 500 mμ long
Influenza	RNA	Helical	+	80 -200 mμ
Polio	RNA	Icosahedral	-	28 mμ
MS2 (Phage)	RNA	Icosahedral	-	24 mμ

Viruses with isometric (nearly spherical) capsids were found to have a somewhat different structure. The protein subunits forming the capsids of these viruses are arranged as if located on the surface of a special class of deltahedron, viz., an icosadeltahedron. This is a polyhedron whose faces are all equilateral triangles and with the symmetry of an icosahedron (a twenty-sided solid). Such solids are said to have cubic or icosahedral symmetry (Figure 12.3).

The introduction of negative staining to electron microscopy gave confirmation of the results obtained by X-ray diffraction. The morphological subunits revealed on the surface of virions by negative staining are termed capsomeres. The number of capsomeres present varies greatly between different viruses, e.g., virions of the phage Ø X 174 have 12 capsomeres whereas the virion of herpes simplex (a human virus) has 162. It seems probable that in most cases a single capsomere consists of more than one structure unit (protein molecule) the latter being the subunits demonstrated by X-ray diffraction. The arrangement of the nucleic acid molecule within the capsid of icosahedral virions is not yet clear. Negative staining has also shown that some animal viruses, although more or less spherical, nevertheless have capsids with helical symmetry. In such viruses (e.g., influenza virus, Newcastle disease virus), the helical capsid is wound up like a ball of wool inside an envelope of lipoprotein. The capsid

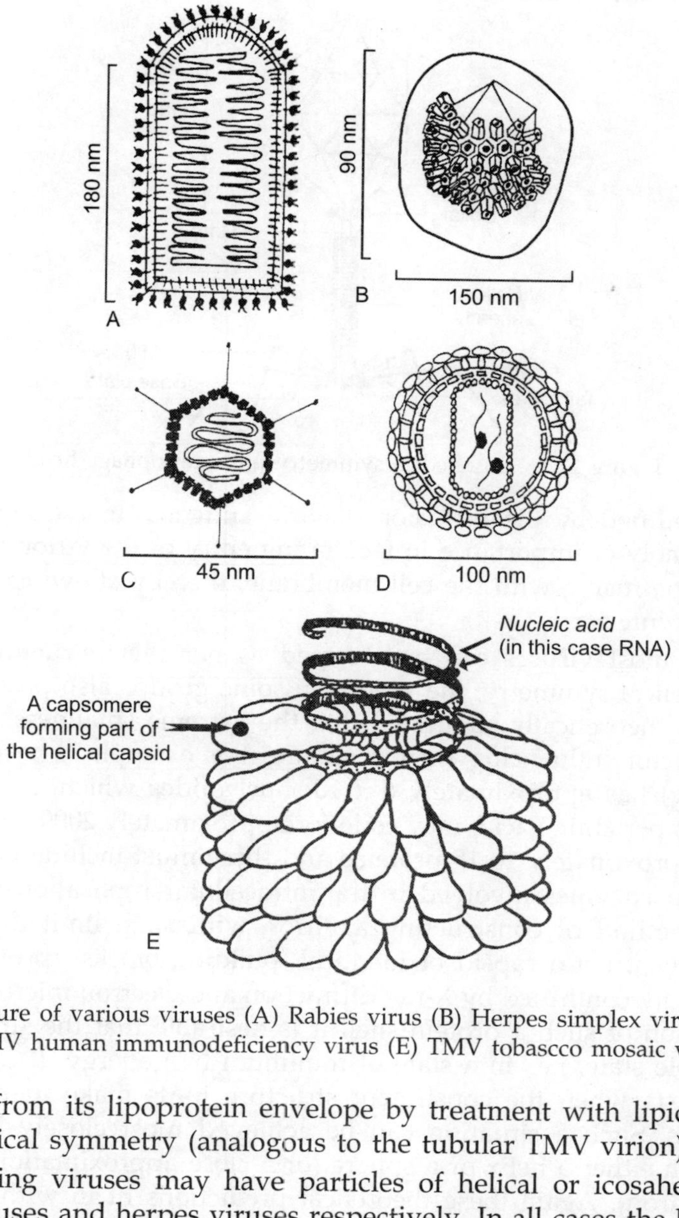

Figure 12.2 Structure of various viruses (A) Rabies virus (B) Herpes simplex virus (C) Adenovirus (D) HIV human immunodeficiency virus (E) TMV tobacco mosaic virus.

can be released from its lipoprotein envelope by treatment with lipid solvent, hence revealing the helical symmetry (analogous to the tubular TMV virion).

Lipid containing viruses may have particles of helical or icosahedral symmetry, e.g., the myxoviruses and herpes viruses respectively. In all cases the lipid is found as a lipoprotein envelope surrounding the nucleocapsid (nucleic acid core plus protein capsid). The viral lipoprotein component has an interesting origin. There is strong evidence to suggest that the lipoprotein is derived directly from the host cell membranes. For example, in the case of herpes simplex virus the virion acquires an envelope of host nuclear membrane as it passes from the nucleus into the cytoplasm. In the myxoviruses, e.g., influenza, the envelope is acquired from the cytoplasmic membrane as the virions bud off from the cell surface. Hence, viral envelopes are of host cell origin although

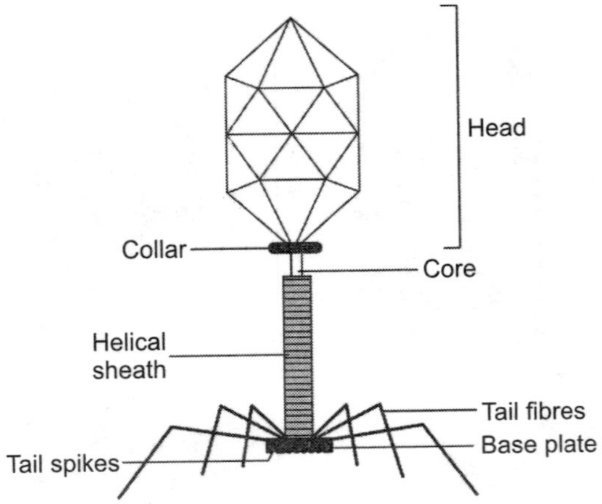

Figure 12.3 Icosahedral symmetry of bacteriophage head.

they may be modified by the inclusion of viral antigens. In the animal viruses, the envelope is probably of importance in facilitating entry of the virion into the host cell; the virus envelope fusing with the cell membrane, thereby allowing the capsid direct access to the cell interior.

Accordingly, most viruses can be grouped as per their structure, having either icosahedral or helical symmetry; the virions in some groups also possess a lipoprotein envelope. It was theoretically suggested that the extreme smallness of viral genomes was the main factor influencing their structure. For example, a virion of poliovirus (RNA containing) has approximately 6×10^3 nucleotides which, assuming a code of three nucleotides per amino acid, may code for approximately 2000 amino acids. This is equivalent to approximately 10-15 proteins and these must include not only structural proteins but also enzymes involved in the intracellular replication of the virus. The most efficient method of constructing a virion with such limited information and materials is to construct a capsid of identical 'building blocks' (protein molecules); a fact which is amply confirmed by X-ray diffraction and electron microscopy. Moreover, in the construction of such a protein shell it is desirable that the structure should be in the most stable state, i.e., in a state of minimum free energy. It can be shown that such a state exists when the constituent structure units make identical bonds with their neighbours. Such a situation can be achieved most closely by arranging the structure units in either a helix or a sphere (or a close approximation to a sphere, i.e., an icosadeltahedron). Again these theoretical predictions fit in with the observations made on virus fine structure by X-ray diffraction and electron microscopy.

Although most viruses appear to fit neatly into the helical or icosahedral categories, there are at least two virus groups which have a more complex structure. The poxviruses (a group including smallpox virus) are large brick shaped or ovoid viruses, the virion being composed of a central DNA containing body (the nucleoid) surrounded by a series of protein and lipid membranes. Many of the larger phages have a complex structure which is clearly correlated with the mechanism by which the viral nucleic acid is injected into the host cell. These phages possess a head and tail (Figure 12.1 and 12.3), the head contains a tightly packed core of nucleic acid (usually DNA) within a thin protein coat, the latter structure being similar to the capsid of the simpler isometric viruses. The head is spherical, ellipsoidal or hexagonal in shape and sizes range from 25 to 100 mμ in diameter. Structural distinction between the nucleic acid and protein components is well illustrated when phage heads are emptied of their DNA by osmotic shock, leaving hollow-headed protein "ghosts". The phage tail is narrower than the head and may be up to 100 mμ in length. It consists of several distinct proteins but its structure varies in complexity with different phages. In the T even phages (T_2, T_4, T_6, a group of phages attacking *Escherichia coli*), the tail consists of an inner tube (possibly having helical symmetry) and an outer contractile sheath. At the head end the sheath is attached to a collar and at the opposite end, to a hexagonal plate (Figure 12.4). Fibres radiating from this terminal plate serve to attach the phase to the host bacterium prior to infection. It is thought that these phages represent a combination of icosahedral and helical symmetry in the structure of the head and tail respectively. Of all viruses, it is perhaps in the structure of phages, the T even in particular, that structure is seen to be correlated with function most elegantly.

12.5 VIRUS REPLICATION

The use of the term 'replicate' infers that the process of virus multiplication is different from that of microorganisms and tissue cells which divide by binary fission with or without mitotic division of their genetic components. Whilst the mode of entering the host cell varies from virus to virus the mode of replication is considered to be similar for all and has been most completely worked out for bacteriophage. The viral nucleic acid upon entering the cell takes over control of the cellular metabolic processes and codes for the separate synthesis of viral nucleic acid and protein which later combine to form the mature virus particle. The virus yield from a cell infected with a single virus particle varies widely but often ranges from 10 to 100 particles.

The host cell must be capable of supporting this sequence of steps in viral replication. Many viruses have a single or limited host cell requirement; others may replicate in a range of different host cells but the quantity of virus produced in each cell type may differ widely (Figure 12.5).

Viruses may be propagated in susceptible animals, plants or microorganisms, or in tissue cultures made from animal or plant tissues. When using animals it is necessary to consider: (1) their natural susceptibility to infection (Table 12.5) or immune status to the virus; (2) the possibility of latent infection with the same or other virus (often the challenge of another virus stimulates a latent virus to become active, as occurs with herpes simplex in man, the cause of the common cold sore on the lips which often erupts when the patient is challenged by a common cold virus); (3) the most suitable route of inoculation which is usually related to the affinity of the virus for particular tissues (Table 12.5). Infection is recognized by characteristic signs and symptoms of disease.

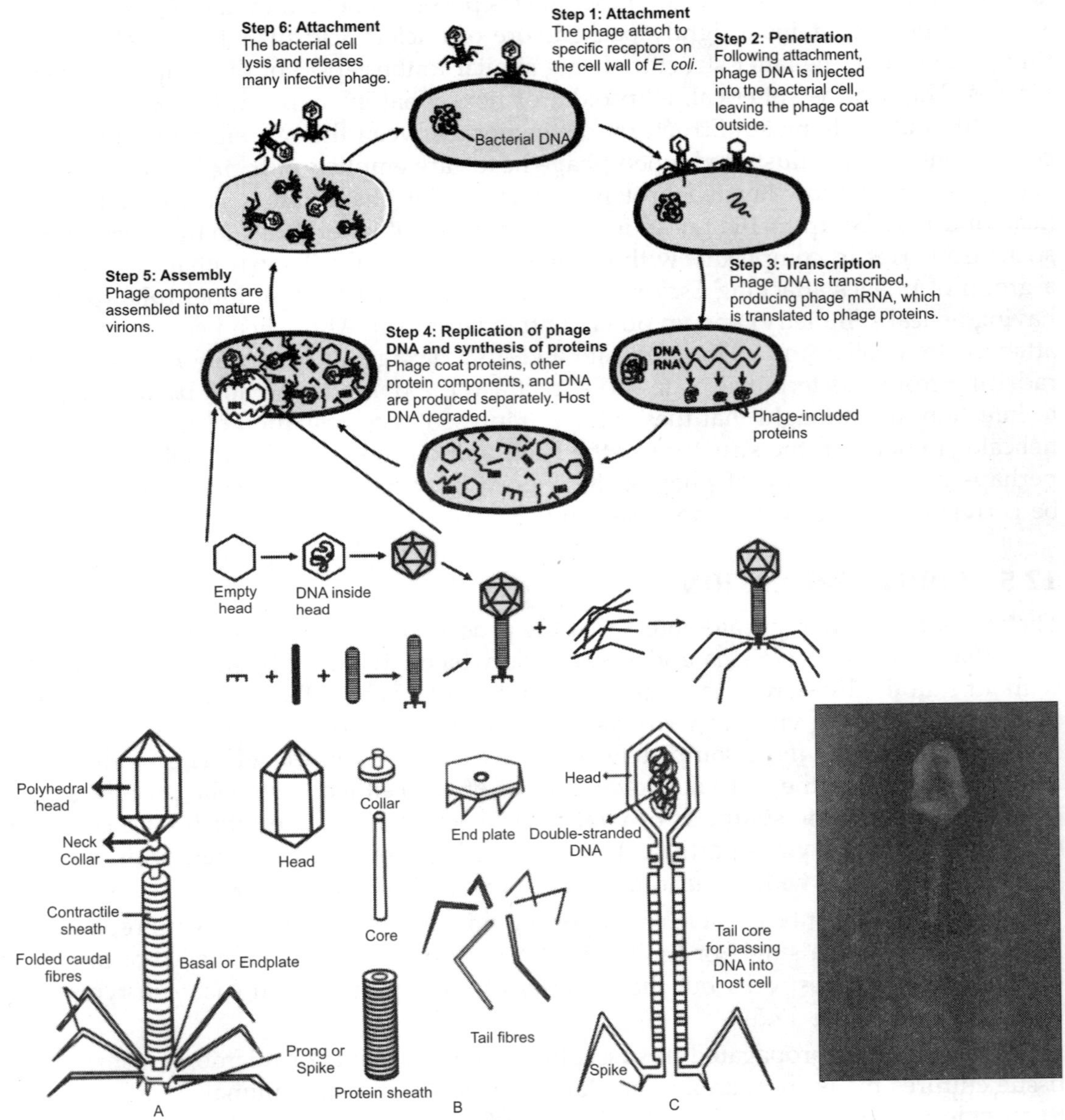

Figure 12.4 Structure and multiplication of T even bacteriophage.

A widely used host for animal virus propagation is the developing chick embryo. The tissues and associated membranes support the replication not only of fowl viruses but also of mammalian viruses to which the hatched chick and adult bird are not susceptible. Whilst maternal antibodies against fowl viruses may be present in small amounts in the yolk of the fertile egg the chick embryo does not possess an antibody-forming mechanism. Evidence of virus replication may be indicated by pock formation in the chorioallantoic membrane, by lesions and inclusion bodies in particular tissues,

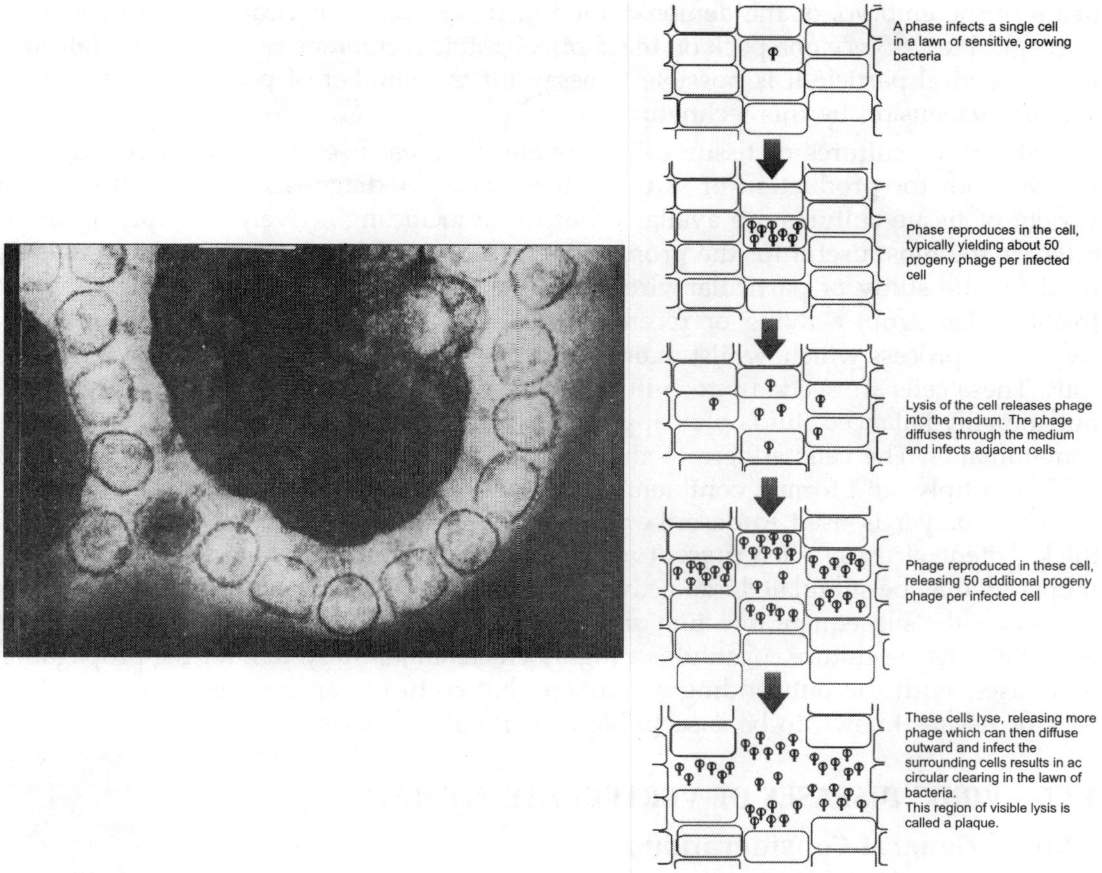

Figure 12.5 Bacteriophage attachment, penetration of genetic material and multiplication.

Table 12.5 The natural and experimental hosts of some common animal viruses

Virus	*Natural Host*	*Animal*	*Route of inoculation*
Arthropod -borne Encephalitides (Arboviruses)	Man and other mammals	Mouse	Cerebral
Canine Distemper	Dog	Ferret	Nasal
Foot and mouth	Cattle animals	Suckling Mice, Guinea pigs,	Hind footpads, Cerebral
Influenza	Man	Chimpanzee, Ferret	Nasal
Mumps virus	Man	Monkey	Parotid gland
Polio virus	Man	Monkey	Cerebral, Peritoneal
Rabies	Dog, Man	Mpuse	Cerebral

death of the embryo, or the demonstration of the presence of viral antigen by specific serological tests. Since one pock on the chorioallantoic membrane can arise from infection with one viral particle it is possible to assay for the number of pock forming units in a virus suspension by this technique.

Laboratory cultures of tissue cells provide the most useful means of propagating viruses both for production of virus antigens and for diagnosis of virus diseases. A variety of tissue cultures are available but those including actively multiplying tissue cells are the most useful for the propagation of viruses. Monolayer cultures are widely used for the study of particular virus-cell interactions and cytopathic effects. Tissues freshly taken from a living or recently dead animal are treated with enzymes (e.g., trypsin), a process which, whilst maintaining cell viability, causes disaggregation of the cells. These cells are seeded into culture tubes or bottles containing physiologically and nutritionally balanced fluids plus appropriate antibiotics to inhibit bacterial and fungal contaminants. The cells adhere to the lower surface of the vessel when incubated at 37°C, multiply, and form a confluent tightly-packed sheet of cells within 2-4 days; the term monolayer is used to describe the cell sheet which is rarely more than one cell thick. Latent viruses, when present in the animal tissue from which the cultures were prepared, may be stimulated into active replication by either the enzymic disaggregation process or the subsequent infection of the tissue culture with another virus. Despite this possibility, tissue culture constitutes a highly efficient *in vitro* system for the propagation of viruses, with the outstanding advantage that cultures can be prepared from most cells or tissues known to be susceptible to particular viruses.

12.6 VIRUS DISEASES OF VERTEBRATE ANIMALS

12.6.1 General Considerations

Unlike the majority of plant virus diseases most animal virus diseases cannot be diagnosed solely on their signs and symptoms. Viruses do not produce exotoxins and the diseases they cause are the direct result of their primary and secondary replication cycles within the various tissue cells of the animal body. An understanding of how this replication occurs in tissue cells provides an understanding of the disease processes taking place in the animal body as a whole. The replication of viruses in tissue cells leads to the biological malfunctioning of those cells and if large numbers of cells are involved malfunctioning of the organ generally follows death of the animal.

Some animal viruses replicate in a limited range of tissue cells, e.g., influenza virus replicates only in cells of the respiratory tract, while others replicate in a wide variety of tissue cells, e.g., smallpox virus in cells of the skin, lungs, and other internal tissues. The latter category of viruses can therefore spread to other susceptible tissues by blood-borne dissemination from a primary site of infection. The presence of virus in the blood stream is not necessarily an indication of wide tissue susceptibility since viruses with a limited tissue range may 'spill-over' into the blood stream following replication in the susceptible tissues.

As with other microbial diseases, the severity of a virus disease depends upon the size of the infecting dose, the state of health of the animal, its age, sex, and degree of immunity.

12.6.2 In vitro Infections

12.6.2.1 Cell susceptibility

There is for each virus a range of susceptible (sensitive) tissue cells in which it will readily replicate. Other types of cells, in which replication is difficult or impossible, constitute the resistant (insensitive) cells. Cell susceptibility is related to such factors as the ability of the virus to adsorb to the cell plasma membrane, which, in turn, depends on the presence of virus 'receptors' on the membrane, as well as the provision by the cell of the appropriate biosynthetic processes for the support of replication of the invading virus.

12.6.2.2 Replication sequence

Tissue culture studies have revealed the nature of the replication sequence which occurs in the infected host cell. This sequence includes: (1) **adsorption** of the virion to the cell plasma membrane; (2) **penetration** of the virion into the cell through active pinocytosis (i.e., engulfment in a manner similar to phagocytosis but on a smaller scale) by the host cell itself or by fusion of the virus envelope with the cell membrane; (3) **eclipse** of the virion, due to enzymically controlled decapsidation processes which release the nucleic acid 'core', intact virus being no longer found in the cell; (4) **the latent period**, when the biosynthesis of new virus components is undertaken; (5) **the assembly period**, when early, immature virus forms are first appearing in the cell; (6) **maturation** of progeny virions from the early immature forms; and (7) **release** of the progeny virions from the infected cells by either budding-off processes (e.g., influenza virus) or cell lysis (e.g., poliovirus). With some animal viruses the complete sequence from (1) to (7) may take as little as 7-8 hours.

12.6.2.3 Cytopathic effects

Following virus replication within the host cell observable cytopathic effects (CPE) may result. These include: (1) **hyperplasia**, or cellular proliferation (e.g., herpes simplex virus); (2) **polykaryocytosis**, or giant cell formation (e.g., measles virus and ectromelia virus); (3) **inclusion body formation**, when inclusions are formed either intracytoplasmically (e.g., poxviruses) or intranuclearly (e.g., herpesviruses) but only occasionally at both sites at the same time (e.g., measles, distemper and rinderpest viruses); (4) **malignant transformation**, when the host cells acquire tumour cell-like characters and may give rise to tumours *in vivo* (e.g., papova viruses); (5) **vacuolation**, of either the cytoplasm (e.g., SV (simian virus) and VA (vacuolating agent) groups of viruses) or the nucleus (e.g., pig pox virus); (6) **necrosis**, which has as its initial signs a 'rounding-off' (e.g., polioviruses) or a 'tailing off' (e.g., adenoviruses) of the host cells and as its terminal signs a shrinking and a contraction in cell volume with heavy nuclear staining (pyknosis), observable in stained preparations (e.g., polioviruses). With some viruses a single CPE is observed in the host cell; with others more than one are found. When necrosis is one of these it is always the terminal cytopathic effect.

12.6.2.4 *In vivo infections*

With a few exceptions the signs and symptoms of virus diseases in animals are insufficient to be diagnostic. However, they may be sufficient to permit a presumptive diagnosis to be made, while awaiting the laboratory identification of the causal virus.

12.6.2.4.1 Signs and symptoms

The signs and symptoms presented by a particular virus disease reflect closely the primary and secondary replication sites in the body and the ensuing CPEs. *Poliovirus* infects man principally via the oral route and is insensitive to the low pH of the stomach. The primary replication sites are the tonsils and the various lymph nodes of the digestive tract. During this stage, virus is found in abundance in the faeces and diarrhoea may be present. Both virulent and occasionally avirulent strains pass via the intestinal wall into the blood stream (viraemia) by which the virus is disseminated throughout the tissues. The viraemia results in an elevated temperature and may induce an invasion of the central nervous system (CNS). Secondary replication sites next become infected including the CNS, lymphatic structures and brown fat and, at this stage of the disease, overt paralysis is observed. Since the virus induces necrosis in its host cells it follows that widespread necrosis of neurons in the CNS, a tissue system with a poor regenerative capacity, results in a permanent paralysis which varies in degree with the extent of the necrosis.

Smallpox virus in man usually infects via the respiratory route, although infection via the dermal tissues may sometimes occur. The enteric route is much less likely because of the susceptibility of the virus to a low pH, such as would be encountered in the stomach. Replication of the virus may produce transient signs of respiratory disease. From this primary site the virus passes to the regional lymph nodes and thence, via the blood stream, to the liver, spleen, and other tissues where the secondary replication takes place. From these tissues the virus proceeds via the blood stream to the skin and mucous membranes where the typical focal rash is formed. Both viraemias will produce temperature elevations, the second being most severe (*ca.* 104-105°F). During the second viraemia the patient suffers a toxaemic illness of about 4 days' duration with fever, systemic symptoms and prostration, often with accompanying headache, backache, pains in the limbs, and vomiting. The typical focal rash is seen on the third or fourth day, involving the buccal and pharyngeal mucosa and the skin, principally of the face, forearms, and lower legs, although the trunk is involved. The rash, at first macular, soon becomes papular, vesicular, and finally pustular (8-9 days) when crusts commence to form.

From these two examples it is evident that viruses are individualistic in respect of their replication sequence and CPE in the animal body. Similar signs, symptoms, and lesions are, however, not necessarily evidence for common patterns of replication. For example, both smallpox and papilloma (wart) viruses give lesions in the skin, but whereas smallpox virus replicates in internal tissues the papilloma viruses are confined entirely to the skin tissues. Also lesions in the CNS are encountered in poliomyelitis, rabies, and the various encephalomyelitides, although each virus has a different pattern of replication in the body.

12.6.2.4.2 Laboratory diagnosis

Laboratory investigations are usually necessary for the diagnosis of virus diseases. These depend upon sampling the right clinical specimens from the patient at the optimum stage of the disease, and upon their transport to the laboratory under conditions which will ensure maximum survival of the virus. Clinical specimens may be used for anyone of three purposes: (1) for virus isolation: (2) for serological examination; (3) for direct examination.

Virus Isolation

The specimens, or suitable extracts of them, are inoculated into appropriate tissue cultures and/or susceptible animals. With fast multiplying and adaptable viruses (e.g., poliovirus and some myxoviruses) only 4-10 days may be needed for their isolation and identification; with slow multiplying and less adaptable viruses (e.g., adenoviruses, measles virus) up to 1-2 months may be required. Sometimes one or more 'blind' sub-cultivations will be required in the tissue cultures before a CPE is obtained. Positive isolation enables some form of definitive diagnosis to be made; negative isolation, however, does not necessarily indicate the absence of a virus. Upon isolation, it is usually possible to make a presumptive identification based on the observed CPE of the tissue culture cells. Confirmation of this identification rests on specific serological tests, e.g., complement fixation, neutralization, haemagglutination-inhibition, haemadsorption-inhibition, interference-inhibition, and plaque-reduction tests.

Serological Examinations

It requires two samples of serum, an acute serum at the time of onset of the disease and a convalescent serum at least 14 days later. These are treated by suitable serological tests *(vide supra),* using appropriate standard virus preparations as antigens, and the level of circulating antibodies in each serum is determined. A fourfold rise in antibody titre in the convalescent compared with the acute serum is usually accepted as significant and diagnostic. A single serum sample is worthless for diagnosis by serological means, although it may be of use in epidemiological surveys which set out to determine mean antibody levels to a particular virus.

Direct Examination

It is made of specimens obtained from localized lesions. For instance, the scab overlying an old cutaneous lesion or preferably, because of the difficulty in softening a dried scab, scrapings from beneath it, are usually a rich source of virus material. These specimens are smeared on microscopic slides, air-dried and methyl alcohol fixed. The smear is then ready for staining and microscopical examination. Conjunctival scrapings from herpetic and in elusion conjunctivitis can be similarly tested. Emulsified scab material is often screened under the electron microscope for viruses of the poxvirus or herpesvirus groups. By these examinations attempts are made to identify either virus particles or virus inclusions.

At postmortem, various tissue specimens may be collected for direct examination for evidence of a virus CPE. Unlike the lesions produced by bacteria, little or no pus is formed in virus lesions. The tissue reaction is therefore never as vigorous, and may be limited to infiltration of the affected site by mononuclear cells or other leucocytes of the blood. Tissue specimens are examined by conventional histopathological means; the CPEs most usually observed are inclusion body formation, hyperplasia, and cellular necrosis.

Inclusions, when found, are so characteristic as to be diagnostic. Their importance in diagnosis was so widely recognized that it became the custom at one time to name them after their discoverer (e.g., Negri bodies in the rabid brain). They are acidophilic in staining reaction and most are homogeneous in composition. Inclusions most frequently indicate the original 'factory sites' of virus replication within the cell and are not usually composed of aggregates of virions. The characters of virus inclusions are generally similar but specifically different.

Hyperplasia is the characteristic CPE of oncogenic (i.e., tumour-forming) viruses. Cellular proliferation may be so rapid as to give tumour-like masses in the body, e.g., the Shope papilloma and the Shope fibroma of rabbits, and the Rous sarcoma of chickens. Less localized proliferation may give rise to the diffuse swellings seen in rabbit myxomatosis. Herpes virus has been observed in association with the proliferating lymphoid cells of Burkitt's lymphoma.

Cellular necrosis, when observed, is usually focal in nature. Pus is never observed at these sites except when there is secondary bacterial infection, e.g., smallpox pocks on the skin. Necrosis in a vital organ or system, e.g., liver, kidney, or CNS, may lead to malfunction. For example, the characteristic jaundice of yellow fever is the result of malfunctioning of the liver.

12.6.2.5 Interferons in Cytoimmunity

In contrast to other microbial infections, antibody immunity is not now considered to be a major factor in recovery from virus infections; whilst formed against virus antigens, antibodies are ineffective intracellularly and do not, therefore, inhibit virus multiplication. There is increasing evidence that interferon is one of the major factors in this recovery. **Interferons** are soluble proteins, with a molecular weight of about 30,000 and their production in tissue cells is induced by many viruses and some non-viral agents. Inactivated or avirulent viruses are usually more active inducers than infective, virulent viruses since the latter rapidly inhibit cellular protein and RNA synthesis, both of which are required for interferon production. Interferons exhibit species specificity and inhibit the multiplication of a variety of viruses only in the same cells or cells from the same animal species.

There is little uptake of interferon by treated cells and it has no effect on the virus extracellularly. It does not interfere with virus adsorption, penetration or assembly of progeny virus particles. Inside the cell, interferon induces the formation of an inhibitor protein via DNA-dependent RNA synthesis and the inhibitor protein interferes with transcription of the information on the virus genome by preventing polysome formation. The infected cell still dies but no progeny virus is synthesized or released. Studies on the mode of action of interferons have yielded considerable information on virus-cell interactions.

12.6.2.6 Slow viruses

Certain agents, called 'slow viruses' are known to produce chronic infections. These include Chronic Infectious Neuropathic Agents (termed CHINA in the abbreviated form) of which scrapie in sheep is one example. Another slow virus is the agent of Aleutian disease of mink. A **slow virus** is a virus, or a virus like agent, etiologically associated with a disease, having a long incubation period of months to years and then a gradual onset of symptoms which progress slowly but irreversibly and terminate in a severe compromised state or, more commonly, death. A **slow virus disease** is a disease that, after an extended period of latency, follows a slow, progressive course spanning months to years, frequently involving the central nervous system and ultimately leading to death. Examples include the Visna-Maedi virus, in the genus Lentivirus (family Retroviridae), that causes encephalitis and chronic pneumonitis in sheep, and subacute sclerosing panencephalitis which is apparently caused by the measles virus. Many of the 'slow viruses' have very unusual properties, e.g., the scrapie agent is less than 25 mμ in size, is extremely resistant to heat and chemicals (e.g., formalin) and contains little or no nucleic acid. Such agents clearly pose problems when compared with more conventional forms of viruses.

12.7 VIRUS DISEASES OF PLANTS

Plant viruses are usually distinguished either by the visible symptoms they produce in the host or by serological methods. Viruses causing disease in animals induce the formation of antihodies in the host just as bacterial pathogens do. In plants such a response does not occur, but specific antibodies are produced when plant viruses are injected into animals. The reaction between antisera produced in this way and sap from plants infected with the same virus is useful in providing a rapid means of detecting and identifying some plant viruses. The symptoms of virus diseases are, however, of primary importance in the identification of viruses as well as being a measure of the harm sustained by the host.

12.7.1 Symptoms of Virus Diseases of Plants

12.7.1.1 External symptoms

A virus causing a plant disease may be systemic, that is, distributed throughout the plant, or localized, producing only local lesions (necrotic spots) on leaves or other parts of the host. The same virus may produce a systemic infection on one host and only local lesions on another strains of a virus may give markedly different symptoms on a particular host. Thus, the strains of tobacco mosaic virus (TMV) from tobacco *(Nicotiana tabacum)* are systemic in this host but produce only local lesions on leaves of *Nicotiana glutinosa*. Strains from tomato, however, produce only local effects on the tobacco plant. When infection is systemic the whole plant may show the characteristic symptoms of the disease or these may be restricted to certain parts, as with tobacco mosaic which infects tobacco systemically but produces the typical mosaic symptoms only on young shoots developing after primary infection has taken place. The most common symptom of plant virus disease is a reduction in size and cropping power of the plant or part of it, as in 'reversion' of blackcurrant where the berries are abnormally small, lettuce mosaic where the diseased plants are smaller than healthy ones, or Potato virus diseases where the tubers are few and small. Such reduction in size is usually accompanied or preceded by other symptoms, notably mosaic, mottling, striping or spotting of the leaves, wrinkling or curling of leaves or a general chlorosis of the green parts of the plant. Wilting and premature leaf fall may occur, as with potato Y virus. The shape of leaves may be altered, as with reversion of blackcurrant and the 'fern' leaf effect produced by cucumber mosaic virus in tomato leaves. Outgrowths (enations) from the leaf surface also occur. Striping or mottling of petals is frequent, as with the well-known 'broken' or Rembrandt tulips. The fruit may be distorted or altered, as in swollen shoot of cacao where the pods of diseased trees are abnormally short, in cucumber mosaic where warty outgrowths occur on the fruit, or certain virus diseases of tomato which cause spotting or cankering of the fruit. Shoot elongation may be inhibited by the presence of a virus as with nettlehead of the hop, proliferation of lateral buds may cause a 'witches broom' effect, or local excess formation of xylem may give rise to canker as in swollen shoot of cacao. Roots may also be affected, often as a result of damage to the phloem as in elm phloem necrosis, or they may die from starvation resulting from reduction in leaf area or chlorophyll content of the shoot.

The recognition and identification of a virus disease is complicated by the production of different symptoms on the same host by a particular virus under different conditions, by the existence of strains of a particular virus often differing in virulence and consequently in effects, and by the fact that the same virus may attack a range of often unrelated hosts producing strikingly different symptoms. Some diseases once thought to be due to viruses are now considered to be caused by mycoplasmas.

12.7.1.2 Effect of environment on symptoms

Factors such as temperature, light and the water supply available to the host profoundly influence the symptoms of virus diseases. It is well known that, in glasshouse experiments, the severity of a virus disease may alter according to the season, some diseases being more severe in summer and others in winter. The former are mainly diseases causing the accumulation of the products of photosynthesis in the leaves, as with leaf roll of potato, and hence the effect is increased with greater light intensity.

High temperature reduces the severity of the symptoms of a number of virus diseases but the limiting temperature varies for different viruses. Thus, while a short period at 20°C suppresses symptoms of crinkle and mosaic of potato, 35°C is necessary to suppress symptoms of tobacco mosaic. Low temperature also may mask the presence of a virus, as with tobacco mosaic which produces typical symptoms only over a range of 10-30°C, and strawberry yellow edge which does not show at temperatures below 16°C although susceptible host plants may become infected at temperatures too low for the development of visible symptoms. Over the range permitting the development of typical symptoms an increase in temperature or in light intensity usually reduces the incubation period, i.e., time from inoculation to the appearance of visible symptoms. The intensity of the response may also increase. The effect of plant nutrients and water supply on the symptoms of virus diseases are usually correlated with their effects on rate of growth of the host, symptoms being generally more severe when growth is rapid.

12.7.1.3 Internal symptoms

Virus diseases often cause the destruction or modification of tissues or cell contents. Phloem necrosis is common in virus-infected plants and leads to abnormal accumulation of carbohydrates in the leaves. In mosaic diseases chloroplasts are altered or destroyed. Some viruses induce abnormalities in meiosis leading to failure to produce viable seeds. Mature particles of certain viruses may be demonstrated in the host tissues with the aid of the electron microscope. Intracellular inclusions have long been noted in the cytoplasm of cells infected with certain plant viruses. These may be amorphous (X-bodies) or may consist of crystals of characteristic shape. Electron micrographs suggest that both types may be aggregates of the virus. One virus, tobacco etch, causes the formation of intranuclear crystalline inclusions in tobacco.

12.7.2 Transmission of Plant Viruses

12.7.2.1 Natural transmissions

A large proportion of plant viruses are carried from a diseased plant to a healthy one by vectors (most commonly insects). Blackcurrant reversion is carried by a mite (itself the cause of 'big bud' of blackcurrant); nematodes have also been shown to act as vectors of some viruses; the chytrid *Olpidium brassicae* is able to transmit tobacco necrosis virus and big vein virus to lettuce plants; mop top virus of potato is transmitted by *Spongospora subterranean* (a member of the Plasmodiophorales). Some viruses can be transmitted only by a specific vector, some may be transmitted by a vector and also by other means, and some cannot be insect transmitted. Many viruses are transmitted mechanically by diseased plants rubbing against healthy ones in windy weather (e.g., potato X virus) and others by root contact or through the soil. The more resistant viruses (e.g., TMV) may survive in the dead remains of diseased plants in the soil, others (e.g., *Arabis* mosaic) are known to survive in weed seeds. Crop plants may become infected from these sources. Virus diseases of crop plants may be spread by common horticultural practices such as grafting, pruning, disbudding, or picking flowers. Viruses are not generally carried by true seeds and it is possible that infected seeds are commonly rendered non-viable and thus are incapable of spreading the disease. However, a few virus diseases, such as lettuce mosaic and *Arabis* mosaic are seed-borne. Vegetative propagating units, such as bulbs, corms, rhizomes or tubers frequently carry viruses. A notable example is the spread of potato viruses by the use of infected 'seed' tubers. Since the tubers of virus-infected plants are smaller than those of healthy ones they tend to be selected for planting. Pollen from diseased barley plants carries barley false stripe virus to healthy ones.

12.7.2.2 *Experimental transmission*

Inoculation of healthy plants with juice from one suspected of containing a virus is one method of identifying viruses. Inoculations are made by grafting, by mechanical means or by the use of suitable vectors. Where it is possible to graft a shoot of a diseased plant on to a healthy one, or vice versa, transmission usually follows. However, only fairly closely related species can be grafted to one another. A parasitic plant, the dodder, has been used to transmit viruses from one host plant to another but again this method is limited in application.

The simplest method of transmission is by rubbing infected sap on to a healthy plant but many viruses cannot be transmitted in this way. Even with those which can be so transmitted care must be taken that the amount of damage to the plant tissues is at a minimum or the virus will not become established. Carborundum powder or other abrasives may be added to the virus extract to produce uniform slight abrasions of the leaf when rubbed; buffers should also be added, to control the pH of the plant sap.

Insects are bred experimentally and used for transmission of those viruses which cannot be successfully established by other means. This involves a special technique of handling insects and host plants and the use of cages to confine the insects on the experimental plants.

12.7.3 Control of Plant Virus Diseases

A plant systemically infected with virus can seldom be cured. This can sometimes be done by heat treatment, however, where there is a sufficient favourable margin between the temperature at which the virus is destroyed and that which damages the plant. The most successful form of heat treatment is that used for the control of the flower-distorting virus (Aspermy virus) in chrysanthemums. This is carried out commercially by a number of cutting producers and has resulted in a great improvement in chrysanthemum flower quality. Other examples are 'phony' disease of peach which is eradicated if infected trees are exposed to a temperature of 35°C for a short period (obviously practicable only with trees growing under glass) and some sugar cane viruses which can be eliminated from cuttings by treatment in a hot water bath at 50°C. Control of the vectors of virus diseases is difficult but effective, and modification of agricultural techniques sometimes prevents spread of disease.

Virus-free clones of certain crop plants, e.g., potato, have been raised from tissue cultures of the meristems. Growing points of shoots are usually free of virus and cultures from these thus give disease-free plants.

Control of most plant virus diseases must, however, be sought through the breeding of resistant varieties, general plant hygiene, removal of source of infection, and the prevention of spread of a virus within a crop.

12.8 INSECT VIRUSES

12.8.1 Introduction

Two major types of virus disease of insects are the inclusion body diseases and non-inclusion diseases. In inclusion body diseases the virus particles are occluded in polyhedral protein crystals (polyhedra), in granules (capsules), or in spindle-shaped bodies. These constitute the inclusion bodies that are to be seen under the light microscope in the tissues of the diseased insects. Polyhedral inclusion bodies may lie either in the nuclei or in the cytoplasm of the host's cells. In relation to this the corresponding viruses are divided into those that cause nuclear and cytoplasmic polyhedroses respectively. These two sorts of virus differ not only in their sites of multiplication but also in their morphology and in the symptoms they produce. The non-inclusion viruses differ strikingly from the foregoing types in that they lie quite free in the tissues of the hosts. Viruses of invertebrates have for long been regarded as a separate and rather special group of viruses. This is not only because their hosts are distinctive but also because viruses that have polyhedral and granular inclusion bodies are not known in vertebrates or plants. The gap between insect virology and general virology has, however, narrowed greatly with the discovery of an increasing number of non-inclusion viruses in insects and other invertebrates, and with the recognition of two other distinct and important categories of insect viruses. These are the arboviruses (arthropod-borne viruses) that infect both vertebrates and blood sucking arthropods, and an analogous group of viruses that multiply in both plants and the insects that serve as vectors.

12.8.2 Types of Insect Viruses

12.8.2.1 Viruses of nuclear polyhedroses

Nuclear polyhedroses occur most commonly in larvae of Lepidoptera (the moths and butterflies) but are known also in Hymenoptera and Diptera. The viruses multiply within nuclei in chiefly the skin, blood, fat-body, and tracheae. The virus particles contain DNA and are rod-shaped (*ca.* 20-50 mμ diameter, 200-400 mμ long), and become enclosed either singly or in bundles within paracrystalline protein which forms microscopically visible polyhedra. These enlarge and rupture the nuclei in which they form, and ultimately the cell bursts. As the disease progresses the tissues of the host disintegrate and liquefy and the skin, which by now has become very fragile, ruptures. The liberated fluid contains myriads of virus containing polyhedra. These are insoluble in water and resistant to ordinary bacterial decay. If diseased insects are left to decay in water, the polyhedra eventually form a layer at the bottom of the vessel. In nature, the polyhedra become ingested with food plants and dissolve in the alkaline juices of the gut of susceptible insect species. Though they are insoluble in water, polyhedra may dissolve if the pH falls below 5 or rises above 8.5. It appears that their dissolution in the gut and the consequent liberation of occluded virus particles is not enzymic. Analyses of polyhedra reveal traces of silicon (e.g., 0.1-0.3%) solution of silicates incorporated in the structure of the polyhedra by alkaline agents in the gut may account for dissolution of the polyhedra. From the gut the viruses enter susceptible tissues and disease develops usually 10-12 days after ingestion of the virus. In addition to infection via the mouth there is much evidence to suggest that nuclear polyhedrosis viruses may be transmitted through the eggs. This may be possible because infection does not always lead to acute disease and viable eggs may be laid by diseased females.

12.8.2.2 Viruses of cytoplasmic polyhedroses

Many cytoplasmic polyhedroses are known. Most affect only Lepidoptera. As in nuclear polyhedroses the virus particles are occluded in polyhedral protein crystals, but these do not dissolve completely in weak alkali. A rather sponge-like residue remains which is pitted with the sockets in which the nearly spherical RNA-containing virus particles were formerly located. The individual virus particles are regular twenty-sided (icosahedral) bodies 60-65 mμ diameter.

The development of cytoplasmic polyhedroses is different from that of nuclear polyhedroses. The viruses multiply only in cells of the gut, usually the epithelial cells of the mid-gut, from which they spread to fore and hind guts. The skin is not attacked, and as a result the body does not become fragile and burst. The polyhedra appear in the cytoplasm of infected cells. They vary much in size, even within the one insect, but are clearly visible with the light microscope. Though the polyhedra themselves are located in the cytoplasm, autoradiographic studies have indicated that the viral RNA may in fact be synthesized within the nucleoli of infected cells. From thence it presumably passes into the cytoplasm where the protein coat of the virus particle is formed. The polyhedra in the gut wall frequently show through the dorsal integument of the host as pale yellow or whitish areas. Infected larvae develop more slowly than do healthy ones, and thus are smaller; they have reduced appetites, and sometimes disproportionately large heads or long bristles. When the disease is far advanced large numbers of polyhedra are liberated in the lumen of the gut from which they may be regurgitated or voided in faeces. Besides spread in this way there is almost certainly hereditary transmission of virus via the egg.

12.8.2.3 Viruses of granuloses

Granuloses have so far been encountered only in larvae, and very occasionally pupae, of Lepidoptera. The diseases are characterized by the appearance in the tissues of millions of granules, so small (*ca.* 300-500 mμ diameter) as to be barely visible beneath the light microscope. Each granule is a capsule that invests one or occasionally two rod-shaped virus particles which contain DNA. The granules occur in both nuclei and cytoplasm of the host's cells of which those of the fat body seem to be the main site of virus development. As the disease progresses other tissues are affected, and normally these include the skin. In this event the marked liquefaction of the internal tissues after death, usually 4-20 or so days after infection, is followed by rupture of the fragile body wall, and the symptoms of the granulosis then closely resemble those of nuclear poly hedroses.

Like polyhedra the granules consist of protein arranged in a paracrystalline lattice, and usually within each capsule a solitary virus particle lies inside two membranes. The capsule, which is resistant to ordinary putrefactive processes, is soluble in various acids and alkalis. The function of the capsule appears to be the same as that of polyhedra, namely, the protection of the otherwise labile virus particle(s) after release from the host body. Granulosis viruses in general appear to remain infective for up to two years and even may survive in the soil for this time with little deterioration. Percolating water does not readily remove them from the soil. A granulosis virus has been found to remain infective on leaves for up to four months. Spread of granuloses in nature among individuals of the same generation is mainly by ingestion of contaminated food material, but transmission from one generation to the next can be achieved by way of the eggs.

12.8.2.4 Spindle viruses

Spindle viruses have been discovered in larvae of the cockchafer *Melolontha melolontha.* They are characterized by the formation in the cytoplasm in susceptible host cells of inclusion bodies of which some are spindle-shaped and 0.5-12 μ long and others ovoid and up to 20 μ long.

12.8.2.5 Viruses of non-inclusion diseases

The number of non-inclusion or free viruses known in insects and other invertebrates has increased rapidly after their discovery, and is now rather more than a dozen. The majority of these are located in the cytoplasm of host cells. Only two, densonucleosis of *Galleria mellonella* and the flaccidity virus of *Antheraea eucalypti* are known to have an affinity with the nucleus. Other free viruses include several that infect bees, for example, acute (ABPV) and chronic (CBPV) paralysis viruses, sacbrood virus (SBV), and one concerned in the causation of European foulbrood; transparency or 'wassersucht' virus of beetles; paralysis virus of crickets; the virus causing 'clear heads' disease of silkworms; lethargy virus in cockchafer; Malaya disease of Indian rhinoceros beetles; a virus disease of *Cirphis unipuncta;* 'iridescent viruses' in *Tipula* (TIV), *Chilo* (CIV), *Sericesthis* (SIV), and mosquitoes (MIV); and hereditary sigma virus of *Drosophila.* Paralysis diseases of crabs and mites are also caused by free viruses.

Non-inclusion viruses are a rather heterogeneous group and vary among other respects in the shape of the virus particles and the sort of nucleic acid contained. The majority are more or less spherical, but CBPV particles are slightly elongated and variable in size, and the Malaya disease virus is rod shaped in its mature form but spherical at an earlier stage of organization of the particles. Densonucleosis virus, TIV, SIV, and CIV contain DNA, but transparency, *Antheraea* flaccidity, CBPV, ABPV, sacbrood, and European red mite viruses contain RNA. The sort of nucleic acid has not been determined for all.

TIV has been investigated thoroughly. It causes a disease of larvae of the crane fly *(Tipula paludosa)* in which the body fluid becomes brilliantly iridescent owing to the presence of microcrystals that are built up of crystalline arrays of rather large (130 mμ diameter) virus particles. The latter in addition to DNA contain lipid and chiefly protein. Though DNA containing, they lie in the cytoplasm of the host's cells. At a late stage of the disease a quarter of the dry weight of the insect may be virus. This is the highest proportion known for an animal disease. Cross-inoculation tests with this distinctive virus have shown that it is able to infect other Diptera as well as Lepidoptera and Coleoptera.

Drosophila flies infected with Sigma virus become permanently paralysed if exposed to carbon dioxide in doses that merely anaesthetize uninfected flies. This virus is passed on from generation to generation.

12.8.2.6 Arboviruses

Arboviruses are acquired by blood-sucking arthropods (mosquitoes, sandflies, certain gnats and ticks) when they feed on vertebrates in whose blood there is a sufficiently high level of virus. The viruses multiply, become established in the salivary gland, and subsequently are transmitted back to susceptible vertebrates, mainly wild animals including rodents, birds and reptiles. Several arboviruses had been catalogued, and of them cause disease in man. The nucleic acid in all arboviruses for which information is available is RNA, and the shape of the majority of arboviruses is spherical. Their antigenic relationships allow about three-quarters of known arboviruses to be distributed among a number of antigenic groups, more than half of which include human pathogens. The human pathogens are found mainly in groups A and B.

Typical arboviruses are generally believed to produce no cellular damage in the arthropod host. In susceptible vertebrates the symptoms of infection range from influenza-like ailments to severe disorders of the central nervous system and from mild to severe haemorrhagic disease. Arbovirus diseases in man include St. Louis encephalitis, Western encephalitis, Venezuelan equine encephalitis (that also affects horses), dengue, and yellow fever. Epidemics of arbovirus diseases sometimes cause considerable public concern. Though vaccination has proved effective in controlling yellow fever, it is possible that for arbovirus diseases generally, the best means of control is reduction of populations of the appropriate arthropods. The effective reservoirs of arboviruses are not generally known, but might include ticks, in which life-long infections may occur, and in temperate regions certain vertebrates such as hibernating rodents.

12.8.2.7 Plant pathogenic viruses in insects

Insects serve as vectors for many viruses that cause diseases in higher plants. Some insects act in this way merely by becoming contaminated about the mouth parts whilst biting the plant. Others act by introducing virus-containing fluid into the plant. Some of these insects merely transmit the virus particles that they have acquired from infected plants on which they have fed, but some leaf hoppers and aphids actually become infected by the viruses that cause disease in plants and within their bodies the viruses multiply. The viruses of wound tumour, maize mosaic, maize rough dwarf, and pea enation mosaic are examples. In all these, virus particles have actually been seen in cells of the insect with the aid of the electron microscope. In a number of instances the infections have been shown to cause cytopathic changes, or to reduce the fecundity or longevity of the insect.

12.8.2.8 General aspects of insect viruses

Two different viruses may concurrently infect the one insect. It has been found in the laboratory that administration of even heat-inactivated granulosis viruses may enhance the virulence of nuclear polyhedrosis viruses, though no marked effect has been found with the converse arrangement. The feeding to an insect of a virus not known to occur in it can stimulate a latent virus to activity. This can be a complicating factor in experiments on cross infection. Latent viruses are extremely frequent in insects, and a variety of factors ('stressors') which can induce a state of stress in the insect are effective in activating them. Such stressors include unfavourable temperatures or humidities, overcrowding, unsuitable food, feeding with certain chemicals, and inoculation with a foreign virus. Stressors probably act indirectly through the metabolism of the insect rather than directly upon the latent virus. The physical state of the latent virus for example whether it is in a provirus or fully formed condition is unknown. Viruses which normally form inclusion bodies probably do not do so in their latent state.

12.8.2.9 Natural and applied control of insect pests

Spontaneous virus diseases are commonly responsible for severe outbreaks of disease in insect populations in the field and their recurrence each season may contribute to pest control. Though outbreaks may result from acquisition of new infections it is possible that latent infections that become activated by stress are of more importance. There are several examples of the successful use by man of viruses to control insect pests, chiefly those which feed openly on the host plant.

A number of circumstances serve to heighten the chance of success in any given instance. The virus used should be highly virulent and should kill the insect quickly enough to prevent it doing much damage before it dies. Because nuclear polyhedroses and most granuloses result in liquefaction of the body contents and disintegration of the skin, they offer the best chance of further liberation and spread of the virus. It is advantageous if the insect to be controlled is one which is gregarious and feeds openly on the foliage. As in all infectious diseases, high host population density favours spread of the disease. Inclusion body viruses offer the advantage that the inclusion body protects the virus and so allows its longer storage and extends its retention of infectivity once distributed in the environment. Virus preparations have been applied as dusts or sprays. Administration of a number of insect viruses to man and other mammals have revealed no harmful effects, but in view of the host specificity known among viruses that infect mammals it would be unwise to generalize from the limited data available. The very existence of arboviruses argues for caution. Although some insect viruses have been successfully produced in tissue culture, current technology virtually requires that viruses for control purposes must be produced by rearing and infecting host populations in the laboratory.

12.9 BACTERIOPHAGE

12.9.1 Introduction

Enormous numbers of different bacteriophages (phages) are known since almost all species of bacteria, including actinomycetes, are susceptible to infection by specific ones. Bacteriophages exhibit high specificity of action, each being able to infect only one group of closely related organisms, usually a single species (although the T phages of *Escherichia coli* will also attack *Shigella sonnei*), or occasionally only a few strains of a single species. This highly selective activity provides a very useful method for typing bacterial isolates, and phage typing methods are widely used in modern epidemiological studies. Phages together with their host bacteria provide a model which is convenient for the intensive study of host-parasite relationships and virus multiplication. In particular extensive studies of the T even coli phages (T_2, T_4 and T_6) have contributed greatly to present day knowledge of the genetics of microorganisms.

12.9.2 Demonstration of Bacteriophages

Bacteriophages can be isolated from many natural sources such as faeces, sewage, and polluted water. Like all viruses, phages multiply only within living cells and since they can be seen directly only by electron microscopy, phage activity must usually be demonstrated by indirect means. When a drop of a bacteria-free filtrate of a fluid containing phage particles is added to a young broth culture of susceptible bacteria, the culture will clear visibly within a few hours due to lysis of the bacterial cells. The bacteria-free filtrate can be used to infect a second broth culture and this process may be continued indefinitely. The lytic activity of phage was first demonstrated on solid media where, on a culture plate seeded or carpeted with susceptible bacteria, phage growth and activity is indicated by cleared areas known as plaques (Figure 12.6).

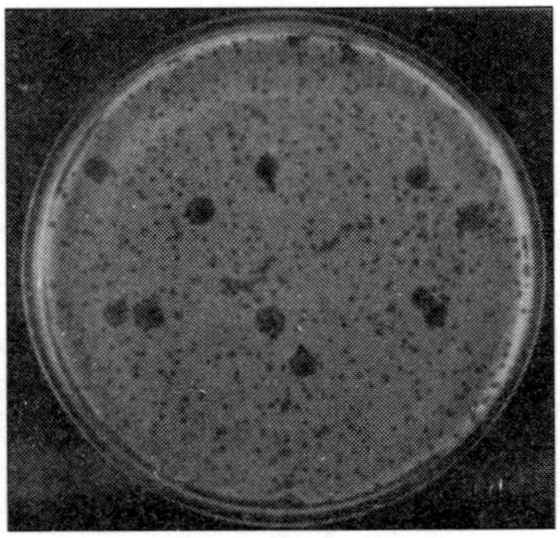

Figure 12.6 Bacteriophage plaques of different sizes due to lysis of bacterial growth.

These plaques behave like colonies; they may show differences in size and general morphology and can be 'picked off' by touching with an inoculating needle and transferred to a fresh culture of susceptible bacteria. Moreover, a single phage particle produces one plaque by multiplication and liberation of many phages which attack and lyse neighbouring bacteria. Hence, fluids containing phage particles can be titrated by plating measured volumes of suitable dilutions of the fluid with sensitive bacteria. The plaque count is analogous to a colony count in bacterial counting methods.

To study the physical and chemical properties of a phage, the preparation needs to be as free as possible from host cell material. Phages are usually concentrated and purified by growth in liquid cultures of host bacteria; the virus particles are then separated from bacterial debris by differential centrifugation and repeated washing. However, the presence of phage is not always evident since latent infection may occur without bacterial lysis. Accordingly, phages are subdivided into two main groups by the relationship they establish with their bacterial hosts. *Virulent* or *lytic* phages always produce lysis of the host cells. *Temperate* or *symbiotic* phages may lyse the host, but sometimes their genetic material becomes associated with the host cell as *prophage* and is transferred to daughter cells on division. Thereafter a small proportion (10^{-2} to 10^{-7} per cell generation) of the progeny cells lyse to release free phage. Hence, bacterial cultures which carry temperate phages are termed *lysogenic* and the process by which the phage enters the prophage state is known as *lysogenization*.

12.9.3 Characteristics of Bacteriophage

Although spherical and filamentous phages are known, most larger phage particles are tadpole shaped, consisting of a head and a tail (Figure 12.3). The proteins which make up the head membrane, tail core, sheath, and fibrils are all distinct from each other and are antigenic (Figure 12.4). Phage preparations exhibit high serological specificity without cross-reactions with host antigens. Complement-fixing antigens have been recognized on the head protein but antibody to this protein does not neutralize the infectivity of phage. Some antigens have been demonstrated in the tail, which produce neutralizing antibodies that are thought to inactivate the phage irreversibly by blocking a specific site on the phage tail, but they do not destroy the phage.

Phage particles are organized structures, exhibiting general stability of type, but since their properties are controlled by genes, they are liable to change through mutation. Phage mutation may yield mutants which differ from the parent in host range, plaque type, and many other properties. When a bacterium is infected by two closely related phages, *genetic recombination* can occur in which some of the new phage particles liberated at the end of the reproductive cycle may possess the combined properties of the two original phages. These properties will then subsequently be passed on to new phages during further reproductive cycles.

12.9.3.1 Virulent bacteriophage

There may be so many points of attack on each cell if a very large number of phage particles attack a susceptible bacterial culture that the cells may disrupt; the contents bursting out through the numerous holes in the cell wall, made by the phages. This is described as lysis. When the number of phages is few, they are able to infect the bacterial cells without causing immediate lysis. Then, instead of the bacterial cell synthesizing its own DNA, its metabolic machinery produces components for phage DNA and protein structures. After a short interval of 15-30 minutes, many new phage particles are assembled inside the cell and are released by the cell bursting; this is 'lysis from within'. The stages in this infective process are similar to the multiplication cycles of other viruses, and may be considered as follows: (1) adsorption of free phage; (2) penetration of phage nucleic acid; (3) intracellular development; (4) maturation and release of new phage (Figure 12.7).

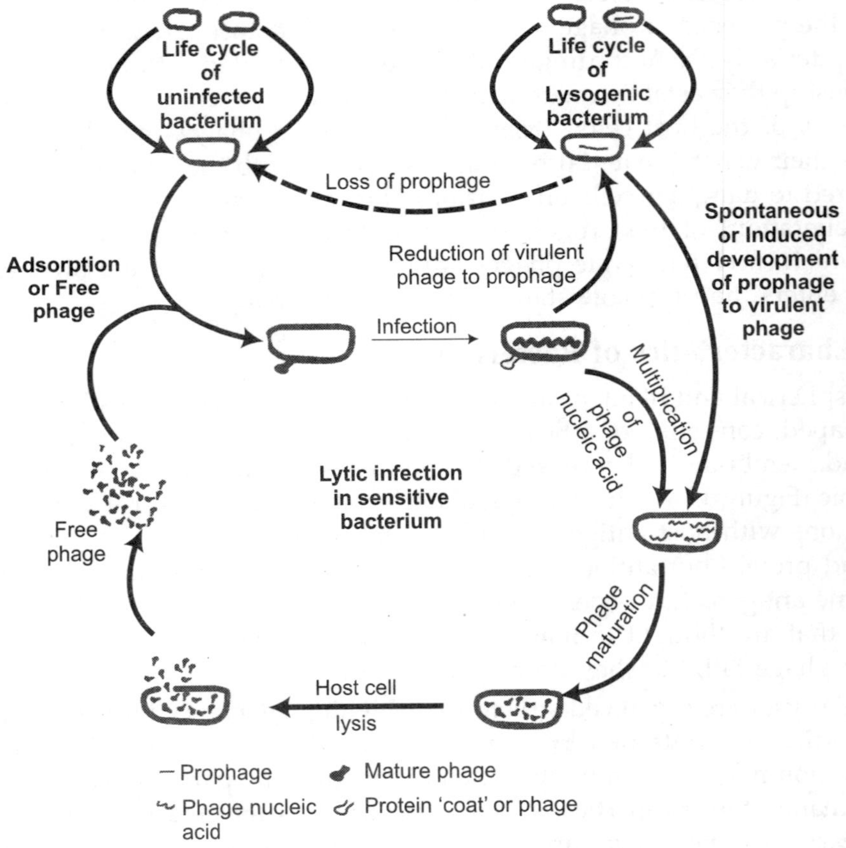

Figure 12.7 Lytic cycle of bacteriophage.

Study of the characteristic sequence of events during a viral multiplication cycle entails making various determinations at precise times after initial infection. To do this, the infection must be *synchronized;* otherwise infection of different cells is spread over a long period of time during which new virus particles, released when some infected cells lyse, will infect other cells. Synchronization can be achieved by allowing adsorption of virus to bacteria for only a few minutes and then removing the unadsorbed virus.

When it is desired to study the multiplication cycle in cells infected by single virus particles, 'one-step conditions' are employed (Figure 12.8); once adsorption is complete, the virus cell mixture is diluted into a large volume of nutrient broth so that progeny viruses released later are very unlikely to become adsorbed to the uninfected cells because they are so sparse.

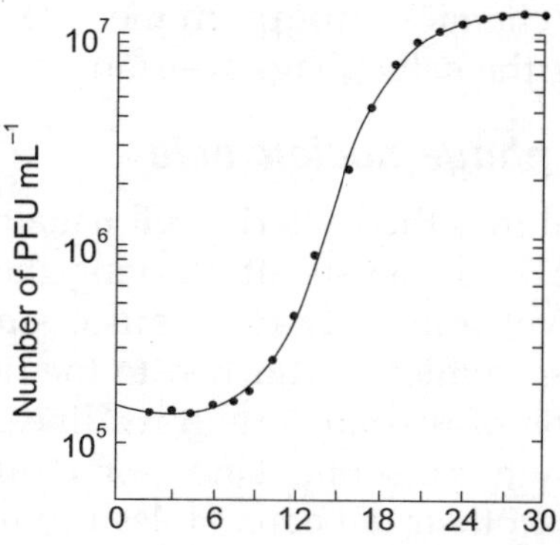

Figure 12.8 One step growth curve of bacteriophage.

12.9.3.2 Adsorption of free phage

For adsorption to take place, the phage particles must first collide with the cell, then bonds must be established between the two surfaces. Under optimum conditions, this process is very rapid, almost all phages being adsorbed within a few minutes. Adsorption is usually irreversible but can sometimes be reversed by varying conditions such as pH and temperature. Frequency of collision between phages and bacteria depends mainly on the concentrations of virus particles and cells. When there is an excess of phage particles, as many as 300 may adsorb to one bacterial cell. Frequency of collision is also affected by the size and mobility or the virus and the mutual attraction or repulsion brought about by electrostatic forces. The formation of bonds after collision is dependent on the chemical structure of the two surfaces. As with the union between an antigen and antibody the presence of complementary chemical groups on the two surfaces is essential, e.g., with coli T_1 amino groups on the viral surface bond with receptors containing acidic groups on the cell surface. The receptor sites of the bacterial cell render it highly specific for a small range of phages and cannot be found in the walls of bacteria resistant to those phages. The receptors may be situated in different layers of the wall, *e.g.*, receptors for phages T_2 and T_6 exist in the lipoprotein layer of the wall whereas phages T_3 and T_4 adsorb to receptors in the lipopolysaccharide layer. Receptor substances have been separated from cell wall preparations by gentle chemical extraction and have been shown to combine specifically with phage active against the whole cell. The phage surface must also be modified before adsorption can occur; this may be by attachment of positively charged cations (e.g., Mg^{++} or in some cases, L-tryptophane. The extent of modification varies from one phage to another but each phage is specific with regard to the co-factor required for adsorption. Some phages (T-even phages) are equipped with fibrils which extend from the base plate at the tip of the tail. These phages adsorb to receptors on the bacterial cell by attachment of their fibrils to the cell surface leaving the virus particle in a characteristic position at right angles to the cell wall and with the head sticking out. The receptors for the very small phages f1 and f2 are only found on the surface of F^+ cells; these phages attach to the specific pili which confer the male character on the F^+ cells. Bacterial mutation can change the nature of the chemical groups on the receptors and by so doing prevent phage adsorption, rendering the cells phage resistant.

12.9.3.3 Penetration of phage nucleic acid

Phage nucleic acid is injected into the bacterial cell usually by insertion of the tail core in the cell wall and contraction of the sheath, leaving the empty protein coat outside. The T-even phages are known to hydrolyse the mucopeptide of the bacterial cell wall by means of a phage lysozyme which is attached to the tip of their tails. The dissolved hole may help in the insertion of the tail core in readiness for injection of nucleic acid, but such a hole is apparently not essential since some other phages which possess no lysozyme are still able to insert their tail core. Holes produced by phage lysozyme may cause a leakage of cell contents; normally the holes are repaired fairly quickly but if a very large number of phages produce holes in the same cell within a short time, there may be so much leakage that the cell lyses without virus multiplication.

Electron microscopy of T-even phages has shown that irreversible adsorption of a phage always results in contraction of the tail sheath. The base plate and fibrils are held firmly against the cell surface and contraction of the sheath pulls the head towards the cell, causing the hollow tube or core to penetrate the cell wall reaching through to the cell membrane but probably not piercing this. Rather like primitive muscle, the sheath contains the source of energy for inserting the tube through the wall and does not depend on an energy supply from the bacterium.

Other phages which are not equipped with contractile sheaths are still able to effect penetration with their tube but at a much slower rate than the T even phages and are dependent on the host cell for the energy required to achieve entry of their tube or, in the case of the tiny spherical phages, of their capsid.

Once the cell membrane is reached, the release of phage DNA is triggered off and the contents of the head are expressed on the membrane through which the nucleic acid molecules are able to pass into the cell. The empty protein head and tail structures remain attached to the bacterial cell wall but they take no further part in the infective process.

12.9.3.4 Intracellular development

The presence of phage cannot be demonstrated by plaque formation on a susceptible plate culture if the bacterial cells are disrupted soon after the injection of phage nucleic acid. This short period before phage can once more be demonstrated is called the **eclipse phase**.

Synthesis of a new protein required by the phage for multiplication commences immediately after injection of phage DNA into the host cell. The first products, often referred to as 'early proteins', include certain enzymes which are required for later construction of the complicated molecules peculiar to the phage, e.g., kinases for the formation of nucleotide triphosphates, and a phage DNA polymerase. Subsequently during the eclipse phase, 'late proteins' appear which include the protein sub-units for the phage head and tail structures. Construction of new phage DNA is accomplished with some components from the medium and some from degraded host DNA. Genetic recombination, in which phage particles undergo random exchange of genetic material, can occur at this stage.

12.9.3.5 Maturation and release of new phages

The assembly of the newly-synthesized components to form a complete or mature, infectious virus particle is known as maturation. Maturation of phages involves assembly of many components and therefore occurs in several steps. Viral DNA molecules are first condensed into large particles surrounded by a membrane and resembling phage heads. Phage heads are completed by further condensation of capsid monomers around the DNA condensate, but the heads are not yet stable since premature lysis of the cells at this stage by artificial means results in the DNA escaping from the protein.

Some temperate phages can confer new properties on the host cell and this is known as **phage conversion**. The prophage behaves as a host gene, each cell receiving the prophage acquiring the new property. For example, virulent toxin-producing strains of *Corynebacterium diphtheriae* have been isolated from cultures of avirulent non toxin-producing strains as a result of infection with specific phages.

12.9.4 Medical Applications

12.9.4.1 Phage-typing

The success of phage-typing depends on the very high specificity of phages for their host bacteria. Phage-typing methods were first introduced in 1925 to distinguish strains of salmonellae. As a result of subsequent development of the methods, the type distribution of *Salmonella typhi* is now known for most of the civilized world and routine typing contributes greatly to our knowledge of the epidemiology of typhoid fever. The phages employed for *Staphylococcus typhi* are active only against freshly-isolated strains possessing the Vi-antigens and the method is termed 'Vi phage typing'. Phage-typing is widely used for distinguishing strains of *Staphylococcus aureus.* This enables the sources of a particular phage type of staphylococcus responsible for an outbreak of sepsis, to be traced and eliminated. Staphylococcal phage-typing presents quite different problems from the typing of salmonellae. Whereas there are only a few hundred typhoid carriers in England, there are many millions of staphylococcal carriers and thus a much greater variety of types are found among staphylococci than typhoid bacilli. Despite the fact that staphylococcal phages are not as stable as those of *Staphylococcus typhi,* the method does permit a fairly reliable identification of individual strains which is very valuable in short-term investigations.

Phage-typing of *Staphylococcus aureus* is technically more difficult than the typing of salmonellae; the staphylococcal phages are derived from lysogenic strains of *Staphylococcus aureus* and selected on the basis of their range of host specificities. Twenty-two phages are included in the basic set of typing phages; the phages are numbered by international agreement and are divisible into four groups of antigenically related phages. Almost all strains derived from human sources belong to groups I, II, or III. Anyone strain of *Staphylococcus aureus* is likely to be susceptible to several phages, usually of the same antigenic group. Some strains of *Staphylococcus aureus* are resistant to all the phages used and cannot be typed but new phages are added to the basic set from year to year, thus increasing the range and value of the typing method. The phages are prepared for use from filtrates of infected susceptible host cultures known as propagating strains. The filtrates are purified and then titrated to determine the dilution required to just produce confluent lysis when spotted on to a culture plate seeded with the host bacterium; this dilution is referred to as *Routine Test Dilution* (or R.T.D.) and is the dilution normally employed in the typing method. A culture plate is seeded with the staphylococcus to be typed, and then each of the phages of the standard set is dropped on to the agar surface in a standard pattern, often within a grid marked on the culture plate. After drying, the plates are incubated, usually at 30°C. The phage type of the test strain is then designated by the numbers of the phages which have produced lysis, under the standardized conditions of the test, e.g., phage type 84/85 is a strain of *Staphylococcus aureus* which has recently caused several serious outbreaks of hospital infection, and which is lysed only by phages 84 and 85. Some strains which are not lysed by phages at their routine test dilution, may be typable when more concentrated phage suspensions are employed (usually 1000 × R.T.D.).

Phage-typing is a very valuable method but since it is an empirical procedure, it is open to many sources of error. For example, if phage lysates are too concentrated, non-specific lysis may occur. Moreover, since most staphylococcal propagating strains are lysogenic, some harbouring as many as five different temperate phages at a time, concentrated lysates are likely to contain both the propagated typing phage and some of the temperate phages. These temperate phages may themselves become lytic or they may alter the susceptibility of the test organism to the typing phage.

12.9.4.2 Phage therapy

Because phages kill bacteria but are harmless to man, d'Herelle regarded the presence of phages as an important factor in recovery from infectious disease and urged its therapeutic administration in a variety of infectious processes. Phage therapy has been attempted on many occasions, e.g., oral administration in enteric infections, instillation into the urinary bladder in cases of cystitis, local application on boils, etc., but the results have been disappointing. The great variety of phage-specific strains of bacteria within a species and the rapidity with which phage-resistant forms can develop are probably important limiting factors. Hence, with the possible exception of cholera, it appears that phage therapy is of little value.

12.10 BACTERIOCINS

The bacteriocins are a class of bacterial products which exercise a very specific bactericidal activity; they have some relationship with bacteriophages and possibly also with sex factors. They were discovered and first studied in the intestinal Gram-negative organisms but many other bacteria, both Gram-negative and Gram-positive, produce them. They take their specific names from the organisms producing them, e.g., colicins from *Escherichia coli*, pyocins from *Pseudomonas pyocyanea (aeruginosa)*, megacins from *Bacillus megaterium.*

All true bacteriocins are proteins but some may become associated with the lipopolysaccharide of the cell wall of the producer and form a complex which can be split only by drastic methods. Some bacteriocins, particularly pyocins, in electron micrographs resemble bacteriophage tail structures. They may be regarded as products of defective phage genomes, able to code for only part of the phage structure. The initiation of bacteriocin synthesis kills the producer cell. Like bacteriophages they are adsorbed to specific receptors on, or possibly under, the bacterial surface from which they exercise a lethal action. The target of the lethal effect varies between bacteriocins. They may degrade the cell's DNA, or inhibit DNA and RNA synthesis, or interfere with protein synthesis. Unlike bacteriophages they are not regenerated in the cells to which they have been adsorbed and, as a rule, they do not lyse them in the process of killing.

While adsorption of the bacteriocin 'particles' takes place very rapidly, there is an interval between adsorption and the irreversible bactericidal action during which the cell may be rescued by inactivation of the bacteriocin already adsorbed to the cell with trypsin. Most, though not all, bacteriocins are good antigens and evoke in rabbits the formation of neutralizing antibodies. Precipitating antibodies are produced also, but it is not entirely clear whether they react with the bacteriocin itself or with an associated component.

The potential or actual synthesis of bacteriocin is due to genetic determinants, the bacteriocinogenic factors, which resemble prophages but are probably not integrated into the bacterial chromosome. Cells carrying such factors are immune to the effect of the corresponding bacteriocin. As with a prophage, the synthesis of the lethal product specified by a bacteriocinogenic factor is repressed in most cells harbouring the factor and is active in only a few of them; but the synthesis of bacteriocin can be induced, like the synthesis of phage components, by U.V. irradiation, mitomycin C and other agents. The massive amount of bacteriocin produced after induction is due mainly to a great increase in the proportion of organisms producing it rather than to an increase in the amount produced by individual cells.

Some colicinogenic factors (genetic determinants for colicins) can be transferred from producer to non-producer strains. The factor for colicin Ib (col I b), acts as a sex factor directing the formation of pili which allow cells to conjugate and thus to pass on the col factor. Other col factors are not able to cause formation of their own pili, but rely for transmission on col I b, the F factor, or on transduction by bacteriophages.

The specificity of action of bacteriocins has led to their use for bacterial typing. Whereas most phage typing schemes are based on the sensitivity of the test strain to phage, most bacteriocin typing schemes are based on the activity spectra of bacteriocins produced by the test strain. Typing of *E. coli, Shigella sonnei, Proteus* spp., and *Pseudomonas aeruginosa* has been done with encouraging results.

12.11 VIRUS DISEASES OF FUNGI

Until the early 1960s no disease of a fungus had been proved to be due to a virus, although viruses had been suggested as the cause of certain abnormalities in fungus fruit-bodies and mycelium. Indeed several authors had cited the supposed lack of virus diseases as a character distinguishing fungi from actinomycetes.

Mushroom growers, however, had long been concerned about poor cropping and the production of abnormal fruiting bodies under conditions of intensive cultivation. It has now been shown conclusively that a number of well-known disorders of cultivated mushrooms, such as 'watery stipe' and 'die-back', are due to a virus or a complex of viruses. Electron microscopy shows that virus particles of at least two and probably three distinct types (viz., small and larger spherical particles and elongated ones) may be present in watery stipe fruit-bodies. These particles are morphologically comparable with those of certain plant viruses and are distinct from those of bacteriophages. Serological tests and density-gradient centrifugation gave similar results for all the types of virus particle observed. Spread of the virus along infected hyphae and through anastomoses between diseased and healthy ones has been demonstrated.

Polyhedral virus particles, 25-30 mμ in diameter and containing RNA, have been isolated from a slow-growing strain of *Penicillium stoloniferum*. The viral RNA from this fungus stimulates the production of interferon in animals. Cultures of *P. stoloniferum* free of virus were obtained from heat-treated conidia of the infected strain. There is strong evidence that some other examples of slow growth and abnormal form of molds is due to virus infection.

12.12 ALGAL VIRUSES

12.12.1 Eukaryotic Algal Viruses

Viruses of eukaryotic algae are incredibly diverse, sweeping generalizations about their ecology are rare. These obligate parasites infect a range of algae and their diversity can be illustrated by considering that isolates range from small particles with ssRNA genomes to much larger particles with 560 kb dsDNA genomes. Molecular research has also provided clues about the extent of their diversity especially considering that genetic signatures of algal viruses in the environment rarely match cultivated viruses. One general concept in algal virus ecology that has emerged is that algal viruses are very host specific and most infect only certain strains of their hosts; with the exception of viruses of brown algae, evidence for interspecies infectivity is lacking. Although some host–virus systems behave with boom-bust oscillations, complex patterns of intraspecies infectivity can lead to host–virus coexistence obfuscating the role of viruses in host population dynamics. Within the framework of population dynamics, host density dependence is an important phenomenon that influences virus abundances in nature. Variable burst sizes of different viruses also influence their abundances and permit speculations about different life strategies, but as exceptions are common in algal virus ecology, life strategy generalizations may not be broadly applicable. Gaps in knowledge of virus seasonality and persistence are beginning to close and investigations of environmental reservoirs and virus resilience may answer questions about virus inter-annual recurrences. Studies of algal mortality have shown that viruses are often important agents of mortality reinforcing notions about their ecological relevance, while observations of the surprising ways viruses interact with their hosts highlight the immaturity of our understanding. Considering that just two decades ago algal viruses were hardly acknowledged, recent progress affords the optimistic perspective that future studies will provide keys to unlocking our understanding of algal virus ecology specifically, and aquatic ecosystems generally. Viruses have been associated with rapid changes in marine algal populations. Lysis of *Chlorella* strains has been recorded and virus particles reported and designated as 'chlorellophages' by Russian workers. The active lysis was only possible in the presence of bacteria although these had no harmful effect on the growth of the *Chlorella*.

12.12.2 Cyanophages (Blue-Green Algal Viruses)

Cyanobacteria (Blue-green algae) are widespread in the aquatic environment and often occur in mass blooms related to eutrophication of water bodies. The economic and nuisance effects of blue-green algae on water quality and in causing fish intoxications have been amply documented. Virus infections of cyanobacterial cells have been recognized and are best known as cyanophages. They were first isolated in 1963 from infected cells of *Lynghya* cultured from a waste stabilization pond. It also infects species of *Plectonema* and *Phormidium* and hence has been coded as LPP-l. This virus is stable over a pH range from 5 to 11 which is similar to that within which the host *Plectonema boryanum* grows. In culture, plaques vary in size between 0.1 to 8 mm in diameter and this is considered to be due to strain variation (Figure 12.9). The virus contains double-stranded DNA with a molecular base composition of 55 per cent guanine plus cytosine.

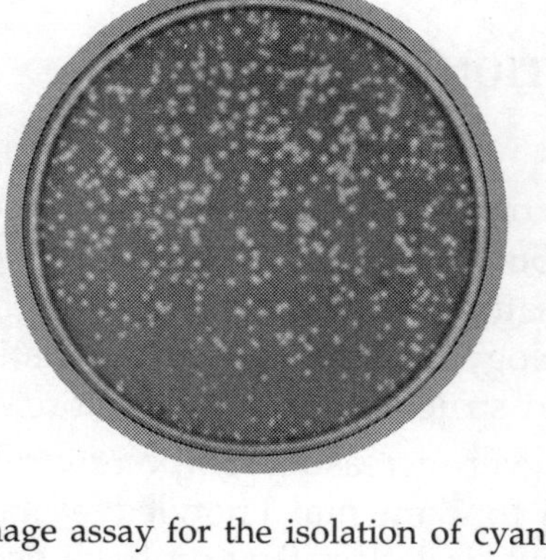

Figure 12.9 Phage assay for the isolation of cyanophage AS-1.

The probable influence of cyanophages on the geographical and seasonal distribution, as well as population dynamics, of cyanobacteria, and their possible use in biological control, added a further impetus to investigations of the cyanophages. The cyanophages isolated so far have been named according to their known hosts. Accordingly, cyanophage groups were designated by the initials of the generic names of the hosts, to which arabic numerals have been added to designate serological subgroups; different isolates of similar types were marked by a suffix consisting of the first letter of the provenance. LPP-1, LPP-2, infect *Lyngbya, Plectonema* and *Phormidium* with 1 and 2 serotype, AS-1 infects *Anacystis nidulans* and *Synecococcus cedrorum*, SM-1 infects *Synecococcus elongates* and *Microcystis aeruginosa* and N-1 infects *Nostoc muscorum* belonging to serotype 1. They are all double stranded DNA containing viruses very similar to T even phages except for the multiplication which is photosynthetic depended.

The cyanophages cause a lateral displacement of the photosynthetic lamellae and the virus particles remain attached to the lamellae after the whole cell has lysed. The technique of growing the virus within algae growing on agar plates revealed that the virus particles had long 'tails'. It is suggested that the virus may develop in virogenic stroma in the nucleoplasm or associated with 'polyhedral bodies'. 'Ghost' virus particles can be seen attached to the cell wall after infection.

Lysis of cells occurs randomly along the filaments of *Plectonema boryanum* and eventually the filaments fragment into separate cells. Cyanophages exhibit a one-step growth curve typical of bacterial viruses.

The growth cycle follows with a virus particle attaches to the outer wall of the host cell by its tail; remains on the outside but injects its DNA into the cell as in some phages; then multiplies in the nucleoplasm where fine threads are observable in infected cells but not in uninfected ones, and is then thought to migrate to the photosynthetic lamellae. As virus formation continues, empty spaces appear and from these the helices move into the virogenic stroma formed where the photosynthetic lamellae are displaced. Here the helices become enclosed in a protein coat and then compressed into the mature particles.

The naturally occurring virus is widespread, having been found world from freshwater to marine environment.

12.13 THE CLASSIFICATION OF VIRUSES

12.13.1 Introduction

Early attempts at viral taxonomy were thwarted by the paucity of fundamental information about viruses. Such systems of virus classification were based largely on the more easily observed features of virus diseases, e.g., tissue tropisms, host range, symptomatology, and pathology, rather than characteristics of the virus particles, such as chemical composition and structure, which are more difficult to determine.

In more recent years the ever-increasing tempo of virological research has produced a concomitant expansion in fundamental knowledge about viruses. Hence, the more recently proposed systems of viral classification have been centred on the more basic and immutable characters of the virion, e.g., type of nucleic acid and fine structure.

An internationally agreed system of virus classification and nomenclature has been formulated which include a system embracing all types of viruses (plant, bacterial, vertebrate, and invertebrate) and in which viruses are given Latinized binomial names; the genera, etc., being defined by a non-phylogenetic hierarchy of characters (type of nucleic acid, virion fine structure, etc.). In opposition to these proposals, others have suggested a binomial nomenclature based on the present vernacular names plus an informative cryptogram which for any particular virus could be amended as more facts are obtained. This 'vernacular name and cryptogram' nomenclature was designed to be non-prejudicial with respect to virus classification but its proponents favour the eventual adoption of the Adansonian type of classification . These systems of classification have been chosen is a matter for speculation. Meanwhile, as the debate proceeds, virologists must continue to use the existing nomenclature (largely based on the vernacular names of virus diseases), together with the systems of virus grouping which have evolved out of necessity over the past few years.

Viruses can be conveniently divided into two main sections, the DNA and RNA viruses respectively. The former includes most bacteriophages and the larger animal viruses; the latter most plant viruses so far investigated, together with the smaller animal and some phages. Other characters currently used in working classifications are capsid symmetry, presence of envelope, site of replication in the cell, reaction to ether treatment, number of capsomeres (for viruses with icosahedral symmetry), diameter of helix (for viruses with helical symmetry), particle size. Individual viruses are more often identified by serological means. The following groups of animal viruses have been defined using the above criteria.

12.13.2 DNA-containing Animal Viruses

Parvovirus group (also known as picodnaviruses). A group of small (Latin, *parvus* = small), non-enveloped icosahedral viruses with particles approximately 22 mμ in diameter. Members of the group include several viruses isolated from rodents, e.g., Kilham's rat virus, and the Virus of feline panleucopenia.

Papovavirus group: Named after three of the group members: the *papllioma* viruses (causing warts), *polyoma* virus (causing tumours in mice) and *vacuolating* agent (found in 'normal' monkey kidneys), The virions are non-enveloped, icosahedral and 40-55 mμ in diameter. Nearly all members of the group are oncogenic (tumour producing).

Adenovirus group: First isolated from the adenoidal glands of man. Members of the group are non-enveloped, icosahedral viruses, 70-80 mμ in diameter. Most contain a common group antigen as well as a type specific antigen, Adenoviruses are generally associated with respiratory diseases and often produce latent infections. Man, ox, dog, mouse, and chicken are among the species from which adenoviruses have been isolated. Some of these isolates produce tumours when inoculated into newborn hamsters.

Herpesvirus group: The icosahedral virions of this group are approximately 110 mμ in diameter and are readily recognized by their surrounding lipoprotein envelope. The integrity of this envelope appears to be essential for infectivity. The latter is therefore destroyed by ether treatment. The group takes its name from the virus causing herpes simplex (cold sores) in man. Other herpesviruses include those of varicella-zoster (chicken pox-shingles), pseudorabies, bovine rhinotracheitis, and infectious laryngotracheitis (fowl). Many herpes infections show latency.

Poxvirus group: These are the largest (230 x 300 mμ) and most complex of the vertebrate viruses. Replication occurs in the cytoplasm of the host cell; this feature is unique since all other DNA viruses replicate in the nucleus. The various poxvirus subgroups depend largely on serological relationships e.g., viruses of the vaccinia subgroup (vaccinia, cowpox, mousepox, smallpox, etc.), and are sufficiently closely related to give common neutralizing antibodies. Some poxviruses produce tumours, e.g., rabbit fibroma and myxoma, others either produce local skin lesions (pocks) or more generalized infections with a rash, e.g., smallpox.

12.13.3 RNA-containing Animal Viruses

Picornavirus group: These small (pico = small), non-enveloped, icosahedral viruses may be subdivided according to their acid sensitivity. Acid stable picornaviruses include poliovirus, the coxsackieviruses and the echoviruses (Enteric Cytopathic Human Orphan viruses). Acid labile members include the rhinoviruses (common cold viruses) and the virus of foot and mouth disease. Infections with picornaviruses are often inapparent (hence the term 'orphan' viruses without a disease). Diseases produced by different members of the group include infections of the gut (sometimes spreading to the central nervous system as in poliomyelitis), upper respiratory tract, meninges, skin, and heart.

Reovirus group. Non-enveloped, icosahedral viruses, 70-75 mμ diameter. Reoviruses have been isolated from the respiratory and enteric tract of man and other animals (Respiratory Enteric Orphan viruses). As their name suggests they are not definitely associated with any known diseases.

Arbovirus group: This group is somewhat unsatisfactory since it contains a large number of viruses, many of which have different physicochemical properties. However, some members are known to have icosahedral symmetry and possess some form of envelope. The group derives its name from the fact that members are *Arthropod-borne,* i.e., arboviruses can multiply in arthropods and then be transmitted by bite to vertebrates (including man, domestic animals, birds, bats, and reptiles). The group contains more than 150 viruses including the equine encephalitis viruses and the virus of yellow fever.

Myxovirus group: The group name is derived from the fact that members have receptors for certain mucins. Hence, myxoviruses are able to agglutinate red blood cells (haemagglutination). The particles are roughly spherical, 80-200 mμ in diameter and have a coiled-up tubular nucleocapsid (80-90 Å in diameter) with helical symmetry. Surrounding the latter is an envelope containing haemagglutinin. The particles also contain the enzyme neuraminidase which causes elution of virus from red blood cells. There are three main serological subgroups; type A viruses cause disease in man, birds, horses, and swine, while type B and C have been recovered only from man.

Paramyxovirus group: These viruses are similar in some respects to the myxoviruses, e.g., some produce haemagglutination and contain neuraminidase. However, the particles and helical nucleocapsids are somewhat larger, 100-300 mμ and 18 mμ in diameter, respectively. This group includes the parainfluenza viruses, mumps virus, and Newcastle disease virus. The serologically related viruses of measles, distemper, and rinderpest have been included largely on morphological grounds. Some members of the group produce mild infections of the upper respiratory tract while others produce more severe generalized diseases with rashes.

Rhabdovirus group: Group of bullet-shaped viruses (Greek, rhabdos = a rod) 60 × 225 mμ in size. The virions are enveloped and have helical symmetry. Members of the group include the viruses of rabies and vesicular stomatitis (a bovine disease).

Other viruses

Other viruses (plant, bacterial, and insect), have been grouped according to the principles similar to those used above for the vertebrate viruses. While no generally accepted group names have been established, some of the plant viruses can be grouped according to clearly defined structural and biological characters, e.g., the tobacco mosaic (TMV) group distinguished by the rod-shaped virion, its relative stability outside the host plant, serological characteristics, and absence of transmission by vectors; the turnip yellow mosaic (TYMV) group, differing from TMV in shape of particle, transmission, and serology. Much more work is required, however, before all plant viruses can be satisfactorily grouped in this manner.

Experiments

12.14 ISOLATION OF BACTERIOPHAGE FROM SEWAGE

Visual evidence of lysis is demonstrated by mixing a culture of bacteria with phage and growing the mixture on nutrient agar. Areas where the phages are active will show up as clear spots called plaques. The most thoroughly studied bacterial viruses are those that parasitize *Eschereichia coli*. They are collectively called coliphages. They are readily isolated from raw sewage.

12.14.1 Materials

Raw sewage, phage broth, pure culture of *E. coli*, incubator, centrifuge, membrane filtration unit, Erlenmeyer flask, test tubes, nutrient broth, soft nutrient agar, hard agar, Petri dishes, graduate and 5 mL pipettes, water bath.

Media

Nutrient broth (1 g Peptone, 0.5 g yeast extract, 0.25 g NaCl, 0.8 g potassium phosphate, dibasic in 100 mL distilled water), 10 × strength nutrient broth (Peptone 20 g, yeast extract 10 g, NaCl 5 g, potassium phosphate-dibasic 16 g, in 200 mL distilled water), warm nutrient agar plates (6,100 × 15 mm plates) (12–15 g agar/liter nutrient medium). Tubes containing 3 mL each of warm, top agarose (one per plate) (7.5 g agarose/litre nutrient broth, molten, cooled to 45°C).

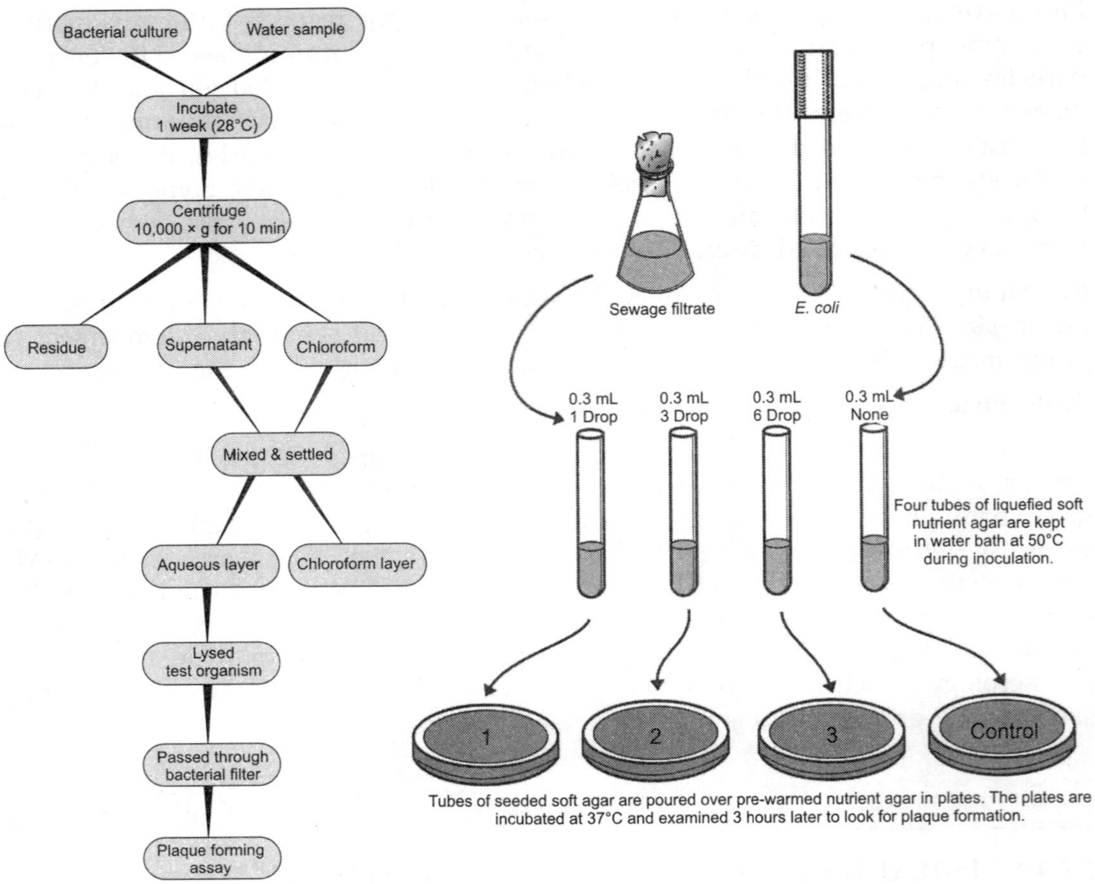

Figure 12.10 Schematic procedure for the isolation of bacteriophage.

12.14.2 Protocol

12.14.2.1 First step filtration and amplification of bacterial viruses (figure 12.10)

1. Collect the raw sewage from sewage treatment.
2. Mix the sewage (100 mL) with 10 mL of 10 × strength nutrient phage broth and 10 mL of *E. coli* suspension.
3. Incubate the mixture at 37°C for 24 hours.
4. Centrifuge the sewage at 2500 rpm for 10 min so that the bacteria and other solid matrix are separated. Decant the supernatant.
5. Repeat the process three times to obtain the filter.
6. Filter the supernatant through the membrane filter (0.45 μm) to remove the bacterial cells. It is quite possible that sometimes, filter gets clogged or choked. It is, therefore, desirable to change the filter and pass the filtrate through the new filter.

7. Connect the filtration assembly through vacuum pump to quick the filtration process.
8. Keep vacuum pump on so that filtrate may accumulate the flask of filtration unit.
9. Transfer aseptically the final filtrate so obtained into a pre-sterilized Erlenmeyer flask.
10. Incubate both cultures at 37°C, shaking for 24 hours.

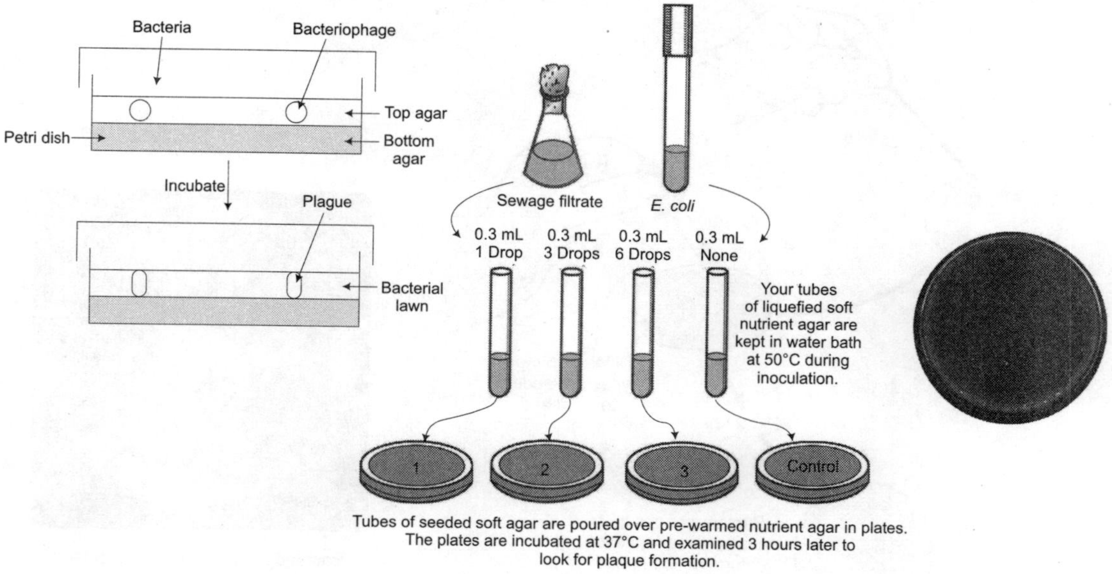

Figure 12.11 Double agar layer method of seeding *Escherichia coli* culture with phage.

12.14.2.2 Second step: Bacteriophage isolation and plating

1. Procure 4 tubes of soft nutrient agar of jelly-like consistency and keep on water bath at 50°C. Label them as 1, 2, 3 and 4.
2. Pour 1 mL filtrate in tube 1, 2 mL filtrate in tube B and 4 mL in tube C and keep tube D as blank having no filtrate.
3. Inoculate all the tubes with 0.5 mL in *E. coli* and pour into Petri dishes previously containing hard agar and label them 1, 2, 3, and 4 respectively to the tubes.
4. After the medium is solidified, invert the plates, keep them in the incubator for several hours and examine for plaque formation. (Figure 12.11)

12.14.3 Results and Observations

Record the results in triplicate and take the average of the readings as plaque forming units

Phage Typing

Phage typing is a method used for detecting single strains of bacteria. It is used to trace the source of outbreaks of infections. The viruses that infect bacteria are called bacteriophages ("phages" for short) and some of these can only infect a single strain of bacteria. These phages are used to identify different strains of bacteria within a single species.

A culture of the strain is grown in the agar and dried. A grid is drawn on the base of the Petri dish to mark out different regions. Inoculation of each square of the grid is done by a different phage. The phage drops are allowed to dry and are incubated: The susceptible phage regions will show a circular clearing where the bacteria have been lysed, and this is used in differentiation (Figure 12.12).

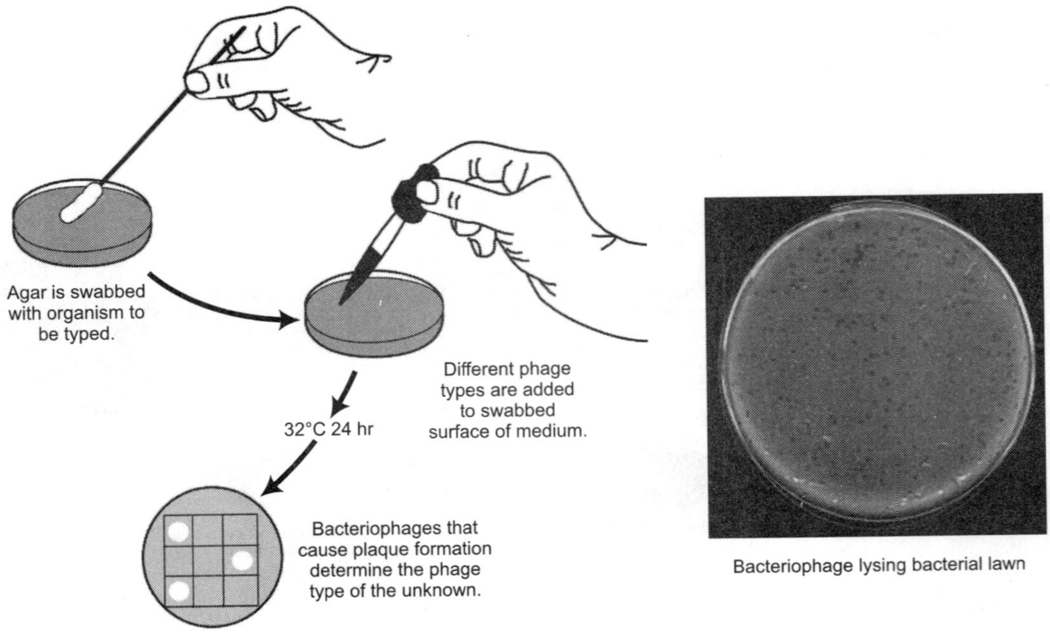

Figure 12.12 Demonstration for phage typing.

12.15 PREPARATION OF HIGH TITRE OF BACTERIOPHAGE OF ISOLATED BACTERIAL VIRUS

The titre is defined as high titre of phage which is required for plaque bioassay. In this procedure initially a single plaque is taken from a plate by sterilized inoculating needle or sterilized toothpicks and suspended in tryptone broth. After vortex for 2-3 minutes, it is transferred to an overnight grown culture of specific host bacteria and the lysate is subjected to plaque forming assay as described in the earlier experiment.

12.15.1 Materials

Troptone broth, host bacterial culture, nutrient agar plates with plaque, sterile needle or sterile toothpicks, sterile pipettes, incubator shaker.

12.15.2 Protocol

1. Procure a plate containing plaque.
2. Scoop out a plaque from the plate by sterilized inoculating needle or sterilized toothpicks.

3. Meanwhile, prepare an overnight grown bacterial culture of the host cell.
4. Mix 0.1 mL plaque broth in 10 mL culture of the host cell.
5. Incubate the mixture at 37°C for 6-8 hours.
6. Centrifuge the mixture at 10,000 rpm for 20 min to remove the bacterial cells.
7. Add 1-2 drops of chloroform for lysing the bacterial cell, if any, present in the suspension.
8. Pour the suspension (1-2 mL) on each plate containing nutrient agar.
9. Incubate the plates at 37°C for 24 hours in the inverted position.
10. Incubate the plates at 37°C for 24 hr in the inverted position.
11. Observe the plate for plaque formation.

12.15.3 Results and Observations

Numerous plaques of uniform size are visible against the dark background of agar.

12.16 ONE STEP GROWTH CURVE OF COLIPHAGE AND DETERMINE THE NUMBER OF PHAGE PARTICLES RELEASED BY THE LYSIS OF THE SINGLE INFECTED BACTERIAL CELL (BURST SIZE)

The growth cycle of the bacteriophage begins with the adsorption of the phases on the surface of the sensitive bacterial host cells. The phage genome is injected into the bacterial cells with the functions of the tail of the phages piercing the cell wall. Inside the cell, the phage organizes its synthetic capacity by utilizing the host metabolism and rapidly reproduces itself. Proteins of the host cells are rapidly reduced. Proteins are synthesized to assemble the viral genome giving rise to proper viral particle. The host cell then lyses releasing up to several hundred bacteriophages. The period of time from the adsorption of the phage to its host cell, until the beginning of the release of progeny, is called 'latent period or eclipse period. The period between the end of the latent period and the *plateau phase* (where maximum PFU plaque forming unit number is observed and remained constant) is called 'rise period.' The number of phage particles released per cell at the end of this period is called 'burst size' (Figure 12.13). Multiplicity of infection (MOI) is a frequently used the term in virology which refers to the number of virions that are added per cell during infection. If one million virions are added to one million cells, the MOI is one. If ten million virions are added, the MOI is ten. Add 100,000 virions, and the MOI is 0.1. The concept is straightforward. High MOI is used when the experiment requires that every cell in the culture is infected. By contrast, low MOI is used when multiple cycles of infection are required. However, it is not possible to calculate the MOI unless the virus titer can be determined, for example, by plaque assay or any other method of quantifying infectivity. In other words, it may be defined as the ratio of the number of infectious virus particles to the number of target cells present in a defined space.

Many bacteriophages require divalent cations such as Mg^{++} and Ca^{++} for attachment to bacterial host cells. All host cells do not recognize the viral particles and some viral

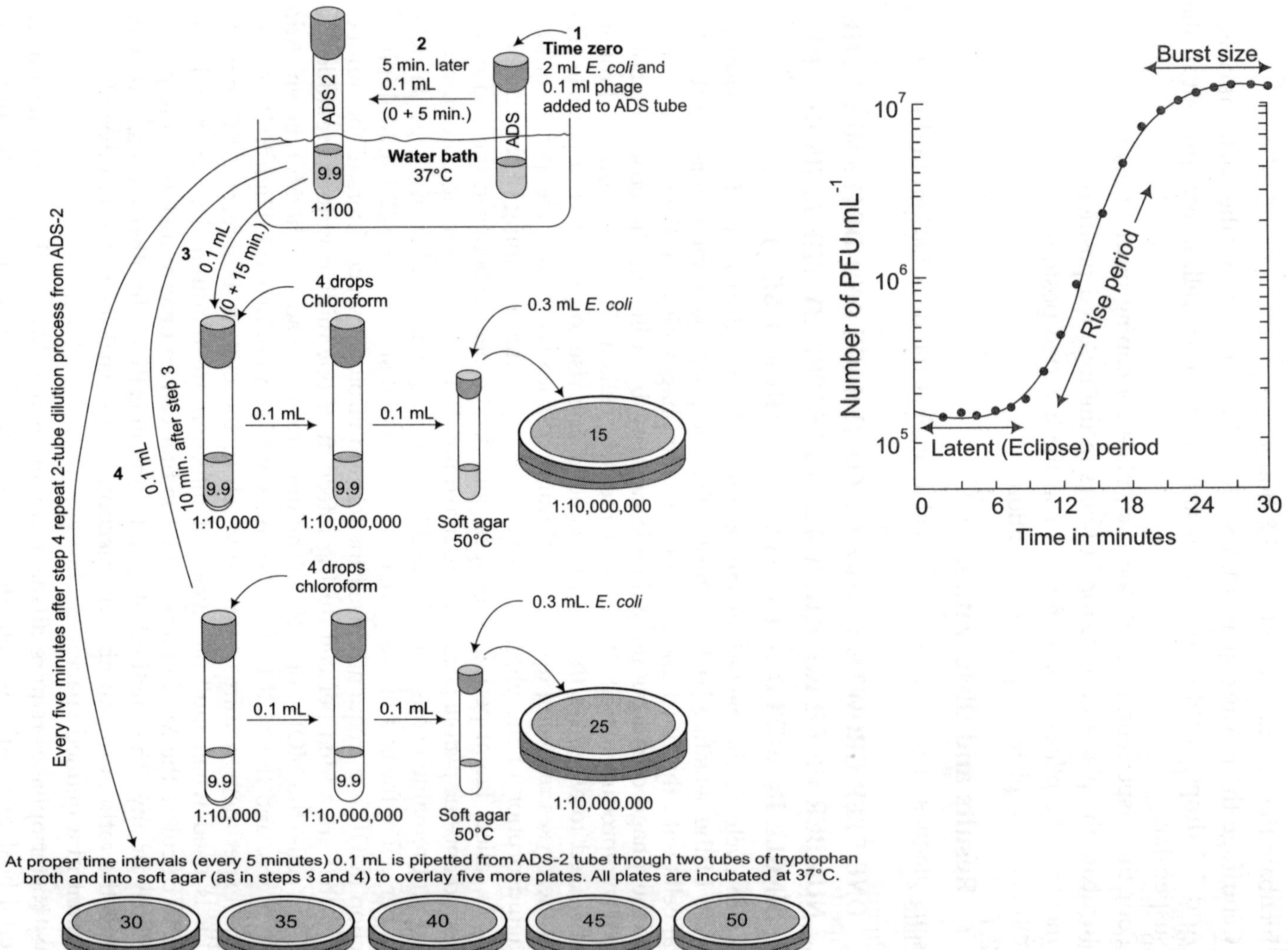

Figure 12.13 Single step growth curve of coliphage number of plaque forming unit as a function of time in minutes.

particles are aborted due to overinfection of the single cell in the adsorption mixture. The number of viral particles injecting the host cells in real term is defined as the input multiplicity.

12.16.1 Materials

Raw sewage, phage broth, pure culture of *E.coli*, incubator, centrifuge, membrane filtration unit, Erlenmeyer flask, test tubes, nutrient broth, soft nutrient agar, hard agar, Petri dishes, graduate and 5 mL pipettes, water bath, semilog graph papers pH meter, test tube stand, beef extract, peptone, agar, *E.coli* culture, bacteriophage lysate.

12.16.2 Protocol

12.16.2.1 Bacteriophage production burst size determination (single step growth curve)

1. Plate out an *E. coli* strain B culture at dilutions of 10^{-5}, 10^{-6}, and 10^{-7}.
2. Pipette 0.9 mL of *E. coli* into a sterilised test tube.
3. Pipette out 0.1 mL of a bacteriophage suspension in 0.9 mL of *E. coli* wait for 5 minutes so that the phage particles are absorbed by 80% of the bacterial cells.
4. Pipette the phage and *E. coli* mixture in a sterile centrifuge tube and centrifuge at 5000 rpm for 20 min. Decant the supernatant, resuspend the pellet of cells in 0.9 of nutrient broth and mix well by swirling. This removes unabsorbed phage particles.
5. Add 0.1 mL of the above cell suspension to 9.9 mL of nutrient broth and mix well by swirling.
6. Inoculate 0.1 mL of the above diluted cell suspension to 9.9 mL of nutrient broth. Mix well by swirling. Incubate at 37°C in the water bath provided.
7. Melt five tubes of regular nutrient agar and five tubes of soft nutrient agar (0.7% agar) temper the agar to 45-50°C and pour a tube of regular nutrient agar into each of five Petri plates, and allow to harden.
8. Temper the soft agar tubes to 45°C, add 3-4 drops of original *E. coli* strain B culture to each tube.
9. Consider the step 3 to be zero time. Add 0.1 mL sample from diluted culture of step 6 to a tube of the inoculated soft agar at intervals of 20, 25, 30, 35 and 40 minutes. Mix each tube and pour on the surface of one of the nutrient agar plates prepared in step 7. Allow the soft agar to solidify and incubate the plates (inverted) at 37°C.
10. Determine the count of *E. coli* from the plates made in step 1. Examine the plates from step 8 and count and record number of plaques at the various time intervals.
11. Determine the *E. coil* count from the step 2.
12. Examine the plaque number and calculate PFU in the original infected culture taking into consideration of dilution factor.

13. Plot the number of PFU versus time (use semi-log graph sheet). Calculate the burst size of the bacteriophage in one step growth.

12.16.3 Results

Calculation for example Assuming that the 0.1 mL of the phage added contained 10^8 PFU. Therefore, the PFU in the 2 mL of the bacterial culture would contain 5×10^7 PFU/mL. Consider the bacterial population (in 2 mL) in the beginning of the experiment is 1×10^8/mL, but the number of bacterial cells infected by the phage is only 5×10^7 cells/mL (IM – Input multiplicity).

$$\text{Therefore, the multiplicity of infection (MOI) is} = \frac{\text{Number of phages infecting bacterial cells}}{\text{Number of host cells}}$$

$$= \frac{5 \times 10^7}{1 \times 10^8} = 0.2$$

This means that, every two cells out of ten bacterial cells are infected with the phages.

If the total PFU obtained at the plateau is 5×10^9 PFU/mL, then the burst size can be calculated as follows:

$$\text{The burst size} = \frac{\text{Number of PFU released at the plate}}{\text{Number of infected bacterial cells}} = \frac{\text{Final PFU} - \text{intial PFU}}{\text{Number of interial cells}}$$

$$= \frac{5 \times 10^9 - 5 \times 10^7}{5 \times 10^7} = \frac{4.95 \times 10^9}{5 \times 10^7} = 99$$

The burst size is 99 indicating that each cell after lysis releases about 99 phage particles.

12.17 DEMONSTRATION OF LYSOGENY AND ISOLATION OF RHIZOBIUM PHAGE (RHIZOPHAGE)

With the invasion of bacterial cell by the viral DNA, one of the two things occur; lysis or lysogeny. In the event that lysis occurs, the metabolism of the bacterial cell becomes reoriented to the synthesis of new viral DNA and protein to produce mature phage particles. Once all the cellular material is used up, the cell bursts to release phage virions that, in turn, are prepared to invade other cells. Phage that causes lysis is said to be virulent. If the phage does not cause lysis; however, it is known as temperate and establishes a relationship with the bacterial cell known as lysogeny. In these cells, the DNA of the phage becomes an integral part of the bacterial chromosome becoming a prophage and behaves as if it is a bacterial chromosome. Lysogenic bacteria grow normally, but their cultures always contain some phage. Periodically, however, phage virion are released by lysogenized cells in lytic bursts similar to that seen in the lytic cycle. Upon exposure to physical or chemical mutagens such as UV exposure or mitomycin c, NTG treatment, the viral genome excises and reverts back.

12.17.1 Material

Rhizobial culture, indicator strain of rhizobia, incubator, shaker, sterilized pipettes, Petri dishes, test tubes and media.

Medium A (Yeast sucrose phage broth) g/L in distilled water Sucrose 2.5, $K_2HPO_4.7H_2O$ 0.5, $MgSO_4.7H_2O$ 0.2, $CaSO_4$ 0.16, NaCl 0.10, $FeCl_3.6H_2O$ 0.02, yeast extract 0.5, distilled water to 1 litre.

Medium B (Yeast sucrose agar plates) Yeast sucrose broth with 7 g agar per litre.

Medium C (defined agar medium)

Solution 1 g/L in distilled water: Mannitol 10, $Na_2HPO_4.12H_2O$ 0.45, $Na_2SO_4.10H_2O$ 0.6, KNO_3 0.6, Thiamine HCl 0.1, biotin 0.5, agar 7.5 and distilled water to 1 litre.

Solution 2 g/L in distilled water: $MgCl_2.6H_2O$ 0.1, $CaCl_2.6H_2O$ 0.1 and distilled water to 1 litre.

Sterilize solution 1 and 2 separately and pour 0.5 mL solution 2 in to Petri dishes. After the medium is solidified, pour 5 mL of melted agar solution 1 over the surface of solution 2.

Medium D (diluting broth) g/L in distilled water: Sucrose 2.5, K_2HPO_4 0.5, NaCl 0.1 and distilled water to 1 liter.

12.17.2 Protocol

1. Prepare medium a, Bor the defined medium C, inoculate 5 mL of any of the medium with desired rhizobial strain and incubate for 12 hours.
2. Pour 5 mL of the inoculated media on a pre-sterilized Petri dish (9 cm diameter), incubate to get translucent background of colonial growth.
3. Inoculate a flask containing 100 mL sterilized medium 2 with a loopful of young culture and incubate at 26°C in a shaker until it contains 2 × 10^8 rhizobia/mL. Pour 1 mL of medium over the first one and incubate as, confluent growth of bacterium is obtained.
4. Mix 1 mL of 24 hours old culture of the same bacterium with 0.5 mL indicator suspension in 5 mL of melted agar medium, add to the Petri dishes to form the second layer of agar.
5. Incubate the Petri dish for 48 hr at 26°C and, then observe the development of the plaques.

12.17.3 Results and Observations

Note the plaque size. Free phage particles from plaques (i.e., spontaneous lysis or rhizobia) are obtained due to the induction of the tested culture by the indicator strain demonstrating lysogeny.

Avoid too heavy bacterial growth. Use low salt medium D to check complete lysis followed by rapid growth due to aggregation.

12.18 ISOLATION OF CYANOPHAGE LPP FROM POND WATER

Cyanophages are the viruses that attack the members of cyanobacteria. They are known to be abundant and dynamic component in the aquatic microbial communities that can regulate the biomass production and species composition of varied microorganisms and other aquatic life forms like bacteria and phytoplankton. They are also known to mediate gene transfer between microorganisms in aquatic ecosystems. Cyanophages are known to play a significant role in the ecology of cyanobacteria and that they are very possibly affecting natural control of cyanobacterial blooms. Distributions of cyanophages are affected immensely by the change in environmental conditions. Temperature, salinity, pH, solar radiation, depth, and seasonal variation greatly affect the attachment and multiplication of cyanophage population.

Cyanophages are capable of entering into a lysogenic relationship with their host. However, induction of the lytic state under lysogenic conditions appears to be more difficult with cyanophage than with bacteriophage. High multiplicity of infection is known to favour lysogeny. Also, the nutritional status of the cells is also an important factor. In nutrient-rich media, the lytic course predominates, whereas growing cells in poor media increases the chances of the lysogenic pathway after single infection. Cells raised in nutrient rich media have a higher concentration of the host global regulator RNase III, which leads to elevated rates of expression of the protein N favouring lytic growth. In carbon-starved cells, on the other hand, RNase III and consequently N concentrations are low. Under these conditions, N translation is repressed. This reduction of N concentration would reduce Q expression to a level that provides more opportunity for lysogenic response. When the phage DNA enters the host cell, then there are options for the operation of both lytic and the lysogenic cycle. Both the processes require the synthesis and also the expression of the immediate early and the delayed early genes. But after this point, there occurs the point of deviation of both the processes: lytic process occurs if the late genes are expressed and lysogeny happens if the repressor gets expressed.

The cyanobacterial-cyanophage association represents an ideal model system for studying the mechanism of oxygenic photosynthesis under viral infection and also the correlation between the viral abundance and the surrounding environment. The cyanophages also infect marine cyanobacteria. Due to the resemblance between cyanophages and the bacteriophages, as well as the presence of common type of photosynthetic activity in cyanobacteria and higher plants, provide an ideal and excellent model system to study the photosynthetic activity under viral infection. The cyanophages show extreme diversity in their structure, habitat and host range. On the structural basis, they are differentiated into three major groups.

12.18.1 Material

Fresh culture of cyanobacteria (Lyngbya, Plectonema, Phormidium), modified Chu 10 agar slants, pond water, chloroform, sterile Petri dishes, illuminated incubator, sterile culture tubes, centrifuge, sterile membrane filtering assembly, membrane filter.

12.18.1.1 Modified Chu 10 medium

Modified Chu 10 medium g/L in distilled water: $Ca(NO_3)_2.4H_2O$ 0.04, $K_2HPO_4.3H_2O$ 0.01, $MgSO_4.7H_2O$ 0.025, $NaCO_3$ 0.02, $NaSiO_3.9H_2O$ 0.025, Ferric citrate 0.003, Citric acid 0.003, Glass distilled water to 1 litre.

To the above medium add 1 mL/L of Fogg's micronutrient solution is added. The composition of the solution: $MnCl_2.4H_2O$ 0.01, $Na_2MoO_42H_2O$ 0.01, H_3BO_3 0.01, $CuSO_4.5H_2O$ 0.001, $ZnSO_4.7H_2O$ 0.001, distilled water to 100 mL. pH of this medium should be adjusted to 7.5 before autoclaving.

1.5% agar is added for preparing agar slants for culturing cyanobacteria.

12.18.1.2 Magnesium saline composition g/L in distilled water

Magnesium saline composition g/L in distilled water 0.2 $MgCl_2.6H_2O$, 5.85 NaCl and distilled water to 1 litre.

12.18.2 Protocol

1. Collect two litres of water sample from ponds, add 2-3 drops of chloroform, shake vigourously and filter through glass filter paper (Whatman GF/C) to remove extraneous matter.
2. The filtrate is concentrated down to 20 mL through dialysis against a solution of high molecular weight (20,000) polyethylene glycol (PEG) in magnesium saline. Such a solution virtually soaks up water in the filtrate.
3. The concentrate is then filtered through an ultrafine sintered glass filter which allows only virus size particles to pass through. Prepare dilution as per Figure 12.14.
4. 5 mL of this ultrafiltrate is incubated with the host culture for 10-15 days. There is complete lysis of the cyanobacterial culture with the disappearance of green colour in the culture flask.

12.18.2.1 Plaque forming unit (PFU) assay

5. Plaque counts are determined on plates in which 5 mL of an inoculated agar had been evenly distributed over a 15 mL solidified layer of 1.5% modified Chu 10 modified agar medium.
6. The surface layer is prepared in test tubes, which is consisted of 0.5 mL of an appropriately diluted virus suspension and 2 mL concentrated cyanobacterial filament of 3 to 4 weeks old culture of host cells and 2.5 mL of melted agar medium (medium with 1% agar) and plated on agar plates having a base of 0.8% agar containing medium (30 mL in 100 × 17 mm Petri plates).
7. The inverted plates are incubated in an illuminated incubation chamber at 30°C. Plaques start appearing after 4 days of incubation (Figure 12.14 and Figure 12.15).
8. The clonal population of the virus is raised from a single plaque isolate and lysate is always filtered through a sintered glass filter (2.5 cm GF/C; Whatman, England) before being used in the experiment.

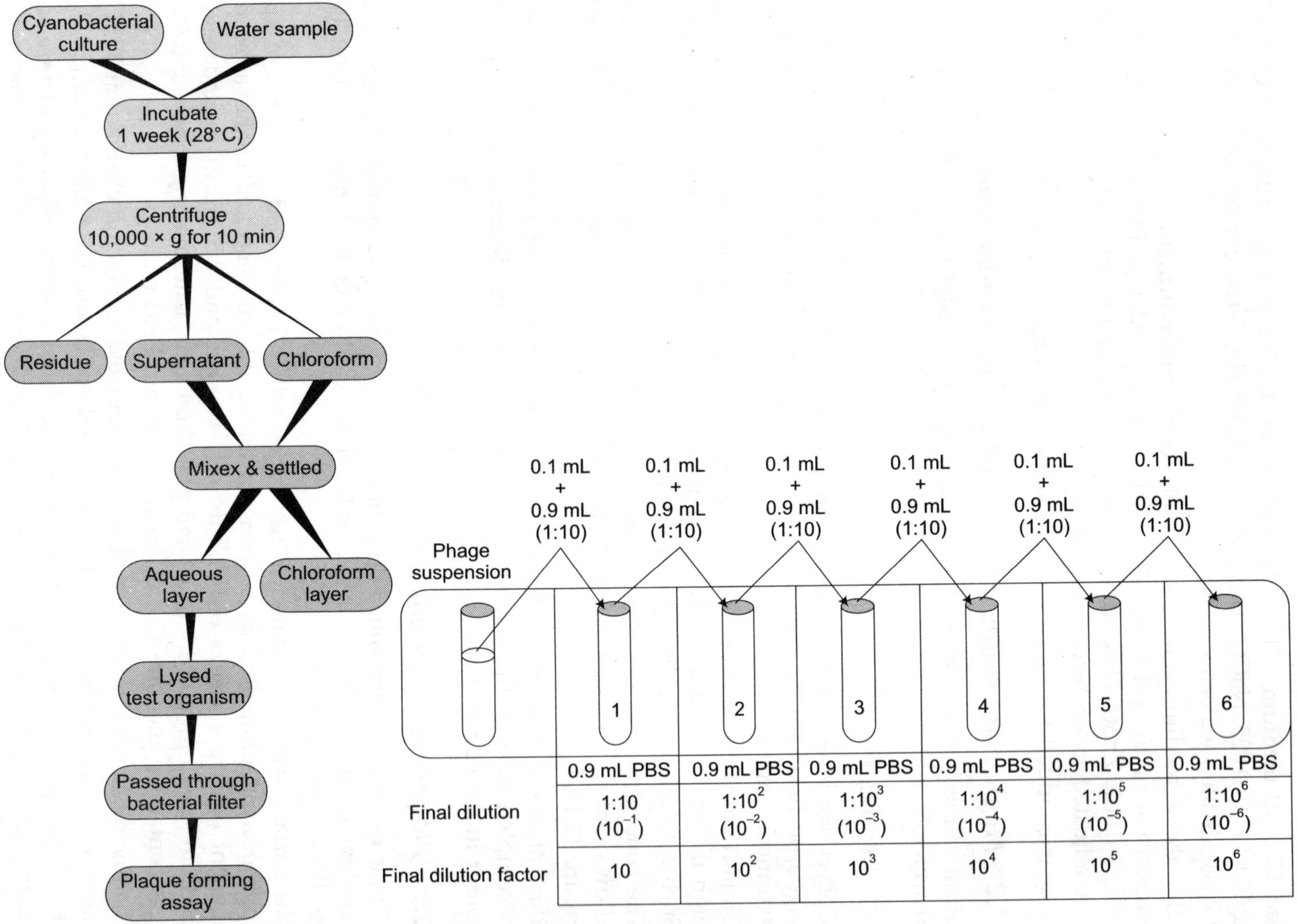

Figure 12.14 Schematic demonstrations for the isolation of cyanophage.

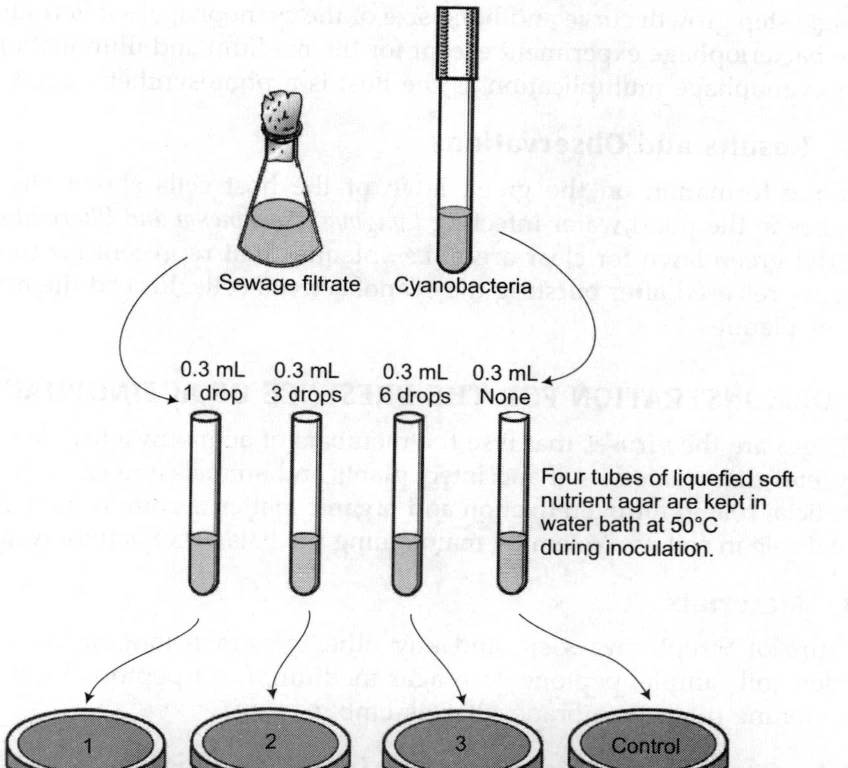

Figure 12.15 Procedure for Plaque Forming Unit (PFU) Assay of cyanophage.

9. Single step growth curve and burst size of the cyanophage is determined similar to the bacteriophage experiment except for the medium and illumination is essential for cyanophage multiplication as the host is a photosynthetic autotroph.

12.18.3 Results and Observations

Clear plaque formation on the green lawn of the host cells shows the presence of cyanophages in the pond water infecting *Lyngbya, Plectonema and Phormidium.* Examine the bacterial green lawn for clear areas, i.e., plaques that represent for the presence of cyanophages released after bursting the cyanobacterial cells. Record the number, shape and size of plaques.

12.19 DEMONSTRATION FOR THE PRESENCE OF ACTINOPHAGES IN SOIL

Actinophages are the viruses that lyse the members of actinomycetes. There are several actinomycetes that reside in soil and infect plants and animals and cause disease besides their beneficial role in nitrogen fixation and organic matter decomposition. Actinophages play a vital role in soil ecosystem by maintaining the balance of actinomycte population.

12.19.1 Materials

Pure culture of Streptomyces sp. and any other genera belonging to actinomycetes, 20 g garden soil sample, peptone corn agar medium, meat peptone broth, Erlenmeyer flask, membrane filter, membrane filter assembly.

12.19.1.1 Peptone-corn agar medium in distilled water

Glucose 10, NaCl 5, $CaCO_3$ 0.5, peptone 5, Corn extract 5, agar 20, and distilled water to 1 liter. pH of the medium is adjusted to 7-7.2.

12.19.1.2 Beef extract peptone broth in distilled water

Beef Extract 3, peptone 5, and distilled water to 1 liter. pH of the medium is adjusted to 7-7.2.

12.19.2 Protocol

1. Prepare corn agar plates and inoculate with the pure culture of *Streptomyces* sp.
2. Collect 20 g garden soil sample and transfer into a 70 mL beef extract - peptone broth in a 250 mL conical flask.
3. Incubate the flask at 25-30°C for 24 hr on a shaker at low speed.
4. Decant the water and filter through the membrane filter aseptically under vacuum and store the filtrate in a sterilized screw cap tubes.
5. Place 1-2 drops of filtrate in the centre of young actively growing cultures of Streptomyces sp and tilt the plate so as to spread the filtrate over a definite part of the agar surface.
6. Incubate the culture at 25 - 26°C for 24 - 72 hrs.

12.19.3 Results and Observations

Observe the growth of Streptomyces sp. in the plate. Growth of Streptomyces sp. is inhibited in filtrate inoculated areas and plaques (small clear lysed areas) are also observed. It can also be demonstrated by subculturing the active agent. Prepare the fresh culture of Streptomyces sp. again. Cut a small piece of agar from plaques/inhibition zone and spread it over the surface of pure culture with spatula. Incubate the plates as above. Appearance of several distinct plaques confirms the presence of actinophages.

12.20 VIRUS CULTURE IN EMBRYONATING CHICKEN EGGS

Before the development of cell culture, many viruses were propagated in embryonated chicken eggs. This method is most commonly used for growth of the virus. The excellent yield of virus from chicken eggs has led to their widespread use in research laboratories and for vaccine production. Avian-embryo culture is economical and convenient compare to laboratory animals. This is commonly used for the isolation, identification, titrating and maintenance of many animal viruses as well as for the production of vaccines.

12.20.1 Materials

Chicken eggs, candling apparatus, syringe, sterile normal saline solution, incubator, sterile pipettes, disinfectant solution.

12.20.2 Protocol

1. Candle two embryonated eggs with the long axis in the horizontal plane. Mark the location of the air sac. Swab the marked area with the iodine-alcohol disinfectant. After sterilizing an 18-gauge needle by dipping it into the disinfectant solution and flaming, use the needle to puncture the shell of an egg at the upper extremity of the shell over the air sac (Figure 12.16).
2. Using a 1 mL syringe, with a 27 gauge, 3/4" needle, inoculate the allantoic cavity with the dilution of disease New-Castle viral disease or any other virus by

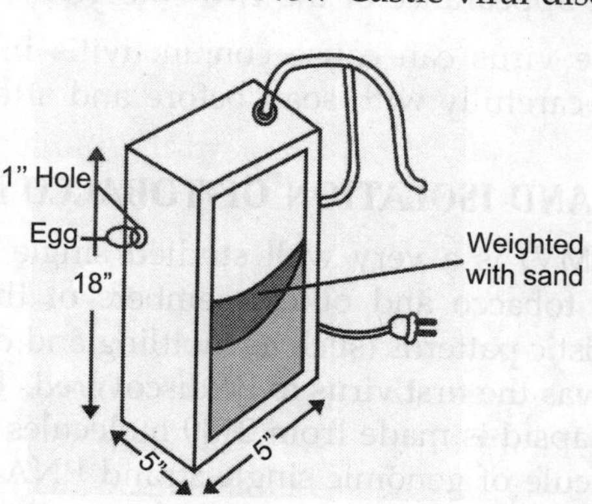

Figure 12.16 Candling an egg.

inserting the needle perpendicularly through the puncture in the shell, passing the needle into the egg along its full length parallel to the long axis of the egg. Inject 0.2 mL of the virus preparation and remove the syringe needle. Seal the puncture hole with duco cement or any other suitable cell cementing material (Figure 12.17).

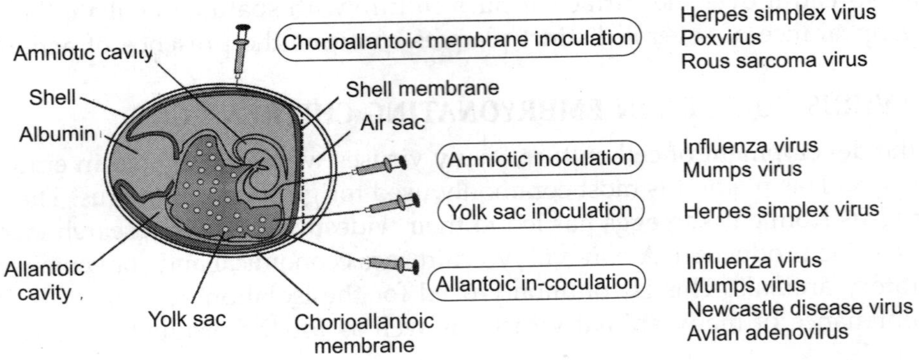

Figure 12.17 Inoculation of virus into allantoic cavity.

3. As step 2 inoculate the second embryonated egg with 0.2 mL of sterile saline using the above steps.
4. Incubate the inoculated eggs at 37°C in an incubator having coster trays in order to maintain the humidity.
5. Candle the inoculated eggs each succeeding laboratory period; examining for the death of embryos as evidenced by the cessation of the movement or disappearance of veins from the egg shell. New castle disease virus causes death of embryo with in 3 or 4 days following inoculation.
6. Crack the shell and empty the contents into a Petri plate when the embryo is dead. Compare the appearance of the two embryos.

Note: New-Castle disease virus can cause conjunctivitis in humans and hence it is advisable to wash hands carefully with soap before and after the experiment.

12.21 EXTRACTION AND ISOLATION OF TOBACCO MOSAIC VIRUS (TMV)

Tobacco mosaic virus (TMV) is a very well studied single stranded RNA virus that infects plants, especially tobacco and other members of the family solanaceae. The infection causes characteristic patterns (such as mottling and discoloration) on the leaves (hence, the name). TMV was the first virus to be discovered. Tobacco mosaic virus has a rod-like appearance. Its capsid is made from 2130 molecules of coat protein (see image to the left) and one molecule of genomic single strand RNA 6400 bases long. The coat protein self-assembles into the rod-like helical structure (16.3 proteins per helix turn) around the RNA which forms a hairpin loop structure.

12.21.1 Material

Diseased (TMV) tobacco leaves, waring blender, sterilized water, low speed centrifuge, ultracentrifuge, cheese cloth, test tubes, buffer solution, sodium borate (0.02 M) pH 8.2, sodium ascorbate (0.01 M). Triton X-100, Tris HCl (10 mM) pH 8.0, EDTA (1 mM), sucrose solutions of various concentrations (10% - 65%), syringe, dialysis tubing.

12.21.2 Protocol

12.21.2.1 Extraction and pathogenicity test

1. Take a few virus-infected leaves; grind the diseased leaves in a mixture of 0.02 M borate buffer pH 8.2 containing 0.01 M sodium ascorbate as a reducing agent to avoid any oxidation of the virus in a waring blender.
2. Filter the homogenate through several layers of cheese cloth for collection of sap with virus (Figure 17.16).
3. Filter the particles by pouring the contents of the blender through a filter paper.
4. Pass the filtrate through a membrane filter (pore size 0.45 μm) using membrane filtration assembly.
5. Keep a healthy tobacco plant beside and select the leaves, which are to be inoculated. Wash them with carborundum using a wash bottle.
6. Rub the leaf with a piece of cotton that has been soaked in the filtrate from step 4.
7. Wash the carborundum and excess virus from the leaf with distilled water from the wash bottle.
8. Keep aside, a control set.
9. Check the plants upto 3 weeks. Look for both localised and systemic symptoms.

Tobacco Mosaic Virus can also be isolated from some commercial products. Russian and Turkish brand of cigarettes are good sources of these viruses (Figure 12.18).

12.21.2.2 Purification of TMV virus

1. Take the filtrate from step 3, add 0.5% Triton X-100 and stir for one hour.
2. Centrifuge the sap at low speed (10,000 rpm) for 10 minutes, discard pellet, and recover the supernatant.
3. Layer the supernatant over cushion of 5 mL of 30% sucrose in 10 mM Tris HCl, 1 mM EDTA (pH 8.0) and centrifuge at 27,000 rpm for 3 hr.
4. Discard the supernatant and collect the tiny pellets of virus sediments at the bottom of the tube.
5. Resuspend the virus pellet in the buffer and keep for overnight.
6. Repeat step 2 to 5 two or three times and further purification of the virus is subjected to density gradient centrifugation.
7. Further, the semi-purified preparation is submitted to a linear density sucrose gradient (10-40%), and ultracentrifugation at 80,000 × g for 2.5 h or prepare

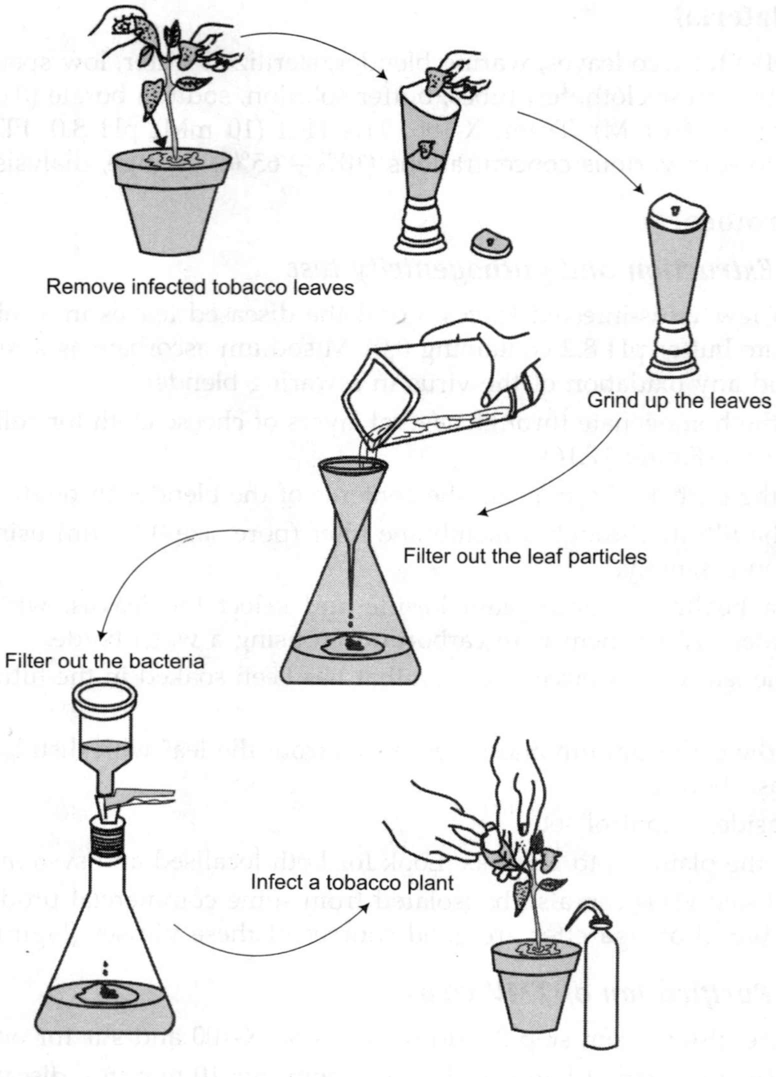

Figure 12.18 Steps in the infection with Tobacco mosaic virus.

sucrose solution gradient (10%, 20%, 30%, 45%) in centrifuge tubes for layering (Figure 12.19) of virus suspension on sucrose gradient.

8. Spin the tubes at high speed at 80,000 × g for 2.5 h in an ultracentrifuge in order to form a band of virus particles at the top of the centrifuge tube.
9. Collect virus band as separate fraction through a puncture in the bottom of the centrifuge tube or by injecting a syringe into plastic centrifuge tube.
10. Transfer virus fraction in dialysis tubing and remove impurities by dialysis against extraction buffer (0.02 M sodium borate pH 8.2 with 0.01 M sodium ascorbate) or against water for 24 hr at 4°C.

12.21.3 Results and Observations

Observe the dialysis tubing for virus. Pure virus particles are obtained in the dialysis tubing which can be used for further studies.

12.22 CULTIVATION OF VIRUSES IN TISSUE CULTURE

As the viruses do not reproduce independent of the living host cells; they cannot be cultured in the same as bacteria and eukaryotic microorganisms. Tissue culture is the *in vitro* propagation of cells. A convenient source for tissue is the chick embryo. *In vitro* cultivation of animal viruses has eliminated the need to kill the animals. This technique has become possible by the development of growth media for animal cells and by the availability of antibiotics which prevent bacterial and fungal contaminations in cultures. Some viruses cause the formation of plaques in tissue culture procedure. The clonal cell lines suspended in suitable media are infected with any desired virus which replicates inside the multiplying cells. If the viruses are virulent, they cause lysis of cells and virus particles are released in the surrounding medium. These newly produced virus particles (virions) infect the adjacent cells. As a result, localized areas of cellular destruction and lysis (called **plaques**) often are formed. Plaques may be detected if stained with dyes, such as neutral red or trypan blue that can distinguish living from dead cells. Viral growth does not always result in the lysis of cells, to form a plaque. Animal viruses, in particular, can cause microscopic or macroscopic degenerative changes or abnormalities in host cells and in tissues called **cytopathic effects,** cytopathic effects may be lethal, but plaque formation from cell lysis does not always occur.

12.22.1 Study of Morphology of Chicken Fibroblast Monolayers

12.22.1.1 Protocol

1. Place the chicken embryo in phosphate buffered saline (PBS) with the following composition:
   (a) NaCl 8 g, KCl 0.2 g, $CaCl_2.2H_2O$ 0.13 g, $MgCl_2.6H_2O$ 0.1 g and distilled water to 800 mL.
   (b) $Na_2HPO_4$1.15 g, KH_2PO_4 0.2 g, and distilled water to 200 mL. Autoclave separately and mix when cool
2. Cut tissue into pieces of about 1 mm in size
3. Wash fragmented sections several times in fresh phosphate buffered saline.
4. Mix the fragments in Trypsin to break the clumps into individual cells.
5. Allow the cell clumps to pass through gauze fitted on a centrifuge tube, add serum and centrifuge at 1000 rpm for 10 minutes.
6. Decant the supernatant and resuspend the cells in a measured amount of tissue culture medium.
7. Calculate the number of cells per mL of the above suspension using a haemocytometer.

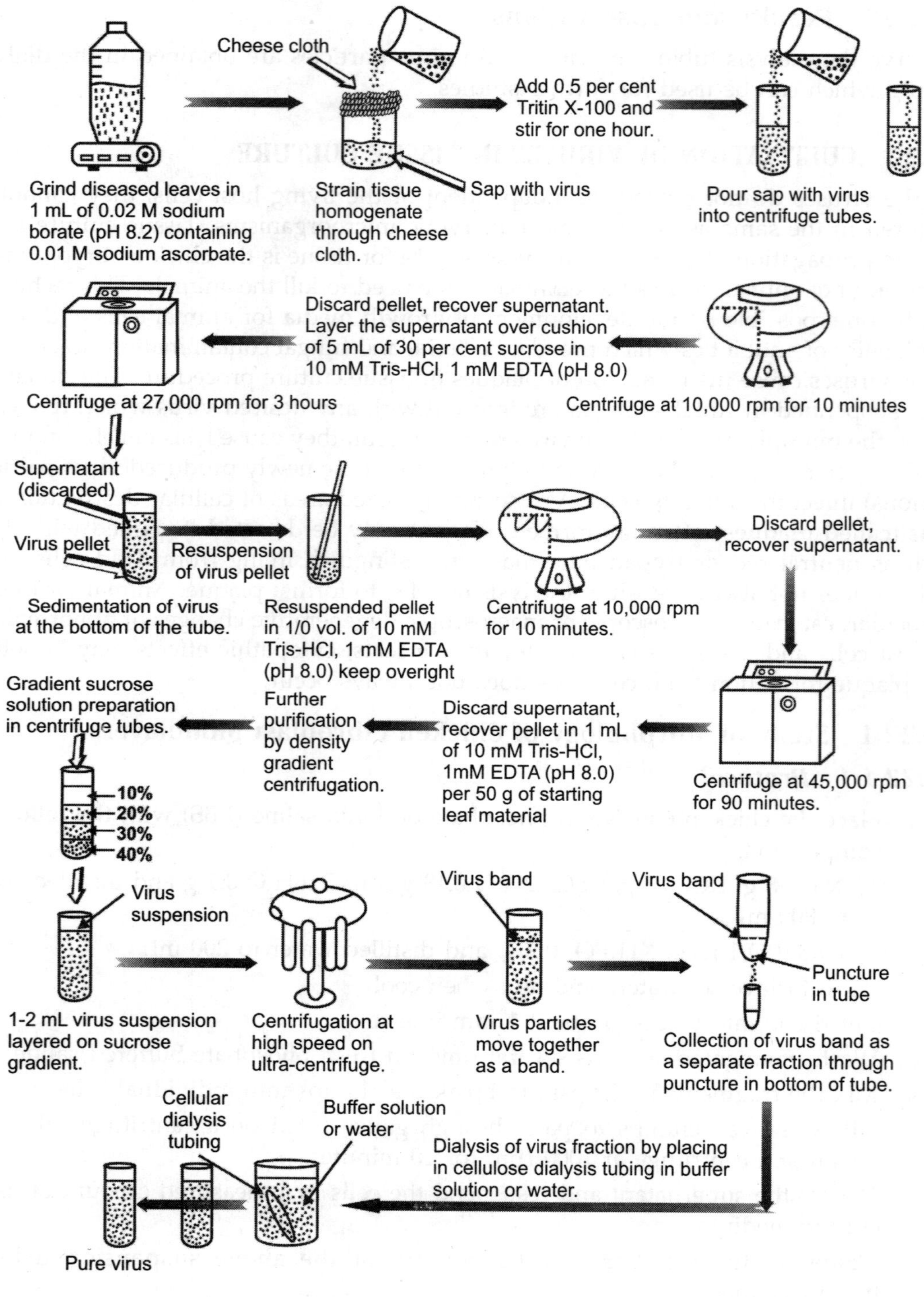

Figure 12.19 Schematic illustration of TMV viral purification.

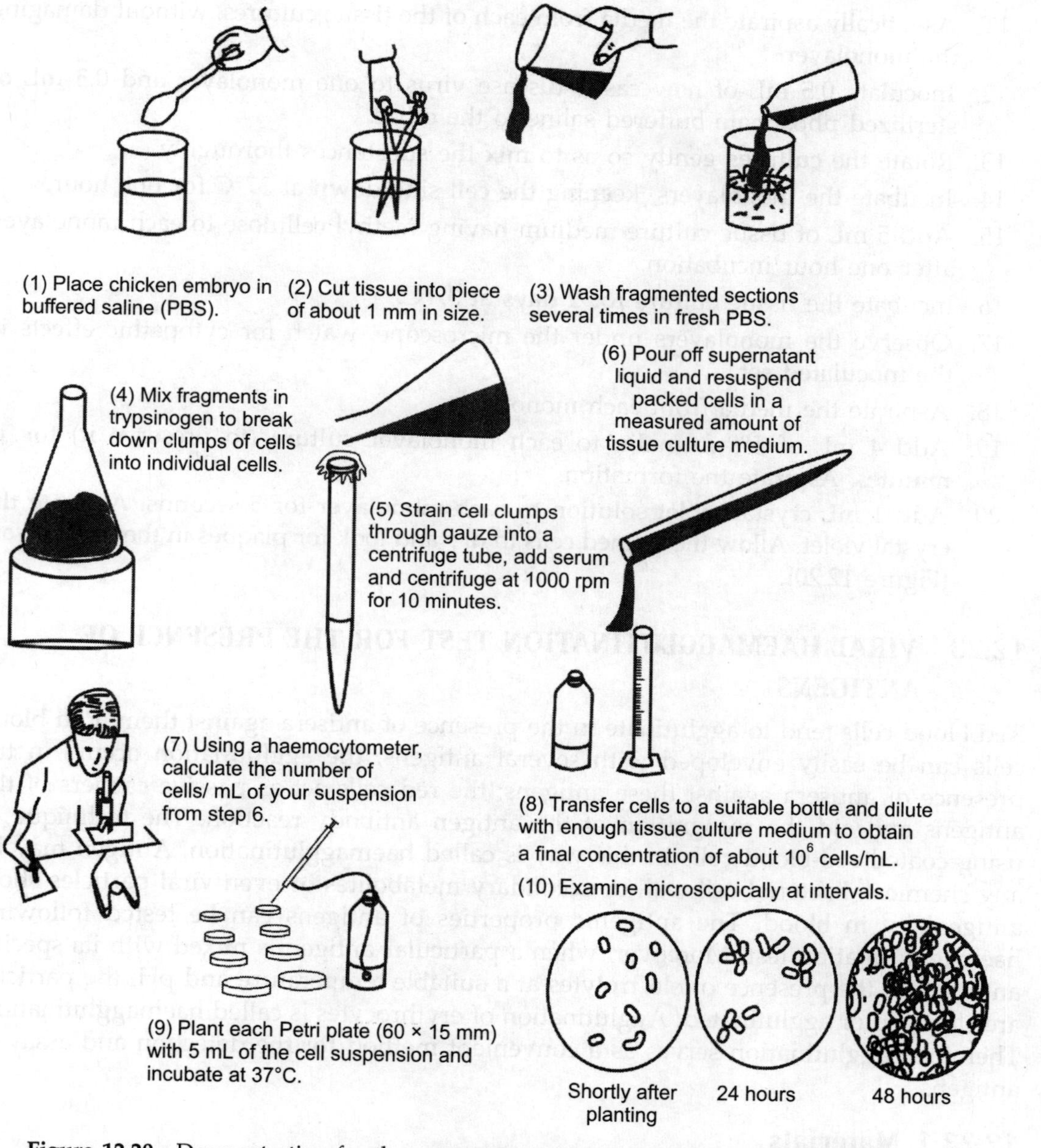

Figure 12.20 Demonstration for the preparation of primary tissue for monolayer cultures Experiment Study of morphology of chicken fibroblast monolayers.

8. Transfer the cells to a flask and dilute it with enough tissue culture medium to obtain a final concentration of about 10^6 cells/mL.
9. Inoculate each Petri plate with 5 mL of the cell suspension and incubate at 37°C.
10. Examine the plates microscopically at different intervals.

11. Aseptically aspirate the media from each of the tissue cultures, without damaging the monolayer.
12. Inoculate 0.5 mL of new castle disease virus to one monolayer and 0.5 mL of sterilized phosphate buffered saline to the other.
13. Rotate the cultures gently so as to mix the substances thoroughly.
14. Incubate the monolayers, keeping the cell side down at 37°C for one hour.
15. Add 5 mL of tissue culture medium having methyl cellulose to each monolayer, after one hour incubation.
16. Incubate the tissue culture for 2 days at 37°C.
17. Observe the monolayers under the microscope, watch for cytopathic effects in the inoculated set.
18. Aspirate the media from each monolayer.
19. Add 4 mL of 5% formalin to each monolayer culture (to preserve it) for 10 minutes. Aspirate the formation.
20. Add 1 mL crystal violet solution to each monolayer for 5 seconds. Aspirate the crystal violet. Allow the stained cells to dry and look for plaques in the monolayers (Figure 12.20).

12.23 VIRAL HAEMAGGLUTINATION TEST FOR THE PRESENCE OF ANTIGENS

Red blood cells tend to agglutinate in the presence of antisera against them. Red blood cells can be easily enveloped with several antigens, the agglutination occurs in the presence of antisera against these antigens; the red cells act as passive carriers of the antigens and aid the recognition of the antigen antibody reaction. The technique of using coated red-blood cells in this way is called haemagglutination. Antigen may be any chemical, bacterial cell wall or secondary metabolites or even viral particles show antigenicity in blood. The antigenic properties of antigens can be tested following haemagglutination test. Moreover, when a particular antigen is mixed with its specific antibody in the presence of electrolytes at a suitable temperature and pH, the particles are clumped or agglutinated. Agglutination of erythrocytes is called haemagglutination. Therefore, agglutination serves as a convenient method for the detection and assay of antigens.

12.23.1 Materials

Permanent Lucite 'U' or 'V' plates, microtiter dropper (0.025 mL), microtiter diluters (0.025 mL), diluents (such as complement fixation test diluents is suitable for influenza), antigen (e.g., haemagglutinin), serum (heated to 56°C for 30 min and suitably treated to remove non-specific inhibitors; filter sterilized), RBCs of optimum concentration of a mammal like sheep, goat, mouse, etc.

12.23.2 Protocol

1. Place one drop of diluents in about 11 wells of the microlitre plate. Use the last well as control.
2. Place one drop of antigen and 1 drop of diluents in the first well so as to get 1:2 dilution.
3. Further serially dilute the antigen in the subsequent wells and discard the last drop (0.025 mL) from the test well.
4. Add 1 drop of diluents to each well used.
5. Add one drop of RBCs to each well including the control well.
6. Incubate the well at appropriate temperature so as to settle the cells of control well to the bottom.

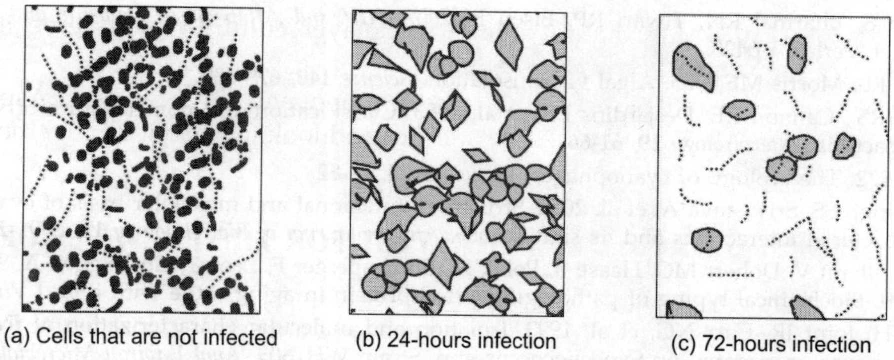

Figure 12.21 Normal and infected chicken fibroblast cells.

12.23.3 Results

The tests should be recorded as follows:

1. Cells distributed evenly in the bottom of the well-complete agglutination (+)
2. Cells compacted in a small red button at the bottom-no agglutination (–)
3. Non-agglutinated cells overlying a thin layer of agglutinated cells-partial agglutination (+)

Read the titre of the antigen as the highest dilution that gives haemagglutination (Figure 12.21).

Suggested Readings

Abedon ST, Duffy S, Turner PE 2009. Bacteriophage Ecology. *Encyclopedia Microbiol* 42-57.

Ackermann HW, DuBow MS. 1987. *Viruses of Prokaryotes: General properties of bacteriophages*, Vol. I. Boca Raton, Florida, CRC Press.

Ackermann HW, DuBow MS. 1987. *Viruses of Prokaryotes: Natural Groups of bacteriophages*, Vol. II. Boca Raton, Florida, CRC Press.

Adams MH. 1959. *Bacteriophages* New York, Wiley-Interscience.

Alcami A, Koszinowski UH.2000. Viral mechanisms of immune evasion. *Trends in Microbiology* 8, 410–418.

Bailey S., Clokie MR, Millard A, et al. 2004. Cyanophage infection and photoinhibition in marine cyanobacteria. *Res Microbiol* **155**, 720-725.

Baker AC., Goddard VJ, Davy J, et al. 2006. Identification of a diagnostic marker to detect freshwater cyanophages of filamentous cyanobacteria. *Appl Environ Microbiol* **72**, 5713-5719.

Bisen PS, Audholia S, Bhatnagar AK. 1985. Mutation to resistance for virus AS-1 in the cyanobacterium *Anacystis nidulans*. *Microbiol Lett* **29**, 7-13.

Desjardins PR, Olson GB.1983. Viral Control of Nuisance Cyanobacteria (Blue-Green Algae). II. Cyanophage Strains, Stability on Phages and Hosts, and Effects of Environmental Factors on Phage-Host Interactions; *California Water Resource Center*, Davis, CA.University of California.

Hawker, LE, Linton AH. 1974. *Microorganisms : Function, Form and Environment*. Edward Arnold (Publishers) Ltd. London 727p.

Holland JJ. ed. 1992. *Genetic Diversity of RNA Viruses*. Springer-Verlag, New York.

Kartikeyan S, Bharmal RN, Tiwari RP, Bisen PS.2007. *HIV and AIDS Basic Elements and Priorities*, Springer Verlag, pp428.

Safferman RS, Morris ME.1963. Algal virus: isolation. *Science* **140**, 679-680.

Safferman RS., Cannon RE, Desjardins PR, et al. 1983. Classification and nomenclature of viruses of cyanobacteria. *Intervirology* **19**, 61-66.

Shilo M. 1972. The ecology of cyanophages. *Bamidgeh* **24,** 76-82.

Singh P, Singh SS, Srivastava A, et al. 2012. Structural, functional and molecular basis of cyanophage-cyanobacterial interactions and its significance. *African Journal of Biotechnology* **11,** 2591-2608.

Tester S, Juillerat V, Doherr MG, Haase B, Polak M, Ehrensperger F, Leeb T, Zurbriggen A, Seuberlich **T.** 2009. Biochemical typing of pathological prion protein in aging cattle with BSE. *J Virol* **6,** 64.

Wilson WH, Joint IR, Carr NG, et al. 1993. Isolation and molecular characterization of five marine cyanophages propagated on Synechococcus spp. Strain WH7803. *Appl Environ Microbiol* **59**, 3736-3743.

Young R, Wang IN, Roof WD. 2000. Phages will out: strategies of host cell lysis. *Trends Microbial* **128**, 8.

Young R. 1992. Bacteriophage lysis: mechanism and regulation. *Microbiol Rev* **56,** 430-481.

Chapter **13**

Identification of Microbes

13.1 *ESCHERICHIA COLI*

13.1.1 Introduction

E. coli is the abbreviated version of *Escherichia coli,* a bacterium found in the lower intestines of mammals and birds discovered by Theodor Escherich in 1885. Although *E. coli* is a necessary intestinal bacterium that helps with the digestion of food, it can be very dangerous if ingested. If a person inadvertently ingests the bacteria, the resulting infection is called *E. coli* enteritis, which causes the small intestine to become inflamed. People can contract an *E. coli* infection by drinking contaminated water, eating fruit or vegetables that have been watered with contaminated water, drinking unpasteurized milk, or eating undercooked ground meat. The *E. coli* infection can also be caught by coming into contact with others who are infected or by working in environments where one might come into contact with human or animal faeces, such as farms, day-care centres, nursing homes, or hospitals. The most common way to contract an *E. coli* infection is by eating food that are not fully cooked. The symptoms of *E. coli* infection are primarily acute diarrhoea that may or may not be bloody, severe stomach cramps, bloating, and gas. While these are the most prevalent symptoms, many people infected with *E. coli* might also experience continuous abdominal pain, loss of appetite, fever, and in rare cases, vomiting.

E. coli are Gram-negative rods 2 to 4 μ by 0.4 MT Symbol commonly seen as coccobacillary form rarely as air filamentous form. These have 4-8 peritrichate flagella and are sluggishly motile. They do not form any spore. The facultative anaerobes develop into a circular raised and smooth colony emitting a faecal odour. Biochemically they are oxidase (–) catalase (+), capable of hydrolysing gelatin, reducing nitrate and can produce indole. It is incapable of utilising citrate and urea.

13.1.2 Isolation

The genus *Shigella* and *Escherichia* are closely related except for the inability of former to utilise Lactose and bromothymol blue agar is used as a selective medium.

13.1.3 Medium Composition (g/L)

Peptone 10, NaCl 5, yeast extract 5, agar 20, distilled water to 1L, pH 7. Autoclave and add the following sterilized solutions.

5% maranil solution 1 mL (addition of maranil (dodecylbenzolsulfonate) at a concentration of 0.005% inhibits swarming of *Proteus*), 50% sodium thiosulfate 2 mL, 1 % bromothymol blue 10 mL, 33% Lactose 25 mL, 33% Glucose 1.2 mL.

(i) Prepare serial dilution of 5 mL of the collected sample.
(ii) Inoculate 0.1 mL of highest dilution on the agar plate.
(iii) Isolate the colonies appearing on the agar plates.

13.1.4 Standard Analysis of Water

It consists of 3 steps: (i) Presumptive test (ii) Confirmed test (iii) Completed test

13.1.4.1 Presumptive test

1. Inoculate 10 mL portions of each water sample collected to 10 mL of lactose broth (double strength)'
2. Incubate at 37°C for 48 hrs.
3. Observe after 24 hrs, presence of gas showing a positive presumptive test; appearance of bubble in next 24 hrs shows a doubtful presumptive test; absence of gas formation indicates a negative confirmed test (Figure 13.1).

Composition of Lactose Broth: Beef extract 6 g, peptone 10 g, distilled water 1 L.

13.1.4.2 Confirmed test

1. Streak, using a very little inoculum, a plate of Eosine Methylene Blue agar Endo agar.
2. Inoculate about one loopful of culture from positive tube to Brilliant Green Lactose Broth.
3. Incubate at 37°C for 48 hrs.

Gas formation in lactose broth shows a positive confirmed test. The appearance of characteristic small, dark greenish or large pinkish mucoid colonies also confirm presence of coliform in the sample.

Eosine Methylene Blue Agar (EMB)

Peptone 10 g, lactose 5 g, Sucrose 5 g, K_2HPO_4 2 g, Eosin Y 0.4 g, Methylene blue 0.06 g, agar 15 g and distilled water to 1 L.

Endo Agar

Peptone 10 g, lactose 10 g, K_2HPO_4 3.5 g, Na_2SO_4 5 g, Basic fuchsin, agar 15 g and distilled water to 1 L.

Brilliant Green Lactose Bile Broth (BGLB)

Peptone 10 g, oxgall 20 g, lactose 10 g, brilliant green 13.3 mg and distilled water to 1 L final pH 7.2.

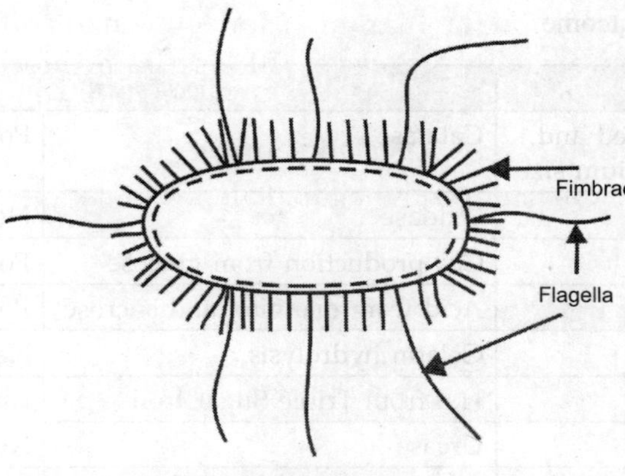

A cell of *Escherichia coli*

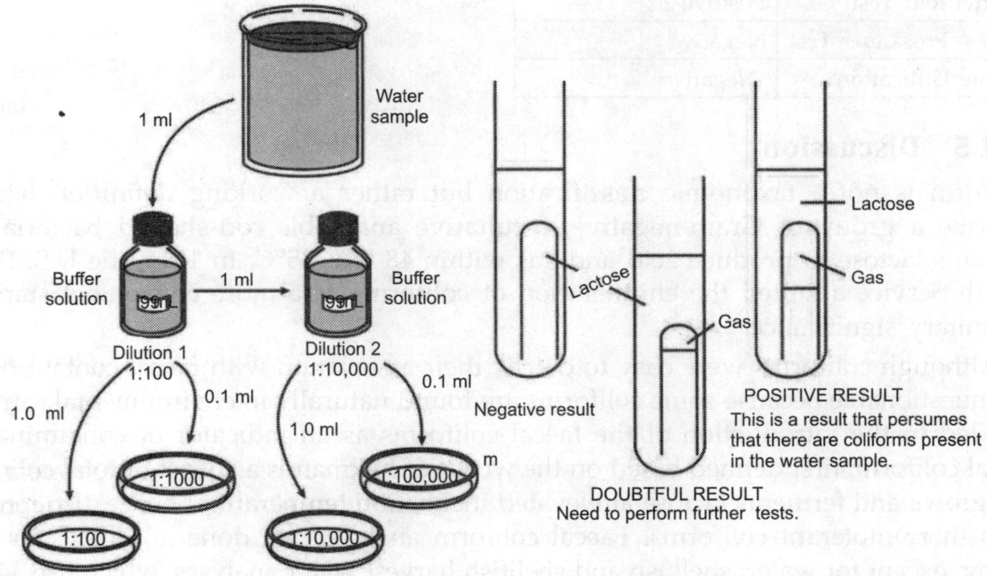

Figure 13.1 Flow diagram for dilution preparation and the presumptive test.

13.1.4.3 Completed test

1. Inoculate a loopful of bacterial cultural from EMB or ENDO AGAR PLATE to Lactose broth.
2. Streak a loopful of culture on to nutrient agar slant.
3. Incubate at 37°C for 48 hrs.
4. Gas formation in lactose broth and the appearance of Gram-negative bacteria on the plates shows a positive completed test (Figure 13.2).

Identification

The isolate tested Gram-negative and picked from nutrient agar slant should be subjected to following tests (Table 13.1).

Table 13.1 Summary of expected outcome

Morphological Test		*Biochemical*	
Colony:	Circular, raised and smooth, medium size	Catalase	Positive
Flagella stain	Positive	Oxidase	Negative
Capsular stain	Negative	Gas production from glucose	Positive
Gram stain:	Negative	Acid from mannitol and sucrose	Positive
Endospore	Negative	Gelatin hydrolysis	Negative
Motility	Positive	H_2S from Triple Sugar Iron	Negative
IMVIC Reactions		Urease	Negative
Indole production	Positive	Nitrate	Negative
Methyl Red Test:	Positive		
Voges- Proskauer Test	Negative		
Citrate Utilization	Negative		

13.1.5 Discussion

Coliform is not a taxonomic classification but rather a working definition used to describe a group of Gram-negative, facultative anaerobic rod-shaped bacteria that ferments lactose to produce acid and gas within 48 h at 35°C. In 1914, the U.S. Public Health Service adopted the enumeration of coliforms as a more convenient standard of sanitary significance.

Although coliforms were easy to detect, their association with faecal contamination was questionable because some coliforms are found naturally in environmental samples. This led to the introduction of the faecal coliforms as an indicator of contamination. Faecal coliform, first defined based on the works of Eijkman is a subset of total coliforms that grows and ferments lactose at elevated incubation temperature, hence also referred to as thermotolerant coliforms. Faecal coliform analyses are done at 45.5°C for food testing, except for water, shellfish and shellfish harvest water analyses, which use 44.5°C. The faecal coliform group consists mostly of *E. coli* but some other enterics such as *Klebsiella* can also ferment lactose at these temperatures and therefore, be considered as faecal coliforms. The inclusion of *Klebsiella* sp. in the working definition of coliforms diminished the correlation of this group with faecal contamination. As a result, *E.coli* has re-emerged as an indicator, partly facilitated by the introduction of newer methods that can rapidly identify *E. coli.*

Currently, all 3 groups are used as indicators but in different applications. Detection of coliforms is used as an indicator of sanitary quality of water or as a general indicator of sanitary condition in the food-processing environment. Faecal coliforms remain the standard indicator of choice for shellfish and shellfish harvest waters; and *E. coli* is used to indicate recent faecal contamination or unsanitary processing. Almost all the methods

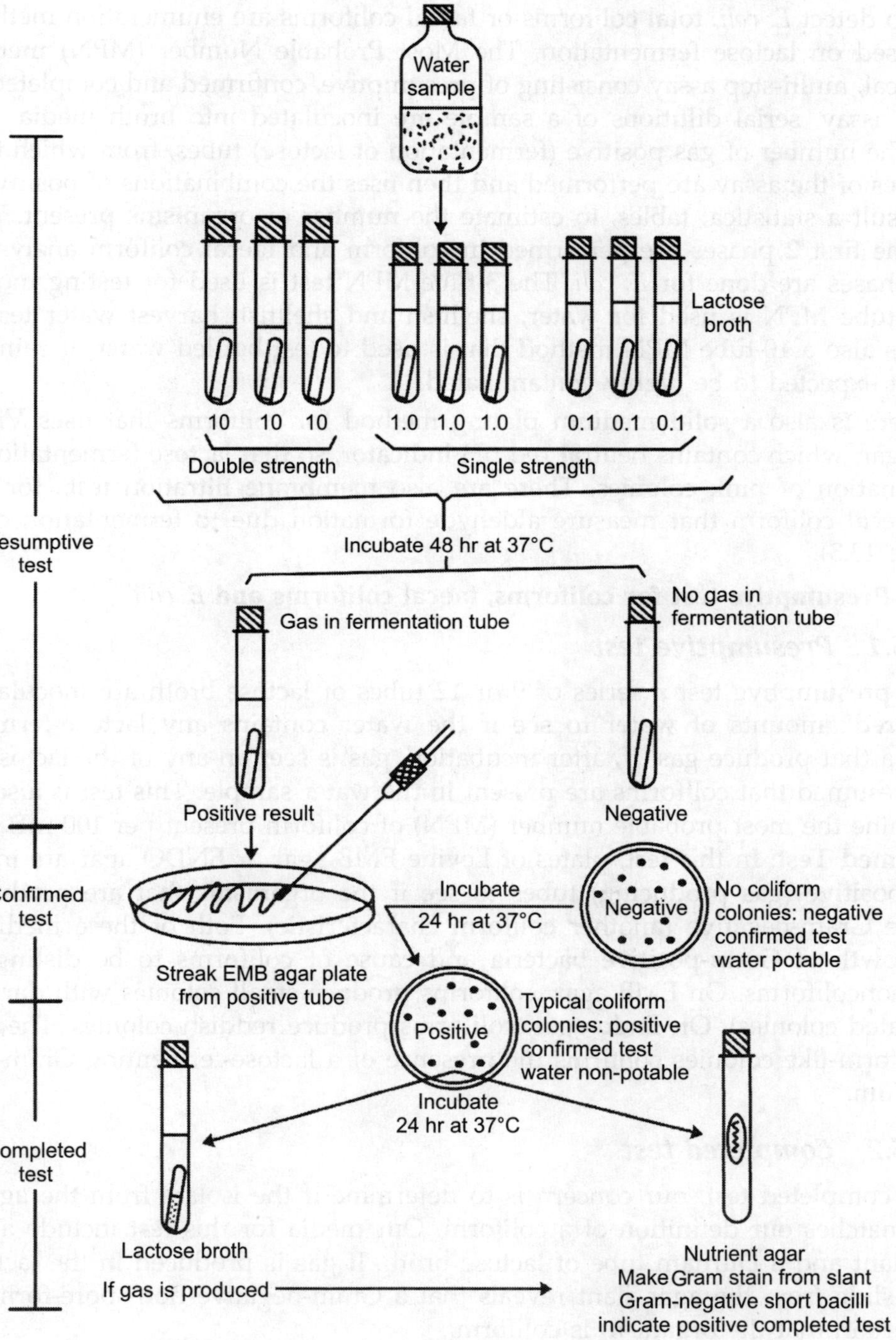

Figure 13.2 Flow diagram for the standard analysis of water.

used to detect *E. coli*, total coliforms or faecal coliforms are enumeration methods that are based on lactose fermentation. The Most Probable Number (MPN) method is a statistical, multi-step assay consisting of presumptive, confirmed and completed phases. In the assay, serial dilutions of a sample are inoculated into broth media. Analysts score the number of gas positive (fermentation of lactose) tubes, from which the other 2 phases of the assay are performed and then uses the combinations of positive results to consult a statistical tables, to estimate the number of organisms present. Typically only the first 2 phases are performed in coliform and faecal coliform analysis, while all 3 phases are done for *E. coli*. The 3-tube MPN test is used for testing most foods. The 5-tube MPN is used for water, shellfish and shellfish harvest water testing and there is also a 10-tube MPN method that is used to test bottled water or samples that are not expected to be highly contaminated.

There is also a solid medium plating method for coliforms that uses Violet Red Bile Agar, which contains neutral red pH indicator, so that lactose fermentation results in formation of pink colonies. There are also membrane filtration tests for coliform and faecal coliform that measure aldehyde formation due to fermentation of lactose (Figure 13.3).

MPN - Presumptive test for coliforms, faecal coliforms and *E. coli.*

13.1.5.1 Presumptive test

In the presumptive test a series of 9 or 12 tubes of lactose broth are inoculated with measured amounts of water to see if the water contains any lactose-fermentation bacteria that produce gas. If, after incubation, gas is seen in any of the lactose broths, it is presumed that coliforms are present in the water sample. This test is also used to determine the most probable number (MPN) of coliform present per 100 mL of water. **Confirmed Test:** In this test, plates of Levine EMB agar or ENDO agar are inoculated from positive (gas producing) tubes to see if the organisms that are producing the gas are Gram-negative (another coliform characteristic). Both of these media inhibit the growth of Gram-positive bacteria and cause of coliforms to be distinguishable from noncoliforms. On EMB agar, coliforms produce small colonies with dark centres (nucleated colonies). On Endo agar, coliforms produce reddish colonies. The presence of coliform-like colonies confirms the presence of a lactose-fermenting Gram-negative bacterium.

13.1.5.2 Completed test

In the completed test, our concern is to determine if the isolate from the agar plates truly matches our definition of a coliform. Our media for this test include a nutrient agar slant and a Durham tube of lactose broth. If gas is produced in the lactose tube and a slide from the agar slant reveals that a Gram-negative non-spore-forming rod, it is certain that the organism is coliform.

The completion of these three tests with positive results establishes that coliforms are present; however, there is no certainty that *E. coli* is the coliform present. The

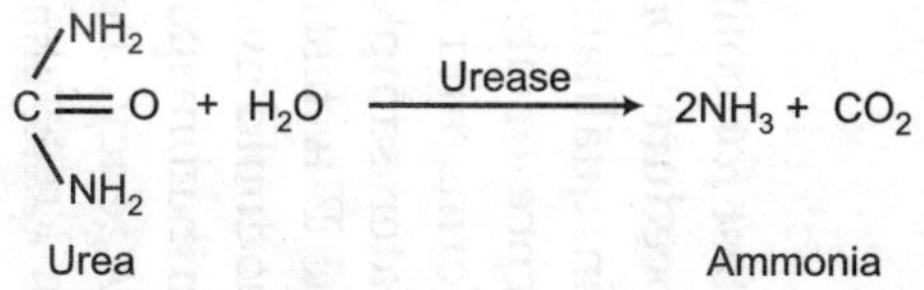

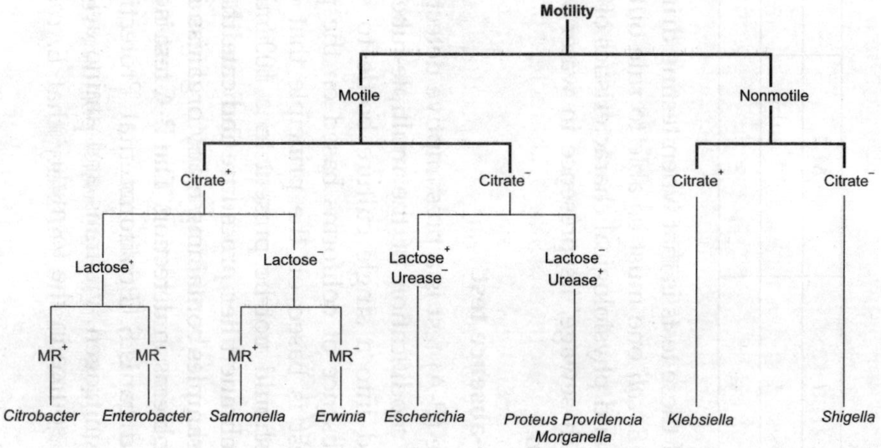

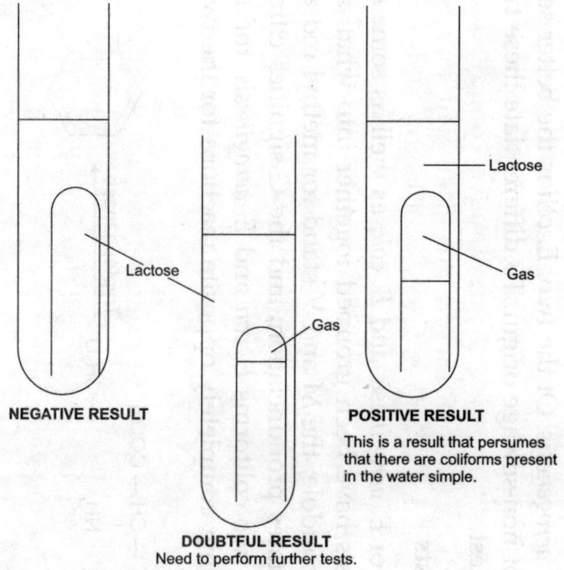

Figure 13.3 Separation outline for different Gram-negative rods and cocci.

organism might be *E. aerogenesis*. Of the two, *E. coli* is the better sewage indicator since *E. aerogenesis* can be of non-sewage origin. To differentiate these two species, one must perform the **IMViC** test.

13.1.5.2.1 IMViC tests

In the differentiation of *E. aeroginesis* and *E. coli,* as well as some other related species, four physiological tests have been grouped together into what are called the **IMViC** tests. The I stands for indole; the M and V stand for methyl red and Voges-Proskauer tests: (i) simply facilitates pronunciation; and the C signifies citrate utilization. In the differentiation of the two coliforms *E. coli* and *E. aeroginesis,* the test results appear as charted below, revealing completely opposite reactions for the two mechanisms on all tests.

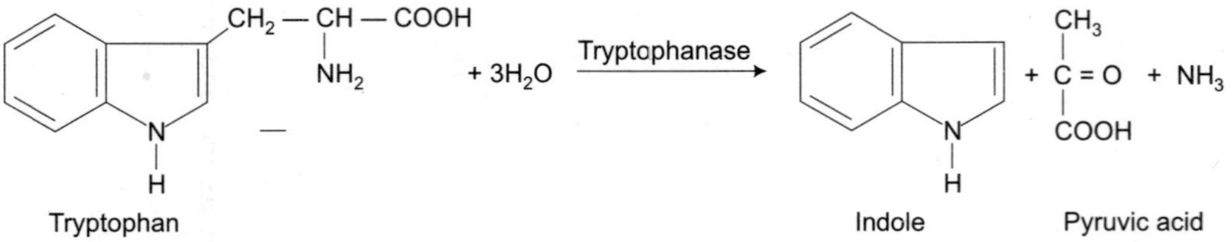

	I	*M*	*V*	*C*
E. coli	+	+	–	–
E. aerogenes	–	–	+	+

The significance of these tests is that when testing drinking water for the presence of the sewage indicator *E. coli* one must be able to rule out *E. aerogenes,* which has many of the morphological and physiological characteristics of *E. coli*. Since *E. aerogenes* is not always associated with sewage, its presence in water would not necessarily indicate sewage contamination.

13.1.5.2.2 Presence-absence test

The Presence-Absence (P-A) test is a presumptive detection test for coliforms in water. The test is a simple modification of the multiple-tube procedure. One test sample, 100 mL, is inoculated into a single culture bottle to obtain qualitative information on the presence or absence of coliforms based on the presence or absence of lactose fermentation. This test is based on the principle that coliforms and other pollution indicator organisms should not be present in a 100 mL water sample. Comparative studies with the membrane filter procedure indicate that the P-A test may maximize coliform detection in samples containing many organisms that could overgrow coliform colonies and cause problems in detection. The P-A test is described in standard methods for water testing and by U.S. Environmental Protection Agency. Beef extract and peptones provide the nitrogen, vitamins and amino acids in Presence-Absence Broth. Lactose is the carbon source in the formula.

13.2 *PSEUDOMONAS*

13.2.1 Introduction

Pseudomonas is a very large and important group of non-fermenting, Gram-negative bacteria which is found in substantial numbers as free living saprophytes in soils, fresh water, and many other natural habitats. Some species are pathogenic to humans, animals or plants. The basic morphological characters common to *Pseudomonas* are the rod-shaped cells; usually motile, the presence of flagella, a Gram-negative reaction and the absence of spores. *Pseudomonas* is one of the few microorganisms which can grow in simple mineral media at the expense of large varieties of low molecular weight carbon compounds. Identification is based on the common morphological and biochemical reactions, motility, the types of flagella, oxidase test, catalase test, presence of extracellular amylase, gelatinase and caseinase, oxidation fermentation test, nitrate reduction and production of pigments.

13.2.2 Isolations

13.2.2.1 Enrichment technique

(i) Pre-incubate the source material at 20°C for 24 hrs. Solid sources, e.g., egg, fish or poultry should be placed in phosphate buffer dilution tanks, shaken and incubated at 20°C for 24 hrs. This facilitates their distribution during streak plating.

(ii) Following pre-incubation, streak the source material on novobiocin-penicillin cycloheximide agar medium.

13.2.2.1.1 Composition of novobiocin-penicillin cycloheximide

Novobiocin 45 µg, penicillin G 44.9 mg (75,000 units), cycloheximide 75 mg.

Mix 3 mL of absolute ethanol dilute with 50 mL of sterilized distilled water. Add to 1 litre sterilized, tempered *Psedomonas* agar F. *Psedomonas* produce pyocyanin (water soluble) and fluorescein pigment. The latter fluoresce when exposed to UV light. These pigments are produced on P agar and F agar medium.

13.2.2.1.2 *Pseudomonas* nutrient agar F medium

Peptone 20 g, K_2HPO_4 1.5 g, Magnesium sulfate 1.5 g, Agar 15 g, distilled water to 1 L final pH 7.2 +/– 0.2 (at 25°C).

Store prepared media below 8°C, protected from direct light. Store dehydrated powder, in a dry place, in tightly-sealed containers.

3.2.2.1.3 *Pseudomonas* nutrient agar P medium

Peptone 20 g, $MgSO_4.7H_2O$ 1.4 g, potassium sulfate 10 g, glycerol 10 mL, agar 15 g, distilled water to 1 L adjust pH to 7.2.

Peptone provides the carbon and nitrogen necessary for bacterial growth. It enhances the production of pyocyanin by the addition of magnesium chloride and potassium sulfate and inhibits the formation of fluorescein. Both pigments diffuse from *Pseudomonas* colonies into the medium on which they grow. Some Pseudomonas strains elaborate both pigments, while others elaborate only one of the two. Pyocyanin on this medium is a blue colour. Glycerol serves as an energy source and also helps to promote pyocyanin production. It is not intended to be used as an isolation medium for pathogens from clinical specimens, but may be used for differentiation of isolates obtained from clinical specimens on other media.

Optional CFC Selective Supplement (Fluka 53477) can be added to make the media selective. Prepare dilution series of the sample material (dilution factor 10). Pipet two 1 mL aliquots from each dilution step and inoculate the plate using pour plate method. Incubate the plates up to 72 hours at 20-25°C.

13.2.2.1.4 *Pseudomonas* cetrimide agar

Enzymatic digest of gelatin 20 g, $MgCl_2$ 1.4 g, potassium sulfate 10 g, glycerol 10 mL, cetrimide (cetyltrimethylammonium bromide) 0.3 g, agar 15 g, distilled water to 1L, Final pH 7.2 ± 0.2 at 25°C.

13.2.3 Procedure

(i) Inoculate 1 g of compost soil in about 10 mL of sterilised distilled water.

(ii) Prepare serial dilution and spread 0.1 mL of highest dilution on nutrient agar plate.

(iii) It is advisable that the sample must first be inoculated on *Pseudomonas* P agar, *Pseudomonas* F agar, *Pseudomonas* cetrimide agar, and then transferred to nutrient agar to check the contamination.

Determine the count of the fluorescing bacteria under the UV lamp and the total microbial count.

DICHOTOMY CHART FOR IDENTIFICATION OF *Psedomonas* sp.

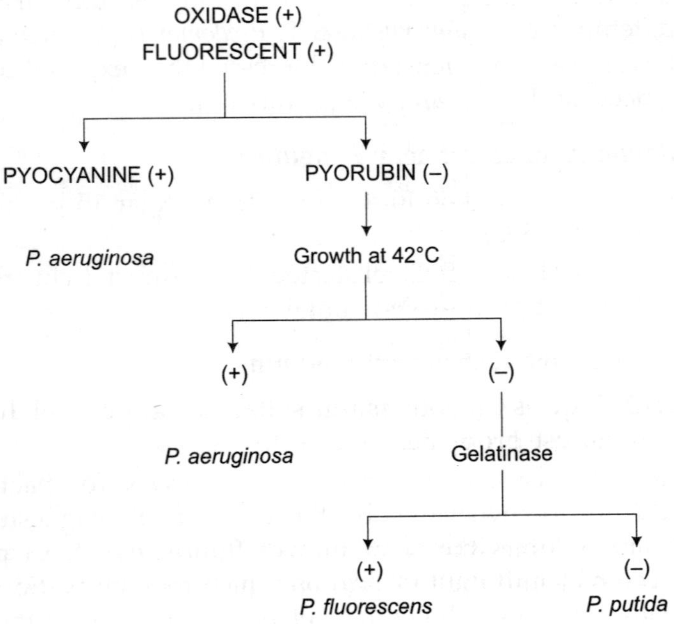

1. Suspend above ingredients in 1L distilled water and add 10 g of glycerol.
2. Heat the medium to boiling in order to dissolve completely.
3. Dispense in tubes or flasks and autoclave.
4. Allow to cool in slanted position.
5. Dispense sterilised medium onto sterilised Petri plates.
6. Inoculate tubes and plates both by surface streaking.

Composition of cetrimide agar (for isolation of *P. aeruginosa*)

13.2.3.1 Method

1. Suspend above ingredients in 1 litre distilled or de-ionised water. Add 10 mL glycerol and heat to boiling to dissolve.
2. Dispense in tubes or flasks.
3. Autoclave for 15 minutes at 15 psi.

13.2.4 Results

The round shaped (small dot) smooth margined colony isolated from nutrient agar plate should be Gram-negative, motile with rod-shaped cells, devoid of spore, sheath or flagella, positive catalase and oxidase reactions (Table 13.2).

Table 13.2 Summary of expected outcome

Morphological Test		*Biochemical*	
Colony:	Smooth, Round	Catalase	Positive
Flagella stain	Positive	Oxidase	Positive
Capsular stain	Negative	Caseinase	Negative
Gram stain:	Negative	Gelatinase	Negative
Endospore	Negative	Amylase	Negative
Motility	Positive	Gas production from glucose broth	Negative
Pigment Production	Positive	Acid from mannitol and sucrose	Positive
		Oxidation/Fermentation reaction	Oxidation
		Growth at 42°C	Positive
		H_2S from Triple Sugar Iron	Negative
		Urease	Negative
		Nitrate	Negative

The colony morphology is smooth margin, round pale brown or cream colour negative reaction with glucose broth, negative gelatinase, caseinase and H_2S production.

13.2.5 Discussion

The genus *Pseudomonas* is characterised by morphological properties such as the rod-shaped cells which are either straight or slightly curved in one plane, the presence of polar flagella Gram-negative reaction, the absence of spores, sheaths and prosthecas (Figure 13.4). They display chemo-organotrophic metabolism with a typical respiratory system having no fermentation and photosynthesis activity, inability to fix dinitrogen, and capacity for growth at the expense of a large variety of organic substances. Catalase positive reaction and glucose broth with bromo thymol blue indicator and Durham tube (no change in the dye-colour by acid fermentation products and gas production) indicate that the isolate has an aerobic obligately respiratory type of metabolism with oxygen as the terminal electron acceptor. The presence of cytochrome C in the electron transport chain is ascertained by positive oxidase reaction. "O" reaction in the oxidase/fermentation test indicates that the isolates are able to catabolise the glucose by oxidation. The production of fluorescent pigment is a definite reaction for *Pseudomonas*. The function of these pigments is to chelate iron in the environment for the use of the cells, and its overproduction is stimulated when the culture is transferred to iron poor medium. Out of 100 known species of *Pseudomonas* only 8 produce fluorescent pigment. These 8 species are P. *aeruginosa, P. aureofaciens, P. choloraphis, P. fluorescens, P. putida* and among plant pathogens *P. syringae, P. viridiflava* and *P. cichorii*. According to Bergey's Manual, *P. putida* has synonyms such as *P. convexa, P. eisenbergii, P. incognita, P. ovalia, P. rugosa* and *P. striata.* They were isolated from soil and water having 62.5 mole %. G + C content.

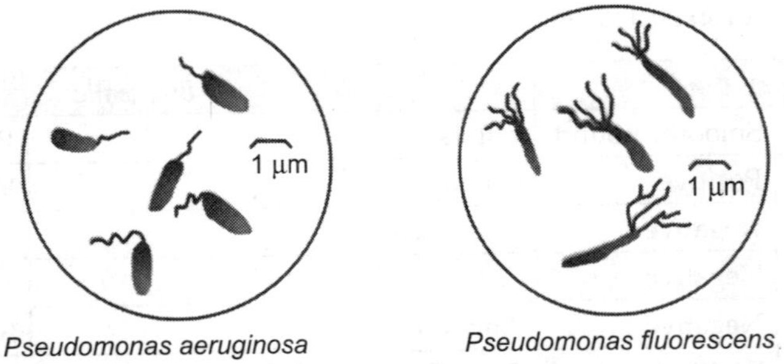

Figure 13.4 *Pseudomonas* morphology.

13.3 *PROTEUS*

13.3.1 Introduction

Proteus is a genus of Gram-negative Proteobacteria, a member of Enterobacteriaceae. *Proteus bacilli* are widely distributed in nature as saprophytes, being found in decomposing animal matter, in sewage, in manure soil, and in human and animal faeces. They are opportunistic pathogens, commonly responsible for urinary and septic infections, often nosocomial. *Proteus mirabilis* is part of the normal flora of the human gastrointestinal tract. It can also be found free living in water and soil. When this organism, however, enters the urinary tract, wounds, or the lungs it can become pathogenic. *Proteus mirabilis* commonly causes urinary tract infections and the formation of stones. It is a small Gram-negative bacillus and a facultative anaerobe.

Proteus sp. is characterized by its swarming motility. They have the ability to elongate themselves and secrete a polysaccharide when in contact with solid surfaces, making it extremely motile. The ability to swim over the surface of agar media, led to its name *Proteus* (a Greek deity, *Proteus* which could change its shape). *Proteus* is characterised by motile, Gram-negative bacilli, highly pleomorphic with spreading colonies (spreaders) actively proteolytic, ability to ferment maltose and other carbohydrates with acid and gas production but inability to ferment lactose. They have the dimension of 0.5 μ × 0.2 μ as short rods while up to 20 μ as filamentous forms. Spore and capsule absent, fimbriated with a facultative aerobic nature.

13.3.2 Isolation

1. Take 5 mL of sewage sample.
2. Prepare serial dilutions and spread 0.1 mL each of the above dilutions on to the previously made nutrient agar plates.
3. Transfer any swarming colony if visible to deoxycholate citrate agar medium.
4. Observe the plate for isolated colonies after 48 hrs incubation at 37°C. Swarming should stop by now.
5. Check the contamination by re-streaking on nutrient agar plates. The colonies should now gain their swarming nature.

13.3.2.1 Composition of deoxycholate citrate agar medium

Peptone 10 g, lactose 10 g, sodium deoxycholate 1 g, NaCl 5 g, K_2HPO_4 2g, ferric citrate 1 g, sodium citrate 1 g, neutral red 0.03 g, agar 15 g and distilled water 1 L

13.3.2.2 PPA slope medium

DL-phenylalanine 2 g, meat extract 3 g, Na_2HPO_4 1 g, NaCl 5 g, agar 1.2 g and distilled water to 100 mL.

13.3.3 Procedure

Sterilize tube (slants) at 10 psi for 10 mins and solidify as a long slant. Inoculate the slant heavily and incubate overnight. Add 4-5 drops of 10% (w/v) ferric chloride slope and allow running over the surface. Observe the production of dark green colour.

The isolate is subjected to following tests for identification.

1. PPA Slope Test (Phenyl-Pyruvic Acid Test) Henrickson (1950)

Phenylanine Solution: Dissolve 0.2 g DL-phenylalanine in 100 mL distilled water. Add 0.1 g merthiolate and 4 mL of 0.01% of phenol red. Turn it to pink colour (if necessary by adding 0.1 M sodium carbonate).

Method: Suspend the growth of an overnight agar culture in normal saline, transfer to a tube add 0.5 mL of phenylalanine solution. Shake the mixture well and incubate the tube (pH 2). Add about 6 drops of half saturated ferric ammonium sulfate Fe

$(NH_4SO_4)_2$ shake thoroughly and observe for the development of a deep green colour, when the test is positive. The depth of the colour indicates the degree of the positivity of the reaction (Figure 13.5).

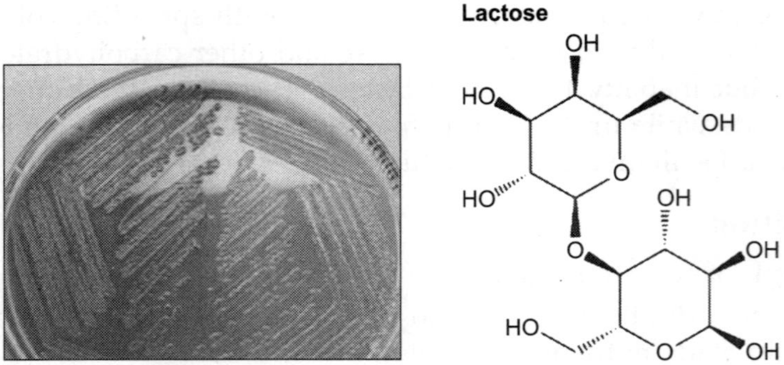

Figure 13.5 *Proteus vulgaris* MacConkey agar plate with an active bacterial culture.

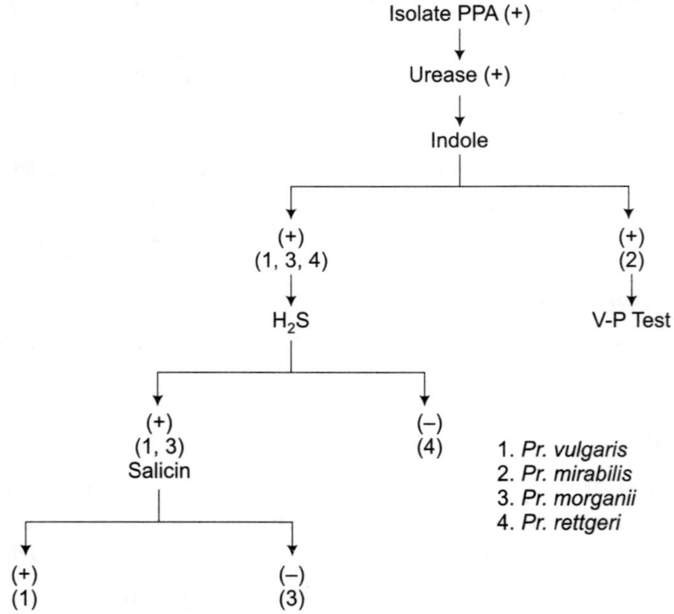

13.3.4 Results

Irregularly shaped, swarming colony turning to small isolated ones if isolated from sewage or manure can be taken as bacterium *Proteus*. These are small, Gram-negative, motile with small rod-like structure. No spore sheath or capsule is seen. Each cell is however fimbriated, flagellar staining is positive. Genus is a non-lactose fermenter but can produce acid and gas from glucose, maltose, salicin, mannitol, etc., depending upon species. Spreading on nutrient agar is stopped in presence of antiseptics, 5% alcohol (Table 13.3 and Table 13.4).

Table 13.3 Summary of expected outcome

Morphological Test		*Biochemical*	
Colony:	Pleomorphic	Catalase	Positive
Flagella stain	Negative	Oxidase	Negative
Capsular stain	Negative	Gas production from glucose	Positive
Gram stain:	Negative	Lactose	Negative
Endospore	Negative	Matose	Positive
Motility	Positive	Salicin	Positive
IMVIC Reactions		PPA Slope Test	Positive
Indole production	Positive	Gelatinase	Negative
Methyl Red Test:	Positive	H_2S from Triple Sugar Iron	Positive/ Negative
Voges-Proskauer Test	Negative	Urease	Positive
Citrate Utilization	Negative	Nitrate	Negative

Table 13.4 Non-lactose fermentation

Organism	*Indole*	*Mannito*	*H_2S*	*Salicin*	*VP*	*Maltose*	*Citrate*	*Gelatin*	*Glucose*
Pr. vulgaris	+	–	+	AG	–	AG	–	+	AG
Pr. mirabilis	–	–	+	AG	+	–	+	+	AG
Pr. morganii	+	–	+	–	–	–	–	–	AG
Pr rettgeri	+	–	–	AG	–	–	+	–	A

A Acid, AG-Acid and Gas, G-Gas

13.3.5 Discussion

The genus *Proteus* is characterised by morphological properties such as rod-shaped cells, presence of flagella, swarming nature, a Gram-negative response, absence of spores, sheathes, prosthecae or capsules and having chemoorganotropnic nature, ability to ferment glucose, mannose and maltose but inability to produce acid and gas from lactose. The properties of the four species commonly found species *viz. Pr. vulgaris, Pr. mirabilis, Pr. morganii,* and *Pr. rettgeri* are very distinct and the four can be distinguished using simple biochemical tests like indole production, urease, H_2S production, salicin fermentation, etc.

Key characters of *Proteus* sp. PPA +, Urease +, Differences in properties of *Pr. vulgaris, Pr. mirabilis, Pr. morganii, Pr. Rettgeri, Pr. mirabilis* cannot produce acetyl methyl carbinol and indole, the other 3 can break tryptophan to yield indole.

13.4 *SHIGELLA*

13.4.1 Introduction

Shigella is a large group of Gram-negative, facultative anaerobic, nonspore forming, non-motile, slender, rod-shaped genus of bacteria of about 0.5 μ × 2.5 m size, fimbriae present, pathogenic to humans and animals and cause bacillary dysentry closely related to *Salmonella*. Genus was isolated by Kiyoshi Shiga (1898) from a case of acute bacillary dysentery in Japan. They are unable to fix nitrogen, non-photosynthetic aerobic or facultative anaerobic, fermentative in nature. Isolation is achieved through enrichment method while identification is based on general biochemical tests such as Gram reaction, motility, oxidase catalase reaction, acid and gas production on glucose, lactose, sucrose and salicin, hydrolysis of gelatin and urea, utilization of citrate, and other IMViC reactions.

13.4.2 Materials and Methods

13.4.2.1 MacConkey agar

Dissolve 20 g of peptone, 5 g of sodium chloride, 5 g of sodiumtauro cholate in 1L distilled water by heating and adjust the pH to 8 with 40% NaOH. Solidify the medium by adding 15 g of agar and dissolve it by steaming. Clarify egg white at 55°C (white of 2 eggs suspended in 50 mL of water added to one litre of medium and steamed for 2 hrs) filter adjust pH to 7.4. Distribute in 100 mL quantities in 250 mL Erlenmeyer flasks. Melt agar before use and add 10% lactose solution in water, previously sterilised, add 1% final concentration of neutral red (dissolved in 50% alcohol) and pour into sterilised petri dishes under aseptic condition.

13.4.2.2 Selenite - F broth

Sodium hydrogen (acid) selenite 4 g, peptone 5 g, lactose 5 g, Na_2HPO_4 9.5 g, NaH_2PO_4 0.5 g, distilled water to 1 L adjust the pH to 7.2.

Distribute in aliquots of 10 mL in test tubes and sterilize by steam for at 10 psi 30 mins. A slight amount of precipitate may occur but it will not affect the medium.

13.4.2.3 Direct isolation

1. Inoculate about 1 mL of sewage sample in 10 mL of distilled water.
2. Prepare a serial dilution of the suspension.
3. Spread 0.1 mL of highest dilution on the MacConkey/EMB agar plate.
4. Incubate for 2 days at 37°C.
5. Restreak on to nutrient agar plate to check contamination.
6. Repeat the procedure till a single type of colony is obtained.

13.4.2.3.1 Enrichment I

1. Repeat procedure 1 to 2 of direct isolation.
2. Inoculate 0.1 mL of above dilution in selenite - F broth
3. Incubate for 2 days at 37°C.
4. Take a loopful and streak on MacConkey agar plate.
5. Repeat procedure 4 till a single type of colony is obtained.

13.4.3 Identification of Species

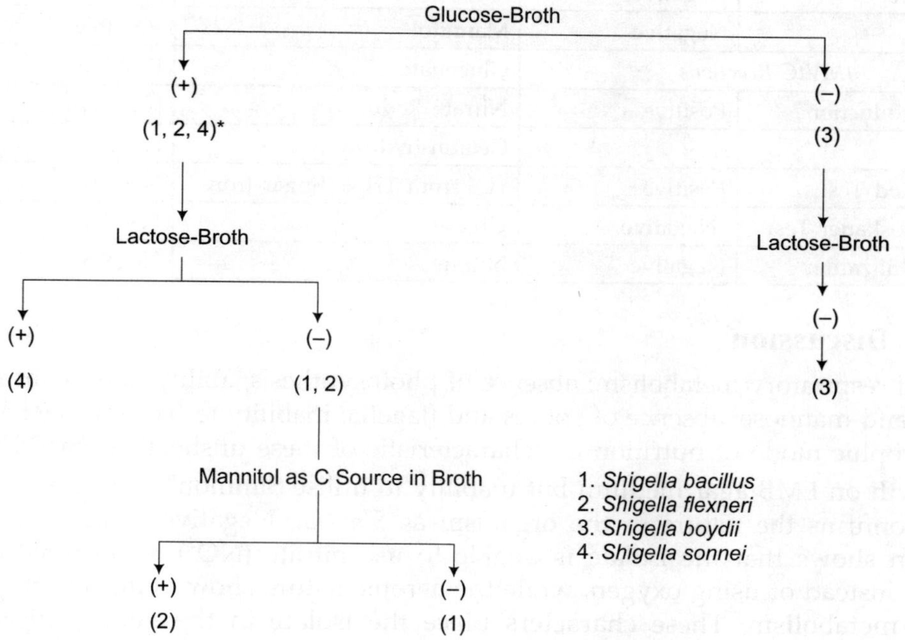

Dichotomous chart for identification of species

13.4.4 Results

A translucent colony picked up from MacConkey agar (a selective media for *Shigella)* bearing rod-shaped cells without spores and flagella stand as a prominent candidate for the *Shigella* genus. The positive response to catalase and oxidase test, inability to form acetyl methyl carbinol (V-P negative), utilise citrate, unable to ferment salicin are some of the important characters of above genus (Figure 13.6, Table 13.5).

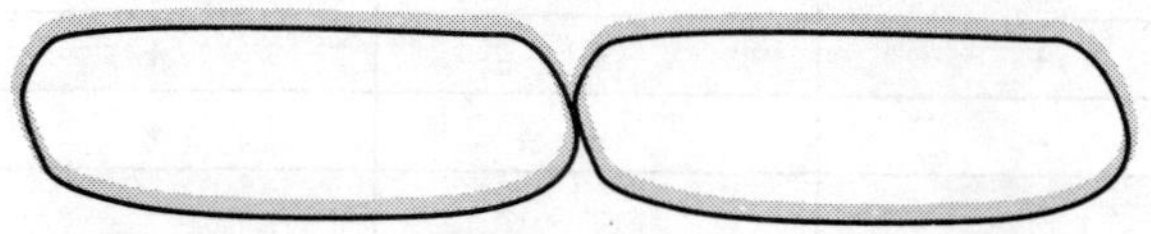

Figure 13.6 Division of *Shigella*.

Table 13.5 Summary of expected outcome

Morphological Test		*Biochemical*	
Colony:	Large, translucent	Catalase	Positive
Flagella stain	Negative	Oxidase	Negative
Capsular stain	Negative	Gas production from glucose	Positive
Gram stain:	Negative	Lactose	Variable

Contd...

Contd...

Endospore	Negative	Matose	Positive
Motility	Negative	Mannitol	Variable
IMVIC Reactions		Gluconate	Positive
Indole production	Positive	Nitrate Reduction	Negative
		Gelatin hydrolysis	Negative
Methyl Red Test:	Positive	H_2S from Triple Sugar Iron	Negative
Voges-Proskauer Test	Negative	Urease	Positive
Citrate Utilization	Negative	Nitrate	Negative

13.4.5 Discussion

A typical respiratory metabolism; absence of photosynthesis; ability to ferment sucrose, glucose and mannose; absence of spores and flagella; inability to fix dinitrogen; chemo-organotrophic mode of nutrition are characteristic of these unsheathed bacilli *Shigella.*

Growth on EMB agar medium but inability to utilise Simmon's citrate and ferment salicin confirms the nature of the organism as *Shigella.* Negative response to nitrate reduction shows that the isolate is unable to use nitrate (NO_3^-) as terminal electron acceptor instead of using oxygen, while the aerobic nature show obligately respiratory type of metabolism. These characters place the isolate in the genus *Shigella.* The identification of species is based on the ability to ferment various carbon sources since *Shigella bacilli* can ferment only glucose while *Shigella sonei* can ferment all the three glucose, lactose and mannose. *Shigella boydii* is very distinct in having tendency to ferment indole and glucose broth while *Shigella flexneri* is limited to glucose and mannitol (Table 13.5 and Table 13.6).

Table 13.6 Key characters of *Shigella* sp.

Carbon	*Shigella flexneri*	*Shigella boydii*	*Shigella sonei*	*Shigella bacilli*
Glucose	+	–	+	+
Mannitol	+	+	+	–
Lactose	–	–	+	–
Indole	–	+	–	–

The G + C content ranges from 49-53 mole %. Among the 4 species of *Shigella* each has its serotype.

13.5 *SALMONELLA*

13.5.1 Introduction

Salmonella (genus *Salmonella*), group of rod-shaped motile, flagellated variant, Gram-negative, facultatively anaerobic, nonsporing and devoid of capsules forming bacteria belonging to family Enterobacteriaceae with 1,400 different strains.Their principal habitat is the intestinal tract of humans and other animals. Some species exist in animals without causing disease symptoms; others can result in any of a wide range of mild to serious infections termed salmonellosis in humans. Common ailments caused by *Salmonella* infection include paratyphoid, typhoid fever, and food poisoning. It is transmitted through contaminated food, egg, and dairy products. The symptoms of stomach pain, diarrhea, nausea, fever, and vomiting begin in 6 to 72 hours after infection. An enormous amount of contaminated food products, especially poultry products, increases the incidence of salmonellosis. The patient is given plenty of fluids to prevent dehydration. Preventive measures include thorough cooking of animal-derived foodstuffs, refrigerating foods, and avoiding raw eggs. Most human infections with *Salmonella* result from the ingestion of contaminated food or water. This pathogen has several serotype which are given a specific name evoking the role S. *typhi* (agent of typhoid), S. *paratyphi* (para typhoid bacterium), S. *abortusovis* (agent of abortion in sheep) and S. *typhimurium* (agent of disease of mouse). The biochemical tests are fairly constant in this group unlike *Proteus*. These are non-lactose and non-sucrose fermenters. Acid and gas is usually produced with a wide range of sugar. Indole is not formed and gelatin and urea are not decomposed. Aerobic or facultatively anaerobic by nature, forms an initial smooth colony which changes to rough on prolonged incubation. Gram-negatives non-sporing (2.5 μ × 0.5 μ) actively motile peritrichate flagella.

13.5.2 Isolation

1. Take about 5 mL of sewage sample.
2. Prepare serial dilutions.
3. Inoculate 0.1 mL of the highest dilution on MacConkey agar and deoxycholate citrate agar plates.
4. Incubate at 37°C for 2 days.
5. Colourless, translucent, pale, non-lactose fermenting colonies will develop.

13.5.2.1 MacConkey agar

Ingredients per litre of deionized water

Bacto peptone 17 g, proteose peptone 3 g, lactose 10 g, bile salt mixture 1.5 g, sodium chloride 5 g, agar 15 g, neutral red 0.03 g and crystal violet 0.001 g.

13.5.2.2 Composition of deoxycholate citrate agar medium

Ingredients per litre of deionized water.

Peptone 10 g, lactose 10 g, sodium deoxycholate 1 g, NaCl 5 g, K_2HPO_4 2 g, ferric citrate 1 g, sodium citrate 1 g, neutral red 0.03 g, agar 15 g

13.5.2.3 Enrichment method

(i) Sample can also be inoculated in Wilson and Blair medium which inhibits growth, of *E. coli* while *Salmonella* develops first as greenish colony with black centre which enlarges to fill the entire colony after 24-48 hrs of incubation. The colonies are round, about 2 mm in diameter black and surrounded by metallic sheen.

(ii) In, S-S agar, it grows as a black smooth colony turning to rough on ageing.

13.5.2.4 Composition of Wilson and Blair's medium

(a) Dissolve 30 g of bismuth-ammonium citrate scales in 100 mL distilled water. To this, add 100 g anhydrous sodium sulphite. Make final volume 1000 mL with distilled water and add 21 g of sodium phosphate cryslals (Na_2HPO_4. $12H_2O$). Boil and cool. This mixture can be stored for months.

(b) Iron citrate brilliant green mixture: Mix 1% solution of ferric citrate with 1% brilliant green and make volume with distilled water to 25 mL. The mixture will keep for months.

To 1 litre of sterilised nutrient agar aseptically add 70 mL of (A) and 4 ml of (B). Mix and plate.

13.5.2.5 Composition of SS-agar

Beef extract 5 g, lactose 10 g, bile salt mixture 8.5 g, sodium citrate 8.5 g, sodium. thiosulfate 8.5 g, ferric citrate 1 g, distilled water to 1 L and adjust pH to 7.

13.5.3 Identification

Identification of genus is quite easy owing to its coliform nature supplemented by their inability to ferment lactose, sucrose, dextrose and manitol and motile in nature.

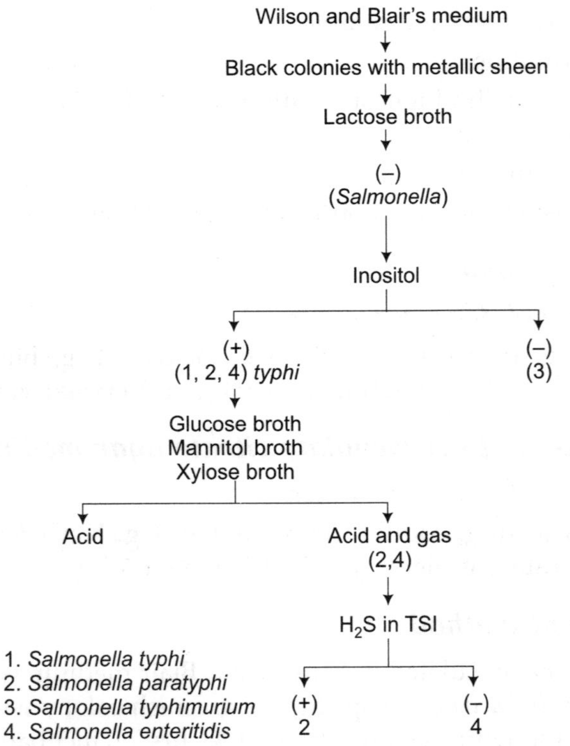

Dichotomous chart for the identification of species

13.5.4 Results

Smooth colonies with black centres during early periods of incubation turning to rough with metallic sheen on S-S agar and growing as white translucent mass of bacteria on MacConkey agar medium, while developing, as black colony on Wilson and Blair's medium can be taken as *Salmonella*.

The usual biochemical reactions alone are not sufficient to establish the identity of *Salmonella*. The negative urease reaction, no evolution of gas or acid production in lactose broth, however, are the strict *Salmonella* characteristics (Figure 13.7 and Table 13.7).

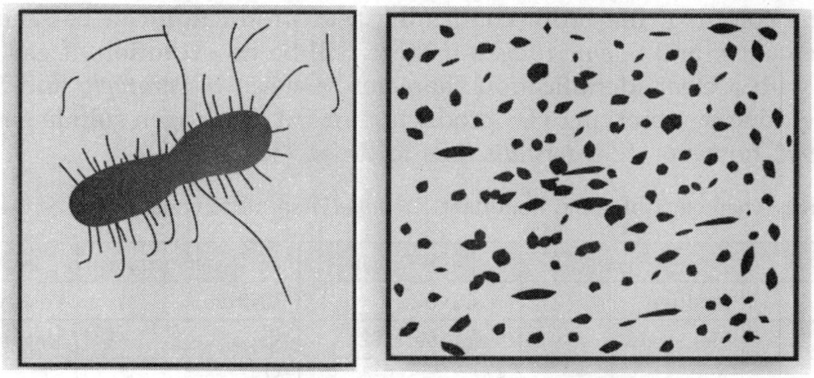

Figure 13.7 Culture of *Salmonella typhi* on agar plate.

Table 13.7 Summary of expected outcome

Morphological Test		*Biochemical*	
Colony on:		Glucose Broth	Positive
MacConkey Agar	White	Mannitol Broth	A/AG
SS-Agar	Black	Xylose Broth	A/AG
Wilson and Blair's Medium	Black	Inositol Broth	+/–
Flagella stain	Positive	Lactose Broth	Negative
Capsular stain	Negative	Mannitol	Variable
Gram stain:	Negative	Gelatin hydrolysis	+/–
Endospore	Negative	H_2S from Triple Sugar Iron	+/–
Motility	Positive	Urease	Negative
IMVIC Reactions		Methyl Red test	+/–
Indole production	Negative	A indicates for acid only; AG acid and gas both; (+/–) variable	
Citrate Utilization	Negative		
Voges-Proskauer Test	Negative		

13.5.5 Discussion

Principal characters of *Salmonella* are peritrichous flagella, absence of spores, prosthecae capsules, Gram-negative reaction, chemoorganotrophic mode of nutrition, ability to ferment glucose and mannose; strictly non lactose fermenter. Microscopic observations and the sudden change in colony texture from smooth to rough, with the increasing length of incubation an obligately respiratory metabolism with oxygen as the terminal electron acceptor do not leave any doubt that the isolate is *Salmonella.*

The establishment of species depends on the basis of ability to ferment glucose, sucrose, xylose and inositol. The indicated sugars are broken down with evolution of acid by S. *typhi,* S. *paratyphi* and S. *enteritidis* but not by S. *typhimurium* species. A change in the colour of the broth confirming acid production can be seen in case of S. *typhi,* S. *paratyphi* and S. *enteritidis* but there will be no evolution of gas in S. *typhii.* This leaves with a clear identification character between *S. paratyphi* and *S. enteritidis* which can be done by observing H_2S production in TSI. Hydrogen sulfide gas is evolved by S. *paratyphi,* however, S. *enteritidis* fails to do so (Table 13.8).

Table 13.8 Key characters of some important *Salmonella* sp. differentiation

Test	*S. typhi*	*S. paratyphi*	*S. typhimurium*	*S. eneteritidis*
Motility	Positive	Positive	Positive	Positive
Lactose	Negative	Negative	Negative	Negative
Glucose	A	AG	AG	AG
Mannose	A	AG	AG	AG
Dulcitol	Negative	AG	AG	AG
Xylose	A	AG	AG	AG
Inositol	Negative	AG	AG	Negative
H_2S	+/-	Positive	Positive	Negative
Urease	Negative	Negative	Negative	Negative

13.6 *VIBRIO*

13.6.1 Introduction

Vibrio is fairly large and important group of Gram-negative, comma-shaped rods 1.5 µ to 3 µ in length by 0.2 µ to 0.4 µ in breadth arranged singly or in pairs. It is actually motile, showing darting or scintillating movement with the single large flagellum. In old culture involution forms occur in the form of S or spirillium. The aerobes can grow on ordinary media and develop into small colonies within 24 hrs. They can ferment a variety of sugars but remain passive towards lactose. The prominent species is *Vibrio cholerae,* the casual organism of cholera. Oxidase positive reaction shows presence of oxygen as an essential factor for final transport of the electron through ETC. Indole production and urea hydrolysis are positive reactions however, gelatin liquefaction is not performed.

13.6.2 Isolation

13.6.2.1 Composition of medium

Thiosulfate – citrate – bile salt – sucrose (TCBS agar medium) Kobayashi *et al.,* 1963.

Yeast extract 5 g, peptone 10 g, sucrose 20 g, sodium thiosulfate 10 g, sodium citrate 3 g, Ox-gall 5 g, sodium chloride 10 g, ferric citrate 1 g, thymol blue 0.04 g, bromothymol blue 0.04 g.

The medium is heated to dissolve the ingredients cooled to 45°C and poured into Petri dishes. It should not be autoclaved.

13.6.2.2 Direct isolation

1. Take 1 g of sludge sample.
2. Dissolve it in distilled water and prepare serial dilutions.
3. Spread 0.1 mL of highest dilution on the TCBS agar plate.
4. Observe the development of colonies.

13.6.2.3 Reed's modification of Wilson-Blair's medium

It is Wilson-Blair's medium to which 1% sodium chloride and mannose are added and final pH is adjusted to 9.2. It has the advantage that it inhibits the growth of non-agglutinable *Vibrios.*

Composition of Wilson-Blair's medium has been described for the isolation of *Salmonella.*

13.6.2.4 Pellicle formation in peptone water identification

The identification of *Vibrio,* a member of family Spirillaceae is accomplished using the usual biochemical reactions.

13.6.3 Results

The translucent, smooth, bluish colonies on nutrient agar developing into a pellicle like growth on peptone water which can be taken as an aspirant for the genus *Vibrio.* Each cell bearing a spiral rod-shaped structure with a single long flagellum confirms the morphological characters of *Vibrio.* The positive oxidase reaction, indole production, Simmon's citrate utilization, gelatin hydrolysis, glucose fermentation, etc., are some key characters of *Vibrio* group (Figure 13.8 and Table 13.9).

Table 13.9 Summary of expected outcome

Morphological Test		*Biochemical*		*Fermentation*	
Colony;	Cream coloured	Gelatin hydrolysis	Positive	Glucose Broth	AG
Gram stain	Negative	Oxidase	Positive	Sucrose Broth	Negative
Flagella stain	Positive	Inositol Broth	+/-	Salicin	Negative
Endospore	Negative	Lactose Broth	Negative	Inositol Broth	AG

Contd...

Contd...

Capsular stain	Negative	Mannitol	Variable	Lactose Broth	Negative
Motility	Positive	Citrate Utilization	Positive (Delayed)	Methyl Red test	+/-
IMViC Reactions		H_2S from Triple Sugar Iron	Negative	A indicates for acid only; AG acid and gas both; (+/–) variable	
Indole production	Negative	Urease	Negative		
Citrate Utilization	Negative	Growth in peptone water (a) +8% NaCl (b) –NaCl: (c) Growth at 42°C:	Pellicle Positive Negative Positive		
Voges-Proskauer Test	Negative				

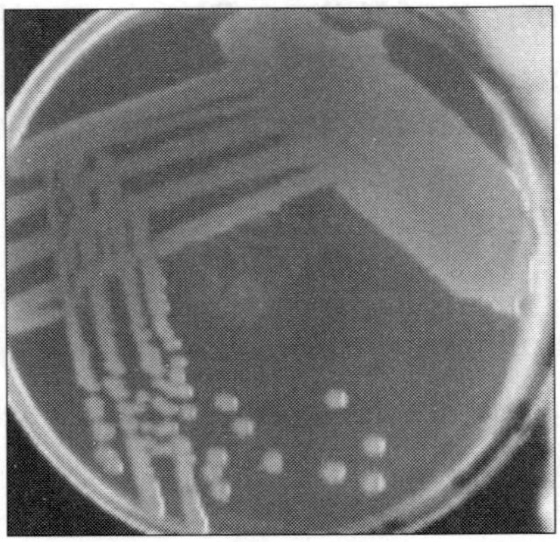

Figure 13.8 24 h old colonies of *Vibro parahaemolyticus* on TCBS.

13.6.4 Discussion

The rod-shaped cells, with a long single flagellum, thin cell wall, low peptidoglycan content, if observed in an isolate, it should be taken as *Vibrio*. The positive oxidase reaction confirms oxygen as an ultimate electron acceptor in the ETC the production of H_2S in TSI, negative response to the urease and a positive gelatinase reaction. Inability to ferment inositol and salicin, formation of pellicle in peptone water broth confirms that the isolate is *Vibrio*. The enrichment media used is TCBS agar, gelatin agar and the peptone water broth. The G + C content varies from 40-50 mole %.

13.7 *STAPHYLOCOCCI*

13.7.1 Introduction

The genus *Staphyloccous* is a group of Gram-positive, spherical bacteria that forms irregular cellular aggregation or grape-like clusters but seldom, form chains when grown in liquid medium (Figure 13.9). They produce characteristic pigment ranging in colour from orange to golden yellow to pink to white. All pathogenic strains are coagulase positive while non-pathogenic species are not. They are also β-haemolytic. *S. aureus* is pathogenic. In human beings disorders caused by pathogenic *Staphylococci* are cellulitis, pustules, boils, carbuncles, impetigo and secondary infections. One of the most common types of food poisoning is *Staphylococcal* food poisoning caused by enterotoxin excreted in food. Among the most serious staphylococcal infections are septicemia, meningitis, pneumonia and osteomyelitis. *Staphylococci* can be differentiated from *Micrococci* on the basis of anaerobic fermentation of glucose. All pathogenic species are mannitol fermenters.

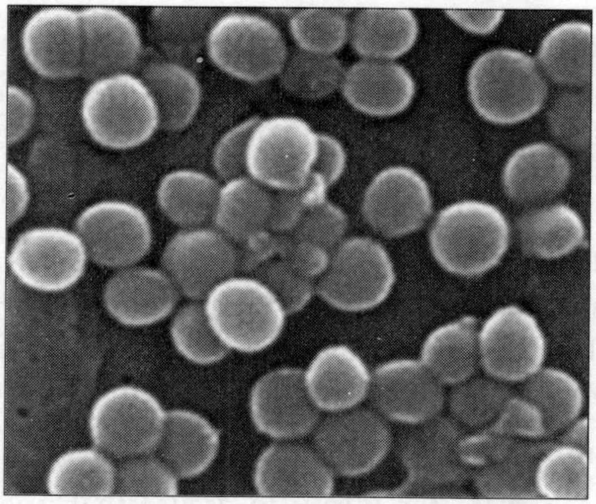

Figure 13.9 Electron micrograph of *Staphylococci.*

13.7.2 Isolation

They are found on skin and mucous membranes of warm blooded animals and in some food and curing brines.

Simplest enrichment procedure for *Staphylococcus* is to culture the source material anaerobically in a medium containing 7% NaCl. The high osmotic concentration of this medium inhibits osmotically sensitive organisms. The culture is purified by streak plating from the enrichment broth and selecting a colony composed of organisms that are catalase positive, Gram-positive cocci and ferment glucose anaerobically. It produces golden yellow to orange pigment on plates.

13.7.3 Procedure

13.7.3.1 Enrichment

(i) Inoculate the source material in 7% NaCl broth and seal the tube with vaspar or vaseline.

(ii) Incubate at 30°C for 2 days.

(iii) Streak culture from enrichment broth onto the plate count agar plate and incubate at 30°C for 2 days.

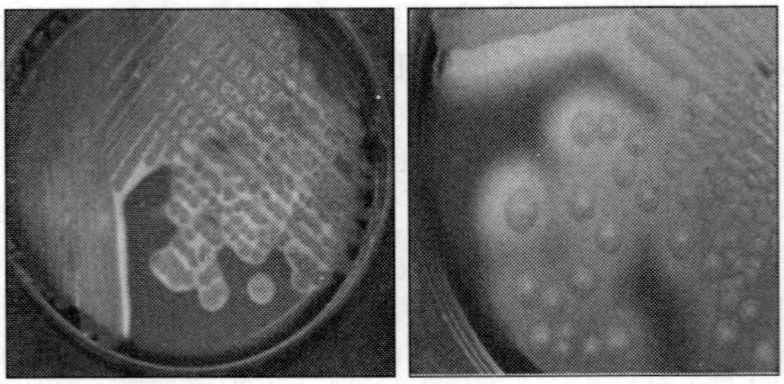

Figure 13.10 Blood agar plates and hemolysis steps in the coagulase test enlarged view on hemolysis.

13.7.3.2 Composition of media

13.7.3.2.1 Plate count agar media (ingredients per litre of deionized water)

Tryptone 5 g, Yeast extract 2.5 g, Glucose 1 g, Agar 15 g

13.7.3.3 Isolation

1. Examine the isolated colonies on streak plate for catalase test.
2. Prepare a wet mount for the study of the morphology of the bacterium.
3. Emulsify a small bit of colony in sterile broth and streak on a plate count agar plate. Incubate both at 30°C for 24 hrs.

13.7.3.4 Confirmation

1. Examine the streak plate for purity to assure that a single colony type was transferred.
2. Prepare a Gram stain.
3. Inoculate a brmocresol purple glucose broth from the overnight broth culture and incubate at 30°C for 5 days.

13.7.3.5 Biochemical tests

13.7.3.5.1 Coagulase activity

Principle: Coagulase positive *Staphylococci* synthesize an enzyme coagulase that forms a clot from normal blood plasma. Clot formation constitutes a positive test.

Method

1. The blood used is prevented from clotting by addition of citrate or oxalate or versene buffers, and the blood cells in suspension are sedimented out by centrifugation. The plasma is employed in test and add 0.5 mL of versenated/ buffered plasma to test tube.
2. Now add enough cells of the exponential culture under study to give a turbid suspension.
3. Incubate at 37°C for 1 hr and examine (Figure 13.11).

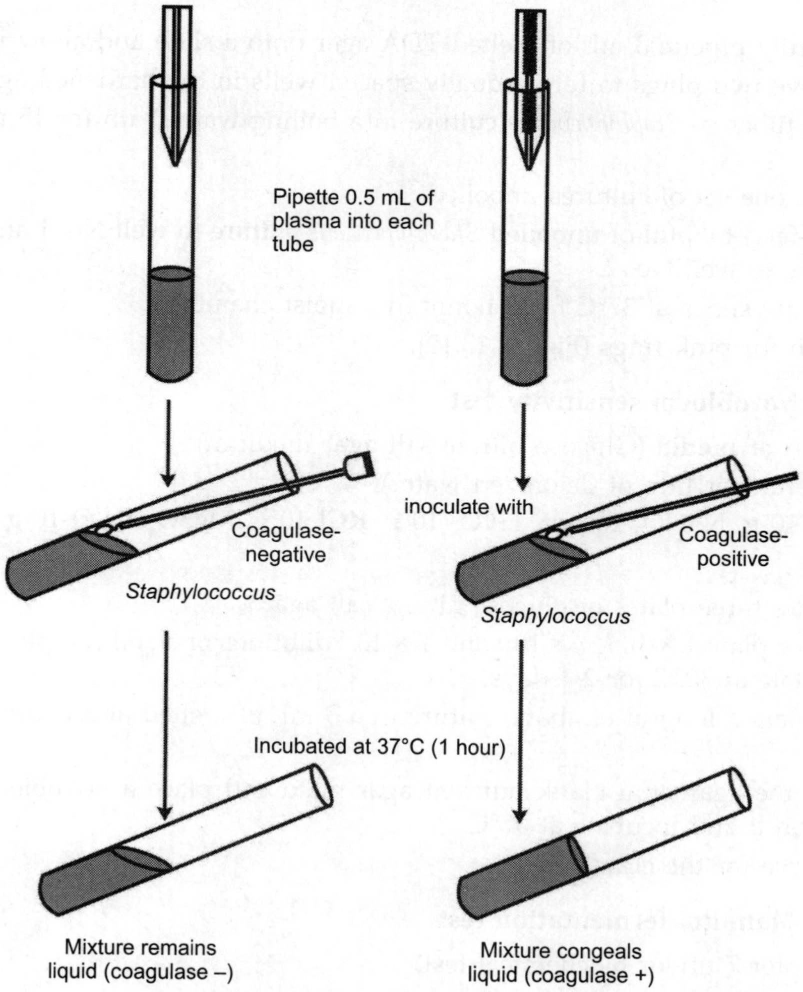

Figure 13.11 Steps in the coagulase test.

13.7.3.5.2 DNase test

In addition to coagulase reaction most infectious *Staphylococcus* produce a thermostable enzyme nuclease that depolymerase DNA. This enzyme, the DNase, can be tested using agar containing dissolved DNA and a dye, toluidine blue O which is exposed to a heated culture of organism. Presence of nuclease develops pink zones where DNA has been hydrolysed.

Composition of Tuluidine blue DNA agar medium (TDA)

(Ingredients per litre of deionized water)

DNA 0.3 g, $CaCl_2$ (0.01 M) 1 mL, NaCl 1 g, agar 10 g in 0.05 M tris buffer pH 9.0. After boiling the mixture to completely dissolve the DNA and agar, add 3 mL of 0.1 M toluidine blue O.

Method

1. Carefully pipette 3 mL of melted TDA agar onto a slide and allow it to harden.
2. Remove two plugs to form equally spaced wells in the hardened agar.
3. Place tubes of *Staphylococcus* culture in a boiling water bath for 15 minutes and cool.
4. Leave one set of cultures unboiled.
5. Transfer a loopful of unboiled *Staphylococcus* culture to well No. 1 and the boiled culture to well No. 2.
6. Incubate slides at 37°C for 2 hours in a moist chamber.
7. Watch for pink rings (Figure 13.12).

13.7.3.5.3 Novobiocin sensitivity test

Composition of media (Glucose nitrate salt agar medium)

(Ingredients per litre of deionized water)

Glucose 10 g, $NaNO_3$ 10 g, K_2HPO_4 10 g, KCl 10 g, $MgSO_4.7H_2O$ 10 g, agar 15 g.

Method

1. Prepare three plates of glucose nitrate salt agar.
2. Surface plate 1×10^{-3}, 1×10^{-4} and 1×10^{-5} dilutions of a soil sample and incubate the plate at 30°C for 2-4 days.
3. Inoculate a loopful of above culture in a 3 mL of melted and tempered nutrient agar.
4. Pour the agar on a blank nutrient agar plate and place a novobiocin antibiotic disc on it and incubate at 37°C.
5. Observe for the clear zones.

13.7.3.5.4 Mannitol fermentation test

(See chapter 7 under biochemical test)

13.7.3.5.5 Haemolysis test

Composition of media (Blood agar medium)

(Ingredients per litre of deionized water)

Tryptone 10 g, beef extract 3 g, NaCl 5 g, agar 15 g

Procedure

Sterilize after cooling to 45°C add 5% fresh defibrinated blood.

(i) Melt a tube (15 mL) of blood base agar and maintain it at 45°C.

(ii) Temper a tube of sterile blood in water bath at 45°C.

(iii) Transfer aseptically 1 mL of blood to the cooled tube of agar and mix with the pipette.

(iv) Mix by rotating the plate, avoiding any bubble formation.

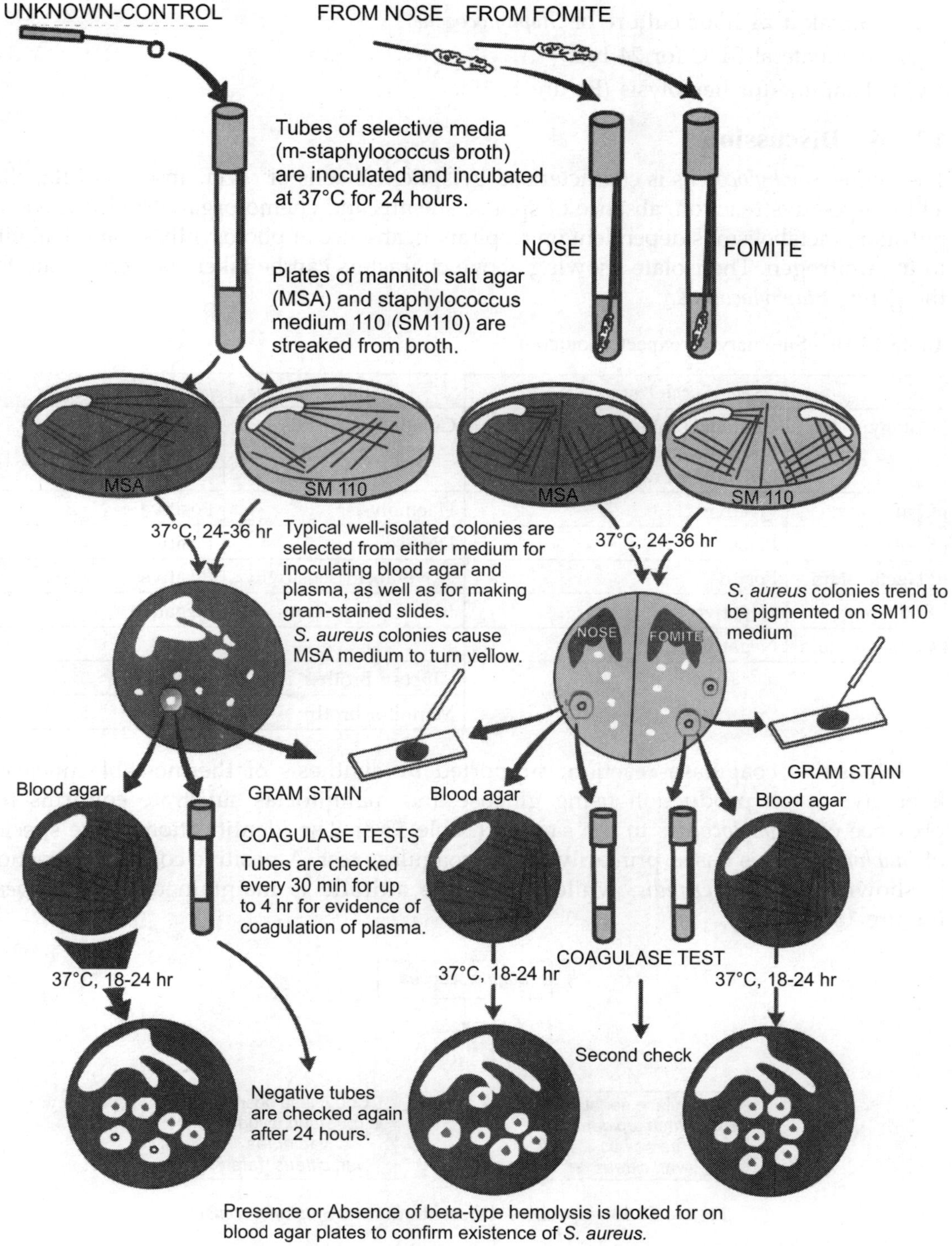

Figure 13.12 Flow diagram for presumptive identification of staphylococci.

(v) Streak a 24 hour culture of *Staphylococcus.*
(vi) Incubate at 31°C for 24 hrs.
(vii) Examine for hemolysis (Figure 13.10).

13.7.4 Discussion

The genus *Staphylococcus* is characterised by spherical cells or cocci, absence of flagella, a Gram-positive reaction, absence of spores, sheaths and chemo-organotrophic mode of nutrition, metabolism is dependent on respiration, absence of photosynthesis and inability to fix dinitrogen. The isolate showing above characters can be taken as a candidate for the genus *Staphylococcus.*

Table 13.10 Summary of expected outcome

Morphological Test		*Biochemical*	
Colony:	Circular, pinheaded, convex, Entire, translucent colourless On nutrient agar plate	Gelatin hydrolysis	Positive
Gram stain	Positive	Haemolysis:	Positive
SS-Agar	Black	DNAse	Positive
Flagella stain	Positive	Novobiocin sensitivity	Positive
Endospore	Negative	Indole production	Negative
Capsular stain	Negative	*Fermentation*	
		Glucose Broth	Positive
		Mannitol Broth	Positive

A positive coagulase reaction, supported by synthesis of thermostable nuclease haemolysis, gas production using glucose and mannitol as substrate confirms the presence of *Staphylococcus* in the sample (Table 13.8). The identification of the species of *Staphylococcus* is based primarily on the coagulase test. A positive coagulase reaction is shown by *S. epidermidis,* while a negative attitude is maintained by S. *pyogens* (Figure 13.13).

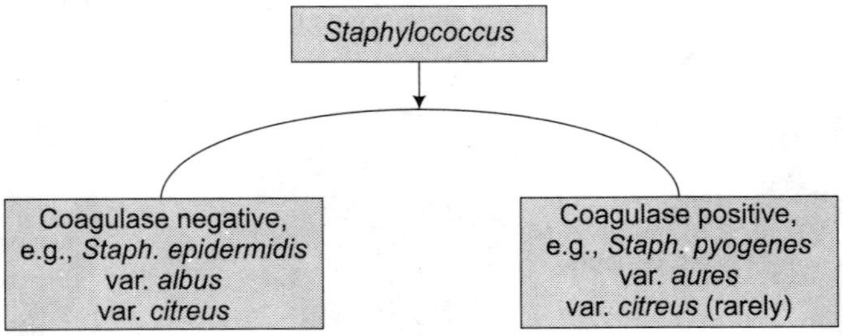

Figure 13.13 Specifies differentiation of *Staphylococcus.*

13.8 *STREPTOCOCCI*

13.8.1 Introduction

Streptococci differ from staphylococci in that they are arranged primarily in chains rather than in clusters. In addition to causing many mixed infection with staphylococci, the streptococci can also, separately, cause diseases such as sore throat, scarlet fever, impetigo (bacterial endocarditis), rheumatic fever and *Glomerulonephritis*. Several species are normal inhabitants of the pharynx. They can also be isolated from the surface of the teeth, saliva, skin, colon, rectum, and vagina. The streptococci of greatest medical significance are *S. pyogenes, S. agalactiae,* and *S. pneumoniae*. Of lesser importance are *S. faecalis, S. faecium,* and *S. bovis*. The genus *Streptococci* is characterised by spherical or ovoid cells; which form chains and are non-motile and non-sporing (Figure 13.14). The growth is scanty in absence of native protein in the media. They produce characteristic hemolytic changes in blood. Only acid is produced by carbohydrate fermentation. *S. pyogenes* is also called pathogenic hemolytic *Streptococci* of lancefield group A and the most common infective syndrome is an acutely inflammed throat with or without exudate called tonsilitis. There is acute inflammation often with oedema and fever, wound, bums, chronic skin lesions may become infected with S. *pyogens,* which may extend and cause cellulitis or get into blood stream.

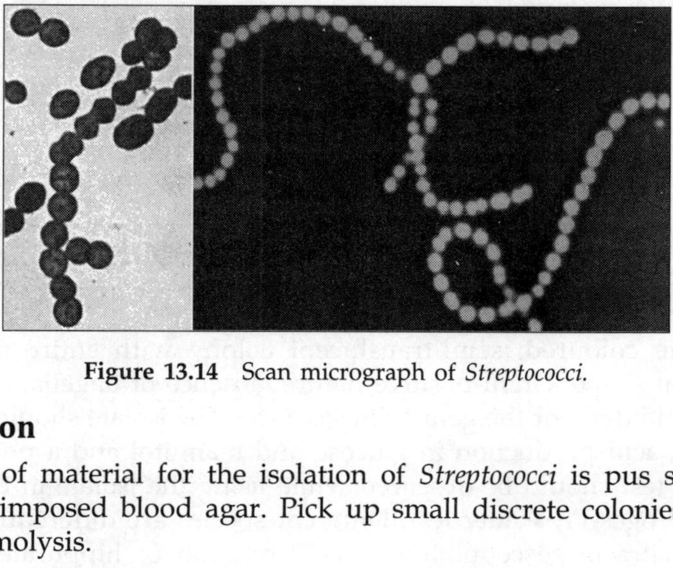

Figure 13.14 Scan micrograph of *Streptococci.*

13.8.2 Isolation

The best source of material for the isolation of *Streptococci* is pus sample. Plate the sample on superimposed blood agar. Pick up small discrete colonies surrounded by clear zone of hemolysis.

13.8.2.1 Identification

13.8.2.1.1 Composition of media

Bile solubility test

Grow the isolate in meat infusion broth.

Meat infusion broth (Ingredients per litre of deionized water)

Lean meat 500 g, peptone 10 g, NaCl 5 g, pH adjusted to 7.4.

1. Add minced meat to 500 mL of water and extract for 24 hrs in the cold. Strain and boil filtrate at 100°C for 15 min, filter and add peptone and NaCl and make up the volume to 1L. Adjust pH to 7.4.
2. Take about 1 mL broth in each of the two test tubes.
3. To one add (0.1 mL) of sodium taurocholate solution (0.5%) and to the other add 0.1 mL of saline.
4. Incubate at 37°C for 30 min and observe for lysis.
5. Confirm by negative staining.

13.8.2.2 Results

If the isolate is *Streptococci,* it should develop as small, smooth margined, circular, colony semi-translucent with a low convex surface and glistening appearance.

It should produce acid from glucose besides a negative response towards methyl red, bile solubility test, catalase and Voges-Proskauer Test (Table 13.11).

Table 13.11 Summary of expected outcome

Morphological Test		*Biochemical*	
Colony:	Cream coloured	Catalase	Negative
Gram stain	Positive	Haemolysis:	Positive
Flagella stain	Negative	Glucose Both	Positive/Acid
Endospore	Negative	Mannitol Broth	Positive/Acid
Capsular stain	Negative	Nitrate Reduction	Positive
Motility	Negative	Bile Solubility	Negative
		Methyl red test	Negative
		Voges-Proskauer Test	Negative

13.8.2.3 Discussion

A circular, cream coloured, semi-translucent colony with entire margin, each cell showing spherical shape, Gram-positive nature, absence of flagella, capsule and spore is a promising candidate for the genus *Streptococcus.* The isolate should show a negative catalase reaction, acid production in glucose and mannitol and a positive nitrate test. A strict negative test should be observed while using the isolate in the bile solubility, methyl red and Voges-Proskauer test. Different species are differentiated on the basis of hemolysis, bacitracin susceptibility, CAMP reaction or hippurate hydrolysis, SXT sensitivity, bile hydrolysis, tolerance to 6.5% NaCl, optochi susceptibility and bile solubility described in Figure 13.15 and Table 13.11.

13.9 *XANTHOMONAS*

13.9.1 Introduction

Xanthomonades are rod shaped, aerobic, polarly flagellated monotrichous bacteria. They have an antagonistic association with higher plants and develop into yellow

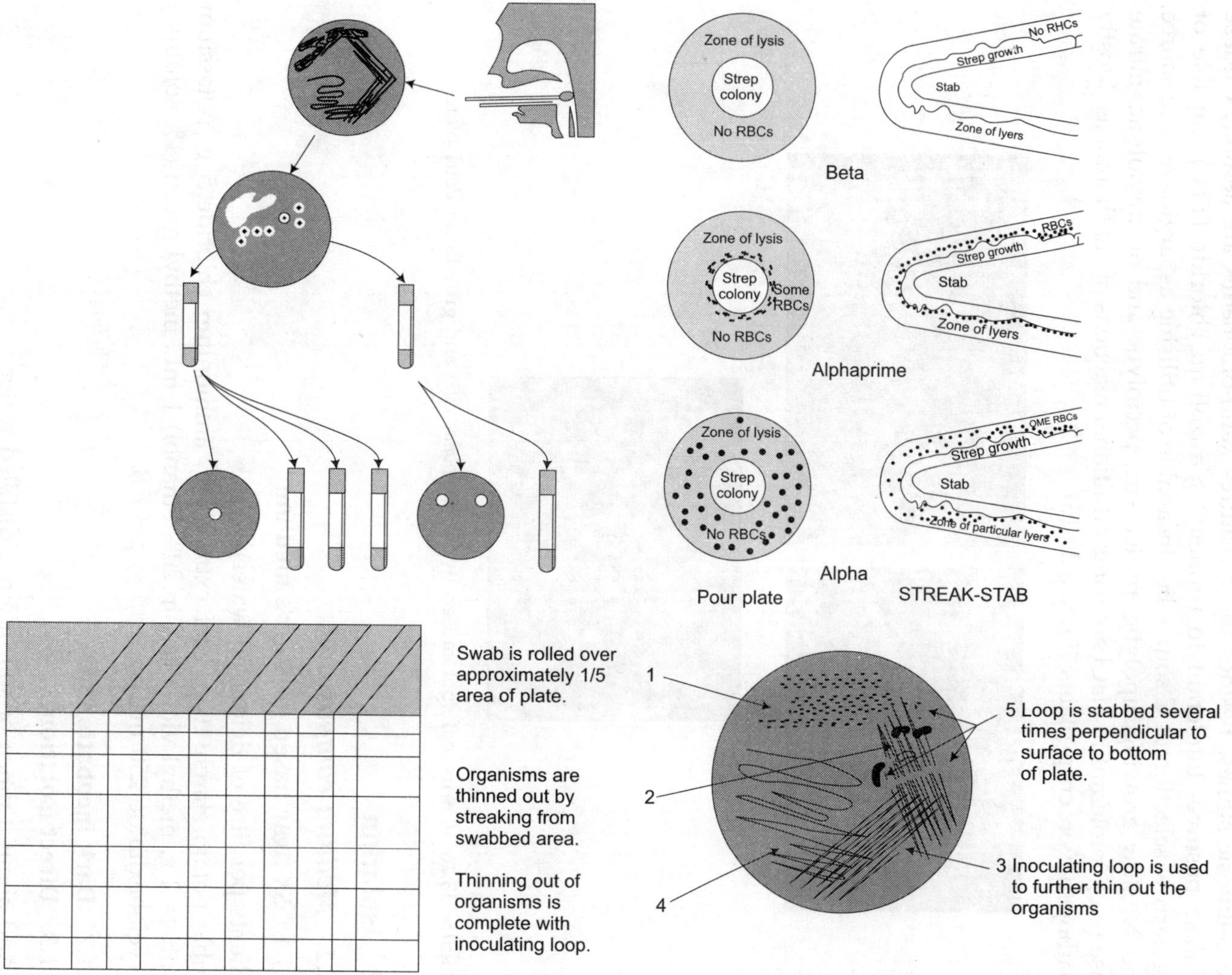

Figure 13.15 Flow chart for the identification *Streptococci* sp., streak-stap procedure of blood agar inoculation, comparison of hemolysis types on pour plates and streak-stab plates, blood agar inoculation techniques for the CAMP; bacitracin, and SXT tests and physiological test for the Streptococcal differentiation.

coloured mucoid colonies on sugar containing media. This characteristic feature is the unique property of the brominated, aryl-polyene esters (xanthomonadian) pigments and the cellulosic slimes. Besides these features xanthomonades are aerobic, oxidase and catalase positive. Intolerant to triphenyl tetrazolium chloride (TTC), capable of utilising narrow selection of C source but incapable of utilising aspargine as sole source of C and N, slow growing, lipolytic, proteolytic, pectolytic and or amylolytic, nitrate reductase (-ve), intolerant to NaCl are some additional reactions of xanthomonas. Mostly phytopathogenic on crop plants (Figure 13.16).

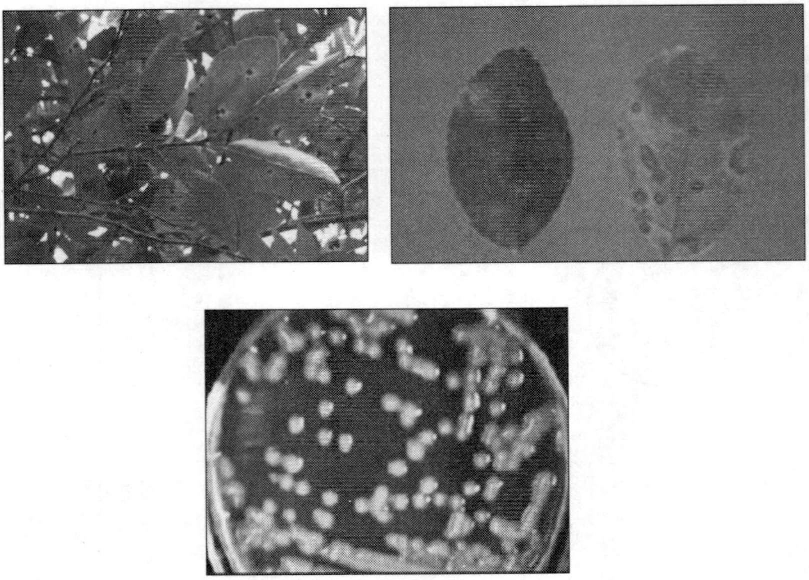

Figure 13.16 Citrus leaf spot caused by *Xanthomonas citri* and growth on Petri plates.

13.9.2 Isolation

13.9.2.1 Medium composition

13.9.2.1.1 SX Agar based on Starr's medium

(Ingredients per litre of deionized water)

Soluble potato starch 10 g, beef extract 1 g, ammonium chloride 5 g, potassium diphosphate 2 g, methyl violet (1% in 20% ethanol) 1 mL, methyl green (1% solution) 2 mL, cycloheximide 250 mg, agar 15 g, pH 6.8.

13.9.2.1.2 Dark incubation

13.9.2.1.3 Direct isolation

1. Add about a pinch of soil in 5 mL distilled water.
2. Make serial dilutions.

3. Spread 0.1 mL of highest dilution on SX agar plates.
4. Incubate for 3 days at 30°C and observe the colony morphology.
5. Starch digestion can be observed after 5 days of growth.

13.9.2.2 Morphological and biochemical tests

After 3 days of incubation small transluscent colonies surrounded by clear zone of 4.5 mm diameter as a result of starch digestion within 5 days. The colonies grow into 3 mm diameter mucoid, glistening, convex circular with entire margin, translucent with bluish purple centres. The starch digestion zone has a diameter of nearly 8-10 mm.

13.9.3 Results

Xanthomonas are expected to develop into 4-5 mm diameter zone of starch digestion. Within 5 days of incubation, they develop into mucoid colonies (Figure 13.16) with entire margins and translucent with bluish purple centres. The starch digestion zone is of around 8-10 mm in diameter. It is a Gram-negative, polarly flagellated with positive catalase and oxidase reactions. Nitrate reduction should be negative.

13.9.4 Discussion

Xanthomonas can be identilied by morphological properties such as rod-shaped cells, presence of polar flagella a Gram-negative reaction, slime formation by the unique property of xanthomonadian pigment, brominated aryl-polyene esters, absence of photosynthesis and inability to fix nitrogen.

Growth on selective SX agar medium, positive reactions towards catalase (+ve), oxidase (+ve), amylase (+ve), caseinase (+ve). Absence of colour change in glucose broth with bromocresol purple indicator and durham tube confirms its aerobic nature. Nitrate reductase (–ve), ammonia evolution (–ve), amylolytic (+ve), pectolytic activity (+ve), with intolerance to NaCl assign the isolate to be the genus of *Xanthomonas* (Table 13.12) with G + C content ranging from 63-71 as per Bergey's manual.

Table 13.12 Summary of expected outcome

Morphological Test		*Biochemical*	
Colony:	Yellow	Catalase	Positive
Gram stain	Positive	Oxidase	Positive
Flagella stain	Positive	Caseinase	Positive
Endospore	Negative	Gelatinase	Positive
Capsular stain	Negative	Amylase	Positive
Motility	Positive	Pectinase	Positive
Pigment Production	Positive	Nitrate reduction	Negative
		2% NaCl Tolerance	Negative

13.10 *SERRATIA*

13.10.1 Introduction

Serratia is a member of Enterobacteriacae having synonymus like Chromobacterium. Micrococcus found in water, air, bread, milk, potatoes, etc. They are motile with 2-4 peritrichate flagella. Colonies are 1-2 mm in diameter after a growth of 48 hrs with low convex smooth, translucent surface and vary from colourless to bright red in colour. Some species and biotypes produce a non-diffusible red pigment digiosin which is 2 methyl-3 amyl-methoxy prodigiosene.

13.10.2 Identification

Identification of genus is based on usual biochemical tests. Presence of pigments and use of selective media play an important role in this genus. Five species have been defined, these are *S. marcescens, S. proteamaculans, S. plymutheca, S. marinorubra* and *S. odorifera.*

Each bacterium is a small, Gram-negative rod, oval or coccobacillary, 0.7 μ to 1 μ by 0.7 μ with rounded ends 2-4 peritrichate flagella, devoid of spores. *Serratia* can be identified by usual biochemical coliform reactions with some specific tests like gluconate, iodoacetate, xylosidase tests. Production of extracellular gelatinase, lecithinase and DNase are special characteristics of *Serratia.*

13.10.2.1 Isolation

Pigmented species give pink to red colonies on nutrient agar. While the non-pigmented ends appear as opaque, whitish, mucoid or transparent, smooth colonies.

The salt tolerance and relatively low growth temperature of all *Serratia* may help to devise a means of enriching media. A nutrient broth containing 4% NaCl and incubated at 15-20°C will allow multiplication of *Serratia* but not of *Enterobacter cloacae* or *Aeromonas hydrophila.*

Capryllate Thallus (CT) Agar medium has been devised for the selective isolation of all *Serratia* species and biotypes. Thus CT agar derived from M-70 minimal medium is made as follows (Starr et al., 1976).

13.10.2.1.1 CT agar for selective isolation of *Serratia*

Preparation of trace element solution (Ingredients per litre of deionized water)

H_3PO_4 1.96 g, $FeSO_4.4H_2O$ 0.0556 g, $MnSO_4$ $4H_2O$ 0.0223 g, $CuSO_4$ $5H_2O$ 0.025 g, CO $(NO_3)_2$ $6H_2O$ 0.03 g, H_3BO_3 0.0062 g. Store at 4°C.

Solution A (Ingredients per 500 mL of deionized water)

$CaCl_2.2H_2O$ 0.0147 g, MgSO4.7H_2O 0.123 g, KH_2PO_4 0.0680 g, K_2HPO_4 0.610 g, Trace element solution (see above) 10 mL, Caprylic acid 1.1 mL, Yeast extract (5% wt/vol) 2 mL, Thallous sulfate 0.25 g, Adjust the pH to 7.2 using NaOH. Autoclave at 110°C (or 120°C for 20 min)

Solution B (Ingredients per 500 mL of deionized water)

NaCl 7 g, $(NH_4)_2$ SO_4 1 g, agar 15 g.

Adjust pH to 7.2. Autoclave and mix solution A and B aseptically and the resulting media is poured in thick layers into sterile plastic Petri dishes (25-30 mL for Petri dishes 9-10 cm in diameter). The media is no longer effective if it is re-melted but plates of CT agar keeps well for several weeks at 4°C if contamination and desiccation are prevented.

13.10.2.1.2 Tests for identification

For the identification of species some specific tests should be performed. These are:

(i) Growth on 5 mg/mL tetracycline (Tc)

(ii) Tetrathionate reduction

(iii) Add 5 mg/mL of Tc to nutrient agar and observe growth.

Tetrathionate Reductase (Le Minoreta 1970)

The liquid medium is composed of potassium tetrathionate 5 g, bromothymol blue (0.2% aqueous solution) 25 mL.

Peptone water (Ingredients per litre of deionized water)

Peptone 10 g, NaCl 5 g

Adjust pH to 7.4, filter sterilize and dispense into small Durham tubes. Inoculate and observe after 1-2 days.

C source utilization test

C source utilization test is perfomed **by** omitting the yeast extract, thallus sulphate and caprylate from solution A from the CT agar medium. Neutralised C sources are added in solution A to give a 0.1% final concentration in the complete medium and 0.2% in the case of carbohydrates.

13.10.2.1.3 Specific tests

(a) Gas production in glucose agar

Prepare glucose agar medium consisting of nutrient agar supplemented with 1% glucose and dispense into tubes. Autoclave, cool and inoculate before the agar sets. Examine the tubes for bubbles of gas production up to 3 days. The reaction is considered positive when a light green to rusty-yellow colour develops after 2 days incubation. Development of a deep blue colour is considered a negative test.

(b) Iodoacetate test (Lysenko 1961) (Ingredients per litre of deionized water)

Tryptone 1 g, Yeast extract 1 g, NaCl 5 g, K_2HPO_4 0.3 g, glucose 10 g, bromothymol blue (1.5 % aqueous solution) 2 mL, pH 7.2

Above medium is sterilized by boiling for 30 minutes on each of 3 successive days. Sodium iodoacetate (filter sterilised) is added to a final concentration of 10^{-3} M (208 mg/L). The inoculated medium is examined for acid production each day for 5 days. Controls without iodoacetate are also inoculated

(c) β-xylosidase

Aqueous (1% wt/vol.) p nitrophenyl β-xyloside is dispensed into the wells in 0.05 mL amounts followed by 0.05 mL of a fresh bacterial suspension in 0.25 M phosphate buffer pH 7 and examine the plates after 24 hours.

(d) Susceptibility to 5-Fluoro cytosine

For determining susceptibility to 5-FC on M-70 minimal agar plate containing 0.2 (wt/vol.) glucose is inoculated with few drops of culture in broth spread on the plate. After drying for 15 minutes at 17°C, a paper disc impregnated with 100 mg 5-FC is placed on the plate. After overnight incubation, the plate is examined for an inhibition zone.

The species identification is described in flow sheet (Figure 13.17).

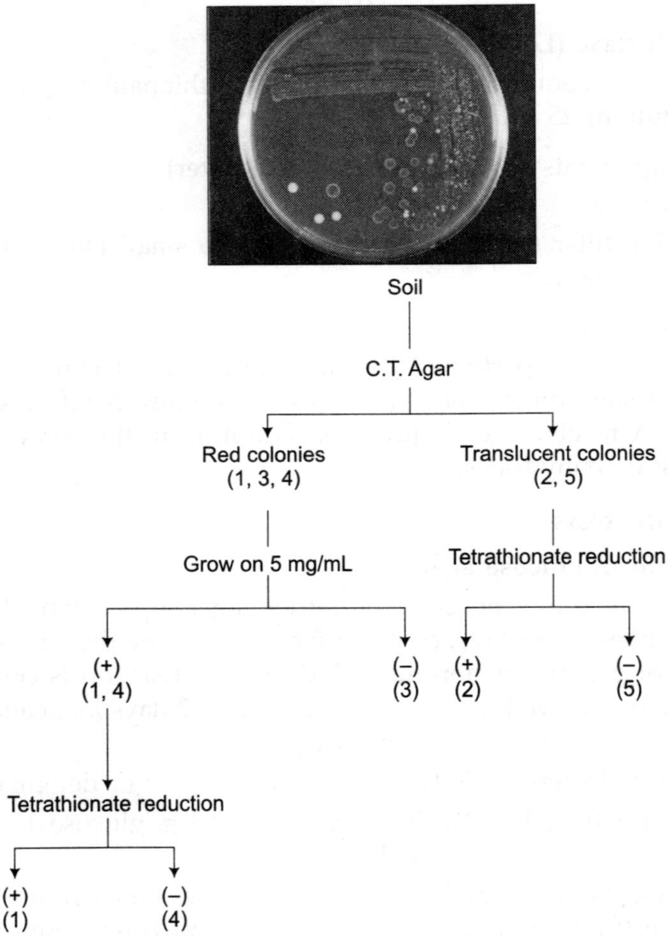

Figure 13.17 Flow chart of *Serratia* for differentiation of different species (1) *Serratia marcescens*, (2) *Serratia proteamaculans*, (3) *Serratia plymuthica*, (4) *Serratia marinorubra*, (5) *Serratia odorifera*.

13.10.2.2 Results

An aerobically growing *Serratia* produces red or white translucent colonies on CT agar plates depending on the species. It can tolerate 5% NaCl solution and is also able to grow at low temperatures. Citrate utilization and Voges-Proskauer's tests are positive. They produce acid from glucose and maltose and can also hydrolyse lipid and protein. Besides xylosidase activity, iodoacetate, gluconate, tetrathionate reductase tests are all positive (Table 13.13).

Table 13.13 Summary of expected outcome from Figure 13.17

	Test	1	2	3	4	5
1.	Red pigment	+	–	+	+	–
2.	Musty vegetable odour	+	–	–	+	–
3.	Growth at 5°C	–	+	+	–	+
4.	Growth at 40°C	+	–	–	+	–
5.	Growth with KCN	+	+	–	+	+
6.	Indole production	–	–	–	–	+
7.	Tetra thionate reduction	+	+	–	–	+
8.	Gas from glucose agar	–	+	+	–	–
9.	Acid from sucrose	+	+	+	+	+
10.	Chitin hydrolysis	+	+	+	+	–
11.	Tween - 80 hydrolysis	+	+	+	+	–
12.	Growth with 5 mg/mL Tc	–	–	–	–	–
C.	**Source Utilization**					
13.	Cellobiose	+	+	+	+	–
14.	Lactose	+	+	+	+	–
15.	Sucrose	+	+	+	+	+
16.	H_2S	–	–	–	–	–
17.	V.P. Test	+	–	+	–	–
18.	Gelatinase	+	+	+	+	+
19.	Amylase	+	+	+	+	+
20.	Growth with 4% NaCl	+	+	+	+	+
21.	Citrate utilization	+	+	+	+	+

13.10.2.3 Discussion

The genus *Serratia* is characterised by morphological properties such as the rod-shaped cells which are either straight or slightly curved in one plane. They are chemo-organotroph, presence or absence of spores, sheaths or prosthecas with respiratory metabolism and absence of photosynthesis and fermentation and ability to utilise various

C sources. Very few coliforms are pigmented and if a pigmented isolate can grow on C.T. agar with salt tolerance, low temperature and ability to hydrolyse protein, starch and lipid but lacks the potential to use urea assigns the bacterium genus *Serratia.*

Moreover, a positive test with 5 mg/mL tetracycline/tetrathionate reduction, accompanied with utilization of range of C compounds, positive gluconate, iodoacetate and β-xylosidose leaves no doubt of the isolate being *Serratia.*

The species can be confirmed through gas production in glucose and tetrathionate reduction as a promising reaction without ambiguity (Table 13.13).

The results if properly fitted to the key characters place the isolate in the genus *Serratia* according to Bergey's manual. It has G + C content of 40-53 mole %.

13.11 *RHIZOBIUM*

13.11.1 Introduction

The rhizobia are soil microorganisms that inhabit the rhizosphere of legumes and other plants. They are a rather more diverse group of organisms than might be supposed, but are united by their ability to produce nodules on legumes. A proper understanding of the Rhizobium legume association requires knowledge of the relationship of rhizobia one to another and to other members of the Rhizobiaceae. Rhizobia are short to medium, Gram-negative rods (0.5-0.9 × 1.2-3 μm) older cells bear capsule, and are sporulated, flagella are arranged in polar, subpolar or peritrichous fashion. The legumes are important in agriculture not only for the fact that they fix nitrogen and can thus conserve nitrogen fertilizers, but also because the plants and seeds are high in protein. Legumes also have ecological significance as they provide nitrogen inflow to natural habitats such as tropical forests. Some tropical legume trees are grown for their wood, and the nitrogen-enriched soil is exploited when the wood is removed.

The growth of most *Rhizobia* is water clear, white or cream coloured, with transparent or translucent areas differentiated into regions of greater opacity owing to the presence of a strain or substrain with in a species. Identification of *Rhizobium* is based on common biochemical tests such as: Gram reaction, motility, flagellar staining, no H_2S production, litmus milk reaction, inability to utilise citrate, and intolerance to 2% NaCl on YMA.

13.11.2 Isolation

1. **Composition of Yeast Extract Mannitol Agar (YMA) medium** (Ingredients per litre of deionized water)

   K_2HPO_4 0.5 g, $MgSO_4.7H_2O$ 0.2 g, NaCl 0.1 g, mannitol 10 g, yeast extract power 0.4 g, agar 15 g
2. **Composition of Glucose-Peptone Agar** (Ingredients per litre of deionized water) Glucose 10 g, peptone 20 g, NaCl 5 g, agar 15 g, pH 7.2

13.11.2.1 Direct isolation

(i) Take 1 gm of soil (collected from legume inhabiting field) and suspend in 10 mL of distilled water.

(ii) Prepare serial dilutions as discussed in cultivation of bacteria.

(iii) Inoculate about 0.1 mL of the highest dilution on YMA plates.

(iv) Examine the plates daily for 5 days at 26°C for the development of typical white colonies either:

(a) With fast grower 2.5 mm viscous or with free flowing gum (exo-polysaccharide);

(b) A slow grower detectable after five days and not much greater than 1-1.5 mm. On 10 days incubation these will be gum resistant and adherently viscous.

Note: Inclusion of congo red dye may help recognition of relatively uncoloured rhizobial colonies avoiding the chance of mistaking *Agrobacterium* (deep pink/red) for *Rhizobia* which gives all the biochemical tests designed for *Rhizobia.*

Addition of 0.002% of cycloheximide eliminates the fungal contaminants, if any.

13.11.2.2 Nodule formation

The principle is to grow sterile host plants in sterilised nitrogen free medium with and without the *Rhizobium* strain under test and to observe whether nodules are formed in presence of bacteria. The test can also show whether or not nitrogen is fixed in nodules. Small seeded plants (e.g., *Trifolium pratense)* can be grown wholly enclosed with in glass tubes fitted with cotton wool plugs to protect from contamination.

Clear uniform seed is rinsed with 95% ethanol and immersed for 3 min in 0.2% (w/v) $HgCl_2$ solution. The seed is then washed thoroughly with at least 10 changes of sterile water and planted directly on agar slopes in tubes or set to germinate on moist filter paper or water agar (1.2%) in Petri dishes at an appropriate temperature.

Composition of seedling agar (Ingredients per litre of deionized water)

CaH_2PO_4 1 g, HPO_4 0.2 g, $MgSO_4.7H_2O$ 0.2 g, NaCl 0.2 g, $FeCl_3$ 0.1 g, agar 10 g, pH 6.8

Inoculate the slopes with bacteria.

Nodules develop when the plants are 2-4 weeks old and first plant leaves open.

13.11.3 Results

Large white colonies, isolated from soil on yeast extract mannitol agar with Gram-negative character, ability to grow at pH 9.5 and utilise biuret with a positive flagella staining reaction and nodule formation (Table 13.14).

13.11.4 Discussion

The genus *Rhizobium* is characterized by morphological properties such as the rod-haped cells presence of flagella in either polar, subpolar or peritrichorus fashion. A Gram-negative reaction, absence of spores in young culture, however, old cultures show spores. Chemoorganotrophic mode of nutrition, ATP generation through glucose break down, aerobic nature, ability to fix dinitrogen are characters of *Rhizobium.*

The principle species of Rhizobium are *R. trifoli, R phaseoli, R. meliloti.* The key characters of these species are shown in dichotomous chart (Figure 13.18). G + C content of *Rhizobium* is 59.1-65.5 mole % as per Bergey's manual.

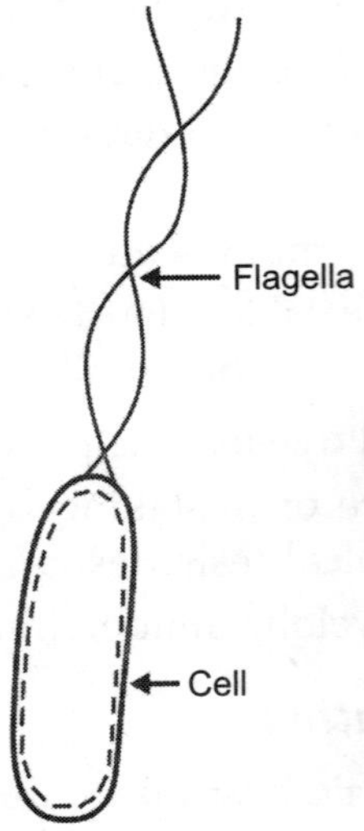

Rhizobium showing two-polar flagella (diagrammatic).

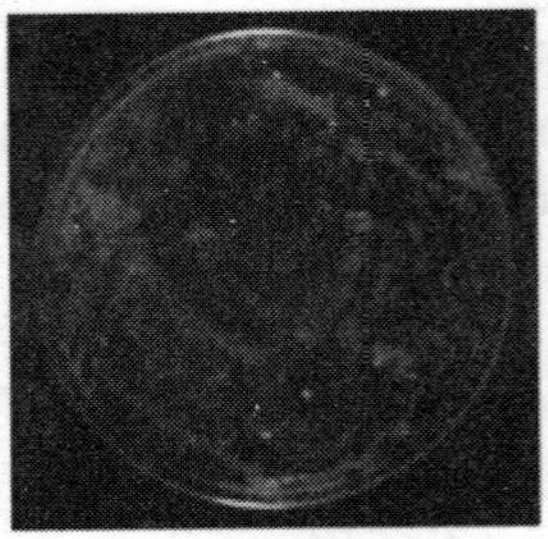

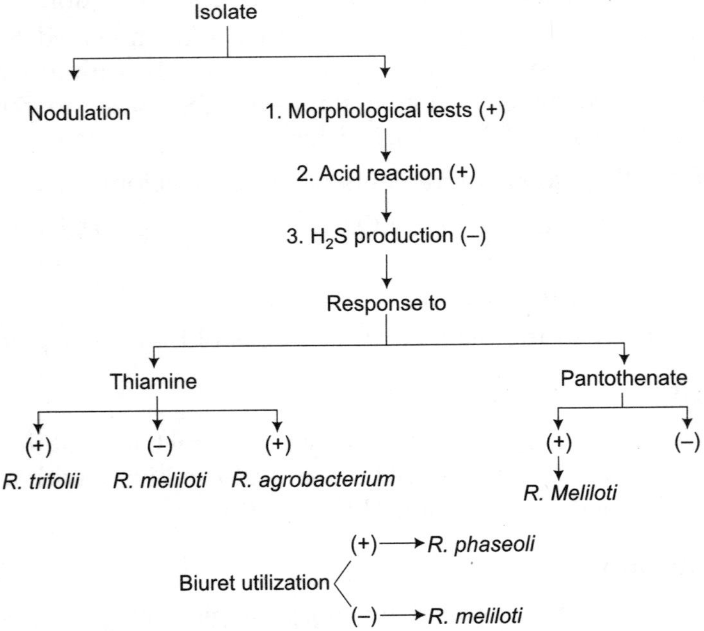

Figure 13.18 Flow chart for the identification of *Rhizobia.*

13.12 *BACILLUS*

13.12.1 Introduction

Members of the genus bacillus are Gram-positive, rod-shaped, catalase producing spore-formers that requires oxygen, motile (flagellated) bacterium. However, this is a very diverse group of organisms and some species are actually Gram-negative or facultative. Smith *et al.* (1952) subdivided the genus into 3 groups on the basis of spore shape and sporangial swelling. The identification and isolation *B. megaterium, B. subtilis, B. circulans, B. calidolaclis* species are strictly soil inhabitants. *B. cereus* is widespread in nature and readily found in soil, where it adopts a saprophytic life cycle; germinating, growing and sporulating in this environment (Figure 13.19). Other species within this genus include *B. anthracis, B. mycoides, B. thuringiensis, B. pseudomycoides and B. weihenstephanensis* are normally food contaminant/toxin producers. Genomic sequencing data has shown *B. anthracis, B. cereus* and *B. thuringiensis* to be very closely related with their 16S rRNA gene sequence sharing more than 99% similarity. Spores are able to survive harsh environments including normal cooking temperatures. Spores are more resistant to environmental stress than vegetative cells due to their metabolic dormancy and tough physical nature. The diarrhoeal syndrome is caused by diarrhoeal toxins produced during growth of the bacteria in the small intestine.

Table 13.14 Summary of expected outcome

Morphological Test		*Biochemical*	
Colony:	Large smooth	Growth on Glucose	Peptone Agar: Slight
Gram stain	Negative	Litmus-Milk Test	Negative
Flagella stain	Positive	Growth on 2% NaCl +YMA	Negative
Endospore	Positive	H_2S production	Negative
Capsular stain	Positive	Citrate utilization	Negative
Motility	Positive	Nodule formation	Positive
		Growth at pH 9.S	Negative
		Ability to utilize dulcitol and raffinose	+/-
		Acid production	+/-
		Biuret utilization	2% Biuret in YMA
		Response to thiamine and pantothenate	+/-

13.12.2 Isolation

Nutrient agar medium composition (Ingredients per litre of deionized water)

Beef extract 3 g, peptone 5 g, NaCl 3 g, agar 15 g, pH 7.2

13.12.2.1 Direct isolation

1. Inoculate 1 g of soil in 10 mL of distilled water.
2. Prepare serial dilutions.
3. Spread 0.1 mL of highest dilution on nutrient agar plates.
4. Incubate at 37°C for 2 days.
5. Restreak the isolated colonies on nutrient agar plates

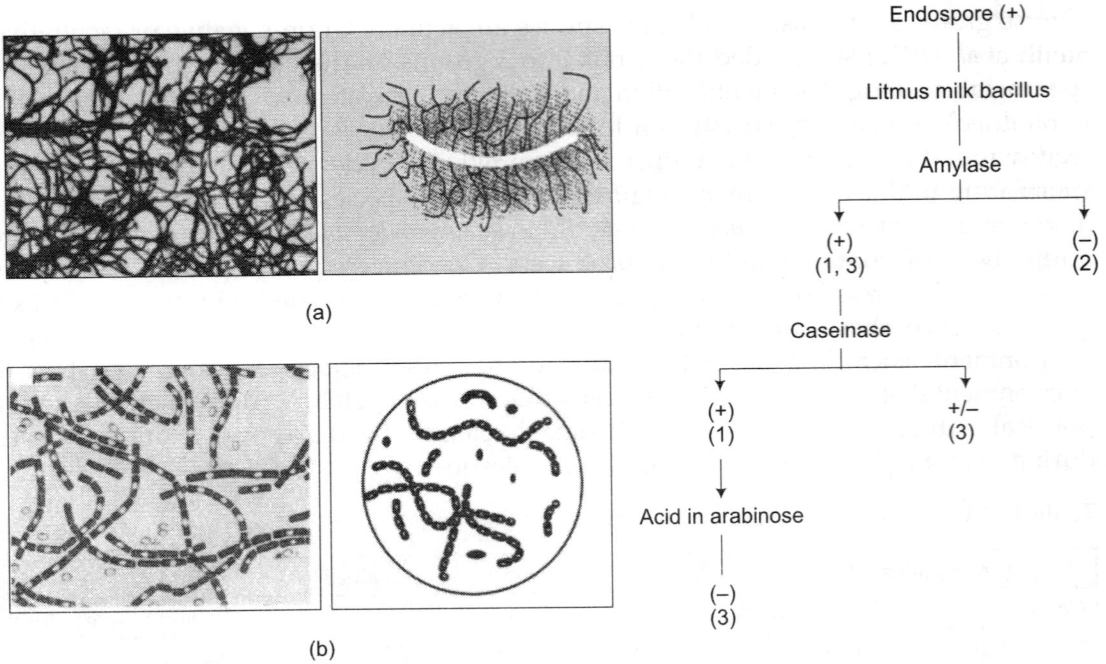

Figure 13.19 Flow chart for the identification of *Bacillus* where (a) Chain of Gram stain cells with flagella, (b) chain of bacteria with their central endospore 1 indicates *Bacillus subtilis*, 2 *Bacillus circulans*, 3 *Bacillus calidolactis*.

13.12.2.2 Selective isolation

Composition of soil extract agar (Gorden and Regnear, 1963): To 400 g air dried, sieved garden soil rich in organic matter in a 2 litre conical flask, add 960 mL of tap water. Autoclave at 121°C for 60 min and allow to cool in the unopened autoclave overnight. Then filter the extract through paper and reautoclave the filtrate at 121°C for 20 minutes, and allow to stand at room temperature for at least 2 weeks. The clear suspension can be used in making soil extract agar.

Composition of soil extract agar medium

Peptone 5 g, beef extract 3 g, agar 15 g, tap water 750 mL, soil extract 250 mL.

With new batches of soil try small lots of medium between 10 and 100% v/v amount of soil extract. The concentration producing best sporulation should be used for the bulk medium once isolated. Bacillus can be grown routinely on nutrient agar medium.

Spread 0.1 mL of the highest dilution on soil extract agar plates.

13.12.3 Identification Results

The most important identifying character of bacillus is presence of endospore. Besides this, they are Gram-positive and catalase producing bacteria.

13.12.4 Discussion

Bacilli are medium size, Gram-positive, centrally placed endospore, catalase positive and unchanged litmus milk confirms that the isolate belongs to genus Bacillus. Typical respiratory metabolism with chemo-organotrophic mode of nutrition and inability to photosynthesize. The identification of species is based on starch hydrolysis which is accomplished by *B. subtilis* and *B. calidolactis* while *B. circulans* is unable to do so. The casein hydrolysis differentiates between *B. calidolactis* and *B. subtilis.* This is further confirmed by acid production in arabinose (Figure 13.19 and Table 13.15).

Table 13.15 Summary of expected outcome

S. No.	*Tests*	*Group I*	*Group II*	*Group III*
A.	**Morphological**			
1.	Colony			
2.	Flagella stain	+	+	+
3.	Capsular stain	–	–	–
4.	Endospore stain	–	–	–
5.	Gram stain	+	+	+
6.	Motility test	–	–	–
B.	**Biochemical**			
1.	Amylase	+	–	–/+
2.	Gelatinase	+/–	–	+/–
3.	Caseinase	–	–	–
4.	Catalase	+	+	+
5.	Litmus milk test	–	–	–
		Acid production from		
a.	Arabinose	+/–	–	–
b.	Mannitol	+/–	+	–
c.	Lactose.	–	–	+/–
d.	Xylose	+/–	+	+/–
e.	Cellobiose	+/–	+	+/–

13.13 *LACTOBACILLUS*

13.13.1 Introduction

Lactobacillus is a very large group of homofermentative as well as heterofermentative. Gram-positive bacteria which appear to come from two very distinct type of nutrient habitat *viz.*

(i) Man and animals
(ii) Plant materials

In man and animals they dominate the flora, in the upper gastro-intestinal tract and as in the oral cavity and lower intestine they may form only a minority of the total, strictly anaerobic flora. On plant materials, lactobacilli occur in very small number being dominated by the more successful leuconostocs. *Leuconostocs* are traditionally found in association with plant matter, fermenting vegetables, milk, dairy products, and wines and meats.

The basic morphological characteristics are the rod-shaped usually non-motile, non-spore forming. It produces lactic acid lowering final pH of the medium to 2.2 - 3.4, resistant to acids and is micro-aerophillic. They grow well in milk. The identification of genus is based on usual morphological and biochemical tests such as Gram reaction, motility, presence of flagella, oxidase test, catalase reaction, presence of extracellular gelatinase and nitrate reduction.

13.13.2 Isolation

SL-Medium Composition (Rogosa, Mitchelie and Weismann, 1951)

(Ingredients per litre of final volume in deionized water)

Trypticase 10 g, yeast extract 5 g, KH_2PO_4 6 g, diammonium citrate 2 g, $MgSO_4.7H_2O$ 0.58 g,

$MnSO_4.4H_2O$ 0.28 g, glucose 10 g, arabinose 5 g, sucrose 5 g, Tween-80 1 mL, sodium acetate $5H_2O$ 2.5 g, agar 15 g. Glacial acetic acid 99.5% required to titrate pH of medium to 5.4

Dissolve agar separately by steaming in 500 mL distilled water, dissolve all the other ingredients except acetate and acetic acid in 300 mL distilled water without heating, add this to melted agar and steam for further 5 minutes excessive heating must be avoided. Dissolve the sodium acetate in about 15 mL distilled water without heating add acetic acid as a 10% v/v aqueous solution to adjust pH to 5.4. Make up volume to 20 mL. Add this buffer mixture to hot based medium in convenient amount in sterilised screw capped bottles. The medium should be clear light straw coloured giving a firm gel. Avoid repeated melting and cooling. The addition of arabinose sucrose and glucose allows the growth of strains that preferably ferment these sugars.

1. Take about 5 mL of curd and make a suspension in 5 mL of distilled water.
2. Prepare serial dilutions.
3. Plate 0.1 mL of highest dilution on SL agar plates.
4. Incubate at 37°C for 2 days.
5. Observe the colonies.
6. Restreak on to nutrient agar plates to check contamination (Figure 13.20).

13.13.3 Results

The isolate should grow as a small, smooth margined, white or off white on SL medium. Each member is Gram-positive, immotile rod, without flagella, or spore or sheath and catalase and oxidase positive. The confirmation for axenic culture should be made by streaking the isolate on to the nutrient agar plate.

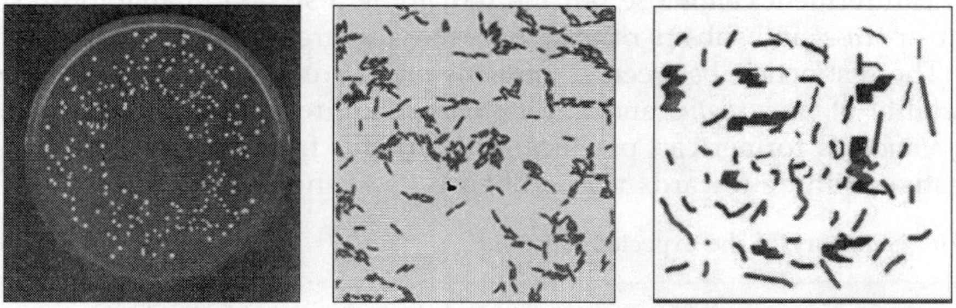

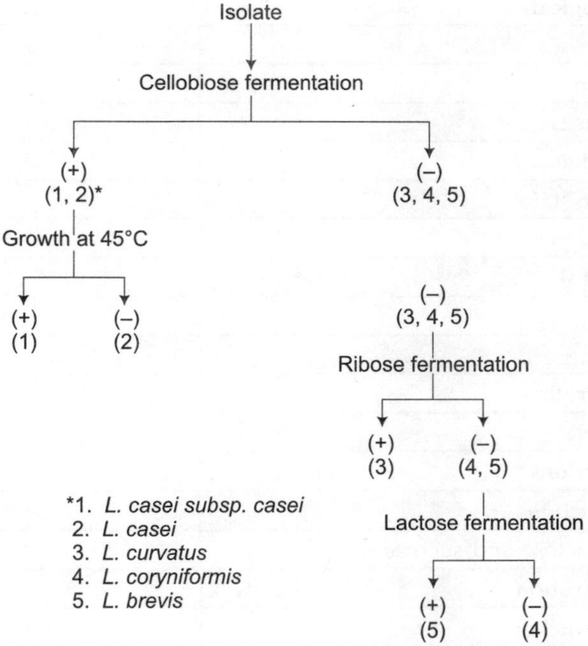

Figure 13.20 Flow chart for the species differentiation of *Lactobacilli*.

13.13.4 Discussion

The genus Lactobacillus is characterised by morphological properties such as the rod-shaped cells, absence of flagella, a Gram-positive reaction, absence of spores, sheathes, capsules, and by physiological properties such as a fermentative mode of nutrition, absence of photosynthesis and nitrogen fixing ability and a potent producer of lactic acid. The isolate from SL medium bearing above characters should be considered belonging to genus *Lactobacillus*.

The change in the colour of glucose, lactose or maltose broth accompanied by down ward displacement of the liquid due to the production of gas are the promising features of *Lactobacillus,* inability to reduce nitrate and a positive oxidase test shows that the bacteria require oxygen as a terminal electron acceptor. Among the four species *L. casei* sub-sp. *rhamnosus,* shows the quite enormous distinction. Since both the sub-sp.

of *L. casei* can ferment cellobiose but the remaining 3 sp. are unable to do so. Among *L. casei* sub-sp. *casei* and sub-sp. *rhamnosus* former can grow at 45°C while latter is unable to do so. The distinction between *L. curvatus* and *L. coryniformis* is made good owing to the inability of *L. curvatus* and *L. coryniformis* to grow at 45°C. Among *L. curvatus* and *L. coryniformis* former can produce acid and gas from ribose while the latter has strict negative attitude towards ribose (Figure 13.20 and Table 13.16).

Table 13.16 Summary of the expected outcome

S. No.	*Tests*	*Results*
A.	**Morphological**	
1.	Colony	White
2.	Gram stain	+
3.	Flagella stain	–
4.	Capsule stain	–
5.	Endospore stain	–
6.	Motility	–
B.	**Biochemical**	
1.	Oxidase	+
2.	Catalase	+
3.	Glucose Broth	+
4.	Gelatinase	+
5.	**T.S.I. reactions**	
a.	H_2S production	+
b.	Glucose, lactose and sucrose	+
c.	Acid production	+
6.	Nitrate reduction	+
7.	Litmus milk test	+
8.	Growth at 45°C	+/–
9.	Cellobiose fermentation	+/–
10.	Ribose fermentation	+/–

According to Bergey's manual, *Lactobacillus* were isolated from milk. Rogosa (1970) has divided the genus into 3 groups, viz., Thermobacterium, Streptobacterium and Betabacterium on the basis of optimal growth temperatures and germenative end products. Thermobacterium and Streptobacterium are homofermentative, the difference lies in the ability of former to grow at 45°C and inability of the latter to do so. All the species discussed here except *L. brevis* belongs to Streptobacterum. *L. brevis* falls under Betabacterium. It is a heterofermentative bacterium requires temperature less than 45°C. The G + C content varies from 30-40 mole%.

13.14 *AZOTOBACTOR*

13.14.1 Introduction

Soil contains several microorganisms which may live freely as well as in association with some plants. *Azotobacter* is a genus of free-living diazotrophic bacteria having the highest metabolic rate compared to any other microorganisms. They fix atmospheric nitrogen into ammonia which is utilized by plants as a nitrogen source.

Azotobacter is Gram-negative, polymorphic, motile, oval or spherical bacteria that form thick-walled cysts and may produce large quantities of capsular slime. Old population of bacteria includes encapsulated forms and have enhanced resistant to heat, desiccation and adverse conditions. The cyst germinates under favourable conditions to give rise to vegetative cells. They also produce polysaccharides. *Azotobacter* sp. is sensitive to acidic pH, high salts and temperature above 35ºC. There are four important species of *Azotobacter* viz. *A. chroococcum, A. agilis, A. paspali* and *A. vinelandii* which are commonly found in our soils.

13.14.2 Isolation

Azotobacter sp. can be isolated by the soil dilution method, direct isolation or soil plate method using any of the medium, viz., Ashby's medium, Jensen's medium, Beijerinck's medium. In direct isolation, lumps of soil are spread on a nitrogen free medium. After incubation of 3 days at 28ºC, *Azotobacter* colonies appear on the agar medium. *Azotobacter* is capable of nitrogen fixation, utilizing mannitol as a carbon source and atmospheric nitrogen as nitrogen source. It does not require organic nitrogen source. Hence, a medium comprising of mineral salts and an easily metabolisable sugar mannitol allows selective growth of *Azotobacter* sp. A selective enrichment step prior to the isolation is always recommended.

Composition of Ashbys Mannitol Agar (Ingredients per litre in deionised water)

Medium is used for cultivation of *Azotobacter* species that can use mannitol and atmospheric nitrogen as source of carbon and nitrogen respectively.

Mannitol 20 g, K_2HPO_4 0.2 g, $MgSO_4.7H_2O$ 0.2 g, NaCl 0.2 g, K_2SO_4 0.1 g, $CaCO_3$ 5 g, agar 15 g, final pH (at 25°C) 7.4 ± 0.2

Jensen's Medium (Ingredients per litre in deionised water)

Sucrose 20 g, K_2HPO_4 1 g, $MgSO_4.7H_2O$ 0.5 g, NaCl 0.5 g, $FeSO_4.7H_2O$ 0.1 g, Na_2MoO_4 0.005 g, $CaCO_3$ 2 g, agar 15 g.

Sterilize by autoclaving at 15 lbs pressure (121°C) for 15 minutes. Mix well and pour into sterile Petri plates. Due to presence of calcium carbonate, the prepared medium forms opalescent solution with white precipitate

Beijerinck's medium (Ingredients per litre in deionised water)

K_2HPO_4 1 g, $MgSO_4.7H_2O$ 0.2 g, $FeSO_4.7H_2O$ 0.05 g, $CaCl_2$. $2H_2O$ 0.1 g, $Na_2MoO_4.2H_2O$ 0.001 g, glucose 10 g, adjust pH to 7.2. Sterilize by autoclaving at 15 lbs pressure (121°C) for 15 minutes. Glucose is to be sterilized separately.

If this medium is to be used to isolate *Azotobacter* from soil, sterilization is not necessary.

13.14.3 Procedure

1. Prepare 50 mL nitrogen free mannitol broth.
2. Inoculate 1 mL of finely ground and sieved soil sample.
3. Incubate for 7 days at room temperature.
4. A thin film of growth appears on the surface of the broth. Touch a sterile loop to lift a little amount of the film and transfer it on a sterile nitrogen free agar plates.
5. Streak for isolation of the culture.
6. Incubate the plate in an inverted position for 3 days at room temperature.

13.14.4 Observation

Azotobacter sp. produces small, watery, brown black or colorless colonies.

1. Observe the colonies and study the characteristics and Gram nature of the *Azotobacter* sp.
2. Identify the species using the key for identification for different species (Table 13.17 and Figure 13.21).

Table 13.17 Differential characters in members of Azotobactereaceae

Microbes	*Water-Soluble Pigments*								
	cysts	*motility*	*Long filaments*	*Brown-black*	*Brown-black to red-violet*	*Red violet*	*green*	*Yellow-green Fluorescent*	*Blue-white fluorescent*
Azotobacter									
A. chroococcum	+	+	–	–	–	–	–	–	–
A. vinelandii	+	+	–	–	–	+	+	+	–
A. beijerinckii	+	–	–	–	–	–	–	–	–
A. nigricans	+	–	–	+	+	+	–	–	–
A. armeniacus	+	+	–	–	+	+	–	–	–
A. paspali	+	+	+	–	–	+	–	+	–
Azomonas									
A. agilis	–	+	–	–	–	–	–	+	+
A. lnsignis	–	+	–	+	–	+	–	+	–
A. macrocytogenes	–	+	–	–	–	–	–	+	+

13.15 *AZOSPIRILLUM*

13.15.1 Introduction

Azospirillum is cosmopolitan in distribution and occurs in a large number, in rhizospheric soils in association with the roots of a variety of C3 and C4 plants. The *Azospirillum*

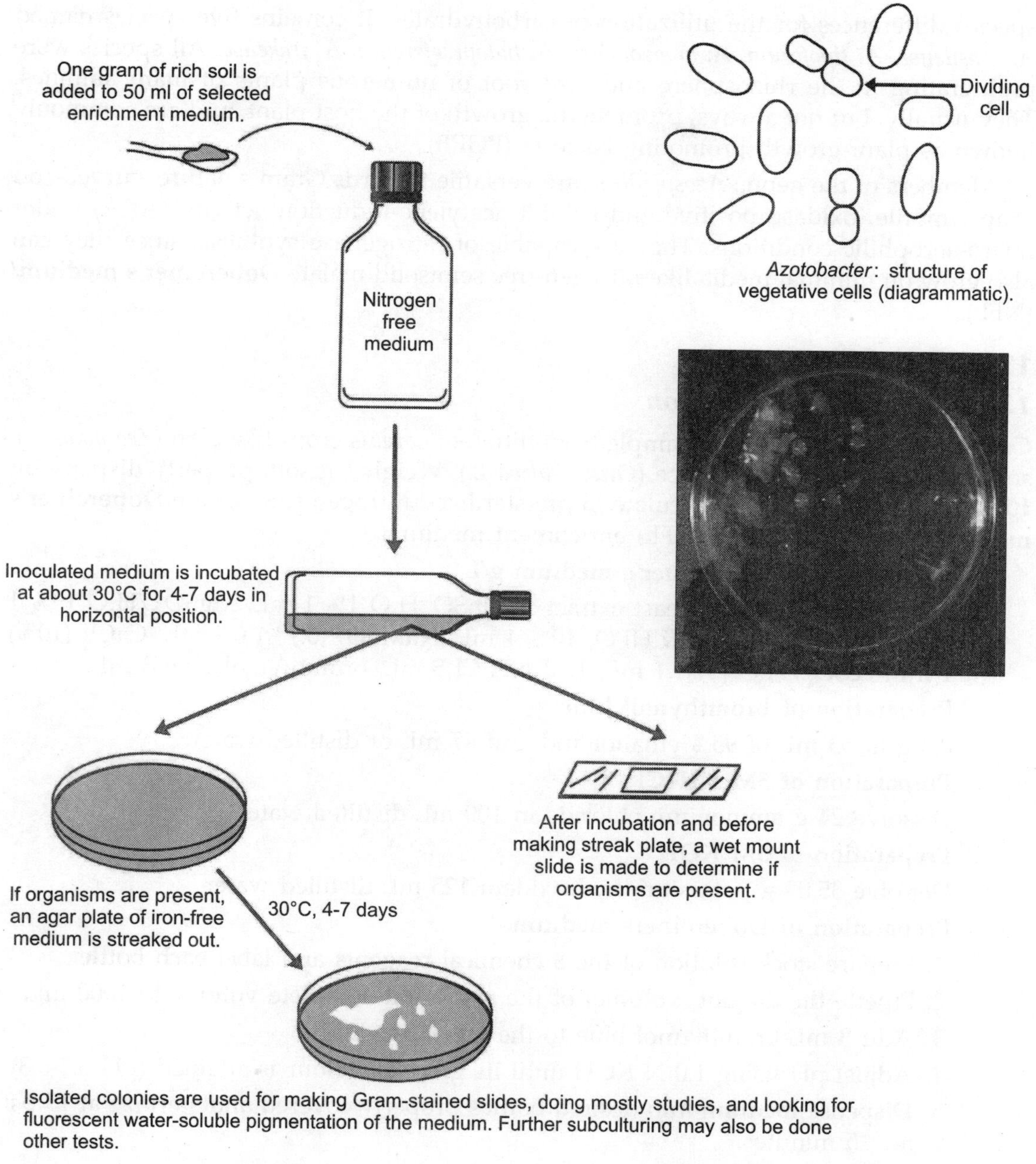

Figure 13.21 Flow chart for the isolation of *Azotobacter* sp.

are common habitant of tropical and temperate regions. *Azospirillum* have well-adapted and highly versatile metabolism that allow them to proliferate in carbon rich, nitrogen poor rhizosphere, rhizoplane and the endorhizosphere environments. *Azospirillum* prefers organic acids like malate, succinate and lactate. There are significant interspecies

species differences for the utilization of carbohydrates. It contains five species named *A. brasilense, A. lipoferum, A. amazonense, A. halopraeference, A. irakense.* All species were proliferating in the rhizosphere and host root of numerous plants of many families. They usually, but not always, promote the growth of the host plant, and are commonly known as plant-growth promoting bacteria (PGPB).

Members of the genus *Azospirillum* are versatile towards Gram's nature, curved-rod shape, motile, oxidase positive and exhibit acetylene-reduction activity (ARA) under micro-aerophilic conditions. They are capable of nitrogenase synthesis since they can also grow on minimal media like nitrogen-free semisolid malate **Dobereiner's medium/** (NFb).

13.15.2 Isolation

13.15.2.1 Sample collection

Collect the soil and host root sample from different cereals crops like corn (*Zea mays L.*), sorghum (*Sorghum* sp.) and rice (*Oryza sativa L.*). Weigh 1 g soil, properly dispense in 10 mL distilled water and inoculate in pre-sterilized nitrogen-free malate **Dobereiner's medium/** semisolid malate (NFb) enrichment medium.

1. **N-free malate Dobereiner's medium g/L**

   Malic acid 5, KOH 4, yeast extract 5, $MnSO_4.H_2O$ 1% 1 mL, $MgSO_4.7H_2O$ 10% 1 mL, NaCl 10 % 2 mL, K_2HPO_4 10% 4 mL, $NaMoO_4$ (0.1%) 0.2 mL, $CaCl_2$ (10%) 1 mL, $FeSO_4.7H_2O$ (5%) 1 mL, 1 M NH_4Cl 5 mL, bromthymol blue 3 mL.

   Preparation of bromthymol blue

   0.5 g in 53 mL of 95% ethanol and add 47 mL of distilled water.

   Preparation of 5M NH4Cl

   Dissolve 24 g ammonium chloride in 100 mL distilled water

   Preparation of 5M KOH

   Dissolve 35.05 g potassium hydroxidein 125 mL distilled water

   Preparation of Dobereiners medium

   1. Prepare stock solution of the 8 chemical reagents and label each bottle.
   2. Pipette the aliquot (volume) of the stock and complete volume to 1000 mL.
   3. Add 3 mL bromthymol blue to the liquid medium,
   4. Adjust pH using 1.0 N KOH until its greenest colour is attained (pH 6.5-6.8).
   5. Dispense medium into desired bottles properly covered and sterilize at 15 psi for 15 minutes.

2. **N-free malate Nfb Medium g/L**

   Malic acid 5, K_2HPO_4 0.5, $MgSO_4.7H_2O$ 0.2, NaCl 0.1, $CaCl_2$ 0.2, trace element 2 mL. 0.5% Alcoholic Bromophenolthymol blue (0.5% alcoholic solution) 2 mL, NH_4Cl 1, agar 1.8 pH 6.9-7.3.

   Preparation of Trace Element Solution g/L

   $Na_2MoO_4.2H_2O$ 0.232 , $MnSO_4$ 0.2 , H_3BO_3 0.2, $CuSO_4.7H_2O$ 0.24

- Bromthymol blue (0.5% aqueous) 2.0 mL solution (dissolved in 0.2 N KOH)
- Fe EDTA (1.64% solution) 4 mL
- Vitamin solution 1 mL

13.15.3 Maintenance of Isolates

- Semi-solid-N free Dobereiner's medium
- Solid slants
- NA (Nutrient agar) slants + mineral oil
- Tryptic soy agar slants

13.15.4 Results and Observations

Development of white or pink irregular, wrinkled, dense and undulating fine pellicle on the semisolid malate medium is very characteristic of *Azospirillum*. It is polymorphic, varying in size, Gram-positive, containing β-hydroxybutyrate granules. Catalase positive, oxidase positive, gelatinase and urease positive, citrate utilization positive, H_2S negative, Voger-Proskaur negative, indole production negative. Cells have one half spiral turn and show spiral movement. Lateral and single polar flagella are present.

LABORATORY EXCERCISES ON FUNGI

13.16 FUNGAL STAINING

In contrast to bacteria, heat-fixed smears of fungi are generally useless for microscopic examination. It is thus necessary to examine either stained or unstained wet mounts of fungi. Furthermore, because of their size, it is generally unnecessary to use an oil-immersion objective to view preparations of fungi. Use of stains such as Gram's stain and Ziehl Neelsen stain, used principally in bacteriology, is somewhat limited in mycology, Gram's stain can be used to co-visualise yeasts but the results are often unsatisfactory. Methylene blue can also be used to stain yeasts but lactophenol cotton blue, a combination of a stain and a mountant, is the preparation of choice for the microscopic examination of yeasts. Fungal spores can be stained using a modification of the Ziehl Neelsen stain, developed to stain mycobacteria. Strong carbol fuchsin is introduced into the spores by heating a fixed preparation, and the stain is removed from other structures by treatment with 5% sulphuric acid. Methylene blue or malachite green is used as counterstains. Spores stain red, and other structures take up die colour of the counterstain. Ascospores can be demonstrated in this manner (Figure 13.22).

The most common method used to visualise fungi is to use lactophenol cotton blue. Light green is an alternative stain to cotton blue that can be used to stain fungal structures. However, it does not stain fungi so intensely, and it tends to fade with time. It is therefore not suitable for preparing mounts that are to be kept for long periods. Cotton blue stains hyphae blue, particularly young hyphae, and the intensity of staining

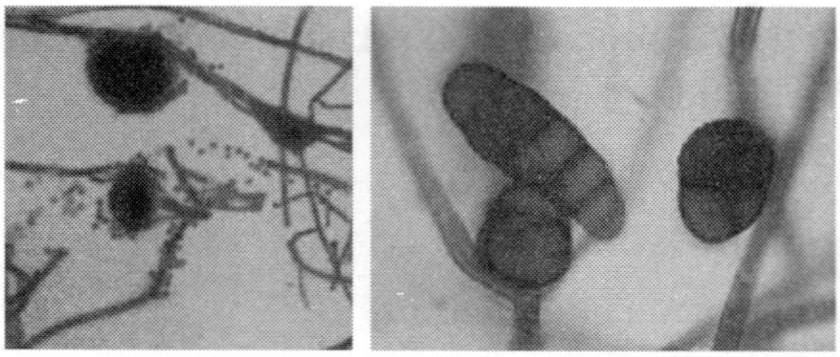

Figure 13.22 Fungal mycelia staining with Lactophenol cotton blue stain.

increases with time. Old mounts can become overstained. Lactophenol is prepared by dissolving phenol crystals, lactic acid and glycerol in distilled water. When it is used on its own as a mountant, lactophenol can be employed to observe the natural colours of fungal structures.

Composition of Lactophenol Cotton Blue Stain (Fungal stain)

Phenol crystals	20 g
Lactic acid	20 mL
Glycerol	40 mL
Cotton blue	0.05 g

Dissolve the phenol crystals in the order ingredients by heating the mixture gently under a hot water tap.

To visualise fungi microscopically, a portion of the fungal growth is removed from the culture using a sterile needle or wire hook. The sample is then placed in a drop of alcohol on the surface of a clean microscopic slide. The alcohol is used to drive out air that becomes trapped between the fungal structures. A drop of lactophenol cotton blue is then added to the preparation. Yeast cultures are generally emulsified directly in lactophenol cotton blue and the culture material is more easily manipulated with a wire loop. Care should be taken not to add too much mountant to the preparation, and if this accidentally happens, then the excess should be gently removed using blotting paper. The edge of a coverslip is then placed onto the drop of mountant carefully and the coverslip is lowered over the preparation, avoiding the introduction of air bubbles.

With certain fungi it can be difficult to remove sufficient material from the culture. This is particularly the case with tough, compact colonies. Alternatively, powdery colonies may yield only spores. In such cases it is often advantageous to remove a little of the agar substrate along with the fungal material. With any preparation from a mold culture, it is usually necessary to tease out the fungal growth on the slide to make visualisation easier. This is generally done in the alcohol drop with two sterile mounted needles. It is often necessary to add more alcohol to prevent the fungus from drying out during this procedure. Mounts prepared in lactophenol can be kept for several weeks or months. More permanent preparations can be made by sealing the edges of the coverslip. This is achieved by painting on several layers of clear nail varnish around the edge of the coverslip. Each layer of nail varnish must be allowed to dry before application of the next layer. If this is to be successful, then it is important that the edges of the coverslip are clean and dry before applying the nail varnish. Properly sealed, such preparations can be kept for many years.

13.17 ISOLATION OF FUNGI

The isolation of fungi in laboratory culture requires the growth of colonies on solid media, since in liquid cultures yeasts form either sediment or a pellicle and molds generally grow to form a fungal ball. In liquid cultures, sporulating structures are apparent only if the fungus grows as a surface mat. Such structures are difficult to handle and it is hard to make preparations for microscopic examination from them. These problems are overcome by the use of solid media.

Cultures of fungi on solid media can be made either in Petri dishes or in tubes, where an agar slope is inoculated with the culture to be examined. Tube cultures take up far less room than Petri dish cultures and are less prone to aerial contamination with spores present in the laboratory air. However, because Petri dishes have a greater surface area, cultures are provided with more room for colony spread and development and the fungal growth is much more easily visualised and manipulated. Petri dish cultures also have the advantage of better aeration than tube cultures. This is an important consideration when cultivating obligate aerobes. Fungal cultures are, in general, maintained for much longer than bacterial cultures. Many bacteria give rise to colonies after overnight incubation, but fungal cultures can take up to three weeks to develop. During this time, loss of moisture from the solid medium in a Petri dish can be a problem. To overcome the difficulty of the medium drying out, an adequate quantity of agar medium must be poured into the plate. Typically, between 35 and 45 mL per 9 cm Petri dish is used for a fungal culture, whereas bacterial cultures rarely require more than 20 mL in a similar sized plate. A further measure to cut down water loss from Petri dish cultures is to seal the plate with an oxygen-permeable adhesive tape or parafilm. The problem of aerial contamination is also reduced using this measure.

Great care must be taken when handling fungal cultures, particularly when molds are being grown. Some pathogenic fungi must always be grown in tubes because of the very high risk of infection and the serious threat to the health of laboratory workers that they pose. Cultures should be opened only when absolutely necessary and be kept open for as short a time as possible. Not only does this help to reduce aerial contamination of that culture, it also helps to prevent the spread of spores from the colonies being examined. Observation of this precaution also protects the laboratory worker. Although many molds may not be infectious, their spores can be highly **allergenic.** These can give rise to severe **allergic** reactions, especially if inhaled. One should never sniff a mold culture.

Many fungi grow at relatively low temperatures, with the optimum growth often occurring in the range 20-30°C, although some fungi that are associated with warm-blooded animals including humans and thermophilic fungi may also grow at higher temperatures 35°C. Although incubators are used to cultivate fungi, many cultures will grow quite happily, although perhaps more slowly, if maintained at room temperature.

Stock cultures are often maintained in screw-top McCartney bottles that contain agar slopes. Such bottles are also used for the transportation of fungi. It is very important that the caps are kept loose. Although the rubber stoppers in such bottles are permeable to oxygen, the atmosphere inside a bottle that has a tightly screwed cap soon becomes depleted of oxygen resulting to either in the cessation of growth, or in abnormal growth and development of the culture. Cultures that are exposed to such stresses are frequently **pleomorphic:** that is, they display unusual structures or produce only vegetative growth and none of their characteristic spores.

13.18 FUNGAL CULTURE TECHNIQUES

In contrast to many bacteriological media, which have a neutral or slightly alkaline pH, fungi tend to grow best on an acid medium, typically pH 5.0-6.0. The pH of many fungal media is selective against many bacteria, but the selective effect may be enhanced by the addition of antibacterial antibiotics such as chloramphenicol and rose bengal to the growth medium. Fungal growth media also tend to have a much higher carbohydrate content than do bacteriological media. This imposes osmotic stresses that again select against the growth of many bacteria. Because of the low pH of fungal media, agar tends to become hydrolysed at high temperatures. For this reason, sterilization temperatures are not usually raised above 115°C.

13.19 MEDIA USED IN THE CULTIVATION OF FUNGI

One of the most commonly used media for the cultivation of fungi in the medical laboratory is Sabouraud's medium. This is a peptone-based medium having a concentration of glucose between 2-4% is significantly higher than those media used for bacteriological culture. Sometimes additional natural sources of complex carbohydrates are used to enhance growth or other characteristics of the culture. Complex media such as these also contain vitamins and other growth factors that enhance growth and proper development of the fungal culture.

Common examples of the other complex mycological media incorporating natural substrates include rice starch agar, used to stimulate the formation of mycelia and chlamydoconidia by *Candida albicans,* potato starch agar, malt extract agar and corn meal agar. Extracts of corn meal enhance the production of penicillin by members of the genus *Penicillium.* It was this discovery, made during the early 1940s, that enabled the first large-scale production of an antibiotic that has revolutionised clinical practice, even though the antibacterial properties of these molds were described by Joseph Lister in the nineteenth century, 60 years before Alexander Fleming published his observations on the antibacterial effects of this fungus. When penicillin was first used on a patient, a policeman who was dying of overwhelming staphylococcal sepsis, the antibiotic was in such short supply that it had to be re-extracted from the patient's urine. The benefits of using corn meal can thus be appreciated. During the period 1911-13, the laboratory attendant in the Botany Laboratory of the University of Cambridge reported to collect unused cultures of *Penicillium glaucum* to use as a salve to treat skin infections. This folk remedy had been used in his family for a very long time. This was at least 16 years before Alexander Fleming described the activity of penicillin.

Some of the media used to cultivate fungi are synthetic and do not use complex organic components. Czapek Dox medium contains mineral salts with added sucrose. Like rice starch agar, this is also used to stimulate mycelial growth and the formation of chlamydoconidia by *Candida albicans.* This is enhanced by lowering the oxygen tension in the medium, and this can be achieved by placing a sterile coverslip over the edge of the inoculum. Some fungi grow less well on defined media such as Czapek Dox agar than on complex media, although Czapek Dox agar is better for the induction of spore production in many fungi.

13.19.1 Semi-Synthetic Media

1. **Corn Meal Agar g/L in distilled watewr**
   Corn meal 20, Peptone 20, Dextrose 15, Agar 15
2. **Oat meal Agar g/L in distilled water**
   Oats 100, Agar 15
3. **Potato Dextrose Agar g/L in distilled water**
   Peeled potato 250, Dextrose 20 g, Agar 15 g, pH 6-6.5
4. **Soil Extract Agar g/L in distilled water**
   Soil extract 100 mL, Glucose 1, K_2HPO_4 0.5, Agar 15, pH 7-7.2.
   Preparation of Soil Extract
   Steam 1000 g garden soil in one L of tap water in an autoclave at 15 psi pressure for 30 min. Add a small quantity of calcium carbonate and filter the soil suspension through double layers of filtre papers. The filtrate is sterilized in a flask and stocked, for future use.
5. **Malt Extract Agar g/L in distilled water**
   Malt Extract 25, Agar 15 g
   Ideal for the growth of wood destroying and many other fungi.
6. **Martin's Rose Bengal streptomycin Agar g/L in distilled water**
   Dextrose 10, Peptone 5, KH_2PO_4 1, $MgSO_4.7H_2O$ 0.5, Rose Bengal (1 part in 30,000 parts of medium), Streptomycin 0.03, Agar 15.
7. **Waksman's medium g/L in distilled water**
   Specially used for counting soil actinomycetes
   Dextrose 10, Peptone 5, KH_2PO_4 1, $MgSO_4.7H_2O$ 0.5, Agar 15, pH adjusted to 4.0.
8. **Sabraud's Medium g/L in distilled water**
   Medium is suitable for many fungi including human or animal pathogens.
   Glucose (or Maltose) 40, Peptone 10, Agar 15.

13.19.2 Synthetic Media

1. **Czapek's solution g/L in distilled water (for soil inhabiting fungi)**
   Sucrose 30, Sodium nitrate 2, K_2HPO_4 1, $MgSO_4.7H_2O$ 0.5, KCl 0.5, $FeSO_4$ 0.01,
2. **Coon's medium g/L in distilled water (For *Fusarium*)**
   Dextrose 3.6, Saccharose 7.2, $MgSO_4.7H_2O$ 1.23, KH_2PO_4 2.72, KNO_3 2.02.
   Add 25 ppm of malachite green or 40 ppm Gention Violet.
3. **Richards solution (for *Aspergillus*) g/L in distilled water**
   KNO_3 10, KH_2PO_4 5, $MgSO_4.7H_2O$ 1.23, $FeCl_3$ 0.02, Sucrose 50 and pH adjusted to 6,6 – 7.2

13.19.3 Special Techniques Used for Fungal Culture

One of the principal aims of laboratory culture is to identify the fungi present in a sample, and a commonly employed taxonomic technique is to observe the spore structures that are produced by a mold. The preparation of a slide mount obviously causes considerable disruption to the structure of the fungus to be examined. This may cause severe difficulty in visualising all the necessary morphological features of the specimen. To overcome this difficulty, there are methods of visualising structures *in situ* without disturbing the fungal growth. The commonest of these are *slide culture, cellophane culture* and *bait technique.*

13.19.3.1 Slide culture

Slide culture depends upon the hyphae and sporophores attaching themselves to the glass of a coverslip or microscope slide, and thus the technique is not suitable for the examination of all fungi. To prepare a slide culture, a sterile glass slide is placed on a bent glass rod in a Petri dish. The glass rod raises the slide above a sterile, moist pad, used to keep the culture hydrated. A block about 1 centimetre square is cut from a solid growth medium that is known to support the growth of the fungus under examination. The block is then transferred aseptically to the centre of the slide, and each edge of the square is inoculated with the fungus. A sterile coverslip is placed over the inoculated block, and the culture is then incubated. When sufficient growth has occurred, the coverslip is removed and a drop of alcohol is placed on the coverslip to dispel any air from the fungal material. The preparation is then stained with a drop of lactophenol cotton blue. The coverslip is then inverted over a clean slide and the culture is ready for examination under the microscope.

An advantage of the slide culture method is that two preparations can be made from one culture. The second preparation is made by removing the agar block from the supporting microscope slide, treating with alcohol, staining the preparation as before and covering the culture material with a clean coverslip (Figure 13.23).

13.19.3.2 Cellophane culture

Cellophane cultures exploit a similar process. Small pieces of sterile cellophane, about 1.5 centimetres square are placed on the surface of a solid growth medium, and are

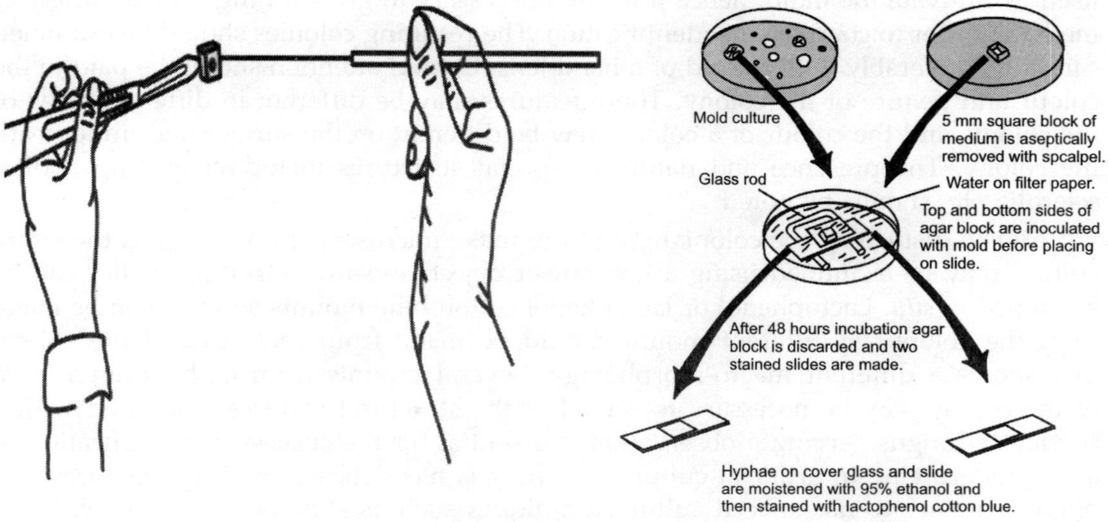

Figure 13.23 Slide culture technique.

inoculated with the fungus to be studied. Fungi generally grow well on the cellophane support, obtaining their nutrients by diffusion through it. When sufficient growth has occurred, the cellophane square is removed and placed on a slide. Alcohol and then lactophenol cotton blue are added as above, and a clean coverslip is placed on top of the mount. It is best to use only a small inoculum when making cellophane preparations, since it is not possible to make preparations from large colonies grown in this way. Ideally a mixture of mycelia and spores is used to initiate the culture.

13.19.3.3 Bait technique

With certain aquatic fungi, agar cultures can be produced in the conventional manner, but spores are formed only in natural waters. To induce spore formation in these fungi, the culture has to be introduced into natural water. This is performed by using 'bait', commonly a mixture of small seeds. Cress, onion or hemp seeds are first sterilised by autoclaving, and their seed coat removed with sterile forceps. These are added, together with the fungus to be studied, to a sterile mixture of coarsely filtered pond water and distilled water. The fungus will then grow on the bait, and may be easily retrieved from the culture for examination when the spore structures have had time to develop.

Fungi may be cultured from natural environments using baiting techniques. For example, fungi able to attack keratin substrates can be cultivated from soil by burying hair or even tennis balls in the earth. Similarly, leather, foodstuffs and other natural substrates can be used as bait to attract specific fungi. Indeed, it is not an uncommon practice to use leather and sterilised dung to cultivate fungi in the laboratory.

13.19.4 Identification of Molds

The identification of a mycelial fungus involves the study of both macro- and micromorphology. These will vary depending upon the growth medium and temperature used to cultivate the mold, hence it is often necessary to grow a fungus on a variety of media in order to complete its identification. The resulting colonies should be examined carefully, preferably with the aid of a hand-lens. Careful attention should be paid to the colour and texture of the colony. These features may be different in different areas of the colony, and the colour of a colony may be different on the surface and underneath the colony. The presence and nature of special structures including fruiting bodies, sclerotia etc. should be noted.

Prior to disturbing the colonial growth to make microscopic preparations the entire culture may be examined using a low-power objective, so that fruiting bodies can be examined *in situ*. Lactophenol or lactophenol cotton blue mounts should then be made from the colony. A different mount should be made from each area of the colony that shows a different macro-morphology. Several mounts from each different area of the colony may be necessary to see all of the structural features associated with a particular fungus. Accurate observation is essential for the successful identification of a fungus growing in artificial culture. If a fungus has delicate sporing structures it is often necessary to use special culture techniques such as slide or cellophane cultures that allow the micro-morphology of the fungus to be examined undisturbed.

It may be difficult to differentiate fruiting structures such as pycnidia. The best way to identify these structures is to squash them to see what comes out, perithecia, cleistothecia and sclerotia from each other. Sometimes sclerotia are difficult or impossible to squash. Sclerotia yield numerous oil droplets. Large numbers of conidia are released from squashed pycnidia, and asci and ascospores are released from cleistothecia and perithecia. Ascospores and asci tend to ooze out of their fruiting structures.

When all the observations necessary have been made, and the structures and their dimensions carefully noted, then the culture may be identified after reference to an identification key. There are several books that contain identification keys for fungi.

13.19.4.1 Isolation and identification of fungi

1. Prepare 6 plates of potato dextrose agar media composition described earlier.
2. Label them as A, B, C, D, E and F.
3. Inocu1ate A and B with soil, C and D with water and expose the remaining ones to air.
4. Each sample should be diluted serially.
5. Incubate at 30°C for 48 hrs.
6. Examine the plate cultures for gross morphology.

13.19.4.2 Materials

Soil sample, potato dextrose agar medium, nutrient agar medium, distilled water, pipette, sterile Petri plates, sterile test tube, inoculating loop, distilled water, bunsen burner, wax marking pencil, etc.

13.19.4.3 Protocol

1. Prepare nutrient agar medium and Potato Dextrose Agar medium in sterile flask.
2. Separate rhizospheric soil from 4 to 5 roots and collect in Petri plates.
3. Add about 1 g soil in 10 mL distilled water in a test tube and mark it as 10° or crude.
4. Now serially dilute the soil sample by transferring 1 mL of previous soil solution from test tube to 9 mL distilled water in next tube.
5. Prepare dilution upto 10^{-6} (Figure 13.26).
6. Transfer 1 mL of soil suspension from each dilution to sterile Petri plates.
8. Now pour the molten and cooled Potato Dextrose Agar medium and Nutrient agar medium in various plates.
9. Incubate the plates for 24 - 48 h at 37°C for bacterial growth and 2-6 days at 27°C for fungal growth (Figure 13.24 and Figure 13.25).

Observe the plates for growth and note different type of colonies on the plates. Record the colony morphology and identify the organisms by following the litreature (Figure 13.27).

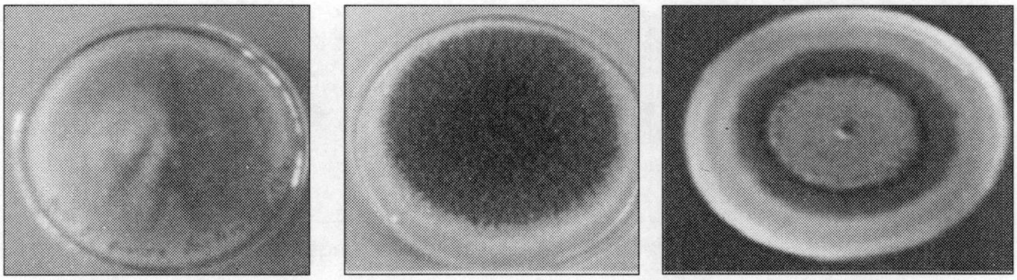

Sectoring in culturing of fungi in *Fusarium sp.*, *Curvularia* sp. and *Aspergillus* sp.

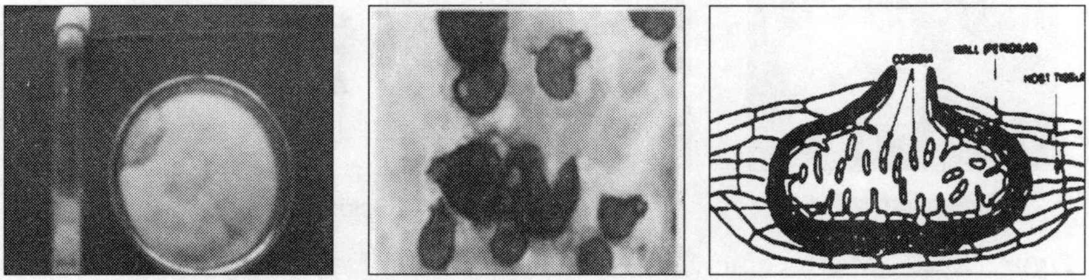

Phoma herbarum | *Pycnidia of Phoma herbarum* | *Conidia and conidiophore*

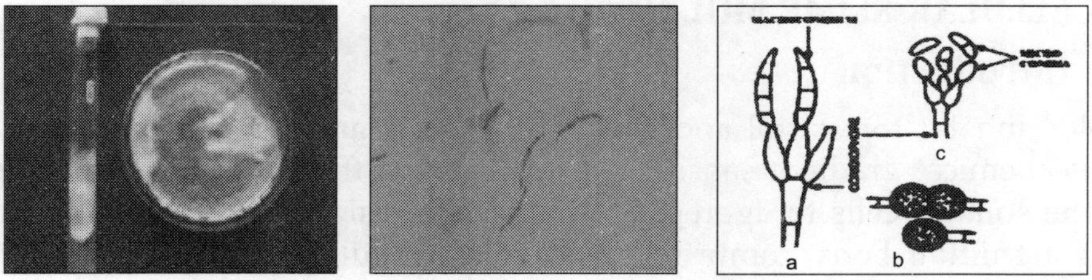

Fusarium roseum | *Micro and Macroconidia* | *Macro, Micro and Chlamydospore*

Figure 13.24 Microscopic view of some common fungi.

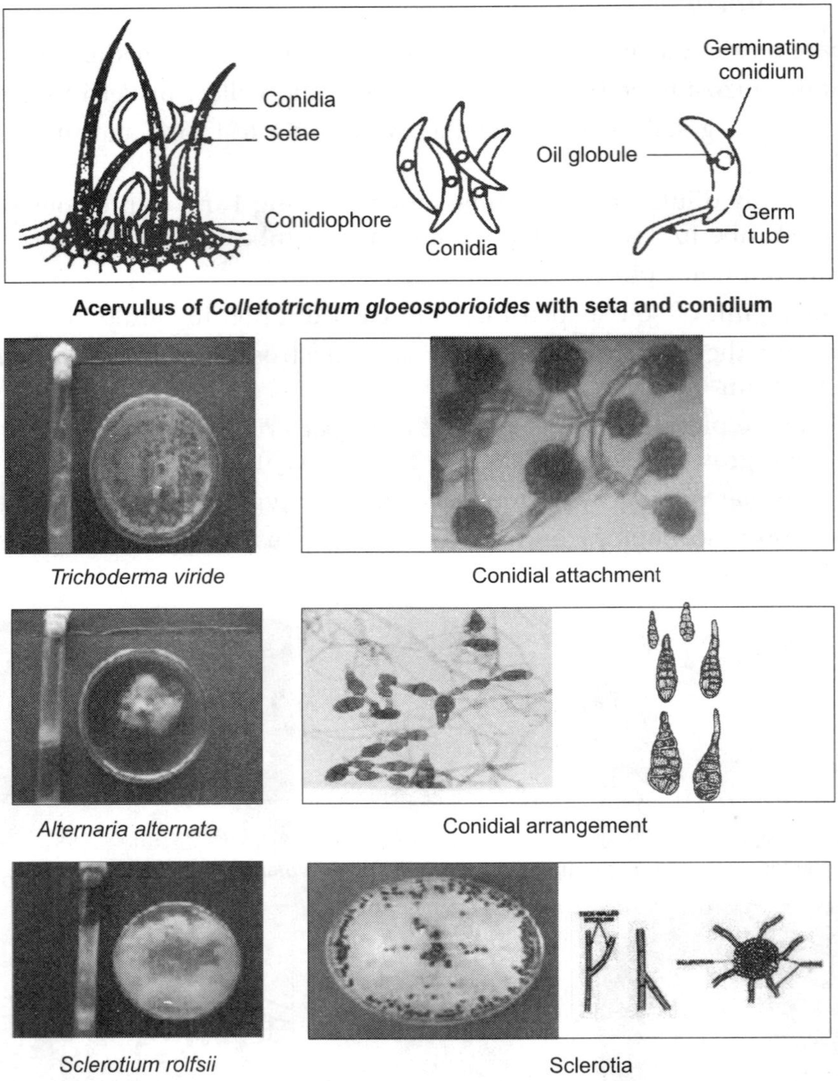

Figure 13.25 Microscopic views of some common Deuteromycete members.

13.20 CELLULAR SLIME MOLDS

13.20.1 Introduction

Slime molds inhabit forest soil and consume bacteria and yeast, which they track by chemotaxis-chemical gradient sensing (of folic-acid in this case). Starvation, however, prompts the solitary cells to aggregate and develop as a true multicellular organism, producing a fruiting body comprised of cellular, cellulosic stalk supporting a bolus of spores. Thus, slime mold has evolved mechanisms that direct the differentiation of a homogeneous population of cells into distinct cell types, regulate the proportions between tissues and orchestrate the construction of an effective structure of the dispersal

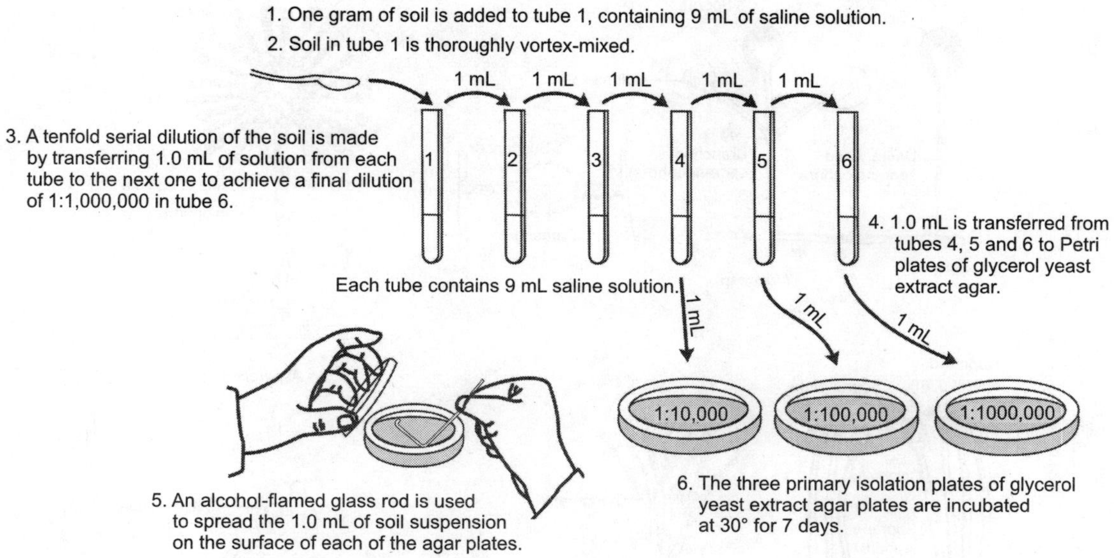

Figure 13.26 Diagrammatic view for preparing serial dilution of soil sample.

of spores. Many of the genes necessary for these processes in slime mold were inherited by Metazoa (animal with specialized cells) and fashioned through evolution for use within many different modes of development. Analysis of the proteins shows that slime mold diverged from the animal-fungal lineage after the plant-animal split, but it seems to have retained more of the diversity of the ancestral genome than have plants, animals or fungi, i.e., it possesses a level of complexity that is greater than the yeast, but much simpler than plants or animals. It represents one of the earliest branches from the last common ancestor of all eukaryotes. As described earlier, cellular slime molds are not true fungi. A typical example of this group is *Dictyostelium discoidium.* It starts its life cycle with the dispersal of spores from a sporangium (Figure 13.28). Upon the absorption of water, the wall of the spore cracks open and a single amoeba like cell, a *Myxamoeba,* emerges. These cells live independently, as they move about feeding on bacteria, they grow and reproduce by fission. This, free living, feeding stage ends following depletion of nutrients, the myxamoeba stream to an aggregation centre and forms a *Pseudoplasmodium.* The *Pseudoplasmodium* migrates as a unit; leaving behind a trail of slime. When the migration ceases, the cells of the *Pseudoplasmodium* differentiate to form a sporangium with spores. The cells from the front, or apical, region of the slug organise to form a stalk. Cells from posterior region, move up the stalk as a ball of cells and become tip of the stalk. These cells transform into spores, each capable of initiating another life cycle of this slime mold.

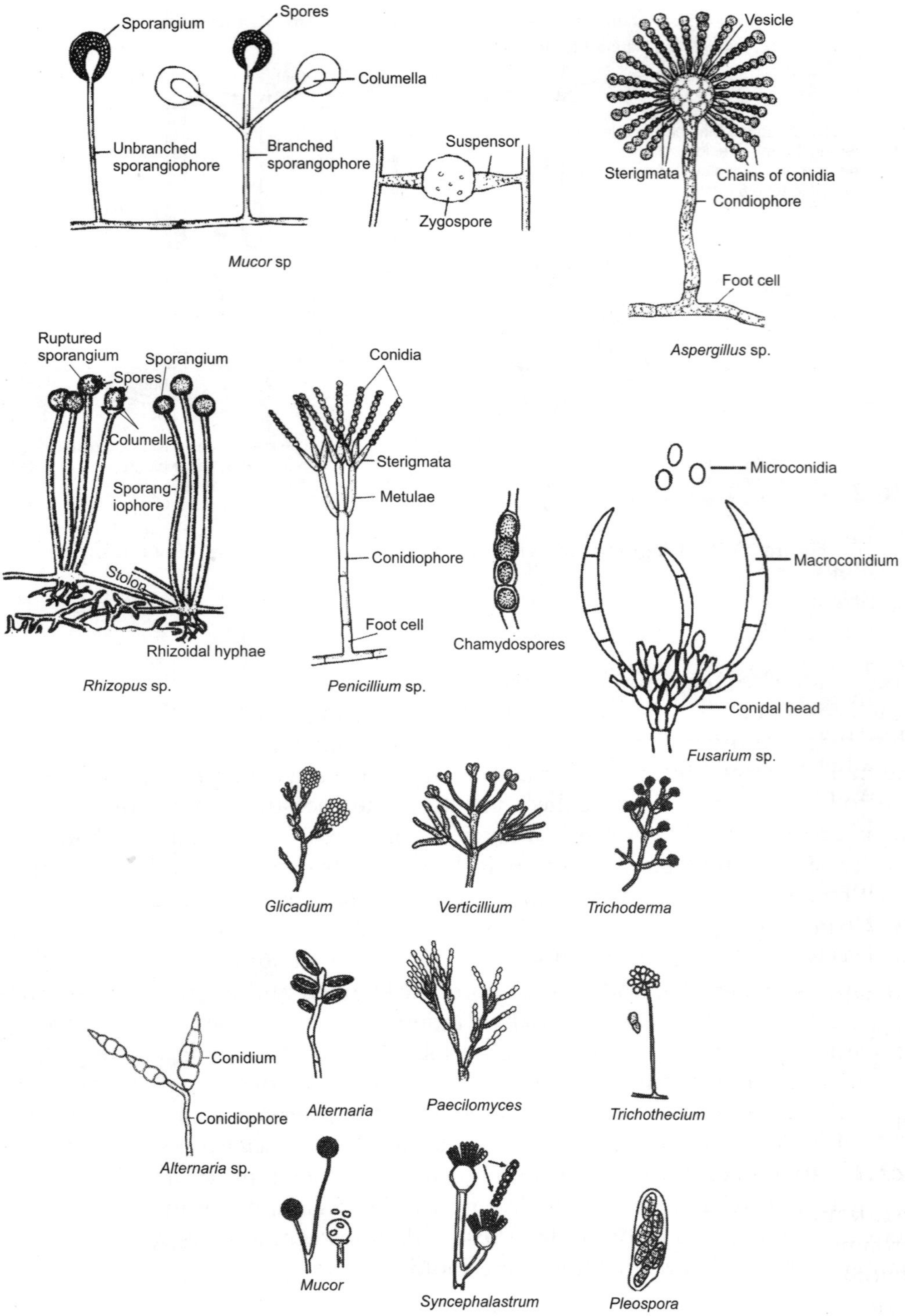

Figure 13.27 Diagrammatic view of some common soil fungi.

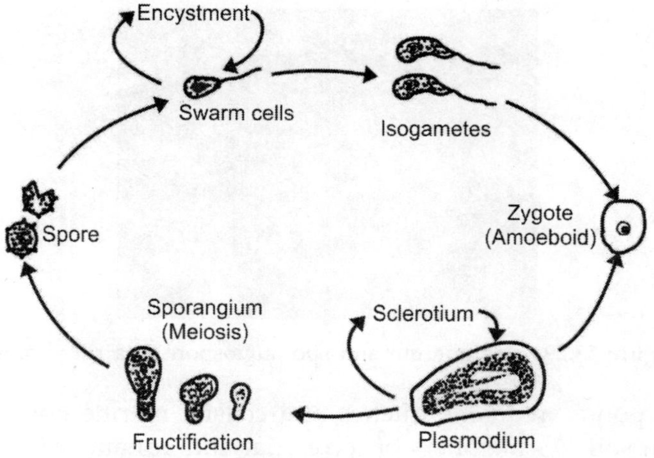

Figure 13.28 Life cycle of slime mold.

13.20.2 Procedure

1. Prepare a nutrient agar plate. Using a wax pencil draw two lines forming a right angle on the bottom of the agar plate.
2. Streak a loopful of *E. coli* on the agar surface along each of the lines marked on the bottom of the plate.
3. Inoculate a loopful of *Dictyostelium discoideum* at one end of each bacterial streak. Incubate the plate at room temperature for 2 days.
4. After incubation, examine the growth on the plate with the naked eye and with a hard lens. It is advisable to observe the plate on the stage of a microscope and examine under the 4 and 16 mm objectives. If an aggregation is visible disrupt the aggregate with a needle to observe the component cells. Repeat the procedure with a *Pseudoplasmodium*. If no fruiting bodies are detected, reincubate the culture for 2 days, and then re-examine.
5. *Dictyostelium discoideum* can be isolated from the sample using following *Dictyostelium* medium (Figure 13.29).

Composition of *Dictyostelium discoideum* medium g/L in distilled water

Proteose peptone 10, Yeast extract 5 , Glucose 10 , Sodium acetate* 0.35, KH_2PO_4 0.35, pH 6.4-6.6. * In place of sodium acetate 2.3 mL of glacial acetic acid can be used.

13.21 ISOLATION OF CHYTRID

13.21.1 Introduction

Chytridiomycetes are true fungi that require water to disperse throughout their environment. Unlike ascomycetes, basidiomycetes, or zygomycetes, chytrids, as they are commonly called, produce flagellated, asexual reproductive cells known as zoospores. Chytrid zoospores are readily dispersed in the presence of free water but this need for water does not restrict them solely to aquatic habitats. Besides being commonly found

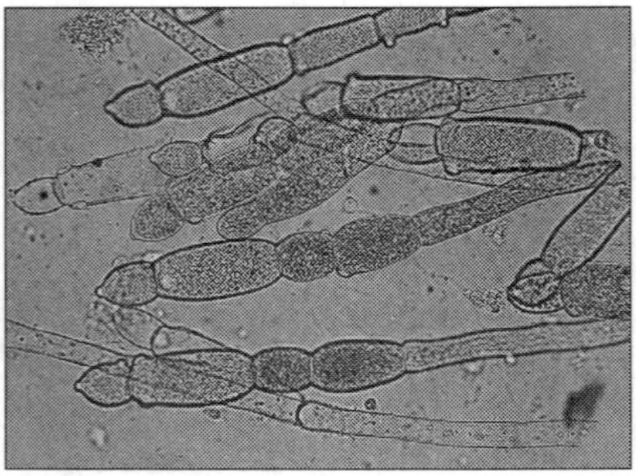

Figure 13.29 Sporangium and sporangiospore in a slime mold.

in lakes, streams, ponds, roadside ditches and coastal marine environments, chytrids also are present in soil. As members of terrestrial and aquatic microbial communities, chytrids play an important ecological role in the degradation of recalcitrant materials like keratin, chitin, and cellulose. Chytrids live saprobically or as parasites in, or on, a number of different organisms and substrates such as pollen grains, insect exoskeletons, protists and small invertebrates, amphibian skin, other fungi, pieces of plants, fruits, and waterlogged twigs.

Chytrides can be isolated by baiting in soil and water samples. When baiting water samples, place the sample into a dish with some of the accompanying organic material and then add a few pieces of bait. The type of bait used will depend on the kind of chytrid you are seeking. The size of the bait should be big enough to be retrieved but small enough to be examined on a microscope slide (about this big). For soil samples, to avoid the culture from being overrun by bacteria add only one to two grams of soil to a dish half filled with sterilized water. The type of baits commonly used include purified shrimp exoskeleton (chitin), snake skin, defatted blond baby hair (keratin), cellophane, lens paper, boiled or autoclaved dialysis tubing, white onion skin, hemp rope, corn straw (cellulose), and pollen (pine pollen, spruce pollen, sweetgum pollen). Another way to bait for chytrids is to use bait bags. Bait bags are made from mosquito netting and can be used to determine what chytrids are active in aquatic habitats. Several pieces of bait and two or three marbles are put into the bags. The bags are then placed into the lake, pond, stream, bog, or ditch for a week or more before being re-collected. The length of time the bait bags remain depends on the time of year and a good rule of thumb is a week when water temperature is above 20°C and longer at lower temperatures. The baits are then examined microscopically for chytrids back in the laboratory.

Another good thing to do is to record the temperature of the water. This is important later when culturing because some cultures may grow better at the temperature of the environment from which they were isolated versus room temperature. Allow a couple of weeks before discarding them as some chytrids take longer time to appear than others. This is especially true with members of the *Nowakowskiella* clade, which can take up to three weeks to appear on onion skin and corn straw baits (Figure 11.12).

13.21.2 Isolating into Pure Culture

If you see mature sporangia of something you want, isolate it immediately before it becomes overgrown by other undesirable organisms. Two main ways of isolation involve harvesting zoospores from pieces of bait placed in depression slides and scraping sporangia from bait onto agar media.

With the depression slide method, cut out the part of the bait with the greatest number of mature sporangia and place it in the depression with a drop or two of distilled water. Watch and wait for zoospores to be released. Once zoospores are swimming around, pick them up with a clean Pastuer pipette and transfer the zoospore suspension to an agar plate. Make a mark with the tip of the pipette on the agar surface where the drop of water will be placed for later reference when checking inverted plates with the compound microscope. Then pipette all of the liquid onto the agar and tip the plate so the water can spread. Allow the free water on the surface of the agar to dry before wrapping the plate junction with laboratory film (e.g., Parafilm "M" ™ American National Can).

The other method of isolation involves scraping sporangia from the bait onto agar medium. While looking through the dissecting microscope, a sharpened needle is used to move sporangia from the piece of bait onto the surface of an agar plate. The sporangia are then picked up with the needle and dragged through the agar to clean them of bacteria, yeast, etc. When cleaning sporangia by dragging through agar, it is easier to move the plate than the needle. One can also repeatedly pick up and put down the sporangium on fresh agar as a way to remove bacteria. Once sporangia are cleaned, they are placed individually onto a new agar plate. It is a good idea to circle the location of each sporangium on the bottom of the plate as this makes them easier to find when looking under the microscope.

Common media used for isolation include PMTG, mPMTG, corn meal agar, and 1/4 strength YPSS. Isolation plates should be examined daily for growth of chytrids. Chytrids usually grow slower than the competing (other fungi, Oomycetes, bacteria, etc.) contaminants and should be removed by cutting out them aseptically to prevent any further contamination. Removing contaminants (weeding) allows the chytrid time to develop before it is moved to a clean plate. Maintaining chytrid cultures involves transferring to new nutrient agar slants, preferably in screw top test tubes, every six weeks at room temperature or every three months if cultures are kept at 4°C. But as with isolating chytrids some cultures require different ways of keeping them going, so do not be afraid to experiment to see if something else will work. You never know until you try!

13.22 IDENTIFICATION OF YEASTS

Yeasts are relatively simple structures and, paradoxically; this makes their identification more difficult than that of other fungi. Identification of yeasts is founded upon their limited morphological differences and on their biochemical properties. Morphological features used in the identification of yeasts include such things as cell size, shape and the presence or absence of a capsule. Some yeast possess the ability to produce pseudohyphae, whilst a minority can produce a true septate mycelium identical with that found in molds. Some yeast is capable of sexual reproduction to produce ascospores. A small number of yeasts reproduce sexually to produce basidiospores. The capsulated yeast *Cryptococcus neoformans* that causes meningitis, particularly in patients with AIDS, is yeast that belongs to the Basdiomycotina. *Candida albicans* is an important commensal and opportunist pathogen of humans. It can be identified from clinical specimens by its ability to produce a mycelial **germ-tube** when incubated for a period of 1-2 hours in serum at 37°C. The biochemical tests involve observing the ability of the isolate to assimilate various compounds, mainly sugars and nitrogen sources, in addition to the ability to ferment different sugars. The identification of yeasts has been standardised and there are commercially available strips that can be inoculated to test for carbon and nitrogen assimilation. Depending upon the pattern of substrate usage an identification profile can be generated, and the identity of the yeast obtained either by reference to a manual or by using a computer database.

13.22.1 Isolation of Yeast from Soil Ripened Grape

13.22.1.1 Introduction

Yeast is a group of unicellular fungi, which are cosmopolitan in distribution. Most yeast are placed in Ascomycotina of the kingdom Mycota. They have been isolated from a natural substance like leaves, flower, sweet fruits, grains, bakery product and soil. Yeast plays a major role in many industry, viz., fermentation and food industries (*Saccharomyces cerevisiae*), production of single cell protein (*Candida succiphila*) animal feed industries (*Candida utilis*) and natural decomposition of organic matter in soil and water. While it is common colonizer of human mucosal surfaces and considered being nonpathogenic for immunocompetent hosts and may cause infections particularly in immunocompromised patients. Morphologically, yeast are Gram-positive, nonfilamentous, unicellular, fungi that are typically spherical or oval in shape. They reproduce by budding, a process in which new cells are formed as protuberance (bud) from the parent cell and separates from it upon reaching maturity. Sometimes daughter cells fail to detach themselves, a short chain of cells is formed which is known as pseudomycellium or pseudohypha (Figure 13.30).

13.22.1.2 Materials

Glucose yeast extract (GYE) agar medium, sterile Petri plates, sterile test tube, distilled water, glass rod, Bunsen burner, inoculating loop, wax marking pencil, etc.

Composition of GYE agar medium g/L

Yeast extract 5, Glucose 20, agar 15 pH adjusted to 7.6 at room temperature

13.22.1.3 Protocol

1. Place a ripened grape into sterile distilled water and crush with the help of sterile glass rod.

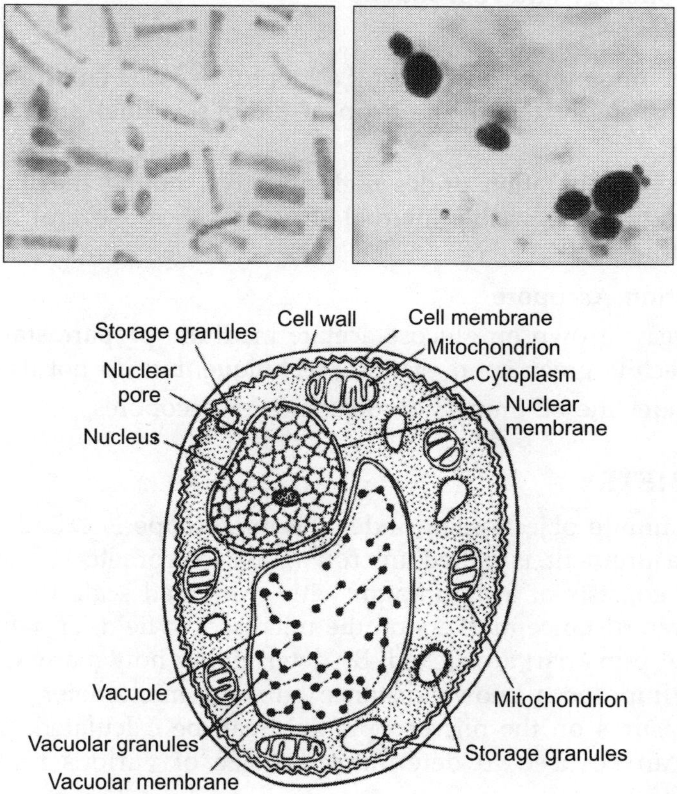

Figure 13.30 Yeast cells stained on microslide and growth in media.

2. Perform Gram's staining of the suspension and observe for yeast cell in smear.
3. Take loopful suspension and streak on GYE agar plate, incubate the plate at 27°C for 24 to 48 h.
4. After incubation observe for colonies of yeast and note down cultural characteristics.
5. Pick up a well isolated colony of yeast and prepare the suspension in sterile distilled water.
6. Perform the Gram's staining of the suspension and verify the purity of the yeast culture.
7. If necessary, preserve the culture by transferring the culture on GYE agar slant.

13.22.1.4 Results and observations

Since, yeasts are unicellular microorganisms, their growth resembles a bacterial colony rather than fungal colony. Growth is rapid, and colony is formed within 24-48 hrs. Colony size ranges from 2-3 mm and have flat, slightly raised or convex elevation. Most colonies are smooth, dull and have soft pasty consistency with characteristic alcoholic smell; the colony appears to have white, creamy, dirty white or light brown pigments.

13.22.1.4.1 Morphology and reproduction

Vegetative cell

1. Prepare wet amounts by suspending a loopful of yeast culture in a water iodine solution (3 drops water to one drop of Gram's iodine) and layer each with a cover slip.
2. Observe each preparation under high-objective, noting the morphology of the yeasts, including any visible internal structures. Scan several microscopic fields to find budding cells.

Sexual Reproduction Ascopore

1. From the yeast grown on glucose acetate medium, prepare stains of the culture by the malachite green spore staining technique, but do not use heat.
2. Observe under the oil-immersion objective for ascopores.

13.23 MICROMETRY

Measurement of minute objects seen under the microscope is called Micrometry. The unit of such measurement is called micron (μ or micrometer). The ocular or eye-piece micrometer consists of a diaphragm with graduated scale unit. The ruled lines superimpose certain distance markers on the microscopic field, on which parallel lines approximately 0.01 mm apart are etched. By determining how many units of the ocular micrometer superimposed a known distance on stage micrometer, the exact distance each division measures on the microscopic field can be calculated. Once caliberated, the micrometer can be used to determine the size of various microscopic objects (Figure 13.31).

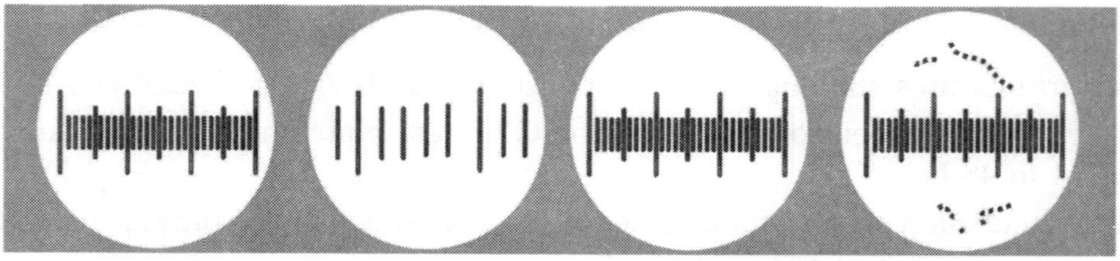

(a) Ocular micrometer in microscope eyepiece divisions arbitrary.

(b) Stage micrometer distance absolute; each division equals 10 micrometer.

(d) The ocular micrometer super-imposed on the stage micrometer. In this lens system what is the absolute length in micrometers covered by 10 ocular divisions?

(e) Stage micrometer (scale b) is now replaced by a slide. From your calculations in C, what is the length of the chain of streptococcl?

Figure 13.31 Standardization of the ocular micrometer and its use in measuring bacterial cells (1 micron = 1 μm = 1 × 10^6 metres).

13.23.1 Procedure

1. Insert the ocular micrometer in the ocular lens. Replace the ocular lens in microscope and observe.
2. Mount the stage micrometer on microscope stage and micrometer with low power objective observe under oil immersion objective.
3. Rotate the ocular so as to superimpose the lines of ocular micrometer upon those of stage micrometer. Begin with the lines of two micrometers coinciding at one end of the field, count the spaces of each micrometer to a point at which lines of micrometer coincide again. This finishes the standardisation.
4. Replace the stage micrometer with the object and calculate the results in micrometers.

13.24 MUSHROOM IDENTIFICATION

13.24.1 Introduction

Mushrooms can be found in forests around the world. Given the proper environment, mushrooms will grow and can offer a good source of natural vitamins and minerals. Mushrooms can also bring illness and even death to people who are unaware of certain types of wild mushrooms. Cultivated mushrooms are therefore the preferred and most reliable source of supply. Mushrooms are commonly used for various dishes in different shapes and forms. The most commonly and easily cultivated mushrooms around the world are white button *(Agaricus brunnescens (bisporus)*, oyster mushrooms (*Pleurotus ostreatus*), ear mushrooms (*Auricularia polytricha*), and straw mushrooms (*Volvariella volvacea*), Shiitake *(Lentinula edodes)* Reishi mushroom *(Ganoderma lucidum)*. *Macrocybe* sp., *Agrocybe* sp. types can also be cultivated successfully but will require more attention and knowledge. Mushrooms (often refers to the fruiting body of the gill fungi), yeasts and algal foods are frequently mentioned as alternative sources of food. Of these, mushrooms are the most preferred. Modem mushroom culture produces more protein per unit area of land than any other kind of agriculture and technology at present available. Mushroom farming is becoming successful because of its very low inputs. It is estimated that about 300 million tonnes of fresh mushroom can be produced from just one-fourth of world's annual yield of straw (2325 million tonnes). It was calculated that approximately 317 million metric tonnes of fresh mushroom could be produced annually that would provide 197 g of fresh mushrooms daily to each person in the world.

13.24.2 Morphology

Mushrooms can be defined as "a macro-fungus with distinctive fruiting bodies, epigeous or hypogeous, large enough to be seen with naked eyes and picked up by the hands". The mushroom fruiting body may be umbrella like or of various other shapes, size and colour. Commonly it consists of a cap or pileus and a stalk or stipe but others have additional structures like veil or annulus, a cup or volva. Cap or pileus is the expanded portion of the carpophore (fruit body) which may be thick, fleshy, membranous or corky. On the underside of the pileus, gills are situated. These gills bear spores on their surface and exhibit a change in colour corresponding to that of the spores. The attachment of the gills to the stipe helps in the identification of the mushroom (Figure 13.32).

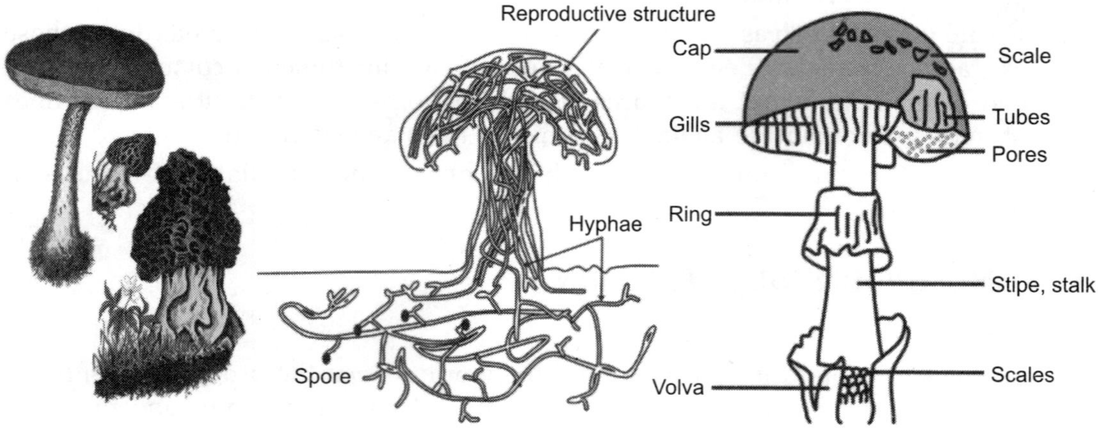

Figure 13.32 Morphology of mushrooms.

Mushroom as a vegetable has become popular due to its food value in terms of protein and medicinal contents. They also show potential for use in waste management. However, the successful operation of mushroom cultivation depends on the proper knowledge of laboratory techniques and practices involved in spawn production.

Mushrooms are the fleshy fungi which constitute a major group of lower plant kingdom. The mushroom is a common fungal fruit body that produces basidiospores at the tip of club-like structures, called *basidia,* which are arranged along the gills of the mushroom. Beneath the mushroom, in the soil, is the mold colony itself, consisting of a mat of intertwined hyphae, sometimes several feet in diameter. The mushrooms first appear as white tiny balls consisting of short stem (stipe) and a cap (pileus), which begin to open up like an umbrella (Figure 13.32 and Figure 13.33). The delicate membrane or veil (velum) enveloping the cap tears off, if allowed to develop fully, and lamellae (gills) radiating from the stalk into the cap become visible. These gills become darkened as the basidiospores (seeds) develop into millions and fall to the ground for starting their life cycle once again for second generation of mushrooms. Since mushrooms grow independently of sunlight so they can be grown in complete darkness but the darkness is not an essential prerequisite. They are relatively fast growing, do not require fertile soil, since grown on composted or uncomposted agro-wastes and their culture can be concentrated within a relatively small space. In addition to floor, air space is also utilized resulting in higher production. It is a labour intensive indoor activity which can help the landless, small and marginal farmers to raise their income, diversify economic activity and can create gainful employment especially for unemployed/under-employed youths, weaker section of the society and women folk.

Agaricus campestris is one among several types of mushroom seen growing wild (Table 13.18). It is popularly called temperate mushroom or *'Khumb'* and grows on dead organic matter under suitable environmental conditions. It derives its carbonaceous food by decomposing lignin, cellulose and hemicellulose present in agro-wastes with the help of extracellular enzymes secreted by the mycelium. Microbic protein available in compost is the chief source of organic nitrogen for its assimilation.

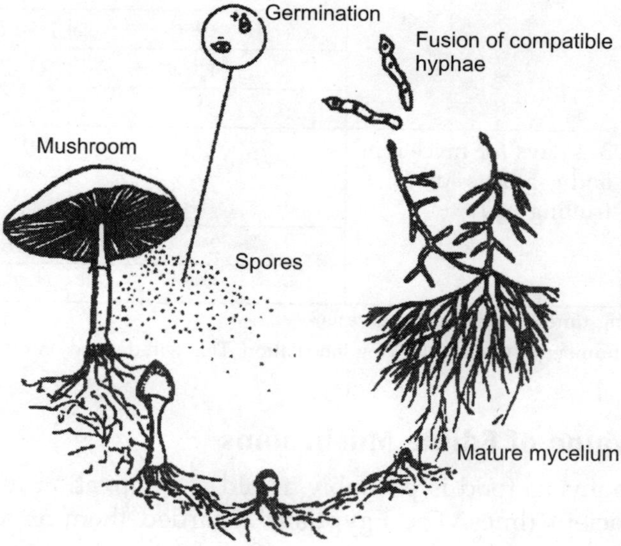

Figure 13.33 Life cycle of mushroom in natural environment.

Table 13.18 Incubation period for the cultivation of few selected tropical mushrooms

Type of mushroom	*Incubation time (weeks)*	*Mushroom Flushes**	*Production time** (weeks)*
Oyster mushrooms (Pleurotus *ostreatus*)	4	1st	5
		2nd	8
		3rd	11
		4th	14
		5th	17
Ear mushrooms (*Auricularia polytricha*)	4	1st	4–5
		2nd	8–9
		3rd	11–12
		4th	14–15
		5th	Beware of mites
Lentinus squarrosulus	4–5 or more	1st	5–6
		2nd	8–9

Contd...

Contd...

		3rd	11–12
		4th	14–15
		5th	17–18
Lentinus polychrous	4-5 or more	1st	5–6
		2nd	8–9
		3rd	11–12
		4th	14–15
		5th	17–18
Straw mushrooms (*Volvariella volvacea*)	3–4 days for mycelium and 4–5 days for fruiting body.	1st	7–9 days
		2nd	14-16 days
		3rd	21-23days

*Flushes means harvesting time or number of harvests

**Production time is the number of weeks following inoculation. This will depend on the season and to the amount of care given by farmer.

13.24.3 Food Value of Edible Mushrooms

The use of mushrooms as food is probably as old as civilization itself. They have been a delicacy since ancient times. The Egyptians regarded them as a food for Pharaohs. The Greeks and Romans described them as *'food for the Gods"* and were served only on festive occasions. These were earlier preferred for their flavour and taste while their nutritive value was recognized later. Mushrooms provide a rich addition to the diet in the form of proteins, carbohydrates, minerals and vitamins. The protein content of fresh mushrooms (3g/100 g) is about 3.7 per cent. It is twice as high as that in most vegetables except green peas, Brussels sprouts and pulses and is much lower to meat, egg, fish and cheese. They have a high percentage of all the nine essential amino acids with traces of sugar and without cholesterol having low caloric value (less than 35 K cal per 100 g), richer in vitamins (B1, B2, Niacin, B12, pantothenic acid and vitamin C) than most vegetables and almost free from fat (O.2g / 100 g), a good source of minerals P, K, Fe and Cu recognizing them as a valuable source of nutritive and protective food.

13.24.4 Cultivated Edible Mushrooms

Presently, about a dozen fungi are cultivated in over 100 countries with a production of 2.2 million tones. Five genera, *Agaricus, Lentinus, Volvariella, Pleurotus* and *Fiammulina,* contribute about 91 per cent of total production. White button (*Agaricus brunnescens* (*bisporus*) has the largest share (56%) followed by shiitake (*Lentinus edodes*) (14%), paddy straw (*Volvariella volvacea*) (8%), Oyster (*Pleurotus* spp.) (7.7%) and others (13%). USA produces about one-fourth of total white button mushroom production (0.35 million metric tonnes) followed by China (0.31), France (0.2), Netherlands (0.15), UK (0.12) and Taiwan (0.062). It is now grown in almost all the continents. Technology for its cultivation has advanced to a state in which this mushroom can be grown profitably in locations with climatic conditions quite different from those in Europe and America where mushroom culture had its beginning and greatest development.

Studies show that certain types of mushrooms have a direct impact on body activities (Table 13.19).

Table 13.19 Mushrooms as nutraceutical

Termitomite sp.	*Good for brain and memory*
Volvariella volvaceae	Heal wounds
Coprinus sp.	Help the digest and decrease phlegm
Auricularia sp.	Clean lungs
Agaricus sp.	Increase mother's milk
Hericium erinacius	Heal wounds in intestine
Pleurotus sp.	Decrease muscle malpighia
Lentinula edodes	Good for baby's cartilage
Flammulina velutipes	Good for liver
Agrocybe cylindraceae	Good for kidney and urine
Schizophyllum commune	Decrease leucorrhea
Dictiophora sp.	Cure dysentery and decrease rotting
Trimella fuciformis	Good for sperm, semen and kidney
Lentinus sp.	Control the whole body system

13.24.5 Important Medicinal Mushrooms

Mushrooms have a long history of use in traditional Chinese medicine. In fact it is estimated that in China more than 270 species of mushrooms are believed to have medicinal properties with 25% of them thought to have antitumour capability. *Ganoderma lucidum* (Reishi mushroom), *Coriolus versicolour* (turkey tail), *Grifola frondosa* (Maitake), *Lentinula edodes* (Shiitake), *Cordyceps* species (Keera ghas), *Tremella fuciformis, Poria cocos* and *Pleurotus* species (Oyster or Dhingri) have gained importance in modern medicine for their various pharmacological, immunological, anticancerous and other medicinal values.

13.24.6 Poisonous Mushrooms

The order Agaricales, commonly called gill fungi contains over 270 genera. These include the mushrooms, the toadstools and the boletes. The edible species are called 'mushrooms' and the poisonous ones, toadstools. These fungi are incapable of causing infectious diseases but produce toxic substances that poison a person who ingests them. These poisonous substances are collectively known as mycotoxins (myco = fungus + toxin = poison) and result in mycetismus (mushroom poisoning) following ingestion of poisonous mushrooms. These mushrooms contain lethal substances that destroy liver cells and excite the nervous system. The most deadly mushrooms are the **death cap**—*Amanita phalloides*, the closely related **destroying angles**—*Amanita virona*, the **fool's cap**—*Amanita verna* and the **fly-agaric**—*Amanita muscaria*. In the **death cap**, toxic principle is a mixture of α and β **amanitin** and **phalloidin**—both are complex cyclic polypeptides containing sulfur. Cooking does not destroy the toxin nor it is affect by the human digestive juices. Symptoms of poisoning appear only after 8-24 hours of ingestion and by the time the toxin is absorbed by the body; neither vomiting nor a stomach pump can help then. Eating poisonous mushrooms may result in various types of reaction like nervous disorder, gastric disorder, haemolytic disorder and muscular disorder. Some time even edible mushrooms can cause indigestion in healthy people and some people may be allergic to a species which is harmless to others. Causes of discomfort and indigestion may be due to eating too much, or eating mushrooms with indigestible food, or having been incorrectly cooked. Mushrooms may also cause illness if they are taken with alcohol.

13.24.7 Cultivation of Mushrooms

The most common mushrooms that can be cultivated using straw as the principal ingredient are *Agaricus brunnescens* (= *A. bisporus*) and *Volvariella volvacea*. Other edible species that can be artificially cultivated using straw as main substrate are mostly species of *Pleurotus* and *Stropharia*. Another quite popular mushroom, *Lentinus edodes* can be grown on freshly cut wood of several tree species. Usually, the wheat and rice straw are used for the cultivation of species of *Agaricus* and *Volvariella*, respectively.

13.24.8 Identification of Mushrooms by Spore Print Method

Several methods are used to identify mushroom and one of the most common method adopted is the spore colours. One of the generic feature of mushroom is the colour of spores which varies in different genera, e.g., white, green, orange, brown, cream, etc. Though *Agaricus* and *Amanita* look alike from outer appearance, yet the spores of *Agaricus* are chocolate brown and that of *Amanita* white to pale. Spores are produced on basidia formed in gills of the basidiocarp.

The spore print is a simple method used for identification of gilled fungi. When the cap of mushroom is held stationary on a paper in a closed chamber for a long period, the impression of pattern of gills is printed, and the spores get accumulated onto the paper and invisible spores appear coloured (Table 13.20).

Table 13.20 Specific spore prints of mushrooms

Mushroom Type	*Genus*	*Spore colour*
Gilled mushroom	*Amanita*	White to pale
	Chlorophyllum	Green
	Clitocybe	Fleshy colour
	Hebeloma	Rusty to yellow brown

Contd...

Contd...

	Hygropharus	Cream
	Inocybe	Brown or cinnamon
	Lepiota	Green
	Naematoloma	Purple brown
	Panaceolus	Black
	Pleurotus	Lavender
	Russula	White, cream or yellow
	Schizophyllum	Pink, salmon or flesh
Tubed mushroom	*Boletus*	Yellow to olive brown
	Cyrodon	Yellow to yellow brown
	Suillus	Yellow brown to cinnamon
Puffball mushroom	*Scleroderma*	Purple black

13.24.8.1 Method

1. Take a fresh mushroom for preparing the spore print for identification. Carefully cut the gill with the help of sharp knife.
2. Place gill side down of the cap on a half black and half white paper in such a way that half of gill should be towards black side and half towards white side.
3. Place a drop of water on the cap of mushroom and cover it by a glass bowl.
4. Leave this set up for about 8 hours or overnight undisturbed (Figure 13.34).

Gently remove the glass bowl and cap from the paper. This area shows the print of gills and spore deposits. Carefully study the gill pattern and spore colour and identify mushrooms studied by this method. Match colour with the colour dictionary and identify the mushroom.

13.24.9 Isolation and Maintenance of Mushroom Culture

13.24.9.1 Method

1. Take a fresh mushroom for tissue culture, clean the surface with 70% aqueous ethyl alcohol, transfer a small piece with the help of inoculation needle under laminar air flow.
2. Clean both hands and bottles with alcohol and insert hands into the laminar air cabinet.
3. Hold needle with 2 fingers in a 45° angle, flame needle to disinfect until the needle turns red. Make sure it does not touch any surface after flaming.
4. While needle cools down (15-20 seconds - hold needle not to touch anything or place it on the clean surface of a glass).
5. Using other fingers, tear mushroom lengthwise (do not use knife to cut).

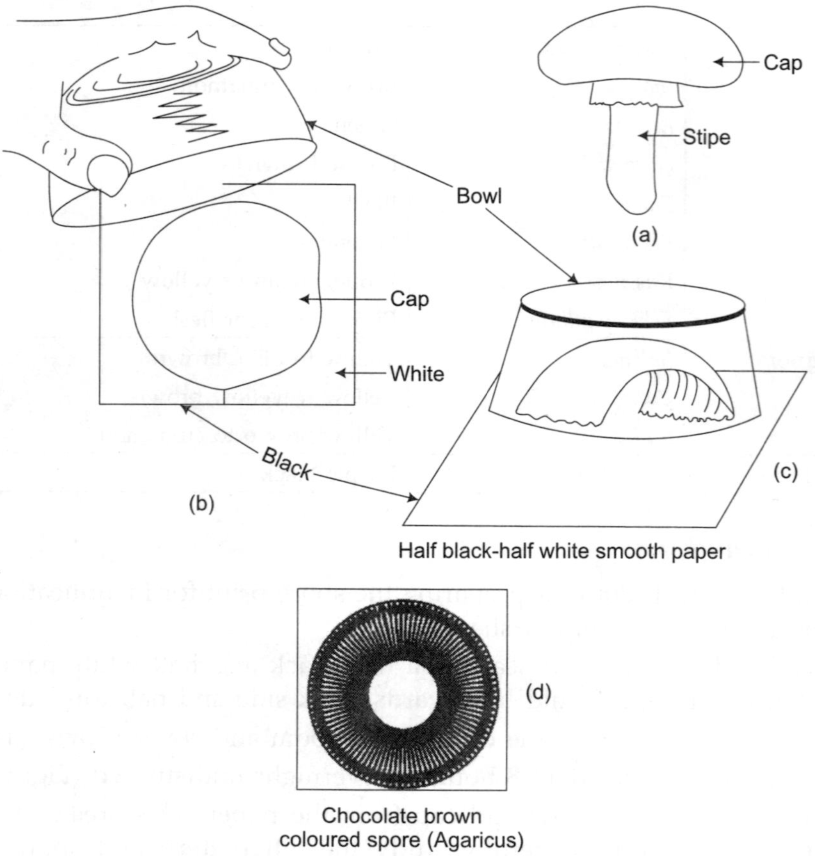

Figure 13.34 Flow diagram explaining spore print method for identification of mushroom.

6. With the needle, cut a small piece (2 mm × 2 mm) of fleshy tissue from inside the mushroom (in the middle between the cap and the stalk). Make sure that it is clean and does not touch the outside of the mushroom.
7. Flame around the PDA plate. Remove lid of PDA plate in front of flame to secure against contamination using other fingers. Inoculate the mushroom flesh on to the plates. Close the plate immediately near the flame (Figure 13.35).
8. Insert the needle in the bottle and inoculate by placing small piece of cut mushroom in the middle of the PDA's surface. Make sure the piece of mushroom does not touch anything before entering the PDA bottle close the mouth of the bottle before the flame with cotton plug. Label plates and indicate: Date, type of mushroom, culture #.
9. Incubate plates and culture bottles at 28 ± 20°C in an incubator for 7 days and observe growth.
   Observe cottony radial growth on agar media.

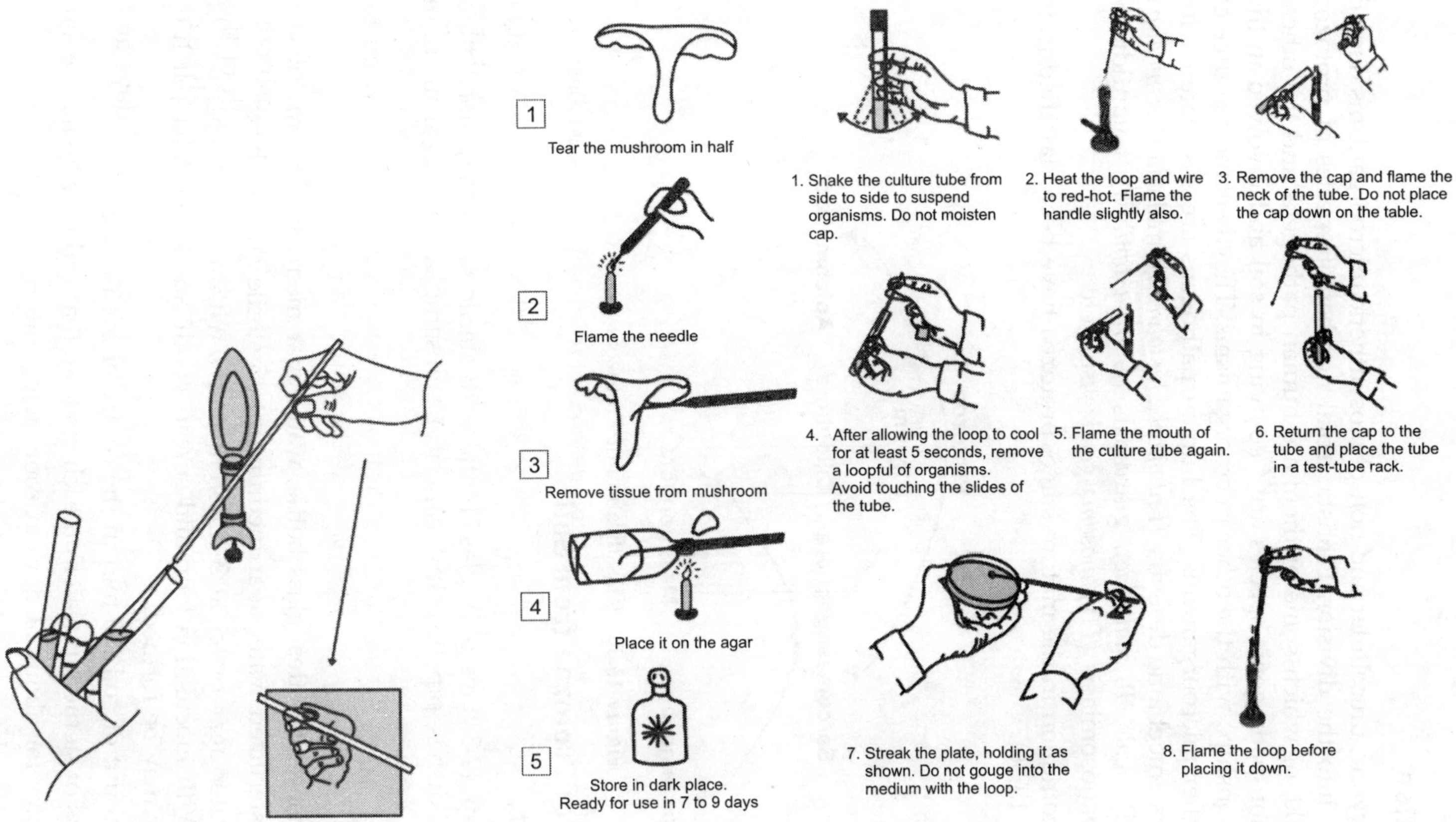

Figure 13.35 Isolation of the mushroom culture.

13.25 PROTOZOA

13.25.1 Introduction

Protozoa are the eukaryotic, unicellular and non-photosynthetic microorganisms devoid of cell wall. They fall into the division Protista. Total number of species of protozoa account for nearly 4000, of which some of them are human pathogens and the others non-pathogens. The non-pathogenic species are free-living in soil and involved in the establishment of relationships with the other microorganisms. They become a source of nutrient release for other microorganisms. The human pathogens are associated with certain diseases such as amoebiotic desentry (*Entamoeba histolytica),* malaria *(Plasmodium vivax, P. falciparum, P. ovale, P. malariae*), giardiasis (*Giardia lamblia*), balantidiasis (*Balantidium coli),* trypanosomiasis (*Trypanosoma gambiensis*), etc.

On the basis of locomotion in the mature stage, protozoa have been classified in to three phylum (Figure 13.36).

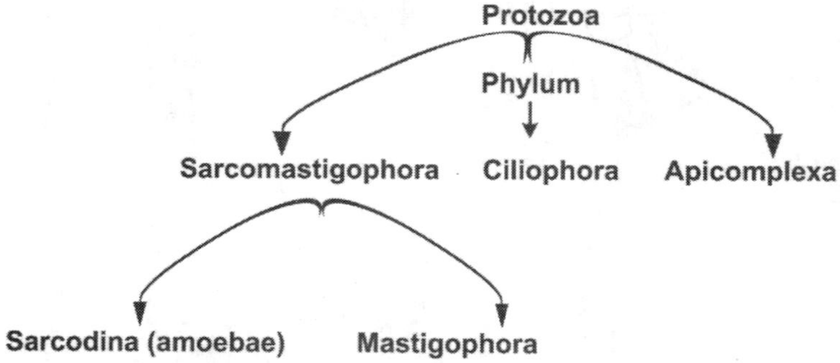

Figure 13.36 Classification of Protozoa.

13.25.2 Isolation of Protozoa from Soil

13.25.2.1 Materials

Culture *of Aeromonas* sp. (2-3 days old), glass bottle with sterile tap water, inoculation loop, microscope, Petri dishes, pipettes (1mL capacity), soil sample, water agar medium (2%).

13.25.2.2 Protocol

1. Sterilize Petri dishes, pipettes, glass bottle, water agar medium 2% agar, etc
2. Pour 20 mL of sterilized water agar medium into 3 sterile Petri dishes.
3. Inoculate the water agar plates at 3 different places with a thick suspension of *Aeromonas* sp. With inoculation loop and spread gently so as a circular patch of 2.5 cm diameter may be formed.
4. Add a small amount of soil (50 mg) in the centre of each circle.
5. Pour a few drops of sterile tap water on each circle so that soil may be moistened.
6. Incubate the plates for about 14 days at room temperature.
7. Add more water on the soil to keep it always moist, as and when required.

13.25.2.3 Results and observations

Observe the plates under the microscope for the presence of protozoa which move on an agar surface (Figure 13.37).

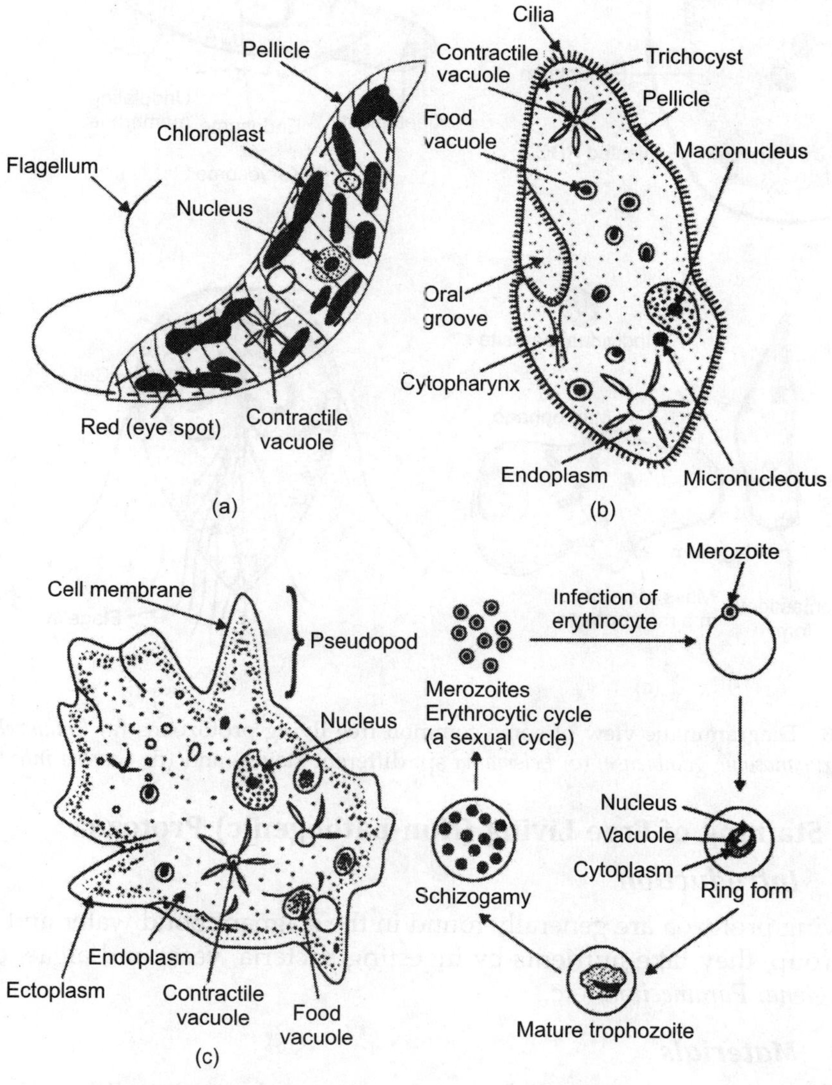

Figure 13.37 Diagrammatic view of some common protozoans: (a) *Euglena* with a whip-like flagellum for movement; (b) *Paramecium* with cilia over its surface; (c) *Amoeba* with pseudopods (lob-like projections); (d) *Plasmodium sporozoites* (non-motile) are released into the blood when an infected mosquito bites a person.

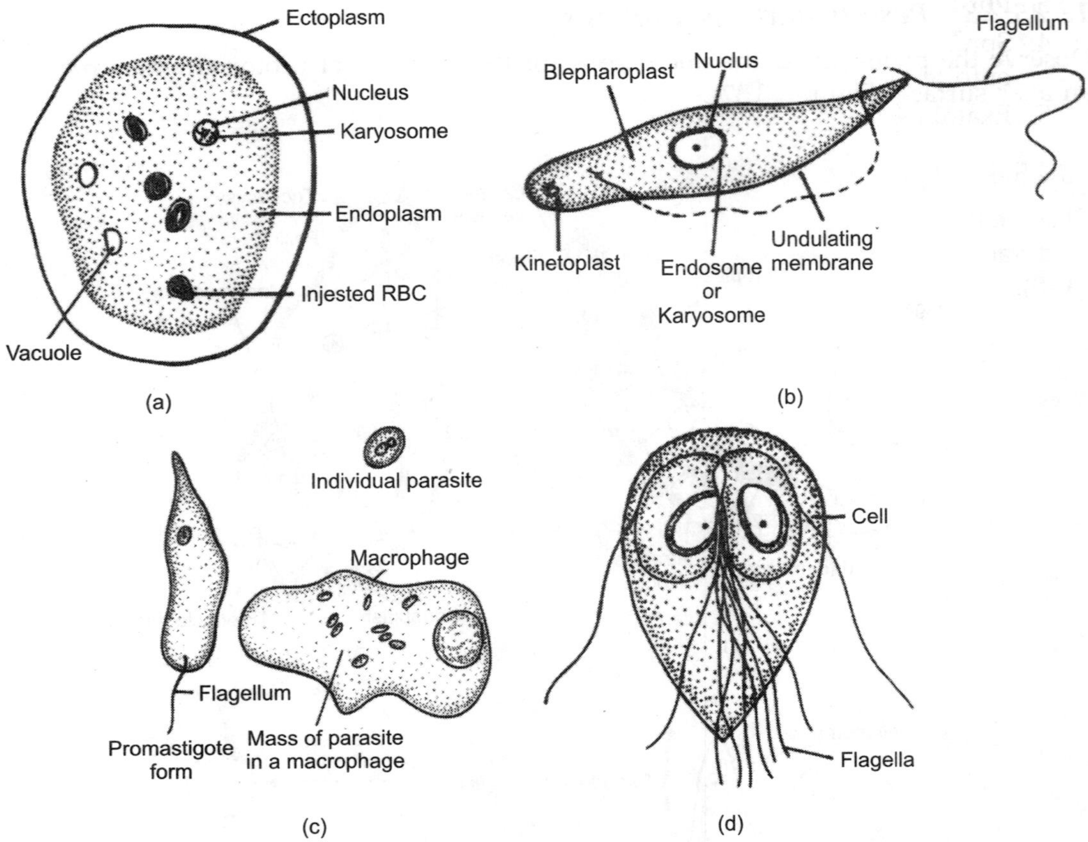

Figure 13.38 Diagrammatic view of some common free living protozoans (a) *Entamoeba histolytica,* (b) *Trypanosoma gambiense,* (c) *Leismania* sp. different stages, and (d) *Giardia intestinalis.*

13.25.3 Staining of Free Living (Non-pathogenic) Protozoa

13.25.3.1 Introduction

The free living protozoa are generally found in the stagnant pond water and soil. Unlike parasitic group, they take nutrients by ingesting bacteria, yeast and algae, e.g., *Amoeba proteus, Euglena, Paramecium,* etc.

13.25.3.2 Materials

Stagnant pond water, methyl cellulose, microscope, glass microslides, pipettes (1 mL capacity).

13.25.3.3 Protocol

1. Take a drop of pond water from the bottom of sample bottle and place in the center of a clean slide.
2. Pour a drop of methylcellulose sample (methylcellulose slow down the movement of the protozoa).

3. Put a clean coverslip in such a way that no bubble forms.
4. Spread the sample by the edge of the coverslip and gently lower the cover slip onto the slide.
5. Examine this slide under the microscope.

13.25.4 Results and Observations

Observe the slide under the microscope and note the types of locomotory parts, nucleus, food vacuoles, flagella and cilia etc and identify following standard litreature (Figure 13.38).

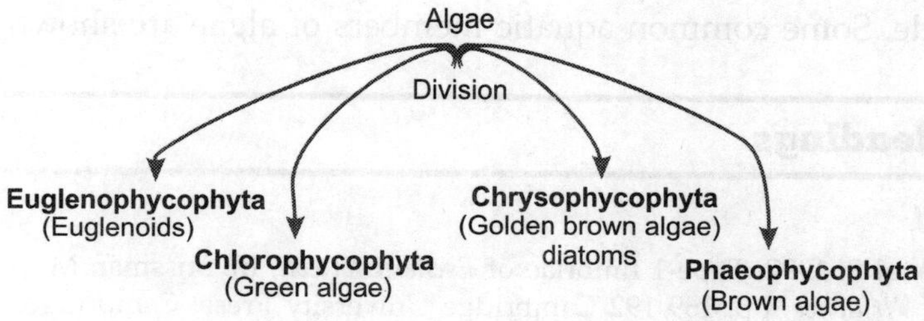

Figure 13.39 Classification of algae.

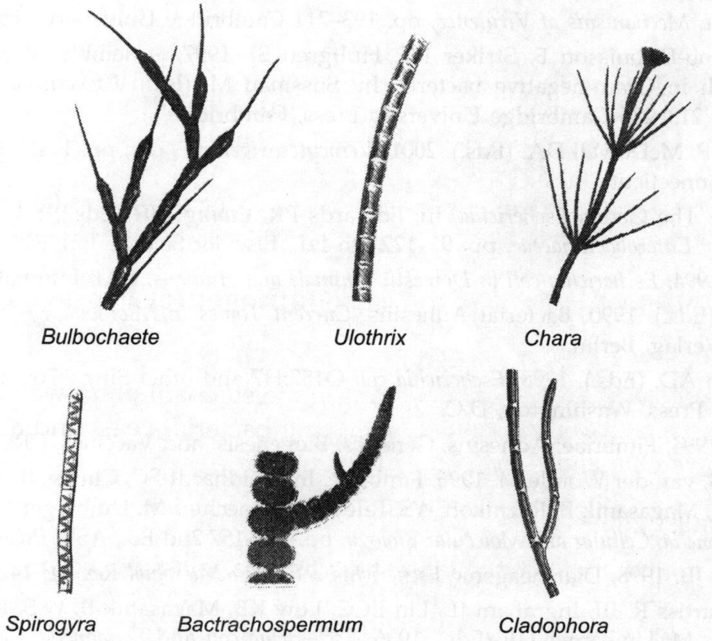

Figure 13.40 Schematic diagram of different forms of algae.

13.26 OBSERVATION OF ALGAE FROM POND WATER

13.26.1 Introduction

Algae are all photosynthetic differing from plant in that they lack tissue differentiation. They may be unicellular, colonial form or filamentous. Undifferentiated algal structures are referred to as a thallus. They are universally present and are classified on the basis of pigments present in them (Figure 13.39).

13.26.2 Protocol

They can be seen in a soil or water sample directly under a compound microscope on a microscopic slide. Some common aquatic members of algae are shown in Figure 13.40.

Suggested Readings

Escherichia coli

Abraham SN, Jaiswal S. 1997. Type-1 fimbriae of *Escherichia coli*. In: Sussman M. (Ed.), *Escherichia coli: Mechanisms of Virulence*, pp. 169 192 Cambridge University Press, Cambridge.

Cowan ST, Steel KJ, Barrow GI, Feltham RKA. (Editors). 1995. *Cowan and Steel' s Manual for the Identification of Medical Bacteria*, University of Cambridge, Cambridge.

de Graaf FK, Gaastra W. 1997. Fimbriae of enterotoxigenic *Escherichia coli*. In: Sussman M. (Ed.), *Escherichia coli: Mechanisms of Virulence*, pp. 193–211 Cambridge University Press, Cambridge.

Dodson KW, Jacob-Dubuisson F, Striker RT, Hultgren SJ. 1997. Assembly of adhesive virulence-associated pili in Gram-negative bacteria. In: Sussman M. (Ed.), *Escherichia coli: Mechanisms of Virulence*, pp. 213–236 Cambridge University Press, Cambridge.

Duffy G, Garvey P, McDowell DA. (Eds.). 2001. *Verocytotoxigenic E. coli*, pp. 1–457 Food and Nutrition Press Inc., Connecticut.

Ewing WH. 1986. The Genus *Escherichia*. In: Edwards PR, ***Ewing WH*** (Eds.), Edwards and Ewing' s *Identification of Enterobacteriaceae*, pp. 93–122 4th Ed., Elsevier Science Publishing Co., New York.

Gyles CL. (Ed.). 1994. *Escherichia coli in Domestic Animals and Humans*, CAB International, Wallingford.

Jann K, Jann B. (Eds.). 1990. Bacterial Adhesins. *Current Topics in Microbiology and Immunology*, 151 pp.Springer-Verlag, Berlin.

Kaper JB, O'Brien AD. (Eds.). 1998. *Escherichia coli* O157:H7 and other Shiga Toxin-Producing Strains, 465 pp. ASM Press, Washington, D.C.

Klemm P. (Ed.). 1994. Fimbriae: Adhesins, Genetics, Biogenesis, and Vaccines, CRC Press, Boca Raton.

Low D, Braaten B, van der Woude M. 1996. Fimbriae. In: Neidhardt FC, Curtiss III R, Ingraham JL, Lin ECC Low KB, Magasanik B, Reznikoff WS, Riley M, Schaechter M, Umbarger HE (Eds), *Escherichia coli* and *Salmonella*: *Cellular and Molecular Biology*, pp. 146–157 2nd Ed., ASM Press, Washington, D.C.

Nataro JP, Kaper JB. 1998. Diarrheagenic *Escherichia coli*. *Clin Microbiol Rev* **11:** 142–201.

Neidhardt FC, Curtiss R, III, Ingraham JL, Lin ECC, Low KB, Magasanik B, W.S. Reznikoff WS, Riley M, Schaechter M, Umbarger HE (Eds). 1996. *Escherichia coli* and *Salmonella*: *Cellular and Molecular Biology*, 2nd Ed., ASM Press, Washington, D.C.

Sussman M. (Ed.). 1997. *Escherichia coli*: *Mechanisms of Virulence*, 639 pp University of Cambridge Press, Cambridge.

Pseudomonas

Blazevic DJ, Köpcke MH, Matsen JM. 1973. Incidence and identification of Pseudomonas fluorescens and *Pseudomonas putida* in the clinical laboratory. *Appl Microbiol* **25,** 107-110.

Brodsky MH, Nixon MG. 1973. Rapid method for detection of *Pseudomonas aeruginosa* on Mac Conkey-Agar under ultraviolet light. *Appl Microbiol* **26,** 219-220.

Formiga LCD.1985. New possibilies for the laboratory diagnosis of diphtheria. *Brazilian J Med Biol Res* **18**, 401-402.

Gaby WL, Free E. 1953. Occurrence and identification of nonpigmented strains of *Pseudomonas aeruginosa* in the clinical laboratory. *J Bacteriol* **65**, 746–746.

King EO, Ward MK, Raney DE. 1954. Media for the demonstration of pyocyanin and fluorescein, *J Lab Clin Med* **44**, 301-307.

Leifson E. 1951. Staining, shape and arrangement of bacterial flagella. *J Bacteriol* **62**, 377–389.

Leifson E. 1958. Identification of Pseudomonas, Alcaligenes and related bacteria. *Zentralbl Bakteriol Orig* **173**, 487–488.

Leifson E. 1966. Bacterial taxonomy: a critique. *Bacteriol Rev* **30**, 257–266.

MacFaddin JF. 1985. Media for isolation-cultivation-identification-maintenance of medical bacteria, vol. 1. Williams & Wilkins, Baltimore, Md.

Moore HB, Pickett MJ. 1960. The Pseudomonas-Achromobacter group. *Can J Microbiol* **6**, 35–42.

Murray PR, Baron EJ, Jorgensen JH, Pfaller MA, Yolken RH, (Eds.), 2003. *Manual of Clinical Microbiology*, 8th Ed. ASM, Washington, D.C.

Redfearn MS, Palleroni NJ, Stanier RY. 1966. A comparative study of *Pseudomonas pseudomallei* and *Bacillus mallei.* J Gen Microbiol **43**, 293–313.

Sherris JC, Shoesmith JG, Parker MT, Breckon D. 1959. Tests for the rapid breakdown of arginine by bacteria: their use in the identification of pseudomonads. *J Gen Microbiol* **21**, 389–396.

Stanier RY, Palleroni NJ, Doudoroff M. 1966. The aerobic pseudomonads: a taxonomic study. *J Gen Microbiol* **43**,159–271.

Sutter VL, Hurst V, Landucci AO. 1966. Pseudomonads in human saliva. *J Dent Res* **45**, 1800–1803.

Sutter VL, Hurst V, Lane CW. 1967. Quantification of *Pseudomonas aeruginosa* in feces of healthy human adults. *Health Lab Sci* **4**, 245–249.

Vongraevenitz A. 1965. *Pseudomonas stutzeri* isolated from clinical specimens. *Am J Clin Pathol* **43**, 357–360.

Wahba AH, Darrell JH. 1965. The identification of atypical strains of *Pseudomonas Aeruginosa*. *J Gen Microbiol* 1965 **38**, 329–342.

Proteus

Ewing WM, Jardetzky WS, Press F. 1957. Elastic Waves in Layered Media. McGraw-Hill, New York.

Henrickson SD. 1950. A comparison of the phenylpyruvic acid reaction and the urease test in the differentiation of Proteus from other enteric organisms. *J Bacteriol* **60**, 225-231.

O'Hara C M, Brenner FW, Miller J M Classification. 2000. Identification, and Clinical Significance of *Proteus, Providencia,* and *Morganella Clin Microbiol Rev* **13**, 534–546.

O'Hara C M, Steigerwalt A G, Green D, McDowell M, Hill B C, Brenner D J, Miller J M. 1999. Isolation of *Providencia heimbachae* from human feces. *J Clin Microbiol* **37**, 3048–3050.

Robinson A, McCarter Y S, Tetreault J.1995. Comparison of Crystal Enteric/Nonfermenter system, API 20E system, and Vitek Automicrobic system for identification of gram-negative bacilli. *J Clin Microbiol* **33**, 364–370.

Ronald A R.1994. Urethritis and cystitis. In: Hoeprich P D, Jordan M C, Ronald A R, eds. Infectious disease: a treatise of infectious processes. pp. 564–570. 5th ed. J. Lippincott, Philadelphia, Pa

Rustigian R, Stuart C A. 1941. Decomposition of urea by *Proteus. Proc Soc Exp Biol Med* **47,** 108–112.

Rustigian R, Stuart C A. 1943. Taxonomic relationships in the genus *Proteus. Proc Soc Exp Biol Med* **53,** 241–243.

Singer J, Bar-Chay J. 1954. Biochemical investigation of Providence strains and their relationship to the *Proteus* group. *J Hyg* **52,** 1–8.

Skirrow M B. 1969. The Dienes (mutual inhibition) test in the investigation of *Proteus* infections. *J Med Microbiol* **2,** 471–477.

Traub W H, Craddock M E, Raymond E A, Fox M, McCall C E.1971. Characterization of an unusual strain of *Proteus rettgeri* associated with an outbreak of nosocomial urinary-tract infection. *Appl Microbiol* **22,** 278–283.

Trüper H G. 1999. Replacement of strain NCTC 4175, since 1963 the neotype strain of *Proteus vulgaris,* with strain ATCC 29905—opinion 70. *Int J Syst Bacteriol* **49**, 1949.

Watanakunakorn C, Perni S C. 1994. *Proteus mirabilis* bacteremia: a review of 176 cases during 1980–1992. *Scand J Infect Dis* **26,** 361–367.

Wenner J J, Rettger L F. 1919. A systematic study of the *Proteus* group of bacteria. *J Bacteriol* **4,** 331–353.

Winslow C-E A, Kligler I J, Rothberg W. 1919. Studies on the classification of the colon-typhoid group of bacteria with special reference to their fermentative reactions. *J Bacteriol* **4,** 429–503.

Yao J D C, Moellering R C., Jr . 1999. Antibacterial agents. In: Murray P R, et al., eds. *Manual of clinical microbiology* pp. 1474–1504. 7th ed. American Society for Microbiology Washington, D.C.

Shigella

Adler B, Sasakawa C, Tobe T, Makino S, Komatsu K, Yoshikawa M. 1989. A dual transcriptional activation system for the 230 kb plasmid genes coding for virulence-associated antigens of *Shigella flexneri. Mol Microbiol* **3,** 627-635.

Al-Hasani K, Henderson IR, Sakellaris H, Rajakumar K, Grant T, Nataro JP, Robins-Browne R, Adler B. 2000. The *sigA* gene which is borne on the *she* pathogenicity island of *Shigella flexneri* 2a encodes an exported cytopathic protease involved in intestinal fluid accumulation. *Infect Immun* **68,** 2457-2463.

Ashida H, Toyotome T, Nagai T, Sasakawa C. 2007. *Shigella* chromosomal IpaH proteins are secreted via the type III secretion system and act as effectors. *Mol Microbiol* **63**, 680-693.

Bahrani FK, Sansonetti PJ, Parsot C. 1997. Secretion of Ipa proteins by *Shigella flexneri*: inducer molecules and kinetics of activation. *Infect Immun* **65,** 4005-4010.

Blocker A, Gounon P, Larquet E, Niebuhr K, Cabiaux V, Parsot C, Sansonetti P. 1999. The tripartite type III secreton of *Shigella flexneri* inserts IpaB and IpaC into host membranes. *J Cell Biol* **147,** 683-693.

Boquet P, E. Lemichez E. 2003. Bacterial virulence factors targeting Rho GTPases: parasitism or symbiosis? *Trends Cell Biol* **13,** 238-246.

Buchrieser C, Glaser P, Rusniok C, Nedjari H, D'Hauteville H, Kunst F, Sansonetti P, Parsot C. 2000. The virulence plasmid pWR100 and the repertoire of proteins secreted by the type III secretion apparatus of *Shigella flexneri*. Mol Microbiol **38,** 760-771.

Cherla RP, Lee SY, Tesh VL. 2003. Shiga toxins and apoptosis. *FEMS Microbiol Lett* **228,** 159-166.

Demers B, Sansonetti PJ, Parsot C. 1998. Induction of type III secretion in *Shigella flexneri* is associated with differential control of transcription of genes encoding secreted proteins. *EMBO J* **17,** 2894-2903.

Ewing WH. 1949. *Shigella* nomenclature. *J Bacteriol* **57,** 633-638.

Gorden J, Small PL. 1993. Acid resistance in enteric bacteria. *Infect Immun* **61,** 364-367.

Hilbi H. 1999. Host responses to secreted *Shigella* virulence factors. *Curr Opin Infect Dis* **12,** 221-228.

Hsu BM, Wu SF, Huang SW, Tseng YJ, Ji DD, Chen JS, Shih FC. 2010. Differentiation and identification of Shigella spp. and enteroinvasive Escherichia coli in environmental waters by a molecular method and biochemical test. *Water Res* **44**, 949-955.

Izard T, Tran Van Nhieu G, Bois PR. 2006. *Shigella* applies molecular mimicry to subvert vinculin and invade host cells. *J Cell Biol* **175,** 465-475.

Jennison AV, Verma NK. 2004. *Shigella flexneri* infection: pathogenesis and vaccine development. *FEMS Microbiol Rev* **28,** 43-58.

Mavris M, Page AL, Tournebize R, Demers B, Sansonetti P, Parsot C. 2002. Regulation of transcription by the activity of the *Shigella flexneri* type III secretion apparatus. *Mol Microbiol* **43,** 1543-1553.

Niyogi SK. 2005. Shigellosis. *J Microbiol* **43,** 133-143.

Ogawa M, Sasakawa C. 2006. Intracellular survival of *Shigella*. *Cell Microbiol* **8,** 177-184.

Schroeder GN, Hilbi H. 2008. Molecular Pathogenesis of *Shigella* spp.: Controlling Host Cell Signaling, Invasion, and Death by Type III Secretion. *Clin Microbiol Rev* **21**,134–156.

Salmonella

Ackers ML, Puhr ND, Tauxe RV, Mintz ED.2000. Laboratory-based surveillance of *Salmonella* serotype typhi infections in the United States: antimicrobial resistance on the rise. *JAMA* **283**, 2668-2673.

Agarwal SC.1962. Chloramphenicol resistance of *Salmonella* species in India, 1956-61. *Bull Wld Hlth Orgn* **17**, 331-335.

Cho JC, Kim SJ. 1999. Viable, but non-culturable, state of a green fluorescence protein-tagged environmental isolate of Salmonella typhi in groundwater and pond water; *FEMS Microbiology Letters* **170**, 257-264.

Crump JA, Barett TJ, Nelson JI, Angulo FJ. 2003. Reevaluating fluoroquinolone breakpoints for *Salmonella enterica* serotype Typhi and for non-Typhi salmonellae. *Clin Infect Dis* **37**, 75–81.

Crump JA, Luby SP, Mintz ED. 2004. The global burden of typhoid fever. *Bull World Health Org* **82,** 346-353.

Effa EE, Bukirwa H.2008. Azithromycin for treating uncomplicated typhoid and paratyphoid fever (enteric fever) *Cochrane Database Syst Rev* 4, CD006083.

Gorden J, Small PL. 1993. Acid resistance in enteric bacteria. *Infect Immun* **61**, 364-367.

Jay S, Grau FH, Smith K, Lightfoot D, Murray C, Davey G.R. 1997. *Salmonella* In: *Food borne microorganisms of public health importance*, pp 169-232 5th Edition. AIFST (NSW Branch), Sydney, Australia

Jesudason MV, John TJ. 1992. Plasmid mediated multidrug resistance in *Salmonella typhi. Ind J Med Res* **95**, 66-67.

Lynch MF, Blanton EM, Bluens S, Polyak C, Vojdani J, Stevenson J, Medalla F, Barzilay E, Joyce K, Barrett T, Mintz ED. 2009. Typhoid fever in the United States, 1999-2006 *JAMA* **302**, 859-865.

Mourad AS, Matwally M, Nour Ei Deen A, Threlfall EJ, Rowe B, Mapes T, Hedstrom R, Bourgeois AL, J. R. Murphy JR. 1993. Multiple drug-resistant *Salmonella typhi. Clin Infect Dis* **17**, 135-136.

Nath G, Tikoo A, Manocha H, Tripathi AK, Gulati AK. 2003. Drug resistance in *Salmonella typhi* in North India with special reference to ciprofloxacin. *J Antimicrobial Chemother* **46**, 145-153.

Pfeifer Y, Matten J, Rabsch W.2009. *Salmonella enterica* Serovar Typhi with CTX-M beta-lactamase, Germany. *Emerg Infect Dis* 15, 1533–1535.

Ryan CA, Hargrett-Bean NT, Blake PA. 1989; Salmonella typhi infections in the United States, 1975-1984: increasing role of foreign travel. *Re Infect Dis* **11**,1-8.

Sanghavi SK, Mane MP, Niphadkar KB.1999. Multidrug resistance in *Salmonella* serotypes. *Ind J Med Microbiol* **17,** 88-90.

Sinha A, Sazawal S, Kumar R, Sood S, Reddaiah VP, Singh B, Rao M, Naficy A, Clemens JD, Bhan MK. 1999. Typhoid fever in children aged less than 5 years *Lancet* 354, 734-737.

Tankhiwale SS, Agrawal G, Jalgaonkar SV. 2003. A Preliminary Report on Current Antibiogram of *Salmonella enterica* serotype *typhi* in Nagpur. *Ind J Med Microbiol* **21**: 292

Threlfall EJ, Ward LR .2001. Decreased susceptibility to ciprofloxacin in *Salmonella* enterica serotype *typhi,* United Kingdom. *Emerg Infect Dis* **7**, 448-450.

Wait DA, Sobsey MD. 2001; Comparative survival of enteric viruses and bacteria in Atlantic ocean seawater. *Wat Sci Tech* **43**,139-142.

Watson KC. 1978. Laboratory and clinical investigation of recovery of *Salmonella typhi* from blood. *J Clin Microbiol* **7**, 122-126.

Watson KC.1954. Clot culture in typhoid fever *J Clin Pathol* **7**, 305-307.

Wayne, PA. Clinical and Laboratory Standards Institute .2006. *Clinical and Laboratory Standards Institution: Methods for Dilution Antimicrobial Susceptibility Tests for Bacteria that Grow Aerobically.* Approved standard, CLSI document M7-A7.

Vibrio

Alice AF, Naka H, Crosa JH. 2008. Global gene expression as a function of the iron status of the bacterial cell: influence of differentially expressed genes in the virulence of the human pathogen *Vibrio vulnificus.* Infect *Immun* **76,** 4019-4037.

Bisharat N, Cohen DI, Harding RM, Falush D, Crook DW, Peto T, Maiden MC. *2005.* Hybrid *Vibrio vulnificus. Emerg Infect Dis* **11,** 30-35.

CDC. 1993. *Vibrio vulnificus* infections associated with raw oyster consumption. *Morb Mortal Wkly Rep* **42,** 405-407.

CDC. 2006. Preliminary FoodNet data on the incidence of infection with pathogens transmitted commonly through food-10 states, United States, 2005. *Morb Mortal Wkly Rep* **55,** 392-395.

CDC. 2007. Preliminary FoodNet data on the incidence of infection with pathogens transmitted commonly through food—10 states, United States, 2006. *Morb Mortal Wkly Rep* **56**, 336-339.

Chuang YC, Yuan CY, Liu CY, Lan CK, Huang AH. 1992. *Vibrio vulnificus* infection in Taiwan: report of 28 cases and review of clinical manifestations and treatment. *Clin Infect Dis* **15,** 271-276.

DePaola A, Capers GM, Alexander D. 1994. Densities of Vibrio vulnificus in the intestines of fish from the U.S. Gulf Coast. *Appl Environ Microbiol* **60**, 984-988.

Gulig PA, Bourdage KL, Starks AM. 2005. Molecular pathogenesis of *Vibrio vulnificus. J Microbiol* **43,** 118-131.

Kreger A, Lockwood D. 1981. Detection of extracellular toxin(s) produced by *Vibrio vulnificus. Infect Immun* **33**, 583-590.

Lee JH, Rhee JE, Park U, Ju HM, Lee BC, Kim TS, Jeong HS, Choi SH. 2007. *Identification and functional analysis of Vibrio vulnificus SmcR, a novel global regulator. J Microbiol Biotechnol* **17,** 325-334.

Melissa KJ, Oliver JD. 2009.*Vibrio vulnificus*: Disease and Pathogenesis. *Infect. Immun* **77,** 1723-1733.

Oliver JD. 2005. Wound infections caused by *Vibrio vulnificus* and other marine bacteria. *Epidemiol Infect* **133**, 383-391.

Oliver JD. 2006. *Vibrio vulnificus,* In: Thompson FL, Austin B, Swing J (eds.), *The biology of vibrios.* p. 349-366 ASM Press, Washington, DC.

Strom MS, Paranjpye RN. 2000. Epidemiology and pathogenesis of *Vibrio vulnificus. Microb Infect* **2,** 177-188.

Wright AC, Simpson LM, Oliver JD. 1981. Role of iron in the pathogenesis of *Vibrio vulnificus* infections. *Infect Immun* **34**, 503-507.

Yoshida S, Ogawa M, Mizuguchi Y. 1985. Relation of capsular materials and colony opacity to virulence of Vibrio vulnificus. *Infect Immun* **47**, 446-451.

Yu HN, Lee YR, Park KH, Rah SY, Noh EM, Song EK, Han MK, Kim BS, Lee SH, Kim JS. 2007. Membrane cholesterol is required for activity of Vibrio vulnificus cytolysin. *Arch Microbiol* **187,** 467-473.

Zuppardo AB, Siebeling RJ. 1998. An epimerase gene essential for capsule synthesis in *Vibrio vulnificus. Infect Immun* **66,** 2601-2606.

Zuppardo AB, Siebeling RJ. 1998. An epimerase gene essential for capsule synthesis in *Vibrio vulnificus. Infect Immun* **66,** 2601-2606.

Staphylococcus

Aires de Sousa M, Bartzavali C, Spiliopoulou I, Sanches IS, Crisostomo MI, de Lencastre H. 2003. Two international methicillin-resistant *Staphylococcus aureus* clones endemic in a university hospital in Patras, Greece. *J Clin Microbiol* **41,** 2027-2032.

Centers for Disease Control and Prevention 2003. Outbreaks of community-associated methicillin-resistant *Staphylococcus aureus* skin infections-Los Angeles County, California 2002-2003. *Morb Mortal Wkly Rep* **52,** 88.

Easmon CSF, Adlam C.1983. *Staphylococci and staphylococcal infections.* Vols 1 and 2. Academic Press, London.

Foster TJ, McDevitt D. 1994. Molecular basis of adherence of staphylococci to biomaterials. In: Bisno AL, Waldvogel FA (eds): Infections Associated with Indwelling Medical Devices, p. 31, 2nd Edition. American Society for Microbiology, Washington, D.C.

Gillet Y, Issartel B, Vanhems P, Lina G, Vandenesch F, Etienne J, Floret D. 2001. Severe staphylococcal pneumonia in children. *Arch Pediatr* **8,** 742s-746s.

Huebner J, Goldmann DA.1999. Coagulase-negative staphylococci: role as pathogens. *Annu Rev Med* **50,** 223-236.

Huletsky A, Giroux R, Rossbach V, Gagnon M, Vaillancourt M, Bernier M, Gagnon F, Truchon K Bastien M, Picard FJ, van Belkum A, Ouellette M, Roy PH, Bergeron MG. 2004. New real-time PCR assay for rapid detection of methicillin-resistant *Staphylococcus aureus* directly from specimens containing a mixture of Staphylococci. *J Clin Microbiol* **42,** 1875-1884.

Kaufmann ME. 1998. Molecular bacteriology: protocols and clinical applications, p. 33-50. Human Press, Totowa, NJ.

Prevost G, Couppie P, Prevost P, Gayet S, Petiau P, Cribier B, Monteil H, Piemont Y. 1995. Epidemiological data on *Staphylococcus aureus* strains producing synergohymenotropic toxins. *J Med Microbiol* **42,** 237-245.

Prevost G, Mourey L, Colin DA, Menestrina G. 2001. Staphylococcal pore-forming toxins. *Curr Top Microbiol Immunol* **257,** 53-83.

Robinson DA, Kearns AM, Holmes A, Morrison D, Edwards G, O'Brien FG, Tenover FC, McDougal LK, Monk AB, Enright MC. 2005. Re-emergence of early pandemic *Staphylococcus aureus* as a community-acquired methicillin-resistant clone. *Lancet* **365,** 1256-1258.

Yamaguchi T, Nishifuji K, Sasaki M, Fudaba Y, Aepfelbacher M, Takata T, Ohara M, Komatsuzawa H, Amagai M, Sugai M. 2002. Identification of the *Staphylococcus aureus etd* pathogenicity island which encodes a novel exfoliative toxin, ETD, and EDIN-B. *Infect Immun* **70,** 5835-5845.

Streptococcus

Alam S, Brailsford SR, Whiley RA, Beighton D. 1999. PCR-based methods for genotyping viridans group streptococci. *J Clin Microbiol* **37,** 2772-2776.

Balter S, Benin A, S. Pinto SWL, Teixeira LM, Alvim GG, Luna E, Jackson D, LaClaire L, Elliott J, Facklam R, Schuchat A. 2000. Epidemic nephritis in Nova Serrana, ***Brazil Lancet*** **355,** 1776-1780.

Beall B, Facklam RR, Elliott JA, Franklin AR, Hoenes T, Jackson D, LaClaire L, Thompson T, Viswanathan R. 1998. *Streptococcal emm* types associated with T-agglutination types and the use of conserved emm gene restriction fragment patterns for subtyping group A streptococci. *J Med Microbiol* **47,** 1-6.

Beighton D, Russell RRB, Whiley RA. 1991. A simple biochemical scheme for the differentiation of *Streptococcus mutans* and *Streptococcus sobrinus. Caries Res* **25,** 174-178.

Bert F, Lambert-Zechovsky N. 1997. Septicemia caused by *Streptococcus canis* in a human. *J Clin Microbiol* **35,** 777-779.

Bisno AL, Collins CM, Turner JC. 1996. M proteins of group C streptococci isolated from patients with acute pharyngitis. *J Clin Microbiol* **34,** 2511-2515.

Bridge PD, Sneath PHA. 1982. *Streptococcus gallinarum* sp. nov. and *Streptococcus oralis* sp. nov. *Int J Syst Bacteriol* **32,** 410-415.

Bruckner DA, Colonna P. 1997. Nomenclature for aerobic and facultative bacteria. *Clin Infect Dis* **25,** 1-10.

Collins MD, Aguirre M, Facklam RR, Shallcross J, Williams AM. 1992. *Globicatella sanguis* gen. nov., a new Gram-positive catalase-negative bacterium from human sources. *J Appl Bacteriol* **73,** 433-437.

Collins MD, Hutson R, Falsen E, Nikolaitchouk N, LaClaire L, Facklam RR. 2000. An unusual Streptococcus from human urine, *Streptococcus urinalis* sp. nov. *Int J Syst Evol Microbiol* **50,** 1173-1178.

Collins MD, Hutson RA, Falsen E, Inganas E, Bisgaard M. 2002. *Streptococcus gallinaceus* sp. Nov., from chickens. Int *J Syst Evol Microbiol* **52,** 1161-1164.

Coykendall AL. 1989. Classification and identification of the viridans streptococci. *Clin. Microbiol Rev* **2,** 315-328.

Facklam RR, Elliott J, Pigott N, Franklin AR. 1995. Identification of *Streptococcus porcinus* from human sources. *J Clin Microbiol* **33,** 385-388.

Facklam RR. 1977. Physiological differentiation of viridans streptococci. *J Clin Microbiol* **5,** 184-201.

Higgins R, Gottschalk M. 1990. An update on Streptococcus suis identification. *J Vet Diagn Investig* **2,** 249-252.

Schleifer KH. 1986. Gram-positive cocci, p. 999-1002. In: Sneath PHA, Mair NS, Sharp ME, Holt JG (ed.), *Bergey's manual of systematic bacteriology,* vol. 2. The Williams and Wilkins Co., Baltimore, Md.

Sherman JM. 1937. The streptococci. *Bacteriol Rev* **1,** 3-97.

Taranta A, Moody MD. 1971. Diagnosis of streptococcal pharyngitis and rheumatic fever. Pediatr. *Clin North Am* **18,** 125-143.

Vela AI, Fernandez E, Lawson PA, Latre MV, Falsen E, Dominguez L, Collins MD, Fernandez-Garayzabal JF. 2002. Streptococcus entericus sp. nov., isolated from cattle intestine. *Int J Syst Evol Microbiol* **52,** 665-669.

Von Baum H, Klemme FR, Geiss HK, Sonntag HG. 1998. Comparative evaluation of a commercial system for identification of gram-positive cocci. *Eur J Clin Microbiol Infect Dis* **17,** 849-852.

Willcox MDP, Zhu H, Knox KW. 2001. *Streptococcus australis* sp. nov., a novel oral Streptococcus. *Int J Syst Evol Microbiol* **51,** 1277-1281.

Zhu H, Willcox MDP, Knox KW. 2000. A new species of oral Streptococcus isolated from Sprague-Dawley rats, *Streptococcus orisratti* sp. nov. *Int J Syst Evol Microbiol* **50,** 55-61.

Xanthomonas

Ezuka A, Kaku H. 2000. A historical review of bacterial blight of rice. *Bull Natl Inst Agrobiol Resour Japan* **15**, 61–74

Goto M. 1990. *Fundamentals of Bacterial Plant Pathology*. Academic Press, San Diego.

Hopkins CM, White FF, Choi SH, Guo A, Leach JE. 1992. Identification of a family of avirulence genes from *Xanthomonas oryzae* pv. oryzae *Mol Plant Microbe Interact* **5,** 451–459.

Jenkins CL, Starr MP. 1982. The brominated aryl-polyene (xanthomonadin) pigments of *Xanthomonas juglandis* protect against photobiological damage. *Curr Microbiol* **7**, 323-326.

Leach JE, White FF. 1996. Bacterial avirulence genes *Annu Rev Phytopathol* **34**, 153–179.

Lee SW, Choi SH, Han SS, Lee DG, Lee BY. 1999. Distribution of *Xanthomonas oryzae* pv. oryzae strains virulent to Xa21 in Korea. *Phytopathology* **89**, 928–933.

Ray SK, Rajeshwari R, Sonti RV. 2000. Mutants of *Xanthomonas oryzae* deficient in general secretory pathway are virulent deficient and unable to secrete xylase *Mol Plant Microbe Interact* **13**, 394–401.

Rudolf K. 1993. *Xanthomonas.* Chapman & Hall, London.

Starr MP, Jenkins CL, Bussey LB, Andrewes AG. 1977. Chemotaxonomic significance of the xanthomonadins, novel brominated aryl-polyene pigments produced by bacteria of the genus *Xanthomonas. Arch Microbiol* **113**, 1-9.

Starr MP, Stephens WL. 1964. Pigmentation and taxonomy of the genus *Xanthomonas. J Bacteriol* **87**, 293-302.

Stephens WL, Starr MP. 1963. Localization of carotenoid pigment in the cytoplasmic membrane of *Xanthomonas juglandis. J Bacteriol* **86**, 1070-1074.

Serratia

Carbonell GV, Della Colleta HHM., Yano T, Darini ALC, Levy CE, Fonseca BA L. 2000. Clinical relevance and virulence factors of pigmented *Serratia marcescens. FEMS Immunology & Medical Microbiology 28*, 143-149.

Coulthurst SJ, Williamson NR, Harris AK, Spring DR, Salmond GP. 2006. Metabolic and regulatory engineering of Serratia marcescens: mimicking phage-mediated horizontal acquisition of antibiotic biosynthesis and quorum-sensing capacities. *Microbiology* 152, 1899–1911.

Engel HJ, Collignon PJ, Whiting PT, Kennedy KJ. 2009. Serratia sp. bacteremia in Canberra, Australia: a population-based study over 10 years. *European Journal of Clinical Microbiology & Infectious Diseases 28*, 821-824.

Filloux A, Hachani A, Bleves S. 2008. The bacterial type VI secretion machine: yet another player for protein transport across membranes. *Microbiology* 154,1570–1583.

Grimont F, Grimont PAD. 1992. The genus Serratia. *The Prokaryotes 3*, 2822–2848.

Hejazi A, Falkiner FR. 1997. Serratia marcescens. *J Med Microbiol* 46, 903–912.

Katara G, Hemvani N, Chitnis S, Chitnis V, Chitnis, DS. 2008. Surface disinfection by exposure to germicidal UV light. *Indian Journal of Medical Microbiology* **26**, 241-242.

Lysenko, 0. 1961. Pseudomonas-an attempt at a general classification. *J Gen Microbiol* **25**, 379-408.

Pitt TL. 1982. State of the art: typing of *Serratia marcescens. J Hosp Infect* **3**, 9-14

Public Health Agency of Canada. 2004. In: Best M, Graham ML, Leitner R, Ouellette M, Ugwu K. (Eds.), *Laboratory Biosafety Guidelines* (3rd ed.). Canada: Public Health Agency of Canada.

Sleigh JD. 1983. Antibiotic resistance in *Serratia marcescens. BMJ* **287,** 1651-1653.

Starr MP, Grimont PAD, Grimont F, Starr PB. 1976. Caprylate-thallous agar medium for selectively isolating *Serratia* and its utility in the clinical laboratory. *Journal of Clinical Microbiology* **4**, 270–276.

Van Houdt R, Givskov M., Michiels CW. 2007. Quorum sensing in Serratia. *FEMS Microbiology Reviews 31*, 407-424.

Williams RF, Gott CL, Qadri SMH. 1971. Induction of pigmentation in nonproliferating cells of *Serratia marcescens* by addition of single amino acids. *J Bacteriol* **106,** 444-448.

Rhizobium

Aarons SR, Graham PH. 1991. Response of *Rhizobium leguminosarum* bv. *phaseoli* to acidity. *Plant Soil* **134**, 145–151.

Abdel-Wahab AM, Abd-Alla MH. 1995. Nodulation and nitrogenase activity of *Vicia faba* and *Glycine max* in relation to rhizobia strain, form and level of combined nitrogen. *Phyton* **35**, 177–187.

Abdel-Wahab HH, Zahran HH. 1979. Salt tolerance of *Rhizobium* species in broth culture. *Z Allg Mikrobiol* **19**, 681–685.

Abu-Gharbia M A. 1996. Growth of rhizobia, nodulation and nitrogen fixation of some legumes under pesticide treatment. *Sohag Pure Appl Sci Bull Fac Sci South Valley Univ Egypt* **12**, 87–104.

Albrecht S L, Bennett J M, Boote K J. 1994. Relationship of nitrogenase activity to plant water stress in field grown soybeans. *Field Crop Res* **8**, 61–71.

Bayoumi HEA, Biro B, Balazsy S, Kecskes M.1995. Effects of some environmental factors on *Rhizobium* and *Bradyrhizobium* strains. *Acta Microbiol Immunol Hung* **42**, 61–69.

Bordeleau L M, Prevost D. 1994. Nodulation and nitrogen fixation in extreme environments. *Plant Soil* **161**, 115–124.

Bottomley P. 1991. Ecology of *Rhizobium* and *Bradyrhizobium*. In: Stacey G, Burris RH, Evans HJ (eds.). *Biological nitrogen fixation*. pp. 292–347. Chapman & Hall; New York.

Burns RC, Hardy RWF. 1975. *Nitrogen fixation in bacteria and higher plants*. Springer-Verlag, New York.

Gouffi K, Pica N, Pichereau V, Blanco C. 1999. Disaccharides as a new class of nonaccumulated osmoprotectants for *Sinorhizobium meliloti*. *Appl Environ Microbiol* **65**, 1491–1500.

Igual JM, Rodriguezbarrueco C, Cervantes E. 1997. The effect of aluminum on nodulation and nitrogen fixation in *Casuarina cunninghamiana Miq. Plant Soil* **190**, 41–46.

Kirda C, Danso SKA, Zapata F. 1989. Temporal water stress effects on nodulation, nitrogen accumulation and growth of soybean. *Plant Soil* **120**, 49–55.

Kishinevsky BD, Sen D, Weaver RW. 1992. Effect of high root temperature on *Bradyrhizobium*-peanut symbiosis. *Plant Soil* **143**, 275–282.

McGrath JW, Hammerschmidt F, Quinn JP. 1998. Biodegradation of phosphonomycin by *Rhizobium huakuii* PMY1. *Appl Environ Microbiol* **64**, 356–358.

Moawad H, Beck D. 1991. Some characteristics of *Rhizobium leguminosarum* isolates from uninoculated field-grown lentil. *Soil Biol Biochem* **23**, 917–925.

O'Hara GW, Glenn AR. 1994. The adaptive acid tolerance response in root nodule bacteria and *Escherichia coli*. *Arch Microbiol* **161**, 286–292.

Ramos MLG, Gordon AJ, Minchin FR, Sprent JI, Parsons R. 1999. Effect of water stress on nodule physiology and biochemistry of a drought tolerant cultivar of common bean (*Phaseolus vulgaris* L.) *Ann Bot* **83**, 57–63.

Sangakkara UR, Hartwig UA, Noesberger J. 1996. Soil moisture and potassium affect the performance of symbiotic nitrogen fixation in faba bean and common bean. *Plant Soil* **184**, 123–130.

Smith SR. 1997. *Rhizobium* in soils contaminated with copper and zinc following the long-term application of sewage sludge and other organic wastes. *Soil Biol Biochem* **29**, 1475–1489.

Tate RL. 1995. Soil microbiology (symbiotic nitrogen fixation) pp. 307–333, John Wiley & Sons, Inc. New York.

Viera-Vargas MS, Souto CM, Urquiaga S, Boddey RM. 1995. Quantification of the contribution of N_2 fixation to tropical forage legumes and transfer to associated grass. *Soil Biol Biochem* **27**, 1193-1200.

Walsh KB. 1995. Physiology of the legume nodule and its response to stress. *Soil Biol Biochem* **27**, 637–655.

Zitzer SF, Archer SR, Boutton TW.1996. Spatial variability in the potential for symbiotic N_2 fixation by woody plants in a subtropical savanna ecosystem. *J Appl Ecol* **33**, 1125–1136.

Zou N, Dort PJ, Marçar NE. 1995. Interaction of salinity and rhizobial strains on growth and N_2^- fixation by *Acacia ampliceps*. *Soil Biol Biochem* **27**, 409–413.

Bacillus

Agata N, Ohta M, Yokoma K. 2002. Production of *Bacillus cereus* emetic toxin (cereulide) in various foods. International *Journal of Food Microbiology* **73**, 23–27

Ankolekar C, Rahmati T, Lebbe RG. 2009. Detection of toxigenic *Bacillus cereus* and *Bacillus thuringiensis* in US rice. *International Journal of Food Microbiology* **128**, 460–466.

Arnesen SLP, Fagerlund A, Granum PE. 2008. From soil to gut: *Bacillus cereus* and its food poisoning toxins. *FEMS Microbiology Reviews* **32**, 579–606

Augustin JC. 2011. Challenges in risk assessment and predictive microbiology of foodborne spore-forming bacteria. *Food Microbiology* **28**, 209–213

Becker H, Schaller G, Von Wiese W, Terplan G. 1994. *Bacillus cereus* in infant foods and dried milk products. *International Journal of Food Microbiology* **23**, 1–15.

Dagmar F, Pukall R. 2001. Reclassification of bioindicator strains *Bacillus subtilis* DSM 675 and *Bacillus subtilis* DSM 2277 as *Bacillus atrophaeus. Int J Sys and Evol Microbio* **51**, 35–37.

Eglezos S, Huang B, Dykes GA, Fegan N. 2010. The prevalence and concentration of *Bacillus cereus* in retail products in Brisbane, Australia. *Foodborne Pathogens and Disease* **7**, 867–870

Ehling-Schulz M, Guinebretière M, Monthan A, Berge O, Fricker M, Svensson B. 2006. Toxin gene profiling of enterotoxic and emetic *Bacillus cereus. FEMS Microbiology Letters* **260**, 232–240.

ESR. 2010. *Bacillus cereus.* Minstry for Primary Industries, New Zealand. http://www.foodsafety.govt.nz/elibrary/industry/Bacillus_Cereus-Spore_Forming.pdf. Accessed 16 August 2012

FDA. 2012. Bad bug book: Foodborne pathogenic microorganisms and natural toxins handbook, 2nd ed. US Food and Drug Administration, Silver Spring, p. 93-96. http://www.fda.gov/Food/FoodborneIllnessContaminants/CausesOfIllnessBadBugBook/ucm2 006773.htm. Accessed 27 March 2013

Fermanian C, Lapeyre C, Fremy J, Claisse M. 1997. Diarrhoeal toxin production at low temperatures by selected strains of *Bacillus cereus. Journal of Dairy Research* **64**, 551–559

Finlay WJJ, Logan NA, Sutherland AD. 2002. *Bacillus cereus* toxin production in cooked rice. *Food Microbiology* **19**, 431–439.

Ghelardi E, Celandroni F, Salvetti S. 2002. Identification and charaterization of toxigenic *Bacillus cereus* isolates responsible for two food poisoning outbreaks. *FEMS Microbiology Letters* **208**,129–134.

Gordon RE, Haynes WC, Pang CHN. 1973. *The Genus Bacillus. Agriculture Handbook No. 427.* United States Department of Agriculture, Washington, DC.

Lim E, Lopez L, Borman A, Cressey P, Pirie R. 2012. Annual report concerning foodborne disease in New Zealand 2011. Ministry for Primary Industry, New Zealand. http://www.foodsafety.govt.nz/science-risk/human-health-surveillance/foodborne-diseaseannual-reports.htm. Accessed 11 April 2013

Montville TJ, Matthews KR.2005. Food Microbiology: An Introduction. ASM Press, Washington D.C.

Pielaat A, Fricker M, Nauta MJ, van Leusden FM. 2005. Biodiversity in *Bacillus cereus.*

RIVM report 250912004/2005. *National Institute for Public Health and the Environment,* The Netherlands.

Priest FG, Goodfellow M, Shute LA, Berkeley RCW. 1987. *Bacillus amyloliquefaciens* sp. nov., nom. rev. *Int J Syst Bacteriol* **37**, 69-71.

Roberts MS, Nakamura LK, Cohan FM. 1994. *Bacillus mojavensis* sp. nov., distinguishable from *Bacillus subtilis* by sexual isolation, divergence in DNA sequence, and differences in fatty acid composition. *Int J Syst Bacteriol* **44**, 256-264.

Roberts MS, Nakamura LK, Cohan F M. 1996. *Bacillus vallismortis* sp. nov., a close relative of *Bacillus subtilis,* isolated from soil in Death Valley, California. *Int J Syst Bacteriol* **46**, 470-475.

Scallan E, Hoekstra RM, Angulo FJ, Tauxe RV, Widdowson M, Roy SL, Jones JL, Griffin PM. 2011. Foodborne illness acquired in the United States - Major pathogens. *Emerging Infectious Diseases* **17**, 7–11

Schneider KR, Parish ME, Goodrich RM, Cookingham T. 2004. Preventing Foodborne Illness: *Bacillus cereus* and *Bacillus anthracis*. http://edis.ifas.ufl.edu/pdffiles/FS/FS10300.pdf. Accessed 21 March 2010

Schoeni JL, Wong ACL. 2005. *Bacillus cereus* food poisoning and its toxins. *Journal of Food Protection* **68,** 636–648

Senesi S, Ghelardi E. 2010. Production, secretion and biological activity of *Bacillus cereus* enterotoxins. *Toxins* **2**, 1690–1703

Slaten DD, Oropeza R, Werner SB. 1992. An outbreak of *Bacillus cereus* food poisoning; are caterers being supervised sufficiently? *Public Health Reports* **107**, 477–480

Smith NR, Gordon RE, Clark FE. 1952. *Aerobic Sporeforming Bacteria. Agriculture Mongraph No. 16*. United States Department of Agriculture, Washington, DC.

Sutherland AD, Limond AM. 1993. Influence of pH and sugars on the growth and production of diarrhoeagenic toxin by *Bacillus cereus*. *Journal of Dairy Research* **60,** 575–580

Toh M, Moffitt MC, Henrichsen L, Raftery M, Barrow K, Cox JM, Marquis CP, Neilan BA. 2004. Cereulide, the emetic toxin of *Bacillus cereus,* is putatively a product of nonribosomal protein synthesis. *Journal of Applied Microbiology* **97**, 992–1000.

Vilain S, Luo Y, Hildreth M, Brözel V (2006) Analysis of the life cycle of the soil saprophyte *Bacillus cereus* in liquid soil extract and in soil. Applied and Environmental Microbiology 72:4970–4977

Zigha A, Roesnfield E, Schmitt P, Duport C. 2006. Anaerobic cells of *Bacillus cereus* F4430/73 respond to low oxidoreduction potential by metabolic readjustments and activation of enterotoxin expression. *Archives of Microbiology* **185**, 233

Lactobacillus

Alander M, Satokari R, Korpela R, Saxelin M, Vilpponen-Salmela T, Mattila-Sandholm T, von Wright A. 1999. Persistence of colonization of human colonic mucosa by a probiotic strain, Lactobacillus rhamnosus GG, after oral consumption. *Appl Environ Microbiol* **65,** 351-354.

Fuller R. 1989. Probiotics in man and animals. *J Appl Bacteriol* **66,** 365-378.

Guarner F, Schaafsma GJ. 1998. Probiotics. *Int J Food Microbiol* **39,** 237-238.

Jay JM. 1996. *Modern food microbiology,* 5th ed. Chapman and Hall, New York, N.Y.

Lilley DM, Stillwell RH. 1965. Probiotics: growth promoting factors produced by microorganisms. *Science* **147,** 747-748.

MacFaddin JD.1985. Media for isolation-cultivation-identification-maintenance of medical bacteria, vol. 1. p. 678-680. Williams & Wilkins, Baltimore, M.D.

Mercenier A. 1999. Lactic acid bacteria as vaccines, p. 113-127. In: Tannock GW (ed.), *Probiotics: a critical review*. Horizon Scientific Press, Wymondham, United Kingdom.

Reid G. 1999. The scientific basis for probiotic strains of lactobacilli. *Appl Environ Microbiol* **65,** 3763-3766.

Rogosa M, Mitchell JA, Wiseman RF. 1951 A selective medium for the isolation and enumeration of oral and fecal lactobacilli. *J Dental Res* **30,** 682-689.

Rogosa M. 1970. Characters used in the classification of lactobacilli. *Int J Syst Bacteriol* **20**, 519-533.

Sanders ME. 2003. Probiotics: considerations for human health. *Nutr Rev* **61**,91-99.

Tannock GW. (ed.). 2002. *Probiotics and prebiotics: where are we going?* Caister Academic Press, Wymondham, United Kingdom.

Tannock GW. 2003. Probiotics: time for a dose of realism. *Curr Issues Intest Microbiol* **4**,33-42.

Tannock GW. 2004. A Special Fondness for Lactobacilli Appl Environ Microbiol 70, 3189-3194.

Walter J, Heng NCK, Hammes WP, Loach DM, Tannock GW, Hertel C. 2003. Identification of *Lactobacillus reuteri* genes specifically induced in the mouse gastrointestinal tract. *Appl Environ Microbiol* **69**, 2044-2051.

Azotobacter

Bandyopadhyay, S, Naik SG, O'Carroll IP, Huynh BH, Dean DR, Johnson MK, Dos Santos PC. 2008. A proposed role for the Azotobacter vinelandii NfuA protein as an intermediate iron-sulfur cluster carrier. *J Biol Chem* **283**, 14092-14099.

Bertsova, YV, Bogachev AV, Skulachev VP. 2001. Noncoupled NADH:ubiquinone oxidoreductase of Azotobacter vinelandii is required for diazotrophic growth at high oxygen concentrations. *J Bacteriol* **183**, 6869-6874.

Dingler C, Kuhla J, Wassink H, Oelze J. 1988. Levels and activities of nitrogenase proteins in *Azotobacter vinelandii* grown at different dissolved oxygen concentrations. *J Bacteriol* **170,** 2148–2152.

Frazzon J, Dean DR. 2002. Biosynthesis of the nitrogenase iron-molybdenum-cofactor from *Azotobacter vinelandii. Met Ions Biol Syst* **39,** 163-186.

Hill S. 1988. How is nitrogenase regulated by oxygen? *FEMS Microbiol Rev* **34**, 11–130.

Jünemann S, Wrigglesworth JM. 1995. Cytochrome bd oxidase from *Azotobacter vinelandii. J Biol Chem* **270**, 16213-16220.

Moshiri F, Crouse B R, Johnson M K, Maier R J. 1995. The "Nitrogenase-protective" FeSII protein of *Azotobacter vinelandii*: overexpression, characterization, and crystallization. *Biochemistry* 34, 12973–12982.

Oelze J. 2000. Respiratory protection of nitrogenase in Azotobacter species: is a widely held hypothesis unequivocally supported by experimental evidence? *FEMS Microbiol Rev* **24,** 321-333.

Robson R L, Postgate J R. 1980. Oxygen and hydrogen in biological nitrogen fixation. *Annu Rev Microbiol* **34**, 183–207.

Storz G, Imlay J. 1999. Oxidative stress. *Curr Opin Microbiol* **2**, 188–194.

Vignais PM, Billoud B, Meyer J. 2001. Classification and phylogeny of hydrogenases. *FEMS Microbiol Rev* **25**, 455-501.

Vignais PM. 2008. Hydrogenases and H(+)-reduction in primary energy conservation. *Results Probl Cell Differ* **45**, 223-252.

Wang Z-C, Burns A, Watt G D. 1985. Complex formation and O_2 sensitivity of *Azotobacter vinelandii* nitrogenase and its component proteins. *Biochemistry* **24**, 214–221.

Azospirillum

Bashan, Y, Holguin G, De-Bashan LE. 2004. Azospirillum-plant Relationships: Physiological, Moecular, Agricultural, and Environmental Advances 1997-2003 *Canadian Journal of Microbiology* **50,** 521-527.

El-Shanshoury AR. 1995. Interaction of *Azotobacter chroococcum, Azospirillum brasilense* and *Streptomyces mutabilis,* in relation to their effect on wheat development. *J Agron Crop Sci* **175**,119-127.

Haghighi H, Sam Daliri M, Mobaser HR, Abbas Moosavi A. 2011. Effect of different nitrogen and potassium fertilizer levels on quality and quantity yield of flue-cured tobacco (Coker 347). *World Applied Sci J* **15**, 941-946.

Hegazy NA, Neimela S .1976. A note on the estimation of Azotobacter densities by membrane filter technique. *J Appl Bacteriol* **41**,311.

Holguin, G, Patten CL, Glick BR. 1999. Genetics and Molecular Biology of Azospirillum. *Biology and Fertility of Soils* **29**, 10-23.

Karaivazoglou NA, Tsotsolis NC, Tsadilas CD. 2007. Influence of liming and form of nitrogen fertilizer on nutrient uptake, growth, yield, and quality of Virginia (flue- cured) tobacco. *Field Crops Res* **100**, 52-60.

Mrkovacki N, Milic V. 2001. Use of *Azotobacter chroococcum* as potentially useful in agricultural application. *Ann Microbiol* **51**, 145-158.

Narula N, Kumar V, Singh B, Bhatia R, Lakshminarayana K. 2005. Impact of biofertilizers on grain yield in spring wheat under varying fertility conditions and wheat-cotton rotation. *Arch Agr Soil Sci* **51**, 79–89.

Nugroho L H, Verpoorte R. 2002. Secondary metabolism in tobacco. *Plant Cell Tissue Org Cult* **68**,105–125.

Okon, Y. 1994. *Azospirillum/Plant Associations.* CRC Press, Boca Raton.

Steenhoudt O, Vanderleyden J. 2000. Azospirillum, a free-living nitrogen-fixing bacterium closely associated with grasses: genetic, biochemical and ecological aspects. *FEMS Microb Rev* **24,** 487-506.

Tien TM, Gaskins MH, Hubbell DH. 1979.Plant Growth Substances Produced by *Azospirillum brasilense* and Their Effect on the Growth of Pearl Millet (*Pennisetum americanum* L.) *Appl Environ Microbiol* **37, 1016-1024.**

Vessey JK .2003. Plant growth promoting rhizobacteria as biofertilizers. *Plant Soil* **255**, 571-586.

Fungi, Algae and Protozoa

Alexopoulos, C.J., C.W. Mims, and M. Blackwell. 1996. *Introductory Mycology*, 868. 4th ed. New York: John Wiley and Sons.

Alexander, M. 1961. *Introduction to Soil Microbiology*,314. John Wiley and Sons, New York and London,

Alexander, M. 1965. Most-Probable-Number Method for microbial population. In *Methods of soil analysis*, ed. C. A. Black, 1467-1472 Part 2 Academic Press, New York.

Bakker, A.W. and B. Schippers. 1987. Microbial cyanide production in the rhizosphere in relation to potato yield reduction and *Pseudomonas* spp. mediated plant growth simulation. *Soil Biol. Biochem* 19: 451-457.

Barron, G.L. 1968. The genera of hyphomycetes from soil. 364 Williams and Wilkins, Baltimore.

Bartha, R. and D. Pramer. 1965. Features of a flask and method for measuring the persistence and biological effects of pesticides in soil. *Soil Science* 100: 68-70.

Benson, H.J. 1990. *Microbiological Applications*-A *laboratory manual in general microbiology*. 459 Win. C. Brown Publishers, Iowa.

Bric, J.M., R. M. Bostock, and S. E. Silversone. 1991. Rapid *in situ* assay for indole acetic acid production by bacteria immobilized on a nitrocellulose membrane. *Appl Environ Microbiol* 57: 535-538.

Brockwell, J. 1963. Accuracy of plant infection technique for counting populations of *Rhizobium trifolii*. *Appl. Microbiol* 11: 377-383.

Bushby, H.V.A. 1981. Quantitative estimation of rhizobia in non-sterile soil using antibiotics and fungicides. *Soil Biol Biochem* 13 : 237-239.

Cappuccino, J.G. and N. Sherman, N. 1991. *Microbiology: A Laboratory manual*. 462. The Benjamin/ Climmings Publishing Company, Amsterdam, The Netherlands.

Carr, N. G. and B. A. Whitton. 1982. The biology of cyanobacteria, 688. University of California Press, CA,

Casida, L.E. 1968. Methods for the isolation and estimation of activity of soil bacteria. In *The ecology of soil bacteria*. eds. T. R. G. Gray and D. Parkinson, Liverpool University Press, Liverpool.

Casida, L.E., Jr., Klein, D.A. and Sautoro, T. 1964 Soil dehydrogenase activity *Soil Sci* 98**:** 371-376.

Cochran, W.G. 1950. Estimation of bacterial densities by means of the most probable number *Biometrics* 6: 105-116.

de Lorenzo, V., M. Herrero, U. Jakubzik, et al. 1960. Mini-Tn5 tranposon derivatives for insertion mutagenesis, promoter probing, and chromosomal insertion of cloned DNA in gram negative Eubactria. *J Bacteriol* 172: 285-294.

Dick, W.A. and M. A. Tabatabai. 1992. Significance and uses of soil enzymes. In *Hand Book of Soil Sciences Properties and Processes*, eds. P. M. Huang, Y. Li, M.E. Sumner, 2nd ed. 95-127, CRC Press, New York.

Domsch, K.H., W. Gams, and T.H. Anderson. 1980. Compendium of Soil Fungi. Academic Press, New York.

Drahos, D.J., B. C. Hemming, and S. McPherson. 1986. Tracking recombinant organisms in the environment: β-galactosidase as a selectable, non-antibiotic marker for fluorescent pseudomonads. *Bio/Technology* 4: 439-443.

Dye, D.W. 1962. The inadequacy of the usual determinative tests for identification of *Xanthomonas* spp. *New Zealand J Science* 5: 393-416.

Elliot, L.F. and J. M. Lynch. 1984. *Pseudomonas* as a factor in the growth of winter wheat (*Triticum aestivum L.*) *Soil Biol. Biochem* 16: 69-71.

Fred, E.B., I. L. Baldwin, and F. Mc Coy. 1932. *Root nodule bacteria and leguminous plants* University of Wisconsin Press, Madison, Wisconsin.

Gibson A.H. 1963 Physical environment and symbiotic nitrogen fixation. I. The effect of root temperature on recently nodulated *Trifolium subterraneum L.* plants. *Aust J Biol Sci* 16: 28-42.

Glick, B.R. 1995. The enhancement of plant growth of free living bacteria. *Can J Microbiol* 41: 109-117.

Gordon A.S. and R. P. Weber. 1951. Colorimetric estimation of indole acetic acid. *Plant Physiol* 26: 192-195.

Gould, W.D., C. Hagedorn, T. R. Bardinelli, and R. M. Zablotowiez. 1985. New selective medium for enumeration and recovery for fluorescent pseudomonads from various habitats. *Apl Environ Microbiol* 49: 28-32.

Hagedorn, C. and J. G. Holt. 1975. Ecology of soil arthrobactrers in Clarion-Websters toposequence of Iowa. *Appl. Microbiol* 29: 211-218.

Halvorson, H.O. and N. R. Ziegler. 1933 Application of statistics to problems in bacteriology: I. A means of determining bacterial population by the dilution method. *J Bacteriol* 25: 101-121.

Hardy R.W.F., R. C. Burns and G. W. Parshall. 1971. The biochemistry of N_2 fixation *In: Bioinorganic Chemistry. Advances in Chemistry,* 219-247. American Chemical Society. Series 100.

Hardy R. W. F., R. D. Holsten, E. K. Jackson, et al. 1968 The acetylene-ethylene assay for N_2 fixation : laboratory and field evaluation. *Plant Physiol* 43: 1185-1207.

Hartel, P.G., J. J. Fuhrmann, Jr. W. F. Johnson, et al. 1994. Survival of a *lacZY* containing *Pseudomonas putida* strain under stressful abiotic soil conditions. *Soil Sci Soc Am J* 58: 770-776.

Hattemer-Frey, H.A., E. J. Brandt, and C. C. Travis. 1990. Small-scale field test of the genetically engineered *lacZY* marker. *Reg Toxicol Pharmacol* 11: 253-261.

Hendricks, C.W. and N. Pascoe. 1988. Soil microbial biomass estimates using 2450 MHz microwave irradiation. *Plant and Soil* 110:39-47.

Horwath, W. R. and E. A. Paul. 1994. Microbial biomass. In *Methods of soil analysis,* eds. R.W. Weaver, S. Angle, P. Bottomley, et al. 753-773. Part 2. SSSA. (Soil Science Society of America Series, 5), Madison, WI.

Kirk, P.M., P. F. Cannon, D. W. Minter, and J. A. Stalpers. 2008. Ainsworth and Bisby's Dictionary of Fungi. CAB International, Wallingford, Oxon, UK.

Herrero, M., V. de Lorenzo, and K. T. Timmis. 1990. Transposon vectors containing non-antibiotic resistance selection markers for cloning and stable chromosome insertion of foreign genes in gram-negative bacteria. *J Bacteriol* 172: 6557-6567.

Hleyn, J., M. Bicknell, and M. Gilstrap. 1995. *Microbiology experiments: A health science perspective.* 354. Wm. C. Brown Publishers, London.

Kapulnik, Y. 1996. Plant growth promoting rhizosphere bacteria, *In*: *Plant Roots the Hidden Half.* 769-781 eds. Y. Waisel, A. Eshel and U. Kafkafi, Marcel Dekker, New York.

Kloepper, J. W. 1993. Plant growth-promoting rhizobacteria as biological control agents. In *Soil Microbial Ecology: Applications in Agricultural and Environmental Management.* 255-274. ed. F. B. Metting, Jr., Marcel Dekker Inc., New York,

Jackson, M.L. 1967. *Soil Chemical Analysis,* Prentice Hall of India Pvt Ltd., New Delhi.

Jensen, H.L. 1942. Nitrogen fixation in leguminous plants. I. General characters of root-nodule bacteria isolated from species of *Medicago* and *Trifolium* in Australia. *Proc Linn Soc NSW* 66: 98-108.

Jensen, H.L. 1954. The Azotobacteriaceae. *Bact Rev* 18: 195-214.

Johnson, L.F. and E. A. Curl. 1972. Methods for research on the ecology of soil-borne plant pathogens. 247. Burgess Publishing Co., Minneapolis, MN.

Johnston, A.W.B. 1975 Identification of *Rhizobium* strains in pea root nodules using genetic markers. *J Gen Microbiol* 87: 343-350.

Josey, D.P. J. L. Beynon, A. W. B. Johnston, et al. 1979. Strain identification in *Rhizobium* using Intrinsic antibiotic resistance. *J Appl Bacteriol* 46: 333-350.

King, E.O.K., M. K.Ward, and D. E. Rancy. 1954. Two simple media for the demonstration of pyocyanin and fluorescein. *Journal of Laboratory and Clinical Medicine* 44: 301-307.

Klein, D.A., T. C. Loh, and R. L. Goulding. 1971. A rapid procedure to evaluate dehydrogenase activity of soils low in organic matter. *Soil Biol Biochem* 3: 385-387.

Kloepper, J.W, J. Leong, M. Teintze, et al. 1980. Enhanced plant growth by siderophores produced by plant growth promoting rhizobacteria. *Nature* (London) 286:885-886.

Laguerre, G., M. R. Allard, F. Revoy, et al. 1994. Rapid identification of rhizobia by restriction fragment length polymorphism analysis of PCR amplified 16S rRNA genes. *Appl Environ Microbiol* 60: 56-63.

Mishustin, E.N. and V. K. Shilinkova. 1969. *The Biological fixation of atmospheric nitrogen by free-living bacteria.* Review of Research, Soil Biology, UNESCO.

Mullis, K.B. 1990. The unusual origin of the polymerase chain reaction. *Sci Am* 262: 56-61.

Nautiyal, C. S. (1999). An efficient microbiological growth medium for screening phosphate solubilizing microorganisms. FEMS Microbiol. Let. 170: 265-270.

Norris, J.R. and H. M. Chapman. 1968. Classification of *Azotobacter.* In *Identification methods for microbiologists,* 19-27. eds. B. M. Gibbs, and D. A. Shapton, Academic Press, London and New York.

Okon, Y., S. L. Albrecht, and R. H. Burris. 1977. Methods for growing *Spirillum lipoferum* and for counting it in pure culture and in association with plants. *Appl Environ Microbiol* 33: 85-87.

Pagan, J.D., J. J. Child, W. R. Scowcroft et al. 1975. Nitrogen fixation by *Rhizobium* cultured on a defined medium *Nature* 256: 406-407.

Pikovskaya, R.E. 1948. Mobilisation of phosphorus in soil in connection with vital activity of some microbial species. *Mikrobiologiya* 17: 362-370.

Powlson, D.S. 1976. The effect of biological treatment on metabolism in soil. V.A method of measuring soil biomass. *Soil Biol. Biochem* 8: 209-213.

Riddell, R.S. 1950. Permanent stained mycological preparations obtained by slide culture. Mycologia 42:265-270

Rodriguez-Caceres, E.A. 1982. Improved medium for isolation of *Azospirillum sp. Appl Environ Microbiol* 44: 990-991.

Kormanik, P. P. and A. C. McGraw. 1982. Quantification of Vesicular-arbuscular Mycorrhizae in Plant Roots. In *Methods and Principles of Mycorrhizal Research*, 37-36. ed. N.C. Schenck, The American Phytopathological Society.

Phillips, J. M. and D. S. Hayman. 1970. Improved procedures for clearing roots and staining parasitic and vesicular-arbuscular mycorrhizal fungi for rapid assessment of infection. *Trans Brit Mycol Soc* 55: 158-161.

Schollhorn, R. and Burris, R.H. 1967 Acetylene as a competitive inhibition of N_2 fixation. *Proc Natl Acad Sci USA* 58: 213-216.

Schwyn, B. and J. B. Neilands. 1987. Universal chemical assay of the detection and determination of siderophore. *Anal Biochem* 60: 47-56.

Seeley, H.W., Jr. and P. J. Denmark. 1975. *Microbes in action: A Laboratory manual of microbiology*, 361. D.B. Taraporevala Sons & Co. Pvt. Ltd., Mumbai, India,

Shende, S.T., R. G. Apte, and T. Singh. 1977. Influence of *Azotobacter* on germinatin of rice and cotton seeds. *Curr Sci* 46: 675-676.

Smith, S.E. and D.J. Read. 1997. Mycorrhizal symbiosis. 2nd ed. *Soil Science Society of America* (SSSA book series) 5, 605, Academic Press, Madison, WI.

Stewart W,D.P., G. P. Fitzgerald, and R. H. Burris. 1967. *In Situ* studies on N_2 fixation using the acetylene reduction technique. *Proc Natl Acad Sci USA* 58: 2071-2078.

Sylvia, D.M. 1994. Vesicular-arbuscular mycorrhizal (VAM) fungi. In *Methods of soil analysis* 351–378. 5, Madison, WI.

Tabatabai, M.A., 1994. Soil enzymes. In: *Microbiological and Biochemical Properties*. eds. R.W. Weaver, J.S. Angle, and P.S. Bottomley Part 2., *Soil Science Society of America* (SSSA book series) 775–833. Madison, WI

Vance, E.D., P. C. Brookes, and D. S. Jenkinson. 1987. A extraction method for measuring soil microbial biomass C. *Soil Boil Biochem* 19: 703-707.

Vela. G.R. and O. Wyss. 1964. Improved stain for visualization of *Azotobacter* encystment. *J Bacteriol* 87: 476-477.

Vincent, J.M. 1970 *A manual for the practical study of the root nodule bacteria*.1-13. Blackwell Scientific Publication, Oxford and Edinburgh.

Voorhorst W.G.B., R.I.L. Eggen, E. J. Luesink, et al. 1995. Characterization of the *cel* /B gene coding for β-glucosidase from the hyper thermophicli archaeon *Pyrococcus furiosus* and its expression and site directed mutation in *Escherichia coli*. *J Bacteriol* 177: 71051-7111.

Watanabe, T.1994. Pictorial atlas of soil and seed fungi : morphologies of cultured fungi. Lewis Publishers, Boca Raton, FL.

Willems, A. and M. D. Collins. 1993. Phylogenetic analysis of rhizobia and agrobacteria based on 16S ribosomal RNA gene sequences. *Int J Sys Bacteriol* 43: 305-313.

Woese, C.R. 1987 Bacterial evolution. *Microbiol Rev* 51: 221-271.

Zuberer, D.A. 1994. Recovery and enumeration of viable bacteria. Pp. In *Methods of Soil Analysis. Microbiological and biochemical properties*. R.W. Weaver, J.S. Angle, and P.S. Bottomley Part 2, *Soil Science Society of America* (SSSA book series) 119-144. Madison, WI.

Chapter 14

Microbial Genetics

14.1 INTRODUCTION

Microbiology began to merge with genetics after the 1928 experiments of Griffith who showed that traits could be transferred between bacteria, but he could not identify the transferring substances. Another 16 years passed before Avery and his group pinpointed the substances as deoxyribonucleic acid (DNA) and the pace of discovery quickened, and biochemists redefined genes as segments of the DNA. With the epic announcement of the structure of DNA by Francis and Crick in 1953, a theory emerged as to how genes could function in the synthesis of protein. There began a period of intense research in molecular genetics that continues to the present day. Microbes especially bacteria stand at the forefront of genetic research for several reasons. They have only a single chromosome and, thus, are far easier to work with than such as Mendel's pea plants or Morgan's fruit flies (where two genes exist as a trait). Also, bacteria can be cultivated conveniently in test tubes, and they produce the next generation much more rapidly than fruitflies, which take several weeks, or pea plants, which require a growing season of several months.

14.2 BACTERIAL GENETICS AND GENE TECHNOLOGY

Bacterial genetics is concerned with the gene changes that take place in a bacterium's chromosome and plasmids, changes that reflect themselves in the morphology, physiology, and pathology of the organism. The changes can take place by mutation and by recombination.

Mutation is a permanent change in the cellular DNA, occurring spontaneously in nature or by induced methods in the laboratory. Such things as ultraviolet light, chemicals and base analogs may induce mutations. Deletions or insertions involving the chromosome as well as transposons and insertion sequences may also be mutagenic.

Recombinations imply an exchange of DNA between bacteria and thus, an acquisition of genes. In one form of recombination called transformation, a "competent" bacterium takes up DNA from the local environment; this DNA has been left behind by a disrupted bacterium. The new DNA displaces a segment of equivalent DNA in the recipient cell, and new genetic characteristics are assumed. In another form, conjugation, a live donor cell contributes a portion of its DNA to a recipient cell. Plasmids are commonly transferred but chromosomal DNA may also be contributed. In the third form of recombination, transduction, a virus enters a bacterium and later replicates within it. During replication, the virus incorporates some bacterial DNA to its protein coat and transports that DNA to a new bacterium. In the specialized form of transduction, the virus first attaches to, then detaches from the bacterial chromosome taking a segment of bacterial DNA with it. Thus, all three forms of recombination are characterized by the introduction of new genes to a recipient bacterium.

Genetic engineering is an outgrowth of studies in bacterial genetics. In one type of this technology, plasmids are isolated from a bacterium, spliced with foreign genes, and then inserted to fresh bacteria where the foreign genes are expressed as protein. Bacteria are thus used as the biochemical factories for the synthesis of such proteins as insulin, interferone, human growth hormones, and others. But genetic engineering is only one branch of modern biotechnology. This technology utilizes DNA-based techniques to perform diagnoses, identify individuals in forensic medicine, detect and treat genetic and spark innovative approaches to agriculture. The practical benefits of bacterial genetics have helped revolutionize myriad fields of human endeavour. Much of our knowledge of bacterial genetics has come from the detailed study of *Escherichia coli K12*, other bacteria such as *Salmonella typhimurium, Bacillus subtilis, Streptomyces* spp., *Pseudomonas aeruginosa* and *Saccharomyces cerevisiae* from eukaryotes have also attracted interest. Multiply marked strains of these microbes have been successfully produced and isolated. Systems of genetic exchange have been studied, and chromosome maps constructed.

14.3 THE BACTERIAL CHROMOSOME

Most of the genetic information in a bacterial (or prokaryotic) cell is contained within the chromosome, a single molecule of DNA arranged as a double helix in a closed loop (Figure 14.1).

The chromosome exists freely in the cytoplasm without the surrounding membrane or protein support found in eukaryotic cells. It occupies about 10 per cent of the total volume of a bacterial cell, and when extended its full length, it is about 1 millimetre (mm) long. This is approximately 1000 times the length of the bacterium that contains it. The tight packing accounts in part for the explosive release of DNA when the membrane of the cell is broken. The comparison of eukaryotic chromosome with prokarytic chromosomme is shown in Table 14.1.

The chromosome of the intestinal bacterium *Escherichia coli* has probably studied more thoroughly than any other single chromosome. Distributed around the chromosome are individual sites to which genetic activity can be traced. Each site, called a locus, consists of several genes for that activity. The chromosome of *E. coli* has about 4000 genes. Some viruses, by contrast, have as few as seven genes, while human chromosomes have a total of over 100,000.

14.3.1 Replication of the Chromosome

The bacterial chromosome replicates during the process of binary fission. At the beginning of the replication sequence, the DNA is usually anchored to particular

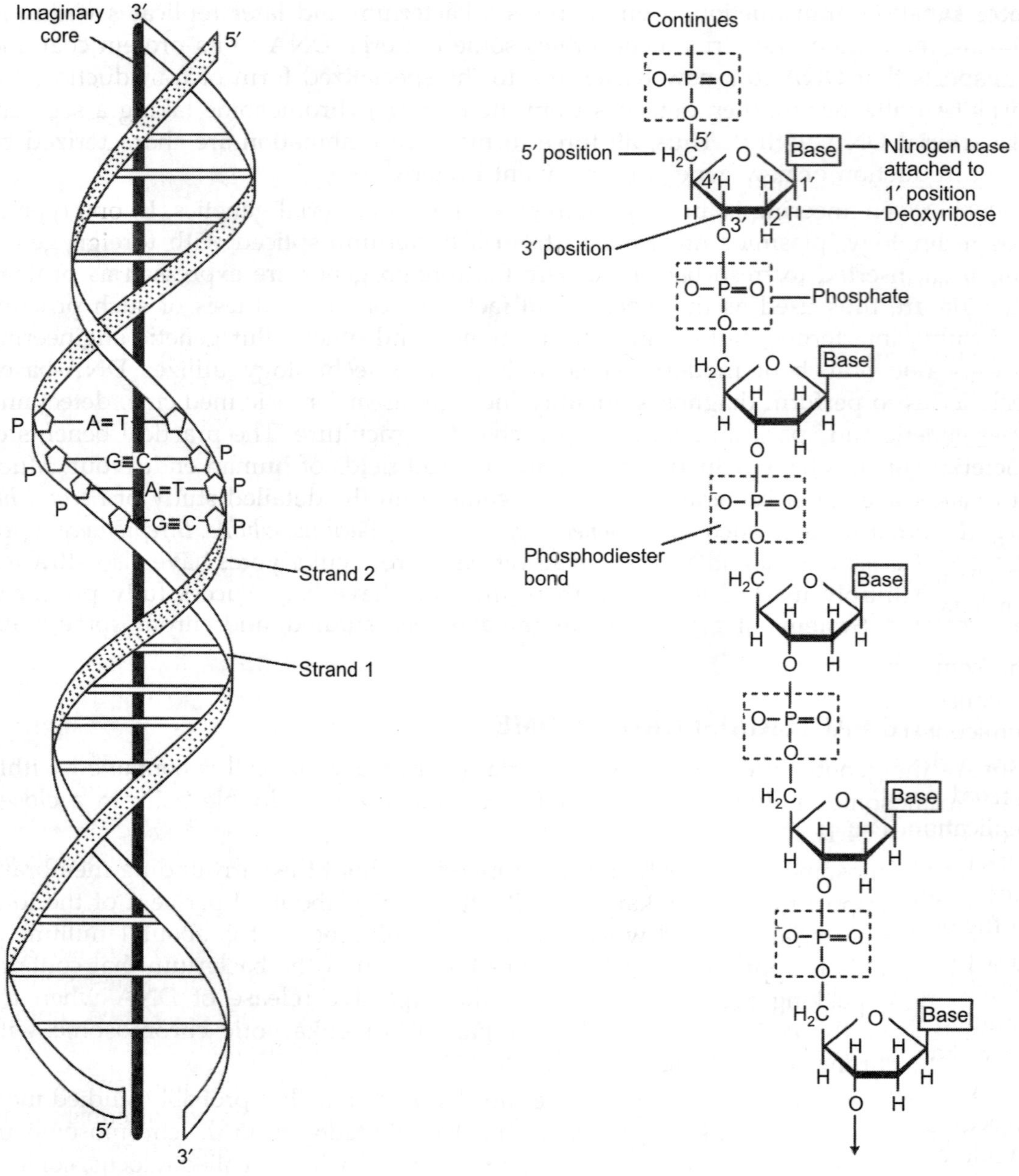

Figure 14.1 Double helix structure of DNA.

point on the bacterial cell membrane. The double helix then unwinds and enzymes synthesize a new strand of DNA for each of the original strand. This is accomplished by joining together nucleotides with DNA polymerase. Each new strand then combines with a parent strand and the replicated DNA twists to reform the double helix. This combination of new and parent strands was first observed in *E. coli* and is called

Table 14.1 Some important differences in chromosomes of prokaryote and eukaryote

Prokaryote bacterial chromosomes	*Eukaryote chromosome*
Replicates during binary fission	Replicates during mitosis
Single molecule of DNA per genetic trait (haploid)	Two molecules of DNA per genetic trait (some organisms haploid)
Closed loop	Linear
No protein present	Histone protein present
No dominance of recessive genes	Genes are dominant or recessive
Organized at the nucleoid	Organized in the nucleus with a nuclear membrane
No introns present	Introns present
DNA also in plasmids	DNA also in mitochondria
Mutations occur in DNA	Mutations occur in DNA
Genetic recombination occur	No genetic recombination occur
4000 genes in E. coli chromosome	100,000 genes in human chromosomes
About 1 mm in length	Several hundreds of millimetres in length
Replicates by a semi conservative mode or rolling circle method	Replicates by semi-conservative mode

the semiconservative method of replication because one strand of the parent DNA is conserved in the new DNA molecule and one strand is newly synthesized. The semiconservative method accounts for DNA replication, but it does not shed light on how a closed loop chromosome replicates. Cairns in 1962 showed that DNA unwinds at a fixed point, whereupon an enzyme nicks the closed loop at a site called the origin of replication. The two strands now separate, or "unzip" to establish a V-shaped replicating fork (Figure 14.2). Synthesis of DNA then occurs along the sides of the fork. Along one side of the fork DNA is manufactured by continuous assembly of nucleotides, beginning at the origin. However, along the other side, the synthesis begins at the point of forking and proceeds in a discontinuous fashion. Here the DNA is synthesized in a series of segments that later join with the help of an enzyme called DNA ligase. The segments came to be known as Okazaki fragments, after Reiji Okazaki who discovered them in 1968. Each of the new DNA strands then combines with a parent strand.

A second type of DNA replication is called the rolling circle mechanism. This process takes place in bacteria undergoing the sexual mating process of conjugation. While one strand of DNA remains in a closed loop, an enzyme nicks the other strand. The broken strand then "rolls off" the loop and serves a template for synthesis of a DNA strand complementary to itself. When the two strands combine, a double helix reforms. Meanwhile the intact loop revolves degrees and serves as a template for a strand of DNA complementary to itself. The loop combines with its new strand to form a second double helix. The bacterium now has two chromosomes, one of which will be used in mating.

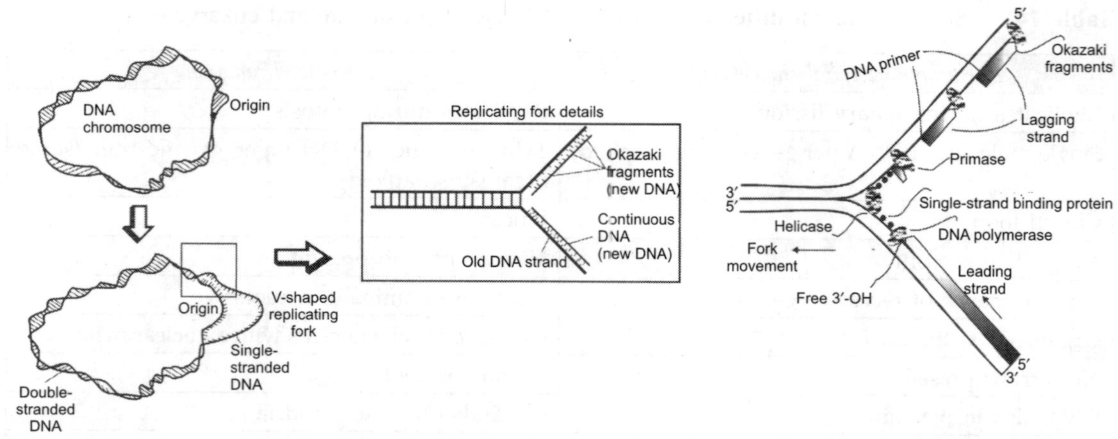

Figure 14.2 Replication of the closed looped chromosomes.

14.4 NATURE AND REPRODUCTION OF GENETIC MATERIAL

The only parts of the eukaryote nucleus that persist through cell division are the chromosomes, which consist of deoxyribonucleic acid (DNA) and a complex basic protein (histone). The interphase nucleus also contains varying amounts of ribonucleic acid (RNA), but this cannot be a candidate for the role of genetic material because it is largely confined to the nucleolus, which disappears during cell division.

Because DNA and histone replicate and divide together it is technically difficult to determine which is the genetic material. The first clue was indirect; it was found that a simple protein-protamine replaces histone in sperm, thus pointing to DNA as the only substance with genetic continuity. This was confirmed when it was discovered that pure DNA could transfer information from one bacterial strain to another; since then refined experiments, using isotypically labelled DNA and protein precursors, have shown beyond doubt that only the DNA is transferred intact from one generation to the next in organisms.

14.4.1 Division of the Genome

14.4.1.1 Viruses

The amount of genetic information carried by viruses is insufficient for them to initiate and complete their own replication. Because of this, they take control of the infected cell's metabolic systems to energize their replication. For example, when abacteriophage infects a bacterium, synthesis of bacterial DNA, RNA, and protein stops within five minutes; then the bacterial system begins to reproduce the phage DNA and latter to form phage coat protein. Similarly, when a tobacco cell is infected with tobacco mosaic virus (TMV), it first synthesizes virus RNA and then virus coat protein. In other words, viruses do not divide; they force other systems to copy them.

14.4.1.2 Bacteria

DNA exists as a close ring in bacteria which replicates by sequential division of the two single strands in this ring. Circular DNA has not been observed in *Bacillus subtilis,* although it is possible that the DNA of this organism is in circular form during replication. DNA replication in bacteria does not occur at the same time as cell division, so that there may be more than one copy of the chromosome in a cell; the mean number of chromosomes per bacterium in a culture of *E. coli* varies between one (in resting populations) and four (during exponential growth).

14.4.2 Mitosis

The genome replication in prokaryotes is a continuous process. In more complex cells of higher organisms, where the DNA is organized into more than one chromosome, there is usually a regular sequence of replication the mitotic cycle. Immediately after the mitotic division there is usually a period (G1) when RNA and protein, etc., are synthesized. Then in the synthetic (S) phase the DNA and histone of the chromosomes doubles. Following S there is another growth phase, G2, and only then is the cell capable of a further mitosis.

In mitotic prophase, the chromosome become visible as double strands (chromatids) which are attached to each other only at appoint of constriction, the centromere. At the end of the prophase the nuclear membrane breaks down and a spindle is formed. The centromeres become attached to the spindle fibres and come to lie on the equator of the cell at metaphase. Anaphase is marked by the sudden division of the centromeres, whose daughters, with their attached chromatids, travel or are pulled towards opposite poles of the spindle, At telophase the daughter chromatids coalesce and nuclear membranes are formed around the two new nuclei. In this way identical genomes are passed to the two daughter cells (Figure 14.3).

14.4.3 Meiosis

Underlying the many different life cycles of eukaryotes, there is a basic alternation of haploid cells containing one set of genetic information and diploid cells containing two sets. Meiosis is the process by which one diploid nucleus gives rise to four haploid nuclei (fusion of two haploid nuclei results in formation of a zygote and restores the diploid condition).

The meiotic process differs from mitosis in two ways – there are two successive nuclear divisions without intervening synthesis of DNA, and the nuclei produced are not identical to each other or to the parent nucleus. Because of the complexity of meiotic prophase it is convenient to divide it into a number of stages.

When the chromosomes first become visible in leptotene, they appear as single strands although DNA synthesis is complete. During zygotene, homologous chromosomes from the two parental strains (shown as heavy and light lines in Figure 14.4) come to lie together, twining around each other so that in pachytene, when pairing is complete, it is impossible to distinguish them. These pairs of homologous chromosomes are now referred as bivalent. In diplotene, the bivalents open out and each chromosome is seen to consist of two chromotides; at various points along the bivalent pairs of chromatids are seen to have exchanges partners. These exchanges or chiasmata (singular chiasma)

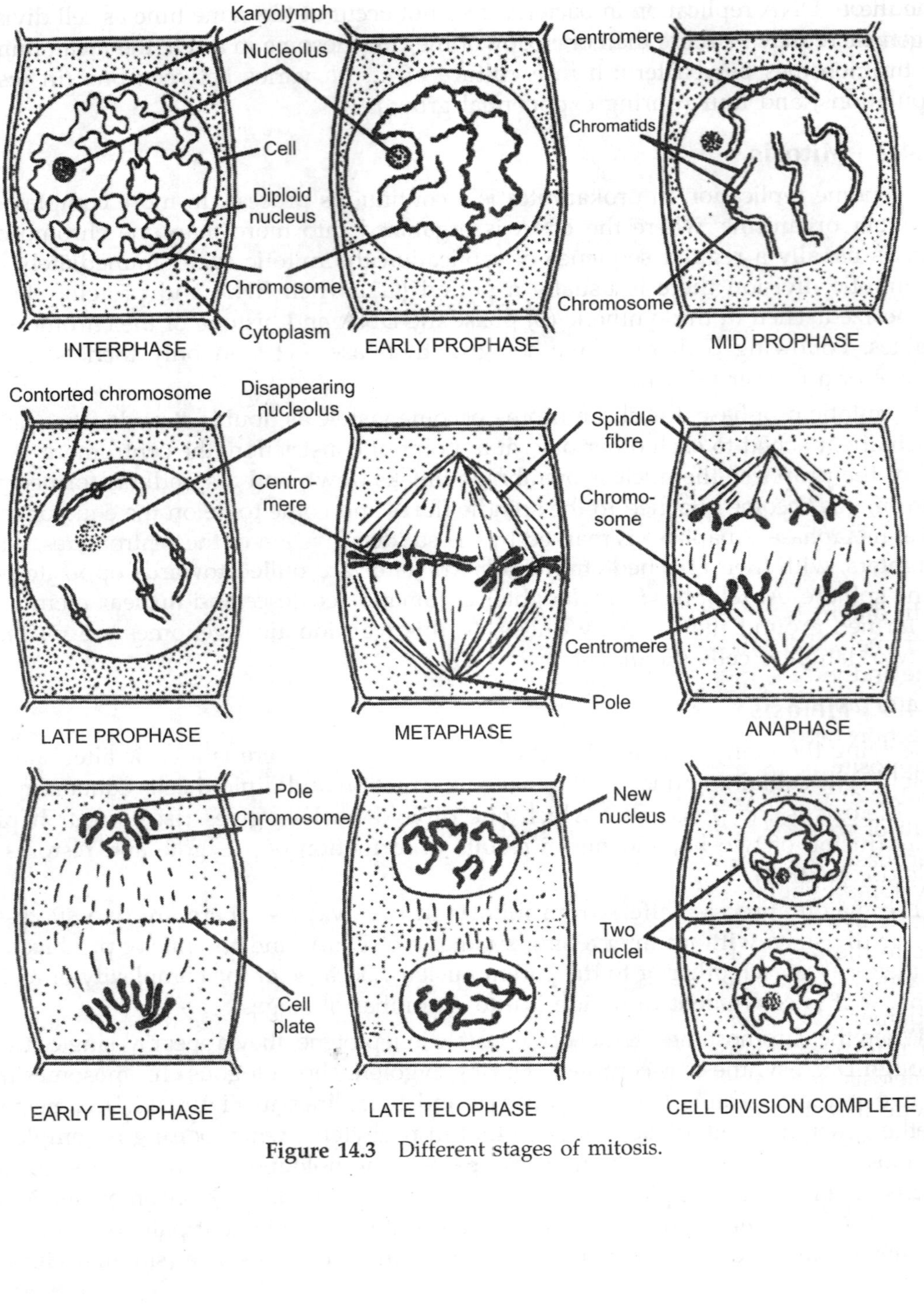

Figure 14.3 Different stages of mitosis.

are the products of breakage and reunion of chromatids during pachytene, at which time there is a small amount of DNA synthesis, and they are the cytological expression of genetic crossing over. The homologous undivided centromeres repeal each other but for the chiasmata would pull the bivalents apart.

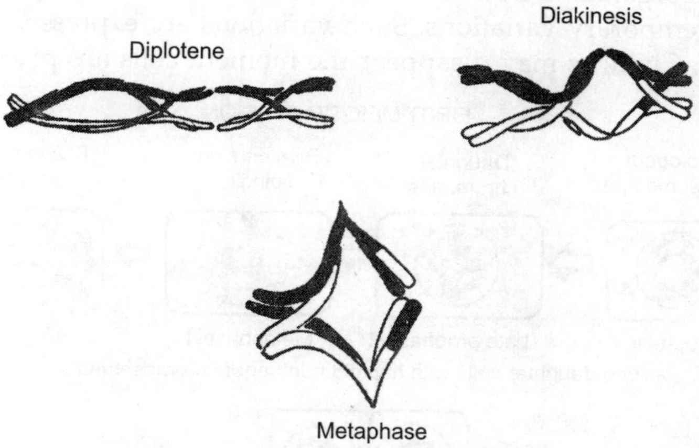

Figure 14.4 Progressive changes in a bivalent with two chiasmata from diplotene to metaphase I of meiosis.

In diakinesis, the nuclear membrane breaks down and a spindle is organized. The centromeres become attached to spindle fibres and jockey for position until at metaphase I the individual bivalents are positioned with their chiasmata more or less on the equator and their centromeres randomly oriented on either side. At anaphase I the attraction between paired chromatids lapses and the still undivided centromeres pass to opposite poles of the spindle to give two telophase I nuclei each with the haploid number of undivided chromosomes. The sister chromatids of these chromosomes as a result of crossing over are not genetically identical. At this stage there may be a short interphase, but more usually the chromosomes pass directly to prophase II and two new spindles are organized. This time the centromeres come to lie on the equator of the spindle by metaphase II and divide normally at anaphase II. The daughter chromatids pass to the four poles to form four haploid nuclei (Figure 14.5).

Although superficially akin to mitosis, this second nuclear division differs from it fundamentally because of the random orientation of centromeres and the crossing-over which took place during prophase I. These two factors ensure that all four products of meiosis are different. In the Euascomycetes, which have been important in many genetic investigations, the tetrad is arranged linearly and each member undergoes a further genuine mitosis to form an ascus containing a row of eight ascospores.

14.5 MUTATION

The phenomenon of mutation is fundamental to genetics; it is a prerequisite of the evolution and, thence, the recognition of new genes. A mutation can be defined as any permanent change in the quality or quantity of an organism's heritable information.

Not all this information is confined to the chromosomes; for example, both plastids and mitochondria are known to have genetic continuity and contain their own DNA. The existence of nuclear genes controlling plastid development, however, indicates that this specific DNA is not sufficient for plastid autonomy. Environmental factors play an important role in reorganizing DNA to adjust the bacterial cell in new environment and this is known as temporary variations. Such variations are expressed morphologically or physiologically. Changes may disappear the moment cells are placed in the normal

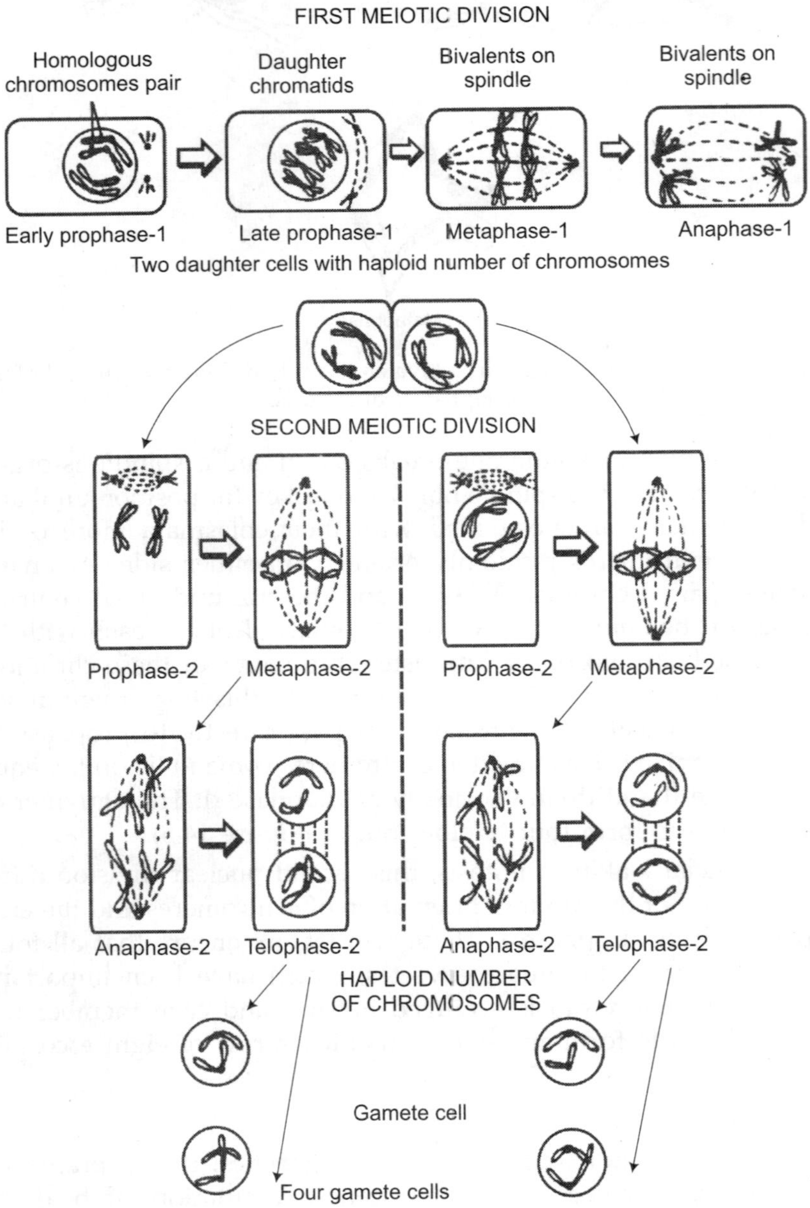

Figure 14.5 Different stages of I and II meiotic division.

environment. For example, an overgrown old culture of *E. coli* in a nutrient depleted medium, the shape is changed and they become coccoid, on re-inoculation into the fresh medium regain the original form. DNA is when permanently reorganized with alteration in gene then it is of permanent nature and designated as permanent variations. Such variations are known as mutations, i.e., a heritable change in the base sequence of the DNA of an organism. A mutant by definition will differ from its parental strain in genotype in addition to the visible properties of the mutant, its phenotype. They are able to survive on permanent transfers in new environment. Such mutations may be physical or chemical in nature. Physical agents for the mutations may be UV light, temperature shift, pH shift or nutrient deficiencies. Chemical mutations are due to some chemicals which may be designated as chemicals, mutants like NTG, antibiotics, heavy metals, etc. Variations of this type occur spontaneously. Some mutations are also caused by the transfer of DNA from one organism to another, either directly or by conjugation or indirectly by some vector like phages (Table 14.2).

Table 14.2 Kinds of mutants

Description	*Nature of change*	*Detection of mutant*
Non-motile	Loss of flagella; non-functional flagella	Compact colonies instead of flat, spreading colonies
Non-capsulated	Loss or modification of surface capsule	Small, rough colonies instead of larger, smooth colonies
Rough colony	Loss or change in lipopolysaccharide	Granular, irregular colonies instead of smooth. layer glistening colonies
Nutritional	Loss of enzyme in biosynthetic pathway	Inability to grow on medium lacking the nutrient
Sugar (Fermentation)	Loss of enzyme in degradative pathway	Do not produce colour change on agar fermentation containing sugar and a pH indicator
Drug resistant	Impermeability to drug or drug target is detoxified	Growth on medium containing a growth-inhibitory concentration of the drug
Virus resistant	Loss of virus receptor	Growth in presence of large amounts of virus
Temperature senstive	Alteration of any essential protein so that it is more heat sensitive	Inability to grow at a temperature normally supporting growth (e.g., 37°C) but still growing at a lower temperature (e.g., 25°C)
Pigmentation	Loss of enzyme in biosynthetic pathway leading to loss of one or more pigments	Detect visually by mutant being different colour or colourless

Contd...

Contd...

Cold sensitive	Alteration in an essential protein so that it is inactivated at low temperature	Inability to grow at a low temperature (e.g., 25°C) that normally supports growth

14.6 NUCLEAR AND EXTRANUCLEAR MUTATION

Since the genome is replicated and divided exactly between daughter cells, chromosome mutation can be detected almost immediately in haploid organisms and will appear in specific proportions in diploids following meiosis and fertilization. On the other hand, if one plastid mutates in a cell containing twenty, a number of cell generations must elapse before a cell is formed which by chance contains none but mutant plastids. There is no reason to suspect that the mutational events in extranuclear particles are in any way different to those within the chromosomes, but the technical difficulties of their isolation and characterization have so far led to concentration of research on nuclear genes.

Most spontaneous mutations occur at a frequency of the order of 10^{-7} to 10^{-9} per cell generation, and their behaviour can usually be interpreted as a change in the nucleotide sequence specifying the phenotype studied. Some genes, however, mutate at much higher frequencies, and it is sometimes necessary to invoke ingenious explanations for these unstable genes. The following account is restricted to the better understood mechanisms of nucleic acid mutation and of structural changes in chromosomes.

14.7 MUTAGENIC AGENTS AND THEIR EFFECTS

The frequency of mutational events can be increased by treating cells with mutagenic (mutation inducing) agents. This is called induced mutation, and the resulting mutant cells are called induced mutants. Chemical, physical, and biological agents can be used as mutagens. The mechanisms by which they produce their effects are being discussed with some selected examples in the following paragraphs (Table 14.3).

Table 14.3 Mutagenic agents and their mode of action

Mutagen	*Mode of action*
5-Bromouracil (BU)	Replaces thymine by pairing with guanine
2-Aminopurine (AP)	Replaces adenine by pairing with cytosine
Nitrous acid	Deamination, strand crossing over
Hydroxylamine	Hydroxylation of cytosine
Ethylmethane sulphonate	Alkylation of purines, transitions
N-methyl-N′-nitro-N-nitroso-guanine	Synthesis of methyl-guanine during replication, transitions and multiple mutations
Acridine orange	Frameshift mutations by intercalation
Ultraviolet radiation (UV) 254 nm	Pyrimidine dimmers, repair errors
X-rays 5 nm wavelength	Breakage of single- and double-stranded DNA

14.7.1 Nucleic Acid Mutation

14.7.1.1 Substitution

Substitution mutation is of two types: Transition mutations involve the replacement of a purine by a purine or of a pyrimidine by pyrimidine, e.g., when A is replaced by G in DNA; transversion mutations are those in which a purine is replaced by a pyrimidine or vice-versa (e.g., A by T). Both types usually lead to a simple change of one amino acid in the gene product (mis-sense mutation; Figure 14.6). if, however, a single base change generates a nonsense triplet (e.g., CAG to UAG), translation of m-RNA is terminated at the site of the mutation and the resultant polypeptide is shorter than its wild-type analogue (nonsense mutation).

DNA (sense strand)	CAT	CGT	TTA	CAT	CA*T	TTA
RNA (messenger)	GUA	GCU	AAU	GUA	GU*A	AAU
Polypeptide	–Val	–Val	–Asn–	–Val	–VAl*	–Asn–

Figure 14.6 Replacement of alanine by valine following a G to A transition (*); mis-sense mutation.

14.7.1.2 Addition or deletion

Unlike the single amino acid changes induced by substitution, deletion or addition of a single nucleotide leads to a one-base shift in the frame of reading of the code. Such a frameshift mutation produces a completely new polypeptide and is therefore likely to alter the phenotype much more profoundly than a simple mis-sense mutation. Sometimes a more drastic type of mutation occurs in which more than one nucleotide is deleted in a single mutational event; these multisite or deletion mutatnts, in which the base sequence corresponding to an entire polypeptide is occasionally lost, are extremely useful in genetic mapping (Figure 14.7).

DNA	AGC	GCG	ATG	ACG	A	AGC	GCG	TGA	ACG	A
RNA	UCG	CGC	UAC	UGC	U	UCG	CCU	ACU	GCU	U
Polypeptide	- Ser	- Arg	- Tyr -	Cys -		- Ser	- Pro* -	Thr* -	- Ala*	-

Figure 14.7 New polypeptide chain (*) produced by deleting a single base from a triplet coding for arginine: frameshift mutation. It is assumed in this figure that the m-RNA is read 'left' to 'right'.

14.7.1.3 Chromosome mutation

There are different types of aberration of chromosome that can occur, viz., inversions, duplications, deletions, and translocations. They can all be placed in one of the two categories; those which affect the viability of the cell's immediate progeny, and those with a delayed effect on the organism's fertility. The molecular mechanisms of mutation are almost certainly similar in the chromosomes of bacteria and those of higher organisms, but the complex structure of the latter makes it difficult to recognize any alteration apart from gross structural changes.

14.7.1.3.1 Viability

A break in a chromosome has no effect on the cell containing it so long as no division takes place. At division, however, those chromosome fragments which lack a centromere usually fail to reach the telophase nuclei, which therefore lose some genetic information. In a diploid nucleus, the other homologue may be able to supply the lost information and there may be no immediate effect, but in haploids loss of information can lead to early death.

14.7.1.3.2 Fertility

Exchange of chromosome parts within or between chromosomes usually has no immediate physiological effect on the cell and its mitotic progeny since there is no change in the total information content. At meiosis, however, crossing-over between paired aberrant chromosomes frequently leads to the formation of unbalanced gametes with consequent reduction of fertility.

14.7.1.4 Induced mutations

Most of the information gathered on mutation has come from induced mutations. Induced mutations are those in which the cause can be identified. Occasionally, these mutations occur by accident. More often they result from planned experiments in which laboratory personnel subject bacteria to chemical or physical agents. The agents are referred to as mutagens. Experiments in molecular genetics reveal that a broad variety of induced mutations may arise, depending on the type of mutagen used.

Mutations arising from treatment with ultraviolet light are among the best understood. Ultraviolet light is a form of energy not perceived by human eye. When this energy is absorbed by DNA, it induces adjacent thymine molecules to link together. The DNA thus loses its ability to insert adenine bases in mRNA molecules during protein synthesis. Ultraviolet light is sometimes used for sterilization purposes because it quickly kills bacteria. In Gram-negative rod *Serratia marcescens,* ultraviolet light induces mutations that prevent the organism from forming its normal red pigment. Nitrous acid will also cause an induced mutation. This chemical mutagen converts DNA's adenine molecules to hypoxanthine molecules. Adenine will normally slot thymine into position during DNA replication, but hypoxanthine slot cytosine into place. Later, when protein synthesis takes place, the new DNA codes for guanine in the mRNA instead of the normal adenine (Figure 14.8).

14.7.1.5 Chemical mutagens

Chemical mutagens are perhaps of greater intrinsic interest than physical mutagens because it is sometimes possible to predict their effects from a knowledge of their properties. They fall into three groups, not always mutually exclusive: nucleotide

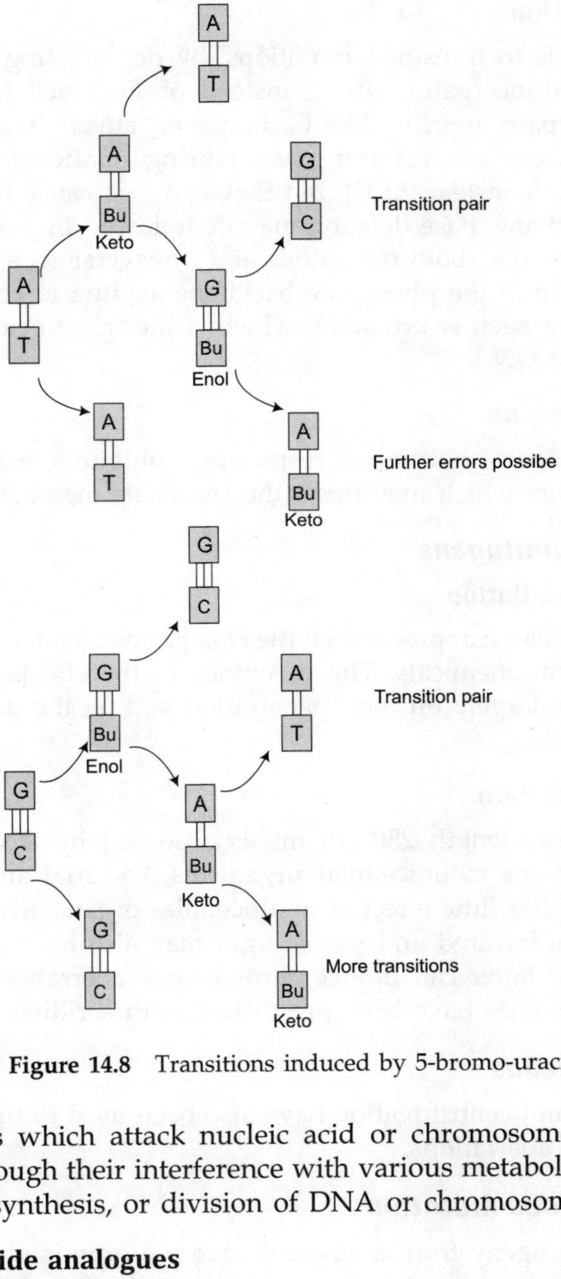

Figure 14.8 Transitions induced by 5-bromo-uracil.

analogues; products which attack nucleic acid or chromosomes directly; and those products which, through their interference with various metabolic processes, indirectly affect the stability, synthesis, or division of DNA or chromosomes.

14.7.1.5.1 Nucleotide analogues

5-bromo-uracil (BU), which in its normal keto form can pair with adenine and in its rare enol form with guanine, can lead to AT→GC transitions. Similarly 2-amino-purine (AP), which normally pairs with thymine, can occasionally pair with cytosine without changing its tautomeric state, again leading to AT→GC transitions.

14.7.1.5.2 Direct action

Nitrous acid also leads to transition mutations, by deaminating A, C, and G to give respectively hypoxanthine (pairs with C instead of T), uracil (pairs with A instead of G) and xanthine (pairs with C, like G; hence no effect) (Figure 14.9). Alkylating agents induce transitions by upsetting base-pairing relationships (e.g., methylated guanine can pair with T instead of C), but they can also cause loss of bases from the sugar phosphate backbone. Base deletion may be followed by replacement by either a purine or a pyrimidine (i.e., both transitions and transversions are induced) or it may be followed by rupture of the phosphate backbone leading to chromosome breakage. Another important mutagen is proflavin, which is thought to cause both deletions of and short additions to DNA.

14.7.1.5.3 Indirect action

This group includes drugs, hormones, respiratory inhibitors, excess O_2, pH changes, and all the other factors which may upset the cell or its metabolism in any way.

14.7.1.6 Physical mutagens

14.7.1.6.1 Ionizing radiation

Ionizing rays and particles can produce all the changes outlined above and are therefore much less specific than chemicals. The variations in the effects of different types of radiation depend very largely on their penetration and on the density of the induced track of excited radicals.

14.7.1.6.2 Other radiation

Ultraviolet light of wavelength 250-270 mμ is absorbed by DNA and is frequently used to induce mutations in unicellular organisms, bacterial and fungal spores and pollen grains, etc. UV has little effect on multicellular organisms, however, because of its lack of penetration, infrared and visible light may also be mutagenic under rather special conditions; the latter can induce chromosome aberrations in the presence of oxygen if the chromosomes have been pre-treated with acridine orange in the dark.

14.7.1.7 Miscellaneous

Ultrasonic treatment and centrifugation have also been used to upset cell division and produce chromosome aberrations.

14.7.1.8 Spontaneous mutation

If large numbers of progeny from a single source are examined it is always possible to find spontaneous mutants, but one cannot be certain that these have not resulted from mutagenic effects in the background. Since, it is possible to select strains with high mutation rates, and since some genes are more likely to mutate than others, this is clearly not a simple situation. Apart from the mutagenic background, there are cellular control mechanisms, such as DNA repair enzymes, which could affect mutability in different strains.

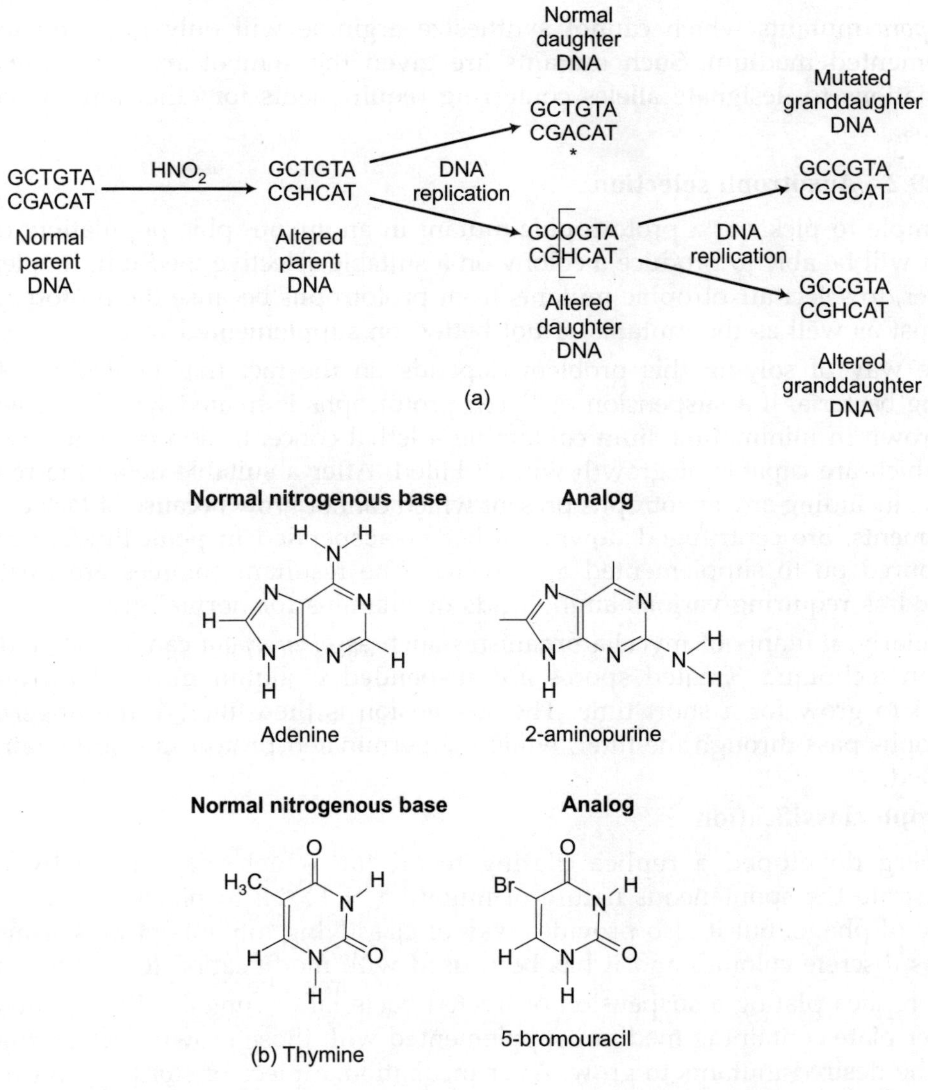

Figure 14.9 Mutation in bacteria (a) Nitrous acid causes induced mutations in bacteria (b) base analogs induce mutations by substituting for nitrogenous bases in the synthesis of DNA.

14.7.1.9 Biochemical mutants

14.7.1.9.1 Prototrophs and auxotrophs

Prototrophic (effectively = wild type, symbol⁺) mL can often be grown on simple media. *Chlamydomonas* required only inorganic salts to grow normally in liquid culture on an agar plates in the presence of light. *E. coli* will grow on a similar medium supplemented with an energy source such as glucose, while *Neurospora crassa* requires the vitamin biotin in addition.

Auxotrophic mutants are unable to grow on the minimal media which will support prototrophs because they cannot synthesize particular compounds. For example,

Neurospora mutants which cannot synthesize arginine will only grow on arginine-supplemented medium. Such mutants are given the symbol *arg*; there are similar abbreviations to designate alleles conferring requirements for other amino acids and vitamins.

14.7.1.9.2 Auxotroph selection

It is simple to pick out a prototrophic mutant in an auxotrophic population; only the mutant will be able to produce a colony on a suitable selective medium. It is less easy, however, to select auxotrophic mutants from prototrophs because the prototrophs will grow just as well as the mutants, if not better, on supplemented medium.

One way of solving this problem depends on the fact that penicillin kills only growing bacteria. If a suspension of *E. coli* prototrophs is treated with a mutagen and then grown in minimal medium containing a lethal concentration of penicillin, all the cells which are capable of growth will be killed. After a suitable delay the remaining bacteria, including any auxotrophs present which cannot grow because of lack of various supplements, are centrifuged down, washed, resuspended in penicillin-free medium, and poured on to supplemented agar plates. The resultant colonies are treated and classified as requiring various amino acids or vitamins for normal growth.

Similarly, mutants of mycelia organisms such as *Neurospora* can be isolated by the filtration technique. Treated spores are suspended in liquid minimal medium and allowed to grow for a short time. The suspension is then filtered; the ungerminated auxotrophs pass through the filter, while the germinated prototrophs are retained and discarded.

Auxotroph classification

Lederberg developed a replica plating technique which was originally used to demonstrate the spontaneous nature of mutation of *E. coli* to phage resistance in the absence of phage, but it also provides way of classifying mutants of organisms which grow as discrete colonies and it has been used with modification for *Neurospora* too.

For replica plating a suspension of treated bacterial or fungal cells is poured on to a master plate containing medium supplemented with those growth factors which will allow the desired mutants to grow. After incubation, a piece of sterile velvet mounted on the end of a cylindrical block is pressed gently on to the plate; this 'replicator' picks up a few cells from each colony on the master plate, and it is pressed successively on to a series of plates containing minimal medium alone with various supplements. After further incubation, colonies appear on these replica plates in the same pattern as on the master plate; absence of a particular colony from a plate is presumptive evidence that it requires the supplement missing from that plate (Figure 14.10).

14.7.1.10 Biochemical pathways

One of the first biosynthetic pathways worked out was that of on arginine requiring mutants of *Neurospora crassa* which fall into a number of groups including:

1. Those which grew on ornithine, citrulline, or arginine.
2. Those which grew on citrulline, or arginine but not on ornithine.

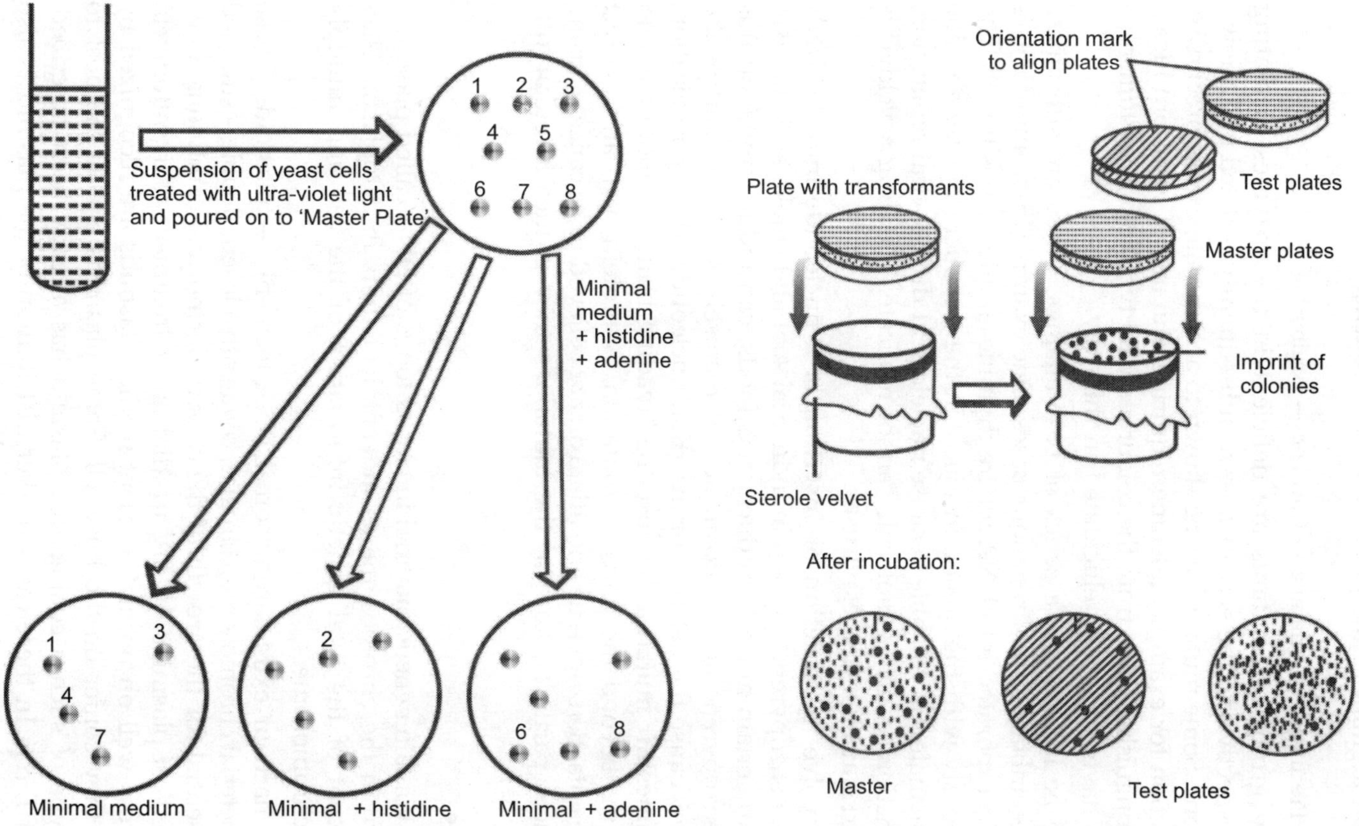

Figure 14.10 Selection of adenine and histidine requiring mutants by replica plating.

3. Those which grew only on arginine.
4. This suggests that arginine is the product of the sequence:

$$\text{Precursor} \xrightarrow{1} \text{Ornithine} \xrightarrow{2} \text{Citruline} \xrightarrow{3} \text{Arginine}$$

And that the three groups of mutants are deficient in the enzymes mediating reactions 1, 2, and 3 respectively. The enzymes were subsequently isolated and shown to catalyse these reactions. Some mutants can be shown to accumulate excess precursors before the blocked reaction for example, adenine-requiring mutants of yeast produce a red pigment which distinguishes them from the colourless wild type colonies. Numerous biochemical pathways have since been elucidated in this way.

Mutations can also be induced by a series of base analogs. These are substances that bear a chemical resemblance to nitrogenous bases. One example, 5-bromouracil, is taken up by cells and incorporated into DNA where thymine should be positioned. The new DNA functions poorly with the analog in place. In the treatment of diseases due to DNA viruses, a base analog is valuable because nucleic acid directs viral replication, and a virus with a functionless DNA molecule cannot replicate. The drug acyclovir is a base analog that works against herpes viruses.

Other mutations involve the **deletion** or **insertion** of a nucleotide into the DNA molecule. For example, benzopyrene, present in industrial soot and smoke and aflatoxin a fungal toxin found in certain animal products and foods, can induce mutation this way. Substances as benzopyrene and aflatoxin do not themselves become part of the DNA molecule, but they cause the deletion or insertion nucleotides during replication. This may lead to a **frameshift mutation** because the "reading frame" is modified, and the wrong mRNA forms beyond the point of mutation. A protein with an incorrect amino acid sequence results. Benzopyrene and aflatoxin are believed to be cancer-causing (carcinogenic) substances partly because of the mutations they induce in eukaryotic cells.

14.7.1.11 Plasmids

Some microorganisms contain genetic material in closed loops of DNA called plasmids. Plasmids exist apart from the chromosome as independent units in cytoplasm. They contain about 2 per cent of the total genetic information of the cell and multiply independently of the chromosome.

Many bacteria can harbour extra-chromosomal DNA elements – plasmids. These small (relative to the bacterial chromosome), circular, covalently closed, double - stranded DNA units are not essential for the growth of the cells under normal conditions. Cells which have been 'cured' of plasmids, by UV irradiation or treatment with mitomycin C or acridine dyes, grow well on normal nutrient media. Plasmids are recognized by the special characters they confer on the host cell. Some plasmids enable the host to conjugate. The prototype of such a conjugative plasmid has already been described, namely the F factor of *E coli*. In these cases further distribution of the plasmid is also facilitated because of the direct contact between cells. Other plasmids can transiently integrate into the bacterial chromosome rather like a prophage, and thus lose their autonomous state for a period.

Plasmids can mobilize the host chromosome and can therefore contribute to exchange and recombination of genetic material under natural conditions. In genetic engineering and gene technology, plasmids are used as vectors of foreign DNA, the methods applied are analogous to natural processes (Figure 14.11). Plasmids can also be found in eukaryotes, but have been far less intensively studied up to now.

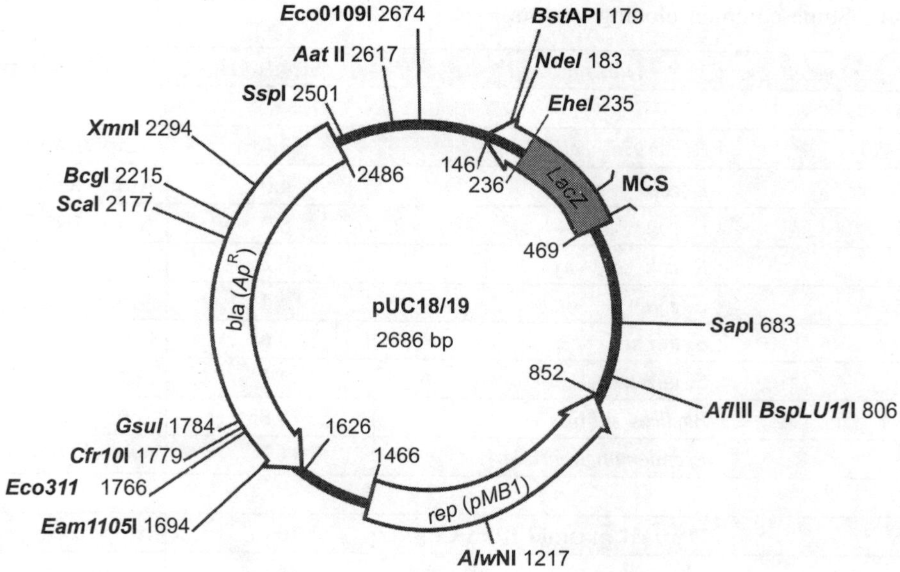

Figure 14.11 Physical map of pUC.

Their number varies from one per cell to as many as several hundred in each bacterial cell. This is referred as copynumber. A number of host properties are specified by plasmids, such as antibiotic and heavy metal resistance, nitrogen fixation, pollutant degradation, bacteriocin, and toxin colicin. The small size of plasmids and the presence of antibiotic resistance make them ideal vectors for cloning. They are not easily broken by physical manipulation or by cycles of freeze–thawing during storage in the laboratory. Plasmid vectors can carry DNA fragments ranging from hundreds of base pairs to much longer fragments of many thousand base pairs.

Naturally occurring plasmids often lack several important features required for a high-quality cloning vector like a small size, unique restriction endonuclease recognition sites, and one or more selectable genetic markers for identifying recipient cells that carry the cloning vector-insert DNA construct. They can be modified by *in vitro* digestion and ligation reactions. Most plasmid vectors containing multiple cloning sites (MCSs) are in common use. They have recognition sequences for many different REs. It is almost always possible to find a vector carrying restriction sites that are compatible with the terminus of a particular fragment of foreign DNA (Figure 14.12).

The F factor can exist in two states in the cell: either autonomous (independent of the chromosome, as in F^+ or integrated (as part of the chromosome – Hfr). This property is also shown by temperate phages. Although the F factor and temperate phages have such different physiological functions, both can be lost and re-acquired by infection, and both can pick up pieces of chromosomal material and carry these to other cells (F′ factors and λ gal^+ particle). Some common cloning vectors are described in Table 14.4.

Table 14.4 Some common cloning vectors

Cloning vector	*Natural occurrence*	*Size (Kb)**	*Selective marker***
Plasmid			
pACYC 177	*Escherichia coli*	3.7	Amp^r, Kan^r
pBR 322	*E. coli*	4.0	Amp^r, tct^r
pBR 324	*E. coli*	8.3	Amp^r, tel^r, EI imm.
pMB9	*R. coli*	8.3	Tcf
pRK 646x	*E. coli*	3.4	Amp^r
pCI94	*S. aureus*	3.6	Ery^r
pSA 0501	*S. aureus*	4.2	Str^r
pBS 161-1	*Bacillus subtilis*	3.65	Tct^r
pWWO	*Pseudomonas putida*	11.7	Kan^r
Cosmids			
pJC 74	Derived plasmid from Col EL	16	Amp^r, EI imm.
pJC 720	-do	24	Elimm./Rif^r
pHC 79	Derivative ofpBR 322	6	Amp^r, Tet^r
Viruses			
SV40	Mammalian cells	5.2	–
Phage M13′	*E. coli*	6.4	–
Phage[1]	*E. coli*	4.9	–

, I Kb (Kilobase pairs) = 1.000 base pairs = 0.66 mega dalton;", Resistance to ampicillin (Amp′), tetracycline (Tel), erythrdrnycine (EryO streptomycine (Str^r), Kanamycin (Kan′). rifampicin (Rif^r), and colicin production (EL imm.)

14.7.1.12 *Transposable genetic elements*

The method of inducing mutagenesis by transposons is becoming a standard technique in bacterial genetics. Transposons (Tn) elements are short sections of double stranded DNA, consisting of more than 2000 base pairs. They usually code for resistance to one or sometimes several antibiotics. Transposons are able to move or jump within the genome, even between a bacterial chromosome and a plasmid, and they are able to become integrated in a number of different sites on the genome. Insertion of a transposon within a structural genus thus interrupts the normal nucleotide sequence of the gene so that it can no longer deliver the information for the synthesis of the normal, functionally potent polypeptide. Transposons thus serve primarily for production of insertion mutants.

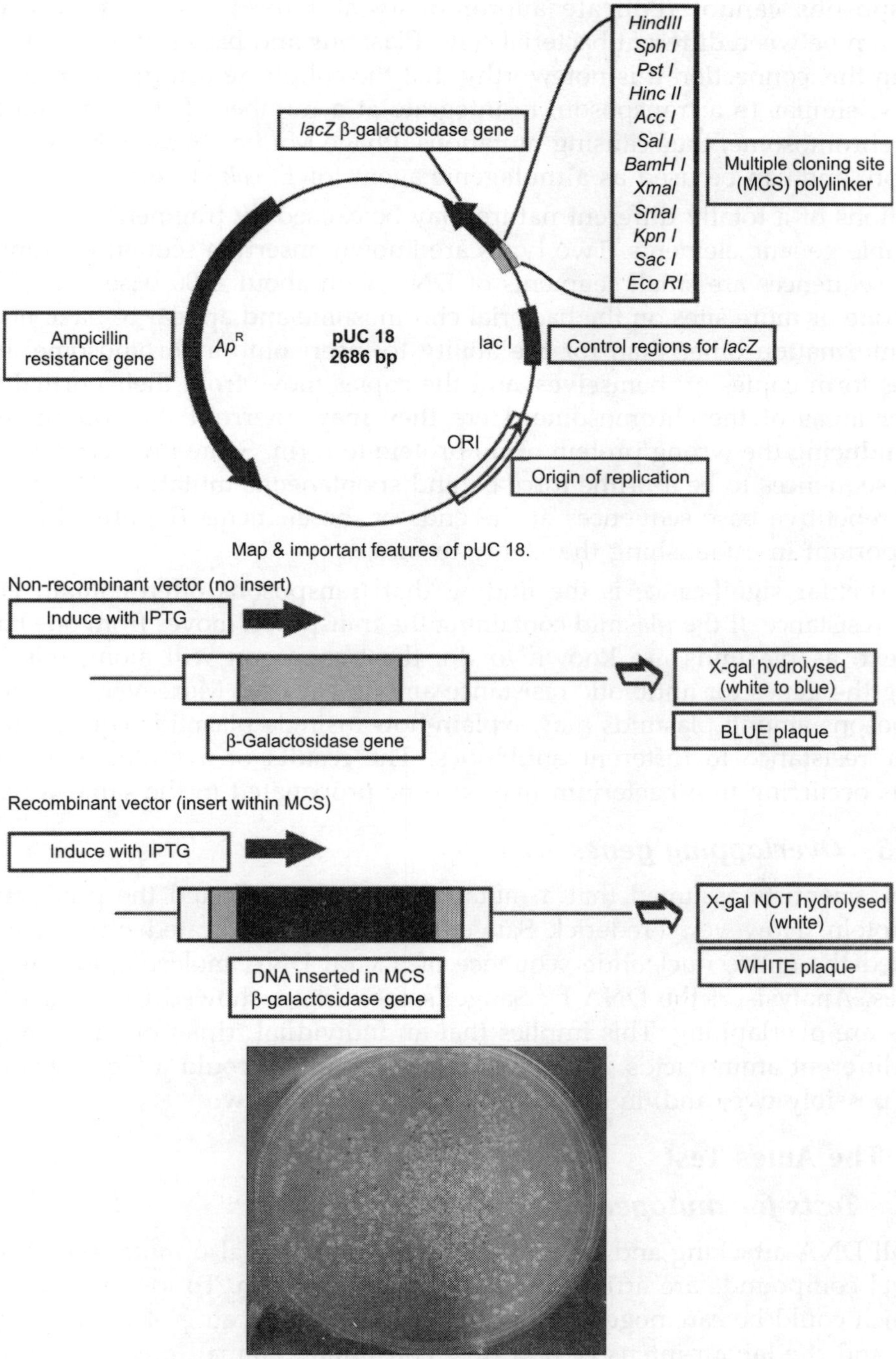

Fiigure 14.12 Plasmid technology showing blue and white recombinant bacterial colonies.

Transposons cannot replicate auotomously and need a carrier or vector for transmission between different bacterial cells. Plasmids and bacteriophages are suitable vectors. In this connection it is noteworthy that the coliphage Mu (mutator phage) has the ability, similar to a transposon, to integrate at a number of different sites on the bacterial chromosome, thus causing mutations. Phage Mu has become known as 'giant transposon' and can be used as a mutagenic agent for *E. coli* strains.

Mutations of a totally different nature may be caused by fragments of DNA called transposable genetic elements. Two types are known: insertion sequences transposons. Insertion sequences are small segments of DNA with about 1000 base pairs. They are found at one or more sites on the bacterial chromosome and appear to have no specific genetic information other than for the ability to insert onto a chromosome. Insertion sequences form copies of themselves, and the copies move from their normal position into other areas of the chromosome. Here they may interrupt the coding sequence, thereby inducing the wrong protein or no protein to form. Some investigators consider insertion sequences to be a prime force behind spontaneous mutation. The presence of inverted repetitive base sequences at the ends of the elements (Figure 14.13) appears to be important in establishing the ability to move.

Of particular significance is the finding that transposons often contain genes for antibiotic resistance. If the plasmid containing the transposon moves from one bacterium to the next, as plasmids are known to do, the transposon will along with it, thus spreading the genes for antibiotic resistance among bacteria. Moreover, the movement of transposons among plasmids may explain how a single plasmid acquires numerous genes for resistance to different antibiotics. The results of a mutation or series of mutations occurring in a bacterium may also be propagated in the same way.

14.7.1.13 Overlapping genes

For years, scientists assumed that a mutation in a gene affected the production of a single protein. However, Frederick Sanger's 1977 work indicated otherwise. Sanger determined the entire nucleotide sequence of a viral DNA molecule and mapped its 5386 bases. Analysis of the DNA by Sanger's co-workers showed that at least four of the genes are overlapping. This implies that an individual triplet of bases might code for two different amino acids. Thus, a mutation at a base could affect not one amino acid but possibly two, and, in effect, not one protein, but two.

14.7.2 The Ames Test

14.7.2.1 Tests for mutagenicity

Almost all DNA-attacking and carcinogenic substances are also mutagens. About 500-1000 novel compounds are artificially produced per annum. To identify those among them which could be carcinogenic is a difficult problem because of the large numbers involved and the labour-intensive and time consuming animal tests needed to gain a

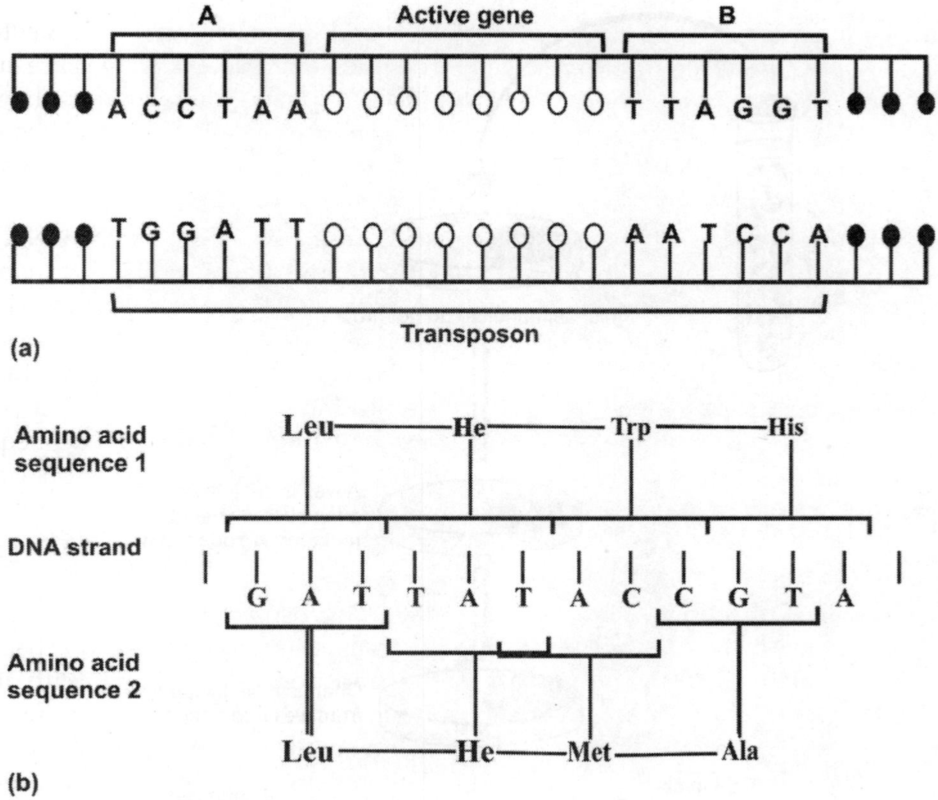

Figure 14.13 (a) Structure of a typical transposon (b) the effect of overlapping genes. The single stranded DNA a provides the genetic code for two amino acid sequences. Below the DNA strand, the base sequence of the third amino acid (Met) includes thymine, a base from the previous base sequence.

degree of certainty. Bruce Ames of the University of California developed a procedure to determine whether a chemical can induce a bacterial mutation and thus be a potential agent of cancer. The procedure, called the Ames test, is widely used as well as relatively inexpensive, rapid, and accurate. The principle of this test is illustrated in Figure 14.14.

The original Ames test has been modified and expanded to increase its reliability and sensitivity. Test organisms have been developed which carry multiple mutations that make them hypersensitive towards mutagens. This is because they lack the DNA repair mechanisms which would normally correct some of the damage. Release of a prophage in a lysogenised bacterium by the substance under test is also used as an indicator of the mutagenicity. This testing procedure, which has drastically lowered the number of animal experiments, is now applied on a worldwide scale. The reliability is very high; of 100 substances with proven mutagenicity, 84 registered positive in a test system with *Salmonella typhimurium* and 91 with *Escherichia coli.*

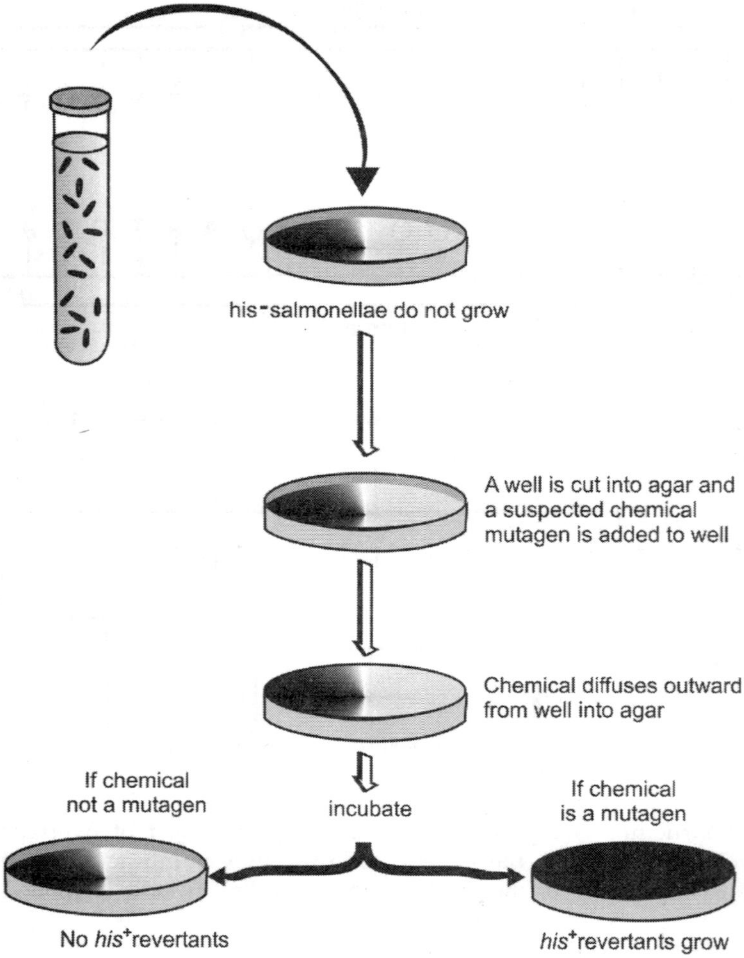

Figure 14.14 Ames test for the mutagenicity of a substance.

The test bacterium is a histidine requiring mutant of *Salmonella typhimurium* with a very low reversion frequency. The cells are mixed with rat liver extract and plated out on a mineral agar which cannot support their growth. The function of the liver extract is – as in the living organism – to supply enzymes for metabolic conversions of the test substance that could produce mutagenicity. A filter paper is saturated with the substance under test and placed on the nutrient medium. If the test substance has mutagenic effects, some mutations would restore ability of the test bacterium to synthesize histidine, and thus enable such mutants to grow, expecially in the diffusion zone of the mutagen.

14.7.3 DNA Repair

DNA is permanently exposed to cellular enzyme activities which can lead to alteration (alkylation, depurination, bond breakage) as well as energy-rich and energy-poor radiations (light, radioactive isotopes in rocks, buildings, food). Life has developed over billions of years in the presence of such radiations. It would be surprising, therefore, if no mechanisms for DNA repair had developed in cells. Such DNA repair mechanisms are mostly inducible. Organisms which have been exposed to sublethal doses of DNA-damaging agents are commonly more resistant than organisms without such exposure. Of the many known mechanisms of DNA repair only some of the important are described below.

14.7.3.1 Photo- and dark-reactivation

The discovery that DNA damage is partially repairable was made during UV light treatment of bacteria. If treatment of bacterial suspensions with high doses of UV irradiation is followed by incubation in the dark, only a very small percentage of the cells can form colonies. However, if the cells are exposed to light of longer wavelength (<320-384 nm) the percentage of survivors increase by orders of magnitude. The photo-reactivation involves an enzyme called photolyase. This gains energy from the longer wavelength light which it uses to split thymine dimmers into monomers. In this way the original state is regained and this system of DNA repair works error free.

Another, similarly reliable mechanisms of irradiation repair do not involve light. Dark reactivation involves the excision of defective sections of the DNA strands (excision repair) and their replacement by newly synthesized nucleotides. The process is catalyzed by a correction endonuclease which represents an enzyme complex of the gene products *uvr A, B,* and *C.* The extent to which such repair of radiation damage is effective varies in different bacterial strains. Radiation resistance of some bacteria is due to a highly effective repair mechanism.

14.7.3.2 SOS repair

When DNA is substantially damaged by high doses of mutagenic agents or radiation, or replication is inhibited by certain antibiotics like novobiocin or mitomycin C, an alarm reaction is elicited in the cell, called the SOS response. Extensive modification takes place. Among others, the release of temperate phage delay of cell division, decreased respiration, higher protein turnover, and finally, the induced synthesis of repair enzymes. This process is regulated by two proteins, LexA and RecA. RecA normally plays a role in the genetic recombination. LexA is a DNA binding protein (a repressor protein) which binds at the operator region and inhibits transcription into mRNA. LexA acts as repressor of a number of cellular functions under SOS conditions, signaled by an increase in single stranded DNA. LexA loses its repressor function through proteolytic cleavage which abolishes its ability to bind DNA. This proteolysis is stimulated by the RecA protein, which apparently has no proteolytic activity itself. The SOS induced repair system differs from the constitutive repair system in its high error roate, hence error prone repair. In this way many mutations, i.e., UV-induced mutations, occur as a consequence of incorrect repair. This is an important point in the context of cancer chemotherapy, since the agents used to inhibit DNA replication and therefore cell division. But they can also cause considerable damage because of their mutagenic effects.

14.8 BACTERIAL RECOMBINATION

The second general method for altering the genetic material of a bacterium is through the process of recombination. Although mutation may be considered a form of recombination, here, recombination in the more restricted sense to mean the genetic process in which two organisms are involved: a donor cell and a recipient cell. The donor cell contributes chromosomal DNA or plasmid DNA to the recipient cell. If plasmids are obtained, they exist independently in the recipient's cytoplasm and begin to multiply and encode proteins immediately. If the recipient obtains chromosomal DNA, the new DNA pairs with a complementary region of recipient DNA and replaces it. Thus, there is no change in quantity of the recipient's DNA but there may be a substantial change in its quality. There are three methods identified for bacterial recombinant, viz., transformation, conjugation, and transduction.

14.8.1 Transformation

Gene transfer by soluble DNA, which has been extracted, or otherwise liberated from a donor bacterium, to a recipient bacterium is called transformation. This is the longest-known and historically most important kind of gene transfer in bacteria.

In 1928, Frederick Griffith published the results of an interesting set of experiments with *Streptococcus pneumoniae.* This organism is a major cause of bacterial pneumonia; the bacterium is often referred to as a pneumococcus (plural: pneumococci). The results were considered unusual, but in retrospect, he gave the first clues to gene activity.

Pneumococci occur in different strains. There is an encapsulated strain designated S, because the organisms grow in smooth colonies. This strain causes pneumonia. There is also an unencapsulated strain, designated R because the colonies are rough. Organisms in this strain are harmless. In an early experiment, Griffith showed that mice injected with live S strain pneumococci died, while those injected with live R strain pneumococci lived. This was what he expected. Also, he showed that mice injected with dead S strain orgnisms lived (Figure 14.15), Again, this result was not unusual. He mixed dead S strain bacteria with live R strain bacteria and injected the mixture into mice. The mice died. Griffith questioned how a mixture of live harmless bacteria (R) and dead pathogenic bacteria (S) could kill the mice. His answer came when he autopsied the animals; they were full of live S strain pneumococci. Apparently the live R strain bacteria had been transformed to live S strain bacteria. Five years later, James L. Alloway (1933) of the Rockefeller Institute confirmed Griffith's work. Alloway used fragments and debris from the S strain cells to transform the R strain cells. However, it was not until 1944 that Oswald T. Averv and his associates Colin M. MacLeod and Maclyn N. McCarty, also of the Rockefeller Institute, purified the transforming substance. These investigators found that the substance was not protein, as had been anticipated, but a then obscure organic compound called deoxyribonucleic acid (DNA).

The majority of scientists was blind to Avery's discovery and was reluctant to accept DNA as a hereditary substance. Geneticists of that period were not trained as chemists, and Avery's experiments were difficult to repeat. Also, many scientists believed that

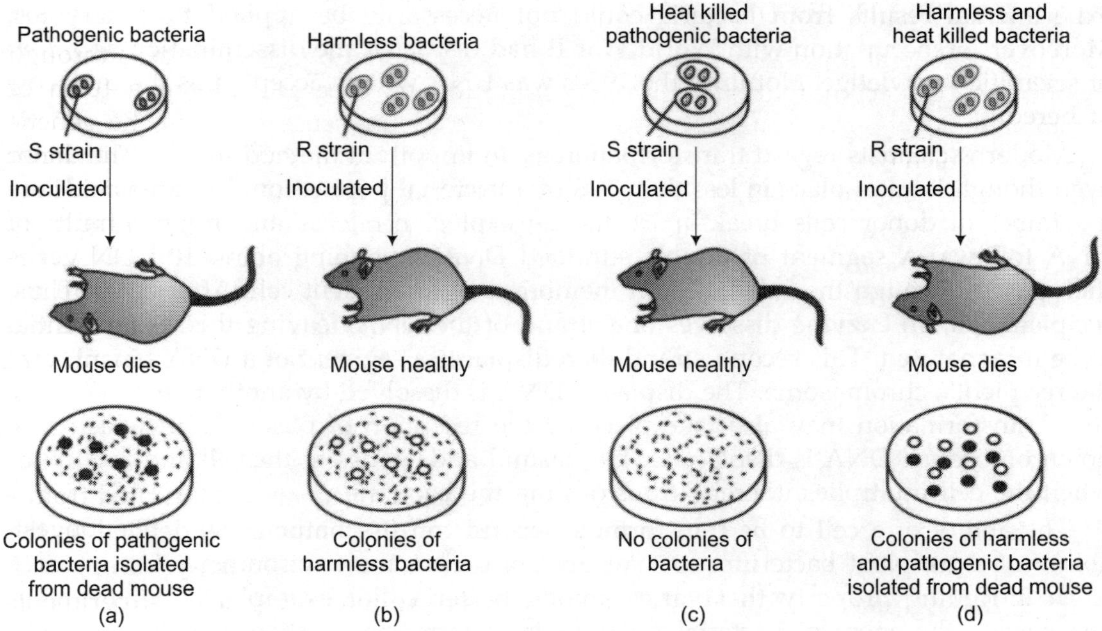

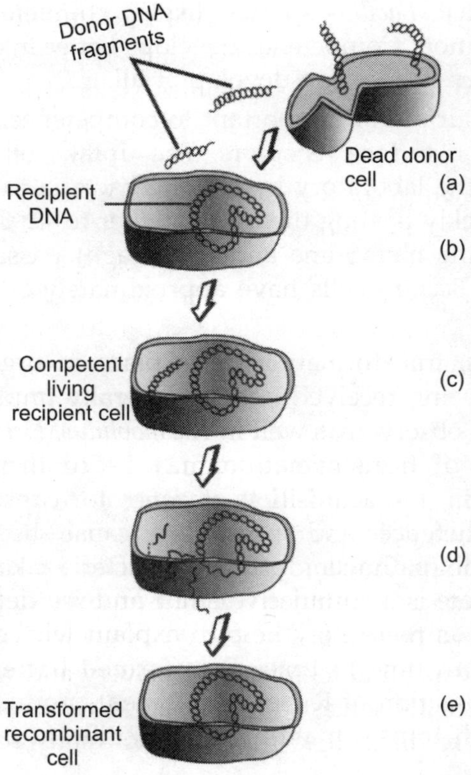

Figure 14.15 Bacterial transformation.

experimental results from bacteria could not necessarily be applied to eukaryotes. Moreover, preoccupation with World War II had restricted the dissemination and flow of scientific knowledge. Not until the 1950s was DNA widely accepted as the molecule of heredity.

Modern scientists regard transformation as an important method for recombination even though it takes place in less than 1 % of a bacterial population. In transformation, a number of donor cells break apart and an explosive release and fragmentation of DNA follows. A segment of double stranded DNA containing about 10 to 20 genes then passes through the cell wall and membrane of a recipient cell. After entry to the recipient cell, an enzyme dissolves one strand of the DNA, leaving the second strand to be incorporated. This second strand then displaces a segment of a DNA strand from the recipient's chromosome. The displaced DNA is dissolved by another enzyme in the cell. Transformation may also take place by the reception of plasmids, in which case no chromosomal DNA is displaced. The plasmid adds genes to the cell and multiplies when the cell multiplies, thereby transforming the bacterium.

The ability of a cell to be transformed depends on its competence, defined as the ability of a recipient bacterium to take up DNA from the environment. Competence is an intriguing property that varies among bacteria. For example, in *Streptococcus pneumoniae,* the agent of bacterial pneumonia, competence is displayed in the entire cell population late in the logarithmic phase of growth when a competence-provoking factor is produced. By contrast, *Bacillus* species display competence during the early events leading to spore formation. Competence develops better in certain growth media than in others, and sometimes it does not develop at all.

Factors affecting the cell surface are important to competence, particularly changes in membrane permeability or surface receptors. The uptake of DNA by *E. coli,* for example, can be induced in the laboratory by chilling bacteria to 4°C, in the presence of calcium chloride, then quickly heating them to together to 42°C with the DNA. This treatment apparently alters the membrane and encourages passage of DNA strands. Competent streptococci and *Bacillus* cells have approximately 50 binding sites where DNA uptake can occur.

Under natural conditions, transformation takes place in organisms whose DNA is very similar to the DNA being received, which generally implies cells of the same species. The process has been observed as well in *Haemophilus, Neisseria,* and *Azotobacter* species. One of the effects of transformation may be to increase an organism's pathogenicity, In pneumococci, the acquisition of genes for capsule formation allows an organism to avoid body defences and survive and cause disease. It has also been demonstrated that when mildly pathogenic strains of bacteria take up DNA from other mildly pathogenic strains, there is a cumulative effect and the degree of pathogenicity increases. Observations such as these may help to explain why especially pathogenic strains of bacteria appear from time to time. Transformed bacteria may also display drug resistance from the acquisition of R factors. Pigment production may be another characteristic derived through transformation. How significant is transformation as a means of genetic recombination under natural conditions? No one knows for sure, but it appears certain that transformation occurs regularly where bacteria exist in very crowded conditions, such as in the human intestine.

14.8.2 Conjugation

In the recombination process called conjugation, two live bacteria come together, and the donor cell transfers genetic material to the recipient cell. This process was first postulated in 1946 by Joshua Lederberg and Edward Tatum in a series of experiments with *E. coli.* Lederberg and Tatum mixed two different strains of bacteria and found that genetic traits could be transferred among them as long as contact occurred.

Experiments in the 1950s by Francois Jacob and Elie L. Wollman established that conjugating bacteria were of two mating types. Certain "male" type donated their DNA. These were designated F^+ cells. Other "female" type received the DNA and came to be known as F^- cells. Jacob and Wollman found that F^- cells (recipients) became F^+ cells (donors) when they acquired a small amount of DNA. This eventually led William Hayes to discover the F (fertility) factors present in the donor cell. Although the male and female nomenclature is still used today, the process of bacterial conjugation is fundamentally different from sexual mating, since the mechanisms of DNA incorporation are very different.

In contemporary microbiology, the F factors are known to be plasmid, the double-stranded loops of DNA existing apart from the chromosome. The factors (plasmids) contain about 20 genes, most of which are associated with conjugation. These genes code for enzymes that replicate and move DNA during conjugation and for enzymes and structural proteins needed to synthesize special pili at the cell surface. Known as F pili, these hair-like fibers contact the recipient bacteria, then retract so that the surfaces of the donor and recipient are very close or touching one another. At the area of contact, a channel or conjugation bridge is believed to form (Figure 14.16).

Once contact has been made, the F factor begins replicating by the rolling circle mechanism. A single strand of the factor then passes over or through the channel to the recipient. When it arrives, enzymes synthesize a complementary strand, and a double helix forms once again. The double helix bends to a loop and reforms an F plasmid, thereby completing the conversion of the recipient from F^- cell to F^+ cell. Meanwhile, back in the donor cell, a new strand of DNA is form to complement the leftover strand of the F plasmid.

The transfer of F factors involves no activity of the bacterial chromosome; therefore the recipient does not acquire new genes other than those on the factor. A type of conjugation that accounts for the passage of chromosome material does exist in bacteria, however. Strains of bacteria that display the ability to donate chromosomal genes are called high frequency of recombination, or Hfr strains (Figure 14.16). Such strains were discovered in the 1950s by William Hayes in *E. coli.*

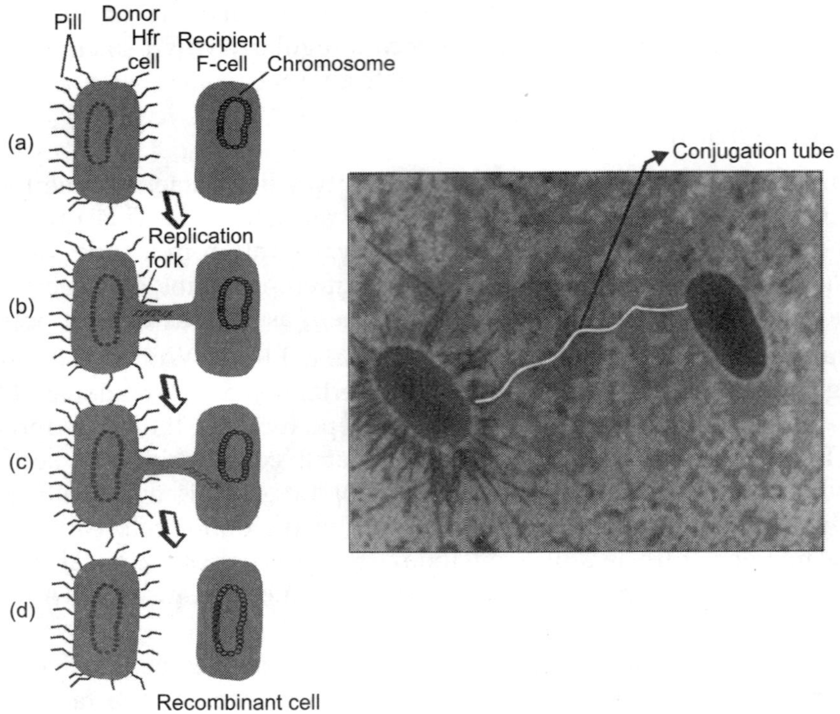

Figure 14.16 Bacterial conjugation.

In Hfr strains, the F factor attaches to the chromosome. This integration is a rare event and requires that an insertion sequence be present on the chromosome to recognize the F factor. At the point of attachment, the chromosome opens and a copy of one strand of the chromosomal DNA is made by the roll circle mechanism. A portion of single-stranded DNA then passes via the channel into the recipient cell, where it replaces the complementary region in the recipient's chromosome. The first genes to enter the recipient are F factor genes, but these are not the ones that control the donor state. Instead, the last segment enter the recipient are the F factor genes for the donor state. These rarely enter the recipient, however, because conjugation is often interrupted by such things as movement that breaks the attachment. (An estimated 100 minutes is required for a complete transfer of an *E. coli* chromosome.) Thus the F^- cell usually remains a recipient, but it has some new genes acquired from the donor.

In certain cases the entire F factor is transferred to the recipient. When this happens, the factor usually detaches from the recipient's chromosome, and enzymes synthesize a strand of complementary DNA. The factor now forms a loop to assume an existence as a plasmid, and the recipient becomes a donor (an F^+ cell).

On occasion, the F factor breaks free from the chromosome of an Hfr cell before conjugation takes place and resumes an independent status. The Hfr cell then reverts to an F^+ cell. Sometimes when the F factor leaves the chromosome, it takes along a segment of chromosomal DNA. The factor with its extra DNA now called an **F′ factor** (pronounced F-prime). When the F′ factor is transferred during a subsequent conjugation, the recipient acquires chromosomal genes from the donor, a process known as **sexduction.** Sexduction results in a recipient with its own genes for a particular process as well as additional genes from the donor for that same process. In the genetic sense, the recipient is a partially diploid organism since there are two genes for a given function.

Conjugation has been demonstrated among various genera of bacteria, in contrast to transformation, which appears to occur only among cells of the same species. For example, conjugation is known to occur between such Gram-negative bacteria as *Escherichia* and *Shigella, Salmonella* and *Serratia,* and *Escherichia* and *Salmonella.* Intergenic transfer has great significance in the transfer of antibiotic resistance genes carried on plasmids. When the F factors are plasmids, the transfer occurs readily. Moreover, when the genes are attached to transposons, the transposons may "jump" from ordinary plasmids to F factors, after which transfer may occur.

Gram-positive bacteria also appear capable of conjugation. Microbiologists have experimented extensively with *Streptococcus mutans,* a common cause of dental cavities. In this organism, conjugation appears to involve only plasmids especially those carrying genes for antibiotic resistance. Similar, observations have been made in *Bacteroides* and *Clostridium* species. However, the mechanism of plasmid transfer in Gram-positive bacteria is poorly understood, and chromosomal transfer has not been demonstrated (Table 14.5).

Table 14.5 Differences between genetic recombinations in bacteria

Principal mode	*Transformation*	*Conjugation*	*Transduction*
DNA transfer method	Across wall and membrane of recipient	Through channel after cell-to-cell contact	By a vector
Amount of DNA transfer	Few genes	Few genes to entire chromosome	Few genes
Plasmid transferred	Yes	Yes	Not likely
Entire chromosome transferred	No	Sometimes	No
Virus required	No	No	Yes
Live bacteria required	Yes	Yes	Yes
Cell debris required	Yes	No	No
Used to acquire antibiotic resistance	Yes	Yes	Not likely

14.8.3 Transduction

The first observation of transduction — the transfer of genetic material from donor to recipient by a phage vector — was made by Zinder and Lederberg in 1952 in *Salmonella typhimurium.* They prepared a U-tube at the bottom of which was a filter imperpeable to bacteria; on one side of the filter was placed a histidine-requiring culture, and on the other a tryptophane-requiring culture that was lysogenized with phage P22. The tube was incubated, gentle suction being applied to each side alternately to allow mixing of the culture supernatants through the filter. Prototrophs (*his*$^+$ *trp*$^+$) appeared at a frequency of about 10^{-7} on the side initially containing the tryptophane-requiring lysogen, showing that phage P22 particles had migrated to the side containing the histidine-requiring strain, picked up *trp*$^+$ markers from it, migrated back, and converted the *trp*$^-$ strain to *trp*$^+$. They ruled out conjugation and transformation because the cells were separated by a thin membrane and DNA was absent in the extracellular fluid. Eventually, they discovered a virus in the fluid and learned the details of what was taking place. The virus that participates in transduction is called a bacteriophage, or simply phage.

In the life cycle of a bacteriophage, the phage interacts with bacteria in two ways. In one way the phage invades a bacterium, reproduces within the bacterium, and then destroys the bacterium as new phages are released. The cycle is called the lytic cycle of infection because at the cycle's conclusion the invaded bacteria lyse, or rupture. Phages that cause lysis are known as virulent or lytic phages.

The second way that phages interact with bacteria also involves invasion of the bacterium. In this case, however, the phage DNA codes for a repressor protein that prevents replication. The phage DNA may remain in the cytoplasm as a plasmid or it may form a closed circle, align next to the bacterial chromosome, and integrate into the bacterial chromosome, as the F factor does in Hfr strains. This process of non-replication is called lysogeny. The phage that enters into lysogeny is known as a temperate phage and the integrated viral DNA is a prophage. Because there are two forms of phage interaction with bacteria, there are two ways in which phages can be involved in the transfer of bacterial genes: generalized transduction and specialized transduction.

14.8.3.1 Generalized transduction

Process is carried out by virulent phages that have a lytic cycle of infection (Figure 14.17). During phage replication a long, linear chain of DNA is first produced. This long DNA is then enzymatically cut into sections that will become phage DNAs. During this process, segments of bacterial DNA may accidentally get caught up in the cutting process, and bacterial segments may end up incorporated to phages. This event is considered rare (1 in every 100,000 phage progeny may have bacterial DNA). Such a phage is fully formed but carries no phage genes. Even though it carries DNA, it cannot encode its' own replication. Now the transduction takes place. The transducing phage is released and attaches to a new (recipient) bacterium and injects its bacterial DNA. Once released in the recipient, the new genes may pair with a section of the recipient's DNA and replace the section. The recipient has now been transduced (changed) using genes from the donor bacterium and the phage intermediary. Two examples of transducing phages are phage P1 that infects *E. coli* and P22 that infects *Salmonella typhimurium.*

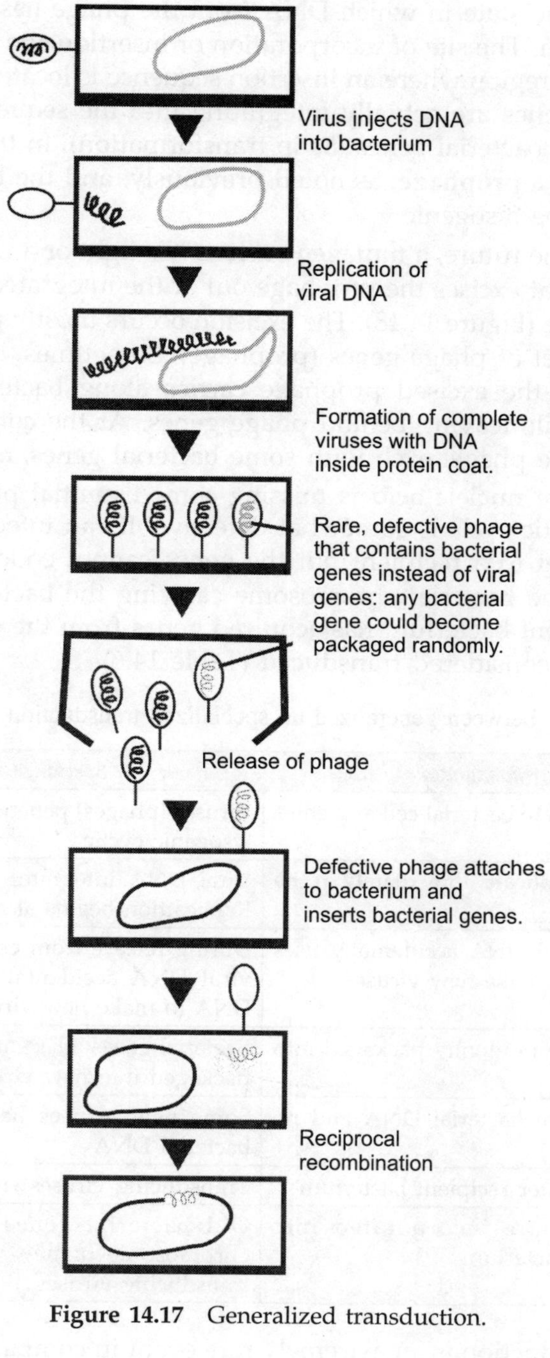

Figure 14.17 Generalized transduction.

14.8.3.2 Specialized transduction

Specialized transduction involves the same genes each time, unlike generalized transduction that can involve any bacterial genes. Specialized transduction occurs as a result of **lysogeny,** the state in which DNA from the phage has incorporated into the DNA of the bacterium. The site of incorporation or insertion site may vary for different phages and is often a region where an insertion sequence is located. When incorporation takes place, phage genes are actually integrating into the sequence of bacterial genes rather than replacing bacterial genes (as in transformation). In the integrated state, the set of phage genes is a **prophage,** as noted previously, and the bacterium carrying the prophage is said to be lysogenic.

At some time in the future, a mutagen such as UV light or a DNA inhibitor activates an enzyme system that excises the prophage out of the integrated state and causes it to re-enter the lytic cycle (Figure 14.18). The excision occurs mostly precisely and results in release of an intact set of phage genes (prophage). Sometimes, however, an **imprecise excision** occurs and the excised prophage carries along bacterial genes adjacent to the insertion site while leaving behind phage genes. At the conclusion of replication, multiple copies of the phage, each with some bacterial genes, are produced.

Because the phage nucleic acid is missing some essential phage genes, the phage resulting from the lytic cycle is somewhat defective. It can infect another and transfer the defective gene set to a recipient but the genes cannot code a lytic cycle. Instead, they integrate into the bacterial chromosome carrying the bacterial genes with them. As before, the recipient bacterium has acquired genes from the original bacterium and the recipient is now considered transduced (Table 14.6).

Table 14.6 Differences between generalized an specialized transduction

Generalized transduction	*Specialized transduction*
Viruses (phages) penetrate bacterial cell and enter lytic cyclic	Viruses (phages) penetrate bacterial cell and enter lysogenic cyclic
Viral DNA begins to replicate immediately in the bacterial cytoplasm	Viral DNA integrates into DNA of bacterium. Replicattion begins at a later time.
During replication, viral DNA accidentally uses some bacterial DNA to make new viruses	During release from chromosome in replication, viral DNA accidentally excises some bacterial DNA to make new viruses
Any bacterial genes are randomly packaged into new viruses	Bacterial genes adjacent to incorporated virus are packaged into new viruses
Some new viruses have bacterial DNA and no viral genes	Some new viruses have both viral DNA and bacterial DNA
Transducing viruses enter recipient bacterium	Transducing viruses enter recipient bacterium
Old bacterial genes are incorporated into chromosome of new bacterium	Old bacterial genes are incorporated into chromosome of new bacterium together with transducing viruses.

Specialized transduction is an extremely rare event in comparison to the generalized form, because genes do not easily break free from the bacterial chromosomes. However, the potential for recombination is substantial because lysogeny with prophages has been well established in clinical microbiology. For example, the diphtheria bacillus, *Corynebacterium diphtheriae* harbours a prophage that provides the genetic code for a toxin that destroys organ cells. Other toxins believed to be coded by prophages include

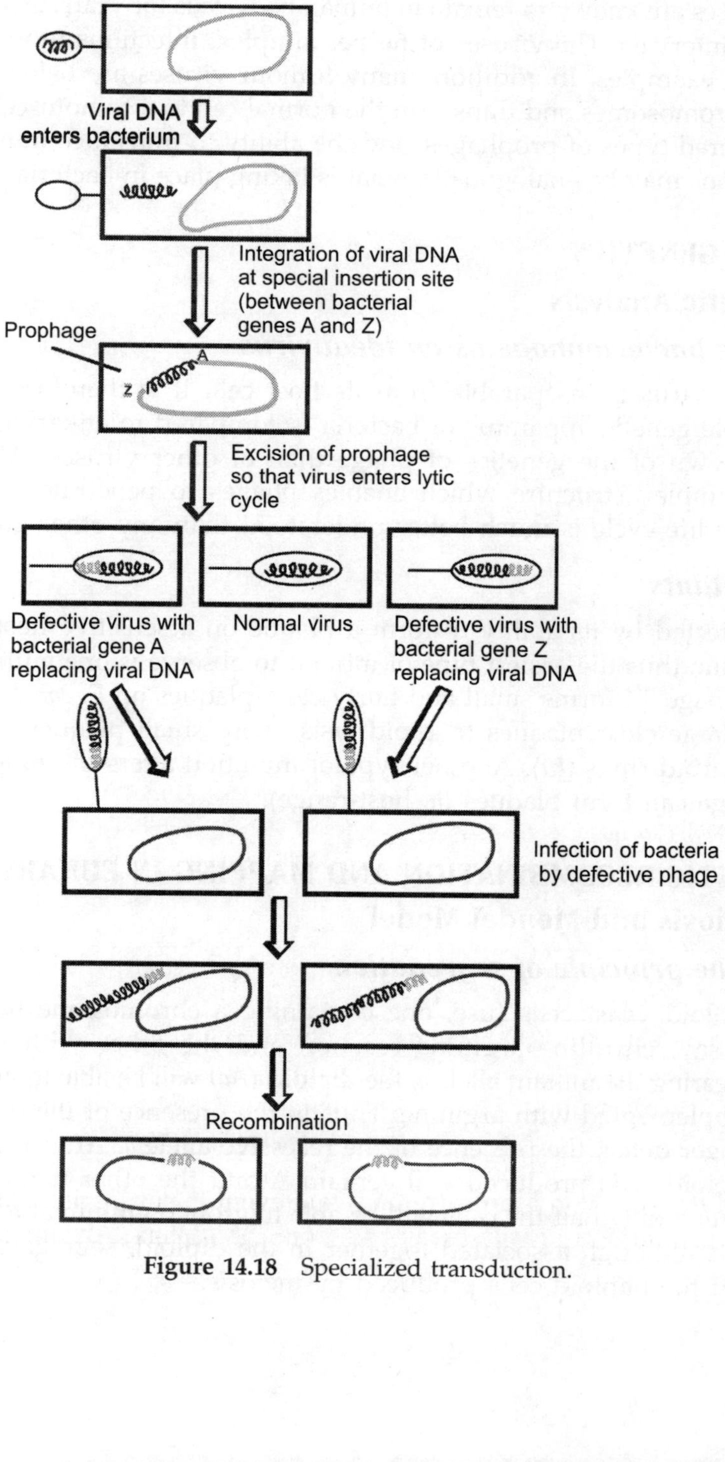

Figure 14.18 Specialized transduction.

staphylococcal entero toxin in food poisoning, clostridial toxins in some forms of animal botulism, and streptococcal toxins in scarlet fever. Also, *Salmonella* cells carry prophages for lipopolysaccharides. These lipopolysaccharides provide the basis for separating *Salmonella* into serological types (serotypes) rather than species.

Certain viruses are known to remain in human body cells for years, express themselves at unspecified intervals. The viruses of herpes simplex, infectious mononucleosis, and chickenpox are examples. In addition, many tumour viruses are believed to associate with human chromosomes and transform the normal cell to a tumour cell. Such viruses may be considered types of prophages, and the ability to transduce human cells under clinical conditions may be analogous to what is taking place in bacterial recombination.

14.9 VIRAL GENETICS

14.9.1 Genetic Analysis

14.9.1.1 The bacteriophage as an ideal virus

The study of a virus is inseparable from its host cell. It is therefore not surprising, when the simple genetic apparatus of bacteria is compared to eukaryotes, to find that far more is known of the genetics of phage than of other viruses. Also, despite the superficially complex structure which enables phages to penetrate the bacterial cell wall, the phage life cycle is much better understood than any other group of viruses.

14.9.1.2 Mutants

A phage is detected by its ability to form a plaque on a sensitive host strain layered on solid medium; thus the easiest type of mutant to observe is one with altered plaque morphology. Phage T4 forms small and fairly clear plaques on *E. coli B;* mutant strains may produce large clear plaques (*r,* rapid lysis), very small plaques (*m,* minute) and plaques with turbid rings (*tu*). Another type of mutation alters the range of strains on which the phage can form plaques (*h,* host-range).

14.10 GENETIC RECOMBINATION AND MAPPING IN EUKARYOTES

14.10.1 Meiosis and Mendel Model

14.10.1.1 The principle of segregation

When two haploid yeast cells fuse, one containing a chromosome bearing the gene A mediating, say, citrullin→arginine reaction and the other with its homologous chromosome bearing the mutant allels *a,* the diploid *(Aa)* will be able to grow on minimal medium unsupplemented with arginine. Thus in the presence of the dominant gene A one can no longer detect the presence of the recessive allele *a.* After meiosis, however, two of the haploid cells produced will contain A and the other two *a,* so that in the haploid generation only half the cells will be able to grow on minimal medium. In other words A and *a,* although associated together in the diploid, segregate independently and unchanged the haploid cells produced by meiosis.

Mendel originally produced this principle of segregation by observing the products of fertilization in pea plants, which are diploid; which can be directly observed in yeast which is having a haploid generation. Following random fusion of haploid yeast cells carrying the two genes, yielding diploids, one can distinguish between the genotypic ration of 1 *AA*: 2*Aa*:1 *aa* and the phenotypic ration of three prototrophs to one auxotrophs.

14.10.1.2 The principle of independent assortment

Consider a second gene pair *B* and *b*, mediating, say, biotin production versus requirement, carried on a second pair of homologous chromosomes, one can again get the genotypic ratio of 1 *BB* : 2 *Bb* : 1 bb and the phenotypic ratio of three biotin producers to one biotin require, following random fusion of cells carrying either *B* or *b*. When observed the inheritance of both sets of genes, however, the haploid progeny of the *AaBb* diploids includes four types of cells in appropriately equal numbers:

1. Grow on minimal medium; hence genotype *AB*
2. Require biotin; hence genotype *Ab*.
3. Require arginine; hence genotype *aB*
4. Require both arginine and biotin; hence genotype *ab*.

14.10.1.3 Chromosome mapping

14.10.1.3.1 The exception: linkage

Look at the haploid progeny of an *AaBb* diploid in which both pairs of genes are carried on the same pair of chromosomes, however, the principle of independent assortment no longer hold good. *A* and *B* still segregate from *a* and *b*, but no longer at random.

If *A* and *B* were originally on one of the parental chromosomes and *a* and *b* on the other, then *Ab* and *aB*, the recombinants, will only be formed if crossing over has taken place between *A* and *B*. When *A* and *B* are placed at equal distances along the chromosomes, this means that if there is only a single chiasma per bivalent then only – of the tetrads will include the recombinant haploids *Ab* and *aB*. There will therefore be an excess of the parental types (*AB* and *ab* derived from the original haploids) and the two genes are said to be linked. Further, if a number of genes are found to be linked in this way, they are said to constitute a linkage group, implying that they all belong to the same chromosome. The two recombinant classs, *Ab* and *aB*, will be present in approximately equal number, since both parents contribute similar haploid chromosome complements to the cross.

14.10.1.3.2 Random spore analysis

The information gained from the study of linked genes can be used to determine the order and degree of separation of the genes along the chromosome. One third of the tetrad are 1 *AB* : 1 *Ab* : 1 *aB* : 1 *ab* while the other two-thirds total 4 *AB* : 4 ab; that is, one-sixth of the haploids are recombinant. This $\frac{1}{6}$ expressed as a percentage gives a map distance of 16.67 units (known as *centimorgans*) between *A* and *B*. If we then find a third gene *C* to be 12 units from *B* and 29 units from *A* (i.e., 58 per cent of tetrads include the results of chiasmata between *A* and *C*), we can deduce that the linear order of the genes is:

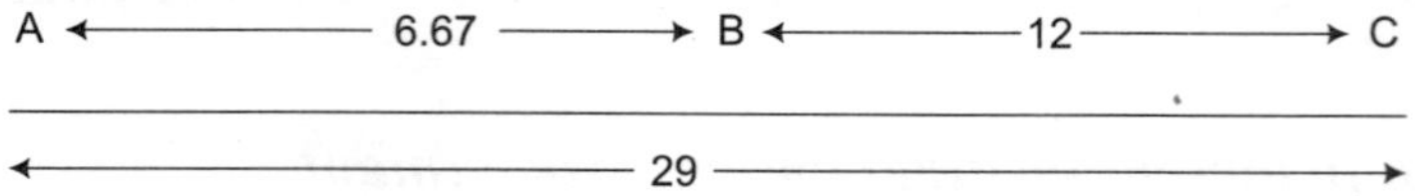

14.10.1.3.3 Gene order

Although this method of genetic mapping is slightly more complex than we have made it appear, it is possible to build up a linkage map of all the genes in a chromosome in this way, using the so-called *three-point test cross*, on the principle that the probability of recombination between any two genes is proportional to the distance between them.

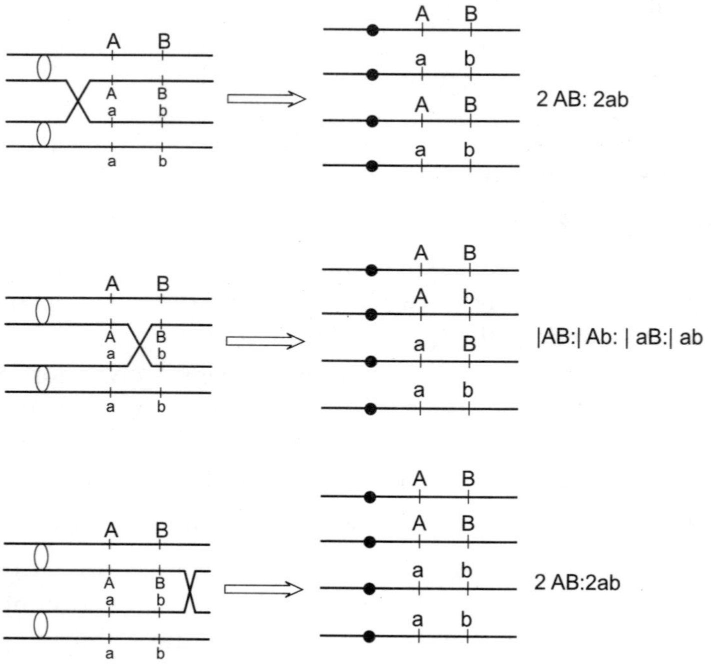

Figure 14.19 Relationship of chiasma position to the segregation of two linked genes A and B.

14.10.1.3.4 Interference

The three-point test cross reveals occasional double cross-overs, which are the products of independent chiasmata occurring in the two regions *AB* and *BC*. Now, the chance of the two occurring together in the same bivalent is the product of the chances of either event occurring singly – 16.67 per cent × 12 per cent, i.e., 2 per cent, in this example. But the actual number of double cross-overs observed is often significantly lower than the expected number as calculated in this way, since a chiasma in one region is found to reduce the chance of another occurring in an adjacent region. This phenomenon, known as interference, is explained by the mechanical restraints imposed on crossing over by the rigidity of the chromatids. Interference is usually expressed as the coincidence, i.e., the ratio of expected to observed double crossovers. As we should expect, the degree of interference usually increases as the distance between crossovers decreases.

14.10.1.3.5 Tetrad analysis

In ascomycetes such as *Neurospora* and *Sordaria* where linear tetrads are formed the products of individual meioses can be isolated mechanically and anlysed. When spore colour mutants are studied this analysis can be carried out by direct microscopic observation of the tetrads can be used to determine the map distance of the gene from the centromere. For example, the asco mutant of *Neurospora* has colourless instead of black spores, and the distribution of black and colourless spres in the asci of a heterozygote depends on whether or not a chiasma has occurred between the gene and the centromere. Thus 20 second division asci occur for every 80 first division asci, and a chiasma occurs between the centromere and the gene in 20 per cent of the asci (Figure 14.20).

Note that in tetrad analysis, count crossover asci, not crossover spores containing crossover chromatids. This is important to remember because in a second division

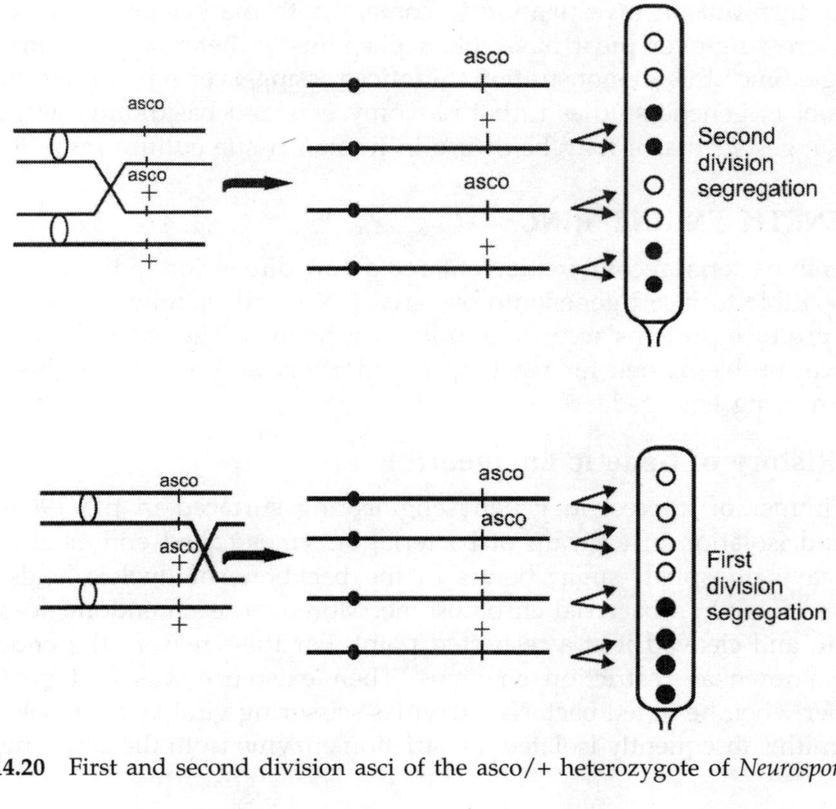

Figure 14.20 First and second division asci of the asco/+ heterozygote of *Neurospora crassa.*

ascus only half the spores contain chromosomes derived from the crossover chromatids, Because of this, if we wish to equate gene-centromere distances obtained in this way with the map distances obtained by random spore analysis, we must divide the gene-centromere distance by two. In this example the map distance between the gene and the centromere is:

$$\frac{\text{Second division asci}}{\text{Total asci}} \times \frac{100}{2} = \frac{20}{100} = \frac{100}{2} = 10 \text{ Units}$$

Similarly, if we obtain linkage data from tetrad analysis, the scores must always be divided by two to equate them with those obtained from random spore analyses.

14.10.1.3.6 Mitotic crossing-over

Although meiosis has not been observed in the imperfect fungus *Aspergillus niger*, Pontecorvo has discovered a parasexual system in this species which allows gene recombination. If strains of the same mating type are allowed to grow together, they form hybrid mycelia (heterokaryons) in which the nuclei of both strains coexist in the same cytoplasm; the haploid nuclei, however, occasionally fuse to form diploids which continue to divide with the haploids in the same mycelium. By chance sectors arise containing only the heterozygous diploid nuclei, and sometimes non-meiotic heploidization occurs (*c.* 10^{-3} per mitosis). Some of the haploid conidia derived from these sectors turn out on investigation to contain both marker genes, thus providing that mitotic crossing-over must have taken place in the heterozygous diploid nuclei at some stage. Since this demonstration, mitotic crossing-over has turned out to be an important tool in genetic studies in both ascomycetes and basidiomycetes, and it has even been suggested that it may be of use in human tissue culture investigations.

14.11 GENETIC ENGINEERING

Experiments in bacterial recombination entered a new dimension in the late 1970s when it became possible to insert genes into bacterial DNA and thereby establish a cell line that would produce proteins according to the instructions. The use of basic research to solve practical problems had far reaching ramification, and an entirely new industry, genetic engineering emerged.

14.11.1 History of Genetic Engineering

The first glimmer of interest in genetic engineering surfaced in the 1960s with the discovery and isolation of a group of bacterial enzymes called endonucleases. These enzymes cleave phosphate-sugar bonds in the backbone of nucleic acids and they could be used to open a bacterial chromosome. Moreover, each endonuclease scanned a DNA chain and cleaved it at a restricted point. For this reason, the endonucleases came to be known as restriction enzymes. Their existence was first postulated by Warner Arber when he noted bacterial enzymes scissoring viral DNA at selected spots. Hamilton Smith subsequently isolated a restriction enzyme from the Gram-negative rod

Haemophilus influenza. In 1971, Daniel Nathans used Smith's enzyme to split the DNA of simian virus 40 (SV40), a cause of tumour in monkeys. In 1978, the Noble Prize in Physiology or Medicine was awarded to the three scientists. By that time over 100 different restriction enzymes had been isolated and characterized.

Among the first scientists to attempt a genetic manipulation was Paul Berg of Stanford University. In 1971, Berg and his co-workers opened the DNA molecule from the SV40 virus and spliced it to a bacterial chromosome. In doing so, they constructed the first recombinant DNA molecule. However, the process was extremely tedious because the bacterial and viral DNAs had blunt ends. Berg therefore had to utilize exhaustive enzyme chemistry to form staggered ends that would combine easily. Nevertheless, he achieved a momentous first and was later honoured as a co-recipient with Sanger of the 1980 Nobel Prize in Chemistry.

While Berg was performing his experiments in 1971, an important breakthrough came from Herbert Boyer and his group at the University of California. Boyer isolated a restriction enzyme that would nick a chromosome and leave it with mortise like staggered ends. The bits of single-stranded DNA extending out from the chromosome easily attached to a new fragment of DNA in recombinant experiments. Scientists quickly dubbed the single-stranded "sticky ends."

During that same period, Stanley Cohen of Stanford was accumulating data on the plasmids of *Escherichia coli* (Figure 14.21). Cohen found that he could isolate **plasmids** from the bacterium and insert them into fresh bacteria by suspending the organisms in calcium chloride and heating them suddenly to achieve a transformation. Once inside *E. coli* cells, the plasmids multiplied independently and produced copies of themselves in succeeding generations. Cohen's data indicated that scientists need not work with the larger, less manageable chromosome. The final link to the process was provided by the DNA ligases. These enzymes, known since the 1960s, function during the replication of DNA and the repair of broken DNA molecules. Essentially, they operate in a manner opposite to the endonucleases; that is, they seal together DNA fragments. Progress came rapidly. Working together, Boyer and Cohen isolated plasmids from *E. coli* and opened them with restriction enzymes. Next they inserted a segment of foreign DNA using DNA ligase. Then they implanted the plasmids into fresh *E. coli*. By 1973, they had successfully spliced genes from *Staphylococcus aureus* into *E. coli*. The recombined plasmids were termed chimeras from the mythical lion-goat-serpent monster of Greek literature.

Genetic engineering experiments intrigued the scientific community because genes from widely divergent species could be spliced together. However, some scientists saw the specter of disaster if certain experiments involving cancer genes were performed. They therefore called for a meeting to discuss the ramifications of genetic engineering experiments, and in 1975, an international group met at Asilomar, California, to map out guidelines to be overseen by the National Institutes of Health (NIH). As the years unfolded, scientists realized that their fears were exaggerated, and most NIH restrictions were relaxed. Only the most risky ones remain under federal regulation.

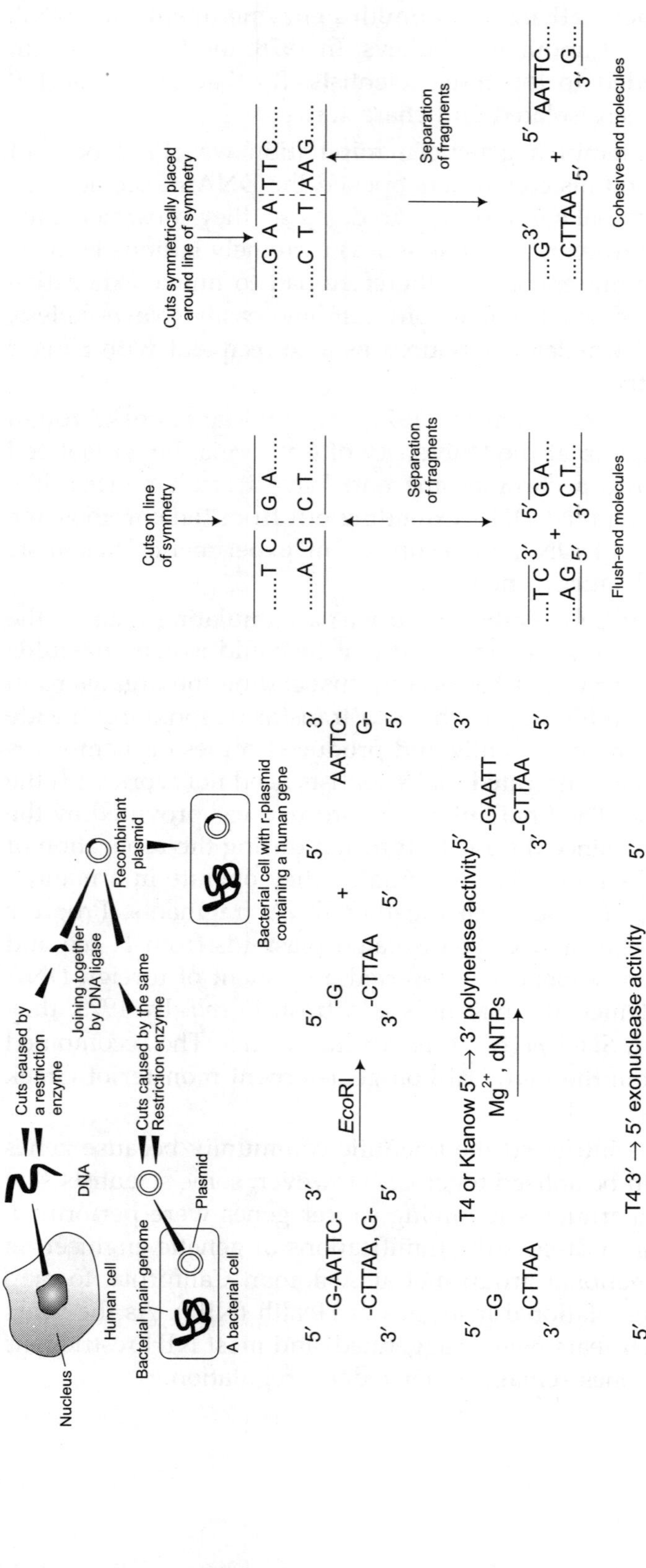

Figure 14.21 Genetic engineering tools.

There is also the legal problem of who owns genetically altered bacteria. In the late 1970s, researchers engineered a strain of *Pseudomonas* to dissolve oil rapidly. They hoped to use the new bacteria to clean up oil spills. The question arose as to whether a patent could be obtained for the organism. Testimony was presented in the courts and, finally, in 1980, the U.S. Supreme Court ruled by 5 to 4 vote that a patent could, indeed, be issued for the new organism.

14.11.2 Modern Applications

By the 1990s, thousands of companies worldwide were working on the applications of genetic engineering and DNA technology. Some were research companies with special units for these studies, while others were established solely to pursue and develop new products by gene-splicing techniques.

The pharmaceutical products of DNA are numerous and diverse. In 1986 the Eli Lilly company began marketing Humulin, a bacteria-produced form of insulin. For decades, insulin had been obtained from the pancreas tissues of animals, a factor that accounted for occasional allergic reactions. With the advent of genetic engineering, the insulin could be produced by inserting "insulin genes" to bacterial plasmids and using bacteria such as *E. coli* as chemical factories (Figure 14.22).

Also during the 1980s, a genetically engineered form of interferon came into use. This antiviral drug is produced by genetically altered *E. coli* cells. Interferon has been approved by the federal Food and Drug Administration (FDA) for use against a form of leukemia as well as for Kaposi's (a cancer occurring in AIDS patients), malignant myeloma, and multiple myeloma. Research is continuing on the use of antisense molecules as therapeutic agents. Antisense molecules are fragments of nucleic acids that unite with and neutralize mRNA molecules carrying the genetic message for protein synthesis. To treat AIDS, for example, an antisense molecule would block the mRNA for synthesis of new HIV particles.

DNA technologists have also used the new biochemical methods to human growth harmone. In 1986, the genetically engineered form of this hormone became available as Protropin to treat pituitary dwarfism. (Child suffering from this disease grew five inches in a year.) By 1994, a type of synthetic Factor VIII was available for use in patients with hemophilia, and a protein called tissue plasminogen activator (TPA) was in use to dissolve clots. To prevent infectious diseases, there was a genetically engineered vaccine for hepatitis B, and research was continuing on an AIDS vaccine using components produced in bacterial cells. Optimism was also high for innovative vaccines for cholera, malaria, and herpes simplex (Figure 14.23).

Genetic engineering has extended into many realms of science. In agriculture, for example, genes for herbicide resistance have been transplanted from bacteria into tobacco plants demonstrating that plants can be engineered to better tolerate the herbicides used for weed control. For tomato growers, a notable advance was made in 1988 when researchers at Washington University spliced genes from a pathogenic virus into tomato plant cells and demonstrated that the cells would produce viral proteins at their surface. The viral proteins then blocked viral encroachment and lent resistance to the tomato plants.

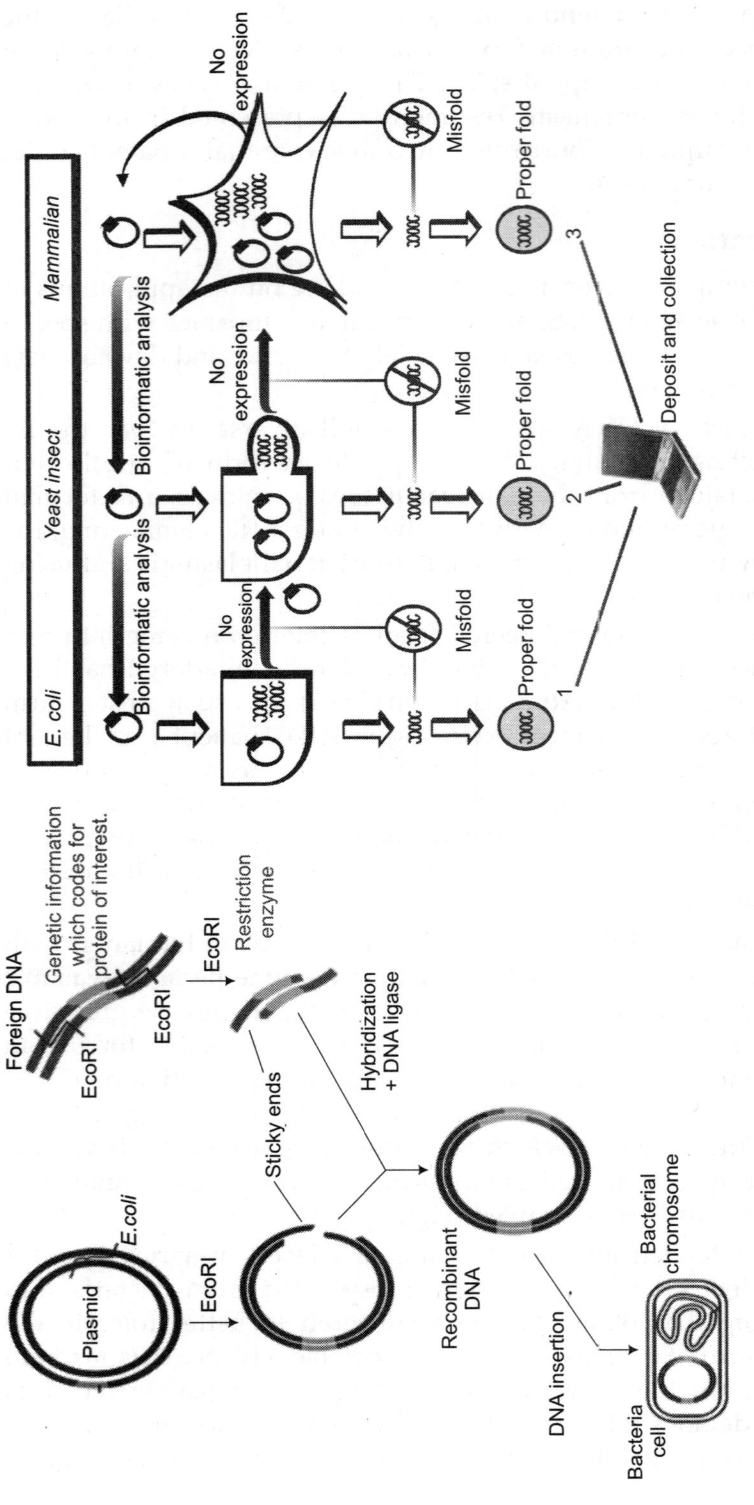

Figure 14.22 Scope of genetic engineering.

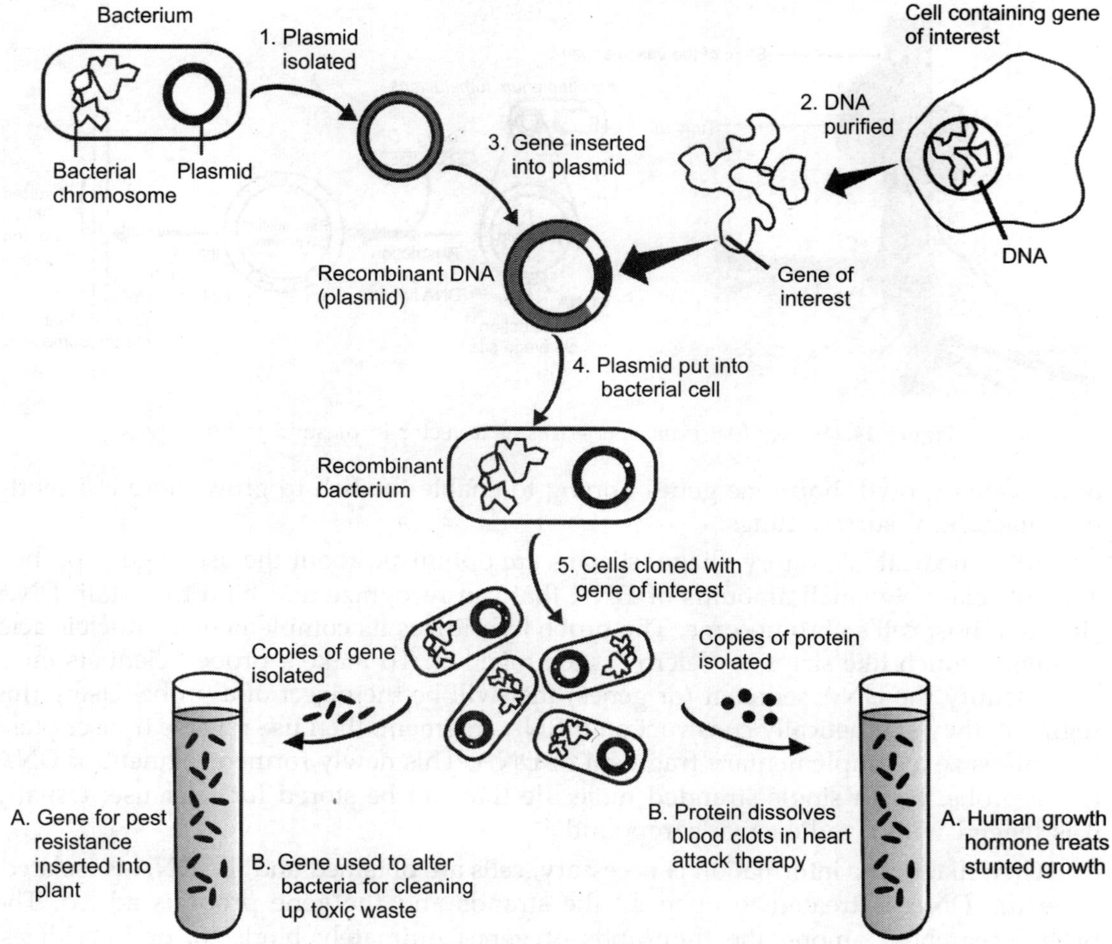

Figure 14.23 Product development employing gene technology.

For gene transfer experiments in plants, the mechanism often used for transfer is a plasmid from *Agrobacterium tumefaciens*. This organism causes a plant tumour called crown gall when DNA from the bacterium inserts itself into the plant cell's chromosomes during lysogeny. Researchers can therefore splice the desired gene into the plasmid and allow the bacterium to infect the plant. In 1989, scientists successfully removed the tumour-inducing genes from the bacterial plasmid so it can now serve as a gene carrier without causing plant damage (Figure 14.24).

The dairy industry may be the first to feel the dramatic effect of the new DNA technology. In the 1980s, for example, researchers at Cornell University injected dairy cows with bacteria-produced bovine growth hormone and reported a 41 per cent increase in milk from the experimental cows. Also being researched is a pig with more meat and less fat, a product of genetically engineered porcine growth hormone. And in 1989, scientists at Auburn University endowed young carp with extra copies

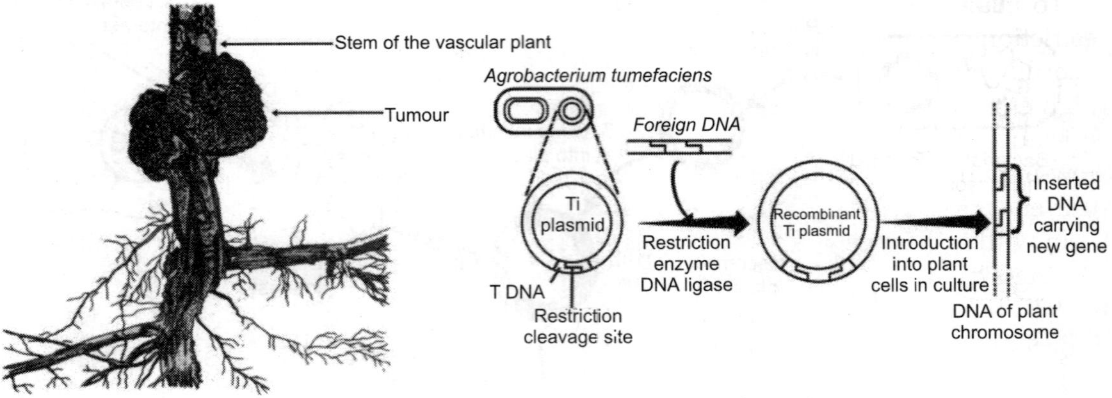

Figure 14.24 *Agrobacterium tumefaciens* as a vector in plasmid technology.

of activated growth hormone genes hoping to enable the fish to grow more efficiently in aquacultural surroundings.

In the medical laboratory, diagnosticians are optimistic about the use of gene probes, a set of relatively small amounts of DNA that can recognize and bind to certain DNA sites in a host cell's chromosome. The probe hybridizes its complementary nucleic acid sequence, much like strips of Velcro™ stick together. To make a probe, scientists must first identify the DNA segment (or genes) that will be the object of a probe. Using this segment, they synthetically construct an mRNA fragment, then use reverse transcriptase to synthesize a complementary fragment of DNA. This newly formed segment of DNA is the probe. It is a single-stranded molecule that can be stored for later use. Usually it is tagged with a radioactive compound.

When diagnostic information is necessary, cells are obtained and the DNA is isolated. Now the DNA is treated to open all the strands and the gene probe is added. The probe "searches" among the thousands of genes ultimately binds to, or hybridizes, the complementary gene. In doing so it concentrates the radioactivity at that site, and when the accumulated radioactivity is detected it can be assumed that a match has been made. The genes for such human diseases as cystic fibrosis, Huntington's chorea, and Alzheimer's disease can now be identified using specific gene probes. Indeed, gene probes are currently used to detect a variety of genetic disorders as well as to locate and identify the bacteria, viruses, and other pathogens of numerous human diseases.

Detection of a different sort can be accomplished through the law enforcement technology of DNA fingerprinting, derived from observations reported by Alec J. Jeffreys in 1985. Jeffreys noted that short, repetitive segments of DNA of unknown function exist between the body's functional genes. He found that the segments appear in all people, but the number of times they are repeated changes per person: the segments may be repeated two times in person A, 20 times in person B, and 200 times in person C. And since the segments often follow one another in a chromosome, the length of a chromosomal fragment varies from person to person.

To make a DNA fingerprint, human cells are obtained, the DNA is extracted, and restriction enzymes scissor the DNA into fragments. The fragments are then separated according to size by a process called electrophoresis. An electric current drives the fragments through a gel in a narrow channel. The shortest fragments meet the least resistance and travel the fastest, moving to the far end of the groove; the longer, slow-moving fragments meet more resistance remain closer to the starting point (the near end). Radioactive gene probes are then used to seek out the fragments and mark them with radiation. In this way the fragment positions can be revealed as dark bands, looking similar to a bar code. DNA fingerprinting was introduced to the legal system in 1987, but although accepted in many courts, the technique has been rejected by some judges because results can be inconclusive.

One of the most ambitious projects in the history of molecular genetics, indeed the history of biology, was launched in January 1989. That month a group of biologists, ethicists, industrial scientists, computer experts, and engineers began a monumental effort to map the human genome, that is, identify the nitrogenous bases and their sequence in the 100,000 odd human genes. Their goal: to spell out for the world the entire genetic message hidden in the code of bases in human DNA. With the National Institutes of Health supporting the project, a staggering $3 billion may be expended during the 15 years estimated for completion. There are about three billion pairs of bases that need to be sequenced, so the project should cost $1 per base pair. Most genes consist of about 10,000 to 150,000 base pairs, and between them are endless stretches of bases that appear to have no meaning, at least in the views of contemporary scientists. The human genome to be deciphered has been equated to a rope two inches in diameter and 32,000 miles long, all neatly arranged in a structure (the nucleus) about the size of a domed stadium.

Determining the nature of the human genome will require a multistep approach beginning with assigning genes to particular chromosomes, continuing with locating genes on a chromosome, and ending with determining the sequence of base pairs in the DNA chain. Each step will require technology not yet invented and the imagination of scientists who are still in grade school.

14.12 EXPERIMENTS ON MICROBIAL GENETICS

14.12.1 Mutant Isolation by Gradient Plate Method

An excellent way to determine the ability of organisms to produce mutants that are resistant to antibiotics is to grow them on a gradient plate of a particular antibiotic. Such a plate consists of two different wedge-like layers of media: a bottom layer of plain nutrient agar and a top layer of nutrient agar with the antibiotic. Since the antibiotic is only in the top layer, it tends to diffuse into the lower layer, producing a gradient of antibiotic concentration from low to high.

In this exercise, a gradient plate using streptomycin in the medium. *E. coli*, which is normally sensitive to this antibiotic, will be spread over the surface of the plate and incubated for 4 to 7 days (Figure 14.25). Any colonies that develop in the high concentration area are streptomycin-resistant mutants.

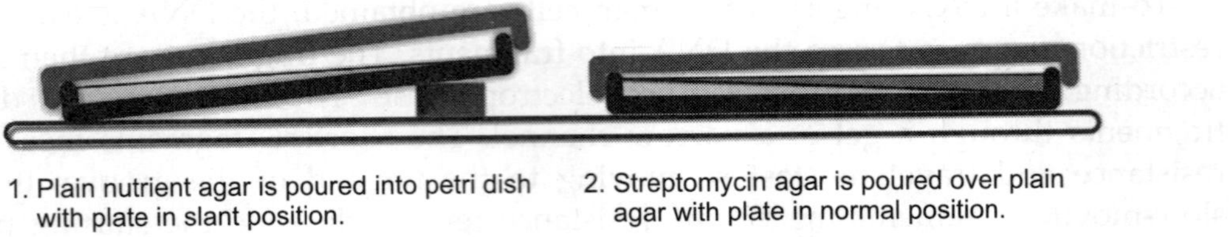

Figure 14.25 Gradient plate method for the isolation of mutant.

14.12.1.1 *Materials*

Sterile Petri plate, nutrient agar (10 mL per tube), tube of streptomycin solution (1%) 1 wood spacer ($^1/_8$″ × ½″ × 2″), 1 mL pipette

14.12.1.2 *Protocol*

14.12.1.2.1 Plate preparation

The gradient plate should have a high concentration of 100 mcg of streptomycin per milliliter of medium. The concentration is 10 times the strength used in sensitivity disks in the Kirby-Bauer test method. Prepare a gradient plate as follows:

1. Liquefy two pours of nutrient agar and cool to 50°C.
2. With wood spacer under one edge of Petri plate, pour contents of one agar pour into plate. Let stand until solidifications has occurred.
3. Remove the wood spacer from under the plate.
4. Pipette 0.1 mL of streptomycin into second agar pour, mix tube between palms, and pour contents over medium of the plate that is now resting level on the table.
5. Label the low and high concentration areas on the bottom of the plate.

14.12.1.2.2 Inoculation

The inoculation procedure is illustrated in Figure 14.26. The technique involves spreading a measured amount of culture on the surface of the medium with a glass bent rod to provide optimum distribution.

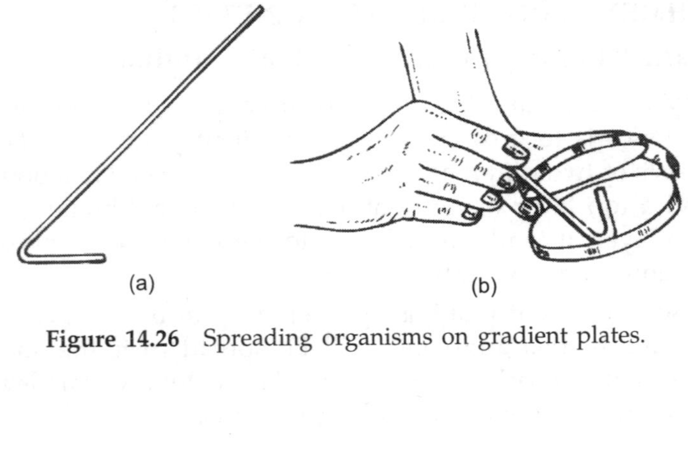

Figure 14.26 Spreading organisms on gradient plates.

14.12.1.3 Results and observations

Beaker with 95% ethanol, glass rod spreader, nutrient broth culture of *E. coli,* 1 mL pipette.

14.12.1.3.1 Protocol

1. Pipette 0.1 mL of *E. coli* suspension onto the surface of medium in Petri plate.
2. Sterilize glass spreading rod by dipping it in alcohol first and then passing it quickly through the flame of a Bunsen burner. Cool the rod by placing against sterile medium in plate before contacting organisms.
3. Spread the culture evenly over the surface with the glass rod.
4. Invert and incubate the plate at 37°C for 4 to 7 days in a closed canister or plastic bag. Unless incubated in this manner, exercise dehydration might occur.

14.12.1.3.2 First evaluation

After 4 to 7 days, look for colonies of *E. coli* in the area of high streptomycin concentration. Count the colonies that appear to be resistant mutants and record your count.

Select a well-isolated colony in the high concentration area and, with a sterile loop, smear the colony over the surface of the medium toward the higher concentration portion of the plate. Do this with two or three colonies. Return the plate to the incubator for another 2 or 3 days.

14.12.1.3.3 Final evaluation

Examine the plate again to note what effect the spreading of the colonies had on their growth.

14.12.2 Demonstration of Streaking Plates Using Loop/Toothpick

Bacterial culture streaking allows bacteria to reproduce on a culture medium in a controlled environment. The process involves spreading bacteria across an agar plate and allowing them to incubate at a certain temperature for a period of time. Bacterial streaking can be used to identify and isolate pure bacterial colonies from a mixed population.

14.12.2.1 Materials

Mixed culture plate, autoclave sterilized tooth picks, Bunsen burner, laminar flow, fresh Petri plates containing nutrient agar medium, glass marker.

14.12.2.2 Protocol

1. While wearing gloves, sterilize an inoculating loop by placing it at an angle over a flame. The loop should turn orange before you remove it from the flame. A sterile toothpick may be substituted for the inoculating loop. Do not place toothpicks over a flame.

2. Remove the lid from a culture plate containing the desired microorganism.
3. Cool the inoculating loop by stabbing it into the agar in a spot that does not contain a bacterial colony.
4. Pick a colony and scrape off a little of the bacteria using the loop. Be sure to close the lid.
5. Using a new agar plate, lift the lid just enough to insert the loop.
6. Streak the loop containing the bacteria at the top end of the agar plate moving in a zig-zag horizontal pattern until 1/3 of the plate is covered.
7. Sterilize the loop again in the flame and cool it at the edge of the agar away from the bacteria in the plate that you just streaked.
8. Rotate the plate about 60 degrees and spread the bacteria from the end of the first streak into a second area using the same motion in step 6.
9. Sterilize the loop again using the procedure in step 7.
10. Rotate the plate about 60 degrees and spread the bacteria from the end of the second streak into a new area in the same pattern.
11. Sterilize the loop again.
12. Replace the lid and invert the plate. Incubate the plate overnight at 37°C (98.6° F).
13. Observe bacterial cells growing in streaks and in isolated areas.

14.12.2.2.1 Streaking plates

When picking and streaking lots of bacterial colonies it is often quicker to use sterile toothpicks and sticks instead of using a wire loop that must be sterilized between each colony by passing it through a flame. Using the technique shown below, eight colonies can be streaked for isolation on a single plate (Figure 14.27) (Save the used toothpicks and sticks to be re-autoclaved).

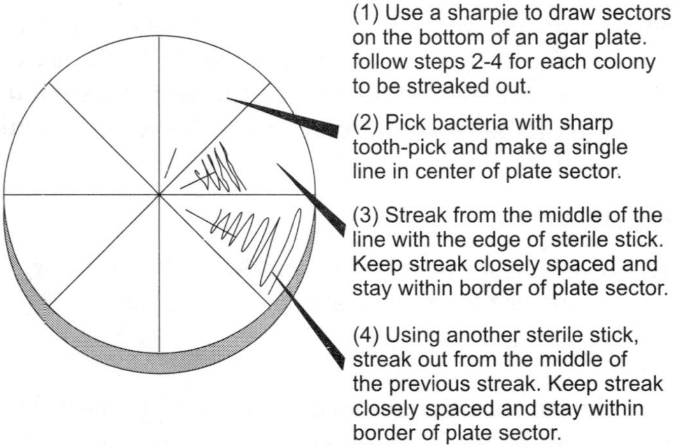

Figure 14.27 Spreading on plates by toothpick.

14.12.2.2.2 Picking and patching colonies

When checking out bacterial strains, it is often useful to patch many colonies on a single Petri plate so they can be tested simultaneously by replica printing onto a series of media. To do this, place a fresh master plate over a "patching grid". Touch the top of each colony with a sterile toothpick and draw a small × on the new master plate. (Only use each toothpick once and save the used toothpicks to be reautoclaved). Many patches can be placed on a single plate. After patching, incubate the plate overnight to let the patches grow up. The next day this plate can be used as a master for replica-printing. Always mark each plate at the top of the patch grid since the patch grids are symmetrical (Figure 14.28).

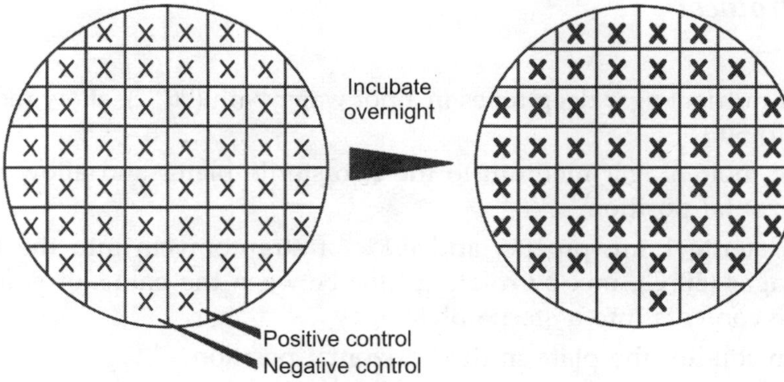

Figure 14.28 Picking and patching colonies.

14.12.3 Isolation of Streptomycin-Resistant Mutant by Replica Plating Method

E. coli could develop mutant strains that are streptomycin/ampicillin-resistant. The replica plating technique is applied in order to test the nature of mutation in the bacterium, i.e., whether induced by the antibiotic or occurred spontaneously. To prove whether or not such a colony exists on a plain agar plate having between 500 and 1000 colonies could be a difficult task as the transfer of organisms from each colony to medium supplemented with streptomycin would be required. The task of transferring several hundred colonies to test one mutant is simplified by the use of replica plating technique in one step (Figure 14.29).

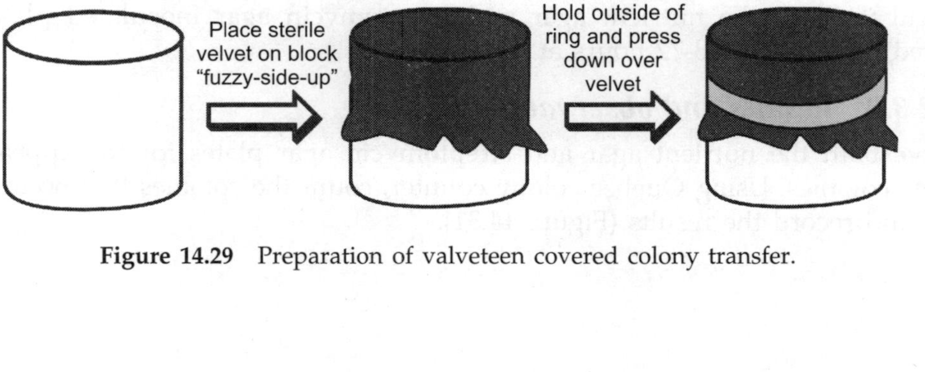

Figure 14.29 Preparation of valveteen covered colony transfer.

This experiment deals with the ampicillin-resistant mutant isolation of *E. coli* by the replica plating technique, where simply a velveteen covered colony transfer device is used for making the transfer on nutrient agar medium supplemented with and without ampicillin (Figure 14.30).

14.12.3.1 Materials

24 hours old actively growing broth of *E. coli* culture; 10 mL nutrient agar deep tubes; streptomycin sulfate (10 mg/ 100 mL of sterile water); sterile Petri dishes; sterile velveteen colony carrier; beaker with 95% ethanol, bent glass rod; 1 mL serological pipette; Quebec colony counter.

14.12.3.2 Protocol

First Day

1. Melt the nutrient agar deep tubes in a hot water bath (96°C), allow molten medium to cool to 50°C.
2. Pour the molten agar medium to the two sterile plates and allow solidifying in the horizontal position.
3. Using a sterile 1 mL pipette, add 0.1% of streptomycin into the third tube of molten agar (50°C), mix by rotating tube between the palms of your hands, and pour the contents into a sterile plate.
4. Allow to solidify the plate in the horizontal position.
5. Using the sterile 1-mL pipette, add 0.2 mL of the *E. coli* test culture to the surface of nutrient agar plate.
6. Using an alcohol dipped and flamed bent glass rod, spread the inoculums over the entire agar surface by rotating the plate.
7. Incubate the plate in an inverted position for 24-48 hours at 37°C.

Observe the incubated plate for the growth of *E. coli* culture on the agar surface.

Second Day

Carefully lower the sterile velveteen colony carrier on to the colonies of *E. coli* growing on the nutrient agar plates.

Inoculate the nutrient agar plate by lightly pressing the carrier onto the medium.

Without returning the carrier to the original culture plate, inoculate the streptomycin agar plate by following the procedure described above.

Incubate both the nutrient agar and streptomycin agar inoculated plates in an inverted position for 48-72 hours at 37°C.

14.12.3.3 Results and observations

Observe both the nutrient agar and streptomycin agar plates for the appearance of discrete colonies. Using Quebec colony counter, count the colonies that occur on both plates and record the results (Figure 14.31).

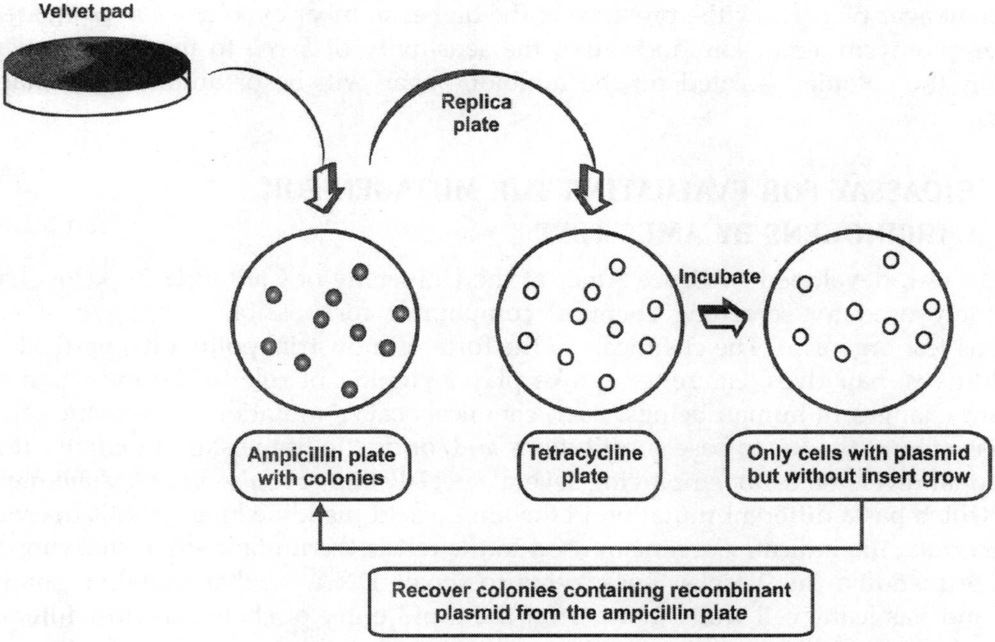

Figure 14.30 Replica plating to detect recombinant plasmids.

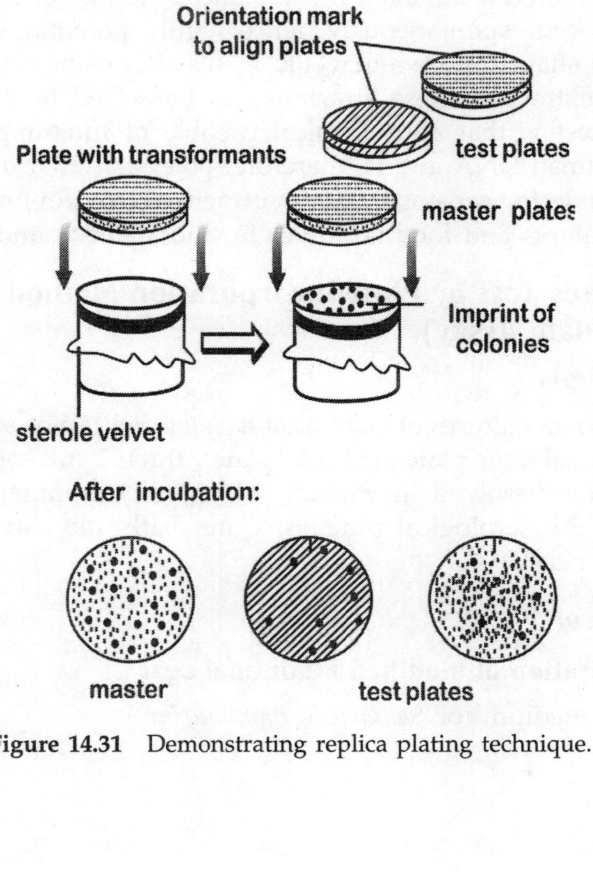

Figure 14.31 Demonstrating replica plating technique.

Nutrient agar plate show the presence of the higher number of colonies as compared to the streptomycin agar plate, indicating the sensitivity of *E coli* to these antibiotics. Moreover, the colonies isolated on the antibiotic agar will be of antibiotic resistant mutants.

14.13 BIOASSAY FOR EVALUATING THE MUTAGEN FOR CARCINOGENS BY AMES TEST

The Ames test, developed by Bruce Ames at the University of California-Berkeley, has been widely used for screening chemical compounds for possible carcinogen using bacteria as test organism. The chemicals in the form of industrial pollutants, pesticides, food additives, hair dyes, cigarettes smoke play a significant role in the induction of malignant changes in human beings. Such chemicals cause mutations in somatic cells. Mutations are mostly due to base substitutions and/or due to frame shift mutations. It is a mutational reversion assay employing several special stains of *Salmonella typhimurium,* each of which has a different mutation in the amino acid histidine biosynthesis operon. S. *typhimurium,* the indicator organism used in the test is the mutant strain that cannot synthesize histidine (his^-) lacks the enzyme to repair DNA, so that mutation shows rapidly and has leaky cell walls that permit the rapid entry of chemicals. In addition, his^- *S. typhimurium* strain carries mutation-enhanced plasmid, the combined effects of these three characters increase the sensitivity of the organism to mutagenic treatments. A test agent is considered a mutagen if it enhances the rate of back-mutation beyond levels that would occur spontaneously. Since highly potential carcinogens, such as benzanthracene and aflatoxins are mutagenic agents after being acted on by mammalian liver enzymes, an extract of these enzymes can be added to the test medium. The premise of Ames test is that any chemical capable of mutating bacterial DNA can similarly mutate human DNA and is, therefore, potentially hazardous. The Ames test has proved invaluable for screening an assortment of environmental (e.g. pesticides) and industrial pollutants and food additives for mutagenesis and carcinogenicity.

14.13.1 The Ames Test by Plate Incorporation Method (Pour Plate Assay)

14.13.1.1 Materials

24 hours actively grown cultures of *Salmonella typhimurium* Ames strain TA98 or TA1538 (ATCC), three minimal agar plates (30 mL/plate), three 2 mL top agar plates (2 mL/tube), 2-nitrofluorene dissolved in ethanol (10 μg/mL), commercial hair dye, sterile Pasteur pipettes, 1 mL serological pipettes, water bath, mechanical pipetting device, glassware marking pencil.

14.13.1.2 Protocol

14.13.1.2.1 Preparation of modified E minimal agar

(Standard minimal medium for *Salmonella typhimurium*):

A stock solution of 50× E medium is prepared by dissolving each of the following ingredients in the order given: Distilled water 670 mL, $Mg.SO_4.7H_2O$ 10 g, Citric acid 100 g, K_2HPO_4 (anhydrous) 500 g, $NaHPO_4.4H_2O$ 175 g.

Store the 50× stock solution over chloroform and sterilize when diluted to 1×. Add 2% glucose and 1,5% agar per litre of 1× E medium.

14.13.1.2.2 Top agar

NaCl 5 g, Agar 6.5 g, Distilled water 1000 mL.

Dissolve agar in 500 mL of distilled water and add NaCl dissolved in 100 mL and make volume to 1 litre. Dispense 100 mL into bottles, autoclave at 121°C for 15 minutes. Add 10 mL of a sterile stock mixture of L- histidine-HCl (0.05 mM i.e., 10.5 mg/100 mL) to 100 mL top agar and maintain at 45°C until used. Pour 2 mL of top agar in sterile test tubes (10 × 75 mm).

1. Prepare E minimal agar plates by pouring 30 mL of the medium into each Petri plate.
2. Label the bottoms of the three minimal E agar plates as follows: Positive control, negative control and unknown.
3. Liquify three tubes of top agar and maintain the molten agar at 45°C.
4. To each molten top agar tube, aseptically add 0.1 mL of an overnight grown culture of *S. typhimurium His*$^-$ auxotroph.
5. Thoroughly mix the bacterium into the top agar by rotating the tube between the palms of your hands.
6. Add 0.5 mL of the test compound (dissolved in water, DMSO, ethanol or formaldehyde depending upon its solubility – 10 µg/mL).
7. Mix the contents immediately by rotating the tube between the palms and pour to contents over the surface of a minimal E glucose agar plate (unknown).
8. Tilt the plate all sides for even distribution of the top agar layer. The entire operation (steps 4-7) should be carried out within 20 seconds.
9. Repeat steps 4-7 for each of the other two tubes of top agar, one plate with control plates of bacteria alone (negative control) and other plate with bacteria plus the mutagen (i.e., 2-nitrofluorene dissolved in alcohol, 10 µg/mL) (positive control).
10. Allow the agar to harden in the dark for a few minutes.
11. Incubate the plates in an inverted position in the dark at 37°C for 24 to 48 hours.

14.13.1.3 Results and observations

Examine the negative control plate for the appearance of the colonies of histidine revertants (phototrophs) and the test plate for the his$^+$ colonies induced by the chemical. Count the number of colonies in all the three plates and record the data in a tabular form (Figure 14.32).

The positive test plate, i.e., added with the known and strong mutagenic agent will have a high density of revertants i.e., his$^+$ colonies induced by the chemical. The negative control plate will either have no or very few revertants. It is due to spontaneous back mutations (i.e., *his*$^-$→*his*$^+$) which always occurs. By comparing the number of colonies growing on the control plate with the test plate, the degree of mutagenicity of the chemical agent can be calculated. Chemicals that produce an increased incidence of back mutation are considered carcinogens. (i.e., more number of *his*$^+$ revertants) are considered carcinogens.

14.13.2 The Ames Test by Spot Method

In this method, the mutagen is omitted from the top agar mixture and is applied after the top agar containing the bacterial inoculum has been poured and hardened. The mutagen can be added directly as a few crystals or 6 mm sterile filter disc impregnated with the test chemical dissolved either in water, alcohol or DMSO solution is placed in the centre of the test plate and the plates are incubated. Following diffusion of the test compound from the disc, a concentration gradient of chemicals is established. A qualitative indication of the test compound can be determined by noting the number of colonies present on the plate. Spot test is widely used for screening of chemical compounds for mutagenicity.

14.13.2.1 Materials

24 hours actively grown cultures of *Salmonella typhimurium* Ames strain TA98 or TA1538 (ATCC), three minimal agar plates (30 mL/plate), three 2 mL top agar plates (2 mL/tube), 2-nitrofluorene dissolved in ethanol (10 µg/mL), commercial hair dye, sterile Pasteur pipettes, 1 mL serological pipettes, water bath, mechanical pipetting device, glassware marking pencil, Sterile filter paper discs, forceps.

14.13.2.2 Protocol

1. Label the bottoms of three minimal E glucose agar plates as negative control, positive control and hair dye.
2. Molt the three tubes of top agar in a water bath and maintain at 45°C.
3. Aseptically add 0.1 mL of *Salmonella typhimurium* test culture into molten top agar tube and thoroughly mix the contents by rotating the test tube between the palms of your hands or by vertexing for three seconds.
4. Immediately pour the top agar cultures onto the E minimal agar plate (control positive) and allow to solidity.
5. Repeat steps 3 and 4 for each of the other two top agar tubes.
6. With a sterile forceps, dip each disc into its respective test chemical solution and drain by touching the side of the container.

   OR

   Position a filter paper disc on the edge of a minimal agar plate with sterile forceps. Saturate the disc with a test compound.
7. Place the chemical impregnated disc in the center of the respectively labelled E minimal glucose agar plates. Gently press down on the discs.

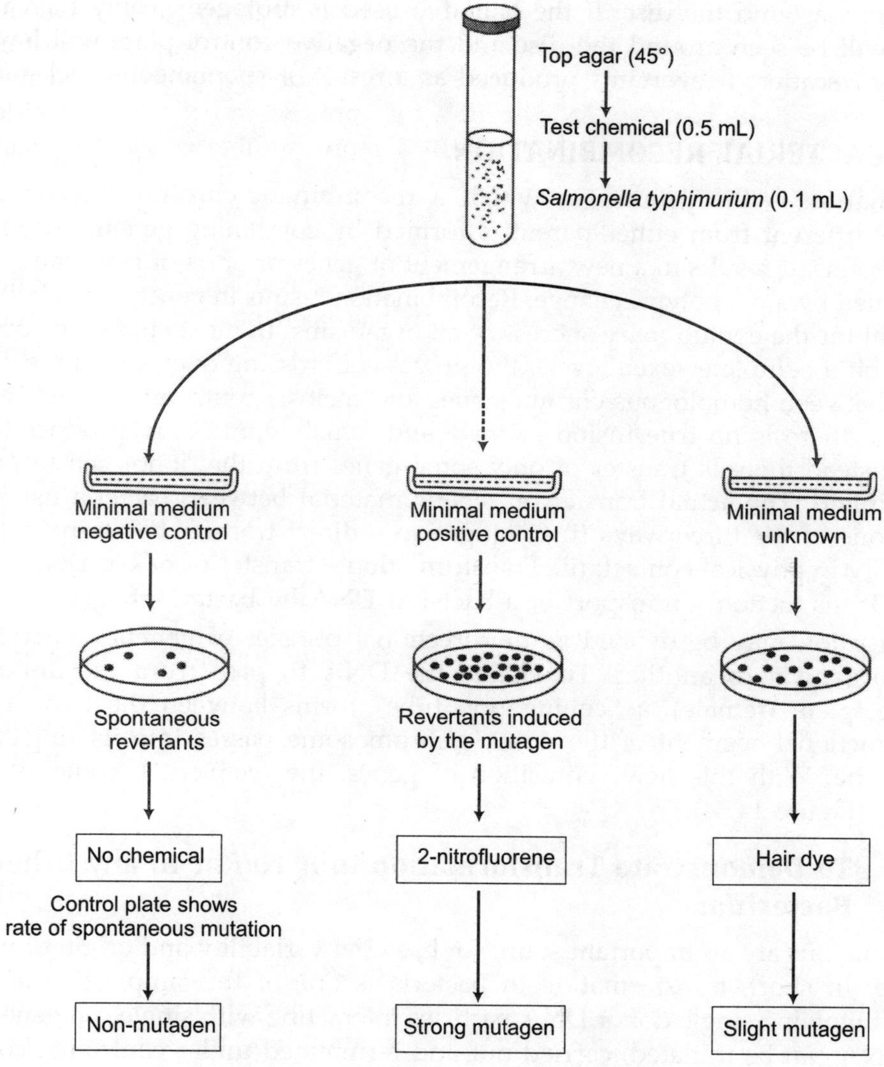

Figure 14.32 Pour plate method for Ames test.

8. Insert a sterile disc dipped in sterile water on the negative control plates.
9. Incubate all the three plates for 24-48 hours at 37°C.
10. Examine all the three plates for the appearance of colonies around the disc. Count the number of colonies in each plate and record the result in a tabular form (Figure 14.33).

14.13.2.3 Results

On examination after 48 hours of incubation, the positive test plates *i.e.* the plates having the disc impregnated with 2 - nitrofluroene, a strong mutagen, will have a high density of revertants around the disc. If the hair dye used is mutagenic, only then revertant colonies will be seen around the disc and the negative control plate will have either no or a few scattered revertants produced as a result of spontaneous back-mutations.

14.14 BACTERIAL RECOMBINATION

Recombination is the process in which a recombinant chromosome, one with a genotype different from either parent is formed by combining genetic material from two organisms. It results in a new arrangement of genes or parts of genes and normally accompanied by a phenotypic change. Recombination results in genetic variability, which is essential for the evolutionary success of all organisms. In most eukaryotic organisms, who exhibit a complete sexual cycle, the process of crossing over, exchange of genetic material between homologous chromosomes and meiosis contribute to this variability. In bacteria there is no true fusion of male and female gametes to produce a diploid zygote, instead there is transfer of only some genes from the donor cell to produce a partial diploid. The actual transfer of genetic material between bacteria usually takes place in one of the three ways (i) Conjugation - direct transfer between two bacteria temporarily in physical contact; (ii) Transformation - transfer of naked DNA fragment; and (iii) Transduction - transport of a bacterial DNA by bacteriophages.

Conjugation may be defined as unidirectional transfer of genetic material (DNA) by one organism to another. To enable the DNA to pass from the donor (male) to the recipient (female), a "conjugation tube" forms between the two organisms. Only a fractional segment of the donor's chromosome passes into its single, circular chromosome. With this new acquisition of genes, the recipient becomes a different organism (Figure 14.34).

14.14.1 To Demonstrate Transformation in *E. coli* or in any Other Bacterium

Transformations are an important source of bacterial variability and origin of new types in nature. In short, transformation in bacteria is one of the simplest recombination systems, involving single donor DNA particles interacting with single cell genomes. The donor DNA can be initiated, carried out and terminated under controlled conditions. The process of transformation essentially includes removal of deoxyribonucleic acid from one strain of a particular bacterium and addition to another strain of that bacterium. Some of the cells in the second strain are able to transmit characteristics of the first to their progeny (Figure 14.35).

The different steps involved in the process of transformation are as follows:

1. Donor DNA segment is absorbed by the recipient cell.
2. Absorbed DNA becomes irreversibly bound.
3. The DNA begins the process of integration. A part of the donor DNA, possibly only one strand, is integrated, probably by substitution into the genome of the recipient. Demonstration of the process of transformation includes two steps (i) DNA extraction and (ii) Transformation.

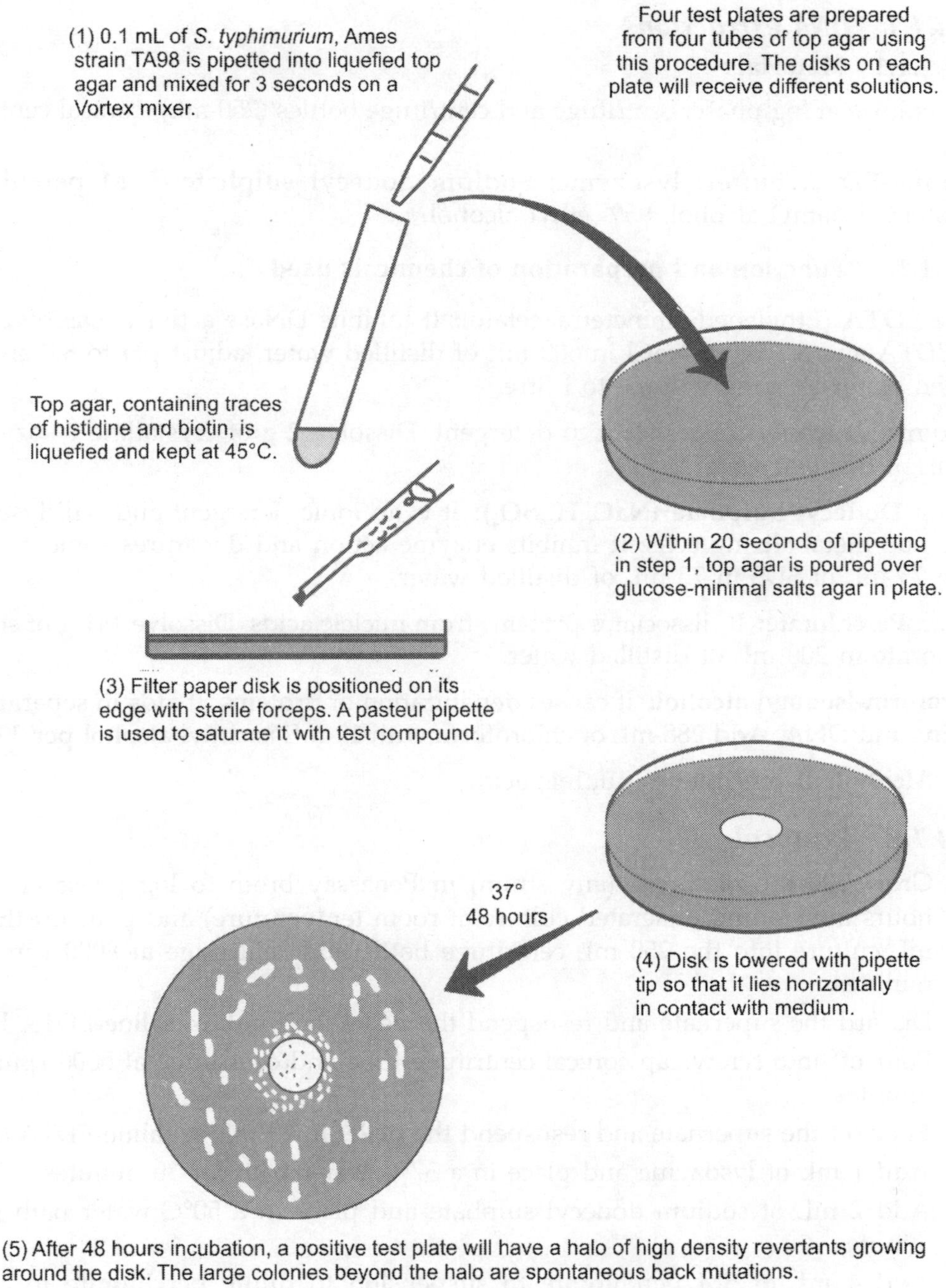

Figure 14.33 Protocol for Ames test by spot method.

14.14.1.1 DNA extraction

14.14.1.1.1 Materials

E. coli culture at log phase; centrifuge and centrifuge bottles (250 mL); conical centrifuge tubes;

Saline-EDTA buffer; lysozyme; Sodium dodecyl sulphate; 5 M perchlorate; chloroform-isoamyl alcohol; 95% ethyl alcohol.

14.14.1.1.2 Function and preparation of chemicals used

Saline EDTA (Ethylenediaminetetraaeetate)**:** It inhibits DNase activity. Dissolve 37.22 g of EDTA and 8.77 g of NaCl in 600 mL of distilled water, adjust pH to 8.0 and add distilled water to bring volume to l litre.

Lysozyme: It lyses cells resistant to detergent. Dissolve 2 g of crystalline lysozyme in 100 mL of distilled water.

Sodium Dodecyl Sulphate ($NaC_{12}H_{26}SO_4$): It is an ionic detergent and will lyse most of the non-metabolizing cells. It inhibits enzyme action and denatures some proteins. Dilute 25 mL of SDS in 75 mL of distilled water.

Sodium Perchlorate: It dissociates proteins from nucleic acids. Dissolve 141 g of sodium perchlorate in 200 mL of distilled water.

Chloroform-lsoamyl alcohol: It causes denaturation of proteins. It aids in separation of proteins and DNA. Add 288 mL of chloroform with 12 mL isoamyl alcohol per 300 mL.

Ethyl Alcohol: It precipitates nucleic acid.

14.14.1.2 Protocol

1. Grow 200 mL of *E. coli* (any strain) in Penassay broth to log phase (12 to 24 hours in standing unaerated culture at room temperature) and pour off this 200 mL culture into the 250 mL centrifuge bottle and centrifuge at 8000 rpm for 5 minutes.
2. Discard the supernate and resuspend the pellet in 25 mL of saline-EDTA buffer.
3. Pour off into screw-cap conical centrifuge tube and centrifuge at 5000 rpm for 5 minutes.
4. Pour off the supernate and resuspend the pellet in 13 mL of saline-EDTA buffer.
5. Add 1 mL of lysozyme and place in a 37°C water bath for 30 minutes.
6. Add 2 mL of sodium dodecyl sulphate and place in a 50°C water bath for 10 minutes.
7. Add 4 mL of 5M perchlorate to suspension to dilute perchlorate to a final concentration of 1M.
8. Add 15 mL of chloroform-isoamyl alcohol.
9. Tighten the cap and shake vigorously.
10. Centrifuge at 5000 rpm for 3 minutes.

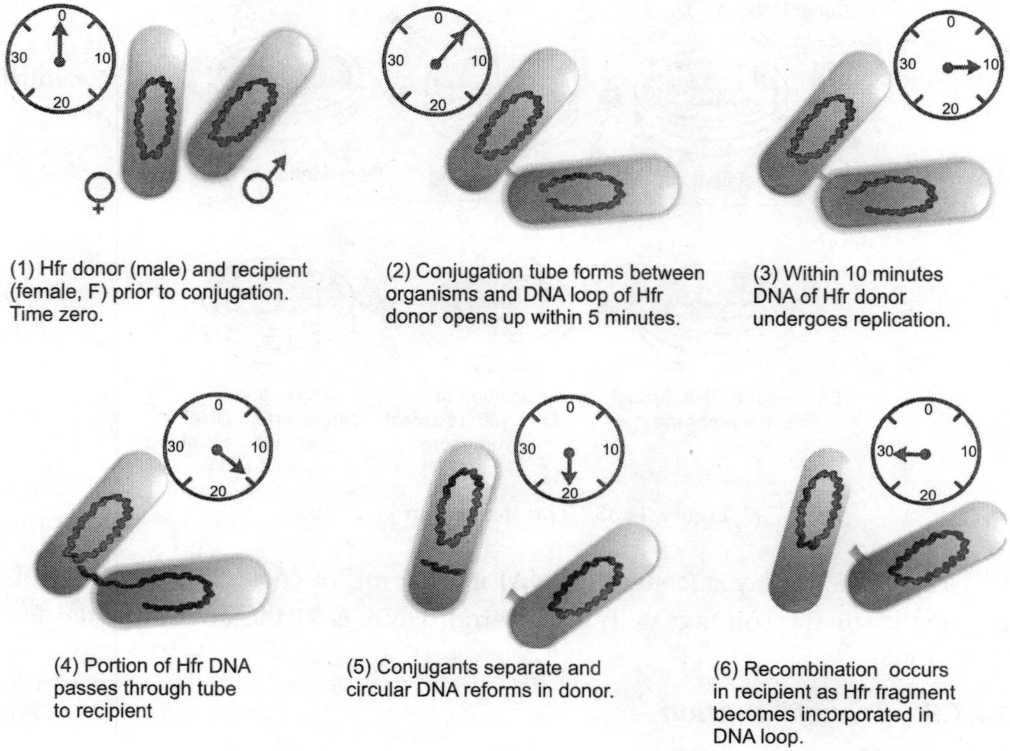

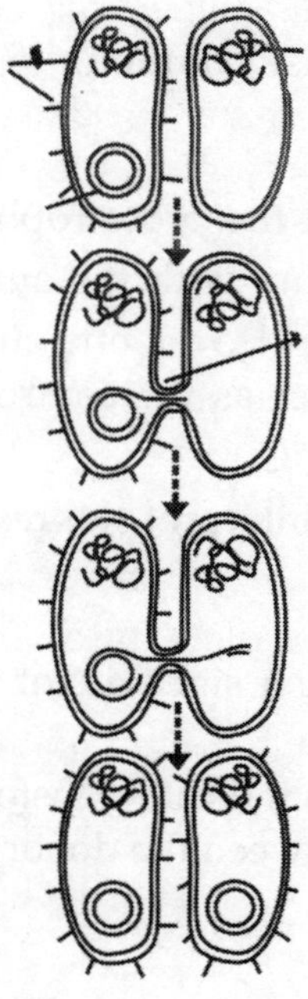

Figure 14.34 Demonstrating mechanism of bacterial conjugation.

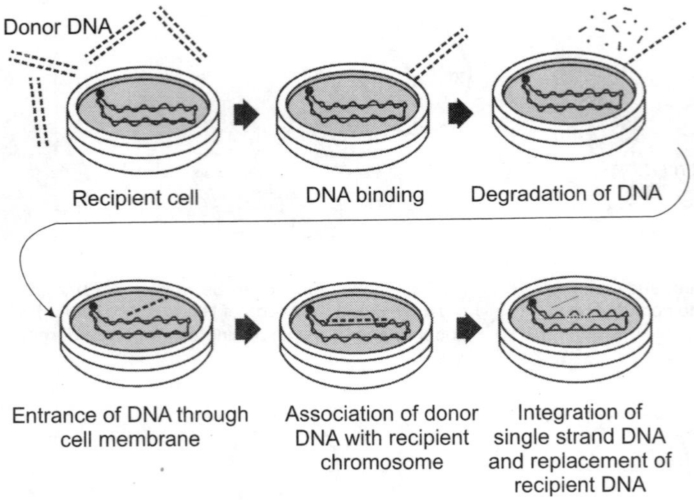

Figure 14.35 Transformation in bacteria.

11. Decant the top layer (contains DNA) into 30 mL of cool 95% ethyl alcohol.
12. Gently stir the solution with a glass rod. DNA is at the bottom in the form of strands.

14.14.1.3 Transformation

14.14.1.3.1 Materials

Solution of DNA from a prototrophic strain of *E. coil; E. coil* culture of auxotrophic strain for valine and leucine, i.e., Val^- Leu^-. DNase solution.

14.14.1.3.2 Protocol

1. Add DNase solution to the DNA extract of the prototrophic strain.
2. Inoculate two plates having minimal medium with the auxotrophic strain.
3. Spread DNA, solution (100 µg/mL) of step 1 over one cultured plate of step 2 with a glass spreader. The other plate serves as the control.
4. Incubate both the plates for 48 hours at 37°C.
5. Count the colonies in both the plates and interpret the results.

14.14.1.4 Results and observations

(i) No colonies are expected in the control dish since auxotrophs cannot grow on minimal medium.

(ii) Presence of colonies in the experimental dish indicate transformation of auxotrophs to prototrophs by recombination of genes between the donor DNA and auxotrophic genome.

14.14.2 Calcium Chloride ($CaCl_2$) Induced Transformation in *E. coli*

A genetic alteration in a cell resulting from the introduction of free DNA from the environment across the cell membrane is known as transformation. Genetic transformation is a process by which a cell takes up naked DNA from the surrounding medium and incorporates it to acquire an altered genotype that is heritable. Bacteria are the only organisms (except yeast) known to transform in nature although cells of higher organisms including that of mammals are transformable by artificial techniques. The overall process in naturally transformable organisms may be divided into five common steps: (i) development of competence (ii) binding DNA (iii) integration of DNA (iv) formation of pre-integration complex and (v) integration of DNA into the recipient cell chromosomes or exist as an extra-chromosomal DNA (plasmid).

The competence of a bacterial cell for transformation could be artificially induced by exposing cells to calcium chloride prior to the addition of DNA. It involves the exposure of growing bacterial cells to a hypotonic solution of calcium chloride at 0°C, causing cells to swell (spheroplast formation). DNA added to the transformation mixture forms a DNase resistant complex of hydroxyl-calcium phosphate that adheres to the cell surface. This complex can be taken up by the cell during a brief 42°C heat pulse. After a few hours of growth in rich medium to allow the spheroplast to recover and the transformed genes to be expressed, transformants can be isolated by plating on selective medium. For example, cells transformed by pBR322 DNA can be selected on medium containing the antibiotics ampicillin and tetracycline. With minor modifications, this basic transformation protocol has been made to work for a variety of bacteria that do not take up exogenous DNA normally. It is suggested further that magnesium ions may play an important role in the stability of the DNA during uptake, and many transformation methods now include $MgCl_2$ during the treatment of the bacteria (Figure 14.36).

14.14.2.1 Materials

Conical flasks, test tubes, centrifuge, pipettes, incubator, Petri plates, autoclave, eppendorf tubes, colorimeter, tryptone, yeast extract, sodium chloride, NaOH, distilled water, $CaCl_2$, DNA, Tris EDTA buffer, ampicillin, and tetracycline, distilled water, *E. coli,* plasmid DNA (pBR 322).

LB medium g/L in distilled water: Casein enzymic hydrolysate 10, Yeast extract 5, Sodium chloride 5, Final pH (at 25°C) 7.0 ± 0.2

Tris-EDTA buffer: 10 mM Tris 1 mM EDTA dissolved in distilled water.

14.14.2.2 Protocol

1. From the overnight grown culture take 0.1 mL and inoculate into the 25 mL of fresh LB medium in 100 mL flask.
2. Allow it to grow till the O.D reaches 0.5 - 0.6 at 540 nm.
3. Chill the culture by keeping it in ice for 30 min.

4. Spin the culture at 5000 rpm for 10 min and discard the supernatant.
5. Suspend the pellet in 5 mL of 50 mM or 10 mM $CaCl_2$.
6. Keep it in the ice (0°C) for 30-60 min.
7. Spin down the culture and discard the supernatant.
8. Suspend the pellet in 1 mL of 10 mM $CaC1_2$, take 0.2 mL of culture and add 50 mg to 500 mg of DNA in Tris-EDTA buffer and keep in the ice for 30 min.
9. In the control experiment, to 0.2 mL of the culture, add only buffer without DNA.
10. Give a heat shock at 42°C for 2 min and transfer it to ice and keep again for 5 min.
11. Now add 0.8 mL of sterile LB broth and incubate at 37°C for 45 min. For the cells to multiply and express the marker.
12. Take an aliquot of 0.1 mL each of these cells and plate on LB agar containing Ampicillin (50 mg/mL) or Tetracycline (15 mg/mL) or both and incubate the plates at 37°C for overnight.
13. Also, plate 0.1 mL of the control culture (step 9) on the selective plate and incubate at 37°C along with plates of step 12 (after incubation no colonies should appear in the control plates).

14.14.3 Demonstration of Genetic Recombination in Bacteria by Conjugation

In this experiment, an attempt has been made to demonstrate the existence of the phenomenon of genetic recombination by using two auxotrophs of *E. coli*. An auxotroph is a bacterial mutant that requires one or more growth factors that the wild strain, or prototroph, can synthesize. Auxotrophs are produced in the laboratory by subjecting prototrophs to mutagenic agents, such as ultraviolet irradiation or mitomycin C. One of the auxotrophs used in this experiment is able to synthesize methionine, but not threonine, from a minimal medium containing only glucose, ammonia, and inorganic salts. In other words, this auxotroph has lost the gene that exists in the prototroph that would enable it to synthesize threonine in the presence of a minimal medium. Thus, it cannot grow on such a medium. Its genetic composition is designated as thr^-met^+. The other auxotroph in this experiment is able to synthesize threonine, but not methionine, and also is unable to grow on the minimal medium. It is designated as thr^+met^-. If these two auxotrophs are able to conjugate to produce a strain that is thr^+met^+ to have a genetic combination, like the original prototroph, that will grow on the minimal medium. The experiment demonstrates that the conjugation can occur to bring such a genetic recombination of genes. The donor in this experiment is thr^+met^-. The recipient is thr^- met^+.

14.14.3.1 Materials

12 hours actively growing (10^8 cells per mL) broth of auxotroph strain of *E. coli* (thr^+ met^-) Male (M) strain 1, 12 hours actively growing (10^8 cells per mL) broth of auxotroph strain of *E. coli* (thr^- met^+) Female (F) strain 2, Two Petri plate of Trypticase soy agar

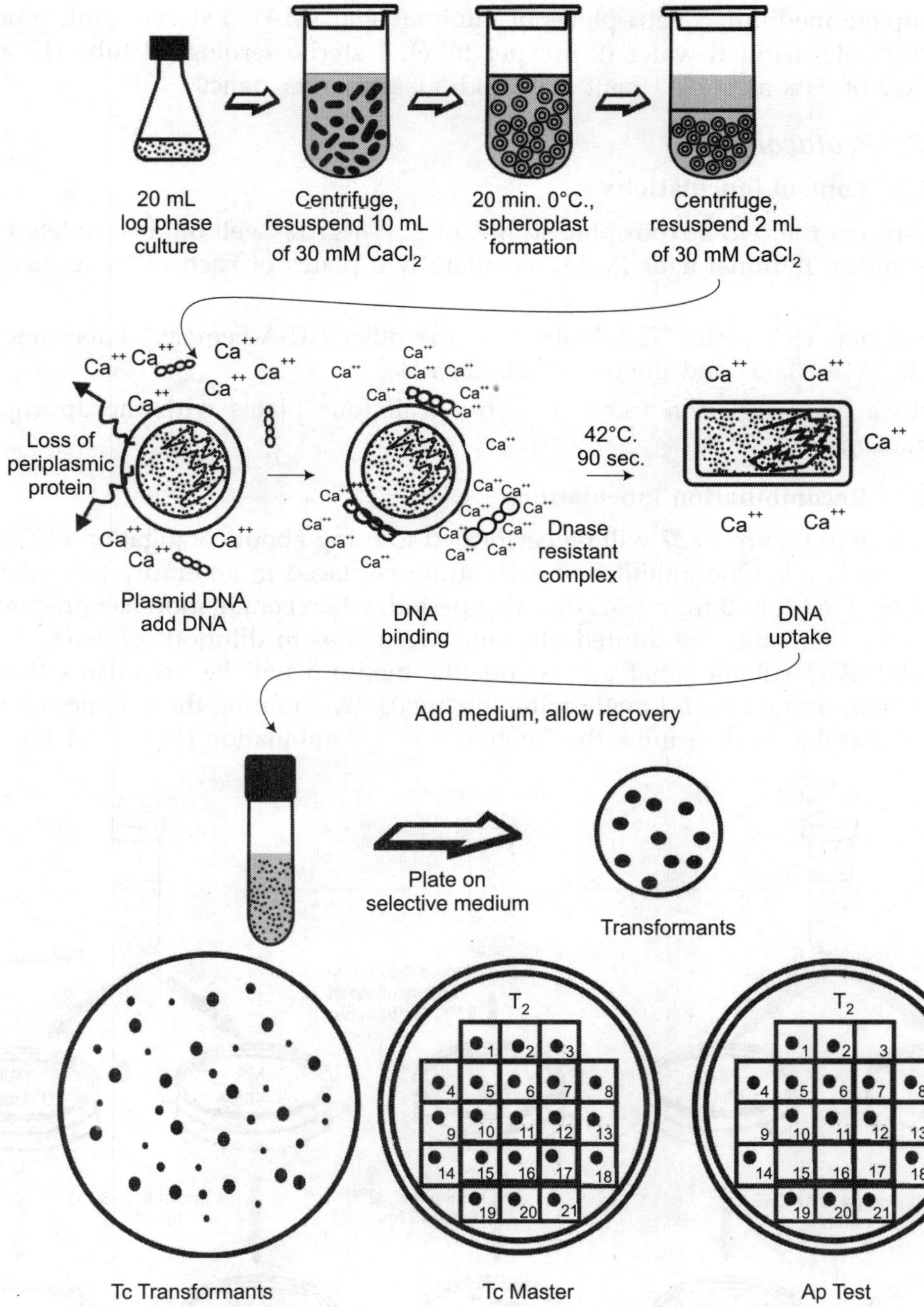

Figure 14.36 Diagrammatic representation of the $CaCl_2$-induced transformation procedure for *E. coli*.

(TSA) complete medium, 5 Petri plates of minimal agar (MA), 3 sterile 1 mL pipettes, 3 tubes of sterile distilled water (9 mL per tube), 1 sterile serological tube (13 × 100 mm), beaker of 95% alcohol, 1 bent glass rod, glass marker pencil.

14.14.3.2 Protocol

14.14.3.2.1 Control inoculations

To establish that the two auxotrophic strains of *E. coli* grow well on a complete (TSA) but not at all on minimal agar (MA), inoculate two plates of each of these media as follows:

1. Label one TSA plate "TSA-Male" and the other "TSA-Female." Label one MA plate "MA-Male" and the other "MA-Female."
2. With a good isolation technique, streak all four plates with the appropriate organisms.

14.14.3.2.2 Recombination inoculations

The procedure in Figure 14.37 will be performed to bring about conjugation of the two auxotrophs of *E. coli*. One millilitre of each strain is placed in an empty test tube and is allowed to stand for 30 minutes. After this period, when conjugation has presumably occurred, the conjugants are diluted out onto MA plates in dilutions of 1:100, 1:1000, and 1:10,000. Any colonies that appear on this medium will be organisms that can synthesize both threonine and methionine (thr^+met^+). By counting the colonies on these plates, it is possible to determine the frequency of recombination (Figure 14.37).

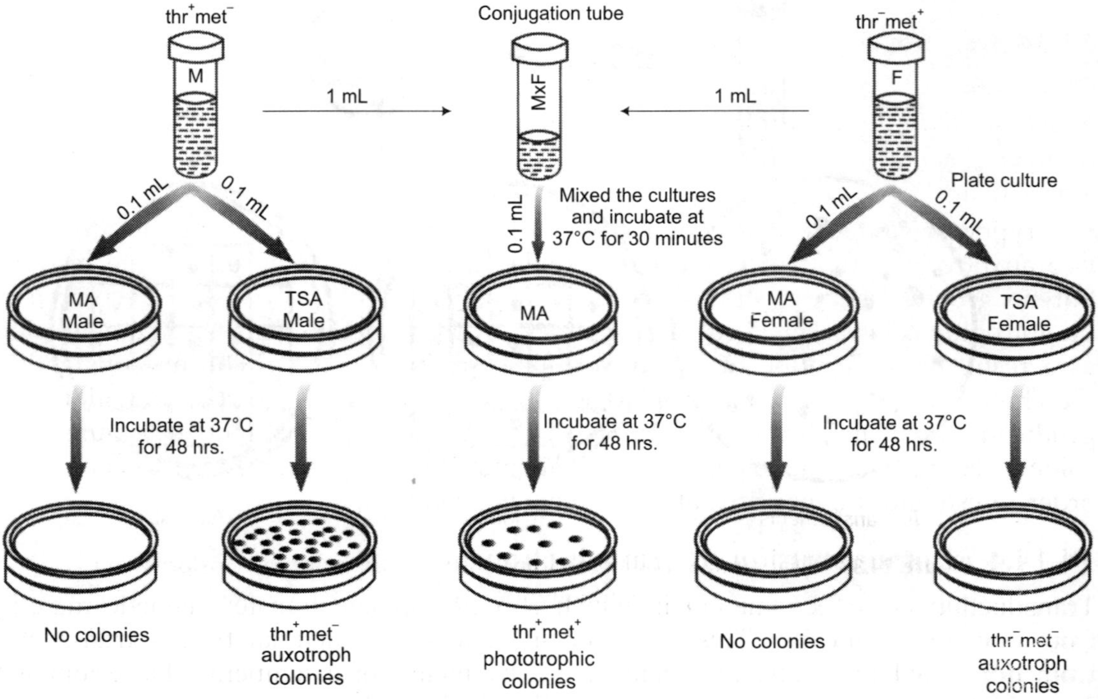

Figure 14.37 Demonstrating conjugation in bacteria using single auxotrophs of *E. coli*.

1. Label a sterile serological tube "M × F." This will be used for conjugation of the two strains of *E. coli*.
2. With separate sterile 1 mL pipettes, aseptically transfer 1 mL from each culture into a sterile 13 × 100 mm test tubes labeled as conjugation tube. Let it stand for 30 minutes.
3. Mix the two cultures by gently rotating the tube between the palms of hands.
4. While conjugation is taking place, label the three water blanks: I, II, and III. Also, label the three plates of minimal agar: I-1:100, II-1:1000, and III-1:10,000.
5. With a sterile pipette, transfer 1 mL from the conjugation tube into water blank I. Mix the contents in tube I by drawing the dilution up into the pipette three times and discharging slowly into the tube.
6. With the same pipette, transfer 0.1 mL from tube I to plate I. Do not spread at this time.
7. Still using the same pipette, transfer 1 mL from tube I to tube II and mix contents in tube II as above. Now transfer 0.1 mL from tube II to plate II. Do not spread at this time.
8. Still using the same pipette, transfer 1 mL from tube II to tube III and repeat all above procedures to inoculate plate III from tube III with 0.1 mL of diluted organisms.
9. With a sterile bent glass rod, spread the organisms over each plate. Be sure to sterilize the rod before spreading organism on each plate.
10. Incubate the two TSA and five MA plates, inverted, for 2 to 4 days at 30°C.

14.14.3.3 Results and observations

Observe the two TSA and two minimal agar control plates inoculated with the individual *E. coli* strain as well as minimal plate inoculated with the mixed culture. Count the number of colonies on all the five plates and record the results. Examine the control streak plates.

Trypticase soy agar (TSA) control plates are expected to exhibit the growth of the two auxotrophic strain of *E. coli* (i.e. $thr^{+}met^{-}$ and $thr^{-}met^{+}$) on a complete medium but no growth of these strains on the minimal agar (control) plates. Recombinant phototrophic colonies of *E. coli* are observed on the experimental minimal agar plate as a result of conjugation of two auxotrophic strains of *E. coli*. During conjugation, the chromosomes of two auxotrophs get associated and undergo recombination thus producing a recombinant strain with genetic composition of thr^{+} met^{+}. The number of colonies counted on the experimental minimal agar plate be compared to controls in order to estimate the per cent of recombinants produced.

14.14.4 Demonstration of Transduction

Transduction is a phenomenon in which a bacteriophage transfers genetic material from one cell to another. This is a process in which a bacterial trait is transmitted from donor cell to a recipient bacterial cell by means of a temperate bacteriophage. Transducing phage may originate from cells which are lysed or may originate as the prophage particle from lysogenized cells following induction. Transduction system is of two types. Generalized and specialized. In the former, the phage may exchange genetic material at any site on the bacterial genome, and in the latter the phage shows specificity for a single site on the bacterial genome.

14.14.4.1 Materials

Test tubes, cooling centrifuge, conical flasks, pH meter, pipettes, Petri-plates, incubator, tryptone, yeast extract, sodium chloride, NaOH, $MgSO_4.7H_2O$, $CaCl_2.2H_2O$, HCl, Na_2HPO_4, KH_2PO_4, NH_4Cl, agar, distilled water, bacterial culture of *E. coli* RR_1 (Pro^-); *E. coli* (C 600, Pro^+) P1 phage lysate.

LB agar plates (g/ L): Tryptone 10 g, agar 10 g, yeast extract 5 g, NaCl 10 g, pH adjusted to 7.2 – 7.4 with 10 N NaOH. Agar medium is autoclaved and cooled medium to pour into Petri plates. Soft agar plates are prepared with 0.5% agar.

Minimal agar medium (M63) Na_2HPO_4 5.8 g, Agar 15 g, KH_2PO_4 3g, NaCl 0.5 g, NH_4Cl 1 g.

Dissolve the above chemicals in one litre of distilled water and pH is adjusted to 7.0 and sterilized.

The following sterilized ingredients are added just before use to complete the medium: 10% glucose or glycerol 50 mL, 1 M $MgSO_4.7H_2O$ 1 mL, 0.01 M $FeCl_3$ 1 mL, 1.5% agar and poured to plates.

MC buffer $MgSO_4.7H_2O$ 0.1 M, $CaCl_2.2H_2O$ 0.005.

TM buffer 10 mM Tris-HCl (pH 8) containing 10 mM $MgSO_4.7H_2O$

14.14.4.2 Protocol

1. Grow the *E.coli* RR_1 Pro-culture in LB medium (5 mL) overnight.
2. Centrifuge and pellets are suspended in 5 mL of MC buffer.
3. Aerate the suspension at 37°C for 15 min.
4. Add 0.8 mL of the suspended cells to each of '6' test tubes and label them .
5. Add to each tube 0.2 mL of diluted (10^{-1} to P1 lysate obtained from *E. coil* C 600 Pro^+. Add only buffer in one of the tube as control which has no lysate.
6. Allow for pre-absorption of phage particles on cells by incubating the tubes at 37°C for 20 min.
7. An aliquot (0.1 mL) of the above is mixed with 10 mL of top agar kept at 45°C.
8. Mix well and pour the mixture on LB bottom agar plates.
9. Incubate at 37°C for 24-48 h.
10. Observe the appearance of colonies on LB agar.

14.14.4.3 Results

Note the colonies on LB agar plates. Then replicate the colonies on minimal agar and score for prototroph. Calculate the transduction frequency, express it in terms of the number of prototrophs (transductant) per recipient cell.

14.15 FUNGAL GENETICS

Neurospora is an Ascomycetes fungus and is commonly called bread mold. The fungus is being widely used as an eukaryotic model to unravel the questions in eukaryotic genetics. It is haploid eukaryotes with seven chromosomes. It can be grown on a chemically defined medium and has a brief morphologically distinct asexual life cycle producing billions of conidia with identical genotype. Controlled genetic analysis is possible in the sexual phase of its life cycle using the mating type alleles A and a. The fusion of nuclei of the two mating type result in the formation of a diploid nucleus that undergoes meiosis producing eight ascospores after mitotic division of each of the meiotic products (Figure 14.38).

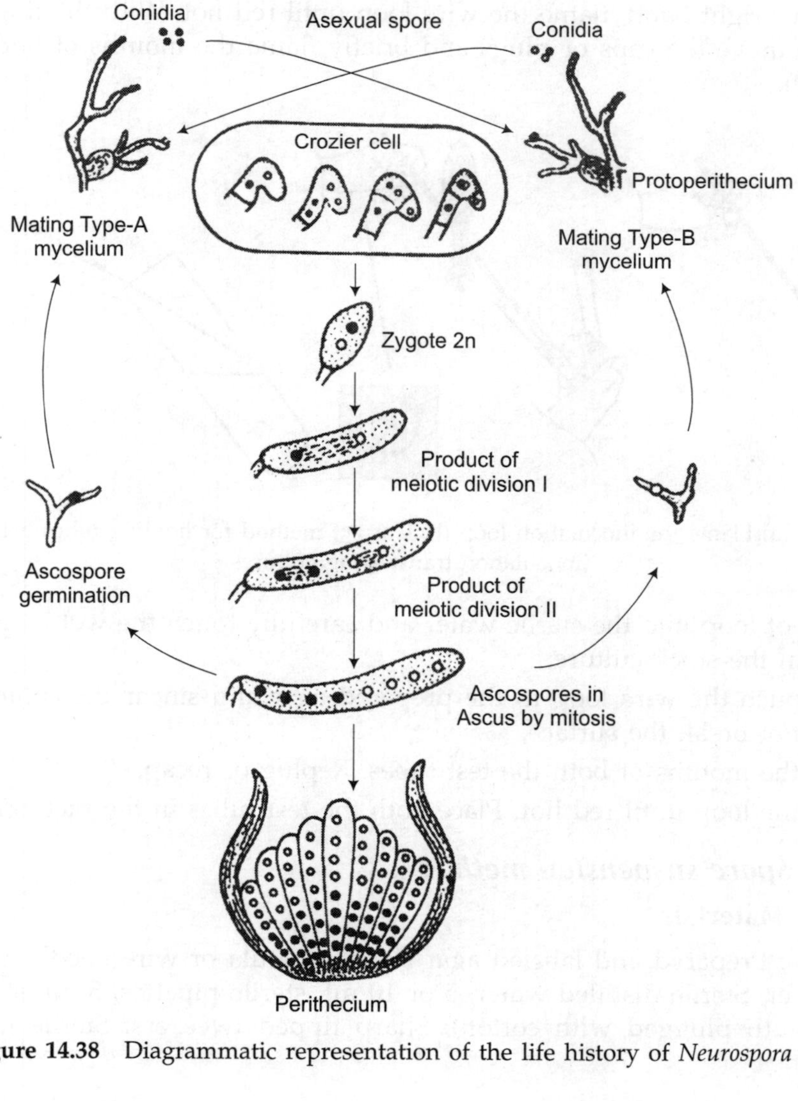

Figure 14.38 Diagrammatic representation of the life history of *Neurospora* sp.

14.15.1 Harvesting of *Neurospora*

14.15.1.1 Flamed-loop method

14.15.1.1.1 Materials

Stock culture; Prepared and labeled agar slants; Wire loops; Bunsen burner; Sterile distilled water.

14.15.1.1.2 Protocol

Place the stock cultures and the prepared agar slants on your left and an empty rack on your right.

Hold one stock culture and one agar slant in your left hand. Loosen caps and plugs.

Using your right hand, flame the wire loop until red hot. With the fingers of your right hand remove the caps or plugs and briefly flame the mouths of both test tubes (Figure 14.39).

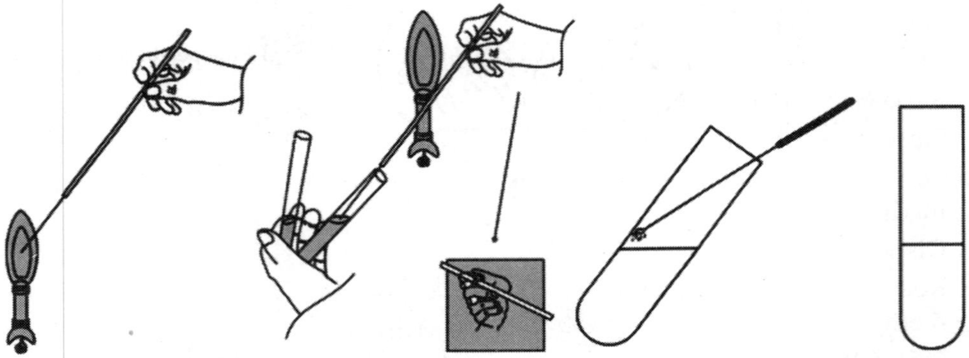

Figure 14.39 (a) Flame the inoculation loop (b) Correct method for holding tubes before and after inoculation transferring cultures.

Dip the hot loop into the sterile water and carefully touch the wet loop to some of the conidia in the stock culture.

Lightly touch the wire loop to the prepared agar and smear the conidia over the surface. Do not break the surface.

Reflame the mouths of both the test tubes. Replug or recap.

Reflame the loop until red hot. Place both the test tubes in the rack on your right.

14.15.1.2 Spore suspension method

14.15.1.2.1 Materials

Stock culture; Prepared and labeled agar slants; Spatula or wire needle; 95% alcohol; Bunsen burner; Sterile distilled water, 5 or 10 mL sterile pipettes; 5 mL sterile reverse pipettes (mouth plugged with cotton); Sharp tipped tweezers; Sterile test tubes or dropper bottles.

Protocol

1. Place the tip of the spatula or wire needle and the tweezers into the alcohol for sterilization.
2. Pour a small amount of sterile water into the flasks (or test tubes) of stock cultures, wetting as much of the conidia as possible. Rotate the stock culture to wet the conidia.
3. Flame the spatula to burn off the alcohol and then cool it by immersing the tip in the water in the flask.
4. Use the spatula to break up the conidia, remove the spatula and place in the alcohol.
5. Place the reverse or plugged end of the reverse pipette into the flask and withdraw the conidial suspension.
6. Remove the pipette from the flask.
7. Flame the tweezers and remove the plug from the pipette with the help of tweezers and place both in the alcohol.
8. Release the liquid suspension into a clean and sterile test tubes or dropper bottle.
9. Repeat the procedure for each culture (Figure 14.40).
10. Store in a cool place until needed for inoculation.
11. To inoculate, add 1 or 2 drops of well shaken conidial suspension to prepared media.
12. Disperse the conidia by rotating the prepared media.
13. Keep the inoculated media (by spore suspension method or by flamed-loop method) in an incubator at 30 to 34°C until the desired amount of growth has been achieved.
14. Autoclave samples for 30 minutes at 20 psi to kill spores and mycelia when the experiment is completed.

[**Note:** *Methods of study of Sordaria, Aspergillus and Saccharomyces (Yeast) are similar to Neurospora experiments. They all belong to the same group of fungus, Ascomycetes as Neurospora having similar life cycles.]*

14.16 TO STUDY THE ORDERED (TETRAD) AND RANDOM ASCOSPORE ANALYSIS IN *NEUROSPORA* sp.

The sexual spores (ascospores) are produced from the fusion of two nuclei of opposite mating types through exchange of conidia between the two mating types. Fertilization is followed by immediate meiotic division and the four products of meiosis undergo one mitotic division resulting in 4 pairs of ascospores which are linearly arranged in an ascus. The members of each pair are genetically identical and each pair represents one meiotic product. The position of ascospores in the ascus reflects the behaviour of the chromosomes during meiosis. The asci are enclosed in a cup-like structure called the perithecium. The ripened ascospores escape through an opening called the ostiole. The formation of the eight ascospores is represented as:

Zygote → meiosis → 4 products → mitosis → 8 ascospores

Ascospore analysis in *Neurospora* is of three distinct type (i) ordered or linear tetrad analysis, (ii) unordered or non-linear tetrad analysis, and (iii) random spore analysis. Ordered tetrad analysis is based upon the first and second division segregation of alleles as detected by the spore position in the ascus.

First division segregation for any particular gene indicates no crossover between that gene and its respective centromere. This results in 4: 4 ascospore patterns. If a gene for black colour having an allele causing albino spore are present on two homologous chromosomes and there is no crossover at meiosis 1, the pattern of spore is top four black and bottom four white or top four white and bottom four black, depending upon the pole to which the chromosome with the coloured allele passes at first anaphase. The chances for the two arrangements are the same, so the two cases occur with equal frequency (Figure 14.41 (a) and (b)). Second division segregation reflects a crossover between a particular gene and its centromere, thus producing a variety of ascospore arrangements (Figure 14.41 (c), (d), (e) and (f)). If a crossover between the locus of the spore colour factor and the centromere occurs, both chromosomes have one chromatid with one allele and the other chromatid with the other allele. Depending upon the orientation at second anaphase, one of the 4 types is produced. All these types have equal probability.

For an ordered tetrad analysis, the perithecia (of 10 to12 days old cross) are broken open. The asci are separated, and the ascospores are carefully leased out of the ascus. Each ascosopores, in order, is very carefully implanted on an agar block; (placed in

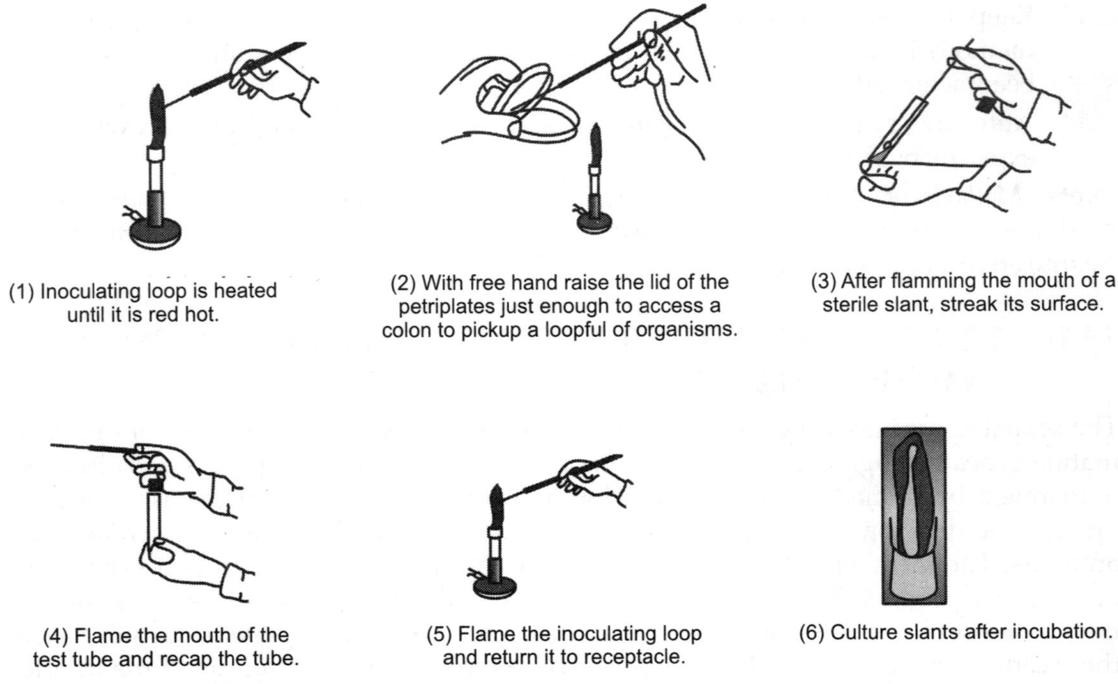

Figure 14.40 Demonstration of complete inoculation procedure.

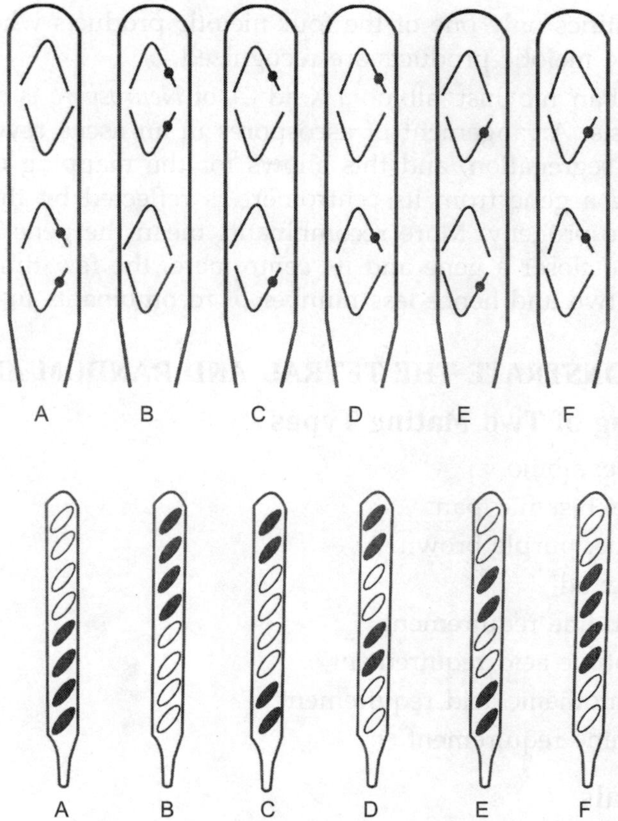

Figure 14.41 Linearly arranged second anaphase of meiosis with linearly arranged ascospores of Neurospora. A, B First division segregation, C, D, E, F different arrangements of ascospores resulting from second division segregation of chromosomes.

a tube) which is then placed at 60°C for 30 minutes (heat shocked). This initiates germination and colony formation.

Tetrad analysis reveals that crossing-over takes place at the four-strand stage of meiosis. It also has helped in determining the gene order on the chromosomes and the distribution of the homologous chromosomes in meiosis I and II.

Ordered and unordered tetrad analysis does not differ significantly. Mathematical analysis is necessary for unordered tetrad since the type of segregation for any given ascus cannot be determined. If tetrad analysis is not possible, random single ascospores are isolated for study. Perithecia of one month old cross eject the ascospores from the asci on to the wall of the cross tube. Ascospores of many perithecia can be pooled, diluted and plated on the appropriate medium and placed in separate tubes for growth and subsequent analysis. The class-data can be pooled and analysed statistically. Such analysis is used mainly for calculating the recombination values allowing the study of linkage relationships between genes. A two-factor cross or a three-factor cross is used for such linkage studies. This type of analysis has limited application where genetic analysis is needed; which is possible in the case of an ordered tetrad analysis. Random spore analysis identifies only one of the four meiotic products whereas with ordered tetrad analysis, all 4 meiotic products are recognized.

Here, the cross Pan⁻ nic⁻ hist⁻ albino a X ad⁻ A of *Neurospora* is described for tetrad and random analysis. Arrangement of ascospores in an ascus reveals the occurrence of second-division segregation, and this allows for the mapping of the centromeres. Distance separating a gene from its centromere is reflected by the frequency of the recombinants in the progeny. More recombinants mean the gene and its centromere are farther part, the closer a gene and its centromere, the few times a crossover will occur between the two and hence less number of recombinants in the progeny.

14.17 TO DEMONSTRATE THE TETRAD AND RANDOM ANALYSIS

14.17.1 Crossing of Two Mating Types

Strain a: Phenotype: albino
Genotype: $hist^-nic^-pan^-$

Strain A: Phenotype: purple brown
Genotype: ad^-
$hist^-$: histidine requirement
nic^-: nicotinic acid requirement
pan^-: pantothenic acid requirement
ad^-: adenine requirement

14.17.2 Materials

Inoculating loops, absolute alcohol, sterlized water, crossing tube (25 × 150 mm) containing a strip of Whatman No. 1 filter paper (2.5 × 7.5 cm), one tube containing Neurospora strain.

14.17.2.1 Protocol

1. Heat the loop until red hot or dip in alcohol, then flame to sterilize.
2. Cool the loop (if it is heated), dip it briefly into the tube of sterile water. Remove the cap of the cross tube containing strain A.
3. Lightly brush the loop against the conidia of strain A and gently remove the loop and replace the cap.
4. Place the loop with the conidia into the cross tube and swirl the loop in the medium to disperse the conidia.
5. Flame the loop until red hot, dip it into alcohol and flame again.
6. Repeat step 2 to 5 for strain a.
7. Store the crossing tube at 25°C.

14.18 TETRAD ANALYSIS

After 12 days of the cross.

14.18.1 Materials

Dissecting microscope, wire loop, razor blade, glass slide, 4% agar in Petri dish, agar block 2.5 by 5 cm cut when needed from the 4% agar, 24 tubes each containing 2 mL of solidified Vogel's medium N plus 2% glucose plus supplements pan, nic, hist and ad; cross tube of cross of section A.

14.18.2 Protocol

1. Remove a plump, dark, fertile perithecium and place it on an agar block. Add 1 or 2 drops of household diluted bleach to kill contaminant asexual conidia.
2. With the sharp needles, slice open the perithecium under the dissecting microscope. Perithecium can also be squeezed with the tips of forceps.
3. After squeezing, a bunch of asci, looking like a bunch of bananas comes out. It normally has 10 asci with 8 spores inside.
4. Separate the asci carefully with needles, scissors etc.
5. Three asci can be selected (one may select more) and transfer them (by picking with the tip of the needle) to fresh agar block on the glass slide (Figure 14.42).
6. When the block is dry, squeeze the ascospores from the asci, one at a time, and move the ascospore to clear areas of the agar block. Using a sterile forceps or needle, pick up the agar under the ascospore and place into individual tubes with the medium. Label the tubes and keep them in order.
7. Heat shock the 24 culture (3 asci × 8 spores = 24) in 60°C water bath for 30 minutes. Then place them at 25°C for 48 hours so that enough conidia should be formed.

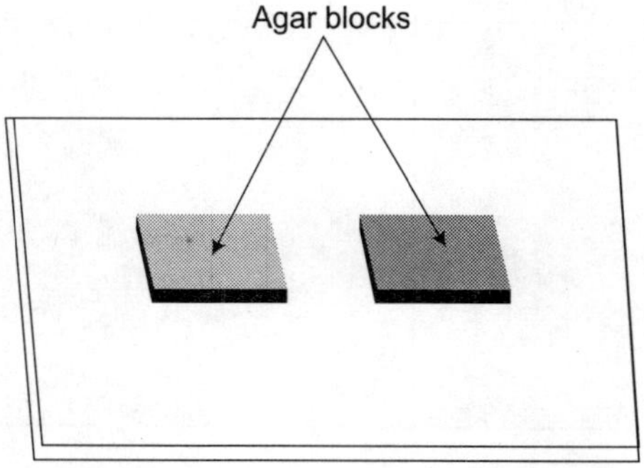

Figure 14.42 Agar blocks on a side.

14.19 PIGMENTATION AND NUTRITIONAL REQUIREMENT ANALYSIS

14.19.1 Materials

One Petri plate each of the following special media:

Vogel's minimal + 1% sorbose + 0.2% sucrose + one of the following supplement combination:

hist, nic, ad; hist, nic, pan; hist, ad, pan; ad, pan, nic.

(Histidine hydrochloride 200 mg/L, adenine sulphate 200 mg/L, nicotinamide 20 mg/L, pantothenic acid 20 mg/L, adjust pH of all media to 5 to 6 and autoclave for 8 minutes).

24 separate cultures of step 7 section B, sterile inoculating needle

14.19.2 Protocol

Observe each of the 24 cultures for pigment formation.

The expected pigment types are:

Albino – parental; purple – brown – parental; albino – purple – recombinant; wild - recombinant; wild type pigment is orange. Orange + purple = brown colour; white + purple = albino-purple colour.

Prepare results in the form of Table 14.7.

Table 14.7 Tetrad analysis (pigment nutritional requirement)

Ascus	*Pigment*	*Spore Position*	*Growth or colony on*				*Cenetype*
			ad	*pan*	*hist*	*nic*	
1.	1						
	2						
	3						
	4						
	5						
	6						
	7						
	8						
2.	1						
	2						
	3						
	4						
	5						
	6						
	7						
	8						
3.	1						
	2						
	3						
	4						
	5						
	6						
	7						
	8						

Place Petri plates of special media on the corresponding places shown in Figure 14.43.

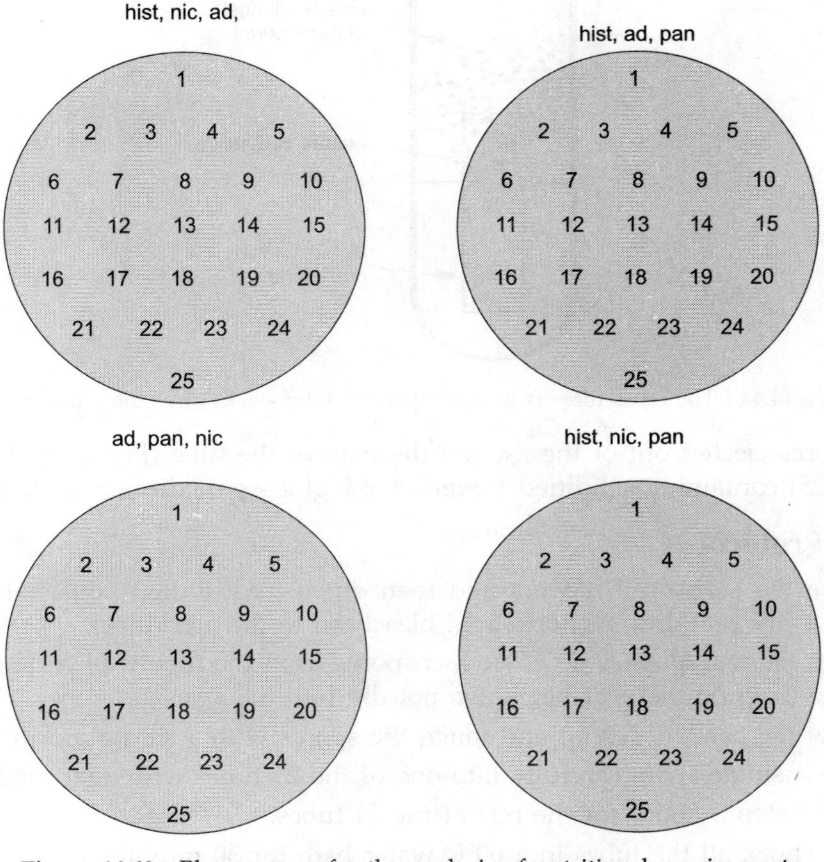

Figure 14.43 Plate pattern for the analysis of nutritional requirements.

Take tube number 1 from 24 tubes and place conidia from this tube on spot number 1 of all the four plates. Repeat this for the rest of the 23 tubes placing conidia on corresponding points.

Incubate plates for 48 hours at 30°C. Score growth as positive (+) or negative (–) and record data in the form of Table 14.8.

14.20 RANDOM SPORE ANALYSIS

Should be done one month after the cross.

14.20.1 Materials

Agar block on slides as shown in Figure 14.44.

Sterile tube containing 1 mL of sterile water, crossing tube (25 × 150 mm) containing the original cross of part A, with a black dust on the walls. These black spores are

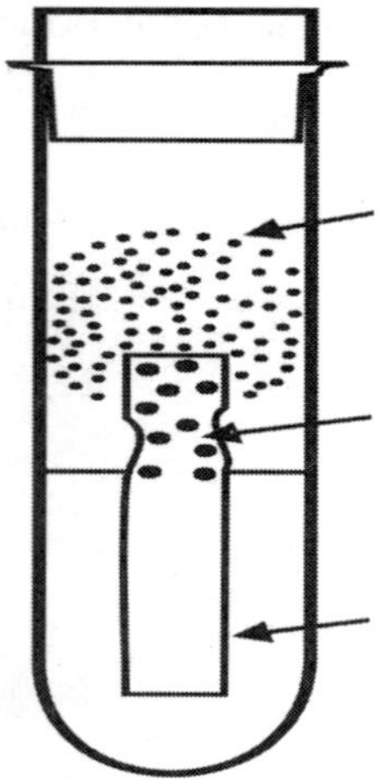

Figure 14.44 The cross tube as it looks three to four weeks after the cross is made.

the ascospores ejected out of the asci on the wall of the tube (Figure 14.44). 25 tubes (label 1 to 25) containing solidified Vogel's + 2% glucose medium + hist, nic, ad, pan.

14.20.2 Protocol

1. Flame the loop until red hot and then dip it into diluted household bleach to cool. Place one drop of household bleach on to the agar block.
2. Using the wet loop, scrap some ascospores from the tube wall and spread them in the drop on the agar block. Do not dig into the agar.
3. Allow the agar to dry up and touch the spores with a sterile picking needle.
4. Place a single spore carefully into one of the 25 tubes with agar medium.
5. Repeat step 3 and 4 for the rest of the 24 tubes.
6. Heat shock all the tubes in a 60°C water bath for 30 minutes.
7. Place tubes at 25°C for 1 week.

14.20.3 Scoring Tubes for Pigmentation

Observe the tubes of step 7 of the preceding section and record the observations in the form of Table 14.8.

Table 14.8 Random spore-pigmentation analysis

Colour	*Number*
Anbino	
Purple-brown	
Albino-purple	
Wild-type	
No growth	
Total	

14.21 TEST FOR NUTRITIONAL REQUIREMENT

14.21.1 Materials

One plate each of the four special media listed in section 14.22.

14.21.2 Protocol

1. Place the first Petri plate as per Figure 14.43.
2. With a sterile and cool needle, pick a few conidia from tube 1 and place them on to the agar on spot 1. Small amount of conidia should be placed.
3. Repeat steps 1 and 2 for the other three Petri plates.
4. Repeat steps 1, 2 and 3 for the rest of the 24 tubes inoculating all corresponding spots in the four Petri plates. Needle should be sterilized each time it is placed in the tube.
5. Leave the four plates at 30°C for two days.

14.21.3 Scoring for Nutritional Requirement

Observe the four plates for growth and record the observations in the form of Table 14.9. Write symbol + (positive) for a colony and – (negative) for no growth or poor growth.

Table 14.9 Random score analysis (nutritional requirement)

Ascus	*Pigment*	*Spore Position*	*Growth or colony on*				*Cenetype*
			ad	*pan*	*hist*	*nic*	
1							
2							
3							
4							
5							
6							
7							
•							
•							
•							
•							
•							
•							
25							

14.21.4 Analysis

Tetrad analysis from Table 14.7.

1. Are asci 1, 2 and 3?
   (i) A tetratype (four genotypes present)?

(ii) A parental ditype (two genotypes, both parental)?

(iii) Non-parental ditype (two genotypes, both recombinants)?

Determine the nature of each ascus based upon the above three types.

2. Are the parental ditypes equal number of non-parental ditypes?
3. Make the diagrams for the chromatid exchanges that must have occurred to result in the type of asci obtained in this experiment.
4. If any of the ascospores failed to germinate, infer their genotypes.
5. Did ascus 1 or 2 or 3 show second-division segregation for any of the genetic markers that are considered in this cross? Calculate the centromere distance for each marker in the cross using the following formula:

$$\%\ \text{cross over} = \frac{\text{Second-division segregation frequency}}{} \times 100$$

$$1\% = 1\ \text{map unit}$$

Random spore analysis from Table 14.9.

Try to find out the following:

1. Number of genotypic classes present in the asci.
2. Number of each genotype present.
3. Number of genotypic classes missing.
4. Genotypes of each class missing.
5. The most frequent and the rarest genotype occurring in the asci.
6. Since the original cross was pan⁻ nic⁻ hist⁻ al a X ad⁻ A, taking genes pair-wise, determine the recombination frequency between them. (For example, for nic and hist, the parental types are nic, hist ad^{++}). The recombination frequency can be calculated as:

$$\%\ \text{recombinants} = \frac{(\text{number of nic}^{+}) + (\text{number of hist}^{+})}{\text{Total number of ascospores}} \times 100$$

Percentage recombination can be calculated similarly for other gene pairs. Enter data in the form of Table 14.10.

7. Find out the genes showing maximum segregation (do not show linkage).
8. Also find out the genes that move together as a unit (show linkage).

Table 14.10 Linkage (Percentage crossing over between)

	pan	*nic*	*hist*	*ad*
pan				
nic				
hist				
ad				

14.22 SUGGESTED MEDIA FOR *NEUROSPORA, SORDARIA* AND *ASPERGILLUS* EXPERIMENTS

14.22.1 Vogel's Medium

Trace elements solution: Distilled water 95 mL, citric acid.H_2O 5 g, $ZnSO_4.7H_2O$ 5 g, $Fe(NH_4)_2(SO_4)_2.6H_2O$ 1 g, $CuSO_4.5H_2O$ 0.25 g, $MnSO_4.H_2O$ 0.05 g, H_3BO_3 0.05 g, $Na_2MoO_4.2H_2O$ 0.05 g, Preservative (Chloroform) 0.5 mL.

14.22.2 Biotin Solution

Distilled water 500 mL, 95% absolute ethyl alcohol 500 mL, biotin 10 mg.

14.22.3 Medium N Stock Solution (50×)

Sodium citrate.$2H_2O$ 125 g, KH_2PO_4 anhydrous 250 g, NH_4NO_3 anhydrous 100 g, $MgSO_4.2H_2O$ 10 g, $CaCl_2.2H_2O$ 5 g, Trace element solution 5 mL, Biotin solution (0.1 mg/mL) 2.5 mL.

Add the ingredients successively. Each should dissolve before the next is added. Add 3 mL of chloroform as a preservative.

14.22.4 Medium N Final Solution

To prepare 1 litre of medium N

Medium N stock solution 20 mL; sucrose 20 g; Difco bacto agar 20 g; distilled water 980 mL. Autoclave the solution for 15 minutes at 20 psi. Cool to 50°C, pour into test tubes and prepare slants. Allow the agar to harden.

14.22.5 Wastegaard-Mitchell Crossing Medium

Trace element solution

$Na_2B_4O_7.10H_2O$ 88.4 mg, $CuCl_2.2H_2O$ 268 mg, $FeCl_3.6H_2O$ 970 mg, $MnCl_2.4H_2O$ 72 mg, $(NH_4)_6.6Mo_7O_{24}.4H_2O$ 37 mg, $ZnCl_2$ 42 mg, distilled water to 1 litre.

14.22.6 Biotin Solution

95 % absolute ethyl alcohol 500 mL; distilled water 500 mL; biotin 10 mg.

Dissolve the biotin in alcohol, and then add water.

14.22.7 Wastegaard-Mitchell Crossing Medium Stock Solution (20×)

KNO_3 20 g; KH_2PO_4 20 g; $MgSO_4.7H_2O$ 20 g; NaCl 2 g; $CaCl_2$ 2 g; Trace element solution 20 mL; Biotin solution 10 mL; distilled water to 1 litre.

The components are added successively with constant stirring. Do not autoclave or heat. To preserve, add 2 mL of chloroform.

14.22.8 Crossing Medium Final Solution

To prepare 1 litre of the crossing medium, use the following:

Distilled water 950 mL; stock solution 50 mL; sucrose 15 g; NaOH enough to adjust pH to 6.5; Difco bacto agar 12 g.

Follow the same procedure as given for the preparation of 1 litre of medium N.

14.23 STUDY OF HETEROKARYOSIS IN *NEUROSPORA* FROM THE NUTRITIONALLY DEFICIENT MUTANTS

Classically, genes are studied, using three methods: Recombination, mutation and analysis of gene products. In this experiment, an attempt is made to explain what a gene is by utilizing a technique for gene analysis called heterokaryosis or complementation.

Heterokaryosis is an act of complementation or the act of coming together to make whole or complete. In genetics, this term is used when two mutant genomes are brought together in a common cytoplasm, allowing for growth or an increase in a specific gene function over that of either mutant type. Thus, the complementary action of homologous sites of genetic material results in wild or near-wild phenotype.

The two genomes may come together in the same cell or cytoplasm by several means. These are: the fusion of germ cells or gametes and the formation of dikaryons in fungi.

The terms complementation and recombination are often confused. Recombination is a physical exchange of genetic material between homologous chromosomes resulting in the recombination of genes and physical establishment of gene order. The recombinant individual has the total amount of genetic content equal to the original content, the quality of genetic content being changed by the act of recombination. Complementation is the expression of a wild or near-wild phenotype by an organism containing two mutant genomes. It can be used to determine the order of the mutational sites within a gene and is based on functional rather than recombinational tests. In complementation test two "different mutant strains of the same mating type were when jointly inoculated onto to minimal medium, their hyphae would fuse (but not their nuclei), and the mixing of the cytoplasm takes place. Such a mixed strain is called a heterokaryon. (Hetero=more than one, karyon = nucleus). If two different non-allelic nutritionally requiring auxotrophic mutants, ad^- (adenine requiring) and pro^- (proline requiring), form a heterokaryon. It is described as ad^- pro^+, ad^+ pro^-, because the genome that is ad^- carries the wild type pro^+ allele and the genome pro^- carry the wild-type ad^+ allele. This heterokaryon is capable of growing on a minimal medium because the nuclei that are ad^+ synthesize adenine and the nuclei that are pro^+ synthesize proline. This way, the mutant nuclei are compensated for in a common cytoplasm, and the complementation of the nuclei results in a near-wild growth. Non-allelic mutants that code for specific enzymes necessary in a biochemical pathway also show complementation. For example, two mutant strains of Neurospora requiring riboflavin ($rib\text{-}l^-$ and $rib\text{-}2^-$) form heterokaryon, which grows on minimal medium, since each of the two strains has a different enzyme block in the pathway for the production of riboflavin

$$A \rightarrow B \rightarrow C \rightarrow \text{riboflavin}$$

rib-1 rib-2

The definition of the gene as the smallest unit of mutation, recombination and function is changed as a result of data obtained from intergenic complementation studies.

(Intragenic – within a gene, Intrallelic – between alleles, Intergenic – between different genes)

"Cistron" is now used to describe a functional genetic unit and is formed from the words cis and trans. In the cis configuration, both mutations are positioned on the same chromosome and are obtained as a recombination product from a cross between two mutants. The alternate trans configuration contains both mutations, but are one on each homologue of the diploid.

A________________ B	A________________ b
A________________ b	a________________ B
cis position	trans position

In this diagram, a and b are mutant alleles of gene A and B.

14.23.1 Complementation between Genes or Non-Allelic Mutants

Since separate genes are responsible for the coding of separate enzymes or polypeptides, when two genomes come together in a common cytoplasm the wild type allele of one gene compensates for its homologous mutant gene. A second form of complementation results when each gene is responsible for the coding of a specific polypeptide of one enzyme and both polypeptides interact or polymerize in the cytoplasm to form an enzyme or other protein (haemoglobin). Thus, in either case, each mutant can produce at least one normal product, and a form of complementation occurs when the two genomes come together. The phenomenon of "allelic complementation" (interallelic) has been demonstrated for a number of loci in Neurospora.

In short, the act of coming together of two mutants and compensating for the mutant genes of each other is described as complementation and the formation of hyphae with two different nuclei is known as heterokaryosis.

To obtain stable heterokaryons, it is necessary to inoculate the two strains together on a medium which will permit neither strain to grow by itself. Thus, if one strain requires riboflavin and the other lysin for growth; and the two are inoculated together on minimal medium, neither strain can grow alone, but the two can co-operate and grow together since each can synthesize the compound required by the other. Heterokaryons produced in this way are called "balanced' or 'forced' heterokaryons.

If conidia from a 'balanced' heterokaryon between two auxotrophs (nutrition requiring mutants) are harvested and plated on minimal medium, no colonies will appear because the conidia are uninucleate and the heterokaryon is dissociated into its two components. Synthesis of new heterokaryons is possible if harvested conidia are plated together on minimal medium in high density. Sometimes a diploid colony, heterozygous for the nutritional characters of the two parents appears and hence is prototrophic.

14.23.1.1 Materials

One Petri dish containing complete medium; eight Petri dishes containing minimal medium; long tipped pipettes; container for used pipettes; conidial suspension of two mutants of the hits⁻ nic⁻ pan⁻ = phenotype-albino = A, ad⁻ = phenotype-purple brown = B, 37°C incubator.

14.23.1.2 Protocol

1. Place the rack of conidial suspension on your left and another empty rack on your right.
2. Using your right hand grasp a sterile pipette near its mouth (wide end).
3. Using your left hand pick up the test tube containing suspension A.
4. Remove the cap of the test tube by grasping it between the palm and little finger. Do not place the cap on the desk, it may lead to contamination.
5. Using the pipette remove a small portion of the conidial suspension. Recap the test tube and place the test tube in the container on your right.
6. Using your left hand remove the top of the Petri dish with complete medium long enough for the inoculation.
7. Inoculate the Petri dish at different spots. Place the used pipette in the wash container.
8. Select a fresh sterile pipette and repeat step 3 to 7 for suspension B.
9. After the liquid from the suspension has dried into the agar, carefully place the plates in a 37°C incubator for 24 to 48 hours.
10. After a sufficient incubation period, round patches of mycelia growth and tufts of conidia appear. Place the plates in a cold room.
11. Transfer small pieces of mycelial mat from the sections showing mixed conidia (white and purple) to plates containing minimal medium at various spots.
12. Incubate the plates at 37°C for 72 hours.
13. Select the mycelial sectors showing thick growth, with mixed conidia of both 'types of pigmentation and transfer these to fresh minimal medium plates.
14. Incubate the plates at 37°C for 48 hours.
15. Fast growing mycelia represent heterokaryons.

Suggested Readings

Benson, H.J. 1990. *Microbiological Applications-A laboratory manual in general microbiology*. Iowa, Win. C. Brown Publishers.

Bisen PS.2012. *Laboratory Protocols in appled life sciences*. 1st ed. Boca Raton, CRC Press.

Bisen, P.S. and A. Sharma. 2012. *Introduction to instrumentation in Life Sciences*, 1st ed. Boca Raton, CRC Press.

Bisen, P.S., G. B. K. S. Prasad and M. Debnath. 2012. *Microbes Concepts and Applications*, 699. 1st ed. New York, John Wiley and Sons.

Burrus V, Waldor MK. 2004. Shaping bacterial genomes with integrative and conjugative elements. *Res Microbiol* **155**, 376-386.

Cappuccino, J.G. and N. Sherman, N. 1991. *Microbiology: A Laboratory manual.* The Amsterdam, The Netherlands Benjamin/Climmings Publishing Company,

Chen I, Christie PJ, Dubnau D. 2005. The ins and outs of DNA transfer in bacteria. *Science* 310, 1456-1460.

Dubnau D. 1999. DNA uptake in bacteria. *Annu Rev Microbiol* 53, 217-244.

Hleyn, J., M. Bicknell, and M. Gilstrap. 1995. *Microbiology experiments: A health science perspective.* 354. London., Wm. C. Brown Publishers.

Lewin B. 1997. *Genes* VI. Oxford University Press.

Marinšek, L. R., and M. Vodovnik. 2007. The applications of microbes in environmental monitoring *In: The applications of microbes in environmental monitoring. Current Research and Educational Topics and Trends in Applied Microbiology* Edited by A. Méndez-Vilas, ©FORMATEX.

Miller JH. 1992. *A short course in bacterial genetics: a laboratory manual and handbook for Escherichia coli and related bacteria.* Cold Spring Harbor Laboratory, Cold Spring Harbor, N.Y.

Pollack R. A., L. Findlay, W. Mondschein, R. Ronald Modesto. 2004. *Laboratory Exercises in Microbiology,* 280 pages New York, John Wiley and Sons.

Prsecott, L. M., J. P. Harley, and D. A. Klein. 2005. *Microbiology.* 6th Ed. 992. New York, McGraw Hill.

Sambrook J, Russell DW. 2001. *Molecular cloning: a laboratory manual,* 3rd ed. Cold Spring Harbor Laboratory, Cold Spring Harbor, N.Y.

Seeley, H.W., Jr. and P. J. Denmark. 1975. *Microbes in action: A Laboratory manual of microbiology,* 361. Mumbai, India, D.B. Taraporevala Sons & Co. Pvt. Ltd.

Shirling E. B., and D. Gottlieb. 1966. *Methods, classification, identification and description of genera and species. Vol. 2,* Baltimore, The Williams and Wilkins Company.

Thomas CM, Nielsen KM. 2005. Mechanisms of, and barriers to, horizontal gene transfer between bacteria. *Nature Rev Microbiol* **3**, 711-721.

Tortora, G. J., B. R. Funke and C. L. Case. 2009 *Microbiology: An Introduction,* Eighth 10th Ed. 960. San Fransisco, CA, Benjamin Cummings

Tortosa P, Dubnau D. 1999. Competence for transformation: a matter of taste. *Curr Opin Microbiol* **2**, 588-592.

Willets N, Skurray R. 1980. The conjugation system of F-like plasmids. *Ann Rev Genet* **14**, 41-76.

Chapter **15**

Soil Microbiology

15.1 INTRODUCTION

The microbial population in soil is responsible for maintaining the geochemical cycle and ensuring continued fertility with the removal of natural litter from the surface of the earth. Many of the transformations are cyclic systems comprising sequential reactions of compounds containing a common element such as carbon, nitrogen, sulphur and phosphorus, etc. However, it is rare to find even the majority of the steps in a cycle occurring in a particular situation at any one time. Some stages in these systems result in an amelioration of the soil, whereas others decrease its fertility, thus both the encouragement and control of these stages is agriculturally desirable. In some instances, stages in a cycle are mediated only by organisms which are extremely sensitive to environmental conditions, others take place as a result of the action of a wide variety of microorganisms and proceed under a wide range of conditions.

15.2 CARBON CYCLE

Carbon is required for the building of all organic compounds. Carbon in the form of carbon dioxide (CO_2) is obtained from the atmosphere and transformed into a usable organic form by organisms. The reservoirs for the carbon cycle are the atmosphere, where carbon dioxide exists as a free gas, fossil organic deposits (such as oil and coal), and durable organic materials like cellulose. Mineral carbonates, such as limestone, are a significant geological sink for carbon. During the process of carbon fixation, carbon dioxide is taken up from the atmospheric reservoir (or from biocarbonates dissolved in water) by plants, photosynthetic bacteria, and algae and is "fixed" into organic substances. For the biological links, the carbon cycle comes full cycle when carbon is released by either plants and animals as they respire or after life as they decompose. Organisms respire carbon dioxide as a waste product from the breakdown of organic molecules as their cells derive energy from oxidizing the molecules containing "fixed" carbon. The burning of organic material such as wood or fuels also results in the release of carbon dioxide from organic carbon (Figure 15.1).

Quantitatively the most important series of transformation in which soil organisms are concerned is that involving carbon compounds. Organic residues added to the soil contain about 50% carbon which is eventually converted to carbon dioxide as a result of the action of the soil population. The microflora is responsible for the evolution of 95% of the gas produced and animals for only 5%. Carbon dioxide is fixed mainly by

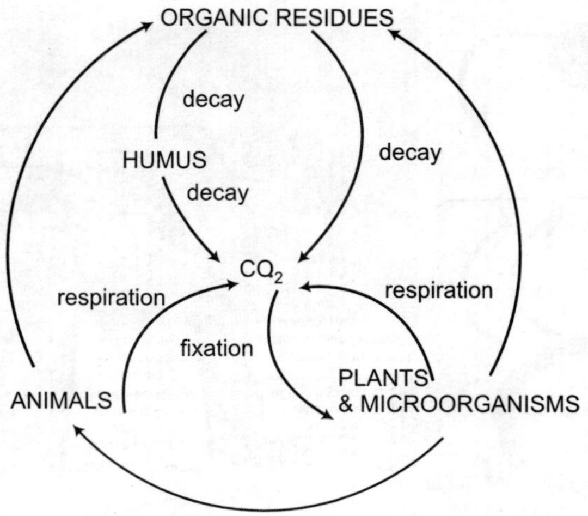

Figure 15.1 The carbon cycle.

green plants but all heterotrophic organisms also use small quantities of this gas and the rate of process is such that it has been estimated that the atmospheric supply would be exhausted in twenty-five to thirty years if it were not replenished in this way. A much simplified explanation of the cycle is given in Figure 15.2.

Photosynthesis is a process that converts carbon dioxide into organic compounds, especially sugars, using the energy from sunlight. Photosynthesis is also the source of the carbon in all the organic compounds within organisms' bodies, the process always begins when energy from light is absorbed by proteins called photosynthetic reaction centres that contain chlorophylls. In plants, these proteins are held inside organelles called chloroplasts, while in bacteria they are embedded in the plasma membrane. Some of the light energy gathered by chlorophylls is stored in the form of adenosine triphosphate (ATP). The rest of the energy is used to remove electrons from a substance such as water. These electrons are then used in the reactions that turn carbon dioxide into organic compounds (Figure 15.2).

15.2.1 Decomposition of Organic Residues

When plant residues are returned to the soil, various organic compounds undergo decomposition. Decomposition is a biological process that includes the physical breakdown and biochemical transformation of complex organic molecules of dead material into simpler organic and inorganic molecules. The continual addition of decaying plant residues to the soil surface contributes to the biological activity and the carbon cycling process in the soil. Breakdown of soil organic matter and root growth and decay also contribute to these processes. Carbon cycling is the continuous transformation of organic and inorganic carbon compounds by plants and micro- and macroorganisms between the soil, plants and the atmosphere. Soluble molecules account for only a very small proportion of the total. In plants the main types of high molecular

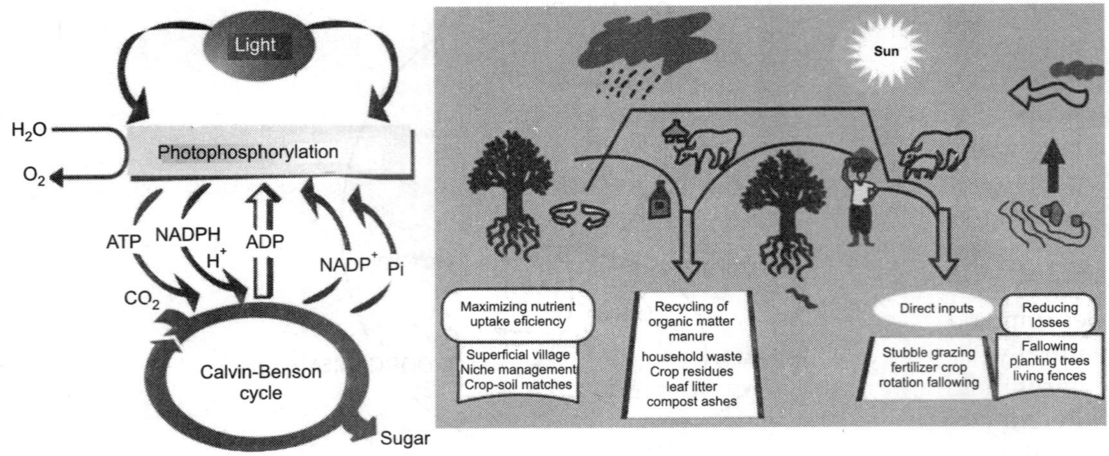

Figure 15.2 Systhesis of organic residue through photosynthesis and its disposal.

weight compounds which occur are cellulose, hemicelluloses, and lignin synthesized through photosynthesis (Figure 15.2), together with smaller quantities of fats, waxes and oils, proteins, and nucleic acids.

On the surface of the soil, residues are attacked by bacteria, fungi, and animals, the latter usually eating remains that have been partly digested by the microflora. The predominant microbial types involved and the rate of the attack depend on the chemical nature of the residue, the environmental conditions and the nature of the underlying soil. Of the environmental conditions, moisture is particularly important and in dry conditions the attack will be slow. Temperature also affects the rate of degradation and the composition of the attacking flora. This will also be affected by the nature of the underlying soil which determines to a large extent the population available for the degradation.

The animal population in the soil also plays a role in the decay process. The activities of springtails, bark lice and earwigs for example, increase the surface area of the residue available to the microflora and various worms eat the partially decomposed material and transport it into the soil proper. Further decay occurs in the gut as a result of the action of the microflora. The conditions are more favourable within the soil for the microflora to increase degradation rate. In particular, the role of bacteria becomes more important in most types of soil and a zymogenous response occurs involving an increase in the numbers of most of the heterotrophic soil bacteria. At every point in this process, the initial attack on the high molecular weight polymers is by enzymes liberated into the soil or acting at the outside surface of the producing cell. Environmental factors may alter the rate of decay by their effect both on the microflora and on the enzymes outside the cell.

Most of the carbon contained in organic residues is oxidized to carbon dioxide but some is incorporated in microbial tissues which themselves will eventually be decomposed. The remainder is incorporated into humus. This is the dark coloured amorphous organic material which is as important for soil fertility as is evidenced by the infertility of soils from which it is absent. The processes whereby it is formed are not understood, although it may be that soil animals are indispensable since in soils where there is no soil fauna humus is not found. It is not a single chemical substance and its composition varies but it is essentially polyphenolic in nature and contains some amino acids and amino sugars.

15.2.2 Decomposition of Humus

Decomposition of organic matter in soil is largely a biological process that occurs naturally. Its speed is determined by three major factors: soil organisms, the physical environment and the quality of the organic matter. In the decomposition process, different products are released: carbon dioxide (CO_2), energy, water, plant nutrients and resynthesized organic carbon compounds. Successive decomposition of dead material and modified organic matter results in the formation of a more complex organic matter called humus. This process is called humification. Humus affects soil properties. As it slowly decomposes, it colours the soil darker; increases soil aggregation and aggregate stability; increases the CEC (the ability to attract and retain nutrients); and contributes N, P and other nutrients. The amount of humus in a mature soil varies little from year to year provided it is not depleted as a result of the cultural practices. Humus is relatively resistant to degradation either because it is intrinsically stable or because it is adsorbed on to clay particles, but whatever the reason, only a small fraction of the total is broken down in anyone year. As with the degradation of plant and animal residues added to the soil, the rate of breakdown is influenced by environmental factors such as aeration, temperature, pH and in addition the amount of organic matter in the soil; when the amount is large the rate of humus decomposition is high. This is probably because the zymogenous flora attacks the humus fraction in addition to the added organic residues. Normally, however, humus degradation is achieved mainly by the autochthonous flora.

The waste products produced by microorganisms are also soil organic matter. This waste material is less decomposable than the original plant and animal material, but it can be used by a large number of organisms. By breaking down carbon structures and rebuilding new ones or storing the C into their own biomass, soil biota plays the most important role in nutrient cycling processes and, thus, in the ability of a soil to provide the crop with sufficient nutrients to harvest a healthy product. The organic matter content, especially the more stable humus, increases the capacity to store water and store (sequester) C from the atmosphere (Figure 15.3).

15.2.3 Methane Formation and Oxidation

In permanently waterlogged anaerobic soils; methane is emitted from the soil mainly due to the activity of methanogenic bacteria. Anaerobic bacteria may convert organic matter to organic acids, carbon dioxide, and hydrogen which are in turn converted by certain strict anaerobic bacteria to methane. Only a few bacteria are capable of

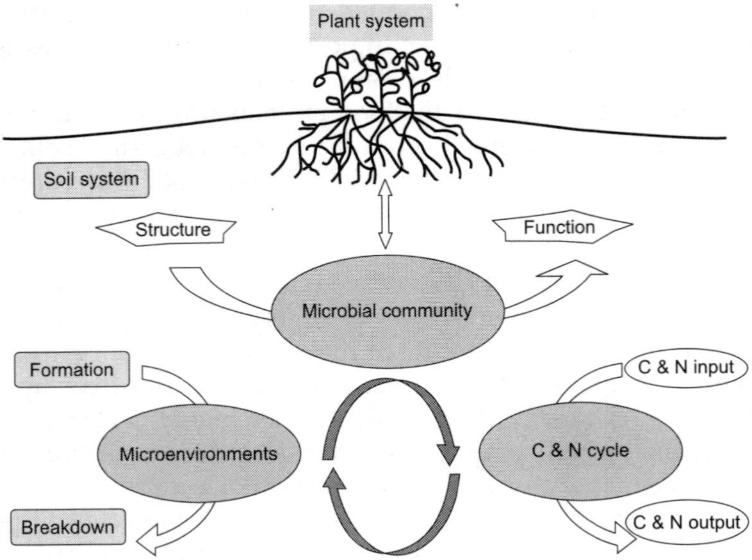

Figure 15.3 Flow chart explaining breakdown of natural residues.

methane production which, although they are morphologically dissimilar, is included in the family Methanobacteriaceae. This process is of no significance in agricultural soils which are temporarily waterlogged but a combination of poor drainage and high organic matter content may vary rarely produce conditions such that methane is produced and liberated at the surface as gas. In waterlogged anaerobic paddy field soils where methane generation occurs, a methane oxidizing flora exists at the surface and converts most of the gas to carbon dioxide. Methane oxidizing bacteria also occur in several other soil types including well-drained ones in which they are more numerous in the lower layers of the soil.

15.3 NITROGEN CYCLE

Nitrogen in the air becomes a part of biological matter mostly through the actions of bacteria and algae in a process known as nitrogen fixation. Legume plants such as clover, alfalfa, and soybeans form nodules on the roots where nitrogen fixing bacteria take nitrogen from the air and convert it into ammonia, NH_3. The ammonia is further converted by other bacteria first into nitrite ions, NO_2^-, and then into nitrate ions, NO_3^-. Plants utilize the nitrate ions as a nutrient or fertilizer for growth. Nitrogen is incorporate in many amino acids which are further reacted to make proteins.

Ammonia is also made through a synthetic process called the Haber Process. Nitrogen and hydrogen are reacted under great pressure and temperature in the presence of a catalyst to make ammonia. Ammonia may be directly applied to farm fields as fertilizer. Ammonia may be further processed with oxygen to make nitric acid. The reaction of ammonia and nitric acid produces ammonium nitrate which may then be used as a fertilizer. Animal wastes when decomposed also return to the earth as nitrates.

To complete the cycle other bacteria in the soil carry out a process known as denitrification which converts nitrates back to nitrogen gas. A side product of this reaction is the production of a gas known as nitrous oxide, N_2O. Nitrous oxide, also known as "laughing gas" – mild anesthetic, is also a greenhouse gas which contributes to global warming. An important difference between the two cyclic systems is that nitrogen turnover determines productivity in most agricultural situations, hence the activity of those soil organisms metabolizing nitrogen compounds in such a way as to remove nitrogen in a form assimilable by higher plants from the soil, may be very important with respect to loss of soil fertility (Figure 15.4).

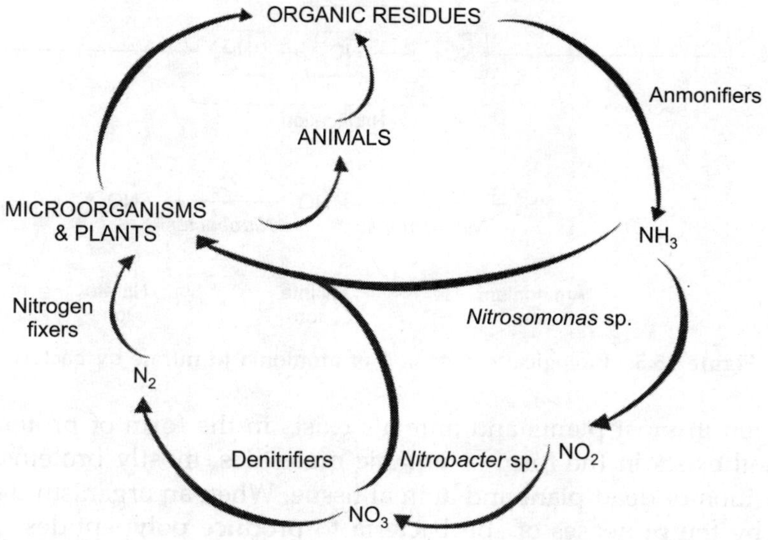

Figure 15.4 N_2 cycle.

15.3.1 Soil Ammonification

The process of the production of ammonia from organic compounds is called ammonification. Nitrogen is essential for the synthesis of structural and functional protein for living organisms. Protein is decomposed to amino acids that, in turn, are deaminated to liberate ammonia when the organisms die. The process of production of ammonia from organic compounds is called ammonification. Majority of bacteria in soil participate in the process and is a very important step for bacteria and plants to assimilate it. The process of the production of ammonia from organic compounds is called ammonification (Figure 15.6). In addition to the ammonification of amino acids, other compounds such as nucleic acids, urea, and uric acid go through the ammonification process. The bacteria that accomplish it (*Bacillus, Clostridium, Proteus, Pseudomonas,* and *Streptomyces*) are called ammonifying bacteria. Ammonification of organic compounds is a very important step in the cycling of nitrogen in soils, since most autotrophs are unable to assimilate amino acids, nucleic acids, urea, and uric acid and use them for their own enzyme and protoplasm construction.

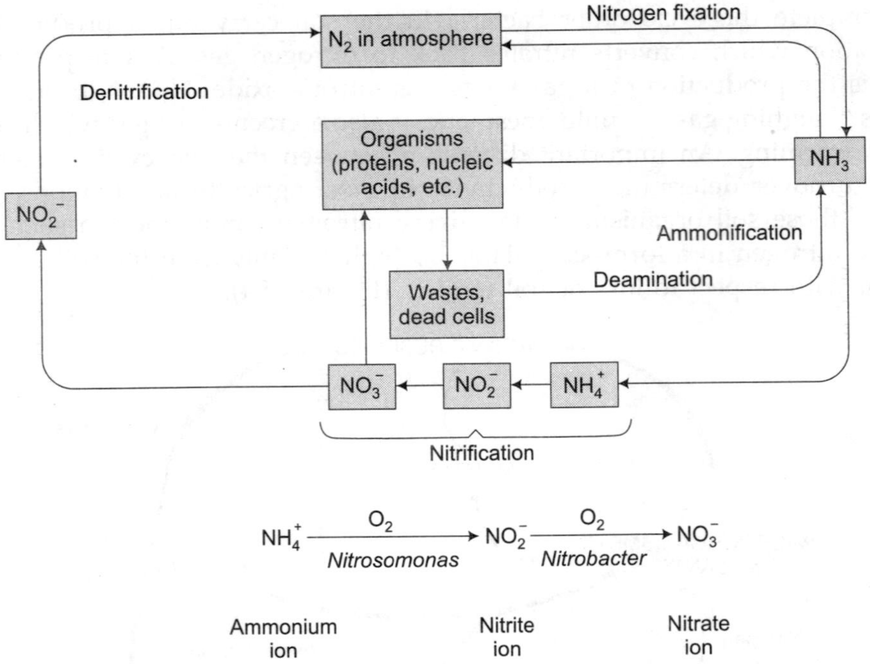

Figure 15.5 Biological conversion of ammonia to nitrate by bacteria.

The nitrogen in most plants and animals exists in the form of protein. Most of the nitrogen in soil exists in the form of organic molecules, mostly proteins derived from the decomposition of dead plant and animal tissue. When an organism dies, its proteins are attacked by the proteases of soil bacteria to produce polypeptides (peptones) and amino acids($C_2H_4NO_2$.R). This process is called peptonization. Then, the amino groups on the amino acids are removed by a process called deamination, producing ammonia (NH_3). In most soils, the ammonia dissolves in water to form ammonium ions (NH^{4+}) (Figure 15.5).

Ammonification is a natural cycle which makes the earth's supply of nitrogen available to organisms which need it, such as plants. Like many other natural cycles, the nitrogen cycle can be disrupted by human activities which can lead to imbalances at various stages, sometimes causing environmental problems.

Ammonification can be a major problem area in the nitrogen cycle when human intervention occurs, as build-ups of ammonia can cause health problems and environmental issues. One of the most elementary of the ammonification reactions is the oxidation of the simple organic compound, urea ($CO(NH_2)_2$), to ammonia through the action of a microbial enzyme known as urease (two units of ammonia are produced for every unit of urea that is oxidized). Urea is a commonly utilized agricultural fertilizer, used to supply ammonia or ammonium for direct uptake by plants, or as a substrate

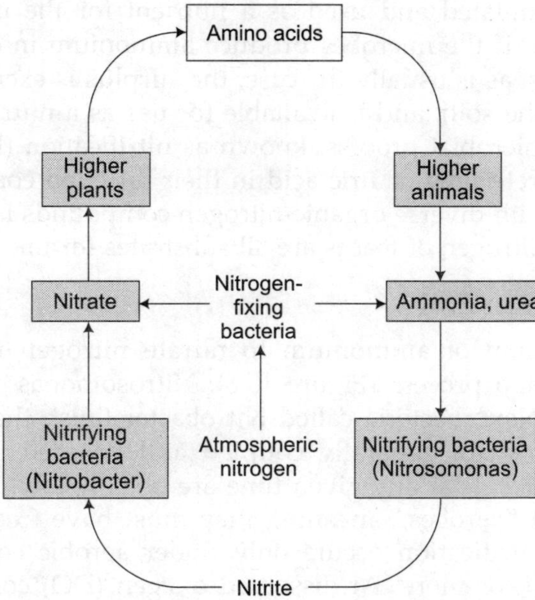

Figure 15.6 Various processes of peptonization.

for the microbial production of nitrate through nitrification. Ammonium is a suitable source of nitrogen uptake for many species of plants, particularly those that live in acidic soils and waters. However, most plants that occur in non-acidic soils cannot utilize ammonium very efficiently. Animals receive their supply of nitrogen through the foods they eat, but plants must assimilate inorganic forms of this nutrient from their environment. However, the rate at which the environment can supply inorganic nitrogen is limited and usually small in relation to the metabolic demands of plants. Therefore, the availability of inorganic forms of nitrogen is frequently a limiting factor for the productivity of plants. This is a particularly common occurrence for plants growing in terrestrial and marine environments, and to a lesser degree, in fresh waters (where phosphate supply is usually the primary limiting nutrient, followed by nitrate).

The dead biomass of plants, animals, and microorganisms contain large concentrations of organically bound nitrogen in various forms, such as proteins and amino acids. The process of decomposition is responsible for recycling the inorganic constituents of the dead biomass and preventing it from accumulating in large unusable quantities (Figure 15.6). Decomposition is, of course, mostly carried out through the metabolic functions of a diverse array of bacteria, fungi, actinomycetes, other microorganisms, and some animals. Ammonification occurs under oxidizing conditions in virtually all ecosystems and is carried out by virtually all microorganisms that are involved in the decay of dead organic matter. In situations where oxygen is not present, a condition referred to as anaerobic, different microbial decay reactions occur, and these produce nitrogen compounds known as amines. The microbes derive some metabolically useful energy from the oxidation of organic nitrogen to ammonium. In addition, much of the ammonium is assimilated and used as a nutrient for the metabolic purposes of the microbes. However, if the microbes produce ammonium in quantities that exceed their own requirements, as is usually the case, the surplus is excreted into the ambient environment (such as the soil), and is available for use as a nutrient by plants, or as a substrate for another microbial process, known as nitrification (Figure 15.7). Animals, in contrast, mostly excrete urea or uric acid in their nitrogen containing liquid wastes (such as urine), along with diverse organic-nitrogen compounds in their feces. The urea, uric acid, and organic nitrogen of feaces are all substrates for microbial ammonification.

15.3.2 Nitrification in Soil

The biological conversion of ammonium to nitrate nitrogen is called nitrification. Nitrification is a two-step process (Figure 15.8). Nitrosomonas convert ammonia and ammonium to nitrite. Next, bacteria called Nitrobacter finish the conversion of nitrite to nitrate. The reactions are generally coupled and proceed rapidly to the nitrate form; therefore, nitrite levels at any given time are usually low. These bacteria known as "nitrifiers" are strict "aerobes," meaning they must have free dissolved oxygen to perform their work. Nitrification occurs only under aerobic conditions at dissolved oxygen levels of 1mg/L or more. At dissolved oxygen (DO) concentrations less than 0.5mg/L, the growth rate is minimal. Nitrification requires a long retention time, a low food to microorganism ratio (F:M), a high mean cell residence time (measured as MCRT or Sludge Age), and adequate buffering (alkalinity). The nitrification process produces acid. This acid formation lowers the pH of the biological population and can cause a reduction of the growth rate of nitrifying bacteria. The optimum pH for *Nitrosomonas* and *Nitrobacter* is between 7.5 and 8.5; most treatment plants are able to

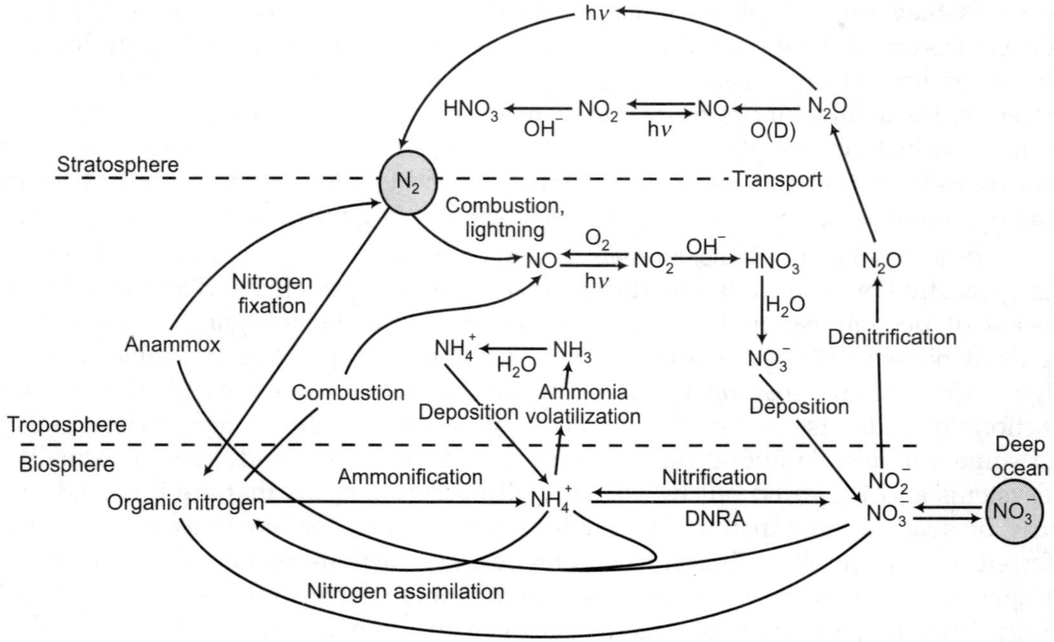

Figure 15.7 Nitrogen assimilation and ammonification in the soil.

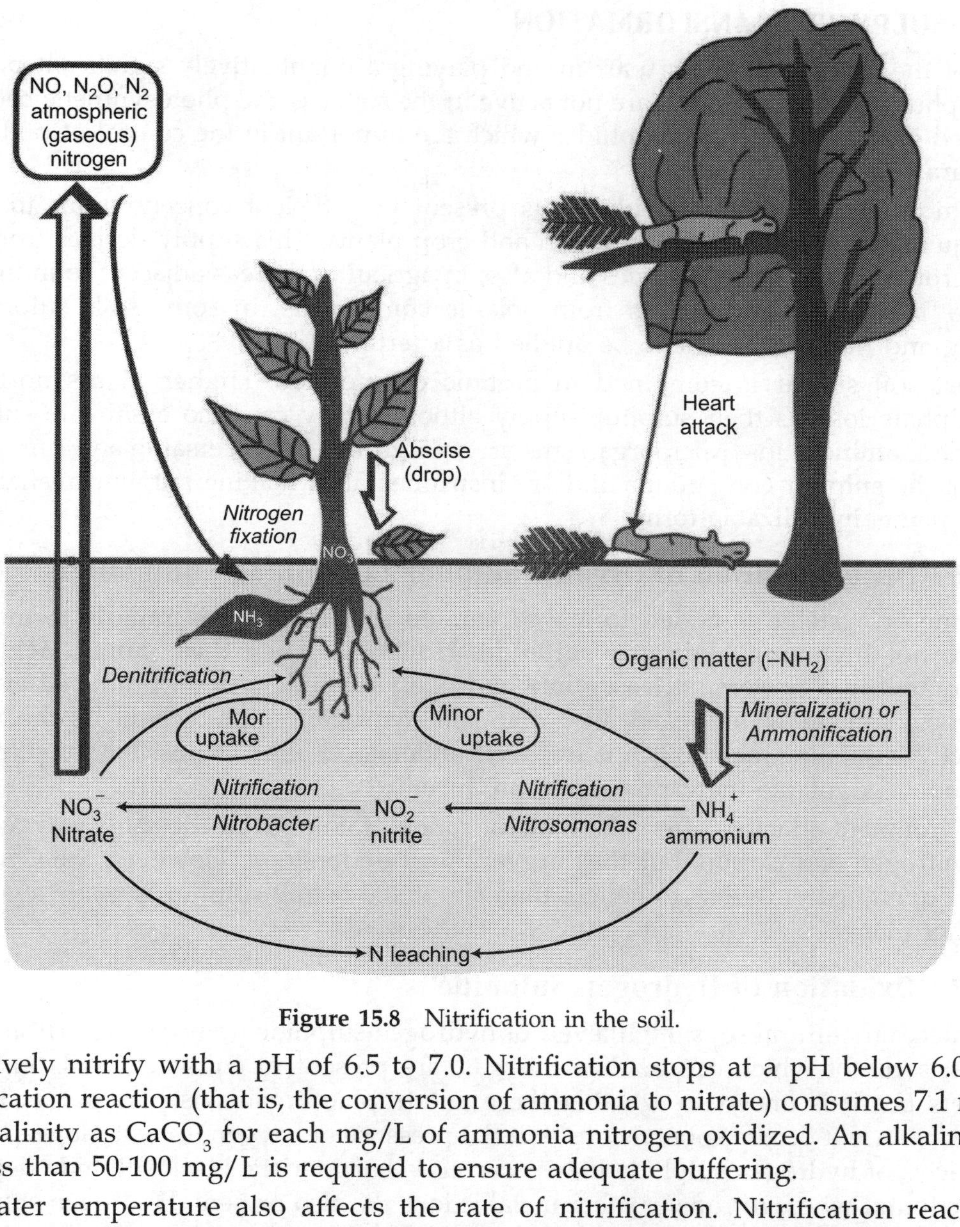

Figure 15.8 Nitrification in the soil.

effectively nitrify with a pH of 6.5 to 7.0. Nitrification stops at a pH below 6.0. The nitrification reaction (that is, the conversion of ammonia to nitrate) consumes 7.1 mg/L of alkalinity as $CaCO_3$ for each mg/L of ammonia nitrogen oxidized. An alkalinity of no less than 50-100 mg/L is required to ensure adequate buffering.

Water temperature also affects the rate of nitrification. Nitrification reaches a maximum rate at temperatures between 30 and 35°C (86°F and 95°F). At temperatures of 40°C (104°F) and higher, nitrification rates fall to near zero. At temperatures below 20°C, nitrification proceeds at a slower rate, but will continue at temperatures of 10°C and less. However, if nitrification is lost, it will not resume until the temperature increases to well over 10°C. Some of the most toxic compounds to nitrifiers include cyanide, thiourea, phenol and heavy metals such as silver, mercury, nickel, chromium, copper and zinc. Nitrifying bacteria can also be inhibited by nitrous acid and free ammonia (Figure 15.8).

15.4 SULPHUR TRANSFORMATION

Most of the organisms taking part in and playing a quantitatively significant part in the sulphur cycle (Figure 15.9) are not active in the soil, e.g. the photosynthetic bacteria involved in the oxidation of sulphides which are important in the control of pollution in natural waters.

In most agricultural soils sulphur is present in sufficient concentrations to meet the requirements of the soil population and crop plants. This supply derives from the parent rock, from organic residues and also, in agricultural areas adjacent to industrial regions, in substantial amounts from volatile compounds. In some soils sulphur is limiting and may, with benefit, be applied as a fertilizer.

Most soil sulpur is combined in organic compounds. Higher plants normally use sulphate ions as their sulphur supply although they can also assimilate sulphur containing amino acids. Microorganisms are as a group, more versatile and can utilize most of the sulphur compounds and are instrumental in making sulphur available to higher plants in utilizable form.

15.4.1 Decomposition of Organic Sulphur Containing Compounds

If cystine or cysteine is added to a well aerated soil, sulphate is rapidly formed by a route not involving hydrogen sulphide. However, when these amino acids are incorporated in a protein, at least some hydrogen sulphide will be produced even in an aerobic soil. In waterlogged anaerobic soil hydrogen sulphide will be the major product. Methionine metabolism is less well understood but it is possible for sulphate, sulphide, or a volatile mercaptan to be produced.

Environmental factors affect the mineralization of sulphur in the same way as they affect nitrogen and carbon, but they are less well understood. However, the C:S ratio of a residue apparently has to be less than about 200 before sulphur is made available to higher plants.

15.4.2 Oxidation of Hydrogen Sulphide

In aerated agricultural soils, high levels of hydrogen sulphide do not occur, *Thiobacillus* species, some of which can oxidize this gas, are present in cultivated soils in small numbers but it is uncertain whether these autotrophic organisms play a part in the oxidation since it occurs spontaneously in the presence of oxygen. In waterlogged soils, high levels of hydrogen sulphide may occur and characteristic black sulphide deposits may be found where ferrous or other metallic ions are also present. Hydrogen sulphide may be oxidized by filamentous colourless sulphur bacteria such as *Beggiatoa* where there is sufficient oxygen, possibly at the surface of the mud.

15.4.3 Oxidation of Hydrogen Sulphur

The addition of sulphur to soil induces the development of a large population of *Thiobacillus* species and the lowering of the soil pH due to the production of sulphuric

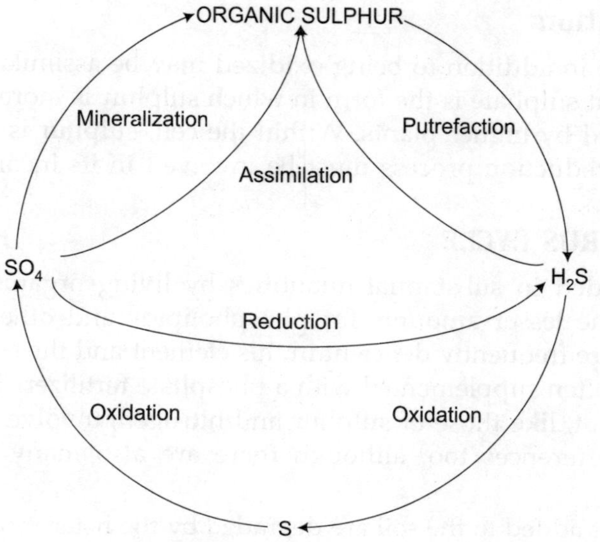

Figure 15.9 The sulphur cycle.

acid by these organisms, one of which, *Thiobacillus thiooxidans*, can tolerate a pH of below 0.5. The pathway of sulphur oxidation by these bacteria is not clear although it seems certain that several sulphur compounds of intermediate oxidation starts synthesizing between sulphur and sulphate are involved.

15.4.4 Reduction of Sulphate

The organisms chiefly responsible for reducing sulphate in the soil are members of the genus *Desulphovibrio.* These organisms are widespread in nature and occur in a wide range of environmental conditions. They can be found growing both heterotrophically, using such compounds as carbohydrates, organic acids, and alcohols as electron donors, and also autotrophically, using molecular hydrogen as the electron donors.

In agricultural soils, these organisms will not play a very important part, since the aeration conditions will be such that they will not be significantly active. However, in waterlogged paddy field soils where anaerobic conditions prevail, they will be responsible for the generation of hydrogen sulphide which may damage the roots of rice plants growing there although evidence of such a toxic effect is not conclusive. Apart from this, sulphate-reducing organisms are economically important in a number of ways. They are responsible for causing extensive corrosion to underground iron pipes by the removal, and use as an electron donor, of hydrogen which normally forms a protective layer around these pipes in the absence of air; further, corrosion then occurs. It is also thought that they may play a part in liberating oil from oil bearing shales. In the past they probably took part in the formation of sulphur deposits by producing hydrogen sulphide which was then oxidized by microorganisms or by purely chemical processes.

15.4.5 Assimilation

Hydrogen sulphide in addition to being oxidized may be assimilated as such by some microorganisms, but sulphate is the form in which sulphur is more usually taken up by these organisms and by higher plants. Within the cell, sulphur is mainly in a reduced form and hence a reduction process must be involved in its incorporation.

15.5 PHOSPHORUS CYCLE

Phosphorus is needed in substantial quantities by living organisms for nucleic acid synthesis and in the lesser amounts for phospholipids and other organic phosphate compounds. Soils are frequently deficient in this element and the supply from plant and animal residues is often supplemented with a phosphate fertilizer. The transformation of phosphorus does not, like those of sulphur and nitrogen, involve autotrophic bacteria. There are other differences too, although there are also many similarities between those processes.

Organic residues added to the soil are degraded by the heterotrophic microorganisms, the rate of process depends on the environmental conditions as do the rates of sulphur and nitrogen mineralization. In fact, the conditions for rapid ammonification are the same as those for speedy phosphate release. Substantial amounts remain bound in organic material. The phosphorus content of the added residue determines whether phosphate is made available to crop plants or whether it is incorporated into microbial tissue. The level above which phosphate is made available is about 0.2%. Phosphate is also made available to plants as a result of the liberation by microorganims of organic acids which dissolve insoluble inorganic phosphate compounds in the soil (Figure 15.10).

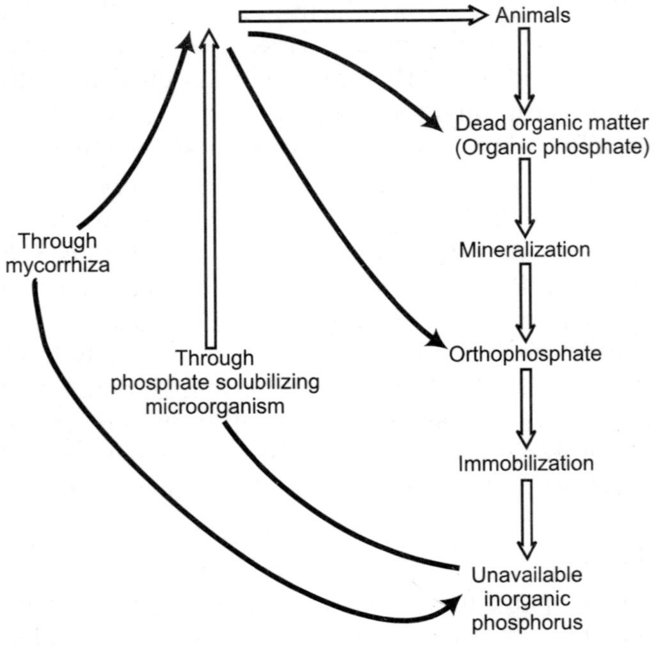

Figure 15.10 Phosphorus mineralization.

15.6 OTHER TRANSFORMATION

In addition to the major elements described above, organisms require smaller amounts of several others. The fungal flora appears to be less sensitive to iron and manganese availability than are the bacteria. *Bacillus* and *Arthrobacter* have a particularly high requirement for iron. A much greater, although incomplete, knowledge exists of the transformation of the various compounds of these two elements and their effect on the availability of the ions to plants. In soils of normal pH, transformation of manganese is the result of biological activity. Only at pH values in excess of 8 is this element converted to the insoluble manganic form, usually manganic oxide (MnO_2), by chemical means. Manganese exists predominantly as the divalent manganous form at pH below 5.5 which it is often stated is the form utilized by plants. There is, however, now some evidence that certain manganese oxides also can act as a source of assimilable manganese. But as a result of the activity of the manganese-oxidizing flora, soils may become deficient in available forms of this element and a reduction in the crop yield may occur. Bacteria oxidizing manganese are common only in the upper layers of heavy soils but are more widely distributed in sandy soil – a reflection of the aeration of the two soil – types presumably. Certain heterotrophic bacteria, such as *Pseudomonas, Corynebacterium*, and *Flavobacterium*, can oxidize manganese as can be by *Metallogenium* which is widely distributed in soil but apparently only develops in association with a fungus. *Pedomicrobium,* which resembles closely with *Hypomicrobium* may according to species, oxidize not only manganese but also iron. Reduction of manganese compounds to a form readily utilizable by plants can result from the production of acid and/or by a reduction of the redox potential of the soil as a result of the activity of the microflora. However, the importance of bacteria in making manganese available to plants is uncertain since the return of aerobic and be neutral pH conditions results in its re-oxidation.

In contrast to the role of the soil microflora in manganese transformation, the part played by bacteria in the conversion of the iron compounds in neutral and alkaline soils is unimportant. Some heterotrophic bacteria may cause deposition of iron from organic complexes by using the organic compounds as an energy source and it is possible, but not certain, that iron deposited in this way may play a part in the formation of the iron pans which occur in some soils. Certain other heterotrophic bacteria can also reduce ferric to ferrous iron under anaerobic conditions. Where sulphide ions occur this will be precipitated as ferrous sulphide. Most of the iron bacteria including the filamentous forms such as *Sphaerotilus* and the unicellular forms such as *Gallionella* are active in water as are several of the manganese bacteria. The availability of several other elements including calcium, magnesium, potassium, zinc, copper, molybdenum, cobalt, and boron is affected by biological activity and arsenic and selenium can be converted into volatile compounds by the action of certain soil fungi.

15.7 RHIZOSPHERE

Unseen soil microbes respond to and influence global climate change. The respiration of soil microbes returns to normal after a number of years under heated conditions and consume much of the available food that future levels of decomposition were reduced because of food scarcity. Some soil microbes are also adapted to the changed environment and reduce their respiration accordingly. Various types of organisms and microorganisms live in soil. Some microorganisms also burrow and channel through soil, which improves soil structure and aggregation, while other microorganisms have the ability to break down resistant organic matter such as lignin, toxins, and pesticides. Microorganisms also have the ability to protect plants from antagonistic pathogens and some can dissolve minerals, making nutrients available to plants. Fungi are able to break down resistant materials such as cellulose, gums, and lignin. They dominate in acidic, sandy soils and fresh organic matter. Actionomycetes also are able to decompose resistant substances in soil. One type, *Frankia*, help plants to get nutrients needed from the air by breaking triple bond nitrogen down into ammonium that plants can use. Antibiotics are reported to be produced by soil actinomycetes. Bacteria decompose a wider range of earth material than any other microbe group. Heterotrophs gain their energy and carbon from other organisms while autotrophs synthesize their own energy from light or by chemical oxidation. Some bacteria can fix nitrogen into forms that plants can use. How quickly decomposition of dead organic matter occurs, depends on soil temperature and soil moisture. Without the microorganisms, dead organic matter would pile high on the earth's surface.

15.7.1 Microbial Distribution

Soil is an excellent culture media for the growth of many types of organisms. This includes bacteria, fungi, algae, protozoa and viruses. In addition, various nematodes, insects, etc., are also present. A spoonful of soil contains billions of microorganisms. In general, the majority of microbial population is found in the upper six to twelve inches of soil and the number decreases with depth. Higher number occurs in the organically rich surface layers than in the underlying mineral soils. Particularly, high numbers of microbes occur in association with plant roots. Fungal populations are favoured in soils of low pH, and bacteria tend to occur in higher numbers at higher pH value. There is about two and a half times more carbon in the soil than in the atmosphere. The soil population is responsible for transformation in the soil which are important for its continued fertility and which ensure the removal of natural litter from the surface of the earth. Soil organisms, including microorganisms, use soil organic matter as food. As they break down the organic matter, any excess nutrients (N, P and S) are released into the soil in forms that plants can use. This release process is called mineralization (Figure 15.11).

Plant roots exude an enormous range of potentially valuable small molecular weight compounds into the rhizosphere. Some of the most complex chemical, physical, and biological interactions experienced by terrestrial plants are those that occur between roots and their surrounding environment of soil (i.e., the rhizosphere). Interactions involving plants' root in the rhizosphere include root-root, root-insect, and root-microbe interactions. The most intense interactions between microbes and plants take

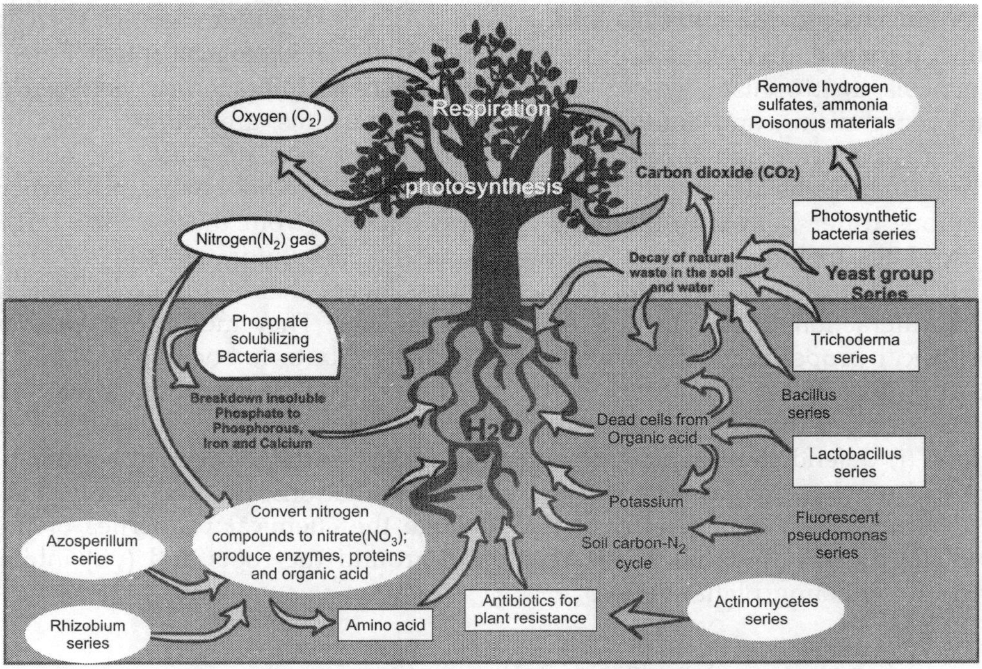

Figure 15.11 Distribution of microorganisms in the soil and their functions.

place at the rhizosphere, which is the interface between plant roots and the soil. The influence of plants on microbial population structure and function in the rhizosphere has important ecological implications for soil function, including biogeochemical cycles. Similarly, soil microbes have a tremendous influence on plant health and productivity. One straightforward and visible benefit for the plant is a better supply of and access to nutrients.

15.7.2 Rhizosphere Microflora

The region in the vicinity of roots can be distinguished into many microhabitats. The term rhizosphere was introduced in 1904 by the German scientist Hiltner to denote that region of the soil which is subject to the influence of plant roots. Rhizosphere is characterized by greater microbiological activity than the soil away from plant roots. Rhizosphere is the region of the soil immediately surrounding the roots of a plant. Some authors include the root surface rhizoplane also under rhizosphere. From an operational viewpoint the rhizosphere is the zone of the soil surrounding a plant root where the biology and chemistry of the soil are influenced by the root. This zone is about 1 mm wide, but has no distinct edge. Rather, it is an area of intense biological and chemical activity influenced by compounds exuded by the root, and by microorganisms feeding on the compounds. As plant roots grow through soil they release water soluble compounds such as amino acids, sugars and organic acids that supply food for the microorganisms. The food supply means microbiological activity in the rhizosphere which is much greater than in soil away from plant roots. In return, the microorganisms provide nutrients for the plants. All this activity makes the rhizosphere the most dynamic environment in the soil. Practically all ecological interactions, such as symbiosis, syntrophism, synergism, commensalism and antagonism between plants and microorganisms, and among different microorganisms, are found in this region. Root secretes a wide range of organic materials (root exudates) into the soil which greatly influence the development of a rhizosphere microbial community. Microbial populations differ both quantitatively and qualitatively from those in the bulk soil. Root exudates include amino acids, organic acids, carbohydrates, sugars, vitamins, mucilage and proteins. The exudates act as messengers that stimulate biological and physical interactions between roots and soil organisms. They modify the biochemical and physical properties of the rhizosphere and contribute to root growth and plant survival (Figure 15.12). However, the fate of the exudates in the rhizosphere and the nature of their reactions in the soil remain poorly understood. The exudates have several functions (i) defend the rhizosphere and root against pathogenic microorganisms; (ii) attract and repel particular microbe species and populations; (iii) keep the soil around the roots moist; (iv) obtain nutrients; (v) change the chemical properties of the soil around the roots; (vi) stabilize soil aggregates around the roots and (vii) inhibit the growth of competing plant species.

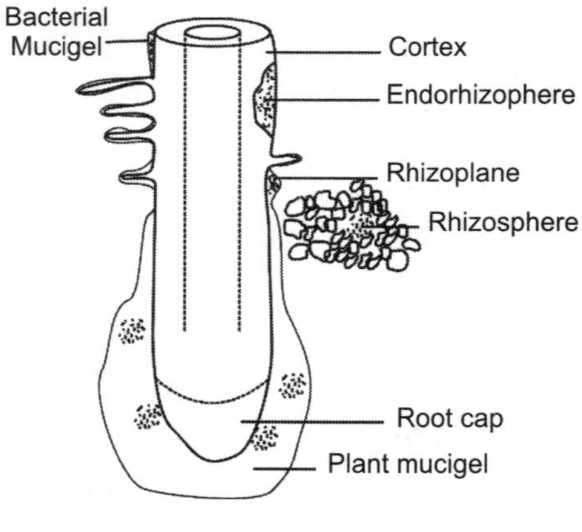

Figure 15.12 Distribution of microorganisms in the rhizosphere.

15.7.3 Nutrient Uptake

The plant roots, which the rhizosphere is associated with, can affect the physical environment of the rhizosphere. As plants transpire water with more force during the day than during the night, they change the soil water potential immediately near their roots and so the rhizosphere undergoes fluctuations that the bulk soil avoids. Movement of organic matter away from the root as well as bacteria colonizing new locations occurs more readily in sandy soils than clayey soils. Sand has larger pores between each granule allowing microorganisms and exudates to travel. Therefore, the larger the granule size, the further the rhizosphere and microorganisms associated with it, will extend into the surrounding soil.

Plant roots compact the soil on the short term as they grow, but once they die and decay, can actually leave soil more porous. Several factors can lower the pH in the rhizosphere. Respiration leads to carbon dioxide (and eventually to bicarbonate/carbonic acid) generation. In addition to respiration of the roots themselves, the rhizosphere is very rich in carbon results in other organisms from prokaryotes to fungi to small animals living and respiring in the rhizosphere more than in the bulk soil. Fungi are found in more acidic soils than alkaline, and bacteria have a very broad pH spectrum where they can survive. The influencing effect of pH in the rhizosphere is critical in supporting a biologically diverse microbial community. Plant-derived compounds are responsible for providing the additional carbon that allows the rhizosphere to host a large variety of organisms. These compounds fall into five categories: exudates, secretions, mucilages, mucigel, and lysates. Exudates include surplus sugars, amino acids, and aromatics that diffuse out of cells to the intercellular space and surrounding soil. Due to their diffusive nature, exudates are limited to compounds of low molecular weights. Secretions are byproducts of metabolic activity. Because they are actively released from the cell, secretions can be of both low and high molecular weight. When an epidermal root cell dies and is broken open, lysates from within the cell become available to the surrounding microbial community. Mucilages are cells sloughed off the root cap as the root grows. Abrasive forces of the root against soil particulate matter are responsible for the removal of cells. These cells consist of cellulose, pectin, starch, and lignin. Mucigel is a slime coating the surface of a root that increases the connectivity between plant roots and the surrounding soil. It is more common on the main body of the root and root hairs than the tip. During dry spells, mucigels are responsible for allowing plants to continue to uptake water and nutrients. However, accumulated evidence suggests that root exudates have a major role in determining outcomes of interactions in the rhizosphere and, ultimately, plant and soil community dynamics. Microbes also indirectly aid nutrient uptake bacteria of the Azospirillum genus, promote increased root mass and more efficient nitrogen uptake from the soil in response to the plant hormone indole-3-acetic acid. Using these bacteria and fungi could provide significant environmental benefits as they would allow a reduction in the application of nitrogen and phosphorus fertilizers.

15.7.4 Plant Protection

In addition to enhancing the nutrient supply of plants, microbes also confer a degree of protection against plant diseases. In particular, various bacteria and fungi, especially of the genera *Pseudomonas, Bacillus* and *Trichoderma,* produce a range of metabolites against other phytopathogenic fungi. Such biocontrol agents are already being used efficiently in the field, but they have not yet attained the degree of efficacy and consistency that is needed for large scale commercialization. But there is the potential for improvement and with development; such microbes could become a realistic alternative to the heavy fungicide regimens used in agriculture at present.

In addition to these direct effects on plant growth, rhizobacteria exert another health-promoting effect on the plants with which they interact. This phenomenon is known as induced systemic resistance and arises when interactions with non-pathogenic bacteria confer better disease resistance on plants. Induced systemic resistance differs from acquired systemic resistance in plants; the later occurs in response to localized microbial attack. Induced resistance is triggered by microbial interactions in the rhizosphere, but also enhances resistance of remote aerial plant parts against pathogens.

This systemic change in plant physiology bears conceptual similarities to other phenotypes, such as improved stress, drought or disease resistance, that have been linked to plant–fungal associations. Little is known about why bacteria induce this condition, which is clearly beneficial to the plant. Although induced systemic resistance is generally ascribed to interactions with bacteria, there is emerging evidence that some fungi, particularly endophytic fungi, may also be able to induce a similar response in plants.

The interactions of rhizosphere microbes with plants depend on the establishment of intimate associations between the two partners. This intimate cooperation between plant and bacteria displays a high level of host specificity. There is also a growing body of evidence suggesting that many other associations between plants and microbes show similar degrees of specificity; different plant species, and even different cultivars of the same plant species, establish distinct microbial populations in their rhizospheres when grown in the same soil.

The formation of these communities depends, at least in part, on the activation of specific programmes of gene expression in the microbe in response to chemical signals secreted from the plant. A pertinent example is the induction of nodulation genes in receptive rhizobia, which are triggered by the production and secretion of particular flavonoids by the plant. In the case of the rhizobia–legume interaction, the plant also responds to bacterial signals, and it is likely that this type of chemical cross talk is typical of other microbe plant interactions. Other examples of plant derived signals that influence microbial gene expression include phenolics exuded from plant wounds, which induce expression of virulence genes in pathogenic *Agrobacterium* spp., and compounds that mimic the quorum sensing signals used by bacteria to regulate gene expression.

15.7.5 Free Living Dinitrogen Fixation

Eighty percent (80%) of the atmosphere is nitrogen gas (N_2). Unfortunately N_2 is unusable by most living organisms. Plants, animals, and microorganisms can die of nitrogen deficiency, if surrounded by N_2 they cannot use. All organisms use the ammonia (NH_3) form of nitrogen to synthesize amino acids, proteins, nucleic acids, and other nitrogen-containing components necessary for life. Biological nitrogen fixation (BNF) process changes inert N_2 to useful NH_3. This process is mediated in nature only by bacteria and certain species of actinomycetes. In the free-living system, plants benefit when the bacteria die and release nitrogen to the environment, or when the bacteria are loosely associated with the roots of plants. In legumes and a few other plants, the bacteria live in small club-like growths on the roots called nodules. Within these nodules,

N_2 fixation occurs, and the NH_3 produced is directly absorbed by the plant. Nitrogen fixation by legumes is a close/symbiotic relationship between a *Rhizobium* bacterium and a legume host plant. Biological nitrogen fixation takes many forms in nature from (1) symbiotic forms including blue-green algae (*Nostoc*); (2) lichens, actinomycetes, non legume; (3) associative symbiosis and the (4) free-living soil bacteria. These types of N_2 fixation contribute significant quantities of NH_3 to natural ecosystems. Nitrogen fixation by legumes can be in the range of 25-75 pounds of N per acre per year in a natural ecosystem and several hundred pounds in a cropping system. Symbiotic N_2 fixation occurs through associations of plant roots with nitrogen-fixing bacteria. The symbioses are: (1) between many leguminous species and *Rhizobium* or *Bradyrhizobium*, forming nodules on roots; (2) between a small number of non-leguminous genera and *Frankia*. The utilization often associative BNF technology in grass and cereal crops was found to be useful in the development of profitable agriculture technologies. The existing associations of tropical grasses with nitrogen-fixing bacteria under favourable conditions, may be contributing significantly to the N economy of these plants. *Azospirillum* as a "biofertilizer" is particularly important in agricultural systems where fertilizer inputs are either impractical (range lands), undesirable (organic farming), or not possible (subsistence agriculture). Experiments on inoculation of crops with *Azospirillum* or other diazotrophs often resulted in enhanced plant growth or nitrogen content under environmental conditions, improve nutrient assimilation, alter root size and function. Numerous studies have shown greater N_2 fixation activities in inoculated plants than in uninoculated controls. In addition to N_2 fixation, inoculation with *Azospirillum* results in the following benefits:

1. promotion of root hair development and branching;
2. increased uptake of N, P, K and microelements;
3. improved water status of plants; and
4. increased dry matter accumulation and grain yield.

Inoculated plants when examined under the electron microscopes revealed invasion of the cortical layer. Several nitrogen-fixing bacteria were isolated from grasses and cereal crops which when used as inoculants were found capable of enhancing shoot growth, root density and yield of rice, corn and sugar cane. The bacteria were found to possess at least 57% of the characteristics of the genus *Azospirillum*. These associative N_2 fixing bacteria were found to be capable of producing growth regulators like gibberellins and cytokinins, which were thought to contribute to the stimulated plant growth. The bacteria belong to the genus *Azospirillum* and are the most promising microorganisms that colonize roots of economically important grasses and cereals.

1. *Azospirillum* spp. are highly adaptable, being able to grow under:
2. Anaerobic conditions (nitrate used as eletron acceptor)
3. Microaerobic (elemental or ammonia used as N source)
4. Fully aerobic conditions (ammonia, nitrate, amino acid or combined N only)

15.7.6 Associative Nitrogen Fixation

The development of nitrogen fixation in response to a deficiency of available nitrogen may well be an integral part of the nitrogen cycle of natural ecosystem and low input farming system thereby maintaining a nitrogen balance in the environment. Associative nitrogen fixation is carried out by a number of bacterial species living on the roots of non leguminous plants. Nitrogen-fixing enteric bacteria of the genera *Klebsiella* and *Enterobacter* have frequently been isolated from the roots of various plants and are known to be able to adhere to the roots of grasses and cereals *in vitro*.

Fimbriae are involved in enterobacterial adhesion to grass roots. Two types of fimbria occur on *Klebsiella* associated with plants: the so-called type 3 and type 1 fimbriae. The two fimbrial types differ morphologically and serologically and in their binding specificities: the type 1 fimbriae bind to mannosides, but the receptor structure for the type 3 fimbria is not known. In associative nitrogen fixation, certain nitrogen fixing bacteria establish themselves close to the root or can be found on the leaf surface. They use root exudates, secretions and sloughed cells as an energy source. Some bacteria like *Acetobacter* can occupy internal root tissues of sugarcane. The amount of nitrogen fixed by these microbes is indirect, as 90% of the nitrogen only becomes available when the bacteria die.

An example is *Azotobacter paspali* which grows as a sheath around the root of the sand grass *Paspalum notatum*. No nodules are formed, but the plant is benefited greatly. Survival of *A. paspali* away from the plant is poor. The acetylene test shows that the *A. paspali* and *P. notatum* association fixes as much nitrogen as a moderate legume crop. Whether this also takes place under natural conditions is yet to be established. *A. paspali* associated with only the five tetraploid ecotypes of *P. notatum* out of 33 stimulated. *A. paspali* lives outside the roots in the rhizosphere. This environment is ideal for heterotrophic nitrogen fixing bacteria. It is almost anaerobic, a factor of importance to the association, which though aerobic is very sensitive to oxygen. The energy and carbon requirements of the heterotrophic free living nitrogen fixers are met with by the root cell and soluble root exudates. These organisms occur in the rhizospheres of a variety of plants. Other examples of associative symbiosis are rice with *Beijerinckia*, and the grass *Digitaria decumbens* with *Spirillum lipoferum*.

Mycorrhizal fungi and the symbiotic nitrogen fixing actinomycete Frankia may occupy inter- and intracellular spaces inside the roots. They may establish a highly integrated mutualistic association with their plant hosts. Helically lobed bacteria have been reported in the collapsed epidermal cells of *Ammophila arenaria* and within roots of *Zea mays*. They appear to enter by intercellular penetration of the epidermal cells.

Associative nitrogen fixing bacterium is defined as the bacterium that not only lives on rhizospherial environment, but also fixes N_2 from the atmosphere and contributes passively to the plant growth. Nitrogen fixing bacteria in rice root system were classified as in Table 15.1. The common associative N_2 fixing bacteria in rice rhizosphere are *Alcaligenes faecalis, Enterobacter cloacae, Klebsiella oxytoca, Klebsiella planticola, Azospirillum brasilense* and *Azospirillum lipferum*. Some strains have been isolated from *Alcaligenes faecalis* and *Azospirillum brasilense*. But only *Azospirillum, Enterobacter cloacae, Alcaligenes faecalis* and *Klebseilla pneumoniae* have been proved as safe strains used for biofertilizer. Generally, the number of associative N_2 fixing bacteria in paddy is larger than that in dry land, as large as 103-107 cells/g soil. Associative N_2 fixing bacteria live mainly in the rhizosphere. Types and number of soil microorganisms depend primarily on the components of root exudates and chemical characteristics of root residues. In terms of rhizobacteria, the carbon sources needed for survival of rhizobacteria must be provided by plant roots because the capacity of degrading organic matter for rhizobacteria is very weak. In addition, there are some growth regulators and antibiotics in root exudates, which regulate the growth of associative N_2 fixing bacteria. On the other hand, plant roots have selectivity for types of microbes, such an effect is surveyed on different varieties of rice and difference in nitrogen fixation activity of one strain of associative nitrogen fixing bacteria inoculated to different varieties of rice. Rice root exudates and residues supply associative nitrogen fixing bacteria with organic acids and sugars for carbon source and growth regulators such as GA3. While IAA is determined if rice roots were incubated with associative nitrogen fixing bacteria. Associative N_2 fixing bacteria belong to PGPR (Plant Growth-Promoting Rhizobacteria) (1) Associative N_2 fixing bacteria provided rice with N sources such as ammonia exudates; (2) Associative N_2 fixing bacteria enhance the growth of rice roots by exudation of growth regulators such as GA3 and IAA; (3) Most associative N_2 fixing bacteria have nitratase, which goes into plant roots after inoculation and assist in nitrate reduction in plant and increase the N level, hence enhancing N_2 fixation; (4) Associative N_2 fixing bacteria can enhance plant mineral uptake; (5) Associative N_2 fixing bacteria can enhance the growth of lateral roots.

Table 15.1 Classification of associative nitrogen fixing bacteria in rice root system

Mode of nutrition/ Type of Organism	*Name of Organism*
1. Autotroph Photosynthetic N-fixing bacteria	*Rhodobacter(Rhodopseudomonas)* *Rhodospirillum*
2. Heterotroph	
• Autofixing Bacteria/aerobic	*Azotobacter, Azotomonas*
• Autofixing Bacterial slightly aerobic	*Derxia, Methylomonas, Bacillus*
• Autofixing Bacterial anaerobic-aerobic	*Clostridium, Desulfotomaculum, Desulfovibrio*
• Associative nitrogen fixing bacterialaerobic	*Beijerinckia*
• Associative nitrogen fixing bacteria slightly anaerobic	*Alcaligenes, Arthrobacter, Azospirillum, Flavobacterium, Pseudomonas*
• Associative nitrogen fixing bacteria oxidative-reductive	*Enterobacter, Klebsiella*

15.7.7 Nematodes and Protozoa

Protozoa and nematodes that graze on bacteria are also more abundant in the rhizosphere. Thus, much of the nutrient cycling and disease suppression needed by plants occurs immediately adjacent to roots. Their feeding stimulates microbial activity and enhances the turnover of nutrients, particularly N, which would otherwise become immobilized in microbial tissue. In experimental systems the presence of microbivorous nematodes or protozoa (grazers) enhances N mineralization and the uptake of N by plant. Plant root exudates stimulate microbial growth which in turn, encourages grazing activity. The activity of grazers in the rhizosphere, therefore, is directly relevant to the availability of N to the growing plants.

15.8 HUMUS ESTIMATION IN SOIL/MANURE

Humus is a complex mixture of heterogenous organic compounds derived from plant and animal residues and is formed by the reaction of decomposition, synthesis and polymerization. It is an integral part of soil organic matter and can be extracted by neutral salts or alkali separating humic and non-humic fraction.

15.8.1 Materials

Extracting solution

Dissolve 44.6 g of sodium pyrophosphate and 4 g NaOH in 1 litre water to maintain 0.1 M strength of the solution.

15.8.2 Protocol

1. Take 20 g soil or manure in a conical flask, add 100 mL of extractant (1:5 ratio). Shake well and allow standing for 18 hrs.
2. Filter through Whatman Filter paper No. 1. (The residue portion on filter paper is called humin and the filtrate is humus. It is dark brown in colour with pH 13).
3. To get rid of sodium, the humus is dialyzed in dialysis tubes or parchment paper against running tap water for 24 hrs.
4. It is free of sodium. Measure the volume and a known volume (10 mL) is taken in a pre-weighed 100 mL beaker. The beaker is dried in an oven at 70°C to a constant weight. The difference in weight gives the amount of humus in 10 mL. Accordingly, it is calculated on a percentage basis.

15.9 DETERMINATION OF MICROBIAL CARBON BIOMASS

The content of soil organic matter affects the stability of agro-ecosystems and is controlled by many factors such as cultivation. Over short periods, changes in microbial biomass C can be a sensitive index of changes in the content of soil organic matter. The soil microbial biomass acts as the transformation agent of the organic matter in soil. As such, the biomass is both a source and sink of the nutrients C, N, P and S contained in the organic matter. It is the center of the majority of biological activity in soil. Soil microorganisms are important for the release and retention of nutrients and soil microbial biomass does play an important role while assessing nutrient and energy flow in the soil system. If a soil is fumigated, the respiration rate of soil immediately after removal of the fumigant will be less than that of non-fumigated soil. After a time, the respiration rate of the fumigated soil will increase sharply to a value exceeding that of the non-fumigated soil and still than subside. There is a temporary flush in CO_2 evolution in the previously fumigated soil. This flush of decomposition (CO_2 evolution) is the result of the decomposition of microbial cells killed by the fumigation. The measurement of this flush in respiration could indicate the approximate size of the soil biomass. Fumigation-incubation method helps in determining the soil microbial biomass carbon.

15.9.1 Materials

Soil sample, desiccators, vacuum pump, Millipore filtration assembly, filter paper, centrifuge, test tubes, beakers, conical flasks, chloroform, glass beads.

Reagents freshly prepared (quantities suitable for about 60 samples):

2M KCl solution 1 litre.

50% Analytical grade ethanol in 1 litre distilled water.

4N Acetate buffer (544 g sodium acetate to 400 mL water and stir while heating until dissolved). Cool to room temperature and add 100 mL of glacial acetic acid and make up to 1 litre; check the pH is 5.51 ± 0.03.

Ninhydrin reagent dissolve 2 g ninhydrin and 0.2 g hydrindantin in 50 mL 2-methoxyethanol, add 50 mL acetate buffer and mix; store in a dark glass reagent bottle.

2.5×10^{-4} M leucine solution prepare a stock solution of 2.5×10^{-2} M by dissolving 0.3279 g of L-leucine in 100 mL 2M KCl. For the assay, dilute 1 mL of stock solution with 99 mL of 2M KCl; this standard contains 7 µg N per 2 mL.

2.5×10^{-4} M ammonium sulphate solution prepare a stock solution of 2.5×10^{-2} M by dissolving 0.1652 g analytical grade ammonium sulphate in 100 mL 2M KCl. For the assay, dilute 1 mL of stock solution with 99 mL of 2M KCl; this standard contains 7 µg N per 2 mL solution.

15.9.2 Protocol

15.9.2.1 Sample collection

In the field, soil should be sampled with a spade from the top 10 – 10 cm, after the clearance of dead wood and leaf litter. The number of samples taken should be sufficient to calculate confidence limits for the data produced, i.e., at least 5 samples per transect, but two per 5 m section is a better number, making 16 in total, or two bulked samples per section, making a total of 8. At each sampling point, about 500 g of soil should be placed in a clean plastic bag and taken immediately to the laboratory and brushed firmly through a coarse sieve to remove stones and pieces of wood. The soil should be then air-dried for about 24 hours, subsampled for the determination of moisture content (by evaporation to dryness at 105°C) and rebagged for further processing, which should begin as soon as possible, and not more than 48 hours later. Soil waiting for processing can be stored at 4°C, but must not be frozen or oven-dried.

15.9.2.2 Analysis

The first step is to determine water-holding capacity (WHC). This need not be done on more than one or two samples per transect, as long as the site is pedologically homogeneous.

1. Place about 50 g of soil in a large filter funnel plugged loosely with glass wool and clamp to stand in a beaker. Pour about 100 mL of water onto the soil and leave overnight.
2. Weigh the moist soil and funnel, then dry in an oven at 105°C for 24 hours and weigh again.

   Remove the dry soil from the funnel and weigh this alone.

   $$\text{WHC} = \frac{\text{Weight of wet soil-Weight of dry soil}}{\text{Weight of dry soil}} = \text{g } H_2O/\text{g dry soil}$$

   WHC is then used to adjust all soil samples subsequently weighed out to 40% WHC (= 8.0 g H_2O per 20 g soil), taking account of the moisture already present (from the earlier determination of moisture content). 20 g subsamples adjusted in this way are then ready for fumigation and/or extraction.
3. Weigh two subsamples of 20 g each of air-dried soil into small glass beakers and adjust to 40% WHC by the addition of water. One subsample will be fumigated with chloroform and then extracted with KCl solution, and the other, which is a control, will be extracted with KCl solution without exposure to chloroform. Mark each beaker accordingly using pencil and tape (not felt-tip pens).
4. Place the samples for fumigation in a large vacuum desiccators containing a beaker with about 50 mL of chloroform. Add a few glass beads to the chloroform, close the desiccators and evacuate (using either an electrical or water-jet vacuum pump) until the chloroform starts boiling. Seal the desiccators and leave in the dark for exactly 10 days at 25°C. Leave the control soil samples, loosely covered with foil, under the same conditions, but without exposure to chloroform.
5. After 10 days, extract each soil sample with 63 mL of 2M KCl for one hour, using an end-over-end (or rotary) shaker. Centrifuge for 2-5 min at 2500 × g to separate soil and supernatant. Carefully remove 10 mL of supernatant with a graduated plastic syringe, fit a Millipore prefilter (AP20 013 00) to the syringe and expel the supernatant through the prefilter into a clean 10 mL screw-cap plastic or glass vial. This filtrate can then be frozen pending further processing and analysis, but note that until this point is reached the timetable of the procedure must be adhered to strictly.
6. Use the standard leucine and ammonium sulphate solutions to check the linearity of the colour development with ninhydrin and calibrate the reagent as follows:

   Make up four tubes containing 1 mL of solution with 0, 0.7, 1.4, 2.1 and 2.8 μg N, respectively (i.e., use 0, 0.2, 0.4, 0.6 and 0.8 mL of diluted stock and make up to 1 mL with 2M KCl). Put 1 mL of each of these standards in each of three tubes. Add 1 mL of freshly prepared ninhydrin to each, mix with a vortex and boil in a water bath for 15 min. Cool the tubes in cold water and add 5 mL of 2M KCl to each. Seal with parafilm, shake vigorously for 5 min and immediately read optical density at 570 nm against a blank tube of 2M KCl. Average OD for the three replicates.

7. For each subsample filtrate, prepare three tubes:

A blank containing 1 mL 2M KCl.

Sample 1, containing 0.5 mL of filtrate from the fumigated samples + 0.5 mL 2M KCl.

Sample 2, containing 0.5 mL of filtrate from the unfumigated samples + 0.5 mL 2M KCl.

Add 1 mL of freshly prepared ninhydrin to each, mix with a vortex and boil in a water bath for 15 min. Cool the tubes in cold water and add 5 mL of 2M KCl to each. Seal with parafilm, shake vigorously for 5 min and immediately read optical density at 570 nm against a blank tube of 2M KCl.

15.9.2.3 Results and observations

Microbial biomass carbon (MBC) is defined as 21× the release of ninhydrin reactive N from soils fumigated for 10 days at 25°C. The calculation of results is, therefore:

MBC = ([t_{10}abs – $t_0$abs – y intercept]) × (sample dilution) × (extraction volume/dry soil mass) × (21). It should be expressed as µg C per g dry weight of soil.

15.10 ESTIMATION OF ALGAL (PHOTOSYNTHETIC) BIOMASS IN SOIL CORES

Among the cellular constituents that can be used as a measure of biomass, pigments, primarily chlorophylls are more widely accepted. All chlorophylls are extractable in 80-90% acetone and can be measured spectrophotometrically in the spectrum of 665 nm. Addition of DMSO (Dimethylsulfoxide) is known to improve the extent of extraction, especially in soil samples. Soil chlorophyll estimation provides a good indicator of growth rate of algal populations and has been extensively utilized in soil ecological studies.

15.10.1 Materials

Fresh soil cores, glass vials, acetone, DMSO (dimethylsulfoxide), centrifuge.

15.10.2 Protocol

1. Fresh soil cores (0–3 cm or 0–5 cm) are collected with the help of tube auger, and placed in 55 mL glass vials and tightly sealed using subaseal stoppers.
2. Acetone: DMSO (1:1) mixture is added @ 4mL g^{-1} soil and vortexed to allow proper mixing of soil and extractant.
3. The vials are stored in dark, at room temperature until all the pigments get extracted (48–96 hr).

4. The samples are thoroughly shaken, and the coloured solvent removed, centrifuged to remove debris and OD values taken at 665 nm with a spectrocolorimeter against a blank and expressed in terms of per g soil.

15.10.3 Results and Observations

Quantification of Chlorophyll-a is made by employing the absorption coefficient 12.5.

Concentration of chlorophyll-a (μg/mL) = $12.5 \times A_{665}$ where A_{665} = Absorption at 665 nm

15.11 SOIL DECOMPOSITION AND MICROBIAL COMMUNITY STRUCTURE (WINOGRADSKY TECHNIQUE)

Soil decomposition is a necessary and important process in every ecosystem. It is the process of recycling nutrients that all organism need for survival. Bacteria and fungi are the main components of this process that involves the breaking down of detritus and dead organic materials. Humus (decomposed material) is the final step in this process that supply soil with nutrient such as calcium, phosphate, potassium, and other ions. In any given condition, natural environment teems with microorganisms, to provide a specific combination of nutrients and oxygen that allows only certain microorganisms to survive. One can use their own designed composed media, and inoculum from any waterbody to create and study a mini-ecosystem, called a Winogradsky column, in the laboratory. The objective of this experiment is to demonstrate the impact of eutrophication and plant nutrients on soil decomposition and microbial community construction in the forest and agriculture environments using Winogradsky technique.

15.11.1 Materials

Water sample, agriculture soil, forest soil, mixing container, grass cuttings, pine needles, calcium sulfate, calcium carbonate.

15.11.2 Protocol

1. Obtain a clean and clear container and label your initial, date, and soil type.
2. Place about 200 mL compost soils (agriculture or forest soil) to a mixing container.
3. Add pond water and stir until it is about the consistency of apple sauce.
4. Add 5 g grass cutting for the agriculture soil and 5 g pine needle cutting for the forest soil as a carbon source (carbon source in the form of cellulose).
5. Add an equal amount of calcium carbonate and calcium sulfate and mix until the mixture become drier (source of carbon and sulfur).
6. Pour or spoon this mixture into the decomposition chamber to approximately 3-4 cm in depth.
7. Mix it well with a spoon or stirring rod to remove any air pockets.
8. Add plain agriculture or forest soil to the respective chamber until the depth of the soil mixture reaches between 6 and 8 cm. DO NOT STIR!

9. Add pond water, leaving about 4 cm of headspace.
10. Insert a thermometer into the soil and record the starting temperature in °C.
11. The soil column in each decomposition chamber should be covered with aluminium foil, seal with parafilm, and place next to the window (Figure 15.13).

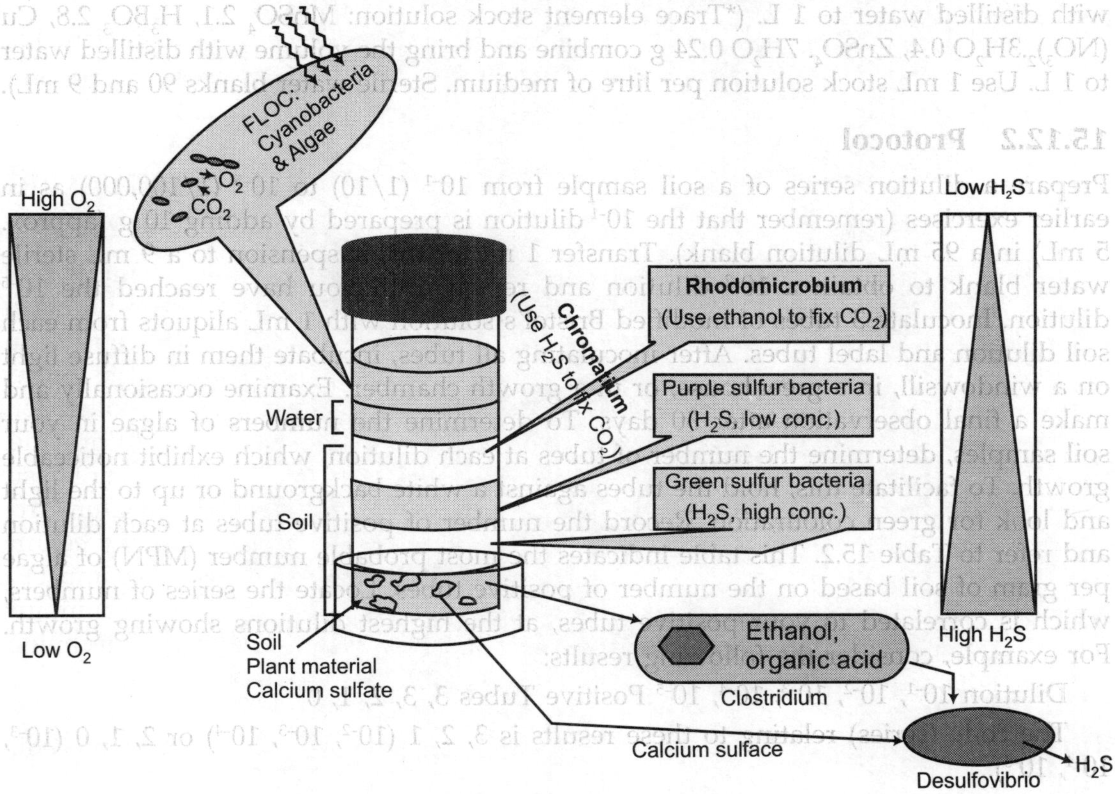

Figure 15.13 Demonstration of Winogradsky technique.

15.11.3 Results and Observations

Create data sheet for weekly observations and discussion of the developed soil layers.

15.12 SOIL ALGAE ENUMERATION

Since algae are photoautotrophs, they can be grown on selective culture media devoid of organic carbon. This eliminates competing microbes, which would outgrow the algae and possibly inhibit their development. One can make these media further selective for cyanobacteria (blue-green algae) by eliminating all sources of fixed nitrogen (ammonium or nitrate, etc.). Enumeration of soil algae is most frequently accomplished by a dilution technique involving inoculation of replicate tubes of an appropriate liquid medium with subsamples from the dilution tubes. This technique is known as the most probable number technique and represents a statistical probability approach to counting microorganisms. It is employed when the group of organisms being investigated is not readily cultured on solid media.

15.12.1 Materials

Modified Bristol's solution (9 mL per test tube) g/L in distilled water: $NaNO_3$ 0.25, $CaCl_2$, 0.025, $MgSO_4$ $.7H_2O$ 0.075 g, K_2HPO_4 0.075, KH_2PO_4 0.018 g, NaCl 0.025, $FeCl_3$ 0.01, $NaMoO_4$ $.2H_2O$ 0.002 and trace minerals*, 1 mL combined and bring the volume with distilled water to 1 L. (*Trace element stock solution: $MnSO_4$ 2.1, H_3BO_3 2.8, Cu $(NO_3)_2.3H_2O$ 0.4, $ZnSO_4$. $7H_2O$ 0.24 g combine and bring the volume with distilled water to 1 L. Use 1 mL stock solution per litre of medium. Sterile water blanks 90 and 9 mL).

15.12.2 Protocol

Prepare a dilution series of a soil sample from 10^{-1} (1/10) to 10^{-5} (1/100,000) as in earlier exercises (remember that the 10^{-1} dilution is prepared by adding 10 g (approx. 5 mL) in a 95 mL dilution blank). Transfer 1 mL of this suspension to a 9 mL sterile water blank to obtain a 10^{-2} dilution and repeat until you have reached the 10^{-5} dilution. Inoculate 3 tubes of modified Bristol's solution with 1 mL aliquots from each soil dilution and label tubes. After inoculating all tubes, incubate them in diffuse light on a windowsill, in a greenhouse, or in a growth chamber. Examine occasionally and make a final observation after 30 days. To determine the numbers of algae in your soil samples, determine the number of tubes at each dilution, which exhibit noticeable growth. To facilitate this, hold the tubes against a white background or up to the light and look for green colouration. Record the number of positive tubes at each dilution and refer to Table 15.2. This table indicates the most probable number (MPN) of algae per gram of soil based on the number of positive tubes. Locate the series of numbers, which is correlated to your positive tubes, at the highest dilutions showing growth. For example, consider the following results:

Dilution 10^{-1}, 10^{-2}, 10^{-3}, 10^{-4}, 10^{-5} Positive Tubes 3, 3, 2, 1, 0

The code (series) relating to these results is 3, 2, 1 (10^{-2}, 10^{-3}, 10^{-4}) or 2, 1, 0 (10^{-3}, 10^{-4}, 10^{-5}).

Refer to Table 15.2 and find these "codes" by reading across the columns. MPN number for 3, 2, 1 is 1.50 and for 2, 1, 0 is 0.15. The MPN number associated with the proper code is multiplied by the reciprocal of the center dilution of the series (series B in this table), to obtain the most probable number per gram of the original soil. Thus the 3, 2, 1 code yields 1.5×10^3/g and 2, 1, 0 yields 0.15×10^{-4} or 1.5×10^3/g also. It should be stressed here that the 3-tube MPN provides minimal accuracy and is used here primarily to introduce the concept and the technique. After you have calculated the most probable number of algae in your soil samples, observe the contents of some tubes (low dilution and high dilution) microscopically using simple wet mounts. Note the pigmentation and morphology of the organisms. Try to classify one or more of these using Prescott's book *How to Know the Fresh Water Algae* available in the laboratory. The most probable number (MPN) technique is a statistical approach to quantification and, as in the plate count technique, involves the preparation of a decimal dilution series. The sample must be diluted to extinction. An aliquot of each dilution (usually 1.0 or 0.1 mL) is then added to a series of tubes containing a growth medium. The number of replicate tubes employed dictates the degree of accuracy. Tables are available for 3, 5, 8, and 10 tube determinations. The table below is a three-tube table. As an example, consider the following: a sample is decimally diluted to extinction. Three 1 mL aliquots from each dilutions is dispensed to three tubes of nutrient broth. After incubation, growth is determined by turbidity. Any turbidity is considered a positive test. At a dilution of 10^{-2} all three tubes are turbid, 10^{-3} has two turbid tubes, 10^{-4} has one turbid tube and 10^{-5} has none. The combination is 3, 2, 1, 0 or 2, 1, 0. Referring to the table, the combination 3, 2, and 1 gives a MPN value of 1.50 for the centre dilution and 2, 1, 0 gives a value of 0.15. The MPN value is multiplied by the reciprocal of the dilution. In these examples, the MPN is (1.5×10^3).

Table 15.2 Three-tubes Most Probable Number (MPN) table

# Of Positive Tubes				*# Of Positive Tubes*			
Series A	*Series B*	*Series C*	*MPN*	*Series A*	*Series B*	*Series C*	*MPN*
0	0	0	<0.03	2	0	0	0.09
0	0	1	0.03	2	0	1	0.14
0	0	2	006	2	0	2	0.20
0	0	3	0.09	2	0	3	0.26
0	1	0	0.03	2	1	0	0.15
0	1	1	0.06	2	1	1	0.20
0	1	2	0.09	2	I	2	0.27
0	1	3	0.12	2	I	3	0.34
0	2	0	0.06	2	2	0	0.21
0	2	1	0.09	2	2	1	0.28
0	2	2	0.12	2	2	2	0.35
0	2	3	0.16	2	2	3	0.42
0	3	0	0.09	2	3	0	0.29
0	3	1	0.13	2	3	1	0.36
0	3	2	0.16	2	3	2	0.44
0	3	3	0.19	2	3	3	0.53
I	0	0	0.04	3	0	0	0.23
I	0	1	0.07	3	0	1	0.39
I	0	2	0.11	3	0	2	0.64
1	0	3	0.15	3	0	3	0.95
I	1	0	0.07	3	I	0	0.43
1	1	1	0.11	3	1	I	0.75
1	1	2	0.15	3	1	2	1.20
I	1	3	0.19	3	1	3	1.60

Contd...

Contd...

1	2	0	0.11	3	2	0	0.93
I	2	I	0.15	3	2	I	1.50
1	2	2	0.20	3	2	2	2.10
I	2	3	0.24	3	2	3	2.90
I	3	0	0.16	3	3	0	2.40
I	3	I	0.20	3	3	I	4.60
I	3	2	0.24	3	3	2	11.00
1	3	3	0.29	3	3	3	>24.00

Soil Metabolic Assessment

15.13 ENZYME-PHOSPHATASE ACTIVITY

While enzymes in soil are sometimes associated with roots, many are of microbial origin. Soil enzymes may retain their activity long after their release from microbial cells primarily because the enzymes form complexes with humic and clay colloids. Complexing with these colloids renders the enzymes highly resistant to denaturation and degradation. Thus, soil can have significant enzymatic activity independent of a native microbial community.

Phosphatase mediates the generalized reaction:

$$C\text{-}PO_4 + H_2O \xrightarrow{\text{Phosphatase}} C\text{-}OH + HPO_4$$

Phosphatase activity in soil is estimated by using *p*-nitrophenyl phosphate as a substrate. As the phosphate is cleaved, *p*-nitrophenol is left. This hydrolyzed product is light sensitive. Samples should, therefore, be kept covered with aluminum foil, to prevent photolysis.

15.13.1 Materials

Soil samples (agriculture or forest), 2 mL Eppendorf microcentrifuge tubes, *p*-nitrophenyl-phosphate solution (100 mM) g/L = 1M, 0.7422 g/20 mL = 100 mM, CaCl2 solution (0.5 M).

NaOH solution (0.5 M), *p*-nitrophenol standard stock solution (100 µM), 0.013911 g/100 mL = 1 mM solution to make 100 mL of 1 µM, take 10 mL of 1 mM solution + 90 mL water. Use the 1 µM solution to make five standard solutions between 0 and 10 µM (10, 5, 2.5, 1.25, and 0).

Microplate reader (vertical path length photometer with 405 nm interference filter), Two flat bottom 300-µL per wells 96-well microplate reader rack, aluminium foil, micropipettes and micropipette tips, spatula, marking pen, and tape.

15.13.2 Protocol

1. Label the microcentrifuge tubes as enzyme assay treatment and control.
2. Wrap each microcentrifuge with aluminium foil.

3. Weigh out 0.25 g soil (dry mass basis).
4. Add 1 mL distilled H_2O to the soil and swirl gently.
5. Initiate the reaction by adding 0.25 mL of 100 mM *p*-nitrophenyl-phosphate to the treatment microcentrifuge tube but NOT to control.
6. Place microcentrifuge tubes in the rack and shake at 175 rpm for 30 min.
   Note: The algae MPN assay should be started during this incubation time.
7. After the incubation period, add 0.25 mL of the *p*-nitrophenyl-phosphate solution to control.
8. Terminate the reaction by adding 0.25 mL of each 0.5 M CaCl2 and 0.5M NaOH to all tubes.
9. Place the treatment and the control tubes of each soil opposite each other in the microcentrifuge and centrifuge at 10,000 rpm for 10 min.
10. Remove tubes from the microcentrifuge and cover with aluminium foil.
11. Run samples against of standard solutions of 0 to 10 µM of *p*-nitrophenol.
12. Samples may require dilution to fit the standard curve. Use dH_2O to make 1:3 dilutions (0.1 mL sample to 0.2 mL distilled H_2O).
13. Place sample wells in the microplate reader rack and mark their locations.
14. The Microplate Program Manager will calculate the concentration of *p*-nitrophenol in each well based on the standard curve.

15.13.3 Results and Observations

Subtract the control *p*-nitrophenol concentration from the treatment *p*-nitrophenol concentration to get the net reaction result. Because each mole of *p*-nitrophenol produces one mole of phosphate, the final data can be directly expressed as µmol phosphate released. Multiply your net results by the dilution factor to express the total as µmol phosphate released g^{-1} soil h^{-1}.

15.14 DEHYDROGENASE ACTIVITY OF THE SOIL

Biological oxidation of organic compounds is generally through a dehydrogenation process, and there are many dehydrogenases (enzymes catalyzing dehydrogenation), which are highly specific. The overall process of dehydrogenation may be presented as follows:

$$XH_2 + A \rightarrow X - AH_2$$

Where XH_2 is an organic compound (hydrogen donor) and A is a hydrogen acceptor. Tetrazolium salts are representatives of a unique class of compounds. These compounds have combinations of desirable properties. They are quaternary NH_4^+ salts and, as such, possess a high degree of water solubility. The method described is based on extraction with methanol and colorimetric determination of TPF produced from the reduction of TTC in soils. The transformation occurs through the rupture of the ring of TTC.

15.14.1 Materials

1. **3% solution of 2, 3, 5-Triphenyltetazolium chloride (TTC)** Dissolve 3 g of TTC in about 80 mL of water and finally make up the volume to 100 mL with water.
2. **Tiphenyl formazan (TPF) standard solution** Dissolve 100 mg of TPF in about 80 mL of methanol and adjust the volume to 100 mL with methanol, mix thoroughly.

15.14.2 Protocol

1. Weight 6 g soil sample for each of the three sterile screw cap tubes.
2. Add 2.5 mL sterile distilled water in two tubes and 1.0 mL of 3% aqueous solution of triphenyl-tetazolium chloride.
3. To the third tube (control) add 3.5 mL sterile distilled water.
4. Mix thoroughly with a mixer and incubate at 30°C for 24 hrs.
5. Methanol extraction: Remove the sample from tubes by shaking with methanol onto a Buchner funnel fitted with Whatman No. 5 paper. Wash the soil with methanol until no more colour could be extracted.

   [Precaution: During this procedure it is necessary to keep the sample wet al all times until extraction is complete to avoid air drawn through the soil].
6. Pour the filtrate in a 100 mL volumetric flask and make up to volume with methanol.
7. After mixing, determine the absorbance at 485 nm.
8. If suspended soil particles are present in tubes, centrifuge the filtrate at 14,000 rpm for 10 minutes before reading absorbance.

15.14.3 Results and Observations

Calculate the amount of TPF produced by reference to a calibration graph prepared from TPF standards. (To prepare this graph, dilute 10 mL of TPF standard solution o 100 mL with methanol (100 µg of TPF/mL). Pipette 5, 10, 15 and 20 mL aliquots of this solution into 100 with methanol and mix thoroughly. Measure the intensity of the red colour of TPF as described for samples. Plot the absorbance readings against the amount of TPF in 100 mL standard solutions).

15.15 RESPIRATION/BIOMASS TESTS

Respiration of microorganisms in soil is one of the earliest and still is one of the most frequently, used indices of soil microbial activity. Measurements of respiration have been found to be well correlated with other parameters of microbial activity such as nitrogen or phosphorus transformations, metabolic intermediates, and average microbial numbers.

Carbon comprises about 45 to 50% of the dry matter of plant and animal tissues. When these tissues or residues are metabolized by microorganisms, O_2 is consumed, and CO_2 is evolved in accordance with the following generalized reaction:

$$(CH_2O)_X + O_2 \longrightarrow CO_2 + H_2O + \text{intermediates} + \text{cell material} + \text{energy}$$

In this reaction, all of the organic carbon should eventually be released as CO_2. In actual practice, under normal aerobic conditions, only 60 to 80% of the carbon is evolved as CO_2 because of incomplete oxidation and synthesis of cellular and intermediary substances. The quantities of CO_2 evolved depend on the type of carbon substrate, the environmental conditions, and the types and numbers of microorganisms involved. When a portion of the soil is fumigated (or otherwise sterilized, e.g., microwave irradiation), the difference in CO_2 evolved between fumigated and non-fumigated soil can be used to estimate microbial biomass. The experiment introduces a simple titrimetric method for measuring CO_2 evolution from soils, and demonstrates the effects that different carbonaceous substrates have on the rate of CO_2 evolution (microbial respiration). In addition, comparisons between the fumigated and non-fumigated samples will provide data for estimating the microbial biomass of the original soil.

15.15.1 Materials

Soil samples, glucose, cellulose, NaOH, NH_4NO_3, beaker, $BaCl_2$.

15.15.2 Protocol

15.15.2.1 Response to carbon addition

Treatments are of agriculture soil (alone, with glucose, or with cellulose additions) and forest soil (alone, with glucose, or with cellulose additions).

Weigh 50 g (dry mass basis) of soil (soil should be at ca. 10% moisture) and place in 1 pint Mason jar. Add 2 mL of NH_4NO_3 solution, 250 mg glucose plus NH_4NO_3 or 250 mg cellulose powder plus NH_4NO_3 (note, add dry ingredients and mix before adding N-solution). Place a 50 mL beaker on the soil. Add exactly 10 mL of 1 N NaOH to the 50 mL beaker and seal with the Mason jar lid. CO_2 produced by the soil is absorbed by the alkali. After 1 week of incubation, remove NaOH with a syringe, place in the beaker, and recharge beaker with fresh NaOH for the second week reading. To determine the amount of CO_2 produced:

1. Add 1 mL of phenolphthalein and 1 mL of 50% $BaCl_2$ to the beaker of NaOH to precipitate the carbonate as an insoluble barium carbonate.
2. Titrate to neutralize the unused alkali with 1 N HCl and record the exact used acid volume.

15.15.3 Results and Observations

Calculate the amount (mg) of C or CO_2 = $(B\text{-}V)\ NE$

Note: $(B\text{-}V)$ is a simplification of $(T\text{-}V)\text{-}(T\text{-}B)$

where, T = Total volume of NaOH at the start of the experiment, V = Volume (mL) of acid to titrate the alkali in the CO_2 collectors from treatments (i.e., with glucose and cellulose) to the endpoint, B = Volume (mL) of acid to titrate the alkali in the CO_2 collectors from controls (i.e., without C addition) to the endpoint, N = Normality of the acid, and E = Equivalent weight.

If data are expressed in terms of carbon, E=6; if expressed as CO_2, E=22. Express the results as mg carbon dioxide produced per g soil.

15.16 DEMONSTRATE THE BIOMASS BY A BIOMETER VESSEL

15.16.1 Materials

Soil sample, Erlenmeyer flask, microwave oven, NaOH, HCl

15.16.2 Protocol

Weigh 50 g dry mass of well mixed soil (soil should contain about 10% moisture) and place in Erlenmeyer flask of a biometer vessel. One vessel from each soil is microwaved for 30 seconds (high power, 700 W, 2450 MHz microwave). Note this is sufficient to kill approximately 90% of the cell. After microwaving, close vessel with rubber stopper. A similar vessel is assembled with non-microwaved soil. Charge side tubes with exactly 10 mL of 1.0 N NaOH and seal. After 10 days, titrate the NaOH as above to determine, the CO_2 evolved (Figure 15.14).

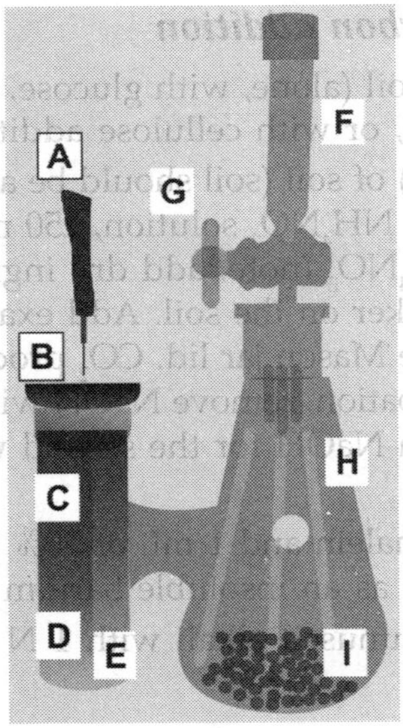

Figure 15.14 Biometer vessel (a) rubber policeman (b) 15-gauge needle (c) 50-mL tube (d) alkali solution (e) polyethylene tubing (f) Ascarite filter (g) stopcock (h) 250-mL flask (i) 50-g soil sample.

15.16.3 Results and Observations

Calculate the mg C biomass by the formula:

(HCl used in control - HCl used in microwaved soil)/0.5 = mg C biomass (assuming 50% conversion of biomass to C in 10 days).

Phosphate Solubilizing Bacteria

Phosphorus (P) is one of the major essential macronutrients for plants and is applied to soil in the form of phosphatic fertilizers. However, a large portion of soluble inorganic phosphate applied to the soil as chemical fertilizer is immobilized rapidly and becomes unavailable to plants. Microorganisms are involved in a range of processes that affect the transformation of soil phosphorus and are thus an integral part of the soil P cycle. Soil bacteria are effective in releasing P from inorganic and organic pools of total soil P through solubilization. Genus *Bacillus, Rhodococcus, Arthrobacter* are some important genus of bacteria which are actively involved in phosphate solubilization (Figure 15.15 and 15.16).

15.17 ISOLATION OF PHOSPHATE SOLUBILIZING BACTERIA FROM SOIL

Phosphorus is one of the vital nutrients for microorganism next to nitrogen. Several species of bacteria such as *Bacillus* sp. and *Pseudomonas* sp. degrade and solubilise the insoluble phosphates into soluble forms through the mechanism of secretion of organic acids such as glycolic acid, acetic acid etc. This acid decreases pH and causes dissolution of bound form of phosphate. The source of phosphates in soil is both minerals and organic phosphates. For isolation of these bacteria, a special type of Pikovskaya's medium is used.

15.17.1 Materials

Soil sample, sterile Petri plates, sterile test tube, inoculating loop, distilled water, Bunsen burner, glassware marking pencil, etc.

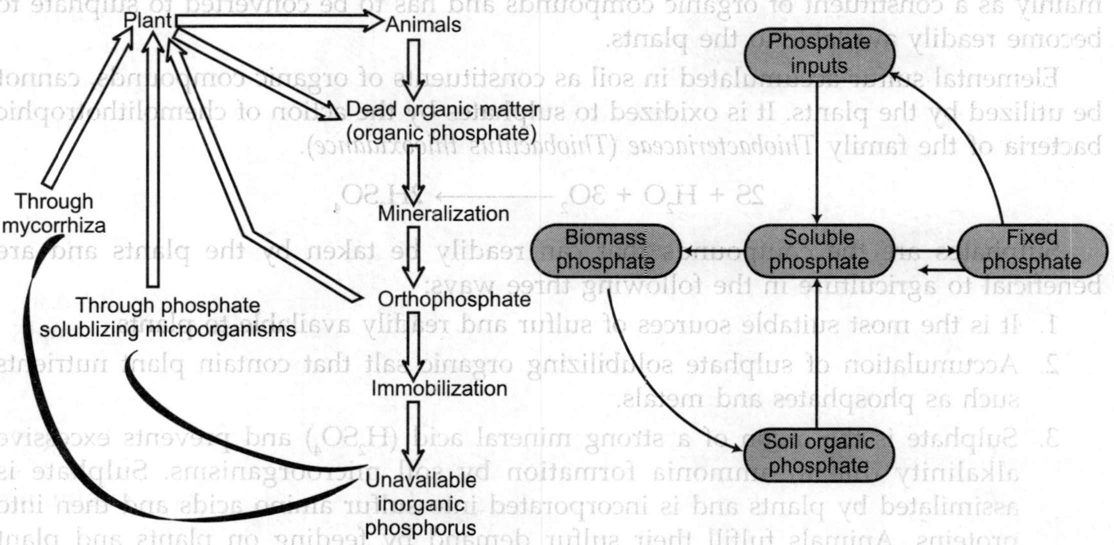

Figure 15.15 Phosphorus cycle.

Pikovaskaya medium g/L in distilled water

Glucose 10, yeast extract 0.5, $Ca_3(PO_4)_2$ 5*, $(NH_4)_2SO_4$ 0.5, KCl 0.2, NaCl 0.2, $MgSO_4.7H_2O$ 0.1, Yeast extract 0.5, $FeSO_4.7H_2O$ Trace, $MnSO_4$ Trace, Agar 15.0g, pH 7.0 ± 0.2

[*Stock suspension of 2.5% $Ca_3(PO_4)_2$ was prepared in distilled water and was autoclaved for preparation of plates or broth, 10 mL of stock suspension was added aseptically to the 90 mL of sterilized medium.]

15.17.2 Protocol

Prepare soil sample by serial dilution method up to the dilution of 10^{-4} described earlier, transfer 1 mL of suspension from 10^{-4} dilution into agar plates containing **Pikovskaya's medium** supplemented with phosphate; incubate in Petri plates at 25°C for 4 to 5 days.

15.17.3 Results and Observations

Transparent zones of clearing around the colonies of microorganisms indicate that phosphate has been solubilised in clearing zones, isolate the culture and identify.

15.18 ISOLATION OF SULFUR OXIDIZING BACTERIA FROM SOIL

Sulfur is considered the fourth major plant nutrient after nitrogen, phosphorus and potassium and is one of the sixteenth nutrient elements which are essential for the growth and development of plants, especially in the agricultural crop production. The importance of sulfur is equal to that of nitrogen in terms of protein synthesis. Plant roots take sulfur in the form of sulphate (SO_4), which undergoes a series of transformation prior to its incorporation into the original compounds. The soil bacteria like *Thiobacillus thioxidans* play a major role in sulfur transformation. Sulfur is accumulated in the soil mainly as a constituent of organic compounds and has to be converted to sulphate to become readily available to the plants.

Elemental sulfur accumulated in soil as constituents of organic compounds, cannot be utilized by the plants. It is oxidized to sulphates by the action of chemolithotrophic bacteria of the family *Thiobacteriaceae* (*Thiobacillus thiooxidance*).

$$2S + H_2O + 3O_2 \longrightarrow 2H_2SO_4$$

Sulphates are the compounds that can readily be taken by the plants and are beneficial to agriculture in the following three ways:

1. It is the most suitable sources of sulfur and readily available to plants.
2. Accumulation of sulphate solubilizing organic salt that contain plant nutrients such as phosphates and metals.
3. Sulphate is the anion of a strong mineral acid (H_2SO_4) and prevents excessive alkalinity due to ammonia formation by soil microorganisms. Sulphate is assimilated by plants and is incorporated into sulfur amino acids and then into proteins. Animals fulfill their sulfur demand by feeding on plants and plant products.

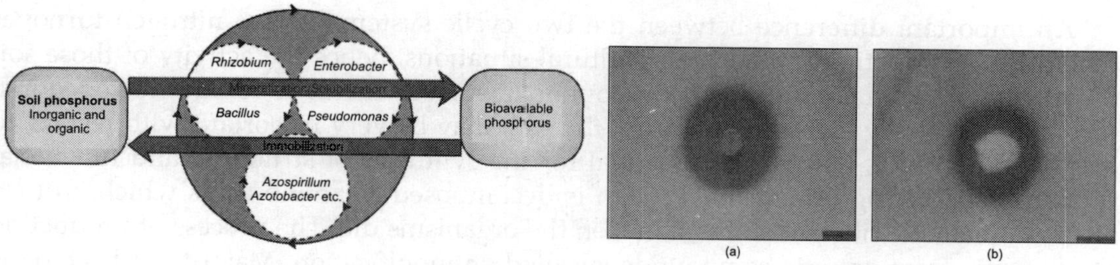

Figure 15.16 Schematic illustration of phosphate mineralization in soil by different bacterial genus and the phosphate-solubilizing zone formed on assay medium containing $Ca_3(PO_4)_2$ (a) and lecithin (b).

15.18.1 Materials

Soil sample, Thiosulphate mineral salt medium, flask, sterile Petri dish, sterile test tube, distilled water, Bunsen burner, inoculating loop, wax marking pencil, etc.

Thiosulphate mineral salt medium (MSM) in distilled water g/L KNO_3 2, NH_4Cl 1, KH_2PO_4, 2, $NaHCO_3$ 2, $MgSO_4.7H_2O$ 0.8, $Na_2S_2O_3.5H_2O$ 5 and 1.0 mL trace element solution with the pH adjusted to 6 with 1N KOH. The trace element solution contained in 1L of distilled water g/L: Na_2-EDTA 50, $CaCl_2.2H_2O$ 7.34, $FeSO_4.7H_2O$ 5, $MnCl_2.4H_2O$ 2.5, $ZnSO_4.7H_2O$ 2.2, $(NH_4)_6Mo_7O_{24}.4H_2O$ 0.5, $CaSO_4.5H_2O$ 0.2 and 11.0 g NaOH. 15 g of agar was added to solidly the medium.

15.18.2 Protocol

Five mL of the soil/wastewater samples were added to a shaking flask containing 50 mL of the MSM and incubated in a rotary shaker (100 rpm) at 30°C for 7 days. The development of turbidity in the medium was assumed to be due to bacterial growth and any flask that also showed a pH drop indicating the growth of thiobacilli was chosen for purification by streaking onto the thiosulfate MSM solid medium also incubated at 30°C for 7 days. The purified cultures were initially screened by growing in the MSM containing Na_2S instead of $Na_2S_2O_3$.

15.18.3 Results and Observations

Observe pH reduction of thiosulphate mineral salt broth and on thiosulphate mineral salt agar medium. Observe for round, smooth, white colonies with an average diameter of 0.5 mm. Microscopic and biochemical examination of sulfur oxidizing bacteria are Gram-negative, nonmotile, short rod catalase and oxidase positive.

15.19 NITROGEN CYCLE

The activity of the organisms involved in the series of processes which constitute the nitrogen cycle results in nitrogen being made available to plants in an assimilable form. These organisms, therefore, act in an analogous way to those involved in the carbon cycle and which make carbon dioxide available as a result of oxidation of carbon-containing compounds.

An important difference between the two cyclic systems is that nitrogen turnover determines productivity in most agricultural situations, hence the activity of those soil organisms metabolizing nitrogen compounds in such a way as to remove nitrogen in a form assimilable by higher plants from the soil, may be very important with respect to loss of soil fertility. Nitrogen is essential for the synthesis of structural and functional protein for all living organisms. Protein is decomposed to amino acids which, in turn are deaminated to liberate ammonia when the organisms die. The process of production of ammonia from organic compounds is called ammonification. Majority of bacteria is soil participate in the process and is a very important step, for bacteria and plants to assimilate it.

In addition to the ammonification of amino acids, other compounds such as nucleic acids, urea, and uric acid go through the ammonification process. The bacteria that accomplish it (*Bacillus, Clostridium, Proteus, Pseudomonas,* and *Streptomyces*) are called ammonifying bacteria. Ammonification of organic compounds is a very important step in the cycling of nitrogen in soils, since most autotrophs are unable to assimilate amino acids, nucleic acids, urea, and uric acid and use them for their own enzyme and protoplasm construction.

15.19.1 Demonstration of Ammonification in Soil

To demonstrate the existence of the ammonification process, inoculate peptone broth with a sample of soil, incubate it for a week, and test for ammonia production. After a total of 7 days incubation it will be tested to see if the amount of ammonia has increased.

15.19.1.1 Materials

Peptone broth, garden soil, inoculation needle, test tubes, Nessler's reagent (commercially available), spot plate, bromothymol blue and indicator chart.

15.19.1.2 Protocol

1. Weigh out 0.1 g of soil.
2. Add the soil to a tube of peptone broth (ammonia forming broth).
3. Label the tube with the soil description, incubate at room temperature with the cap loosened for 1 week.
4. After 7 days, test the medium for ammonia with the following procedure.
5. Deposit a drop of Nessler's reagent into a depression of a spot plate (Figure 15.17).
6. Add a drop of the inoculated peptone broth/soil culture to the Nessler's reagent in the depression. Repeat steps using uninoculated peptone broth.
7. Check the pH of the two tubes by placing 50 μL of each in separate depressions on the spot plate and adding one drop of bromothymol blue to each one and compare the colour with a colour chart or set of indicator tubes to determine the pH.

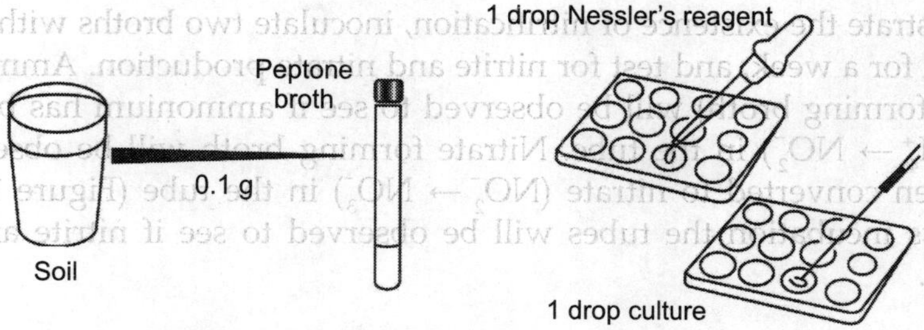

Figure 15.17 Test for ammonification.

15.19.1.3 Results and observations

Interpret the amount of ammonia present as follows:

(a) No color-no ammonia –

(b) Pale orange clour-small amount of ammonia +

(c) Orange-more ammonia ++

(d) Brown precipitate-large amount of ammonia +++

Compare results with the uninoculated control. Record results.

	Incubated soil culture	Uninoculated broth
Ammonia produced		

15.19.2 Demonstration of Nitrification in Soil

Some microbes can utilize nitrates as electron acceptors and metabolize organic substances without the need for oxygen. This process reduces nitrates to nitrites and free nitrous oxide and nitrogen gases ($NH_4^+ \rightarrow NO_2^- \rightarrow NO_3^-$). The nitrate released into the soil is highly soluble and easily assimilated by photoautotrophic bacteria, algae, and plants that convert it into the amino acids needed for their own enzyme and protoplasm construction. This process is called nitrification. The chemoautotrophic bacteria that accomplish it (*Nitrobacter, Nitrococcus, Nitrosococcus,* and *Nitrosomonas*) are called nitrifying bacteria. These chemoautotrophs use the energy of reduced nitrogen compounds, such as ammonium and nitrite, as an energy source for the autotrophic production of organic compounds. Nitrification requires oxygen as a reactant and occurs in aerobic soils. *Nitrosomonas* and *Nitrobacter* are autotrophic and appear to be the principal organisms in soil that can perform these reactions. There are some heterotrophs that can produce nitrites and nitrates, but the amount is insignificant. These organisms are very small Gram-negative rods that do best in a completely inorganic medium using carbon dioxide as their carbon source. Many kinds of organic matter are toxic to them, especially the substances that have free amino groups.

To demonstrate the existence of nitrification, inoculate two broths with a soil sample, incubate them for a week, and test for nitrite and nitrate production. Ammonium sulfate broth (nitrite forming broth) will be observed to see if ammonium has been converted to nitrite ($NH_4^+ \rightarrow NO_2^-$) in the tube. Nitrate forming broth will be observed to see if nitrite has been converted to nitrate ($NO_2^- \rightarrow NO_3^-$) in the tube (Figure 15.18). After a total of 7 days incubation the tubes will be observed to see if nitrite and nitrate has been released.

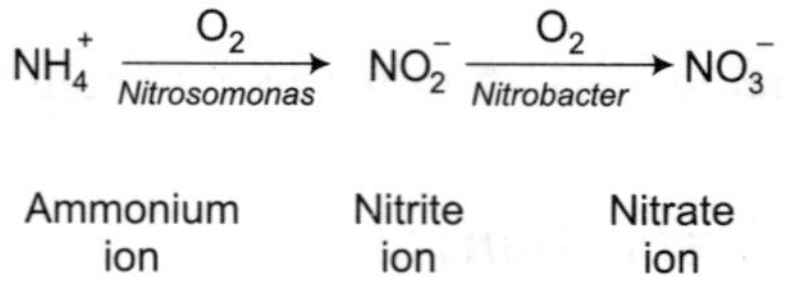

Figure 15.18 Nitrification process by bacteria in soil.

15.19.2.1 Materials

Composition of Medium (g/L)

Ammonium sulfate [$(NH_4)_2$ SO_4]	2	–
Sodium nitrite [$NaNO_2$]	–	1
Magnesium sulphate [$MgSO_4$, H_2O]	0.5	0.5
Ferrous sulphate [$FeSO_4$, $7H_2O$]	0.03	003
Sodium chloride [NaCl]	0.3	0.3
Magnesium carbonate [$MgCO_3$]	10	–
Sodium carbonate [Na_2CO_3]	–	1
Dipotassium pbosphate [K_2HPO_4]	1	1

Reagents

Trommsdorf's Solution: Add slowly, with constant stirring, a boiling solution of 20 g of zinc chloride in 100 mL of distilled water to a mixture of 4 g of starch in water. Continue heating until the starch is dissolved as much as possible, and the solution is nearly clear. Dilute with water and add 2 g of zinc iodide (potassium iodide will do). Dilute to 1 litre and filter. Alternatively, the reagent is commercially available.

Nessler's reagent *(for ammonia)*: Dissolve 50 g of KI in the smallest possible quantity of cold water (50 mL). Add a saturated solution of mercuric chloride (about 22 g in 350 mL of water will be needed) until an excess is indicated by the formation of a precipitate. Then add 200 mL of 5 N NaOH and dilute to 1 litre. Let settle, and draw off the clear liquid. Alternatively, the reagent is commercially available.

Diphenylamine: Dissolve 0.2 g in 100 mL of concentrated sulfuric acid.

Ammonium sulphate and sodium nitrate is separately sterilized by bacterial filter or at low pressure 5-8 psi for 30 minutes so as to protect the chemical nature of salt and added to maintain the required concentration.

Ammonium ion containing medium as a nitrogen source is to be used to detect the presence of nitrite producing (*Nitrosomonas)* bacteria and a medium containing nitrate is to be used for isolating nitrate producing *(Nitrobacter)* bacteria.

15.19.2.2 Protocol

Test for *Nitrosomonas*

Take 25 mL broth of the medium in 6 oz flat bottle separately for *Nitrosomonas* and *Nitrobacter* bacterial isolation, inoculate with 100-200 µL soil suspension and incubate at room temperature for one week and measure the nitrite production in ammonium containing medium and nitrate production in nitrite medium (Figure 15.19).

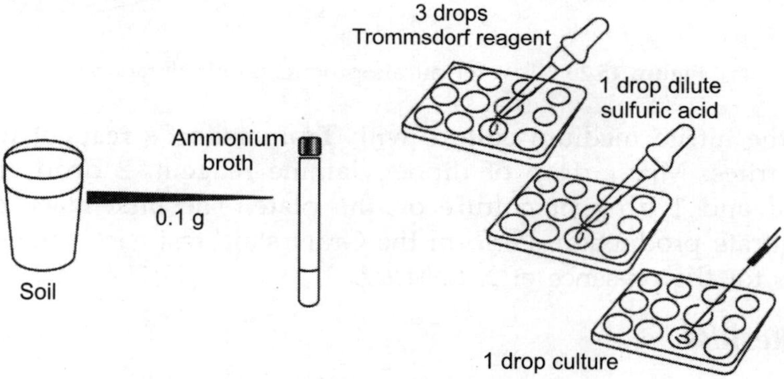

Figure 15.19 Test for nitrification (nitrite production by *Nitrosomonas*).

Measurement of Nitrite Production

Mix 3-5 drops of **Trommsdorf's reagent** with 1 drop of dilute sulphuric acid (1:3 conc) on a spot plate and transfer 2-3 drops of the culture from the ammonium medium to this reagent, mix to avoid false result. Appearance of a blue black colour confirms the production of conversion of ammonium to nitrite. Subsequently perform the test of the residual ammonium ions in the medium by **Nessler's reagent**. Negative test for absence of ammonium ions confirm the complete conversion of ammonium to nitrite. Perform the Gram stain test each time. Gram-negative cocci confirms for the presence of Nitrosomonas.

Test for *Nitrobacter*

Diphenylamine test is performed for the production of nitrate by *Nitrobacter* bacteria. This reagent produces a blue black colour in the presence of nitrate or nitrites. It is, therefore, necessary to make sure that no nitrite is present when it is used as a test for nitrates (Figure 15.20).

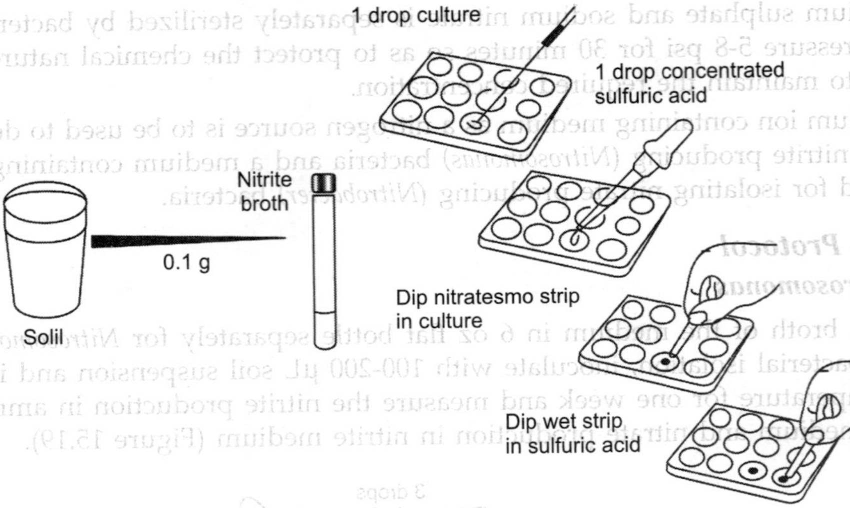

Figure 15.20 Test for nitrate production by *Nitrobacter*.

First test the nitrite medium culture with **Trommsdorf's reagent** to establish the absence of nitrites. Mix 1 drop of diphenylamine reagent, 2 drop of concentrated sulphuric acid and 1 drop of culture on the plate. The blue black colour will be evidence of nitrate production. Perform the Gram stain test each time Gram-negative cocci confirms for the presence of *Nitrobacter*.

15.19.2.3 Results

	Incubated soil culture	Uninoculated broth
Nitrite		
Nitrate		

15.20 MOST PROBABLE NUMBER METHOD FOR *NITROSOMONAS* AND *NITROBACTER*

The soil bacteria that oxidize ammonium to nitrite and further to nitrate are chemoautotrophic. They are able to use an inorganic material as their sole source of energy and carbon dioxide as their sole source of carbon. This property is taken to enumerate the nitrifying bacteria. In the case of *Nitrosomonas,* dilutions of soil are inoculated into an inorganic medium containing ammonium as the source of nitrogen. If *Nitrosomonas* is present in viable form in the inoculums, growth will occur, and nitrite will be produced. Hence, a positive test for nitrite in the inoculated medium but not in the controls indicates the presence of *Nitrosomonas.* To enumerate and isolate the *Nitrobacter* group, the medium employed is free of organic material and contains nitrite as an energy source. The presence of *Nitrobacter* in the inoculums is indicated by a negative test for nitrite will have been oxidized to nitrate if *Nitrobacter* group are present but not if they are absent.

15.20.1 Materials

Erlenmeyer flasks 250 mL capacity, culture tubes, shaker, sterile pipettes, Griess-Illosvay reagent (Merck).

Medium for *Nitrosomonas* and *Nitrobacter* g/L in distilled water $(NH_4)SO_4$, 0.235; KH_2PO_4, 0.200; $CaCl_22H_2O$, 0.040; $MgSO_4.7H_2O$, 0.040; $FeSO_4.7H_2O$, 0.005; NaEDTA.$7H_2O$, 0.005; phenol red, 0.005 pH adjusted to 8.0. The same medium is used for both *Nitrobacter* and *Nitrosomonas*, but for the cultivation of *Nitrobacter*, along with the above ingredients, $NaNO_2$ (247.0 mg/mL) is added.

Griess-Illosvay reagent (a) dissolve 0.6 g sulfanilic acid in 70 mL hot (90 + C) distilled water; cool the solution, add 20 mL of concentrated hydrochloric acid (HCl), dilute the mixture to 100 mL with distilled water, and mix; (b) dissolve 0.6 g of alpha naphthylamine in 10 to 20 mL of the distilled water containing 1 mL of concentrated hydrochloric acid (HCl); dilute to 100 mL with distilled water and mix, and (c) dissolve 16.4 g of sodium acetate (CH_3COONa ~$3H_2O$) in distilled water, dilute to 100 mL with distilled water and mix.

Zinc-copper-manganese dioxide mixture: Mix together 1 g of powdered zinc metal (Zn), l g of powdered manganese dioxide (MnO_2), and 0.1 g of powdered copper (Cu).

15.20.2 Protocol

1. Collect the soil sample, and prepare serial dilutions.
2. From the highest serial dilution prepared, transfer 1 mL aliquots to each of five tubes containing sterile medium for *Nitrobacter* and *Nitrosomonas.*
3. Make a similar set of inoculations using the four next lower and consecutive serial dilutions.
4. Incubate the incubated tubes for 3 weeks at 30°C. Also, include a set of uninoculated controls.
5. After incubation, test each tube using Griess-Illosvay reagent. Development of pink colour indicates the presence of nitrite.
6. To all those tubes that yield a negative test, add a small pinch of Zn–Cu–MnO_2 mixture. Find the most probable number from MPN table (Table 15.2).

15.20.3 Results and Observations

The presence of *Nitrobacter* in the inoculums is indicated by a negative test for nitrite will have been oxidized to nitrate if *Nitrobacter* group are present but not if they are absent.

15.21 DEMONSTRATION OF MICROBIAL DENITRIFICATION IN SOIL

Some microbes can utilize nitrates as electron acceptors and metabolize organic substances without the need for oxygen. This process reduces nitrates to nitrites and free nitrous oxide and nitrogen gases ($NO_3^- \rightarrow NO_2^- \rightarrow N_2O \rightarrow N_2$). The gases diffuse out the soil and into the atmosphere removing their nitrogen from the nitrogen exchange pool. This process is called denitrification (Figure 15.21).

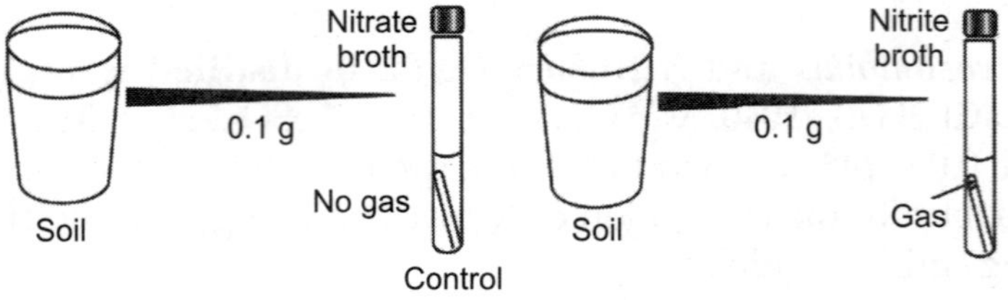

Figure 15.21 Test for denitrification.

The bacteria that accomplish it (*Alcaligenes, Bacillus, Paracoccus, Pseudomonas*) are called denitrifying bacteria. Denitrification happens most frequently in waterlogged, anaerobic soils. It can be minimized by aeration of the soil through cultivation.

To demonstrate the existence of this process, inoculate nitrate broth in a Durham tube with a sample of soil, incubate it for a week, and test for gas production. After a total of 7 days incubation it will be observed to see if nitrogen gas has collected in the Durham tube.

15.21.1 Materials

Two tubes of nitrate broth (nitrogen gas forming broth), Durham tube (containing an inverted vial), soil sample.

15.21.2 Protocol

1. Weigh 0.1 g of soil, add into a tube containing nitrate broth (Nitrogen Gas Forming Broth).
2. Label the tube with the soil description. If any gas is present in the inverted vial, remove it by inverting the entire tube and tapping lightly.
3. Incubate at room temperature with the cap tight for 1 week.
4. After 7 days, test the medium for nitrogen gas with the following procedure.

15.21.3 Results and Observations

1. Observe the inoculated and incubated nitrate broth/soil culture Durham tube. Look for a gas bubble inside the inverted vial.
2. Repeat step 1 using uninoculated nitrate broth Durham tube.
3. Interpretation of nitrogen gas presence is as follows:
   (a) Gas present in inverted vial +
   (b) No gas in inverted vial –

Compare results with the uninoculated control. Record results as per following tabulated form.

	Incubated Soil Culture	Uninoculated Broth
Nitrogen Gas		

15.22 ISOLATION AND MICROSCOPIC EXAMINATION OF FUNGI

15.22.1 Materials

Martin's agar medium: KH_2PO_4 1 g, $MgSO_4.7H_2O$ 0.5 g, peptone 5 g, dextrose 10 g, Rose Bengal (1% aqueous solution) 3.3 mL, streptomycin/streptopenicillin 30 mg (to be added after autoclaving and just before pouring), agar agar 15 g and distilled water to 1 litre. The medium is also known as Peptone-dextrose Rose Bengal agar medium.

Soil sample, pipettes of 10 mL and 1 mL capacity, Erlenmeyer flasks 250 mL capacity, incubator, cotton blue, lactophenol, syringe with needle, spirit lamp/ Bunsen burner.

15.22.2 Protocol

1. Collect soil at random in sterilized polythene bags, make a composite sample by mixing the above samples.
2. Add 0.005 to 0.15 g (approximately) air-dried soil each in 5 sterile Petri plates with the help of a sterilized cooled loop or transfer needle.
3. Add 15-20 mL of melted, cooled (45°C) medium, supplemented with streptomycin/ streptopenicillin and Rose Bengal, to each soil inoculated Petri plate.
4. Dispense soil particles throughout the medium by gentle rotation of the Petri dishes.
5. Allow the plates to solidify.
6. Incubate the plates at 25°C in an inverted position for 15 days.

15.22.3 Results and Observations

Examine selected colonies using the dissecting microscopes and low power objective of the compound microscope. Look especially for fruiting structures (e.g., spores). Aseptically remove small portions of fungal and an actinomycete culture from dilution plates and make 'squash' mounts in various stains. Observe the plates for the appearance of the fungal colonies after 3 days of incubation and continue till 15 days. Purify various fungal isolates for identification up to a specific level and future use (Figure 15.22).

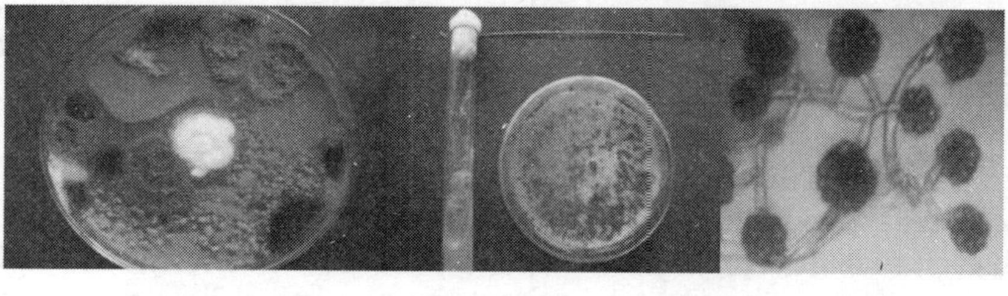

Figure 15.22 Mixed culture of fungi from soil in a Petri plate, *Trichoderma viride* on culture plate and slants and microscopic examination of conidial attachment of *Trichoderma* viride.

15.23 ISOLATION OF MICROORGANISMS FROM SOIL

Microorganisms in soil are important because they affect the structure and fertility of different soils. Soil microorganisms exist in large numbers in the soil as long as there is a carbon source for energy. A large number of bacteria in the soil exist, but because of their small size, they have a smaller biomass. Actinomycetes are a factor of 10 times smaller in number but are larger in size, so they are similar in biomass to bacteria. Fungus population numbers are smaller, but they dominate the soil biomass when the soil is not disturbed. Bacteria, actinomycetes, and protozoa are hardy and can tolerate more soil disturbance than fungal populations, so they dominate in tilled soils while fungal and nematode populations tend to dominate in untilled or no-till soils. Despite this diversity, bacteria, including the mold like actinomycetes and fungi are the most prevalent as shown in Table 15.3.

Table 15.3 Relative number and biomass of microbial species at 0-6 inches (0-15 cm) depth in soil microbes are in the mixed population and should be separated into individual species for study. Several techniques are applied to isolate and study microorganisms in pure culture. Different media are employed to support the growth of these microbes.

Microorganisms	*Number/g of soil*	*Biomass (g/m²)*
Bacteria	10^8 _ 10^9	40 - 500
Actinomycetes	10^7 - 10^8	40 - 500
Fungi	10^5 _ 10^6	100 1500
Algae	10^4 _ 10^5	1 50
Protozoa	10^3 - 10^4	Varies
Nematodes	10^2 - 10^3	Varies

15.23.1 Materials

Though various methods are available to isolate microbes from soil, the serial dilution agar plating method or viable plate count method is one of the commonly used one.

Organisms	*Media*
1. Bacteria	Nutrient agar
2. Actinomycetes	Glycerol Yeast extract agar
3. Cyanobacterial Algae	Chu 10/A & A agar
4. Molds	Glucose Peptone acid agar/ Potato dextrose agar with Rose Bengal and antibiotics to check bacterial growth 10^2 – 10^3

15.23.2 Protocol

The method is based upon the principle that when material containing microorganisms is cultured, each viable microorganism will develop into a colony, hence the number of colonies appearing on the plates represent the number of living organisms present in the sample. In serial dilution agar-plate method, a known amount (1 mL or 1 g) of material is suspended or agitated in a known volume of sterile NaCl water blank (9 mL to make the total volume to 10 mL) to make a microbial suspension. Serial dilutions 10^{-2} to 10^{-7} are made by pipetting measured volumes (1 mL - 9 mL) of sterile water. Finally, 0.1 mL aliquot of various dilutions is added to sterile Petri dishes to which 15 mL of sterile, cool, molten media is added to each one. If it is fungi or actinomycetes the glycerol yeast agar and Sabourauds agar supplemented with 10 µg of chlortetracycline per millimetre of medium to inhibit the growth of bacteria., The dilutions of 10^{-2} to 10^{-5} are selected for enumeration of fungi, 10^{-3} to 10^{-6} for actinomycetes and 10^{-4} to 10^{-7} for bacteria are selected. Upon solidification the plates are incubated, in an inverted position for 3-7 days at 25°C (Figure 15.23).

15.23.3 Results and Observations

The number of colonies appearing on dilution plates are counted, averaged and multiplied by the dilution factor to find the number of cells/spores per gram of the sample

Number (average of 3 replicates) of No. of cells/mL or g

$$= \frac{\text{Colonies} \times \text{dilution factor}}{\text{Dry wt. of soil}}$$

Dilution factor = Reciprocal of the dilution (e.g., $10^7 = 10^7$)

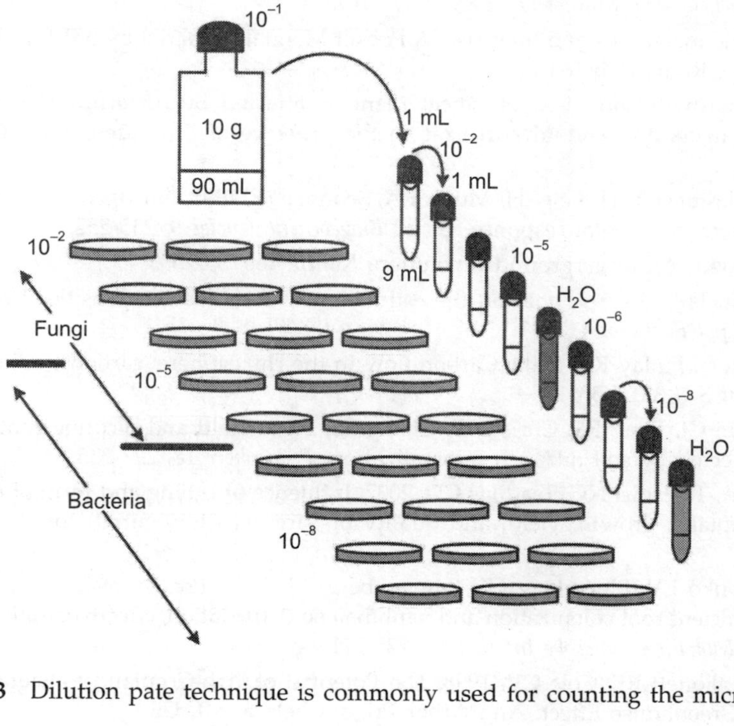

Figure 15.23 Dilution pate technique is commonly used for counting the microorganism.

Suggested Readings

Azcon R, Barea Jm. 1976. Synthesis Of auxins, gibberrellins and cytokinins by *Azotobacter vinelandii* and *Azotobacter beijerinckii* related to effects produced on tomato plants. *Plant Soil* **43**, 609–613.

Barker AV, Bryson GM. 2007. Nitrogen. In: Barker AV, Pilbeam DJ (Eds.). *Handbook of Plant Nutrition* p. 21-51 CRC Press. Taylor & Francis Group, Boca Ratan.

Benfey PN, Bennett M, Scheifelbein J. 2010. Getting to the root of plant biology: impact of the Arabidopsis genome sequence on root research. *The Plant journal* **61**, 992-1000.

Berg G, Smalla K. 2009. Plant species and soil type cooperatively shape the sturcutre and function of microbial communities in the rhizosphere. *FEMS Microbiol Ecol* **68**, 1-13.

Burgmann H, Meier S, Bunge M, Widmer F, Zeyer J. 2005. Effect of model root exudates on structure and activity of a soil diazotroph community. *Environ Microbiol* **7**, 1711-1724.

Calvaruso C, Turpault M, Frey-Klett P. 2006. Root-associated bacteria contribute to mineral weathering and to mineral nutrition in trees: a budgeting analysis. *Appl Environ Microbiol* **72**, 1258-1266.

Canada Soil Survey Committee, 1978. The Canadian System of Soil Classification, Agriculture Canada, Queen's Printer, Ottawa.

Cheng W, Johnson DW, Fu S. 2003. Rhizosphere effects on decomposition: Control of plant species, phenology and fertilization *Soil Sci Soc Am J* **67**, 1418-1427.

Collins HP, Blevins RL, Bundy LG, Christenson DR, Dick WA, Muggins DR, Paul EA. 1999. Soil carbon dynamics in corn-based agroecosystems: Results from carbon-13 natural abundance. *Soil Sci Soc Am J* **63**, 584-591.

Dagley S. 1967. The microbial metabolism of phenolics. p. 287-317. *In* McLaren AD, Peterson GH. (eds.), *Soil Biochemistry*. Marcel Dekker, New York.

FAO, 1984. Legume inoculants and their use. A Pocket Manual Prepared by NiFTAL Project and FAO, USA and FAO, Rome Italy, 63 pp

Fares CN. 1997. Growth and yield of wheat plant as affected by biofertilisation with associative, symbiotic N2-fixers and endomycorrhizae in the presence of the different P-fertilizers. *Ann Agr Sci* **42**, 51-60.

Galloway JN, Schlesinger WH, Levy H, Muchat A, Schnoor JL. 1995. Nitrogen fixation: Anthropogenic enhancement-environmental response. *Global Biogeochem. Cycles* **9**, 235-252.

Gewin V. 2010. Food: An underground revolution. *Nature* **466**, 552-553.

Hegazy NA, Neimela S. 1976 A note on the estimation of Azotobacter densities by membrane filter technique. *J Appl Bacteriol* **41**, 311.

Jones DL, Nguyen C, Finlay RD. 2009. Carbon flow in the rhizosphere: carbon trading at the soil-root interface. *Plant Soil* **321**,5-33.

Ju XT, Chao FC, Li CJ, Jiang RF, Christie P, Zhang FS. 2008. Yield and nicotine content of flue-cured tobacco as affected by soil nitrogen mineralization. *Pedosphere* **18**, 227-235.

Karaivazoglou NA, Tsotsolis NC, Tsadilas CD. 2007. Influence of liming and form of nitrogen fertilizer on nutrient uptake, growth, yield, and quality of Virginia (flue- cured) tobacco. *Field Crops Res* **100**, 52-60.

Kuiper I, Kravchenko LV, Bloemberg GV, Lungenberg BJJ. 2002. *Pseudomonas putida* strain PCL 1444, selected for efficient root colonization and naphthalene degradation, effectively utilizes root exudates components. *Mol Plant Microbe Interact* **15**, 734-741.

Lal R, Kimble JM, Follett RF, Cole CV. 1998. The Potential of U.S. Cropland to Sequester Carbon and Mitigate the Greenhouse Effect. Ann Arbor Press, Chelsea, MI, USA.

Maillet F, Poinsot V, Andre O, Puech-Pages V, Haouy A, Gueunier M, Cromer L, Giraudet D, Formey D, Niebel A, Andres Martinez E, Driguez H, Cecard G, Denarie J. 2011. Fungal lipochitooligosaccharide symbiotic signals in arbuscular mycorrhizal. *Nature* **469**, 58-64.

Martin F, Duplessis S, Ditengou F, Lagrange H, Voiblet C, Lapeyrie F. 2001. Developmental cross talking in the ectomycorrhizal symbiosis: signals and communication genes. *New Phytologist* **151**, 145-154.

Millar CE, L.M. Turk LM, Foth HD, 1965. Fundamentals of Soil Science, 4th ed., Wiley, New York. 491 pp.

Narula N, Kumar V, Singh B, Bhatia R, Lakshminarayana K. 2005. Impact of biofertilizers on grain yield in spring wheat under varying fertility conditions and wheat cotton rotation. *Arch Agr Soil Sci* **51**, 79–89.

Postgate J. 1998. *Nitrogen fixation*, 3rd Ed., Cambridge University Press, Cambridge, U. K.

Raich JW, Mora G. 2005. Estimating root plus rhizosphere contributions to soil respiration in annual crop lands. *Soil Sci Soc Am J* **69**, 634–639.

Sifola MI, Postiglione L. 2003. The effect of nitrogen fertilization on nitrogen use efficiency of irrigated and non-irrigated tobacco (*Nicotiana tabacum* L.). *Plant and Soil* **252**, 313-323.

Simonson RW. 1959. Outline of a generalized theory of soil genesis. *Soil Sci Soc Amer Proc* **23**, 152-156.

Sjoblad RD, Bollag JM. 1981. Oxidative coupling of aromatic compounds by enzymes Volume 5, Marcel Dekker, New York.

Stevenson FJ. 1986. *Cycles of Soil: Carbon, Nitrogen, Phosphorus, Micronutrients*. Wiley Interscience, New York.

Vessey JK. 2003. Plant growth promoting rhizobacteria as biofertilizers. *Plant Soil* **255**, 571-586.

Von Bülow JF, Döbereiner J. 1975. Potential for nitrogen fixation in maize genotypes in Brazil. *Proc Natl Acad Sci USA* **72**, 2389–2393.

Zhao G, Liu Jingshuang L, 2009. Effects of elevated CO2 concentration and nitrogen supply on biomass and active carbon of freshwater marsh after two growing seasons in Sanjiang Plain, Northeast China. *Journal of Environmental Sciences* **21**, 1393–1399.

Chapter 16

The Aquatic Environment

16.1 INTRODUCTION

Water is essential for the well-being of all living organisms and is crucial for the growing of our food and the operation of all our industries. All microbes live in an aqueous environment and their distribution is extremely diverse. Without water they either perish or become inactive. Rain, snow, or hail remove large numbers of microorganisms from the air. Hence, water reaching the earth by precipitation is not sterile. Over most of the earth's surface rainwater collects microorganisms from the surfaces of plants, buildings, or soil on which it falls. Large numbers are acquired from the soil and pass out with the drainage water into streams and rivers and thence into freshwater lakes and eventually into the ocean or inland seas. The number and type of microorganisms in surface water varies according to the source of the water, its organic and inorganic content and with geographical, biological, and climatic factors. Microbes have evolved that can live in saturated salt (sodium chloride) solutions at temperatures from below freezing to >150°C; they thrive in waters full of toxic substances like copper, cyanide, lead, silver, gasoline, oil, benzene, and a plethora of other noxious natural and man-made substances. Water from natural springs or artesian wells is usually relatively free of organisms owing to the filtering effect of natural percolation through the surface rocks. The largest forms of heterotrophic organisms exist on the bottoms and banks of rivers and lakes where organic matter predominates. It is a near natural law that where liquid water, an energy source and basic nutrients exist, you will find living microbes. This is why the scientific community is so interested in finding water on Mars or any other planet/moon in our solar system. 70% of the earth is covered with water. However, the threat of human activities by contamination of water supplies with sewage is a prime concern of everyone. The three descriptive terms in front of the word water changes the entire picture. These words are: potable, polluted and fresh. In running water the effect of the current is important. Most naturally occurring water contains nutrients and other substances in solution, together with colloidal and particulate matter. Pure water probably never occurs in nature but the concentration of dissolved substances varies from negligible amounts in some upland waters to progressively higher concentrations in fresh water which has collected by drainage of agricultural and industrial areas, brackish waters of estuaries, the saline waters of the oceans and the extremely saline waters of some inland lakes in regions of excessive evaporation, e.g., the Dead Sea. The most concentrated of these is still a relatively dilute solution, but the range is sufficient to exert a selective effect on the organisms. Some organisms inhabiting tidal estuaries are able to tolerate a wide range of salinity but few, if any, extend over the whole range. In fresh water the concentration of nutrients does not determine the species present but influences the productivity or number of individuals developing. Productive waters are termed eutrophic, unproductive ones, oligotrophic.

16.2 AQUATIC MICROORGANISMS

16.2.1 Primary Producers

Algae, cyanobacteria, protozoans and microscopic invertebrtaes are the primary producers of organic matter. The algae and cyanobacteria are the main producers of organic matter in aquatic habitats through photosynthesis although they occupy the water column only down to the point where photosynthetically available light penetrates. This may be to the bottom of a shallow lake or to the seabed near the coast but is only a relatively thin 'veneer' of water in the open ocean. The factors affecting the growth of algae and the means of study are similar in both marine and freshwater habitats. The organisms in the two environments are, however, almost entirely distinct, few if any species being common to both. The same genus may occur, but not the same species. There are two major spheres in the aquatic environment; the benthic, which encompasses a vast range of habitats all associated with solid/liquid interfaces and therefore usually with the bottom, e.g., the surfaces of rocks, silt, plants, animals, etc. (Figure 16.1), and the planktonic, which encompases all the niches within the water mass itself and within which the organisms must either float or be capable of swimming. Study of the biology of the organisms in these aquatic spheres is the province of biological limnology and oceanography, though these are intimately interconnected with physical and chemical limnology and oceanography which studies the framework within which the organisms grow.

The algal flora tends to be quite discrete in each habitat. In the benthos two distinct life forms are found, one attached and non-motile, the other unattached and capable of horizontal and vertical movements. The attached species (Figure 11.42 and 11.43) grow on rock or stone surfaces (epilithic) either as single cells or in groups, and include many diatoms (e.g., *Synedra, Meridion, Licmophora)*, the cyanobacteria *Chamaesiphon,* creeping filaments or semicircular mucilage colonies (e.g., *Hildenbrandia, Rivularia, Chaetophora),* or filamentous outgrowths attached by basal holdfast cells (e.g., *Cladophora, Sphacelaria* and many small Rhodophytes). Many of these tend to be very firmly 'glued' to the substrate, some even penetrating it (e.g., the lime boring species of cyanobacteria). Some produce considerable amounts of mucilage in which they are embedded whilst others precipitate calcium carbonate around them and build up nodular growths some of which become fused with the rock. 'Beach rock' is formed in a similar manner in many tropical regions by the cementing of sand grains and calcium carbonate amongst the algal cells. Very much smaller algal cells are attached amongst the coating of bacteria on both freshwater and marine sand grains (epipsammic flora); the majority belong to the Bacillariophyta, though some cyanobacteria (Figure 10.45 and 10.47) and chlorophyta (Figure 11.44) are also reported. A very dense flora develops in some situations; the

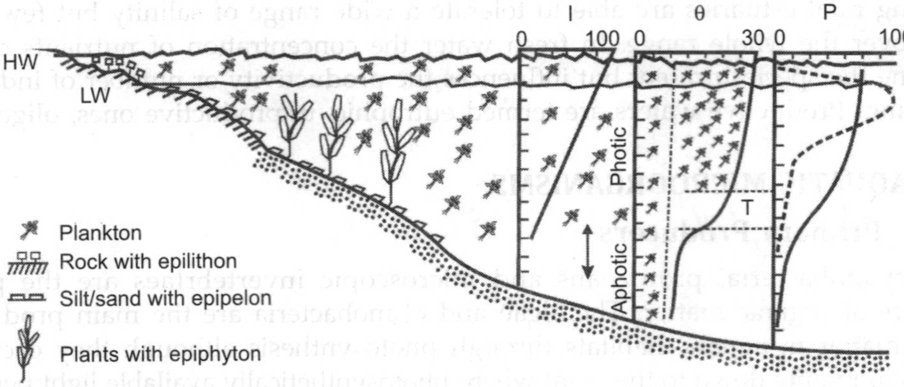

Figure 16.1 Diagrammatic representation of an ideal lake depicting the shore into deep water in either a fresh water lake or in the sea. The distribution of algal associations is shown and three graphs illustrating the decreasing illumination (I) with depth against 100% illumination on the surface; winter temperature is depicted in dotted lines in °C(θ) and summer temperature in solid lines; position of thermocline (T); photosynthesis (P), without surface inhibition (solid line) and with inhibition (broken line); maximum photosynthesis represented as 100%. On the temperature diagram the plankton is shown distributed down the complete profile during winter and confined to the water above the thermocline in summer. HW—high water, LW—low water.

cells are usually attached by mucilage pads or short mucilage stalks and they tend to be so small that they are protected by the slight concavities of the grains. A similar attached flora occurs on the undersurface of sea ice and even penetrates into the interstitial water.

The second large group of attached species occurs on larger plants (epiphytic), and on animals (epizoic). The exact relationship between these associations and those on inanimate matter is not clear, though it is certain that many species occur in only one or the other habitat. The same morphological types are present as in the lithophilic associations but there is a greater degree of host specificity, e.g., on some large marine algae a single diatom species such as *Isthmia* or *Licmophora,* is dominant, whilst on others a large number occurs (Figure 11.45).

The former is often the situation also when small crustacea are found with epizooic algal populations. On many freshwater plants a seasonal succession of epiphytes occurs. The relationship between the growth of the host, the seasonally changing environments, and the epiphytes is not yet clear. It is also becoming apparent that the metabolic activities of the host affect the epiphyte populations by the excretion of nutrients and metabolic products. Owing to the intensive concentration of the epiphytic biomass at a convenient surface it forms one of the most important 'grazing grounds' for protozoa, etc., and this flora is grazed often almost exclusively by the vegetarian fish, particularly of tropical waters. Amongst the primarily attached forms of the true epiphyton occur numerous other unattached 'organisms; algae such as desmids, coccoid and flagellate green algae, motile diatoms, etc., collectively termed metaphyton.

It is also extremely rich in bacteria, fungi, and protozoa (Figure 11.30, Figure 11.33 and 11.37).

Wherever sediments are deposited in water, shallow enough to receive adequate radiation, a rich motile algal flora grows (epipelic). The prime necessity for life in this habitat is that the species are able to maintain themselves on the surface, either because they are motile (e.g., flagellates, diatoms, and cyanobacteria), or by forming flocculent masses which sediment more slowly than do the inorganic particles. These algae frequently form growths obvious to the naked eye and some mat together and may often be seen rising to the surface buoyed up by the bubbles of oxygen formed during photosynthesis. This habitat abounds in species of diatoms which have raphe systems on the two valves and many of this display an innate diurnal rhythm of motility. The motility rhythm is even more pronounced where this association occurs on inter-tidal sediments. In this latter habitat the algae (diatoms and *Euglena* species) move beneath the silt at sunset and move up on to the surface during the early morning. Tidal cover during the daylight period induces a downward movement before the tide actually reaches the algae; this rhythm is one which can often be demonstrated under constant laboratory conditions, showing that it is under cellular control and is not impressed by the environment.

The same diatom genera are often present on both freshwater and marine sediments but the species are different e.g., *Caloneis, Diploneis, Navicula, Pinnularia, Amphora, Cymatopleura, Campylodiscus, Nitzschia,* and *Surrirella* are all common, but whereas *Diploneis* and *Amphora* are represented in the freshwater epipelon by four of five species there are up to 100 or more in the marine epipelon. Cyanobacteria (e.g., *Aphanothece, Merismopedia, Anabaena, Oscillatoria)* and desmids (e.g., *Closterium),* are more common in fresh water than in the sea, where the other components tend to be dinoflagellates and occasional Euglenoids.

The plankton does not support such a variety of life forms since the habitat is isotropic. One small variant is formed by the algae, bacteria, and protozoa which live at the water/air interface (neustonic); they have been little studied. The only common motile forms are flagellates some of which are capable of vertical migration especially under relatively calm conditions. The other components tend to sink slowly and are kept in suspension only by water movement, e.g., the bulk of the diatoms (*Asterionella, Fragillaria, Cyclotella, Stephanodiscus,* and *Melosira* in fresh water and *Asterionella, Chaetoceros, Rhizosolenia, Bacteriastrium, Guinardia, Leptocylindrus,* and colonial *Nitzschia* species in the marine habitat). A few species have buoyancy mechanisms, e.g., gas vacuoles in cyanobacteria *(Microcystis, Anabaena, Aphanizomenon),* oil globules *(Botryococcus)* and unknown mechanisms in forms such as *Halosphaera.* Even the forms without apparent buoyancy have some subtle mechanism which aids their maintenance in the plankton since death is always followed by rapid sedimentation. Some species appear to be able to maintain themselves at given depths, e.g., some dinoflagellates are 'shade plants' only found in the sub-surface ocean water, others are 'sun plants' and predominate at the surface (Figure 11.44). The mechanism for maintaining this is not known. All these organisms are large forms easily collected by drawing fine silk nets through the water. However, between these large particles and passing through most collecting nets are a whole assemblage of minute organisms (less than 5 μ in diameter) often collectively called nano-plankton with the smaller species sometimes separated off as μ-plankton. These are algal and bacterial components and they may be numerous although in fresh water their total biomass is probably less than the larger species. In the oceans many of these forms are flagellates (e.g., *Chrysochromulina)* and many are extremely delicate and very difficult to collect and preserve. Even less is known of their distribution and biology than for the larger forms, since techniques involving direct observations are not possible and although plating techniques reveal some, this is a selective procedure. In nature the species comprising the plankton are probably never randomly dispersed either horizontally or vertically. Swarms of species occur in a complex interaction between growth rates, grazing, sexual reproduction, nutrient depletion, etc. In large lakes and the oceans even more extensive patches occur in the various current systems.

16.2.2 Role of Radiation Flux

Three interrelated major parameters affect the whole aquatic microbiological system, viz., the radiation flux reaching the sea or lake surface, the thermal properties of the water, and the nutrient flux within the water. Primary fixation of carbon by algae is dependent on the radiation flux and this varies from a very high figure at 0° latitude, where two minima and two maxima occur every twelve months, to the single maximum towards the poles where, for a short summer period, the daily flux exceeds that of the tropics but where at all other times light is absent or severely limiting to algal growth. Most algae grow intermittently as do annual land plants and the commencement of growth may be related to the light climate. Additionally radiation is absorbed and its spectral composition modified as it penetrates water, thus even in pure water (which does not exist in nature) some 2% of blue light and 4% of the yellow is absorbed in 1 m. Since particulate matter, including the suspended algae, bacteria, etc., reduces the amount transmitted these figures are approached only in the purest oceanic water whilst in coastal water 40% or more may be absorbed in the first metre. The effect of this light absorption can be readily shown if the algae living on the sediments from the shore are studied down a depth profile. These can be estimated either as cell numbers, cell volume, or by estimation of some cell component (e.g., chlorophyll-*a)*. In a relatively clear lake this population has virtually disappeared between 6-8 m but in a very small lake with turbid water a comparable population exists only down to 2 m. Measurement of a planktonic population in this way is not possible since water currents continuously mix the population through all depths and hence light limitation of growth is better determined by inoculating plankton into stoppered bottles which can be hung at each depth. Growth may be measured over short intervals by measuring photosynthesis either by the change in oxygen content in the bottles or by uptake of lie supplied in a soluble form. However, the highly illuminated surface water often has a lower rate of carbon fixation than has the sub-surface; this is related to inhibition of carbon fixation by high light intensity. In general the red end of the spectrum is absorbed rapidly and the blue penetrates to the highest depths. Thus algae at various depths are exposed to light of different wavelengths as well as to different amounts. The second major effect of radiation entering the water is its heating effect, which if the water was pure and undisturbed would show an exponential fall with depth. However, owing to cooling of the surface by evaporation and movement via wind-induced currents, a very characteristic pattern develops for all except very shallow waters. The radiation heats the surface waters and this is transmitted to lower depths by wind-induced currents. When heating becomes excessive in early spring the resultant reduction in density of the surface heated water makes it difficult to mix this water into the denser cold layer below (rather like pouring water on to syrup and blowing on the surface to mix the two liquids). The layer of warm water then rests all summer on the cold lower layer and the region of rapid temperature drop between the two is known as the thermocline; the water mass is said to be thermally stratified and active circulation is then confined to the upper layer. In the autumn, surface cooling and increased wind stress eventually break down the stratification, sometimes within 24 hours, and the water then circulates to a greater depth, which may be to the bottom in many lakes or through 100 m or more in the open ocean. Thus, during the most unfavourable time of the year the algae are circulating throughout a greater depth and therefore spend a longer period in the region where photosynthesis is impossible. In the summer, however, they are trapped in the warm surface water which is also well illuminated.

16.2.3 Nutrient Distribution and Uptake in Nature

This stratification has two major effects on the nutrient supply to the microorganisms, firstly absorption of nutrients is excessive in the surface water above the thermocline and nutrient concentrations may thus be reduced to limiting amounts. Replenishment is slow, since transport across the thermocline is hampered by relative lack of water movements. Secondly, in the lower water, oxygen is being depleted by respiration of all the inhabitants, both micro- and macroorganisms, and is neither being replaced by downward movement, since the thermocline is a barrier, nor by photosynthesis, since it is below the limit of light penetration. Various degrees of de-oxygenation result and these are accompanied by a steady build-up of nutrients in this region. In lakes, these tend to diffuse out of the sediments when the surface changes from an oxidized to a reduced state. Thus the surface few metres of freshwater lakes and the upper 100 m or so of the ocean may be rich in organisms but depleted of nutrients whilst below this depth is a rich reservoir of nutrients unavailable to the surface crop until mixed by some agency which breaks down the thermocline. In the sea this nutrient-rich water upwells in some coastal areas and these results in tremendous algal productivity. The breakdown of thermal stratification in fresh water allows the deep nutrient-rich water to mix with the surface water but this usually occurs in autumn by which time light intensity is often too low to allow much growth before the spring.

The factors affecting the growth of microorganisms in nature have been studied for only a few species. Nutrient depletion has been shown to be responsible for the decline of the growth rate of such diatoms as *Asterionella* and *Stephanoducus*. Since diatoms have an absolute requirement for silica and this substance is only supplied slowly, via inflows or by recycling from the lower waters, it can be easily exhausted. This effect is pronounced on organisms in the plankton but as might be expected less obvious amongst the benthic forms which are living close to a supply of silica in the sediments or possibly diffusing from some plant stems. Other algae which require carbon and not silica for their cell walls are unlikely to be limited by lack of wall material since carbon is continuously supplied via solution of carbon dioxide and the interaction of pH, carbonates, and bicarbonates. Instead nutrients such as nitrogen and phosphorus become limiting to many algae especially in mid-summer when supply from the lower waters is limited. There is other evidence, however, to show that some algae, such as *Dinobryon,* grow in lakes when the concentration of the major ions has been reduced and this is coupled with further experimental evidence that high phosphate concentrations are deleterious.

The onset of growth of vernal species is not related to a build-up of nutrients having high concentration throughout the winter, but is triggered by increase in radiation flux. Temperature is still very low and although it does affect the growth rate is not the prime factor. Other species (e.g., *Oscilatoria rubescens* and *Melosira italica)* behave as winter or cold-water shade plants. Both these species tend to disappear at least from the surface waters of lakes in spring. *M. italica* has been shown to remain alive on the sediments for several years and its sudden appearance in the plankton in the autumn is due to mixing of this material upwards by autumn gales. It is a heavy species and if ice cover occurs it sinks owing to the reduction of wind-induced turbulence. In spring, as the temperature increases and stratification begins, it sinks out of the surface water.

Since all species in the aquatic environment have patterns of seasonal growth it is obvious that the problems are complex, and, in addition, there are well authenticated examples of biological interactions which influence the primary fluctuations correlated with the cycle of physico-chemical variables. Indirect effects of the production of antibiotics have been demonstrated in culture, but although these are undoubtedly important, it is extremely difficult in the natural habitat to separate effects due to antibiotics from those of other factors. Much more drastic, however, are the effects of parasitism, e.g., of the chytrid *Rhizophidium* on planktonic diatoms. Algal viruses/ cyanophages also have a profound effect on cyclic appearance and disappearance of algal/ cyanobacterial populations. They are known to be abundant and dynamic component in the aquatic microbial communities that can regulate the biomass production and species composition of varied microorganisms and other aquatic life forms like bacteria and phytoplankton (Chapter 12). They are also known to mediate gene transfer between microorganisms in aquatic ecosystems.

An imperfectly understood aspect of the chemical composition of the environment is its effect on the species composition of all the algal associations. No two water masses are alike in their species complement but certain broad features can be discerned particularly where there is a single overriding factor (e.g., saline lakes tend to be dominated by certain cyanobacteria, such as *Spirulina;* very acid, low-calcium lakes by desmids and chrysophyceae; organically rich waters by euglenoids). However, almost all algae can be cultured in defined media containing the common elements plus the vitamins thiamin, cyanocobalamin and biotin, all of which are generally present in natural waters. It is now clear that certain species are abundant in low-calcium/acidic waters (e.g., the diatoms, *Frustulia, Stenopterobia, Actinella)* whilst others occur only in higher calcium/alkaline waters (e.g., *Gyrosigma, Cymatopleura, Epithemia).* Similar correlations can be made for almost every group of algae. In the marine environment, however, there is no such easy distinction, since the waters are all alkaline and salinity varies only slightly, at least in the open oceans which appear to have a similar flora throughout the world, though ecological races do occur related to temperature and salinity. Brackish waters of varying degrees are an exception and have characteristic floras.

16.3 HETEROTROPHIC MICROORGANISMS

Heterotrophic microorganisms are usually present in natural waters in direct proportion to the amount of organic matter available. Bacteria, aquatic fungi, and protozoa are associated with the various communities of algae described above. By their activities they break down organic compounds and thus return plant nutrients to the water.

Bacteria are found wherever sufficient organic matter is present and vary in number with the amount of such food material available. They are numerous around submerged vegetation and in and just above the mud layer at the bottom of both fresh water and the sea. In the absence of algae, competition for oxygen is intense on the bottom sediments and oxygen becomes limiting for aerobic species. Anaerobic bacteria and a few fungi with an unusually low oxygen requirement then predominate.

The number of bacteria present in freshwater plankton is directly correlated with the amount of organic matter present and hence is higher in eutrophic waters. In waters relatively unpolluted by intestinal organisms, both pigmented and non-pigmented bacteria occur, including species of *Pseudomonas, Chromobacterium, Achromobacter, Flavobacterium,* and *Micrococcus.* Organisms washed into water from soil are mainly spore-bearing bacilli, including species of *Bacillus* (aerobic) and *Clostridium* (anaerobic), but also include members of the coli-aerogenes group (*Eschericheae*) normally found on plants and decaying vegetation. Nitrifying bacteria and species of *Streptomyces* are frequently present. In water polluted with animal or human excreta, many organisms characteristic of the intestinal canal survive for considerable lengths of time and include *Escherichia coli, Streptococcus faecalis, Proteus vulgaris,* and *Clostridium perfringens.*

Occasionally, intestinal pathogens, such as *Salmonella typhi,* and, in certain parts of the world, *Vibrio cholerae* and other water-borne pathogens may be found. Bacteria, including species of *Pseudomonas, Vibrio, Flavobacterium,* and *Achromobacter,* are present in the surface layers of the sea but their numbers differ widely with conditions. They are most numerous where the organic content of the water is high, that is near coasts, particularly where land drainage occurs, in harbours and in areas of highly productive plankton. Various counts show decreases from more than 10^6 per mL in harbours and around 4×10^5 per mL near the shore to as low as 10^1 to 2.5×10^3 per mL a few miles offshore. A large proportion of marine bacteria are pigmented and thus protected from the sun's rays. Some, such as the purple sulphur bacteria, may be sufficiently numerous to impart a red tinge to the water. Many are phosphorescent. Species of actinomycetes and yeasts have also been recorded. Marine bacteria tend to be smaller than freshwater ones, but sulphur bacteria and the spirochaetes are exceptions. All are salt-tolerant (halophilic) and most are relatively intolerant of high temperatures. Marine bacteria play similar role in the nitrogen, sulphur, phosphorus, and carbon cycles in the sea to those played by other species in the soil. Some bacteria may be troublesome by eroding metal (e.g., ships' plates) immersed in sea water.

A wide range of aquatic fungi is found as in moving water or in epiphytic algal communities, e.g., chytrids parasitic on algae, protozoa and other aquatic fungi, where oxygen is available. Many are saprophytic on organic matter. Stems of such freshwater aquatic plants as *Phragmites* and species of *Scirpus* are parasitized by members of aquatic ascomycota. A few brown seaweeds are parasitized by species of *Mycosphaerella,* some Pyrenomycetes and some fungi imperfecti. Pyrenomycetes (e.g., species of *Ceriosporopsis)* also attack and erode wood submerged in the sea. Decaying leaves of deciduous trees in well-aerated streams and lakes support a characteristic flora of aquatic hyphomycetes. These are taxonomically unrelated forms and although most of them resemble one another in the production of conidia with projecting arms or spines or consisting of a curved or branched row of cells, these superficially similar spores arise by quite different processes. Their shape, however, is well adapted to their habitat and they readily become entangled in submerged leaves which they colonize after germination. Many of the aquatic ascomycetes have ascospores of a similar shape to the conidia of the aquatic hyphomycetes. *Saprolegnia* and related species and *Monoblepharis* (found on submerged sticks) are also limited to well-aerated water. These produce motile spores (zoospores) and resting spores (oospores) which are non-motile and relatively thick walled.

Where oxygen is scarce, as in deposits of leaves or other organic matter at the bottom of stagnant ponds or where excessive growth of bacteria has depleted the supply, the fungus flora is quite different from that of similar substrata in well-aerated waters. Certain hyphomycetes, such as species of *Clathrosphaerina* and *Helicodendron,* which have distinctive coiled spores may be present as hyphae but usually produce their conidia only when removed from the water and placed in humid air. They have been described as 'aero-aquatic' fungi. Members of the Blastocladiales may also be present. It has been shown that *Blastocladia pringsheimiana* is almost anaerobic and that it produces its resting spores only under conditions of high carbon dioxide concentration.

Most water molds of the family Saprolegniaceae are less common in strongly eutrophic waters than in rather less productive ones, but the related *Leptomitus lacteus* (the so-called sewage fungus), increases with the organic content of the water and is particularly common in water polluted by sewage or organic industrial effluents where it may become a serious nuisance by actual mechanical blocking of channels and by exhaustion of the available oxygen supply.

Aquatic protozoa are common in both fresh and salt waters.

Planktonic protozoa (ciliates, flagellates, and amaeboid) and other animals, particularly the grazing species, also fluctuate more or less with the phytoplankton. They are numerous in marine plankton, and include flagellates, foraminifera (members of which sink and build up thick calcareous deposits on the ocean floor, e.g., the globigerina ooze), radiolaria (which similarly contribute to siliceous deposits, e.g., radiolarian ooze) and ciliates. A transparent, complex microorganisms that, to the inexperienced appear as protozoan (Chapter 11). In most instances, these moving "monsters" are "rotifers", in some cases they are copepods, daphnis or any one of the other forms. A few invertebrates, such as *Dugesia* and *Hydra*, are macroscopic in adult form but they are microscopic when immature. Coelentrates are almost exclusively marine. The only common freshwater form is hydra. In addition there are a few less common freshwater genera similar to the marine hydroids. The hydra is very common in ponds and attached to rocks, twigs or other substrata. Around the mouth at the free end are five tentacles of various lengths, depending on the species. Smaller organisms such as Daphnia are grasped by the tentacles and conveyed to the mouth (Figure 1.17).

Other small animals and larval forms are also present. Grazing of the phytoplankton by these microscopic animals and by the young fishes may be a factor in the fluctuation of the algae and even be a contributory cause of the frequent disappearance of phytoplankton under conditions of good nutrition.

16.4 WATER POLLUTION, SEWAGE DISPOSAL AND WATER SUPPLY

16.4.1 Pollution of Natural Waters

The increasing use of fertilizers and the disposal of wastes in streams are resulting in rapidly increasing nutrient content (eutrophication) in many natural waterbodies, and enormous algal crops result, followed by their decay giving massive pollution of shores. In spite of considerable study few algicides have proved effective and non-toxic to other organisms. Sodium pentachlorophenate has been used but it is toxic to fish. Quinone derivatives, e.g., 2,3-dichloronaphthoquinone (dichlone), are more effective and apparently less harmful. Similar problems of excessive algal growth arise in industrial plants, e.g., in power station cooling towers, and ponds. Here the above chemicals and quaternary ammonium compounds and tetrachlorobenzoquinone (chloronil) have been used successfully. The potent fish-killing alga *Pyrmnesium parvum* has been controlled in Israeli fish ponds by treatment with ammonium sulphate. The degree of pollution of both fresh and marine waters can be determined by studying the algal growth. Benthic algae are very sensitive indicators and in the most extreme polluted zones only algae such as *Oscillatoria chlorina* and *Spirulina jenneri* occur and in the slightly less polluted zones the diatoms *Nitzschia palea* and *Gamphonema parvulum* appear. Pollution often reduces the diversity of the algal flora but within limits increases enormously the growth of the resistant species, owing to the removal of competitors and predators. This is particularly noticeable in marine habitats where removal of molluscs, etc., results in intense growth of diatoms and allows the sporelings of green algae such as *Enteromorpha* to become established. A further deleterious effect of algae is their growth on all objects placed in water or which conduct water for industrial processes. Ships are particularly prone to colonization by algae and in spite of the use of algicidal paints there is still no completely satisfactory solution to this expensive problem. The recent growth in number of nuclear power stations, from which a large amount of excess heat has to be dissipated, is giving cause for concern over 'heat pollution' of rivers and estuaries. If the waters of these are used for cooling not only is there direct danger to fish, but interference with the numbers and types of microorganisms may disrupt the food chain or favour harmful species.

16.4.2 Sewage Disposal

In highly industrialized countries, the disposal of industrial and domestic waste displays an enormous problem. The easiest means of sewage disposal, still practised by coastal and estuarine towns, is simply to run the raw sewage into the sea or nearby river. The larger inland communities have been forced, by lack of convenient dumping grounds, to develop sewage purification systems whose end-products are mainly pure water and some harmless solids.

Wastewater is the used water supply of a community and consists of domestic waste water or sewage, including human excrement and wash waters that is drained into a sewerage system, industrial waterborne wastes such as acids, oil, greases and animal and vegetable matter discharged by factories, and ground, surface and atmospheric waters that enter the sewerage system. Domestic waste water or sewage consists of approximately 99.9% water and 0.2 to 0.03% suspended solids, and other soluble organic and inorganic substances. Wastewater and sewage treatment involves a large-scale use of microorganisms for bioconversion on an industrial scale. Wastewater treatment is necessary before wastewater can be disposed of without producing significant undesirable or even harmful effects. Disposal of inadequately treated wastewater leads to greater possibility for dissemination of pathogenic microorganisms and toxic chemicals that endanger ecosystems and threaten public health. Wastewater and sewage treatment from individual dwellings or households can be accomplished by anaerobic digestion and/or by aerobic metabolic processes. One of the commonly used structures is the septic tank, which is an aerobic digesting system.

A septic tank (Figure 16.2) is an enclosed sewage settling tank designed to retain the solids of the sewage entering the tank long enough to permit adequate decomposition of the sludge (collected particulate matter). As sewage enters this type of tank, sedimentation occurs from the upper portion, permitting a liquid with fewer suspended solids to be discharged from the tank. The sedimented solids are subject to degradation by anaerobic bacteria, hence the end products are still very high in BOD and odorous. The effluent from the septic tank is distributed under the soil surface through a disposal field.

In open sewage oxidation ponds green flagellates and coccoid algae grow in distinct sequence and are essential to the purification process since they supply the oxygen for the bacterial growth. Unfortunately they have to be removed, by the ton, which itself provides problems (Figure 16.3).

They can be processed into cattle food by dilution with other feeding stuffs; undiluted algae from these plants are too concentrated a protein source for direct consumption. The initial engineering problems of pond design, recovery of algae (by centrifugation

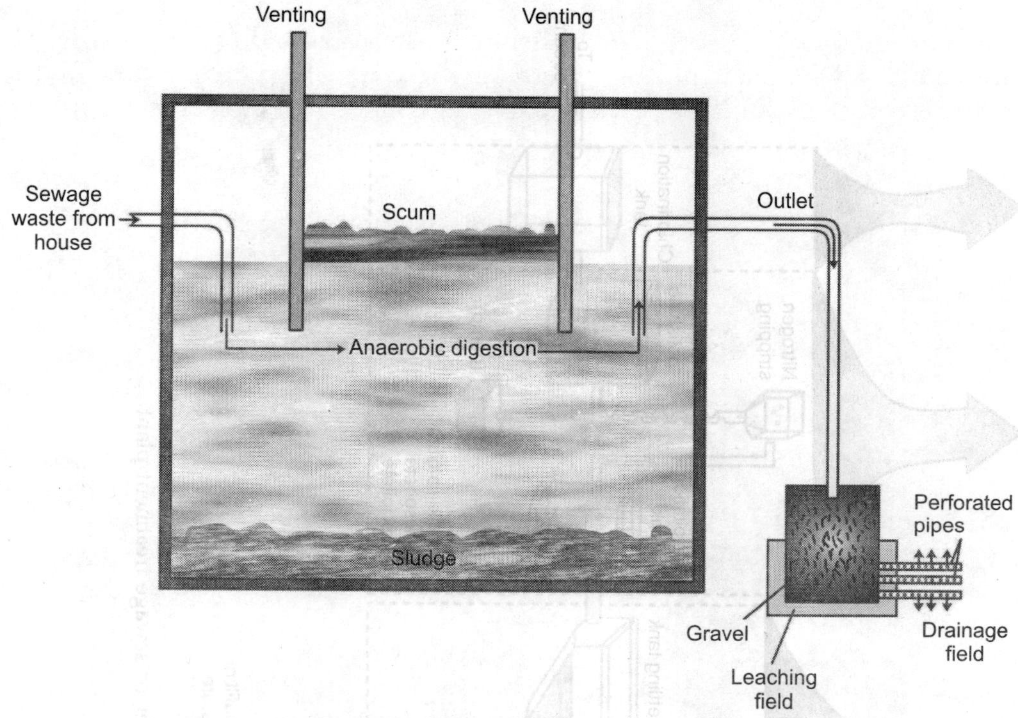

Figure 16.2 Cross section of a septic tank.

or flocculation), dewatering (centrifugation, drum drying or sand bed drying), and processing have already been overcome. The incidental or intended effect of the algal growth in these processes is the removal of vast quantities of inorganic salts from the waters.

The disposal of trade wastes presents many problems. Some industrial liquors, such as the effluents from flax retting, from factories dealing with sugar beet or milk, or from slaughterhouses or breweries, can be mixed with sewage or treated in similar disposal plants. Other trade wastes may contain poisonous organic material (e.g., effluents from factories processing leather, cellulose, textiles, or glue) or poisonous inorganic substances (e.g., effluents from coke ovens, which contain inorganic cyanides, from metal works, where acids and metallic compounds will be present, or from lead or copper mines). These cannot be dealt with by ordinary methods of sewage disposal without preliminary treatment to remove the toxic substances. A different type of pollution is that from china-clay mines where a fine deposit of clay covers the bottom of streams is continually replenished, thus preventing growth of most microorganisms. This is only a local effect and is not a danger to public health.

Sewage consists mainly of water containing organic and inorganic dissolved and suspended substances along with many microorganisms. A typical domestic sewage would have the composition shown in Table 16.1.

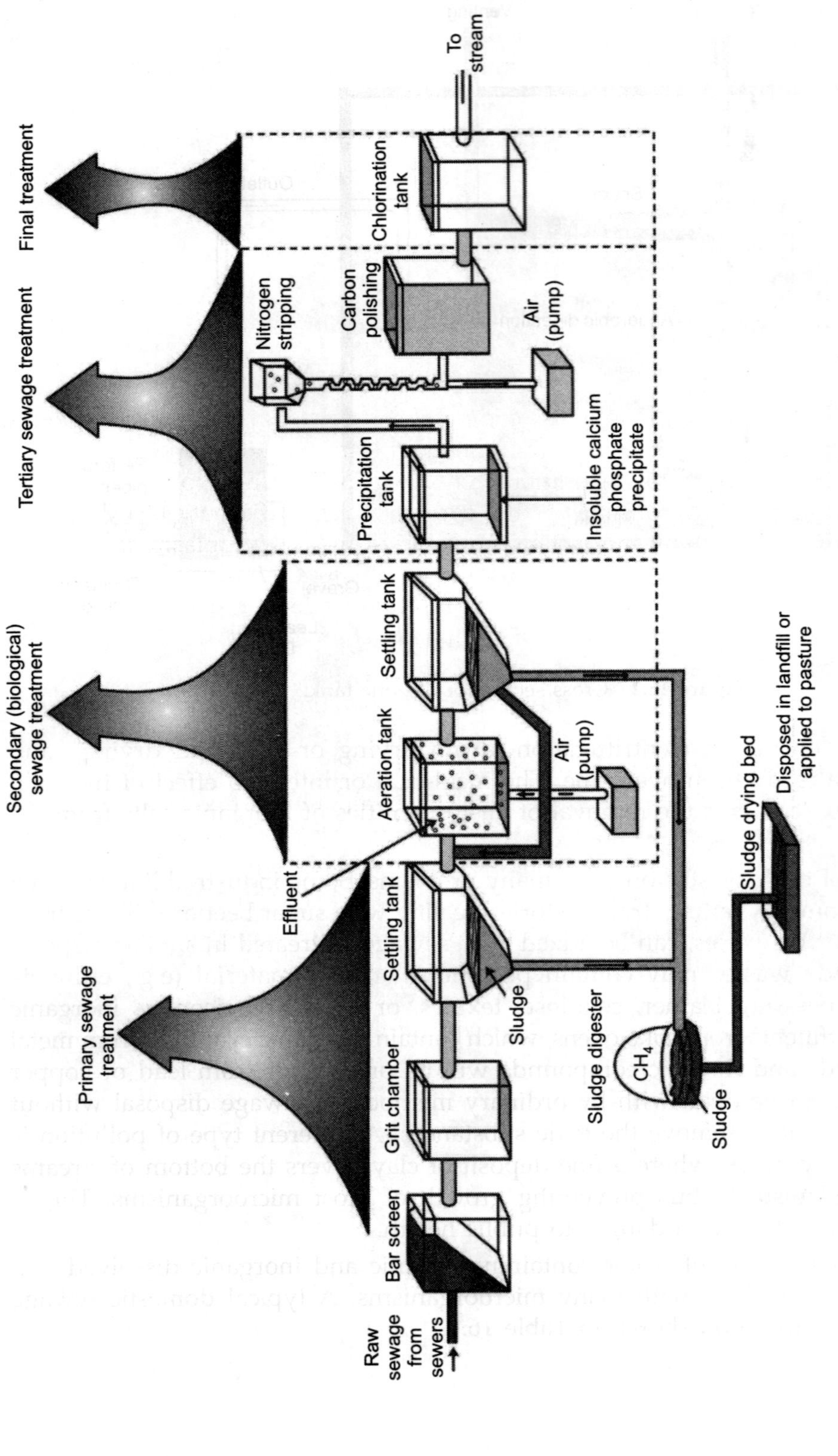

Figure 16.3 Municipal waste water or sewage treatment plant.

Table 16.1 Composition of typical domestic sewage

Constituents	*% organic carbon*
Fatty acid	22.8
Carbohydrates	17.7
Proteins	10.0
Fatty acid esters	9.1
Soluble acids	6.8
Anionic surface active agents	4.5
Amino acids	1.6
Creatinine	1.1
Amino sugars	0.6
Amides	0.5

One of the important means of observing the quality of sewage is by the measurement of its biochemical oxygen demand (BOD). This is defined as the weight of dissolved oxygen (usually in mg) required by a definite volume of liquid (usually 1 litre) during five days incubation at 20°C recognized as the international standard temperature for incubation. The aim of purifying sewage is to reduce the BOD so that if effluent waters are run into a stream or river, the indigenous flora and fauna will not die for lack of oxygen.

When sewage is received at a processing plant, there is a preliminary screening to remove solid matter. Grit and stones are allowed to settle out from a slow flowing stream of sewage. There is a regular removal of putrefactive sludge to prevent the onset of anaerobic digestion and the remaining material which is removed after varying periods of sedimentation consists of liquid with suspended floes of organic matter. This may be treated in one of several ways.

16.5 TREATMENT OF WATERY EFFLUENT

16.5.1 Activated Sludge Treatment

This process is used at large sewage works and consists of running the sewage into tanks where it is aerated either by compressed air or stirring. This vigorous agitation continues from 5 to 15 hours, and during this time flocculation of the organic matter takes place. The effluent is subsequently run into settling tanks, where the sediment is recycled back to provide an inoculum for the incoming sewage, whilst the water is sufficiently pure to be discharged into a river. When the sewage contains an excess of certain substances, e.g., carbohydrates of brewery effluent, it is necessary to add nitrogen to keep the balance correct. The dominant floc-forming bacterium is *Zooglea ramigera.* Nitrifying bacteria are also present, and while these do not compete with the heterotrophs for nutrients, there is considerable competition for oxygen. Fungi do not become well established, and if *Geotrichum* species appear it usually indicates an upset in the balance of the process. Protozoa are present both as the free-swimming forms and the stalked species, the latter being dominant. Their presence is most important because they feed on bacteria, and effluent from this process lacking protozoa is often cloudy.

Some of the known chemical conversions are shown below:

1. Metabolism of urea.

$$CO(NH_2)_2 + H_2O \rightarrow CO_2 + 2NH_3$$

2. Nitrification as carried out by *Nitrosomonas europaea.*

$$2NH_4^+ + 3O_2 \rightarrow 2H^+ + 2H_2O + 2NO_2^-$$

$$2NO_2^- + O_2 \rightarrow 2NO_3^-$$

3. Nitrate reduction.

$$2NO_3^- \rightarrow 2NO_2^- \rightarrow 2(HNO) \rightarrow 2NH_2.OH \rightarrow 2NH_3$$

$$2(HNO) \rightarrow N_2 + H_2O$$

The activated sludge process will remove 85-95 per cent of BOD and the same percentage of suspended materials. Thus, it enables the solids to be spread on the land without danger to health, and water to be returned to the rivers without causing pollution.

16.5.2 Biological Filters

Biological filters consist of a circular tank, above ground to a height of about 4 ft, packed with inert material such as coke that acts as a filter bed. Over the top of the bed, four sparge arms slowly rotate distributing effluent evenly over the whole surface. The microorganisms become stratified within the filter bed, different types growing at a level where the oxygen concentration and composition of the effluent suits their particular requirements best. In the upper layers such heterotrophs as *Zooglea ramigera* are to be found and as the organic matter diminishes in the lower layers autotrophs such as *Nitrobacter* and *Nitrosomonas* grow. In addition, there are such organisms as *Fusarium* and *Geotrichium* growing on the surface, molds being more numerous than in the activated sludge process. Protozoa abound, the motile species grow in the upper layers whilst the stalked kinds grow lower down the beds. A macrofauna of small animals graze upon the solid matter that gets caught in the interstices of the bed, thus preventing the plant from becoming blocked. Examples of the chemical changes that take place and some of the organisms responsible for them are shown below.

1. NH_4 → *Nitrosomonas* sp. → NO_2 → *Nitrobacter* sp. → NO_3
2. Phenolic substance → *Vibrio* sp., *Pseudomonas* sp., *Actinomyces* → CO_2 + H_2O
3. CNS → *Thiobacillus sp* → $CO_2 + NH_3 + SO_4^-$

$$S_2O_3 \rightarrow \textit{Thiobacillus sp} \rightarrow SO_4^-$$

Reduction of BOD is about the same as that of the activated sludge process, whereas the removal of suspended solids is slightly less, varying from 70-90%. Synthetic detergents have been added to sewage in recent years. These are surface active agents that cause considerable foaming and also reduce oxygen transfer by about 20% in the activated sludge process.

It has been shown that if the benzene ring contains an alternative substituent, biological attack takes place through this group. Furthermore, the straight-chain compounds are more readily attacked than branched chain ones. The tendency now is to produce detergents containing a higher proportion of straight-chain compounds, and some countries now have legislation prohibiting the manufacture of detergents resistant to biological degradation.

16.5.3 Oxidation Ponds

Settled sewage is run into ponds and held for about 30 days. The action of bacteria produces carbon dioxide and ammonia. The presence of these substances encourage the growth of algae which releases oxygen during photosynthesis. These conditions permit aerobic bacteria to grow which decompose the organic matter further, leaving a minimum of residue (Figure 16.3).

16.5.4 Purification of Water Supplies

16.5.4.1 Storage

When water is stored in a reservoir the amount of suspended food material normally tends to decrease, both by sedimentation and as a result of the activities of microorganisms. This, in turn, leads to a reduction in the number of microorganisms and in particular of pathogenic bacteria which are unable to survive in competition with saprophytic species. Populations of planktonic and benthic algae similar to those in natural water occur in water supply reservoirs. These algae are controlled by similar factors to those in nature and the same principles apply with the added complication of control by man, e.g., thermal stratification may be prevented by inflow design, pumping, etc. The algae are here normally beneficial since they absorb certain nutrients, supply oxygen to the water, have some anti-bacterial effect, assist filtration, etc., but these activities can readily become deleterious when the algal growth becomes too great. Water reservoirs receiving acid drainage water with low nutrient concentrations rarely give trouble, but the same water impounded in lowland reservoirs or river water containing large amounts of dissolved chemicals can produce enormous algal crops (the so-called water blooms). These may decay and produce very unpleasant by-products, causing taints or odours. If the water is being drawn off during such a growth the beds of sand through which the water is filtered in the treatment plants become clogged within a few hours and have to be drained and the surface layer of sand removed. Algae, such as *Cladophora,* may grow actually on the filters and then must be removed. Control of the algae in the reservoirs may be achieved by adding small amounts of copper sulphate to the water, but this does not always have the desired effect since the removal of one alga is often followed by the growth of another and as is often the case after copper sulphate treatment this may be a smaller and even more troublesome species. Reservoir management, including the installation of jets at the inlets to cause the water to circulate, switching to alternative reservoirs, drawing off the water at the most suitable level from a stratified reservoir, and the regulation of the storage period to give the maximum reduction of bacteria with the minimum increase of algae, is important in overcoming the problem of excessive algal growth.

16.5.4.2 Filtration

Even if no additional pollution occurs during the process, storage of water in open lakes and reservoirs is insufficient to render it safe for drinking; hence artificial methods of purification must also be employed. The water is first piped to a filter plant which may be a slow sand filter suitable for reasonably pure waters and for use where adequate space is available for the large filters needed, or a rapid sand filter, suitable for turbid waters or where insufficient land is available.

16.5.4.3 Slow sand filters

The natural process of filtration through rocks is imitated by allowing the waters to percolate slowly through clean sand. Purification does not, however, depend only on mechanical straining of the water by the sand but is achieved mainly by the activity of the microorganisms present in the system. After filtration has proceeded for several days a slimy gelatinous mass composed of bacteria, protozoa, and algae accumulates, particularly in the upper layers of the sand. This layer brings about biological oxidations and reductions, slowly closes up the pores of the filter thus rendering it more effective by slowing down the rate of filtration and provides predatory protozoa which feed on bacteria. The net result is that the water emerging from the filter is chemically and biologically purer than before filtration. Eventually, the slime layer becomes too thick and slows filtration too much. The filter bed is then cleaned, re-laid with fresh sand and filtration is renewed. A newly prepared filter bed is of low efficiency and the water from it cannot be used until a new slime layer has developed.

16.5.4.4 Sand filters

It differs from the slow filters in having a smaller filter area and in the rapidity and entirely mechanical nature of the process. The organic matter in the water is first precipitated by chemicals such as ammonium aluminium sulphate, and the sticky, bulky precipitate is allowed to sediment in large settling basins. Bacteria as well as any colouring matter in the water are readily adsorbed on to the precipitate and are thus carried down with it. The clean supernatant fluid is finally run over a sand filter to remove any precipitate still in suspension and the purified water is collected at the bottom of the filter (Figure 16.4).

Most microbial diseases of animals and man arise from an exogenous source. One important method of disease control is achieved, therefore, by treating or removing the source of infection to prevent pathogenic microorganisms from reaching susceptible hosts.

16.6 LAND TREATMENT

Settled sewage is run onto arable land where accumulated organic matter is removed by microbiological oxidation. This system is not used today as a major method for dealing with effluent, but is usually applied to the cleaning of effluents from other processes such as activated sludge.

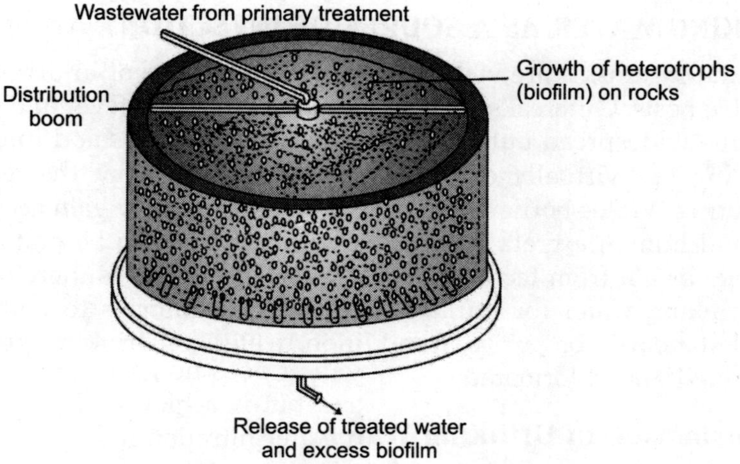

Figure 16.4 The trickling filter.

16.6.1 Treatment of Solids

Sludge of this type is derived from the initial screening of raw sewage, from the settled deposits of activated sludge or as humus sludge from percolating filters. These are combined and placed in a 'digester'. This is the part of the sewage that may contain pathogens and it is, therefore, essential that this material is rendered innocuous before disposal. The digester, in reality, is nothing more than a large-scale septic tank whose contents are stirred and heated. The most common temperature range is 30-35°C which is applied for about one month. Thermophilic digestion using a temperature of 50°C is sometimes practised, but is less usual than the mesophilic digestion which is more easily controlled. During digestion, methane, with a calorific value of about 700-750 BTU/cu ft, is evolved and in a large sewage plant this is used as a source of heat and power. Organisms producing methane belong to several genera as illustrated by the following list: *Methanobacterium formicum, Methanobacillus omelianskii, Methanosarcina barkerii* (Chapter 21). They utilize a variety of substrates including alcohols, fatty acids, CO_2, and H_2 to produce methane and may be highly substrate specific. Experiments using labelled carbon have shown that the CO_2 carbon becomes the methane carbon. Other changes taking place during digestion are the liquefaction of solids and the breakdown of carbohydrates, fats, and proteins. The digestion process can give rise to very obnoxious odours since indole, skatole, and compounds containing mercaptans are produced. When sludge digestion is complete, the excess liquid is returned to the digester and the residual solid material is dried by vacuum filtration, by heat or in underdrained beds. The remaining dry solid can then be disposed of by incineration or applied to the land as manure.

16.7 DRINKING WATER AS A SOURCE OF INFECTION

Some pathogenic organisms are water-borne and when present in drinking water may infect susceptible hosts. Outbreaks of diseases, such as typhoid fever and cholera in man, were at one time widespread but the control of water supplies and improved hygiene in food handling has virtually eradicated these diseases from the technically more advanced countries. Water-borne pathogenic organisms usually gain access to the water from animal and human excreta and sewage. Water may also be polluted by poisons of non-microbial origin from factory wastes. Constant control is therefore necessary to ensure that drinking water for human consumption conforms to both chemical and bacteriological standards of purity. In addition, water for drinking purposes may be artificially purified and chlorinated.

16.7.1 Chlorination of Drinking Water

Neither natural purification nor filtration renders water safe for drinking, and disinfection is finally achieved by the addition of chlorine. Chlorination destroys all water-borne pathogenic bacteria and also pathogenic intestinal protozoa, such as *Entamoeba histolytica,* the cause of amoebic dysentery. When chlorine is added to water a portion combines with reduced organic material in the water and is no longer effective. This 'chlorine demand' varies considerably with different waters according to the amount of organic matter present and consequently the amount of chlorine that must be added to give a safe residual amount of free chlorine (0·2 to 0·4 part per million) also varies. The small amount of free chlorine necessary to render water safe is harmless and usually cannot be detected, but waters with an abnormally large 'chlorine demand' may have an unpleasant flavour after chlorination, due to the presence of such chlorine compounds as chlorophenols.

16.7.2 Bacteriological Standards of Water Supplies

Bacterial pollution of water may originate either from individuals with clinical symptoms of disease or from symptomless carriers of enteric (i.e., intestinal) pathogens such as typhoid bacilli. Pathogens are difficult to detect in water for a number of reasons. They may be present only sporadically owing to intermittent excretion and dilution and, unless the water is being continuously polluted from an infected source, the pathogens usually disappear before the disease is recognized in someone infected by drinking the contaminated water. The isolation of pathogens necessitates the concentration of large volumes of the suspected water and the use of selective media, techniques which are complex and unsuitable for routine examinations. In contrast, harmless members of the normal faecal flora are constantly being excreted and may persist much longer in polluted water. Bacteriological tests to determine the suitability of water for drinking are designed therefore to detect the presence in the water of organisms of the normal flora of the gut. The detection of these organisms is relatively simple but their presence is indicative of faecal contamination which renders the water potentially dangerous. Simple routine tests for faecal pollution, including viable counts and the demonstration of organisms of known faecal origin, are therefore employed at regular intervals to screen supplies of water used for drinking purposes (Chapter 13).

16.7.2.1 Viable counts

Counts are made from dilutions of the water sample on a standard medium after incubation at 22°C and 37°C: Most saprophytic bacteria grow at 22°C and all organisms able to grow on the standard medium at this temperature will be included in this count. Since the number of saprophytes present is likely to be proportional to the amount of organic matter available for their nutrition, this count gives a rough indication of the relative amount of organic material present in the water. The majority of organisms growing at 37°C will be parasites or potential parasites of man and animals, derived from soil, excreta, or sewage. A high count at 37°C, relative to the count at 22°C almost always indicates pollution with animal or human excreta. In unpolluted water the ratio of the count at 22°C to that at 37°C is usually greater than 10:1; in polluted water the ratio may be 1: 1.

16.7.2.2 Demonstration of microorganisms of faecal origin

The water is further examined for the presence of specific bacteria of known intestinal origin. Certain species, as *Escherichia coli, Streptococcus faecalis,* and *Clostridium perfringens,* usually present in the human intestine and frequently in those of other animals, are useful indicators of pollution with excreta. Many coliform bacteria are found on vegetation and in soil but *E. coli* are found only transiently outside the alimentary canal and its presence in water is thus almost certain evidence of recent faecal pollution. The demonstration of even a few *E. coli* in water is sufficient to condemn it as unfit for human consumption even though no pathogens have been found in it (Figure 16.5).

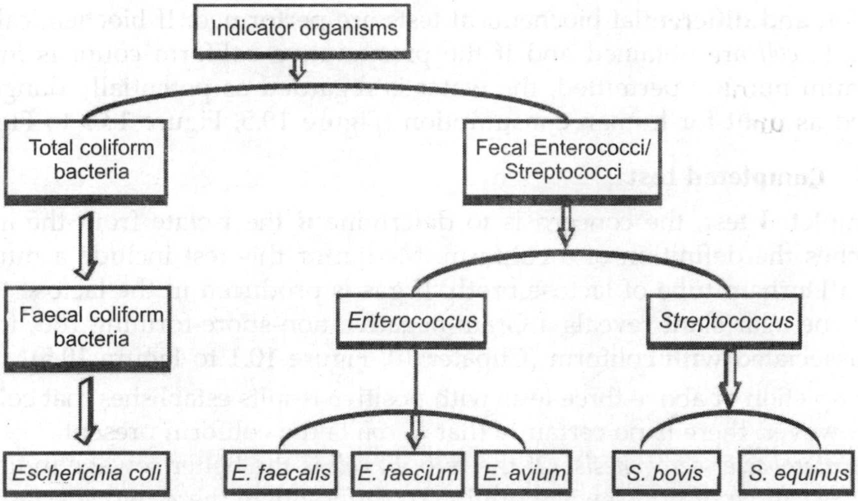

Figure 16.5 Illustration of various coliform bacteria as water pollution.

Two tests are used to detect *E. coli*: (1) the presumptive coliform count and (2) the identification of isolated strains by differential biochemical tests.

16.7.2.2.1 Presumptive coliform count

This count is performed in a selective nutrient fluid medium to which bile salts, lactose; and a pH indicator are added. The bile salts inhibit the growth of most non-intestinal organisms but not that of *E. coli* and allied organisms. The presence of coliform bacilli under these conditions is demonstrated by the production of acid and gas by fermentation of lactose (Figure 10.5, Figure 13.1 to Figure 13.3). Detailed methods have been discussed in Chapter 10 and 13.

For statistical reasons a range of volumes and dilutions of the water under test is added to appropriate volumes of the bile-lactose medium. The greater the number of *E. coli* present and the larger the sample of water examined, the higher will be the chance that *E. coli* will be contained in any particular sample and vice versa. The distribution of coliform positive reactions in the range of volumes tested from each water sample is referred to probability tables and the probable number of coliform bacilli in 100 mL of the original water sample is deduced. The standards required vary for different types of water but, in general, an unchlorinated water with more than 5 coliform bacilli per 100 mL would be regarded as unfit for drinking. Chlorinated water would obviously be expected to have no viable coliforms present.

16.7.2.2.2 Differential coliform tests

Fermentation of lactose in the presumptive coliform count may be due to anyone of a number of coliform bacilli and not necessarily to *E. coli*. It is essential, therefore, to determine whether the organism detected in the count is the type known to be parasitic in the gut. The organism is first isolated on solid medium from one of the tubes showing fermentation and differential biochemical tests are performed. If biochemical reactions typical for *E. coli* are obtained and if the presumptive coliform count is in excess of the maximum number permitted, the water is regarded as potentially dangerous and condemned as unfit for human consumption (Figure 10.5, Figure 13.1 to Figure 13.3).

16.7.2.2.3 Completed test

In the completed test, the concern is to determine if the isolate from the agar plates truly matches the definition of a coliform. Media for this test include a nutrient agar slant and a Durham tube of lactose broth. If gas is produced in the lactose tube and a slide from the agar slant reveals a Gram-negative non-spore-forming rod, it is certain that it is associated with coliform (Chpater 10, Figure 10.1 to Figure 10.5).

The completion of above three tests with positive results establishes that coliforms are present; however, there is no certainty that *E. coli* is the coliform present. The organism might be *Enterobacter aerogenesis*. Of the two, *E. coli* is the better sewage indicator since *E. aerogenesis* can be of non-sewage origin. To differentiate these two species, one must perform the **IMViC.**

IMViC Tests

In the differentiation of *E. aeroginesis* and *E. coli,* as well as some other related species, four physiological tests have been grouped together into what are called the **IMViC** tests. The I stands for indole; the M and V stand for methyl red and Voges-Proskauer tests: i simply facilitates pronunciation; and the C signifies citrate utilization. In the differentiation of the two coliforms *E. coli* and *E. aeroginesis,* the test results appear as charted (Chapter 13, Section 13.1.5.2.1), revealing completely opposite reactions for the two mechanisms on all tests. The significance of these tests is that when testing drinking water for the presence of the sewage indicator *E. coli* one must be able to rule out *Enterobacter aerogenes,* which has many of the morphological and physiological characteristics of *E. coli.* Since *E. aerogenes* is not always associated with sewage, its presence in water would not necessarily indicate sewage contamination.

Concept of indicator organism

Indicator organisms are used to measure potential faecal contamination of environmental samples. The presence of coliform bacteria, such as *E. coli,* in surface water is a common indicator of faecal contamination. Coliform bacteria in water samples may be quantified using the most probable number (MPN) method, a probabilistic test which assumes cultivable bacteria meet certain growth and biochemical criteria. If preliminary tests suggest that coliform bacteria are present at numbers in excess of an established cut-off, faecal contamination is suspected and confirmatory assays such as the Eijckman test are conducted. Coliform bacteria selected as indicators of faecal contamination must not persist in the environment for long periods of time following efflux from the intestine, and their presence must be closely correlated with contamination by other faecal organisms. Indicator organisms need not be pathogenic. Certain criteria should be followed before an indicator organism can be considered reliable in predicting a health risk:

1. The organism must be exclusively of faecal origin and consistently present in fresh faecal waste.
2. It must occur in greater numbers than the associated pathogen.
3. It must be more resistant to environmental stresses and persist for a greater length of time than the pathogen.
4. It must not proliferate to any great extent in the environment.
5. Simple, reliable, and inexpensive methods should exist for the detection, enumeration, and identification of the indicator organism.

Organisms that fit these criteria include the coliform bacteria, faecal streptococci (enterococci) and the sulfite-reducing clostridia i.e. *Clostridium perfringens.* Gastrointestinal pathogens known to have caused outbreaks of enteric disease are largely from the systematically defined family, Enterobacteriaceae, and include *Salmonella, Shigella, Yersinia enterocolitica, Klebsiella pneumoniae, Enterobacter* and enterotoxigenic *Escherichia coli* (*E. coli*). *Vibrio cholerae* and *Campylobacter jejuni* are two other enteric pathogens often found in contaminated water. These organisms are spread by water contaminated with faecal material from humans and other warm-blooded animals (Chapter 10).

Coliform is not a taxonomic classification but rather a working definition used to describe a group of Gram-negative, facultative anaerobic rod-shaped bacteria that ferments lactose to produce acid and gas within 48 h at 35°C. Although coliforms were easy to detect, their association with faecal contamination was questionable because some coliforms are found naturally in environmental samples. This led to the introduction of the faecal coliforms as an indicator of contamination. Faecal coliform, first defined based on the works of Eijkman is a subset of total coliforms that grows and ferments lactose at elevated incubation temperature, hence also referred to as thermotolerant coliforms. Faecal coliform analyses are done at 45.5°C for food testing, except for water, shellfish and shellfish harvest water analyses, which use 44.5°C. The faecal coliform group consists mostly of *E. coli* but some other enterics such as *Klebsiella* can also ferment lactose at these temperatures and therefore, be considered faecal coliforms. The inclusion of *Klebsiella* sp. in the working definition of coliforms diminished the correlation of this group with faecal contamination. As a result, *E. coli* has reemerged as an indicator, partly facilitated by the introduction of newer methods that can rapidly identify *E. coli*.

Currently, all 3 groups are used as indicators but in different applications. Detection of coliforms is used as an indicator of sanitary quality of water or as a general indicator of sanitary condition in the food-processing environment. Faecal coliforms remain the standard indicator of choice for shellfish and shellfish harvest waters; and *E. coli* is used to indicate recent faecal contamination or unsanitary processing. Almost all the methods used to detect *E. coli*, total coliforms or faecal coliforms are enumeration methods that are based on lactose fermentation. The Most Probable Number (MPN) method is a statistical, multi-step assay consisting of presumptive, confirmed and completed phases. In the assay, serial dilutions of a sample are inoculated into broth media. Analysts score the number of gas positive (fermentation of lactose) tubes, from which the other 2 phases of the assay are performed and then uses the combinations of positive results to consult a statistical tables, to estimate the number of organisms present. Typically only the first 2 phases are performed in coliform and faecal coliform analysis, while all 3 phases are done for *E. coli*. The 3-tube MPN test is used for testing most foods. The 5-tube MPN is used for water, shellfish and shellfish harvest water testing and there is also a 10-tube MPN method that is used to test bottled water or samples that are not expected to be highly contaminated.

There is also a solid medium plating method for coliforms that uses Violet Red Bile Agar, which contains neutral red pH indicator, so that lactose fermentation results in formation of pink colonies. There are also membrane filtration tests for coliform and faecal coliform that measure aldehyde formation due to fermentation of lactose.

16.7.2.3 Membrane Filter (Mf) method

The MF technique is an effective, accepted technique for testing fluid samples for microbiological contamination. It involves less preparation than many traditional methods, and is one of a few methods that will allow the isolation and enumeration of microorganisms. The MF technique also provides presence or absence information within 24 hours. The Membrane Filter (MF) technique was introduced in the late 1950s as an alternative to the Most Probable Number (MPN) procedure for microbiological analysis of water samples. The MF Technique offers the advantage of isolating discrete colonies of bacteria, whereas the MPN procedure only indicates the presence or absence of an approximate number or organisms. The various steps are described below.

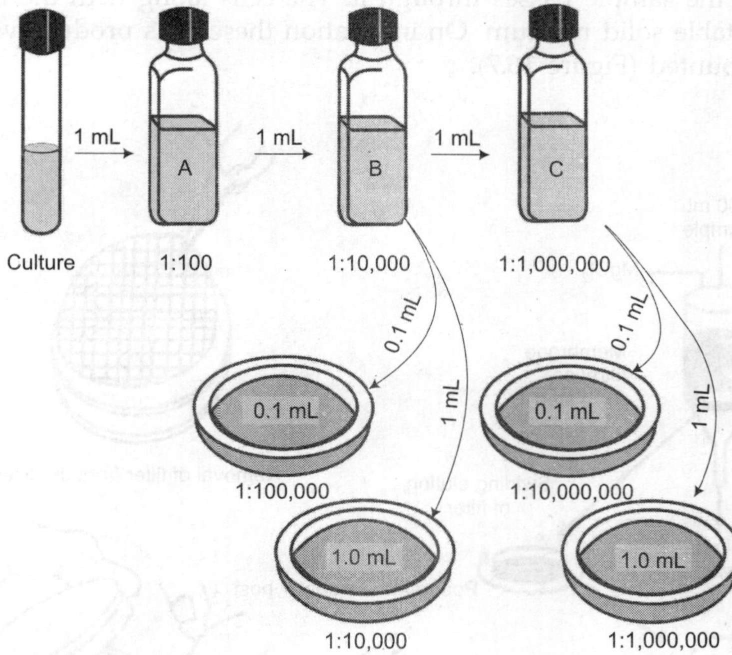

Figure 16.6 Method of preparing proper dilution samples.

(i) Collect the sample and make necessary dilutions (Figure 16.6).

(ii) Select the appropriate nutrient or culture medium. Dispense the broth into a sterile Petri dish, evenly saturating the absorbent pad.

(iii) Flame the forceps, and remove the membrane from the sterile package.

(iv) Place the membrane filter into the funnel assembly.

(v) Flame the pouring lip of the sample container and pour the sample into the funnel.

(vi) Turn on the vacuum and allow the sample to draw completely through the filter.

(vii) Rinse funnel with sterile buffered water. Turn on vacuum and allow the liquid to draw completely through the filter.

(viii) Flame the forceps and remove the membrane filter from the funnel.

(ix) Place the membrane filter into the prepared Petri dish.

(x) Incubate at the proper temperature and for the appropriate time period.

(xi) Count the colonies under 10-15 × magnification.

(xii) Confirm the colonies and report the results.

When the number of indicator organisms in water is very low, direct inoculation on solid media is not practicable and other methods must be used by which large volumes can be examined. Other, methods of examination include membrane filtration and multiple tube tests. Membrane filter techniques are widely used for the enumeration of bacteria from water sample. In this method the bacterial cells are filtered through a membrane as the sample passes through it. The cells along with the membrane are placed on a suitable solid medium. On incubation these cells produce visible colonies which can be counted (Figure 16.7).

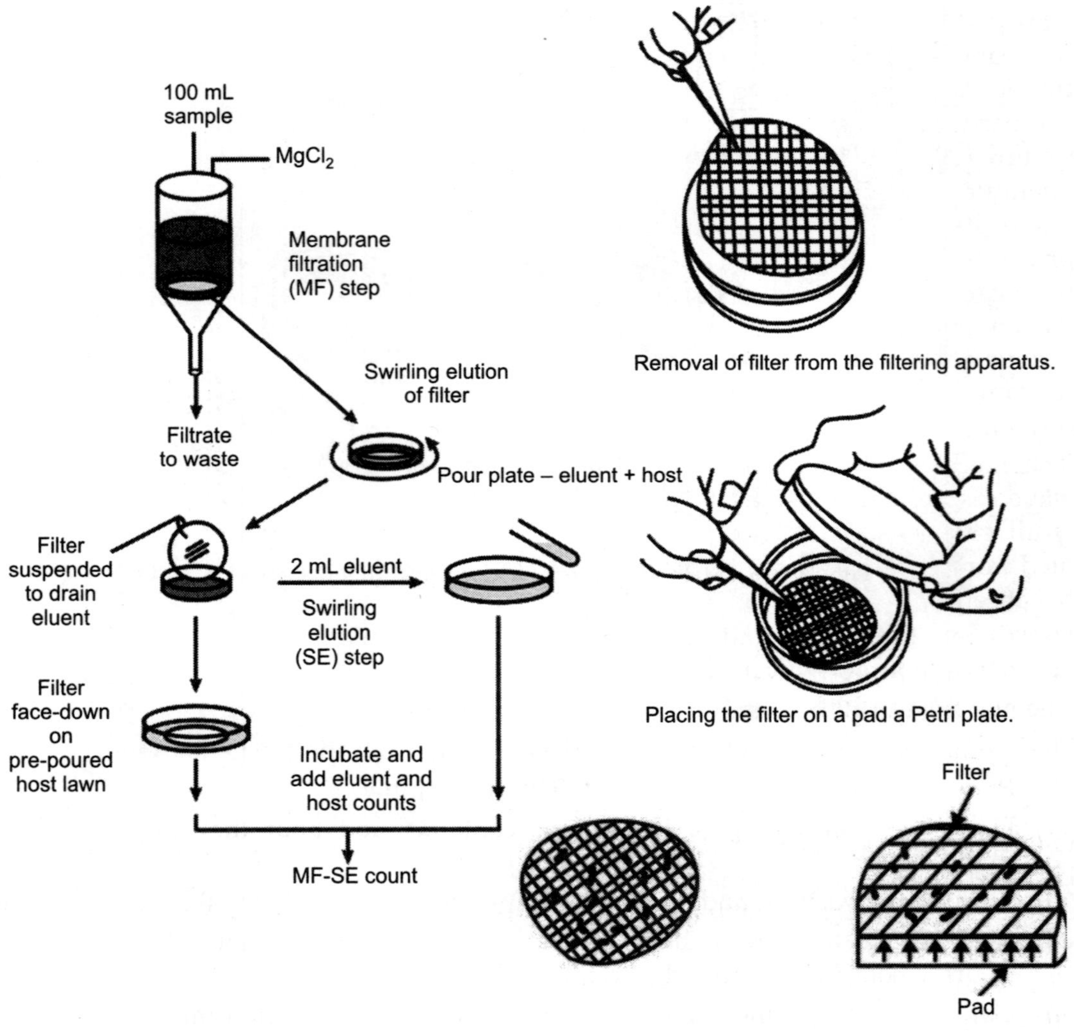

Figure 16.7 Membrane filtration technique.

16.7.2.3.1 Counting of faecal coliform

Compute the density from the sample quantities that produced MF counts within the desired range of 20 to 60 faecal coliform colonies. This colony density range is more restrictive than the 20 to 80 total coliform range because of larger colony size on M-FC medium. Record densities as faecal coliforms per 100 mL. Compute the count by the equation: (Total) Coliform colonies/100 mL = (Coliform colonies counted x 100) / (mL sample filtered)

16.7.2.3.2 Counting of faecal *Streptococci*

The *Streptococci* have been under consideration as indicators of faecal pollution for many years. Their poor acceptance as a measure of faecal pollution from human and warm-blooded animal excreta has been due in part to the relatively low recovery rates in comparison to coliform densities in polluted waters; the multiplicity of detection procedures; poor agreement between the various methods for their quantitative enumeration; and the lack of detailed and systematic studies on the sources, survival, and interpretation of streptococci in various types of waters. Furthermore, undue emphasis has been placed on the *Streptococcus faecalis* group (enterococci) with little or no regard for other streptococcal strains present in the gut of humans and warm blooded animals or birds. The predominating species of streptococci may vary markedly in various animal excreta. Some detection methods yield excellent results with human faecal samples but give poor quantitative recovery of the streptococci present in pig or cow faeces.

The faecal streptococcus group consists of a number of species of the genus *Streptococcus,* such as *S. faecalis, S. faecium, S. avium, S. bovis, S. equinus,* and *S. gallinarum.* They all give a positive reaction with Lancefield's Group D antisera and have been isolated from the faeces of warm-blooded animals. In addition, *S. avium* sometimes reacts with Lancefield's Group Q antisera. *S. faecalis* subsp. *liquefaciens* and *S. faecalis* subsp. *zymogenes* are differentiated based on the ability of these strains to liquefy gelatin and hemolyze red cells. However, the validity of these subspecies is questionable.

The normal habitat of faecal *Streptococci* is the gastrointestinal tract of warm-blooded animals. *S. faecalis* and *S. faecium* once were thought to be more human-specific than other *Streptococcus* species. Other species have been observed in human feces but less frequently. Similarly, *S. bovis, S. equinus,* and *S. avium* are not exclusive to animals, although they usually occur at higher densities in animal faeces. Certain streptococcal species predominate in some animal species and not in others, but it is not possible to differentiate the source of faecal contamination based on the speciation of faecal streptococci.

The faecal streptococci have been used with faecal coliforms to differentiate human faecal contamination from that of other warm-blooded animals. A ratio greater than four is considered indicative of human faecal contamination, whereas a ratio of less than 0.7 was suggestive of contamination by nonhuman sources. The value of this ratio has been questioned because of variable survival rates of faecal streptococcus group species. *S. bovis* and *S. equinus* die off rapidly, once exposed to aquatic environments, whereas *S. faecalis* and *S. faecium* tend to survive longer. Furthermore, disinfection of wastewaters appears to have a significant effect on the ratio of these indicators, which may result in misleading conclusions regarding the source of contaminants. The ratio is affected also by the methods for enumerating faecal streptococci. The membrane filter procedure has a false-positive rate ranging from 10 to 90% in marine and fresh waters. For these reasons, the FC/FS ratio cannot be recommended, and should not be used as a means of differentiating human and animal sources of pollution.

16.7.2.3.3 Clostridium perfringens

Clostridium perfringens (formerly known as *C. welchii*) is a Gram-positive, rod-shaped, anaerobic, spore-forming bacterium of the genus *Clostridium. C. perfringens* is ubiquitous in nature and can be found as a normal component of decaying vegetation, marine sediment, the intestinal tract of humans and other vertebrates, insects, and soil. The *Clostridium perfringens* enterotoxin (CPE) mediating the disease is heat-labile (dies at 74°C) and can be detected in contaminated food, if not heated properly, and feces.

Incubation time is between 6 and 24 (commonly 10-12) hours after ingestion of contaminated food. Often, meat is well prepared but too far in advance of consumption. Since *C. perfringens* forms spores that can withstand cooking temperatures, if let stand for long enough, germination ensues and infective bacterial colonies develop (Figure 16.8). Symptoms typically include abdominal cramping and diarrhea; vomiting and fever are unusual. The whole course usually resolves within 24 hours. Very rare, fatal cases of clostridial necrotizing enteritis (also known as Pig-Bel) have been known to involve "Type C" strains of the organism, which produce a potently ulcerative β-toxin. This strain is most frequently encountered in Papua New Guinea.

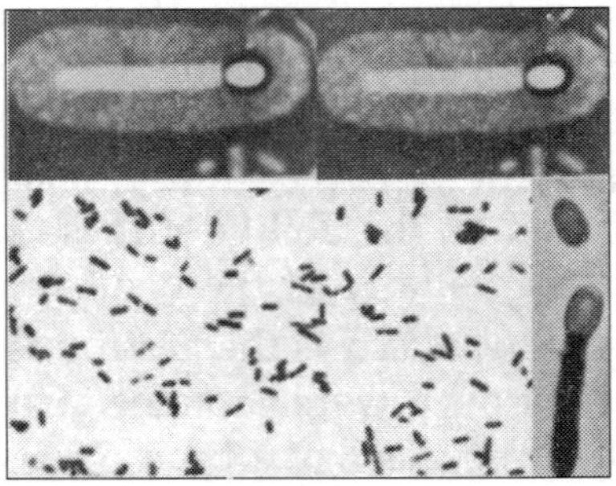

Figure 16.8 *Clostridium perfringens*

It is likely that many cases of *C. perfringens* food poisoning remain subclinical, as antibodies to the toxin are common among the population. This has led to the conclusion that most of the population has experienced food poisoning due to *C. perfringens.* On blood agar plates, *C. perfringens* grown anaerobically produces β-haemolytic, flat, spreading, rough, translucent colonies with irregular margins. A Nagler agar plate, containing 5-10% egg yolk, is used to identify strains that produce α-toxin, a diffusible lecithinase that interacts with the lipids in egg yolk to produce a characteristic precipitate around the colonies. One-half of the plate is inoculated with antitoxin to act as a control in the identification.

16.7.2.3.4 Heterotrophic Plate Count (HPC)

The Heterotrophic Plate Count (HPC) is a procedure used to estimate the number of live heterotrophic bacteria that are present in a water sample. A sample of water is put on a plate that contains nutrients that the bacteria need to survive and grow. The nutrient media that is most often used for this test is called R2A Agar, which is a gelatine-like substance that is best suited to the needs of water bacteria. After 5-7 days, the number of small spots on the plate, called colonies, is counted, and a measure of how many bacteria are present in each millilitre of water can be determined (Figure 16.9). The HPC results are generally reported as CFU/mL or Colony Forming Units per millilitre. Each colony-forming unit represents an initial single, live bacterium that was capable of multiplying until it could be observed on the plate. This test can provide useful information about water quality and supporting data on the significance of coliform test results. High concentrations of the general bacterial population may hinder the recovery of coliforms.

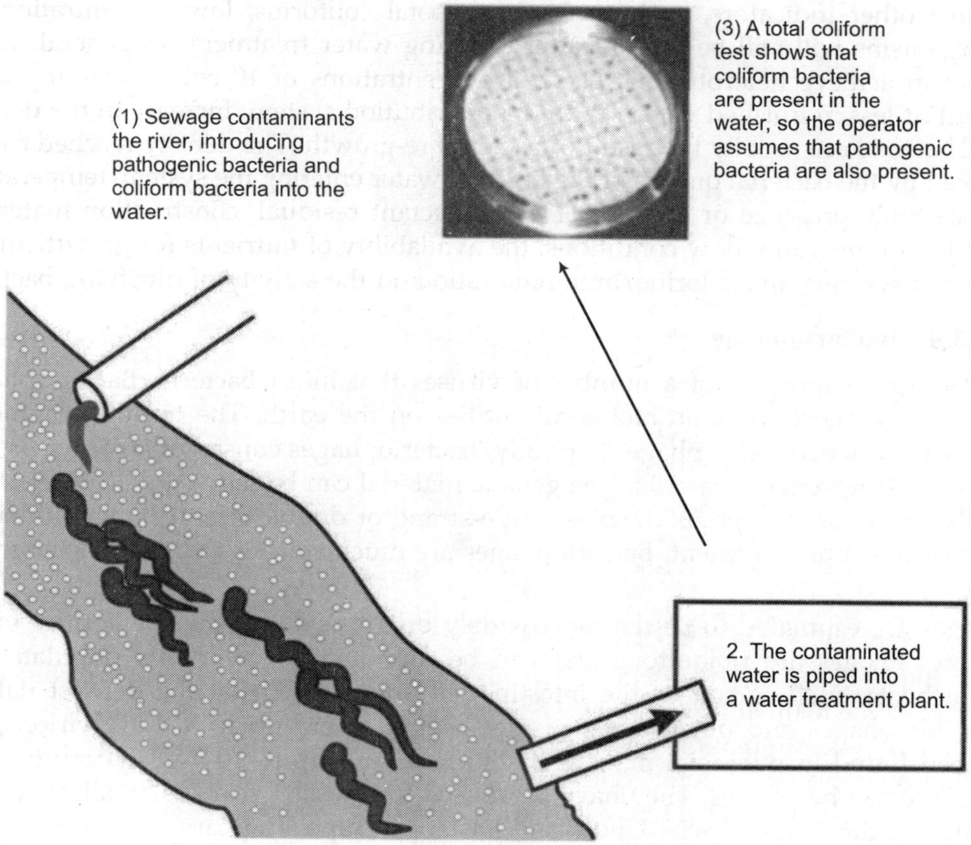

Figure 16.9 Coliform test.

It is important to understand that the colony count, alone, does not allow one to draw conclusions about the risks to public health. However, it currently serves as a relatively easy way to measure filtration and disinfection efficiency, as well as the estimated numbers of bacteria in areas that have the potential for increased contamination.

Growth of bacteria following drinking water treatment is normally referred to as re-growth. This type of growth is typically reflected in higher HPC values measured in water samples. Re-growth generally occurs in areas of distribution or plumbing systems where the water may remain stationary for a longer amount of time, in bottled water, water softeners or carbon filters. In order to ensure that re-growth of bacteria is kept to a minimum, general water safety practices such as maintenance protocols, regular cleaning, temperature management and maintenance of a disinfectant residual (e.g. chlorine) should be in place. If the HPC points exceed recommendations, one should consider looking at whether the system has been adequately cleaned, whether the disinfectant residual is effective, and the efficiency of temperature management. A failing in any of these areas could lead to elevated HPC levels.

Unlike other indicators, such as *E. coli* or total coliforms, low concentrations of HPC organisms will still be present after drinking water treatment. In general, water utilities can achieve heterotrophic bacteria concentrations of 10 colony-forming units (cfu/mL) or less in finished water. Within a distribution system, increase in the density of HPC bacteria are usually the result of bacterial re-growth. The density reached can be influenced by the bacterial quality of the finished water entering the system, temperature, residence time, presence or absence of a disinfectant residual, construction materials, surface-to-volume ratio, flow conditions, the availability of nutrients for growth and in chlorinated systems, the chlorine/ammonia ratio and the activity of nitrifying bacteria.

16.7.2.3.5 Bacteriophage

Bacteriophage is any one of a number of viruses that infect bacteria. Bacteriophages are among the most common biological entities on the earth. The term is commonly used in its shortened form, phage. Typically, bacteriophages consist of an outer protein capsid enclosing genetic material. The genetic material can be ssRNA, dsRNA, ssDNA, or dsDNA ('ss-' or 'ds-' prefix denotes single-strand or double-strand) long with either circular or linear arrangement. Bacteriophages are much smaller than the bacteria they destroy.

Phages are estimated to be the most widely distributed and diverse entities in the biosphere. Phages are ubiquitous and can be found in all reservoirs populated by bacterial hosts, such as soil or the intestines of animals. One of the densest natural sources for phages and other viruses is sea water, where up to 9×10^8 virions/mL have been found in microbial mats at the surface, and up to 70% of marine bacteria may be infected by phages. They have been used for over 60 years as an alternative to antibiotics in the former Soviet Union and Eastern Europe. They are seen as a possible therapy against multidrug resistant strains of many bacteria.

Bacteriophages may have a lytic cycle or a lysogenic cycle, and a few viruses are capable of carrying out both. With lytic phages such as the T4 phage, bacterial cells are broken open (lysed) and destroyed after immediate replication of the virion (Figure 16.10). As soon as the cell is destroyed, the new phages can find new hosts.

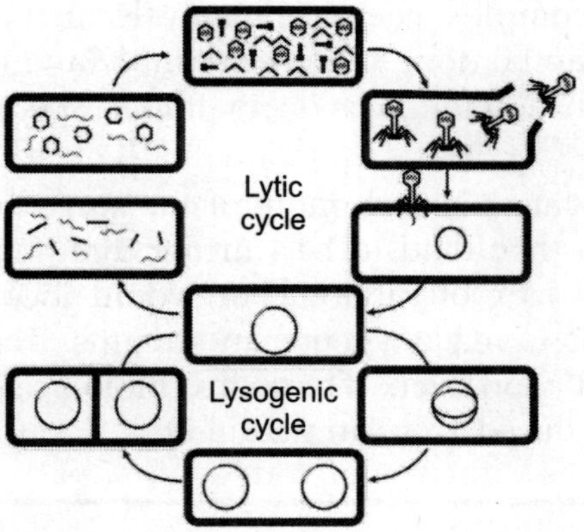

Figure 16.10 Lysogenic and lytic cycle of bacteriophage.

In contrast, the lysogenic cycle does not result in immediate lysing of the host cell. Those phages able to undergo lysogeny are known as temperate phages. Their viral genome will integrate with host DNA and replicate along with it fairly harmlessly, or may even become established as a plasmid. The virus remains dormant until host conditions deteriorate, perhaps due to depletion of nutrients, then the endogenous phages (known as prophages) become active. At this point they initiate the reproductive cycle, resulting in lysis of the host cell. As the lysogenic cycle allows the host cell to continue to survive and reproduce, the virus is reproduced in all of the cell's offspring. Sometimes prophages may provide benefits to the host bacterium while they are dormant by adding new functions to the bacterial genome in a phenomenon called lysogenic conversion (Figure 16.10). A famous example is the conversion of a harmless strain of *Vibrio cholerae* by a phage into a highly virulent one, which causes cholera. This is why temperate phages are not suitable for phage therapy.

Bacteriophages attach to specific receptors on the surface of bacteria, including lipopolysaccharides, teichoic acids, proteins, or even flagella. This specificity means that a bacteriophage can only infect certain bacteria bearing receptors that they can bind to, which in turn determines the phage's host range. Host growth conditions also influence the ability of the phage to attach and invade bacteria. As phage virions do not move independently, they must rely on random encounters with the right receptors when in solution (blood, lymphatic circulation, irrigation, soil water, etc). Complex bacteriophages use a hypodermic syringe-like motion to inject their genetic material into the cell. After making contact with the appropriate receptor, the tail fibres bring the base plate closer to the surface of the cell. Once attached completely, the tail contracts, possibly with the help of ATP present in the tail (Prescott, 1993), injecting genetic material through the bacterial membrane.

16.7.2.3.6 Other indicator organism

All organisms need certain conditions to survive and multiply. Since each aquatic organism has specific tolerances of chemical and physical conditions, the presence or absence of particular organisms can tell us a lot about the body of water we are studying. Compared to complex chemical analysis, using surveys of organisms to determine water quality can be done relatively simply and effectively. A species which is normally present in an aquatic ecosystem under specific conditions is called an indicator organism.

Ponds and other still waters absorb more of the sun's heat energy and retain it for longer periods of time, so they tend to be warmer than fast flowing streams. Because of this heat retention and lack of motion that would incorporate oxygen, ponds are usually much lower in dissolved oxygen than streams. This critical difference affects every aspect of the pond food web. Therefore, biological indexes designed for fast flowing streams cannot be used to accurately assess the water quality of ponds.

Suggested Reading

American Public Health Association. 1999. Standard Methods for the Examination of Water and Wastewater, American Water Works Association, Washington, DC: Water Environment Federation,

Baker Jr., M. B., P. F. Ffolliott, L. F. Debano, D. G. Neary. 2004. Riparian areas of the Southwestern United States, Hydrology, Ecology, and Management: Boca Raton, London: Lewis Publishers, CRC Press.

Bisen PS, Khare P.1991. Mitigating effect of physicochemical factors on Ni^{2+} , Hg^{2+} and Cu^{2+} toxicity in *Cylindrospermum* IU 942. *Environ Technol Lett* **12**, 297-301.

Bisen PS, Mathur S.1993. Adaptive response of wild and mutant type *Synechococcus cedrorum* to a polychlorinated pesticide Endosulfan. *Biomed Environ Sc* **6**, 265-272.

Bisen PS, Shukla HD, Gupta A, Bagchi SN.1987. Preliminary characterization of a novel *Synechococcus* isolate showing mercury, cadmium and lead tolerance. *Environ Technol Lett* **8**, 427-432.

Garte, S. J.(Ed). 1994. Molecular Environmental Biology, Lewis Publishers, Boca Raton, London:CRC Press.

Irini Koukkou Anna (Ed). 2011.Microbial Bioremediation of Non-metals: Current Research, Norfolk, Caister Academic Press.

Mayers SP. 2000. Developments in aquatic microbiology. *Int Microbiol* **3**, 203-211.

McCrary KJ, Harclerode Case CL, Gentry TJ, Aitkenhead-Peterson JA. 2013. *Escherichia coli* regrowth in disinfected sewage effluent: effect of DOC and nutrients on regrowth in laboratory incubations and urban streams. *Water Air Soil Pollut* **224**, 1412-1421.

Nierenberg W. (Ed.). 2012. Encyclopedia of Environmental Biology Vol. 1,2,3, London: Elsevier.

Villahermosa D, Corzo A, Gonzalez JM, Portillo MC, Garcia-Robledo E, Papaspyrou S. 2013. Reduction of net sulfide production rate by nitrate in wastewater bioreactors, kinetics and changes in the microbial community. *Water, Air & Soil Pollution* **224**, 1738.

Zimmermann J, Portillo MC, Serrano L, Ludwig W, Gonzalez JM. 2012. Acidobacteria in fresh water ponds at Donana National Park, Spain. *Microbial Ecology* **63**, 844-855.

Chapter 17

Air Microbiology

17.1 INTRODUCTION

Air microbiology deals with the study of living microbes which are suspended in the air. Air mainly acts as dispersal or transport medium for microorganisms occurring in relatively small numbers compared to soil or water. When compared to soil or water, air contains less number of microorganisms. Air is the simplest one of all the environments and it occurs in a single phase gas.

The air contains gases, dust particles, dried vapour droplets, in addition to these; air also contains more number of microorganisms. The relative quantities of various gases in air, by volume percentage are nitrogen 78%, oxygen 21%, argon 0.9%, carbon dioxide 0.03%, hydrogen 0.01 % and other gases in trace amounts. The air has vegetative cells and spores of bacteria, fungi, algae, protozoan cyst and viruses (Figure 17.1).

Component	Symbol	Volume	
Nitrogen	N_2	78.084%	99.998%
Oxygen	O_2	20.947%	
Argon	Ar	0.934%	
Carbon dioxide	CO_2	0.033%	
Neon	Ne	18.2 parts per million	
Helium	He	5.2 parts per million	
Krypton	Kr	1.1 parts per million	
Sulfur dioxide	SO_2	1.0 parts per million	
Methane	CH_4	2.0 parts per million	
Hydrogen	H_2	0.5 parts per million	
Nitrous oxide	N_2O	0.5 parts per million	
Xenon	Xe	0.09 parts per million	
Ozone	O_3	0.07 parts per million	
Nitrogen dioxide	NO_2	0.02 parts per million	
Iodine	I_2	0.01 parts per million	
Carbon monoxide	CO	Trace	
Ammonia	NH_3	Trace	

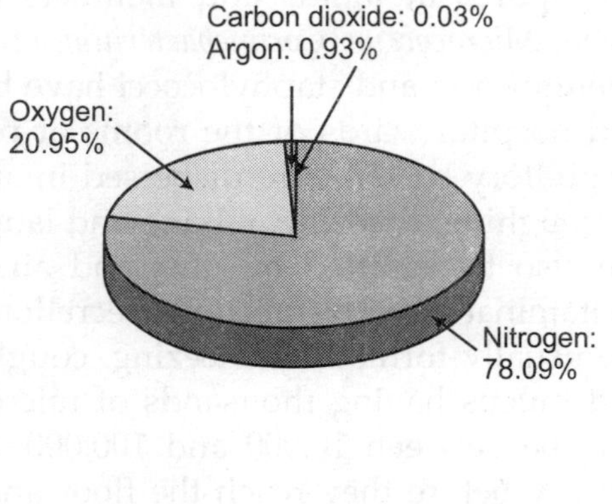

Figure 17.1 Gas composition of air.

Atmospheric layers can be recognized up to a height of about 1000 km. The layer nearest to the earth is called troposphere which is extended up to 11 km in temperate regions whereas in tropics up to about 16 km. The atmosphere as a habitat is characterized by high light intensities, extreme temperature variations, low amount of organic matter and a scarcity of available water making a non hospitable environment for microorganisms for their survival.

Substantial number of microbes is found in the lower regions of the atmosphere. The temperature of the atmosphere varies near the earth's surface. However, there is a steady decrease of about 1°C per 150 metre until the top of the troposphere. Above the troposphere, the temperature starts to increase. This troposphere is characterized by a heavy load of microorganisms. Since air is often exposed to sunlight, it has a higher temperature and less moisture. So, if not protected from desiccation, most of the available microbial forms will die.

17.2 MICROBIAL DISTRIBUTION IN AIR

No microbes are native to the atmosphere rather they represent allochthonous populations transported from aquatic and terrestrial habitats into the atmosphere. Microbes of air within 300 – 1,000 or more feet of the earth's surface are the organisms of soil that have become attached to fragments of dried leaves, straw or dust particles, being blown away by the wind. Species vary greatly in their sensitivity to a given value of relative humidity, temperature and radiation exposures. More microbes are found in air over land masses than far at sea. Commonly found fungal spores belonging to *Curvularia, Alternaria, Cladosporium, Penicillium* and *Aspergillus, Rhizopus,* are more in number than other forms over sea within about 400 miles of land in both polar and tropical air masses at all altitudes up to about 10,000 feet.

Microbes found in air over populated land areas below altitude of 500 feet in clear weather include spores of *Bacillus* and *Clostridium,* ascospores of yeasts, mycelium and spores of molds and members of Streptomycetaceae, pollen, protozoan cysts, algae, *Micrococcus, Corynebacterium,* etc. Microbes such as tubercle bacilli, streptococci, pneumococci and staphylococci have been demonstrated in the dust and air of schools and hospital wards or the rooms of persons suffering from infectious diseases. These respiratory bacteria are dispersed in air in the droplets of saliva and mucus produced by coughing, sneezing, talking and laughing. Respiratory and some enteric tract viruses are also transmitted by dust and air. Pathogens primarily derived from the objects contaminated with infectious secretions after drying become infectious dust. Droplets are usually formed by sneezing, coughing and talking. Each droplet consists of saliva and mucus having thousands of microbes. The number of bacteria in a single sneeze may be between 10,000 and 100,000. Small droplets in a warm and dry atmosphere are dry before they reach the floor and thus quickly become droplet nuclei.

Many plant pathogens can be predicted by measuring the concentration of air-borne spores which are responsible transporting them from one field to another through air and cause infection (Table 17.1). Air-borne diseases such as diphtheria, meningitis, pneumonia, tuberculosis and whooping cough are caused by human bacterial pathogens (Table 17.2).

Table 17.1 Some common air-borne plant diseases

Plant Disease	*Pathogens*
Late Blight of Potato	*Phytopthora infestans*
Brown and Black Rust of wheat	*Puccinia graminis*
Loose Smut of wheat	*Ustilago tritici*
Dutch Elm Disease	*Ceratocystic ulmi*
Downy mildew	*Pseudoperonospora humili*
Maize rust	*Puccinia sorghi*
Southern com leaf blight	*Helminthosporium maydis*
Powdery mildew of barley	*Erisiphe graminis*

Table 17.2 Some common air-borne human diseases

Human disease	*Pathogens*
Bacterial diseases	
Pulmonary tuberculosis	*Mycobacterium tuberculosis*
Pneumonia	*Klebsiella pneumonia*
Pulmonary anthrax	*Bacillus anthracis*
Legionellosis	*Legionella sp.*
Whopping cough	*Bordetella pertussis*
Diptheria	*Corynebacterium diptheriae*
Fungal disease	
Aspergillosis	*Aspergillus fumigatus*
Coccidioidomycosis	*Coccidioides immitis*
Viral disease	
Influenza	Influenza virus
Hunta virus pulmonary	Huntavirus
Hepatitis	Hepatitis virus
Chickenpox	Herpesvirus
Common cold	Picornavirus
Dengu fever	Flavavirus

17.2.1 Sources of Microorganisms in Air

Air does not have an indigenous flora of its own as it doesn't contain enough moisture and nutrients to support the growth and reproduction of microbes. One of the most common sources of air microflora is the soil. Man-made actions like digging or ploughing the soil releases soil-borne microbes into the air. Soil microorganisms are disturbed by the wind blow, liberated into the air and remain suspended there for a long period of time. Similarly, microorganisms found in the water are also released into the air in the form of water droplets or aerosols. Splashing of water by wind action or tidal action may also produce droplets or aerosols.

Air currents may bring the microorganisms from plant or animal surfaces into air. These organisms may be either commensals or plant or animal pathogens. Plant pathogenic microorganisms are spread over very long distances through air. Fungal rust caused by *Puccinia graminis* and smut disease caused by *Ustilago tritici* of wheat are the classical examples of travelling of spores over a thousand kilometres. The main source of air-borne microorganisms in human beings is surface flora which may be shed at times and may be disseminated into the air. However, the transmission of animal diseases is not usually important in outside air.

Similarly, the commensal as well as pathogenic flora of the upper respiratory tract and the mouth are constantly discharged into the air by activities like coughing, sneezing, talking and laughing. The microbes are discharged out in three different forms which are grouped on the basis of their relative size and moisture content. They are droplets, droplet nuclei and infectious dust. The microflora of air can be classified under two headings outdoor and indoor microflora.

17.2.1.1 Outdoor microflora

The air in the atmosphere, which is found outside the buildings, is referred to as outdoor air. The dominant micro flora of outside air is fungi. Common genera of fungi are *Curvularia, Alternaria, Cladosporium and Sporobolomyces*. Besides these genera, other genera found in air are *Aspergillus, Penicillium, Phytophthora* and *Erysiphe*. The outdoor air also contains basidispores, ascospores of yeast, fragments of mycelium and conidia of molds. Among the bacterial genera *Bacillus* and *Clostridium, Mycobacterium, Sarcina, Micrococcus, Corynebacterium, Serratia, Psudomonas* and *Achromobacter* are widely found in the outside air. The number and kind of microorganisms may vary from place to place, depending upon the human population densities.

17.2.1.2 Indoor microflora

The air found inside the building is referred as indoor air. The common genera of fungi in indoor air are *Penicillium, Aspergillus, Mucor, Rhizopus, Alternaria, Curvularia,* and of bacteria are *Staphylococci, Bacillus* and *Clostridium.* In case of occupants being infected, the composition slightly varies with latitude and to a lesser extent with altitude.

17.2.1.3 Droplet

Droplets are usually formed by sneezing, coughing or talking. Each consists of saliva and mucus. Droplets may also contain hundreds of microorganisms which may be pathogenic if discharged from diseased persons. Pathogens will be mostly of respiratory tract origin. The size of the droplet determines the time period during which they can remain suspended. Most droplets are relatively large, and they tend to settle rapidly in still air. When inhaled these droplets are trapped on the moist surfaces of the respiratory tract. Thus, the droplets containing pathogenic microorganisms may be a source of infectious disease. It has been estimated that the number of bacteria in a single sneeze may be between 10,000 and 100,000.

17.2.1.3.1 Droplet nuclei

Small droplets in a warm, dry atmosphere tend to evaporate rapidly and become droplet nuclei. Thus, the residue of solid material left after drying up of a droplet is known as droplet nuclei. These are small, 1-4 μm, and light. They can remain suspended in air for hours or days, travelling long distances. They may serve as a continuing source of infection if the bacteria remain viable when dry. Viability is determined by a set of complex factors including, the atmospheric conditions like humidity, sunlight and temperature, the size of the particles bearing the organisms, and the degree of susceptibility or resistance of the particular microbial species to the new physical environment. Thus, droplet nuclei may act as more potential agents of infectious diseases than droplets.

17.2.1.3.2 Infectious dust

Large aerosol droplets settle out rapidly from air on to various surfaces and get dried. Nasal and throat discharges from a patient can also contaminate surfaces and become dry. Disturbance of this dried material by bed making, handling a handkerchief having dried secretions or sweeping floors in the patient's room can generate dust particles which add microorganisms to the circulating air. Most dust particles laden with microorganisms are relatively large and tend to settle rapidly. Some droplets are of such size that complete evaporation occurs in a warm, dry climate, and before they reach the floor quickly become droplet nuclei. These are small and light, and may float about for a relatively long period. Air-borne diseases are transmitted by two types of droplets, depending upon their size. Droplet infection remains localized and concentrated, whereas air-borne infection may be carried long distances and is in diluted form.

Microorganisms can survive for relatively longer periods in dust. This creates a significant hazard, especially in hospital areas. Infective dust can also be produced during laboratory practices like opening the containers of freeze dried cultures or withdrawal of cotton plugs that have dried after being wetted by culture fluids. These pose a threat to the people working in laboratories.

17.3 TECHNIQUES FOR MICROBIOLOGICAL ANALYSIS OF AIR

There are several methods, which require special devices, designed for the enumeration of microorganisms in air. The most important ones are solid and liquid impingement devices, filtration, sedimentation, centrifugation, electrostatic precipitation, etc. However, none of these devices collects and counts all the microorganisms in the air sample tested. Some microbial cells are destroyed and some entirely pass through in all the processes. Some of the methods are described below.

17.3.1 Impingement on Liquids

In this method, the air drawn is through a very small opening or a capillary tube and bubbled through the liquid. The organisms get trapped in the liquid medium. Aliquots of the liquid broth are then plated to determine its microbial content.

17.3.2 Impingement on Solids

In this method, the microorganisms are collected, or impinged directly on the solid surface of agar medium. Colonies develop on the medium where the organism impinges. Several devices are used, of which the **settling-plate technique** is the simplest. In this method the cover of the Petri dish containing an agar medium is removed, and the agar surface is exposed to the air for several minutes. Certain number of colonies develops on incubation of the Petri dish. Each colony represents particle carrying microbes. Since the technique does not record the volume of air actually sampled, it gives only a rough estimate. However, it does give information about the kind of microorganisms in a particular area. Techniques where in a measured volume of air is sampled have also been developed. These are sieve and slit type devices. A sieve device has a large number of small holes in a metal cover, under which is located a Petri dish containing an agar medium.

A measured volume of air is drawn, through these small holes. Air-borne particles impinge upon the agar surface. The plates are incubated and the colonies counted. In a **slit device** the air is drawn through a very narrow slit onto a Petri dish containing agar medium. The slit is approximately the length of the Petri dish. The Petri dish is rotated at a particular speed under the slit. One complete turn is made during the sampling operation.

17.3.2.1 Limitations of slit sampling

When sitting and operating the sampler inside a room major source of bacterial contamination in the air is the dust from the skin and clothing's of the operating person liberated by the body, movements. Therefore unnecessary movements should be avoided while sampling.

17.3.2.2 Advantages of slit sampler

This is a highly efficient device which can collect up to about 95% of the water droplet particles sprayed into air. Even respiratory secretion droplet nuclei of 0.2 μm diameter can be collected in large numbers.

17.3.2.3 Disadvantages of slit sampler

Slit sampler is a cumbersome and noisy equipment. When the vacuum pump is enclosed with acoustic insulation noise can be reduced to some extent. Quieter and portable samplers are also available but they are less efficient in collecting sample particles.

17.3.2.4 Sieve sampler

This is a mechanically simpler form of impinger. The instrument is more or less similar to that of slit sampler with an enclosed chamber. The particles containing microorganisms are distributed over the plate as separate air jets through several holes. Upon incubation these particles form colonies which can be counted. For more efficient sampling and size grading of particles a multistage sieve device has been developed in which several impingers with holes of different sizes are arranged in series.

17.3.3 Electrostatic Precipitation

Electrostatic precipitation is an efficient method of removing particles from air. The air is allowed to pass through the electrodes in this system. The charged particles fall on a rotating disc which is fed with collecting fluid at a rate of 10 mL/minute. Air is sucked into the chamber by a rotating fan at the bottom. The low resistance of the system enables high rates of air flow. They are suitable for large volumes of air.

A modified sampler has been devised which contains two removable covers. Each unit has one upper electrode and one lower electrode. In one unit the upper electrode is negative and the lower electrode is positive and in the other unit the electrical condition is reversed. Air is drawn at equal rates in both the units. Charged microorganisms are collected in the Petri dishes placed on opposite electrodes.

17.3.4 Filtration

This is a simple method for collecting particles from air. The filter can be made of any fibrous or granular material like sand, glass fibre and alginate wool (in phosphate buffer). However, recovery of organisms for culture is not so easy.

17.3.4.1 Millipore filter

This type of filter is made of pure and biologically inert cellulose ethers. They are prepared as thin porous, circular membranes of about 150 μm thickness. The filters have different porosity with different grades varying from 10 nm to 8 nm. The assemblage contains a funnel-shaped inlet and a tube-like outlet. In between these two the filter is fitted. The outlet may be connected to a vacuum pump to suck known amount of air. After collecting required volume of air through the filter, it can directly be placed onto the surface of a solid medium. After incubation colonies formed can be counted. The method is similar in principle to that described for water sampling (Figure 17.2).

17.4 SIGNIFICANCE OF AIR MICROFLORA

Air microflora is very low in number when compared with the microorganisms of other environments, however, playing a very significant role. This is due to the fact that the air is in contact with almost all animate and inanimate objects. Louis Pasteur (1861) provided the first evidence that microorganisms could occur as air-borne contaminants. He used special cotton in his air sampler onto which the microorganisms were deposited. He microscopically demonstrated the presence of microorganisms in the cotton. In his famous swan necked flask experiment, he showed that growth could not occur in sterile media unless air-borne contamination had occurred.

17.4.1 Air Microflora and Human Health

The significance of air microflora in human health relies on the fact that air acts as a medium for the transmission of infectious agents. An adult man inhales about 5 m^3 of air per day. Although most of the microorganisms present in air are harmless saprophytes and commensals, less than 1% of the air-borne bacteria are pathogens. Even though the

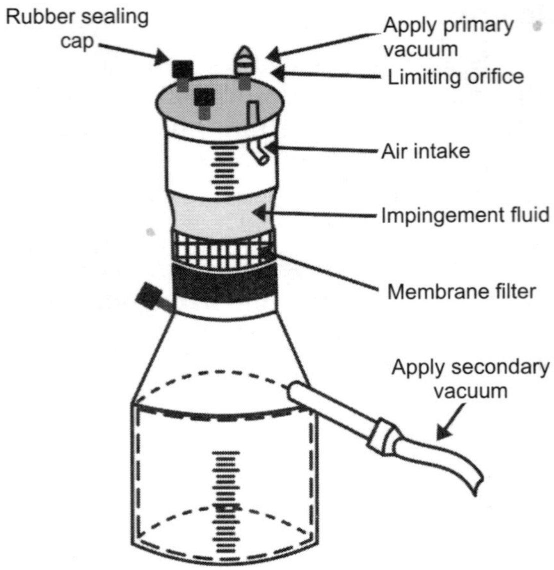

Figure 17.2 Millipore filter assembly.

contamination level is very low, the probability of a person becoming infected will be greater if he is exposed to a high concentration of air-borne pathogens. Carriers, either with the manifestation of corresponding symptoms or without any apparent symptoms, may continuously release respiratory pathogens in the exhaled air. *Staphylococcus aureus* is the most commonly found pathogen in air since the carriers are commonly present. The number of *S. aureus* in air may vary between 0-1/m^3 and 50/m^3.

Practically speaking, outdoor air doesn't contain disease causing pathogen in a significant number to cause any infection. The purity of outdoor air, however, is an essential part of man's environment. Dispersion and dilution by large volume of air is an inherent mechanism of air sanitation in outside air. In the case of indoor air, chances for the spread of infectious disease are more, especially in areas where people gather in large numbers. For example, in theatres, schools, etc.

17.4.2 Air Microflora and Hospitals

Although hospitals are the war fields for combating against diseases, there are certain occasions in which additional new infectious diseases can be acquired during hospitalization. Air within the hospital may act as a reservoir of pathogenic microorganisms which are transmitted by the patients.

Infection acquired during the hospitalization is called nosocomial infections and the pathogens involved are called nosocomial pathogens. Infections, manifested by the corresponding symptoms, after three days of hospitalization can be regarded as nosocomial infection. Nosocomial infection may arise in a hospital unit or may be brought in by the staff or patients admitted to the hospital. The common microorganisms associated with hospital infection are *Haemophilus influenzae, Streptococcus pneumoniae, Staphylococcus aureus, Pseudomonas aeruginosa,* members of Enterobacteriaceae and respiratory viruses. Development of high antibiotic resistance is a potential problem among nosocomial pathogens. Nosocomial pathogens may cause or spread hospital outbreaks. Nosocomial pneumonia is becoming a serious problem nowadays and a number of pathogens have been associated with it. Frequent agents are Staphylococcus aureus, *Streptococcus pneumoniae, Pseudomonas aeruginosa, Enterobacter, Klebsiella, Escherichia coli* and *Haemophilus influenzae*. Other less frequent agents are enterococci, streptococci other than *S. pneumoniae, Serratia marcescens, Citrobacter freundii, Acinetobacter* sp. and *Xanthomonas* sp.

In addition *Legionella, Chlamydia pneumoniae* and *Mycobacterium tuberculosis* have also been reported. Nosocomial transmissions of tuberculosis from patients to patients and from patients to healthcare workers have also been well documented. There are two main routes of transmission for nosocomial pathogens, contact (either direct or indirect) and air-borne spread. Air-borne spread is less common than the spread by direct or indirect contact. The source may be either from persons or from inanimate objects.

In case of spread from persons the droplets from mouth, skin scales from nose, skin exudates and infected lesion transmit diseases such as measles, tuberculosis, pneumonia, staphylococcal sepsis and streptococcal sepsis. Talking, coughing and sneezing produce droplets. Skin scales are shed during wound dressing or bed making. In case of inanimate sources particles from respiratory equipment and air-conditioning plant may transmit diseases. These include Gram-negative respiratory infection, Legionnaire's disease and fungal infections.

17.4.3 Air Microflora and Industries

As long as microorganisms remain in the air, they are of little importance. When they come to rest they may develop and become beneficial or harmful. Knowledge of the microorganisms in air is of importance in several aspects. The microbes are identified with the type of industries.

17.4.4 Food Manufacture

Microorganisms that have been transported through the air and have settled on, or in, the material are involved in various fermentation products. Production of alcoholic beverages, vinegar, sauerkraut, ensilage, dairy products, etc., is often due to microbial activity.

17.4.5 Spoilage of Foods and Fermentation Products

Microorganisms are often troublesome in the home and in industry where foods and other fermentation products are prepared. In industrial processes, where particular organisms are to be grown, sterile air free from contaminating organisms is to be supplied to maintain the process.

17.5 AIR-BORNE DISEASES

There are two main sources of microorganisms in air. These are saprophytic soil organisms raised as dust, and organisms from body tissues introduced into the air during coughing, sneezing, talking, and singing. Droplets expelled during coughing, sneezing, etc., consist of saliva and mucus, and each of them may contain thousands of microorganisms.

17.5.1 Major Diseases Transmitted by Air

Among various modes of disease transmission, air is one of the important routes and a number of diseases have been shown to be transmitted through air. Since human beings and animals are continuously inhaling the external air, the chances for an air-borne microorganism to find a host and cause infection are more. Most of the respiratory tract infections are acquired by inhaling the air containing the pathogen. Microorganisms in droplets and infectious dusts and spores can be easily disseminated through air. Since air enters the body through the respiratory tract and since such diseases frequently localize in the nose and throat, they are called respiratory diseases as a group. Some of the respiratory diseases having an air-borne mode of transmission, are briefly described as follows.

17.5.1.1 Air-borne bacterial diseases

- **Brucellosis:** It is mainly an occupational disease among veterinarian, butcher and slaughter house workers. *Brucella* sp. is the causative organism.
- **Pulmonary Anthrax:** *Bacillus anthracis* is the causative agent. Transmission is mainly by inhaling the dust contaminated by animal products.
- **Diseases Caused by *Streptococcus pyogenes*:** Several diseases are caused by *Streptococcus pyogenes* which is mainly transmitted through air. Diseases caused by *Streptococcus pyogenes* occur in the throat, skin, and systemically.
- **Rheumatic Fever:** This is upper respiratory tract infection by *S. pyogenes* characterized by inflammation and degeneration of heart valves.
- **Pneumococcal pneumonia:** ***Streptococcal pneumonia*** is of major occurrence among the bacterial pneumonia.
- **Meningitis:** *Haemophilus influenzae* causes meningitis in children between 6 weeks and 2 years of age.
- **Diptheria:** Diphtheria is mainly contracted by children. Infection of the tonsils, throat and nose and generalized toxemia are the symptoms. The causative agent is *Corynebacterium diphtheria.*
- **Tuberculosis**: Pulmonary tuberculosis is a severe respiratory disease. Loss of appetite, fatigue, weight loss, night sweats and persistent cough are some of the symptoms. Causative agent is *Mycobacterium tuberculosis.*
- **Legionellosis:** It is a type of branchopneumonia. *Legionella pneumophila* is the causative agent. It occurs in natural water. At times it enters and proliferates in cooling tower, air cooler and shower bath. Spraying and splashing of water containing pathogen may produce aerosols which are disseminated in air.
- **Meningococcal Infections** *Neisseria meningitidis* is the cause of adult bacterial meningitis and the clinical manifestations of the disease can be quite varied, ranging from transient fever and bacteremia to fulminant disease with death ensuing within hours of the onset of clinical symptoms.
- **Whooping Cough (Pertussis):** Whooping cough or pertussis is a highly contagious, primarily childhood respiratory disease caused by *Bordetella pertussis* which is small Gram-negative, aerobic coccobacillus.

17.5.1.2 *Air-borne fungal diseases*

- **Cryptococcosis:** Leads to mild pneumonitis. Causative agent is the yeast *Cryptococcus neoformans.* It is a soil saprophyte. Infection is acquired by inhalation of soil particles containing the causative agent.
- **Blastomycosis:** Formation of supportive and granulomatous lesions in any part of the body. *Blastomyces dermatitis* is the causative agent. It is a soil-borne fungus and hence inhalation of soil particles containing the fungus produces the infection.
- **Coccidiodomycosis:** Infection may not be apparent but in severe cases it is fatal. Usually, infection leads to self-limited influenza fever known as valley fever or desert rheumatism. Causative agent of the disease is a soil fungus, *Coccidioides immitis.* Inhalation of dust containing arthrospores of the fungus leads to infection.
- **Aspergillosis:** It is an opportunistic disease of humans. Causative agent is *Aspergillus fumigatus.* Infection occurs through inhalation of spores.

17.5.1.3 *Air-borne viral diseases*

- **Common Cold:** It is the most frequent of all human infections. Characteristic symptom includes running noses. Rhinovirus is the causative agent. Droplets with nose and throat discharges from infected persons are the source.
- **Influenza:** Symptoms of influenza are nasal discharge, headache, muscle pains, sore throat and general weakness. Causative agents are orthomyxo virus.
- **Measles:** Measles is the most common communicable human disease mainly affecting children. Symptoms are fever, cough, cold and red, blotchy skin rash. Causative virus is morbillivirus. Source of infection is respiratory tract secretions in the form of droplets.
- **Mumps:** It is a communicable disease and is a common childhood disease. It is characterized by painful swelling of parotid glands and salivary glands. Mumps virus causes the disease. Droplets containing infected saliva are the main source.
- **Adeno Viral Diseases:** Adenoviruses cause acute self-limiting respiratory and eye infections. Adenoviruses are transmitted by air-borne mode. Diseases include acute febrile pharyngitis, acute respiratory disease and adenovirus pneumonia.

17.6 CONTROL MEASURE

17.6.1 Physical Techniques

17.6.1.1 *Dust control*

As stated earlier, the dust particles are laden with large number of microbial agents that cause diseases and allergies. Since dust occurs in homes, schools, factories, laboratories and hospitals, its suppression to lower down microbial content is, therefore, extremely important. Use of dry-vacuum pickup followed by the application of suitable disinfectant detergents solution has strongly been recommended to control dust in indoor environments. In addition, oiling floors, bedclothes, and other textiles has proved a highly effective dust control device.

17.6.1.2 Ultraviolet radiation

This technique has great potential value for reducing indoor air-borne microbial population. Various types of germicidal lamps are used for this purpose. These lamps emit radiations in 250-260 nm regions, the most effective bactericidal region. These radiations prove effective only when they come in direct contact with the particles laden with microbes; it is because these radiations have little penetrating power. Since these radiations are irritating to human eyes and skin, skillful installation of the lamps is required to avoid injury to persons using those rooms.

17.6.1.3 Laminar airflow system

This is a technique recommended for controlling indoor microbial population. This technique represents unidirectional airflow system in which the air passes high-efficiency particulate air (HEPA) filters and the entire body of air moves with uniform velocity along parallel flow lines. The HEPA filters consist of cellulose acetate plated around aluminium foil and can remove particles as small as 0.3 mm. The laminar airflow system is, therefore, suitable device in electronic and aerospace industries where an extremely high degree of cleanliness is required for product reliability. In laminar flow, air is drawn in from the atmosphere and passes through a 5µm filter with 95 percent efficiency. About 80% of the air in the room is recycled through this filter. Before the air is delivered to the operating site it is passed through high efficiency particulate air (HEPA) filters with 0.3 µm pore size and 99.97 percent efficiency. Thus, bacterial contamination is prevented. Coarse filters should be changed every 3-4 months depending upon the climatic conditions while HEPA filters should be changed every six months.

17.6.2 Chemical Agents

17.6.2.1 Bactericidal vapours

Indoor air-borne microbial population can be effectively reduced by vaporizing or spraying certain chemical substances into the air. Some of such chemical substances are propylene glycol, triethylene glycol, resorcinol, hypochlorous acid and b-propiolactone. Polyethylene glycol and triethylene glycol are colourless, tasteless, non-irritating, and non-toxic chemicals. Nearly all microorganisms present in a litre of heavily contaminated air can be killed by vapour from only 0.5 mg of polyethylene glycol; triethylene glycol is nearly ten times effective in comparison to polyethylene glycol. However, great care is required for safe and efficient application of chemical agents and, therefore, they may be used only when rapid control of air-borne microbial content is essential.

17.6.2.2 Tube sampler

This is one of the oldest devices for collecting and enumerating microorganisms in the air. It consists of a tube with an inlet at the top and an outlet at the bottom which is narrower than the top end. Near the bottom there is a filter of wet sand which is supported by a cotton plug below. The entire device can be sterilized. After sterilization, the air to be sampled is allowed to pass through the sand and cotton. Microorganisms as well as dust particles containing microorganisms in the air are deposited in the sand filter as the air passes through it. Later the sand is washed with broth and a plate count is made from the broth by taking aliquotes of the broth.

17.7 DETERMINATION OF AIR MICROFLORA

The numbers and kinds of microorganisms contaminating the air are determined by the sources of contamination in the environment. Several disease causing microorganisms are present in air involved in causing various diseases such as influenza, cold and some time mycoses. Thus, the isolation of these organisms is important for determining the composition of causal organism present in air. The sampling of air to determine its content requires special apparatus and devices such as solid impingement devices and liquid impingement devices. By these methods spores of microorganism fallen on the medium can be detected and the composition of air can be determined.

17.7.1 Procedures

Prepare Czapek dox agar medium and nutrient agar medium with a pinch of antibacterial and antifungal agent in the medium, respectively, for fungal and bacterial isolation separately, pour molten agar medium on 4 to 5 plates separately, allow the plates to solidify. Open plates in air for 5 to 8 minutes by removing its top cover, incubate the exposed plates at 25°C for 48 to 72 h for Czapek dox agar plates and at 37°C for 24 to 48 h for nutrient agar plates. Observe the plate for microbial growth. Determine the distribution by using colony counter and identify different types of bacteria in the plates and record the results. Fungal members may be identified by microscopical observations.

17.8 EXPERIMENTS ON AIR MICROBIOLOGY

17.8.1 Isolate Microorganisms from Air

17.8.1.1 Principle

The numbers and kinds of microorganisms contaminating the air are determined by the sources of contamination in the environment. Thus the isolation of these organisms is important for determining the composition of causal organism present in air. By these methods spores of microorganism fallen on the medium is detected and the composition of air can be determined.

17.8.1.2 Requirements

Nutrient agar medium, Czapek dox agar medium, colony counter, sterile Petri dish, sterile test tube, inoculating loop, distilled water, Bunsen burner, disinfectant for desktop, sponge, wax marking pencil etc.

17.8.1.3 Protocol

1. Prepare Czapek dox agar medium and nutrient agar medium with a pinch of antibacterial and antifungal agent, respectively.

2. Pour both the above molten agar medium on 4 to 5 plates.
3. Allow the plates to solidify.
4. Allow the plates to open in air for 5 to 8 minutes by removing its top cover.
5. Incubate the exposed plates at 25°C for 48 to 72 h for Czapek dox agar plates and at 37°C for 24 to 48 h for Nutrient agar plates.

17.8.1.4 Results and observations

1. Observe the plate for microbial growth. Determine the distribution by using colony counter.
2. Identify the different type of bacteria in the plates and record the results.

17.8.2 Isolation of Bacteria from Phyllosphere and Phylloplane

17.8.2.1 Principle

Leaf surface microbes may perform an effective function in controlling the spread of air-borne microbes inciting plant diseases. Presence of a fungal spore on the surface of leaves incite the formation of a chemical substance referred to as phytoalexin which are active in host defence mechanisms. The phyllosphere and phylloplane microflora are of special interest as some of them have antagonistic effect against fungal parasites, some degrade plant surface wax and cuticle and produce phytoalexins. There is a continuous diffusion of plant metabolites from the leaves which support the microbial growth and, in turn, these microbes protect the plant from pathogens.

17.8.2.2 Requirement

Fresh leaves, Czapek dox agar medium, Nutrient agar medium, distilled water, pipettes, polythene bags, Petri plates.

17.8.2.3 Protocol

1. In a sterile container collect the leaves and take it in laboratory.
2. Cut small pieces of leave with sterile cork borer.
3. Transfer the leave discs in 10 mL distilled water.
4. Now prepare serial dilution upto 10^{-6} in test tube.
5. Transfer 1 mL from each dilution in separate plates and pour molten and cooled medium on it.
6. Incubate at 27°C for 2 to 6 days in Czapek dox agar plates and at 37°C for 24 to 48 h for Nutrient agar plates.

17.8.2.4 Results and observations

Observe the plates after incubation for growth and identify the organisms on leaf surface (phylosphere). Note the results for further studies.

Suggested Readings

Al-Dagal M, Fung DY. 1990. Aeromicrobiology--a review. *Critical Reviews in Food Science and Nutrition* **29**, 333-340.

Amato, Pierre. "Clouds Provide Atmospheric Oases for Microbes." Microbe Magazine: n. pag. American Society for Microbiology. Web. 1 Apr. 2012.

Brandl H, von Daniken A, Hitz C, Krebs W. 2008. Short-term dynamic patterns of bioaerosol generation and displacement in an indoor environment *Aerobiologia* **24,** 203-209.

Burrell R. 1991. Microbiological agents as health risks in indoor air. *Environ Health Perspect* **95**, 29–34.

Castellan RM, Olenchock SA, Kinsley KB, Hankinson JL. 1987. Inhaled endotoxin and decreased spirometric values. An exposure-response relation for cotton dust. *N Engl J Med* **317**, 605–610.

Dutkiewicz J, Jabłoński L, Olenchock SA. 1988. Occupational biohazards: a review. *Am J Ind Med* **14,** 605–623.

Heyden S, Pratt P. 1980. Exposure to cotton dust and respiratory disease. Textile workers, 'brown lung,' and lung cancer. *JAMA* **244**,1797–1798.

Kaufmann AF, McDade JE, Patton CM, Bennett JV, Skaliy P, Feeley JC, Anderson DC, Potter ME, Newhouse VF, Gregg MB, Brachman PS. 1981. Pontiac fever: isolation of the etiologic agent (*Legionella pneumophilia*) and demonstration of its mode of transmission. *Am J Epidemiol* **114**, 337–347.

McCormick A, Loeffler J, Frank Ebel F. 2010. *Aspergillus fumigatus*: Contours of an opportunistic human pathogen. *Cellular Microbiology* **12,** 1535-1543.

McDade JE, Shepard CC, Fraser DW, Tsai TR, Redus MA, Dowdle WR. 1977. Legionnaires' disease: isolation of a bacterium and demonstration of its role in other respiratory disease. *N Engl J Med* **297**, 1197–1203.

Proctor BE, Parker BW. 1938. Microbiology of Upper Air. *J Bacteriol* **4**, 363–375.

Robinson M, Stilianakis N, Drossinos Y. 2012. Spatial Dynamics of Airborne Infectious Diseases. *Journal of Theoretical Biology* **297**, 116-126.

Chapter **18**

Microbes and Plants

18.1 INTRODUCTION

Interaction between plants and microbes is of varied type, ranging from the usually effects of soil microorganisms on soil fertility, and hence on plant growth, to destructive parasitism, resulting in damage or death of the plant. Symbioses cover all examples of the living together of different species, including the destruction of one organism by another. Often it is used in a limited sense to denote a state of controlled or reciprocal parasitism in which each partner benefits to some extent from the presence of the other. The terms commensalism or mutualism are sometimes preferred in describing this condition but not commonly used by plant pathologists. The activities of microorganisms in the soil, already described in chapter on rhizosphere, necessarily affects plant. A closer relationship is that between plant roots and the microflora of the rhizosphere of layer of soil surrounding them. Here we discuss in depth the association of microflora with the aerial surface of the plants (epiphytes) and controlled parasites (symbiosis) and destructive plant parasites (plant diseases).

18.2 MICROORGANISMS EPIPHYTIC ON PLANTS

Numerous microorganisms occur on the aerial surface of the plants, subsisting on traces of nutrients exuding from the plants or present in dust falling upon them. These epiphytic species do not penetrate the plant and except in wet weather when molds may grow to an extent sufficient to reduce the light reaching the leaves, do not harm it. They are able to survive under severe climate conditions, many of the bacteria being protected by a mucous sheath. Comparatively few species are represented, but the number of individual organisms may be very large. Estimates of the number of bacteria (mainly non-sporulating rods) on newly harvested grain vary from 500,000 to 1,330,000 per gram. This is probably higher than the numbers on the actively growing parts of the plants, since spread on to the floral parts is encouraged by sugary liquids produced there. *Cladosporium herbarum* is readily isolated from dead leaves of many species and from twigs and the bark of trees. The numbers of individuals and of species increase rapidly in wet weather and also on leaves covered with the sticky 'honey dew' resulting from aphis attack. Leaves so attacked may become completely covered with a sooty layer of spores and mycelium. The flora is then known as 'sooty mold'. The constituents of sooty molds are external to the plant but may seriously harm it by blocking stomata, thus interrupting transpiration and aeration, and by screening the leaves from light thereby reducing photosynthesis. Although, plant surfaces have to be considered a hostile environment to microbes, more than 85 different species of microorganisms in 37 genera have already been reported in the phyllosphere of different crops.

18.3 MICROORGANISMS ENDOPHYTIC ON PLANTS

Fungi, bacteria and yeast can live symbiotically in the tissues of host plants without causing visible harm. Such microorganisms are known as endophtes. They spend part or all of their life cycle colonizing either inter- or intra-cellularly the healthy tissues of plants without causing any immediate overt effects. They have been isolated from liverworts, hornworts, mosses, lichens, equisetopsids, ferns, sea grasses and seed plants from arctic to the tropics. They are known to promote plant growth and yield, act as biocontrol agents and are known to produce secondary metabolites, such as enzymes, growth hormones, antimicrobial, antifungal, or anticancerous substances. It is also known that the endophytes apart from being enormous source of secondary metabolites with varied properties also display the ability to reduce metals producing nanoparticles. Among all microorganisms, endophytic fungi have been more extensively studied group of organisms.

18.4 CONTROL PARASITISM (SYMBIOSIS)

Plant parasites include some organisms mainly external to the host but deriving nutrition from it by houstoria which penetrate the epidermal cells, and others which spend all or part of their life cycles within the tissues or even inside the cells. The effects of such parasites vary from death of the host to minor damage which may be negligible, or, as with the so called symbiotic relationships, may actually be balanced by beneficial effects.

Lichens, the root nodules of leguminous plants and the mycorrhiza of plants and orchids are the classical examples of symbiotic relationships where both the partners are benefited and neither are seriously harmed. However, in all these cases, the balance between the partners is a delicate one and that if it is upset by a change in environmental conditions favouring one partner this may then damage or destroy the other. The relationship is thus one of controlled parasitism in which normally the mutual benefits are greater than the mutual disadvantages.

18.4.1 Plant Root Symbionts

Awareness of the advantageous effects of leguminous crops on soil fertility can be traced back to biblical times. However, it was not until much later that Atwater, in 1885, and Hellriegel and Wilfarth, in 1888, demonstrated that such plants as peas and clover can utilize nitrogen from the air when they are nodulated; plants without nodules requiring nitrate for growth. In 1888 Beijerinck isolated a bacterium, now included in the genus Rhizobium, from nodule material and from soil. It was concluded that as a result of this association of microorganisms in the nodules nitrogen was fixed. However, rigorous demonstration of this was obtained only when ^{15}N labelled nitrogen was used.

Neither of the partners in this symbiotic association is capable of nitrogen fixation in the absence of the other but they are not dependent on each other for existence. The legume is capable of using combined nitrogen in the form of nitrate and the bacteria exist independently even in soils which have not supported a legume crop. The legumes are important in agriculture not only for the fact that they fix nitrogen and can thus conserve nitrogen fertilizers but also because the plants and seeds are high in protein. The seeds of grain legumes are important nutritionally for humans, and the forage legumes (clovers, lucerne or alfalfa, and vetches) also supply a high-protein diet to livestock. The population in soil is, however, larger when the macrosymbiont has been grown there. A proper understanding of the Rhizobium legume association requires knowledge of the relationship of rhizobia one to another and to other members of the Rhizobiaceae. The leguminosae is a large family of about 750 genera and 20,000 species (Figure 18.1). A proper understanding of the Rhizobium legume association requires knowledge of the relationship of rhizobia one to another and to other members of the Rhizobiaceae. However, not all species or genera associate with Rhizobium to form root nodules. Legumes also have ecological significance as they provide nitrogen inflow to natural habitats such as tropical forests. Some tropical legume trees are grown for their wood, and the nitrogen-enriched soil is exploited when the wood is removed. There are several other uses of legumes, but even so only a tiny fraction of the total number of species in this family is exploited by man. Because there is such a wide range of plants in this family, nodule form and modes of infection will differ in detail according to the host species and rhizobial strain.

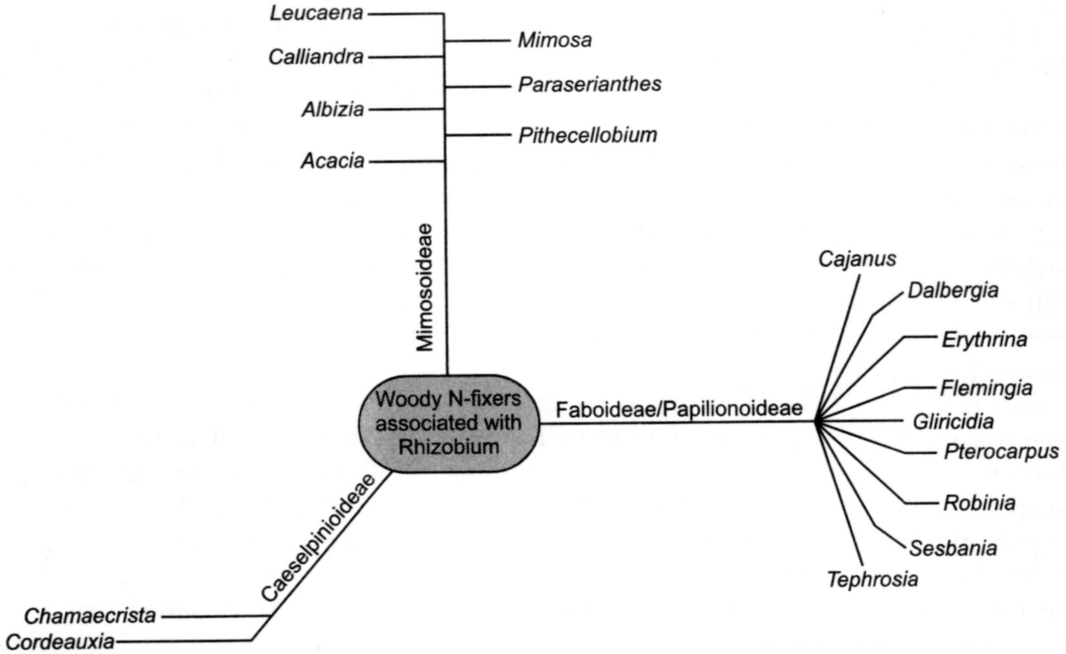

Figure 18.1 Woody N-fixers associated with Rhizobium.

18.4.1.1 Rhizobium legume association

All peas and beans are legumes. Nitrogen fixing plants are called legumes. The plants that work together with nitrogen fixing bacteria called Rhizobia, to "fix" nitrogen. The Rhizobia chemically convert the nitrogen from the air to make it available for the plant. Leguminous plants live in a symbiotic relationship with the nitrogen fixing bacteria. Rhizobia live in nodules in the plant's root. This way the plant can look after its own nitrogen needs and fertilizer is not required. In addition, when the crop is harvested and the plant cut back to ground level, the root nodules should release all the valuable fixed nitrogen.

The grain legumes themselves contribute 20% of nitrogen fixed and considering that most of the fixation in grasslands will be by forage legumes and a proportion of the fixation in forest and woodlands will be by tree legumes with a conservative estimate of 50% of natural nitrogen fixation is accomplished by the Rhizobium-legume association. Nitrogen fixing leguminous plants has the unique ability to fix atmospheric nitrogen in the ground and make their own fertilizers. Plants owe partial credit for this effort to their symbiotic relationship with nitrogen fixing bacteria. The leguminous plants provide nutrients to the bacteria in return for which the bacteria fix atmospheric nitrogen through anaerobic processes. The primary function of nitrogen fixing bacteria is 'survival' and in their efforts to survive, they enter into a symbiotic relationship with leguminous plants or some survive on their own. Legumes nodulation by Rhizobium is a complex process requiring their coordinated expression of both bacteria and plant genes. The rhizobia are soil microorganisms that inhabit the rhizosphere of legumes and other plants. A proper understanding of the Rhizobium legume association requires knowledge of the relationship of rhizobia one to another and to other members of the Rhizobiaceae.

18.4.1.1.1 The rhizobium

There are two main types of Rhizobia, distinguished by a number of differences but referred to according to their growth rate on laboratory media as 'fast growers' and 'slow growers'. The slow grower rhizobia have been placed in a separate genus *Bradyrhizobium*. The division of Rhizobium into species is based on the interaction with plants; those bacteria which nodulate clovers, for instance, are put in *Rhizobium trifolii* and those that nodulate peas and vetches are put in *Rhizobium leguminosarum*. Although this classification is still used, it has been realized for a long time that it is not satisfactory. There is a group which has such a wide specificity that it has not been assigned to a species in slow growing rhizobia, but is called the cowpea group after one of the many hosts. Since the classification was first accepted, more and more exceptions have been found with regard to species host specificity. Some strains can be put into more than one species but generally retain the name from the host from which they were first isolated. Agrobacterium is a member of the Rhizobiaceae and this genus also interacts with plants, viz. *A. tumifaciens, A rubi* form crown galls and *A. rhizogenes* causes proliferation of adventitious roots or hairy root disease in dicotyledons.

Agrobacterium has been reported to be more similar to fast growing rhizobia. Studies using the techniques of numerical taxonomy, where every character is given equal weight, have used over 100 different characters in their analysis and reported that two types of rhizobia are distinct from one another and sufficiently distinct to put them into separate genera. The closeness of *Agrobacterium* to the fast growing rhizobia has also led to the suggestion that they should be included in the same genus, *Rhizobium*. The placing of Agrobacterium with Rhizobium may seem odd because they have such different effects on plants. However, both *Agrobacterium* and *Rhizobium* have shown the characters concerned with the plant interaction are coded on a plasmid. The pathogenic property of the *Agrobacterium* is plasmid borne and a part of the plasmid DNA from *Agrobacterium* is transferred to the plant to become incorporated into the plant chromosomes. Subsequently, it was discovered that the nodulating properties of the fast growing strains of Rhizobium were coded on plasmid genes, as were also the nitrogen fixing enzymes. Plasmids bearing the nodulating and nitrogen fixing genes have not so far been found in the slow growing rhizobia. In this group the genes are probably therefore on the chromosomes. Rhizobia are recognised to form root nodules, but some strains are capable of forming nodules on stems, although it is root primordial on the stems that become infected. *Sesbania rostrata* and *Aeschynomene afraspera* are two species that can be so nodulated. Some of the strains of *Rhizobium* which form stem nodules are able to fix sufficient nitrogen to support growth in the free living state. In order to do this, however, the oxygen concentration must be low, as they have no means of oxygen protection. The slow growing rhizobia in general are able to fix nitrogen in culture but are unable to fix sufficient nitrogen to grow. The fast growing rhizobia have not so far been found to fix nitrogen outside the root nodule. Some strains of Rhizobium can form effective nodules with some tropical non-legumes, species of *Parasponiam,* a genus in the family Ulmaceae (Figure 18.1). Average amounts of nitrogen added to the soil by a leguminous crop are frequently in excess of 100 lb per acre per year.

Rhizobia are short to medium, Gram-negative rods (0.5–0.9 × 1.2–3 µm) older cells bear capsule, and are sporulated, flagella are arranged in polar, subpolar or peritrichous fashion. Identification of *Rhizobium* is based on common biochemical tests such as: Gram reaction, motility, flagellar staining, no H_2S production, litmus milk reaction, inability to utilise citrate, and intolerance to 2% NaCl on YMA.

18.4.1.1.2 Recognition

Plant defence mechanism normally prevents the entry of microorganisms that may cause disease, so that there must therefore be some means of mutual recognition between the symbiotic partners to enable infection to occur and the association to develop. However, when it is growing with adequate mineral nitrogen the symbiosis is not an advantage to the plant because of the high cost to the plant of carbon substrates required to develop and maintain the symbiotic condition. Legumes thus resist infection and nodulation is inhibited if the nitrogen content of the soil is high. When the soil is depleted of nitrogen, rhizobia are recognized and infection occurs. The recognition process is thus controlled by the nitrogen status of the environment through its interaction with the plant. Rhizobia have a restricted host range with a high degree of specificity in the recognition process. The basis of recognition must be the molecules on the surface of the partners and these must be sufficiently diverse if specificity is to be attained. Suitable molecules are proteins and polysaccharides. The range of sugars in the surface polysaccharides of both plant and bacterium is wide enough to provide sufficient diversity. If the sugars are the recognition signals, then there must be a mechanism to distinguish specific sugars and to mediate the recognition response between plant and bacterium. Lectins have several binding sites for sugars and can thus attach the bacterium to the root surface by binding specific sugars on the plant surface and the bacterial surface through binding sites on the same lectin molecule (Figure 18.2).

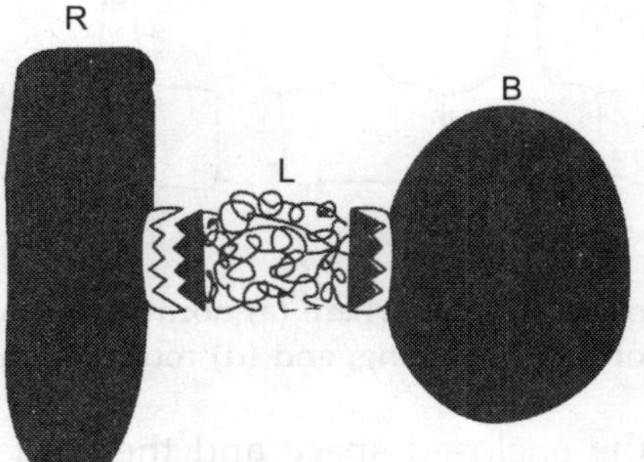

Figure 18.2 Lectin binding mechanism where L-lectin, R-root, B. bacterium infection.

18.4.1.1.3 Infection

The morphological alteration of root hair is the next process of infection. Root hairs become elongated, curled and branched (Figure 18.3). There is an interchange of small diffusible molecules between the host and the bacterium. A polysaccharide fraction, isolated from the rhizobial growth medium, has been shown to cause root hairs to curl moderately and also to branch. However, the presence of active rhizobia is required to induce the strong root-hair curling that takes place on infection. Within the rhizosphere of the legume the growth of rhizobia is stimulated by root exudates which also promote the growth of competitors. The bacterium stimulates root hair elongation by indole acetic acid and then root hair deformation results from another Rhizobium product possibly in conjunction with IAA which alone does not cause deformation, root hair is next induced to form polygalacturonase (PG) by bacterial polysaccharide material.

The bacteria at this stage are to the root hair attached and become trapped in folds formed as a result of the deformation. This enables high IAA and PG concentration to develop and invagination of the root hair wall will then occur. This invagination becomes the infection thread. A curled root hair is not essential for infection; it is thought to facilitate infection as the bacteria become enclosed by the root hair walls. This produces a closed environment in which enzymes involved in root hair penetration are prevented from diffusing away. The root hair cell wall is dissolved away at a localized site and the bacterium can then enter. On entry, new wall is formed, cutting the bacterium off from the host cell's contents. At its edge this wall is connected to the root hair wall and it thus appears that the root hair cell wall invaginates at this point.

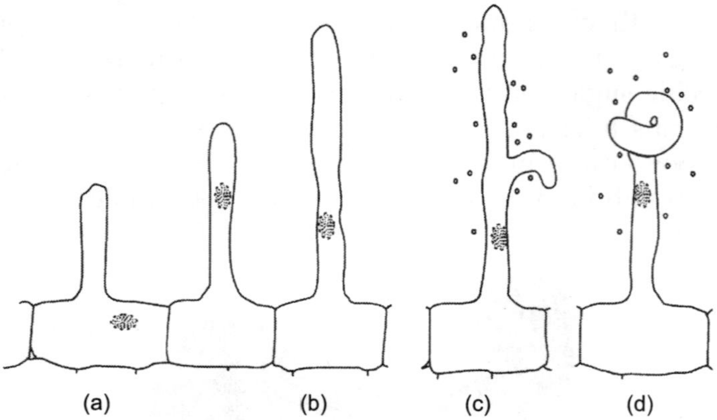

Figure 18.3 Root hair deformation types where (a) normal root hair; (b) root hair elongation; (c) root hair branching; and (d) root hair curling.

The bacteria divide in the enclosed space and the wall enclosing them extends to grow as a tube towards the base of the root hair cell. This tube is termed the infection thread. Although they are within the root, as the infection thread wall is plant cell wall and it, in turn, is surrounded by the plasmalemma, the bacteria are extracellular (Figure 18.4).

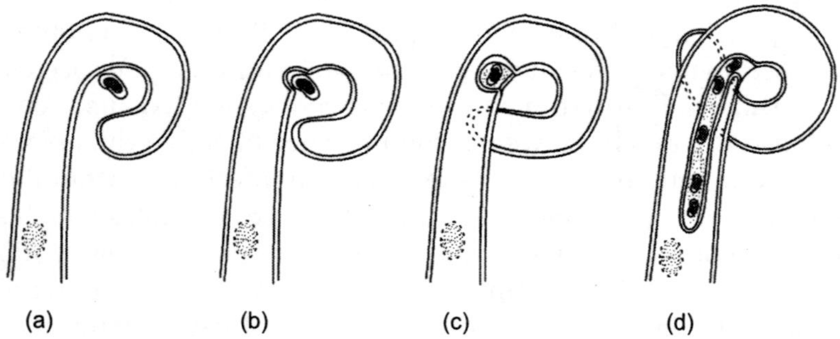

Figure 18.4 Types of root hair deformations after advancing infection by *Rhizobium* (a) normal root hair; (b) during root hair elongation; (c) during root hair branching; (d) during root hair curling.

18.4.1.1.4 Nodule development

The infection thread continues to grow beyond the root hair cell and penetrates the cortex of the root. When the infection thread has penetrated some way into the cortex, the cells in the inner cortical region are stimulated to divide. The dividing cells are tetraploid, and it was once thought that it was only tetraploid cells that were caused to come into division. However, it is known that diploid cells divide and become tetraploid by duplication of the chromosome material. The initial division is, more often than not, opposite the xylem poles and the nodules emerge in the same position as the lateral roots. They are not, however, modified lateral roots as they are initiated in the cortex and not in the pericycle as are true lateral roots.

At this stage nodule development may vary depending upon the host. The main types of nodule show either determinate growth, as found on soya beans, or indeterminate growth found on peas. When once formed, the determinate nodules are not capable of further growth by cell division, whereas the intermediate types maintain a meristem and are capable of continued growth over long periods.

18.4.1.1.5 Determinate nodules

In these nodules the bacteria are released into dividing cells, as a result of which the bacteria are distributed through the central cell mass of the nodule. The cells finally cease to divide, so that further increase in nodule size occurs only through cell enlargement. The cortex contains vascular strands that connect with the root vascular system. Within the cortex is an 'endodermis' layer with cells which have suberized walls, which may or may not have Casparian bands. This layer has no intercellular spaces and thus presents a diffusion barrier to gases passing into and out of the nodule.

18.4.1.1.6 Indeterminate nodules

Indeterminate nodules differ in a number of ways from determinate nodules. The bacteria are released from the infection threads into newly divided cells that do not divide again. This means that the bacteria must be distributed within the nodule by the infection thread which continues to grow and branch so that cells behind the dividing cells become infected. As the nodule enlarges still further, the meristem itself may divide and give rise to branched nodules.

The continued growth of the nodule means that the nodule can be divided into zones of different age with oldest cells nearest to the axis of the root (Figure 18.5). At the apex of the nodule is the meristem which produces some diploid cells on the outer side, giving rise to the nodule cortex and tetraploid cells towards the inside. Next is a zone of differentiating tetraploid cells which become enlarged following infection with rhizobia, and this is followed by a region of mature cells containing bacteroids and in which nitrogen fixation takes place. The pink colour of this region is due to the leghaemoglobin which is contained within the infected cells.

Finally, if the nodule is old enough, there is a region of senescing nodule cells. These can be distinguished as the leghaemoglobin breaks down to form a green pigment.

18.4.1.1.7 Bacteroids formation

The bacteria are released from the infection thread at a point where it is free of host cell wall. The bacteria migrate first from the centre of the infection thread, where they are encased in extracellular polysaccharides, to the periphery of the thread, and are

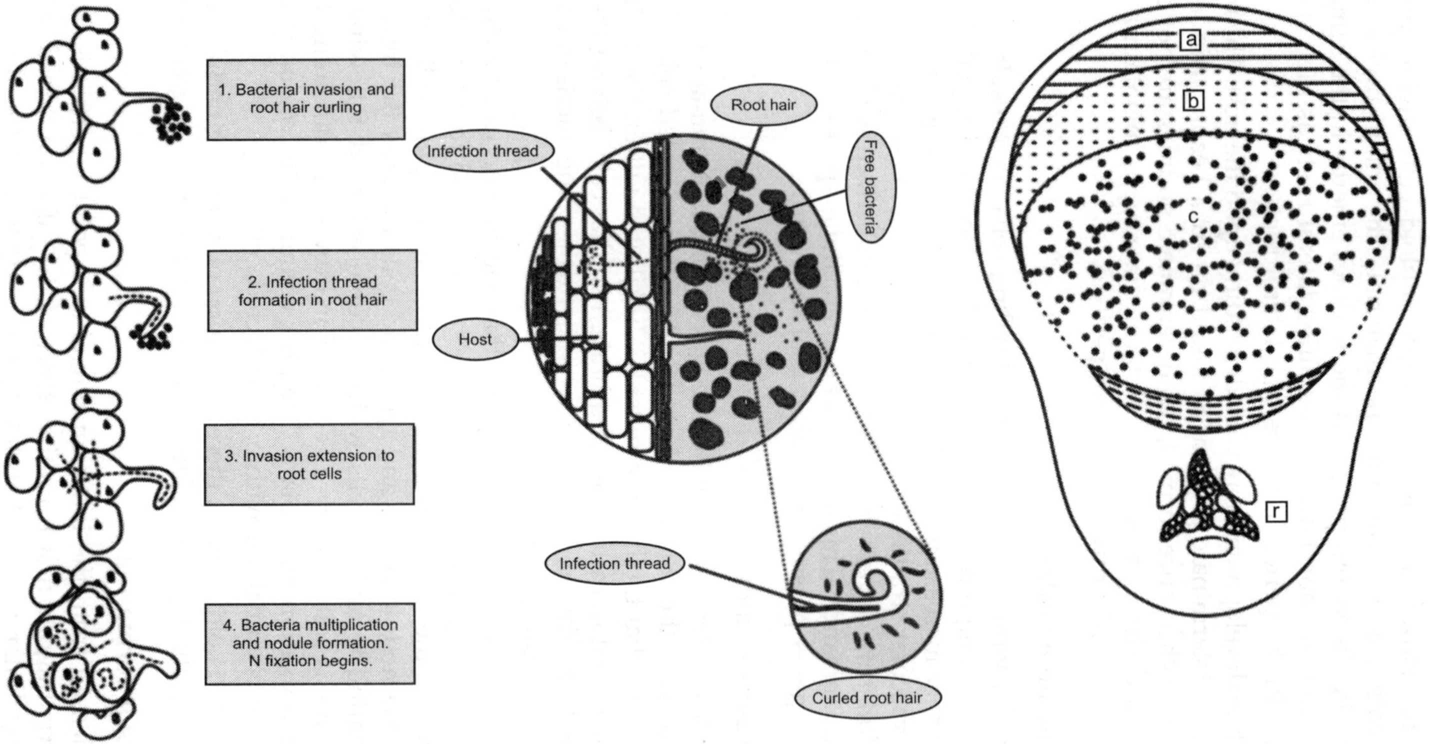

Figure 18.5 Infection process and indeterminate nodule showing different zones (a) meristematic zone; (b) cell infection zone with differentiation; (c) active nitrogen fixing tissue; (d) senescent cells; (r) Root.

released into the plant cell cytoplasm and surrounded by the plasmalemma by a process of endocytosis. This membrane is now termed as peribacteroid membrane. The contents of the host cell are seen to be undamaged when they contain bacteria. The bacteria divide along with the peribacteroid membrane so that each bacterium is surrounded by its own membrane envelope. In some cases as in soy bean, the peribacteroid membrane does not divide as frequently as the bacteria and as a result each membrane envelope may contain several bacteria. When the cytoplasm is almost full of bacteria, the bacteria enlarge and change shape and become bacterioids. The bacteroids of soya bean differ little in morphology from the bacteria although they may be slightly larger. However, in other plants such as clover and pea, the bacteroids have typical shapes. In pea, the bacteroids often have a Y-shape while clover contains pear shaped bacteroids. They are much larger than the bacteria-pea bacteroids showing about a 40 fold increase in volume. The host's organelles are pushed to the periphery of the cell and become concentrated at the corners of the cell next to the intercellular spaces. There is always a change in bacterial morphology to form bacteroids with a difference in the biochemical pathways; cytochrome components to cope with the low oxygen concentrations within the nodule; also, of course, the synthesis of the nitrogen-fixing system takes place. The most noticeable change in the host cell at this stage is the synthesis of leghaemoglobin.

18.4.1.1.8 Effective nodules

Effective nodules can be defined as those nodules which fix sufficient nitrogen to be of benefit to the host plant. Not all host-Rhizobium strain combinations are equally effective — some fix more nitrogen than others. The reasons for this are complex and effectiveness may depend upon environmental factors. For example, one strain may be most effective if the roots are maintained at a low temperature; at higher temperatures another strain may be the more effective. Effectiveness may also have a more straight forward basis: for example, strains that possess the uptake hydrogenase may be more effective than similar strains which lack this enzyme.

18.4.1.1.9 Ineffective nodules

The formation of the nodule can be seen to be a complex process. It may break down at any stage, and if it does so, an ineffective nodule is formed. Ineffective nodules either do not fix nitrogen, or else fix so little that the plant does not derive any benefit. Although a great nuisance to agriculture, strains which form ineffective nodules are of value to workers who seek to understand the underlying processes of nodule formation. There is a wide range of defects in ineffective nodules, varying from the inability of the bacteria to emerge from the infection thread to the production of nodule tissue with bacteroids of nitrogen-fixing ability in which the fixing cells senesce prematurely so that they do not last long enough in an active state to be productive.

18.4.1.2 Biochemical pathways

18.4.1.2.1 Leghaemoglobin

Leghaemoglobins are monomoric proteins which will reversibly bind oxygen, and are similar in many ways to myoglobin, the red oxygen carrying protein bound in mammalian muscle. They have one haem group and can bind one oxygen molecule. They have a very high affinity for oxygen and are half saturated at oxygen concentration of around 10-20 nM. This compares with values of 220 nM for horse myoglobin and 126 nM for the β-chain of human haemoglobin.

Legume root nodules may contain more than one form of laghaemoglobin; soya bean has four. These forms differ in amino acid sequence and composition but have a number of critical sequences that are conserved and are common to all the leghaemoglobin. These conserved sequences will be associated with the functional parts of the molecule which bind the haem and provide the correct environment to the haem which influences the oxygen binding.

18.4.1.2.2 Synthesis of leghaemoglobin

The synthesis of leghaemoglobin is interesting from the point of view of the formation of the functional symbiosis. It involves the cooperation of both partners in the synthesis of a compound peculiar to, and necessary for, the efficient functioning of the association. The polypeptide, globin, is coded for on the plant chromosomes and synthesized on the plant ribosomes. It was demonstrated that this was a plant protein when it was found that different species of legume possessed different leghaemoglobin molecule even though they were inoculated with the same strains of *Rhizobium.* It has been known for a long time that the rate of nitrogen fixation within the nodule is closely correlated with the concentration of leghaemoglobin. The close linkage of leghaemoglobin with nitrogen fixation even led at one time to the suggestion that leghaemoglobin was the nitrogen binding agent. If reversible bind to the oxygen and its function would be related to this.

18.4.1.2.3 Carbon metabolism

Carbon is transported from the shoot to the root nodule in the form of sugars. In the nodule, the sugars have to be transformed into the carbon skeletons of the nitrogen transport compounds and also have to be respired to produce ATP and reducing power for nitrogen fixation. The sugars can be metabolized by the glycolytic pathway until phosphoenolpyruvate is produced which then can be carboxylated to oxalacetate. Some of this acid will be used to form aspartic acid and asparagines for nitrogen transport, but the rest will be reduced by malic dehydrogenase to malate. This can occur in cyatoplasm and does not need the intervention of the mitochondria (Figure 18.6).

18.4.1.2.4 The rhizobium-parasponia association

Parasponia belonging to family Ulmaceae also nodulate with rhizobia. The association is interesting, as the nodules fix nitrogen and share some features with legume nodule and some with actinorhizal nodules. The first stage is the modification of the root hairs, as it is in both the legume-Rhizobium association and in the actinorhizal association. The root hairs in this case are modified somewhat differently. Instead of the root hairs being composed of a single cell they become multicellular. Below the modified hairs, cell divisions occur in the outer layers of the cortex. It is thought that this production

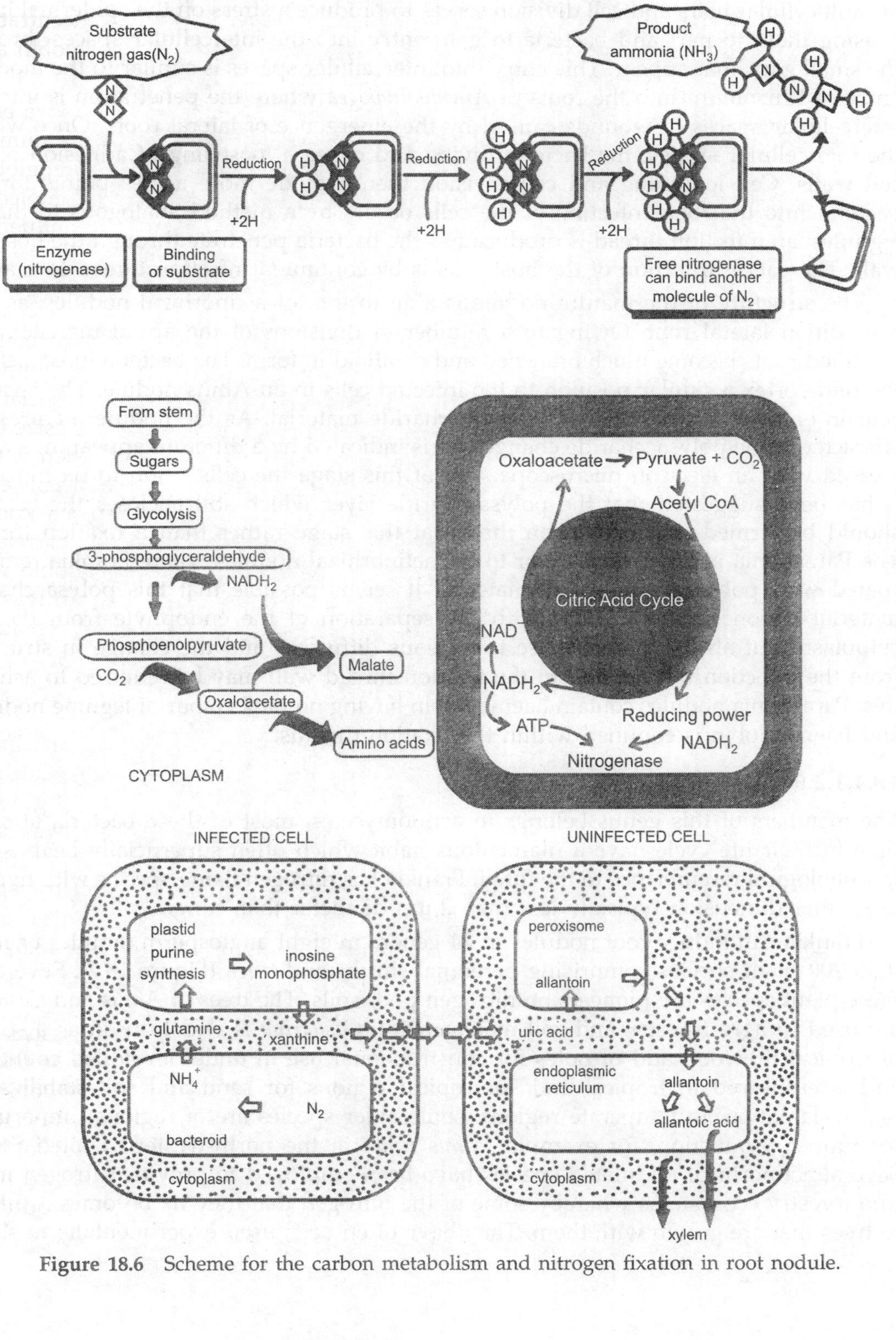

Figure 18.6 Scheme for the carbon metabolism and nitrogen fixation in root nodule.

of multicellular hairs and cell division serves to produce a stress on the epidermal hairs, causing them to part and bacteria to gain entry into the intercellular spaces through the small gaps that appear. This entry into intercellular spaces is similar to the mode of entry of Rhizobium into the roots of *Arachis hypogea* where the penetration is into the intercellular spaces of wounds caused by the emergence of lateral roots. Once within the intercellular spaces, the bacteria divide and cause a loosening of adhesion of the cell walls. Cell loosening and cell division then provide more access points for the bacteria into the root. Infection of the cells occurs by a method analogous to that in legumes: an infection thread is produced as the bacteria penetrate through the host cell wall. Thereafter, infection of the host cells is by continued infection thread formation.

The structure of the mature nodule is akin to that of actinorhizal nodules, as it is a modified lateral root. Owing to a number of divisions of the apical meristem the modified roots become much branched and coralloid in form. The bacteria infect cells in the mid-cortex a similar position to the infected cells in an Alnus nodule. The bacteria remain enclosed within cell-wall polysaccharide material. As the nodule matures the character of the polysaccharide changes, as is indicated by a different appearance when viewed with an electron microscope, and at this stage the cells begin to fix nitrogen. It has been suggested that the polysaccharide layer which still encloses the bacteria should be termed as an infection thread at this stage rather than a fixation thread. The Parasponia nodules are similar to the actinorhizal nodules where Frankia remains coated with polysaccharide wall material. It seems possible that this polysaccharide material is concerned not only with the separation of the endophyte from the cell cytoplasm but also with resistance to gaseous diffusion, and the change in structure from the infection thread wall to the fixation thread wall may be required to achieve this. Parasponia nodules contain haemoglobin having no pink colour of legume nodules, and haemoglobin is confined within the fixation threads.

18.4.1.2.5 Actinorhizal root nodules

The members of this genus belongs to actinomycetes, most of these bacteria at some time in their life cycle have a filamentous habit which often superficially bears some morphological resemblance to the fungi. Frankia has nitrogen fixing species with hyphae of smaller dimension typically less than 2 μm diameter than fungi.

Frankia can induce root nodules in 24 genera in eight angiosperm families of more than 200 plant species comprising perennial shrubs and trees (Figure 18.7). Several of these plant species are pioneers on nitrogen poor soils. The trees of *Alnus* and *Casurina* are used in agro forestry and soil management. *Casuarina* is widely used as a source of tropical firewood and of poles for constructional use in underdeveloped countries, and is employed in tropical and subtropical regions for sand and soil stabilization and reclamation. In temperate regions, some alder species are of regional importance for timber production, for example, *Alnus rubra* in the north-western United States. Several actinorhizal nodulated species have been employed to provide nitrogen input into forestry ecosystems whereby some of the nitrogen that they fix becomes available to trees that are grown with them. They have often been used experimentally in short-rotation 'energy' plantations, because of their ability to fix nitrogen. Many species of actinorhizal plants are pioneers and colonize denuded sites that are nitrogen-poor. This ability has led to the use of some of these associations for amenity purposes to restore derelict land. Although the host range of *Frankia* is wide, it is limited to eight families within which there is specificity for infection of different genera, and therefore the problem of recognition has to be overcome. It is probable that this recognition involves lectin binding as has been suggested for Rhizobium-legume recognition. The binding of Frankia to the host cell walls has been observed, but the host lectins that may be involved have not as yet been isolated. In contrast to rhizobia, Frankia fixes nitrogen in the free living state by forming specialised vesicles where the nitrogenase enzyme is protected from oxygen. The membranes of these vesicles contain high level of triterpine hopanoids believed to be involved in O_2 barrier formation.

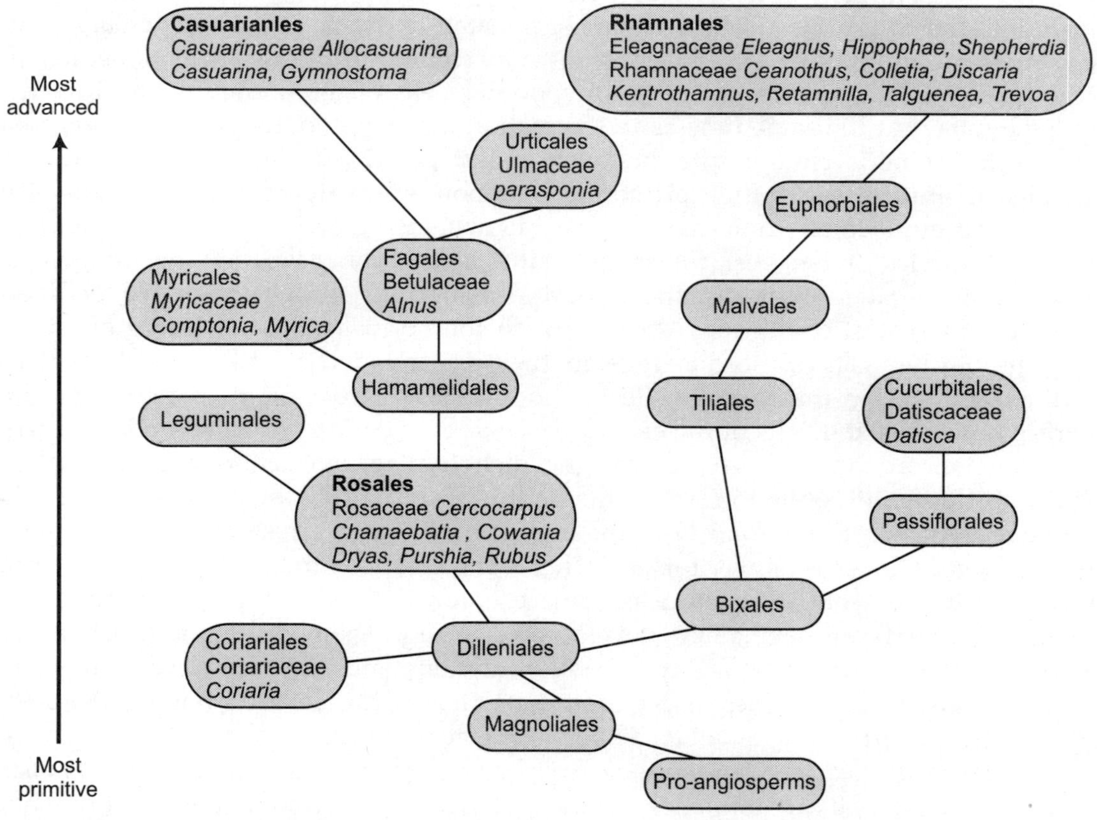

Figure 18.7 Relationship between the woody dicotyledonous angiosperm between the known orders of Actinorhizal plants based on the phylogenetic system of Huchinson (1973) (After Dixon and Wheeler, 1986).

Like *Rhizobium, Frankia* usually infects the plant via root hair deformation stimulating cell proliferation in the host's cortex via the intercellular spaces of the epidermis. The infection remains intercellular and infection threads are not formed until the host cells are invaded. The penetrating hypha is encapsulated in cell-wall-like material and can undergo further differentiation to produce vesicles in which nitrogen fixation occurs. The mechanism by which this occurs has not been studied in actinorhizal plants, but an interesting co-operation of a third organism in this process has been described. Some pseudomonads have the capability also to deform root hairs, although this is not specific to the hosts of *Frankia*. In Alnus roots inoculated with *Frankia* together with *Pseudomonas cepacia*, the number of deformed root hairs is considerably increased compared with the numbers formed following inoculation with *Frankia* alone. The increase in deformed root hairs then increases the number of sites suitable for infection and subsequent nodule formation. *Frankia* enters root hairs at the same type of site as *Rhizobium* and the hyphae entering the root hair are also separated from the cytoplasm by cell wall, but in Frankia infections the cell wall adheres closely to the hyphae (Table 18.1). After entering the deformed root hair cell, the hyphae may branch, and they then grow down into the cortex. Host cells there are stimulated to divide and also to enlarge, but they remain diploid. This cell division causes a swelling in the root which, because it is not the nodule proper, is called the prenodule. Cells of the prenodule are invaded by the hyphae. At the same time as the formation of the prenodule, nodule primordia develop in the pericycle opposite the xylem poles. Cell divisions in the pericycle and endodermis initiate the nodule primordia but soon afterwards the cortical cells also divide. The number of primordia formed and whether they branch or not depends upon the species of the host. Since they form in the pericycle, the development is characteristic of lateral root formation, and the actinorhizal nodules resemble modified roots structurally, although they arise between the normal points of origin of lateral roots proper. As the primordia develop they are invaded by hyphae which then continue growing in the cortex while the nodule grows from the meristem at its tip. Meristematic cells and cells of the central vascular cylinder are, however, not invaded. The meristem at the apex divides causing dichotomous or trichotomous branching which produces the coralloid form of the nodule. Two morphological types of nodule can be recognized. In the *Alnus* type, the nodule meristem is arrested so that a globose mass of nodule lobes is slowly formed. This is the most common morphological form of nodule. In the Myricaceae and Casuarinaceae, however, after a period of arrest the nodule lobe meristem begins new development to form the nodule root, a negatively geotropic structure which is always free of endophyte and which is of importance in oxygen supply to the nodules under conditions of poor soil aeration (Figure 18.8). All *Frankia* strains form vesicles when fixing nitrogen in culture, however, vesicle form and structure *in vivo* is not a suitable primary taxonomic characteristic for *Frankia* since the host plant can influence its morphology. Furthermore, *Frankia*, does not form bacteriods. *Frankia* hyphae or vesicles are not released into the plant cytoplasm, but the bacteria remain enveloped by the invaginated host plasmalemma membrane, embedded in a matrix of cell wall material. The diffusion barrier present in the cortex of legume nodules is not necessary for actinorhizal nodules where the vesicles provide protection from oxygen. The reason for the variable amounts of haemoglobin in the nodules of different actinorhizal plant genera is of interest that the largest quantities of haemoglobin are to be found in *Casuarina* nodules, where the typical vesicle structure is either absent or rarely present (Table 18.1).

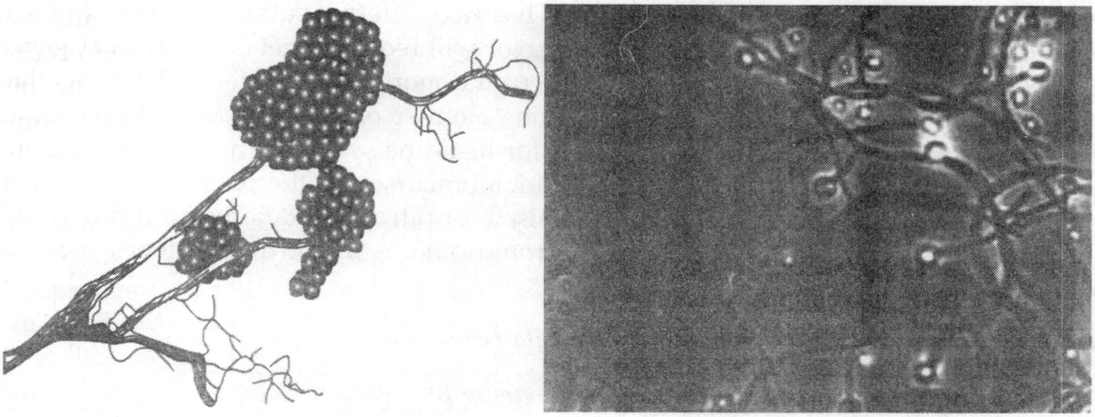

(a) (b)

Figure 18.8 Symbiotic relationship of Frankia with Actinorhizal plants.

Table 18.1 Comparison of some aspects of legume root nodules and actinorhizal root nodules.

Nodule infection and development	*Legumes nodules*	*Actinorhlzal nodules*
1. Mutual recognition by lectius	Cortex	?
2. Organism bound	+	+
3. Deformation of root hairs	+	+
4. Infection through root hairs	+	+
5. Organism separated from the plant cell contents by plant cell wall and plasmalemma	+	+
6. Release of organism from enclosure by cell wall	+	–
7. Development of specific nitrogen-fixing form	Bacteroids	Vesicles
8. Haemoglobin present in infected tissue	+	+
9. Origin of the nodule	Cortex	Pericycle
10. Increase in ploidy of host cells	+	–
11. Only polyploid cells infected	+	–
12. Vascular tissue as strands in the cortex	+	–
13. Vascular tissue as central cylinder	–	+
14. Diffusion barrier 'endoderm is' present in the cortex	+	–

18.4.2 Genetics of Nitrogen Fixation

Nitrogenase protein is composed of component 1 and component 2. The component 1 is made up of MoFe protein and component two of Fe protein. Biochemistry of nitrogen fixation reveals that several genes are necessary to code for the nitrogen fixing system. The Fe protein is composed of two subunits but as each of these is the same, one gene will code for this protein. The MoFe protein has two different sub-units, each of which will require one gene. The molebdenum cofactor will require a gene, and further genes will be necessary to code for any special electron donors in the system. The genes that code for these proteins are all adjacent to one another on the *Klebsiella* chromosome. They lie between the *his* gene, which codes for histidine synthesis, and the *shi* A gene, which codes for the synthesis of shikimic acid, a precursor on the pathway of aromatic ring biosynthesis. 17 genes have been recognised for nitrogen fixation accounting for all the DNA on this stretch of the *Klebsiella* chromosome. A list of the nif genes together with their functions is given in Table 18.2.

Table 18.2 Functions of the nif genes of *Klebsiella pneumoniae*

Gene	*Function of the gene or gene product*
H	Codes for the subunit of the Fe protein
D	Codes for the a subunit of the FeMo protein
K	Codes for the a subunit of the FeMo protein
M	Activation of the Fe protein
B	Synthesis and insertion of the iron molebdenum cofactor, FeMoCo
N	As for B
E	As forB
V	As tor B
F	Codes for flavodoxin
J	Codes for pyruvate:flavodoxin oxidoreductase
A	Codes for an activator molecule for the other nif genes
L	Codes for a repressor molecule for the other nif genes
Q	Possibly concerned with molybdenum uptake
S	Possibly concerned with processing the FeMo protein
U	AsS
X	Unknown
y	Unknown

18.5 CYANOBACTERIAL SYMBIOSES

The range of plants with which cyanobacteria can form a symbiosis is very wide indeed. It encompasses diatoms, fungi, mosses, liverworts, ferns, cycads and angiosperms. The cyanobacterial symbiont has not always been identified with certainty, but is usually a species of *Anabaena* or *Nostoc*, both of which are heterocystous forms. More often than not the proportion of heterocysts to vegetative cells is much higher in the symbiotic form than in the free-living cyanobacteria. Studies on the differentiation of heterocysts in the filaments have shown that this is determined by the nitrogen status. Nitrogen is fixed by the heterocysts and diffuses into the vegetative cells. As these cells divide the distance of the distal cell from the heterocyst increases and the available nitrogen is depleted by the demands for growth. When this reaches a low level a new heterocyst is developed. The nitrogen compound that is 'sensed' to indicate the nitrogen status is thought to be glutamine and the concentration of this compound controls the development of the cell into a heterocyst. In the symbiosis where the products of nitrogen fixation are excreted for use by the host, combined nitrogen levels along the filament become depleted sooner than in free-living forms thus resulting in a greater proportion of heterocysts in symbiotic systems. As with other symbioses, there are morphological adaptations of the host and also physiological adaptations to cater for the special demands of nitrogen fixation (Figure 18.9).

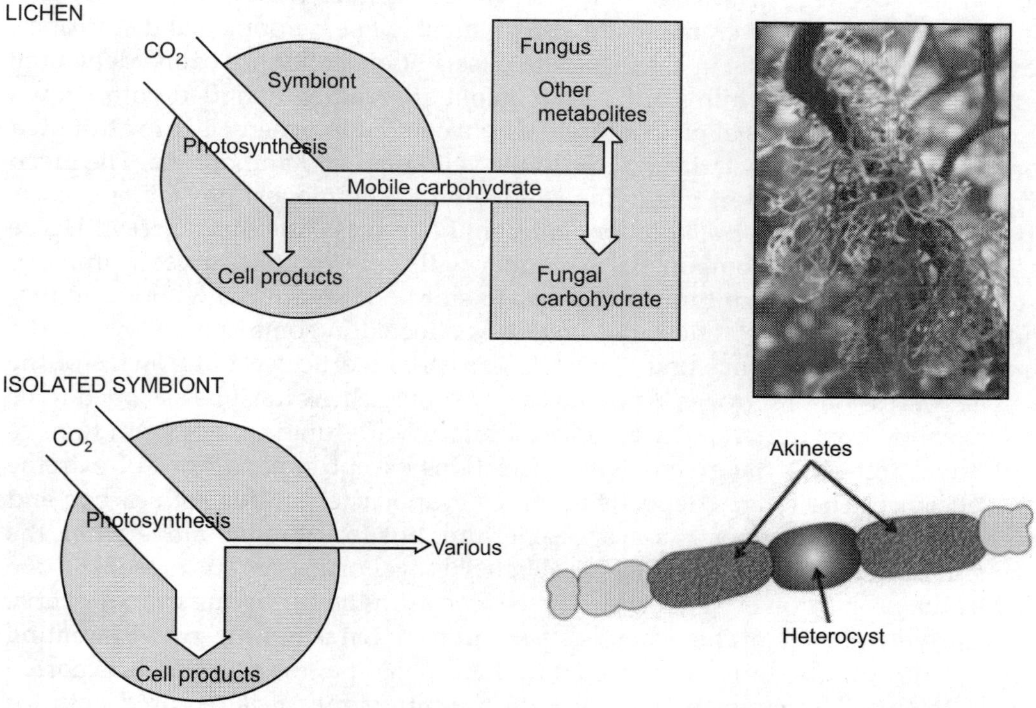

Figure 18.9 Diagrammatic representation of symbiotic association with Lichen and cyanobacterial heterocyst and akinete differentiation in *Anabaena variabilis*.

Lichens are symbioses of fungi, ascomycetes and basidiomycetes with algae. By far the highest proportion of lichen species is associations of fungi with green algae, but about 25 genera have cyanobacteria as the 'algal' symbiont. There are also a few species, such as *Peltigera aphthosa* and *Stereocaulon vesuvianum,* that have a triple association. The main part of the thallus is composed of an association of the fungus with a green alga and special structures, cephalodia, contain cyanobacteria which fix nitrogen for the thallus as a whole. The structure of the lichen thallus varies from Collema, where the fungal hyphae and cyanobacterial filaments intermingle through the thallus, to *Peltigera canina,* where the cyanobacteria are confined to a thin layer beneath the upper cortex of fungal hyphae. In the lichens the form is determined by both the fungus and the alga, so that it is not possible to ascertain whether any particular change in form is concerned with the cyanobacterial association or to nitrogen fixation. The fungal partner is known as the mycobiont and the photosynthetic partner as the photobiont or phycobiont. The morphology, physiology and biochemistry of lichens are very different from those of the isolated fungus and cyanobacteria in culture. Lichens occur in some of the most extreme environments on earth arctic tundra, hot deserts, rocky coasts, and toxic slag heaps. However, they are also abundant as epiphytes on leaves and branches in rain forests and temperate woodland, on bare rock, including walls and gravestones, and on exposed soil surfaces (e.g., *Collema*) in otherwise mesic habitats. Lichens are widespread and may be long-lived; however, many are also vulnerable to environmental disturbance, and may be useful to scientists in assessing the effects of air pollution, ozone depletion, and metal contamination. Cultures of the phycobiont are readily identified with known free-living algae. Out of 26 algal genera found as phycobiont in lichens, 17 are attributed to Chlorophyta, 8 to Cyanobacteria and one to the yellow-green Xanthophyta. The green algae *Trebouxia* (rarely found in free-living state) are the commonest phycobiont genus in temperate regions; *Trentepophlia* predominates in the tropics. Attempts to resynthesize lichens from their isolated components have met with only limited success, probably owing to the need for exact cultural conditions to suit both organisms without unduly favouring one of them. Vegetative reproduction is effected in some species by soredia (loose masses) of fungal hyphae and algal cells) or isidia (corticated outgrowths of the thallus). The possession of nitrogen-fixing ability by some lichens must be assumed to be an advantage; however the cyanobacterial associations with fungi do not seem to have any great advantage over the green algal associations except in conditions of extreme nitrogen deficiency. The fungus depends upon the cyanobacterium for both carbon and nitrogen when the sole partner is a cyanobacterium, but for nitrogen alone when the cyanobacteria are confined to cephalodia. When the cyanobacteria are present in the main thallus the proportion of heterocysts is the same as in the free-living cyanobacteria. This is because the growth of the lichen is slow and nitrogen is not the growth-limiting factor. When the cyanobacteria are confined to the cephalodia, the nitrogen is exported to the main thallus. This forms a large sink for the nitrogen as it is required both for the fungus and its green algal partner. As the cephalodia form a small proportion of the whole organism, the nitrogen is depleted and as a consequence the proportion of heterocysts is high. The enzyme concerned with the incorporation of nitrogen in free-living cyanobacteria is glutamine synthetase. The amount of this enzyme in the endophytic form of the cyanobacterium is considerably reduced so that it is unable to incorporate much of the nitrogen that it has fixed and ammonia is exported to the fungus. Amounts of other nitrogen-metabolizing enzymes concerned with the further processing of nitrogen compounds for amino-acid synthesis are also reduced in the endophyte. The mechanism by which this reduction in nitrogen-metabolizing enzymes is achieved is not as yet known. Lectins from the fungus have been found which bind specifically to the cyanobacterial partner. Many lichens accumulate extracellularly a wide variety of unique chemical substances. These are useful in delimiting species. Most species grow very slowly. Many occur in exposed situations on soil, rocks, walls, tree trunks, and branches, under conditions where neither partner could survive alone (Figure 18.10). Others grow luxuriantly in damp habitats such as tropical forests or where the atmosphere is humid from sea spray and mists. Lichens behave as hydrophilic gels, having no mechanism for controlling water absorption or loss. In the dry state (2-15% water content) they are tolerant of extreme temperature. Survival of particular species at –216°C and of others at 100°C has been reported. Saturation after a period of drought leads to a rapid rise in the respiration rate but the rate of photosynthesis rises slowly. Lichens have also been used in making dyes and perfumes, as well as in traditional medicines. Thallus is the body of the lichen. The fungal hyphae (filaments), branch and then fuse together (anastomose) to form a mesh of hair-like threads. The top surface is normally a layer of tightly packed hyphae known as 'cortex' covering below the algal layer where the photobiont lives. Below this is the medulla an area of loose hyphae in which nutrients are stored. Sometimes a lower cortex exists, in others the medulla rests on the surface. Filamentous lichens consist of chains of algal cells wrapped around with fungal hyphae. Cladonia are a successful group of lichens which have a primary thallus and a secondary thallus. The primary thallus is small and clings closely to the substrate while the secondary thallus is a shrubby growth like **Fruticose lichens**. Once the lichen is established the primary thallus often dies off. Lichens have been classified on the basis of their morphology, symbiotic relationship and the type of structure as (1) **Crustose lichens** on the basis of forming a crust on the surface of the substrate on which they are growing. This crust can be quite thick and granular or actually embedded within the substrate. In this latter case the fruiting bodies still rise above the surface. In many crustose lichens the surface of the thallus breaks up into a cellular, crazy-paving like pattern. Crustose lichens tend to grow out from their edges and have their fruiting bodies in their centre. Crustose lichens are very difficult to remove from their substrates. (2) **Squamulose lichens** have a portion of their thallus lifted off the substrate to form 'squamules'. They are otherwise similar to crustose lichens in that they possess an upper cortex but no lower cortex. (3) **Foliose lichens** have an upper and lower cortex. They are generally raised to some extent above the substrate but connected to it by rhizines (specialised root-like hyphae). They are easier to remove from their substrate (4) **Fruticose lichens** are attached to their substrate by a single point and rise, or more usually, dangle from this are also known as shrubby lichens. Some foliose lichens can be stubby like fruticose lichens, however, close examination will reveal that the algal part exists only on one side of the flattish thallus whereas in fruticose lichens it exists as a ring around the thallus, even when it is flattened as in *Ramalina* sp. (5) **Leprose lichens** are an odd group of lichens which have never been observed to produce fruiting bodies and have not yet been identified properly, or at least not yet given full scientific names. They not only lack an inner cortex, but also lack an outer one, i.e. no cortex, only an algal cell layer and sometimes a weakly defined medulla (Figure 18.10).

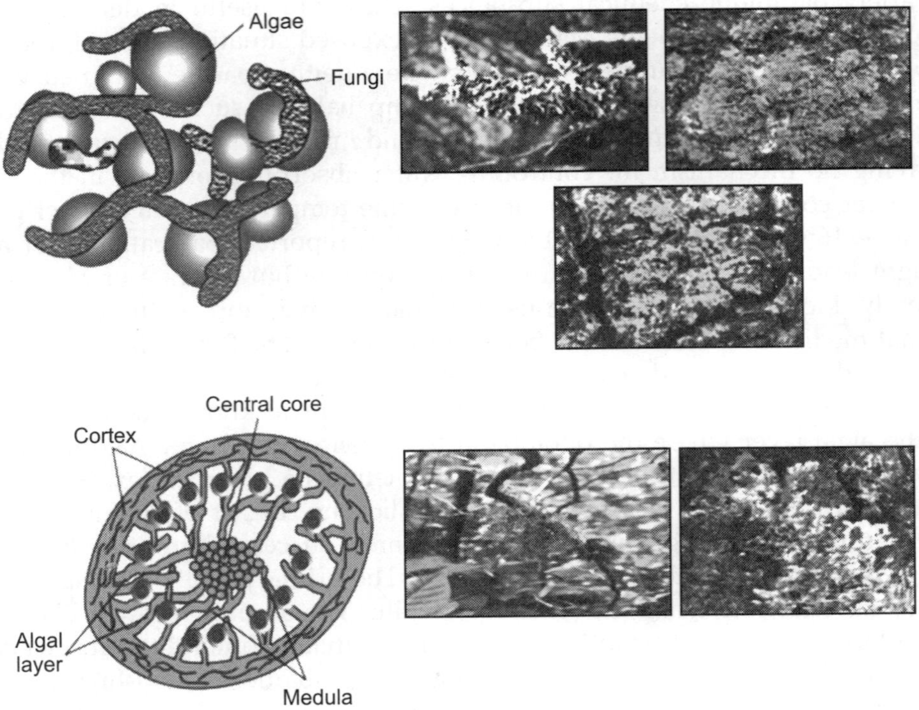

Figure 18.10 Lichens of various forms.

18.5.1 Bryophytes

Cyanobacteria form a variety of associations with bryophytes ranging from the almost accidental to very close symbioses. *Nostoc* is a widespread cyanobacterial genus that is a common symbiont and the association between *Nostoc* and hornworts is an excellent example of a close symbiosis. Hormogonia that fragment from a parent cyanobacterium can move, by gliding, over a wetable surface and so disperse. The mobile, filamentous hormogonia are the means by which *Nostoc* enters the bryophytes, the immobile parental filaments being unable to do so. Hormogonia can form new heterocysts and so begin the creation of new cyanobacterial colonies. Within a hornwort thallus there are *Nostoc* cavities which hold *Nostoc* cells. Studies of the hornwort *Anthoceros punctatus* have shown that when starved of nitrogen the hornwort release hormogonia-inducing chemicals to their surroundings and induce hormogonia formation in any nearby free-living *Nostoc* colonies. The released hormogonia-attracting chemicals help to guide the hormogonia to attach into the appropriate cavities. Experimental evidence indicates that the bryophytes release chemical hormogonia-repressing factors into the *Nostoc* cavities to suppress hormogonia and promote heterocyst creation instead. In free-living cyanobacteria the heterocysts make up 2-5% of all cells. Within the *Nostoc* cavities up to 60% of the cyanobacterial cells are heterocysts. In at least the case of *Anthoceros punctatus* about 80% of the nitrogen fixed by the heterocysts is released to the bryophyte. There is no evidence of photosynthesis by the cyanobacterial cells within the cavities so the necessary carbohydrates must be supplied by the photosynthesizing bryophyte partner. Overall, there is good evidence that the bryophyte partner regulates the behaviour of the *Nostoc* (Figure 18.11).

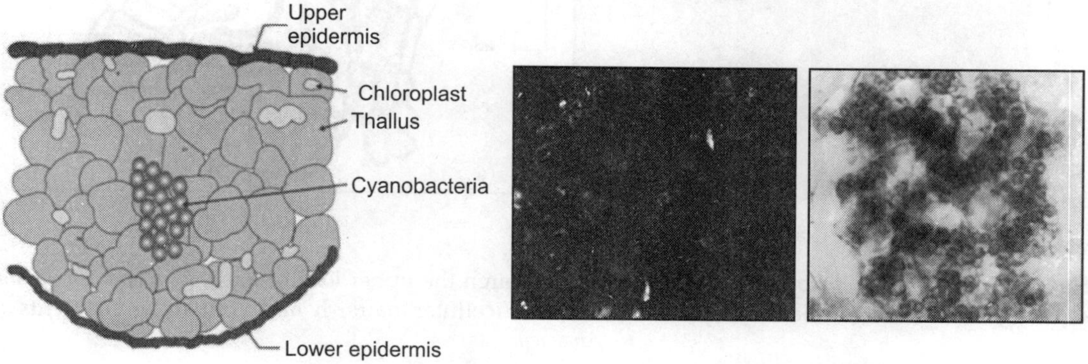

Figure 18.11 VTS of *Anthoceros punctatus* thallus showing cyanobacterial zone, thallus of *Anthoceros punctatus* detailed view of thallus showing loose filament of cyanobacteria.

Looser associations between cyanobacteria and bryophytes have been found in various habitats. Examples are the mosses *Bryum algovicum* (in dune valleys), *Ceratodon purpureus* and *Funaria hygrometrica* (burnt areas), species of *Grimmia* and *Racomitrium* (lava fields) and the liverworts *Anthelia juratzkana* and *Porella navicularis* (on trees). In these associations the cyanobacteria had the low percentage of heterocysts typical of free-living forms. While the bryophytes took up nitrogen, in most cases any benefits to the cyanobacteria were unclear. Cyanobacteria have also been found growing on *Sphagnum* plants in *Sphagnum* bogs and in some cases even inside the empty cells in *Sphagnum* leaves.

18.5.2 Azolla

Azolla is a small fern which grows on the surface of water. It has a prostrate branching stem on which the deeply bilobed leaves are alternately arranged. The lower leaf lobe floats on the surface of the water and thus enables the upper lobe to remain in an aerial position above the water. The adventitious roots, through which mineral uptake occurs, hang down into the water. The plants are usually found in very slow-moving or still water as they are fairly fragile, and although they can regrow if fragmented, overall growth is slowed down. *Azolla* is a tropical genus and thus cannot withstand frosts. In hot weather it is prone to insect attack and disease. However, the sporophyte has to be kept in cultivation during the hot summer in order to produce the following crop as methods of inducing sporulation, harvesting and germinating the spores have yet to be developed. This makes the management of *Azolla* difficult and time consuming. In spite of this, it is used extensively in the Far East for the provision of nitrogen for rice crops (Figure 18.12).

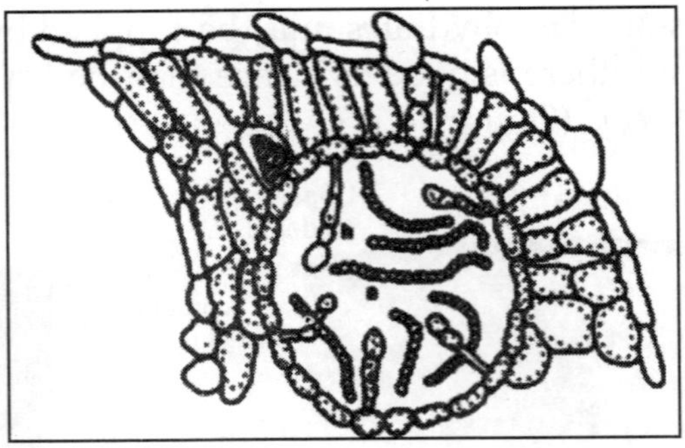

(a) (b)

Figure 18.12 (a) Azolla sp. (b) A vertical section through the upper lobe of a leaf of *Azolla caroliniana* showing the mucilage-filled cavity lined with multicellular hairs, h and containing filaments of *Anabaena*, a.

Azolla is applied to rice in two ways. It can be grown separately and ploughed in as a green manure, or it can be applied to the paddy fields at planting time and grown until it is shaded out by the rice. It can also be applied in both ways to a rice crop. Increases in yield of rice of about 20% can be obtained when *Azolla* is applied.

The plant grows vigorously and under good conditions has a doubling time of about two days. A good crop of *Azolla* can accumulate as much as 10 kg N/ha/day but the amount of nitrogen that can be fixed is limited, since the growing season is restricted to the cooler winter or spring due to the disease and insect problems brought by the hot weather. Because of this, it is managed so that it will be shaded out by the growing rice crop in early summer. Part of the time will be spent in the development of *Azolla*, so that although fixation rates are high the amount fixed per season is not as high as 10 kg N/ha/day would suggest. However, its usefulness is indicated by the fact that *Azolla* has extended to cover over 1.5 million hectares in China and Vietnam.

The endophyte of *Azolla* is *Anabaena azollae*. Filaments of *Anabaena* which have no heterocyst and are thus not nitrogen-fixing are associated with the apical meristems of the main rhizome and branches of *Azolla*. They are enclosed in the lobes of a differentiated leaf which protects both the *Anabaena* filaments and the meristem from the environment. In the leaves which develop from the meristem, a small depression forms in the basal surface of the upper leaf lobe which has associated with it a branched hair that grows towards the apical *Anabaena* filaments. The hair has a transfer structure with deeply infolded walls which increase the surface area of the cell membrane. This suggests that it is concerned with the transfer of metabolites to and from the endophyte. The *Anabaena* cells remain near the branched hair and the epidermal cells of the leaf enlarge to enclose the cavity caused by the depression in the leaf surface. Thus, both branched hair and *Anabaena* filaments become enclosed within the upper leaf lobe. At this time a further branched hair develops in the cavity and then a number of simple hairs. These all have transfer cell structure. The branched hairs are adjacent to the leaf traces and the simple hairs more randomly arranged, and on the basis of their distribution it has been suggested that the simple hairs are concerned with the transfer of sugars to the endophyte while the branched hairs are more concerned with the absorption of ammonia released by the endophyte into the cavity. As the epidermis closes over the cavity, heterocysts develop on the *Anabaena* filaments and nitrogen fixation commences.

Modifications in the *Azolla* host plant involve the production of a leaf with a cavity and modified cells, the hairs, which facilitate the interchange of molecules between the partners (Figure 18.12). *Anabaena* does not undergo any morphological changes other than to produce an increased proportion of heterocysts. However, in this symbiosis, biochemical changes are evident which enable the host to benefit from the symbiosis. Nitrogen fixed by the endophyte is released into the cavity as ammonia where it is taken up and transferred to the growing apex of the host, and also to the growing *Anabaena* at the stem tip which also depends upon this nitrogen for growth. As in the lichens the release of the newly fixed nitrogen is caused by the fact that the endophyte cannot use most of it for itself. The normal route of nitrogen assimilation in *Anabaena* is by glutamine synthetase. The endophyte has very little, if any, of this enzyme, and over 90% of glutamine synthetase activity in the symbiotic association is found in the host. Thus, the ammonia that the *Anabaena* cannot use will leak into the cavity where it is available for the host. The *Anabaena* filaments in the leaf activity have a high frequency of heterocysts which lack photosystem II and thus cannot fix carbon. Sucrose must therefore be obtained from the host, and this can be metabolized to supply reductant for the high nitrogen fixation rates.

18.5.3 Cycadaceae

The cycad association has a number of points of special interest. The symbiosis does not seem to be very specific with regard to the cyanobacterial partner as even with one species of cycad, *Macrozamia communis,* both *Nostoc* and *Anabaena* have been isolated from the root nodules. The root nodules like those of the actinorhizal symbiosis, are composed of modified roots, and the gross morphology of the nodules is rather similar. The modified roots become apogeotropic and grow up towards the surface of the soil, branching dichotomously or trichotomously to form coralloid clusters. Cross-sections of the root nodules show that the cortex is in two parts with a distinct inner and outer cortex. The endophyte is confined to the intercellular spaces between the cortical layers. Anatomical investigation has shown that the outer cortex is in fact a persistent root cap. The root cap does not slough off as in normal roots, but remains and grows with the root. The inner layer of the root cap cells does not keep pace in elongation with the cells of the inner cortex, so that intercellular spaces are formed which are later infiltrated with the endophyte, and the endophyte is thus confined to one layer within the root (Figure 18.13). It appears that cyanobacteria penetrate the cortex at a break in the epidermis of a young root before the periderm has laid down suberized cells. Whether these are accidental breaks, caused by unequal plant growth, or are caused by the cyanobacteria, has not been determined. Having penetrated into the cortex, the cyanobacteria migrate through the intercellular spaces or may digest their way through the middle lamella to reach the spaces between the root cap, outer cortex, and the root inner cortex.

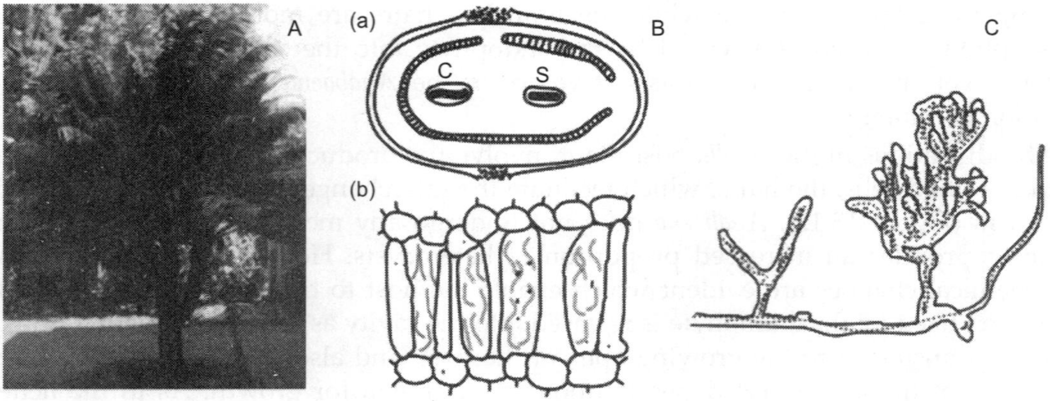

Figure 18.13 A. Cycas plant B. (a) T. S. of corolloid root showing, (b) zone occupying the mucilage filled intercellular space, C. Corolloid root.

In the lichens, *Azolla* and most cycad systems, the cyanobacteria exist outside the host's cells, but in one species of cycad, *Macrozamia communis*, the cyanobacteria have been found inside the cells. This is similar to the symbiosis with *Gunnera*. The cyanobacterial cells are inter- or intracellular in the cycad symbiosis, they still retain pigment, even when the nodules are below the soil surface so rendering photosynthesis impossible. The cyanobacteria thus live heterotrophically on carbohydrate supplied by the plant. Around 5 to 10% of the vegetative cells of a filament can transform into heterocysts when cyanobacteria are deprived of both nitrate and ammonia. Heterocysts are rounded; seemingly empty cells, usually distributed regularly along a filament or at one end of a filament. They develop as a result of the differentiation of the vegetative cells and are the sole sites of nitrogen fixation in heterocystous cyanobacteria. At the time of transformation of a vegetative cell into heterocyst, the former synthesizes a very thick new wall, reorganizes its photosynthetic membranes, discards its photosystem-II and phycobiliproteins, and synthesizes the nitrogen fixing enzyme nitrogenase. Photosystem-I still operate and generates ATP (Figure 18.14 and Figure 10.50).

18.5.4 Gunnera

Gunnera is a genus of about 50 species in the Haloragaceae, the water milfoil family. It is distributed in the southern hemisphere, mainly in the tropical and subtropical regions. The species vary from those of about 6 cm in overall height to those having large palmate leaves over one metre across. The cyanobacterial symbiont is Nostoc, and this symbiosis is of interest as it is the only known cyanobacterial association with an angiosperm. The symbiosis is present in all species of *Gunnera* so far examined, and it is probable that all species in this genus are associated with cyanobacteria.

The genus does not contain any plants of economic importance. Some species are grown in gardens for decorative purposes. The flowers are very small but *G. manicata* is a very handsome plant (Figure 18.15). The amount of nitrogen fixed has been estimated on a few occasions and estimates vary from 20 to 76 kg N/ha/year.

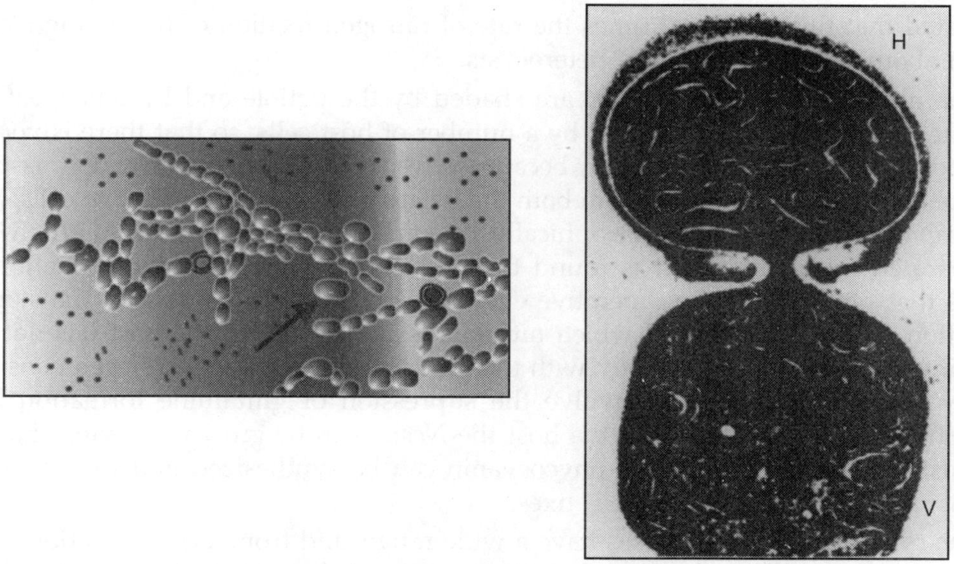

Figure 18.14 Metabolic interaction of heterocyst (H) and vegetative cells (V).

Figure 18.15 A full grown Gunerra plant.

The *Nostoc* enters glands found on the stem in the axils of the leaves. There are three such glands associated with each petiole base: one situated centrally on the stem above the emergence of the petiole and the other two lying laterally on either side of the petiole. The *Nostoc* grows into the stem tissue in a fan-like form so that cells of Nostoc invade host cells well within the stem tissue. The infection process has not so far been investigated. The filaments invade the host cells and become intracellular. Inside the cells they are separated from the host cell's contents by the plasmalemma. After having invaded the host cell, further growth occurs and a large part of that growth involves formation of heterocysts. The development of heterocysts continues through the life of the plant cell until cell senescence, when about 60% of the *Nostoc* cells are heterocysts. However, at this late stage nitrogen fixation has ceased.

In the earlier stages, when heterocysts form 20-40% of the *Nostoc* cells, the high proportion of nitrogen-fixing cells provides very active nitrogen fixation, and it has been estimated that this is several times the rate of nitrogen fixation of free-living *Nostoc* in which about 5% of the cells are heterocysts.

The glands that contain *Nostoc* are shaded by the petiole and by other leaf bases, and the endophyte is also covered by a number of host cells, so that there is very little light for photosynthesis. However, because phycocyanin, a pigment that is associated with photosystem-II, is absent from both the heterocysts and the vegetative cells, *Nostoc,* in symbiosis, must live heterotrophically on carbon fixed by the host. When *Nostoc* is separated from the host it is found that light can stimulate nitrogen fixation. This shows that photosystem-I is operative even though it cannot be used in the symbiotic condition. The mechanism by which nitrogen is transferred to the host has not so far been investigated, but by analogy with the other symbioses that have been considered, one would expect that it to involve the supression of glutamine formation by the endophyte. When isolated from the host the *Nostoc* can be grown, showing that when the host's influence is removed phycocyanin can be synthesized and that it can then also assimilate the nitrogen that it fixes.

The cyanobacterial symbioses have a wide range, and from a consideration of some of the associations it can be seen that there is a great diversity of interactions, ranging from the lichens where the cyanobacterial partner photosynthesizes and provides the host with both carbon and nitrogen, through the association with *Azolla* where photosynthesis takes place but carbon is supplemented by the host, to the situation in cycads and in *Gunnera* where the endophyte is dependent wholly upon the host for carbon.

18.6 MYCORRHIZAS AND MYCOTROPHIC PLANTS

Mycorrhiza literally means "root fungus" refers to the symbiotic association that exists between plant roots and fungi. Living together is a common situation in the underground environment, where the roots of 90% of the land plants join some soil fungi to create the most widespread type of mutualistic symbiosis, namely, mycorrhizae. They are obviously not harmful and are often beneficial. If there is one thing developed beyond doubt with mycorrhizae, it is that they increase uptake of phosphorus from soil (a pipeline between the root and the soil). Mycorrhizal plants not only grew better than those non- infected, but were also more resistant to numerous forms of both biotic and abiotic stress. Such complexes of root and fungus are termed mycorrhizas (fungus roots). Investigation of mycorrhizae thus appeared to open the way to the adoption of agronomic and forestry techniques based on respect for the natural microflora, a less indiscriminate use of artificial fertilisers, and the rehabilitation of rundown areas through the use of more natural microbiological techniques. The role of mycorrhizae for sustainable agriculture is a widely accepted norm and thrown these ideas in the scientific limelight.

They are in fact a necessary starting point for the elaboration of a new underground ecology and can be seen as a fundamental link for the flow of nutrients between plant and soil. This in turn can be translated into new approaches to the management of forests, the control of soil erosion, as well as recolonization and production of plants with economic interest. With many forest trees penetration of the roots takes place to only a slight extent and the mycelium is mainly external (Ecotrophic mycorrhizas) in which one fungus is dominated on the root surface. In many other mycorrhizas the fungus penetrates into the cortical cells or intercellular spaces of the host root (Endotrophic mycorrhizas) (Figure 18.16). In some plants the endophytic fungus is not confined to the roots but permeates the aerial parts also. Such plants are said to be mycotrophic. Mycorrhizae can be broadly divided into two catagories.

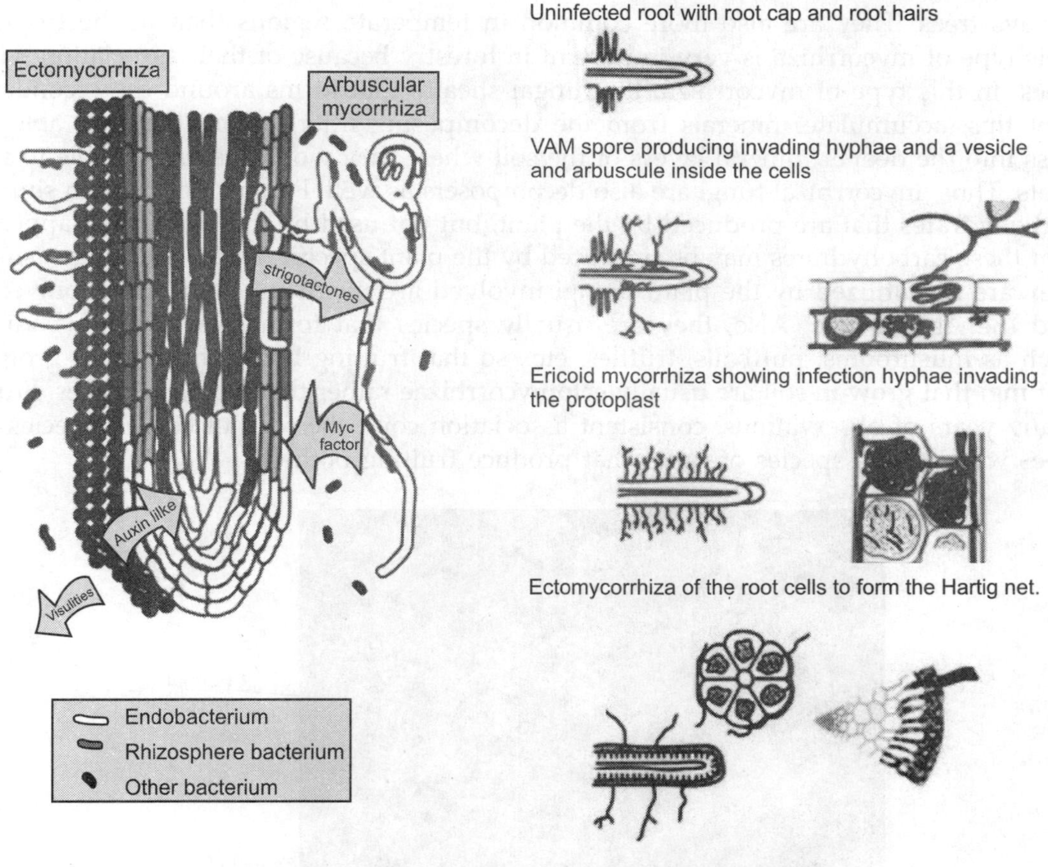

Figure 18.16 Major forms and association of mycorrizas.

18.6.1 Ectomycorrhizae

Ectomycorrhizae are characterized by forming an external sheath of mycelium around the root tips and hyphal cells do not penetrate the cell walls (intercellular) although they may go between cells in the cortex (Hartig Net— a hyphal network, that extends into the root penetrating between epidermal and cortical cells; this network is a site of nutrient exchange between the fungus and the host plants; Hartig net is typical for ectomycorrhiza). This category of mycorrhiza is very uniform in appearance, and biologically identical despite having literally thousands of different species of fungi, in the Ascomycota and Basidiomycota. For this reason, it is not subdivided into further subcategories as in endomycorrhizae. It is referred to as "ecto-" because the fungal symbiont does not invade the cell protoplasm. However, the fungus does form a thick sheath around the root tip and mycelium also grows between the cells of the cortex. The infected roots are very distinctive, forming short, paired, branches (Figure 18.17). While there are a large number of fungi that are ectomycorrhizae, plants that have ectomycorrhizae are restricted to only a few families of plants, and these plants are always trees. They are also more common in temperate regions than in the tropics. This type of mycorrhiza is very important in forestry because of their association with trees. In this type of mycorrhiza, the fungal sheath, that forms around the secondary root tips, accumulate minerals from the decomposing litter, before they are able to pass into the deeper mineral layers of the soil where they would be unavailable to the roots. Thus, mycorrhizal fungi are also decomposers as well. Fungus does obtain simple carbohydrates that are produced by the plant, but not used by the plant. So it appears that these carbohydrates may be produced by the plant specifically for the fungus since they are not utilized by the plant. Fungi involved are members of the Basidiomycota and the Ascomycota. Also, they are usually species that form large fruiting bodies, such as mushrooms, puffballs, truffles, etc., so that fruiting bodies from these groups of fungi that grow in soil are usually ectomycorrhizae rather than being saprobes. From many years of observations, consistent association could be seen of certain species of trees with certain species of fungi that produce fruiting bodies.

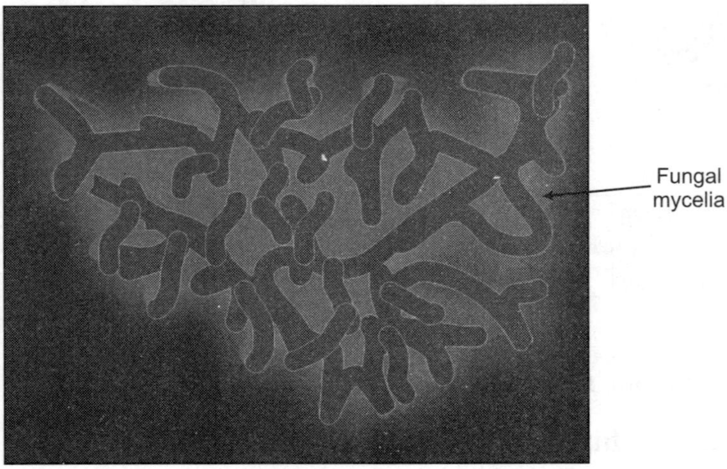

Figure 18.17 Infected roots from pine, note the distinctive, short, pairs of branches.

Although the mycelium of many ectomycorrhizae fungi can be grown in an agar based synthetic/semi-synthetic medium, their growth is slow compared to in soil and with no fructification. This was the reason why "cultivation" of truffles, e.g., *Tuber melanosporum*, which is mycorrhizal, requires planting of the host trees that have been inoculated with the fungus in order to obtain fruiting bodies. It has been demonstrated that unknown growth factors exuded by the roots seems to stimulate mycelial growth. There are undoubtedly many more factors involved, with regards to growth of the fungi which are not known. The ectomycorrhizal root that is formed has a morphology that is distinct from that of uninfected roots. One distinctive characteristic of the infected root tips is that they lack root hairs. This is unusual because root hairs are normally present, in abundance, in the young root. This morphology is in part due to the fungus secreting auxin, a plant hormone, that acts upon the root development and causing it to have branching. Branching of the root system differs with different plant families.

18.6.1.1 Economic relevance

Plants that are involved in ectomycorrhizae are always trees and are found only in a few families. They include the Betulaceae, Beeches and Alders, Casuarinaceae, Ironwood, Fagaceae Oaks, Myrtaceae Eucalyptus and Pinaceae Pines, Douglas Firs, Firs, etc. Most of these are utilized as a source of lumber, and in the case of the Pine family, millions of trees are used annually as Christmas trees. When planting these trees, it is a routine practice, in forestry, to inoculate the seedling with a mycorrhizal fungus. This group of mycorrhiza also protects plants against root pathogens by developing a fungal sheath of the ectomycorrhizal fungus covering the root tips. Fungal root pathogens would be unable to gain entry into the root system of the host.

18.6.2 Endomycorrhizae (Endotrophic)

They are characterized by the lack of an external sheath around root tip and the penetration of cortical cells (intracellular) by the fungus mycelium. This category of mycorrhiza is far more common than the ectomycorrhizal type. Generally, it can be said that plants that do *not* form ectomycorrhizae will be the ones that form endomycorrhizae. However, because of the absence of macroscopic fruiting bodies, the presence of endomycorrhizae is more difficult to demonstrate. Because of the lack of visibility, this group was considered to be rare until a method was devised that could readily detect such fungi in the soil and demonstrate that they are in fact very common (Figure 18.18). The only common feature that they all share is that the mycelium of the fungal symbiont will gain entry *into* the host, root cells by cellulolytic enzymes. Unlike the ectomycorrhizae, roots which are infected with mycorrhizal fungi do not differ morphologically from those that are not infected, i.e., root hairs are present and sheath is not formed around the root tip. However, the type of association that is formed between the host and fungus vary a great deal in the different categories of endomycorrhizae.

18.6.2.1 Vesicular-arbuscular mycorrhizae (VAM)

This category of mycorrhiza can be found throughout the world, but more abundant in the tropics than in temperate regions, and is associated with more plants than any of the other categories of mycorrhizae. These are the 'arbuscular vesicular' or phycomycetous mycorrizas. The endophytes are of phycomycetous type with few or no septa and commonly produce large, globose, or irregular vesicles in the intercellular spaces or

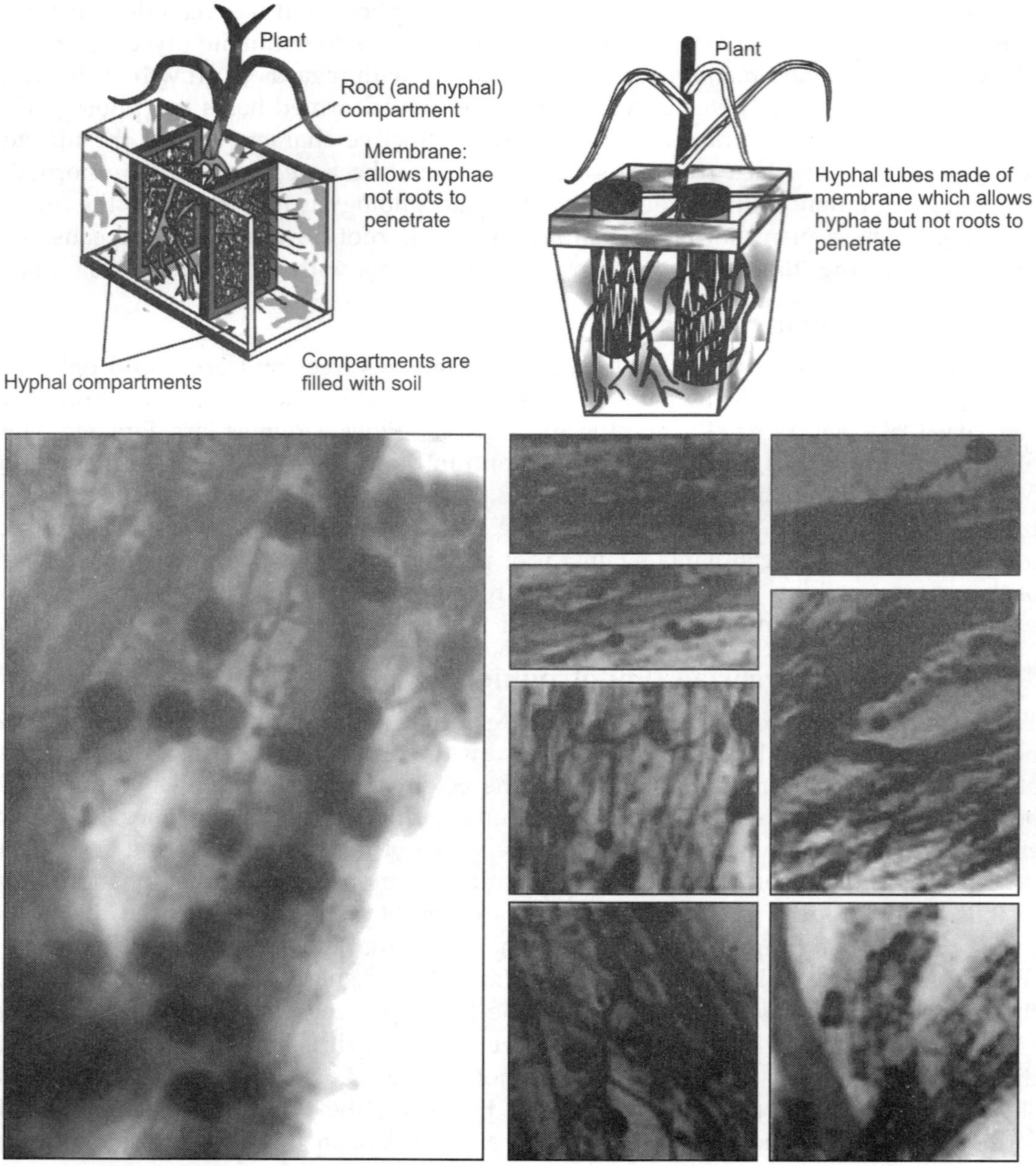

Figure 18.18 Experiment demonstrating association of mycorrhizal growth of *G. mosseae* on vascular plants.

less commonly within the cells of the root cortex, and much branched, haustorium-like, intracellular structures, termed arbuscules. The arbuscules finally disintegrate and are probably digested by the host. There is evidence that the aerial parts of some species of the host plant, such as the wild garlic, *Allium ursinum*, produce an antibiotic substance which prevents the invasion of bulbs and aerial shoots by the root fungus.

The name of this type of mycorrhizae comes from the distinct structures that can be seen inside the cells of the infected roots, the rounded **vesicles** and the branched tree-like **arbuscules**. There is also extensive mycelium in the soil, but none of it is organized in any fashion. The vesicles and arbuscules contain the stored minerals that are needed by the plant. These structures lyse in the root cells and in this way the minerals become available to the plant. The group of fungi involved is always a member of the Zygomycota. The mycelium under the microscope is coenocytic (nonseptate), which is expected from fungi in this division. There are only a few genera of fungi involved, but because of the lack of specificity of these genera to specific host plants, they have been found to have largest host range of any mycorrhizal group.

The VAM fungi normally produce assorted types of spores which can be used in the identification of these fungi, i.e., zygospores, chlamydospores and azygospores. It was once thought that these fungi were nothing more than a rare curiosity. However, this was only because a technique was needed, which could more efficiently find VAM spores, than by simply sifting through the soil. Once this technique was found, this type of mycorrhiza was found to be the most common in nature. It is because VAM have a broad host range they were once considered to be a future tool in agriculture, i.e., fertilizer substitute. However, because these fungi cannot be grown in the absence of a host plant, individual inoculations would have to be done for each plant. This would be impractical for any grains grown as well as for most crops, but have been utilized in planting of fruit trees which are planted individually (Figure 18.19).

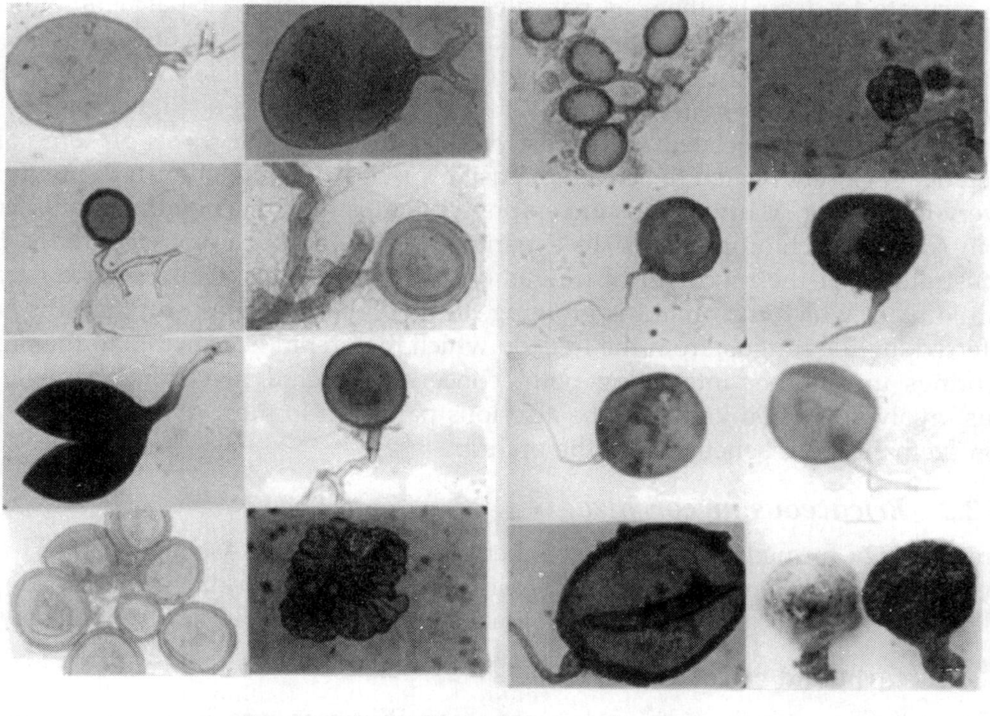

Figure 18.19 Different types of VAM spores.

18.6.2.2 Orchid mycorrhizae

The best-known example of an endotrophic mycorrhiza is that of certain orchids. It is a well documented horticultural fact that the seeds of many orchids, which are very small and relatively poorly differentiated, fail to germinate in sterilized soil but will do so in the presence of soil in which the same species has been previously grown. This is due to the presence of mycorrhizal fungus in such soil. The germinating seeds become infected at an early stage. In the roots of older plants coils (pelotones) of fungal hyphae are present in certain cortical cells, the exact layer occupied varying with the host species. The meristematic tissue of the root tip is normally not invaded and the infected zone does not usually extend throughout the root, the part just below the junction with the aerial parts of the plant being free of infection. The tubes of most tuberous species and the aerial shoots of both green and chlorophyll-less ones are infected and it has been shown that certain tubers produce an antibiotic substance which inhibits the growth of the mycorrhizal fungus. The root cells containing healthy hyphae are usually adjacent to others in which the hyphae are obviously disintegrating (digestive cells). It is assumed that the fungus obtains nutrients from the orchid and that the latter benefits by the synthesis of organic substances by the fungus which become available by the digestion of the hyphae. It is now known that some orchid seeds germinate in the absence of mycorrhizal fungus if sucrose is added to the culture medium. The chlorophyll-less orchids are special case since they are unable to carry out photosynthesis. These were formerly described as saprophytes, but it is probably more accurate to describe them as parasites on their mycorrhizal partners, although the exact nature of the relationship is unknown. The Japanese orchid *Gatrodia elata,* which largely consists of a colourless tuberous structure, is unable to flower unless it is invaded by *Armillaria mellea,* normally a destructive parasite of higher plants. The mycorrhizal fungus associated with most orchids is *Rhizoctonia repens* or some allied species, again closely related to actively parasitic species. Some relationships are unique and very interesting. Many orchids are epiphytes, that is they live on other plants rather than in soil, and achlorophyllous. In experiments with orchid epiphytes, it has been demonstrated that the mycorrhizal fungus on the orchid roots also acts as a parasite upon the plant which the orchid is growing. In this type of relationship, food is being transferred, by the fungus, from the tree, on which the orchid is growing, to the orchid. This brings up another interesting point concerning orchid mycorrhizal fungi. The fungus involved is often known to be a serious pathogen to most plants, but for some reason seems to be a benefactor to the orchid.

18.6.2.3 Ericaceous mycorrhizae

The mycorrhiza formed in this group is between fungi in the Ascomycota, and more rarely in the Deuteromycota, and species in the family's Epacridaceae, Ericaceae and Pyrolaceae. Three subcategories are recognized, arbutoid, ericoid and monotropoid.

18.6.2.3.1 Arbutoid mycorrhizas

They are of the ect-endo type and, among the Ericaceae, are found in the genera *Arbutus, Arctostaphylos* and *Arctous*. They are also found in *Pyrola* of the Pyrolaceae which is also in the Ericales [N.B. Some authors recognise this as a separate type of mycorrhiza]. There is a well-developed fungal sheath and a Hartig net restricted to the outer layers of cells. Extensive penetration of the outer cortical cells occurs and hyphal coils fill the cells. This type of mycorrhiza resembles the ectomycorrhizas formed with forest trees but differs from them because of the cell penetration. The fungi involved are Basidiomycetes and mainly species that form ectomycorrhizas with forest trees. The presence of the fungal sheath, as in ectomycorrhizas, insulates the host from the soil. Hence, all minerals absorbed by mycorrhizal roots must pass through this sheath. Arbutoid mycorrhizal associations are variants of ectomycorrhiza found in certain plants in the Ericaceae characterised by hyphal coils in epidermal cells. *Gaultheria* and *Kalmia* have ericoid mycorrhizas as well as arbutoid associations.

18.6.2.3.2 Ericoid mycorrhizae

This group is probably the most important, with respect, to its potential applications. Ericoid mycorrhizae have evolved in association with plants that are continually stressed by factors within the soil. The soil is typically extremely acid, low in available minerals because mineralization is inhibited. Plants with ericoid mycorrhizae seem to have a high tolerance to these stresses and there is good reason to believe that this is related to the presence of the mycorrhizal fungus and that the survival of the host is dependent upon the fungus.

18.6.2.3.3 Monotropoid mycorrhizae

One of the characteristics is that they have chlorophyll and can produce their own food through the process of photosynthesis. However, this is not true of all plants. The Monotropaceae and Pyrolaceae are two families of plants which are achlorophyllous. Thus, plants in these families are more dependent upon their mycorrhizal partners than plants which can carry out photosynthesis. The means by which food is obtained by these plants is similar to that of the epiphytic orchids described above. However, morphologically, they are very different. The achlorophyllous host has mycorrhizae roots that appear to be formed by an ectomycorrhizal fungus, but the epidermal and outer cortical cells are penetrated by the fungus. The fungus also forms an ectomycorrhizal relationship with a tree which is capable of photosynthesis. So, as in the case of the epiphytic orchids, the photosynthetic tree indirectly provides carbohydrates to these achlorophyllous plants, as well as to the fungus. Both hosts probably obtain their mineral requirements through the fungus.

The roots of wide range of higher plants, including many ferns, frequently contain endophytic fungi of a different type from those of the orchids. These are the 'arbuscular vesicular' or phycomycetous mycorrizas. The endophytes are of phycomycetous type with few or no septa and commonly produce large, globose, or irregular vesicles in the intercellular spaces or less commonly within the cells of the root cortex, and much branched, haustorium-like, intracellular structures, termed arbuscules. The arbuscules finally disintegrate and are probably digested by the host.

Most common endophytes is one of a number of a species of the curious zygomycetous genus Endogone. These produce hypogeous fruit-bodies containing large chlamydospores, zygospores or a mixture of both types of spores. Such fruit-bodies have been shown to be in direct association with the hyphae in the roots and moreover, the endophytic association has been induced by inoculation of soil with the appropriate fruit-bodies. Isolated spores of size and structure similar to that of the chlamydospores of Endogone have been demonstrated in soil and have also induced the development of endophytes in the roots of seedlings under experimental conditions. Other fungi have been claimed as the endophyte in arbuscular-vesicular mycorrhizas, but of these only a species of Pythium isolated from such roots has induced the condition in seedlings, fern prothalli and liverworts grown under strictly aseptic conditions. It could well be that Endogone, Pythium and possibly other fungi are the fungal partners of the same plant species according to the environmental, and particularly soil, conditions. If there is one thing developed beyond doubt with mycorrhizae, it is that they increase uptake of phosphorus from soil (a pipeline between the root and the soil).

Because of the fundamental importance of arbuscular mycorrhizas in key ecosystem, an understanding of the symbiosis is an essential prerequisite for the development of sustainable agriculture systems and as guidelines for the protection of natural ecosystems.

18.6.3 Classification of Arbuscular Mycorrhiza

The classification of arbuscular fungi presented here is that of Schüßler et al. (2001) with modifications of Oehl and Sieverding (2004), Walker and Schüßler (2004), Sieverding and Oehl (2006), Spain et al. (2006), Walker et al. (2007a, b), and Palenzuela et al. (2008). In this system of classification, fungi forming or considered to form arbuscular mycorrhizae are placed in four orders, i.e., Archaeosporales, Diversisporales, Glomerales, and Paraglomerales, comprising ten families and thirteen genera, belonging to the class Glomeromycetes of the phylum Glomeromycota (Oehl and Sieverding 2004; Palenzuela et al. 2008; Schüßler et al. 2001; Sieverding and Oehl 2006; Spain et al. 2006; Walker and Schüßler 2004; Walker et al. 2007). *Geosiphon pyriformis* of the family Geosiphonaceae (Archaeosporales) does not form arbuscular mycorrhizae. It forms endocytosymbioses with cyanobacteria (*Noctos* sp.) and was included into the Glomeromycota based on only its close molecular relationship.

GLOMEROMYCOTA C. Walker & Schuessler
- **Glomeromycetes** Cavalier-Smith
  - **Archaeosporales** C. Walker & Schuessler
    - **Ambisporaceae** C. Walker, Vestberg & Schuessler
      - *Ambispora* Spain, Oehl & Sieverd.
    - **Archaeosporaceae** J.B. Morton & D. Redecker emend. Oehl & Sieverd.
      - *Archaeospora* J.B. Morton & D. Redecker
      - *Intraspora* Oehl & Sieverd.
    - **Geosiphonaceae** Engler. & E. Gilg emend. Schuessler
      - *Geosiphon* (Kütz.) F. Wettst.

Diversisporales C. Walker & Schuessler
Acaulosporaceae J.B. Morton & Benny
Acaulospora Gerd. & Trappe emend. S.M. Berch
Kuklospora Oehl & Sieverd.
Acaulospora Gerd. & Trappe emend. S.M. Berch
Diversisporaceae C. Walker & Schuessler
Diversispora C. Walker & Schuessler
Otospora Oehl, J. Palenzuela & N. Ferrol
Entrophosporaceae Oehl & Sieverd.
Entrophospora R.N. Ames & R.W. Schneid. emend. Oehl & Sieverd.
Gigasporaceae J.B. Morton & Benny
Gigaspora Gerd. & Trappe emend. C. Walker & F.E. Sanders
Scutellospora C. Walker & F.E. Sanders
Pacisporaceae C. Walker, Blaszk., Schuessler & Schwarzott
Pacispora Oehl & Sieverd.
Glomerales J.B. Morton & Benny
Glomeraceae Piroz. & Dalpé
Glomus Tul. & C. Tul.
Paraglomerales C. Walker & Schuessler
Paraglomaceae J.B. Morton & D. Redecker
Paraglomus J.B. Morton & D. Redecker

Molecular investigations showed that fungi of the phylum are more closely related to members of the phyla Ascomycota and Basidiomycota than to those of the phylum Zygomycota, as, e. g., Gerdemann and Trappe (1974), Lawrynowicz (1979), and Morton and Benny (1990), Morton and Msiska (2010).

The fungi of the order Archaeosporales form endocytosymbioses with photoautotrophic prokaryotes (*Geosiphon pyriformis*) or produce mycorrhizae with arbuscules, with or without vesicles. Their spores are colourless and do not react in Melzer's reagent. Glomoid spores (identical to those of fungi of the genus *Glomus*) form singly or in clusters on or under the soil surface. Acaulosporioid spores (similar to those of members of the genus *Acaulospora*) develop singly in the soil. The fungi differ from other arbuscular fungi by the possession of the rRNA SSU gene signature YCT**AT**CYKYCTGGTGAKRCG, corresponding to homologous position 691 of the *Saccharomyces cerevisiae* SSU rRNA sequence J01353, with the highlighted nucleotides being specific for the taxon. The order Archaeosporales contains three families, Archaeosporaceae with the genera *Archaeospora* and *Intraspora*, Ambisporaceae with the genus *Ambispora*, and Geosiphonaceae with the genus *Geosiphon*.

Members of the order Diversisporales form mycorrhizae with arbuscules, frequently lacking vesicles, with or without auxiliary cells. Spores develop either inside (entrophosporioid spores of the genera *Entrophospora* and *Kuklospora*) or laterally on the neck of a sporiferous saccule (acaulosporioid spores of the genus *Acaulospora* and *Otospora*), from a bulbous base on the sporogenous hypha (gigasporioid spores of the genera *Gigaspora* and *Scutellospora*), or blastically at the tip of a sporogenous hypha (glomoid spores of the genera *Diversispora* and *Pacispora*). They differ from other arbuscular fungi by the possession of the rRNA SSU gene sequence signature YV**R**RYW/1-5/NGY**Y**YGB, corresponding to homologous position 658 of the *S. cerevisiae* SSU rRNA sequence J01353 SSU rRNA, GTYARDYH**M**HYY/2-4/GRADRKKYGWCRAC, corresponding to homologous position of the *S. cerevisiae* SSU rRNA sequence position 1346 of the *S. cerevisiae* SSU rRNA sequence J01353, TTATCGGTTR**A**ATC, corresponding to homologous position 650 of the *S. cerevisiae* rRNA SSU sequence J01353, and A**CT**GAGTTMATYT, corresponding to homologous position 1481 of the *S. cerevisiae* rRNA SSU sequence J01353 with the coloured nucleotides being specific for the taxon. The order Diversisporales is represented by five families, Diversisporaceae with the genera *Diversispora* and *Otospora*, Acaulosporaceae with the genera *Acaulospora* and *Kuklospora*, Entrophosporaceae with the genus *Entrophospora*, Gigasporaceae with the genera *Gigaspora* and *Scutellospora*, and Pacisporaceae with the genus *Pacispora*.

Fungi of the order Glomerales usually are hypogeous, rarely epigeous. They produce mycorrhizae with arbuscules, vesicles, and spores. Spores form either blastically at the tip of a sporogenous hypha or intercalary inside them. Spores occur singly, in clusters or sporocarps having a peridium. They differ from other arbuscular fungi by the possession of the rRNA SSU gene sequence signature.

YTRRY/2-5R**YYA**RGTYGNCARCTTCTTAGAGGGACTATCGG**T**GTYTA**A**CCGRTGG, corresponding to homologous position 1353 of the *S. cerevisiae* SSU rRNA sequence J J01353, with the coloured nucleotides being specific for the taxon. The order Glomerales includes one genus, *Glomus*.

Species of the order Paraglomerales form arbuscular mycorrhizae, rarely with vesicles. Spores are glomoid and colourless. The fungi differ from other arbuscular fungi by the possession of the rRNA SSU gene sequence signature G**C**GAA**G**CGTCATG**G**CCTTAAC**C**GG**C**CGT, corresponding to homologous position 703 of the *S. cerevisiae* SSU rRNA sequence J01353, with the coloured nucleotides being specific for the taxon. The order Paraglomerales is represented by one family, Paraglomeraceae, containing one genus, *Paraglomus*.

At present, the number of described species of arbuscular fungi is ca. 200. According to Morton et al. (1994), the number of species of this group of fungi may come up to 2700.

18.7 SYMBIOTIC ASSOCIATIONS WITH ANIMALS

The number of symbiotic associations between animals and microorganisms is immense. The rumen symbiosis in ruminants represents one of the few cases where the advantages to both partners are fairly obvious. In the trophosomes of *Riftia pachyptila*, H_2S oxidising bacteria grow as endosymbionts. They are supplied with oxygen and H_2S by the blood of the host and serve as the sole food for the worm. They are also reported to contain methane-oxidising bacteria as endosymbionts.

Protozoa, both free living and those ciliates, flagellates, and amoebae that inhabit the intestinal tracts of ruminants, termites and cockroaches, very often are themselves colonised by bacterial ecto- and endo-symbionts. Many protozoa are hosts to methanogenic bacteria, cyanobacteria and unicellular algae. Among the anaerobic ciliates and amoebae found in sea- and freshwater sediments (and in the rumen) some contain methanogenic bacteria as endosymbionts. These symbionts can be easily demonstrated by fluorescence microscopy because of the fluorescence of their coenzyme (F_{420} and methylpterin). The symbiosis rests on the ability of some protozoa to evolve molecular hydrogen which can be converted to methane by the methanogenic bacteria. The green algae are called zoochlorellae and the yellow and brown algae zooxanthellae on the basis of the sapropelic protozoa which contain phototrophic organisms recognised by their pigments. The flagellates *Cyanophora* harbours symbiotic cyanobacteria, known as cyanellae.

Many insects harbour ciliates, yeasts and bacteria as symbionts, either in the intestinal tract or in special intestinal appendices. The microorganisms are present as ectosymbionts or in the cells of specific tissues as endosymbionts. The function of these microorganisms can be easily surmised in the case of those insects, such as wood eating termites, that feed on poorly degradable materials (wood cellulose), or in the case of many aphids and heteroptera that have an unbalanced diet (plant juices). The symbionts either carry out digestive functions or supply required accessory nutrients (steroids, vitamins, amino acids, etc). In many insects the organs that harbour the endosymbionts (mycetomes) are derived from continuation or appendices of the hind gut. In contrast, the rumen in ruminants is an organ of the foregut. Hind-gut appendices are also found in mammals. In many herbivorous mammals the digestive action supported by microorganisms takes place in hind-gut appendages. Since the host cannot reap sufficient advantage from this location of the symbiosis, it becomes understandable that some animals eat their own excreta (coprophagy) in order to benefit from the basic and accessory nutrients provided by the microbes. The mutualistic character of the association between humans and their intestinal flora becomes obvious when the relationship is disturbed and the bacterial population is eliminated by antibiotic treatment or chemotherapeutic measures. The function of the intestinal flora is especially clearly recognisable in animals reared under sterile conditions. Although the animals can develop normally provided that they are fed an appropriate diet, their sensitivity to infections is drastically enhanced. It seems that a normal bacterial flora also has an important protective function against pathogenic and opportunistic microorganisms.

Similar relations exist between humans and the skin flora. Human skin has a characteristic bacterial flora, consisting primarily of mycobacteria, streptococci, and the propionibacteria, deriving their nutrients from perspiration. The normal skin flora does not have any adverse consequences for man, except for the production (under some circumstances) of odorous substances. The useful functions of the skin flora, again, are recognised when topical application or excessive internal medication with antibiotics has a growth inhibitory effect on the skin bacteria. Many marine fish have organs of luminescence. These are usually highly differentiated morphologically and are colonised by luminescent bacteria (*Vibrio fischeri*).

18.8 DESTRUCTIVE PLANT PARASITES (PLANT DISEASES)

Symptoms of disease in plants can arise from a variety of causes and in some instances are due to external factors in the plant's environment. These include an excess or deficiency of mineral nutrients in the soil and adverse climatic conditions such as drought, frost or excessive solar radiation. Again, damage to plants sometimes follows the application of toxic chemicals, or results from drift of such sprays applied to neighbouring plants. Most commonly, however, plant disease symptoms are due to attack by pathogenic organisms including fungi, bacteria, viruses , parasitic nematodes and insect pests. The symptoms of disease can take many forms and are the reflection of changes at a cellular level, themselves arising from physico-chemical reaction between host and parasite. In a number of diseases the structure of the pathogen constitutes the most prominent symptoms of the disease. Several of such symptoms are described below.

18.8.1 Fungal Diseases of Plants

18.8.1.1 Mildew

Mildews are plant diseases in which the pathogen is seen as a growth on the surface of the host. They appear as white, grey, brownish, or purplish patches of varying size on leaves, herbaceous stems, or fruits. In downy mildews the superficial growth is a tangled cottony or downy layer, while in the powdery mildews enormous number of spores are formed on superficial growth of the fungus giving a dusty or powdery appearance. Black minute fruiting bodies may also develop in the powdery mass.

18.8.1.2 Rusts

These are diseases with rusty symptoms. The rusts appear as relatively small pustules of spores, usually breaking through the host epidermis. The pustules may be either dusty or compact, and red, brown, yellow, or black in colour (Figure 18.20).

18.8.1.3 Smuts

The word smut means a sooty or charcoal-like powder. In plant disease known as smuts affected parts of the plant show a black or purplish-black dusty mass. These symptoms usually appear on floral organs, particularly the ovulary, and the pustules are usually considerably larger than those of the rust. Smut symptoms may also be found on stem, leaves, as well as roots (Figure 18.21).

18.8.1.4 White blisters

On leaves of cruciferous and other plants there may be found numerous white blister-like pustules which break open and expose powdery masses of spores. Such diseases have been commonly known as white rusts. Since there is nothing common between them and the true rusts they may be more appropriately called white blisters.

18.8.1.5 Scab

The term scab refers to a roughened or crustes-like lesion or to a freckled appearance of the diseased organ. In some diseases of this type the parasite appears at a certain

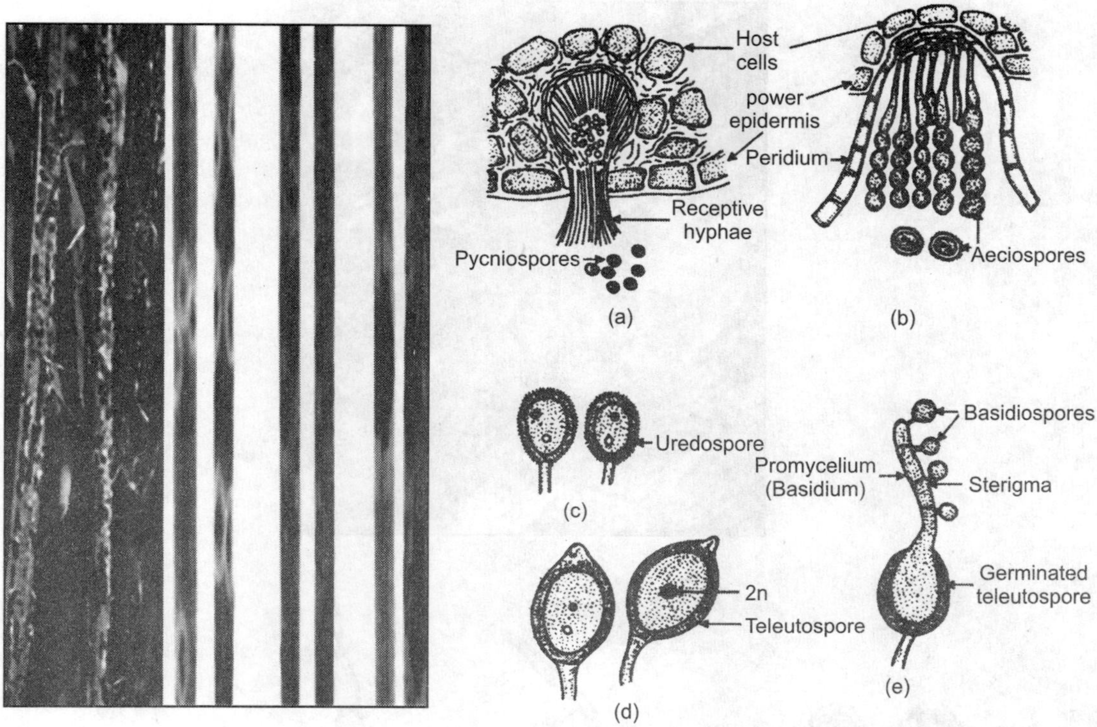

Figure 18.20 Brown rust of wheat caused by *Puccinia graminis.*

stage, in others it is never seen. Thus, this term may be listed in both the major groups of symptoms (Figure 18.22).

18.8.1.6 Sclerotia

A sclerotium is a compact, often hard, mass of dormant fungus mycelium. In some cases, as in ergot of grasses, the sclerotium assumes a characteristic shape; in others the shape may be more variable. Sclerotia are most often black, or they may be buff or dark brown or purplish in colour. They cause root rot in many cereal and vegetable crops and the causal organism is *Rhizoctonia solani* and *Sclerotium rolfssi* (Figure 18.23).

18.8.1.7 Leaf spots, leaf blights and anthracnoses

Fungal leaf spots, leaf blights and anthracnoses are caused by a large number of fungi imperfecti (Deuteromycota). Leaf spots and leaf blights resemble each other considerably in the expression of symptoms as well as mechanism of pathogenesis. However, in leaf blight a larger area of leaf or the whole leaf is involved. In certain cases whole plant is involved. In leaf spots, fungal pathogens generally produces its reproductive structures in the centre of the lesion in the dead areas (Figure 18.24 and Figure 18.25).

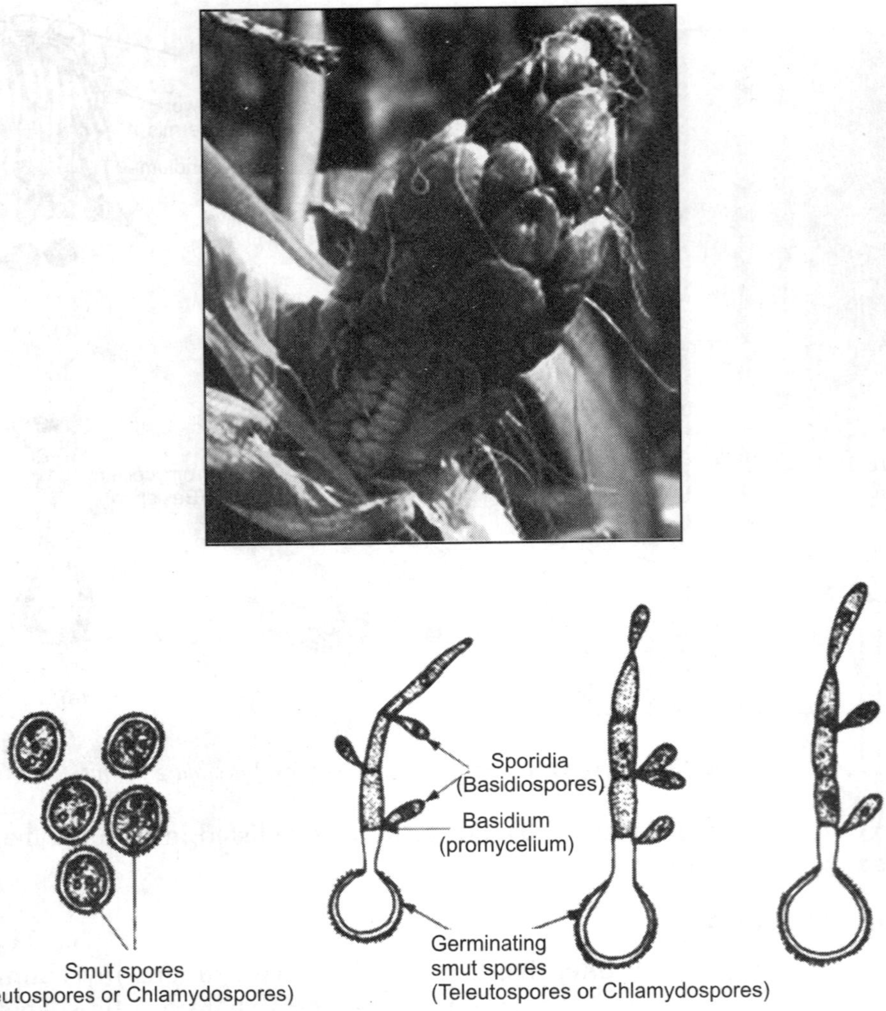

Figure 18.21 Smut of maize caused by *Ustilago maydis*.

18.8.1.8 Blotch

This symptom consists of a superficial growth giving the fruit a blotched appearance as in sooty blotch and fly-speck disease of apple fruits.

18.8.1.9 Wilt

Wilt disease is any number of diseases that affect the vascular system of plants. Attacks by fungi, bacteria and nematodes can cause rapid killing of plants, large tree branches or even entire trees. Causal organisms invade the vascular vessels and cause the xylem to fail to transport water to the foliage, thus causing wilting of stems and leaves Figure 18.26.

Figure 18.22 Scab of potato caused by *Streptomyces scabies*, wart disease of potato caused by *Synchytrium endobioticum* (a) and late Blight of potato caused by *Phytophthora infestans* (b & c).

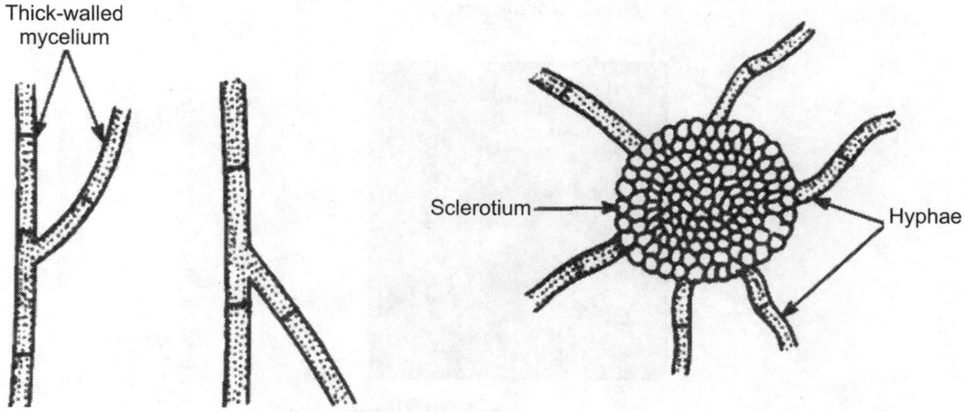

Figure 18.23 Sclerotium of causal organism of root rot *Sclerotium rolfsii*.

18.8.1.10 Fruiting bodies

The wood rooting fungi develop relatively large spore-bearing structures (sporophores) which are either fleshy or woody. The causal fungus can be identified by means of the characteristic features of these sporophores.

18.8.1.11 Exudation

In several bacterial diseases, such as in bacterial blight of paddy and fire blight of pome fruits, masses of bacteria ooze out to surface of the affected organ where they may be seen as drops of various size or as thin smear over the surface.

18.8.1.12 Tar spots

These are somewhat raised, black coated fungus bodies with the appearance of a flattened out drop of tar on leaves.

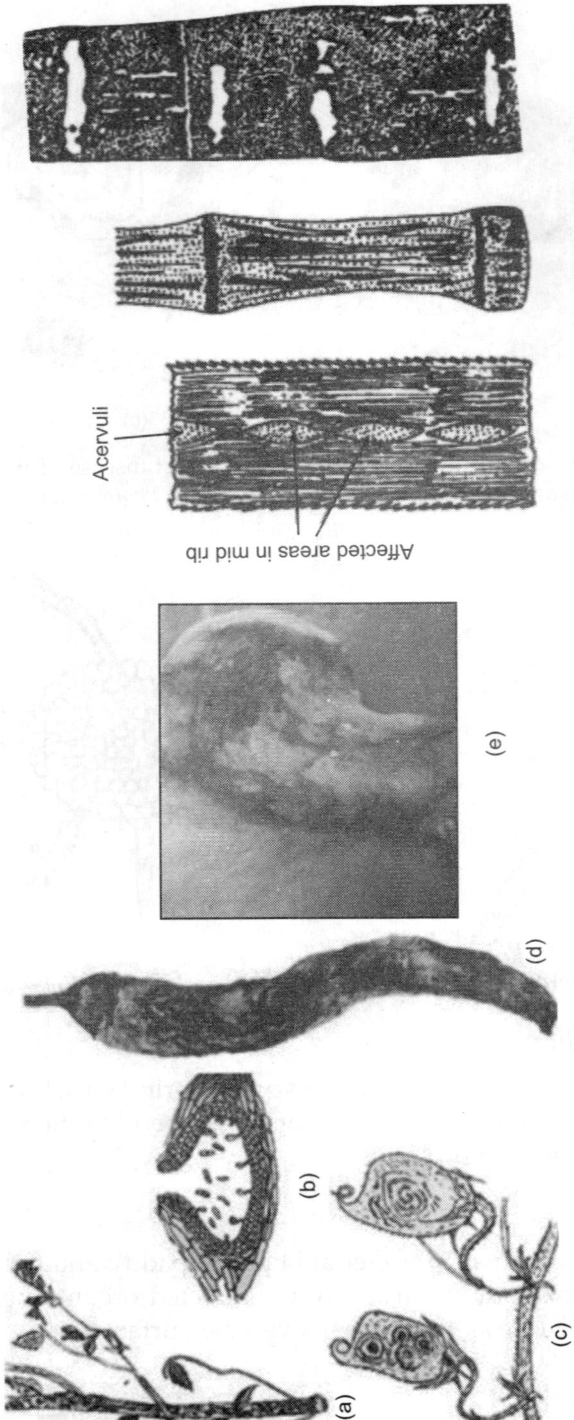

Figure 18.24 Blight of Chickpea *Cicer arietinum* caused by *Ascochita rabiei* (a) A diseased plant (b) Pycnidium showing conidiophores and conidia, (c) Diseased pod showing lesions (d) Anthracnose in ripe fruit of *Capsicum annuum* (e) Anthracnose of Papaya (*Carica papaya*) caused by *Colletotrichum gloeosporoides* and red rot of sugar cane caused by *Colletotrichum* sp.

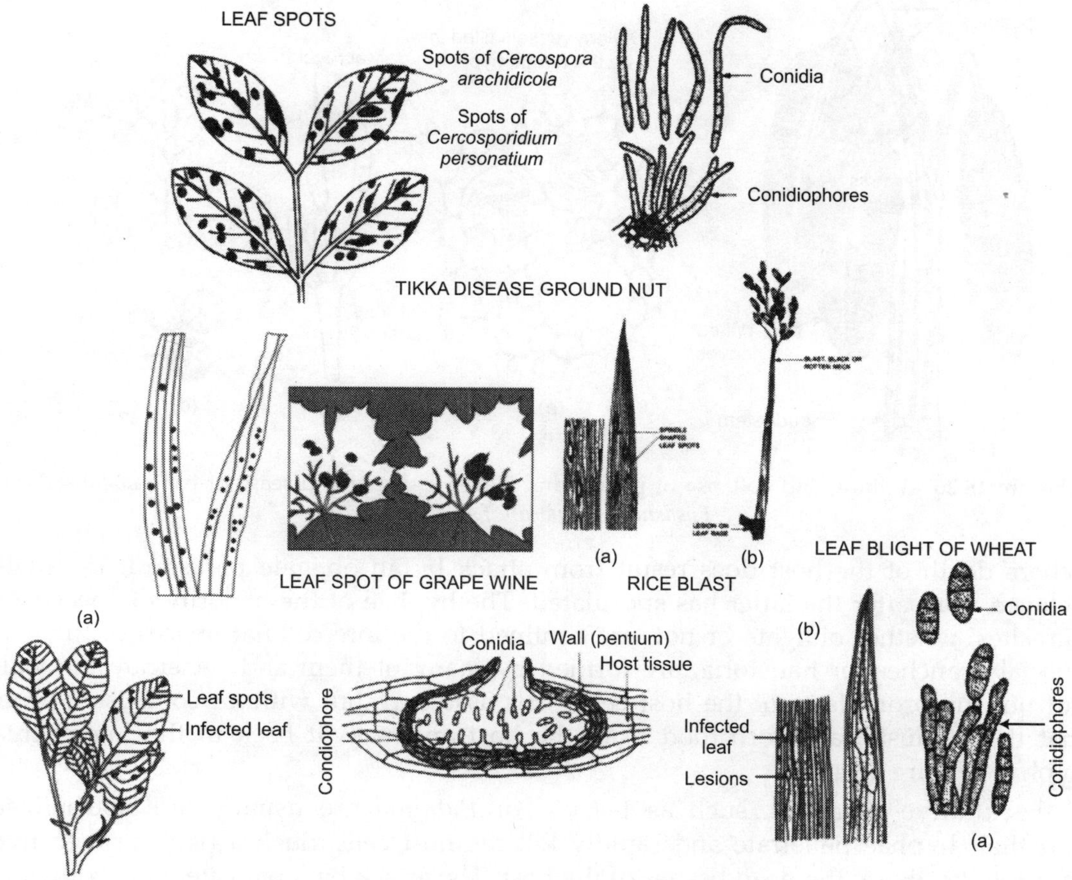

Figure 18.25 (a) Blast of rice (*Oryza sativa*) caused by *Pyricularia grisea* (b) Leaf spot of *Vinifera* sp. caused by *Septoria* sp.

18.8.2 Host-Parasite Relationship

The degree of dependence of the parasite on the host varies enormously. The rust fungi, the downy mildew and powdery mildew fungi and a few species from other groups have been considered to be strictly obligate parasite, i.e., the organisms which have not yet been grown in axenic culture. Many saprophytes which are part of the normal flora of the soil or of other habitats are able to attack certain plants under suitable conditions. These facultative parasites, which include such fungi as *Botritis cinerea* and many others, are usually non-specialized parasites on a relatively wide range of host plants. They are in marked contrast to the highly specialized obligate parasites which have a strictly limited host range. The obligate parasite is entirely dependent upon the host for its food supply and its physiology must be closely adapted to that of the host. Obligate parasite may induce considerable hypertrophy of the host but usually do not kill it rapidly. Many rust are perennial within the host and cause minimum of damage.

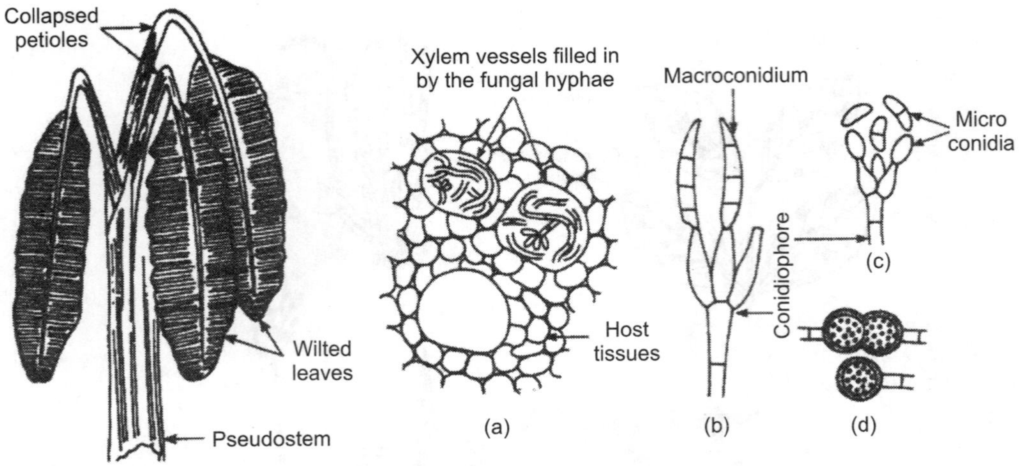

Figure 18.26 Wilting and collapse of petiole and leaves in Panama disease of banana caused by *Fusarium oxysporum f.sp. cubense.*

Where death of the host does result from attack by an obligate parasite it is usually delayed until after the latter has sporulated. The hyphae of the majority of specialized parasites, whether obligate or not, are confined to the intercellular spaces of the host. Special branches, or haustoria, are formed by many of them and penetrate the walls but not the protoplasts of the host cells. It is assumed, but without conclusive proof, that these haustoria absorb food from the host and pass it back to the intercellular hyphae (Figure 11.4).

Facultative parasites, such as *Botrytis* or *Pythium* are usually less specialized, and their hyphae penetrate and rapidly kill the host cells. Such a parasite then lives saprophytically on the dead tissues of the host. The attack by a parasite may be divided into three phases: the pre-penetration phse, covering the dispersal of the parasite, its arrival on a suitable new host and such further growth and development of the dispersal unit as may be essential before penetration of the host can take place; the penetration phase, that is, the actual entry into the host, and the post-penetration phase, which is the phase of the establishment of the parasite within the host and the subsequent exploitation of the tissues after penetration has been accomplished.

18.8.2.1 Phases in Infection

18.8.2.1.1 Pre-penetration phase

The methods by which a parasitic organism reaches a new host plant or a new part of the original host are many and varied. It seems likely that many soil inhabiting fungi parasitic on plants are capable of growing at least short distances through the soil and thus reach the host roots in the form of active mycelium. Some produce resistant sclerotia on the surface of penetrating parts of plant root. On germination the sclerotia produce hyphae which as with *Rhizoctonia solani* produces flat sclerotia on potato tubers, spread to and attack the susceptible young shoots of the host. Many plant diseases are seed borne. The fungus may be present as dormant mycelium within the seed, as spores adhering to the surface or as a spore bearing structure like pycnidia in *Phoma* and *Septoria* or as Sclerotia. Seed borne fungi are in a position to attack on germination of the seed and cause pre-emergence damping off of seedlings after emergence and a veriety of foot-rots or anthracnose diseases caused by *Colletotrichum gloeosporioides* in which the plant is weakened even if it survives.

The aerial parts of plants become infected by infection units (propagules) carried to them by the agency of wind, water (including dew and rain splashes), insects or other animals or by contact with infected plants. The infection unit of phytopathogenic bacteria is the bacterial cell itself and raindrops or insects are the commonest dispersal agents. With fungi the infection unit is usually a spore and here the problem is threefold; namely, the release of the spore from the parent mycelium or fruit body (spore discharge), its transport to the host plant (spore dispersal) and the alighting of the spore on the host surface. Weather is of importance in determining the production of spores or other propagules, their successful transport to the new host and their development upto the moment of initial penetration of the host tissues. Many fungal spores will germinate only when submerged in water. Those of the powdery mildews are an exception in germinating best in humid air. For the majority a thin film of rain or dew on the host surface will suffice, but if this dries up before penetration takes place the germ tube may die. Temperature also influences the rate of germination. The germination of spores or other propagules may also be influenced by secretions from the host plant. Thus, germination of conidia of *Botritis cinerea* is stimulated or inhibited by volatile or diffusible substances given off by various plants. Most fungal spores contain sufficient essential nutrients to permit germination, but growth of germ tubes soon ceases in the absence of an external supply of food. Some spores, however, either fail to germinate or germinate only slowly in the absence of an external food supply. The amount of the infective material "inoculum potential" is also important and many fungi are able to attack only when this is sufficiently large.

18.8.2.1.2 Penetration phase

Bacteria and some fungi enter the host only through natural openings, such as stomata or lenticels, or through wounds. Other fungi are able to penetrate the intact cuticle. Hyphae seem to reach the stomata by chance but, once there, reactions take place, such as the production of an appressorium over the stoma by rust hyphae, which suggest that some influence emanates from the stoma and that entry may then be guided by such a stimulus. Chemotaxis may be involved in the swimming of zoospores towards plant rootlets. When fungal spores or bacteria alight on the newly cut surface of a vascular bundle they may actually be sucked into the vessel and there is evidence that some disease organisms, such as the apple canker fungus enter mechanically in this way.

Penetration of the intact cuticle by certain fungi is almost certainly mechanical and is independent of chemical stimuli or enzyme action. Such fungi attach themselves firmly to any hard surface and may form an appressorium at the point of contact (Figure 11.4 and Figure 11.5). From this, fine penetration hyphae grow out and bore through the cuticle and epidermal cell wall. This shows that penetration is mechanical, since if enzymes played a part one would expect hardness of a membrane to delay penetration, but not prevent it. The mechanism by which such mechanical penetration takes place is unknown. There is now evidence that some fungi can secrete wall degrading enzymes capable of digesting cutin playing a role in pathogenesis.

18.8.2.1.3 Post-penetration phase

After the fungus has gained access to the host by one means or another, the hyphae advance through the host tissues. In some diseases, e.g., leaf spots, their advance is restricted; in others, progressive lesions result. The development of symptoms, as already mentioned, is a reflection of changes brought about by the fungus. The nature of the host-parasite interaction is considered more fully below.

The physiological and biochemical nature of pathogenicity

Toxic products of fungi: Various categories of toxic compounds are secreted by fungi which are involved in pathogenesis; they are chemically diverse and are classified as follows:

(i) Extracellular enzymes, pectolytic and cellulolytic, which attack specific plant cell wall structures such as cellulose and pectin in the middle lamella region.

(ii) Non-enzymic toxin, which may be of high and low molecular weight. High molecular weight compounds are polypeptide or bacterial polysaccharides. Componds of low molecular weight include fumaric acid, fusaric acid, alterneric acid, oxalic acid etc. The term 'vivotoxin " has been used for toxins produced by a pathogen, the host, or both, as distinct from those secreted *in vitro* for which no role in pathogenicity has been proved.

(iii) Growth substances, of which gibberellins secreted by *Gibberella fujikuroi* afford a good example. These compounds are responsible for the excessive elongation of the internodes of rice seedlings infected by the fungus.

The action of groups (ii) and (iii) is characterised by modification of host metabolism rather than structural degradation.

Effect on the host at the cellular and sub-cellular levels: There are some physical changes in the host during infection which are:

(i) Structural effects, resulting from enzyme action, are frequently observed in the cell wall area due to the activity of pectolytic, cellulolytic, proteolytic enzymes, lipase enzymes in the cell membrane.

(ii) Physiological effects, due to toxin or the secondary effects of enzyme attack, also occur. The respiration rate generally rises, and there is also a shift to the pentose phosphate pathway, while the rate of photosynthesis generally falls. Electrolyte leakage from cell membrane leads to cell death. In the case of viral and at least some fungal infections, there are drastic changes in the proteins synthesized by the host cells; levels of sugars, amino acids and nucleic acids are frequently altered.

18.9 BACTERIAL DISEASES

Unlike the fungi, most of the phytopathogenic bacteria are incapable of direct penetration of cuticularized walls or layers of cork cells of the host. The main points of entry are non-cutinized areas such as root hairs, stigmas, etc. In addition, the majority of infections by bacteria occur through wounds or punctures caused during cultural operations or by insects, nematodes, etc. The bacterial diseases are mainly disseminated through the agency of seed, soil, air and water.

Like fungal disease the bacterial diseases also produce different types of symptoms depending upon the species, host-parasite relationship, and environments. These diseases may be grouped in the following categories.

18.9.1 Blights

The invasion by the bacterium leads to very rapid and extensive necrosis of the affected plant parts resulting in scorched appearance of the host.

18.9.2 Soft Rots

The major effect is softening of the tissues due to disintegration of cells and dissolution of middle lamella, as a result of enzyme action. Very often a dirty liquid oozes out of the affected parts. In many cases, the disintegration is preceded by change of colour.

18.9.3 Leaf Spots

The necrosis of the tissue around the substomatal spaces results in the appearance of necrotic areas on the lamina surface after a specific bacterium invades the leaf through stomata. The dead tissues appear brown or water-soaked and the spots usually remain restricted in growth.

18.9.4 Tumours and Galls

In many bacterial diseases the effect of invasion by the pathogen is hyperplasia and hypertrophy of invaded tissues. As a result, tumours develop on the affected organs. The formation of root nodules in the legume crops in response to bacterial action is also a similar effect. Crown gall (*Agrobacterium tumefaciens*) is the best known and most extensively studies diseases in this group. It affects many plant species (Figure 18.27).

18.9.5 Cankers

Cankers or warty outgrowths are formed on leaves, twigs, and fruits. They result from necrosis of the tissues and reaction of undamaged tissues to produce cork cells. Such diseases are mostly localized (Figure 18.28).

18.9.6 Vascular Diseases

In some of the bacterial leaf spot diseases the organism moves into the vascular system and becomes systemic. In others, the invasion is concentrated in the vascular tissues

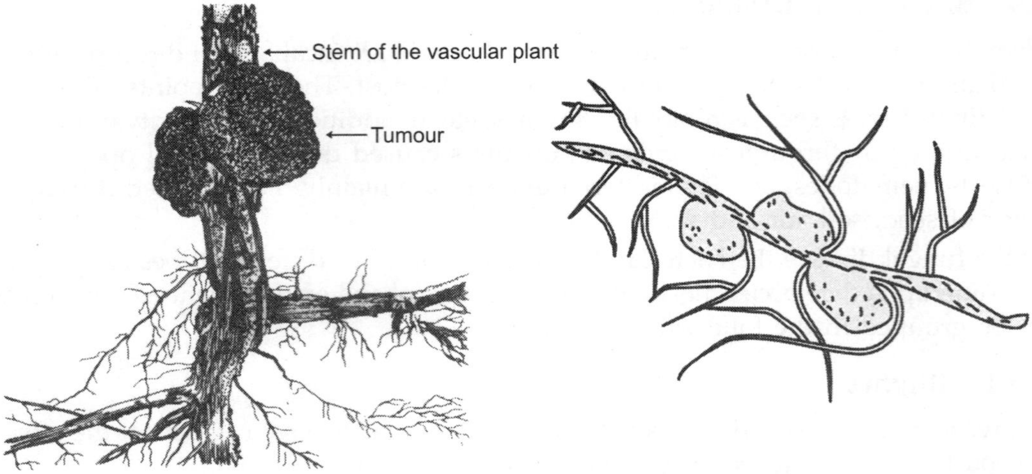

Figure 18.27 Tumour formation in vascular plants after infection with *Agrobacterium.*

Figure 18.28 Early stage of bacterial canker in citrus leaves caused by *Xanthomonas.*

causing typical wilt of the plant by plugging the water conducting vessels and by production of toxin. Common examples are bacterial wilt of tomatoes, brown rot of potatoes, bacterial wilt of cucurbits, bean, maize, etc.

18.10 PLANT VIRAL DISEASES

The virus affected plants show such histopathological changes as hypertrophy, hyperplasia, hypoplasia, and necrosis. The physiology of the plant is also affected. There is generally decreased photosynthetic activity, increased rate of respiration, accumulation of soluble nitrogen compounds particularly amides, increased activity of polyphenoloxidases and accumulation of oxidised polyphenol derivatives and decreased activity of growth regulating substances. The external symptoms which are visible to the naked eye are usually the manifestation of these internal abnormalities. The external symptoms which are visible on leaves, stem, flowers, fruits, seeds and roots may be of one or more of the following types depending upon the nature of the virus.

18.10.1 Chlorosis

Leaf chlorosis resulting into whitening or yellowing of part or the entire leaf (Figure 18.29).

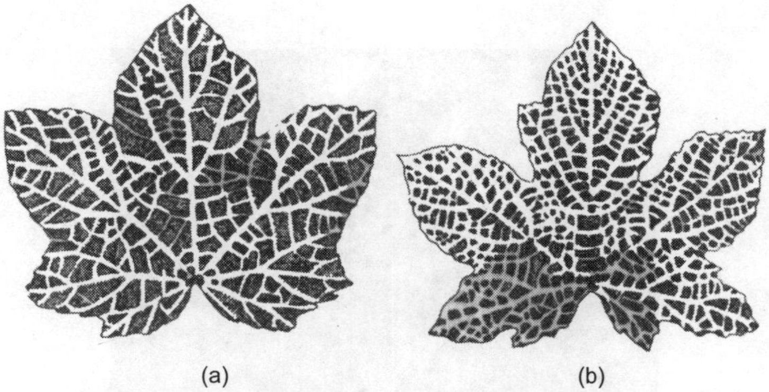

Figure 18.29 Yellow vein mosaic of okra (*Abelmoschus esculentus*).

18.10.2 Mottle or Mosaic

A pattern of green and white or green and yellow in which either colour may be predominant. In monocotyledons, such as sugar cane mosaic, this pattern may take the shape of definite streaks or stripes (Figure 18.30).

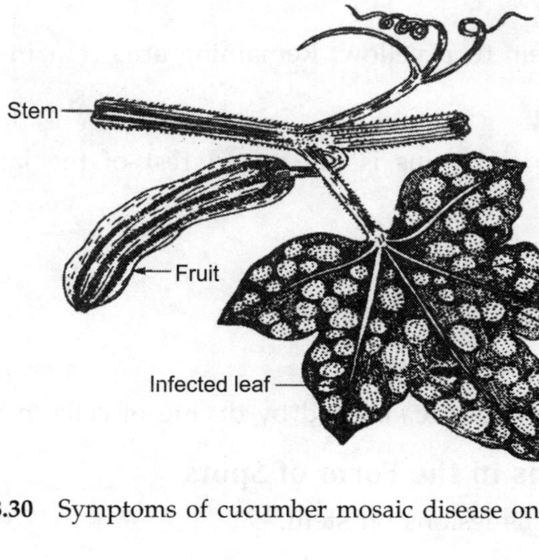

Figure 18.30 Symptoms of cucumber mosaic disease on different parts.

18.10.3 Puckering or Pouch

They are developed as pouch or puckering like in green parts of leaves.

18.10.4 Burnt Appearance

Burnt-like appearance or scorching of the entire leaf or its margins.

18.10.5 Curling, Wrinkling or Rolling

Appear on leaves (Figure 18.31). Sometimes leaves may become thick and leathery, full of starch.

Figure 18.31 Papaya leaf curl with viral mosaic symptoms.

18.10.6 Deformities

Such as filiform, fern-like appearance of leaves and reduction in the size of the leaves.

18.10.7 Vein-clearing

The tissues close to the vein turn yellow. Remaining area remains green.

18.10.8 Vein-banding

The parenchyma close to the veins is green and rest of the lamina surface shows chlorosis.

18.10.9 Galls

Development of the galls on the leaf surface.

18.10.10 Different Types of Spots

Such as ring spots, black spots, etc., caused by drying of cells in a particular fashion.

18.10.11 Stem Lesions in the Form of Spots

Stripes or cankers appear as lesions on stem.

18.10.12 Bunchy Top

Spindle tuber, spindle shoots, and similar abnormal growth of stems. Internal or external **necrosis** of tissues of the stem.

18.10.13 Spike-like

Spike like growth of the entire stem.

18.10.14 Abnormal Thinning

Appear on the leaf lamina and of the stem.

18.10.15 Late Germination

Late germination of the cuttings used for propagation.

18.10.16 Loss of Germinability

18.10.17 Stunting, Wilting, Necrosis and Ring Spots

Suggested Readings

Agrios GN. 2005. *Plant Pathology*, 5 edition, Academic Press, New York.

Appleby CA. 1984. Leghemoglobin and *Rhizobium* respiration. *Annual Review of Plant Physiology* **33**, 443-478.

Beijerinck MW. 1901. Über oligonitrophile Mikroben. *Zbl Backt* **7**, 561-582.

Benson DR, Silvester WB. 1993. Biology of *Frankia* strains, actinomycete symbionts of actinorhizal plants. *Microbiological Reviews* **57**, 293-319.

Bisen PS, Debnath M, Prasad GBKS. 2012. *Microbes Concepts and Applicatins*. Wiley-Blackwell, Hoboken, New Jersey, USA.

Bisen PS, Dev A, Gour RK, Jain RK, Sengupta LK.1995. Study of Vesicular-arbuscular mycorrhizal fungus *Glomus mosseae* in soil samples of Bhopal. In: *Mycorrhizal Biofertilizers For the Future* (Ed. Alok Adholeya and Sujan Singh), TERI, New Delhi pp 73-76.

Bisen PS, Gour RK, Jain RK, Dev A, Sengupta LK.1996. VAM colonization in tree species planted in Cu, Al, and coal mines of Madhya Pradesh with special reference to *Glomus mosseae*. *Mycorrhiza News* {T E R I New Delhi} 8: 9-11.

Dixon ROD, Wheeler CT. 1986. *Nitrogen fixation in Plants* Chapman and Hall, New York.

Elkan GH. 1987. *Symbiotic Nitrogen Fixation Technology*. Marcel Dekker, Inc. N.Y.

Fay P. 1992.Oxygen relation of nitrogen fixation in cyanobacteria. *Microbiol Rev* **56**, 340-373.

Gerdemann JW, Trappe JM. 1974. The *Endogonaceae* of the Pacific Northwest. *Mycologia Memoir* **5**, 1–76.

Herridge DF, Peoples MB, Boddey, RM. 2008. *Global inputs of biological nitrogen fixation in agricultural systems*. *Plant Soil* **311**, 1-18.

Hutchinson J. 1973. *The Families of Flowering Plants*. Clarendon Press, 3rd edn. Oxford, UK

Ingram D. 1999. *Plant Disease*. Collins, Glasgo, UK

Lane C. 2012. *Fungal Plant Pathogens*. CABI Publishing, Oxford. UK

Lawrynowicz M. 1979. *Endogonales,* klebiankowate. In: Skirgiello A., Zadara M., Lawrynowicz M. Grzyby X. Warszawa-Kraków, pp. 273-295.

Leben C. 1965. Epiphytic Microorganisms in Relation to Plant Disease. *Ann Rev Phytopathol* **3**, 209-230.

Morton JB, Benny GL. 1990. Revised classification of arbuscular mycorrhizal fungi (*Zygomycetes*): a new order, *Glomales,* two new suborders, *Glomineae* and *Gigasporineae,* and two families, *Acaulosporaceae* and *Gigasporaceae,* with an emendation of *Glomaceae. Mycotaxon* **37**, 471–491.

Morton JB, Msiska Z. 2010. Phylogenies from genetic and morphological characters do not support a revision of *Gigasporaceae* (*Glomeromycota*) into four families and five genera. *Mycorrhiza* **20**, 483–496.

National Research Council. 1994. Biological Nitrogen Fixation: Research Challenges, National Academy Press, Washington, DC.

Oehl F, Sieverding E. 2004. Pacispora, a new vesicular arbuscular mycorrhizal fungal genus in the Glomeromycetes. *Journal of Applied Botany* **78**, 72-82.

Palenzuela J, Ferrol N, Boller T, Azcón-Aguilar C, Oehl F. 2008. *Otospora bareai,* a new fungal species in the *Glomeromycetes* from a dolomitic shrub-land in the National Park of Sierra de Baza (Granada, Spain). *Mycologia* **100,** 282–291.

Peoples MB, Faizah AW, Rerkasem B, Herridge DF 1989. Methods for Evaluating Nitrogen Fixation by Nodulated Legumes in the Field. *ACIAR Monograph* No. **11**, 76 p., Canberra.

Schüßler A, Schwarzott D, Walker C. 2001. A new fungal phylum, the Glomeromycota: phylogeny and evolution. *Mycological Research* **105**, 1413-1421.

Shah VK, Brill WJ. 1977. Isolation of an iron-molybdenum cofactor from nitrogenase. *Proc Natl Acad Sci USA* **74**, 3249–3253.

Sieverding E, Oehl F. 2006. Revision of *Entrophospora* and description of *Kuklospora* and *Intraspora,* two new genera in the arbuscular mycorrhizal *Glomeromycetes. Journal of Applied Botany and Food Quality* **80**, 69–81.

Silver WS, Postgate JR. 1973. Evolution of asymbiotic nitrogen fixation. *J Theor Biol* **40**, 1–10.

Spain JL, Sieverding E, Oehl F. 2006. *Appendicispora,* a new genus in the arbuscular mycorrhizal-forming *Glomeromycetes,* with a discussion of the genus *Archaeospora. Mycotaxon* **97**, 163–182.

Tyler B. 1978.Regulation of the assimilation of nitrogen compounds. *Annu Rev Biochem* **47**,1127–1162.

Vadakattu G, Paterson J. 2006. Free-living bacteria lift soil nitrogen supply. *Farming Ahead* **169**, 40.

Walker C, Schüßler A. 2004. Nomenclatural clarifications and new taxa in the Glomeromycota. *Mycological Research* **108,** 981-982.

Walker C, Vestberg M, Demircik F, Stockinger H, Saito M, Sawaki H, Nishmura I, Schüßler A. 2007. Molecular phylogeny and new taxa in the *Archaeosporales* (*Glomeromycota*): *Ambispora fennica* gen. sp nov., *Ambisporaceae* fam. nov., and emendation of *Archaeospora* and *Archaeosporaceae. Mycological Research* **111**, 137-153.

Chapter 19

Medical Microbiology and Immunology

19.1 INTRODUCTION

The skin and mucous membranes of animals are constantly exposed with environmental organisms and become readily colonized by various microbial species. These contaminants include saprophytes, pathogens, and many intermediate forms. The mixture of organisms regularly found at any anatomical site is referred to as the normal flora. Most organisms fail to establish themselves in any particular site on the body, partly as a result of the self-sterilizing action of many tissues and partly owing to competition between microorganisms for the same site. Certain species, not normally pathogenic, are particularly suited to colonization of various sites; some are present in a particular site under almost all conditions, others may be there only temporarily. The normal flora of humans consists of a few eukaryotic fungi and protists, but bacteria are the most numerous and obvious microbial components of the normal flora. Thus, in man *Escherichia coli* is a normal intestinal parasite whereas *Staphylococcus aureus* is present on the skin of only some individuals and at certain times but most have a special capacity to spread from host to host, establish themselves, and later invade deeper tissues from the site initially colonized.

19.2 NORMAL FLORA

Environment significantly influences the types of microorganisms present as the normal microbial flora. Nevertheless, there are certain bacterial species which are invariably found in particular sites in the healthy animals. Thus, it is possible to give general descriptions of normal flora for a given animal species. The normal flora of corresponding sites in different animal hosts varies widely and is profoundly influenced by such factors as body temperature and diet. *E. coli* is the best known bacterium that regularly associates itself with humans, being an invariable component of the human intestinal tract. Even though *E. coli* is the most studied of all bacteria, and we know the exact location and sequence of 4288 genes on its chromosome, we do not fully understand its ecological relationship with humans. Sometimes the relationship between a member of the normal flora and its host cannot be deciphered. Such a relationship where there is no apparent benefit or harm to either organism during their association is referred to as a commensal relationship. Many of the normal flora that are not predominant in their habitat, even though always present in low numbers, are thought of as commensal bacteria.

19.3 NORMAL FLORA OF THE HUMAN SKIN

Human skin is not a particularly rich place for microbes to live. The skin surface is relatively dry, slightly acidic and the primary source of nutrition is dead cells. This is an environment that prevents the growth of many microorganisms, but a few have adapted to life on our skin. The normal flora of the skin and sebacious glands is restricted because of the natural disinfecting powers of the skin. Microbes able to survive on the skin must be able to withstand the antimicrobial properties associated with sweat and oil secretions. It consists of a few species of bacteria including Gram-positive cocci (mainly *Staphylococcus albus* and *Sarcina* sp.) and diphtheroids. These are all non-pathogenic but sometimes the pathogenic species, *Staphylococcus aureus,* is found on the face and hands, particularly in individuals who are nasal carriers. These organisms persist even after intensive washing. *Propionibacterium acnes* is a Gram-positive bacterium that inhabits the skin. *P. acnes* are anaerobes and they lives in pores and glands where oxygen levels are lower. As the name implies *P. acnes* causes the common skin condition called acne. Although acne outbreaks can result in emotional and physical discomfort the infection is not life threatening. A point complemented by *P. acnes* performing an important role through occupying niches that might otherwise be colonized by more dangerous pathogens. Many other organisms, not generally considered to be part of the normal flora, may be present temporarily in numbers largely determined by the standard of personal hygiene, especially in moist crevices of the skin and surrounding the body orifices, such as the anus. In the fatty and waxy secretions of skin, lipophilic yeasts are common; *Pityrosporum ovalis* occurring most commonly on the scalp and *P. orbiculare* on the glabrous skin. In the secretions of ears and genitalia, saprophytic acid-fast bacilli (e.g., *Mycobacterium smegmatis)* and diphtheroids are common. The tears secreted on the conjunctiva of the eyes contain the enzyme lysozyme, which dissolves bacteria on the eye, and this keeps the number of organisms in this site at a low level. Some common bacteria reported from human skin are aerobic micrococci, *Proteus* sp., *Candida* sp., *Pseudomonas* sp., *Hemophilus* sp., *Neisseria* sp., *Trichophyton sp., Bacillus* sp., and *Corynebacterium* sp.

19.4 NORMAL FLORA OF THE HUMAN RESPIRATORY TRACT

The human nose is home to the infamous Gram-positive bacterium *Staphylococcus aureus,* best known for its role in hospitals where it is a major cause of surgical wound and systemic infection. *Staphylococcus aureus* is often present in large numbers in the nose, this being the main carrier site of this important pathogen. Diphtheroids and *Staphylococcus albus* are found in limited numbers in the nose. It is often carried in the noses of healthcare workers and transmitted from patient to patient. In addition to these organisms, larger numbers of many more species of bacteria are to be found in the nasopharynx; predominantly mixed viridans streptococci (e.g., *Streptococcus salivarius)* and *Neisseria* spp. (e.g., *N. pharyngis)*. The nasal sinuses are normally sterile in a healthy person. Occasionally pathogens, such as pneumococci, *Haemophilus influenzae, Neisseria meningitidis, Klebsiella pneumoniae,* and allied organisms, may be harboured in the nasopharynx of apparently healthy individuals. The trachea, bronchi, and pulmonary tissues are, in the healthy state, almost free of microorganisms. Air is inhaled through the tortuous nasal passages which act as effective filters. Bacteria adhering to particulate matter are removed at this site. Most organisms that pass this filter are trapped on the mucous surfaces of the nasopharynx and trachea, being swept upwards by ciliated epithelial linings of the tract, and are subsequently removed by expectoration or swallowing. There is thus no normal flora of the mucosa of trachea and bronchi.

19.5 NORMAL FLORA OF THE GENITAL TRACT OF THE HUMAN FEMALE

Relative to other microbial populations of the human body little is known about the normal flora of the vaginal tract. The predominant bacterial species are *Lactobacillus*. It is thought that *Lactobacillus* may prevent the growth of yeast by producing hydrogen peroxide, a by-product of bacterial metabolism. In the sexually mature human female the vagina is acidic, owing to the production of lactic acid from glycogen in the epithelial linings, a defense against other bacteria. Consequently the main flora consists of mixed acid-tolerant organisms, including the predominant *Lactobacillus acidophilus,* together with corynebacteria and anaerobic streptococci. Mycoplasmas have been isolated on a number of occasions. Before puberty and after menopause the vaginal secretions are alkaline and contain normal skin organisms, e.g., staphylococci, streptococci, coliforms, and diphtheroids. The organs of the urinary tract are generally free of bacteria, possibly as a result of the flushing action of the urine.

19.6 NORMAL FLORA OF THE HUMAN ALIMENTARY TRACT

The presence of particles of food, epithelial debris, and secretions makes the mouth a nutritionally favourable habitat for a great variety of bacteria. The flora is subject to great variation but always includes very large numbers of α and non-haemolytic streptococci and, in addition, non-pathogenic species of *Neisseria,* anaerobic spirochaetes, *Fusobacterium* (particularly between gums and teeth) and lactobacilli (in the acidic cavities of teeth, especially in the presence of dental caries). The yeast-like fungus *Candida albicans,* the cause of 'oral thrush' and other diseases, is also normally present but causes disease only in persons, particularly infants, suffering from malnutrition or in adults with debilitating diseases, e.g., diabetes, or in whom the normal bacterial flora is disturbed by the use of antibiotics, steroids and cytotoxic drugs.

The majority of vegetative organisms, other than acid-fast bacilli, originating in food or from the mouth are destroyed by the hydrochloric acid of the gastric juices so that the stomach, when empty, is relatively free of living microorganisms. In the small intestine the influence of bile secretions becomes gradually apparent; the highly buffered foods revert to an alkaline pH and the numbers of bacteria increase, reaching a maximum in the caecum and colon. The flora includes coliform bacilli (particularly *Escherichia coli*), enterococci, clostridia (mainly *Clostridium perfringens*), species of *Bacillus*, Gram-negative pleomorphic non-sporing anaerobes of the genus *Bacteriodes* (this genus constitutes the predominant group of organisms), saprophytic acid-fast bacilli and many Gram-negative rods of varied types. Large numbers of organisms die during their passage down the gut and approximately one-third of the dry weight of faeces consists of bacteria.

Diet greatly influences the predominant flora. This is illustrated by the changes in flora in a child as a result of variations in diet during early life. In the breast-fed infant the flora consists predominantly of *Lactobacillus bifidus* which, during the early days of life, may form 90 per cent of the intestinal flora. With change in diet other organisms, characteristic of the adult alimentary canal, gradually become predominant. A high meat diet results in an increase in proteolytic organisms.

19.6.1 Origin of the Alimentary Flora

The alimentary tract of the foetus is sterile but at birth becomes rapidly and progressively infected by way of the mouth and anus. Immediately after birth, organisms commonly found in the mouth of the new-born are of the same type as those in the vagina of the mother. A few days later other organisms, characteristic of the normal mouth flora of adults, develop and it is thought that these originate from the naso-pharynx of the mother. The intestinal flora of infants gradually changes with time as other organisms gain access until a complex mixture of organisms resembling that in the adult is attained.

19.6.2 The Importance of the Mammalian Intestinal Flora

Considerable variation in intestinal flora is found among mammalian species. *Escherichia coli*, for example, is far less frequently found in herbivores than in carnivores. In guinea pigs lactobacilli comprise approximately 80 per cent of the flora, the remainder consisting of organisms from soil and air. Consequently the importance of the intestinal flora must be considered in relation to the host species. Unlike the flora of herbivores, the flora of carnivores is probably not essential to life. Animals, including guinea pigs and chicks, have been successfully reared on a balanced and sufficient diet under completely sterile conditions without showing stunted growth. With chicks any lack of normal ability to digest their food completely was made good by eating more. In these animals, therefore, life has been shown to be possible in the absence of a gut flora, although there is an indication that the enzymic activities of the bacteria assist in the digestion of the food. However, the administration of broad spectrum antibiotics over a long period may result in vitamin deficiency, presumably because these vitamins are normally synthesized by intestinal organisms.

In contrast to carnivorous or omnivorous animals, which depend wholly upon the secretion of enzymes into their alimentary canal to digest their food, herbivorous animals rely upon the enzymic activities of microorganisms to bring about the initial digestion. In these animals the gut flora is essential to life and a symbiotic relationship exists between the host and the microbial flora of the gut. The breakdown of vegetable matter requires a large chamber where the food can be retained and slow transit through the digestive tract to allow sufficient time for microbial degradation to take place. The alimentary tract of herbivorous animals is modified to include an enlarged portion or portions of the tract, such as the caecum and colon in the horse, or an enlarged portion of the oesophagus, including rumen and reticulum in addition to the true stomach, in ruminants (Figure 19.1). In the latter the diet is subjected to two fermentations, one preceding and the other following the conventional mammalian digestive process.

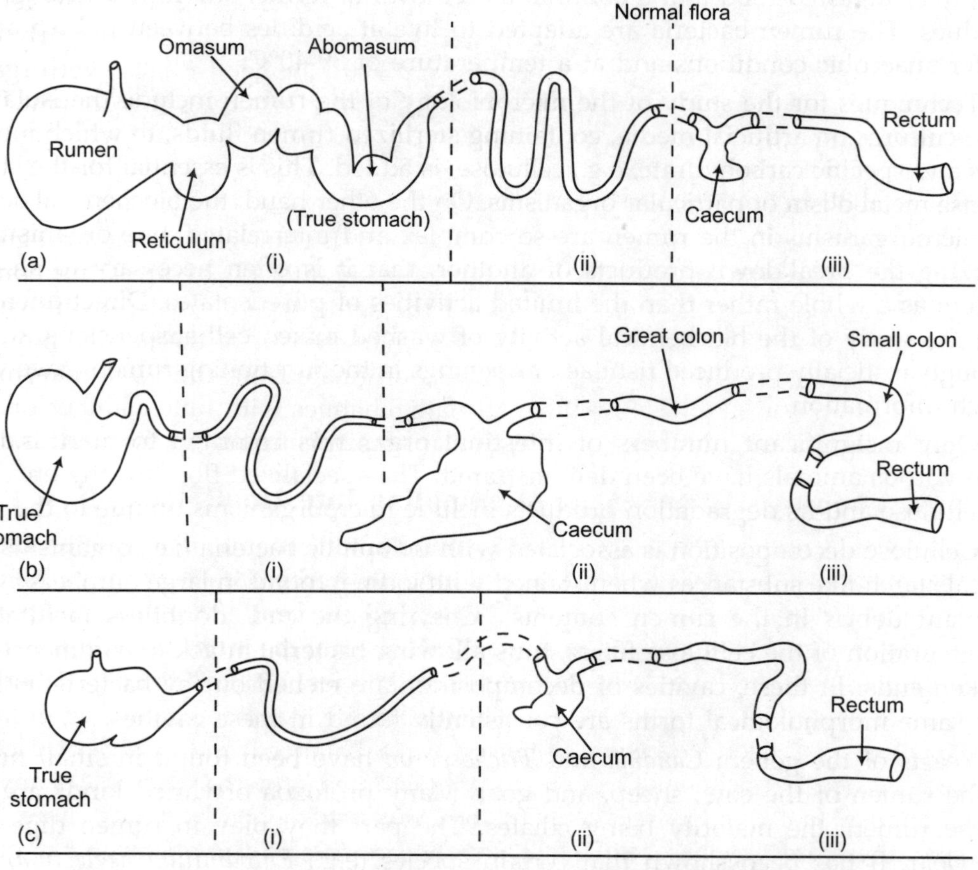

Figure 19.1 The digestive tracts of different animals showing how ruminants differ from other animals. (a) Herbivorous ruminant, e.g., cow. (b) Herbivorous non-ruminant, e.g., horse. (c) Omnivorous animal, e.g., Rat. (i) pre-absorptive part of gut, (ii) small intestine, (iii) large intestine.

19.7 MICROBIOLOGY OF THE RUMEN

The food of ruminants is complex, containing carbohydrates (mainly cellulose), proteins, fats, numerous organic compounds, and minerals. Cellulose, derived from plant cell walls, is broken down by the enzyme cellulase secreted by bacteria. This energy-yielding carbohydrate is converted into microbial cells and metabolic by-products including carbon dioxide, methane, ammonia, and acetic, propionic, lactic, and butyric acids. The bacterial proteins and their products of metabolism, including vitamins, are food for the host.

A complex mixture of microorganisms is involved in the breakdown of food. An extremely dense population of many species of bacteria and protozoa results from the favourable conditions provided in the rumen. The rumen consists of a system not greatly different from that of a laboratory steady-state continuous culture with a regular supply of ingested food and a continuous removal of fermentation products and food residues. The rumen bacteria are adapted to live at acidities between pH 5.5 and 7.0, under anaerobic conditions and at a temperature of 39-40°C.

Techniques for the study of the microbiology of the rumen include the isolation of pure cultures on artificial media, containing sterilized rumen fluids, to which inorganic salts and specific carbohydrate, e.g., cellulose, is added. This is essential for the study of precise metabolism of particular organisms. On the other hand, the biochemical activities of microorganisms in the rumen are so complex and interrelated, one organism often utilizing the breakdown products of another, that it is often necessary to study the system as a whole rather than the limited activities of pure isolates. Direct microscopy and the study of the biochemical activity of washed mixed cell suspensions, obtained through artificially produced fistulae or openings in the functioning rumen, have yielded much information.

Only insignificant numbers of intestinal organisms common to man and other omnivorous animals have been demonstrated. The specialized flora for the breakdown of cellulose and its degradation products include microorganisms unique to the rumen.

Cellulose decomposition is associated with iodophilic bacteria (i.e., organisms which reveal starch-like substances when stained with iodine) found in large numbers attached to plant debris in the rumen contents. 'Chewing the cud' doubtless facilitates the disintegration of the cellulose fibres, thus allowing bacterial attack to commence at the broken ends. In these, cavities of decomposition are etched out by bacterial enzymes. The same morphological forms are consistently found in these cavities.

Yeasts of the genera *Candida* and *Trichosporon* have been found in small numbers in the rumen of the cow, sheep, and goat. Many protozoa of varied kinds are found in the rumen, the majority being ciliates. The part they play in rumen digestion is not clear. It has been shown that certain species (e.g., *Entodinitum neglectum)* fail to grow in artificial culture if cellulose is omitted from the medium. It does not follow, however, that cellulose is digested by the protozoa since in all cultures there is active fermentation by cellulolytic bacteria. The necessity for cellulose in the medium could therefore be due to a requirement by the protozoa for the metabolic products of the bacteria digesting the cellulose. It is generally thought that protozoa play no vital role in rumen metabolism.

The diversity of bacteria in the rumen is striking. Some types are found regularly in the rumen flora; others are highly host specific. The flora varies according to the chemical composition; some bacteria have a broad and others a narrow range of biochemical activity. Some of the typed species of rumen bacteria are considered according to the substrate attacked.

19.7.1 Cellulose Digestion

Three kinds of rod-shaped cellulolytic bacteria present in large numbers in the rumen are *Bacteriodes succinogenes, Butyrivibrio fibrisolvens,* and *Clostridium lochheadii. Bacteroides succinogenes* is a Gram-negative, crescent-shaped organism with migratory and biochemical activities similar to the myxobacterium *Cytophaga succinicans*. It ferments glucose and cellobiose in addition to cellulose; from cellulose, succinic acid is the most important product. Cellulolytic cocci include *Ruminococcus flavefaciens,* which grows in chains, is Gram-positive and produces yellow colonies, and *Ruminococcus albus,* a Gram-negative coccus which characteristically produces white colonies.

19.7.2 Starch Digestion

Amylolytic, but non-cellulolytic, rumen bacteria include *Streptococcus bovis, Bacteroides amylophilus,* and *Selenomonas ruminantium,* a crescent-shaped flagellate cell. The fermentation products vary among strains but include formic, acetic, propionic, butyric, and lactic acids.

19.7.3 Hemicellulose Digestion

Hemicelluloses are plant carbohydrates insoluble in water, which constitute a large part of the forage consumed by ruminants. Most of the cellulolytic and amylolytic bacteria attack this carbohydrate. Other important rumen bacteria include *Veillonella alcalescens,* which is concerned with fat metabolism and lactate utilization, and *Peptostreptococcus elsdenni,* also an important lactate fermenter which follows the acrylic pathway of propionate formation.

19.7.4 Synthesis of Vitamins of the B Complex

The amount of B vitamins in rumen contents is found to be several times greater than that expected on the basis of the vitamin content of the food ingested. This is the result of synthesis by rumen bacteria but not protozoa. The vitamin concentration in the rumen is not influenced by the amount present in the diet; metabolic control of bacterial synthesis maintains a steady level. This explains the fact that the concentration of these vitamins in ruminant's milk remains steady irrespective of fluctuations in the diet. Although the synthesis of vitamins by bacteria in the alimentary canal is not limited to ruminants, it is probably of particular importance to these animals since for the larger part of the year they graze on dried pastures of poor quality which are deficient in vitamins.

19.7.5 Protein Synthesis

The most common nitrogen source for ruminants is provided as protein in forage. The proteins are broken down by ruminant microorganisms into peptides, amino acids, and ammonia, these products being used in different ways for the synthesis of microbial protein. This in turn becomes available to the host when bacteria die and ate digested by the ruminant. Protein is therefore used by the host in the form of rumen bacteria and this constitutes the major part of the ruminant's nitrogenous food. This knowledge has opened up a new approach to nutritional problems by adding urea or ammonium salts to diets deficient in protein, with the effect of allowing cheaper high-fibre feeding stuffs (e.g., straw) to be economically utilized. The microbial degradation of protein in the rumen inevitably means loss through ammonia and urea. Efforts are therefore being made to protect food proteins (e.g., by formalin) so that they may enter the abomasum before undergoing digestion.

19.7.6 The Influence of Antibiotics on Alimentary Flora

Considerable changes in faecal flora result from the administration of sulphoriamides and broad spectrum antibiotics. Antibiotics, often in combination, are sometimes prescribed prior to alimentary surgery in order to render the bowel virtually free of living bacteria, with a view to avoiding septic complications. This has met with mixed success. In a proportion of patients, the normal bowel flora is superseded by antibiotic-resistant staphylococci, resulting in a staphylococcal enteritis. Overgrowth of other antibiotic-resistant microorganisms, such as *Candida* spp., may also produce complications. The addition of small amounts of antibiotics to the feed of poultry and swine has sometimes been shown to increase food conversion and rate of weight gain. Antibiotics have also been tried with ruminants but only young animals have shown an advantage in food conversion. The ruminants' nutrition depends on rumen microorganisms the addition of antibiotics at low levels to the food has negligible effect, but in higher amounts may interfere with digestion. The mechanism of antibiotic growth stimulation is not fully understood but it is thought that the antibiotics check the growth of toxic organisms during the early days of life when the gut of the new-born animal, which is thinner and absorbs nutrients more efficiently than the adult gut, is first colonized by microorganisms. There are, however, reasons why this practice of feeding antibiotic supplemented foods should be discontinued. Even sub-inhibitory levels of antibiotics are selective for antibiotic-resistant organisms, and if these organisms carry the genetic determinants of infective drug resistance transfer to enteric pathogens may occur; as a result the pathogens would become resistant to the antibiotics being used and would not subsequently yield to antibiotic therapy.

19.8 COPROPHILOUS FUNGI

Coprophilous fungi (*dung-loving* fungi) are a type of saprobic fungi that grow on animal dung. The dung of various herbivorous animals supports a large fungus flora. The hardy spores of coprophilous species are unwittingly consumed by herbivores from vegetation, and are excreted along with the plant matter. The distribution of coprophilous fungi is closely linked to the distribution of the herbivores on which they rely, such as rabbits, deer, cattle, horses and sheep. Some species rely on a specific species for dung; for instance, *Coprinus radiatus* and *Panaeolus campanulatus* grow almost exclusively on horse faeces, while others, such as *Panaeolus sphinctrinus*, can grow on any faeces or even just particularly fertile soil (Figure 19.2). Further, some species (*Conocybe rickenii*) can be found in large numbers in areas where manure has been used as a soil fertilizer, such as in gardens. Some coprophilous fungi are also known to grow from the dung of omnivores (*Chaetomium globisporum* from rat droppings) or even carnivores (*Chaetomium rajasthanense*, from tiger feces).

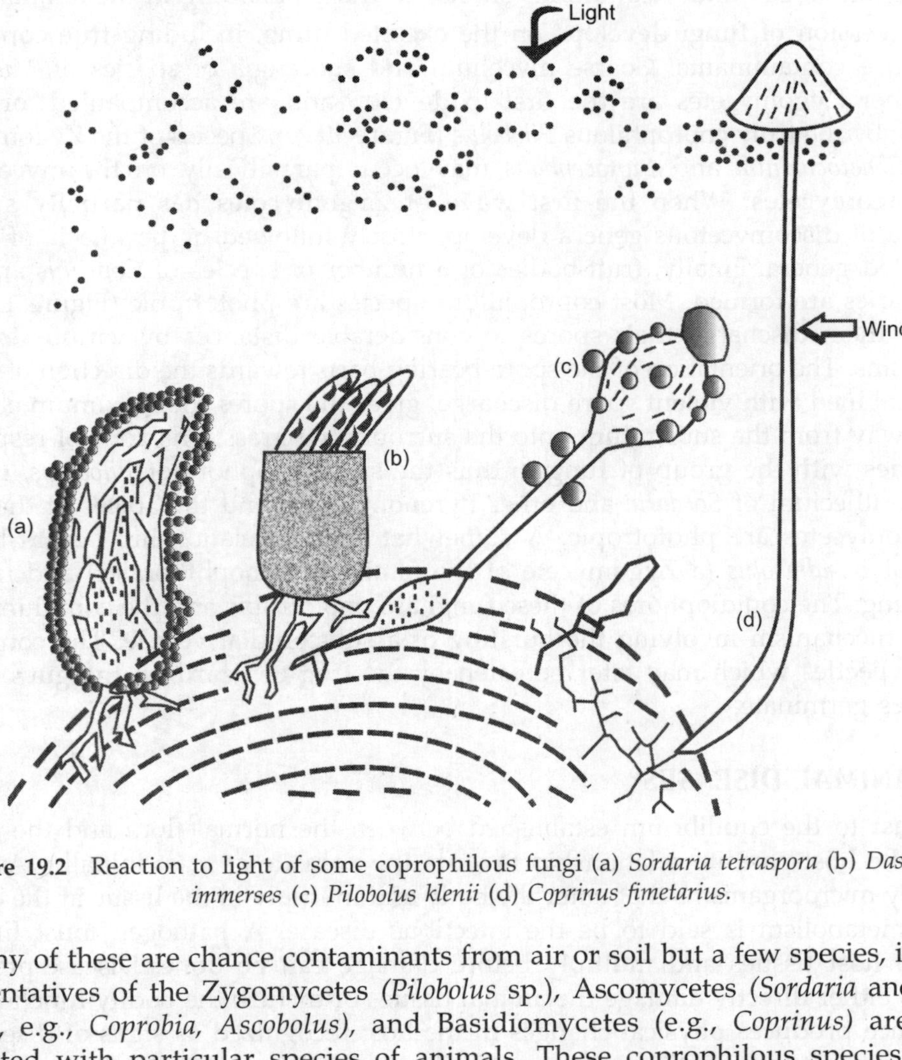

Figure 19.2 Reaction to light of some coprophilous fungi (a) *Sordaria tetraspora* (b) *Dasyobolus immerses* (c) *Pilobolus klenii* (d) *Coprinus fimetarius.*

Many of these are chance contaminants from air or soil but a few species, including representatives of the Zygomycetes *(Pilobolus* sp.), Ascomycetes *(Sordaria* and related genera, e.g., *Coprobia, Ascobolus),* and Basidiomycetes (e.g., *Coprinus)* are always associated with particular species of animals. These coprophilous species may be taxonomically unrelated but are biologically similar in their adaptation to the habitat. Their spores are ingested with the grass or other animal food and pass through the alimentary canal of the animal. Many of them germinate during passage through the gut, thus gaining a start which enables them to compete with faster growing species, such as *Mucor* spp., which reach the dung only after excretion. It is not known whether these germinated spores play any part in the breakdown of cellulose in the gut, but their part would certainly be a minor one compared with that of the more numerous rumen and intestinal cellulolytic bacteria. The spores of many coprophilous species are thick-walled or enveloped in mucilage, some contain black or purple pigments and many germinate only after exposure to relatively high temperatures or treatment with digestive juices, i.e., under conditions similar to those obtaining in the animal's gut.

A succession of fungi develops on the excreted dung, including true coprophiles and chance contaminants. Coarse mycelium and sporangia of species of *Mucor* and some other Zygomycetes are the first to develop and are accompanied, or closely followed, by the truly coprophilous *Pilobolus* (Figure 19.2c). Species of the Zygomycetous genera *Chaetocladium* and *Piptocephalis* may occur parasitically on the mycelium of other Zygomycetes. When the first wave of Zygomycetes has partially subsided apothecia of discomycetous genera develop, closely followed by perithecia of *Sordaria* and related genera. Finally, fruit-bodies of a number of species of *Coprinus* and some other agarics are formed. Most coprophilous species are phototropic (Figure 19.2) and many of them discharge their spores to considerable distances by various explosive mechanisms. The orientation of the spore-bearing parts towards the direction of greatest light, combined with violent spore discharge, gives the spores the maximum chance of falling away from the substratum onto the surrounding grass. The zone of response to light varies with the group of fungus; thus the sporangiophore of *Pilobolus,* the neck of the perithecium of *Sordaria* and other Pyrenomycetes, and the apices of the asci of the Discomycetes are phototropic. A somewhat similar relationship occurs between species of *Basidiobolus* (a Zygomycete of the family Entomophthorales) and frog and lizard dung. The conidiophores of these fungi are phototropic and the conidium is shot off by a mechanism involving the bursting of a sub-conidial vesicle. The conidia are eaten by beetles which may later be eaten by the frog or lizard, in the gut of which the spores germinate.

19.9 ANIMAL DISEASES

In contrast to the equilibrium established between the normal flora and the host, in disease the host-parasite relationship favours the parasite at least initially. A disease caused by microorganisms that enter the body and multiply in the tissue at the expense of host metabolism is said to be the infectious disease. A pathogen must first gain access to host tissues and multiply before damage can be done. Disease-producing parasites either directly damage the animal tissues or disturb the bodily function. Such disturbance produces physical changes in the host recognized as signs of disease and these range from minor tissue damage to death. Infectious diseases of animals may be caused by bacteria, fungi, protozoa, or viruses. A few parasitic bacteria and fungi are able to propagate in nature outside the animal host and assume a pathogenic role when introduced into the body. Historically, bacteria have been the cause of some of the most deadly diseases and widespread epidemics of human civilization. Smallpox and malaria, diseases caused by other microbes, have killed more humans than bacterial diseases, but diseases such as tuberculosis, typhus, plague, diphtheria, typhoid, cholera, dysentery and pneumonia have taken a large toll of humanity. At the beginning of the twentieth century, most bacterial diseases of humans and their etiologic agents have been identified, although variants continue to evolve and sometimes new ones emerge (Table 19.1).

Table 19.1 The discoveries of the main bacterial pathogens

1877	Anthrax	*Bacillus anthracis*	Koch, R.
1878	Suppuration	*Staphylococcus*	Koch, R.
1879	Gonorrhea	*Neisseria gonorrhoeae*	Neisser, A.L.S.
1880	Typhoid fever	*Salmonella typhi*	Eberth, C.J.
188J	Suppuration	*Streptococcus*	Ogston, A.
1882	Tuberculosis	*Mycobacterium tuberculosis*	Koch, R.
1883	Cholera	*Vibro cholera*	Koch, R.
1883	Diphtheria	*Corvnebacterium diphtheria*	Klebs, T.A.E.
1884	Tetanus	*Clostridium tetani*	Nicolaier, A.
1885	Diarrhea	*Escherichia coli*	Escherich, 'I'.
1886	Peneumonia	*Streptococcus pneumonia*	Fraenkel, A.
1887	Meningitis .	*Neisseria meningitides*	Weichselbaum, A.
1888	Food poisoning	*Salmonella enteritidis*	Gaertner, A.A.H.
1892	Gas gangrene	*Clostridium botulinum*	Welch, W.H.
1894	Plague	*Yersinia pestis*	Kitasato, S., Yersin, A.J.E. (independently)
1896	Botulism	*Clostridium botulinum*	Van Ermengem, E.M.P.
1898	Dysentery	*Shigella dysenteriae*	Shiga, K.
1900	Paratyphoid	*Salmonella paratyphi*	Schottmuller, H.
1903	Syphilis	*Treponema pallidum*	Schaudinn, F.R. and Hoffman, E
1906	Whooping cough	*Bordetella pertussis*	Border, J., and Gengou, O.

Aspergillus fumigatus is common1y present in soil and on plants or plant products such as grain or straw, but when the spores, in sufficient quantity, are inhaled by birds, or occasionally by other animals including man, the respiratory disease aspergillosis may result. In contrast most pathogens, whilst able to survive for varied periods of time outside the animal host, normally do not multiply away from the animal. The majority of pathogenic bacteria and fungi have been grown under laboratory conditions devised to resemble the host environment. Many grow relatively easily; others only with great difficulty because they require highly nutritious media and prolonged incubation. Thus *Mycobacterium johnei,* the causal pathogen of chronic enteritis in cattle and sheep, will grow only on egg media containing extracts of other mycobacteria and on primary isolation takes upwards of two months to produce visible growth. A few have not so far been grown in artificial culture or, if so, have not retained full virulence, e.g., *Mycobacterium leprae,* the causal pathogen of leprosy, and *Treponema pallidum,* the causal pathogen of syphilis. Several clinical tests are performed to help identify an organism on the basis of different biochemical reactions can be measured. Most important tests are summerized in Table 19.2 and details of biochemical test are described in Chapter 7.

Table 19.2 Important clinical diagnostic tests for bacteria

Carbohydrate fermentation	*Acid and/or gas during fermentative growth with sugars or sugar alcohols*	*Both medium with carbohydrate and phenol red as pH indicator; inverted tube for gas*	*Enteric bacteria differentiation (also several other genera or species separations with some individual sugars.*
Catalase	Enzyme decomposes hydrogen peroxide, H_2O_2	Add drop of H_2O_2 to dense culture and look for bubbles	*Bacillus* (+) from *Clostridium* (–); *Streptococcus* (–) from *Mircococcusi, Staphylococcus* (+)
Citrate utilization	Utilization of citrate as; sole carbon source, results in alkalinization of medium	Citrate medium with bromothymol blue as pH indicator, look for intense blue colour (alkaline pH)	*Klebsiella-Enterobacter* (+) from *Escherichia* (–), *Edwardsiella* (–) from *Salmonella* (+)
Coagulase	Enzyme causes clotting of blood plasma	Mix dense liquid suspension of bacteria with plasma, incubate, and 'look for sign of clot in few minutes	*Staphylococcus aureus* (+) from *S. epidermidis* (–)
Decarboxylases (lysine, ornithine arginine)	Decarboxylation of amino acid releases CO_2 and amine	Medium enriched with amino acid. Bromcresol purple pH indicator. Alkaline pH if enzyme action, indicator becomes purple	Aid in determining bacterial group among the enteric bacteria
β-Galactosidase (ONPG) Test	Orthonitrophenyl β-galactoside (ONPG) is an artificial substrate for the enzyme. When hydrolyzed, nitrophenol (yellow) is formed	Incubate heavy suspension of lysed culture with ONPG; look for yellow colour	*Citrobacter* and *Arizona* (+) from *Salmonella* (–). Identifying some *Shigella* and *Pseudomonas* species
Gelatin liquefaction	Many proteases hydrolyze gelatine and destroy the gel	Incubate in broth with 12% gelatine. Cool to check for gel formation. If gelatine hydrolyze, tube remains liquid upon cooling	To aid in identification of *Serratia, Pseudomonas, Flavobacterium, Clostridium*

Contd...

Contd...

Hydrogen sulphide (H_2S) production	H_2S produced by break down of sulphur amino acid or reduction of thiosulfate	H_2S detected in iron-rich medium from formation of black ferrous sulphide (many variants: Kliger's iron agar triple sugar iron agar, also	In enteric bacteria, to aid in identifying *Salmonella Arizona, Edwardsiella,* and *Proteus*
Indole test	Tryptophan from proteins converted to indole.	Detect indole in culture medium with dimethyl-aminobenzaldehyde (red colour)	To distinguish *Escherichia* (+) from *Klebsiella-Enterobacter* (–); *Edwardsiella* (+) from *Salmonella* (–)
Methyl red test	Mixed-acid fermenters produce sufficient acid to lower pH below 4.3	Glucose-broth medium. Add methyl red indicator to a sample after incubation	To differentiate *Escherichia* (+ culture red) from *Enterobacter* and *Klebsiella* usually –, culture yellow.
Nitrate	Nitrate as alternate electron acceptor, reduced to NO_2^- or N_2	Broth with nitrate. After incubation, detect nitrite with α-niphtylamine-sulfanilic acid (red colour). If negative, confirm that NO_2^- still present by adding zinc dust to reduce NO_3^- to NO_2^-. If no colour after zinc, then NO_2^-? N_2.	To aid in identification of enteric bacteria (usually +)
Oxidase test	Cytochrome *c* oxidizes artificial electron acceptor: tetramethyl (or dimethyl)-*p-phenylenediamine*	Broth or agar. Oxidase-positive colonies on agar can be detected by flooding plate with reagent and looking for blue or brown colonies	To separate *Neisseria* and *Moraxella* (+) from *Acinetobacter* (–). To separate enteric bacteria (all –) from *Pseudomonads* (+). To aid in identification of *Aeromonas* (+)
Oxidation-fermentation (O/F) test	Some organisms produce acid only when growing aerobically	Acid production in top part of sugar-containing culture tube; soft agar used to restrict mixing during incubation	To differentiate *Micrococcus* (aerobic acid production only) from *Staphylococcus* (acid anaerobically). To characterize *Pseudomonas* (aerobic acid production)
Phenylalanine deaminase test	Deamination produces phenylpyruvic acid, which is detected in a colourimetric test	Medium enriched in phenylalanine. After growth, add ferric chloride reagent, look for green colour	To characterize the genus *Proteus* and the *Providencia* group

Contd...

Contd...

Starch hydrolysis	Iodine-iodide gives blue colour with starch	Grow organism on plate containing starch. Flood plate with Gram's iodine look for clear zones around colonies	To identify typical starch hydrolyzcrs such as *Bacillus* spp.
Urease test	Urea (H_2N-CO-NH_2) split to $2NH_2+CO_2$	Medium with 2% urea and phenol red indicator. Ammonia release raises pH, intense pink-red colour	To distinguish *Klebsiella* (+) from *Escherichia* (–). To distinguish *Proteus* (+) from *Providencia* (-)
Voges-Proskauer test	Acetoin produced from sugar fermentation	Chemical test for acetoin using α-napthol	To separate *Klebsiella* and *Enterobacter* (+) from *Escherichia* (–). To characterize members of genus *Bacillus*.

The survival of a pathogenic microorganism depends on many interdependent factors. Under natural conditions, if an obligate parasite is to survive, it must be able to infect a susceptible host in which it can propagate. A high degree of infectivity is essential to the survival of those organisms which do not have a resistant phase. Other factors, such as the numbers of susceptible hosts available, also limit the survival of the pathogen. In island populations and other isolated communities, where the number of susceptible hosts is limited, certain diseases, such as measles, may flourish for a time then rapidly die out owing to this limiting factor. An organism which seriously damages its host may not survive, since if the host is killed the organism perishes unless it can first reach another susceptible individual. On the other hand, an organism of low virulence may fail to survive owing to inability to overcome the body defences of the host. Among the most successful of the common pathogens are those which have a high degree of infectivity but relatively low virulence under natural conditions, e.g., the virus of the common cold.

19.10 DISPERSAL OF PATHOGENIC MICROORGANISMS

19.10.1 Escape of Pathogens from Infected Host

The successful survival of a parasite depends on its ability to escape from one host and be transmitted to and infect another susceptible animal. Pathogenic microorganisms usually escape from the living animal in excreta from the alimentary tract, in secretions from mucous membranes of the respiratory tract, in organic debris such as desquamated skin and hair, and in pus from infected wounds. Coughing and sneezing assist the escape of organisms from the respiratory tract.

19.10.2 Transmission of Pathogens Between Hosts

This usually occurs between hosts of a single animal species but sometimes infection arises between hosts of different species. Examples of the latter are the infection of man by dermatophytic fungi from infected domestic animals, by *Pasteurella multocida* from a cat bite or scratch, or infection of a dog with the tubercle bacillus derived from its master.

Certain diseases are contracted by direct association with an infected host. This may occur by inhalation of droplets from an individual with a respiratory disease (e.g., tuberculosis, common cold virus), by direct contact with an infected host either venereally (e.g., syphilis and gonorrhoea in man or vibriosis in cattle) or generally (e.g., ringworm infection).

Contamination of specific sites may result in infection. For instance, infections of the human external ear may be caused by *Aspergillus niger* and *A. fumigatus;* following mastoid operation the resultant cavity, in which debris collects due to impaired drainage, frequently becomes infected by any of the following: *Candida* spp., *Aspergillus niger, A. fumigatus, A. flavus,* and *A. terreus.* Bacteria also often infect these sites.

Many diseases are contracted indirectly by ingestion of food contaminated at source from an infected animal or contaminated subsequently by polluted water or by excreta from an infected patient or healthy carrier (e.g., food poisoning, typhoid, tuberculosis, brucellosis, etc.). Horses may be indirectly infected with anthrax by feeding on hay grown the previous summer on anthrax contaminated soil, since the spores survive throughout the winter months. Another method of indirect infection is by penetrating the skin with a contaminated instrument, as often occurs with tetanus, or by contamination of an abrasion, as in anthrax in man.

Some pathogens are transmitted to a new host by arthropod vectors. This is the mode of transmission of the plague bacillus and is very common with many virus, protozoan, and rickettsial infections. The rickettsiae of typhus fever, transmitted by lice, infect the susceptible host by insect excreta entering either by the bite of the vector or through an abrasion caused by the patient's scratching.

The length of time during which pathogenic organisms survive outside their hosts varies enormously from species to species. Some, like the organisms responsible for human venereal diseases *(Neisseria gonorrhoeae* and *Treponema pallidum)* which are mainly transmitted during coitus, rapidly die outside the body. Many pathogens of the respiratory tract are more hardy but are favoured by fairly rapid transmission in cough spray, although infection may persist in contaminated dust and may be inhaled later; this has often been shown to occur in cross-infection in fever hospital wards. In general, pathogens of the intestinal tract have much greater resistance and can survive for many days, or even months, outside the body. Spore-bearing pathogens, such as the causal organisms of tetanus, gas-gangrene, and anthrax, survive as spores in a fully virulent condition in manured soils for years.

19.10.3 Routes of Infection

Many fully virulent microorganisms are able to cause disease in the animal body only if they enter by specific routes. *Clostridium tetani*, the cause of tetanus, is frequently found without harmful effects in the intestinal tract, especially of domesticated animals (horse, pigs, etc.), to which it gains access via the mouth. It assumes a pathogenic role usually if introduced into a deep wound, either following surgical manipulation or accidental penetration, in which anaerobic conditions essential for its growth are found. Many alimentary infections, on the other hand, are unable to establish themselves when introduced into abrasions but readily infect via the mouth. The same organism may show varying degrees of virulence when introduced into the same host by different routes; thus anthrax bacilli and staphylococci are more virulent when they enter through skin abrasions than when ingested; the tubercle bacillus is more virulent when it enters by the respiratory than by the alimentary route, probably owing to the differences in the defence mechanisms.

Closely linked with the portal of entry is the affinity of many organisms for particular tissues, e.g., the pneumococcus for lung tissue, the leprosy bacillus for skin tissue, the meningococcus for the meninges of the brain, leptospirae for the liver and kidney, and *Vibrio cholerae* for the intestine.

19.11 PATHOGENICITY OF MICROORGANISMS

Since Koch's observation in 1878 on the characteristic symptoms and anatomical lesions in animals caused by six morphologically distinct microorganisms, it has been well established that each pathogenic organism produces a specific range of disease. Certain pathogens produce the same symptoms in a variety of animals hosts, e.g., paralysis resulting from the toxin of the tetanus bacillus. Others are able to produce different forms of disease in the same host. For instance, *Streptococcus pyogenes* in man can cause tonsillitis; scarlet fever, erysipelas, or wound infections. Some organisms are highly pathogenic to certain hosts and non-pathogenic to others, as, for instance, *Corynebacterium diphtheriae* which is pathogenic naturally to man and experimentally to the guinea pig but to which the rat is completely resistant. Other pathogens produce disease of varying severity in different animal hosts, as illustrated by *Bacillus anthracis* which produces an acute, usually fatal septicaemic disease in cattle, sheep, and horses, inflammatory oedema of the pharynx in pigs and dogs, and a malignant pustule at the site of entry through the skin in man. Only those strains of known pathogenic species able to overcome the resistance of the host and propagate in the animal body are virulent; those which cannot do this are termed avirulent. There is, however, no clear distinction between the two, since virulence is a relative term dependent upon many factors which can be altered under a variety of environmental influences. The virulence of many bacterial and viral pathogens may be diminished by such procedures as artificial culture, culture at temperatures above the optimum, or passage through animals in which the disease is not normally found. These methods are used to produce attenuated vaccines for artificial immunization. Frequently loss of virulence is associated with a change from the smooth colonial form to the 'rough'. Selection of smooth virulent colonies from a rough avirulent culture is not readily accomplished, but in certain instances can be achieved simply by animal passage when the few virulent organisms present in the culture will assume a pathogenic role and later be isolated from pathological material in pure culture. Rough avirulent pneumococci can be transformed into a virulent form by growing them in the presence of killed suspensions of whole, smooth pneumococci, or simply in the presence of the genetically important transforming substance extracted from smooth pneumococci. Griffith in 1928, reported that injected mice with live non-capsulated and heat killed capsulated pneumococci; neither of these proved fatal when injected separately, but when both were injected together, some of the mice, died and living capsulated bacteria were recovered from their blood. This same conversion of non-capsulated to capsulated pneumococci was later demonstrated *in vitro,* and it was found that a cell-free extract of the capsulated strain could carry out the transformation (Figure 14.15). Griffith at first supposed that the agent *(transforming principle)* that transformed the non-capsulated strain was the capsular polysaccharide of the killed culture, but later work has identified it unequivocally with DNA. Transformation has since been demonstrated in many bacteria, although work has been concentrated on the pneumococcus, *Haemophilus influenzae,* and *Bacillus subtilis;* it is also found that the transforming principle can carry any donor gene.

Virulent toxin producing strains of the diphtheria bacillus have been isolated from cultures of avirulent non-toxin-producing strains after exposure to particular bacteriophages, indicating that in this species there is an association between lysogeny and virulence; however, not all lysogenic strains of the diphtheria bacillus are virulent.

The properties which distinguish a virulent from an avirulent bacterium are twofold, namely, 'invasiveness', i.e., the ability of the organism to establish itself and propagate in the host tissues, and 'toxicity', i.e., the ability of the organisms to form toxins which destroy or damage tissues or impair their physiological functions.

19.11.1 Invasiveness

Apart from those organisms which are introduced into the body through wounds, most microorganisms initially lodge on the skin or mucous membranes through which they pass in order to invade the underlying tissues. Most mucous membranes, particularly those of the nasopharynx and intestine, are liberally supplied with lymphatic channels near to the surface. Organisms impinging on the membranes of the nasopharynx are constantly being removed by ciliated epithelial cells or engulfed by wandering phagocytic cells which carry them away to the lymph channels. If the organisms escape these defences and multiply, infection may be initiated. Local multiplication of organisms such as *Staphylococcus aureus* may occur in crypts of the skin, in hair follicles, or sebaceous glands, thereby setting up local damage followed by penetration into the sub-epithelial tissues. Skin secretes substances which are bactericidal and these must be overcome by microorganisms before they can multiply or reach a penetrable surface. Many pathogens are thus destroyed before penetration is achieved. After penetration the subsequent course of the infection varies with the type of organism and the nature of the internal defences of the host. The infection may be confined to local abscess formation, as is common with staphylococci, or the organism may penetrate through subcutaneous tissues to give rise to spreading skin lesions, as in streptococcal skin infections, e.g., erysipelas in man caused by *Streptococcus pyogenes,* or a clinically similar condition in swine caused by *Erysipelothrix rhusiopathiae.* Some pathogens after local multiplication spread via the lymph channels to the lymph nodes, which rapidly become enlarged owing to the stimulation of defence mechanisms in these glands. These may check the spread of infection; otherwise organisms continue to spread either in the lymph or the bloodstream. Organisms entering the bloodstream are rapidly circulated to all parts of the body (bacteraemia) and they may set up local foci of infection in various organs from which pus may escape into the bloodstream (pyaemic spread). Progressive multiplication of the organism in the bloodstream is termed septicaemia.

19.11.2 Toxicity

The ability of different pathogens to invade the body varies considerably. Many with very little power to invade are, however, extremely harmful through their ability to produce highly potent toxins. The tetanus bacillus, for instance, has no invasive powers but, when introduced into a deep wound, multiplies at the local site and generates a powerful toxin which has a specific affinity for nerve cells and brings about destruction of those parts of the central nervous system which it reaches, resulting in symptoms of paralysis. The diphtheria bacillus does not invade beyond the tissues of the throat, nasopharynx, and bronchi but its toxin, produced *in situ,* is absorbed and circulated around the body in the bloodstream to produce the various generalized symptoms of diphtheria, in particular, damage to heart muscle tissue. Many strongly invasive organisms, such as streptococci and the gas-gangrene anaerobic sporing bacilli, are producers of potent toxins which result in gross tissue destruction. The various bacterial toxins are divided into two classes, namely, endotoxins and exotoxins.

19.11.2.1 Endotoxins

Endotoxins are structural components of bacterial cells released after death and disintegration of the cell, and are then found in bacteria-free filtrates of cultures. In the body endotoxin release occurs when the bacteria are phagocytosed or otherwise lysed by natural components of the body fluids (e.g., complement in the presence of antibody. Endotoxins are produced mainly by Gram-negative pathogens and non-pathogens (e.g., species of *Salmonella, Escherichia, Proteus, Shigella, Brucella,* and *Neisseria).* Irrespective of the organism from which the endotoxin came the symptoms produced in the host are similar and include fever, diarrhoea, local haemorrhage, especially in the visceral tissues, fall of blood pressure and shock. The severity of the effect depends on the dose of endotoxin and the sensitivity of the animal to it; sensitivity is variable. Endotoxins have been chemically extracted from a number of bacterial species and have been found to consist of complex molecules of protein, lipid, and polysaccharide, corresponding to the O-somatic antigens of the bacterial cell walls. Serum antibodies are produced following immunization with bacterial endotoxins; these combine with the O-somatic antigens in the whole cell producing antibacterial effects (e.g., bacteriolysis) but the antibodies do not completely neutralize the toxic action of the endotoxins when these are released from the cell. Endotoxins are more heat stable than exotoxins but do not act at such high dilution. They produce their effects on the host by damaging small blood vessels by a hypersensitive type of reaction, the sensitivity of the patient arising from past exposure to the same antigens which induced the state of hypersensitivity.

19.11.2.2 Exotoxins

In contrast to endotoxins, they are readily separable from the living producer cells and may be obtained by growing the bacteria in a suitable liquid medium and separating the toxin from the cells by centrifugation or filtration. Those which have been isolated as pure chemical substances (e.g., tetanus and diphtheria toxins) are simple proteins. They are frequently highly toxic for the host, the lethal dose of pure crystalline tetanus and botulinus toxins for mice being of the order of 0.0000001 mg. They act as enzymes and as such are generally unstable to chemicals (e.g., formalin slowly converts them to toxoid), to oxygen (e.g., oxygen-sensitive haemolysins), and to heat. Sensitivity to heat varies widely; diphtheria toxin is sensitive to 65°C, *Clostridium perfringens* α-toxin to 100°C, and staphylococcal enterotoxin and the toxin of botulinus to higher temperatures. Some exotoxins exhibit specific affinity for tissues; for instance, tetanus bacilli and other bacteria produce neurotoxins which destroy cells of the central nervous system, the symptoms of paralysis being produced several days after the toxin is released into the body. Other toxins are rapid in action and are named according to the effects they produce; haemolysins, as produced by streptococci, lyse red blood cells; leucocidins, e.g., the leucotoxin of *Staphylococcus aureus*, destroy leucocytes, the wandering phagocytes of the blood stream (Figure 19.3).

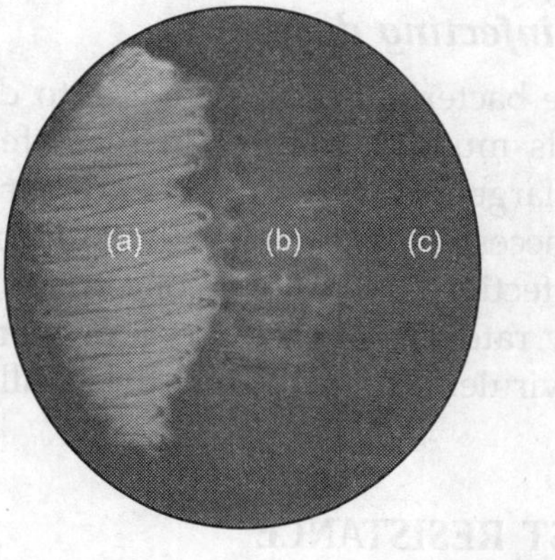

Figure 19.3 (a) β-hemolysis (b) α-hemolysis (c) γ-hemolysis.

The necrotoxins exhibit more general tissue destruction, such as the general destruction of tissue by the gas-gangrene group of Clostridia. The same toxin may produce a number of effects. For instance, the α-toxin of *Staphylococcus aureus* acts as a haemolysin, a leucocidin, a necrotoxin, and a lethal toxin. In most instances the mechanism of action is unknown. The α-toxin of *Clostridium perfringens* acts as a lecithinase, a lecithin-splitting enzyme which can attack lecithin-containing cell membranes such as those of red blood cells. The toxin of diphtheria appears to be the protein moiety of the bacterial cytochrome *b* enzyme and may act by interfering with synthesis of cytochrome *b* in the host's tissues. Some organisms are able to produce toxins resulting in a tissue rash (e.g., *Streptococcus pyogenes,* which produces an erythrogenic toxin causing the rash of scarlet fever). Specific neutralizing antibodies (antitoxins) can be demonstrated in the bloodstream of an animal following infection with an exotoxin-producing bacterium, or these may be stimulated by artificially immunizing the animal with a toxoid.

Recent advances in rapid diagnostics have made it possible to isolate, identify and determine the antibiotic sensitivity of most routinely encountered pathogenic bacteria within 48 hours of sampling. The physician on the careful examination of the patient, may suspect that an infectious disease is present. Samples of the infected tissues or fluids are then collected for microbiological, immunological and molecular biological analysis (Figure 19.4).

19.11.2.3 *Other cell products*

Some bacterial cells produce substances which influence the course of an infection in ways other than by toxic action. For instance, pathogenic staphylococci produce the enzyme coagulase which induces the coagulation of blood plasma, thereby setting up a barrier to the circulating body defences and assisting the organism to establish itself in the host body. In contrast, streptococci produce the enzymes streptokinase, which dissolves fibrin clots, and hyaluronidase, which breaks down hyaluronic acid (a mucopolysaccharide which holds together the mesodermal structures of subcutaneous tissues) thereby facilitating spread of the organism from the primary focus.

19.11.2.4 *Size of the infecting dose*

It is very rare for a single bacterial cell to give rise to disease. More usually a critical minimum number of cells must be present in the infecting dose before invasion of the host can occur. Very large numbers of spores of certain pathogenic fungi, such as *Aspergillus fumigatus,* are necessary to establish an infection. Even when microorganisms are able to set up an infection, the rate of onset of symptoms, the severity of the disease, and the mortality rate, are often directly proportional to the size of the initial infecting dose. The more virulent the pathogen, the smaller the minimum dose required to produce infection.

19.12 GENERAL HOST RESISTANCE

The health of an individual can be influenced by many environmental factors. Before an organism can reach the body tissues, the epithelial coverings, e.g., skin and mucous membranes must be penetrated. In a healthy system, when these natural barriers are intact, many pathogenic microorganisms are removed before they can invade the body. Any factor able to damage these natural barriers (e.g., minor abrasions of the skin,

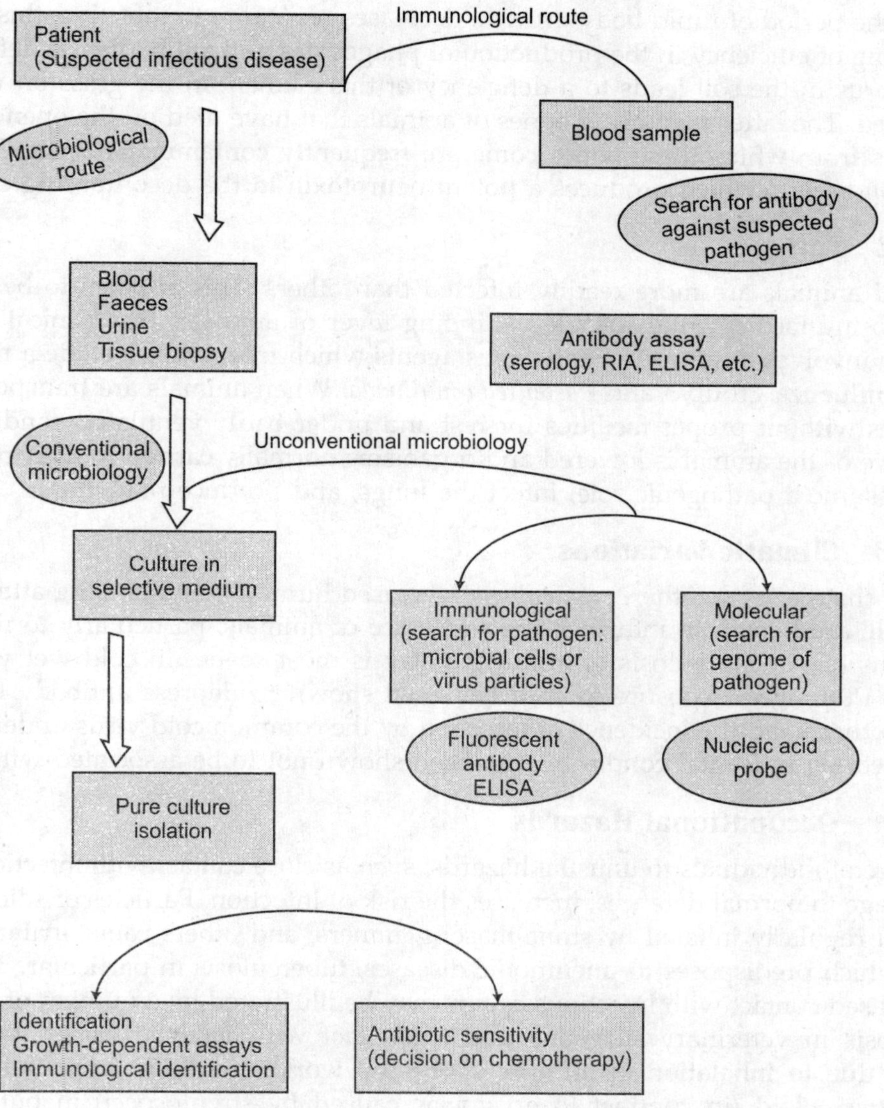

Figure 19.4 Protocol used for investigation.

or burns by acids or heat) deprives the system of this protection. Several diseases of childhood rarely occur in adult life, the early occurrence of the disease establishing a lasting immunity in the individual. This acquired immunity is the result of either or both of two mechanisms of defence against microbial infection, namely, antibody-dependent or humoral immunity and cell-mediated immunity.

19.12.1 Diet

The influence of diet on resistance is a very complex problem interconnected with many other factors. The deficiency of vitamins A and C, and of proteins, particularly during the period of rapid body growth, decreases resistance to infection; this is due to a lowering of efficiency in the production of phagocytes and antibodies. A deficiency of phosphorus in the soil leads to a deficiency of this element in the grass on which the cattle feed. The latter then chew bones of animals that have died on the open field. The carcasses from which these bones come are frequently contaminated with *Clostridium botulinum* type C which produces a potent neurotoxin in the decomposing carcass.

19.12.2 Fatigue

Fatigued animals are more readily infected than others. This appears to be the main predisposing factor in the so-called shipping fever of animals. This clinical condition possibly involves a number of infectious agents which frequently include a myxovirus of parainfluenza group 3 and *Pasteurella multodda.* When animals are transported long distances without proper facilities for rest and under badly ventilated conditions, the resistance of the animal is lowered and organisms normally carried in their respiratory tracts assume a pathogenic role, infect the lungs, and produce pneumonia.

19.12.3 Climatic Variations

Sudden changes of weather, particularly increased humidity and chilling after a warm, dry spell, are known to influence the resistance of animals, particularly to respiratory infections (e.g., aspergillosis of young poultry is most severe in cold wet weather in spring). Prolonged exposure to cold has been shown to depress antibody formation. On the other hand the incidence of infection by the common cold virus under carefully controlled experimental conditions has been shown not to be associated with chilling.

19.12.4 Occupational Hazards

Exposure of individuals to unusual hazards, such as close contact with infectious agents or damage to normal defences, increases the risk of infection. Particles of silica or other material regularly inhaled by stone-masons, miners, and others cause irritation of the lungs which predisposes to pneumonic diseases, tuberculosis in particular. The effects of increased contact with infectious agents may be illustrated by a number of examples; brucellosis in veterinary surgeons due to contact with infected cattle; woolsorter's disease due to inhalation of anthrax spores by workers sorting untreated imported wool; erysipeloid (in contrast to erysipelas caused by streptococci) in butchers and abattoir workers as a result of infection with *Erysipelas rhusiopathiae* through abrasions caused by contaminated pig bones; Q-fever by inhalation of dust (contaminated with rickettsias) by workers in close contact with infected cattle.

19.12.5 Irradiation

The effect of radiation on the resistance is dependent on the degree of exposure. Small doses of X-rays have found to increase resistance, probably owing to a stimulation of the immunity producing centres of the body. On the other hand, the large doses of ionizing radiations, X-rays or atomic radiations, decrease the resistance of the host by damaging the lymphoid tissue so that it can no longer produce antibody in response to infection. Radiation may also damage bone marrow tissue, thereby impairing the production of red blood cells and leucocytes in the adult mammal.

19.12.6 Presence of Other Parasites

Damage caused to the host by one organism may increase the risk of infection to a second, as in black disease of sheep, in which infection of the liver with *Clostridium oedematiens*, a blood-borne infection from the gut, occurs if liver flukes have previously caused necrosis of the liver, thereby producing suitable anaerobic conditions for the rapid propagation of the *Clostridium*.

19.12.7 Age and Sex

Susceptibility to infection varies considerably with the age. Newborns are frequently very susceptible to organisms, such as those of the normal gut flora, which are not usually pathogenic to older one. Infection often occurs during the critical period whilst the body is acquiring its gut flora and before it has built up a resistance against these organisms. Thus, young children succumb to infection by specific strains of *Escherichia coli;* young lambs frequently, become infected with *Clostridium perfringens* from the soil and develop lamb dysentery. Newborn may, however, be more resistant to certain infections than are slightly older ones if they have received a temporary immunity conferred on them from the mother. Once this temporary immunity is lost, young animals are frequently more susceptible to certain infections than are mature ones. For instance, babies in their first year of life are usually immune to measles because of passive immunity transferred from the mother but in the next few years are highly susceptible. *Brucella abortus* is unable to establish itself in calves but readily invades the uterus and udder in the cow and the testicles in the bull due to the sex hormones or other growth promoting substances, present only in the adult animal, are essential for the establishment of this pathogen.

19.12.8 Barriers

19.12.8.1 Physical barriers

Skin

No microbe can invade skin when intact. The low pH of skin due to lactic and fatty acids is not congenial for the growth of microbes. The epidermis forms thin outer layer with tightly packed epidermal cells and keratin (water proofing) and is completely renewed every 15-30 days. The dermis is thicker inner layer that contains sebaceous glands associated with hair follicles. The sebum produced consists of lactic and fatty acids.

Mucous membranes

Membrane lining on gastrointestinal, urogenital and respiratory tracts consists of ciliated epithelial cells and collectively represent a surface area equivalent to that of a basket ball court. Various secretions of body such as saliva, tears and mucous secretions along with cilia of epithelial cells play an important role in outward movement of microbial pathogens that gained entry through any of these entry points.

19.12.8.2 Physiologic barriers

Temperature

Normal body temperature inhibits growth of most microorganisms and elevated body temperature (fever) can have a direct effect on pathogenic microorganisms.

pH

Acidic pH of stomach, skin, and vagina inhibits microbial colonization and growth.

19.12.8.3 Chemical factors

A number of chemical factors also participate in defence against day to day encountered pathogens and include: fatty acids, lactic acid, lysozyme (able to cleave the petidoglycan layer of the bacterial cell wall), cryptidins and α-defensins (produced in base of crypts of small intestine – damage cell membranes) β-defensins (produced within skin, respiratory tract – also damages cell membranes) surfactant proteins A and D (present in lungs – function as opsonins which enhance the efficiency of phagocytosis), interferons (group of proteins produced by cells following viral infection, which then bind to nearby cells and ignite mechanisms which inhibit viral replication), complement (a group of serum proteins that circulate in an inactive proenzyme state which can be activated by a variety of specific and nonspecific molecules resulting in destruction of pathogenic organisms).

19.12.8.4 Endocytic and phagocytic barriers

Endocytosis

It is a process by which pathogen derived soluble antigens or macromolecules present in the extracellular tissue fluids get internalized by cells. Endocytosis takes place through pinocytosis or receptor-mediated endocytosis.

Phagocytosis

Ingestion of particulate material including whole pathogenic microorganisms is referred to as phagocytosis. The plasma membrane expands around the particulate material to form large vesicles called phagosomes (10-20 times larger than endosome). Once particulate matter is ingested into phagosomes, the phagosomes fuse with lysosomes and the ingested material is then digested by a process similar to that seen in endocytosis. Only specialized cells are capable of phagocytosis, whereas endocytosis is carried out by virtually all cells.

Macrophages and neutrophils are considered 'professional phagocytes' while fibroblasts and epithelial cells are known as "non-professional" phagocytes. The latter cells can be induced to become phagocytic under certain circumstances such as during intense inflammation.

19.13 GENERAL CELLULAR RESISTANCE

When bacterial cells pass through the natural barriers of the primary defences of the body they are treated in the same way as inert particles by the cellular defences.

Specialized cells, known as phagocytes, rapidly engulf the organisms and unless killed by the toxins of the bacteria, begin to digest and destroy them. These phagocytic cells may be fixed in the liver, spleen, and lymph glands and include macrophages or histiocytes; other phagocytes may be circulating in the blood and consist mainly of large mononuclear leucocytes and polymorphonuclear leucocytes, the so-called white cells of the blood.

Many virulent organisms kill phagocytes, resulting in the formation of pus, and continue to multiply. The subsequent course of the infection then depends on whether the system is stimulated to produce sufficient phagocytes to overcome the rapidly multiplying bacteria. In this the cellular response is aided by the other forms of immunity and any artificial assistance that may be provided by chemotherapeutic treatment. Frequently virulent organisms are able to resist phagocytosis by means of special properties of their capsule which give them some degree of protection.

19.14 SPECIFIC IMMUNE RESPONSES

Humoral immunity depends on the formation of substances termed antibodies which react with and neutralize the infective properties of the invading microorganism. In addition to its importance as a mechanism of immunity, the specificity of antibody reactions has provided a means of demonstrating differences and similarities between substances capable of stimulating antibody production and has led to the emergence of the science of immunology. This is widely used in the identification and classification of microorganisms, the diagnosis of disease, the immunochemical analysis of components of microorganisms and other organic substances (Table 19.3).

Table 19.3 Some clinical immunological procedures for diagnosis of infectious diseases

Pathogen of disease	*Antigen*	*Serological procedure**
Streptococcus (group A) *Neisseria meningitidis*	Streptolosin O (exotoxin) Dnase (extracellular protein) Capsular polysaccharide	Neutrilization hemolosis Neutrilization enzyme Passive hemagglutination (N. *meningitidis* polyasaccharide adsorbed to red cells)
Salmonella *Vibrio cholerae*	*N. meningitidis* cells O or H antigen O antigen	Indirect fluorescent antibody Agglutination (widal test) ELISA Agglutination Bactericidal (in presence of complement) ELISA
Brucella *Corynebacterium diphtheria* *Mycobacterium tuberculosis*	Cell wall antigen Toxin Tuberculin (Partially Purified bacterial Proteins, PPD)	Agglutination ELISA Skin test (Schick test) Skin test (Tuberculin test)
Syphilis *Treponema pallidum*)	Cardiolipin-lecithin-cholesterol	ELISA Flocculation [Venereal Disease Research laboratory (VDRL) test]

Contd...

Contd...

Rickettsial diseases (Q fever, typhus, Rocky Mountain Spotted fever)	*T. Pallidum* antigens *T. Pallidum* cells Killed rickettsial cells	ELISA Indirect fluorescent autibody (FTA) Complement-based assay or cell agglutination test
Influenza virus	Influenza virus suspensions Nasopharynx cells containing influenza virus Human Immunodeficiency Virus (HIV)	ELISA Complement-based assay Immunofluorescence FLISA Western blot

*Excent for the skin tests. the serum of the natient is assayed for antibody against the specific antigen by the methods shown.

19.14.1 Antigens (Ags)

An **antigen** is any molecule that stimulates an immune response. Antigens are molecules on the surface of pathogens, viruses, fungi, or bacteria. Nonliving substances such as bacterial toxins, chemicals, drugs, and foreign particles may also function as antigens. Most antigens are proteins or polysaccharides, though small molecules coupled to carrier proteins (haptens) can also be antigenic. Certain polysaccharides and polypeptides may also be antigenic. Other substances, such as lipids and nucleic acids, are antigenic when conjugated with proteins, the latter possibly operating by retaining the antigenic determinants in the body for a sufficiently long time to elicit an antibody response. Antigens have a large molecular weight of at least 5000, i.e., they are too large to be disposed of by the host's normal physiological functions; they can be soluble or particulate and are essentially foreign to the host. Microorganisms and many of their products are therefore strongly antigenic. The segment of an antigenic molecule to which its cognate antibody binds is termed an antigenic determinant or epitope.

Antigens may be classified as immunogens if they stimulate an immune response, as haptens if they induce an immune response only when coupled to an immunogenic carrier molecule, or as toleragens if they cause cells of the immune system to become selectively unresponsive to re-exposure to the same molecules. Epitope is the smallest part of an antigen that is "seen" by somatically generated B and T cell receptors. Antigens contain one or more epitopes or antigenic determinants, the basic units recognized by immune receptors and antigenic specificity depends on epitopes (Figure 19.5).

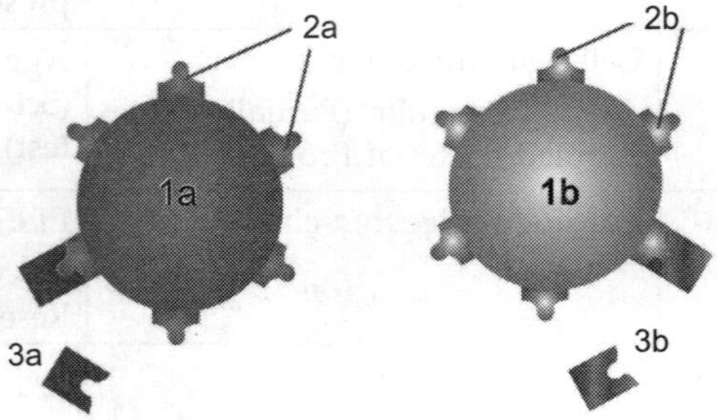

Figure 19.5 Antigens and antigenic determinant.

19.14.2 Antibodies (Abs)

An antigen may enter the animal host and gain access to the tissues during the course of a natural microbial infection or it may be introduced artificially. Under this stimulus antibodies are formed principally by cells of the lymphoid tissue, especially plasma cells, which are distributed throughout the reticulo-endothelial system of the animal body. Plasma cells are not normal body cells but develop from small lymphocytes under an antigenic stimulus. Antibodies are serum globulins having low solubility in water and are completely precipitated by half-saturated ammonium sulphate (Table 19.5). By the technique of electrophoresis (i.e., under the influence of an electric field), the negatively charged serum globulins migrate to the positive cathode. The slowest diffusing fraction of an immune serum, the gamma globulin, contains most antibody (Table 19.4 and Table 19.5). These immunoglobulins can be separated by chemical and physical means into at least five major components, IgG, IgA, IgM,IgD and IgE (Table 19.6).

Table 19.4 Five different classes of immunoglobulins.

Ig class	*II-Chain*
JgG	Gamma (γ)
IgA	Alpha (α)
IgM	mu (μ)
1gD	delta (δ)
IgE	epsilon (ε)

Table 19.5 The classification of the human immunoglobulins

Class	*Serum mg/mL*	*Mol. Wt*	*Light chain*	*Heavy chain*	*Structure**
IgG	12	150 000	k or λ	γ	$k_2\gamma_2$
IgA	3	>300 000	k or λ	α	$(k_2\alpha_2)_n$
IgM	1	950 000	k or λ	μ	$(k_2\mu_2)_s$
IgD	0.1	160 000	k or λ	δ	$k_2\delta_2$
IgE	0.001	200 000	k or λ	ε	$k_2\varepsilon_2$

*Alternative structures are possible with the k chains replaced by λ chains

Table 19.6 Physico-chemical properties of human immunoglobulins

Properly	*IgG**	*IgM*	*19A***	*IgD*	*IgE*
Heavy chain	γ	μ	α_1	δ	ε
Mean serum (mg/mL)	9	1.5	3.0	0.03	0.00005
% total antibody	80-85	5-10	5-15	<1	0.002-0.05
Valency	2	5 (10)	2-4	2	2

Contd...

Contd...

Mass of heavy chain (kDa)	51	65	56	70	72
Total molecular mass (kDa)	146	970	160***	184	188
Placental transfer	+	–	–	–	–
Half-life in serum (days)****	23	5	6	–	–
Complement activation Classical pathway	++	+++	–		
Alternative pathway	–	–	+	–	–
Induction of mast cell degranulation	–	–	–	–	+
Major characteristics	Abundant Ig in body fluids; neutralizes toxins, opsonizes bacteria, activates complement, maternal antibody	First to appear after antigen stimulation; very effective agglutination; expressed as membrane bound antibody on B cells.	Secretory antibody: protects external surfaces	Present in B-cell surface; B-cell recognition of antigen	Anaphylactic mediating antibody; resistance to helminthes
% carbohydrate	3	7-10	7	12	11

*Properties of lgG subclass I; ** Properties of IgA subclass 1; ***sIgA = 360-400 kDa:

****Time required for half of the antibodies to disappear.

Each is antigenically distinct and the dominant component in the serum at any time depends upon a number of factors. For instance, IgM is chiefly formed in young animals and some lower animals, such as Elasmobranchs. (More primitive vertebrates and invertebrates do not possess the capacity to produce antibody.) A primary antigenic stimulus results in IgM formation whilst subsequent stimulation by the same antigens produces IgG. Only IgG can pass from mother to foetus by the placental circulation and thus only IgG is found in the serum of the new-born. Particulate antigens tend to stimulate IgM production. The properties of various class of immunoglobulin are described below

IgG

IgG is a monomer and is the most abundant immunoglobulin and represents 75% of serum immunoglobulins. It is the only class of Ig that crosses the placenta except for IgG2. It can fix complement (except IgG4) and binds to Fc receptors (except IgG2 and IgG4) on blood cells including phagocytic cells. Hence, IgG is considered 'opsonin'.

IgM

IgM is the third most common serum Ig and exists generally as a pentamer (19S immunoglobulin) but it can also exist as a monomer. The monomeric form is present on the surface of B-cells and mediates antigen recognition. The membrane bound IgM contains an additional stretch of 20 amino acids at C-terminal end and is noncovalently associated with two additional membrane bound proteins called Ig-alpha and Ig-beta with long cytoplasmic tails. These additional proteins function as signal transducing molecules upon antigen interaction. IgM is the first Ig to be made by the foetus and the first Ig to be produced in primary immune response. The pentameric form contains an additional chain called 'J' chain which helps in polymerization of monomeric IgM. Single molecules of pentameric IgM can fix complement whereas a minimum of two adjacent IgGs are required for complement fixation to take place. It is an efficient agglutinator.

IgA

IgA is the second most abundant serum immunoglobulin and exists as a monomer in serum and as a dimer in secretions. IgA is the major class of Ig in secretions — tears, saliva and colostrums plays an important role in local (mucosal) immunity. The secretory IgA is associated with two other peptides, viz., a J chain and a secretory piece or T piece. The secretory piece helps IgA to be transported across mucosa and also protects it from degradation in the secretions. Normally, IgA does not fix complement, unless aggregated.

IgD

IgD is primarily found on B cell surfaces and found in low levels in serum. The function of serum IgD is uncertain while membrane bound IgD functions as a receptor. IgD does not bind complement. The membrane bound IgD, like IgM, contains a hydrophobic stretch of amino acids and is associated with signal transducing Ig-alpha and Ig-beta chains.

IgE

IgE exists as a monomer and has an extra domain in the constant region. It is present in very low concentrations in serum as it can bind effectively to Fc receptors on basophils and mast cells even before interacting with antigen. It is the mediator of Type I hypersensitivity (allergy) reactions and is elevated significantly in allergy and parasitic helminth infections. IgE does not fix complement. But eosinophils have Fc receptors for IgE and binding of eosinophils to IgE-coated helminths results in killing of the parasite by a process known as antibody dependent cell mediated cytotoxicity (ADCC). Details about the physico-chemical properties of human immunoglobulin are described in Table 19.6.

19.14.3 General Functions

19.14.3.1 Antigen binding

Antibody binds to a specific antigenic determinant that is complementary to the CDR region. The primary function of immunoglobulins is antigen binding which may result in neutralization or elimination from the host.

19.14.3.2 Effector functions

Frequently the binding of an antibody to an antigen may not result in elimination; rather, the antigen-antibody complexes may bind to phagocytic cells via Fc receptors and get phagocytosed. Antibody is thus considered 'opsonin'. The effector functions of immunoglobulins include:

1. ***Complement activation*** Antibody with bound antigen can fix and activate Complement proteins by 'Classical pathway'. The Complement activation leads to lysis or breakdown of bacterial and other pathogens and release of biologically active substances that mediate inflammation.
2. ***Binding to various cell types*** Antibodies may bind to phagocytic cells, lymphocytes, platelets, mast cells, and basophils which have Fc receptors. The binding of free antibodies or antigen-antibody complexes to these cell types result in a variety of effector functions such as 'opsonization', mast cell degranulation (IgE) and antibody dependent cell mediated cytotoxicity (ADCC). Some immunoglobulins such as IgG bind to receptors on placental trophoblasts, which facilitate transfer of the IgG across the placenta and these results in transplacental transfer of maternal immunity to the fetus and newborn.

19.14.4 Haptens

Haptens are substances, which cannot induce the formation of antibodies (not immunogenic) by themselves, but can react with specific antibodies, e.g., capsular polysaccharide of pneumococci, cardiolipin. Penicillin is a hapten. It can break down and bind to serum proteins to form a combined molecule that can produce an allergic reaction such as rash and fever, rarely death. Haptens may be converted to complete antigens by combining with a large inert carrier molecule.

$$\mathbf{Ab + Ag \longrightarrow (Ab\text{-}Ag)_n}$$

This reaction is highly specific, and the antibody is bound by close range non-covalent Van der Waals' forces to a small site on the antigen known as the antigenic determinant. A single antigenic molecule may contain a number of different antigenic determinants and bovine serum albumin (BSA) has several hundred antigenic determinants per molecule (Figure 19.6).

Antibodies are also specific in that they normally react only with foreign substances and not with the plasma and tissues of the individual. This ability to recognize 'self' as opposed to 'non-self' is extremely important, and problems occur when antibodies are made against one's own tissue giving rise to autoimmune diseases such as rheumatoid arthritis and ulcerative colitis.

19.15 IMMUNITY

The immune response is how the body recognizes and responds to the entry of microbial pathogens and altered self molecules such as tumour cells that appear foreign to the

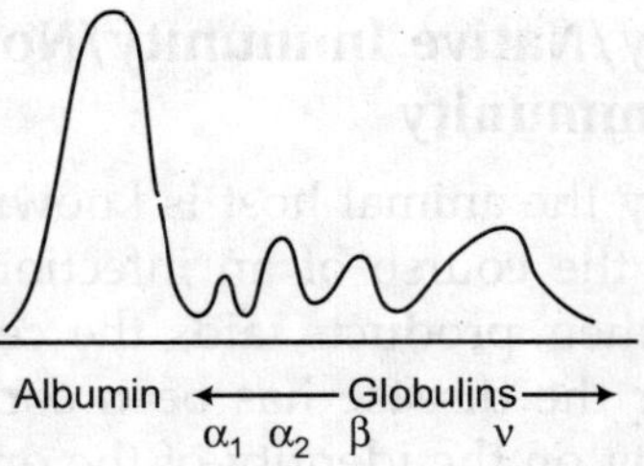

Figure 19.6 Scan of a stained cellulose acetate electrophoresis strip of human serum protein.

body. The immune system recognizes the pathogen as a whole or the substances known as antigens released by the pathogen and responds appropriately. Antigens are molecules on the surface of pathogens, viruses, fungi, or bacteria. The immune system recognizes and destroys these antigens.

Anatomically, the immune system is composed of specialized organs and several interdependent cell types that collectively protect the body from the invasion of foreign pathogens and altered self cells. Many of these cell types have specialized functions and are specialized in engulfment of bacteria and other microbes, kill parasites, kill tumour cells, or viral-infected host cells. The immune system can functionally be divided into two wings, viz., innate immune system (non-specific immunity) and acquired or adaptive immune system (specific immunity) and details are elaborated in Figure 19.7.

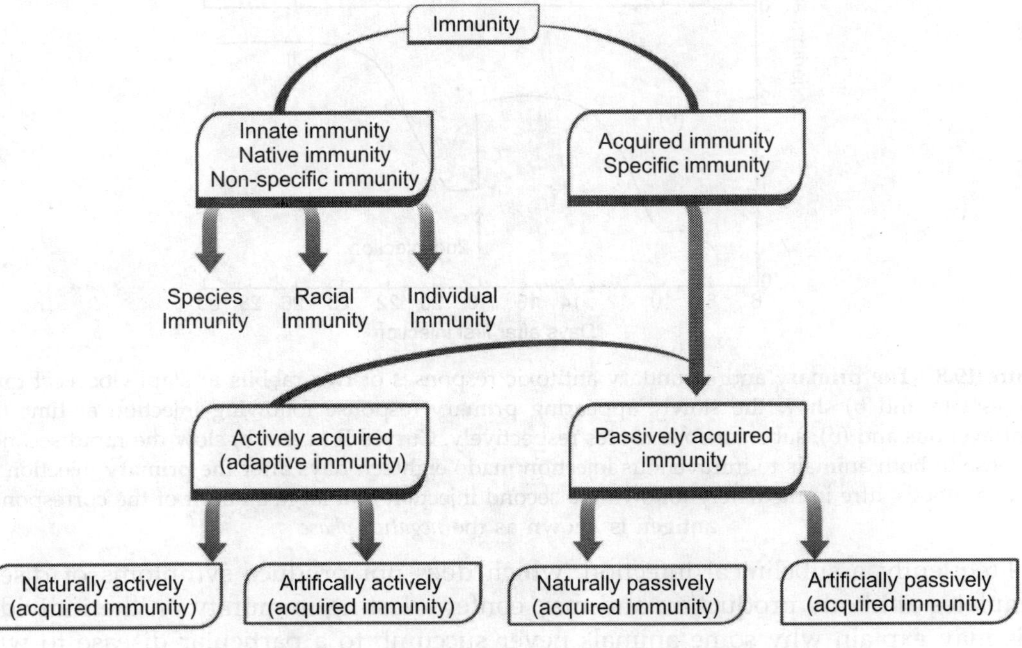

Figure 19.7 Demostration of immune system.

19.15.1 Innate Immunity/Native Immunity/Nonspecific Immunity or Natural Active Immunity

The production of antibody by the animal host is known as active immunity. The titre of antibody increases during the course of an infection and, by combining with the infecting microorganisms or their products, aids the cellular defences of the body to function with impunity. After the disease has been controlled, the titre remains high for months or years, depending on the identity of the original infection, but eventually begins to fall. When, however, the animal, which is specifically immune, is exposed to another attack by the same pathogen, antibody production occurs rapidly and only rarely does the infection establish itself a second or subsequent time. This sequence of events may be compared with antibody production following a second or subsequent 'booster' injection as practised in immunization (Figure 19.8).

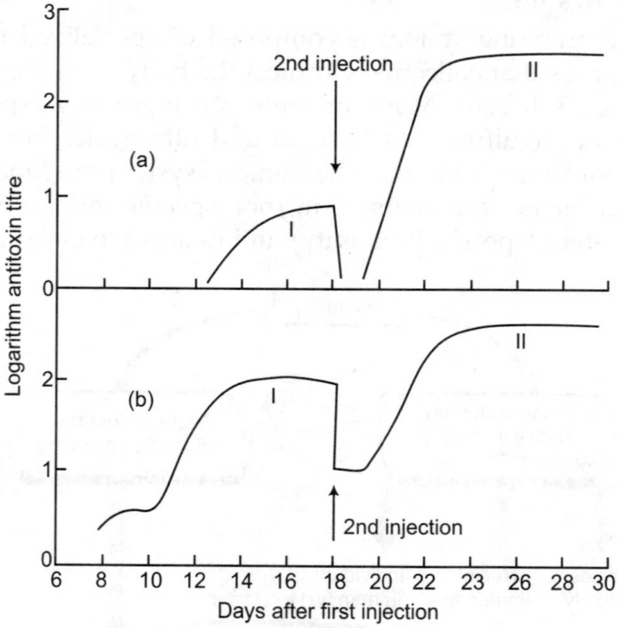

Figure 19.8 The primary and secondary antitoxic responses of two rabbits to staphylococcal toxoid. Curves I (a and b) show the slowly appearing primary response following injection at time 0, by a, intravenous and (b), subcutaneous routes respectively. Curves II (a and b) slow the rapid secondary response in both animals to intravenous injection made eighteen days after the primary injection. The fall in antibody titre immediately following a second injection of a large amount of the corresponding antigen is known as the *negative phase.*

Frequently a subclinical infection, which does not produce symptoms of disease, stimulates antibody production and may confer a lasting immunity on the individual. This may explain why some animals never succumb to a particular disease to which others of the same host species fall ready victims. It is the resistance that an individual possesses by virtue of his/her genetic and body constitution. Thus, it is in-born.

19.15.1.1 *Species immunity*

Resistance to infection varies with the species. Species immunity is the resistance exhibited by all members of a species towards a pathogen. Basic anatomical, physiological and biochemical characteristics of a species can determine whether a microorganism can be pathogenic for that species, or not.

19.15.1.2 *Racial immunity*

With a species, different races may exhibit differences in their resistance, due to genetic factors. Such as, races with sickle cell anaemia prevalent in Mediterranean coast are resistant to infection by the malaria parasite, *Plasmodium falciparum.*

19.15.1.3 *Individual immunity*

Different individuals in a race exhibit differences in innate immunity and is known as individual immunity. Such as, some people have many more colds during the winter than do others. Several factors such as age, malnutrition, and hormonal disturbance influence the level of innate immunity.

19.15.2 Acquired Immunity/Specific Immunity

The resistance that an individual acquires during life is known as acquired immunity. It is of two types – actively acquired and passively acquired. The difference between the two are elaborated in Table 19.7.

Table 19.7 Differences between active and passive immunity

Actively acquired immunity	*Passively acquired immunity*
1. Produced by active functioning of the immune system	1. Received passively; no involvement of immune system
2. Stimulated by infections, antigens (vaccines, allergens)	2. Acquired by administration of antibodies of sensitized T- cells.
3. Provides effective and long lasting immunity	3. Immunity is less effective and transient.
4. There is a lag phase (latent period) required for the formation of antibodies	4. Offers immediate protection
5. Immunological memory is present. Subsequent contact with antigen will have a booster effect	5. No memory, subsequent administration of antibodies is less effective due to immune elimination.
6. Negative phase may occur due to the combination of antigen with preexisting antibody	6. No negative phase
7. Not applicable in immunodeficient hosts.	7. Applicable in immunodeficient hosts.
8. Active immunization is mainly employed in prophylaxis	8. Passive immunization has therapeutic and prophylactic uses.

Actively acquired immunity may be natural or artificial. Similarly, passively acquired immunity may be natural or artificial. Thus, there are four ways to acquire acquired immunity.

19.15.2.1 Active natural immunity

It is due to natural exposure to antigen from infections (disease or subclinical infections). For example, persons who recover from viral diseases such as smallpox, chickenpox, and measles will develop lifelong natural active immunity.

19.15.2.2 Active artificial immunity

When an antigen is deliberately introduced into an individual to stimulate his/her immune system, the immunity developed is known as active artificial immunity. Active artificial immunity is the preferred method of acquiring immunity because it produces long lasting immunity without any disease such as vaccination.

The methods by which artificial active immunity may be achieved simulates that resulting from natural exposure to infection. Antibody production can be induced artificially by administering relevant antigens to the patient in the form of vaccines of dead or living microorganisms or their products (Table 19.8). It is extremely important that vaccines be prepared from organisms antigenically similar to prevalent members of the same species in order that antibodies corresponding to the virulent pathogens may be formed. Only occasionally are fully virulent organisms used as vaccines because of the danger of initiating the disease in susceptible individuals. Immunization against avian infectious laryngotracheitis utilizes the fully virulent virus, but this is introduced at an unusual site on the mucous membrane of the bursa of Fabricius, a pouch of the cloaca, and care must be taken to ensure that the virus does not gain access to the respiratory system. A harmless inflammatory reaction results but immunity against the virulent respiratory disease is achieved.

Table 19.8 Some agents used in artificial immunization

Process	*Disease*	*Agents*
Living agents		
1. Unmodified 2. Attenuated	Rubella Tuberculosis Poliomycelitis Brucellosis Measles Smallpox Anthrax Rabies Yellow fever	Controlled exposure to natural infection B.CG. (Bacillus of Clamette Guerin), poliovirus (sabin) Strain 19 in cattle Measles virus Vaccina virus Spore vaccine of B. *anthracis* Rabies virus (goat, rabbit or egg-adopted) Yellow fever virus

Contd...

Contd...

Killed inactivated agents		
1. Microorganisms	Enteric fever (e.g., typhoid) Whooping cough Poliomyelitis Cholera Influenza Measles Leptospirosis Foot and mouth disease Canine distemper	*Bordetella pertussis* Poliovirus (salk) *Vibrio cholera* Influenza virus Measles virus *Leptospira* sp. Foot and mouth virus Distemper virus
2. Toxoids	Diphtheria Tetanus Infection by anaerobic spore bearing bacilli in animals	Tetanus toxoids Anaerobic bacilli and toxoid

(b) Passive immunity		
	Disease	*Agents*
Diptheria	Diphtheria	Antitoxin
Gas gangrene	*Clostridial myonecrosis*	Antitoxin
Tetanus *Canine distemper*	*Clostridium tetani* *canine distemper*	Antitoxin Antitoxin
Animal infection with sporing anaerobes	Virus	Antitoxin
Rubella (Gelman measles)	*Rubella virus*	Human gamma globulin (pooled)

19.15.3 Passively Acquired Immunity

Passive immunity occurs when another person or animal develops antibodies and the antibodies are transferred to a non-immune individual, who did not produce antibodies, and there is no involvement of the immune system. Passive immunity is preferred in some situations when immediate protection is required.

19.15.3.1 Passive natural immunity

Immunity results when antibodies are transferred from mother to child. During her life, the mother has been exposed to many antigens. These antibodies protect the mother and the developing foetus against disease. The antibody IgG can cross the placenta from mother to the foetus and provides natural passive immunity to the foetus. If the mother nurses her child, IgA in the mother's milk, particularly in colostrum provides natural passive immunity to the child.

19.15.3.2 *Passive artificial immunity*

It is achieved by administration of specific antibodies or sensitized T-lymphocytes, e.g., tetanus antitoxin (anti-tetanus serum, ATS), diphtheria antitoxin, pooled human gamma globulin against Hepatitis B, convalescent sera and antirabic serum. Passive immunization may also be used to suppress active immunity in Rh negative women with Rh positive baby by the administration of anti-Rh antibody (lgG). Achieving passive artificial immunity usually begins with vaccinating an animal such as horse. After the animal's immune system responds to the antigen, antibodies (sometimes T-cells) are removed from the animal and injected into the individual requiring immunity. In some cases, a human who has developed immunity is used as a source. However, this technique provides only temporary immunity because the antibodies are metabolized and eliminated (immun6e elimination).

19.16 ORGANS OF THE IMMUNE SYSTEM

19.16.1 Primary Lymphoid Organs

The primary lymphoid organs give rise to the birth of various cells associated with immune response. There are two primary lymphoid organs.

19.16.1.1 *Bone marrow*

All the cells of the immune system are initially derived from the bone marrow through a process called hematopoiesis. During hematopoiesis, bone marrow-derived stem cells differentiate into mature cells either in bone marrow or into precursors of cells that migrate out of the bone marrow to continue their maturation elsewhere. The bone marrow generates B cells, natural killer cells, granulocytes and immature thymocytes, besides red blood cells and platelets.

19.16.1.2 *Thymus*

Immature thymocytes, also known as prothymocytes, leave the bone marrow and migrate into the thymus. In presence of thymic growth factors and stromal cells, the thymocytes get differentiated into T cells with specialized functions. Through a remarkable maturation process sometimes referred to as thymic education, T cells that are beneficial to the body are spared, while those T cells that might evoke a detrimental autoimmune response get eliminated. The mature T cells leave the thymus into the bloodstream.

19.16.1.3 *Secondary lymphoid organs*

The secondary lymphoid organs provide battleground where the interaction between cells of lymphoid system and microbial derived antigens, activation of T-cells and B-cells takes place resulting in generation of appropriate immune responses.

19.16.1.4 Spleen

The spleen filters of the blood-borne pathogens. It consists mainly of red blood cells, macrophages, dendritic cells, B cells, T cells and natural killer cells. In addition to direct capturing of foreign invaders (antigens) from the bloodstream, migratory macrophages and antigen presenting cells bring antigens to the spleen via the bloodstream. An immune response is initiated when the macrophage or dendritic cells present the antigen to the appropriate B or T cells. This organ can be considered a battleground for cells of the immune system and foreign invaders. In the spleen, B cells get activated and differentiated to effector plasma cells in presence of T-cell derived cytokines and produce large amounts of specific antibodies. Also, the spleen serves as site for destruction of aged red blood cells.

19.16.1.5 Lymph nodes

The lymph nodes filters of tissue-borne pathogens that enter the lymphatic system. Lymph nodes are found throughout the body and composed mainly of dendritic cells and macrophages, B cells and T cells. The macrophages and dendritic cells dwelling in lymph nodes capture antigens, process them and present these processed antigens to T and B cells, consequently initiating an appropriate immune response.

19.16.1.6 Mucosa associated lymphatic tissue

This is less organized and consists of loosely organized clusters of lymphoid cells. These cells play a key role in dealing with mucosal pathogens.

19.17 CELLS OF IMMUNE SYSTEM

19.17.1 Cells of Innate Immune System

19.17.1.1 Granulocytes or polymorphonuclear (pmns) leukocytes

The polymorphonuclear leukocytes consist of three cell types, viz., neutrophils, eosinophils and basophils, based on their staining characteristics with certain dyes. These cells play a major role in the elimination of bacteria and parasites from the body. They engulf these foreign organisms and degrade them in antigen non-specific fashion.

19.17.1.2 Macrophages

Macrophages are key cells in the regulation of adoptive immune responses. They are often referred to as scavengers or antigen-presenting cells (APC) because they pick up and ingest foreign materials and have the capability to present these antigens to T cells of the adoptive immune system. This is one of the important prerequisites in the initiation of adaptive immune response. Macrophages activated by certain cytokines exhibit increased levels of phagocytosis and also secrete large amounts of monokines required for generation of inflammatory response.

19.17.1.3 Natural killer cells

Natural killer (NK) cells are similar to CD8+ cells and function as effector cells that directly kill viral-infected cells, most notably herpes and cytomegalovirus-infected cells. NK cells also act on certain tumours such as melanomas, lymphomas. Unlike the CD8+ T cells, NK cells kill their targets without prior activation in antigen non-specific fashion. However, in presence of cytokines from CD4+ T cells, NK cells kill their tumour or viral-infected host cells more effectively.

19.17.1.4 Dendritic cells

Dendritic cells function as antigen presenting cells (APC) and are considered superior to macrophages in antigen presentation to helper T- cells. These cells are found in the bloodstream and other tissues of the body such as skin, lungs, etc. The dendritic cells capture antigen and bring it to the nearby lymphoid organs where an immune response is initiated. The dendritic cells are known to bind high amount of HIV, and are believed to serve as a reservoir of HIV that may be transmitted to CD4+ T cells.

19.17.2 Cells of Adaptive Immune System

19.17.2.1 T-cells

T-lymphocytes are divided into many subsets that are functionally and phenotypically different. The T-helper subset, also called the CD4+ T-cell, is a pertinent coordinator of immune regulation. The main function of the T-helper cell is to augment immune responses by the secretion of specialized factors called cytokines that activate other cells of the immune system to fight off infection.

Another important subset of T-cell is CD8+ T-cell which is involved in direct killing of certain tumour cells, viral-infected cells and allogenic cells. Another important type of T-cell is regulatory T-cells which play an important role in regulation of immune responses. All types of T-cells are found throughout the body and they often depend on the secondary lymphoid organs (the lymph nodes and spleen) as sites for activation to take place, but they are also found in other tissues of the body, most conspicuously the liver, lung, blood, and intestinal and reproductive tracts.

19.17.2.2 B-cells

The major function of B-lymphocytes is the production of specific antibodies in response to foreign proteins of bacteria, viruses, and parasites. Antibodies are specialized defence proteins that specifically recognize and bind to specific antigenic epitopes. Antibody binding to a foreign substance or antigen, serves as a means of signalling phagocytic cells to engulf, kill or remove that substance from the body.

19.18 ANTIGEN-ANTIBODY REACTIONS

19.18.1 Antigen-Antibody Interaction

The hypervariable regions of heavy and light chains of Fab portion of the antibody form antigen combining site of an antibody. The CDRs of one H-chain and one L-chain constitute one antigen binding site of an antibody. The antigenic epitopes nestles in a cleft formed by the combining site of the antibody and held together by the non-covalent bonds, viz., hydrogen bonds, electrostatic bonds, Van der Waals forces and hydrophobic bonds. Multiple bonding between the antigen and the antibody ensures that the antigen will be bound tightly to the antibody, although the interaction is reversible. The antigen-antibody interactions are affected by factors such as, affinity and avidity of the antibody towards antigen.

Specificity refers to the ability of an individual antibody to bind to only one antigenic determinant or the ability of a population of antibodies to react with only one antigen. Antibodies can distinguish differences in the primary structure of an antigen, isomeric forms of an antigen, secondary and tertiary structure of an antigen and thus highly specific for a given epitope.

Cross reactivity refers to the ability of an individual antibody to react with more than one antigenic determinant or the ability of a population of antibodies to react with more than one antigen. Cross reactions arise because of shared epitopes.

When an antigen is mixed with antiserum containing its homologous antibody, physico-chemical forces induce the antigen and antibody to combine together. Each antibody molecule possesses two combining sites of equal specificity, thus one molecule can bridge two molecules of antigen. Since antigens are multivalent the result is the formation of a lattice structure which is often macroscopically visible as in the agglutination of red blood cells or bacteria.

The detection of an antigen-antibody reaction is only possible if demonstrable reactions take place. These include agglutination, precipitation, complement fixation, killing and lysis of whole cells, capsule swelling, phagocytosis, and neutralization of the physiological activity of the antigen. Antibodies used to be named according to the type of reaction they produced, such as agglutinin, precipitin, lysin, and opsonin. It is now appreciated that the same antibody may produce a number of effects depending on the environmental circumstances. An agglutinating antibody, for instance, may render cells susceptible to phagocytosis and killing or lysis in the presence of complement, in addition to agglutinating the particulate antigen. On the other hand, some antibodies may be deficient in certain properties, as when an agglutinating antibody fixes complement but poorly.

19.18.1.1 Agglutination reaction

This reaction occurs between homologous antibody and particulate antigen in the presence of electrolyte. When a suspension of cells (bacteria or red blood cells) is mixed with homologous antiserum a visible clumping or agglutination occurs (Figure 19.9). Agitation of the mixture by shaking or stirring speeds up the reaction which also occurs more rapidly at higher temperatures. By use of a standard suspension of cells and a series of dilutions of antiserum the antibody can be titrated. This test is widely used for blood grouping and the identification and diagnosis of bacteria and their infections. Agglutinins, in addition to agglutinating bacteria, may also immobilize motile ones, an observation used in the Treponema Immobilization Test for the diagnosis of syphilis.

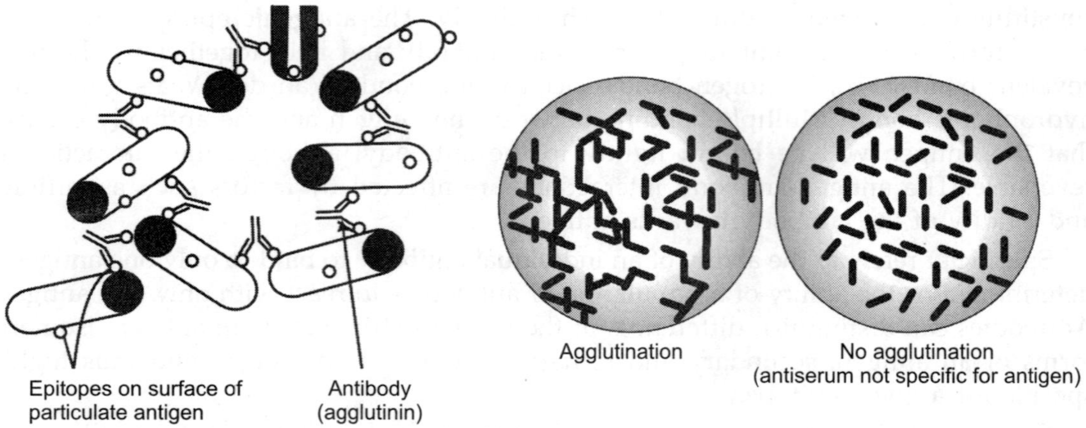

Figure 19.9 Agglutination reaction.

Soluble antigens may be adsorbed on to carrier particles, e.g., bacterial polysaccharide on to red blood cells (sometimes treated with tanning agents, e.g., bis-diazo-benzidine, to aid the adsorption) and these cells agglutinate when mixed with antiserum against the bacterial polysaccharide.

19.18.1.2 Precipitin reactions

Soluble antigens, such as serum and bacterial polysaccharides, when mixed with homologous antibody in suitable proportions, precipitate in the presence of electrolyte. Excess of either reagent results in soluble non-precipitating complexes being formed. For good precipitation reactions it is therefore necessary to determine the optimum conditions for a visible reaction to take place. The reaction may be used as a ring test, where a ring of precipitate forms at the fluid interphase when antigen is layered onto an antiserum; as a flocculation test when floccules of antigen-antibody complex are formed in a mixture of both reagents; as a gel diffusion test (often called the Ouchterlony test after the first worker to describe it), in which both reagents diffuse towards each other in an agar gel (Figure 19.10a). In gel diffusion, lines of precipitate, corresponding to each antigen-antibody reaction, are formed in the gel at points where optimal levels of reactants meet. This technique is sometimes superimposed on a primary separation of immunoglobulins by immuno-electrophoresis. Antigens are caused to move through an agar gel under the influence of an electric current by which separation is accomplished. Antiserum, containing antibody to the various antigens, is later diffused from a source at right angles to the prior movement of the antigens and lines of precipitate form at the points reached by the various antigens (Figure 19.10b).

19.18.2 Lysis by Complement

Antibody may combine with living bacteria and usually bring about their agglutination, but the organisms still remain viable and fully pathogenic. However, a component of normal serum, known as complement, will combine with bacterial cells treated with

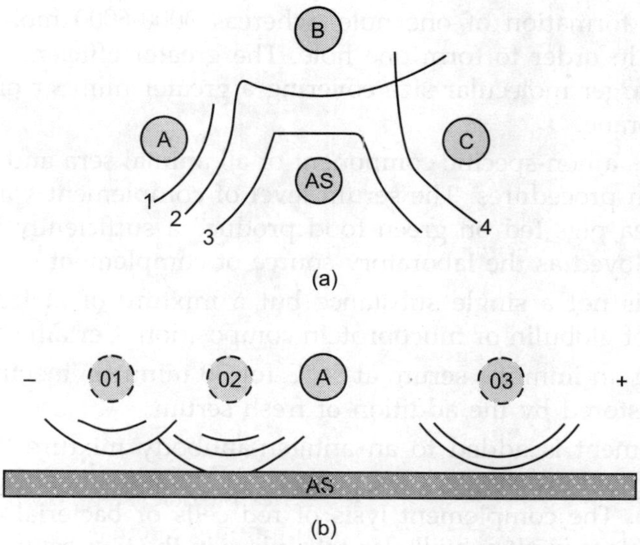

Figure 19.10 Precipitation reactions in agar plate.

(a) Lines of precipitation laid down in an agar gel when three antigen complexes (A, B, and C) are set up in an immunodiffusion test against an antiserum (AS). 1. Antigen common to complex A only 2. Antigen common to complexes A and B (hence joined) 3. Antigen common to complexes A, B, and C. Antigen common to complex C only (hence crossing line 2).

(b) Immunoelectrophoresis Antigen A, placed in an agar cup, has been subjected to an electric field B which separates fractions a1; a2; and a3. Each of these fractions includes a number of antigens which migrates in the electric field at the same rate. Subsequent normal diffusion in the agar gel against a composite antiserum (AS) reveals the constituent antigens.

homologous antibody bringing about their death and, in some species, their lysis. Both death and lysis frequently occur with the Gram-negative bacteria, as with *Vibrio cholerae* originally described by Pfeiffer, but no lysis occurs with Gram-positive bacteria. This is thought to be due to the greater thickness of cell walls in Gram-positive bacteria, the antibody not being able to make close contact with the cell membrane. Complement combines with most antigen-antibody systems with certain exceptions, e.g., toxin-antitoxin complexes.

Red blood cells are also lysed in the presence of both homologous antibody and complement; this has been used as a model system for the elucidation of the mechanisms of complement lysis. Indeed, red cell lysis, first described by Bordet and Gengou in 1901, is now extensively used in the laboratory for detecting the presence of free complement *in vitro* and especially in the complement fixation test.

Electron microscope studies have shown that complement lysis of red cells is the result of a large number of holes being produced in the cell membrane (80-100 mμ diameter) through which the haemoglobin leaks out. The initial addition of homologous antibody to the red cells did not give this lysis; only the final addition of complement produced the holes. Further, qualitative studies showed that one molecule of IgM was sufficient for the formation of one hole, whereas 3000-6000 molecules of IgG were required per cell in order to form one hole. The greater efficiency of IgM is thought to be due to its larger molecular size covering a greater number of denaturation sites on the cell membrane.

Complement is a non-specific component of all animal sera and is not increased by any immunization procedures. The serum level of complement varies from animal to animal, but guinea pigs fed on green food produce a sufficiently high level for their serum to be employed as the laboratory source of complement.

Complement is not a single substance but a mixture of at least nine substances (C_1, C_2, C_3, etc.) of globulin or mucoprotein composition. Certain components are heat labile and heating an immune serum at 56°C for 30 minutes 'inactivates' complement; activity can be restored by the addition of fresh serum.

When complement is added to an antigen-antibody mixture the components are adsorbed in a defined order (i.e., C_1, C_4, C_2, $C_{3,5,6,7,8,9}$) at a site adjacent to the antigen-antibody junction. The complement lysis of red cells or bacterial cells, which can be observed to take place *in vitro*, 'will also take place in the tissues of the immune animal. For instance, complement participates in haemolytic anaemia of the new-born, bringing about the haemolysis of red blood cells in the presence of specific antibody. A further possible role of complement is in assisting polymorphs to lyse ingested bacterial cells. It should be noted that the neutralizing antibody, which constitutes the basis of viral immunity, can neutralize the infectivity of virus particles in the absence of complement.

19.18.3 Complement Fixation

The lysis of red blood cells by complement is the basis of a widely used serological test known as the complement fixation test. Many antigen-antibody reactions, involving viral or soluble antigens, are not visible, but it is possible to demonstrate the reaction by superimposing the visible lysis of red blood cells in the presence of complement.

To detect specific antibody in a patient's serum, a standard volume of the neat or diluted serum (heated at 56°C for 30 minutes to destroy its natural complement) is mixed with a standard amount of the specific antigen. To this is added a minimal quantity of titrated fresh normal serum (usually guinea pig) as a source of complement. This combines with the antigen-antibody complex if the patient's serum contains antibody. To demonstrate that this reaction has occurred, sensitized sheep red blood cells (i.e., red blood cells treated with homologous antiserum) are added. These cells will only lyse in the presence of free complement and hence act as an indicator of the experimental system. If antibody is present in the patient's serum, it will combine with the antigen and fix complement—the consequence will be no lysis of the red blood cells. On the contrary, if no antibody is present, the complement will be free and available to lyse the sensitized red blood cells. These reactions are illustrated in Figure 19.11.

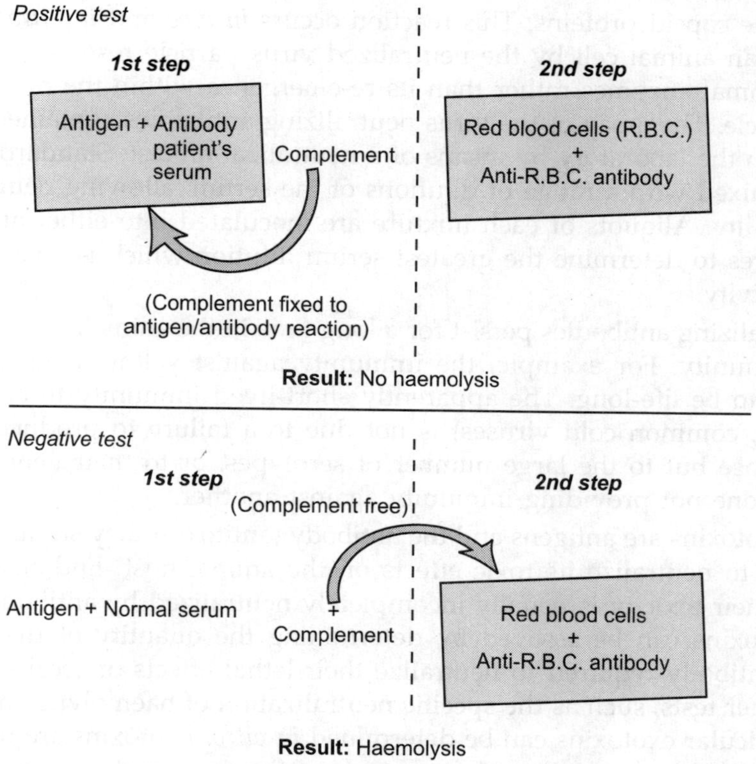

Figure 19.11 The complement fixation test.

19.18.4 Phagocytosis

Foreign particles found in the body are phagocytosed by macrophages and monocytes in the tissues and polymorphonuclear leucocytes in the peripheral blood of the host. In contrast to inert particles, virulent bacteria are removed only after prior modification usually by antibody. In the presence of complement, the antibody treated bacteria are killed, after which they are removed as inert particles. Components of normal serum function to a lesser degree. For instance, opsonins, both IgM and IgG of normal gamma globulin, probably act by neutralizing the negative charge on the bacterial cell so that they are repulsed to a lesser degree by the negatively charged phagocytes. A measure of the opsonic index, i.e., the ability of a patient's serum to render bacterial cells susceptible to phagocytosis, was used at one time to measure the patient's natural resistance to infection.

19.18.5 Neutralization

Antibody is usually able to neutralize the toxic or infective power of a microorganism. Neutralizing antibodies constitute the basis of animal immunity against virus diseases. These antibodies are stimulated by the protein components of the virus capsid, and the neutralization of virus infectivity entails the firm adsorption of the homologous antibody molecules to the capsid proteins. This reaction occurs *in vivo* and *in vitro*. Subsequent penetration of an animal cell by the neutralized virus particle results in its digestion by phagolysosomal enzymes rather than its re-emergence within the cytoplasm as an infectious particle. The assay of the virus neutralizing antibodies contained in a serum is carried out in the laboratory by means of a neutralization test. Standard amounts of the virus are mixed with a range of dilutions of the serum, allowing neutralization to proceed to finality. Aliquots of each mixture are inoculated into either animal tissues or tissue cultures to determine the greatest serum dilution which is able to neutralize the viral infectivity.

Virus neutralizing antibodies persist for a long time in the animal, thereby providing prolonged immunity. For example, the immunity against yellow fever and German measles tends to be life-long. The apparently short-lived immunity to certain viruses (e.g., influenza, common cold viruses) is not due to a failure to produce a sufficient immune response but to the large number of serotypes, or to their genetic variation, infection with one not providing immunity against another.

Bacterial exotoxins are antigens and the antibody (antitoxin) they stimulate combines with the toxin to neutralize its toxic effects on the animal host. Endotoxins are poor antigens and their toxicity is usually incompletely neutralized by antibody stimulated by them. Exotoxins can be assayed by determining the quantity of the more stable homologous antibody required to neutralize their lethal effects or local skin reactions in animals. Other tests, such as the specific neutralization of haemolytic and lecithinase activity of particular exotoxins can be determined *in vitro*. Exotoxins are not destroyed by antitoxins but may be recovered from toxin-antitoxin complexes, and, because of this, complexes of these reagents are unstable as immunizing agents.

19.18.6 Haemagglutination Inhibition

Certain viruses or products of their intracellular replication (haemagglutinins) are able to agglutinate red blood cells of various animal species *in vitro*. This reaction is termed haemagglutination (HA) and, since under controlled conditions the degree of agglutination is proportional to the quantity of virus, the HA test can be used as an assay method for the virus. The different degrees of agglutination are readily detected by allowing the agglutinated red cells to settle out in round-bottomed tubes or in the cups of perspex plates, giving different settling patterns. The sensitivity of the HA test is not great; with influenza virus one haemagglutinating unit (i.e., the quantity of virus which will agglutinate 50 per cent of the red cells in the tube) may be 10^6 virus particles. Further, since both infectious and non-infectious particles haemagglutinate, the test is a measure of the total virus in a preparation, not of infectivity (Figure 19.12).

Following infection with a haemagglutinating virus, antibodies are formed which inhibit the HA reaction, and these are termed haemagglutination-inhibiting antibodies. These may be assayed in the laboratory by means of the haemagglutination-inhibition (HAI) test. Standard amounts of virus are mixed with a range of dilutions of the serum, and, after a suitable incubation period, standard amounts of a 1 per cent red cell suspension are added to each mixture. The settling patterns of the red cells in the tubes or cups determine the HAl endpoint of the serum. Serum from a convalescent patient gives a fourfold or greater rise in titre compared with serum taken from the same patient during the acute stage of the illness.

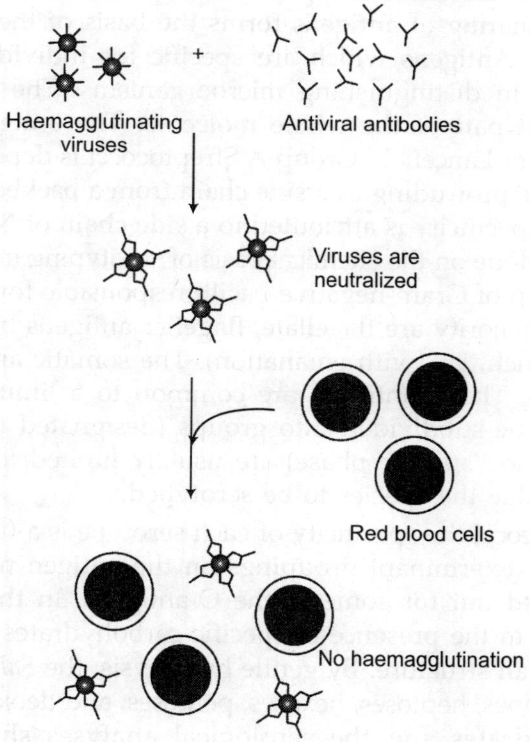

Figure 19.12 Haemagglutination inhibition (HI).

19.19 SPECIFICITY OF ANTIGEN-ANTIBODY REACTIONS

Antigen-antibody reactions are highly specific, making it possible to recognize differences between proteins and other antigens of closely related plant or animal origin and even between tissues within the same host. This specificity depends primarily on the nature of the chemical groupings of opposite charge on molecules attracting each other. In addition, the spatial arrangement of the molecules, dependent upon the *ortho-*, *meta-*, or *para-* positions of chemical groups and stereoisometric spatial arrangements *(leavo-, dextro-*, or *meso-)*, imprint their influence upon the shape of the molecules and enable the antigen and antibody to combine together in close proximity, thereby allowing shortrange inter-molecular forces to operate. These include van der Waal's forces, electrostatic forces, and hydrogen bonding. The union of an antigen with its antibody is usually very firm but the avidity (i.e., the strength of the attracting forces between antigen and antibody) varies from system to system and in + some cases the union is reversible. No profound changes occur in either molecule as a result of combination, as is seen in antitoxin-toxin reactions from which the unchanged toxin can be fairly readily released.

The intact bacterial cell possesses multiple antigens on the cell wall surface in addition to those of flagella, fimbriae, and capsules. Many antigens are common to different species of the same genus and occasionally different genera, as for instance, antigens common to the rickettsiae of typhus and typhus-like fevers and certain strains of *Proteus,* where the sharing of antigens forms the basis of the Weil-Felix diagnostic test for these diseases. Antigens which are specific for individual species or strains are those most useful in distinguishing microorganisms. The determinant group is often a relatively small part of the whole molecule. The antigenic specificity of the polysaccharide capsule of Lancefield Group A Streptococci is dependent on the presence of N-acetyl glucosamine protruding as a side chain from a backbone of rhamnose in the polymer; in Group C, specificity is attributed to a side chain of N-acetyl galactosamine. Similar work has been done on the chemical basis of serotyping in *Salmonella.* This genus consists of a large group of Gram-negative bacilli responsible for a range of diseases of man and animal. The majority are flagellate, flagellar antigens being called H-antigens (from the German, Hauch, Le., with emanation). The somatic antigens of the cell wall are termed O-antigens. The O-antigens are common to a limited number of species enabling *Salmonella* to be subdivided into groups (designated by capital letters). The H-antigens (when in their specific phase) are usually limited to one or a few species and these antigens enable the species to be serotyped.

As with the streptococci the specificity of each serotype is a function of the chemical structure of particular determinant groupings on the antigen molecule, the nature of which has been worked out for some of the O-antigens. In these the specificity has been shown to be due to the presence of specific carbohydrates on the polysaccharide molecules of the cell wall structure. By gentle hydrolysis, the *Salmonella* polysaccharide can be split into hexamines, heptoses, hexoses, pentoses, and deoxy-sugars. Comparisons between the carbohydrates and the serological analyses show that the terminal dideoxyhexoses (of which only five have been identified) determine the specificity of the antigenic determinants. Each *Salmonella* species examined has been shown to possess only one of these sugars which is characteristic of the serological group based on O-antigens (Table 19.9). For instance, paratose is characteristic of O-antigen 2 in species falling in Group A, abequose of antigen 4 in Group B, tyvelose of antigen 9 in Group D and colitose of antigen 35 in Group O. The specific sugar always occupies the terminal position but the antigen specificity may be modified by the way the sugar is linked to the molecule or by the sugar inmediately next to it on the side chain.

19.20 HYPERSENSITIVITY

Contact with an infectious disease usually results in an increased resistance by the host to subsequent exposure to the same infectious microorganism. Occasionally the converse occurs, the animal becoming abnormally sensitive to the infectious agent or other environmental antigens. Many types of hypersensitivity are encountered which may be divided according to whether the host response to contact with the sensitizing agent is immediate or delayed. In the former the host reaction is the result of antigen-antibody reactions; in the latter the sensitivity is mediated by host cells specifically sensitized to the antigen.

Table 19.9 Somatic antigen and carbohydrate units of specific O-antigens of selected salmonella groups A, B, D and O

Salmonella species	*Group*	*O-antigens*	*Hexoses*			*6-deoxy hexose Rham.*	*Terminal sugar 3.6 dideoxyhexoses*			
			Gaf.	*Glu.*	*Man.*		*Abequose*	*Colitose*	*Paratose*	*Tyvelose*
S. paratyphi A	*A*	1, **2**, 12	+	+	+	+	–	–	+	+
S. paratyphi B	*B*	1, **4**, 5, 12	+	+	+	+	–	–	–	–
S. typhimurlum	*B*	1, 4, 5, 12	+	+	+	+	+	–	–	–
S. abortus equi	*B*	**4**, 12	+	+	+	+	+	–	–	–
S. budapest	*B*	1, **4**, 12	+	+	+	+	+	–	–	–
S. stanley	*B*	**4**, 5, 12	+	+	+	+	+	–	–	–
S. salunatus	*D*	**4**, 12	+	+	+	+	+	–	–	–
S. typhi	*D*	**9**, 12, vi	+	+	+	+	–	–	–	+
S. enteritidis	*D*	1, **9**, 12	+	+	+	+	–	–	–	+
S. gallinarum	*D*	1, **9**, 12	+	+	+	+	–	–	–	+
S. adelaide	*O*	**35**	+	+	–	–	–	–	–	–
S. monschaui	*O*	**35**	+	+	–	–	–	–	–	–

Group specific O-antigens are in heavy type.

19.20.1 Immediate Hypersensitivity

Immediate hypersensitivity has been classified into three types diagrammatically represented in Figure 19.13 Type 1 occurs in patients previously sensitized to a particular foreign antigen. The antibody produced is thought to be absorbed to the surface of tissues cells. The interaction of antigen-antibody at cell surfaces results in tissue damage and the release of pharmacologically active substances such as histamine when a second injection of the same antigen is given. These substances cause the smooth muscle of particular organs to contract which results in profound shock. The classical example is *anaphylaxis* (meaning 'against protection') which may follow a second injection of a foreign serum, as in serum therapy or by exposure to antibiotics. In small animals anaphylactic shock may result in death; in man and large domesticated animals symptoms of shock include bronchial spasm, pulmonary oedema (outpouring of fluids into the tissues), and urticaria (skin rashes). When the antigen is introduced locally, as the inhalation of pollen or dust, etc., in hay fever and asthma, respectively, the symptoms of excessive secretion of the mucous membranes or bronchial spasm occur locally although the patient may be generally sensitive.

Type 2 occurs when circulating antibody reacts with an antigenic component of tissue cells bringing about their damage. This may occur in acute glomerular nephritis, a disease of the kidneys, and rheumatic fever. Both these conditions follow infection with specific types of streptococci. Antibodies produced against the streptococci cross-react with the basal membrane of kidney substance and heart muscle respectively.

Since antibodies begin to appear in the serum after about ten days, the symptoms of these diseases do not arise until sometime after the streptococcal infection has cleared up (Figure 19.13).

Type 3 is demonstrated by the toxic symptoms which follow an antigen-antibody reaction in the blood in the presence of excess antigen as in serum sickness which frequently follows injection of large volumes of serum into certain hosts. Antibodies produced against the foreign serum are demonstrated in the patient's serum after the normal period of about 10 days (Figure 19.13).

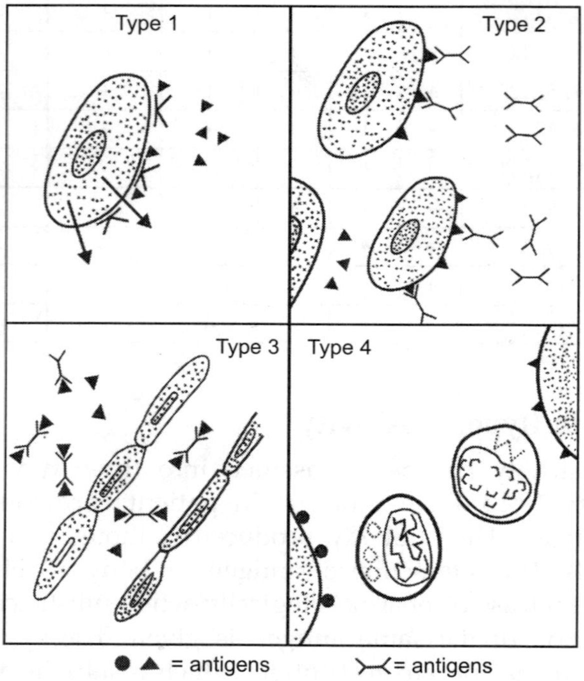

Figure 19.13 Illustrating the four types of allergic reaction which may be deleterious to the tissues and harmful to the host. Type 1. Free antigen reacting with antibody absorbed to the cell surface results in the release of histamine and other pharmacologically active substances form the cell. Type 2. Antibody reacting either with cell surface of with antigens or haptens which becomes attached to cell surfaces: complement plays a major destructive role. Type 3. Antigen and antibody reacting in antigen excess forming complexes which are toxic to cells. Type 4. Specifically modified mononuclear cells.

19.20.2 Delayed Hypersensitivity

These react with excess antigen (the foreign serum) still circulating in the blood and form toxic complexes which produce symptoms of serum sickness. Delayed hypersensitivity (Figure 19.13 Type 4) occurs in patients with chronic infections such as tuberculosis. In these diseases cell-mediated immunity plays a greater role than antibody immunity. As a result of long exposure to the bacterial antigens, lymphocytes and other mononuclear cells become sensitized to products of the bacterial cells (e.g., tuberculin), with which they react specifically when these substances are introduced locally either by natural infection or by injection. The area becomes red and swollen owing to the infiltration of cells into the local site and the reaction reaches a peak after 48-72 hours; hence the term delayed hypersensitivity. This reaction is used as a clinical test to determine if a patient has been previously, or is at present, infected. No tissue reaction occurs in a patient who has had no previous exposure to the tubercle bacillus. In animals the test is known as the Tuberculin test and in man, the Heaf test. A Heaf-negative patient can be converted to a Heaf-positive patient by BCG vaccination.

Cell-mediated immunity, stimulated by chronic infection, provides another mechanism of host resistance. It can be passively transferred by living lymphoid cells but not by serum and, as with passive antibody immunity, passive cell mediated immunity is of relatively short duration. Not all cell-mediated immune responses are advantageous to the host. Exposure to foreign tissues induces this type of immune response and homografts (i.e., tissue transplants from another host of the same species not genetically identical) are rejected after about two weeks by infiltration of specifically sensitized lymphocyte like cells.

19.21 IMMUNITY IN VIRAL INFECTIONS

19.21.1 Humoral Components

19.21.1.1 Nonspecific components

A number of humoral components of the nonspecific immune system function in resistance to viral infections.

Interferons

Interferon is one of the first lines of defence against viruses because it is induced early after virus infection before any of the other defence mechanisms appear (*e.g.* antibody, Tc cells etc.). There are three types of interferon, IFN α (also known as leukocyte interferon), IFN β (also known as fibroblast interferon) and IFN γ (also known as immune interferon). IFN α and IFN β are also referred to as Type I interferon and IFN γ as Type II. There are approximately 20 subtypes of IFN-α but only one IFN-β and IFN γ (Table 19.10). Transcription of the IFN genes occurs only after exposure of cells to an appropriate inducer such as viruses, double stranded RNA, LPS. Among the viruses, the RNA viruses are the best inducers while DNA viruses are poor IFN inducers.

Table 19.10 Types and properties of interferons

Interferon			
Property	*Alpha*	*Beta*	*Gamma*
Genes	>20	1	1
Inducers	Viruses (RNA>DNA) dsRNA	Viruses (RNA>DNA) dsRNA	Antigens, Mitogens
Principal source	Leukocytes, Epithelium	Fibroblasts	Lymphocytes

Complement

Most viruses do not fix complement by the alternative route. However, the interaction of a complement-fixing antibody with a virus infected cell or with an enveloped virus can result in the lysis of the cell or virus. Thus, by interfacing with the specific immune system, complement also plays a role in resistance to viral infections.

Specific responses

Virus specific antibody can directly neutralize virus infectivity by preventing the attachment of virus to receptors on host cells or entry of the virus into the cell. Antibodies can also prevent uncoating of virus by interfering with the interaction of viral proteins involved in uncoating. Complement fixing antibodies can assist in the lysis of viral infected cells or enveloped viruses. Antibodies can also act as opsonins and augment phagocytosis of viruses either by promoting their uptake via Fc or C3b receptors or by agglutinating viruses to make them more easily phagocytosed. Antibody coated virus infected cells can be killed by K cells thereby preventing the spread of the infection. Table 19.10 shows in brief the various antiviral activities of antibody.

Opsonization of viruses with antibody can enhance their uptake by phagocytic cells (Table 19.11). If the virus is able to survive in the phagocyte, this allows for the spread of virus infection. HIV is an example of viruses that can survive in macrophages.

Table 19.11 Mediation of anti-viral activities of antibody

Target	*Agent*	*Mechanism*
Free virus	Antibody alone	Blocks binding to cell Blocks uncoating of virus
	Antibody + Complement	Damage to virus envelope Opsonisation of virus
Virus-infected cell	Antibody bound to infected cells	ADCC by K cells, NK cells and/ or macrophages

19.21.2 Cellular Components

In addition to the barriers and humoral components involved in resistance to and recovery from viral infections, there are several different cells that play a role in our antiviral defences (Table 19.12).

Table 19.12 Virus mediated immunosuppression

Host defence affected	*Virus*	*Virus product*	*Mechanism*
Interferon	EBV	EBERS (small RNAs)	Blocks protein kinase
	Vaccinia	elF-2alpha homolog	Prevents phosphorylation of elF-2alpha by protein kinase
Antibody	HSV-I	gE/gI	Binds Fe-gamma and blocks function

Contd...

Contd...

	Myxoma	IFN-gamma receptor homolog	Competes for I FN-gamrna and blocks function
	Shope fibroma virus	TNF receptor	Competes for TNF and blocks function
Cytokines	EBV	IL-IO homolog	Reduces fFN-gamma function
MHCClass I	CMV	Early protein	Prevents transport of peptide-loaded MHC
	Adenovirus	E3	Blocks transport of MHC to surface
Apoptosis	Adenovirus	I4.7K	Inhibits capsases
	EBV	BcI-2 homolog	Anti-apoptotic

19.21.2.1 *Nonspecific*

Macrophages Macrophages contribute to antiviral defences in a number of ways.

Intrinsic antiviral activity: Macrophages help limit viral infections by virtue of their intrinsic ability to prevent replication of viruses. However some viruses are able to replicate or at least survive in macrophages and thus can be spread by macrophages.

Extrinsic antiviral activity: Macrophages are also able to recognize virus infected cells and to kill them. Thus, macrophages also contribute to antiviral defences by virtue of their cytotoxic activity.

ADCC: Virus infected cells that are coated with IgG antibodies can be killed by macrophages by ADCC.

IFN production: Macrophages are a source of IFN.

NK CELLS NK cells act by recognizing and killing virus infected cells. The recognition of virus infected cells is not MHC-restricted or antigen specific. Thus, NK cells will kill cells infected with many different viruses. NK cells can also mediate ADCC and can kill virus infected cells by this mechanism. The activities of NK cells can be enhanced by IFN-γ and IL-2.

19.22 CHEMOTHERAPY AS AN AID TO RESISTANCE

Antibody immunity aids an animal either by neutralizing bacterial toxins so that they cannot damage the host or by inactivating the parasites in the presence of complement and can be removed by the phagocytic cells in the same manner as inert particles. Chemotherapeutic agents, by their bacteriostatic or bactericidal actions, similarly aid the body defences to deal with the invading organisms. Satisfactory chemotherapeutic agents are selectively active against specific parasites without being toxic to the host and are extremely valuable in controlling infections already established in the body. In particular, the antibiotics are highly selective inhibitors, each one showing a defined spectrum of activity against microorganisms owing to their specialized modes of action. The clinically useful antibiotics can be grouped according to their range of antibacterial activity. Some are active against Gram-positive bacteria and pathogenic Gram-negative cocci but relatively inactive against Gram-negative bacilli; a second group is active to a greater or lesser degree against most species of bacteria, the larger viruses and rickettsiae, and are referred to as broad spectrum antibiotics; a third group is particularly active against the Gram-negative bacilli (Table 19.13).

Table 19.13 Antibacterial agents used in clinical practicle

Group 1 *Active against* *Gram-positive bacteria and Gram-negative cocci*	*Group 2* *Active against:* *Gram-positive and Gram-negative bacteria, Bedsoniae and rickettsiae*	*Group 3* *Active against* *Gram-negative bacilli*
Penicilli	Semi-synthetic penicillins	← Streptomycin
Benzy penicillin (G)		
Phenoxymethyl penicillin(V)	Ampicillin	← Kanamycin
Cephalosporin →	Carbenicillin	
Erthromycin	Tetracyclines	← Neomycin
Lincomycin	Chloramphenicol	
Novobiocin	Sulphonamides	Polymixin (colistin)
Fucidin		
Vancomycin		
Ristocctin		
Bacitracin		

The arrows indicate a broader spectrum of activity by particular antibacterials than those organisms listed in the group.

A very few, e.g., streptomycin, are active against the tubercle bacillus. The use of chemotherapy during the early stages of an infection will prevent the production of antibody immunity since the stimulating microorganisms will be killed and removed from the body. Chemotherapy may not therefore always be immediately desirable since it may prevent development of lasting immunity. Certain bacteria, such as *Streptococcus pyogenes*, pneumococci, and meningococci, are consistently sensitive to penicillin. Other bacterial species, normally susceptible to a particular drug, frequently produce resistant variants; with these organisms the choice of antibiotic for treatment of an infection must be based on prior tests of their sensitivity *in vitro*.

Resistance to antibiotics may arise by a number of possible routes.

1. Mutations within the bacterial population may result in resistant variants being formed from the sensitive organisms. When this occurs during treatment the sensitive population may be killed by antibiotic therapy leaving the few resistant mutants to proliferate and cause a recurrence of the disease. This type of complication can often be avoided by the use of high dosage of the antibiotic or combined therapy, a second drug dealing with mutants resistant to the first.

2. Within a bacterial species a number of strains are frequently found to be naturally resistant. This is characteristic of *Staphylococcus aureus,* many strains of which produce varying quantities of the enzyme penicillinase which destroys penicillin. Exogenous infection with such a strain means that the pathogen will not respond to penicillin and consequently another suitable drug, such as one of the penicillinase-resistant semi-synthetic penicillins, e.g., methicillin, must be used. Within a closed community (e.g., hospitals), sensitive strains of staphylococci are readily killed by widespread use of antibiotic therapy but this results also in a build-up by selection of resistant strains (hospital strains) within an isolated community. Both staff and patients may become carriers of resistant staphylococci in their noses and on their skin and cross-infection among patients entering hospital rapidly occurs.
3. Infective drug resistance differs from other forms of resistance in that the resistance can be transferred to sensitive cells by conjugation. The resistance factor which is transferred is a self-replicating extrachromosomal particle carrying genetic information. Infective drug resistance is transferred between Gram-negative bacteria of both the same and different genera and species. This fact is of great epidemiological importance since sensitive pathogens may receive the resistance factor from non-pathogens in the alimentary canal. The resistance factor may carry resistance to only a single antibiotic but resistance of a multiple type may occur. Any environmental influence, such as the feeding of sub-inhibitory antibiotic food supplements to animals, which will select for resistant members of the normal non-pathogenic flora, is potentially dangerous since this resistance may be subsequently transferred to pathogens during an infection and these, in turn, will not respond to antibiotic therapy. Should these resistant pathogens be transferred to man the antibiotics against which the organisms are resistant will no longer be useful in treatment. For instance, many *Salmonella typhimurium* infections of man derived from cattle (especially calves) carry multiple resistance to antibiotics.

19.23 EXPERIMENTS

19.23.1 Isolation and Presumptive Identification of Gram-negative Intestinal (Enteric) Bacteria

The digestive tract represents a very attractive environment. It provides shelter from the outside world, yet is easily accessible, provides nutrients in great number and variety, and is less likely to trigger an immune response. It is therefore not surprising that many bacteria live in the human guts, most of them harmlessly, with some even playing useful functions for their host. However, some species of bacteria can be highly pathogenic when they enter and colonize the human digestive tract. Such enteric bacterial pathogens typically cause gastroenteritis, but can also have life threatening consequences. The enteric pathogenic bacteria *Salmonella* and *Shigella* are of prime medical concern. They cause enteric fever, food poisoning, and bacillary dysentery. *Salmonella typhi*, which causes typhoid fever, is by far the most significant pathogen of the salmonella group. In addition to the typhoid organisms, there are 10 other distinct salmonella species and over 2200 serotypes. The shigella, which are the prime causes of human dysentery, comprise four species and many serotypes within genera are organisms of similar biochemical characteristics that can most easily be differentiated by serological typing. Lactose fermentation separates the Salmonella and Shigella from most of the other members of Enterobacteriaceae. Final differentiation of the two enteric pathogens from Proteus relies on motility, hydrogen sulfide production and urea hydrolysis. The details about the isolation and identification of the two enteric bacteria are elaborated in Figure 13.8 and Figure 13.9.

There are several excellent selective differential media that have been developed for the isolation of these pathogens. Various inhibiting agents such as brilliant green, bismuth sulfite, sodium desoxycholate and sodium citrate are included in them. For, Salmonella typhi, bismuth sulfite agar appears to be the best medium. Colonies of *S typhi* on this medium appear black due to the reduction of sulfite to sulfide. Other widely used media are MacConkey agar, Hektoen Enteric agar (HE), and Xylose Lysine Desoxycholate (XLD) agar. These media may contain bile salts, and/or sodium desoxycholate to inhibit Gram-positive bacteria. To inhibit coliform and other nonenterics, they may contain a citrate. All of them contain lactose and a dye so that if an organism is a lactose fermenter, its colony will take on a colour characteristics of the dye present.There are five steps that are used to prove the presence of these pathogens in a sample (1) enrichment (2) isolation (3) fermentation test (4) physiological test and finally (5) serotyping.

Look for the isolated colonies for *Salmonella* and *Shigella*. The characteristics to look for on each medium are as follows:

1. **MacConkey agar** *Salmonella, Shigella* and other non-lactose-fermenting species produce smooth, colourless colonies. Coliforms that ferment lactose produce reddish, mucoid, or dark centered colonies.
2. **Hektoen Enteric (HE) agar** *Salmonella* and *Shigella* colonies are greenish-blue. Some species of Salmonella will have greenish blue colonies with black centres due to H_2S production. Coliform colonies are salmon to orange and may have a bile precipitate.
3. **Xylose Lysine Desoxycholate (XLD) agar** Most *Salmonella* produce red colonies with black centres, a few may produce red colonies that lack black centres. Shigella colonies are red. Coliform colonies are yellow. Some *Pseudomonas* produces false positive red colonies. The detail protocol is shown in the Figure 19.14 and also described in Chapter 13.

19.23.2 Phage Typing

Phage typing is a method used for detecting single strains of bacteria. It is used to trace the source of outbreaks of infections. The viruses that infect bacteria are called bacteriophages ("phages" for short) and some of these can only infect a single strain of bacteria. These phages are used to identify different strains of bacteria within a single species.

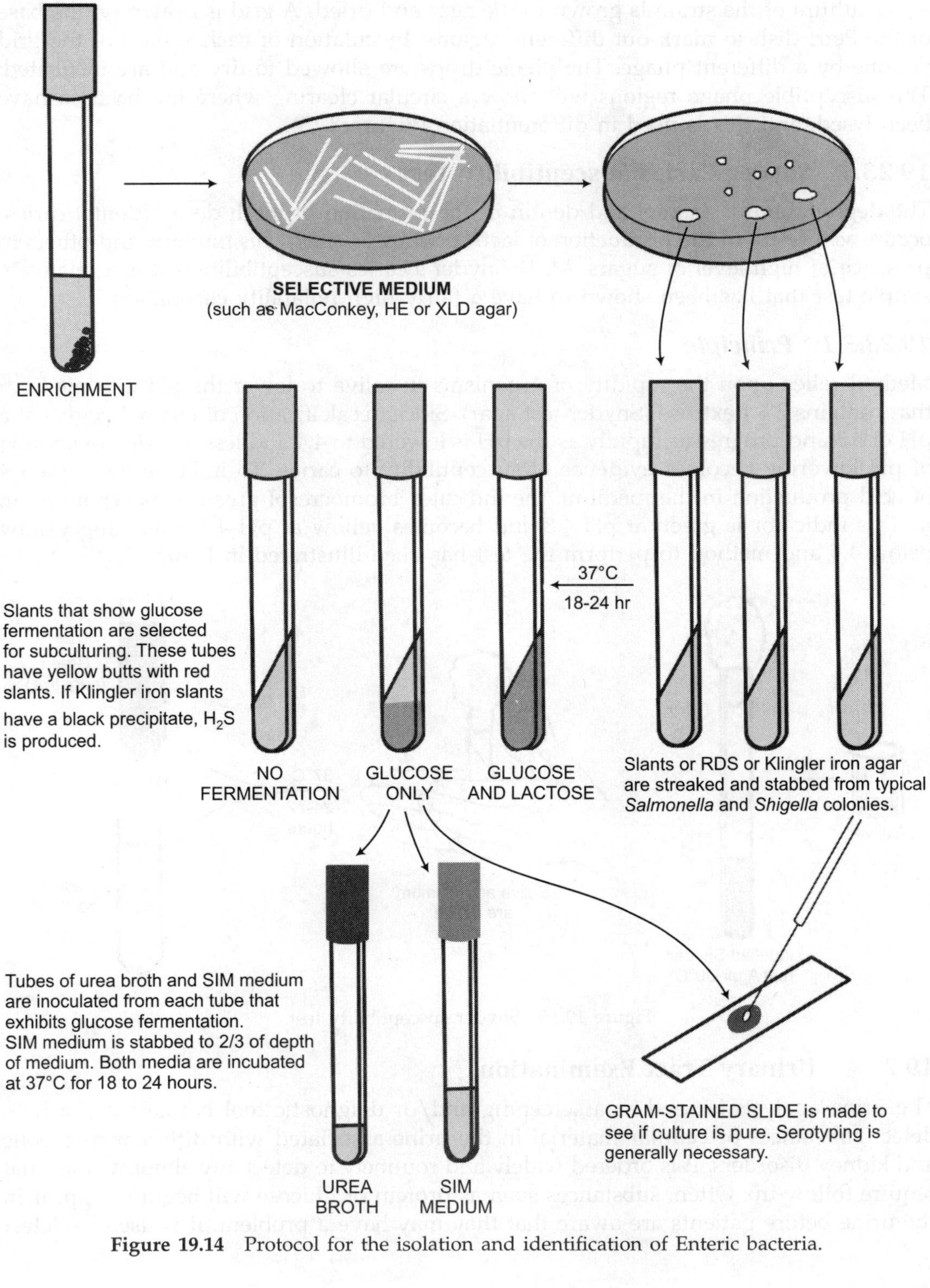

Figure 19.14 Protocol for the isolation and identification of Enteric bacteria.

A culture of the strain is grown in the agar and dried. A grid is drawn on the base of the Petri dish to mark out different regions. Inoculation of each square of the grid is done by a different phage. The phage drops are allowed to dry and are incubated: The susceptible phage regions will show a circular clearing where the bacteria have been lysed, and this is used in differentiation (Figure 8.13).

19.23.3 Snyder Caries Susceptibility Test

The degradation of enamel and dentin in the formation of tooth decay (dental caries) occurs as a result of the production of lactic acid by *Streptococcus* mutants and others in presence of high level of sugars. M. L. Snyder's caries susceptibility test is a relatively simple test that has been shown to have a fairly high reliability correlation.

19.23.3.1 Principle

Method relies upon the rapidity of organisms in saliva to lower the pH in a medium that contains 2% dextrose (Snyder test agar). Since decalcification of enamel begins at a pH of 5.5 and progresses rapidly as the pH is lowered to 4.4 and less, the demonstration of pH lowering becomes evidence of susceptibility to caries. To indicate the presence of acid production in the medium, the indicator bromocresol green is incorporated in it. This indicator is green at pH 4.8 and becomes yellow at pH 4.4 remaining yellow below 4.4 and method to perform the test has been illustrated in Figure 19.15.

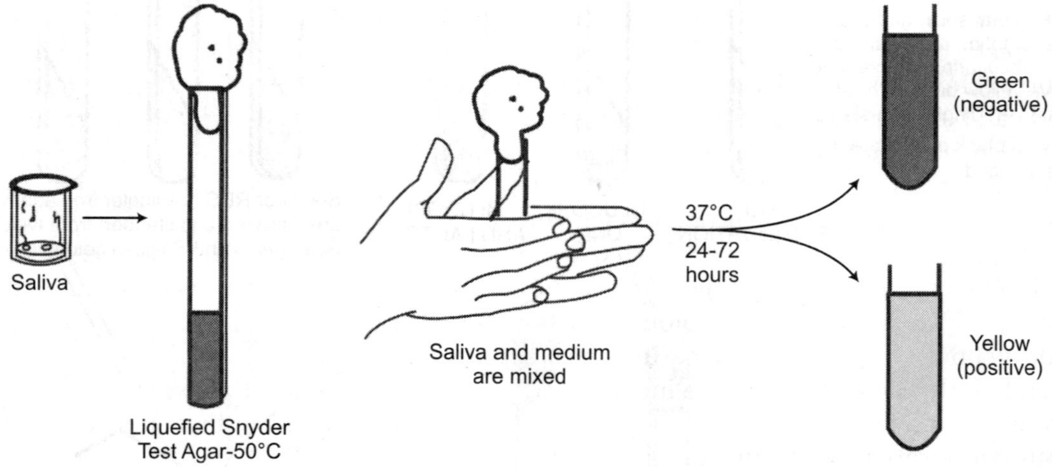

Figure 19.15 Snyder susceptibility test.

19.23.4 Urinary Tract Examination

The urine analysis is used as a screening and/or diagnostic tool because it can help detect substances or cellular material in the urine associated with different metabolic and kidney disorders. It is ordered widely and routinely to detect any abnormalities that require follow-up. Often, substances such as protein or glucose will begin to appear in the urine before patients are aware that they may have a problem. It is used to detect urinary tract infections (UTI) and other disorders of the urinary tract. In patients with acute or chronic conditions, such as kidney disease, the urinalysis may be ordered at intervals as a rapid method to help monitor organ function, status, and response to treatment. Urinalysis results can have many interpretations. Abnormal findings are a warning that something may be wrong and should be evaluated further. Generally, the greater the concentration of the atypical substance, such as greatly increased amounts of glucose, protein, or red blood cells, the more likely it is that there is a problem that needs to be addressed. But the results do not tell the doctor exactly what the cause of the finding is or whether it is a temporary or chronic condition. A normal urinalysis does not guarantee that there is no illness. Some people will not release elevated amounts of a substance early in a disease process, and some will release them sporadically during the day, which means that they may be missed by a single urine sample. In very dilute urine, small quantities of chemicals may be undetectable. Chronic or acute infections of the urinary tract may involve the kidneys, ureters, bladder or urethra. Such infections of this tract enter by way of the urethra; very few originate in the blood. Following examinations are normally performed for urine analysis.

19.23.4.1 The visual examination

Urine's colour, clarity, and concentration are examined first. Urines most often have variety of shades of yellow, from very pale or colourless to very dark or amber. Unusual or abnormal urine colours can be the result of a disease process, some medications, or the result of eating certain foods. For example, some people excrete red-coloured urine after eating beets. The colour is from the natural pigment of beets and is not a cause for worry. However, red-coloured urine can also occur when blood is present in the urine and can be an indicator of disease or damage to some part of the urinary system. Blood can also be a contaminant that gets into the urine unintentionally during collection, such as from hemorrhoids or from a woman's menstruation. Once this contaminating blood is in the urine, it will be detected during the chemical phase of a urinalysis, and your doctor will initially assume that it came from the urinary tract. The importance of blood in urine is discussed further in the chemical and microscopic examination sections. The depth of urine colour is also a crude indicator of urine concentration; (a) pale yellow or colourless urine indicates a dilute urine where lots of water is being excreted; (b) dark yellow urine indicates concentrated urine and the excretion of waste products in a smaller quantity of water, such as is seen with the first morning urine, with dehydration, and during a fever.

Urine clarity refers to how clear the urine is. Usually, laboratories report the clarity of the urine using one of the following terms: clear, slightly cloudy, cloudy, or turbid. "Normal" urine can be clear or cloudy. Substances that cause cloudiness but that are not considered unhealthy include mucus, sperm and prostatic fluid, cells from the skin, normal urine crystals, and contaminants such as body lotions and powders. Other substances that can make urine cloudy, like red blood cells, white blood cells, or bacteria, indicate a condition that requires attention. Urine colour and clarity can be a sign of what substances may be present in urine. However, confirmation of suspected substances is obtained during the chemical and microscopic examinations.

19.23.4.2 The chemical examination

To perform the chemical examination, most clinical laboratories use commercially prepared test strips. These are narrow plastic strips that hold small squares of paper called test pads, arranged in a row. The test pads have chemicals impregnated into them. When a strip is briefly, but completely, dipped into urine, the test pads absorb the urine and a chemical reaction changes the colour of the pad within seconds to minutes and compare the colour change for each reaction pad to a colour chart provided with the test strips to determine the result for each test. Each reaction pad must be evaluated at the appropriate time for that chemical. If too little time or too much time has passed since the reaction you may get incorrect results. To reduce timing errors and eliminate variations in colour interpretation, automated instruments are frequently used to "read" the reaction colour on each test pad. The degree of colour change on a test pad can also give an approximation of the amount of substance present. For example, a slight colour change in the test pad for protein may indicate a small amount of protein present in the urine whereas a deep colour change may indicate a large amount. The most frequently performed chemical tests using reagent test strips are; specific gravity, pH, protein, glucose, ketones, blood, leukocyte esterase, nitrite, bilirubin, and urobilinogen. Some reagent test strips also have a test pad for ascorbic acid (vitamin C).

Specific Gravity (SG)

The first test, specific gravity, is actually a physical characteristic of the urine, a measure of urine concentration that can be determined using a chemical test.

There are no "abnormal" specific gravity values. This test simply indicates how concentrated the urine is. Specific gravity measurements are actually a comparison of the amount of solutes (substances dissolved) in urine as compared to pure water. If there were no solutes present, the SG of urine would be 1.000 – the same as pure water. Since all urine has some solutes a urine SG of 1.000 is not possible. If a person drinks excessive quantities of water in a short period of time or gets an intravenous (IV) infusion of large volumes of fluid, then the urine SG may be as low as 1.002. The upper limit of the test pad, an SG of 1.035, indicates concentrated urine, one with many solutes in a limited amount of water. Knowing the urine concentration helps healthcare providers decide if the urine specimen they are evaluating is the best one to detect a particular substance. For example, if they are looking for very small amounts of protein, a concentrated morning urine specimen would be the best sample.

pH

As with specific gravity, there are typical but not "abnormal" pH values. The kidneys play an important role in maintaining the acid-base balance of the body. Therefore, any condition that produces acids or bases in the body such as acidosis or alkalosis, or the ingestion of acidic or basic foods, can directly affect urine pH. Diet can be used to modify urine pH. A high-protein diet or consuming cranberries will make the urine more acidic. A vegetarian diet, a low-carbohydrate diet, or the ingestion of citrus fruits will tend to make the urine more alkaline.

Some of the substances dissolved in urine will precipitate out to form crystals when the urine is acidic; others will form crystals when the urine is basic. If crystals form while the urine is being produced in the kidneys, a kidney stone or "calculus" can develop. By modifying urine pH through diet or medications, the formation of these crystals can be reduced or eliminated.

Protein

The protein test pad measures the amount of albumin in the urine. Normally, there will not be detectable quantities. When urine protein is elevated, you have a condition called proteinuria; this can be an early sign of kidney disease. Albumin is smaller than most other proteins and is typically the first protein that is seen in the urine when kidney dysfunction begins to develop. Other proteins are not detected by the test pad but may be measured with a separate urine protein test. Other conditions that can produce proteinuria include: (a) disorders that produce high amounts of proteins in the blood, such as multiple myeloma, (b) conditions that destroy red blood cells, (c) inflammation, malignancies (cancer), or injury of the urinary tract, for example, the bladder, prostate, or urethra and (d) vaginal secretions that get into urine

Glucose

Glucose is normally not present in urine. When glucose is present, the condition is called glucosuria. It results from either: (a) An excessively high glucose concentration in the blood, such as may be seen with people who have uncontrolled diabetes mellitus; and (b) a reduction in the "renal threshold." When blood glucose levels reach a certain concentration, the kidneys begin to excrete glucose into the urine to decrease blood concentrations. Sometimes the threshold concentration is reduced and glucose enters the urine sooner, at a lower blood glucose concentration. Some other conditions that can cause glucosuria include hormonal disorders, liver disease, medications, and pregnancy. When glucosuria occurs, other tests such as blood glucose are usually performed to further identify the specific cause.

Ketones

Ketones are not normally found in the urine. They are intermediate products of fat metabolism. They can form when a person does not eat enough carbohydrates (for example, in cases of starvation or high-protein diets) or when a person's body cannot use carbohydrates properly. When carbohydrates are not available, the body metabolizes fat instead to get the energy it needs to keep functioning. Ketones in urine can give an early indication of insufficient insulin in a person who has diabetes. Severe exercise, exposure to cold, and loss of carbohydrates, such as with frequent vomiting, can also increase fat metabolism, resulting in ketonuria.

Blood (Haemoglobin)

This test is used to detect haemoglobin in the urine (haemoglobinuria). Haemoglobin is a oxygen-transporting protein found inside red blood cells (RBCs). Its presence in the urine indicates blood in the urine (known as haematuria). The small number of RBCs normally present in urine (see microscopic examination) usually results in a "negative" test. However, when the number of RBCs increases, they are detected as a "positive" test result. Even small increase in the amount of RBCs in urine can be significant. Numerous diseases of the kidney and urinary tract, as well as trauma, medications, smoking, or strenuous exercise can cause haematuria or haemoglobinuria.

This test cannot determine the severity of disease nor be used to identify where the blood is coming from. For instance, contamination of urine with blood from haemorrhoids or vaginal bleeding cannot be distinguished from a bleed in the urinary tract. This is why it is important to collect a urine specimen correctly and for women to tell their healthcare provider that they are menstruating when asked to collect a urine specimen. Sometimes a chemical test for blood in the urine is negative, but microscopic examination shows increased numbers of RBCs. When this happens, test the sample for ascorbic acid (vitamin C), because vitamin C has been known to interfere with the accuracy of urine blood test results, causing them to be falsely low or falsely negative.

Leukocyte Esterase

Leukocyte esterase is an enzyme present in most white blood cells (WBCs). Normally, a few white blood cells (see microscopic examination) are present in urine and this test is negative. When the number of WBCs in urine increases significantly, this screening test will become positive. When the WBC count in urine is high, it means that there is inflammation in the urinary tract or kidneys. The most common cause for WBCs in urine (leukocyturia) is a bacterial urinary tract infection (UTI), such as a bladder or kidney infection.

Nitrite

This test detects nitrite and is based upon the fact that many bacteria can convert nitrate to nitrite in your urine. Normally, the urinary tract and urine are free of bacteria. When bacteria find their way into the urinary tract, they can cause a urinary tract infection (UTI). A positive nitrite test result can indicate a UTI. However, since not all bacteria are capable of converting nitrate to nitrite, one can still have a UTI despite a negative nitrite test.

Bilirubin

Bilirubin is not present in the urine of normal, healthy individuals. Bilirubin is a waste product that is produced by the liver from the haemoglobin of RBCs that are removed from circulation. It becomes a component of bile, a fluid that is secreted into the intestines to aid in food digestion. In certain liver diseases, such as biliary obstruction or hepatitis, bilirubin leaks back into the bloodstream and is excreted in urine. The presence of bilirubin in urine is an early indicator of liver disease and can occur before clinical symptoms such as jaundice develop.

Urobilinogen

Urobilinogen is normally present in urine in low concentrations. It is formed in the intestine from bilirubin, and a portion of it is absorbed back into the bloodstream.

Positive test results help detect liver diseases such as hepatitis and cirrhosis and conditions associated with increased RBC destruction (haemolytic anemia). When urine urobilinogen is low or absent in a patient with urine bilirubin and/or signs of liver dysfunction, it can indicate the presence of hepatic or biliary obstruction

19.23.4.3 The microscopic examination

A microscopic examination may or may not be performed as part of a routine urinalysis. It will typically be done when there are abnormal findings on the physical or chemical examination. It is performed on urine sediment – urine that has been centrifuged to concentrate the substances in it at the bottom of a tube. The fluid at the top of the tube is then discarded and the drops of fluid remaining are examined under a microscope. Cells, crystals, and other substances are counted and reported either as the number observed "per low power field" (LPF) or "per high power field" (HPF). In addition, some entities, if present, are estimated as "few," "moderate," or "many," such as epithelial cells, bacteria, and crystals.

Red Blood Cells (RBCs)

Normally, a few RBCs are present in urine sediment. Inflammation, injury, or disease in the kidneys or elsewhere in the urinary tract, for example, in the bladder or urethra, can cause RBCs to leak out of the blood vessels into the urine. RBCs can also be a contaminant due to an improper sample collection and blood from hemorrhoids or menstruation.

White Blood Cells (WBCs)

The number of WBCs in urine sediment is normally low. When the number is high, it indicates an infection or inflammation somewhere in the urinary tract. WBCs can also be a contaminant, such as those from vaginal secretions.

Epithelial Cells

Normally, in men and women, a few epithelial cells from the bladder (transitional epithelial cells) or from the external urethra (squamous epithelial cells) can be found in the urine sediment. Cells from the kidney (kidney cells) are less common. In urinary tract conditions such as infections, inflammation, and malignancies, more epithelial cells are present. Determining the kinds of cells present helps the healthcare provider pinpoint where the condition is located. For example, a bladder infection may result in large numbers of transitional epithelial cells in urine sediment. Epithelial cells are usually reported as "few," "moderate," or "many" present per low power field (LPF).

Microorganisms (bacteria, trichomonads, yeast)

In health, the urinary tract is sterile; there will be no microorganisms seen in the urine sediment. Microorganisms are usually reported as "none," "few," "moderate," or "many" present per high power field (HPF). Bacteria from the surrounding skin can enter the urinary tract at the urethra and move up to the bladder, causing a urinary tract infection (UTI). If the infection is not treated, it can eventually move to the kidneys and cause pyelonephritis. Less frequently, bacteria from a blood infection (septicaemia) may move into the urinary tract. This also results in a UTI. Special care must be taken during specimen collection, particularly in women, to prevent bacteria that normally live on the skin or in vaginal secretions from contaminating the urine. A urine culture may be performed if a UTI is suspected. In women (and rarely in men), yeast can also be present in urine. They are most often present in women who have a vaginal yeast infection, because the urine has been contaminated with vaginal secretions during collection. If yeast are observed in urine, then tests for a yeast (fungal) infection may be performed on vaginal secretions.

Trichomonads

Trichomonads are parasites that may be found in the urine of women or men (rarely). As with yeast, the trichomonads are actually infecting the vaginal canal and their presence in urine is due to contamination. If these are found during a urinalysis, then follow-up testing for *Trichomonas vaginalis* may be performed to look for a vaginal infection.

Casts

Casts are cylindrical particles sometimes found in urine that are formed from coagulated protein secreted by kidney cells. They are formed in the long, thin, hollow tubes of the kidneys known as tubules and usually take the shape of the tubule (hence the name). Under the microscope, they often look like the shape of a "hot dog" and in healthy people they appear nearly clear. This type of cast is called a "hyaline" cast (Figure 19.16). When a disease process is present in the kidney, other things such as RBCs or WBCs can become trapped in the protein as the cast is formed. When this happens, the cast is identified by the substances inside it, for example, as a red blood cell cast or white blood cell cast. Different types of casts are associated with different kidney diseases, and the type of casts found in the urine may give clues as to which disorder is affecting the kidney. Some other examples of types of casts include granular casts, fatty casts, and waxy casts. Normally, healthy people may have a few (0–5) hyaline casts per low power field (LPF). After strenuous exercise, more hyaline casts may be detected. Cellular casts, such as RBC and WBC casts, indicate a kidney disorder.

Crystals

Urine contains many dissolved substances (solutes) – waste chemicals that your body needs to eliminate. These solutes can form crystals, solid forms of a particular substance, in the urine if; (a) the urine pH is increasingly acidic or basic; (b) the concentration of dissolved substances is increased; and (c) the urine temperature promotes their formation.

Crystals are identified by their shape, colour, and by the urine pH. They may be small, sand-like particles with no specific shape (amorphous) or have specific shapes, such as a needle. Crystals are considered "normal" if they are from solutes that are typically found in the urine. Some examples of crystals that can be found in the urine of healthy individuals include (a) amorphous urates, (b) crystalline uric acid, (c) calcium

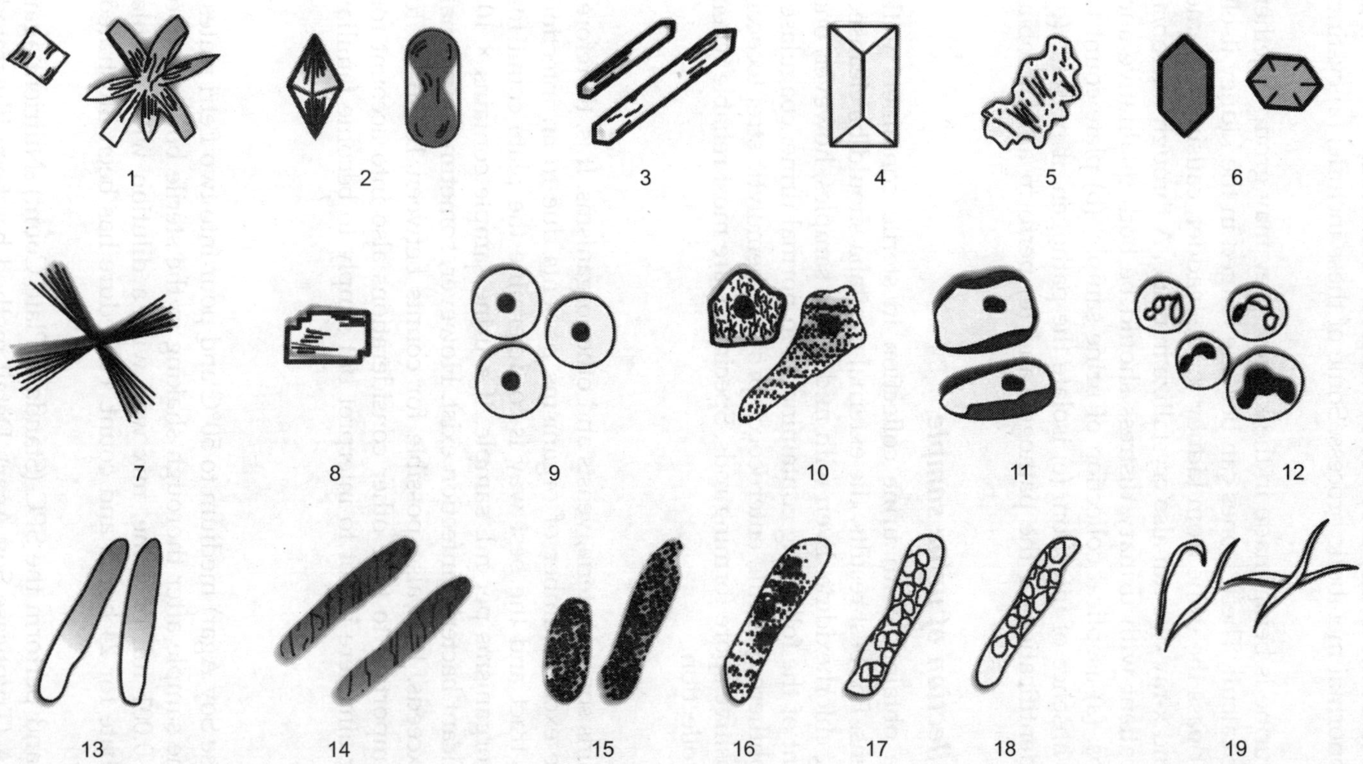

Figure 19.16 Microscopic elements in urine.

oxalates, (d) amorphous phosphates, (e) calcium carbonate If the crystals are from solutes that are not normally in the urine, they are considered "abnormal." Abnormal crystals may indicate an abnormal metabolic process. Some of these include; (a) cystine, (b) tyrosine, (c) leucine.

When crystals form as urine is being made in the kidney, they may group together to form kidney "stones" or calculi. These stones can become lodged in the kidney itself or in the ureters, tubes that pass the urine from kidney to the bladder, causing extreme pain. Medications, drugs, and x-ray dye can also crystallize in urine. A through microbial analysis of urine from a patient with urinary distress should be both qualitative and quantitative following steps: (a) aseptical collection of urine sample, (b) plate count to determine the presence or absence of infection, (c) isolate the pathogen, if present, (d) analysis of presumptive identification of the pathogen and (e) perform antimicrobial sensitivity test.

19.23.4.4 Aseptical collection of urine sample

Meaningful results can be obtained with urine collection in sterile containers with midstream voided specimens. For best results, the external genitalia should be cleansed with liquid soap containing chlorhexidine. Even with midstream samples, however, one can expect to find low count of the following contaminants in normal urine: coagulase-negative staphylococci, diphtheroid bacilli, enterococci, proteus, hemolytic streptococci, yeast and aerobic Gram-positive spore forming rods. Specimens are most reliable when plated immediately after collection.

19.23.4.5 Plate count

Normal urine always contains some bacteria, yeasts and other organisms. It is, therefore, necessary to determine the excess number of organisms presents due to an infection somewhere in the urinary tract and the best way is to determine the plate count for evaluating the number of organisms per mL sample. If a urine sample contains $\times 10^5$ one may assume the significant bacterial infection exist. However, sometime in normal individual, this number exceeds. It is also possible for counts between 10^3 to 10^5 to be significant. Thus, it is important to take other considerations also into account for evaluating the results. Our aim here is not to interpret but simply to become familiar with the basic procedures.

19.23.4.6 Procedure

Liquefy two TSA (Trypticase Soy Agar) medium to 50°C and pour into two Petri plates, cool and inoculate the urine sample after thorough shaking in the sterile bottle to two tubes separately 0.01 and 0.001 mL of urine, mix well with a dilution with inocula from a urine sample, incubate for 24 hours and count. Procedure has been illustrated in Figure 19.17.

Take the urine sample and perform the SPC (Standard Plate Count). Nutrient agar medium is replaced by TSA (Trypticase Soy Agar). Detail method has been illustrated in following Figure 19.18.

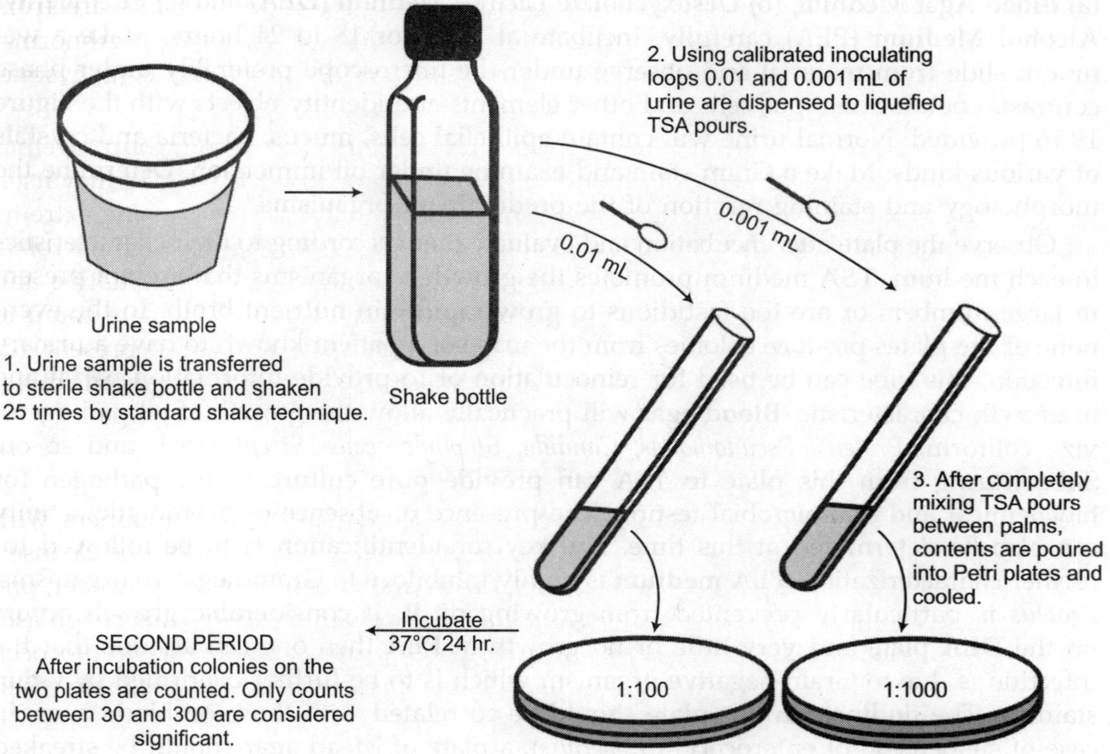

Figure 19.17 Plate count of urine sample.

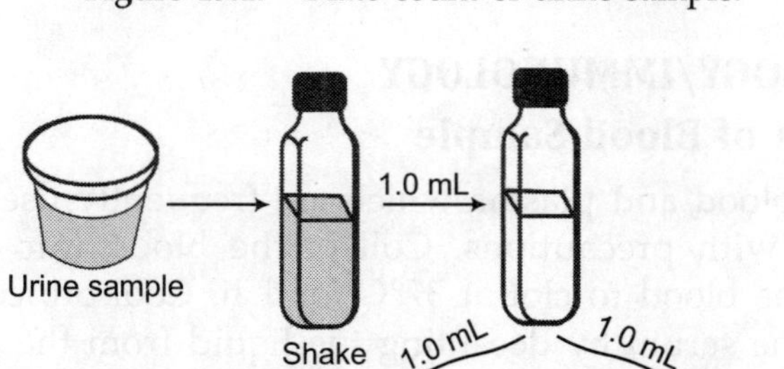

Figure 19.18 Standard plate count for urine sample.

Once it is established that an infection exists, the next step is to isolate the pathogen and identify it. Concentrate the sample of urine by centrifugation in a sterile screw cap centrifuge tubes for 10 minutes at 2000 rpm. Decant all but the 0.5 mL from the tube in the bottom and resuspend with the help of wire loop. Inoculate a tube of thioglycolate medium with a loopful of sediment. Strek a loopful of the sediment on three agar plates

(a) **Blood Agar Medium,** (b) **Desoxycholate Lactose Medium (DLA)** and (c) **Phenylethyl Alcohol Medium (PEA)** carefully. Incubate at 37°C for 18 to 24 hours. Make a wet mount slide from material and observe under the microscope preferably under phase contrast. Look for cast, pus cells and other elements and identify objects with the Figure 19.16 provided. Normal urine will contain epithelial cells, mucus, bacteria and crystals of various kinds. Make a Gram stain and examine under oil immersion. Determine the morphology and staining reaction of the predominant organisms.

Observe the plate after incubation and evaluate them according to their characteristics in each medium. **TSA** medium promotes the growth of organisms that are not present in large numbers or are too fastidious to grow rapidly in nutrient broth. In the event none of the plates produce colonies from the urine of a patient known to have a urinary infection, this tube can be used for reinoculation or to provide information pertaining to growth characteristic. **Blood agar** will practically allow the growth of all pathogens, viz., coliform, *Proteus, Pseudomonas, Candida, Staphylococcus, Streptococcus* and so on. Subculturing from this plate to TSA can provide pure culture of the pathogen for biochemical and antimicrobial testing. The presence or absence of haemolytic activity can also be determined at this time. The key for identification is to be followed for further characterization. **PEA medium** is highly inhibitory to Gram-negative organisms. *Proteus* is particularly prevented from growing on it. If considerable growth occurs on the **DLA** plate and very little or no growth in **PEA** then one can assume that the infection is due to Gram-negative organism which is to be further confirmed by Gram staining. The findings on this plate should be correlated with those on blood agar. In case of suspension of enterococci (*S. faecalis*), a plate of **Mead agar** should be streaked and incubated at 37°C. Enterococci produce pink colonies on this medium.

19.24 HAEMATOLOGY/IMMUNOLOGY

19.24.1 Collection of Blood Sample

The collection of the blood and plasma which are frequently used in the experiments must be carried out with precautions. Collect the blood into a glass, not plastic, container and leave the blood to clot at 37°C for 1 h. Cool on ice to allow the clot to retract and separate the serum by decanting the liquid from the clot and centrifuging on a bench centrifuge. Store the serum at 0°C until required then warm to room temperature just before use. Immunochemical assays are now used extensively in the biological sciences, and some examples of those currently in use are described herein as the series of experiments.

19.24.2 Determination of ABO Blood Group by Slide Agglutination Test

The interaction between antibody and a particulate antigen results in visible clumping called agglutination. Antibodies that produce such reactions are called agglutinins. Agglutination reactions are similar in principle to precipitation reactions; they depend on the crosslinking of polyvalent antigens. Just as, an excess of antibody inhibits precipitation reactions, such excess can also inhibit agglutination reactions; this inhibition is called the prozone effect. Because prozone effects can be encountered in many types of immunoassays, understanding the basis of this phenomenon is of general importance. Agglutination reactions are routinely performed to type red blood cells (RBCs). In typing for the ABO antigens, RBCs are mixed on a slide with antisera to the A or B blood-group antigens. If the antigen is present on the cells, it agglutinate, forming a visible clump on the slide. Determination of which antigens are present on donor and recipient RBCs is the basis for matching blood types for transfusions (Figure 19.19 and Table 19.14). When red cells are mixed with various reagent antisera (soluble antibody), agglutination will occur on the slides containing cells positive for (possessing the antigen) the corresponding antigen. No agglutination will occur when the red cells do not contain the corresponding antigen. One primary application of this principle is blood typing (Figure 19.21).

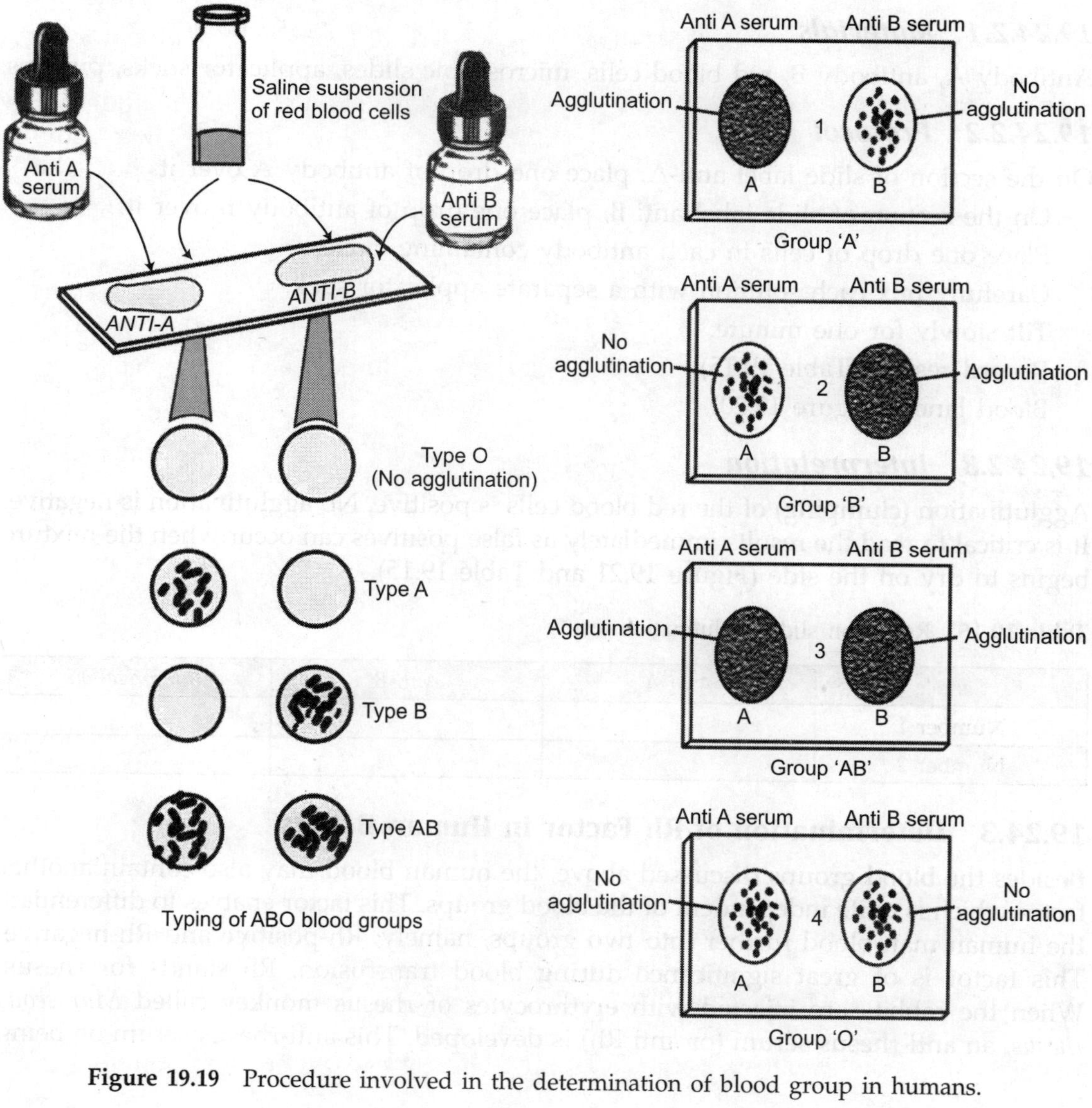

Figure 19.19 Procedure involved in the determination of blood group in humans.

Table 19.14 Blood grouping

Blood group	*Antigen on RBC*	*Antibodies in plasma*	*Can donate to recipient*	*Can receive from Donor*
A	A	b (anti B)	A, AB	O, A
B	B	a (anti A)	B, AB	O, B
AB	A and B	nil	AB,	A, B, AB, O
O	nil	a and b	A, B, AB, O	O

Figure 19.20 Blood lancet.

19.24.2.1 Materials

Antibody A, antibody B, red blood cells, microscopic slides, applicator sticks, pipettes

19.24.2.2 Protocol

On the section of slide label anti-A, place one drop of antibody A over it.

On the section of slide label anti-B, place one drop of antibody B over it.

Place one drop of cells in each antibody containing circle.

Carefully mix each solution with a separate applicator stick.

Tilt slowly for one minute.

Record results (Table 19.15).

Blood lancet (Figure 19.20).

19.24.2.3 Interpretation

Agglutination (clumping) of the red blood cells is positive. No agglutination is negative. It is critical to read the results immediately as false positives can occur when the mixture begins to dry on the side (Figure 19.21 and Table 19.15).

Table 19.15 Result on slide agglutination test

Sample	*Anti-A*	*Anti-B*	*Interpretation*
Number 1			
Number 2			

19.24.3 Determination of Rh Factor in Human Beings

Besides the blood groups discussed above, the human blood may also contain another factor which is quite independent of the blood groups. This factor enables to differentiate the human man blood further into two groups, namely, Rh-positive and Rh-negative. This factor is of great significance during blood transfusion. Rh stands for rhesus. When the rabbits are injected with erythrocytes of rhesus monkey called *Muraceous rhesus,* an anti-rhesus serum (or anti Rh) is developed. This anti-rhesus serum on being

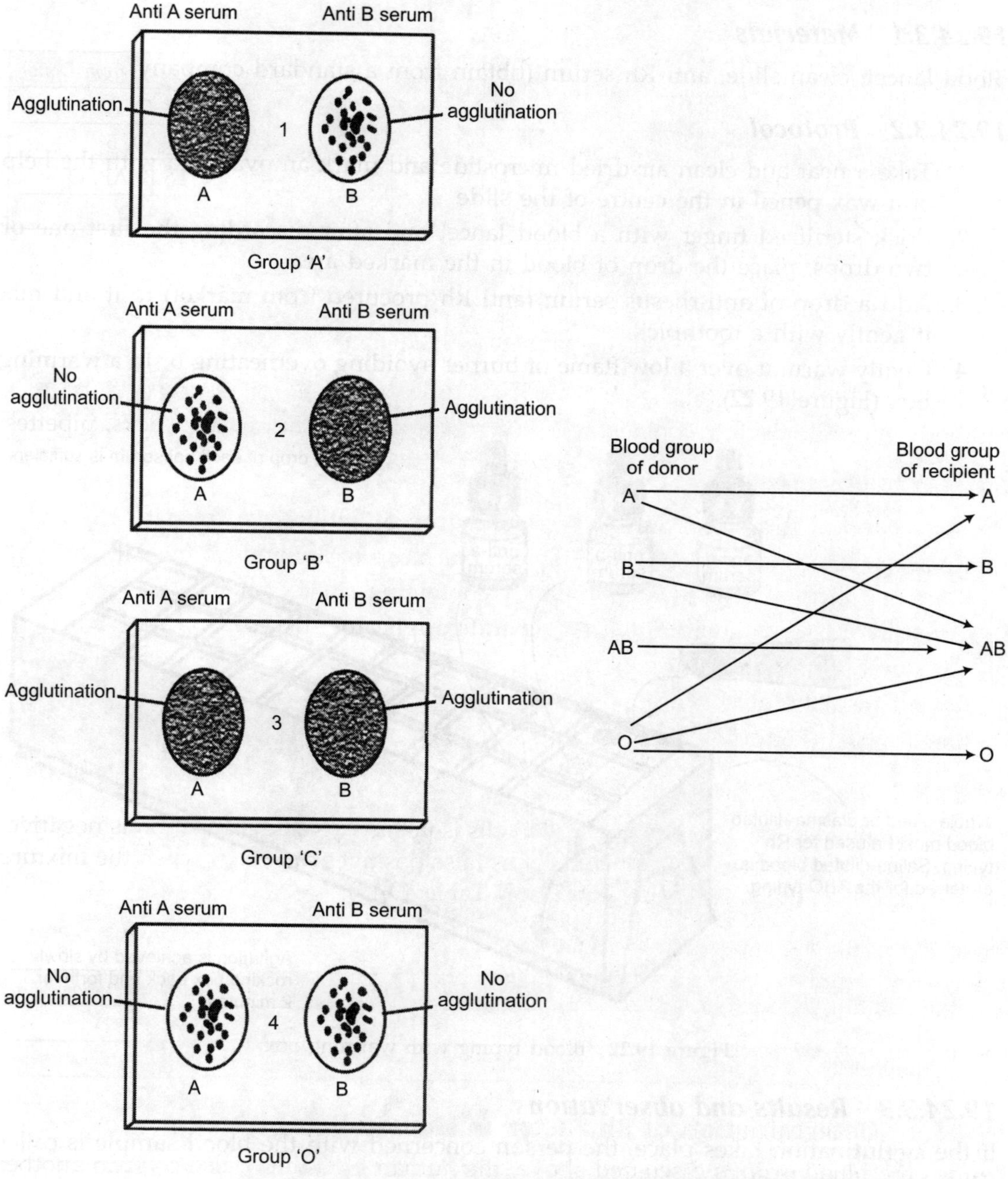

Figure 19.21 Compatible blood groups for transfusion.

tested brings about the agglutination of the erythrocytes of monkey as well as a great percentage of human population. About 97% Indians possess Rh positive factor, and the rest Rh negative factor.

19.24.3.1 Materials

Blood lancet, clean slide, anti Rh serum (obtain from a standard company).

19.24.3.2 Protocol

1. Take a neat and clean air-dried microslide and mark an oval area with the help of a wax pencil in the centre of the slide.
2. Prick sterilized finger with a blood lancet and after discarding the first one or two drops, place the drop of blood in the marked area.
3. Add a drop of anti-rhesus serum (anti Rh procured from market) to it and mix it gently with a toothpick.
4. Gently warm it over a low flame of burner avoiding overheating or in a warming box (Figure 19.22).

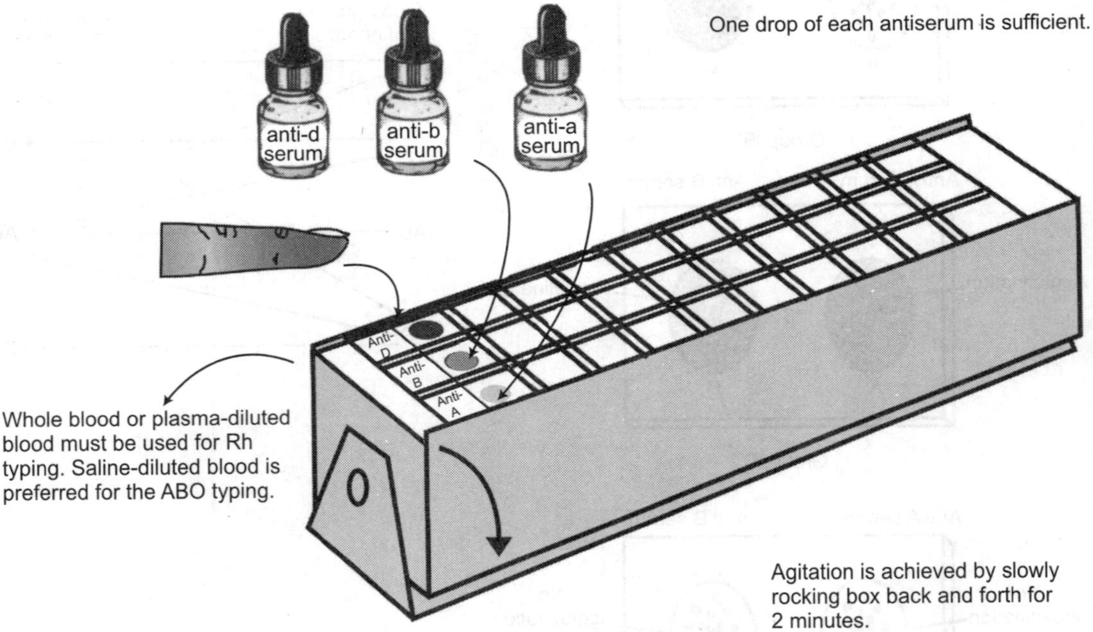

Figure 19.22 Blood typing with warming box.

19.24.3.3 Results and observations

If the agglutination takes place, the person concerned with the blood sample is called (Rh^+) otherwise (Rh^-).

19.24.4 Demonstration of Agglutination Reaction of Unknown Bacterial Culture by Slide Agglutination Technique

Many closely related microorganisms that are identical physiologically can be differentiated only by determining their antigenic nature. The method of determining the presence of specific antigen in a microorganism is called serological typing (serotyping). Organisms of different species not only differ in morphology and physiology, but they also differ in protein makeup. The different proteins of bacterial cells that are able to stimulate antibody production when injected into an animal are antigens. The antigen structure of each species of bacteria is unique to that species and, like the fingerprint of an individual, can be used to identify the organism. Many closely related microorganisms that are identical physiologically can be differentiated only by determining their antigenic nature. Serotyping consists of applying suspension of the organism to antiserum, which is specific for the known antigens. If the antigens are present, the antibodies in the antiserum will combine with the antigens, causing agglutination or clumping of the bacterial cell. Serotyping is playing a key role in vaccine research against several infectious agents including viruses and bacteria. Method used for the serotyping is illustrated below in Figure 19.23.

19.24.4.1 Materials

Slide/plate containing 3 wells, serological tube, phenolised saline (0.85% sodium chloride, 0.5% phenol), *Salmonella* 0 polyvalent serum, *Salmonella* 0 antigen.

19.24.4.2 Protocol

1. Take a clean depression slide or spot plate and label the 3 depressions as A, B and C.
2. Prepare a phenolized saline suspension of each culture in separate serological tubes by transferring one or more loopfuls of organisms in one mL of phenolized saline. Mix it thoroughly and observe that it should be turbid.
3. Transfer 0.05 mL from the phenolized saline suspension of one tube to cavity A and B.
4. Add one drop of *Salmonella* antiserum inside each of the two depressions B and C.
5. Add one drop of phenolized saline to depression A.
6. Add one drop of antigen to depression C.
7. Separately mix the cells with serum in each depression with an unused end of the application stick.
8. Gently rock the slide back and forth, holding it horizontally for 2-3 minutes.
9. Repeat this process on another depression slide for the other organisms. To cavity B and C add 1 drop of Salmonella 0 polyvalent serum.
10. Compare all the three mixtures for agglutination reaction.

19.24.4.3 Results and observations

If agglutination appears in cavity B, the bacterial sample is *Salmonella.* Slide agglutination technique is shown in Figure 19.21.

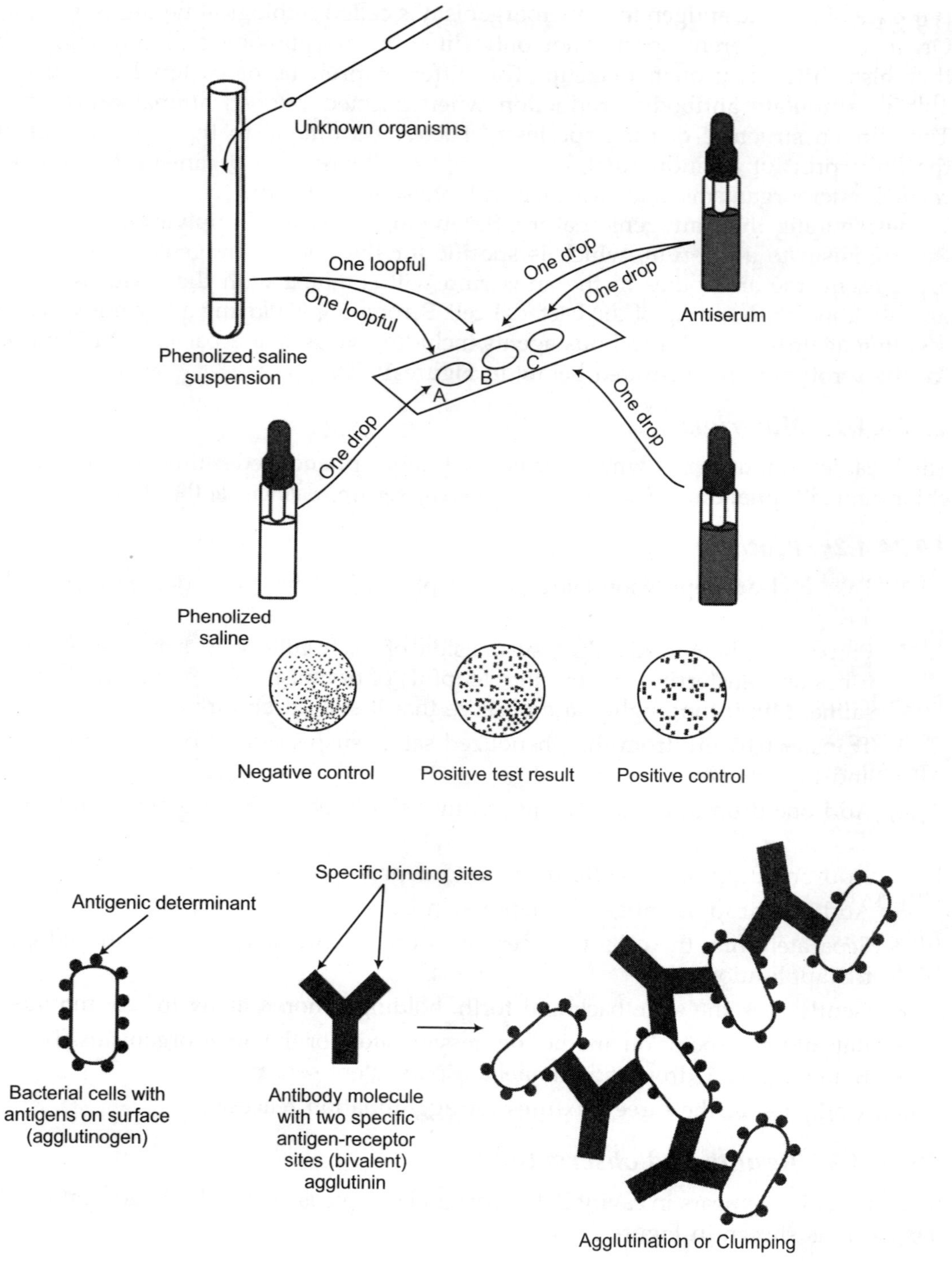

Figure 19.23 Slide agglutination test for serotyping.

19.24.5 Estimation of Total Leukocyte Count by Haemocytometer

The glacial acetic acid lyses the red cells while the gentian violet slightly stains the nuclei of the leukocytes. The blood specimen is diluted 1:20 in a WBC pipette with the diluting fluid and the cells are counted under low power of the microscope by using a counting chamber. The number of cells in undiluted blood is reported per cu mm (µL) of whole blood.

Increase in total leukocyte count of more than 10,000/cu mm (µL) is known as leukocytosis and decrease of less than 4 000 cu mm (µl) as leukopenia.

19.24.5.1 Causes of leukocytosis

Pathological

Leukocytosis is a common disease when there is a transient period of infections. The degrees of the rise in leukocytes depend on the type and severity of the infection and the response of the body. The infection may be 1) bacterial 2) viral 3) protozoal (malaria) or 4) parasitic (filaria, hookworm infection). Leukocytosis is also observed in severe hemorrhage and in leukaemia.

Physiological

Age: At birth the total leukocyte count is about 18,000/cu mm (µL). It drops gradually at adult level.

Pregnancy: At 'full term' the total count tends to be about 12,000 to 15,000/cu mm (µL). It rises soon after delivery and then gradually returns to normal.

High temperature.

Severe pain.

Muscular exercise.

19.24.5.2 Causes of leukopenia

Certain viral and bacterial infections (typhoid) lead to leukopenia rather than leukocytosis.

Infections

Bacterial (typhoid, paratyphoid, tuberculosis, etc.)

Viral (hepatitis, influenza, measles, etc.)

Protozoal (malaria)

Some cases of leukaemia.

Primary bone marrow depression (Aplastic anaemia).

Secondary bone marrow depression (due to drugs, radiation, etc.)

Anemia (Iron deficiency Megaloblastic etc)

19.24.5.3 Normal values

- Adults 4,000-10,000/cu mm (µL)
- At birth 10,000-25,000/cu mm (µL)

- 1 to 3 years 6,000-18,000/cu mm (μL)
- 4 to 7 years 6,000-15,000/cu mm (μL)
- 8 to 12 years 4,500-13,500/cu mm (μL)

19.24.5.4 Materials

Microscope

Improved Neubauer Chamber (Figure 19.24)

WBC pipette (Figure 19.25).

WBC diluting fluid containing Glacial acetic acid 2.0 mL, 1% (w/v) gentian violet: 1.0 mL and distilled water 97 mL. This solution is stable at room temperature (25°C ± 5°C). A pinch of thymol may be added as a preservative.

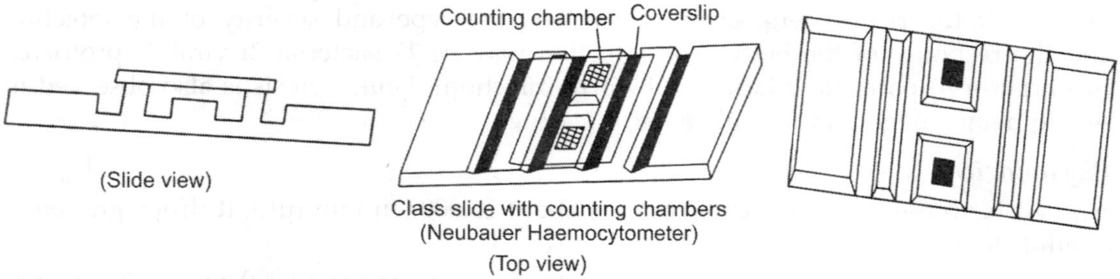

Figure 19.24 Haemocytometer with Neubauer counting chamber with coverslip over it.

19.24.5.5 Specimens

Double oxalated or EDTA blood

Capillary blood (specimen need not be a fasting sample)

19.24.5.6 Protocol

1. Draw blood up to 0.5 mark of a WBC pipette.
2. Carefully, wipe excess blood outside the pipette by using cotton. Draw diluting fluid up to 11 mark.
3. Mix the contents in the pipette and after five minutes by discarding few drops, fill the counting chamber and allow the cells to settle for two to three minutes.
4. Since bulb pipettes are not recommended following procedure is performed,
5. Make a 1:20 dilution of blood with diluting fluid by with 20 μL of blood in a glass tube (10 × 75 mm). Cork the tube tightly and mix the suspension by rotating in a cell-suspension mixer for at least 1 minute. Fill the Neubauer counting chamber by means of a Pasteur pipette or glass capillary.
6. Focus on one of the 'W' marked areas (each having 16 small squares) by turning objective to low power. Count cells in all four W marked corner.

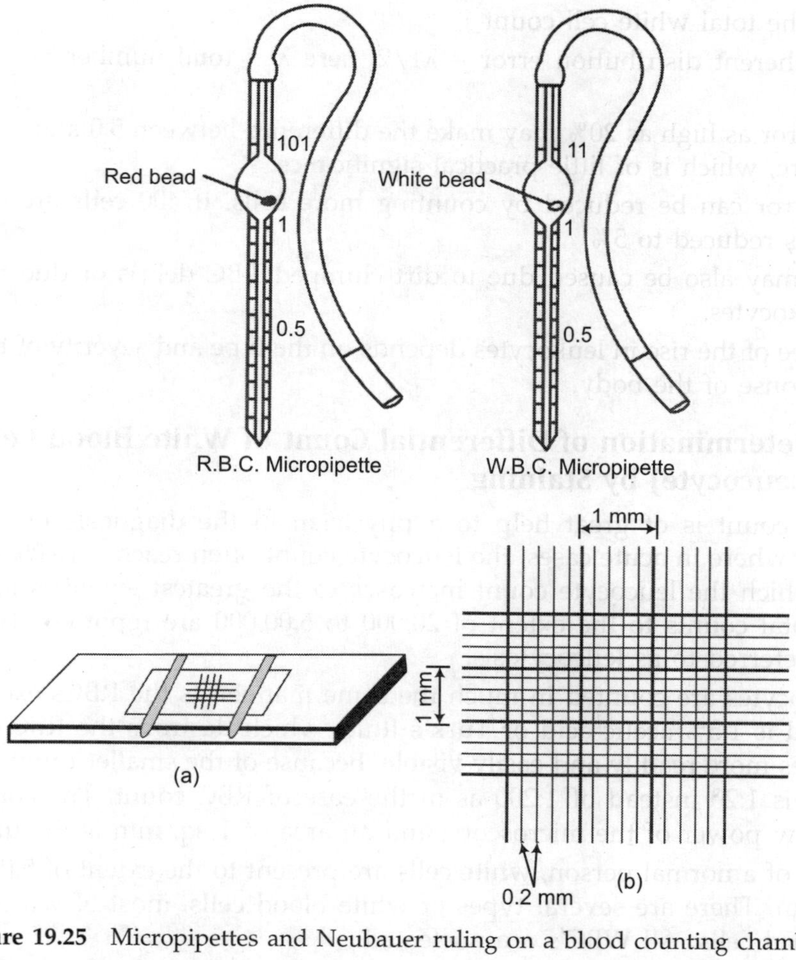

Figure 19.25 Micropipettes and Neubauer ruling on a blood counting chamber.

19.24.5.7 Calculations

Number of white cells/cu mm (μL) of whole blood = (number of white cells counted × dilution)/(area counted × depth of fluid) (Where: Dilution = 20)

Area counted 4 × 1 sq.mm = 4 sq mm

Depth of fluid = 0.1 mm (constant)

Hence, number of white cells per cu mm (μL) of whole blood = (No. of cells counted × 20)/(4 × 0.1) = No of cells counted × 50

Note: The precautions taken are exactly the same as for REC counting technique. The sources of error are also same as for RBC counting technique. However, in the case of WBC counting extra care is taken during the preparation and storage of WBC diluting fluid. It should be perfectly free from dust particles and yeast cells; otherwise falsely high counts are obtained due to the presence of yeast cells and dust particles.

Error of the total white cell count

1. The inherent distribution error = λ1/2, here λ = total number of cells in each area.
2. The error as high as 20% may make the difference between 5.0 and 6.0 × 10^9 cells per litre, which is of little practical significance.
3. The error can be reduced by counting more cells. If 400 cells are counted the error is reduced to 5%.
4. Error may also be caused due to dirt clumped RBC debris or due to clumping of leukocytes.

The degree of the rise in leukocytes depends on the type and severity of the infection and the response of the body.

19.24.6 Determination of Differential Count of White Blood Cells (Leucocyte) by Staining

A leucocyte count is of great help to a physician in the diagnosis of disease, like appendicitis, where in acute cases, the leucocyte count often reaches 15,000-20,000. The disease in which the leucocyte count increases to the greatest extent is leukaemia in which the total counts to the extent of 20,000 to 5,00,000 are reported. This disorder of blood is referred to as leucocytosis.

The leucocytes are counted in much the same manner as the RBCs except that the diluting fluid is 1.5% acetic acid or Turk's fluid, which destroys the RBC's and make the leucocytes more readily and easily visible. Because of the smaller number of WBCs, the dilution is 1:20 instead of 1:200 as in the ease of RBC count. The count is made under the low power of the microscope and an area of 1 sq. mm is the unit counted. In the blood of a normal person, white cells are present to the extent of 5,000 to 10,000 per cubic mm. There are several types of white blood cells, most of which are larger than red blood cells. All WBC's are nucleated.

Blood platelets are normally present to the extent of 3,50,000 per cubic mm. They are much smaller than red cells and are of irregular shape. They are devoid of nuclei, and in a stained preparation, they are visible in groups or clusters (Figure 19.26), shows the general appearance, relative size and other characteristics of different types of cells as revealed in the microscopic examination.

Hypersegmented Neutrophil

A hypersegmented neutrophil is a mature neutrophil. The nucleus of a hypersegmented neutrophil is divided into six or more segments or lobes.

Eosinophil

Eosinophils aid in detoxification. They also break down and remove protein material. The cytoplasm of an eosinophil contains numerous coarse, reddish-orange granules, which are lighter coloured than the nucleus.

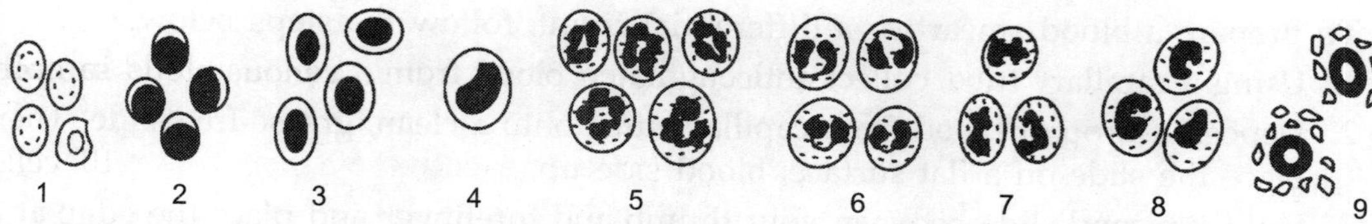

Figure 19.26 1. RBC 2. Small lymphocyte 3. Large lymphocyte 4. Monocyte or large mono nuclear 5. Polymorphonuclears (Neutrophils) 6. Eosinophils 7. Basophils (Mast cells) 8. Transitionals 9. RBC surrounded by blood platelets.

Basophil

The function of basophilic cells is unknown. It is believed, however, that basophilic cells keep the blood from clotting in inflamed tissue. Scattered large, dark-blue granules that are darker than the nucleus, characterize the cell as a basophil. Granules may overlay the nucleus as well as the cytoplasm.

Lymphocyte

The function of lymphocytes is also unknown, but it is believed that they produce antibodies and destroy the toxic products of protein metabolism. The cytoplasm of a lymphocyte is clear sky blue, scanty, with few unevenly distributed, azurophilic granules with a halo around them. The nucleus is generally round, oval, or slightly indented, and the chromatin (a) network of fibres within the nucleus) is lumpy and condensed at the periphery.

Monocyte

The monocyte, the largest of the normal white blood cells, destroys bacteria, foreign particles, and protozoa. Its colour resembles that of a lymphocyte, but its cytoplasm is a muddy grey-blue. The nucleus is lobulated, deeply indented or horseshoe-shaped, and has a relatively fine chromatin structure. Occasionally, the cytoplasm is more abundant than in the lymphocyte.

19.24.6.1 Protocol

The procedure for the differential white cell count is completed in 4 steps: Step 1: Making the blood smear, Step 2: Staining the cells, Step 3: Counting the cells, and Step 4: Reporting the count.

19.24.6.1.1 Preparation of blood smears

The simplest way to count the different types of white cells is to spread them out on a glass slide. The preparation is called a blood smear. There are two methods of making a blood smear: the slide method and the cover glass method.

It is very important to make a good blood smear. If it is made poorly, the cells may be so distorted that it will be impossible to recognize them. Prepare at least two smears for each patient as the additional smear should be examined to verify any abnormal findings.

To prepare a blood smear for a differential count, follow the steps below:

1. Using a capillary tube, collect anticoagulated blood from a venous blood sample.
2. Deposit a drop of blood from capillary tube onto a clean, grease-free slide. Then place the slide on a flat surface, blood side up.
3. Hold a second slide between your thumb and forefinger and place the edge at a 23-degree angle against the top of the slide that holds the drop of blood. Back the second slide down until it touches the drop of blood. The blood will distribute itself along the edge of the slide in a formed angle.
4. Push the second slide along the surface of the other slide, drawing the blood across the surface in a thin, even smear. If this is done in a smooth, uniform manner, a gradual tapering effect (or "feathering") of the blood will occur on the slide (Figure 19.27).

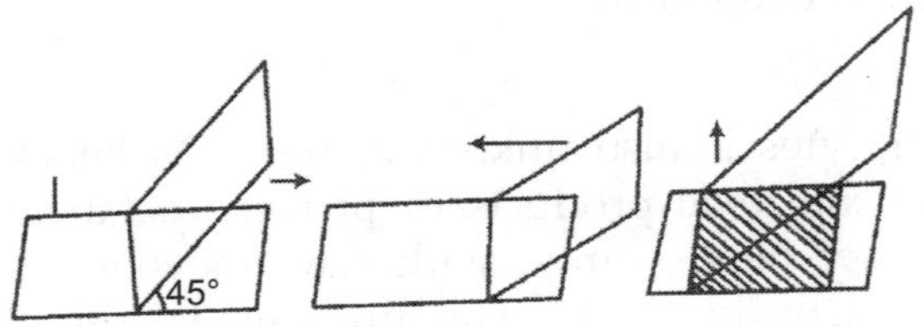

Figure 19.27 Preparation of blood smear.

5. A small drop of blood is placed about ¾ inch away from one end of the slide. The drop should not exceed 1/8 diameter.
6. The spreader slide is moved in the direction of arrow, allowing drop of blood to spread along slide's back edge.
7. The spreader slide is pushed along the slide, dragging the blood over the surface of the slide.
8. A pencil is used to mark off both ends of the smear to retain the staining solution on the slide.

19.24.6.1.2 Staining the cells

1. Fix for at least 30 seconds in absolute methanol.
2. Remove methanol by tilting the slide or by simply removing from the fixing jar.
3. Apply staining solution I freshly diluted with an equal part of buffer for 5 minutes on a horizontally positioned slide or in a jar.
4. Transfer slide from the jar without washing (or remove staining solution by holding slide vertically) into staining solution II that has been freshly diluted with 9 parts buffer for 10 to 15 minutes.
5. Transfer slide to jar with buffer for 1 rinse after removing stain.
6. Wash slide with ample water.
7. Transfer slide to a jar containing water for 2 to 5 minutes.
8. Dry the slide in a tilted position; do not blot dry.

(i) Mount a coverglass if desired. (ii) Observe under oil immersion lens different types of cells (Figure 19.28).

Examine the slide for WBC's and record the number of each type

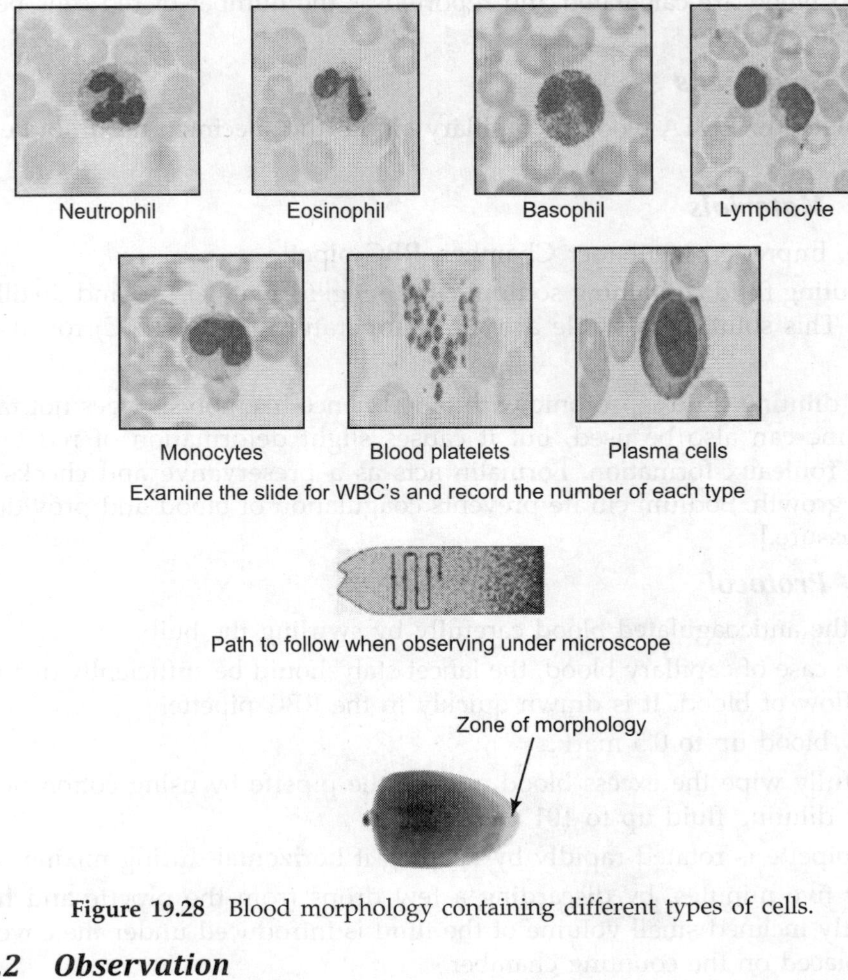

Figure 19.28 Blood morphology containing different types of cells.

19.24.6.2 Observation

Record results as per Table 19.16.

Table 19.16 Differential count of WBC

Cells	*Counts*
% Segmented neutrophils	
% Banded neutrophils	
% Eosinophil	
% Lymphocytes	
% Monocytes	

19.24.7 Determination of Total Erythrocyte Count by Haemocytometer

The blood specimen is diluted 1:200 with the RBC diluting fluid and cells are counted under high power (40 × objective) by using a counting chamber. The number of cells in undiluted blood are calculated and reported as the number of red cells per cu mm (μL) of whole blood.

19.24.7.1 Specimens

Double oxalated or EDTA blood or capillary blood (the specimen need not be a fasting sample).

19.24.7.2 Materials

Microscope, Improved Neubauer Chamber, RBC pipette.

RBC diluting fluid containing sodium citrate 3 g, formalin 1 mL and distilled water to 100 mL. This solution is stable at room temperature (25°C ± 5°C) for at least one year.

[*Note:* RBC diluting fluid is isotonic with blood hence haemolysis does not take place. Normal saline can also be used, but it causes slight deformation of red blood cells and allows rouleaux formation. Formalin acts as a preservative and checks bacterial and fungal growth. Sodium citrate prevents coagulation of blood and provides correct osmotic pressure.]

19.24.7.3 Protocol

1. Mix the anticoagulated blood carefully by swirling the bulb.
2. In the case of capillary blood, the lancet stab should be sufficiently deep to allow free flow of blood. It is drawn quickly in the RBC pipette.
3. Draw blood up to 0.5 mark.
4. Carefully wipe the excess blood outside the pipette by using cotton or a gauze, draw diluting fluid up to 101 mark..
5. The pipette is rotated rapidly by keeping it horizontal during mixing.
6. After five minutes, by discarding a few drops from the pipette and holding it slightly inclined small volume of the fluid is introduced under the covers which are placed on the counting chamber.
7. Allow the cells to settle for 2 to 3 minutes.
8. Place the counting chamber on the stage of the microscope.
9. Switch to low power objective. Adjust light and locate the large square in the centre with 25 small squares. Now switch to high power objective.

19.24.7.4 Calculations

The red blood cells in the four corner squares and in the centre square are counted.

Use following formulae for the calculation of red blood cells.

• Total red blood cells per litre of blood = RBCs/cu mm (μL) × 10^6 Or use the following formula – Red cell count (per liter) = [(no. of cells counted)/(volume counted μL)] × dilution × 10^6

Total red blood cells/cu mm (μL) = [number of red cells counted × dilution]/[area counted × depth of fluid]. Where (1) Dilution = 1:200 (i.e., 200)

Area counted = 80/400 = 1/5 sq mm

Since cell are counted in 5 bigger squares and such square is further divided into 16 small squares.

Each small square = 1/400 sq mm

Hence, area of (5 × 16) = 80 such areas = 80/400 sq mm = 1/5 sq mm

Depth of fluid = 1/10 mm

Number of red cells counted = N

Hence, total red blood cells, cu mm = [(N × 200)]/[(1/5) × (1/10)] = N × 200 × 50 = N × 10000

19.24.7.5 Observations

Additional information

For dilution of blood, instead of using an RBC pipette follow the procedure given below:

1. By using a 20 μL (standardized) pipette, mix 20 μL of blood with 4 mL of RBC diluting fluid in a glass test tube (10 × 75 mm). It is mixed well and then, introduce small volume of this fluid into a counting chamber.
2. The errors of blood counting by visual means are considerable due to the random distribution of the cells in the counting chamber. The movement of the cells in the chamber during the filling process causes them to collide, and this influences their distribution. The inherent distribution error can be reduced by counting more cells.
3. To confirm specific results, it is better to repeat a count using a second chamber.

Sources of Error

Falsely high counts

1. Due to collection of blood from the area where there is haemo concentration.
2. Due to inadequate wiping of the pipette.
3. Due to improper pipetting of blood as well as the fluid.
4. Because of improper mixing.
5. Due to uneven distribution in the counting chamber.
6. Due to errors in calculations.

Falsely low counts

1. Blood dilution with tissue fluid due to oedema or squeezing.
2. Improper pipetting and dilution (when blood drawn is less and if dilution is above the mark).
3. Dilution of the contents in the pipette by saliva.

4. Uneven distribution of the cells in the counting chamber.
5. Error in calculations.
6. Use of improperly standardized counting chamber and bad adjustment of the coverslip

19.24.7.6 Clinical significance

At birth, the total erythrocyte count varies from 6.5 million to 7.25 million per cubic mm (cu mm). There is the steady decline after a few hours and at the end of 15 days to one month there is a small rise to normal adult levels. An increase in total erythrocyte count is observed in conditions such as haem concentration due to burns, cholera, etc., in central cyanotic states as seen in chronic heart disease, conditions of decreased lung function such as emphysema and in polycythemia. Decrease in erythrocyte count is observed in old age, during pregnancy and group of diseases classified under anaemia.

Normal values

Male: 4.5 to 6.0 × 106 cells/cu mm (μL).
Female: 4.0 to 4.5 × 106 cells/cu mm (μL).

19.24.7.7 Results and observations

Report results in view of above points

19.24.8 Slide Agglutination Test (Serotyping)

Microorganisms, viruses, or cells are classified together based on their cell surface antigens. Serotyping plays an essential role in determining species and subspecies. Several microorganisms which are identical physiologically can be differentiated only by determining their antigenic nature. The antigenic properties are responsible for stimulating antibody production when injected into animals. The antigenic structure of each species of bacteria/virus is unique to that species and like the fingerprint of an individual which can be used for the identification of organisms. The method of determining the serotype is known as **serotyping** which is based on a variety of factors, including virulence, lipopolysaccharides (LPS) in Gram-negative bacteria, presence of an exotoxin, plasmids, phages, genetic profile (such as determined by polymerase chain reaction), or other characteristics which differentiate two members of the same species, allowing the epidemiologic classification of organisms to the subspecies level. A group of serovars with common antigens is called a **serogroup**. It consists of adding a suspension of the organisms to antiserum, which contains antibodies that are specific for the known antigens. If the antigens are present, the antibodies in the antiserum will combine with the antigens, causing agglutination, or clumping of the bacterial cells. In the identification of the various serotypes of two different genera, the use of antisera is generally performed after basic biochemical tests. The *Salmonella* genus of bacteria, for example, has been determined to have over 4400 serotypes, including *Salmonella enterica* serovar Typhimurium, *S. enterica* serovar Typhi, and *S. enterica* serovar Dublin. *Vibrio* *cholerae*, the species of bacteria that causes cholera, has over 200 serotypes, based on cell antigens. Only two of them have been observed to produce the potent enterotoxin that results in cholera: 0:1 and 0:139.

19.24.9 Role in Organ Transplantation

The immune system is capable of discerning a cell as being 'self' or 'non-self' according to that cell's serotype. In humans, that serotype is largely determined by human leukocyte antigen (HLA), the human version of the major histocompatibility complex. Cells determined to be non-self are usually recognized by the immune system as foreign, causing an immune response, such as haemagglutination. Serotypes differ widely between individuals; therefore, if cells from one human (or animal) are introduced into another random human, those cells are often determined to be non-self because they do not match the self-serotype. For this reason, transplants between genetically non-identical humans often induce a problematic immune response in the recipient, leading to transplant rejection. In some situations this effect can be reduced by serotyping both recipient and potential donors to determine the closest HLA match. Serotyping is playing a key role in vaccin research against several infectious agents including viruses and bacteria. Method used for the serotyping is illustrated below in Figure 19.29.

19.24.10 Latex Fixation Test (or Latex Agglutination Test)

This test is based on agglutination of particles and has been used as diagnostic tool for many years. Agglutination is defined as the clumping of particles to form insoluble aggregates. In a latex fixation test, a sample is mixed with latex beads coated with antibodies. If the antigen is present, it will react with the antibodies causing the latex beads to clump. These tests employ latex particles, gelatin beads, colloidal particles, or preserved mammalian or avian blood cells to facilitate visualization of agglutination. The components of an agglutination test consist of a test card or microscope slide and an agglutination solution with analyte-specific molecules that are attached to particles. A specimen is placed on a microscope slide or in a microtest plate well, and then the carrier particles are added. The mixture is then stirred or agitated and incubated or allowed to settle for a few minutes.

The results are determined by visually observing agglutination of the carrier particles as they bind to the specific analyte(s) in the sample. In strong reactions, the interpretation of results is simple, but in weak or marginal reactions, the interpretation can be difficult, and accurate results are possible only after much experience and practice. In the case of microwell plate assays, the pattern of particles that have settled needs to be interpreted to determine a result (Figure 19.30). Advantages of agglutination methods include a low individual test cost, ability to obtain semiquantitative results, and a relatively short time to obtain results. Semiquantitative results can be obtained by performing 2- to 10-fold specimen dilutions. Disadvantages include the need to carefully interpret marginal results and problems with specificity due to interfering substances in many assays. The technique is used to detect antibodies, such as those produced in response to the rubella virus or the rheumatoid factor. May be used for Rickettsia,

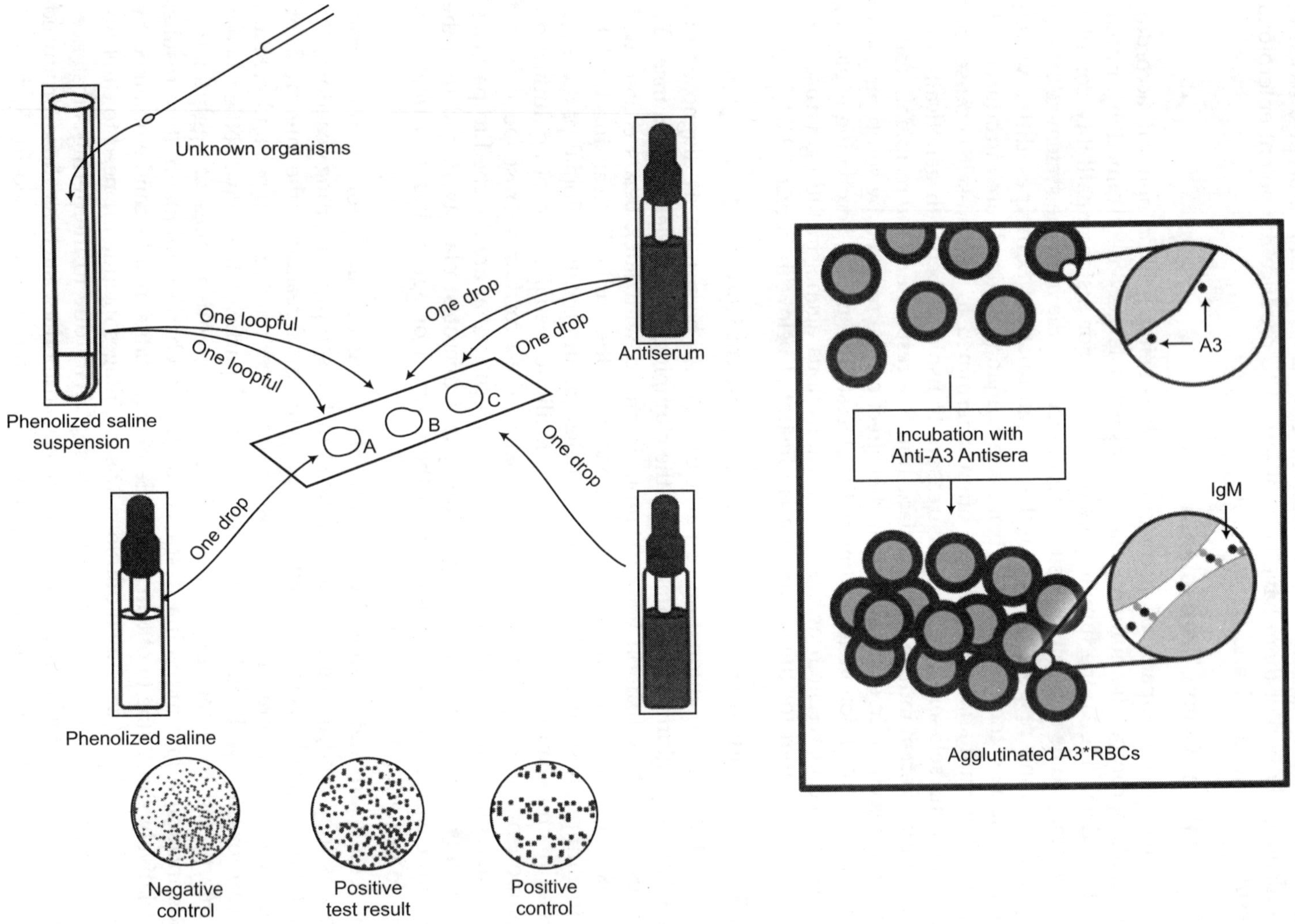

Figure 19.29 Slide agglutination test for serotyping; agglutination of HLA A3 positive red blood cells (RBCs) with anti-A3 alloreactive antisera containing anti-A3 IgM.

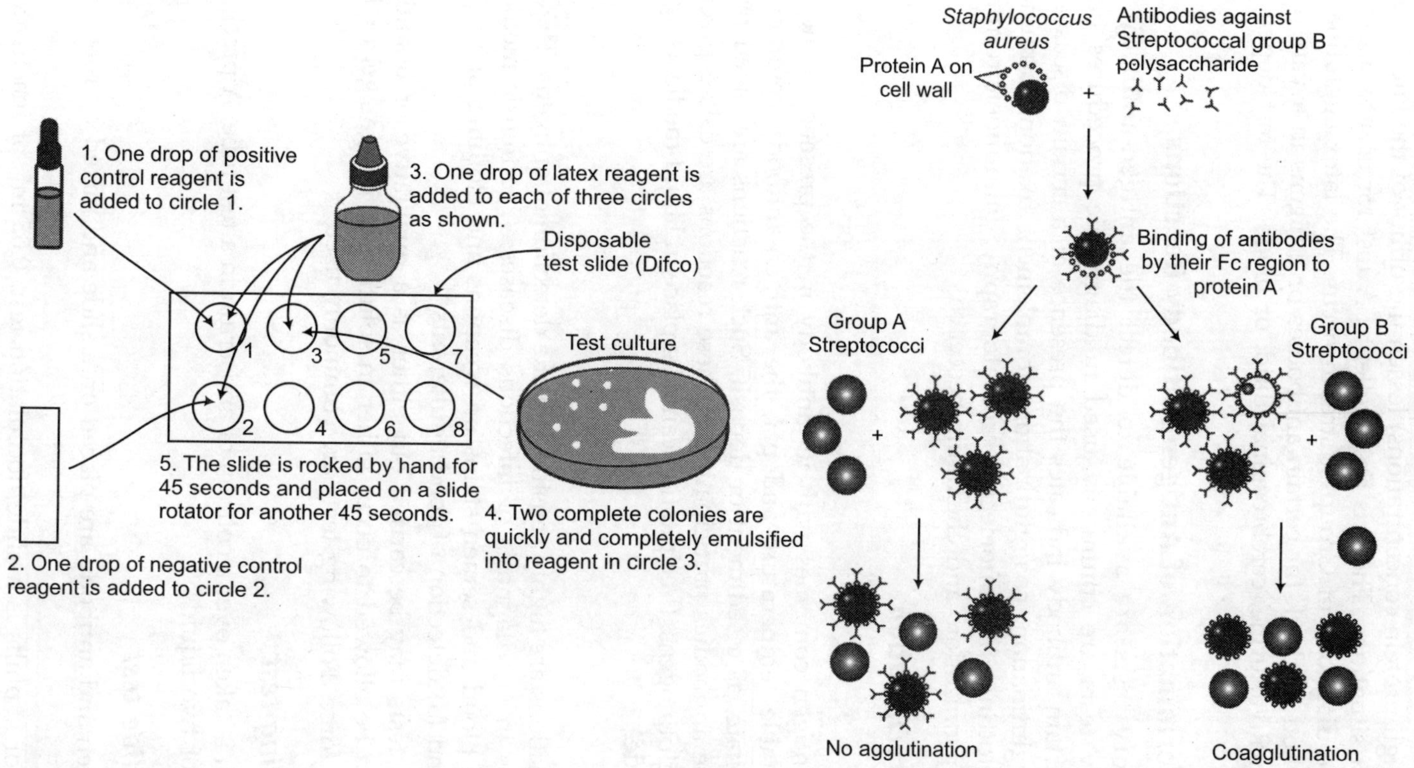

Figure 19.30 Slide agglutination test for the presence of coagulase and/or protein A for *S. aureus*.

coccidioidomycosis, though numerous sources of interferences (e.g., large immune complexes and high protein concentrations) lessen the utility of the latex agglutination (LA) test in most situations. There is an excellent LA test for cryptococcus. It is also used in definitive diagnosis of Group A streptococcus. The latex agglutination test is a laboratory method to check for certain antibodies or antigens in a variety of bodily fluids including saliva, urine, cerebrospinal fluid, or blood. The test depends on type of samples.

19.24.11 Determination of Antigen-Antibody Reactions

Several laboratory tests are available to detect the antigen-antibody reactions. These laboratory tests are commonly used in diagnostic procedures because the presence of a certain antibody indicates the presence of a certain disease. These are precipitation, agglutination, coagglutination, complement fixation, immunolabelling, immunoelectroblotting, immunoelectrochromatography, immunoelctronmicroscopy, neutralization, opsonization, and chemiluminescence.

19.24.11.1 Precipitation

Definition

When soluble antigen combines with its antibody in the presence of normal saline (NaCl) at a suitable temperature and pH, the antigen-antibody complex forms an insoluble precipitate, as a sediment or deposit. Such reactions are called precipitation reactions and the antibodies participating in these reactions are called precipitins and the antigens, agglutinogens. Precipitation can take place in liquid media as agar, agarose or polyacrylamide.

Applications

Precipitation reactions are highly sensitive for the detection of antigens. Therefore, they find applications in (1) diagnosis of infectious diseases; (2) forensic medicine in the identification of blood and semen stains on weapons and clothing or in other crime investigations and (3) detection of food adulterants.

Precipitation tests may be carried out either as a qualitative or quantitative test. Precipitation can be allowed to occur either in solution or in agar gel or in agar gel with an electric force. Following tests are commonly used.

19.24.11.2 Ring test

Antigen solution is taken over a column of antiserum in a test tube. A precipitate forms at the junction of two liquids.

19.24.11.3 Slide test

A drop of antigen and antibody are placed on a slide and mixed by rotation, floccules appear.

VDRL test for syphilis is a slide flocculation test. 0.05 mL of inactivated patient's serum is taken on special slide with depression (VDRL slides). One drop of freshly prepared cardiolipin antigen is added with a syringe that delivers 60 drops per mL. The slide is rotated at 180 rpm, in a VDRL, rotator for 4 min and examined under the microscope with low power objective. Uniform distribution of crystals in the drop indicates that the serum is nonreactive while the formation of floccules indicates that the serum is reactive. With reactive sera, serial dilutions can be tested to obtain the reactive titre.

RPR card test utilizes the cardiolipin antigen coated with fine carbon particles. It makes the test result clearer and evident to the naked eye, and there is no need of microscope. RPR test can be done with unheated serum or plasma. A drop of patient's serum or plasma and a drop of antigen is added on a card in the circle and rotated for a few minutes. If black particles are aggregated in the centre, the test is nonreactive; if the black particles are seen throughout the circle, the test is reactive.

19.24.11.4 Tube test

The antigen solution and antiserum (patient's serum) are taken in a test tube, mixed, incubated and then observed for precipitation. Usually, several tubes with different dilutions of antigen are set up since precipitation may be inhibited in the post zone region.

Kahn test is a tube flocculation test used for the diagnosis of syphilis. 0.15 mL of inactivated patient's serum is taken in 3 tubes and mixed with different amounts (0.05, 0.25, 0.0125 mL) of freshly prepared antigen, shaken in the Kahn shaker at 280 oscillations per minute and examined after addition of saline. Floccules appear in a positive test. Negative test shows uniform opalescence. In Kahn verification test, the test is observed at both 1 and 37°C. The antibody due to syphilis gives a strong reaction at 37°C and the nonspecific antibody at 1°C.

19.24.11.5 Immunodiffusion (or) precipitation in gel (or) gel precipitation

When precipitation is allowed to occur in a gel, the reaction is visible as a distinct band, which is stable and can be stained for preservation. An agar concentration of 0.3 to 1.5% allows for diffusion of most reactants. Agarose gel may be preferred to agar because agar has a strong negative charge while agarose has no charge. The rate of diffusion is affected by the size of particles, temperature, gel viscosity, amount of hydration and interactions between the matrix and reactants. As each antigen-antibody reaction gives rise to a line of precipitation, the number of different antigens in the sample can be made out.

Immunodiffusion also indicates identity, cross-reaction and non-identity between different antigens. Gel precipitation can be done by several ways.

Single diffusion in one dimension (Oudin method)

In this method, antibody is incorporated into gel in a test tube and the soluble antigen is directly pored over the antibody-containing gel. The antigen diffuses into the gel and forms a line of precipitation (Figure 19.31a).

Double diffusion in one dimension (Oakley-Fulthrope method)

In this method, the antibody is incorporated into gel in a test tube, above which a layer of plain agar gel is placed. The soluble antigen is then poured over on top of this plain gel. This results in the diffusion of both soluble antigen and antibody towards each other into intervening layer of plain agar and interact to form a precipitation ring (Figure 19.31b).

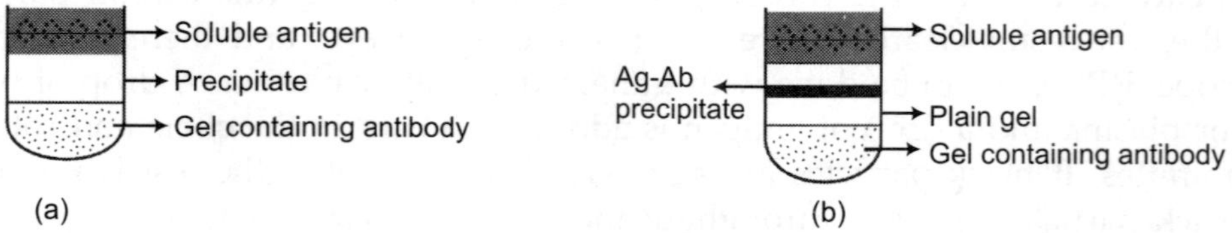

Figure 19.31 (a) Single diffusion in one dimension (Oudin method) (b) Double diffusion in one dimension (Oakley method).

Single diffusion in two dimensions (Mancini method)

It is a quantitative immunodiffusion test used to quantitate serum proteins. Immunoglobulins, complement factors, measure antibodies to influenza virus in the sera, and estimate serum transferring and alpha fetoprotein (Figure 19.32).

Monospecific antibody is added to molten agar and poured on slides or Petri dishes and allowed to set. Wells are cut in the gel and filled with known amounts of various dilutions of reference standard antigen solutions. The unknown test antigen is added to a separate well and allowed to diffuse. A ring of precipitate forms around each well and the diameter of the ring is proportional to the antigen concentration in the well. The wider the ring, the greater will be the antigen concentration. The diameters of the rings formed by the standard antigen concentrations are measured and plotted as a function of antigen concentration. Concentrations values for the test antigens are then read from the plotted graph.

Double diffusion in two dimensions (Ouchterlony method)

It is used to detect enterotoxin in cases of suspected food poisoning, to detect antibodies against specific pathogens in the diagnosis of infectious diseases, widely used for the diagnosis of small pox. Elek's test for toxigenicity of diphtheria bacilli is a special variety of double diffusion in two dimensions, to identify fungal antigens and to detect antibodies to extractable nuclear antigens (Figure 19.33).

The test is performed by pouring molten agar onto glass slides or Petri dishes and allowing it to settle. The antiserum is placed in the central well, and different antigens are added to wells surrounding the central well, punched in agar. The antigen and antibody diffuse from the wells independently through the gel in two dimensions, horizontally and vertically. After incubation for 12-48 hours in a moist chamber, lines of precipitate are formed at the sites of antigen-antibody combination. If the two adjacent precipitation lines fuse forming an arc (V-shaped), it indicates reaction of identity of the antigens. If the two adjacent precipitation lines fuse and form a spur or

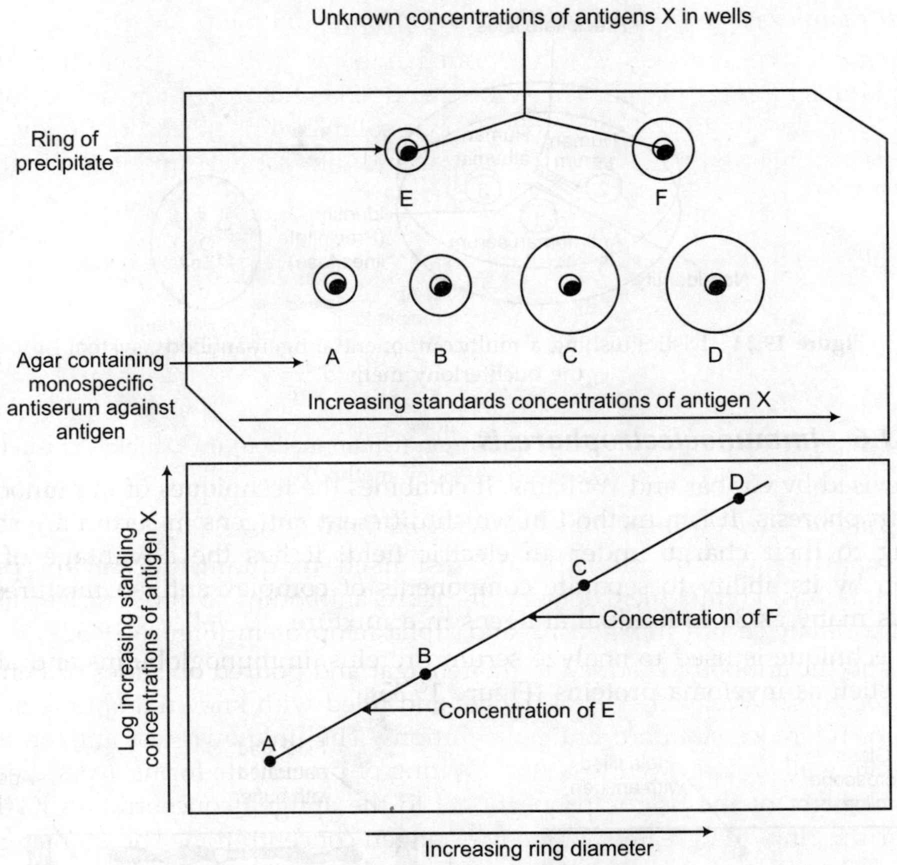

Figure 19.32 Single radial immunodiffusion test (Mancini method).

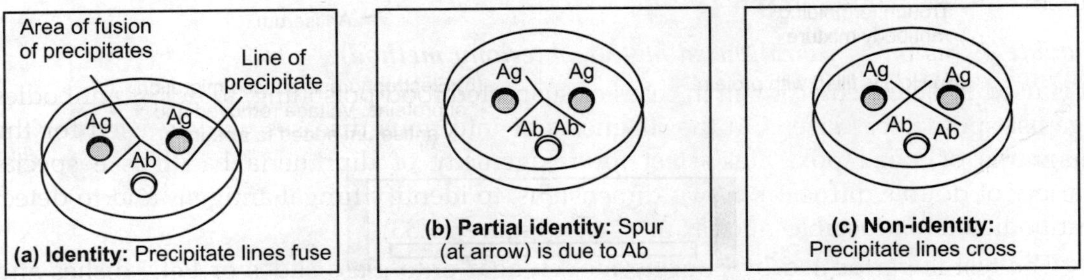

Figure 19.33 Ouchterlony test.

projection (Y-shaped, the stem of Y called spur), it indicates reaction of partial identity (cross-reactivity) of the antigens. If the two adjacent precipitation lines cross each other (X-shaped), it indicates reaction of non-identity (Figure 19.34).

Uses: It is used for the comparison of various antigens and antisera or for comparison of several antigens at a time on one agar gel plate.

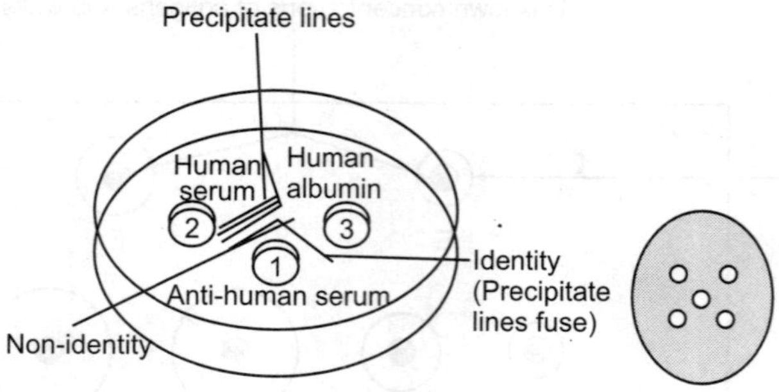

Figure 19.34 Distinguishing a multicomponent antigen-antibody system by the ouchterlony method.

19.24.11.6 Immunoelectrophoresis

It was devised by Grabar and Williams. It combines the techniques of immunodiffusion and electrophoresis. It is a method in which different antigens in serum are separated according to their charge under an electric field. It has the advantage of greater resolution by its ability to separate components of complex antigen mixtures. It can resolve as many as 39 individual antigens in a mixture.

This technique is used to analyze serum proteins, immunoglobulins and abnormal proteins such as myeloma proteins (Figure 19.35).

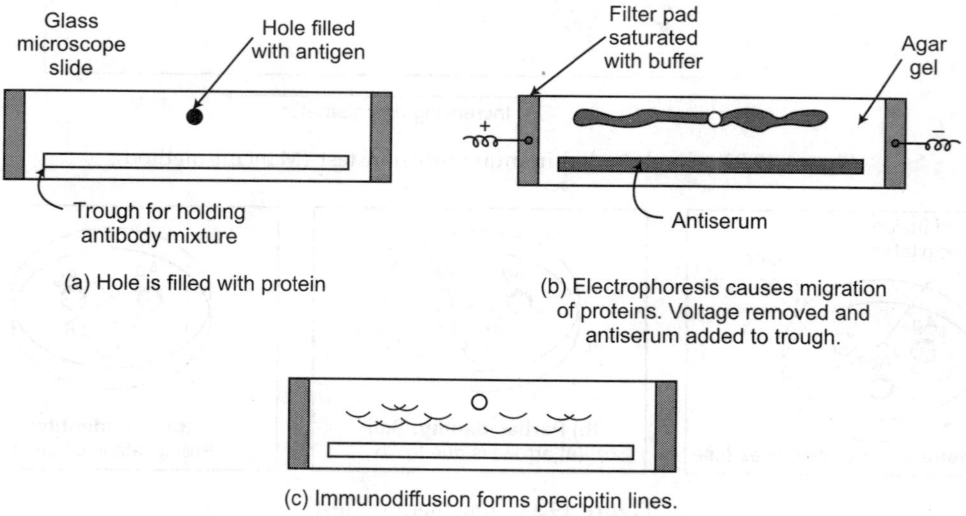

Figure 19.35 Arrangement for immune electrophoresis.

The test is performed in two stages:

(i) Agar gel is taken on a slide, and a well is cut and filled with an antigen mixture such as human serum. Electric current is applied in an electrophoresis chamber.

As a result, antigens are separated and distributed as separate spots based on their electrical charge. Positively charged proteins move to the negative electrode and negatively charged proteins move towards the positive electrode.

(ii) A rectangular trough is then cut parallel to the electrophoretic separation and filled with appropriate antibody, and incubated. The antigens and antibodies diffuse towards each other into the gel and form a series of complex pattern of are shaped precipitation bands that can be better visualized by staining.

19.24.12 Electroimmunodiffusion (Electroimmuno Assay)

There are several methods to perform these tests.

19.24.12.1 Counter current immunoelectrophoresis (CIEP)

This test depends on the diffusion of antigens towards anode and antibodies towards the cathode through the gel under electric field. The test is performed on a glass slide with agarose gel in which a pair of wells are punched. One well is filled with antigen solution and the other with antibodies, and electric field applied in electrophoresis chamber. The antigen and antibody, which have opposite charge, diffuse towards each other in an electric field, by forming a line of precipitation where they meet in equivalent amounts in about 30-60 minutes (Figure 19.36).

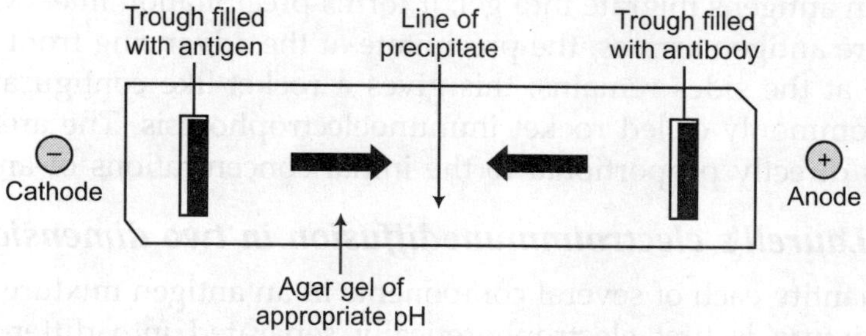

Figure 19.36 Counter immune electrophoresis.

Uses: It is a rapid and highly specific method for detection of both antigens and antibodies in serum, CSF and other body fluids in the diagnosis of many infectious diseases including bacterial, viral, fungal and parasitic. It is commonly used for detection of HBsAg in patient serum, alpha fetoprotein in serum, *Cryptococcus* antigens in CSF, meningococcal antigens in CSF, and detection of hydatid and amoebic antigens in serum.

19.24.12.2 Rocket electrophoresis

This method is an adaptation of radial immunodiffusion. It is used for the quantitative estimation of human plasma antigens and viral antigens. It combines the speed of electrophoresis with quantification of antigen of radial immunodiffusion (Figure 19.37).

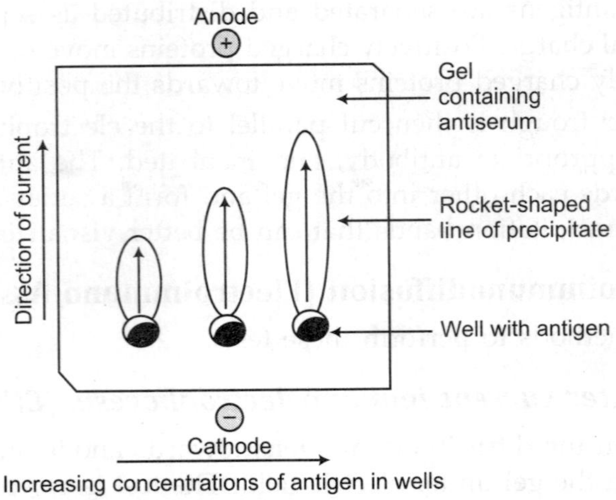

Figure 19.37 Rocket electrophoresis.

The antibody is added to molten agar gel and poured on a glass slide and allowed to set; wells are cut and filled with antigen solution (standard concentrations and unknowns) and current is applied in an electrophoresis chamber, allowing little time for diffusion. When antigens migrate into gel, it forms precipitation lines with antibodies in the gel. As more antigen arrives, the precipitate at the advancing front dissolves, while the precipitate at the sides remains; this gives a rocket-like configuration. Hence, the procedure is commonly called rocket immunoelectrophoresis. The area of each rocket or its height is directly proportional to the initial concentrations of antigen.

19.24.12.3 Laurell's electroimmunodiffusion in two dimensions

It is used to quantity each of several components in an antigen mixture. In this method, the antigen mixture is first electrophoretically separated into different components, followed by immunodiffusion into gel containing antibody so that rocket like precipitation lines are produced. Thus, this technique is similar to immunoelectrophoresis and is a variant of rocket electrophoresis (Figure 19.38).

1. Antigens are separated by their electrophoretic mobility on the first run.
2. The second run at right angles to the first drives the antigens into the antiserum containing gel to form precipitation rockets.

19.24.13 Agglutination Reactions

Agglutination means clumping of the particles. In most of the agglutination reactions, the antigen is particulate, and antibody is in the soluble form. However, if antibodies are coated on particles, agglutination can also occur with soluble antigens.

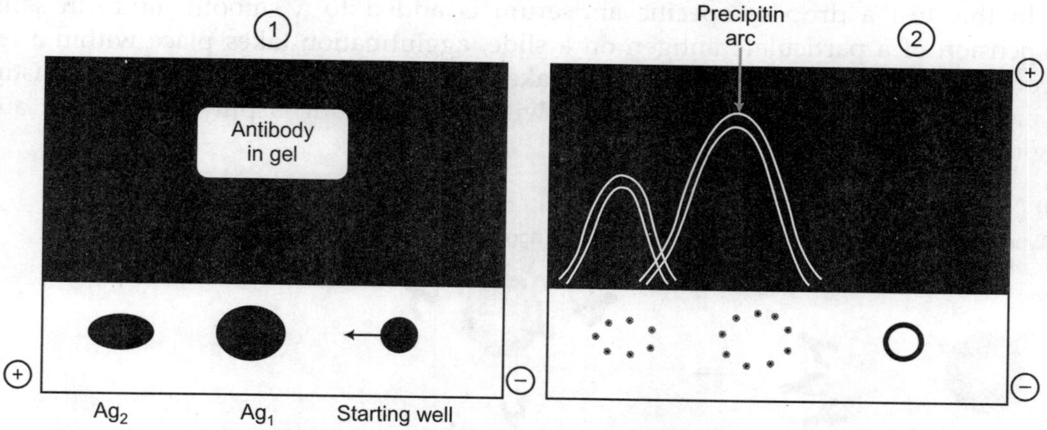

Figure 19.38 Laurell's two-dimensional rocket electrophoresis.

Definition: When a particulate antigen combines with its corresponding antibody in the presence of an electrolyte at a suitable temperature and pH, the antigen-antibody complexes form large clumps/aggregates/agglutinates that are visible and can be seen with the unaided eye. Such reactions are called agglutination reactions (Figure 19.39). The antibodies participating in these reactions are called **agglutinins** and antigens referred to as **agglutinogens.** Agglutination tests are more sensitive for both either qualitative or quantitative assay of antibodies.

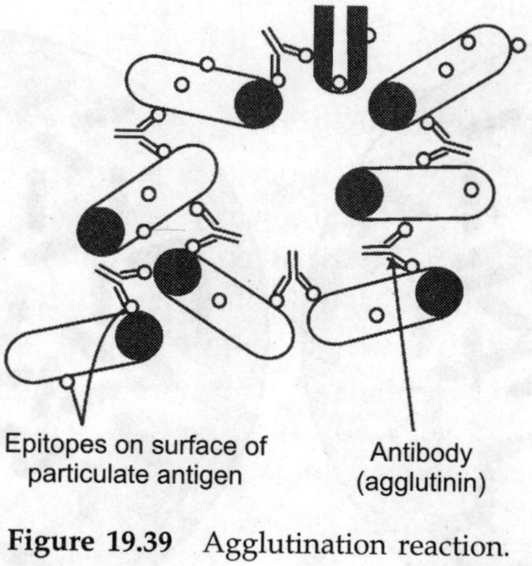

Figure 19.39 Agglutination reaction.

19.24.13.1 *Slide agglutination test*

This is the direct agglutination reaction, mainly utilized for the identification of bacterial isolates from clinical specimens as well as for blood grouping (to type RBC) and cross-matching.

In this test a drop of specific antiserum is added to a smooth, uniform saline suspension of a particulate antigen on a slide, agglutination takes place within a few seconds which is usually visible to the naked eye (Figure 19.40). A control consisting of antigen suspension in normal saline without antiserum is put up to check auto agglutination of the antigen.

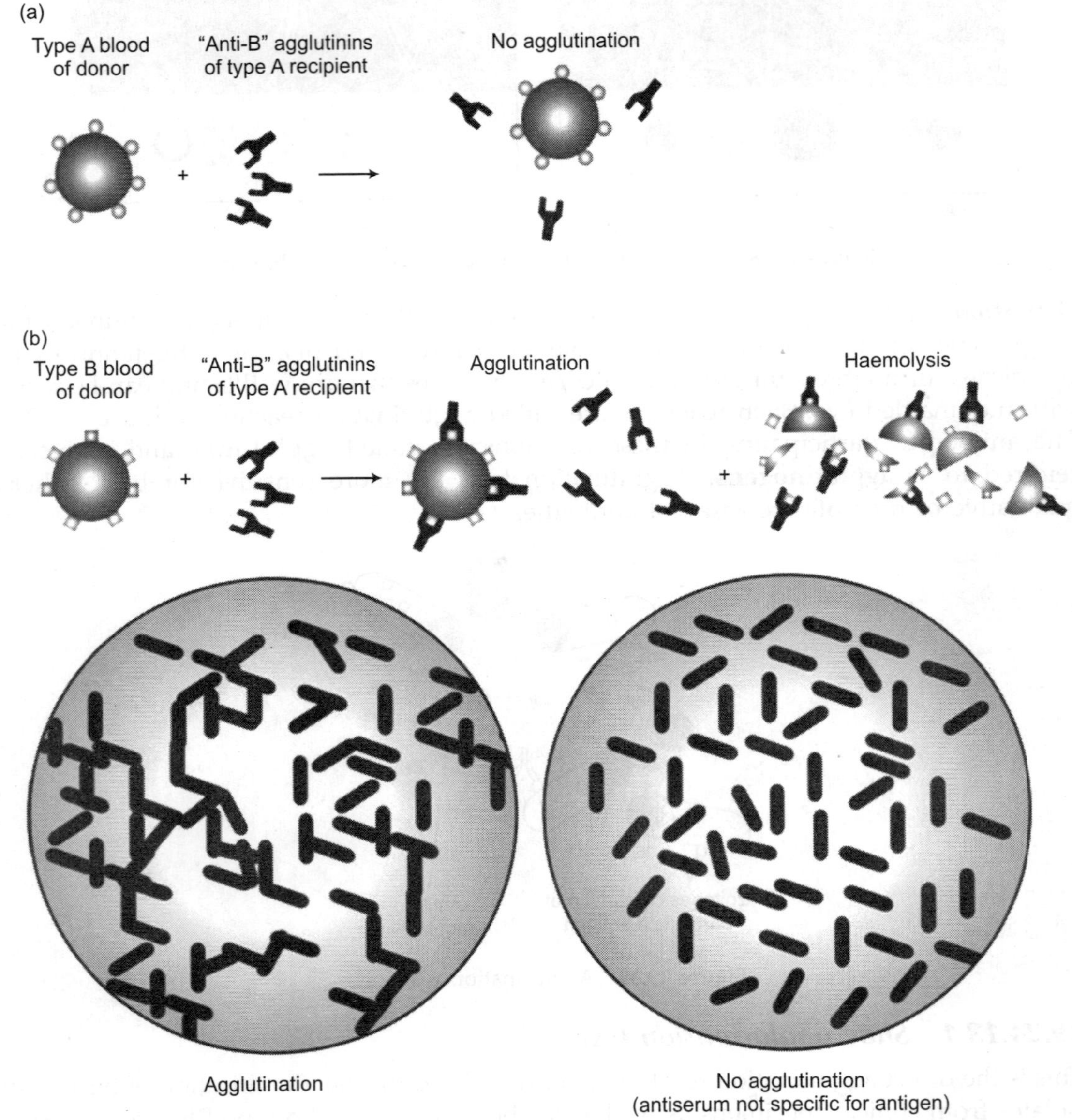

Figure 19.40 Illustration of slide agglutination test.

This is the most widely used standard quantitative method for the estimation of antibodies in serum specimens.

The test is performed by adding a fixed amount of particulate antigen to a series of tubes containing doubling dilutions of the antiserum (patient's serum); for example, tube 1 = , tube 2 = , tube 3 = , tube 4 = , tube 5 = and so on. The reciprocal of the highest dilution in which agglutination occurs is taken as the titer of the antiserum (Figure 19.41) The test is used viz. Widal test for the serodiagnosis of typhoid and paratyphoid fevers (enteric fever), tube agglutination test for brucellosis, Weil-Felix reaction is a heterophile agglutination test for the serodiagnosis of typhus fevers, Paul-Bunnel test is a heterophile agglutination test for the serodiagnosis of infectious mononucleosis, and cold agglutination and *Streptococcus* MG agglutination tests are heterophile agglutination tests for the serodiagnosis of primary atypical pneumonia.

19.24.13.2 Passive (indirect) agglutination test

In this test, soluble antigens are detected by first being made to adhere to the surfaces of inert particulate carrier substances such as latex beads, bentonite particles and RBC. When specific antiserum is added to such antigen-coated particles, large clumps or aggregates are formed that are easily visible to the naked eye. Thus, a precipitation test can be converted into an agglutination test (Figure 19.42). Passive agglutination tests are more sensitive for detecting antibodies.

19.24.13.3 Latex agglutination test

Test is widely employed for the detection of ASLO, CRP, RA factor etc. Latex particles are cheap, relatively stable and do not cross react with other antibodies.

19.24.13.4 Rose-Waaler test is a passive haemagglutination test

(Sensitized Sheep Rbc Agglutination Test)

Test is used to detect RA factor. In passive agglutination, soluble antigens are first made to adhere to particulate carriers before reaction with antibodies. When the carrier is red blood cells, the method is termed passive haemagglutination.

19.24.13.5 Indirect haemagglutination (IHA) test

When RBCs are coated with antigen to detect antibodies in the serum, the test is called the IHA. This is most commonly used for the serodiagnosis of many parasitic diseases such as extra-intestinal amoebiasis, hydatid disease and toxoplasmosis.

When the antigen is a natural constituent of a particle (e.g., RBC, bacterial cell), the agglutination reaction, is referred to as direct agglutination. When the agglutination reaction takes place between antibodies and soluble antigen that has been attached to an insoluble particle, the reaction is referred to as passive agglutination.

19.24.13.6 Latex agglutination test for pregnancy

The modern pregnancy test detects the increased levels of human chorionic gonadotropin (HCG) hormone that occurs in female urine and blood early in pregnancy.

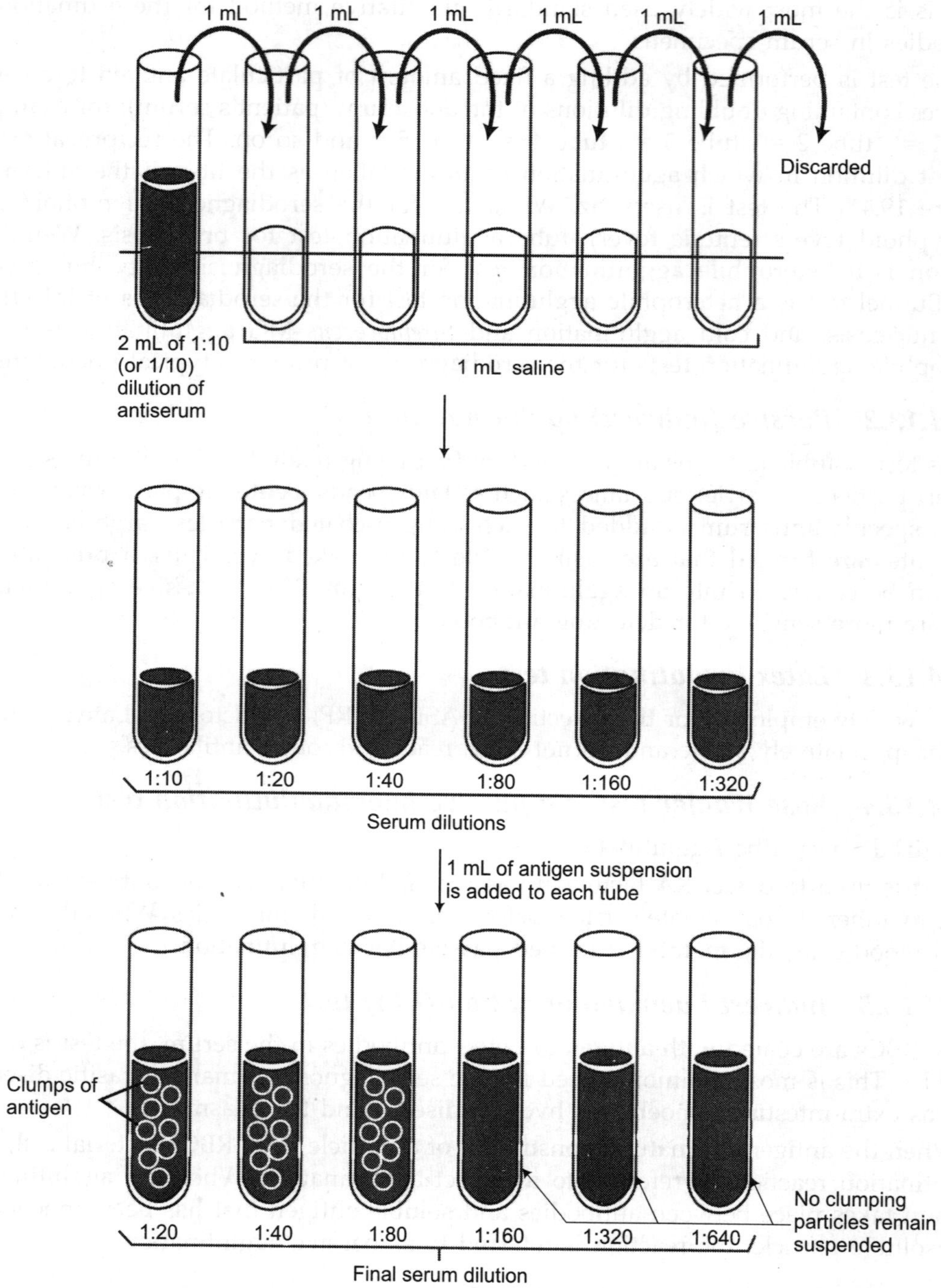

Figure 19.41 Protocol for preparing doubling dilution for tube agglutination test.

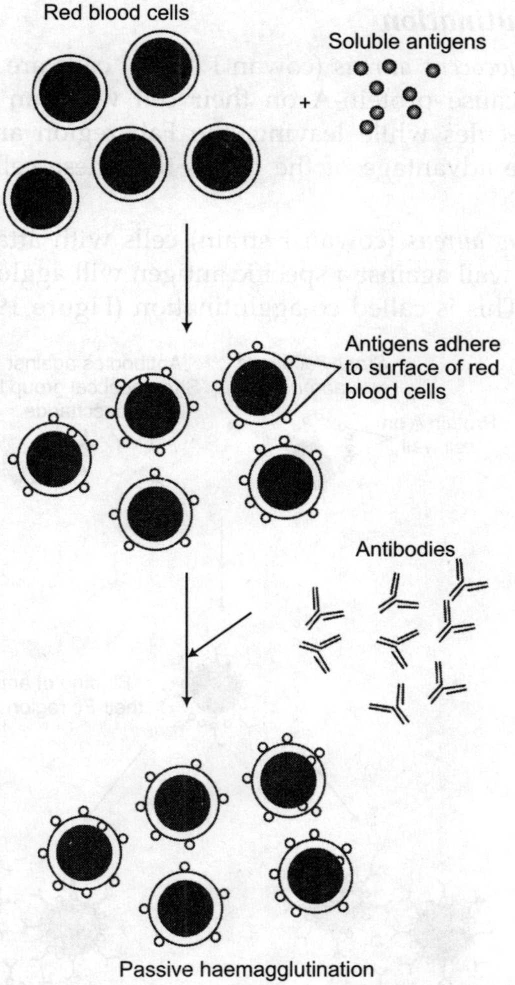

Figure 19.42 Passive agglutination test.

The urine from the female containing HCG is mixed with an antibody specific for HCG. Then, latex particles coated with HCG are added. If HCG is present, it binds to HCG specific antibodies, thereby preventing them from agglutinating the latex particles coated with HCG. Therefore, in a positive test there will be no agglutination. In a negative test, the latex particles coated with HCG are agglutinated by antibody to HCG.

19.24.13.7 Reverse passive agglutination test

When instead of the antigen, antibodies are attached to the RBCs to detect antigens, it is known as reverse passive haemagglutination (RPHA), e.g., detection of HBsAg in patient's serum.

19.24.13.8 Coagglutination

In this method *Staphylococcus aureus* (cowan I strain) cells are used as carrier particles to coat antibodies because protein-A on their cell wall can react with F_c region of immunoglobulin molecules while leaving, the Fab region are free to combine with a specific antigen. The advantage of the test is that these cells show greater stability than latex particles.

Thus, *Staphylococcus aureus* (cowan I strain) cells with attached IgG antibodies on their protein-A of cell wall against a specific antigen will agglutinate bacteria that have that specific antigen. This is called co-agglutination (Figure 19.43).

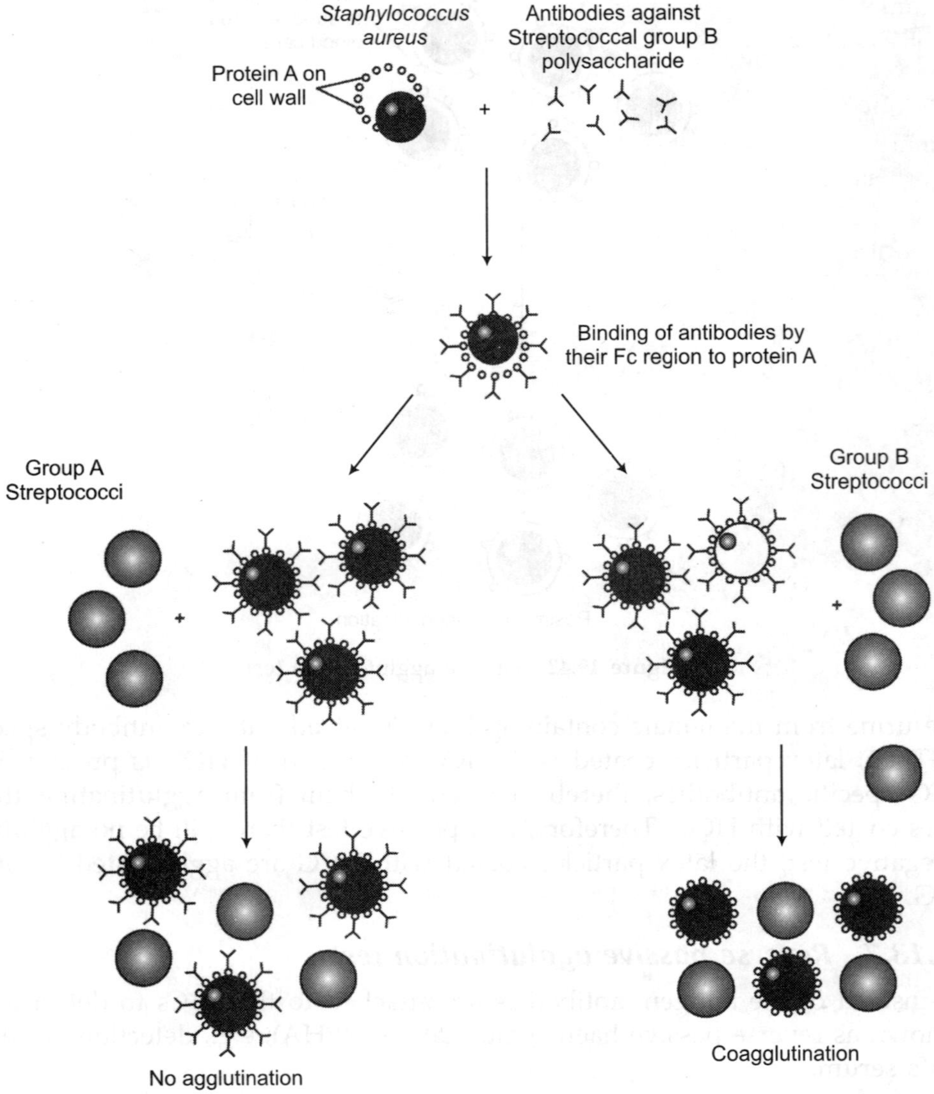

Figure 19.43 Illustrating coagglutination.

It has several important diagnostic applications such as streptococcal grouping, gonococcal and mycobacterial typing, detection of typhoid antigen from enteric fever, detection of cryptococcal antigen from CSF in the diagnosis of cryptococcal meningitis, detection of amoebic and hydatid antigens in the serum for the diagnosis of amoebiasis and hydatid disease respectively.

19.24.14 Coombs Test (Antiglobulin Test)

It is used to detect nonagglutinating, antierythrocyte (Anti-Rh IgG) antibodies. The major applications of Coombs test are in evaluating haemolytic disease of the new born, diagnosis of autoimmune haemolytic anaemia, and erythrocyte (RBC) typing in blood banks.

19.24.14.1 Principle

Anti-Rh IgG antibodies when react with RBC (coated on RBC) fail to produce agglutination. The addition of an antiglobulin antiserum (Coombs serum) produced in heterologous species (e.g., Rabbit antihuman-globulin) produces marked agglutination.

The test may be direct or indirect (Figure 19.44). The direct Coombs test measures bound antibody, whereas the indirect Coombs test measures serum antibody.

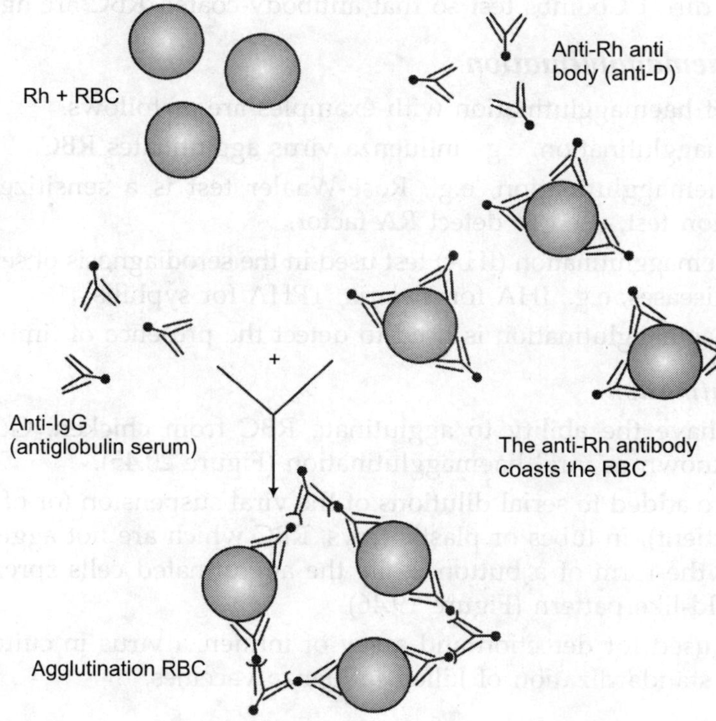

Figure 19.44 Illustrating antiglobulin (Coombs) test.

19.24.14.2 Direct Coombs test

In the direct Coombs test, anti-immunoglobulins are added to RBC that are suspected of having antibodies bound to antigens on their surfaces. For example, a newborn baby is suspected of having haemolytic disease of the newborn caused by maternal anti-Rh IgG antibodies that are bound to the baby's RBC. Then, the addition of anti-immunoglobulin to suspension of baby's RBC causes binding of the anti-immunoglobulin to the maternal IgG on the surface of the RBC leading to agglutination.

19.24.14.3 Indirect Coombs test

It is a two-stage reaction for detection of incomplete (anti-Rh IgG) antibodies in patient's serum. The indirect Coombs test is used to detect the presence of antibodies in the serum, specific to antigens on RBC. The serum antibodies (Anti-Rh IgG), when added to the RBC, may fail to cause agglutination because of the zeta potential. The subsequent addition of anti-immunoglobulin will cause agglutination. A common application of the indirect Coombs test is to detect anti-Rh IgG antibodies in the blood of a Rh negative woman.

The woman's (patient's) serum is first incubated with Rh positive RBC so that antibody coats RBC. In the second step, Coombs antiglobulin serum (anti-immunoglobulin) is added, as in the direct Coombs test so that antibody-coated RBC are agglutinated.

19.24.14.4 Haemagglutination

Different types of haemagglutination with examples are as follows.

1. Viral haemagglutination, e.g., influenza virus agglutinates RBC.
2. Passive haemagglutination, e.g., Rose-Waaler test is a sensitized sheep RBC agglutination test, used to detect RA factor.
3. Indirect haemagglutination (IHA) test used in the serodiagnosis of several parasitic, bacterial diseases, e.g., IHA for malaria, TPHA for syphilis.
4. Fimbrial haemagglutination is used to detect the presence of fimbriae.

Viral Haemagglutination

Certain viruses have the ability to agglutinate RBC from chicken, Guinea pig and humans and is known as viral haemagglutination (Figure 20.45).

When RBC are added to serial dilutions of the viral suspension (or of serum from a virus-infected patient), in tubes or plastic trays, RBC which are not agglutinated settle at the bottom in the form of a button while the agglutinated cells spread as an even coating in a shield-like pattern (Figure 19.46)

Uses: Test is used for detection and assay of influenza virus in culture fluids and for titration and standardization of killed influenza vaccines.

Haemagglutination inhibition (HI)

The antibodies against viruses that can cause haemagglutination can bind to viruses and prevent them from agglutinating RBC. This inhibition phenomenon is called viral

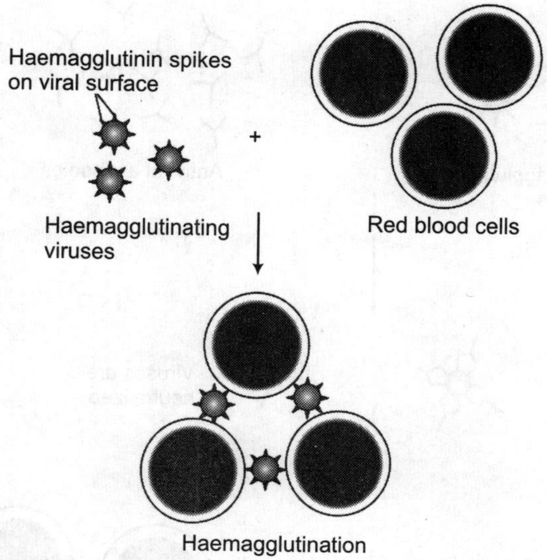

Figure 19.45 Illustration for viral haemagglutination test.

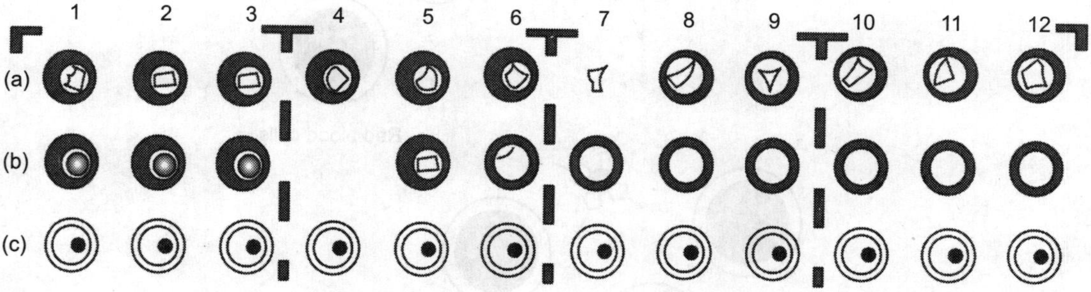

Figure 19.46 Haemagglutination reactions seen in the wells of a microtiter plate. Top row (a) complete haemagglutination has occurred in all of the wells, middle row, (b) only wells 3 through 6 show complete haemagglutination. Botton row (c) no haemagglutination in any of the wells.

neutralization, and the test based on it is called a viral haemagglutination inhibition (HI) test, which is used for detection and quantitation of antibody against virus (Figure 19.47).

19.24.15 Complement Fixation Test (CFT)

CFT detects the presence of complement fixing antibodies (IgM and IgG) present in a patient's serum. Complement-fixation tests are used in the diagnosis of certain viral, fungal and bacterial diseases, e.g., Wassermann test for the diagnosis of syphilis.

19.24.15.1 Principle

The ability of antigen-antibody complexes to fix complement is made use of in CFT. The test requires five reagents: 1) Antigen may be soluble or particulate, 2) Antibody (usually patient's serum), 3) Complement (C): The source of complement is Guinea pig

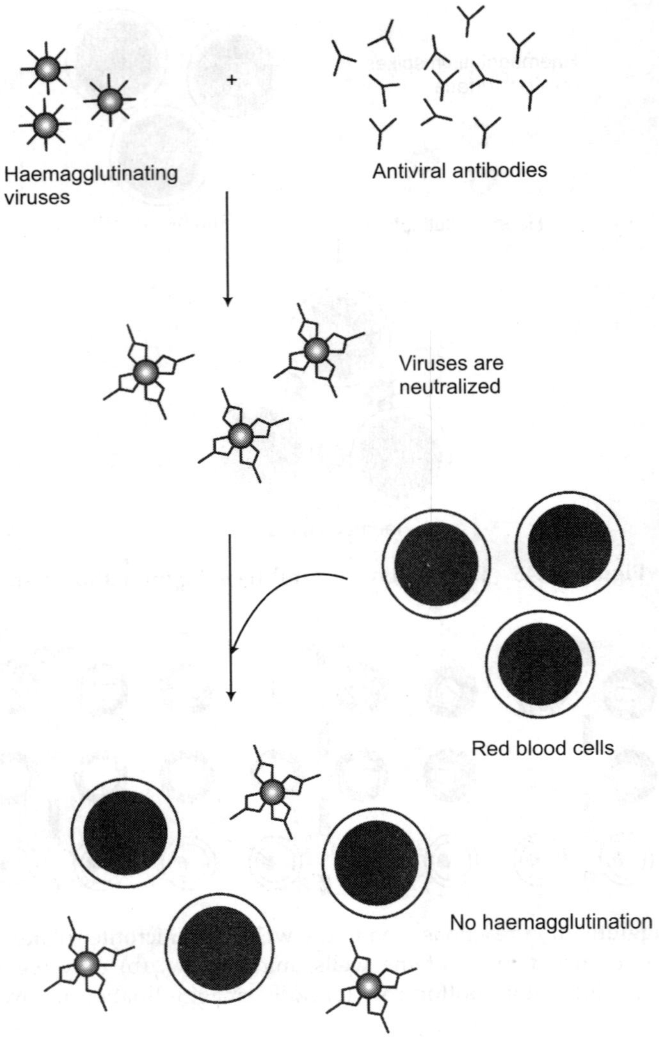

Figure 19.47 Haemagglutination inhibition (HI).

serum. As C is heat labile, the serum should be freshly drawn or preserved either in lyophilized or frozen state or by Richardson's method, 4) sheep RBC, 5) amboceptor, an antibody to sheep RBC raised in rabbits. Sheep RBC and amboceptor together are known as haemolytic system.

19.24.15.2 Procedure

The test may be carried out in small tubes, in plastic microtitration plates or even in the automated system (Figure 19.48).

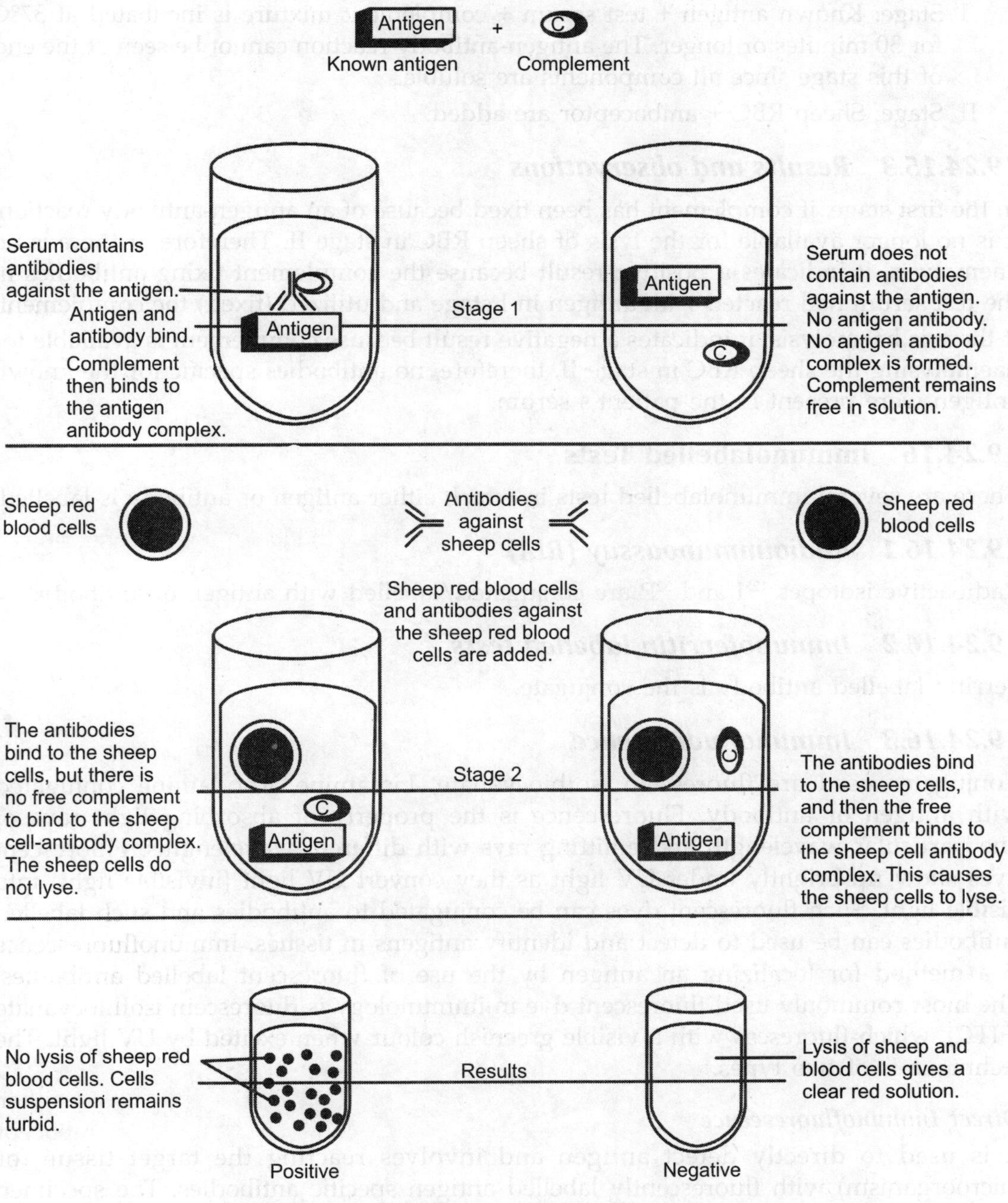

Figure 19.48 Complement fixation test (CFT).

The test is performed in two stages:

The test serum from the patient must first be inactivated at 56°C for 30 minutes to destroy its complement activity.

I Stage: Known antigen + test serum + complement mixture is incubated at 37°C for 30 minutes or longer. The antigen-antibody reaction cannot be seen at the end of this stage since all components are soluble.

II Stage: Sheep RBC + amboceptor are added.

19.24.15.3 Results and observations

In the first stage, if complement has been fixed because of an antigen-antibody reaction, it is no longer available for the lysis of sheep RBC in stage II. Therefore, if there is no haemolysis, it indicates a positive result because the complement fixing antibodies in the test serum had reacted with antigen in I stage and utilized (fixed) the complement. If there is haemolysis, it indicates a negative result because complement is available for haemolysing the sheep RBC in stage II, therefore, no antibodies specific for the known antigen were present in the patient's serum.

19.24.16 Immunolabelled Tests

There are several immunolabelled tests in which either antigen or antibody is labelled.

19.24.16.1 Radioimmunoassay (RIA)

Radioactive isotopes ^{125}I and ^{32}P are conjugated/labelled with antigen or antibody.

19.24.16.2 Immunoferritin labelled tests

Ferritin labelled antibody is the conjugate.

19.24.16.3 Immunofluorescence

Conjugates used are fluorescein isothiocyanate, Lissamine, Rhodamine conjugated with antigen or antibody. Fluorescence is the property of absorbing light rays of one particular wavelength and emitting rays with different wavelengths. Fluorescent dyes show up brightly under UV light as they convert UV light (invisible light) into visible light. Such fluorescent dyes can be conjugated to antibodies and such labelled antibodies can be used to detect and identify antigens in tissues. Immunofluorescence is a method for localizing an antigen by the use of fluorescent labelled antibodies. The most commonly used fluorescent dye in immunology is fluorescein isothiocyanate (FITC), which fluoresces with a visible greenish colour when excited by UV light. The technique is of two types.

Direct Immunofluorescence

It is used to directly detect antigen and involves reacting the target tissue (or microorganism) with fluorescently labelled antigen specific antibodies. The specimen (cells or microorganisms) containing the antigen are fixed on to a slide. Then specific fluorescein labelled antibodies are added to the slide and incubated. The slide is washed to remove any unbound antibody and examined with the fluorescence microscope for a yellow-green fluorescence, which reveals the antigen location. Direct fluorescent antibody (DFA) technique is usually used to identify a particular kind of microorganism in a clinical specimen, e.g. *Streptococcus pyogenes* in a throat swab from a patent. Direct immunofluorescence is used to identify antigens found on the surface of Group A Streptococci and to diagnose EPEC, *Neisseria meningitides, Salmonella typhi, Shigella sonnei, Listeria monocytogenes, Haemophilus influenza* and the Rabies virus.

Direct immunofluorescence is also used clinically for identifying lymphocytic subsets and for demonstrating the presence of specific protein deposition in certain tissues such as kidney and skin in cases of SLE. The major disadvantage of the direct immunofluorescence is the preparation of separately labelled antibody for each pathogen.

Indirect immunofluorescence

In this technique, a known antigen is fixed on to a slide. The test antiserum (patient's serum) is then added, and if the specific antibody is present, it reacts with the antigen to form a complex. When fluorescent-labelled anti-immunoglobulin is added, it binds to the specific antibody that has already reacted with the antigen in the smear. After incubation and washing, the slide is examined with the fluorescence microscope. The occurrence of fluorescence shows that antibody specific to the test antigen is present in the serum (Figure 19.49).

The indirect fluorescent antibody (IFA) is used to detect the presence of antibodies to a specific microorganism in the body fluid of patient, e.g., (1) the presence of IgG antibodies to an oral spirochete in the serum or saliva of a patient with gum disease; (2) *Treponema pallidum* antibodies in the diagnosis of syphilis (FTA-ABS, FTA-200) and antibodies to other microorganisms. The IFA is more sensitive and has the advantage of using a single labelled antiglobulin (antibody to IgG) reagent to detect many different specific antigen-antibody interactions.

19.24.17 ELISA

Enzyme-linked immunosorbent assay (ELISA) is highly sensitive, highly specific and less expensive technique used in serology, to detect antigens or antibodies. This test involves the linking of various label enzymes to either antigens or antibodies. Horse radish peroxidase, alkaline phosphatase is labelled with either antigen or antibody. The ELISA is used for the detection of many infectious viruses, bacteria, fungi, parasites (protozoa) and drugs, toxins, food allergens, etc. ELISA is based on the following two principles:

1. Antibodies and some antigens can absorb to a solid support such as polystyrene, polyvinyl or polycarbonate wells or membranes (discs) of polyacrylamide paper or plastic.
2. Antigens and antibodies can be linked to enzymes, with the resulting complexes still fully functional, both immunologically and enzymatically.

Enzyme activity is used to measure the quantity of antigen or antibody present in the test sample. Enzymes used in ELISA include β-galactosidase, glucose oxidase, peroxidase, and alkaline phosphatase. The test is usually performed by using microtiter plates (96-well) suitable for automation. There are three basic ELISA techniques, viz., double-antibody sandwich technique, indirect immunosorbent technique and competitive ELISA.

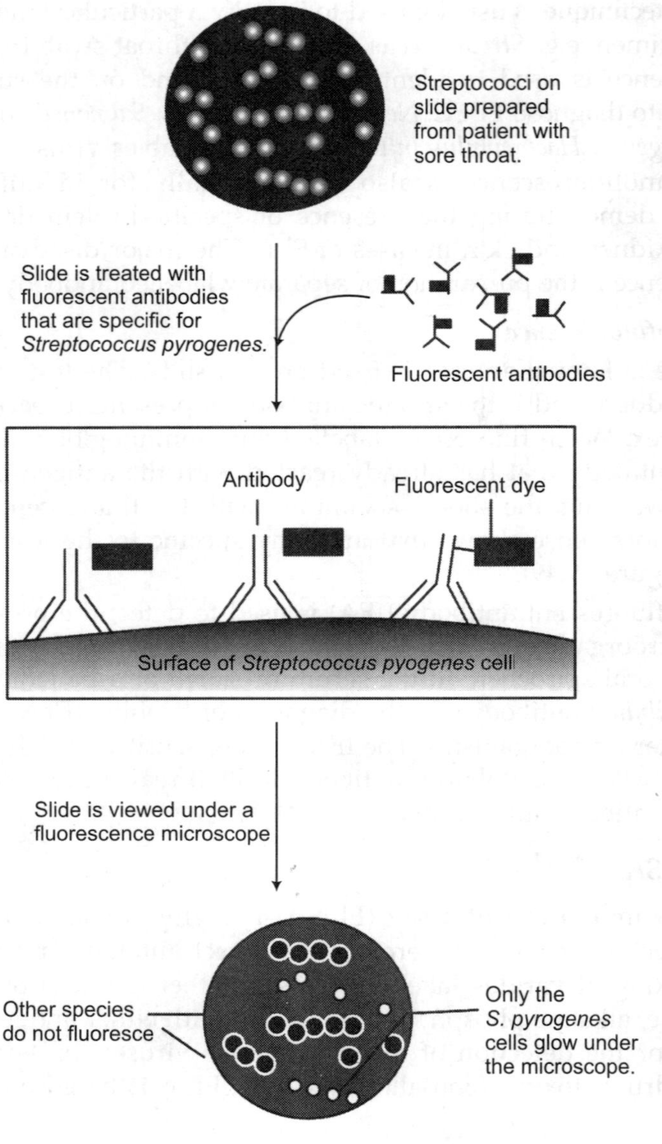

Figure 19.49 Direct fluorescent antibody (DFA) test.

19.24.17.1 Double antibody sandwich assay (sandwich ELISA)

It is used for detection of antigens (e.g., HBsAg in patient's serum, rotavirus and enterotoxin of E. *coli* in feces) (Figure 19.50). In this technique, specific antibody is placed in wells of a microtiter plate (or it may be attached to a membrane) and the antibody adheres to the inner surface of each well. The test antigen is added to each well. If the antigen reacts with the antibody, the antigen is retained when the well is washed to remove unbound antigen (Figure 19.51).

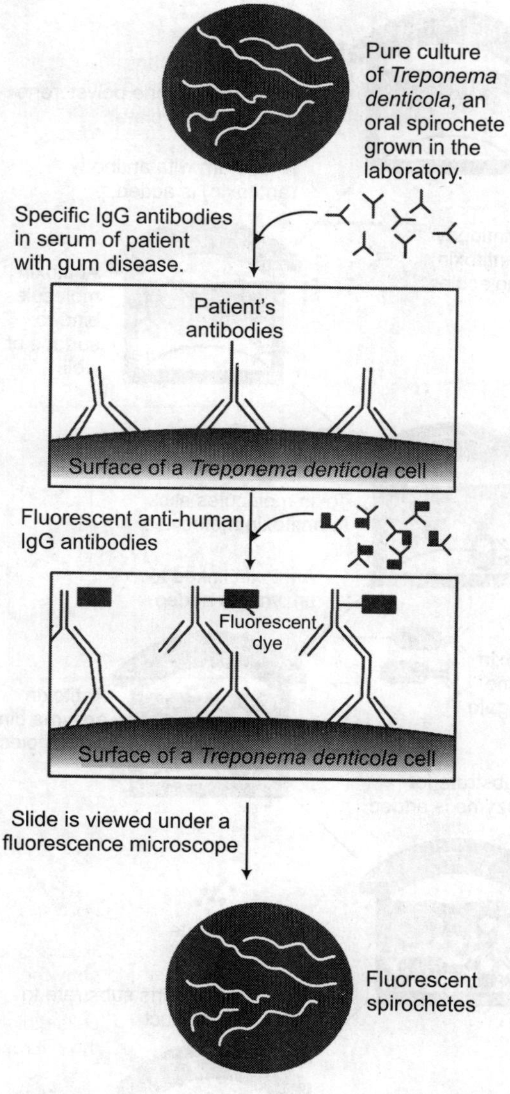

Figure 19.50 Indirect Fluorescent Antibody (IFA) technique.

An antibody-enzyme conjugate specific for the antigen is then added; it will bind to the antigen already fixed by the first antibody, forming an antibody (with enzyme)-antigen-antibody sandwich.

A substrate that the enzyme will convert to a coloured product is then added, and any resulting product is quantitatively measured by optical density scanning with an ELISA Reader.

If the antigen has reacted with the adsorbed antibodies in the first step, the ELISA test is positive. If the antigen is not reacted with the adsorbed antibody, the ELISA test is negative.

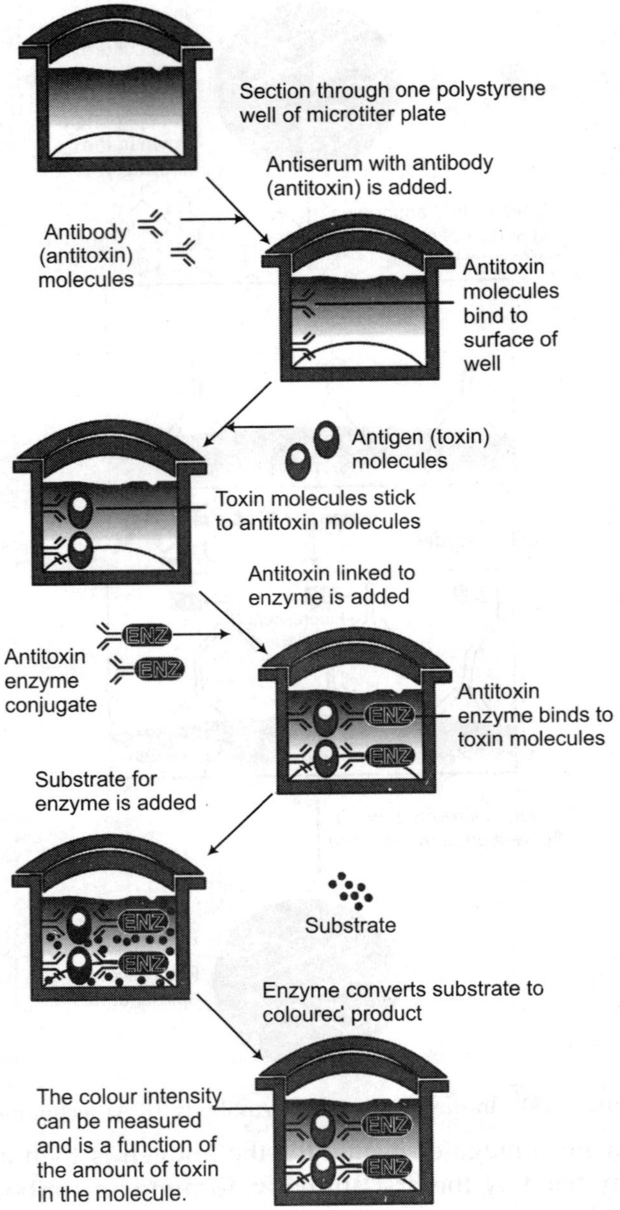

Figure 19.51 Double antibody sandwich ELISA technique for the detection and measurement of antigen (against *Staphylococcus* enterotoxin in a food extract). The presence of as little as 0.4 ng of toxin per mL of food extract can be detected

19.24.17.2 Indirect immunosorbent assay (Indirect ELISA)

It is used for the detection of antibodies (e.g., antibodies to HIV, Dengue, Japanese encephalitis and Rubella virus in the serum sample) (Figure 19.52).

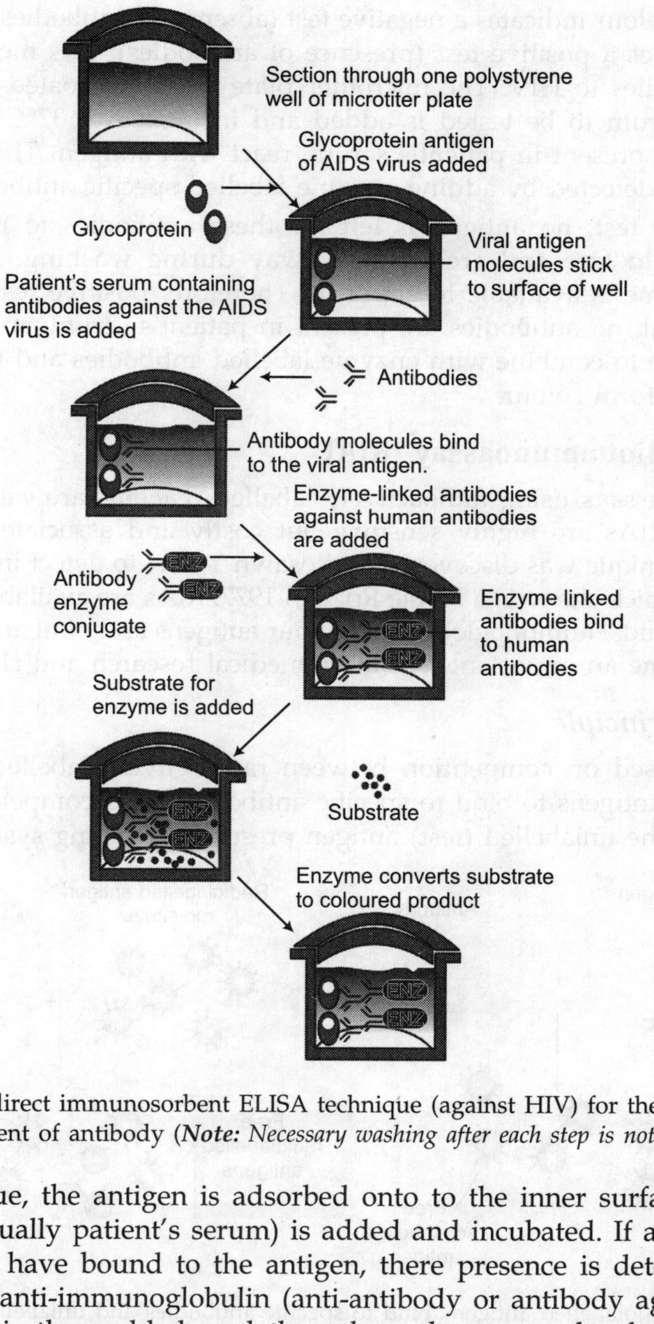

Figure 19.52 Indirect immunosorbent ELISA technique (against HIV) for the detection and measurement of antibody (***Note:** Necessary washing after each step is not shown*).

In this technique, the antigen is adsorbed onto to the inner surface of well wall. Test antiserum (usually patient's serum) is added and incubated. If any antibodies in the test antiserum have bound to the antigen, there presence is detected by adding an enzyme-linked anti-immunoglobulin (anti-antibody or antibody against antibody). Enzyme substrate is then added, and the colour change can be detected by ELISA Reader in terms of optical density.

19.24.17.3 Competitive ELISA

In competitive ELISA, two specific antibodies, one conjugated with enzyme and other present in patient's serum are used. Two antibodies compete for the same antigen.

Appearance of colour indicates a negative test (absence of antibodies) while the absence of colour indicates a positive test (presence of antibodies). It is most commonly used to detect antibodies to HIV. The microtiter plate wells are coated with HIV antigen. The patient's serum to be tested is added and incubated at 37°C and then washed. If antibodies are present in patient's serum, react with antigen. This antigen-antibody complex can be detected by adding enzyme labelled specific antibodies to HIV.

In a positive test, no antigen is left for these antibodies to react. Hence, these antibodies remain free and are washed away during washing. When substrate is added, no enzyme is available to act on it. Therefore, positive test shows no colour. In a negative test, no antibodies are present in patient's serum, antigen in the coated wells is available to combine with enzyme labelled antibodies and the enzyme acts on the substrate to form colour.

19.24.18 Radioimmunoassay (RIA)

Immunological assays using radioactively labelled reagents are called Radio immune assays (RIAs). RIAs are highly sensitive but costly and associated with radioactive hazard. The technique was discovered by Rosalyn Yalow to detect insulin concentration in plasma, for which she won a Noble Prize in 1977. RIAs are available for all hormones, drugs in body fluids, immunoglobulins, tumour antigens and viral antigens (e.g. HBsAg) and it has become an important tool in biomedical research and clinical practice.

19.24.18.1 Principle

The assay is based on competition between radioactively labelled antigen (e.g. ^{125}I) and unlabelled antigens to bind to specific antibodies. This competition is determined by the level of the unlabelled (test) antigen present in reacting system (Figure 19.53).

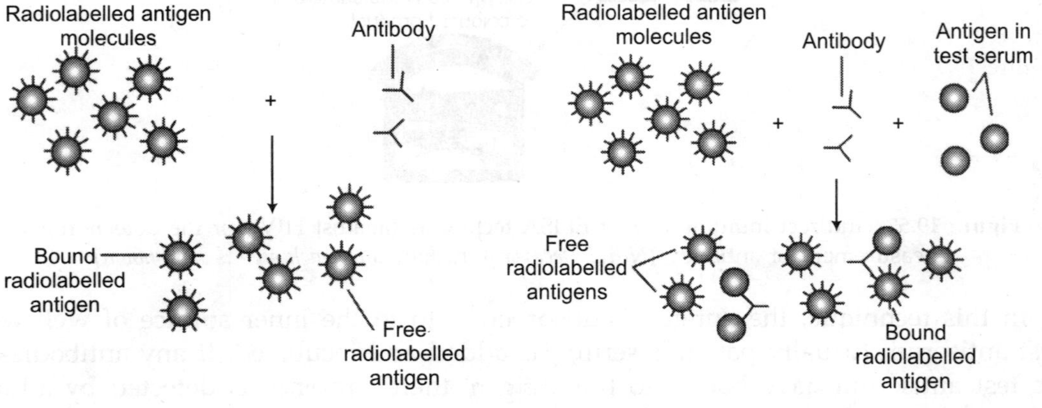

Figure 19.53 Radiolabelled antigens bind to specific antibodies and unlabelled antigen (variable) amount in test sample) competes with labelled antigen (fixed amount) for antibody-combining sites (fixed amount of antibodies).

19.24.18.2 Procedure

Measured quantities of radiolabelled antigen (of the same kind being tested) and antibody (specific to the antigen being tested) are mixed and incubated, one mixture with and one without added test sample. At increasing concentrations of test antigen, (unlabelled antigen), more labelled antigen is displaced or prevented from binding to the antibody molecules. The antigen-antibody complex is washed to remove unbound radiolabelled antigen from the mixture. The radioactivity associated with the antibody is then detected by means of radioisotope analyzers and autoradiography (photographic emulsions that show areas of radioactivity). A little amount of bound radioactivity indicates that there is a large amount of antigen, and a large amount of bound radioactivity indicates that there is little antigen in the sample.

A standard calibration curve is plotted on the graph by taking the ratio of bound radiolabelled antigen to free radiolabelled antigen against varying known amounts of unlabelled antigen and the antigen concentration of the test sample can be read from this curve (Figure 19.54).

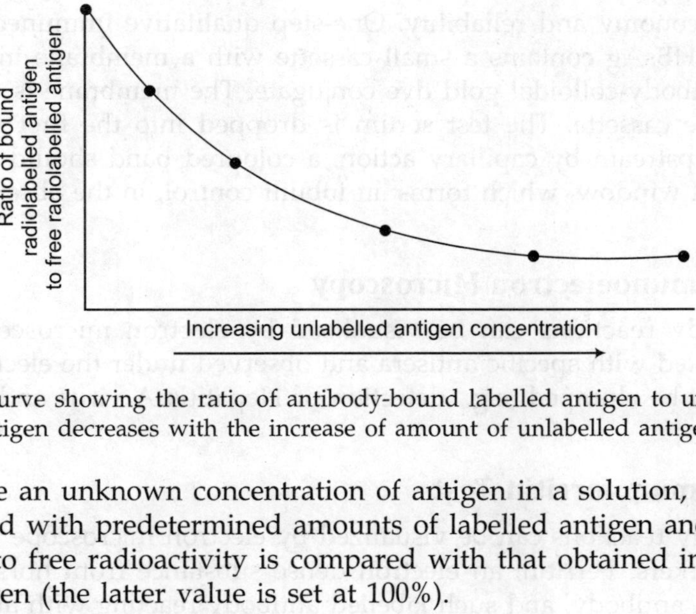

Figure 19.54 Curve showing the ratio of antibody-bound labelled antigen to unbound labelled antigen decreases with the increase of amount of unlabelled antigen.

To determine an unknown concentration of antigen in a solution, a sample of the solution is mixed with predetermined amounts of labelled antigen and antibody. The ratio of bound to free radioactivity is compared with that obtained in the absence of unlabelled antigen (the latter value is set at 100%).

19.24.19 Immuno Electroblotting

In immunoblot techniques, antibodies detect individual proteins (antigens) in complex mixtures. The technique combines the sensitivity of enzyme immunoassay with much greater specificity, viz., Western blot.

19.24.19.1 Western blotting

It is used as a confirmatory test for the serodiagnosis of HIV infection. This technique is used to identify proteins (antigens) by using specific antibodies as analytical probes.

The sources of antibody probes may be serum from patients with known infections or purified polyclonal antibody preparations or monoclonal antibodies.

Protein antigen mixture is first separated into its components by polyacrylamide and then blotted onto to the nitrocellulose membrane. It is then incubated with specific antibody (patient's serum). This primary antibody binds to its target protein antigen and is immobilized on the nitrocellulose blotting support. This antigen-antibody complex can then be identified with radioactive or enzyme labelled antibodies (anti-antibodies or antihuman gamma globulin) to primary antibodies, and can then be visualized by autoradiography. If required, the labelled anti-antibody can be removed, and the protein antigens can be studied further.

When enzyme tagged antihuman gamma globulin is added to detect primary antibody, enzyme substrate is added subsequently. The substrate changes colour in the presence of enzyme and permanently stains the nitrocellulose paper.

19.24.20 Immunochromatographic Technique

Immunochromatographic technique has wide applications in serodiagnosis due to its simplicity, economy and reliability. One-step qualitative immunechromatographic test system for HBsAg contains a small cassette with a membrane impregnated with anti HBsAg antibody-colloidal gold dye conjugate. The membrane is exposed at three windows on the cassette. The test serum is dropped into the first window. As the serum travels upstream by capillary action, a coloured band should appear in every case at the third window, which forms an inbuilt control, in the absence of which the test is invalid.

19.24.21 Immunoelectron Microscopy

Antigen-antibody reactions can be visualised by electron microscope. When viral particles are mixed with specific antisera and observed under the electron microscope, they are seen to be clumped, e.g., detection of Hepatitis A virus and Rotavirus from fecal samples.

19.24.22 Immunoferritin Test

Antigen-antibody reactions can be visualized by electron microscope with the help of appropriate markers. Ferritin, an electron dense substance from horse spleen can be conjugated with antibody, and such labelled antibody reacting with an antigen can be visualized under the electron microscope.

19.24.23 Immunoenzyme Test

Antigen-antibody reactions can be visualized by electron microscope with the help of appropriate markers such as enzymes, e.g., peroxidases, glucose oxidase, phosphatases, tyrosinase etc.

Some stable enzymes, such as peroxidases can be conjugated with antibodies. Tissue sections carrying the corresponding antigens are treated with peroxidase labelled antisera.

The peroxidase bound to the antigen can be visualized under the electron microscope, by treatment with enzyme substrate.

19.24.24 Neutralization

Neutralization tests are antigen-antibody reactions that determine whether the activity of a toxin or virus has been neutralized by antibody. Laboratory animals or tissue culture cells are used as indicator systems in these tests. The toxin or virus to be assayed has known effects on the indicator system. The effect in the animals might be death, paralysis or skin lesions.

19.24.24.1 Toxin neutralization tests

The test is based on the principle that the biological action of toxin is neutralized on reacting with specific neutralizing antibodies called antitoxins, e.g., Nagler reaction and Schick test. For example, when the exotoxin of *Clostridum botulinum* is suspected of causing food poisoning in a person, a sample of the suspect food or the serum, stools or vomits of the patient is collected. Two groups of indicator mice are used. The control group receives the botulinum antitoxin and the experimental group does not. Filtrates of the samples are injected into both groups of mice. If the toxin is present, all mice except those receiving the antitoxin will die, and the test is positive for botulism.

Neutralization of viruses by their specific antibodies is called **virus neutralization** tests. Viral neutralization is frequently used to detect viral infections. Suspected blood serum containing viral antibodies can be introduced into tissue culture cells or embryonated eggs. If antibodies are present against the virus, viral neutralization will occur and prevent the virus from infecting the culture cells. No cytopathic effect is seen. Viral haemagglutination inhibition (HI) test is an example of virus neutralization test commonly used in the diagnosis of viral infections such as influenza, mumps and measles. The mechanism of neutralization of toxins is illustrated in Figure 19.55.

19.24.25 Opsonization

Opsonization (Greek; Opson = prepare victims for) is a process in which microorganisms or other particles are coated by antibody and/or complement, thereby being prepared for recognition and ingestion by phagocytic cells.

Opsonins are substances that make an antigen more susceptible to phagocytosis. Antibodies (IgG) act as opsonins by connecting to the antigen at $F(ab)_2$ site and to the macrophages by FC receptor. This binding forms a bridge between the phagocyte and the antigen. The macrophage then phagocytoses the antigen and antibody (Figure 19.56).

When antigen-antibody complexes activate complement pathway, C3b fragment assembles on the surface of microorganisms. Both neutrophils and macrophages possess surface receptors for C3b. A bridge is formed between microorganisms and phagocytic cells via complement that aids in phagocytosis. If both antibody and C3b opsonize, binding is greatly enhanced. The opsonic index is defined as the ratio of the phagocytic activity of patient's blood for a given bacterium, to the phagocytic activity of blood from

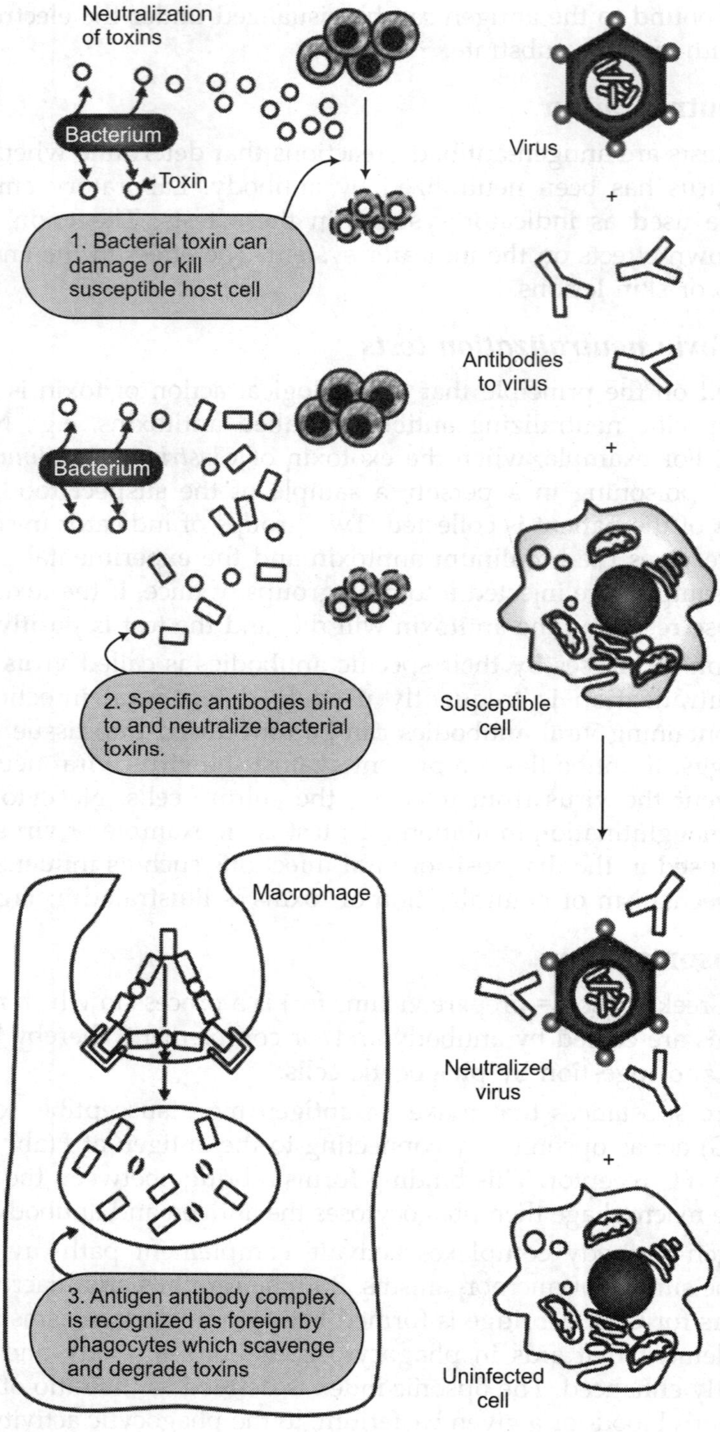

Figure 19.55 Flow diagram showing neutralization of toxin.

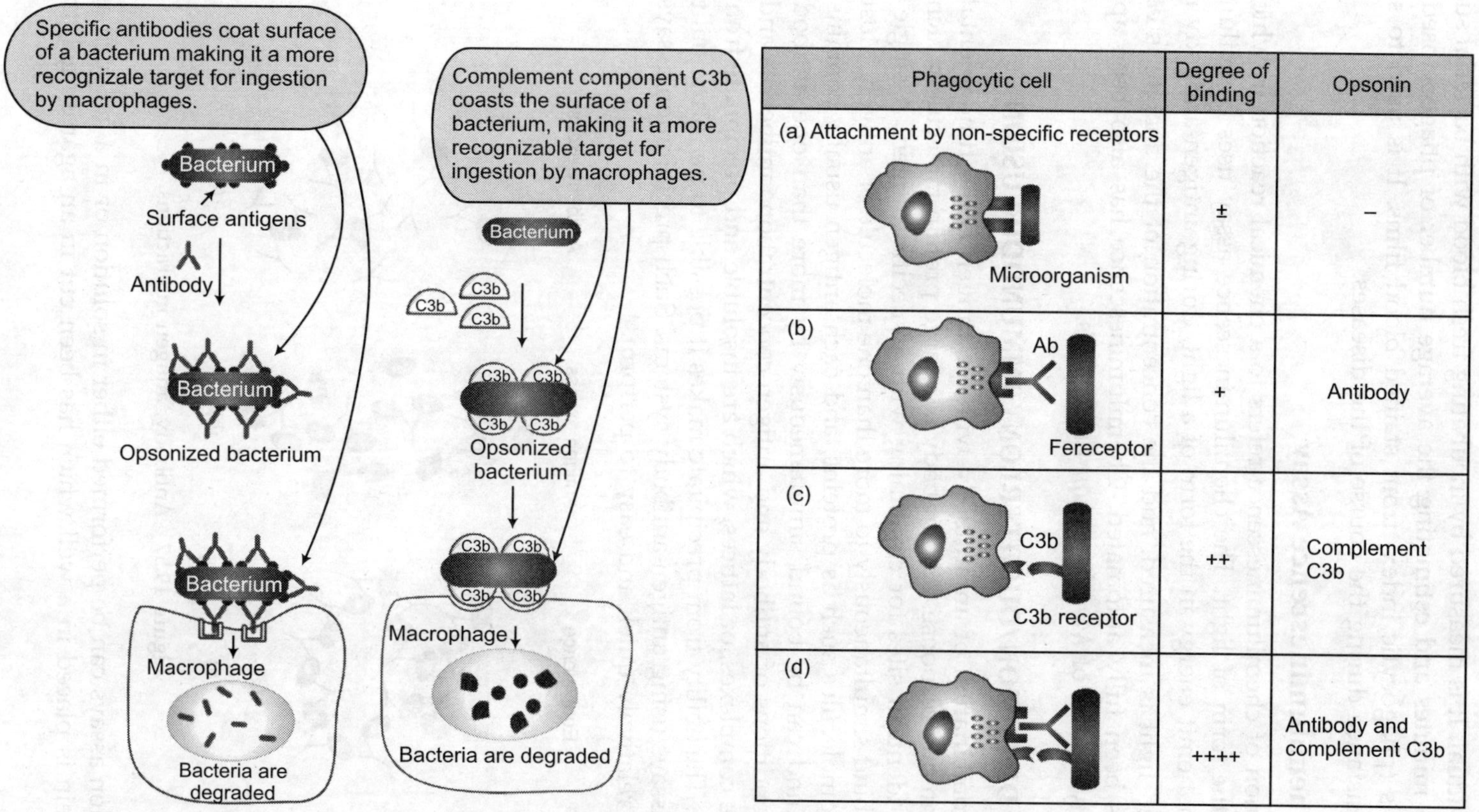

Figure 19.56 Opsonization for neutralization of toxin and bacterium.

a normal individual. It is measured by incubating fresh blood with bacterial suspension at 37°C for 15 minutes and estimating the average number of phagocytosed bacteria per neutrophils (phagocytic index) from stained blood films. It is used to study the progress of resistance during the course of the diseases.

19.24.26 Chemiluminescence Assay

The phenomenon of chemiluminescence refers to a chemical reaction in which energy is emitted in the form of light. The chemiluminescence assay uses chemiluminescent compounds that emit energy in the form of a light, during antigen-antibody reactions.

The emitted light is measured, and the concentration of the analyte is calculated. The assay has been fully automated. Chemiluminescence has also been applied for drug sensitivity testing of *Mycobacterium tuberculosis.*

19.25 GEL DIFFUSION/OUCHTERLONY IMMUNODIFFUSION

The specific combination of an antibody with an antigen is the fundamental reaction of humoral immune response. An antibody is a large protein molecule which contains two or more, identical sites for combining with a specific antigen. A single antibody molecule can bind simultaneously to more than one molecule of antigen. Most antigens are also macromolecules such as proteins, and each antigen usually contains multiple binding sites and can be bound simultaneously by more than one antibody. Under appropriate conditions of antibody and antigen concentration, antibodies and antigens can form large complexes, or lattices, which are insoluble and precipitate from solution (Figure 19.57). The ability from precipitate makes it possible to perform qualitative and quantitative assays using antigen-antibody systems. Such precipitation assays are very sensitive and relatively quick and easy to perform.

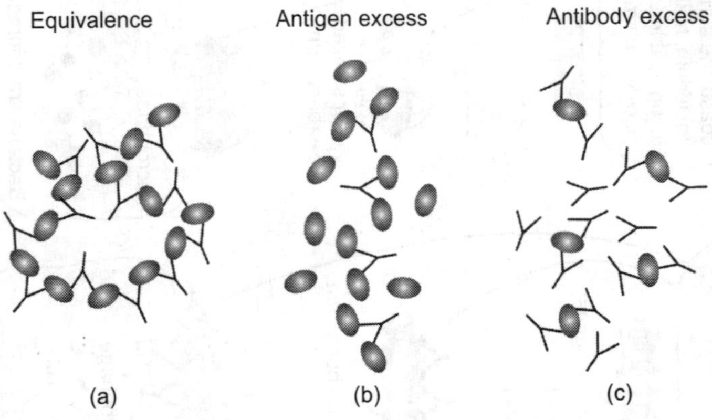

Figure 19.57 Antibody, antigen interactions.

Precipitation assays can be performed either in solution or in a gel such as agarose. When a protein is placed in a well which has been cut in an agarose gel, the protein will diffuse away from the well, through the gel, in all directions, to create a gradient of the molecule where the concentration decreases with increasing distance from the well.

This occurs because proteins can move freely in the gel, which microscopically is more like a matrix with many holes than like a solid mass, and the kinetic energy of the molecules will cause them to move randomly. If antibodies and antigens are placed in wells in different areas of a gel, they will move toward each other, as well as all other directions, and will form opaque bands of precipitate (precipitation lines) where diffusion fronts meet. Maximum precipitation will occur in the area of gel where concentrations of antibody and antigen molecules are at an optimal ratio. Closer to the antibody well, antibody concentration will be too high for precipitate formation (antibody excess zone). Closer to the antigen well, the antigen concentration will be too high (antigen excess zone).

A major advantage of this type of gel diffusion assay is that more than one antigen-antibody system in a mixture can be detected. It is important to remember that when antibodies are produced by injecting an antigen into an animal, many different antibody molecules reacting with different parts of the antigen (e.g. different proteins in a mixture, or antigenically different parts of a single protein) will be found in the serum of the animal. It is this serum, referred to as antiserum, which is usually used as a source of antibodies in experiments such as these. A single antigen will combine with the antibody it induces (its homologous antibody) to from a single precipitin line in a gel diffusion assay. When two antigens are present, each behaves independently of the other. The positions of the precipitin lines will depend primarily on the sizes of the antigens and their rates of diffusion in the gel. Thus, the number of precipitin lines indicates the number of antigen-antibody systems present. This is actually the minimum number of antigen-antibody systems since different precipitin lines often develop near each other and appear as a single line (Figure 19.58).

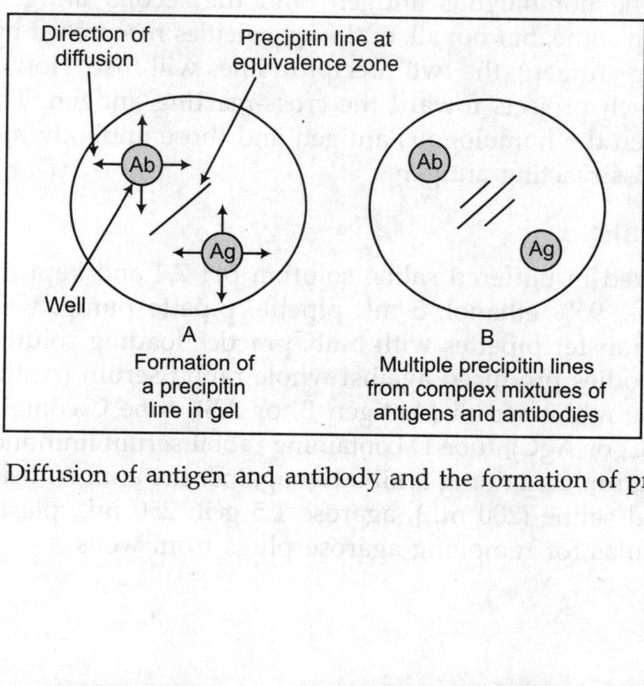

Figure 19.58 Diffusion of antigen and antibody and the formation of precipitin lines.

Double diffusion, referring to the process described above in which both antigen and antibody molecules diffuse from a well, is a simple but very useful procedure invented by the Swedish scientist, Ouchterlony. It is used to compare antigens for the number of identical or cross-reacting determinants. If a solution of antigen is placed into two adjacent wells, and the antibody is placed equidistant from the two antigen wells, the two precipitin lines formed will join at their closest ends and fuse (Figure 19.59).

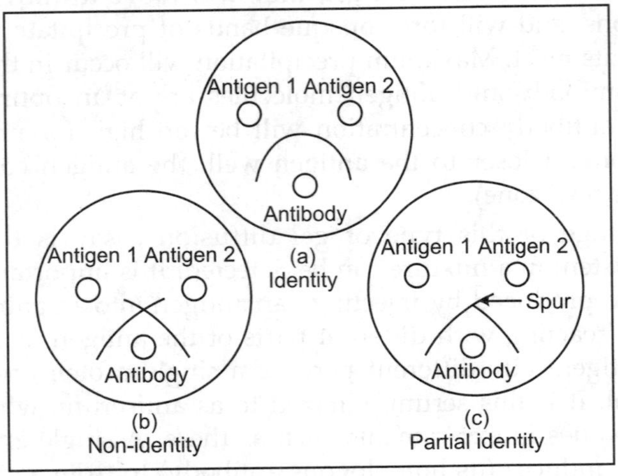

Figure 19.59 Detection of antigenic identity by the double diffusion.

This is known as a reaction of identity. When unrelated antigens are placed in adjacent wells, and the centre well is filled with antibody reacting with both antigens, the precipitin lines will form independently of each other and will cross. This is known as a reaction of non-identity. If one of the antigens was used to produce the antibody sample (i.e. it is the homologous antigen) and the second antigen shares with the homologous antigen some, but not all, of the specificities recognized by the antibody (i.e. it is a cross-reacting antigen), the two precipitin lines will fuse. However, an additional spur will form which projects toward the cross-reacting antigen. The spur represents the reaction between the homologous antigen and those antibody molecules which do not bind to the cross-reacting antigen.

19.25.1 Materials

1% agarose dissolved in buffered saline solution pH 7.4 and kept in 60°C water bath, 4 microscope slides, 95% ethanol, 5 mL pipette, pipette pump, well cutter, template for cutting wells, transfer pipettes with bulb, practice loading solution (optional), tube A containing antibodies produced against whole rabbit serum (Antibody, or Ab), tube B containing whole rabbit serum (Antigen B, or AB), tube C containing rabbit serum albumin (Antigen C, or AgC), tube D containing rabbit serum immunoglobulin (Antigen D, or AgD). Antiserum containing antibodies against the antigen being tested, antigen, Phosphate-buffered saline (200 mL), agarose 1.5 g in 250 mL, plastic or lid for pans, toothpicks or spatulas for removing agarose plugs from wells.

19.25.2 Staining and Destaining Solution

0.1M Tris buffer pH 7.5, destaining solution 50 : 40 : 10 of water : methanol : acetic acid, staining solution 0.1% Coomassie blue in 50 : 40 : 10 of water : methanol acetic acid.

Preparation of solutions

1. For phosphate-buffered saline, dissolve the following in 150 mL distilled water:
2. 19 g NaCl, 1.28 g $KH_2PO_{4,}$ 1.63 g Na_2HPO_4 $7H_20$ adjust pH to 7.2, add distilled water to bring final volume to 250 mL.
2. 1% agarose solution
   (a) Add 180 mL phosphate-buffered saline to 250 mL flask. Add 1.8 g agarose. Swirl to disperse large clumps.
   (b) Heat to dissolve agarose using a hot plate to bring mixture to boiling with occasional careful swirling. Continue heating until all particles have dissolved, and the final solution is clear.
   (c) Keep melted agarose in 60°C water bath to prevent solidification, if the agarose does solidify, remelt it in a hot water bath or in a microwave oven.
3. Tris buffer: Tris solution (0.01 M.), dissolve 1.21 g Tris base in 700 mL distilled water. Adjust pH to 7.5 and add distilled water to bring the volume to 1000 mL.

19.25.3 Protocol

Preparation of slides with sample wells

1. Clean all slides by placing a small amount of alcohol. After cleaning, do not touch the clean surface with fingers.
2. Label the slides to be used for the experiment at one end and number them consecutively 1, 2, 3. Be careful not to touch the clean surface with fingers.
3. Using a 5 mL pipette, carefully pipette 3.5 mL of melted agarose, allow to cool slightly, onto each slide by moving the pipette tip along the centre length of the slide and dispensing the agarose. The melted agarose should spread evenly over the surface of the slide, without running over the edges. Repeat with the remaining two slides.
   [*Note:* If you are not successful at covering the slide with agarose, wipe off the slide, clean the slide again with alcohol as described in step 1, relabel the slide, and try again.]
4. Allow the agarose to solidify. This will take approximately 10 min at which time the gel will appear slightly opaque.
5. For the three slides to be used for the experiment, place the template under one of the slides so that the pattern is in the centre of the slide. The distance between the wells is important. Try to follow the template as accurately as possible (Figure 19.60).
6. Cut the five wells with the well cutter, using a gentle punching motion. Remove the agarose plugs by lifting the edges with a toothpick or spatula.

[*Note:* If well placement is not accurate, there should be enough room on the slide to re-cut the wells after repositioning the template under the slide.]

7. Repeat steps 5 and 6 with the remaining two slides.

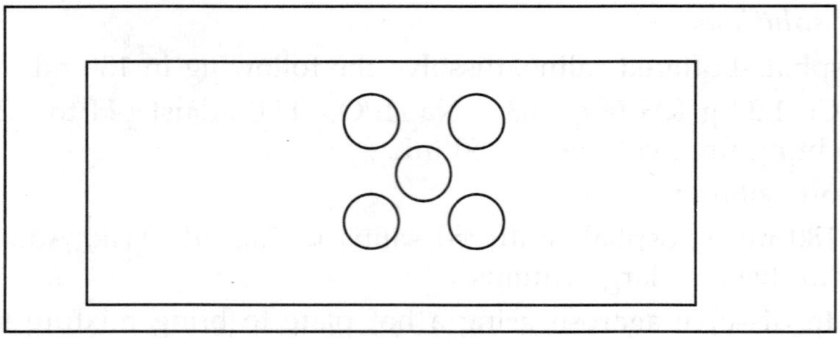

Figure 19.60 Template for cutting wells.

Loading the samples

1. **Load samples as:** Squeeze the pipette bulb to draw slowly, the liquid up into the pipette. The sample should remain in the lower portion of the pipette.
   Place the pipette tip just over, not inside, the sample well. Maintain steady pressure on the pipette bulb to prevent sample from being drawn back up into the pipette,
   Slowly squeeze the pipette bulb to eject three drops of sample. The well should appear full. Be careful not to overfill the well and cause spillage on the agarose surface. Such spillage would affect your results (Figure 19.61).
2. Fill the centre wells of all three slides with 3 drops (approximately 30 μL) of antiserum (Ab, tube A), using a transfer pipette.
3. Fill the outer wells with 3 drops of antigen (tubes B, C and D), using a clean transfer pipette for each antigen as follows in Figure 19.60 to Figure 19.62.
4. Carefully place the slides in the incubation chamber on top of the wet paper towel layer. Cover the chamber with plastic wrap and let allow precipitin lines to form. The precipitin lines should be visible in 24 hours.
5. Predict the patterns that will be formed by the precipitin lines, based on the relationships between the antigens and the information given the introduction. Sketch your predictions on the diagrams of the slides above.

(Optional) Staining slides

1. Cover the slides with Iris buffer. 0.01 M pH 7.5 and incubate at room temperature for at least a hours, change the Iris buffer once.
2. Remove buffer and cover the slides with 0.1% Coomassie blue stain and let sit 30-60 min.

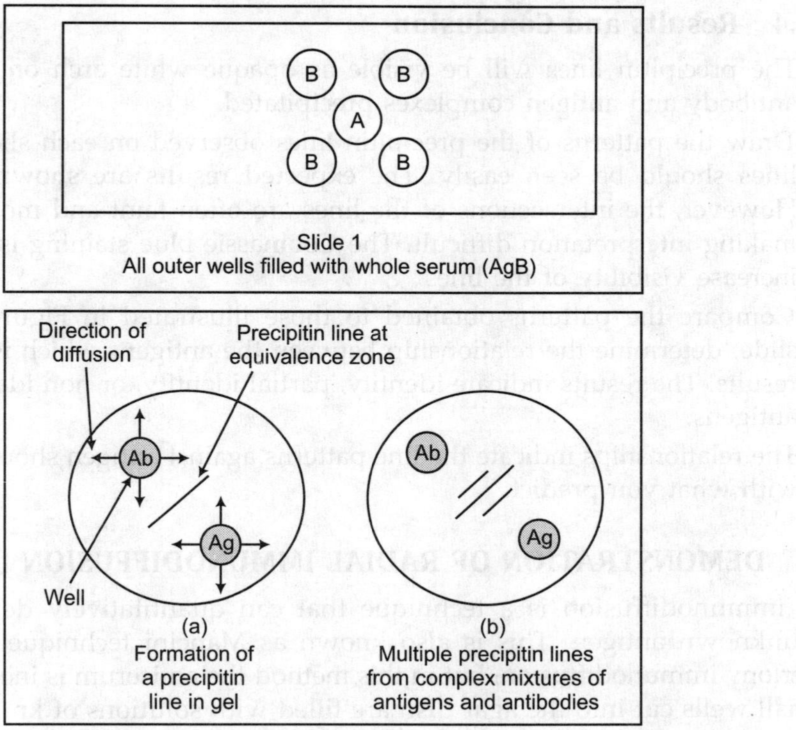

Figure 19.61 Filling of well.

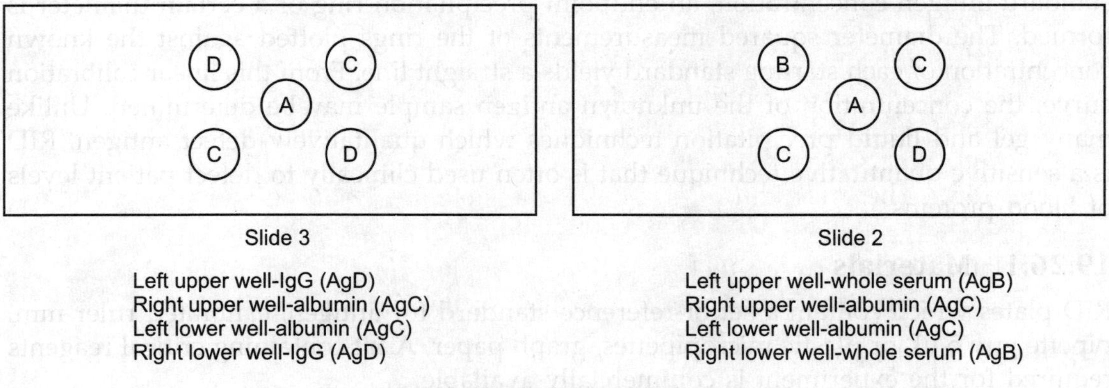

Figure 19.62 Complete test.

3. Remove stain and cover slides with destain. Change destain when it becomes blue (about one hour), add fresh destain, and repeat until the gel is clear enough to visualize blue precipitin lines.
4. Remove destain, cover dish and store slides in the refrigerator until results are observed.

19.25.4 Results and Conclusion

1. The precipitin lines will be visible as opaque white arch on each slide where antibody and antigen complexes precipitated.
2. Draw the patterns of the precipitin lines observed on each slide. The precipitin lines should be seen easily. The expected results are shown in Figure 19.63. However, the intersections of the lines are often faint and more difficult to see, making interpretation difficult. The Coomassie blue staining is recommended to increase visibility of the lines.
3. Compare the patterns obtained to those illustrated in Figure 19.63. For each slide, determine the relationship between the antigens which is indicated by the results. The results indicate identity, partial identity, or non-identity between the antigens.
4. The relationships indicate the line patterns against antigen should be is consistent with what you predict.

19.26 DEMONSTRATION OF RADIAL IMMUNODIFFUSION (RID)

Radial immunodiffusion is a technique that can quantitatively determine the level of an unknown antigen. This is also known as Mancini technique and is similar to Ouchterlony immunodiffusion, but in this method the antiserum is incorporated into the gel. Small wells cut into the agar dish are filled with solutions of known concentration antigen and an unknown solution sample. The antigen solution diffuses outwards front the well in a circular pattern. In the endpoint or Mancini method used in this experiment, antibody is present in excess amount and diffusion continues until a stable ring of antigen-antibody precipitate forms, generally within 24 to 48 hours. For each standard antigen concentration, an endpoint precipitation ring of a certain diameter is formed. The diameter squared measurements of the rings plotted against the known concentration of each starting standard yields a straight line. From this linear calibration curve, the concentration of the unknown antigen sample may be determined. Unlike many gel and liquid precipitation techniques which qualitatively detect antigen, RID is a sensitive quantitative technique that is often used clinically to detect patient levels of blood proteins.

19.26.1 Materials

RID plates which contain a set of reference standard for antigen, calibrated ruler mm, pipette gun 5 μL or plastic micropipettes, graph paper. A kit containing critical reagents required for the experiment is commercially available.

19.26.2 Protocol

1. Perform the test using the samples, following the directions provided with the kit, add 5 μL of antigen solution standards to each of four wells. Rinse micropipettes or change tips. To the fifth well add 5 μL of the unknown antigen solution. Snap

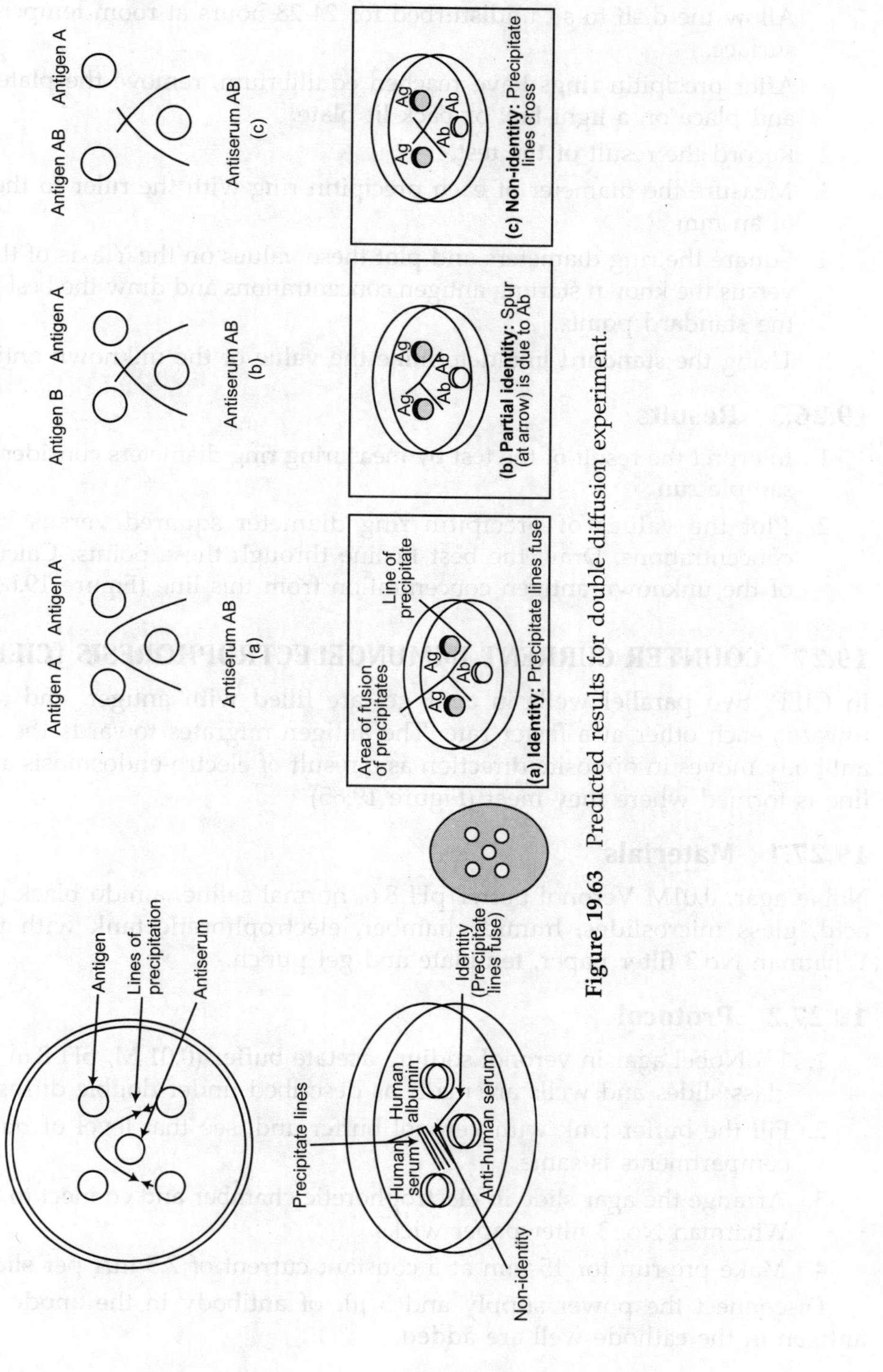

Figure 19.63 Predicted results for double diffusion experiment.

the cover of the Petri dish back on. Place the dish into a zip lock bag and seal. Allow the dish to sit undisturbed for 24-28 hours at room temperature on a flat surface.

After precipitin rings have reached equilibrium, remove the plate from the bag and place on a light box or back lit plate.

2. Record the result of the test.
3. Measure the diameter of each precipitin ring with the ruler to the nearest tenth of an mm.
4. Square the ring diameters and plot these values on the Y axis of the graph paper versus the known starting antigen concentrations and draw the best lit line through the standard points.
5. Using the standard line determine the value of the unknown antigen.

19.26.3 Results

1. Interpret the result of the test by measuring ring diameters considering any control sample run.
2. Plot the values of precipitin ring diameter squared versus known antigen concentrations. Draw the best fit line through these points. Calculate the value of the unknown antigen concentration from this line (Figure 19.64).

19.27 COUNTER CURRENT IMMUNOELECTROPHORESIS (CIEP)

In CIEP, two parallel wells in agar gel are filled with antigen and antibody move towards each other at a faster rate. The antigen migrates towards the anode and the antibody moves in opposite direction as a result of electro-endosmosis and a precipitin line is formed where they meet (Figure 19.65).

19.27.1 Materials

Noble agar, 0.01M Veronol buffer pH 8.6, normal saline, amido black (1%), 2% acetic acid, glass microslides, humid chamber, electrophoretic tank with power supply, Whatman No.3 filter paper, template and gel punch.

19.27.2 Protocol

1. 1% Nobel agar in veronol sodium acetate buffer (0.01 M, pH 8.6) is layered over glass slides and wells are made as described under double diffusion technique.
2. Fill the buffer tank with veronol buffer and, see that level of buffer in the two compartments is same.
3. Arrange the agar slide in electrophoretic chamber and connect to the buffer with Whatman No. 3 filter paper wicks.
4. Make pre run for 15 min at a constant current of 7.5 mA per slide.

Disconnect the power supply and 5 μL of antibody in the anode well and 5 μL antigen in the cathode well are added.

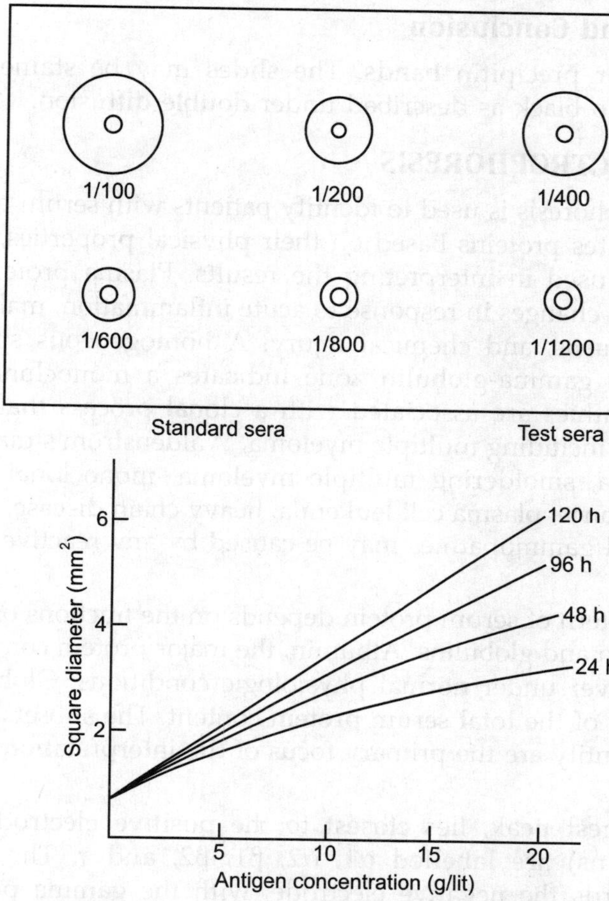

Figure 19.64 Albumin measurement by single radial diffusion meth.

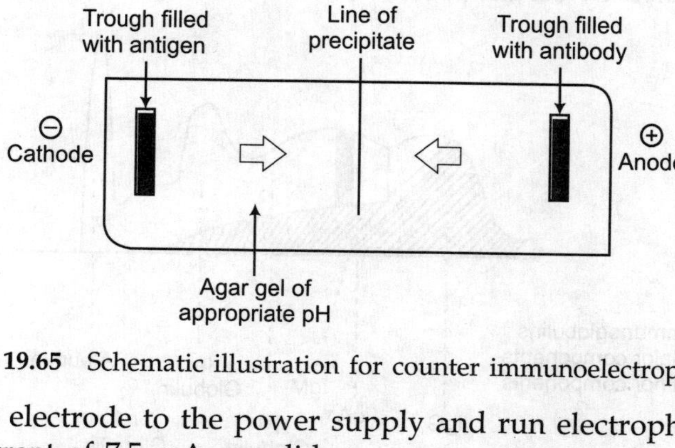

Figure 19.65 Schematic illustration for counter immunoelectrophoresis.

6. Connect the electrode to the power supply and run electrophoresis for 1 hr at a constant current of 7.5 mA per slide.
7. After the run, disconnect the power supply and incubate the slide in a humid chamber at room temperature for 12 hrs.

19.27.3 Results and Conclusion

Observe the slides for precipitin bands. The slides may be stained for improved sensitivity in 1% amido black as described under double diffusion.

19.28 SERUM ELECTROPHORESIS

Serum protein electrophoresis is used to identify patients with serum protein disorders. Electrophoresis separates proteins based on their physical properties, and the subsets of these proteins are used in interpreting the results. Plasma protein levels display reasonably predictable changes in response to acute inflammation, malignancy, trauma, necrosis, infarction, burns, and chemical injury. A homogeneous spike-like peak in a focal region of the gamma-globulin zone indicates a monoclonal gammopathy. Monoclonal gammopathies are associated with a clonal process that is malignant or potentially malignant, including multiple myeloma, Waldenström's macroglobulinemia, solitary plasmacytoma, smoldering multiple myeloma, monoclonal gammopathy of undetermined significance, plasma cell leukemia, heavy chain disease, and amyloidosis. In contrast, polyclonal gammopathies may be caused by any reactive or inflammatory process.

Electrophoretic pattern of serum protein depends on the fractions of two major types of the protein: albumin and globulins. Albumin, the major protein component of serum, is produced by the liver under normal physiologic conditions. Globulins comprise a much smaller fraction of the total serum protein content. The subsets of these proteins and their relative quantity are the primary focus of the interpretation of serum protein electrophoresis.

Albumin, the largest peak, lies closest to the positive electrode. The next five components (globulins) are labelled α1, α2, β1, β2, and γ. The peaks for these components lie towards the negative electrode, with the gamma peak being closest to that electrode. Figure 19.66 shows a typical normal pattern for the distribution of proteins as determined by serum protein electrophoresis.

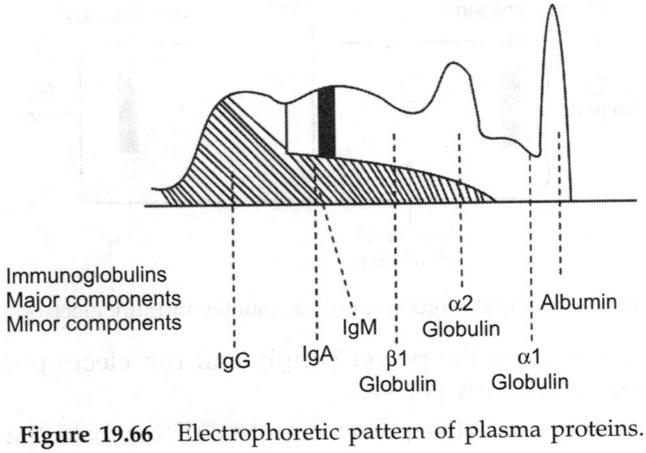

Figure 19.66 Electrophoretic pattern of plasma proteins.

19.28.1 Materials

Glass slides, agarose, electrophoresis apparatus, power pack,

1 X immune buffer

Prepared from 10×: 1mL 10× immunobuffer + 9 mL distilled water.

Veronal Buffer 0.1 M pH 8.6

(a) Na-barbitone = 10 g/L (5 g/500 mL)

(b) Na-Acetate = 6.5 g/L (3.28 g/500 mL)

(c) HCl 0.1 N = 34.2 µL/L (17.1 µL/500 mL)

19.28.2 Protocol

1. Prepare 1% Agarose in 1× immuno buffer.
2. Prepare Gel plate by pouring 5-6 mL of agarose on to the slide.
3. After the solidification, cut a trough on the gel with the help of blade.
4. Take an electrophoretic tank and add Veronal buffer.
5. Transfer agarose slide to the electrophoretic chamber.
6. Connect the slides to anodic and cathodic buffer chambers with Whatmann filter paper wicks.
7. Make pre run for 5 min at 7.5 mA.
8. Load 20 µL serum containing few drops of Bromo phenol blue dye in the trough.
9. Run electrophoresis for about 180 min.
10. Remove slides from electrophoretic chamber.
11. Wash slides in normal saline for about 30 min.
12. Immerse slides in 2% acetone solution for about 15 min.
13. Dry the slides in an incubator at 40°C.
14. Stain the slides in 1% amido black or Coomassie blue stain.

19.28.3 Results and Conclusion

Observe the slides for bands.

19.29 DEMONSTRATION OF IMMUNOELECTROPHORESIS

In immunoelectrophoresis, two methods are combined, namely, agar-gel electrophoresis and immunodiffusion. Method is used in both clinical and research laboratories for separating and identifying proteins on the basis of their electrophoresis behaviour and their immunological properties. Immunoelectrophoresis is used to separate and characterize a mixture of serum proteins as well as to examine the specificity of the antigen-antibody interaction.

In the clinical laboratory, this technique is used for diagnosis. It is utilized in examining certain serum abnormalities especially those involving immunoglobulins, and also semi-quantitative proteins analysis of urine, cerebrospinal fluid, pleural fluids, etc. In the research laboratory, this technique may be used to monitor antigen and/or antibody purifications, to detect impurities, to analyze soluble antigens of plant and animal tissues and microbial extracts.

In immunoelectrophoresis, the proteins are first subjected to an electric field and separated by zone electrophoresis on the basis of their different charge-to-mass ratios. The electric field is then shut off, and antibodies to the proteins are introduced into the system. Diffusion of both antigen and antibody takes place and at a particular locus, the equivalence point is reached resulting precipitation. In classical immunoelectrophoresis, the antibody is added in a trough cut out of the electrophoretic gel after electrophoresis.

19.29.1 Materials

2 g agarose, 500 mL TAE (Tris-acetate-EDTA) buffer pH 8.0, 10 mL whole serum, 200 mL goat anti-rabbit IgG, horizontal electrophoresis apparatus

TAE (50 × stock, 1 litre) 24.2 g Tris base, 57.1 mL glacial acetic acid, 37.2 g Na_2EDTA. H_2O pH 8.5.

Stock solution dissolve 12.1 g trishydrasymethyl-amino-methane (Tris) and 17 g of ethyleliediaminetetraacetate, disodium salt (EDTA) in 450 mL of ionized water. Adjust pH to 8.5 with glacial acetic acid (25 mL). Make up to 500 mL with water.

Working buffer dilute stock 50 times

Dilute antiserum in TAE if necessary

Prepare agarose gel Mix 0.5 g of agarose and 50 mL of diluted TAE buffer in a 250 mL Erlenmeyer flask. Dissolve by boiling in a microwave oven or on a hot plate. Allow to cool to 50°C. Temperature of molten gel is critical to be at 50°C.

19.29.2 Protocol

1. Coat a glass plate by applying 2% agar solution with a small brush.
2. Cast the gel on to the glass plate by pouring the molten gel to a thickness of 1-2 mm and allow cooling.
3. Cut a small hole/trough in the gel by punching the agarose gel with a Pasteur pipette.
4. Transfer the gel to the electrophoresis apparatus and add enough buffers, to just cover the gel.
5. Add 2 drops of a bromophenol blue solution to the serum.
6. Load whole serum into the well with a micropipette or capillary tubing.
7. Elevate slide 2-3 mm above electrophoresis - buffer solution. Make contact between the buffer and agarose by using a wick of filter paper pre-soaked in electrophoresis buffer. One end of the wick should overlap the end of the gel by 3-4 mm and the other end submerged in electrophoresis buffer.
8. Electrophorese at 50 V and 30 mA for 1 hour.

9. Carefully remove the gel from the electrophoresis chamber. Cut two troughs, parallel to electrophoretic run. Use a sharp blade to delineate the troughs and a Pasteur pipette tip to remove the cut gel.
10. Load the left through with goat anti-rabbit whole serum
11. Place the gel in a closed humidifying chamber containing moistened paper towels and allow diffusion to take place over a 24 hour period or until a visible precipitate is formed. (Figure 19.67).

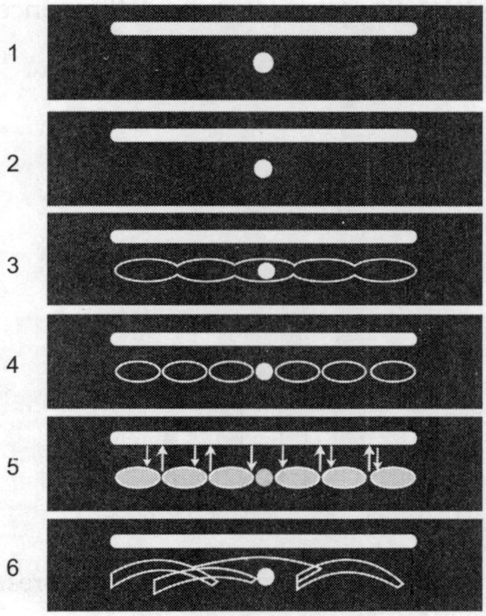

Figure 19.67 Flow diagram illustrating immunoelectrophoresis.

19.29.3 Results

1. Note the formation of arcs of precipitate in the gel.
2. Identify the number of proteins in the rabbit whole serum from the number of arcs of precipitate.
3. The actual number of arcs of precipitate will depend on the number of antigens in serum and the efficiency of separation.

19.30 DETERMINATION OF ALBUMIN BY LAURELL ROCKET IMMUNOELECTROPHORESIS

Rocket immunoelectrophoresis is very similar to radial immunodiffusion in that the antigen is incorporated into the agar-gel. This time, however, the samples move through the gel by electrophoresis rather than by simple diffusion.

The antiserum is incorporated into the agar, and holes are cut along the origin, to hold the test material and dilutions of a standard antigen for calibration. When potential is applied across the gel, the antigen migrates towards the anode while the antibody in the gel moves towards the cathode and immune complex forms where they meet. Initially, there is an excess of antigen so that the immune complex is soluble but as the experiment proceeds the antigen continues to migrate and becomes more dilute as more antigen is present in the immune complex. Later on equivalence is reached, and an insoluble immune complex is formed. The complex itself migrates but eventually becomes stationary when the entire antigen is present in the immune complex. The precipitate is rocket shaped and, for a given antibody concentration, there is a linear relationship between the distance moved by the precipitate and the concentration of the antigen (Figure 19.68).

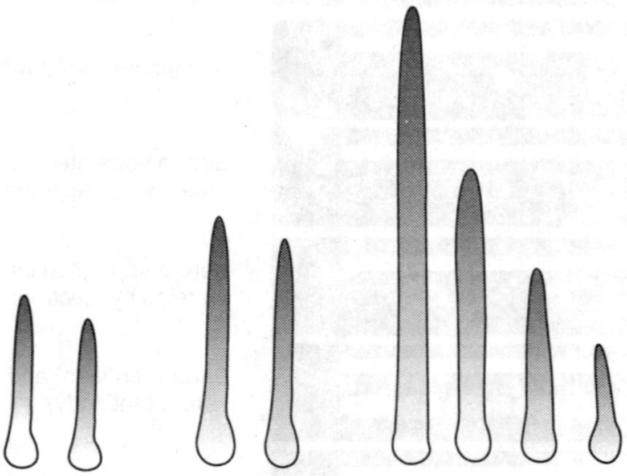

Figure 19.68 Demonstration of Laurell rocket immunoelectrophoresis.

19.30.1 Materials

Materials described for immunoelectrophoresis and immunodiffusion in earlier experiments, Coomassie blue (0.25 per cent w/v in methanol:acetic acid: water, 1 L 5:1:5), destaining solution (methanol:acetic acid:water, 5:1:5), hair drier.

19.30.2 Protocol

Prepare the plates as for the immunoelectrophoresis but add 0.6 μL of rabbit anti-human albumin per cm^2 of the plate area. Cut 12 wells into the gel in a straight line 1.5 cm from one edge of the plate and place in the electrophoresis apparatus. Apply the diluted standards to four of the wells and 1/400 dilutions of the tests to the other wells. All dilutions are made in saline as in the previous experiments. Pipette 3μL of the samples into each well as quickly as possible so as to minimize the amount of radial diffusion. When all the samples have been added to the plate, which should be set up in duplicate, place the lid of the tank on the apparatus and turn on the power. Carry out electrophoresis at 5-10 V/cm with the cooling system activated. At the end of this time, turn off the power and remove the plate. Place the slides under a number of filter papers with a light weight on top, leave for 10 min then remove the filter papers.

Dry the plate completely in a stream of hot air from a hair drier then immerse the dry plate in Coomassie blue for 10 min. Remove the excess stain with the destaining solution and measure the heights of the rockets from the anodal edge of the well to the peak of the rocket.

19.30.3 Results and Conclusion

Plot the rocket height against albumin concentration to produce a standard curve, and then use this curve to estimate the albumin concentrations in the eight serum samples.

19.31 PURIFICATION OF IMMUNOGLOBULINS BY ION EXCHANGE CHROMATOGRAPHY

When human serum is added to an ion exchange column such as DEAE-Sephadex, most of the serum proteins become bound to the column. However, by using a buffer with a pH greater than 6.5, IgG is not absorbed and is eluted in the first fractions. The other proteins are then removed from the column by a stepwise elution with increasing concentrations of salt. The isolated fractions are then analysed by cellulose acetate electrophoresis and immunoelectrophoresis.

19.31.1 Materials

DEAE-Sephadex A50-120, disposable syringes (10 mL), Tris-HCl buffer (10 mmol/L, pH 8.0), buchner filtration apparatus, NaOH (1 mol/L), HCl (1 mol/L), Glass wool or plastic sinter, Peristaltic pumps that can pump less than 60 mL/h, human serum, UV spectrophotometers or column monitoring equipment, solid polyethyleneglycol.

Buffered saline solutions (Prepare the different concentrations of 250 mL saline in the Tris-HCl buffer (a) 20 mmol/L, (b) 50 mmol/L, (c) 100 mmol/L, (d) 150 mmol/L, (e) 200 mmol/L, (f) 300 mmol/L.

Electrophoresis horizontal electrophoresis apparatus, power pack, barbitone buffer (0.07 mol/L pH 8.6), ponceau S stain (2 g/L in 30 g/ L TCA), acetic acid (5% v/v), cellulose acetate strips, Whatman No. 3 MM paper, methanol: water (3 : 2), citrate buffer (0.1 mol/L pH 6.8).

Immunoelectrophoresis 2 g agarose, 500 mL TAE (Tris-acetate-EDTA) buffer pH 8.0, 10 mL whole serum, 200 mL goat anti-rabbit IgG, horizontal electrophoresis apparatus

TAE (50× stock, 1 litre) 24.2 g Tris base, 57.1 mL glacial acetic acid, 37.2 g Na_2EDTA. H_2O pH 8.5.

Stock solution dissolve 12.1 g trishydrasymethyl-amino-methane (Tris) and 17 g of ethyleliediaminetetraacetate, disodium salt (EDTA) in 450 mL of deionized water. Adjust pH to 8.5 with glacial acetic acid (25 mL). Make up to 500 mL with water.

Working buffer dilute stock 50 times. Dilute antiserum in TAE if necessary.

Prepare agarose gel Mix 0.5 g of agarose and 50 mL of diluted TAE buffer in a 250 mL Erlenmeyer flask. Dissolve by boiling in a microwave oven or on a hot plate. Allow to cool to 50°C. Temperature of molten gel is critical to be at 50°C.

19.31.2 Protocol

Preparation of column Add 2 g of DEAE-Sephadex A50-120 to 70 mL of water and leave until swollen. Degas the suspension under vacuum and leave to equilibrate to room temperature. Place some glass wool in the bottom of a 10 mL disposable syringe as a support and prepare a column of the ion exchange material. Wash the column with 1 mol/L NaOH then repeatedly with water until the pH falls to about 7. Wash the saline then 1 mol/L HCl and finally with the tris-HCl buffer until the eluate registers pH 8.0. Alternatively the material can be washed on a Buchner funnel.

Separation of immunoglobulins Load 0.5-1.0 mL serum on to the DEAE-Sephadex and collect five fractions of about 2.5 mL each by pumping the tris-HCl buffer through the column. In the meantime prepare 25 mL of a series of fluids containing sodium chloride at the following concentrations in the buffer: 20 mmol/L, 50 mmol/L, 100 mmol/L, 150 mmol/L, 200 mmol/L, 300 mmol/L. Pump these solutions through the column in turn starting with the lowest salt concentration and collect fractions until all the protein has been eluted with that solution, then move on to a higher concentration of salt. Carefully monitor the proteins eluted by measuring the absorbance of each fraction at 280 nm and plot the extinction fraction number.

19.31.3 Results

Concentrate each fraction between 5 and 10 times by dialysing against solid polyethyleneglycol for about 3 h and examine the protein concentrates by cellulose acetate electrophoresis, immunodiffusion and immunoelectrophoresis. Compare the electrophoresis pattern of fractions.

19.32 PREGNANCY TESTING BY IMMUNOLOGICAL METHODS

Pregnancy is a stage of conception in women that can be detected in the early stage by using the immunological methods. However, this test is based on the presence of a hormone *"human chorionic gonadotropin'* (hCG). It is a glycoprotein produced by trophoblastic cells of the placenta that begin about 10 days after the conception, and found in the urine of only pregnant women. The hCG is a dimer consisting of two sub-units alpha and beta. The molecular structure of hCG is similar to the other glycoprotein hormones found in urine such as LH (luteinising hormone), FSH (follicle stimulating hormone) and TSH (thyroid stimulating hormone).

After five weeks of LMP (last menstrual period), the concentration of hCG rises sharply in urine and attains peak level at 10 weeks of gestation (LMP). During the first trimester, it reaches 1, 00,000 IU/mL. The IU (international unit) is more accurate and related to specific gonadotropic activity of 0.1 mg of dried standard kept at the National Institute for Medical Research (London). There are several methods developed for a pregnancy test, but the most accurate and rapid ones are the immunological methods. Such as latex particle agglutination inhibition test, radioimmunoassay for the a-subunit of hCG, ELISA test, etc.

19.32.1 Latex Agglutination Inhibition Test

The presence of hCG in urine depends on the specificity of an antigen-antibody reaction. The heat-coagulation test of proteinuria is performed before proceeding for this test. A false pregnancy test can be anticipated if this test is positive. Standard technique for latex haemagglutination inhibition test is given in this experiment. The immunological test can be performed in two ways. It differs in carrier used for an external source of hCG. A positive and a negative control should be prepared.

19.32.1.1 Materials

Urine of pregnant women (collected in a clean glass container), glass slides, dropper, Anti- hCG serum, a wooden stick, antigen.

19.32.1.2 Protocol

1. Collect urine of pregnant women in a thoroughly clean glass container free from any residual soap or detergent.
2. Take anti- hCG serum in a clean dropper and put on a clean glass slide.
3. Add one drop of urine to anti-hCG serum and mix thoroughly.
4. Add 1-2 drop of antigen (hCG-coated latex particle suspension).
5. Mix with a clean wooden stick.
6. Rock the slide very slowly and gently for no longer than 90 seconds and observe agglutination.

19.32.1.3 Results and observations

The added antiserum will not be neutralized if the urine does not contain hCG. Still active antiserum reacts with hCG causing agglutination when the hCG coated latex suspension is added. Evenly suspended latex particles show a positive result and the clumping formation shows a negative result. In case of proteinuria, a false positive result may occur.

[*Note:* In some of the methods the latex particles are coated with hCG antibody, therefore, reversal of interpretation and the result should be considered, i.e., agglutination implies for a positive result.]

19.32.2 Haemagglutination Inhibition Test

In this test anti-hCG serum and sensitized red cells are used. The anti-hCG serum is neutralized having no effect of the sensitized red cells. The red cells are gathered at the bottom of the tube giving the positive test, and if the hCG is absent in urine the reactions will not take place.

19.32.2.1 Materials

Small size test tubes, urine sample, sensitised red cells, anti-hCG serum.

19.32.2.2 Protocol

1. Collect the urine sample in a clean glass vial and appropriately dilute it with distilled water.
2. Take 1-2 drops of anti-hCG serum in the glass tube.
3. Add 1-2 drops of urine with anti-hCG serum and mix properly.
4. Add 1-2 drops of sensitized red cells in the suspension.

19.32.2.3 Results and observations

The anti-hCG serum will be neutralized if urine contains hCG and the sensitized red cells will not be affected. The red cells will gather at the bottom of the tube and form a ring giving the positive evenly dispersed layer in the test tube showing the negative result. This test is not affected by proteinuria.

19.32.3 Pregnancy Testing Through Commercially Available Kits

There are several pregnancy test kits available commercially such as beta Visipreg latex agglutination pregnancy test kit (sensitivity = 0.5 IU/mL), VISIPREG STRIP pregnancy test kit (sensitivity = 25 IU/mL), VISIPREG CARD pregnancy test kit, etc.

19.32.3.1 Visipreg strip method

The VISIPREG STRIP is a two-site sandwich immune-assay in which antiserum, i.e., anti-hCG (purple colour) is used both in conjugate as well as in native forms. The conjugates are dried on the strip (Figure 19.69). Its working is based on the principle of ELISA which determines hCG qualitatively present in the pregnant women. When the strip is dipped in urine, the hCG if present migrates through capillary action and reacts with dried conjugate (anti-hCG) and form a complex 'anti-hCG - conjugate hCG' complex. This complex again migrates forward on the membrane which is pre-dispersed with antiserum (anti-hCG). Antibodies of this serum can bind the hCG molecules present in 'anti-hCG - conjugate-hCG' complex. Wherever the antibodies of hCG serum are present, it forms a visible purple band along the exact location of immobilised anti-hCG conjugate resulting in a positive test. If no hCG is present in urine, the anti-hCG conjugate will pass through immobilised anti-hCG band giving no coloured line. This shows the negative test.

Also, the strip contains a control region near the holding end. It consists of a band of immobilized antiserum. This band will bind all encountered free conjugate and forms a coloured line irrespective of whether hCG is present in urine or not. Appearance of control line indicates the correct testing procedure only.

The VISIPREG STRIP kept in a pouch should be stored at 4-8°C so that the activity of antiserum may be preserved. The testing work should be done at room temperature around 25°C.

Materials

Urine sample, VISIPREG STRIP.

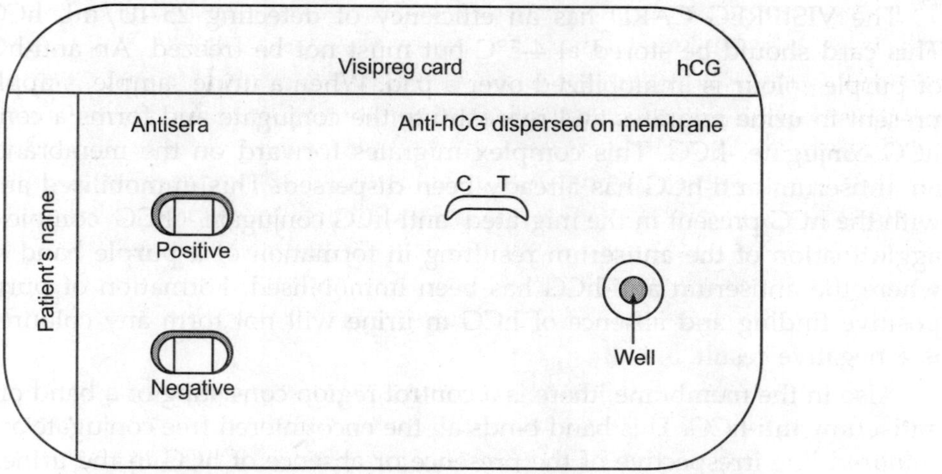

Figure 19.69 Diagrammatic illustration of VISIPREG card used for pregnancy testing.

Protocol

1. Collect urine in a clean container at any time in a day and store at 2-8°C for about 72 hours before testing.
2. Bring both urine sample and pouch of the VISIPREG STRIP at room temperature.
3. Take 0.5 mL urine in a clean vial.
4. Hold the strip at unprinted end and dip the printed end in urine up to the level marked 'MAX' (Figure 19.69) (do not dip the strip in urine above the level marked 'MAX' as the conjugate may be dissolved into the urine instead of migrating up).
5. Wait for 5 minutes and see colour development.

Results and observations

Appearance of purple line near the dipping end of the strip indicates a positive result. Absence of this line indicates the negative result. The control line will appear purple near the holding end in case the finding is negative. However, it should be noted that in a positive test sometimes two lines also appear (i.e., the positive test line and the control line). It is because when the concentration of hCG is very high in urine, only the positive line will appear. The control line may appear very faint which sometimes may not be visible. If none of these lines appears, the test is wrong. Repeat the test taking the other VISIPREG STRIP.

19.32.3.2 Visipreg card test

VISIPREG CARD pregnancy test is a qualitative laboratory immunoassay device used for determination of hCG in urine. The VISIPREG CARD is a two-site sandwich assay where antiserum anti-hCG is dispersed on a membrane both in conjugate form as well as in native form. It is based on the technological innovation over the principles of ELISA.

The VISIPREG CARD has an efficiency of detecting 25 IU/mL hCG or greater. This card should be stored at 4-5°C but must not be freezed. An anti-hCG conjugate of purple colour is immobilized over a pad. When a urine sample is applied, the hCG present in urine migrates and reconstitute the conjugate and forms a complex of anti-hCG conjugate -hCG. This complex migrates forward on the membrane onto which an antiserum anti-hCG has already been dispersed. This immobilised anti-hCG binds with the hCG present in the migrated 'anti-hCG conjugate – hCG' complex. This causes agglutination of the antiserum resulting in formation of a purple band along the site where the antiserum anti-hCG has been immobilised. Formation of purple band is a positive finding and absence of hCG in urine will not form any coloured band. This is a negative result.

Also in the membrane, there is a control region consisting of a band of immobilised antiserum anti-hCG. This band binds all the encountered free conjugate and develops a coloured line irrespective of the presence or absence of hCG in the urine. Appearance of this coloured line represents the correct procedure and viability of the reagents in the test device.

Materials

Test device, dropper, desiccants pouch (not for use in the test).

Protocol

1. Collect urine of pregnant women in a clean glass vial. The urine collected in morning contains the highest concentration of hCG, but it may be collected any time in a day.
2. Before testing bring the urine sample, test device, control/reference materials to the room temperature, 25-30°C.
3. Write the patient's name or reference number on the space marked on the card.
4. Put 3 drops (about 0.1 mL) of urine into the centre of the sample well with the help of a dropper holding vertically. Care should be taken to allow each drop to soak before adding the next drop. Wait for 5-10 minutes.

Results and observations

Appearance of two purple coloured lines (one in the result region and the second in the control region) is suggestive of a positive result (Figure 19.69). However, appearance of only one purple coloured line in the control region is indicative of the negative result. If there are no coloured lines neither in the test region nor in the control region, does this show the improper handling of the test or due to improper storage and exposure of the test device after opening the pouch and/or expiry of the card?

19.33 HIV DOT TEST OF AIDS PATIENTS

HIV 1 and HIV 2 are the causal agents of AIDS (acquired immunodeficiency syndrome). After entering in the blood of humans, HIV disrupts the immune system leading to multiple complex diseases called 'disease syndrome'. HIV kits are now available commercially. The HIV DOT kit has in-built procedural control combined test for HIV

1 and HIV 2. The HIV DOT test is a rapid (requires 3-5 minutes) visual screening test against human serum or plasma. It is an accurate immunoassay of HIV 1 and HIV 2 antibodies in human serum/plasma and does not require any instrumentation. All reagents are ready to be used in the kit. This test is quite cheap, absolutely safe and extremely reliable. Each reagent is individually packed in dropper vials eliminating cross contamination. The test device is such that it contains a porous membrane having two visible circles. The membrane is supported with absorbent. The HIV peptide antigens (toxin) are immobilised onto a porous membrane. The test sample of a patient and solutions can pass through the membrane and absorb into the underlying absorbent.

19.33.1 Materials

Colloidal gold conjugate solution, Protein A, HIV antibodies, Membrane, Absorbent

Suspected HIV patient, a clean and dry sterile vial, centrifuge, the test device, buffer solution, wash solution, colloidal gold conjugate, positive control (no risk of infection synthetic recombinant peptide used), sample dropper (disposable pipette).

19.33.2 Protocol

1. Collect blood from the suspected HIV patient(s) in a clean, dry and sterile glass vial and allow to clot.
2. Centrifuge the blood at room temperature to separate the serum assay the serum immediately, however, it may be stored at 2-8°C.
3. Bring all the solution, test device and sample to room temperature before the start of the test.
4. Add 2 drops of buffer solution into the inner circle of the device (Figure 19.70) and wait for a few second to get it soaked.

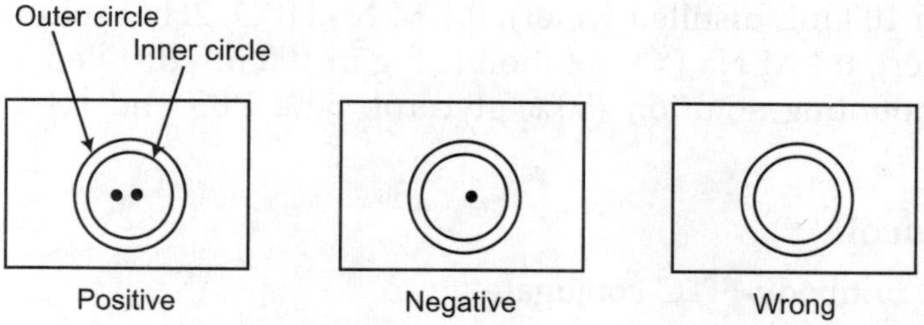

Figure 19.70 HIV dot test of the suspected AIDS patient.

5. Add 1 drop of the patient's serum/plasma with the help of a dropper provided with the kit and wait to get the serum soaked by the membrane.
6. Add 2 drops of buffer solution and allow it to soak in.
7. Again add 2 drops of wash solution and wait to soak in the membrane.
8. Add 2 drops of colloidal gold conjugate solution and wait to soak in.
9. Finally, add 3 drops of wash solution and record the result.

19.33.3 Results

If the patient's serum contains the HIV antibodies, it will bind to the HIV peptide antigens (toxin). The conjugate complex will bind to the Fe portion of the HIV antibodies to give a distinct red colour against a white background. Therefore, the positive result will show two visible dots in the inner circle. However, if only one dot is visible, it shows that the patient's serum does not contain detectable HIV antibodies giving the negative result. On the other hand if none of the dots appears, it shows the procedural error or deterioration of reagents. In such cases, re-test the specimen with a new test device.

For the assessment of positive procedure, a control dot appears in the inner circle. It is because a rabbit IgG is immobilized on the membrane surface. This control verifies that the test reagents (colloidal gold conjugate, wash buffer and the test device) are functioning well, and reagents have been added in the correct order.

19.34 DEMONSTRATION OF IMMUNOFLUORESCENCE ASSAY

Target antigens in parasite, tissue or cells against specific antibodies can be localized by incubating them directly with fluorochrome labelled specific antibodies or directly using second antibodies labelled with fluorochromes. Fluorochromes emit fluorescence when excited with ultraviolet radiation. Fluorescein is the most widely used fluorochromes. The protocol describes indirect fluorescent antibody test (IFAT) on intact filarial parasite.

19.34.1 Materials

Brugia malayi infective (L3) larvae/ microfilariae (mf)/ adult worms, anti-human IgG, test sera, fluorescein isothiocyanate (FITC), PBS 0.01 M, pH 7.4 and 0.05 M pH 7.4, 0.145 M sodium chloride solution (850 mg in 100 mL distilled water), 0.2 M $Na_2HPO_4.2H_2O$ solution (3.5 g in 100 mL distilled water), 0.1 M Na_2HPO_4 $2H_2O$ solution (1.75 g in 100 mL distilled water), 0.1 M Na_3PO_4 solution (3.8 g in 100 mL distilled water), sephadex G-25 column, mounting solution (50% glycerol, 50% PBS and 0.1% sodium azide), Evan's Blue (0.1%).

19.34.2 Protocol

Anti-human IgG antibody-FITC conjugate

1. Take 250 µL (2.5 mg protein) of anti-human IgG antibodies in plastic vial and dilute to 1 mL with 0.145 M sodium chloride solution.
2. Add 500 µL of 0.2 M Na_2HPO_4 dropwise while stirring.
3. Add 1 mL FITC solution (12.5 µg FITC/mg protein), in 0.1 M Na_2HPO_4.
4. Adjust the pH to 9.5 using 0.1 M Na_2PO_4.
5. Incubate the mixture at 25°C for 45 min.
6. Apply the conjugate mixture and pass through sephadox G- 25 column (2 × 30 cm) and elute with 0.01 M PBS (pH 7.2) at a flow rate of about 30 mL/mm. Collect the first elute yellow coloured fractions that contain anti-human IgG antibodies conjugated with FITC.

7. Concentrate FITC conjugate by ultra-membrane filtration, add preservative sodium azide (0.1%) and store at 4°C.

Immunofluorescence assay

1. Incubate about 10 L3 larvae/100 microfilariae (mf)/2 adult worms with 200 µL of optimally diluted (1:10) test sera in PBS (0.05 M) in a conical vial at room temperature for 1 hr.
2. Wash the parasites with PBS two times using a centrifuge at 500 RPM for 2 min.
3. Add 20 µL of optimally diluted (1:10 in 0.05 M PBS) anti-human IgG-FITC conjugate and incubate for 30 min at 40°C.
4. Wash the parasite with 0.05 M PBS once (500 rpm for 2 min).
5. Add 5 µL of 0.1% Evan's blue.
6. Wash the parasites with 0.05 M PBS once (500 rpm for 2 min).
7. Add 5 µL mounting buffer, transfer the parasite onto a glass slide, place cover slip and seal with nail polish.

19.34.3 Results and Observations

Examine under a fluorescence microscope and compare the yellow-green fluorescence with the parasite incubated with positive sera compared to those incubated with normal sera.

19.35 IDENTIFICATION OF BACTERIA BY FLUORESCENT ANTIBODY TECHNIQUE (FAT)

The antibodies produced by bacteria combine with antigen and form antigen-antibody complex. This complex can be observed after fluorescing it against a dark background with UV light by using a fluorochrome, i.e., fluorescent dye.

19.35.1 Materials

Microslide, coverslip, spirit lamp, Petri dish, Whatman No, 1 filter paper, lint-free paper, fluorescent microscope, PBS (phosphate buffer saline) (pH 7.4 – 7.6), specific conjugate, distilled water, low fluorescence immersion oil, mounting fluid (9 mL glycerine plus 1 mL of PBS pH 9), fluorochromes (such as rhodamine, fluoroscein isothiocyanate, acridine orange, auramine, etc).

19.35.2 Protocol

1. Inoculate the bacterial colonies into a few drops of PBS having pH 7.4 – 7.6 and make the suspension.
2. Clean slide and put a drop of suspension on the test area. Prepare smear, air dry and heat fix the bacterial cells on a glass slide by gently warming. Do not make the slide hot over the flame of the spirit lamp.

3. Place a drop of the bacterial conjugate over the smear and distribute it evenly over the smear. Care should be taken not to dry the conjugate otherwise it will give a false result.
4. Place the slide in Petri dish containing a moist filter paper for one hour or incubate in a humid chamber at 37°C.
5. Rinse the slide in a jar containing PBS for 10 minutes. Finally, rinse the slide with distilled water.
6. Gently blot dry the slide using lint-free paper. For each test use separate slide.
7. Place a drop of mounting fluid and apply the cover slip.
8. Clean the objective lenses and condenser lenses of the microscope using lens paper.
9. Place 1-2 drops of low-fluorescence immersion oil on the condenser after lowering it.
10. Place the slide on the stage of the microscope and put a drop of immersion oil on the cover slip. Adjust the microscope to view the required area.
11. Examine the field under low power and concentrate the light in a small area by adjusting the condenser.
12. Now, switch over to oil immersion objective and observe under the microscope using UV light (Figure 19.71).

19.35.3 Results

The bacterial cells that produce specific antibodies will show fluorescence.

19.36 IMMUNOFLUORESCENCE LABELLING OF CULTURED CELLS

The following is a basic "generic" method for localizing proteins and other antigens by indirect immunofluorescence. The method relies on proper fixation of cells to retain cellular distribution of antigen and to preserve cellular morphology. After fixation, the cells are exposed to the primary antibody directed against the protein of interest, in the presence of permeabilizing reagents to ensure antibody access to the epitope. Following incubation with the primary antibody, the unbound antibody is removed, and the bound primary antibody is then labelled by incubation with a fluorescently tagged secondary antibody directed against the primary antibody host species. For example, incubation with a mouse IgG primary antibody might be followed by incubation with a RITC (rhodamine isothiocyanate-labelled goat anti-mouse IgG secondary antibody. After removal of the secondary antibody, the specimen is ready for viewing on the fluorescence microscope. Once the conditions for observing specific immunolocalization have been identified for a given antibody and cell type, double labelling with two antibodies can be employed to compare localizations. To do this, primary antibody incubation can contain two antibodies generated in two species (e.g., mouse and rabbit), followed by incubation with two secondary antibodies coupled to different fluorophores.

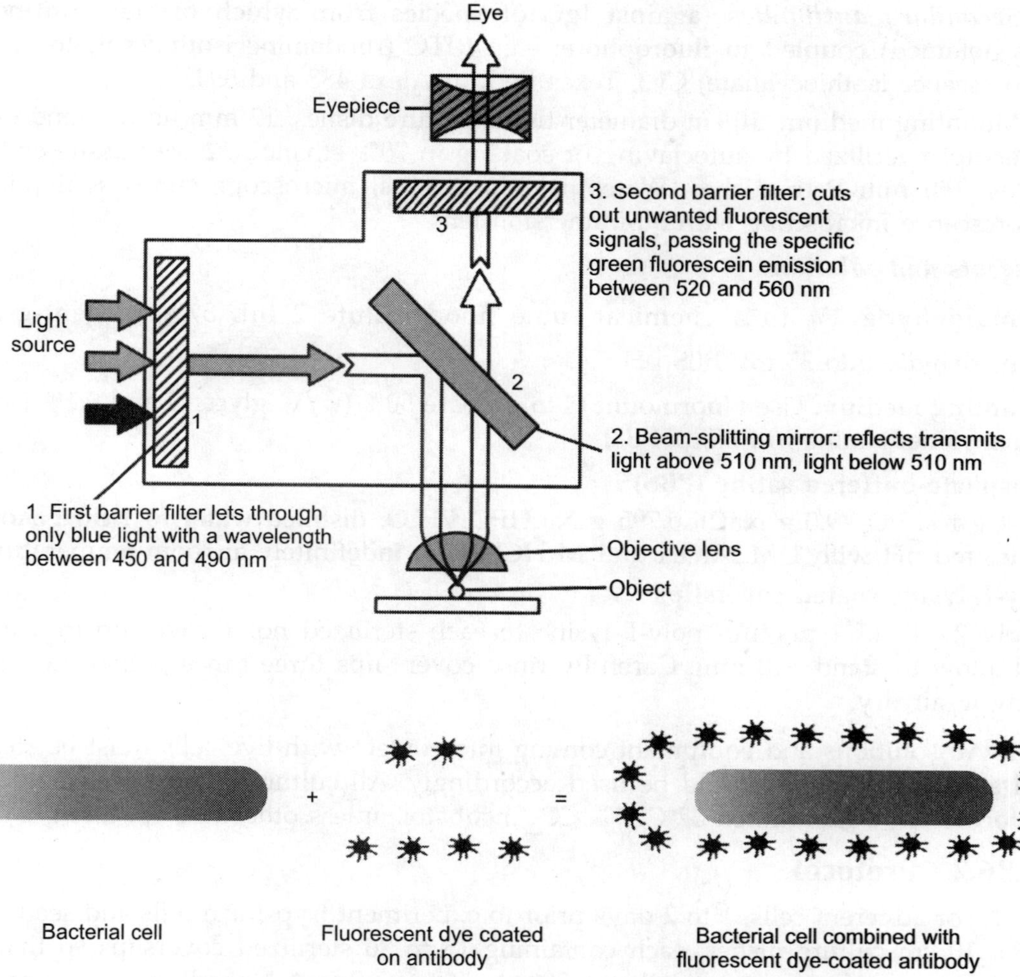

Figure 19.71 Fluorescence staining technique and microscopy.

Care should be taken, however that the two antibody combinations, especially the secondary antibodies, should not cross-react.

19.36.1 Materials

Cells of interest growing in tissue culture, Trypsin/EDTA solution, 2% formaldehyde, Phosphate-buffered saline pH 7.4, PBS/FBS: PBS, pH 7.4, containing 10% fetal bovine serum (FBS), primary antibody.

0.1% (w/v) Saponin in PBS/FBS Prepare fresh from 10% (w/v) saponin stock solution

Controls Preimmune serum (if using rabbit polyclonal antibody) or antigen add in excess to the primary antibody

Secondary antibodies (against IgG of species from which primary antibody was obtained) coupled to fluorophore: e.g., RITC (rhodamine isothiocyanate), FITC (fluorescence isothiocyanate) Cy3, Texas Red, or Alexa 488 and 594.

Mounting medium, 10 cm diameter tissue culture dishes, 12 mm no. 1 round glass coverslips, sterilized by autoclaving or soaking in 70% ethanol, 12 well tissue culture plates, 150 mm Petri dishes, Watchmaker's forceps, microscope slides, nail polish, Fluorescence microscope with oil-immersion lens.

Reagents and solutions

Formaldehyde 2% In a chemical fume hood, dilute 2 mL 37% reagent-grade formaldehyde into 35 mL PBS pH 7.4.

Mounting medium Use Fluormount G to prepare 50% (w/v) glycerol and 0.1% (w/v) *p*-phenylenediamine in PBS, pH 8.0.

Phosphate-buffered saline (PBS)

0.144 g KH_2PO_4, 9.0 g NaCl, 0.795 g $Na_2HPO_4 \cdot 7H_2O$, distilled water to 1 litre. Adjust to desired pH with 1 M NaOH or 1 M HCl. Store indefinitely at room temperature.

Poly-L-lysine-coated coverslips

Apply 25 µL of 1 mg/mL poly-L-lysine to each sterilized no. 1 coverslip in a hood and allow to stand ~10 min. Carefully rinse cover slips three times with water, then allow to air dry.

Note: All solutions and equipment coming into contact with live cells must be sterile, and aseptic technique should be used accordingly. All culture incubations should be performed in a humidified 37°C, 5% CO_2 incubator unless otherwise specified.

19.36.2 Protocol

1. For adherent cells, 1 to 2 days prior to experiment trypsinize cells and seed onto 10 cm culture dishes, each containing 15 to 20 sterilized coverslips so that on the day of experiment cells are 20% to 50% confluent. Nonadherent cells can be coaxed into adhering to the coverslips by precoating the coverslip with poly-L-lysine. Apply 10 to 20 µL of suspended cells to each cover slip, let sit 10 min, then proceed with fixation (step 3). Alternatively, cells can be attached to cover slips using a cytocentrifuge by following the manufacturer's instructions.
2. On the day of experiment, transfer each coverslip individually to a well of a 12-well tissue culture dish containing 1 mL culture medium. Subject cells to the desired experimental conditions (e.g., treat with various drugs, inhibitors, or temperatures prior to fixation and immunostaining).
3. Aspirate medium and add 1 mL of 2% formaldehyde to each well. Allow cells to fix at room temperature for 10 min.
4. Aspirate the formaldehyde fixative and wash cover slips twice, each time by adding 1 mL PBS, pH 7.4, let it stand for 5 min, then aspirating the PBS. Add 1 mL PBS/FBS to the fixed coverslips and let stand 10 to 20 min, to block nonspecific sites of antibody adsorption.

[*Note:* Throughout the procedure, do not let the cells dry out.]

5. In 1.5 mL microcentrifuge tubes dilute primary antibodies in 0.1% saponin/PBS/FBS. Typically affinity-purified antibodies are diluted in the range of 1 to 10 μg/mL, and rabbit antisera are diluted between 1:100 and 1:1000. If using a commercial antibody, follow suggested dilutions from manufacturer. During initial characterization, it is wise to try a range of dilutions of antibody.
6. Prepare controls containing only 0.1% saponin/PBS/FBS or (if available) containing preimmune antiserum (if rabbit polyclonal antibody is being used) or specific (primary) antibody with the antigen added in excess. Controls are often the most important part of an immunofluorescence experiment.
7. Micro centrifuge antibody dilutions and control solutions 5 min at maximum speed, room temperature, to bring down aggregates in pellet. Pipette the antibody solution from above the aggregate pellet.
8. Place a 10 × 10 cm piece of parafilm in the bottom of a 150 mm Petri dish. In a grid pattern that replicates the 12 wells used to incubate the coverslips, label the appropriate place on the parafilm for each coverslip with a marker.
9. Apply a 25 μL drop of appropriate primary antibody solution to each numbered section. Carefully remove each coverslip from the 12-well plate with watchmaker's forceps, blot the excess fluid by touching the edge to a kimwipe, then invert the coverslip over the appropriate 25 μL drop, making sure that the side with the cells is facing down. Place the top on the Petri dish and incubate 1 hr at room temperature.

   Sometimes proper labelling will require a longer incubation time or the Petri dish will be incubated >1 hr for convenience. For longer incubations, add some wetted kimwipes to the dish to maintain a humid atmosphere. The incubation can be extended overnight at 4°C, if necessary.

   [*Note:* It is important to be always aware of which side of the cover slip the cells are on. Cells should be facing up when in the 12 well plate, but facing down when placed on the antibody. Picking up cover slips with the forceps is awkward at first, but becomes easier with practice.]
10. Carefully pick up each inverted coverslip and flip it over so that it is cell-side-up, then place in a well of a 12 well plate. Wash each coverslip three times to remove unbound antibody, each time by adding 1 mL PBS/FBS, letting stand 5 min, then aspirating the solution.
11. Dilute fluorophore-conjugated secondary antibodies in 0.1% saponin/PBS/FBS. Mix, then microfuge as in step 7 to remove aggregates. Typically, commercial preparations are diluted between 1:100 and 1:500.
12. Prepare an incubation chamber as in step 8. Apply 25 μL of appropriate secondary antibody solution to each numbered section and invert coverslip over drop as in step 8. Cover Petri dish and protect from light with aluminum foil or place chamber in drawer. Incubate 1 hr at room temperature.

13. Wash coverslips as in step 10. After removal of last PBS/FBS wash, add 1 mL PBS.
14. Label slides and place 1 drop of mounting medium onto slide. Pick up coverslip from well, gently blot off excess PBS by touching the edge to a kimwipe, then invert coverslip, cell-side-down, onto drop. Gently blot mounted coverslip with paper towel, then seal edge of coverslip onto the slide by painting the edge with a rim of nail polish. Let it dry. The fixed, mounted, and nail polish sealed coverslips can be stored in the dark for 6 months to 1 year at 4°C.

19.36.3 Results and Observations

View specimen on fluorescence microscope using a 63 × oil immersion objective.

19.37 ISOLATION OF IgG FROM SERUM

IgG is the main antibody isotype found in blood and extracellular fluid allowing it to control infection of body tissues by binding many kinds of pathogen-representing viruses, bacteria, and fungi-IgG protects the body from infection. IgG is secreted as a monomer that is small in size allowing it to easily perfuse tissues. It is the only isotype that can pass through the human placenta, thereby providing protection to the foetus in the uterus. Along with IgA secreted in the breast milk, residual IgG absorbed through the placenta provides the neonate with humoral immunity before its own immune system develops. Colostrum contains a high percentage of IgG, especially bovine colostrum. In individuals with prior immunity to a pathogen, IgG appears about 24-48 hours after antigenic stimulation.

Isolation of IgG class of immunoglobulin is carried out in two steps. In the first phase A. Immunoglobulin-rich fraction is prepared by salt precipitation and in the second phase B. IgG is separated from other classes of immunoglobulins by ion-exchange chromatography.

19.37.1 Isolation of Immunoglobulin-rich Fraction from Serum by Salt Precipitation

19.37.1.1 Principle

Different proteins can be gradually precipitated out from their aqueous solutions using highly soluble salts such as ammonium sulphate. As the concentration of the salts is increased they compete with protein for water molecules and lead to their gradual precipitation based upon their requirement for water molecules to be in the soluble form. By convention, the final concentration of ammonium sulphate required for precipitation is expressed in term of percentage of saturation. Immunoglobulin-rich fraction can be precipitated out from serum by 33% saturation with ammonium sulphate.

19.37.1.2 Materials

Serum, 0.01 M phosphate buffer pH 7.2, saturated ammonium sulphate solution (add about 75 g of ammonium sulphate to 100 mL distilled water and stir. Heat the solution and filter through Whatman No.1 filter paper while it is hot. After the solution comes to room temperature, adjust the pH to 6.8 using concentrated ammonium hydroxide and store. Ammonium sulphate crystals should be formed in the container after few hours), magnetic stirrer, refrigerated centrifuge.

19.37.1.3 Protocol

1. Add 5 mL of saturated ammonium sulphate solution gradually to 10 mL of serum under constant stirring at ice cold temperature.
2. Keep the suspension in ice for 15 minutes with occasional stirring.
3. Centrifuge at 3000 RPM for 30 min at 4°C.
4. Discard the supernatant and wash the precipitate twice with 40% ammonium sulphate solution (mix thoroughly with a glass rod and centrifuge at 3000 rpm for 30 min at 4°C).
5. Reconstitute the precipitate in small volume of 0.01 M saline phosphate buffer (pH 7.2) and dialyze against the same buffer for 24 hr with frequent changes of buffer.

Removal of salt from protein fraction

The ammonium sulphate is removed from protein fraction by dialysis and the protein concentrated by filtered by ultrafiltration.

Principle

Dialysis is the separation process which depends on the differential transport of solutes of different sizes across a porous barrier separating to liquid where the driving force is concentration gradient only. It is usually used to separate solutes too large to diffuse through the barrier from those small enough to diffuse through it (Figure 19.72).

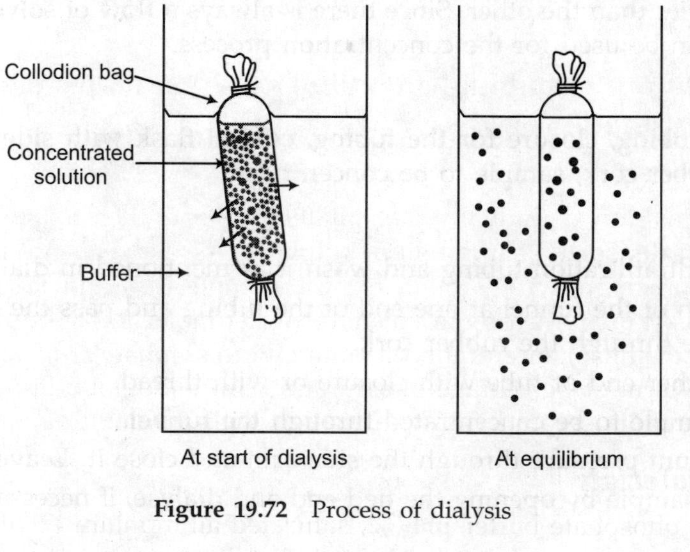

Figure 19.72 Process of dialysis

19.37.1.4 Materials

Dialysis membrane, closures for dialysis membrane, SPB (0.01M sodium phosphate buffer pH 7.2), antibody to be desalted (Human Ig), magnetic stirrer and magnetic bar.

19.37.1.5 Protocol

Cut the required length of dialysis membrane and place in distilled water.

Open the membrane by gently seizing from one end to another and wash the inner side of membrane by passing the distilled water gently without damaging the membrane. Close one end of the membrane using closure or thread. Pour the antibody solution through another end and close the same. Place the filled bag in a beaker containing 0.01 M SPB and magnetic bar and keep the beaker on a magnetic stirrer and adjust the speed (at 4°C). Change the pre-cooled buffer after every 4-6 hr for 3 times (1 lit each time). Collect the antibody.

[*Note:* Choice of the pore size of the membrane depends on the substance to be dialysed. Extensive cleaning of the membrane is necessary before use. The membrane should not dry because it may damage the membrane. It is advisable to test the system for leaks using distilled water before adding the protein solution to the tubing.]

19.37.1.6 Ultra membrane filtration

Principle

Ultrafiltration is a device which can withstand high pressure and concentrates a protein solution using selective permeable membranes. The function of the membrane is to let the water and small molecules pass through while retaining the protein. Ultrafiltration is sometimes called reverse osmosis. It is a process in which the solutes and solvent, up to a certain critical size are forced to the barrier by high pressure on one side of the porous barrier than the other. Since there is always a flow of solvent moving along the solute, it can be used for the concentration process.

Materials

Ultrafiltration tubing, closure for the tubing, conical flask with side arm, funnel and single hold rubber cork, sample to be concentrated.

Protocol

Cut 10-15 cm ultrafiltration tubing and wash it as mentioned in dialysis.

Insert the tip of the funnel at one end of the tubing and pass the end of the funnel with membrane through the rubber cork.

Close the other end of tube with closure or with thread.

Pour the sample to be concentrated through the funnel.

Apply vacuum pressure through the side arm and close it. Leave the flask at 4°C.

Collect the sample by opening the tied end and dialyse, if necessary (Figure 19.73).

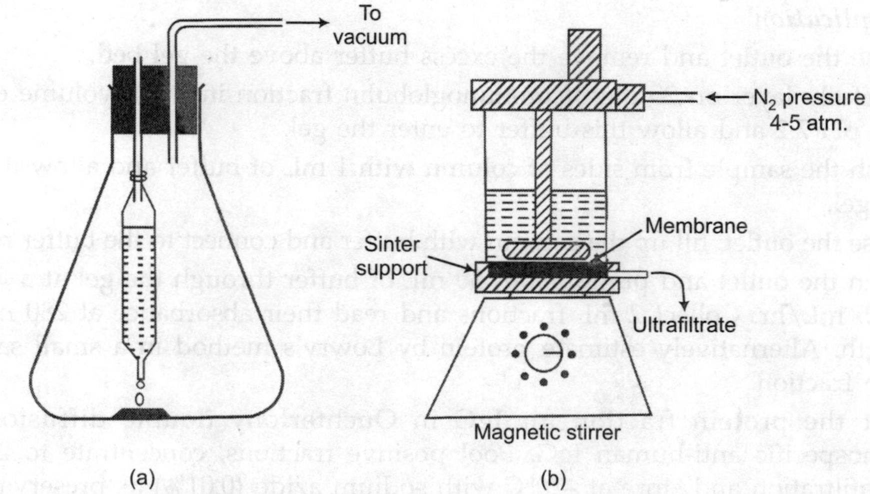

Figure 19.73 Ultrafiltration of IgG solution in dialysing sac.

19.37.1.7 Isolation of IgG from immunoglobulin-rich fraction

Principle

Fractionation of protein by ion exchange chromatography exploits differences in their net charges. The immunoglobulin-rich fraction prepared by ammonium sulphate precipitation is applied on DEAE cellulose column at pH 7.2. At this pH, DEAE cellulose is positively charged while all classes of immunoglobulins except IgG carry negative charges. This situation allows only IgG to be eluted out while the other classes of immunoglobulins interact with the matrix and hence are retained in the column.

Materials

DEAE cellulose (Whatman DE-52), 0.01 M SBP, pH 7.2, Whatman filter paper No. 1, chromatography set up, spectrophotometer, Immunoglobulin (Ig) (γ- globulin fraction)

Protocol

1. Suspend about 3 gm of DEAE cellulose in 0.01 M SPB (pH 7.2) and wash the slurry by changing the buffer 5 or 6 times to remove fine particles.
2. Take a glass column (10 × 2 cm) with sintered glass disc or glass wool plug at the bottom and mount it vertically.
3. Inject about 10 mL of 0.01 M SPB, pH 7.2 through the outlet of column with the help of the syringe and avoid trapping of air bubbles below the sintered glass disc or glass wool plug.
4. Close the outlet and pour thick slurry of DEAE cellulose along the sides of the column and allow the buffer to run about 25 mL/hr. Disturb the settled bed slightly with a glass rod if some slurry is to be added.
5. Cover the gel bed with a Whatman No. 1 filter paper circle (having a diameter slightly lesser than that of the column) and equilibrate the column by pumping 0.01 M SPB, pH 7.2 at a flow rate of about 25 mL/hrs.

Sample application

1. Close the outlet and remove the excess buffer above the gel bed.
2. Carefully layer on 25 mg of immunoglobulin fraction in 1 mL volume of 0.01M SPB pH 7.2 and allow this buffer to enter the gel.
3. Wash the sample from sides of column with 1 mL of buffer and allow it to enter the gel.
4. Close the outlet, fill up the column with buffer and connect to the buffer reservoir.
5. Open the outlet and pump about 30 mL of buffer through the gel at a flow rate of 25 mL/hr. Collect 2 mL fractions and read their absorbance at 280 nm wave length. Alternatively estimate protein by Lowry's method in a small sample of each fraction.
6. Test the protein fraction for IgG in Ouchterlony double diffusion using monospecific anti-human IgG. Pool positive fractions, concentrate to 2 mL by ultrafiltration and store at –20°C with sodium azide (0.01%) as preservative.

19.38 PURIFICATION OF IgM

IgM forms polymers where multiple immunoglobulins are covalently linked together with disulfide bonds, mostly as a pentamer but also as a hexamer. IgM is primarily found in serum; however, because of the J chain, it is also important as a secretory immunoglobulin. Many IgM antibodies will precipitate when dialyzed against distilled water. The IgM produced by this method is of sufficient purity to be used for fragmentation, fluorescein isothiocyanate (FITC) or biotin labelling, or radiolabelling. The protocol involves centrifugation of ascites fluid or monoclonal antibody supernatant, followed by simple dialysis against water. IgM is separated from contaminating proteins (such as a normal mouse proteins or proteins from fetal bovine serum) by use of size-exclusion (SE) chromatography. The purity of IgM is checked by SDS-PAGE, and the pure IgM is stored in borate buffer.

19.38.1 Materials

Ascites fluid or monoclonal antibody (MAb) supernatant, PBS, centrifuge, dialysis tubing, 26 × 900 mm column with appropriate size-exclusion (SE) resin

19.38.2 Protocol

1. Clarify ascites fluid by centrifugation and remove lipid. If using MAb supernatant, centrifuge 30 min at 15,000 × g (11,000 rpm in SS-34 rotor), room temperature or 4°C, to remove cell debris. Save the supernatant.
2. Transfer clarified ascites fluid or MAb supernatant to dialysis tubing and dialyze against ≥10 times the sample volume of distilled water for 24 hr at 4°C.
3. Centrifuge dialyzed material 1 hr at 15,000 to 20,000 × g (11,000 to 13,000 rpm in SS-34 rotor), room temperature or 4°C, and discard supernatant.

4. Prepare a 26 × 900 mm column with SE resin and equilibrate in PBS. Dissolve pellet from step 3 in ≤5 mL PBS and apply to column. Elute protein with PBS and collect 100 fractions (1% of column volume). IgM will be in the first protein peak to elute from the column.
5. Assess IgM concentration spectrophotometrically at A_{280}. Pool the eluates containing pure IgM.
6. Check the purity of the IgM on a 10% SDS-polyacrylamide gel under reducing conditions.

   Load 1 to 80 μL on the gel, depending on the apparatus used. On a reduced gel, IgM shows two bands one at 78 kDa, representing the heavy chain, and the other at 25 kDa, representing the light chain.
7. Store IgM at 1 to 20 mg/mL in PBS at 4°C. IgM is stable under these conditions for several months. Some IgM preparations will come out of solution more easily than others; this is unavoidable. Each IgM must be observed for precipitation.

Suggested Readings

Brewer, J. M., A. J. Pesce and R. B. Ashworth. 1974. *Experimental Techniques in Biochemistry,* New Jersey, Printice-Hall.

Dabbas, D. J. 2002. *Diagnostic immunochemistry.* New York, Churchill Livingstone.

Hayat, M. A. 2002. *Microscopy, Immunochemistry and Antigen Retrieval Methods: For Light and Electron Microscopy,* London. Plenum Press.

Hermanson, G. T. 1996. *Bioconjugate Technique.* London, Academic Press.

Janeway, C. A., P. Travers P, M. Walport, et al. 2005. *The Immune System in Health and Disease,* 6th edn. Oxford, Garland Science.

Johnstone, A. P., and R. Thorpe. 1996. *Immunochemistry in Practice,* 3rd edn. Oxford, Blackwell Science.

Johnstone, A. P.,and M. W. Turner. (eds.). 1997. *Immunochemistry - A practical Approach,* Oxford, IRL Press.

Lachmann, P. I. , D. R. Peters, and P. S. Rosens PS (eds.). 1992. *Clinical Aspects of Immunology,* 5th edn. Oxford, Blackwell Science.

Plummer, D. T. 1988. *An Introduction to Practical Biochemistry,* New York. McGraw-Hill.

Roitt, I. M. 2001. *Essential Immunology,* 10th edn. Oxford, Blackwell Science.

Tiwari, R. P., A. Jain, Z. Khan, et al. 2013. Development of novel antigenic peptide cocktail for the detection of antibodies to HIV 1/2 by ELISA. *J. Immunological Methods* 387: 157-166.

William, E., and W. E. Paul. 2008. *Fundamental immunology,* New York, Lippincott Williams & Wilkins.

Wilson, K., and J. Walker. 2003. *Practical Biochemistry;* Principle and techniques. 5th edition, Cambridge, Cambridge University Press.

Chapter

20

Microbes in Waste Management

20.1 INTRODUCTION

Major environmental pollutants are the products of large scale manufacturing, processing and handling of industrial waste discharged from chemical and petroleum industry which lead to serious surface and subsurface soil contamination. A wide variety of hazardous and toxic hydrocarbons such as polychlorinated biphenyls (PCBs), trichloroethylene (TCE), polyaromatic hydrocarbons (PAHs) and pesticides are designated as xenobiotics because of their relative recalcitrance to biodegradation and these compounds have the potential to bioaccumulation in living system. Microbial removal of contaminants from the environment often takes place without human intervention. This has been termed as "intrinsic bioremediation". In certain cases, where the contamination does not pose an immediate health threat and remains localized, intrinsic bioremediation could be opted for a choice of decontamination. The waste management strategy for the future should meet the benefits of the humanity, respect principles of planet ecology, and compatibility with other habitability systems. For these purpose the waste management technologies, relevant to the application of the physicochemical properties of bacteria are of great value. The biological treatment method is based upon the biodegradation of organic substances by various microorganisms. The advantage of the biodegradation waste management in general allows: (i) to diminish the volume of organic wastes (ii) the biological hazard of the wastes is controlled and (iii) this system may be compatible with the other systems. The basis of bioremediation is the enormous natural capacity of microorganisms to degrade organic compounds. This capacity could be improved by applying the GMMs (genetically modified microorganisms).

20.2 BIOREMEDIATION

Bioremediation is defined as the process whereby organic wastes are biologically degraded under controlled conditions to an innocuous state, or to levels below concentration limits established by regulatory authorities. Process is defined as the use of living organisms for the recovery/cleaning up of a contaminated medium (soil, sediment, air, water). The process of bioremediation might involve introduction of new organisms to a site, or adjustment of environmental conditions to enhance degradation rates of indigenous fauna. Bioremediation can be applied to recover brownfields for development and for preparing contaminated industrial effluents prior to discharge into waterways. Bioremediation technologies are also applied to contaminated wastewater, ground or surface waters, soils, sediments and air where there has been either accidental or intentional release of pollutants or chemicals that pose a risk to human, animal or ecosystem health. Studies in molecular biology and ecology offer opportunities for more efficient biological processes. The microorganisms may be indigenous to a contaminated area or they may be isolated from elsewhere and brought to the contaminated site. Contaminant compounds are transformed by living organisms through reactions that take place as a part of their metabolic processes. Biodegradation of a compound is often a result of the actions of multiple organisms. Microorganisms are often imported to a contaminated site to enhance degradation by a process known as bioaugmentation and the microorganisms performing the function of bioremediation are known as bioremediators. For bioremediation to be effective, microorganisms must enzymatically attack the pollutants and convert them to harmless products. As bioremediation can be effective only where environmental conditions permit microbial growth and activity, its application often involves the manipulation of environmental parameters to allow microbial growth and degradation to proceed at a faster rate (Figure 20.1). Like other technologies, bioremediation has its limitations. Some contaminants, such as chlorinated organic or high aromatic hydrocarbons, are resistant to microbial attack. It is not easy to predict the rates of clean-up for a bioremediation exercise as the clean-up operation is either slow or not at all. There are no rules to predict if a contaminant can be degraded. Bioremediation techniques are typically more economical than traditional methods such as incineration, and some pollutants can be treated on site, thus reducing exposure risks for clean-up personnel, or potentially wider exposure as a result of transportation accidents. Most bioremediation systems are run under aerobic conditions, but running a system under anaerobic conditions may permit microbial organisms to degrade otherwise recalcitrant molecules.

Bioremediation technologies can be generally classified as *in situ* or *ex situ*. *In situ* bioremediation involves treating the contaminated material at the site while *ex situ* involves the removal of the contaminated material to be treated elsewhere (Figure 20.2). Some examples of bioremediation technologies are bioventing, landfarming, bioreactor, composting, bioaugmentation, rhizofiltration, and biostimulation. Bioremediation can occur on its own (natural attenuation or intrinsic bioremediation) or can be spurred on via the addition of fertilizers to increase the bioavailability within the medium (biostimulation). Advancements have also proven successful via the addition of matched microbe strains to the medium to enhance the resident microbe population's ability to break down contaminants.

On the basis of removal and transportation of wastes for treatment, basically there are two methods: *in situ* bioremediation and *ex situ* bioremediation.

20.2.1 *In Situ* Bioremediation

In situ bioremediation is the clean-up approach which directly involves the contact between microorganisms and the dissolved and adsorbed contaminants for biotransformation. Biotransformation in the surface environment is a very complex process. Potential advantages of *in situ* bioremediation methods include minimal site

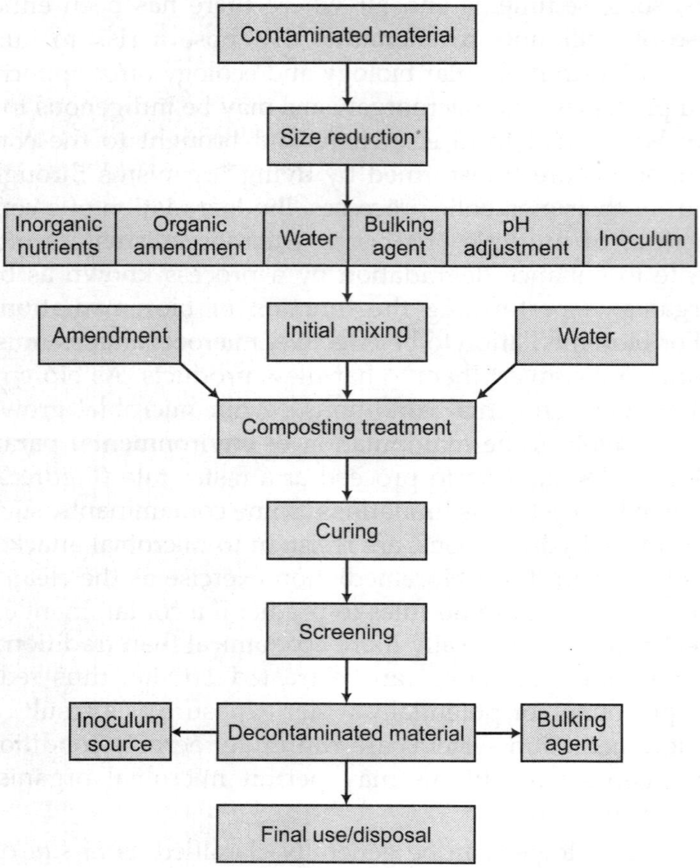

Figure 20.1 Flow chart explaining bioremediation process at a contaminated site.

disruption, simultaneous treatment of contaminated soil and groundwater, minimal exposure of public and site personnel, and low costs. But the disadvantages are *(i)* time consuming method as compared to other remedial methods, *(ii)* seasonal variation of microbial activity resulting from direct exposure to prevailing environmental factors, lack of control of these factors, and *(iii)* problematic application of treatment additives (nutrients, surfactants and oxygen). The microorganisms act efficiently well only when the waste materials help them to generate energy and nutrients to build up more cells. When the native microorganisms lack biodegradation capacity, genetically engineered microorganisms (GEMs) may be added to the surface during *in situ* bioremediation. But stimulation of indigenous microorganisms is preferred over addition of GEMs. There are two types of *in situ* bioremediation: intrinsic and engineered *in situ* bioremediation.

20.2.1.1 Intrinsic bioremediation

Conversion of environmental pollutants into the harmless forms through the innate capabilities of naturally occurring microbial population is called intrinsic bioremediation.

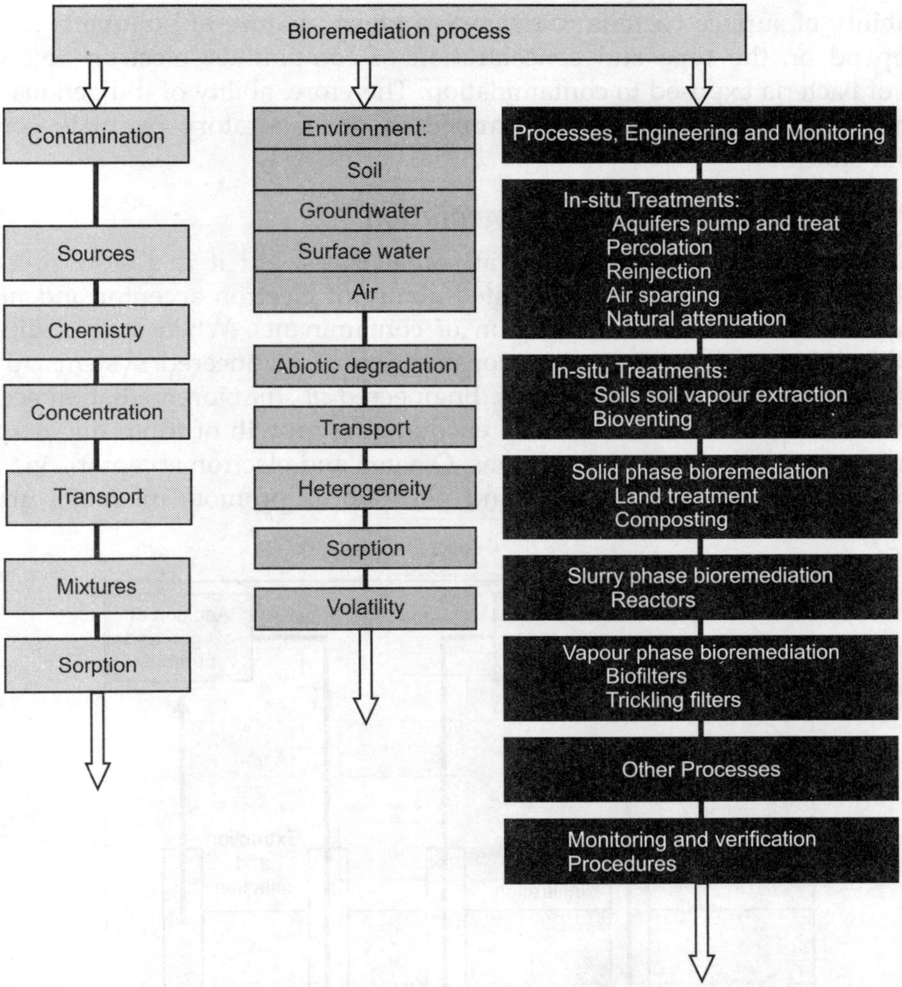

Figure 20.2 Bioremediation technology.

However, there is increasing interest on intrinsic bioremediation for control of all or some of the contamination at waste sites. The intrinsic, i.e., inherent capacity of microorganisms to metabolize the contaminants should be tested at laboratory and field levels before use for intrinsic bioremediation. The progress of the site monitoring programme of intrinsic bioremediation should be recorded time to time. The conditions of the site that favour intrinsic bioremediation are groundwater flow throughout the year, carbonate minerals to buffer acidity produced during biodegradation, supply of electron acceptors and nutrients for microbial growth and absence of toxic compounds. The other environmental factors such as pH, concentration, temperature and nutrient availability determine whether or not biotransformation takes place. Bioremediation of waste mixtures containing metals such as Hg, Pb, As and cyanide at toxic concentration can create a problem.

The ability of surface bacteria to degrade a given mixture of pollutants in groundwater depend on the type and concentration of compounds, electron acceptor and duration of bacteria exposed to contamination. Therefore, ability of indigenous bacteria degrading contaminants can be determined in the laboratory by plate count and microcosm studies.

20.2.1.2 Engineered in situ bioremediation

Intrinsic bioremediation is satisfactory at some places, but it is a slow process due to poorly adapted microorganisms, limited ability of electron acceptor and nutrients, cold temperature and high concentration of contaminants. When site conditions are not suitable, bioremediation requires construction of engineered systems to supply materials that stimulate microorganisms. Engineered *in situ* bioremediation accelerates the desired biodegradation reactions by encouraging growth of more microorganisms via optimizing physicochemical conditions. Oxygen and electron acceptors, viz., nitrate and sulphate and nutrients, nitrogen and phosphorus promote microbial growth in surface (Figure 20.3).

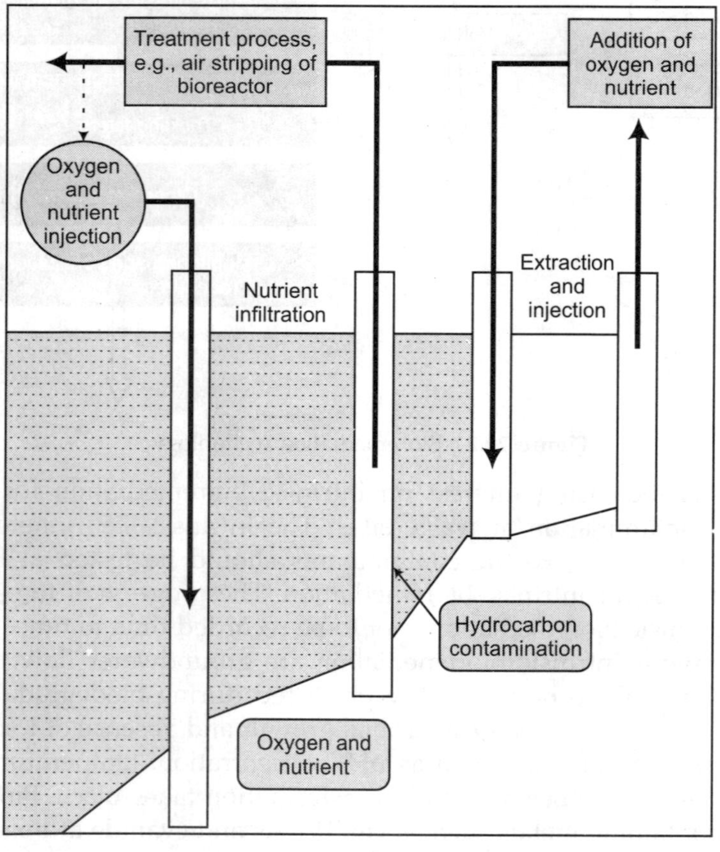

Figure 20.3 Surface treatment using above-ground reactor, injection of oxygen, acid nutrient and extraction wells.

20.2.2 *Ex Situ* Bioremediation

Ex situ bioremediation involves removal of waste materials and their collection at a place to facilitate microbial degradation. *Ex situ* bioremediation technology includes most of the disadvantages and limitations. It also suffers from costs associated with solid handling process, e.g., excavation, screening and fractionation, mixing, homogenizing and final disposal. On the basis of phases of contaminated materials under treatment *ex situ* bioremediation is classified into two: (i) solid-phase system (including land treatment and soil piles) i.e., composting, and (ii) slurry-phase systems (involving treatment of solid-liquid suspensions in bioreactors).

20.2.2.1 Solid-phase treatment

Solid-phase system includes organic wastes, (e.g., leaves, animal manures and agricultural wastes), and problematic wastes, (e.g., domestic and industrial wastes, sewage sludge and municipal solid wastes). The traditional clean-up practice involves the informal processing of the organic materials and production of composts which may be used as a soil amendment. ***Composting***: Composting is a self-heating, substrate-dense, managed microbial system. It is a solid-phase biological treatment technology, suitable the treatment of a large amount of contaminated solid materials. However, many hazardous compounds are resistant to microbial degradation due to complex chemical structure, toxicity and compound concentration that hardly support growth. Microbial growth is also affected by moisture, pH, inorganic nutrients and particle size. Because composting of hazardous wastes typically involves the bioremediation of contaminated substrate-sparse soils, support of microbial self-heating needs incorporation of the proper amount of supplements. The hazardous compounds reported to disappear through composting includes aliphatic and aromatic hydrocarbons and certain halogenated compounds. The possible routes leading to disappearance of hazardous compounds include volatilization, assimilation, absorption, polymerization and leaching.

Composting can be done in open system, i.e., land treatment, and in a closed system. The *open land system* can be an inexpensive treatment method, but the temperature fluctuates from summer to winter. Therefore, rate of biodegradation of waste materials declines. Secondly, land treatment system may become oxygen limited, depending on the amount of substrate, depth of waste, application, etc. However, efficiency of open treatment system can be increased by passing air (Figure 20.4). This approach is referred to as *engineered soil piles* and *forced aeration treatment*. The *closed treatment system* is preferred over the open land treatment system because controlled air is supplied to maintain the microbial activity. As a result of microbial growth and volatilization of hazardous compounds, internal temperature gradually rises. Therefore, use of blowers for air circulation and exhaust for removal of toxic volatiles are set up in closed treatment system (Figure 20.5). Ventilators supply oxygen and remove heat through evaporation of water.

A pH range of 5.0-7.8 promotes the maximum rate of degradation of hazardous wastes, however, lignin degradation is rapid between pH range of 3.0-6.5 suggesting

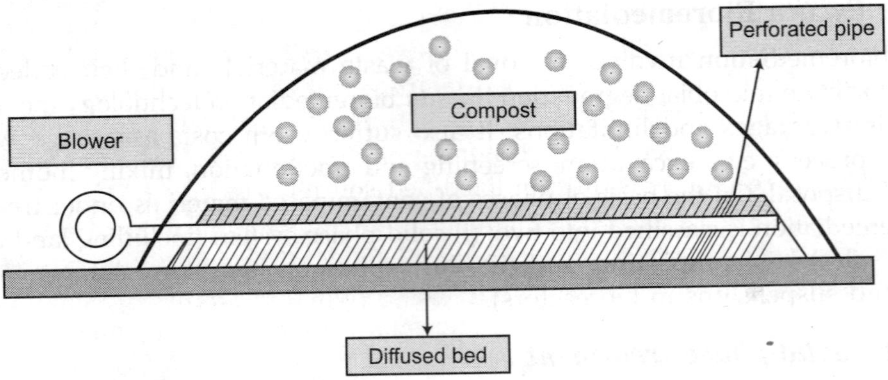

Figure 20.4 Physical configuration of an open composting side view.

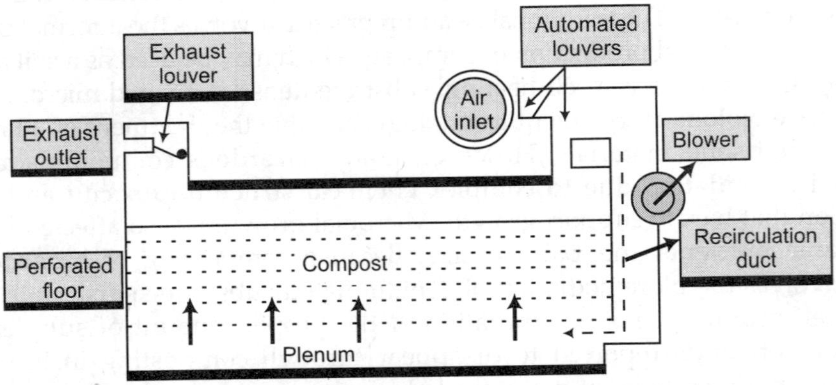

Figure 20.5 Physical configuration of a closed composting system with forced ventilation side view.

that the optimal pH levels can be species, site and waste specific. Gradual colonization of organic materials is occupied by indigenous microflora, but hazardous chemicals may inhibit microbial growth. Therefore, *bioaugmentation* (i.e., use of commercial or GEMs (Genetically engineered microorganisms) of wastes is recommended.

20.2.2.2 Slurry-phase treatment

The contaminated solid materials (soil, degraded sediments, etc), microorganisms and water formulated into slurry are brought within a bioreactor, i.e., fermenter. Thus slurry-phase treatment is a triphasic system involving three major components: water, suspended particulate matter and air. Here water serves as suspending medium where nutrients, trace elements, pH adjustment chemicals and desorbed contaminants are dissolved. Suspended particulate matter includes a biologically inert substratum consisting of contaminants (soil particles) and biomass attached to the soil matrix or free in suspending medium. Air provides oxygen for bacterial growth. Slurry-phase reactors are new design in bioremediation. The objectives of bioreactor designing are to *(i)* alleviate microbial growth limiting factors in the soil environment such as substrate, nutrients and oxygen availability, *(ii)* promote suitable environmental conditions for bacterial growth such as moisture, pH, temperature and *(iii)* minimize mass transfer limitations and facilitate the desorption of organic material from the soil matrix. Biologically there are three types of slurry-phase bioreactors: (i) aerated lagoons, (ii) low-shear airlift reactor, and (iii) fluidized-bed soil reactor.

Aerated Lagoon It is used for treatment of small common municipal waste water. Nutrients and aeration are supplied to the reactor. Mixers are fitted to mix different components and form slurry, whereas surface aerators provide air required for microbial growth. The process may be used as single-stage or multistage operation. If the waste contains volatiles, this reactor is not appropriate (Figure 20.6).

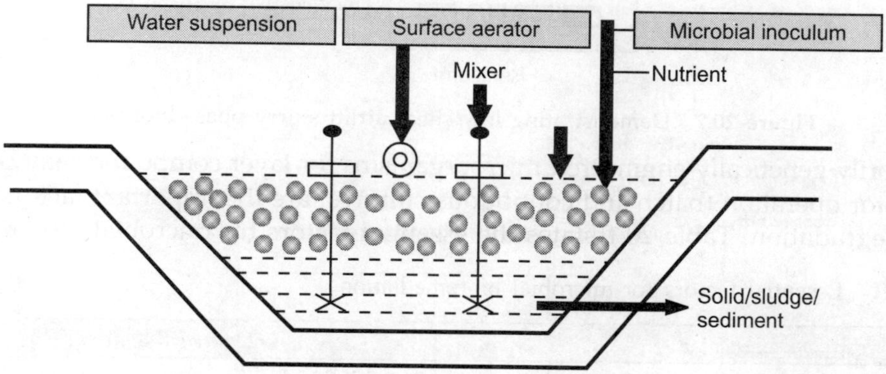

Figure 20.6 Slurry-phase lagoon.

Low-shear Airlift Reactors (LSARs) The LSARs are useful when waste contains volatile components. The tight process control and increased efficiency of bioreactors are required.

LSARs are cylindrical tanks which are made up of stainless steel. In this bioreactor pH, temperature, nutrient addition, mixing and oxygen can be controlled as desired. Shaft is equipped with impellers. It is driven by motor set up at the top. The rake arms are connected with blades which are used for resuspension of coarse materials that tend to settle on the bottom of the bioreactor. Air diffusers are placed radially along the rake arm. Airlift provides to bottom circulation of contents in reactor. Baffles make the hydrodynamic behaviour of slurry-phase bioreactors (Figure 20.7). Pre-treatment process includes size fractionation of solids, soil washing, milling to reduce particle size and slurry preparation. Certain surfactants such as anthracene, pyrene, perylene, etc., are added to enhance the rate of biodegradation.

20.2.2.2.1 Factors affecting slurry-phase biodegradation

pH (optimum 5.5-8.5), moisture content, temperature (20-30°C), oxygen (aerobic metabolism preferred), aging, mixing (mechanical and air mixing), nutrients; N, P and micronutrients, microbial population (naturally occurring microorganisms are

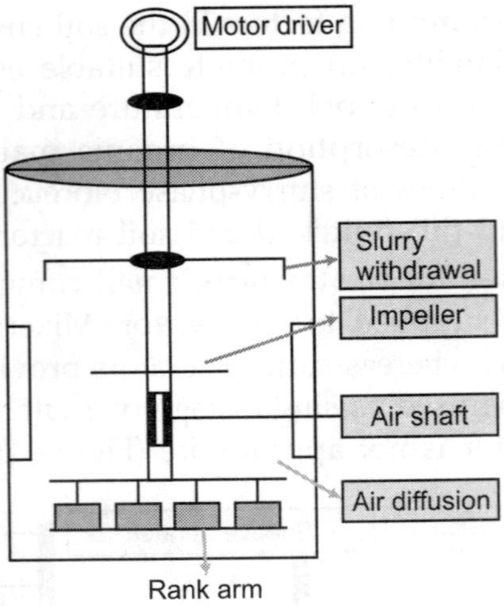

Figure 20.7 Demonstrating low shear airlift slurry phase bioreactor.

satisfactorily genetically engineered microorganisms for layer compound may be added), and reactor operation (batch and continuous cultures) are the important factors affecting the biodegradation. Table 20.1 states the essential factors for microbial bioremediation.

Table 20.1 Essential factors for microbial bioremediation

Factor	*Desired Conditions*
Microbial population	Suitable kinds of organisms that can biodegrade all of the contaminants
Oxygen	Enough to support aerobic biodegradation (about 2% oxygen in the gas phase or 0.4 mg/litre in the soil water)
Water	Soil moisture should be from 50-70% of the water holding capacity of the soil
Nutrients	Nitrogen, phosphorus, sulphur, and other nutrients to support good microbial growth
Temperature	Appropriate temperatures for microbial growth 20-30°C
pH	Best range is from 6.5 to 7.5

20.3 MICROORGANISM AND ORGANIC POLLUTANTS

Oil is insoluble in water and is less dense. It floats on water surface and forms slicks. In bulk storage tank, microbial growth is not possible provided water, and air is supplied. The microorganisms which are capable of degrading petroleum include pseudomonads, various corynebacteria, mycobacteria and some yeasts. However, there are two methods for bioremediation of hydrocarbons/oil spills, by using a mixture of bacteria, and using genetically engineered microbial strains. Many factors control biodegradability of a contaminant in the environment. Before attempting to employ bioremediation technology, one needs to conduct a thorough characterization of the environment where the contaminant exists, including the microbiology, geochemistry, mineralogy, geophysics, and hydrology of the system. Many chemical contaminants in the environment can be readily degraded because of their structural similarity to naturally occurring organic carbon, the amounts added may exceed the carrying capacity of the environment. Many pollutants are currently or were in the past used as pesticides. Others are used in industrial processes and in the production of a range of goods such as solvents, polyvinyl chloride, and pharmaceuticals. Microbes remove ammonia from wastewater treatment systems, clean up toxic waste and reduce phosphorus in lakes. Each strain of bacteria consumes a very limited range of hydrocarbons. Methods have, therefore, been devised to introduce a mixture of bacteria. Moreover, mixture of bacteria has successfully been used to control oil pollution in water supplies or oil spills from ships. The rate of degradation could not be accelerated without human intervention. Artificially, well characterized mixture of the bacterial strain along with inorganic nutrients such as phosphorus and nitrogen are pumped into the ground or applied to oil spill areas as required. This increases the rate of bioremediation significantly.

20.4 PROCESS OF BIODEGRADATION

The process of biodegradation may be different for different substances but in general, biodegradable substances are decomposed into carbon dioxide, methane, and water as the final products. Biodegradation is the chemical breakdown of materials by a physiological environment. The term is often used in relation to ecology, waste management and environmental remediation (bioremediation). Organic material can be degraded aerobically with oxygen, or anaerobically, without oxygen. A term related to biodegradation is biomineralisation, in which organic matter is converted into minerals. Biosurfactant, an extracellular surfactant secreted by microorganisms, enhances the biodegradation process.

Biodegradable matter is generally organic material such as plant and animal matter and other substances originating from living organisms, or artificial materials that are similar enough to plant and animal matter to be put to use by microorganisms. Some microorganisms have the astonishing, naturally occurring, microbial catabolic diversity to degrade, transform or accumulate a huge range of compounds including hydrocarbons (e.g., oil), polychlorinated biphenyls (PCBs), polyaromatic hydrocarbons (PAHs), pharmaceutical substances, radionuclides and metals. Major methodological breakthroughs in microbial biodegradation have enabled detailed genomic, metagenomic, proteomic, bioinformatic and other high-throughput analyses of environmentally relevant microorganisms providing unprecedented insights into key biodegradative pathways and the ability of microorganisms to adapt to changing environmental conditions. Biodegradation can be measured in a number of ways. The activity of aerobic microbes can be measured by the amount of oxygen they consume or the amount of carbon dioxide they produce. It can be measured by anaerobic microbes and the amount of methane or alloy that they may be able to produce. Biodegradable plastics are of two main types in the market: hydro-biodegradable plastics (HBP) and oxo-biodegradable plastics (OBP). Both have to undergo chemical degradation by hydrolysis and oxidation respectively resulting into their physical disintegration and a drastic reduction in their molecular weight. These smaller, lower molecular weight fragments are then amenable to biodegradation (Figure 20.8).

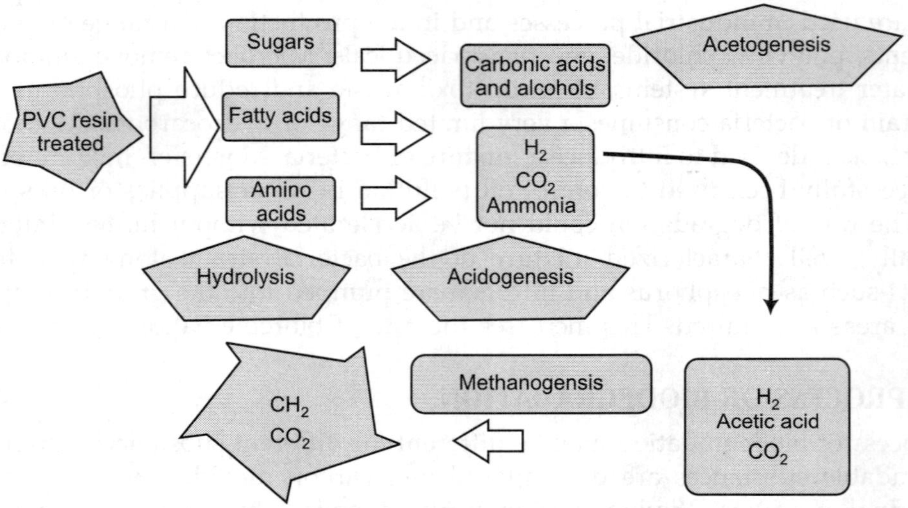

Figure 20.8 Biodegradation of PVC resin.

OBPs are made by adding a small proportion of compounds of specific transition metals (iron, manganese, cobalt and nickel are commonly used) into the normal production of polyolefins such as polyethylene (PE), polypropylene (PP) and polystyrene (PS). The additives act as catalysts to speed up the normal oxidative degradation, increasing the overall process by up to several orders of magnitude (factors of 10). It generally takes a small amount of catalyst, relative to the amount of material being catalyzed, to perform the required function. Also, by definition, the catalyst is not consumed in the reaction.

The products of the catalyzed oxidative degradation of the polyolefins are precisely the same as for conventional polyolefins because, other than a small amount of additive present, the plastics are conventional polyolefins. Many commercially useful hydrocarbons (e.g., cooking oils, polyolefins, many other plastics) contain small amounts of additives called antioxidants that prevent oxidative degradation during storage and use. Antioxidants function by 'deactivating' the free radicals that cause degradation. Lifetime (shelf life + use life) is controlled by antioxidant level and the rate of degradation after disposal is controlled by the amount and nature of the catalyst.

Since there are no existing corresponding standards that can be used directly in reference to plastics that enter the environment in other ways other than compost, i.e., as terrestrial or marine litter or in landfills, OBP technology is often attacked by the HBP industry which is unable to live up to the standards (which are actually the standards for composting). It has to be understood that composting and biodegradation are not identical. OBP can however be tested according to ASTM D6954, and (as from 1.1.2010) UAE 5009:2009. HBPs tend to degrade and biodegrade somewhat more quickly than OBP, but they have to be collected and put into an industrial composting unit. The end result is the same — both convert to carbon dioxide (CO_2), water (H_2O) and biomass. OBP are generally less expensive, possess better physical properties and can be made with current plastics processing equipment. HBP emits methane in anaerobic conditions, but OBP does not.

Polyesters play a predominant role in hydro-biodegradable plastics due to their potentially hydrolysable ester bonds. HBP can be made from agricultural resources such as corn, wheat, sugar cane, or fossil (petroleum-based) resources, or blend of the two. Some of the commonly-used polymers include PHA (polyhydroxyalkanoates), PHBV (polyhydroxybutyrate-valerate), PLA (polylactic acid), PCL (polycaprolactone), PVA (polyvinyl alcohol), PET (polyethylene terephthalate), etc. It would be misleading to call these "renewable" because the agricultural production process burns significant amounts of hydrocarbons and emits significant amounts of CO_2. OBPs (like normal plastics) are made from a by-product of oil or natural gas, which would be produced whether or not the by-products were used to make plastic. HBP technology claims to be biodegradable by meeting the ASTM D6400-04 and EN 13432 standards. However, these two commonly quoted standards are related to the performance of plastics in a commercially managed compost environment. They are not biodegradation standards. Both were developed for hydro-biodegradable polymers where the mechanism including biodegradation is based on reaction with water and state that in order for a production to be compostable, the following criteria need to be met:

Disintegration, the ability to fragment into non-distinguishable pieces after screening and safely support bio-assimilation and microbial growth; **Inherent biodegradation**, conversion of carbon to carbon dioxide to the level of 60% and 90% over a period of 180 days for ASTM D6400-04 and EN 13432 respectively. There is therefore little or no carbon left for the benefit of the soil, but the CO_2 emitted to atmosphere contributes to climate-change.

Safety, that there is no evidence of any eco-toxicity in finished compost and soils can support plant growth; and **toxicity**, that heavy metal concentrations are less than 50% regulated values in soil amendments.

20.4.1 Relationship Between Contaminant Structure, Toxicity and Biodegradability

Some type of contaminant structure can lead to low degradation even if the contaminant structure is similar to naturally occurring molecules. Microbial degradation is thought to be the primary mechanism for PAH degradation in soil and is dominated by members of the *Sphingomonas, Burkholderia, Pseudomonas,* and *Mycobacterium* taxonomic groups. Microbial communities that are capable of digesting specific compounds will proliferate exponentially in response to digestible contamination. However, some higher order PAHs are not accessible to microbes as food due to their absorption inside organic particles or location in small pores that are inaccessible for bacteria. The degree to which these "etches" accelerate degradation is unclear, and the addition of nitrogen and phosphorous-containing soil amendments is the best way to facilitate the growth of biofilms, given supportive pH. The presence of lower molecular weight PAHs supports microbial communities that may be needed for the metabolization of higher molecular weight PAHs, which puts older, more weathered sites at a disadvantage in the absence of amendments.

Actions and interactions between the three fundamental sets of ecotoxicological factors: contamination, abiotic, and biotic factors is depicted in Figure 20.9. Pesticides reaching to the soil are acted upon by several physical, chemical, and biological forces. However, physical and chemical forces are acting upon/degrading the pesticides to some extent, microorganisms plays a major role in the degradation of pesticides. Many soil microorganisms have the ability to act upon pesticides and convert them into simpler non-toxic compounds. This process of degradation of pesticides and conversion into non-toxic compounds by microorganisms is known as "biodegradation". Not all pesticides reaching to the soil are biodegradable and such chemicals that show complete resistance to biodegradation are called "recalcitrant".

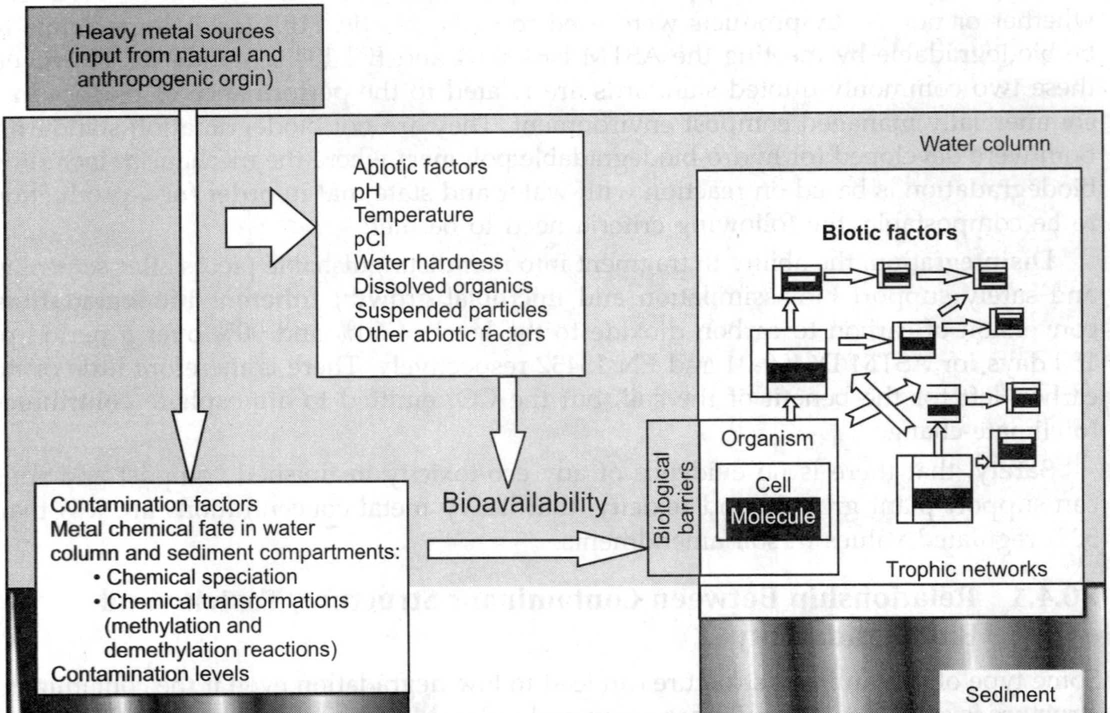

Figure 20.9 Flow representation for ecotoxicological approach to metal bioaccumulation and transfers at ecosystem level.

20.4.2 Conversion of the Pesticide Molecule to a Non-toxic Compound

Detoxification is not synonymous with degradation. Since a single chance in the side chain of a complex molecule may render the chemical non-toxic. It is the conversion of non-toxic substrate into a toxic molecule, e.g., herbicide, 4-butyric acid (2, 4-DB) and the insecticide Phorate are transformed and activated microbiologically in soil to give metabolites that are toxic to weeds and insects. Organisms must have necessary catabolic activity required for degradation of contaminant at a fast rate to bring down the concentration of contaminant, the target contaminant must be bioavailable, soil conditions must be congenial for microbial /plant growth and enzymatic activity and cost of bioremediation must be less than other technologies of removal of contaminants.

20.5 SIGNIFICANCE OF GENETICALLY ENGINEERED BACTERIAL STRAINS

The technique was introduced for the first time in 1979 by Dr. Anand Mohan Chakrabarty, an India born American scientist. He isolated a strain of *Pseudomonas putida* that contained the XYL and NAH plasmid as well as a hybrid plasmid derived by recombination parts of CAM and OCT (these are incompatible and cannot co-exist as separate plasmids in the same bacterium). This strain could grow rapidly on crude oil because it was capable of metabolizing hydrocarbons more efficiently than any other single plasmid. In 1990, the USA Government allowed him to use this superbug for cleaning up of an oil spill in water of State of Texas. Superbug was produced on a large scale in the laboratory, mixed with straw and dried. The bacteria laden straw can be stored until required. When the straw was spread over oil slicks, the straw soaked up the oil and bacteria broken up the oil into non-polluting and harmless products.

20.5.1 Bioremediation of Industrial Wastes

A variety of pollutants is discharged in the environment from a large number of industries/mills. Textile industry alone contributes a significant amount of pollutants to waterbodies such as enzymes, acids, alkali, alcohols, phenols, dyes, heavy metals, viz., zinc, cadmium, mercury, copper, chromium and lead in traces, radionuclides, etc. Actinomycetes show a higher capacity to bind metal ions compared to fungi and bacteria. The living microbial cells accumulate metals intracellularly at a higher concentration, whereas dead cells precipitate metals in and around cell walls by several metabolic processes. Dead biomass immobilized on polymeric membrane absorbs uranium well, and immobilized *Aspergillus oryzae* cells on reticulated foam particles are used for Cd removal. *Aspergillus niger* biomass contains up to 30 per cent of chitin and glucans. Chitin phosphate and chitosan phosphate of fungi absorb a greater amount of uranium than Cu, Cd, Mn, Co, Mg and Ca.

20.5.2 Bioremediation of Dyes

Maximum number of dyes undergo degradation through reduction by azo-reductase of *Pseudomonas* strains in the chemostat culture. This enzyme catalyses azo-linkage of the dye. During the degradation process of azo, NAD(P) acts as electron donor for the degradation of black liquor pulp mill effluents by the strains of *Pseudomonas putida*. Some anaerobic bacteria, *Streptomyces* and fungi (e.g., *Phaenerochaete, Chrysosporium)* have been characterized for decolouration of chromogenic dyes. The enzymes involved in dye degradation are lignases (lignin peroxidase), Mn (II) dependent peroxidase and glyoxal oxidase. These enzymes are well associated with lignin degrading system.

20.6 MICROBES AND METAL POLLUTANTS

High concentrations of heavy metals in soil have an adverse effect on microorganisms and microbial processes. Toxic effects of heavy metals on soil microorganisms have been extensively studied in the past and almost every group of organisms has been studied in this respect. Fungi and bacteria constitute the main components of the soil microbial biomass. It has often been stated that fungi are more tolerant to heavy metals as a group than bacteria. Heavy metals are naturally present in some ecosystems, their industrial use leads to serious environmental problems. The use of metal-resistance bacteria can help to remove metal from contaminated environments. Heavy metal contaminated soil/ water problem has already become the global focus and understanding the regulation of heavy metal resistance could be useful for biological waste treatment and estimating the impact that industrial activity may have on natural ecosystems.

20.6.1 Metals Defined

Microbes encounter metals and metalloids of various kinds in the environment and it is, therefore, not surprising that they should interact with them, sometimes to their benefit, at other times to their detriment. Of particular practical interest are the base metals including vanadium, chromium, manganese, iron, cobalt, nickel, copper, zinc, molybdenum, silver, cadmium and lead; the precious metals gold and silver; and the metalloids arsenic, selenium, and antimony. In nature, these metals and metalloids exist mostly as cations, oxyanions, or both in aqueous solution, and mostly as salts or oxides in crystalline (mineral) form or as amorphous precipitates in insoluble form. A few, like iron, copper, and gold, may also exist in the metallic state in nature. All microbes, whether prokaryotic or eukaryotic, employ metal species for structural functions and/ or catalytic functions. The alkali metals Ca and Mg serve structural as well as catalytic functions. The metals V, Cr, Mn Fe, Co, Ni, Cu, Zn, Mo, and W and the metalloid Se may participate in catalytic functions. For such uses, low environmental concentrations are sufficient.

Some prokaryotes can employ metal species that can exist in more than one oxidation state, among them Cr, Mn, Fe, Co, Cu, As, and Se, as electron donors or acceptors in their energy metabolism. For this function, the metals or metalloids must occur in sufficiently high concentration locally to meet the organisms' demand. Microbial interactions with small quantities of metals or metalloids do not exert a major impact on metal or metalloid distribution in the environment, whereas interactions with larger quantities, as are required in energy metabolism for instance, have a noticeable impact. Some of the latter interactions are commercially exploited or have a potential for it.

A number of microbes are able to use some metals or metalloids as electron donors or acceptors in energy metabolism. They include eubacteria and archaea. As energy sources, oxidizable metals or metalloids may satisfy the entire energy demand of an organism (chemolithotrophs). For example, the eubacteria *Thiobacillus ferrooxidans* and *Leptospirillum ferrooxidans* and the archaea *Acidianus brierleyi* and *Sulfolobus acidocaldarius* are able to obtain all their energy for growth from the oxidation.

20.6.2 Metal Toxicity Effect on Microbial Cells

It is apparent that metal toxicity can be heavily influenced by environmental conditions. Binding of metals to organic materials, precipitation, complexation, and ionic interactions are all important phenomena that must be considered carefully in laboratory and field studies. It is also obvious that microbes possess a range of tolerance mechanisms, most featuring some kind of detoxification. Many of these detoxification mechanisms occur widely in the microbial world and are not only specific to microbes growing in metal-contaminated environments.

20.6.3 Mechanism of Microbial Metal Resistance and Detoxification

Microbial cells have resistances to essentially all of the toxic heavy metals of the Periodic Table. In bacterial cells, the genetic determinants of these resistances are frequently found on small extra chromosomal plasmids and transposons. Sometimes the resistances are associated with detoxifying enzymes. This is true for the $Hg^{2+} \rightarrow Hg^0$ reductase, the $As^{3+} \rightarrow As^{5+}$ oxidase and the $Cr^{6+} \rightarrow Cr^{3+}$ reductase. In other cases, such as As^{5+}, Ag^+ and Cd^{2+}, no change in redox state occurs but, rather, uptake and transport differences accompany resistance determinants.

Some metals, including iron, zinc, copper, manganese, etc., are micronutrients used in the redox processes, regulation of the osmotic pressure, and also enzyme components. Other metals are not essential. However, even essential metals such as zinc and copper are toxic at high concentrations and can cause damage at intracellular level. High metal concentration damage DNA and membrane and loss of enzyme function. Bacteria utilize a variety of resistance mechanisms to protect themselves from toxic metals concentrations that involve permeability barriers, intra- and extracellular sequestration, efflux pumps, enzymatic detoxification and reduction. Many microorganisms demonstrate resistance to metals in water, soil and industrial waste. Genes located on chromosomes, plasmids, or transposons encode specific resistance to a variety of metal ions. There are five basic mechanisms that convey an increased level of cellular resistance to metals (Figure 20.10); they are active transport efflux of the toxic metal from the cell; enzymic conversion; intra- or extracellular sequestration; exclusion by a permeability barrier; reduction in sensitivity of cellular targets.

Bacterial plasmids contain genetic determinants for resistance systems for Hg^{2+} (and organomercurials), Cd^{2+}, AsO_2, $AsO_4{}^{3-}$, $CrO_4{}^{2-}$, $TeO_3{}^{2-}$, Cu^{2+}, Ag^+, Co^{2+}, Pb^{2+}, and other metals of environmental concern. In some cases, there is the potential for using genetically engineered microbes for bioremediation. Recombinant DNA analysis has

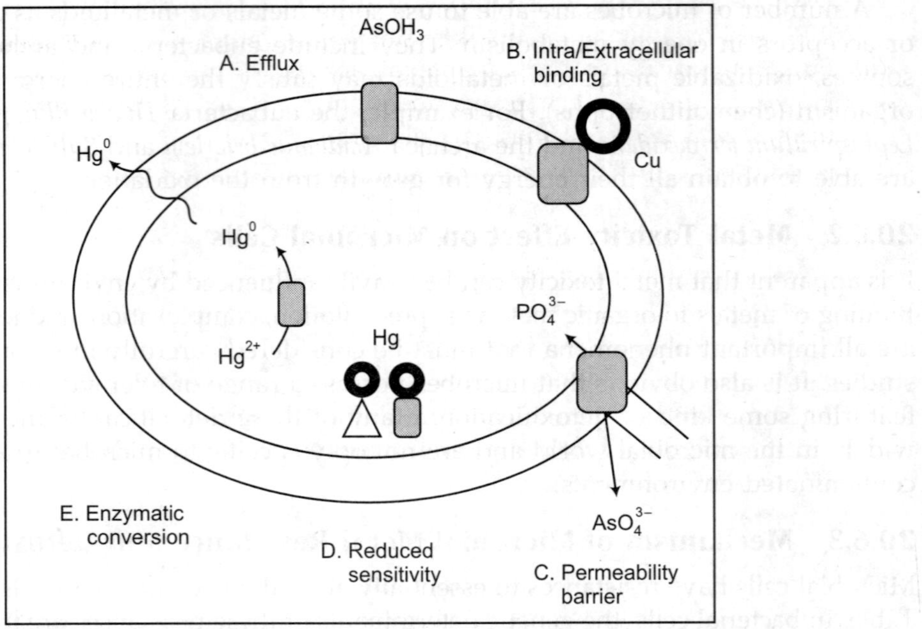

Figure 20.10 Methods of metal resistance.

been applied to mercury, cadmium, zinc, cobalt, arsenic, chromate, tellurium and copper resistance systems. The eight mercury resistance systems that have been sequenced all contain the gene for mercuric reductase, the enzyme that converts toxic Hg^{2+} ions to less toxic volatile metallic Hg°. Four of these systems also determine the enzyme organomercurial lyase, which cuts the HgC bond and thus detoxifies methylmercury and phenylmercury. Two sequenced Cd^{2+} resistance determinants govern cellular efflux of Cd^{2+} assuring a low level of intracellular Cd^{2+}: not an obvious candidate for bioremediation. Cadmium accumulation by bacterial metallothionein or phytochelatin is a potentially useful process, but only preliminary reports have appeared on bacteria producing polythiol polypeptides. For arsenic resistance, a unique efflux ATPase maintains low intracellular As levels. A bacterial AsO_2^- oxidase has been reported, with the potential of converting more toxic As (III) into less toxic As (V), but this system has not been studied in recent years. For chromate, resistance results from reduced cellular uptake. However, both soluble and membrane-bound Cr (VI) reductase bacterial activities convert more toxic Cr (VI) to less toxic Cr (III) in different bacteria.

20.6.4 Bioremediation of Heavy Metals

Bacteria, algae, fungi, actinomycetes and higher plants accumulate high amount of heavy metals in their cells.

(i) **Algae:** The species of *Chlorella, Anabaena inaequalis, Anacystis nidulans, Cylindrospermum* sp., *Westiellopsis prolifica, Stigeoclonium tenue, Synechococcus* sp. etc. tolerate heavy metals. However, several species of *Chlorella, Anabaena,* marine algae

have been used for the removal of heavy metals. But the operational condition limit the practical application of these organisms.

(ii) **Fungi:** Fungi also are capable of accumulating heavy metals in their cells. However, several mechanisms operate in them for removal of heavy metals from the solution, a few of these are elaborated below :

(a) **Metabolism-independent accumulation:** The positively charged ions in the solution are attracted to negatively charged ligands in cell materials. Biosorption of metal ion occurs on microbial cell surface. But composition of biomass and other factors affect biosorption. For example, in *Rhizopus arrhizus* adsorption depends on ionic radius of Li^{3+}, Mn^{2+}, Cu^{2+}, Zn^{2+}, Cd^{2+}, Ba^{2+}, Hg^{2+} and Pb^{2+}. However, binding of Hg^{2+}, Ag^{2+}, Cd^{2+}, $A1^{3+}$, Ni^{2+}, Cu^{2+} and Pb^{2+} strongly depends on the concentration of yeast cells.

(b) **Metabolism-dependent accumulation:** In fungi and yeast, heavy metal ions are transported into the cells through cell membrane. However, as a result of metabolic processes ions are precipitated around the cells and synthesize intracellularly metal-binding proteins. Energy-dependent uptake of Cu^{2+}, Cd^{2+}, Co^{2+}, Ni^{2+}, Zn^{2+} by fungi has been demonstrated. Moreover, intracellular uptake is influenced by certain external factors such as pH, anions, cations and organic materials, growth phase, etc. Metal uptake by growing batch culture was found maximum during the lag phase and early log phase in *Aspergillus niger, Penicillium spinulosum* and *Trichoderma viride*.

(c) **Extracellular precipitation and complexation**. Fungi produce several extracellular products which can complex or precipitate heavy metals. For example, several fungi and yeast release high affinity Fe-binding compounds that chelate iron. These are called siderophores. The Fe^{3+} chelates which are formed outside the cell wall are taken up into the cell. In *Saccharomyces cerevisiae* removal of metals is done by their precipitation as sulphides, e.g., Cu^{2+} is precipitated as CuS.

20.6.5 Bioremediation Through VAM Fungi

Among the soil microorganisms, arbuscular mycorrhizal fungi provide a direct link between soil and roots, thereby enhancing the uptake of nutrient by the host plant. They are reported to be present on metal-contaminated sites and therefore to be of great importance in metal uptake by plant roots. Arbuscular mycorrhizal fungi are soil microorganisms that establish mutual symbiosis with the majority of higher plants, providing direct links between fungi and roots. Selected VAM fungi are introduced through plants in coal mine areas. Extensive infection of most plant species colonizing coal waste has been observed. It has been found that VAM fungi improved the growth and survival of desirable revegetation species. Increased growth of red maple, maize, alfalfa and several other plants inoculated with VAM fungi growing in coal mine soil has been recorded. Host plants that have evolved metal tolerance normally can do so without their mycorrhizal symbionts and it appears that being in symbiosis may help them to fare better. Further, the effect of AM fungi on the heavy metal uptake of the host plant depends on the physical and chemical properties of the contaminated soil and on the heaviness and duration of the contaminated load on the host plant and the fungal species.

20.7 BIOREMEDIATION OF XENOBIOTICS

Many of the artificially made complex compounds, i.e., xenobiotics persist in environment and do not undergo biological transformation. Microorganisms play an important role in the degradation of xenobiotics, and maintaining of steady state concentrations of chemicals in the environment. Detoxification is not synonymous with degradation. It is the conversion of non-toxic substrate into a toxic molecule, e.g., herbicide, 4-butyric acid (2, 4-D B) and the insecticide Phorate are transformed and activated microbiologically in soil to give metabolites that are toxic to weeds and insects. Organisms must have necessary catabolic activity required for degradation of contaminant at a faster rate to bring down the concentration of contaminant. The target contaminant must be bio-available and soil conditions must be congenial for microbial/plant growth and enzymatic activity with cost effective technology (Figure 20.11).

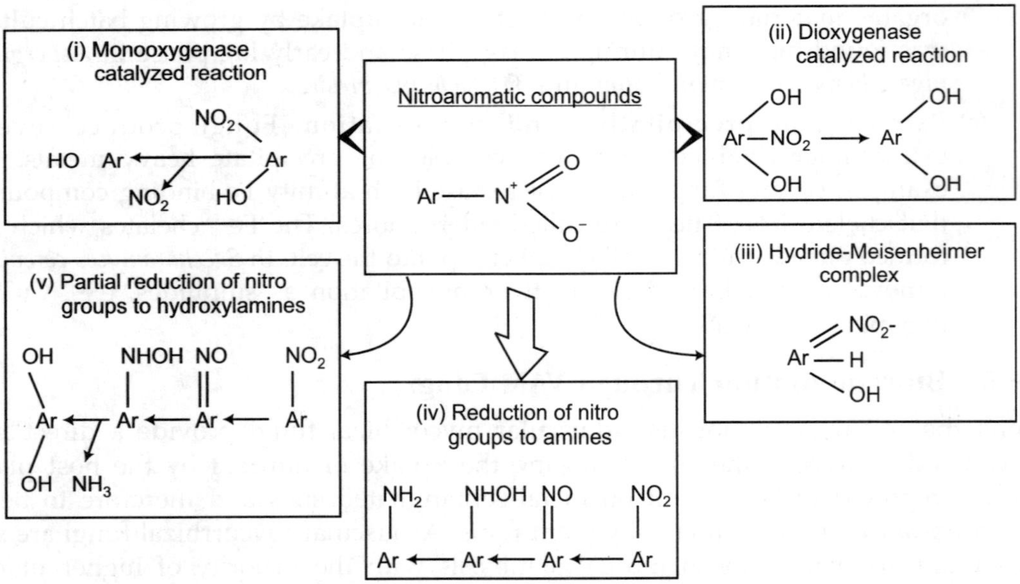

Figure 20.11 Microbial strategies for remediation of nitro-aromatic compounds by aerobes (i) (ii) (iii) and (iv) and anaerobes (v).

The complete degradation of a pesticide molecule to its inorganic components that can be eventually used in an oxidative cycle removes its potential toxicity from the environment. Understanding the characters of microbes to provide clues to the evolution of degradative pathways and makes the task of gene manipulation easier to construct the genetically engineered microbes capable of degrading the pollutants.

20.7.1 Microbial Degradation of Xenobiotics

Biodegradation of pesticides occurs by aerobic soil microbes. Pesticides are of wide varieties of chemicals, e.g., chlorophenoxyalkyl caboxylic acid, substituted urea, nitrophenols, triazines, phenyl carbamates, organochlorines, organophosphates, etc. Duration of persistence of herbicides and insecticides in soil is given in Table 20.2. Organophosphates, e.g., diazion, methyl parathion and parathion are perhaps the most extensively used insecticides under many agricultural systems. Biodegradation through hydrolysis of *p-o*-aryl bonds by *Pseudomonas diminuta* and *Flavobacterium* are considered the most significant steps in the detoxification of organophosphorus compounds. Organomercurials, (e.g., Semesan, Panodrench, Panogen) have been practised in agriculture since the birth of fungicides. Several species of *Aspergillus, Penicillium* and *Trichoderma* have been isolated from Semesan treated soil which have shown ability to grow over 100 ppm of fungicide *in vitro.* The major fungicides used in agriculture are water soluble derivatives such as Ziram, Ferbam, Thiram, etc. All these are degraded by microorganisms.

Table 20.2 Duration of persistence of insecticides and herbicides in soil

Biocides	*Time taken for 75-100% disappearance*
A. Chlorinated insecticides	
DDT (1,1,I-trichloro-2,2-bis-(p-chlorophenyl) ethane	4 years
Aldrin	3 years
Chlordane	5 years
Heptachlor	2 years
Lindane (hexachloro-cyclohexane)	3 years
B. Organophosphate insecticides	
Diazinon	12 years
Malathion	1 week
Parathion	1 week
C. Herbicides	
2,4-D (2,4-dichlorophenoxyacetic acid)	4 weeks
2,4,5-T	30 weeks
Atrazine	40 weeks
Simazine	48 weeks
Propazine	1.5 years

Source: Madigan et al. (1997).

Steps of PCP degradation has been shown in Figure 20.12. Pentachlorophenol (PCP) is a broad spectrum biocide which has been used as the fungicide, insecticide, herbicide, algicide, disinfectant and antifouling agent. Bioreactors containing alginate immobilized + Polyurethane foam immobilized PCP degrading *Flavobacterium* (ATCC39723) cells have

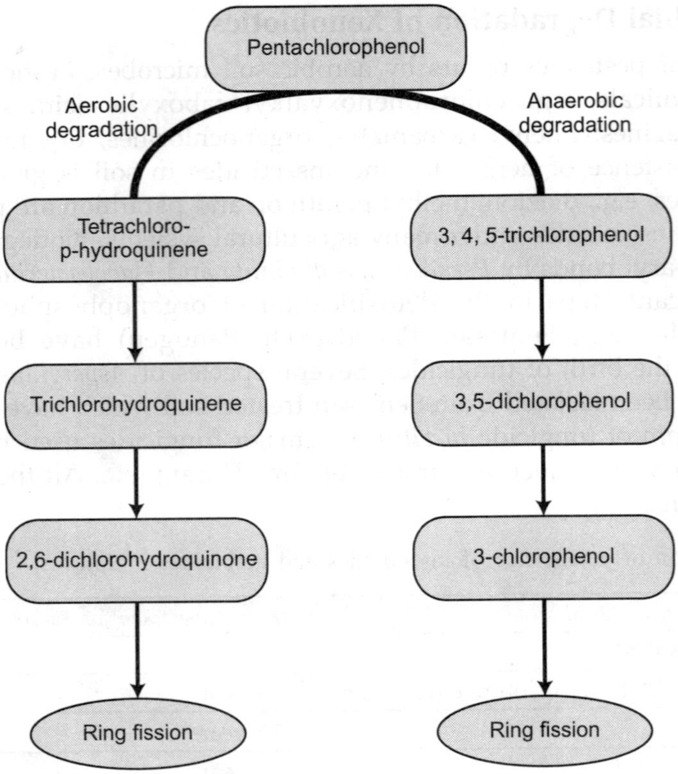

Figure 20.12 Outline of aerobic and anaerobic degradation of pentachlorophenol.

been used to remove PCP from contaminated water. Absorption of PCP by Polyurethane immobilized matrix plays a role in reducing the toxicity of PCP. Flavobacterium removes and detoxify PCP.

DDT ((l,l,l-trichlofo-2,2-bis (p-chlorophenyl) ethane)) is an insecticide that persists in soil for four years. Degradation pathway of DDT involves an initial dechlorination of the trichloromethyl group to form 1,1-dichloro 2-ethane which then undergoes a further dechlorination, oxidation and decarboxylation to form bis methane. Subsequent cleavage of one of the normal aromatic rings yields p-chlorophenyl acetic acid, which may also undergo ring cleavage. Microorganisms associated with DDT degradation are *Aspergillus flavus, Fusarium oxysporum, Mucor aternans, Phanerochaete chrysosporium, Trichoderma viride,* etc. Environmental factors including pH, temperature, bioavailability, nutrient supply and oxygen availability affect biodegradation of pesticides.

20.7.2 Gene Manipulation of Pesticide-degrading Microorganisms

Day-by-day the number of xenobiotic-degrading microorganisms is increasing. However, pesticide-degrading genes of only a few microorganisms have been characterized. Most of genes responsible for catabolic degradation are located on the chromosomes, but in a few cases these genes are found on plasmids or transposons. Camphor and naphthalene

(NAH) degrading genes of *Pseudomonas putida* was isolated from the plasmid. Discovery of these genes made it possible to construct a new genetically engineered strain of *P. putida* that alone was potent to degrade camphor, naphthalene, xylene, toluene, octanes and hexanes. Two genes involved in early steps of Y-HCH degradation in UT26 was cloned and sequenced. The *lin*A gene encodes Y-HCH dehydrochlorinase which converts Y-HCH to 1,2,4TCB via Y-PCCH and 1,4-TCDN. The *lin*B gene encodes 1,4-TCDN chlorohydrolase which converts 1,4-TCDN to 2,4-DDOL via 2,4,5-DNOL. This gene is a member of the haloalkane dehalogenase family with a broad range specificity for substrate. The genetically engineered *P. putida* comprises both *lin*A and *linB* genes.

The *pod* gene, initially isolated from *Flavobacterium* sp. ATCC27551 and *Pseudomonas diminuta* has been well characterized. It is associated for degradation of pesticides such as parathion, methyl parathion, etc. A recombinant DNA plasmid containing *pod* gene has been successfully transferred into a fungus, *Gliocladium virens* with expression though at low level. The recombinant *G. virens* is a useful soil saprophyte fungus showing a strong mycoparasitic activity against many fungal pathogens. Strains of *G. virens* are potential for use in the bioremediation of contaminated soil. Optimization of *pod* expression in bioremedially useful organism such as *P. chrysosporium, G. virens,* etc., holds a great promise in reducing the pesticide pollution.

20.8 MICROBIAL LEACHING

Future sustainable development requires measures to reduce the dependence on non-renewable raw materials and the demand for primary resources. New resources for metals must be developed with the aid of novel technologies. In addition, improvement of already existing mining techniques can result in metal recovery from sources that have not been of economical interest until today. Metal-winning processes based on the activity of microorganisms offer a possibility to obtain metals from mineral resources not accessible by conventional mining. Microbes such as bacteria and fungi convert metal compounds into their water-soluble forms and are biocatalysts of these leaching processes. Additionally, applying microbiological solubilisation processes, it is possible to recover metal values from industrial wastes which can serve as secondary raw materials.

Microbial leaching is a process described as being "the dissolution of metals from their mineral source by certain naturally occurring microorganisms" or "the use of microorganisms to transform elements so that the elements can be extracted from a material when water is filtered through it". Microbial leaching has emerged as an important branch of biotechnology in recent years. Microbial technology renders help in case of recovery of ores which cannot be economically processed with chemical methods, because they contain low grade metals. Therefore, large quantity of low grade ores are produced during the separation of high grade ores. The low grade ores are discarded in waste heaps which enter in the environment. The low grade ores contain a significant amount of nickel, lead, and zinc ores which could be processed by microbial leaching. The technology for microbial leaching of uranium and copper has been widely commercialized. But large scale leaching process may cause environmental problems when dump is not managed properly resulting in the seepage of leach fluids containing large quantity of metals and low pH into nearby natural water supplies and ground water.

Thus, biomining is economically sound hydrometallurgical process with lesser environmental problem than conventional commercial application. However, it is an interdisciplinary field involving metallurgy, chemical engineering, microbiology and molecular biology having tremendous practical application.

20.8.1 Microorganisms Used for Leaching

The most commonly used microorganisms for bioleaching are *Thiobacillus thiooxidans* and *T. ferrooxidans.* The other microorganisms may also be used in bioleaching, viz., *Bacillus licheniformis, B. luteus, B. megaterium, B. polymyxa, Leptospirillum ferrooxidans, Pseudomonas fluorescens, Sulfolobus acidocaldarius, Thermothrix thioparus, Thiobacillus thermophilica,* etc.

20.8.2 Chemistry of Microbial Leaching

Thiobacillus thiooxidans and *Thiobacillus ferrooxidans* have always been found to be present in a mixture on leaching dumps. *Thiobacillus* is the most extensively studied Gram-negative bacillus bacterium which derives energy from oxidation of Fe^{2+} or insoluble sulphur. In bioleaching process, there are two following reaction mechanisms:

20.8.2.1 Direct bacterial leaching

In direct bacterial leaching, a physical contact exists between bacteria and ores and oxidation of minerals takes place through several enzymatic catalyzed steps. For example, pyrite is oxidized to ferric sulphate as below:

$$2FeS_2 + 7O_2 + 2H_2 \xrightarrow{\text{T.ferroxidans}} 2FeSO_4 + 2H_2SO_4$$

20.8.2.2 Indirect bacterial leaching

In indirect bacterial leaching, microbes are not in direct contact with minerals but leaching agents are produced by microorganisms which oxidize them.

$$FeS_2 + Fe_2(SO_4) \longrightarrow 3FeSO_4 + 2S^{\circ}$$

$$2S^{\circ} + 3O_2 + 2H_2O \longrightarrow 2H_2SO_4$$

Oxidation of ferrus (Fe^{2+}) to ferric (Fe^{3+}) by *T. ferrooxidans* at low pH is given below:

$$4FeSO_4 + 2H_2SO_4 + O_2 \xrightarrow{\text{T.ferroxidans}} 2Fe_2(SO_4)_3 + 2H_2O$$

20.8.2.3 Leaching process

There are three commercial methods used in leaching:

(i) **Slope Leaching:** About 10,000 tonnes of ores are ground first to get fine pieces. It is dumped in large piles down a mountain side leaching dump. Water containing the inoculum of *Thiobacillus* is continuously sprinkled over the pile. Water is collected at bottom. It is used to extract metals and generate bacteria in an oxidation pond.

(ii) **Heap Leaching:** The ore is dumped in large heaps called leach dump. Further steps of treatment are as described for slope leaching.

(iii) ***In situ* Leaching:** In this process ores remain in its original position in the earth. Surface blasting of rock is done just to increase permeability of water. Thereafter, water containing *Thiobacillus* is pumped through drilled passage to the ores. Acidic water seeps through the rock and collects at bottom. Again from bottom water is pumped, mineral is extracted and water is reused after generation of bacteria.

20.8.3 Bioleaching of Copper

Copper leaching plants have been widely used for many years throughout the globe. It is operated as simple heap leaching process or combination of both heap leaching and *in situ* leaching process. Dilute sulphuric acid (pH 2) is percolated down through the pile. The liquid coming out of the bottom of pile reach in mineral. It is collected and transported to precipitation plant; metal is reprecipitated and purified. Liquid is pumped back to top of pile and cycle is repeated. For the removal of copper, the ores commonly used are chalcocite (Cu_2S), chalcopyrite ($CuFeS_2$) or covellite (CuS). Several other metals are also associated with these ores. Chalcocite is oxidized to the soluble form of copper (Cu^{2+}) and covellite by *T. ferrooxidans.*

$$Cu_2S + O_2 \longrightarrow CuS + Cu^{2+} + H_2O$$

Covellite is oxidized to copper sulphate chemically or by bacteria.

$$2CuFeS_2 + 8\tfrac{1}{2}\ O_2 + H_2SO_4 \longrightarrow 2CuSO_4 + Fe_2(SO_4)_3 + H_2O$$

Thereafter, strictly chemical reaction occurs which is the most important reaction in copper leaching.

$$CuS + 8Fe^{3+} + 4H_2O \longrightarrow Cu^{2+} + SFe^{2+} + SO_4^{2-} + 8H^+$$

Copper is removed as below:

$$Fe^{2+} + Cu^{2+} \longrightarrow Cu° + Fe^{2+}$$

Fe^{2+} is transferred in oxidation pond

$$Fe^{2+} + 1/4O_2 + H^+ \xrightarrow{\text{T.ferroxidans}} Fe^{3+} + 1/2O_2$$

The Fe^{3+} ions produced are an oxidation of ores; therefore, it is pumped back to pile. Sulphuric acid is added to maintain pH. An outline of microbial leaching of copper is shown in Figure 20.13. Microbial leaching of copper has been widely used in the USA, Australia, Canada, Mexico, South Africa and Japan.

20.8.4 Bioleaching of Uranium Leaching

Uranium leaching is more important than copper, although less amount of uranium is obtained than copper. For getting one tonne of uranium, a thousand tonnes of uranium ore must be handled. *In situ,* uranium leaching is gaining vast acceptance. However, uranium leaching from ore on a large scale is widely practised in USA, South Africa, Canada and India.

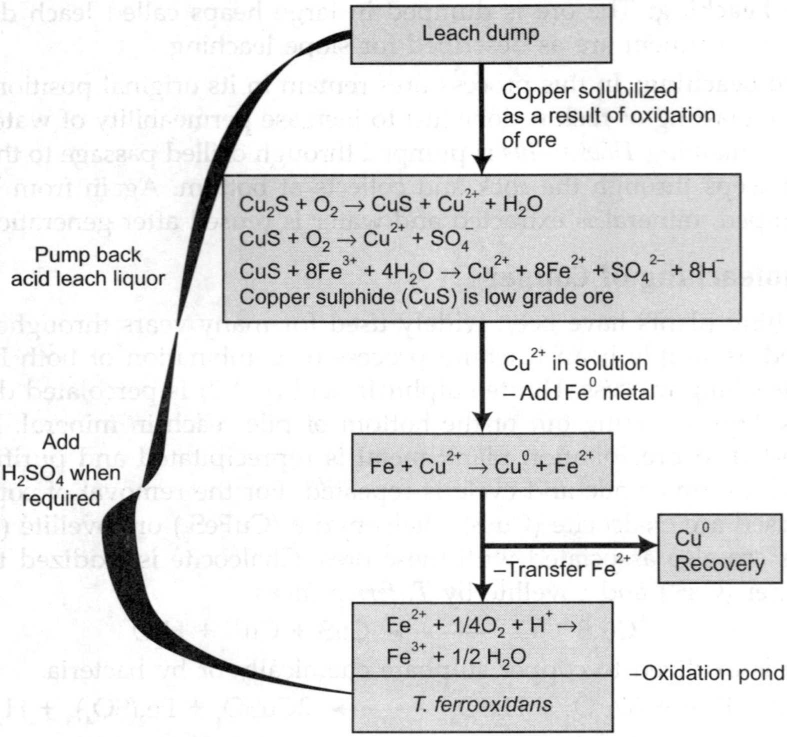

Figure 20.13 Flow diagram of microbial leaching of copper.

Insoluble tetravalent uranium is oxidized with a hot H_2SO_4/Fe^{3+} solution to make soluble hexavalent uranium sulfate at pH 1.5-3.5 and temperature 35°C.

$$UO_2 + Fe_2(SO_4)_3 \longrightarrow UO_2SO_4 + 2FeSO_4$$

Uranium leaching is an indirect process. *T. ferrooxidans* do not directly attack on uranium ore, but on the iron oxidant. The pyrite reaction is used for the initial production of Fe^{3+} leach solution.

20.8.5 Bioleaching of Gold and Silver

Microbial leaching of precious metal ores to enhance gold and silver recovery is one of the most promising applications. Gold is obtained through bioleaching of arsenopyrite/pyrite ore and its cyanidation process. Silver is more readily solubilized than gold during microbial leaching of iron sulfide.

20.8.6 Bioleaching of Silica Leaching

Magnesite, bauxite, dolomite and basalt are the ores of silica. *Bacillus licheniformis,* an isolate from magnesite ore deposits is found to be an excellent microbe for silican uptake and concomitant minerals. The silican uptake is restricted to adsorption on bacterial cell surface rather than internal uptake through the membrane.

20.9 UTILIZATION OF AGRO WASTE AND SEWAGE

20.9.1 Cellulose Decomposition in Soil

The most abundant renewable natural resource on this planet is the cellulose. Its annual worldwide production is estimated as much as 22 × 10^{10} tonnes through the photosynthetic fixation of CO_2 and 4 × 10^9 tonnes are available for further processing by various conventional and non-conventional methods. The interest in the biological degradation of cellulose has arisen mainly due to the need for conservation of the resources through their reuse to meet the expanding requirements of the raw materials as well as for the removal of pollutants from the soil environment by recycling. Most of the microbes conserve and recycle this biomass in the form of agricultural and forest waste by converting it into biofuels and feed protein. In nature, cellulose is usually found to be associated with either polysaccharides like starch, pectin and a variety of hemicelluloses. The chemical composition of raw material is an important criterion before their utilization for the production of valuable products. The importance of these raw materials is based on the total cellulose percentage (Figure 20.14). Cellulose is a linear homopolymer of cellobiose units connected by β-(1-4) glucosidic linkage. It is an important constituent of cell wall of plants. Cellulose is mainly composed of two regions (i) amorphous region and (ii) crystalline region. The susceptibility of cellulose mainly depends on the amorphous region. The crystalline region of cellulose fibre is impermeable to water molecules because of its compact nature.

Many fungi, bacteria, actinomycetes from soil origin are able to degrade cellulose into different metabolic products. Fungal members have been found to be the most active forms in soil to degrade the lignocellulosic waste and help to convert it in the form of compost. They also help to maintain the carbon cycle in nature. Most of the cellulolytic fungi are reported to be of thermophylic in nature. Some mesophyles are also reported to degrade the cellulose (Figure 20.15).

20.9.2 Cellulase Complex

In general, degradation of insoluble cellulose to soluble sugars *in vitro* by microbial systems involves a complex enzyme system or systems which act together to bring about the hydrolysis of cellulose. Exoglucanase or cellobiohydrolase is the major cellulase produced by cellulolytic fungi. The purified cellobiohydrolase exhibits a strong activity towards crystalline cellulose but it is competitively inhibited by cellobiose and glucose. Almost all cellobiohydrolase are glycoproteins and sugar content in them varies from as high as 50% to 3% depending upon the strains and type of fungi. Endoglucanase attacks cellulose polymers internally to produce glucose, cellobiose and other water soluble cellodextrins. Endoglucanase consists of two group of enzymes differentiated on the basis of their molecular weight. Endoglucanases are described as glucoproteins in which sugar contents are around 10%. β-glucosidase or cellobiase is important in regulation of cellulolysis and induction of cellulose. *In vivo,* cellobiase hydrolyzes exogenous cellobiose to glucose therby influencing cellulose biosynthesis through an

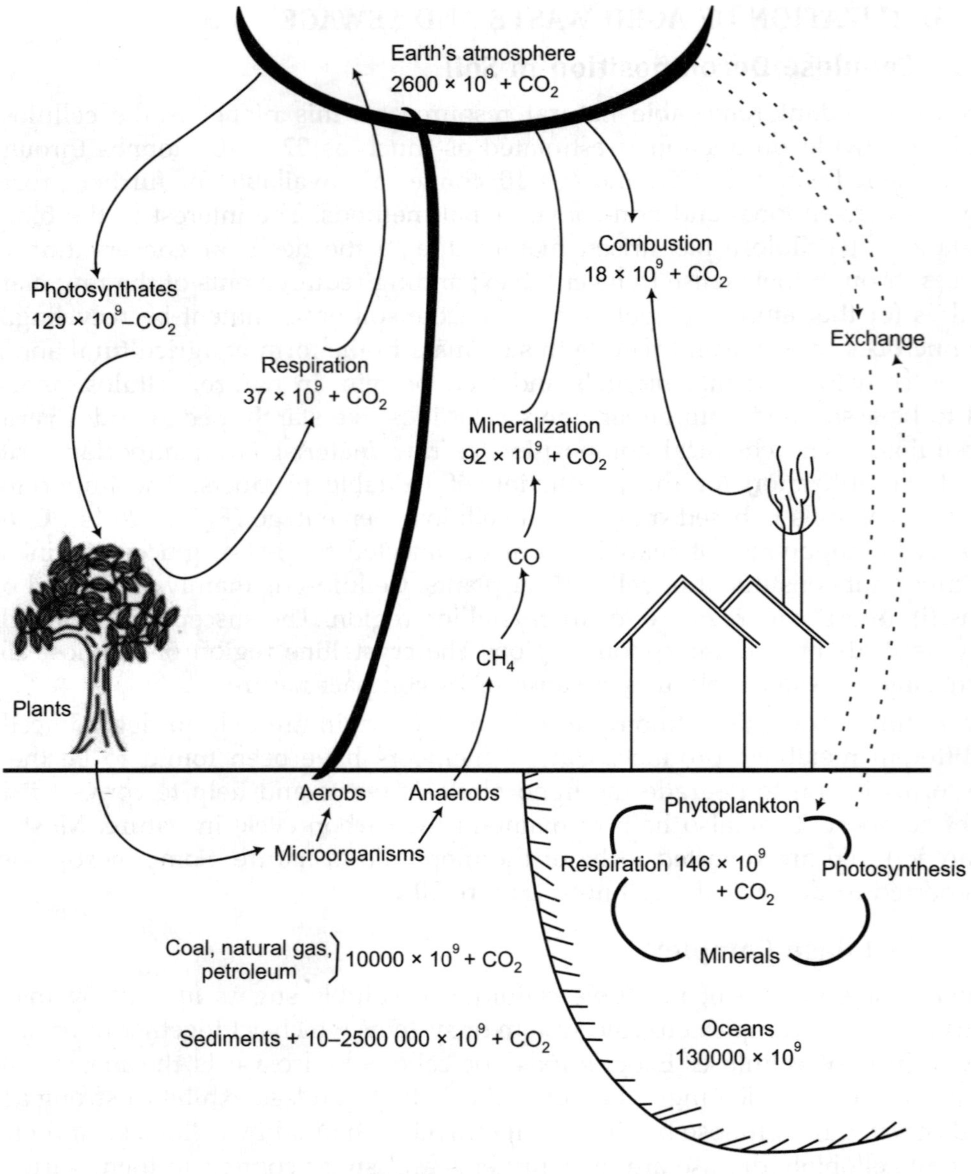

Figure 20.14 Carbon cycle in the biosphere.

induction repression mechanism. They have been reported to exist intracellularly and extracellularly in microbes under stationary phase and also for their use for enhancing the cellulolysis. The two types of β-glucosidase include β-glucosidase and aryl-β-glucosidase. Both of them are considered to be different due to their differences in substrate specificity, genetic control of biosynthesis and physical properties.

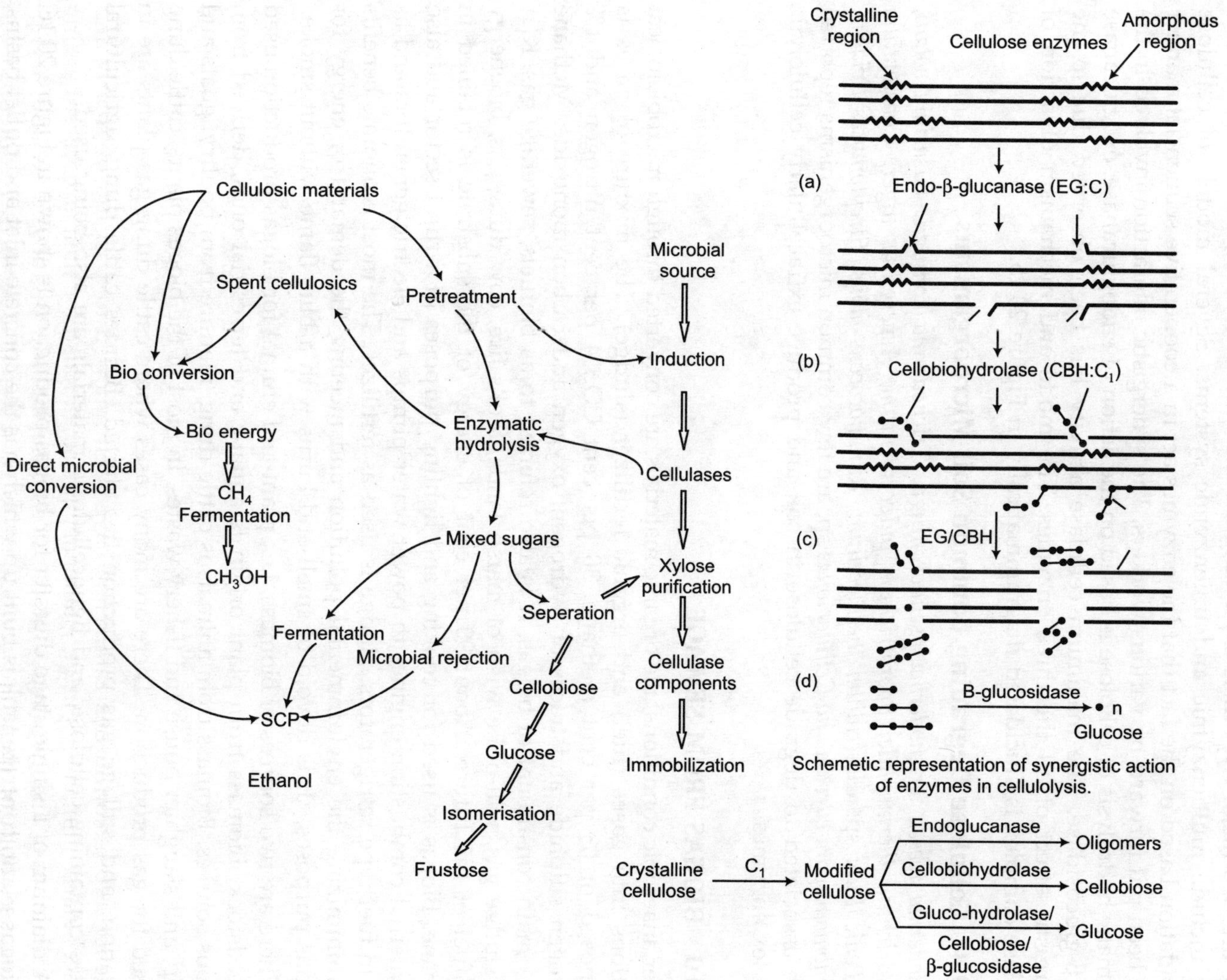

Figure 20.15 Schematic representation of various processes of microbial degradation of cellulose.

20.9.3 Mode of Action

There are two major hypotheis explaining the mechanism of cellulose biodegradation. They include multienzymic and unienzymic systems. Several models of cellulose degradation based on the fact that the enzymes act in a cooperative sequencial manner have been put forward by various workers. The synergistic interaction involved in the enzymatic hydrolysis of cellolose is based on the action of endoglucanase, exoglucanase and β-glucosidase. The maximum cellulase activity has been observed due to the synergistic effect of all the three enzymes. The mode and mechanism of action of cellulase complex is elucidated diagrammatically in Figure 20.15.

20.9.4 Cellulose Degrading Common Soil Microorganisms

Trichoderma reesei, Trichoderma pseudokoningii, Trichoderma koningii, Fusarium solani, Sporotrichum thermophile, Sporotrichum pulverulentum, Aspergillus niger, Chaetomium sp., *Penicillium* sp. *Clostridium thermocellum, Ruminococcus albus, Streptomyces lividens, Thermomonospora curvata and Cellulomonas* sp. are few common microorganisms reported to be associated to degrade cellulose in soil and produce extracellularly cellulolytic complex enzymes.

20.10 BIOGAS FROM SEWAGE

Under anaerobic conditions, the organic materials are converted through microbiological reactions into gases (fuel) and organic fertilizer (sludge). The mixture of gases is composed of 63 per cent methane, 30 per cent CO_2, 4 per cent nitrogen and 1% hydrogen sulphide and traces of hydrogen, oxygen and carbon monoxide. Methane is the main constituent of biogas. It is also referred to as biofuels, sewerage gas, Klar gas, sludge gas, will-o-the wisp of marsh lands, fool's fire, cow dung gas, bioenergy and fuel of the future. About 90 per cent of energy of the substrate is retained in methane. Biogas is used for cooking and lighting purposes in a rural sector and also for internal combustion engines to power water pumps and electric generators and as fuel in fuel type refrigerators. Sludge is used as fertilizer. The most economic benefits are minimizing the enviornmental pollution and meeting the demand of energy for various purposes. It is devoid of smell and burns with a blue flame without smoke.

There are two sources of biomass, i.e., plant and animal for biogas production used as feedstock. Biomass from plant origin is aquatic and terrestrial ones, derived from various sources. Biomass from animals is cattle dung, manure from poultry, goats and sheep, and slaughter house and fishery wastes. In most of the biogas plants, cattle dung is used for gas production. There are many cases where cattle dung gas plants are in operation, and sullage gas production has started. Besides cattle dung, agricultural wastes containing cellulosic and lignocellulosic materials are also being used.

A diagram of a single stage digester for biogas production is shown in Figure 20.16. In biogas plant, concrete tank is built up which has the concrete inlet and outlet basins. Fresh cattle dung is deposited into a charging pit, which leads into the digestion tank.

Dung remains in the tank. After 50 days, sufficient amount of gas is accumulated in the gas tank, which is used for household purposes. Digested sludge is removed from the basin and is used as organic fertilizer. Usually, digesters are buried in the soil in order to benefit from insulation provided by soil. In cold climate, digester can be heated by the installations provided from composting for the agricultural wastes. Anaerobic digestion is accomplished in 3 stages, solubilization, acidogenesis and methanogenesis. These stages are characterized by 3 groups of bacteria.

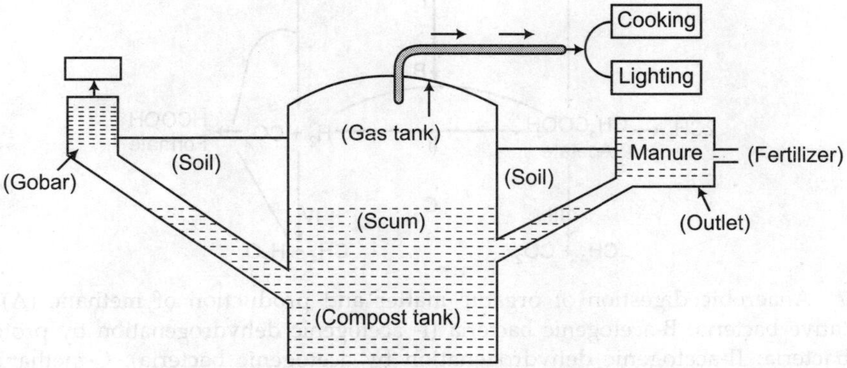

Figure 20.16 Illustration of cattle dung gas plant.

20.10.1 Solubilization

It is the initial stage, when feedstock is solubilized by water and enzymes. The feedstock (cattle dung and other organic polymers) is dissolved in water to make the slurry. The complex polymers are hydrolyzed into organic acids and alcohols by hydrolytic fermentative bacteria which are normally anaerobes (Figure 20.17a).

20.10.1.1 Acidogenesis

During this stage, the second group of bacteria, i.e., facultative anaerobic and H_2 producing acidogenic bacteria convert the simple organic material via oxidation/reduction reactions into acetate, H_2 and CO_2. These substances serve as food for the final stage. Fatty acid is converted into acetate, H_2 and CO_2 via acetogenic dehydrogenation by obligate H_2 producing acetogenic bacteria. There is other group of acetogenic bacteria which produce acetate and other acids from H_2 and CO_2 via acetogenic hydrogenation (Figure 20.17b I and II).

20.10.1.2 Methanogenesis

This is the final stage of anaerobic digestion where acetate and H_2 plus CO_2 are converted by methane producing bacteria (methanogens) into methane, carbon dioxide, water and other products (Figure 20.17c III and IV).

Different species of methanogens are involved in break-down of complex organic matter into acetate or other organic acids. Acetate is one of the substrates of

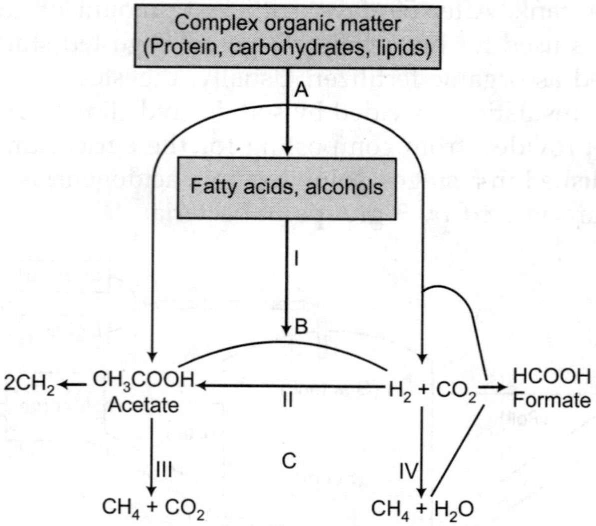

Figure 20.17 Anaerobic digestion of organic matter and production of methane (A) Hydrolytic and fermentative bacteria; B-acetogenic bacteria (I- acetogenic dehydrogenation by proton reducing acetogenic bacteria; II-acetogenic dehydrogenation by acetogenic bacteria); C-methanogenesis by acetoclastic methanogens (i.e., acetate respiratory bacteria) (III, and hydrogen oxidizing methanogens IV)

methanogenic bacteria. Hydrogen with CO_2 is a general substrate for methanogenesis. Numbers of these bacteria differ with the type of substrates.

Methanogens. Methanogens are a unique group of bacteria. They are obligate anaerobes and have slow growth rate. They play a major role in the breakdown of substrate into gas form. They are the only organisms which can anaerobically catabolize acetate and H_2 to gaseous products in the absence of exogenous electron acceptors other than CO_2 or light energy. In their absence, effective degradation would cease because of the accumulation of non-gaseous, reduced fatty acid and alcohol products of the fermentative and other H_2 using bacteria that have almost the same energy content as the original organic matter. In morphology, they are of different types such as cocci, bacillii, spirilli and sarcinea and have been classified as below:

Order 1. Methanobacteriales

Family. Methanobacteriaceae

Methanobacterium formicicum

M. bryantii

M. thermoautotrophicum

Methanobrevibacter ruminantium

M. arboriphilus

M. smithii

Order 2. Methanococcales

Family. Methanococcaceae

Methanococcus vanielii

M. voltae

Order 3. Methanomicrobiales

Family 1. Mathanomicrobiaceae

Methanomicrobium mobile

Methanogenium cariaci

M. marisnigri

Methanospirillum hungatei

Family 2. Methanosarcinaceae

Methanosarcina barken

They all require H_2 and formate (except *M. bryantii, M. thermoautotrophicum* and *M. arboriphilus)* for growth and methane production, whereas *M. barkeri* requires (besides H_2) methanol (CH_3OH), methyl amine (CH_3NH_2) and acetate for their growth. Thus, the methanogens are either autotrophs or utilize simple organic compounds as formate, acetate, and methyl amine and occupy the terminal position in anaerobic food chain.

Mechanism of Methane Formation. Metabolically the methanogens are very peculiar. Carbon dioxide fixation, Calvin cycle, serine or hexulose pathway are absent in them. The mechanism of methane formation is not well understood. Several coenzymes, which are not present in any other group of bacteria, are involved in the process. These coenzymes are methyl coenzymes M, hydroxymethyl coenzyme M, coenzyme F420, coenzyme F430, component B, corrinoids, methanofuran or carbon dioxide reducing factor and methanopterin and formaldehyde activating factor. Primary reaction in which carbon monoxide takes part is as below:

$$CO + H_2O \longrightarrow CO_2 + H_2$$

The secondary reaction takes place in the presence of sufficient hydrogen:

$$CO_2 + 4H_2 \longrightarrow CH_4 + 2H_2O$$

Other reactions showing methane formation from various substrates are given below:

$$4CH_3OH \longrightarrow 3CH_4 + CO_2 + 2H_2O$$

(Methanol)

$$4HCOOH \longrightarrow CH_4 + 3CO_2 + 2H_2O$$

(Formate)

$$CH_3COOH \longrightarrow 12\ CH_4 + 12\ CO_2\text{-}$$

(Acetate)

20.10.1.3 Factors affecting methane formation

Following are the factors affecting methane production:

Slurry Proper solubilization of organic materials (the ratio between solid and water) should be 1:1 when it is the household things.

Seeding In the beginning seeding of slurry with the small amount of sludge of another digester activates methane evolution. Sludge contains acetogenic and methanogenic bacteria.

pH For the production of sufficient amount of methane, optimum pH of digester should be maintained between 6-8 as the acidic medium lowers methane formation.

Temperature Fluctuation in temperature reduces methane formation, because of inhibition in growth of methanogens. In the case of mesophilic digestion, temperature should be between 30°C and 40°C but in case of thermophilic ones, it should be between 50°C and 60°C.

Nitrogen Concentration Excess amount of nitrogen inhibits growth of bacteria, and thereby lowers methane production. Therefore, use of such materials should be discouraged.

Carbon-nitrogen (C: N) ratio Improper C:N ratio lowers methane production. Maximum digestion occurs when C:N ratio is 30:1. Amendment of nitrogen or carbon substrates should be done exogenously according to the chemical nature of substrates used in fermentation.

Creation of Anaerobic Conditions It is obvious that methane production takes place in strictly anaerobic condition; therefore; the digesters should be totally airtight. In Indian conditions, digesters are buried in the soil.

20.10.1.4 Biogas production from different feedstocks

Although, there are various types of biomaterials differing in chemical composition are cow dung, cattle waste, agricultural wastes, municipal wastes, industrial wastes and aquatic biomass.

Water Hyacinth (*Eichornia crassipes* and several aquatic weeds)

This weed can be used for the production of laboratory and domestic fuel. Its decomposition rate is higher than that of cow dung for gas production. Water hyacinth is totally decomposed within 3 days in summer while cow dung takes 8 days. The ratios between total gas evolved by water hyacinth and cow dung under the identical conditions in summer and winter in India are 5:3 and 5:1, respectively. Gas produced from water hyacinth by methanogens in an airtight tank contains 70 per cent methane and 30 per cent CO_2. Carbon dioxide can be removed before its use for cooking, heating, lighting and source of power for engines, tractors and cars. Increase in recreational value of waterbodies from which hyacinth is to be removed can be an additional advantage of exploiting it for gas production. The hyacinth biomass contains substantial amount of nutrients, the removal of which may keep a waterbody clean and less eutrophic.

20.11 MICROORGANISMS USED AS SINGLE CELL PROTEIN (SCP)

Many groups of microorganisms are used as sources of proteins. Some of the microorganisms with their carbon and energy sources are given in Table 20.3.

Table 20.3 Single cell protein (SCP) produced on the selected substrates

Microgialmicroorganisms groups		*Protein (% 100 g)., on dry weight basis*	*Substrates*
Algae	*Chlorella pyrenoidosa*	–	CO_2 (10%). light
	Scenedesmus acinus	–	CO_2 sunlight
	Spirulina maxima	53	CO_2 (5%). combustion gases, bicarbonate, sunlight (in pond)
Bacteria	*Achromobacter delvacvate*	–	Diesel oil in fermenter
	Bacillus megaterium	–	Collagen meat packing waste in fermenter
	Cellulomonas sp.	87	Bagasse
	Methylomonas clara	13	Methanol
	Pseudomonas sp.	–	n-alkanes fuel oil
Actinomycete Fungi	*Nocardia* sp.	–	n-alkanes
	Thermomonospora fusca	5.6	Cellulose pulp
I. Yeasts	*Candida lipolytica*	65-69	n-alkanes
	C. utilis	54	Potato starch waste
	B. utilis	53	Sulphite liquor
	Saccharomyces cerevisiae	45	Molasses
	Saccharomyces cerevisiae	54	Beer
	Saccharomyces fragilis		Milk whey
	Rhodotorula glutinis	–	Domestic sewage
	Torulopsis sp.	–	Methanol
II. Molds	*Aspergillus niger*	50	Molasses
	Trichoderma viride	64	Straw. starch
	Paecilomyces varioti	55	Sulphite waste liquor
III. Mushrooms	*Agaricus campestris*	36-45	Glucose, palnt wastes
	Morchella crassipes	31	Glucose, cheese whey, sulphite. liquor.

20.12 MUNICIPAL SEWAGE

Human activities generate a tremendous volume of sewage and wastewater that require treatment before discharge into waterways. Often this wastewater contains excessive amounts of nitrogen, phosphorus, and metal compounds, as well as organic pollutants that would overwhelm waterways with an unreasonable burden. Wastewater also contains chemical wastes that are not biodegradable, as well as pathogenic microorganisms that can cause infectious disease.

20.12.1 Solid Waste

The rapid urbanization and change in lifestyle has increased the waste load and thereby pollution loads on the urban environment to unmanageable and alarming proportions. The existing waste dumping sites are full beyond capacity and under unsanitary conditions leading to pollution of water sources, proliferation of vectors of communicable diseases, foul smell and odours, release of toxic metabolites, anaesthetic ambience and eyesore, etc. It is difficult to get new dumping yards and open dumping is prohibited by law. This is particularly true for countries, with severe constraints of land availability, dense population, environmental fragility and expectation for management of solid wastes relies on an overly centralized approach. In earlier days, municipal wastes, comprised mainly of biodegradable matter, did not create much problem to the community as the quantity of wastes generated was either recycled/reused directly as manure or was within the assimilative capacity of the local environment. The biodegradable waste of the urban centres was accepted by the suburban rural areas for composting in the agricultural fields. With increasing content of plastics and non-biodegradable packaging materials, municipal wastes became increasingly unacceptable to cultivators. As a result, the excessive accumulation of solid wastes in the urban environment poses serious threat.

Microorganisms are the agents which bring about the conversion of these wastes into useful products like fuel gases, fuel alcohol and also compost which can be used as manure. The gases produced in landfills due to decomposition by anaerobic organisms also can be used as a source of energy. The major problem today in producing energy from waste is the cost factor. Efforts are being made to produce genetically modified organisms which will produce energy from wastes more efficiently and at least cost.

This decomposition process may be aerobic or anaerobic. Solid waste decomposition is carried out by bacteria decomposing complex organic materials to simple water soluble organic compounds. These are then converted to CO_2 and H_2O aerobically, or to CH_4 anaerobicaly. Fungi are mostly aerobic and feed on decaying organic matter. Soil fungi play a vital role in stabilizing solid wastes in composting and land filling processes by decomposing plant tissues like cellulose and lignin. Protozoa are predators on bacteria. They are found wherever bacteria are prevalent. Thus they help to maintain the equilibrium of microbial flora in solid waste disposal systems.

The solid waste is placed on slow moving conveyer belts. Materials like corrugated paper are handpicked and then ferrous materials are removed by magnetic separation. Thus the materials which are not easily biodegradable are separated. The waste is then ground in hammer mills and converted to pulp. Then it is mixed with nutrient source, water and fillers. Nutrient sources like sewage sludge, night soil or animal manure is used. Wood chips or ground corn cobs may be used as fillers. Water is added to maintain 50% moisture. Many Indian compost cycles do not include fillers material and other nutrient sources.

20.12.2 Wastewater Teatment and Filling

The chemical and biological waste in sewage and water must be broken down before it is deposited to the soil and environment. This breakdown can effectively be controlled by managing the microbial population in waters and encouraging microorganisms to digest the organic matter. The water must then be purified before it is considered fit to drink. Water taken from ground sources must also be treated before consumption.

To purify water for drinking, a number of processes are conducted to reduce the microbial population and maintain that population at a safe level. First, the solid matter is allowed to settle out in a sedimentation tank. Flocculating materials such as alum are used to drag microorganisms to the bottom of the tank.

Then the filtration process is begun. Water is filtered through either a slow sand filter or a rapid sand filter. These processes remove 99 per cent of the microorganisms. The slow sand filter is composed of finer grains of sand, and the filtration process takes longer than in the rapid sand filter, where larger grains are used.

Many communities then purify the water by chlorination. When added to water, chlorine maintains the low microbial count and ensures that the water remains safe for drinking purposes. Chlorine gas or hypochlorite (NaOCl) is used for chlorination purposes. The water is chlorinated until a slight residue of chlorine remains. Sewage treatment involves a more complex set of procedures than are needed for water purification because the volume of organic matter and the variety of microorganisms are much greater.

The first treatment, or primary treatment, of sewage and wastewater involves the removal in settling tanks of particulate matter such as plant waste. The solids that sediment are strained off, and the sludge is collected to be burnt or buried in landfills. Alternatively, it can be treated in an anaerobic sludge-digesting tank.

During the secondary treatment of wastewater and sewage, the microbial population of liquid and sludge waste is reduced. In the anaerobic sludge digester, microorganisms break down the organic matter of proteins, lipids, and cellulose into smaller substances for metabolism by other organisms. Results of these breakdowns include organic acids, alcohols, and simple compounds. Methane gas is produced in the sludge tank, and it can be burnt as a fuel to operate the waste treatment facility. The remaining sludge is incinerated or buried in a landfill, and its fluid is recycled and purified (Figure 20.18).

Primary treatment is represented by the steps preceding secondary treatment, and tertiary treatment is performed in the chlorination tank at the conclusion of the process.

In aerobic secondary sewage treatment, the fluid waste is aerated and then passed through a trickling filter. In this process, the liquid waste is sprayed over a bed of crushed rocks, tree bark, or other filtering material. Colonies of bacteria, fungi, and protozoa grow in the bed and act as secondary filters to remove organic materials. The

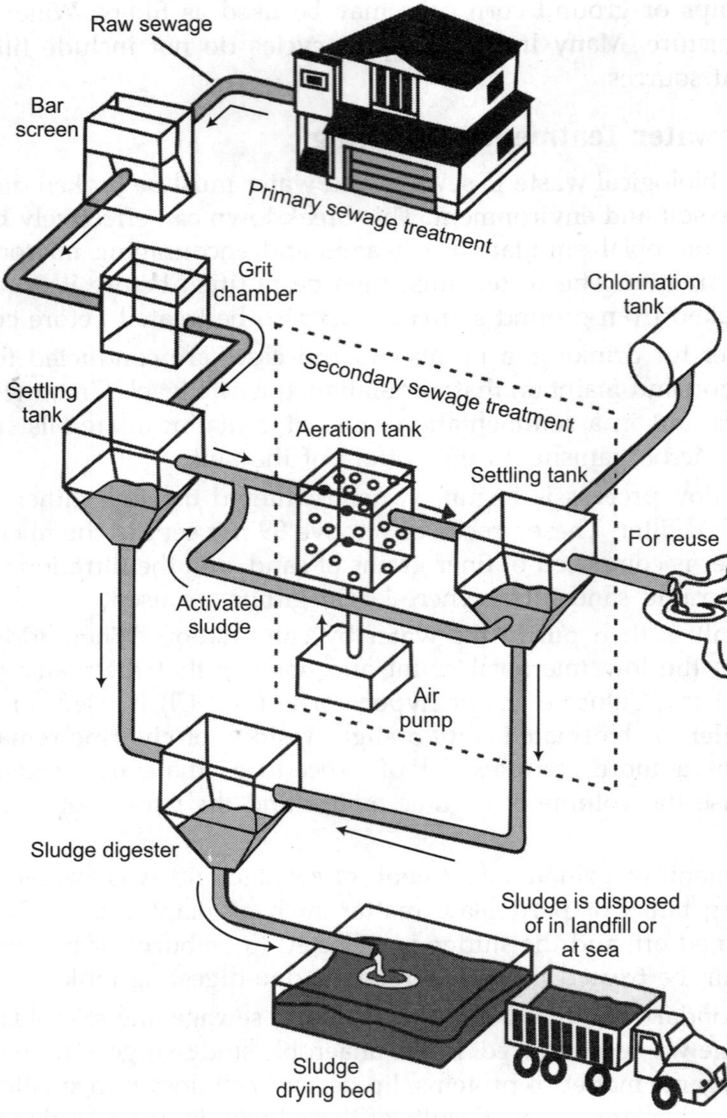

Figure 20.18 Methods used in sewage treatment in a large municipality.

microorganisms metabolize organic compounds and convert them to carbon dioxide, sulfate, phosphates, nitrates, and other ions. The material that comes through the filter has been 99% cleansed of microorganisms.

Liquid waste can also be treated in an activated digester after it has been vigorously aerated. Slime-forming bacteria form masses that trap other microorganisms to remove them from the water. Treatment for several hours reduces the microbial population significantly, and the clear fluid is removed for purification. The sludge is placed in a landfill or at sea.

In the tertiary treatment of sewage, the fluid from the secondary treatment process is cleansed of phosphate and nitrate ions that might cause pollution. The ions are precipitated as solids, often by combining them with calcium or iron, and the ammonia is released by oxidizing it to nitrate in the nitrification process. Adsorption to activated charcoal removes many organic compounds such as polychlorinated biphenyls (PCBs), a chemical pollutant.

The home septic system is a waste treatment facility on a small scale. In a septic tank, household sewage is digested by anaerobic bacteria, and solids settle to the bottom of the tank. Solid waste is carried out of the outflow apparatus into the septic field beneath the ground. The water seeps out through holes in tiles and enters the soil, where bacteria complete the breakdown processes. A similar process occurs in cesspools, except that sludge enters the ground at the bottom of the pool and liquids flow out through the sides of the pool.

20.12.3 Vermicomposting

Vermicomposting is the phenomenon of compost formation by earthworms. Obviously, earthworms play an important role in the cycling of plant nutrients, turnover of organic matter and maintenance of soil structure. They can consume 10-20 per cent of their own biomass per day. The most important effect of earthworms in agro-ecosystems is the increase in nutrient cycling, particularly nitrogen. They ingest organic matter with a relatively wide C:N ratio and convert it to earthworm tissue with a lower C:N ratio. Food consumption varies from 36 to 69 mg/g live worm/day. Cast production ranges from 4 to 6 mg/g live worm/day. Some of the known and potential waste decomposer (such as *Drawida nepalensis*, etc.) earthworms may be introduced in such places where they are absent. *D. nepalensis* is a slow growing vermicomposting species and also shows parthenogenesis; life cycle is presented in Figure 20.19.

20.13 HAZARDS OF ENVIRONMENTAL ENGINEERING

The genetically modified microorganisms which are purposefully released into the environment for bioremediation, plant growth stimulation or biocontrol of pathogens and pests of plants and animals either do well as hoped, influence the other harmful microorganisms or themselves changed are still in debate. However, some of performance of GMMs have been discussed in this context. There are several types of risks associated with release of GMMs into the environment (Table 20.4).

Table 20.4 Possible types of risks associated with the release of GMMs.

Type	*Mechanism*	*Results*
A. Risk for humans, animals, plants	Transfer of drug resistance gene into clinical and/or symbiotic strains,	New forms of known diseases
	GMMs useful for food production	Toxins and/or unwanted physiologically active substances in food

Contd...

Contd...

B. Risk for environment	GMMs will overgrew the indigenous strain	Threat to biodiversity
	Disturbance of ecological balance	Activation of earlier unknown pathogen
	Massive transfer of foreign genes into indigenous strains	Formation of new pathogen
C. Social risk	Beneficial usage of the GMMs in industrial countries	Widen gaps in economical and social diferences between industrial and developing countries
	GMMs release for military and/or improper purpose	Industrial countries will use territories of developing countries for field tests of GMMs release
D. Ethical risk	Commercial secrecy concerning information about release of GMMs	Violation of consumers right

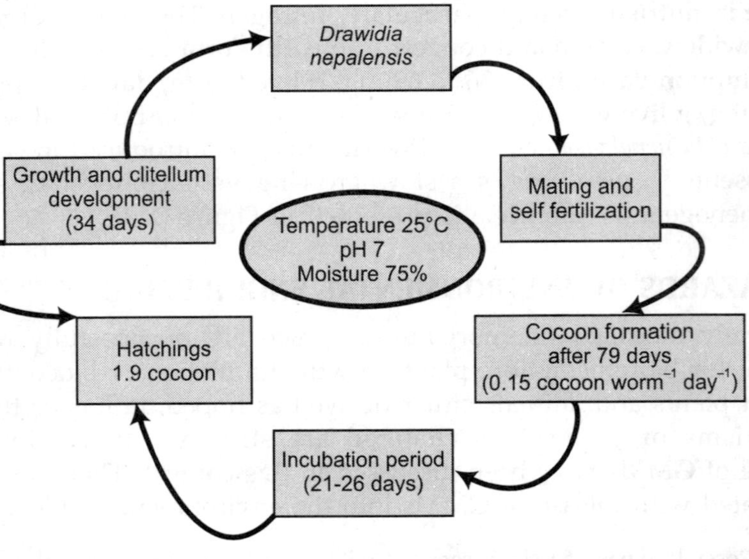

Figure 20.19 Life cycle of a vermicomposting earthwarm maintained on cow manure.

20.13.1 Survival of Released Genetically Modified Microorganisms (GMMs) in the Environment

The GMMs after release into the environment are much influenced by multiple factors. Before release GMMs live in very ideal conditions, but after release their survival is affected by both abiotic and biotic factors. Due to presence of poor quality of available nutrients their growth and multiplication declined. Substrates accessible to microorganisms in open nature are so limiting that within one year a typical soil microbe could go through 1-36 generations of growth. Therefore, under nutrient starvation conditions, non-differentiating microorganisms enter in a dormant viable state. During this course of time certain changes in redox potential, energy status and composition of proteins occur. It has been found that starvation induced proteins (unfoldases and chaperones), that protect non-growing cells of *E. coli* are produced. After some days of starvation, microorganisms do not require energy to remain viable but enter into a viable but not culturable (VNC) state. During this period they can remain viable for years, and retain their plasmid. According to physiological survival strategy of non-differentiating bacteria, there is a very high probability of entering of released GMMs into VNC state. In addition to physiological strategy of survival, the GMMs have genetical strategy of survival.

20.13.2 Adaptive Mutagenesis in GMMs

The GMMs released into the environment do not grow even in the presence of potential substrates. It seems that the non-growing cells 'adapt' themselves to the potential substrates present in the environment. It may be assumed that in populations of microorganisms growing and persisting in open environment the random mutagenesis is increasing their biodiversity, adaptive mutagenesis 'adapt' the non-growing cells to the substrates potentially present in the environment. However, if foreign genes from GMMs spread into the indigenous microbes it may result hazardous effects.

20.13.3 The GMMs Participation in Gene Transfer

There are four ways of gene transfer in microbial ecosystem, viz., transformation by free DNAs, transduction by bacteriophage DNAs, conjugation by plasmids, and conjugal transfer by conjugative transposons. Conjugal gene transfer is of the most considerable ecological importance. In an experiment the GMMs and natural soil strains were released in sterile soil. Conjugal gene transfer from GMMs to soil strains occurred with a frequency of about $3 \times 10^{-6} - 3 \times 10^{-5}$ of transconjugants per donor. The addition of glucose and/or tryptone enhanced up to 2×10^{-4}. However, in non-sterile soil natural microorganisms inhibit conjugative gene transfer. On the other hand in the rhizosphere rich in nutrients, effective gene transfer among *Agrobacterium*, *Pseudomonas* and *Rhizobium* has been observed.

There are abundance of plasmids in aquatic microbial systems. In some lake waters up to 46 per cent of heterotrophs contain plasmids, but in river water 10-15 per cent. Majority of these plasmids are conjugative. The frequency of plasmid transfer in aquatic systems varies from 5×10^{8} to 2.5×10^{-3} conjugants per donor. Moreover, very effective plasmid transfer was obtained between GMMs and strains of epilithon (aquatic microbial community associated with stones), 10^{-2} transconjugants per recipient.

20.13.4 Gene Transfer via Conjugative Transposons

The conjugative transposons are self-transmissible elements that normally are integrated into a chromosome or plasmid but excise themselves and transfer by conjugation to recipients. The conjugative transposons are capable of transferring themselves as well as driving the transmission of other elements. For example, conjugative transposons of *Bacteroides* can mobilize co-resident plasmids either in *cis* or *trans* position. The indigenous strains have an efficient capability to transfer their plasmids into GMMs released in soil or aquatic ecosystems.

20.13.5 Effect of Environmental Factors on Gene Transfer

Soil and water polluted with heavy metals and organic xenobiotics are characterized by the presence of many strains with plasmids carrying genes encoding resistance to heavy metals and biodegradation. In such polluted ecosystem selective pressure could be stimulative for conjugative transfer between GMMs and natural strains. However, the risk of spread of foreign genes by conjugation could be reduced if GMMs do not contain any conjugative plasmids.

20.13.6 Ecological Impact of GMMs Released into the Environment

The GMMs released into the environment disturb the ecological balance by inhibiting/promoting growth of indigenous mycroflora or replacing the natural strains.

20.13.7 Growth Inhibition of Natural Strains

A genetically modified strain of *Klebsiella planticola* constructed to produce ethanol from organic wastes was found *(a)* to destroy mycorrhizal fungi, *(b)* to reduce plant growth, and (c) to increase the population of parasitic nematodes of plants. A genetically modified strain of *P. putida* pR103 degraded in soil microcosm the herbicide 2,4-D (2,4-dichlorophenoxyacetic acid) and resulted in accumulation of 2,4-DCP (2,4-dichlorophenol) which in turn is toxic metabolite. The 2,4-DCP was more toxic than 2,4-D as the former reduced the rate of CO_2 evolution of soil microflora as compared to 2,4-D.

20.13.8 Growth Stimulation of Indigenous Strains

The GMMs strain of *Erwinia carotovora* affected indigenous bacterial community with the significant increase in total bacterial density. This increase is attributed to an inoculum nutrient effect. The inoculated cells of *E. carotovora* died and became a source of nutrients for indigenous soil microorganisms. More interesting finding is that the genetically modified *Pseudomonas cepacia* released in soil microcosm containing the pollutant increased the taxonomic diversity of soil microorganisms. This is because the intermediate metabolites of genetically modified *P. cepacia* produced after biodegradation of pollutant which stimulated the growth of static cells. This has resulted in increased taxonomic diversity. Stimulation of plasmid transfer and recombination might cause an increase in genetic diversity.

20.13.9 Replacement of Natural Strains

A strain of *Pseudomonas* PCI released in a wheat root system replaced efficiently the indigenous pseudomonads in their ecological niche. A genetically modified strain of

Agraobacterium tumifaciens, with the deleted genes of tumour formation replaced the initial wild strain and protected plants from tumour formation. This ability of GMMs is useful in agricultural biotechnology for protection of plants against pathogens.

20.13.10 Monitoring of GEMs in the Environment

In a microbial community, diverse group of microorganisms are found. It is now possible to monitor individual microbe in a community. Sensitive and selectable trackling tools, e.g., *ice, lacZY, xylE, lux* systems are used in monitoring studies. Nucleic acid hybridization is also useful method where rRNA specific probes can identify the specific microbial strains of interest. Bioluminiscent reporter gene has potential to identify the specific microbe, for example, enzyme luciferase acts on substrate luciferin. In addition, genes capable of giving resistance against certain antibiotics have also been used in environmental monitoring as these detoxify antibiotics such as gentamycin, ampicillin, tetracyclin, kanamycin, steptomycin, etc.

20.13.11 Risk Assessment of the GMMs Released into the Environment

When GMMs are released in the environment, they get dispersed and it becomes difficult to eliminate them. Relatively a few small scale field tests have been conducted with genetically modified bacteria (10 strains) and genetically modified viruses (5 strains). Most of the European countries, USA and Japan have environmental regulations that ban the deliberate release of GMMs. But the USA has permitted deliberate release of some of the GMMs. No country has defined what steps will be required to obtain permission for commercial use of GMMs for *in situ* bioremediation.

In July 1995, European parliament passed a resolution saying that "a legally binding international biosafety protocol is necessary". It must be negotiated by States partly on the United Nations' convention on biological diversity. Therefore, resolution of deliberate release of GMMs are being carried out in many developing countries for safe use of GMMs. Otherwise the whole biosphere will be at risk. India has formulated Biosafety Rules and Regulations for GMMs. There are two principal uncertainties of the assessment of risks associated with possible ecological consequences of release of GMMs: *(i)* total number of hazardous strains present in the environment as only 5 per cent of total species are known and rest of 95 per cent unknown, and *(ii)* absence of definite criteria for evaluation of similarity between microorganisms. However, estimates of similarity made by different experts may vary drastically. Moreover, there are certain questions unasked which may possibly help in assessment of risk arising from release of GMMs in the environment (Velkov, 1996).

20.14 EXPERIMENTS

The importance of biodegradation is established, and a number of methods for monitoring the chemical disappearance of contaminants and the activities of microorganisms in bioremediation are described. The monitoring tools needed to manage and to design more effective bioremediation technologies which are explained through following experiments.

20.14.1 Isolation of Diesel Degrading Bacterial Isolates from Contaminated Soil Sample

In natural environment, microbes occur almost always in a mixed population composed of a multitude of different strains and species. For analyzing the properties of a defined organism out of such a mixed population, a pure culture is required. One of the promising ways is the biological degradation of oil by bacteria. The bacteria metabolize the oil in much the same way humans convert food into energy. Like food, oil is a compound rich in carbon. The following experiment can be used to demonstrate that some types of bacteria can degrade oil. Apart from direct isolation of strains by diluting and plating, enrichment cultures with hydrophobic substrates are very promising for the isolation of diesel degrading microbes.

The principle of enrichment culture is to provide growth conditions that are very favourable for the organisms of interest and as unfavourable as possible for competing organisms. Hence, the microbes of interest are selected and enriched.

20.14.1.1 Materials

1. Diesel contaminated soil samples
2. Bushnell Hass broth as Enrichment medium supplemented with diesel oil (10%v/v).
3. Nutrient agar medium

***Bushnell Hass Broth* (g/L)** Monopotassium phosphate 1.0; Dipotassium phosphate 1.0; Ammonium nitrate 1.0; Ferric chloride 0.5; Magnesium sulphate 0.2 g; Calcium chloride 0.02; distilled water to 1 litre. pH adjusted to 7.0

***Nutrient agar media* (g/L):** Peptone 5.0; Beef extract 3.0; Sodium chloride 5.0; Agar 15.0; distilled water to 1 litre. pH adjusted to 7.0

20.14.1.2 Procedure

1. Collect diesel contaminated soil samples from the depth of 0-15 cm with the help of sterilized spatula in sterilized polythene bags and bring to the laboratory under ice cold condition.
2. Add 10 g soil sample in 100 mL Bushnell Haas (BH) broth supplemented with diesel 10% (v/v).
3. Incubate at 30ºC at 170 rpm in a shaker for two weeks.
4. After two weeks, transfer 1 mL of grown media into freshly prepared enrichment media and incubate at the same conditions as described in step 3.
5. Prepare serial dilutions (1/10) from the third enrichment process and plate out onto BH agar plates, which is covered with 100 µl of diesel oil and incubate at 30ºC.
6. Streak the single colony onto nutrient agar plates, incubate at 30°C overnight and store at 4°C for further use.

Interpretation **The bacterial isolates exhibiting growth on diesel containing media can be considered diesel utilizing microorganism and can be used for remediation of diesel** contaminated sits (Figure 20.20).

Diesel contaminated site

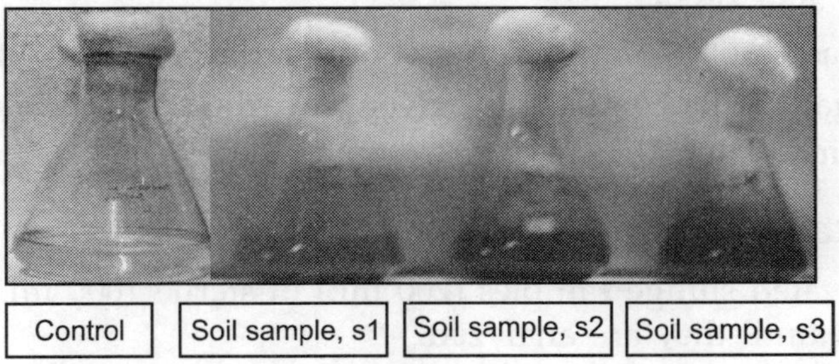

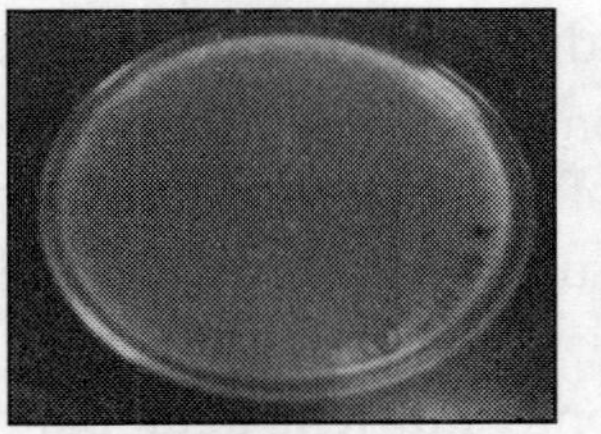

Isolation of diesel degrading bacteria on BH agar plate supplemented with 10% diesel

Figure 20.20 Isolation of diesel degrading bacteria from diesel contaminated soil sample. Enrichment of diesel degrading bacteria in Bushnell Hass broth supplemented with 10% diesel

20.14.2 Isolation of Diesel Degrading Bacterial Isolates from Wastewater Sample

The petrochemical industry is normally found in tropics, where a great amount of wastewaters are released on waters and soils. Although these wastewaters are treated to reduce contaminant organic charge, they still contain high contaminant concentrations that are discharged on the swampy zones. The wastewaters released by oil-processing and petrochemical enterprises contain large amounts of toxic derivatives, such as polycyclic and aromatic hydrocarbons, phenols, sulfides, and heavy metals. The major problems faced by oil refineries are the safe disposal of oily sludge generated during the processing of crude oil. Improper disposal of oily sludge leads to environmental pollution, and represents a serious threat to groundwater. Many of the constituents of oily sludge are carcinogenic and potent immunotoxicants. In such areas, the wastewaters remain for large periods where they make possible the development of a great quantity of native strains acclimatable to the toxic derivatives, being potentially capable of degrading hydrocarbons. This approach to reclaiming contaminated land reduces the threat to groundwater and enhances the rate of biodegradation. Many bacterial strains, each capable of degrading a specific compound, are needed commercially for bioremediation.

20.14.2.1 Matrerials

1. Activated sludge sample
2. Plate count agar (PCA),
3. Bushnell Hass broth composition given in earlier experiment.
4. Nutrient agar media composition given in earlier experiment.
5. Normal saline water (0.85%)

20.14.2.2 Procedure

1. Collect activated sludge samples (200 mL) in sterile 1000 mL bottles and keep it refrigerated until they are analyzed.
2. Remove 10 mL of activated sludge sample and homogenize in 90 mL of 0.85% (w/v) saline solution and serially dilute (10-fold).
3. Pour aliquots (0.1 mL) on centre of Petri plates and add PCA, and BH medium separately for isolation of diesel degrading bacterial isolates.
4. Count the total number of colonies on each medium, after 7 days incubation at 30°C.
5. Streak a single colony onto nutrient agar plates for isolation of pure bacterial isolates, incubate at 30°C overnight and store at 4°C for further use.

S.No.	*Wastewater samples*	*Total heterotrophic bacterial count*	*Total hydrocarbon utilizlng bacterial count*
1.			
2.			
3.			
4.			
5.			

Interpretation The total number of hydrocarbons utilizing bacterial count in wastewater sample can be used to determine the extent of occurrence of bacterial isolates having potential of biodegradation and isolation of pure culture would help in their use in bioremediation of the contaminated sites.

20.14.3 Isolation of Diesel Degrading Bacteria from Air (Syal et al., 2008)

Air is rich in microorganisms, many of which are important to humans. In the present study, a method of isolating diesel-degrading bacteria from air is devised, which is a simpler, faster and more efficient method of isolating these microorganisms compared to the conventional method of isolation from diesel contaminated soil or water through enrichment culture techniques.

20.14.3.1 Materials

1. Bushnell–Hass medium
2. Nutrient agar media

21.14.3.2 Procedure

1. For isolation of diesel-degrading bacterial isolates from air, the experiment is set up in 250 mL flasks containing 100 mL Bushnell–Hass broth, supplemented with increasing concentrations of diesel (2–8%).
2. Leave it open for isolating bacterial strains from air against a blank which is well covered.
3. The flasks are left open for 10 days and manual shaking is done at regular intervals.
4. Prepare serial dilutions (1/10) and plate out on nutrient agar medium and incubate at 30°C overnight.
5. Streak the single colony onto nutrient agar plates, incubate at 30°C overnight, obtain pure cultures and store at 4°C for further use.

Interpretation The bacterial isolates exhibiting growth on diesel containing media can be considered diesel utilizing microorganism (Figure 20.21).

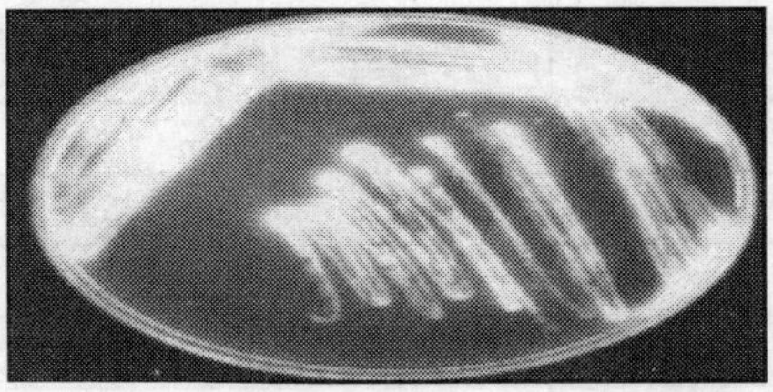

Figure 20.21 Diesel degrading bacterial isolate from air.

20.14.4 Determination of Growth Potential of Diesel Degrading Bacterial Isolates at Different Concentrations of Diesel Oil

An increase in diesel degradation corresponded to an increase in cell number during the degradation processes demonstrate the isolate's ability to utilize diesel as an energy source. Optical density, measured in a spectrophotometer, can be used as a measure of the concentration of bacteria in a suspension. As visible light passes through a cell suspension the light is scattered. Greater scattering indicates the presence of higher bacterial cell concentration.

20.14.4.1 Materials

1. Bacterial strains
2. Bushnell-Hass broth
3. Nutrient broth
4. 0.85% normal salin (sterilized)
5. Commercial diesel oil

20.14.4.2 Procedure

1. Inoculate pure bacterial isolates in nutrient broth and incubate it for 24 hr at 37°C.
2. Centrifuge the broth at 10,000 rpm for 10 min.
3. Discard the supernatant and wash the cell pellet twice with 0.85% normal saline.
4. Resuspend the cell pellet in BH broth until A_{600} is equivalent to 1.2.
5. Use this inoculum for inoculation of BH broth supplemented with diesel oil (5%, 10%, 15%, 20% v/v). Prepare two control flasks one devoid of bacterial inoculum and another containing bacterial inoculum but without supplement of a diesel as carbon and energy source.
6. Incubate all these flasks at 30°C undershaking (120 rpm) for 10 days
7. Determine growth potential of each isolate by taking absorbance at 600 nm from Day 0 upto Day 10.

Observation Table:

S.No	*Incubation days*	*Bacterial isolate A*	*Bacterial isolate B*	*Bacterial isolate C*	*Bacterial isolate D*	*Bacterial isolate E*
1.	Day 0					
2.	Day 1					
3.	Day 2					
4.	Day 3					
5.	Day 4					
6.	Day 5					
7.	Day 6					
8.	Day 7					
9.	Day 8					
10.	Day 9					
11.	Day 10					

Interpretation The bacterial isolate exhibiting maximum growth in diesel could be a potential candidate for bioremediation of diesel contaminated sites.

20.14.5 Determination of Degradation Potential of Bacterial Isolates in Different Concentrations of Diesel Oil

Oil eating microbes (OEMs) which are strains of bacteria that naturally use oils in the environment as their food source. Oil is made up mostly of hydrocarbon, which OEMs consume. The basic process of cleaning up an oil spill involves applying a special OEMs containing solution to the spill. First, the solution breaks down the oil to molecule size, thus increasing the source area. The increase into surface area begins the oxygenation process; this revives dormant microbes so they will begin feeding on the hydrocarbons. Nutrients in the OEMs solution help these activated microbes to survive. There are three basic types of hydrocarbons, straight chains, branched chains and six-member ring. The OEMs break down all three of these hydrocarbons into fatty acids or carboxylic acid, which are then further broken down for energy and carbon atoms. They are used in the citric acid cycle to generate energy. Thus, oil is broken down into basic, non-toxic elements, carbon-dioxide and water. OEMs require air, water and a source of nutrients such as oils.

20.14.5.1 Materials

1. Bacterial strains
2. Bushnell-Hass broth, Nutrient broth, 0.85% Normal saline
3. Commercial diesel fuel
4. Solvents; Chloroform, Dichloromethane, Hexane

20.14.5.2 Procedure

1. Inoculate pure bacterial isolates in nutrient broth and incubate it for 24 hr at 37°C.
2. Centrifuge the broth at 10,000 rpm for 10 min.
3. Discard the supernatant and wash the cell pellet twice with 0.85% normal saline.
4. Resuspend the cell pellet in BH broth until A_{600} is equivalent to 1.2.
5. Use this inoculum for inoculation of BH broth supplemented with diesel oil (5%, 10%, 15%, 20% v/v). Prepare two control flasks one devoid of bacterial inoculum and another containing bacterial inoculum but without supplementation of a diesel as carbon and energy source.
6. Incubate all these flasks at 30°C undershaking (120 rpm) for 10 days
7. Determine degradation potential of each isolate by analysis of residual diesel from Day 0 upto Day 10 through partition gravimetric method.

Residual diesel analysis

(a) Centrifuge the culture grown broth at 10,000 rpm for 10 min and separate the aqueous phase.

(b) Acidify the aqueous phase with 0.1N HCl upto 2 or below.

(c) Shake the aqueous phase with equal volume of each solvent (Chloroform, dichloromethane and hexane). Shake vigorously for 5 min and extract the solvent layer from separatory funnel to flask. (Repeat it with all the three solvents.)

(d) Carry out the distillation in distilling flask for all the three solvents in water bath at 70°C for 15 min.

(e) After the solvent has evaporated, if the residue contains visible water, add 2 mL acetone, evaporate on water bath and repeat the addition and evaporation step until all visible water has been evaporated.

(f) Allow the residue to cool down for 3 min and calculate the percentage of diesel degraded with following formula.

$$\% \text{ diesel on degraded} = \frac{\text{Wt. of diesel degraded}}{\text{Original weight of diesel oil introduced}} \times 100$$

(g) Where, the weight of diesel oil degraded is determined as original weight of diesel oil-weight of oil remaining.

Observation Table:

S.No.	*Bacterial Isolate*	*% diesel degradation*							
		5%		*10%*		*15%*		*29%*	
		0 day	10th day	0 day	10th day	0 day	10th day	0 day	10th day
1.									
2.									

Interpretation The bacterial isolate degraded maximum degradation of diesel could be the potential for bioremediation of contaminated sites.

20.14.6 Immobilization of Diesel Degrading Bacteria by Sodium Alginate

Immobilized cells have been used and studied extensively for the production of useful chemicals, for the treatment of wastewaters and for the bioremediation of contaminated soil caused by numerous toxic chemicals. Immobilization not only simplifies separation and recovery of the immobilized bacteria and the binding agent, but it also makes the application reusable, which reduces the overall cost. Immobilized materials, furthermore, have comparatively longer operating lifetimes due to an enhanced stability of the macromolecules or cells and, consequentially, protection from adverse conditions. Immobilization of bacterial cells significantly enhance the biodegradation rate of crude oil compared to free-living cells.

Immobilization is known to reduce competition with indigenous microorganisms, offer protection from predation, extremes of the pH and toxic compounds in the contaminated soil. There is also evidence of increased biological stability including plasmid stability in immobilized cells. Immobilization also provides the added benefit of reducing the cells' exposure to the elements, thus preventing them from being washed away, particularly in an open water system.

20.14.6.1 Materials

1. ***Tryptic soya broth*** (g/L) Tryptone 17 g, Peptone 3 g, Sodium chloride 5 g, Di potassium hydrogen phosphate 2.5 g, Dextrose 2.5 g, Distilled water 1000 mL pH adjusted to 7.0
2. 0.85% Normal saline
3. Sterile sodium alginate solution (4%)
4. Sterile 0.2 M $CaCl_2$

20.14.6.2 Procedure

1. Grow the bacterial isolates in separate culture flasks using 200 mL tryptic soya broth at 37°C and 150 rpm for 18-24 hours.
2. Harvest the cells by centrifugation (10.000 rpm for 10 minutes), rinse three times with sterile saline solution (0.85% NaCl), and resuspend in 150 mL sterile sodium alginate (4%), use this suspension as the inoculum for the immobilization matrix.
3. Test two different cell concentrations by diluting 50 mL (1.8×10^{10}cfu/mL) or 100 mL (3.7×10^{10} cfu/mL) in 200 mL of 4% (w/v) sterilized alginate solution.
4. Stir the suspension until a homogenous mixture is formed. Drip the resulting alginate/cell mixtures into ice cold, sterile $CaCl_2$ solution (0.2 M) under constant stirring. Beads of approximately 2 mm diameter are formed during the process.
5. Resuspend the beads in fresh $CaCl_2$ solution with gentle agitation for 2 h.
6. Add 10 g of the immobilized bacteria to the 250 mL of Bushnell-Hass media containing 5% diesel as substrate.
7. Incubate at 37°C for 150 rpm up to 30 days.
8. After incubation, wash the alginate beads with sterile 0.85% saline solution thrice.
9. The degradation process is identical to one used during the fermentation.

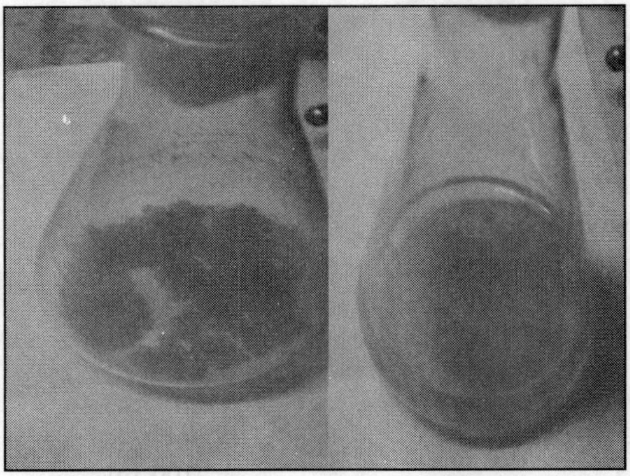

Figure 20.22 Immobilized diesel oil degrading bacterial cells.

10. After incubation determine the residual diesel oil level by using a modified partition gravimetric method.
11. To test the stability of biodegradation capacity overtime, subject the immobilized hydrocarbon degrading bacterial cells to repeated cycles of fermentation (a series of five 30 day periods) (Figure 20.22).
12. Calculate the per cent diesel degradation as described in earlier experiment

Interpretation

The ability to maintain the capacity for biodegradation for extended periods of time and during repeated use suggest that Ca^{2+} alginate immobilization provides a certain level of membrane stabilization and increased cell permeability, which is conceivably conducive to the bacterial degradation capacity on account of improved cell protection.

Observation Table:

No. of cycles	*I Cycle*	*II Cycle*	*III Cycle*	*IV Cycle*	*V Cycle*
0/0 degradation of diesel					

20.14.7 Assay of Biofilm Formation for Diesel Degrading Bacteria Using Microtiter Plate

This experiment system, commonly referred to as the 96-well plate assay, is a simple high-throughput method used to monitor microbial attachment to an abiotic surface. In brief, cells are grown in microtiter dishes for a desired period of time, and after incubation, the wells are washed to remove any planktonic bacteria. Cells remaining adhered to the wells are subsequently stained with a dye that allows visualization of the attachment pattern. This surface-associated dye can also be solubilized for semiquantitative assessment of the biofilm formed.

20.14.7.1 Materials

Bacterial isolate

Bushnell-Hass medium

0.1 M Phosphate Buffer

Sodium dihydrogen phosphate	15.601 g
Disodium hydrogen phosphate	14.196 g
Distilled water	1000 mL
pH	7.0

***Saline Water* (0.85%)**

Sodium chloride	0.85 g
Distilled water	100 mL

***Crystal Violet Solution* (25%)**

Crystal violet	25 g
Distilled water	100 mL

33% glacial acetic acid,
95% v/v ethanol for solubilizing dye and biofilm biomass
96-well microtiter plates (with lids)
Commercial diesel oil

20.14.7.2 Procedure

1. Inoculate bacterium of interest in a 3-5-mL Bushnell-Hass broth supplemented with diesel (1% v/v) and incubate for 3 days at 30°C under shaking at 120 rpm.
2. Pipette 50 µl of each culture into wells in a microtiter plate containing 200 µl Bushnell Hass medium supplemented with diesel (5% v/v).
3. Cover the plate and incubate at 30°C temperature for two weeks.
4. After two weeks remove the medium carefully with a micropipette.
5. Remove planktonic bacteria from each microtiter dish rinsing it with 100 µl Phosphate buffer saline and briskly shaking the dish out over the waste tray. To wash wells, submerge the plate in the water tray and then vigorously shake out the liquid over the waste tray. Replace buffer when it becomes cloudy.
5. Add 100 µl of 25% crystal violet solution to each well. Stain 10 min at room temperature.
6. Shake each microtiter dish out over the waste tray to remove the crystal violet solution.
   This step will remove any crystal violet that is not specifically staining the adherent bacteria.
7. Invert each microtiter dish and vigorously tap on paper towels to remove any excess liquid. Allow plates to air-dry. At this stage, the staining is stable and the dried plates may be stored at room temperature for at least several weeks.
8. Add 100 µl of 33% of glacial acetic acid to each stained well. Allow dye to solubilize by covering plates and incubating 10 to 15 min at room temperature.
9. Measure the optical density (OD) of each of these samples at a wavelength of 600 nm in an ELISA reader.

Interpretation of Result

The microtiter plate biofilm assay is repeated on three separate plates for all bacterial strains, and the averages and standard deviations for each strain are calculated. Based on OD value, the investigated strains are classified into categories as following; strains that do not produce biofilm, weak, moderate and strong biofilm producers (Figure 21.23). Cut off OD value is defined as three standard deviations (SD) above the mean OD of negative control (uninoculated medium); ODc= average OD of negative control + 3 × SD of negative control. Final OD value of tested strain is expressed as average

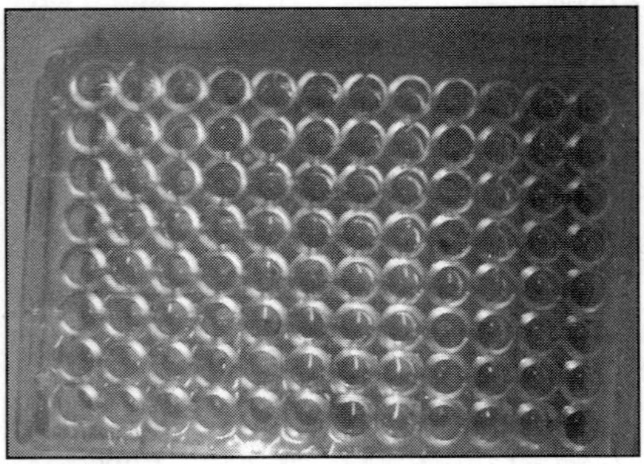

Figure 20.23 Microtiter plate showing formation of biofilm by bacterial isolates.

OD of a strain reduced by ODc value (OD= average OD of a strain-ODc). ODc value is calculated for each microtiter plate separately. The strains are classified as follows.

1. Strains that do not produce a biofilm OD≤ODc
2. Weak biofilm producers ODc<OD<(2XODc)
3. Moderate biofilm producer (2XODc)<OD≥(4XODc)
4. Strong biofilm producer (4XODc)<OD

20.14.8 Isolation of Azo Dye Degrading Bacteria from Textile Effluent

Many of the known bacterial isolates that could decolorize azo dyes can only do so under static or anaerobic conditions. In addition, growth and decolorizing ability are often a two-stage process. Azo dyes are the largest class of commercially available synthetic dyes and have found wide applications in textiles, food, cosmetics, plastic, laboratories, leather, paper printing, colour photography, pharmaceutical and toy industries. Their widespread use, especially in textile and dyestuff industries, coupled with the fact that azo dyes are not readily degraded in conventional aerobic treatment systems makes this class of xenobiotic a significant environmental problem. The employment of physico-chemical methods, although effective, is cost prohibitive and often lead to extra solid wastes. As a viable alternative, biological processes have received increasing interest owing to their cost effectiveness. Over the past decades, many microorganisms that are capable of degrading azo dyes, including bacteria have been documented.

20.14.8.1 Materials

Activated sludge sample, azo dye

Growth medium (g/L)

NaH_2PO_4 3.5, K_2HPO_4 5, $(NH_4)_2SO_4$ 2.5, $MgSO_4.7H_2O$ 0.3, $FeSO_4$ 0.05 mg, $CuSO_4$0. 01 mg, $ZnSO_4$ 0.005 mg, $CoCl_2$ 0.005 mg, $MnCl_2$ 0.005 mg, $CaCl_2$ 0.005 mg, Na_2MoO_4 0.005 mg, Peptone 5, Yeast extract 5, NaCl 5.

20.14.8.2 Procedure

Screening of azo-dye degrading bacteria

1. Collect the activated sludge sample from wastewater treatment plant in textile factory as a source of high performance bacterial decolourizer.
2. Acclimatize the enrichment bacterial culture from the sludge samples for 1 month in the presence of azo dye and use it as a stock culture.
3. Isolate the pure bacterial cultures in a screening medium supplemented with azo dye (50 mg/L) using pour method.
4. Collect the colonies around which clean zones expanded for further experiments.
5. Cultivate the single strains in 500 mL flasks containing 150 mL growth medium at 30°C in a rotary shaker for 24 hours.

Reduction of azo dye by bacterial strain

1. Grow aerobically to the stock culture for 24 hours at 30°C in a rotary shaker.
2. Transfer the aerobically grown culture to plastic tubes (15 mL).
3. Add the azo dye at different concentrations of azo dye ranging from 100 to 1500 mg and fill the tubes completely with the bacterial culture sealed with screw caps and incubate at 30°C.
4. Centrifuge the grown culture at 10,000 g for 15 min, decant the supernatant and use it for determination of dye concentration.
5. Determine the concentration of dye in the supernatant by reading absorbance at 520 nm with UV-VIS spectrophotometer. Compare this absorbance with standard curve plotted using different concentrations of the dye. Conduct all assays in triplicates with non-inoculated controls.

Interpretation

Interpret data on the basis of results obtained.

Observation Table:

S.No	*Initial concentration of azo dye (mg/mL)*	*Decolorization (%)*		
		After 24 h of incubation	*After 48 h of incubation*	*After 72 h of incubation*
1.				
2.				
3.				
4.				
5.				
6.				
7.				
8.				

20.14.9 Screening of Biosurfactant Producing Bacterial Isolates from Diesel Contaminated Soil

"Biosurfactants are microbially produced surface-active compound. They are amphipilic compounds with both hydrophilic and hydrophobic region causing them to aggregate at interphases between fluids with different polarity such as water and hydrocarbon." Biosurfactants enhance the emulsification of hydrocarbons, have the potential to solubilise hydrocarbon contaminants and increase their availability for microbial degradation. The use of chemicals for the treatment of a hydrocarbon polluted site may contaminate the environment with their by-products, whereas biological treatment may efficiently destroy pollutants, while being biodegradable themselves. Biosurfactant producing microorganisms, therefore, may play an important role in accelerating bioremediation of hydrocarbon from contaminated sites.

20.14.9.1 Materials

Mineral salt media (Bodour and Miller-Maier, 1998): The MSM was a mixture of solution A and solution B. Solution A contained (in g/L): 2.5 of $NaNO_3$, 0.4 of $MgSO_4\,7H_2O$, 1.0 of NaCl, 1.0 of KCl, 0.05 of $CaCl_2 2H_2O$, and 10 mL of concentrated phosphoric acid (85%). This solution was adjusted to pH 7.2 with KOH pellets. Solution B contained (in g/L): 0.5 of $FeSO_4 7H_2O$, 1.5 of $ZnSO_4 7H_2O$, 1.5 of $MnSO_4 H_2O$, 0.3 of K_3BO_3, 0.15 of $CuSO_4 5H_2O$, and 0.1 of $Na_2MoO_4 2H_2O$. One millilitre of solution B was added to 1000 mL of solution A to form the MSM. The broth cultures were incubated with shaking (200 rpm) for 7 to 9 days at 23°C. The cell suspensions were then tested for the presence of surfactant by using the qualitative drop-collapse method.

Hickey-Trenser medium (1952) (in g/L): Dextrin 10, 1g/L meat extract 1, yeast extract 1, trypton 2, $CoCl_2$ 0.02, agar 15, pH 7.2

20.14.9.2 Procedure

1. Inoculate one loopful of bacteria of interest in a 250 mL flask containing 50 mL of production medium and incubate at 30°C with reciprocal shaking at 200 rpm for 5 days.
2. After incubation, fermentation broth is subjected to four different screening method: Drop collapse test, oil spread test, cell surface hydrophobicity test and emulsification assay.

(a) Drop collapse test

This assay relies on the destabilization of liquid droplets by surfactants. Therefore, drop of a cell suspension or of culture supernatant are placed on an oil coated, solid surface. If the liquid does not contain surfactants, the polar water molecules are repelled from the hydrophobic surface and the drops remain stable. If the liquid contains surfactants, the drop spreads or even collapse because the force or inteıfacial tension between the liquid drop and the hydrophobic surface is reduced. The stability of drop is dependent on surfactant concentration and correlates with surface and interfacial tension. The drop collapse assay is rapid and easy to carry out, requires no specialized equipment and just a small volume of sample. In addition, it can be performed in microplates. This assay has been applied several times for screening purposes. However, the test displays a relative low sensitivity since a significant concentration of surface active compounds must be present in order to cause a collapse of the aqueous drops on the oil or glass surfaces.

Procedure

1. Add 2 µl of diesel oil to each well of a 96 well microtiter plate lid.
2. The lid is equilibrated for 1 hr at room temperature, and then add 5µl of the culture to the surface of oil.
3. Observe the shape of the drop on the surface of oil after 1 min.
4. Biosurfactant producing cultures yield flat drops with the scoring system ranging from + to ++++ corresponding to the partial to complete spreading on the oil surface. Cultures giving rounded drops are scored as negative indicative of the lack of the biosurfactant production.
5. Aliquots from a culture of each strain is analysed on two separate microtiter plate.

(b) Oil spread Method

The oil spreading method is rapid and easy to carry out, requires no specialized equipment and can be carried out with just a small volume of sample. The method is useful with low activity and low quantity of biosurfactant. The oil spreading technique is a reliable method to detect biosurfactant production by diverse microorganisms.

Procedure

1. Use 24 h old bacterial inoculum grown in Bushnell–Hass medium.
2. Fill Petri plates base with 50 mL of distilled water.
3. On this water spread 20 µl of diesel oil uniformly.
4. Further add 10 µl of culture at different spots on the diesel oil which is coated on water surface.
5. Occurrence of clear zone is an indication of biosurfactant producing strains.
6. Further add 10 µl of culture at different spots on the diesel oil which is coated on water surface.
7. Occurrence of clear zone is an indication of biosurfactant producing strains.

S.No	*Bacterial isolates*	*Drop collapse Test*	*Oil spread Test*	*Cell surface hydrophobicity test*	*Emnlsification assay*
1.					
2.					
3.					
4.					
5.					

Interpretation The strains with higher cellular hydrophobicity tended to approach the hydrophobic carbon source (diesel) located on the upper surface of the medium, contributing to the floating behaviour Therefore the bacteria with higher cell surface hydrophobicity, emulsification activity through production of biosurfactant and wide range of hydrocarbon utilization effectively help in the degradation of oil.

(c) Cell surface hydrophobicity test

The bacterial adhesion to hydrocarbons (BATH) assay is used to determine changes in cell surface hydrophobicity during growth on mineral salt medium with 2% diesel.

Procedure

1. Harvest bacterial cells by centrifugation at 8,000 × g from growth cultures for 10 min at 4°C, wash it twice, and suspend in PUM buffer (22.2 g $K_2HPO_4.4H_2O$; 7.26 g KH_2PO_4; 1.8 g urea and 0.2 g $MgSO_4.7H_2O$ in 1 litre distilled water, pH 7.2) to an initial absorbance at 400 nm to 1.0.
2. Vortex diesel oil (0.5 mL) and cell suspensions (2.0 mL) in a test tube for 2 min and equilibrate for 15 min.
3. Remove the bottom aqueous phase carefully with a Pasteur pipette
4. Measure optical density at 400 nm.
5. The adherence is expressed as the percentage decrease in optical absorbance of the lower aqueous phase following the mixing procedure, compared with that of the cell suspension before mixing (Figure 20.24).

Figure 20.24 Biosurfactant causing emulsification and foaming on Bushnell-Hass media containing diesel oil.

Cellular hydrophobicity is defined as follows:

$$\%\text{Hydrophobicity} = \frac{1 - OD_{400} \text{ of a aqueous phase of cellular activity after mixing with hydrocarbon}}{OD_{400} \text{ of inital culture prior to hydrocarbon addition}} \times 100$$

20.14.10 Respiration/Dehydrogenase Assay of Bacterial Isolates in Presence of Different Concentrations of Alkanes

Several methods have been developed to study cell viability and proliferation in cell populations. This miniaturization method (96 well microplate formate) allows to analyze many samples rapidly and simultaneously. The microplate format also reduces the amount of culture medium and cells required and is economical. Colorimetric assays allow to measure samples directly in the microplate with an ELISA plate reader. The microtiter plate method is a selective method for detection of diesel degrading microorganisms with rapidity and convenience. The dehydrogenases are enzymes that oxidize their substrate by removing hydrogen from the substrate and passing it to a suitable acceptor. They differ from oxidases in that the suitable acceptor cannot be atmospheric oxygen; in almost all cases, it is a coenzyme-like nicotinamide adenine dinucleotide (NAD), nicotinamide adenine dinucleotide phosphate (NADP), or a flavoprotein (FAD). The hydrogen is then passed either through a hydrogen transport system to react ultimately with oxygen or to an enzymatic process in which it is used for biosynthetic reactions. which is (1) the substrate that is to be oxidized by the dehydrogenase, (2) any cofactor that is required as the immediate hydrogen acceptor of the dehydrogenase activity (e.g., NAD or NADP), (3) any activators of the enzyme that may be known from biochemical or histochemical studies, and (4) a tetrazolium salt that acts as the trapping agent for the hydrogen and in becoming reduced by yielding a red coloured precipitate of formazan.

20.14.10.1 Materials

Bushnell-Hass broth, Tetrazolium salt (TTC) solution (0.1% in 100 mM phosphate buffer or glycyl glycine at pH 7.8.), Cyclohexane, Octane, Hexadecane, octadecane, and diesel oil.

20.14.10.2 Procedure

1. Preparation of inoculum
   (a) Inoculate pure bacterial isolates in nutrient broth and incubate for 24 hr at 37°C.
   (b) Centrifuge the grown media at 10,000 rpm for 10 min. Remove the supernatant and wash the cell pellet twice with 0.85% sterilized saline.
   (c) Resuspend cells in BH broth until A_{600} is equivalent to 1.2.
2. Use 50 µl of bacterial inoculum to inoculate individual wells of a 96-well microtiter plate containing 200 µl sterile BHMS medium supplemented with various alkanes (cyclohexane, octane, hexadecane, octadecane, and diesel fuel) at different concentrations (0 – 4% v/v).
3. Incubate the microtiter plates at 30°C for two weeks.
4. Add 50 µl of triphenyl tetrazolium chloride (TTC) to each well.
5. Incubate the plate at room temperature for 24h for colour development.
6. Read the optical density at 430 nm on ELISA reader.

Inerpretation

Since the proliferative cells are metabolically more active than non-proliferating (resting) cells, the assays are suitable not only for the determination of cell viability and factor mediated cytotoxicity but also for the determination of cell activation and proliferation.

Observation Table:

Bacteialisolate	Optical density at 430 nm																								
	Cyclohexune					Octane					Hexadecane					Octadecan,					Diesel				
	0%	1%	2%	3%	4%	0%	1%	2%	3%	4%	0%	1%	2%	3%	4%	0%	1%	2%	3%	4%	0%	1%	2%	3%	4%
1.																									
2.																									
3.																									
4.																									
5.																									

20.14.11 Laboratory Scale Bioremediation of Diesel Contaminated Soil by Indigenous Bacterial Consortium

Bioremediation is the transformation or degradation of contaminants into non-hazardous or less hazardous chemicals. There are following three different sets for defining bioremediation:

Biotransformation the alteration of contaminant molecules into less or nonhazardous molecules.

Biodegradation the breakdown of organic substances into smaller organic or inorganic molecules.

Mineralization complete biodegradation of organic materials into inorganic constituents such as CO_2 or H_2O.

All the three processes of bioremediation can occur either *in situ* (at the site of contamination) or *ex situ* (contaminant taken out of the site of contamination and treated elsewhere).

20.14.11.1 Materials

1. Petroleum contaminated soil (450 g)
2. Aluminium pans- 441cm^2 surface area and 1764 cm^2 volume
3. Nutrient supplements
   $(NH_4)_2SO_4$– 250 mg /kg
   K_2HPO_4– 100 mg /kg
4. Sterile distilled water – 20 mL
5. Bacterial consortium preparation – 40 mL (2.6×10^8 cells mL^{-1})

20.14.11.2 Procedure

1. Preparation of microcosm

   A. Place the contaminated soil (450 g) in a set of aluminum pans with a surface area of 441 cm^2 and a volume of 1764 cm^2.

(a) Prepare control by sterilizing soil through autoclaving three times at 121°C for 30 min.

(b) Cover the microcosm with aluminium foil and keep at room temperature (27°C).

2. Mix the soil properly to provide sufficient air and O_2.
3. Moistened the soil by addition of 20 mL of sterile distilled water every week until the end of the experiment to maintaine the proper moisture.

Add nutrient supplements $(NH_4)_2SO_4$ – 250 mg/kg, K_2HPO_4 – 100 mg/kg.

4. Add 40 mL 2.6×10^{-8} cells mL^{-1} of bacterial consortium, isolated previously from the contaminated soil samples.
5. ***Biodegradation analysis***

(a) Sample the contaminated soils from microcosm at 0,1,2,3,4,5 and 6 weeks for chemical and microbiological analysis.

(b) Determine the total bacterial count (viable cell count) of the soil on the onset and after the end of the experiment with standard dilution plating technique using sterile nutrient agar medium.

(c) Extract total petroleum hydrocarbon from 10 g of soil by sequentially extraction with 100 mL of hexane, methylene chloride and chloroform.

6. Pool all three extracts at room temperature and dry evaporate the solvents at 40°C on water bath.
7. After evaporation of the solvents determine the amount of the residual TPH gravimetrically.

Calculate the percentages of degradation by following the expression:

$$\% \text{ degradation} = \frac{(\text{TPH control-TPH treatment})}{\text{TPH control}} \times 100$$

20.14.12 Demonstration of Biodegradation of Polycyclic Hydrocarbon

Polycyclic aromatic hydrocarbons (PAHs) are ubiquitous pollutants found in soil at wood preservation plants, gas works, oil refineries, runoff from asphalt pavements, and combustion processes. Their physicochemical properties, which include low water solubility and high adsorption coefficient, make soils and sediments environmental sinkers for PAHs. PAHs are genotoxic and carcinogenic and represent considerable environmental concern. Their mutagenicity varies with the number of aromatic rings. PAHs released into the environment could be removed by many processes, including volatilization, photo-oxidation, chemical oxidation, bioaccumulation, and adsorption on soil particles. However, the principal process for successful removal and elimination of PAHs from the environment is microbial transformation and degradation. In PAHs contaminated soil, microorganisms capable of utilizing and degrading hydrocarbons could be employed for PAH elimination.

20.14.12.1 Materials

Prepare **Mineral basal salt medium (MBSM)** containing (g/L in distilled water) 1.0 $(NH_4)_2SO_4$, 0.1 $CaCl_2.2H_2O$, 0.8 K_2HPO_4, 0.2 KH_2PO_4, 0.2 $MgSO_4 \cdot 7H_2O$, and trace elements made up of 12 mg of $FeSO_4 \cdot 7H_2O$, 3 mg of $MnSO_4 \cdot 7H_2O$, 3 mg of $ZnSO_4 \cdot 7H_2O$, 1 mg of $CoSO_4 \cdot 7H_2O$, and 1 mg of $(NH_4)_6Mo_7O_{24} \cdot 4H_2O$.

Prepare stock solution of phenanthrene in a brown bottle at a concentration of 1.0 mg mL^{-1} acetone at 4°C, and wrap the bottle to avoid any light illumination before dilution.

20.14.12.2 Procedure

Isolation of phenanthrene degrading bacterial isolates

Add 25 g of fresh soil sample immediately after field collection in 500 mL sterilized MBSM containing 50 mg mL^{-1} phenanthrene in Erlenmeyer flasks, and shaken in an orbital shaker at 150 rpm at 30°C in the dark.

Two weeks later, transfer 10 mL aliquots to 100 mL of fresh MBSM containing 50 mg mL^{-1} phenanthrene, and shake the flasks for another two weeks. Repeat this step three times to isolate and purify microbial strains to degrade phenanthrene.

Biodegradation of Phenanthrene

1. For biodegradation studies, inoculate bacterial strains into 100 mL of MBSM containing 50 mg mL^{-1} phenanthrene in Erlenmeyer flasks.
2. Incubated flasks for 24 h at room temperature, while shaking at 150 rpm, in dark. Then, use 10 mL of each pre-culture as inoculum, which contained 1.5-2.0 × 10 cells mL^{-1} of the isolated bacterial strains.
3. Determine the phenanthrene biotransformation by the bacteria in parallel flasks by using 2 litre Erlenmeyer flasks containing 500 mL of sterile liquid MBSM plus four different concentrations of phenanthrene including 25, 50, 100, and 200 mg L^{-1}. Incubate the flasks in three replicates in darkness at 30°C, and maintain initial pH 7.0 ± 0.2 during the experiment.
4. Also run uninoculated medium flasks and autoclaved controls in duplicate for each treatment.
5. The growth of bacteria on phenanthrene was quantified daily by measuring absorbance at 660 nm. After six days, the media from bacterial and control flasks were extracted for the residual phenanthrene.
6. The residual phenanthrene in the media was extracted with diethyl ether two times, and the extract was dried with anhydrous Na_2SO_4. Then, the extract was analyzed by GC-FID.

20.14.13 Isolation of Phenol Degrading Bacterial Isolates from Contaminated Environmental Samples

Phenolic compounds are priority pollutants due to their toxic and harmful effects even at a low concentration. In addition to the biological sources such as oil mill effluents and animal wastes; the source of phenol contaminants are mostly from oil refineries, coal conversion. Phenol, cresols, dimethylphenols and resorcinols are considered major pollutants in the oil-shale semi-coke dump leachates that contaminate the surrounding soils and pose a threat to the groundwater. Microbial bioremediation has proved to be the most effective method for phenol removal from environment.

20.14.13.1 Materials

Contaminated soil sample, Oil refinery waste water sample, Phenol broth media, Phenol agar medium.

***Phenol broth media* (g/L in distilled water)**

KH_2PO_4 2.25, KH_2PO_4 2.25, $(NH_4)_2SO_4$ 1.0, $MgCl_2.6H_2O$ 0.2, NaCl 0.1, $FeCl_3.6H_20$ 0.02, Bromothymol blue 0.06, $CaCl_2$ 0.01 pH adjusted to 7.0

20.14.13.2 Procedure

1. Add portions of soil (1 g) or wastewater (1 mL) sample to the Erlenmeyer flasks containing 100 mL phenol broth supplemented with phenol as a sole carbon source at a concentration of 3 mM.
2. Incubate it at 30°C for 7 days on a rotary shaker operating at 180 rpm.
3. Add 5 mL aliquot of grown culture to the fresh phenol broth medium (100 mL) and incubate again as mentioned above.
4. Repeat it for four further subculturing (each after two days at 30°C, 180 rpm).
5. Streak out inoculum from the flasks and purify phenotypically different colonies on phenol agar media.
6. Utilization of phenol is accompanied by decrease in pH and thus a colour change from green to yellow of the bromothymol blue.

Growth and Phenol Removal Assay

1. Growth potential of bacterial isolates is routinely assayed by turbidity measurement (O.D at 600 nm) by UV-Visible spectrophotometer.
2. The phenol removal assay is carried out by using 2,4 dichloro-quinon-4-chloroimide dyes (Gibb's reagent).
3. Centrifuge the 150 µl grown culture at 6000 g for 10 min, decant the supernatant add 30 µl Na_2HCO_3 and 20 µl Gibb's reagent to it.
4. Determine the reduction of phenol in samples by measuring the absorbance of the colour developed in the supernatant of the sample at 630 nm with UV-VIS spectrophotometer.
5. Determine the phenol concentration from absorbance calibration curve of standard solutions.

Suggested Readings

Bagchi SN, Karamchandani A, Bisen PS. 1985. Isolation and preliminary characterization of cadmium and lead tolerant strains of *Lyngbya* IU 487. *Microbios Lett* **29**, 65-68.

Bartha R, Atlas RM. 1977. The microbiology of aquatic oil spills. *Adv Appl Microb* **22**, 225-266.

Berreck M, Haselwandier K. @001. Effect of arbuscular mycorrhizal symbiosis upon uptake of cesium and other cations by plants. *Mycorrhiza* **10**, 275-280.

Bisen PS, Debnath M, Prasad GBKS. 2012. *Microbes Concepts and Applicatins*. Wiley-Blackwell, Hoboken, New Jersey, USA.

Bisen PS, Dev A, Gour RK, Jain RK, Sengupta LK.1995. Study of Vesicular-arbuscular mycorrhizal fungus *Glomus mosseae* in soil samples of Bhopal. In: *Mycorrhizal Biofertilizers For the Future* (Ed. Alok Adholeya and Sujan Singh), TERI, New Delhi pp 73-76.

Bisen PS, Gour RK, Jain RK, Dev A, Sengupta LK.1996. VAM colonization in tree species planted in Cu, Al, and coal mines of Madhya Pradesh with special reference to *Glomus mosseae*. *Mycorrhiza News* {T E R I New Delhi} 8: 9-11.

Bisen PS, Khare P.1991. Mitigating effect of physicochemical factors on Ni^{2+} , Hg^{2+} and Cu^{2+} toxicity in *Cylindrospermum* IU 942. *Environ Technol Lett* **12**, 297-301.

Bisen PS, Mathur S.1993. Adaptive response of wild and mutant type *Synechococcus cedrorum* to a polychlorinated pesticide Endosulfan. *Biomed Environ Sc* **6**, 265-272.

Bisen PS, Shukla HD, Gupta A, Bagchi SN.1987. Preliminary characterization of a novel *Synechococcus* isolate showing mercury, cadmium and lead tolerance. *Environ Technol Lett* **8**, 427-432.

Bodour A, Miller-Maier RM. 1998. Application of a modified drop collapse technique for surfactant quantification and screening of biosurfactant-producing microorganisms. *Journal of Microbiological Methods* **32**, 273-280.

Bogue RW. 2003. Biosensors for monitoring the environment. *Sensor Review* **23**, 302-310.

Borgne SL, Paniagua D, Vazquez-Duhalt R. 2008. Biodegradation of organic pollutants by halophilic bacteria and archaea. *J Mol Microbiol Biotechnol.* **15**, 74-92.

Boudou A, Ribeyre F. 1997. Aquatic ecotoxicology: from the ecosystem to the cellular and molecular levels. *Environ Health Perspects* **105(Suppl 1)**, 21-35.

Brown EJ, Braddock JF. 1990. Sheen screen, a miniaturized Most-Probable-Number method for enumeration of oil degrading microorganisms. *Appl Environ Microbiol* **56**, 3895-3896.

Bushnell LD, Haas HF. 1941. The utilization of certainhydrocarbons by microorganisms. *J Bacteriol* **41**, 653-673.

Chakrabarty AM, Gunsalus IC. 1969. Defective phage and chromosome mobilization in Pseudomonas putida. *Proc Natl Acad Sci U S A* **64**, 1217–1223.

Chakrabarty AM, Gunsalus IC. 1979. Transduction and genetic homology between Pseudomonas species putida and aeruginosa. *J Bacteriol* **103**, 830–832.

Chakrabarty AM. 1972. Genetic basis of the biodegradation of salicylate in Pseudomonas. *J Bacteriol* **112**, 815–823.

Colberg PJS, Young LY. 1995. Anaerobic Degradation of Nonhalogenated Homocyclic Aromatic Compounds Coupled with Nitrate, Iron, or Sulfate Reduction. In: *Microbial Transformation and Degradation of Toxic Organic Chemicals, Wiley-Liss, New York,* 307-330.

Delong EF, Pace NR. 2001. Environmental Diversity of Bacteria and Archaea. *Syst Biol* **50**, 470-478.

Dev A,Gour RK,Jain RK, Bisen PS, Sengupta LK. 1997. Effect of Vesicular Arbuscular Mycorrhiza-*Rhizobium* (VAM-*Rhizobium*) interaction on heavy metal (Cu, Zn, Fe) uptake in soyabean (*Glycine max,* var. JS-335) under variable P doses. *Int J. Trop Agr* **15,** 75-79.

Dopson M, Baker-Austin C, Koppineedi PR, Bond PL. 2003. Growth in sulfidic mineral environment : metal resistance mechanisms in acidophilic microorganisms. *Microbiology* **149,** 1959-1970.

Doyle A, Griffith JB.1998. *Biotechnology* pp77-79 Wiley, New York,

Ehrlich HL. 2002. How microbes mobilize metals in ores: A review of current understandings and proposals for further research. *Minerals & Metallurgical Processing* **19,** 220-224.

Friedrich B, Schwartz E. 1993. Molecular biology of hydrogen utilization in aerobic chemolithotrophs. *Annu Rev Microbiol* **47,** 351-383.

G. Endo, Narita M, C.-C. Huang CC, Silver S. 2003. Microbial heavy metal resistance transposons and plasmids: potential use for environmental biotechnology. *J Environ Biotech* **2,** 71-82.

Gadd GM, Griffiths AJ. 1978. Microorganisms and heavy metal toxicity. *Microbial Ecol* **4,** 303-317.

Gothalwal R, Bisen PS. 1998. Degradation of Chlorinated Hydrocarbon Insecticides by *Pseudomonas* spp. *Int J Trop Agric* **16,** 189-193.

Gothalwal R, Bisen PS. 2001. Bioremediation. *In*: *Encyclopedia of Pest Management* (eds. David Pimentel) Marcel Dekker, Inc. pp.89-93.

Gothalwal R,Bisen PS. 1993. Isolation and physiological characterization of *Synechococcus cedrorum* 1191 strain tolerant to heavy metals and pesticides. *Biomed and Environ Sc* **6,** 187-194.

Hickey RJ, Tresner HD. 1952. A cobalt-containing medium for sporulation of *Streptomyces* species. *J Bacteriology* **64,** 891-92.

Jannasch HW. 1989. Chemosynthetically sustained ecosystems in the deep sea. In: *Autotrophic Bacteria(Eds Schlegel HG, Bowien B.) Springer, New York,* 147-166.

Jeffery WW, Simon S. 1984. Bacterial resistance and detoxification of heavy metals. *Enzyme and Microbial Technology* **49,** 1-112.

Juwarkar A, Khirsagar DG. 1991. Emulsification and oil degradation by marine bacteria. *Indian J Mar Sci* **20,** 78-79.

Karlsen DA, Larter SR. 1991. Analysis of petroleum fractions by TLC-FID: applications to petroleum reservoir description. *Org Geochem* **17,** 603-617.

Kulkarni M, Chaudhari A. 2007. Microbial remediation of nitro-aromatic compounds: An overview. *Journal of Environmental Management* **85,** 496-512.

Leyval C, Turnau K, Haselwandter K. 1997. Effect of heavy metal pollution on mycorrhizal colonization and function: physiological, ecological and applied aspects. *Mycorrhiza* **7,** 139-153.

Marquez Rocha FJ, Hemandez- Rodriguez V, Lamela MT. 2001. Biodegradation of diesel oil in soil by a microbial consortium. *Water Air Soil Pollu* **128,** 313-320.

Mehta R, Shukla HD, Bisen PS. 1992. Physiological and genetical characterization of a novel *Synechococcus* sp. isolate showing resistance against cyanide, heavy metals and drugs. *Environ Tech Lett* **13,** 253-258.

Mueller JG, CernigliaCE, Pritchard PH. 1996. Bioremediation of Environments Contaminated by Polycyclic Aromatic Hydrocarbons. In Bioremediation: *Principles and Applications, Cambridge University Press, Cambridge,* 125-194.

Nagatta Y, Nariya T, Ohtomo R, Fukudo M, Yano K, Takagi M. 1993. Cloning and sequencing of a dehalogenase gene encoding an enzyme with hydrolase activity involved in the degradation of gamma-hexachlorocyclohexane in Pseudomonas paucimobilis. *J Bacteriol* **175,** 6403-6410.

Nisha S and Meera V. 2005. Bioremediation of phenol contaminated soil. *Poll Res* **24,** 481-485.

Peng JF, Song YH, Yuan P, Cui CY, Qiu GL. 2009. The emediation of heavy metals contaminated sediment. *J Hazard Mater* **161,** 633-640.

Pilon-Smits E, Pilon M. 2002. Phytoremediation of Metals Using Transgenic Plants. *Critical Reviews in Plant Sciences* **21,** 439–456.

Ruggiero CE, Neu MP, Matonic JH, Reilly SD. 2000. Dissolution of plutonium (IV) hydroxide by desferrioxamine siderophores and simple organic chelators. *Inorg Chem* **41**, 3593-3595.

Samanta KS, Singh OV, Jain RK. 2002. Polycyclic aromatic hydrocarbons: environmental pollution and bioremediation. *Trends Biotechnol* **20**, 243 – 248.

Sharma SK, Bisen PS. 1992. Hg^{2+} and Cd^{2+} induced inhibition of light induced proton efflux in the cyanobacterium *Anabaena flos-aquae. Biometals* **5,** 163-167.

Sharma SK, Singh DP, Shukla HD, Ahmad A,Bisen PS. 2001. Influence of sodium ion on heavy metal-induced inhibition of light-regulated proton efflux and active carbon uptake in the cyanobacterium *Anabaena flos-aquae. World J Microbiol Biotech* **17**, 707-711.

Shokrollahazdeh S, Azizmoheni F, Golmohammad F, Shokouhi H, Khademhighat F. 2008. Biodegradation potential and bacterial diversity of petrochemical wastewater treatment plant in Iran. *International Journal of Biodegradation and Biodeterioration* **45,** 234-236.

Singh C, Lin J. 2008. Isolation and characterization of diesel oil degrading indigenous microorganism in Kwazulu-Natal, South Africa. *African Journal of Microbiology* 1, 1972-1982.

Singh DP, Gothalwal R, Bisen PS. 1989b. Toxicity of sodium diethyl-dithiocarbamate (NaDDC) on photoautotrophic growth of *Anacystis nidulans* IU 625. *J Basic Microbiok* **29**,685-694.

Singh DP, Khare P, Bisen PS. 1989a. Effect of Ni^{2+}, Hg^{2+} and Cu^{2+} on growth, oxygen evolution and photosynthetic electron transport in *Cylindrospermum* IU 942. *J Pl Physiol* **134,** 406-412.

Singh DP, Sharma SK,Bisen PS. 1993. Differential action of Hg^{2+} and Cd^{2+} on the photosynthetic apparatus of *Anabaena flos-aquae. Biometals* **6,** 125-132.

Suzuki I, Lee D, Mackay B, Harahuc L, Oh JK. 1999. Effect of Various Ions, pH, and Osmotic Pressure on Oxidation of Elemental Sulfur by *Thiobacillus thiooxidans. Appl Environ Microbiol* **65,** 5163-5168.

Velkov VV. 1996 Environmental Genetic Engineering: Hope or Hazard? *Curr Sci* **70,** 823–832.

Velkov VV. 1996. Stress-induced evolution and the biosafety of genetically modified microorganisms released into the environment. *J BioSc* **26**, 667-683.

Vidali M. 2001. Bioremediation. An overview. *Pure Appl Chem* **73**, 71163-71172.

White PJ, Broadley MR. 2000. Mechanisms of caesium uptake by plants. *New Phytol* **147**, 241-256.

Chapter 21

Microbes in Food and Dairy Production

21.1 INTRODUCTION

The production of safe food, largely free from pathogenic and spoilage organisms, is vitally important with the rapid changes in the pattern of food distribution now taking place in highly industrialized countries. Cities are so large that the inhabitants are becoming increasingly dependent on regular supplies of reasonably priced food, pre-packaged and available all the year round. Such foodstuffs, marketed over a wide area, must be microbiologically safe. Only companies with large capital resources and wide networking can afford the necessary amount of research and quality control to produce food to these high standards. Further, in order to keep the large food factories running economically, constant supply of animals, vegetables and fruits are needed, of standard size and quality. This, in turn, has led to intensive farming practices with extremely high standards of hygiene and scientific principles of nutrition with rigid microbiological control at all stages. The same pattern will be repeated with fish, crustaceans, and mollusks. Rising world population continue to pressure us to search and to make new foods, either entirely microbiological in origin, or bland starch and protein, suitably modified in flavour and nutrient status by the action of microorganisms.

Food microbiologists are concerned with the level of spoilage organism in the raw materials, often of animal and plant origin, the standard of hygiene during processing, the efficiency of preservation and storage methods, as well as the incidence of organisms responsible for food poisoning.

21.2 THE NATURAL MICROFLORA

The natural microflora of food or beverage can consist of three main components, i.e., those associated with the raw material, those acquired during processing, and those surviving preservation and storage. They can be further subdivided into harmless organisms, producing either desirable or undesirable flavour changes in the food, and pathogens forming dangerous enterotoxins.

Microorganisms spoiling the aroma, taste, colour, or texture of the product may be molds, yeasts, or bacteria. The precise flora that is present at any stage will depend on the nutrient status of a food, its temperature, pH, water content, etc., as well as on the nature of the organisms themselves. Thus, many molds flourish below pH 3.8 in the presence of air, even at low levels of available water. Aerobic yeasts are similarly acid-tolerant and can spoil pickles, dry wines, and ciders. Some yeast has special properties, thus can be found in many habitats with *Debaryomyces* sp. are aerobic, salt-tolerant, and can utilize nitrite as a sole source of nitrogen; commonly found on the cut surfaces of cured meats, low water activity, such as seawater, from which it was initially isolated, cheese, wine, beer, fruit and soil as well as in high-sugar products. Fermenting yeasts can grow anaerobically at low pH and are thus associated with spoiled fruit juices; osmophilic yeasts are contaminants of sugar; highly sugared or salted products. Similarly, bacteria can be tolerant of heat (thermophilic), cold (psychrophilic), low pH (acid-tolerant), salt (halophilic) etc. Even foods without very marked characteristics carry an association of bacteria, e.g., fresh meat and fish have *Pseudomonas* and *Achromobacter* species that give way to a micrococci/lactobacilli association when the flesh cured. Often the natural flora is succeeded by a factory flora, as in cheese and cider making. Many foods and beverage are stable only because of the early development of particular bacteria reduce the pH by producing lactic acid and thus inhibit the development of food pathogens. Again, foods are often rendered unpalatable by spoilage organisms before pathogens have had time to develop in sufficient numbers.

Food pathogens include salmonellae, *Clostridium botulinum*, *C. Perfringens*, and some of the staphylococci and streptococci. All non-spore forming pathogens are killed by pasteurization of the food, but preformed enterotoxins are not destroyed by the temperature used in pasteurization. Where the raw material contains pathogenic organisms, there is a danger of re-infection of the pasteurized product. Staphylococci are more salt-tolerant than many spoilage organisms, and this can give them an advantage in some preserved foods. Faecal streptococci (*Streptococcus faecalis* and other group D streptococci) and coli forms, including *Escherichia coli*, are common in foodstuffs; but in this situation are not reliable indicators of faecal contamination, though they may be so in water supplies. These organisms grow readily on food residues in badly maintained processing lines. Thus their numbers in food are generally considered giving an indication of the degree of factory hygiene. They are normally absent from liquids pasteurized and processed in a closed system. Massive numbers (10^6 to 10^7/g) in other foods would be regarded as a potential hazard to health, since, in such circumstances, they have been suspected of causing food poisoning. The general food poisoning bacteria grow optimally at 37°C and little or no multiplication occurs under refrigeration, a technique widely used to restrict bacterial multiplication.

21.3 METHODS OF FOOD PRESERVATION

Any product, whether solid or liquid, can be sterilized if heated long enough and/or treated with a suitable concentration of germicide. Usually, the treatment with substrates of neutral pH is so drastic that accompanying changes in the flavour, texture, and colour render the product unacceptable. Techniques are designed to inhibit the multiplication of the spoilage flora for such type of product prior to consumption of the food. The treatments available are numerous, the choice depending not only on the nature of micro-flora, but also on the chemical and enzymatic constitution of the product (Figure 21.1).

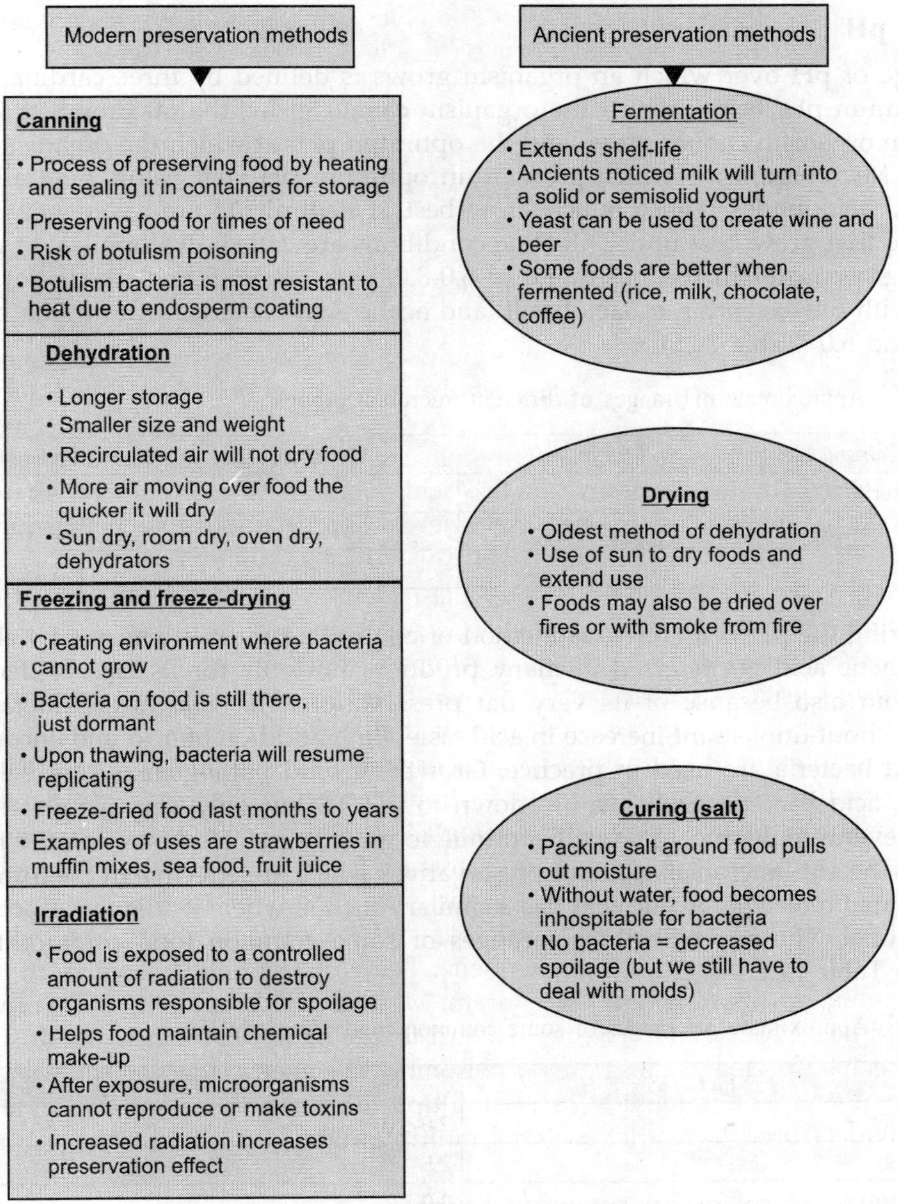

Figure 21.1 Modern and ancient methods of preservation.

21.3.1 Chemical Inhibition

The use of herbs and spices in the middle ages moderate or mask the effects of food spoilage is well known. Essential oils, e.g., clove has been shown to have only mild anti-bacterial properties, so that they are now used only as flavouring agents. The details of the methods of ancient and modern food preservation are shown in Figure 21.1.

21.3.2 pH

The range of pH over which an organism grows is defined by three cardinal points: the minimum pH, below which the organism cannot grow, the maximum pH, above which the organism cannot grow, and the optimum pH, at which the organism grows the best. Microorganisms which grow at an optimum pH well below neutrality (7.0) are called acidophiles. Those which grow best at neutral pH are called neutrophiles and those that grow best under alkaline conditions are called alkalophiles. In general, bacteria grow faster in the pH range of 6.0-8.0, yeasts 4.5-6.5 and filamentous fungi 3.5-6.8, with the exception of lactobacilli and acetic acid bacteria with optima between pH 5.0 and 6.0 (Table 21.1).

Table 21.1 Approximate pH ranges of different microbial groups

Organisms	*Minimum*	*Optimum*	*Maximum*
Most Bacteria	4.5	6.5-7.5	9.0
Yeasts	1.5-3.5	4.0 6.5	4.0 6.5
Molds	1.5-3.5	4.5 6.8	8.0 11.0

Lowering the pH is an obvious method of controlling microorganisms by chemical means. Lactic acid is preferred in many products, not only for its lack of distinctive flavour, but also because of its very flat preservation. This allows considerable pH change without unpleasant increase in acid taste. Both addition of acid and encouraging lactic acid bacteria are used in practice. Growth of food pathogens is rare below pH 5.5; lactic acid bacteria grow readily down to pH 3.8 but only very slowly down to pH 3.0. Several mold and yeast will continue to grow at pH 2.3. Lowering the pH also improves the anti-microbial action of preservatives which are effective in the form of the undissociated molecule. Acetic acid has a similar function where its flavour is acceptable or traditional. The approximate pH ranges of some common food commodities are shown in Table 21.2.

Table 21.2 Approximate pH ranges of some common food commodities

Product	*PH*
Citrus fruits	2.0-5.0
Soft drinks	2.5-4.0
Apples	2.9-3.3
Banana	4.5-4.7
Beer	3.5-4.5
Meat	5.6-6.2
Vegetables	4.0-6.5
Fish (most spp)	6.6-6.8
Milk	6.5-6.8

Contd...

Contd...

Wheat flour	6.2-6.8
Egg white	8.5-9.5
Fermented shark	10.5-11.5

21.3.3 Water Activity (a_w)

Water is often the major constituent in foods. Even relatively 'dry' foods like bread and cheese usually contain more than 35% water. The state of water in a food can be most usefully described in terms of water activity.

Water activity of a food is the ratio between the vapour pressure of the food, when in a completely undisturbed balance with the surrounding air, and the vapour pressure of pure water under identical conditions. Water activity, in practice, is measured as Equilibrium Relative Humidity (ERH) and is given by the formula:

$$\text{Water Activity } (a_w) = \text{ERH}/100$$

Water activity is an important property that can be used to predict food safety, stability and quality. The various applications of water activity includes; maintaining the chemical stability of foods, minimizing non-enzymatic browning reactions and spontaneous autocatalytic lipid oxidation reactions, prolonging the desired activity of enzymes and vitamins in foods, optimizing the physical properties of foods such as texture.

Water activity scale extends from 0 (bone dry) to 1.00 (pure water). But most foods have a water activity in the range of 0.2 for very dry foods to 0.99 for moist fresh foods. Based on regulations, if a food has a water activity value of 0.85 or below, it is generally considered non-hazardous. This is because below a water activity of 0.91, most bacteria including the pathogens such as *Clostridium botulinum* cannot grow. But an exception is *Staphylococcus aureus* which can be inhibited by water activity value of 0.91 under anaerobic conditions but under aerobic conditions, it requires a minimum water activity value of 0.86. Most molds and yeasts can grow at a minimum water activity value of 0.80. Thus a dry food like bread is generally spoiled by molds and not bacteria. In general, the water activity requirement of microorganisms decreases in the following order: Bacteria > Yeast > Mold. Below 0.60, no microbiological growth is possible. Thus the dried foods like milk powder, cookies, biscuits, etc., are more shelf stable and safe as compared to moist or semi-moist foods.

Factors that affect water activity requirements of microorganisms include the following: kind of solute added, nutritive value of culture medium, temperature, oxygen supply, pH, inhibitors. Each microorganism has a minimal water activity for growth as shown in Table 21.3.

Water activity of some foods and susceptibility to spoilage by microorganisms is shown in Table 21.4.

Table 21.3 Minimum water activity values of spoilage microorganisms

Microbial group	*Minimum a_W*	*Examples*
Most bacteria	0.91	*Salmonella* sp. *Clostridium botulinum*
Most yeasts	0.88	*Torulopsis* sp.
Most molds	0.80	*Aspergillus flavus*
Halophilic bacteria	0.75	*Wallemia sebi*
Xerophilic molds	0.65	*Aspergillus echinulatas*
Osmophilic yeasts	0.60	*Saccharomyces bisporus*

Table 21.4 Water activity of some foods and susceptibility to spoilage by microorganisms

Water activity	*Microorganisms grow at this a.and above*	*Food examples*
1.00-0.95	*Pseudomonas. E. coli, Proteus, Shigella, Klebsiella, Bacillus, Clostridium perfringens* & some yeasts	Highly perishable fresh foods & canned fruits, vegetables, meat, fish, milk. eggs: foods containing up to 40% {w/w} sucrose or 7% NaCl.
0.95-0.91	*Salmonella, Vibrio parahemolyticus, Clostridium botulinum. Serratia, Lactobacillus, Pediococcus*; some molds and yeasts	Some cheeses (cheddar, Swiss), Cured meats; some fruit juice concentrates; bread; high moisture prunes; foods containing 55%(w/w) sucrose or 12% NaCl
0.91-0.87	Many yeasts like *Candida, Torulopsis, Hansenula; Micrococcus*	Fermented sausages; sponge cakes; dry cheese; margarine: foods containing 65%(w/w) sucrose (saturated) or 15% NaCl
0.87-0.80	Most molds, *Staphylococcus aureus*, most *Saccharomyces* sp., *Debaromyces*	Most fruit juice concentrates, sweetened condensed milk, flour, rice, pulses containing 15-17% moisture, salami
0.80-0.75	Most halophilic bacteria, Mycotoxigenic *Aspergilli*	Jam, Marmalade, Glace fruits, Soy sauce
0.75-0.65	Xerophilic molds, *Saccharomyces bisporus*	Rolled oats containing 10% moisture; Fudge; Marshmallows; Jelly; Some dried fruits: Nuts, Peanut Butter
0.65-0.60	Osmophilic yeasts, Few molds	Dried fruits containing 15-20% moisture' Honey
0.50	No microbial proliferation	Pasta containing 12% moisture; Spices containing 10% moisture
0.40	No microbial proliferation	Whole egg powder containing 5% moisture
0.30	No microbial proliferation	Cookies, biscuits crackers, bread crusts, etc., containing 3-5% moisture

Contd...

Contd...

0.20	No microbial proliferation	Whole milk powder containing 2-3% moisture. Dried vegetables containing 5% moisture. Corn flakes containing 5% moisture, Instant coffee.

Water acts as an essential solvent that is needed for most biochemical reactions by the microorganisms. Water activity of the foods can be reduced by several methods: by the addition of solutes or hydrophilic colloids, cooking, drying and dehydration: (e.g., egg powder, pasta), or by concentration, (e.g., condensed milk) which restrict microbial growth so as to make the food microbiologically stable and safe. A wide variety of foods are preserved by restricting their water activity.

21.3.3.1 Dried or low moisture foods

These contain less than 25% moisture and have a final water activity between 0.0 and 0.60. e.g., dried egg powder, milk powder, crackers, and cereals. These products are stored at room temperature without any secondary method of preservation. These are shelf stable and do not spoil as long as moisture content is kept low.

21.3.3.2 Intermediate moisture foods

These foods contain between 15% and 50% moisture content and have a water activity between 0.60 and 0.85. These foods normally require added protection by secondary methods such as pasteurization, pH control, refrigeration, preservatives, but they can also be stored at room temperature. These include dried fruits, cakes, pastries, fruit cake, jams, syrups and some fermented sausages. These products are usually spoiled by surface mold growth.

21.3.3.3 Chemical preservatives

Organisms can also be inhibited specifically by the addition of small quantities of a chemical preservative. The ideal properties of an antimicrobial food preservative are summarized below:

1. It is preferable that the preservative should kill rather than inhibit the microorganisms. Having killed them it should itself decompose into innocuous products.
2. A bacteriostatic preservative would be equally satisfactory if it is destroyed only during final cooking, but if it is to be used in conjunction with thermal control processes it would need to have adequate heat resistance.
3. The range of specificity should correspond with the range of microorganisms able to develop in the food, i.e., it must inhibit both food poisoning and spoilage organisms.
4. Any preservative used as a supplement to thermal processing should give similar protection against *Clostridium botulinum* as that given by the standard thermal processing alone, i.e., a reduction in viable spores by a factor of 10^{12}.

5. A preservative should not be inactivated or removed by chemical reaction with the food, by some specific inhibitor in the food, or by-products of microbial metabolism.
6. The preservative should not stimulate the development of resistant strains, and should be avoided totally if the same is also used therapeutically or as an additive to animal feeds.
7. There should be a chemical/analytical method for analyzing the effective portion of the preservative.

No known antibacterial food preservative has all these properties, and there is a dearth of suitable compounds active in the pH range 5.0 to 7.0, a range which is important especially for wet foodstuffs of high nutritive value such as fish meat, and milk.

Food preservatives can be divided into two groups on the basis of their mode of action. In the first, which includes acids, esters, and phenols, the compound is absorbed by the solid components of the bacterial cell and, if of high lipid solubility, is concentrated on the cell membrane and various cell structures. The metabolic effects of salicylic acid, for example, are related to its influence on ATP formation in the mitochondria. The action of such preservatives becomes increasingly less effective with an increase in the lipid and solid content of the food to which it is added. The antimicrobial effect of the second group, quinines and nitrofurans, depends primarily on their ability to penetrate cell membranes. Quinones are able to penetrate any type of cell, whereas nitrofurans have a selective action on bacteria which they can penetrate, but are unable to penetrate yeasts. Once inside the cell, excessive amounts of coenzyme are required to produce the corresponding hydroquinones and aminofuran compounds, ultimately disrupting electron transport in the cell. The role of preservatives in the prevention of spore germination is also important. No known preservative will prevent germination of *Bacillus cereus* spores, but at minimum inhibitory concentrations nisin, subtilin, diethyl pyrocarbonate, and sodium nitrite prevent growth immediately after germination. Spores that shed their spore wall are prevented from elongating into vegetative cells by sodium benzoate, whereas tylosin, sodium sorbate, sodium metabisulphite, and sodium chloride allow some increase in length but prevent cross-wall formation. At greater concentrations, all preservatives prevent any development after germination. Nisin is used to prevent gas formation by clostridia in cheese, but attempts to use it in other foodstuffs have not been entirely successful; part of its activity is lost during heating or curing and on storage it is eventually lost completely so that any bacterial spores present can then develop.

Knowledge of a range of organisms sensitive to a preservative is most important. Thus, growth of yeasts and molds follows suppression of bacteria by the use of tetracyclines. Attempts to use tetracyclines for delaying spoilage in poultry carcasses led to the selective growth of salmonellae, which may be relatively resistant to antibiotics.

Chemically, it is easier to conserve acid foodstuff and beverages. Sulphurous acid (sulphur dioxide) controls the acid-tolerant bacteria of ciders and wines, but molds and fermenting yeasts, common to such environments, are highly resistant. Sorbic acid is used in cheese wrapping to inhibit surface mold growth. It has some activity against yeasts and in France it is permitted in sweet white wines, together with sulphur dioxide, which is antibacterial and an antioxidant. Benzoic acid, used as a preservative in soft drinks, is effective against many types of yeast, but *Saccharomyces acidifacienscan* metabolizes it to some extent in the presence of sugar. Diethyl pyrocarbonate is active against yeasts, particularly in the presence of alcohol, so that it would be more effective in wines than in freshly pressed fruit juices. All substances suggested for use as antimicrobial food preservatives must be submitted to a statutory two-year evaluation programme that includes feeding to rat and dogs, before being permitted in foods and beverages.

Chemical inhibition of growth can also be obtained with gases. Thus, with the original Boehi process non-sterile apple juice was impregnated with eight atmospheres pressures of carbon dioxide and stored at ambient temperature. The process had to be modified, since contaminating lactic acid bacteria (acid-tolerant and microaerophilic) grew and spoiled the product. The juice is now concentrated to a third or fourth of its original volume, saturated with the same percentage of carbon dioxide (0.8% wt) and held at 2°C. There is a cheaper process in France for the sterile storage of flash-pasteurized juice at ambient temperature under a blanket of nitrogen gas. Stringent inspection is required since traces of oxygen would encourage the growth of any mold and yeast contaminants; gas production by the latter would burst the tanks.

Smoking meat, fish, cheese, etc. (normally in conjunction with salting), is an ancient method of food preservation brought up to date. Essentially, it is a vapour absorption process, the solid particles playing only a minor role. The active components of the vapour include volatile fatty acids which have both a bactericidal and a residual mild bacteriostatic effect. Usually, micrococci and staphylococci are inhibited, leaving a flora consisting mainly of lactic acid bacteria.

21.3.3.4 Dehydration and use of concentrated solutions

A wide range of foods is preserved by dehydration or the use of concentrated solutions of salt or sugar. The lower the moisture content of a product, the less liable it is to support microbial growth. A similar effect is obtained the higher the osmotic pressure of food or beverage, whether this is due to added salt or sugar. Hence, the terms xerophile, halophile, and osmophile are used to describe organisms likely to be found in extremes of such environments.

Moisture requirements of microorganisms for survival or growth are expressed as water activity (a_w), i.e., equal to one-hundredth part of the corresponding relative humidity. Water activity can also be related to osmotic pressure and absolute temperature.

Each species of microorganism has its own characteristic optima and range of a_w values. Bacteria normally exist only between 0.995 to0.990 a_w, although staphylococci can exist döwn to 0.86 a_w, and a few halophiles down to as low as 0.75 (saturated sodium chloride solution). Yeasts can withstand drier conditions than bacteria; 'osmophilic' species such as *Zygosaccharomyces barkeri* (*S. rouxii* var. *polymorphus*) can exist at a_w as low as those tolerated by most molds. However, some molds can exist at lower a_w values than any other microorganisms, some species tolerating an a_w as low as 0.62. Reducing the available water below the optimum serves merely to increase the lag phase and decrease the rate of growth; at a_w values 0.65 to 0.62 mold spoilage would be unlikely to become serious in less than one-and-a-half to two years.

It is possible to compare the susceptibility of all types of foods and beverages to microbial growth by using the concept of available water. Thus the a_w in frozen foods is given by the ratio of the vapours pressure of ice to that of water at a temperature under consideration. At –5°C, –10°C, and –15°C the corresponding a_w values are 0.9526, 0.9074, and 0.8642, so that the inability of bacteria to grow on frozen food below –5°C could be due to the limitations of unsuitable a_w. The ability of specific molds to grow below this temperature is due to their tolerance both of low temperature and low a_w values. Some specific effects are also due to the toxic effect of ions; for an osmophilic yeast these are in the order of toxicity $K^+ < Na^+ < Mg^{++} < Ca^{++}$ Li+; $< Cl^- < SO_4^-$. It must not be forgotten that enzyme reactions can still continue even in a dry substrate at 0.35 a_w albeit very slowly indeed and irrespective of whether water is a reactant (hydrolytic enzymes) or not oxidizing enzymes).

21.3.4 Temperature

21.3.4.1 High temperature

In practice the terms 'sterilization' and 'pasteurization' are used to differentiate different levels of heat treatment. The second usually implies that only some of the spoilage organisms or food pathogens are destroyed. With both processes, it is first necessary to know the amount of heat required to inactivate an organism or, if spore-forming, its spores. This will vary with the chemical nature of the food, e.g., its pH, water activity (a_w), and the presence of inhibitors. Secondly, heat penetration studies must be carried out on the food in its actual container during the heating process. The rate of penetration is modified by the volume and shape of the container, and the viscosity of the contents, presence of solid particles, etc. Thirdly, since the death/time curve approaches zero only at infinity, it is never possible to kill every organism in every container in every batch produced. Hence, some statistical limit must be agreed initially, e.g., a reduction in the number of viable spores of *Clostridium botulinum* by a factor of 10^{12} in the canning of certain foods, such as wet food at neutral pH.

Cases of botulism have been reported in the past, mainly in the USA, following the consumption of vegetables, mushrooms, etc., canned or bottled in the home. The heat treatment given was sufficient to destroy the spoilage that would normally keep *Clostridium botulinum* in check, but left spores of the latter undamaged. If the subsequent storage temperature were sufficiently high the spores grew and produced toxin, without altering the physical appearance of the food. These infections have now disappeared where governmentally sponsored and tested recipes are used in conjunction with domestic pressure cookers. With acid foods and beverages a relatively mild heat treatment will give satisfactory results, since *Clostridium botulinum* will not develop at low pH.

The amount of heat required to inactivate bacterial spores (units of lethal heat) is usually calculated in terms of F values, or the number of minutes at 121°C. This assumes instantaneous heating and cooling but, as this is impossible in practice, the equivalent heating effects occurring during the heating and cooling processes must also be calculated from standard sets of tables as graphs. The values would be inconveniently small if used for calculating heat requirements in pasteurization, which is carried out below 100°C. Instead, Pasteurization Unit (P.U.) is used; for pickles the reference temperature is usually 82°C, while 60°C has been proposed for brewing. It is often difficult to determine the effectiveness of sterilization or pasteurization, except by incubating large numbers of samples. With clear liquids, membrane filtration of the contents of a specified number of containers is a routine measure, while chemical determination of the remaining amount of the enzyme phosphatase is a suitable test for milk and beer. For milk, there is also the short Methylene Blue test carried out 24 hours after heating, to determine the amount of post-pasteurization infection.

Normally, packaged goods other than acid or cured products are processed to at least the *Clostridium botulinum* spores standard but difficulties are sometimes experienced with other heat-resistant organisms, e.g., the spores of thermophilic bacteria. The problem is greater with very viscous products of neutral pH, like packaged rice pudding, whose flavour and appearance would be spoiled by the excessive heating needed to kill extreme thermophiles. Whether the contents of the can remain sound will depend on the minimum growth temperature of the organisms remaining after heating and the temperature at which cans are stored. Canners now have rigid standards for the numbers of thermophiles they will accept in sugar, starch, and other additives. Staphylococcal food poisoning from consumption of commercially packaged vegetables has now been eliminated following recognition that the source of the infection was the operatives handling the newly processed cans. The cans are sterile when leaving the cooker, but 2 to 3% of cans with good commercial seams will leak temporarily during cooling. Hence, infection liquid on the outside of the same could be drawn into the cans while it was cooling. The amount of water entering is 300×10^{-4} per mL for large leaks and 8 to 30×10^{-4} per mL for small. The packaging industries have adopted following measures to solve above problems (1) rigid control of cooling water chlorination, (2) no mechanical damage to the can seam while it is in motion, (3) keeping the wet runways between cooler and dryer to a minimum (surface count on the runway must be < 500 organisms/4 sq. inch), (4) keeping any subsequent runways dry, (5) rigid sterilization programme for the equipment, (6) no manual handling of wet cans. Typhoid outbreaks, such as the one in Aberdeen in 1964, associated with imported packaged meats have usually originated from the use of non-chlorinated river water, contaminated with sewage, for cooling the cans after heating.

Considerable developments are now taking place in sterile canning. The product is flash-heated in a heat exchanger to the required temperature, then filled into sterilized cans that are sealed for the required time before cooling. Basically, the same process is used for the U.H.T. process for milk (138°C for a few seconds) and the flash-pasteurization of carbonated beer or cider under pressure prior to bottling.

21.3.4.2 Low temperature

Food and beverages may be chilled to between 0 and 5°C to delay spoilage while awaiting sale or consumption, or to –18°C for extended storage. Three main groups of organisms are important in this respect in deep frozen foods (i) food poisoning bacteria, (ii) psychrophiles, and (iii) molds. Psychrophiles are defined as organisms that grow well on solid media at 0°C, forming colonies visible to the naked eye in one week. It is important to note that this is not their optimum temperature. Most psychrophilic bacteria are strains of the genus *Pseudomonas,* with some from the genera *Flavobacterium, Achromobacter, Alcaligenes, Escherichia,* and *Aerobacter*. They are very rare amongst Gram-positive bacteria. Psychrophilic yeasts usually belong to the genera *Candida* and *Rhodotorula.* Vegetative forms of molds are more sensitive to cold than are spores. Spores of certain species of the genera *Monilia, Chaetostylum, Cladosporium, Aspergillus, Fusarium, Mucor, Thamnidium,* and *Botrytis,* are particularly resistant. Some of these are able to grow slowly at low levels of water activity and prevention of growth therefore requires the addition of CO_2 or exclusion of air from the package.

It is difficult to determine accurately the minimum temperatures at which organisms will grow. The freezing of the medium alters the ability of the organism to grow, but the addition of antifreezes of low osmotic pressure, (e.g., glycerol) or super-cooling without crystallization are beginning to show that some organisms can be induced to grow at –5° to –7°C. Without these techniques, minimum growth temperatures are normally determined by extrapolation of the growth rate or the reciprocal of the lag period. The lag phase before growth occurs may be so long that the results of the experiment depend more on its duration than on the experimental conditions. With these provisos it is generally accepted that (1) the lowest temperature for bacterial growth is –10°C, (2) growth of *Staphylococcus aureus* and *Salmonella* sp. has not been reported at +6.7°C (Enterotoxin production by staphylococci is unknown below + 18°C), (3) *Clostridium botulinum* (types A, B, C, and D) do not grow or produce toxin below + 10°C, and (4) *Clostridium botulinum* type E can grow and produce toxin at +3.3°C after prolonged incubation, but there is no recorded case of botulism arising from this type of activity at low temperatures.

Frozen foods are not sterile and most spoilage is due to improper handling prior to freezing. Freezing is not generally bactericidal unless it is carried out slowly and the produce subsequently stored at a comparatively high temperature (0 to –10°C). Fluctuating storage temperatures within this range also reduce the bacterial load; food pathogens are found to die more rapidly between these limits (maximal at –2°C) than they do below –17°C. Hence the commercial freezing temperature cannot be guaranteed to destroy pathogens, particularly with modern freezing techniques using *rapid* freezing at –35°C followed by storage at 18°C. The cryogenic technique of extremely rapid freezing, by spraying with liquid nitrogen, has hardly any effect on bacterial numbers at all. Hence, prevention of food poisoning from frozen foods depends on the use of good quality foodstuffs, on maintaining impeccable hygienic standards during processing and on the avoidance of contamination and multiplication of microorganisms in the thawed product. All vegetables and some meat and fish products are heat treated before freezing. The cooling period between the two processes should be as short as possible to avoid bacterial growth. Products containing uncooked meat, especially imported frozen boned meat (readily contaminated during the boning procedure) are particularly troublesome. Salmonellae, usually from infected animals, are liable to be present and even with some heat pre-treatment there is a danger of cross-infection between raw and cooked meat. Similarly, egg products are likely to contain salmonellae and must be pasteurized at 65°C for 2.5 minutes and not re-infected afterwards.

After a frozen food is thawed it does not then spoil any more rapidly than it would have done had it not been frozen. But microbial control must be exercised during thawing; this is done either in cold rooms or by using dielectric heating, the latter being too rapid for bacterial growth to occur. Normally, even if unheated frozen vegetables are thawed and stored above the minimum growth temperature, the food is damaged by obvious taints from the harmless spoilage flora rather than by clostridia. The latter organisms appear to be inhibited by these psychrophiles. Similarly, food being defrosted reaches the minimum temperature for growth of the psychrophilic microflora before that of the toxigenic staphylococci. Danger would arise in products containing fairly high concentrations of salt, since staphylococci may then grow faster than less salt-tolerant spoilage bacteria. The ability of *Salmonella typhimurium* to grow in the presence of a competing microflora appears to depend on the nature of the food. No frozen food should contain salmonellae, irrespective of whether it will be cooked before being eaten or not. This is perhaps an ideal situation, practicable only in pasteurized products and frozen vegetables. Faecal indicators and enterococci rarely grow below 5°C but their presence in substantial numbers in a food thawed above this temperature could falsely indicate insanitary processing conditions.

Two industrial processes based on both freezing and heating have been developed. The first is the dielectric defrosting of blocks of frozen meat, fish, and crustacea, which takes 20 to 30 minutes compared with 25 to 30 hours needed for cold room defrosting. The average temperature after dielectric defrosting is still just below 0°C, consequently there is very little change in the bacterial count. The second is freeze-drying, when a frozen food is dried by microwave heating under vacuum at a temperature just below the incipient melting point. These products, however, need the same care in their pre-freezing and re-constitution phases as do other frozen products

21.3.4.3 Miscellaneous methods

Clear liquids can be clarified and then filtered free of all microorganisms. By this process organisms are removed from fluids and may be separated from their soluble products of metabolism, e.g., exotoxins. Filters are made from many materials including unglazed porcelain, diatomaceous earth, asbestos fibres, sintered glass and synthetic membranes of cellulose nitrate (collodion) and cellulose acetate. All types are available in a range of grades of different pore size of particle which is to be removed. Membrane filters are now made in a wide range of pore sizes including many much smaller than the smallest bacterium. Many millions of gallons of wines, beers, ciders, clear fruit juices, soft drinks, sugar syrups, etc., are sterilized annually in this manner. In contrast to these well-tried method, efforts have been made over the past several years to use atomic or ionizing radiation for food preservation. Unfortunately, off-odours and taints tend to form at the dose levels needed to sterilize foods.

21.4 SPECIFIC FOODS AND BEVERAGES

The common microbiological spoilage defects that occur in different foods with some examples are shown in Table 21.5 and are being discussed below in brief for different food.

Table 21.5 Types of food spoilage with some examples of casual organisms

Foods	*Spoilage type*	*Associated microorganisms*
Bread	Moldy Ropy	*Rhizopus nigricans, Penicillium,* *Bacillus subtilis*
Pickles, Sauerkraut	Film/Pink Yeasts	*Rhodotorula*
Fresh Meats	Putrefaction	*Alcaligenes, Clostridium, Proteus vulgaris, Psedomonas fluorescens*
Cured Meats	Moldy Souring Greening, Slime	*Rhizopus, Penicillium, Aspergillus* *Pseudomonas, Micrococcus* *Lactobacillius, Leuconostoc*
Fish	Discolouration	*Pseudomonas, Alcaligenes, Flavobacterium*
Poultry, Eggs	Slime, Odour Colourless and green rots	*Alcaligenes, Pseudomonas* *Proteus, Pseudomonas fluorescens, Alcaligenes, Pseudomonas*
Concentrated juicies, fresh fruits and vegetables	'Off' flavour soft rots	*Acetobacter; Lactobacillus, Leuconostoc, Rhizopus, Erwinia*
Milk/Bitter/Cream/Cheese	Gray and Black mold Ropiness	*Botritis, Aspergillus niger; Alcaligenes viscolactis, Bacillus, Streptococcus, Lactobacillus, Micrococcus*
	Fat Decomposers	*Yeasts*, molds, *Alcaligenes, Proteus, Pseudomonas. Micrococcus, Bacillus, Clostridium*
	Alkali Formers	*Alcaligenes viscolactis, Pseudomonas fluorescens*
	Flavour change	*Leuconostoc, Lactobacillus, Streptococcus*

21.4.1 Milk

Milk is synthesized by cells within the mammary gland and is virtually sterile when secreted into the alveoli of the udder. Beyond this stage of milk production, bacterial contamination can generally occur from three main sources; within the udder, outside the udder, and from the surface of equipment used for milk handling and storage. Cow health, environment, milking procedures and equipment sanitation can influence the level of microbial contamination of raw milk. Equally important is the milk holding temperature and length of time milk is stored before testing and processing that allow bacterial contaminants to multiply. All these factors will influence the total bacteria count (SPC) and the types of bacteria present in raw bulk tank milk.

Milk drawn aseptically from the udder shows a predominance of staphylococci (the great majority coagulase-negative) and diphtheroids (mainly heat sensitive corynebacteria), both groups being part of the cow's normal skin flora. Some udders harbour mastitis streptococci, coagulase-positive staphylococci have become the commonest cause of mastitis. However, thermoduric organisms are uniformly absent from milk collected aseptically. Under usual milking conditions further contamination occurs from milking utensils, the dust of the dairy, the udder of the animals, and from the milking operatives. Of these the most important source is the milking utensils which, if unsterilized, become coated with large numbers of bacteria and contribute enormous numbers of organisms to the milk.

Not unnaturally, single farm samples show wide variations both qualitatively and quantitatively, ranging from those in which udder organisms are predominant to those with a denser and more complex flora. Counts of thermoduric bacteria (micrococci, corynebacteria, aerobic spore-forming bacilli, and thermophilic strains of *Streptococcus faecalis)* above 10^2-10^3 / mL are indicative of heavy infection from contaminated equipment. Bulk samples, particularly from non-refrigerated tanks, carry heavy contamination and give evidence of bacterial growth, particularly of *Streptococcus lactis* and *Streptococcus kefir*. Holding the raw milk unchilled (10 to 20°C) allows an active growth of the two last-named species, coagulase-negative staphylococci and Gram-negative rods *(Alcaligenes viscolactis* and fluorescent and non-fluorescent pseudomonads). Finally, the flora is likely to be dominated by *Streptococcus lactis,* resulting in souring owing to fermentation of lactose. Lower temperatures favour the growth of Gram-negative rods which turn the milk alkaline and eventually putrid by their proteolytic activity.

Under laboratory conditions, milk pasteurized by being held at 63°C for three minutes succumbs to deterioration by *Bacillus cereus* (including *Bacillus mycoides)*. In commercial practice, whether flash (High Temperature Short Time - HTST) or in-bottle pasteurization methods are used, *Bacillus cereus* again overgrows the slower growing coryneforms, but is itself overtaken by Gram-negative rods derived from recontamination. These in turn are outgrown by the milk-souring organisms, *Streptococcus lactis* and *Streptococcus cremoris. Bacillus cereus* causes not only rapid sweet curdling but also 'bitty' cream and 'ropiness'. 'Sterile' milk has long been produced commercially by heating the bottled milk above 100°C. It is now replaced by Ultra-High Temperature treated (UHT) milk, filled aseptically into sterile bottles and Tetrapaks.

21.4.2 Butter

Butter is made from separated cream, which is usually soured, since butter made from soured cream keeps better than that made from sweet. The souring of the cream is brought about by organisms naturally present in the milk, or the milk may first be pasteurized and specific starters then added. Organisms concerned in the making of butter include *Streptococcus lactis* and *Streptococcus cremoris,* which ferment lactose and sour the cream, and two other capsulated streptococcal-like organisms, *Leuconostoc citrovorum* and *L. dextranicum*. The latter organisms attack citric acid, a by-product of lactose fermentation, to produce diacetyl which imparts a buttery aroma to the cream. Many other identified strains are also present and add their particular flavours to the butter.

The temperature at which the cream is soured is very important. Temperatures around 20°C are low enough to prevent growth of thermophilic spoilage organisms, which have survived pasteurization, and yet are sufficiently high to allow growth of the desirable streptococci. In butter-making the soured cream is churned. This causes the fat globules to aggregate to form butter, leaving the buttermilk which can be drained off.

Bacteria are restricted to the moisture droplets throughout the butter. Washing of butter to replace these droplets of buttermilk by water deprives bacteria of nutrients and limits their growth, thus improving the keeping quality of the butter. 'Working' of butter to break the moisture into smaller droplets also improves the keeping quality by limiting the available nutrients within a droplet. This does not apply to molds, which can penetrate the surrounding wall of fat and move to other droplets.

21.4.3 Cheese

Cheese is one of the most important milk products. The widely used Cheddar cheese has a close texture, firm mellow body, and a mild nutty flavour. The raw milk is heated to 68°C for 15 seconds to destroy most vegetative organisms, including coagulase-positive staphylococci (which have increased with the wider use of milking machines) as well as the udder-derived coagulase-negative strains. It is important that the milk be pasteurized without delay to ensure that any strains of *Staphylococcus aureus* which may be present would not have had time to produce enterotoxins. The pasteurized milk is not sterile and will contain thermoduric organisms. A starter is required to produce lactic acid *in situ* in the curd; this can be either *Streptococcus lactis, Streptococcus cremoris,* or a mixture of these, with or without *Leuconostoc cremoris,* etc. Great care must be taken not to build up specific bacteriophages in cheese rooms, particularly where a single strain of bacterium is always employed. Single strains are used when waxy, bland-flavoured cheese is favoured, whereas multi-strains commonly produce the more mellow, nutty flavour. Antibiotics used to control mastitis cause problems in cheese making and penicillinase-producing strains of coagulase-negative staphylococci have been used both to remove the antibiotic and because their lipolytic action can contribute to the cheese flavour. The starter rapidly gives a low redox potential, produces protein degradation product and a pH of 5.0, all conditions essential to the development of the lactobacilli and unfavourable to other bacteria; clostridia are inhibited by the antibiotic nisin produced by some strains of streptococci. If the streptococci fail, so also do the lactobacilli and the cheese develops such faults as taints and gas produced by coliforms and clostridia, and faulty colour by a variety of organisms. Toxigenic staphylococci would also find the conditions favourable.

Renneting with calf rennet follows immediately after the starter which also produces the optimum pH for casein decomposition and clotting. The coagulum is cut into small pieces to release the whey, the degree of syneresis being dependent on the temperature and the rate of acid production by the bacteria. Finally, the curd is consolidated into blocks under light pressure until they become plastic, when they are broken into pieces, sprinkled with salt and pressed into molds. The shaped blocks are first dipped in molten paraffin wax to exclude air and thus to prevent external mold growth and weight loss, and then ripened at 13°C for quick process cheese or 5.5°C for the old slow-ripening type. The optimum period of maturing varies from 2 to 14 months, according to the previous treatment and type of cheese required. During cheese ripening, the numbers of lipolytic organisms are low; the lactic streptococci used as starters are soon inhibited while the numbers of lactobacilli increase progressively during the later stages of ripening of all cheeses, especially the hard varieties. The main flora consists of homofermentative types *(Lactobacillus casei* and *Lactobacillus plantarum); Lactobacillus brevis* and *Lactobacillus lactis* occur much less frequently. The total number of lactobacilli and streptococci, both alive and dead, is extremely large and probably accounts for as much as 0·1% of the weight of the cheese. In Camembert cheese, *Streptococcus lactis* is dominant during the first phase of growth, *Penicillium* and *Oidium* in the second and *L. casei* in the third phase. Blue cheeses are made from open textured cheese through the action of varieties of the mold *Penicillium roqueforti.* The drying cheese coat supports the growth of molds, yeasts *(Torulopsis* sp.), streptococci, and lactobacilli. The microflora of the coat of Stilton cheese is characteristic of the factory producing it. Clammy rind on Dutch cheese is caused by the growth of corynebacteria that develop only at pH 5.3 and is favoured by the prior development of yeasts (*Debaryomyces, Candida, Trichosporon* sp.) that decompose lactic acid and form ammonia from proteins, thus altering the pH in favour of the corynebacteria. On the other hand, surface ripened cheeses such as Liederkrantz, Trappist, Tilsiter, etc., that have a brown surface smear of aerobic bacteria, have antibiotic activity due to *Brevibacterium linens.*

21.4.4 Fermented Milks

Many different strains of lactobacilli, either alone or in combination with lactose fermenting yeasts and streptococci, are used to produce characteristic fermented beverages. In Europe and the Middle East yoghurt is made from milk thickened by heating or the addition of dried milk solids and fermented with *Lactobacillus bulgaricus* and *Streptococcus thermophilus.* It has custard-like consistency and is often flavoured with ground fruit or nuts. Kefir is a fizzy beverage made from fermented mares' milk using 'grains' that contain yeasts, streptococci, lactobacilli, and micrococci. The grains are strained off after their action is complete and added to the next batch of milk. Koumiss, also made from mares' milk, is a greyish-white liquid with uncurdled casein, often bottled and allowed to carbonate by the action of the microorganisms. It normally contains 1 – 2.5% v/v alcohol. Acidophilus milk has been a fashionable drink for several decades since it was thought that *Lactobacillus acidophilus* would survive in the gut and replace so-called harmful bacteria. However, it is doubtful whether this organism would survive without a high carbohydrate diet rich in growth factors.

The most common usage for bacteria in food preparation is with dairy fermentations. Yogurt and cheeses have been made for centuries using bacteria. The ancients may not have known exactly what kind of bacteria that was needed or if what were needed was, indeed, bacteria. All they knew was that the previous batch was required to make a new one. Many people lack the ability to break down and absorb lactose, the sugar molecule in milk. As a result, it enters the gut, producing acid and gas, causing pain and diarrhoea. Fermented milk products metabolize lactose into lactic acid, which is more tolerable for many people. It is found in two isomer form D and L (Figure 22.4). The most common fermented milk product is yogurt. The lactobacilli used in the making of many yogurts, however, may not be the same type as found within the common flora of humans as there are many different strains (**Probiotics**). Some of the bacteria used in the diary industry are listed in Table 21.6.

Table 21.6 Genera of lactic acid bacteria

Genus	*Cell Morphology*	*Fermentation route*	*Growth at 10°C 45°C*		*Growth in NaCl at: 6.5% 18%*		*Growth at 4.4 pH: 9.6*		*Lactic acid Isomer*
Lactobacillus	rods	horno/hetero	±	±	±	–	±	–	D, L, DL
Camobacterium	rods	homo/hetero	+	–	–	–	–	–	L
Lactococcus	cocci	homo	+	–	–	–	±	–	L
Streptococcus	cocci	homo	–	+	–	–	–	–	L
Enterococcus	cocci	homo	+	+	+	–	+	+	L
Vogococcus	cocci	homo	+	–	–	–	±	–	L
Leuconostoc	cocci	hetero	+	–	±	–	±	–	D
Oenococcus	cocci	hetero	+	+	±	–	±	–	D
Pediococcus	coccittetrads)	homo	±	±	±	–	+	–	D, L, DL
Tetragenococcus	cocci(tetrads)	homo	+	–	+	+	–	+	L
Aerococcus	cocci(tetrads)	homo	+	–	+	–	–	+	L
Weissella	coccoid	hetero	+	–	±	–	±	–	D, L, DL

Lactic acid bacteria are a heterogenous family of mainly low G+C Gram-positive anaerobic, non-sporulating and acid tolerant bacteria. They can ferment various nutrients through a homo-or heterofermentative route into primarily lactic acid, but also into by products such as acetic acid, formic acid, ethanol and carbon dioxide. They contribute to rapid acidification of food products, but also to flavour, texture and nutrition. The genome sequencing of food and health related LAB has been booming. Around forty LAB genomes have been fully sequenced, including the genera *Lactobacilli, Lactococci, Streptococci, Pediococcus, Oenococcus* and *Leuconotoc* (Figure 21.2).

Acidophilus milk is made with *Lactobacillus acidophilus.* **Butter** is made from pasteurized cream, to which a lactic acid starter has been added. The starter contains, for example, *Streptococcus cremoris* or *S. lactis,* but requires *Lactobacillus diacetylactis* to give it its characteristic flavour and odour. **Cheese** is often made with *Streptococcus* and *Lactobacillus* bacteria. Fermentation lowers the pH, thus helping in the initial coagulation of the milk protein, as well as giving characteristic flavours (Figure 21.3).

In such Swiss cheeses, the typical flavour is the result of the use of *Propionibacterium.* Cheese can be classified within two groups - ripened and unripened. Unripened cheeses consist of cottage cheese, cream cheese, and Mozzarella, for example. These are soft cheeses and are made by the lactic acid fermentation of milk. Many different bacteria are used to produce the various cheeses, but *Lactococcus lactis* and *Leuconostoc cremoris* are used most often. Soft cheeses can take one to five months to ripen; hard cheeses, three months to a year or more; and very hard cheeses, like Parmesan, can take twelve to eighteen months. The blue veins found in cheeses, are caused by growth of *Penicillium roqueforti,* which is deliberately added to cheese. Originally, it was found as a natural contaminant of the areas where it was made. The holes in Swiss cheese are the result of *Propionibacterium shermanii.* The surfaces of Camembert and Brie are innoculated with *Penicillium camembertii,* which then develops in a skin on the surface. Limburger is soaked in brine to encourage the growth of *Brevibacterium linens* (it should come as no surprise that this is the same bacteria isolated from smelly feet!). **Kefir** includes many different microbes, including yeasts, lactobacilli, lactococci, and leuconostocs. Depending on geographical locations, the precise types of microbes will vary. **Yogurt** usually requires the addition of *Lactobacillus bulgaricus, Lactococcus thermophilus,* and/ or *Streptococcus thermophilus* to the milk. Lactic acid bacteria comprise an ecologically diverse group of microorganisms with the formation of lactic acid as the primary metabolite of sugar metabolism. These bacteria utilize sugars by either homo- or heterofermentative pathways (Figure 21.4).

21.5 CONTROL OF FOOD SUPPLIES

The increase in urban populations and improvements in methods of food preservation have led to large-scale transport of basic foods from the producer to the consumer areas. This has inevitably increased the risk of infection of many people from a common food source. This risk can be considerably reduced by suitable precautions.

21.5.1 Milk Supplies

Consumer-milk is usually transported by bulk-collection services; it is therefore most essential that adequate measures of control are observed since contaminated milk from one source may be mixed with a large volume of clean milk. The first requirement is good animal husbandry and dairy technique to produce a clean product of high quality.

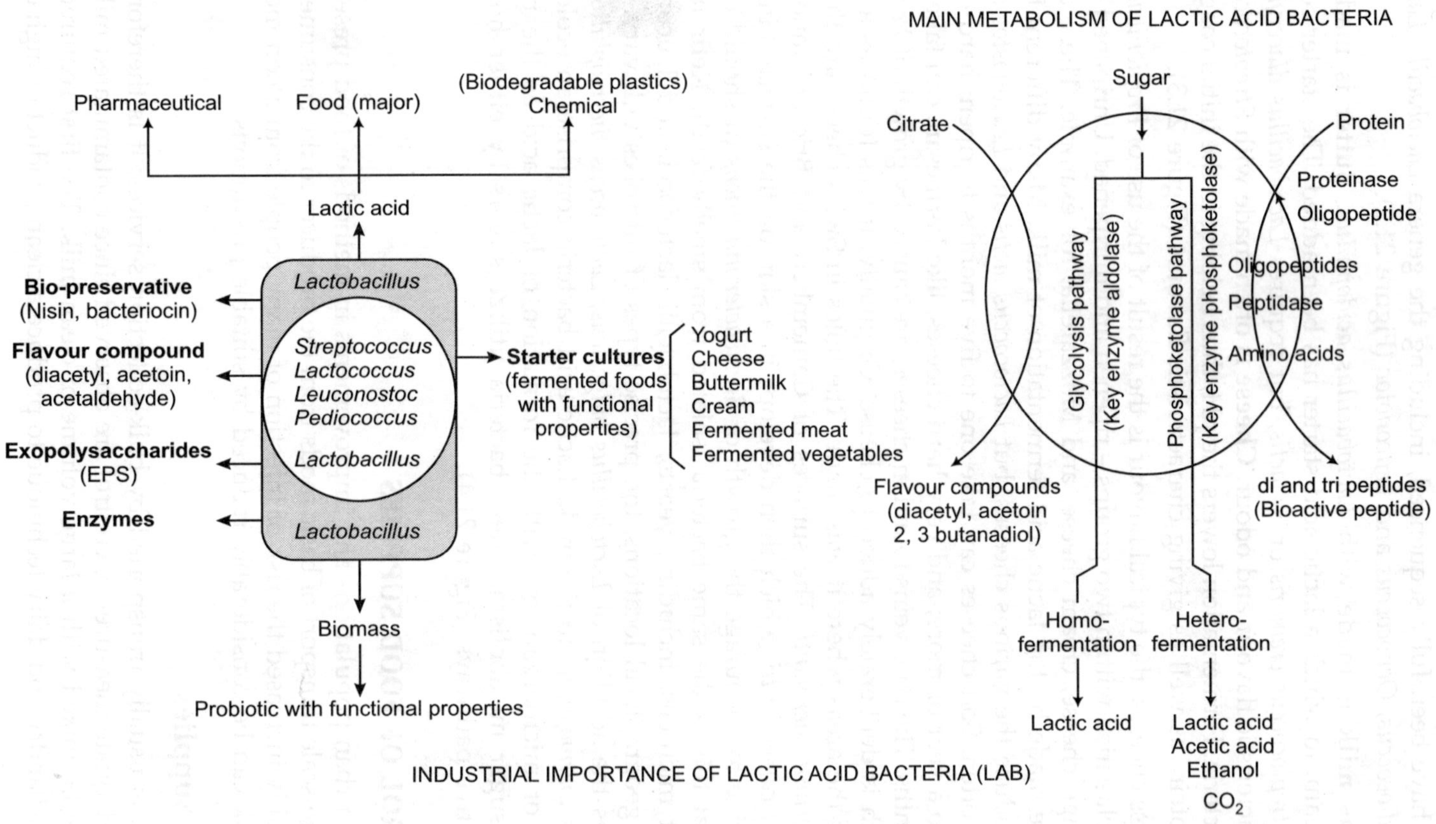

Figure 21.2 Salient feature of lactic acid bacteria.

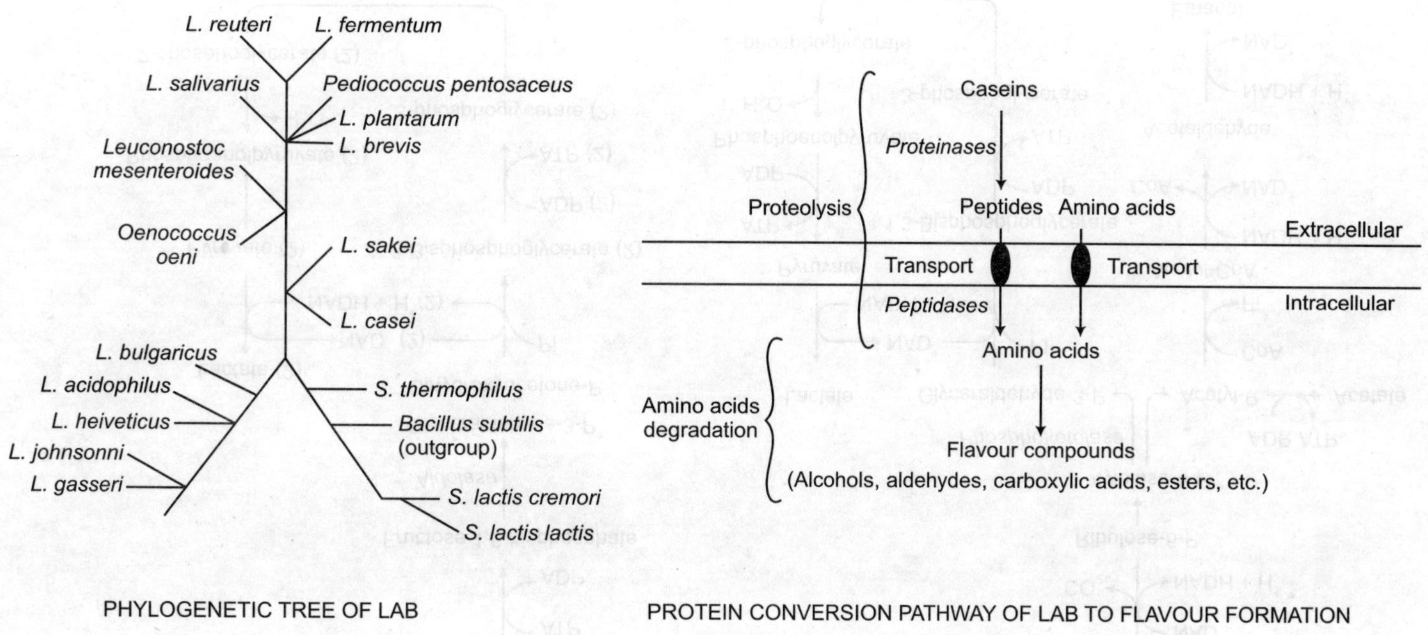

Figure 21.3 Phylogenetic tree and important commercial importance of LAB.

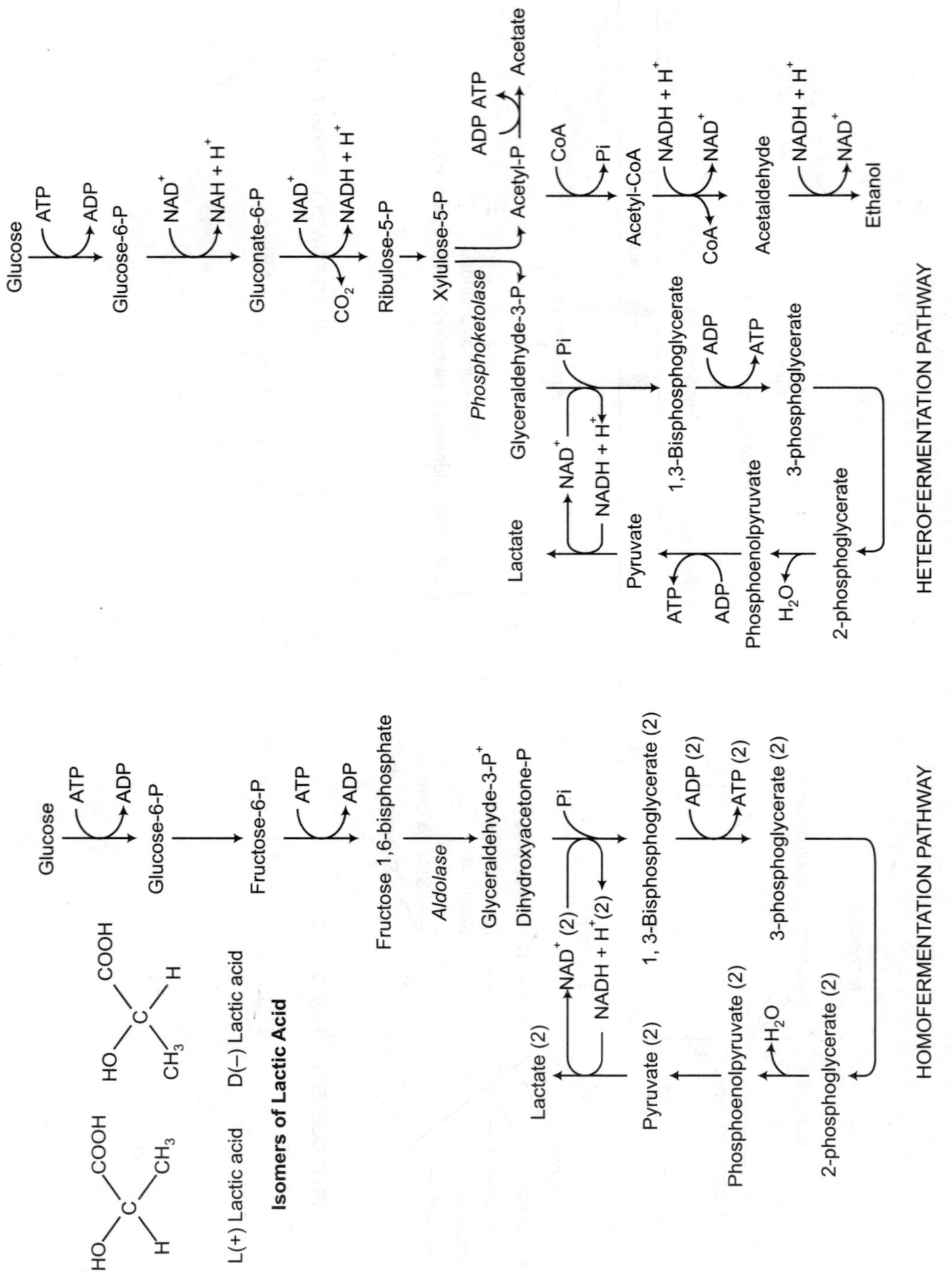

Figure 21.4 Homo- and heterofermentative pathways of LAB.

As an additional safeguard most milks are heat-treated to kill pathogenic bacteria which may be present and at the same time to reduce the number of contaminants thereby improving keeping quality.

21.5.1.1 Methods of heat treatment of milk

Heat treatment is the most widely used method of preservation of milk. Three methods are currently used.

Pasteurization This method of partial destruction of the microbial population by heat was first introduced by Pasteur to kill contaminating organisms which interfered with the fermentation processes in the manufacture of wine. Its application to milk was first used in Denmark to safeguard pigs against infection from bovine pathogens, but its widest industrial application today is in treating milk for human consumption. By holding the milk for a defined time at a standard temperature, as, for instance, 15 seconds at 161°F (71.7°C) in the High Temperature Short Time process, most non-sporing organisms, including all nonsporing pathogens, are killed. This renders the milk safe for drinking and extends its keeping quality.

Boiling Greater numbers of microorganisms are killed when milk is held at a temperature not less than 100°C for a period, which ensures that it will comply with the turbidity test, and the bottles are sealed immediately afterwards. This process imparts a caramelized flavour to the milk and homogenization occludes a visible cream line. Milk subjected to this process is often called sterilized but total sterility is not achieved and it will not keep indefinitely at normal temperatures.

Ultra heat treated By this method, the milk is exposed to a temperature of not less than 270°F (132.2°C) for at least one second. Usually this treatment renders it sterile and therefore gives it excellent keeping qualities. There is no cream line, the cream being dispersed throughout the milk as in homogenized milk. From 10 to 20 per cent of vitamins are destroyed and a slight flavour is imparted, but this becomes less upon storage. Because of the excellent keeping quality of the milk it is likely to become very popular since less frequent deliveries to the consumer will be possible and export to other countries is facilitated.

21.5.1.2 Statutory standards for milk

The one official test to which an untreated milk must comply is the *Methylene Blue Test.* A standard solution of methylene blue is added to a sample of the milk on the day after it left the farm and the sample is then incubated at 37°C. Satisfactory milk will not reduce the dye to the colourless leuco-form within 30 minutes. Methylene blue is a redox potential indicator and is reduced by the microbial activity in badly contaminated samples.

Pasteurized milk is required to pass two official tests prescribed by the Ministry of Health. A *Methylene Blue Test* similar to the one described for untreated milk is performed on the milk 24 hours after it has been pasteurized. This determines the bacterial cleanliness of the milk at the time when it would normally reach the consumer, and measures contamination which may arise from improperly cleaned bottles subsequent to processing. The other test, the *Phosphatase Test,* is a quantitative method of detecting the amount of the natural enzyme, phosphatase, normally present in milk. In pasteurized milk, most of this enzyme is destroyed and only a minimal amount is detectable. This test checks that the milk has been adequately pasteurized.

Sterilized milk must conform to the *Turbidity Test.* Proteins are denatured by boiling, but if the milk has received inadequate heating, soluble proteins remain and may be detected by the turbidity test. The milk is first saturated with ammonium sulphate and filtered; a tube of the clear filtrate is then plunged into boiling water and kept in it for 5 minutes. The formation of a turbidity, due to the heat denaturation of proteins, indicates that the milk had been insufficiently heated. Ultra Heated Milk is tested by a *Colony Count.* A standard loopful of the milk is cultured in 5 mL of yeastral milk agar at 37°C for 48 hours. Not more than 10 colonies are permitted if the milk is to pass the test. Neither the phosphatase nor the turbidity test can be applied to Ultra Heated Milk since phosphatase and undernatured proteins remain after the ultraheat treatment.

21.6 PROBIOTICS

Probiotics are defined as foods containing live microorganisms, which actively enhance health of consumers by improving the balance of microflora in the gut when ingested live in sufficient numbers (FAO/WHO 2001). Functional food is defined as food product that contain one or more functional compounds based on scientific evidence having certain physiological function and health benefit, and has been proved to be safe (Figure 21.5). The recommended doses of probiotics is 10^8–10^{11} CFU per serving with one to several servings per day.

These days consumers are concerned about the synthetic chemicals used as preservatives in food, and there is resulting trend towards less processed food. The untreated foods can harbour dangerous pathogens which are able to multiply under refrigeration and without oxygen. A solution to this dilemma is the use of antimicrobial metabolites of fermentative microorganisms. Many antimicrobial chemicals have been in use for a long time without any known adverse effects. Many of the organic compounds which have stirred interest are antimicrobial metabolites of bacteria used to produce, or associated with fermented foods. In fermentation, the raw materials are converted by microorganisms (bacteria, yeast and molds) to products that have acceptable qualities of food. In common fermented products such as yogurt, lactic acid is produced by the starter culture bacteria to prevent the growth of undesirable microorganisms. Food fermentations have a great economic value and it has been accepted that these products contribute in improving human health. LAB has contributed in the increased volume of fermented foods worldwide especially in foods containing probiotics or health promoting bacteria.

The recent advances in biotechnology have significantly increased the production of high quality, nutritious and tasteful foods that remain fresh for long time and are

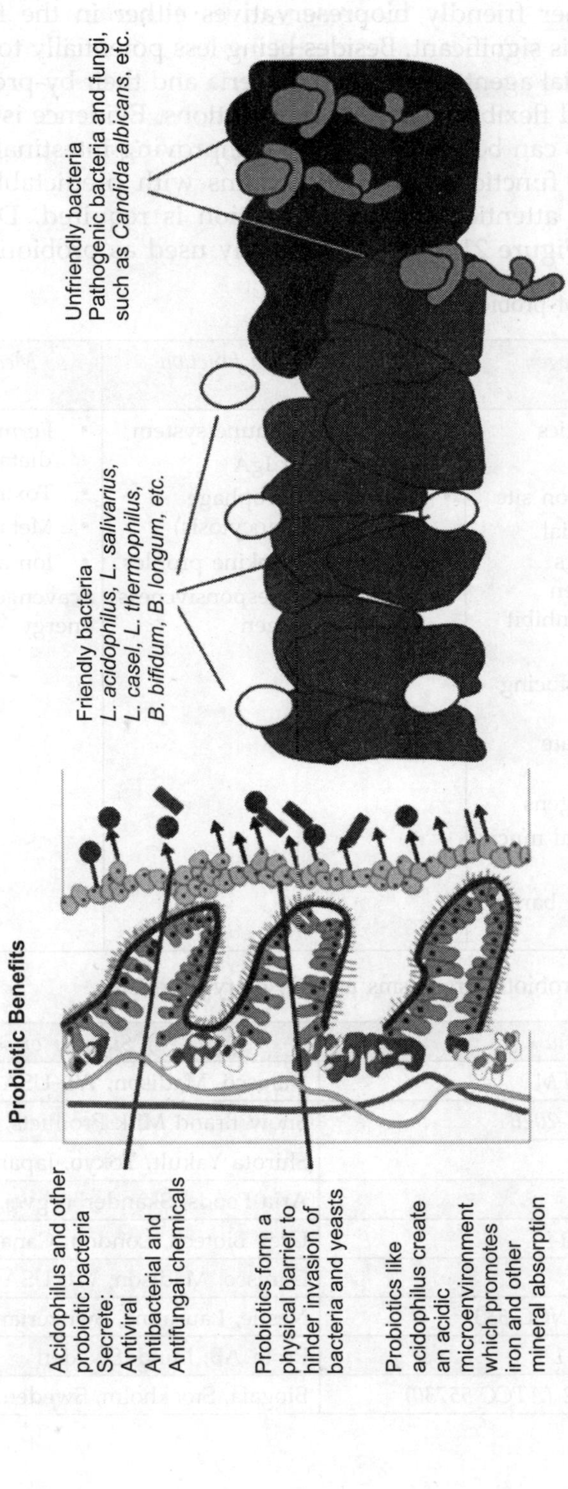

Figure 21.5 Intestinal association of probiotics.

completely safe and less reliant on artificial additives. The potential application of bacteriocins as consumer friendly biopreservatives either in the form of protective cultures or as additives is significant. Besides being less potentially toxic or carcinogenic than current antimicrobial agents, lactic acid bacteria and their by-products have shown to be more effective and flexible in several applications. Evidence is accumulating that confirms that probiotics can benefit the host by improving intestinal well-being (Table 21.7). In order to have functional probiotic strains with predictable and measurable beneficial effects, strict attention to strain selection is required. Different species of Lactobacillus listed in Figure 21.8 has been widely used as probiotics.

Table 21.7 Mechanisms of probiotics

Protective function pathogen displacement	*Immune system function*	*Mechanism of action*

• Colonization of probiotics
• Compete for nutrients
• Competition for adhesion site
• Secretion of antimicrobial substances by probiotics (organic acids, hydrogen peroxide, bacteriocin) inhibit pathogen
• Alter local pH (by producing short chain fatty acids and lactic acids) to create an unfavourable local environment for pathogens
    • Stimulate epithelial mucin production
    • Enhance intertinal barrier function
	• Stimulate immune system
	• Induction of IgA
	• Increase macrophage activity (Phagocytosis)
	• Modulate cytokine profiles
	• Induce hyporesponsivcness to food antigen
		• Ferment nondigestible dietary residue
		• Toxin inactivation
		• Metabolize cholesterol
		• Ion absorption control
		Scavenge superoxide radicals. energy

Table 21.8 Commercial probiotic organisms used in dairy products

Organism	*Supplier or source*
Lactobacillus acidophilus NCFM	Danisco, Madison, WI, USA
Lactobacillus acidophilus SB7-2026	Snow Brand Milk Products, Tokyo. Japan
Lactobacillus casei strain	Shirota Yakult, Tokyo, Japan
Lactobacillus casei F 19	Aria Foods, Skanderborgvej, Denmark
Lactobacillus fermentum RC-14	Urex Biotech, London, Canada
Lactobacillus gasseri ADH	Danisco, Madison, WI, USA
Lactobacillus johnsonii KAI (NCC 533)	Nestle, Lausanne, Switzerland
Lactobacillus plantarum 29911	Probi AB, Lund, Sweden
Lactobacillus reuteri SD 2112 (ATCC 55730)	Biogaia, Stockholm, Sweden

Contd...

Contd...

Lactobacillus rhamnosus GR-I	Urex Biotech, London, Canada
Lactobacillus rliamnosus GG (.4TCC 53103)	Valio Ltd., Helsinki, Finland
Lactobacillus salivarius UCCI18	University College. Cork, Ireland
Bifidobacterium longum SBT 2928	Snow Brand Milk Products, Tokyo, Japan
Bifidobacterium longum BB536	Morinaga Milk Industry, Zama City, Japan
Bifidobacterium breve strain Yakult	Yakult, Tokyo, Japan
Lactobacillus lactis	Local Manufacturers
Lactobacillus delbreuckii supsp. *bulgaricus*	Local Manufacturers
L. lactis subsp. *cremoris*	Local Manufacturers

The basic morphological characteristics are the rod-shaped usually non-motile non-spore forming. It produces lactic acid lowering final pH of the medium to 2.2 - 3.4, resistant to acids and is microaerophillic. They grow well in milk. The identification of genus is based on usual morphological and biochemical tests such as Gram reaction, motility, presence of flagella, oxidase test, catalase reaction, presence of extracellular gelatinase and nitrate reduction. The acceptable manufacturing process of probiotics has been defined in Figure 21.6.

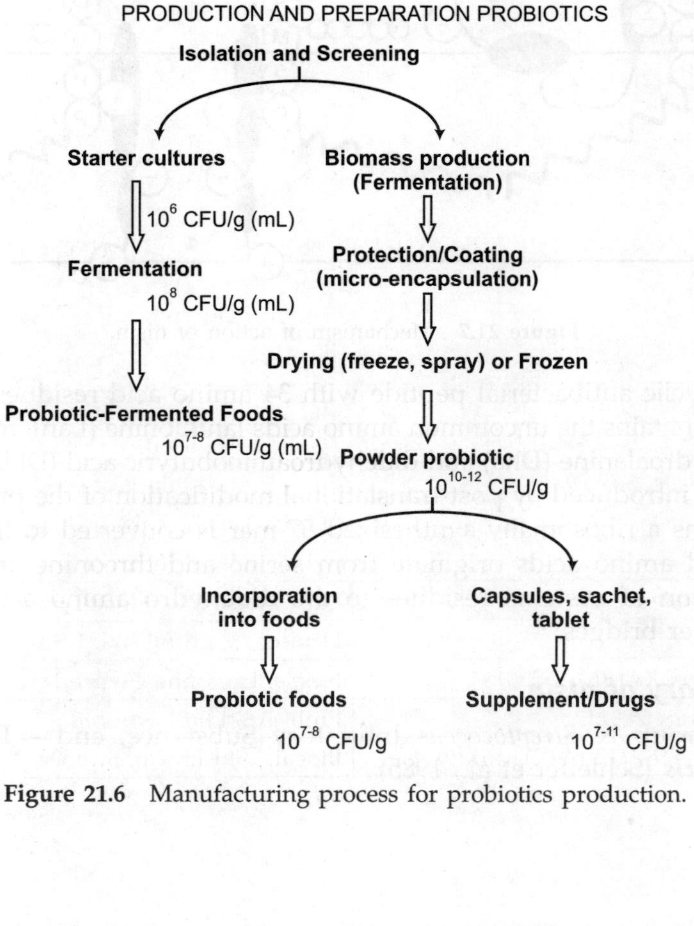

Figure 21.6 Manufacturing process for probiotics production.

21.6.1 Nisin

Nisin is bacteriocins which are ribosomally synthesized closely related antimicrobial peptides or proteins produced by strains of *Lactococcus lactis* subsp. *lactis* showing a narrow or broad antimicrobial activity spectrum against Gram-positive bacteria sensitive to proteolysis enzymes. They are heat stable and are active in a wide pH range (maximum stability at pH_3).

Nisin preparation consists of nisin and sodium chloride with an activity of not less than 900 units per mg. Nisin may be produced in a sterilized medium of non-fat milk solids or of a non-milk-based fermentation source, such as yeast extract and carbohydrate solids. Nisin can be recovered from the fermentation medium by various methods, such as injecting sterile, compressed air (froth concentration); membrane filtration; acidification; salting out; and spray-drying. Bacteriocins act by destabilization and permeabilization of the cytoplasmic membrane through the formation of transitory poration complexes or ionic channels that cause the reduction or dissipation of the proton motive force (PMF) (Figure 21.7).

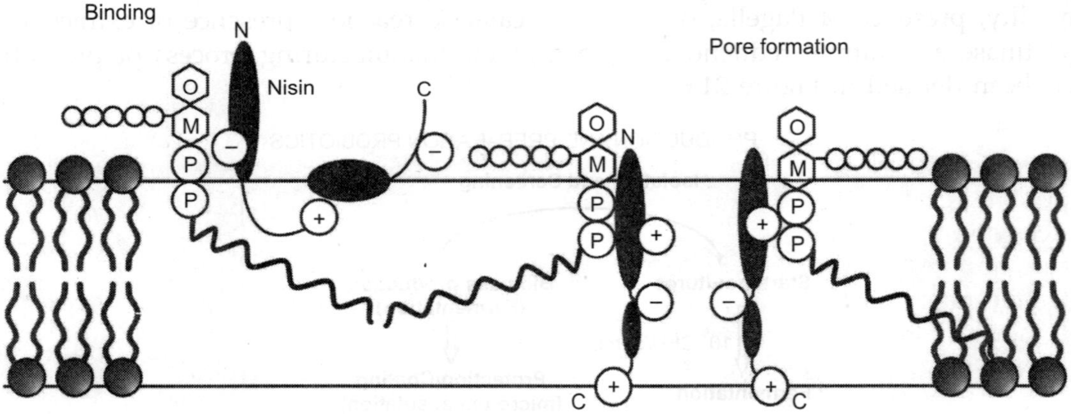

Figure 21.7 Mechanism of action of nisin.

It is a polycyclic antibacterial peptide with 34 amino acid residues used as a food preservative. It contains the uncommon amino acids lanthionine (Lan), methyllanthionine (MeLan), didehydroalanine (Dha) and didehydroaminobutyric acid (Dhb). These unusual amino acids are introduced by post-translational modification of the precursor peptide. In these reactions a ribosomally synthesized 57-mer is converted to the final peptide. The unsaturated amino acids originate from serine and threonine, and the enzyme-catalysed addition of cysteine residues to the didehydro amino acids result in the multiple thioether bridges.

21.6.1.1 History of nisin

Produced by group N *Streptococcus* Inhibitory Substance, end – IN (revised into ***Lactococcus lactis*** (Schleifer et al., 1985)

- 1928 – 1947: research continued on nisin
- 1953 – Commercial production started on nisin
- 1969 – joint expert FAO/WHO approved nisin as bio-preservative
- 1989 – nisin obtained GRAS status by US FDA
- Nisin is adopted by BPOM as one of food preservatives
- 30% (6 of 20) of human milk samples contains nisin-producing bacteria suggests that humans may have a long history of consuming nisin-producing bacteria. Nisin-producing *L. lactis* may protect mothers (mastitis) and infants (toxication) from pathogenic skin flora, such as *Staphylococcus aureus* (Beasley and Saris, 2004).

Nisin is produced by fermentation using the bacterium *Lactococcus lactis*. Commercially, it is obtained from the culturing of *Lactoccus lactis* on natural substrates, such as milk or dextrose, and is not chemically synthesized. It is used in processed cheese, meats, beverages, etc., during production to extend shelf life by suppressing Gram-positive spoilage and pathogenic bacteria (Table 21.9). While most bacteriocins generally inhibit only closely related species, Nisin is a rare example of a "broad-spectrum" bacteriocin effective against many Gram-positive organisms, including lactic acid bacteria (commonly associated with spoilage), *Listeria monocytogenes* (a known pathogen), etc. However, when coupled with the chelator EDTA, Nisin has also been known to inhibit Gram-negative bacteria, as well. Nisin is soluble in water and can be effective at levels nearing the parts per billion ranges. In foods, it is common to use Nisin at levels ranging from ~1-25ppm, depending on the food type and regulatory approval. Due to its naturally selective spectrum of activity, it is also employed as a selective agent in microbiological media for the isolation of Gram-negative bacteria, yeast, and molds. Subtilin and epidermin are related to nisin. All are members of a class of molecules known as lantibiotics. Following two methods illustrate the production of yogurt (Figure 21.8).

Table 21.9 Level of nisin in food applications

Food Application	*Typical target organisms*	*Level of nisin (mg/Kg or mg/L)*
Processed cheese	*Clostridium* sp. *Bacillus* sp.	5-15
Pasteurized milk and milk products	*Clostridium* sp. *Bacillus* sp.	0.25-10.0
Pasteurized chilled soups	*B. cereus* *C. pasteurianum*	2.5-6.25
Crumpets	*B. cereus*	4-6.25
Canned foods (high acid)	*Clostridium botulinum* and *Clostridium thermosaccharolyticum*	2.5-5.0
Ricotta cheese	*Listeria monocytogenes*	2.5-5.0
Continental type cooked sausage	Lactic acid bacteria, *Brochothrix thermosphacta*, *Listeria monocytogenes*	5-25

Contd...

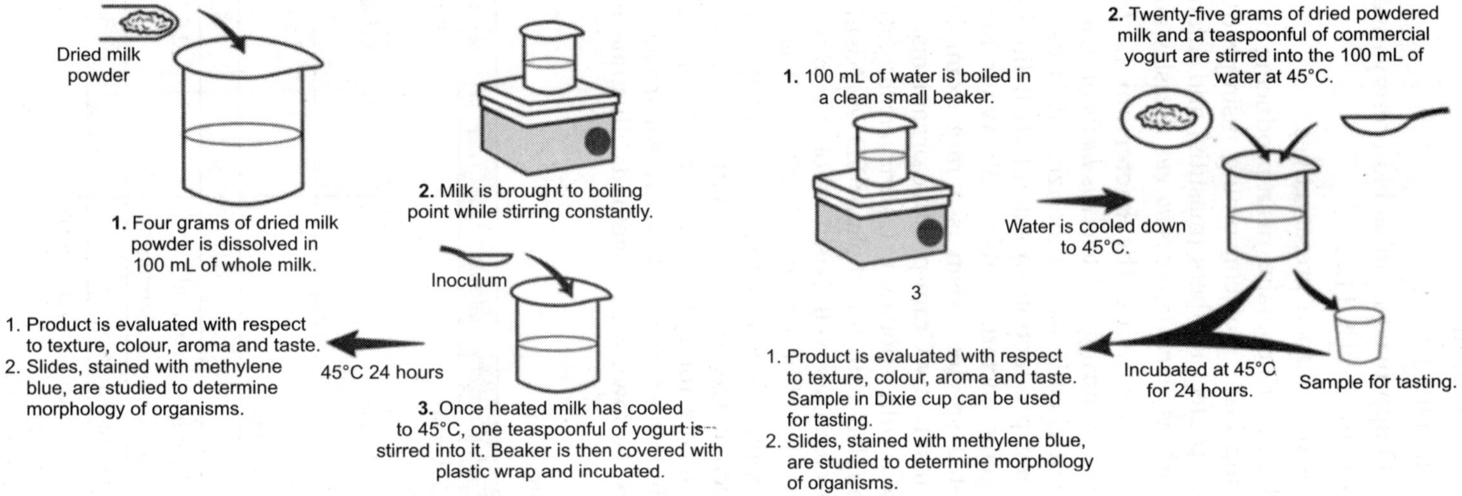

Figure 21.8 Common methods of preparing yogurt.

Dipping sauces	*Lactic acid bacteria*	1.25-6.25
Salad dressings	*Lactic acid bacteria,*	25.0-37.5
Beer: pitching yeast wash	*Lactic acid bacteria, e.g., Lactobacillus, Pediocococcus*	25.0-37.5
Post-fermentation		0.25-1.25

21.7 MEAT

The microflora of freshly dressed meat from healthy animals is derived only slightly from the flesh (mainly the lymph nodes) but mainly from dirt on the animal, its faeces, the personnel and instruments in the abattoir, and the air flora of the chill rooms (Figure 21.9). Animals must not be fatigued or distressed before slaughter, otherwise *rigor mortis* sets in early and the meat putrefies rapidly. The high muscle glycogen content of a rested animal ensures the continued presence of some lactic acid, which is unfavourable to spoilage bacteria in the meat. Efficient stunning suspends the heart action. Otherwise contaminants from the knife or from the cut area circulate around the body during the bleeding process. Strains of *Serratia, Achromobacter* and *Pseudomonas,* which are resistant to the bactericidal action of fresh mammalian blood, are particularly important in this respect. Immobilization of pigs with carbon dioxide prior to bleeding is helpful for both these reasons; its use could well be extended to cattle and sheep. During evisceration it is essential to remove the intestines from the carcass cutting area immediately; this reduces the amount of contamination by bacteria from the gut. When the carcasses are hanging in the chill room (2° to 3°C for 2 to 3 days) before cutting up, the microflora can include bacteria, molds, and yeasts. Most of the bacteria are mesophiles and tend to die out during chilled storage; only a few are associated with any specific defect or spoilage of stored meat. Molds and yeasts may then form between 1 to 10% of the flora. Growth on the uncut surfaces which are covered with a layer of fat and connective tissue is very limited.

During chilled storage any of the following changes may take place, depending on the original microbial 'load':

1. Psychrophilic bacteria grow readily on moist surfaces of the meat, thus producing individual colonies that coalesce, especially if there is condensation, forming slime and off-odours. *Pseudomonas* sp. are virtually the sole slime formers when the meat is stored at l0°C, while at 15°C approximately equal amounts of *Micrococcus* and *Pseudomonas* occur.
2. Putrefactive spoilage can occur in the deep tissue of large thick pieces. *Clostridium navy;* and other putrefactive anaerobes have been implicated with bone taint, while salt-tolerant bacteria, able to grow between 0 to 3°C in bone marrow, are responsible for ham souring.
3. The mold *Cladosporium herbarum* can grow as black spots on chilled meat during transhipment. Carcasses of home-killed meat, held for long periods in chilled rooms may become infected with *Cladosporium, Rhizopus, Mucor,* and *Thamnidium*

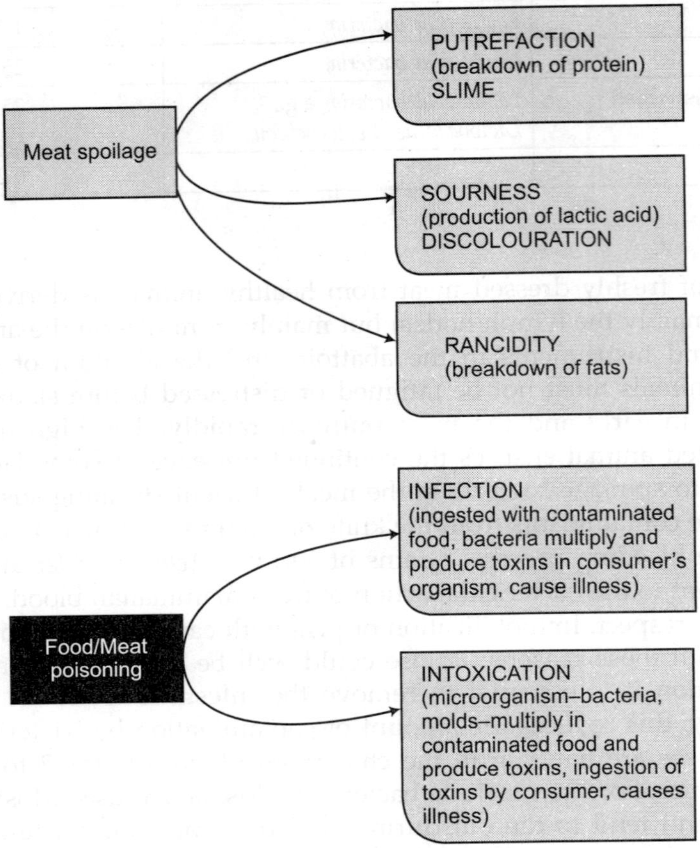

Figure 21.9 Meat and food spoilage factors.

species but these molds rarely develop with the shorter storage periods and damper conditions now in vogue.

4. Rancidity can be caused by lipolytic bacteria and yeasts.

After a chilling period the carcasses are transported, cut and re-chilled. Each cut re-distributes the organisms and adds to the total number of bacteria, so that minced meat, which is cut most of all, has the highest count. The microfloras of chilled carcasses and cut meat are mainly aerobic surface contaminants with very few anaerobic spore-formers. Good quality meat has an initial load of *ca* 10^3-10^4 bacteria/cm^2 and should not become slimy before 14 days at 0°C. When slime does occur the count is usually 3 to 10 × 10^7 bacteria/cm^2, mainly *Pseudomonas* and *Achromobacter* species.

When meat has lost its freshness it becomes offensive and is not likely to be eaten. The consumer, however, may be unable to judge whether apparently fresh meat has come from a diseased animal and he is protected against this risk by inspection of animals slaughtered for food. Qualified inspectors examine the animal both before and after death and only satisfactory meat is released for human consumption. Premises where meat is sold or prepared are regularly inspected. The prevention of cross-infection from meat is the complete eradication of certain diseases from the domesticated animal population within a country.

21.8 BACON

Originally bacon is prepared by packing split pig carcasses in dry salt for a period, giving a brown tough product. It is now customary to inject the meat with brine containing sodium chloride, potassium nitrate, sodium nitrite, polyphosphate, etc., followed by steeping in a similar solution for several days. The brine tanks contain large number of bacteria (10^9/mL), mainly salt-tolerant micrococci and some lactobacilli, part of whose function is the reduction of nitrate to nitrite. The sliced bacon is vacuum-packed in oxygen-impermeable plastic wrapping; the exclusion of air prevents the attractive pink colour of nitrosomyoglobin from fading. The bacterial flora of micrococci (10^5 to 10^6/g) is fairly stable at low temperature, but it changes rapidly at higher storage temperatures. Slices with a normal salt content (5 to 7 per cent w/v in the aqueous phase) spoil after about 15 days at 20°C with the scented sour smell derived mainly from lactic acid bacteria. At 30°C the slices putrefy from the proteolytic action of the mesophilic coagulase-negative staphylococci. Increasing the salt content in the bacon to 8 to 12% reduces the number of lactic acid and proteolytic bacteria, thereby increasing the storage life at 30°C. Sliced bacon can support the growth of *Staphylococcus aureus* if the storage temperature is 30°C and the numbers of competing organisms are low. Packs of sliced bacon should therefore be kept cool during sale and in the home. The same is also true of cooked or smoked hams.

21.9 FISH

The flesh of healthy fish is largely free of organisms, but their external surfaces carry an appreciable bacterial flora (skin 10^2 to 10^7/ cm^2, gill tissue and intestines 10^3 to 10^8 /g). The aerobic bacteria of marine fish are largely psychrophiles, species of *Pseudomonas, Achromobacter*, coryneforms, *Flavobacterium*, micrococci, *Vibrio*, and possibly *Alcaligenes*. The percentage composition of the flora differs in fish from different parts of the world, with larger numbers of mesophiles on fish from warmer seas. The intestines usually contain some strictly anaerobic bacteria, similar to those found in deposits on the sea bottom. In certain parts of the world there has been a progressive increase in the incidence of botulism caused by fish contaminated with *Clostridium botulinum* Type E, an organism able to grow at low temperatures. Intestines of fish caught in sewage polluted areas commonly contain typical coliform and food poisoning bacteria. Fish caught by deep-sea trawlers is eviscerated and packed in ice, that caught inshore may be so treated after landing. Evisceration contaminates the flesh with intestinal organisms and a low storage temperature is essential to prevent these from multiplying with consequent spoilage of the fish. As on meat, *Pseudomonas* species are the most active spoilage organisms at 0°C.

Cooling usually ceases once the fish reach the market leading to an increase in the flora already present. Further contamination is likely from fish boxes, filleting knives and boards, etc. The nature and amount of the microflora developing depends upon the temperature during transit and sale. Many fish are brined and/or smoked, with numerous permutations of treatment on both unsplit and split fish. The overall effect of brining is to reduce the Gram-negative bacteria and to increase Gram-positive types such as micrococci and coryneforms. Cold smoking, while reducing the total bacterial load considerably, has little effect on the composition of the flora. Molds never occur on fresh fish but they are a problem on smoked fish, the spores being derived from the sawdust used in smoke production. In contrast, hot-smoked fish, (i.e., processed at temperatures of 65-75°C for 30 minutes or longer) are usually sterile unless *Clostridium botulinum* was present originally and the salt concentration is below 3 per cent. Probably one of the safest fish products, as far as food poisoning organisms are concerned, is fish sticks. For these, frozen rectangular blocks of cod and haddock fillets are cut or sawn to size, dipped in batter, fried at 205 to 260°C for 2 minutes, packed and re-frozen. The bacterial counts remain at a very low level since the interior of the sticks remain frozen throughout.

21.10 FERMENTED EASTERN FOODS AND BEVERAGES

Fermentations of bland starch (cereal products) and protein (particularly fish) with molds, yeasts, and bacteria have been practised for centuries in the Orient to add flavour, improve nutrient quality, or prevent spoilage. Some of these fermented foods and beverages are an important source of minerals and vitamins in the diet of large populations. Table 21.10 lists a few of the many preparations known. Fermented drinks, prepared from millet or other local cereals, are also an important source of vitamins in the Japan, diet of many African populations.

Table 21.10 Some ethnic fermented food

Product	*Raw Material*	*Microorganisms used*	*Country of origin*
Katsuobushi	Bonito (fish)	*Aspergillus glaucus*	Japan, Thailand.
Nam-pla (fish sauce)	Fish	Halophilic bacteria	SEAsia
Shoyu	Wheat and Soybean	*Aspergillus oryzae,* yeast and *Lactobacillus* sp.	Japan
Natto	Soybean	*Bacillus subtilis*	Japan
Tempeh	Soybean	*Rhizopus oligosporus*	Indonesia
Sufu	Soybean curd	*Actinomucor elegans*	China
Hamanatto	Soybean	*Aspergillus oryzae*	Japan
Tao-Si	Soybean	*Aspergillus oryzae*	Phillinines
Tu-Su	Soybean	*Asoergillus oryzae*	China

Contd...

Contd...

Ontjom	Peanut press cake	*Neurospora sitophila*	East Indies
Sake	Rice	*Lactobacillus* sp. and *Saccharomyces cerevisiae*	Japan
Indian idli	Soaked rice and mung beans are soaked separately, ground and incubated.	*Leuconostoc mesenterodies* and *Streptococcus faecalis*	Common in Southern India
African acid porridges	Starchy cereal	*Corynebacterium manihot* and *Geotrichum candidum*	Common in African countries
Gundruk	Leafy brassica vegetable	Fermented and dried sour; high vitamin B complex	All Nepalese, Sikkim and Bhutanese

21.10.1 Lactic Pickles

Pickling is the process of preserving food by anaerobic fermentation in brine or vinegar. The resulting food is called a pickle. This procedure gives the food a salty or sour taste. In South Asia, edible oils are used as the pickling medium with vinegar. They are made from certain individual varieties of vegetables and fruits that are chopped into small pieces and cooked in edible oils like sesame oil or brine with many different Indian spices like asafoetida, red chili powder, turmeric, fenugreek, and plenty of salt. Some regions also specialize in pickling meats, mushroom and fish. Vegetables can also be combined in pickles to make mixed vegetable pickle. Some varieties of fruits and vegetables are small enough to be used whole. When both salt concentration and temperature are low, *Leuconostoc mesenteroides* dominates, producing a mix of acids, alcohol, and aroma compounds. At higher temperatures *Lactobacillus plantarum* dominates, which produces primarily lactic acid. Many pickles start with *Leuconostoc*, and change to *Lactobacillus* with higher acidity. Another distinguishing characteristic is a pH less than 4.6, which is sufficient to kill most bacteria. Pickling can preserve perishable foods for months. Antimicrobial herbs and spices, such as mustard seed, garlic, cinnamon or cloves, are often added. If the food contains sufficient moisture, pickling brine may be produced simply by adding dry salt. For example, German sauerkraut and Korean kimchi are produced by salting the vegetables to draw out excess water. Natural fermentation at room temperature before sunlight, by lactic acid bacteria, produces the required acidity. Other pickles are made by placing vegetables in vinegar. Unlike the canning process, pickling (which includes fermentation) does not require that the food be completely sterile before it is sealed. The acidity or salinity of the solution, the temperature of fermentation, and the exclusion of oxygen determine which microorganisms dominate, and determine the flavour of the end product. Pickles produced in traditional method contain *Lactobacillus*, produced by fermentation in brine (salt and water). Pickles produced using vinegar does not contain *Lactobacillus*. *Lactobacillus* makes traditional pickles probiotic.

While vinegar is commonly used in the Western world in the preparation of pickles, many vegetables are preserved by fermentation with lactic acid bacteria. Thus in the preparation of Sauerkraut, the solid, round-headed cabbages are washed to remove the undesirable Gram-negative bacteria on the outer leaves *(Pseudomonas, Flavobacterium,* and *Achromobacter* sp.) and then sliced and mixed with brine. Sugar, mainly sucrose, nutrients, and mineral salts are leached into the liquid which is then fermented by a succession of three bacterial species, derived from the innermost leaves. The heterofermentative *Leuconostoc mesenteroides,* and *Lactobacillus brevis* produce lactic and acetic acids, carbon dioxide, alcohol, mannitol, and dextran while the homofermentative *Lactobacillus plantarum* produces lactic acid only. The optimum temperatures and salt concentrations are 13-18°C and 1.8 – 2.25% salt. At the upper ends of these limits, the homofermentative species *Streptococcus faecalis* and *Pediococcus cerevisiae* tend to replace the less salt-tolerant *Leuconostoc mesenteroides.* At 3% salt the product is tough and can develop a pink colouration with the growth of carotenoid-producing yeasts. At too low a salt concentration the product is unpleasantly soft due to the growth of bacteria capable of digesting pectin and cellulose. Pure culture inoculations have never produced the succession of species that gives the best flavoured product. Olives and small cucumbers are also brined and preserved by the action of lactic acid bacteria.

21.11 COCOA AND COFFEE

Fermentation is also used to prepare certain products for processing, as with cocoa and coffee beans, where removal of unwanted outer layers is facilitated by the action of microorganisms. Coffee and chocolate require *Erwinia dissolvens, Leuconostoc,* and *Lactobacillus* species plus the yeasts of the genus *Saccharomyces* to remove the tough outer coats. The microbes do not affect the taste of coffee but are necessary to confer the characteristic taste to cocoa and chocolate. The bacteria *S. napoli* and *S. eastbourne* often use chocolate as a vector. It is thought that the chocolate provides protection for the bacterium as it passes through the acidic environment of the stomach. This was observed when higher incidents of illness were reported in children.

21.12 EGGS

The contents of some 90% of freshly laid hens' eggs are sterile; their shells and the interior of the remainder contain Gram-positive bacteria of ovarian origin. However, the predominant species inside rotten eggs are Gram-negative rods and extra-genital in origin, derived from the nesting material and not the faeces. The predominant species are *Alcaligenes faecalis, Aeromonas liquefaciens, Proteus vulgaris,* non-proteolytic strains of *Cloaca* sp., *Citrobacter* sp., and *Pseudomonas fluorescens.* These organisms possess combinations of proteolytic and lecithinase activity, pigment and hydrogen sulphide production. The progress of bacterial rotting follows a definite pattern. First there is contamination and penetration of the shell, followed by growth in the cell membranes. The extent of growth is controlled by the antimicrobial action of the conalbumin in the egg albumen. Eventually the yolk makes contact with the shell membrane and the organisms are able to utilize the glucose of the albumen and infect the complete contents of the egg.

Salmonellae can proliferate in eggs without producing any visible symptoms other than a faint turbidity in the albumen. *Salmonella typhimurium, S. anatum,* and *S. enteritidis* are frequently encountered even in eggs from clinically normal birds and the importance of duck eggs as a source of human salmonellae is well established. Humans can become infected by salmonellae not only from fresh whole eggs but also from frozen, liquid, or spray-dried products made from lower grade eggs. Because of the lack of obvious spoilage symptoms, salmonella infections are not apparent and the contents of such eggs are added to the main bulk for processing. This is in addition to any contamination from the hands of the process workers or from the containers into which the eggs are poured.

The control of infection from egg sources relies on reduction of the incidence of infection in the flock, reduction of contamination of the shells by faeces, special attention to the sites where eggs are laid, the sale of only clean, unwashed eggs, and the education of kitchen staff to encourage adequate cooking of food containing eggs or its products.

21.13 BREAD

In the conventional bread making process, a dough of flour, water, salt, etc., is allowed to ferment with a culture of the yeast *Saccharomyces cerevisiae,* originally overnight, now for 3 hours. The dough is worked, allowed to rise, then reworked, placed into tins, allowed to rise for a further 50 minutes (final proof) and baked. This is now being replaced by the high-energy system (Chorleywood Bread Process) in which the dough is worked mechanically. The amount of yeast is increased by 50% and the addition of a fast-acting oxidizing agent and fat are essential parts of the process. Production time is decreased by some 60%, much less space at controlled humidity is required in the bakery, the loaf has a lower staling rate, and the yield of bread is increased 4-5 per cent. Yeasts produced by controlled hybridization are now available and are particularly suitable for modern methods of bread making.

Both compressed yeast and bakery products are subject to contamination. Bacteria contaminating compressed yeasts are mainly lactobacilli, normally considered harmless; only heavy infections of the slime-forming *Leuconostoc mesenteroides* are liable to affect the baking properties of the yeast. Nowadays contamination with wild yeasts is rare, but before rigid sanitation programmes were instituted in yeast factories the presence of such aerobic yeasts as *Candida krusei, C. mycoderma, C. tropicalis, Trichosporon cutaneum, T. candida,* and *Rhodotorula mudlaginosa* was not uncommon. The widespread use of moisture-impervious wrappings has made control of molds such as *Aspergillus* and *Penicillium* species and *Oidium (Oospora) lactis* of special importance, since they can form unsightly blotches on the surface of the yeast block beneath the wrapping. Acid calcium phosphate and acetic acid are commonly used as preservatives in bakery products to prevent the development of bacterial ropiness. The addition of propionates in bread, sorbic acid and its salts in flour, and both these in baked products are being allowed to use for the prevention of mold growth. Reducing the available water content (a_w) to as low a level as possible without spoiling texture and appearance helps to delay mold growth but, in the absence of fungicides, cannot prevent it. Strict cleanliness within the bakery is absolutely essential to ensure that any waste flour is quickly removed, otherwise it becomes moldy and spores are discharged into the atmosphere.

Sour dough bread is made with a mixture of *Lactobacillus brevis* (provides aroma), *L. plantarum* (provides crumb elasticity), and yeast (for raising power). Increasing the holding temperature decreases the percentage of yeast in the mix, until at 35°C it is practically yeast-free.

21.14 BEER AND LAGER

An alcoholic beverage is a drink that contains ethanol, commonly known as alcohol (although in chemistry the definition of "alcohol" includes many other compounds). Beer has been a part of human culture for 8000 years. In Germany, Ireland, the United Kingdom, and many other European countries, drinking beer (and other alcoholic beverages) in a local bar or pub is a cultural tradition. Non-alcoholic beverages are drinks that usually contain alcohol, such as beer and wine, but contain less than 0.5% alcohol by volume. This category includes low-alcohol beer, non-alcoholic wine, and apple cider.

Wines are made from a variety of fruits, such as grapes, peaches, plums or apricots. The most common wines are produced from grapes. The soil in which the grapes are grown and the weather conditions in the growing season determine the quality and taste of the grapes which, in turn, affects the taste and quality of wines. When ripe, the grapes are crushed and fermented in large vats to produce wine. Beer is also made by the process of fermentation. A liquid mix, called wort, is prepared by combining yeast and malted cereal, such as corn, rye, wheat or barely. Fermentation of this liquid mix produces alcohol and carbon dioxide. The process of fermentation is stopped before it is completed to limit the alcohol content. The alcohol so produced is called beer (Figure 21.10). It contains 4 to 8 per cent of alcohol.

Beer is made by fermenting a hop-flavoured extract of barley malt with a top-fermentation strain of *Saccharomyces cerevisiae.* The bottom-fermenting yeast, *S. carlsbergensis,* is required for lager. Barley is converted into malt by allowing the soaked grain to germinate under controlled conditions of temperature and humidity, leading to the formation of α-and β-amylases, and the breakdown of the protein, hordein, to amino acids. The process is stopped by raising the temperature and removing the rootlet and plumules. The exact treatment depends on the type of malt required. In some countries, the protein is fully converted into amino acids, lager malts still contain appreciable quantities of protein, for beer manufacture. Stout malts are dried at 105°C to give a dark-coloured extract; malts from distilleries and vinegar breweries receive no final kilning and still contain limit-dextrinase.

An extract or wort is now prepared from the ground malt, the exact conditions of extraction again being varied according to the type of beer. In some countries starch is hydrolysed by the action of the amylases to glucose, maltose, malto-triose and -tetrose, and oligosaccharides; glucose and fructose are also present in the extract. Worts from lager also need extensive proteolytic action during mashing. Sometimes flaked barley

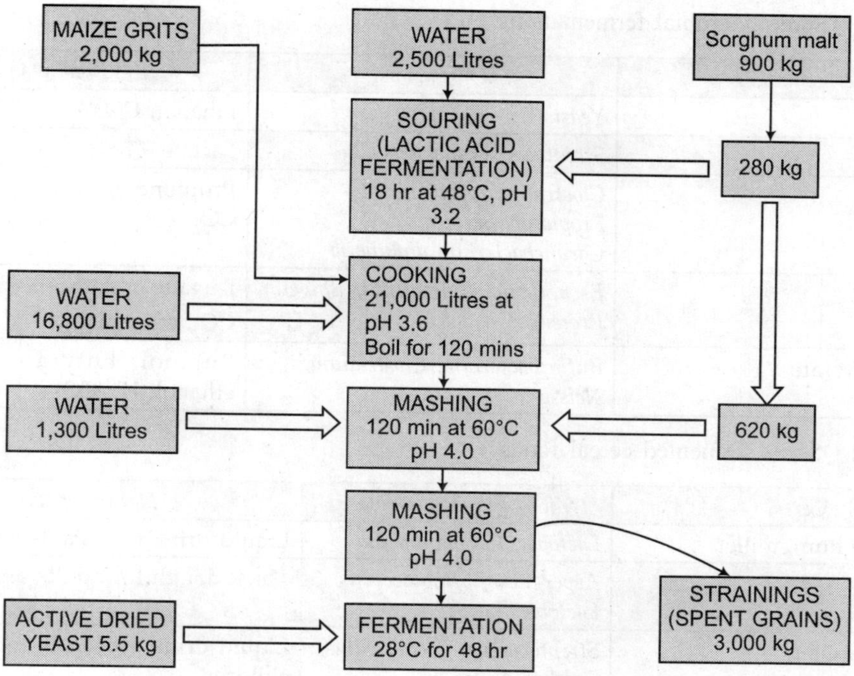

Figure 21.10 Flow chart for the production of beer.

or maize, or gelatinized unmalted cereal grits are added to supplement the starch supply (Table 21.11 and Table 21.12). When extraction is complete the wort is boiled under pressure with hops, the tannins from which co-precipitate with any remaining proteins. A proportion of the humulone and lupulone complexes are converted into their corresponding isomers during boiling, ultimately giving bitterness to the beer and conferring some resistance to Gram-positive bacteria. The cooled and filtered wort is 'pitched' with yeast and allowed to ferment. Distillery and vinegar brewers' worts are not boiled (so that further saccharification due to the enzyme limit-dextrinase can take place during fermentation), neither are they hopped. Bacterial infection is often restricted by inoculating part of the wort with *Lactobacillus delbrueckii* and, when acidified sufficiently, boiling this before returning it to the remainder of the wort ready for yeasting; sulphuric acid may be used as an alternative method of reducing the pH. Again, some power-alcohol distilleries sterilize the wort chemically with 1000 ppm ammonium fluoride and use a fluoride-adapted yeast.

The two types of brewing yeasts differ not only as to species but also in other properties. *Saccharomyces cerevisiae* has round cells, most of the crop rises to the surface at the end of the fermentation and, characteristically, when supplied with raffinose ferments only one-third of the molecule, leaving melibiose. *Saccharomyces carlsbergensis* is cylindrical, forms a deposit and can ferment the whole of raffinose. There are strains of both yeasts with intermediate morphology and brewing properties; the latter can be modified of course by the brewing system employed. Beer wort is yeasted with between 2 and 6 g moist yeast/L, depending on its original specific gravity. After 12-

Table 21.11 Some microbial fermentations

Fermentation	*Organisms*	*Products*
Ethanol	*Yeasts*	Ethanol, CO_2
Lactate	*Streptococcus, Lactobacillus*	Lactate
Propionate	*Clostridium propionicum, Propionibacterium, Corynebacterium diphtheria*	Propionate, acetate, succinate, CO_2
Mixed acid	*Escherichia, Salmonella, Shigella, Proteus*	Lactate, acetate, succinate, H_2, CO_2, ethanol
Butanol-butyrate	*Butyribacterium, Clostridium, Neisseria*	Butanol, butyrate, acetone, ethanol. H_2, CO_2

Table 21.12 Some fermented cereal foods

Cereal	*Microorganism involved*	*Nature of use*
Maize, sorghum, millet	*Lactobacillus plantarum*	Liquid drink for infants and young
Maize	*Aspergillus, Streptococcus, Lactobacillus*	Thick dough for adults and children
Sorghum, millet	*Streptococcus, Lactobacillus, Saccharomyces*	Liquid drink especially for sick children
Maize	Lactic acid bacteria	Liquid drink for adults and children
Maize, sorghum	Lactic bacteria yeasts and molds	Paste for infants and young children

18 hours at 15°C a rapid fermentation ensues, the temperature is raised by 3 to 7°C and the pH falls from 5.2 to 4.1. The thick yeast head that collects after 21 days is skimmed off, fermentation slackens and the beer, still containing some sugar, is ready for clarification and natural conditioning in tanks before sale. Some modern breweries use a non-flocculant yeast that ferments very rapidly; the fermentation is terminated at the desired sugar content by centrifugation. Traditionally, part of the yeast crop (often a balanced collection of strains) was collected and used for the next fermentation. Now a days pure cultures are being used increasingly and are replaced after 12 to 20 brews, depending on the standard of hygiene. Lager has always been made with a pure culture, the crop being collected from the bottom of the tank after the main fermentation, subsidiary crops in the pre-fermentation and conditioning tanks being discarded. The pitching temperature, 5 to 9°C, is lower than with beer; fermentation at 12°C is more protracted (7 to 14 days), and storage at 2°C may take 6-40 weeks before the lager is considered sufficiently conditioned and stabilized for sale. Distillery worts of high specific gravity are pitched at 21°C with a heavy inoculum of a yeast specially bred for alcohol tolerance; fermentation is rapid, the temperature rising to 25-30°C during the process. Either a fresh culture is used for each fermentation or the previous crop is acid-washed and reused.

21.15 CIDER AND WINE

Successful brewing is dependent not only on the biochemical properties of a yeast strain but also on the physical properties of the mass culture. In contrast, cider and wine were made by controlling a natural flora of yeasts, molds, and bacteria, so that only organisms with desirable properties were allowed to grow. The microbiology of cider is being discussed here, that of wine is similar in many respects. In the traditional cider-making countries, the raw material consists of special varieties of apples, high in tannin and low in acidity. Nowadays these are supplemented with dessert and culinary apples graded as unsuitable for the fresh fruit market; other countries use these latter sorts as their sole source of raw material. The fruit as it arrives at the factory carries both an internal and an external microflora. Some of the common molds are *Candida pulcherrima, Kloeckera apiculata, Torulopsis jamata, Aspergillus niger, Botritis cinerea, Penicillium* sp. together with the 'black yeast' *Aureobasidium pullulans* and occasionally rhodotorulae, *Candida,* and *Torulopsis* sp. *Saccharomyces* sp. and bacteria are rare in sound fruit but mold-damaged specimens carry large populations, not only of mold spores but also of acetic acid bacteria and fermenting yeasts. Both the molds and acetic bacteria produce sulphite binding compounds that are of great significance at later stages of processing. Hence, sound, preferably washed, fruit is essential. The apples are milled to a pulp and, pressed hydraulically in cloth envelopes between wooden drainage racks. Formerly the racks and cloths were washed very infrequently, consequently they soon supported a dense population of *Saccharomyces* species; these were the yeasts mainly responsible for the subsequent fermentation. At first *Candida* sp. and other poorly fermenting yeasts are most numerous but they are rapidly overgrown by *Kloeckera* sp. after a few degrees drop in specific gravity. The latter then co-exist and in turn, are overgrown by *Saccharomyces* sp. while in the fully fermented dry ciders film yeasts and acetic bacteria become more important. The whole process is improved by treatment of the freshly pressed juice with sulphur dioxide, the actual amount being varied according to the pH. Below pH 3.3, 100 ppm is sufficient, while 150 ppm is desirable between 3.3 and 3.8; to be effective more SO_2 is required above 3.8 than is allowed by law (200 ppm), but most factories process a mixture of apples whose juice pH is generally 3.5-3.6. It is vitally important that the concentration of sulphite binding compounds be kept low in the juice, otherwise no free sulphur dioxide remains in the solution and this is the only effective moiety (sometimes referred to as undissociated sulphurous acid). After treatment only *Saccharomyces* sp. remain, so that in effect sulphiting selects a pure culture from the original microflora. However, with higher standards of cleanliness in the press rooms, the juices have a flora similar to that of the fruit, i.e., virtually free of fermenting yeasts. Hence, the addition of pure cultures of bottom-fermenting wine yeasts, following sulphiting, is becoming almost universal in cider factories. It also means that both the time needed for fermentation and the flavour of the cider become more predictable, since there is no longer any need to rely on what were virtually factory contaminants.

A second fermentation of malic acid to lactic acid and carbon dioxide, due to lactic acid bacteria, nearly always occurs during either the yeast fermentation or storage of the cider. The organisms that can bring about this change include both homo- and heterofermentative lactic rods and cocci, including the organisms responsible for producing polysaccharide slime or ropiness. The bacterium found most frequently is *Lactobacillus pastorianus* var. *quinicus.* If citric acid is present, as in perry but not cider, diacetyl, lactic, and acetic acids are also formed and the flavour is thereby spoiled. The source of these bacteria is still in some doubt since they appear to be absent from many sulphited juices. Ciders stored in contact with air, especially those held in small wooden containers, soon show surface growths of *Acetobacter xylinum* and the film yeasts *Pichia membranaefaciens* and *Candida mycoderma.* The problem is now rare in large factories where juice sulphiting and very large storage tanks are normal practice.

Basically, wines are made by a similar process but many other fermentation treatments are possible, due to the much wider range of climatic conditions in which grapes are grown. Thus grapes can be 'raisinified' by being left on the vine or by being dried in the sun; wines made from their high gravity juices are fermented with osmophilic yeasts such as *Saccharomyces rouxii.* Sauterne is made from grapes naturally infected with the mold *Botrytis cinerea* that dehydrates the grapes, metabolizes some of the acid and produces glycerin and the mild antibiotic *botryticin.* Consequently the subsequent fermentation of the very sugary juice ceases prematurely, leaving a soft, sweet, luscious-flavoured wine. Climate also has an effect on the type of yeasts found on the grapes. The greater the ambient temperature, the greater the concentration of alcohol-tolerant *Saccharomyces* yeasts and sporing apiculate yeasts *(Hanseniaspora valbyensis)* and the fewer there are of their non-sporing counterparts, (e.g., *Kloeckera magna* and *K. apiculata).* Finally, a number of *Saccharomyces* species have been isolated from both cider and wine. These are *S. acidifaciens, S. carlsbergensis, S. cerevisiae, S. delbrueckii, S. elegans, S. florentinus, S. fructuum, S. oviformis, S. rosei, S. rouxii,* and *S. uvarum.* In addition, there are other *Saccharomyces* spp. apparently characteristic of each beverage. *Brettanomyces* sp. have been isolated from cider and wine, in the Bordeaux area and in dry wines made within a 40-mile radius of Cape Town. There is one wine for which no parallel exists in cider-making, namely, sherry. The yeasts responsible for the characteristic fermentation of alcohols and organic acids, sometimes called *S. beticus, S. cheresiensis,* etc., develop as a veil on the surface of the dry wine when it is exposed to air. Unlike C. *mycoderma* and other spoilage film yeasts, the sherry yeasts form acetaldehyde and other characteristic flavours from the alcohol; the remaining process is a complicated blending system (the solera) also used for the sun-baked wine, Madeira.

21.16 FOOD AS A SOURCE OF INFECTION AND DISEASE

The two most common types of food-borne illness are intoxication and infection. A food-borne infection is caused by the ingestion of food containing pathogenic microorganisms (i.e., bacteria, virus or parasite) which must multiply within the gastrointestinal tract, producing widespread inflammation. The food-borne illness may be due to *Campylobacter jejuni, Clostridium botulinum, Clostridium perfringens, Cyclospora cayetanensis, Escherichia coli,* (0157:H7), *Shigella, Salmonella,* and *Staphylococcus aureus. Campylobacter jejuni* can be transmitted to humans via unpasteurized milk, contaminated water, and raw or undercooked meats, poultry, and shellfish. *C. botulinum,* which causes botulism, is the most deadly of all food pathogens. Toxins, not bacteria, cause illness. Bacteria that produce toxins include *S. aureus* and *C. botulinum*. Symptoms may include nausea, vomiting, diarrhoea, abdominal pain or distress, and fever. Pathogens may be introduced by flies or, most important of all, by the handling of foods by human carriers of food poisoning organisms. Most risk is entailed with foods prepared for consumption the following day, the organisms multiplying if the food is kept at a temperature conducive to growth.

21.16.1 Food Poisoning

A special group of diseases contracted by the ingestion of food are designated 'food poisoning'. Outbreaks of these are usually explosive in character and may be related to ingestion of a specific meal or common food source. Food poisoning may be due to poisons derived from plant or animal sources, or to chemicals added inadvertently or as preservatives at too high concentrations, or to the presence of harmful microorganisms or toxins produced by them. The symptoms produced by different poisons may be similar and the final diagnosis thus depends upon laboratory tests. The time from the consumption of the poisonous food to the onset of symptoms varies from 10 minutes to 2 hours with some forms of chemical poisoning, to around 6 hours with bacterial toxins, 12 to 72 hours when poisoning is due to living bacteria and 2-3 days when poisonous toadstools are the cause.

21.16.1.1 Bacteria causing food poisoning

The type of foods involved in food poisoning are those favouring growth of the causative organisms, so that large numbers of bacteria or their products are present in the food at the time it is ingested. Outbreaks of food poisoning frequently involve a large proportion of the people who have eaten a particular food. Two groups of agents are recognized: pathogenic bacteria which can infect the alimentary canal and produce symptoms of gastroenteritis, and bacteria (or their toxins) which do not infect the gut but produce various toxic symptoms when ingested with food.

21.16.1.1.1 Infective food poisoning

Salmonellosis

Salmonellosis is a form of food infection that may result when foods containing Salmonella bacteria are consumed. In this form of food poisoning the infective agent is able to establish itself in the gut, multiply rapidly and produce symptoms of disease over a number of days, often for about a week or longer. This form of poisoning is mainly· limited to species of *Salmonella.* When infection is established, endotoxins, resulting from the breakdown of dead bacterial cells, cause irritation to the gut linings resulting in symptoms of acute gastroenteritis. The Salmonella group includes many hundreds of types of serologically related organisms which are found in almost every species of animal, including birds and reptiles. In Great Britain meat, milk, synthetic cream, and eggs (particularly duck eggs) are the foods most commonly contaminated with salmonellae. Meats from cattle and pigs have been found to be a frequent source of infection, particularly when made up into processed foods such as brawns, sausages or pies, which may be incompletely cooked or subsequently contaminated during handling. Ideally, salmonella should be absent from food but if adequately cooked, these pathogens are destroyed and there is no risk of food poisoning, since there are no residual toxins to produce food poisoning symptoms. Symptoms of salmonella food poisoning vary in severity but include acute gastroenteritis accompanied by severe headaches followed by nausea, vomiting, abdominal pain, and diarrhoea. Fever frequently occurs. These symptoms usually arise between 12 and 72 hours after the meal has been consumed, by which time the organisms are established in the gut. Symptoms of toxic food poisoning may arise earlier if sufficiently large numbers of organisms are ingested. In favourable cases the symptoms subside within a week, but in a small percentage of patients organisms invade the tissues from the gut and may cause death. Even when symptoms disappear, the patient may continue to excrete the pathogens and thereby remain a potential source of infection for susceptible individuals and of contamination of food. If this continues for a long period after recovery, the patients are termed carriers. It is most important to ensure that such people are not employed in food handling.

Cholera

Cholera is a severe water-borne diarrhoeal disease caused by *Vibrio cholerae*, a Gram-negative curved bacillus bacterium, transmitted almost exclusively via contaminated water, food, etc. *V. cholerae* and *V. parahaemolyticus* are pathogens of humans. Both produce diarrhoea, but in ways that are entirely different. *V. parahaemolyticus* primarily affects the colon while *V. cholerae* is noninvasive and affects the small intestine by secreting an enterotoxin.

V. cholerae produces cholera toxin, the model for enterotoxins, whose action on the mucosal epithelium is responsible for the characteristic diarrhoea of the disease cholera. The clinical description of cholera begins with sudden onset of massive diarrhoea. This results from the activity of the cholera enterotoxin which activates the adenylate cyclase enzyme in the intestinal cells, converting them into pumps which extract water and electrolytes from blood and tissues and pump it into the lumen of the intestine. This loss of fluid leads to dehydration, anuria, acidosis and shock. The watery diarrhoea is speckled with flakes of mucus and epithelial cells and contains enormous numbers of vibrios. The loss of potassium ions may result in cardiac complications and circulatory failure. The immediate treatment of the disease is the oral rehydration therapy with NaCl plus glucose to estimate water uptake by the intestine. Antibiotics such as tetracyclin, trimethoprim-sulfamethoxazole or ciprofloxacin are commonly used for treatment.

Camphylobacteriosis and Helicobacteriosis

Campylobacteriosis is caused by consuming food or water contaminated with *Campylobacter jejuni*. *C. jejuni* generally inhabits intestinal tracts of healthy animals (especially chickens) and in untreated surface water. Raw and inadequately cooked foods and non-chlorinated water are the main sources of human infection. The organism grows best in a reduced oxygen environment, is easily killed by heat. Diarrhoea, nausea, abdominal cramps, muscle pain, headache and fever are common symptoms. Onset usually occurs two to five days after eating contaminated food. Duration is two to seven days, but can be weeks with such complications as urinary tract infections and reactive arthritis.

Infection of gastric epithelial cells with *Helicobacter pylori* induces strong pro-inflammatory responses by activating nuclear transcription factors NF-KB and AP-1. Several reports indicate that multiple bacterial factors and cellular molecules are involved in this signaling.

Listeriosis

Listeriosis is a dangerous infection caused by eating food contaminated with bacteria called *Listeria monocytogenes,* small Gram-positive rods and non-spore forming. The *L. monocytogene* infections are common in wild animals, domesticated animals, and in soil and water. Raw milk or products made from raw milk may carry these bacteria. The bacteria most often cause a gastrointestinal illness. In some cases, it may lead to septicaemia or meningitis. The bacteria may cross the placenta and infect the developing baby. Infections in late pregnancy may lead to stillbirth or death of the infant within a few hours of birth. About half of infants infected at or near term will die.

21.16.1.1.2 Bacterial cells or their products as cause of food poisoning

In this form of food poisoning no infection of the gut occurs but bacterial cells or exotoxins of bacterial origin produce symptoms of food poisoning when ingested in foods. Since these toxins are absorbed by the gut wall they are called enterotoxins. They are preformed in the food by growth of the pathogenic organism and although the vegetative cells may be killed by heating the food prior to eating, many of the toxins, being moderately resistant to heat, remain and cause symptoms of food poisoning when ingested. The most important bacterial toxic food poisonings are caused by the toxins of specific strains of *Staphylococcus aureus* and *Clostridium botulinum,* the latter frequently causing a fatal form of food poisoning known as botulism. Strains of *Clostridium perfringens* also cause toxic food poisoning and certain non-pathogenic bacteria, such as *Proteus* sp., not generally considered to be toxigenic, produce acute gastroenteritis when consumed in sufficiently large numbers.

Staphylococcal Food poisoning

Most outbreaks of this type of food poisoning are associated with milk and its products (e.g., cream fillings in cakes and milk deserts such as custards) and prepared meat (e.g., pies or brawns). Staphylococcal food poisoning is due to enterotoxin B and the symptoms are characterised by severe vomiting, diarrhoea, abdominal pain and cramps. Recovery usually occurs within 6 to 24 hours. These foods favour the rapid multiplication of staphylococci (and consequent toxin production) in warm weather if they are not refrigerated. In non-pasteurized milk the organisms may come from an infection of the cow's udder, but since staphylococci are frequently found on human skin and in the nasopharynx, there are many opportunities for contamination of the food during preparation. Fortunately, only a few types of staphylococci (generally, phage types of group III) are able to produce the enterotoxin which is formed in significant amounts only when the bacterial count reaches 10^5 to 10^6 per gram of food. Symptoms of staphylococcal food poisoning may arise after only 1 hour but more usually after about 12 hours from ingestion of the toxin, and include violent vomiting, diarrhoea, and prostration but there is no fever and recovery is rapid, often within 24 hours. This type of poisoning is only occasionally fatal.

Botulism

Clostridium botulinum is a spore-bearing anaerobe occurring in soil. It may be ingested with food but does not multiply in the body. Inadequate sterilization of foods originally contaminated from the soil and stored under anaerobic conditions conducive to the growth of this organism are responsible for the majority of outbreaks. There are 7 types (type A, B, C, D, E, F, and G) of these neurotoxins recognised on the basis of specificity. The neurotoxin of *C. botulinum* is a protein and a dose as low as 0.01 mg is said to be fatal to human beings. A and B producing strains often found on fruits and vegetables and honey. Symptoms appear 12-36 hours after ingestion and include nausea and vomiting (B and E) visual impairment: blurred, ptosis, dilated pupils, loss of mouth and throat function (A and B) dry mouth, throat, tongue, sore throat fatigue and loss of coordination respiratory impairment abdominal pain and either diarrhoea or constipation in general, cranial nerve first affected and then descend. Sixty to seventy per cent cases of botulism results to death. Neurotoxin (BoTox) is water soluble, produced as a single polypeptide 150,000 MW (progenitor), cleaved by a protease to form two polypeptides (H and L chains) which then become S-S bonded. All serotypes block the exocytotic release of acetylcholine from synaptic vesicles at peripheral motor nerve terminals. No case of botulism is known to have arisen as a result of eating fresh foods. The preformed toxin, when ingested, is absorbed through the mucosa of the stomach and upper part of the intestine. The symptoms differ from all other forms of food poisoning in that the toxin acts directly upon the central nervous system. Nausea and vomiting usually first occur within 24 hours and are followed several days later by paralysis of specific muscles due to damage of the nerve centres controlling them. Thus, double vision results from paralysis of the eye muscles, difficulty in swallowing follows paralysis of the throat muscles. If paralysis of the muscles of the respiratory tract occurs, death from respiratory failure results.

Non-specific bacterial causes of food poisoning

Occasionally, outbreaks of food poisoning have been reported in which the foods involved were heavily contaminated with species of *Proteus, Escherichia, Streptococcus,* or *Bacillus.* It is generally thought that large numbers of these organisms have an irritating effect on the gastrointestinal mucosa and set up symptoms of toxic food poisoning.

This emphasizes the fact that gross contamination of food with organisms generally considered to be non-pathogenic, followed by storage at temperatures conducive to bacterial propagation, presents a potential hazard to the consumer.

21.16.1.2 Poisonous fungi

Even edible species of fungi are liable to prove indigestible and to cause minor illness if eaten in excess and individuals may be allergic to particular species. A few agarics, however, are very poisonous. The most deadly species in Britain is the death cap, *Amanita phalloides*. The consumption of even a small piece of a fruit-body of this species may be fatal and at best will cause a long and serious illness. Attempts have been made to prepare a serum against this species but this is effective only if it is known at once that the death cap has been eaten. More often the victim is unaware that he has eaten a poisonous variety and the symptoms develop after only one or two days by which time treatment is useless. The best safeguard is the ability to recognize this and a few other poisonous species. The toxins of *A. phalloides,* which cause degeneration of liver and kidneys, and progressive paralysis of the central nervous system, are not destroyed by cooking, but those of some other species are inactivated by heat, so that fruit-bodies which are dangerous when eaten raw are safe after adequate cooking. For instance, the Ascomycete, *Gyromitra esculenta* produces a heat labile toxin, *helvolic acid,* which acts on red blood cells.

The sclerotia of the Pyrenomycete *Claviceps purpurea* the ergot of rye, are of medicinal value, owing to the presence of an alkaloid which is extracted and used to stop excessive bleeding, particularly in childbirth. They are also poisonous; rye grass infected with ergot can cause abortion in sheep and cattle, and flour prepared from grain admixed with the 'ergots' causes serious disease (formerly known as St. Anthony's Fire) in persons consuming it. Another Pyrenomycete, *Gibberella zeae,* infects the ears of cereals and renders the grain poisonous to man, pigs, and dogs but not to sheep or cattle. The darnel grass fungus, a non-sporing systemic endophyte of *Lolium temulentum,* is said to render the seeds of this grass poisonous to stock, but this has not been proved and it has been suggested that in the instances reported contamination with ergot had occurred.

Aflatoxins are toxic metabolites produced by certain fungi in/on foods and feeds. They are probably the best known and most intensively researched mycotoxins in the world. Aflatoxins have been associated with various diseases, such as aflatoxicosis, in livestock, domestic animals and humans throughout the world. The occurence of aflatoxins is influenced by certain environmental factors; hence the extent of contamination will vary with geographic location, agricultural and agronomic practices, and the susceptibility of commodities to fungal invasion during preharvest, storage, and/or processing periods. Aflatoxins have received greater attention than any other mycotoxins because of their demonstrated potent carcinogenic effect in susceptible laboratory animals and their acute toxicological effects in humans. As it is realized that absolute safety is never achieved, many countries have attempted to limit exposure to aflatoxins by imposing regulatory limits on commodities intended for use as food and feed.

21.17 LABORATORY DIAGNOSIS OF BACTERIAL FOOD POISONING

Confirmation of clinical diagnosis rests either on laboratory isolation of the causal organism, principally from faeces, or demonstration of the toxin either in the patient or in the incriminated food, if some is still available. Since in staphylococcal food poisoning the causal organism is rarely present in the faeces of patients, isolation and subsequent bacteriophage typing is attempted from food, vomit, etc. Often the staphylococci have been killed by subsequent heating of the food and then confirmation of this type of food poisoning is only presumptive. Whilst many animal tests to detect staphylococcal enterotoxin in food and culture filtrates have been explored, the only reliable method is to feed small amounts of culture filtrates to human volunteers in an attempt to reproduce food poisoning symptoms. The organism causing botulism is spore-bearing and its spores are highly heat-resistant. It may survive in cooked food, from which it may be isolated. As a further test, the toxin, obtained as a sterile filtrate from a suspension of the food or gut contents, is injected intraperitoneally into mice. If mice, protected against botulism by the injection of specific antiserum survive whilst nonprotected ones die, proof of the presence of botulism toxin is established.

21.17.1 Tools and Methods to Detect Microbes in Food

Many of these molecular tools have been accepted and implemented in standard protocols for detection and quantification of the most important pathogenic bacteria, such as *Salmonella* sp. and *Listeria monocytogenes*. Despite rapid diffusion of molecular tools in microbiology laboratories, there are still many drawbacks and obstacles concerning specificity, reproducibility and reliability of nucleic acids and antibody based technologies for microbial detection. That led, to some extent, to an underestimation of those methods as a permanent alternative to conventional culture-based detection techniques that sill represent the "golden standard" for microbial diagnostic. The complexity of food and water matrices and the cross reaction of some molecular probes to target sites of innocuous bacteria closely related to pathogenic ones are the main enthralling challenges which researchers are still arguing with. In order to maintain food safety standards, conventional microbiological methods are still being used to detect bacteria and other organisms in food (Table 21.13). However, these techniques are not ideal, as often it can be many days before results are known which may be of particular economic importance for those foods with a short shelf-life. The introduction of newer technology, such as nucleic acid probe and related amplification technology in other fields, has transformed the detection of many organisms. The Polymerase Chain Reaction (PCR) allows nucleic acid probes, with their inherent specificity, to be used to detect organisms present in very low numbers within a short period of time. However, at present, in food microbiology, there are technical problems with using the PCR, as certain components in food interfere with the reaction.

21.17.2 Commercial Food Safety Diagnostics

The implementation of food safety should be seen as an ongoing process, which is influenced by environmental, socio-economical, political and cultural factors. Food safety issues need to be addressed on a continuous basis, from a regional, national, European and global point of view. New, flexible tools are required for evaluating and managing new food safety challenges (Figure 21.11). To guarantee the safety of foodstuffs producers have therefore shifted their focus towards the use of food safety management tools, most important HACCP (Hazard Analysis Critical Control Point), and the consequent application of hygienic measures, based on Good Manufacturing/ Hygienic practice (GMP/GHP). Food safety management tools use input of scientific information to identify critical contamination points in the food chain and the production process, and design measures to control them. However, the lack of reliable data is often limiting the usefulness of this approach and therefore data collection is one of the priorities for future food safety strategies. In the absence of relevant data, other strategies might still have to be used to control food hazards. The Food and Drug Administration (FDA) has been involved in the regulation of *in vitro* diagnostic devices (IVDs or laboratory tests) since the introduction of the Medical Device Amendments of 1976. IVDs developed as kits or systems intended for use in multiple laboratories require review by the FDA before being marketed to ensure appropriate performance and labeling.

Table 21.13 Characteristics of microorganisms responsible for food-borne illness

Microorganism	*Food-borne illness*	*Symptoms*	*Common food sources*	*Incubation*
Bacillus cereus	Intoxication	Watery diarrhoea and cramps, or nausea and vomiting	Cooked product that is left uncovered milk, meats, vegetables fish, rice, and starchy foods	0.5-15 hours
Campylobacter jejuni	Infection	Diarrhoea, perhaps accompanied by fever, abdominal pain, nausea, headache, and muscle pain	Raw chicken, other foods contaminated by raw chicken, unpasteurized milk, untreated water	2-5 days
Clostridium botulinum	Intoxication	Lethargy, weakness, dizziness, double vision, difficulty speaking, swallowing, and/ or breathing; paralysis; possible	Inadequately processed, home-canned foods; sausages; seafood products; chopped bottled garlic; kapchunka; molona; honey	18-36 hours
Escherichia coli group	Infection	Watery diarrhoea, abdominal cramps, low-grade fever, nausea, malaise	Contaminated water, undercooked ground beef, unpasteurized apple juice and cider, raw milk, alfalfa sprouts, cut melons	12-72 hours

Contd...

Contd...

Listeria monocytogenes	Infection	Nausea, vomiting, diarrhoea; may progress to headache, confusion. loss of balance	Ready-to-eat foods contaminated with bacteria, including raw milk, cheeses, ice cream, raw	Unknown may range from a few days to 3 weeks
Salmonella	Infection	Abdominal cramps, diarrhoea,	Foods of animal origin; other foods contaminated through contact with feces,	12-72 hours
Shigella	Infection	Fever, abdominal pain and	Faecally contaminated foods	12-48
Staphylococcus aureus	Intoxication	Nausea, vomiting,	Foods contaminated by improper handling and holding temperature esmeats and meat	1-12 hours
Hepatitis A	Infection	Jaundice, fatigue, abdominal	Raw or undercooked molluscan	15-50 days
Norwalk-type viruses	Infection	Nausea, vomiting, diarrhoea,	Shellfish grown in faecally contaminated water;	12-48 days
Giardia lamblia	Infection	Diarrhoea, abdominal cramps,	Water and foods	1-2 weeks
Trichinella spiralis	Infection	Nausea, diarrhoea, vomiting	Raw and undercooked pork	1-2 days

So, what is there in existence to manage quality and safety, and how do they relate to each other? Below are listed the most well known methods to manage quality and/ or safety, and these will be briefly discussed individually and then how they integrate with each other. The food safety rules and their relationship are explained in Figure 21.11 and Figure 21.12.

Good Hygienic Practices / Good Manufacturing Practices

The terms GHP and GMP refers to measures and requirements which any establishment should meet to produce safe food. These requirements are prerequisites to other and more specific approaches such as HACCP, and are often now called prerequisite programmes. In recent years the term Standard Sanitary Operating Procedures (SSOP) has also been used in the US to encompass basically the same issues, i.e., best practices.

Hazard Analysis Critical Control Point (HACCP)

Hazard Analysis Critical Control Point (HACCP) is a systematic approach which identifies, evaluates, and controls hazards which are significant for food safety (CAC,

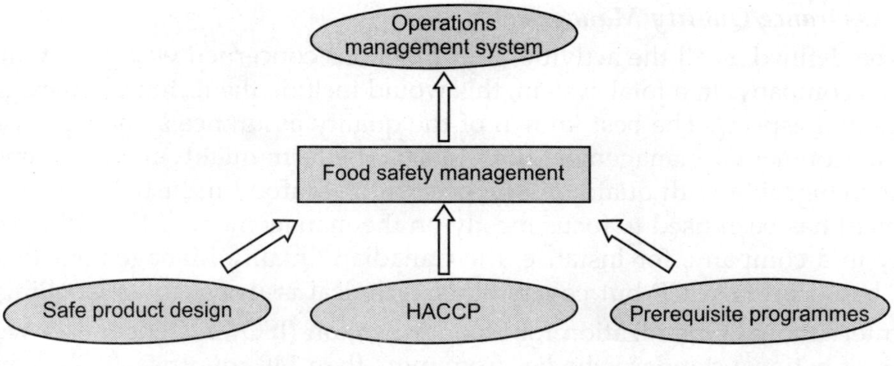

Figure 21.11 Procedure for food safety management tools.

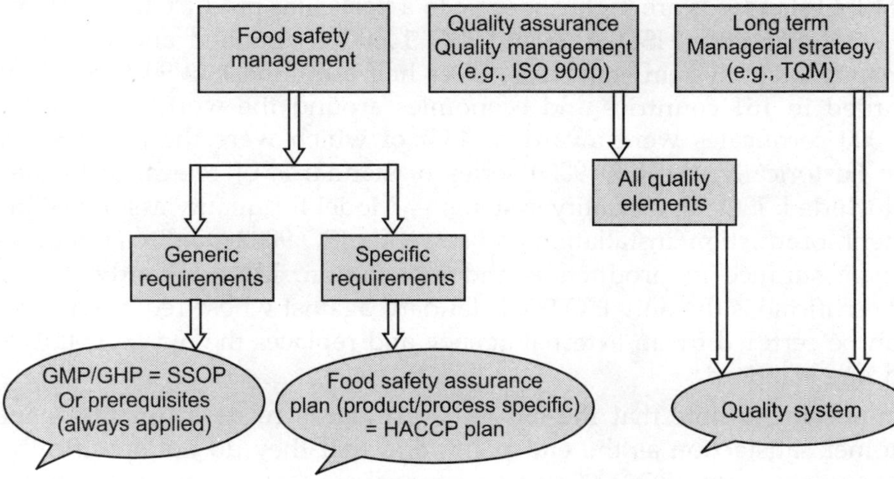

Figure 21.12 Food safety rules and their relationship.

1997). HACCP ensures food safety through an approach that builds upon foundations provided by good manufacturing practice. It identifies the points in the food production process that require constant control and monitoring to make sure the process stays within identified limits. Statistical Process Control systems are relevant to this operation. HACCP is legislated in many countries, including the USA and the European Union. The combination of GHP/GMP and HACCP is particularly beneficial in that the efficient application of GHP/GMP allows HACCP to focus on the true critical determinants of safety.

Quality Control

It is an important subset of any quality assurance system and is an active process that monitors and, if necessary, modifies the production system so as to consistently achieve the required quality. It can be argued that QC is used as part of the HACCP system, in terms of monitoring the critical control points in the HACCP plan. However, traditional QC is much broader than purely this focus on critical control points for safety systems.

Quality Assurance/Quality Management

This can be defined as all the activities and functions concerned with the attainment of quality in a company. In a total system, this would include the technical, managerial and environmental aspects. The best known of the quality assurance standards is ISO 9000 and for environmental management, ISO 14000. The term quality management is often used interchangeably with quality assurance. In the seafood industry, the term quality management has been used to focus mostly on the management of the technical aspects of quality in a company, for instance, the Canadian Quality Management Programme which is based on HACCP but covers other technical issues such as labelling.

The International Organization for Standardization (ISO) in Geneva is a worldwide federation of national standards bodies from more than 140 countries. ISO's work results in international agreements which are published as International Standards. The vast majority of ISO standards are highly specific to a particular product, material, or process. However, two standards, ISO 9000 and ISO 14000, mentioned above, are known as generic management system standards. Over half a million ISO 9000 certificates have been awarded in 161 countries and economies around the world and in 2001 alone over 100 000 certificates were awarded, 43% of which were the new ISO 9001:2000 certificate. Historically, the ISO 9000 series of standards of relevance to the seafood industry included: ISO 9001 Quality systems – Model for quality assurance in design/ development, production, installation and servicing ISO 9002 Quality systems – Model for quality assurance in production and installation. More recently, the new ISO 9001:2000 certificate is the only ISO 9000 standard against whose requirements a quality system can be certified by an external agency and replaces the old ISO 9001, 9002 and 9003 with one standard.

It is important to note that the ISO 9000 standards relate to quality management with customer satisfaction as the end point, and that they do not specifically refer to technical processes only. ISO 9000 gives an assurance to a customer that the company has developed procedures (and adheres to them) for all aspects of the company's business.

ISO 14000 is primarily concerned with environmental management. Introduced much later than the ISO 9000 series, there are now over 35 000 ISO 14000 certificates awarded in 112 countries or economies of the world. During 2001, nearly 14,000 certificates were awarded, around 40% of the total awarded since the introduction of the standard.

In most countries, implementation of ISO 9000 quality management systems or ISO 14000 environmental systems are voluntary.

Quality Systems

This term covers organizational structure, responsibilities, procedures, processes and the resources needed to implement comprehensive quality management (Jouve et al., 1998). They are intended to cover all quality elements. Within the framework of a quality system, the prerequisite programme and HACCP provides the approach to food safety.

Total Quality Management (TQM)

TQM is an organization's management approach, centred on quality and based on the participation of all its members and aimed at long-term success through customer satisfaction and benefits to the members of the organization and to society (Jouve et al., 1998). Thus TQM represents the organizations' "cultural" approach and together with the quality systems provides the philosophy, culture and discipline necessary to commit everybody in the organization to achieve all the managerial objectives related to quality.

21.17.3 Business Outlook for Food Diagnostics

Consumer demands safe food today. The commercial food industry's ability to identify bacterial pathogens and unsafe residues has resulted in an almost fivefold increase in food recalls by major manufacturers since 1988. New technology allows Government regulatory agencies to identify a bacterial pathogen and trace it back to its source more rapidly (Figure 21.13). The key to this new technology is the availability of rapid food safety diagnostics.

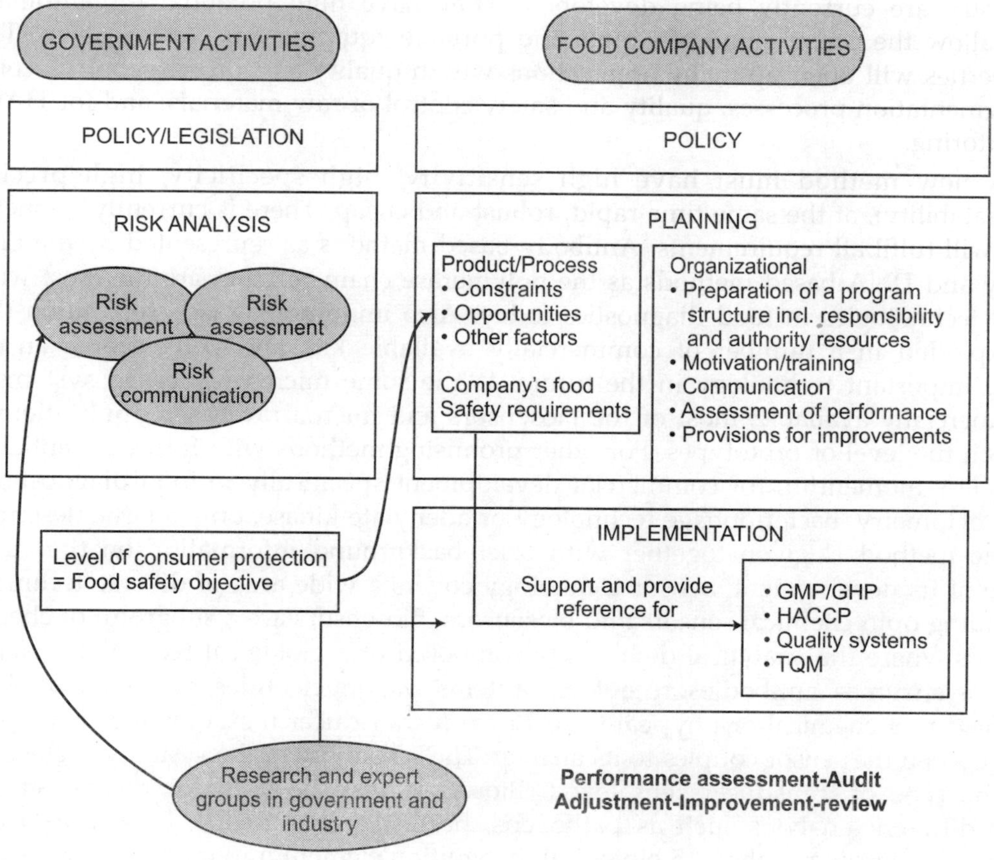

Figure 21.13 New policies and activities of the government and food company for food safety.

These rapid food safety diagnostics provide a quick, "positive or negative" answer before the food product enters the distribution system. Tests take up to 30 hours to complete because of the requirement for a bacterial growth enrichment period. This growth enrichment period is necessary to increase the total number of bacteria so they can be detected using current technology. A negative answer means that the product does not contain that particular bacteria or toxin and no further testing is required. A positive answer means that further testing is needed at a reference laboratory using standard laboratory methods to confirm the actual presence of the bacteria or toxin. For example, rapid screening for *Salmonella typhimurium* requires approximately 24 hours to complete.

21.17.4 Next Generation Diagnostics Industry

The most promising breakthroughs of the development of online or onsite, sensitive, low-cost, rapid methods for routine use are expected to be made in the area of sensor technology. Many prototypes for food diagnostic application in the food and drink industry are currently being developed. They have high potential for automation and allow the construction of simple and portable equipment for fast analysis. These properties will open up many applications within quality and process control, control of fermentation processes, quality and safety control of raw materials, and for HACCP monitoring.

A new method must have high sensitivity, high specificity, high precision (repeatability), at the same time rapid, robust and cheap. There is currently no method that will fulfil all requirements. Antibody based methods as represented by the ELISA assay and DNA based methods as the polymerase chain reaction are the most widely used technologies in food diagnostics today. Also immunomagnetic separation (IMS) is exploited in a number of commercially available kits and will become an even more important technology in the future. While some microarray based systems are commercially available, most of the biosensors and microarray based applications are still on the level of prototypes. For other promising methods with future potential, but with less momentum for commercial development specifically in food diagnostics, as flow cytometry, bacteriophage technology or adenylate kinase, only a basic description of the method is given together with brief background information on the current stage of its development. Sensor technology covers a wide area of diverse techniques, including opto-chemical sensors and biosensors. Biosensors are a subgroup of chemical sensors where the analytical devices are composed of a biological recognition element such as enzymes, antibodies, receptors, proteins, oligonucleotides, or even a whole cell coupled to a chemical or physical transducer. A transducer measures the changes that occur when the sensor couples to its analyte. The sensitivity of the system is determined by the type of transducers employed. Biosensors can be used for the detection of very different analytes such as pathogens, pesticides and toxins. Biosensors can be grouped according to their 15 biological recognition element into immunosensors using antibodies and hybrid sensors using DNA or RNA probes. There have been many sensors developed for the detection of food-borne pathogens with the goal to overcome problems associated with traditional microbiological detection techniques such as being time and labour intensive (Baeumner, 2003). In fact, biosensor advancements have greatly improved our ability to detect minute quantities of analytes as research into biosensors has mainly focused on detection platforms with very low detection limits (Rider *et al.*, 2003). It has been estimated that 38% of reported pathogen biosensors in the past 20 years were designed for the food industry. However, only a limited amount of methods are combined and currently exploited for their use in food diagnostics. As recognition elements, bioaffinity based receptors that use the selective interaction between ligand and receptor, antibody or nucleic acid are most widely used. As transducers, electrochemical and optical systems have gained practical importance. As nanobiotechnology progresses, sensors to detect pathogens or their constituents become smaller and more sensitive. Owing to the nature of these nanoscale sensors, the sample size from which the detection is being made is typically a microlitre or smaller. Therefore, the challenge for scientists developing detection methods for pathogens in foods is in the sample preparation. Although the sample preparation requirements will vary from one food product to another, research into this step is required to bridge the emerging field of nanosensors with the food industry. Thus, while the organism with the largest number of diagnosed cases may fluctuate from year to year, the food industry will always be looking for detection systems that will help identify all pathogens of concern in its food products (Rider et al., 2003).

Optical biosensors have been developed for rapid detection of contaminants in foods, including pathogens, and several have evolved into commercial prototype systems. The analyte in the food interacts with the bioactive molecule, usually an antibody. Antibodies can be immobilized directly on the fibre, either on the blunt end or along the sides of a fibre tip. The binding of antibody and analyte is detected as a change in an optical signal measured through the fibre-optic assembly. The light from a laser travels to the fibre tip and penetrates into the area outside the tip. A fluorescently labelled complex binds to the antibodies on the tip. The fluorescent signal then radiates in all directions, and some of it travels back up the fibre tip to the detector. Detection of molecules in solution can be made either by direct binding to the biosensor coating molecules, or by competition binding with soluble capture molecules added together with the sample. Future applications might include protein quality and the detection of allergens genetically modified proteins, BSE prions, pathogens and biocide residues.

21.18 EXPERIMENTS

21.18.1 Bacterial Count

The bacterial count in milk is the most reliable and accepted indication of sanitary quality. Although, the human pathogens may not be in a high count it may indicate a diseased udder, unsanitary handling of milk. In general, a high count means that there is a possibility of disease transmission. On the other hand, it is necessary to avoid the wrong interpretation of low plate counts, since it is possible to have pathogens such as brucellosis and tuberculosis with permissible number of bacterial count. Following methods are normally used for microbiological examination of milk.

21.18.1.1 Direct microscopic count of organisms

This method is especially useful to determine the milk quality in much shorter time than is possible with a standard plate count. In addition to being much faster than the SPC, the direct microscopic count has two other distinct advantages. First of all, it will reveal the presence of bacteria that do not form colonies on an agar plate at 35°C; thermophiles, psychrophiles, and dead bacteria would fall in this category. Secondly, the presence of excessive number of leukocytes and pus-forming streptococci on a slide will be evidence that the animal that produced the milk has an udder infection (mastitis).

Direct microscopic count is accomplished by staining a measured amount of milk that has been spread over an area of one square centimetre on a slide. The slide is examined under oil and all of the organisms in an entire microscopic field are counted. Several fields are counted to get average field counts in order to maintain accuracy. It is necessary to measure the field area in order to translate the readings into organisms per mL.

High quality milk will have very few organisms per field, necessitating the examination of many fields. A slide made of poor quality milk, on the other hand, will reveal large numbers of bacteria per field, thus requiring the examination of fewer fields. In view of all these advantages, it is apparent that the direct microscopic count has real value in milk testing. It is widely used for testing raw milk in creamery receiving station and for diagnosing the types of contamination and growth in pasteurized milk products. Milk that has been separated, blended, homogenized, and pasteurized will lack leukocytes and normal flora.

Shake the milk sample for nearly 50-60 times so as to break the large bacterial clumps, transfer 0.01 mL of the milk to one square on the slide using micropipette. Allow the slide to air and then place it over a beaker of boiling water for 5 minutes dry to steam fix it. Flood the slide with xylol to remove fat globules (Figure 21.14). Remove the xylol from the slide by flooding the slide with 95% ethyl alcohol. Gently immerse the slide into a beaker of distilled water to remove alcohol under running water. The milk film will wash off. Stain the smear with methylene blue for 15 seconds and dip the slide again in water to remove the excess stain. Decolorize the smear to pale blue with 95% alcohol and dip in water to stop decolorization. Allow the slide to completely air-dry before examination.

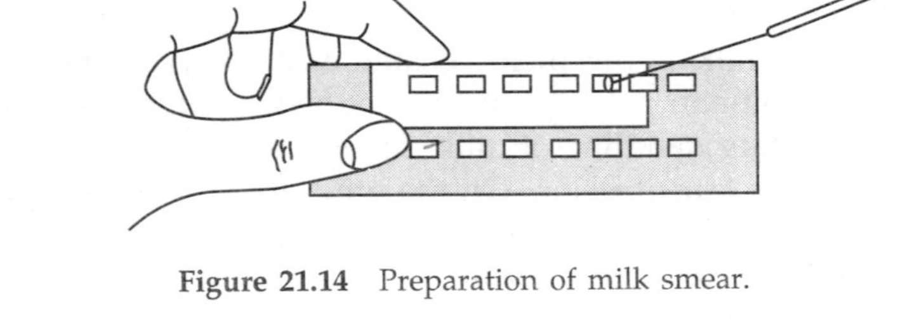

Figure 21.14 Preparation of milk smear.

Least Count of Microscopic Measurement

The relationship of the field to a millilitre is the microscopic factor (MF). Stage micrometer is to be used to calculate the microscopic field. пr² formula is used for measuring the area. Suppose 0.01 mL milk is used in the smear and the area of the slide is 1 cm^2. MF can easily be calculated. Measurement will require to focus the stage micrometer on the microscope under oil immersion keeping in mind that each space is equal to 0.01 mm. Field area is calculated in square millimetres employing пr² formula where π is equal to 3.14. Square millimetres are converted to square centimetre by dividing with 100. Calculate the number of fields in one square centimetre by dividing one square centimetre by the area of the field in square centimetres. To get the part of a millilitre that is represented in a single field (MF), multiply the number of fields by 100. The value should be around 500,000. Therefore, a single field represents 1/500,000 of each mL of milk.

It is observed that the number per millilitre is higher than a standard plate count (SPC) but a clum count is closed to the SPC. Both methods, viz., individual cells and clumps are normally employed. The details of the methods are depicted in Figure 21.15.

21.18.1.2 Litmus milk test

Milk is a complex nutritional source that contains proteins (mainly casein) in an aqueous solution of lactose and minerals. Bacterial enzymes alter the media and may bring about various changes. Litmus is added to the medium to detect pH changes that may occur as a result of these enzymatic reactions. Litmus is blue at above pH 8.3 while below a pH of 4.5 it is red. By producing acid from the fermentation of lactose present in milk a bacterium may also cause the milk to curdle or clot in the bottom of the tube. Litmus acts as an electron acceptor thus becoming reduced by bacterial metabolism. This reaction is sometime observed as a white colour in the medium. **Litmus milk test** is used to distinguish between different types of bacteria. The lactose (milk sugar), litmus (pH indicator), and casein (milk protein) contained within the medium can all be metabolized by different types of bacteria. The test itself tells whether the bacterium can ferment lactose, reduce litmus, form clots, form gas, or start peptonization (Figure 21.16).

21.18.1.3 Qualitative screening of milk quality (methylene blue test)

Milk having enough amount of microbial population will quickly consume dissolved oxygen resulting to reduce the oxidation reduction potential. Methylene blue loses its colour when it is reduced. It is also known as reductase test. In this test, 1 mL of 1:25000 diluted aqueous methylene blue solution is added in freshly collected 10 mL milk sample. The tube is sealed with a rubber stopper and slowly inverted 3-4 times to mix well, incubated at 35°C in a water bath and examined after every 30 minutes for up to 6 hours. The time taken for methylene blue to become colourless is the methylene blue reduction time (MBRT). The shorter the MBRT, the poorer the quality

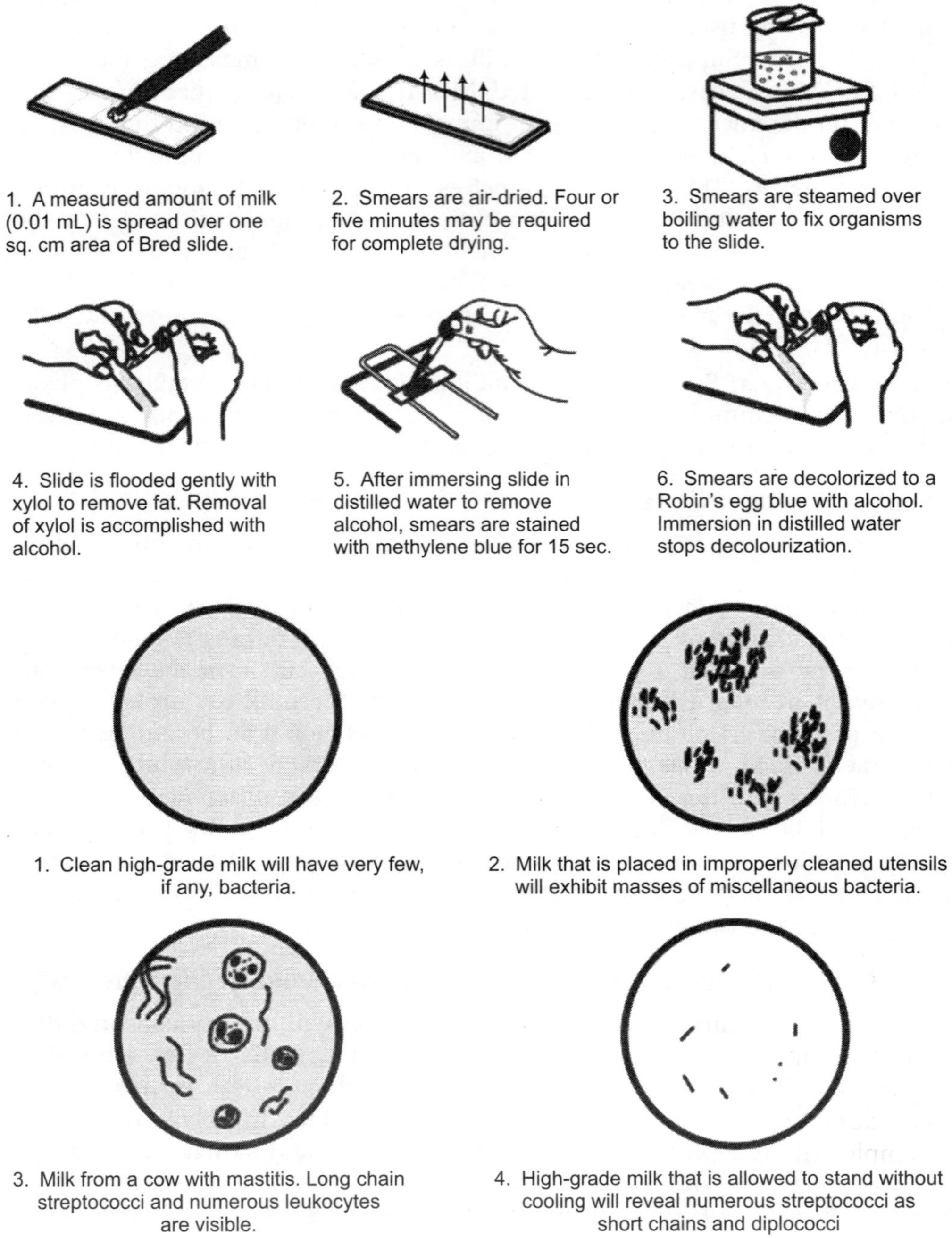

Figure 21.15 Microscopic measurement of bacterial count in milk.

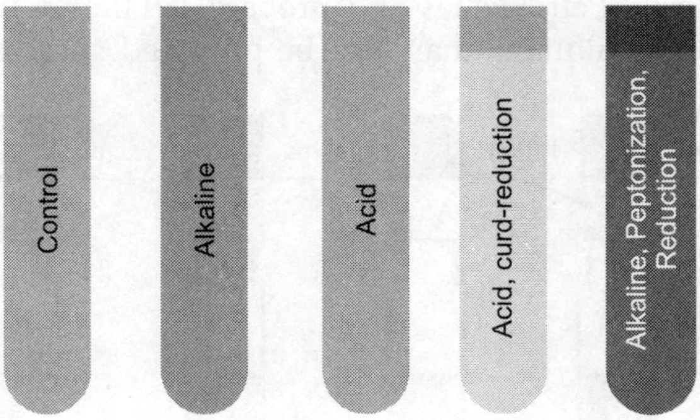

Figure 21.16 Litmus milk test.

of milk. An MBR of 30 minutes is considered good quality of milk. Raw milk primarily contains *Streptococcus lactis, Escherichia coli* which are strong reducers. The test is used for screening the milk at milk collection centre. However, the test will not identify the psychrophiles and thermophiles which are in large numbers in milk (Figure 21.17).

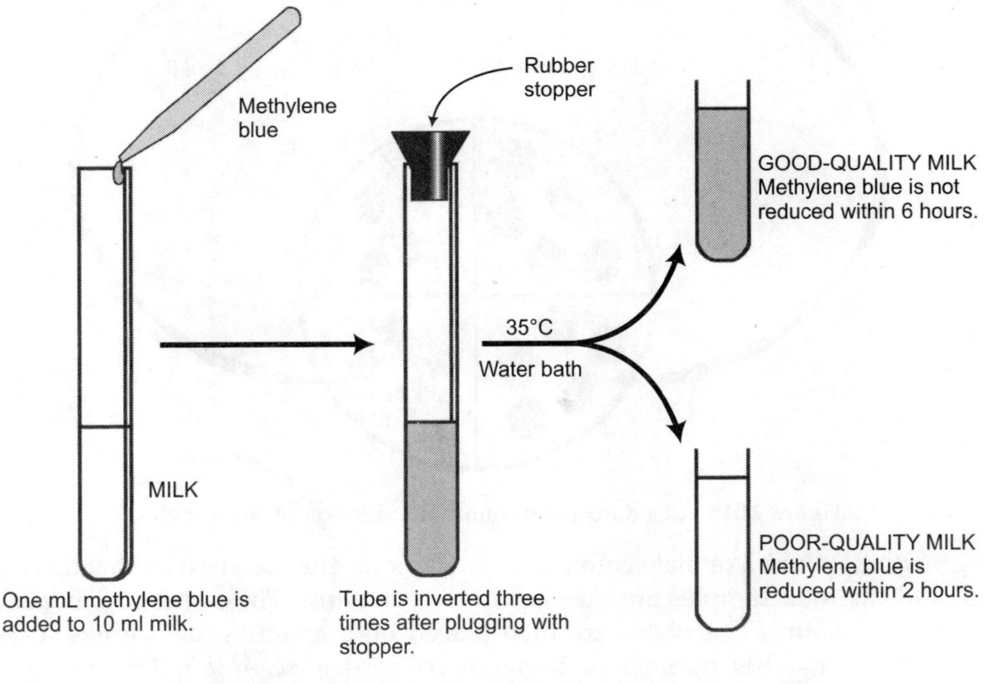

Figure 21.17 Demonstrating methylene blue test for milk.

21.18.1.4 Standard plate count: a quantitative test

This test determines the total number of bacteria in a specified amount of milk, generally a millilitre. This is used for grading of milk. Under aseptic conditions, one mL of milk is added to 99 mL of distilled water or buffer. One mL and one-tenth mL samples are then transferred to sterile Petri dishes and prepare 1:100 and 1: 1000 dilutions of the milk, respectively. Other dilutions may also be prepared successively (Figure 21.18).

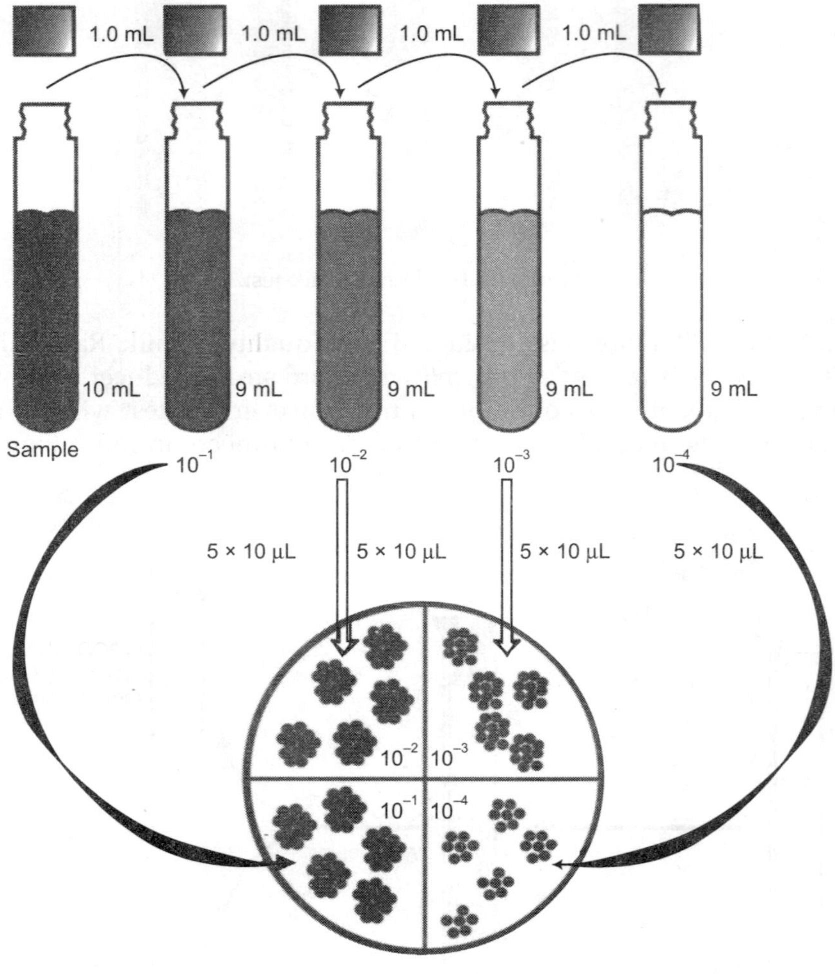

Figure 21.18 Standard plate count after dilution of the samples.

A growth medium like plate count agar or tryptone glucose yeast extract agar is then added, and the milk samples are mixed with the medium. The dishes are incubated at 37°C for 24-48 hours. The plates are then placed on a counting device like a Quebec colony counter and the number of bacterial colonies is recorded. The colony count falling between 30 and 300 is selected and multiplied by the reciprocal of the dilution factor to obtain the bacterial count per mL of milk. If 248 colonies appeared on the 1: 100 plate and 16 colonies on, the 1:1000 plate, 248 would be selected and multiplied by 100, giving 24,800 total bacteria per mL of milk sample.

The standard plate count (SPC), also referred to as the aerobic plate count or the total viable count, is one of the most common tests applied to indicate the microbiological quality of food. The significance of SPCs, however, varies markedly according to the type of food product and the processing it has received. When SPC testing is applied on a regular basis it can be a useful means of observing trends by comparing SPC results over time.

Three levels of SPC are listed below based on food type and the processing/handling the food.

Level 1 applies to ready-to-eat foods in which all components of the food have been cooked in the manufacturing process/preparation of the final food product and, as such, microbial counts should be low.

Level 2 applies to ready-to-eat foods which contain some components that have been cooked and then further handled (stored, sliced or mixed) prior to preparation of the final food or where no cooking process has been used.

Level 3 SPCs not applicable. This applies to foods such as fresh fruits and vegetables (including salad vegetables), fermented foods and foods incorporating these (such as sandwiches and filled rolls). It would be expected that these foods would have an inherent high plate count because of the normal microbial flora present.

[*Note:* An examination of the microbiological quality of dairy and food products should not be based only on SPCs. The significance of high (unsatisfactory) SPCs cannot truly be made without identifying the microorganisms that predominate or without other microbiological testing for pathogenic and nonpathogenic organisms].

21.19 FOOD MICROBIOLOGY

The detection of pathogenic bacteria is key to the prevention and identification of problems related to health and safety. Legislation is particularly tough in areas such as the food industry, where failure to detect an infection may have terrible consequences. Throughout the world, food production, preparation and distribution have become increasingly complex, and raw materials are often sourced globally. Changes in food processing techniques, food distribution and the emergence of new food pathogens have changed the epidemiology of food-borne diseases. Food-borne microorganisms are continuously changing due to their inherent ability to evolve and their amazing capacity to adapt to different forms of stress. New primary production technologies and food manufacturing practices are introduced all the time; food consumption patterns and the demographic structure of many countries continue to change.

Approximately 1/3rd of all food manufactured in the world is lost due to spoilage. On the basis of the shelf-life, food can be classified as non-perishable foods (pasta), semiperishable foods (bread) and perishable foods (eggs). The conditions for spoilage relate to water, pH, physical structure, oxygen and temperature. Various intrinsic factors are also responsible like composition, pH, presence and availability of water, oxidation-reduction potential, altered by cooking, physical structure and presence of antimicrobial substances. Microbial content of foods (microbial load) can be evaluated by determining bacterial titer.

The intestinal tract of warm-blooded animals contains a variety of bacterial species. When an animal is butchered, these bacteria may be spread from the gut to the edible portions of the carcass. While such bacteria generally are not harmful to humans (beef is most often contaminated with *E. coli*, a normal inhabitant of the human intestine), some types of bacteria (such as *Salmonella*, found frequently in chickens) can cause food poisoning. It is, therefore, essential that meats be thoroughly cooked prior to eating. The non-pathogenic varieties of bacteria in foods may cause chemical changes which alter the flavour and texture of food (Figure 21.19). Because of this, meats are refrigerated or frozen to retard the growth of bacteria. Since some bacteria can grow over a wide range of temperatures (from less than 10°C to greater than 45°C for *E. coli*), refrigeration is adequate for only short term storage of meats.

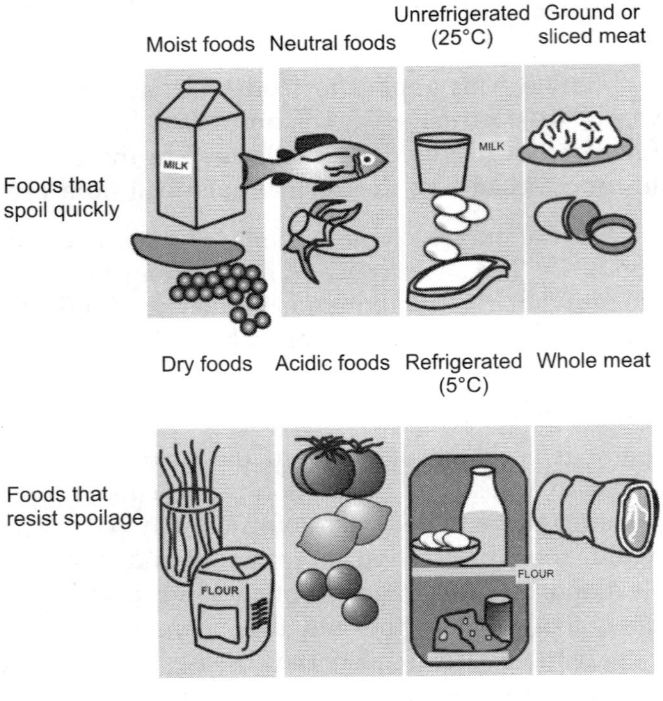

Figure 21.19 Microbial food spoilage.

21.19.1 Microbial Spoilage of Canned Food

Canned foods may spoil either due to biological or chemical reasons. Biological spoilage of canned food occurs due to the action of various microorganisms. Spore forming bacteria, e.g., *Clostridium, Bacillus* represent the most important group of canned food spoiling microorganisms because of their heat resistant nature (thermophilic nature). In addition, there are other microorganisms which are not heat resistant (mesophilic) but enter through the leakage of the container during cooling and spoil the food. In this way, we can divide biological spoilage of canned food into following two categories.

21.19.1.1 Biological spoilage by thermophilic bacteria

Canned foods under processing result in spoilage by thermophilic bacteria, the bacteria that grow best at temperature of 50°C or higher. Five types of this spoilage can be recognized.

Flat sour spoilage In canned foods, production of acid and no gas is referred to as flat sour spoilage because the food becomes sour, but the can shows no evidence of food spoilage because no gas is produced, i.e., the can remains flat. Thus, the spoilage cannot be detected unless the can is opened. The spoilage is caused by *Bacillus* spp. such as *B. coagulans* and *B. stearothermophilus* resulting in sour, abnormal odour, sometimes cloudy liquor in food content of the can.

Thermophilic anaerobic (TA) spoilage *Clostridium thermosaccharolyticum,* an obligate thermophile, causes spoilage. The can swells and may burst due to production of CO_2 and H_2. The food becomes fermented sour, cheesy, and develops butyric odour.

Sulfide spoilage *Clostridium nigricans* is involved in this spoilage. It produces H_2S gas which is absorbed by the food product. The latter becomes usually blackened and gives "rotten egg" odour.

Putrefactive anaerobic spoilage *Clostridium sporogenes* causes spoilage through putrefaction. The can swells and may burst. Putrefaction may result from partial digestion of the food. The latter develops typical "putrid" odour.

Aerobic sporeformer's spoilage *Bacillus* spp., the aerobic bacteria, causes spoilage. If the canned food is cured meat, swelling of the can is observed.

21.19.1.2 Biological spoilage by mesophilic microorganisms

Bacillus spp., *Clostridium* spp., *yeast,* and other molds which are mesophilic (an organism growing best at moderate temperature range of 25 to 40°C) are mainly responsible for this type of canned food spoilage. As stated earlier, these organisms enter through the leakages of the container during cooling. *Clostridium butyricum* and *C. pasteurianum* result in butyric acid type of fermentation in acidic (tomato juice, fruits, fruit juices, etc.) or medium acidic (corn, peas, spinach, etc.) food with swelling of the container due to the production of CO_2 and H_2. *Bacillus subtilis* and *B. mesenteroides* have been reported as spoiling canned sea-foods, meats, etc. Other mesophilic bacteria which have been reported in cans are *Bacillus polymixa, B. macerans, Streptococcus* sp., *Pseudomonas, Proteus,* etc. Yeasts and molds have also been found present in canned foods. Yeasts result in CO_2 production and swelling of the cans.

21.19.1.3 Microbial spoilage of refrigerated meat

Lipases are class of enzymes, which catalyze the hydrolysis of long chain triglycerides. These are glycerol ester hydrolases that also catalyses esterification, interesterification, acidolysis and aminolysis in addition to the hydrolytic activity on triglycerides. Beside these activities, there are two more classes of enzymes, mostly secreted by microorganisms and these are the oxygenases and the reductases which also bring about changes in lipids.

21.19.1.4 Bacterial counts

SPC (Standard Plae Count), milipore filter and multiple tube tests is normally used much in the same manner that they are used on water and milk to determine the total counts and the presence of coliforms. However, to get the organisms in suspension the food blender is used for blending meat, dry fruits, frozen foods, etc., as described in Figure 21.20. Fruit juice and other beverages are used in the manner as we do for milk and water.

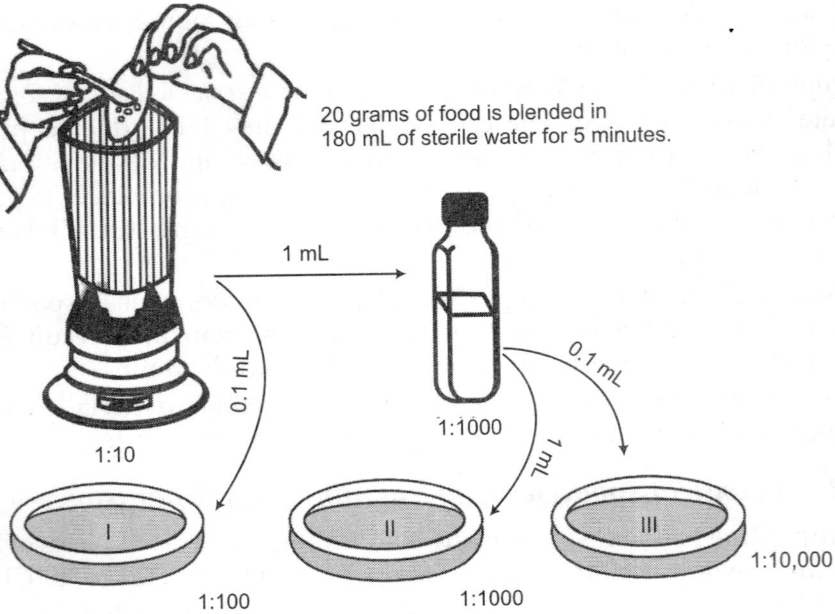

Figure 21.20 Blending and SPC procedure for food.

Canned Food

Microbial spoilage of commercially canned food and heat processed is confined to heat resistant endospore. Canning normally involves heat exposure for long period at temperature that is adequate to kill spores of most bacteria especially to the processing of low acid foods in which *Clostridium botulinum* can thrive to produce botulism food poisoning.

Spoilage occurs when the heat processing fails to meet accepted standards. This is because of several reasons (1) usually in case of home canning, (2) carelessness in handling the raw materials before canning, resulting in an unacceptably high level of contamination that ordinary heat processing may be inadequate to control, (3) malfunctioning of equipment resulting to undetected underprocessing and (4) defective containers that permit the entrance of organisms after the heat process. There are three types of food spoilage caused by heat resistant bacteria (a) Flat sour – pertains to spoilage in which acids are formed with no gas production; results sour food in cans that have flat ends (b) T.A. Spoilage - caused by thermophlic anaerobes which produce acid and gas (CO_2 and H_2 but not H_2S) in low acid foods (c) Stinker Spoilage – Cans swell to various degrees, sometime bursting due to the spore formers that produce H_2S and blackening of the can and contents. Blackening is due to the reaction of H_2S with the iron in the can to form iron sulfide. To perform the experiments on microbial spoilage follow the Figure 21.21 with the inoculation of cans with following 5 cultures viz. *Bacillus stearothermophilus, Bacillus coagulans, Clostridium sporogenes, Clostridium thermosaccharolyticum* and *Escerichia coli* sepeartely. Incubation temperature for *Clostridium thermosaccharolyticum* and *Bacillus stearothermophilus* be 55°C, *Clostridium sporogenes* and *Bacillus coagulans* at 37°C and 30°C for *Escerichia coli.* If cans begin to swell during incubation they should be placed in refrigerator. After incubation place the cans in hood to open with the help of punch type can opener. Hold an inverted plastic funnel over the can in case the can is swollen during perforation to minimize the effect of any explosive release of contents. Remove about 10 mL of the liquid through the opening, pour it into a small plastic beaker, cover with parafilm and later use for making stained slides for Gram staining and endospore staining and examine under brightfield oil immersion.

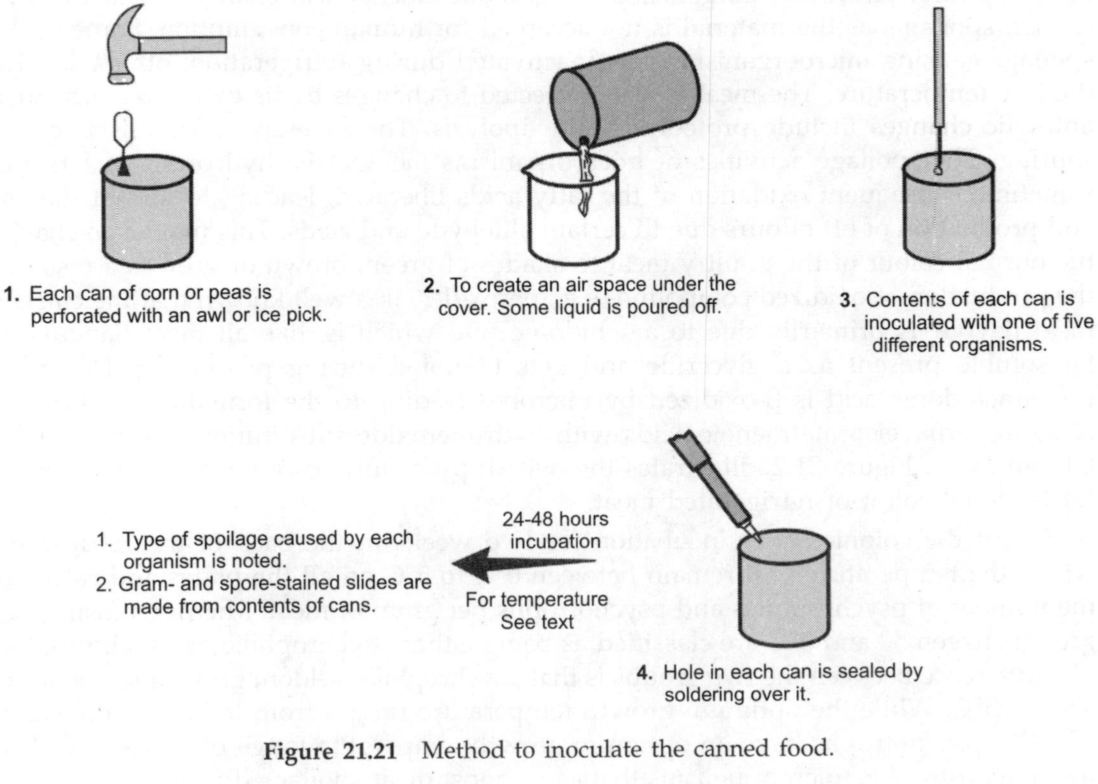

Figure 21.21 Method to inoculate the canned food.

Refrigerated Meat

The widespread use of refrigeration to store and preserve food stuffs, provides great diversity of nutrient rich habitat for psychrophilic and psychrotolerant food spoilage microorganisms. The lipases produced by certain organisms are significant to the food industry in that they improve the traditional chemical processes of food manufacture. However, certain microorganisms like *Arthobacter* sp., *Pseudomonas fragi, Pseudomonas fluorescence* and *Serratia marcesans* which produce cold active lipases, were isolated from refrigerated milk, meat products and other spoiled food samples.

Dressed poultry are highly susceptible to spoilage by many microorganisms. The flesh and other parts like the liver, has hardly any microorganisms in any living animals; therefore, most of the contamination comes during the slaughtering processes like bleeding, defeathering, removal of viscera, washing and subsequent handling. In order to prevent spoilage by these microorganisms, meat is refrigerated usually at –10 to –15°C. However, during selling these are stored at –2 to –5°C over the counter. Sometimes it has been observed that the temperature of the counter reaches 0 to 2°C and the material is held in this temperature for more than 8 h. Such sort of practice leads to increase in aerobic microbial count especially like those of the psychrophiles. These microorganisms bring about biochemical changes, which can be termed spoilage, as the material is not accepted for human consumption. Some of the spoilage causing microorganisms gets inactivated during refrigeration; others survive the low temperature. The meat is also subjected to changes by its own enzymes. Such autolytic changes include proteolysis and lipolysis. The excessive autolysis is called souring. The spoilage activities of microorganisms include the hydrolysis of fats and sometime subsequent oxidation of the fatty acids liberated, leading to loss of flavour and production of off odours due to certain aldehyde and acids. This may even change the normal colour of the poultry meat to shades of green, brown or grey as a result of the production of oxidized compound, e.g., peroxides. It is well known that the chicken meat flavour is primarily due to arachidonic acid which is, like all meat flavours, is fat soluble present as a glyceride and gets liberated during processing. However, this arachidonic acid is β-oxidized by microbes leading to the formation of different N-hydroperoxyeicosatetraenoic acids with hydroperoxide substitution at C5, C8, C11, C12 and C15. Figure 21.22 illustrates the overall procedure to perform the experiment for bacterial count of refrigerated meat.

Count the colonies after incubation for two weeks in the back of the refrigerator where the temperature will remain between 0°C to 5°C on all the plates and calculate the number of psychrophiles and psychotrophs per gram of meat. Microorganisms that grow between 50 and 0°C are classified as being either psychrophilic or psychotrophic. The difference between the two groups is that psychrophiles seldom grow at temperature above 25°C. While the optimum growth temperature ranges from for psychrophiles is 15-18°C, psychotrophs have an optimum growth temperatue range of 25°C–30°C. It is the psychrotrophic microorganisms that cause most meat spoilage during refrigeration. Select a colony from the plates and prepare a Gram stain, endospore stain and flagella staining, etc., and observe under oil immersion with the aid of bright field microscope.

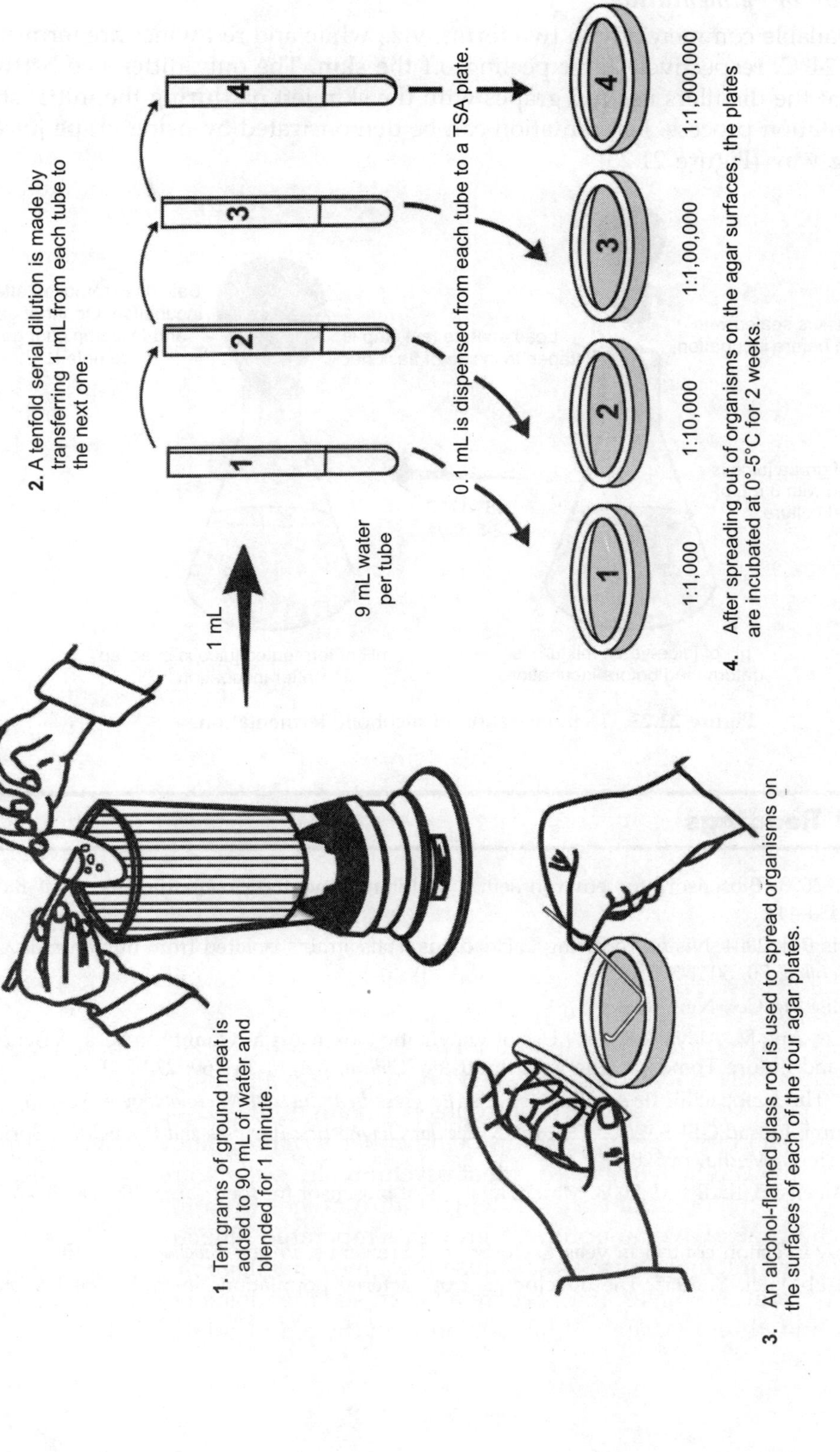

Figure 21.22 Method describing bacterial counting of refrigerated meat.

Demonstration of Fermentation

Wines are available commercially in two forms, viz., white and red wines are fermented at 13°C and 24°C, respectively after peeling off the skin. The only difference between the two is that the distillers use red grapes with the skin left on during the initial stage of the fermentation process. Fermentation can be demonstrated by using grape juice in the following way (Figure 21.23).

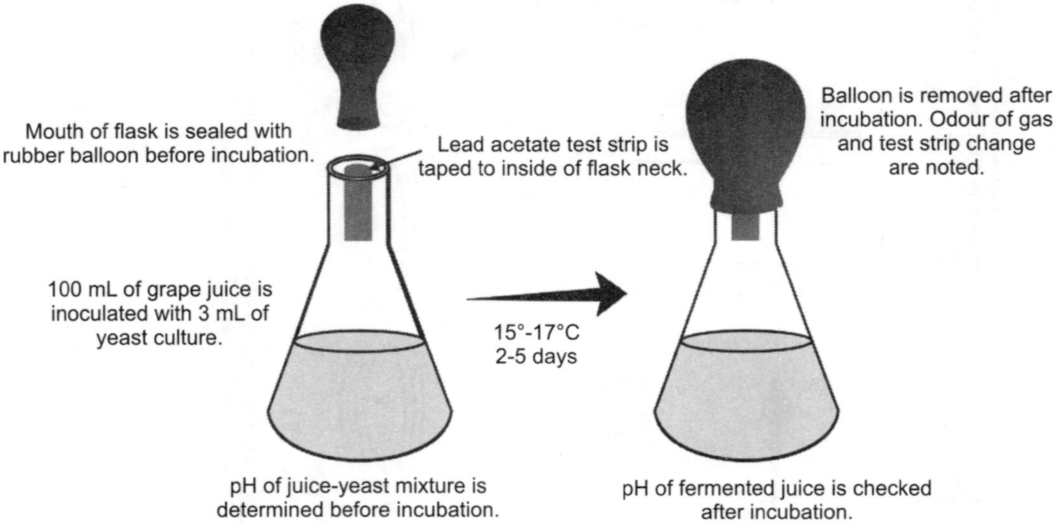

Figure 21.23 Demonstration of alcoholic fermentation.

Suggested Readings

Baeumner, A.J. 2003. Biosensors for environmental pollutants and food contaminants. *Anal Bioanal Chem* 377, 434-445.

Beasley SS, Saris PEJ. 2004. Nisin-producing Lactococcus lactis strains isolated from human milk. *Appl Environ Microbiol* 70, 5051-5053.

Chemical Publishing Co., New York,

Cock SL, Zetty A, Ana M, Alaya AA. 2009 Use of Enzymatic Biosensors as Quality Indices: A Synopsis of Present and Future Trends in The Food Industry. *Chilean J Agric Res* 69, 270-280.

Davis JG. 1963. The lactobacilli. II. Applied aspects. *Progress in Industrial Microbiology* **4**, 95-136.

Debnath Mousumi, Prasad GBKS, Bisen PS.2010. *Molecular Diagnostics: Promises and Possibilities*, Springer Science+Business Media, pp520.

Evangelyn C, Alocilja S, Radke M .2003. Market analysis of biosensor for food safety. *Biosens Bioelectron* 18,841-846.

Fowell RR. 1967. Infection control in yeast factories and breweries. *Process Biochem* **2,** 11-15.

Gibson T, Abd-El-Malek Y. 1957. The development of bacterial populations in milk. *Can J Microbiol* 3, 203-213.

Herson AC, Hulland ED. 1964. *Canned Foods: an introduction to their microbiology.* 291 pp.

Jouve JL, Stringer MF, Baird-Parker AC. 1998. *Food safety Management Tools.* ILSI Europe Risk Analysis in Microbiology, Brussels, Belgium.

Mello LD, Kubota LT. 2002. Review of the use of biosensors as analytical tools in the food and drink industries. *Food Chem* 77, 237-256.

Patel PD. 2006. Overview of affinity biosensors in food analysis. *J AOAC Int* 89, 805-818.

Rasooly A, Herold KE. 2006. Biosensors for the analysis of food- and waterborne pathogens and their toxins. *J AOAC Int* 89, 873-883.

Rider TH, Petrovick MS, Nargi FE, Harper JD, Schwoebel ED, Mathews RH, Blanchard DJ, Bortolin LT, Young AM, Chen J, Hollis MA. 2003. A B-cell based sensor for rapid identification of pathogens. *Science* 301, 213-215.

Schleifer KH, Kraus J, Dvorak C, Kilpper-Bälz R, Collins MD, Fischer, W. 1985. Transfer of *Streptococcus lactis* and related streptococci to the genus *Lactococcus gen. nov. Syst Appl Microbiol* 6, 183–195.

Wang H L, Hesseltine C W. 1979. Mold-modified foods. In: *Microbial Technology (Eds HJ Peppler and D Perlman) 2nd ed. Academic Press Inc, New York,* 95-129.

Chapter **22**

The Industrial Uses of Microorganisms

22.1 INTRODUCTION

Microorganisms were used in industrial processes even before their existence was known. The production of fermented beverages and vinegar, and the leavening of bread are all traditional processes which have come down to us from time immemorial. The discovery of microorganisms with their multiplicity of highly specific biochemical activities has stimulated a steady growth of industrial fermentation processes. Perhaps the most famous of all industrial fermentations is that of acetone-butanol production by *Clostridium acetobutylicum*. During the First World War, acetone, sorely needed for the manufacture of cordite, was produced by this fermentation.

Between the wars, the industrial use of microorganisms was extended, but it was not until after the Second World War that the fermentation industries developed on the massive scale that we know today. The major stimulus was the need to produce antibiotics. Just prior to the war, some of the larger American companies had developed microbial fermentations to produce fumaric acid and this was one of the first examples of the use of neutral conditions. Hitherto, most industrial fermentations had been conducted under acid conditions where sterility was not highly critical. With the advent of neutral fermentations, there were new bio-engineering problems to be solved in order to maintain pure cultures on a large scale. The neutral conditions and the special sterile precautions required for the production of fumaric acid were exactly those required for the mass cultivation of antibiotic-producing microorganisms in the post-war period.

Although the industrial processes using microorganisms are too numerous to mention, many have declined and some have disappeared altogether. With the enormous amounts of crude oil being distilled for petroleum, there are many useful hydrocarbons in other fractions unsuitable for fuels. These form the starting materials for such organic solvents as acetone and butanol which can be produced more efficiently and cheaply from these sources.

Thus, the need to use such organisms as *Clostridium acetobutylicum* has disappeared. In spite of the expansion of the petrochemical industries, there are some substances that are still most easily produced microbiologically and these may be conveniently discussed under the following headings: (1) Substances produced as a major end-product of microbial metabolism, (2) The use of microorganisms to reshape molecules that do not participate in their normal metabolism, (3) Mixed culture fermentations, and (4) Fermentations for the production of microorganisms.

The development of fermentation process to their present degree of efficiency was a lengthy and costly procedure. It is, therefore, not surprising that commerical firms are reluctant to disclose their manufacturing methods. Indeed, some of the methods are such closely guarded secrets that they are not even patented. Thus, while the broad outlines of industrial fermentations are well known, the details often remain undisclosed.

22.2 SUBSTANCES PRODUCED AS A MAJOR END-PRODUCT OF MICROBIAL METABOLISM

22.2.1 Antibiotics

22.2.1.1 The penicillins

One of the most dramatic fermentation stories of the century started in the late twenties when Fleming recorded the effect of a stray mold, *Penicillium notatum,* upon some of his pathogenic bacteria. This observation was not pursued further until the pressure of wartime needs stimulated its fuller investigation in the 1940s and led to the founding of a new industry and the saving of many lives; a remarkable example of the application of microbiology on the industrial scale. When *Penicillium notatum* was originally grown as a surface organism on liquid media, levels of only 10 units/mL of penicillin were produced (an international unit of benzyl pencillin is 0.5988 μgm or 1,670, i.e.,/mg). With improved media this was increased to 200 units/mL. The introduction of the more slowly metabolized lactose, instead of glucose, together with corn steep liquor as a growth medium increased yields even further.

This was again improved by the use of *Penicillium chrysogenum* followed by the selection of many mutants produced artificially with the aid of such mutagens as ultra-violet light. Nowadays, this antibiotic is produced under submerged aerated conditions in stainless steel vats of 20,000 gallons capacity with the control of such variables as pH and glucose concentration (Figure 22.1).

Glucose has supplanted lactose because it is now added in controlled doses giving the same slow carbohydrate metabolism. Amounts in excess of 7000 units/mL of penicillin are attained under modern fermentation conditions. Many other organisms are known to produce penicillins, including *Aspergillus, Trichophyton, Epidermophyton,* and *Cephalosporium* species. It was also discovered that penicillin made in Britain differed from the American product. Investigations showed that these two penicillins had the same basic molecular structure with different substituent groups. The American preparation is benzyl penicillin or penicillin G, while the British equivalent at that time was pentenyl penicillin or penicillin F.

During the course of chemical analyses on fermentation liquors, it was found that there was often more penicillin present than could be accounted for by the measurable antimicrobial activity. This led to the discovery of 6-amino penicillanic acid (6APA)

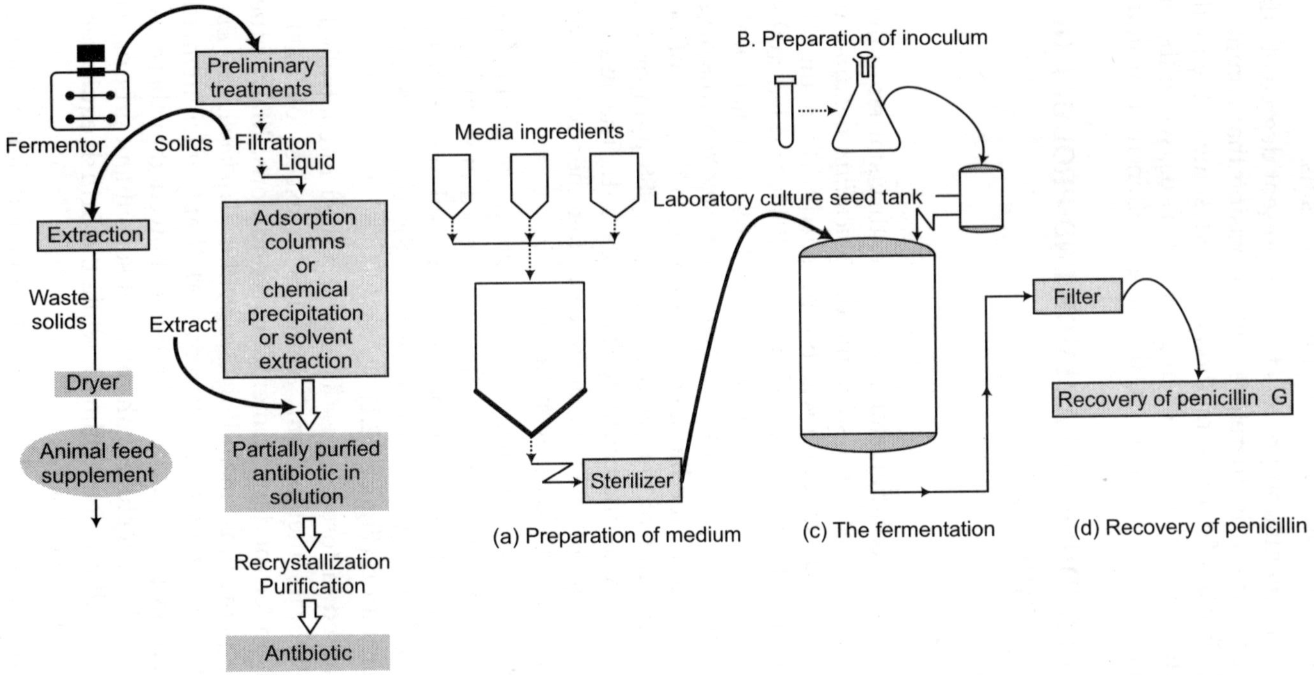

Figure 22.1 Flow chart showing production of antibiotics for the production of penicillin.

which has an NH_2 group on the β lactam ring and hence no side chain. Although 6APA is biologically inactive, side chains can be substituted by acylation either chemically or enzymically. A wide range of organisms that contain penicillin acylases including *Escherichia coli* and *Alcaligenes faecalis* is now known. These will link specific side chains to 6APA producing additions to the penicillin family with novel properties. One such semi-synthetic penicillin, Cloxacillin has been formulated which is stable to the action of μ lactamase (penicillinase) and to acid, the latter property allowing administration by mouth.

22.2.1.2 The tetracyclines

Oxytetracycline was first obtained from *Streptomyces rimosus* and chlortetracycline from *Streptomyces aureofaciens.* The latter is normally used for commercial production and may be grown on a variety of media including corn steep liquor, groundnut or soy bean meal. For the production of chlortetracycline, lard oil is also added while arachis oil is used in the manufacture of oxytetracycline. Sucrose or starch are the most suitable carbohydrate sources for this organism. Like many antibiotics, the tetracyclines are toxic to the organisms that produce them, and since the toxicity is thought to be due to sequestration of trace metals a plentiful supply of calcium or magnesium ions is added to overcome this effect. These organisms are favoured by a neutral pH and a temperature of 28-33°C. Aeration is essential.

22.2.1.3 Other antibiotics

Table 22.1 illustrates the sources and the name of some more important common antibiotics. It should be remembered that although all the antibiotics discussed here are derived from fungi or actinomycetes, others such as subtilin and bacitracin are produced by bacteria. The usefulness of the latter is limited by their greater toxicity.

Table 22.1 Sources and name of some common antibiotics

Name of the antibiotics	*Source*
Griseofulvin	*Penicillium petulum*
Streptomycin	*Streptomyces griseus*
Chloramphenicol (Now made synthetically fonnarly known as chloromycetin)	*Streptomyces venezualae*
Erythromycin	*Streptomyces erythreus*
Streptomycin	*Streptomyces griseus*
Penicillin	*Penicillium crysogenum, P. notatum*

22.2.2 Dextran

Dextran is a polysaccharide composed of glucose molecules and may be found in slimy masses wherever sucrose is used. Sugar refineries and ham curing factories are prone to trouble from dextran-forming bacteria. Dextran can also block pipelines, make floors slippery, render food uneatable and beverages undrinkable. It has, however, an important use in blood transfusion where it is employed as a blood volume expander. Its application in this way is possible because it is relatively inert, neither causing pyrogenic (i.e., raised body temperatures) nor allergic reactions and will remain sufficiently long in the circulation to allow time for protein renewal in the blood plasma. It is stable to autoclaving and can be stored for long periods at room temperature without deterioration. One particular advantage is that it can be prepared with a known mean molecular weight. The heterofermentative lactic coccus *Leuconostoc mesenteroides* and *L. mesenteroides* is normally used for the production of dextran. This is normally maintained in sucrose agar and is gradually worked up through intermediate volumes to a 1000 gallon scale for commercial production. The medium used is sucrose plus yeast extract supplemented with ammonium and potassium phosphates. Incubation temperature is 23°C, aeration is unnecessary since the organism is microaerophilic. Additions of alkali are made to keep the pH above 5.0 because below this point dextran production is reduced. The overall reaction for dextran production is as follows:

$$n \text{ sucrose} \rightarrow \text{dextran/sucrose} \rightarrow (\text{glucose})\ n + n \text{ fructose}$$

The fructose is subsequently used as an energy source by the bacteria, while the glucose units are joined together into chains mainly with 1-6 linkages but sometimes 1-4 and occasionally 1-3. Branched chains vary in amount according to the particular strain of bacterium used. Hence *L. mesenteroides* is suitable for commercial production because it gives 95 per cent straight-chain molecules, these being the most acceptable for clinical use.

Dextran is removed from the culture fluid by acetone precipitation, and this so-called (native dextran is then ready for the final stage of degrading to the correct molecular size. The methods available for this molecular sizing operation are as follows:

1. ***Acid hydrolysis:*** Dextran is heated with acid, and the average molecular size is checked by viscosity measurement. This method has the advantage that it can be performed without separating the dextran from the fermentation liquor.
2. ***Heating under pressure:*** Degradation may be brought about by heating the dextran under pressure at 160°C in the presence of sodium sulphite to prevent oxidation, and calcium carbonate, to keep the pH at 7.
3. ***Enzymic degradation:*** Various microorganisms contain dextranases, the one most frequently used being *Cellvibrio fulva.*
4. ***Mechanical breakdown:*** High frequency waves emitted by a piezoelectric quartz crystal when stimulated electrically will break down high molecular chains and give a product of considerable homogeneity. Viscosity is the main criterion used for estimating molecular size.

22.2.3 Acid Fermentation

Several acids are being produced microbiologically on a commercial scale. These are acetic, lactic, citric, fumaric, itaconic, and gluconic acids, some of these being produced from molds which require high aeration rates in the submerged conditions employed (Figure 22.2). Only some of them are discussed here.

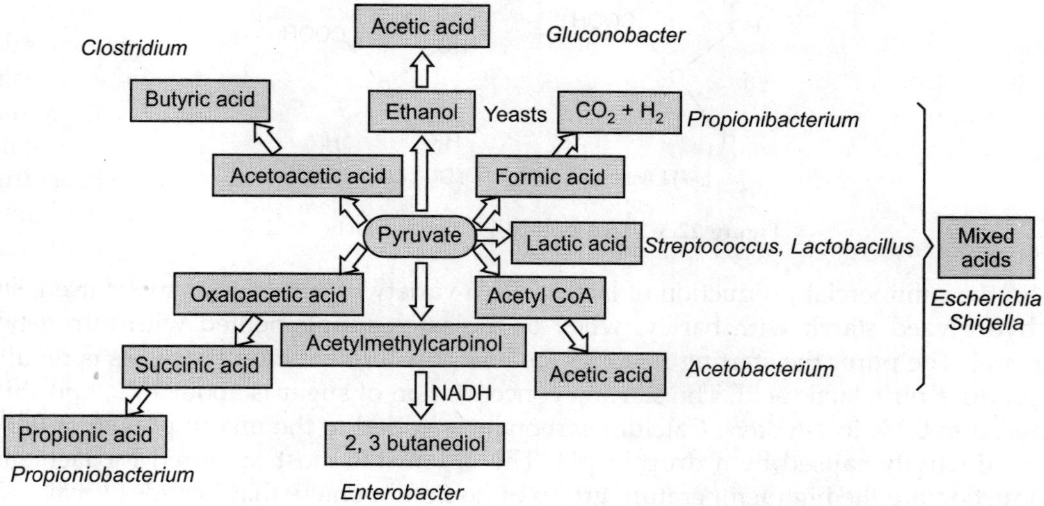

Figure 22.2 Schematic representation of production of various organic acids by different bacteria.

22.2.3.1 Lactic acid

This is one of several acids produced microbiologically on a commercial scale. Lactic acid, in contrast, is produced by microaerophilic bacteria belonging to the genus *Lactobacillus*. This acid is an odourless, colourless liquid having an acid flavour. It is used in a variety of ways finding applications in the drinks industry as a flavouring, in the preservation of food as an adjunct or substitute for vinegar, and in the form of calcium lactate as a convenient means of getting calcium into the body. Other uses are in the plastics and leather industries. A number of grades are produced varying in purity and concentration (22-85%) according to the purpose for which they are required. Broadly, the grades may be classified as pharmaceutical, edible, and technical. Irrespective of grades, however, all types are made by the same process, and it is the rigour of the purification processes that decides the final grade. The most common organisms used in lactic acid production are the homofermentative lactobacilli. In these, as with yeasts, the Embden-Meyerhof-Parnas (EMP) pathways is a main energy-gaining mechanism. The bacterial pathway, however, differs from that of the yeasts because there is no pyruvic decarboxylase present, therefore, there is no loss of CO_2 and no subsequent acceptance of hydrogen by acetaldehyde. Instead, pyruvic acid performs this role in these bacteria and, by accepting hydrogen, is reduced to lactic acid as shown below.

CH3.CO.COOH + 2NADH → CH3.CH.OH.COOH + 2NAD+

Pyruvic acid Lactic acid

This type of lactobacillus produces a single end-product, namely, racemic lactic acid. However, by varying temperature and type of medium, the relative proportions of the two optical isomers can be varied (Figure 22.3).

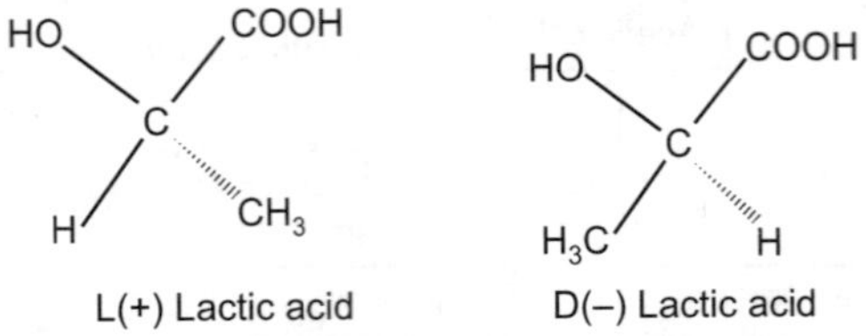

Figure 22.3 Two optical isomers of lactic acid.

In the commercial production of lactic acid, a variety of substrates may be used, such as hydrolyzed starch with barley, whey or molasses supplemented with nitrogenous material. The purer the starting material the less post-fermentation processes is required to produce pure lactic acid. The starting concentration of sugar is about 15%, and this is reduced to 0.1% in 4-6 days. Calcium carbonate is added to the mix to prevent a fall-off in productivity caused by a drop in pH. The organisms most favoured for lactic acid production are the high temperature group of homofermenters that includes *Lactobacillus delbrueckii.* This allows a working temperature of 50°C which has the advantage of suppressing possible contaminants with lower optimum growth temperatures. To maintain this unusually high temperature in vats that may be of 30,000 gallons capacity, heating coils are fitted and the medium is stirred to ensure uniform temperature and composition. Aeration is not applied, since this is an anaerobic process. As with all large scale batch fermentations, it is necessary to build up the inoculum by passing it through a series of tanks of increasing volume until finally the size is sufficient to initiate a rapid fermentation in the large vats used for commercial production (Figure 22.4). When fermentation is complete, the yield is about 85 per cent of the fermentable hexose. The liquor is then heated to 80°C to kill the organisms and coagulate the protein. The end-product, present as the calcium salt, is acidified with sulphuric acid which frees the lactic acid and precipitates the calcium as insoluble sulphate. Heavy metals, at one time removed by treatment with sodium sulphide, are now taken out with ion exchange resins. Lactic acid is decolorized with powdered charcoal and is extracted from other constituents by the use of iso-propyl ether. The acid is then extracted from the ether by using water. Both of these processes can be operated continuously in counter-current towers. For the very pure grades, such as those required in the manufacture of plastics, a lactic acid ester is prepared, the impurities distilled off, and the lactic acid and alcohol recovered.

22.2.3.2 *Citric acid*

Citric acid ($CH_2COOH\ COH\ COOH\ CH_2\ COOH$), a tricarboxylic acid, is a natural constituent of a variety of fruits (Figure 22.5). Citric acid is a commodity chemical, and more than a million tonnes are produced every year by fermentation. It is used mainly as an acidifier, as a flavoring, preservative, and as a chelating agent. In this production technique, which is still the major industrial route to citric acid used today, cultures of *Aspergillus niger* are fed on a sucrose or glucose-containing medium to produce citric acid. The source of sugar is corn steep liquor, molasses, hydrolyzed corn starch

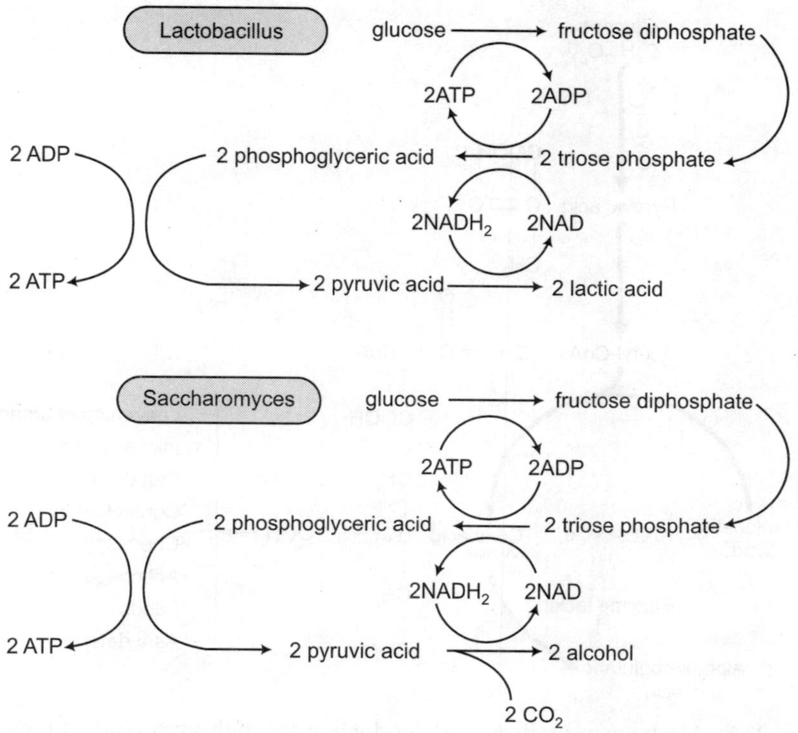

Figure 22.4 Lactic acid and alcohol fermentation.

or other inexpensive sugary solutions. After the mold is filtered out of the resulting solution, citric acid is isolated by precipitating it with lime (calcium hydroxide) to yield calcium citrate salt, from which citric acid is regenerated by treatment with sulfuric acid (Figure 22.6).

Cane molasses obtained from sugar mills is used as an inexpensive carbon substrate. Cane molasses contains water 20%, sugar contents 62%, non-sugar contents 10%, and inorganic salts (ash contents) 8%, making a blackish homogenous liquid with high viscosity. Ash contents include ions such as Mg, Mn, Al, Fe and Zn in variable ratio. Sugar content is diluted to about 25% sugar level. The molasses solution, after adding 35 mL of 1N H_2SO_4 per litre, is boiled for half an hour, cooled, neutralized with lime-water (CaO) and is left to stand overnight for clarification. The clear supernatant liquid is diluted to 15% sugar level. When molasses is used, it is diluted to a sugar concentration of 15-20% with dilute sulfuric acid and to a pH 5.5 – 6.5. Other nutrients for fungal growth are then added and the mixture is sterilized for 30 min to reduce or eliminate the inhibitory action of metals. Stainless steel fermenters are employed for citric acid fermentation. The incubation temperature is kept at 30 ± 1°C throughout the fermentation period. Agitation speed of the stirrer is kept constant while aeration rate is maintained. Sterilized silicone oil is used to control foaming during fermentation. Critic acid is precipitated as calcium citrate using 1 part of hydrated lime for every

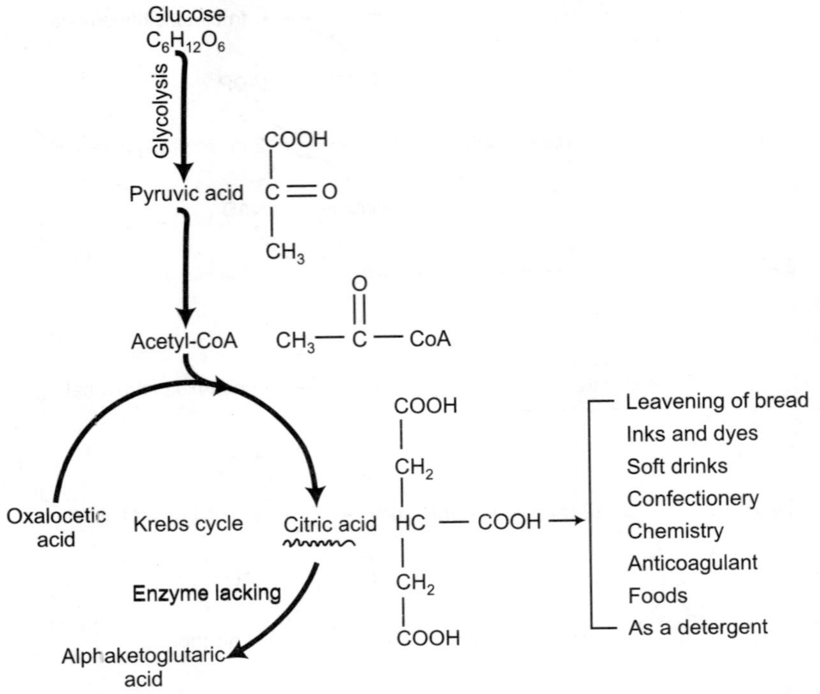

Figure 22.5 Mechanism of citric acid production through glycolytic pathway and Krebs cycle in *Aspergillus niger.*

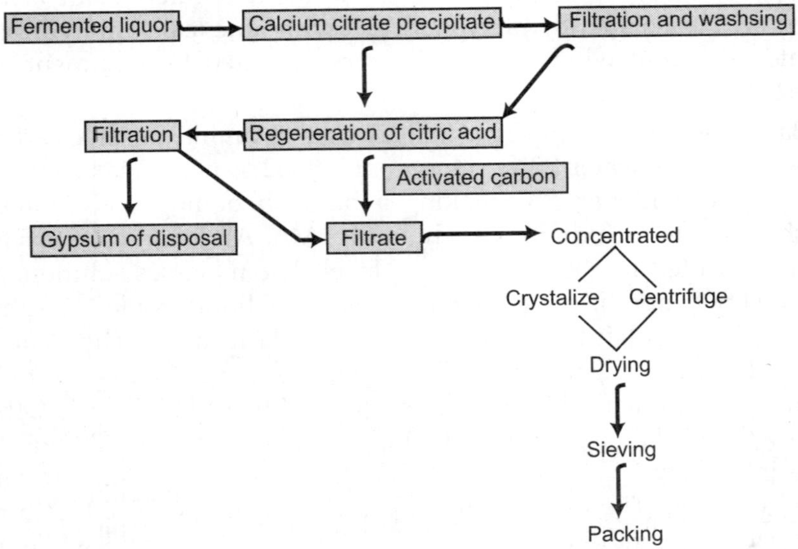

Figure 22.6 Flow chart for the downstream processing of citric acid production by *Aspergillus niger.*

2 parts of liquor added over a one hour period, while the temperature raised to 95°C. Precipitation of citric acid can be carried at a temperature of 50°C for 20 min. The precipitated calcium citrate is filtered, washed with water several times and transfer to acidulate (Figure 22.6).

22.2.4 Amino Acids

In the 1950s *Corynebacterium glutamicum* was found to be a very efficient producer of L-glutamic acid. Since this time biotechnological processes with bacteria of the species *Corynebacterium* developed to be among the most important in terms of tonnage and economical value. L-glutamic acid and L-lysine are bulk products nowadays. L-valine, L-isoleucine, L-threonine, L-aspartic acid and L-alanine are among other amino acids produced by *Corynebacteria*. Applications range from feed to food and pharmaceutical products. The growing market for amino acids produced with *Corynebacteria* led to significant improvements in bioprocess and downstream technology as well as in molecular biology. During the last decade big efforts were made to increase the productivity and to decrease the production costs.

At present, the amino acid industry has come to occupy an important role in world chemical industries. China is an agricultural country and has a very large population. Annual demand for amino acids used in feed additives and pharmaceutical products is huge. Most of the amino acids are currently manufactured in China. But the industrial production process has not been set up for a few limited kinds of amino acids such as L-tryptophan, L-histidine and L-arginine. The extraction method is still an industrial process for L-cysteine. However, the extraction method depends on the availability of natural protein-rich resources such as hair keratin, feather and the production process is not environmental friendly as unpleasant odours produced and problems of waste treatment. So there is an urgent need for establishing the production process in order to meet the demand for these amino acids.

These substances are used for various clinical purposes, such as post-operative therapy, as supplements to low protein diets, and as food flavourings. A common ingredient of certain dehydrated foods such as soups is mono-sodium glutamate. Amino acids can be prepared by chemical synthesis, but this has the disadvantage of producing both optical isomers. In contrast, biosypthesis produces the L-isomer which is the form found in most biological systems. Sometimes, a combination of chemical synthesis and biosynthesis is employed as in the production of lysine from diamino pimelic acid. The production of the diaminopimelic acid is purely chemical, while the final step to lysine is biological. The main amino acids prepared by fermentation are glutamic acid and lysine (Table 22.2). In the preparation of the former, the carbohydrate (usually glucose or sucrose) is present at a concentration of 10-20 per cent. Urea or $NH4^+$ salts provide the nitrogen; magnesium, manganese, and iron, the metallic constituents, and such substances as pelargonic acid are added as stimulators. The pH is maintained between 7 and 8 by the addition of urea or ammonia gas. The concentration of biotin is most important because it controls the yield.

Table 22.2 Organisms and methods for amino acid production

Amino acids	*Method*	*Organisms*
Aspartic acid	Transamination from oxalacetate or fumaric acid and ammonia	*E. coli, Serratia mercescens, Bacillus* sp.
L-homoserine	This organism produces L-lysine with L-homoserine. Varying the concentration of biotin changes the ratio of the yields of glutamate, lysine, and homoserine.	*Micrococcus glutamicus*
L-threonine	Made by a strain which has growth requirement for diaminopimelic acid or one which also has requirement for methionine. Mannitol or sorbitol instead of glucose increases	*E. coli*
L-methionine	The organism is cultured in a medium containing mecapto-α-hydroxy butyric acid	*Pseudomonas* sp.
L-valine	Strain (a), (b), (c) and (d) all accumulate L-valine in the medium, strain (e) is an auxotrophic mutant requiring L-leucine or L-isoleucine. Biotin favours L-valine production.	(a) *Aerobacter cloacae* (b) *Paracolabacterium coliforme* (c) *E. coli* (d) *Aerobacter aerogenes* (c) *Micrococcus glutainicus*
L-omithin	This organism accumulates ornithin and requires either L-arginine or L-citruline as a precursor	*Micrococcus glutamicus*
DL-alanine	This is transamination reaction involving pyruvic acid and glutamic acid and a racemization	*Corynebacterium gelatinosum*
L-phenylalanine	Phenylalanine is required to accumulate L-tyrosine, Shikimic, α-hydroxyphenyl pyruvic, and D-quinic acid or tryptophane stimulate L-tyrosine production	*Micrococcus glutamicus*
L-tryptophane	Strain (a), (b) and (c) will convert indole in submerged cultute to L-tryptophane. Strain (d) and (e) will convert 3-indole pyruvic acid to L-tryptophane by reductive deamination. Strain (I) produces L-tryptophane from anthranilic acid.	(a) *Claviceps purpuria* (b) *Rhizopus oryzae* (c) *Ustilago avenue* (d) *Micrococcus* sp. (e) *Serratia* sp. (f) *Hansenula anomala*
L-lysine	Organism use sugar cane syrup to convert	*Corynebacterium glutamicum*
Glutamic acid	Organism use sugar cane syrup to convert	*Corynebacterium glutamicum*

The organism commonly used in the manufacture of glutamic acid is a *Corynebacterium* sp. but strains of the genera *Brevibacterium, Micrococcus, Microbacterium,* have also been used. Removal of glutamic acid from the growth medium is easy because it is insoluble in acid conditions. High yielding media are treated as follows: the organisms are removed, the filtrate concentrated and acidified to pH 3.2, and the glutamic acid, thus precipitated, is recovered by filtration. Lower yielding media are treated with a suitable ion exchange resin, and this method not only separates but also purifies the glutamic acid (Figure 22.7). Lysine preparation is similar to that of glutamic acid, the usual organism being a mutant of *Micrococcus glutamicus*. Biotin is less critical than for the production of glutamic acid, whereas the concentrations of homoserine, threonine, and methionine have a profound effect upon yield. Lysine is more difficult to separate from the growth medium than glutamic acid, but ion exchange resins offer the best means of extraction. In the final stages, it is usually obtained as lysine monohydrochloride and the pure amino acid produced by recrystallization.

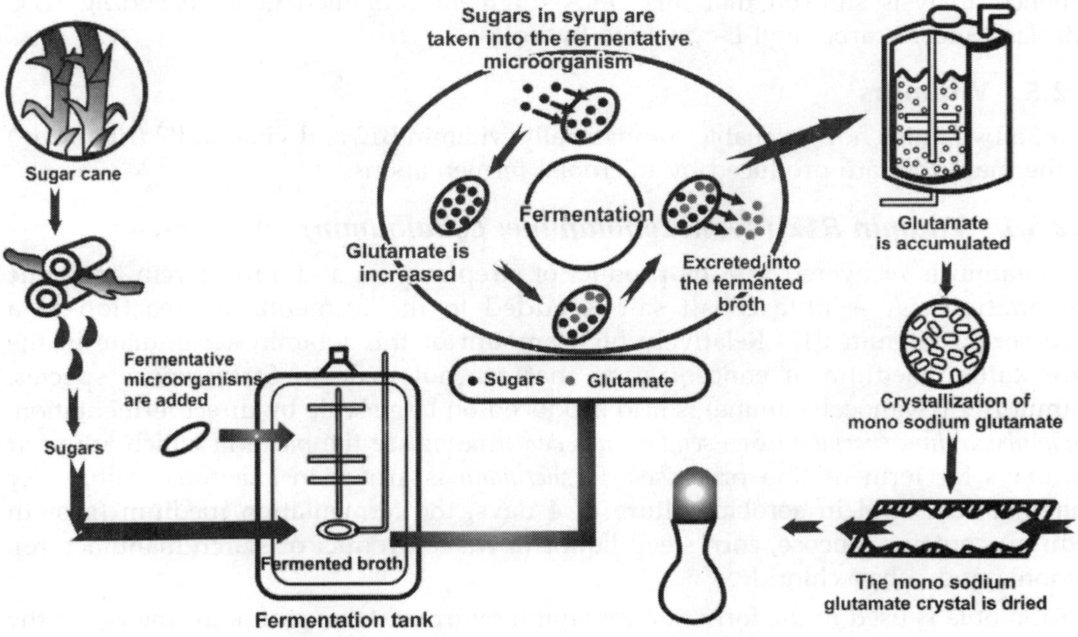

Figure 22.7 Production of monosodium glutamate by fermentation.

Many other amino acids can be made in smaller quantities as summarized in Table 22.2. *Corynebacterium glutamicum, Corynebacterium crenatum* and *Corynebacterim pekinese* are used in amino acid production process. But the mechanism of amino acids accumulation of those mutants has not been extensively investigated. So far the strain improvement has mainly been carried out by an alternate process of mutagenesis and screening. It is difficult to increase the production yield further by these methods.

22.2.4.1 Microbial production of L-tryptophan

A microorganism with amino acid racemase activity was screened and identified as *Pseudomonas putida*. The amino acid racemase has low substrate specificity. It has activity toward most aliphatic amino acids. A technique of L-tryptophan production was established by coupling the reactions catalyzed by amino acid racemase and tryptophan synthase. Optimization of cell cultivation and transformation conditions was carried out. In addition, two distinct amino acid racemase genes from *P. putida* were cloned. Only alanine racemase activity was found in recombinant *E. coli* TG1 containing DadX gene. Another amino acid racemase encoded by alr gene showed low substrate specificity.

22.2.4.2 *Enzymatic synthesis of L-cysteine*

DL-ATC is a precursor of L-cysteine synthesis. A microorganism having ability to utilize DL-ATC as a sole carbon and nitrogen source and accumulating L-cysteine was screened and identified as *Pseudomonas* sp. A 6 Kb DNA fragment isolated from the genome DNA of *Pseudomonas* sp. was shown to be involved in conversion of DL-ATC. Sequence analysis showed that this DNA fragment contained genes encoding ATC hydrolase and N-carbomoyl-L-cystenine hydrolase.

22.2.5 Vitamins

Out of all vitamins now available commercially, vitamin B12 and vitamin B2 (riboflavin) are the main that are produced by microbial fermentations.

22.2.5.1 *Vitamin B12 (Cyanocobalamine; Cobalbumin)*

This vitamin is recovered as a by-product of streptomycin and aureomycin antibiotic fermentations. A soluble cobalt salt is added to the fermentation reaction as a precursor to vitamin B12. Relatively high amount of this vitamin accumulate in the fermentation medium at concentration that are not toxic to *Streptomyces* species. Vitamin B12 (Cyanocobalamine) is also produced on large scale by direct fermentation. *Propionibacterium shermanii* or *Pseudomonas denitrificans* are the bacteria which are used nowadays for fermentation processes. *P. shermanii* is grown in anaerobic culture for 3 days at 30°C and in aerobic culture of 4 days; the fermentation medium (growth medium) contains glucose, corn steep liquor (a waste product of starch manufacture), ammonia and cobalt chloride.

Ammonia is used in the form of ammonium hydroxide that maintains the pH of the medium at 7. *Pseudomonas denitrificans* is grown for 2 days in aerobic culture medium containing sucrose, betaine, glutamic acid, cobalt chloride, 5, 6 dimethylbenzimidazol, and salts. The vitamin resulted in the fermentation are retained within the cells. The cells are, therefore, collected by high-speed centrifugation and the vitamin B12 is recovered by releasing the vitamin from the cells by treating with acid, heating, etc. The vitamin thus released from the cell is absorbed on ion-exchange resin IRC-50 or charcoal. It is then purified further by phenol and water. The vitamin is finally crystallized from aqueous-acetone solutions. The normal yield is 23 mg/litre to 30 mg/litre when *Propionibacterium* is used.

22.2.5.2 Vitamin B2 (Riboflavin)

Riboflavin (vitamin B2) is recovered as a by-product of acetone butanol fermentation and is produced by various *Clostridium* species. But, *Eremothecium ashbyii* and *Ashnua gossypii*, the yeasts that are used to commercially produce this vitamin by direct fermentation. The requirement nutrient medium (growth medium) for fermentation contains glucose, corn oil, soy oil, and glycine. When *Ashbya gossypii* is used, the fermentation is normally carried out for 4-6 days at 36°C in aerobic conditions, and at pH 6-7.5. The recovery of the vitamin is at the rate of 4.25 g/litre. Later on, the yeast cells are recovered and used as a feed supplement for domestic animals to supply riboflavin required by them.

22.3 THE USE OF MICROORGANISMS TO RESHAPE MOLECULES

One process, already described earlier in antibiotics which might also be included here is the application of acylase enzymes in conjunction with 6APA to produce the new penicillins. In this type of process, the organisms employed do not metabolize the substances presented to them but merely have the enzymic capacity to change certain groupings within the molecule. Steroids have frequently been treated in this way and there are a large number of organisms available to carry out a wide range of chemical changes.

Two well known examples of biotransformation are: use of a zygomycetous fungus, *Rhizopus nigrinans* to hydroxylate progesterone (progesterone is obtained by a series of chemical steps from stigmosterol, extracted from soybeans) for the systhesis of cortisone hormone. This conversion involves three reactions. The first reaction is a typical microbial transformation carried out by *Rhizopus nigricans* and the other two reactions are performed chemically as shown below:

$$\text{Progesterone} \xrightarrow[\textit{Rhizopus nigrinans}]{\text{Biotransformation}} 11\alpha\text{-hydroxyprogesterone}$$

$$\xrightarrow{\text{Chemical reaction}} \text{Hydrocortisone} \xrightarrow{\text{Chemical reaction}} \text{Cortisone}$$

Steroids are compounds of extreme importance in the economy of the animal body. Steroid hormones are secreted by the adrenal cortex and are responsible for the regulation of carbohydrate and mineral metabolism. Other well-known examples of steroid-secreting organs are the ovaries and testes that regulate the animal's sexual activities. Recently, steroid compounds have been administered for therapeutic reasons. Examples of such applications are the use of cortisone as an anti-inflammatory agent, the menstrual cycle regulators incorporated into the 'pill' for contraceptive use, and more recently their administration as postoperative immune reaction suppressors in organ transplant patients. Steroids are complicated molecules with the basic structure as illustrated in Figure 22.8.

To synthesize a steroid from a suitable starting material takes about 32 steps and even to move an oxygen from C_{11} to C_{12} may take up to 12 steps. These lengthy chemical

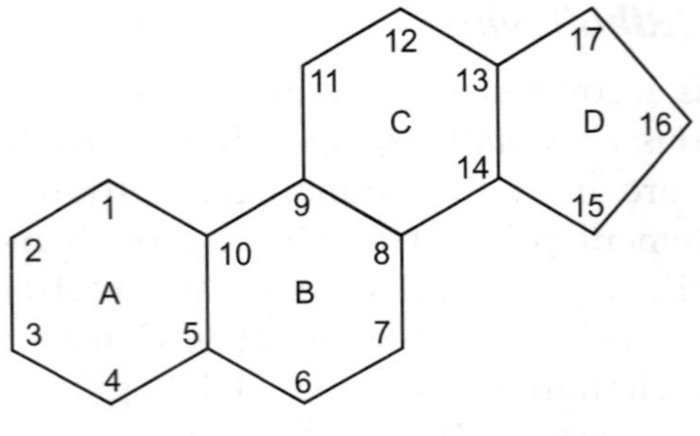

Figure 22.8 Chemical structure of steroid.

syntheses are not practicable for producing steroids on a commercial scale, and new methods were therefore sought. The first microbiological steroid transformation was made in 1937 by Mamoli and Vercellone, and since that time a whole range of steroids have been changed in chemical structure by an equally wide range of microorganisms. The advantage of the microbiological method is that highly specific changes in the steroid molecule can be effected in a single step. The known range of chemical changes includes hydroxylation, dehydrogenations, etc. The organisms concerned include many molds, particularly species of Zygomycetes and Deuteromycetes (Figure 22.9) and Actinomycetes and a few species of yeasts and unicellular bacteria. Although the first microbial transformation of a steroid was made as long ago as 1937, one factor that discouraged such processes on a large industrial scale was the relative insolubility

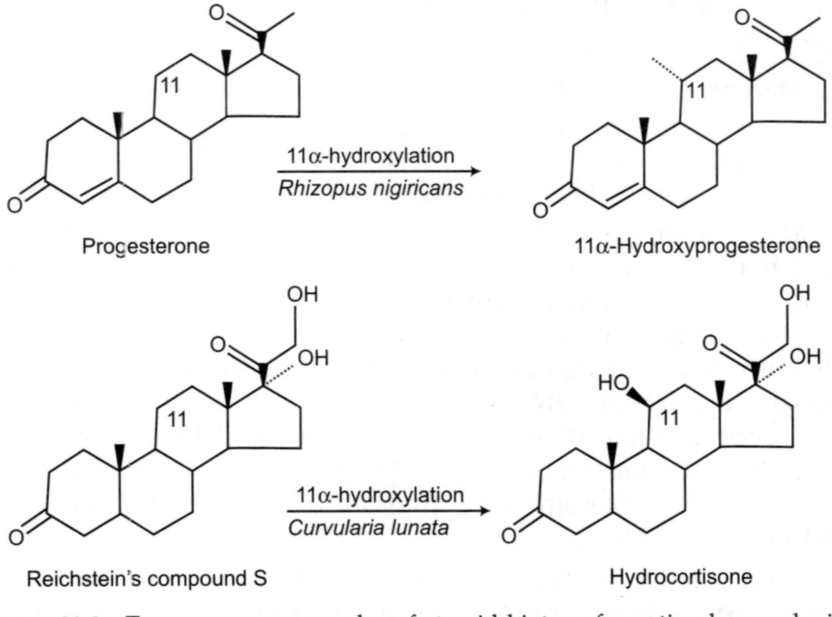

Figure 22.9 Two common examples of steroid biotransformation by employing *Rhizopus nigricans* and *Curvularia lunata*.

of these complex compounds. One of the factors that stimulated industrial-scale steroid transformations was the discovery that oxygen at the C_{11} position imparted important anti-inflammatory properties. This gave rise to a steady demand for such compounds as cortisone and stimulated new ways of manufacturing this and related compounds on a large scale. Thus, one of the early discoveries was the transformation of compound S (11 deoxycortisol) to cortisol and smaller amounts of cortisone by the mold *Cunninghamella blakesleeana.*

Such transformations as these are now usually being performed on a 30,000 gallon scale. A suitable growth medium such as glucose, molasses, or corn steep liquor, with yeast extract added as a nitrogen source are provided for the transforming organism. The large-scale tanks are agitated and aerated, the organisms being allowed to reach maximum growth. Towards the end of the growth phase, the appropriate precursor is added in a non-toxic hydrophilic organic solvent such as ethanol, acetone, or propylene glycol. The chemical change can be followed chromatographically, and when complete the newly formed steroid is extracted with organic solvents. Although, owing to the insolubility of steroids, the total weight per batch of culture is small, losses are also small and a 95 per cent recovery is quite common.

There seems little doubt that the transformation of steroid compounds by microorganisms is a process that will not be easily supplanted by chemical methods. The ease with which the substituent groupings of this complicated molecule are changed in a way that cannot be achieved by chemical means is a factor that appeals to manufacturers. In spite of the large range of transforming organisms known, little has been done to select high yielding mutants as has been successfully done in the antibiotics industry.

22.4 MIXED CULTURE FERMENTATIONS

There are a number of industrial fermentation processes that use mixed cultures. These include the production of fermented vinegar making, beverages, and ensilage of grass for cattle feed.

22.4.1 The Manufacture of Vinegar

This is prepared as the result of two microbiological processes, namely: alcoholic fermentation and acetification mediated respectively by yeasts and by bacteria of the genus *Acetobacter*. It is interesting to note that different countries have a preference for certain substrates. In Britain, malt wort is commonly used, whereas in France and other continental countries grape is the usual starting material, whilst in the U.S.A. apple juice is the preferred substrate.

The acetic acid (vinegar) fermentation is one of the oldest known to mankind, occurring naturally as an unwanted spoilage of wine. The active ingredient in vinegar is acetic acid, which is produced by acetic acid bacteria oxidizing an alcohol containing fruit juice. Vinegar is actually produced in two stages: the first utilizing a yeast such as *Saccharomyces cerevisiae* to produce ethanol from plant sugar. The second stage involves an aerobic fermentation carried out by acetic acid bacteria such as *Acetobacter* and *Gluconobacter* (Figure 22.10). These bacteria oxidize ethanol through acetaldehyde to acetic acid as shown below:

$2C_2H_5OH + O_2$ Alcohol dehydrogenase $\rightarrow CH_3CHO + H_2O$
Acetaldehyde

$CH_3CHO + O_2$ Aldehyde dehydrogenase $\rightarrow CH_3COOH$
Acetic acid

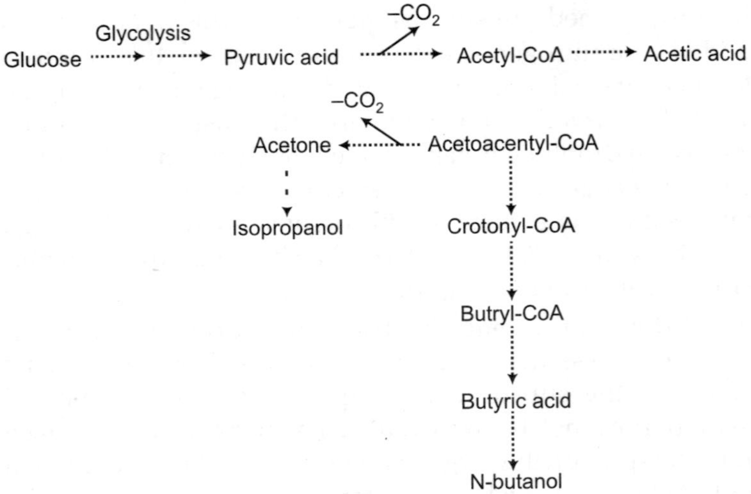

Figure 22.10 Schematic diagram showing production of n-butanol, acetone, isopropanol, acetic acid during acetone butanol fermentation.

All *Acetobacter* species are aerobic and usually found in the wild state growing as a pellicle at the air/liquid interface of alcoholic liquors. This is utilized in the old-fashioned French system called the Orleans process. In this, acetic acid bacteria are allowed to grow on the surface of wine in partly filled casks. When the wine has been slowly converted to vinegar, part is drawn off at the base of the cask, the original volume being re-established by replenishing with new wine. Although this is a slow process, it produces vinegar of the highest quality. Perhaps the most widely used method for vinegar production is the so-called quick or trickling process. The apparatus consists of a large wooden, tower packed with beechwood shavings, birch twigs or other non-compressible material. A sparge at the top of the tower distributes alcoholic liquor over the packing, which provides a support for the acetic acid bacteria. The liquor gradually acetifies as it trickles down the tower to a reservoir at the base from which it is recycled until the process is complete. Acetification being exothermic, the heat produced causes the tower to act as a chimney drawing cold air in at the base and discharging warm at the top. A series of vents around the base of the tower can be opened or closed to regulate air flow, while condensers at the exit trap all volatiles entering in the warm air. This method, although slow, is simple in operation and requires the minimum of attention and mechanical maintenance.

The third method of vinegar manufacture is by cultivating acetic acid bacteria submerged in a suitable substrate. Constant and intense aeration is required to keep the organisms in a rapid phase of acetification and provision must be made for efficient dissipation of heat. This is usually effected by the presence of cooling coils in the acetifier that are supplied with coolant by way of a solenoid valve actuated by the temperature of the vinegar. The shape and size of vessels used for submerged acetification may vary, but the commonest is a tall narrow vat which will accommodate froth that accumulates on the surface of the liquor. These vessels may be of wood or stainless steel. The submerged system is not only used as a batch culture but also continuously when a rotating stirrer may also be incorporated to induce more thorough mixing and aeration. Such systems have the advantage of producing large quantities of vinegar in a relatively short time in a minimum of vat space. They do require, however, complex control mechanisms and suffer from the disadvantage that if the air supply is interrupted, even momentarily, the culture may take several days to recover.

Dilute acetic acid produced by synthetic means is sold as a substitute for vinegar. In Britain, at least, the condiment may not be called vinegar unless it has been produced by one of the microbiological processes here described. Furthermore, it is not recognized as vinegar unless it contains a minimum of 4% of volatile acid.

22.4.2 Beverages

An alcoholic beverage is a drink that contains ethanol, commonly known as alcohol (although in chemistry the definition of "alcohol" includes many other compounds).

Beer has been a part of human culture for 8,000 years. In Germany, Ireland, the United Kingdom, and many other European countries, drinking beer (and other alcoholic beverages) in a local bar or pub is a cultural tradition. Non-alcoholic beverages are drinks that usually contain alcohol, such as beer and wine, but contain less than 0.5% alcohol by volume. This category includes low-alcohol beer, non-alcoholic wine, and apple cider.

Wines are made from a variety of fruits, such as grapes, peaches, plums or apricots. The most common wines are produced from grapes. The soil in which the grapes are grown and the weather conditions in the growing season determine the quality and taste of the grapes which, in turn, affects the taste and quality of wines. When ripe, the grapes are crushed and fermented in large vats to produce wine. Beer is also made by the process of fermentation. A liquid mix, called wort, is prepared by combining yeast and malted cereal, such as corn, rye, wheat or barely. Fermentation of this liquid mix produces alcohol and carbon dioxide. The process of fermentation is stopped before it is completed to limit the alcohol content. The alcohol so produced is called beer (Figure 22.11). It contains 4 to 8 per cent of alcohol.

Whisky is made by distilling the fermented juice of cereal grains such as corn, rye or barley (Figure 22.12). Scotch whisky was originally made in Scotland. The word

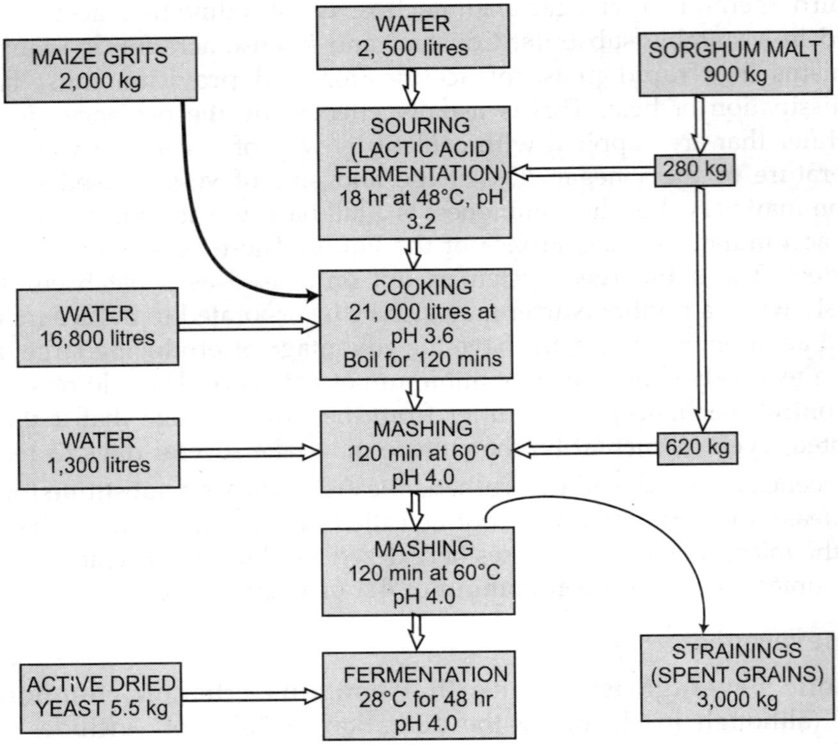

Figure 22.11 Production of beer.

"Scotch" has become almost synonymous with whisky of good quality. Rum is a distilled beverage made from fermented molasses or sugarcane juice and is aged for at least three years. Caramel is sometimes used for colouring. Brandy is distilled from fermented fruit juices. Brandy is usually aged in oak casks. The colour of brandy comes either from the casks or from caramel that is added. Gin is a distilled beverage. It is a combination of alcohol, water and various flavours. Gin does not improve with age, so it is not stored in wooden casks. Liqueurs are made by adding sugar and flavouring such as fruits, herbs or flowers to brandy or to a combination of alcohol and water. Most liqueurs contain 20-65 per cent alcohol. They are usually consumed in small quantities after dinner.

22.5 ENZYMES

Microbial enzymes are most widely used in the food and beverage industries and to a lesser extent in clinical and analytical laboratories and as protease detergents in washing powders. The most economical and convenient method of producing these enzymes is by microbial fermentation. *Bacillus stearothermophilus* produces amylases as secondary metabolites, but most other microbes produce enzymes as primary metabolites, during exponential growth. The major commercial utilization of microbial

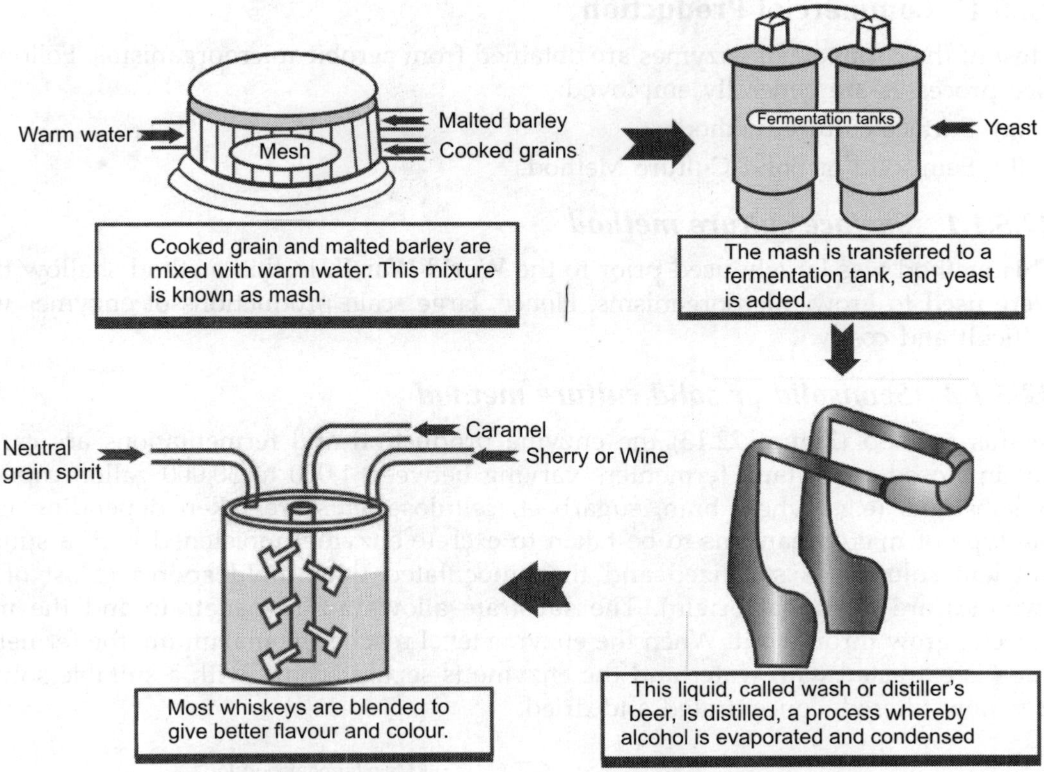

Figure 22.12 Production of whiskey.

enzymes is in the food and beverage industries (Jeenes et al., 1991), in pharmaceutical textile and leather industries, in analytical processes as well as in detergent and dairy industry. Most enzymes are synthesized in the logarithmic phase of the batch culture and may, therefore, be considered primary metabolites. However, some, for example, amylases of *Bacillus stearothemophilus* are produced by idiophase, type cultures and may be considered equivalent to secondary metabolites (Manning and Campbell, 1961). Enzymes may be produced from animals and plants as well as microbial sources, but the production by microbial fermentation is the most economical and convenient method. For example, microbial rennin is being used instead of calf rennin and proteases used in detergents are produced by microbes. It is now possible to engineer microbial cells to produce animal or plant enzymes. Most of the enzymes in industrial use are extracellular proteins produced by *Aspergillus* sp. or *Bacillus* sp. and include α-amylase, β-glucanase, cellulase, dextranase, lactase, lipase, pectinase, proteases and others. The extent of purification required for most of these enzymes is minimal, and they can be produced in tonnes without serious problems.

22.5.1 Commercial Production

Most of the commercial enzymes are obtained from aerobic microorganisms. Following two processes are generally employed:

(a) Surface culture method;

(b) Semisolid or Solid Culture Method.

22.5.1.1 Surface culture method

This method was largely used prior to the World War II. In this method shallow trays were used to grow microorganisms. Hence, large scale productions of enzymes were difficult and costly.

22.5.1.2 Semisolid or solid culture method

In this method (Figure 22.13) the enzyme production and fermentations are carried out in closed, deep tank fermenters varying between 1,000 to 30,000 gallon capacity. A substrate, (e.g., wheat bran, sugarbeet, cellulose, etc., are taken depending upon the type of microorganisms to be taken to excrete enzyme) moistened with a suitable nutrient solution is sterilized and then inoculated with mold spores (most of the enzymes are of fungal origin). The substrate allows air to penetrate and the mold mycelia grow throughout. When the enzyme level reaches its maximum, the fermented mass is extracted with water and the enzyme is separated out with a suitable solvent. It is now filtered, concentrated and dried.

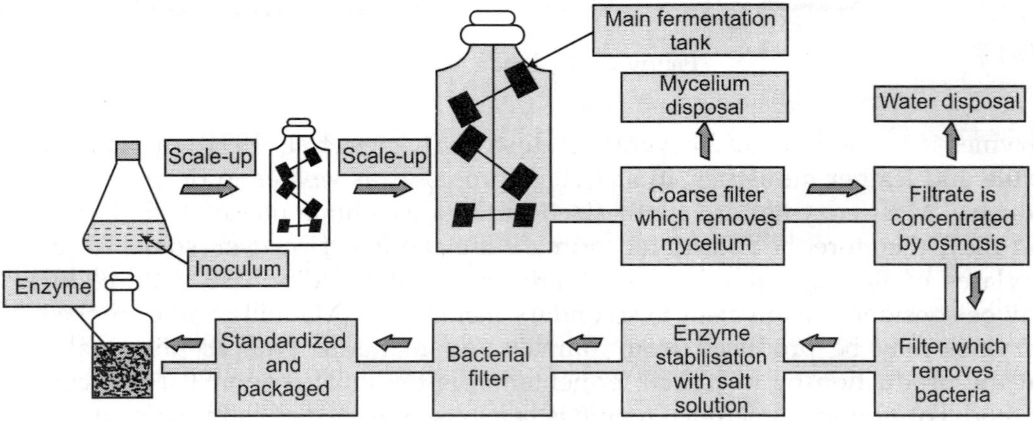

Figure 22.13 Schematic representation of enzyme production

Microbial enzymes are also used for production of synthetic polymers. Plastic industry mostly uses chemical methods for producing alkene oxides used in the production of plastics. It is now possible to synthesise alkene oxides by using microbial enzymes and genetically engineered strains would make commercial production feasible. Some microbial enzymes and their applications are shown in Table 22.3. The synthesis of alkene oxides from alkenes requires sequential action of three enzymes: pyranose-2-oxidase from the fungus *Oudmansiella mucida*, a haloperoxidase from the fungus *Caldariomyces*, and an epoxidase from a *Flavobacterium* sp.

Table 22.3 Some important microbial enzymes and their applications

Source	*Enzymes*	*Reaction*	*Application*
Bacillus	α-Amylase	Starch hydrolysis	Converts starch to glucose or dextran in food industry
	Proteases	Protein digestion	Help laundering
Escherichia	Penicillin Acylase	Benzoyl cleavage	Production of 6-APA
	L-Asparaginase	Removal of L-asparagine involved in tumour growth	Leukaemia/cancer treatment
Aspergillus	Amyloglucosidase	Dextrin hydrolysis	Glucose production
	β-Galactosidase	Lactose hydrolysis	Lactose hydrolysis in milk or whey
	Aminoacylase	Hydrolysis of acylated L-amino acids	Resolution of racemic mixtures
	Glucose oxidase	Oxidation of glucose	Glucose detection in blood
Streptomyces	Glucose isomerase	Conversion of glucose to fructose	Production of high fructose syrups
Several marine bacteria	Luciferase	Bioluminescence	Assay for ATP
Several bacteria and cyanobacteria	Nucleases (restriction endonucleases)	Hydrolysis of phosphodiester bonds in nucleic acids	Genetic engineering

22.6 SOLVENTS

Although the production of acetone and butanol by *Clostridium* strains was a thriving industrial fermentation process, it is no longer competitive with the chemical synthesis of solvents and has been discontinued. However, studies on the molecular biology of *Clostridium* strains suggest that genetic engineering for improved solvent production is feasible, and could result in the revival of the industrial fermentation process. Acetone and butanol were historically produced through fermentation of carbohydrate raw materials. Conventional feedstocks such as grain and molasses, and the energy required to recover products by distillation, are too costly for traditional batch fermentation to compete with petrochemical synthesis. Common uses for organic solvents are in dry cleaning, (e.g., tetrachloroethylene), as a paint thinner, (e.g., toluene, turpentine), as nail polish removers and glue solvents (acetone, methyl acetate, ethyl acetate), in spot removers, (e.g., hexane, petrol ether), in detergents (citrus terpenes), in perfumes (ethanol), and in chemical synthesis. The use of inorganic solvents (other than water) is typically limited to research chemistry and some technological processes. An important property of solvents is boiling point. This also determines the speed of evaporation. Small amount of low-boiling solvents like diethyl ether, dichloromethane, or acetone will evaporate in seconds at room temperature, while high-boiling solvents like water or dimethyl sulfoxide need higher temperatures, an air flow, or the application of vacuum for fast evaporation.

Low boilers: boiling temperature below 100°C (boiling point of water)

Medium boilers: between 100°C and 150°C

High boilers: above 150°C

Most organic solvents have a lower density than water, which means they are lighter and will form a separate layer on top of water. Most of the halogenated solvents like dichloromethane or chloroform will sink to the bottom of a container, leaving water as the top layer.

22.7 ENSILAGE

The process of ensilage is an alternative to the drying of green crops for winter feeding of cattle and sheep. The resulting silage is stored in a succulent form. The process involves the controlled fermentation of green crops such as grass, maize, and occasionally sugar-beet tops or kale, and has the advantage of being largely independent of weather conditions. Processing can thus be done at the optimum time, the most nutritious product being achieved with younger, rather than mature crops. Success depends on adequate temperature control and exclusion of oxygen. The green crop is packed and compressed in containers (silos), in pits in the ground or in stacks above-ground enclosed in polythene. Molasses is sometimes added to promote fermentation and increase palatability. As the fermentation proceeds the temperature rises and the product becomes more acid. The acidity prevents the growth of many putrefactive organisms and, as the oxygen is used up, molds and strictly aerobic bacteria are also inhibited. Only facultative and strict anaerobes continue to grow. At first, species of *Escherichia* and *Aerobacter* predominate, but as the pH falls these are replaced by lactobacilli and streptococci, which ferment the plant material to produce lactic, butyric, propionic, and acetic acids and esters giving flavours much relished by cattle. Active fermentation continues for three or four weeks, after which the temperature falls, and a fairly stable product results.

22.8 FERMENTATION FOR THE PRODUCTION OF MICROORGANISMS

22.8.1 Baker's Yeast

All the processes so far considered have been concerned with the production of an end-product from a substrate. The conversion of part of the substrate to biomass is incidental in all these processes, but once the optimum amount of end-product has been achieved the organisms are merely a waste product. In the preparation of bakers' yeast, the roles are reversed, for it is the aim of the producer to achieve the maximum quantity of microorganisms from the substrate. Pasteur in 1876 first showed that aerating

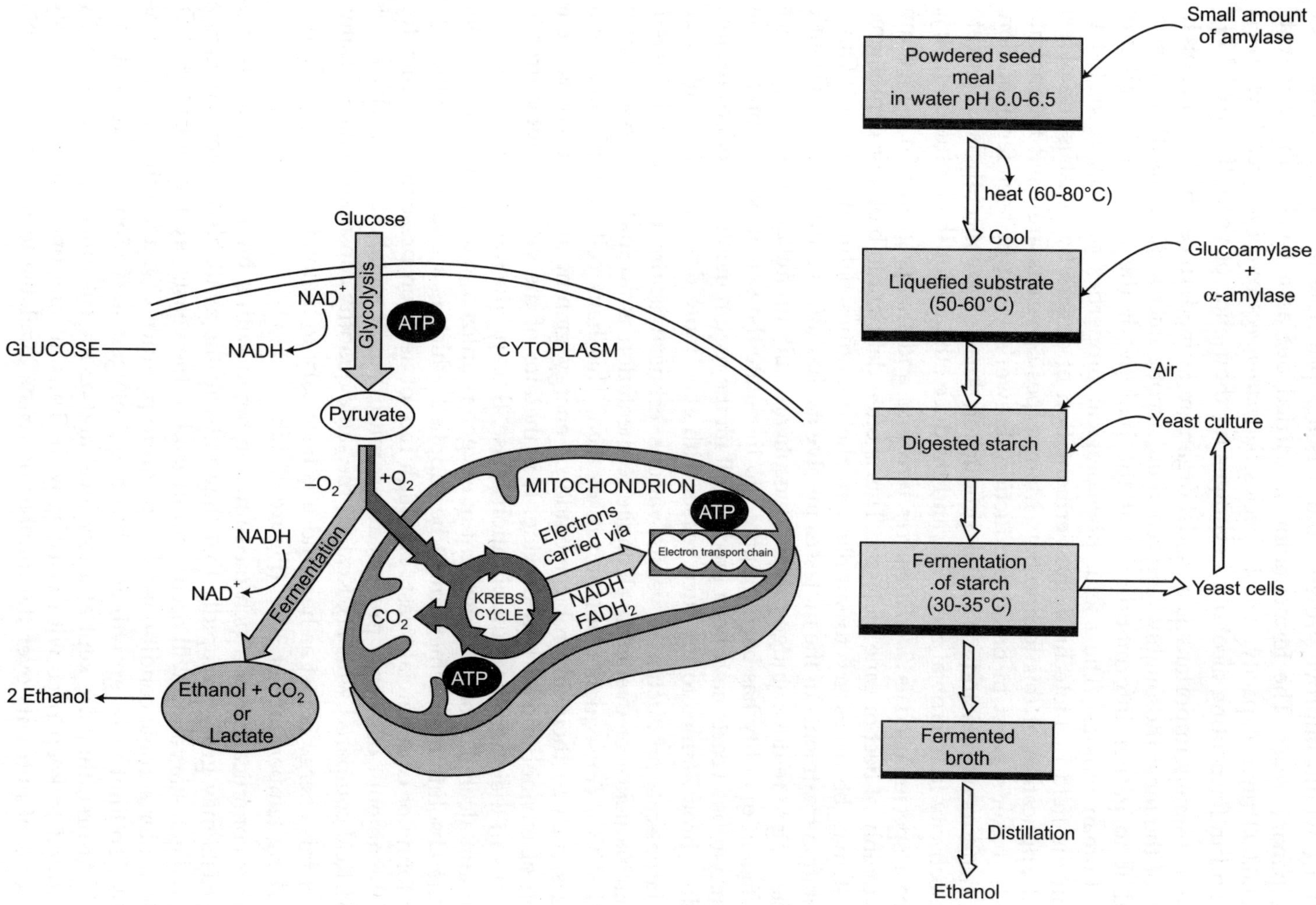

Figure 22.14 Ethanol production for biofuel industry by yeast.

a yeast culture would improve growth, and it is this principle that is used today for producing panary yeast. The fermentation is carried out as a batch culture in vats of 50,000-200,000 titres capacity. Yeast species *Saccharomyces cerevisiae* has been used in baking and in fermenting alcoholic beverages for thousands of years (Figure 22.14).

It is also extremely important as a model organism in modern cell biology research, and is one of the most thoroughly researched eukaryotic microorganisms. Researchers have used it to gather information about the biology of the eukaryotic cell and ultimately human biology. The useful physiological properties of yeast have led to their use in the field of biotechnology. Fermentation of sugars by yeast is the oldest and largest application of this technology. Many types of yeasts are used for making many foods: baker's yeast in bread production; brewer's yeast in beer fermentation; yeast in wine fermentation and for xylitol production. Yeasts have recently been used to generate electricity in microbial fuel cells, and produce ethanol for the biofuel industry.

The yeast species used is *Saccharomyces cerevisiae* and this is grown in a medium consisting mainly of beet or cane molasses. This is often deficient in biotin or pantothenic acid, and it may be necessary to supplement the quantities naturally present. Other prefermentation treatments of the molasses involve its dilution, treatment with sulphuric acid and heating to remove sludge protein. Inocula are built up initially from single cells until a sufficient quantity has been prepared to inoculate the vats used for commercial production. Various yeasts have been bred to full fill the requirements of modern baking, and breeding programmes to produce yeasts with even more desirable characteristics.

In the large-scale production of yeasts, various technical factors must be considered. The normal sequence of events in alcoholic fermentation is a rapid multiplication of yeast cells with a consequent depletion of oxygen. This halts cell multiplication and the yeasts switch to their anaerobic method of energy gaining with the consequent production of ethanol and carbon dioxide. It is the aim of the yeast manufacturer to keep the cells in their aerobic multiplication phase, thus producing maximum biomass with minimum alcohol. About 12 per cent of oxygen is utilized in aeration so that about 40 times the absolute requirement must be supplied which is a total of 330,000 cu. ft. of air per tonne of commercial yeast. The high rate of aeration presents problems, for example, the amount of energy produced per unit of sugar metabolized aerobically is about 1384 Kcal compared with 117 Kcal for the same quantity consumed anaerobically.

Some of this energy is lost as heat and it is necessary to maintain steady conditions and a working temperature of 30°C (Figure 22.15).

A further complication of a rich medium combined with a high rate of aeration is that the yeasts may grow so rapidly that they outstrip the available oxygen. In such a situation, oxygen starvation will occur and instead of the yeasts producing CO_2 + H_2O, as they do under aerobic conditions, they begin to produce CO_2 and ethanol, which is undesirable. To limit yeast growth and thus avoid oxygen starvation, the nutrients are controlled. At first, they are kept at a low level and as the cells increase exponentially they are added to keep pace with yeast growth. Thus, at any one time there is only a small excess of nutrients over the number of yeasts available to use them. Another

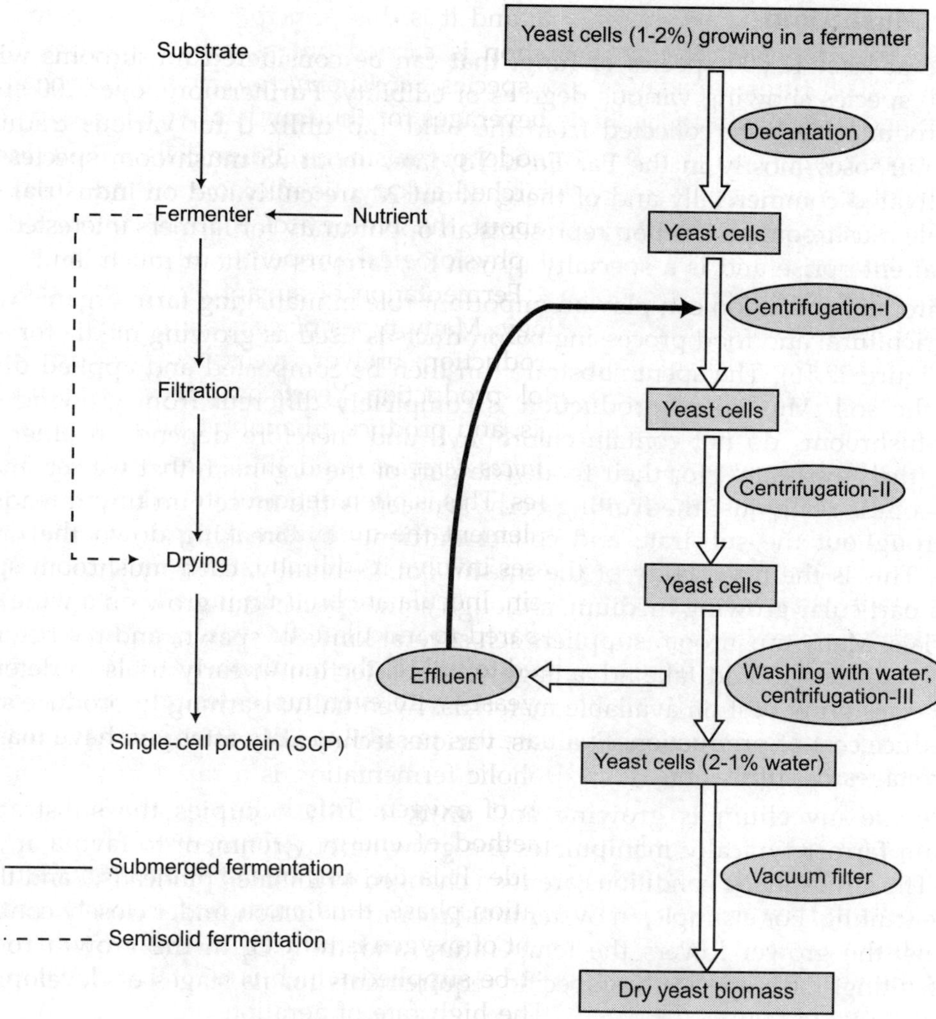

Figure 22.15 Production of single cell protein and dry yeast biomass.

variable requiring control is pH. Since nitrogen is often added in the form of ammonium sulphate the removal of the ammonium ions tends to create acid conditions which are adjusted by periodical additions of alkali to keep the pH in the range 4.0-5.5.

At the end of yeast growth, when the nutrients are exhausted, aeration is continued for 30-60 minutes to 'ripen' the yeast. It has been found that this part of the process stabilizes the yeast and imparts better keeping qualities. After removal from the vats, the yeast is centrifuged and washed several times then chilled to 2-4°C. Water is removed by vacuum filter dehydrators and the yeast packaged. Packing material can be of importance in the preservation of baker's yeast, and contamination can only be prevented by strict controls throughout the process.

22.8.2 Mushroom

There are at least 12,000 species of fungi that can be considered mushrooms with at least 2000 species showing various degrees of edibility. Furthermore, over 200 species of mushroom have been collected from the wild and utilized for various traditional medical purposes mostly in the Far East. To date, about 35 mushroom species have been cultivated commercially and of these, about 20 are cultivated on industrial scale. Small-scale mushroom production represents an opportunity for farmers interested in an additional enterprise and is a specialty option for farmers without much land.

Mushroom production can play an important role in managing farm organic wastes when agricultural and food processing by products is used as growing media for edible fungi (Figure 22.16). The spent substrate can then be composted and applied directly back to the soil. Mushroom production is completely different from growing green plants. Mushrooms do not contain chlorophyll and therefore depend on other plant material (the "substrate") for their food. The part of the organism that we see and call a mushroom is really just the fruiting body. Unseen is the mycelium tiny threads that grow throughout the substrate and collect nutrients by breaking down the organic material. This is the main body of the mushroom. Generally, each mushroom species prefers a particular growing medium, although some species can grow on a wide range of materials. Many mushroom suppliers sell several kinds of spawn, and the beginning mushroom farmer should take advantage of this selection in early trials to determine which species grow best on available materials. Eventually, learning to produce spawn might reduce cost of production. Evaluate this possibility only after you have mastered the later stages of cultivation.

While the mycelium is growing and until it fully occupies the substrate the mushroom farmer typically manipulates the growing environment to favour mycelial growth. The atmospheric conditions are then changed to initiate "pinheads," and then to complete fruiting. For example, in oyster mushroom production under closely controlled conditions, the grower lowers the temperature and the CO_2 in the growth room to initiate fruiting. Each species has specific requirements for its stages of development.

22.9 EXPERIMENTS

22.9.1 Isolation of Antibiotic Produced from Soil

Several soil microorganisms are known to produce antibiotics and are of medical significance. The experiment described herein the isolation of antibiotic producing actinomyces from soil.

Prepare the soil dilution in order to thin out the microbial load in the dilution as indicated in Figure 22.17 and follow the procedure described. Prepare a series of six dilution tubes to produce a final dilution of 10^{-6}.

22.9.2 Isolation of Industrially Important Microorganisms

Several properties of microorganisms are increasingly important in industry which is being exploited in a variety of ways, viz., as a source of high quality food and other

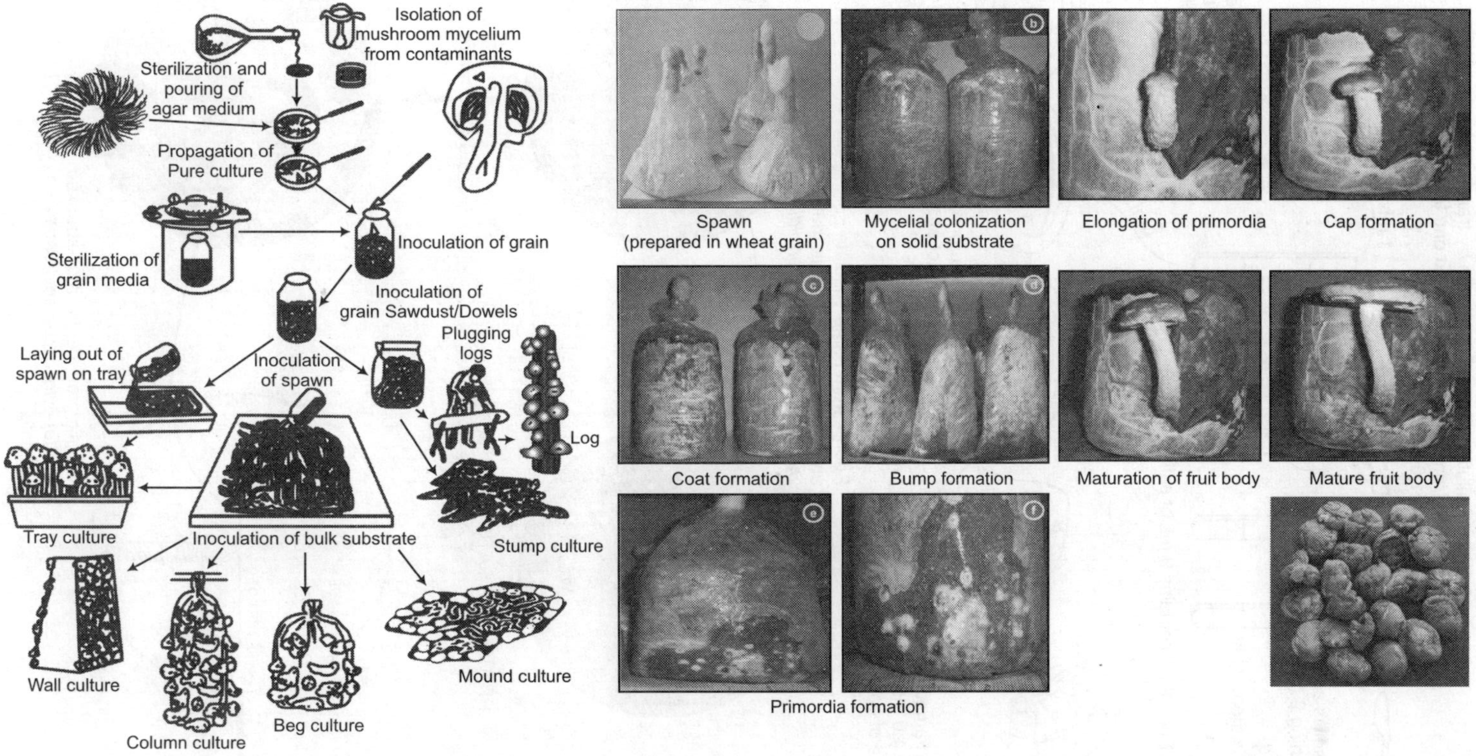

Figure 22.16 Commercial production of mushrooms.

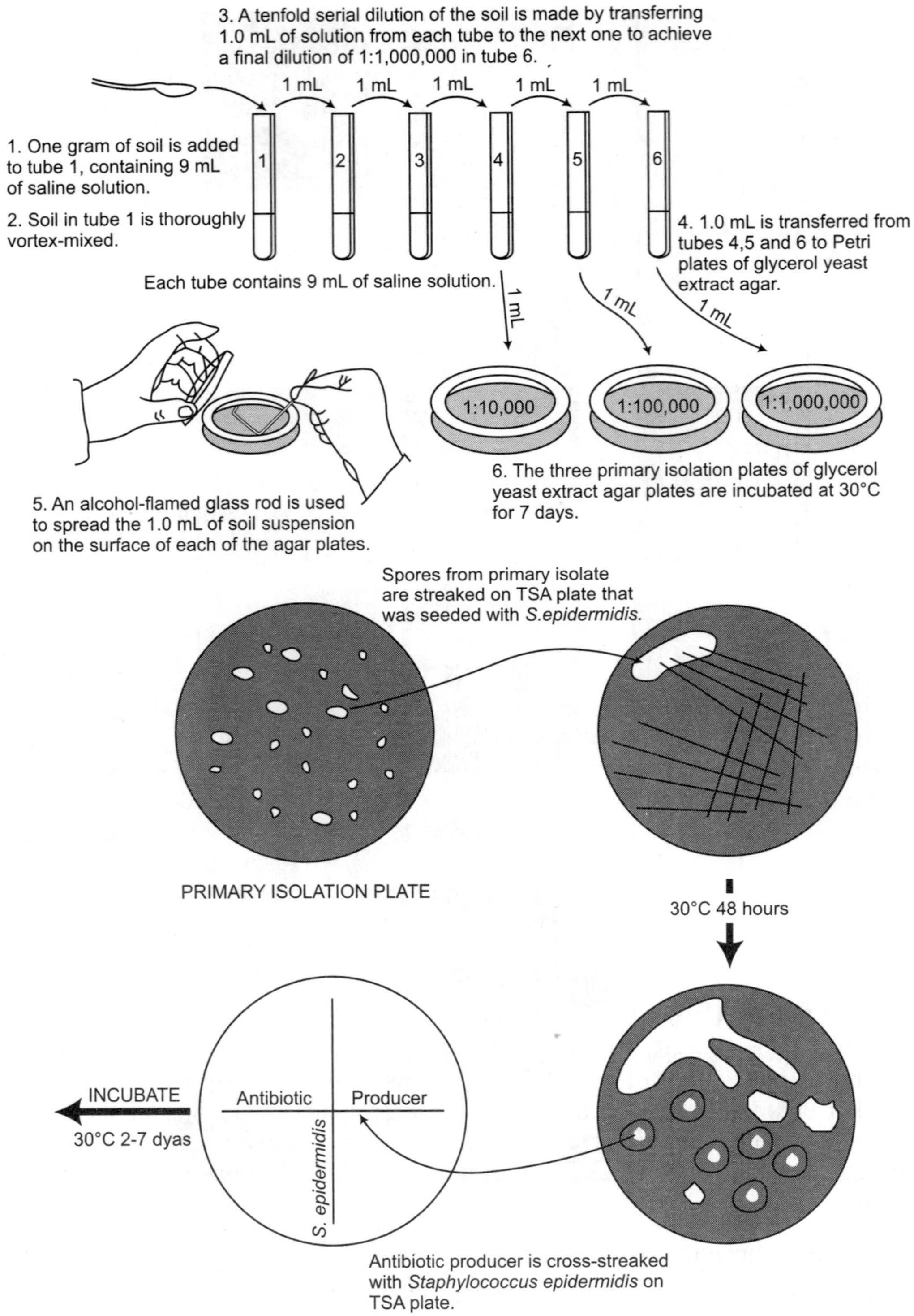

Figure 22.17 Isolation procedure for antibiotic producing *Actinomyces* from soil.

products such as enzymes for the industrial applications degrading pollutants and toxic wastes to clean environment, as vectors to treat disease and enhance agricultural productivity. In industry, microorganisms are used to make products such as antibiotics, vaccines, steroids, alcohol and other solvents, vitamins, amino acids and enzymes. Microorganisms can even leach valuable minerals from low-grade ores.

22.9.2.1 *Materials*

Shake flasks, Petri plates, agar medium, yeast extract.

22.9.2.2 *Protocol*

Isolation of pure or mixed culture involves following criteria for selecting a microorganism of industrially important processes:

1. **The nutritional characteristics of the organism**: It is frequently required that a process should be carried out by employing an inexpensive medium or a pre-determined one.
2. **The optimum temperature of the organisms:** The use of organisms having an optimum temperature above 40°C considerably reduce the cooling costs of a large-scale fermentation and therefore, the use of such a temperature in the isolation procedure may be beneficial.
3. The reaction of the organisms with the equipment to be employed and the suitability of the organism to the type of process to be used.
4. The stability of the organism and its amenability to genetic manipulation,
5. The productivity of the organism, measured in its ability to convert substrate into product and to give a high yield of product per unit time.
6. **The ease of product recovery from the culture:** The ideal isolation procedure commences with an environmental source (frequently soil) which is highly probable to be rich in the desired types, as to favour the growth of those organisms possessing the industrially important characteristic. Alternatively, the isolation procedure may be designed to include certain microbial 'weeds' to encourage the growth of more novel types.

22.9.3 Isolation Methods Utilizing Selection of the Desired Characteristics as Selective Factors

Enrichment culture is a technique which involves a mixed population and providing conditions either suitable for the growth of the desired type or unsuitable for the growth of the other organisms, e.g., by the provision of particular substrates or the inclusion of certain inhibitors prior to the culture stage it is often advantageous to subject the environment source (normally soil) to conditions which favours the survival of the organisms.

22.9.3.1 *Enrichment liquid culture*

1. Frequently carried out in shake flasks.

2. Growth of the desired type from a mixed inoculum result in the modification of the medium and therefore changes the selective force which may allow the growth of other organisms, viable from the initial inoculum.
3. The selective force may be re-established by inoculating the enriched culture into identical fresh medium (sub-culturing). Such sub-culture may be repeated several times.
4. The time of sub-culture in an enrichment process is critical and should correspond to the point at which the desired organisms are dominant.
5. The problems of time of transfer and selection on the basis of maximum specific growth rate in batch culture may be overcome by the use of a continuous process where fresh medium is added to the culture at a constant rate.
6. The effect of substrate concentration on the specific growth rate of a microorganism In continuous culture, growth rate is controlled by the dilution rate and is related to the limiting substrate concentration by the equation (Figure 22.18).

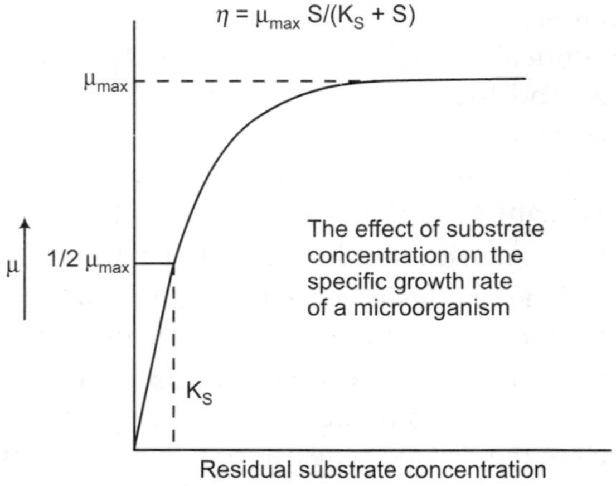

Figure 22.18 The effect of substrate concentration on the specific growth rate of a microorganism.

7. In continuous culture the specific growth rate is determined by the substrate concentration which is equal to the dilution rate, so that at dilution rates below point Y strain B would be able to maintain a higher growth rate than strain A, whereas rates above Y strain A would be able to maintain a higher growth rate (Figure 22.19).
8. Thus the organisms which are isolated by continuous enrichment culture will depend on the dilution rate employed.
9. The enrichment medium employs organic carbon as a sole source of carbon and should be carbon limited. The inclusion of other organic carbon sources, such as vitamins or yeast extract may result in the isolation of strains adapted to using these, rather than the principal carbon source, as energy sources.

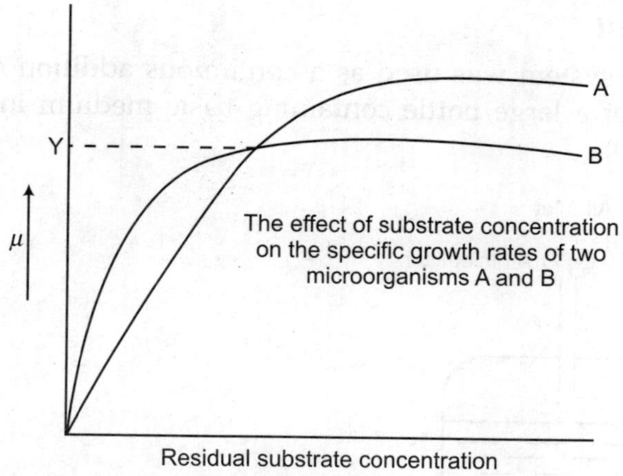

Figure 22.19 The effect of substrate concentration on the specific growth rate of two microorganisms A and B.

22.9.3.2 Disadvantages

The main difficulty in using a continuous enrichment process is the washout of the inoculum before an adopted culture is established. It is, therefore, advisable to start the isolation process in batch culture using 20% inoculum and as soon as growth is observed, the culture should be transferred to fresh medium and the subsequent purification and stabilization of the enrichment is performed in continuous culture.

22.9.3.3 Solutions to the problem of early washout

Use a turbidostat and/or two-stage chemostat. In the turbidostat, the cell density of the culture is continually monitored photometrically and whenever the turbidity of the culture reaches a critical point, the addition of fresh medium to the growth chamber is triggered. In the chemostate, fresh medium is continuously being added to the culture chamber and the rate of growth of the microorganism being cultured is regulated by the rate of flow of fresh medium into the vessel. Growth is regulated by the concentration of a limiting nutrient in this system:

22.9.3.4 Turbidostat

1. It is a continuous-flow system provided with photoelectric cell to determine the turbidity of the culture and maintain the turbidity between set points by initiating or terminating the addition of medium.
2. Thus, washout is avoided as the medium supply will be switched on/off if the biomass falls below the lower fixed point.
3. The use of turbidostat will result in selection on the basis of maximum specific growth rate as it operates at high levels of limiting substrate.

22.9.3.5 Chemostat

The first stage of the system was used as a continuous addition of inoculum and the 2nd stage consisted of a large bottle containing basic medium inoculated with a soil infusion (Figure 22.20).

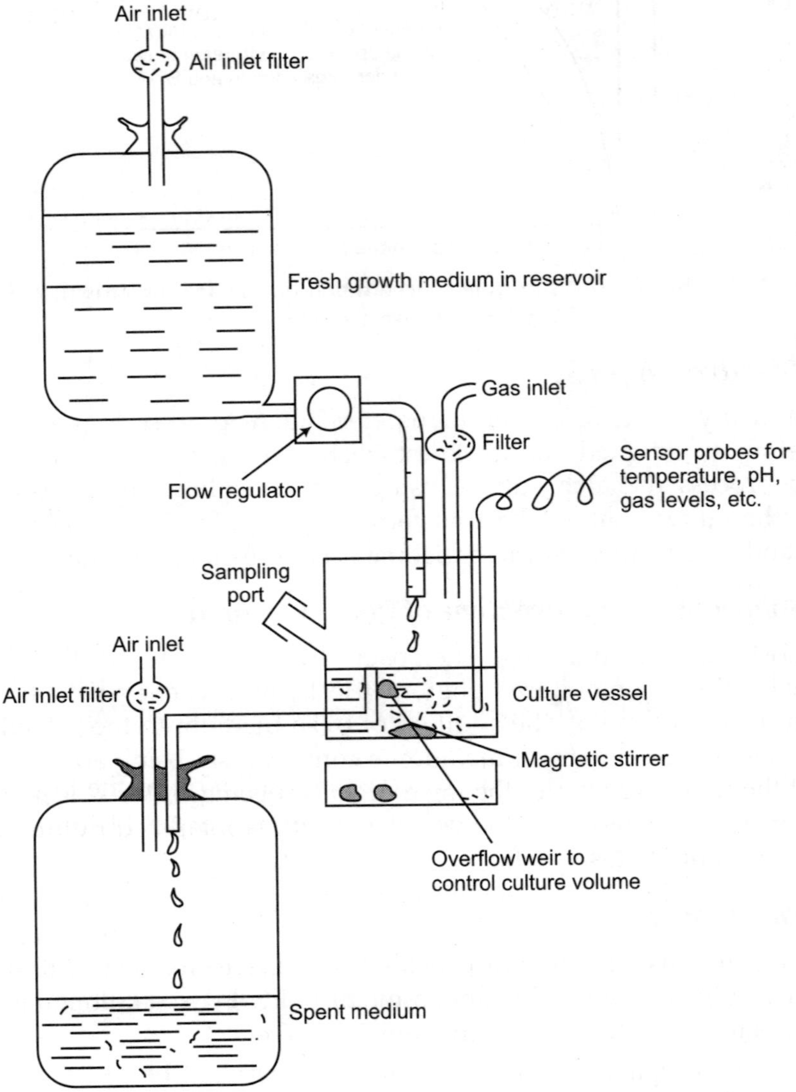

Figure 22.20 Continuous culture system (Chemostate).

22.9.3.6 Enrichment cultures using solidified media

1. Soils of various pH were used as the initial inoculum and to a certain extent, the number of isolates correlated with the alkalinity of the soil sample.

2. The soil samples were pasteurized to eliminate non-sporulating organisms and then the sample is spread onto the surface of agar media at pH 9-10, containing an insoluble protein.
3. The size of the clearing zone could not be used quantitatively to select high produces as there was not an absolute correlation between the size of the clearing zone and the production of alkaline protease in submerged culture.

22.9.4 Isolation Methods Not Utilizing Selection of the Desired Characteristic

The synthesis of some products does not give the producing organism any selective advantages which may be exploited directly in an isolation procedure. Examples include the production of antibiotics and growth promoters. Therefore, a pool of organisms has to be isolated and subsequently tested for the desired characteristic. A major problem faced by industrial microbiologists in this situation is the resolution of strains which have already been screened many times before.

1. Developing procedures to favour the isolation of unusual taxa which are less likely to have been screened previously.
2. Identifying selectable features correlated with the undetectable industrial trait thus enabling an enrichment process to be developed.
3. The incorporation of particular antibiotics into isolation media may result in the selection of the resistant taxa.

22.9.4.1 Screening methods

1. The evolution of antibiotic screens serves as an excellent illustration.
2. Antibiotics were initially detected by growing the potential producer on an agar plate in the presence of an organism against which antimicrobial action was required.
3. Production of the antibiotic was detected by inhibition of the test organisms.
4. Alternatively, the microbial isolate could be grown in liquid culture and the cell-free broth is tested for activity.
5. Antibiotics that inhibit cell-wall biosynthesis and the mode of action resulted in a very significant increase in the discoveries of new β–lactose antibiotics. Increasing frequency of penicillin and cephalosporin resistance amongst clinical bacteria led to the development of mechanism based screens for the isolation of more effective antibacterial antibiotics.

22.9.4.2 Screening mechanism

The progress in molecular biology, genetics and immunology has also contributed extensively to the development of innovative screens by enabling the construction of specific detector strains, increasing the availability of enzymes, receptors and developing extremely sensitive assays. Molecular probes for particular gene sequences may enable the detection of organisms capable of producing certain product groups. The development of reporter gene assays has been used in the search for metabolites that disrupt viral replication.

22.9.5 Determination of Thermal Death Point (TDP) and Thermal Death Time (TDT)

Temperature is one of the most important physical factors affecting microbial growth. Bacteria are different from higher plants and animals as they lack homeostatic mechanism and cannot regulate against heat generated by metabolism and are, therefore, directly and readily affected by temperature. Over a limited temperature range, there is a twofold increase in the rate of enzyme catalysed reactions for every 10°C rise in temperature.

Bacteria may be divided into three major groups with respect to their temperature requirements: (i) ***Psychrophiles***, those with optimum temperature between 0°C and 20°C, (ii) ***Mesophiles,*** those with optimum temperature between 20°C and 40°C; and (iii) Thermophiles, with optimum temperature between 40°C and 80°C. Thermophiles are of two types (i) ***Facultative Thermophiles*** with an optimum temperature of growth between 45°C and 60°C and (ii) ***Obligate Thermophiles*** with optimum temperature above 60°C. Temperature in the range between 50°C and 100°C are normally in the lethal range for bacterial cells and spores. The range of temperature preferred by bacteria is genetically determined resulting in enzyme expression with different temperature requirements.

Each organism grows within a particular temperature range (i.e., cardinal temperature points). The minimum growth temperature is the lowest temperature at which growth of a species will occur. The highest temperature at which a species can attain its maximum growth temperature and a species grows faster at its optimum growth temperature.

22.9.5.1 *Effect of incubation temperatures on growth*

Precautions

Culture incubated at higher temperatures (55°C and above) should be kept in moist chamber in order to avoid drying of the slants.

Materials

Cultures of *Trichoderma viridi, Fusarium solani, Eshcherichia coli* and *Bacillus cereus,* Nutrient agar slants (Bacterial culture), Czapek-Dox agar plates (fungal culture), Thermometer, Glass marker, Pipettes, Spreader, Water bath, Bunsen burner, Inoculating loop, Inoculating needle, Refrigerator set at 4°C, Four incubators set at 15°C, 35°C, 55°C and 75°C.

Protocol

1. Label nutrient agar slants with the test organisms and the temperature of incubation (0, 4, 15, 35, 55, 75°C).
2. Inoculate six nutrient agar slants each (Label 0, 4, 15, 35, 55, 75°C) with *E.coli* and *Bacillus cereus* by streaking the surface of each slope.

3. Repeat step 2 for inoculating the remaining cultures.
4. Label each of the Czapek-Dox agar plates with *Fusarium solani* and *Trichoderma viridi* and the temperature of incubation (0, 5, 15, 35, 55, 75°C).
5. Inoculate six Czapek-Dox agar plates with *Fusarium solani* and other six plates with *Trichoderma viridi* by transferring a disc on the centre of each plate.
6. Incubate one inoculated slant and plate from each set at 0°C (ice box of refrigerator), 4°C (refrigerator) for 10-14 days and other slants and plates at 15, 35, 55 and 75°C (incubators set at required temperature) for 24-48 hours (bacterial) and 7-14 days (fungal cultures).

Results and observations

Examine cultures after two days of incubation (incubated at 15, 35, 55, and 75°C) and after 10-14 days of incubation (at 0°C and 4°C) for the presence (+) or absence (–) of growth and the degree of growth (i.e., minimal growth (+), moderate growth (2+) heavy growth (3+), very heavy (maximum) growth (4+). Express the results in the form of a table and determine the minimum, maximum and optimum temperature range for each microorganism and classify each organism as ***Psychropile, Mesophile*** and ***Thermophile*** as per Table 22.4.

Table 22.4 Minimun, maximum and optimum temperatures of different microorganisms

Organism	*Temperature in °C*						*Cardinal temperature*			*Classification as T, M and P*
	0	*14*	*15*	*35*	*55*	*75*	*Min.*	*Max.*	*Opt.*	
T. viridi										
F. solani										
E. coli										
B. cereus										

22.9.5.2 Lethal effects of temperature thermal death point (TDP)

Introduction

Time of exposure is a vital factor in assessing the lethal effect of high temperature on bacterial cell. For this purpose two methods are useful: (i) the thermal death point (TDP); the temperature at which an organism is killed in 10 minutes of exposure; and (ii) thermal death time (TDT), is the time required to kill cells/ spores suspension at a given temperature. It is necessary to compare the susceptibility of different organisms to rising temperatures. However, some factors such as pH, moisture, composition of media and age of cells influence the TDP. Enzymatic activities of any organism operate well at their optimum temperature which varies with the organism. Increase/decrease in temperature influences the microbial growth and survival leading to death also. The degree of tolerance is measured by exposing the microbes to gradually increasing temperature for a given period, e.g., 10 minutes and determining their survival (Figure 22.21).

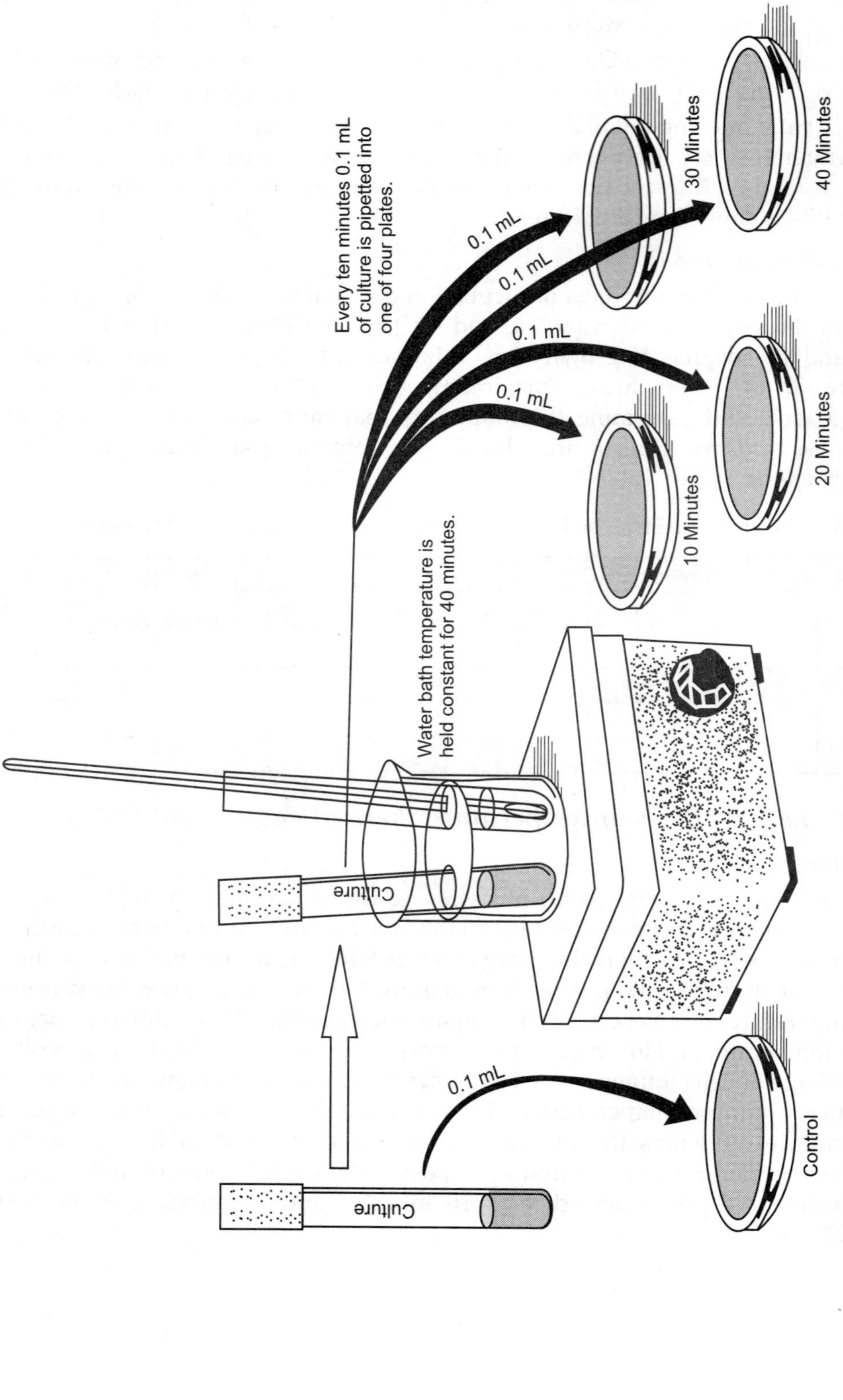

Figure 22.21 Flow chart for determining thermal death point.

Precautions

1. The inoculum should be at least 1-2 cm away from the plate centre where the streak sector converges on Petri plates.
2. Constant attention should be given to the temperature of water bath so that adjustment can be made before a serious fluctuation in temperature occurs.

Materials

Nutrient broth cultures of *Escherichia coli* and *Bacillus subtilis* and *Staphylococcus aureus*. (30-48 hours old culture in flasks), hot water bath, nutrient agar in Erlenmeyer flask 100 mL, sterile empty tubes with cap (15), sterile cotton-plugged 10 mL pipettes (3) culture tube with 10 mL water (1), centigrade thermometer, wax glass marking pencil, culture tube stand, inoculating loop/needle, bunsen burner. Set the water bath up to 60°, 70°, 80°, 90° and 100°C.

Protocol

1. Liquefy the nutrient agar medium and cool to 50°C.
2. Label five Petri plates: control, 10 min, 20 min, 30 min and 40 min.
3. Shake the culture of organisms and transfer 0.1 mL of organisms with a 1 mL pipette to the control plate.
4. Place the culture and a tube of sterile nutrient broth into the water broth. Remove the cap from the tube of nutrient broth and insert a thermometer into the tube.
5. As soon as the temperature of the nutrient broth reaches the desired temperature, record the time. Watch the temperature carefully to make sure it does not vary appreciably.
6. After 10 minutes have elapsed, transfer 0.1 mL from the culture to the 10 min plate with a fresh 1 mL pipette. Repeat this operation at 10 minute intervals until all the plates have been inoculated. Use fresh pipette each time and be sure to shake the culture before each delivery.
7. Pour liquefied nutrient agar (50°C) into each plate, rotate and cool.
8. Incubate at 37°C for 24 to 48 hours. After evaluating plates, record results as per Table 22.5.

Table 22.5 Growth of different microorganisms at different temperatures

Organism	*Temperature in °C*				
	60	*70*	*80*	*90*	*100*
Bacillus megaterium					
Escherichia coli					
Staphylococcus aureus					

Observations and Result

Examine the plates for presence (+) or absence (–) of growth in all the sectors and compare with control.

Represent your results in a tabular form and find the thermal death point (at which death of an organism occurs) of each organism.

22.9.5.3 Determination of thermal death time (TDT)

Introduction

TDT is the time required to kill the cells/spores/propogules of an organism at a given temperature. TDT differs from organism to organism. The degree of heat tolerance by a microorganism is determined first by exposing cells, etc., at a gradually increasing time, and then assessing its survival by sub-culturing on respective nutrient medium (Figure 22.22).

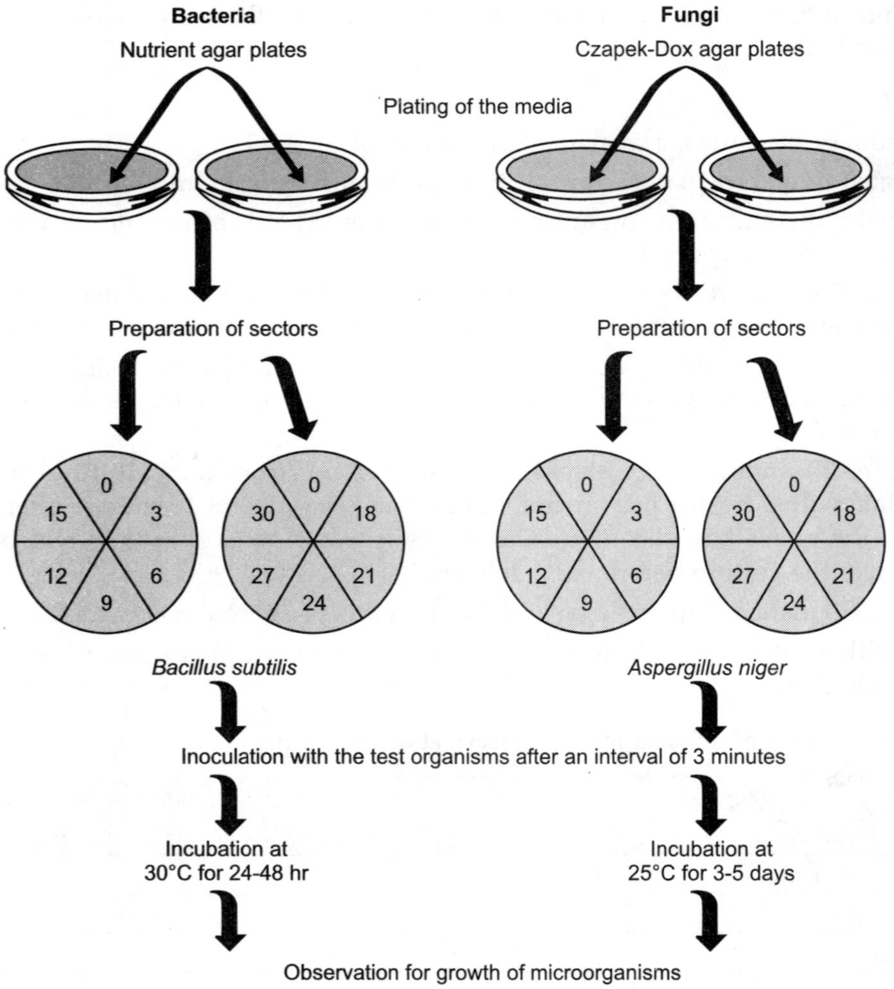

Figure 22.22 Flow chart for determination of thermal death time (TDT) of microorganisms.

Materials

Nutrient broth cultures of *Escherichia coli* and *Bacillus subtilis* (30-48 hours old culture in flasks), hot water bath, nutrient agar in Erlenmeyer flask 100 mL, sterile empty tubes with cap (15), sterile cotton-plugged 10 mL pipettes (3), culture tube with 10 mL water (1), centigrade thermometer, wax glass marking pencil, culture tube stand, inoculating loop/needle, bunsen burner.

Protocol

Prepare nutrient broth, pour 2 mL into 11 small sized tubes, plug properly and autoclave at 121°C for 20 min.

2. Cool down and inoculate each tube with a loopful suspension of the bacterial culture and incubate at 30°C for 24 hours.
3. Simultaneously, prepare nutrient agar medium and pour sterilize and cool down medium into two Petri dishes.
4. Take nutrient agar plates and divide into six sectors on the back of the lower half with the help of a glass marker (pen or pencil).
5. Label the sector of one plate with time intervals to be treated as 0, 3, 6, 9, 12, 15 and the other with 0, 18, 21, 24, 27 and 30 min where, 0 refers for control.
6. Inoculate the 0 sector with the test bacterial suspension.
7. Note the TDP of the test bacterium from the experiment as described in the earlier experiment.
8. Prepare a water bath and maintain the temperature at 60°C.
9. Place 10 tubes containing 24 hours old bacterial culture into the water bath maintained at 60°C.
10. After an exposure to 3 minutes, take out one tube and quickly cool under running tap water.
11. Make streak inoculation of the suspension on sector 3 of the plate.
12. Repeat the process at every 3 minutes interval, i.e., exposure to 6, 9, 12, 15, 18, 21, 24, 27, 30 minutes and make streak inoculated on sector 6 to 30, respectively.
13. Incubate and inoculate plates at 30°C for 24 – 48 hours.
14. Tabulate the results as per Table 22.6.

Table 22.6 Determination of thermal death time (TDT) of the microorganisms

Test Organisms	*Time of exposure (min) at 60 °C*											*Thermal death time TDT (minutes)*
	0	3	6	9	12	15	18	21	24	27	30	
Escherichia coli												
Bacillus subtilis												
Staphyllococcus aureus												

Results and observations

Observe the inoculated plates for the presence of growth in various sectors and compare with the '0' time sector and record the growth as + (growth present) and - (no growth)

and determine the TDT of each organism, i.e., the time to kill the test organism at the given temperature.

[*Note:* Identical method can be adopted to determine the TDT of fungi by using conidial/hyphal suspension. Medium used for fungal should be Czapex Dox agar and Czapek Dox broth. The inoculated plates should be incubated at 28°C for about a week].

22.9.6 Determination of Ethanol Production Using Different Substrates

Fermentation is the conversion of a carbohydrate such as sugar into an acid or an alcohol through microbes. Fermentation occurs naturally in many different foods given the right conditions, and humans have intentionally made use of it for several years. Fermentation is a biological and enzymatic process of conversion of carbohydrates into alcohols, lactic acid, carbon dioxide, etc. by some bacteria and fungi. These organisms use fermentation as a method of obtaining energy in the form of adenosine triphosphate (ATP). It is a cellular process that transfers the energy in glucose bonds to bonds in ATP.

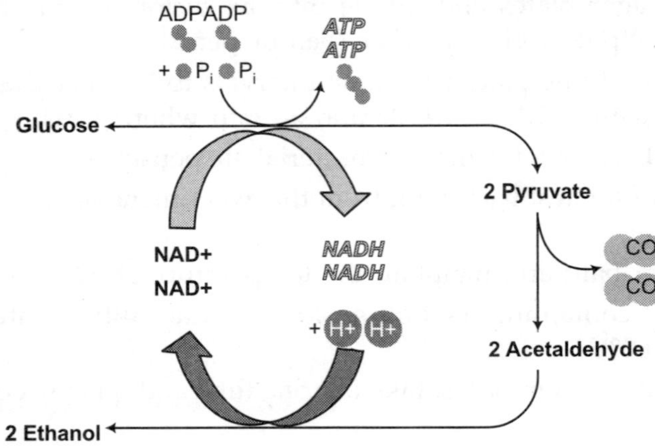

$$C_6H_{12}O_6 \quad 2CH_3CH_2OH + 2CO_2$$

A series of intermediate steps involving enzymes result in the breakdown of $C_6H_{12}O_6$ to pyruvic acid which is then converted to ethyl alcohol and CO_2.

$$CH_3CCOCOH \xrightarrow{\text{Pyruvate decarboxylase}} CH_3CHO + CO_2$$

$$CH_3CHO + NADH_2 \xrightarrow{\text{Alcohol dehydrogenase}} CH_3CH_2OH + NAD$$

The three main by-products of alcohol industry are CO_2 yeast and distillery effluent.

22.9.6.1 Materials

Grape juice broth culture of *Saccharomyces cerevisiae var. ellipsoideus*, pasteurized commercial grape juice (500 mL), 1% phenolphthalein solution, 0.1 N sodium hydroxide, sucrose 20 g, one litre sterile Erlenmeyer flask, rubber stopper (fitted with a 5 cm glass tube plugged with cotton), sterile 10 mL pipettes, burette, spatula, alcoholometer, distillation apparatus, wire gauze, glass beads, ring stand, Bunsen burner.

22.9.6.2 Protocol

1. Take 500 mL of the grape juice into a 1 litre sterile flask.
2. Add 20 mL of sucrose and 50 mL of culture of the yeast (10% starter culture) to the flask. Size of the inoculum ranges from 8-10% with an average of 4% volume by volume.
3. High degree of aeration and agitation is necessary for preparing the inoculum since this favours rapid growth of yeast, producing a large amount of cells mass. pH is maintained at 4.8 to 5 while temperature at 28°C - 30°C.
4. Close the flask with the rubber stopper.
5. Inoculate the fask at 25°C for 21 days and determine total acidity, volatile acidity and alcohol at different intervals.
6. Add 20 g of sucrose to the fermenting juice after 2 and 4 days of incubation.
7. Determine total acidity and volatile acidity by titration method at 1 to 7 day interval, as described below:
   (a) Take 10 mL of the fermenting juice to a flask.
   (b) Add 10 mL of distilled water and 5 drops of 1% phenolphthalein solution as indicator.
   (c) Mix the contents.
   (d) Titrate against 0.1 N sodium hydroxide taken in a burette to the first persistent pink colour.
   (e) Note the volume of sodium hydroxide used.
   (f) Repeat steps *a* through *e* taking fermenting juice after 14 and 21 days.
   (g) Repeat steps *a* to *e* using uninoculated (control) grape juice.
8. Determine the % alcohol by dipping alcoholometer in the fermenting juice at 7 day intervals.
9. Assemble the distillation apparatus as illustrated in the Figure 22.23.
10. Pour 100 mL of fermented grape juice in the distilling flask, add a few glass beads and put thermometer in the hole.
11. Start circulation of water through the condenser.
12. Light the Bunsen burner and allow the liquid to drip in the receiver flask.
13. Adjust the heat in such a way that the drops come at the rate of one per second.

22.9.6.3 Results and observations

Record and note the appearance of odour and presence of turbidity or sediment of the content of the flask.

Total acidity and volatile acidity can be calculated by applying the following formula:

$$\text{Total acidity (expressed as \% tartaric acid)} = \frac{\text{Volume (mL) of alkali used} \times \text{Normality of alkali} \times 7.5}{\text{Weight of sample in gm (1 mL = 1 gm)}}$$

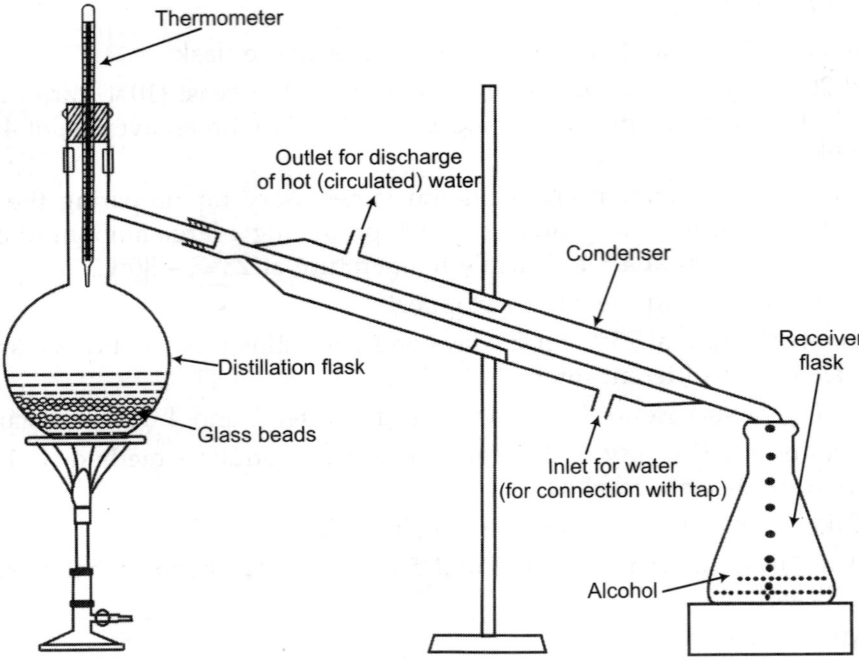

Figure 22.23 Distillation assembly for recovery of alcohol.

$$\text{Volatile acidity (expressed as \% acetic acid} = \frac{\text{Volume of alkali} \times \text{Normality of alkali} \times 60}{\text{Weight of sample in gm}}$$

Record total acidity, volatile acidity and per cent alcohol of the fermenting juice at 7 day intervals. Determine the amount of alcohol per 100 mL of the grape juice produced under laboratory conditions at 12 hr, 24 hr or 7 day intervals.

22.9.7 Sugar Fermentation by Yeast (*Saccharomyces cerevisiae*) for Ethanol and Cell Biomass Production

Yeast can metabolize sugar in two ways, aerobically, with the aid of oxygen, or anaerobically, without oxygen. When the yeast respires aerobically, oxygen gas is consumed at the same rate that CO_2 is produced; there would be no change in the gas pressure in the test tube. When yeast ferments the sugars anaerobically, however, CO_2 production will cause a change in the pressure of a closed test tube, since no oxygen is being consumed. The use of this pressure change monitors the fermentation rate and metabolic activity of the organism.

22.9.7.1 Materials

A small laboratory grade fermentor of 2 litre capacity, 50 mL centrifuge tubes, glucose reagent, yeast inoculum, distilled water (Figure 22.24).

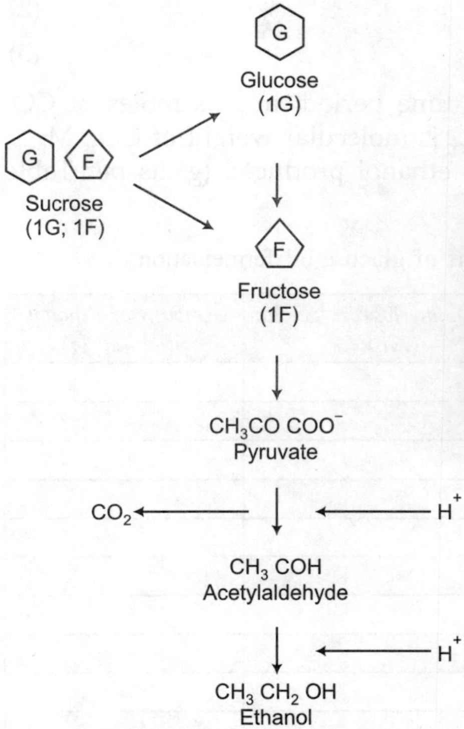

Figure 22.24 Fermentation reaction. The visible colour in the media shows high cell density of *Saccharomyces cerevisiae.*

22.9.7.2 Protocol

1. Take two centrifuge tubes and add glucose to each centrifuge tubes. Add 1, 2, 3 and 4 g glucose to achieve 5, 10, 15 and 20% w/v glucose concentrations, respectively.
2. Add distilled water to make up the volume to 20 mL.
3. Add 150 μL yeast inoculum to glucose solution. Yeast inoculum is prepared by incubating, active dry yeast with distilled water, for 20 min in 32°C water bath at 120 rpm.
4. Weigh the tubes. This is the zero hour reading.
5. Loosen the cap of centrifuge tubes. Keep them in 32°C water bath with shaking at 100 rpm.
6. Weigh the tubes again and record the time.
7. Next day, weigh the tubes after 24 hrs (according to the time you started the fermentation). Introduce the procedures for the laboratory.

Necessary Data, see Table below

Necessary Equations

$$W_{CO2} = W_t - W_0, \qquad \text{(refer Table 22.7)} \qquad (1)$$

$$m_{EtOH} = m_{CO2} = W_{CO2}/M_{CO2} \quad (2)$$

$$A_{EtOH} = m_{EtOH} * M_{EtOH} \quad (3)$$

W_{CO2} is mass of CO_2 produced at a particular time period; m_{CO2} is moles of CO_2 produced; m_{EtOH} is moles of ethanol produced; M_{CO2} is molecular weight of CO_2; M_{EtOH} is molecular weight of ethanol; A_{EtOH} is amount of ethanol produced (g) as per Table 22.7.

Table 22.7 Result on the effect of different concentrations of glucose oil fermentation

Glucose Concentration, % wlv	*Time, hr*	*Weight of Centrifuge Tube, (W_{time})*	*Calculate loss in weight, (W_t-W_0)*	*Amount of Ethanol produced (g)*
5	0	W_0=		
	8	W_8=		
	24	W_{24}=		
	40	W_{40}=		
10	0	W_0=		
	8	W_8=		
	24	W_{24}=		
	40	W_{40}=		
	0	W_0=		
15	8	W_8=		
	24	W_{24}=		
	40	W_{40}=		
20	0	W_0=		
	8	W_8=		
	24	W_{24}=		
	40	W_{40}=		

22.9.7.3 *Data analysis procedure*

1. Determine amount of ethanol produced (in gram) at respective time periods for four different glucose concentrations in the fermentation media.
2. Plot amount of ethanol produced versus fermentation time for four glucose concentrations.
3. Discuss correlation of total ethanol produced with glucose concentrations.

22.9.8 Measurement of Yeast Biomass Using Haemocytometer

22.9.8.1 *Materials*

Dry yeast, distilled water, methylene blue solution.

22.9.8.2 Protocol

1. Yeast inoculum is prepared by adding 5 g dry yeast and 25 mL distilled water.
2. This solution is incubated in 32°C water bath at 120 rpm for 20 min.
3. After incubation, yeast solution is diluted 1000 times.
4. To 1000 μL of diluted yeast solution, methylene blue dye is added. Dead yeast cells become blue in colour while live cells retain original mustard yellow colour.
5. Pour 100 μL of the above solution into hemocytometer well.
6. Count yeast cells in the four corner squares of the hemocytometer.
7. Yeast concentration will be determined based on dimensions of the hemocytometer square and cells counted.

22.9.8.3 Calculations

$$\text{Total live cell concentration, cells/mL} = \frac{\text{Number of live cells* Dilution Factor}}{\text{Volume of corner square}}$$

$$\text{Total dead cell concentration, cells/mL} = \frac{\text{Number of dead cells* Dilution Factor}}{\text{Volume of corner square}}$$

Volume of the corner square, mL = 16* volume of smallest square.

Determining Cell Concentration

(1) **Zeroing the spectrophotometer:** Set the wavelength to 630 nm. Using the knob on the left, set the reading to 0% transmission when the chamber is empty; using the knob on the right, set the reading to 100% transmission when the chamber contains a test tube with about 4 mL of pure medium.

(2) **Determining cell concentration:** First flush out the sample tube by taking an 8-10 mL fermenting sample. Take about 4 mL from sample, mix by shaking the tube and transfer it to a glass test tube. Clean the outside of the test tube with ethanol, insert it into the spectrophotometer, and record the absorbance reading. If the absorbance reading is greater than 0.25, a typical limit of linear correlation between the absorbance and cell mass concentration, dilute the sample with a known amount of pure medium, and measure the absorbance again to check if the absorbance reading is on the linear portion of the calibration curve. Record the time you take the sample along with the absorbance reading in the linear range as well as the dilution factor.

Glucose measurement

Take undiluted yeast sample tube and insert the thin plastic sample tubing from the glucose analyzer into it. If the analyzer is in the "Standby" mode, hit the "Run" key; after the machine has calibrated, hit the "Sample" key. If the analyzer is not in the "Standby" mode, just hit the "Sample" key. The glucose concentration in the medium from sample will print out in about one minute. Record this value.

Ethanol assay

(1) Take 0.5 mL from undiluted sample, pipette it into an Eppendorf tube, and spin at 14,000 rpm for 5 minutes.
(2) Add 2 mL of Ethanol Assay Reagent to each of three cuvettes.
(3) Add 10 μL of pure medium to the first cuvette; this is Blank. Cover with parafilm and mix gently.
(4) Add 10 μL of the Ethanol Standard (.08% w/v) to the second cuvette; this is mL Standard. Cover with parafilm and mix gently.
(5) Add 10 μL of sample (from the supernatant from step a above) to the third cuvette; this is your sample. Cover with parafilm and mix gently.
(6) Incubate cuvettes at room temperature for 10 minutes.
(7) Clean the outside of the cuvette, and then read the absorbance of each at 340 nm; use the blank to zero the spectrophotometer. If the sample absorbance is > 1.7, dilute the supernatant from sample 1:4 and perform the assay again using this diluted supernatant. Don't redo the blank and standard.
(8) Calculate the ethanol concentration for sample using the following formula:

$$[\text{Ethanol}]\,(\text{g/L}) = \frac{\text{A (Sample)}}{\text{A (Standard)}} \times 0.8 \text{ (for the undiluted samples)}$$

Record this value.

22.9.9 Lactic Acid and Cell Biomass Production by *Streptococcus thermophilus*

Fermentation is an anaerobic (without oxygen) process; cellular respiration is aerobic (utilizing oxygen). All living organisms, including bacteria, protists, plants, and animals, produce ATP in fermentation or cellular respiration and then use ATP in their metabolism. The process of fermentation followed by *Streptococcus thermophilus* is homolactic fermentation in which one mole of glucose gets converted into two moles of lactic acid.

$$C_6H_{12}O_6 \rightarrow 2\ CH_3CHOHCOOH$$

Before the fermentation starts molecule of glucose must split into two moles of pyruvate and the process is known as glycolysis. Fermentation and cellular respiration involve redox reactions (oxidation-reduction reactions). Redox reactions are defined in terms of electron transfers, being reduction the gain of electrons and oxidation the loss of electrons. In cellular respiration, two hydrogen atoms are removed from glucose (oxidation) and transferred to a coenzyme called nicotinamide adenine dinucleotide (NAD^+), reducing this compound to NADH.

22.9.9.1 *Materials*

A small laboratory grade fermenter, balance, autoclave, cooling centrifuge, pH meter, hot plate.

Media Composition:

Ingredients		g/litre
Universal peptone	–	10
Meat extract	–	5
Yeast extract	–	5
Glucose	–	20
Dipotassium hydrogen phosphate	–	2
Diammonium hydrogen citrate	–	2
Sodium acetate	–	5
Magnesium sulfate	–	0.1
Manganous sulfate	–	0.05

Final pH 6.5 +/–0.2 at 37°C

22.9.9.2 Protocol

1. Prepare 1 litre MRS broth and pour into fermenter vessel
2. Autoclave media at 121 psi for 20 min in vessel and allow it to cool overnight
3. Inoculate actively growing culture of *S. thermphilus* to fermenter vessel and initiate fermentation process
4. Allow the process to run for 24 hours
5. Harvest the culture from the fermenter by centrifugation at 4000 rpm

Fermentation process

The inoculation of *Streptococcus thermophilus* should always be done carefully under total aseptic conditions through the inoculation port on the top plate of fermenter with the help of inoculation syringe (Figure 22.25).

After inoculation pH should be adjusted to 7, allow the fermentative process to run for 24 hours for complete cell multiplication leading to high cell density or mass production. The required conditions for *S. thermophilus* to grow in a uniform manner are:

- Temperature should not be less than 36°C and not more than 41°C.
- pH of the media should be in between 6.5 to 7.2
- Very low amount of oxygen can be supplied as the organism is facultative anaerobe

The pH of the medium starts dripping as the microbe enters the log phase of the bacterial growth curve because of the production of lactic acid which is an end product of the homo-fermentative process of the microbe to gain energy for survival.

22.9.10 Microbial Production of Citric Acid Using *Aspergilus niger*

Citric acid (CH_2COOH COH COOH CH_2 COOH), a tricarboxylic acid, is a natural constituent of a variety of fruits. Citric acid is a commodity chemical, and more than a million tonnes are produced every year by fermentation. It is used mainly as an

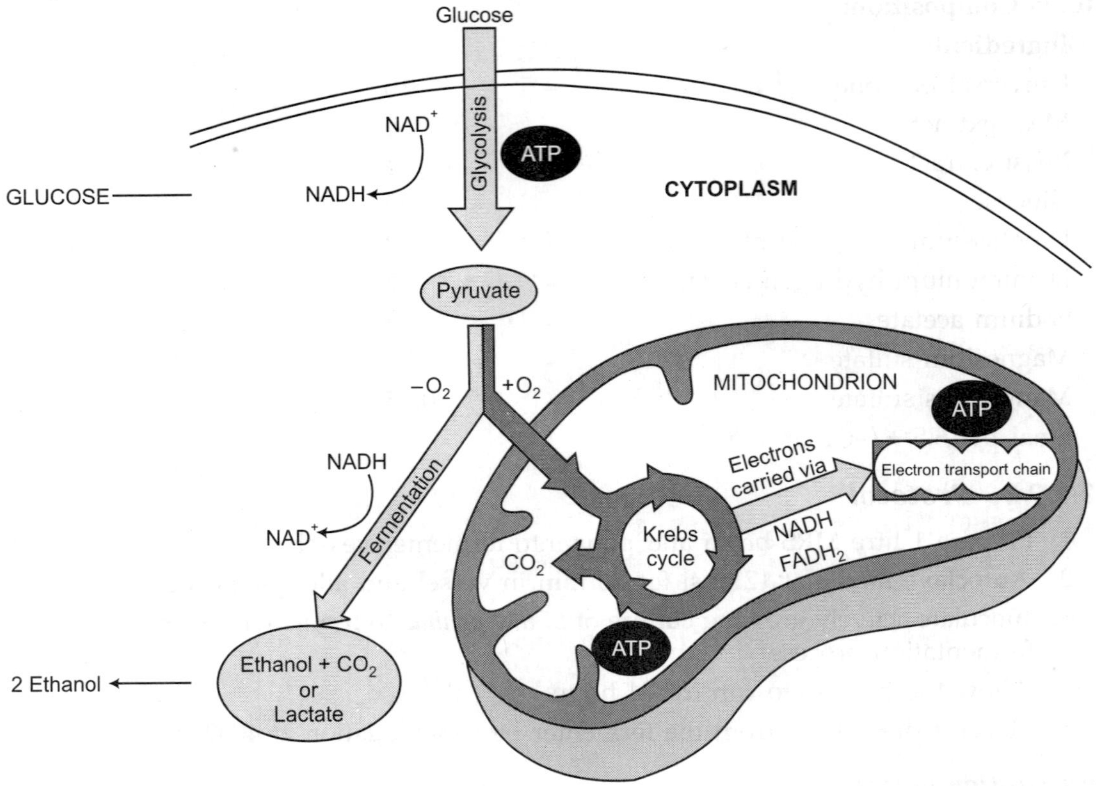

Figure 22.25 Stages of cellular respiration and fermentation in a laboratory grade 2 litre fermenter with MRS broth.

acidifier, as a flavouring, preservative, and as a chelating agent. In this production technique, which is still the major industrial route to citric acid used today, cultures of *Aspergillus niger* are fed on a sucrose or glucose-containing medium to produce citric acid. The source of sugar is corn steep liquor, molasses, hydrolyzed corn starch or other inexpensive sugary solutions. After the mold is filtered out of the resulting solution, citric acid is isolated by precipitating it with lime (calcium hydroxide) to yield calcium citrate salt, from which citric acid is regenerated by treatment with sulfuric acid.

22.9.10.1 Materials

Organism and culture maintenance Soil isolate of *Aspergillus niger* is used. The strain is developed by alternate treatment of ultraviolet irradiations (1.6×10^2 J/m^2/S) and N- methyl-N -nitro-N-nitrosoguanidine (NTG - 100 mg/mL) for different time intervals (5 - 45 min) and maintained on sterilized potato dextrose agar medium (diced potato 200 g/L, dextrose 20 g/L and Agar 15 g/L), pH 4.5 and stored at 5°C in the refrigerator. All the culture media, unless otherwise stated, are sterilized in autoclave at 15-lbs/inch^{2}pressure (121°C) for 15 min.

Preparation of fermentation medium for citric acid production

Sugar (Molasses/Sucrose)	4 -15%
$(NH_4)_2SO_4$, NH_4NO_3, $NaNO_3$, KNO_3, urea, etc.	0.25%
KH_2PO_4	0.10 – 0.15%
$MgSO_4 .7H_2$	00.02 – 0.025%

Trace element iron 0.2 ppm, copper sulphate 4.7 mg% zinc 1-2 mM

pH Molasses medium	5 to 6
Sucrose medium	2 to 3

Pre-treatment of molasses Cane molasses obtained from sugar mills is used as an inexpensive carbon substrate. Cane molasses contains water 20%, sugar contents 62%, non-sugar contents 10%, and inorganic salts (ash contents) 8%, making a blackish homogenous liquid with high viscosity. Ash contents include ions such as Mg, Mn, Al, Fe and Zn in variable ratio. Sugar content is diluted to about 25% sugar level. The molasses solution, after adding 35 mL of 1N H_2SO_4 per litre, is boiled for half an hour, cooled, neutralized with lime-water (CaO) and is left to stand overnight for clarification. The clear supernatant liquid is diluted to 15% sugar level. When molasses is used, it is diluted to a sugar concentration of 15-20% with dilute sulfuric acid and to a pH 5.5 – 6.5. Other nutrients for fungal growth are then added and the mixture is sterilized for 30 min to reduce or eliminate the inhibitory action of metals.

Inoculum Preparation 100 mL of molasses medium (Sugar 15%, pH 6.0) containing glass beads, in 1L cotton wool plugged conical flask is sterilized. One mL of conidial suspension (6.5×10^6 conidia) from the slant culture is aseptically transferred. The conidial count is made by Haemocytometer Slide Bridge. The flask is then incubated at 30°C in an incubator shaker at 200 rpm for 24 hours.

22.9.10.2 Method

Fermentation technique Stainless steel fermenter of 15 L capacity with working volume of 9-L (60%) is employed for citric acid fermentation. Prepared inoculum is transferred to the production medium at a level of 5% (v/v). The incubation temperature is kept at 30 ± 1°C throughout the fermentation period of 144 hours. Agitation speed of the stirrer is kept at 200 rpm while aeration rate is maintained at 1.0-4.0 l/min[1]. Sterilized silicone oil is used to control foaming during fermentation (Figure 22.26).

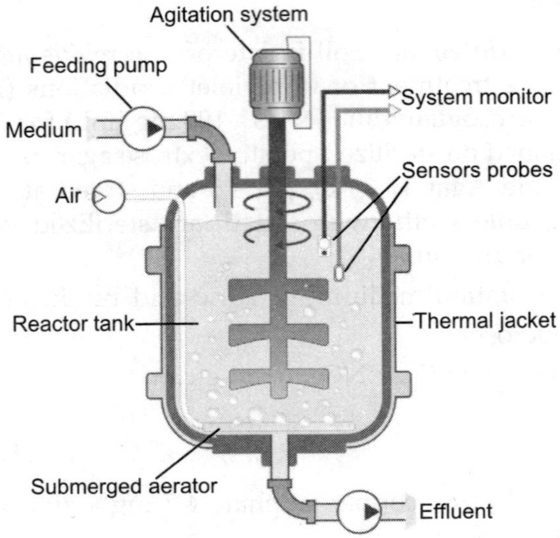

Figure 22.26 External view of an ideal stainless steel fermenter of 15 L capacity

Product Recovery

The crude fermented liquor containing citric acid, obtained either from surface or submerged, or solid state is subjected to the following method for product recovery:

1. Filter crude fermented mother liquor to remove mycelia or other suspended impurities. Heat the mother liquor to 80°C – 90°C add of small amount of hydrated lime to precipitate oxalic acid.
2. Critic acid is then precipitated as calcium citrate using 1 part of hydrated lime for every 2 parts of liquor added over a one hour period, while the temperature raised to 95°C. Precipitation of citric acid can be carried at a temperature of 50°C for 20 min.
3. The precipitated calcium citrate is filtered, washed with water several times and transfer to acidulate. Treat with H_2SO_4, filter the solution again to remove $CaSO_4$ cacitrate $+H_2SO_4 \rightarrow CaSO_4$ + Citrate. Mother liquor containing citric acid is decolorized by passing through ion exchange column with activated charcoal.
4. Concentrate the liquor under vacuum and finally run into low temperature crystallizer. Citric acid is crystallized as monohydrate.

5. Quality of citric acid can further be improved by treating with metal free Ca $(OH)_2$. Also add 9-12% calcium ferrocyanide to mother liquor at 95° - 97°C for 5-8 min to improve the quality of citric acid (Figure 22.6).

22.9.10.3 Results

Citrate obtained from *Aspergilus niger* recovered by this method is recommended for food, drug and cosmetic uses.

22.9.11 Production of Spawn for White Button Mushroom *(Agaricus brunnescens* Syn. *A. bisporus)*

For the successful cultivation of any mushroom on a small scale or commercial scale, one of the most important requirements, is the seed of that species/variety. *Spawn-a pure culture of the mycelium grown on a special medium is* the mushroom seed, comparable to the vegetative seed in crop plants. The production of spawn is done by professionals in the laboratory under controlled conditions of temperature, light and humidity. Spawn can be produced either by germinating basidiospores or by culturing small pieces of vegetative mycelium of a mushroom on a suitable substrate. The success of mushroom cultivation and its yield depend to a large extent on the purity and quality of the spawn used. Mushrooms for spawn production can be grown on sterilised cereal grain (wheat, rye, sorghum or bajra), but usually grain colonised with mycelium *(grain-spawn)* is used as an inoculum for composts. Grain is preferred as a substrate for mushroom spawn because grain gives a large number of inoculation sites, each with a high inoculum potential derived from the nutrient base the outgrowing fungi can utilise. This helps to ensure that the compost is rapidly permeated, which is important for the exclusion of competitors as well as for the rapid production of fruit-bodies. The media used for maintenance, multiplication and preservation of mushroom culture are: Potato Dextrose agar (PDA), Yeast Potato Dextrose agar (YPDA), Malt extract agar and Rice bran decoction medium.

Spawn production mainly consists of three steps: *(i)* substrate preparation; *(ii)* substrate inoculation; and *(iii)* incubation of the inoculated substrate for spawn production or growth of the mycelium on the substrate. Preferably fresh spawn should be used for mixing with compost for better results. Protocol for the spawn production of *Agaricus bisporus* is given in Figure 22.27.

22.9.11.1 Materials

Organism: Pure culture of *Agaricus brunnescens* syn. ***A. bisporus,* cereal grains** (sorghum, wheat, rye, etc.), bottles (flask type), cotton (gauze), paper squares 7 cm × 7 cm, calcium sulphate (gypsum), calcium carbonate (chalk), glucose bottles/milk bottles/polypropylene bags, cotton, alkathene sheets, autoclave, laminar flow cabinet, Incubator/storage room, wire guage balance, Bunsen burner, water.

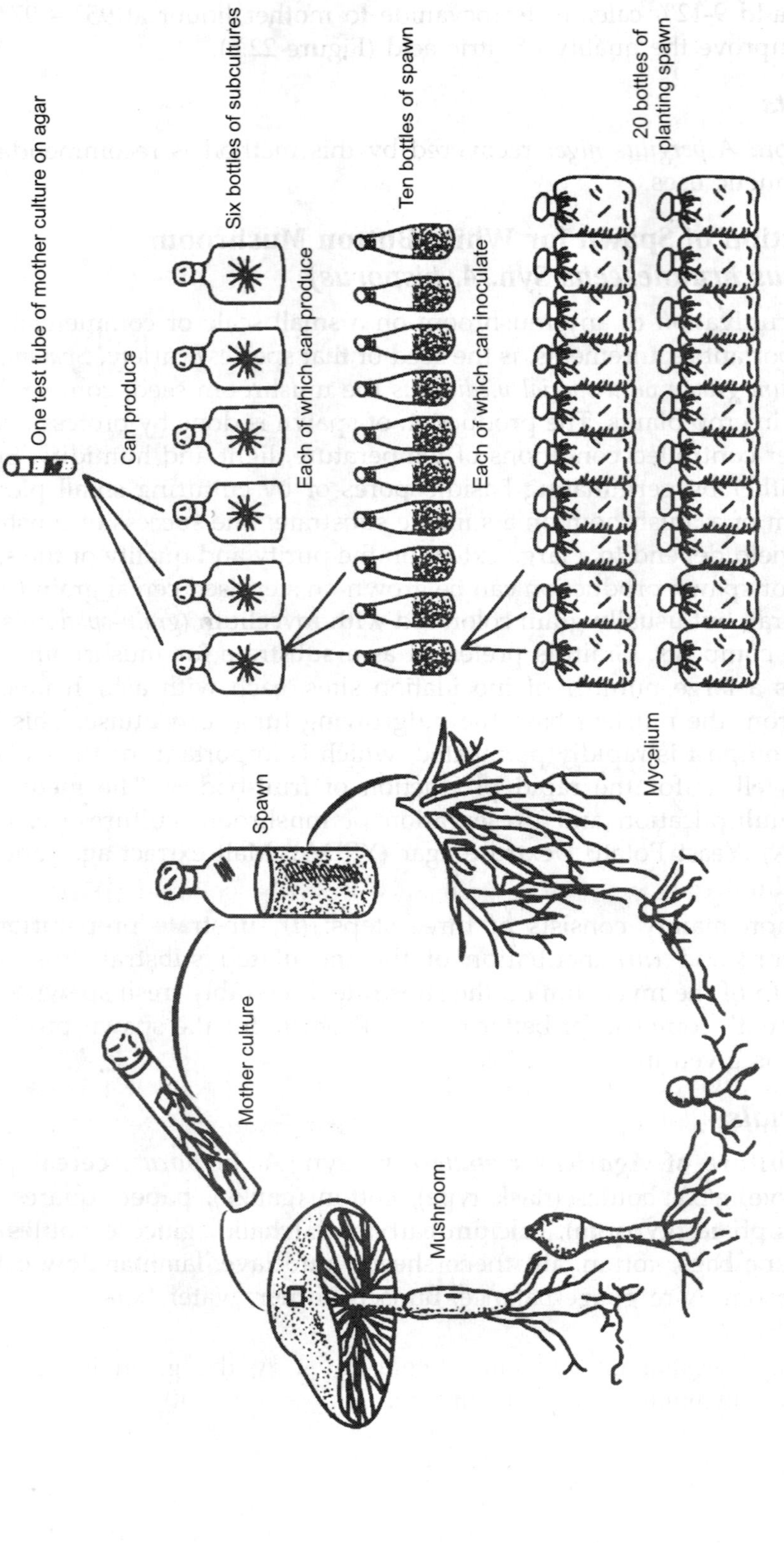

Figure 22.27 Life cycle from mushroom to spawn and multiplication of spawn

22.9.11.2 Method

Substrate Preparation

1. Soak cereal grains for one night; 2 litres of water per 1 kg of grain, wash and strain cereal grains to remove all water.
2. Steam sorghum seeds for 30-45 minutes to soften grains, drain water and spread cereal grains to cool down and decrease moisture.
3. 5 Fill 3/4 of bottle with cereal grains, carefully prepare cotton plug, tightly plug mouth of bottle with cotton and leave out for ventilation.
4. Allow the grain to surface dry by spreading over alkathene sheets, in shade, for a few hours.
5. Mix the grain thoroughly with chemicals, (e.g., calcium sulfate and calcium carbonate at 2% and 0.5% respectively on dry weight basis of the grain), to adjust pH of the grain at 7 - 7.8. The grain must not be coagulated at this stage.
6. Fill the grain-chemical mixture in 500 mL glucose/milk bottles/polypropylene bags (300-350 g boiled grains/container). However, the first generation spawn (master spawn) must be prepared in glass bottles due to their convenience in handling for further subculturing.
7. Plug the bottles/containers with non-absorbent cotton (Figure 22.28 and Figure 22.29).
8. Sterilize the substrate by autoclaving at 121°C (15 psi) for 30 minutes.
9. Repeat the process of sterilization after 24 hours of first autoclaving.
10. Allow the substrate container to come to room temperature for making the substrate ready for inoculation.

Inoculation of Substrate

1. Inoculate the substrate (grain in containers) with the mycelium of the mushroom grown on a specific medium by transferring mycelium in agar on the grain under aseptic (sterile) conditions.
2. Shake the containers, after plugging, to distribute fragments of the mycelium.

Incubation

1. Store (incubate) the inoculated containers at 20-25°C in darkness for 3 weeks.
2. Shake the containers for an even distribution of mycelium, after a few days of incubation or as soon as mycelium is visible on grain.

At the time of boiling of grain in water, observe the grain for their intact nature. Observe the inoculated grain during incubation at regular intervals for the appearance of white mycelium on the grain as well as for the appearance of contaminants.

22.9.11.3 Results

Appearance of silky whitish growth completely covering the grain indicates the preparation of spawn of white button mushroom.

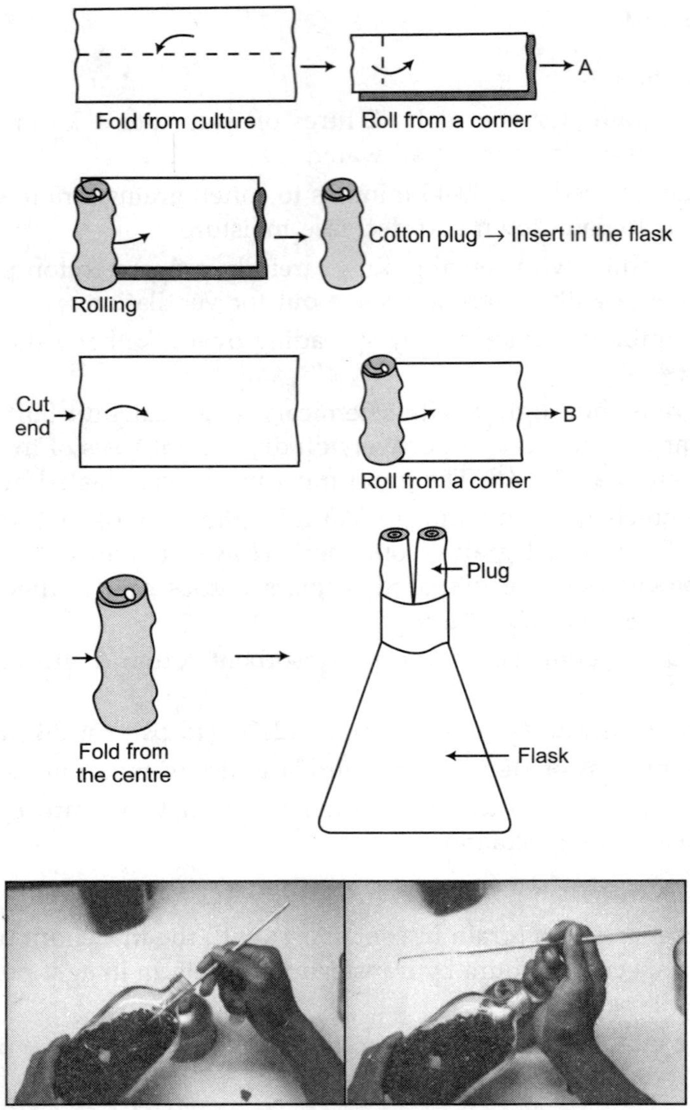

Figure 22.28 Plug preparation and method of inoculation for spawn preparation.

Storage of Spawn

Store the spawn, if not needed immediately, at 0-4°C in a refrigerator for a maximum period of 6 months. Spawn, if stored at low temperature, should be allowed to attain room temperature (25°C) before being used for spawning the compost.

Transportation of Spawn

Transport the spawn in refrigerated vans as higher temperature (above 32°C) is detrimental to mushroom mycelium.

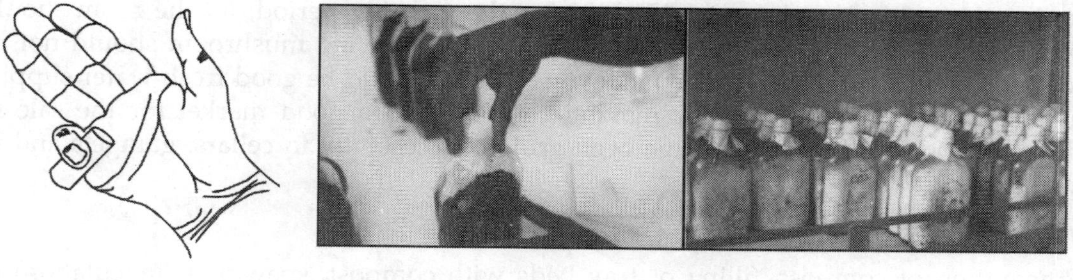

Figure 22.29 Plugging after inoculation.

22.9.11.4 Precautions

1. Always use healthy and whole grain for spawn making.
2. Remove the damaged grain, weed seeds and inert material before boiling of grain.
3. Never boil the grain for a longer period.
4. Check the master spawn for contamination; contaminated as well as suspected to be contaminated bottles should be discarded immediately.
5. Always keep a check for microbial contamination of the mushroom culture during incubation.
6. Always use cotton plugs to maintain aerobic conditions in the spawn containers.

22.9.12 Cultivation of White Button Mushroom (*Agaricus brunnescens = A. bisporus*)

Agaricus brunnescens = ***A. bisporus*** is a temperate mushroom which can grow best only under temperate conditions and 24°C is the optimum temperature required for its development and spread of mycelium in the compost. However, for the formation of fruit-bodies it requires temperature ranging from 14-18°C. Temperature, moisture, ventilation and good spawn (seed) are the essential conditions for the optimal growth of the mushroom. They require nearly a saturated atmosphere with moisture yet the direct application of water on beds is harmful to the growing crop. Good ventilation maintains congenial environmental conditions for the exchange of toxic gas with the introduction of adequate fresh air. Carbon dioxide level of 0.10 to 0.15 per cent volume is necessary during crop production that can be achieved by giving 4 to 6 air changes/hour or introducing 10 cubic ft fresh air/sq ft bed area/hour. Spawn, a pure culture of the mycelium from a selected mushroom grown on a convenient medium, should be of good quality; i.e., it should be a strain originating from a single specimen or of a perfect crop; the substratum must be covered with the white mycelium, it must be uniform and at the time of removal from the container, be absolutely free from molds and other microbial contaminants. Commercial and amateur mushroom cultivation is done indoors. The various installation required for mushroom cultivation vary with the size of a mushroom house. The space requirements necessary for successful production of *A. bisporus* are: *(i)* the location should be easily accessible so that the manure and casing soil can be brought in and removed conveniently; *(ii)* the room should be well ventilated; *(iii)* direct sunlight should not fall on the mushroom bed; *(iv)* the room temperature should not exceed 15°C during the growing period; *(v)* the room should be fitted with heating facility; and *(vi)* location for growing mushroom should not be too moist and should be at such a site where there should be good fresh water supply, availability of the fertilizers for making compost and a good market for the sale of the mushrooms. Mushrooms have been grown successfully in cellars, garage's and in abandoned rooms.

The protocol for cultivation of A. bisporus

Preparation of compost, filling of tray beds with compost, spawning (inoculation) of beds, casing, watering of beds, harvesting of mushrooms, storage.

22.9.12.1 Materials

Cement concrete platform (14 × 7 metres) for preparing compost, tray beds (wooden trays made of pinus or deodar wood, 100 × 50 × 22 cm) (Figure 22.30), compost ingredients, spawn (mushroom seed) of *A. bisporus* 10 bottles/25 beds, casing soil, sprayer with a fine nozzle, box for gathering mushrooms (Figure 22.31), steriliziation facility, wooden mold. (It consists of three wooden boards: one end board, and two side-boards, fastened with a clamp with notches at the top, wooden board (25 × 12 cm) with handle for compressing the compost, filter paper sheets, lime or carbonate of lime, pH meter.)

22.9.12.2 Protocol

Preparation of Compost

Compost is the substrate on which mushroom grows. The microbial degradation of organic wastes by several microorganisms makes the substrate selective for the growth of *A. bisporus.* During composting, distinct changes occur in the physical, chemical and biological characteristics of the straw, all of which influence the productivity of *A. bisporus* subsequently. Synthetic compost, made out of wheat straw for growing *A. bisporus* consists of the following constituents for 25 tray beds:

Chopped wheat straw (3-6 cm long) 300 kg, wheat bran 15 kg, calcium ammonium nitrate (or ammonium sulfate) 6 kg, superphosphate 7.5 kg, urea 2.4 kg, potassium sulfate 3 kg, gypsum 30 kg, saw dust 10-12 kg.

Procedure

1. Wet the saw dust by spraying water or leaving it overnight after mixing all the constituents except wheat straw.
2. Spread the wheat straw over the cement floor on the following day and wet it thoroughly by sprinkling water.
3. Spread the pre-mixed constituents over the wheat straw surface and mix thoroughly.
4. Stack this mixture into a pile of 1.30 metre wide and 1.30 metre high, using indigenously fabricated wooden mold.

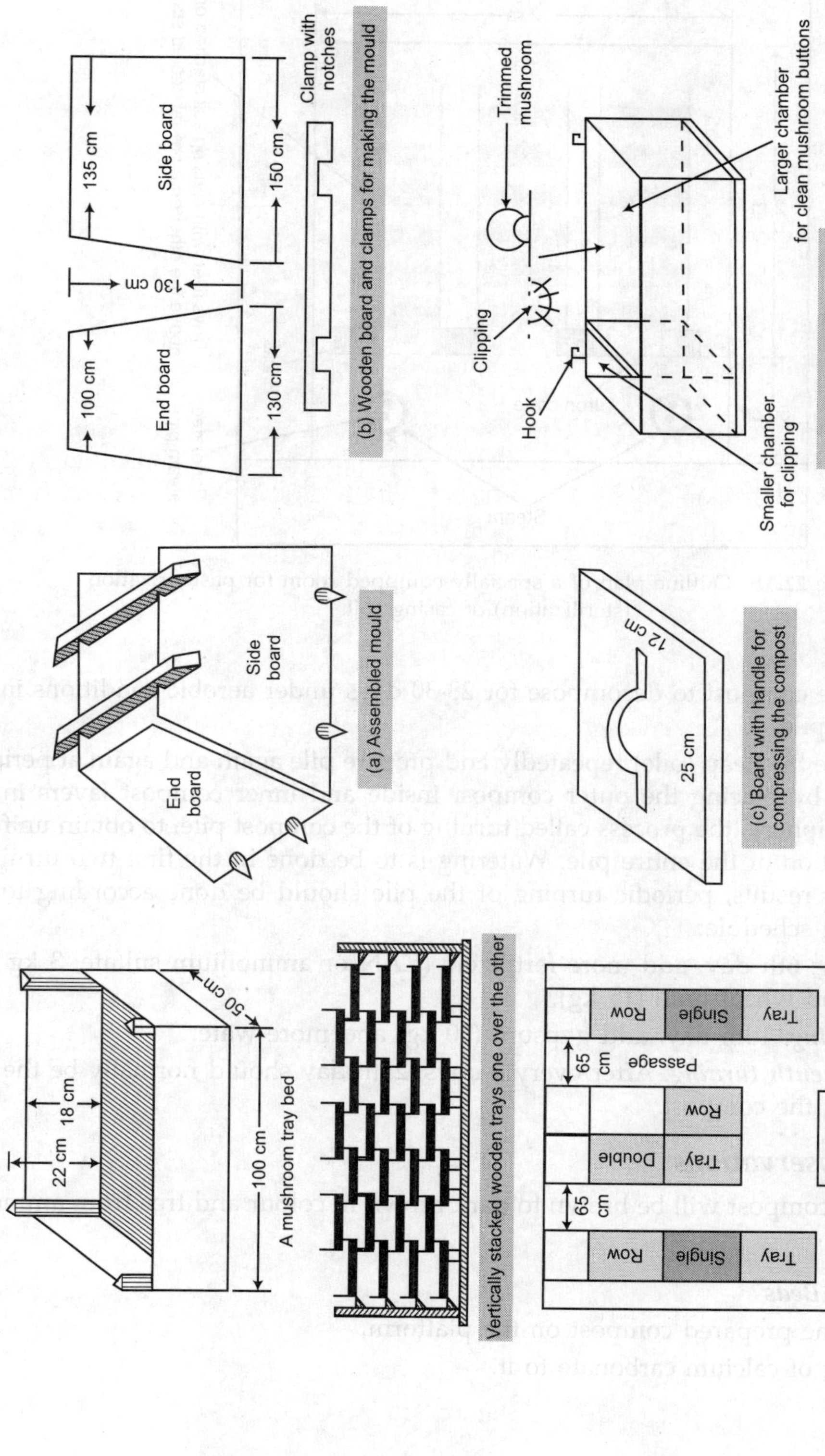

Figure 22.30 Details of mushroom tray bed and mode of stacking of trays and appliances for cropping of white button mushroom.

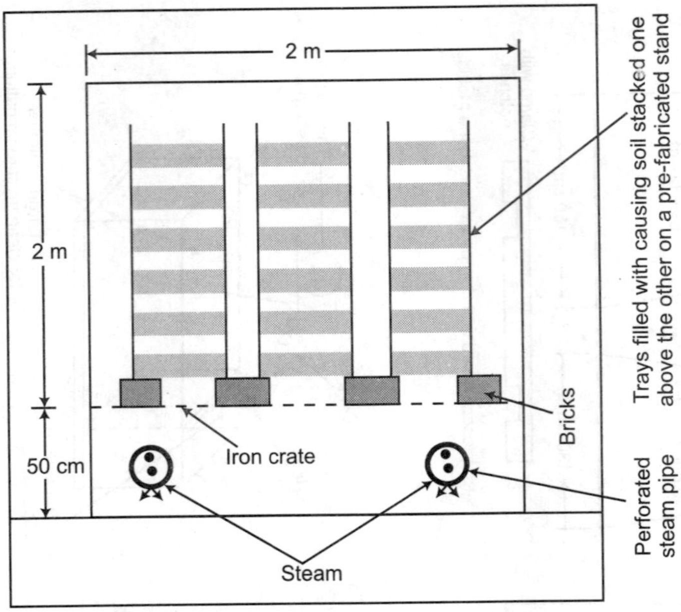

Figure 22.31 Outline plan of a specially equipped room for pasteurization (sterilization) of casing soil.

5. Allow the compost to decompose for 28-30 days under aerobic conditions in the compost pile.
6. Dismantle the heap (pile) repeatedly and prepare pile again and again at periodic intervals by placing the outer compost inside and inner compost layers in the outer periphery, the process called turning of the compost pile, to obtain uniform fermentation of the entire pile. Watering is to be done in the first two turnings. For good results, periodic turning of the pile should be done according to the following schedule:

First turning: 6th day, add more fertilizers (CAN or ammonium sulfate, 3 kg and urea, 1.2 kg) and wheat bran (15 kg).

Second turning: 10th day, add gypsum (30 kg) and more water.

Third to seventh turning: After every 3 days. 25th day should normally be the last day for turning the compost.

22.9.12.3 Observations

Well prepared compost will be brown to dark brown in colour and free from ammonia in colour.

Filling of Tray Beds

(a) Spread the prepared compost on the platform.
(b) Mix 3 kg of calcium carbonate to it.

(c) Fill the compost in all corners and edges of a tray.
(d) Compress fairly the compost in the tray using a wooden board leaving 1 cm clear space on the top of the tray.

Spawning

Spawning means planting mushroom mycelium, growing on a suitable substrate, in the compost. *A. bisporus* spawn is the mycelium especially grown on wheat, sorghum or other cereal grains.

(a) Perform the spawning by spreading the spawn on tray beds when half filled with compost and again after the tray is filled completely. During spawning, the spawn is gently mixed with forefingers and pressed uniformly each time.
(b) Cover the trays with filter paper/newspaper sheets.
(c) Sprinkle water on newspaper sheets, to provide humidity.
(d) Stack the inoculated trays vertically, one over the other, depending on the height of the room as shown in the Figure
(e) Continue water spraying twice a day or less depending upon available humidity in the atmosphere throughout the spawn running and cropping period.

Maintain temperature of the room between 24 and 25°C for 12-15 days for running of the spawn, i.e., formation of mycelium strands all over the tray beds.

22.9.12.4 Observations

Observe for the white cottony mycelium over the compost surface and colour of the compost that changes from dark to light brown which is indicative of successful completion of spawn running period.

Casing

Casing means covering the compost with a thin layer of soil or soil-like material after the spawn has spread in the compost (spawn run). Soil has been the universal casing material. But all soils as such cannot be used as 'casing soil' with advantage, so it is especially prepared soil that can be used for casing.

(a) well rotten cow dung, mixed with light soil in 3 : 1 ratio;
(b) soil and sand in the ratio of 1: 1;
(c) farm yard manure and gravel, ratio 4: 1; *(d)* farm yard manure and loam ratio 1 : 1;
(e) soil peat mixture 2 : 1; and
(f) spent compost, sand and slaked lime (4: 1: 1) and nematicide mixture. Of these, the last one is most often used.

Preparation of Casing Soil

(a) Mix four parts of spent compost with one part of sand to which 5 kg of slaked lime per cubic metre of compost is added.

(b) Treat the mixture with nemagon, a nematicide, by spraying or sprinkling it at the rate of 5 mL per cu. metre.
(c) Leave the material in a pile (1.20 M × 1.0 M) under a tree shade for a period of one year, turning it every 4 months.
(d) Sieve the spent compost, which turns into black soil in one year, for removal of undecomposed lumps.

22.9.12.5 Observations

Observe the casing soil for black colour and pH. A pH between 8.0 and 8.5 is the most suitable for mushroom yield. Adjust pH, if desired, by the addition of lime or carbonate of lime or free stone to the casing mixture.

Sterilization of Casing Soil

Sterilize the casing material either by the chemicals (formalin, chloropicrin, methyl bromide) or by heating or by steam from boiler through perforated pipes (Figure 22.31) and temperature raised to 70-75°C and maintained for 6 hours. The purpose of sterilization is to kill harmful microorganisms (fungi, nematodes, insects).

Casing the Beds

(a) Remove the newspaper sheets from the spawned beds after 3 weeks.
(b) Gently press the compost with the help of wooden board.
(c) Cover the beds with 2-2.5 cm thick layer of sterilized casing material.

Watering of Beds

Spray the beds over the casing soil with a fine nozzle of a sprayer to maintain relative humidity between 70 and 80 per cent.

22.9.12.6 Observations

Observe the beds for mushroom crop which can be expected after 5 to 20 days. Mushrooms mostly appear *in-"flushes"* and at a temperature of 15°C; it generally takes 7 to 8 days to come to the button stage from the first appearance of the formation of a pin-head. There is an interval of 8 to 10 days between flushes.

22.9.12.7 Harvesting of mushrooms

Harvest the crop everyday (in the morning and evening depending upon the market demand) to get a good quality of mushrooms (the cap still being tight to its stalk is the right stage to harvest the mushrooms). Harvesting is done by holding the cap with forefingers slightly pressed against the soil and twisted out. The mycelial strands and soil particles clinging to the base of the stalk (stem) are cut off with a knife. Specially designed wooden box is used for collection of mushrooms from multistoried trays, each provided with the hooks for resting it against the side board of the mushroom tray.

22.9.12.8 Storage

Store the mushrooms at 4°C in a refrigerator for a few days to avoid the deterioration, because mushrooms are a highly perishable commodity. The white colour turns brown and then black in a couple of hours at high summer temperature. Soon after water oozes out and the mushrooms become unfit for cooking.

22.9.12.9 Precautions

1. Always prepare compost with wheat-straw over 300 kg in quantity, as the heat generated in smaller piles will not be sufficient for fermentation.
2. The compost pile should not become hard pressed to avoid anaerobic conditions as decomposition takes place only under aerobic conditions.
3. If ammonia smell is noticeable, more turnings of the compost should be done to free the compost from ammonia smell.
4. Care should be taken to wet the composting material to the extent that water drops do not ooze out, if the compost is squeezed in the palm of hand.
5. Water should never be poured over the spawn to avoid the failure of the crop.
6. Proper care should be taken not to allow the temperature to rise above 75°C so that useful microorganisms present in the soil are not killed.
7. Care should be taken to leave at least one metre clear space in between the last tray and the ceiling, so that air circulation is not impaired.

22.9.13 Cultivation of *Ganoderma lucidum* (Ling Zhi)

Ganoderma lucidum (Ling Zhi) is a basidiomycete white rot macrofungi which has been used extensively as "the mushroom of immortality" in China, Japan, Korea and other Asian countries for 2000 years. A great deal of work has been carried out on therapeutic potential of *Ganoderma lucidum*. The basidiocarp, mycelia and spores of *Ganoderma lucidum* contain approximately 400 different bioactive compounds, which mainly include triterpenoids, polysaccharides, nucleotides, sterols, steroids, fatty acids, proteins/peptides and trace elements which has been reported to have a number of pharmacological effects including immunomodulation, anti-atherosclerotic, anti-inflammatory, analgesic, chemopreventive, antitumour, chemo- and radio-protective, sleep promoting, antibacterial, antiviral (including anti-HIV), hypolipidemic, anti-fibrotic, hepatoprotective, anti-diabetic, anti-androgenic, anti-angiogenic, anti-herpetic antioxidative and radical-scavenging, anti-aging, hypoglycemic, estrogenic activity and anti-ulcer properties. *Ganoderma lucidum* has now become recognized as an alternative adjuvant in the treatment of leukemia, carcinoma, hepatitis and diabetes. The macrofungus is very rare in nature rather not sufficient for commercial exploitation for vital therapeutic emergencies, therefore the cultivation on solid substrates, stationary liquid medium or by submerged cultivation has become an essential aspect to meet the driving force towards the increasing demands in the international market. It can be grown under solid substrate and submerged fermentation and cultivation methods are described, keeping in mind that *G. lucidum* is a therapeutic fungal biofactory (Figure 22.32 and Figure 22.33).

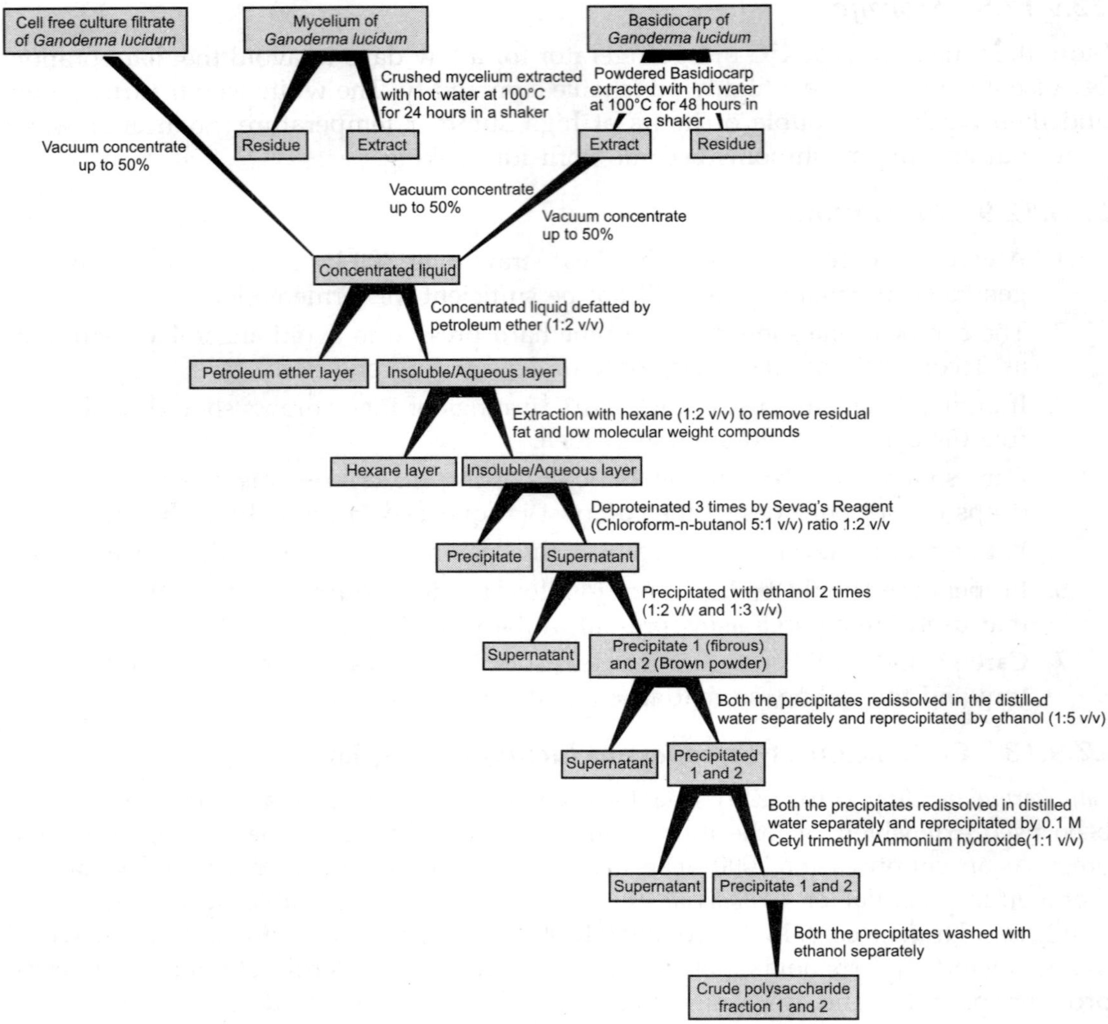

Figure 22.32 Flow chart for the isolation and characterization of polysaccharides from *Ganoderma lucidum.*

22.9.13.1 Materials

Organism Pure culture of *Ganoderma lucidum* (Ling Zhi), cereal grains (sorghum, wheat, rye, etc.), bottles (flask type), cotton (gauze), sawdust of the broad leaf plant *Tectona grandis* (Teak), narrow leaf plant *Acacia* spp. as substrate, wheat bran, rice bran, $CaSO_4$, $CaCO_3$, plastic bag, autoclave, laminar flow cabinet, Incubator/storage room, Wire guage balance, Bunsen burner, water.

Culture maintenance: The media used for maintenance, multiplication and preservation of mushroom culture are: Potato Dextrose agar (PDA), Yeast Potato Dextrose agar (YPDA).

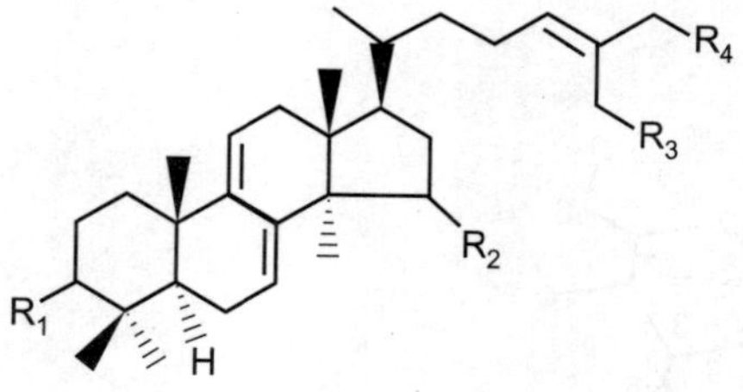

Ganoderiol F — $R_1 = O$, $R_2 = H$, $R_3 = R_4 = OH$
Ganodermadiol — $R_1 = \beta\text{-OH}$, $R_2 = R_3 = H$, $R_4 = OH$
5a-lanosta-7,9 (11), 24-triene-15a-26-dihydroxy-3-one — $R_1 = O$, $R_2 = \alpha\text{-OH}$, $R_3 = H$, $R_4 = OH$

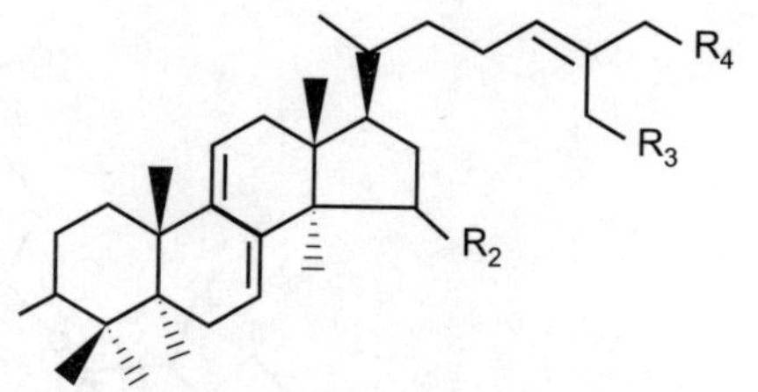

Ganoderiol F — $R_1 = O$, $R_2 = H$, $R_3 = R_4 = OH$
Ganodermadiol — $R_1 = \beta\text{-OH}$, $R_2 = R_3 = H$, $R_4 = OH$
5a-lanosta-7,9 (11), 24-triene-15a-26-dihydroxy-3-one — $R_1 = O$, $R_2 = \alpha\text{-OH}$, $R_3 = H$, $R_4 = OH$

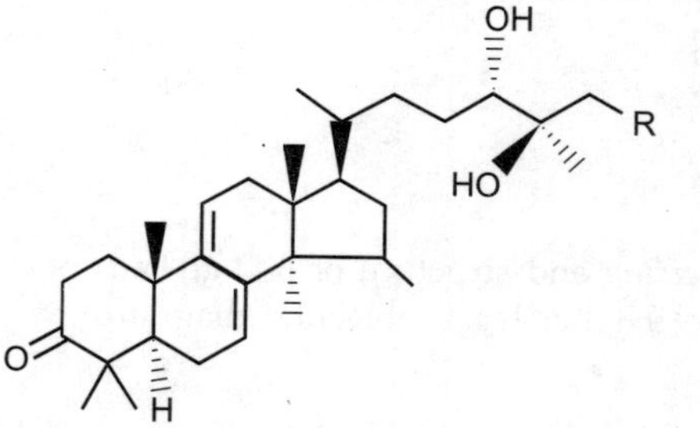

Ganodermanontriol $R = OH$
Ganodermanondiol $R = H$

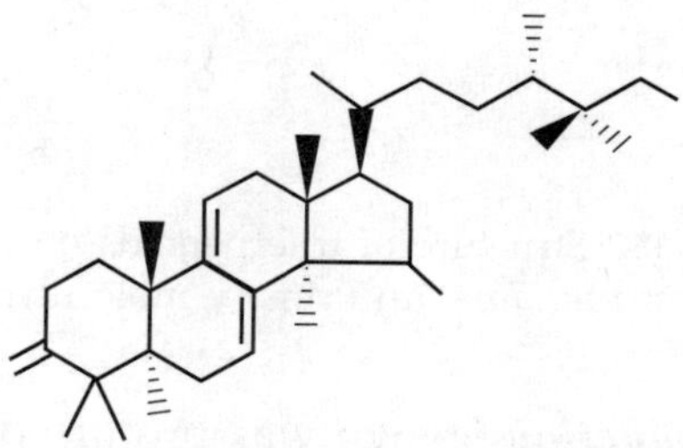

Ganodermanontriol $R = OH$
Ganodermanondiol $R = H$

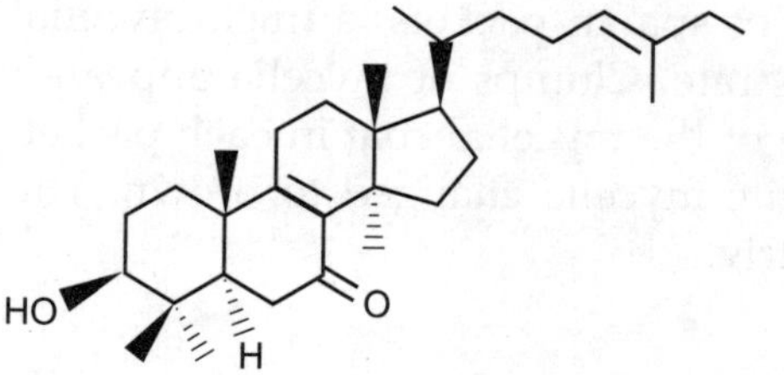

Lucidadiol $R = OH$
Lucialdehyde C $R = O$

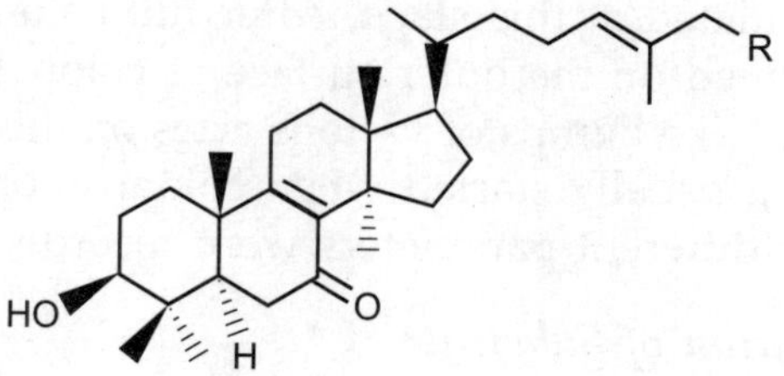

Lucidadiol $R = OH$
Lucialdehyde C $R = O$

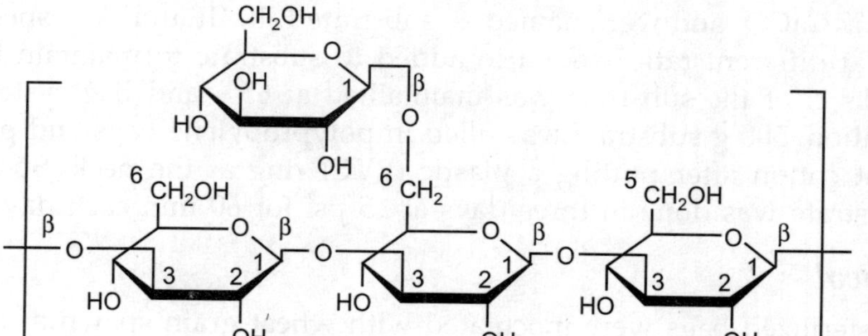

(a)

Contd...

Contd...

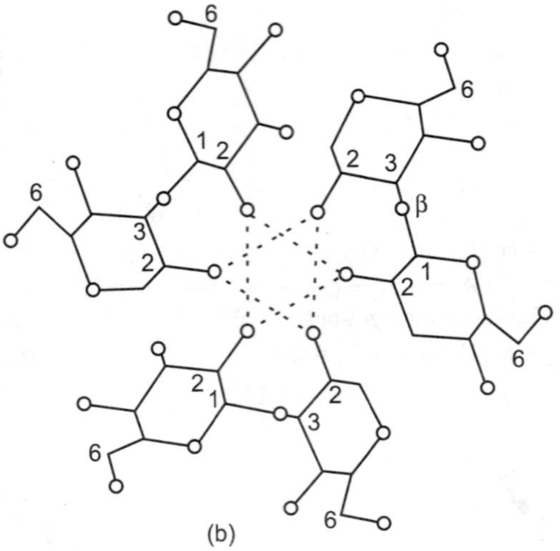

(b)

Figure 22.33 Structure of triterpenoids of *Ganoderma lucidum* and structure of β-D-glucans of *Ganoderma lucidum*, (a) Primary molecular diagram; (b) Higher level molecular diagram.

Spawn production Spawn was produced and maintained in spawn bottle using wheat grain with calcium sulfate and calcium carbonate at 2% and 0.5%, respectively on dry weight basis of the grain, to adjust pH of the grain at 4.5 to 5.0 pH. The grain must not be coagulated at this stage. After full colonization of spawn packets, a thick mycelial coat formed on the outer surface of colonized substrate. Clumps of mycelia appeared as blister like bumps of various sizes on the surface of the mycelial coat in each packet. Bumping usually started when colonization of white mycelia changed to brown. The data on different parameters were recorded regularly.

Preparation of Substrate

The substrate prepared contains saw dust (teak) + 20% wheat bran + 2% $CaSO_4$ + 0.6% $CaCO_3$; saw dust (teak) + 20% rice bran + 2% $CaSO_4$ + 0.6% $CaCO_3$; saw dust (*Acacia* spp.) + 20% wheat bran + 1.5% $CaSO_4$ + 0.5% $CaCO_3$; saw dust (*Acacia* spp.) + 20% rice bran + 1.5% $CaSO_4$ + 0.5% $CaCO_3$ and were named as substrate I, II, III and IV respectively. $CaSO_4$ and $CaCO_3$ at different ratio were also added to substrate to maintain the pH 4.5. The moisture level of the substrate was maintained at 65% and Bag system was adopted for cultivation. 500 g substrate was filled in polypropylene bags and plugged with non-absorbent cotton after putting a plastic (PVC) ring at the neck. Successive sterilization of substrate was done in three days at 15 psi for 60 min each day.

22.9.13.2 Protocol

Upon cooling the sterilized bags were inoculated with wheat grain spawn at the rate of 4% and incubated in dark. After complete mycelial colonization (bags were white all over) the bags were opened and rolled back to expose upper surface and proper conditions (temperature, R.H. and light intensity) for different stages of primordia and fruit-body formation were optimized and provided. After harvesting the first flush, conditions for primordia formation were again switched on (i.e., temp.- 28 ± 2°C, R.H.- 95%, light- 800 lux) for starting and completing the second flush and the same procedure has been repeated for the succesive flush. The optimum temperature for mycelial colonization was 30 ± 2°C.

22.9.13.3 Results and observations

For the complete mycelial colonization substrate- I normally take 18 days; substrate - II take 16 days; substrate – III take 26 days and substrate- IV take 23 days. Further, primordia is formed in 12-14, 10-12, 16-19 and 14-16 days in substrate-I, substrate-II, substrate- III and substrate- IV respectively, soon after mycelial colonization of *G. lucidum*. The optimum light intensity (800 lux), R.H. (95%) and temperature (30 ± 2°C) are suitable for primordia formation (Figure 22.34).

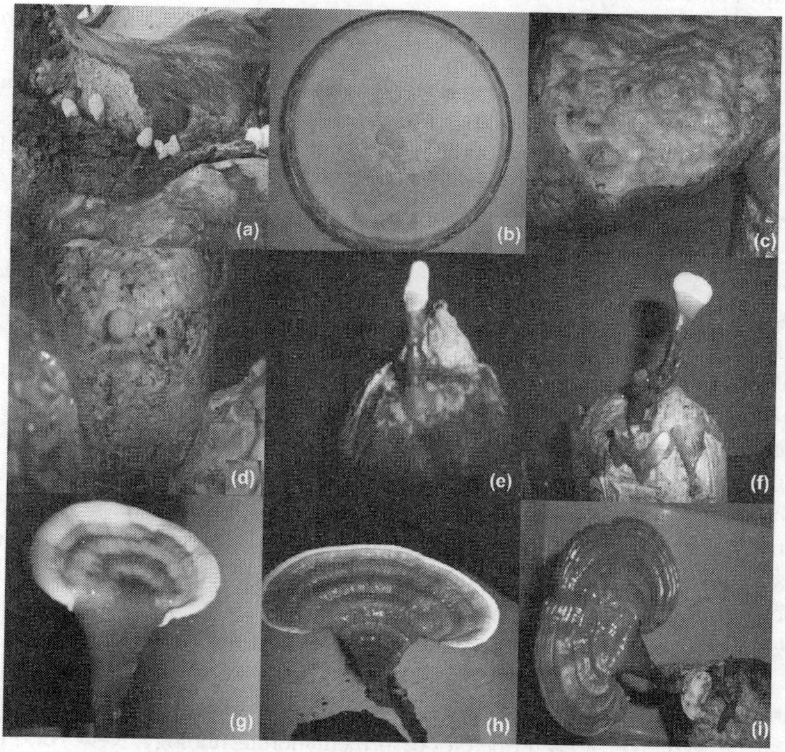

Figure 22.34 Different growth stages of *Ganoderma lucidum* during artificial cultivation under solid substrate state fermentation on polypropylene bags; (a) Source of strain isolation; (b) Culture plate (growing in potato dextrose agar medium); (c) Mycelial colonization on solid substrate; (d) Primordia formation; (e) elongation of primordia; (f) Cap formation; (g) Flattening and growth of cap; (h) Thickening of cap; (i) Maturation of fruit body.

The following data are taken into account for record during the experiment.

(a) **Days for completion of running of mycelium:** Time is recorded in days for completion of spawn running and mycelial cover on each substrate.

(b) **Days for primordia formation:** Record the time in days for primordia formation on each substrate.

(c) **Fruiting body formation:** Record the time in days for fruiting body formation on each substrate.

Suggested Readings

Arora, D. 1986. Mushrooms Demystified. Berkeley, CA: Ten Speed Press.

Bisen, P. S., R. K. Baghel, B. S. Sanodiya, G. S. Thakur, G. B. K. S. Prasad. 2010. *Lentinus edodes*: A Macrofungus with Pharmacological Activities. *Current Medicinal Chemistry* 17: 2419-2430.

Christopher, H. 2002.Medicinal Mushrooms (Herbs and Health Series), Redondo Beach, CA: Botanica Press.

Doran, P.M. 2012. *Bioprocess Engineering Principles*, Waltham, MA, USA: Academic /Elsevier.

Gyun, Jong, Kim, Chan Jeong, Joo Soo Kwang, Yoo Young Je Kim. 2011. Biotechnology and Bioprocess Engineering 16 Gene Cloning and Expression of a 3-ketovalidoxylamine C-N-lyase from Flavobacterium saccharophilum IFO 13984. *Biotechnology and Bioprocess Engineering* 16 :366-373.

Jayani RS, Saxena S, Gupta R. 2005. "Microbial pectinolytic enzymes: a review." *Process Biochem* 40, 2931-2944.

Jones DT, Woods DR. 1986. Acetone-butanol fermentation revisited. *Microbiol Rev* 50, 484-524.

Preuss, H. G. and S. Konno. 2002. Maitake Magic, Crawford Ave, Evanston, IL: Freedom Publishing Company (IL).

Rao, D.G. 2010. *Introduction to Biochemical Engineering*, New Delhi: Tata McGraw-Hill Education.

Sanodiya, B. S., G. S. Thakur, R. K. Baghel, G. B. K. S. Prasad, P. S. Bisen.2009. *Ganoderma lucidum*: A potent pharmacological macrofungus. *Current Pharmaceutical Biotechnology* 10: 717-742.

Saue M, Porro D, Mattanovich D, Branduardi P. 2008. Microbial production of organic acids: expanding the markets. *Trends Biotechnol* 26, 100-110.

Shuler ML., Kargi F. 2012.*Bioprocess Engineering:Basic Concepts*, Pearson Education, New Delhi.

Stamets, P. 1983. The Mushroom Cultivator: A Practical Guide to Growing Mushrooms at Home, Seattle: Agarikon Press.

Stamets, P. 1996. Psilocybin Mushrooms of the World: An Identification Guide, Berkeley, CA: Ten Speed Press.

Stamets, P. 2000. Growing Gourmet and Medicinal Mushrooms. Berkeley, CA:Ten Speed Press.

Stanbury, P.F., and A. Whitaker. 1984. Principles of fermentation technology. New York: Pergamon Press.

Walton MT, Martin JL. 1979. Production of butanol-acetone by fermentation. In: *Microbial Technology. (Eds HJ Peppler and D Perlman) 2nd ed. Academic Press Inc, New York,* 187-209.

Glossary

ABO blood group system The classification of red blood cells based on the presence or absence of A and B carbohydrate antigens.

Abscess A localized accumulation of pus.

Acellular Vaccine A vaccine consisting of antigenic parts of cells.

Acid A substance that dissociates into one or more hydrogen ions (H^+) and one or more negative ions.

Acid-Fast Stain A differential stain used to identify bacteria that are not decolorized by acid-alcohol.

Acidic Dye A salt in which the color is in the negative ion; used for negative staining.

Acidophile A bacterium that grows below pH 4.

Acid Precipitation Rain, snow, sleet, hail, or drizzle containing acidic oxides of sulfur, nitrogen, etc. that are released into the atmosphere, usually from the combustion of fossil fuels, and return to Earth dissolved in precipitation.

Acquired Immune Deficiency The inability, obtained during the life of an individual, to produce specific antibodies or T cells, due to drugs or disease.

Acquired Immunity The ability, obtained during the life of the individual, to produce specific antibodies.

Activated Macrophage A macrophage that has increased phagocytic ability and other functions after exposure to mediators released by T cells after stimulation by antigens.

Activated Sludge System A process used in secondary sewage treatment in which batches of sewage are held in highly aerated tanks; to ensure the presence of microbes efficient in degrading sewage, each batch is inoculated with portions of sludge from a precious batch, called activated sludge.

Activation Energy The minimum collision energy required for a chemical reaction to occur.

Active Site a region on an enzyme that interacts with the substrate.

Active Transport Net movement of a substance across a membrane against a concentration gradient; requires the cell to expend energy.

Acute Disease A disease in which symptoms develop rapidly but last for only a short time.

Adenocarcinoma Cancer of glandular epithelial tissue.

Adenosine Diphosphate (ADP) The substance formed when ATP is split and energy is released.

Adenosine Triphosphate (ATP) An important intracellular energy source.

Adherence Attachment of a microbe or phagocyte to another's plasma membrane or other surface.

Adhesin A carbohydrate-specific binding protein that projects from procaryotic cells; used for adherence, also called ligand.

Aerobe An organism requiring oxygen (O_2) for growth.

Aerobic Respiration Respiration in which the final electron acceptor in the electron transport chain is oxygen (O_2).

Aerotolerant Anaerobe An organism that does not use molecular oxygen (O_2) but is not affected by its presence.

Aflatoxin A carcinogenic toxin produced by *Aspergillus flavus*.

Agar A complex polysaccharide derived from a marine seaweed and used as a solidifying agent in culture media.

Agglutination A joining together or clumping of cells.

AIDS (acquired immunodeficiency syndrome) An infectious disease caused by the human immunodeficiency virus (HIV) in which the virus destroys the immune system and leaves the body vulnerable to opportunistic infections and malignancies.

Airborne Transmission The spread of pathogens farther than 1 meter in air from reservoir to susceptible host.

Alcohol An organic molecule with the functional group -OH.

Alcohol Fermentation A catabolic process, beginning with glycolysis, that produces ethyl alcohol.

Aldehyde An organic molecule with the functional group.

Alga (plural: **algae**) A photosynthetic eucaryote; some algae are included in the Kingdom Protista and some in the Kingdom Plantae.

Algal Bloom An abundant growth of microscopic algae, producing visible colonies in nature.

Algin a sodium salt of mannuronic acid ($C_6H_8O_6$); found in brown algae.

Allergen An antigen that evokes a hypersensitivity response.

Allergy *See* hypersensitivity.

Allograft A graft between people who are not identical twins.

Allosteric Inhibition The process in which an enzyme's activity is changed because of binding on the allosteric site.

Allosteric Site The site on an enzyme at which a noncompetitive inhibitor binds.

Amanitin a polypeptide mushroom toxin that causes liver and nerve damage.

Ames Test A procedure using bacteria to identify potential carcinogens.

Amination The addition of an amino group.

Amino Acid An organic acid containing an amino group and a carboxyl group.

Amino Acid Sequencing Determination of the order of the amino acids in a protein.

Aminoglycoside An antibiotic consisting of amino sugars and an aminocyclitol ring; for example, streptomycin.

Amino Group $-NH_2$.

Ammonification The release of ammonia from nitrogen-containing organic matter by the action of microorganisms.

Amoeba An organism belonging to the Kingdom Protista that moves by means of pseudopods.

Amphibolic Pathway A pathway that is both anabolic and catabolic.

Amphitrichous Having tufts of flagella at both ends of a cell.

Anabolism All synthesis reactions in a living organism; the building of complex organic molecules from simpler ones.

Anaerobe An organism that does not require oxygen (O_2) for growth.

Anaerobic Respiration Respiration in which the final electron acceptor in the electron transport chain is an inorganic molecule other than oxygen (O_2); for example, a nitrate or sulfate.

Anaerobic Sludge Digester Anaerobic digestion used in secondary sewage treatment.

Anal Pore A site in certain protozoa for elimination of waste.

Analytical Epidemiology Comparison of a diseased group and a healthy group to determine the cause of the disease.

Anamnestic Response *See* memory response.

Anaphylaxis A hypersensitivity reaction involving IgE antibodies, mast cells, and basophils.

Animalia The kingdom composed of multicellular eucaryotes lacking cell walls.

Anion An ion with a negative charge.

Anoxygenic Not producing molecular oxygen; typical of cyclic photophosphorylation.

Antagonism Active opposition; for example, between two drugs or two microbes.

Antibiotic An antimicrobial agent produced naturally by a bacterium or fungus.

Antibody A protein produced by the body in response to a foreign substance, or antigen, and capable of combining specifically with that antigen, usually leading in vivo to its destruction or inactivation; also called immunoglobulin.

Antibody-Dependent Cell-Mediated Cytotoxicity (ADCC) The killing of antibody-coated cells by natural killer cells and phagocytes.

Antibody Titer The amount of antibody in serum.

Anticodon The three nucleotides by which a tRNA recognizes an RNA codon.

Antigen Any substance that, when introduced into the body, causes antibody formation and reacts only with its specific antibody; also called immunogen.

Antigen-Antibody Complex The combination of an antigen with the antibody that is specific for it; the basis of immune protection and many diagnostic tests.

Antigen-Binding Sites a site on an antibody that binds to an antigenic determinant.

Antigenic Determinant A specific region on the surface of an antigen against which antibodies are formed; also called epitope.

Antigenic Drift A minor variation in the antigenic makeup of influenza viruses that occurs with time.

Antigenic Shift A major genetic change in influenza viruses causing changes in H and N antigens.

Antigen-Presenting Cell (APC) A macrophage or dendritic cell that engulfs an antigen and presents fragments to T cells.

Antigen Receptors An antibody-like molecule on B and T cells that enables them to recognize and bind to their specific antigen.

Antihuman Immune Serum Globulin (anti-HISG) An antibody that reacts specifically with human antibodies.

Antimicrobial Drug A chemical that destroys pathogens without damaging body tissues.

Antimicrobial Peptide An antibiotic, derived from an animal, that is fast-acting, bactericidal, and has a broad spectrum of activity.

Antisense Agent An experimental type of antimicrobial drug; a short segment of DNA that binds to a target area on mRNA blocks the synthesis of a pathogenic protein by a pathogenic microorganism.

Antisense Strand (- strand) Viral RNA that cannot act as mRNA.

Antisepsis A chemical method for disinfection of the skin, mucous membranes, or other living tissues; the chemical is called an antiseptic.

Antiserum A blood-derived fluid containing antibodies.

Antitoxin A specific antibody produced by the body in response to a bacterial exotoxin or its toxoid.

Antiviral Protein (AVP) A protein made in response to interferon that blocks viral multiplication.

Apoenzyme The protein portion of an enzyme, which requires activation by a coenzyme.

Apoptosis The natural programmed death of a cell; the residual fragments are disposed of by phagocytosis.

Aquatic Microbiology The study of microorganisms and their activities in natural waters.

Arbuscule Fungal mycelia in plant root cells.

Archaea Procaryotic organisms lacking peptidoglycan; members of the Domain Archaca.

Arthrospore An asexual fungal spore formed by fragmentation of a septate hypha.

Arthus Reaction Inflammation and necrosis at the site of injection of foreign serum, due to immune complex formation.

Artificially Acquired Active Immunity The production of antibodies by the body in response to a vaccination.

Artificially Acquired Passive Immunity The transfer of humoral antibodies formed by one individual to a susceptible individual, accomplished by the injection of antiserum.

Ascospore A sexual fungal spore produced in an ascus, formed by the Ascomycetes class of fungi.

Aseptic Surgery Surgery preformed in microbe-free conditions

Ascus A saclike structure containing ascospores; found in the ascomycetes.

Asepsis The absence of contamination by unwanted organisms; aseptic means free of pathogens.

Aseptic Packaging Commercial food preservation by filling sterile containers with sterile food.

Asexual Spore A reproductive cell produced by mitosis and cell division (eucaryotes) or binary fission (actinomycetes).

Atom The smallest unit of matter that can enter into a chemical reaction.

Atomic Number The number of protons in the nucleus of an atom.

Atomic Weight The total number of protons and neutrons in the nucleus of an atom.

Attenuated Whole-Agent Vaccine A vaccine that uses as antigen living but attenuated (weakened) microorganisms. See also inactivated whole-agent vaccine.

Autoclave Equipment for sterilization by steam under pressure, usually operated at 15 psi and 121°C.

Autograft A tissue graft from one's self.

Autoimmune Disease Damage to one's own organs due to action of the immune system.

Autotroph An organism that uses carbon dioxide (CO_2) as its principal carbon source; also called lithotroph.

Auxotroph A mutant microorganism with a nutritional requirement that is absent in the parent.

Axial Filament The structure for motility found in spirochetes; also called endoflagellum.

Bacillus (plural: **bacilli)** (1) Any rod-shaped bacterium. (2) When written as a genus, refers to rod-shaped, endospore-forming, facultatively anaerobic, gram-positive bacteria.

Bacteremia A condition in which there are bacteria in the blood.

Bacterial Species A population of cells with similar characteristics. See also eucaryotic species; viral species.

Bacteriology The scientific study of bacteria

Bacterium (plural: bacteria) All living organisms with procaryotic cells.

Bacterial Growth Curve A graph indicating the growth of a bacterial population over time.

Bactericide A substance capable of killing bacteria.

Bacteriocin A toxic protein produced by bacteria that kills other bacteria.

Bacteriophage (phage) A virus that infects (multiplies in) bacterial cells.

Bacteriostasis A treatment capable of inhibiting bacterial growth.

Bacteroid An enlarged *Rhizobium* cell found in root nodules.

Base A substance that dissociates into one or more hydroxide ions (OH_2) and one or more positive ions.

Base Analog A chemical that is structurally similar to the normal nitrogenous bases in nucleic acids but with altered base-pairing properties.

Base Pairs The arrangement of nitrogenous bases in nucleic acids based on hydrogen bonding; in DNA, base pairs are AT and GC; in RNA, base pairs are AU and GC.

Base Substitution The replacement of a single base in DNA by another base, causing a mutation; also called point mutation.

Basic Dye A salt in which the color is in the positive ion; used for bacterial stains.

Basidiospore A sexual fungal spore produced in a basidium, characteristic of the basidiomycetes.

Basidium A pedestal that produces basidiospores; found in the basidiomycetes.

Basophil A granulocyte (white blood cell) that readily takes up basic dye and is not phagocytic; has receptors for IgE Fc regions, which react with antigens, causing the release of histamine and other factors that result in an allergic reaction.

B cell A type of lymphocyte that acts as one of the main components of the humoral immune system; differentiates into antibody-secreting plasma cells and memory cells.

BCG Vaccine A live, attenuated strain of *Mycobacterium bovis* used to provide immunity to tuberculosis.

Benthic zone The sediment at the bottom of a body of water.

Bergey's Manual *Bergey's Manual of Systematic Bacteriology*, the standard taxonomic reference on bacteria; also refers to *Bergey's Manual of Determinative Bacteriology*.

Beta Oxidation The removal of two carbon units from a fatty acid to form acetyl CoA.

Biguanide A group of antimicrobial drugs, including chlorhexidine, especially useful on skin and mucous membranes.

Binary Fission Bacterial reproduction by division into two daughter cells.

Binomial Nomenclature The system of having two names (genus and specific epithet) for each organism; also called scientific nomenclature.

Biochemical Oxygen Demand (BOD) A measure of the biologically degradable organic matter in water.

Bioconversion Changes in organic matter brought about by the growth of microorganisms.

Biofilm A microbial community that usually forms as a slimy layer on a surface.

Biogenesis The concept that living cells can arise only from preexisting cells.

Biogeochemical Cycle The recycling of chemical elements by microorganisms for use by other organisms.

Biological Transmission The transmission of a pathogen from one host to another when the pathogen reproduces in the vector.

Bioluminescence The emission of light from the electron transport chain of certain living organisms.

Biomass Organic matter produced by living organisms and measured by weight.

Bioreactor A fermentation vessel with controls for environmental conditions, e.g., temperature and pH.

Bioremediation The use of microbes to remove an environmental pollutant.

Biotechnology The industrial application of genetic engineering, usually to produce products such as pharmaceuticals.

Biovar A subgroup of a serovar based on biochemical or physiological properties; also called biotype.

Blade A flat leaflike structure of multicellular algae.

Blastospore An asexual fungal spore produced by budding from the parent cell.

Blood-Brain Barrier Cell membranes that allow some substances to pass from the blood to the brain but restrict others.

Brightfield Microscope A microscope that uses visible light for illumination; the specimens are viewed against a white background.

Broad-Spectrum Antibiotic An antibiotic that is effective against a wide range of both gram-positive and gram-negative bacteria.

Broth Dilution Test A method of determining the MIC by using serial dilutions of an antimicrobial drug.

Bubo An enlarged lymph node caused by inflammation.

Budding (1) Asexual reproduction beginning as a protuberance from the parent cell that grows to become a daughter cell. (2) Release of an enveloped virus through the plasma membrane of an animal cell.

Budding Yeast Following mitosis, a yeast cell that divides unevenly to produce a small cell (bud) from the parent cell.

Buffer A substance that tends to stabilize the pH of a solution.

Bulking A condition arising when sludge floats rather than settles in secondary sewage treatment.

Burst Size The number of newly synthesized bacteriophage particles released from a single cell.

Burst Time The time required from bacteriophage attachment to release.

Calvin-Benson Cycle The conversion of CO_2 into reduced organic compounds; used by autotrophs.

Capnophile A microorganism that grows best at relatively high CO_2 concentrations.

Capsid The protein coat of a virus that surrounds the nucleic acid.

Capsomere A protein subunit of a viral capsid.

Capsule An outer, viscous covering on some bacteria composed of a polysaccharide or polypeptide.

Carbapenems Antibiotics that contain a ß-lactam antibiotic and cilastatin.

Carbohydrates An organic compound composed of carbon, hydrogen, and oxygen, with the hydrogen and oxygen present in a 2:1 ratio; carbohydrates include starches, sugars, and cellulose.

Carbon Cycle The series of processes that converts carbon dioxide (CO_2) to organic substances and back to carbon dioxide in nature.

Carbon Fixation The synthesis of sugars by using carbons from CO_2 gas. *See also* Calvin-Benson cycle.

Carbon Skeleton The basic chain or ring of carbon atoms in a molecule; for example,

Carboxyl Group –COOH.

Carboxysome A procaryotic inclusion containing ribulose 1,5-diphosphate carboxylase.

Carcinogen Any cancer-causing substance.

Casein Milk protein.

Catabolic Repression Inhibition of the metabolism of alternate carbon sources by glucose.

Catabolism All decomposition reactions in a living organism; the breakdown of complex organic compounds into simpler ones.

Catalase An enzyme that catalyzes the breakdown of hydrogen peroxide to water and oxygen.

Catalyst A substance that increases the rate of a chemical reaction but is not altered itself.

Cation A positively charged ion.

Cell Culture Animal or plant cells grown in vitro.

Cell-Mediated Immunity An immune response that involves T cells binding to antigens presented on infected cells; T cells then differentiate into several types of effector T cells, including helper and cytotoxic.

Cellular Respiration *See* respiration.

Cell Wall The outer covering of most bacterial, fungal, algal, and plant cells; in eubacteria, it consists of peptidoglycan.

Center for Disease Control and Prevention (CDC) A branch of the U. S. Public Health Service located in Atlanta, Georgia that serves a central source of epidemiological information.

Central Nervous System (CNS) The brain and the spinal cord. *See* also peripheral nervous system.

Centriole A structure consisting of nine microtubule triplets, found in eucaryotic cells.

Cephalosporin An antibiotic produced by the fungus *Cephalosporium* that inhibits the synthesis of gram-positive bacterial cell walls.

Cercaria A free-swimming larva of trematodes.

Chancre A hard sore, the center of which ulcerates.

Chemical Bond An attractive force between atoms forming a molecule.

Chemical Element A fundamental substance composed of atoms that have the same atomic number and behave the same way chemically.

Chemical Energy The energy of a chemical reaction.

Chemically Defined Medium A culture medium in which the exact chemical composition is known.

Chemical Reaction The process of making or breaking bonds between atoms.

Chemiosmosis A mechanism that uses a proton gradient across a cytoplasmic membrane to generate ATP.

Chemistry The science of the interactions between atoms and molecules.

Chemoautotroph An organism that uses an inorganic chemical as an energy source and carbon dioxide (CO_2) as a carbon source.

Chemoheterotroph An organism that uses organic molecules as a source of carbon and energy.

Chemokine A cytokine that induces, by chemotaxis, the migration of leukocytes into infected areas.

Chemotaxis Movement in response to the presence of a chemical.

Chemotherapeutic Agent A chemical(drug) used in the treatment of disease.

Chemotherapy The treatment of a disease with chemical substances.

Chemotroph An organism that uses oxidation-reduction reactions as its primary energy source.

Chimeric Monoclonal Antibody A genetically engineered antibody made of human constant regions and mouse variable regions.

Chlamydospore An asexual fungal spore formed within a hypha.

Chloramphenicol A broad-spectrum bacteriostatic chemical.

Chloroplast The organelle that performs photosynthesis in photoautotrophic eucaryotes.

Chlorosome Plasma membrane folds in green sulfur bacteria containing bacteriochlorophylls.

Chromatin Threadlike, uncondensed DNA in an interphase eucaryotic cell.

Chromatophore An infolding in the plasma membrane where bacteriochlorophyll is located in photoautotrophic bacteria; also known as thylakoid.

Chromosome The structure that carries hereditary information, chromosomes contain genes.

Chronic Disease An illness that develops slowly and is likely to continue or recur for long periods.

Cilium (plural: **cilia)** A relatively short cellular projection that moves in a wavelike manner.

Ciliate A member of the protozoan phylum Ciliophora that uses cilia for locomotion.

Cistern A flattened membranous sac in endoplasmic reticulum and the Golgi complex.

Cladogram A dichotomous phylogenetic tree that branches repeatly, suggesting the classification of organisms based on the time sequence in which evolutionary branches arose.

Class A taxonomic group between phylum and order.

Clonal Deletion The elimination of B and T cells that react with self.

Clonal Selection The development of clones of B and T cells against a specific antigen.

Clone A population of cells that are identical to the parent cell.

Clue Cells Sloughed-off vaginal cells covered with *Gardnerella vaginalis*.

Coagulase A bacterial enzyme that causes blood plasma to clot.

Coccobacillus (plural: coccabacilli) A bacterium that is an oval rod.

Coccus (plural: cocci) A spherical or ovoid bacterium.

Codon A sequence of three nucleotides in mRNA that specifies the insertion of an amino acid into a protein.

Coenocytic Hypha A fungal filament that is not divided into uninucleate cell-like units because they lack septa.

Coenzyme A nonprotein substance that is associated with and that activates an enzyme.

Coenzyme A (CoA) A coenzyme that functions in decarboxylation.

Coenzyme Q *See* ubiquinone.

Cofactor (1) The nonprotein component of an enzyme. (2) A microorganism or molecule that acts with others to synergistically enhance or cause disease.

Coliforms Aerobic or facultatively anaerobic, gram-negative, nonendospore-forming, rod-shaped bacteria that ferment lactose with acid and gas formation within 48 hours at 35°C; used as an indicator organism for tests of bacteriological purity.

Collagenase An enzyme secreted by several species of *Clostridium* that beaks down collagen and facilitates the spread of gas gangrene.

Collision Theory The principle that chemical reactions occur because energy is gained as particles collide.

Colony A clone of bacterial cells on a solid medium that is visible to the naked eye.

Colony Hybridization The identification of a clone containing a desired gene by using a DNA probe that is complementary to that gene.

Colony-Stimulating Factor (CSF) Asubstance that induces certain cells to proliferate of differentiate.

Commensalism A symbiotic relationship in which two organisms live in association and one is benefited while the other is neither benefited nor harmed.

Commercial Sterilization A process of treating canned goods aimed at destroying the endospores of *Clostridium botulinum.*

Communicable Disease Any disease that can be spread from one host to another.

Competence The physiological state in which a recipient cell can take up and incorporate a large piece of donor DNA.

Competitive Inhibitor A chemical that competes with the normal substrate for the active site of an enzyme. See also noncompetitive inhibitor.

Complement A group of serum proteins involved in phagocytosis and lysis of bacteria.

Complementary DNA (cDNA) DNA made in vitro from an mRNA template.

Complement Fixation The process in which complement combines with an antigen–antibody complex.

Complex Medium A culture medium in which the exact chemical composition is not known.

Complex Virus A virus with a complicated structure, such as a bacteriophage.

Composting A method of solid waste disposal, usually plant material, by encouraging its decomposition by microbes in piles or containers, converting it into a smaller final product; often used as a soil conditioner.

Compound A substance composed of two or more different chemical elements.

Compound Light Microscope (LM) An instrument with two sets of lenses that uses visible light as the source of illumination.

Compromised Host A host whose resistance to infection is impaired.

Condensation Reaction A chemical reaction in which a molecule of water is released; also called dehydration synthesis.

Condenser A lens system located below the microscope stage that directs light rays through the specimen.

Congenital Immune Deficiency The inability, due to an individual's genotype, to produce specific antibodies or T cells.

Conidiophore An aerial hypha bearing conidiospores.

Conidiospore An asexual spore produced in a chain from a conidiophore.

Conjugated Vaccine A vaccine consisting of the desired antigen and other proteins.

Conjugation The transfer of genetic material from one cell to another involving cell-to-cell contact.

Constitutive Enzyme An enzyme that is produced regardless of how much substrate is present.

Contact Inhibition The cessation of animal cell movement and division as a result of contact with other cells.

Contact Transmission The spread of disease by direct or indirect contact or via droplets.

Contagious Disease A disease that is easily spread from one person to another.

Continuous Cell Line Animal cells that can be maintained through an indefinite number of generations in vitro.

Corepressor A molecule that binds to a repressor, enabling the repressor to bind to an operator.

Cortex The protective fungal covering of a lichen.

Counterstain A stain used to give contrast in a differential stain.

Covalent Bond A chemical bond in which the electrons of one atom are shared with another atom.

Crisis The phase of a fever characterized by vasodilation and sweating.

Crista (plural: cristae) A foldings of the inner membrane of a mitochondrion.

Crossing Over The process by which a portion of one chromosome is exchanged with a portion of another chromosome.

Culture Microorganisms that grow and multiply in a container of culture medium.

Culture Medium The nutrient material prepared for growth of microorganisms in a laboratory.

Curd The solid part of milk that separates from the liquid in the making of cheese, for example.

Cutaneous Mycosis A fungal infection of the epidermis, nails, or hair.

Cuticle The nonliving outer covering of helminths.

Cyanobacteria Oxygen-producing photoautotrophic procaryotes; also called blue-green algae.

Cyclic AMP (cAmp) A molecule derived from ATP, in which the phosphate group has a cyclic structure; acts as a cellular messenger.

Cyclic Photophosphorylation The movement of an electron from chlorophyll through a series of electron acceptors and back to chlorophyll; purple and green bacterial photophosphorylation.

Cyst A sac with a distinct wall containing fluid or other material; also, a protective capsule of some protozoa.

Cysticercus An encysted tapeworm larva.

Cystitis Inflammation of the urinary bladder.

Cytochrome Oxidase An enzyme that oxidizes cytochrome c.

Cytochromes A proteins that functions as an electron carrier in cellular respiration and photosynthesis.

Cytokine A small protein released from human cells in response to bacterial infection; directly or indirectly may induce fever, pain, or T-cell proliferation.

Cytolysis The destruction of cells, resulting from damage to their cell membrane, that causes cellular contents to leak out.

Cytopathic Effect (CPE) A visible effect on a host cell, caused by a virus, that may result in host cell death (cytocidal effect) or damage (noncytocidal effect).

Cytoplasm In a procaryotic cell,, everything inside the plasma membrane; in a eucaryotic cell, everything inside the plasma membrane and external to the nucleus.

Cytoplasmic Streaming The movement of cytoplasm in a eucaryotic cell.

Cytoskeleton Microfilaments and microtubules that provide support and movement for eucaryotic cytoplasm.

Cytostome The mouthlike opening in some protozoa.

Cytotoxic T (T_C) cells A specialized T cell that destroys infected cells presenting antigens.

Cytotoxin A bacterial toxin that kills host cells or alters their functions.

Darkfield Microscope A microscope that has a device to scatter light from the illuminator so that the specimen appears white against a black background.

Dark (light-independent) Reaction The second stage of photosynthesis, in which electrons and energy from ATP are used to reduce CO_2 to sugar. *See also* Calvin-Benson cycle.

Deamination The removal of an amino groups from amino acids to form ammonia. *See also* ammonification.

Death Phase The period of logarithmic decrease in a bacterial population; also called logarithmic decline phase.

Debridement Surgical removal of necrotic tissue.

Decarboxylation The removal of carbon dioxide (CO_2) from an amino acid.

Decimal Reduction Time (DRT) The time (in minutes) required to kill 90% of a bacterial population at a given temperature; also called D value.

Decolorizing Agent A solution used in the process of removing a stain.

Decomposition Reaction A chemical reaction in which bonds are broken to produce smaller parts from a large molecule.

Deep-Freezing Preservation of bacterial cultures at 250°C to 295°C.

Definitive Host An organism that harbors the adult, sexually mature form of a parasite.

Degeneracy Redundancy of the genetic code; that is, most amino acids are signaled by several codons.

Degerming The removal of microorganisms in an area of an injection, usually by alcohol swab; also called degermation.

Degranulation The release of contents of secretory granules from mast cells or basophils during anaphylaxis.

Dehydration Synthesis *See* condensation reaction.

Dehydrogenation The loss of hydrogen atoms from a substrate.

Delayed-Type Hypersensitivity Cell-mediated hypersensitivity.

Delayed Hypersensitivity T (T_D) cells A specialized T cell that produces lymphokines in type IV hypersensitivities.

Denaturation A change in the molecular structure of a protein, usually making it nonfunctional.

Dendritic Cell A type of antigen-presenting cell characterized by long fingerlike extensions; found in lymphatic tissue and skin.

Denitrification The reduction of nitrates to nitrites or nitrogen gas.

Dental Plaque A combination of bacterial cells, dextran, and debris adhering to the teeth.

Deoxyribonucleic Acid (DNA) The nucleic acid of genetic material.

Deoxyribose A five-carbon sugar contained in DNA nucleotides.

Dermatomycosis A fungal infection of the skin; also known as tinea or ringworm.

Dermatophyte A fungus that causes a cutaneous mycosis.

Dermis The inner portion of the skin.

Descriptive Epidemiology The collection and analysis of all data regarding the occurrence of a disease to determine its cause.

Desensitization The prevention of allergic inflammatory responses.

Desiccation The removal of water.

Dichotomous Key An identification scheme based on successive paired questions; answering one question leads to another pair of questions, until an organism is identified.

Differential Interference Contrast (DIC) microscope An instrument that provides a three-dimensional, magnified image.

Differential Medium A solid culture medium that makes it easier to distinguish colonies of the desired organism.

Differential Stain A stain that distinguishes objects on the basis of reactions to the staining procedure.

Differential White Blood Cell Count The number of each kind of leukocyte in a sample of 100 leukocytes.

Dimorphism The property of having two forms of growth.

Dioecious Referring to organisms in which organs of different sexes are located in different individuals.

Diplobacilli (singular: **diplobacillus**) Rods that divide and remain attached in pairs.

Diplococci (singular: **diplococcus**) Cocci that divide and remain attached in pairs.

Diploid Cell A cell having two sets of chromosomes; diploid is the normal state of a eucaryotic cell.

Diploid Cell Line Eucaryotic cells grown in vitro.

Direct Agglutination Test The use of known antibodies to identify an unknown cell-bound antigen.

Direct Contact Transmission A method of spreading infection from one host to another through some kind of close association of the hosts.

Direct FA Test A fluorescent-antibody test to detect the presence of an antigen.

Direct Microscopic Count Enumeration of cells by observation through a microscope.

Disaccharide A sugar consisting of two simple sugars, or monosaccharides; sucrose (table sugar) is a disaccharide.

Disease An abnormal state in which part or all of the body is not properly adjusted or is incapable of performing normal functions; any change from a state of health.

Disinfection Any treatment used on inanimate objects to kill or inhibit the growth of microorganisms; a substance used is called a disinfectant.

Disk-Diffusion Method An agar-diffusion test to determine microbial susceptibility to chemotherapeutic agents; also called Kirby-Bauer test.

d-isomer A stereoisomer.

Dissimilation A metabolic process in which nutrients are not assimilated but are excreted as ammonia, hydrogen sulfide, etc.

Dissimilation Plasmid A plasmid containing genes coding for the production of enzymes that trigger the catabolism of certain unusual sugars and hydrocarbons.

Dissociation The separation of a compound into positive and negative ions in solution. *See also* ionizations.

Disulfide Bond A covalent bond that holds together two atoms of sulfur.

Division A phylum; used in botany and microbiology.

DNA Base Composition The percentage of guanine plus cytosine in an organism's DNA.

DNA Fingerprinting Analysis of DNA by electrophoresis of restriction enzyme fragments of the DNA.

DNA Ligase An enzyme that covalently bonds a carbon atom of one nucleotide with the phosphate of another nucleotide.

DNA Probe A short, labeled, single strand of DNA or RNA used to locate its complementary strand in a quantity of DNA.

DNA Sequencing A processes by which the nucleotide sequence of DNA is determined.

Domain A taxonomic classification for related kingdoms.

Donor Cell A cell that gives DNA to a recipient cell during genetic recombination.

DPT Vaccine A combined vaccine used to provide active immunity, containing diphtheria and tetanus toxoids and killed *Bordetella pertussis* cells.

Droplet Transmission The transmission of infection by small liquid droplets carrying microorganisms.

d Value *See* decimal reduction time.

Dysentery A disease characterized by frequent, watery stools containing blood and mucus.

Eclipse Period The time during viral multiplication when complete, infective virions are not present.

Edema An abnormal accumulation of fluid in body parts or tissues, causing swelling.

Electron A negatively charged particle in motion around the nucleus of an atom.

Electron Acceptor An ion that picks up an electron that has been lost from another atom.

Electron Donor An ion that gives up an electron to another atom.

Electronic Configuration The arrangement of electrons in shells or energy levels in an atom.

Electron Microscope A microscope that uses a flow of electrons instead of light to produce an image.

Electron Shell A region of an atom corresponding to an energy level.

Electron Transport Chain A series of compounds that transfer electrons from one compound to another, generating ATP by oxidative phosphorylation.

Electroporation A technique by which DNA is inserted into a cell using an electrical current.

Elementary Body An infectious form of *Chlamydia.*

ELISA (Enzyme-Linked Immunosorbent Assay) A group of serological tests that use enzyme reactions as indicators.

Emerging Infectious Disease (EID) A new or changing disease that is increasing or has the potential to increase in incidence in the near future.

Emigration The process by which phagocytes move out of blood vessels; also called diapedesis.

Embden–Meyerhof pathway *See* Glycolysis.

Emulsification The process of mixing two liquids that do not dissolve in each other.

Encephalitis Infection of the brain.

Encystment Formation of a cyst.

Endemic Disease A disease that is constantly present in a certain population.

Endergonic Reaction A chemical reaction that requires energy.

Endocarditis Infection of the lining of the heart (endocardium).

Endocytosis The process by which material is moved into a eucaryotic cell.

Endoflagellum. *See* axial filament.

Endolith An organism that lives inside rock.

Endoplasmic Reticulum (ER) A membranous network in eucaryotic cells connecting the plasma membrane with the nuclear membrane.

Endospore A resting structure formed inside some bacteria.

Endosymbiotic Theory A model for the evolution of eucaryotes proposing that organelles arose from procaryotic cells living inside a host procaryote.

Endotoxic Shock *See* Septic shock.

Endotoxin Part of the outer portion of the cell wall (lipid A) of most gram-negative bacteria; released on destruction of the cell.

End-Product Inhibition *See* feedback inhibition.

Enrichment Culture A culture medium used for preliminary isolation that favors the growth of a particular microorganism.

Enteric The common name for a bacterium in the family Enterobacteriaceae, which inhabit the intestinal tracts of humans and other animals.

Enterotoxin An exotoxin that causes gastroenteritis, such as those produced by *Staphylococcus, Vibrio,* and *Escherichia.*

Entner–Doudoroff Pathway An alternate pathway for the oxidation of glucose to pyruvic acid.

Envelope An outer covering surrounding the capsid of some viruses.

Enzyme A protein that catalyzes biochemical reactions in a living organism.

Enzyme Induction The process by which a substance can cause the synthesis of an enzyme.

Enzyme-Linked Immunasorbent Assay. *See* ELISA.

Enzyme–Substrate Complex A temporary union of an enzyme and its substrate.

Eosinophil A granulocyte whose granules take up the stain eosin.

Epidemic Disease A disease acquired by many people in a given area in a short time.

Epidemiology The science that studies when and where diseases occur and how they are transmitted.

Epidermis The outer portion of the skin.

Epitope *See* antigenic determinant.

Equilibrium The point of even distribution.

Equivalent Treatments Different methods that have the same effect on controlling microbial growth.

Ergot A toxin produced in sclerotia by the fungus *Claviceps purpurea* that causes ergotism, a disease characterized by hallucinations and the contraction of arteries and uterine muscles.

Ethambutol A synthetic antimicrobial agent that interferes with the synthesis of RNA.

Ethanol A flammable, intoxicating liquid produced by microorganisms, especially yeasts; ethyl alcohol.

Etiology The study of the cause of a disease.

Eubacteria Procaryotic organisms possessing peptidoglycan cell walls; members of the Domain Eubacteria.

Eucarya All eucaryotes (animals, plants, fungi, and protists); members of the Domain Eucarya.

Eucaryote A cell having DNA inside a distinct membrane-enclosed nucleus.

Eutrophication The addition of organic matter and subsequent removal of oxygen from a body of water.

Exchange Reaction A chemical reaction that has both synthesis and decomposition components.

Excision Repair The repair of DNA involving the removal of defective nucleotides and replacement with functional ones.

Exergonic Reaction A chemical reaction that releases energy.

Exon A region of a eucaryotic chromosome that codes for a protein.

Exotoxin A protein toxin released from living, mostly gram-positive bacterial cells.

Experimental Epidemiology The study of a disease using controlled experiments.

Exponential Growth Phase *See* log phase.

Extracellular Polysaccharide (ESP) A glycocalyx, composed of sugars, that permits bacteria to attach to various surfaces.

Extreme Halophile An organism that requires a high salt concentration for growth.

Extreme Thermophile An organism whose optimum growth temperature is at least 80°C; also called hyperthermophile.

Extremophile A microorganism that lives in environmental extremes of temperature, acidity, alkalinity, or salinity.

Eyespot A pigmented area capable of detecting the presence of light.

Facilitated Diffusion The transfer of a substance across a plasma membrane from an area of higher concentration to an area of lower concentration, mediated by carrier proteins (permeases).

Facultative Anaerobe An organism that can grow with or without molecular oxygen (O_2).

Facultative Halophile An organism capable of growth in, but not requiring, 1–2% salt.

Family A taxonomic group between order and genus.

Feedback Inhibition Inhibition of an enzyme in a particular pathway by the accumulation of the end-product of the pathway; also called end-product inhibition.

Fermentation The enzymatic degradation of carbohydrates in which the final electron acceptor is an organic molecule, ATP is synthesized by substrate-level phosphorylation, and oxygen (O_2) is not required.

Fever An abnormally high body temperature.

F Factor (Fertility Factor) A plasmid found in the donor cell in bacterial conjugation.

Fibrinolysin A kinase produced by streptococci.

Filtration The passage of a liquid or gas through a screenlike material; a 0.45-μm filter removes bacteria.

Fimbria (plural: **fimbriae**) An appendage on a bacterial cell used for attachment.

Fission Yeast Following mitosis, a yeast cell that divides evenly to produce two new cells.

Five Kingdom System A classification system in which organisms are assigned to one of five kingdoms: Animalia, Plantae, Fungi, Protista, or Monera.

Fixed Macrophage A macrophage that is located in a certain organ or tissue (e.g., liver, lungs, spleen, or lymph nodes); also called a histiocyte.

Fixing (1) In slide preparation, the process of attaching a specimen to a slide. (2) Regarding chemical elements, combining elements so that a critical element can enter the food chain. *See also* Calvin-Benson cycle; nitrogen fixation.

Flagellum (plural: **flagella**) A thin appendage arising from one or more locations on the surface of a cell; used for cellular locomotion.

Flagellate A member of the protozoan subphylum Mastigophora that uses flagella for locomotion.

Flaming The process of sterilizing an inoculating loop by holding it in an open flame.

Flat Sour Spoilage Thermophilic spoilage of canned goods not accompanied by gas production.

Flatworm An animal belonging to the phylum Platyhelminthes.

Flavin Adenine Dinucleotide (FAD) A coenzyme that functions in the removal and transfer of hydrogen ions (H^+) and electrons from substrate molecules.

Flavin Mononucleotide (FMN) A coenzyme that functions in the transfer of electrons in the electron transport chain.

Flavoprotein A protein containing the coenzyme flavin; functions as an electron carrier in cellular respiration.

Flocculation The removal of colloidal material during water purification by adding of a chemical that causes the colloidal particles to coalesce.

Flow Cytometry A method of separating and counting cells using a flow cytometer, which detects cells by the presence of a fluorescent tag on the cell surface.

Fluid Mosaic Model A way of describing the dynamic arrangement of phospholipids and proteins comprising the plasma membrane.

Fluke A flatworm belonging to the class Trematoda.

Fluorescence The ability of a substance to give off light of one color when exposed to light of another color.

Fluorescence-Activated Cell Sorter (FACS) A modification of a flow cytometer that sorts cells in addition to separating and counting them.

Fluorescence Microscope A microscope that uses an ultraviolet light source to illuminate specimens that will fluoresce.

Fluorescent-Antibody (FA)technique A diagnostic tool using antibodies labeled with fluorochromes and viewed through a fluorescence microscope; also called immunofluorescence.

Fluoroquinolone A synthetic antibacterial agent that inhibits DNA synthesis.

Focal Infection A systemic infection that began as an infection in one place.

Folliculitis An infection of hair follicles, often occurring as pimples.

Fomite A nonliving object that can spread infection.

Forespore A structure consisting of chromosome, cytoplasm, and endospore membrane inside a bacterial cell.

Frameshift Mutation A mutation caused by the addition or deletion of one or more bases in DNA.

Freeze-Drying *See* lyophilization.

FTA-ABS Test An indirect fluorescent-antibody test used to detect syphilis.

Functional Group An arrangement of atoms in an organic molecule that is responsible for most of the chemical properties of that molecule.

Fungus (plural: fungi) An organism that belongs to the Kingdom Fungi; a eucaryotic absorptive chemoheterotroph.

Fusion The merging of plasma membranes of two different cells, resulting in one cell containing cytoplasm from both original cells.

Gamete A male or female reproductive cell.

Gametocyte A male or female protozoan cell.

Gamma Globulin *See* immune serum globulin.

Gastroenteritis Inflammation of the stomach and intestine.

Gas Vacuole A procaryotic inclusion for buoyancy compensation.

Gel Electrophoresis The separation of substances (such as serum proteins) by their rate of movement through an electrical field.

Gene A segment of DNA (a sequence of nucleotides in DNA) that codes for a functional product.

Gene Library A collection of cloned DNA fragments created by inserting restriction enzyme fragments in a bacterium, yeast, or phage.

Gene Therapy Treating a disease by replacing abnormal genes.

Generalized Transduction The transfer of bacterial chromosome fragments from one cell to another by a bacteriophage.

Generation Time The time required for a cell or population to double in number.

Genetic Code The mRNA codons and the amino acids for which they code.

Genetic Engineering Manufacturing and manipulating genetic material in vitro; also called recombinant DNA technology.

Genetic Recombination The process of joining pieces of DNA from different sources.

Genetics The science of heredity.

Genetic Screening Techniques for determining which genes are in a cell's genetic material.

Genotype The genetic makeup of an organism.

Genus (plural: genera) The first name of the scientific name (binomial); the taxon between family and species.

Germicide A substance capable of killing microorganisms.

Germination The process of starting to grow from a spore or endospore.

Germ Theory of Disease The principle that microorganisms cause disease.

Global Warming The warming of the atmosphere caused by increasing accumulations of CO_2, which absorbs infrared radiation from sunlight and slows its escape from Earth's surface; also known as the greenhouse effect.

Globulin A class of proteins having certain solubility characteristics; widespread in animal tissue. *See also* immunoglobulin.

Glycocalyx A gelatinous polymer surrounding a cell.

Glycolysis The main pathway for the oxidation of glucose to pyruvic acid; also called Embden-Meyerhof pathway.

Golgi Complex An organelle involved in the secretion of certain proteins.

Graft-Versus-Host (GVH) Disease A condition that occurs when a transplanted tissue has an immune response to the tissue recipient.

Gram-Negative Bacteria Bacteria that lose the crystal violet color after decolorizing by alcohol; they stain pink after treatment with safranin.

Gram-Negative Cell Wall The cell wall of gram-negative bacteria, a peptidoglycan layer surrounded by a lipopolysaccharide outer membrane.

Gram-Positive Bacteria Bacteria that retain the crystal violet after decolorizing by alcohol; they stain dark purple.

Gram-Positive Cell Wall The cell wall of most gram-positive bacteria, consisting of peptidoglycan and teichoic acids.

Gram Stain A differential stain that classifies bacteria into two groups, gram-positive and gram-negative.

Granulocyte A leukocyte with visible granules in the cytoplasm; includes neutrophils, basophils, and eosinophils.

Griseofulvin A fungistatic antibiotic.

Group Translocation In procaryotes, active transport in which a substance is chemically altered during transport across the plasma membrane.

Gumma A rubbery mass of tissue characteristic of tertiary syphilis.

Halogen One of the following elements: fluorine, chlorine, bromine, iodine, or astatine.

Haploid Cell A eucaryotic cell or organism with one of each type of chromosome.

Hapten A substance of low molecular weight that does not cause the formation of antibodies by itself but does so when combined with a carrier molecule.

Helminth A parasitic roundworm or flatworm.

Helper T (T_h) cell A specialized T cell that often interacts with an antigen before B cells interact with the antigen.

Hemagglutination The clumping of red blood cells.

Hemoflagellate A parasitic flagellate found in the circulatory system of its host.

Hemolysin An enzyme that lyses red blood cells.

Herd Immunity The presence of immunity in most of a population.

Hermaphroditic Having both male and female reproductive capacities.

Heterocyst A large cell in certain cyanobacteria; the site of nitrogen fixation.

Heterolactic Describing an organism that produces lactic acid and other acids or alcohols as end-products of fermentation; e.g., *Escherichia.*

Heterotroph An organism that requires an organic carbon source; also called organotroph.

Hfr Cell A bacterial cell in which the F factor has become integrated into the chromosome; Hfr stands for high frequency of recombination.

High-Efficiency Particulate Air (HEPA) Filter A screenlike material that removes particles larger than 0.3 μm from air.

HIGH-TEMPERATURE SHORT-TIME (HTST) Pasteurization Pasteurizing at 72°C for 15 seconds.

Histamine A substance released by tissue cells that causes vasodilation, capillary permeability, and smooth muscle contraction.

Histocompatibility Antigen An antigen on the surface of human cells.

Histone A protein associated with DNA in eucaryotic chromosomes.

Holdfast The branched base of an algal stipe.

Holoenzyme An enzyme consisting of an apoenzyme and a cofactor.

Homolactic Describing an organism that produces only lactic acid from fermentation (for example, *Streptococcus*).

Host An organism infected by a pathogen. *See also* definitive host; intermediate host.

Host Range The spectrum of species, strains, or cell types that a pathogen can infect.

Hot-Air Sterilization Sterilization by the use of an oven at 170°C for approximately 2 hours.

Human Leukocyte Antigen (HLA) complex Human cell surface antigens. *See also* major histocompatibility complex.

Humoral Immunity Immunity produced by antibodies dissolved in body fluids, mediated by B cells; also called antibody-mediated immunity.

Hyaluronidase An enzyme secreted by certain bacteria that hydrolyzes hyaluronic acid and helps spread microorganisms from their initial site of infection.

Hybridoma A cell made by fusing an antibody-producing B cell with a cancer cell.

Hydrogen Bond A bond between a hydrogen atom covalently bonded to oxygen or nitrogen and another covalently bonded oxygen or nitrogen.

Hydrolysis A decomposition reaction in which chemicals react with the H^+ and OH_2 of a water molecule.

Hydroxyl Radical A toxic form of oxygen (OH•) formed in cytoplasm by ionizing radiation and aerobic respiration.

Hyperacute Reflection Very rapid rejection of transplanted tissue, usually in the case of tissue from nonhuman sources, caused by the development in humans of antibodies against distantly related animals.

Hyperbaric Chamber An apparatus to hold materials at pressures greater than 1 atmosphere; used to oxygenate gangrenous tissue.

Hypersensitivity An altered, enhanced immune reaction leading to pathological changes; also called allergy.

Hyperthermophile *See* extreme thermophile.

Hypertonic (Hyperosmotic) Solution A solution that has a higher concentration of solutes than an isotonic solution.

Hypha A long filament of cells in fungi or actinomycetes.

Hypotonic (Hypoosmotic) Solution A solution that has a lower concentration of solutes than an isotonic solution.

ID_{50} The bacterial concentration required to produce a demonstrable infection in 50% of the test host population.

Idiophase The period in the production curve of an industrial cell population in which secondary metabolites are produced; a period of stationary growth following the phase of rapid growth. See also trophophase.

IgA The class of antibodies found in secretions.

IgD The class of antibodies found on B cells.

IgE The class of antibodies involved in hypersensitivities.

IgG The most abundant class of antibodies in serum.

IgM The first class of antibodies to appear after exposure to an antigen.

Imidazoles An antifungal drug that interferes with sterol synthesis.

Immune Complex A circulating antigen–antibody aggregate capable of fixing complement.

Immune Serum Globulin The serum fraction containing immunoglobulins (antibodies); also called gamma globulin.

Immunity The body's defense against particular pathogenic microorganisms; also called specific resistance.

Immunization *See* vaccination.

Immunodiffusion Test A test consisting of precipitation reactions carried out in an agar gel medium.

Immunoelectrophoresis The identification of proteins by electrophoretic separation followed by serological testing.

Immunofluorescence *See* fluorescent-antibody technique.

Immunogen *See* Antigen.

Immunoglobulin A protein (antibody) formed in response to exposure to an antigen; provides specific immunity to that antigen if encountered again. *See also* globulin.

Immunological Escape Resistance of cancer cells to immune rejection.

Immunological Surveillance The body's immune response to cancer.

Immunology The study of a host's specific defenses to a pathogen.

Immunosuppression Inhibition of the immune response.

Immunotherapy Making use of the immune system to attack tumor cells, either by enhancing the normal immune response or by using toxin-bearing specific antibodies. *See also* immunotoxin.

Immunotoxin An immunotherapeutic agent consisting of a poison bound to a monoclonal antibody.

Inactivated Whole-Agent Vaccine A vaccine that uses as antigen killed, but otherwise intact, microorganisms. *See also* attenuated whole-agent vaccine.

Inapparent Infection *See* subclinical infection.

Incidence The fraction of the population that contracts a disease during a particular period of time.

Inclusion Material held inside a cell, often consisting of reserve deposits.

Inclusion Body A granule or viral particle in the cytoplasm or nucleus of some infected cells important in the identification of viruses that cause infection.

Incubation Period The time interval between the actual infection and first appearance of any signs or symptoms of disease.

Indicator Organism A microorganism, such as a coliform, whose presence indicates conditions such as fecal pollution of food or water.

Indirect (Passive) Agglutination Test An agglutination test using soluble antigens attached to latex or other small particles.

Indirect Contact Transmission The spread of pathogens by fomites (nonliving objects).

Indirect FA Test A fluorescent-antibody test to detect the presence of specific antibodies.

Inducer A sustance that acts to induce the transcription of a gene.

Induction The process that turns on the transcription of a gene.

Infection The invasion or growth of microorganisms in the body.

Infection Thread An indentation in a root hair of a plant that allows *Rhizobium* to infect the root.

Infectious Disease A disease in which pathogens invade a susceptible host and carry out at least part of their life cycle in the host.

Inflammation A host response to tissue damage characterized by redness, pain, heat, and swelling; and sometimes loss of function.

Innate Resistance The resistance of an individual to diseases that affect other species and other individuals of the same species.

Inoculum A culture medium in which microorganisms are implanted.

Inorganic Compound A small molecule that does not usually contain carbon.

Insertion Sequence (IS) The simplest kind of transposons.

Interferon (IFN) An antiviral protein produced by certain animal cells in response to a viral infection.

Interleukins A chemicals that causes T-cell proliferation. *See also* cytokine.

Intermediate Host An organism that harbors the larval or asexual stage of a helminth or protozoan.

Intoxication A condition resulting from the ingestion of a microbially produced toxin, or effects from the formation of toxin by bacterial growth in the body.

Intron A region in a eucaryotic gene that does not code for a protein or mRNA.

Invasin A surface protein produced by *Salmonella typhimurium* and *Escherichia coli* that rearranges nearby actin filaments in the cytoskeleton of a host cell.

Iodophor A complex of iodine and a detergent.

Ion A negatively or positively charged atom or group of atoms.

Ionic Bond A chemical bond formed when atoms gain or lose electrons in the outer energy levels.

Ionization The separation (dissociation) of a molecule into groups of atoms with electrical charges.

Ionizing Radiation High-energy radiation with wavelength less than 1nm; causes ionization. X rays and gamma rays are examples.

Ischemia Local loss of blood supply.

Isograft A tissue graft between identical twins.

Isomer One or two molecules with the same chemical formula but different structures.

Isoniazid (INH) A bacteriostatic agent used to treat tuberculosis.

Isotonic (Isomatic) Solution A solution in which, after immersion of a cell, osmotic pressure is equal across the cell's membrane.

Isotope A form of a chemical element in which the number of neutrons in the nucleus is different from the other forms of that element.

Kelp A multicellular brown alga.

Keratin A protein found in epidermis, hair, and nails.

Kinase (1) An enzyme that removes a (P) from ATP and attaches it to another molecule (2) A bacterial enzyme that breaks down fibrin (blood clots).

Kingdom A taxonomic classification above phylum and division.

Kinins A substance released from tissue cells that causes vasodilation.

Kirby–Bauer Test. *See* disk-diffusion method

Koch's Postulates Criteria used to determine the causative agent of infectious diseases.

Krebs Cycle A pathway that converts two-carbon compounds to carbon dioxide (CO_2), transferring electrons to NAD^+ and other carriers; also called tricarboxylic acid (TCA) cycle or citric acid cycle.

Lagging Strand During DNA replication, the daughter strand that is synthesized discontinuously.

Lag Phase The time interval in a bacterial growth curve during which there is no growth.

Larva The sexually immature stage of a helminth or arthropod.

Latent Disease A disease characterized by a period of no symptoms when the pathogen is inactive.

Latent Infection A condition in which a pathogen remains in the host for long periods without producing disease.

LD_{50} The lethal dose for 50% of the inoculated hosts within a given period.

Leading Strand During DNA replication, the daughter strand that is synthesized continuously.

Lepromin Test A skin test to determine the presence of antibodies to *Mycobacterium leprae*, the cause of leprosy.

Leukocidins A substances produced by some bacteria that can destroy neutrophils and macrophages.

Leukocyte A white blood cell.

Leukotriene A substance produced by mast cells and basophils that causes increased permeability of blood vessels and helps attach phagocytes to pathogens.

Lichen A mutualistic relationship between a fungus and an alga or a cyanobacterium.

Ligand *See* adhesine.

Light (Light-Dependent) Reactions The first stage of photosynthesis, in which light energy is used to convert ADP and phosphate to ATP. *See also* photophosphorylation.

Light-Repair Enzyme An enzyme that splits thymine dimers in the presence of visible light.

Limnetic Zone The surface zone of an inland body of water away from the shore.

Limulus Amoebocyte Lysate (LAL) Assay A test to detect the presence of bacterial endotoxins.

Lipase An enzyme that breaks down fats into their component glycerol and fatty acids.

Lipid A molecule composed of glycerol and fatty acids.

Lipid A A component of gram-negative outer membrane; endotoxin.

Lipopolysaccharide (LPS) A molecule consisting of a lipid and a polysaccharide, forming the outer layer of gram-negative cell walls.

l-isomer A stereoisomer; the amino acids found in proteins are always l-isomers.

Littoral Zone The region along the shore of an inland body of water where there is considerable vegetation and where light penetrates to the bottom.

Local Infection An infection in which pathogens are limited to a small area of the body.

Localized Anaphylaxis An immediate hypersensitivity reaction that is restricted to a limited area of skin or mucous membrane; for example, hayfever, a skin rash, or asthma. *See also* systemic anaphylaxis.

Logarithm Decline Phase *See* death phase.

Log Phase The period of bacterial growth or logarithmic increase in cell numbers; also called exponential growth phase.

Lophotrichous Having two or more flagella at one end of a cell.

Luciferase An enzyme that accepts electrons from flavoproteins and emits a photon of light.

Lymphangitis Inflammation of lymph vessels.

Lymphocyte A white blood cell involved in specific immune responses.

Lyophilization Freezing a substance and sublimating the ice in a vacuum; also called freeze-drying.

Lysogenic Conversion A change in the characteristics of a microorganism due to carrying a prophage in which the bacterial cell is immune to infection by the same type of phage.

Lysogenic Cycle Stages in viral development that result in the incorporation of viral DNA into host DNA.

Lysis (1) Destruction of a cell by the rupture of the plasma membrane, resulting in a loss of cytoplasm. (2) In disease, a gradual period of decline.

Lysogeny A state in which phage DNA is incorporated into the host cell without lysis.

Lysosome An organelle containing digestive enzymes.

Lysozyme An enzyme capable of lysing bacterial cell walls.

Lytic Cycle A mechanism of phage multiplication that results in host cell lysis.

Macrolide An antibiotic that inhibits protein synthesis; for example, erythromycin.

Macromolecule A large organic molecule.

Macrophage A phagocytic cell; an enlarged monocyte.

Macrophage Activation Factor Increases macrophages' efficiency at destroying ingested cells.

Macrophage Chemotactic Factor Attracts macrophages to infection site.

Macrophage Migration-Inhibiting Factor Prevents macrophages from leaving infection site.

Macule A flat, reddened skin lesion.

Magnetosome An iron oxide inclusion, produced by some gram-negative bacteria, that acts like a magnet.

Major Histocompatibility Complex (MHC) The genes that code for histocompatibility antigens; also known as human leukocyte antigen (HLA) complex.

Malolactic Fermentation The conversion of malic acid to lactic acid by lactic acid bacteria.

Malt Germinated barley grains containing maltose and amylase.

Malting The germination of starchy grains resulting in glucose and maltose production.

Margination The process by which phagocytes stick to the lining of blood vessels.

Mast Cell A type of cell found throughout the body that contains histamine and other substances that stimulate vasodilation.

Maximum Growth Temperature The highest temperature at which a species can grow.

Mechanical Transmission The process by which arthropods transmit infections by carrying pathogens on their feet and other body parts.

Medulla A lichen body consisting of algae (or cyanobacteria) and fungi.

Membrane Attack Complex Complement proteins C5–C9, which together make lesions in cell membranes that lead to cell death.

Membrane Filter A screenlike material with pores small enough to retain microorganisms.

Memory Cells A long-lived B or T cell responsible for the memory, or secondary, response.

Memory Response A rapid rise in antibody titer following exposure to an antigen after the primary response to that antigen; also called anamestic response or secondary response.

Meningitis Inflammation of the meninges, the three membranes covering the brain and spinal cord.

Merozoite A trophozoite of *Plasmodium* found in infected red blood cells or liver cells.

Mesophile An organism that grows between about 10°C and 50°C; a moderate-temperature–loving microbe.

Mesosome An irregular fold in the plasma membrane of a procaryotic cell that is an artifact of preparation for microscopy.

Messenger RNA (mRNA) The type of RNA molecule that directs the incorporation of amino acids into proteins.

Metabolic Pathway A sequence of enzymatically catalyzed reactions occurring in a cell.

Metabolism The sum of all the chemical reactions that occur in a living cell.

Metacercaria The encysted stage of a fluke in its final intermediate host.

Metachromatic Granule A granule that stores inorganic phosphate and stains red with certain blue dyes; characteristic of *Corynebacterium diphtheriae*. Collectively known as volutin.

Methane The hydrocarbon CH_4, a flammable gas formed by the microbial decomposition of organic matter; natural gas.

Microaerophile An organism that grows best in an environment with less molecular oxygen (O_2) than is normally found in air.

Microinjection The introduction of material, such as DNA, directly into a cell, usually with a glass pipette.

Micrometer (µm) A unit of measurement equal to 10^{-6} m.

Microorganism A living organism too small to be seen with the naked eye; includes bacteria, fungi, protozoa, and microscopic algae; also includes viruses.

Microtubule A hollow tube made of the protein tubulin; the structural unit of eucaryotic flagella and cilia.

Microwave An electromagnetic wave with wavelength between 10^{-1} and 10^{-3} m.

Minimal Bactericidal Concentration (MBC) The lowest concentration of a chemotherapeutic agent that will kill test microorganisms.

Minimal Inhibitory Concentration (MIC) The lowest concentration of a chemotherapeutic agent that will prevent growth of the test microorganisms.

Minimum Growth Temperature The lowest temperature at which a species will grow.

Miracidium The free-swimming, ciliated larva of a fluke that hatches from the egg.

Missense Mutation A mutation that resulting in the substitution of an amino acid in a protein.

Mitochondrion (plural: **mitochondria)** An organelle containing the respiratory ATP-synthesizing enzymes.

Mitosis A eucaryotic cell division process in which the chromosomes are duplicated, followed by division of the cytoplasm of the cell.

MMWR Morbidity and Mortality Weekly Report; a CDC publication containing data on notifiable diseases and topics of special interest.

Mole An amount of a chemical equal to the atomic weights of all the atoms in a molecule of the chemical.

Molecular Biology The science dealing with DNA and protein synthesis of living organisms.

Molecular Weight The sum of the atomic weights of all atoms making up a molecule.

Molecule A combination of atoms forming a specific chemical compound.

Monobactam A synthetic antibiotic with a ß-lactam ring that is monocyclic in structure, in contrast to the bicycle ß-lactam structure of the penicillins and cephalosporins.

Monera *See* Procaryotae.

Monoclonal Antibody A specific antibody produced by an in vitro clone of B cells hybridized with cancerous cells.

Monocyte A phagocyte that is the precursor of a macrophage.

Monoecious Having both male and female reproductive capacities.

Monomer A small molecule that collectively combines to form polymers.

Mononuclear Phagocytic System A system of fixed macrophages located in the spleen, liver, lymph nodes, and bone marrow.

Monosaccharide A simpler sugar consisting of 3–7 carbon atoms.

Monotrichous Having a single flagellum.

Morbidity (1) The incidence of a specific disease. (2) The condition of being diseased.

Morbidity Rate The number of people affected by a disease in a given period of time in relation to the total population.

Mordant A substance added to a staining solution to make it stain more intensely.

Mortality The number of deaths from a specific notifiable disease.

Mortality Rate The number of deaths resulting from a disease in a given period of time in relation to the total population.

Most Probable Number (MPN) method A statistical determination of the number of coliforms per 100 ml of water or 100 g of food.

Motility The ability of an organism to move by itself.

M Protein A heat- and acid-resistant protein of streptococcal cell walls and fibrils.

Mucous Membranes Membranes that line body openings, including the intestinal tract, open to the exterior; also called mucosa.

Mutagen An agent in the environment that brings about mutations.

Mutation Any change in the nitrogenous base sequence of DNA.

Mutation Rate The probability that a gene will mutate each time a cell divides.

Mutualism A type of symbiosis in which both organisms or populations are benefited.

Mycelium A mass of long filaments of cells that branch and intertwine, typically found in molds.

Mycology The scientific study of fungi.

Mycorrhiza A fungus growing in symbiosis with plant roots.

Mycosis A fungal infection.

Mycotoxin A toxin produced by a fungus.

Nanometer (nm) A unit of measurement equal to 10^{-9} m, 10^{-3} μm.

Natural Killer (NK) cells A lymphoid cells that destroys tumor cells and virus-infected cells.

Natural Penicillins The first agents of the penicillin family that were produced; penicillins G and V are example. *See also* semisynthetic penicillins.

Naturally Acquired Active Immunity Antibody production in response to an infectious disease.

Naturally Acquired Passive Immunity The natural transfer of humoral antibodies; transplacental transfer.

Necrosis Tissue death.

Negative (Indirect) Selection The process of identifying mutations by selecting cells that cannot function, using replica plating.

Negative Staining A procedure that results in colorless bacteria against a stained background.

Neurotoxin An exotoxin that interferes with normal nerve impulse conduction.

Neutralization An antigen–antibody reaction that inactivates a bacterial exotoxin or virus.

Neutron An uncharged particle in the nucleus of an atom.

Neutrophil A highly phagocytic granulocyte; also called polymorphonuclear leukocyte (PMN) or polymorph.

Nicotinamide Adenine Dinucleotide (NAD^+) A coenzyme that functions in the removal and transfer of hydrogen ions (H^+) and electrons from substrate molecules.

Nicotinamide Adenine Dinucleotide Phosphate ($NADP^+$) A coenzyme similar to NAD^+.

Nitrogen Cycle The series of processes that converts nitrogen (N_2) to organic substances and back to nitrogen in nature.

Nitrogen Fixation The conversion of nitrogen (N_2) into ammonia.

Nitrosamine A carcinogen formed by the combination of nitrite and amino acids.

Nomenclature The system of naming things.

Noncommunicable Disease A disease that is not transmitted from one person to another.

Noncompetitive Inhibitor A chemical that does not compete with the substrate for an enzyme's active site. *See also* allosteric inhibition; competitive inhibitor.

Noncyclic Photophosphorylation The movement of an electron from chlorophyll to NAD^+; plant and cyanobacterial photophosphorylation.

Nonionizing Radiation Short-wavelength radiation that does not cause ionization; ultraviolet (UV) radiation is an example.

Nonsense Codon A special terminator codon that does not code for any amino acid.

Nonsense Mutation A base substitution in DNA that results in a nonsense codon.

Nonspecific Resistance Host defenses that afford protection against any kind of pathogen. *See also* specific resistance.

Normal Microbiota The microorganisms that colonize an animal without causing disease; also called normal flora.

Nosocomial Infection An infection that develops during the course of a hospital stay and was not present at the time the patient was admitted.

Notifiable Disease A disease that physicians must report to the U. S. Public Health Service; also called reportable disease.

N (Neuraminidase) Spikes Antigenic projections from the outer lipid bilayer of influenza virus.

Nuclear Envelope The double membrane that separates the nucleus from the cytoplasm in a eucaryotic cell.

Nuclear Pore An opening in the nuclear envelope through which materials enter and exit the nucleus.

Nucleic Acid A macromolecule consisting of nucleotides; DNA and RNA are nucleic acids.

Nucleic Acid Hybridization The process of combining single complementary strands of DNA.

Nucleic Acid Vaccine A vaccine made up of "naked" DNA, usually in the form of a plasmid; also called DNA vaccine. Injected into muscle, DNA stimulates the production of proteins that stimulate an immune response.

Nucleoid The region in a bacterial cell containing the chromosome.

Nucleolus (plural: **nucleoli)** An area in a eucaryotic nucleus where rRNA is synthesized.

Nucleoside A compound consisting of a purine or pyrimidine base and a pentose sugar.

Nucleotide A compound consisting of a purine or pyrimidine base, a five-carbon sugar, and a phosphate.

Nucleus (1) The part of an atom consisting of the protons and neutrons. (2) The part of a eucaryotic cell that contains the genetic material.

Nutrient Agar Nutrient broth containing agar.

Nutrient Broth A complex medium made of beef extracts.

Objective Lenses In a compound light microscope, the lenses closest to the specimen.

Obligate Aerobe An organism that requires molecular oxygen (O_2) to live.

Obligate Anaerobe An organism that is unable to use molecular oxygen (O_2).

Obligate Halophile An organism that requires high osmotic pressures such as high concentrations of sodium chloride.

Ocular Lens In a compound light microscope, the lens closest to the viewer; also called eyepiece.

Oligodynamic Action The ability of small amounts of a heavy metal compound to exert antimicrobial activity.

Oncogene A gene that can bring about malignant transformation.

Oncogenic Virus A virus that is capable of producing tumors; also called oncovirus.

Oocyst An encysted sporozoan zygote in which cell division occurs to form the next infectious stage.

Operator The region of DNA adjacent to structural genes that controls their transcription.

Operon The operator and promoter sites and structural genes they control.

Opportunistic Pathogen A microorganism that does not ordinarily cause a disease but can become pathogenic under certain circumstances.

Opsonization The enhancement of phagocytosis by coating microorganisms with certain serum proteins (opsonins); also called immune adherence.

Optimum Growth Temperature The temperature at which a species grows best.

Order A taxonomic classification between class and family.

Organelle A membrane-enclosed structure within eucaryotic cells.

Organic Compound A molecule that contains carbon and hydrogen.

Organic Growth Factor An essential organic compound that an organism is unable to synthesize.

Osmosis The net movement of solvent molecules across a selectively permeable membrane from an area of higher concentration to an area of lower concentration.

Osmotic Pressure The force with which a solvent moves from a solution of lower solute concentration to a solution of higher solute concentration.

Oxidation (1) The removal of electrons from a molecule. (2) The addition of oxygen to a molecule.

Oxidation Pond A method of secondary sewage treatment by microbial activity in a shallow standing pond of water, in which aeration is promoted by wave action or mechanical agitation.

Oxidation-Reduction A coupled reaction in which one substance is oxidized and one is reduced; also called redox reaction.

Oxidative Phosphorylation The synthesis of ATP coupled with electron transport.

Oxygenic Producing oxygen, as in plant and cyanobacterial photosynthesis.

PABA Para-aminobenzoic acid; a precursor for folic acid synthesis.

Pandemic Disease An epidemic that occurs worldwide.

Parasite An organism that derives nutrients from a living host.

Parasitology The scientific study of parasites (protozoa and parasitic worms).

Parasitism A symbiotic relationship in which one organism (the parasite) exploits another (the host) without providing any benefit in return.

Parenteral Route A portal of entry for pathogens by deposition directly into tissues beneath the skin and mucous membranes.

Pasteurization The process of mild heating to kill particular spoilage microorganisms or pathogens.

Pathogen A disease-causing organism.

Pathogenesis The manner in which a disease develops.

Pathogenicity The ability of a microorganism to cause disease by overcoming the defenses of a host.

Pathology The scientific study of disease.

Pellicle The flexible covering of some protozoa.

Penicillins A group of antibiotics produced either by *Penicillium* (natural penicillins) or by adding side chains to the ß-lactam ring (semisynthetic penicillins).

Pentose Phosphate Pathway A metabolic pathway that can occur simultaneously with glycolysis to produce pentoses and NADH without ATP production; also called hexose monophosphate shunt.

Peptide Bond A bond joining the amino group of one amino acid to the carboxyl group of a second amino acid with the loss of a water molecule.

Peptidoglycan The structural molecule of eubacterial cell walls consisting of the molecules N-acetylglucosamine, N-acetylmuramic acid, tetrapeptide side chain, and peptide side chain; also known as murein.

Perforin Protein that makes a pore in a target cell membrane, released by T_C cells.

Pericarditis Inflammation of the sac around the heart (pericardium).

Period of Convalescence The recovery period, when the body returns to its predisease state.

Peripheral Nervous System (PNS) The nerves that connect the outlying parts of the body with the central nervous system.

Periplasmic Space The region of a gram-negative cell wall between the outer membrane and the cytoplasmic membrane.

Peritrichous Having flagella distributed over the entire cell.

permease A carrier protein in the plasma membrane.

Peroxidase An enzyme that breaks down hydrogen peroxide.

Peroxide Anion An oxygen anion consisting of two atoms of oxygen.

Peroxygen A class of oxidizing-type sterilizing disinfectants.

Persistent Viral Infection *See* slow viral infection.

pH The symbol for hydrogen ion (H^+) concentration; a measure of the relative acidity or alkalinity of a solution.

Phage *See* bacteriophage.

Phage Typing A method of identifying bacteria using specific strains of bacteriophages.

Phagocyte A cell capable of engulfing and digesting particles that are harmful to the body.

Phagocytosis The ingestion of solids by cells.

Phagolysosome A digestive vacuole.

Phagosome A food vacuole of a phagocyte; also called phagocytic vesicle.

Phase-Contrast Microscope A compound light microscope that allows examination of structures inside cells through the use of a special condenser.

Phenol A disinfectant often used as a standard against which others are measured; also called carbolic acid.

Phenolic A synthetic derivative of phenol.

Phenotype The external manifestations of an organism's genotype, or genetic makeup.

Phosphate Group A portion of a phosphoric acid molecule attached to some other molecule:

Phospholipid A complex lipid composed of glycerol, two fatty acids, and a phosphate group.

Phosphorus Cycle The various solubility stages of phosphorus in the environment.

Phosphorylation The addition of a phosphate group to an organic molecule.

Photoautotroph An organism that uses light as its energy source and carbon dioxide (CO_2) as its carbon source.

Photoheterotroph An organism that uses light as its energy source and an organic carbon source.

Photophosphorylation The production of ATP by photosynthesis.

Photosynthesis The conversion of light enery from the sun into chemical energy; the light-fueled synthesis of carbohydrate from carbon dioxide (CO2).

Phototaxis Movement in response to the presence of light.

Phototroph An organism that uses light as its primary energy source.

Phylogeny The evolutionary history of a group of organisms; phylogenetic relationships are evolutionary relationships.

Phylum A taxonomic classification between kingdom and class.

Phytoplankton Free-floating photoautotrophs.

Pilus (plural: **pili)** An appendage on a bacterial cell used for the transfer of genetic material during conjugation.

Pinocytosis The engulfing of fluid by infolding of the plasma membrane.

Plankton Free-floating aquatic organisms.

Plantae The kingdom composed of multicellular eucaryotes with cellulose cell walls.

Plaque A clearing in a bacterial lawn resulting from lysis by phages. *See also* dental plaque.

Plaque-Forming Units (pfu) Visible viral plaques counted.

Plasma The liquid portion of blood in which the formed elements are suspended.

Plasma Cell A cell that activated a B cell differentiates into; plasma cells manufacture specific antibodies.

Plasma (Cytoplasmic) Membrane The selectively permeable membrane enclosing the cytoplasm of a cell; the outer layer in animal cells, internal to the cell wall in other organisms.

Plasmid A small circular DNA molecule in bacteria that replicates independently of the chromosome.

Plasmodium (1) A multinucleated mass of protoplasm, as in plasmodial slime molds. (2) When written as a genus, refers to the protozoan agent of malaria.

Plasmolysis Loss of water from a cell in a hypertonic environment.

Plate Count A method of determining the number of bacteria in a sample by counting the number of colony-forming units on a solid culture medium.

Pleomorphic Having many shapes, characteristic of certain bacteria.

Pneumonia Inflammation of the lungs.

Point Mutation *See* Base substitution.

Polar Molecule A molecule with an unequal distribution of charges.

Polyene Antibiotic An antimicrobial agent that alters sterols in eucaryotic plasma membranes and contains more than four carbon atoms and at least two double bonds.

Polymer A molecule consisting of a sequence of similar molecules, or monomers.

Polymerase Chain Reaction (PCR) A technique using DNA polymerase to make multiple copies of a DNA template in vitro.

Polymorphonuclear Leukocyte (PMN) *See* neutrophil.

Polypeptide (1) A chain of amino acids. (2) A group of antibiotics.

Polysaccaride A carbohydrate consisting of 8 or more monosaccharides joined through dehydration synthesis.

Porins A type of protein in the outer membrane of gram-negative cell walls that permits the passage of small molecules.

Portal of Entry The avenue by which a pathogen gains access to the body.

Portal of Exit The route by which a pathogen leaves the body.

Positive (Direct) Selection A procedure for picking out mutant cells by growing them.

Pour Plate Method A method of inoculating a solid nutrient medium by mixing bacteria in the melted medium and pouring the medium into a Petri dish to solidify.

Precipitation Reaction A reaction between soluble antigens and multivalent antibodies to form visible aggregates.

Precipitin Ring Test A precipitation test performed in a capillary tube.

Predisposing Factor Anything that makes the body more susceptible to a disease or alters the course of a disease.

Prevalence The fraction of a population having a specific disease at a given time.

Primary Cell Line Human tissue cells that grow for only a few generations in vitro.

Primary Infection An acute infection that causes the initial illness.

Primary Metabolite A product of an industrial cell population produced during the time of rapid, logarithmic growth. See also secondary metabolite.

Primary Producer An autotrophic organism, either chemotroph or phototroph, that converts carbon dioxide into organic compounds.

Primary Response Antibody production to the first contact with an antigen. *See also* Memory response.

Primary Sewage Treatment The removal of solids from sewage by allowing them to settle out and be held temporarily in tanks or ponds.

Prion A self-replicating protein with no detectable nucleic acids.

Privileged Site (Tissue) An area of the body (or a tissue) that does not elicit an immune response.

Procaryote A cell whose genetic material is not enclosed in a nuclear envelope.

Procaryotae The kingdom to which all procaryotic organisms belong; also called Monera.

Prodromal Period The time following the incubation period when the first symptoms of illness appear.

Profundal Zone The deeper water under the limnetic zone in an inland body of water.

Proglottid A body segment of a tapeworm containing both male and female organs.

Promoter The starting site on a DNA strand for transcription of RNA by RNA polymerase.

Prophage Phage DNA inserted into the host cell's DNA.

Prostaglandins A hormonelike substance that is synthesized in many tissues and circulates in the blood; exact function is unknown.

Prostheca A stalk or bud protruding from a procaryotic cell.

Protease Inhibitor A chemical that inhibits proteolytic enzymes (proteases).

Protein A large molecule containing carbon, hydrogen, oxygen, and nitrogen (and sulfur); some proteins have a helical structure and others are pleated sheets.

Protista The kingdom to which protozoa belong; unicellular, eucaryotic organisms.

Proton A positively charged particle in the nucleus of an atom.

Protoplast A gram-positive bacterium or plant cell treated to remove the cell wall.

Protoplast Fusion A method of joining two cells by first removing their cell walls; used in genetic engineering.

Protozoan (plural: **protozoa)** A unicellular eucaryotic organism belonging to the Kingdom Protista.

Provirus Viral DNA that is integrated into the host cell's DNA.

Pseudohypha A short chain of fungal cells that results from the lack of separation of daughter cells after budding.

Pseudopod An extension of a cell that aids in locomotion and feeding.

Psychrophile An organism that grows best at about 15°C and does not grow above 20°C; a cold-loving microbe.

Psychrotroph An organism that is capable of growth between about 0°C and 30°C.

Purines The class of nucleic acid bases that includes adenine and guanine.

Pus An accumulation of dead phagocytes, dead bacterial cells, and fluid.

Pyocyanin A blue-green pigment produced by *Pseudomonas aeruginosa.*

Pyrimidines The class of nucleic acid bases that includes uracil, thymine, and cytosine.

Quaternary Ammonium Compound (quat) A cationic detergent with four organic groups attached to a central nitrogen atom; used as a disinfectant.

Quinine An antimalarial drug derived from the cinchona tree and effective against schizogomy stage in red blood cells.

Quinolone An antibiotic whose mode of action is to inhibit DNA replication by interfering with the enzyme DNA gyrase.

Radioimmunoassay (RIA) A method of measuring the amount of a compound using radioactively labeled antibodies.

Rapid Plasma Reagin (RPR) test A serological test for syphilis.

r-determinant A group of genes for antibiotic resistance carried on R factors.

Receptor An attachment for a pathogen on a host cell.

Recipient Cell A cell that receives DNA from a donor cell during genetic recombination.

Recombinant DNA A DNA molecule produced by recombination.

Recombinant DNA Technology *See* genetic engineering.

Recombinant Vaccine A vaccine made by recombinant DNA techniques.

Redia A trematode larval stage that reproduces asexually to produce cercariae.

Redox Reaction *See* oxidation-reduction.

Red Tide A bloom of planktonic dinoflagellates.

Reducing Medium A culture medium containing ingredients that will remove dissolved oxygen from the medium to allow the growth of anaerobes.

Reduction (1) The addition of electrons to a molecule. (2) The gain of hydrogen atoms in a molecule.

Refractive Index The relative velocity with which light passes through a substance.

Rennin An enzyme that forms curds as part of any dairy fermentation product; originally from calves' stomachs, now produced by molds and bacteria.

Replica Plating A method of inoculating a number of solid minimal culture media from an original plate of complete medium; mutant colonies that do not grow on the minimal media can be selected from the original plate.

Replication Fork The point where DNA strands separate and new strands will be synthesized.

Repression The process by which a repressor protein can stop the synthesis of a protein.

Repressor A protein that binds to the operator site to prevent transcription.

Reservoir Of Infection A continual source of infection.

Resistance The ability to ward off diseases through nonspecific and specific defenses.

Resistance (R) Factor A bacterial plasmid carrying genes that determine resistance to antibiotics.

Resistance Transfer Factor A group of genes for replication and conjugation on the R factor.

Resolution The ability to distinguish fine detail with a magnifying instrument; also called resolving power.

Respiration (1) An ATP-generation process in which chemical compounds are oxidized and the final electron acceptor is usually an inorganic molecule. (2) The process by which living organisms produce carbon dioxide (CO_2).

Restriction Enzyme An enzyme that cuts double-stranded DNA, leaving sticky ends.

Retort A device for commercially sterilizing canned food by using steam under pressure; operates on the same principle as an autoclave but is much larger.

Reverse Transcriptase An RNA-dependent DNA polymerase; an enzyme that synthesizes a complementary DNA from an RNA template.

Reversible Reaction A chemical reaction in which the end-products can readily revert to the original molecules.

RFLP Restriction fragment length polymorphism; a fragment resulting from restriction-enzyme digestion of DNA.

Rh Factor An antigen on red blood cells of rhesus monkeys and most humans; possession makes the blood Rh+.

Rh Blood Group System The classification of red blood cells based on the presence or absence of Rh antigens.

Rhizine A rootlike hypha that anchors a fungus to a surface.

Ribonucleic Acid (RNA) The class of nucleic acids that comprises messenger RNA, ribosomal RNA, and transfer RNA.

Ribose A five-carbon sugar that is part of ribonucleotide molecules and RNA.

Ribosomal RNA (rRNA) The type of RNA molecule that forms ribosomes.

Ribosomal RNA (rRNA) Sequencing Determination of the order of nucleotide bases in rRNA.

Ribosome The site of protein synthesis in a cell, composed of RNA and protein.

ribozyme An enzyme consisting of RNA that specifically acts on strands of RNA by removing sections and splicing together the remaining pieces.

Rifamycin An antibiotic that inhibits bacterial RNA synthesis.

Ring Stage A young *Plasmodium* trophozoite that looks like a ring in a red blood cell.

RNA Primer A short strand of RNA used to start synthesis of the lagging strand of DNA, and to start the polymerase chain reaction.

Root Nodule A tumorlike growth on the roots of certain plants containing symbiotic nitrogen-fixing bacteria.

Rotating Biological Contactor A method of secondary sewage treatment in which large disks are rotated while partially submerged in a sewage tank; they accumulate a film of sewage-metabolizing microbes that are exposed to aerobic conditions as the disks rotate.

Rough ER Endoplasmic reticulum with ribosomes on its surface.

Roundworm An animal belonging to the phylum Nematoda.

S (Svedberg unit) Notes the relative rate of sedimentation during ultra-high-speed centrifugation.

Salt A substance that dissolves in water to cations and anions, neither of which is H^+ or OH^-.

Sanitization The removal of microbes from eating utensils and food preparation areas.

Saprophyte An organism that obtains its nutrients from dead organic matter.

Sarcina (plural: sarcinae) (1) A group of eight bacteria that remain in a packet after dividing. (2) When written as a genus, refers to gram-positive, anaerobic cocci.

Sarcoma A cancer of fleshy, nonepithelial tissue or connective tissue.

Saturation The condition in which the active site on an enzyme is occupied by the substrate or product at all times.

Saxitoxin A neurotoxin produced by some dinoflagellates.

Scanning Electron Microscope (SEM) An electron microscope that provides three-dimensional views of the specimen magnified 1000–10,000x.

Schick Test A skin test to detect the presence of antibodies to diphtheria.

Schizogony The process of multiple fission, in which one organism divides to produce many daughter cells.

Scientific Nomenclature *See* binomial nomenclature.

Sclerotia The compact mass of hardened mycelia of the fungus *Claviceps purpurea* that fills infected rye flowers; produces the toxin ergot.

Scolex The head of a tapeworm, containing suckers and possibly hooks.

Secondary Infection An infection caused by an opportunistic microbe after a primary infection has weakened the host's defenses.

Secondary Metabolite A product of an industrial cell population produced after the microorganism has largely completed its period of rapid growth and is in a stationary phase of the growth cycle. *See also* primary metabolite.

Secondary Response *See* memory response.

Secondary Sewage Treatment Biological degradation of the organic matter in wastewater, following primary treatment.

Secretory Vesicle A membrane-enclosed sac produced by the ER; transports synthesized material into cytoplasm.

Selective Medium A culture medium designed to suppress the growth of unwanted microorganisms and encourage the growth of desired ones.

Selective Permeability The property of a plasma membrane to allow certain molecules and ions to move through the membrane while restricting others.

Selective Toxicity The property of some antimicrobial agents to be toxic for a microorganism and nontoxic for the host.

Self-Tolerance The ability of an organism to recognize and not make antibodies against self.

Semiconservative Replication The process of DNA replication in which each double-stranded DNA molecule contains one original strand and one new strand.

Semisynthetic Penicillins Modifications of natural penicillins by introducing different side chains that extend the spectrum of antimicrobial activity and avoid microbial resistance.

Sense Codon A codon that codes for an amino acid.

Sense Strand (+ strand) Viral RNA that can act as mRNA.

Sepsis A toxic condition resulting from the growth and spread of bacteria in blood and tissue.

Septa A cross-wall in a fungal hypha.

Septate Hypha A hypha consisting of uninucleate cell-like units.

Septicemia The proliferation of bacteria in the blood, accompanied by fever; sometimes causes organ damage.

Septic Shock A sudden drop in blood pressure induced by toxins produced by bacterial proliferation in the blood, usually endotoxins from gram-negative bacteria.

Serial Dilution The process of diluting a sample several times.

Seroconversion A change in a person's response to an antigen in a serological test.

Serological Testing Techniques for identifying a microorganism based on its reaction with antibodies.

Serology The branch of immunology that studies blood serum and antigen–antibody reactions in vitro.

Serovar A variation within a species; also called serotype.

Serum The liquid remaining after blood plasma is clotted; contains antibodies (immunoglobulins).

Sexual Dimorphism The distinctly different appearance of adult male and female organisms.

Sexual Spore A spore formed by sexual reproduction.

Shuttle Vector A plasmid that can exist in several different species; used in genetic engineering.

Sign A change due to a disease that a physician can observe and measure.

Simple Diffusion The net movement of molecules or ions from an area of higher concentration to an area of lower concentration.

Simple Stain A method of staining microorganisms with a single basic dye.

Singlet Oxygen Highly reactive molecular oxygen (O_2).

Slide Agglutination Test A method of identifying an antigen by combining it with a specific antibody on a slide.

Slime Layer A glycocalyx that is unorganized and loosely attached to the cell wall.

Slime Mold A fungus like protist.

Slow Viral Infection A disease process that occurs gradually over a long period; also called persistent viral infection.

Sludge Solid matter obtained from sewage.

Smear A thin film of material containing microorganisms, spread over the surface of a slide.

Smooth ER Endoplasmic reticulum without ribosomes.

Solute A substance dissolved in another substance.

Solvent A dissolving medium.

Southern Blotting A technique that uses DNA probes to detect the presence of specific DNA in restriction fragments separated by electrophoresis.

Specialized Transduction The process of transferring a piece of cell DNA adjacent to a prophage to another cell.

Species The most specific level in the taxonomic hierarchy; also called specific epithet. *See also* bacterial species; eucaryotic species; viral species.

Spectrum of Microbial Activity The range of distinctly different types of microorganisms affected by an antimicrobial drug; a wide range is referred to as a broad spectrum of activity.

Spheroplast A gram-negative bacterium treated to damage the cell wall, resulting in a spherical cell.

Spicule One of two external structures on the male roundworm used to guide sperm.

Spike A carbohydrate-protein complex that projects from the surface of certain enveloped viruses.

Spirillum (plural: **spirilla)** (1) A helical or corkscrew-shaped bacterium. (2) When written as a genus, refers to aerobic, helical bacteria with clumps of polar flagella.

Spirochete A corkscrew-shaped bacterium with axial filaments.

Splicing A process by which an RNA molecule cuts out introns and rejoins exons to produce a molecule of mRNA.

Spontaneous Generation The idea that life could arise spontaneously from nonliving matter.

Spontaneous Mutation A mutation that occurs without a mutagen.

Sporadic Disease A disease that occurs occasionally in a population.

Sporangiophore An aerial hypha supporting a sporangium.

Sporangiospore An asexual fungal spore formed within a sporangium.

Sporangium A sac containing one or more spores.

Spore A reproductive structure formed by fungi and actinomycetes.

Sporogenesis *See* sporulation.

Sporozoite A trophozoite of *Plasmodium* found in mosquitoes, infective for humans.

Sporulation The process of spore and endospore formation; also called sporogenesis.

Spread Plate Method A plate count method in which inoculum is spread over the surface of a solid culture medium.

Staphylococci (singular: **Staphylococcus)** A grapelike cluster or broad sheet of spherical cells.

Stationary Phase The period in a bacterial growth curve when the number of cells dividing equals the number dying.

Stem Cell A fetal cell that gives rise to bone marrow, blood cells, and B and T cells.

Stereoisomers Two molecules consisting of the same atoms, arranged in the same manner but differing in their relative positions; mirror images; also called D-isomer and L-isomer.

Sterile Free of microorganisms.

Sterilization The killing of all microorganisms, including endospores.

Steroid A specific group of chemical substances, including cholesterol and hormones.

Stipe A stemlike supporting structure of multicellular algae and basidiomycetes.

Strain A group of cells derived from a single cell.

Streak Plate Method A method of isolating a culture by spreading microorganisms over the surface of a solid culture medium.

Streptobacilli (singular: **Streptobacillus)** Rods that remain attached in chains after cell division.

Streptococci (plural: **streptococcus)** Cocci that remain attached in chains after cell division.

Structural Gene A gene that determines the structure of a protein.

Subacute Disease A disease with symptoms that are intermediate between acute and chronic.

Subclinical Infection An infection that does not cause a noticeable illness; also called inapparent infection.

Subcutaneous Mycosis A fungal infection of tissue beneath the skin.

Substrate Any compound with which an enzyme reacts.

Substrate-Level Phosphorylation The synthesis of ATP by direct transfer of a high-energy phosphate group from an intermediate metabolic compound to ADP.

Subunit Vaccine A vaccine consisting of an antigenic fragment, often produced by genetic engineering.

Sulfhydryl Group –SH.

Sulfonamide A bacteriostatic compound that interferes with folic acid synthesis by competitive inhibition; also called sulfa drug.

Sulfur Cycle The various oxidation and reduction stages of sulfur in the environment, mostly due to the action of microorganisms.

Superantigen An antigen that activates many different T cells, thereby eliciting a large immune response.

Superficial Mycosis A fungal infection localized in surface epidermal cells and along hair shafts.

Superinfection The growth of a target pathogen that has developed resistance to an antimicrobial drug being used; the growth of an opportunistic pathogen.

Superoxide Dismutase An enzyme that destroys superoxide free radicals.

Superoxide Free Radical A toxic form of oxygen (O_2^-) formed during aerobic respiration.

Suppressor T Cell (T_S) A T cells that is thought to end an immune response after an antigen is no longer present.

Surface-Active Agent Any compound that decreases the tension between molecules lying on the surface of a liquid; also called surfactant.

Susceptibility The lack of resistance to a disease.

Symbiosis The living together of two different organisms or populations.

Symptom A change in body function that is felt by a patient as a result of a disease.

Syncytium A multinucleated giant cell resulting from certain viral infections.

Synergism (1)The effect of two microbes working together that is greater than the effect of either acting alone. (2) The principle whereby the effectiveness of two drugs used simultaneously is greater than that of either drug used alone.

Synthesis Reaction A chemical reaction in which two or more atoms combine to form a new, larger molecule.

Synthetic Drug A chemotherapeutic agent that is prepared from chemicals in a laboratory.

Systemic Anaphylaxis A hypersensitivity reaction causing vasodilation and resulting in shock; also called anaphylactic shock.

Systemic (Generalized) Infection An infection throughout the body.

Systemic Mycosis A fungal infection in deep tissues.

Tachyzoite A rapidly growing trophozoite form of a protozoan.

T Antigen An antigen in the nucleus of a tumor cell.

Tapeworm A flatworm belonging to the class Cestoda.

Target Cell An infected body cell to which defensive cells of the immune system bind.

Taxis Movement in response to an environmental stimulus.

Taxonomy The science of the classification of organisms.

T Cell A type of lymphocyte, which develops from a stem cell processed in the thymus gland, that is responsible for cell-mediated immunity.

T-Dependent Antigen An antigen that will stimulate the formation of antibodies only with the assistance of helper T cells; usually a protein, such as those found on viruses, bacteria, and toxins.

Teichoic Acid A polysaccharide found in gram-positive cell walls.

Temperature Abuse Improper food storage at a temperature that allows bacteria to grow.

Terminator The site on a DNA strand at which transcription ends.

Tertiary Sewage Treatment A method of waste treatment that follows conventional secondary sewage treatment; nonbiodegradable pollutants and mineral nutrients are removed, usually by chemical or physical means.

Tetracyclines A broad-spectrum antibiotic that interferes with protein synthesis.

Tetrad A group of four cocci.

Thallus The entire vegetative structure or body of a fungus, lichen, or alga.

Thermal Death Point (TDP) The temperature required to kill all the bacteria in a liquid culture in 10 minutes.

Thermal Death Time (TDT) The length of time required to kill all bacteria in a liquid culture at a given temperature.

Thermoduric Heat resistant.

Thermophile An organism whose optimum growth temperature is between 50°C and 60°C; a heat loving microbe.

Thermophilic Anaerobic Spoilage Spoilage of canned foods due to the growth of thermophilic bacteria.

Thylakoid A chlorophyll-containing membrane in a chloroplast. A bacterial thylakoid is also known as a chromatophore.

Tincture A solution in aqueous alcohol.

T-Independent Antigen An antigen that will stimulate the formation of antibodies without the assistance of helper T cells; usually a substance with repeated sequences, such as a polysaccharide.

T-Independent Antigen B cells do not require T_H cells to respond to this type of antigen. *See also* T-dependent antigen.

Ti Plasmid An *Agrobacterium* plasmid carrying genes for tumor induction in plants.

Titer An estimate of the amount of antibodies in an antibody-containing fluid, such as serum.

Total Magnification The magnification of a microscopic specimen, determined by multiplying the ocular lens magnification by the objective lens magnification.

Toxemia The presence of toxins in the blood.

Toxigenicity The capacity of a microorganism to produce a toxin.

Toxin Any poisonous substance produced by a microorganism.

Toxoid An inactivated toxin.

Trace Element A chemical element required in small amounts for growth.

Transamination The transfer of an amino group from an amino acid to an organic acid.

Transcription The process of synthesizing RNA from a DNA template.

Transduction The transfer of DNA from one cell to another by a bacteriophage. *See also* generalized transduction; specialized transduction.

Transferrin A human iron-binding protein that reduces iron available to a pathogen.

Transfer RNA (tRNA) The type of RNAmolecule that brings amino acids to the ribosomal site where they are incorporated into proteins.

Transformation (1) The process in which genes are transferred from one bacterium to another as "naked" DNA in solution. (2) The changing of a normal cell into a cancerous cell.

Transient Microbiota The microorganisms that are present in an animal for a short time without causing a disease.

Translation The use of mRNA as a template in the synthesis of protein.

Transmembrane Channel A pore in a target cell membrane produced by complement.

Transmission Electron Microscope (TEM) An electron microscope that provides high magnifications (10,000–100,000x) of thin sections of a specimen.

Transposon A small piece of DNA that can move from one region of the DNA molecule to another.

Triazole An antifungal antibiotic, related to the imidazoles, used to treat systemic fungal infections.

Trickling Filter A method of secondary sewage treatment in which sewage is sprayed out of rotating arms onto a bed of rocks or similar materials, exposing the sewage to highly aerobic conditions and sewage-metabolizing microbes.

Triplex Agent An experimental type of antimicrobial drug; a short segment of DNA that binds to a target area on a double strand of DNA blocks the synthesis of a pathogenic protein by a pathogenic microbe.

Trophophase The period in the production curve of an industrial cell population in which the primary metabolites are formed; a period of rapid, logarithmic growth. *See also* idiophase.

Trophozoite The vegetative form of a protozoan.

Tuberculin Skin Test A skin test used to detect the presence of antibodies to *Mycobacterium tuberculosis*.

Tumor Necrosis Factor (TNF) A polypeptide released by phagocytes in response to bacterial endotoxins; induces shock; also called cachectin.

Tumor-Specific Transplantation Antigen (TSTA) A viral antigen on the surface of a transformed cell.

Turbidity The cloudiness of a suspension.

Turnover Number The number of substrate molecules metabolized per enzyme molecule per second.

12D Treatment A sterilization process that would result in a decrease of the number of *Clostridium botulinum* endospores by 12 logarithmic cycles.

Ubiquinones A low–molecular weight, nonprotein carrier in an electron transport chain; also called coenzyme Q.

UDP-N-acetylglucosamine (UDPNAc) A compound necessary for the biosynthesis of peptidoglycan.

Ultra-High-Temperature (UHT) treatment A method of treating food with high temperatures (140–150°C) for very short times to make the food sterile so that it can be stored at room temperature.

Uncoating The separation of viral nucleic acid from its protein coat.

Undulating Membrane A highly modified flagellum on some protozoa.

Use-Dilution Test A method of determining the effectiveness of a disinfectant using serial dilutions.

Vaccination The process of conferring immunity by administering a vaccine; also called immunization.

Vaccine A preparation of killed, inactivated, or attenuated microorganisms or toxoids to induce artificially acquired active immunity.

Vacuole An intracellular inclusion, in eucaryotic cells, surrounded by a plasma membrane; in procaryotic cells, surrounded by a proteinaceous membrane and containing gas.

Valence The combining capacity of an atom or a molecule.

Vancomycin An antibiotic that inhibits cell wall synthesis.

Variolation An early method of vaccination using infected material from a patient.

Vasodilation Dilation or enlargement of blood vessels.

VDRL Test A rapid screening test to detect the presence of antibodies against *Treponema pallidum*, the cause of syphilis. (VDRL stands for Venereal Disease Research Laboratory.)

Vector (1) A plasmid or virus used in genetic engineering to insert genes into a cell. (2) An arthropod that carries disease-causing organisms from one host to another.

Vegetative Referring to cells involved with obtaining nutrients, as opposed to reproduction or resting.

Vehicle Transmission The transmission of a pathogen to a large number of people by an inanimate reservoir.

Vibrio A curved or comma-shaped bacterium; when written as a genus, a gram-negative, motile, facultatively anaerobic curved rod.

Vibrioid A curved, helical bacterium that does not make a complete turn or twist.

Viral Hemagglutination The ability of certain viruses to cause the clumping of red blood cells in vitro.

Viral Hemagglutination Inhibition Test A neutralization test in which antibodies against particular viruses prevent the viruses from clumping red blood cells in vitro.

Viremia The presence of viruses in the blood.

Virion A complete, fully developed viral particle.

Viroid An infectious piece of "naked" RNA.

Virology The scientific study of viruses.

Virulence The degree of pathogenicity of a microorganism.

Virus A submicroscopic, parasitic, filterable agent consisting of a nucleic acid surrounded by a protein coat.

Volutin Stored inorganic phosphate in a procaryotic cell. *See also* metachromatic granule.

Wandering Macrophage A macrophage that leaves the blood and migrates to infected tissue.

Wassermann Test A complement fixation test used to diagnose syphilis.

Western Blotting A technique that uses antibodies to detect the presence of specific proteins separated by electrophoresis.

Wheal A small, burning or itching swelling on the skin, such as that resulting from a skin test or a mosquito bite.

Whey The fluid portion of milk that separates from curd.

Xenodiagnosis A method of diagnosis based on exposing a parasite-free normal host to the parasite and then examining the host for parasites.

Xenograft A tissue graft from another species; also called xenotransplant.

Zone Of Inhibition The area of no bacterial growth around an antimicrobial agent in the disk-diffusion method.

Zoonosis A disease that occurs primarily in wild and domestic animals but can be transmitted to humans.

Zygospore A sexual fungal spore characteristic of the zygomycetes.

Zygote A diploid cell produced by the fusion of two haploid gametes.

Index

C

P

R

S

W

X

Y